ORGAN SYSTEMS OF THE HUMAN BODY

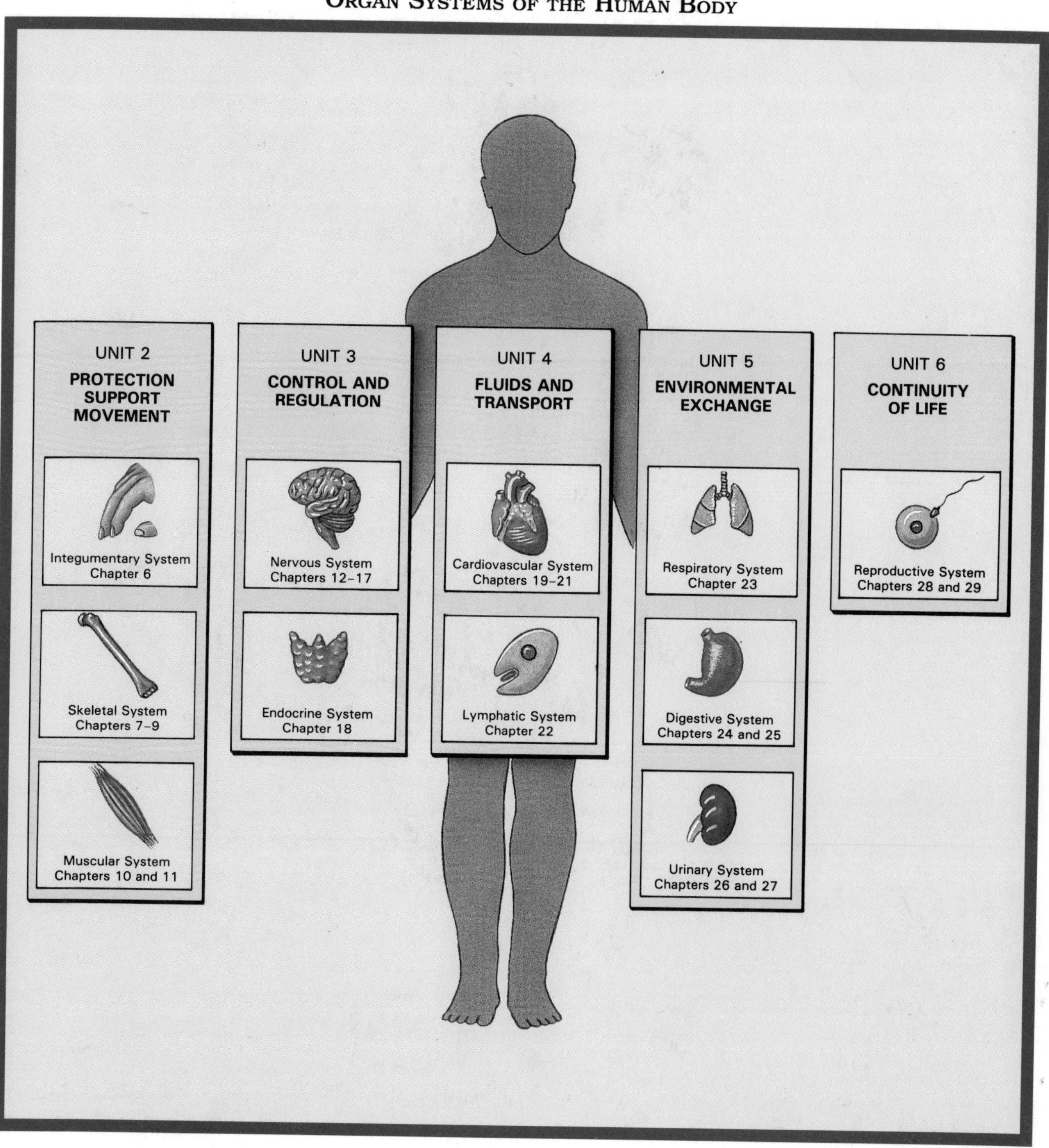
UNIT 2
PROTECTION SUPPORT MOVEMENT
Integumentary System
Chapter 6
Skeletal System
Chapters 7–9
Muscular System
Chapters 10 and 11
UNIT 3
CONTROL AND REGULATION
Nervous System
Chapters 12–17
Endocrine System
Chapter 18
UNIT 4
FLUIDS AND TRANSPORT
Cardiovascular System
Chapters 19–21
Lymphatic System
Chapter 22
UNIT 5
ENVIRONMENTAL EXCHANGE
Respiratory System
Chapter 23
Digestive System
Chapters 24 and 25
Urinary System
Chapters 26 and 27
UNIT 6
CONTINUITY OF LIFE
Reproductive System
Chapters 28 and 29

Fundamentals of Anatomy and Physiology

Text and Illustration Team

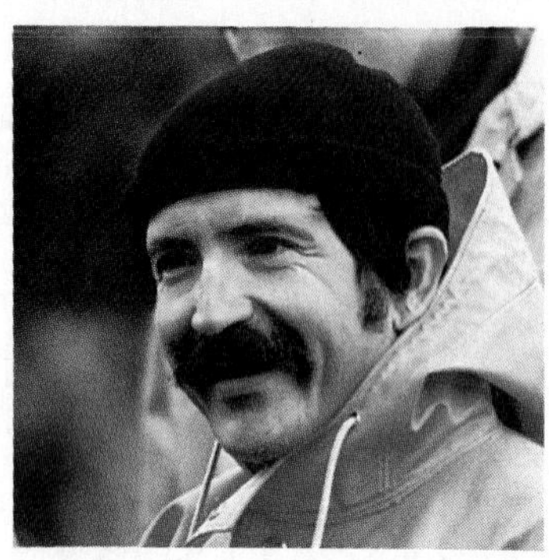

Frederic H. Martini received his Ph.D. from Cornell University in Comparative and Functional Anatomy. He has broad interests in vertebrate biology, with special expertise in anatomy, physiology, histology, and embryology. Dr. Martini's research publications include articles and technical reports. He has also written magazine articles and a book for the amateur naturalist, EXPLORING TROPICAL ISLES AND SEAS, that was a Book of the Month Club selection in 1984.

Dr. Martini has been teaching undergraduate courses in anatomy and physiology (comparative and/or human) since 1970. Until 1989 he taught courses on human anatomy and physiology at Maui Community College each winter and spent his summers at Cornell University's Shoals Marine Laboratory. Dr. Martini now devotes his winters to developing new approaches to anatomy and physiology education for *Fundamentals of Anatomy and Physiology* and related projects. He continues his summer teaching at SML, where he lectures on comparative vertebrate anatomy, physiology, ecology, and evolution. Dr. Martini is a member of the Human Anatomy and Physiology Society, the National Association of Biology Teachers, the American Society of Zoologists, and the Western Society of Naturalists.

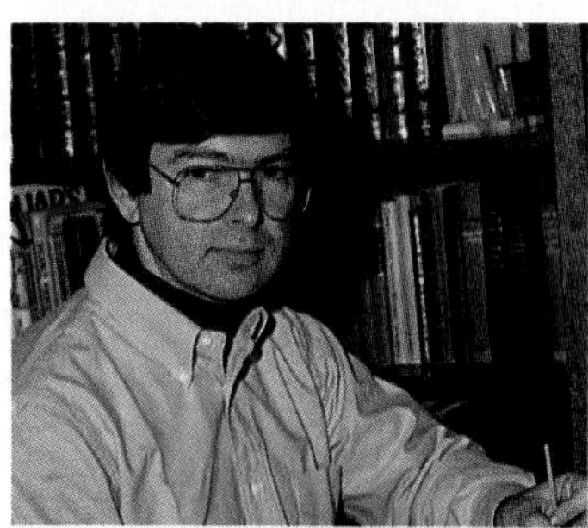

Dr. William C. Ober (art coordinator and illustrator) received his undergraduate degree from Washington and Lee University and his M.D. from the University of Virginia in Charlottesville. While in medical school he also studied in the Department of Art as Applied to Medicine at Johns Hopkins University. After graduation Dr. Ober completed a residency in family practice, and is currently on the faculty of the University of Virginia as a Clinical Assistant Professor in the Department of Family Medicine. He is also part of the Core Faculty at Shoals Marine Laboratory, where he teaches biological illustration in the summer program. Dr. Ober now devotes his full attention to medical and scientific illustration.

Claire W. Garrison, R.N., (illustrator) practiced pediatric and obstetric nursing for nearly 20 years before turning to medical illustration as a full-time career. Following a five-year apprenticeship, she has worked as Dr. Ober's associate since 1986. Ms. Garrison is also a Core Faculty member at Shoals.

Texts illustrated by Dr. Ober and Ms. Garrison have received national recognition and awards from the Association of Medical Illustrators (Award of Excellence), American Institute of Graphics Arts (Certificate of Excellence), Chicago Book Clinic (Award for Art and Design), Printing Industries of America (Award of Excellence), and Bookbuilders West. They are also recipients of the Art Directors Award.

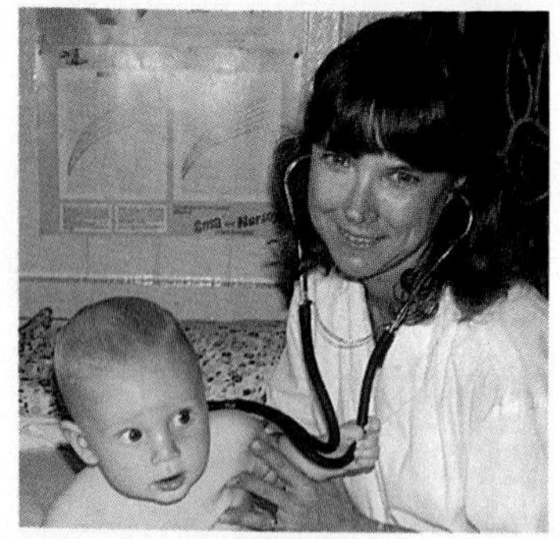

Dr. Kathleen Welch (clinical consultant) received her M.D. from the University of Washington in Seattle and completed her residency at the University of North Carolina at Chapel Hill. For two years she served as Director of Maternal and Child Health at the LBJ Tropical Medical Center in American Samoa, and subsequently was a member of the Department of Family Practice at the Kaiser Permanente Clinic, Lahaina, Hawaii. She is a member of the American Academy of Family Practice, the Hawaii Medical Association, and the Human Anatomy and Physiology Society.

second edition

Fundamentals of Anatomy and Physiology

Frederic Martini, Ph.D.

with

William C. Ober, M.D.
Art coordinator and illustrator

Claire W. Garrison, R.N.
Illustrator

Kathleen Welch, M.D.
Clinical consultant

Prentice Hall, Englewood Cliffs, New Jersey 07632

Library of Congress Cataloging-in-Publication Data

Martini, Frederic,
Fundamentals of anatomy and physiology/Frederic Martini;
illustration coordinators, William C. Ober, Claire W. Garrison.
2nd. ed.
P. cm.
Includes index.
ISBN 0–13–334590–4:
1. Human physiology. 2. Human anatomy. I. Title.
[DNLM: 1. Anatomy. 2. Physiology. QS M3855f]
QP34.5.M27 1992
612—dc20
DNLM/DLC
for Library of Congress 91–31947
CIP

Acquisition Editor: *David Kendric Brake*
Development Editor: *Dan Schiller*
Production Editor: *Debra A. Wechsler*
Marketing Manager: *Kelly Albert*
Copy Editor: *Margo Quinto*
Interior Designer: *Lorraine Mullaney*
Cover Designer: *Bruce Kenselaar/Anne Bonanno*
Design Director: *Florence Dara Silverman*
Prepress Buyer: *Paula Massenaro*
Manufacturing Buyer: *Lori Bulwin*
Page Layout: *Lorraine Mullaney, Maureen Eide, Meryl Poweski,* and *Janet Schmid*
Photo Research: *Stuart Kenter*
Photo Editor: *Lorinda Morris-Nantz*
Illustrators: *William C. Ober, M.D.; Claire W. Garrison, R.N.; Tina Sanders; MediVisuals; Ron Ervin;* and *Craig Luce*
Production Assistants: *Joanne Jimenez* and *Nancy Bauer*
Editorial Intern: *Elizabeth Pfaffenroth*
Editorial Assistant: *Mary DeLuca*
Proofreader: *Marie DeFelice*

Cover: Jess Curtis, a dancer with the San Francisco-based performance company, Contraband. Photo by Bonnie Kamin.

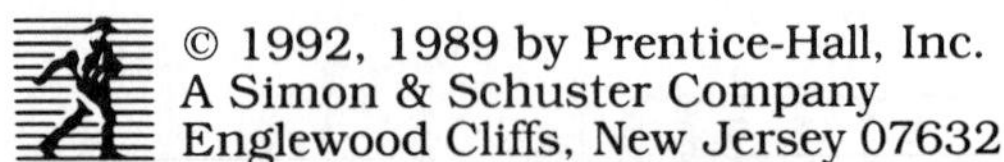

Printed in the United States of America
10 9 8 7 6 5 4 3

ISBN 0-13-334590-4

Printice-Hall International (UK) Limited, *London*
Prentice-Hall of Australia Pty, Limited, *Sydney*
Prentice-Hall Canada Inc., *Toronto*
Prentice-Hall Hispanoramericana, S.A., *Mexico*
Prentice-Hall of India Private Limited, *New Delhi*
Prentice-Hall of Japan, Inc., *Tokyo*
Simon & Schuster Asia Pte. Ltd., *Singapore*
Editora Prentice-Hall do Brasil Ltda., *Rio de Janerio*

Contents in Brief

■ UNIT 1 Levels of Organization

■ UNIT 2 Support and Movement

■ UNIT 3 Control and Regulation

■ UNIT 4 Fluids and Transport

■ UNIT 5 Environmental Exchange

■ UNIT 6 Continuity of Life

To my students,
who suggested that I start this project,
and to my friends and family,
whose understanding and support
enabled me to complete it.

Contents

UNIT 3 Control and Regulation

UNIT 4 Fluids and Transport

UNIT 5 Environmental Exchange

23 The Respiratory System 747

24 The Digestive System 783

25 Metabolism and Energetics 827

26 The Urinary System 859

27 Fluid, Electrolyte, and Acid-Base Balance 895

UNIT 6 Continuity of Life

28 The Reproductive System 917

Preface

Pick up a newspaper or turn on the television, and you will find that the human body is in the news. Medical breakthroughs and epidemics; sports records and sports injuries; famines and artificial sweeteners; contraception, surrogate motherhood, and the abortion debate; cholesterol and carotene; laser surgery and drug abuse; airbags and deodorant ads; air pollution, water pollution, the ozone hole—these are just a few of the topics that either affect or reflect some aspect of human anatomy or physiology.

The increase in available, useful information concerning the structure and function of the body has the potential to affect every aspect of our lives. There have never been greater opportunities for employment in applied health-related fields, from nursing to sports training, from dietetics to occupational safety. Yet, paradoxically, this same information explosion in the health sciences has also had negative impacts on society. One of the most serious has been the creation of a "knowledge gap" that sharply divides the population. A large segment of the populace not only lacks the educational background necessary to take advantage of these opportunities, but is unable to apply the information available on health issues to their personal lives. Without the ability to evaluate health care and nutritional options, people make inappropriate and even dangerous decisions.

This textbook introduces basic concepts and principles important to an understanding of the human body. It has two primary goals:

1. Building a foundation for later courses dealing with specific topics in human anatomy or physiology;
2. Providing a framework for the organization, interpretation, and application of related information obtained outside of the classroom.

The aim has been to present essential information clearly, with suitable emphasis on the concrete, applied aspects of each topic. Those pursuing careers in the medical or allied health sciences will thus acquire a context into which additional knowledge obtained in subsequent classes can be integrated. For those who seek careers outside the biomedical fields, the perception that anatomical and physiological processes are understandable, relevant, and logical should remain intact and valuable long after the origin and insertion of the *latissimus dorsi* muscle have been forgotten.

■ About the New Edition

THEMES

To succeed in this course, students must do more than develop a large technical vocabulary and retain a large volume of detailed information. They must relate the information learned in each chapter to data introduced in earlier chapters, and understand how these specific details affect the overall functioning of the body. In other words, anatomy and physiology students must develop their capacity for critical thinking, abstraction, and concept integration. These skills are important for everyone, but they are especially vital for those pursuing careers in the health sciences. Recent educational research has demonstrated that memorization of information does not improve diagnostic abilities *unless the individual learns the material in a logical framework that stresses concept organization.*

Unfortunately, introductory students are often unprepared for this type of course because their previous courses stressed rote memorization and recital of discrete blocks of information. An anatomy and physiology course thus gives the student a chance to develop or improve the ability to think clearly and logically. This text has evolved to enable this process to be as easy and enjoyable as possible.

The cornerstone of the second edition is the concept that the human body functions as an integrated unit. That integration exists at all levels, from cell to cell, tissue to tissue, organ to organ, and system to system. Homeostasis is maintained through interactions that occur on each of these levels, as well as between levels. When homeostasis breaks down, symptoms of disease appear. These basic concepts are introduced early in the text, and reinforced in all subsequent chapters.

DEVELOPMENT

This text has evolved in response to the concerns of today's anatomy and physiology instructors and students. Planning of the second edition began after on-campus meetings with faculty and students across the country. The writing of the book was greatly aided by the comments and suggestions of those using the first edition. Drafts of the new edition were further modified in the light of reviewer feedback, together with detailed input from focus groups and reviewer conferences sponsored by Prentice Hall.

While this edition was in preparation, many novel methods of organizing and presenting scientific concepts were being discussed at national meetings, such as those of the National Association of Biology teachers, the Human Anatomy and Physiology Society, and the American Society of Zoologists. These new approaches were largely based on the research of science educators, whose work has focused attention on how and why students learn. Many of these innovative concepts have been adapted to provide new solutions to such problems as the wide diversity of students' learning types, backgrounds, degrees of preparation, and intellectual abilities, as well as limited instructional time and resources.

The resulting text differs in many ways from the first edition, and from other anatomy and physiology texts. Each new feature, detailed in later sections, addresses one or more of the specific challenges that we have all encountered in our courses. The design and implementation of these features reflects the combined efforts of the many instructors who were willing to offer advice, ideas, and suggestions for this edition.

GENERAL FEATURES

Important new features of the second edition include:

Expanded Treatment, in Text and Art, of Physiological Mechanisms

Coverage of muscle physiology, neurophysiology, the autonomic nervous system, sensory mechanisms, cardiovascular physiology, immunology, urinary function, and acid-base balance has been enhanced and revised.

Extensive Use of Concept Maps and Flow Charts

The new illustration program provides visual summaries of organizational and functional relationships within and between vital systems.

Increased Use of Figures Showing the Relationships between Macroscopic and Microscopic Structure

The first edition contained more photomicrographs, spread through more chapters, than any competing text. However, introductory students are most familiar (and most comfortable) with the highest levels of organization, those of the individual or organ system. They are much less familiar with, and considerably more apprehensive of, events at the molecular or cellular level. The new figures for the second edition bridge the gap between the familiar, macroscopic world and the unfamiliar world of cells and tissues.

An Emphasis on Applied Topics, Bridging the Gap between Theoretical, Abstract Principles and Concrete, Real-life Examples

The running text makes frequent reference to concrete, real-life examples that drive home the impact of relatively abstract material. The first edition included boxed material and essays concerning clinical conditions and diagnostic patterns, which were included to enhance student interest and provide useful applied information. This material has been reorganized in the second edition. In addition to boxes dealing with clinical conditions, this edition includes boxed material dealing with diagnostic procedures, topics related to sports and fitness, health news, and information on career options. (These features are described in more detail in a later section.)

Expanded Pedagogy

The pedagogical elements in the second edition represent one of the major changes in this revision. Realizing that not all students comprehend and internalize information in the same way, the pedagogical structure of each chapter now offers more help to the diverse types of students who take this course. Every element from the chapter-opening vignette to the Three Level Review system that ties together the end of chapter material, as well as several key ancillaries to the text, are indicative of a sensitivity to the ways students learn and retain information.

Boxed Summaries of Embryological and Fetal Development in Chapters Dealing with Individual Systems

These summaries are provided for instructors who opt to cover basic morphogenesis on a system-by-system basis. For instructors who prefer to cover developmental topics exclusively in Chapter 29, that chapter contains a summary table that includes cross-references to appropriate portions of the text.

A Discussion of the Effects of Aging on Each System

The field of geriatrics is becoming increasingly important as the baby-boomer generation grows older.

Systems Integrators

These figures, found near the end of each systems chapter, reinforce the concept that the body functions as an integrated unit rather than as a set of relatively isolated, independent systems.

Many of these new features have been specifically designed to address the problems students encounter with this material. The extensive use of concept maps and flow charts provides a visual overview that should assist visual learners in mastering difficult material and relationships. The macro- to micro-figures bring the microscopic world into perspective and make structural and functional relationships easier to understand. The emphasis on applied topics maintains student interest; the expanded pedagogy enables them to keep track of their progress. Cross-referencing of text and figures, coupled with special system-integration figures, helps students develop an integrated perspective on the functioning of the body.

INTERNAL CHAPTER ORGANIZATION

Each of the 29 chapters in the second edition has the same basic layout:

Opening Chapter Vignette

Each chapter opens with an intriguing full page photograph and a narrative vignette that briefly highlights the relationship between the image in the photograph and the focus of that particular chapter. In the selection of each photo and the development of each vignette, an attempt was made to avoid intimidating clinical and scientific information. Rather, this feature is designed to stimulate student interest, generate questions, and establish lines of thought that will propel students into the chapter with anticipation instead of anxiety.

Chapter Objectives

Every chapter begins with a numbered list of manageable learning objectives that focus on the key concepts of that chapter. It is my belief that mastering the chapter objectives will provide students with a foundation to build upon. These objectives were developed knowing that many students cannot hope to synthesize and apply important concepts or develop critical thinking skills until they have sufficiently internalized the most basic concepts in a chapter. The learning objectives also play a significant role in the Three Level Learning System discussed below.

Body of the Chapter

The main text of the chapter follows, broken down by numerous subheadings. As in the first edition, key terms are presented in boldface, and pronunciation guides and word roots are indicated where appropriate. Special features of each chapter include the following:

Integrated and Comprehensive Illustrations The art program and the text evolved together, and the layout helps the reader correlate the information provided by the text and the illustrations.

Concepts Maps, Flow Charts and Summary Tables When practical, functional relationships are highlighted through concept maps and flow charts. As well, information presented in a block is summarized in easy-to-follow tables.

The Use of Analogies Whenever possible, basic physiological principles are related to familiar physical principles or events in everyday life. This helps to create a mental picture that enhances comprehension.

Additional Vocabulary Development Several chapters contain tables that summarize information concerning the proper use of terms dealing with anatomical orientation, directional terms, and descriptive terms used when dealing with the skeletal or muscular systems.

Concept Checkpoints Two to four questions are placed near the end of major sections in each chapter. These questions are intended to provide a quick means of checking reading comprehension and improving the ability to integrate the information contained in blocks of text material.

Cross-referencing A concept links (∞) icon and page reference will be found wherever the development of a new concept builds on material presented earlier in the text.

Boxed Material The text flows around several different types of boxed discussions. These boxes, located immediately adjacent to the relevant narrative, provide useful insights into the relevance or application of important concepts. Because the material is both boxed and categorized by topic, it can be read, if assigned, or ignored without disturbing the flow of essential information in the running text. There are four basic types of boxes:

Clinical Comments: Clinical Comments boxes contain relevant clinical material. Each box describes a limited number of disorders that illustrate the consequences of homeostatic imbalances. The discussion is directly tied to the normal physiological or anatomical material in the adjacent text.

The number of Clinical Comments varies from chapter to chapter. Although few instructors assign all of the clinical material, *these boxes, together with the supplemental material in the Clinical Manual, address virtually all of the major clinical conditions and problems affecting each system.* In addition to demonstrating the relevance of text material, the Clinical Comments are valuable as references. Students and instructors may refer to the Clinical Comments for information about personal or family health concerns. Because they are intended as references, care has been taken to ensure that the information contained is current, concise, and accurate.

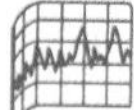

Diagnostics: Diagnostics boxes focus attention on laboratory or other clinical procedures important to the diagnosis of disease or useful in patient evaluation. This information will be of particular interest to those interested in careers as medical technicians, or to readers attempting to interpret the results of clinical tests.

Sports and Fitness: Sports and Fitness boxes deal with medical topics related to athletic activities, such as exercise schedules, cardiovascular fitness, androgen abuse or blood doping, and to sports-related injuries. These boxes will be of special interest to those interested in physical education, physical therapy, or kinesiology.

Health News: The Health News boxes consider high-visibility issues that often make headlines or stimulate controversy. Genetic engineering, fad diets, and surrogate motherhood are examples of topics considered in these sections.

Chapter Review Each chapter ends with an extensive Chapter Review comprising several integrated features, all of which are designed to work together to help students study, review, apply and integrate the new material of the chapter into the general framework of the course. Each module contains the following elements:

Review of Selected Clinical Terms
This section reviews the definitions and pronunciations of the most important clinical terms discussed in the chapter text and boxed material. In addition to defining major terms, the review section includes an index to all of the clinical information in the chapter.

Study Outline
The Study Outline reviews the major concepts and topics in summary fashion. Relevant page numbers are indicated for major headings, and related key terms are noted. Students able to recognize the statement, understand the concept, and define the key terms should have a firm grasp of essential material. If they encounter an unfamiliar term or phrase, the adjacent page reference should make the review process much easier.

Review Planner
This important pedagogical aid offers the student a suggested strategy for reviewing the material in the chapter. This strategy is based upon a **Three Level Review System** that affords each student the opportunity to review material in increasing levels of difficulty.

Level One involves a restatement of that chapter's learning objectives (the basic concepts to be mastered) combined with a listing of specific review questions from the text that drill and reinforce those objectives.

A **Level Two** icon is followed by a list of question numbers from the text that take the student to the next level of difficulty. Level two questions encourage students to combine, integrate and relate the basic concepts or objectives mastered in level one.

A **Level Three** icon is likewise followed by a list of question numbers from the text. Level three questions promote critical thinking skills at a point where such skills are most effectively developed—after mastering level one and level two material.

C M In addition to the three level review system, the Review Planner also includes icons that highlight additional clinical material and applications that can be found in the Clinical Manual (see supplements list). An icon for the Prentice Hall/ABC video library is also present in many chapters; this icon will be followed by a list of brief, informative videos that the student may elect to review.

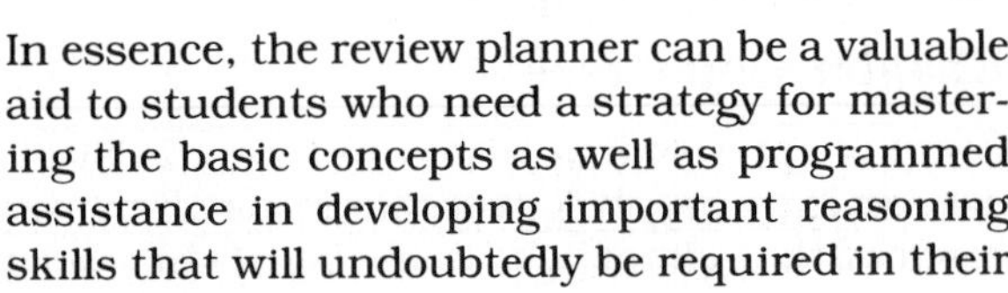

In essence, the review planner can be a valuable aid to students who need a strategy for mastering the basic concepts as well as programmed assistance in developing important reasoning skills that will undoubtedly be required in their future careers.

Review Questions
This section includes a series of questions that are segregated by level of difficulty and question type (multiple-choice, diagram or table completion, true-false, matching, fill-ins, short essay). As already mentioned, the Review Planner refers to these questions in the context of the Three Level Review System.

Career Close-ups Found at the end of selected chapters, boxed career profiles introduce the student to former anatomy and physiology students who are now enjoying active careers in a number of related areas.

OTHER USEFUL PEDAGOGICAL FEATURES

Systems Overview

Chapter 5, the last of the introductory chapters, is followed by a brief Interchapter titled *Systems Overview* that introduces each of the body's organ systems by outlining its structure and functions. These striking combinations of images, tables, and text integrate concepts introduced in Chapter 3 (cytology) and Chapter 5 (histology) while providing an organizational framework for later chapters dealing with specific systems.

Endpapers

The endpapers of this text contain information that most readers would consider essential reference material.

- The front endpapers (1) summarize the hierarchy of structural levels and (2) introduce the basic organization of the text by unit, chapter, and organ system.
- The back endpapers (1) summarize the most commonly used conversions for weights and measures, (2) define anatomical directions and planes of section, (3) provide a quick index to the in-text boxes, and (4) list all the topics treated in the Clinical Manual by chapter. The back endpaper folds out, so that the directional terms can be kept in view while reading the text.

Appendices

The appendices contain material that most instructors will use at some time in the course.

- Appendix I provides a review of the important systems of weights and measures used in the text. Students are usually advised to review this material while completing the introductory chapter, to avoid confusion and distress later in the text.
- Appendix II includes the periodic table of the elements, for instructors who would like to cover the material in Chapter 2 in greater detail.
- Appendix III details the structural classes of amino acids.
- Appendix IV discusses biological solutions and the expression of physiological concentrations.
- Appendix V contains reference tables that report normal physiological values for body fluids.
- Appendix VI describes the detailed pathways of aerobic respiration, including the TCA cycle, for potential reference in Chapters 4 and/or 25.

Answers to Concept Checks

These answers let students monitor their progress and their abilities to deal with the various types of questions encountered in each chapter.

Eponyms, Abbreviations, and Glossary

The glossary provides pronunciations and definitions of important terms. It is preceded by lists of the eponyms (commemorative names) and abbreviations used in the text. These lists allow easy reference by the reader.

■ Supplements

New York Times Contemporary View A program designed to enhance student access to current information of relevance in the classroom. Through this program, the core subject matter in the text is supplemented by a collection of time-sensitive articles from one of the world's most distinguished newspapers, THE NEW YORK TIMES. These articles demonstrate the vital, ongoing connection between what is learned in the classroom and what is happening in the world around us.

To enjoy the wealth of information of THE NEW YORK TIMES daily, a reduced subscription rate is available. For information, call toll-free: 1-800-631-1222. PRENTICE HALL and the NEW YORK TIMES are proud to co-sponsor A CONTEMPORARY VIEW. We hope it will make the reading of both textbooks and newspapers a more dynamic, involving process.

ABC Videos Approximately 23 video segments from such award-winning ABC News programs as 20/20, Nightline and The Health Show. Each video reinforces the relationship between concepts in the text and applications in the real world. These videos are brief, instructional and perfect for the student who wants to probe a topic just a bit further. All ABC videos are referenced in the Review Planner at the end of each chapter.

Anatomy and Physiology Laser Disc This valuable instructional resource offers hundreds of still images (photo micrographs and color line art) from the text as well as selected video images from our ABC and Lifetime libraries. In addition, this disc offers numerous frames containing critical thinking questions and concept maps that allow instructors to highlight the key concepts that they feel are important.

Tutorial Software This software, which follows the same three level review format found in the text, the study guide and the test item file, allows the student to encounter material in increasing levels of difficulty, from basic concept review to critical thinking problems. On screen graphics make this review and skill development tool even more valuable.

Clinical Manual and Anatomy and Physiology Toolkit A handy collection of additional clinical and diagnostic discussions that further highlight material in the text. Additionally, this supplement offers helpful articles on how to study, how to use and develop concept maps and how to improve writing and thinking skills. A perfect companion for the student who needs a little extra support or desires to probe the material in the text a bit deeper.

Pronunciation Tapes A series of useful audio tapes which drill students on vocabulary terms and word roots important in the anatomy and physiology course. These tapes will be especially valuable to the busy student who commutes or who can otherwise find the time to enhance their skills by listening at their own pace and convenience.

Transparency Pack (acetates and masters) 150 full color acetates taken from the text, selected especially for their instructional value in the classroom. Approximately 50 masters containing key charts, graphs and concept maps from the text are also included. (Available as slides as well.)

Dissection Slides Approximately 70 full color cadaver dissection slides that fully depict key body regions and systems. Recommended for laboratory and classroom use.

Course Organizer/Resource Guide—Created by Dr. Harry E. Peery. A user's manual for the entire Martini learning system (text and ancillaries). This manual will help the instructor organize lectures and integrate ancillaries such as slides, acetates, tutorial software, videos, quizzes and laser disc into the classroom presentation. In addition, this valuable manual contains suggested classroom activities designed to appeal to and challenge various types of learners.

Test Item File—Written by George Karleskint of St. Louis Community College. A bank of over 3,000 questions that include multiple choice, true/false, matching, short answer/essay questions. All questions have been developed according to the three level review system which presents questions in increasing levels of difficulty; this format allows the instructor to create comprehensive tests or a variety of pretests and quizzes targeted at varying degrees of student ability.

Prentice Hall Test and Data Manager A versatile program available in IBM and MacIntosh formats which gives instructors the ability to create tests, quizzes and tutorial programs.

Telephone Testing Service This is a service to instructors who would like Prentice Hall to prepare and print testing material for them. The instructor calls Prentice Hall toll free and chooses the questions from the *Test Item File*. Prentice Hall will prepare the test with no charge to the instructor. Tests will be prepared and mailed to instructors in 48 hours, along with an answer key, and an answer sheet for students.

Study Guide—Developed by Charles Seiger, Atlantic Community College. A unique review tool organized around the three level review system. Students can use this study guide to review and internalize the basic objectives (concepts) in each chapter and then progress to higher cognitive challenges that involve concept synthesis and critical thinking. Numerous kinds of questions and exercises, including concept mapping, clustering and brief writing activities, enable students of all abilities to enjoy flexible use of this important ancillary.

Laboratory Manuals Written by Roberta Meehan of the University of Northern Colorado, both the cat and pig version of the manual offer class tested experiments that require only the most basic equipment to conduct. The author emphasizes interaction in the laboratory environment and includes writing and concept mapping activities designed to increase student comprehension of the key anatomical and physiological concepts that support each exercise. (Available with an Instructor's Guide.)

■ Acknowledgments

This textbook represents a group effort, rather than being the product of any single individual. Foremost on the list stand the reviewers, whose advice, comments, and collective wisdom helped to shape the text into its final form. Their interest in the subject, their concern for the accuracy and method of presentation, and their experience with students of widely varying abilities and backgrounds made the review process an educational experience. To these individuals, who carefully recorded their comments, opinions, and sources, I would like to express my sincere thanks and best wishes.

The following individuals devoted large amounts of time reviewing the second edition manuscript:

Debra Joan Barnes, *Contra Costa College*
Steven Bassett, *Southeast Community College*
Cynthia Bottrell, *Scott Community College*
C. David Bridges, *Purdue University*
William M. Chamberlain, *Indiana State University*

O. D. Cockrum, *Texas State Technical College*
John Dziak, *Community College of Allegheny*
Jeff Gerst, *North Dakota State University*
Ernest Joe Harber, *San Antonio College*
Ann Harmer, *Orange Coast College*
Donna Hoel, *Stark Technical College*
Elvis J. Holt, *Purdue University*
Susan Lustick, *San Jacinto College North*
Richard Mostardi, *Akron University*
Elizabeth Naugle, *Mission College*
Mark Paulissen, *Slippery Rock University*
Robert A. Sinclair, *San Antonio College*
Jay Templin, *Widener University*
Michael Wood, *Del Mar College*

Participants in the summer 1990 focus group included:

Maxine A'Hearn, *Prince George's Community College*
Charles Biggers, *Memphis State University*
Ann Funkhouser, *University of the Pacific*
George Karleskint, *St. Louis Community College*
Bob McDonough, *DeKalb College*
Roberta Meehan, *University of Northern Colorado*
Charles Seiger, *Atlantic Community College*

The following individuals assisted me by reviewing specific chapters, performing classroom testing, doing comparative reviews, or responding to detailed questionnaires concerning their experiences and opinions after teaching with the first edition:

Maxine A'Hearn, *Prince George's Community College*
William M. Chamberlain, *Indiana State University*
William D. Chapple, *University of Connecticut*
Gerald R. Dotson, *Front Range Community College*
Kathleen M. Gorczyca, *North Shore Community College*
Jean A. S. Helgeson, *Collin County Community College District*
Vickie S. Hennessy, *Sinclair Community College*
Ann Miller, *Middlesex Community College*
Robert L. Preston, *Illinois State University*
Carolyn C. Robertson, *Tarrant County Junior College District*
David S. Smith, *San Antonio College*
Thomas S. Spurgeon, *Colorado State University*

Dan Schiller, Senior Development Editor at Prentice Hall, also played a vital role in the shaping of the second edition. His keen eye for detail was a great help in keeping the text organization, general tone, and level of presentation consistent throughout. The accuracy and currency of the clinical material in this edition primarily reflects the detailed clinical reviews performed by my wife, Kathleen Welch, M.D. Without her expertise, encouragement, and optimism the second edition process would have been far more onerous.

Virtually without exception, reviewers stressed the importance of accurate, integrated, and visually attractive illustrations in aiding the students to understand essential material. The revision of the first edition art program was primarily directed by Bill Ober, M.D. with the collaboration of Claire Garrison, Tina Sanders, and the artists of Vantage Art, Inc., and MediVisuals, Inc. In many respects Dr. Ober also served as a reviewer, for he took the time to discuss the manuscript goals before making decisions about the accompanying figures. His suggestions concerning topics of clinical importance, presentation sequence, and revisions to the proposed art were of incalculable value to me and to the project in general.

Many of these figures include color photographs or micrographs collected from a variety of sources. Much of the work in tracking down these materials was performed by Stuart Kenter, whose efforts are greatly appreciated. Many of the light micrographs prepared by the author used commercially available slides and photomicrographic equipment generously provided by the Shoals Marine Laboratory of Cornell University. Mr. Howard Edgerton, Manager, Histology Slide Department, Carolina Biological Supply, earned my heartfelt thanks by allowing me to explore their extensive slide collection in Burlington, NC. Thanks are also due to Mr. George Ahn, Manager of the Histological Slides Department of Wards Scientific, Rochester, NY, who personally selected additional slides needed for specific figures.

The author also wishes to express his appreciation to the editors and support staff at Prentice Hall who made the entire project possible and who kept the text, art, and production programs on schedule and in relative harmony. Special thanks are due to Ray Mullaney, Editor in Chief, College Book Editorial Development, who gave Dan Schiller extra support and latitude; to Tim Bozik, Editor in Chief for Science and Mathematics, for his support of the project; to David K. Brake, Biology Editor, for being the driving force and project coordinator; to Lorraine Mullaney, for designing the second edition; and to Debra Wechsler, Production Editor, for somehow managing to keep people, text, and art moving in the proper direction at appropriate times. I am also greatly indebted to Bette Weinstein Kaplan, Elizabeth Pfaffenroth, Gauri Bhatia, and Alice Fugate for their work on the Career Close-ups.

My gratitude is extended to the many people

who have provided suggestions, comments, and support while the second edition was under development. Special thanks are extended to the faculty and staff of the Shoals Marine Laboratory, who tolerated the presence of an occasionally distracted associate each summer, and to the students and staff of Maui Community College for their support and encouragement during the winter months.

No one person could expect to produce a flawless textbook of this scope and complexity. Any errors or oversights are strictly my own, rather than those of the reviewers, artists, or editors. In an effort to improve future editions, I would ask that readers with pertinent information, suggestions, or comments concerning the organization or content of this textbook send their remarks addressed to: David Brake, Biology Editor, College Book Division, Prentice Hall, Englewood Cliffs, NJ 07632. Any and all comments and suggestions will be deeply appreciated, and carefully considered in the preparation of the third edition.

Frederic Martini
Haiku, HI 1991

Fundamentals of Anatomy and Physiology

Every day we perform a balancing act. It is an amazingly complex performance and it goes on continuously, without break or intermission. Every part of our physical being is involved in this balance, from individual atoms to the largest organs within us. Yet, unless we study the human body at work, we miss almost the whole show, for we are hardly ever aware of what our bodies are doing each instant.

Our bodies face constant challenges: injury, disease, stresses both mental and physical. Every change, whether it occurs inside us or outside, threatens this delicate balance. Adjusting to these changes is crucial, for if the body loses its balance for very long it will plunge out of control—into malfunction, illness, even death. As you read this book you will become aware of how fragile, yet how resilient our bodies are as they work to maintain the intricate balance of life.

CHAPTER 1

An Introduction to Anatomy and Physiology

Chapter Objectives

After reading this chapter, you will be able to:

1 Describe the basic functions of living organisms.
2 Define the various specialties of anatomy and physiology.
3 Identify the major levels of organization in living organisms.
4 Explain the significance of homeostasis.
5 Describe important mechanisms involved in homeostatic regulation.
6 Use anatomical terms to describe body sections, body regions, and relative positions.
7 Identify the major body cavities.
8 Compare important radiological procedures in terms of their methods, advantages, and disadvantages.

Introduction

The world around us contains an enormous diversity of living organisms that vary widely in appearance and lifestyle. One aim of **biology**, the study of life, is to discover the unity and pattern that underlie this diversity. Biologists have found that despite their obvious differences, all living things perform the same basic functions:

1. They **respond** to changes in their immediate environment. You move your hand away from a hot stove, your dog barks at approaching strangers, fish are scared by loud noises, and tiny amoebas glide toward potential prey.
2. They show **adaptability**. In other words, their responses can vary over time as they adjust to their environment. Spending hours in the tropical sun after a long winter will give most light-skinned individuals a terrible sunburn. But by the end of an active summer, the same degree of exposure might not have a noticeable effect on the skin. This is because the skin adapts to increased sunlight by producing pigments that absorb damaging solar radiation and provide a measure of protection.

3. Over a lifetime, organisms **grow** and **reproduce**, creating subsequent generations of similar organisms.
4. They are capable of producing **movement**. That movement may be internal (transporting food, blood, or other materials inside the body) or external (moving through the environment).

Responsiveness, adaptability, growth, reproduction, and movement are active processes, and organisms must expend energy to perform them. The energy used must continually be replaced or the organism will become unable to function. The need for energy necessitates several additional functions:

5. Organisms **absorb** materials from the environment. They use some of the absorbed substances for growth and break the rest down to provide energy.
6. They usually absorb and consume oxygen, an atmospheric gas. This process is called **respiration**. Oxygen is required because most organisms "burn" nutrients in order to obtain energy.
7. Each living organism generates useless or even harmful waste products that are discharged into the environment in the process of **excretion**.

These are the basic characteristics of living things, both plant and animal. Several additional functions can be distinguished when you consider animals as complex as fish, cats, or human beings. For very small organisms, absorption, respiration, and excretion involve the movement of materials across exposed surfaces. But creatures larger than a few millimeters seldom absorb nutrients directly from their environment. For example, human beings cannot absorb steaks, apples, or ice cream without processing them first. That processing, called **digestion**, occurs in specialized areas where complex foods are broken down into simpler components that can be absorbed easily. Respiration and excretion are also more complicated for large organisms. Humans have specialized structures responsible for gas exchange (lungs) and waste elimination (kidneys). Finally, because absorption, respiration, and excretion are performed in different portions of the body there must be an internal transportation system, or **circulation**.

Biology includes a number of subspecialties, each with a slightly different perspective. This text considers two biological subjects, *anatomy* (a-NAT-o-mē) and *physiology* (fiz-ē-OL-ō-jē). In the course of these 29 chapters you will become familiar with the basic anatomy and physiology of an unusually large and interesting type of animal, the human being.

The Sciences of Anatomy and Physiology

The word *anatomy* has its origins in Greece, as do many other anatomical terms and phrases. A literal translation would be "to cut open." **Anatomy** is the study of internal and external structure and the physical relationships between body parts. **Physiology,** another adopted Greek phrase, is the study of how living organisms perform the vital functions listed above. The two disciplines are closely integrated both theoretically and practically. Anatomical information provides clues about probable functions, and physiological mechanisms can be explained only in terms of the underlying anatomy. This observation leads to a very important concept: *all specific functions are performed by specific structures*. Whether one starts with known anatomy and tries to determine function, or begins with a known function and then looks for a structural basis, the linkage always exists.

Consider a simple nonbiological example. Automobiles are often compared to animals, although cars are much simpler. If an anatomist and a physiologist were examining a Ford pickup truck, the anatomist would begin by measuring and photographing it. The next step would be to take it apart completely and then put it back together. At this point the structural relationships would be well understood. On anatomical grounds, it would be apparent that rotating the steering wheel will turn the front tires. The physiologist would then review the anatomical information and proceed with a detailed functional analysis. How much force must be applied to the steering wheel? Does the amount of force vary when traveling at different speeds? What is the turning radius of the truck under several different road conditions?

In this example the anatomical work was completed first, but that is not always the case. The physiologist could have taken the truck for a test drive and answered these questions with only a superficial understanding of the anatomy. The anatomist would then disassemble the truck to determine the physical relationships that convert rotation of the steering wheel into movement of the tires.

Examples of each sequence can be found in the study of the human body. The anatomical structure of the heart and brain were clearly described in the fifteenth century. But almost 200 years passed before the pumping action of the heart was demonstrated, and many aspects of brain function remain a mystery. On the other hand, important principles of cell physiology were understood decades before the electron microscope revealed the anatomical basis for known functions.

This text will provide a familiarity with anatomical structures and an appreciation of the physiological processes that make human life possible. This information should enable you to understand many kinds of disease processes. It will also provide a perspective that should prove useful for making intelligent decisions about your own life. To this end, we have included a number of boxes titled *Health News*; these deal with recent findings that may help you to protect your personal health and the well-being of those around you.

GROSS ANATOMY

Our discussions will involve several anatomical specialties. **Gross anatomy**, or **macroscopic anatomy**, considers features visible without a microscope. There are many ways to approach gross anatomy, and the strategies have been given titles of their own. **Surface anatomy** refers to the study of general form and superficial markings. Someone interested in **regional anatomy** would examine all of the superficial and internal features in a specific area of the body, such as the head and neck, the trunk, or the arms and legs.

Systemic anatomy considers the structure of major *organ systems*, such as the respiratory or digestive systems. *Organs* are anatomical units with specific functions. The heart, liver, brain, and kidneys are examples of organs. Organ systems are groups of organs that function together to produce coordinated effects. For example, the heart, blood, and blood vessels form the *cardiovascular system*, which distributes oxygen and nutrients throughout the body.

Several other gross anatomical specialties are important in medical diagnosis. **Medical anatomy** focuses on anatomical features that may undergo characteristic changes during illness. **Radiographic anatomy** involves the study of anatomical structures as they are visualized by X-rays, ultrasound scans, or other specialized procedures performed on an intact body. **Surgical anatomy** studies anatomical landmarks important for surgical procedures.

There is considerable overlap between these specialties, but a simple example should illustrate the practical differences. An individual arrives at a physician's office complaining of severe headaches. The physician, using a knowledge of medical anatomy, notes certain abnormalities, including a droopy eyelid, a dilated pupil, and limp, sagging muscles, all on the same side of the face. On the physician's advice, the patient then sees a radiologist who uses a procedure that permits the visualization of blood vessels in the head. This technique reveals a large mass inside the brain; the symptoms can be attributed to pressure on adjacent nerves. On that basis a surgical team is assembled, and a physician skilled in the surgical anatomy of the head opens the skull and removes the mass.

DEVELOPMENTAL ANATOMY

Developmental anatomy examines the changes in form that occur during the period between conception and physical maturity. The most extensive structural changes occur during the first two months of development, for over this period the basic body plan and the foundations of the major organ systems become established. The study of **embryology** involves a detailed examination of these early developmental processes.

Chapter 29 discusses developmental processes in detail. In addition, special sections called *Embryology Summary* provide an overview of selected topics in development. These sections provide a link between early embryological form and the developmental changes that shape the adult body. Developmental anatomy is important in medicine because many structural abnormalities can result from errors in development. When these abnormalities are severe enough to produce clinical symptoms, they are called **congenital defects** (kon-JEN-i-tal; *con-*, with + *gennan*, to produce).

Developmental anatomy considers a tremendous range of sizes, from an adult person to a single cell. It therefore provides a bridge between the realms of macroscopic anatomy and *microscopic anatomy*.

MICROSCOPIC ANATOMY

The boundaries of **microscopic anatomy**, or *fine anatomy*, are established by the limits of the equipment used. A simple hand lens shows details that escape the naked eye, while an electron microscope demonstrates structural details that are only a few billionths of an inch across. As we proceed through the text, we will be considering details at all levels, from macroscopic to microscopic. (Readers unfamiliar with the terms used to describe weights and measurements over this size range should consult the reference tables in Appendix I.)

Microscopic anatomy can be subdivided into specialties that consider features within a characteristic range of sizes. **Cytology** (sī-TOL-o-jē) analyzes the internal structure of individual cells, the smallest units of life. Living cells are composed of chemical substances in various combinations, and our lives depend on the chemical processes occurring in trillions of individual cells. Chapter 2 briefly summarizes the nature of the important chemicals and the mechanisms involved in their interactions. Cytological anatomy is detailed in Chapter 3, while Chapter 4 considers cell functions and their control.

Histology (his-TOL-o-jē) takes a broader perspective and examines groups of cells that form distinctive units. These groups, known as **tissues**, have specific functional roles. Tissues are the focus of Chapter 5. Tissues in combination form **organs**, anatomical units with multiple functions. Many organs are easily examined without a microscope, and at the organ level we cross the boundary into gross anatomy.

PHYSIOLOGY

Physiology considers the function of anatomical structures. A physiologist attempts to determine the physical and chemical processes responsible for vital functions. **Human physiology** is the study of the functions of the human body. These functions are complex and much more difficult to examine than most anatomical structures. As a result, there are even more specialties in physiology than in anatomy.

The cornerstone of human physiology is **cell physiology**. Cell physiology is the study of the functions of living cells. Cell physiology includes events at the chemical and molecular levels: both chemical processes within cells and chemical interactions between cells. Chapters 2–4 focus on the chemical structure, internal organization, and control mechanisms of living cells.

Aspects of **histophysiology**, the study of tissue function, are presented in Chapter 5. **Special physiology** is the study of the physiology of specific organs. Examples of special physiology include *renal physiology* (kidney function), *cardiac physiology* (heart function), and *hepatic physiology* (liver function). **System physiology** considers all aspects of the function of specific organ systems. *Cardiovascular physiology*, *respiratory physiology*, and *reproductive physiology* are examples of system physiology. **Pathological physiology** studies the effects of diseases on organ or system functions. (The Greek word *pathos* refers to disease.) Modern medicine depends on an understanding of both normal and pathological physiology; the practitioner must know not only what is wrong but how to correct it. You will find extensive information on these topics in the *Clinical Comment* and *Diagnostics* boxes scattered through subsequent chapters.

There are also special topics in physiology that address specific functions of the human body as a whole. These specialties focus on physiological interactions between multiple organ systems. For example, *high-altitude physiology* explores the effects of changes in atmospheric pressure, such as one encounters on high mountains or in space, on body functions. *Diving physiology* deals with the effects of swimming or working under water. *Exercise physiology* studies the physiological adjustments to exercise; *sports physiology* considers the effects of participating in various athletic activities such as tennis, football, or marathon racing. Many of these applied topics in physiology are discussed in boxes titled *Sports and Fitness* in later chapters.

■ Levels of Organization

The preceding discussions of anatomy and physiology began with chemicals and cells and proceeded through increasing levels of complexity. Chemicals interact to form cells, cells are organized in tissues, tissues combine to form organs, and organs work together in organ systems that make human life possible. We will use Figure 1-1 to follow the structural

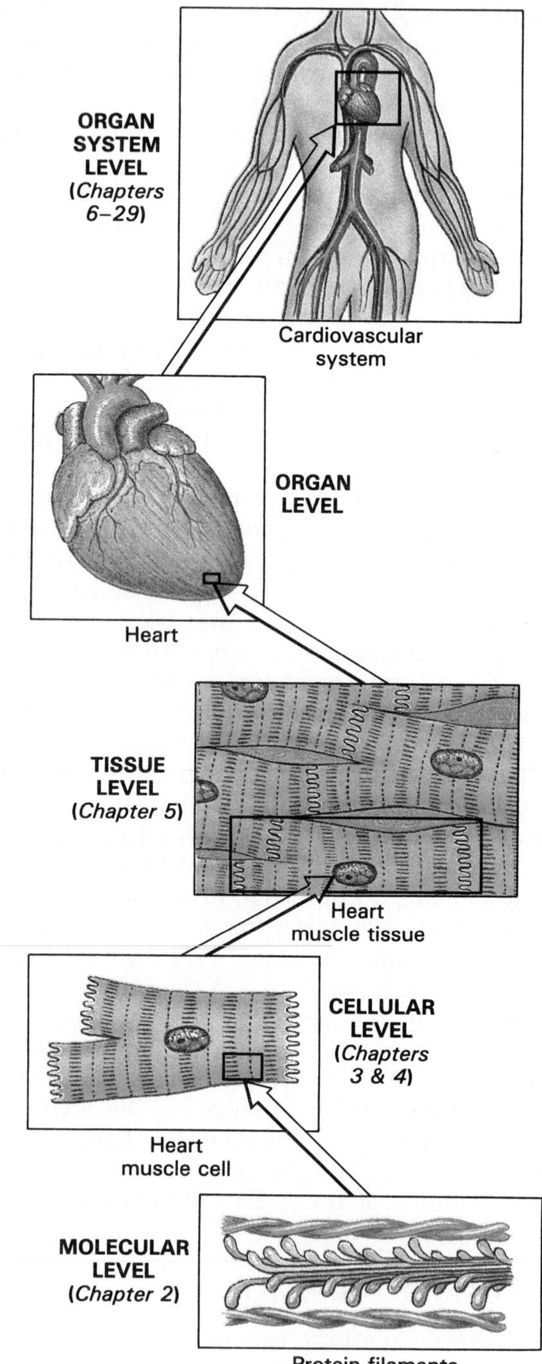

FIGURE 1-1
Levels of Organization. Interacting molecules form heart muscle cells. These cells interlock to make up heart muscle tissue, which constitutes most of the walls of a three-dimensional organ, the heart. The heart is one component of the cardiovascular system, which also includes the blood and blood vessels.

and functional repercussions of these *levels of organization.*

We begin at the *chemical, or molecular, level* of organization, where chemicals interact to form complex molecules with distinctive properties. In this case the molecules form a *muscle cell* in the heart. Muscle cells are unusual because they can contract. These cells are tied together to form a distinctive *muscle tissue*. Muscle tissue, in turn, forms the bulk of the wall of the *heart*, a hollow, three-dimensional *organ*.

Muscle contractions involve chemical interactions between structures inside individual muscle cells. Because adjacent muscle cells are linked together, their contractions are coordinated, and these coordinated contractions produce a heartbeat. When that beat occurs, the internal anatomy of the heart enables it to function as a pump. Each time it contracts, the heart pushes blood into the *circulatory system*, a network of blood vessels. Together the heart, blood, and circulatory system form an *organ system*, the *cardiovascular system*.

Each level is totally dependent on the others. For example, damage at the cellular, tissue, or organ level will affect the entire system. A chemical change

FIGURE 1-2
An Introduction to Organ Systems

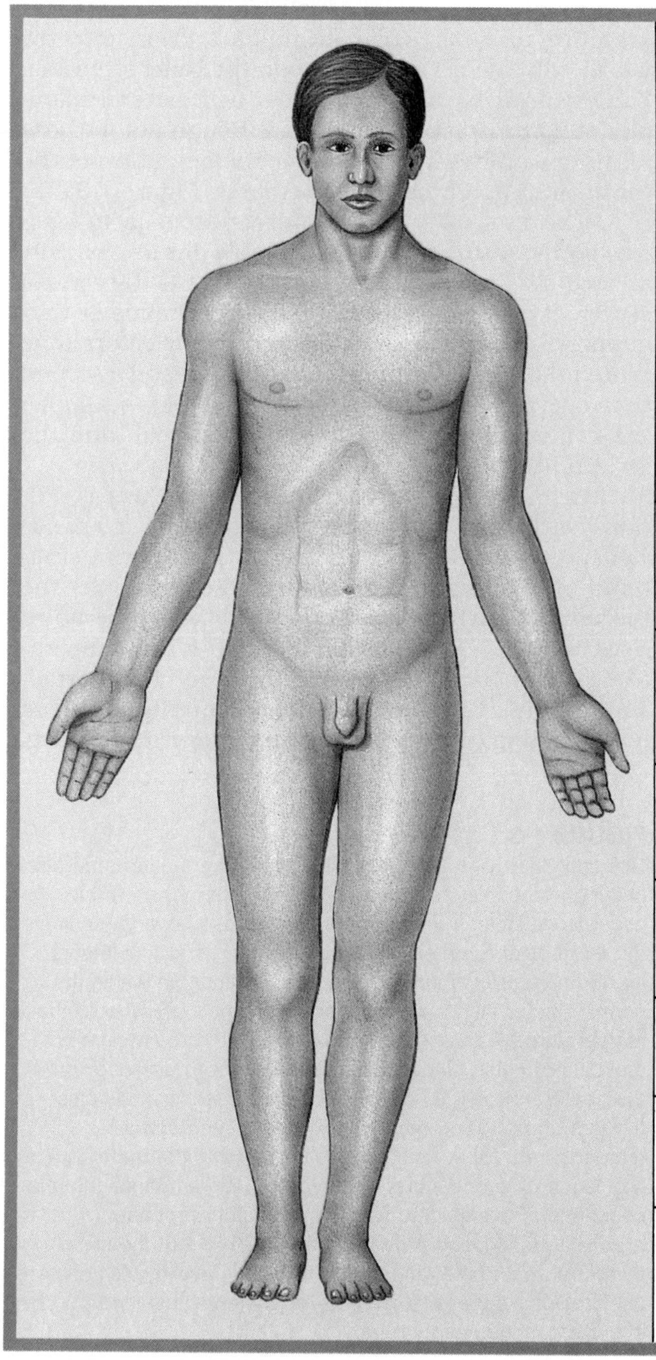

Organ System	Major Functions
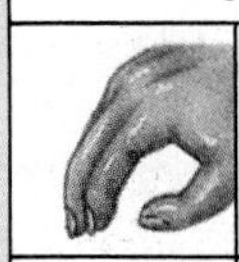 Integumentary system	Protection from environmental hazards, temperature control
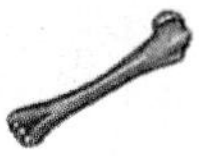 Skeletal system	Support, protection of soft tissues, mineral storage, blood formation
 Muscular system	Locomotion, support, heat production
 Nervous system	Directing immediate responses to stimuli, usually by coordinating the activities of other organ systems
 Endocrine system	Directing long-term changes in the activities of other organ systems
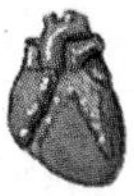 Cardiovascular system	Internal transport of cells and dissolved materials, including nutrients, wastes, and gases
 Lymphatic system	Defense against infection and disease
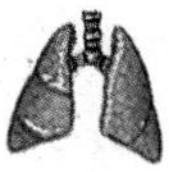 Respiratory system	Delivery of air to sites where gas exchange can occur between the air and circulating blood
 Digestive system	Processing of food and absorption of nutrients, minerals, vitamins, and water
 Urinary system	Elimination of excess water, salts, and waste products
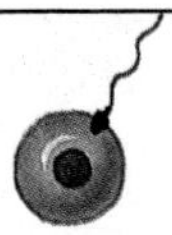 Reproductive system	Production of future generations

in heart muscle cells can disrupt the pattern of contraction and even stop the circulation. Physical damage to the muscle tissue, as in a chest wound, can make the heart ineffective even when most of the heart muscle cells are intact and uninjured. An inherited abnormality in heart structure can make it an ineffective pump, although the muscle cells and muscle tissue are perfectly normal.

Finally, it should be noted that something that affects the *system* will ultimately affect all of its components. For example, the heart cannot pump blood effectively after a massive blood loss, because there may not be enough blood to fill the circulatory system. If the heart cannot pump and blood cannot flow, oxygen and nutrients cannot be distributed. In a very short time, the heart muscle cells will die from oxygen and nutrient starvation, and the tissue will begin to break down.

Of course these changes will not be restricted to the cardiovascular system; all of the cells, tissues, and organs in the body will be damaged. This brings us to another, higher level of organization, that of the entire human being. This level reflects the interactions between organ systems. All are vital; every system must be working properly and in harmony with every other system, or survival will be impossible.

Figure 1-2 introduces the 11 organ systems in the human body and indicates their major functions. Figures similar to this are used throughout the text to summarize the interactions between organ systems.

√ **A histologist investigates structures at what level of organization?**

√ **A researcher studies the factors that cause heart failure. What names might be used for her specialty?**

Homeostasis and System Integration

Organ systems are interdependent, interconnected, and packaged together in a relatively small space. The cells, tissues, organs, and systems of the body live together in a shared environment, like the inhabitants of a large city. City dwellers breathe the city air and eat food from local restaurants; cells in the human body absorb oxygen and nutrients from fluids that surround them. All living cells are in contact with blood or some other body fluid. If the composition of these fluids changes, it will affect cells in some way. For example, suppose there are changes in the temperature or salt content of the blood. The effect on the heart could range from a minor adjustment (heart muscle tissue contracts more often, and the heart rate goes up) to a total disaster (the heart stops beating altogether).

A variety of physiological mechanisms act to prevent potentially disruptive changes in the environment inside the body. The tendency for physiological systems to stabilize internal conditions is called **homeostasis** (hō-mē-ō-STĀ-sis; *homeo-*, unchanging + *stasis*, standing). The term **homeostatic regulation** refers to the adjustments in physiological systems that are responsible for the preservation of homeostasis.

HOMEOSTATIC REGULATION

Homeostatic regulation usually involves a **receptor** sensitive to a particular stimulus and an **effector** whose activity has an effect upon the same stimulus. You are probably already familiar with several examples of homeostatic regulation, although not in those terms. As an example, consider the operation of the thermostat in a house or apartment (Figure 1-3a,b).

The goal of this regulatory system is to keep the temperature within acceptable limits, usually around 72° F (22.2° C). The thermostat is the control center. It consists of a receptor (a thermometer that monitors room temperature) and two effectors (one switch that operates the air conditioner and a second that operates the heater). The principle is simple: the heater turns on if it becomes too cold, and the air conditioner turns on if it becomes too warm.

Regulatory systems such as this one keep conditions within an acceptable range, not at a precise value. Although it may be set at 72° F, minor fluctuations in temperature are ignored. For example, the thermostat might not respond until the temperature rises above 73° F or falls below 71° F. If the temperature rises above 73° F (22.7° C), however, the thermostat responds by turning on the air conditioner. This has a chilling effect, and the room temperature starts

FIGURE 1-3
Control of Room Temperature. (a) A thermostat controls heating and cooling systems to keep temperatures within acceptable limits. If the temperature climbs above those limits, the thermostat turns on the air conditioner, and this lowers the temperature. When the temperature is again within the normal range, the air conditioner is switched off. Should the temperature fall below the normal range, the thermostat will turn on the heater. Once the temperature has returned to normal, the heater will be shut off. In each case the initial stimulus triggers an opposing response that restores normal temperatures. (b) A flow chart that diagrams the thermostatic regulation of temperature. This type of diagram highlights the sequence of events in a complex regulatory process. (c) A graphical description provides information about the room temperature and the range of thermostatic activity. Graphs are useful because they provide a visual summary that incorporates important ranges and values.

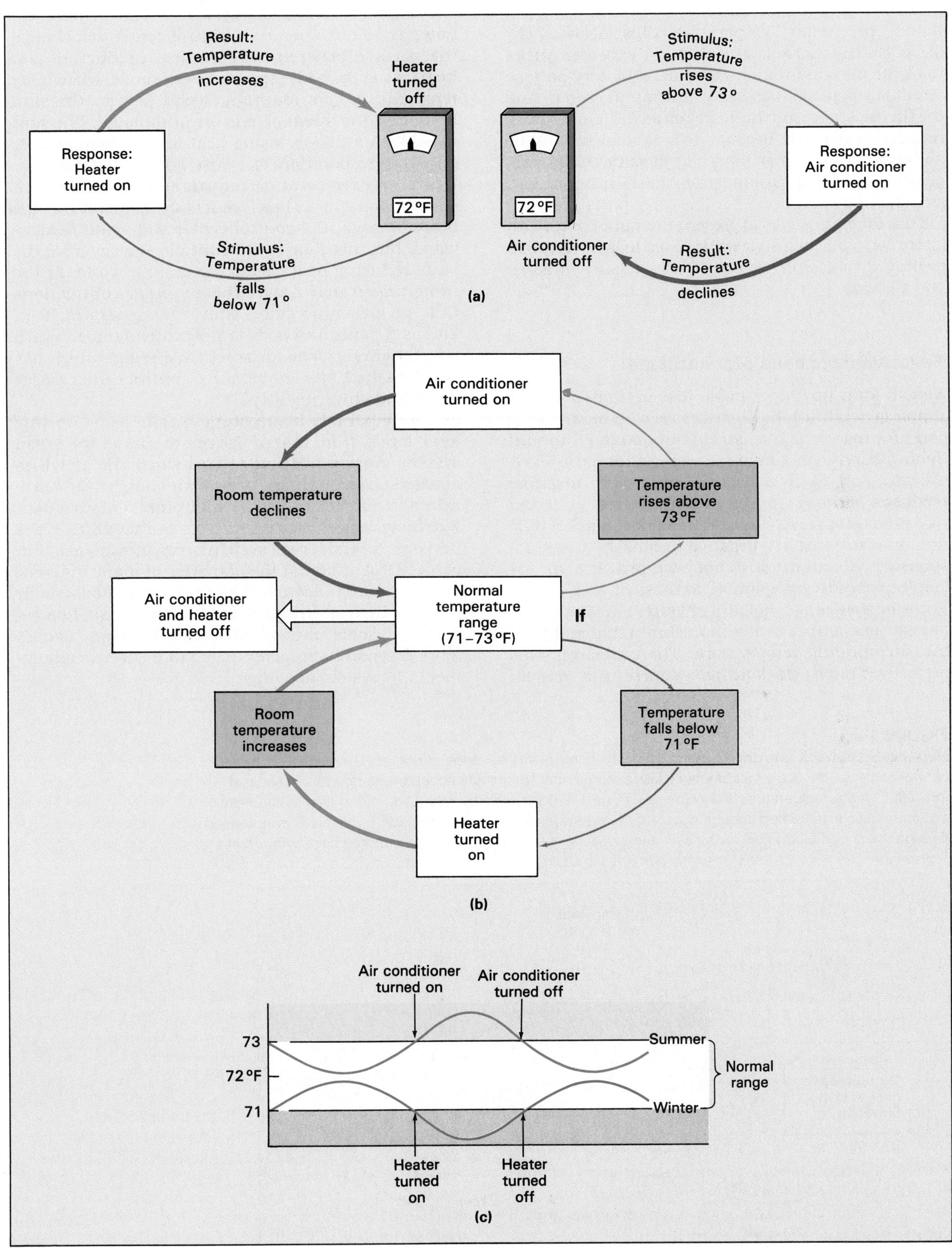
Result:
Temperature
increases
Heater
turned
off
Response:
Heater
turned on
72°F
Stimulus:
Temperature
falls
below 71°
72°F
Air conditioner
turned off
Stimulus:
Temperature
rises
above 73°
Response:
Air conditioner
turned on
Result:
Temperature
declines
(a)
Air conditioner
turned on
Room temperature
declines
Temperature
rises above
73°F
Air conditioner
and heater
turned off
Normal
temperature
range
(71–73°F)
If
Room
temperature
increases
Temperature
falls below
71°F
Heater
turned
on
(b)
Air conditioner
turned on
Air conditioner
turned off
73
72°F
71
Summer
Winter
Normal
range
Heater
turned
on
Heater
turned
off
(c)

to drop. As it approaches 72° F the thermostat shuts the air conditioner off again.

If the room temperature dips below 71° F (21.6° C), the thermostat turns on the heater rather than the air conditioner. But the regulatory pattern remains the same: the heater warms the room, and the thermostat turns the heater off as the temperature returns to normal. Because this system maintains a normal *range* rather than a fixed value, room temperature oscillates around the "ideal" temperature, as indicated in Figure 1-3c. The three parts of Figure 1-3 use different styles to present the same basic regulatory concept; each style will be used in later chapters dealing with the physiology of cells, tissues, organs, and systems.

Negative Feedback Mechanisms

We can sum up the example just presented with a simple generalization. Regardless of whether the temperature rose or fell, *a variation outside of normal limits triggered an automatic response that corrected the situation.* Such a mechanism is called **negative feedback** because the correction involves an action that directly opposes the variation. Most homeostatic mechanisms involve negative feedback. Your responses to fluctuations in body temperature, for example, are quite comparable to those of your house. The similarities are noted in Figure 1-4. The "thermostat" is a control center in the brain that monitors the surrounding temperature. The center remains perfectly content as long as temperatures remain close to 98.6° F (37° C). If body temperature rises above 99° F (37.2° C) or falls below 98° F (36.7° C), however, the temperature control center will change the balance between internal heat production and heat loss at the body surface. For example, when body temperature gets too high, blood flow to the skin increases and sweat glands are stimulated. The skin acts like a radiator, losing heat to the environment, and the evaporation of sweat speeds the process. When body temperature returns to normal, the control center switches itself off. If body temperature dips below normal, the control center will again be activated, but this time it will shift blood away from the skin, reducing heat loss to the outside world. At the same time it may trigger the muscular contractions that we call "shivering." Shivering generates heat, and as it continues body temperature climbs toward normal levels. Once an acceptable temperature has been reached, the temperature control center closes down and shivering stops.

Comparable homeostatic mechanisms operate at all levels, from that of the cell to that of the organ system. **Autoregulation** occurs when the activities of a cell, tissue, organ, or system change automatically when faced with some environmental variation. **Extrinsic regulation** results from the activities of the nervous or endocrine systems, organ systems that can control or adjust the activities of many different systems simultaneously. Extrinsic regulation usually occurs when autoregulation fails to maintain homeostasis in some part of the body. It usually causes more extensive and potentially more effective adjustments in system activities.

FIGURE 1-4
Negative Feedback and the Control of Body Temperature. Body temperature is regulated by a control center in the brain that functions like a thermostat. It normally accepts a temperature range of 98°–99° F. If the temperature falls below 98° F, heat is conserved by restricting blood flow to the skin, and more heat is generated through shivering. If the temperature climbs above 99° F, heat loss is increased through enhanced blood flow to the skin and sweating. In each instance a *variation outside of normal limits triggers an automatic response that corrects the situation.* This mechanism is called negative feedback.

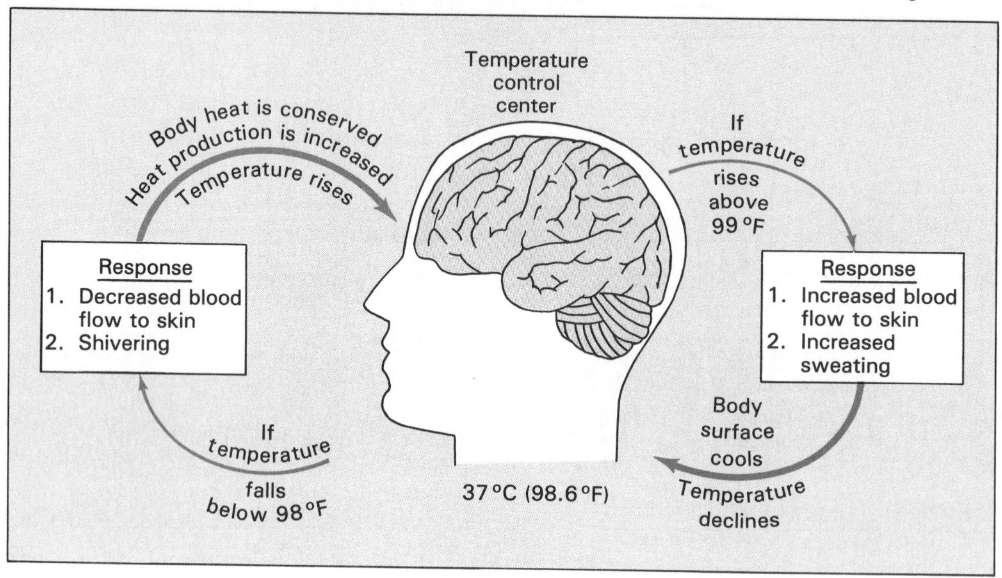

In general, the nervous system performs crisis management by directing rapid, short-term, and very specific responses. When you accidentally set your hand on a hot stove, the rising temperature produces a painful, localized disturbance of homeostasis. The nervous system responds by ordering the contraction of specific muscles that will pull the hand away from the stove. The effects last only as long as the neural activity continues, usually a matter of seconds.

By contrast, the endocrine system releases chemical messengers, called *hormones*, that affect tissues and organs throughout the body. The responses may not be immediately apparent, but when the effects appear they often persist for days or weeks. Examples of endocrine function include the long-term regulation of blood volume and composition and the adjustment of organ system function during starvation or stress. Despite their many functional differences, both systems are usually controlled by negative feedback mechanisms.

Positive Feedback Mechanisms

In a relatively few instances homeostatic regulation involves **positive feedback** mechanisms. In positive feedback *the initial stimulus produces a response that exaggerates the stimulus.* The urge to eat an entire bowl of potato chips after sampling just one might be considered an example of positive feedback in a behavioral sense. In physiological systems, positive feedback occurs in cases where a particular process must proceed swiftly to completion.

One example of positive feedback is diagrammed in Figure 1-5. *Blood clotting* is a chain reaction that occurs when the circulatory system has been damaged. Clotting temporarily repairs damage to the walls of blood vessels and limits the loss of blood. Vessel damage is the important first step that triggers the clotting process. Once it begins to form, the growing blood clot releases chemicals that accelerate the clotting process. As a result, the process rushes to completion and patches the vessel wall.

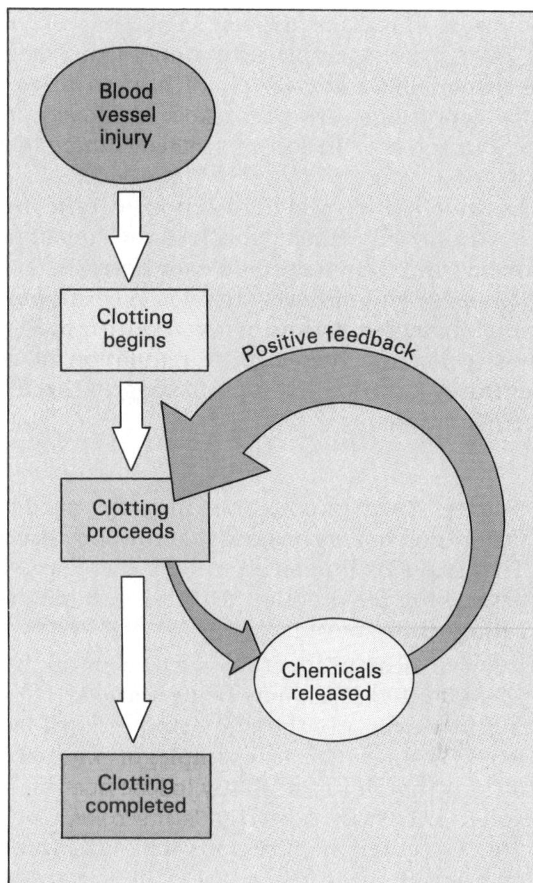

FIGURE 1-5
Positive Feedback. In positive feedback, a stimulus produces a response that reinforces the original stimulus. Positive feedback is important in accelerating processes that must proceed to completion rapidly. In this example, positive feedback enhances the clotting process, which seals breaks in blood vessel walls and prevents blood loss.

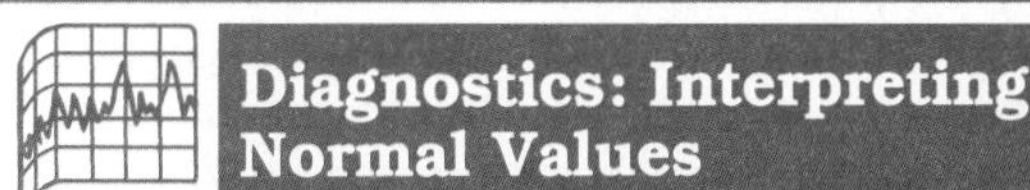

Diagnostics: Interpreting Normal Values

Homeostatic mechanisms normally ignore minor variations that occur within an individual. Moreover, individuals themselves differ in their homeostatic "set points." It is therefore impossible to define "normal" homeostatic conditions. By convention, physiological values are reported either as averages, the average value obtained by sampling a large number of individuals, or as a range that includes 95 percent or more of the sample population. For instance, 5 percent of normal adults have a body temperature outside the "normal" range (below 36.3° C or above 37.1° C). But these temperatures are perfectly normal *for them*, and the variations have no clinical significance. This variability should be kept in mind when reviewing a lab report or clinical discussion, since an unusual value—even one outside the normal range—may represent individual variation rather than a homeostatic malfunction.

HOMEOSTASIS AND DISEASE

Physiological mechanisms do a remarkably good job of maintaining a constant internal environment, regardless of our ongoing activities. But when homeostatic regulation fails, organ systems begin to malfunction and the individual experiences the symptoms of illness, or **disease**. The chapters that follow devote considerable attention to the mechanisms responsible for a variety of human diseases. Specific conditions are presented in boxed form throughout the text, in sections called *Clinical Comment.*

An understanding of normal homeostatic mechanisms can usually enable you to draw conclusions about what might be responsible for observed symptoms. Suppose you discover that you have an abnormally high body temperature. If you return to Figure 1-4 and review the homeostatic regulation of body temperature, you should be able to suggest three possible explanations:

1. Internal heat production may be occurring faster than heat is being lost at the body surface. This happens during strenuous exercise, when active muscles produce heat faster than perspiration can remove it.
2. Internal heat production may be normal, but adequate heat loss may be prevented. Remaining in a sauna bath or wearing a down jacket in the summertime are examples of external conditions that limit the ability to lose heat. In *heat stroke*, otherwise known as "sunstroke," the regulatory center malfunctions and fails to order appropriate heat loss at the body surface. As a result, body temperature climbs to dangerous or even fatal levels.
3. The thermostat may have been reset, so that it accepts a higher temperature as normal. This often occurs in an infection or other illness, when specific chemical messengers in the circulating blood affect the temperature control center to cause the elevated temperature we know as *fever.*

This type of logical analysis makes it possible to ask specific questions and make predictions about normal or abnormal physiological mechanisms. As a result, it also provides a framework for clinical diagnosis and decision making. Subsequent chapters of this book will emphasize major organizational and functional patterns that are relevant to clinical practice.

The analysis just completed is an example of *critical thinking*. In this process information is organized and examined objectively in such a way that conclusions can be evaluated and tested. Critical thinking is essential for understanding science, especially the medical sciences, and this topic is explored further in the Health News section on p. 23. To help you improve your ability to analyze problems and practice critical thinking, each chapter contains several concept questions. These questions, placed near the related text, will help you learn *and apply* important concepts. The review questions at the end of each chapter also include questions that deal with the application of principles to clinical situations.

- √ **Why is homeostatic regulation important to human beings?**
- √ **What happens to the body when homeostasis breaks down?**
- √ **Why is positive feedback helpful in blood clotting but unsuitable for regulation of body temperature?**

A Frame of Reference for Anatomical Studies

If you discovered a new continent, how would you begin collecting information so that you could report your findings? Just strolling around jotting down observations wouldn't be very effective. Comments like, "Over there I saw a tree," or "Monsters found here," won't mean anything to a person who doesn't know where "here" and "there" are. You would have to construct a detailed map of the territory. The completed map would contain (1) prominent landmarks, such as mountains, valleys, or volcanoes; (2) the distance between them; and (3) the direction you traveled to get from one place to another. The distances might be recorded in miles, and the directions recorded as compass bearings (north, south, northeast, southwest, and so on). With such a map you could go back to civilization, make speeches, raise money, and do other things that famous explorers do in their spare time. Even a person who had never heard of *you* could use your map to go directly to a specific location on that continent.

Early anatomists faced similar communication problems. Stating that a bump is "on the back" does not give very precise information about its location. So anatomists created maps of the human body. The landmarks are prominent anatomical structures, and distances are measured in centimeters or inches. Because compass bearings aren't useful, anatomists use special directional terms. In effect, anatomy uses a special language that must be learned almost at the start. It does take some time and effort to develop a working anatomical vocabulary. However, it is absolutely essential if one is to avoid a situation like that depicted in Figure 1-6.

SUPERFICIAL ANATOMY

An understanding of anatomical landmarks and directional references will make subsequent chapters more comprehensible, for with the exception of the integument, none of the organ systems can be seen from the body surface. You must create your own mental maps and extract information from the anatomical illustrations that accompany this discussion.

FIGURE 1-6
The Importance of a Precise Vocabulary. Would you want to be this patient?

Anatomical Landmarks

Important anatomical landmarks are presented in Figure 1-7. The anatomical terms are given in boldface, the common names in plain type, and the anatomical adjectives in parentheses. You should become familiar with all three terms. For example, the term **brachium** refers to the arm, and later chapters will discuss the brachial artery, brachial nerve, and so

FIGURE 1-7
Anatomical Landmarks. The anatomical terms are shown in boldface type, the common names are in plain type, and the anatomical adjectives are in parentheses.

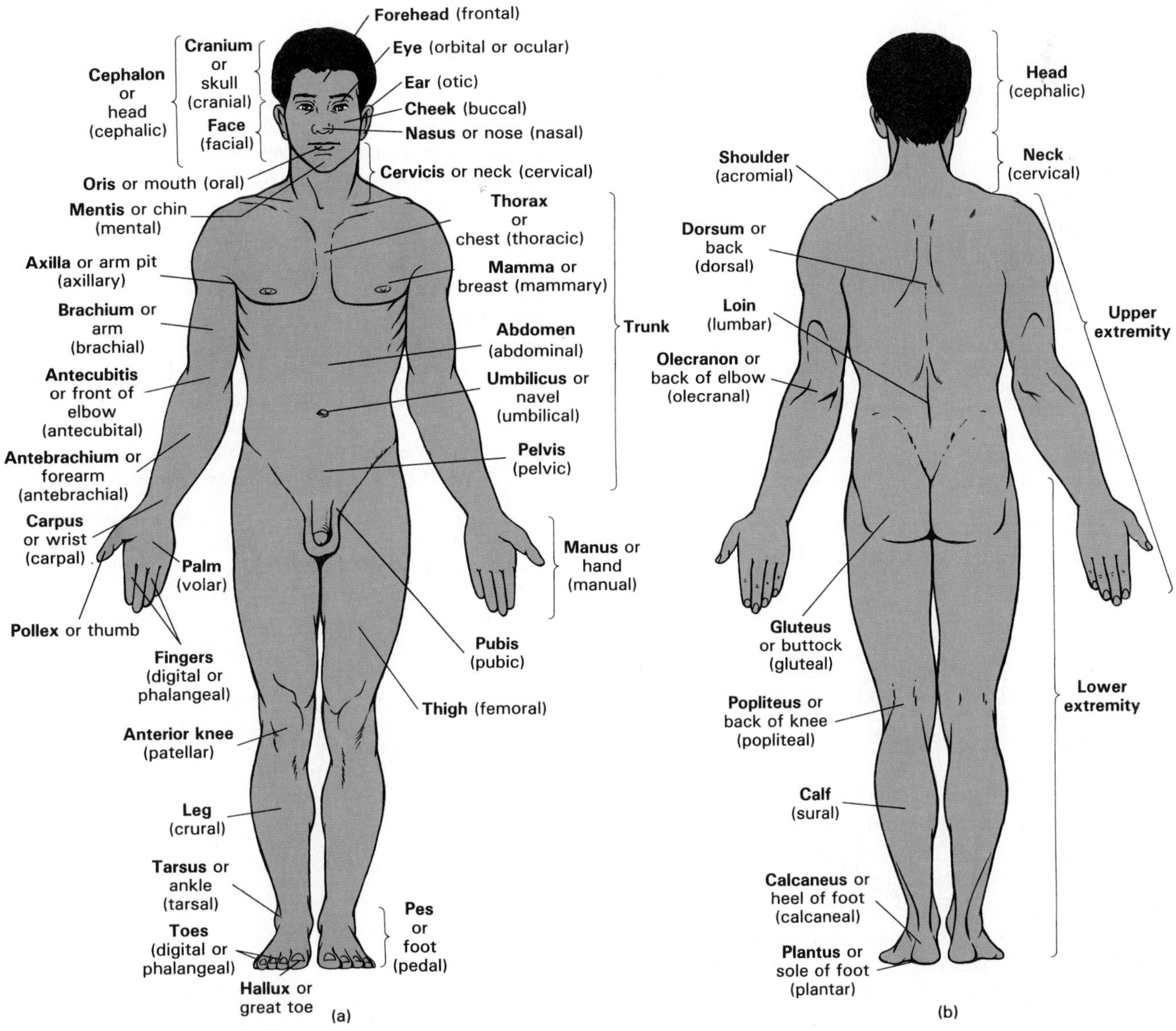

(a)

Cephalon or head (cephalic)
Cranium or skull (cranial)
Face (facial)
Forehead (frontal)
Eye (orbital or ocular)
Ear (otic)
Cheek (buccal)
Nasus or nose (nasal)
Oris or mouth (oral)
Mentis or chin (mental)
Cervicis or neck (cervical)
Axilla or arm pit (axillary)
Brachium or arm (brachial)
Antecubitis or front of elbow (antecubital)
Antebrachium or forearm (antebrachial)
Carpus or wrist (carpal)
Palm (volar)
Pollex or thumb
Fingers (digital or phalangeal)
Thorax or chest (thoracic)
Mamma or breast (mammary)
Abdomen (abdominal)
Umbilicus or navel (umbilical)
Pelvis (pelvic)
Trunk
Manus or hand (manual)
Pubis (pubic)
Thigh (femoral)
Anterior knee (patellar)
Leg (crural)
Tarsus or ankle (tarsal)
Toes (digital or phalangeal)
Hallux or great toe
Pes or foot (pedal)

(b)

Head (cephalic)
Neck (cervical)
Shoulder (acromial)
Dorsum or back (dorsal)
Loin (lumbar)
Olecranon or back of elbow (olecranal)
Upper extremity
Gluteus or buttock (gluteal)
Popliteus or back of knee (popliteal)
Calf (sural)
Calcaneus or heel of foot (calcaneal)
Plantus or sole of foot (plantar)
Lower extremity

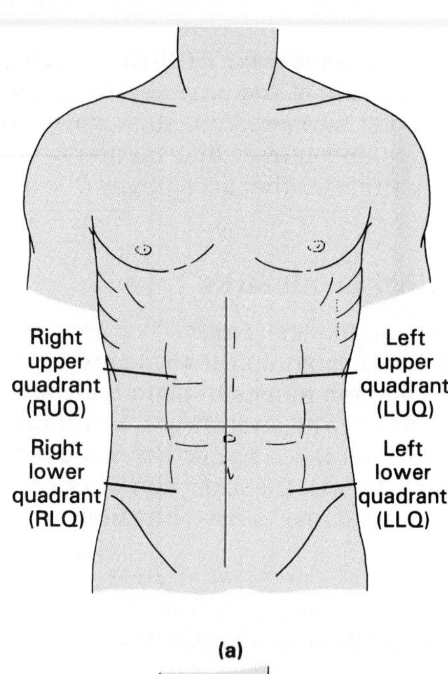

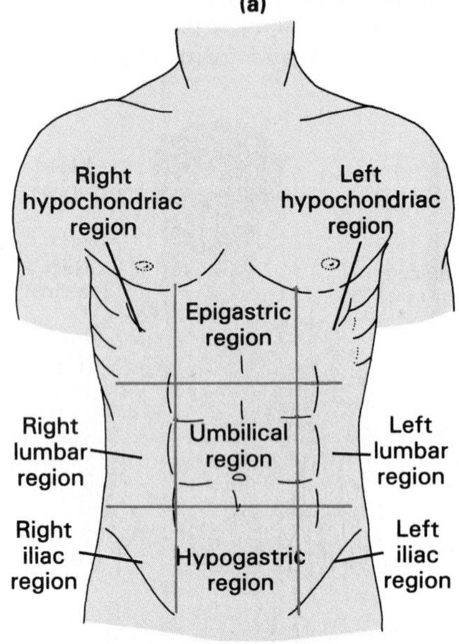

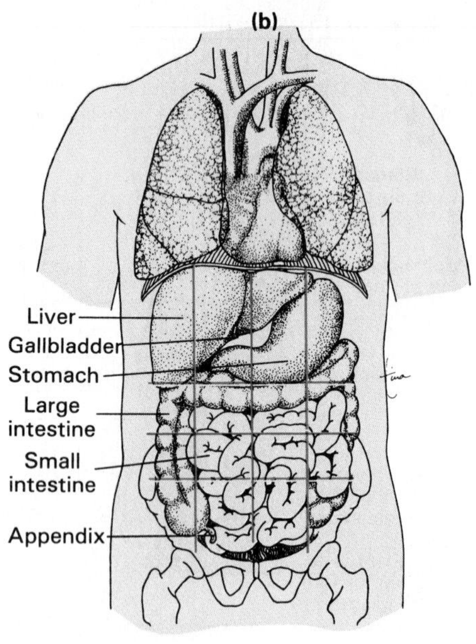

FIGURE 1-8
Abdominopelvic Quadrants and Regions. (a) Abdominopelvic quadrants divide the area into four sections. These terms, or their abbreviations, are most often used in clinical discussions. (b) More precise regional descriptions are provided by reference to the appropriate abdominopelvic region. (c) Quadrants or regions are useful because there is a known relationship between superficial anatomical landmarks and underlying organs.

forth. Understanding the terms and their origins will help you to remember the location of a particular structure, as well as its name.

Standard anatomical illustrations show the human form in the **anatomical position**. In this position the hands are at the sides with the palms facing forward. Figure 1-7 shows an individual in the anatomical position as seen from the front (1-7a) and back (1-7b). A person lying down in the anatomical position is said to be **supine** (SŪ-pīne) when lying face up and **prone** when lying face down.

FIGURE 1-9
Directional References. Important directional terms used in this text are indicated by arrows; definitions and descriptions are included in Table 1-1.

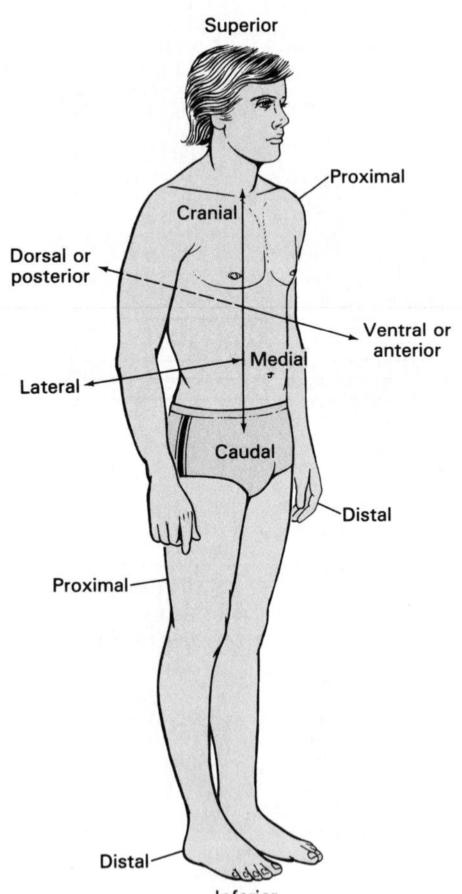

Anatomical Regions

In addition to specific landmarks, anatomists and clinicians often need to use regional terms to describe a general area of interest or injury. Two approaches have developed, both concerned with mapping the surface of the abdominopelvic region. Clinicians refer to the **abdominopelvic quadrants**. The region is divided into four segments using a pair of imaginary lines that intersect at the *umbilicus* (navel). This simple method, shown in Figure 1-8a, provides useful references for the description of aches, pains, and injuries. The location can assist the doctor in deciding the possible cause; for example, tenderness in the right lower quadrant (RLQ) is a symptom of appendicitis, whereas tenderness in the right upper quadrant (RUQ) may indicate gallbladder or liver problems.

Anatomists like to use more precise regional distinctions to describe the location and orientation of internal organs. They recognize nine **abdominopelvic regions** (Figure 1-8b). Figure 1-8c shows the relationship between quadrants, regions, and internal organs.

Anatomical Directions

Table 1-1 and Figure 1-9 show the principal directional terms and examples of their use. There are many different terms, and some can be used interchangeably. Although your instructor may have additional recommendations, the terms that appear frequently in later chapters have been emphasized. When following anatomical descriptions, you may also find it useful to remember that the terms "left" and "right" always refer to the left and right sides of the *subject*, not of the observer.

TABLE 1-1 Regional and Directional References

Region of Body	*Adjective*	*Directional Reference*	*Examples of Descriptive Use*
Front	Ventral or anterior	Ventrally	The navel is on the *ventral* (*anterior*) surface of the trunk.
Back	Dorsal or posterior	Dorsally	The *dorsal* body cavity encloses the brain and spinal cord; moving *dorsally* from the navel you find the muscles of the abdominal wall.
Head	Cranial	Cranially	The *cranial* border of the pelvis; moving *cranially* from the pelvis brings you to the umbilicus.
	Cephalic		(Same usage as cranial)
	Superior		(Same as cranial but you refer to the *superior* surface of the skull.)
	Rostral	Rostrally	(Same as cranial but refers to nose): the eyes are on the *rostral* surface of the head; moving *rostrally* from the back of the skull brings you to the face.
Tail (coccyx)	Caudal	Caudally	The hips are *caudal* to the waist; moving *caudally* from the shoulder brings you to the hips.
	Inferior	Inferiorly	(Same as caudal but the soles are on the *inferior* surfaces of the feet.)
Close to long axis of the body	Medial	Medially	The *medial* surfaces of the thighs may be in contact, moving *medially* across the surface of the chest, you arrive at the sternum.
Away from long axis	Lateral	Laterally	The leg articulates with the *lateral* surface of the pelvis; moving *laterally* from the nose brings you to the eyes.
Toward an attached base	Proximal	Proximally	The wrist is *proximal* to the fingers; moving *proximally* from the wrist brings you to the elbow.
Away from an attached base	Distal	Distally	The fingers are *distal* to the wrist; moving *distally* from the elbow brings you to the wrist.

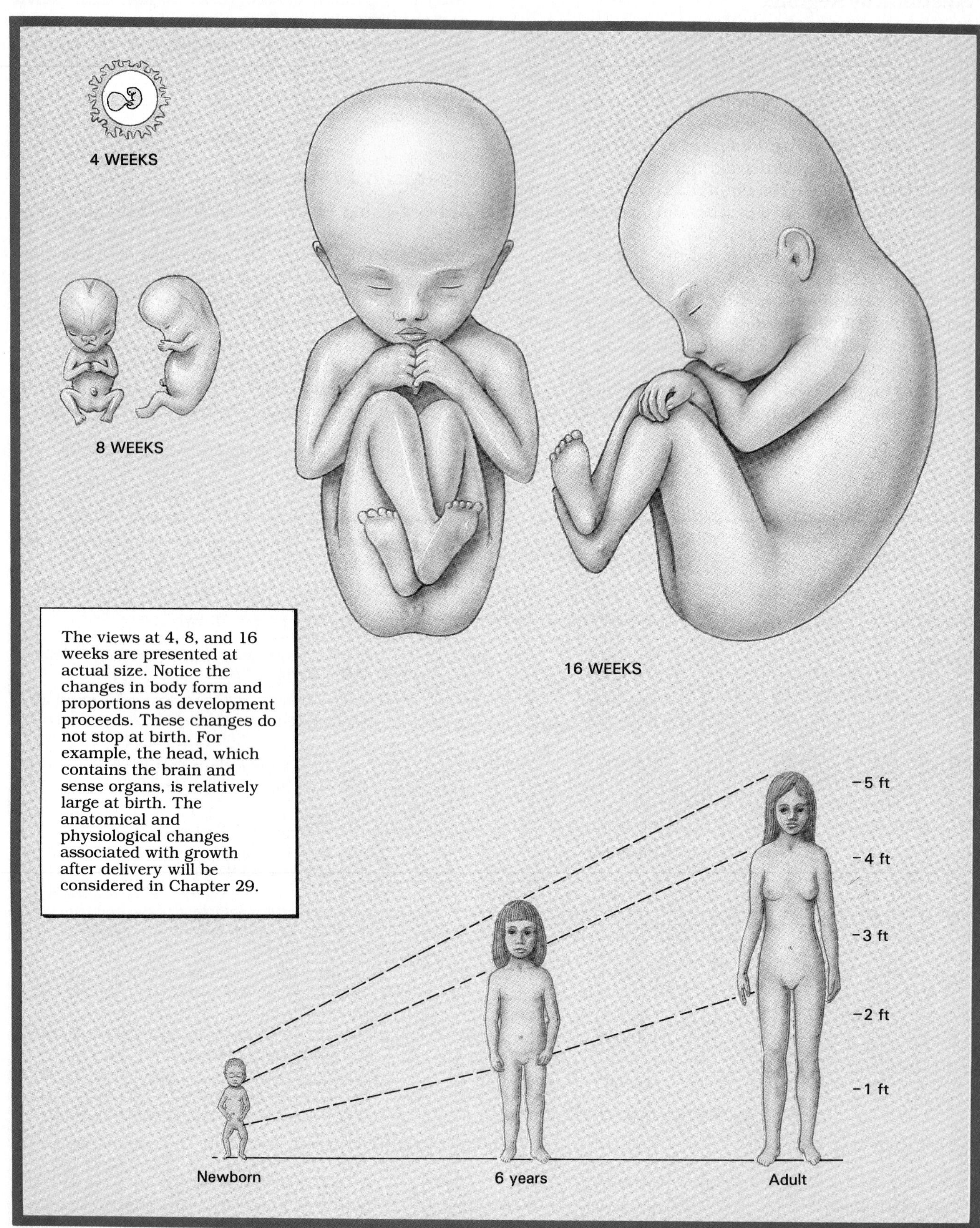

4 WEEKS
8 WEEKS
16 WEEKS
The views at 4, 8, and 16 weeks are presented at actual size. Notice the changes in body form and proportions as development proceeds. These changes do not stop at birth. For example, the head, which contains the brain and sense organs, is relatively large at birth. The anatomical and physiological changes associated with growth after delivery will be considered in Chapter 29.
-5 ft
-4 ft
-3 ft
-2 ft
-1 ft
Newborn
6 years
Adult

Embryology Summary: Growth and changes in body form

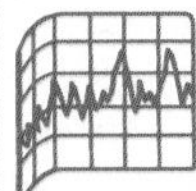

Diagnostics: Anatomy and Observation

Technological innovation has affected modern medicine in many ways, but some of the most important diagnostic procedures have not changed significantly in a thousand years. The next time you visit a physician, pay attention to the way information is obtained. First, your chart is reviewed, giving the doctor information about past or preexisting problems. Then you are asked to describe the reasons for your visit.

As you talk, the doctor not only listens carefully, but watches closely. This is really the first stage in physical assessment, a time for general questions like, "Is the patient moving, speaking, and thinking normally?" The answers will later be integrated with the results of more precise observations.

A physical examination usually follows. There are four basic components to a physical examination:

1. **Inspection:** Inspection is careful observation. A general inspection involves examining body proportions, posture, and patterns of movements. Local inspection is the examination of sites or regions of suspected injury. Of the four components of the physical exam, inspection is often the most important because it provides the largest amount of useful information. Many diagnostic conclusions can be made on the basis of inspection alone; most skin conditions, for example, are identified in this way. A number of endocrine problems and inherited metabolic disorders can produce subtle changes in body proportions that are easily overlooked by the untrained eye.
2. **Palpation:** In palpation the physician uses hands and fingers to feel the body. This procedure provides information on skin texture and temperature, the presence of abnormal tissue masses, the pattern of the pulse, and the location of tender spots. Once again, the procedure relies on an understanding of normal anatomy. A small, soft, lumpy mass in one spot is a salivary gland; in another location it could be a tumor. A tender spot is important in diagnosis only if the observer knows what organs lie beneath it.
3. **Percussion:** Percussion is tapping with the fingers or hand to obtain information about the densities of underlying tissues. For example, the chest normally produces a hollow sound, because the lungs are filled with air. That sound changes in pneumonia, when the lungs contain large amounts of fluid. Of course, to get the clearest chest percussions, the fingers must be placed in the right spots.
4. **Auscultation:** Auscultation (aws-kul-TĀ-shun; *auscultare*, to listen) is listening to body sounds, often using a stethoscope. This technique is particularly useful for checking the condition of the lungs during breathing. The wheezing sound heard in asthma is caused by constriction of the airways, and pneumonia produces a gurgling sound, indicating that fluid has accumulated in the lungs. Auscultation is also important in diagnosing heart conditions. Many cardiac problems affect the sound of the heartbeat or produce abnormal swirling sounds during blood flow.

The entire process of physical examination relies on one fact: the doctor already knows the superficial and deep anatomy of the human body. To detect the abnormal, you must first understand the normal.

SECTIONAL ANATOMY

A presentation in sectional view is sometimes the only way to illustrate the relationships between the parts of a three-dimensional object. An understanding of sectional views has become increasingly important since the development of procedures that enable us to see inside the living body without resorting to surgery.

Planes and Sections

Any slice through a three-dimensional object can be described with reference to three **sectional planes**, indicated in Table 1-2 and Figure 1-10. The **transverse plane** lies at right angles to the long axis of the body, dividing it into **superior** and **inferior** sections. A cut in this plane is called a **transverse section**, or *cross section*. The **frontal**, or **coronal**, **plane**

TABLE 1-2 Terms That Indicate Planes of Section

Orientation of Plane	*Adjective*	*Directional Reference*	*Description*
Parallel to long axis	Sagittal	Sagittally	A *sagittal* section separates right and left portions. You examine a sagittal section, but you section sagitally.
	Midsagittal		In a *midsagittal* section the plane passes through the midline, dividing the body in half and separating right and left sides.
	Parasagittal		A *parasagittal* section misses the midline, separating right and left portions of unequal size.
	Frontal or coronal	Frontally or coronally	A *frontal*, or *coronal*, section separates anterior and posterior portions of the body; coronal usually refers to sections passing through the skull.
Perpendicular to long axis	Transverse or horizontal	Transversely or horizontally	A *transverse*, or *horizontal*, section separates superior and inferior portions of the body.

FIGURE 1-10
Planes of Section. The three primary planes of section are indicated here. Table 1-2 defines and describes them.

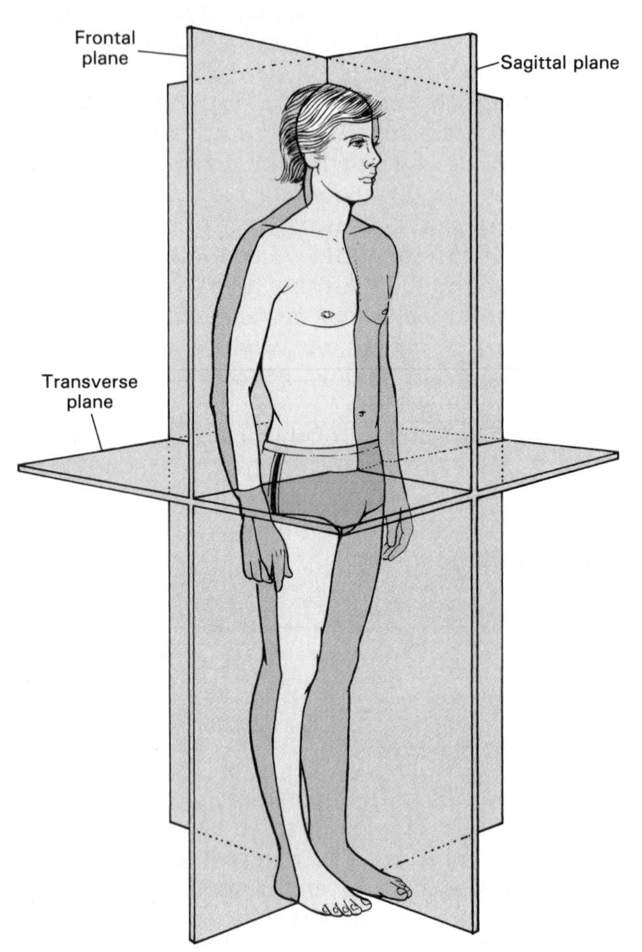

and the **sagittal plane** parallel the long axis of the body. The frontal plane extends from side to side, dividing the body into **anterior** and **posterior** sections. The sagittal plane extends from front to back, dividing the body into *left* and *right* sections. A cut that passes along the midline and divides the body into left and right halves is a **midsagittal section**; a cut parallel to the midsagittal line is a **parasagittal section**.

Sometimes it is helpful to compare the information provided by sections made along different planes. You can experiment with this procedure by mentally sectioning this book, as in Figure 1-11a. (Performing this experiment is not recommended, unless you are dropping the course.) Each sectional plane provides a different perspective on the structure of the book; when combined with observations on the external anatomy, they create a reasonably complete picture.

A more accurate and detailed picture would entail choosing one sectional plane and making a series of sections at small intervals. This process, called **serial reconstruction**, permits the analysis of relatively complex structures. Figure 1-11b shows the serial reconstruction of a familiar object, a piece of elbow macaroni. Serial reconstruction is an important method for studying histological structure and for analyzing the images produced by sophisticated clinical procedures (see Sectional Anatomy and Clinical Technology, below). When anatomical diagrams or clinical procedures present sectional views of the body, the sections are presented as though the observer were standing at the feet and looking toward the head of the subject.

Body Cavities

Viewed in sections, the human body is not a solid object, like a rock, in which all of the parts are fused

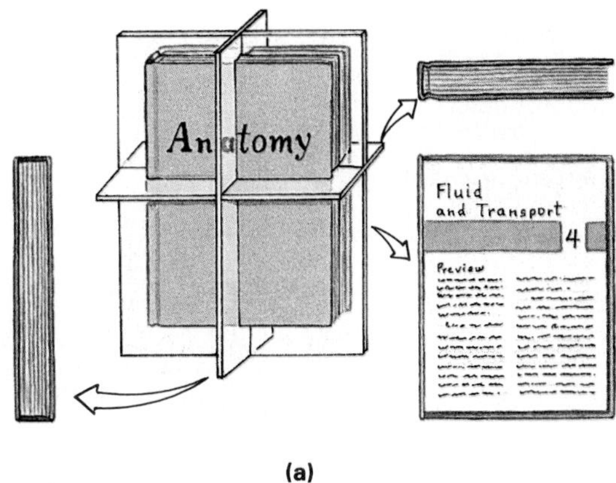

(a)

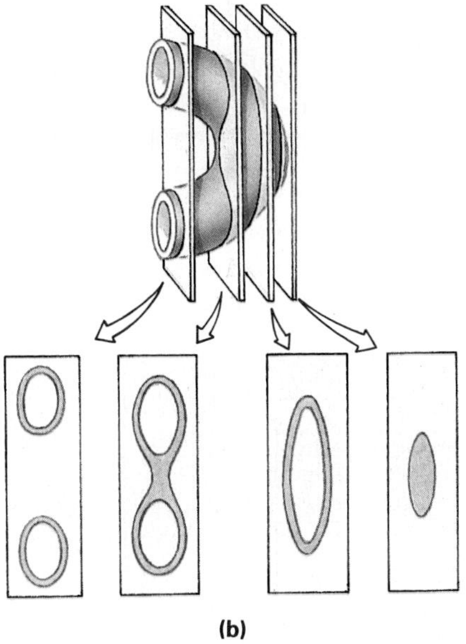

(b)

FIGURE 1-11
Sectional Planes and Visualization. (a) Taking three different sections through a book provides detailed information about its three-dimensional structure. (b) More complete pictures can be assembled by taking a series of sections at small intervals. This process is called serial reconstruction. Notice how the sectional views change; although a simple tube, a piece of elbow macaroni can look like a pair of tubes, a dumbbell, an oval, or a solid, depending on where the section was taken. The effects of sectional plane should be kept in mind when looking at slides under the microscope.

together. Many vital organs are suspended in internal chambers called *body cavities*. These cavities have two essential functions: (1) they protect delicate organs, such as the brain and spinal cord, from accidental shocks, and cushion them from the thumps and bumps that occur during walking, jumping, and running; and (2) they permit significant changes in the size and shape of visceral organs. For example, because they are situated within body cavities, the lungs, heart, stomach, intestines, urinary bladder, and many other organs can expand and contract without distorting surrounding tissues and disrupting the activities of nearby organs.

Two body cavities form during embryonic development. A **dorsal body cavity** surrounds the brain and spinal cord, and a much larger **ventral body cavity**, or **coelom** (SĒ-lom; *koila*, cavity), surrounds developing organs of the respiratory, cardiovascular, digestive, urinary, and reproductive systems. Relationships between the dorsal and ventral body cavities and their various subdivisions can be see in Figure 1-12.

Dorsal Body Cavities The dorsal body cavity (Figure 1-13a) is a fluid-filled space whose limits are established by the **cranium**, the bones of the skull that surround the brain, and the spinal vertebrae. The dorsal body cavity is subdivided into the **cranial cavity**, which encloses the brain, and the **spinal cavity**, which surrounds the spinal cord.

Ventral Body Cavities As development proceeds, internal organs grown and change their relative positions. These changes lead to the subdivision of the ventral body cavity. The formation of the **diaphragm** (DĪ-a-fram), a flat muscular sheet, divides the ventral body cavity into a superior **thoracic cavity**, enclosed by the chest wall, and an inferior **abdominopelvic cavity**, enclosed by the abdomen and pelvic girdle.

By the time of birth, the thoracic cavity has been further subdivided into two **pleural cavities**, each containing a lung, and a **pericardial cavity** that surrounds the heart. Figure 1-13 shows the anatomical relationships of these compartments.

A large central mass of connective tissue, the **mediastinum** (mē-dē-as-TĪ-num or mē-dē-AS-ti-num), surrounds the pericardial cavity and separates the two pleural cavities. In addition to the heart and pericardial cavity, the mediastinum surrounds the trachea, esophagus, and the large arteries and veins attached to the heart.

The heart projects into the pericardial cavity like a fist pushing into a balloon. The base of the heart, corresponding to the wrist, is embedded in the mediastinum, as is the outer wall of the balloon, or **pericardial sac**.

The abdominopelvic cavity, also known as the **peritoneal** (per-i-tō-NĒ-al) **cavity**, has two subdivisions. The **abdominal cavity** extends from the inferior surface of the diaphragm to an imaginary line drawn from the inferior surface of the lowest spinal vertebra to the anterior and superior margin of the pelvic girdle. The portion of the peritoneal cavity inferior to this imaginary line is the **pelvic cavity**.

Sectional Anatomy and Clinical Technology

The use of clinical radioisotopes in scanning procedures is described in Chapter 2. The term **radiologi-**

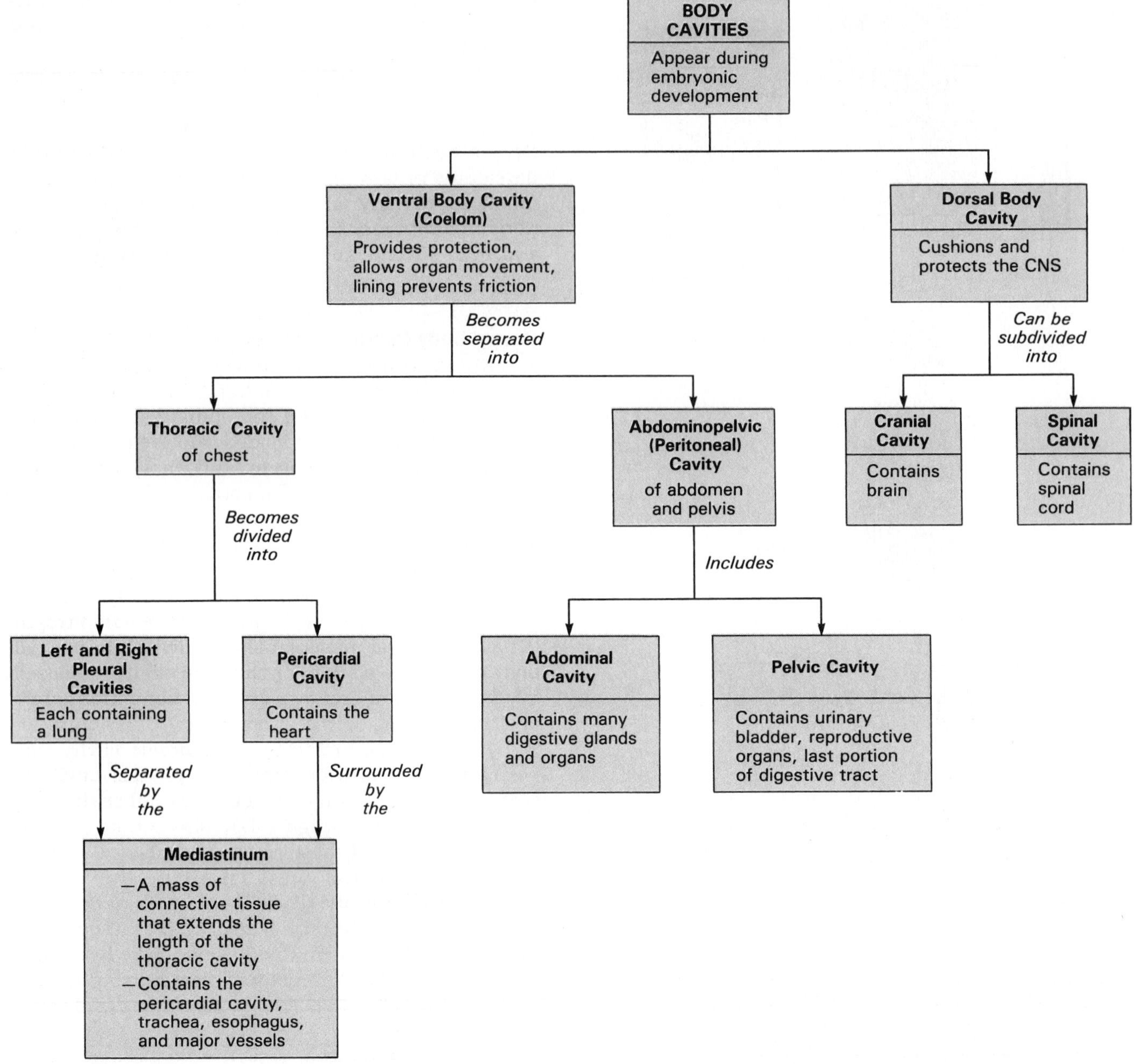

FIGURE 1-12
Relationships of the Various Body Cavities.

FIGURE 1-13
Body Cavities. (a) The dorsal body cavity is bounded by the bones of the skull and vertebral column. The muscular diaphragm divides the ventral body cavity into a superior thoracic cavity and an inferior abdominopelvic cavity. The pericardial cavity is located inside the chest cavity. The heart is suspended within the pericardial cavity like a fist pushed into a balloon. The attachment site, corresponding to the wrist of the hand in the model, lies at the connection between the heart and major blood vessels. (b) An anterior view of the ventral body cavity, showing the central location of the pericardial cavity within the chest cavity. The relationships are seen more clearly in the sectional plane, which shows how the mediastinum divides the thoracic cavity into two pleural cavities.

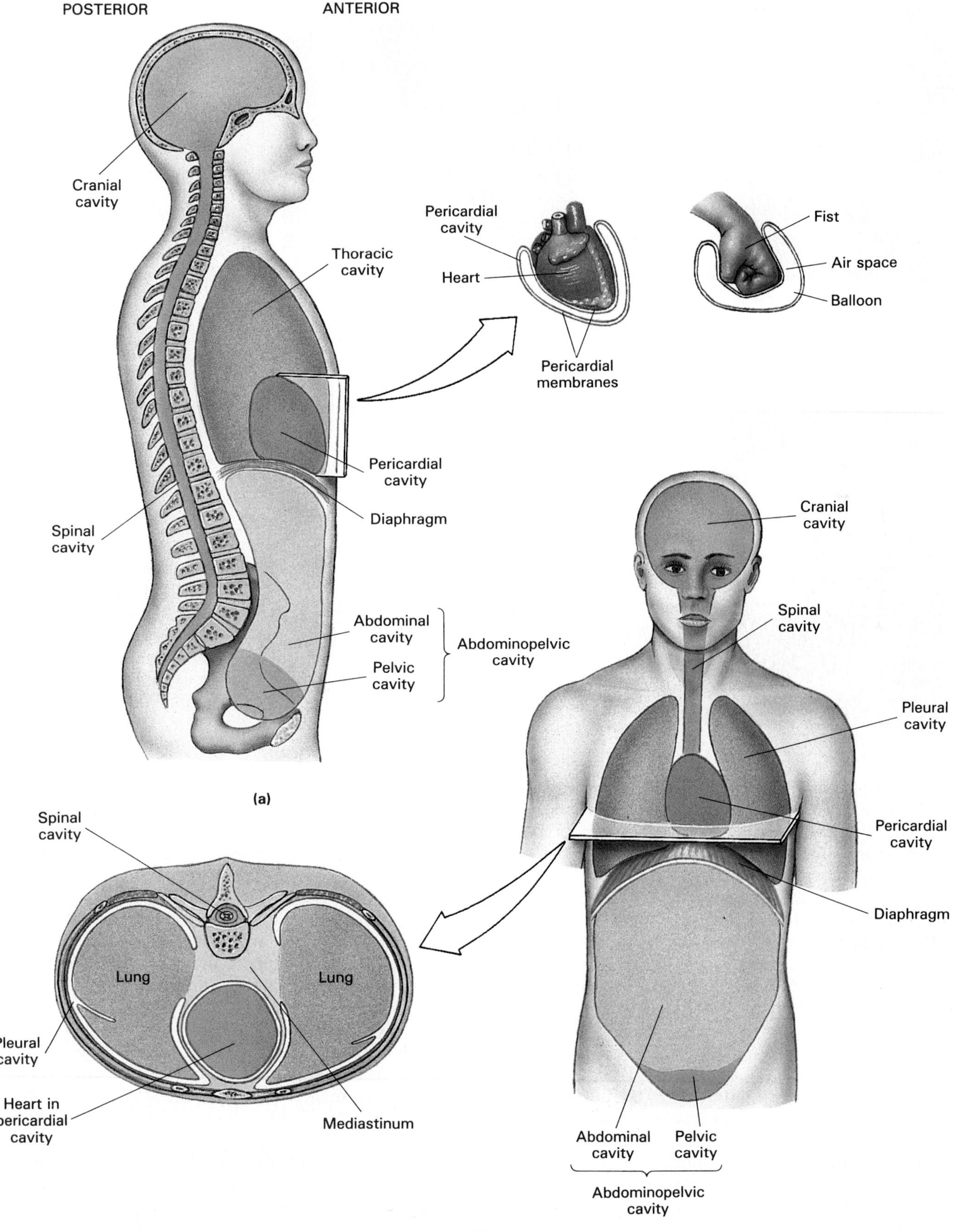
POSTERIOR
ANTERIOR
Cranial cavity
Thoracic cavity
Pericardial cavity
Heart
Pericardial membranes
Fist
Air space
Balloon
Pericardial cavity
Diaphragm
Spinal cavity
Abdominal cavity
Pelvic cavity
Abdominopelvic cavity
(a)
Cranial cavity
Spinal cavity
Pleural cavity
Pericardial cavity
Diaphragm
Spinal cavity
Lung
Lung
Pleural cavity
Heart in pericardial cavity
Mediastinum
Abdominal cavity
Pelvic cavity
Abdominopelvic cavity
(b)

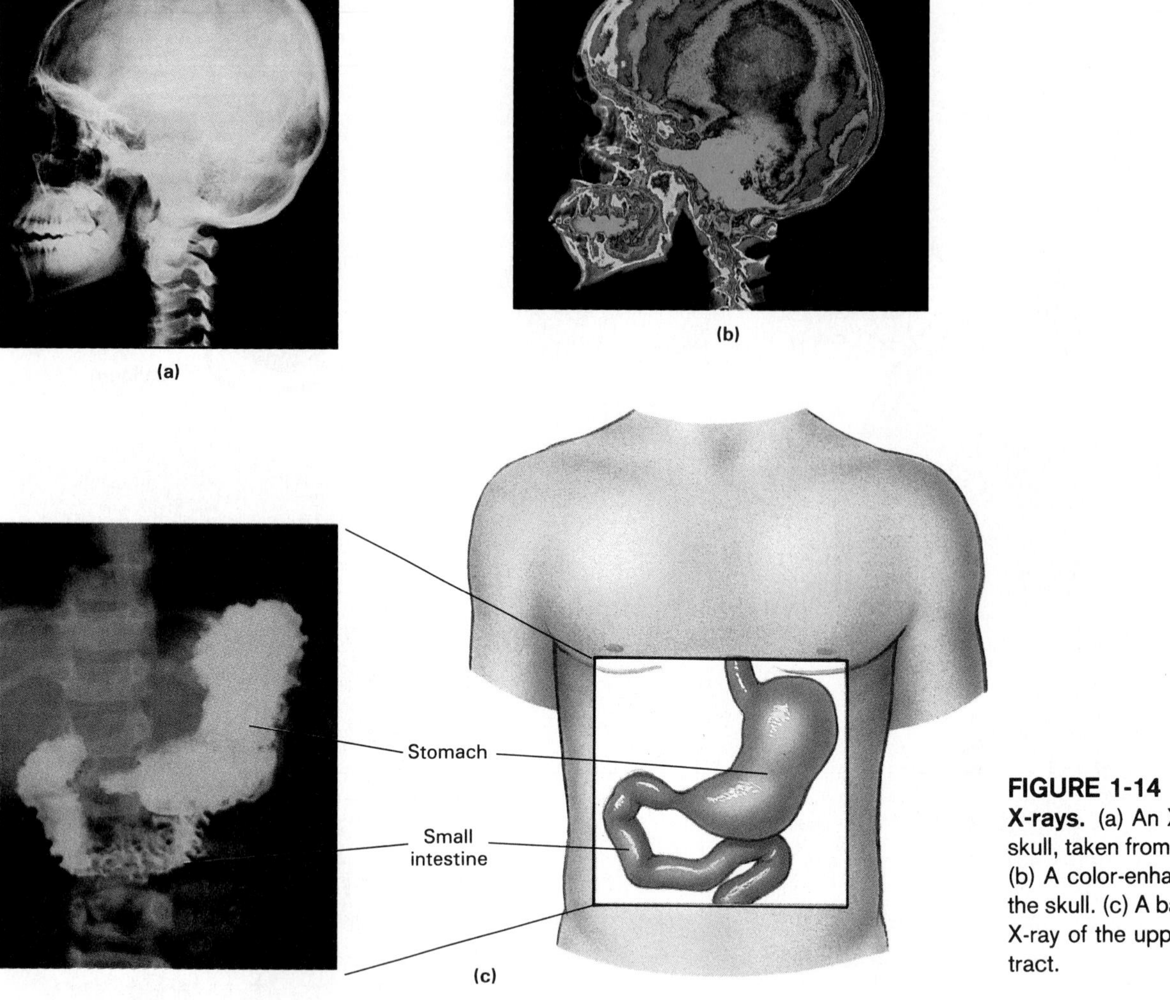

FIGURE 1-14
X-rays. (a) An X-ray of the skull, taken from the right side. (b) A color-enhanced X-ray of the skull. (c) A barium-contrast X-ray of the upper digestive tract.

cal procedures includes not only those scanning techniques that involve radioisotopes but also methods that employ radiation sources outside the body. Physicians who specialize in the performance and analysis of these procedures are called **radiologists**. Radiological procedures can provide detailed information about internal systems. Figures 1-14 and 1-15 compare the views provided by several representative techniques. These figures include examples of X-rays, CT scans, MRI scans, and ultrasound images. Other examples of clinical technology will be found in later chapters.

X-rays **X-rays** are a form of high-energy radiation that can penetrate living tissues. In the most familiar procedure, a beam of X-rays travels through the body and strikes a photographic plate. All of the projected X-rays do not arrive at the film; some are absorbed or deflected as they pass through the body. When X-rays strike something in their path, the higher the atomic weight of the component atoms, the more difficult it is for X-rays to penetrate. The resistance to X-ray penetration is called **radiodensity**. Tissues with a high radiodensity block X-ray penetration, and those with a low radiodensity are essentially transparent to X-rays. In the human body, the order of increasing radiodensity is as follows: air, fat, liver, blood, muscle, bone. Where X-rays can penetrate easily, they strike the photographic plate and expose it; where they cannot penetrate at all, the photographic plate remains unexposed. When the glass plate or film is developed, any unexposed areas are washed clean of photosensitive chemicals and become transparent. Photosensitive chemicals struck by X-rays remain, creating dark areas. The result is a *negative* image, where radiodense tissues, such as bone, appear white, and other, less dense tissues are gray to black.

X-ray images can be produced relatively quickly and easily, compared with many other imaging procedures. Many physicians and dentists have X-ray machines in their offices. They are most often used to differentiate between bruises and broken bones. It takes time and effort to learn to read an X-ray image with confidence, because the picture is a two-dimensional image of a three-dimensional object. For example, the X-ray in Figure 1-14a is a view through the

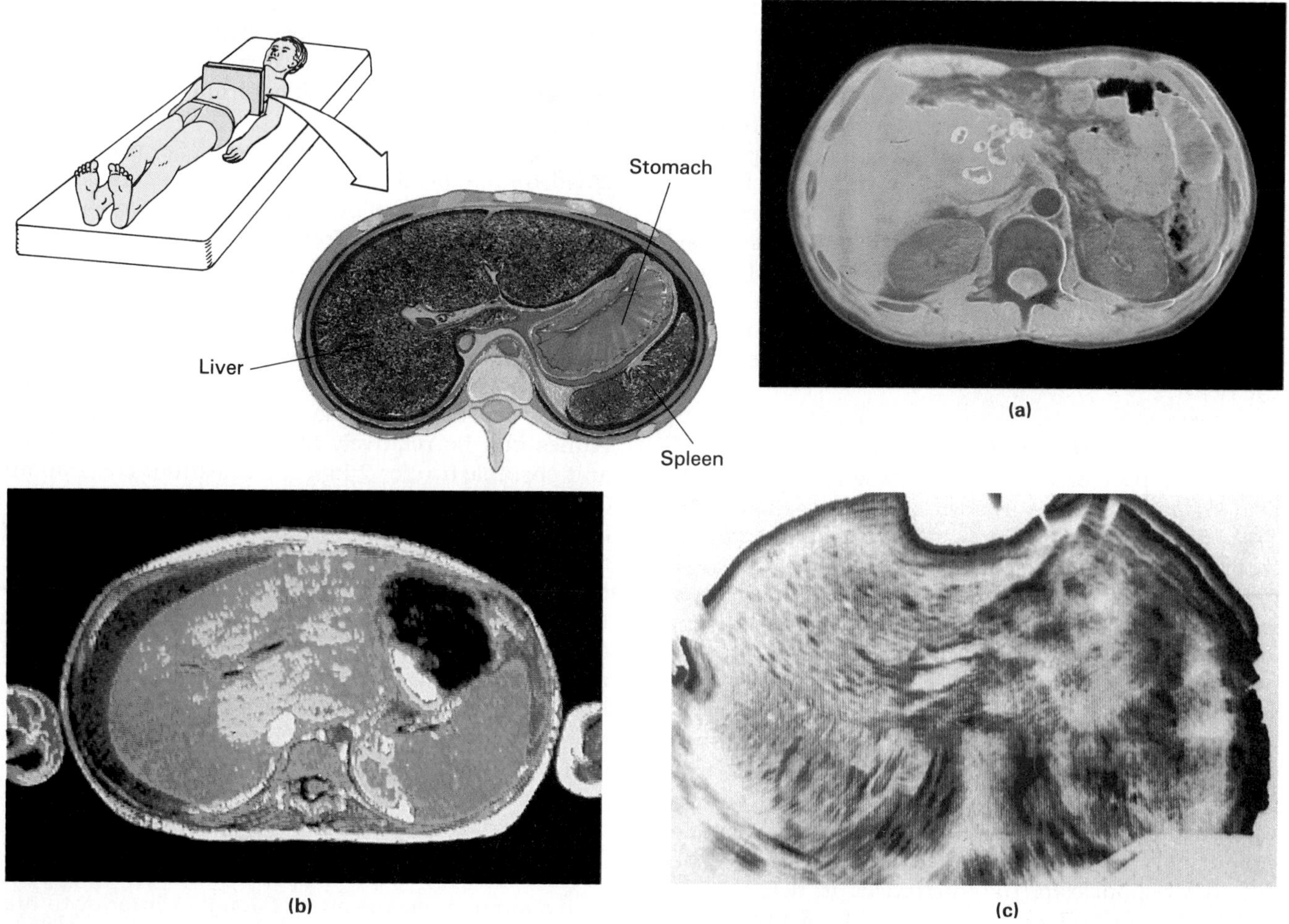

FIGURE 1-15
Scanning Techniques. (a) A color-enhanced CT scan of the abdomen. (b) A color-enhanced MRI scan of the same region. Note the difference in densities of the same organs when viewed in MRI versus CT scans. Depending on the goals of the scan, one procedure or the other may have advantages. (c) An ultrasound scan of the abdomen. The picture is not as crisp as a CT or an MRI scan, but no computer or processing time was required to generate this image on a video screen.

intact skull, as if it had been set on the floor and squashed flat. As a result, it can be difficult to decide whether a particular feature is on the left side (toward the viewer) or on the right side (away from the viewer).

When the goal is the visualization of soft tissues, such as the digestive tract or circulatory system, radiodense dyes or chemical solutions can be used to increase contrast and to highlight special features. For example, to check for ulcers or other stomach or upper digestive tract disorders, X-rays are taken after the patient drinks large quantities of a solution containing barium ions. Barium is very radiodense, and the contours of the gastric and intestinal lining can be seen outlined against the white of the barium solution, as in Figure 1-14c. For examining the large intestine, barium enemas are used instead of "barium milkshakes." To monitor circulatory pathways, radiodense dyes are injected into the circulatory system. This procedure produces an X-ray image known as an **angiogram**.

CT Scans Creating X-ray images simply involves beaming X-rays at a photographic plate. **CT** (**C**omputerized **T**omography), formerly called **CAT** (**C**omputerized **A**xial **T**omography), uses computers to reconstruct sectional views. In this procedure, a single X-ray source rotates around the body. Instead of striking a photographic plate, the X-ray beam strikes a sensor monitored by the computer. The source completes one revolution around the body every few seconds; it then moves a short distance and repeats the process. By comparing the information obtained at each point in the rotation, the computer reconstructs the three-dimensional structure of the body. The result is an image called a **CT scan** that shows a section through the body.

Figure 1-15a shows a color-enhanced CT scan through the abdomen. The color enhancement was done for visual effect; it has little diagnostic value, and radiologists generally prefer to work from sheets of black-and-white film that resemble X-rays. CT scans show three-dimensional relationships and soft tissue structure more clearly than do X-rays. These scans are often used to check for tumors or other tissue abnormalities that would be difficult or impossible to see in a standard X-ray. Another advantage is that the computer can be ordered to perform more complicated reconstructions. The image in Figure 1-15b is a cross section through the body; however, the computer could also generate a sagittal section or even a three-dimensional image using the information provided by a series of scans. CT scans are scattered throughout the text. For example, Figure 20-9 contains several three-dimensional CT scans of the heart.

MRI Scans **MRI** (**M**agnetic **R**esonance **I**maging) surrounds part or all of the body with a magnetic field about 3000 times as strong as that of the earth. This field affects protons within atomic nuclei throughout the body, which line up along the magnetic lines of force like compass needles in the earth's magnetic field. These well-aligned protons are then exposed to brief pulses of radio waves. When struck by a radio wave of the proper frequency, a proton will absorb energy. When the pulse ends, that energy is released, and the source of the radiation is detected by the MRI computers. Each element differs in terms of the radio frequency required to affect its protons. As a result, subtle structural differences can be detected. Figure 1-15b shows an MRI scan of the abdominal region. Note the differences in density between this image, the CT scan, and the X-ray.

This procedure, until recently called **NMR** (**Nu**clear **M**agnetic **R**esonance) has a number of advantages over imaging techniques that use X-rays. For one thing, radio waves are relatively harmless compared with high-energy X-rays. Moreover, by "tuning in" to the frequencies of different elements, MRI can be used to obtain anatomical and physiological information not detectable in CT scans or X-rays.

Ultrasound Compared with other imaging techniques, **ultrasound** is relatively simple and inexpensive. A small transmitter contacting the skin broadcasts a brief, narrow burst of high-frequency sound and then picks up the echoes. The sound waves are reflected by internal structures, and a picture, or **echogram**, can be assembled from the pattern of echoes.

As you can see from Figure 1-15c, the primary disadvantage of ultrasound is that the images lack the striking clarity of other procedures. On the positive side, no adverse affects have been attributed to the sound waves, and fetal development can be monitored without a significant risk of birth defects. Special methods of transmission and processing permit structural analysis of the beating heart, without the complications that can accompany dye injections. Equally important is the fact that ultrasound machines can be relatively inexpensive (under $9000) and portable (under 2 kg). A CT scanner costs about $1 million, and an MRI device nearly twice that amount; neither could be housed in the average living room.

This chapter provided an overview of the locations and functions of the major components of each organ system. It also introduced the vocabulary needed to follow more detailed anatomical descriptions in later chapters. Many of the figures in those chapters contain images produced by the procedures outlined above.

- √ **What type of section would separate the two eyes?**
- √ **If a surgeon makes an incision just inferior to the diaphragm, what body cavity will be opened?**
- √ **Ordinary X-rays do not show the structure of the brain, but MRI scans do. Why?**

This concludes our preview of the major topics covered in this text and the underlying concepts that will guide our discussions. The next four chapters will take you on a tour of the principal levels of organization, from individual atoms to individual human beings.

Health News: The Scientific Method

A great deal of confusion and misinformation exists about just how medical science "works," and people make unwise and even dangerous decisions as a result. Nowhere is this more apparent than when a discussion drifts around to health, nutrition, and cancer. If you are going to be working in a health-related profession, or are even just trying to make sound decisions about your own life, you must learn how to organize information, evaluate evidence, and draw logical conclusions.

Forming a Hypothesis

There is a lot more to science than the collection of information. You could spend the rest of your life carefully observing the world around you, but this won't reveal very much unless you can see some kind of pattern and come up with an idea, or **hypothesis**, that explains your observations.

Hypotheses are ideas that may be correct or incorrect; to evaluate one, you must have relevant data and a reliable method of data analysis. For example, you could propose the hypothesis that radiation emitted by planet X produces immortality. Could anyone prove you wrong? Not very likely, particularly if you didn't specify the location of the planet or the type of radiation. Would anyone believe you? If you were a "leading authority" on something (anything) a few probably would.

That's not as ridiculous as it might seem. For almost 1500 years "everyone knew" that inhaled air went from the lungs through blood vessels to the heart. They knew this because Galen, the famous Roman physician, had said so. Since he was right in several other respects, all of his statements were accepted as true, and contrary opinions were held in low esteem. To avoid making this kind of error, you must always remember to evaluate the hypothesis, not the individual who proposed it!

The evaluation process examines the hypothesis to see if it makes correct predictions about the real world. The steps in this process are diagrammed in Figure 1-16. A valid hypothesis will have three characteristics: it will be *testable*, *unbiased*, and *repeatable*.

A **testable** hypothesis is one that can be studied by experimentation or data collection. Your assertion concerning planet X qualifies as a hypothesis, but it cannot be tested unless we

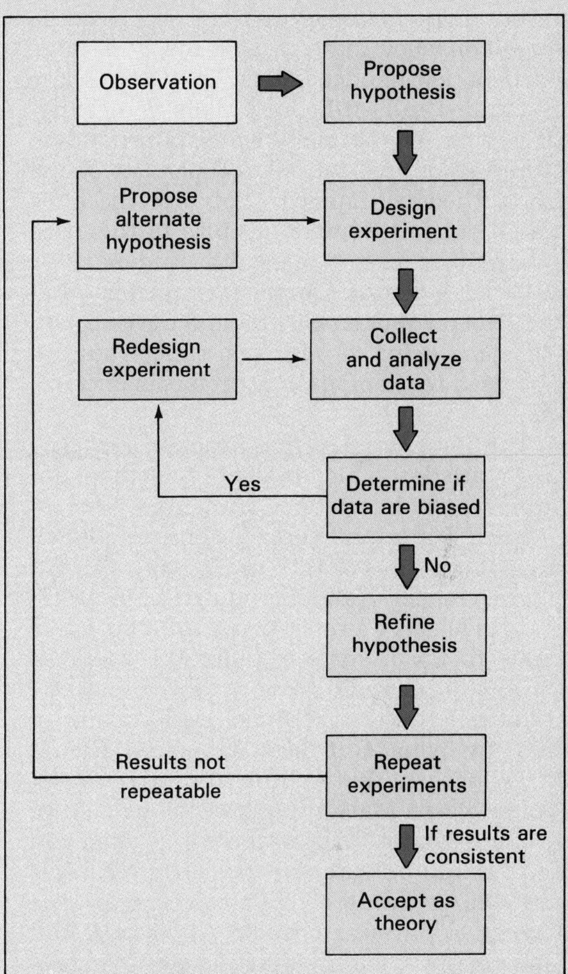

FIGURE 1-16
The Scientific Method. The basic sequence of steps involved in the development of a scientific theory.

find the planet and detect the radiation. An example of a testable hypothesis would be "left-handed airplane pilots have fewer crashes." That's a testable theory because it makes a prediction about the world that can be checked, in this case by collecting and analyzing data.

Avoiding Bias

Suppose, then, that you went out and collected information about all of the crashes in the world, and discovered that 80 percent of all crashed airplanes were flown by right-handed pilots. "Aha!" you might shout, "The hypothesis is correct!" The implications are obvious: ban all right-handed airline pilots, eliminate four-fifths of all crashes, and sit back and wait for your prize from the Air Traffic Safety Association.

Unfortunately, you would be acting prematurely, for your data collection was **biased**. To test your hypothesis adequately, you need to know not only how many crashes involved right-handed or left-handed pilots, but how many right-handed and left-handed pilots were flying. If 90 percent of the pilots were right-handed, but they accounted for only 80 percent of the crashes, then left-handed pilots are the ones to watch out for! Eliminating bias in this case is relatively easy, but in health studies there may be all kinds of complicating factors. Because 25 percent of us will probably develop cancer at some point in our lives, cancer studies will be used to exemplify the problems encountered.

The first example of bias in action concerns cancer statistics, which indicate that there are definite regional variations in cancer rates in the United States and abroad. For example, although the estimated U.S. yearly cancer death rate was 199 per 100,000 population in 1987, the rate in Alaska was only 82 per 100,000, while the rate in the District of Columbia was 315 per 100,000. It would be very easy to assume that this difference is the direct result of country versus city living. But these data alone should not convince you that moving from the District of Columbia to Alaska will lower your risk of developing cancer. To draw that conclusion you would have to be sure that the observed rates were the direct result of just a single factor, the difference in physical location. As you will find in later chapters, many different factors can promote cancer development. To exclude all possibilities other than geography, you would have to be certain that the populations were alike in all other respects. Here are a few possible sources of variation that could affect that conclusion:

- *Different population profiles*: Cancer rates vary between males and females, among racial groups, and among age groups. Therefore, we need to know how the populations of Alaska and the District of Columbia differ in each respect.
- *Different occupations*: Because chemicals used in the workplace are implicated in many cancers, we need to know how the populations of each region are employed and what occupational hazards they face.
- *Different mobilities*: Because the region in which a person dies may not be the same as the one in which he or she lived and developed cancer, we need to know whether people with cancer in Alaska stay in the state or go elsewhere for critical care, and whether people with cancer travel to the District of Columbia to seek treatment at special clinics.
- *Different health care*: Since cancer death rates reflect differences in patterns of health care, we need to know whether residents of Alaska pay more attention to preventive health care and have more regular checkups, whether their medical facilities are better, and whether they devote a larger proportion of their annual income to health services than do residents in the District of Columbia.

You can probably think of additional factors, but the point is that avoiding experimental bias can be quite difficult!

A second example of the problem of bias comes from the motley collection of "miracle cures" that continue to appear and disappear at regular intervals. Pyramid power, pendulum power, crystals, magnetic energy fields, and psychic healers come and go in the news. Miraculous drugs are equally common, whether they are "secret formulas" or South American plant extracts discovered by Mayan colonists from other planets. The proponents of each new procedure or drug report glowing successes with patients who would otherwise have surely succumbed to the disease. And of course all of these remedies are said to have been suppressed or willfully ignored by traditional therapists.

Even accepting that the claims aren't exaggerated, does it prove anything that 1 or 100 or even 1000 patients have been cured? No, it doesn't, for a list of successes doesn't mean very much. To understand why this is so, consider the questions you might pose to an instructor who announced on the first day of class that he or she had given 20 A's last semester. You would want to know how many students were in the class: only 20, or several hundred? You would also want to find out how the rest of the class performed—20 A's and 200 D's might be rather discouraging. You could also check on how the students were selected. If only students with A averages in other courses were allowed to enroll, your opinion should be adjusted accordingly. Finally, you might check with the students, and compare their grades with those given by other instructors teaching the same course.

With just a couple of modifications, the same questions could be asked about a cancer "cure":

- How many patients were treated, how many were cured, and how many died?
- How were the patients selected? If selection depended on wealth, degree of illness, or previous exposure to other therapeutic techniques, then the experimental procedure was biased from the start.
- How many might have recovered regardless of the treatment? Even "terminal" cancers sometimes simply disappear for no apparent reason. Such occurrences are rare, to be sure, but they do happen.
- How do the above statistics compare with those of more traditional therapies when subjected to the same unbiased tests?

The Need for Repeatability

Finally, let's examine the criterion of **repeatability**. It's not enough to develop a reasonable, testable hypothesis and collect unbiased data. Consider the hypothesis that every time a coin is tossed, it will come up heads. You could build a coin-tossing machine, turn it on, and find that in the first experiment of 10 tosses, the coin came up heads every time. Does this prove the hypothesis?

No, it doesn't, despite the fact that it was an honest experiment and the data supported the hypothesis. The problem here is one of statistics, sample size, and luck. The odds that a coin will come up heads on any given toss are 50 percent, or 1 in 2—the same as the odds that it will come up tails. The odds that it will come up heads 10 times in a row are about 1 in 2000—pretty small, but certainly not inconceivable. If that coin is tossed 50 times, however, the chance of getting 50 heads becomes just 1 in 4,500,000,000,000,000, a figure that most people would accept as "vanishingly small." To prove that the hypothesis "a tossed coin always lands heads up" is false, the coin need only come up tails once. So the truth could be revealed by running the experiment with more coin tosses or by letting other people set up identical experiments and toss their own coins.

The point here is that if a hypothesis is correct, anyone and everyone will get the same results when the experiment is performed. If it isn't repeatable, you've got to doubt the conclusions even when you have complete confidence in the abilities and integrity of the original investigator.

If a hypothesis satisfies all these criteria, it can be accepted as a scientific **theory**. The scientific use of this term therefore differs from that used in general conversation; when writers discuss "wild-eyed theories" they are really speaking of untested hypotheses. Hypotheses may be true or false, but by definition theories describe real phenomena, and they make accurate predictions about the world. Examples of scientific theories include the theory of gravity and the theory of evolution. The "fact" of gravity is not in question, and the theory of gravity accounts for the available data. But this does not mean that theories cannot change over time. Newton's original theory of gravity, though used successfully for over two centuries, was profoundly modified and extended by Einstein. Similarly, the theory of evolution originally proposed by Charles Darwin has been greatly elaborated since it was first proposed in the middle of the last century. No one theory can tell "the whole story," and all theories are continually undergoing modification and improvement as we learn more about our universe.

Developing an analytical approach to the world around you takes time and considerable attention to detail. As you continue your education, exercise your imagination. Don't be afraid to ask "what if" or "what about" questions when they occur to you, and watch for conclusions or claims based on simplistic or biased viewpoints. You will probably be fascinated by what you discover!

CHAPTER REVIEW

Review of Selected Clinical Terms

anatomy (a-NAT-o-mē) (*p. 2*)
The study of internal and external structure and the physical relationships between body parts.

embryology (*p. 3*)
The study of structural changes during the first two months of development.

congenital (kon-JEN-i-tal) defect (*p. 3*)
A structural problem resulting from an error in development.

histology (his-TOL-o-jē) (*p. 3*)
The study of tissues.

physiology (fiz-ē-OL-ō-jē) (*p. 4*)
The study of the function of anatomical structures.

disease (*p. 10*)
A malfunction of organs or organ systems resulting from failure of homeostatic regulation.

abdominopelvic quadrant (*p. 13*)
One of four regions of the anterior abdominal surface.

abdominopelvic region (*p. 13*)
One of nine regions of the anterior abdominal surface.

inspection (*p. 15*)
A careful observation of a patient's appearance and actions.

palpation (*p. 15*)
Using hands and fingers to feel the patient's body as part of a physical exam.

percussion (*p. 15*)
Tapping with the fingers or hand to obtain information about the densities of a patient's underlying tissues.

auscultation (aws-kul-TĀ-shun) (*p. 15*)
Listening to a patient's body sounds using a stethoscope.

radiologist (*p. 20*)
A physician who specializes in performing and analyzing radiological procedures.

X-rays (*p. 20*)
High-energy radiation that can penetrate living tissues.

angiogram (*p. 21*)
An X-ray image of circulatory pathways.

CT, CAT (computerized [axial] tomography) (*p. 21*)
An imaging technique that reconstructs the three-dimensional structure of the body.

MRI (magnetic resonance imaging) (*p. 22*)
An imaging technique that employs a magnetic field and radio waves to portray subtle structural differences.

ultrasound (*p. 22*)
An imaging technique that uses brief bursts of high-frequency sound reflected by internal structures.

echogram (*p. 22*)
An image created by ultrasound.

Additional Terms of Clinical Importance

GROSS ANATOMY (*p. 3*): **gross (macroscopic) anatomy, surface anatomy, regional anatomy, systemic anatomy, medical anatomy, radiographic anatomy, surgical anatomy**

DEVELOPMENTAL ANATOMY (*p. 3*): **developmental anatomy**

MICROSCOPIC ANATOMY (*p. 3*): **microscopic anatomy, cytology**

PHYSIOLOGY (*p. 4*): **human physiology, cell physiology, histophysiology, special physiology, system physiology, pathological physiology**

Study Outline

Related Key Terms

Introduction (pp. 1–2)

1. **Biology** is the study of life; one of its goals is to discover the unity and patterns that underlie the diversity of living organisms.
2. All living things perform the same basic functions: they **respond** to changes in their environment; they show **adaptability** to their environment; they **grow** and **reproduce** to create future generations; they are capable of producing **movement**; and they **absorb** materials from the environment. Organisms absorb and consume oxygen during **respiration**, and discharge waste products during **excretion**. **Digestion** occurs in specialized areas of the body to break down complex foods. The **circulation** forms an internal transportation system between areas of the body.

The Sciences of Anatomy and Physiology (pp. 2–4)

1. **Anatomy** is the study of internal and external structure and the physical relationships between body parts. **Physiology** is the study of how living organisms perform vital functions. All specific functions are performed by specific structures.

GROSS ANATOMY (p. 3)

2. **Gross (macroscopic) anatomy** considers features visible without a microscope. It includes **surface anatomy** (general form and super-

Related Key Terms

ficial markings); **regional anatomy** (superficial and internal features in a specific area of the body); **systemic anatomy** (structure of major organ systems); **medical antomy** (features that undergo characteristic changes during illness); **radiographic anatomy** (structures as they are visualized by specialized procedures performed on an intact body); and **surgical anatomy** (landmarks important for surgical procedures.)

DEVELOPMENTAL ANATOMY (p. 3)

3. **Developmental anatomy** examines the changes in form that occur between conception and physical maturity. **Embryology** studies processes during the first few months of development.

congenital defects

MICROSCOPIC ANATOMY (p. 3)

4. The boundaries of **microscopic anatomy** are established by the equipment used. **Cytology** analyzes the internal structure of individual cells. **Histology** examines **tissues** (groups of cells that have specific functional roles.) Tissues combine to form **organs**, anatomical units with multiple functions.

PHYSIOLOGY (p. 4)

5. **Human physiology** is the study of the functions of the human body. It is based on **cell physiology**, the study of the functions of living cells. **Histophysiology** examines tissue function; **special physiology** studies the physiology of specific organs. **System physiology** considers all aspects of the function of specific organ systems. **Pathological physiology** studies the effects of diseases on organ or system functions.

Levels of Organization (pp. 4–6)

1. Anatomical structures and physiological mechanisms are arranged in a series of interacting levels of organization.

Homeostasis and System Integration (pp. 6–10)

1. **Homeostasis** is the tendency for physiological systems to stabilize internal conditions; through **homeostatic regulation** these systems adjust to preserve homeostasis.

HOMEOSTATIC REGULATION (pp. 6–9)

2. Homeostatic regulation usually involves a **receptor** sensitive to a particular stimulus and an **effector** whose activity affects the same stimulus.
3. **Negative feedback** is a corrective mechanism involving an action that directly opposes a variation from normal limits.
4. **Autoregulation** occurs when the activities of a cell, tissue, organ, or system change automatically in response to an environmental change. **Extrinsic regulation** results from the activities of the nervous or endocrine systems.
5. In **positive feedback** the initial stimulus produces a response that exaggerates the stimulus.

HOMEOSTASIS AND DISEASE (pp. 9–10)

6. Symptoms of **disease** appear when failure of homeostatic regulation causes organ systems to malfunction.

A Frame of Reference for Anatomical Studies (pp. 10–22)

SUPERFICIAL ANATOMY (pp. 10–15)

1. Standard anatomical illustrations show the body in the **anatomical position**. If the figure is shown lying down, it can be either **supine** (face up) or **prone** (face down).

brachium

2. **Abdominopelvic quadrants** and **abdominopelvic regions** represent two different approaches to describing anatomical regions of the body.

SECTIONAL ANATOMY (pp. 15–22)

3. The three **sectional planes (frontal** or **coronal plane, sagittal plane,** and **transverse plane)** describe relationships between the parts of the three-dimensional human body.

superior • **inferior**
midsagittal section • **anterior**
parasagittal section • **posterior**
transverse section

4. **Serial reconstruction** is an important technique for studying histological structure and analyzing images produced by radiological procedures.
5. **Body cavities** protect delicate organs and permit changes in the size and shape of visceral organs. The **dorsal body cavity** contains the **cranial cavity** (enclosing the brain) and **spinal cavity** (surround-

cranium • **mediastinum**
pericardial sac

ing the spinal cord). The **ventral body cavity** or **coelom** surrounds developing respiratory, cardiovascular, digestive, urinary, and reproductive organs.

6. During development the **diaphragm** divides the ventral body cavity into the superior **thoracic** and inferior **peritoneal cavities**. By birth the thoracic cavity contains two **pleural cavities** (each containing a lung) and a **pericardial cavity** (which surrounds the heart). The peritoneal or **abdominopelvic cavity** consists of the **abdominal cavity** and the **pelvic cavity**.
7. Important **radiological procedures** (which can provide detailed information about internal systems) include **X-rays, CT scans, MRI,** and **ultrasound.** Each technique has its advantages and disadvantages.

Related Key Terms

radiologists • radiodensity
angiogram • echogram
CAT • NMR

■ Review Planner

	Objective	Level 1
1	Describe the basic functions of living organisms.	**3 27**
2	Define the various specialties of anatomy and physiology.	**1 7 8**
3	Identify the major levels of organization in living organisms.	**10 28**
4	Explain the significance of homeostasis.	**6 14**
5	Describe important mechanisms involved in homeostatic regulation.	**3 9 20 25**
6	Use anatomical terms to describe body sections, body regions, and relative positions.	**11 12 13 15 21 22 24 29 30 34**
7	Identify the major body cavities.	**4 6 16 17 18 19**
8	Compare important radiological procedures in terms of their methods, advantages, and disadvantages.	**2 5 23 32**

Level 2: **26 31 33**

Level 3: **35 36**

■ Review Questions

MULTIPLE CHOICE

1. Physiology is the study of: a) how living organisms perform vital functions; b) internal and external structures in living organisms; c) the physical relationships between body parts; d) the internal structure of individual cells.
2. You're looking at an X-ray image. Areas that are very white are most likely to be: a) air; b) blood; c) fat; d) bone.
3. You spend a summer working in your garden, and develop calluses on areas of your hands that are exposed to chronic pressure from garden tools. This is an example of: a) movement; b) respiration; c) adaptability; d) excretion.
4. The cranial cavity and the spinal cavity comprise the: a) ventral cavity; b) pericardial cavity; c) peritoneal cavity; d) dorsal body cavity.
5. Radiodensity refers to tissues' resistance to ______________: a) X-rays; b) magnetic fields; c) ultrasound echoes; d) CT scans.
6. Failure of homeostasis is likely to produce the state

we call: a) autoregulation; b) disease; c) auscultation; d) feedback.

7. If you studied all the superficial and internal features of the adult foot, you would be studying: a) surface anatomy; b) embryology; c) systemic anatomy; d) regional anatomy.
8. If you studied the changes that occur in the human foot between conception and physical maturity, you would be studying: a) regional anatomy; b) cytology; c) developmental anatomy; d) pathological physiology.
9. Blood clotting in response to injured blood vessels is an example of: a) positive feedback; b) negative feedback; c) disease; d) none of the above.
10. Which of the following lists levels of organization in the correct sequence from least complex to most complex?: a) cells, organ systems, tissues, organs; b) cells, tissues, organs, organ systems; c) tissues, organ systems, organs, cells; d) organ systems, organs, tissues, cells.
11. Exhausted after reading the first chapter of your Anatomy and Physiology text, you assume a supine position. Which of the following are you facing?: a) the ceiling; b) the floor; c) a wall; d) your abdominopelvic quadrants.
12. The eyes are ________ to the nose: a) medial; b) lateral; c) anterior; d) distal.
13. Moving ________ from the ankle, one reaches the knee: a) dorsally; b) proximally; c) distally; d) caudally.
14. The tendency for physiological systems to stabilize internal conditions is called: a) positive feedback; b) homeostasis; c) circulation: d) system physiology.

DIAGRAM

15. Create a table that lists the appropriate adjective(s) and directional reference(s) to describe the following regions of the body: front, back, head, coccyx, close to long axis of the body, away from long axis, toward an attached base, and away from an attached base.

MATCHING QUESTIONS

(Match each structure with the body cavity in which it is found.)

Structures:	**Body Cavities:**
16. Heart	a) Pleural cavity
17. Lung	b) Spinal cavity
18. Spinal cord	c) Pericardial cavity
19. Brain	d) Cranial cavity

TRUE/FALSE

(If the statement is false, your instructor may wish to have you correct it).

20. **T/F:** Stretching a skeletal muscle often triggers a contraction that returns it to its original length. This is an example of positive feedback.
21. **T/F:** The anatomical position shows the human form lying face down.
22. **T/F:** In anatomical descriptions, the terms "left" and "right" refer to the left and right sides of the subject, rather than of the observer.
23. **T/F:** Echograms are more expensive and riskier to the patient than MRI or CT scans, but they provide greater clarity.

SHORT ANSWER/ESSAY

24. A friend informs you that there is a fly crawling on your body. It is on the dorsal surface of your cranium and moving rostrally. Where is the fly and in what direction is it moving?
25. Fill in the blanks: Homeostatic regulation usually involves a(n) ________ that is sensitive to a particular stimulus, and a(n) ________ whose activity exerts an effect upon the same stimulus.
26. What is serial reconstruction, and why is it useful?
27. Describe the basic functions performed by all living organisms.
28. Distinguish between cells, tissues, and organs.
29. What plane would divide the head so that the face remains intact?
30. What plane would divide the trunk into left and right valves?
31. What do the nervous and the endocrine systems have in common? How do their functions differ?
32. Compare the techniques used in X-rays, CT scans, MRI scans, and ultrasound. What are the relative advantages and disadvantages of each?
33. An individual complains of a sharp, stabbing pain in the left lower quadrant and fears it may be appendicitis. Do you think it likely that this diagnosis is correct?
34. If you begin at the elbow and move proximally 25 cm, medially 20 cm, and inferiorly 30 cm, what superficial anatomical landmark will you be near?

CRITICAL THINKING/APPLICATIONS

35. A patient has come to the hospital's X-ray department to have a "barium swallow." She is nervous and asks you what the test involves. What would you tell her? What conditions is this test often used to diagnose?
36. While grocery shopping one day, you read the following label on a package: "In one scientific study subjects who ate this product, in conjunction with a low-fat diet, lowered their blood cholesterol levels by almost 20% within three months." a) Do you detect any bias in this claim? Explain. b) Describe how you could apply the scientific method to test the results of this study.

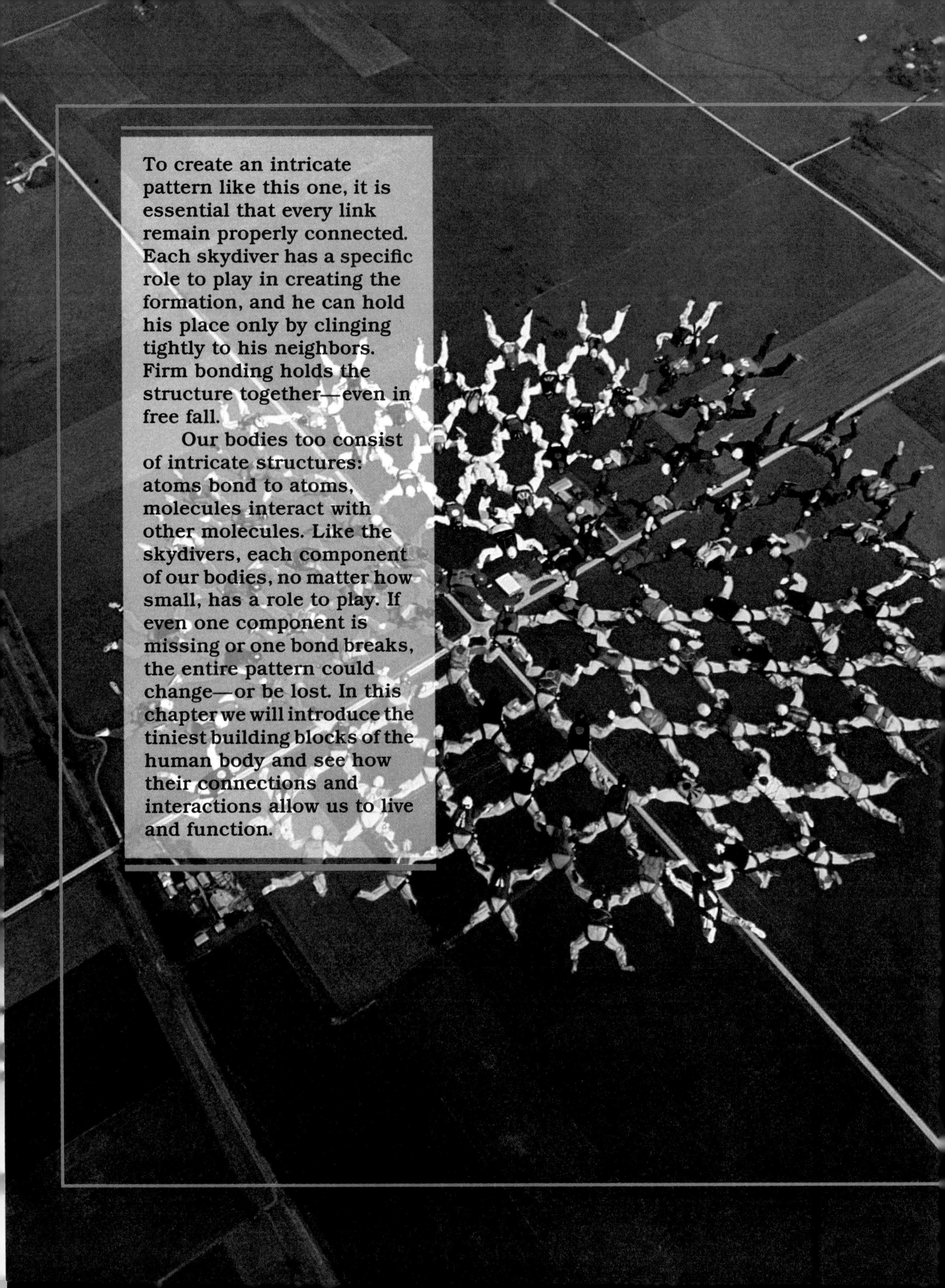

To create an intricate pattern like this one, it is essential that every link remain properly connected. Each skydiver has a specific role to play in creating the formation, and he can hold his place only by clinging tightly to his neighbors. Firm bonding holds the structure together—even in free fall.

Our bodies too consist of intricate structures: atoms bond to atoms, molecules interact with other molecules. Like the skydivers, each component of our bodies, no matter how small, has a role to play. If even one component is missing or one bond breaks, the entire pattern could change—or be lost. In this chapter we will introduce the tiniest building blocks of the human body and see how their connections and interactions allow us to live and function.

CHAPTER 2

The Chemical Level of Organization

Chapter Objectives

After reading this chapter, you will be able to:

1 Describe an atom and how atomic structure affects interactions between atoms.
2 Compare the different ways in which atoms combine to form molecules and compounds.
3 Use chemical notation to symbolize chemical reactions.
4 Distinguish between the three major types of chemical reactions that are important for studying physiology.
5 Explain how the chemical properties of water make life possible.
6 Discuss the importance of pH and the role of buffers in body fluids.
7 Distinguish between organic and inorganic compounds.
8 Describe the physiological roles of inorganic compounds.
9 Discuss the structure and functions of carbohydrates, lipids, proteins, nucleic acids, and high-energy compounds.

■ Introduction

Our examination of the human body begins at the most fundamental level of organization, that of individual atoms and molecules. Air, elephants, oranges, oceans, rocks, and people are all composed of atoms in varying combinations. The unique characteristics of each object, living or nonliving, result from the types of atoms involved and the ways those atoms combine and interact.

In the human body, atoms in combination form vital anatomical structures, and physiological processes depend on the precise control of atomic interactions. If the structures are normal and the interactions are controlled and coordinated, the body functions well. Marathons can be run, chapters read and remembered, songs written and sung. Yet life continues only as long as chemical processes follow an orderly and appropriate sequence. A single structural abnormality or the loss of control over a single chemical process can cause severe illness or even death.

This chapter introduces basic concepts in chemistry, focusing attention on the structure of the human body. This information will establish the foundation needed for later discussions of cellular physiology, tissue organization, and the structure and function of organs and organ systems.

Atoms and Molecules

Atoms are the smallest chemical units of matter—no chemical change can alter their identities. Physicists have blasted atoms apart and identified dozens of different *subatomic particles*, but only three are stable constituents of atomic structure. These three fundamental particles are **protons**, **neutrons**, and **electrons**. Protons and neutrons are similar in size and mass, but *protons* bear a *positive* electrical charge whereas *neutrons* are *neutral*—that is, uncharged. Electrons are lighter, only 1/1836th as massive as protons, and bear a negative electrical charge.

STRUCTURE OF THE ATOM

All atoms contain protons and electrons, normally in equal numbers. The number of protons in an atom is known as its **atomic number**. A chemical **element** is a substance that consists entirely of atoms with the same atomic number. **Hydrogen** is the simplest element, with an atomic number of 1. An atom of hydrogen contains one proton and one electron. The proton is located in the center of the atom and forms the **nucleus**. The electron whirls around the nucleus at high speed, forming an **electron cloud** (Figure 2-1a). For convenience, this structure is often illustrated in the simplified form shown in Figure 2-1b. In this representation the electrons are shown in a spherical **electron shell.**

The dimensions of the cloud determine the overall size of the atom. To get an idea of the scale involved, consider that if the nucleus were the size of a tennis ball the electron cloud would have a radius of six *miles*. In reality, atoms are so small that atomic measurements are most conveniently reported in terms of **nanometers** (NA-nō-mē-ters) (nm) or *angstroms* (Å). A nanometer is 10^{-9} meters (0.000000001 m), and an angstrom is one-tenth of that size (10^{-10} m). The very largest atoms approach 0.5 nm in diameter (0.00000005 cm, or 0.00000002 in.).

Figure 2-2 indicates the important elements found in the human body. Each element has a chemical symbol that represents an abbreviation recognized by scientists everywhere. Most of the symbols are easily connected with the names of the elements, but a few, such as *Na* for sodium, are abbreviations of their Latin names.

FIGURE 2-1
Hydrogen Atoms. (a) The electron cloud of a hydrogen atom is formed by the orbiting of an electron around the nucleus. (b) A two-dimensional model depicting the electron in an electron shell makes it easier to visualize the components of the atom. A typical hydrogen nucleus contains a single proton and no neutrons. (c) A deuterium (^2H) nucleus contains a proton and a neutron. (d) A tritium (^3H) nucleus contains a pair of neutrons in addition to the proton.

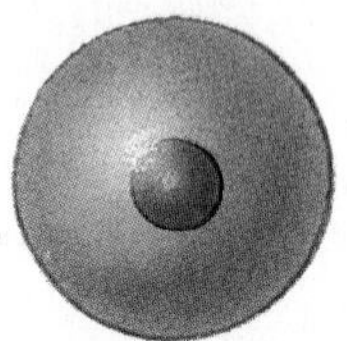

(a) Hydrogen atom (space-filling model)

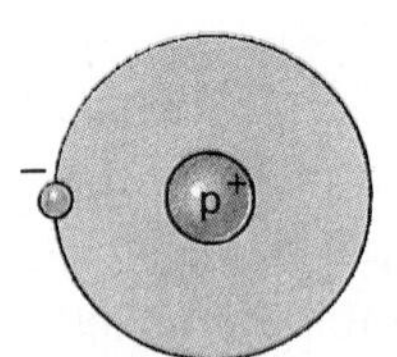

(b) Hydrogen-1

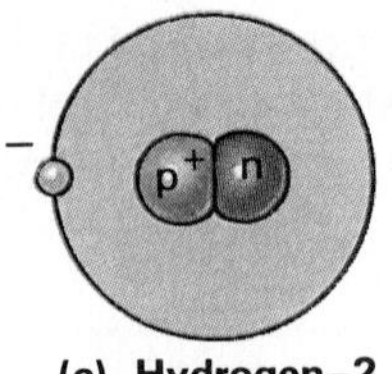

(c) Hydrogen-2 (deuterium)

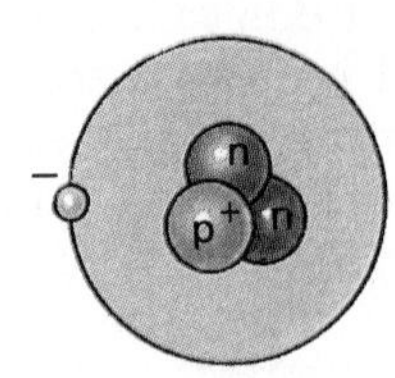

(d) Hydrogen-3 (tritium)

Isotopes

When neutrons are present in an atom they are found in the nucleus together with the protons. Neutrons occur in the atoms of all elements other than hydrogen. In fact, even some hydrogen atoms contain neutrons. Although most hydrogen nuclei consist of a single proton, 0.015 percent also contain one neutron, and a very small percentage contain two.

We can see from this example that the number of neutrons in the nucleus can vary, even among atoms of a single element. **Isotopes** of an element are atoms whose nuclei contain different numbers of neutrons. Because the presence or absence of neutrons has very little effect on the chemical properties of an atom, isotopes are usually indistinguishable except on the basis of weight. The **mass number**—the total number of protons and neutrons in the nucleus—is used to identify a particular isotope. Thus, the three isotopes of hydrogen are designated hydrogen-1, or ^1H; hydrogen-2, or ^2H, also known as *deuterium*; and hydrogen-3, or ^3H, also known as *tritium* (Figure 2-1c,d). Isotopes of larger atoms do not have individual names.

Atomic Weights

An atom of oxygen with an atomic number of 8, contains eight protons and eight neutrons. The mass of this atom is therefore about 16 times that of a

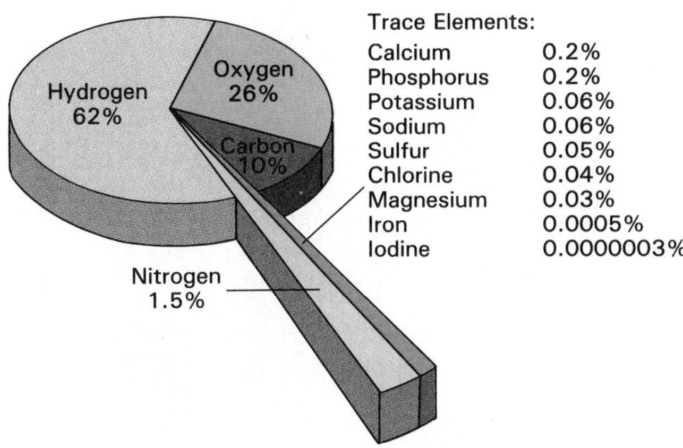

FIGURE 2-2
Elements in the Human Body. The percentages given are estimates of the contribution made by each element to the total number of atoms in the body. Note that just four elements (C, H, O, and N) contribute over 99 percent to the total. Among the elements present in quantities too small to show are silicon (Si), fluorine (F), copper (Cu), manganese (Mn), zinc (Zn), selenium (Se), cobalt (Co), molybdenum (Mo), cadmium (Cd), chromium (Cr), aluminum (Al), and boron (B). The functions of some of these in the body are still poorly understood.

Element	*Significance*
Hydrogen (H)	A component of water and most other compounds in the body.
Oxygen (O)	A component of water and other compounds; oxygen gas essential for respiration.
Carbon (C)	Found in all organic molecules.
Nitrogen (N)	Found in proteins, nucleic acids, and other organic compounds.
Calcium (Ca)	Found in bones and teeth; important for membrane function, nerve impulses, muscle contraction, and blood clotting.
Phosphorus (P)	Found in bones and teeth, proteins, nucleic acids, and high energy compounds.
Potassium (K)	Important for proper membrane function, nerve impulses, and muscle contraction.
Sodium (Na)	Important for membrane function, nerve impulses, and muscle contraction.
Chlorine (Cl)	Important for membrane function and water absorption.
Sulfur (S)	Found in many proteins.
Magnesium (Mg)	A cofactor for several enzymes.
Iron (Fe)	Essential for oxygen transport and energy capture.
Iodine (I)	A component of a hormone (thyroxine) of the thyroid gland.

Clinical Comment: Radioisotopes and Clinical Testing

Radioisotopes are isotopes having unstable nuclei that emit subatomic particles in measurable amounts. **Alpha particles** are generally released by the nuclei of large radioactive atoms, such as uranium. Each alpha particle consists of a helium nucleus: two protons and two neutrons. **Beta particles** are electrons, more often released by radioisotopes of lighter atoms. **Gamma rays** are very high-energy waves comparable to the X-rays used in clinical diagnosis. The **half-life** of any radioactive substance is the time required for a 50 percent reduction in the amount of radiation emitted. The half-lives of radioisotopes range from fractions of a second to thousands of years.

Like X-rays, gamma rays, beta particles, and alpha particles can damage or destroy living tissues. The danger posed by radiation exposure varies, depending on the nature of the emission and the duration of exposure. But radiation also has a variety of beneficial uses in medical research and clinical diagnosis. Weakly radioactive isotopes with short half-lives can sometimes be used to check the structural and functional state of an organ without surgery.

Radioisotopes are useful because they can be incorporated into specific compounds normally found within the body. These labeled compounds, called **tracers**, can be introduced into the body and tracked by the radiation they release. After a labeled compound is swallowed, its uptake, distribution, and excretion can be determined by monitoring the radioactivity of samples taken from the digestive tract, body fluids, and waste products. For example, compounds labeled with radioisotopes of cobalt are used to monitor the intestinal absorption of vitamin B_{12}. Usually cobalt-58 is used, a radioisotope with a half-life of 71 days.

Radioisotopes can also be injected into the blood or other body fluids to provide information on circulatory anatomy and the anatomy and physiology of specific target organs. In **nuclear imaging** the radiation emitted by injected radioisotopes creates an image on a special photographic plate. Such a procedure may be used to identify regions where particular radioactive materials are concentrated or to check the circulation through vital organs. Radioisotopes can also produce pictures of specific organs, such as the liver, spleen, or thyroid, where labeled compounds are removed from the circulation.

The thyroid gland, for example, secretes chemical "messengers" called *hormones* (Chapter 18) that contain iodine atoms. As a result, the thyroid will absorb and concentrate radioactive iodine. This gland, shown in Figure 2-3a, sits below the larynx (voicebox) on the front of the neck. The **thyroid scan** in Figure 2-3b was taken following the injection of iodine-131, a radioisotope with an 8-day half-life. This procedure, called a *thyroid radioactive iodine uptake measurement*, or **RAIU**, can provide information on (1) the size and shape of the gland and (2) the amount of absorptive activity under way. Comparing the rate of iodine uptake with the level of circulating hormones makes it possible to evaluate the functional state of the gland.

PET (**P**ositron **E**mission **T**omography) scans utilize the same principles as standard radioisotope scans, but the analyses are performed by computer. The scans are much more sensitive, and the computers can reconstruct sections through the body that permit extremely precise localization. Among other things, this procedure can analyze blood flow through organs and assess the metabolic activity within specific portions of an organ as complex as the human brain. Figure 2-3c is a PET scan of the brain showing activity at a single moment in time. The scan is dynamic, however, and changing patterns of activity can be followed in real time. PET scans can be used to analyze normal brain function as well as to diagnose brain disorders, and, to date, the technique has served primarily as a research tool. Because the equipment is expensive and bulky, it is unlikely to be available anywhere except in large regional medical centers or universities.

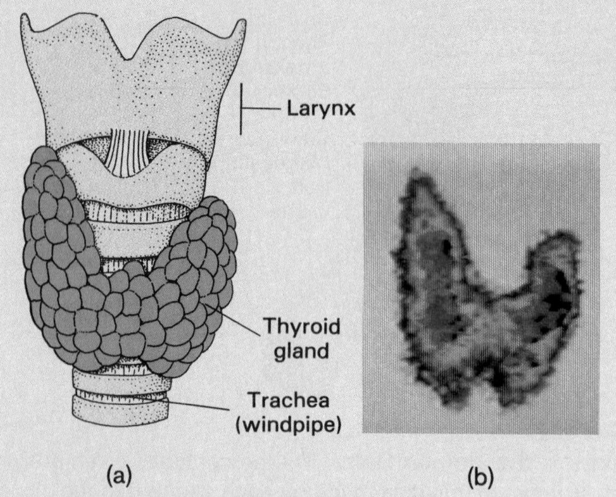

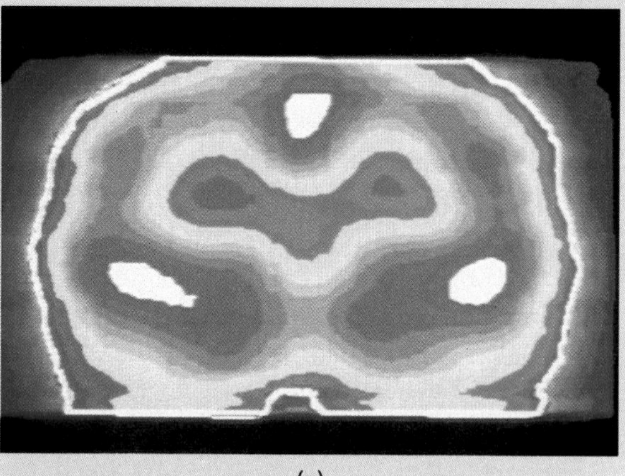

FIGURE 2-3
Imaging Techniques. (a) The position and contours of the normal thyroid gland as seen in dissection. (b) After it has been labeled with radioactive iodine, the thyroid can be examined by special imaging techniques. In this computer-enhanced image, different color intensities indicate differing concentrations of the radioactive tracer. (c) A PET scan of the head as seen in a frontal section. The soft tissue of the brain appears red.

typical hydrogen atom. The **atomic weight** of an element is defined in such a way as to be very close to its mass number. The two values differ because when dealing with large numbers of atoms there must be allowance for (1) the presence of isotopes with different mass numbers, (2) the small difference in mass between protons and neutrons, and (3) the mass of electrons. Mass numbers and atomic weights are included in Table 2-1.

For every element, a quantity that has a mass in grams equal to the atomic weight will contain the same number of atoms. That quantity[1] has been given a special name: a **mole**. Expressing relationships in moles rather than grams makes it much easier to keep track of the relative numbers of atoms in chemical samples and processes. For example, oxygen has an atomic weight of 16.0 and hydrogen has an atomic weight of 1.0, so each oxygen atom weighs 16 times as much as each hydrogen atom. Thus 1 g of hydrogen

[1] The number of atoms, ions, or molecules in a mole—called *Avogadro's number*—is 6.02×10^{23}, or about 600 billion trillion.

will contain the same number of atoms as 16 g of oxygen.

Electrons and Energy Levels

Atoms are electrically neutral because every positively charged proton is balanced by a negatively charged electron. Examine Table 2-1, which shows this pattern clearly. Note that each increase in the atomic number is accompanied by a comparable increase in the number of electrons orbiting around the nucleus. These electrons occupy an orderly series of electron shells, or energy levels.

The number and arrangement of electrons in the outer energy level is extremely important because it determines the chemical properties of every element. Each energy level can accommodate a specific number of electrons. For example, the innermost shell can hold only two electrons. As indicated in Figure 2-4, a hydrogen atom has one electron in this energy level, but a helium atom has two. Lithium has three electrons, so in a lithium atom the first level is filled and the third electron occupies a second energy level. The second level can hold up to eight electrons, and this level is filled in a neon atom with an atomic number of 10. This pattern should be reviewed in Table 2-1 and Figure 2-4. To see how other elements are related to those shown here, consult Appendix II (the periodic table).

CHEMICAL BONDS AND CHEMICAL COMPOUNDS

We mentioned that the characteristics of the outer electron energy level determine the chemical properties of a particular element. An atom with a full outer energy level is very stable. For examples, return to Table 2-1. Helium contains two electrons in its outermost energy level; argon and neon contain eight electrons in their outer energy levels. These elements are called **inert gases** because their atoms neither react with one another nor combine with atoms of other elements.

TABLE 2-1 Atomic Structure of the First Twenty Elements

				Nucleus		Electrons			
Element (Symbol)	*Atomic Number*	*Mass Number*	*Atomic Weight*	*Neutrons*	*Protons*	*First Energy Level*	*Second Energy Level*	*Third Energy Level*	*Fourth Energy Level*
Hydrogen (H)	1	1	1.008	—	1	1			
Helium (He)	2	4	4.003	2	2	2—Inert gas			
Lithium (Li)	3	7	6.941	4	3	2	1		
Beryllium (Be)	4	9	9.012	5	4	2	2		
Boron (B)	5	11	10.810	6	5	2	3		
Carbon (C)	6	12	12.011	6	6	2	4		
Nitrogen (N)	7	14	14.007	7	7	2	5		
Oxygen (O)	8	16	15.999	8	8	2	6		
Fluorine (F)	9	19	18.998	9	9	2	7		
Neon (Ne)	10	20	20.179	10	10	2	8—Inert gas		
Sodium (Na)	11	23	22.990	12	11	2	8	1	
Magnesium (Mg)	12	24	24.305	12	12	2	8	2	
Aluminum (Al)	13	27	26.981	14	13	2	8	3	
Silicon (Si)	14	26	26.086	12	14	2	8	4	
Phosphorus (P)	15	31	30.974	16	15	2	8	5	
Sulfur (S)	16	32	32.060	16	16	2	8	6	
Chlorine (Cl)	17	35	35.453	18	17	2	8	7	
Argon (Ar)	18	40	39.948	22	18	2	8	8—Inert gas	
Potassium (K)	19	39	39.098	20	19	2	8	8	1
Calcium (Ca)	20	40	40.080	20	20	2	8	8	2

Note: For a complete listing of the elements and their atomic numbers and weights, see Appendix II.

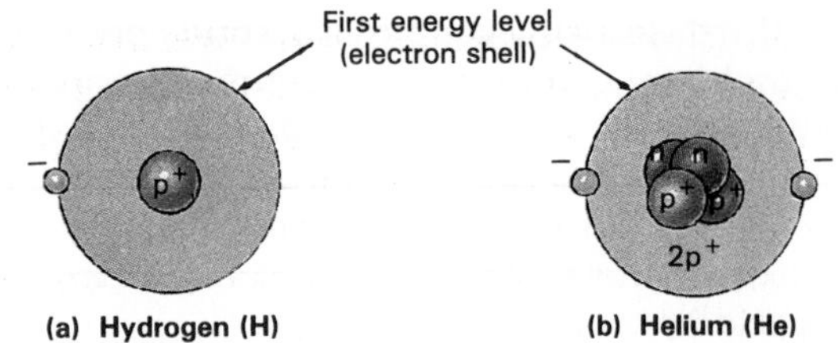

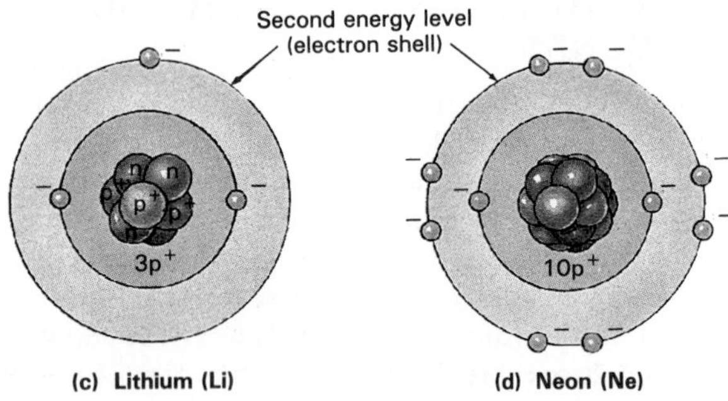

FIGURE 2-4
Atoms and Energy Levels. (a) A typical hydrogen atom has one proton and one electron. The electron orbiting the nucleus occupies the first energy level, diagrammed as an electron shell. (b) An atom of helium has a pair of protons and two electrons. The two electrons orbit in the same energy level. (c) The first energy level can hold only two electrons. In a lithium atom, with three protons and three electrons, the third electron occupies a second energy level. (d) The second level can hold up to eight electrons. A neon atom has 10 protons and 10 electrons; thus both the first and second energy levels are filled. Note that helium, lithium, and neon atoms contain neutrons as well as protons in their nuclei.

Atoms with unfilled outer energy levels are relatively unstable. These atoms can achieve stability by sharing, gaining, or losing electrons. This process usually produces a **molecule**, a chemical structure containing more than one atom. When atoms of different elements combine in this way, the result is a chemical **compound**. A compound is a new chemical substance with properties that can be quite different from those of its component elements. For example, hydrogen and oxygen are highly flammable gases, but combining hydrogen and oxygen atoms produces a compound, water, that can put out fires.

One way that atoms can complete their outer electron shells is by sharing electrons with other atoms. The result is a molecule held together by **covalent** (kō-VĀ-lent) **bonds.**

Covalent Bonds

Individual hydrogen atoms, as diagrammed in Figure 2-4a, are not found in nature. Instead, we find hydrogen molecules. Figure 2-5a diagrams one such molecule. Molecular hydrogen is a gas present in the atmosphere in very small quantities. The two hydrogen atoms share their electrons, with each electron whirling around both nuclei. The sharing of one pair of electrons creates a **single covalent bond.** Note that three different methods can be used to show the structure of a hydrogen molecule: (1) the electron-shell model, which diagrams the positions of the electrons in concentric shells that represent the energy levels; (2) the space-filling model, which shows the molecule

FIGURE 2-5
Covalent Bonds. (a) In a molecule of hydrogen, two hydrogen atoms share their electrons so that each has a filled outer electron shell. This sharing creates a single covalent bond. (b) A molecule of oxygen consists of two oxygen atoms that share two pairs of electrons. The result is a double covalent bond. (c) In a molecule of carbon dioxide, a central carbon atom forms double covalent bonds with a pair of oxygen atoms. Covalently bonded molecules can be depicted in several ways: by *electron-shell diagrams*, which depict the sharing of electrons (note that for simplicity the subatomic particles in the nucleus are not shown); by *space-filling models* that more accurately depict the electron cloud surrounding the nucleus; or by *structural formulas*, in which each shared pair of electrons (covalent bond) is represented by a solid line.

	ELECTRON–SHELL MODEL	SPACE–FILLING MODEL	STRUCTURAL FORMULA
(a) Hydrogen (H_2)			H—H
(b) Oxygen (O_2)			O=O
(c) Carbon dioxide (CO_2)			O=C=O

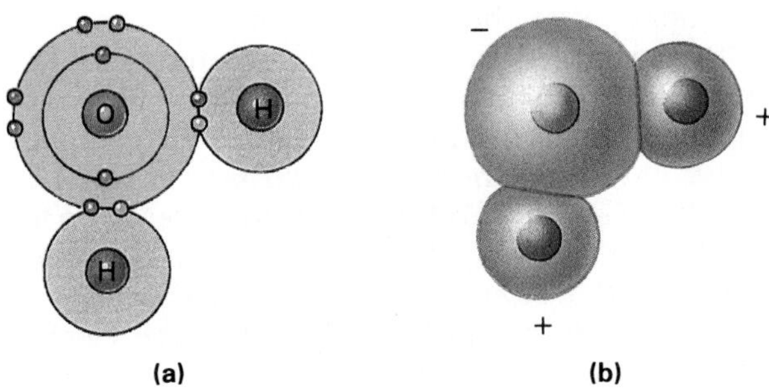

FIGURE 2-6
Polar Covalent Bonds and the Structure of Water. (a) In forming a water molecule, an oxygen atom completes its outer energy level by sharing electrons with a pair of hydrogen atoms. The sharing is unequal because the oxygen atom holds the electrons more tightly than do the hydrogen atoms. (b) Because the oxygen atom has two extra electrons much of the time, it develops a slight negative charge; the hydrogen atoms become weakly positive. The bonds in a water molecule are called polar covalent bonds.

in three dimensions; and (3) the structural formula, which uses a solid line to indicate a shared electron pair.

Oxygen, with an atomic number of 8, has two electrons in its first energy level and six in the second. The oxygen atoms diagrammed in Figure 2-5b attain a stable electron configuration by pooling their resources and sharing two pairs of electrons, forming a **double covalent bond**. Molecular oxygen is an atmospheric gas that is very important to living organisms; our cells would die without a relatively constant supply of oxygen.

The chemical reactions in our bodies that consume oxygen also produce a waste product, **carbon dioxide**. The oxygen atoms in a carbon dioxide molecule form double covalent bonds with the carbon atom, as indicated in Figure 2-5c.

Covalent bonds are very strong because the electrons tie the atoms together. In typical covalent bonds the atoms remain electrically neutral because each shared electron spends just as much time at home as away. (If two people were tossing a pair of baseballs back and forth as fast as they could, on the average, each person would have just one.) Covalent bonds, especially between carbon atoms, create the stable framework of the large molecules that make up most of the structural components of the human body.

Polar Covalent Bonds Many covalent bonds involve relatively equal sharing of electrons. Some, however, do not, because elements differ in how strongly they hold shared electrons. An unequal sharing of electrons creates a **polar covalent bond**. For example, in a molecule of water, diagrammed in Figure 2-6, an oxygen atom forms covalent bonds with two hydrogen atoms. The oxygen atom has a much stronger attraction for the shared electrons than the hydrogen atoms do, so the electrons spend most of their time in the vicinity of the oxygen nucleus. Because it has two extra electrons part of the time, the oxygen atom develops a slight negative charge. At the same time, the hydrogen atoms develop slight positive charges, for their electrons are away part of the time. This unequal sharing makes polar covalent bonds somewhat weaker than other covalent bonds.

Ionic Bonds

In some bonds, one atom has such a strong affinity for electrons that the other atom loses its hold on

FIGURE 2-7
Ionic Bonding. This series of steps diagrams the formation of an ionic bond. Step 1: A sodium atom loses an electron to become a sodium ion with a +1 charge. The full second electron shell now forms the boundary of the ion. A chlorine atom uses the electron to complete its third electron shell, becoming a chloride ion with a −1 charge. Step 2: Because these ions have opposite charges, they are attracted to one another. Step 3: The association of sodium and chloride ions forms the ionic compound sodium chloride, otherwise known as table salt. No electrons are shared; the bond is created by the attraction between oppositely charged ions.

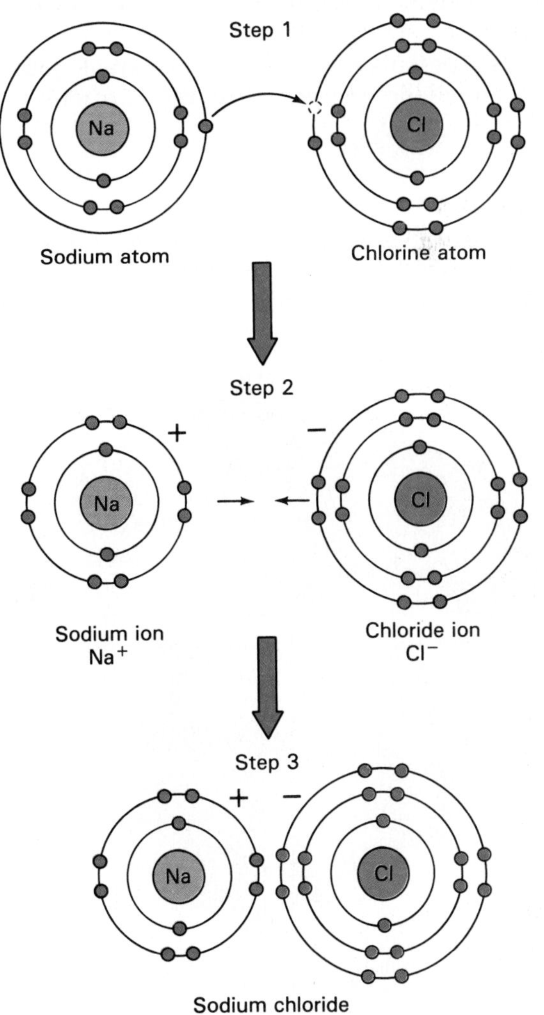

them entirely. When this occurs, an **ionic** (ī-ON-ik) **bond** is formed.

The steps in the formation of an ionic bond are illustrated in Figure 2-7. If you assign a value of +1 to the charge on a proton, the charge on an electron would be −1. As long as the number of protons is equal to the number of electrons the atom will be electrically neutral. If the atom loses an electron, it will exhibit a charge of +1 because there will be one proton without a corresponding electron. Adding an extra electron to the atom will give it a charge of −1. Atoms or molecules that have a (+) or (−) charge are called **ions**. Ions with a positive charge are **cations** (KAT-ī-ons), those with a negative charge are **anions** (AN-ī-ons).

The sodium atom diagrammed in Figure 2-7 (Step 1) has an atomic number of 11, so this atom normally contains 11 protons and 11 electrons. Electrons fill the first and second energy levels, and a single electron occupies level 3. Losing that "extra" electron would give the sodium atom a full outer energy level and produce a sodium ion with a +1 charge. But the electron cannot simply be thrown away; it must be donated to another atom. A chlorine atom has seven electrons in its outer energy level. An additional electron would fill this energy level, and a sodium atom can provide it. In the process (Step 2) the chlorine atom becomes a *chloride ion* with a −1 charge.

Both atoms have now become stable ions with filled outer energy levels. But the two ions do not move apart after the electron transfer, because the positively charged sodium ion is attracted to the negatively charged chloride ion (Step 3). The combination of oppositely charged ions forms the ionic compound *sodium chloride*, otherwise known as table salt. Large numbers of sodium and chloride ions interact to form highly structured crystals, such as the one shown in Figure 2-8. Unlike atoms joined by covalent bonds, atoms in ionic compounds do not share electrons at all; only the attraction of opposite charges holds them together.

FIGURE 2-8
Sodium Chloride. Large numbers of sodium and chloride ions form crystals of sodium chloride that are held together by the attraction between opposite charges.

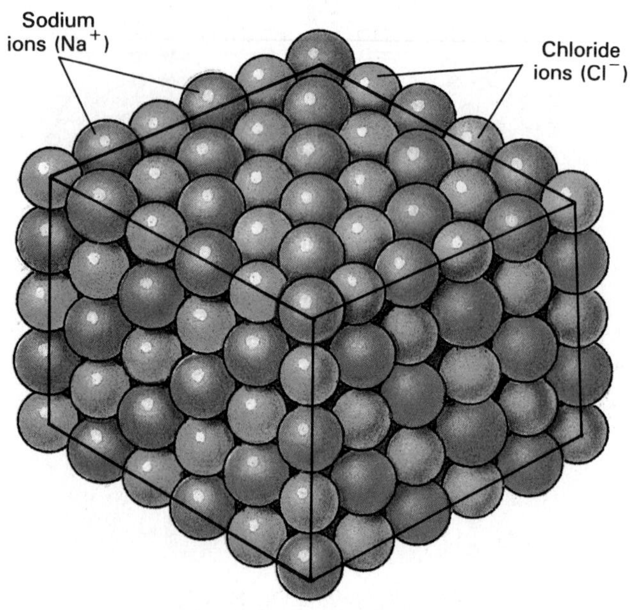

Sodium chloride (NaCl)

FIGURE 2-9
Hydrogen Bonds. (a) The hydrogen atoms of a water molecule have a slight positive charge, and the oxygen atom has a slight negative charge (see Figure 2-6). Attraction between the hydrogen atom of one water molecule and the oxygen atom of another represents a hydrogen bond (indicated by dashed lines). (b) Hydrogen bonding between water molecules at a free surface restricts evaporation and creates surface tension.

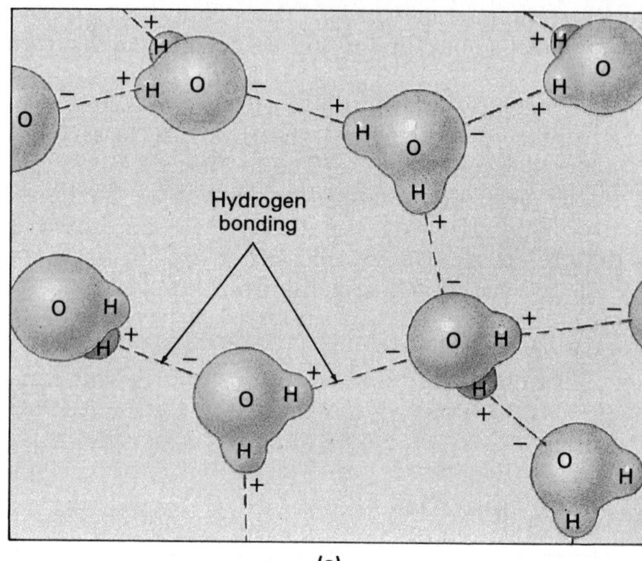

(a)

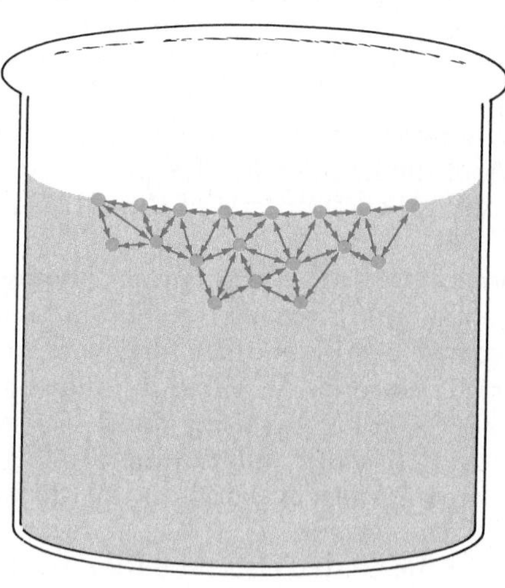

(b)

TABLE 2-2 Molecular Bonds

Bond Type	*Strength*	*Structure*	*Importance*
Covalent bond	Strongest	Electrons shared	Most common, creates stable molecules
Polar covalent bond	Strong	Electrons shared unequally	Common, creates molecules with areas of slight (+) and (−) charge
Ionic bond	Strong	Electrical attraction between opposite charges	Common, forms ionic compounds
Hydrogen bond	Weak	Electrical attraction between a hydrogen atom in a polar covalent bond and a negatively charged atom (often oxygen) in a polar covalent bond of a different molecule or another site on the same molecule	Important effects on the properties of water and the shapes of complex molecules

Hydrogen Bonds

Covalent and ionic bonds tie atoms together in a relatively stable framework. Comparatively weak attractive forces act between atoms in different parts of a large molecule as well as between adjacent molecules. *Hydrogen bonds* are the most important of these attractive forces. Hydrogen bonds affect the shapes and properties of complex molecules, such as proteins; they may also determine the three-dimensional relationships between molecules.

Hydrogen atoms often form polar covalent bonds with the atoms of other elements, such as oxygen or nitrogen. In the process the hydrogen atom develops a slight positive charge and its partner develops a weak negative charge. A **hydrogen bond** is the attraction between such a hydrogen atom and a negatively charged atom in another molecule or at another site in the same molecule. Hydrogen bonds do not create molecules, but they can alter molecular shapes or pull molecules together. For example, hydrogen bonding occurs between water molecules (Figure 2-9a). At the water surface the attraction between molecules slows the rate of evaporation and creates the phenomenon known as surface tension (Figure 2-9b). Surface tension in the tear layer keeps small objects such as dust particles from touching the surface of the eye. If normal surface tension is disrupted (for example, by an airborne droplet of gasoline), this protection is lost. In addition, the rate of evaporation increases dramatically, leading to a dry and irritated eye.

Table 2-2 reviews information concerning hydrogen bonds and the other chemical bonds.

√ **Oxygen and neon are both gases at room temperature. Oxygen combines readily with other elements but neon does not. Why?**

√ **How is it possible for two samples of hydrogen to contain the same number of atoms but have different weights?**

√ **What kind of bond holds atoms in a water molecule together? What attracts water molecules to one another?**

Chemical Notation

Before we can consider the specific compounds found in the human body, we must be able to describe chemical compounds and reactions effectively. Using sentences to describe chemical structures and events often leads to confusion, and a simple form of "chemical shorthand" makes communication much more efficient.

The chemical shorthand we will use is known as chemical notation. Chemical notation enables us to describe complex events in a brief and precise fashion. The rules of chemical notation can be summarized as follows:

1. The abbreviation of an element indicates one atom of that element:

   H = an atom of hydrogen,
   O = an atom of oxygen

You should take the time to learn the abbreviations for the elements included in Figure 2-2, for you will need to recognize them in subsequent chapters.

2. A number preceding the abbreviation of an element indicates more than one atom:

   2 H = two individual atoms of hydrogen
   2 O = two individual atoms of oxygen

3. A subscript following the abbreviation of an element indicates a molecule with that number of atoms:

H_2 = one molecule of hydrogen composed of two hydrogen atoms

O_2 = one molecule of oxygen composed of two oxygen atoms

The symbols for compounds that consist of more than one element follow the same rules.

H_2O = a single molecule of water, made of two hydrogen atoms and one oxygen atom

This notation can be used to indicate both the number of molecules and their structure at the same time.

$2\,H_2O$ = two identical molecules of water, each composed of two hydrogen atoms and one oxygen atom

4. In a description of a chemical reaction, the interacting participants are called **reactants**, and the reaction generates one or more **products**. An arrow indicates the direction of the reaction, from reactants (usually on the left) to products (usually on the right).

Rules 1–4 can be used to describe all of the chemical reactions discussed earlier in the chapter. For example:

a. $H + H \rightarrow H_2$: Two hydrogen atoms (H) combine to produce a single molecule of hydrogen (H_2). (p. 35)

b. $O + O \rightarrow O_2$: Two oxygen atoms (O) combine to produce a single molecule of oxygen (O_2). (p. 35)

c. $2\,H + O \rightarrow H_2O$: Two atoms of hydrogen combine with one atom of oxygen to produce a single molecule of water (H_2O). (p. 36)

d. $Na + Cl \rightarrow NaCl$: An atom of sodium and an atom of chlorine combine to produce the ionic compound sodium chloride (NaCl). (p. 34)

5. A superscript sign following the abbreviation for an element indicates an ion.

Na^+ = one sodium ion

Cl^- = one chloride ion

Ions do not always carry a charge equal to +1 or −1. For example, an atom of calcium (Ca) has two electrons in the third energy level, and it attains a complete outer energy level by "losing" both. The ion that forms has a +2 charge, usually indicated as Ca^{2+}.

6. Chemical reactions neither create nor destroy atoms—they merely rearrange them into new combinations. Therefore, the numbers of atoms of each element must always be the same on both sides of the equation. When this is the case, the equation is **balanced.** For example, this reaction is unbalanced, and therefore misleading:

$$H_2 + O_2 \rightarrow H_2O$$

The equation states that two atoms of hydrogen combine with two atoms of oxygen to produce one molecule of water. It is true that hydrogen and oxygen atoms combine to form water, but here the reactants include two atoms of oxygen, while only one oxygen atom appears in the product. A correct, balanced representation of the reaction would be:

$$2\,H_2 + O_2 \rightarrow 2\,H_2O$$

This equation is balanced because the same number of oxygen and hydrogen atoms are found in the products as in the reactants. Note that each side of the reaction contains four atoms of hydrogen and two of oxygen, although the molecular forms are quite different. Because the equation is balanced, you know that you could take two molecules of hydrogen and one of oxygen and combine them to form a pair of water molecules.

You cannot actually handle individual molecules, nor could you easily count the billions of molecules that take part in ordinary chemical processes in the lab or in the body. Thus the first step in performing such an experiment would be to calculate the **molecular weights** involved. The molecular weight is equal to the sum of the atomic weights of the components. The atomic weight of hydrogen is close to 1, so one hydrogen molecule (H_2) would have a *molecular* weight of 2. Oxygen has an atomic weight of 16, so the molecular weight of an oxygen molecule (O_2) is 32. In practical terms, if you wanted to perform the experiment you would take 4 grams of hydrogen, combine it with 32 grams of oxygen, and produce 36 grams of water. You could also work with ounces, pounds, or tons, as long as the proportions remained the same.

Chemical Reactions

Living cells remain alive and functional by controlling internal chemical reactions. In effect, each cell is a chemical factory. For example, growth, maintenance and repair, secretion, and contraction all involve complex chemical reactions. In a **chemical reaction**, chemical bonds between atoms are broken as atoms are rearranged in new combinations to form different chemical substances.

TYPES OF REACTIONS

Three types of chemical reactions are important to the study of physiology. A **decomposition** reaction breaks a molecule into smaller fragments. You could diagram a typical decomposition reaction as:

$$AB \rightarrow A + B$$

A broken chemical bond releases energy. Our cells can capture some of that energy and use it to power essential functions such as growth, repair, movement, and reproduction.

Synthesis (SIN-the-sis) is the opposite of decomposition. A synthesis reaction assembles larger molecules from smaller components. The discussion of chemical notation included a number of synthetic reactions, such as:

$$Na + Cl \rightarrow NaCl$$

$$H + H \rightarrow H_2$$

These relatively simple reactions could be diagrammed as:

$$A + B \rightarrow AB$$

In the examples above, A and B were individual atoms. In other synthetic reactions they might be molecules that combine to form an even larger product. Synthesis always involves the formation of new chemical bonds, whether the reactants are atoms or molecules. Because it takes energy to create a chemical bond, synthesis usually represents an uphill struggle. Living cells balance their chemical activities, and decomposition reactions provide the energy needed to support synthesis.

In an **exchange reaction** parts of the reacting molecules are shuffled around, as in:

$$AB + CD \rightarrow AD + CB$$

You will notice that there are two products and two reactants. Although the reactants and products contain the same components (A, B, C, and D), they are present in different combinations. In an exchange reaction, the reactant molecules AB and CD break apart (a decomposition) before they interact with one another to form AD and CB (a synthesis). If breaking the old bonds releases more energy than it takes to create the new ones, the exchange reaction will release energy, usually in the form of heat. Such reactions are said to be **exergonic** (*exo-*, outside). If the energy required for synthesis exceeds the amount released by the associated decomposition reaction, additional energy must be provided. Such reactions are called **endergonic** (*endo-*, inside) because they absorb heat.

REVERSIBLE REACTIONS

Chemical reactions are at least theoretically reversible, so that if $A + B \rightarrow AB$, then $AB \rightarrow A + B$. Many important biological reactions are freely reversible. Such reactions can be diagrammed as:

$$A + B \leftrightharpoons AB$$

This equation reminds you that there are really two reactions occurring simultaneously, one a synthesis ($A + B \rightarrow AB$) and the other a decomposition ($A + B \leftarrow AB$). At **equilibrium** (ē-kwi-LIB-rē-um) the two rates are in balance. As fast as a molecule of AB forms, another degrades into A + B. It is possible to predict the result of a disturbance in the equilibrium condition. When the concentration of a reaction participant rises or falls, the rate of reaction will increase or decrease accordingly. In the example above, adding additional AB molecules will increase the rate of conversion to A and B. The concentrations of A and B then rise, and as these concentrations increase so does the rate of AB formation. Eventually an equilibrium is again established.

Not all chemical reactions are easily reversed. The requirements for the two reactions may differ, so that at any given time and place the reaction will proceed chiefly in one direction. For example, the synthesis reaction may occur when the A and B molecules are heated, and the decomposition reaction when AB molecules are placed in water. In that case the reaction would be diagrammed as:

$$A + B \underset{H_2O}{\overset{\text{heat}}{\rightleftharpoons}} AB$$

Many of the reversible chemical reactions that occur within our bodies are controlled by **enzymes**, special compounds that promote chemical processes. These reversible reactions could be diagrammed as:

$$A + B \overset{\text{enzyme \#1}}{\rightleftharpoons} AB$$

Complex reactions proceed in a series of interlocking steps, each step controlled by a different enzyme. Such a reaction sequence is called a **pathway**. A synthetic pathway could be diagrammed as:

$$A + B \underset{\text{Step 1}}{\overset{\text{enzyme \#1}}{\rightleftharpoons}} AB \underset{\text{Step 2}}{\overset{\text{enzyme \#2}}{\rightleftharpoons}} C \underset{\text{Step 3}}{\overset{\text{enzyme \#3}}{\rightleftharpoons}} \text{etc.}$$

Sometimes the steps in the synthetic pathway differ from those of the decomposition pathway, and separate enzymes are involved. The function of enzymes and the regulation of enzymatic pathways will be considered in Chapter 4.

Although the human body is very complex, it contains a relatively small number of elements, as indicated in Figure 2-2. But knowing the identity and quantity of each element will not help you to understand a human being any more than studying the alphabet will help you to understand this text. The rest of this chapter focuses attention on **nutrients** and **metabolites** (me-TAB-o-līts; *metabole*, change). Nutrients are the essential elements and molecules absorbed from food. Metabolites include all of the molecules synthesized or broken down by chemical reactions inside our bodies. Nutrients and metabolites can be broadly categorized as **inorganic** or **organic**. Inorganic compounds do not contain carbon and hydrogen atoms as the primary structural ingredients, whereas carbon and hydrogen form the basis for organic molecules.

√ **In living cells, glucose, a six-carbon molecule, is converted into two three-carbon molecules by a reaction that yields energy. How would you classify this reaction?**

√ **If the product of a reversible reaction is continuously removed, what effect do you think this will have on the equilibrium?**

Inorganic Compounds

Inorganic compounds are usually small molecules held together partially or completely by ionic bonds. The most important inorganic molecules are water and inorganic acids, bases, and salts.

WATER AND ITS PROPERTIES

Water, H_2O, is the single most important constituent of the body, accounting for almost two-thirds of its total weight. A change in body water content can have fatal consequences because virtually all physiological systems will be affected.

Although familiar to everyone, water really has some very unusual properties; Table 2-3 lists those properties that affect human physiology. The properties of water are a direct result of the hydrogen bonding that occurs between adjacent water molecules.

Temperature Effects

The processes that sustain life occur in a fluid environment. Fluids fill the interior of cells and the spaces between them. If these fluids freeze, vital processes stop; if they boil, the cells and tissues are destroyed. Fortunately, the fluid involved is water. Because water molecules are attracted to one another through hydrogen bonding, the temperature must be quite high before individual molecules have enough energy to break free and become water vapor. Consequently water stays in the liquid state over a broad temperature range that roughly corresponds to normal environmental temperatures. Without hydrogen bonding, body fluids would boil at temperatures far below the freezing point of water (0° C), and life would be impossible.

Because water molecules must absorb so much energy to break free of their hydrogen bonds, water carries a great deal of heat away with it when it finally does change from a liquid to a vapor. This accounts for the cooling effect of perspiration on the skin.

Because of hydrogen bonding, water also has a high **heat capacity**. A great deal of energy must be provided to raise the temperature of a volume of liquid water by 1° C. Conversely, a great deal of energy must be lost before a 1° temperature drop will occur. As a result, once a large volume of water has reached a particular temperature, it will change temperature only slowly. This property is called *thermal inertia*. Because water accounts for roughly 66 percent of the weight of the human body, thermal inertia helps stabilize body temperature.

Solvent Properties

A remarkable number of inorganic and organic molecules will dissolve in water, creating a solution. Every **solution** consists of a fluid medium, or **solvent**, in which atoms, ions, or molecules of a dissolved substance, or **solute**, are dispersed. Water is particularly effective as a solvent because it has an unusual chemical structure. The polar covalent bonds in a water molecule are oriented so as to place the hydrogen atoms relatively close together. This creates a **polar**

TABLE 2-3 Important Characteristics of Water

Characteristic	*Significance to Human Physiology*
Percentage of total body weight: 66%	Water gains and losses affect all living cells.
Molecular structure: polar covalent bonds (see Figure 2-6).	Results in hydrogen bonding that accounts for water's distinctive properties.
Freezing point: 0° C (32° F); **boiling point:** 100° C (212° F)	Remains fluid over a wide temperature range.
Capacity to absorb and distribute heat: greatest of all liquids over the temperature range of 0°–100° C.	Prevents rapid changes in body temperature. Rapidly distributes heat from one region of the body to another.
Heat absorbed during evaporation: greatest of all liquids.	Makes perspiration an effective means of losing heat and lowering body temperatures.
Solvent properties: dissolves more inorganic and organic compounds and in greater quantities than any other liquid; ions in solution conduct electrical currents.	Makes water an ideal medium for the absorption and/or transport of inorganic and organic compounds. Conductivity important for the normal functioning of nerve cells and muscle cells.

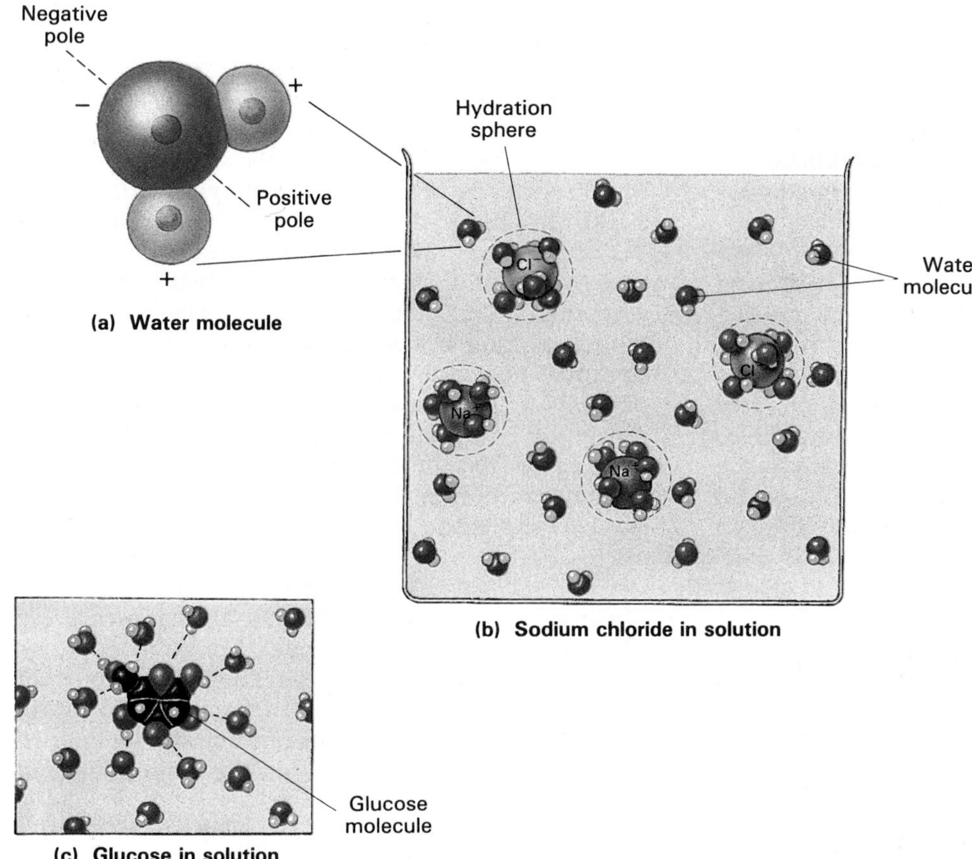

FIGURE 2-10
Water Molecules and Solutions. (a) In a water molecule oxygen forms polar covalent bonds with two hydrogen atoms. Because the hydrogen atoms are positioned toward one end of the molecule, the molecule has an uneven distribution of charges. This creates positive and negative poles. (b) Ionic compounds dissociate in water as the polar water molecules disrupt the ionic bonds. The ions in solution are surrounded by water molecules, creating hydration spheres. (c) Hydration spheres will also form around an organic molecule containing polar covalent bonds. If the molecule is small it will be carried into solution, as shown here with glucose.

molecule, with positive and negative poles, as indicated in Figure 2-10a.

Inorganic compounds are often held together by ionic bonds, and most will undergo **ionization** (ī-on-i-ZĀ-shun), or **dissociation** (di-sō-sē-Ā-shun), in water. In this process, shown in Figure 2-10b, ionic bonds are broken as the ions form hydrogen bonds with water molecules. This produces a mixture of cations and anions surrounded by a **hydration sphere** of polar water molecules. **Electrolytes** (e-LEK-tro-līts) are soluble inorganic molecules whose ions will conduct an electric current in solution. Electrical events involving cell membranes affect the functioning of all cells, and small electric currents carried by ions trigger processes such as muscle contraction and nerve function. Chapters 10 and 12 will discuss these events in more detail.

Sodium ions (Na^+), potassium ions (K^+), calcium ions (Ca^{2+}), and chloride ions (Cl^-) are released by the dissociation of electrolytes in blood and other body fluids. Table 2-4 contains a more complete listing of important electrolytes and the ions released by their dissociation. Alterations in the body fluid concentrations of these ions will disturb almost every vital function. For example, declining potassium levels will lead to a general muscular paralysis, and rising concentrations will cause weak and irregular heartbeats. [**CM:** *Solute Concentrations*]

TABLE 2-4 Important Electrolytes That Dissociate in Body Fluids

NaCl (sodium chloride)	$\rightarrow Na^+$	$+ Cl^-$
KCl (potassium chloride)	$\rightarrow K^+$	$+ Cl^-$
$CaCl_2$ (calcium chloride)	$\rightarrow Ca^{2+}$	$+ 2\ Cl^-$
$NaHCO_3$ (sodium bicarbonate)	$\rightarrow Na^+$	$+ HCO_3^-$
$MgCl_2$ (magnesium chloride)	$\rightarrow Mg^{2+}$	$+ 2\ Cl^-$
$Na_2\ HPO_4$ (disodium phosphate)	$\rightarrow 2\ Na^+$	$+ HPO_4^{2-}$
Na_2SO_4 (sodium sulfate)	$\rightarrow 2\ Na^+$	$+ SO_4^{2-}$

The Most Common Ions in Body Fluids

Cations	*Anions*
Na^+ (sodium)	Cl^- (chloride)
K^+ (potassium)	HCO_3^- (bicarbonate)
Ca^{2+} (calcium)	HPO_4^{2-} (biphosphate)
Mg^{2+} (magnesium)	SO_4^{2-} (sulfate)

Organic molecules often contain polar covalent bonds, which also attract water molecules. The hydration spheres that form then carry these molecules into solution (Figure 2-10c). In this case the molecule is an important soluble sugar, *glucose*. Molecules that readily dissolve in water are called **hydrophilic** (hī-dro-FI-lik; *hydro-*, water + *philus*, loving). **Hydrophobic** (hī-dro-FŌ-bik; *hydro-*, water + *phobos*, fear) molecules have few if any polar covalent bonds. When placed in contact with water molecules, hydration spheres do not form and the molecules do not dissolve. Body fat deposits consist of large, insoluble droplets of hydrophobic molecules, trapped in the watery interior of cells. Gasoline, heating oil, and diesel fuel are examples of hydrophobic molecules not found in the body. (You are probably already aware that such molecules are hydrophobic, for if petroleum products were water-soluble we would not have to clean up the beaches after an oil spill.)

Hydrogen Ion Concentration and the pH of Body Fluids

The concentration of hydrogen ions in body fluids is precisely regulated. Hydrogen ions are extremely reactive in solution; in excessive numbers they can break chemical bonds, change the shapes of complex molecules, and disrupt cell and tissue functions. Some of the hydrogen ions are produced by the dissociation of water; others are produced by the dissociation of solute molecules.

Water itself dissociates to a very slight degree, in a reversible reaction that can be diagrammed as:

$$H_2O \leftrightharpoons H^+ + OH^-$$

The dissociation of a water molecule yields a hydrogen ion (H^+) and a **hydroxyl** (hī-DROK-sil) **group**, OH^-. Very few water molecules dissociate in pure water, and the number of hydrogen and hydroxyl ions is very small. The quantities are usually reported in moles, making it easy to keep track of the relative numbers of hydrogen and hydroxyl ions. A liter of pure water contains around 0.0000001 moles of hydrogen ions and an equal number of hydroxyl ions. In other words, the **concentration** of hydrogen ions in that solution is 0.0000001 moles per liter. This can be written as:

$$[H^+] = 1 \times 10^{-7} \text{ mol}/\ell$$

The brackets around H+ indicate "the concentration of," another example of chemical notation.

The hydrogen ion concentration is so important to physiological processes that a special shorthand is used to express it. The **pH** of a solution is defined as *the negative exponent of the hydrogen ion concentration*[2] expressed in moles per liter. Thus,

[2] A more precise definition is the negative logarithm of the hydrogen ion concentration in moles per liter.

TABLE 2-5 pH and Hydrogen Ion Concentration

Concentration of Hydrogen Ions $[H^+]$		*pH*	
10^0 (1.0)		0	Hydrochloric acid (HCl)
10^{-1} (0.1)		1	Stomach secretions
10^{-2} (0.01)		2	Lemon juice
			Cola drinks
10^{-3} (0.001)	Acidic	3	White wine
10^{-4} (0.0001)		4	Tomato juice
10^{-5} (0.00001)		5	Coffee
10^{-6} (0.000001)		6	Urine
			Saliva
10^{-7} (0.0000001)	NEUTRAL	7	← Distilled water
			Blood, semen
10^{-8} (0.00000001)		8	Bile
10^{-9} (0.000000001)		9	
			Bleach
10^{-10} (0.0000000001)		10	
			Milk of magnesia
10^{-11} (0.00000000001)	Basic	11	
			Ammonia water
10^{-12} (0.000000000001)		12	
10^{-13} (0.0000000000001)		13	
			Drain opener
10^{-14} (0.00000000000001)		14	Lye (NaOH)

instead of using the above expression, we could state that the pH of pure water is $-(-7)$, or 7. This saves space, but you must always remember that the pH number is an exponent. Thus a pH of 6, for example ($[H^+] = 1 \times 10^{-6}$, or 0.000001), means that the concentration of hydrogen ions is ten times as great as it is at a pH of 7 ($[H^+] = 1 \times 10^{-7}$, or 0.0000001). For common liquids the pH scale, included in Table 2-5, ranges from 0 to 14.

Although pure water has a pH of 7, solutions display a wide range of pH values, depending on the nature of the solutes involved. A solution with a pH of 7 is **neutral** because it contains equal numbers of hydrogen and hydroxyl ions. A solution with a pH below 7 is **acidic** (a-SI-dik), meaning that hydrogen ions predominate. A pH above 7 is **basic**, or **alkaline** (AL-kah-lin), with hydroxyl ions in the majority.

The pH of the blood normally ranges from 7.35 to 7.45. Abnormal fluctuations in pH can damage cells and tissues by breaking chemical bonds, changing molecular shapes, and altering cellular function. For example, a blood pH below 7 can produce coma, while a blood pH higher than 7.8 usually causes uncontrollable, sustained muscular contractions. The human body generates significant quantities of acids that threaten homeostasis and must be neutralized. Under unusual circumstances, the loss of acids in body fluids may promote an equally disruptive increase in pH.

- √ **Why does a salt solution conduct electricity but a sugar solution does not?**
- √ **How does an antacid help to decrease stomach discomfort?**
- √ **Why would an extreme change in pH of body fluids be undesirable?**

ACIDS, BASES, AND SALTS

Inorganic Acids and Bases

An **acid** is a solute that dissociates to release hydrogen ions and shift the pH toward acidity. (Because a hydrogen atom that loses its electron consists solely of a naked proton, hydrogen ions are often referred to simply as protons, and acids as "proton donors.") A **strong acid** dissociates completely, and the reaction is essentially one-way. Hydrochloric acid (HCl) is an excellent example, for in water it dissociates completely:

$$HCl \rightarrow H^+ + Cl^-$$

The stomach produces this powerful acid to assist in the breakdown of food. Hardware stores sell HCl, as "muriatic acid," for cleaning sidewalks and swimming pools.

A **base** is a solute that removes hydrogen ions from a solution. Many common bases are compounds that dissociate in solution to liberate a hydroxyl ion. Hydroxyl ions have a strong affinity for hydrogen ions and quickly react with them, tying them up in water molecules, so they satisfy the definition of a base given above. For example, sodium hydroxide, NaOH, is a **strong base**, because in solution it dissociates completely. The reaction that releases sodium and hydroxyl ions can be diagrammed as:

$$NaOH \rightarrow Na^+ + OH^-$$

Strong bases have a variety of industrial and household uses; drain openers and lye are two familiar examples.

Salts

If you add a strong acid to a strong base, an exchange reaction occurs. For example, adding hydrochloric acid to sodium hydroxide results in the reaction:

$$HCl + NaOH \rightarrow H_2O + NaCl$$

This reaction produces water and a salt, in this case sodium chloride (table salt). A **salt** is an inorganic compound created by the reaction of an acid with a base. Salts are held together by ionic bonds, and in water many of them dissociate into a cation and an anion. In the example above, the sodium chloride would immediately dissociate into Na^+ and Cl^- ions. These ions are the most abundant in body fluids, but many others are present in lesser amounts, released by the dissociation of various other compounds. Ionic concentrations in the body are regulated by mechanisms described in Chapter 28.

The dissociation of sodium chloride does not release hydrogen or hydroxyl ions, so NaCl, like many salts, is a "neutral" solute. Other salts, however, indirectly affect the concentration of H^+ and OH^- ions through their interactions with water molecules. Thus they may be slightly acidic or basic in solution.

BUFFERS AND pH CONTROL

Buffers are compounds in body fluids that maintain pH within normal limits (7.35–7.45) by removing or replacing hydrogen ions. One important buffering mechanism depends on the reactions of weak acids and weak bases.

Weak acids and **weak bases** fail to dissociate completely, and a significant number of molecules remain intact. For the same number of molecules in solution, weak acids and weak bases therefore have less of an impact on pH than strong acids and bases. Carbonic acid (H_2CO_3) is an example of a weak acid found in body fluids. Its dissociation releases hydrogen ions and **bicarbonate ions** (HCO_3^-) through the reversible reaction:

$$H_2CO_3 \leftrightharpoons H^+ + HCO_3^-$$

The buffering activity of a weak acid can be demonstrated by adding a strong base. For example, adding sodium hydroxide (NaOH) to carbonic acid (H_2CO_3) produces the reaction:

$$NaOH + H_2CO_3 \rightarrow NaHCO_3 + H_2O$$

This reaction converts a very strong and potentially dangerous base to a relatively harmless weak base and water. The weak base produced is **sodium bicarbonate**, $NaHCO_3$, otherwise known as "baking soda." In solution, sodium bicarbonate reversibly dissociates into a sodium ion and a bicarbonate ion (HCO_3^-) in the reaction:

$$NaHCO_3 \leftrightharpoons Na^+ + HCO_3^-$$

Although this dissociation does not release a hydroxyl ion, it has the same effect on pH because the bicarbonate ion will remove an excess hydrogen ion from the solution. Consider what happens when sodium bicarbonate is combined with hydrochloric acid (HCl). An exchange reaction occurs:

$$NaHCO_3 + HCl \rightarrow NaCl + H_2CO_3$$

This process converts a dangerously strong acid to a relatively harmless weak acid and a salt. Antacids such as Alka-Seltzer®, Rolaids®, and Tums® use weak bases (often sodium bicarbonate) to counteract excess hydrochloric acid in the stomach. Carbonic acid and sodium bicarbonate are two important buffers found in body fluids. We will discuss other buffers and buffer systems in Chapter 27.

Organic Compounds

Organic compounds contain the elements carbon and hydrogen, and usually oxygen as well. Each carbon atom needs four electrons to complete its outer energy level. These electrons are obtained through the formation of single, double, or triple covalent bonds, as indicated in Figure 2-11. Organic molecules can contain long chains of carbon atoms linked by covalent bonds. These carbon atoms often form additional covalent bonds with hydrogen or oxygen atoms, and less often, with nitrogen, phosphorus, sulfur, iron, or other elements.

Many organic molecules are soluble in water. Although the discussion above focused attention on inorganic acids and bases, there are also important organic acids and bases. For example, *lactic acid* is an organic acid generated by active muscle tissues.

There are four major classes of organic compounds: **carbohydrates**, **lipids**, **proteins**, and **nucleic acids**. In addition, the human body contains small quantities of many other organic compounds whose structures and functions will be considered in later chapters.

CARBOHYDRATES

A **carbohydrate** (kar-bō-HĪ-drāt) molecule has carbon, hydrogen, and oxygen in a ratio near 1:2:1. Familiar carbohydrates include the sugars and starches that make up roughly half of the typical American diet. Our tissues can break down most carbohydrates, and

FIGURE 2-11
Carbon Atoms and Organic Molecules. (a) A carbon atom, with four electrons in its outer energy level. (b) Each carbon atom can form four single covalent bonds to complete the outer energy level. Above: the conventional representation of the bond arrangement often used in structural formulas. Below: the actual three-dimensional orientation of these bonds. (c) Many organic molecules consist of long carbon chains. Above: Through covalent bonds each of the carbon atoms shares four electrons. That sharing may involve single, double, or triple covalent bonds with other carbon atoms or atoms of other elements. Below: The presence of double or triple covalent bonds alters the shape of the molecule.

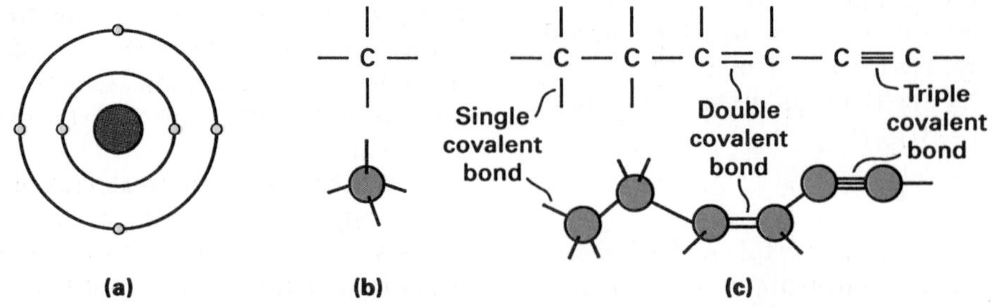

although they may have other functions carbohydrates are most important as sources of energy. There are three major types of carbohydrates: *monosaccharides, disaccharides,* and *polysaccharides.*

Monosaccharides

A **simple sugar**, or **monosaccharide** (mon-ō-SAK-ah-rīd; *mono-*, single + *sakcharon*, sugar), is a carbohydrate containing from three to seven carbon atoms. A triose has three carbons, and there are tetrose (four-carbon), pentose (five-carbon), hexose (six-carbon), and heptose (seven-carbon) molecules as well. The hexose **glucose** (GLOO-kōs), $C_6H_{12}O_6$, is the most important metabolic "fuel" in the body. Figure 2-12 diagrams the structure of a glucose molecule. This figure includes several different representations of the molecule. The straight-chain form (a) shows the component atoms clearly, but in the body the ring form (b) is most common. The actual arrangement of atoms in three dimensions is shown most realistically in a space-filling model (c).

It is important to remember that each organic molecule has a specific three-dimensional shape, and that molecular shape is an important characteristic. **Isomers** are molecules that have the same molecular formula but different structural formulas. That is, they contain the same atoms, but those atoms are bonded together in different arrangements. The body generally treats different isomers as distinct molecules, and they have different properties and fates. For example, Figure 2-13 shows two simple sugars, glucose and **fructose**. Fructose is found in many fruits and in secretions of the male reproductive tract. Although its chemical formula is the same as that of glucose, separate enzymes and reaction sequences control its breakdown and synthesis. Simple sugars such as glucose dissolve readily, and they are rapidly distributed throughout the body by the blood and other body fluids.

FIGURE 2-12
Glucose. (a) The straight-chain structural formula. (b) The ring form that is most common in nature. (c) A space-filling model that shows the actual three-dimensional organization of the carbon atoms in the ring.

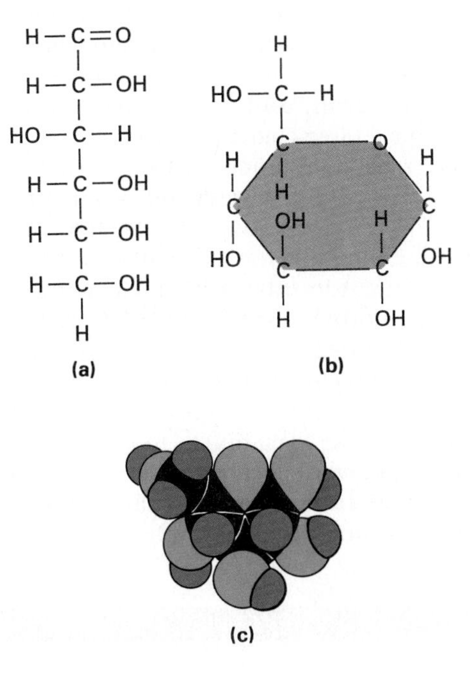

(a) (b)

(c)

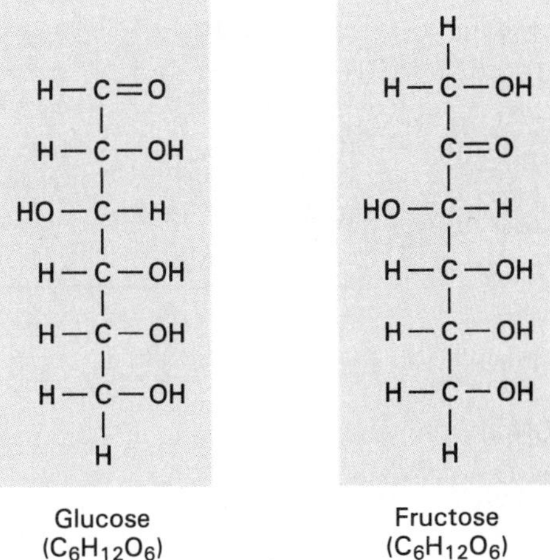

FIGURE 2-13
Isomers. These two molecules have the same chemical formula ($C_6H_{12}O_6$) but different molecular structures and distinct chemical properties.

Disaccharides and Polysaccharides

Two simple sugars joined together form a **disaccharide** (dī-SAK-ah-rīd; *di-*, two). Disaccharides such as **sucrose** (table sugar) have a sugary taste, and they are also quite soluble. Figure 2-14a diagrams the reaction responsible for the formation of sucrose. This is an example of **dehydration synthesis**, the linking together of chemical units by the removal of water to create a more complex molecule. The reverse of this process is **hydrolysis** (hī-DROL-i-sis; *hydro-*, water + *lysis*; dissolution), shown in Figure 2-14b. Hydrolysis breaks a complex molecule into smaller fragments by the addition of a water molecule. Many foods contain disaccharides, but they must be disassembled through hydrolysis before they can be broken down to provide useful energy. Most popular "junk foods," such as candies or sodas, abound in simple sugars (often fructose) and disaccharides such as sucrose.

Dehydration synthesis can build complex molecules from simpler carbohydrates. These large molecules are **polysaccharides** (pol-ē-SAK-ah-rīds; *poly-*, many). Polysaccharide chains may be straight or

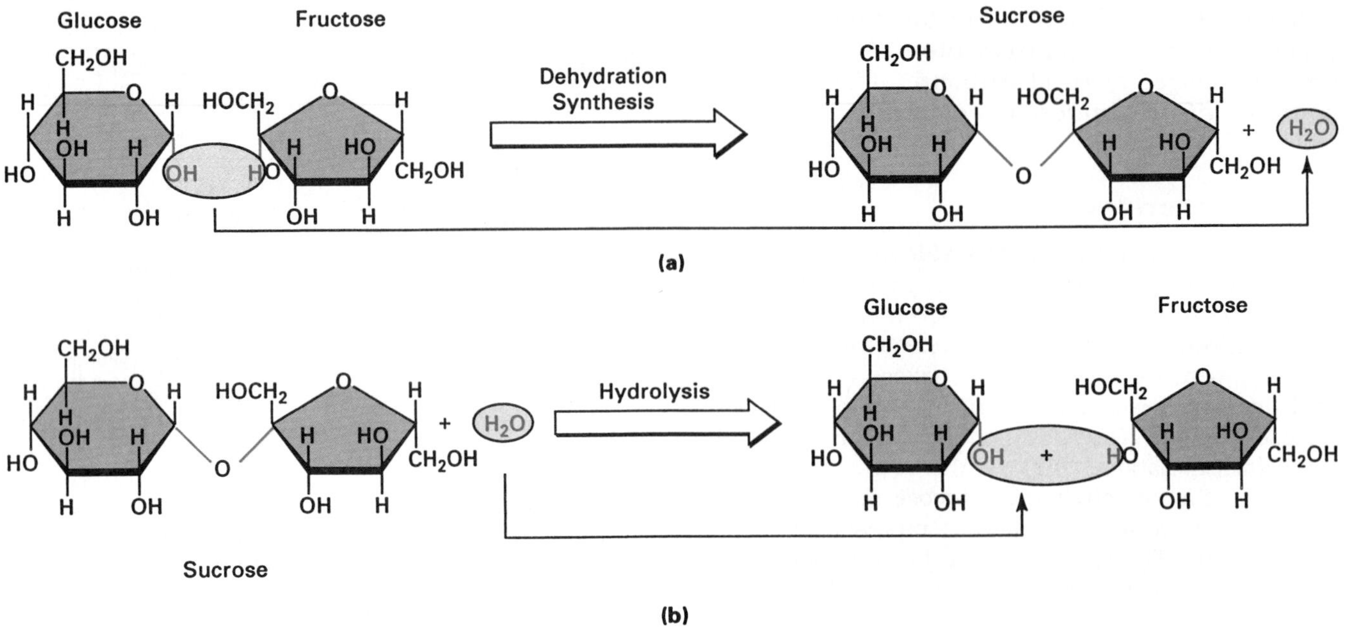

Glucose Fructose Sucrose

Dehydration Synthesis

(a)

Sucrose Glucose Fructose

Hydrolysis

(b)

FIGURE 2-14
The Formation of Complex Sugars. (a) During dehydration synthesis two molecules are joined by the removal of a water molecule. In this example, glucose and fructose are combined to form the disaccharide sucrose. (b) Hydrolysis reverses the steps of dehydration synthesis; a complex molecule is broken down by the addition of a water molecule.

Health News: Artificial Sweeteners

Some people cannot tolerate sugar for medical reasons; others avoid it because recent dietary guidelines call for reduced sugar consumption, or in an effort to lose weight. Thus many people today are using artificial sweeteners in their foods and beverages.

Artificial sweeteners are organic molecules that can stimulate taste buds and provide a sweet taste to foods without adding substantial amounts of calories to the diet. These molecules have a much greater effect on the taste receptors than natural sweeteners, such as fructose or sucrose, so they can be used in minute quantities. For example, *saccharin* is about 300 times sweeter than sucrose. The popularity of this sweetener has declined since it was reported that saccharin may promote bladder cancer in rats. The risk is very small, however, and saccharin continues to be used. There are several other artificial sweeteners currently on the market, including *aspartame* (*NutraSweet®*), *sucralose*, and *acesulfame potassium* (*Ace-K*, or *Surette®*).

Molecules of artificial sweeteners do not resemble those of natural sugars. Saccharin, acesulfame potassium, and sucralose cannot be broken down by the body and have no nutritional value. Aspartame consists of a pair of amino acids. Amino acids are the building blocks of proteins (as discussed later in this chapter), and they can be broken down in the body to provide energy. However, because aspartame is 200 times sweeter than sucrose, very small quantities are needed, so the sweetener adds few calories to a meal. Aspartame does not produce the bitter aftertaste sometimes attributed to saccharin, and thus is used in many diet drinks and low-calorie desserts.

Two new sweeteners, *thaumatin-1* and *monellin*, are proteins extracted from African berries. Thaumatin, roughly 100,000 times sweeter than sucrose, has been approved by the Food and Drug Administration for use in chewing gums. Another artificial sweetener, *cyclamate*, was banned in 1970 after experiments suggested that it caused bladder tumors in laboratory rats. These conclusions have been shown to be incorrect, so cyclamates may soon be reapproved. However, because this sweetener is only about 30 times sweeter than sucrose, it may not have much impact on the marketplace.

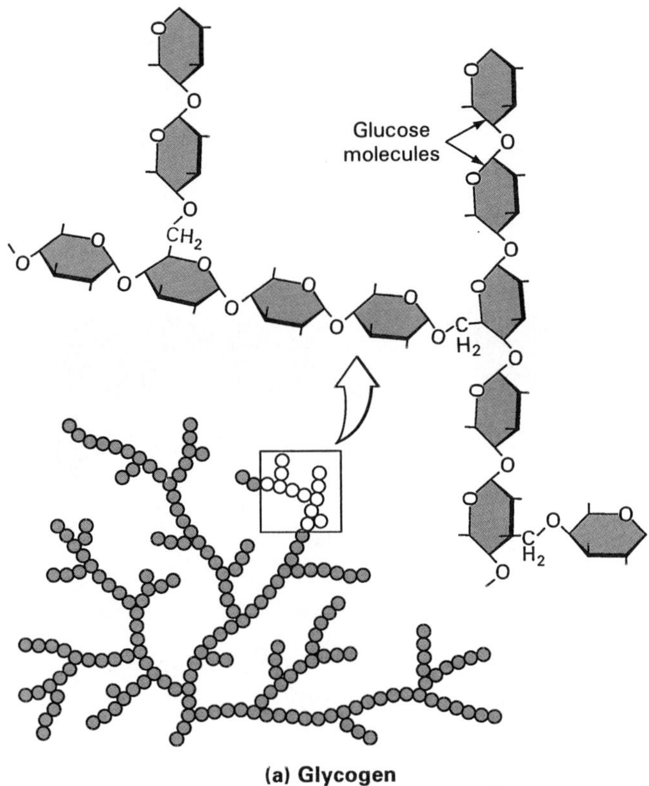

(a) Glycogen

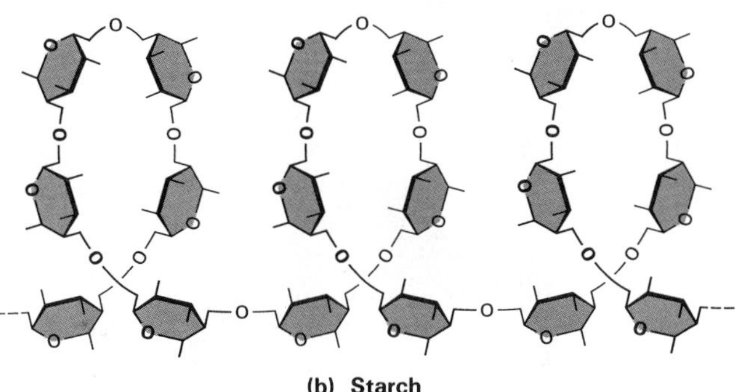

(b) Starch

FIGURE 2-15
Polysaccharides. (a) Liver and muscle cells store glucose in glycogen molecules. (b) Plants produce starches that are glucose-based polysaccharides.

highly branched, as indicated in Figure 2-15. **Glycogen** (GLĪ-ko-jen) is a branched polysaccharide composed of interconnected glucose molecules (Figure 2-15a). Like most other large polysaccharides, glycogen will not dissolve in water or other body fluids. Liver and muscle tissues manufacture and store large quantities of glycogen. When these tissues have a high demand for energy, glycogen molecules are broken down into glucose; when demands are low, the tissues absorb glucose and rebuild glycogen reserves.

The human diet contains large numbers of complex carbohydrates. **Starches** are glucose-based polysaccharides manufactured by plants (Figure 2-15b). The digestive tract can break these molecules into simple sugars, and starches such as those found in potatoes and grains represent a major dietary energy source. **Cellulose**, a structural component of many plants, is a polysaccharide that our bodies cannot digest at all.

Despite their metabolic importance, carbohydrates account for less than 3 percent of our total body weight. Table 2-6 summarizes information concerning the carbohydrates.

LIPIDS

Lipids (*lipos*, fat) also contain carbon, hydrogen, and oxygen, but the ratios do not approximate 1:2:1. In general, a lipid molecule contains much less oxygen than a carbohydrate having the same number of carbon atoms. In addition to carbon, hydrogen, and oxygen, lipids may contain small quantities of other elements, such as phosphorus, nitrogen, or sulfur. Familiar lipids include fats, oils, and waxes. Most lipids are insoluble in water, but special transport mechanisms carry them in the circulating blood.

Lipids form essential structural components of all cells. In addition, lipid deposits are important as energy reserves, for on the average lipids provide roughly twice as much energy as carbohydrates, gram for gram, when broken down in the body. Fat deposits also help maintain body temperature. Many of these deposits lie just under the skin, and heat loss across

TABLE 2-6 Carbohydrates

Structure	*Examples*	*Primary Functions*	*Remarks*
Monosaccharides (simple sugars)	Glucose, fructose	Energy source	Manufactured in the body and obtained from food; found in body fluids.
Disaccharides	Sucrose, lactose	Energy source	Sucrose is table sugar, lactose is present in milk; must be broken down to monosaccharides before absorption.
Polysaccharides	Glycogen, starches, cellulose	Storage of glucose molecules	Glycogen is found in animal cells, starch and cellulose in plant cells.

Clinical Comment: Drugs and Lipids

Lipids account for 10–16 percent of the average body weight, and lipid-soluble drugs accumulate within this mass during treatment. As a result, the symptoms produced by an overdose of a lipid-soluble drug, such as cortisol or other steroids, are slow to subside. Such drugs will continue to diffuse out of the body's lipids even after administration has ceased. The active ingredient in marijuana, cannabinol, is a lipid-soluble molecule; because it slowly diffuses out of body lipids, a drug test can detect marijuana use days after the fact.

On the positive side, the body lipids can be a storehouse for useful compounds. For example, vitamin deficiency diseases do not appear until long after dietary intake of fat-soluble vitamins (A, E, D, or K) has ended, because these vital substances are "stockpiled" in the fats of our bodies.

a layer of lipids is only about one-third of that through other tissues.

All together, lipids normally account for roughly 12 percent of our total body weight. There are many different kinds of lipids in the body. Major lipid types are presented in Table 2-7 and diagrammed in Figures 2-16 and 2-17. We will consider six classes of lipid molecules: *fatty acids, glycerides, prostaglandins, steroids, phospholipids,* and *glycolipids.*

Fatty Acids

Fatty acids are long carbon chains with hydrogen atoms attached. One end of the carbon chain always bears a **carboxylic** (kar-bok-SIL-ic) **acid group**, a structure that can be diagrammed as:

$$\underset{\text{(carbon chain)}}{R\ldots -\overset{\text{OH}}{\overset{/}{C}} = O}$$

The carbon chain is abbreviated by "R" and the carboxylic acid group by COOH. The name carboxyl should help you remember that *carb*on and hydr*oxyl* are the important structural features. Figure 2-16a shows a representative fatty acid, *lauric acid.*

In a fatty *acid* the polar covalent bond between the oxygen and hydrogen atoms of the carboxylic acid

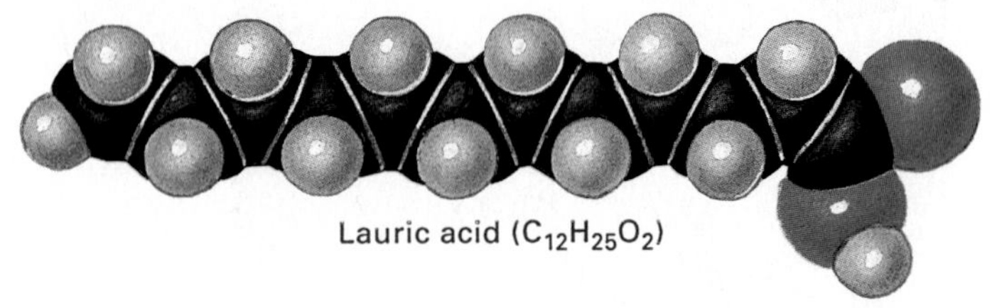

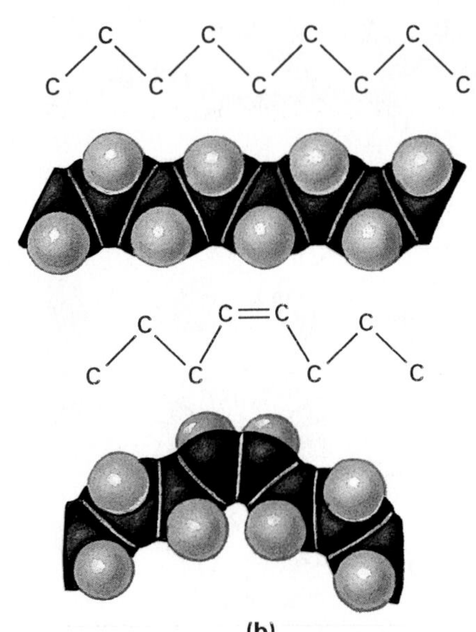

FIGURE 2-16
Fatty Acids. (a) Lauric acid demonstrates two structural characteristics common to all fatty acids: a long backbone of carbon atoms and a carboxylic acid group (–COOH) at one end. (b) A fatty acid may be saturated or unsaturated. Unsaturated fatty acids have double covalent bonds, and their presence causes a sharp bend in the molecule. (c) An omega-3 fatty acid has an unsaturated bond three carbons before the end of the carbon chain.

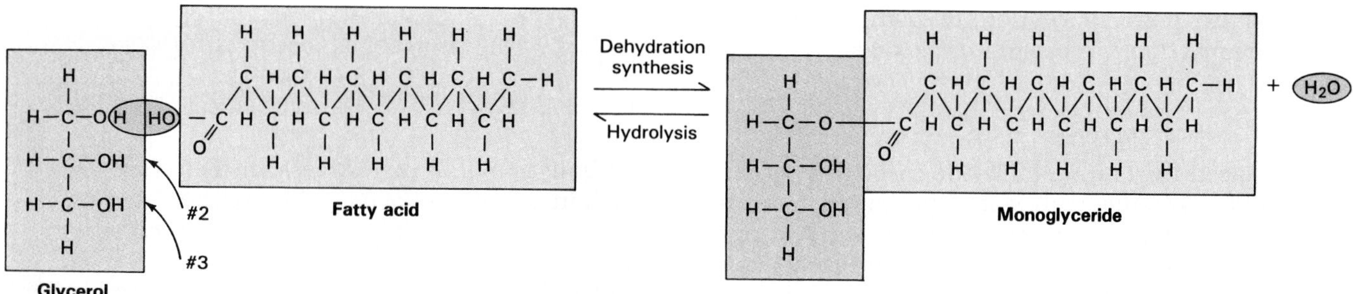

FIGURE 2-17
Glyceride Formation. The formation of a glyceride involves the attachment of fatty acids to the carbons of a glycerol molecule. This example shows the formation of a monoglyceride (glycerol + one fatty acid). Attachment of a second fatty acid at site #2 would create a diglyceride, and addition of a third at site #3 would produce a triglyceride.

TABLE 2-7 Representative Lipids and Their Functions

Lipid Type	*Examples*	*Primary Functions*	*Remarks*
Fatty acids	Lauric acid	Energy sources	Absorbed from food or synthesized in cells; transported in the blood for use in many tissues.
Glycerides	Monoglycerides, diglycerides, triglycerides	Energy source, energy storage, insulation, and physical protection	Stored in fat deposits; must be broken down to fatty acids and glycerol before they can be used as an energy source.
Prostaglandins		Chemical messengers coordinating local cellular activities	Produced in most body tissues.
Steroids	Cholesterol	Structural component of cell membranes, hormones, digestive secretions in bile	All have the same carbon-ring framework.
Phospholipids, glycolipids		Structural components of cell membranes	

Health News: Omega-3 Fatty Acids

Eskimos have lower rates of heart disease than other populations, despite the fact that the Eskimo diet contains high quantities of fats and cholesterol. Their dietary fats have an unusual structure, shown in Figure 2-16c. The carbons in a fatty acid molecule are numbered beginning at the carboxylic acid end; the last carbon in the chain is called the *omega* carbon. The fatty acids in the Eskimo diet have an unsaturated bond three carbons before the omega carbon, a position known as "omega minus 3" or **omega-3**. **Omega-3 fatty acids** are found in fish flesh and fish oils. For unknown reasons the presence of omega-3 fatty acids in the diet reduces the risks of heart disease, rheumatoid arthritis, and other inflammatory diseases. The typical U.S. diet contains large amounts of omega-6 fatty acids and relatively few omega-3 fatty acids. It is generally agreed that increasing fish consumption to two meals per week would have significant health benefits. However, taking large amounts of fish oil supplements cannot be recommended because the risks of high omega-3 fatty acid diets have not yet been fully assessed. One interesting observation is that Eskimos have a much higher incidence of strokes than the general U.S. population. It is not known whether this is related to their consumption of omega-3 fatty acids or whether other environmental or genetic factors may be involved.

group breaks down in solution, releasing a hydrogen ion. This reaction can be summarized as:

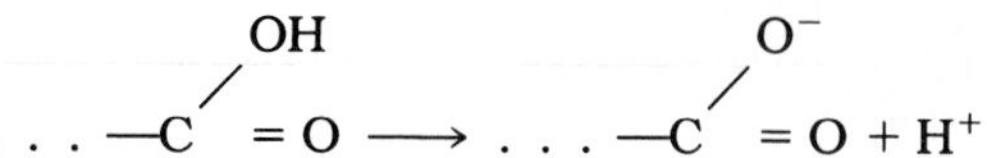

When placed in solution, only the carboxyl end of a fatty acid associates with water molecules, for this is the only polar portion of the molecule. The rest of the carbon chain is hydrophobic, and fatty acids have a very limited solubility in water.

In a **saturated** fatty acid each carbon atom in the hydrocarbon tail has four single covalent bonds. If some of the carbon-to-carbon bonds are double covalent bonds, the fatty acid is **unsaturated**. These terms refer to the number of hydrogen atoms in the fatty acid. Replacing a double bond between carbon atoms with a single bond allows the molecule to accept two more hydrogens; hence a double-bonded chain is said to be unsaturated (see Figure 2-16b). A **monounsaturated fatty acid** has a single unsaturated bond in the carbon tail. A **polyunsaturated** fatty acid contains multiple unsaturated bonds.

Both saturated and unsaturated fatty acids can be broken down for energy, but a diet containing large amounts of saturated fatty acids increases the risk of heart disease and other circulatory problems. Butter, fatty meat, and ice cream are popular dietary sources of saturated fatty acids. Vegetable oils such as olive oil or corn oil contain a mixture of monounsaturated and polyunsaturated fatty acids. Current research indicates that the monounsaturated fats may be more effective than polyunsaturated fats in lowering the risk of heart disease. Increasing the proportion of **oleic acid**, an 18-carbon monounsaturated fatty acid, in cooking oils could therefore yield health benefits.

Glycerides

Individual fatty acids cannot be strung together in a chain by dehydration, as simple sugars can. But they can be attached to another compound, **glycerol** (GLI-se-rol), through a similar reaction. As illustrated in Figure 2-17, dehydration synthesis can produce a **monoglyceride** (mo-nō-GLI-se-rīd) consisting of glycerol plus one fatty acid. Subsequent reactions can yield a **diglyceride** (glycerol + two fatty acids), and then a **triglyceride** (glycerol + three fatty acids). Hydrolysis breaks the glycerides into fatty acids and glycerol. A comparison of Figure 2-17 with Figure 2-14 reveals that dehydration synthesis and hydrolysis operate in the same way, whether the molecules involved are carbohydrates or lipids.

Triglycerides, otherwise known as **neutral fats** have several important functions:

Health News: Fat Substitutes

Although the average American diet is not as rich in fats as that of Eskimos, we still consume more fat than do people in many other parts of the world. Diets high in fat have been linked not only to heart disease but also to certain forms of cancer, and recent recommendations suggest that lowering the percentage of calories we derive from fat would benefit our health. This suggestion has led to an increased interest in the development of possible substitutes for fat.

Fat substitutes provide the texture, taste, and cooking properties of natural fats. One fat substitute, *Simplesse®*, has been approved by the Food and Drug Administration; a second, *Trailblazer®*, is currently under review. Both are made from proteins of egg white and skim milk or whey. The heated proteins are treated to form small spherical masses that have the taste and texture of fats. Simplesse can be used in place of fats in any application other than baking; it is found in low-calorie "ice creams" under the trade name *Simple Pleasures®*. These fat substitutes can be broken down in the body, but they provide less energy than natural fats. For example, ice cream with Simplesse has half the calories of ice cream containing natural fats.

Several other fat substitutes currently being developed are derived from carbohydrates. One of these, *Olestra®*, is made by chemically combining sucrose and fatty acids. The resulting compounds cannot be used by the body, and so contribute no calories at all. Olestra has been approved as an ingredient in margarines and baked goods; its use as a shortening and cooking oil is under review.

1. Fatty deposits in specialized sites represent a significant energy reserve. In times of need the triglycerides are disassembled to yield fatty acids that can be broken down to provide energy.
2. Fat deposits under the skin serve as insulation, preventing heat loss to the environment.
3. A fat deposit around a delicate organ such as a kidney provides a cushion that protects against shocks or blows.

Prostaglandins

Prostaglandins (pros-tah-GLAN-dins) are fatty acids that have five of their carbon atoms joined in a ring (Figure 2-18a). Cells release prostaglandins to coordinate or direct local cellular activities. They are extremely powerful and effective in minute quantities. Almost every tissue in the body contains prostaglan-

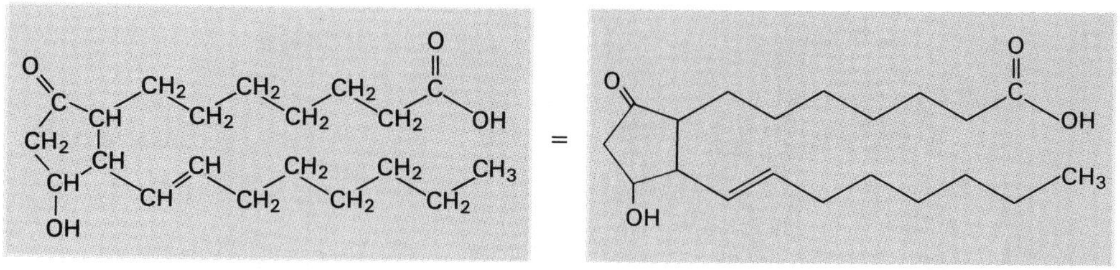

(a)

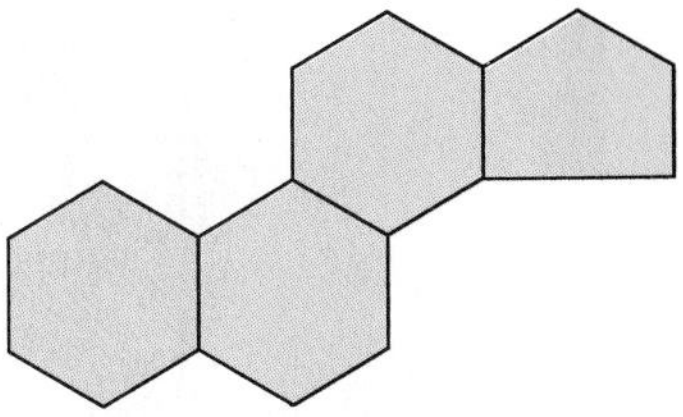

(b)

(c)

FIGURE 2-18
Prostaglandins and Steroids. (a) Prostaglandins are unusual short-chain fatty acids. To appreciate the chain structure, the complete structural formula for a typical prostaglandin (left) should be compared with the abbreviated formula (right), where the molecular framework is shown as a solid line. Each change in the direction of the line indicates the presence of a carbon atom. This diagrammatic method simplifies the chemical formula and makes it easier to visualize the complete molecule. (b) All steroids share this complex four-ring structure. (c) Individual steroids differ in the side chains attached to the carbon rings. This is a molecule of cholesterol.

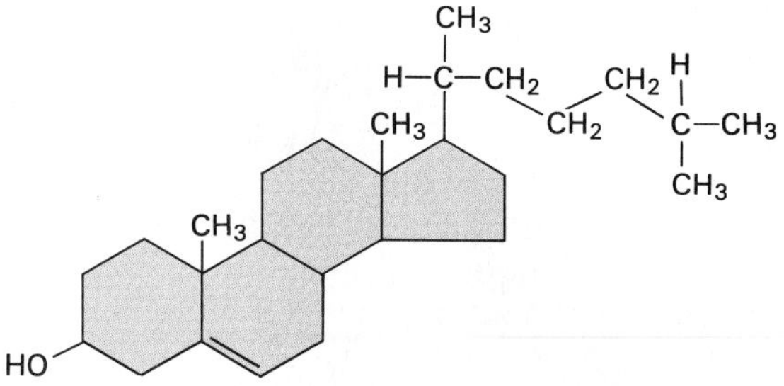

dins; the effects vary depending on the nature of the prostaglandin and the site of release. Consider three examples:

1. Prostaglandins released by damaged tissues stimulate nerve endings and produce the sensation of pain.
2. Prostaglandins released in the uterus help trigger the start of labor contractions.
3. Prostaglandins released in the stomach may help prevent stomach ulcers.

Prostaglandins are often called "local hormones" because hormones are chemical messengers produced at one site to regulate activities under way in a *different* portion of the body. (Hormone structure and function will be considered in Chapter 18).

Steroids

Steroids are large lipid molecules that share the distinctive carbon framework shown in Figure 2-18b. They differ in the carbon chains attached to this basic structure. Figure 2-18c shows the structural formula for the steroid **cholesterol**. All cell membranes contain cholesterol, and a number of hormones, including the male and female sex hormones, are derived from cholesterol.

The cholesterol needed to maintain cell membranes and manufacture steroid hormones comes from two sources. One major source is the diet; meat, cream, and egg yolks are especially rich in cholesterol. The body can synthesize cholesterol as well, and this ability sometimes makes it difficult to control blood cholesterol levels by dietary restriction alone.

Phospholipids and Glycolipids

Phospholipids (FOS-fō-lip-ids) and **glycolipids** (GLĪ-cō-lip-ids) are structurally related, and both can be synthesized from fatty acids. In a *phospho*lipid (Figure 2-19a) a phosphate group (PO_4) serves as a link

Health News: The Cholesterol Controversy

"Do you know your cholesterol level?" This is an important question because a strong link exists between high blood cholesterol concentrations and heart disease. Physicians now recommend maintaining blood cholesterol concentrations at under 200 mg/dℓ (milligrams per 100 mℓ). Current nutritional advice suggests reducing cholesterol intake to under 300 mg per day; this represents a 40 percent reduction for the average American adult. The connection between blood cholesterol levels and heart disease will be examined more closely in Chapters 21 and 25.

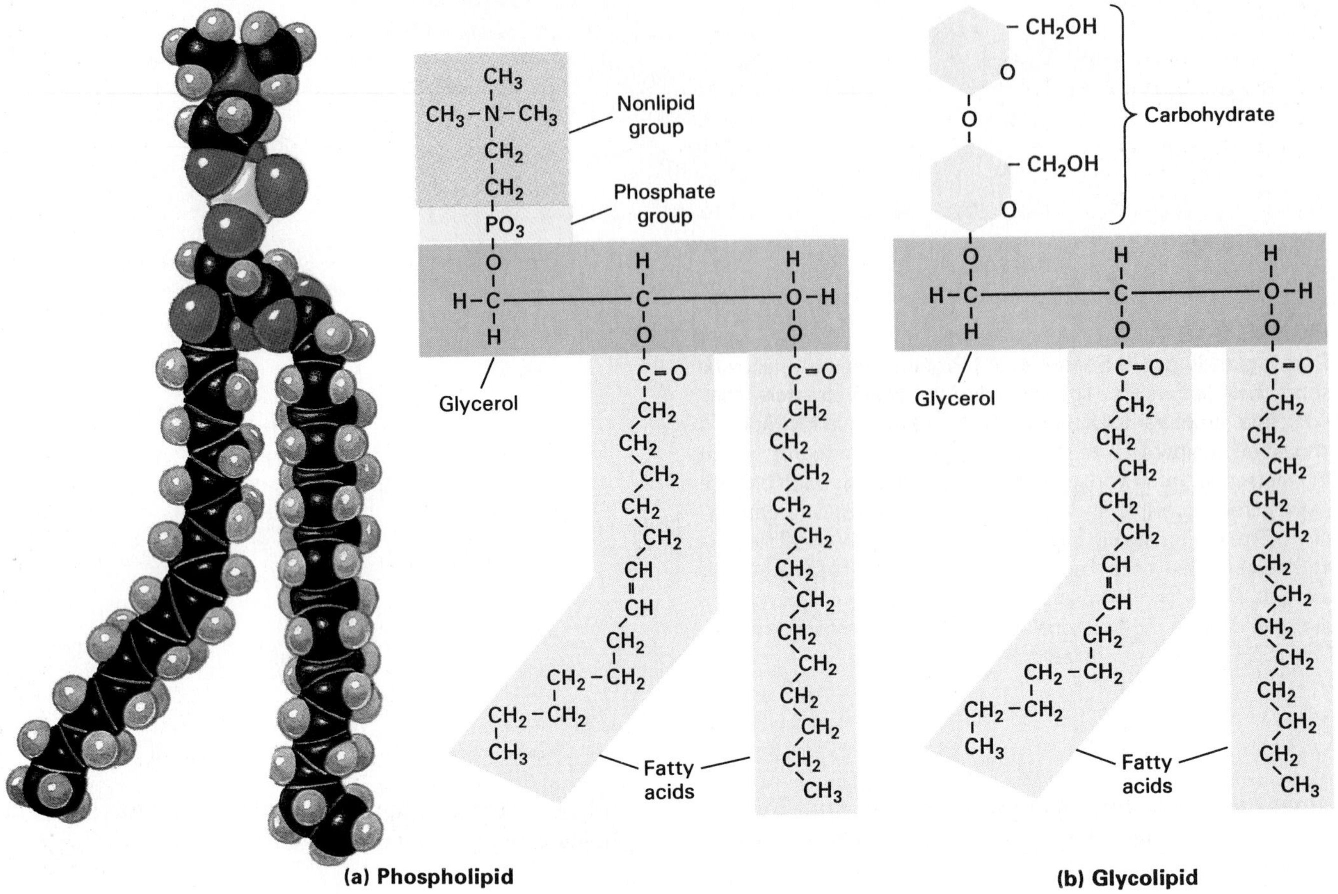

(a) Phospholipid

(b) Glycolipid

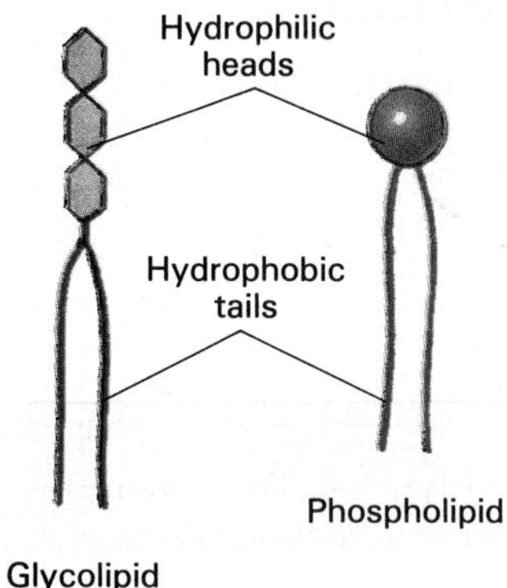

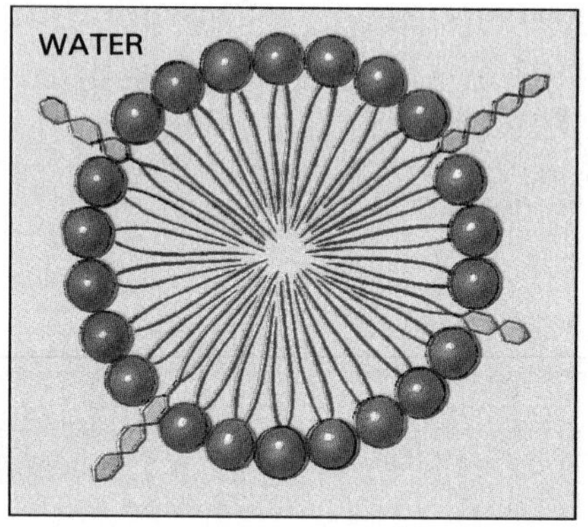

(c) Micelle structure

FIGURE 2-19
Phospholipids and Glycolipids. (a) In a phospholipid a phosphate group links a nonlipid molecule to a diglyceride. This is a molecule of *lecithin*, a representative phospholipid. (b) In a glycolipid a carbohydrate is attached to a diglyceride. (c) When present in large numbers these molecules will form micelles, with the hydrophilic heads facing the water molecules and the hydrophobic tails on the inside of the droplet.

between a diglyceride and a nonlipid group. In a *glycolipid* (Figure 2-19b) a carbohydrate is attached to a diglyceride. Note that placing "-lipid" *last* in these names indicates that the molecule consists primarily of lipid.

The long fatty acid "tails" of these molecules are hydrophobic, but the nonlipid "heads" are hydrophilic. In water, large numbers of these molecules tend to form droplets, or **micelles** (mī-SELLS), with the hydrophilic portions on the outside. Figure 2-19c illustrates this arrangement. Most meals contain a mixture of lipids and other organic molecules, and micelles form as the food breaks down in the digestive tract. In addition to phospholipids and glycolipids, micelles may contain other insoluble lipids, such as steroids, glycerides, and long-chain fatty acids.

Cholesterol, phospholipids, and glycolipids are called **structural lipids** because they form and maintain cellular membranes. Membranes composed of hydrophobic lipids surround each cell, and internal membranes subdivide the interior of the cell. Because they are separated by hydrophobic membranes, each of the intracellular compartments can have a distinctive chemical composition, and the watery interior of the cell is relatively isolated from the watery extracellular environment.

PROTEINS

Proteins are the most abundant organic components of the human body, and in many ways the most important. There are roughly 100,000 different kinds of proteins, and they account for about 20 percent of the total body weight. All proteins contain carbon, hydrogen, oxygen, and nitrogen; smaller quantities of sulfur may also be present.

Proteins perform a variety of essential functions. These include seven major functional categories.

1. *Support*: **Structural proteins** create a three-dimensional framework for the body, providing strength, organization, and support for cells, tissues, and organs.
2. *Movement*: **Contractile proteins** are responsible for muscular contraction; related proteins are responsible for the movement of individual cells.
3. *Transport*: Insoluble lipids, respiratory gases, special minerals, such as iron, and several hormones are carried in the blood attached to special **transport proteins.** Other specialized proteins transport materials from one part of a cell to another.
4. *Buffering*: Proteins provide a considerable buffering action, helping to restrict alterations in pH.
5. *Metabolic regulation*: Special proteins called **enzymes** accelerate chemical reactions in living cells. The sensitivity of enzymes to environmental factors is extremely important in controlling the pace and direction of metabolic operations.
6. *Coordination and control*: Protein hormones can influence the metabolic activities of every cell in the body or affect the function of specific organs or organ systems.
7. *Defense*: The tough, waterproof proteins of the skin, hair, and nails protect the body from environmental hazards. The proteins of the **immune system** known as **antibodies** protect us from disease. Special clotting proteins restrict bleeding following an injury to the circulatory system.

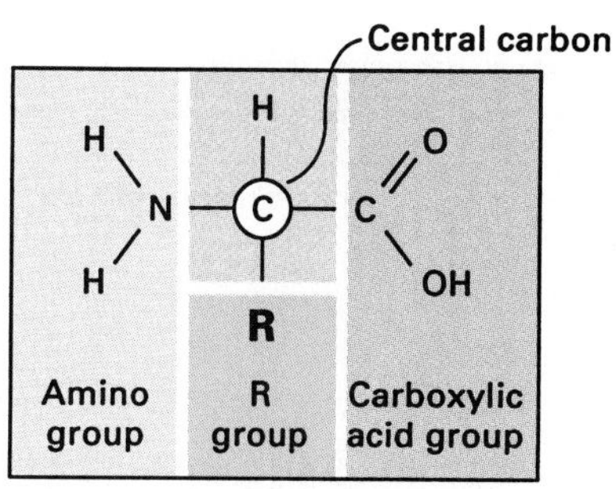

(a) Structure of an amino acid

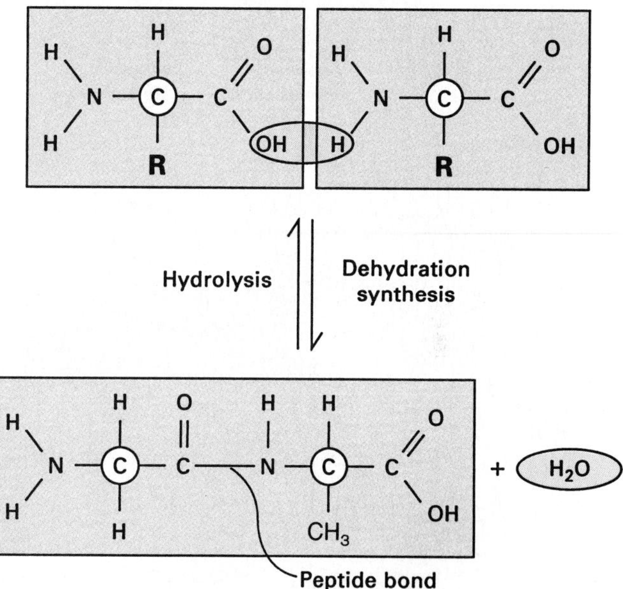

(b) Peptide bond formation

FIGURE 2-20
Amino Acids and Peptide Bonds. (a) Each amino acid consists of a central carbon atom to which four different groups are attached: a hydrogen atom, an amino group ($-NH_2$), a carboxylic acid group ($-COOH$), and a variable group generally designated R. (b) Peptides form as dehydration synthesis creates a peptide bond between the carboxyl group of one amino acid and the amino group of another. In this example glycine and alanine are linked to form a dipeptide.

Structure of Proteins

Proteins are chains of small molecules called **amino acids**. The human body contains significant quantities of 20 different amino acids. A typical protein contains 1000 amino acids, but the largest protein complexes may have a hundred thousand or more. Each amino acid consists of a central carbon atom to which four groups are attached: (1) a hydrogen atom; (2) an **amino group** ($-NH_2$); (3) a carboxylic acid group ($-COOH$); and (4) a variable group, known as an **R group** or *side chain*, which differs from one amino acid to another, giving each its individual chemical properties (Figure 2-20a). The name *amino acid* refers to the presence of the *amino* group and the carboxylic *acid* group. These are hydrophilic groups, and amino acids are relatively small molecules. As a result, they are water-soluble even if the side chains are hydrophobic. Figure 2-20b shows two representative amino acids, **glycine** and **alanine.** (All 20 amino acids commonly found in proteins are diagrammed in Appendix IV.)

As indicated in Figure 2-20b, dehydration synthesis can link two amino acids. This process attaches the carboxylic acid group of one amino acid to the amino group of another. The connection is called a **peptide bond**, and the molecule created is a **dipeptide**.

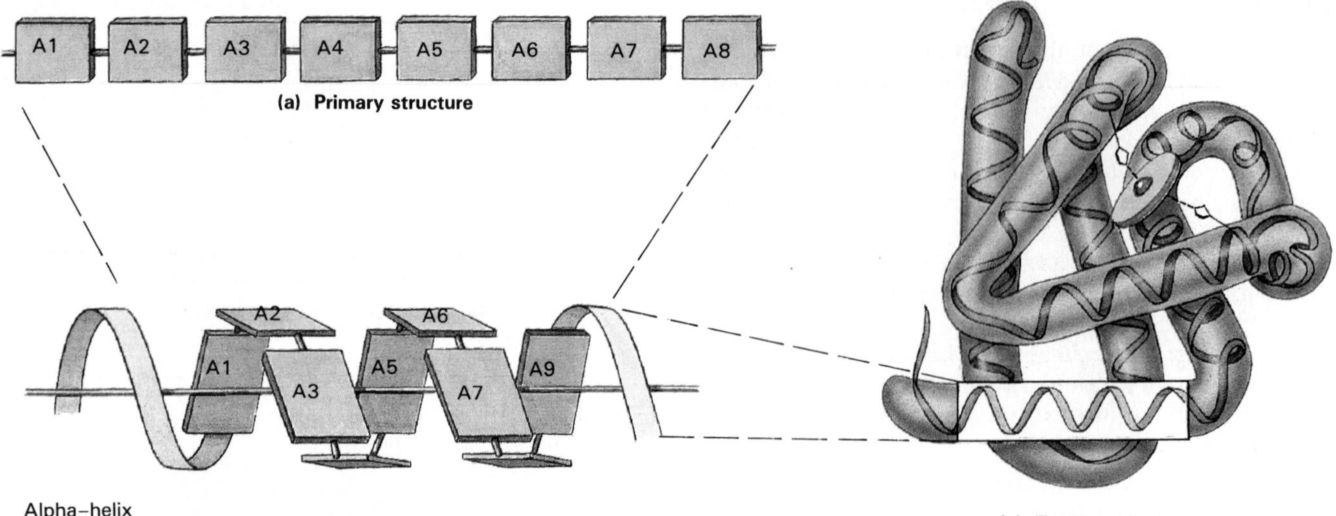

FIGURE 2-21
Protein Structure. (a) The primary structure of a polypeptide is the sequence of amino acids along its length. (b) The secondary structure is created by hydrogen bonding along the length of the polypeptide chain. Such bonding often produces a simple spiral, called an alpha-helix. It may also produce a flattened arrangement known as a pleated sheet. (c) Tertiary structure appears as the completed chain interacts with surrounding water molecules. Attraction between R groups of amino acids in different parts of the chain also play a large role in tertiary structure. This is the structure of myoglobin, a globular protein involved in the storage of oxygen in muscle tissue. In the cylindrical segments the polypeptide chain is arranged in an alpha-helix.

The chain can be lengthened by the addition of more amino acids. Attaching a third produces a **tripeptide**, and there are tetrapeptides, pentapeptides, and so forth. **Polypeptides** containing more than 100 amino acids are usually called proteins. You are probably already familiar with the names of several important proteins, including **hemoglobin** in red blood cells and **keratin** in fingernails and hair.

Shape of Proteins

The characteristics of a particular protein are determined in part by the R groups on its component amino acids. But the properties of a protein are more than just the sum of the properties of its parts, for polypeptides and proteins can have very complex shapes. There are several levels of structural organization:

1. The **primary structure** of a protein is the sequence of amino acids along its length. The primary structure of a short peptide chain has been diagrammed in Figure 2-21a.
2. The **secondary structure** appears as hydrogen bonding occurs between parts of the polypeptide chain. This often creates a simple spiral, known as an **alpha-helix**, shown in Figure 2-21b. Less often, hydrogen bonding produces a flat **pleated sheet**. A single polypeptide chain may have both helical and pleated sections.
3. The **tertiary structure** of a protein is the complex coiling and folding that gives the molecule its unique shape. Tertiary structure results primarily from interactions between the polypeptide chain and the surrounding water molecules, and partly from interactions between the R groups of amino acids in different parts of the molecule. Figure 2-21c shows the tertiary structure of **myoglobin**, a protein found in muscle cells. Myoglobin is an example of a **globular protein**. Globular proteins are compact, usually rounded, and water-soluble.
4. Some proteins contain several polypeptide chains, each with its own secondary or tertiary structure. Interactions between these polypeptides determine the **quaternary structure** of the functional protein. Figure 2-22 shows two im-

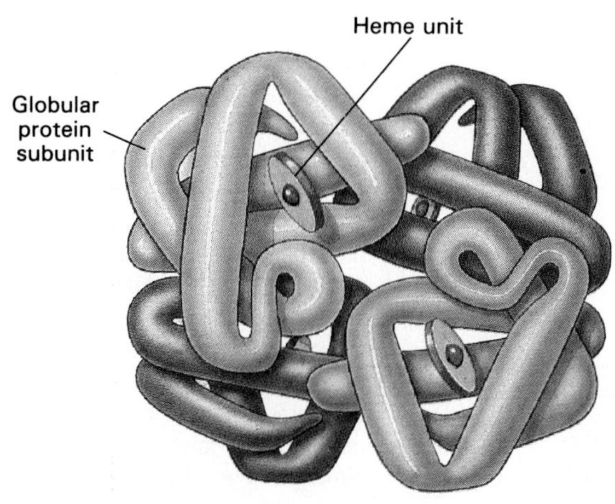

(a) Hemoglobin molecule

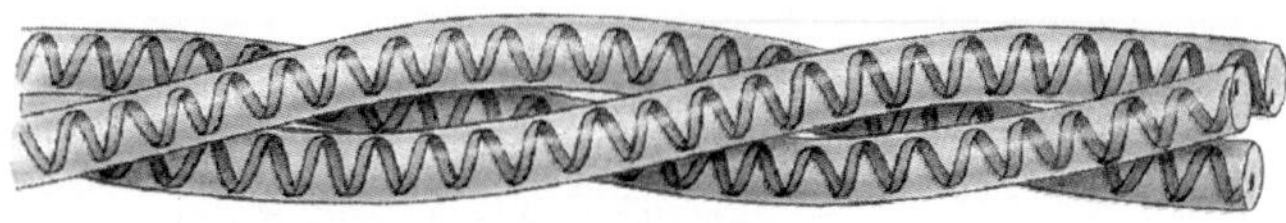

(b) Keratin fiber

FIGURE 2-22
The Quaternary Structure of Proteins. The quaternary structure develops when separate polypeptide subunits interact to form a larger molecule. (a) A single hemoglobin molecule contains four globular subunits, each structurally similar to myoglobin. Hemoglobin transports oxygen in the blood; the oxygen binds to the heme units. (b) In keratin, three fibrous subunits intertwine like the strands of a rope. Keratin is a tough, water-resistant protein found in skin, hair, and nails.

portant examples of quaternary structure. The protein **hemoglobin** (Figure 2-22a) contains four globular subunits, each structurally similar to myoglobin. Hemoglobin is found within red blood cells, where it binds and transports oxygen. In **keratin** (Figure 2-22b) three helical polypeptides are wound together like the strands of a rope. Keratin is an example of a **fibrous protein.** Fibrous proteins are generally insoluble, and they form a strong supporting framework in cells and tissues. Keratin is the tough, water-resistant protein found in skin, nails, and hair.

Denaturation and Protein Function

The shape of a protein determines its functional properties. Small changes in the ionic composition, temperature, or pH of their surroundings can affect protein function by altering tertiary or quaternary structure. Extreme environmental conditions may permanently alter the shape of the protein by breaking the bonds responsible for tertiary and quaternary structure. This irreversible change is called **denaturation**. You see denaturation in progress each time you fry an egg, for the clear egg white contains abundant dissolved proteins. As the temperature rises, the protein structure changes. Eventually the egg proteins become completely denatured, forming an insoluble white mass. Very high body temperatures (over 43° C, or 110° F) will denature tissue proteins. Because denaturation drastically alters the functional characteristics of proteins, high body temperatures cause irreversible damage to organs and organ systems.

Glycoproteins and Proteoglycans

Glycoproteins (GLĪ-kō-prō-tēns) and **proteoglycans** (prō-tē-o-GLĪ-kans) are combinations of protein and carbohydrate molecules. Glyco*protein* molecules are large proteins with small carbohydrate groups attached. Glycoproteins may function as enzymes, antibodies, or protein components of cell membranes. Proteo*glycans* are large polysaccharide molecules connected to short polypeptide chains. Proteoglycan secretions coat the surfaces of the respiratory and digestive tracts, providing lubrication. Proteoglycans dissolved in tissue fluids give them a syrupy consistency.

NUCLEIC ACIDS

Nucleic (noo-KLĀ-ik) **acids** are large organic molecules composed of carbon, hydrogen, oxygen, nitrogen, and phosphorus. Nucleic acids store and process information at the molecular level, inside living cells. There are two classes of nucleic acid molecules, **deoxyribonucleic** (dē-ok-se-rī-bō-noo-KLĀ-ik) **acid**, or **DNA**, and **ribonucleic** (rī-bō-noo-KLĀ-ik) **acid**, or **RNA**.

The DNA in our cells determines our inherited characteristics, such as eye color, hair color, blood type, and so on. It affects all aspects of body structure and function because DNA molecules encode the information needed to build proteins. By directing the synthesis of structural proteins, DNA controls the shape and physical characteristics of our bodies. By controlling the manufacture of enzymes, DNA regulates not only protein synthesis but all aspects of cellular metabolism, including the creation and destruction of lipids, carbohydrates, and other vital molecules.

Several different forms of RNA cooperate to manufacture specific proteins using the information provided by DNA. The functional relationships between DNA and RNA will be detailed in Chapter 4.

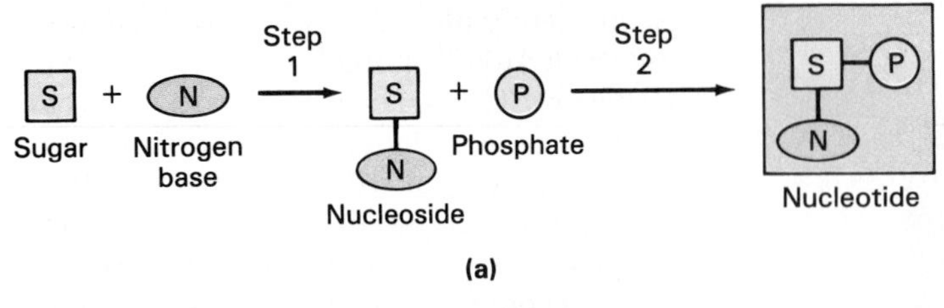

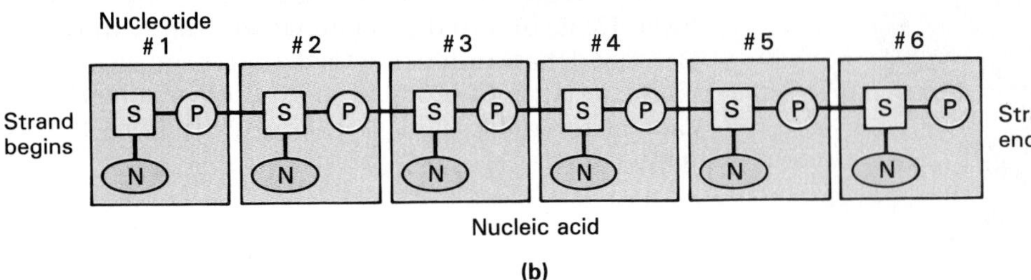

FIGURE 2-23
Nucleic Acid Structure. (a) Step 1: Combination of a sugar and a nitrogen base creates a nucleoside. Step 2: Addition of a phosphate group to a nucleoside produces a nucleotide. (b) Nucleic acids are long chains of nucleotides; the structure of a short chain (six nucleotides) is diagrammed here. Each molecule starts at the sugar-nitrogen base of the first nucleotide and ends at the phosphate group of the last member of the chain.

FIGURE 2-24
RNA and DNA. (a) An RNA molecule consists of a single nucleotide chain. Its shape is determined by the sequence of nucleotides and the interactions between them. (b) A DNA molecule consists of a pair of nucleotide chains linked by hydrogen bonding between complementary base pairs.

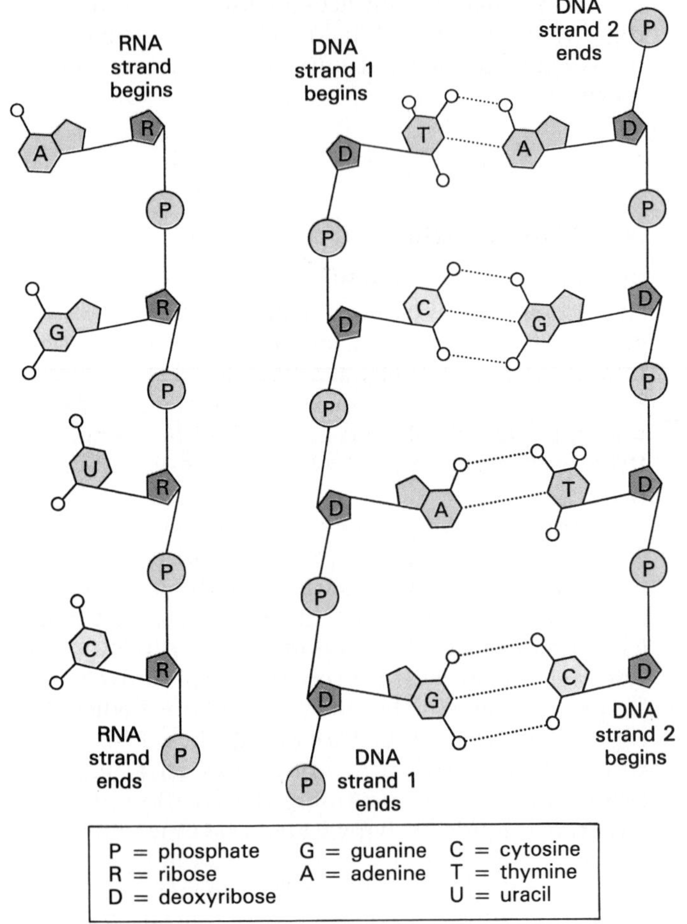

Structure of Nucleic Acids

A nucleic acid consists of a series of **nucleotides** linked by dehydration synthesis. A single nucleotide has three basic components: a sugar, a phosphate group, and a **nitrogen base**. The sugar is always a **pentose** (five-carbon sugar), either **ribose** (in RNA) or **deoxyribose** (in DNA). Each pentose is attached a phosphate group (PO_4) and a nitrogen base. There are five different nitrogen bases: **adenine** (abbreviated A), **guanine** (G), **cytosine** (C), **thymine** (T), and **uracil** (U). Adenine and guanine are two-ringed structures called **purines**; the others are single-ringed **pyrimidines**. Both RNA and DNA contain adenine, guanine, and cytosine. Uracil is found only in RNA, and thymine only in DNA.

Figure 2-23a diagrams the steps in the formation of a nucleic acid. The attachment of a nitrogen base to a carbohydrate (Step 1) creates a **nucleoside** that is named after the nitrogen base. For example, attaching a guanine molecule produces a guanine nucleoside. A nucleotide forms when a phosphate group binds to the carbohydrate of the nucleoside (Step 2). Dehydration synthesis then attaches the phosphate group of one nucleotide to the carbohydrate of another. The nitrogen bases are unaffected, and the "backbone" of a nucleic acid molecule is the sugar-to-phosphate-to-sugar sequence illustrated in Figure 2-23b.

RNA and DNA

There are important structural differences between RNA and DNA. A molecule of RNA consists of a single chain of nucleotides (Figure 2-24a). Its shape depends on the order of the nucleotides and the interactions between them. A DNA molecule consists of a pair of nucleotide chains, as illustrated in Figure 2-24b. Hydrogen bonding between opposing nitrogen bases holds the two strands together. Because of their shapes, adenine can bond only with thymine, and cytosine only with guanine. As a result, adenine/thymine and cytosine/guanine are known as **complementary base pairs**.

The two strands of DNA twist around one another in a **double helix** that resembles a spiral staircase. The stair steps correspond to the nitrogen base pairs. Figure 2-25 presents two three-dimensional views of a DNA molecule. Table 2-8 summarizes the differences between DNA and RNA.

FIGURE 2-25
The Structure of DNA. The molecule is depicted in two ways. (a) The positions of the individual atoms. (b) A more schematic representation highlighting the double helix, or "spiral staircase" organization, of the two strands.

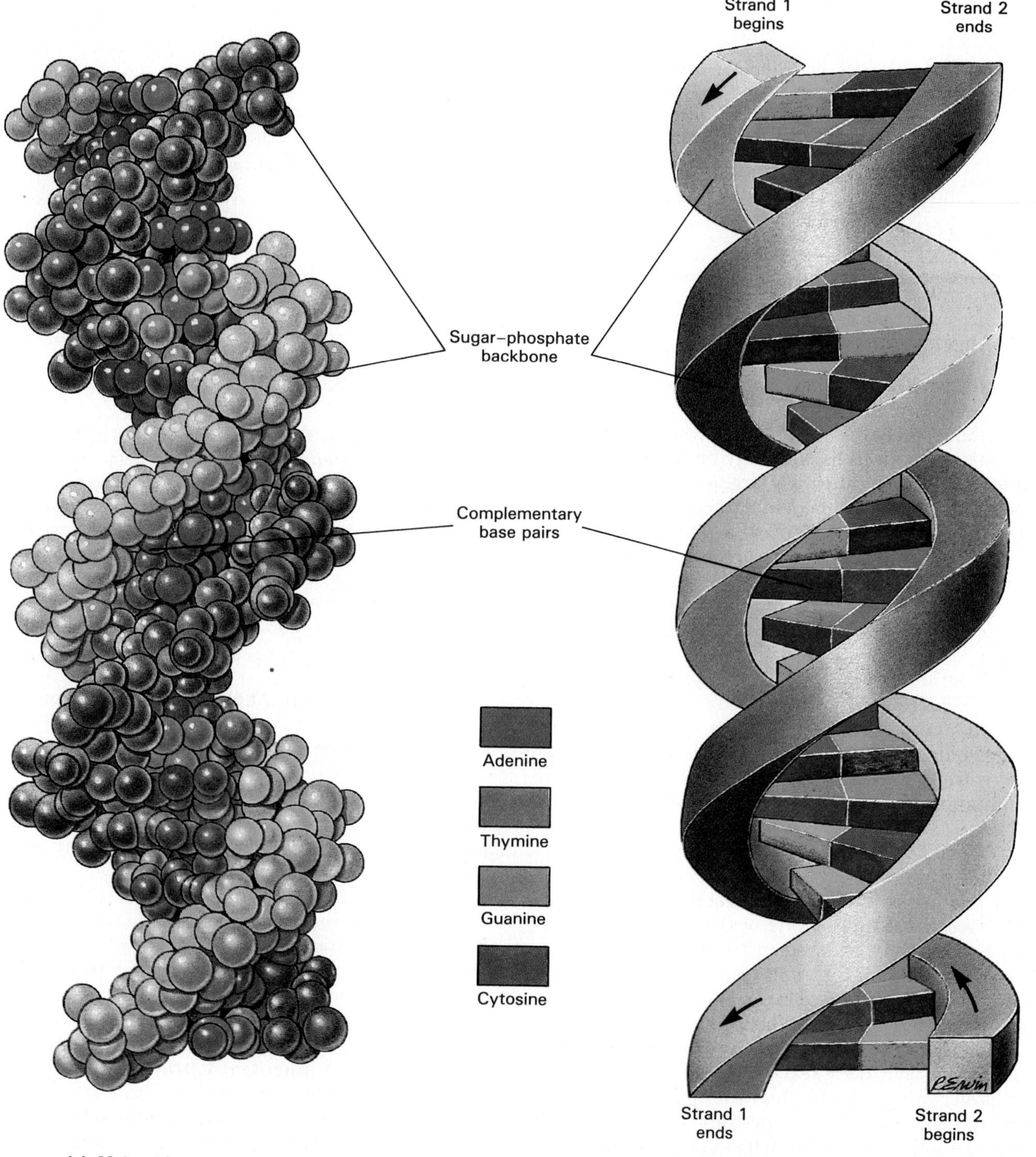

TABLE 2-8 A Comparison of RNA and DNA

Characteristic	*RNA*	*DNA*
Sugar	Ribose	Deoxyribose
N-bases	Adenine Guanine Cytosine Uracil	Adenine Guanine Cytosine Thymine
Number of nucleotides in typical molecule	Varies from under 100 nucleotides to around 50,000	Always over 45 million nucleotides
Shape of molecule	Varies, depending on hydrogen binding along the length of the strand	Paired strands coiled in a double helix
Function	Performs protein synthesis as directed by genetic programming	Stores genetic information

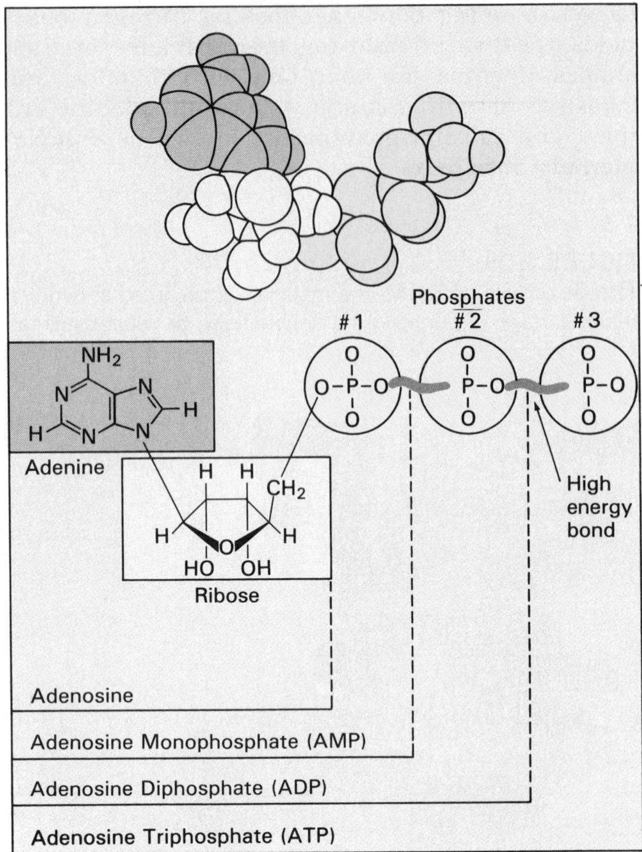

FIGURE 2-26
The Structure of ATP. A space-filling model of ATP (top) has been color-coded to match the structural formula below. Note that the molecule consists of an adenine nucleoside (Figure 2-23) to which three phosphate groups have been joined. Both the second and third phosphates are bound to the molecule by high-energy bonds. Cells most often store energy by attaching a third phosphate group to ADP. Removing the phosphate group releases the energy for cellular work, including the synthesis of other molecules.

HIGH-ENERGY COMPOUNDS

Catabolism releases energy, and living cells can use that energy in constructive ways. The energy released by catabolic reactions is first harnessed to create high-energy bonds. A **high-energy bond** is a covalent bond that stores an unusually large amount of energy. In our cells the formation of a high-energy bond usually involves the attachment of a phosphate group (PO_4) to another molecule. This process is called **phosphorylation** (fos-for-i-LĀ-shun). When that bond is later broken, perhaps in a distant portion of the cell, the energy will be released under controlled conditions.

Phosphorylation creates a high-energy compound, a molecule whose structure includes one or more high-energy bonds. The most important high-energy compound is an adenine nucleoside, adenosine, with three phosphate groups attached. This combination is called **adenosine triphosphate**, or **ATP**.

Figure 2-26 details the structure of ATP. It takes very little energy to attach the first phosphate and create the nucleotide adenosine monophosphate, or AMP. This nucleotide is one of the building blocks of nucleic acids, discussed in the previous section. The phosphate bond in AMP does not contain large amounts of energy, and cells can easily make or break down nucleotide molecules. Attaching a second phosphate to the first produces **adenosine diphosphate**, or **ADP**. This synthetic reaction requires a considerably larger energy input, but the greatest amount of energy is needed to add the third phosphate and produce a molecule of ATP. Within our cells the conversion of ADP to ATP represents the primary method of energy storage, and the reverse reaction provides a mechanism for controlled energy release. The arrangement can be summarized as:

$$\text{ATP} + H_2O \leftrightharpoons \text{ADP} + \text{phosphate group} + \text{energy}$$

When energy sources are available, our cells make ATP from ADP; when energy is required, the reverse reaction occurs.

Although ATP is the most abundant high-energy compound, there are others. These compounds are typically other nucleotides that have undergone phosphorylation. For example, *guanosine triphosphate* **(GTP)** and *uridine triphosphate* **(UTP)** are nucleotide-based high-energy compounds that transfer energy in specific enzymatic reactions. High-energy compounds and their functions will be discussed in more detail in Chapter 4.

Table 2-9 summarizes information concerning the inorganic and organic compounds considered in this chapter.

TABLE 2-9 Structure and Function of Biologically Important Compounds

Class	*Function*	*Building Blocks*
INORGANIC		
Water	Solvent, transport medium for dissolved materials and heat, cooling through evaporation.	Hydrogen and oxygen; absorbed as liquid water or generated via metabolism.
Acids, bases, salts	Important as structural components, buffers, sources of ions.	Obtained from the diet or generated via metabolism.
ORGANIC		
Carbohydrates	Energy source; some structural role when attached to lipids or proteins; energy storage.	C, H, O, and perhaps N; CHO in a 1:2:1 ratio; obtained in diet or manufactured in the body.
Lipids	Energy source; energy storage; insulation; structural components; chemical messengers; physical protection.	C, H, O, and perhaps N or P; CHO not in 1:2:1 ratio; obtained in diet or manufactured in the body.
Proteins	Catalysts for metabolic reactions; structural components; movement; transport; buffers; defense; control and coordination of activities.	C, H, O, N, often S; 20 common amino acids; roughly half can be manufactured in the body; the others must be obtained in the diet.
Nucleic acids	Storage and processing of genetic information.	C, H, O, N, and P; nucleotides composed of phosphates, sugars, and nitrogen bases; absorbed or manufactured.

CHEMICALS AND LIVING CELLS

The human body is more than a collection of chemicals. Biochemical building blocks form functional units called **cells**. Each cell behaves like a miniature organism, responding to internal and external stimuli. A lipid membrane separates the cell from its environment, and internal membranes create compartments with specific functions. Proteins form an internal supporting framework and act as enzymes to accelerate and control the chemical reactions that maintain homeostasis. Nucleic acids direct the synthesis of all cellular proteins, including the enzymes that enable the cell to synthesize a wide variety of other substances. Carbohydrates provide energy for vital activities and form part of specialized compounds, such as proteoglycans or glycolipids. The next two chapters consider the ways these chemicals interact in a living, functional cell.

√ **A food contains organic molecules with the elements C, H, and O in a ratio of 1:2:1. What type of compound is this and what is its major function in the body?**

√ **Why does boiling a protein affect its structure and functional properties?**

CHAPTER REVIEW

Review of Selected Clinical Terms

radioisotopes (*p. 29*)
Isotopes with unstable nuclei that emit subatomic particles in measurable amounts.

tracer (*p. 29*)
A compound labeled with a radioisotope that can be tracked within the body by the radiation it releases.

nuclear imaging (*p. 29*)
A procedure in which an image is created on a photographic plate or video screen by the radiation emitted by injected radioisotopes.

PET (positron emission tomography) scan (*p. 30*)
A nuclear imaging technique in which the emitted radiation is analyzed and the image created by a computer.

mole (M), millimole (mmol) (*p. 30*)
A mole is the number of atoms or molecules in a sample that has a weight in grams equal to the atomic weight (for an element) or the molecular weight (for a molecule.) 1 mmol = 0.001 M. These units are often used to measure solute concentrations.

omega-3 fatty acids (*p. 47*)
Fatty acids abundant in fish flesh and fish oils, that have a double bond three carbons away from the end of the hydrocarbon chain. Their presence in the diet has been linked to reduced risks of heart disease and other conditions.

cholesterol (*p. 49*)
A steroid important in the structure of cellular membranes, that in high concentrations, increases the risks of heart disease.

Additional Terms of Clinical Importance

RADIOISOTOPES AND CLINICAL TESTING (*pp. 29–30*): **alpha particles, beta particles, gamma rays, half-life, thyroid scan, RAIU**

Study Outline

Related Key Terms

Atoms and Molecules (pp. 28–35)

1. **Atoms** are the smallest units of matter; they consist of **protons**, **neutrons**, and **electrons**.

STRUCTURE OF THE ATOM (pp. 28–31)

2. An **element** consists entirely of atoms with the same number of protons (**atomic number**). Within an atom, an **electron cloud** surrounds the **nucleus**. — **hydrogen • nanometers**
3. The **mass number** of an atom indicates the total number of protons and neutrons in its nucleus. **Isotopes** are atoms of the same element whose nuclei contain different numbers of neutrons.
4. Electrons occupy a series of **energy levels** that are often illustrated as **electron shells**. The electrons in the outermost energy level determine an atom's chemical properties. — **atomic weight • mole**

CHEMICAL BONDS AND CHEMICAL COMPOUNDS (pp. 31–35).

5. Atoms can combine to form a **molecule**; combinations of atoms of different elements form a **compound**. Some atoms share electrons to form a molecule held together by **covalent bonds**. — **inert gases**
6. Sharing one pair of electrons equally creates a **single covalent bond**; sharing two pairs equally forms a **double covalent bond**. An unequal sharing of electrons creates a **polar covalent bond**. — **carbon dioxide**
7. An **ionic bond** results from the attraction between **ions**: atoms that have gained or lost electrons. **Cations** are positively charged, **anions** are negatively charged.
8. A **hydrogen bond** is a weak but important force that can affect the shapes and properties of molecules.

Chemical Notation (pp. 35–36)

1. **Chemical notation** allows us to describe reactions between **reactants** that generate one or more **products**. — **molecular weight**, **balanced**

Chemical Reactions (pp. 36–38)

TYPES OF REACTIONS (pp. 36–37)

1. A **chemical reaction** may be classified as a **decomposition**, **synthe-**

Related Key Terms

sis, or **exchange reaction**. **Exergonic** reactions release energy; **endergonic** reactions absorb energy.

REVERSIBLE REACTIONS (p. 37)

2. At **equilibrium** the rates of two opposing reactions are in balance.
3. **Enzymes** control many chemical reactions within our bodies. **Nutrients** and **metabolites** can be broadly classified as **organic** (carbon-based) or **inorganic**.

pathway

Inorganic Compounds (pp. 38–42)

WATER AND ITS PROPERTIES (pp. 38–41)

1. Water is the most important component of the body; its high **heat capacity** helps stabilize body temperature. It is a **polar molecule**.
2. Many inorganic compounds, called **electrolytes**, will undergo **dissociation** in water to form ions.
3. The **pH** of a solution indicates the concentration of hydrogen ions it contains. Solutions can be classified as **neutral**, **acidic**, or **basic (alkaline)** on the basis of pH.

solution • **solvent** • **solute**
ionization
hydration sphere • **hydrophilic**
hydrophobic
hydroxyl group

ACIDS, BASES, AND SALTS (p. 41)

4. An **acid** releases hydrogen ions, while a **base** removes hydrogen ions from a solution. A **salt** produced through the interaction of an acid and a base is an ionic compound that is often neutral.

BUFFERS AND pH CONTROL (pp. 41–42)

5. **Buffers** maintain pH within normal limits in body fluids.

strong acids • **weak acids**
strong bases • **weak bases**
bicarbonate ions
sodium bicarbonate

Organic Compounds (pp. 42–57)

1. **Organic compounds** contain carbon and hydrogen, and usually oxygen as well. They can be classified as carbohydrates, lipids, proteins, or nucleic acids.

CARBOHYDRATES (pp. 42–45)

2. **Carbohydrates** are most important as an energy source for metabolic processes. The three major types are **monosaccharides (simple sugars)**, **disaccharides**, and **polysaccharides**.
3. **Dehydration synthesis** can create complex molecules by joining simpler components, while **hydrolysis** breaks complex molecules into smaller fragments.

glucose • **isomers** • **fructose**

sucrose • **glycogen** • **starches**
cellulose

LIPIDS (pp. 45–50)

4. **Lipids** are water-insoluble molecules that include fats, oils, and waxes. There are six important classes of lipids: **fatty acids**, **glycerides**, **prostaglandins**, **steroids**, **phospholipids**, and **glycolipids**.
5. Fatty acids can be linked to form a **monoglyceride**, **diglyceride**, or **triglyceride** (neutral fat).
6. Prostaglandins and some steroids function as *hormones*.
7. **Cholesterol** is a precursor of steroid hormones and is an important component of membranes.

carboxylic acid group
saturated fatty acid
unsaturated fatty acid
monounsaturated fatty acid
polyunsaturated fatty acid
oleic acid
micelles • **structural lipids**

PROTEINS (pp. 51–53)

8. Proteins perform a great variety of functions in the body. Important types of proteins include **structural proteins**, **contractile proteins**, **transport proteins**, **enzymes**, and **antibodies**.
9. Proteins are chains of **amino acids** linked by **peptide bonds**.
10. Protein and carbohydrate molecules combine to form **glycoproteins** and **proteoglycans**.

immune system
amino group • **R group**
glycine • **alanine** • **dipeptide**
tripeptide • **polypeptides**
hemoglobin • **keratin**
primary structure
secondary structure
tertiary structure
quaternary structure
alpha-helix • **pleated sheet**
myoglobin • **globular protein**
fibrous protein • **denaturation**

NUCLEIC ACIDS (pp. 53–55)

11. **Nucleic acids** store and process information at the molecular level. There are two kinds of nucleic acids: **deoxyribonucleic acid (DNA)** and **ribonucleic acid (RNA)**.
12. Nucleic acids are chains of **nucleotides**. Each nucleotide contains a sugar, a **phosphate group**, and a **nitrogen base**. The sugar is always **ribose** or **deoxyribose**. The nitrogen bases found in DNA are **adenine**, **guanine**, **cytosine**, and **thymine**. In RNA, **uracil** replaces thymine.

purines • **pyrimidines**
nucleoside
complementary base pairs
double helix

HIGH-ENERGY COMPOUNDS (pp. 56–57)

13. A **high-energy bond** stores a large amount of energy; such bonds are often formed through **phosphorylation**. The chief high-energy compound is **adenosine triphosphate (ATP)**, formed by adding a third phosphate group to **adenosine diphosphate (ADP)**.

GTP • **UTP**

CHEMICALS AND LIVING CELLS (p. 57).

14. Biochemical building blocks form functional units called **cells**.

Review Planner

		Level 1	Level 2	35 37 46
1	Describe an atom and how atomic structure affects interactions between atoms.	1 2 4 10 12 23 44		
2	Compare the different ways in which atoms combine to form molecules and compounds.	8 10 13 14 30 32 33	Level 3	47 48 49
3	Use chemical notation to symbolize chemical reactions.	29 36 38 40 41 43		
4	Distinguish between the three major types of chemical reactions that are important for studying physiology.	5 16 29	C M	Solute Concentrations
5	Explain how the chemical properties of water make life possible.	13 18 22 39		
6	Discuss the importance of pH and the role of buffers in body fluids.	17 26 27 30 31 34		
7	Distinguish between organic and inorganic compounds.	7 19		
8	Describe the physiological roles of inorganic compounds.	18 19 42 45		
9	Discuss the structure and functions of carbohydrates, lipids, proteins, nucleic acids, and high-energy compounds.	3 6 9 11 15 20 21 24 25 28 42 45		

Review Questions

MULTIPLE CHOICE

1. The mass number of an atom is: a) the total number of protons in the nucleus; b) the total number of protons and neutrons in the nucleus; c) the total number of particles in the atom; d) the total number of electron shells in the atom.
2. The fundamental subatomic particles are: a) protons, neutrons, and elements; b) neutrons, electrons, and isotopes; c) protons, neutrons, and electrons; d) protons, electrons, and isotopes.
3. Proteins are chains of small molecules called: a) amino acids; b) glycoproteins; c) nucleic acids; d) fatty acids.
4. The atomic number of an element is determined by its: a) electrons; b) energy level; c) neutrons; d) protons.
5. Which of the following is *not* a true statement? a) A synthesis reaction is the opposite of a decomposition reaction; b) In an exchange reaction, larger molecules are exchanged for smaller ones; c) Exchange reactions involve both decomposition and synthesis reactions; d) Exchange reactions can be either exergonic or endergonic.
6. Which of the following is *not* a function of proteins? a) transport; b) support; c) movement; d) insulation.
7. Glucose and fructose are examples of: a) isomers; b) carbohydrates; c) simple sugars; d) all of the above.
8. Chemical bonds that involve sharing electrons equally are called: a) single, double, and polar covalent bonds; b) polar covalent bonds; c) ionic bonds; d) single and double covalent bonds.
9. Most lipids are: a) soluble in water; b) insoluble in water; c) electrolytes.
10. Which of the following is *not* a true statement? a) An electron shell is also known as an energy level; b) The number and arrangement of electrons in the outer energy level determine an element's chemical properties; c) When an atom's outer energy level is full, the atom becomes highly unstable; d) For each element, each energy level holds a specific number of electrons.
11. A diet containing large amounts of ________ increases the risk of heart disease: a) unsaturated fatty acids; b) saturated fatty acids; c) cholesterol; d) all of the above.

DIAGRAMS

12. Draw a diagram of a generalized atom. Label the fundamental subatomic particles and the nucleus.
13. Draw a diagram of a water molecule and label its components.

TRUE/FALSE

(If the statement is false, your instructor may wish to have you correct it.)

14. T/F: The chemical properties of a compound must correspond to those of its component elements.
15. T/F: DNA and RNA are examples of nucleotides.
16. T/F: Exergonic reactions release energy, usually as heat.
17. T/F: The pH of a solution can be defined as the negative exponent (log) of the hydrogen ion concentration.
18. T/F: Water readily dissolves hydrophilic molecules.

MATCHING QUESTIONS

(Match each statement with the most appropriate *term.)*

Statements:

19. Essential inorganic and organic molecules in foods.
20. Lipids that help trigger the onset of labor contractions.
21. Organic molecules that store and process information at the molecular level.
22. Molecules that dissolve readily in water.
23. Negatively charged particles that orbit around the nucleus.
24. The group of organic molecules that includes polysaccharides.
25. Molecules that are present in our fingernails and hair.
26. A solute that dissociates to release hydrogen ions and shifts the pH toward acidity.
27. A substance that can dissolve another substance.
28. A group of organic molecules that includes fats, oils, and waxes.
29. A state of balance that exists when two opposing chemical reactions are proceeding at identical rates.
30. A chemical bond created by the attraction between oppositely charged ions.
31. In healthy human blood, this should range from 7.35 to 7.45.
32. Compounds that do not readily dissolve in water.
33. A chemical bond that involves the unequal sharing of electrons.
34. A substance that is dissolved in a solution.

Terms:

a) Electrons
b) Polar covalent bonds
c) Ionic bond
d) Equilibrium
e) Nutrients
f) Solvent
g) Solute
h) Hydrophilic
i) Hydrophobic
j) pH
k) Acidic
l) Carbohydrates
m) Lipids
n) Prostaglandins
o) Proteins
p) Nucleic acids

SHORT ANSWER/ESSAY

35. Your stomach feels upset, so you take a product that contains sodium bicarbonate and soon you feel better. What chemical reactions are involved?
36. Identify the type of reaction shown in each of the following reactions:
    a) $HCl + NaOH \rightarrow H_2O + NaCl$
    b) $2\,H_2 + O_2 \rightarrow 2\,H_2O$
    c) $NaOH \rightarrow Na^+ + OH^-$
37. Why are oil spills so harmful to oceans and rivers?
38. Are the following equations balanced? If not, explain what is wrong.
    a) $2\,C + O_2 \rightarrow 2\,CO_2$
    b) $2\,KBr + Cl_2 \rightarrow 2\,KCl + Br_2$
    c) $2\,NaCl \rightarrow 2\,Na + Cl$
39. Identify the chemical properties of water that make it so important in the human body.
40. Write the symbol for a sodium ion with a chloride ion, a carboxyl group, a hydroxyl group, and an amino group.
41. Write the symbol for a calcium ion that was formed by "losing" two electrons from its outer energy level.
42. Fill in the blank: amino acids are to proteins as nucleotides are to ________.
43. How would you indicate that the following equation is reversible?
    $PCl_3 + Cl_2$ forms PCl_5
44. One atom has three neutrons; another has four neutrons. Both are atoms of the same element. These atoms are known as ________.
45. The reverse process of dehydration synthesis is ________.
46. Diagram and describe the primary, secondary, tertiary, and quaternary structures of a generalized protein. What causes proteins to assume these shapes? Under what circumstances does a protein form a quaternary structure?

CRITICAL THINKING/APPLICATIONS

47. Mr. Smith is admitted to the hospital with a diagnosis of an enlarged thyroid gland. a) Discuss how radioactive isotopes could be used to investigate physiological conditions that might bear on the cause of the enlarged gland. b) Should the radio isotopes used in this manner be strongly or weakly radioactive? Why?
48. A child is brought to the emergency room with a temperature of 106° F. How does a high fever affect the tissues in this child? Why are these effects irreversible after the body temperature reaches a certain level?
49. Concerned about cholesterol, a friend wants to eliminate all dietary fats. On the basis of what you read in this chapter, would you advise against this? Why or why not?

Many structures, from skyscrapers to patchwork quilts to backyard jungle-gyms, are made up of repeated small, similar components. The human body is built on this same principle, for it is made up of several trillion tiny units: cells. Cells, however, differ from all other building blocks in some remarkable ways. An individual brick can perform few (if any) of the functions of a building. But a cell can perform many of the functions of the organism of which it is part. Indeed, cells are the smallest entities that can perform all the basic life functions discussed in Chapter 1. They can do so because of their intricate structure—for cells are complex, with many specialized components. We'll explore the organization of the cell and its chief constituents in this chapter.

CHAPTER 3

The Cellular Level of Organization: Cell Structure

Chapter Objectives

After reading this chapter, you will be able to:

1 Discuss the basic concepts of the cell theory.
2 Compare the fluid contents of a cell with the extracellular fluid.
3 Explain the structure and importance of the cell membrane.
4 Discuss the ways in which a cell interacts with its environment.
5 Compare the structure and functions of the various cellular organelles.
6 Explain how cells obtain energy to power their operations.
7 Discuss the role of the nucleus as the cell's control center.
8 Describe the cell life cycle and how cells divide.
9 Explain the importance of the transmembrane potential, and describe the mechanisms that maintain it.
10 Compare the ways in which cells attach to other cells and discuss how the methods of attachment affect body tissues.

Introduction

Atoms are the building blocks of molecules; cells are the building blocks of the human body. Cells were first described by the English scientist Robert Hooke, around 1665. Hooke used an early light microscope to examine dried cork. He observed thousands of tiny empty chambers, which he named *cells*. Later that decade other scientists, observing the structure of living plants, realized that in life these spaces were filled with a gelatinous material. Research over the next 175 years led to the cell theory, the concept that cells are the fundamental units of all plant and animal tissues. Since that time the cell theory has been expanded to incorporate several basic concepts relevant to our discussion of the human body.

1. Cells are the building blocks of all plants and animals.
2. Cells are produced by the division of preexisting cells.
3. Cells are the smallest units that perform all vital physiological functions.
4. Each cell maintains homeostasis at the cellular level.

5. Homeostasis at the tissue, organ, system, and individual levels reflects the combined and coordinated actions of many cells.

Cells have a variety of forms and functions. Figure 3-1 gives examples of the range of cell sizes and shapes found in the human body. The relative proportions of the cells in this figure are correct, but all have been magnified roughly 500 times. Together, these and other types of cells create and maintain all anatomical structures and perform all vital physiological functions.

The human body contains trillions of cells, and all our activities, from running to thinking, result from the combined and coordinated responses of millions or even billions of cells. Yet each cell also func-

FIGURE 3-1
The Diversity of Cells in the Human Body. The cells of the body have many different shapes and a variety of special functions. These examples give an indication of the range of forms and sizes; all of the cells are shown with the dimensions they would have if magnified approximately 500 times.

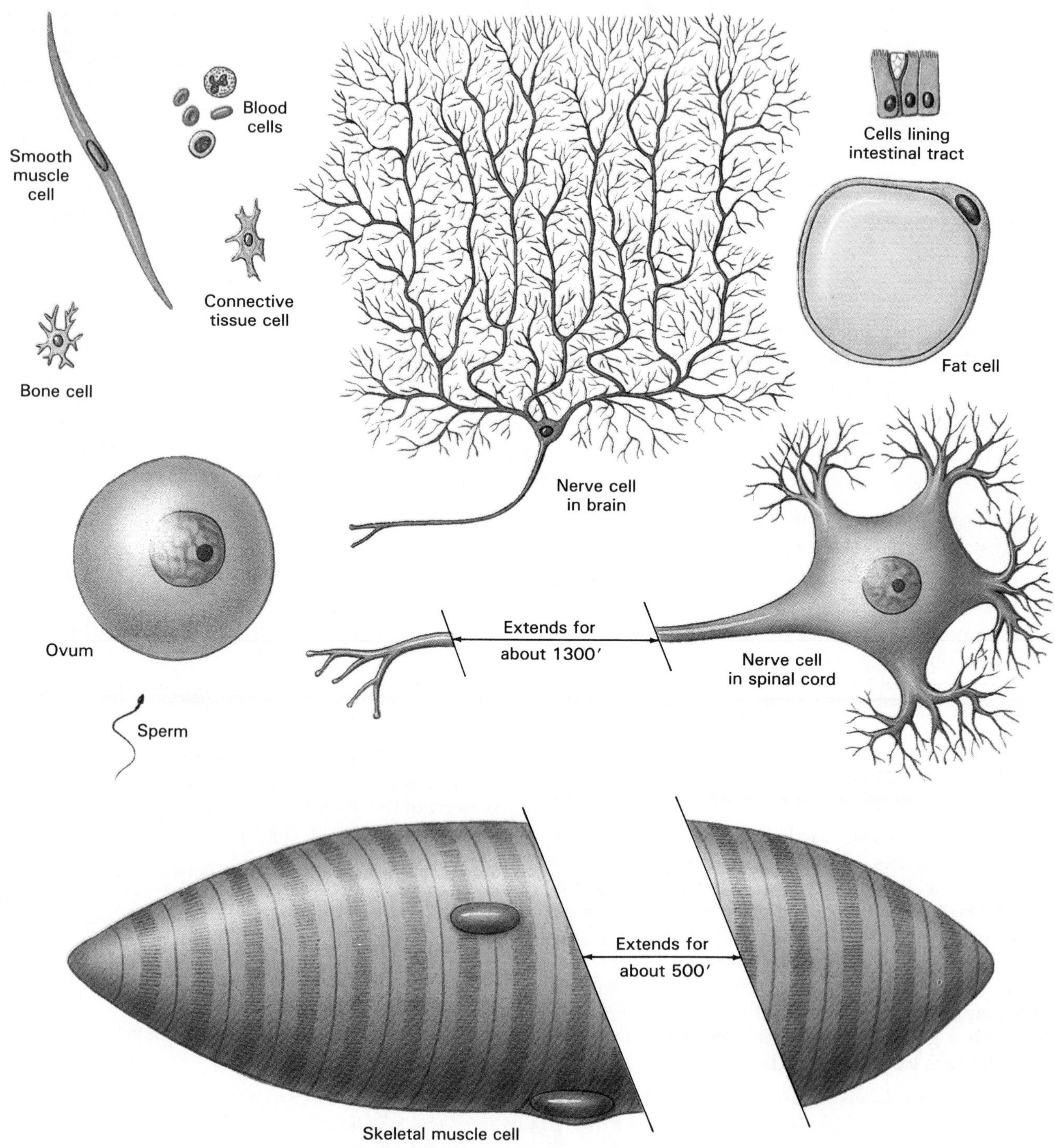

tions as an individual entity, responding to a variety of environmental cues. As a result, anyone interested in understanding how the human body functions must first become familiar with basic concepts in cell biology.

Studying Cells

Cytology (sī-TOL-o-jē; *cyto-*, cell + *-ology*, the study of) is the study of the structure and function of cells. What we have learned over the past 40 years has given us new insights into cellular physiology and the mechanisms of homeostatic control. The two most common methods used to study cell and tissue structure are *light microscopy* and *electron microscopy*.

Before the 1950s most information was provided by light microscopy. A photograph taken through a light microscope is called a light micrograph (LM). Light microscopy can magnify cellular structures about 1000 times and show details as fine as 0.25 μm. (The symbol μm stands for micrometer, often shortened to micron; 1 μm = 0.001 mm, or 0.00004 in.) With a light microscope one can identify cell types, such as muscle cells or nerve cells, and see large intracellular structures. Because individual cells are relatively transparent, thin sections taken through a cell are treated with dyes that stain intracellular structures, making them easier to see.

Although special staining techniques can show the general distribution of protein, lipid, carbohydrate, or nucleic acids in the cell, many fine details of intracellular structure remained a mystery until investigators began using electron microscopy. This technique uses a focused beam of electrons, rather than a beam of light, to examine cell structure. In **transmission electron microscopy** electrons pass through an ultrathin section to strike a photographic plate. The result is a transmission electron micrograph (TEM). Transmission electron microscopy shows the fine structure of cell membranes and intracellular structures. In **scanning electron microscopy** electrons bouncing off exposed surfaces create a scanning electron micrograph (SEM). Although scanning microscopy provides less magnification, it provides a three-dimensional perspective on cell structure.

Many other methods can be used to examine cell and tissue structure, and examples will be found in the pages that follow. Each gives a different perspective, but none tells the entire story. This chapter describes the structure of a typical cell, some of the ways in which cells interact with their environment, and how cells reproduce. Later chapters consider the coordination of cellular activities in physiological systems.

Cellular Anatomy

The "typical" cell is like the "average" person. Any description masks enormous individual variations. Our model cell will share features with most cells of the body without being identical to any. Figure 3-2 shows such a cell, and Table 3-1 summarizes the structures and functions of its parts.

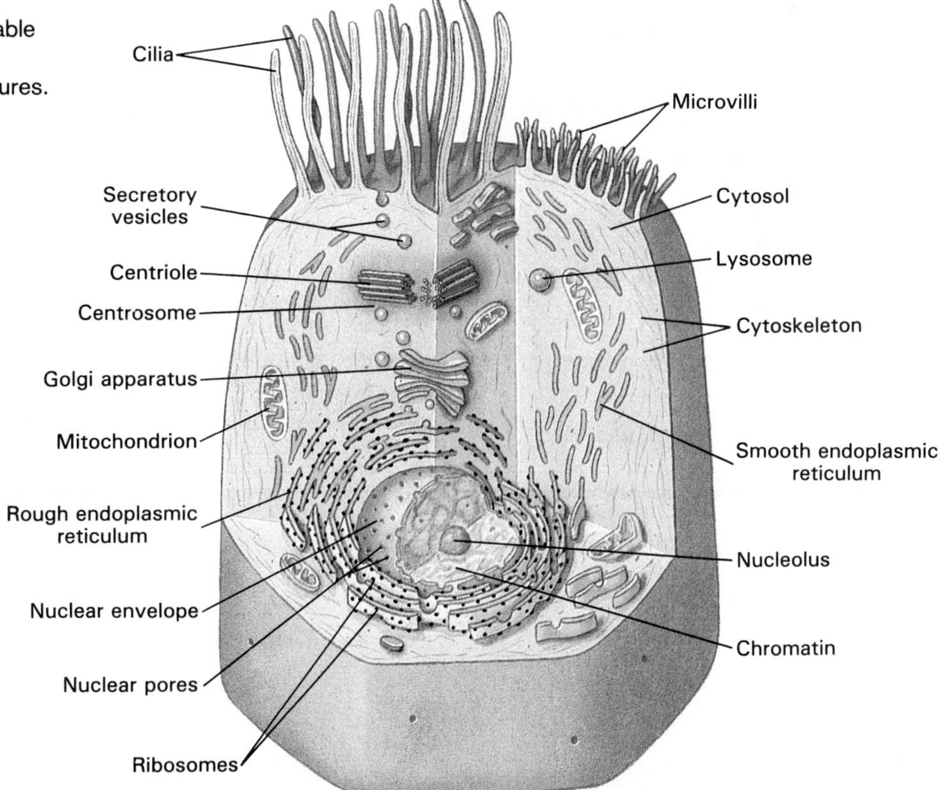

FIGURE 3-2
Anatomy of a Typical Cell. See Table 3-1 for a summary of the functions associated with the various cell structures.

TABLE 3-1 Anatomy of a Typical Cell

Appearance	*Structure*	*Composition*	*Function*
	CELL MEMBRANE	Lipid bilayer, containing phospholipids, steroids, and proteins	Isolation, protection, sensitivity, organization
	CYTOSOL	Fluid component of cytoplasm	Distributes materials by diffusion
	NONMEMBRANOUS ORGANELLES		
	Cytoskeleton: Microtubules, microfilaments	Proteins organized in fine filaments or slender tubes	Strength, movement of cellular structures and materials
	Microvilli	Membrane extensions containing microfilaments	Absorption of extracellular materials
	Cilia	Membrane extensions containing microtubules in 9 × 2 arrangement	Movement of materials over surface
	Centrioles	Two centrioles, at right angles; each composed of microtubles in 9 × 3 array	Movement of chromosomes during cell division
	Ribosomes	RNA + proteins; fixed ribosomes bound to endoplasmic reticulum, free ribosomes scattered in cytoplasm	Protein synthesis
	MEMBRANOUS ORGANELLES		
	Mitochondria	Double membrane, with inner folds (cristae) enclosing important metabolic enzymes	Produce 95% of the ATP required by the cell
	Nucleus	Nucleoplasm containing nucleotides, enzymes, and nucleoproteins; surrounded by double membrane or "nuclear envelope"	Control of metabolism; storage and processing of genetic information
	Nucleolus	Dense region in nucleoplasm	Site of RNA synthesis
	Endoplasmic reticulum	Network of membranous channels extending throughout the cytoplasm	Synthesis of secretory products; intracellular storage and transport
	Rough ER	Ribosomes attached to membranes	Secretory protein synthesis
	Smooth ER	Lacks attached ribosomes	Lipid and carbohydrate synthesis
	Golgi apparatus	Series of stacked, flattened membranes (saccules) containing chambers (cisternae)	Storage, alteration, and packaging of secretory products and lysosomes
	Lysosomes	Vesicles containing powerful digestive enzymes	Intracellular removal of damaged organelles or of pathogens
	Peroxisomes	Vesicles containing degradative enzymes	Neutralization of toxic compounds

TABLE 3-2 A Comparison of Organelles and Organ Systems

Primary Function	*Organelle*	*Organ System*
Protection, isolation from environment	Cell membrane	Integument
Internal support	Cytoskeleton	Skeletal system
Locomotion	Flagella, microfilaments	Muscular system
Regulation of activity		
Long-term	Nucleus	Endocrine system
Short-term	None (cytoplasmic enzymes make immediate short-term adjustments)	Nervous system
Internal transport	Microtubules	Circulatory system
Waste disposal	Exocytosis	Urinary system
Gas exchange (CO_2 and O_2)	None (diffusion at cell membrane)	Respiratory system
Energy source	Mitochondria	Digestive system
Defense against invasion	Lysosomes	Lymphatic system
Reproduction	Centrosome	Reproductive system

A cell survives because the organelles work in an integrated manner. In many ways the parts of a cell function like the physiological systems in the body, and Table 3-2 compares organelles and systems from this perspective.

Figure 3-3 indicates the organizational framework that will be followed as we examine the structure of a typical cell. Our cell floats in a watery medium known as the **extracellular fluid**. A *cell membrane* separates the cell contents, or *cytoplasm*, from the extracellular fluid. The cytoplasm can be further subdivided into a fluid, the *cytosol*, and intracellular structures collectively known as *organelles* (or-gan-ELS; "little organs").

FIGURE 3-3
A Flow Chart for the Study of Cell Structure

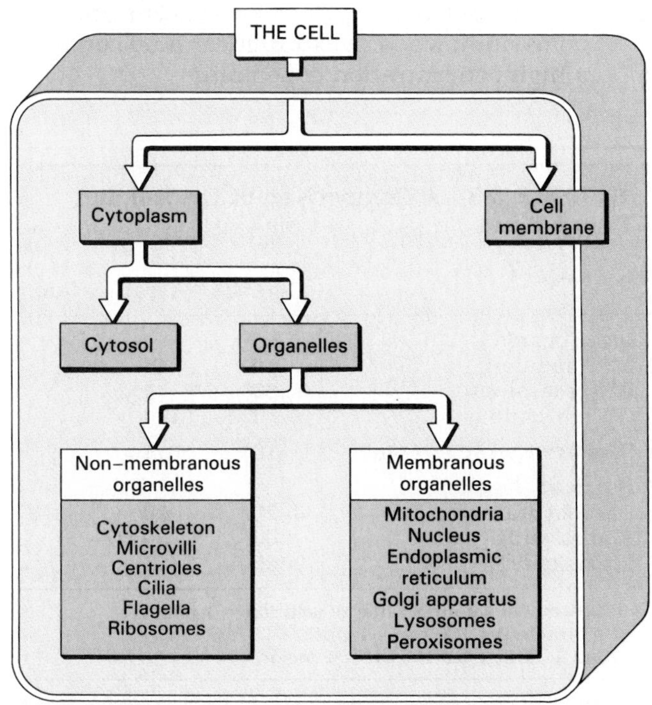

THE CELL MEMBRANE

The **cell membrane** (also called the **plasma membrane**, or *plasmalemma*) forms the outer boundary of the cell. The major components of the cell membrane are phospholipids, proteins, glycolipids, and cholesterol. Figure 3-4a highlights major features of membrane structure.

The cell membrane is called a **phospholipid bilayer** because the phospholipids form two distinct layers. In each layer the phospholipid molecules lie so that the hydrophilic heads are at the surface and the hydrophobic tails are on the inside. Figure 3-4b highlights the similarities in organization between a micelle, described in Chapter 2 (see Figure 2-19c), and the cell membrane. (p. 50) Ions and water-soluble compounds cannot enter the interior of a micelle because the lipid tails of the phospholipid molecules are highly hydrophobic and will not associate with water molecules. For the same reason, these solutes cannot cross the lipid portion of a cell membrane. This makes the membrane very effective in isolating the cytoplasm from the surrounding fluid environment. Such isolation is important because the composition of the cytoplasm is very different from that of the extracellular fluid, and those differences must be maintained.

Peripheral proteins are attached to the inner membrane surface; **integral proteins** are embedded

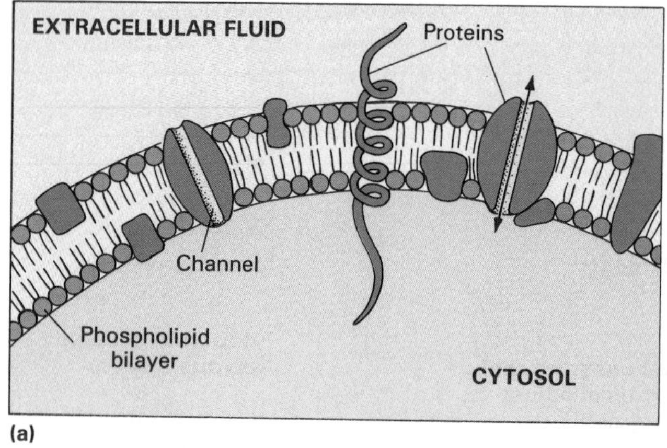

(a)

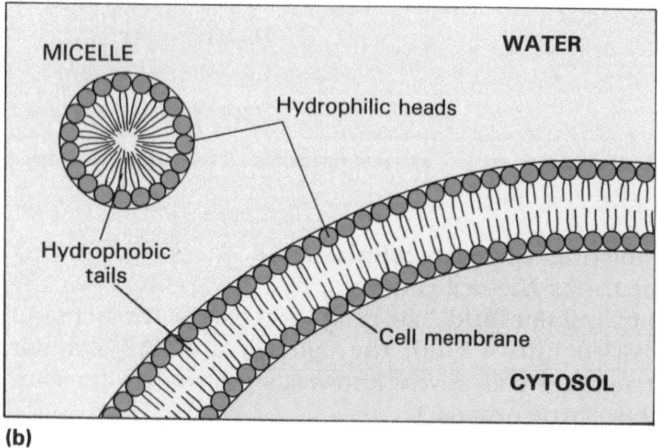

(b)

FIGURE 3-4
The Cell Membrane. (a) Diagrammatic view of membrane structure showing the phospholipid bilayer and associated proteins. (b) The structure of a cell membrane is similar to that of a micelle. The hydrophobic tails of the phospholipid molecules in the cell membrane create a barrier between the watery interior of the cell and the watery extracellular fluids.

in the membrane. Some of the integral proteins form **channels** that let water molecules, small water-soluble compounds, and ions into or out of the cell. Most of the communication between the interior and exterior of the cell occurs via these channels. (We will examine the details of this process later in the chapter.) Proteins, glycoproteins, and glycolipids on the membrane surface also function as receptors. A membrane receptor can bind specific extracellular molecules, triggering a change in cellular activity.

The general functions of the cell membrane include:

1. *Physical isolation*: The cell membrane is a physical barrier that separates the inside of the cell from the surrounding extracellular fluid.
2. *Regulation of exchange with the environment*: The cell membrane controls the entry of ions and nutrients, the elimination of wastes, and the release of secretory products.
3. *Sensitivity*: The cell membrane is the first part of the cell affected by changes in the extracellular fluid. It also contains a variety of receptors that allow the cell to recognize and respond to specific molecules in its environment. Any alteration in the cell membrane may affect all cellular activities.
4. *Structural support*: Specialized connections between cell membranes or between membranes and extracellular materials give tissues a stable structure.

THE CYTOPLASM

Cytoplasm is a general term for the material inside the cell. Cytoplasm contains many more proteins than the extracellular fluid; proteins account for 15–30 percent of the weight of the cell. The cytoplasm includes two major subdivisions:

1. **Cytosol**, or intracellular fluid: The cytosol contains dissolved nutrients, ions, soluble and insoluble proteins, and waste products. The cell membrane separates the cytosol from the surrounding extracellular fluid.
2. **Organelles**: Organelles are structures that perform specific functions within the cell.

The Cytosol

Table 3-3 compares the composition of the cytosol with the composition of tissue fluid, the extracellular fluid that surrounds most of the cells in the body. You should note three important differences:

1. The cytosol contains a high concentration of potassium, whereas extracellular fluid contains a high concentration of sodium.

TABLE 3-3 A Comparison of Cytosol and Tissue Fluid

	Cytosol	*Tissue Fluid*
Ions (mmol/ℓ)[1]		
Na^+ (sodium)	10	145
K^+ (potassium)	160	4
Cl^- (chloride)	3	114
Dissolved proteins (g/dℓ)[2]	16	1.8
Nutrients (mg/dℓ)[3]		
Carbohydrates	0–20	90
Amino acids	200	30
Lipids (g/dℓ)	2–95	0.6

[1] 0.001 mol of solute per liter of solution = 1 mmol/ℓ.
[2] 1 g protein per 100 mℓ = 1 g/dℓ.
[3] 1 mg (0.001 g) per 100 mℓ = 1 mg/dℓ.

2. The cytosol contains a relatively high concentration of dissolved proteins. Many of them are enzymes that regulate metabolic operations, while others are associated with the various organelles. These proteins give the cytosol a consistency that varies between that of thin maple syrup and almost-set gelatin.
3. The cytosol contains relatively small quantities of carbohydrates and large reserves of amino acids and lipids. The carbohydrates are broken down to provide energy, and the amino acids are used to manufacture proteins. The lipids stored in the cell are primarily used as an energy source when carbohydrates are unavailable.

The cytosol sometimes contains masses of insoluble materials known as **inclusions**. Among the most common inclusions are stored nutrients: for example, glycogen granules in liver or skeletal muscle cells and lipid droplets in fat cells.

Organelles

Organelles are found in most cells that are capable of growing and reproducing. Each organelle performs specific functions that are essential to normal cell structure, maintenance, and metabolism. Cellular organelles can be divided into two broad categories. **Nonmembranous organelles** are always in contact with the cytosol. **Membranous organelles** are surrounded by lipid membranes that isolate them from the cytosol, just as the cell membrane isolates the cytosol from the extracellular fluid.

NONMEMBRANOUS ORGANELLES

The cell's nonmembranous organelles include the *cytoskeleton, microvilli, centrioles, cilia, flagella,* and *ribosomes.*

The Cytoskeleton

The **cytoskeleton** is an internal protein framework that gives the cytoplasm strength and flexibility. It has four major components: *microfilaments, intermediate filaments, thick filaments,* and *microtubules.* Their properties are summarized in Table 3-4.

Microfilaments **Microfilaments** are slender protein strands, usually composed of the protein **actin.** In most cells microfilaments are scattered throughout the cytoplasm, and they form a dense layer under the cell membrane. Figure 3-5a shows the superficial layers of microfilaments in an intestinal cell.

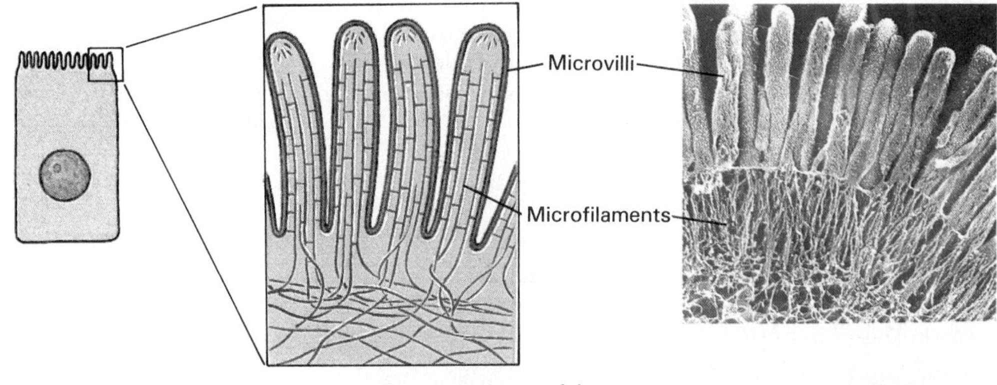

(a)

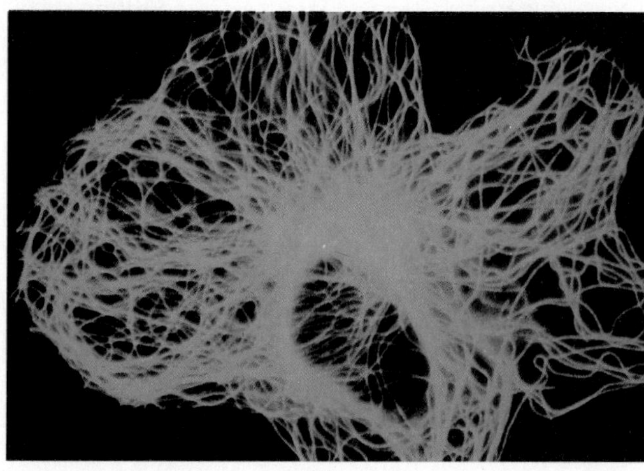

(b)

FIGURE 3-5
Microfilaments and Microtubules. (a) Microfilaments form a network just under the cell membrane. They also extend into microvilli that project above the membrane surface. (b) Microtubules in a living cell. (LM, ×3200)

Microfilaments have two major functions:

1. Microfilaments anchor the cytoskeleton to integral proteins of the cell membrane. This stabilizes the position of the membrane proteins, provides additional mechanical strength to the cell, and firmly attaches the cell membrane to the underlying cytoplasm.
2. Actin microfilaments can interact with filaments composed of another protein, **myosin**, to produce active movement of a portion of a cell or a change in the shape of the entire cell. Such interactions are responsible for the contraction of muscle cells.

Intermediate Filaments **Intermediate filaments,** as their name implies, are defined chiefly by their size (see Table 3-4); their composition varies from one cell type to another. Intermediate filaments (1) provide strength, (2) stabilize the positions of organelles, and (3) transport materials within the cytoplasm. For example, many cells contain intermediate filaments composed of the protein myosin. Interactions between these filaments and actin filaments can change the shapes of these cells. Other specialized intermediate filaments, called **neurofilaments**, are found in nerve cells, where they provide structural support and probably assist in the movement of materials within the cytoplasm.

Thick Filaments **Thick filaments** are relatively massive strands composed of myosin protein subunits. Thick filaments are abundant in muscle cells, where they interact with actin filaments to produce powerful contractions.

Microtubules **Microtubules**, found in all our cells, are hollow tubes built from the globular protein **tubulin**. Figure 3-5b shows the microtubules in the cytoplasm of a representative cell. The number and distribution of microtubules within a cell change over time. A microtubule forms through the aggregation of tubulin molecules; it persists for a time and then disassembles into individual tubulin molecules once again.

Microtubules have a variety of functions:

1. Microtubules form the primary components of the cytoskeleton, giving the cell strength and rigidity and anchoring the position of major organelles.
2. Disassembly of microtubules provides a mechanism for changing the shape of the cell, perhaps assisting in cell movement.
3. Microtubules can attach to organelles and other intracellular materials and move them around within the cell.

The cytoskeleton as a whole incorporates microfilaments, intermediate filaments, and microtubules into a network that extends throughout the cytoplasm. The organizational details are as yet poorly understood, because the network is extremely delicate and thus hard to study in an intact state. Figure 3-6 is based on our current knowledge of cytoskeletal structure.

Microvilli

Microvilli are small, finger-shaped projections of the cell membrane (Figure 3-6a). They are found in cells that are actively engaged in absorbing materials from

TABLE 3-4 The Cytoskeleton

Cytoskeletal Element	*Diameter*	*Protein Composition*	*Location*	*Remarks*
Microfilaments	Under 6 nm	Actin	In bundles beneath the cell membrane and throughout the cytoplasm	Present in most cells; best organized in skeletal and cardiac muscle cells
Intermediate filaments	7–11 nm	Variable	In cytoplasm	Present in most cells; at least five types known
Thick filaments	15 nm	Myosin	In cytoplasm	Found in skeletal and cardiac muscle cells
Microtubules	25 nm	Tubulin	In cytoplasm radiating away from centrosome	Present in most cells

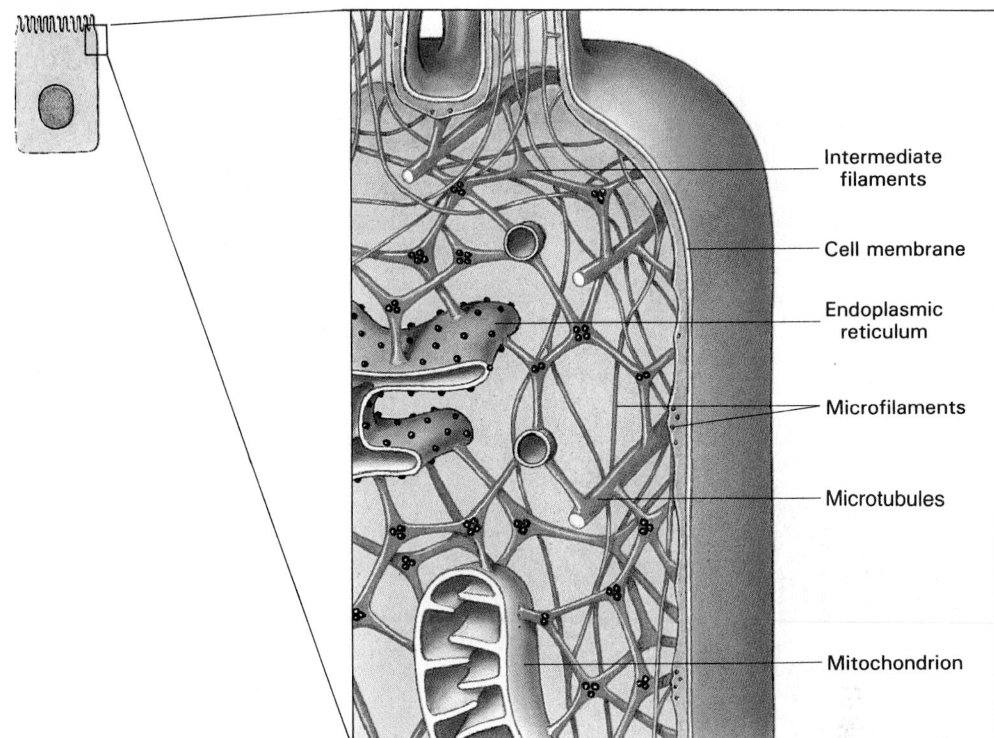

FIGURE 3-6
The Cytoskeleton. The cytoskeleton provides strength and structural support for the cell and its organelles. Interactions between cytoskeletal components are also important in moving organelles and changing the shape of the cell.

the extracellular fluid, such as the cells of the digestive tract and kidneys. A network of microfilaments stiffens each microvillus and anchors it to the underlying cytoskeleton. Interactions between these microfilaments and the cytoskeleton can produce a waving or bending action. Microvilli are important because they increase the surface area exposed to the extracellular environment. Their movements help to circulate fluid around the microvilli, bringing dissolved nutrients into contact with receptors on the membrane surface.

Centrioles, Cilia, and Flagella

The cytoskeleton contains numerous microtubules that function individually. Microtubules can also interact to form more complex structures known as *centrioles*, *cilia*, and *flagella*.

Centrioles A **centriole** (Figure 3-7a) is a cylindrical structure composed of short microtubules. There are nine groups of microtubules, with three in each group. All animal cells that are capable of reproducing themselves contain a pair of centrioles arranged as indicated in Figure 3-2. The **centrosome** is the cytoplasm surrounding this pair. Microtubules of the cytoskeleton usually begin within the centrosome and radiate through the cytoplasm. Centrioles direct the movement of DNA strands during cell division (discussed later in this chapter). Cells that do not divide, such as mature red blood cells and skeletal muscle cells, lack centrioles.

Cilia **Cilia** (singular *cilium*) contain nine pairs of microtubules surrounding a central pair (Figure 3-7b). Cilia are anchored to a compact **basal body** situated just beneath the cell surface. The structure of the basal body resembles that of a centriole. The exposed portion of the cilium is completely covered by the cell membrane. Cilia "beat" rhythmically, as de-

Functions	*Examples*
Provide strength, alter cell shape, bind the cytoskeleton to the cell membrane, tie cells together	Thin filaments in muscle cells
Provide strength, move materials through cytoplasm	Neurofilaments in nerve cells, keratin in skin
Interact with thin filaments to produce muscle contraction	Thick filaments in muscle cells
Provide strength, move organelles	Cilia, centrioles

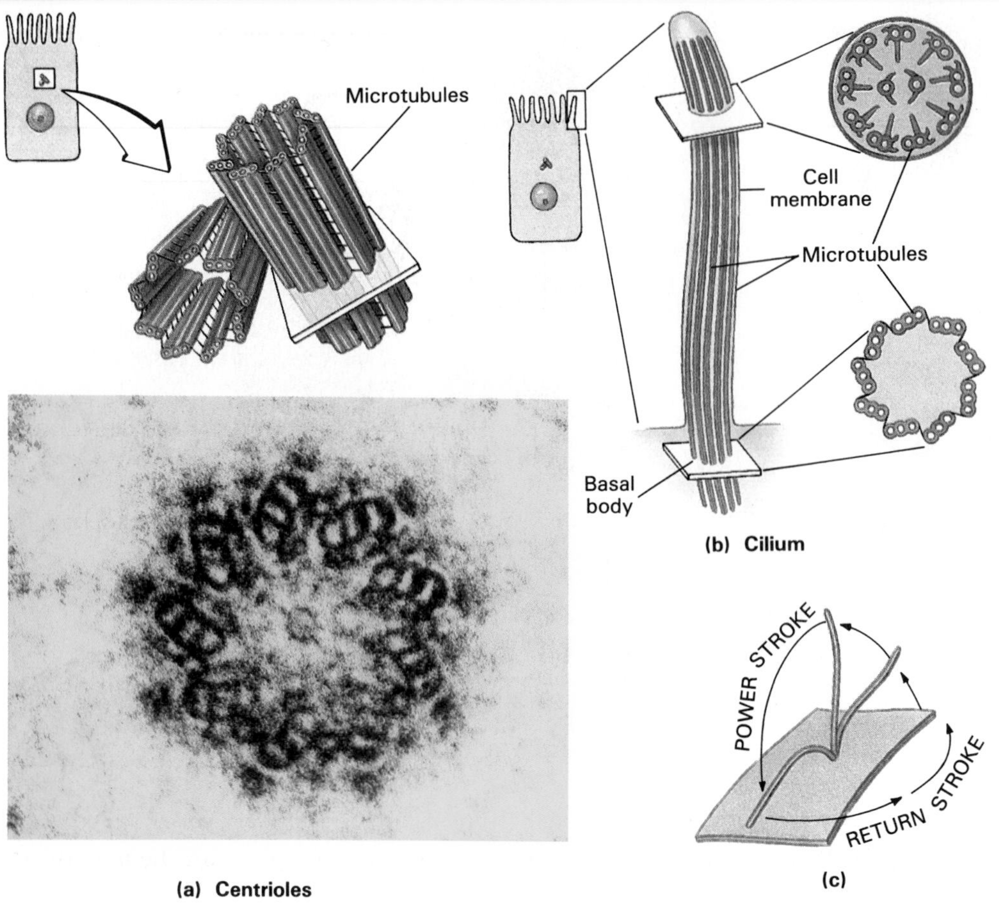

FIGURE 3-7
Centrioles and Cilia. (a) The centrosome contains a pair of centrioles oriented at right angles to one another; note the correlation between the three-dimensional structure and the sectional (TEM, ×54,000) view. (b) A cilium contains nine pairs of microtubules surrounding a central pair. (c) A single cilium swings forward and then returns to its original position. During the power stroke the cilium is relatively stiff, but during the return stroke it bends and moves parallel to the cell surface.

picted in Figure 3-7c, and their combined efforts move fluids or secretions across the cell surface. Cilia lining the respiratory tract beat in a synchronized manner to move sticky mucus and trapped dust particles toward the throat and away from delicate respiratory surfaces.

Flagella **Flagella** (fla-JEL-ah; singular *flagellum*, whip) resemble cilia but they are much larger. Flagella move a cell through the surrounding fluid, rather than moving the fluid past a stationary cell. Flagella on pathogenic (disease-causing) organisms allow them to move through our tissues and body fluids. The sperm cell is the only human cell that has a flagellum, and we will examine its structure in Chapter 28.

Table 3-5 summarizes information concerning centrioles, cilia, and flagella.

Ribosomes

Ribosomes (Figure 3-8) are small, dense structures that cannot be seen clearly with the light microscope. In an electron micrograph, ribosomes are dense granules roughly 25 nm in diameter. They are found in

TABLE 3-5 A Comparison of Centrioles, Cilia, and Flagella

Structure	*Microtubule Organization*	*Location*	*Function*
Centriole	Nine groups of microtubules form a short cylinder	In centrosome near nucleus	Moves DNA molecules during cell division
Cilia	Nine groups of long microtubules form a cylinder around a central pair	At cell surface	Move fluids or solids across cell surface
Flagella	Same as cilia	At cell surface	Move sperm cells through fluid

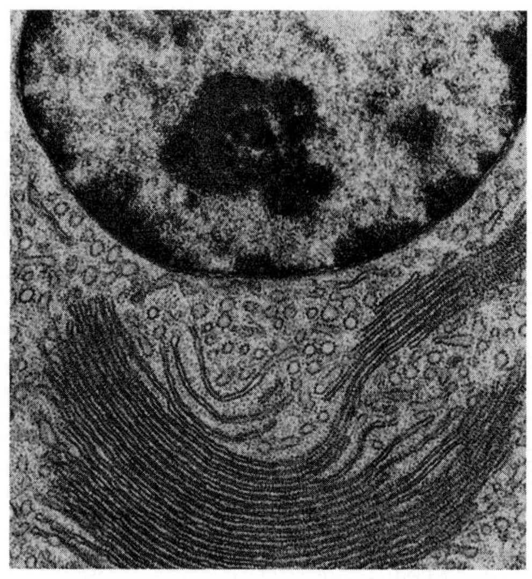

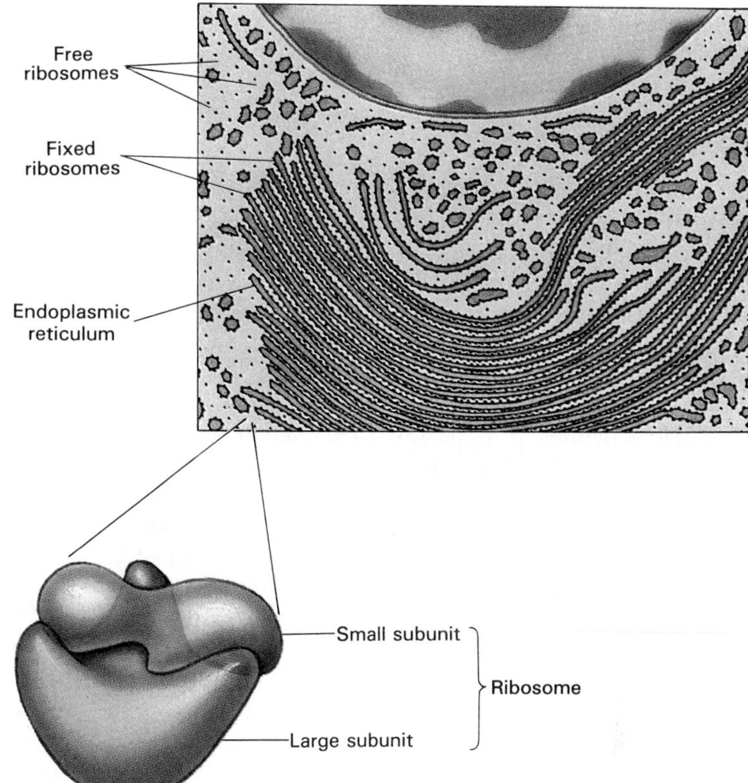

FIGURE 3-8
Ribosomes. Both free and fixed ribosomes can be seen in the cytoplasm of this cell. An individual ribosome has small and large subunits.

all cells, but their number varies depending on the type of cell and its activities. Each ribosome consists of roughly 60 percent RNA and 40 percent protein. At least 80 ribosomal proteins have been identified. These organelles are intracellular factories that manufacture proteins, using information provided by the DNA of the nucleus. (Details of this process will be examined in Chapter 4.)

Clinical Comment: Kartagener's Syndrome

Kartagener's syndrome is an inherited disorder that results from an inability to synthesize normal microtubules. Cilia and flagella containing these abnormal microtubules are unable to move. The primary symptom of this condition is chronic respiratory infection, because the cilia that line the respiratory passageways normally move mucus, dust, and bacteria away from the exchange surfaces of the lungs. Men with this condition are sterile because the flagella on their sperm do not beat, and nonmotile sperm cannot reach and fertilize an egg.

There are two major types of ribosomes: free ribosomes and fixed ribosomes. **Free ribosomes** are scattered throughout the cytoplasm; the proteins they manufacture enter the cytosol. **Fixed ribosomes** are attached to the *endoplasmic reticulum* (*ER*), a membranous organelle. Proteins manufactured by fixed ribosomes enter the ER, where they are modified and packaged for export. These processes are described in the section on the ER later in this chapter.

- √ **Cells lining the small intestine have numerous finger-like projections on their free surface. What are these structures and what is their function?**
- √ **How would the absence of a flagellum affect a sperm cell?**

MEMBRANOUS ORGANELLES

A phospholipid bilayer membrane similar to the cell membrane surrounds each membranous organelle, isolating it from the cytosol. This isolation allows the organelle to manufacture or store secretions, enzymes, or toxins that could adversely affect the cytoplasm in general. Table 3-1 includes six types of membranous organelles: *mitochondria, the nucleus, the endoplasmic reticulum, the Golgi apparatus, lysosomes*, and *peroxisomes*.

Mitochondria

Mitochondria (mī-tō-KON-drē-ah; singular *mitochondrion*; *mitos*, thread + *chondros*, cartilage) are small organelles that have an unusual double membrane (Figure 3-9). An outer membrane surrounds the entire organelle, and a second, inner membrane contains numerous folds, called **cristae**. Cristae increase the surface area exposed to the fluid contents, or **matrix**, of the mitochondrion. The matrix contains metabolic enzymes that perform the reactions that provide energy for cellular functions.

Respiratory enzymes attached to the cristae produce most of the ATP generated by mitochondria. Mitochondrial activity produces about 95 percent of the energy needed to keep a cell alive. Mitochondria produce ATP through the breakdown of organic molecules in a series of reactions that also consume oxygen (O_2) and generate carbon dioxide (CO_2). Chapter 4 will examine mitochondrial energy production in greater detail.

Mitochondria can have a variety of shapes, from long and slender to short and fat. The number of mitochondria in a particular cell vary depending on its energy demands. Red blood cells have none, but mitochondria may account for 20 percent of the volume of an active liver cell.

Mitochondria have an interesting evolutionary history. According to a widely accepted theory, they represent the descendants of bacteria that developed a mutually beneficial relationship with single-celled plants and animals more than a billion years ago. This history explains many unusual features of mitochondria. For example, mitochondria contain DNA, RNA, and the enzymes needed to synthesize proteins. The mitochondrial DNA and the associated enzymes are similar to those found in bacteria, but quite unlike those found in the nucleus of the cell. This synthetic capability enables mitochondria to control their own maintenance, growth, and reproduction. The muscle cells of body builders, for instance, have high rates of energy consumption, and over time their mitochondria respond to the increased energy demands by reproducing. The increased numbers of mitochondria can provide more energy and thus improve muscular performance. [CM: *Mitochondrial DNA, Disease, and Evolution*]

The Nucleus

The **nucleus** is the control center for cellular operations. Most cells contain a single nucleus, but there are exceptions. For example, skeletal muscle cells have many nuclei, and mature red blood cells have none. Figure 3-10 details the structure of a typical nucleus. A **nuclear envelope** surrounds the nucleus and separates it from the cytosol. The nuclear envelope is a double membrane containing a narrow **perinuclear** (*peri-*, around) **space**.

The nucleus directs processes that take place in the cytosol and must in turn receive information about conditions and activities in the cytosol. Chemi-

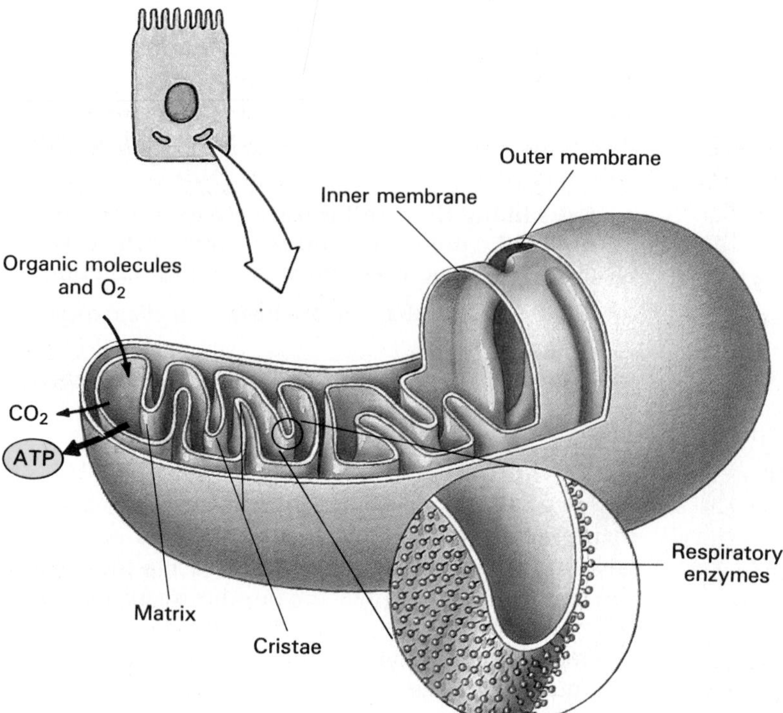

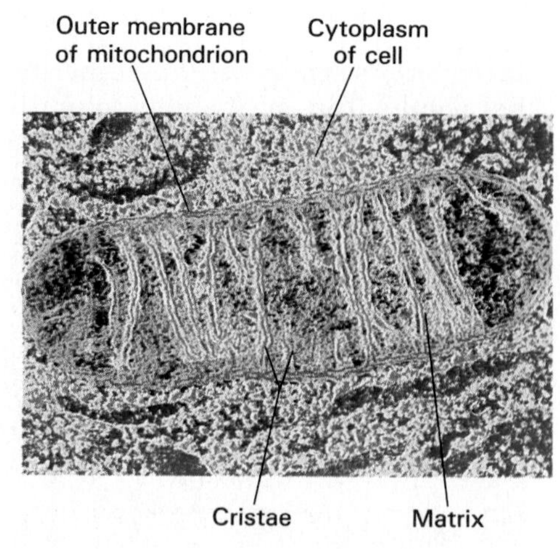

FIGURE 3-9
Mitochondria. This false-color TEM shows a typical mitochondrion in section, and the sketch details its three-dimensional organization. (TEM, ×43,200)

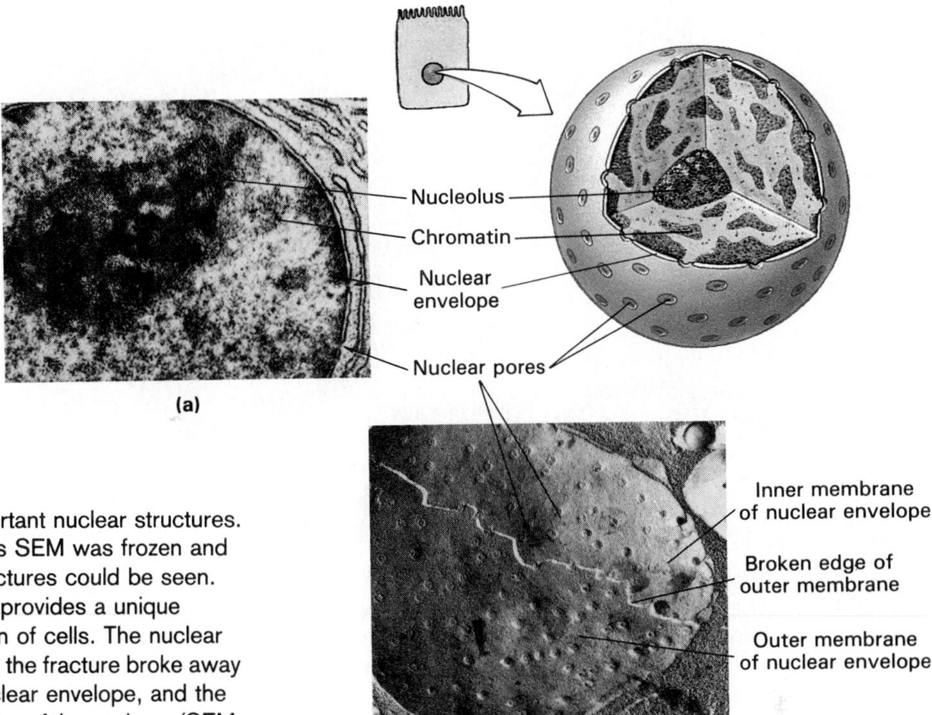

FIGURE 3-10
The Nucleus. (a) TEM showing important nuclear structures. (TEM, ×5700) (b) The cell seen in this SEM was frozen and then broken apart so that internal structures could be seen. This technique, called freeze-fracture, provides a unique perspective on the internal organization of cells. The nuclear envelope and nuclear pores are visible; the fracture broke away part of the outer membrane of the nuclear envelope, and the cut edge can be seen crossing the center of the nucleus. (SEM, ×8250)

cal communication between the nucleus and cytosol occurs through **nuclear pores**. These pores, which cover about 10 percent of the surface of the nucleus, are large enough to permit the movement of ions and small molecules, but too small for the passage of proteins or DNA.

The term **nucleoplasm** refers to the fluid contents of the nucleus. The nucleoplasm contains ions, enzymes, RNA and DNA nucleotides, proteins, small amounts of RNA, and DNA. The DNA strands form complex structures known as **chromosomes** (*chroma*, color). Each chromosome contains DNA strands bound to special proteins called **histones**. Our cell nuclei contain 23 pairs of chromosomes; one member of each pair is derived from our mother and one from our father. The structure of a typical chromosome is diagrammed in Figure 3-11.

At intervals the DNA strands wind around the histones, forming a complex known as a **nucleosome**. The entire chain of nucleosomes may coil around other histones. The degree of coiling determines whether the chromosome is long and thin or short and fat. Chromosomes in a dividing cell are very tightly coiled, and so can be seen clearly as separate structures in light or electron micrographs. In cells that are not dividing, the chromosomes are loosely coiled, forming a tangle of fine filaments known as **chromatin**. Each chromosome may have some coiled regions, and only the coiled areas stain clearly. As a result, the nucleus has a clumped, grainy appearance.

All vital cellular activities involve proteins. The nucleus controls cellular operations through its regulation of protein synthesis; the DNA strands of our chromosomes contain the information needed to synthesize roughly 100,000 different proteins. The details of this process are presented in Chapter 4.

The chromosomes also have direct control over the synthesis of RNA. Most nuclei contain one to four dark-staining areas called **nucleoli** (noo-KLĒ-o-lī; singular *nucleolus*). Nucleoli are nuclear organelles that synthesize the components of ribosomes. A nucleolus contains histones and enzymes as well as RNA, and it forms around a chromosomal region containing the genetic instructions for producing ribosomal proteins and RNA. Nucleoli are most prominent in cells that manufacture large amounts of proteins.

The Endoplasmic Reticulum

The **endoplasmic reticulum** (en-dō-plaz-mik re-TIK-ū-lum), or **ER**, is a network of intracellular membranes. It has three major functions:

1. *Synthesis*: The membrane of the endoplasmic reticulum manufactures proteins, carbohydrates, and lipids.
2. *Storage*: The ER can hold synthesized molecules or materials absorbed from the cytosol without affecting other cellular operations.
3. *Transport*: Materials can travel from place to place in the endoplasmic reticulum.

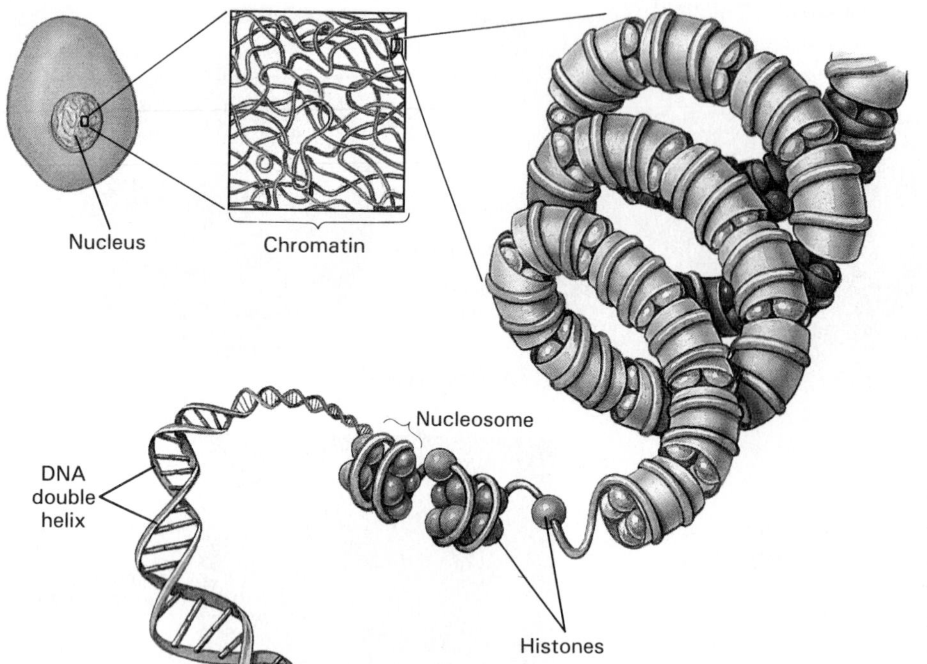

FIGURE 3-11
Chromosome Structure. DNA strands are coiled around histones to form nucleosomes. Nucleosomes form coils that may be very tight or rather loose. In cells that are not dividing, the DNA is loosely coiled, forming a tangled network known as chromatin. When the coiling becomes tighter, as it does in preparation for cell division, the DNA becomes visible as distinct structures called chromosomes.

The endoplasmic reticulum (Figure 3-12) forms hollow tubes, flattened sheets, and round chambers. The chambers are called **cisternae** (sis-TUR-nē; singular *cisterna*, a reservoir for water). There are two distinct types of endoplasmic reticulum, **rough endoplasmic reticulum (RER)** and **smooth endoplasmic reticulum (SER)**.

The RER functions as a combination workshop and shipping depot. It is where many newly synthesized proteins undergo chemical modification and where they are packaged for export to their next destination, the *Golgi apparatus*.

The outer surface of the rough endoplasmic reticulum contains fixed ribosomes. Figure 3-13 diagrams the functional relationship between the ribosomes and the RER. Ribosomes synthesize proteins using instructions provided by a strand of RNA; the mechanism will be detailed in Chapter 4. As the polypeptide chains grow, they enter the cisternae of the endoplasmic reticulum. Inside the ER the protein assumes its secondary or tertiary structure. Some of the proteins are enzymes that will function inside the ER. Other proteins are chemically modified within the ER by the attachment of carbohydrates, creating

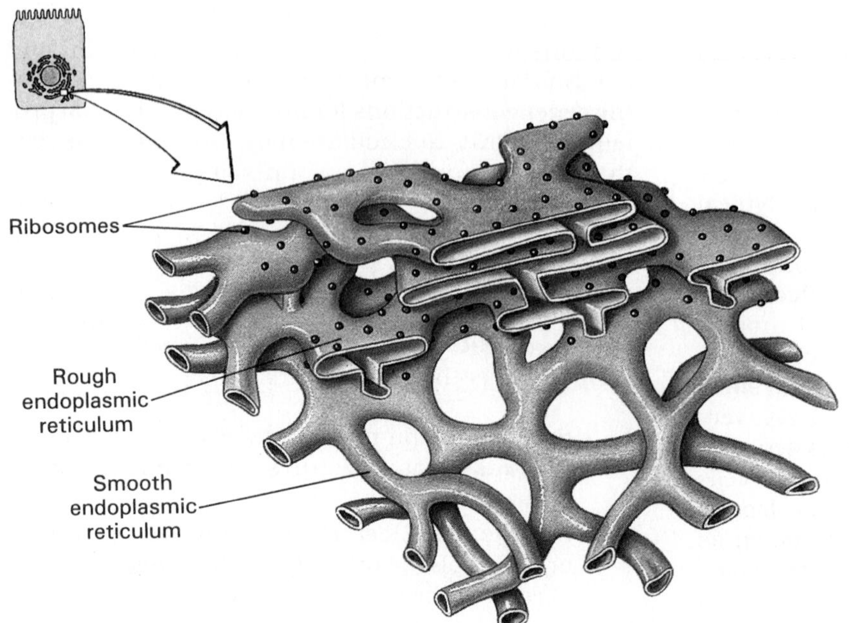

FIGURE 3-12
The Endoplasmic Reticulum. This diagrammatic sketch indicates the three-dimensional relationships between the rough and smooth endoplasmic reticulum.

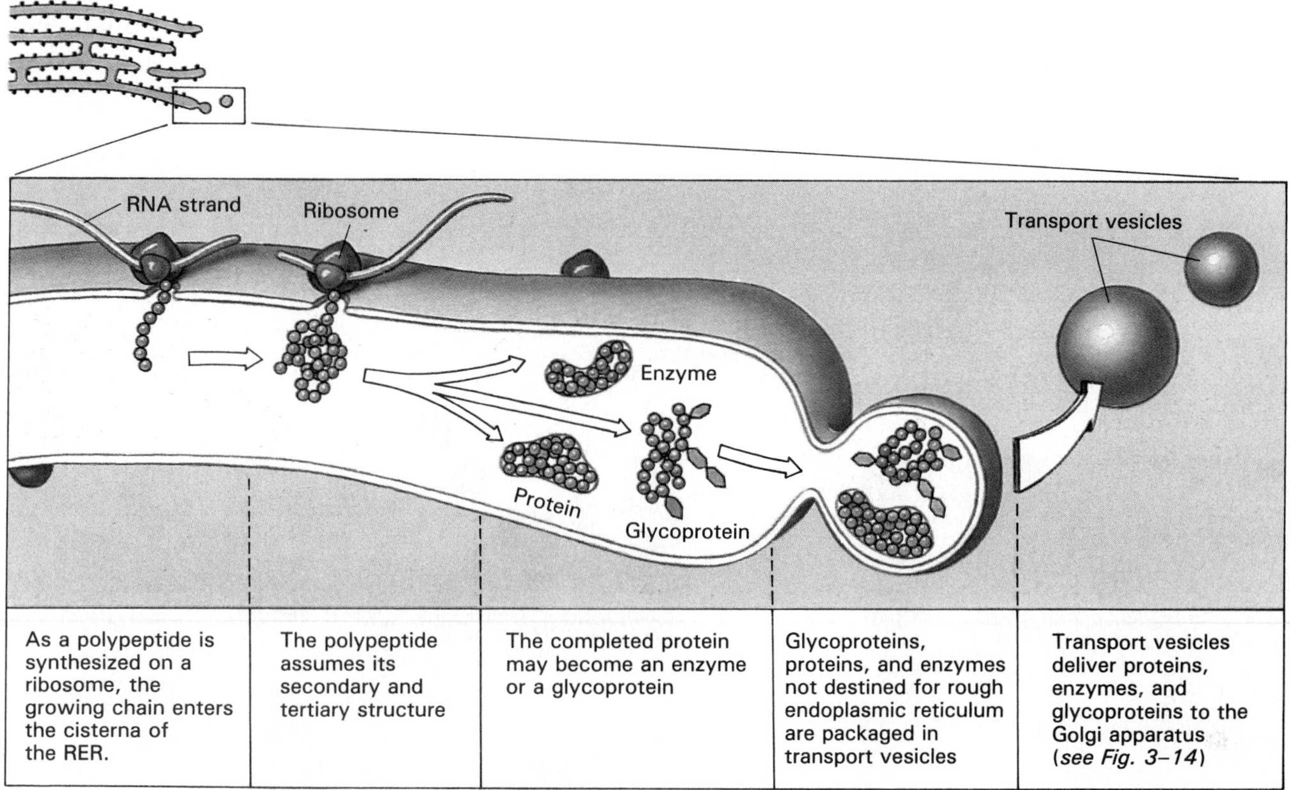

RNA strand
Ribosome
Transport vesicles
Enzyme
Protein
Glycoprotein
As a polypeptide is synthesized on a ribosome, the growing chain enters the cisterna of the RER.
The polypeptide assumes its secondary and tertiary structure
The completed protein may become an enzyme or a glycoprotein
Glycoproteins, proteins, and enzymes not destined for rough endoplasmic reticulum are packaged in transport vesicles
Transport vesicles deliver proteins, enzymes, and glycoproteins to the Golgi apparatus (see Fig. 3–14)

FIGURE 3-13
The Functional Relationship Between Fixed Ribosomes and the Rough Endoplasmic Reticulum

glycoproteins. Most of the proteins and glycoproteins produced by the RER are packaged into small membrane sacs that pinch off the tips of the cisternae. These **transport vesicles** deliver them to the Golgi apparatus.

There are no ribosomes associated with the smooth endoplasmic reticulum. The SER has a variety of functions that center around the synthesis of lipids and carbohydrates. Those functions include:

1. Synthesis of the phospholipids and cholesterol needed for maintenance and growth of the cell membrane, ER, nuclear membrane, and Golgi apparatus in all cells.
2. Synthesis of steroid hormones, such as testosterone (male sex hormone) and estrogen (female sex hormone) in cells of the reproductive organs.
3. Synthesis and storage of glycerides, especially triglycerides, in liver and fat cells.
4. Detoxification or inactivation of drugs in the SER of liver and kidney cells.
5. Synthesis and storage of glycogen in skeletal muscle and liver cells.
6. Removal and storage of calcium ions (Ca^{2+}) or larger molecules from the cytosol. Calcium ions are stored in the SER of skeletal muscle cells, nerve cells, and many other cell types.

The amount of endoplasmic reticulum and the proportion of RER to SER vary depending on the type of cell and its ongoing activities. For example, pancreatic cells that manufacture digestive enzymes contain an extensive RER, and the SER is relatively small. The situation is just the reverse in the cells that synthesize steroid hormones in the reproductive system.

The Golgi Apparatus

The **Golgi** (GOL-jē) **apparatus** (Figure 3-14) consists of flattened membrane discs, called **saccules**. A typical Golgi apparatus consists of five to six saccules; a single cell may contain several sets, each resembling a stack of dinner plates. Most often these stacks lie near the nucleus of the cell.

The major functions of the Golgi apparatus are:

1. Synthesis and packaging of secretions, such as mucus or enzymes,
2. Packaging of special enzymes for use in the cytosol,
3. Renewal or modification of the cell membrane.

The Golgi saccules communicate with the ER and with the cell surface. This communication in-

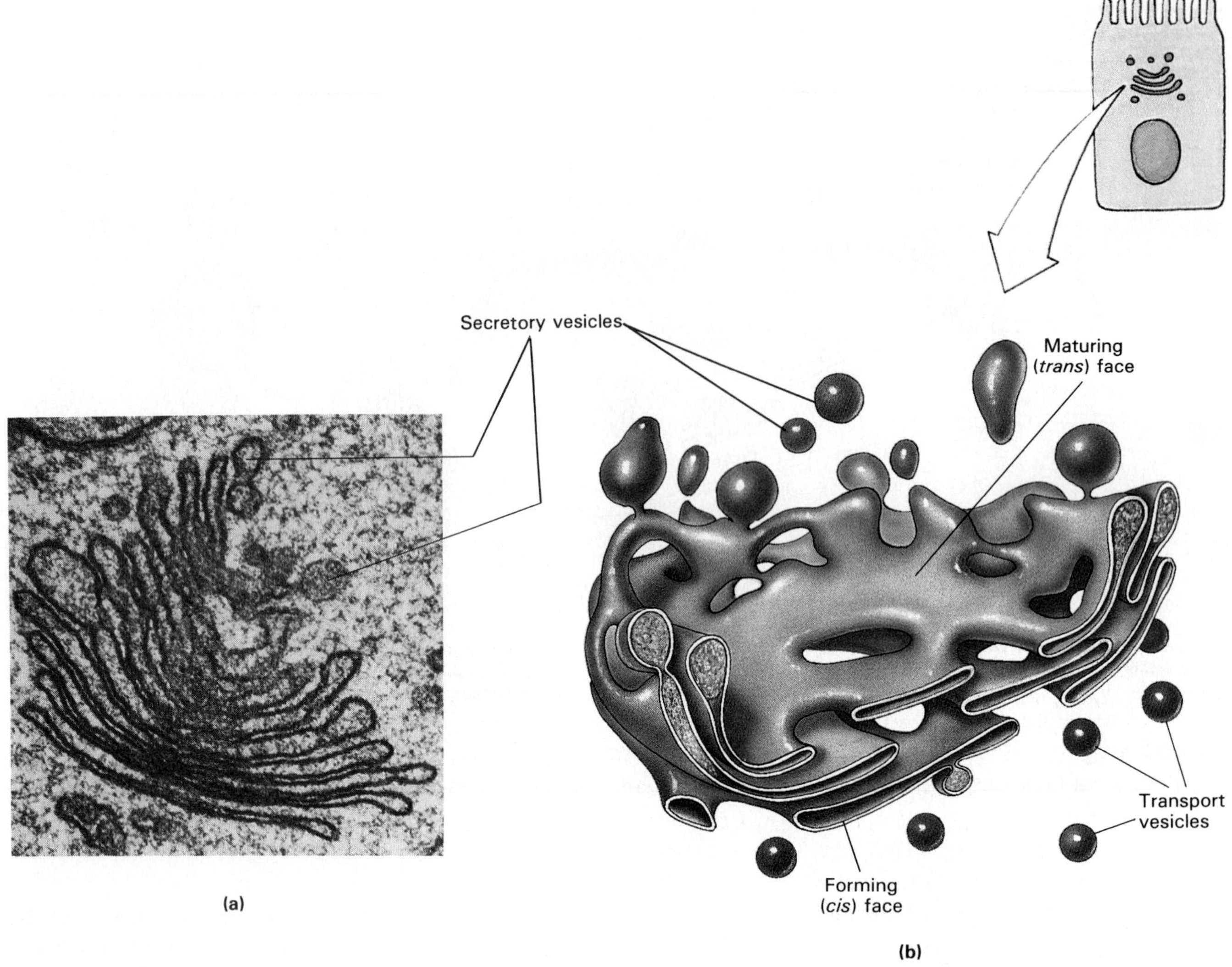

FIGURE 3-14
The Golgi Apparatus. (a) A sectional view of the Golgi apparatus of an active secretory cell. (TEM ×50,000) (b) A three-dimensional view of the Golgi apparatus with a cut edge corresponding to part (a).

volves the formation, movement, and fusion of vesicles.

Vesicles and Secretion Figure 3-15a diagrams the role of the Golgi apparatus in packaging secretions. Protein and glycoprotein synthesis occurs in the RER, and transport vesicles then move these products to the Golgi apparatus. The vesicles usually arrive at a convex saccule known as the *cis* (sis) *saccule*, or *forming face*. The transport vesicles then fuse with the Golgi membrane, emptying their contents into the cisterna. Inside the Golgi, enzymes modify the arriving proteins and glycoproteins. For example, the carbohydrate structure of a glycoprotein may be changed, or a phosphate group, sugar, or fatty acid may be attached to a protein.

Material moves from saccule to saccule by means of small **transfer vesicles**. Ultimately the product arrives at the *trans saccule*, or *maturing face*.

At the trans saccule, vesicles form that carry materials away from the Golgi. Vesicles containing secretions that will be discharged from the cell are called **secretory vesicles**. Secretion occurs as the membrane of a secretory vesicle fuses with the cell membrane. This ejection process, diagrammed in Figure 3-15b, is called **exocytosis** (ek-sō-sī-TŌ-sis). The glycoproteins that cover most cell surfaces, for example, synthesized by the Golgi apparatus, are released through exocytosis.

Membrane Turnover Because the Golgi apparatus continually adds new membrane to the cell surface in this way, it has the ability to change the properties of the cell membrane over time. For example, new glycoprotein receptors can be added, making the cell more sensitive to a particular stimulus; alternatively, receptors can be removed, making the cell less sensitive. Such changes can profoundly alter the sensitivity and functions of the cell.

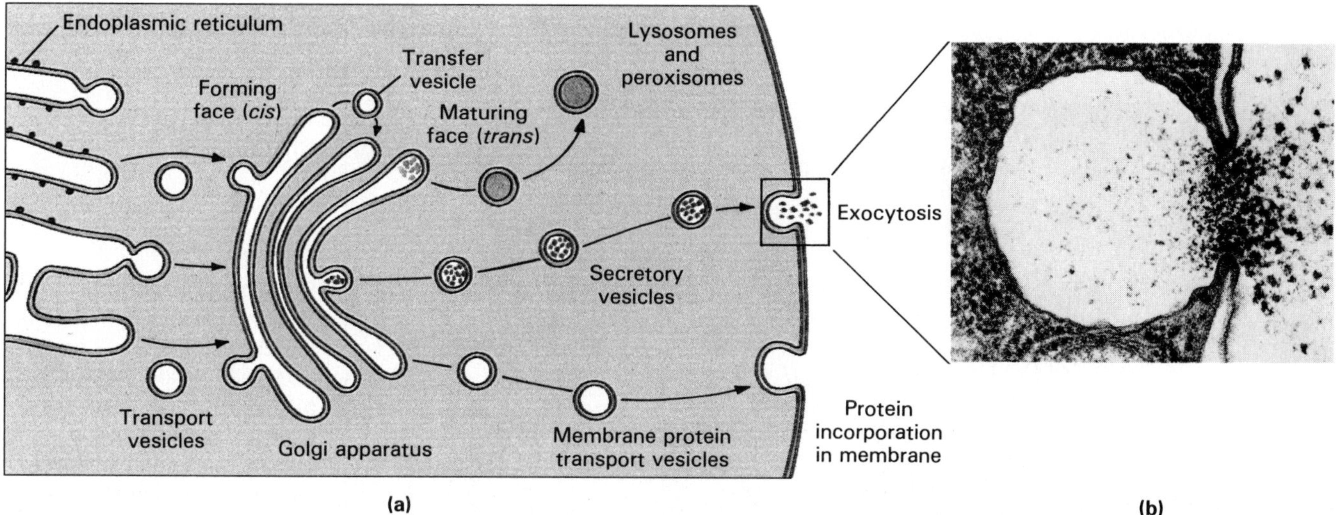

FIGURE 3-15
Golgi Function. (a) This diagram shows the functional link between the ER and the Golgi apparatus. Golgi structure has been simplified to clarify the relationships between the membranes. Transport vesicles carry the secretory product from the endoplasmic reticulum to the Golgi apparatus, and transfer vesicles move membrane and materials between the Golgi saccules. At the maturing face, three functional categories of vesicles develop. Secretory vesicles carry the secretion from the Golgi to the cell surface, where exocytosis releases the contents into the extracellular fluid. Other vesicles add surface area and integral proteins to the cell membrane. Lysosomes and peroxisomes, which remain in the cytoplasm, are vesicles filled with enzymes. (b) Exocytosis at the surface of a cell.

In an actively secreting cell the Golgi membranes may undergo a complete turnover every 40 minutes. The membrane lost from the Golgi is added to the cell surface, and that addition is balanced by the formation of vesicles at the membrane surface. As a result, an area equal to the entire membrane surface may be replaced each hour.

A third class of vesicles produced at the Golgi apparatus never leaves the cytoplasm. These vesicles contain digestive enzymes. The most important are *lysosomes*, whose functions will be detailed in the next section.

The Golgi apparatus has the remarkable ability to produce lysosomes, vesicles containing new membrane proteins, and secretory vesicles simultaneously. It is not known how these products are sorted, packaged, and consigned to their appropriate destinations.

With the exception of the mitochondria, all of the membranous organelles in the cell are either interconnected or in communication through the movement of vesicles. The rough and smooth endoplasmic reticulum are continuous and connected to the nuclear envelope. Transport vesicles connect the ER with the Golgi apparatus, and secretory vesicles link the Golgi apparatus with the cell membrane. Finally, vesicles forming at the exposed surface of the cell remove and recycle segments of the cell membrane. This continual movement and exchange has been called *membrane flow*. Membrane flow is another example of the dynamic nature of cells; functional aspects of membrane flow will be considered in Chapter 4.

Lysosomes

Lysosomes (LĪ-so-sōms; *lyso-*, dissolution + *soma*, body), are vesicles filled with digestive enzymes. *Primary lysosomes* contain inactive enzymes. Activation occurs when the lysosome fuses with the membranes of damaged organelles, such as mitochondria or fragments of the endoplasmic reticulum. This fusion creates a *secondary lysosome*, which contains active enzymes. These enzymes then break down the lysosomal contents. Nutrients reenter the cytosol, and the remaining material is eliminated by exocytosis.

Lysosomes also function in the defense against disease. Cells may remove bacteria, as well as fluids and organic debris, from their surroundings in vesicles formed at the cell surface. This process, described in greater detail later in the chapter, is called **endocytosis** (EN-do-sī-TŌ-sis; *endo-*, within). Lysosomes may fuse with vesicles created in this way, and the digestive enzymes then break down the contents and release usable substances such as sugars or amino acids. In this way the cell at once protects itself against pathogenic organisms and obtains valuable nutrients.

Figure 3-16 summarizes lysosomal functions. Lysosomes perform essential cleanup and recycling functions inside the cell. For example, when muscle

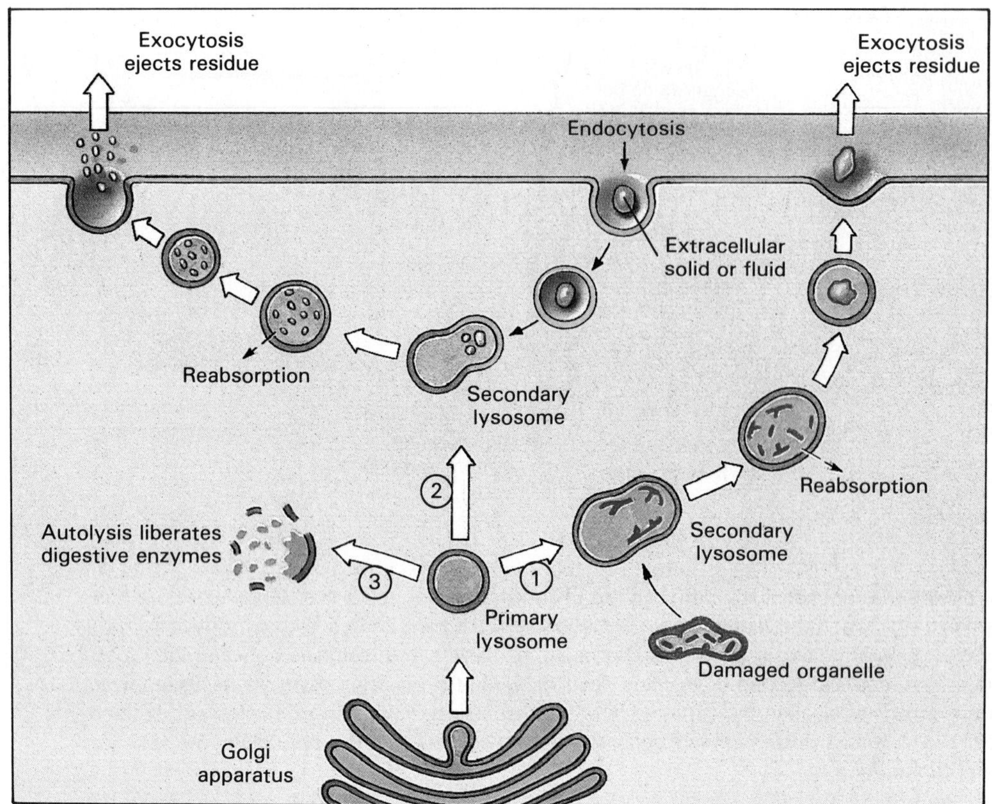

FIGURE 3-16 Lysosomal Functions. Primary lysosomes, formed at the Golgi apparatus, contain inactive enzymes. Activation may occur under three basic conditions: (1) when the primary lysosome fuses with the membrane of another organelle, such as a mitochondrion; (2) when the primary lysosome fuses with an endoyctic vesicle containing fluid or solid materials from outside the cell; or (3) in autolysis, when the lysosomal membrane breaks down following death or injury to the cell.

cells are inactive, lysosomes gradually break down their contractile proteins; if the cells become active once again, this destruction ceases. This regulatory mechanism fails in a damaged or dead cell. Lysosomes then disintegrate, releasing active enzymes into the cytosol. These enzymes rapidly destroy the proteins and organelles of the cell, a process called **autolysis** (aw-TA-li-sis; *auto-*, self). Because the breakdown of lysosomal membranes can destroy a cell, lysosomes have been called cellular "suicide packets." We do not know how to control lysosomal activities, or why the enclosed enzymes do not digest the lysosomal walls unless the cell is damaged. [CM: Lysosomal Storage Diseases]

Peroxisomes

Peroxisomes are smaller than lysosomes and carry a different group of enzymes. In contrast to lysosomes, which are produced at the Golgi apparatus, peroxisomes probably originate at the RER. Peroxisomes absorb and neutralize toxins, such as alcohol or hydrogen peroxide (H_2O_2), that are absorbed from the extracellular fluid or generated by chemical reactions in the cytoplasm. Peroxisomes are most abundant in liver cells, which are responsible for removing and neutralizing toxins absorbed in the digestive tract.

- **Microscopic examination of a cell reveals that it contains many mitochondria. What does this observation imply about the cell's energy requirements?**
- **Cells in the ovaries and testes contain large amounts of smooth endoplasmic reticulum (SER). Why?**

Communication between the Cell and Its Environment

Cells in the body work together to maintain homeostasis at the tissue, organ, and system levels. The essential communication and coordination activities involve the cell membrane, which forms the interface between each cell and its surroundings. The cell membrane regulates the dynamic exchange between the intracellular and extracellular fluids. Regulation is essential because the intracellular and extracellular environments are quite different, and those differences must be maintained to preserve homeostasis. This section explores the role of the cell membrane in maintaining the integrity of the cell and controlling the exchange between the intracellular and extracellular environments.

TABLE 3-6 Composition of the Cell Membrane

Compound	*Percentage of Weight*	*Function*	*Remarks*
Lipids	42	Form barrier to prevent free exchange of water-soluble materials between the intracellular and extracellular fluids	Double layer of phospholipids covers most of the cell surface; cholesterol is dissolved in the phospholipid layer
Phospholipids	25		
Cholesterol	13		
Other lipids	4		
Proteins	55	Form channels that regulate the passage of ions, perform active transport, facilitated transport, and act as receptors for specific extracellular materials	May be partially or totally embeded in the phospholipid bilayer
Carbohydrates	3	Act as receptors that bind specific extracellular materials	Found as components of glycoproteins, glycolipids, and proteoglycans of the glycocalyx

MEMBRANE STRUCTURE

The cell membrane is extremely thin and delicate, ranging from 6 to 10 nm in thickness. Nevertheless, it has a complex structure composed of lipids, proteins, and carbohydrates. Table 3-6 provides details concerning the identities and functions of these organic components.

Membrane Lipids

Figure 3-17 is a diagrammatic sketch that highlights important aspects of cell membrane structure. The phospholipids are organized in two distinct sheets, with the hydrophilic ends (the ends containing the phosphate groups) exposed to the adjacent solution. The hydrophobic lipid tails form the interior of the membrane, in association with cholesterol and small quantities of other lipids. Water-soluble ions and molecules cannot enter this region.

Membrane Proteins

Several types of proteins are associated with the membrane. *Peripheral proteins* are bound to the inner surface of the membrane but are easily separated from it. *Integral proteins* are part of the membrane structure, extending into the hydrophobic region and sometimes crossing it. *Channels* are passageways surrounded by integral proteins. These passages permit the movement of water and ions across the membrane. Some of the channels are called **gated** because

FIGURE 3-17
The Cell Membrane

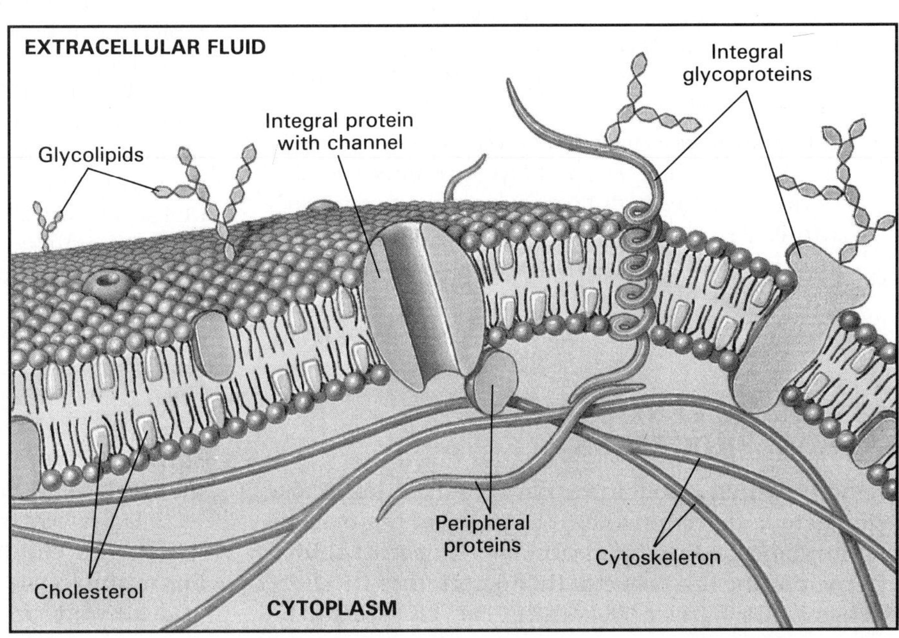

they can open or close to regulate the passage of materials. Channels account for around 0.2 percent of the total membrane surface area.

Membrane structure is not rigid. Some of the integral proteins are anchored to the underlying cytoskeleton, but many embedded proteins drift from place to place across the surface of the membrane like ice cubes in a punch bowl. In addition, the composition of the cell membrane can change over time, because of the removal of membrane surface through endocytosis or membrane addition through exocytosis.

The inner and outer surfaces of the cell membrane differ in protein and lipid composition. For example, cytoplasmic enzymes may be bound to integral proteins along the inner surface of the membrane, while receptors that bind specific materials may be found on its outer surface.

Membrane Carbohydrates

The carbohydrates in the cell membrane are found as components of complex molecules such as proteoglycans, glycoproteins, and glycolipids. The carbohydrate portions of these large molecules extend away from the outer surface of the cell membrane, forming a layer known as the **glycocalyx** (*calyx*, cup). The glycocalyx has a variety of important functions:

1. The glycoproteins and glycolipids form a viscous layer that lubricates and protects the cell membrane.
2. Because the components are sticky, the glycocalyx can help anchor the cell in place, and it also participates in the locomotion of specialized cells.
3. Glycoproteins and glycolipids can function as receptors, binding specific extracellular compounds. Such binding can alter the properties of the cell surface and indirectly affect the behavior of the cell.
4. The characteristics of the glycocalyx are genetically determined. Normal glycoproteins are recognized by the body's immune system as "self" rather than "foreign," and this recognition system keeps the immune system from attacking the tissues of the body.

MEMBRANE PERMEABILITY: PASSIVE PROCESSES

Precisely which substances can enter or leave the cytoplasm is determined by a property of the plasma membrane called its permeability. The **permeability** of a membrane is a property that determines its effectiveness as a barrier. If nothing can cross a membrane, it is described as **impermeable**. If any substance at all can cross without difficulty, the membrane is **freely permeable**. Cell membranes fall somewhere in between and are thus said to be **selectively permeable**. A selectively permeable membrane permits the free passage of some materials and restricts the passage of others. The distinction may be on the basis of size, electrical charge, molecular shape, solubility, or some combination of factors.

The permeability of a cell membrane varies depending on the organization and identity of membrane lipids and proteins. Passage across the membrane may be passive or active. Active processes, discussed later in this chapter, require that the cell draw on an energy source, usually ATP. Passive processes move ions or molecules across the cell membrane without any energy expenditure by the cell. Passive processes include *diffusion*, *osmosis*, *filtration*, and *facilitated diffusion*.

Diffusion

Ions and molecules in solution are in constant motion, bouncing off one another and colliding with water molecules. The result of the continual collisions and rebounds that occur is the process called diffusion. **Diffusion** can be defined as the net movement of material from an area where its concentration is relatively high to an area where its concentration is relatively low. The difference between the high and low concentrations represents a **concentration gradient**, and diffusion takes place until that gradient has been eliminated. Because diffusion occurs from a region of high concentration to one of relatively lower concentration, it is often described as proceeding "down a concentration gradient" or "downhill." When a concentration gradient has been eliminated, molecular motion continues, but it is random; there is no longer a net movement in any direction.

We have all experienced the effects of diffusion, which occurs in air as well as water. The smell of fresh flowers in a vase can sweeten the air in a large room, just as a drop of ink spreads to color an entire glass of water. In each case you begin with an extremely high concentration of molecules in a very localized area. Consider the ink dropped in the water glass, shown in Step 1 of Figure 3-18. Placing that drop in a large volume of clear water establishes a sharp concentration gradient for the dye: the dye concentration is high at the drop and negligible everywhere else. As diffusion proceeds, the dye molecules spread through the solution (Step 2) until they are distributed evenly (Step 3).

Diffusion is important in body fluids because it tends to eliminate local concentration gradients. For example, an active cell generates carbon dioxide and absorbs oxygen. As a result, the extracellular

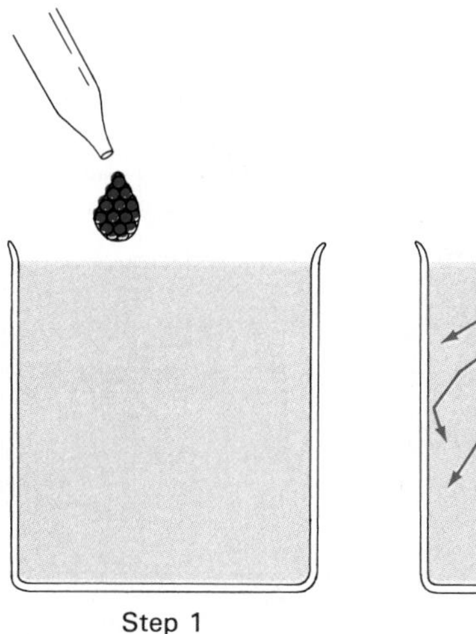

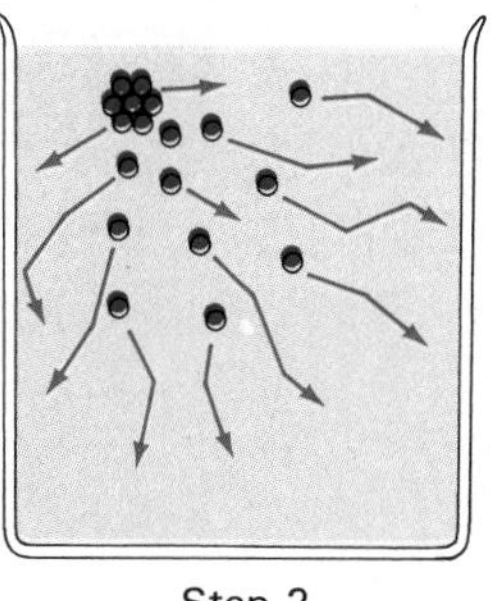

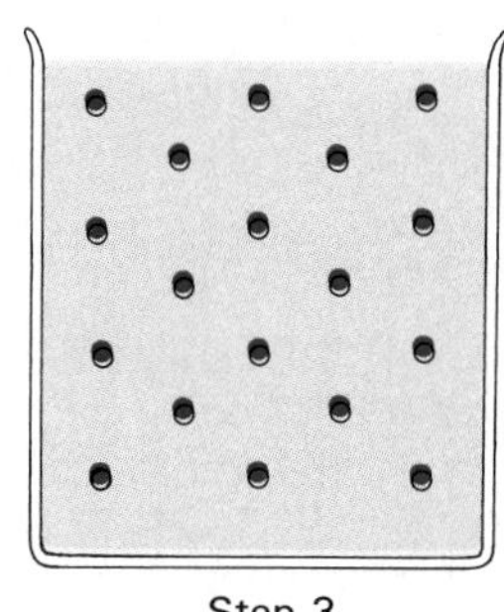

FIGURE 3-18
Diffusion. Step 1: Placing an ink drop in a glass of water establishes a strong concentration gradient because there are many ink molecules in one location and none elsewhere. Step 2: Diffusion occurs, and the ink molecules spread through the solution. Step 3: Eventually diffusion eliminates the concentration gradient, and the ink molecules are distributed evenly.

fluid around the cell develops a relatively high concentration of CO_2 and a relatively low concentration of O_2. Diffusion then distributes the carbon dioxide through the tissue and into the bloodstream. At the same time, oxygen diffuses out of the blood and into the tissue.

To be effective, the rate of diffusion of nutrients, waste products, and dissolved gases must be able to keep pace with the demands of active cells. Important factors that influence diffusion rates include:

1. *Distance*: Concentration gradients are eliminated quickly over short distances. The greater the distance, the longer the time required. In the human body, diffusion distances are usually small. For example, few living cells are farther than 125 μm from a blood vessel.
2. *Size of the gradient*: The larger the concentration gradient, the faster diffusion proceeds. When cells become more active and absorb more oxygen and nutrients, the rate of diffusion from the blood increases.
3. *Molecular size*: Ions and small organic molecules such as glucose diffuse faster than large proteins.
4. *Temperature*: The higher the temperature, the faster the diffusion rate. The human body maintains a temperature of around 37° C, and diffusion proceeds much faster at this temperature than at normal environmental temperatures.

Diffusion Across Cell Membranes In the extracellular fluids of the body, water and dissolved solutes diffuse freely. A cell membrane, however, acts as a barrier that selectively restricts diffusion: some substances can pass through easily, whereas others cannot penetrate the membrane at all. There are only two ways for an ion or molecule to cross a cell membrane: diffuse through one of the membrane channels or diffuse across the lipid portion of the membrane (Figure 3-19). Three major factors determine whether or not a substance can diffuse across a cell membrane: *lipid solubility, channel size,* and *electrical interactions.*

1. **Lipid solubility**. Alcohol, fatty acids, and steroids can enter cells easily because they can diffuse through the lipid portions of the membrane. Dissolved gases such as oxygen and carbon dioxide also enter and leave our cells by diffusion through the lipid bilayer.
2. **The size of the membrane channels**. Water-soluble compounds must diffuse through channels in the membrane. These channels are very small, averaging about 0.8 nm in diameter. Water molecules can enter or exit freely, but even a small organic molecule, such as glucose, is too big to fit through the channels.
3. **The charge on the ion, molecule, or membrane**. Important ions such as sodium (Na^+), chloride (Cl^-), potassium (K^+), and calcium (Ca^{2+}) are small enough to fit through membrane channels. But subtle differences in the diameter and structure of the membrane pores affect the ions' abilities to enter or leave the cytoplasm. For example, like charges (+ and + or − and −) repel one another, so the presence of positive charges in the channel wall can discourage the passage of cations (+) and attract anions (−). Physiologists speak of sodium channels, calcium channels, potassium channels, and so forth, because each ion uses a different channel.

Although the exact mechanisms have yet to be determined, cell membranes control the diffusion rates through each type of channel. The rate of ion movement can change from moment to moment, with some diffusion rates increasing and others declining.

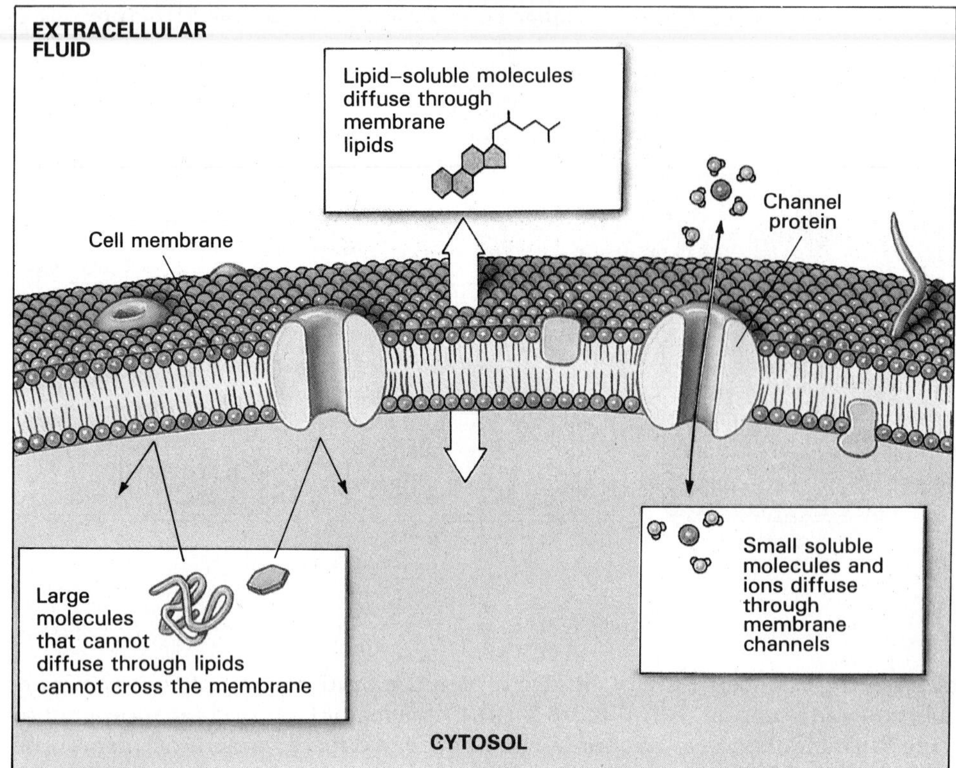
EXTRACELLULAR FLUID
Lipid–soluble molecules diffuse through membrane lipids
Channel protein
Cell membrane
Large molecules that cannot diffuse through lipids cannot cross the membrane
Small soluble molecules and ions diffuse through membrane channels
CYTOSOL

FIGURE 3-19 Diffusion Through Cell Membranes. Small ions and soluble molecules diffuse through membrane channels. Lipid-soluble molecules can cross the membrane by diffusing through the phospholipid bilayer. Large molecules that are not lipid-soluble cannot diffuse through the membrane at all.

The *characteristics* of particular channels are ultimately more important than the mere *number* of channels. For example, sodium channels outnumber potassium channels by about 25:1, but under normal conditions the ratio of sodium entry to potassium loss is only about 1.5:1.

Osmosis

Intracellular and extracellular fluids are solutions that contain a variety of dissolved materials. Each solute tends to diffuse as if it were the only material in solution. For example, the diffusion of sodium ions

Clinical Comment: Drugs and the Cell Membrane

Many clinically important drugs affect cell membranes. Although the mechanism behind the action of general anesthetics, such as *ether*, *chloroform*, and *nitrous oxide*, has yet to be determined, most are lipid-soluble hydrophobic molecules. There is a direct correlation between the potency of an anesthetic and its lipid solubility. Lipid solubility may speed the drug's entry into cells and enhance its ability to block ion channels or alter other properties of cell membranes. The most important clinical result is a reduction in the sensitivity and responsiveness of nerve and muscle cells.

Local anesthetics such as *procaine* and *lidocaine*, as well as *alcohol* and *barbiturate* drugs, are also lipid-soluble. These compounds affect membrane properties by blocking sodium channels in nerve cell membranes. This reduces or eliminates the responsiveness of nerve cells to painful (or any other) stimuli. A very powerful toxin, *tetrodotoxin* (*TTX*), is found in some ocean fish. Eating the internal organs of these fish causes a severe and potentially fatal form of food poisoning, marked by the disruption of normal neural and muscular activities. (Nevertheless, the flesh is considered a delicacy in Japan, where it is prepared by specially licensed chefs and served under the name *fugu*.)

Other drugs interfere with membrane receptors for hormones or chemicals that stimulate muscle or nerve cells. *Curare* is a plant extract that interferes with the chemical stimulation of muscle cell membranes. South American Indians use it to coat their hunting arrows so that wounded prey cannot run away. Anesthesiologists sometimes administer it to patients about to undergo surgery, to prevent reflexive muscle contractions or twitches while the surgery is being performed.

occurs in response to the existence of a concentration gradient for sodium. Changes in the concentration of other dissolved materials will have no effect on the rate or direction of sodium ion diffusion. Some molecules diffuse into the cytoplasm, others diffuse out, and a few, such as proteins, are unable to cross the cell membrane at all. But if we ignore the individual identities, and simply count molecules, we find that the total concentration of dissolved molecules on either side of the cell membrane stays the same.

This state of equilibrium persists because the entire cell membrane is freely permeable to *water*. Whenever a concentration gradient exists, water molecules will diffuse rapidly across the cell membrane until the gradient is eliminated. This movement, which eliminates differences in solute concentrations, occurs in response to a concentration gradient for water molecules. Dissolved solute molecules occupy space that would otherwise be taken up by water molecules. Thus the higher the solute concentration, the lower the water concentration. As a result, water molecules will tend to diffuse across a membrane toward the solution containing a higher solute concentration.

This diffusion of water across a membrane in response to differences in concentration is so important that it is given a special name, **osmosis** (oz-MŌ-sis; *osmos*, thrust). For convenience, we will always use the term *osmosis* when considering water movement and restrict use of the term *diffusion* to the movement of solutes.

Three characteristics of osmosis should be remembered:

1. Osmosis is the diffusion of water molecules across a membrane.
2. Osmosis occurs across a selectively permeable membrane that is freely permeable to water but not freely permeable to solutes.
3. In osmosis water will flow across a membrane *toward the solution that has the highest concentration of solutes*, because that is where the concentration of water is lowest.

Figure 3-20 diagrams the process of osmosis. Step 1 shows two solutions (A and B) with differing solute concentrations separated by a selectively permeable membrane. As osmosis occurs, water molecules cross the membrane until the solute concentrations in the two solutions are identical (Step 2a). Thus the volume of solution B increases at the expense of solution A. The greater the initial difference in solute concentrations, the stronger the osmotic flow. The force of water movement, known as the **osmotic pressure**, can be measured in several ways. For example, a strong enough opposing pressure can prevent the entry of water molecules. Pushing against a fluid generates **hydrostatic pressure**. In Step 2b, hydrostatic pressure in solution B, created by the applied force, balances the osmotic pressure, and no net osmotic flow occurs.

FIGURE 3-20
Osmosis. Step 1: Two solutions containing different solute concentrations are separated by a selectively permeable membrane. Water molecules (small dots) begin to cross the membrane toward the solution with the higher concentration of solutes (larger circles) (solution B). Step 2a: At equilibrium the solute concentrations on the two sides of the membrane are equal. The volume of solution B has increased at the expense of solution A. Step 2b: Osmosis can be prevented by resisting the volume change. The amount of pressure required to stop the osmotic flow provides a measurement of the osmotic pressure.

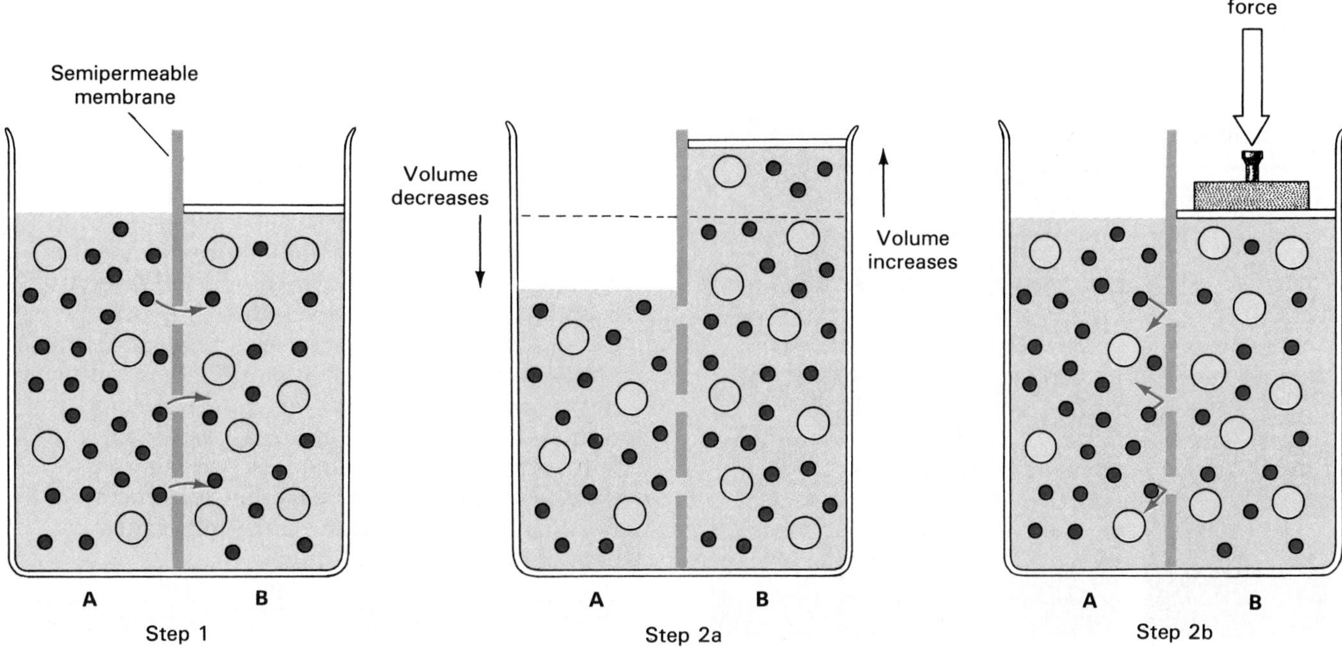

Osmosis eliminates solute concentration differences much more quickly than one would predict on the basis of diffusion rates for other molecules. When water molecules cross a membrane they move in groups held together by hydrogen bonding. So while solute molecules usually diffuse through membrane channels one at a time, water molecules move together in large numbers. This phenomenon is called **bulk flow**.

The total solute concentration in a solution is its **osmotic concentration**, or **osmolarity**. When describing the effects of various osmotic solutions on living cells, the term **tonicity** is used instead. Although it is difficult to see the effects of tonicity in most tissues, Figure 3-21 provides an example. Figure 3-21a shows the appearance of a red blood cell immersed in an isotonic solution. An **isotonic** solution is one that has the same solute concentration as does the cytoplasm, and so will not cause a net movement of water into or out of the cell. A **hypotonic** solution (*hypo-*, below) has a solute concentration lower than that of the cytoplasm. When a cell is placed in contact with a hypotonic solution, water will flow into the cytoplasm by osmosis. If the difference is substantial, the cell will swell up like a balloon, as shown in Figure 3-21b. Ultimately the red blood cell may burst, an event known as **hemolysis** (*hemo-*, blood + *lysis*, dissolution). The cytoplasm then diffuses away, leaving behind cell membranes known as red blood cell "ghosts." Red blood cells are often immersed in hypotonic solutions to obtain ghosts used in membrane research.

A **hypertonic** solution (*hyper-*, above) has a solute concentration higher than that of the cytoplasm. When a cell is placed in contact with a hypertonic solution, water will flow out of the cell and into the surrounding medium by osmosis, and the cell will shrivel and dehydrate. This shrinking of red blood cells, shown in Figure 3-21c, is called **crenation**.

Minor fluctuations in intracellular and extracellular solute concentrations are eliminated in a matter of seconds, but it takes a longer time to make system-wide alterations. After you have drunk a glass of pure water it may take a half-hour for your intracellular and extracellular fluids to become isotonic again. Severe alterations, such as occur in dehydration, are extremely dangerous; they will be considered in later chapters dealing with kidney function and water balance.

FIGURE 3-21
Osmotic Flow Across Cell Membranes. (a) Because these red blood cells are immersed in an isotonic saline solution, no osmotic flow occurs and the cells have their normal appearance. (b) Immersion in a hypotonic saline solution results in the osmotic flow of water into the cells. The swelling may continue until the cell membrane ruptures. (c) Exposure to a hypertonic solution results in the movement of water out of the cells. The red blood cells shrivel and become crenated.

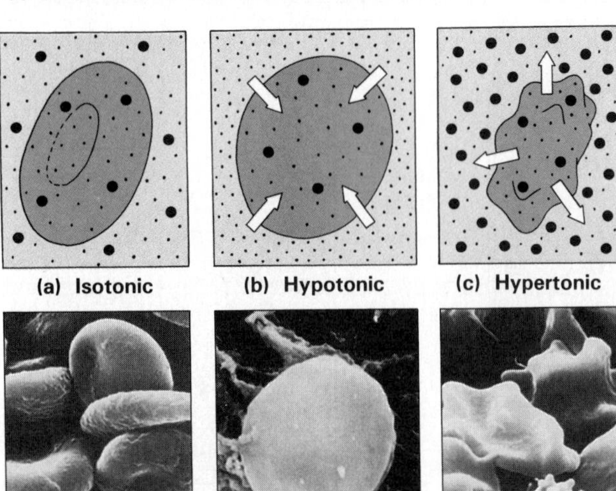

Clinical Comment: Tonicity and Clinical Practice

Although closely related and often used interchangeably, osmolarity and tonicity are not identical. For example, consider a solution that has the same osmolarity as the intracellular fluid but different concentrations of individual ions. If ions can diffuse from the solution into the cell, the osmolarity of the intracellular fluid will increase and that of the solution will decrease. Osmosis will then occur, moving water into the cell. If the process continues, the cell will eventually burst. In this case the solution and the intracellular fluid were equal in osmolarity, but the solution was *not* isotonic.

It is often necessary to give patients large volumes of fluid after a severe blood loss or dehydration. One fluid often administered is a 0.9-percent (0.9 g/mℓ) solution of sodium chloride (NaCl). This solution, which approximates the normal osmotic concentration of the extracellular fluids, is called **normal saline**. It is used because sodium and chloride are the most abundant ions in the extracellular fluid. There is little net movement of either ion across cell membranes; thus normal saline is nearly isotonic with the cytosol of body cells. An alternative treatment involves the use of an isotonic solution containing **dextran**, a carbohydrate that cannot cross cell membranes.

Filtration

In **filtration,** hydrostatic pressure forces water across a membrane, and solute molecules are selected on the basis of size. If the membrane pores are large enough, molecules of solute will be carried along with the water. We can see filtration in action in a coffee machine (Figure 3-22a). Gravity forces hot water through the filter, and the water carries with it a variety of dissolved compounds. The large coffee grounds never reach the pot because they cannot fit through the fine pores in the filter.

In the body, the heart pushes blood through the circulatory system and generates hydrostatic pressure. Filtration occurs across the walls of small blood vessels, pushing water and dissolved nutrients into the tissues of the body (Figure 3-22b). Filtration across specialized blood vessels in the kidneys is an essential step in the production of urine. When filtration occurs across these exceptionally permeable vessels, water moves by bulk flow. Because the filtration pores are very large, the clusters of water molecules carry dissolved ions with them. This accelerates the filtration process.

Facilitated Diffusion

Many essential nutrients, such as glucose or amino acids, are insoluble in lipids but too large to fit through membrane channels. These compounds can be passively transported across the membrane by special **carrier proteins** in a process called **facilitated diffusion**. The molecule to be transported first binds to a **receptor site** on the protein. It is then moved to the inside of the cell membrane and released into the cytoplasm. The process is diagrammed in Figure 3-23.

As in the case of simple diffusion, no ATP is expended in facilitated diffusion, and the molecules move from an area of higher concentration to one of lower concentration. However, facilitated diffusion differs from ordinary diffusion in many other ways, because carrier proteins are involved. The key features of the process include:

1. The carrier proteins involved in facilitated diffusion show *specificity*. That is, there are many different carrier proteins embedded in the cell membrane, and each is selective about what substances it will carry to the cytoplasm. For example, the carrier protein that transports glucose will not transport many other simple sugars.
2. The rate of facilitated diffusion is limited by the availability of carrier proteins, just as enzymatic reaction rates are limited by enzyme concentrations. Consider the transport of glucose across a cell membrane. Initially, as the glucose concentration increases in the fluid surrounding the cell, so does the rate of facilitated diffusion. More and more carrier proteins are called into action, and the "free time" between active cycles de-

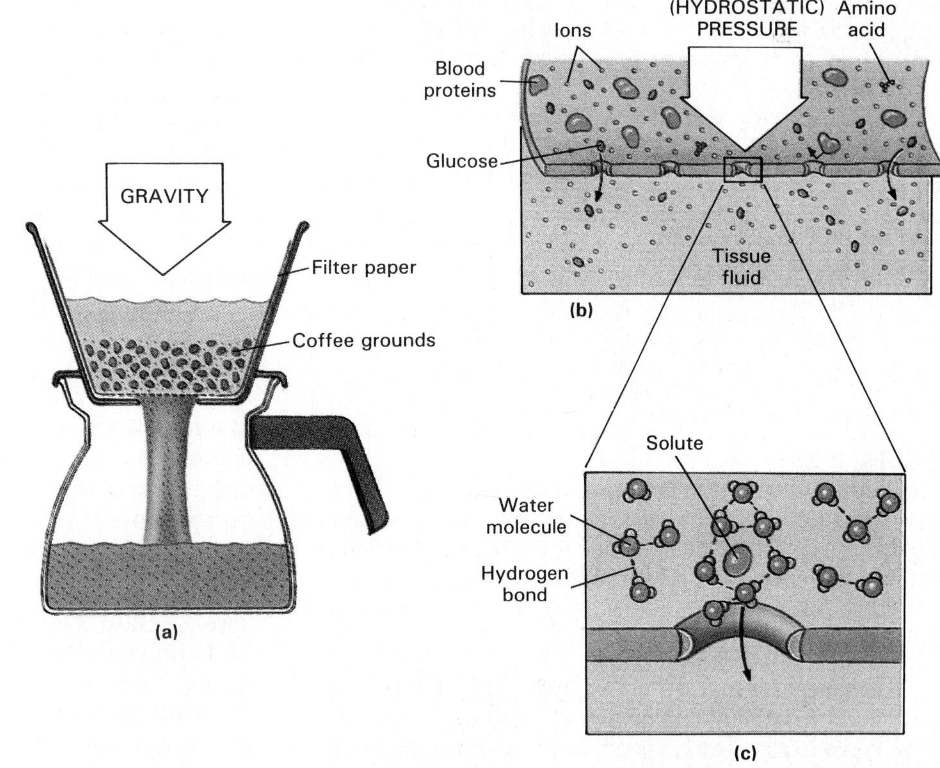

FIGURE 3-22
Filtration and Bulk Flow. (a) In filtration, materials are removed from a solution on the basis of size. In a coffee maker, the paper filter keeps the coffee grounds trapped, but lets smaller, dissolved molecules pass through. Gravity provides the hydrostatic pressure needed to force the coffee water through the filter. (b) In the most delicate blood vessels, blood pressure forces water and dissolved nutrients out of the bloodstream and into the interstitial fluid. Blood proteins are too large to pass through the openings between the cells that line the vessel. (c) Bulk flow occurs because adjacent water molecules are attracted to one another through the formation of hydrogen bonds. Water molecules cross a membrane in groups, and dissolved solutes can be carried along.

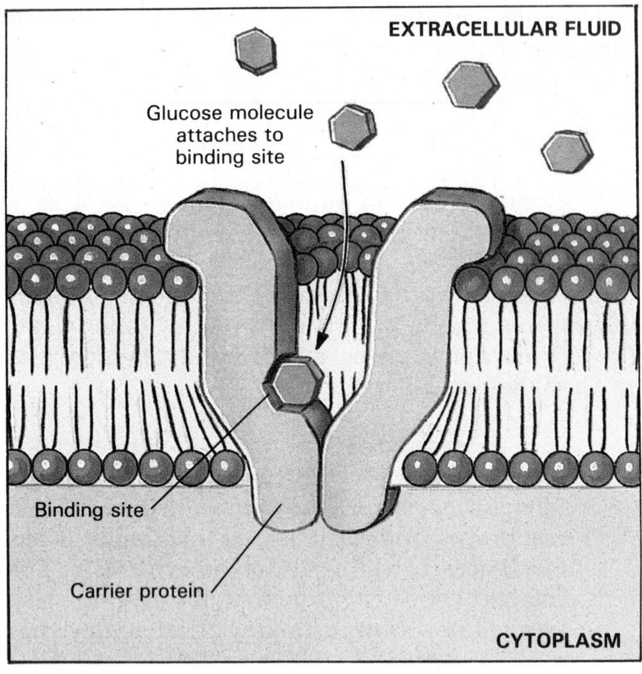

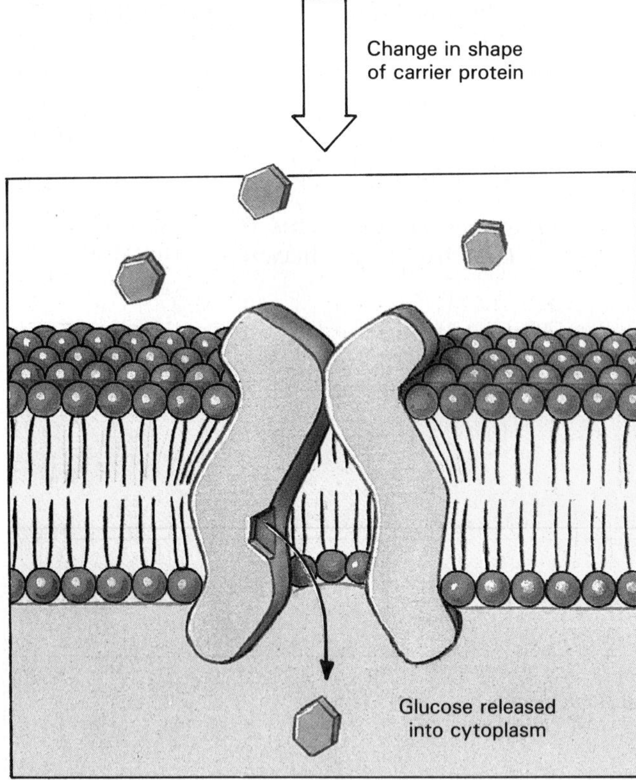

FIGURE 3-23
Facilitated Diffusion. In this process an extracellular molecule, such as glucose, binds to a receptor site on a carrier protein. The binding alters the shape of the protein, which then releases the molecule to diffuse into the cytoplasm.

creases for each carrier molecule. Eventually every carrier molecule is making repeated round-trips at top speed. At this point the carrier system is said to be **saturated**. Because no more carriers are available, and the existing ones cannot work any faster, increasing the extracellular glucose concentration further will not increase the rate of glucose transport into the cell.

3. The rate of carrier protein activity can change in the presence of other molecules, such as hormones. Hormones provide an important means of coordinating carrier protein activity throughout the body. For example, after each meal, blood glucose concentrations rise. This rise triggers the release of the hormone *insulin*. The insulin, in turn, accelerates the movement of glucose molecules into cells by stimulating the activity of the glucose carrier protein. The interplay between hormones and cell membranes will be examined further in chapters dealing with the endocrine system (Chapter 18) and metabolism (Chapter 25).

MEMBRANE PERMEABILITY: ACTIVE PROCESSES

All **active membrane processes** require energy. By spending energy, usually in the form of ATP, the cell can transport substances *against their concentration gradients*. We will consider two active processes: *active transport* and *endocytosis*.

Active Transport

In **active transport** the high-energy bond in ATP provides the energy needed to move ions or molecules across the membrane. The process is complex, and specific enzymes must be present in addition to carrier proteins. Although it has an energy cost, active transport offers one great advantage: it is not dependent on a concentration gradient. As a result the cell can import or export specific materials *regardless of their intracellular or extracellular concentrations*.

All living cells show active transport of the cations sodium (Na^+), potassium (K^+), calcium (Ca^{2+}), and magnesium (Mg^{2+}). Specialized cells can transport additional ions such as iodide (I^-), chloride (Cl^-), and iron (Fe^{2+}). Many of these carrier mechanisms, known as **ion pumps**, move a specific cation or anion in one direction, either in or out of the cell. In a few cases, one carrier protein will move more than one ion at a time. If one ion moves in one direction and the other moves in the opposite direction, the carrier is called an **exchange pump**.

The Sodium-Potassium Exchange Pump Sodium and potassium ions are the principal cations in body fluids. Sodium ion concentrations are high in the extracellular fluids, while sodium concentrations in the cytoplasm are relatively low. The distribution of potassium in the body is just the opposite—low in the extracellular fluids and high in the cytoplasm. As a result, sodium ions slowly diffuse into the cell, and potassium ions leak out. Homeostasis within the cell depends on ejecting sodium ions and recapturing lost potassium ions. This is accomplished through the activity of the **sodium-potassium exchange pump**.

As indicated in Figure 3-24, this ion pump exchanges intracellular sodium for extracellular potassium. On the average, for each ATP molecule consumed three sodium ions are ejected and two potassium ions are reclaimed. Assuming that ATP is readily available, the rate of transport depends on the concentration of sodium ions in the cytoplasm. When that concentration rises, the pump becomes more active.

One indication of the importance of this mechanism for cellular homeostasis is the number of pumps in the membrane. A typical cell membrane has roughly four potassium channels and 100 sodium channels per square micrometer of cell surface. By comparison, that same area will contain around 1000 sodium-potassium exchange pumps. The energy demands of these pumps are impressive; a resting cell may use up to 40 percent of the ATP it produces to power its sodium-potassium exchange pumps. We will consider additional aspects of exchange pump function later in this chapter.

FIGURE 3-24
The Sodium-Potassium Exchange Pump. The operation of the Na^+-K^+ pump is an example of active transport—it depends on energy provided by ATP. For each ATP molecule converted to ADP this ion pump carries three Na^+ ions out of the cell and two K^+ ions into the cell.

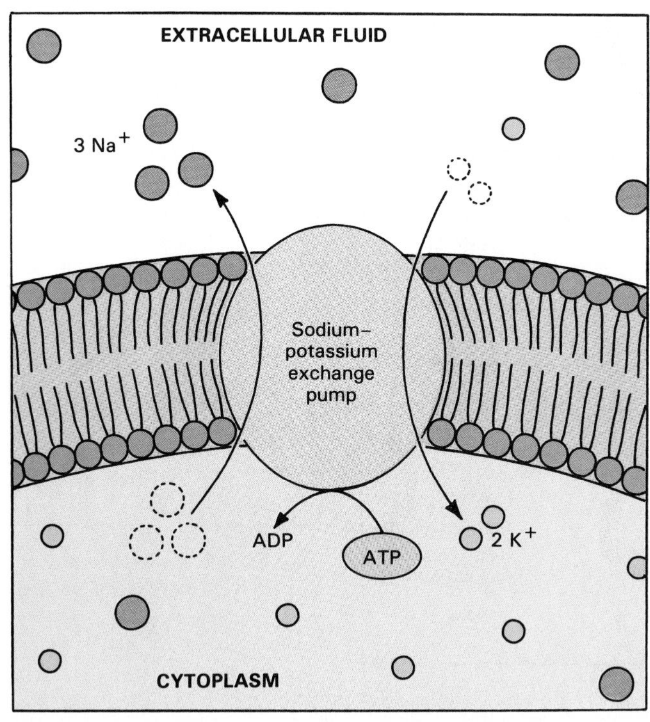

Endocytosis

Endocytosis is the packaging of extracellular materials in a vesicle at the cell surface for importation into the cell. This process, which involves relatively large volumes of extracellular material, is sometimes called *bulk transport*. There are three major types of endocytosis: *pinocytosis*, *receptor-mediated endocytosis*, and *phagocytosis*. All three require energy in the form of ATP and so are classified as active processes. The mechanism is presumed to be the same in each case, but the mechanism itself remains unknown.

All forms of endocytosis produce cytoplasmic vesicles. Once an endocytic vesicle has been formed, however, the contents do not necessarily enter the cytosol. They remain isolated within the vesicle unless they can somehow cross the vesicle wall. This may occur by means of active transport, simple or facilitated diffusion, or the destruction of the vesicle membrane.

Pinocytosis **Pinocytosis** (pi-no-si-TŌ-sis), or "cell drinking," is the formation of vesicles filled with extracellular fluid. In this process, a deep groove or pocket forms in the cell membrane and then pinches off. Once inside the cell, vesicles usually fuse with lysosomes, and the lysosomal enzymes break down organic molecules dissolved in the fluid (Figure 3-16). Nutrients, such as lipids, sugars, or amino acids, then enter the cytoplasm by diffusion or active transport. The membrane of the pinocytotic vesicle then pinches off and returns to the cell surface.

Virtually all cells perform pinocytosis in this manner. In a few specialized cells the vesicles form on one side of the cell and travel through the cytoplasm to the opposite side (Figure 3-25). There they

FIGURE 3-25
Pinocytosis. In specialized cells, the endocytic vesicles travel intact from one side of the cell to another. This is important as a method for the bulk movement of fluid across capillaries.

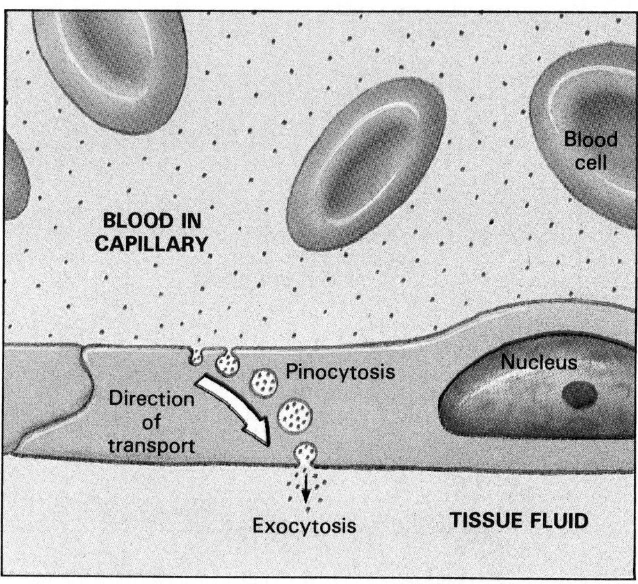

fuse with the cell membrane and discharge their contents. This method of bulk transport is found in cells lining capillaries, the most delicate blood vessels. These cells use pinocytosis to transfer fluid and solutes from the bloodstream to the surrounding tissues.

Receptor-Mediated Endocytosis **Receptor-mediated endocytosis** resembles pinocytosis, but it is far more selective. Pinocytosis produces vesicles filled with extracellular fluid; receptor-mediated endocytosis produces vesicles that contain a specific target molecule in high concentrations. Receptor-mediated endocytosis begins when materials in the extracellular fluid bind to receptors on the membrane surface (Figure 3-26). The receptor molecules are usually glycoproteins, and each has a specific target, such as a transport protein or hormone. These target molecules are known as **ligands**.

Receptors bound to ligands cluster together, creating areas of cell membrane coated with ligands. The coated areas then form grooves or pockets that pinch off to produce **coated vesicles**. Inside the cell the coated vesicles fuse with lysosomes, and lysosomal enzymes free the ligands. The free ligands enter the cytosol by diffusion or active transport, and the vesicle membrane pinches off and returns to the cell surface, its receptors ready to bind more ligands.

Many important substances, including cholesterol and iron ions (Fe^{2+}), are distributed through the body attached to special transport proteins. The proteins are too large to pass through membrane pores, but they can enter the cell through receptor-mediated endocytosis. The transport and absorption mechanisms for specific nutrients, ions, and hormones will be detailed in later chapters.

Phagocytosis **Phagocytosis** (fa-go-si-TŌ-sis), or "cell eating," produces vesicles containing *solid objects* that may be as large as the cell itself. This process is diagrammed in Figure 3-27a. Cytoplasmic extensions called **pseudopodia** (su-do-PŌ-dē-a; *pseudo-*,

FIGURE 3-26
Receptor-Mediated Endocytosis. In this process, specific target molecules called ligands bind to receptors, usually glycoproteins, in the membrane surface. Membrane areas coated with ligand pinch off to form vesicles that fuse with primary lysosomes. The ligands are freed from the receptors and, if necessary, broken down by enzymes before diffusing or being transported into the surrounding cytoplasm. The membrane containing the receptor molecules separates from the membrane of the lysosome and returns to the cell surface to bind additional ligands.

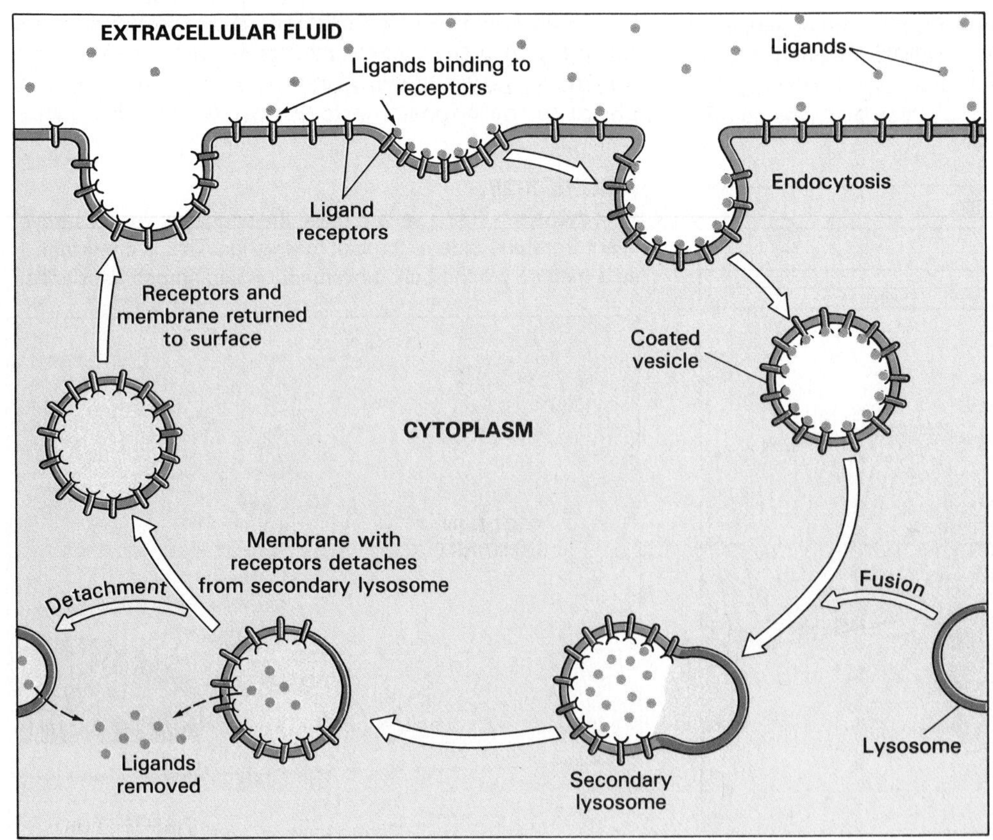

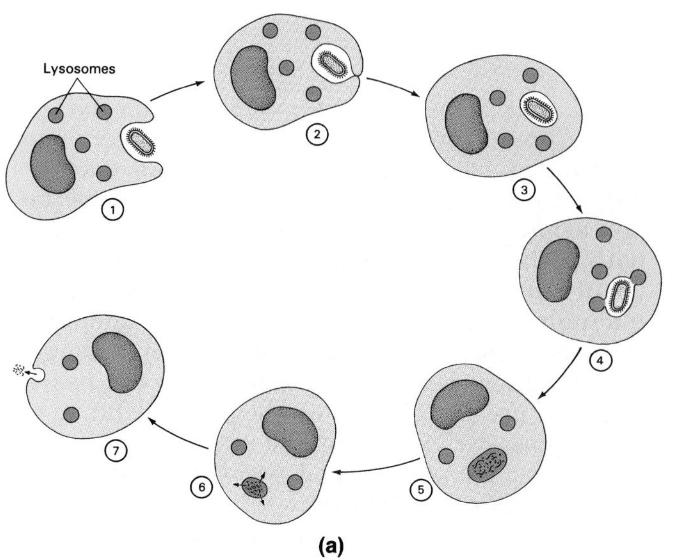

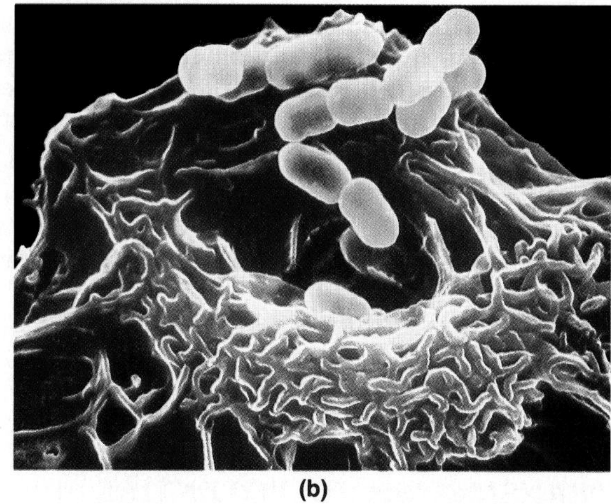

FIGURE 3-27
Phagocytosis. (a) A phagocytic cell first comes in contact with the foreign object and sends cytoplasmic extensions around it (1). The extensions approach one another (2) and then fuse to trap the material within an endocytic vesicle (3). Lysosomes fuse with this vesicle, activating digestive enzymes that gradually break down the structure of the phagocytized material (4–6). Undissolved residue can then be ejected by exocytosis (7). (b) SEM showing a phagocyte engulfing several bacteria.

false + *podon*, foot) surround the object, and their membranes fuse to form a vesicle. The vesicle may then fuse with a lysosome, whereupon its contents are digested by lysosomal enzymes. The SEM in Figure 3-27b gives a three-dimensional view of phagocytosis in action.

Most cells display pinocytosis, but phagocytosis, especially the entrapment of living or dead cells, is performed only by specialized cells of the immune system. Phagocytic cells will be considered in chapters dealing with blood cells (Chapter 19) and the immune system (Chapter 22).

Many different mechanisms can be moving materials in and out of the cell at any given moment. Before proceeding further, review and compare the mechanisms summarized in Table 3-7.

TABLE 3-7 Summary of Mechanisms Involved in Movement Across Cell Membranes

Mechanism	*Process*	*Factors Affecting Rate*	*Substances Involved*
PASSIVE			
Diffusion	Molecular movement of solutes; direction determined by relative concentrations	Size of gradient, molecular size, charge, lipid solubility	Small inorganic ions, lipid-soluble materials (all cells)
Osmosis	Movement of water (solvent) molecules toward high solute concentrations; requires membrane	Concentration gradient, opposing pressure	Water only (all cells)
Filtration	Movement of water, usually with solute, by hydrostatic pressure; requires membrane filter	Amount of pressure, size of pores	Water and small ions (blood vessels)
Facilitated diffusion	Carrier molecules transport down concentration gradient; requires membrane	As above, plus availability of carrier	Glucose and amino acids (all cells)
ACTIVE			
Active transport	Carrier molecules work regardless of any concentration gradients	Availability of carrier, substrate, and ATP	Na^+, K^+, Ca^{2+}, Mg^{2+} (all cells); probably other materials in special cases
Endocytosis	Creation of vesicles containing fluid or solid material	Stimulus and mechanics not understood; requires ATP	Fluids, nutrients (all cells); debris, pathogens (special cells)

✓ **During digestion in the stomach, the concentration of H^+ ions rises to many times the concentration found in the cells of the stomach. What type of transport process could produce this result?**

✓ **Some pediatricians recommend using a 10 percent salt solution for infants with stuffy noses to relieve congestion. What type of solution would this be and how would it affect the cells lining the nasal cavity?**

MEMBRANE PERMEABILITY AND THE TRANSMEMBRANE POTENTIAL

Intracellular and extracellular fluids differ in ionic composition, and the differences persist despite the tendency for diffusion to eliminate them. In particular, the extracellular fluid contains large numbers of sodium ions (Na^+) and chloride ions (Cl^-), whereas the cytoplasm has high concentrations of potassium ions (K^+) and negatively charged proteins (Pr^-). Figure 3-28 diagrams the relative sizes and directions of the concentration gradients for major cations and anions across the cell membrane.

FIGURE 3-28
Concentration Gradients Across the Cell Membrane. The size of the abbreviation for each ion indicates its relative concentration. The arrows indicate the relative abilities of sodium and potassium to cross the membrane.

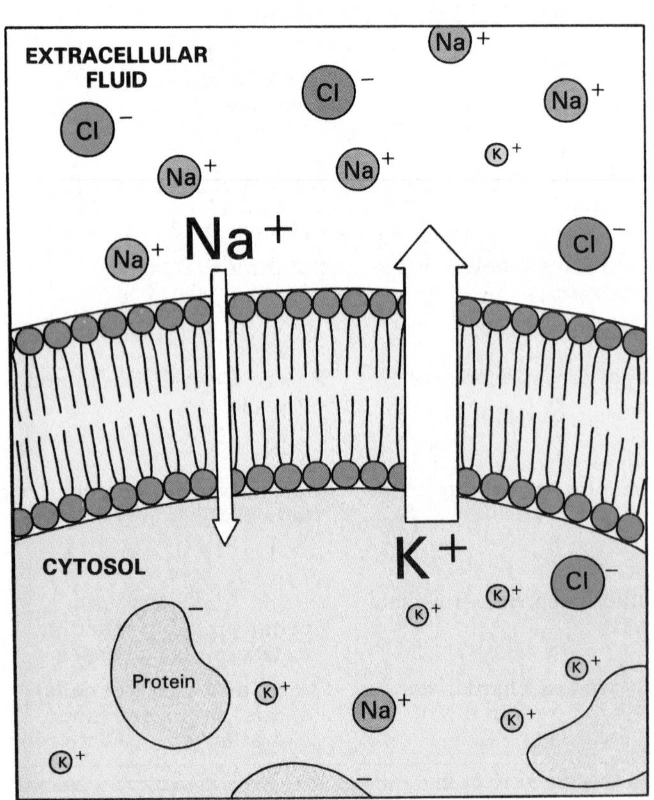

If the cell membrane were freely permeable, diffusion would continue until these ions were evenly distributed across the membrane. This does not happen because living cells have selectively permeable membranes that contain a variety of active transport mechanisms, notably the sodium-potassium exchange pump. However, although cell membranes maintain stable ion concentrations in the cytoplasm, they do not maintain an even distribution of charges. Under normal conditions a cell membrane has an excess of positive charges on the outside and an excess of negative charges on the inside.

Whenever positive and negative ions are kept apart, a **potential difference** exists. The larger the potential difference, the stronger the attraction between positive and negative charges. The unit of measurement for potential difference is the volt (V). Automobile batteries typically maintain a potential difference of 12 V; ordinary flashlight batteries provide 1.5 V.

The two sides of a cell membrane are often compared to the positive and negative terminals of a battery. Because the charges are separated by a cell membrane, the potential difference is called a **transmembrane potential**. The transmembrane potentials of living cells are very small, averaging about 0.070 V for a nerve cell membrane. This is usually expressed as −70 mV, with mV indicating millivolts (thousandths of a volt) and the minus sign signifying that the inside of the cell contains an excess of negative charges.

Creation of the Transmembrane Potential

Two factors, one passive and one active, interact to create and maintain the transmembrane potential. The passive factor is that the membrane permeabilities for sodium and potassium are quite different. Potassium ions can diffuse out of the cell through potassium channels faster than sodium ions can enter the cell through sodium channels. The cell therefore loses positive charges faster than it gains them. Contributing to the imbalance are the cell's many proteins, most of which are negatively charged. Because these molecules are too large to pass through the membrane, the interior of the cell develops an excess of negative charges.

The active factor is the presence of the sodium-potassium exchange pump in the membrane surface. As noted earlier, this ion pump does not operate on a 1:1 basis; it ejects three sodium ions for every two potassium ions reclaimed from the extracellular fluid. At a transmembrane potential of 0 mV, potassium ions diffuse out of the cell much faster than sodium ions enter it, and the exchange pump cannot prevent the net loss of positive charges.

As the interior of the cell becomes increasingly negative, both sodium ions (Na^+) and potassium ions (K^+) are attracted to the inside of the cell membrane.

In the case of sodium ions, the addition of an electrical attraction to the concentration gradient causes an increase in the rate of sodium entry. In the case of potassium ions, the electrical attraction opposes the concentration gradient, and the rate of potassium loss decreases. At a transmembrane potential of −70 mV, three sodium ions enter the cell for every two potassium ions that leave. This rate is precisely balanced by the activity of the sodium-potassium exchange pump, so the transmembrane potential is stabilized at this value.

Functions of the Transmembrane Potential

The transmembrane potential is characteristic of all living cells because it results from the active and passive properties of their cell membranes. The transmembrane potential in an undisturbed cell is called its **resting potential**. Each cell type has a characteristic resting potential between −10 mV and −100 mV—examples include fat cells (−40 mV), thyroid cells (−50 mV), skeletal muscle cells (−85 mV), and cardiac muscle cells (−90 mV).

Although not visible through a microscope, the transmembrane potential is just as important as any structural characteristic or organelle. Many cell functions that involve the cell membrane, such as secretion or ciliary movement, involve changes in the transmembrane potential. But the transmembrane potential itself has significant functions. In addition to the open sodium and potassium channels described above, the cell membrane contains gated ion channels that are closed at the normal resting potential. Because the concentration gradients are large, any stimulus that opens one of these gates will produce a sudden rush of ions into or out of the cell. The amount of ion movement does not depend on the size or nature of the stimulus; the stimulus only opens the floodgates. If it can affect a channel protein, even a relatively weak stimulus can have a significant impact on the cell. Because the transmembrane potential can magnify a stimulus in this way, it greatly increases the cell's sensitivity to its environment.

The ionic movements that occur following stimulation can have a variety of effects on the cell. In muscle cells and nerve cells, such ion movements are essential to their specialized functions. For example, a sudden influx of sodium ions across the membrane of a muscle cell is the first step leading to its contraction. The role of the transmembrane potential in muscle and nerve function will be explored further in Chapters 10 and 12.

CELL ATTACHMENT

Most cells in the body are firmly attached to other cells or to extracellular protein fibers. The attachments occur at cell junctions that are not involved in membrane flow. There are four types of cell junctions: *gap junctions*, *tight junctions*, *intermediate junctions*, and *desmosomes*.

Gap Junctions

In a **gap junction** (Figure 3-29a) the two cells are held together by an interlocking of membrane proteins. Because these are channel proteins, the result is a narrow passageway that lets small molecules and ions pass from cell to cell. Gap junctions are most common in cardiac muscle and smooth muscle tissue; they are also occasionally found between nerve cells.

Tight Junctions

At a **tight junction** (Figure 3-29b) there is a partial fusion of the lipid portions of the two cell membranes. Because the membranes are fused together, tight junctions are the strongest intercellular connections. In addition to providing mechanical strength, tight junctions block the passage of water or solutes between the cells. Tight junctions are often found where cells are exposed to fluids whose composition is very different from that of normal extracellular fluid. For example, tight junctions near the exposed surfaces of cells lining the digestive tract keep enzymes, acids, and wastes from damaging delicate underlying tissues.

Intermediate Junctions

At an **intermediate junction** (Figure 3-29c) the opposing cell membranes, while remaining distinct, are held together by a thick layer of proteoglycans. This proteoglycan layer is called **intercellular cement**; **hyaluronic acid** is the most important proteoglycan involved. The cytoplasm at an intermediate junction contains a dense network of microfilaments that anchor the junction to the cytoskeleton. This arrangement adds strength and helps to stabilize the shape of the cell.

Desmosomes

At **desmosomes** (DEZ-mo-somz; *desmos*, ligament + *soma*, body) there is a very thin proteoglycan layer between the opposing cell membranes, reinforced by a network of intermediate filaments that lock the two cells together (Figure 3-29d). A dense concentration of filaments beneath the cell membrane at a desmosome anchors it to the cytoskeleton. Desmosomes are very strong, and the connection can resist stretching and twisting. These intracellular connections are most abundant between cells in the superficial layers of the skin. The desmosomes create

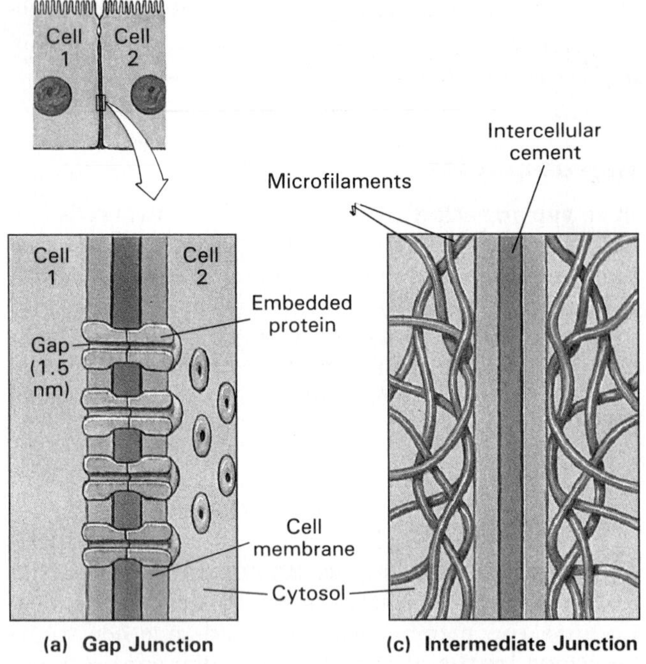

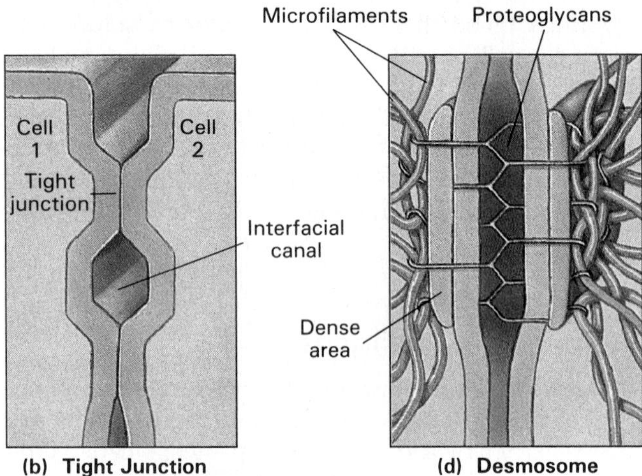

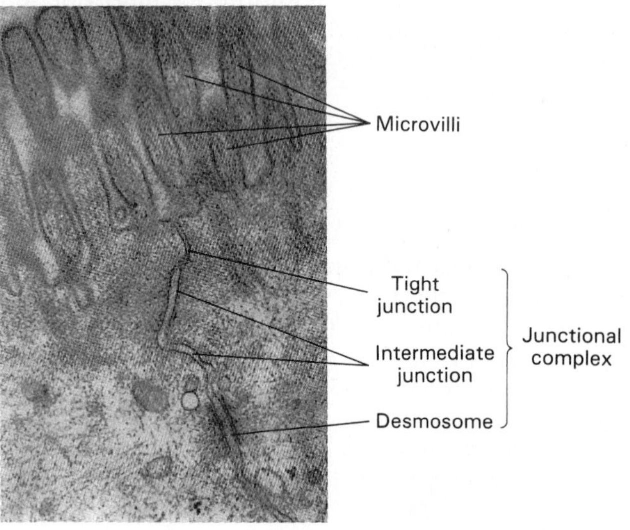

FIGURE 3-29
Cell Attachments. (a) At a gap junction, binding of membrane proteins creates a cytoplasmic connection between two cells. (b) A tight junction is formed by fusion of the outer layers of two cell membranes. (c) At an intermediate junction the membranes are held together by intercellular cement. A network of microfilaments strengthens the region of attachment. (d) A desmosome has a more organized network of microfilaments. Desmosomes attach one cell to another or attach a cell to extracellular structures, such as the protein fibers in connective tissues. (e) A junctional complex consists of a tight junction, an intermediate junction, and a desmosome. The tight junction is closest to the cell surface.

links so strong that even dead skin cells are usually shed in thick sheets rather than individually.

Junctional Complexes

Cells lining the digestive tract, respiratory tract, or other passageways are held together by **junctional complexes**. A single junctional complex consists of a tight junction, an intermediate junction, and a desmosome, with the tight junction closest to the cell surface. Figure 3-29e details the structure of a typical junctional complex.

■ The Cell Life Cycle

Between fertilization and physical maturity a human being goes from a single cell to roughly 75 trillion cells. This amazing increase in numbers occurs through a form of cellular reproduction called **cell division**. The division of a single cell produces a pair of daughter cells, each half the size of the original. For all intents and purposes two new cells have replaced the original one.

Even when development has been completed, cell division continues to be essential to survival. Although cells are highly adaptable, they can be damaged by physical wear and tear, toxic chemicals, temperature changes, or other environmental hazards. In addition, cells, like individuals, are subject to aging. The lifespan of a cell varies from hours to decades, depending on the type of cell and the environmental stresses involved. A typical cell does not live nearly as long as a typical person, so over time cell populations must be maintained by cell division.

Central to cell reproduction is the accurate duplication of the cell's genetic material and its distribution to the two new daughter cells formed by division. This process is called **mitosis** (mī-TŌ-sis). Mitosis occurs during the division of **somatic** (*soma*, body) **cells**. Somatic cells include all of the cells in the body

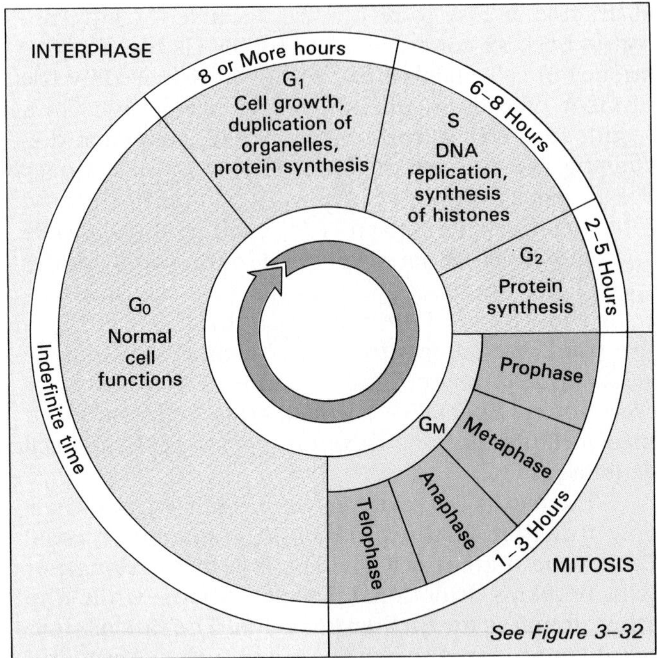

FIGURE 3-30
The Cell Life Cycle

other than the **reproductive cells**, which give rise to sperm or eggs. Production of sperm and eggs involves a distinct process, *meiosis* (mī-Ō-sis), that will be described in Chapter 29.

Most cells spend only a small part of their time actively engaged in cell division. Somatic cells spend the majority of their functional lives in interphase. During **interphase** the cell performs all of its normal functions plus, if necessary, making preparations for division. Figure 3-30 presents the life cycle of a typical cell in greater detail.

INTERPHASE

An interphase cell in the **G_0 phase** is not preparing for mitosis. A cell that is going to divide first enters the **G_1 phase**. In this phase the cell manufactures enough mitochondria, centrioles, cytoskeletal elements, endoplasmic reticulum, ribosomes, Golgi membranes, and cytosol to make two functional cells. In cells dividing at top speed, G_1 may last as little as 8–12 hours. Such cells pour all of their energy into mitosis, and all other activities cease. If G_1 lasts for days, weeks, or months, preparation for mitosis occurs as the cells perform their normal functions. When preparations have been completed the cell enters the **S phase**. Over the next 6–8 hours the cell duplicates its chromosomes.

Throughout the life of a cell, the DNA strands in the nucleus remain intact. DNA synthesis, or **DNA replication**, occurs in cells preparing to undergo mitosis or meiosis. The goal of replication is to copy the genetic information in the nucleus so that one set of chromosomes can be given to each of the two cells produced. Several different enzymes are needed for the process.

As discussed in Chapter 2, a DNA molecule consists of a pair of strands held together by hydrogen bonding between complementary nitrogen bases. Figure 3-31 diagrams the process of DNA replication. It starts when the weak bonds between the nitrogen bases are disrupted, and the strands unwind. As they do so, molecules of the enzyme **DNA polymerase** bind to the exposed nitrogen bases. This enzyme promotes bonding between the nitrogen bases of the DNA strand and complementary DNA nucleotides dissolved in the nucleoplasm.

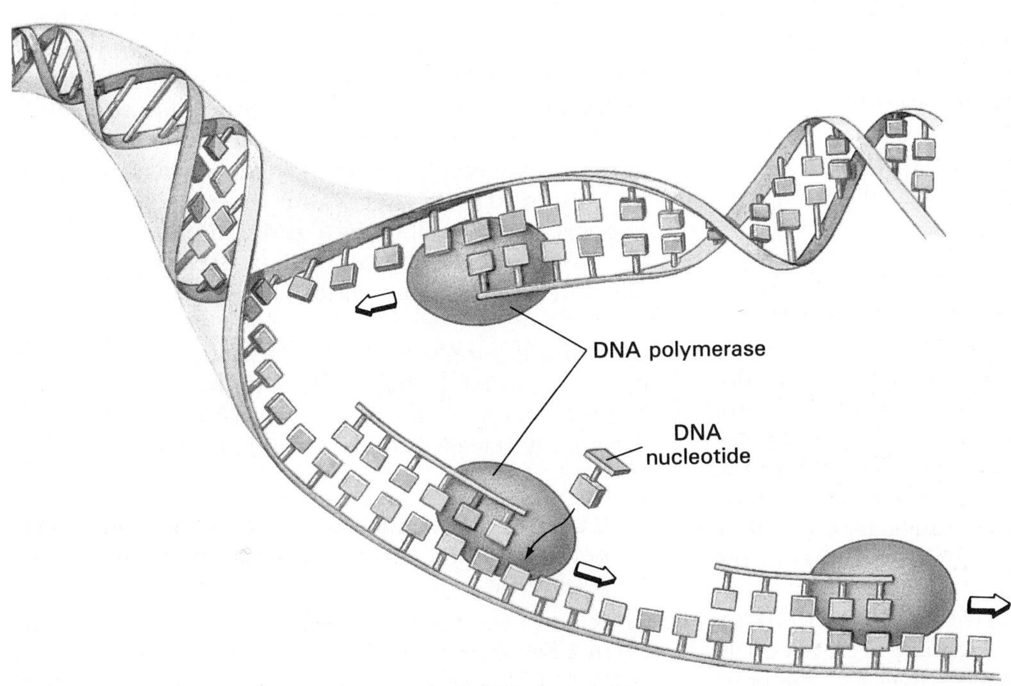

FIGURE 3-31
DNA Replication. In replication the DNA strands unwind and DNA polymerase begins attaching complementary DNA nucleotides along each strand. This produces two identical copies of the original DNA molecule.

Many molecules of DNA polymerase are working simultaneously, along different portions of each DNA strand. This produces short complementary nucleotide chains that are then linked together by enzymes called **ligases** (LI-gās-ez; *liga*, to tie). The final result is a pair of identical DNA molecules.

Once DNA replication has been completed, there is a brief (2–5 hours) **G_2 phase** devoted to last-minute protein synthesis. The cell then enters the **G_m phase**, and mitosis begins.

MITOSIS

Figure 3-32 summarizes the four stages of mitosis.

STAGE 1: **Prophase** (PRŌ-fāz; *pro*, before): Prophase begins when the chromosomes coil so tightly that they become visible as individual structures. As a result of DNA replication during the S phase, there are two copies of each chromosome, called **chromatids** (KRŌ-ma-tids), connected at a single point, the **centromere** (SEN-tro-mer). As the chromosomes appear, the two pairs of centrioles move apart. **Spindle fibers** extend between the centriole pairs; smaller microtubules radiate into the surrounding cytosol. Prophase ends with the disappearance of the nuclear envelope.

STAGE 2: **Metaphase** (MET-a-fāz; *meta*, after): The chromatids now move to a narrow central zone called the **metaphase plate.** A microtubule of the spindle apparatus attaches to each centromere.

STAGE 3: **Anaphase** (AN-uh-fāz; *ana*, back): As if responding to a single command, the chromatid pairs separate and the **daughter chromosomes** move toward opposite ends of the cell.

STAGE 4: **Telophase** (TEL-o-fāz; *telo*, end): This stage is in many ways the reverse of prophase. The nuclear membranes form, the nuclei enlarge, and the chromosomes gradually uncoil. Once the chromosomes disappear, nucleoli reappear and the nuclei resemble those of interphase cells.

Telophase marks the end of mitosis proper, but the daughter cells have yet to complete their physical separation. This separation process, called **cytokinesis** (sī-tō-ki-NĒ-sis; *cyto-*, cell + *kinesis*, motion), usually begins in late anaphase. As the daughter chromosomes near the ends of the spindle apparatus, the cytoplasm constricts along the plane of the metaphase plate. This process continues through telophase, and the completion of cytokinesis marks the end of cell division.

The preparations for cell division that occur between G_0 and the end of the S phase are difficult to recognize in a light micrograph. However, the start of mitosis is easy to recognize, because the chromosomes become condensed and highly visible. The frequency of cell division can thus be estimated by the number of cells in mitosis at any given time. As a result, the term **mitotic rate** is often used when discussing rates of cell division. In general, the longer the life expectancy of a cell type, the slower the mitotic rate. Relatively long-lived cells, such as muscle cells and nerve cells, either never divide or do so only under special circumstances. Other cells, such as those covering the surface of the skin or the lining of the digestive tract, are subject to attack by chemicals, pathogens, and abrasion. They survive only for days or even hours. Special cells called **stem cells** maintain these cell populations through repeated cycles of cell division.

Stem cells are relatively unspecialized, and their only function is the production of daughter cells. Each time a stem cell divides, one of the daughter cells develops functional specializations while the other prepares for further divisions. The rate of stem cell division can vary, depending on the tissue and the demand for new cells. In heavily abraded skin, stem cells may divide more than once a day, but stem cells in adult connective tissues may remain inactive for years.

ENERGETICS AND CELL DIVISION

Dividing cells use an unusually large amount of energy. For example, they must synthesize new organic materials and move organelles and chromosomes within the cell. All of these processes require ATP in substantial amounts. Cells that do not have adequate energy sources cannot divide, and in starvation normal cell growth and maintenance grind to a halt. For this reason, prolonged starvation stunts growth, slows wound healing, lowers resistance to disease, thins the skin, and changes the lining of the digestive tract. The same changes are seen in the late stages of many cancers, because the cancer cells are "stealing" the nutrients that would otherwise be used to support normal cell growth and maintenance.

This chapter began with a description of the parts of a cell. It has ended in a discussion of the relative energy cost of cell division. All cells need a continuous supply of energy, and without it they cannot survive, let alone divide. The next chapter will begin by considering how cells capture and harness the energy released by chemical reactions.

√ **If the cell membrane were freely permeable to Na^+, how would this effect the transmembrane potential?**

√ **What would happen if spindle fibers failed to form in a cell during mitosis?**

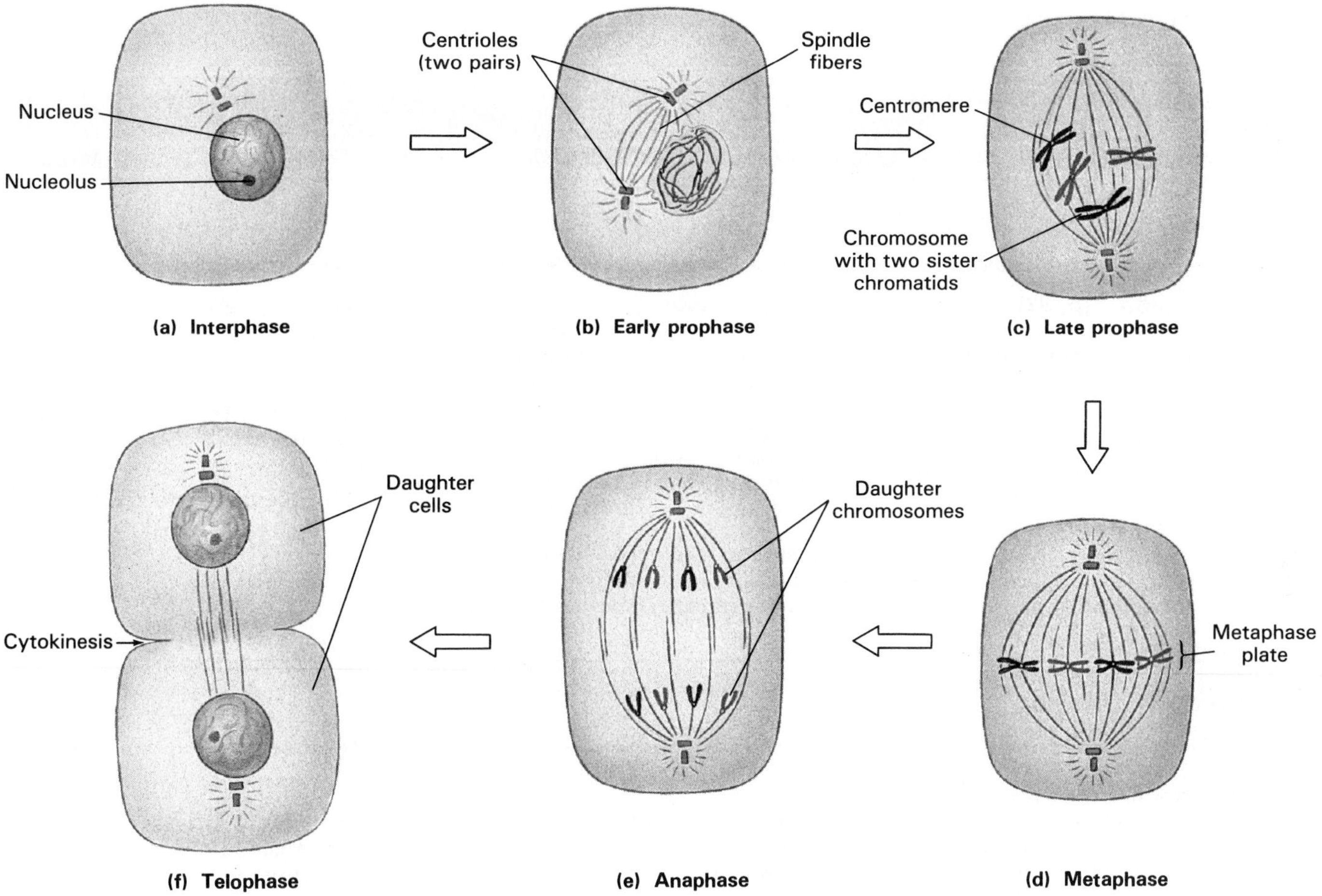

Nucleus
Nucleolus
(a) Interphase
Centrioles (two pairs)
Spindle fibers
(b) Early prophase
Centromere
Chromosome with two sister chromatids
(c) Late prophase
Metaphase plate
(d) Metaphase
Daughter chromosomes
(e) Anaphase
Daughter cells
Cytokinesis
(f) Telophase

FIGURE 3-32
Mitosis

Clinical Comment: Cancer I—Cell Division and Cancer

Mitotic rates are usually well controlled, and in normal tissue the rate of cell division balances cell loss or destruction. When that balance breaks down, the tissue begins to enlarge. A **tumor**, or **neoplasm**, is a mass or swelling produced by abnormal cell growth and mitosis. In a **benign tumor** the cells remain within a connective tissue capsule. Such a tumor seldom threatens an individual's life. Surgery can usually remove the tumor if its size or position disturbs tissue function.

Cells in a **malignant tumor** are no longer responding to normal control mechanisms. These cells divide rapidly, spreading into the surrounding tissues, and they may also spread to other tissues and organs. This spread is called **metastasis** (me-TAS-ta-sis). Metastasis is dangerous and difficult to control. Once in a new location, the metastatic cells produce secondary tumors.

The term **cancer** refers to an illness characterized by malignant cells. Cancer cells gradually lose their resemblance to normal cells. They change size and shape, often becoming unusually large or abnormally small. Organ function begins to deteriorate as the number of cancer cells increases. The cancer cells may not perform their original functions at all, or they may perform normal functions in an unusual way. For example, endocrine cancer cells may produce normal hormones, but in abnormally large amounts. Cancer cells compete for space and nutrients with normal cells. They do not use energy very efficiently, and they grow and multiply at the expense of normal tissues. This accounts for the starved appearance of many patients in the late stages of cancer. We will return to the subject of cancer in Chapters 4 and 5, as well as in later chapters dealing with the specific systems.

CHAPTER REVIEW

Review of Selected Clinical Terms

normal saline (*p. 82*)
A solution that approximates the normal osmotic concentration of extracellular fluids.

dextran (*p. 82*)
A carbohydrate that cannot cross cell membranes; often administered to patients following blood loss or dehydration.

tumor (neoplasm) (*p. 93*)
A mass or swelling produced by abnormal cell growth and division.

benign tumor (*p. 93*)
A mass or swelling in which the cells remain within a connective tissue capsule; rarely life-threatening.

malignant tumor (*p. 93*)
A mass or swelling in which the cells no longer respond to normal control mechanisms, but divide rapidly.

metastasis (me-TAS-ta-sis) (*p. 93*)
The spread of malignant cells into surrounding tissues and organs.

cancer (*p. 93*)
An illness characterized by malignant cells.

Additional Terms of Clinical Importance

KARTAGENER'S SYNDROME (*p. 77*): **Kartagener's syndrome**

Study Outline

Related Key Terms:

Introduction (pp. 59–61)

1. Contemporary cell theory incorporates several basic concepts: (1) cells are the building blocks of all plants and animals; (2) cells are produced by the division of preexisting cells; (3) cells are the smallest units that perform all vital physiological functions; (4) each cell maintains homeostasis at the cellular level; (5) homeostasis at the tissue, organ, system, and individual levels reflects the combined and coordinated actions of many cells.

cytology
transmission electron microscopy
scanning electron microscopy

Cellular Anatomy (pp. 61–76)

1. A cell floats in the **extracellular fluid**. The cell's outer boundary is the **cell membrane** or **plasma membrane**, which is a **phospholipid bilayer**.

THE CYTOPLASM (pp. 64–65)

2. The **cytoplasm** contains a fluid **cytosol** and surrounds **organelles**.

NONMEMBRANOUS ORGANELLES (pp. 65–69)

3. **Nonmembranous organelles** are always in contact with the cytosol. They include the cytoskeleton, microvilli, centrioles, cilia, flagella, and ribosomes.
4. The **cytoskeleton** gives the cytoplasm strength and flexibility. It has four components: **microfilaments**, **intermediate filaments**, **thick filaments**, and **microtubules**.
5. **Microvilli** are small projections of the cell membrane that increase the surface area exposed to the extracellular environment.
6. **Centrioles** direct the movement of DNA molecules during cell division.
7. **Cilia** beat rhythmically to move fluids or secretions across the cell surface.
8. **Flagella** move a cell through surrounding fluid, rather than moving fluid past a stationary cell.
9. **Ribosomes** are intracellular factories that manufacture proteins. There are **free ribosomes** and **fixed ribosomes**.

inclusions
actin • **myosin**
neurofilaments • **tubulin**
centrosome
basal body

MEMBRANOUS ORGANELLES (pp. 69–76)

10. **Membranous organelles** are surrounded by lipid membranes that isolate them from the cytosol.
11. **Mitochondria** are responsible for 95 percent of the ATP production within a typical cell.
12. The **nucleus** is the control center for cellular operations. It is surrounded by a **nuclear envelope**, through which it communicates with the cytosol through **nuclear pores**.
13. The **endoplasmic reticulum (ER)** is a network of intracellular membranes. There are two types: rough and smooth. **Rough endo-**

cristae • **matrix**
respiratory enzymes
perinuclear space
nucleoplasm • **chromosomes**
histones • **nucleosome**
chromatin • **nucleoli**
cisternae • **transport vesicles**

Related Key Terms

plasmic reticulum (RER)** contains ribosomes; **smooth endoplasmic reticulum (SER)** does not.

14. The **Golgi apparatus** packages **lysosomes**, **peroxisomes**, and **secretory vesicles**. Secretions are discharged from the cell in a process called **exocytosis**.
15. **Lysosomes** are vesicles filled with digestive enzymes. The process of **endocytosis** is important in ridding the cell of bacteria and debris.
16. **Peroxisomes** also carry enzymes; they absorb and neutralize toxins.

saccules • transfer vesicles
membrane flow
autolysis

Communication between the Cell and Its Environment (pp. 76–90)

MEMBRANE STRUCTURE (pp. 77–78)

1. **Integral proteins** are part of the membrane itself, while **peripheral proteins** are attached but can separate from it. **Channels** allow water and ions to move across the membrane; some channels are called **gated** because they can open or close.

glycocalyx • permeability
impermeable • freely permeable

MEMBRANE PERMEABILITY: PASSIVE PROCESSES (pp. 78–84)

2. Cell membranes are **selectively permeable**.
3. **Diffusion** is the net movement of material from an area where its concentration is relatively high to an area where its concentration is lower. Diffusion occurs until the **concentration gradient** is eliminated.
4. Diffusion of water across a membrane in responses to differences in concentration is **osmosis**. The force of movement is **osmotic pressure**.
5. In **filtration**, hydrostatic pressure forces water across a membrane; if membrane pores are large enough, molecules of solute will be carried along.
6. **Facilitated diffusion** requires the presence of **carrier proteins**.

hydrostatic pressure
bulk flow
osmotic concentration (osmolarity)
isotonic • tonicity
hypotonic • hypertonic
hemolysis • crenation
receptor site

MEMBRANE PERMEABILITY: ACTIVE PROCESSES (pp. 84–88)

7. All **active membrane processes** require energy. Two important active processes are active transport and endocytosis.
8. **Active transport** mechanisms consume ATP but are independent of concentration gradients. Some **ion pumps** are **exchange pumps**.
9. **Endocytosis** is an active process that can take three forms: **pinocytosis**, **receptor-mediated endocytosis**, and **phagocytosis**.

sodium-potassium exchange pump
ligands • coated vesicles
pseudopodia

MEMBRANE PERMEABILITY AND THE TRANSMEMBRANE POTENTIAL (pp. 88–89)

10. The **potential difference** between the two sides of a cell membrane is a **transmembrane potential**. The transmembrane potential in an undisturbed cell is its **resting potential**.
11. Because it can greatly magnify the effect of a stimulus, the transmembrane potential increases the cell's sensitivity to its environment.

CELL ATTACHMENT (pp. 89–90)

12. Cells can attach to other cells or to extracellular protein fibers in four ways: gap junctions, tight junctions, intermediate junctions, and desmosomes.
13. In a **gap junction** two cells are held together by interlocked membrane proteins, forming a narrow passageway.
14. At a **tight junction** there is a partial fusion of the two cell membranes; these are the strongest intercellular connections.
15. At an **intermediate junction** two cells are held together by a thick layer of proteoglycans called **intercellular cement**.
16. A **desmosome** has a very thin proteoglycan layer between the cell membranes, reinforced by a network of microfilaments.
17. Cells in some areas of the body are linked by **junctional complexes**.

hyaluronic acid

The Cell Life Cycle (pp. 90–93)

1. **Mitosis** refers to the nuclear division of **somatic cells**. **Reproductive cells** (sperm and eggs) are produced by **meiosis**.
2. Most somatic cells spend most of their time in **interphase**.

MITOSIS (p. 92)

3. Mitosis proceeds in four stages: **prophase**, **metaphase**, **anaphase**, and **telophase**.
4. In general, the longer the life expectancy of a cell type, the slower the **mitotic rate**. **Stem cells** undergo frequent mitoses to replace other, more specialized cells.

cell division
G_0 phase • G_1 phase
S phase • DNA replication
DNA polymerase • ligases
G_2 phase • G_m phase
chromatids • centromere
spindle fibers • metaphase plate
daughter chromosomes
cytokinesis

Chapter Review

Review Planner

		Level 1
1	Discuss the basic concepts of the cell theory.	**22**
2	Compare the fluid contents of a cell with the extracellular fluid.	**11 24**
3	Explain the structure and importance of the cell membrane.	**11 15 17 29**
4	Discuss the ways in which a cell interacts with its environment.	**1 9 11 13 21 26 29 30 33 35 37**
5	Compare the structure and functions of the various cellular organelles.	**2 3 4 5 7 8 14 16 19 20 23 25 27 28 31 32 34**
6	Explain how cells obtain energy to power their operations.	**20**
7	Discuss the role of the nucleus as the cell's control center.	**3**
8	Describe the cell life cycle and how cells divide.	**4 6 12 18**
9	Explain the importance of the transmembrane potential, and describe the mechanisms that maintain it.	**13 17 38 40**
10	Compare the ways in which cells attach to other cells and discuss how the methods of attachment affect body tissues.	**10 36**

Level 2 **39–40**

Level 3 **41 42 43**

 C M **Mitochondrial DNA, Disease, and Evolution • Lysosomal Storage Disease**

Review Questions

MULTIPLE CHOICE

1. Facilitated diffusion is a passive process. What feature(s), if any, does it have in common with active transport processes? a) It requires energy from ATP; b) It requires a sodium-potassium exchange pump; c) It requires carrier proteins; d) They have nothing in common.
2. The cytoplasm is: a) a protein framework that strengthens the cell; b) the jellylike substance that surrounds cells in the intestinal tract; c) only the fluid inside the cell; d) everything inside the cell.
3. The organelle that serves as the control center for cellular operations is: a) the brain; b) the mitochondria; c) DNA; d) none of the above.
4. Which of the following is false? a) Chromosomes contain RNA, not DNA; b) Chromosomes contain information for protein synthesis; c) Chromatin refers to chromosomes in cells that are not in the process of dividing; d) Histones are proteins found in chromosomes.
5. Which of the following are not functions of the endoplasmic reticulum: a) It synthesizes proteins, carbohydrates, and lipids; b) It allows materials to travel from place to place; c) It packages secretions to be discharged from the cell; d) It stores molecules absorbed from the cytosol.
6. The nuclear division of somatic cells is called: a) meiosis; b) cell cleavage; c) mitosis; d) interphase.
7. Rough endoplasmic reticulum contains them; smooth endoplasmic reticulum does not. They are a) ribosomes; b) cisternae; c) centromeres; d) centrioles.
8. Which of the following are not functions of vesicles? a) connecting two cells; b) transferring proteins within a cell; c) carrying secretions to be discharged from the cell; d) carrying digestive enzymes that break down bacteria.
9. Water and ions can cross cell membranes through passageways called: a) cristae; b) microtubules; c) high-energy bonds; d) channels.
10. Intercellular cement is the primary "glue" between cells at which type of cell junction? a) desmosome; b) intermediate junction; c) gap junction; d) tight junction.

11. Cell membranes are: a) impermeable; b) freely permeable; c) selectively permeable; d) none of the above.
12. Which is the correct sequence for the stages of mitosis: a) prophase, anaphase, metaphase, telophase; b) prophase, metaphase, telophase, anaphase; c) anaphase, prophase, metaphase, telophase; d) none of the above.
13. All of the following statements about the sodium-potassium exchange pump are true *except*: a) It is important for cell homeostasis; b) It exchanges intracellular sodium for extracellular potassium; c) A typical cell membrane has approximately 1000 of them per square micrometer; d) It exchanges extracellular sodium for intracellular potassium.

DIAGRAMS

14. Sketch a "typical" cell as discussed in this chapter. Illustrate and label the nonmembranous and membranous organelles.

TRUE/FALSE

(If a statement is false, your instructor may wish to have you correct it.)

15. **T/F:** The cell membrane is called a phospolipid trilayer.
16. **T/F:** Some organelles are not essential to normal cellular operations.
17. **T/F:** Each type of cell has a characteristic resting potential.
18. **T/F:** Somatic cells spend most of their functional lives in interphase.

MATCHING QUESTIONS

(*Match each statement with the* most appropriate *term*).

Statements:

19. These structures "beat" to remove particles from our throats and respiratory passages.
20. Mitochondria generate this high-energy compound.
21. It carries a receptor site to which a molecule must bind in order to be transported.
22. These are the smallest units able to perform all vital physiological functions.
23. The sperm cell is the only human cell that possesses this structure.
24. This term describes a solution that has the same concentration of solutes as the cytoplasm.
25. This is a membranous organelle.
26. The "target molecules" to which receptor molecules bind during receptor-mediated endocytosis.
27. These masses of insoluble materials are sometimes found in the cytosol.
28. This process occurs when lysosomes "fall apart," releasing active enzymes into the cell that can destroy its contents.

Terms:

a) Inclusions
b) Flagellum
c) Autolysis
d) Carrier protein
e) Cells
f) Ligands
g) Isotonic
h) Cilia
i) ATP
j) The Golgi apparatus

SHORT ANSWER/ESSAY

29. Define "hydrophilic" and "hydrophobic." Why are the hydrophilic heads and hydrophobic tails of phospholipid molecules important to the cell membrane?
30. Describe phagocytosis. Which cells perform this process? Why is it important?
31. What are nonmembranous organelles? Name the six categories mentioned in this chapter, and discuss the physiological roles of each.
32. Define "membranous organelle" and name the six types discussed in the text. Compare the physiological functions of each.
33. What is hydrostatic pressure?
34. Distinguish between the following terms: microfilaments, mitcrotubules, microvilli, intermediate filaments, neurofilaments, and thick filaments. What function(s) does each serve?
35. Define and compare the transport processes of diffusion, osmosis, filtration, and facilitated diffusion.
36. Distinguish between gap junctions, tight junctions, intermediate junctions, desmosomes, and junctional complexes.
37. What is meant by the terms "sodium channel," "calcium channel," and "potassium channel"?
38. Define the transmembrane potential. Explain its physiological importance.
39. How are cell populations maintained? What health conditions can result if mitotic rates are no longer well controlled?
40. Explain the importance of the sodium-potassium exchange pump.

CRITICAL THINKING/APPLICATIONS

41. An automobile accident victim has lost a great deal of blood. Should paramedics at the accident site administer (1) water, (2) fluid that approximates intracellular fluid, or (3) fluid that approximates extracellular fluid? Explain your answer.
42. One result of regular cigarette smoking is paralysis of the cilia that line the respiratory passageways. What function do these cilia serve? Based on what you have read in this chapter, why is it harmful when they no longer beat? What health problems can result?
43. Two surgeries are being performed. In one case, the anesthetic is methohexital; in the other, thiopental sodium is the anesthetic agent. Thiopental sodium is much more lipid soluble than methohexital. On the basis of what you have read in this chapter, which patient is more likely to go to sleep first? Why?

So far we've learned about the tiny components that make up the human body, and how they interact to keep life going. But what controls this intricate dance of atoms, molecules, organelles, and cells that we call life? Only in the last few decades have we begun to unravel this enormous puzzle. These people are looking at a large part of the answer: DNA.

The DNA in your cells encodes a complete set of instructions for making a human being (rather than a geranium or a jellyfish or a giraffe). More remarkably, they are the directions for making a particular, absolutely unique human being, *you* (and not anyone else). And more remarkable still, although each cell has the complete instruction set for *all* your cells, it "reads" only the ones that apply to it: those that tell it exactly what it must do to play it's tiny but indispensable part in making you—you! We'll explore these and other aspects of how cells control their activities in this chapter.

CHAPTER 4

The Cellular Level of Organization: Control Mechanisms at the Cellular Level

Chapter Objectives

After reading this chapter, you will be able to:

1 Relate basic energy concepts to cell operations.
2 Distinguish between catabolic and anabolic pathways.
3 Explain how metabolic pathways are regulated.
4 Describe the crucial role of enzymes in cellular metabolism.
5 Discuss how cells capture, store, and use energy.
6 Outline the main steps in the metabolic pathway of aerobic respiration.
7 Explain how genes control cell functions.
8 Discuss the mechanisms that control genetic activity.
9 Explain the importance of metabolic turnover.

Introduction

A one-celled organism, such as an amoeba or bacterium drifting about in the ocean leads a relatively independent life. It can absorb nutrients and oxygen from its surroundings and discharge waste products. Under most conditions these activities have no effect on other unicellular organisms or on the local environment, because the ocean is large and individual cells are few and far between.

These relationships do not apply to cells in the human body. A large part of the environment of any body cell consists of other cells, all of which must interact and work together. Moreover, the "ocean" of tissue fluid that they all inhabit is relatively small in relation to the number and volume of body cells. It is hard to imagine the number of cells in the body, because 75 trillion is such a huge number. To get an idea, consider that if you counted one cell every second, it would take over 2 million years to count all of your cells.

Because so many cells are dependent on such a relatively small volume of fluid, their activities must be tightly controlled. If there is any alteration in conditions in the extracellular fluid—a change in pH, or temperature, or nutrient content—the cell must adjust or die. If a hormone or prostaglandin or nerve impulse arrives, the cell must respond appropriately, and that response must be coordinated with the responses of other cells to maintain homeostasis at the levels of the cell, tissue, organ, organ system, and individual.

Chapter 2 introduced the major classes of chemicals in the body, and Chapter 3 examined the cellular structures built from those chemicals.

This chapter deals with control mechanisms that operate at the cellular level. We will consider four major processes:

1. How enzymes affect the chemical reactions under way in living cells;
2. How cells obtain the energy needed to perform vital functions;
3. How genetic information determines the structural and functional properties of cells; and
4. How cells change their activities over time and become specialized to perform specific functions.

Enzymes and Cellular Metabolism

The term **metabolism** (meh-TAB-o-lizm) includes all of the chemical reactions in the body. Through **cellular metabolism** cells capture, store, and use energy to maintain homeostasis and to support essential functions. Most of the chemical reactions involved are regulated by *enzymes* (Chapter 2), special proteins synthesized by the cell. (p. 37)

ENZYMES AND CHEMICAL REACTIONS

Most chemical reactions do not occur spontaneously. Enough energy must be provided to *activate* the molecules before a reaction can begin. **Activation energy** is the amount of energy required to start a reaction. Figure 4-1a diagrams the activation energy needed for a typical reaction. This energy requirement is often quite substantial—many complex organic molecules are very stable. For example, the hydrocarbons in an oil deposit have lasted for millions of years.

Changes in temperature or pH can speed up organic reactions, but such changes would kill cells. To break a disaccharide into a pair of simple sugars, for example, you must boil it in an acid solution. Instead of such violent methods, cells use *enzymes* to speed up the reactions that support life. With the help of enzymes, these reactions occur under conditions normally found in living cells.

Enzymes belong to a class of substances called **catalysts** (KAT-ah-lists; *katalysis*, dissolution): compounds that accelerate chemical reactions without themselves being permanently changed. **Enzymes** are organic catalysts—each enzyme is a protein made by a living cell to promote a specific metabolic reaction.

Like all catalysts, enzymes speed up chemical reactions by lowering the activation energy required for the reaction to occur. Figure 4-1b diagrams the effect of an enzyme on the activation energy of a typical reaction. Keep in mind, however, that lowering the activation energy affects only the *rate* of a reaction, not the direction of the reaction or the products that will be formed. *An enzyme cannot bring about a reaction that would otherwise be impossible.*

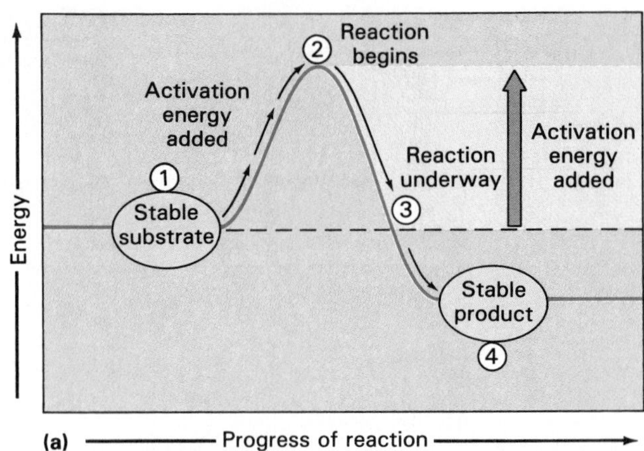

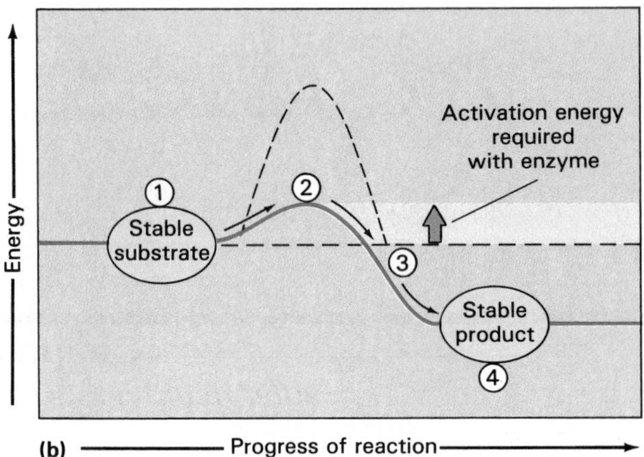

FIGURE 4-1
Activation Energy and Enzyme Function. (a) Before a reaction can begin, considerable activation energy must be provided. In this diagram the activation energy represents the energy required to proceed from Point 1 to Point 2. (b) The activation energy requirement of the reaction is much lower in the presence of an appropriate enzyme. This allows the reaction to take place much more rapidly, without the need for extreme conditions that would harm cells.

TABLE 4-1 Major Classes of Enzymes

Enzyme Class	*Function*	*Examples*
Ligase	Joins two molecules together, requires ATP	Ligases string nucleotides together in nucleic acids, simple sugars in polysaccharides
Lyase	Breaks bonds between carbon atoms or between carbon and nitrogen	Deaminase removes amino group from amino acid
Hydrolase	Breaks large molecules into simpler molecules with the addition of a water molecule	Proteases split proteins into amino acids; amylase splits carbohydrates into simple sugars; lipase breaks apart triglycerides
Transferase	Removes part of one molecule and attaches it to another	Transaminase transfers an amino group from an amino acid
Isomerase	Rearranges the atoms in a molecule without changing the chemical formula	Isomerases are important in preparing some carbohydrates for enzymatic processing
Oxido-reductase	Removes H^+ or electrons from one molecule and gives them to another	Enzymes involved in mitochondrial energy production
Kinase	Attaches a phosphate group with a high-energy bond	Kinases are essential for ATP production and activation of some enzymes

All physiological processes are directly or indirectly dependent on enzyme activity, and thousands of different enzymes are made by living cells. Enzymes are usually named after the reactions they promote; Table 4-1 identifies the major functional classes of enzymes.

ENZYME STRUCTURE AND FUNCTION

Figure 4-2 shows a simple model of enzyme function. The reactants in an enzymatic reaction, called **substrates**, interact to form a specific **product**. Substrate molecules bind to the enzyme at a groove or pocket on the enzyme surface called the **active site**. This binding depends on a combination of physical and chemical "fit." For example, the shape of the substrate may allow it to enter the active site, and hydrogen bonding may hold it there.

The shape of the active site is determined by the tertiary or quaternary structure of the enzyme molecule. It is therefore the unique three-dimensional shape of each enzyme that accounts for its high degree of *specificity*: the ability to accept only one substrate, or a small class of very similar substrates. Once the substrate is bound, the enzyme catalyzes

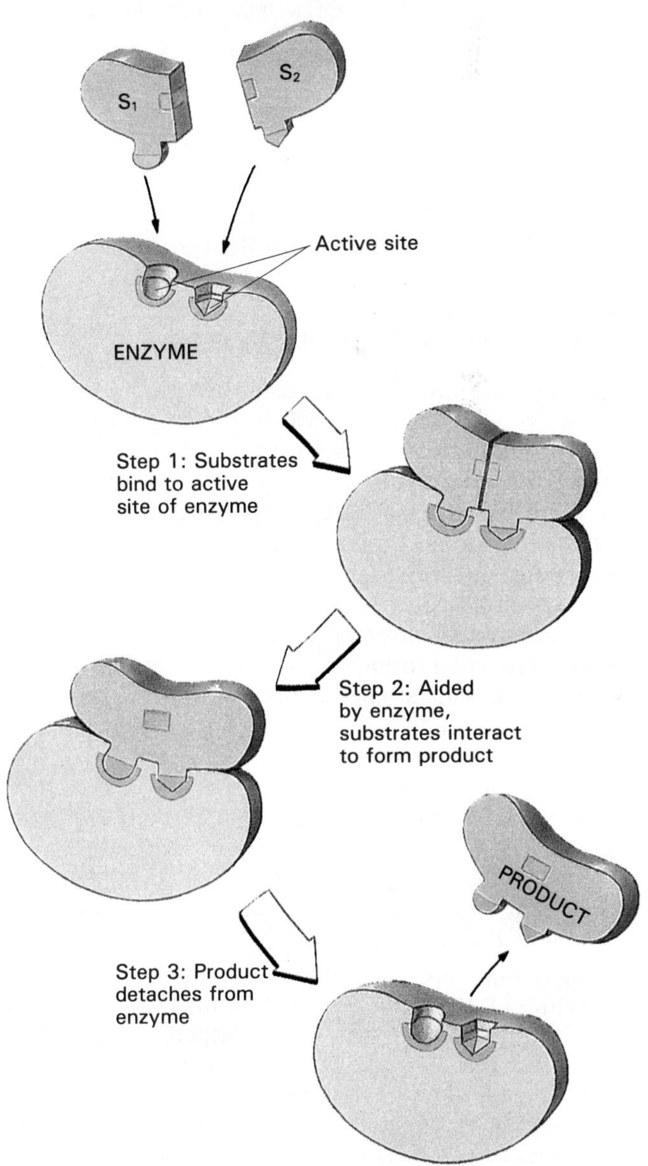
S1
S2
Active site
ENZYME
Step 1: Substrates bind to active site of enzyme
Step 2: Aided by enzyme, substrates interact to form product
PRODUCT
Step 3: Product detaches from enzyme

FIGURE 4-2
Enzyme Structure and Function. Each enzyme contains a specific active site somewhere on its exposed surface. In Step 1 a pair of substrate molecules (S_1 and S_2) bind to the active site. In Step 2 the substrates interact, forming a product that detaches from the active site (Step 3). Because the structure of the enzyme has not been affected, the entire cycle can be repeated.

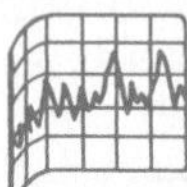

Diagnostics: Isozymes

Isozymes are enzymes that differ in structure but catalyze the same reaction. Different tissues often contain different isozymes. When tissue damage occurs, those enzymes leak out of the injured cells into the tissue fluid and the blood. Thus the identification of a particular isozyme can provide direct evidence of damage to a specific tissue. For example, the enzyme **lactate dehydrogenase (LDH)** catalyzes a reaction that produces *lactic acid* during periods of intense muscular activity. (The reaction sequence and its significance are discussed in Chapter 10.) Cardiac muscle and skeletal muscle have different isozymes of LDH. When a patient reports experiencing severe chest pain, an LDH assay is often performed. If the pain resulted from a heart attack, cardiac LDH levels will be elevated.

Health News: An Introduction to Vitamins

There are two major groups of **vitamins**. *Vitamins A, D, E,* and *K* are called **fat-soluble vitamins** because their hydrophobic structure allows them to dissolve in lipids. We absorb these vitamins with dietary fats and store excess quantities in fatty tissues. These reserves are used when dietary supplies are inadequate. In general, fat-soluble vitamins have a broad range of functions, some of which are not yet thoroughly understood.

The nine **water-soluble vitamins** include the *B vitamins*, common in milk and meats, and *vitamin C*, found in citrus fruits. Most of the water-soluble vitamins are easy to absorb from the diet, but the body does not maintain large reserves. As a result, a reduction in the dietary supply will soon affect metabolic operations, because many of these vitamins are building blocks for coenzymes required by important enzymes. Often the coenzymes involved play key roles in energy production. Several important coenzymes, for example, are derived from B vitamins: *NAD* from *niacin*, *FAD* and *FMN* from *riboflavin*, and *coenzyme A* from *pantothenic acid*. These coenzymes are discussed later in this chapter. Chapter 25 will consider the functions of vitamins and associated nutritional problems in greater detail.

the reaction. The completed product then detaches from the active site and enters the cytosol.

COFACTORS AND COENZYMES

Cofactors are ions or molecules that must attach to the active site before substrate binding can occur. An enzyme activated by an appropriate cofactor is termed a **holoenzyme**. An enzyme without its cofactor is an inactive **apoenzyme** (ĀP-ō-en-zīm). Figure 4-3 diagrams the function of a cofactor. Examples of cofactors include many metal ions such as calcium (Ca^{2+}) or magnesium (Mg^{2+}).

Coenzymes are large organic molecules that function as cofactors. The body converts many **vitamins** into essential coenzymes. Vitamins are structurally related to lipids or carbohydrates, but they have unique functional roles. The human body cannot synthesize most vitamins, and they must be present in the diet. Chapter 25 will present more information concerning nutrition and vitamins.

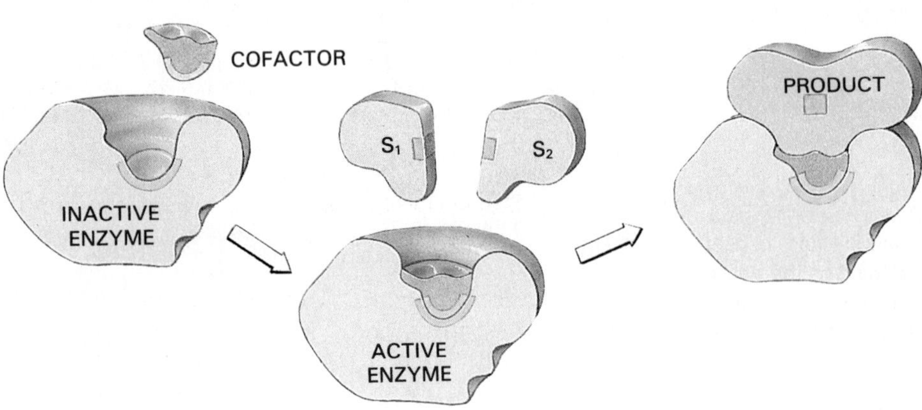

FIGURE 4-3
Cofactors and Enzyme Activity. In some cases a cofactor must bind to the active site before the enzyme can bind substrate molecules and function normally.

CONTROL OF ENZYME ACTIVITY

All cellular structures are formed by chemical reactions, and all cell functions involve chemical reactions. Without enzymes those reactions would occur so slowly that life would be impossible. The factors that control enzyme activity therefore indirectly determine both the structure of the cell and its functional capabilities.

The relationship between enzymes and cofactors is one illustration of how enzyme activity can be controlled. Without its cofactor, the enzyme is nonfunctional; with its cofactor, it can catalyze a specific reaction. A variety of other factors can affect enzyme performance; we will consider a few of the most important.

Temperature and pH

Each enzyme works best at an optimal combination of temperature and pH. For example, sustained high body temperatures (over 40° C) are fatal because enzymes throughout the body undergo denaturation, becoming permanently nonfunctional. Enzymes are equally sensitive to pH changes. *Pepsin*, an enzyme that breaks down proteins in the stomach contents, works best at a pH of 2.0 (strongly acid). The small intestine contains *trypsin*, another enzyme that attacks proteins. Trypsin works only in an alkaline environment, with an optimum pH of 9.5 (somewhat basic).

Enzyme Activation and Inhibition

The precise shape of the active site is extremely important to an enzyme's ability to bind its substrate and catalyze a reaction. Consequently, any factor that affects the tertiary or quaternary structure of an enzyme is likely to have a decisive effect on the enzyme's functional state. Even a small alteration in structure can switch the enzyme "on" or "off." A change from inactive to active state is called **enzyme activation**. One form of activation, mentioned previously, occurs when a cofactor binds to the active site. In other cases, activation may occur when a molecule binds elsewhere on the enzyme. The binding of another molecule can also alter, reduce, or eliminate enzyme activity, an effect known as **enzyme inhibition**. A molecule that interferes with the activity of an enzyme by distorting or blocking access to the active site is called an **enzyme inhibitor**.

Because they produce immediate and often reversible changes in enzyme activity, enzyme activation and enzyme inhibition are important mechanisms for short-term control of the cell's metabolic reactions.

Clinical Comment: Enzyme Inhibition

Chemicals that inhibit enzymes can have powerful effects on cells throughout the body. Several deadly compounds, such as hydrogen cyanide (HCN) and hydrogen sulfide (H_2S), kill cells by inhibiting mitochondrial enzymes involved with ATP production. Many inhibitors with less drastic effects are in clinical use. For example, *warfarin* slows blood clotting by inhibiting liver enzymes responsible for the synthesis of clotting factors. Several of the beneficial effects of *aspirin*, notably the reduction of inflammation, are related to the inhibition of enzymes involved with prostaglandin synthesis. Many important antibiotics, such as penicillin, kill bacteria by competitively inhibiting enzymes that are essential to bacteria but absent from our cells.

ENZYMES, METABOLIC PATHWAYS, AND HOMEOSTASIS

A **metabolic pathway** is a sequence of enzymatic reactions that yields a specific product. Metabolic pathways involve a series of interlocking steps, as shown in Chapter 2. (p. 37) The initial enzymatic step is often controlled by a special *regulatory enzyme*. **Regulatory enzymes**, typically larger and structurally more complex than other enzymes, are sensitive to the action of particular activators or inhibitors. They act as on/off switches—when the regulatory enzyme is inhibited, the entire pathway can be shut down. Moreover, since most metabolic pathways branch and interconnect in various ways, a change in the activity of a single enzyme may affect a single reaction, a group of integrated reactions, a metabolic pathway, or even the metabolic activities of the entire cell.

Regulatory enzymes permit the operation of several different pathways using the same substrate. For example, glucose molecules can (1) be used in the synthesis of glycogen, (2) be broken down to provide energy, or (3) be converted to other compounds, such as other sugars, lipids, or amino acids (Figure 4-4). Different regulatory enzymes control access to these pathways, and the fate of glucose molecules depends on how active each of these enzymes is at a given moment.

The sensitivity of enzymes to substances in their environment provides a mechanism for preserving homeostasis. Each cell contains an assortment of enzymes, and any particular enzyme may be activated

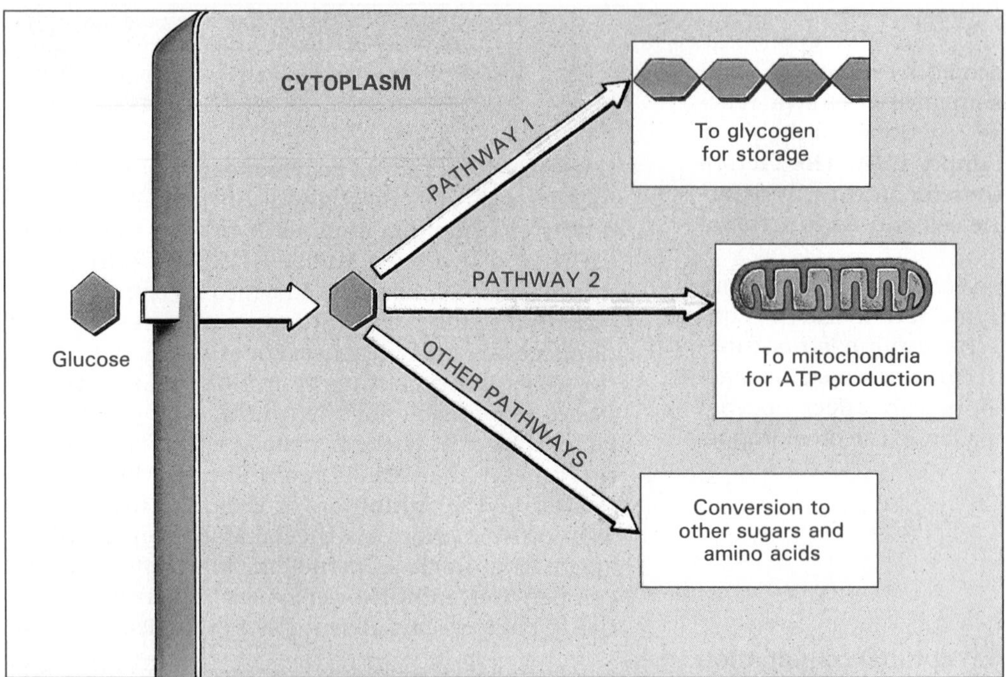

FIGURE 4-4
Metabolic Pathways and Regulatory Enzymes. Typically, a different regulatory enzyme controls the first step in each metabolic pathway or each branch of a pathway. The fate of glucose entering the cell is determined by the activity of regulatory enzymes that control major pathways.

under one set of conditions and inactivated under another. For example, when a muscle cell needs energy it breaks down glucose molecules. When glucose becomes scarce, an enzyme responsible for the breakdown of glycogen becomes active.

■ Cells and Energy

Cells need energy both to maintain homeostasis and to perform special functions that vary from cell to cell. Like most people, they operate on a limited budget, but instead of earning and spending dollars, cells gain and lose energy in the form of ATP. Before examining the ways in which cells capture energy you must become more familiar with the physical principles involved.

BASIC ENERGY CONCEPTS

Most people are familiar with the terms *work*, *energy*, and *heat*. **Work** is performed whenever there is any movement of an object or change in its physical structure. **Energy** is the capacity to perform work; movement or physical change will not occur unless energy is provided. There are two major types of energy: *kinetic energy* and *potential energy*.

Kinetic energy is the energy of motion. When a car hits a tree, it is the car's kinetic energy that causes the damage. **Potential energy** is stored energy. It may result from the position of an object (as when a book sits on a high shelf) or its physical state or structure (as when a spring is stretched or a battery is charged). Kinetic energy had to be used to lift the book, stretch the spring, and charge the battery. The potential energy of these objects is converted back to kinetic energy when the book falls, the spring contracts, or the battery discharges, and the energy released can be used to perform work.

Conversion between potential energy and kinetic energy is never 100 percent efficient. Each time an energy exchange occurs, some of the energy produces **heat**. Heat is an increase in random molecular motion; what we call *temperature* is a measure of that motion.

Living cells perform work in many forms. The contraction of a muscle cell, the movement of organelles, and the synthesis of proteins or nucleic acids are cellular events that require energy. The energy comes from the covalent bonds that hold organic molecules together. Those bonds represent a form of potential energy, and breaking them releases energy that cells can capture. **Catabolism** (kah-TAB-o-lizm; *katabole*, a throwing down) is the decomposition of organic molecules by cells. *This process provides the energy for all cellular activities.*

The organic molecules broken down by cells are normally obtained from the extracellular fluid, as indicated in Figure 4-5. Once they are absorbed into

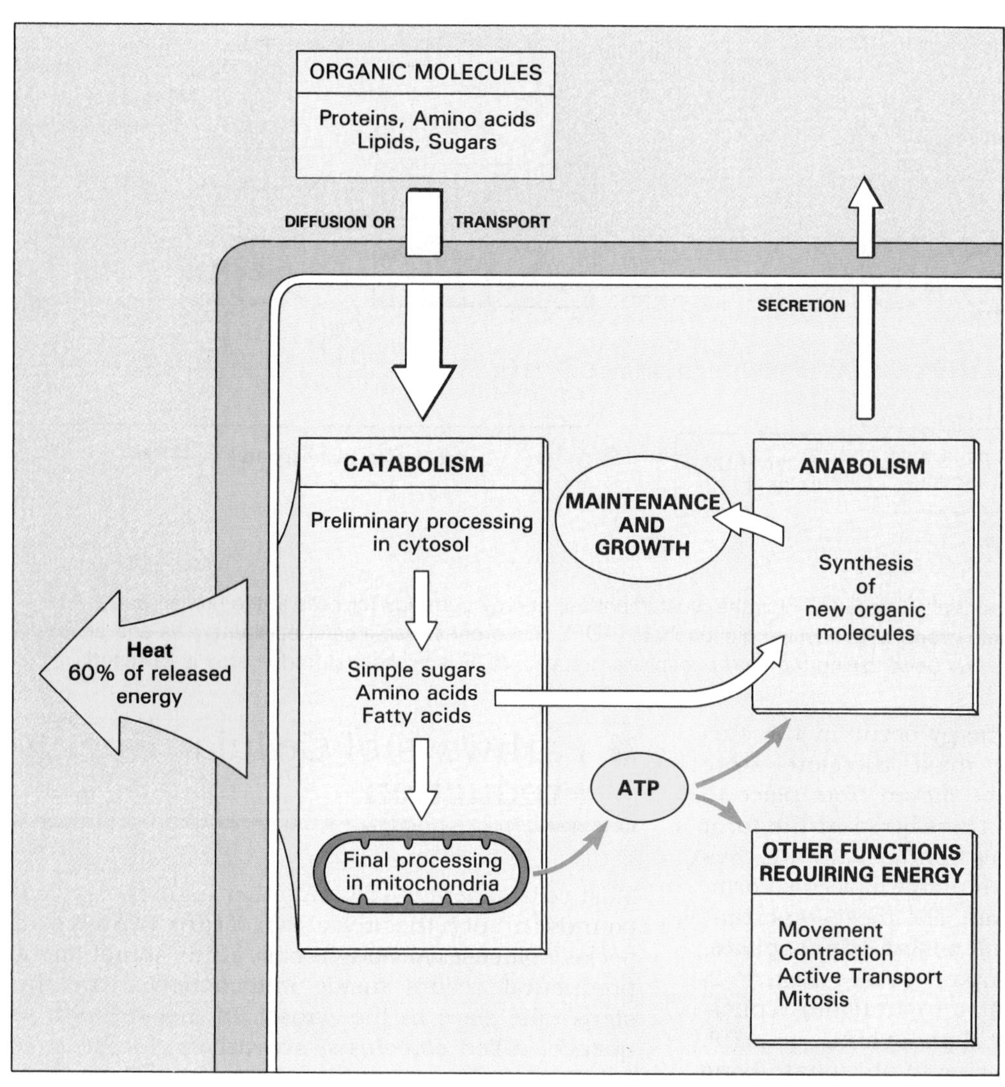

FIGURE 4-5
Cellular Metabolism. The cell obtains organic molecules from the extracellular fluid and breaks them down to obtain ATP. Only about 40 percent of the energy released through catabolism is captured in ATP; the rest is lost as heat. The ATP generated through catabolism provides energy for all vital cellular activities, including anabolism.

the cell, catabolism proceeds in a series of steps. In general, preliminary processing occurs in the cytosol. There, catabolic reactions break down the organized structure of the absorbed molecules. Carbohydrates are degraded to simple sugars or short carbon chains, triglycerides split into fatty acids and glycerol, and proteins are broken down to individual amino acids.

Relatively little ATP is captured in this way. The simple molecules produced can be used to manufacture new organic molecules that suit the needs of the cell, or broken down further for a greater yield of energy. The most important energy-producing steps occur inside mitochondria. Even in these mitochondrial reactions, however, a cell can capture only about 40 percent of the energy released when a covalent bond breaks. The rest escapes as heat that warms the interior of the cell and the surrounding tissues.

Much of the energy captured supports **anabolism** (a-NAB-o-lizm; *anabole*, a throwing upward), the synthesis of new organic molecules. Additional energy is required to power active cell functions, such as contraction, ciliary or cell movement, and active transport.

HIGH-ENERGY COMPOUNDS AND ENERGY TRANSFER

The reactions that release energy and those that use energy often occur in different places and at different times. Cells must therefore store energy in a relatively stable and portable form. **High-energy compounds** store energy for later use. A high-energy compound contains one or more **high-energy bonds**: covalent bonds that, when broken, release large amounts of energy. In our cells the formation of a high-energy bond generally involves the attachment of a phosphate group (PO_4) to an existing molecule. This process, called **phosphorylation** (fos-for-i-LĀ-shun), creates a high-energy compound.

Most of the chemical reactions that release energy occur in the mitochondria, but most of the cellu-

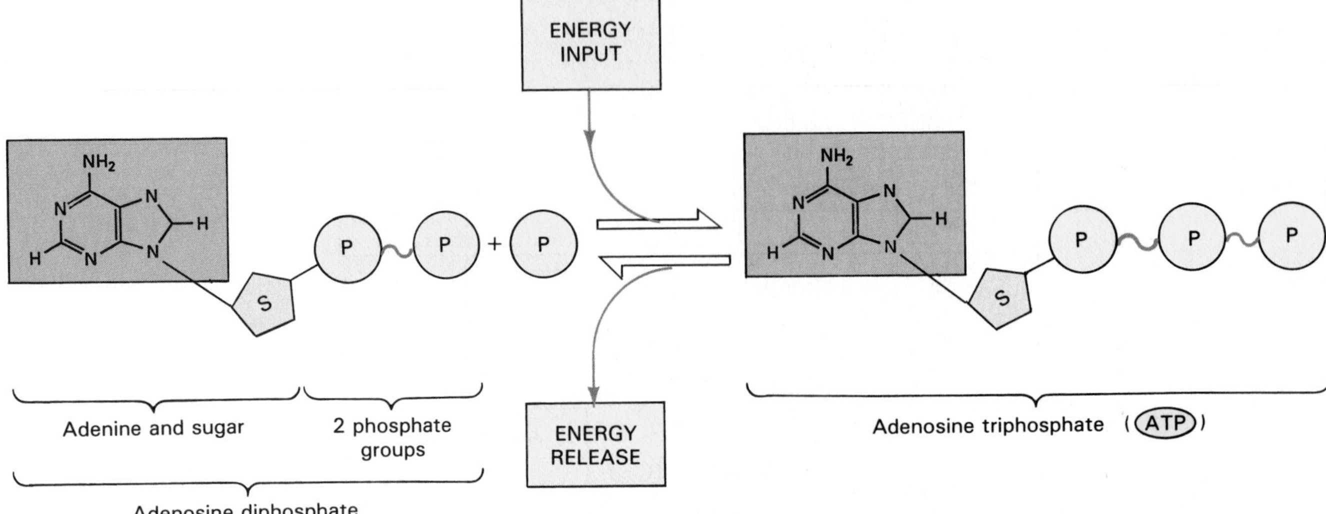

FIGURE 4-6
The Function of ATP. Adenosine triphosphate (ATP) is the most important energy currency for cells in the human body. ATP is created by attaching a phosphate group to adenosine diphosphate (ADP); this process requires special enzymes and an energy source. When the high-energy bond joining the third phosphate group to ADP is broken, stored energy is released.

lar activities that require energy occur in the surrounding cytoplasm. Cells must therefore store energy in a form that can be moved from place to place. Energy is stored and transferred in the form of a **high-energy bond**. Such a bond usually attaches a phosphate group (PO_4) to a suitable molecule, forming a **high-energy compound**. The most important high-energy compound is **adenosine triphosphate**, or **ATP**, discussed in Chapter 2 (see Figure 2-26). (p. 56) ATP is created by attaching a phosphate group to **adenosine diphosphate**, or **ADP**. Living cells break the high-energy phosphate bond under controlled conditions, reconverting ATP to ADP and release energy for their use (Figure 4-6).

Although ATP is the most abundant high-energy compound, there are others. **Guanosine triphosphate (GTP)** and **uridine triphosphate (UTP)**, like ATP, are nucleotide triphosphates. They are essential to specific reaction pathways that cannot accept energy from ATP. GTP provides energy at key steps in the synthesis of proteins, and UTP is required for the synthesis of the pyrimidines (cytosine, thymine, and uracil). **Phosphagens** are high-energy compounds derived from amino acids. **Creatine phosphate (CP)**, which provides energy for skeletal muscle cell contraction, is the most important phosphagen in our bodies.

- ✓ **Why can a prolonged high fever be life-threatening?**
- ✓ **Sugars and fats are "burned" in the body to produce energy, yet these compounds pose little fire hazard in your kitchen. Explain.**

■ Pathways of Cellular Energy Production

Most cells generate ATP and other high-energy compounds through the breakdown of carbohydrates, especially glucose. Although most of the actual energy production occurs inside mitochondria, the first steps take place in the cytosol. In this reaction sequence, called *glycolysis*, six-carbon glucose molecules are broken down into three-carbon fragments. These can be used as building blocks for other compounds or absorbed by the mitochondria for additional energy production.

GLYCOLYSIS

Glycolysis (glī-KOL-i-sis) is the breakdown of glucose in the cytosol (Figure 4-7). A series of enzymatic steps breaks the six-carbon glucose molecule into two three-carbon molecules of **pyruvic acid**. Glycolysis requires (1) glucose molecules, (2) appropriate cytoplasmic enzymes, (3) ATP and ADP, and (4) **NAD** (**n**icotinamide **a**denine **d**inucleotide), a coenzyme that removes hydrogen atoms during one of the enzymatic reactions. If any of these particpants is not present, glycolysis cannot take place.

Glycolysis begins when an enzyme attaches a phosphate group to the last (sixth) carbon atom on a glucose molecule, creating **glucose-6-phosphate**. Although this step "costs" the cell one ATP molecule, it has two important results. First, it traps the glucose molecule within the cell, because phosphorylated glucose cannot cross the cell membrane. Second, it pre-

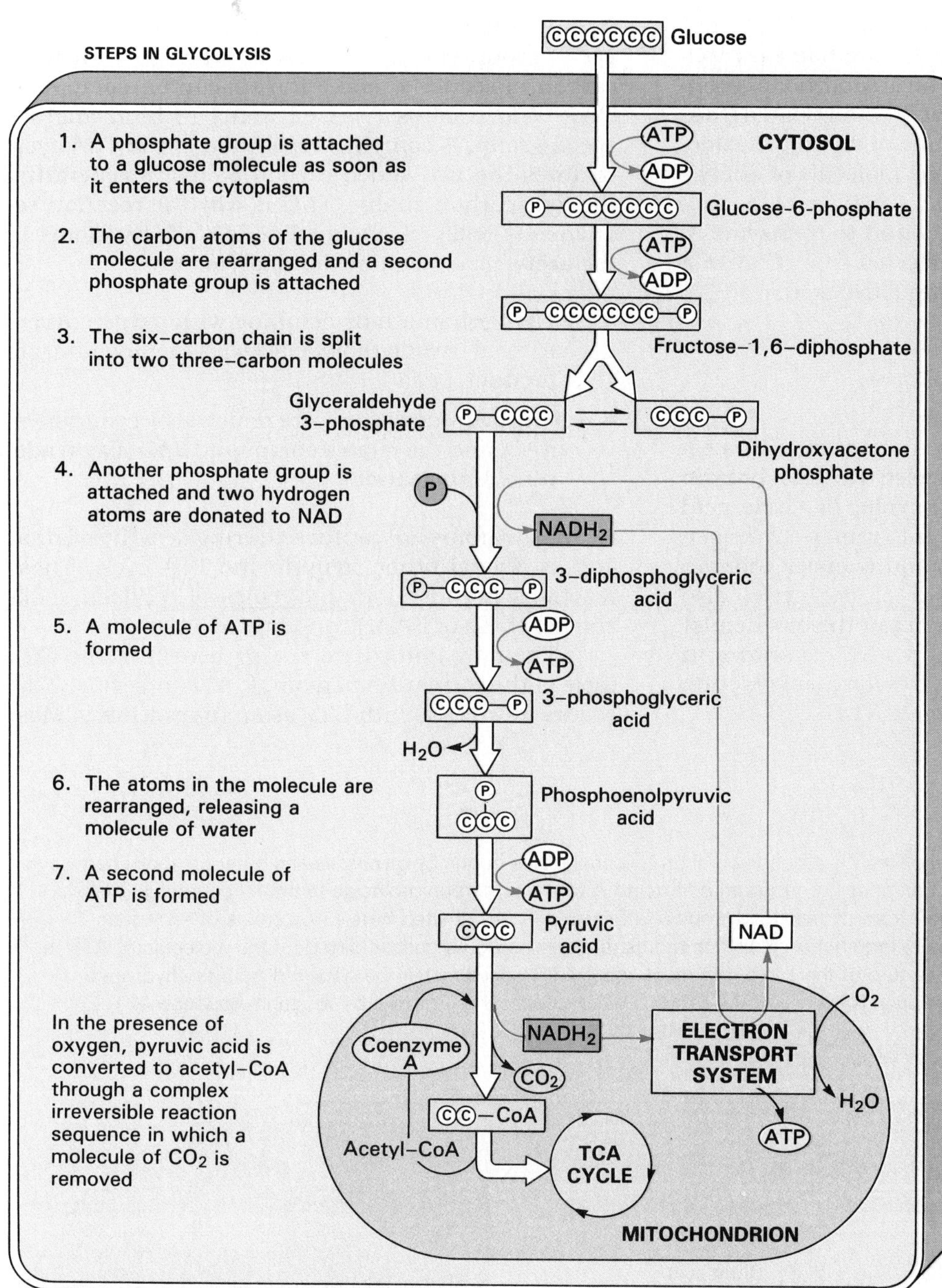
STEPS IN GLYCOLYSIS
Glucose
CYTOSOL
1. A phosphate group is attached to a glucose molecule as soon as it enters the cytoplasm
ATP
ADP
Glucose-6-phosphate
2. The carbon atoms of the glucose molecule are rearranged and a second phosphate group is attached
3. The six-carbon chain is split into two three-carbon molecules
Fructose-1,6-diphosphate
Glyceraldehyde 3-phosphate
Dihydroxyacetone phosphate
4. Another phosphate group is attached and two hydrogen atoms are donated to NAD
NADH2
3-diphosphoglyceric acid
5. A molecule of ATP is formed
3-phosphoglyceric acid
H2O
6. The atoms in the molecule are rearranged, releasing a molecule of water
Phosphoenolpyruvic acid
7. A second molecule of ATP is formed
Pyruvic acid
NAD
O2
In the presence of oxygen, pyruvic acid is converted to acetyl-CoA through a complex irreversible reaction sequence in which a molecule of CO2 is removed
Coenzyme A
CO2
ELECTRON TRANSPORT SYSTEM
H2O
Acetyl-CoA
TCA CYCLE
MITOCHONDRION

FIGURE 4-7
Glycolysis. Glycolysis breaks down a six-carbon glucose molecule into two three-carbon molecules of pyruvic acid. This process involves a series of enzymatic steps. There is a net gain of 2 ATP for each glucose molecule converted to pyruvic acid. In addition, two molecules of the coenzyme NADH are converted to $NADH_2$. Both the pyruvic acid and the $NADH_2$ can still yield a great deal more energy via aerobic respiration. The further catabolism of pyruvic acid begins with its entry into the mitochondrion. There one of its three carbon atoms is removed as carbon dioxide. The other two are attached to a molecule called coenzyme A to form acetyl-CoA, which shuttles them into the TCA cycle (see Figure 4-8).

pares the glucose molecule for further biochemical steps—either catabolic or anabolic.

A second phosphorylation occurs in the cytosol before the six-carbon chain is broken into two three-carbon fragments. Energy benefits begin to appear as these fragments are converted to pyruvic acid. Two of the steps release enough energy to generate ATP. In addition, two molecules of NAD are converted to $NADH_2$. The overall reaction looks like this:

$$\text{Glucose (6-carbon)} + 2\ \text{NAD} + 2\ \text{ATP} \rightarrow$$
$$2\ \text{pyruvic acid (3-carbon)} + 4\ \text{ATP} + 2\ \text{NADH}_2$$

MITOCHONDRIAL ENERGY PRODUCTION

These cytoplasmic reactions yield an immediate net gain of two ATP for the cell. However, a great deal of additional energy is still locked in the chemical bonds of pyruvic acid. The ability to capture that energy depends on the availability of oxygen. If oxygen supplies are adequate, mitochondria absorb the pyruvic acid molecules and break them down completely. The hydrogen atoms are removed by coenzymes; they will ultimately be the source of most of the energy released. The carbon and oxygen atoms are removed and released as carbon dioxide, a process called **decarboxylation** (de-kar-boks-i-LĀ-shun).

Once inside the mitochondrion, each pyruvic acid molecule first loses one carbon atom in a complicated reaction involving NAD and an additional coenzyme called **coenzyme A**, commonly abbreviated **CoA**. This reaction yields one molecule of carbon dioxide, one molecule of $NADH_2$, and one molecule of **acetyl-CoA** (as-e-til-CŌ-ā). Acetyl-CoA consists of a two-carbon **acetyl group** (CH_3CO) bound to coenzyme A. Next, the acetyl group is transferred from CoA to a four-carbon molecule, producing citric acid.

The TCA Cycle

The formation of citric acid is the first step in a sequence of enzymatic reactions called the **tricarboxylic** (trī-kar-bok-SIL-ik) **acid (TCA) cycle**, or **citric acid cycle**. The purpose of the cycle is to remove hydrogen atoms from organic molecules and transfer them to coenzymes. The overall pattern of the cycle, also known as the *Krebs cycle* in honor of the biochemist who described these reactions in 1937, is shown in Figure 4-8. (Those interested in the complete reaction sequence should consult Appendix VI.)

At the start of the TCA cycle, the two-carbon acetyl group carried by CoA is attached to a four-carbon molecule to make the six-carbon compound citric acid. CoA is released intact to bind another acetyl group. A complete revolution of the TCA cycle removes the two added carbon atoms, regenerating the four-carbon chain. (This is why the reaction sequence is called a *cycle*.) The fate of the atoms of the acetyl group can be summarized thus:

- The carbon atoms combine with oxygen atoms to form carbon dioxide (CO_2), a metabolic waste product.
- The hydrogen atoms are removed by coenzymes, NAD, and the related compound **FAD** (**f**lavin **ade**nine **d**inucleotide).

Hydrogen atoms are captured during decarboxylation and at several other steps in the TCA cycle. These reactions produce four molecules of *reduced coenzymes*: three of $NADH_2$ and one of $FADH_2$.

The only immediate energy benefit of the TCA cycle is the formation of a single ATP molecule. This occurs indirectly, with GTP as an intermediary. Most

FIGURE 4-8
The TCA Cycle. The TCA cycle completes the breakdown of organic molecules begun by glycolysis and other catabolic pathways. The cycle begins with the transfer of an acetyl group from coenzyme A to a four-carbon molecule in the mitochondrial matrix. (Details of the formation of acetyl-CoA from pyruvic acid, shown in Figure 4-7, are omitted here.) In a series of enzymatic reactions the two added carbon atoms, together with oxygen atoms, are eliminated as carbon dioxide. One molecule of ATP is produced indirectly (via GTP) for each turn of the cycle, but much more ATP will ultimately be obtained from the hydrogen atoms that are removed by the coenzymes NAD and FAD. (Additional biochemical details will be found in Appendix VI.)

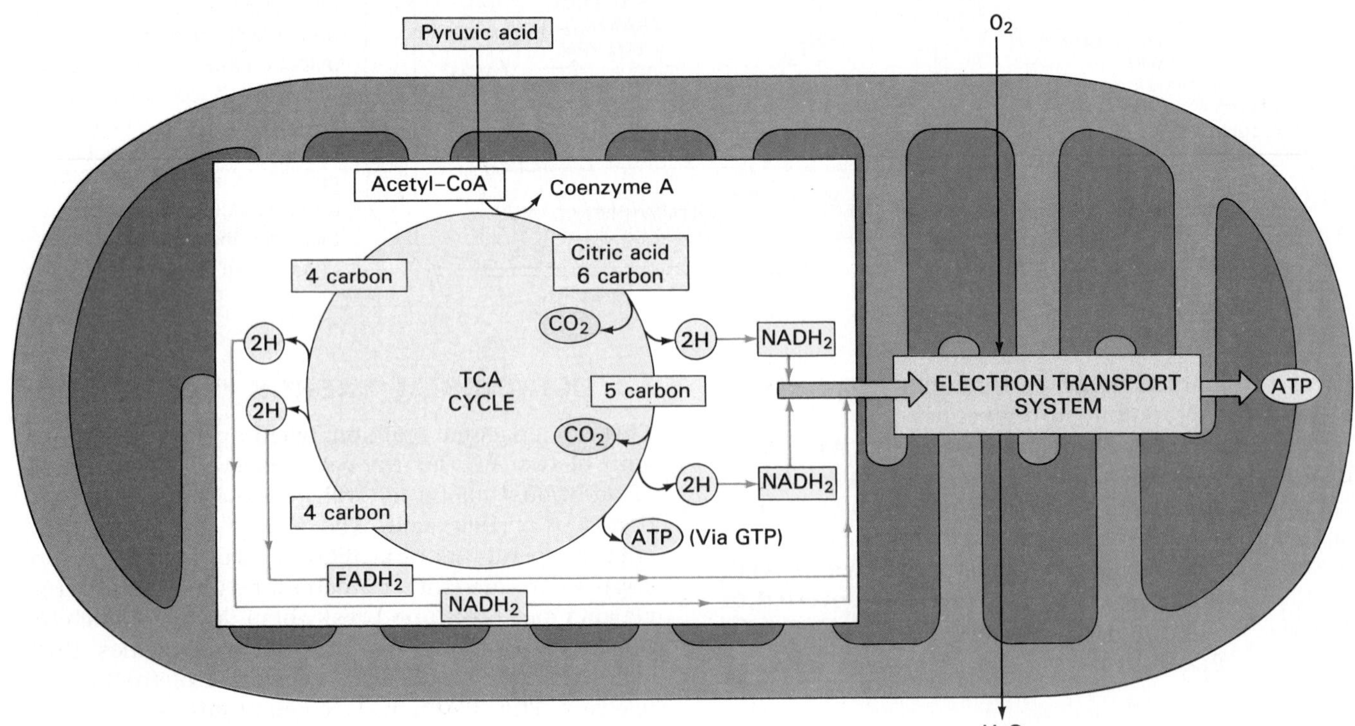

of the energy yield is obtained when the reduced coenzymes transfer their captured electrons to molecules bound to the inner mitochondrial membrane. To understand the mechanism involved, we must take a closer look at the way in which high energy-bonds are created.

Oxidative Phosphorylation and Energy Capture

The goal in cellular energy production is the creation of a high-energy compound, usually ATP. This involves the phosphorylation of ADP. Our cells have two strategies for attaching the phosphate group: *substrate-linked phosphorylation* and *oxidative phosphorylation.*

In **substrate-level phosphorylation** an enzyme uses the energy released by a chemical reaction to transfer a phosphate group to a suitable acceptor molecule, creating a high-energy compound. Most often the acceptor is ADP, and the compound formed is ATP. This process occurs in several of the energy-yielding pathways used by living cells. The ATP produced in glycolysis is obtained in this way, and so is the GTP derived from the TCA cycle. Normally, however, substrate-level phosphorylation provides a relatively small amount of energy compared with oxidative phosphorylation.

Oxidative phosphorylation is the generation of ATP through a reaction sequence that consumes oxygen. This process, which occurs inside mitochondria, produces over 90 percent of the ATP used by our cells. The foundation of this reaction sequence is very simple:

$$2\,H_2 + O_2 \rightarrow 2\,H_2O$$

Living cells can easily obtain the ingredients for this reaction. Hydrogen is a component of all organic molecules, and oxygen is an atmospheric gas. The only problem is that the reaction releases a tremendous amount of energy all at once. In fact, this reaction releases so much energy that it is used to launch rockets into orbit. Cells cannot handle energy explosions; energy release must be gradual, and this is the goal of oxidative phosphorylation. In the process, this powerful reaction proceeds in a series of small, enzymatically controlled steps. Under these conditions energy can be captured and ATP generated.

Oxidation, Reduction, and Energy Transfer The enzymatic steps of oxidative phosphorylation involve *oxidation* and *reduction.* **Oxidation** is the loss of electrons. An electron may be lost by itself or together with a proton—that is, as part of a hydrogen atom (Chapter 2). (p. 28) **Reduction** is the acceptance of electrons or hydrogen atoms. When electrons or hydrogen atoms pass from one molecule to another, the donor is oxidized and the recipient reduced. Electrons are bearers of energy, and in general terms *the reduced molecule gains energy at the expense of the oxidized molecule.*

Oxidative phosphorylation involves a series of such reactions. The initial steps utilize coenzymes that accept and transport hydrogen atoms. The major coenzymes are NAD, FAD, **FMN** (**f**lavin **m**ono**n**ucleotide), and **coenzyme Q** (*ubiquinone*). These molecules are either free in the mitochondrial matrix (NAD, FAD) or bound to the inner mitochondrial membrane (FMN, coenzyme Q).

The Electron Transport System The function of these coenzymes is to feed electrons into the **electron transport system (ETS)**. This system is a sequence of metalloproteins called **cytochromes** (SĪ-tō-krōmz; *cyto-*, cell + *chroma*, color). Each cytochrome has two components: a protein and a pigment. The protein, embedded in the inner mitochondrial membrane, surrounds the pigment complex, which contains a metal ion, either iron (Fe^{3+}) or copper (Cu^{2+}). There are four different cytochromes: **b, c, a,** and **a_3**.

Figure 4-9 summarizes the major steps in oxidative phosphorylation. These steps are:

1. *A coenzyme strips a pair of hydrogen atoms from a substrate molecule.* Different coenzymes are used for different substrate molecules.
2. *Each hydrogen atom ionizes to H^+, donating an electron to the electron transport system.* The hydrogen ions enter the space between the inner and outer mitochondrial membranes.
3. *Electrons are passed along the electron transport system, losing energy in a series of small steps.* During the electron transfers, additional hydrogen ions are pumped across the inner mitochondrial membrane.
4. *At the end of the electron transport system an oxygen atom accepts the electrons, creating an oxygen ion* (O^-). This ion has a very strong affinity for hydrogen ions (H^+); it quickly combines with hydrogen ions from Steps 2 and 3, forming water. Because the reaction has occurred in a series of small steps, the combination of hydrogen and oxygen occurs quietly rather than explosively. The ETS is also called the **respiratory chain** because it consumes oxygen.

ATP Generation The coenzymes of the matrix and the cytochromes of the electron transport chain do not produce ATP directly. Instead, they create the conditions necessary for ATP production. At Steps 2 and 3 above, hydrogen ions were transported across the inner mitochondrial membrane. These ions can-

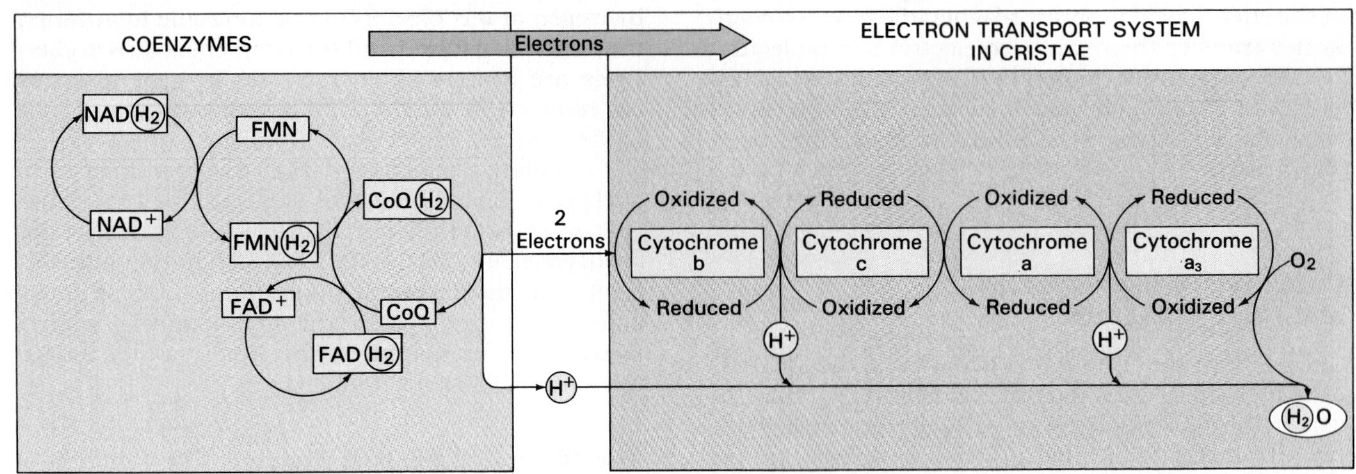

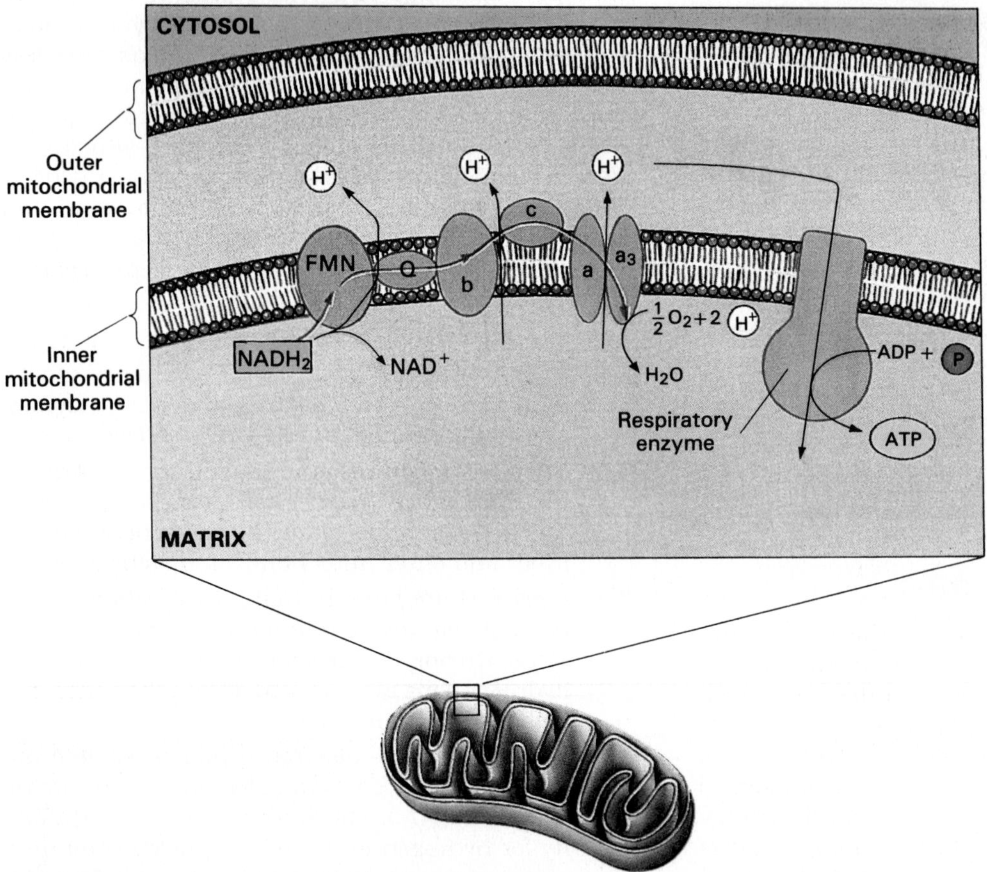

FIGURE 4-9
Oxidative Phosphorylation and Chemiosmosis. Oxidative phosphorylation in mitochondria requires hydrogen and oxygen atoms; it generates ATP and water. The electron carriers of the electron transport system are bound to the inner mitochondrial membrane. Coenzymes in the matrix feed hydrogen atoms into the ETS. As electrons pass down the chain of carriers, hydrogen ions are pumped into the space between the inner and outer mitochondrial membranes. The return of those hydrogen ions through the respiratory complex somehow drives the formation of ATP.

not diffuse through the membrane lipids, so there is a large concentration gradient for hydrogen ions across the inner membrane. In addition to this concentration gradient, there is an electrical gradient, because hydrogen ions bear a positive charge and the electrons that remain inside the membrane are negatively charged. Thus the operation of the ETS creates a large *electrochemical gradient* across the inner mitochondrial membrane.

In the process of **chemiosmosis** (kem-ē-oz-MŌ-sis), ion movement driven by this electrochemical gradient is harnessed to synthesize ATP. Hydrogen ions can reenter the mitochondrial matrix through channels in the *respiratory enzymes* that are integral components of the inner mitochondrial membrane. These enzymes, which look like miniature Tootsie-Roll pops, are visible in Figure 3-9. (p. 70) As hydrogen ions flow back into the matrix through one of the respiratory enzymes, their movement, by some mechanism that has not yet been determined, provides the energy needed to convert ADP to ATP. For each pair of hydrogen atoms removed from a substrate by a coenzyme, six hydrogen ions are pumped across the inner mitochondrial membrane. Their reentry into the matrix provides the energy to generate three molecules of ATP.

Oxidative phosphorylation represents the single most important mechanism for the generation of ATP. A chronic suspension or even a significant reduction in the rate of oxidative phosphorylation will usually kill the cell. If many cells are affected, the individual may die. Hydrogen cyanide gas is sometimes used as a pesticide to kill rats or mice; in some states where capital punishment is legal, it is used to execute criminals. The cyanide ion (CN^-) binds to cytochrome a_3 and prevents the passage of electrons to oxygen. With the last reaction stopped, the entire ETS comes to a halt, like cars at a washed-out bridge. Oxidative phosphorylation can no longer take place, and cells quickly succumb to energy starvation. Because nerve cells have a high demand for energy, the brain is one of the first organs to be affected.

Oxidative phosphorylation ends when the electron transport system gives up electrons to an oxygen atom. It begins when coenzymes accept hydrogen atoms from substrate molecules. The process requires both hydrogen atoms and oxygen, and there must be one oxygen atom for every two hydrogen atoms. *The rate of the entire process is therefore limited by the availability of oxygen and hydrogen atoms in the proper ratio.* Cells obtain oxygen by diffusion from the extracellular fluid. The hydrogen atoms are derived from the breakdown of organic compounds in the cytosol—chiefly glycolysis and the TCA cycle. One revolution of the TCA cycle generates three molecules of $NADH_2$ and one molecule of $FADH_2$. Through the ETS, each molecule of $NADH_2$ yields three molecules of ATP and one of water; the $FADH_2$ yields two molecules of ATP and one of water. As noted earlier, each revolution of the TCA cycle generates a single ATP via GTP. Thus the TCA cycle and oxidative phosphorylation combined can be summarized in the equation:

$$\text{acetyl-CoA} + 3\,O_2 \rightarrow \text{CoA} + 2\,CO_2 + 2\,H_2O + 12\,\text{ATP}$$

A SUMMARY OF AEROBIC RESPIRATION

The complete reaction pathway that begins with glucose and ends with carbon dioxide and water is called **aerobic** (*aer*, air) **respiration**. In most cells, aerobic respiration is the primary method of generating ATP.

Figure 4-10 reviews the entire process of aerobic respiration from an energy standpoint.

- During glycolysis, the cell gains two molecules of ATP directly for each glucose molecule broken down to pyruvic acid. Two molecules of $NADH_2$ are also produced.
- The two pyruvic acid molecules derived from each glucose molecule are fully broken down in mitochondria. Two revolutions of the TCA cycle, each yielding a molecule of ATP, provide a total gain of two more molecules of ATP. However, both the TCA cycle itself and the step leading into it (formation of acetyl-CoA from pyruvic acid) produce additional molecules of reduced coenzymes ($NADH_2$ and $FADH_2$).
- The reduced coenzymes from all these steps feed electrons into the electron transport chain. The total yield is another 32 molecules of ATP.

Summing up, for each glucose molecule processed the cell gains 36 molecules of ATP. All but two of them are produced by the mitochondria.

OTHER CATABOLIC PATHWAYS

Aerobic respiration is relatively efficient and capable of generating large amounts of ATP. It is the cornerstone for normal cellular metabolism, but it has one obvious limitation: the cell must have adequate supplies of both oxygen and glucose. Cells can survive only for brief periods without oxygen; the mechanisms used to deal with low-oxygen emergencies will be considered in detail in Chapter 10. Low glucose concentrations have a much smaller effect on most cells, because they can break down other nutrients to provide substrates for the TCA cycle, as shown in Figure 4-11. Details of the pathways involved will be presented in Chapter 25; for the moment, we will merely note that many cells can switch from one nutrient source to another as the need arises. For example, many cells can shift from glucose-based to lipid-based ATP production when necessary. Skeletal muscles catabolize glucose when actively contracting, but utilize fatty acids when at rest.

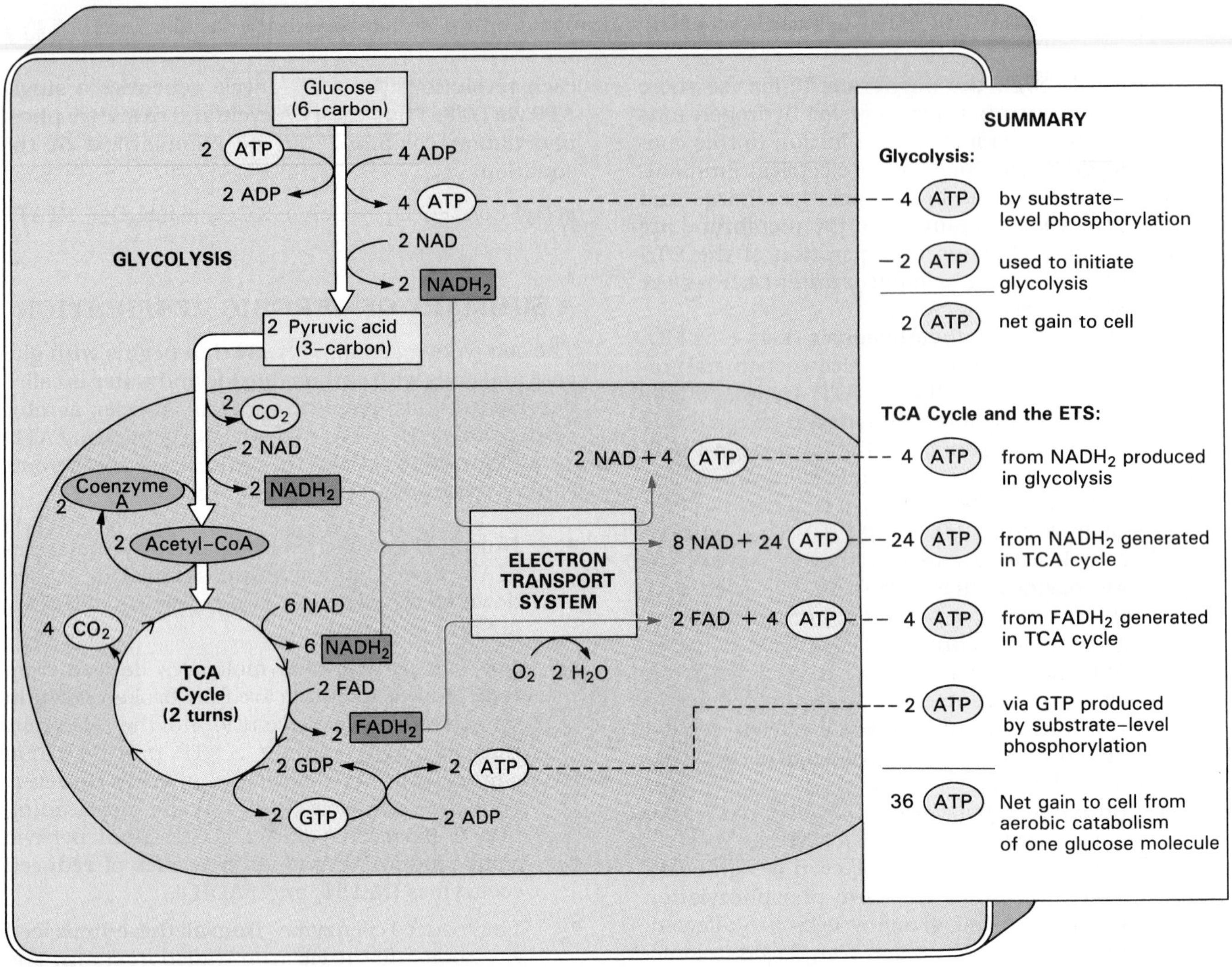

SUMMARY

Glycolysis:

4 ATP	by substrate-level phosphorylation
−2 ATP	used to initiate glycolysis
2 ATP	net gain to cell

TCA Cycle and the ETS:

4 ATP	from $NADH_2$ produced in glycolysis
24 ATP	from $NADH_2$ generated in TCA cycle
4 ATP	from $FADH_2$ generated in TCA cycle
2 ATP	via GTP produced by substrate-level phosphorylation
36 ATP	Net gain to cell from aerobic catabolism of one glucose molecule

FIGURE 4-10
A Summary of the Energy Yield of Aerobic Respiration. For each glucose molecule broken down via aerobic respiration, only two molecules of ATP are produced directly in glycolysis and another two in the TCA cycle. However, glycolysis, the formation of acetyl-CoA, and the TCA cycle all yield molecules of reduced coenzymes ($NADH_2$ and $FADH_2$). Many additional ATPs are produced when electrons from these coenzymes pass down the electron transport chain. Each $NADH_2$ produced inside the mitochondria yields three ATPs. Because the transfer of hydrogens from cytosol to mitochondion costs the cell energy, each $NADH_2$ produced in glycolysis provides only two ATPs. Each $FADH_2$ also yields two ATPs.

Sports and Fitness: Carbohydrate Loading

Although other nutrients can be broken down to provide substrates for the TCA cycle, carbohydrates require the least processing and preparation. It is not surprising, therefore, that athletes have tried to devise ways of exploiting these compounds as ready sources of energy.

Eating carbohydrates *just before* exercising does not improve performance and may actually decrease endurance by slowing the mobilization of existing energy reserves. Runners or swimmers preparing for lengthy endurance contests, such as a marathon or 5-km swim, do not eat immediately before competing, and for 2 hours before the race their diet is limited to drinking water. However, these athletes often eat carbohydrate-rich meals for 3 days before the event. This **carbohydrate loading** increases the carbohydrate reserves of muscle tissue that will be called upon during the competition.

Maximum effects can be obtained by exercising to exhaustion for 3 days before starting the high-carbohydrate diet; this practice is called **carbohydrate depletion/loading**. There are a number of potentially unpleasant side effects to carbohydrate depletion/loading, including muscle and kidney damage, and sports physiologists recommend that athletes use this routine fewer than three times per year.

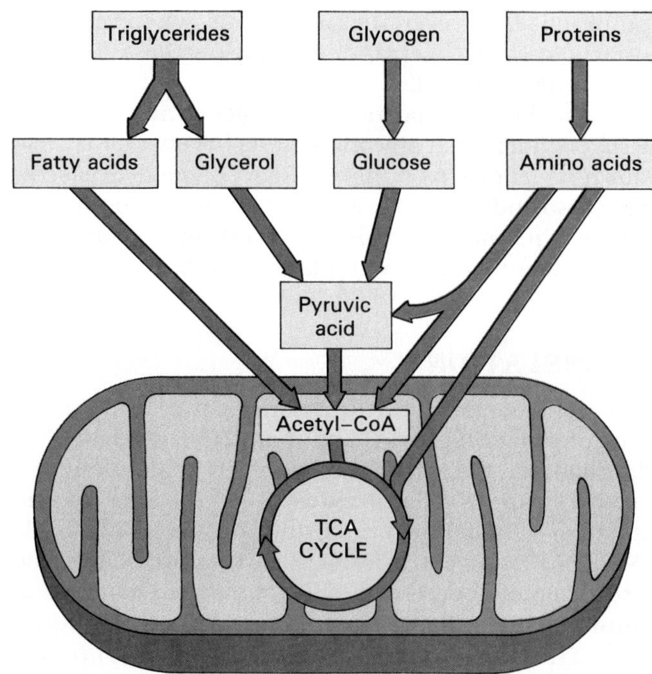

FIGURE 4-11
Alternate Catabolic Pathways.

Cells break down proteins for energy only when lipids or carbohydrates are unavailable, primarily because proteins make up the enzymes and organelles that the cell needs to survive. Nucleic acids are present only in small amounts, and they are seldom catabolized for energy, even when the cell is dying of acute starvation. This restraint makes sense, because it is the DNA in the nucleus that determines all of the structural and functional characteristics of the cell. How it does so will be considered in the next section, which explores the way in which information is stored and retrieved at the cellular level.

√ **Why would you expect to find large amounts of ATP in muscle tissue?**

√ **If the amount of available NAD in a cell were sharply decreased, how would this effect cellular energy production?**

√ **During oxidative phosphorylation, the pH just outside the inner mitochondrial membrane falls. Why?**

■ DNA and Information Storage

The basic structure of nucleic acids was described in Chapter 2 (see Figures 2-23 to 2-25). ∞ (pp. 54, 55) Chapter 3 considered the organization of the nucleus, the organelle that contains the cell's DNA. *The DNA of the nucleus controls the cell by directing the synthesis of specific proteins.* Proteins make up 15–30 percent of the weight of each cell, and through control of protein synthesis virtually every aspect of cell structure and function can be regulated. There are two levels of control involved, one direct and the other indirect:

1. The DNA of the nucleus has *direct* control over the synthesis of structural proteins, such as cytoskeletal components, membrane proteins (including receptors), and secretory products. By issuing appropriate orders, the nucleus can alter the internal structure of the cell, its sensitivity to substances in the extracellular environment, or its secretory functions to meet changing circumstances.
2. The DNA of the nucleus has *indirect* control over all other aspects of cellular metabolism because it regulates the synthesis of enzymes. By ordering the production of appropriate enzymes, the nucleus can regulate all of the metabolic activities and functions of the cell. For example, the nucleus can accelerate the rate of glycolysis by increasing the number of glycolytic enzymes in the cytoplasm.

THE GENETIC CODE

The DNA in a typical nucleus contains the instructions for manufacturing each of the roughly 100,000 different proteins found in the body. This vast quantity of information is stored in the sequence of nitrogen bases (adenine, A; thymine, T; cytosine, C; and guanine, G) along the length of DNA strands. The information storage system used by cells has been termed the *genetic code*.

The genetic code is called a **triplet code** because a sequence of three nitrogen bases can specify the identity of a single amino acid. Each **gene** consists of all the "triplets" needed to produce a specific polypeptide chain. The number of triplets in a gene varies, depending on the size of the polypeptide represented. A relatively short peptide chain might require fewer than a hundred triplets, while the instructions for building a large protein might involve a thousand or more.

Each gene also contains segments responsible for regulating its activity. In effect these are triplets that say "do (or do not) read this message," "message starts here," and "message ends here." The "read me," "don't read me," and "start here" signals form a special control segment at the start of the genetic message.

TRANSCRIPTION

Activated genes do not leave the nucleus, nor do they lose their connections with other genes along the length of the DNA strand. Instead, a messenger carries the instructions from the nucleus to the cytoplasm. The carrier is a single strand of **messenger RNA (mRNA)**. The process of mRNA formation is called **transcription** because the mRNA is "rewriting" the instruction carried by the gene. Figure 4-12 details this process.

As you will recall from Chapter 2, a DNA molecule consists of a pair of strands held together by hydrogen bonding between complementary nitrogen bases. ∞ (p. 54) Each of the DNA strands contains thousands of individual genes. When a specific gene is activated, the weak bonds between the nitrogen bases are temporarily disrupted, and an enzyme, **RNA polymerase**, binds to the initial segments of the gene. This enzyme promotes hydrogen bonding between the nitrogen bases of the gene and complementary nucleotides in the nucleoplasm. The nucleotides involved are those characteristic of RNA, not DNA; RNA polymerase may attach adenine, guanine, cytosine, or uracil, but never thymine. Thus wherever an A occurs in the DNA strand, the polymerase will attach a U rather than a T.

Only a small portion of the gene interacts with RNA polymerase at any one time as the enzyme travels down its length. As it moves, RNA polymerase assembles a strand of mRNA. At the "stop here" command, the enzyme and the mRNA strand detach, and the complementary DNA strands reassociate.

TRANSLATION

The construction of a functional polypeptide using the information provided by an mRNA strand is called **translation**. Each triplet of nitrogen bases along the mRNA strand is called a **codon** (KŌ-don). Every amino acid has at least one unique and specific codon; Table 4-2 includes several examples. During translation the sequence of codons will determine the sequence of amino acids in the polypeptide.

Translation proceeds in three steps: (1) *initiation*, (2) *elongation*, and (3) *termination*. These steps are diagrammed in Figure 4-13.

FIGURE 4-12
Transcription. This diagram shows a small portion of a single DNA molecule, containing a single gene available for transcription. **Step 1:** The two DNA strands separate, and RNA polymerase binds to the control segment of the gene. **Step 2:** The RNA polymerase moves from one triplet to another along the length of the gene. At each site, complementary RNA nucleotides form hydrogen bonds with the DNA nucleotides of the gene. The RNA polymerase then strings the arriving nucleotides together into a strand of mRNA. **Step 3:** Upon reaching the stop signal at the end of the gene, the RNA polymerase and the mRNA strand detach, and the two DNA strands reassociate.

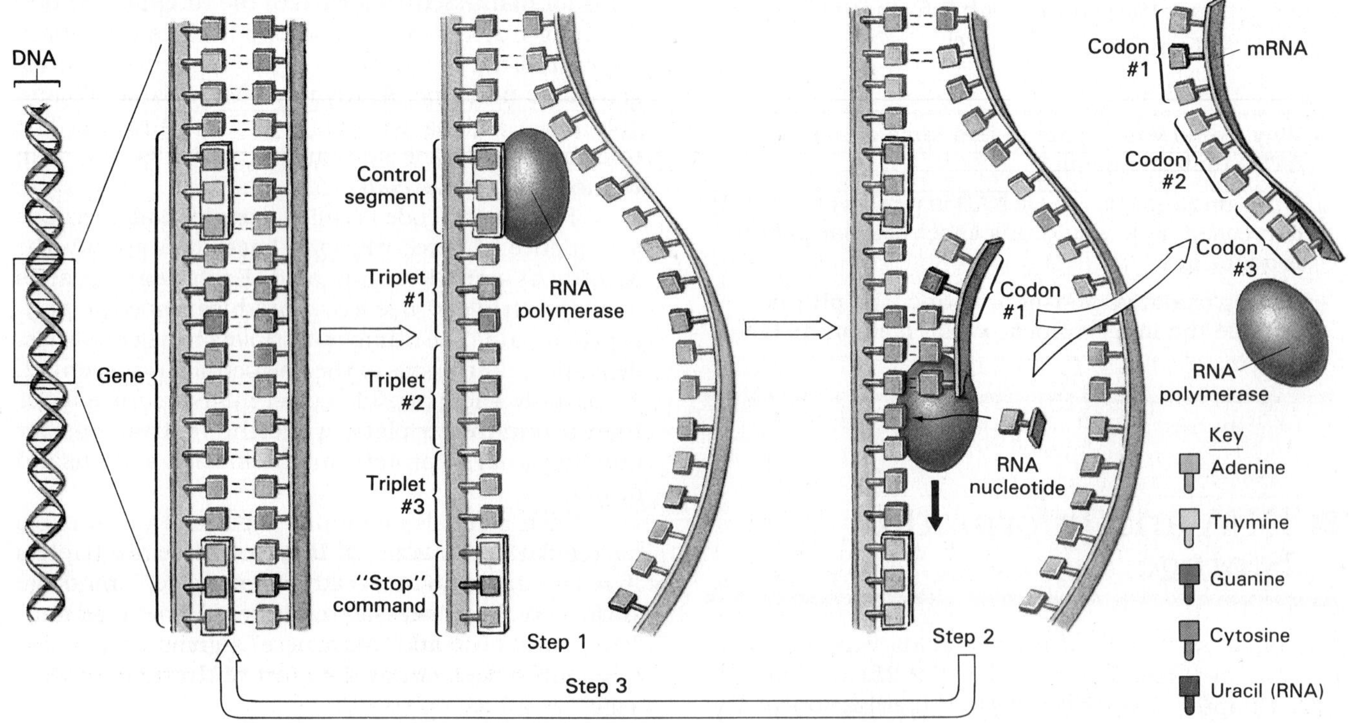

TABLE 4-2 Examples of the Triplet Code

DNA Triplet	*mRNA Codon*	*tRNA Anticodon*	*Amino Acid or Instruction*
AAA	UUU	AAA	Phenylalanine
AAT	UUA	AAU	Leucine
ACA	UGU	ACA	Cysteine
CAA	GUU	CAA	Valine
GGG	CCC	GGG	Proline
CGA	GCU	CGA	Alanine
CGG	GCC	CGG	Alanine
CGC	GCG	CGC	Alanine
TAC	AUG	UAC	Initiator codon
ATT	UAA	[none]	Terminator codon
ATC	UAG	[none]	Terminator codon
ACT	UGA	[none]	Terminator codon

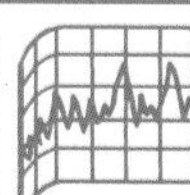

Diagnostics: DNA Fingerprinting

Although each cell type has a different set of active and potentially active genes, all of the nucleated cells in the body carry the same 46 chromosomes and identical DNA segments. All of the DNA in the nucleus does not code for proteins; some segments are involved in the regulation of gene activity, others are responsible for ribosomal RNA (rRNA) or transfer RNA (tRNA) production, and a significant portion of the cell's DNA has no known function. Some of the "useless" segments contain the same nucleotide sequence repeated over and over. The number of segments and the number of repetitions vary from individual to individual. The chances of any two individuals, other than identical twins, having the same pattern of repeating segments is less than one in 9 billion. In other words, it is extremely unlikely that you will ever encounter someone else who has the same pattern of repeating nucleotide sequences present in your DNA.

Individual identification can therefore be made on the basis of a pattern of DNA analysis, just as it can on the basis of a fingerprint. Skin scrapings, blood, semen, hair, or other tissues can be used as a sample source. Information from **DNA fingerprinting** has already been used to convict persons committing violent crimes, such as rape or murder.

Initiation

When the mRNA enters the cytosol it interacts with one or more ribosomes. Each ribosome consists of two subunits that are normally separate and distinct. They differ in size; one is a **light ribosomal subunit** and the other a **heavy ribosomal subunit**, as seen in Figure 3-8. (p. 69) These subunits contain special proteins and a second type of RNA, **ribosomal RNA (rRNA)**. During initiation these subunits interlock around the initial segment of the mRNA strand.

Initiation begins as the mRNA strand binds to the light ribosomal subunit. A molecule of **transfer RNA (tRNA)** then arrives at the first codon. The tRNA molecule is small but highly organized, and it carries a specific amino acid. There are more than 20 different types of transfer RNA, at least one for each amino acid used in protein synthesis. At one place in its structure each tRNA molecule has a different trio of nitrogen bases, known as an **anticodon**, that will bind to a complementary codon on the messenger RNA. *A tRNA molecule that carries a particular amino acid always has an anticodon that is complementary to the mRNA codon specifying that particular amino acid.* This is how the sequence of codons along the mRNA strand determines the sequence of amino acids in a protein.

The first tRNA to arrive bears an anticodon sequence of U-A-C; this tRNA carries the amino acid *methionine*. (This amino acid will ultimately be removed from the finished protein.) The anticodon sequence will bind to the first codon of the mRNA strand, which always has the base sequence A-U-G. When this binding occurs, a heavy ribosomal subunit joins the complex to create a complete ribosome. The mRNA strand nestles in the gap between the light and heavy subunits.

Elongation

A second tRNA now arrives at the adjacent site of the ribosome (Steps 1–2), and its anticodon binds to the next codon of the mRNA strand. Enzymes of the heavy ribosomal subunit then break the linkage between the tRNA molecule and its amino acid. At the same time, they attach the amino acid to its neighbor by means of a peptide bond (Step 3). The ribosome then moves one codon down the mRNA strand. The cycle can now be repeated with the arrival of another molecule of tRNA. The tRNA already stripped of its amino acid drifts away, ready to find another amino acid.

Termination

Elongation continues, adding amino acids to the growing polypeptide chain, until the ribosome reaches the "stop here" signal, or *terminator codon*, at the end of the mRNA strand. The ribosomal subunits now detach, leaving an intact strand of mRNA and a completed polypeptide.

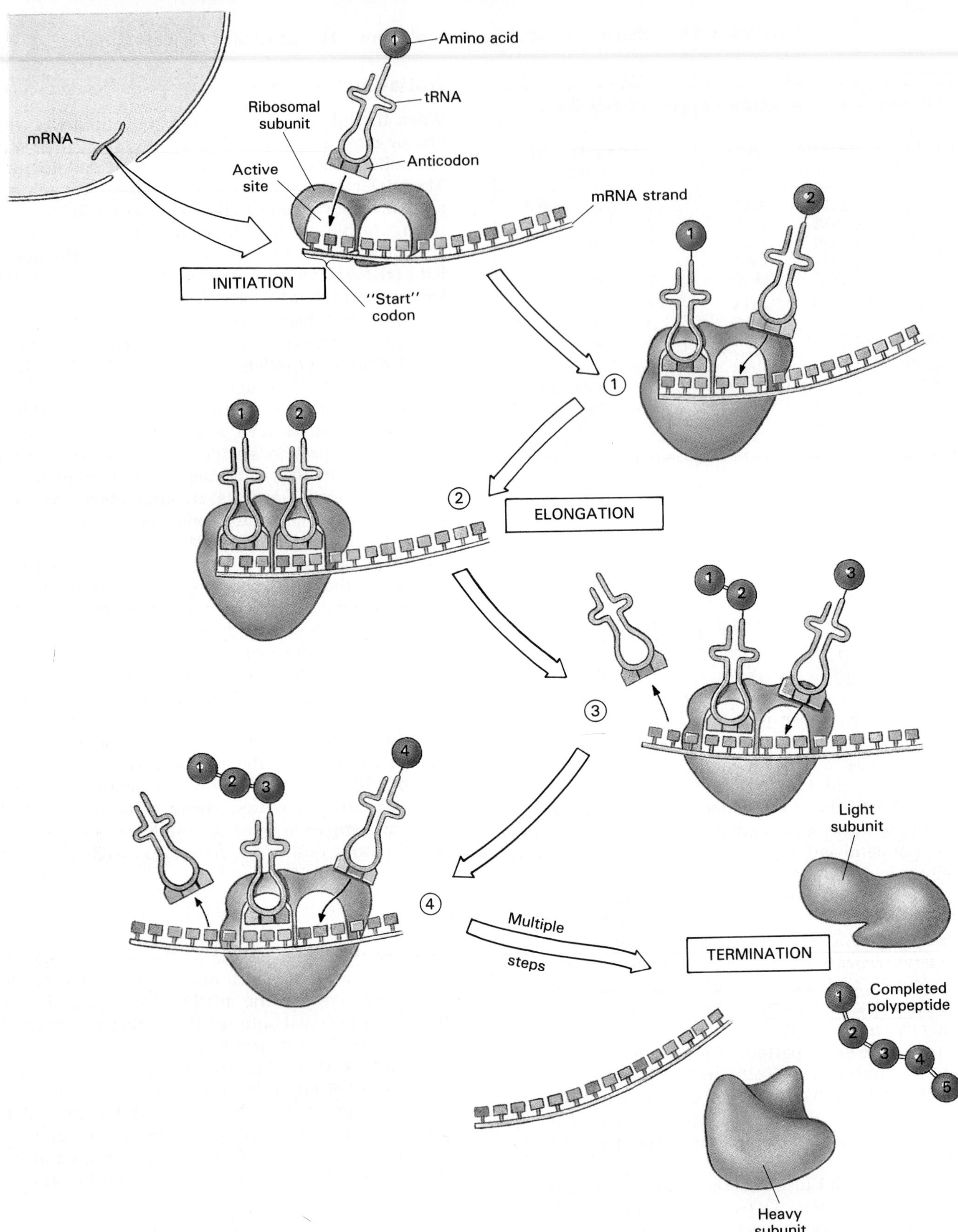

FIGURE 4-13
The Process of Translation. *Initiation*: mRNA binds to the light ribosomal subunit and is joined by the first tRNA, which carries the amino acid methionine. Bonding occurs between complementary base pairs of the codon and anticodon, and the light and heavy ribosomal subunits interlock around the mRNA strand. *Elongation*: A second tRNA arrives at the adjacent site of the ribosome and its anticodon binds to the next mRNA codon. The amino acid that it carries is detached from the tRNA and joined to the first amino acid by a peptide bond. The ribosome moves one codon farther along the mRNA strand; the first tRNA detaches as another tRNA arrives. *Termination*: Elongation continues until the terminator codon is reached; the components then separate.

Translation proceeds swiftly, producing a typical protein in around 20 seconds. The polypeptide begins as a simple linear strand, but secondary and tertiary structures start to appear as elongation proceeds. Energy for translation is provided by GTP rather than ATP; it costs the cell one molecule of GTP to begin the process, two GTP per amino acid incorporated, and one GTP to end it.

Polyribosomes

During translation, only two mRNA codons are involved at any one time. The mRNA strand may contain thousands of codons, and several ribosomes can be translating the same message simultaneously. A **polyribosome** is a series of ribosomes attached to the same mRNA strand. Figure 4-14 shows polyribosomes in a cell actively synthesizing proteins. Polyribosome formation greatly increases the rate of protein synthesis.

Control of Genetic Activity

The regulation of genetic activity involves turning specific genes on or off at appropriate times. For a gene to be available for transcription, the DNA segment must be unwound, the initial segment must be activated, and the appropriate RNA polymerase must be present and activated. Three general regulatory mechanisms are known:

1. **Control by negative feedback**: A gene may stop transcribing mRNA when the concentration of the protein that it codes for reaches some critical level within the nucleoplasm. Transcription may begin again when the concentration declines. This is a very simple, direct mechanism that maintains concentrations within relatively narrow limits.
2. **Control by repressors**: Repressor substances turn off particular genes. Repressors for one gene are often proteins manufactured under the instructions of another gene. Ions or other compounds diffusing into the nucleoplasm can also act as repressors.
3. **Control by inducers**: A gene can be switched on by some other chemical compound that triggers transcription of mRNA at a particular site. For example, most cells catabolize glucose for energy when nutrients are in adequate supply. During starvation, the adrenal gland produces steroid hormones that diffuse through cell membranes and bind to receptors in the nucleus. Their binding triggers the transcription of genes that code for enzymes involved with lipid catabolism. As a result, the cells begin breaking down lipids rather than glucose to obtain energy. In adipose tissue, these hormones stimulate the release of stored lipids. The net result is that the body begins to mobilize and use its lipid reserves for the duration of the crisis.

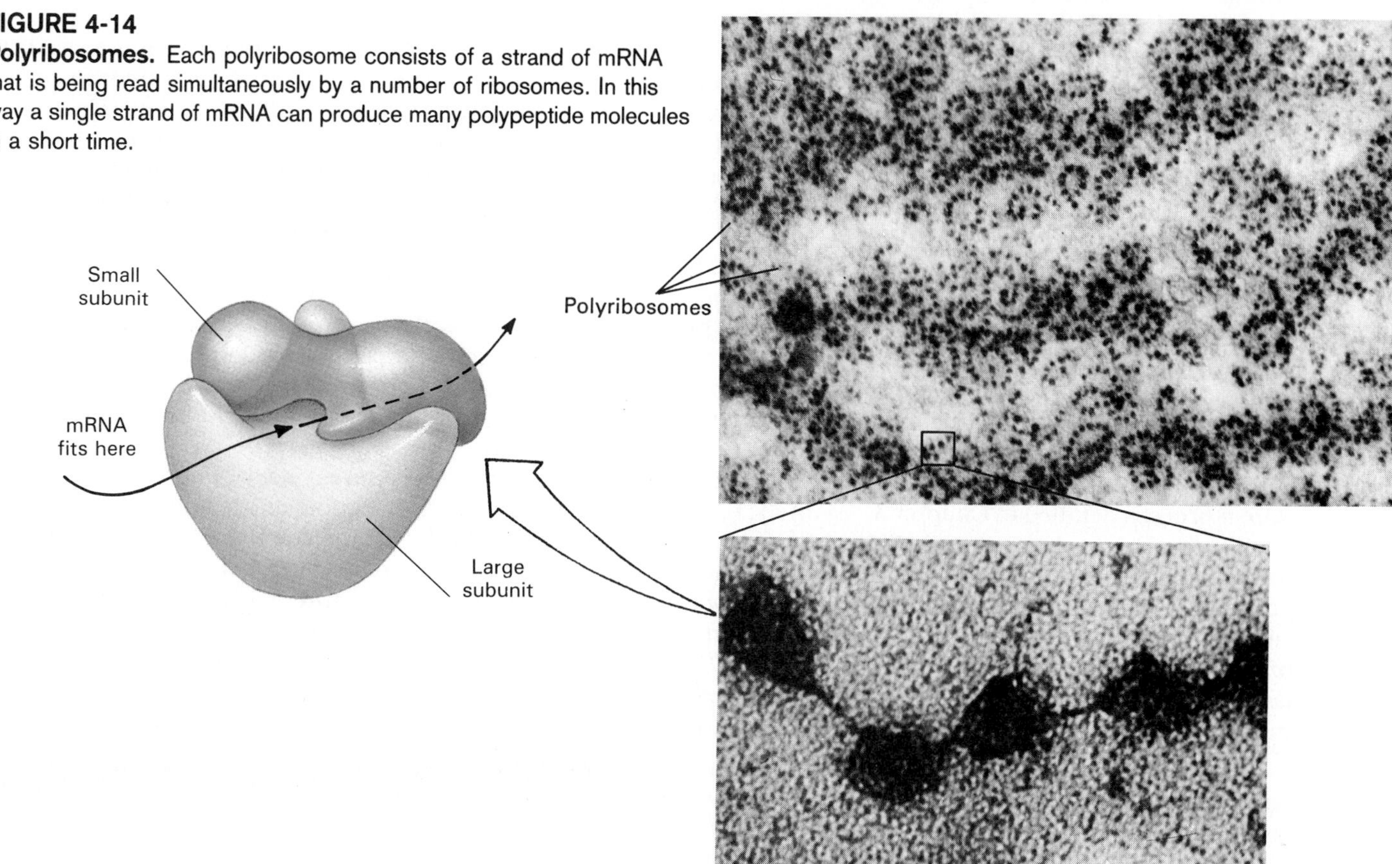

FIGURE 4-14
Polyribosomes. Each polyribosome consists of a strand of mRNA that is being read simultaneously by a number of ribosomes. In this way a single strand of mRNA can produce many polypeptide molecules in a short time.

The effects of inducers can include permanent structural changes as well as temporary regulation of metabolic activities. The steroid sex hormones **estrogen** (female) and **testosterone** (male), for instance, are inducers that activate genes in a wide variety of cells. The differential effects account for the appearance of secondary sexual characteristics, including the distribution of body fat, hair growth, and breast development.

MUTATIONS

Mutations are permanent alterations in a cell's DNA that affect the nucleotide sequence of one or more genes. The simplest is a **point mutation**, a change in a single nucleotide. This affects one codon. In some cases the change will have no apparent effect. The triplet code has some flexibility because several different codons can signal the same amino acid. For example, the amino acid alanine is represented by the mRNA codons GCU, GCC, GCA, or GCG. A point mutation that changed the third nitrogen base from U to C, A, or G would not affect the translated protein. But a point mutation that altered either of the first two bases would produce a codon that specifies a different amino acid, and so change the structure of the completed protein.

With roughly 2.5 billion nucleotides in the DNA of a human cell, a single mistake might seem relatively unimportant. But over 100 inherited disorders have been traced to abnormalities in enzyme or protein structure that reflect alterations in nucleotide sequence. A change in the amino acid sequence of a single structural protein or enzyme can prove fatal. Several cancers and two lethal blood disorders, *thalassemia* and *sickle cell anemia* (p. 614), result from variations in a single nucleotide. [**CM**: *Iron Deficiencies and Excesses*, Chapter 25]

More elaborate mutations can affect chromosomal structure and even break a chromosome apart. Enzymes in the nucleus may attempt repairs, but the fragments may not be joined in the proper sequence, and they may even be attached to the wrong chromosome. Such anomalies may interfere with normal gene control mechanisms.

Mutations most often occur during *DNA replication*, a process detailed in Chapter 3. (p. 91) The likelihood that such an event will occur varies, depending on the gene or chromosome under consideration. In general, mutations are most likely to involve cells undergoing cell division. A single cell, a group of cells, or an entire individual may be affected. The latter will be the case if the changes occur during meiosis or early in development. For example, a muta-

Health News:

Once the mechanics of the genetic code were understood, everyone realized that it would be theoretically possible to change the genetic makeup of organisms—perhaps even of a human being. The popular term for activities related to this goal is **genetic engineering**.

What are some of the key problems confronting genetic engineers? Genes code for proteins; the makeup of each protein is determined by the sequence of codons (nucleotide triplets) in a stretch of DNA. A human cell has 46 chromosomes, two meters of DNA, and roughly 10^9 triplets. Before a specific gene can be modified, its location must be determined with great precision. This involves preparing a map of the appropriate chromosome.

Suppose that the location of a defective gene has been pinpointed. Before attempting to remedy the defect, one would have to determine the nature of the genetic abnormality. For example, the gene could be inactive, overactive, or producing an abnormal protein. It could even be missing entirely.

Finally, it would be necessary to decide how to remedy the defect. Can the gene can be turned on, turned off, modified, or replaced?

Where Is It? Several techniques can be used to create a general map of the chromosomes. **Karyotyping** (KAR-ē-ō-tī-ping; *karyon*, nucleus + *typos*, mark) is the determination of an individual's chromosome complement. Figure 4-15a shows a set of normal human chromosomes. Each chromosome has characteristic banding patterns, and segments can be stained with special dyes. Unusual banding patterns can indicate structural abnormalities. These abnormalities are sometimes linked to specific inherited conditions, such as Down syndrome (Figure 4-15b) and a few cancers, including a form of leukemia.

Such procedures yield *general* information about the location of an abnormal gene. One way to determine the *precise* location of a gene is to work backward, beginning with the protein it produces and deducing the probable sequences of DNA nucleotides responsible.

Despite the logistical difficulties involved, the scientific community in the United States has made the commitment to map the entire

Genetic Engineering and Gene Therapy

(a)

(b)

(c)

FIGURE 4-15
Normal and Abnormal Karyotypes. (a) A micrograph of the normal human chromosome set; the chromosomes have been arranged in this sequence to make comparisons easier. (b) The chromosomes of an individual with Down syndrome. Note the extra copy of chromosome 21. (c) Down syndrome is associated with mental retardation, cardiovascular problems, and a variety of physical abnormalities. The boy (age 9) on the bicycle has the characteristic appearance of an individual with Down syndrome; compare his features with those of his normal 4-year-old sister.

human genome. As of 1989, a gene register maintained by Dr. Victor McKusick and other researchers at Johns Hopkins contained the locations of almost 2000 genes, an impressive number but a minute percentage of the total number. The mapping process, which will take billions of dollars and 10 years or more, will provide basic information about the location of genes on normal human chromosomes.

What's the Problem? This can be a particularly difficult question to answer. Many of the 1500 inheritable genetic disorders are classified according to general patterns of symptoms rather than any specific protein or enzyme deficiency. In some cases the approximate location of the gene has been determined, but the identity of the protein responsible for the clinical symptoms remains a mystery.

What Can Be Done? If the gene is present but overproducing or underproducing, its activity might be controlled by introducing chemical repressors or inducers. Another approach relies upon **gene splicing** to produce a protein missing or present in inadequate quantities in the abnormal individual. Gene splicing (Figure 4-16) begins with the localization of the gene, followed by its isolation. That gene is then "spliced" into the relatively simple DNA strand of a bacterium, creating **recombinant DNA**. Bacteria grow and reproduce rapidly under laboratory conditions, and before long you have a colony of identical bacteria. All of the members of the colony will carry the introduced gene and manufacture the corresponding protein. The protein can be extracted, concentrated, and administered to individuals whose diseases represent deficiencies in the activity of that particular gene.

Gene splicing is also used to obtain large quantities of proteins normally found in very small concentrations. Interferon, an antiviral protein, and human growth hormone are examples of compounds now being produced commercially using gene-splicing technology.

The most revolutionary strategies involve adding normal genes to abnormal cells. In general, this method poses significant targeting problems, for the gene must be introduced into the right kind of cell. For example, placing liver enzymes in fingernails would not correct a metabolic disorder. But when the target cells can be removed and isolated, as in the case of bone marrow, the technique is promising. Actual removal of a defective gene does not appear to be a practical approach, and the focus has been on adding genes that can take over normal functions.

FIGURE 4-16
Gene Splicing. A gene is removed from a human cell nucleus and attached to the DNA in a bacterium, where it directs production of a human protein. Bacterial replication creates a colony of bacteria that share the introduced gene and can yield large quantities of the protein product.

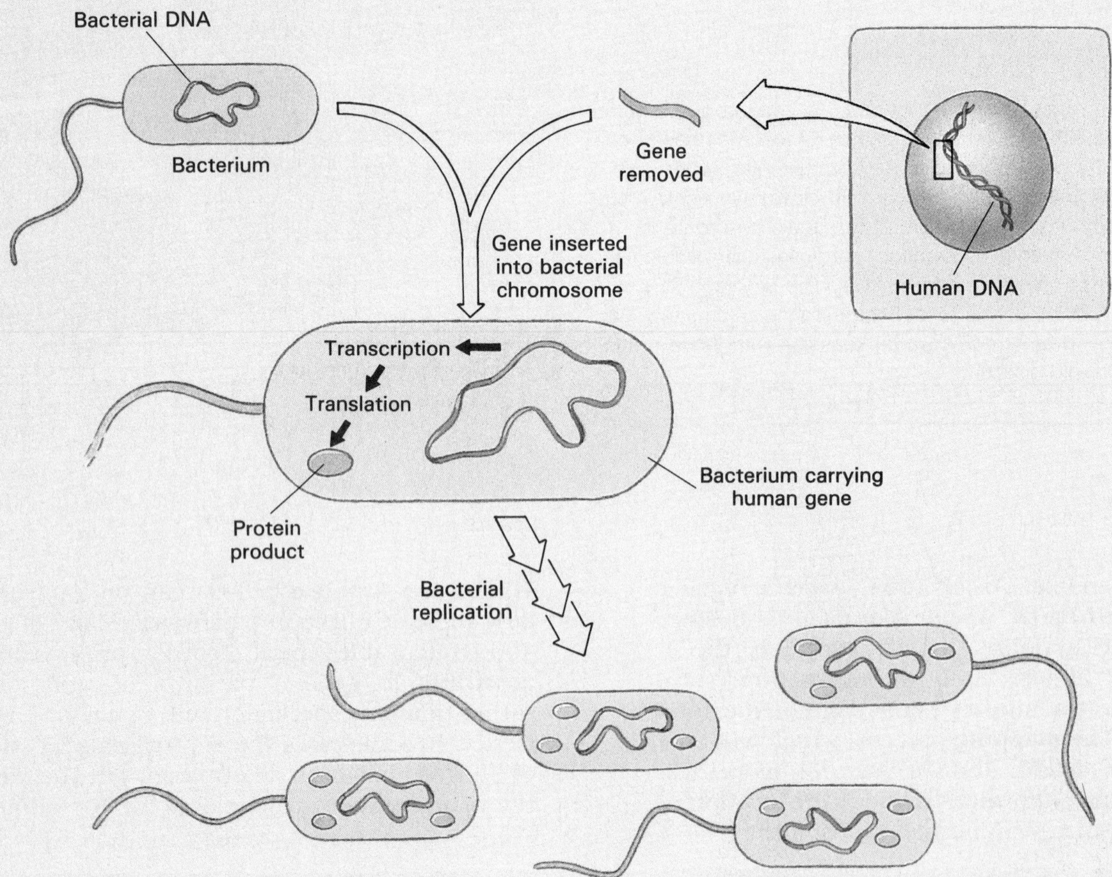

In September 1990, the first gene therapy trials were initiated. The procedure was used to treat a 4-year-old girl afflicted with **adenosine deaminase deficiency (ADA)**. ADA is a rare condition that affects only about 20 children worldwide. Without this enzyme, toxic chemicals build up in cells of the immune system, and as these cells die the body's defenses break down.

ADA results in a complex of symptoms known as **severe combined immunodeficiency disease**, or **SCID**. [CM: *SCID*] SCID can also be caused by other enzyme disorders affecting cells of the immune system. Symptoms include chronic respiratory infections, diarrhea, and a low resistance to viral or bacterial infections. Children with ADA usually die from infections that would pose no threat to normal children. A new drug called *PEG-ADA*, an altered form of the missing enzyme, can prolong life, but it does not cure the condition.

In this clinical trial, blood cells were collected, and cells of the immune system were removed. Short segments of DNA containing the normal gene for adenosine deaminase were then inserted into the nuclei of these cells, and the modified immune cells were returned to the body. Roughly a billion modified cells were reintroduced. Over time these modified cells will undergo mitotic division, producing a large population of normal immune cells. More than a year will pass before there are enough normal cells to improve her disease resistance.

This procedure attempts to relieve the symptoms of disease by inserting genes into defective somatic cells. They do not change the genetic structure of reproductive cells; because the eggs or sperm retain the original genetic pattern, the genetic defect will be passed to future generations. Researchers are much further away from practical methods of changing the genetic characteristics of reproductive cells. Mouse eggs fertilized outside the body have been treated and transplanted into the uterus of a second mouse for development. The gene added was one for a growth hormone obtained from a rat, and the "supermouse" that resulted demonstrated that such manipulations can be performed. The possibilities for manipulating the characteristics of valuable animal stocks, such as cattle, sheep, or chickens, are quite exciting. The potential for altering the genetic characteristics of human beings is somewhat intimidating. Before any clinical variations on *this* theme are tested, our society will have to come to grips with a number of difficult ethical issues!

tion that affects gametes (sperm or eggs) will be inherited by that individual's children.

- √ **Why do you think three nucleotide bases are needed to specify each amino acid in a protein, rather than just one or two?**
- √ **Why can a change in a single nucleotide among the many millions of nucleotides in the nucleus of a cell prove fatal?**
- √ **Why does protein synthesis slow down in starvation?**

Metabolic Turnover and Genetic Regulation

The DNA in the nucleus can exert effective control over cellular operations because cells are dynamic structures. If every protein synthesized by a cell lasted indefinitely, it would be much more difficult to modify cell structure and control metabolic activities. In fact, most cell structures are temporary rather than permanent. Their removal and replacement are part of the process of *metabolic turnover*.

METABOLIC TURNOVER

Cell maintenance involves both destruction and renewal. This cyclical process is called **metabolic turnover**. During metabolic turnover the cell breaks down and replaces its organic contents. Except for the DNA of the nucleus, the organic molecules in the cell are replaced at intervals ranging from hours to months. The average time between synthesis and recycling is known as the **turnover rate**. Table 4-3 indicates the turnover rates for components of various intracellular structures.

All living cells perform metabolic turnover, and it represents a significant energy expense. For example, proteins account for up to 30 percent of the weight of a cell, and they are recycled every week (see Table 4-3). The advantage of this expensive process is that it gives the cell an ability to change its physical or metabolic characteristics over time. These changes would be impossible if structural proteins and enzymes remained intact indefinitely. Such adjustments are essential to maintaining homeostasis at the cellular and systems levels; specific examples will be found in later chapters.

■ TABLE 4-3 Turnover Rates

Cell Type	Component	Half-Life (Days)[a]
Liver	Total protein	5–6
	Enzymes	1 hour to several days, depending on the enzyme
	Glycogen	1–2
	Cholesterol	5–7
Muscle cell	Total protein	30
	Glycogen	0.5–1
Neuron	Phospholipids	200
	Cholesterol	100+
Fat cell	Triglycerides	15–20

[a] Times are average life expectancies. Most values were obtained from studies on mammals other than humans.
(From A. L. Lehninger: *Biochemistry, 2nd ed.* Worth Publishers, New York, 1975, p. 375.)

The catabolic and anabolic events in metabolic turnover can be diagrammed as a closed cycle, as shown in Figure 4-17a. If the organic molecules synthesized are identical to those broken down, cell structure and function will remain unchanged. In reality, the situation is far more dynamic. Consider Figure 4–17b, which diagrams the major anabolic and catabolic events in the cell. There are several

FIGURE 4-17
Metabolic Turnover and Cellular ATP Production. (a) Metabolic turnover can be diagrammed as a closed cycle in which the catabolic steps yield substrates used to support the anabolic steps. (b) A more realistic view considers the integration of metabolic turnover with other cellular functions. Organic molecules released during metabolic turnover may be recycled via anabolic pathways—used to build other large molecules. They can also be broken down for energy production. The cell must provide ATP for both anabolism and other cellular functions. Consequently, it must obtain substrates from the extracellular fluid.

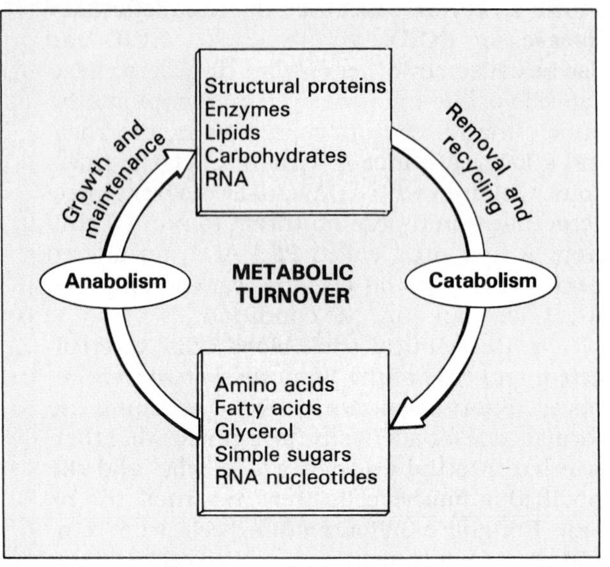

(a)

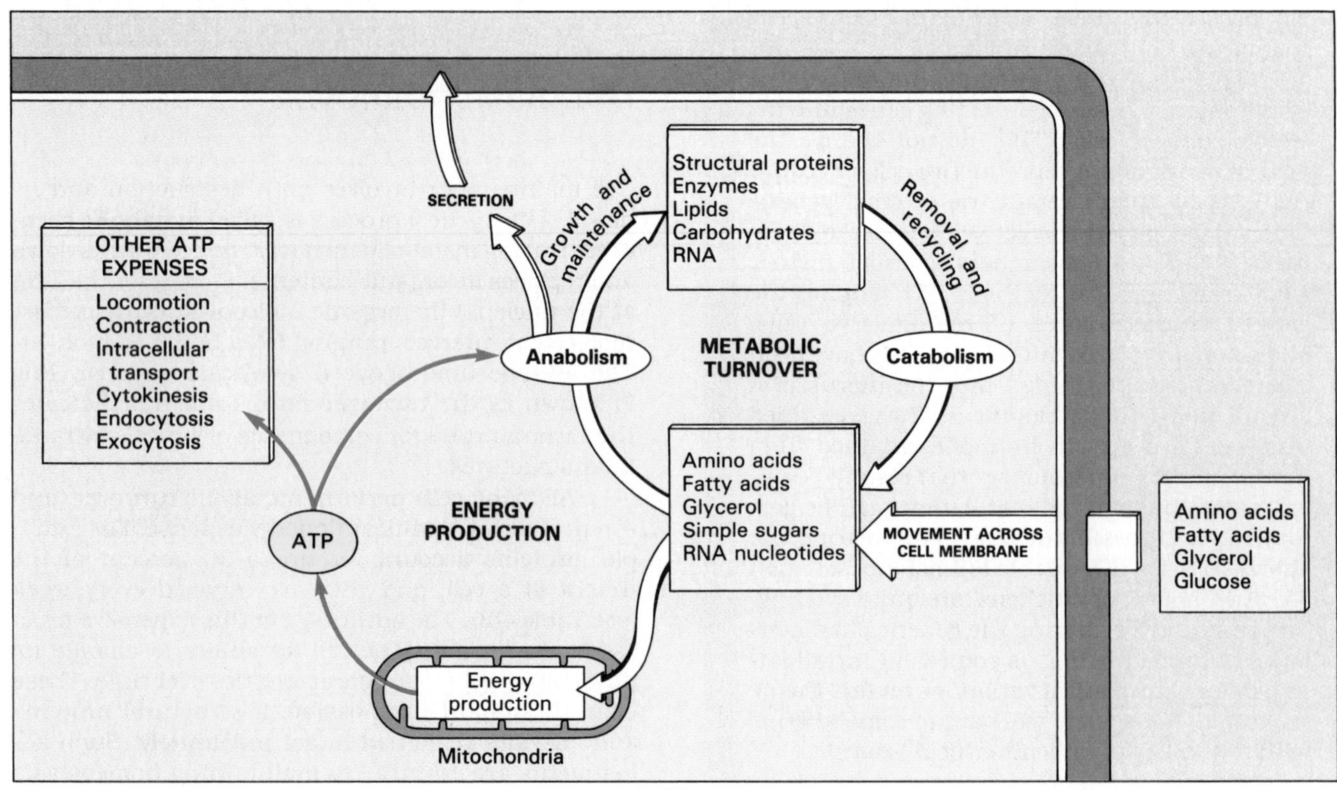

(b)

important points to consider as you examine this diagram:

1. The catabolic and anabolic pathways of metabolic turnover are independently regulated, and the relative rates of anabolism and catabolism can change. If anabolism outpaces catabolism, the cell grows; if catabolism predominates, the cell becomes smaller and less able to perform vital functions.
2. Organic molecules released by catabolic processes are not necessarily used for anabolism; they can also be used by the mitochondria for energy production.
3. Energy is released at each catabolic step, but only part of that energy can be captured by the cell. As a result, it takes more energy to rebuild molecules (anabolism) than the cell gains by breaking them down (catabolism). Moreover, synthesis of molecules is not the only energy expense of the cell. It must also use ATP to perform other vital functions such as transport, secretion, or contraction.
4. This gap in the cell's energy budget makes the cell dependent on an external source of organic compounds that can be catabolized to generate ATP. If the cell cannot obtain enough organic molecules from the extracellular environment, it will start catabolizing organic molecules released by metabolic turnover. The cell will then grow smaller and its functional abilities will be impaired.
5. Because the catabolic and anabolic steps of metabolic turnover are independently regulated, the cell does not necessarily synthesize the same organic molecules that it breaks down. The "decisions" are made by the nucleus, which controls all protein synthesis. By transcribing the genes for specific structural proteins or enzymes, a cell can store or mobilize energy reserves, reorganize the cytoskeleton, synthesize a new secretory product, or shift from glucose to another substrate for ATP production.

THE CONTROL OF CELL FUNCTION

Genes have direct control over the synthesis of enzymes, structural proteins, and RNA within the cell. This section will discuss (1) the relationship between genetic regulation and other, shorter-term regulatory strategies, (2) how the activities of different cells are coordinated, and (3) how differences in genetic activity account for the structural and functional characteristics of different cell types.

Genes and Enzymes: Short-Term versus Long-Term Regulation

Cells remain alive and functional by controlling thousands of chemical reactions in dozens of anabolic and catabolic pathways. Only in this way can they preserve homeostasis at the cellular level of organization. The regulation is dynamic: it must change constantly to meet the demands of a changing environment. These metabolic adjustments may take the form of temporary, short-term modifications in enzyme activity or permanent, long-term changes in enzyme composition or cell structure. Figure 4-18 indicates the general pattern.

FIGURE 4-18
Extracellular Control of Cellular Activities. Changes in the composition of the extracellular fluid, such as the presence of a hormone, prostaglandin, or nutrient, can alter activities within the cell. Enzymes can be activated or deactivated as the result of an extracellular molecule binding at a membrane receptor or entering the cell via diffusion or transport. Genetic activity can also be altered, producing long-term adjustments in the cell's structural or functional characteristics.

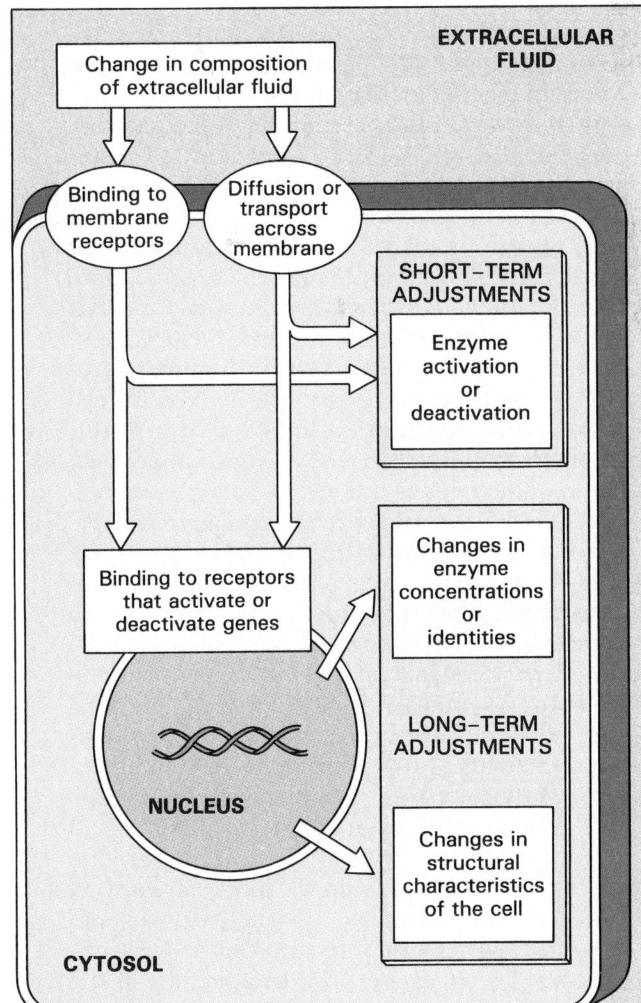

Clinical Comment: Cancer II—Causes

Twenty-five percent of all Americans develop cancer at some point in their lives. It has been estimated that 70–80 percent of these cases involve chemical exposure, environmental factors, or both, and almost half (40 percent) are due to a single stimulus: cigarette smoke. During 1990, almost 500,000 Americans were killed by some form of cancer, making this Public Health Enemy Number 2, second only to heart disease.

A relatively small number of cancers are actually inherited; 18 types have been identified to date, including two forms of leukemia. Most cancers develop through the interaction of genetic and environmental factors, and it is difficult to separate the two completely.

Genetic Factors Two related genetic factors are involved in the development of cancer: *hereditary predisposition* and *oncogene activation*.

An individual born with genes that increase the likelihood of cancer is said to have a hereditary predisposition for the disease. Under these conditions a cancer is not guaranteed, but it is a lot more likely. The inherited genes usually affect tissue abilities to metabolize toxins, control mitosis and growth, perform repairs after injury, or identify and destroy abnormal tissue cells. As a result, body cells become more sensitive to local or environmental factors that would have little effect on normal tissues.

Cancers may also result from somatic mutations that modify genes involved with cell growth, differentiation, or mitosis. As a result, an ordinary cell is converted into a cancer cell. The modified genes are called **oncogenes** (ON-kō-jēns); the normal genes are called *proto-oncogenes*. Oncogene activation occurs by alteration of normal somatic genes. Because these mutations do not affect reproductive cells, the cancers caused by active oncogenes are not inherited.

A proto-oncogene, like other genes, has a regulatory component that turns the gene "on" and "off" and a structural component that contains the triplets that determine protein structure. Mutations in either portion of the gene may convert it to an active oncogene. A small mutation can accomplish this; changing one nucleotide out of a chain of 5000 can convert a normal proto-oncogene to an active oncogene. In some cases, a viral infection can trigger activation of an oncogene. For example, one of the papilloma viruses appears to be responsible for cancer of the cervix (discussion in Chapter 28).

More than 50 oncogenes are known. In addition, a group of anticancer genes has now been identified. These genes, called *tumor-suppressing genes* (*TSG*), or *anti-oncogenes*, suppress division and growth in normal cells. Mutations that alter these genes make oncogene activation more likely. TSG mutation has been suggested as important in promoting several cancers, including several blood cell cancers, breast cancer, and ovarian cancer.

Environmental Factors Many cancers can be directly or indirectly attributed to environmental factors called **carcinogens** (kar-SIN-o-jens). Carcinogens stimulate the conversion of a normal cell to a cancer cell. Some carcinogens are **mutagens** (MŪ-ta-jens)— that is, they damage DNA strands and sometimes cause chromosomal breakage. Radiation is an example of a mutagen that has carcinogenic effects.

There are many different chemical carcinogens in the environment. Plants manufacture poisons that protect them from insects and other predators, and although their carcinogenic activities are relatively weak, many common spices, vegetables, and beverages contain compounds that can be carcinogenic if consumed in large quantities. Animal tissues may also store or concentrate toxins, and hazardous compounds of many kinds can be swallowed in contaminated meals. A variety of laboratory and industrial chemicals, such as coal tar derivatives and synthetic pesticides, have been shown to be carcinogenic. Cosmic radiation, X-rays, and other radiation sources can also cause cancer.

Specific carcinogens will affect only those cells capable of responding to that particular physical or chemical stimulus. The responses vary because differentiation produces cell types with specific sensitivities. For example, benzene can produce a cancer of the blood, cigarette smoke a lung cancer, and vinyl chloride a liver cancer. Very few stimuli can produce cancers throughout the body; radiation exposure is a notable exception. In general, cells undergoing mitosis are most likely to be vulnerable to chemical or radiational carcinogens. As a result, the cancer rates are highest in epithelial tissues, where stem cell divisions occur rapidly and relatively low in nervous and muscle tissues, where divisions do not normally occur.

Changes in Enzyme Activity Regulatory enzymes provide a mechanism for short-term control. Enzymes may be turned on or off by intracellular events, such as a change in the concentration of a particular molecule. Often that molecule is the product of the pathway that the enzyme controls, as described earlier. Extracellular stimuli, such as hormones, can also affect the activities of enzymes within the cell. They bind to receptor proteins in the cell membrane, triggering changes inside the cell that alter the function of particular enzymes. Many important hormones work in this way; insulin, the hormone that stimulates glucose absorption across cell membranes, is an example.

Changes in Enzyme Synthesis The nucleus also controls metabolic activity by regulating the quantity and identity of cytoplasmic enzymes synthesized. By calling for the production of new and different enzymes, the nucleus can change the rate of growth, alter secretory activity, vary the composition of the cell membrane, or affect any other aspect of cell structure or function. Because it takes time to synthesize new enzyme molecules, changing enzyme concentrations is primarily a method for long-term adjustment of cellular activity.

Coordinating Cellular Activities

Genes control activities in the cytoplasm, and feedback from the cytoplasm can activate or inhibit genes. Maintaining homeostasis at the level of the individual *person* involves coordinating the activities of many different cell types. Most regulatory mechanisms rely on the fact that each cell type has a different sensitivity to ions, nutrients, or hormones in the extracellular fluid.

The sensitivity of a particular type of cell may depend on what glycoprotein receptors are present in its membrane. Cells also differ in the specific genes that can be activated by a particular stimulus. The net result is that a single change in the composition of the extracellular fluid can trigger a variety of responses. For example, an adrenal steroid can cause glucose synthesis in liver cells, initiate the release of lipids from fat cells, and leave nerve cells relatively unaffected.

Diversity and Differentiation

Liver cells, fat cells, and nerve cells contain the same genes, but in each case a different set of genes has been turned off *permanently*. In other words, liver cells differ from other cells because they have different genes available for transcription.

When a gene is functionally eliminated, the cell loses the ability to create a particular protein, and thus to perform any functions requiring that protein. Each time another gene switches off, the cell's functional abilities become more restricted. This specialization process is called **differentiation**.

Fertilization produces a single cell with all of its genetic potential intact. There follows a period of repeated mitotic divisions, and as the number of cells increases, differentiation begins. Differentiation is necessary because no one cell could contain the metabolic and structural machinery required to perform all the secretion, absorption, contraction, conduction, storage, and elimination processes in the human body.

Differentiation produces specialized cells with limited capabilities. These cells form organized collections known as **tissues**, each with discrete functional roles. The next chapter examines the structure and function of tissues and considers the role of tissue interactions in the maintenance of homeostasis.

CHAPTER REVIEW

Review of Selected Clinical Terms

isozymes (*p. 98*)
Enzymes that differ in structure but catalyze the same reaction.

vitamins (*p. 98*)
Essential organic nutrients that must usually be obtained from the diet; many function as coenzymes in vital enzymatic reactions.

carbohydrate loading (*p. 108*)
The practice of eating carbohydrate-rich meals for several days to increase the carbohydrate reserves of muscle tissue.

DNA fingerprinting (*p. 111*)
Identifying an individual on the basis of repeating nucleotide sequences in his or her DNA.

genetic engineering (*p. 114*)
The popular term for research and experiments related to changing the genetic makeup of an organism.

karyotyping (KAR-ē-ō-tī-ping) (p. 114)
The determination of an individual's chromosome complement.

recombinant DNA (*p. 116*)
DNA created by splicing a specific gene from one organism into the DNA strand of another organism.

oncogene (ON-kō-jēn) (*p. 120*)
A cancer-causing gene created by a somatic mutation in a normal gene involved with growth, differentiation, or cell division.

carcinogen (kar-SIN-o-jen) (*p. 120*)
An environmental factor that stimulates the conversion of a normal cell to a cancer cell.

mutagen (MŪ-ta-jen) (*p. 120*)
A factor that can damage DNA strands and sometimes cause chromosomal breakage, stimulating the development of cancer cells.

Additional Terms of Clinical Importance

ISOZYMES (*p. 98*): **lactate dehydrogenase (LDH)**

AN INTRODUCTION TO VITAMINS (*p. 98*): **water-soluble vitamin, fat-soluble vitamin**

CARBOHYDRATE LOADING (*p. 108*): **carbohydrate depletion/loading**

GENETIC ENGINEERING AND GENE THERAPY (*p. 114*): **gene splicing, adenosine deaminase deficiency (ADA), severe combined immunodeficiency disease (SCID)**

Study Outline

Related Key Terms

Introduction (pp. 95–96)

1. Homeostasis depends on the tightly controlled interactions of cells in the body.

Enzymes and Cellular Metabolism (pp. 96–100)

1. **Metabolism** refers to all of the chemical reactions in the body. Through **cellular metabolism** cells capture, store, and use energy to maintain homeostasis and support essential functions.

ENZYMES AND CHEMICAL REACTIONS (pp. 96–97)

2. **Activation energy** is the amount of energy required to start a reaction. **Enzymes** are organic **catalysts**—substances that accelerate chemical reactions without themselves being permanently changed.

ENZYME STRUCTURE AND FUNCTION (pp. 97–98)

3. The reactants in an enzymatic reaction, called **substrates**, interact to form a **product** by binding to the enzyme at the **active site**.

COFACTORS AND COENZYMES (p. 98)

4. **Cofactors** are ions or molecules that must attach to the active site before substrate binding can occur. **Coenzymes** are large organic molecules that function as cofactors; many are derived from **vitamins**.

   Related key terms: **holoenzyme** • **apoenzyme**

CONTROL OF ENZYME ACTIVITY (p. 99)

5. A variety of factors can affect enzyme performance. Each enzyme works best at an optimal combination of temperature and pH. **Enzyme activation** and **enzyme inhibition** are important methods for short-term control of reaction rates.

   Related key terms: **enzyme inhibitors**

ENZYMES, METABOLIC PATHWAYS, AND HOMEOSTASIS (pp. 99–100)

6. A **metabolic pathway** is a series of enzymatic reactions that yields a specific product. **Regulatory enzymes** act as "on/off switches" in pathways.

Chapter Review

Related Key Terms

Cells and Energy (pp. 100–102)

1. Cells require energy to maintain homeostasis and to perform special functions.

BASIC ENERGY CONCEPTS (pp. 100–101)

2. **Work** involves movement of an object or a change in its physical structure, and **energy** is the capacity to perform work. There are two major types of energy: kinetic and potential.
3. **Kinetic energy** is the energy of motion. **Potential energy** is stored energy that results from the position or structure of an object. Conversions from potential to kinetic energy are not 100 percent efficient; every energy exchange produces **heat**.
4. Cells gain energy to power their functions by breaking down organic molecules, a process called **catabolism**. Much of this energy supports **anabolism**, the synthesis of new organic molecules.

HIGH-ENERGY COMPOUNDS AND ENERGY TRANSFER (pp. 101–102)

5. Cells store energy in **high-energy compounds** for later use. The most important high-energy compound is **ATP (adenosine triphosphate).** When energy is available, cells make ATP by adding a phosphate group to ADP **(phosphorylation)**. When energy is needed, ATP is broken down to ADP and phosphate.

ADP (adenosine diphosphate)
GTP (guanosine triphosphate)
UTP (uridine triphosphate)
phosphagens
CP (creatine phosphate)

Pathways of Cellular Energy Production (pp. 102–109)

1. Most cells generate high energy compounds by breaking down carbohydrates.

GLYCOLYSIS (pp. 102–103)

2. **Glycolysis** is the breakdown of glucose to pyruvic acid in the cytosol.

pyruvic acid
NAD (nicotinamide adenine dinucleotide)
glucose-6-phosphate

MITOCHONDRIAL ENERGY PRODUCTION (pp. 103–107)

3. Glycolysis captures only a small part of the energy contained in glucose. Much more energy is obtained when the products of glycolysis are broken down further inside mitochondria.
4. The **TCA cycle** is a sequence of reactions in which 2-carbon acetyl groups are catabolized. Hydrogen atoms are removed by coenzymes while carbon and oxygen are released as CO_2.
5. **Oxidative phosphorylation**, the single most important mechanism for generating ATP, produces ATP through a reaction sequence that consumes oxygen. Reduced coenzymes deliver hydrogen ions and electrons from the TCA cycle into the **electron transport system (ETS)**. As the electrons are passed down a sequence of metalloproteins called **cytochromes**, protons are pumped across the inner mitochondrial membrane. As these protons flow back into the matrix, ATP is generated.

decarboxylation
coenzyme A (CoA) • **acetyl-CoA**
acetyl group
FAD (flavin adenine dinucleotide)
oxidation • **reduction**
FMN (flavin mononucleotide)
coenzyme Q • **porphyrin**
respiratory chain
chemiosmosis

A SUMMARY OF AEROBIC RESPIRATION (p. 107)

6. **Aerobic respiration** is the complete reaction pathway that begins with glucose and ends with carbon dioxide and water; in most cells, it is the primary method of generating ATP. Each molecule of glucose broken down yields 36 molecules of ATP: two from glycolysis in the cytosol and the rest from the activity of mitochondria.

OTHER CATABOLIC PATHWAYS (pp. 107–109)

7. Aerobic respiration has the drawback of requiring adequate supplies of oxygen and glucose. Many cells can switch to energy sources other than glucose when the need arises.
8. Lipids are the body's second most important energy source. Proteins can be used to provide energy, but are utilized only in emergencies because they are such crucial components of the cell. Nucleic acids do not contribute significantly to the cell's energy reserves.

DNA and Information Storage (pp. 109–117)

1. The DNA of the nucleus controls the cell by directing the synthesis of specific proteins.

THE GENETIC CODE (p. 109)

2. The cell's information storage system, the genetic code, is called a **triplet code** because a sequence of three nitrogen bases identifies a single amino acid. Each **gene** consists of all the triplets needed to produce a specific polypeptide chain.

TRANSCRIPTION (p. 110)

3. **Transcription** is the process of forming a strand of **messenger RNA (mRNA)**, which carries instructions from the nucleus to the cytoplasm.

RNA polymerase

Chapter Review

TRANSLATION (pp. 110–114)

Related Key Terms

4. During **translation** a functional polypeptide is constructed using the information from an mRNA strand. Each trio of nitrogen bases along the mRNA strand is a **codon**; the sequence of codons determines the sequence of amino acids in the polypeptide. Translation proceeds in three steps: *initiation, elongation, and termination.*
5. **Ribosomal RNA (rRNA)** forms part of the structure of the ribosomes involved in translation. Molecules of **transfer RNA (tRNA)** bring amino acids to the active site of the ribosomal complex. — **light ribosomal subunit** **heavy ribosomal subunit** **anticodon** • **polyribosome**
6. Genetic activity can be regulated by negative feedback, inducers, or repressors.

MUTATIONS (pp. 114–117)

7. **Mutations** are permanent changes in a cell's DNA that affect the nucleotide sequence of one or more genes. A **point mutation** is a change in a single nucleotide that affects only one codon.

Metabolic Turnover and Genetic Regulation (pp. 117–121)

turnover rate

1. DNA can effectively control cellular operations because cells are dynamic.

METABOLIC TURNOVER (pp. 117–119)

2. The cyclical process of destruction and renewal of cellular components is **metabolic turnover**; it requires significant amounts of energy.

THE CONTROL OF CELL FUNCTION (pp. 119–121)

3. Regulation of cell activity may be short- or long-term and involve changes in cell structure, enzyme activity or enzyme synthesis.
4. Maintaining homeostasis at the level of an individual person involves the coordinated activities of many different cell types. Most regulatory mechanisms depend on each cell type's sensitivity to ions, nutrients, and hormones in the extracellular fluid. — **estrogen** • **testosterone**
5. **Differentiation** is the process of specialization that produces cells with limited capabilities. These specialized cells form organized collections called **tissues**, each of which has certain functional roles.

Review Planner

		Level 1
1	Relate basic energy concepts to cell operations.	**9 12 14 18**
2	Distinguish between catabolic and anabolic pathways.	**17 31**
3	Explain how metabolic pathways are regulated.	**1 20**
4	Describe the crucial role of enzymes in cellular metabolism.	**1 7 11 15 16 24 28 32 38**
5	Discuss how cells capture, store, and use energy.	**3 10 19 22 25 34**
6	Outline the main steps in the metabolic pathway of aerobic respiration.	**2 6 8 10 13 26 33 40**
7	Explain how genes control cell functions.	**4 5 15 21 23 27 39 41**
8	Discuss the mechanisms that control genetic activity.	**4 21 23 27 30 35**

Level 2: **36 37 42 43**

Level 3: **44–49**

Genetic Engineering Offers Great Potential for Cures

9	Explain the importance of metabolic turnover.	**8 10 13 29**

Review Questions

MULTIPLE CHOICE

1. Most of the chemical reactions involved in cellular metabolism are regulated by: a) lipids; b) electrolytes; c) enzymes; d) sugars.
2. Glycolysis is the breakdown of: a) pyruvic acid; b) ATP; c) glycogen; d) glucose.
3. The most important high-energy compound in our bodies is: a) ADP; b) GTP; c) ATP; d) H_2O.
4. The instructions for protein synthesis are contained in: a) mitochondria; b) ADP; c) ATP; d) DNA.
5. Transfer ribonucleic acid is also known as: a) TRANS; b) RNA; c) TRA; d) tRNA.
6. Which of the following is *not* produced in aerobic respiration? a) glucose; b) CO_2; c) H_2O; d) ATP.
7. A catalyst ________ a chemical reaction: a) slows down; b) accelerates; c) can either accelerate or slow down; d) does not affect.
8. If the body is unable to catabolize carbohydrates, it is most likely to catabolize: a) proteins; b) lipids; c) nucleic acids; d) none of the above.
9. The energy of motion is called: a) kinetic energy; b) potential energy; c) work; d) activation.

DIAGRAMS

10. Diagram the process of cellular metabolism, showing: catabolism, anabolism, heat, the extracellular fluid, and ATP. Use arrows to illustrate the sequence of events and the relationships between these concepts.
11. Diagram the three-step process in which an enzyme binds two differently shaped substrates, which then form a product and enter the cytosol. Define enzyme, substrate, active site, and product. Include labels on your diagram.

TRUE/FALSE

(If the statement is false, your instructor may wish to have you correct it.)

12. **T/F:** A catalyst promotes a chemical reaction that otherwise would never occur.
13. **T/F:** If a cell is dying and its energy stores are entirely depleted, it will catabolize the DNA in its nucleus.
14. **T/F:** A conversion between potential energy and kinetic energy is not totally efficient.
15. **T/F:** The nucleus performs long-term regulation of metabolic activity.

MATCHING QUESTIONS

(Match each statement with the most appropriate *term.)*

Statements:

16. The reactants in enzymatic reactions.
17. The synthesis of new organic molecules.
18. An increase in random molecular motion.
19. Another name for the TCA cycle.
20. They function as "on/off switches" that control the operation of metabolic pathways.
21. This consists of all the "triplets" necessary to produce a specific polypeptide chain.
22. The loss of electrons from a molecule.
23. A permanent change in a cell's DNA that affects the nucleotide sequence of one or more genes.
24. The location on the surface of an enzyme to which molecules of reactants bind.
25. The acceptance of electrons from another molecule.
26. The removal of carbon and oxygen as carbon dioxide.
27. The process of forming a functional polypeptide using the information provided by mRNA.
28. An ion or molecule that must attach to an enzyme before substrate binding can occur.
29. The cycle during which a cell breaks down and replaces its organic contents.
30. The process through which an mRNA strand is formed.
31. The decomposition of organic molecules within cells.

Terms:

a) Metabolic turnover
b) Citric acid cycle
c) Catabolism
d) Active site
e) Gene
f) Translation
g) Decarboxylation
h) Reduction
i) Substrates
j) Heat
k) Regulatory enzymes
l) Transcription
m) Oxidation
n) Mutation
o) Anabolism
p) Cofactor

Chapter Review

SHORT ANSWER/ESSAY

32. Why are enzymes important to life?
33. Name three major coenzymes that are important to the process of oxidative phosphorylation.
34. Why is it important for cells to store energy in a form that is both stable and mobile? What is this form called?
35. Discuss the three regulatory mechanisms that switch genes on and off.
36. Distinguish between substrate-level phosphorylation and oxidative phosphorylation. Describe the process of each. Which generates more ATP, and why is this process important to life?
37. What is a metabolic pathway? What function do regulatory enzymes serve? Describe the effect of regulatory enzymes on glucose molecules.
38. Discuss several factors that control enzyme activity.
39. Define differentiation, and explain its significance.
40. Discuss the major steps in the TCA cycle. Why is the cycle important?
41. Distinguish between mRNA, rRNA, and tRNA. What role does each play during transcription and translation?
42. In the chapter you read that estrogen and testosterone function as inducers. Explain what this means. For what effects are estrogen and testosterone responsible?
43. Explain why hydrogen cyanide gas can be fatal to humans.

CRITICAL THINKING/APPLICATIONS

44. A patient is rushed to the hospital. His symptoms include radiating chest pain, difficulty in breathing, and rapid pulse rate. A blood test reveals high levels of cardiac lactate dehydrogenase (LDH). a) What is LDH? b) What is an isozyme? c) What is one possible diagnosis for elevated levels of cardiac LDH in the blood? Explain your answer.
45. A marathon runner training for an important race has read about carbohydrate depletion/loading, and he wonders if this would help his performance. He is also thinking about eating a candy bar just before the race to give him a extra "push" of energy. What would you tell him? Why?
46. A woman claims to have been raped by a man who denies the charge. You have been asked to serve as an expert witness at the trial. Explain how you could determine that this suspect is not the rapist.
47. Define "genetic engineering." Explain why determining the locations of human genes is an important aspect of this research.
48. Discuss several techniques that attempt to "correct" genetic activity. What is your opinion of the ethical issues involved? Would you support some of these techniques over others? Explain your answer.
49. Where would a cancer be more likely to start: in the deltoid muscle or in the skin over the same area? Explain your answer.

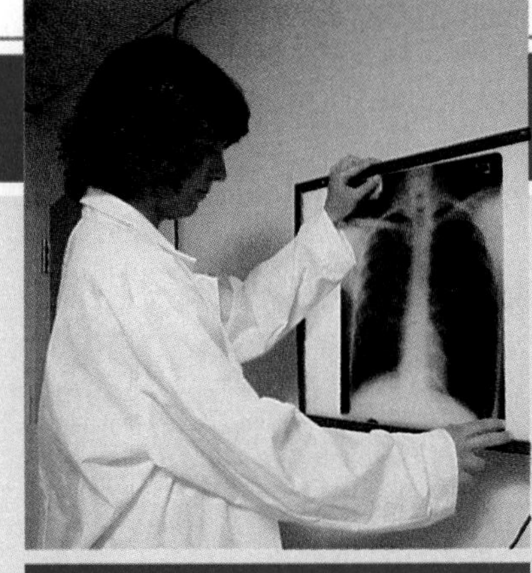

Career Close-up: X-Ray Technologist

Betty Gonzalez is a single parent with two children. When she decided to return to school after six years, she had already worked as a medical assistant in an Obstetrics/Gynecology office and then as a medical assistant for a neurologist. She wanted to be more involved in medicine, but she didn't want to become a doctor or a nurse. She decided that radiology was the right field.

Betty had to take Anatomy and Physiology in her two-year X-ray technology program. "I didn't know what to expect. I certainly didn't know the class would be so demanding." Sometimes she felt overwhelmed by the course and by going back to school. "I often felt like giving up, but I knew that if I did, there would be nowhere for me to go. I couldn't work for low wages forever."

Now that Betty is an X-ray technologist, she finds her job to be an enriching experience. "I just love it. It's really interesting meeting new people, helping to find out what is wrong with them, and being part of helping them get better." She considers the job a good stepping stone for other radiology-related jobs. "There is a great diversity of career possibilities for X-ray technicians. If you don't like diagnostics, you can go into computerized tomography (CT scans), MRI (magnetic resonance imaging), sonography, special procedures like angiograms, nuclear medicine (injecting and tracing radioactive isotopes), or radiation oncology (treating cancer patients with radiation therapy)."

"I now use my anatomy more than my physiology, because I deal mostly with skeletal problems. Actually, in my Anatomy and Physiology course I struggled through the physiology. I had an easier time understanding the systems on the macroscopic level than comprehending activities on the cellular level. I learn through visualizing rather than imagining." Betty uses her visualizing skills all the time when taking an X-ray. "I must look for anatomical landmarks, like a certain bone for a spinal X-ray or the diaphragm for a chest X-ray. It helps me know I have a good piece of diagnostic film when I can visualize the shadows or outlines of muscles and organs."

Betty cares a great deal about the patients she meets. "At a hospital, everything tends to be rush-rush. I really enjoy the times when I can give special attention to people who need it, especially people who don't always get enough respect in the community, like the elderly."

Betty takes regular X-rays, usually of the torso or extremities, and diagnoses diseases ranging from cancer to hernias. To visualize abnormalities that cannot be seen in an ordinary X-ray, she does fluoroscopy in which a radiodense liquid such as barium is injected into the area she wants to visualize on the monitor. "I once had a little boy in the office who was getting a voiding-cystourogram. To do this we fill the bladder with a barium solution so that it shows up clearly in the X-ray. It's not a pleasant procedure, and naturally the boy was terrified, especially because his pregnant mother couldn't be with him. I felt compassion for him when he was crying, probably because he was the same age as my son. I didn't want him to have a bad experience. So I took the star pin I was wearing and said, 'Here. Open your hand so you can think of the stars and make a wish.' He laughed, and I was relieved."

Although X-rays are potentially hazardous, Betty is not afraid. "I know an older technician who was taking X-rays before anything was known about the dangers of radiation, and he now has cataracts and other problems. But today we learn about radiation protection and use caution. Sometimes I have to help hold a patient, which means I am in the room while the X-ray is being taken, so I wear a lead-lined gown. I'm not scared as long as I practice good protection."

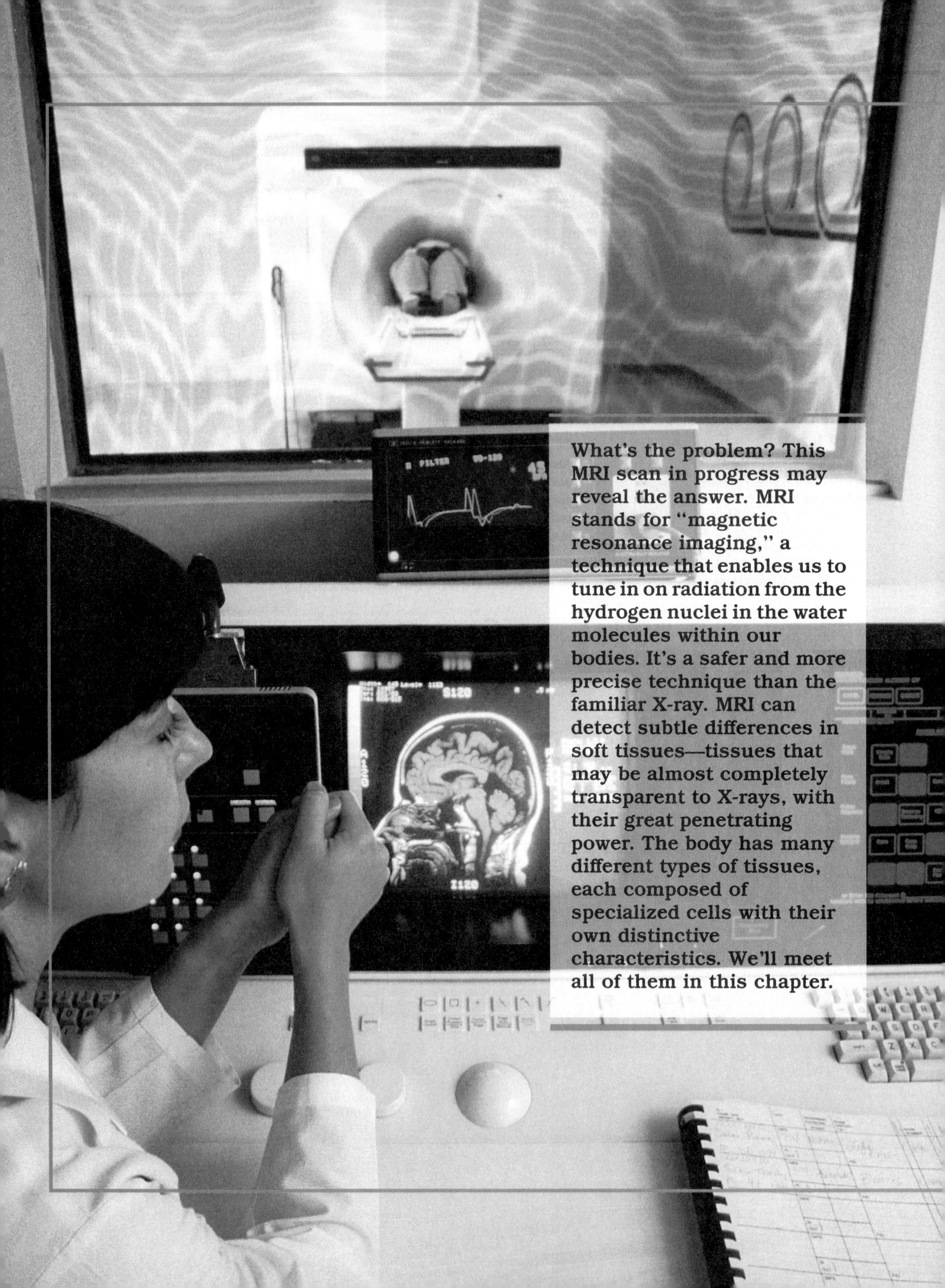

What's the problem? This MRI scan in progress may reveal the answer. MRI stands for "magnetic resonance imaging," a technique that enables us to tune in on radiation from the hydrogen nuclei in the water molecules within our bodies. It's a safer and more precise technique than the familiar X-ray. MRI can detect subtle differences in soft tissues—tissues that may be almost completely transparent to X-rays, with their great penetrating power. The body has many different types of tissues, each composed of specialized cells with their own distinctive characteristics. We'll meet all of them in this chapter.

CHAPTER 5

The Tissue Level of Organization

Chapter Objectives

After reading this chapter, you will be able to:

1. Classify the tissues of the body into four major categories.
2. Discuss the relationship between form and function for each of the primary tissue types.
3. Discuss the types and functions of epithelia.
4. Compare the structures and functions of the various connective tissues.
5. Explain how epithelia and connective tissues combine to form four different types of membranes, and specify the functions of each.
6. Describe the different types of muscle tissue.
7. Discuss the structure and role of neural tissue.
8. Explain how tissues respond in a coordinated manner to maintain homeostasis.
9. Describe how nutrition and aging affect tissues.

Introduction

We have seen in previous chapters how chemical reactions within cells maintain life and support a variety of activities. But no single cell could contain the metabolic machinery and organelles needed to perform all the many functions of the human body. Instead, through the process of differentiation each cell develops a characteristic set of structural features and a relatively restricted range of functions. These structures and functions can be quite distinct from those of nearby cells. Nevertheless, the cells in a given location work together.

A detailed examination of the body reveals a number of patterns at the cellular level. Although there are trillions of cells in the human body, there are only about 200 types of cells. These cell types combine to form **tissues**, collections of specialized cells and cell products that perform a relatively limited number of functions. There are four **primary tissue types**: *epithelia, connective tissue, muscle tissue,* and *neural tissue*. Their basic characteristics are summarized in Figure 5-1.

Epithelia cover exposed surfaces and line internal passageways and body cavities. The surface of the skin is an example of an epithelium. Connective tissues fill internal spaces, providing structural sup-

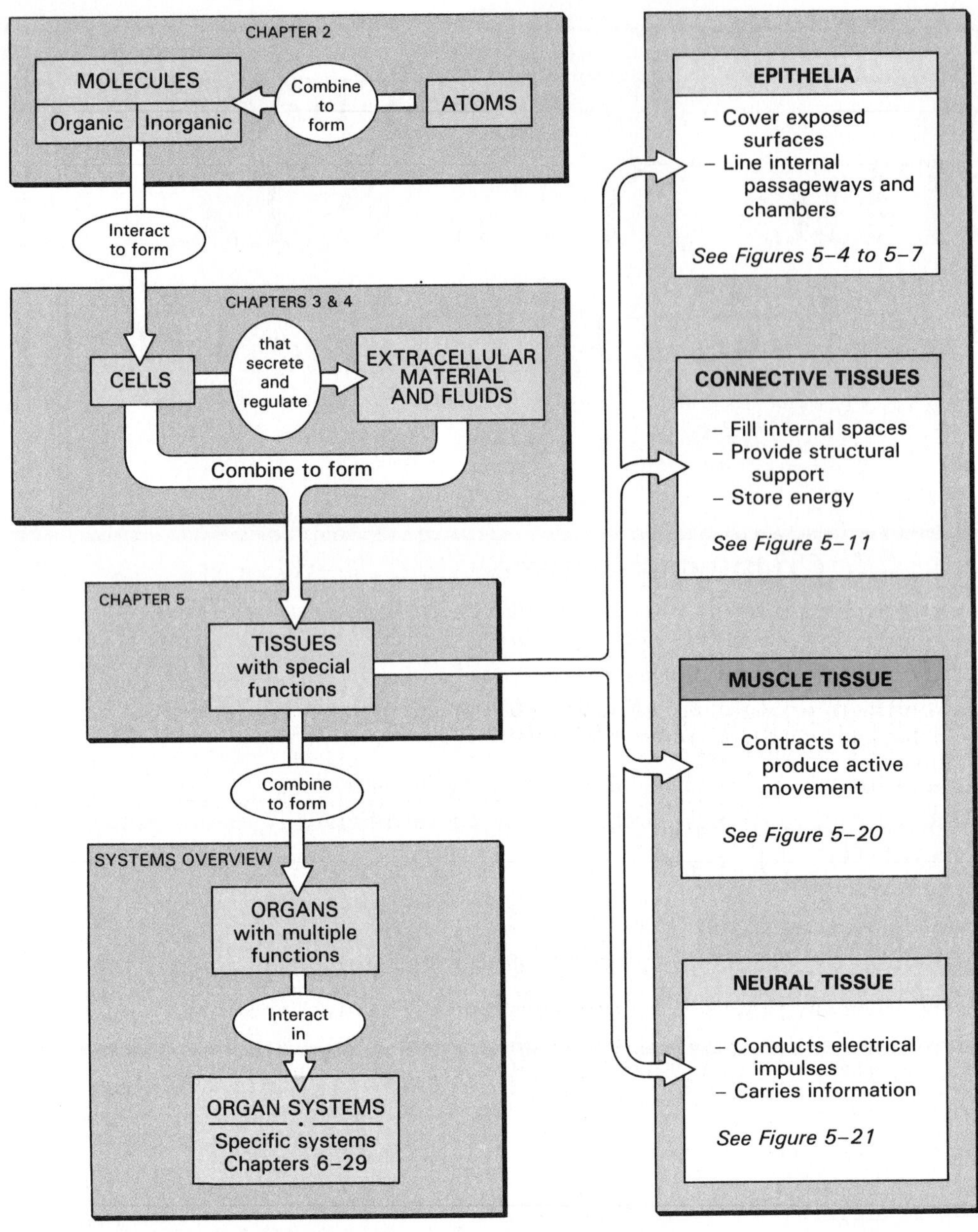

FIGURE 5-1
An Orientation to the Tissues of the Body

port and a framework for communication within the body. They also store energy sources, such as lipids, in bulk quantities. Muscle tissue has the ability to contract and produce active movement. Neural tissue carries information from one part of the body to another in the form of electrical impulses. Each of the organs of the body contains these four types of tissue in varying combinations.

The study of tissues is called **histology** (*histos*, tissue). Such a study provides beautiful examples of the interplay between form and function. Although there is considerable overlap between histology and cytology, they differ in their emphasis: cytology focuses on details of cellular structure, whereas histology considers broad patterns of cellular organization. This chapter will explore those patterns. We will discuss the characteristics of each major tissue type, focusing on the relationship between cellular organization and tissue function. Later chapters will consider the patterns of tissue interaction in various organs and systems in greater detail.

■ Epithelia

An **epithelium** (e-pi-THĒ-lē-um) is a layer of cells that forms a barrier with specific properties. Although firmly attached to underlying connective tissues, an epithelium always has a free surface exposed to the environment or to some internal chamber or passageway. Epithelia consist mainly of cells rather than extracellular materials, and there are few extracellular materials between adjacent epithelial cells. There are no blood vessels in epithelia. Because of this **avascular** (ā-VAS-kū-lar; *a-*, without + *vas*, vessel) condition the epithelial cells must obtain nutrients from deeper tissues or from their exposed surfaces.

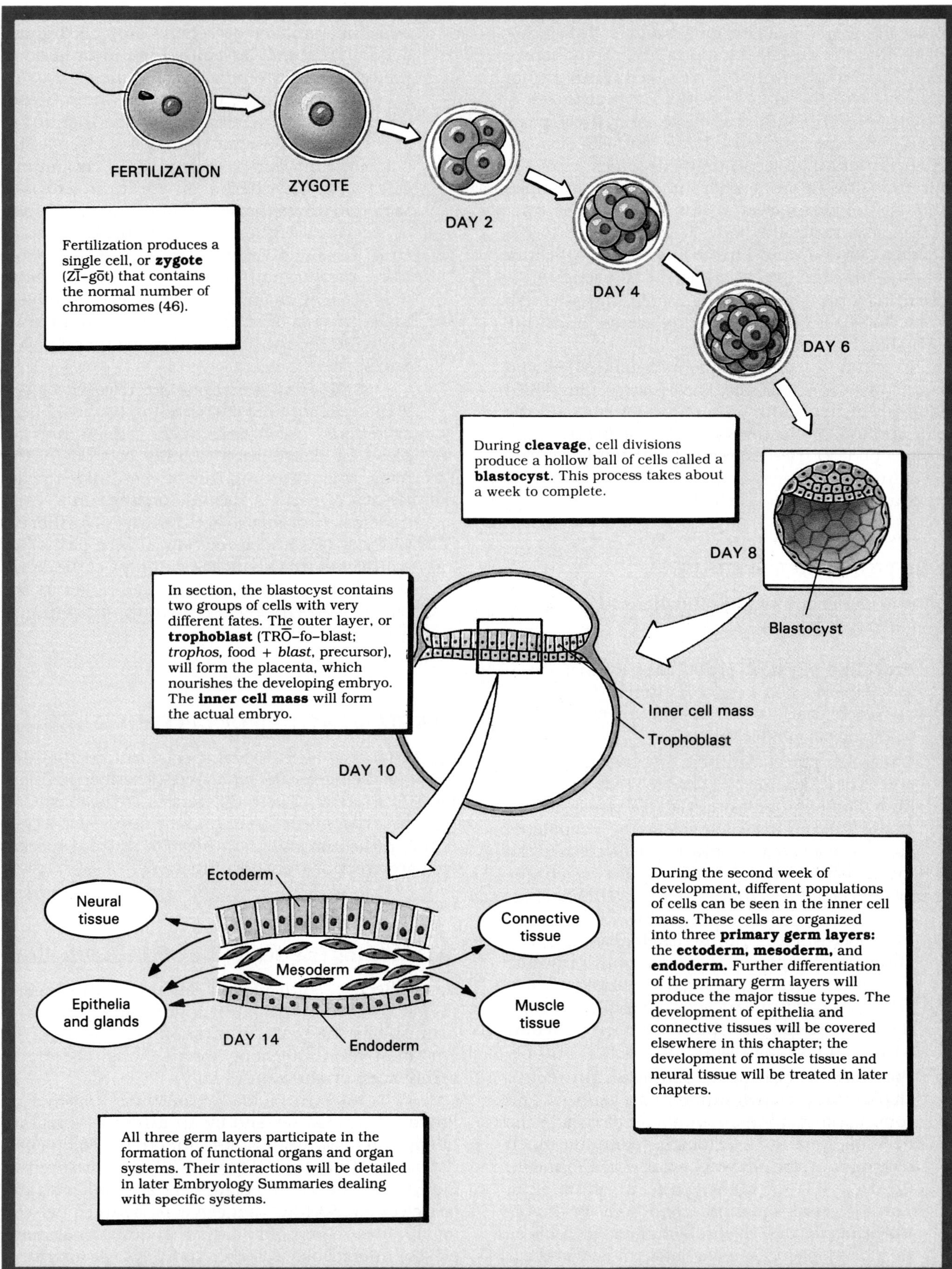

FERTILIZATION
ZYGOTE
DAY 2
DAY 4
DAY 6
Fertilization produces a single cell, or **zygote** (ZĪ-gōt) that contains the normal number of chromosomes (46).
During **cleavage**, cell divisions produce a hollow ball of cells called a **blastocyst**. This process takes about a week to complete.
DAY 8
Blastocyst
In section, the blastocyst contains two groups of cells with very different fates. The outer layer, or **trophoblast** (TRŌ-fo-blast; *trophos*, food + *blast*, precursor), will form the placenta, which nourishes the developing embryo. The **inner cell mass** will form the actual embryo.
Inner cell mass
Trophoblast
DAY 10
Ectoderm
Neural tissue
Connective tissue
Mesoderm
Epithelia and glands
Muscle tissue
DAY 14
Endoderm
During the second week of development, different populations of cells can be seen in the inner cell mass. These cells are organized into three **primary germ layers:** the **ectoderm, mesoderm,** and **endoderm.** Further differentiation of the primary germ layers will produce the major tissue types. The development of epithelia and connective tissues will be covered elsewhere in this chapter; the development of muscle tissue and neural tissue will be treated in later chapters.
All three germ layers participate in the formation of functional organs and organ systems. Their interactions will be detailed in later Embryology Summaries dealing with specific systems.

Embryology Summary: Formation of Tissues

Epithelia cover every exposed body surface. The surface of the skin is an obvious example, but epithelia also line the digestive, respiratory, reproductive, and urinary tracts—internal passageways that communicate with the outside world. Each passageway is lined by an epithelium that has a distinctive appearance and unique functions. These epithelia form selective barriers that separate the deep tissues of the body from the external environment. For example, although the passageway within the digestive tract is physically inside the body, it is connected to the outside world at the mouth and anus. The epithelium that lines the digestive tract absorbs nutrients while preventing unprocessed foods, acids, digestive enzymes, and bacteria from gaining access to delicate internal organs.

In addition, epithelia line internal cavities and passageways, such as the chest cavity, fluid-filled chambers in the brain, eye, and inner ear, and the inner surfaces of blood vessels and the heart. These epithelia prevent friction, regulate the fluid composition of internal cavities, and restrict communication between the blood and tissue fluids.

FUNCTIONS OF EPITHELIA

Epithelia perform essential functions that can be summarized as follows:

1. **Providing physical protection:** Epithelia protect exposed and internal surfaces from abrasion, dehydration, and destruction by chemical or biological agents.
2. **Controlling permeability:** Any substance that enters or leaves the body has to cross an epithelium. Some epithelia are relatively impermeable, whereas others are easily crossed by compounds as large as proteins. Many epithelia contain the molecular "machinery" needed for active transport or facilitated diffusion. The epithelial barrier can be regulated and modified in response to various stimuli. Hormones, for example, can affect the transport of ions and nutrients through epithelial cells. Even physical stress can alter the structure and properties of epithelia—think of the calluses that form on your hands when you do rough work for a period of time.
3. **Providing sensations:** Epithelia can detect changes in the environment they confront and convey information about such changes to the nervous system. For example, pain and touch receptors of the nervous system are found in the deepest layers of the epithelium of the skin. These receptors monitor conditions in the epithelium, alerting the individual to problems as they develop. In the nose, mouth, eye, and ear the sensory receptors are specialized epithelial cells. These cells, which are in direct contact with the nervous system, provide the sensations of smell, taste, sight, and sound.
4. **Producing specialized secretions:** Epithelial cells that produce secretions are called **gland cells**. In a **glandular epithelium** most or all of the cells actively produce secretions.

   Exocrine (*exo-*, outside + *krinein*, to secrete) secretions are discharged onto the surface of the skin or onto an epithelial surface lining one of the internal passageways that communicates with the exterior. In effect, an exocrine secretion leaves the body. There are many kinds of exocrine secretions, and they perform a variety of functions. Enzymes entering the digestive tract, perspiration on the skin, and the milk produced by mammary glands are examples. Exocrine cells often form pockets that are connected to the epithelial surface by tubes, called **ducts.**

   Endocrine secretions are released by the gland cells into the surrounding tissues. These secretions, called **hormones**, diffuse into the blood for distribution to other portions of the body, where they regulate or coordinate the activities of various tissues, organs, and organ systems. (Hormones are discussed further in Chapter 18.) Endocrine cells may be part of an epithelial surface, such as the lining of the digestive tract, or they may be separate, as in the pancreas, thyroid gland, thymus, and pituitary gland.

SPECIALIZATIONS OF EPITHELIAL CELLS

Epithelial cells have several specializations that distinguish them from the typical cell described in Chapter 3. ∞ (p. 61) Those differences are primarily related to (1) the need to maintain the physical integrity of the epithelium and (2) the need to perform specialized secretory or transport functions.

Maintaining the Structure of the Epithelium

To be effective, epithelial cells must remain attached to one another. If the epithelium is damaged or the connections are broken, infection can easily occur. For this reason infections are a serious risk after a severe burn or abrasion.

Cells in epithelia are normally tied together by junctional complexes and by an extensive infolding of opposing cell membranes. Junctional complexes, detailed in Chapter 3 (Figure 3-29), are reviewed in Figure 5-2. ∞ (p. 90) Figure 5-2 also indicates the degree of interlocking that can exist between two cell membranes. The combination of junctional complexes, intercellular cement, and physical interlocking gives the epithelium strength and stability. The extensive connections between cells hold them together and deny access to chemicals or pathogens that may cover their free surfaces.

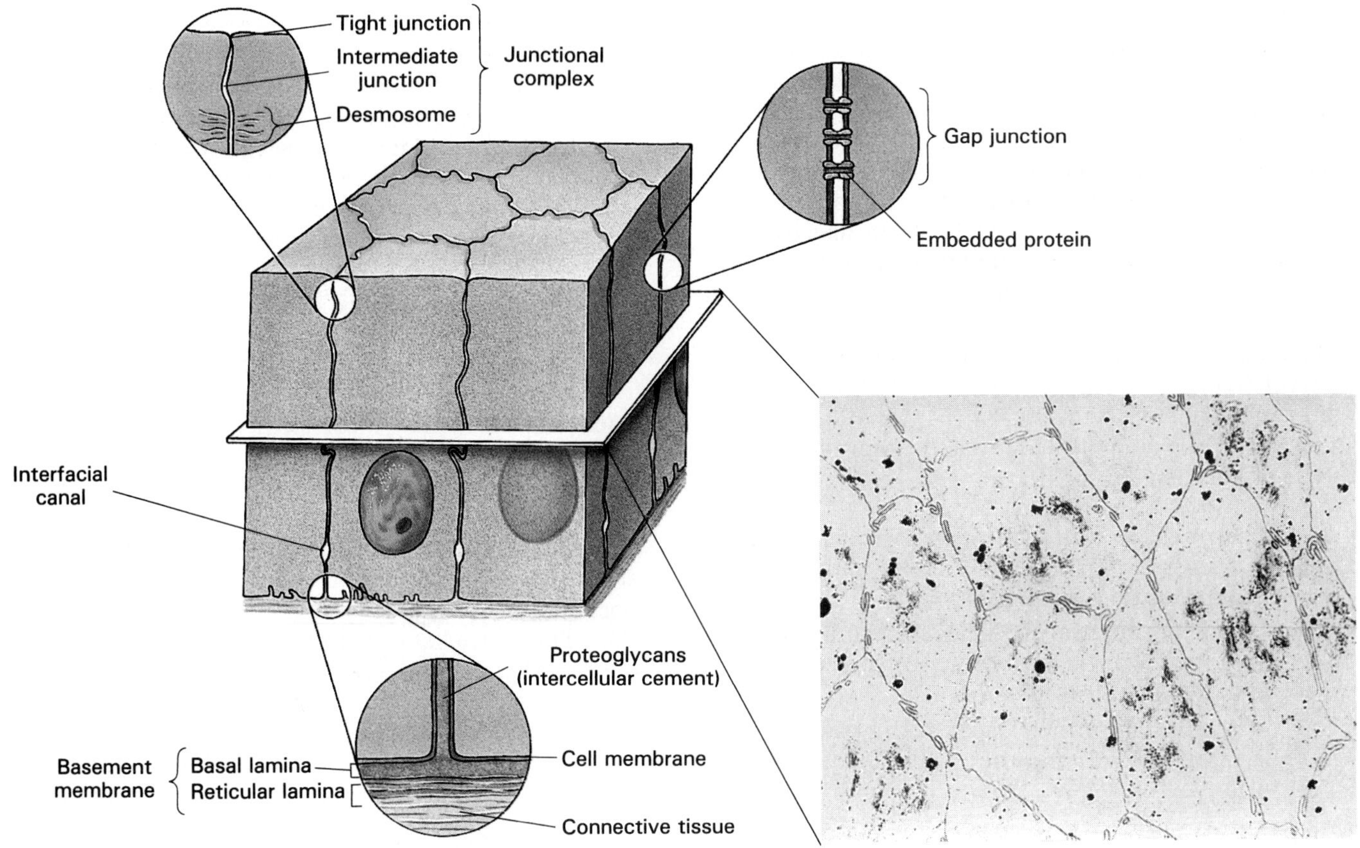

FIGURE 5-2
Organization of Epithelia. The relative positions of epithelial cells are maintained through extensive junctional complexes and intercellular cement. In addition, adjacent cell membranes are often interlocked. The TEM, magnified 2600 times, indicates the degree of such interlocking between columnar epithelial cells. At their inner surfaces epithelia are attached to a basement membrane that forms the boundary between the epithelial cells and the underlying connective tissue.

The tight interconnection of epithelial cells has one noteworthy disadvantage: the interconnections that tie cells together also slow or stop diffusion between cells. This is a problem because epithelia are avascular and therefore depend on diffusion from underlying tissues for nutrient supply and waste removal. Consequently, when an epithelium contains many cell layers (as in the skin), tight junctions often create a network of **interfacial canals** that radiate between the cells and distribute essential nutrients through the layers.

Epithelial cells must not only hold onto one another, they must remain firmly connected to the rest of the body. The inner surface of each epithelium is attached to a special two-part **basement membrane**. The layer closest to the epithelium, called the **basal lamina** (LA-mi-na; *lamina*, thin layer), contains glycoproteins and a network of fine protein filaments. The basal lamina provides a barrier that restricts the movement of proteins and other large molecules from the underlying connective tissue into the epithelium. The deeper portion of the basement membrane, the **reticular lamina**, contains bundles of coarse protein fibers produced by connective tissue cells. The reticular lamina gives the basement membrane its strength. Attachments between the fibers of the basal lamina and those of the reticular lamina hold the two together.

These specializations stabilize the position of the epithelium. In addition, the epithelium must continually repair and renew itself. Epithelial cells lead hard lives, for they may be exposed to disruptive enzymes, toxic chemicals, pathogenic bacteria, or mechanical abrasion. A functional epithelial cell may survive for just a day or two before it is lost or destroyed. The only way the epithelium can maintain its structure over time is through the continual division of stem cells. These stem cells, also known as **germinative cells**, are found in the deepest layers of the epithelium, close to the basement membrane. The role of stem cells in maintaining cell populations was discussed in Chapter 3. (p. 92)

Performing Secretory and Transport Functions

Many epithelial cells are specialized for (1) the production of exocrine or endocrine secretions, (2) the movement of fluids over the epithelial surface, or (3) the movement of fluids through the epithelium itself. These specialized epithelial cells usually show a definite **polarity** along the axis that extends from the basement membrane to the exposed surface of the epithelium. In other words, along this axis the organelles are distributed unevenly. The actual arrangement varies depending on the functional activities of the individual cells. Epithelial polarity also involves a number of more subtle regional differences in cellular organization. For example, there are variations in the distribution of enzymes in the cytoplasm, and the distribution of membrane receptors and carriers on the free surface is different from their distribution on the basal surface of the cell.

The cells shown in Figure 5-3a show a common type of polarity. Notice that:

1. The nucleus is not in the center of the cell; it is closer to the basement membrane than to the free surface.
2. The Golgi apparatus is near the nucleus, oriented toward the surface where secretory products will be released.
3. The cell membrane closest to the basal lamina is highly folded, and there are many mitochondria in this region. The mitochondria provide ATP used to transport materials across the cell membrane.
4. The cell is firmly attached to adjacent cells by junctional complexes. Tight junctions near the free surface block the passage of dissolved materials between the cells. In many epithelia, gap junctions permit extensive chemical communication between adjacent cells.
5. The free surface has microvilli or cilia. Microvilli increase the surface area available for absorption, and cilia move fluids and dissolved substances over the surface of the cell.

Microvilli and Cilia Most epithelial cells have microvilli on their exposed surfaces; there may be just a few, or the entire surface may be carpeted by them. Microvilli, detailed in Chapter 3 (see Figure 3-5), are especially abundant on epithelial surfaces where absorption and secretion take place, such as along portions of the digestive and urinary tracts. ∞ (p. 65) The epithelial cells in these locations are transport specialists, and a cell with microvilli has at least twenty times the surface area of a cell without them.

Typical microvilli are visible in Figure 5-3b. Each microvillus contains a central core of microfilaments that are woven into the microfilament network

FIGURE 5-3
Polarity of Epithelial Cells. (a) Many epithelial cells differ in internal organization along an axis between the free surface and the basement membrane. The free surface frequently bears microvilli; less often, this surface may have cilia or (very rarely) stereocilia. (All three would not normally be found on the same group of cells but are depicted here for purposes of illustration.) Junctional complexes prevent movement of pathogens or diffusion of dissolved materials between the cells. Folds of membrane near the base of the cell increase the surface area exposed to the basement membrane. Mitochondria are typically concentrated in this region, probably to provide energy for the cell's transport activities. (b) An SEM showing the surface of a ciliated epithelium that lines most of the respiratory tract. The small, bristly areas are microvilli found on the exposed surfaces of mucus-producing cells that are scattered among the ciliated epithelial cells. (SEM, × 4284)

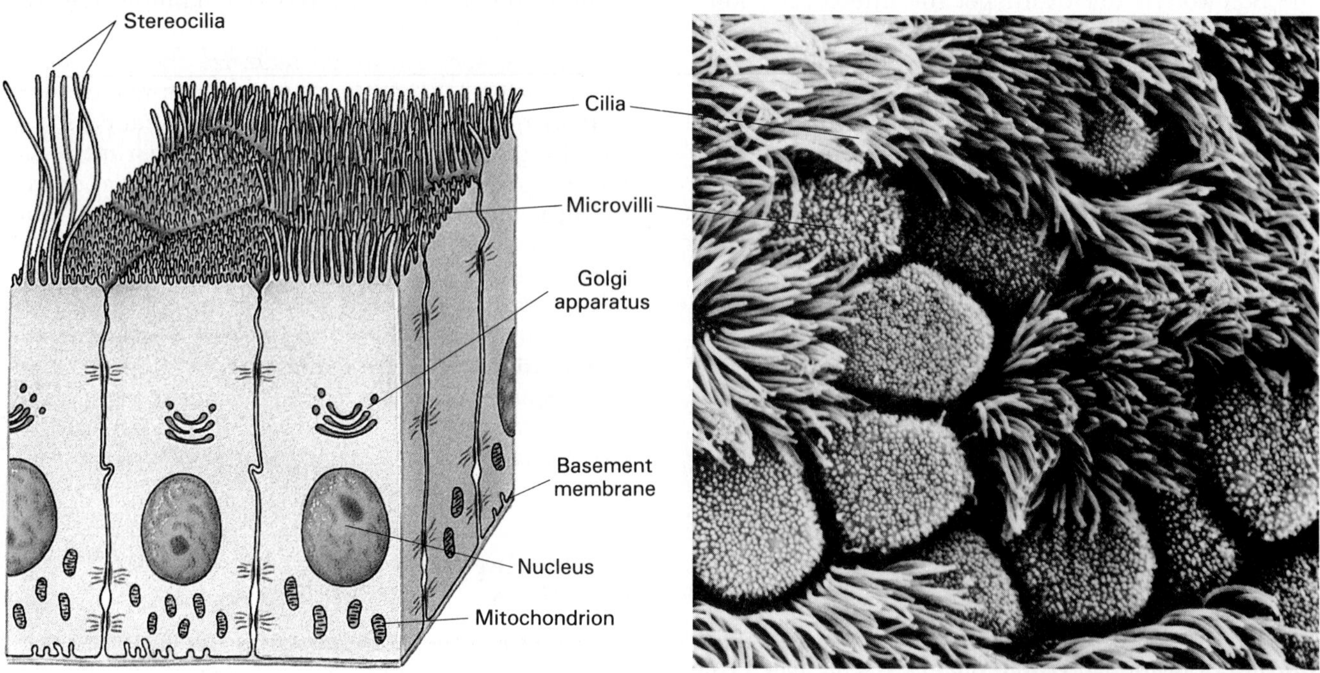

just beneath the cell membrane. Interactions between the microfilaments and other filaments of the cytoskeleton make the microvilli bend or wave. **Stereocilia** are very long microvilli (up to 250 μm) that are incapable of movement. Stereocilia are found only along portions of the male reproductive tract, and they will be discussed further in Chapter 28.

The structure and function of an individual cilium were considered in Chapter 3. (p. 67) Figure 5-3b shows the exposed surface of a **ciliated epithelium**. A typical ciliated cell contains about 250 cilia that beat in a coordinated fashion. Although the biochemical basis for ciliary movement remains a mystery, the process requires ATP and is accompanied by a change in the transmembrane potential.

Materials are moved over the epithelial surface by the synchronized beating of cilia. For example, the ciliated epithelium that lines the respiratory tract moves mucus away from the lungs and toward the throat. The mucus traps irritants and pathogens and carries them away from more delicate surfaces deeper in the lungs. Movement of ions through gap junctions provides the stimulus that coordinates this ciliary activity, and the entire epithelial surface behaves as if it were a single, enormous cell membrane. Injury to the cilia or to the epithelial cells in general can stop ciliary movement and thus halt the protective flow of mucus. This is one effect of smoking; other effects will be considered in later sections.

Epithelial Transport Many epithelial cells perform the directional transport activities described in our discussion of pinocytosis in Chapter 3. In capillaries, along portions of the digestive tract, in the kidney, and elsewhere, simple epithelia transport fluids and dissolved materials from their free surfaces to the basement membrane, or vice versa. This movement is called **epithelial transport**. Figure 3-25 diagrammed one example of epithelial transport, the bulk movement of fluids across capillary endothelia. (p. 85) Other examples will be discussed in later chapters.

CLASSIFICATION OF EPITHELIA

Epithelia are classified according to their appearance. Attention focuses on two variable features, the number of cell layers and the shape of the exposed cells. The classification scheme recognizes two types of layering—*simple* and *stratified*—and three cell shapes—*squamous*, *cuboidal*, and *columnar*.

If there is only a single layer of cells covering the basement membrane, the epithelium is termed **simple epithelium**. Simple epithelia are relatively thin, and because all the cells have the same polarity, the nuclei form a rough line above the basement membrane. Because they are so thin, simple epithelia are also relatively fragile. A single layer of cells cannot

FIGURE 5-4
Squamous Epithelia. (a) A superficial view of the simple squamous epithelium (mesothelium) that lines the peritoneal cavity. The three-dimensional drawing shows the epithelium in superficial and sectional view. (b) A sectional view of the stratified squamous epithelium that covers the tongue.

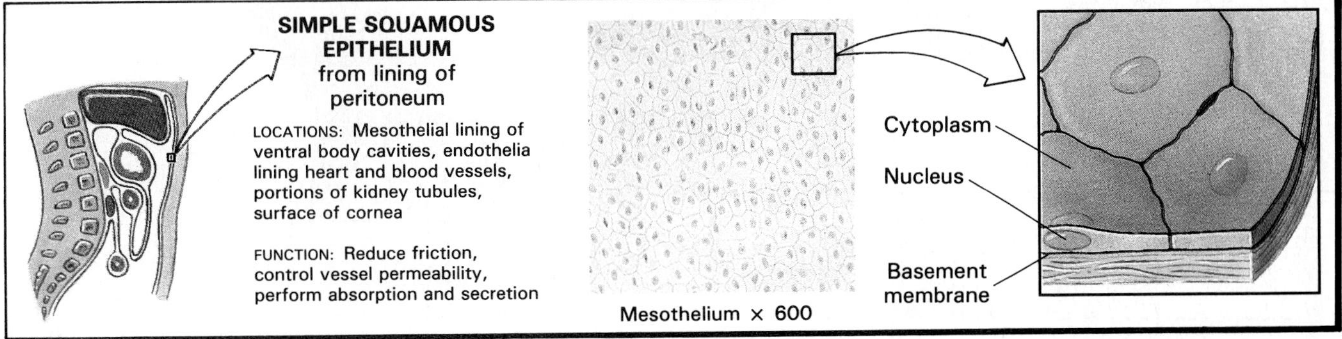
SIMPLE SQUAMOUS EPITHELIUM
from lining of peritoneum
LOCATIONS: Mesothelial lining of ventral body cavities, endothelia lining heart and blood vessels, portions of kidney tubules, surface of cornea
FUNCTION: Reduce friction, control vessel permeability, perform absorption and secretion
Mesothelium × 600
Cytoplasm
Nucleus
Basement membrane

(a)

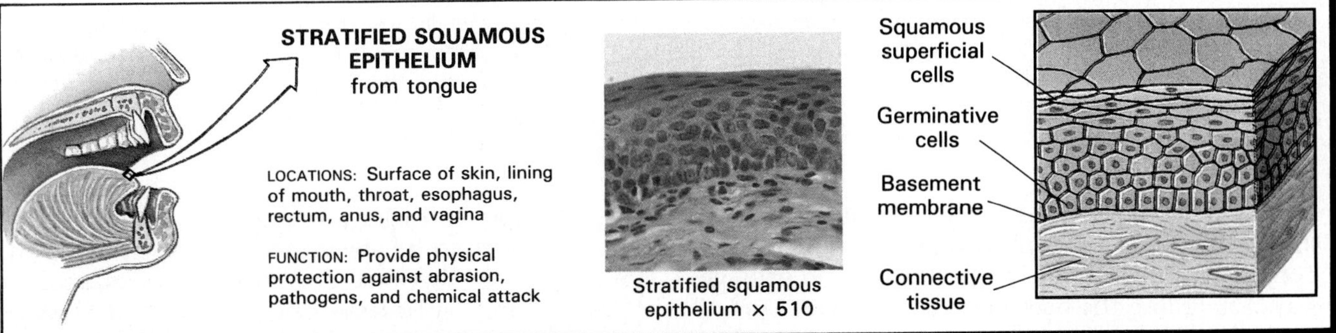
STRATIFIED SQUAMOUS EPITHELIUM
from tongue
LOCATIONS: Surface of skin, lining of mouth, throat, esophagus, rectum, anus, and vagina
FUNCTION: Provide physical protection against abrasion, pathogens, and chemical attack
Stratified squamous epithelium × 510
Squamous superficial cells
Germinative cells
Basement membrane
Connective tissue

(b)

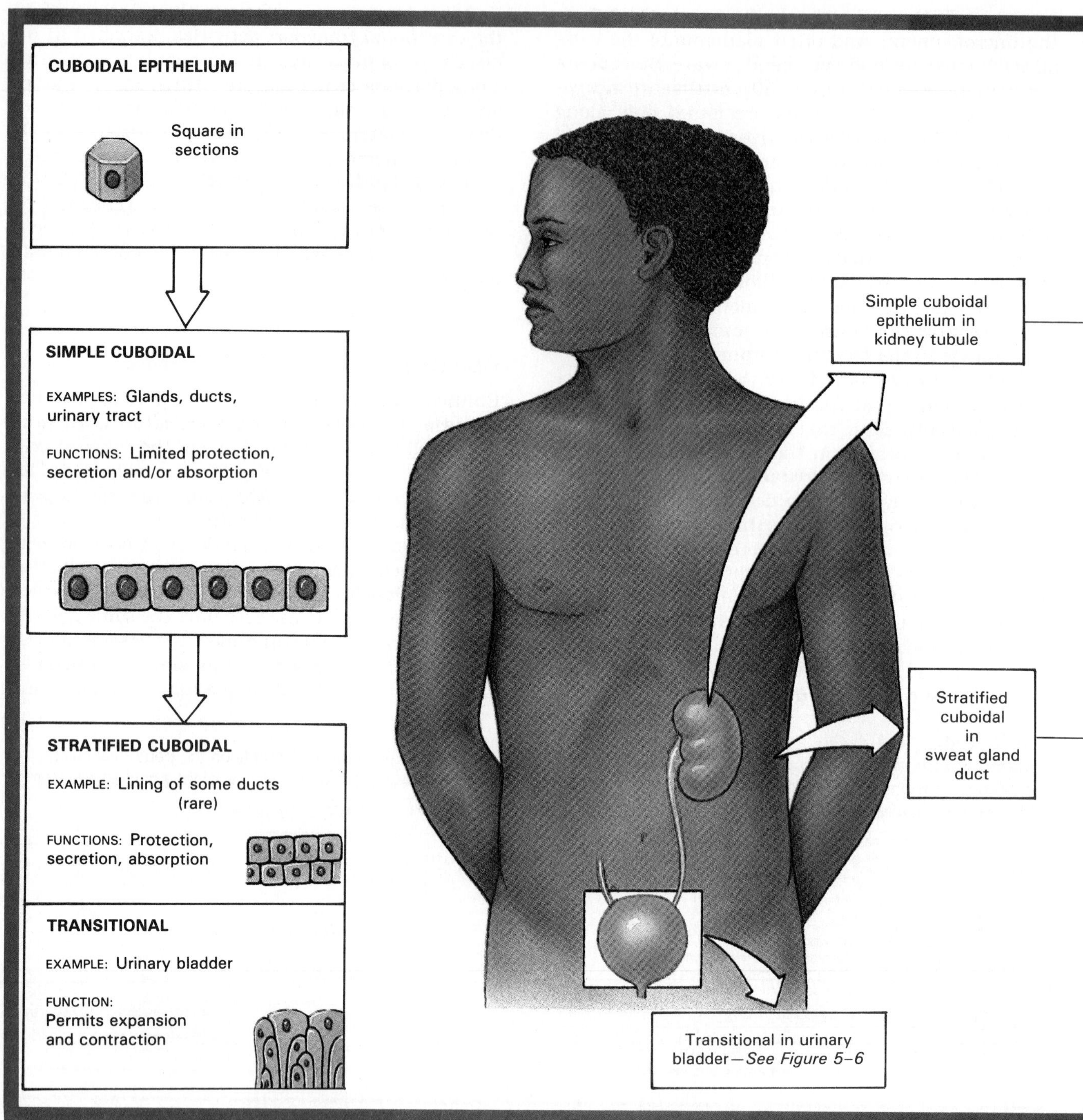

FIGURE 5-5
Cuboidal Epithelia. (a) A section through the cuboidal epithelial cells of a kidney tubule. The diagrammatic view emphasizes structural details that permit the classification of an epithelium as cuboidal. (b) Sectional view of the stratified cuboidal epithelium lining a sweat gland duct in the skin.

provide much mechanical protection, and simple epithelia are found only in protected areas inside the body. They line internal compartments and passageways, including the body cavities, the heart, and all blood vessels.

Simple epithelia are characteristic of regions where secretion or absorption occurs, such as the lining of the digestive and urinary tracts and the gas-exchange surfaces of the lungs. In such places the thinness of simple epithelia is an advantage, for

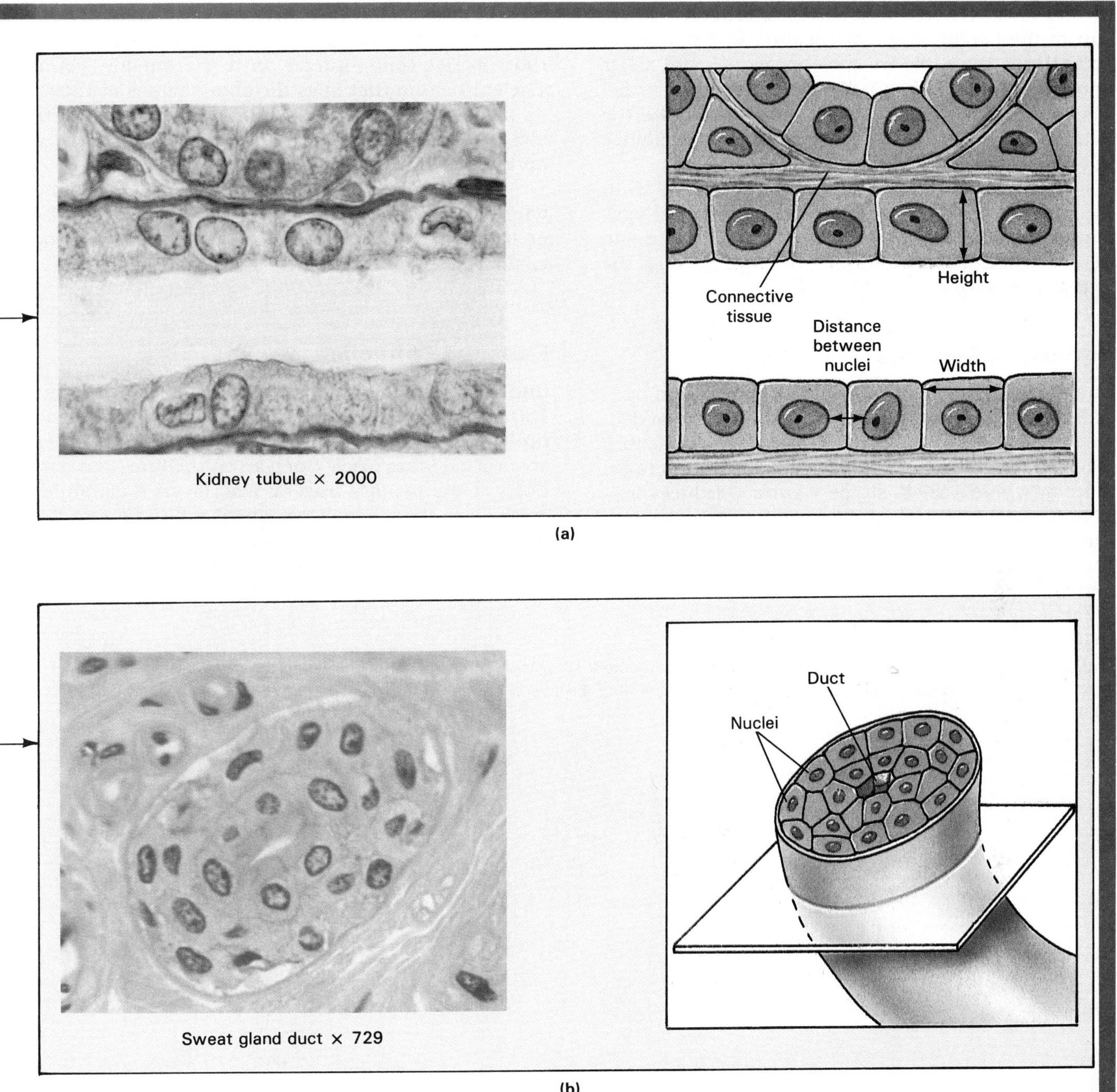

it speeds the passage of materials through or across the epithelial barrier. Because no blood vessels penetrate epithelia, a thick layer of epithelial cells would slow the exchange of materials between the bloodstream and the digestive, urinary, or respiratory tract.

A **stratified epithelium** has several layers of cells above the basement membrane. Stratified epithelia are usually found in areas subject to mechanical or chemical stresses, such as the surface of the skin and the lining of the mouth.

In sectional view, the cells at the surface of the epithelium usually have one of the three basic shapes. In a **squamous epithelium** (SKWĀ-mus; *squama*, plate or scale) the cells are thin and flat. In a sectional view the nucleus occupies the thickest portion of each cell; from the surface, the cells look like fried eggs laid side by side. The cells of a **cuboidal epithelium** resemble little hexagonal boxes; they appear square in typical sectional views. The nuclei are near the center of each cell. They form a neat row, and the distance between adjacent nuclei is roughly equivalent to the height of the epithelium. In a **columnar epithelium** the cells are also hexagonal, but taller and more slender. The nuclei are crowded into a narrow band close to the basement membrane, and the height of the epithelium is several times the distance between two nuclei.

Combining the two basic epithelial layouts (simple and stratified) and the three possible cell shapes (squamous, cuboidal, and columnar) enables one to describe almost every epithelium in the body, as the following examples will show.

Squamous Epithelia

Squamous epithelia are shown in Figure 5-4 on page 129. A **simple squamous epithelium** is the most delicate epithelium in the body. This type of epithelium is found in protected regions where absorption takes place or where a slick, slippery surface reduces friction. Examples include portions of the urinary tract, the respiratory surfaces of the lungs, the lining of body cavities, and the inner surfaces of the circulatory system.

Special names have been given to simple squamous epithelia that line chambers and passageways that do not communicate with the outside world. The epithelium that lines the body cavities is known as a **mesothelium** (mez-ō-THĒ-lē-um; *mesos*, middle). The lining of the heart and blood vessels is called an **endothelium** (en-dō-THĒ-lē-um).

A **stratified squamous epithelium** is found where mechanical stresses are severe. The surface of the skin and the lining of the mouth are good examples.

Cuboidal Epithelia

Cuboidal epithelia are shown in Figure 5-5 on page 130. A **simple cuboidal epithelium** provides limited protection and occurs in regions where secretion or absorption takes place. Such an epithelium lines portions of the urinary tract. It also forms a glandular epithelium that secretes enzymes and buffers in the pancreas and salivary glands, and it lines the passages that transport these exocrine secretions. One endocrine structure, the thyroid gland, contains

FIGURE 5-6
Transitional Epithelia. (a) The lining of the empty urinary bladder, showing transitional epithelium in the contracted state. (LM, × 413) (b) The lining of the full bladder, showing the effects of stretching on the arrangement of cells in the epithelium. (LM, × 486)

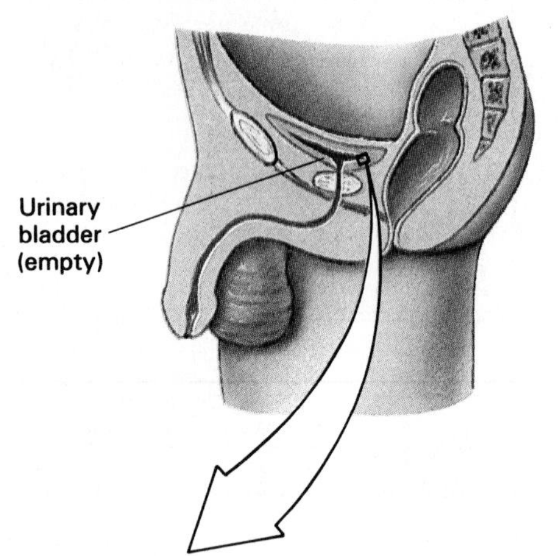

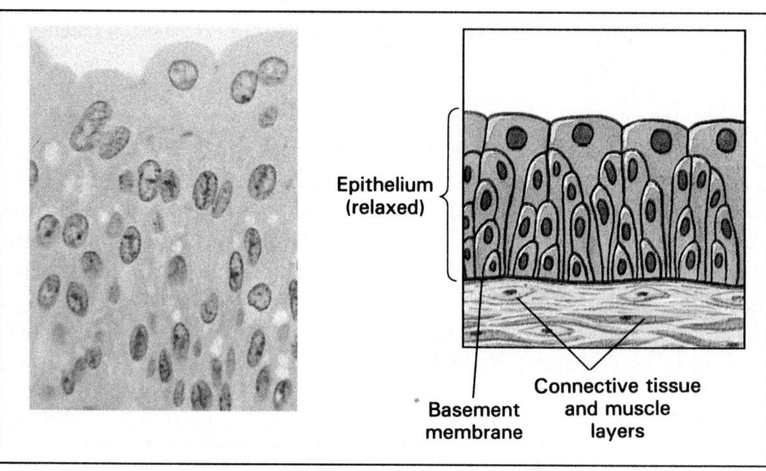

(a)

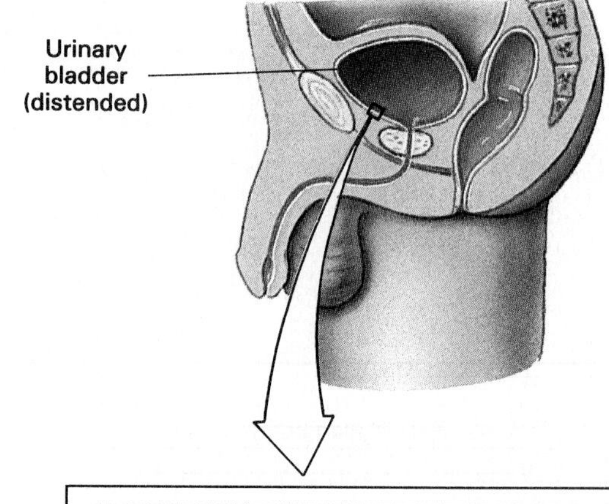

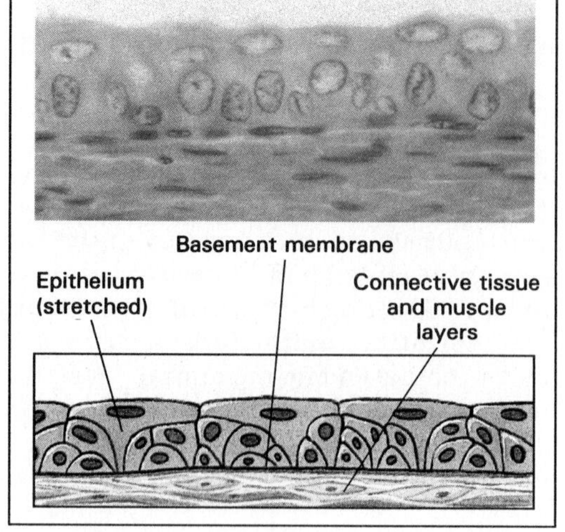

(b)

chambers lined by a cuboidal secretory epithelium that secretes the hormone *thyroxine*. Simple cuboidal epithelia also line kidney tubules involved with the production of urine (Figure 5-5a).

True **stratified cuboidal epithelia** are relatively rare; they are found only along the ducts that drain sweat glands (Figure 5-5b). A **transitional epithelium**, seen in Figure 5-6, resembles a modified form of stratified cuboidal epithelium. This epithelium lines the urinary bladder and other portions of the

FIGURE 5-7
Columnar Epithelia. (a) Micrograph showing the characteristics of simple columnar epithelium. In the diagrammatic sketch, note the relationships between the height and width of each cell; the relative size, shape, and location of nuclei, and the distance between adjacent nuclei. Contrast these observations with the corresponding characteristics of simple cuboidal epithelia shown in Figure 5-5a. (b) The pseudostratified, ciliated, columnar epithelium of the respiratory tract. Note the uneven layering of the nuclei. (c) A stratified columnar epithelium is sometimes found along large ducts, such as this salivary gland duct. Note the overall height of the epithelium and the location and orientation of the nuclei; compare with Figures 5-4b and 5-5b.

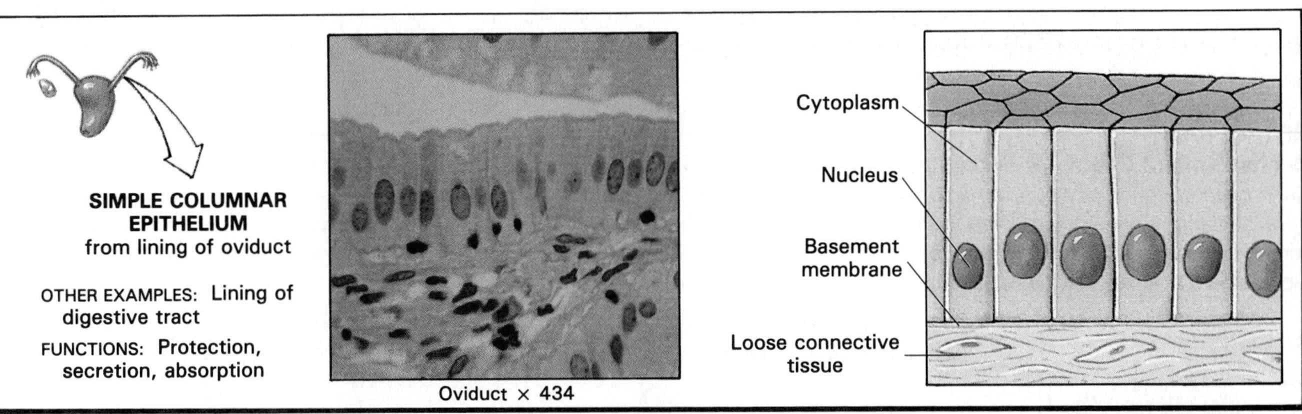

(a)

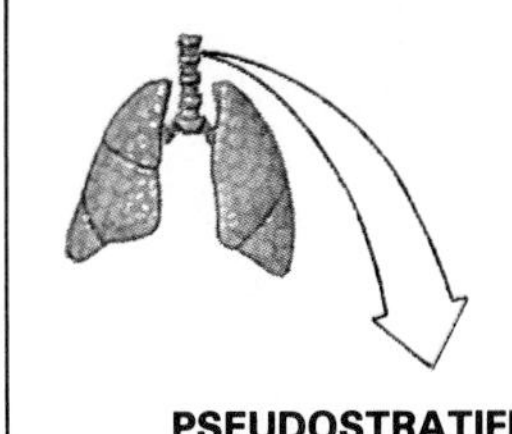

PSEUDOSTRATIFIED COLUMNAR EPITHELIUM
from trachea

FUNCTIONS: Protection, secretion

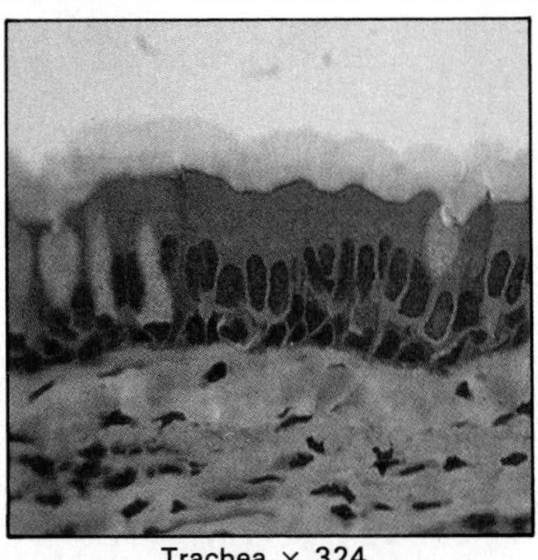

Trachea × 324

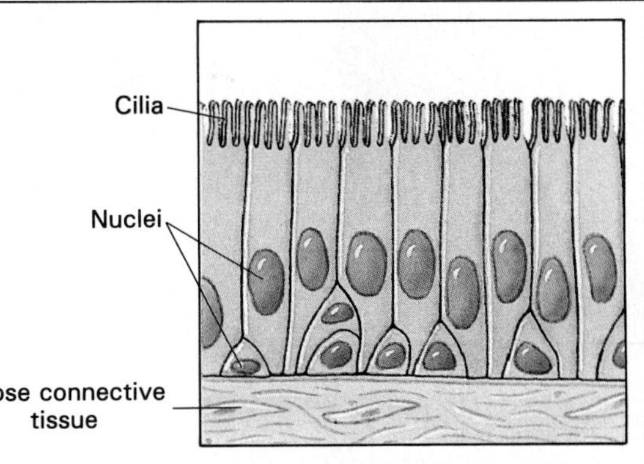

(b)

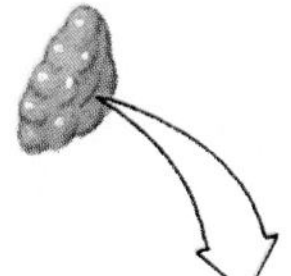

STRATIFIED COLUMNAR EPITHELIUM
from salivary gland duct

OTHER EXAMPLES: Large passageways of respiratory tract

FUNCTION: Protection

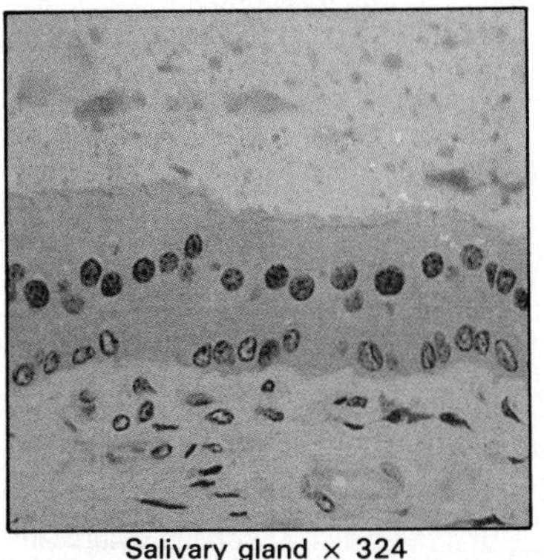

Salivary gland × 324

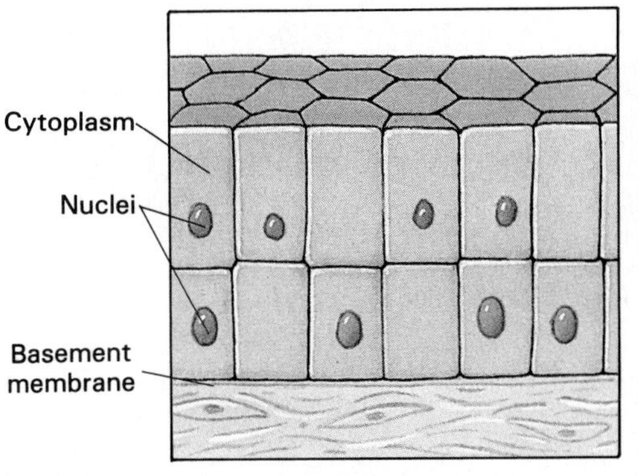

(c)

urinary tract, where great changes in volume occur. In an empty bladder (Figure 5-6a) the epithelium seems to have many layers, and the outermost cells are rounded or cuboidal. The layered appearance results from overcrowding; the actual structure of the epithelium can be seen in the full bladder, when the pressure of the urine has stretched the lining to its natural thickness (Figure 5-6b).

Columnar Epithelia

Columnar epithelia can be seen in Figure 5-7. A **simple columnar epithelium** (Figure 5-7a) provides some protection and may also be encountered in areas where absorption or secretion occurs. This type of epithelium lines most of the digestive tract and many excretory ducts. It protects underlying tissues, and glandular cells intermingled with the other epithelial cells may secrete mucus that lubricates the free surface.

Portions of the respiratory tract contain a columnar epithelium that includes a mixture of cell types. Because their nuclei are situated at varying distances from the surface, the epithelium has a layered appearance. Yet it is not a stratified epithelium because all of the cells contact the basement membrane. As a result, it is known as a **pseudostratified columnar epithelium** (Figure 5-7b). This epithelium lines the trachea, or windpipe, and portions of the male reproductive tract.

Stratified columnar epithelia are relatively rare, providing protection along portions of the respiratory, digestive, and reproductive tracts, as well as along a few large excretory ducts (Figure 5-7c).

FIGURE 5-8
Mechanisms of Glandular Secretion. (a) In merocrine secretion, secretory vesicles are discharged at the surface of the gland cell through exocytosis. (b) Apocrine secretion involves the loss of cytoplasm. Inclusions, secretory vesicles, and other cytoplasmic components are shed in the process. The gland cell then undergoes a period of growth and repair before releasing additional secretions. (c) Holocrine secretion occurs as superficial gland cells break apart. Continued secretion involves the replacement of these cells through the mitotic divisions of underlying stem cells.

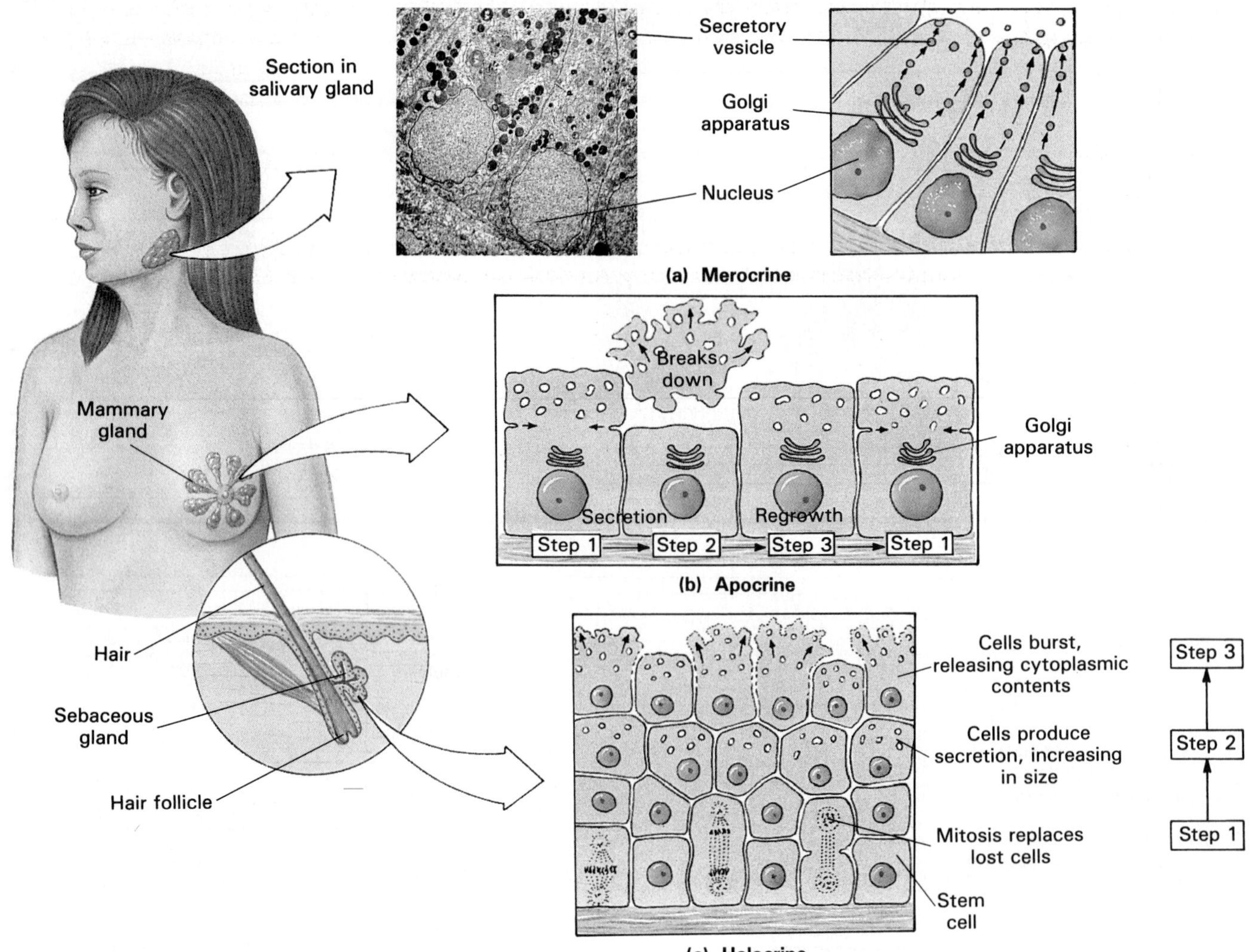

GLANDULAR EPITHELIA

Many epithelia contain gland cells that produce exocrine or endocrine secretions. These cells vary in both how they perform their function and how they are organized.

Methods of Secretion

A glandular epithelial cell may use one of three methods to release its secretions: *merocrine secretion, apocrine secretion,* or *holocrine secretion.*

In **merocrine secretion** (MER-o-krin; *meros,* part + *krinein,* to separate) the product is released through exocytosis, as described in Chapter 3. (p. 74) This method, diagrammed in Figure 5-8a, is the most common mode of secretion. **Apocrine secretion** (A-po-krin; *apo-,* off) involves the loss of cytoplasm as well as the secretory product (Figure 5-8b). The outermost portion of the cytoplasm becomes packed with secretory vesicles before it is shed. Milk production in the breasts and underarm perspiration occur through apocrine secretion.

Merocrine and apocrine secretions leave the cell intact and able to continue secreting. **Holocrine secretion** (HO-lo-krin; *holos,* entire) does not (Figure 5-8c). During holocrine secretion the entire cell becomes packed with secretions and then bursts apart. Thus the product is released but the cell is destroyed. Further secretion depends on gland cells being replaced by the division of stem cells. Sebaceous glands, associated with hair follicles, produce a waxy hair coating by means of holocrine secretion.

Types of Glands

In epithelia that contain scattered gland cells, the individual secretory cells are called **unicellular glands**. **Multicellular glands** are organs containing glandular epithelia that produce exocrine or endocrine secretions.

As noted previously, exocrine glands discharge their products onto some internal or external surface, whereas endocrine glands secrete hormones into the blood or tissue fluids. There are several types of exocrine glands. **Serous glands** secrete a watery solution containing enzymes; **mucous glands** secrete a viscous mucus. **Mixed glands** contain more than one type of gland cell and may produce two different exocrine secretions, one serous and the other mucous. The *submandibular gland,* one of the salivary glands, is an example of a mixed exocrine gland. Some mixed glands even produce both exocrine and endocrine secretions. The *pancreas,* for example, secretes several hormones in addition to digestive enzymes.

As examples of exocrine glands, we will examine three very different glands, one unicellular and two multicellular. (The structure of endocrine glands is considered in Chapter 18.)

Unicellular Exocrine Glands **Goblet cells**, shown in Figure 5-9, are the only example of **unicellular exocrine glands** in the body. Goblet cells are often scattered among other epithelial cells, as along the digestive and respiratory tracts. These gland cells produce a sticky solution containing glycoproteins and proteoglycans, and release it through merocrine secretion. Their secretion, simply called **mucus**, is an effective lubricant and a sticky trap for foreign particles and microorganisms.

Multicellular Exocrine Glands **Multicellular exocrine glands** contain secretory epithelia. In a **secretory sheet** (Figure 5-10a) the glandular cells are exposed to an inner compartment. The mucus-secreting

FIGURE 5-9
Goblet Cells. Goblet cells are unicellular exocrine glands. These secretory cells are often scattered among the simple or pseudostratified columnar epithelial cells of the digestive and respiratory tracts. The micrograph shows a goblet cell in the intestinal epithelium (simple columnar type). Figure 5–3b provides a superficial view of the microvilli that carpet goblet cells in the respiratory epithelium.

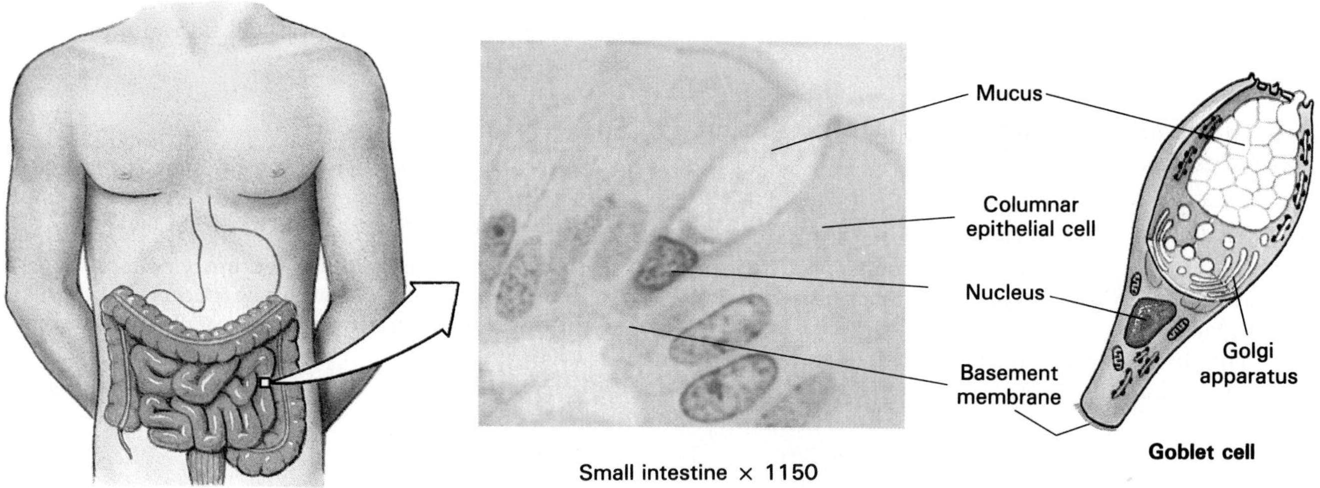

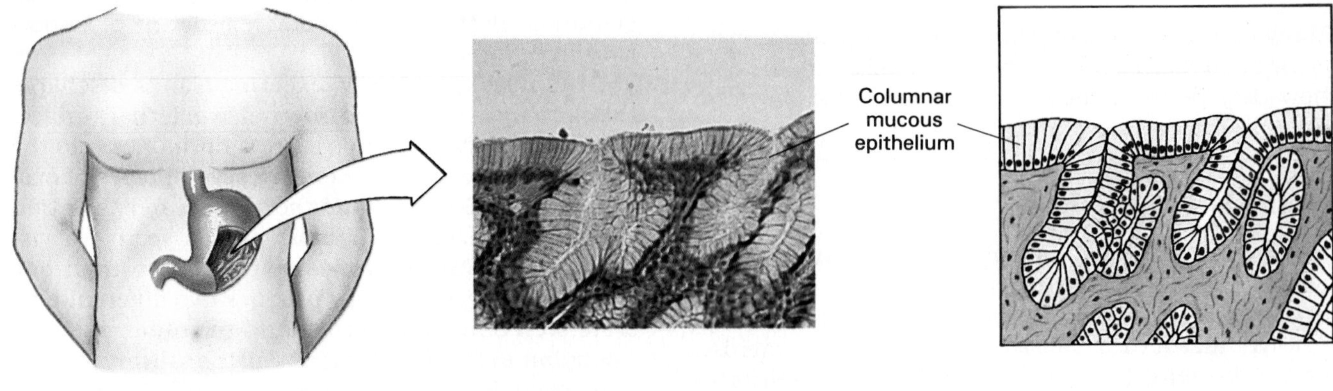

(a) MUCOUS EPITHELIUM

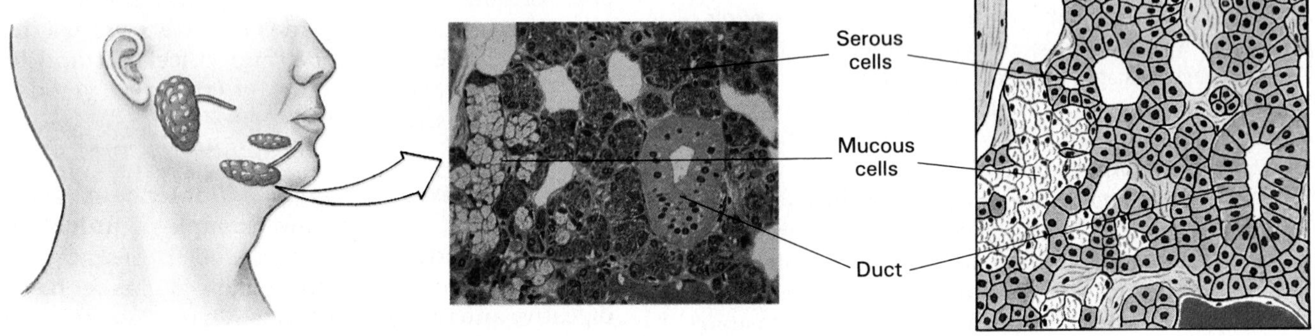

(b) MIXED GLANDULAR EPITHELIUM

FIGURE 5-10
Mucous and Mixed Glandular Epithelia. (a) The interior of the stomach is lined by a mucous epithelium (a secretory sheet) whose secretions protect the walls from acids and enzymes. (The acids and enzymes are produced by glands that discharge their secretions onto the mucous epithelial surface.) (b) The submandibular salivary gland is a mixed gland containing cells that produce both serous and mucous secretions. (LM, × 168)

cells that line the stomach are an example. Their continual secretion protects the stomach from the acids and enzymes it contains. Most other multicellular glands are found in pockets set back from the epithelial surface; their secretory products reach the surface by means of special **excretory ducts**. Figure 5-10b shows an example, a salivary gland that produces mucus and digestive enzymes.

A Structural Classification of Multicellular Exocrine Glands

Many terms are used to describe the physical structure of a complex gland. Two characteristics are used in classifying a gland: (1) the branching pattern of the duct and (2) the shape of the secretory portion of the gland.

The duct is called **simple** if it does not branch, and **compound** if it branches repeatedly. Glands made up of cells arranged in a tube are **tubular**; those made up of cells in a blind pocket are **alveolar** (al-VĒ-ō-lar; *alveolus*, sac) or **acinar** (A-si-nar; *acinus*, chamber). Glands that have a combination of the two arrangements are called **tubuloacinar**. Each glandular area may have its own duct; in the case of **branched** glands several glands share a duct. Specific examples of each gland type will be examined more closely in later chapters.

√ **You look at a tissue under a microscope and see a simple squamous epithelium. Can it be a sample of the skin surface?**

√ **In producing sebum, the secretory cells of sebaceous glands fill with secretions and then rupture, releasing their contents. What kind of secretion is this?**

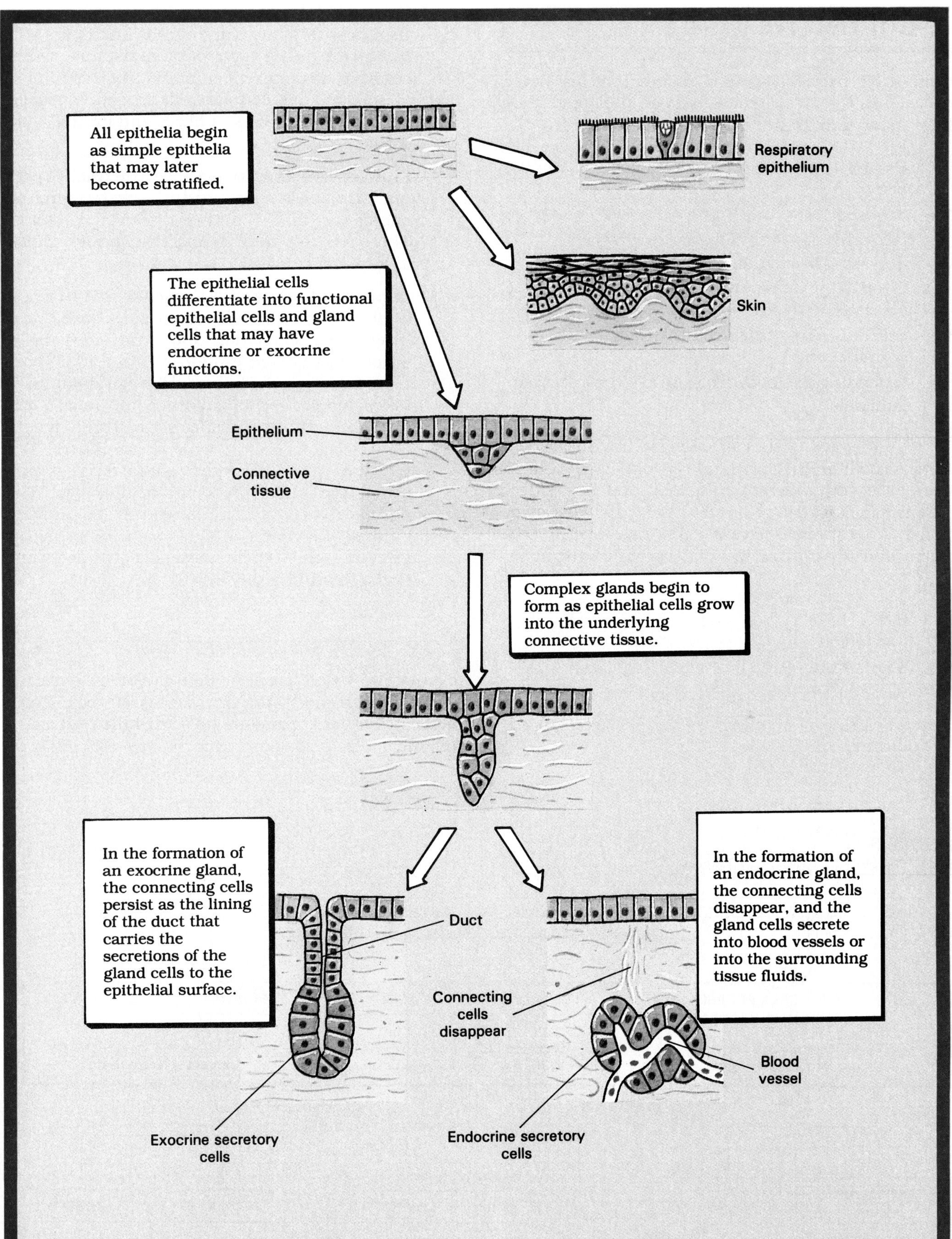

All epithelia begin as simple epithelia that may later become stratified.
Respiratory epithelium
Skin
The epithelial cells differentiate into functional epithelial cells and gland cells that may have endocrine or exocrine functions.
Epithelium
Connective tissue
Complex glands begin to form as epithelial cells grow into the underlying connective tissue.
In the formation of an exocrine gland, the connecting cells persist as the lining of the duct that carries the secretions of the gland cells to the epithelial surface.
Duct
In the formation of an endocrine gland, the connecting cells disappear, and the gland cells secrete into blood vessels or into the surrounding tissue fluids.
Connecting cells disappear
Blood vessel
Exocrine secretory cells
Endocrine secretory cells

Embryology Summary: Development of Epithelia

Connective Tissues

Connective tissues are deep tissues that are never exposed to the environment outside the body. They have many important functions, including:

1. Establishing a structural framework for the body;
2. Transporting fluids and dissolved materials from one region of the body to another;
3. Providing protection for delicate organs;
4. Supporting, surrounding, and interconnecting other tissue types;
5. Storing energy reserves, especially in the form of lipids; and
6. Defending the body from invasion by microorganisms.

Connective tissues are diverse in appearance. Bone, blood, and fat are familiar connective tissues that have very different functions and properties. However, all connective tissues have three basic components: (1) specialized cells, (2) extracellular protein fibers, and (3) a fluid known as the **ground substance**.

CLASSIFICATION OF CONNECTIVE TISSUES

Several classes of connective tissue are recognized, as indicated in Figure 5-11.

- **Connective tissue proper** refers to connective tissues with many types of cells and fibers surrounded by a syrupy ground substance. Other connective tissues differ in terms of the number of cell types they contain and the relative properties and proportions of fibers and ground substance.
- **Fluid connective tissues** have a distinctive population of cells suspended in a watery ground substance that contains dissolved proteins. There are two fluid connective tissues, *blood* and *lymph*.
- **Supporting connective tissues** are of two types, *cartilage* and *bone*. These tissues have a less diverse cell population than connective tissue proper and a dense ground substance that contains closely packed fibers. The combination of fibers and ground substance is known as a **matrix**. The ground substance of cartilage may contain collagen, elastin, or reticular fibers, and the matrix is a gel whose characteristics vary depending on the predominant fiber type. The fibrous matrix of bone is said to be **calcified** because it contains mineral deposits, primarily calcium salts. These minerals give the bone strength and rigidity.

CONNECTIVE TISSUE PROPER

Connective tissue proper contains fibers, a viscous ground substance, and two classes of cells. **Fixed cells** are stationary and are involved with local maintenance, repair, defense, and energy storage. The

FIGURE 5-11
Major Types of Connective Tissue

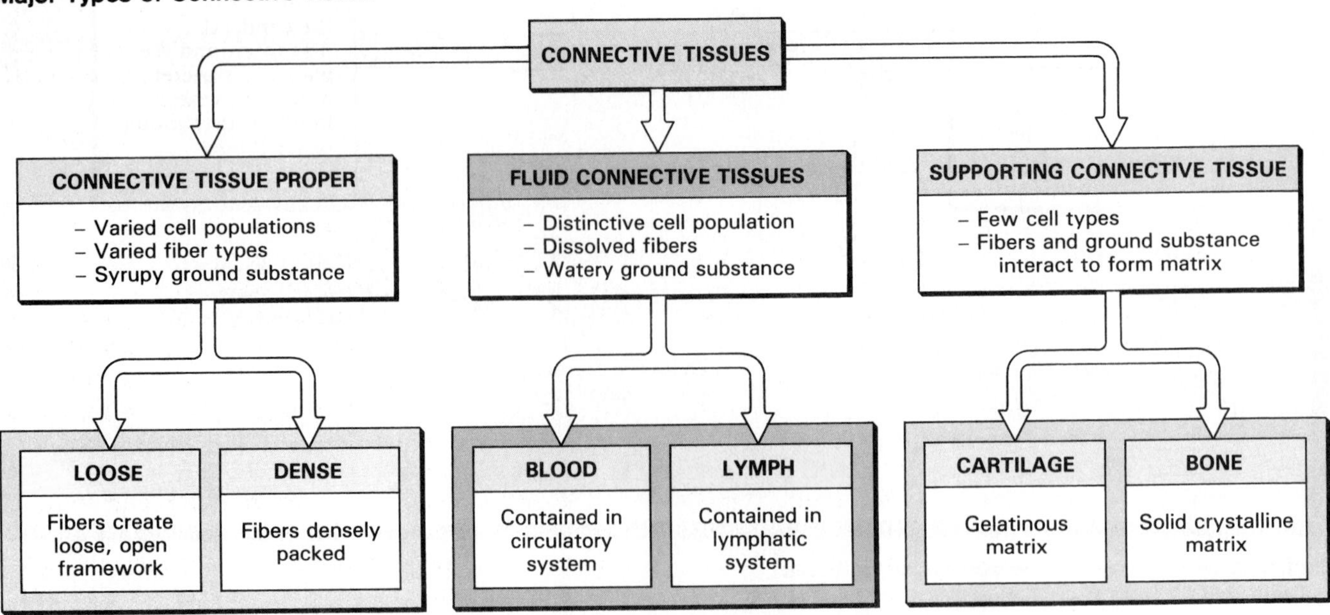

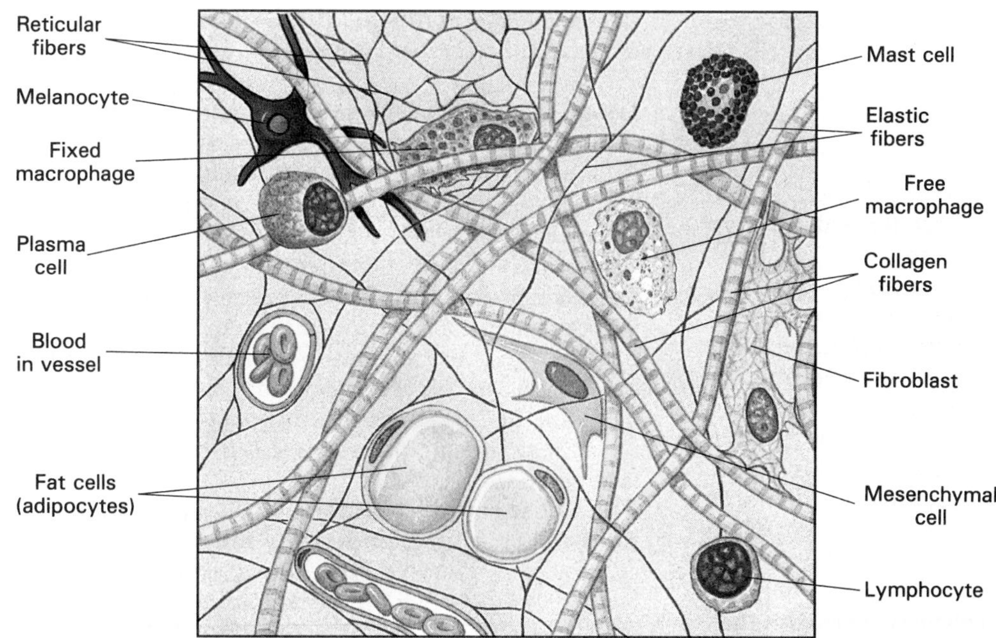

FIGURE 5-12
The Cells and Fibers of Connective Tissue Proper

number of **wandering cells** in an area fluctuates in response to local or regional stimuli. The cells and fibers of connective tissue proper can be seen in Figure 5-12.

Fixed Cells

Fibroblasts (FĪ-brō-blasts) are the most abundant fixed cells in connective tissue proper. These slender or *stellate* (star-shaped) cells are responsible for the production and maintenance of the connective tissue fibers. Each fibroblast manufactures and secretes small protein subunits that interact to form large fibers in the ground substance. In addition, fibroblasts secrete hyaluronic acid, a proteoglycan that gives the ground substance its syrupy consistency.

Fixed macrophages (MAC-rō-fāj; *phagein*, to eat) are scattered among the fibers. These cells phagocytize damaged cells or pathogens that enter the tissue. The engulfed materials are subsequently destroyed by lysosomal enzymes, as detailed in Chapter 3 (Figures 3-16 and 3-27). (pp. 76, 87). Fixed macrophages are very important in policing connective tissues, and when stimulated they release chemicals that mobilize the immune system.

Fat cells are known as **adipose cells**, or simply **adipocytes** (AD-i-po-sīts). A typical adipocyte contains a single, enormous lipid droplet. The nucleus and other organelles are squeezed to one side, making the cell in section resemble a class ring. The number of fat cells varies from one connective tissue to another, from one region of the body to another, and from individual to individual.

Connective tissues may also contain cells that synthesize and store a brown pigment, **melanin** (ME-la-nin); these cells are called **melanocytes** (me-LAN-o-sīts).

Wandering Cells

Free macrophages perform the same function as fixed macrophages, but they are highly mobile. When an infection occurs, macrophages wandering throughout the body are drawn to the affected area. The differences between fixed and free macrophages are not absolute, and during an infection a fixed macrophage can detach and begin roaming around the tissue.

Mast cells are small, mobile connective tissue cells often found near blood vessels. The cytoplasm of a mast cell is packed with granules of **histamine** (HIS-ta-mēn) and **heparin** (HEP-a-rin). These chemicals are released by exocytosis following injury or infection, and they have an immediate effect on local circulation. Histamine dilates blood vessels, thereby increasing the blood flow to the area. Heparin prevents blood clotting that might otherwise oppose the increased flow of blood. In addition to mast cells and free macrophages, **lymphocytes** (LIM-fo-sīts) and **microphages** (special phagocytic cells of the blood) may move through the connective tissue. Their numbers increase markedly if the tissue is damaged, and some of the lymphocytes may then develop into **plasma cells**, the cells responsible for the production of antibodies. As you will recall from Chapter 2, **antibodies** are proteins that destroy invading microorganisms or foreign substances. (p. 51)

In addition to fixed and wandering cells, connective tissues contain small numbers of stem cells, also called **mesenchymal cells**. These cells respond to an injury or infection by dividing to produce daughter cells that differentiate into fibroblasts, macrophages, or other connective tissue cells.

Connective Tissue Fibers

There are three basic types of fibers: *collagen*, *reticular*, and *elastic*. All three types are formed through the aggregation of protein subunits.

Collagen fibers are long, straight, and unbranched. Each collagen fiber contains bundles of protein fibers, and each protein fiber contains three protein subunits, wound together like the strands of a rope. Collagen fibers, the most common fibers in connective tissue proper, are strong but flexible. Although stiff, they will bend when pushed from the side or from either end. However, collagen fibers do not stretch very much, and they are very strong when pulled from either end.

Reticular fibers (*reticulum*, network) contain the same protein subunits as collagen fibers, but they are combined in a different way. These fibers are thinner than collagen fibers and form a branching, interwoven framework that is tough but flexible. Reticular fibers are especially abundant in organs such as the spleen and liver, where they create a complex three-dimensional network that resists forces applied from many directions.

Elastic fibers contain the protein *elastin*. Elastic fibers are branched and wavy, and after stretching they will return to their original length. [CM: *Marfan's Syndrome*]

Ground Substance

Ground substance fills all the spaces between cells and surrounds all the connective tissue fibers. Ground substance in normal connective tissue proper is clear, colorless, and similar in consistency to maple syrup. In addition to hyaluronic acid, it contains a mixture of other proteoglycans and glycoproteins (Chapter 2). (p. 53) The proteoglycans provide lubrication and give the ground substance a sticky texture. Several different glycoproteins attach fixed cells to the connective tissue fibers and guide the movement of mobile cells.

Connective tissue proper is usually divided into loose connective tissues and dense connective tissues on the basis of the relative proportions of cells, fibers, and ground substance.

Loose Connective Tissues

Loose connective tissues are the packing material of the body. These tissues fill spaces between organs, provide cushioning, and support epithelia. Loose connective tissues also anchor blood vessels and nerves, store lipids, and provide a route for the diffusion of materials. There are three types of loose connective tissues: *loose connective (areolar) tissue*, *adipose tissue*, and *reticular tissue*.

Loose Connective Tissue **Loose connective tissue**, or **areolar tissue** (*areola*, little space), is the least specialized connective tissue in the adult body. This tissue, shown in Figure 5-13a on page 142, contains all of the cells and fibers found in any connective tissue proper. Loose connective tissue has an open framework, and ground substance accounts for most of its volume. This syrupy fluid cushions shocks, and because the fibers are loosely organized, loose connective tissue can distort without damage. The presence of elastic fibers makes it fairly elastic, so this tissue returns to its original shape after external pressure is relieved.

Loose connective tissue forms a layer that separates the skin from underlying muscles. In addition to providing padding, the elastic properties of this layer allow a considerable amount of independent movement. Thus, pinching the skin of the arm does not affect the underlying muscle. Conversely, contractions of the underlying muscles do not pull against the skin—as the muscle bulges, the loose connective tissue stretches. Because this tissue has an extensive circulatory supply, drugs are often injected into the loose connective tissue layer under the skin. The tissue distorts to accept the additional fluid volume, and the drug crosses the capillary walls and enters the bloodstream at a slow but relatively constant rate.

In addition to delivering oxygen and nutrients and removing carbon dioxide and waste products,

Clinical Comment: Ground Substance and Resistance to Infection

The proteoglycans in the ground substance give it a thick, syrupy consistency. It is so dense that bacteria have trouble moving through it; for them, it's like swimming in molasses. This density slows the spread of bacteria through the tissue and makes it easier for phagocytes to catch them. Some bacteria, such as *Staphylococcus aureus*, have evolved a way to escape. They secrete the enzyme **hyaluronidase**, which breaks down hyaluronic acid and other proteoglycans. These bacteria are dangerous because they can spread rapidly by liquifying the ground substance of connective tissues and dissolving the intercellular cement that holds epithelial cells together.

the capillaries in loose connective tissue carry wandering cells to and from the tissue. These capillaries may also indirectly support epithelia. Epithelia usually cover a layer of loose connective tissue, and fibroblasts are responsible for maintaining the reticular lamina of the basement membrane. The epithelial cells rely on diffusion across that membrane, and the capillaries in the underlying connective tissue provide the necessary oxygen and nutrients.

Adipose Tissue The distinction between loose connective tissue and fat, or **adipose tissue** (Figure 5-13b), is somewhat arbitrary. Adipocytes account for most of the volume of adipose tissue, but only a fraction of the volume of loose connective tissue. However, tissues that fall between these extremes are common. Adipose tissue provides padding and cushions shocks at least as well as loose connective tissue. It also acts as an insulating blanket that slows heat loss through the skin.

Adipose tissue is common under the skin of the sides, buttocks, and breasts. It fills the bony sockets behind the eyes, surrounds the kidneys, and dominates extensive areas of loose connective tissue in the pericardial and peritoneal (abdominal) cavities. Many television advertisements and popular articles use the term *cellulite* when talking about adipose tissue. Cellulite is adipose tissue containing bands of collagen fibers that extend between the skin and deeper connective tissue layers. It is not common in men, and usually appears in women after age 30. The fibrous attachments restrict the ability of this tissue to stretch, creating a dimpling of the skin surface that is most apparent in overweight individuals.

Adipocytes are metabolically active cells—their lipids are continually being broken down and replaced. When extra nutrients are available, these cells enlarge, a fact with which many of us are all too familiar. Although adipocytes are incapable of dividing, an excess of nutrients can cause the division of mesenchymal cells, which then differentiate into additional fat cells. As a result, areas of loose connective tissue can become adipose tissue in times of nutritional plenty. When nutrients are scarce, adipocytes deflate like collapsing balloons. This is what occurs during a weight-loss program. Because the cells are not killed, merely reduced in size, the lost weight can easily be regained in the same areas of the body.

Reticular Tissue **Reticular tissue** (Figure 5-13c) forms the basic framework and organization for several organs that have a complex three-dimensional structure. This framework, or **stroma**, is found in the liver, the spleen, lymph nodes, and bone marrow. Fixed macrophages and fibroblasts are associated with the reticular fibers, but these cells are seldom visible because the organs are dominated by specialized cells with other functions.

Health News: Liposuction

Liposuction is a surgical procedure for the removal of unwanted adipose tissue. Adipose tissue is flexible but not as elastic as loose connective tissue, and it tears relatively easily. In liposuction, a small incision is made through the skin, and a tube is inserted into the underlying adipose tissue. Suction is then applied, and chunks of tissue containing adipocytes, other cells, fibers, and ground substance are removed. An estimated 115,000 liposuction procedures were performed in the United States during 1990.

This practice has received a lot of news coverage, and many advertisements praise the technique as easy, safe, and effective. In fact, it is not always easy, is sometimes dangerous, and is of limited effectiveness. The density of adipose tissue varies from place to place and from individual to individual, and it is not always easy to suck through a tube. An anesthetic must be used to control pain, and this poses risks. Blood vessels are stretched and torn, and extensive bleeding can occur. Probably the most serious complication is the potential for infection after the liposuction treatment.

Finally, it should be noted that adipose tissue can repair itself, and adipocyte populations recover over time. The only way to ensure that fat lost through liposuction will not return is to adopt a lifestyle that includes a proper diet and adequate exercise. In fact, such a lifestyle could produce the same weight loss *without* liposuction, eliminating the surgical expense and risk.

Dense Connective Tissues

Most of the volume of **dense connective tissues** is occupied by fibers. Dense connective tissues are often called **collagenous** (ko-LA-jin-us) tissues because collagen fibers are the dominant fiber type. In *dense regular connective tissue* the fibers are arranged in an orderly fashion, as in Figure 5-14a,b on page 144. In *dense irregular connective tissue* the fibers form an interwoven meshwork (Figure 5-14c).

Dense Regular Connective Tissue In **dense regular connective tissue** the collagen fibers are packed tightly and aligned with the forces applied to the tissue. Five important examples of this tissue type are

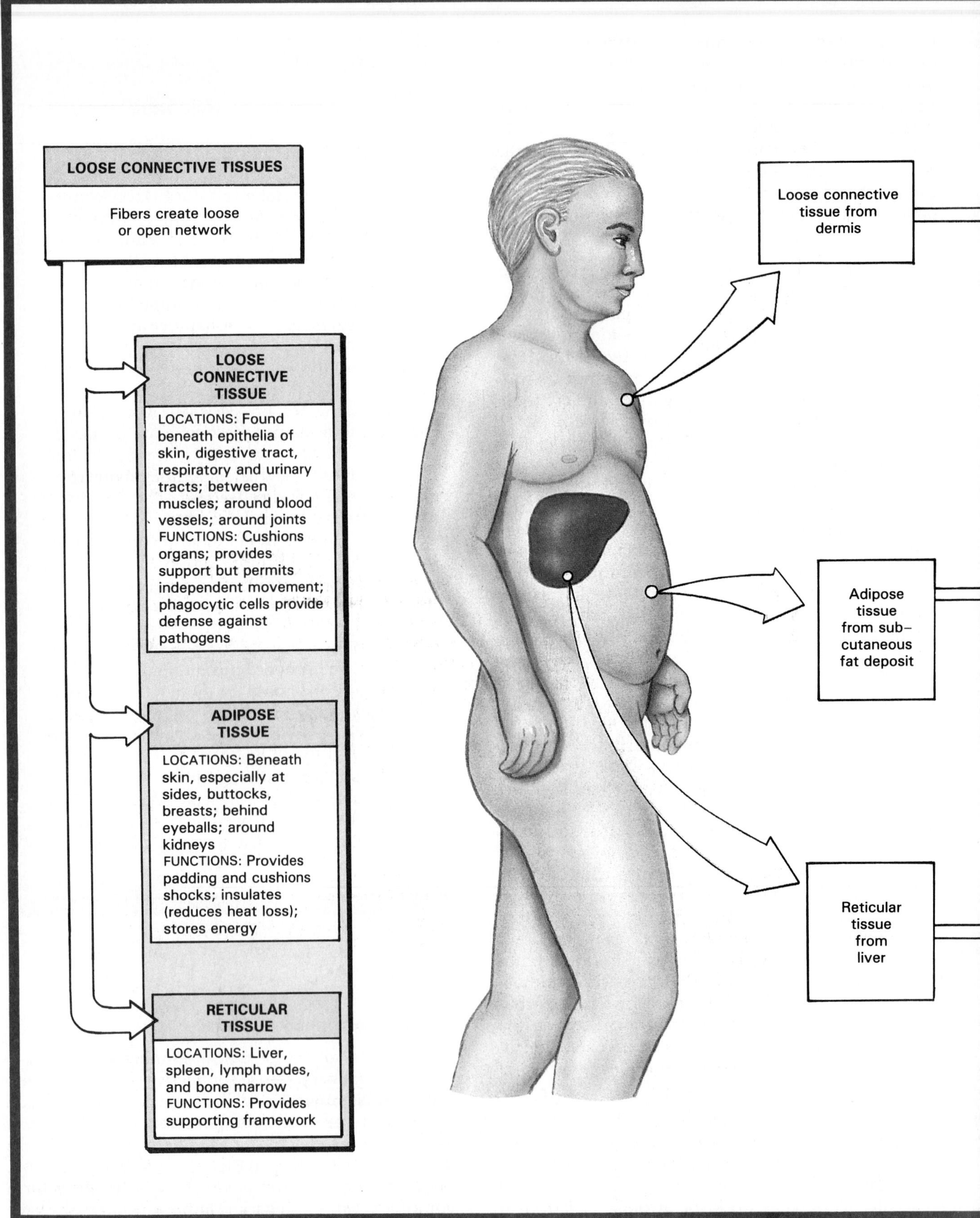

LOOSE CONNECTIVE TISSUES
Fibers create loose or open network
LOOSE CONNECTIVE TISSUE
LOCATIONS: Found beneath epithelia of skin, digestive tract, respiratory and urinary tracts; between muscles; around blood vessels; around joints
FUNCTIONS: Cushions organs; provides support but permits independent movement; phagocytic cells provide defense against pathogens
ADIPOSE TISSUE
LOCATIONS: Beneath skin, especially at sides, buttocks, breasts; behind eyeballs; around kidneys
FUNCTIONS: Provides padding and cushions shocks; insulates (reduces heat loss); stores energy
RETICULAR TISSUE
LOCATIONS: Liver, spleen, lymph nodes, and bone marrow
FUNCTIONS: Provides supporting framework
Loose connective tissue from dermis
Adipose tissue from sub-cutaneous fat deposit
Reticular tissue from liver

FIGURE 5-13
Loose Connective Tissues. (a) Note the open framework of loose connective tissue in this micrograph. All the cells of connective tissue proper are found in loose connective tissue. (b) Adipose tissue is a loose connective tissue that is dominated by adipocytes.

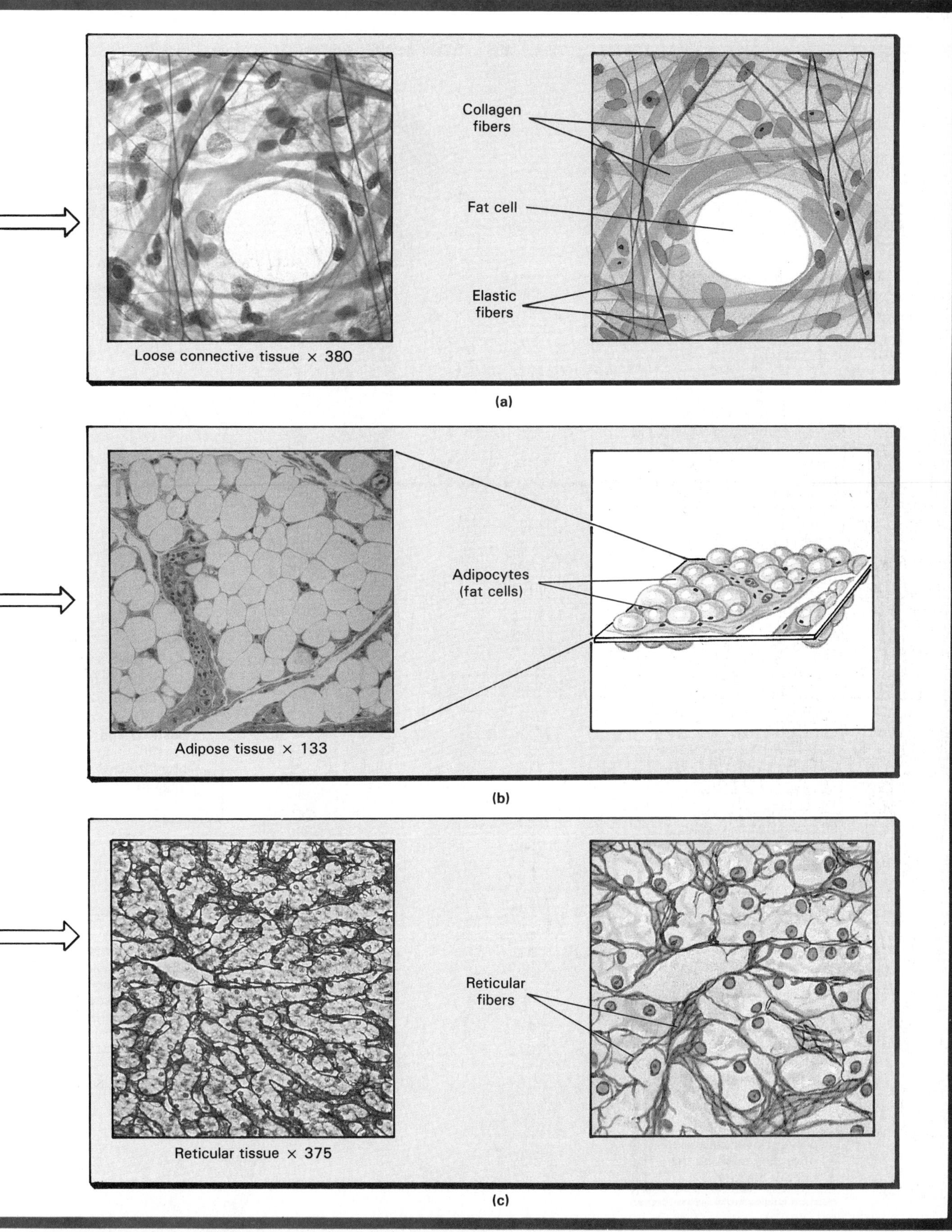

In standard histological preparations the tissue looks empty because the lipids in the fat cells dissolve during the sectioning and staining procedures. (c) Reticular tissue has an open framework of reticular fibers. These fibers are usually very difficult to see because of the large numbers of cells around them.

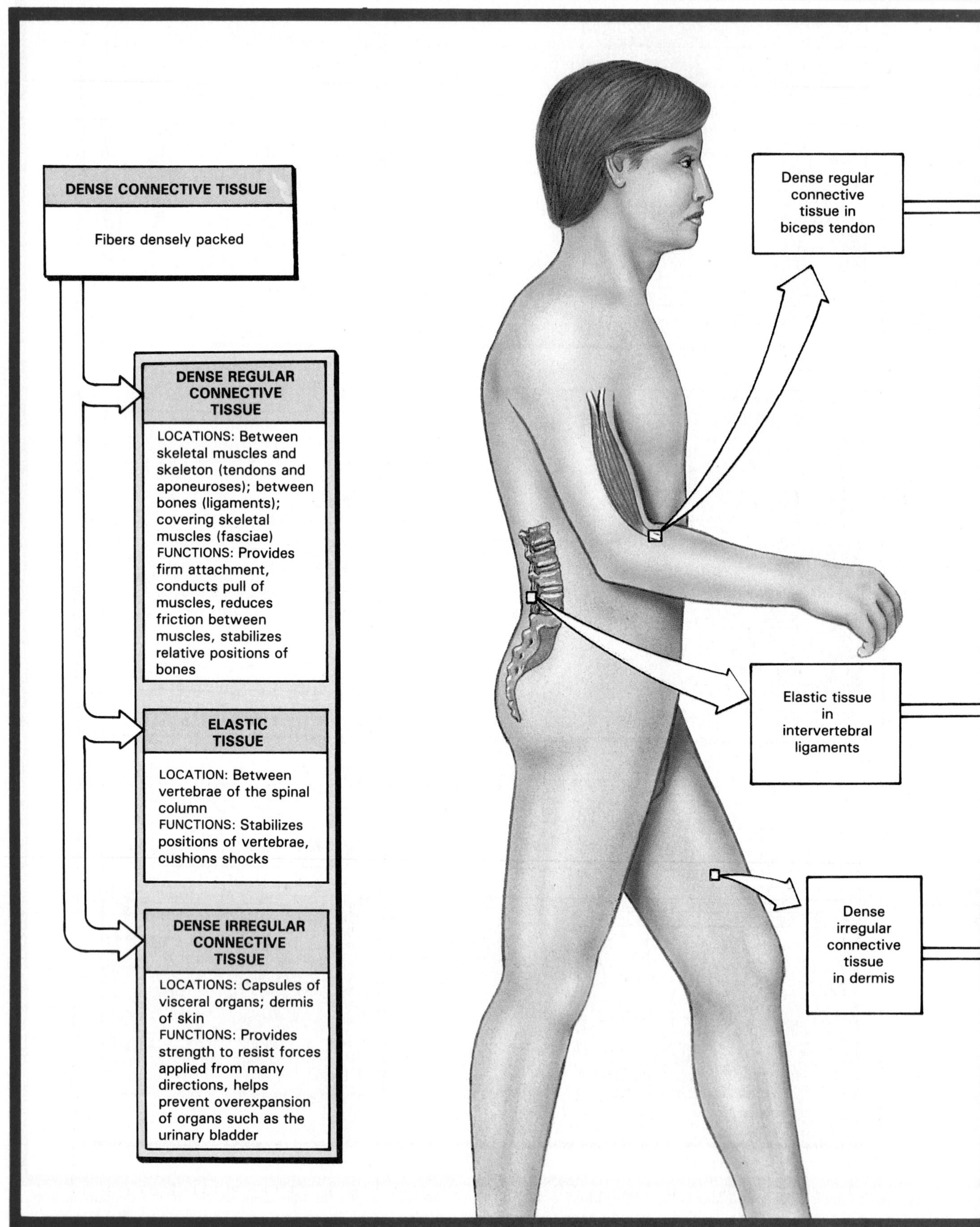

DENSE CONNECTIVE TISSUE
Fibers densely packed
DENSE REGULAR CONNECTIVE TISSUE
LOCATIONS: Between skeletal muscles and skeleton (tendons and aponeuroses); between bones (ligaments); covering skeletal muscles (fasciae)
FUNCTIONS: Provides firm attachment, conducts pull of muscles, reduces friction between muscles, stabilizes relative positions of bones
ELASTIC TISSUE
LOCATION: Between vertebrae of the spinal column
FUNCTIONS: Stabilizes positions of vertebrae, cushions shocks
DENSE IRREGULAR CONNECTIVE TISSUE
LOCATIONS: Capsules of visceral organs; dermis of skin
FUNCTIONS: Provides strength to resist forces applied from many directions, helps prevent overexpansion of organs such as the urinary bladder
Dense regular connective tissue in biceps tendon
Elastic tissue in intervertebral ligaments
Dense irregular connective tissue in dermis

FIGURE 5-14

Dense Connective Tissues. (a) The dense regular connective tissue in a tendon. Notice the densely packed, parallel bundles of collagen fibers. The fibroblast nuclei can be seen flattened between the bundles. (b) An elastic ligament from between the

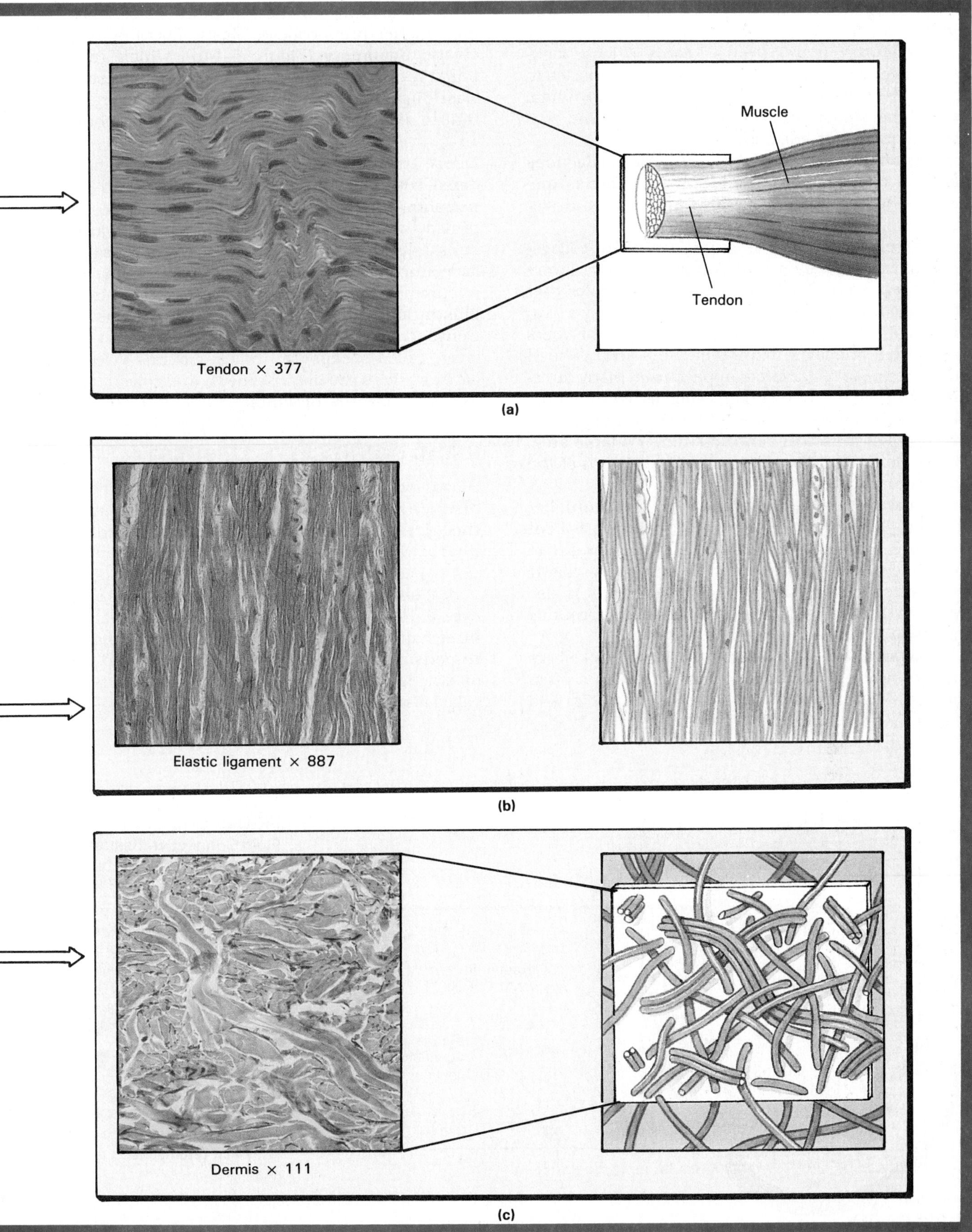

vertebrae of the spinal column. The bundles are fatter than those of a tendon or ligament composed of collagen. (c) The dermis of the skin contains a thick layer of dense irregular connective tissue.

tendons, aponeuroses, fasciae, elastic tissue, and *ligaments.*

Tendons (Figure 5-14a) are cords of dense regular connective tissue that attach skeletal muscles to bones. The collagen fibers run along the longitudinal axis of the tendon and transfer the pull of the contracting muscle to the bone.

Aponeuroses (ap-ō-nū-RŌ-sēz) are collagenous bands or ribbons that resemble broad tendons. Aponeuroses cover the surface of a muscle and assist in attaching the muscle to another structure.

Fasciae (FA-shē-ē; singular *fascia*) are collagenous sheets or bands that contain multiple layers of collagen fibers. Their organization resembles that of plywood: all the fibers in an individual layer run in the same direction, but the orientation of the fibers varies from one layer to another. This arrangement helps fasciae to resist forces applied from many directions. Fasciae surround organs such as skeletal muscles, attach skeletal muscles to bone, and distribute the forces of muscular contraction over a large area. Their role in creating the internal framework of the body is discussed later in this chapter.

In **elastic tissue** elastic fibers greatly outnumber collagen fibers, giving the tissue a springy, resilient structure. Because its ability to stretch allows it to tolerate cycles of expansion and contraction, elastic tissue often underlies transitional epithelia. It is also found in the walls of blood vessels, and surrounds the respiratory passageways.

Ligaments (LIG-a-ments) are bundles of fibers that connect one bone to another. Ligaments often contain elastic fibers as well as collagen fibers, and thus can tolerate a modest amount of stretching. An even higher proportion of elastic fibers is found in **elastic ligaments** (Figure 5-14b), which resemble tough rubber bands. Although uncommon elsewhere, elastic ligaments along the spinal cord are very important in stabilizing the positions of the vertebrae.

Dense Irregular Connective Tissue The fibers in **dense irregular connective tissue** (Figure 5-14c) do not show any consistent orientation. These tissues provide strength and support to areas subjected to stresses from many directions. A layer of dense irregular connective tissue gives skin its strength; a piece of cured leather (animal skin) provides an excellent illustration of the interwoven nature of this tissue. Dense irregular connective tissue also forms a thick layer, called a **capsule**, that surrounds visceral organs, such as the liver, kidneys, and spleen, and encloses the cavities of joints.

FLUID CONNECTIVE TISSUES

Blood and lymph are connective tissues that contain distinctive collections of cells in a fluid matrix. Although many different proteins are dissolved in that matrix, they do not form large insoluble fibers under normal conditions.

A single cell type, the **red blood cell**, or **erythrocyte** (e-RITH-rō-sīt; *erythros*, red), accounts for almost half of the volume of blood. Red blood cells are responsible for the transport of oxygen and carbon dioxide in the blood. The watery ground substance, called **plasma**, also contains small numbers of **leuko-**

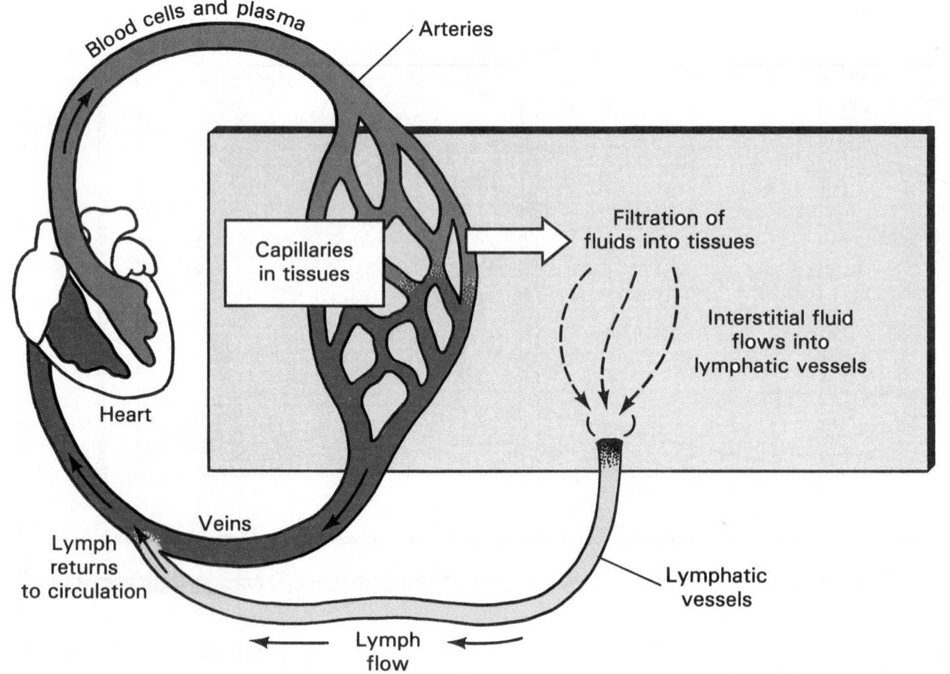

FIGURE 5-15
Fluid Connective Tissues. The functional relationships between blood and lymph. Blood travels through the circulatory system, pushed by the contractions of the heart. In capillaries, hydrostatic (blood) pressure forces fluid and dissolved solutes out of the circulatory system. This fluid mixes with the interstitial fluid already in the tissue. Interstitial fluid slowly enters lymphatic vessels; now called lymph, it travels along the lymphatics and reenters the circulatory system at one of the veins that returns blood to the heart.

cytes (LOO-kō-sīts; *leukos*, white), or **white blood cells**. White blood cells are important components of the immune system, which protects the body from infection and disease. They include phagocytic *microphages*, lymphocytes, and macrophages called **monocytes**. Tiny packets of cytoplasm called **platelets**, which contain enzymes and special proteins, function in the clotting response that seals breaks in the endothelial lining.

The extracellular fluid of the body includes three major subdivisions: *plasma*, *interstitial fluid*, and *lymph*. Plasma is normally confined to the vessels of the circulatory system, and the contractions of the heart keep it in constant motion. A network of **arteries** carries blood away from the heart and toward fine, thin-walled vessels called **capillaries**. **Veins** drain the capillaries and return blood to the heart, completing the circuit. In the tissues, filtration moves water and small solutes out of the capillaries and into the **interstitial fluid** (in-ter-STISH-al; *inter-*, between + *sistere*, to set) that bathes the cells of the body. The chemical composition of the interstitial fluid, also called *tissue fluid*, was noted in Chapter 3 (see Table 3-3). ∞ (p. 64) The major difference between plasma and interstitial fluid is that plasma contains a much higher concentration of dissolved proteins.

Lymph forms as interstitial fluid enters small passageways, or **lymphatics**, that return it to the circulatory system. Figure 5-15 diagrams this flow pattern. Along the way, lymph flows past cells of the immune system that respond to signs of injury or infection. The proportion of cell types in lymph may vary, but ordinarily 99 percent are lymphocytes, and the rest are wandering macrophages or microphages.

SUPPORTING CONNECTIVE TISSUES

Cartilage and bone are called **supporting connective tissues** because they provide a strong framework that supports the rest of the body. In these connective tissues the matrix contains numerous fibers and, in some cases, deposits of insoluble calcium salts.

Cartilage

The matrix of **cartilage** is a firm gel that contains proteoglycans called **chondroitin sulfates** (kon-DROY-tin; *chondros*, cartilage). **Chondrocytes** (KON-drō-sīts) are the only cells found within the matrix; they live in small pockets known as **lacunae** (la-KOO-nē; *lacus*, pool). The physical properties of cartilage depend on the type and abundance of fibers, as well as the proteoglycan content.

Because chondrocytes produce a chemical that discourages the formation of blood vessels, cartilage is avascular. All nutrient and waste product exchange must occur by diffusion through the matrix. The cartilage is set apart from the surrounding tissues by a fibrous **perichondrium** (pe-re-KON-drē-um; *peri-*, around). The perichondrium contains two distinct layers, an outer, fibrous region of dense irregular connective tissue and an inner, cellular layer.

Types of Cartilage The three major types of **cartilage**—*hyaline cartilage*, *elastic cartilage*, and *fibrocartilage*—are shown in Figure 5-16. **Hyaline cartilage** (HĪ-a-lin; *hyalos*, glass) is the most common type of cartilage. The matrix of hyaline cartilage contains closely packed collagen fibers that may account for 40 percent of its weight. Because the fibers do not stain well, they are not apparent when viewed through a light microscope (Figure 5-16a), but the fiber content makes this cartilage tough and somewhat flexible. In the adult, this type of cartilage connects the ribs to the sternum (breastbone), supports the conducting passageways of the respiratory tract, and covers the surfaces of bones within joints.

Elastic cartilage (Figure 5-16b) contains numerous elastic fibers that make it extremely resilient and flexible. Elastic cartilage supports the external flap (*pinna*) of the outer ear, the epiglottis, and the tip of the nose.

Clinical Comment: "Jaws" and the Fight against Cancer

Cartilage cells survive despite the fact that they are crowded together in an avascular matrix. In other tissues, when cells are crowded and active, blood vessels grow into the area and improve oxygen and nutrient delivery. Cartilage cells actively prevent this from happening by secreting a chemical that blocks the growth of blood vessels. This compound has been named **antiangiogenesis factor** (*anti-*, against + *angeion*, vessel + *gennan*, to produce).

One reason cancers can grow so explosively is that blood vessels branch into the developing tumor, delivering supplies to the renegade cells. This growth could theoretically be prevented by antiangiogenesis factor, but the quantities produced in normal human cartilage are extremely small.

Sharks are highly successful marine predators. Although large, their skeletons are cartilaginous, rather than bony. Sharks are now being collected to obtain the antiangiogenesis factor from their cartilages. Because they contain so much cartilage, substantial quantities can be extracted from a single animal.

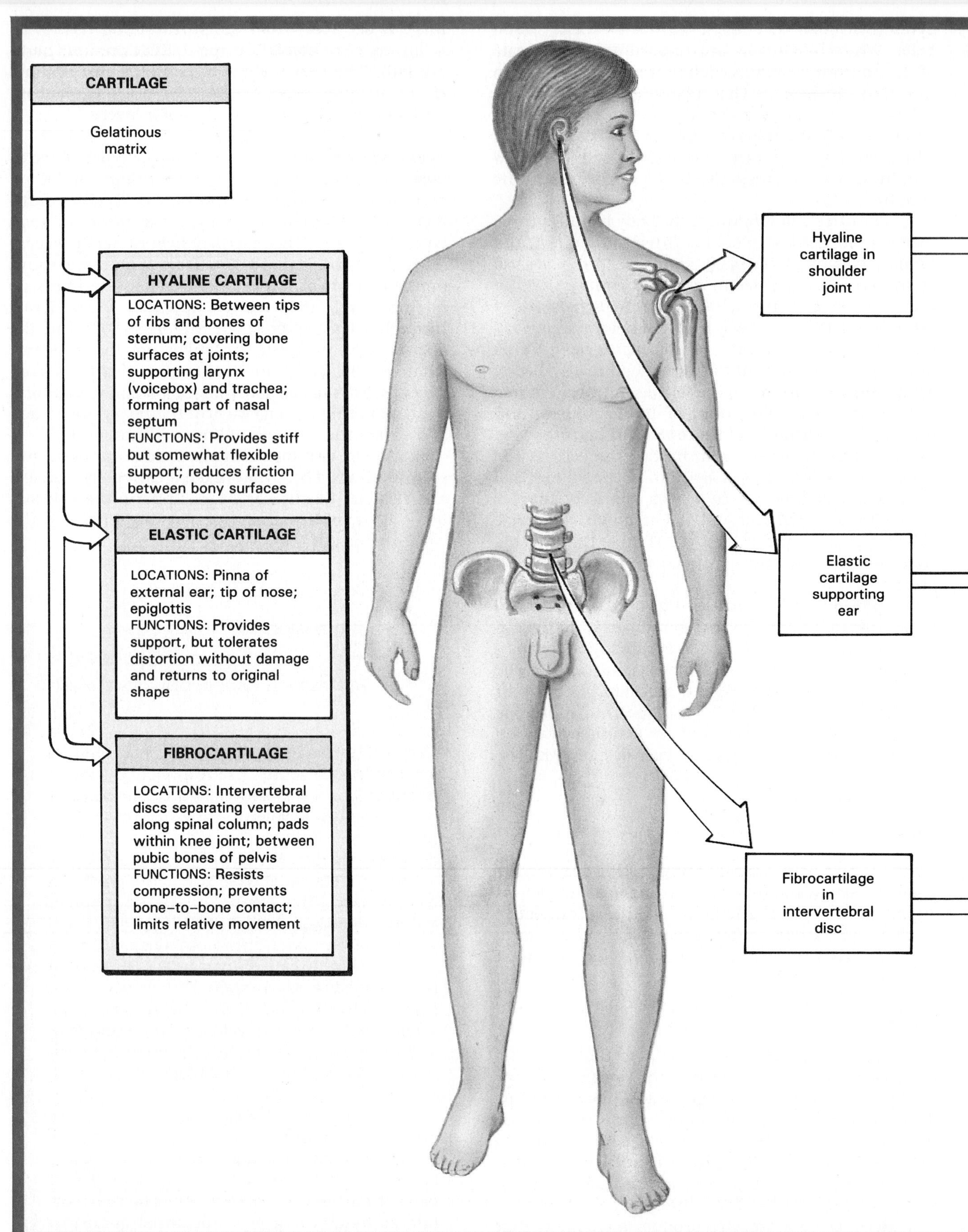

CARTILAGE
Gelatinous matrix
HYALINE CARTILAGE
LOCATIONS: Between tips of ribs and bones of sternum; covering bone surfaces at joints; supporting larynx (voicebox) and trachea; forming part of nasal septum
FUNCTIONS: Provides stiff but somewhat flexible support; reduces friction between bony surfaces
ELASTIC CARTILAGE
LOCATIONS: Pinna of external ear; tip of nose; epiglottis
FUNCTIONS: Provides support, but tolerates distortion without damage and returns to original shape
FIBROCARTILAGE
LOCATIONS: Intervertebral discs separating vertebrae along spinal column; pads within knee joint; between pubic bones of pelvis
FUNCTIONS: Resists compression; prevents bone-to-bone contact; limits relative movement
Hyaline cartilage in shoulder joint
Elastic cartilage supporting ear
Fibrocartilage in intervertebral disc

FIGURE 5-16
Types of Cartilage. (a) Hyaline cartilage: note the translucent matrix and the absence of prominent fibers. (b) Elastic cartilage:

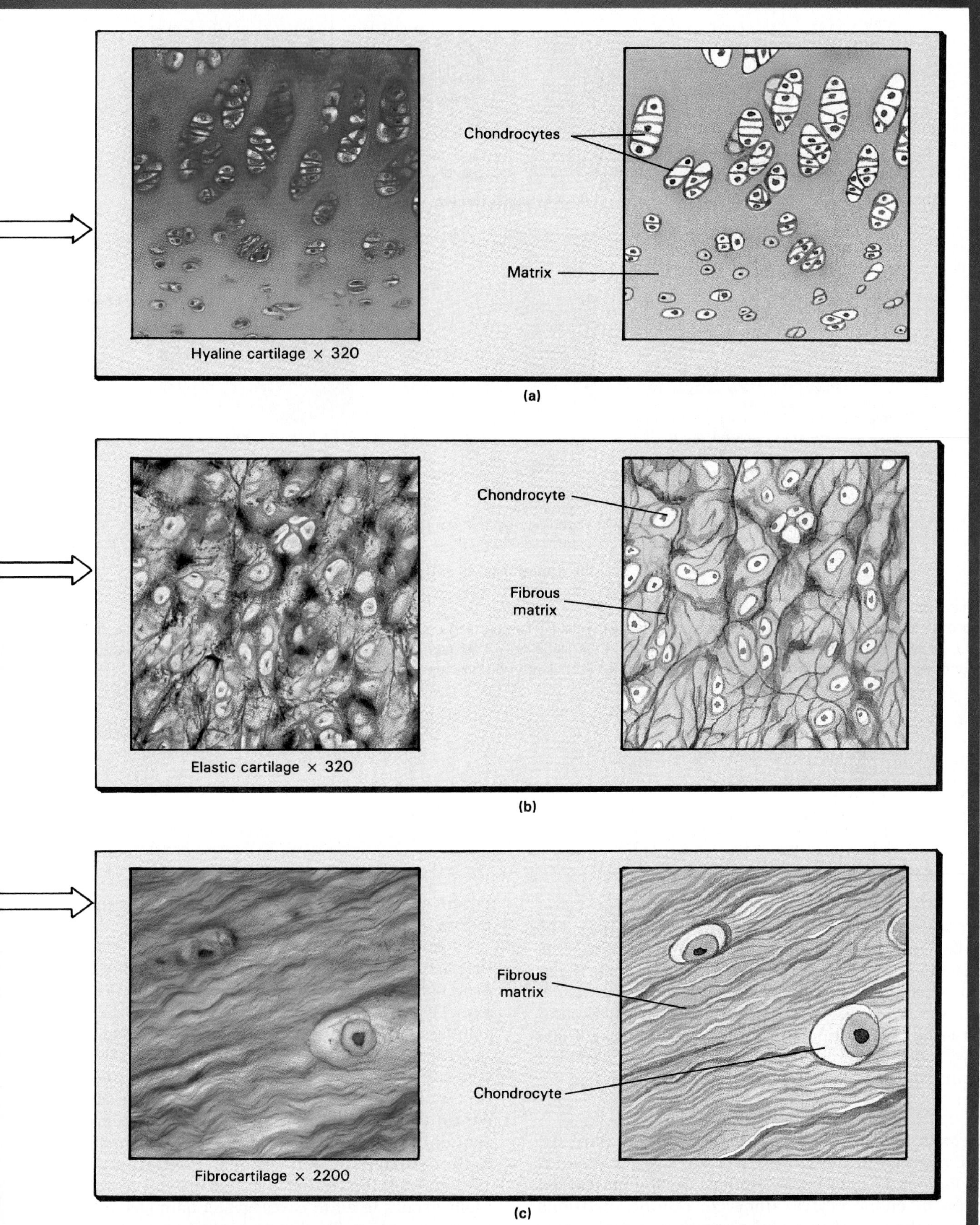

the closely packed elastic fibers are visible between the chondrocytes. (c) Fibrocartilage: the collagen fibers are extremely dense, and the chondrocytes are relatively far apart.

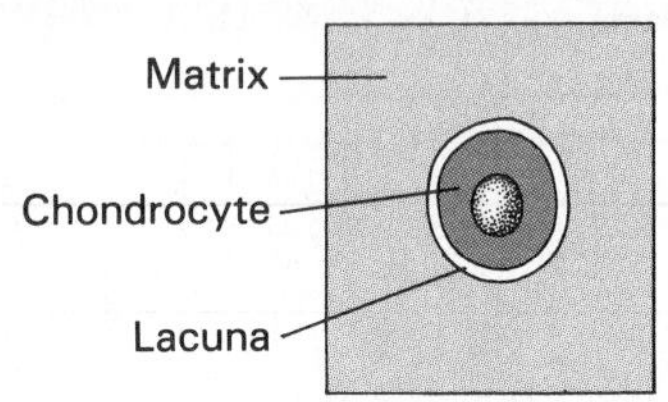

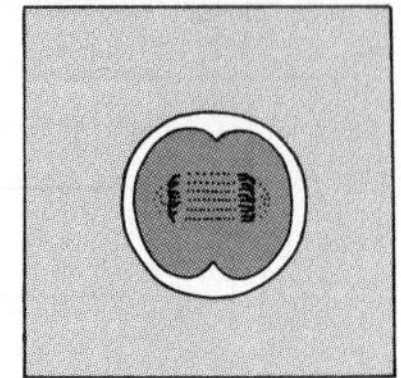

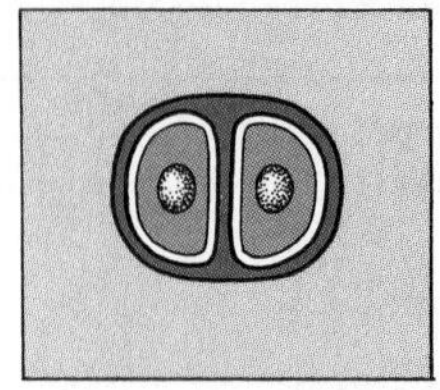

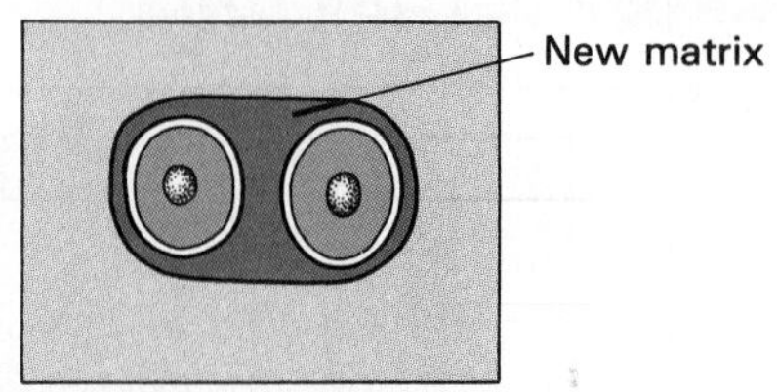

Chondrocyte within a lacuna surrounded by cartilage matrix undergoes division.

As daughter cells secrete additional matrix they move apart, expanding the cartilage from within.

(a) Interstitial Growth

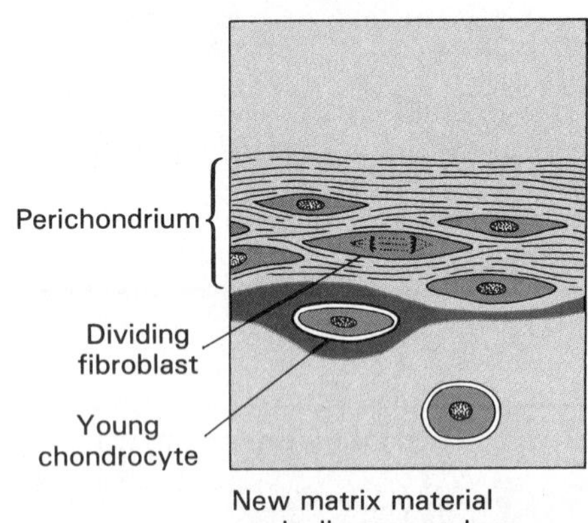

New matrix material gradually surrounds a fibroblast in the cellular layer of the perichondrium.

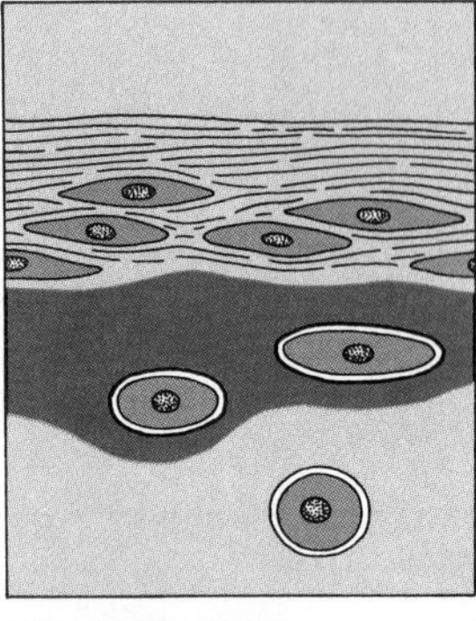

These fibroblasts differentiate into chondrocytes and secrete additional matrix.

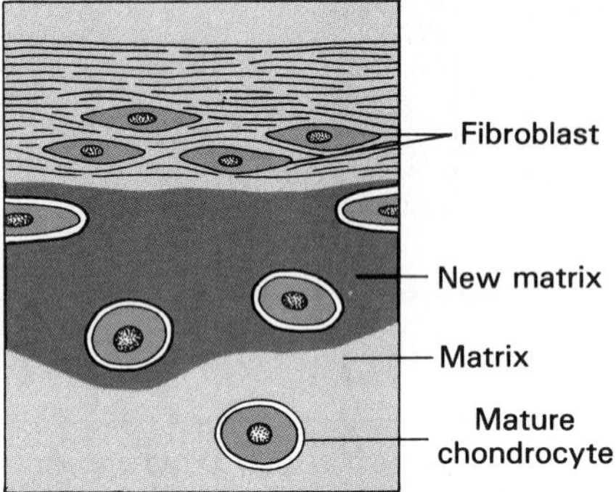

As matrix expands, more fibroblasts are incorporated; they are replaced by divisions of cells in the periosteum.

(b) Appositional Growth

FIGURE 5-17
Formation and Growth of Cartilage. (a) *Interstitial growth.* The cartilage expands from within as chondrocytes divide, grow, and secrete new matrix. (b) *Appositional growth.* The cartilage grows through the addition of cartilage to the outer surface. This process involves the entrapment and differentiation of fibroblasts from the inner, cellular layer of the perichondrium.

Fibrocartilage has little ground substance and an abundance of collagen fibers (Figure 5-16c). The collagen fibers are densely interwoven, making this tissue extremely durable and tough. Fibrocartilaginous pads lie between the vertebrae of the spinal column, between the bones of the pelvis, and around or within a few joints and tendons. In these positions they resist compression, absorb shocks, and prevent damaging bone-to-bone contact.

Growth and Repair of Cartilage Two mechanisms are involved in the growth of a cartilage: *interstitial growth* and *appositional growth.* During **interstitial growth** chondrocytes undergo mitotic divisions within their lacunae, and the new chondrocytes separate as each produces new matrix materials. In this process, diagrammed in Figure 5-17, the cartilage expands from within. Interstitial growth is most apparent during the early phases of cartilage formation, before birth and in early childhood.

Appositional growth begins after the perichondrium has formed, and cartilages can continue to grow by this mechanism even after puberty. In appositional growth (Figure 5-17b), fibroblasts of the inner, cellular layer of the perichondrium become embedded in the matrix, where they differentiate into chondrocytes. The lacunae are small and rather flattened at first, but they enlarge as time passes and the cartilage continues to expand. Each new chondrocyte may subsequently divide and make an additional contribution to the cartilage structure through interstitial growth.

Neither interstitial nor appositional growth normally occurs in adult cartilages. A damaged cartilage heals slowly, as fibroblasts migrate into the injury site and differentiate into chondrocytes. In the case of a severe injury, in which large numbers of chondrocytes are killed, the matrix disintegrates and migrating fibroblasts lay down a network of collagen fibers rather than cartilage. As a result, severe damage to the cartilages of the ear or nose often causes scarring that requires cosmetic surgery.

Under unusual conditions calcium phosphate crystals may precipitate within the matrix, producing **calcified cartilage**. This might seem desirable, as the addition of minerals should make the matrix considerably stronger. Such advantages are short-lived, however, for nutrients cannot diffuse through a calcified matrix any faster than they could through a boulder. Deprived of their nutrient supply, the chondrocytes die and disintegrate, and tissue maintenance ceases. In structures subject to heavy stresses, a different connective tissue, bone, provides the necessary strength.

Sports and Fitness: Cartilages and Knee Injuries

The knee is an extremely complex joint that contains both hyaline cartilage and fibrocartilage. The hyaline cartilage covers bony surfaces, and pads of fibrocartilage within the joint prevent bone contact when movements are under way. Many sports injuries involve tearing of the cartilage pads. This loss of cushioning places more strain on the cartilages within joints and leads to further joint damage. Because cartilages are avascular, they heal poorly, and joint cartilages heal even more slowly than other cartilages. Surgery usually produces only a temporary or incomplete repair.

Over the last 3 years, advances in tissue culture have enabled researchers to grow fibrocartilage in the laboratory. Chondrocytes removed from the knees of injured dogs are cultured in an artificial framework of collagen fibers. They eventually produce masses of fibrocartilage that can be inserted into the damaged joints. Over time the pads change shape and grow, restoring normal joint function. In the future this technique may be used to treat severe knee injuries in humans.

Bone

Because the detailed histology of bone, or **osseous tissue** (OS-ē-us; *os*, bone), will be considered in Chapter 7, this discussion will focus on significant differences between cartilage and bone. Roughly one-third of the matrix of bone consists of collagen fibers. The balance is a mixture of calcium salts, primarily calcium phosphate with lesser amounts of calcium carbonate. This combination gives bone truly remarkable properties. By themselves, calcium salts are strong but rather brittle. Collagen fibers are weaker, but relatively flexible. In bone, the minerals are organized around the collagen fibers. The result is a strong, somewhat flexible combination that is very resistant to shattering. In its overall properties, bone can compete with the best steel-reinforced concrete, at a considerable savings in weight.

The general organization of osseous tissue can be seen in Figure 5-18. Lacunae within the matrix contain bone cells, or **osteocytes** (OS-tē-ō-sīts). The lacunae are often organized around blood vessels that branch through the bony matrix. Although diffusion cannot occur through the calcium salts, osteocytes communicate with the blood vessels and with one

FIGURE 5-18
Bone. The osteocytes in bone are usually organized in groups around a central space that contains blood vessels. For the photomicrograph, a sample of bone was ground thin enough to become transparent. Bone dust filled the lacunae and the central canal, making them appear dark.

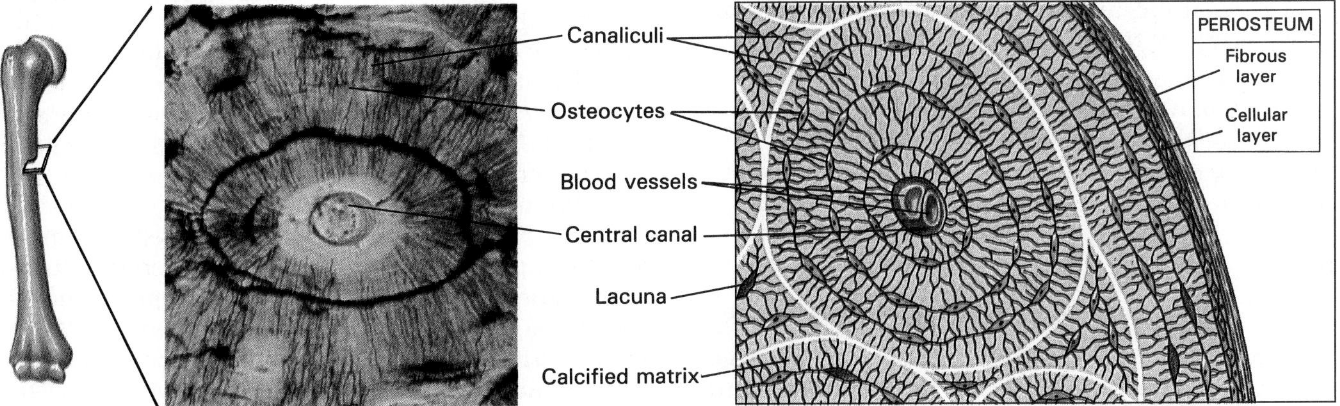

■ **TABLE 5-1 A Comparison of Cartilage and Bone**

Characteristic	*Cartilage*	*Bone*
STRUCTURAL FEATURES		
Cells	Chondrocytes in lacunae	Osteocytes in lacunae
Matrix	Chondroitin sulfate (proteoglycan) dissolved in water	Insoluble crystals of calcium phosphate and calcium carbonate
Fibers	Collagen, elastin, reticular fibers (proportions vary)	Collagen fibers predominate
Vascularity	None	Extensive
Covering	Perichondrium, two-part	Periosteum, two-part
Strength	Limited: bends easily but hard to break	Strong: resists distortion until breaking point is reached
METABOLIC FEATURES		
Oxygen demands	Relatively low	Relatively high
Nutrient delivery	By diffusion through matrix	By diffusion through cytoplasm and fluid in canaliculi
Growth	Interstitial and appositional	Appositional only
Repair capabilities	Limited ability	Extensive ability

another through slender cytoplasmic extensions. These extensions provide a route for the diffusion of materials between the blood vessels and the osteocytes. Early histologists named the passageways through which these cytoplasmic processes ran **canaliculi** (kan-a-LIK-ū-lē; little canals) because they form a branching network within the bony matrix.

Each bone is surrounded by a fibrous **periosteum** (pe-rē-OS-tē-um) that is layered like the perichondrium. Because the matrix is solid rather than fluid, bone cells within the matrix cannot move farther apart. As a result, only appositional growth occurs, and an osteocyte never undergoes division. But unlike cartilage, bone undergoes extensive remodeling on a regular basis, and complete repairs can be made even after severe damage has occurred. Table 5-1 summarizes the similarities and differences between cartilage and bone.

√ **A sheet of tissue has many layers of collagen fibers that run in different directions in successive layers. What type of tissue is this?**

√ **Lack of vitamin C in the diet interferes with the ability of fibroblasts to produce collagen. What effect might this have on connective tissue?**

■ Membranes

Some anatomical terms have more than one meaning, depending on the context. "Membrane" is an excellent example. A review of the anatomical literature would reveal perhaps a dozen different anatomical membranes. For example, at the cellular level, membranes are lipid bilayers that restrict the passage of ions and other solutes—Chapter 3 introduced the cell membrane, the nuclear membrane, and various membranous organelles. (p. 63) At the tissue level, membranes again form a barrier or interface—we have already mentioned the basement membranes that separate epithelia from connective tissues. At still another level, epithelia and connective tissues combine to form membranes that cover and protect other structures and tissues in the body. There are four such membranes: *mucous membranes, serous membranes*, the *cutaneous membrane*, and *synovial membranes*.

MUCOUS MEMBRANES

Mucous membranes (Figure 5-19a on page 154) line cavities that communicate with the exterior, including the digestive, respiratory, reproductive, and urinary tracts. The epithelial surfaces are kept moist at all times; they may be lubricated by mucus produced by goblet cells or multicellular glands, or by exposure to fluids such as urine or semen. The loose connective tissue component of a mucous membrane is called the **lamina propria** (PRŌ-prē-a). Later chapters will consider the organization of specific mucous membranes in greater detail.

Many mucous membranes are lined by simple epithelia that perform absorptive or secretory functions, such as the simple columnar epithelium of the digestive tract. However, other types of epithelia may be involved. For example, a stratified squamous epithelium covers the mucous membrane of the

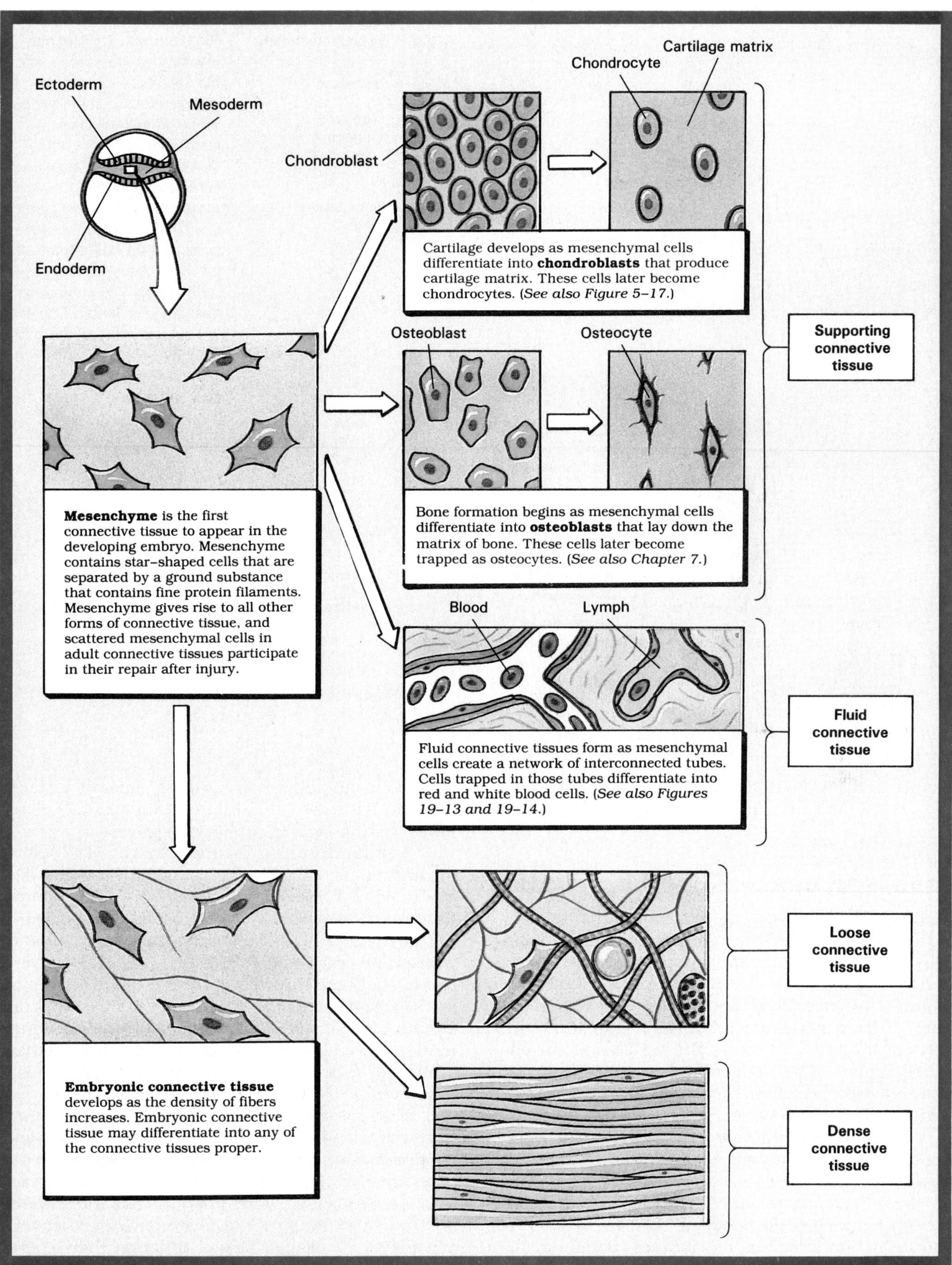

Ectoderm
Mesoderm
Endoderm
Chondroblast
Chondrocyte
Cartilage matrix
Cartilage develops as mesenchymal cells differentiate into **chondroblasts** that produce cartilage matrix. These cells later become chondrocytes. (*See also Figure 5–17.*)
Osteoblast
Osteocyte
Bone formation begins as mesenchymal cells differentiate into **osteoblasts** that lay down the matrix of bone. These cells later become trapped as osteocytes. (*See also Chapter 7.*)
Supporting connective tissue
Mesenchyme is the first connective tissue to appear in the developing embryo. Mesenchyme contains star-shaped cells that are separated by a ground substance that contains fine protein filaments. Mesenchyme gives rise to all other forms of connective tissue, and scattered mesenchymal cells in adult connective tissues participate in their repair after injury.
Blood
Lymph
Fluid connective tissues form as mesenchymal cells create a network of interconnected tubes. Cells trapped in those tubes differentiate into red and white blood cells. (*See also Figures 19–13 and 19–14.*)
Fluid connective tissue
Loose connective tissue
Embryonic connective tissue develops as the density of fibers increases. Embryonic connective tissue may differentiate into any of the connective tissues proper.
Dense connective tissue

Embryology Summary: Origins of Connective Tissues

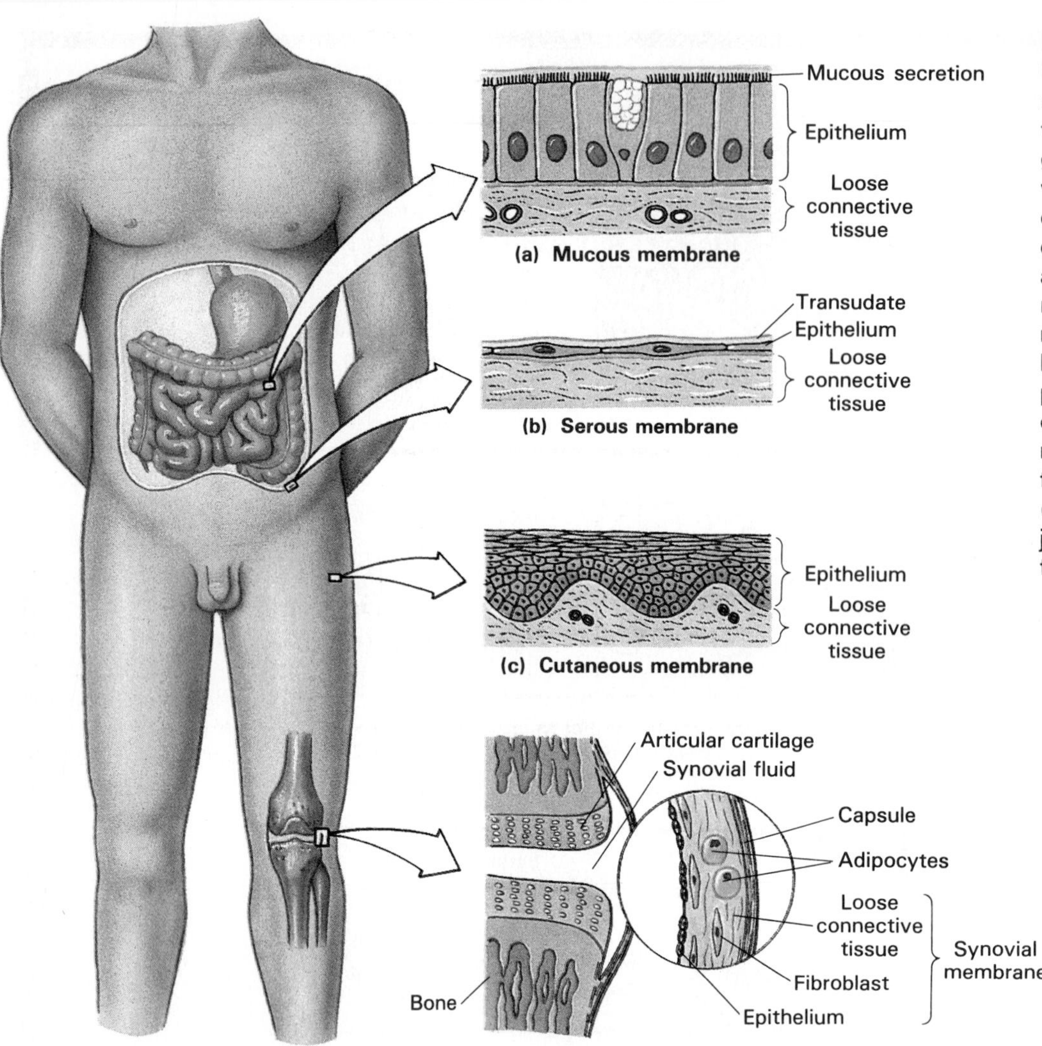

FIGURE 5-19
Membranes. (a) Mucous membranes are coated with the secretions of mucous glands. Mucous membranes with a simple columnar epithelium line most of the digestive and respiratory tracts and portions of the reproductive tract. (b) Serous membranes line the ventral body cavities (the peritoneal, pleural, and pericardial cavities). (c) The cutaneous membrane of the skin covers the outer surface of the body. (d) Synovial membranes line joint cavities and produce the fluid within the joint.

mouth, and a transitional epithelium covers the mucous membrane along most of the urinary tract.

SEROUS MEMBRANES

Serous membranes line the sealed, internal cavities of the body. There are three serous membranes, each consisting of a *mesothelium* supported by loose connective tissue (Figure 5-19b). The **pleura** (PLOO-ra; *pleura*, rib) lines the pleural cavities and covers the lungs. The **peritoneum** (pe-ri-tō-NĒ-um) lines the peritoneal cavity and covers the surfaces of the enclosed organs. The **pericardium** (pe-ri-KAR-dē-um; *peri*, around + *cardium*, heart) lines the pericardial cavity and covers the heart.

The serous membrane that lines the wall of the cavity and the membrane that covers the enclosed organs are in close contact at all times—there are no "wide open spaces" in body cavities. Minimizing friction between these opposing surfaces is the primary function of serous membranes. Serous membranes are extremely permeable, because mesothelia are very thin. Interstitial fluids continually cross the epithelium, keeping the exposed surfaces moist. The squamous epithelial cells have flat, smooth surfaces to begin with, and the fluid coating virtually eliminates friction between opposing surfaces.

The fluid formed on the surfaces of a serous membrane is called a **transudate** (TRANS-ū-dāt; *trans*, across). Specific transudates are called **pleural fluid**, **peritoneal fluid**, or **pericardial fluid**, depending on their source. Ordinarily the total volume of transudate is extremely small, just enough to prevent friction between the walls of the cavities and the surfaces of internal organs. But following an injury or in certain disease states the volume of transudate may increase dramatically, complicating existing medical problems or producing new ones (see the Clinical Comment: Problems with Serous Membranes).

If serous membranes are damaged, the production of transudate may cease, and the parietal and visceral mesothelia will begin to rub against one another, producing abrasion. This mesothelial damage attracts fibroblasts, which migrate into the abraded area and bind the opposing membranes together with a network of collagen fibers. Although their efforts reduce friction, the fibers may compress blood vessels, nerves, or other vital structures in the region. These restrictive fibrous connections are called **adhe-**

sions. Adhesions may occur as a result of surgery, infection, or other injuries that damage serous membranes.

CUTANEOUS MEMBRANE

The **cutaneous membrane** of the skin (Figure 5-19c) covers the surface of the body. It consists of a stratified squamous epithelium and the underlying connective tissues. In contrast to serous or mucous membranes, the cutaneous membrane is thick, relatively waterproof, and usually dry.

The epithelium of the cutaneous membrane must resist abrasion, chemical and bacterial attack, and other stresses. The germinative cell layer is very active, and cell divisions produce daughter cells that replace those lost or shed to the environment. A complex folding of the basement membrane increases the surface area exposed to the loose connective tissue beneath. This folding provides increased space for germinative cell attachment, as well as enlarging the area for diffusion of nutrients and dissolved gases from the connective tissue to epithelial cells.

SYNOVIAL MEMBRANES

Bones of the skeleton contact one another at joints, also called **articulations** (ar-tik-ū-LĀ-shuns). The connective tissues at a joint may be quite restrictive, permitting little if any motion, or the elements may be capable of a significant amount of independent movement. When the bones are mobile, the bony surfaces do not come into direct contact with one another. If they did, abrasion and impacts would damage the opposing surfaces, and smooth movement would be almost impossible. Instead, the ends of the bone are covered with hyaline cartilage and separated by a viscous **synovial fluid**, a product of the synovial membrane that lines the joint cavity.

A **synovial membrane** (sin-Ō-vē-al) (Figure 5-19d) consists of extensive areas of loose connective tissue bounded by a superficial layer of squamous or cuboidal cells. This lining, sometimes called a *synovial epithelium*, differs from other epithelia in two respects: (1) there is no basement membrane and (2) the cellular layer is incomplete, and small spaces exist between adjacent cells. Some of the lining cells are phagocytic and others are secretory. The secretory cells produce a mixture of proteoglycans, primarily hyaluronic acid, and synovial fluid resembles the ground substance of loose connective tissue.

Clinical Comment: Problems with Serous Membranes

Several clinical conditions, including infection and chronic irritation, can cause the abnormal buildup of fluid within one of the ventral body cavities. **Pleuritis**, or **pleurisy**, is an inflammation of the pleural cavities. At first the membranes become rather dry, and the opposing membranes may scratch against one another. This scratching produces a characteristic sound known as a **pleural rub**. Adhesions seldom form between the serous membranes of the pleural cavities. More frequently, continued inflammation and rubbing lead to a gradual increase in the production of fluid to levels well above normal. Fluid then accumulates in the pleural cavities, producing a condition known as **pleural effusion**. Pleural effusion can also be caused by heart conditions that elevate the pressure in blood vessels of the lungs. As fluids build up in the pleural cavities, the lungs are compressed, making it difficult to breathe. The combination of severe pleural effusion and heart disease can be lethal.

Pericarditis is an inflammation of the pericardium. This condition often leads to a **pericardial effusion**: an abnormal accumulation of fluid in the pericardial cavity. When sudden or severe, the fluid buildup can seriously reduce the efficiency of the heart and restrict blood flow through major vessels.

Peritonitis, an inflammation of the peritoneum, can occur following an infection of or injury to the peritoneal lining. Peritonitis is a potential complication of any surgical procedure that requires opening the peritoneal cavity. Liver disease, kidney disease, or heart failure can cause an increase in the rate of fluid movement through the peritoneal lining. The accumulation of fluid, called **ascites** (a-SĪ-tēz), creates a characteristic abdominal swelling. Distortion of internal organs by the contained fluid can result in a variety of symptoms; heartburn, indigestion, and low back pain are common complaints.

These conditions will be discussed further in chapters dealing with the respiratory, cardiovascular, and digestive systems.

Muscle Tissue

Muscle tissue is specialized for contraction. Individual muscle cells are relatively long and slender; as a result, they are usually called **muscle fibers**. A large skeletal muscle fiber may be 100 μm in diameter and 250,000 μm (25 cm) long.

Muscle fiber contraction involves interaction between filaments of *myosin* and *actin*. As you will recall from Chapter 3, these are proteins found in the cytoskeleton of many cells. ∞ (p. 65) However, in muscle fibers they are more numerous and arranged so that their interaction produces a contraction of the entire cell. This contraction process requires large quantities of ATP, so muscle fibers have a high demand for energy.

There are three types of muscle tissue: *skeletal*, *cardiac*, and *smooth*. The contraction mechanism is the same in all of them, but they differ in the organi-

FIGURE 5-20
Muscle Tissue. (a) *Skeletal muscle fibers.* Note the large fiber size, prominent banding pattern, multiple nuclei, and unbranched arrangement. (b) *Cardiac muscle fibers.* Cardiac muscle fibers differ from skeletal muscle fibers in three major ways: size (cardiac muscle fibers are smaller), organization (cardiac muscle fibers branch), and number of nuclei (each cardiac muscle fiber has one central nucleus). Both contain actin and myosin filaments in an organized array that produces the striations seen in both types of muscle fiber. (c) *Smooth muscle fibers.* Smooth muscle fibers are small and spindle-shaped, with a central nucleus. They do not branch, and there are no striations.

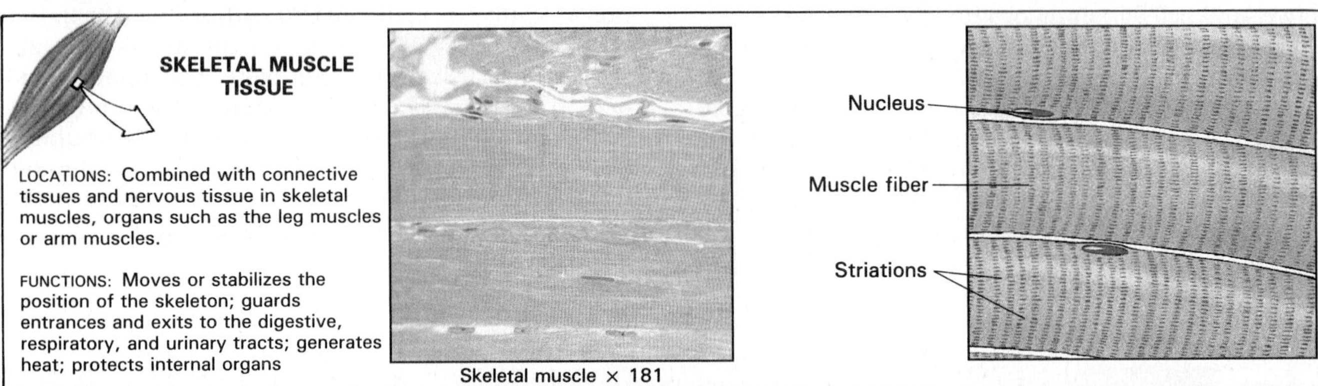
SKELETAL MUSCLE TISSUE
LOCATIONS: Combined with connective tissues and nervous tissue in skeletal muscles, organs such as the leg muscles or arm muscles.
FUNCTIONS: Moves or stabilizes the position of the skeleton; guards entrances and exits to the digestive, respiratory, and urinary tracts; generates heat; protects internal organs
Skeletal muscle × 181
Nucleus
Muscle fiber
Striations

(a)

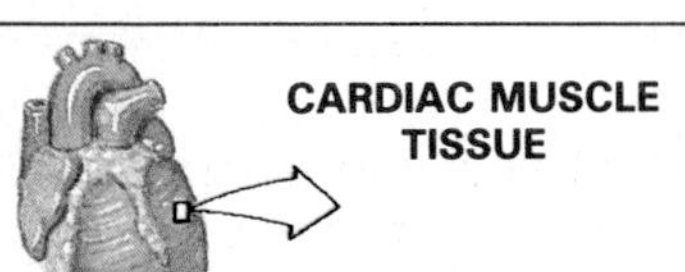

CARDIAC MUSCLE TISSUE

LOCATION: Heart

FUNCTIONS: Circulates blood, maintains blood (hydrostatic) pressure

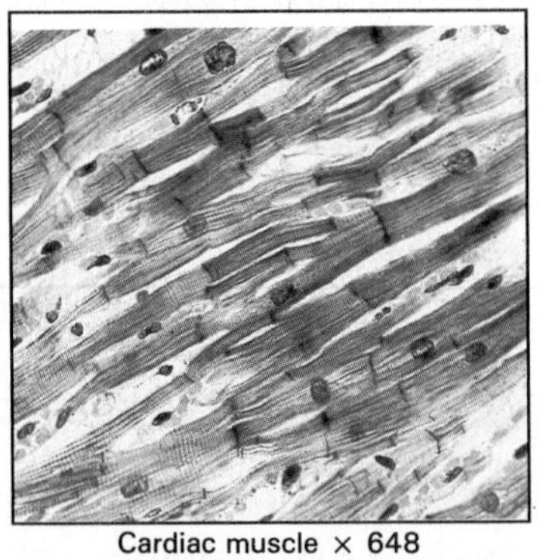

Cardiac muscle × 648

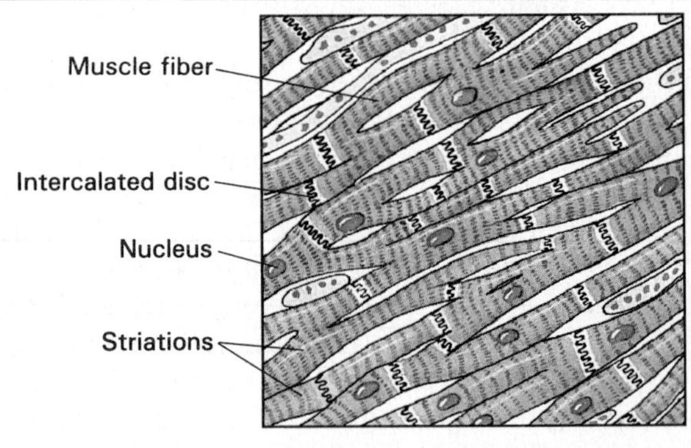
Muscle fiber
Intercalated disc
Nucleus
Striations

(b)

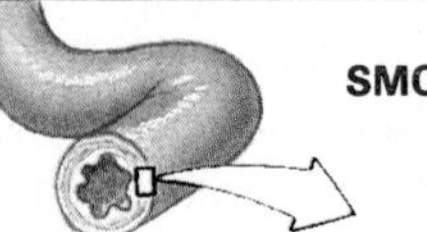

SMOOTH MUSCLE TISSUE

LOCATIONS: Encircles blood vessels; found in the walls of digestive, respiratory, urinary, and reproductive organs

FUNCTIONS: Moves food, urine, and reproductive tract secretions; controls diameters of respiratory passageways; regulates diameter of blood vessels; and contributes to regulation of tissue blood flow

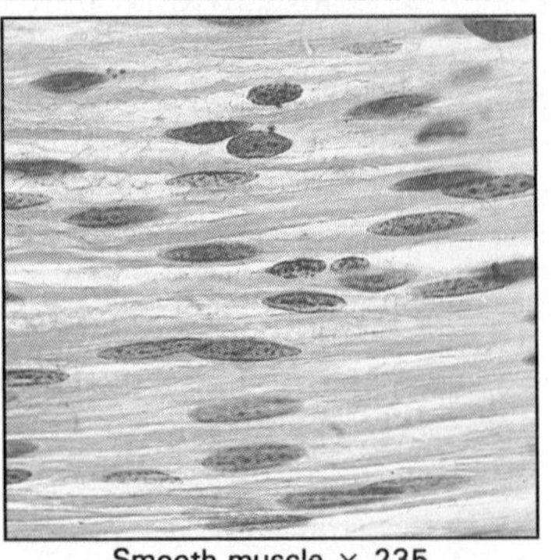

Smooth muscle × 235

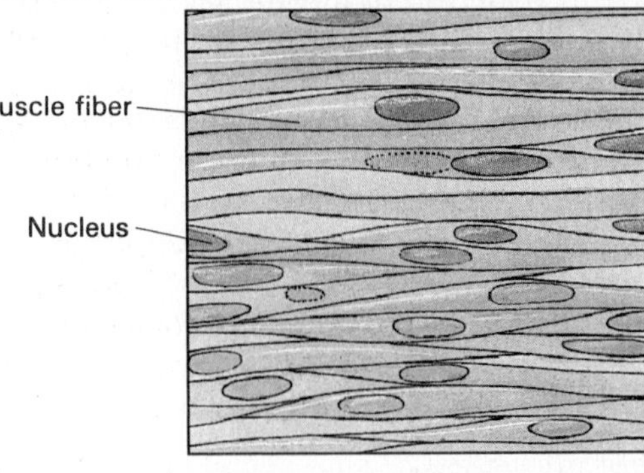
Muscle fiber
Nucleus

(c)

zation of their actin and myosin filaments. Because each will be examined in later chapters (skeletal muscle in Chapter 10, cardiac muscle in Chapter 20, and smooth muscle in Chapter 24) this discussion will focus on general characteristics rather than specific details.

SKELETAL MUSCLE TISSUE

Skeletal muscle tissue contains very large fibers (cells) tied together by loose connective tissue. The collagen and elastic fibers surrounding each cell and group of cells blend into those of a tendon or aponeurosis that conducts the force of contraction, usually to a bone of the skeleton. When the muscle tissue contracts, the bone moves.

Skeletal muscle fibers are very unusual because they may be a foot or more in length, and each cell contains hundreds of nuclei. They are extremely specialized cells that are incapable of undergoing division. However, new muscle fibers can be produced through the division of **satellite cells**, mesenchymal cells that persist in adult skeletal muscle tissue. Thus skeletal muscle tissue can at least partially repair itself after an injury.

Because the actin and myosin filaments are arranged in organized groups, skeletal muscle fibers appear to have a series of bands or stripes running across them. These bands, also known as striations (*striae*), can be seen in Figure 5-20a. Skeletal muscle fibers will not usually contract unless stimulated by nerves, and the nervous system provides voluntary control over their activities. Summarizing these properties, skeletal muscle can be considered as **striated voluntary muscle**.

CARDIAC MUSCLE TISSUE

Cardiac muscle tissue, seen in Figure 5-20b, is found only in the heart. The muscle fibers in cardiac muscle tissue are not as large as those in skeletal muscle tissue, and there is only one nucleus in each cell. Cardiac muscle fibers, or **cardiocytes**, form extensive junctional complexes with one another, creating a branching network that conducts the force and stimulus for contraction from one area of the heart to another. Like skeletal muscle fibers, cardiac muscle fibers are incapable of undergoing division, and this tissue does not contain satellite cells. As a result, cardiac muscle tissue damaged by injury or disease cannot regenerate.

The actin and myosin filaments are organized as they are in skeletal muscle fibers, and the muscle fibers have prominent striations. Unlike skeletal muscle, cardiac muscle fibers do not rely on nerve activity to tell them when to start a contraction. Instead, specialized cardiac muscle fibers, called **pacemaker cells**, establish a regular rate of contraction. Although the nervous system can alter the rate of pacemaker activity, it does not provide voluntary control over individual cardiac muscle fibers. In short, cardiac muscle can be considered as **striated involuntary muscle**.

SMOOTH MUSCLE TISSUE

Smooth muscle tissue, shown in Figure 5-20c, can be found in the walls of blood vessels; around hollow organs such as the urinary bladder; and in layers around the respiratory, circulatory, digestive, and reproductive tracts.

A smooth muscle fiber is small and spindle-shaped, with a single nucleus. These cells are capable of division, and smooth muscle tissue can regenerate after an injury. The actin and myosin filaments in smooth muscle fibers are organized differently from either skeletal or cardiac muscle. Filaments are scattered throughout the cytoplasm, and there are no striations.

Smooth muscle fibers may contract on their own, or their contractions may be triggered by nerve cell activity. The nervous system usually does not provide voluntary control over smooth muscle contractions, and smooth muscle is therefore categorized as **nonstriated involuntary muscle**.

■ Neural Tissue

Neural tissue is specialized for the conduction of electrical impulses that convey information or instructions from one region of the body to another. Most of the neural tissue in the body (98 percent) is concentrated in the brain and spinal cord, the control centers for the nervous system.

Neural tissue contains two basic types of cells: nerve cells, or **neurons** (NOO-rons; *neuro*, nerve), and several different kinds of supporting cells, or **neuroglia** (noo-RŌG-lē-a; *glia*, glue). Neurons transmit the actual signals, which take the form of changes in the transmembrane potential. Neuroglia have four basic functions. They (1) isolate and protect the cell membranes of neurons, (2) provide a supporting framework for neural tissue, (3) regulate the composition of the interstitial fluid, and (4) act as phagocytes that defend neural tissue from pathogens and assist in repair of injuries.

Chapter 12, which considers the properties of neural tissue, provides more detail on its histology and cytology. However, a few general comments will provide background for discussions of the integumentary, skeletal, and muscular systems (Chapters 6–11).

The structure of a representative neuron is diagrammed in Figure 5-21. Neurons are the longest cells in the body, reaching a meter in length; they

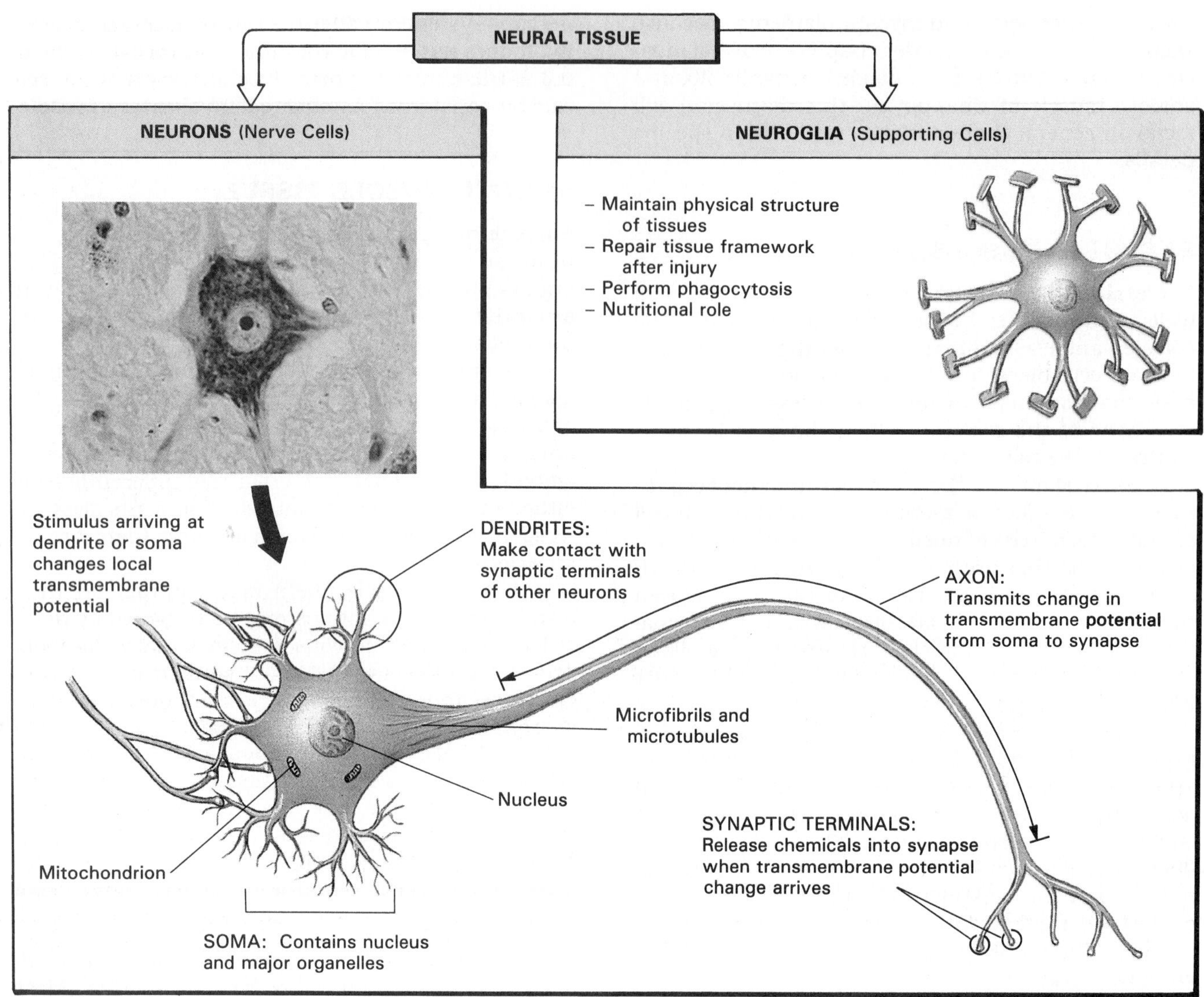

FIGURE 5-21
Neural Tissue

are often called **nerve fibers**, just as muscle cells are called muscle fibers. Most neurons are incapable of undergoing division, and neural tissue has a very limited ability to repair itself after injury.

A typical neuron has a cell body, or **soma**, that contains the nucleus. The stimulus that results in the production of an electrical impulse usually affects the cell membrane of one of the **dendrites** (DEN-drīts; *dendron*, a tree). Stimulation alters the permeability of the cell membrane, eventually producing a massive change in the transmembrane potential that sweeps along the length of the **axon**.

The axon ends at a specialized intercellular junction called a **synapse** (SIN-aps; *syn*, together). On reaching the end of the axon, the change in the transmembrane potential triggers the release of a chemical **neurotransmitter**. This compound diffuses across the synapse and contacts receptors on the opposing cell membrane. The effects of neurotransmitter binding vary depending on the nature of the chemical, the properties of the receptor, and the identity of the stimulated cell.

- √ **The lining of the nasal cavity is normally moist, contains numerous goblet cells, and rests on a layer of connective tissue called the lamina propria. What type of membrane is this?**
- √ **What type of muscle tissue has small, spindle-shaped cells with single nuclei and no obvious banding pattern?**
- √ **Why do you find the same epithelial organization in the pharynx, esophagus, anus, and vagina?**

The Connective Tissue Framework of the Body

Connective tissues create the internal framework of the body. Layers of connective tissue proper connect the organs within the dorsal and ventral body cavities with the rest of the body. These layers (1) provide strength and stability, (2) maintain the relative positions of internal organs, and (3) provide a route for the distribution of blood vessels, lymphatics, and nerves. Outside the body cavities, layers of dense connective tissue surround and separate individual organs, such as skeletal muscles, nerves, or glands. In other areas, regions of loose connective tissue provide padding and flexible support. The entire array of connective tissue layers and wrappings can be divided into three major components: the *superficial fascia*, the *deep fascia*, and the *subserous fascia*. The structural and functional relationships between these layers are diagrammed in Figure 5-22.

The **superficial fascia** is a layer of loose connective tissue that separates the skin from underlying tissues and organs. It has several other names, including the **subcutaneous layer** (*sub*, below + *cutis*, skin) and the **hypodermis** (*hypo*, below + *dermis*, skin). This layer, which will be discussed in further detail in Chapter 6, provides insulation and padding. It also lets the skin or underlying structures move independently.

The **deep fascia** consists of dense connective tissue. The capsules that surround most organs, including the kidneys and the organs in the thoracic and peritoneal cavities, are components of the deep fascia. The perichondrium around cartilages, the periosteum around bones and the ligaments that interconnect them, and the connective tissues of muscle, including tendons and aponeuroses, are all part of the deep fascia. Another layer of dense connective tissue surrounds the body cavities. The dense connective tissue components are interwoven; for example, the deep fascia around a muscle blends into the tendon, whose fibers intermingle with those of the periosteum. This arrangement creates a strong, fibrous network for the body and ties structural elements together.

The **subserous fascia** is a layer of loose connective tissue that lies between the deep fascia and the serous membranes that line body cavities. Because this layer separates the serous membranes from the deep fascia, movements of muscles or muscular organs do not severely distort the delicate lining.

Tissue Repairs

Tissues in the body are not isolated; they combine to form organs with diverse functions. Any injury to the body affects several tissue types simultaneously, and these tissues respond in a coordinated manner. The restoration of homeostasis after an injury involves two related processes. First, the area is isolated while damaged cells or other tissue components are removed and any dangerous microorganisms or harmful chemicals are destroyed. This defense, which involves the coordinated activities of several tissues, is called **inflammation**, or the **inflammatory response**. Inflammation begins immediately

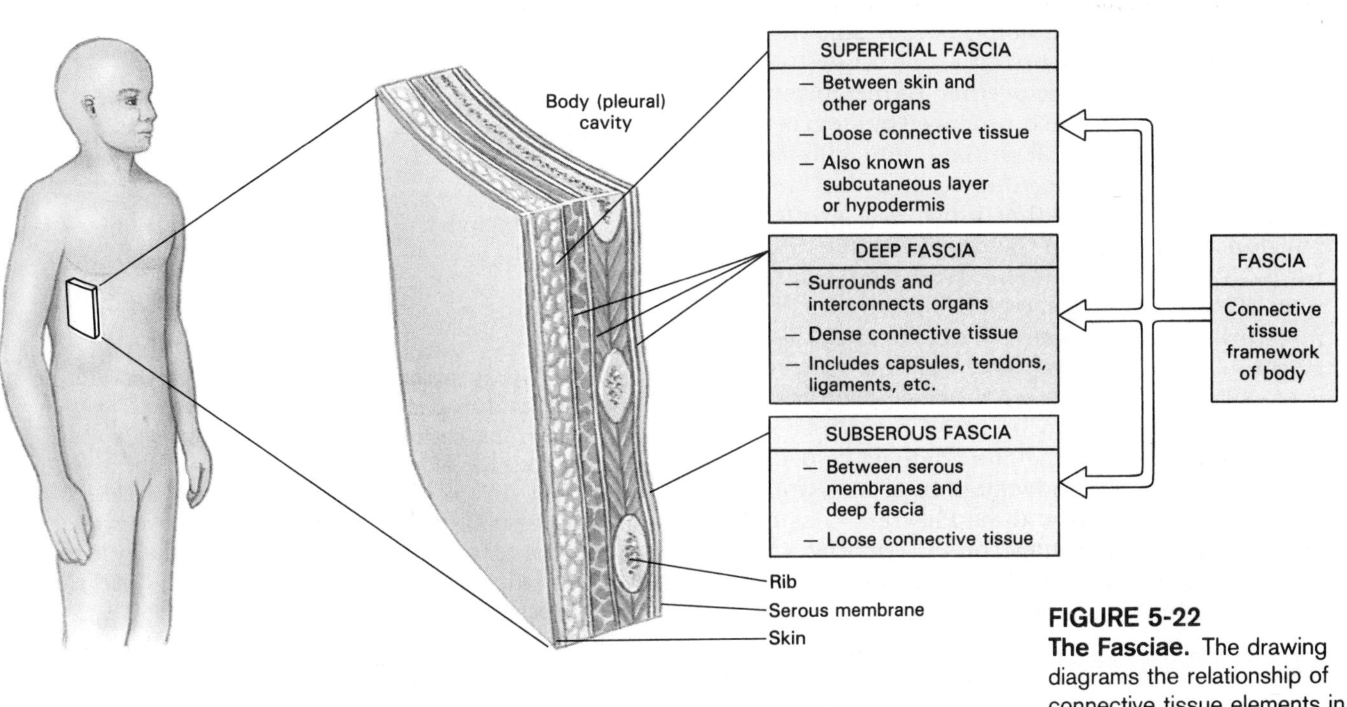
Body (pleural) cavity
SUPERFICIAL FASCIA
– Between skin and other organs
– Loose connective tissue
– Also known as subcutaneous layer or hypodermis
DEEP FASCIA
– Surrounds and interconnects organs
– Dense connective tissue
– Includes capsules, tendons, ligaments, etc.
SUBSEROUS FASCIA
– Between serous membranes and deep fascia
– Loose connective tissue
FASCIA
Connective tissue framework of body
Rib
Serous membrane
Skin

FIGURE 5-22
The Fasciae. The drawing diagrams the relationship of connective tissue elements in the body.

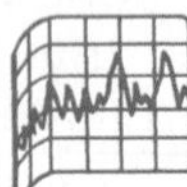

Diagnostics: Tissue Structure and Disease

Pathologists (pa-THOL-o-jists) are physicians who specialize in the diagnosis of disease processes. In their analyses they synthesize anatomical and histological observations to determine the nature and severity of the disease. Pathologists seldom interact with patients directly, and they are more concerned with diagnosis than treatment.

Disease processes affect the histological organization of tissues and organs. Figure 5-23 diagrams the histological changes induced by one relatively common irritating stimulus, cigarette smoke. The first abnormality to be observed is **dysplasia** (dis-PLĀ-zē-a), a change in the normal shape, size, and organization of tissue cells. It is usually a response to chronic irritation or inflammation, and the changes are reversible. In our example, the normal trachea (windpipe) is lined by a pseudostratified, ciliated, columnar epithelium. The cilia move a mucus layer that traps foreign particles and moistens incoming air. The drying and chemical effects of smoking first paralyze the cilia, halting the movement of mucus (Figure 5-23a). As mucus builds up, the individual coughs to dislodge it; this produces "smoker's cough."

Epithelia and connective tissues may undergo more radical changes in structure, caused by the division and differentiation of stem cells. **Metaplasia** (me-ta-PLĀ-zē-a) is a structural change that dramatically alters the character of the tissue. In our example, heavy smoking first paralyzes the cilia, but over time the epithelial cells lose their cilia altogether. As metaplasia occurs, the epithelial cells produced by stem cell division no longer differentiate into ciliated columnar cells. Instead, they form a stratified squamous epithelium that provides a greater resistance to drying and chemical irritation (Figure 5-23b). This protects the underlying tissues more effectively, but it completely eliminates the moisturizing and cleaning function of the epithelium. The cigarette smoke will now have an even greater effect on more delicate portions of the respiratory tract. Fortunately, metaplasia is reversible, and the epithelium gradually returns to normal once the individual quits smoking.

During **anaplasia** (a-na-PLĀ-zē-a) tissue cells change size and shape, often becoming unusually large or abnormally small. In anaplasia (Figure 5-23c), which occurs in smokers developing one form of lung cancer, tissue organization breaks down. Cell divisions are frequent, but not all proceed in the normal way, and many of the tumor cells have abnormal chromosomes. Unlike dysplasia and metaplasia, anaplasia is irreversible.

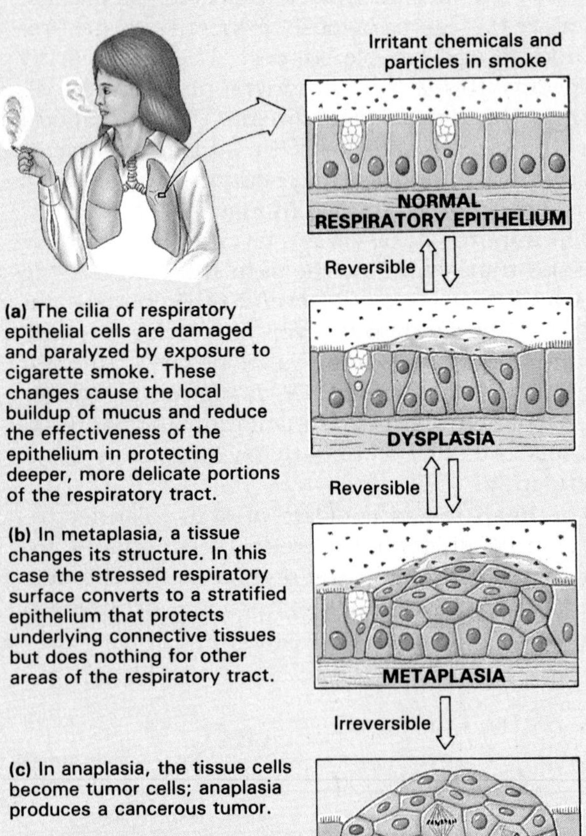

Irritant chemicals and particles in smoke
NORMAL RESPIRATORY EPITHELIUM
Reversible
(a) The cilia of respiratory epithelial cells are damaged and paralyzed by exposure to cigarette smoke. These changes cause the local buildup of mucus and reduce the effectiveness of the epithelium in protecting deeper, more delicate portions of the respiratory tract.
DYSPLASIA
Reversible
(b) In metaplasia, a tissue changes its structure. In this case the stressed respiratory surface converts to a stratified epithelium that protects underlying connective tissues but does nothing for other areas of the respiratory tract.
METAPLASIA
Irreversible
(c) In anaplasia, the tissue cells become tumor cells; anaplasia produces a cancerous tumor.
ANAPLASIA

FIGURE 5-23
Changes in a Tissue under Stress. (a) The cilia of respiratory epithelial cells are damaged and paralyzed by exposure to cigarette smoke. This causes the local buildup of mucus and reduces the protection for more delicate portions of the respiratory tract. (b) The stressed respiratory surface converts to a stratified epithelium that protects underlying connective tissues but does nothing for other areas of the respiratory tract. (c) The tissue cells may then become tumor cells that multiply to produce a cancer.

following an injury and produces several familiar sensations, including swelling, redness, and pain. Second, the damaged tissues are replaced or repaired to restore normal function. This repair process is called **regeneration**.

Inflammation and regeneration are controlled at the tissue level. The two phases overlap; isolation establishes a framework that guides the cells responsible for reconstruction, and repairs are under way well before cleanup operations have ended.

The discussion that follows provides a brief overview of these processes; later chapters, especially Chapter 22, will examine inflammation in more detail. At this time, we will focus on the interaction between different tissues. Our example includes two connective tissues (loose connective tissue and blood), an epithelium (the endothelia of blood vessels), a muscle tissue (smooth muscle in the vessel walls), and neural tissue (sensory nerve endings).

FIRST PHASE: INFLAMMATION

Many stimuli can produce inflammation, including impact, abrasion, distortion, chemical irritation, infection by pathogenic organisms (such as bacteria or viruses), or extreme temperatures (hot or cold). The common factor is that each of these stimuli either kills cells, damages fibers, or injures the tissue in some other way. These changes alter the chemical composition of the interstitial fluid: damaged cells release prostaglandins, proteins, and potassium ions, and the injury itself may have introduced foreign proteins or pathogens.

Immediately after the injury, tissue conditions become even more abnormal. **Necrosis** (ne-KRŌ-sis) refers to the tissue degeneration that occurs after cells have been injured or destroyed. The process begins several hours after the initial event, and the damage is caused by lysosomal enzymes. Lysosomes, described in Chapter 3, break down through autolysis, releasing digestive enzymes that first destroy the injured cells and then attack surrounding tissues. ∞ (p. 75) The accumulation of debris, fluid, dead and dying cells, and necrotic tissue components that may result is known as **pus**. Pus often forms at an infection site in the dermis. An accumulation of pus in an enclosed tissue space is called an **abscess**.

These tissue changes trigger the inflammatory response by stimulating mast cells, connective tissue cells introduced earlier in this chapter. Figure 5-24 follows the events set in motion by the activation of mast cells. Although in this example the injury occurs in a loose connective tissue, the same basic process would occur after injuries to any connective tissue proper. Because these connective tissues are found in all organs, inflammation can occur anywhere in the body.

When an injury occurs that damages fibers and cells, the stimulated mast cells release chemicals (**histamine** and **heparin**). These chemicals stimulate local macrophages, affecting smooth muscle tissue and endothelial cells in the region. The combination of abnormal tissue conditions and chemicals released by mast cells stimulates local neurons almost at once. The pain sensations generated trigger immediate action (such as pulling the affected part of the body away from the source of the injury) that reduces the chances for further damage.

Other responses are diagrammed in Figure 5-24. Three major changes are highlighted.

FIGURE 5-24
Inflammation. A flow chart for the inflammation process.

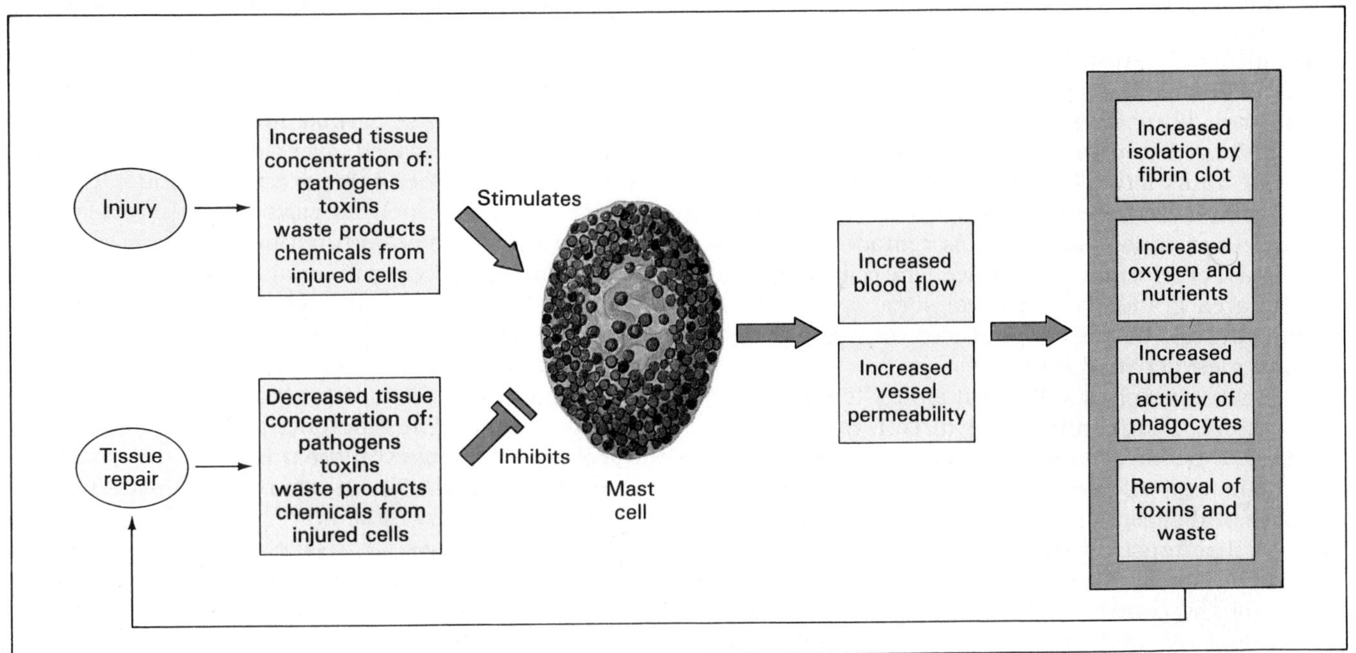

- Histamine relaxes the smooth muscle tissue in the vessel walls, and the vessels enlarge, or **dilate**. This dilation increases the blood flow through the tissue, giving the region a reddish color and making it warm to the touch. The increased blood flow accelerates the delivery of nutrients and oxygen and the removal of dissolved waste products and toxic chemicals. It also brings white blood cells to the region. These cells migrate into the injury site and assist in the defense and cleanup operations.
- Histamine makes the endothelial cells of local capillaries more permeable. Plasma, including blood proteins, now diffuses into the injured tissue, and the area becomes swollen. Once in the tissue, some of the blood proteins combine to form large, insoluble fibers, known as **fibrin**. The heparin released by mast cells prevents the formation of fibrin at the injury site, but a dense fibrous meshwork, called a **clot**, surrounds the damaged area. The clot walls off the inflamed region, slowing the spread of cellular debris or bacteria into adjacent tissues.
- Stimulated macrophages and newly arrived white blood cells defend the tissue and perform cleanup operations. Fixed macrophages, free macrophages, monocytes, and microcytes phagocytize debris and bacteria. Lymphocytes convert to plasma cells, producing antibodies that attack pathogens and foreign proteins. Usually the combination of physical attack, through phagocytosis, and chemical attack, through antibodies, succeeds in cleaning up the region and eliminating the inflammatory stimulus.

SECOND PHASE: REGENERATION

Although the situation is under control and no further injury will occur, many cells in the area have died, either as a result of the initial injury or subsequent regional changes. As tissue conditions return to normal, fibroblasts move into the necrotic area, laying down a network of collagen fibers that stabilizes the injury site. This process, called **fibrosis**, produces a dense, collagenous framework known as **scar tissue**. Over time, scar tissue is remodeled and assumes a more normal appearance. The cell population in the area gradually increases; some cells migrate to the site, and others are produced through the division of mesenchymal cells.

Each organ has a different ability to regenerate after injury. That ability can be directly linked to the pattern of tissue organization in the injured organ. Epithelia and connective tissues regenerate well, smooth and skeletal muscle tissues relatively poorly, and cardiac muscle tissue and neural tissue cannot regenerate at all. The skin, which is dominated by epithelia and connective tissues, regenerates rapidly and completely after injury. This process will be detailed in Chapter 6. In contrast, damage to the connective tissues of the heart can be repaired, but lost cardiac muscle fibers are replaced by scar tissue rather than by functional muscle tissue.

After any injury, the most complete regeneration will occur in young, healthy individuals. Repairs are always slowed by poor health, continuing infection, or an inadequate diet. Age is also a factor, for healing takes more time in elderly patients, even healthy ones.

√ **Why do antihistamines decrease the amount of swelling that occurs during inflammation?**

Tissues, Nutrition, and Aging

Each tissue must be maintained over time, through the process of metabolic turnover (Chapter 4). (p. 117) To maintain tissue homeostasis, tissue cells require adequate amino acids, lipids, energy sources, and vitamins. The precise requirements differ from tissue to tissue. For example, neurons can catabolize only glucose for energy, and they need certain amino acids to synthesize neurotransmitters. Cardiac muscle fibers, which can metabolize lipids or carbohydrates, require different amino acids to build actin and myosin filaments.

Tissues change with age. In general, repair and maintenance activities grow less efficient, and a combination of hormonal changes and alterations in lifestyle affect the structure and chemical composition of many tissues. Epithelia get thinner, and connective tissues more fragile. The thin skin becomes translucent, and the individual bruises easily. Cartilages become stiffer and less resilient, and bones become brittle; joint pains and broken bones are common complaints. Because cardiac muscle fibers and neurons cannot be replaced, over time cumulative losses from relatively minor damage can contribute to major health problems such as cardiovascular disease or deterioration in mental function.

Future chapters will consider the effects of aging on specific organs and systems. Some of these changes are genetically programmed. For example, the chondrocytes of elderly individuals produce a slightly different form of proteoglycan than those of younger people. The difference probably accounts for the observed changes in the thickness and resilience of cartilage. Treating an age-related tissue change of this kind is very difficult.

In other cases the tissue degeneration may be temporarily slowed or even reversed. The age-related reduction in bone strength in women is often caused by a combination of inactivity, low dietary calcium levels, and a reduction in circulating *estrogens* (female sex hormones). A program of exercise, calcium supplements, and hormonal replacement therapies can usually maintain normal bone structure for many years.

This chapter completes the introductory portion of this text. Chapter 5 has introduced the four basic types of tissue found in the human body. In combination these tissues form all of the organs and systems discussed in subsequent chapters. Before proceeding to those chapters take the time to examine the Systems Overview that follows this chapter. The figures in the Overview will help you integrate information contained in the introductory chapters with material covered in later chapters.

Clinical Comment: Cancer III—Development and Treatment

Physicians who specialize in the identification and treatment of cancers are called **oncologists** (on-KOL-o-jists; *onkos*, mass). Pathologists and oncologists classify cancers according to their cellular appearance and their sites of origin. Over a hundred kinds have been described, but broad categories are usually used to indicate the location of the primary tumor. Table 5-2 summarizes information concerning benign and malignant tumors (cancers) associated with the tissues discussed in this chapter. Additional details will be provided in chapters dealing with specific organs and systems.

TABLE 5-2 Benign and Malignant Tumors in the Major Tissue Types

Tissue	*Description*
Epithelia	
Carcinoma	Any cancer of epithelial origin
Adenocarcinoma	Cancers of glandular epithelia
Angiosarcomas	Cancers of endothelial cells
Mesotheliomas	Cancers of mesothelial cells
Connective tissues	
Fibromas	Benign tumors of fibroblast origin
Lipomas	Benign tumors of adipose tissue
Liposarcomas	Cancers of adipose tissue
Leukemias, lymphomas	Cancers of blood-forming tissues
Chondromas	Benign tumors in cartilage
Chondrosarcomas	Cancers of cartilage
Osteomas	Benign tumors in bone
Osteosarcomas	Cancers of bone
Muscle tissues	
Myxomas	Benign muscle tumors
Myosarcomas	Cancers of skeletal muscle tissue
Cardiac sarcomas	Cancers of cardiac muscle tissue
Leiomyomas	Benign tumors of smooth muscle tissue
Leiomyosarcomas	Cancers of smooth muscle tissue
Neural tissues	
Gliomas, neuromas	Cancers of neuroglial origin

Steps in Cancer Formation Cancer develops in a series of steps diagrammed in Figure 5-25. Initially the cancer cells are restricted to a single location, called the **primary tumor** or **primary neoplasm**. Usually, all of the cells in the tumor are the daughter cells of a single malignant cell. At first the growth of the primary tumor simply distorts the tissue, and the basic tissue organization remains intact. Metastasis begins as tumor cells "break out" of the primary tumor and invade the surrounding tissue. When this invasion is followed by penetration of nearby blood vessels, the cancer cells begin circulating throughout the body.

Responding to cues that are as yet unknown, these cells later escape from the circulatory system and establish **secondary tumors** at other sites. These tumors are extremely active metabolically, and their presence stimulates the growth of blood vessels into the area. The increased circulatory supply provides additional nutrients and further accelerates tumor growth and metastasis. Death may occur as a result of compression of vital organs, because nonfunctional cancer cells have killed or replaced the normal cells in vital organs, or because the voracious cancer cells have starved normal tissues of essential nutrients.

Treatment It is unfortunate that the media tend to describe cancer as though it were one disease rather than many. This simplistic perspective fosters the belief that some dietary change, air ionizer, or wonder drug will be found that can prevent the affliction. There is no single, universally effective cure or preventive measure for cancer; there are too many separate causes, possible mechanisms, and individual differences.

The goal of cancer treatment is to achieve **remission**. A tumor in remission either ceases to grow or decreases in size. Basically, the treatment of malignant tumors must accomplish one of the following to produce remission.

1. **Surgical removal or destruction of individual tumors:** Tumors that contain malignant cells can be surgically removed or destroyed by radiation, heat, or freezing. These techniques are very effective if the treatment is undertaken before extensive metastasis has occurred. For this reason early detection is important in improving survival rates for all forms of cancer.
2. **Killing metastatic cells throughout the body:** This is much more difficult and potentially dangerous, because healthy tissues are likely to be damaged at the same time. At present the most widely approved treatments are chemotherapy and radiation.

Chemotherapy involves the administration of drugs that will either kill the cancerous tissues or prevent mitotic divisions. These drugs often affect stem cells in normal tissues, and the side effects are usually quite unpleasant. For example, because chemotherapy slows the regeneration and maintenance of epithelia of the skin and digestive tract, patients lose their hair and experience nausea and vomiting. Several drugs are often administered simultaneously or in sequence, because over time cancer cells can develop a resistance to a single drug. This strategy also tends to reduce the severity of any side effects. Chemotherapy is often used in the treatment of many kinds of metastatic cancer.

Localized radiation is often used to destroy cancers confined to a small region of the body. Massive doses of radiation are necessary if the cancer has metastasized widely. For example, in advanced cases of lymphoma, a cancer of the immune system, enough radiation is administered to kill all of the blood-forming cells in the body. After treatment, new blood cells must be provided by a bone marrow transplant. Chapters 19 and 22 will provide more detailed information about marrow transplants, lymphomas, and other cancers of the blood.

An understanding of molecular mechanisms and cell biology is leading to new approaches that may revolutionize the cancer treatment. One approach focuses on the fact that cancer cells are usually ignored by the immune system. In **immunotherapy**, chemicals are administered that help the immune system recognize and attack the cancer cells. More elaborate experimental procedures involve the creation of customized antibodies using the gene-splicing techniques discussed in Chapter 4. (p. 116) The resulting antibodies are specifically designed to attack the tumor cells in one particular patient. Although this technique shows promise, it remains difficult, costly, and very labor-intensive.

Another new procedure builds upon immunotherapy (and on the radiation concepts treated in Chapter 2). (p. 29) In **boron neutron capture therapy (BNCT)** antibodies made

FIGURE 5-25
The Development of Cancer

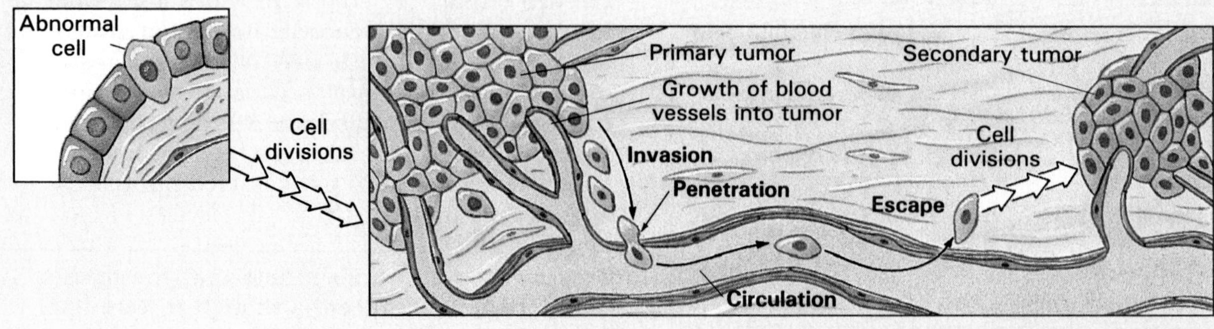

TABLE 5-3 Cancer Incidence and Survival Rates in the United States

Site	Estimated New Cases (1991)	Estimated Deaths (1991)	Five-Year Survival Rates: Diagnosis Date 1960–63	Five-Year Survival Rates: Diagnosis Date 1981–86
DIGESTIVE TRACT				
Esophagus	10,900	9,800	2.5%	7.5%
Stomach	23,800	13,400	9.5%	17%
Colon and rectum	157,500	60,500	36%	53%
RESPIRATORY TRACT				
Lung and bronchus	161,000	143,000	6.5%	13%
URINARY TRACT				
Kidney and Other Urinary	25,300	10,600	37.5%	52%
Bladder	50,200	9,500	38.5%	76%
REPRODUCTIVE SYSTEM				
Breast	175,900	44,800	54%	75%
Ovary	20,700	12,500	32%	38%
Testis	6,100	375	63%	88%
Prostate gland	122,000	32,000	42.5%	71%
NERVOUS SYSTEM	16,700	11,500	18.5%	28%
CIRCULATORY SYSTEM	84,900	47,500	14%[a]	45%[a]
SKIN (MELANOMA ONLY)	32,000	8,500	60%	81%

Data courtesy of the American Cancer Society.
[a] Lymphocytic leukemia only.

to attack cancer cells are labeled with an isotope of boron (B). After these antibodies have been administered and have bound to cancer cells, the patient is irradiated with neutrons. These neutrons do not damage normal tissues. However, the boron atoms absorb neutrons and then release alpha particles (2 neutrons + 2 protons), radiation that kills the cancer cells quite effectively. Because alpha particles do not penetrate far through body tissues, they affect only the cancer cells that have been "tagged" with boron atoms. Other, normal tissues remain unaffected.

Incidence and Survival Rates A statistical profile of cancer incidence and survival rates in the United States is presented in Table 5-3. Interestingly, the profile would be different for other countries, reflecting different environmental conditions and cultural dietary preferences. Bladder cancer is common in Egypt, stomach cancer in Japan, and liver cancer in Africa.

Advances in chemotherapy, radiation procedures, and molecular biology have produced significant improvements in the survival rates for several types of cancer. However, the improved survival rates indicated in Table 5-3 reflect not only advances in therapy but also in early detection. Much of the credit goes to increased public awareness and concern about cancer. In general, the odds of survival increase markedly if the cancer is detected early, especially before it undergoes metastasis. Despite the variety of possible cancers, the American Cancer Society has identified seven "warning signs" that mean it's time to consult a physician. These warning signs are presented in Table 5-4.

TABLE 5-4 Seven Warning Signs of Cancer

Change in bowel or bladder habits
A sore that does not heal
Unusual bleeding or discharge
Thickening or lump in breast or elsewhere
Indigestion or difficulty in swallowing
Obvious change in wart or mole
Nagging cough or hoarseness

CHAPTER REVIEW

Review of Selected Clinical Terms

liposuction (*p. 141*)
A surgical procedure to remove unwanted adipose tissue by sucking it through a tube.

adhesions (*p. 154*)
Restrictive fibrous connections that can result from surgery, infection, or other injuries to serous membranes.

pleuritis (pleurisy) (*p. 155*)
An inflammation of the lining of the pleural cavities.

effusion (*p. 155*)
The accumulation of fluid in body cavities.

pericarditis (*p. 155*)
An inflammation of the pericardium.

peritonitis (*p. 155*)
An inflammation of the peritoneum.

ascites (a-SĪ-tēz) (*p. 155*)
An accumulation of fluid that creates a characteristic abdominal swelling.

pathologists (pa-THOL-o-jists) (*p. 160*)
Physicians who specialize in diagnosing disease processes.

dysplasia (dis-PLĀ-zē-a) (*p. 160*)
A change in the normal shape, size, and organization of tissue cells.

metaplasia (me-ta-PLĀ-zē-a) (*p. 160*)
A structural change that alters the character of a tissue.

anaplasia (a-na-PLĀ-zē-a) (*p. 160*)
An irreversible change in the size and shape of tissue cells.

regeneration (*p. 161*)
The process of repairing injured tissues that follows inflammation.

clot (*p. 162*)
A dense fibrous meshwork that surrounds an injured area of tissue.

oncologists (on-KOL-o-jists) (*p. 163*)
Physicians who specialize in identifying and treating cancers.

primary tumor (primary neoplasm) (*p. 163*)
The site at which a cancer initially develops.

secondary tumor (*p. 163*)
A colony of cancerous cells formed by metastasis, the spread of cells from a primary tumor.

remission (*p. 164*)
A stage in which a tumor stops growing or grows smaller; the goal of cancer treatment.

chemotherapy (*p. 164*)
Administering drugs that either kill cancerous tissues or prevent mitotic divisions.

immunotherapy (*p. 164*)
Administering drugs that help the immune system recognize and attack cancer cells.

Additional Terms of Clinical Importance

GROUND SUBSTANCE AND RESISTANCE TO INFECTION (*p. 140*): **hyaluronidase.**

'JAWS' AND THE FIGHT AGAINST CANCER (*p. 147*): **antiangiogenesis factor.**

PROBLEMS WITH SEROUS MEMBRANES (*p. 155*): **plural rub, pleural effusion, pericardial effusion**

FIRST PHASE, INFLAMMATION (*pp. 159, 161*): **inflammation (inflammatory response), necrosis, histamine, heparin**

SECOND PHASE: REGENERATION (*p. 162*): **fibrosis, scar tissue**

CANCER III—DEVELOPMENT AND TREATMENT: (*p. 164*): **boron neutron capture therapy**

Study Outline

Related Key Terms

Introduction (pp. 123–124)

1. **Tissues** are collections of specialized cells and cell products that are organized to perform a relatively limited number of functions. There are four **primary tissue types**: epithelia, connective tissue, muscle tissue, and neural tissue. **Histology** is the study of tissues.

Epithelia (pp. 124–137)

1. An **epithelium** is an **avascular** layer of cells that forms a barrier with certain properties.

FUNCTIONS OF EPITHELIA (p. 126)

2. Epithelia provide physical protection, control permeability, provide sensation, and produce specialized secretions. **Gland cells** are epithelial cells that produce secretions. — **glandular epithelium**
3. **Exocrine** secretions leave the body; **endocrine** secretions, known as **hormones**, are released by gland cells into the surrounding tissues. — **ducts**

SPECIALIZATIONS OF EPITHELIAL CELLS (pp. 126–129)

4. Epithelial cells are specialized to allow them to maintain the physical integrity of the epithelium and to perform secretory or transport

functions. A network of **interfacial canals** may help distribute essential nutrients through the epithelial layers.

5. The inner surface of each epithelium is connected to a two-part **basement membrane** consisting of a **basal lamina** and a **reticular lamina**. Divisions by **germinative cells** continually replace the short-lived epithelial cells.
6. Epithelial cells may show **polarity**, and often have microvilli.
7. The coordinated beating of the cilia on a **ciliated epithelium** moves materials across the epithelial surface.

CLASSIFICATION OF EPITHELIA (pp. 129–134)

8. Epithelia are classified on the basis of the number of cell layers and the shape of the exposed cells.
9. A **simple epithelium** has a single layer of cells covering the basement membrane, while a **stratified epithelium** has several layers. In a **squamous epithelium** the cells are thin and flat. Cells in a **cuboidal epithelium** resemble little hexagonal boxes; those in a **columnar epithelium** are taller and more slender.

GLANDULAR EPITHELIA (pp. 135–137)

10. A glandular epithelial cell may release its secretions through merocrine, apocrine, or holocrine mechanisms.
11. In **merocrine secretion**, the most common method of secretion, the product is released through exocytosis. **Apocrine secretion** involves the loss of both secretory product and cytoplasm. Unlike the first two methods, **holocrine secretion** destroys the cell; it becomes packed with secretions and finally bursts.
12. In epithelia that contain scattered gland cells, individual secretory cells are called **unicellular glands**. **Multicellular glands** are organs that contain glandular epithelia that produce exocrine or endocrine secretions.

Related Key Terms

stereocilia
epithelial transport

mesothelium • endothelium

serous glands • mucous glands
mixed glands • goblet cells
unicellular exocrine glands
mucus
multicellular exocrine glands
secretory sheet • excretory ducts

Connective Tissues (pp. 138–152)

1. Connective tissues are internal tissues with many important functions: establishing a structural framework; transporting fluids and dissolved materials; protecting delicate organs; supporting, surrounding, and interconnecting tissues; storing energy reserves; and defending the body from microorganisms.
2. All connective tissues have specialized cells, extracellular protein fibers, and a **ground substance**.

CLASSIFICATION OF CONNECTIVE TISSUES (p. 138)

3. **Connective tissue proper** refers to connective tissues that contain varied cell populations and fiber types surrounded by a syrupy ground substance.
4. **Fluid connective tissues** have a distinctive population of cells suspended in a watery ground substance containing dissolved proteins. The two types are blood and lymph.
5. **Supporting connective tissues** have a less diverse cell population than connective tissue proper and a dense ground substance that contains closely packed fibers; the combination of fibers and ground substance is called a **matrix**. The two types of supporting connective tissues are cartilage and bone.

CONNECTIVE TISSUE PROPER (pp. 138–146)

6. Connective tissue proper contains fibers, a viscous ground substance, and two types of cells: **fixed cells** and **wandering cells**.
7. There are three types of fiber in connective tissue: **collagen fibers**, **reticular fibers**, and **elastic fibers**.
8. Connective tissue proper is classified as **loose** or **dense connective tissues.** There are three types of loose connective tissues: **loose connective tissue** or **areolar tissue, adipose tissue,** and **reticular tissue.** Most of the volume in dense connective tissue consists of fibers. There are two types of dense connective tissue: **dense regular connective tissue** and **dense irregular connective tissue.**

fibroblasts • fixed macrophages
adipose cells • adipocytes
melanin • melanocytes
mast cells • histamine
microphages • plasma cells
antibodies • mesenchymal cells
stroma • collagenous • tendons
aponeuroses • ligaments
fasciae • elastic tissue
elastic ligaments • capsule

FLUID CONNECTIVE TISSUES (pp. 146–147)

9. **Blood** and **lymph** are connective tissues that contain distinctive collections of cells in a fluid matrix.
10. Blood contains **red blood cells**, **white blood cells**, and **platelets**. Its watery ground substance is called **plasma**.
11. **Arteries** carry blood from the heart and toward **capillaries**, where water and small solutes move into the **interstitial fluid** of surrounding tissues. **Veins** return blood to the heart.

erythrocyte • leukocytes
monocytes • platelets

Related Key Terms

12. **Lymph** forms as interstitial fluid enters the **lymphatics** which return lymph to the circulatory system.

SUPPORTING CONNECTIVE TISSUES (pp. 147–152)

13. Cartilage and bone are called **supporting connective tissues** because they support the rest of the body. — **lacunae**
14. The matrix of **cartilage** is a firm gel that contains **chondroitin sulfates** (proteoglycans) and cells called **chondrocytes**. A fibrous **perichondrium** separates cartilage from surrounding tissues. There are three types of cartilage: **hyaline cartilage**, **elastic cartilage**, and **fibrocartilage**.
15. Cartilage grows by two different mechanisms, **interstitial growth** and **appositional growth**. — **calcified cartilage**
16. Chondrocytes rely upon diffusion through the avascular matrix to obtain nutrients.
17. **Bone**, or **osseous tissue**, has a matrix consisting of collagen fibers and calcium salts, giving it unique properties.
18. **Osteocytes** depend on diffusion through **canaliculi** for nutrient intake.
19. Each bone is surrounded by a **periosteum** with fibrous and cellular layers.

Membranes (pp. 152–155)

1. Membranes form a barrier or interface. Epithelia and connective tissues combine to form membranes that cover and protect other structures and tissues. There are four types of membranes: *mucous, serous, cutaneous,* and *synovial.*
2. **Mucous membranes** line cavities that communicate with the exterior. Their surfaces are normally moistened by mucous secretions. — **lamina propria**
3. **Serous membranes** line internal cavities and are delicate, moist, and very permeable. — **pleura** • **peritoneum** **pericardium** • **pleural fluid** **peritoneal fluid** • **pericardial fluid** **adhesions**
4. The **cutaneous membrane** covers the body surface. Unlike serous and mucous membranes, it is relatively thick, waterproof, and usually dry.
5. The **synovial membrane**, located at joints or **articulations**, produces **synovial fluid** in joint cavities. Synovial fluid helps lubricate the joint and promotes smooth movement.

Muscle Tissue (pp. 156–157)

1. **Muscle tissue**, consisting of **muscle fibers**, is specialized for contraction. There are three different types of muscle tissue: *skeletal muscle, cardiac muscle,* and *smooth muscle.*

SKELETAL MUSCLE TISSUE (p. 157)

2. **Skeletal muscle tissue** contains very large fibers tied together by collagen and elastic fibers. Skeletal muscle fibers have a striped appearance due to the organization of contractile proteins. The stripes are called **striations.** Because we can control the contraction of skeletal muscle fibers through the nervous system, skeletal muscle can be considered as **striated voluntary muscle**. — **satellite cells**

CARDIAC MUSCLE TISSUE (p. 157)

3. **Cardiac muscle tissue** is found only in the heart. The nervous system does not provide voluntary control over cardiac muscle fibers, or **cardiocytes.** Thus, cardiac muscle is **striated involuntary muscle**. — **pacemaker cells**

SMOOTH MUSCLE TISSUE (p. 157)

4. **Smooth muscle tissue** is found in the walls of blood vessels, around hollow organs, and in layers around various tracts. It is classified as **nonstriated, involuntary muscle**.

Neural Tissue (pp. 157–158)

1. **Neural tissue** is specialized to conduct electrical impulses that convey information from one area of the body to another.
2. Cells in neural tissue are either neurons or neuroglia. **Neurons**, often called **nerve fibers**, transmit information as electrical impulses. There are several kinds of **neuroglia**, but their basic functions include: isolating and protecting the cell membranes of neurons; supporting neural tissue; regulating the composition of the interstitial fluid; and defending neural tissue from pathogens and repairing injuries.
3. A typical neuron has a **soma**, **dendrites**, and an **axon** that ends

at a **synapse**. A change in the transmembrane potential triggers the release of a **neurotransmitter**, which crosses the synapse and contacts receptors on the opposing cell membrane.

Related Key Terms

The Connective Tissue Framework of the Body (p. 159)

1. Internal organs and systems are tied together by a network of connective tissue proper which includes the **superficial fascia** (separating the skin from underlying tissues and organs), the **deep fascia** (dense connective tissue), and the **subserous fascia** (the layer between the deep fascia and the serous membranes that line body cavities).

subcutaneous layer
hypodermis

Tissue Repairs (pp. 159–162)

1. Any injury affects several tissue types simultaneously, and they respond in a coordinated manner. Homeostasis is restored in two processes: inflammation and regeneration.
2. **Inflammation**, or the **inflammatory response**, isolates the injured area while damaged cells, tissue components, and any dangerous microorganisms are cleaned up. **Regeneration** is the repair process that restores normal function.

necrosis • **pus** • **abscess**
dilate • **fibrin** • **clot**
fibrosis • **scar tissue**

Tissues, Nutrition, and Aging (pp. 162–165)

1. Tissues change with age. Repair and maintenance grow less efficient, and the structure and chemical composition of many tissues are altered.

Review Planner

	Objective	Level 1
1	Classify the tissues of the body into four major categories.	8 14 15 16 17 51
2	Discuss the relationship between form and function for each of the primary tissue types.	2 14 15 16 17 25 38
3	Discuss the types and functions of epithelia.	2 14 18 22 23 24 31 33 37 40 47 54 55 56
4	Compare the structures and functions of the various connective tissues.	4 5 6 9 11 13 15 19 26 28 29 35 36 39 41 43 44 45 48 49 57 58 59
5	Explain how epithelia and connective tissue combine to form four different types of membranes, and specify the functions of each.	1 3 12 34 37
6	Describe the different types of muscle tissue.	10 16 20 21 46 50
7	Discuss the structure and role of neural tissue.	17 32 42 52
8	Explain how tissues respond in a coordinated manner to maintain homeostasis.	7 27 30 36
9	Describe how nutrition and aging affect tissues.	20 21 60

Level 2	**53**
Level 3	**61–68**
CM	**Marfan's Syndrome**

Chapter Review

Review Questions

MULTIPLE CHOICE

1. The serous membrane that lines the abdominal cavity is the: a) peritoneum; b) perineum; c) pleura; d) pericardium.
2. Mucus is produced by: a) mesothelia; b) necrosis; c) germinative cells; d) goblet cells.
3. Synovial fluid reduces friction between: a) the parietal and visceral surfaces of serous membranes; b) nonstriated muscle and striated muscle tissue; c) unicellular and multicellular glands; d) bone and cartilage surfaces at joints.
4. The ground substance in blood is called: a) plasma; b) platelets; c) interstitial fluid; d) leukocytes.
5. Canaliculi are found in: a) cartilage; b) bone; c) capillaries; d) dense irregular connective tissue.
6. Internal organs and systems are tied together by a network of _____ tissue: a) muscle; b) nervous; c) connective; d) epithelial.
7. The process of fibrosis produces a thick, collagenous framework at the site of an injury called: a) fibrin; b) fibroblasts; c) clots; d) scar tissue.
8. Which of the following is *not* one of the four primary tissue types? a) muscle tissue; b) fasciae; c) epithelia; d) neural tissue.
9. The dense connective tissue that surrounds most organs is called the: a) superficial fascia; b) skeletal muscles; c) deep fascia; d) organ system.
10. Cardiac muscle tissue is considered: a) striated voluntary muscle; b) striated involuntary muscle; c) nonstriated, involuntary muscle; d) none of the above.
11. Most of the matrix of bone consists of: a) a mixture of calcium salts; b) collagen fibers; c) fibrous periosteum; d) calcified cartilage.
12. The epithelium that lines the heart and blood vessels is called the: a) pericardium; b) mesothelium; c) endothelium; d) myocardium.
13. The combination of fibers and ground substance in connective tissue is called a: a) fibroblast; b) fibrin; c) fibro-ground substance; d) matrix.

DIAGRAMS

14. Draw a chart that illustrates the classification scheme for epithelia. Include descriptive phrases that summarize the characteristics and functions of each feature. Be sure to include all possible combinations.
15. Draw a chart that shows the three major types of connective tissue, and the subgroups below each type. Include descriptive phrases under each type that summarize that type's characteristics and functions.
16. Create a table that compares the following characteristics for the three types of muscle tissue: fiber length, fiber width, nuclei, contractile proteins, and control mechanism(s). Include descriptive phrases that summarize the functions of each type.
17. Draw a diagram that compares the functions (in descriptive phrases) of the two types of cells in neural tissue.

TRUE/FALSE

(If the statement is false, your instructor may wish to have you correct it.)

18. **T/F:** A mixed gland can produce both exocrine and endocrine secretions.
19. **T/F:** Calcified cartilage is a healthy development in adults, since additional calcium makes the cartilage stronger.
20. **T/F:** Cardiac muscle cannot regenerate after an injury.
21. **T/F:** Smooth muscle cannot regenerate after an injury.

MATCHING QUESTIONS

(Match each statement with the most appropriate term.)

Statements:

22. A term that describes epithelial cells that resemble small, square boxes in section.
23. A term that describes epithelial cells that are tall, slender, and rectangular in section.
24. A term used to describe epithelial cells that are thin and flat.
25. Lacking blood vessels.
26. Under normal conditions, the most common type of cell in lymph.
27. Packets of cytoplasm in blood that assist in the clotting response.
28. The connective tissue layer that separates cartilage from surrounding tissues.
29. The connective tissue layer that surrounds bone.
30. An inflammation resulting from pathogens, such as bacteria.
31. A secretion that is discharged onto the surface of the skin or into an internal passage that communicates with the exterior of the body.
32. A nerve cell body.
33. A secretion that is discharged into surrounding tissues.
34. The fluid formed on the surfaces of a serous membrane.

Terms:

a) Transudate
b) Endocrine
c) Exocrine
d) Squamous
e) Polarity
f) Mesenchymal cells
g) Soma
h) Cuboidal
i) Columnar
j) Hyaline
k) Avascular
l) Mesothelium
m) Lymphocyte
n) Platelets
o) Periosteum
p) Perichondrium
q) Infection

35. A type of cartilage.
36. Stem cells in connective tissue that divide to replace lost connective tissue cells.
37. The epithelium that lines body cavities.
38. The uneven distribution of organelles within a cell.

SHORT ANSWER/ESSAY

Identify the following cells. (Give another name for the cell if one was mentioned in the chapter. Also mention the type of tissue in which it is found and its function.)

39. Fat cells
40. Melanin cells
41. Red blood cells
42. Neurons
43. White blood cells
44. Chondrocytes
45. Osteocytes
46. Pacemaker cells

47. Discuss ways in which epithelial cells are specialized and why these specializations are important. What needs do these specializations address?
48. Both blood and bone are connective tissues. Why is one liquid while the other is not?
49. Compare the structures and functions of the three types of cartilage. What category of tissue do they belong to? Where are they found in the body?
50. What is muscle tissue specialized for? Compare the three types of muscle tissue. Where are they found, and can they repair themselves after an injury?
51. Define tissue, and name the four primary tissue types. Compare their basic characteristics and functions.
52. Compare the functions of neurons and neuroglia. To which type of tissue do they belong?
53. Compare/contrast the structure and function of the four types of membranes discussed in this chapter. Where are they located in the body?
54. Distinguish between merocrine, apocrine, and holocrine secretion. Which is the most common? What distinguishes holocrine secretion from the first two processes?
55. What is a ciliated epithelium? What is its role? Give an example.
56. Define "epithelium." Discuss the essential functions of epithelia. Where are epithelia located?
57. Distinguish between collagen fibers, reticular fibers, and elastic fibers. In what type of tissue are they found?
58. Identify the five examples of dense regular connective tissue that the chapter mentions. Compare their structures and functions.
59. What are fixed macrophages and free macrophages? In what type of tissue are they found, and what function(s) do they perform?
60. Discuss the changes in tissues that occur with aging.

CRITICAL THINKING/APPLICATIONS

61. You've always wanted a flatter abdomen, so you go on a diet and lose ten pounds. Two years later you've regained ten pounds and your abdomen looks the way it did before you lost the weight in the first place. Explain why.
62. What is liposuction? Do you think it has long-term benefits?
63. You've cut your hand, and you notice that in the early stages of healing the skin around the cut becomes swollen, red, and warm. Explain the reasons for these symptoms. What is this process called?
64. An elderly woman and a child both slip and fall on the ice. Which one is more likely to be injured, and why? Describe the injuries to tissues which could result.
65. Mrs. J., who smokes a pack of cigarettes a day, suffers from a chronic cough and undergoes a bronchoscopy (examination of her respiratory tract using an illuminated tube). The pathology report shows that the tracheal tissues demonstrate metaplasia. Explain the significance of metaplasia. For what conditions is Mrs. J. at risk?
66. Mike has injured his knee cartilages while playing basketball. He has been told to stop playing basketball for several months, stay off his knee as much as possible, and wait for it to heal. He is growing impatient and asks you why it isn't healing faster. What would you tell him?
67. What is chemotherapy, and what are its side effects? Explain why these side effects occur.
68. Name the warning signals of cancer. For each signal, identify possible site(s) of cancer which that symptom could indicate.

AN ORIENTATION TO THE HUMAN BODY

Systems Overview

Our perspective has gradually changed over the preceding chapters. Atomic structure can only be imagined or indirectly examined through experimental procedures. Cellular details often escape detection unless an electron microscope is used. Tissue structure can be examined with a light microscope, and many conclusions can be drawn on the basis of the texture of the tissue seen with the unaided eye. You may already be familiar with identifying features of some tissues from experience in the laboratory. For example, once you have handled adipose tissue, with its lumpy, greasy texture, it would be difficult to mistake it for any other type of tissue.

Organs are combinations of tissues that perform complex functions. A great deal of information concerning organs and organ structure can be obtained by dissection and direct examination. In organ systems, several organs work together in a coordinated fashion. We can easily observe the functions of intact organ systems as they perform, direct, or moderate the activities of individual human beings. As a result, at the start of this course you probably knew much more about the major organs and systems than you did about cell and tissue structure. The pages that follow are designed to give you an overview of the 11 organ systems in the human body, focusing on their basic structural and functional characteristics. This information will provide a framework for later chapters dealing with specific systems.

Figure SO-1 presents four graphs of the composition of the human body, reflecting the changes in our perspective over the last five chapters. In Chapter 2 the body was treated as a collection of elements (Figure SO-1a) that combine to form molecules, the building blocks of cells (Chapters 3 and 4). Figure SO-1b indicates the composition of the body in terms of water and the major classes of organic compounds that were introduced in Chapter 2 and reappear in many of the discussions in Chapters 3, 4, and 5.

Chapter 5 described the association of roughly 200 types of cells in four body tissues (Figure SO-1c). These tissues combine to form thousands of different organs. Some are quite large and distinctive; the liver, an organ of the digestive tract, weighs about 1.6 kg (3.5 lb), and some skeletal muscles are even larger. Other organs are tiny, and far more numerous; the skin contains about 3 million sweat glands that are barely large enough to see with the unaided eye. Regardless of their size, all of these organs contain all four tissue types, but the proportions vary from organ to organ. For example, all four tissue types contribute extensively to the structure of the stomach, but most of the heart is composed of cardiac muscle tissue. Figure SO-1d analyzes the human body at the organ system level, the focus of the following pages.

Despite their structural and functional differences, all organ systems share certain characteristics:

1. **Specialization** for performance of a limited number of functions. In other words, there is a division of labor among organ systems.
2. **Functional independence** when responding to local environmental stimuli.
3. **Dependence on other organ systems** for nutrient supply, oxygen, and waste removal.
4. **Integration of activity** through neural and hormonal mechanisms.

We will now examine each of the organ systems and introduce the organs associated with them. The boundaries are not always clear-cut, for while each organ is assigned to a particular system on the basis of its primary function, it can also have important secondary functions. For example, the pancreas is primarily an organ of the digestive system because 99 percent of the pancreatic cells are associated with the production of digestive enzymes and buffers. However, because the rest of the pancreatic cell population secretes vitally important hormones, the pancreas is also a component of the endocrine system.

Figures SO-3 through SO-13 detail individual organ systems and summarize the histology of representative organs in each system. These examples should promote integration of new material with the structural and functional patterns described in Chapter 5.

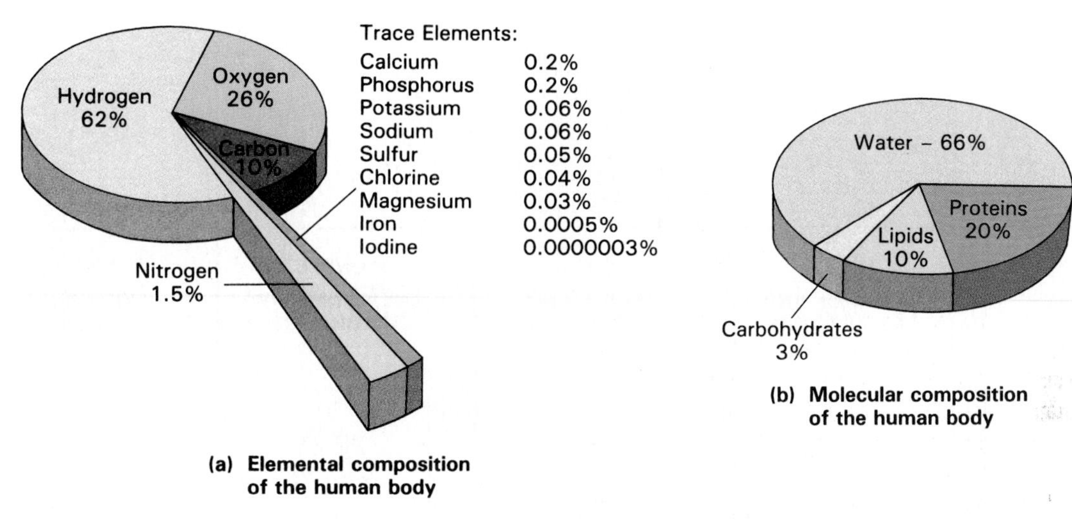

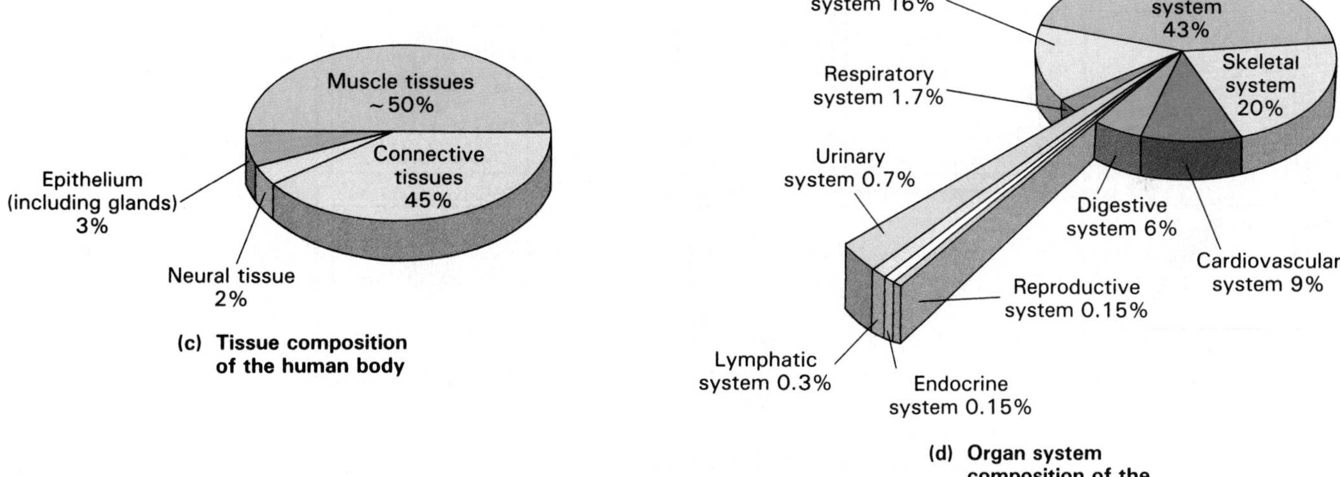

FIGURE SO-1 The Composition of the Human Body from Four Different Perspectives.

(a) The elemental composition of the body.
(b) The major classes of inorganic and organic molecules that constitute the body's cells.
(c) The tissue composition of the body.
(d) The relative sizes of the organ systems.

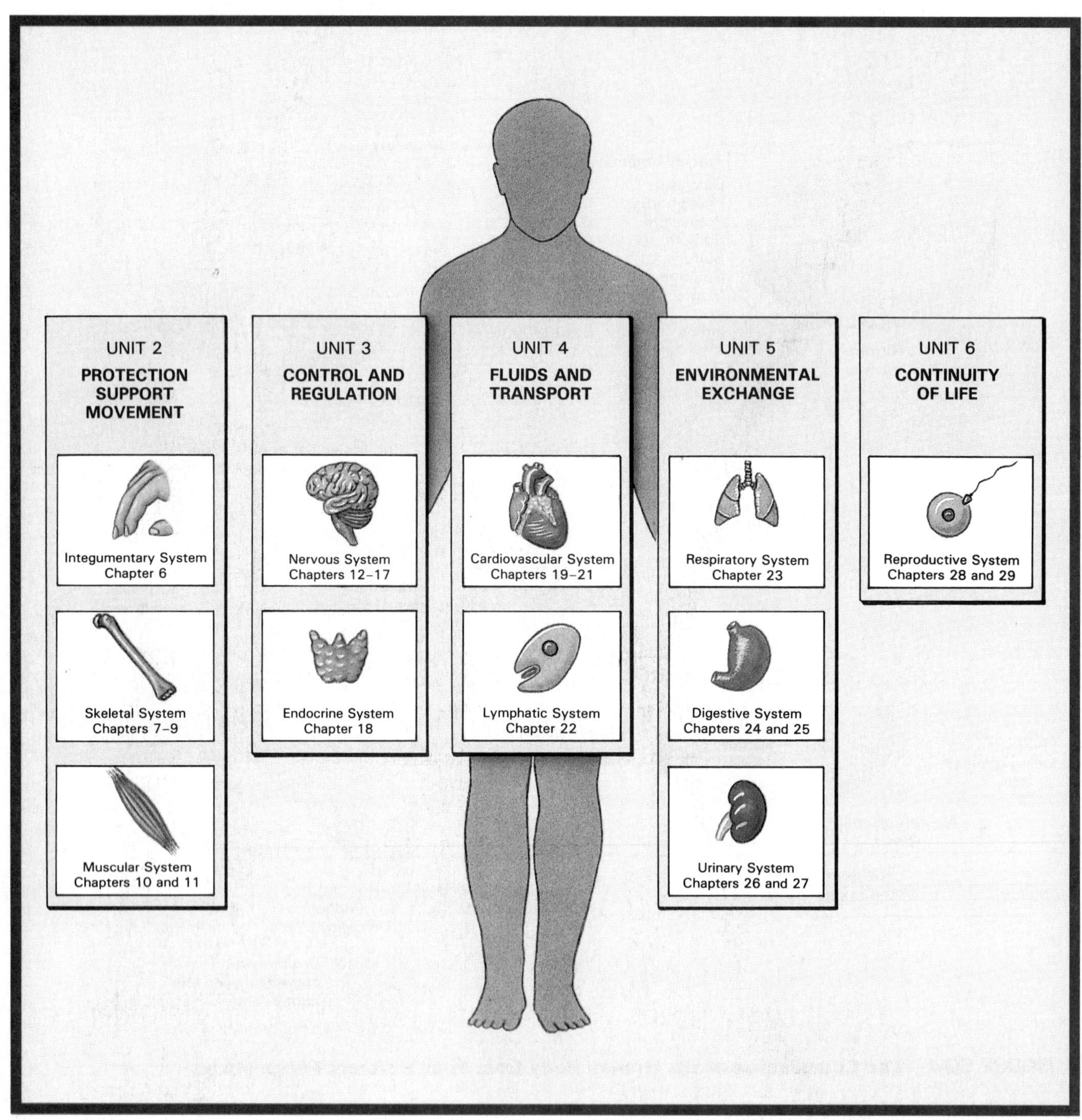

■ FIGURE SO-2 **Organ Systems of the Human Body:** An overview of the next 24 chapters

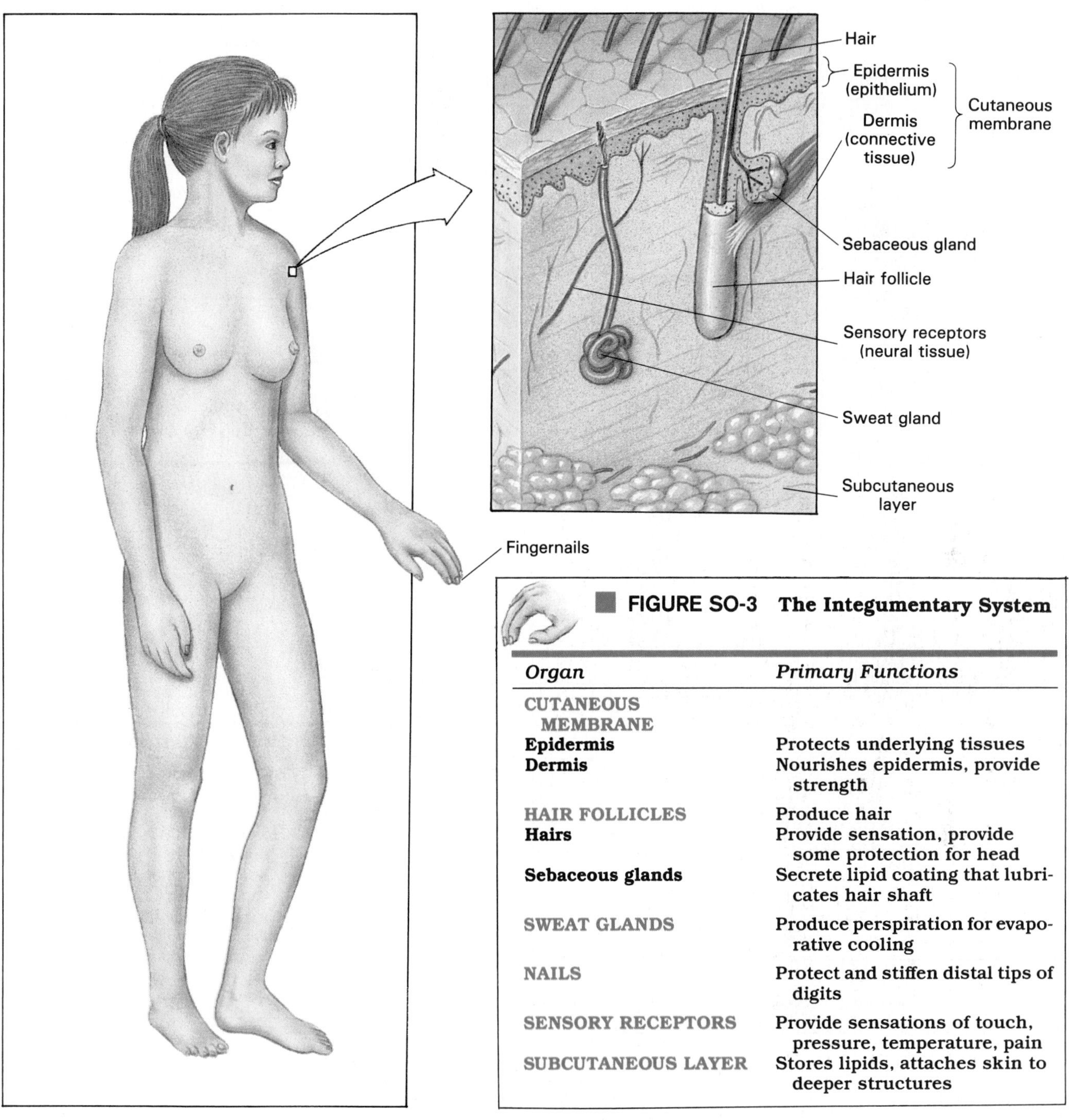

FIGURE SO-3 The Integumentary System

Organ	*Primary Functions*
CUTANEOUS MEMBRANE	
Epidermis	Protects underlying tissues
Dermis	Nourishes epidermis, provide strength
HAIR FOLLICLES	Produce hair
Hairs	Provide sensation, provide some protection for head
Sebaceous glands	Secrete lipid coating that lubricates hair shaft
SWEAT GLANDS	Produce perspiration for evaporative cooling
NAILS	Protect and stiffen distal tips of digits
SENSORY RECEPTORS	Provide sensations of touch, pressure, temperature, pain
SUBCUTANEOUS LAYER	Stores lipids, attaches skin to deeper structures

The **integumentary** (in-teg-ū-MEN-tar-ē; *tegere*, to cover) **system,** or skin, consists of the cutaneous membrane (Chapter 5) and associated structures, such as glands, hair, and nails. This system differs from all others in that it is a structural unit, and for this reason it could be considered an enormous organ rather than an organ system. However, because the functions of the skin are so varied and the component structures so complex, it makes more sense to treat it as an organ system.

Few organ systems are as large, as accessible, as varied in function, and as unappreciated as the skin. The integumentary system accounts for about 16 percent of the total body weight, and its surface is continually abused, abraded, and assaulted by microorganisms, sunlight, and various mechanical and chemical hazards.

The basic components of the integumentary system are shown in Figure SO-3, along with the primary functions of each component. The system as a whole: (1) provides protection from environmental hazards, (2) assists in thermoregulation (as discussed in Chapter 1), (3) provides sensations to the nervous system through stimulation of *sensory receptors*, and (4) assists in the excretion of water and solutes, including organic wastes and ions. Secondary functions include the synthesis of one vitamin (vitamin D) and the storage of lipid reserves in adipose tissue.

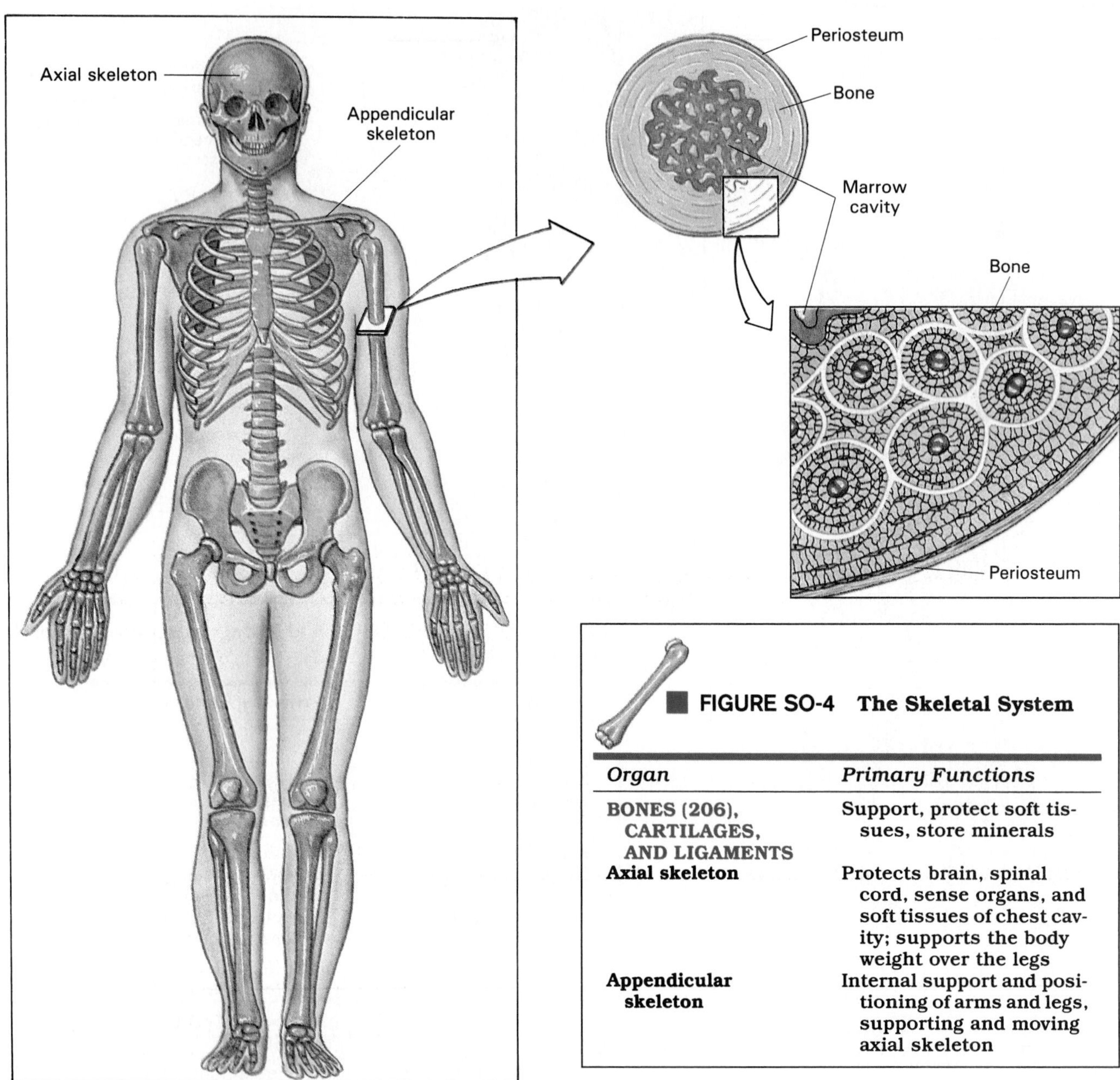

■ **FIGURE SO-4 The Skeletal System**

Organ	*Primary Functions*
BONES (206), CARTILAGES, AND LIGAMENTS	**Support, protect soft tissues, store minerals**
Axial skeleton	**Protects brain, spinal cord, sense organs, and soft tissues of chest cavity; supports the body weight over the legs**
Appendicular skeleton	**Internal support and positioning of arms and legs, supporting and moving axial skeleton**

The **skeletal system** (Figure SO-4) forms an internal supporting framework. It includes approximately 206 bones. (Because bones may develop in tendons and many bones fuse during development, there is some individual variation.) The system also includes a large number of cartilages. Where the skeletal elements interconnect at joints they are stabilized by connective tissues and ligaments that are also considered part of the skeletal system. Tendons and surrounding muscle tissue, too, help to maintain proper skeletal orientation.

The skeletal system can be subdivided into the **axial skeleton** and the **appendicular skeleton**. The axial elements form the vertical axis of the body and enclose the brain and spinal cord. The appendicular skeleton includes the bones of the arms and legs as well as those that connect the limbs to the rest of the body.

The skeletal system supports the body as a whole, as well as individual organs and organ systems. The skeleton directs the forces of skeletal muscle contraction, producing movement. The individual bones store minerals, and they may also protect delicate tissues and organs. For example, in adults, red blood cells are formed in **bone marrow**, a connective tissue framework that fills the spaces, or **marrow cavities**, within many bones. Several bones can also provide protection by creating a bony shield, such as the skull, which surrounds the brain, or a bony framework, such as the rib cage.

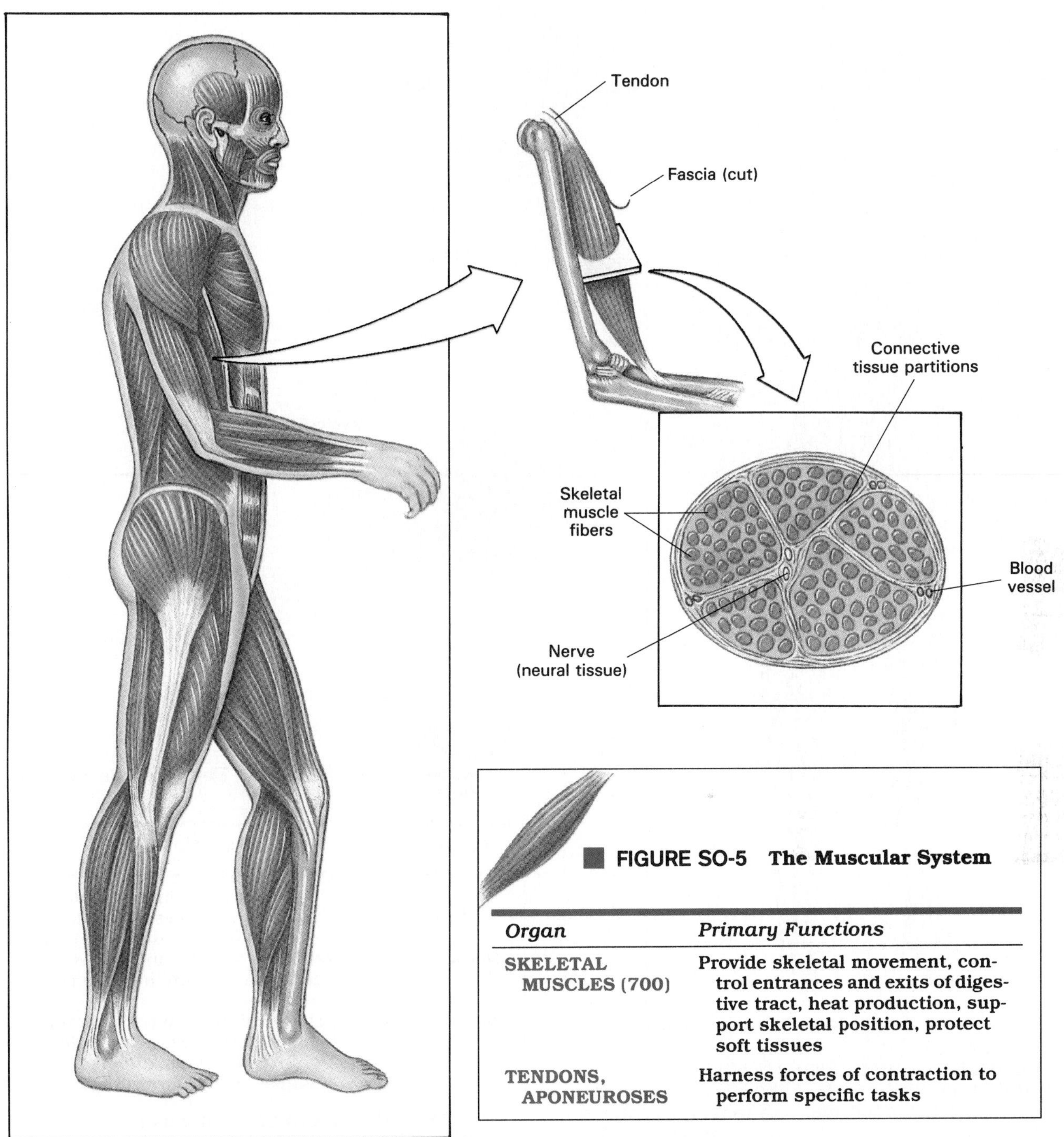

■ FIGURE SO-5 **The Muscular System**

Organ	*Primary Functions*
SKELETAL MUSCLES (700)	**Provide skeletal movement, control entrances and exits of digestive tract, heat production, support skeletal position, protect soft tissues**
TENDONS, APONEUROSES	**Harness forces of contraction to perform specific tasks**

The **muscular system** (Figure SO-5) includes all of the skeletal muscles in the body that are under voluntary control. Of the approximately 700 skeletal muscles in the body, all but about 10 are part of this system. The rest are assigned to other systems; for example, the involuntary skeletal muscles in the pharynx (throat) that assist in swallowing are considered to be components of the digestive system.

Skeletal muscles are organs specialized for contraction. Each skeletal muscle consists of skeletal muscle tissue and the connective tissues (tendons, aponeuroses, and fasciae) that stabilize its position and channel the force of contraction. A muscle also contains blood vessels that provide circulation and neural tissue that monitors and controls its contraction.

Skeletal muscles are directly or indirectly connected to the skeleton. Their contractions can therefore produce skeletal movements that reposition body parts (waving the hand) or the entire body (walking). Skeletal muscles also maintain posture and balance, cradle and support soft tissues, guard the entrances and exits of the digestive tract, and assist in regulating blood flow through vessels. Finally, the heat produced during muscle contraction plays a major role in thermoregulation.

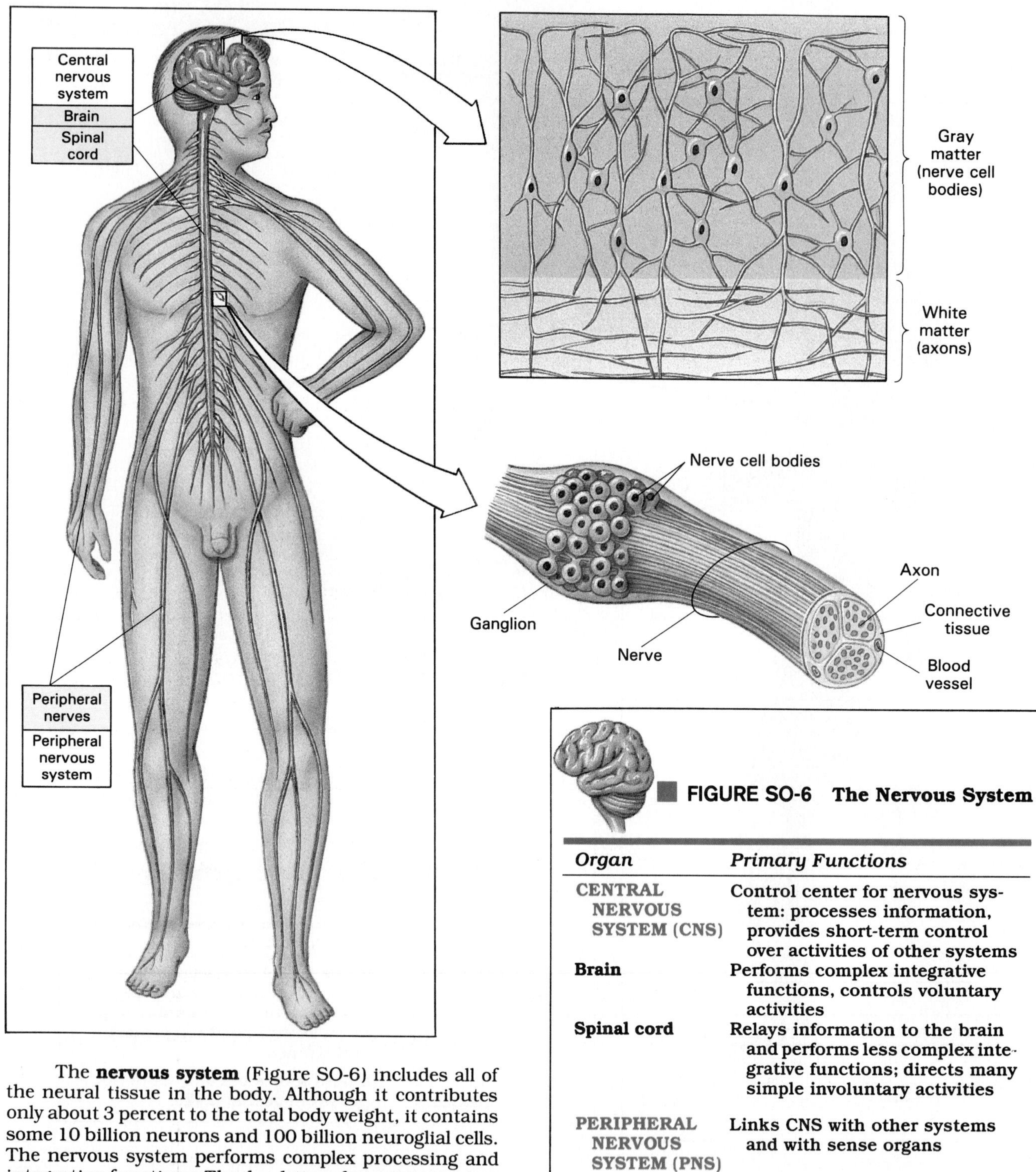

FIGURE SO-6 The Nervous System

Organ	*Primary Functions*
CENTRAL NERVOUS SYSTEM (CNS)	Control center for nervous system: processes information, provides short-term control over activities of other systems
Brain	Performs complex integrative functions, controls voluntary activities
Spinal cord	Relays information to the brain and performs less complex integrative functions; directs many simple involuntary activities
PERIPHERAL NERVOUS SYSTEM (PNS)	Links CNS with other systems and with sense organs

The **nervous system** (Figure SO-6) includes all of the neural tissue in the body. Although it contributes only about 3 percent to the total body weight, it contains some 10 billion neurons and 100 billion neuroglial cells. The nervous system performs complex processing and integration functions. The dendrites of **sensory neurons** are sensitive to physical or chemical changes in peripheral tissues. **Motor neurons** direct or adjust the activities of tissues and organs throughout the body.

The components of the nervous system include the brain, the spinal cord, complex sense organs such as the eye and ear, and the nerves that interconnect these organs and link the nervous system with other systems. The brain and spinal cord form the **central nervous system**, or **CNS**. The CNS integrates sensory information and controls or moderates the activities of other systems. The sensory and motor neurons that bring information to or from the CNS are part of the **peripheral nervous system (PNS)**. The **autonomic nervous system**, or **ANS**, includes both CNS and PNS neurons involved in the automatic regulation of physiological processes, such as heart rate and digestive activities.

Nerve cell bodies are usually clustered together; because these masses are dark and grainy they are known as **gray matter**. Areas of **white matter** have a lighter, reflective appearance; these regions consist largely of axons. In the PNS, axons make up distinct bundles, or **nerves**, while nerve cell bodies form small masses called **ganglia** (singular *ganglion*).

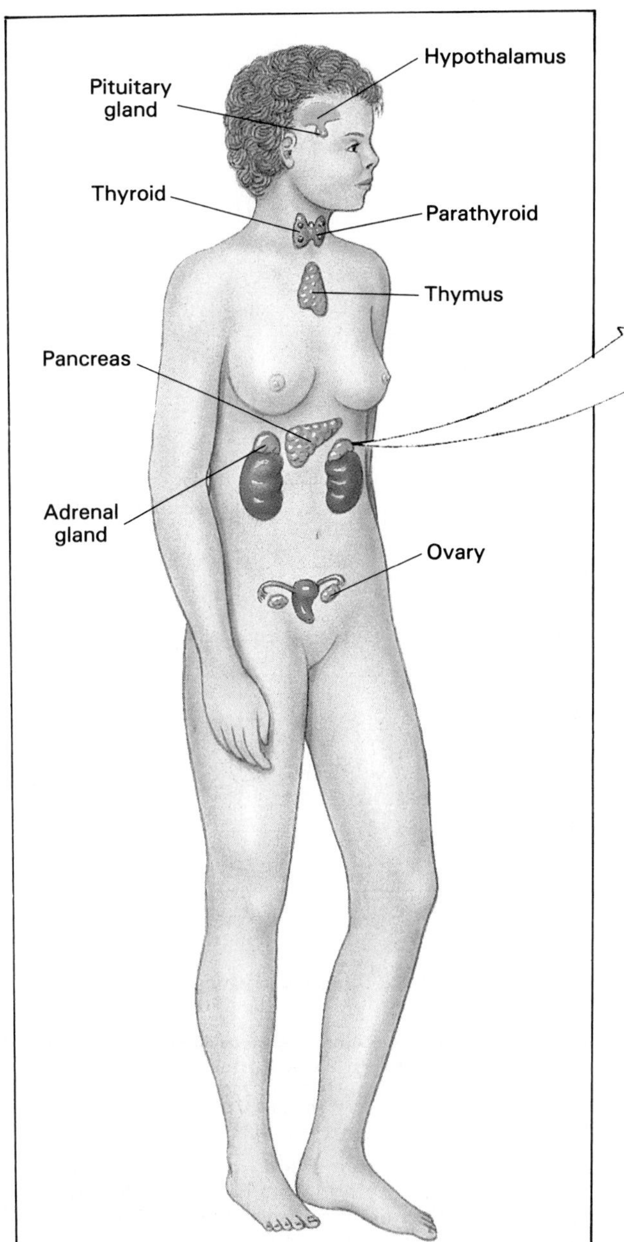

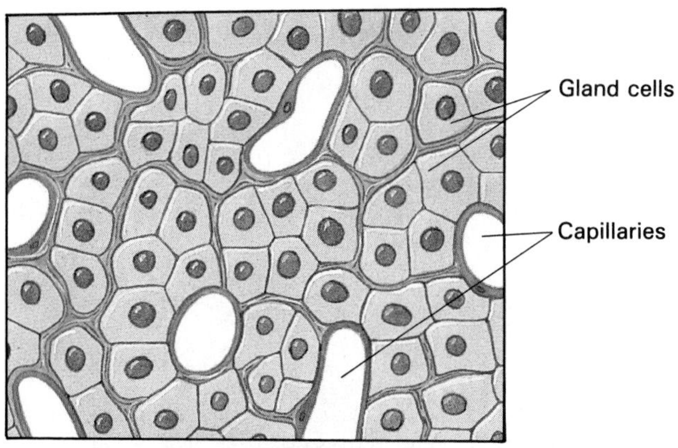

SECTION OF ENDOCRINE GLAND
(Adrenal medulla)

■ **FIGURE SO-7 The Endocrine System**

Organ	*Primary Functions*
PITUITARY GLAND	Controls secretions of other glands, integrates CNS and endocrine activities, regulates growth, secretes hormones involved with fluid balance and smooth-muscle control
THYROID GLAND	Controls tissue metabolic rates and regulates blood calcium levels
PARATHYROID GLAND	Regulates blood calcium levels (with thyroid)
THYMUS	Controls maturation and maintenance of lymphocytes
ADRENAL GLANDS	Control water balance, regulate tissue metabolism of carbohydrates and lipids, increase cardiovascular and respiratory activity in emergencies
KIDNEYS	Control red blood cell production and elevate blood pressure
PANCREAS	Regulates blood glucose levels
HEART	Regulates fluid balance (with pituitary, adrenals, and kidneys)
DIGESTIVE TRACT	Coordinates activities of digestive glands
TESTES	Support sperm production and male sexual characteristics
OVARIES	Support egg production and female sexual characteristics, prepare uterus for embryo and mammary glands for milk production
PINEAL GLAND	Controls body pigmentation and the timing of reproduction

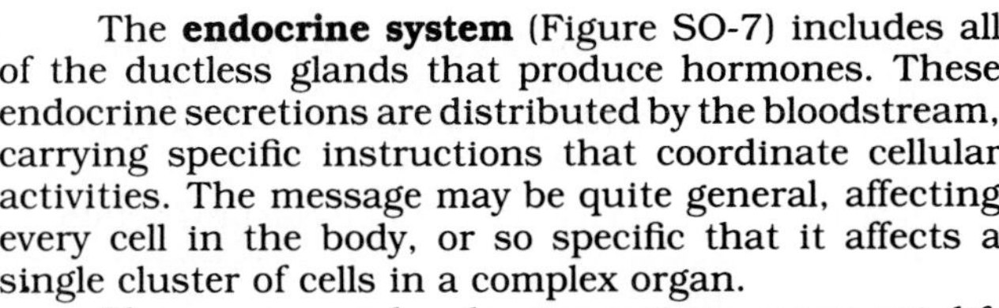

The **endocrine system** (Figure SO-7) includes all of the ductless glands that produce hormones. These endocrine secretions are distributed by the bloodstream, carrying specific instructions that coordinate cellular activities. The message may be quite general, affecting every cell in the body, or so specific that it affects a single cluster of cells in a complex organ.

The nervous and endocrine systems represent different approaches to the regulation and coordination of internal operations. The nervous system specializes in integrating sensory information and providing an immediate, short-term response, often in less than a second after the stimulus arrives. Endocrine regulation proceeds at a more leisurely pace, producing changes in metabolic activity over seconds, minutes, hours, or even years. Examples of endocrine regulation include the control of blood glucose levels and metabolic adjustments to cold, dehydration, or changing electrolyte concentrations. Hormones also regulate sexual characteristics and the development of functional sperm and eggs.

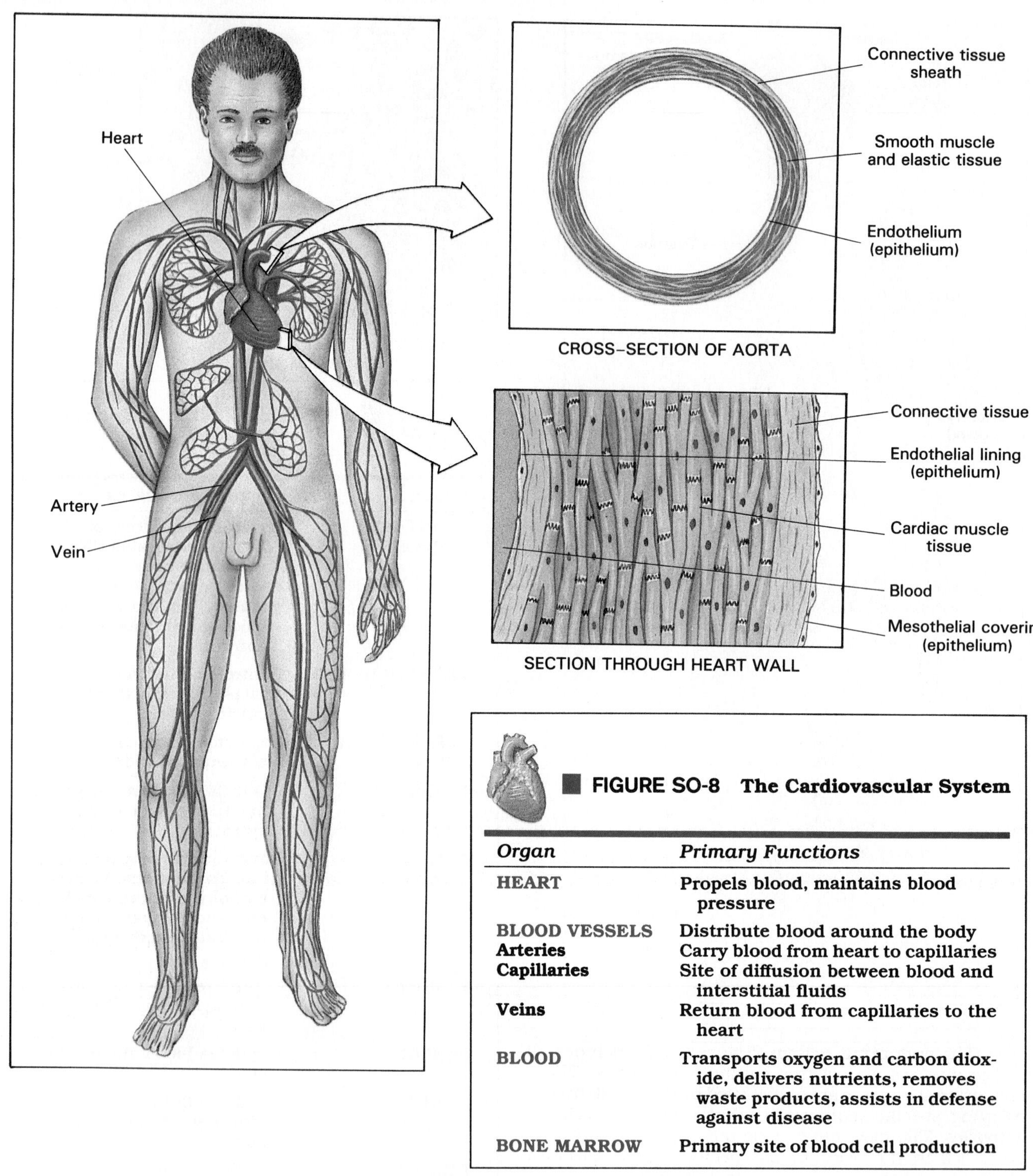

FIGURE SO-8 The Cardiovascular System

Organ	*Primary Functions*
HEART	Propels blood, maintains blood pressure
BLOOD VESSELS	Distribute blood around the body
Arteries	Carry blood from heart to capillaries
Capillaries	Site of diffusion between blood and interstitial fluids
Veins	Return blood from capillaries to the heart
BLOOD	Transports oxygen and carbon dioxide, delivers nutrients, removes waste products, assists in defense against disease
BONE MARROW	Primary site of blood cell production

The **cardiovascular system** (Figure SO-8) consists of the heart, blood, and blood vessels. This system also includes the bone marrow, where red blood cells are formed (see Figure SO-4). The cardiovascular system transports nutrients, dissolved gases, and hormones to tissues throughout the body. It also carries waste products from peripheral tissues to sites of excretion, such as the kidneys. Buffers within the blood help regulate the pH of body fluids. Warm blood carries heat from one location to another, participating in thermoregulation. White blood cells provide defense from disease, while the formation of fibrin restricts the loss of blood through breaks in endothelia or epithelia and slows the spread of pathogens through damaged tissues.

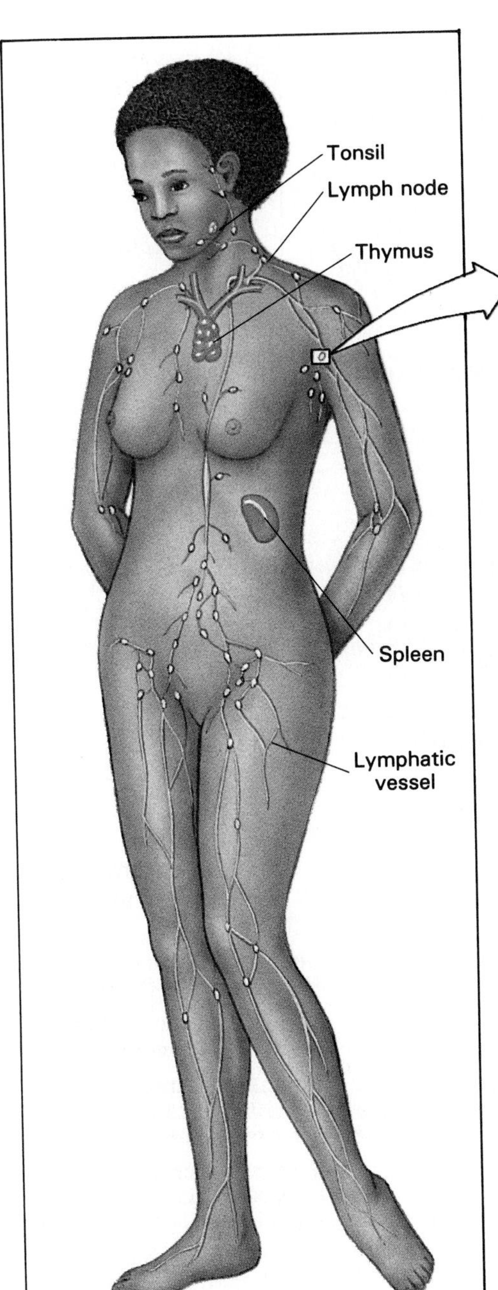

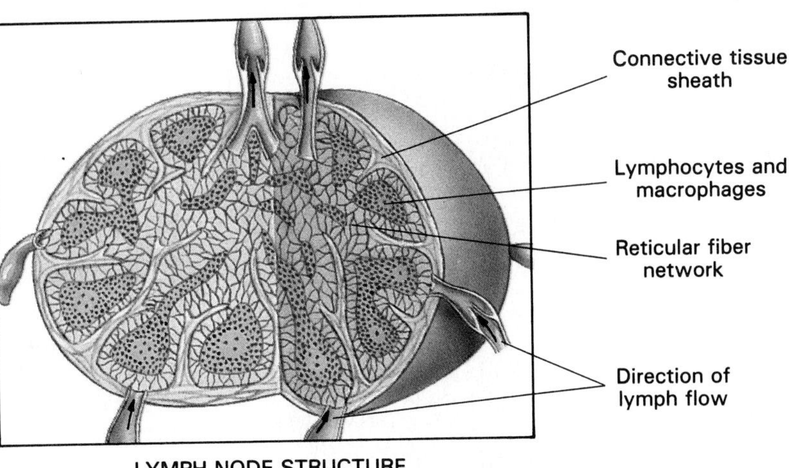

LYMPH NODE STRUCTURE

■ **FIGURE SO-9 The Lymphatic System**

Organ	*Primary Functions*
LYMPHATIC VESSELS	Carry lymph from peripheral tissues to the veins of the cardiovascular system
LYMPH NODES	Monitor the composition of lymph, engulf pathogens, stimulate immune response
SPLEEN	Monitors circulating blood, engulfs pathogens, stimulates immune response
THYMUS	Controls development and maintenance of lymphocytes

The relationship between the cardiovascular system and the **lymphatic system** was described in Chapter 5, in our discussion of blood and lymph. The lymphatic system (Figure SO-9) includes an extensive network of lymphatic vessels that deliver lymph to the circulatory system. In the process lymph transports large molecules such as tissue proteins and lipids that cannot diffuse into capillaries. The lymphatic system also contains **lymphatic organs** that produce or support a large population of lymphocytes, plasma cells, and phagocytes, cell types described in Chapter 5. **Lymph nodes** are small lymphatic organs that contain cells sensitive to changes in the composition of lymph. Examples of larger lymphatic organs include the **thymus gland**, the **tonsils**, and the **spleen**.

The lymphatic system defends the body from disease and toxic substances. Wandering phagocytic cells patrol tissues throughout the body, engulfing debris and pathogens. Fixed macrophages in lymph nodes attack damaged cells or pathogens that have escaped the tissue defenses. Lymphocytes, like macrophages, circulate through the body. When they encounter abnormal cells or pathogens, some lymphocytes physically attack the intruders. Other lymphocytes convert to plasma cells, producing antibodies that provide a biochemical defense.

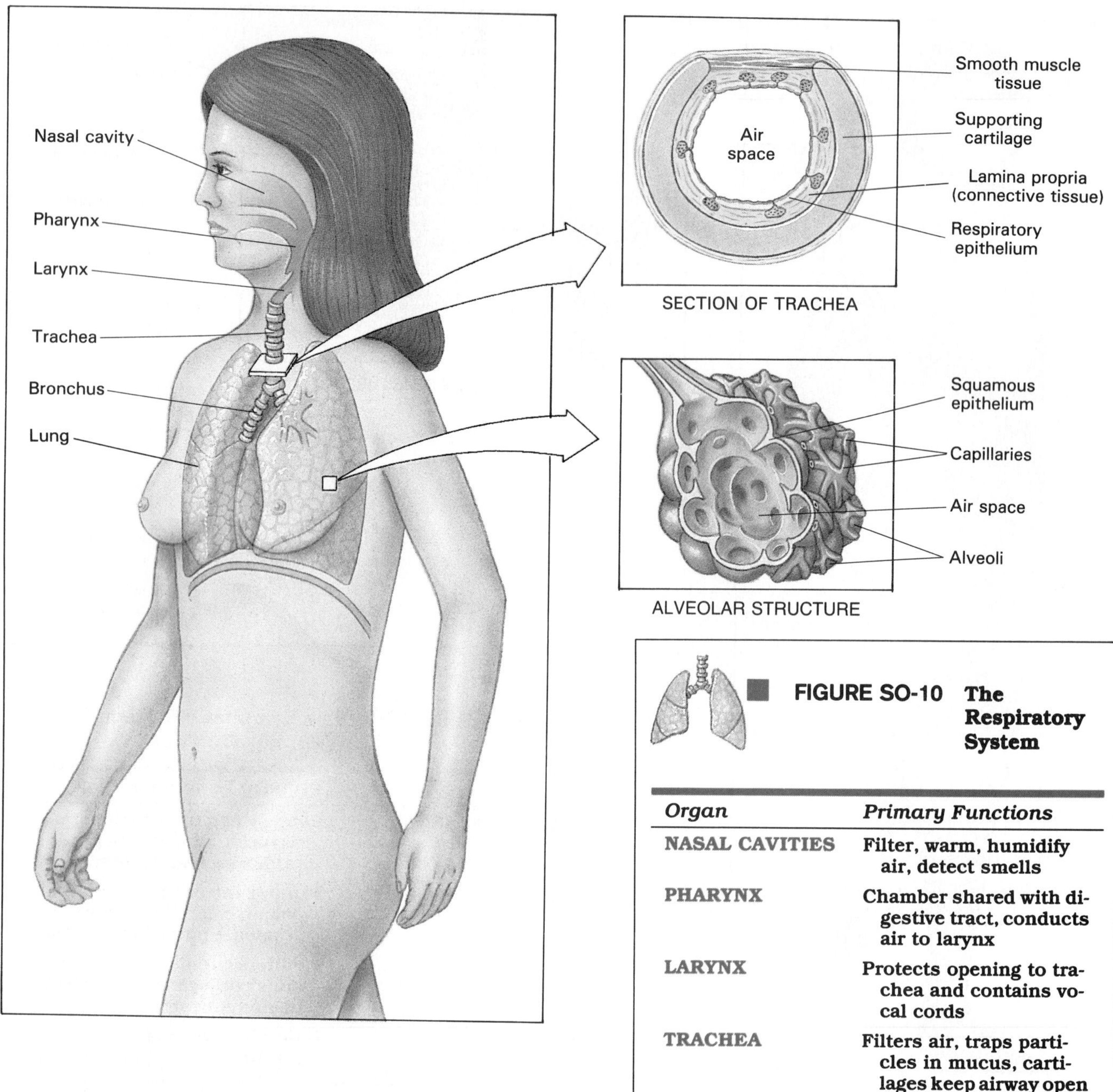

FIGURE SO-10 The Respiratory System

Organ	*Primary Functions*
NASAL CAVITIES	Filter, warm, humidify air, detect smells
PHARYNX	Chamber shared with digestive tract, conducts air to larynx
LARYNX	Protects opening to trachea and contains vocal cords
TRACHEA	Filters air, traps particles in mucus, cartilages keep airway open
LUNGS	Includes airways and alveoli; volume changes responsible for air movement
BRONCHI	Same as trachea
ALVEOLI	Sites of gas exchange between air and blood

The **respiratory system** (Figure SO-10) includes the lungs and the passageways that carry air to them. These passageways begin at the **nasal cavities**, and continue through the **pharynx**, **larynx** (voicebox), **trachea** (windpipe), and **bronchi** (BRONG-kē; singular **bronchus**) before entering the lungs. Within the lungs, the bronchi branch repeatedly, growing ever smaller in diameter. The nasal cavities, trachea, and bronchi are lined by the pseudostratified, ciliated epithelium described in Chapter 5. This epithelium filters, warms, and moistens incoming air, protecting delicate respiratory surfaces. The exchange surfaces themselves, lined by a simple squamous epithelium, form small pockets, called **alveoli** (al-VĒ-ō-lī; singular *alveolus*, a small sac). At these surfaces, gas exchange occurs between the air and the circulating blood.

The primary functions of the respiratory system are the delivery of oxygen and the removal of carbon dioxide. The respiratory system also has a number of important secondary functions. By changing the concentration of carbon dioxide in the blood, the respiratory system helps to regulate the pH of body fluids. This is important because, as noted in Chapter 2, uncontrolled changes in pH can have disastrous effects on cells and tissues. In addition, the evaporation of water in the lungs and along the respiratory passageways helps cool the body. Finally, air forced across the vocal cords produces sounds used in communication.

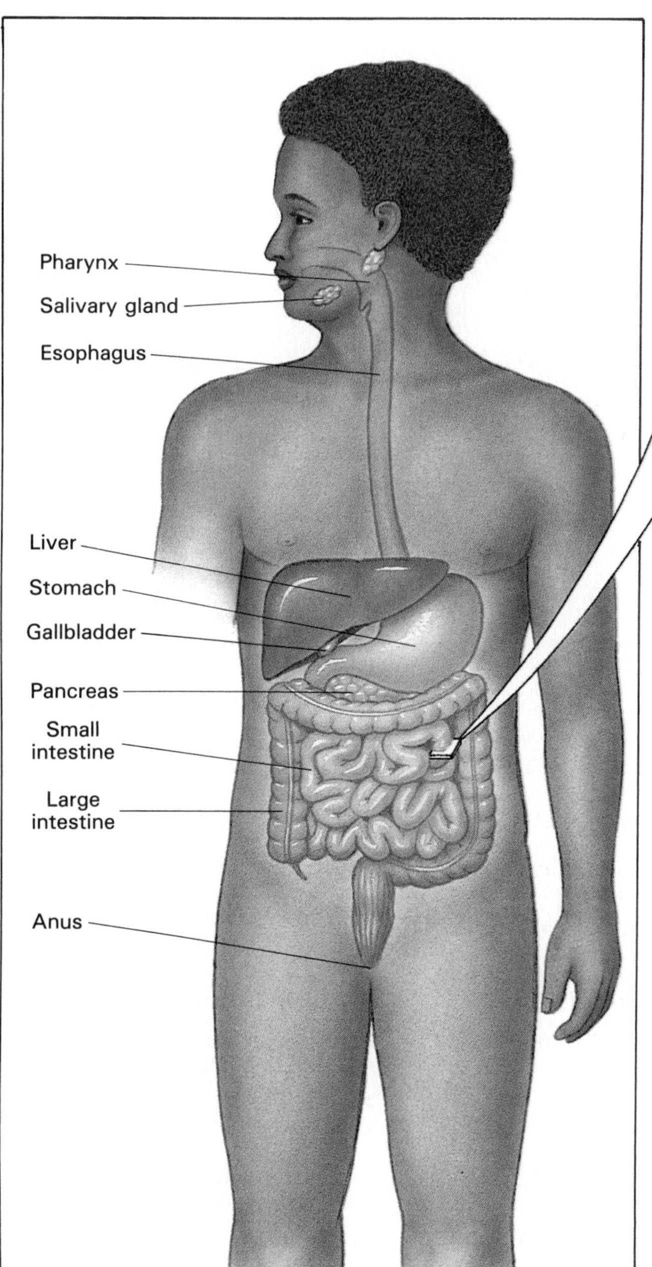

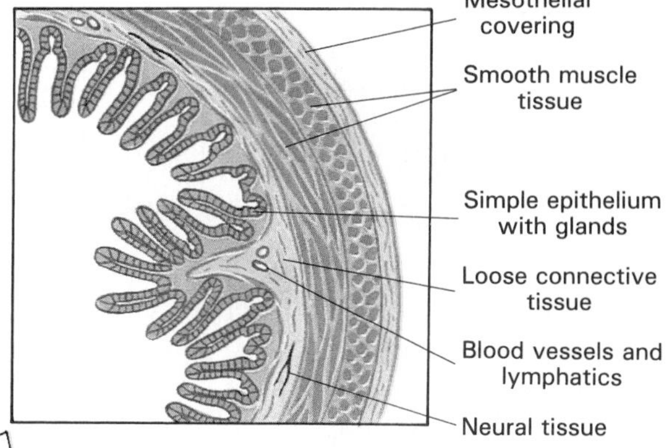

FIGURE SO-11 The Digestive System

Organ	*Primary Functions*
MOUTH	Mixes food with salivary secretions, taste, chewing
SALIVARY GLANDS	Produce buffers and enzymes that begin digestion
PHARYNX	Passageway shared with respiratory system, leads to esophagus
ESOPHAGUS	Delivers food to stomach
STOMACH	Secretes acids and digestive enzymes that break down proteins
SMALL INTESTINE	Absorbs nutrients
LIVER	Secretes bile (important for lipid digestion), regulates nutrient composition of blood, synthesizes blood proteins, stores lipid and carbohydrates reserves
GALLBLADDER	Stores bile for release into small intestine
PANCREAS	Secretes digestive enzymes and buffers into small intestine; contains endocrine cells included in Figure SO-7
LARGE INTESTINE	Removes water from fecal material, stores wastes
ANUS	Opening to exterior for discharge of feces

The central component of the **digestive system** (Figure SO-11) is the digestive tract, a long tube that begins at the mouth and ends at the anus. A number of accessory glands communicate with the tract along its length. The secretions of the **salivary glands**, **stomach**, **small intestine**, and **pancreas** introduce digestive enzymes that assist in the breakdown of food. The **liver**, the largest digestive organ, secretes *bile*, which is essential to the breakdown of lipids. It also regulates the nutrient content of the blood and manufactures important plasma proteins.

The primary function of the digestive system is the breakdown of food for absorption and use by the body. The digestive tract absorbs nutrients, including organic compounds, water, ions, and vitamins. Buffers, ions, and a few metabolic wastes are dumped into the digestive tract by accessory glands, especially the pancreas and liver. These wastes, along with the undigested residue of previous meals, are eliminated from the **large intestine** during the process of **defecation**.

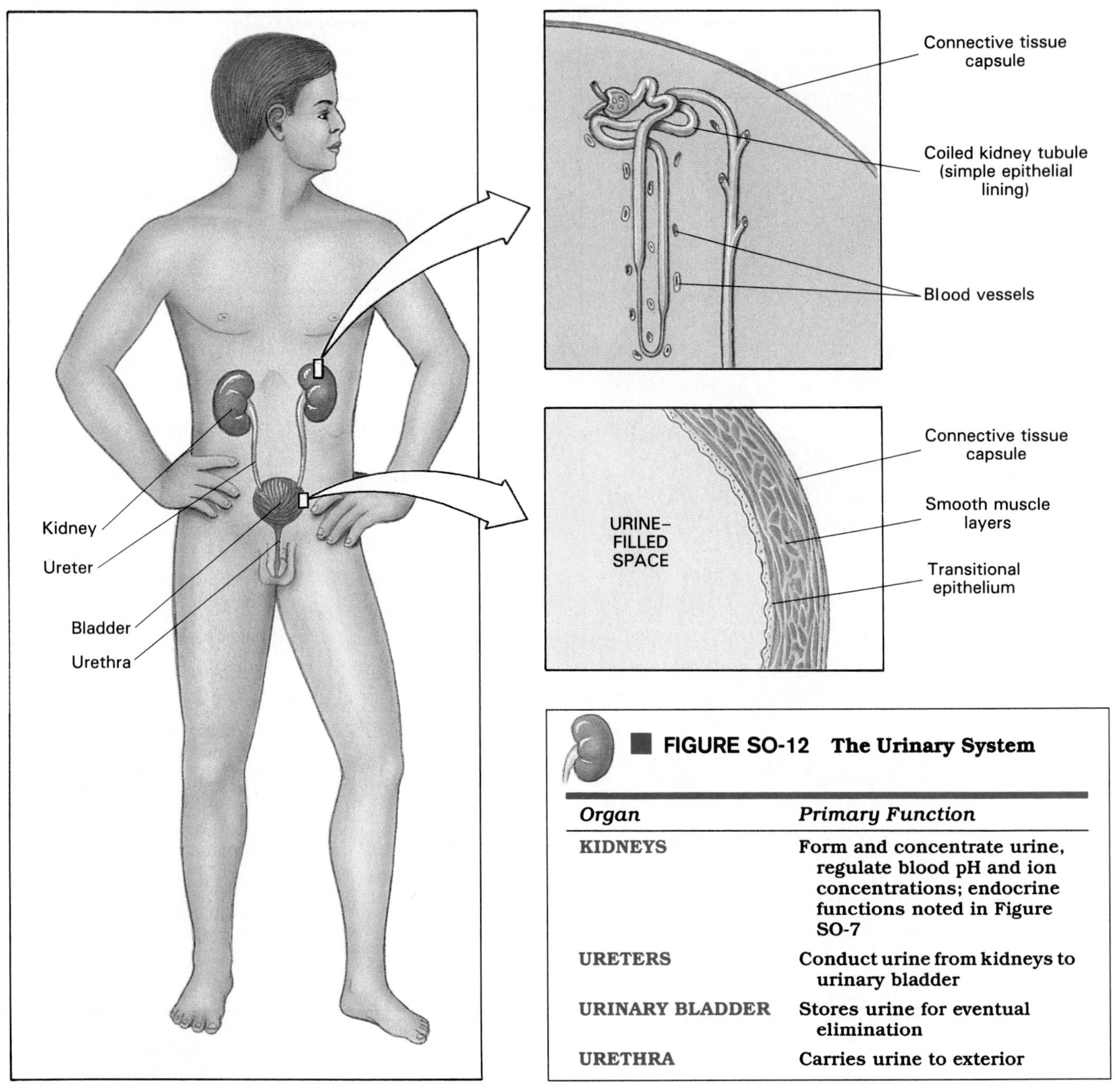

FIGURE SO-12 The Urinary System

Organ	*Primary Function*
KIDNEYS	**Form and concentrate urine, regulate blood pH and ion concentrations; endocrine functions noted in Figure SO-7**
URETERS	**Conduct urine from kidneys to urinary bladder**
URINARY BLADDER	**Stores urine for eventual elimination**
URETHRA	**Carries urine to exterior**

The **urinary system** (Figure SO-12) includes the **kidneys**, where urine production occurs; the **ureters**, which carry urine to the **urinary bladder** for storage; and the **urethra**, which conducts urine from the bladder and the exterior of the body.

The primary function of the urinary system is the selective elimination of waste products, including acids that might otherwise alter the pH of body fluids. It also regulates the fluid and electrolyte composition of the blood. Urine production begins at the kidneys, where filtration occurs across the walls of special capillaries into **kidney tubules**. As this fluid passes along the tubules, its composition changes through osmosis and the transport activities of epithelial cells. The urine that results is stored in the urinary bladder for a variable period before it is eliminated through the process of **urination**.

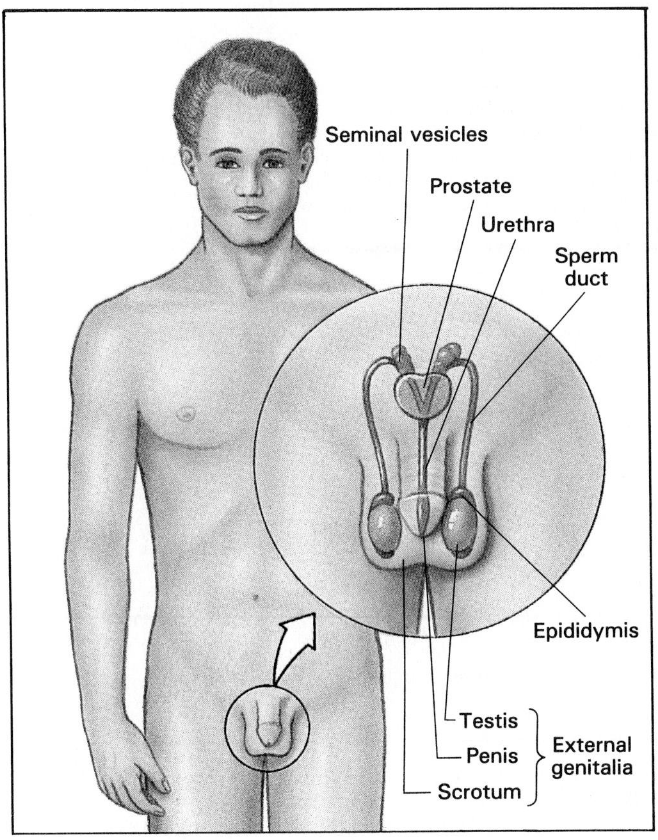

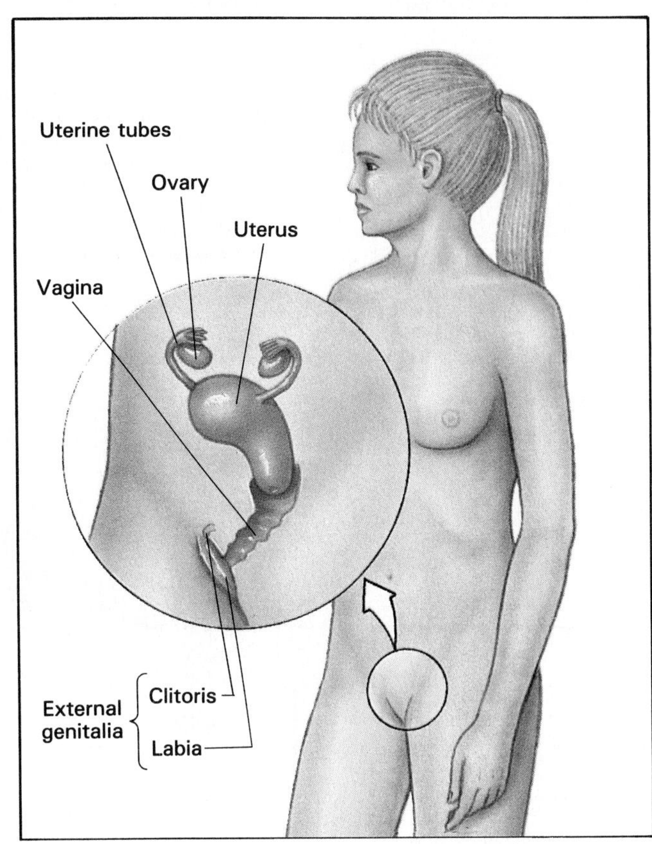

FIGURE SO-13 (a) The Reproductive System of the Male

Organ	*Primary Functions*
TESTES	Produce sperm; see also Figure SO-7
ACCESSORY ORGANS	
Epididymis	Site of sperm maturation
Ductus deferens (sperm duct)	Conducts sperm between epididymis and prostate
Seminal vesicles	Secrete fluid that makes up much of the volume of semen
Prostate	Secretes buffers and fluid
EXTERNAL GENITALIA	
Penis	Erectile organ used to deposit sperm in the vagina of a female
Scrotum	Surrounds the testes

FIGURE SO-13 (b) The Reproductive System of the Female

Organ	*Primary Functions*
OVARIES	Produce eggs; endocrine function in Figure SO-7
UTERINE TUBES	Deliver egg(s) or embryo(s) to uterus; site of normal fertilization
UTERUS	Site of embryonic and fetal development and diffusion between maternal and embryonic bloodstreams
VAGINA	Site of sperm deposition; birth canal at delivery; provides passage of fluids at menses
EXTERNAL GENITALIA	
Clitoris	Erectile organ, produces pleasurable sensations during sexual act
Labia	Contain glands that lubricate entrance to vagina
MAMMARY GLANDS	Produce milk that nourishes newborn infant

The **reproductive system** (Figure SO-13) is responsible for the production of future generations. It also produces hormones that affect the development, growth, and maintenance of many other systems. Sperm or eggs develop within reproductive organs known as **gonads**: the **testes** of a male, and **ovaries** of a female. Reproductive cells, or **gametes**, produced in the gonads travel toward the exterior along ducts that receive the secretions of **accessory organs**. The portions of the reproductive system that are visible at the body surface are known as the **external genitalia**.

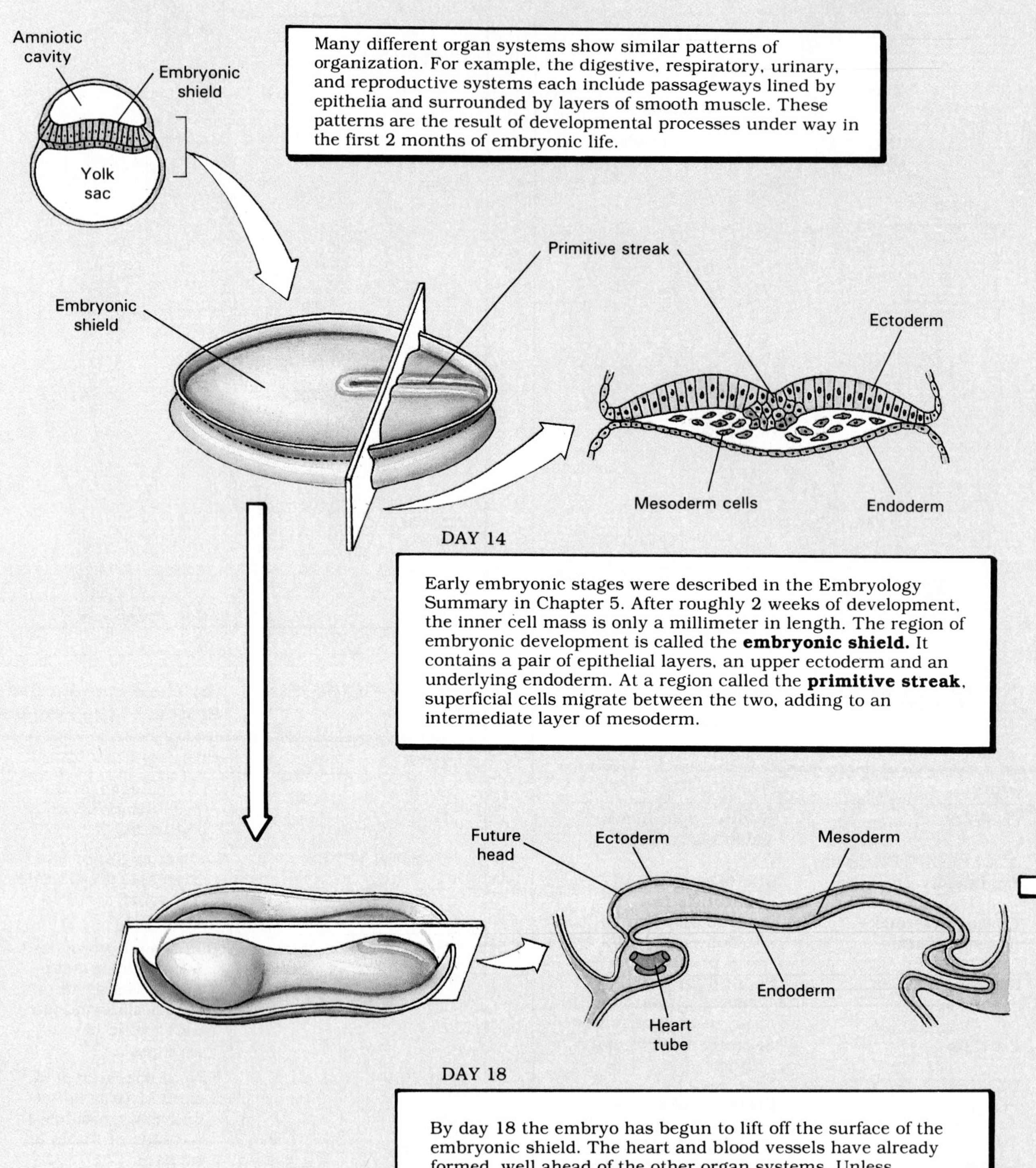

Many different organ systems show similar patterns of organization. For example, the digestive, respiratory, urinary, and reproductive systems each include passageways lined by epithelia and surrounded by layers of smooth muscle. These patterns are the result of developmental processes under way in the first 2 months of embryonic life.

DAY 14

Early embryonic stages were described in the Embryology Summary in Chapter 5. After roughly 2 weeks of development, the inner cell mass is only a millimeter in length. The region of embryonic development is called the **embryonic shield.** It contains a pair of epithelial layers, an upper ectoderm and an underlying endoderm. At a region called the **primitive streak**, superficial cells migrate between the two, adding to an intermediate layer of mesoderm.

DAY 18

By day 18 the embryo has begun to lift off the surface of the embryonic shield. The heart and blood vessels have already formed, well ahead of the other organ systems. Unless otherwise noted, discussions of organ system development in later chapters will begin at this stage.

EMBRYOLOGY SUMMARY: Development of the Organ Systems

DERIVATIVES OF PRIMARY GERM LAYERS	
Ectoderm Forms:	Integumentary system, including hair follicles, nails, and glands communicating with the skin (sweat, milk, and sebum) Lining of the mouth, salivary glands, nasal passageways, and anus Nervous system, including brain and spinal cord Portions of endocrine system (parts of pituitary and adrenal glands)
Mesoderm Forms:	Lining of the body cavities (pleural, pericardial, peritoneal) Muscular, skeletal, cardiovascular, and lymphatic systems Kidneys and part of the urinary tract Gonads and most of the reproductive tract Connective tissues supporting all organ systems Portions of endocrine system (parts of adrenal and endocrine tissues of reproductive tract)
Endoderm Forms:	Most of the digestive system: epithelium (except mouth and anus), exocrine glands (except salivary glands), the liver and pancreas Most of the respiratory system: epithelium (except nasal passageways) and mucous glands Portions of urinary and reproductive systems (ducts only) Portions of endocrine system (thymus, thyroid, pancreas, part of pituitary)

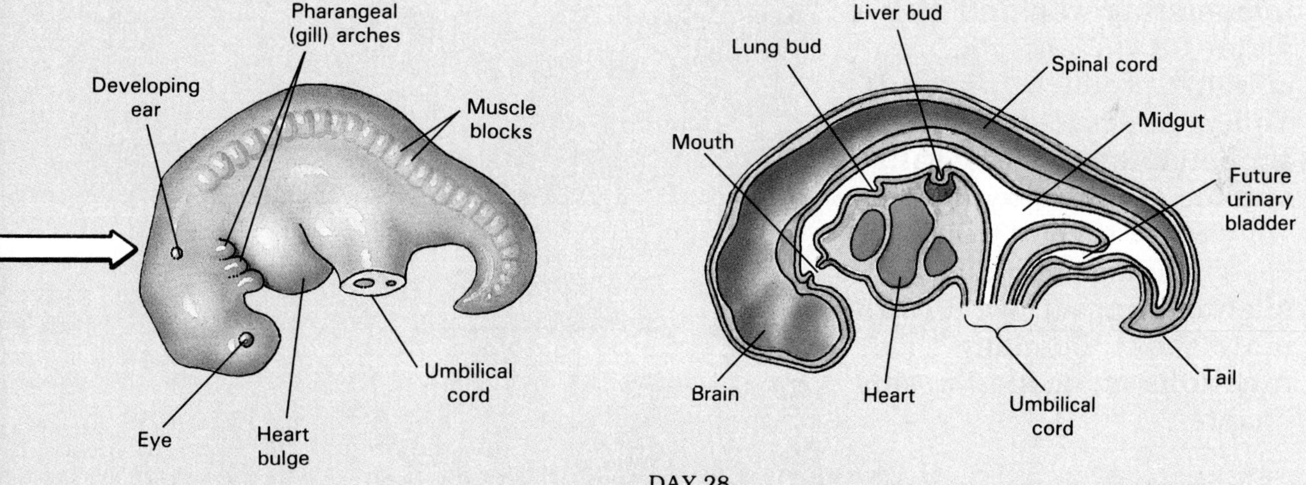

DAY 28

After 1 month you can find the beginnings of all major organ systems. The role of each of the primary germ layers in the formation of organs is summarized in the accompanying table; details will be found in later chapters. The Embryology Summary in Chapter 1 began at this stage and followed subsequent changes in external appearance.

Quick—what's the largest organ system in your body? (Here's a hint—you can see most of it very clearly in this photo!)

The answer surprises many people—it's the skin. The skin is the most over-examined and under-rated component of the human body. Our skin is a versatile covering that does much, much more than just keep the ocean outside and the rest of us inside. As we'll see, it's an active organ that works constantly to maintain homeostasis, adjusting to changes inside and outside the body. Without it your body temperature wouldn't stay a stable 98.6° F and you couldn't produce vitamin D. You would quickly accumulate too many salts and too much water, and your other organs would be easily damaged by the slightest impact. We'll learn more about the many functions of the skin in this chapter.

CHAPTER 6

The Integumentary System

Chapter Objectives

After reading this chapter, you will be able to:

1 Compare the structures and functions of the layers of the skin.
2 Discuss the functions of the skin's accessory structures.
3 Explain what accounts for individual and racial differences in skin, such as skin color.
4 Describe how the integumentary system helps to regulate body temperature.
5 Discuss the effects of ultraviolet radiation on the skin.
6 Describe the mechanisms that produce hair and determine hair texture and color.
7 Explain how the skin responds to injuries and repairs itself.
8 Summarize the effects of the aging process on the skin.

Introduction

This unit considers the integumentary, skeletal, and muscular systems. Together these systems account for almost 80 percent of the body weight. They are united by the fact that among their many other functions, each provides mechanical protection for more delicate internal structures, such as the central nervous system and organs of the thoracic and peritoneal cavities. We begin with the integumentary system, the only complete system we see every day.

Because others see this system as well, we devote a lot of time to the skin and associated structures. Washing the face and hands, brushing or trimming hair, clipping nails, taking showers, and applying deodorant are activities that modify the appearance or properties of the skin. And when something goes wrong with the skin, the effects are immediately apparent. Even a relatively minor skin condition, such as mild acne, will be noticed at once, whereas more serious problems in other systems are often ignored.

This chapter focuses on the important structural and functional relationships in the skin. In the

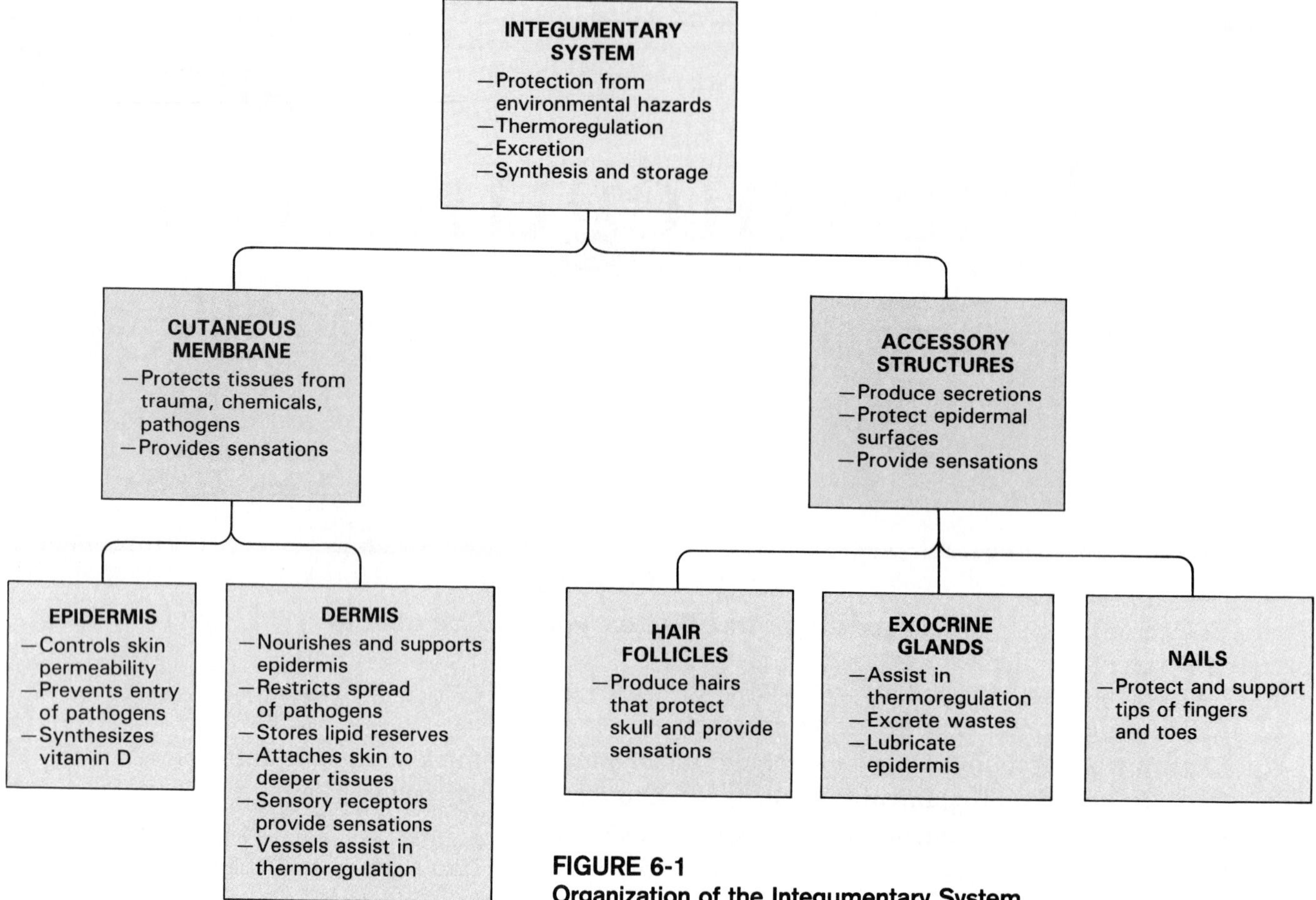

FIGURE 6-1
Organization of the Integumentary System

process, it demonstrates patterns that apply to tissue and organ interactions in other systems.

Integumentary Structure and Function

Figure 6-1 shows the functional organization of the **integumentary system**. It has two major components, the *cutaneous membrane* and the *accessory structures*.

- The **cutaneous membrane** has two components, the superficial epithelium, or **epidermis** (*epi-*, above) and the underlying connective tissues of the **dermis**. Beneath the dermis, the loose connective tissue of the **subcutaneous layer** (*superficial fascia*, or *hypodermis*) attaches the integument to deeper structures, such as muscles or bones.
- The **accessory structures** include hair, nails, and a variety of multicellular exocrine glands.

The integument does not function in isolation. An extensive network of blood vessels branches through the dermis, and sensory receptors that monitor pressure, temperature, and pain provide valuable information to the central nervous system. The general structure of the integument is shown in Figure 6-2; later figures will detail specific organs.

These varied components interact to:

1. Protect underlying tissues and organs;
2. Excrete salts, water, and organic wastes;
3. Maintain normal body temperature;
4. Synthesize vitamin D;
5. Store nutrients; and
6. Detect touch, pressure, pain, and temperature stimuli.

Each of these functions will be explored more fully as we discuss the individual components of the system.

THE EPIDERMIS

The epidermis consists of a stratified squamous epithelium, diagrammed in Figure 6-3. Early histologists gave the epithelial layers Latin names that are still in use.

Epidermis
Dermis
Subcutaneous layer
Hair shaft
Sensory nerve ending
Sebaceous gland
Arrector pili muscle
Sweat gland duct
Hair follicle
Sweat gland
Artery
Vein
Fat

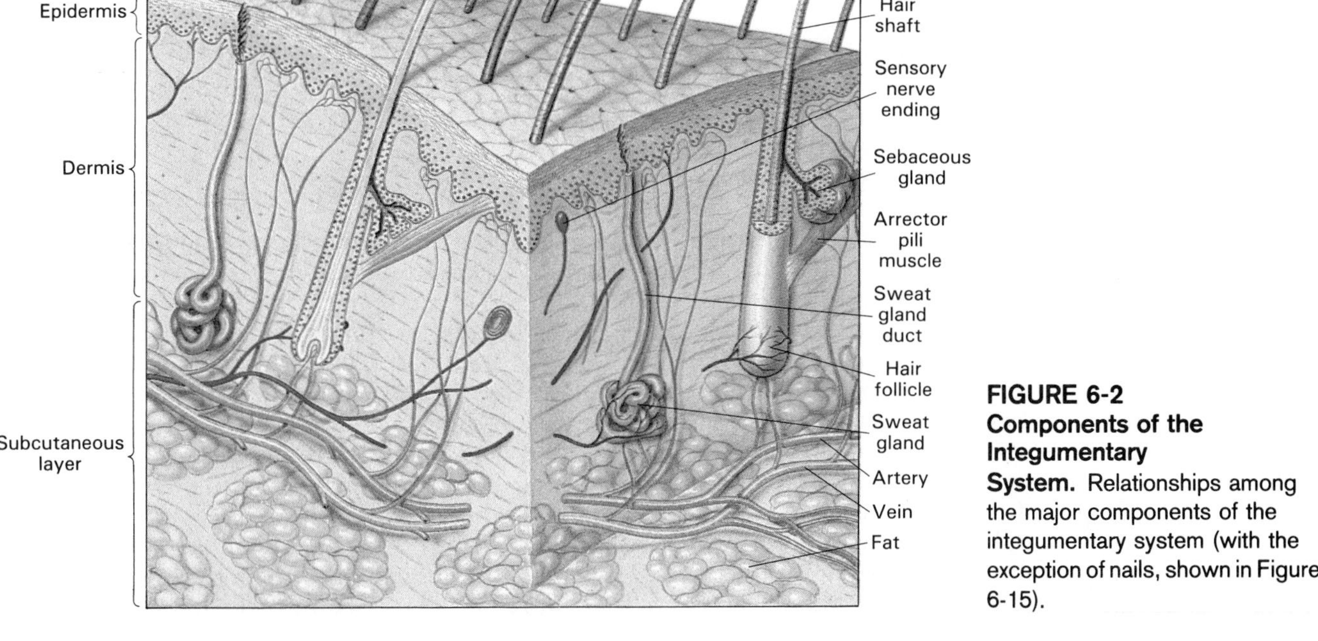

FIGURE 6-2
Components of the Integumentary System. Relationships among the major components of the integumentary system (with the exception of nails, shown in Figure 6-15).

FIGURE 6-3
The Structure of the Epidermis. A light micrograph through a portion of the epidermis, showing the major layers that appear during the maturation of epidermal cells. That maturation involves the production of keratin; stages in this process are indicated at the right. (LM, ×150)

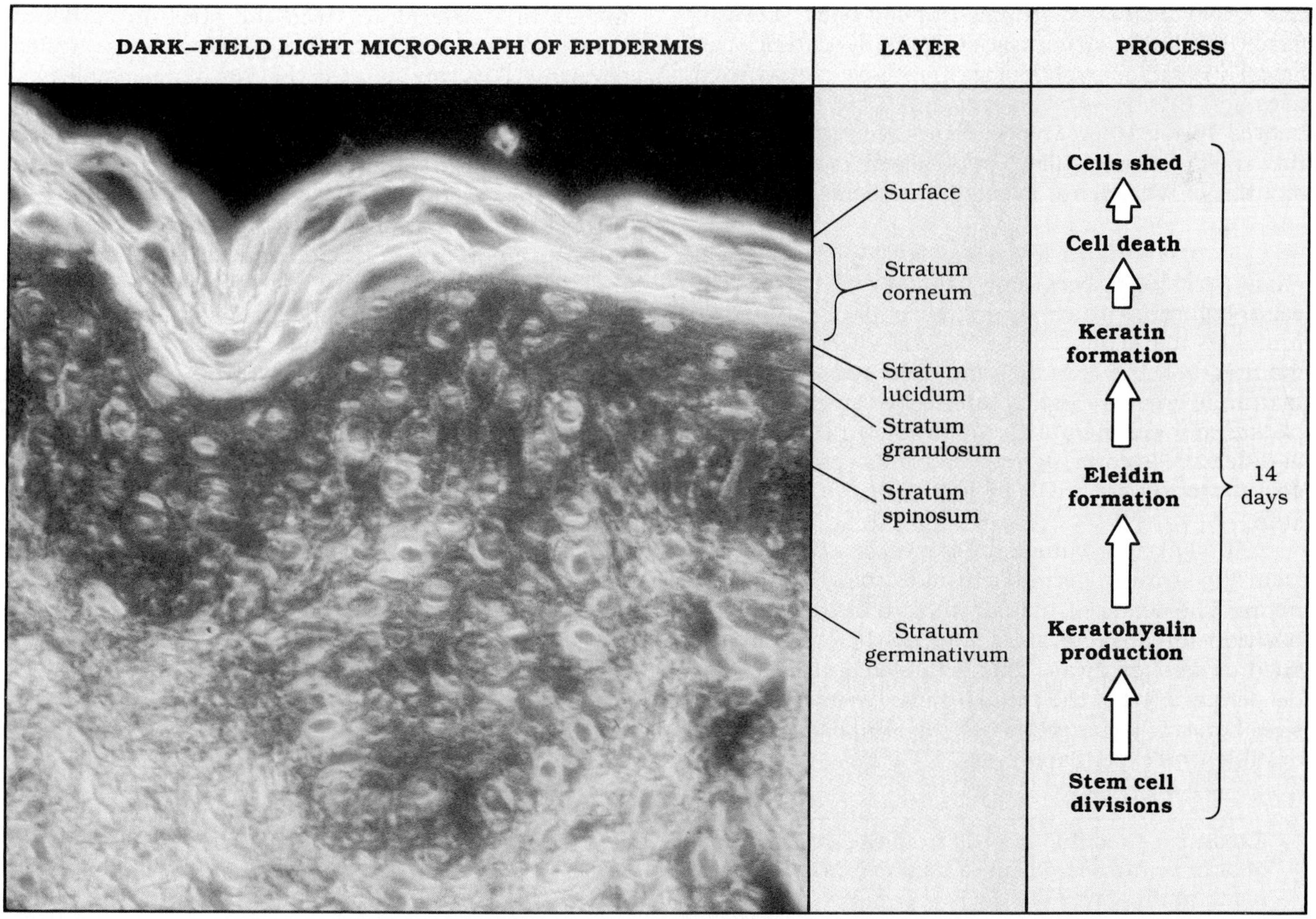

DARK–FIELD LIGHT MICROGRAPH OF EPIDERMIS
LAYER
PROCESS
Surface
Stratum corneum
Stratum lucidum
Stratum granulosum
Stratum spinosum
Stratum germinativum
Cells shed
Cell death
Keratin formation
Eleidin formation
Keratohyalin production
Stem cell divisions
14 days

Layers of the Epidermis

The deepest epidermal layer, the **stratum germinativum** (STRA-tum jer-mi-nā-TĒ-vum), or *stratum basale*, is firmly attached to the basement membrane. Stem cells dominate the stratum germinativum, and their mitotic divisions replace the more superficial cells that are lost or shed at the epithelial surface.

At each division, one of the daughter cells enters the more superficial layer known as the **stratum spinosum** (spiny layer). The stratum spinosum is several cells thick, and the cells are bound together by desmosomes. Standard histological procedures shrink the cytoplasm, but the cytoskeletal elements and desmosomes remain intact, making the cells look like miniature pincushions. Some of these cells undergo further divisions, but mitotic activity ceases by the time they reach the **stratum granulosum** (grainy layer). In this layer the cells begin manufacturing large quantities of the protein **keratohyalin** (ker-a-tō-HĪ-a-lin). At the palms or soles, areas of very thick skin, a glassy **stratum lucidum** (clear layer) covers the stratum granulosum. The cells in this layer are flattened, densely packed, and filled with **eleidin** (el-Ē-i-din), a protein derived from keratohyalin.

Keratohyalin and eleidin represent preliminary steps in the production of the fibrous protein **keratin** (KER-a-tin; *keros*, horn). Keratin, a fibrous protein introduced in Chapter 2, is a biological wonder. (p. 53) It is extremely strong, light, flexible, durable, and water-resistant. In the human body, keratin forms the basic structural component of hair, calluses, and nails, but other animals have more varied uses for this interesting protein. Cow horns and hooves, bird feathers, reptile scales, porcupine quills, the armor of armadillos, and baleen plates in the mouths of whales are examples of the versatility of keratin.

As true keratin fibers are developing, the cells change in shape, becoming thinner and flatter. The cell membranes thicken, and their permeability diminishes markedly. The nuclei and other organelles disintegrate, the cells die, and their subsequent dehydration creates a tightly interlocked layer of keratin packaged in cell membranes. These layers of flattened and dead cells covering the epidermis constitute the **stratum corneum** (KOR-nē-um; *cornu*, horn) of the skin.

It takes approximately 14 days for a cell to move from the stratum germinativum to the stratum corneum. The dead cells usually remain in this exposed position for an additional 2 weeks before they are shed or washed away. This arrangement places the deeper portions of the epithelium and underlying tissues beneath a protective barrier composed of dead, durable, and expendable cells. [**CM:** *Excessive Keratin Production*]

√ **Excessive shedding of cells from the outer layer of skin in the scalp causes dandruff. What is the name of this layer of skin?**

√ **As you pick up a piece of lumber, a splinter pierces the palm of your hand and lodges in the third layer of the epidermis. Identify this layer.**

Keratinization and Epidermal Permeability

An epithelium containing large amounts of keratin is said to be **keratinized** (ker-A-tin-īzed), or **cornified** (KŌR-ni-fīd; *cornu*, horn + *facere*, to make). In comparison with other epithelia, a cornified epithelium is only around 1/100,000th as permeable to water and electrolytes. Normally the stratum corneum is relatively dry; only around 10 percent of the cell weight is water. The dry surface makes it unattractive to many bacteria. Maintenance of this barrier involves coating the surface with the lipid secretions of epidermal glands, a process that resembles waxing a pair of leather sneakers to keep them dry and supple.

Although the stratum corneum is water-resistant, it is not waterproof, and water from the interstitial fluids slowly penetrates the surface, to be evaporated into the surrounding air. Roughly 500 mℓ (about 1 pt) of water per day is lost in this way. This water loss is called **insensible perspiration** to distinguish it from the **sensible perspiration** produced by sweat glands. **Xerosis** (ze-RŌ-sis), or "dry skin," is a common complaint of the elderly and people who live in arid climates. Under these conditions the cell membranes deteriorate, and the stratum corneum becomes more a collection of scales than a unified covering. This can increase the rate of insensible perspiration by 75 times.

When the skin is immersed in water, osmotic forces may move water into or out of the epithelium. (Osmosis was described in Chapter 3. (p. 80)) Sitting in a freshwater bath causes water to move into the epidermis because fresh water is hypotonic (has fewer dissolved materials) compared with body fluids. The epithelial cells may swell to four times their normal size, a phenomenon you may have noticed in the thickly keratinized areas of your palms and soles. Swimming in the ocean reverses the direction of osmotic flow, because the ocean is hypertonic with respect to body fluids. As a result, water leaves the body, crossing the epidermis from the underlying tissues. The process is slow, but long-term exposure to seawater endangers survivors of a shipwreck by promoting dehydration.

Thick and Thin Skin

Cornification occurs everywhere on the skin except over the surfaces of the eyes. Most of the body is covered by **thin skin**, with a total thickness of 1.5–4 mm. In a sample of thin skin, seen in Figure 6-4b, the epidermis averages a mere 0.08 mm in thickness, and the stratum corneum is only a few layers deep. Heavily abraded surfaces such as the palms

Clinical Comment: Transdermal Medication

Drugs in oils or other lipid-soluble carriers can penetrate the epidermis. The movement is slow, particularly through the layers of cell membranes in the stratum corneum, but once a drug reaches the underlying tissues it will be absorbed into the circulation. A useful technique involves placing a sticky patch containing a drug over an area of thin skin. To overcome the relatively slow rate of diffusion, the patch must contain an extremely high concentration of the drug. This procedure, called *transdermal administration*, has the advantage that a single patch may work for several days, making daily pills unnecessary. *Scopolamine*, a drug that affects the nervous system, is administered transdermally to control nausea. Transdermal *nitroglycerin* can be used to improve blood flow within heart muscle and prevent a heart attack.

DMSO (*dimethyl sulfoxide*) is a transdermal drug intended for the treatment of injuries to the muscles and joints of domesticated animals, such as horses or cows. It is a solvent that rapidly crosses the skin, and drugs dissolved in DMSO will be carried into the body at the same time. DMSO has not been tested and approved for the treatment of human patients in the United States, either for joint or muscle injuries or as a transdermal solvent. However, it can be prescribed in Canada and Europe. The long-term risks associated with its use are unknown; reported short-term side effects include nausea, vomiting, cramps, and chills.

of the hands or the soles of the feet, may be covered by 30 or more layers of cornified cells. As a result, the epidermis in those locations may be as much as six times thicker than the epidermis covering the general body surface. This **thick skin** can be seen in Figure 6-4c.

Epidermal Ridges

The deeper layers of the epidermis form **epidermal ridges** that extend into the dermis, increasing the area of contact between the two regions. Dermal projections called **dermal papillae** (singular *papilla*, nipple-shaped mound) extend between adjacent ridges, as indicated in Figure 6-4. The epidermal ridges are monitored by nerves that provide touch, pressure, pain, and temperature information from the epidermal surface.

The contours of the skin surface follow the ridge patterns, which vary from small conical pegs to the complex whorls seen on the skin of the palms and soles. Ridges on the palms and soles increase the surface area of the skin and increase friction, ensuring a secure grip. Ridge shapes are genetically determined: those of each person are unique and do not change in the course of a lifetime. Fingerprints—ridge patterns on the tips of the fingers (Figure 6-5)—can therefore be used to identify individuals, and have been so employed in criminal investigation for over a century.

√ **Why does sitting in fresh water for a long period of time cause swelling of epidermal cells?**

√ **Some criminals sand the tips of their fingers so as not to leave recognizable fingerprints. Would this practice permanently remove fingerprints? Why or why not?**

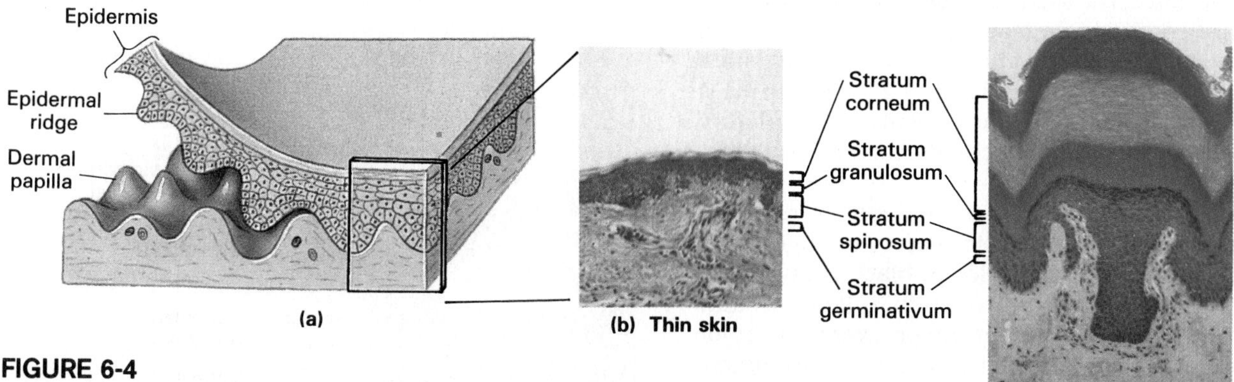

FIGURE 6-4
Thin and Thick Skin. (a) The basic organization of the epidermis. The proportions of the various layers change depending on the location sampled. (b) Thin skin covers most of the exposed body surface. (c) Thick skin covers the surfaces of the palms and soles. Because of its greater thickness, thick skin provides much greater protection against abrasion than thin skin does. (LMs, ×75)

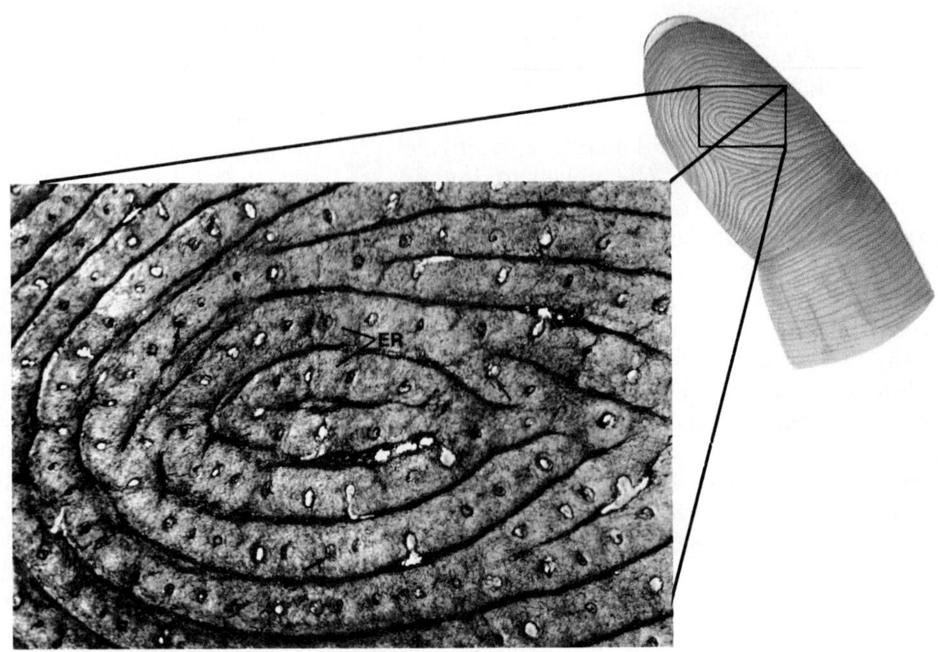

FIGURE 6-5
The Epidermal Ridges of Thick Skin. Fingerprints reveal the pattern of epidermal ridges in thick skin. This scanning electron micrograph shows the ridges at a fingertip. The pits are the openings of sweat gland ducts. (SEM, ×25)

Skin Color

The color of the epidermis is due to an interaction between two basic factors: (1) circulatory supply and (2) pigment composition and concentration.

Blood with abundant oxygen is bright red, and blood vessels in the dermis normally give the skin a reddish tint that is most easily seen in light-pigmented individuals. When those vessels are dilated, as during inflammation (Chapter 5), the red tones become much more pronounced. (p. 159) When the circulatory supply is temporarily reduced, the skin becomes relatively pale; a frightened Caucasian may "turn white" because of a sudden drop in blood supply to the skin. During a sustained reduction in circulatory supply, the blood in the skin loses oxygen and changes color to a much deeper red tone. Seen from the surface, the skin takes on a bluish coloration that is called **cyanosis** (sī-a-NŌ-sis; *kyanos*, blue). Cyanosis is most apparent in areas of thin skin, such as the lips, ears, or beneath the nails. It can be a response to extreme cold or a result of circulatory or respiratory disorders, such as heart failure or severe asthma.

In general, pigment content can overshadow other factors; for example, circulatory changes have less visible effect on skin color in black individuals. But while the skin pigments of black individuals may obscure localized inflammation or cyanosis, color changes can usually be seen through the nails, where the epidermis lacks dark pigments.

Epidermal cells contain variable quantities of two pigments, *carotene* and *melanin*. **Carotene** (KAR-ō-tēn) is an orange-yellow pigment that normally accumulates inside the cells. Carotene pig-

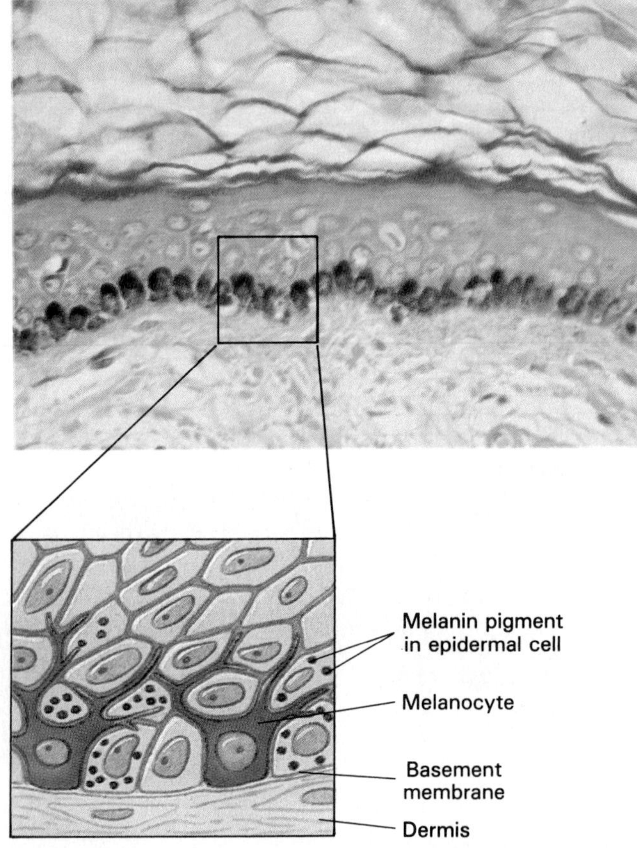

FIGURE 6-6
Melanocytes. The micrograph and accompanying drawings indicate the location and orientation of melanocytes in the deeper layers of the epidermis of a black person.

ments are found in a variety of vegetables; for example, they are responsible for the orange color of carrots. Vegetarians with a special fondness for carrots can actually turn orange from an overabundance of carotenes in the skin.

Melanocytes are pigment cells found in the stratum germinativum, squeezed between the epithelial cells, and within the adjacent connective tissue of the dermis. Melanocytes, shown in Figure 6-6, manufacture and store **melanin**, a yellow-brown, brown, or black pigment. Melanocytes inject melanin into the surrounding epithelial cells, thereby coloring the entire epidermis. The melanin pigments protect these cells from a serious environmental hazard, sunlight.

Sunlight contains **ultraviolet (UV) radiation** that can penetrate the epidermis and reach underlying tissues. A little ultraviolet is useful, for sunlight converts a steroid related to cholesterol into vitamin D, as discussed later in this chapter. But too much ultraviolet radiation produces immediate effects simi-

Health News: Skin Cancers

Almost everyone has several benign tumors of the skin; freckles and moles are examples. Skin cancers are also relatively common. In fact, they are the most common cancers. In 1990, approximately 600,000 cases of skin cancer were diagnosed in the United States, as compared with 840,000 cases involving all other organs. The most common skin cancers are usually caused by prolonged exposure to sunlight.

A **basal cell carcinoma** is a malignant cancer that originates in the germinative (basal) layer. This is the most common skin cancer, and roughly two-thirds of these cancers appear in areas subjected to chronic UV exposure. **Squamous cell carcinomas** are less common, but almost totally restricted to areas of sun-exposed skin. Metastasis seldom occurs in squamous cell carcinomas and almost never in basal cell carcinomas, and most people survive these cancers. The usual treatment involves surgical removal of the tumor, and 95 percent of patients survive 5 years or more after treatment. (This statistic, the 5-year survival rate, is a common method of reporting long-term prognosis.)

Compared with these common and seldom life-threatening cancers, **malignant melanomas** (mel-a-NŌ-mas) are extremely dangerous. In this condition cancerous melanocytes grow rapidly and metastasize through the lymphatic system. In 1990, 27,600 cases of malignant melanoma were diagnosed in the United States, making this a relatively rare disease. The incidence follows the sun: 3900 new cases in California versus 40 in Alaska. The outlook for long-term survival changes dramatically, depending on when the condition is diagnosed. If localized, the 5-year survival rate is 90 percent; if widespread, the survival rate drops to 14 percent.

To catch melanoma at an early stage, it is essential to know what to look for when examining your skin. The key points can be remembered most easily using the mnemonic ABCD.

- *A* is for *asymmetry*: melanomas tend to be irregular in shape. Often they are raised; they may ooze or bleed.
- *B* is for *border*: usually unclear, sometimes notched.
- *C* is for *color*: usually mottled, with many different colors (tan, brown, black, red, pink, white, and/or blue).
- *D* is for *diameter*: a growth more than about 5 mm (1/5 in.) in diameter is dangerous.

A new experimental treatment for melanoma uses genetic engineering technology (discussed in Chapter 4) to manufacture antibodies that target the MSH receptors on the surfaces of melanocytes. ∞ (p. 114) Melanocytes coated with these antibodies are then recognized and attacked by cells of the immune system.

Light-skinned people who live in the tropics are most susceptible to all forms of skin cancer, because their melanocytes are unable to shield them from the ultraviolet radiation. Sun damage can be prevented by avoiding exposure to the sun during the middle hours of the day, and by using a sunblock (*not* a tanning oil or sunscreen)—a practice that also delays the cosmetic problems of sagging and wrinkling. *Everyone* who expects to be out in the sun for any length of time should choose a broad-spectrum sunblock with a sun protection factor (SPF) of at least 15; blonds, redheads, and people with very fair skin are better off with a sun protection factor of 20 to 30.

The use of sun screens may become increasingly important as ozone gas in the upper atmosphere is destroyed by our industrial emissions. Ozone absorbs UV before it reaches the earth's surface, and in doing so it assists the melanocytes in preventing skin cancer. Australia, which is most affected by the depletion of ozone near the south pole (the "ozone hole"), is already reporting an increased incidence of skin cancers.

lar to those of mild or even serious burns. Long-term damage can result from chronic exposure, and an individual attempting to acquire a deep tan places severe stresses on the skin. Alterations in the underlying connective tissues lead to premature wrinkling, and skin cancers can result from chromosomal damage in germinative cells or melanocytes.

Melanin helps prevent these changes by absorbing ultraviolet radiation before it reaches the deep layers of the epidermis and dermis. Within the epidermal cells, melanin concentrates around the outer wall of the nucleus, so it absorbs the UV before it can damage the nuclear DNA. Melanocytes respond to UV exposure by increasing their activity. Unfortunately, the response is not rapid enough to prevent a sunburn the first day at the beach. Melanin synthesis accelerates slowly, peaking around 10 days after the initial exposure.

The ratio between melanocytes and germinative (basal) cells ranges between 1:4 and 1:20, depending on the region of the body surveyed. The observed differences in skin color between individuals and even races do not reflect different *numbers* of melanocytes, merely different levels of synthetic activity. Even the melanocytes of **albino** individuals are distributed normally, although they are incapable of producing melanin. (Albinism, an inherited condition that affects approximately one person in 10,000, will be discussed further in Chapter 29.)

Melanocyte activity normally increases or decreases in response to changes in circulating levels of **melanocyte-stimulating hormone (MSH)**. This peptide, secreted by the pituitary gland, increases the rate of melanin synthesis.

√ **Why does exposure to sunlight or tanning lamps cause the skin to become darker?**

Langerhans Cells and Merkel Cells

Two other cell types, *Langerhans cells* and *Merkel cells*, are scattered among the deeper cells of epidermis, close to the basement membrane. **Langerhans cells** are most common in or below the stratum spinosum. These cells are mobile macrophages that are part of the body's defensive system; their functions will be discussed in Chapter 22. **Merkel cells** are attached to surrounding epidermal cells and to the dendrites of sensory neurons. Together the Merkel cells and neurons provide information about objects touching the skin. Although there are many other kinds of touch receptors, the rest are in the dermis, and are less sensitive than the Merkel cell/neuron combination.

The Epidermis as a Chemical Factory

Although strong sunlight can damage epithelial cells and deeper tissues, limited exposure to sunlight is very beneficial. Epidermal cells in the stratum spinosum and stratum germinativum can convert steroid precursors to vitamin D when exposed to sunlight. Vitamin D is required for normal calcium and phosphorus absorption across the intestinal lining, and an inadequate supply of this vitamin leads to impaired bone maintenance and growth. Children who live in areas where the sky is overcast much of the year can have abnormal bone development. This condition has largely been eliminated in the United States because dairy companies add vitamin D to the milk sold in grocery stores. Chapter 7 will consider vitamin D and its effects on bone growth in greater detail.

THE DERMIS

The **dermis** lies beneath the epidermis. It has two major components, a superficial *papillary layer* and a deeper *reticular layer*. These layers are detailed in Figure 6-7.

Layers of the Dermis

The **papillary layer** (Figure 6-7a) consists of loose connective tissue. This region contains the capillaries and nerves supplying the surface of the skin. The papillary layer derives its name from the dermal papillae that project between the epidermal ridges, as indicated in Figure 6-4.

The deeper **reticular layer** consists of dense, irregular connective tissue. Figure 5-13c was a light micrograph of this portion of the dermis, and Figure 6-7b provides a three-dimensional view. Bundles of collagen fibers leave the reticular layer to blend into those of the papillary layer above, so the boundary line between these layers is indistinct. Collagen fibers of the reticular layer also extend into the subcutaneous layer below (Figure 6-7c). This layer, also known as the superficial fascia, or hypodermis, provides support and attachment for the dermis, but allows flexibility and independent movement. [CM: *Dermatitis*]

Wrinkles and Stretch Marks

The interwoven collagen fibers of the reticular layer provide considerable strength. In addition, an extensive array of elastic fibers enables the dermis to stretch and contract repeatedly during normal movements. Age, hormones, and the destructive effects of ultraviolet radiation reduce the amount of elastin in the dermis, producing wrinkles and sagging skin. The extensive distortion of the dermis that occurs over the abdomen during pregnancy or following a substantial weight gain often exceeds the elastic capabilities of the skin. Although the skin stretches, it does not contract to its original size after delivery or a rigorous diet. The skin then wrinkles and creases, creating a network of **stretch marks**.

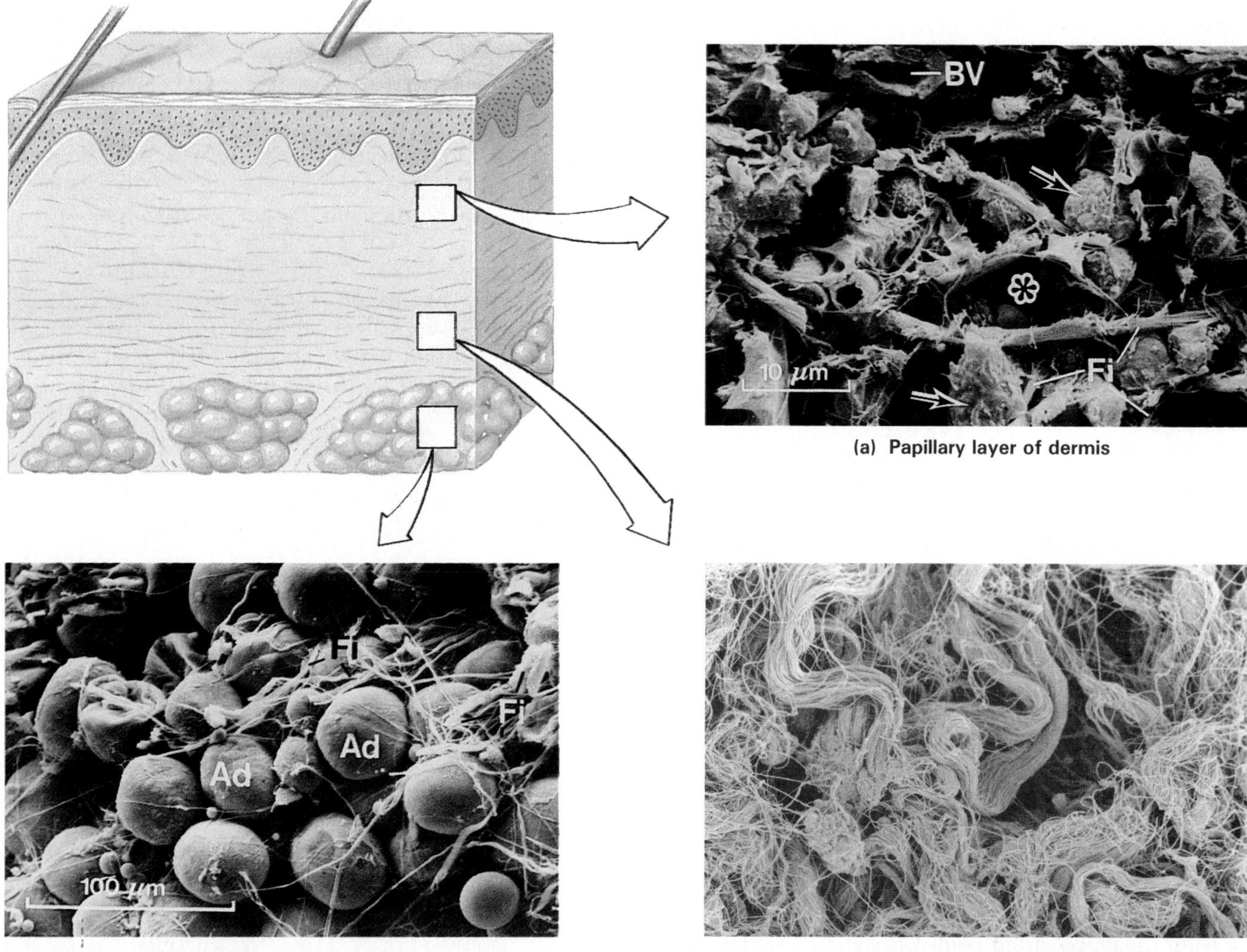

(a) Papillary layer of dermis

(c) Subcutaneous layer

(b) Reticular layer of dermis

FIGURE 6-7
The Structure of the Dermis and Subcutaneous Layer. (a) The papillary layer of the dermis consists of loose connective tissue that contains numerous blood vessels (BV), fibers (Fi), and macrophages (arrows). Open spaces, such as the one marked by an asterisk, would be filled with fluid ground substance. (SEM, ×484) (b) The reticular layer of the dermis contains dense, irregular connective tissue. (SEM, ×1000) (c) The subcutaneous layer contains large numbers of adipocytes (Ad) in a framework of loose connective tissue fibers (Fi). (SEM, ×200)

Tretinoin (*Retin-A*®) is a derivative of vitamin A that can be applied to the skin as a cream or gel. This drug was originally developed to treat acne, but it also increases blood flow to the dermis and stimulates dermal repairs. [CM: *Acne*] As a result, the rate of wrinkle formation decreases, and existing wrinkles become smaller. The degree of improvement varies from individual to individual.

At any one location, the majority of the collagen and elastic fibers are found in parallel bundles. The orientation of these bundles varies depending on the stress placed on the skin during normal movement. The pattern of fiber bundles establishes the **lines of cleavage** of the skin. The typical lines of cleavage are shown in Figure 6-8. A cut parallel to these lines will usually remain closed, while a cut at right angles will be pulled open as the elastic fibers recoil. Surgeons select their incision patterns accordingly, for a neatly closed wound will heal faster and with less scarring.

Other Dermal Components

In addition to protein fibers, the dermis contains a mixed cell population that includes all of the cells of connective tissue proper (Chapter 5). [p. 138] Accessory organs of epidermal origin, such as hair follicles and sweat glands, extend into the dermis. Other systems communicate with the skin via the dermis. For example, the reticular and papillary layers contain a network of blood vessels (cardiovascular system), lymphatics (lymphatic system), and nerve fibers (nervous system). Blood vessels provide nu-

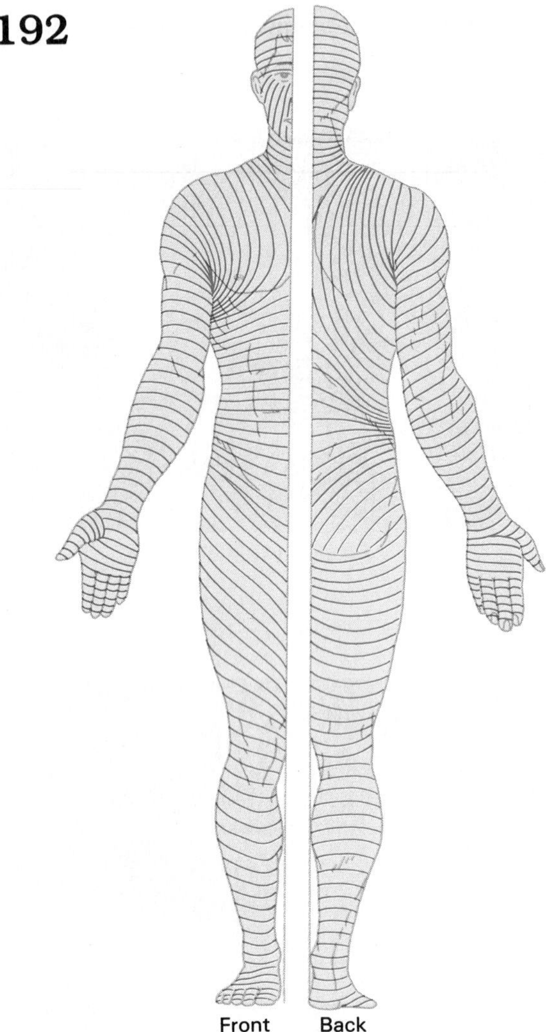

FIGURE 6-8
Lines of Cleavage of the Skin. Lines of cleavage follow lines of tension in the skin. They reflect the orientation of collagen fiber bundles in the dermis.

Clinical Comment: Tumors in the Dermis

Tumors seldom develop in the dermis, and those that do appear are usually benign. Two forms of benign tumors called **hemangiomas** may appear among dermal blood vessels during embryonic development. Viewed from the surface, these form prominent *birthmarks*. A *capillary hemangioma* involves capillaries of the papillary layer. It usually enlarges after birth, but subsequently fades and disappears. *Cavernous hemangiomas*, or "port-wine stains," affect larger vessels in the dermis. Such birthmarks usually last a lifetime; Figure 6-9 shows a typical example.

FIGURE 6-9
A Famous Hemangioma. President Gorbachev, leader of the Soviet Union prior to its collapse, has a prominent cavernous hemangioma, or "port-wine stain," on his forehead.

trients and oxygen and remove carbon dioxide and waste products. Both the blood vessels and the lymphatics monitor conditions in the dermis and assist local tissue defenses and repairs after an injury or infection. The nerve fibers control blood flow, adjust gland secretion rates, and monitor sensory receptors in the dermis and the deeper layers of the epidermis. These receptors, which provide sensations of touch, pain, pressure, and temperature, will be detailed in Chapter 17.

THE SUBCUTANEOUS LAYER

An extensive network of connective tissue fibers attaches the dermis to the subcutaneous layer, or hypodermis. The boundary between these two layers is indistinct, and although the *hypodermis* is not actually a part of the integument it is important in stabilizing the position of the skin in relation to underlying tissues and organs.

As noted in Chapter 5, the subcutaneous layer consists of loose connective tissue with abundant fat cells. (p. 159) Infants and small children usually have a hypodermal blanket of adipose tissue over the entire body. This "baby fat" helps reduce heat loss, for heat diffuses through lipids only around one-third as rapidly as through other tissues. Subcutaneous fat also serves as a substantial energy reserve and a shock absorber for the rough-and-tumble activities of our early years.

As maturation proceeds, the distribution of subcutaneous fat changes, the pattern depending on the sex of the individual. Men accumulate subcutaneous fat at the neck, upper arms, along the lower back, and over the buttocks. In women the breasts, buttocks, hips, and thighs are the primary sites of subcutaneous fat storage. In adults of either sex the hypodermis of the backs of the hands or the upper surfaces of the feet contain few fat cells, whereas distressing amounts of adipose tissue can accumulate in the abdominal hypodermis, producing a prominent "pot belly."

The hypodermis is quite elastic. It contains a limited number of capillaries and no vital organs. This last characteristic makes **subcutaneous injection** a useful method for administering drugs. The familiar term **hypodermic needle** refers to the region targeted for injection.

ACCESSORY STRUCTURES

Accessory structures include hair follicles, sebaceous glands, sweat glands, and nails.

Hair Follicles

Hairs project above the surface of the skin almost everywhere except over the sides and soles of the feet, the palms of the hands, the sides of the fingers and toes, the lips, and portions of the external genital organs. These hairs originate in complex organs called **hair follicles**. Before discussing the functions of hair, we must examine the structure of hair follicles in greater detail.

Structure of Hair Follicles Hair follicles extend deep into the dermis, often projecting into the underlying subcutaneous layer. Key features of follicle structure are illustrated in Figure 6-10. The epithe-

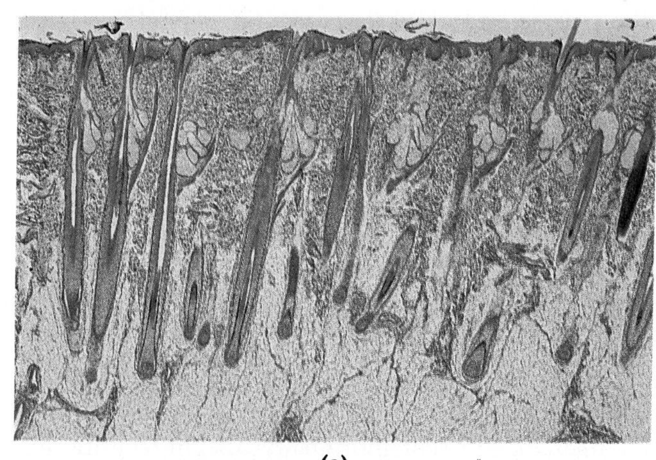

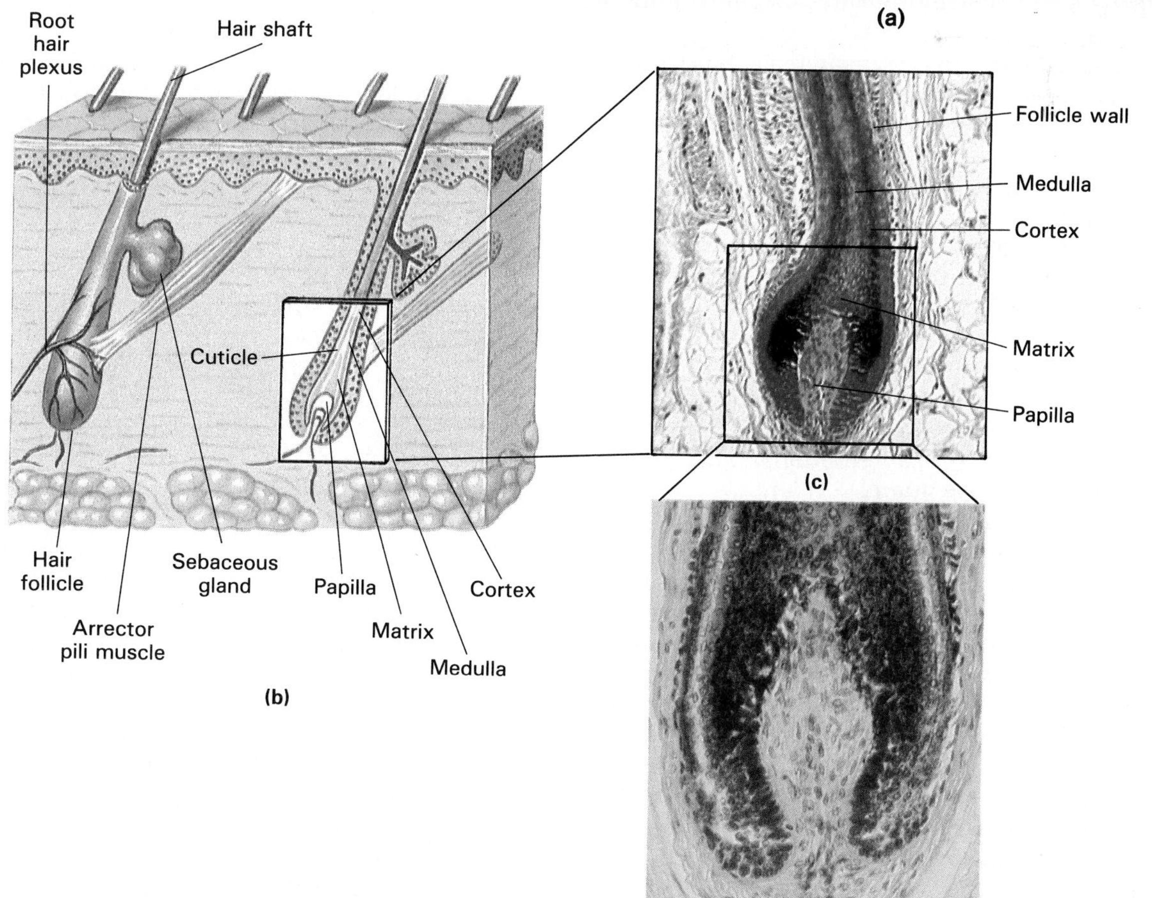

FIGURE 6-10
Hair Follicles. (a) A light micrograph showing the sectional appearance of the skin of the scalp. Note the many hair follicles and the way they extend into the dermis. (LM, ×13) (b) Three-dimensional drawing showing the structure of a hair follicle. (c) A light micrograph showing closer view of the base of the follicle and hair shaft. (LM, ×85); (d) Light micrograph showing the matrix and papilla at the hair root, where hair growth occurs (LM, ×215).

lium at the base of a follicle surrounds a small connective tissue **papilla** containing capillaries and nerves. The **matrix** of the follicle consists of those epithelial cells that surround the papilla. The cells of the matrix are responsible for the formation of the hair itself.

Hair production involves a specialization of the cornification process. Basal cells of the matrix divide, and overlying layers undergo keratinization. Those closest to the center of the papilla form the soft core, or **medulla**, of the hair, while cell divisions farther away produce the **cortex**. The medulla contains flexible **soft keratin** similar to that found in the stratum corneum of the epidermis. **Hard keratin** in the cortex gives the hair its stiffness. Matrix cells near the edge of the papilla form the **cuticle**, a layer of hard keratin that coats the hair.

Each individual hair has a **root** that encloses the matrix and a **shaft** of varying size, shape, and color. Differences result from the size of the follicles, the activity of follicular cells, and the shapes of the hairs. For example, straight hairs are round in cross section, whereas curly ones are rather flattened. **Vellus hairs** are the fine "peach fuzz" hairs found over much of the body surface. **Terminal hairs** are heavy, more deeply pigmented, and sometimes curly. The hair on your head, including your eyebrows and eyelashes, are examples of terminal hairs. **Intermediate hairs** are those hairs of intermediate character, such as the hairs of the arms or legs. Some hair follicles alter the structure of the hairs in response to circulating hormones, and this accounts for many of the changes in hair distribution that begin at puberty.

Functions of Hair In furry mammals, hairs provide an insulating blanket and considerable protection from abrasion. The underlying epidermis is very thin, and is cornified only in exposed areas such as the tip of the snout or the pads of the paws. Special smooth muscles in the dermis, the **arrector pili** (a-REK-tōr PI-li) muscles, can pull on the follicles and elevate the hairs. Contraction may be caused by emotional states, such as fear or rage, or as a response to cold. Unlike the furry mammals, we receive little mechanical protection and no insulating benefits from our complement of hairs except on the tops of our heads. Nevertheless, we have retained the arrector pili muscles, and similar emotional and physical states produce "goose bumps."

The 5 million hairs on the human body do have important functions. A full head of hair protects the scalp from ultraviolet light and can cushion a blow to the head. The hairs guarding the entrances to the nostrils and external ear canals help prevent the entry of foreign particles and insects, and eyelashes perform a similar function for the surface of the eye. Although individual hairs are composed of dead cells, a **root hair plexus** of sensory nerves surrounds the base of each hair. Thus movement of the shaft of a single hair may catch your attention almost immediately, even when the movement is very slight. This provides an "early warning" system that can help prevent injury. For example, you can often detect and swat a mosquito before it reaches the skin surface.

Color of Hair Variations in hair color reflect differences in structure and variations in the pigment produced by melanocytes at the papilla. These characteristics are genetically determined, but the condition of your hair may be influenced by hormonal or environmental factors. As pigment production decreases with age, the hair color lightens toward gray. White hair results from the presence of air bubbles within the hair shaft. Because the hair itself is quite dead and inert, changes in coloration are gradual. Unless bleach is used, it is not possible for hair to "turn white overnight," as some horror stories would have us believe.

Growth and Replacement of Hair Many people are subject to anxiety attacks when they find hairs clinging to their hairbrush instead of to their heads. Such hair loss is not a sign of approaching baldness, but merely a reflection of the **hair growth cycle**, diagrammed in Figure 6-11.

A hair in the scalp grows for 2–5 years, at a rate of around 0.33 mm/day. Variations in the hair growth rate and in the duration of the growth cycle account for individual differences in uncut hair length.

While hair growth is under way, the root of the hair is firmly attached to the matrix of the follicle. At the end of the growth cycle the follicle becomes inactive, and the hair is now a **club hair**. The follicle gets smaller, and over time the connections between the hair matrix and the root of the club hair break down. When another growth cycle begins, the follicle produces a new hair, and the old club hair gets

FIGURE 6-11
The Hair Growth Cycle

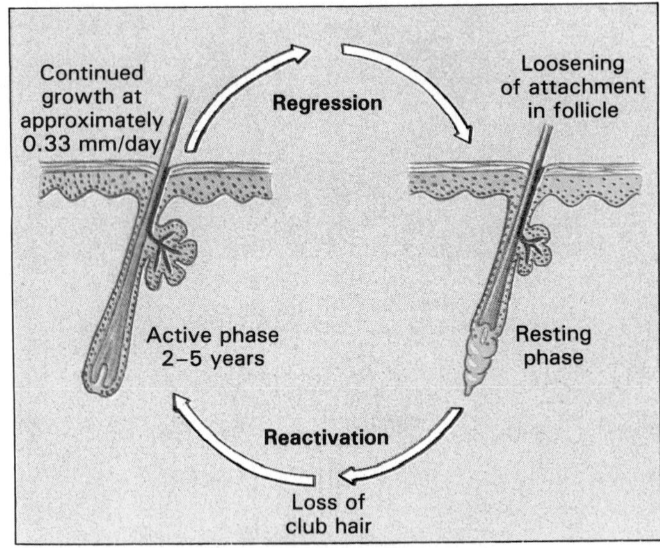

pushed toward the surface. It will then be shed, perhaps onto your brush. (There is a 30-ft beard on display at the Smithsonian National Museum, but it contains the interlocked shafts of many generations of hairs.)

On the average, about 50 hairs are lost each day, but several factors may affect this rate. Sustained losses of over 100 hairs per day usually indicate that something is wrong. Temporary increases in hair loss can result from drugs, dietary factors, radiation, high fever, stress, and hormonal factors related to pregnancy.

In males, changes in the level of circulating sex hormones can affect the scalp, causing a shift from terminal hair to vellus hair production. This alteration is called **male pattern baldness**. [CM: *Baldness and Hirsutism*]

√ What happens to the dermis when it is excessively stretched, as in pregnancy or weight gain?

√ What condition is produced by the contraction of the arrector pili muscles?

√ A person suffers a burn on the forearm that destroys the epidermis and a portion of the dermis. When the injury heals, would you expect to find hair growing again in the area of the injury?

Sebaceous Glands

The skin contains two types of exocrine glands, *sebaceous glands* and *sweat glands* (Figure 6-12). **Sebaceous** (se-BĀ-shus) **glands** are holocrine glands that discharge a waxy, oily secretion into hair follicles.

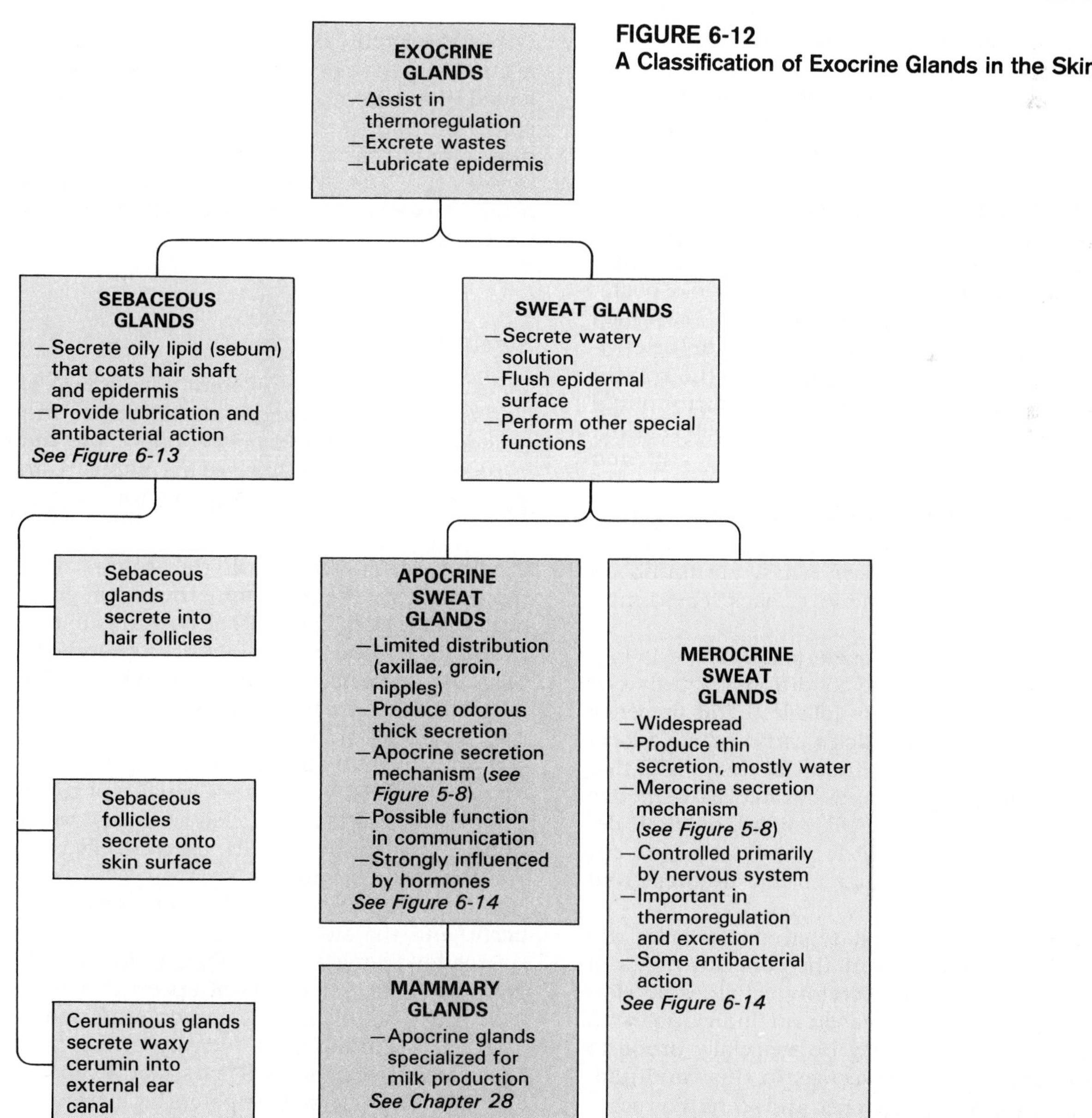

FIGURE 6-12
A Classification of Exocrine Glands in the Skin

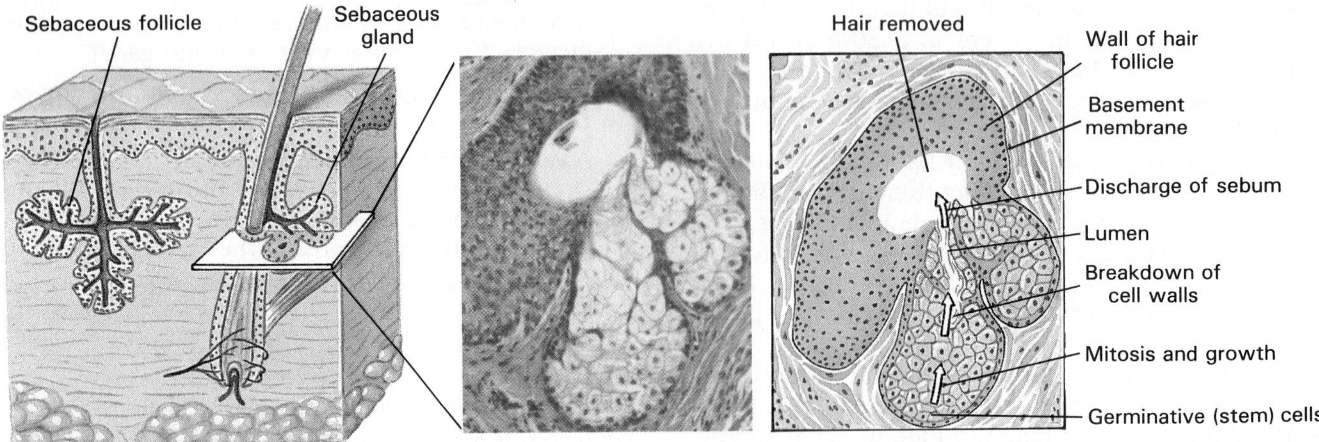

FIGURE 6-13
Sebaceous Glands and Follicles. The structure of sebaceous glands and sebaceous follicles in the skin. (LM, ×110)

These glands can be seen in Figure 6-13. Several sebaceous glands may communicate with a single follicle by means of short ducts. These glands are classified as *simple alveolar glands* because the ducts do not branch and the gland itself resembles a sac (*alveolus*, hollow sac). The gland cells manufacture large quantities of lipids as they mature, and their eventual death releases the lipids into the open passageway, or **lumen**, of the gland. Contraction of the arrector pili muscle that elevates the hair squeezes the sebaceous gland, forcing the waxy secretions onto the surface of the skin. This secretion, called **sebum** (SĒ-bum), provides lubrication and inhibits the growth of bacteria. Keratin is a tough protein, but dead, cornified cells become dry and brittle once exposed to the environment. Sebum lubricates and protects the keratin of the hair shaft and conditions the surrounding skin. Shampooing removes the natural oily coating, and excessive washing can make hairs stiff and brittle.

Sebaceous follicles are large sebaceous glands that communicate directly with the epidermis. These follicles, which never produce hairs, are found on the integument covering the face, back, chest, nipples, and male sex organs.

Although sebum has bactericidal (bacteria-killing) properties, under some conditions bacteria can invade sebaceous glands or follicles. The presence of bacteria in glands or follicles can produce a local inflammation known as **folliculitis** (fo-lik-ū-LĪ-tis). If the duct of the gland becomes blocked, a distinctive abscess called a **furuncle** (FUR-ung-kl), or "boil" develops. The usual treatment for a furuncle is to cut it open, or "lance" it, so that normal drainage and healing can occur.

Sebaceous glands and sebaceous follicles are very sensitive to changes in the concentrations of sex hormones, and their secretory activities accelerate at puberty. For this reason an individual with large sebaceous glands may be especially prone to develop **acne** during adolescence. In this condition, sebaceous ducts become blocked and secretions accumulate, causing inflammation and providing a fertile environment for bacterial infection. [CM: *Acne*]

Seborrheic dermatitis is an inflammation around abnormally active sebaceous glands. The affected area becomes red, and there is usually some epidermal scaling. Sebaceous glands of the scalp are most often involved. In infants, mild cases are called "cradle cap." Adults know this condition as "dandruff." Anxiety, stress, and food allergies can exaggerate the problem.

Sweat Glands

The skin contains two different populations of sweat glands, *apocrine sweat glands* and *merocrine sweat glands*. The differences between apocrine and merocrine secretion were discussed in Chapter 5, and both gland types are shown in Figure 6-14. (p. 135)

Apocrine Sweat Glands In the armpits, around the nipples, and in the groin, **apocrine sweat glands** communicate with hair follicles. These are coiled tubular glands that produce a viscous, cloudy, and potentially odorous secretion. Special **myoepithelial cells** (*myo-*, muscle) surround the secretory epithelia, and their contraction discharges the accumulated secretion into the hair follicles. Myoepithelial cells are not muscle cells, but they are capable of contraction. The secretory activities of gland cells and the contractions of myoepithelial cells are controlled by neural and hormonal stimuli.

Apocrine sweat glands begin secreting at puberty, and the sweat produced is an excellent food source for bacteria that enhance its odor. In other mammals, the secretions of apocrine sweat glands have social significance, conveying information concerning the individual's identity and sexual state. The human sense of smell is less acute, and our society prefers to ignore this potential method of infor-

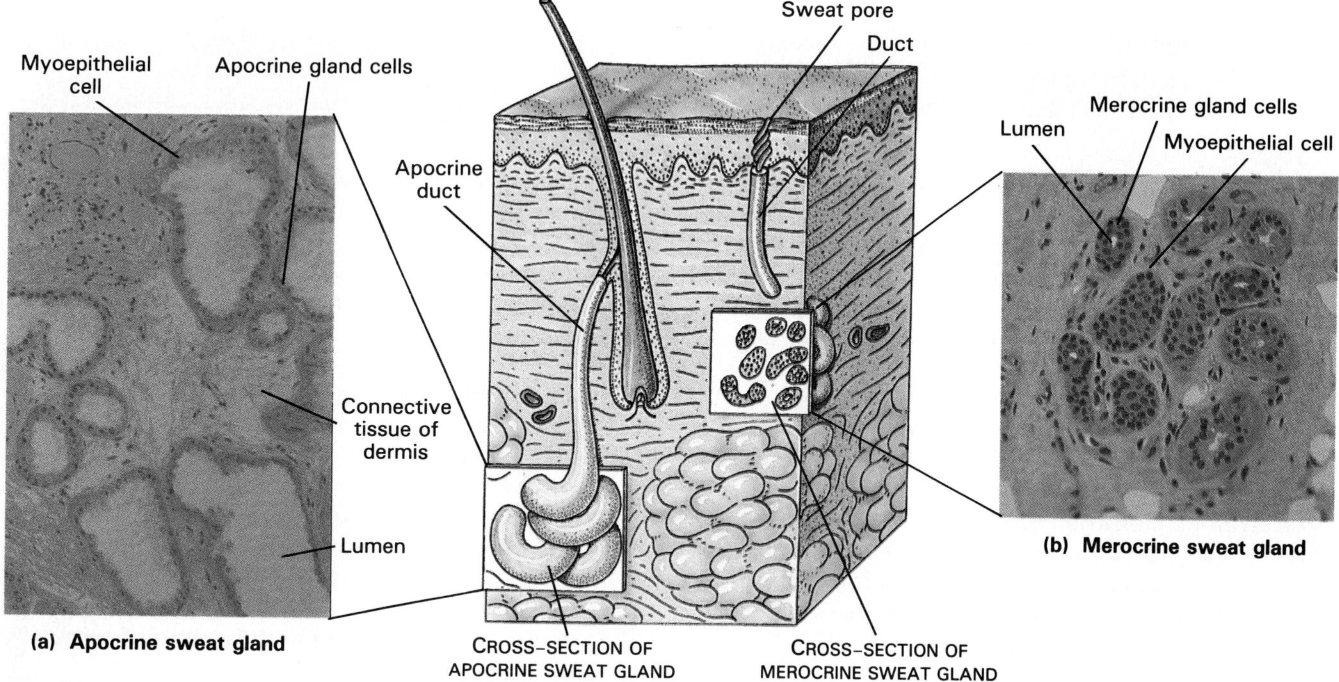

FIGURE 6-14
Sweat Glands. (a) Apocrine sweat glands are found in the axillae (armpits), groin, and nipples. They produce a thick, odorous fluid by apocrine secretion. (LM, ×84) (b) Merocrine (eccrine) sweat glands produce a watery fluid by merocrine secretion. (LM, ×190)

mation transfer.[1] The result is a thriving industry devoted to the production of antiperspirants and hygiene sprays that either inhibit perspiration or mask its odor.

The **mammary glands** of the breasts are structurally and evolutionarily related to aprocrine sweat glands. A complex interaction between sexual and pituitary hormones controls their development and secretion. Mammary gland structure and function will be considered in Chapter 28.

Merocrine Sweat Glands **Merocrine**, or **eccrine** (EK-rin), **sweat glands** are far more numerous and widely distributed than apocrine glands. The adult integument contains around 3 million eccrine glands. Palms and soles have the highest numbers; it has been estimated that the palm of the hand has about 3000 glands per square inch.

Merocrine sweat glands are coiled tubular glands that discharge their secretions directly onto the surface of the skin. They are smaller than apocrine sweat glands, and they do not extend as far into the dermis. The primary functions of the *sensible perspiration* released by the merocrine sweat glands are to cool the surface of the skin and to reduce body temperature. When a person sweats in the hot sun, all the merocrine glands are working together. The blood vessels beneath the epidermis are flushed with blood, and the skin assumes a reddish coloration. The skin surface is warm and wet, and as the moisture evaporates the skin cools. If body temperature falls below normal, perspiration ceases, blood flow to the skin declines, and the cool, dry surfaces release little heat into the environment. The negative feedback mechanisms involved in thermoregulation (temperature control) were described in Chapter 1. (p. 8)

The perspiration produced by merocrine glands is a clear secretion that is more than 99 percent water, but it does contain a mixture of electrolytes, metabolites, and waste products. It is the presence of the electrolytes that gives sweat a salty taste. Merocrine secretions can provide a significant excretory route for these materials, as well as for a number of ingested or administered drugs. When all of the merocrine sweat glands are working at maximum, the rate of perspiration may exceed a gallon per hour, and dangerous fluid and electrolyte losses can occur. For this reason marathon runners and other athletes in endurance sports must pause frequently to drink fluids.

Merocrine secretions also provide protection from environmental hazards by diluting harmful chemicals and discouraging the growth of microorganisms. Microbes must cross the relatively dry, sebum-coated stratum corneum in order to penetrate the epidermis. To begin their attack they must usually attach to the surface, and a periodic flushing with sweat helps to wash these pests away before they make significant progress.

[1] In women, apocrine secretions still undergo cyclic changes in composition tied to the menstrual cycle—clear evidence of their evolutionary origin.

Sports and Fitness: On the Front Line

Regular exercise and participation in sports can have substantial health benefits, many of which will be examined in later chapters. But while internal systems are being slimmed down and toned up, one system—the integument—suffers increased abuse, and may actually deteriorate as a result.

Chafing During repetitive movements, skin surfaces near creases, such as those between the trunk and limbs, rub against one another or against clothing. Friction gradually strips away the protective layers of the stratum corneum, producing damp, sore areas. In addition to being painful, these areas lack the normal epidermal barriers to infection.

Trauma Those engaged in sports are likely to damage the skin through impact with the floor, the ground, or their opponents. For example, "strawberries" are abrasion injuries produced by skidding across a basketball court. Raquetball players diving for a shot or runners sliding into second base can develop similar injuries.

Infection Any break in the skin provides an avenue for infection. Even on intact skin, the combination of warmth and dampness creates an inviting environment for fungal pathogens. Several species may infect the epidermis, most frequently in covered areas where perspiration collects. Infection often occurs between the toes, causing the condition known as "athlete's foot," or in the skin creases of the groin, where it produces symptoms of "jock itch" in men.

Sun Outdoor sports are often associated with cases of severe sunburn. Often the participants are perspiring freely, making it difficult to maintain an effective coating of sunscreen. By the time they begin noticing their skin color, it is already too late. It is no exaggeration to say that a life of exercise and sun exposure can produce a healthy 40-year-old individual who has the physique of a 30-year-old and the skin of someone age 50 or older.

Water Seawater spray, often associated with wind, increases drying by coating the skin surface with a layer of salt. As perspiration dissolves the salt, it creates a concentrated solution that pulls water across the spithelial layers by osmosis. For skin divers, continued exposure to seawater leads to a wrinkling of the skin and a reduction in skin sensitivity, especially to touch and to painful stimuli. Injuries to the skin then become more likely.

Sebaceous and apocrine glands can be switched on or off, but no regional control is possible. When one sebaceous or apocrine gland is activated, so are all the other glands of that type in the body. Merocrine sweat glands are much more precisely controlled, and the amount of secretion and the area of the body involved can be varied independently. For example, when you are nervously awaiting an anatomy and physiology exam, your palms may begin to sweat.

Ceruminous Glands **Ceruminous** (se-ROO-mi-nus) **glands** are modified sweat glands located in the external auditory canal. Their secretions combine with those of nearby sebaceous glands, forming a mixture called **cerumen**, also known as "ear wax." Ear wax, together with tiny hairs along the ear canal, probably helps trap foreign particles or small insects and keep them from reaching the eardrum.

Modified sweat glands are also found in the margins of the eyelid and the tissues of the breasts; these will be considered in later chapters.

Nails

Nails form over the tips of the fingers and toes on the opposite side from the epidermal ridges described earlier. The nails protect the exposed tips of the fingers and toes and help limit their distortion when they are subjected to mechanical stress— for example, in running or grasping objects. The structure of a nail can be seen in Figure 6-15, and you may also use your thumbnail as a reference. The body of the nail covers the **nail bed**, but nail production occurs at the **nail root**, an epithelial fold not visible from the surface. The deepest portion of the nail root lies very close to the periosteum of the bone of the fingertip, or *distal phalanx* (FĀ-lanks).

A portion of the stratum corneum of the fold extends over the exposed nail nearest the root, forming the **cuticle**, or **eponychium** (ep-ō-NIK-ē-um; *epi-*, over + *onyx*, nail). Underlying blood vessels give the nail its pink color, but near the root these vessels may be obscured, leaving a pale crescent known as the **lunula** (LOO-nu-la; *luna*, moon). The **nail body** is recessed beneath the level of the surrounding epithelium, and it is bounded by **nail grooves** and **nail folds**. The **free edge** of the nail extends over a thickened stratum corneum, the **hyponychium** (hi-pō-NIK-ē-um).

Changes in the shape, structure, or appearance of the nails may indicate the existence of a disease process affecting metabolism throughout the body. For example, the nails may turn yellow in patients who have chronic respiratory disorders, thyroid gland disorders, or AIDS. They may become pitted and distorted in psoriasis, and concave in some blood disorders.

Local Control of Integumentary Function

The integumentary system displays a significant degree of functional independence. That is, it responds directly and automatically to local influences without the involvement of the nervous or endocrine systems. For example, when the skin is subjected to mechanical stresses, stem cells in the stratum germinativum divide more rapidly, and the depth of the epithelium increases. That is why calluses form on your palms when you perform manual labor, such as shoveling. A more dramatic display of local regulation can be seen following an injury to the skin.

INFLAMMATION OF THE SKIN

The epidermis provides significant protection from mechanical and chemical hazards because it is relatively thick and covered with keratin. Although the surface of the skin harbors a variety of microorganisms, most of them are harmless as long as they remain outside the stratum corneum. Penetration of the superficial layers may produce familiar conditions such as "athlete's foot" (a fungal infection), warts, and cold sores (both viral infections). To reach the underlying connective tissues, a bacterium must survive the bactericidal components of sebum, avoid being flushed from the surface by the sweat gland secretions, penetrate the stratum corneum, squeeze between the junctional complexes of deeper layers, escape the Langerhans cells, and cross the basement membrane. After such a difficult journey, the papillary layer of the dermis must resemble bacterial heaven, for it is warm, dark, loosely organized, and contains a nutritious ground substance.

If the protective barriers are crossed, or if an injury breaks through the epidermis, mast cells within the dermis respond by triggering a powerful inflammatory response. The process of *inflammation* was introduced in Chapter 5. Inflammation in the skin is important in defending the body against serious injury and disease. [**CM:** *Complications of Inflammation*]

REGENERATION

The skin can regenerate effectively even after considerable damage has occurred, because stem cells persist in both the epithelial and connective tissue components. Germinative cell divisions replace epidermal cells, and mesenchymal cell divisions replace lost dermal cells. Figure 6-16 shows stages in the regeneration of the skin after an injury. When damage extends through the epidermis and into the dermis, bleeding usually occurs. The fibrin clot, or **scab**, that forms at the surface temporarily restores the integrity of the epidermis and restricts the entry of additional microorganisms. Cells of the stratum germinativum undergo rapid divisions and begin to migrate along the sides of the wound in an attempt to replace the missing epidermal cells. [**CM:** *A Classification of Wounds*]

If the wound covers an extensive area or involves a region covered by thin skin, dermal repairs must be under way before epithelial cells can cover the sur-

FIGURE 6-15
Structure of a Nail. These drawings illustrate the prominent features of a typical fingernail as viewed from the surface and in section.

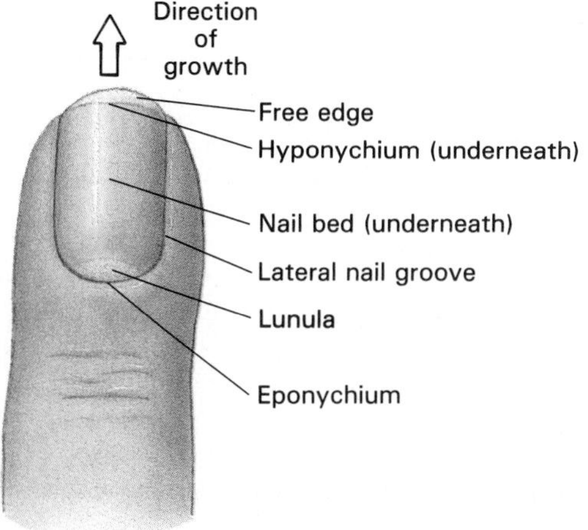

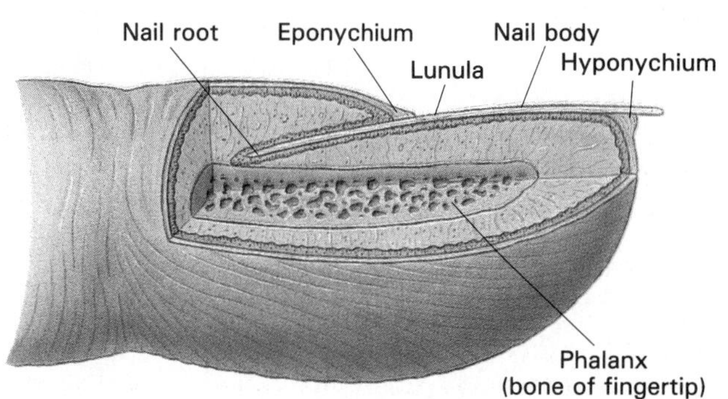

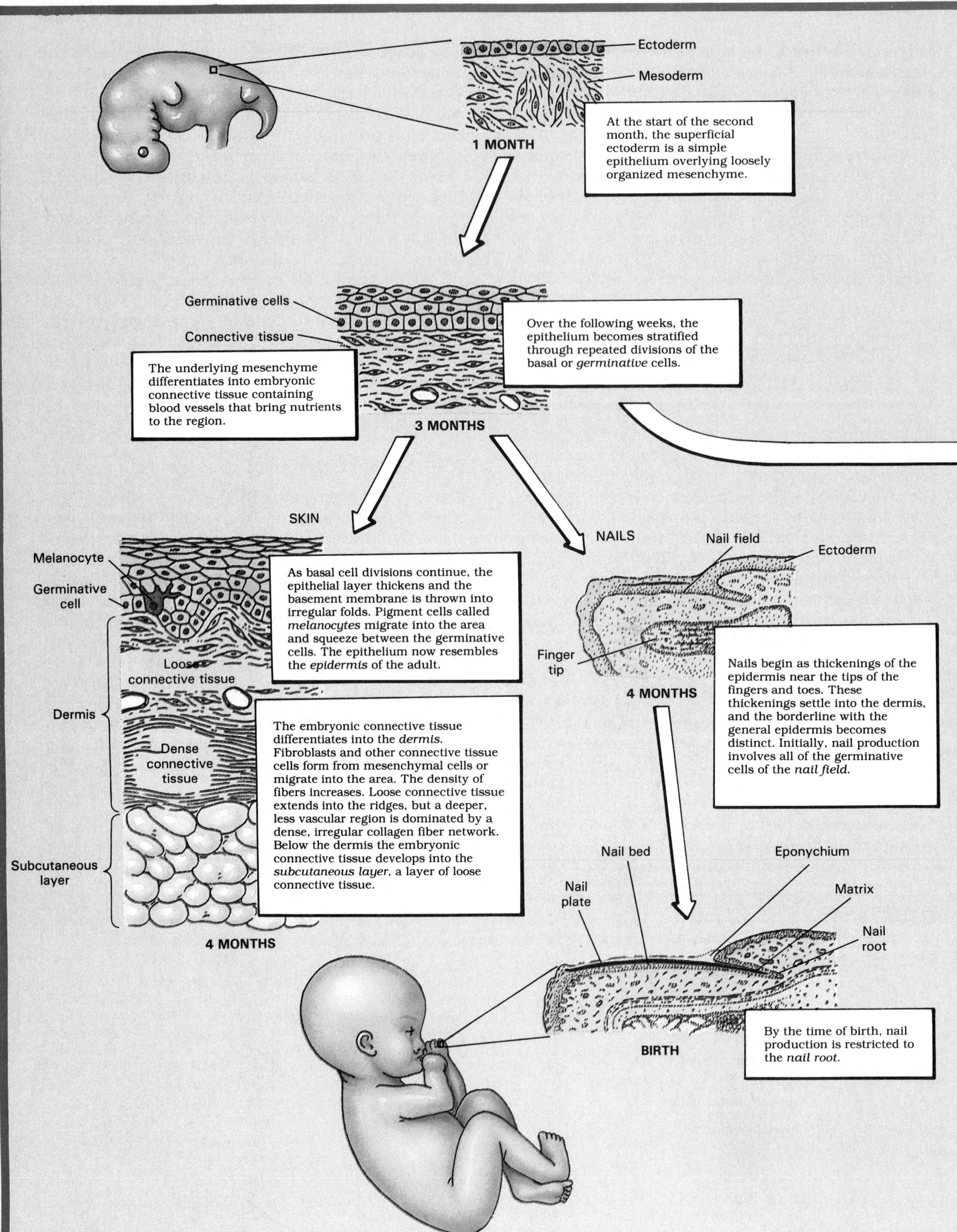

Ectoderm
Mesoderm
1 MONTH
At the start of the second month, the superficial ectoderm is a simple epithelium overlying loosely organized mesenchyme.
Germinative cells
Connective tissue
The underlying mesenchyme differentiates into embryonic connective tissue containing blood vessels that bring nutrients to the region.
Over the following weeks, the epithelium becomes stratified through repeated divisions of the basal or germinative cells.
3 MONTHS
SKIN
Melanocyte
Germinative cell
As basal cell divisions continue, the epithelial layer thickens and the basement membrane is thrown into irregular folds. Pigment cells called melanocytes migrate into the area and squeeze between the germinative cells. The epithelium now resembles the epidermis of the adult.
Loose connective tissue
Dermis
Dense connective tissue
The embryonic connective tissue differentiates into the dermis. Fibroblasts and other connective tissue cells form from mesenchymal cells or migrate into the area. The density of fibers increases. Loose connective tissue extends into the ridges, but a deeper, less vascular region is dominated by a dense, irregular collagen fiber network. Below the dermis the embryonic connective tissue develops into the subcutaneous layer, a layer of loose connective tissue.
Subcutaneous layer
4 MONTHS
NAILS
Nail field
Ectoderm
Finger tip
4 MONTHS
Nails begin as thickenings of the epidermis near the tips of the fingers and toes. These thickenings settle into the dermis, and the borderline with the general epidermis becomes distinct. Initially, nail production involves all of the germinative cells of the nail field.
Nail bed
Eponychium
Matrix
Nail plate
Nail root
BIRTH
By the time of birth, nail production is restricted to the nail root.

Embryology Summary: Development of the Integumentary System

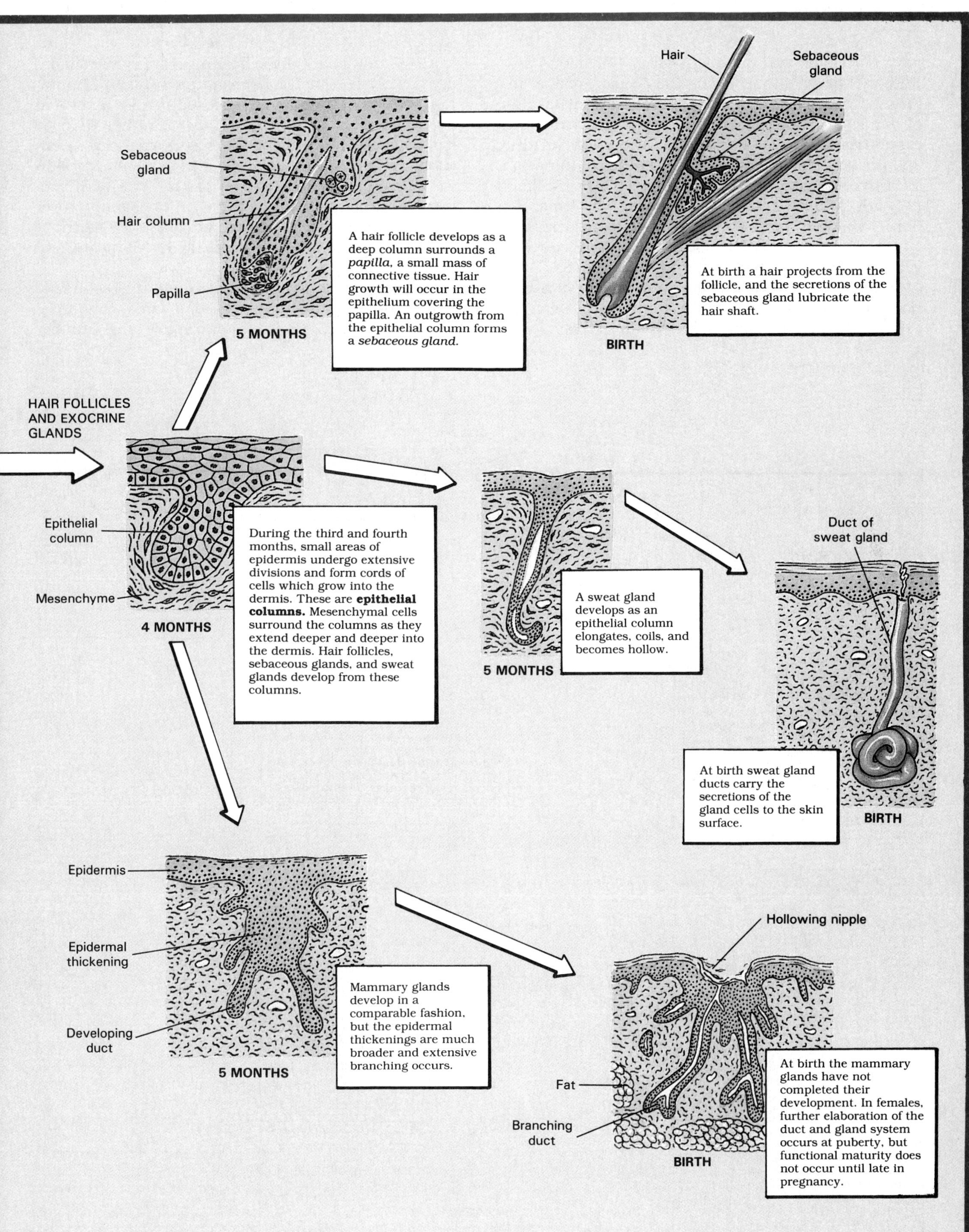
Hair
Sebaceous gland
Sebaceous gland
Hair column
Papilla
A hair follicle develops as a deep column surrounds a *papilla*, a small mass of connective tissue. Hair growth will occur in the epithelium covering the papilla. An outgrowth from the epithelial column forms a *sebaceous gland*.
5 MONTHS
At birth a hair projects from the follicle, and the secretions of the sebaceous gland lubricate the hair shaft.
BIRTH
HAIR FOLLICLES AND EXOCRINE GLANDS
Epithelial column
Mesenchyme
4 MONTHS
During the third and fourth months, small areas of epidermis undergo extensive divisions and form cords of cells which grow into the dermis. These are **epithelial columns.** Mesenchymal cells surround the columns as they extend deeper and deeper into the dermis. Hair follicles, sebaceous glands, and sweat glands develop from these columns.
A sweat gland develops as an epithelial column elongates, coils, and becomes hollow.
5 MONTHS
Duct of sweat gland
At birth sweat gland ducts carry the secretions of the gland cells to the skin surface.
BIRTH
Epidermis
Epidermal thickening
Developing duct
5 MONTHS
Mammary glands develop in a comparable fashion, but the epidermal thickenings are much broader and extensive branching occurs.
Hollowing nipple
Fat
Branching duct
BIRTH
At birth the mammary glands have not completed their development. In females, further elaboration of the duct and gland system occurs at puberty, but functional maturity does not occur until late in pregnancy.

face. Fibroblast and mesenchymal cell divisions produce mobile cells that invade the deeper areas of injury, migrating along fibrin strands. Endothelial cells of damaged blood vessels also begin to divide, and capillaries follow the fibroblasts, eventually uniting and providing a circulatory supply. The combination of fibrin clot, fibroblasts, and an extensive capillary network is called **granulation tissue**. Over time, the fibrin dissolves, and the number of capillaries declines. Fibroblast activity leads to the appearance of collagen fibers and typical ground substance.

While dermal repairs are in progress, **contraction** pulls the edges of the wound closer together. The mechanism of contraction is uncertain, but it is an essential part of the healing process when damage has been extensive. For example, after an amputation, over 90 percent of the exposed surface is covered by contraction of the wound edges, rather than by the divisions and migration of epithelial cells. Contraction distorts the adjacent surface, as if the skin were being stretched to cover the injury site. If contraction and epithelial cell migration cannot cover the wound, **skin grafts** may be required; methods of skin grafting are discussed in the Clinical Comment: Burns and Grafts.

After an injury the integumentary repairs do not restore the integument to its original condition. The repair site contains an abnormally large number

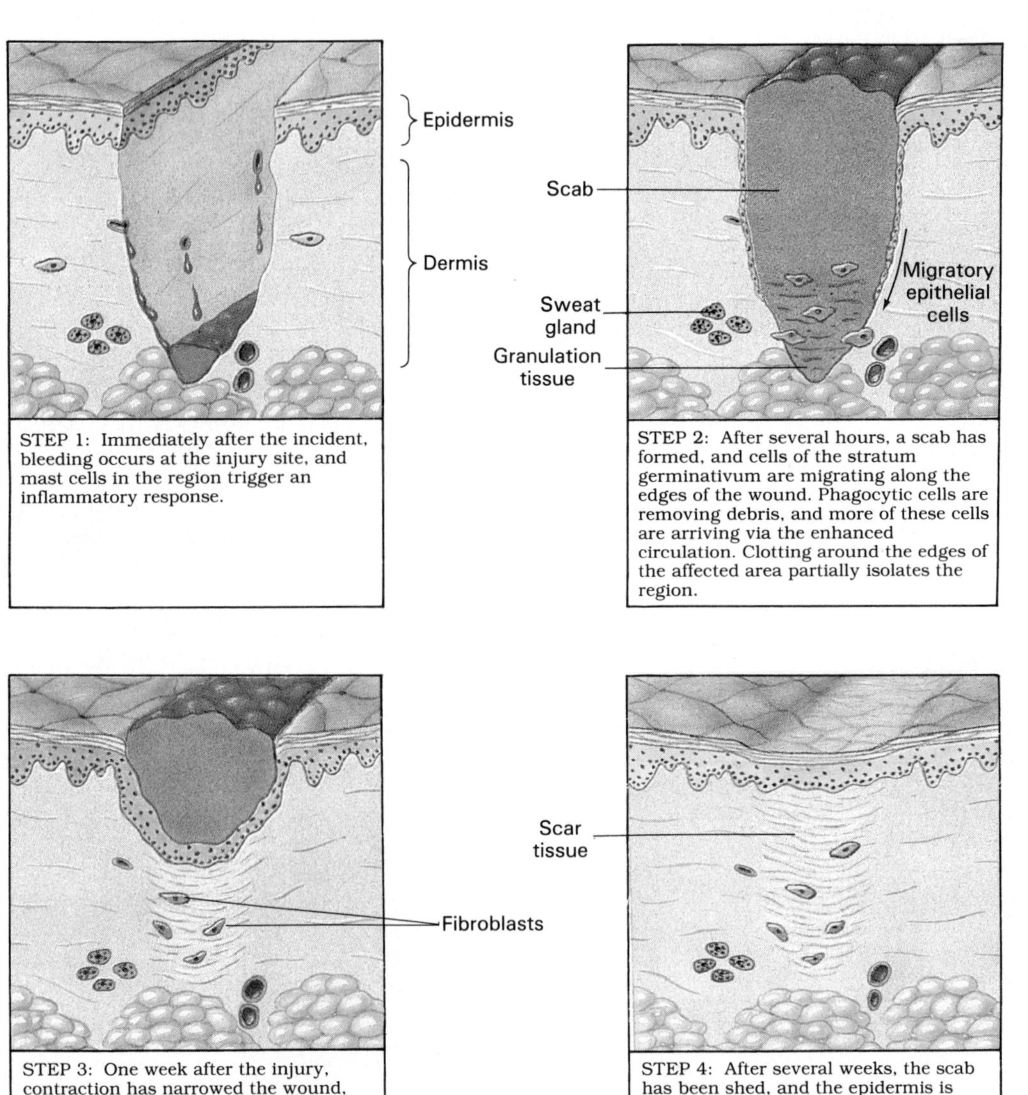

Epidermis
Dermis
Scab
Sweat gland
Granulation tissue
Migratory epithelial cells
STEP 1: Immediately after the incident, bleeding occurs at the injury site, and mast cells in the region trigger an inflammatory response.
STEP 2: After several hours, a scab has formed, and cells of the stratum germinativum are migrating along the edges of the wound. Phagocytic cells are removing debris, and more of these cells are arriving via the enhanced circulation. Clotting around the edges of the affected area partially isolates the region.
Scar tissue
Fibroblasts
STEP 3: One week after the injury, contraction has narrowed the wound, and the scab has been undermined by epidermal cells migrating over the meshwork produced by fibroblast activity. Phagocytic activity around the site has almost ended, and the fibrin clot is disintegrating.
STEP 4: After several weeks, the scab has been shed, and the epidermis is complete. A shallow depression marks the injury site, but fibroblasts in the dermis continue to create scar tissue that will gradually elevate the overlying epidermis.

FIGURE 6-16
Integumentary Repair

of collagen fibers and relatively few blood vessels. Damaged hair follicles, sebaceous or sweat glands, muscle cells, and nerves are seldom repaired, and they too are replaced by fibrous tissue. The formation of this rather inflexible, fibrous, noncellular scar tissue can be considered as a practical limit to the healing process.

It is not known what regulates the extent of scar tissue formation. In some individuals, most often blacks or dark-skinned people, scar tissue formation may continue beyond the requirements of tissue repair. The result is a flattened mass of scar tissue that begins at the injury site and grows into the surrounding dermis. This thickened area of scar tissue,

Clinical Comment: Burns and Grafts

Burns result from exposure of the skin to heat, radiation, electrical shock, or strong chemical agents. The severity of the burn reflects the depth of penetration and the total area affected. The clinical classification shown in Table 6-1 depends on the depth alone. In a **first-degree burn** damage is confined to the superficial layers of the epidermis. The inflammation produced leads to the characteristic **erythema** (er-i-THĒ-ma), or redness, typical of mild sunburns. First-degree burns are mildly painful, and the damage is repaired in a few days. Some separation of the skin layers occurs, and as "peeling" takes place the deeper, less keratinized layers are exposed.

A **second-degree burn** kills epidermal cells down to the stratum germinativum and into the dermis. Inflammation is extensive, and these burns are quite painful. In thin-skinned areas, fluid may accumulate within the epidermis or between epidermis and dermis. These **blisters** often rupture at the surface, providing easy access to microbial invaders. Healing takes 1–2 weeks, and some scar tissue may form.

First- and second-degree burns are also called **partial-thickness** burns because damage is restricted to the superficial layers of the skin. Accessory structures such as hair follicles and glands are usually unaffected. **Full-thickness**, or **third-degree, burns** destroy the epidermis and dermis, extending into subcutaneous tissues. These burns are actually less painful than second-degree burns, because sensory nerves are destroyed along with accessory structures, blood vessels, and other dermal components. Extensive third-degree burns cannot repair themselves, because the extent of the destruction makes it impossible for granulation tissue to form. Without granulation tissue, epithelial cells cannot migrate across the injury site, and the site remains exposed to potential infection.

Roughly 10,000 people die from burns each year in the United States. The larger the area burned, the more significant the effects

TABLE 6-1 A Classification of Burns

Classification	*Damage Report*	*Appearance and Sensation*
First-degree burn	*Killed:* superficial cells of epidermis. *Injured:* deeper layers of epidermis, papillary dermis.	Inflamed, tender.
Second-degree burn	*Killed:* superficial and deeper cells of epidermis; dermis may be affected. *Injured:* damage may extend into reticular layer of the dermis, but many accessory structures unaffected.	Blisters, pain.
Third-degree burn	*Killed:* all epidermal and dermal cells. *Injured:* hypodermal and deeper tissues and organs.	Charred, no sensation at all.

on integumentary function. Figure 6-17 presents a standard reference for calculating the percentage of total surface area involved. Burns that cover more than 20 percent of the skin surface represent serious threats to life because they affect the following functions:

Fluid and Electrolyte Balance Even areas with partial-thickness burns lose their effectiveness as barriers to fluid and electrolyte losses. In full-thickness burns, the rate of fluid loss through the skin may reach five times the normal level.

Thermoregulation Increased fluid loss means increased evaporative cooling. More energy must be expended to keep body temperature within acceptable limits.

Protection from Attack The epidermal surface, damp from uncontrolled fluid losses, encourages bacterial growth. If the skin is broken at a blister or the site of a third-degree burn, infection is likely. Widespread bacterial infecion, or **sepsis** (*septikos*, rotting), is the leading cause of death in burn patients.

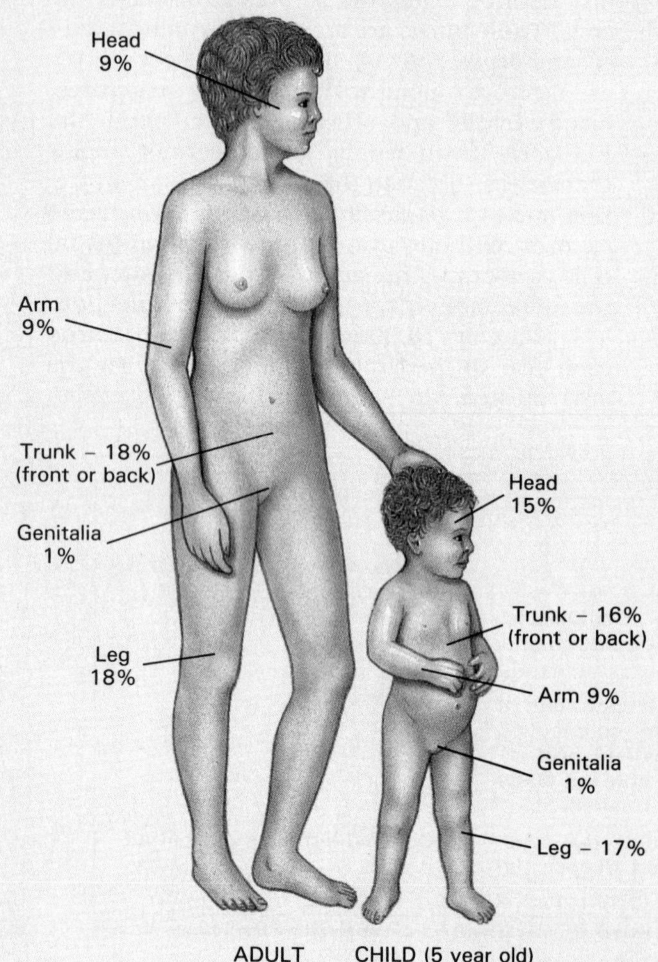

FIGURE 6-17
A Quick Method for Estimating the Percentage of Surface Area Affected by Burns in Adults and Small Children. The method of estimation is called the "rule of nines" because of the surface area proportions in the adult. This rule must be modified in children because their proportions are quite different (see the Embryology Summary in Chapter 1, p. 14).

Effective treatment of full-thickness burns focuses on these procedures:

1. Replacing lost fluids and electrolytes,
2. Providing sufficient nutrients to meet increased metabolic demands for thermoregulation and healing,
3. Preventing infection by cleaning and covering the burn while adminstering antibiotic drugs, and
4. Assisting tissue repairs.

Because full-thickness burns cannot heal unaided, surgical procedures are necessary to encourage healing. In a **skin graft**, areas of intact skin are transplanted to cover the burn site. A **split-thickness graft** takes a shaving of the epidermis and superficial portions of the dermis. A **full-thickness graft** involves the epidermis and both layers of the dermis.

With the development of fluid replacement therapies, infection control methods, and grafting techniques, the recovery rate for severe burns has improved dramatically. At present, young patients with burns over 80 percent of the body have an approximately 50 percent chance of recovery.

Recent advances in cell culture techniques may improve survival rates further. It is now possible to remove a small section of skin and grow it under controlled laboratory conditions. Over time, the germinative cell divisions produce large sheets of epidermal cells that can then be used to cover the burn area. From initial samples the size of postage stamps, square yards of epidermis have been grown and transplanted onto body surfaces. Although questions remain concerning the strength and flexibility of the repairs, skin cultivation represents a substantial advance in the treatment of serious burns. [CM: *Synthetic Skin*] called a **keloid** (KĒ-loyd), is covered by a shiny, smooth epidermal surface. Keloids most often develop on the upper back, shoulders, anterior chest, and earlobes.

Skin repairs proceed most rapidly in young, healthy individuals. For example, it takes 3–4 weeks to complete the repairs to a blister site in an 18–25-year-old. The same repairs at age 65–75 take 6–8 weeks. However, this is just one example of the changes that occur in the integumentary system as a result of the aging process.

Aging and the Integumentary System

Aging affects all of the components of the integumentary system. The major changes include:

1. The epidermis thins as germinative cell activity declines, making the elderly more prone to injury and skin infections.
2. The number of Langerhans cells decreases to around 50 percent of levels seen at maturity. This decrease may reduce the sensitivity of the immune system and further encourage skin damage and infection.
3. Vitamin D production declines by around 75 percent. The result can be muscle weakness and a reduction in bone strength.
4. Melanocyte activity declines, and in Caucasians the skin becomes very pale. With less melanin in the skin, the elderly are more sensitive to sun exposure and more likely to experience sunburn.
5. Glandular activity declines. The skin becomes dry and often scaly because sebum production is reduced; sweat glands are also less active (see 8, below).
6. Hair follicles stop functioning or produce thinner, finer hairs. With decreased melanocyte activity, these hairs are gray or white.
7. The dermis becomes thinner, and the elastic fiber network decreases in size. The integument therefore becomes weaker and less resilient; sagging and wrinkling occurs. These effects are most pronounced in areas exposed to the sun.
8. The blood supply to the dermis is reduced at the same time that sweat glands become less active. This combination makes the elderly less able to lose body heat, and overexertion or overexposure to warm temperatures can cause dangerously high body temperatures.
9. With changes in levels of sex hormones, secondary sexual characteristics in hair and body fat distribution begin to fade. In consequence, people age 90–100 of both sexes and all races look very much alike.
10. Skin repairs proceed relatively slowly, and recurring infections may result.

These changes are summarized in Figure 6-18.

√ **What will happen if the duct of an infected sebaceous gland becomes blocked?**

√ **Deodorants are used to mask the effects of secretions from what type of skin gland?**

√ **Older individuals do not tolerate the summer heat as well as they did when they were young, and are more prone to heat-related illness. What accounts for this change?**

FIGURE 6-18
Changes in the Skin during the Aging Process

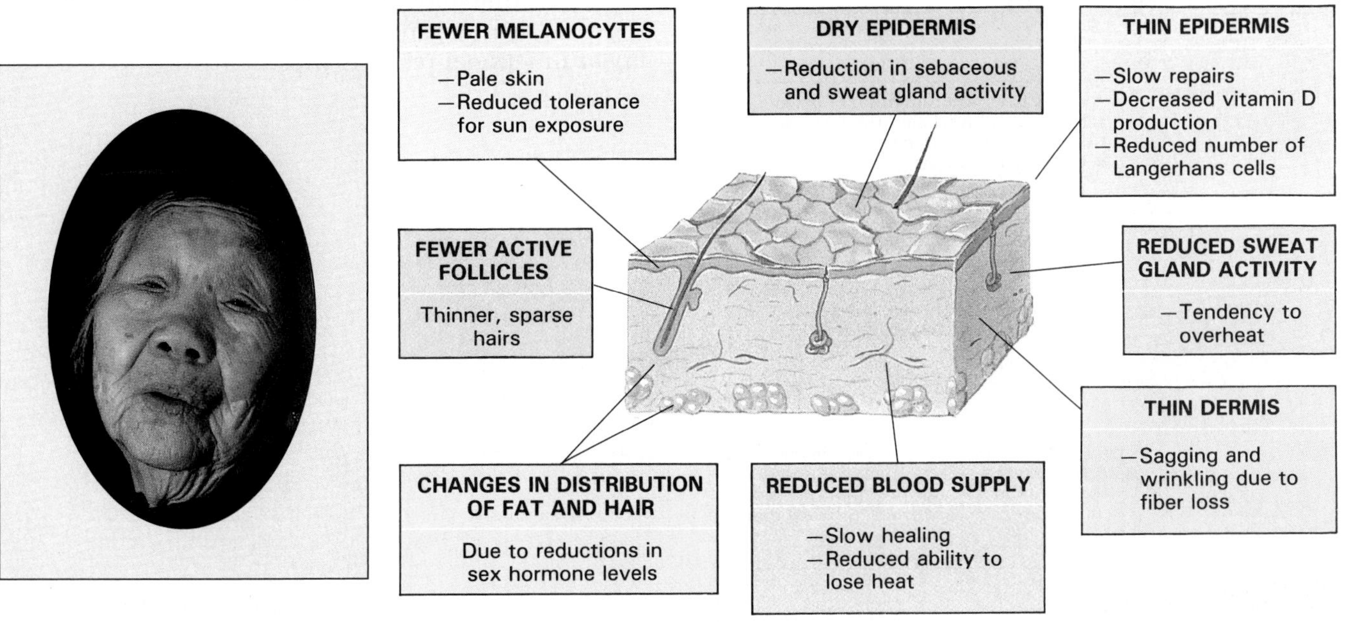

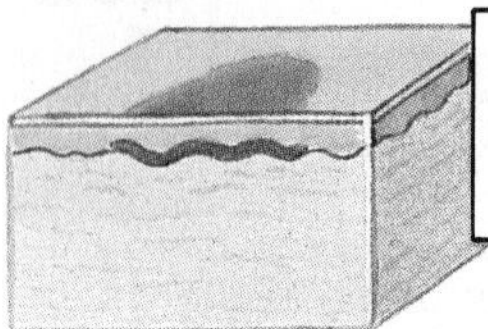
A flat **macule** is a localized change in skin color. Example: freckles

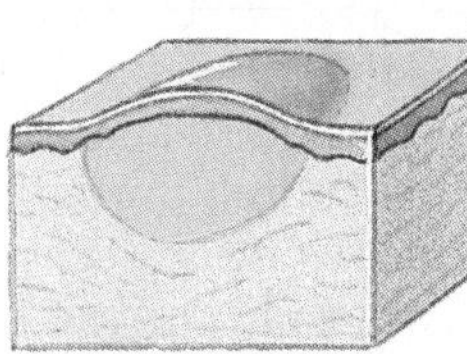
Accumulation of fluid in the papillary dermis may produce a **wheal**, a localized elevation of the overlying epidermis. Example: hives

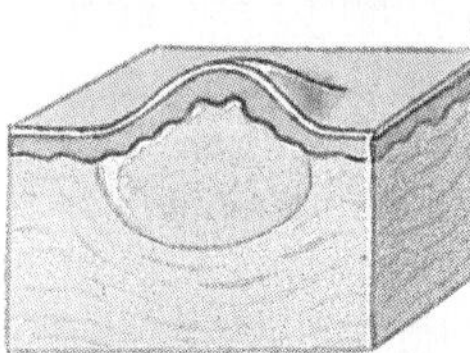
A **papule** is a solid elevated area containing epidermal and papillary dermal components. Example: mosquito or other insect bite

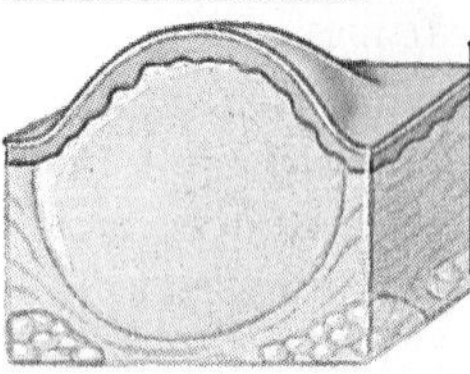
Nodules are large papules that may extend into the subcutaneous layer. Example: cyst

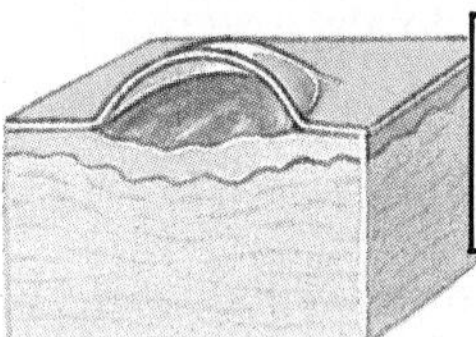
A **vesicle**, or blister, is a papule with a fluid core. A large vesicle may be called a bulla. Example: second–degree burn

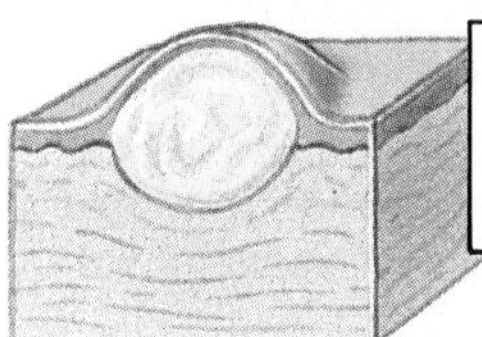
A **pustule** is a papule sized lesion filled with pus. Example: acne pimple

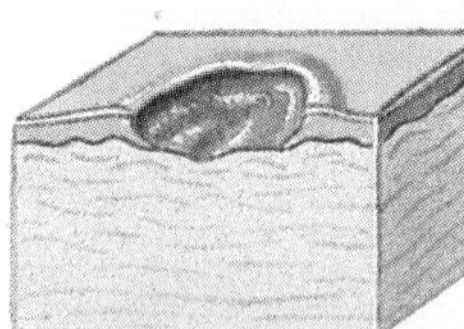
An **erosion**, or *ulcer*, may occur following the rupture of a vesicle or pustule. Eroded sites have lost part or all of the normal epidermis. Example: decubitis ulcer

A **crust** is an accumulation of dried sebum, blood, or interstitial fluid over the surface of the epidermis. Example: seborrheic dermatitis

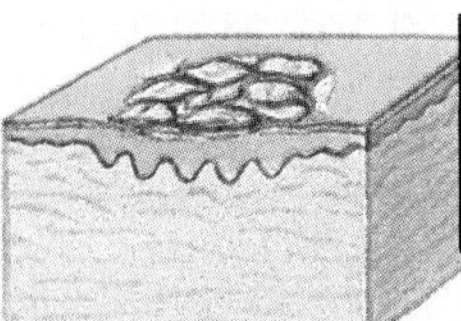
Scales form as a result of abnormal keratinization. They are thin plates of cornified cells. Example: psoriasis

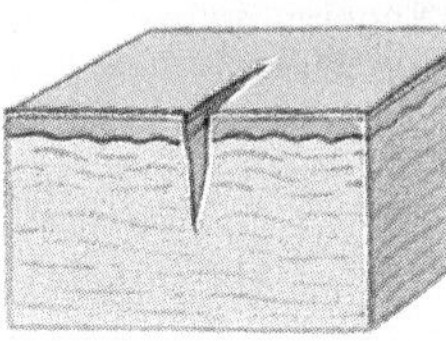
A **fissure** is a split in the integument that extends through the epidermis and into the dermis. Example: athletes foot

FIGURE 6-19
Skin Signs

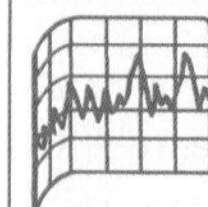

Diagnostics: Symptoms and Signs

Many different skin disorders produce the same uncomfortable sensations. For example, **pruritis** (proo-RĪ-tus), an irritating itching sensation, is an extremely common symptom associated with skin conditions. When investigating skin rashes or inflammations, dermatologists use a combination of investigative interviews ("What have you done?" or "How does it feel?") and physical examination to arrive at a diagnosis. They also pay particular attention to **skin signs**, characteristic abnormalities in the skin surface. Figure 6-19 diagrams the most common skin signs.

The general condition of the skin may also be significant. For example, changes in skin color, flexibility, elasticity, dryness, or sensitivity often appear following the malfunctions of other organ systems. Examples of secondary changes in nail color were given earlier in the chapter.

Integration with Other Systems

Although it can function independently, many activities of the integumentary system are integrated with those of other systems. Figure 6-20 diagrams the major functional relationships.

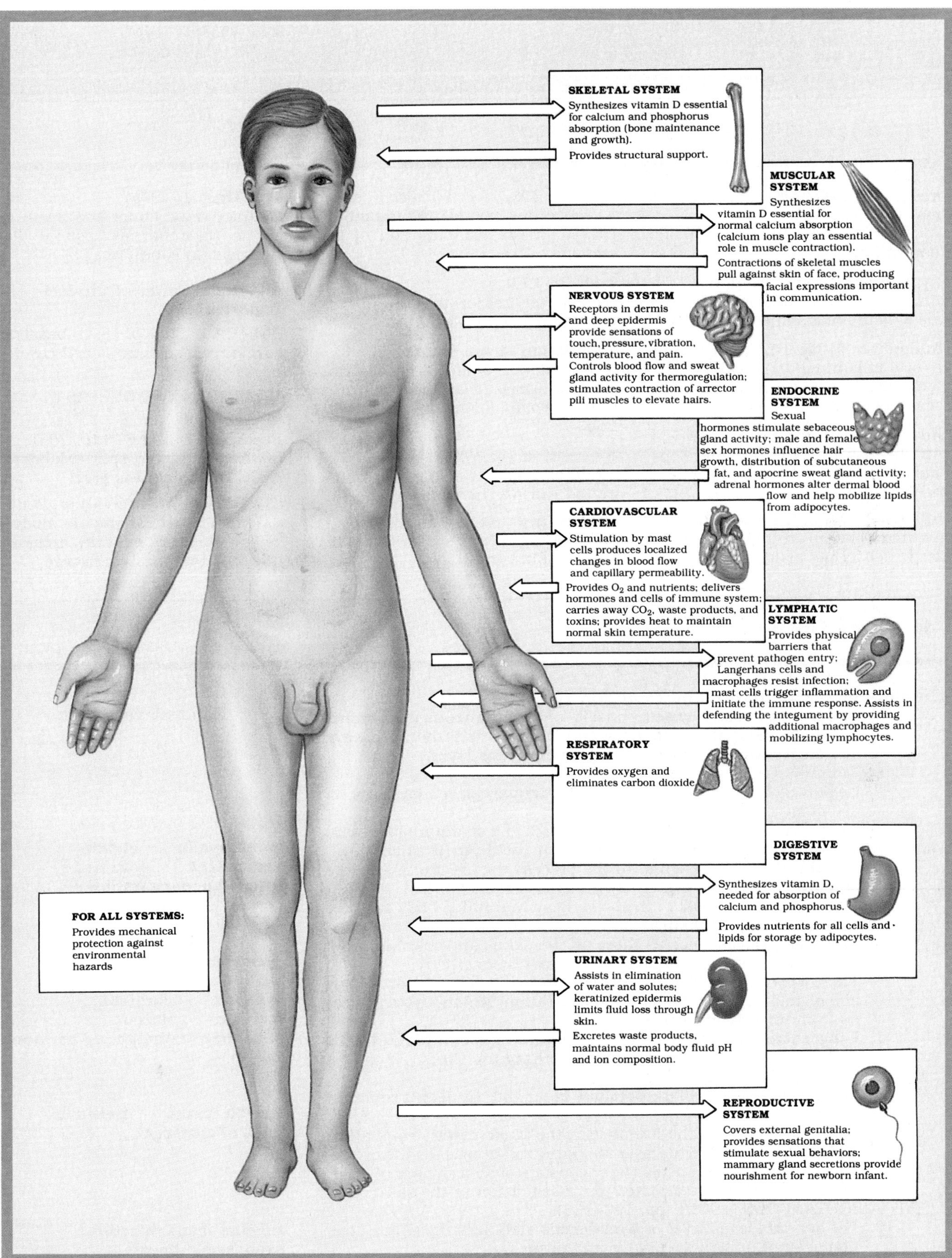

SKELETAL SYSTEM
Synthesizes vitamin D essential for calcium and phosphorus absorption (bone maintenance and growth).
Provides structural support.
MUSCULAR SYSTEM
Synthesizes vitamin D essential for normal calcium absorption (calcium ions play an essential role in muscle contraction).
Contractions of skeletal muscles pull against skin of face, producing facial expressions important in communication.
NERVOUS SYSTEM
Receptors in dermis and deep epidermis provide sensations of touch, pressure, vibration, temperature, and pain.
Controls blood flow and sweat gland activity for thermoregulation; stimulates contraction of arrector pili muscles to elevate hairs.
ENDOCRINE SYSTEM
Sexual hormones stimulate sebaceous gland activity; male and female sex hormones influence hair growth, distribution of subcutaneous fat, and apocrine sweat gland activity; adrenal hormones alter dermal blood flow and help mobilize lipids from adipocytes.
CARDIOVASCULAR SYSTEM
Stimulation by mast cells produces localized changes in blood flow and capillary permeability.
Provides O_2 and nutrients; delivers hormones and cells of immune system; carries away CO_2, waste products, and toxins; provides heat to maintain normal skin temperature.
LYMPHATIC SYSTEM
Provides physical barriers that prevent pathogen entry; Langerhans cells and macrophages resist infection; mast cells trigger inflammation and initiate the immune response. Assists in defending the integument by providing additional macrophages and mobilizing lymphocytes.
RESPIRATORY SYSTEM
Provides oxygen and eliminates carbon dioxide.
DIGESTIVE SYSTEM
Synthesizes vitamin D, needed for absorption of calcium and phosphorus.
Provides nutrients for all cells and lipids for storage by adipocytes.
URINARY SYSTEM
Assists in elimination of water and solutes; keratinized epidermis limits fluid loss through skin.
Excretes waste products, maintains normal body fluid pH and ion composition.
REPRODUCTIVE SYSTEM
Covers external genitalia; provides sensations that stimulate sexual behaviors; mammary gland secretions provide nourishment for newborn infant.
FOR ALL SYSTEMS:
Provides mechanical protection against environmental hazards

FIGURE 6-20
Functional Relationships between the Integumentary System and Other Systems

CHAPTER REVIEW

Review of Selected Clinical Terms

malignant melanoma (mel-a-NŌ-ma) (*p. 189*)
A skin cancer originating in malignant melanocytes.

hypodermic needle (*p. 193*)
A needle used to administer drugs via subcutaneous injection.

folliculitis (fo-lik-ū-LĪ-tis) (*p. 196*)
A local inflammation caused by bacteria invading a sebaceous gland or sebaceous follicle.

furuncle (FUR-ung-kl) (*p. 196*)
An abscess or boil that develops when the duct of a sebaceous gland is blocked.

seborrheic dermatitis (*p. 196*)
An inflammation around abnormally active sebaceous glands.

acne (*p. 196*)
A sebaceous gland inflammation caused by an accumulation of secretions.

scab (*p. 199*)
A fibrin clot that forms at the surface of a wound to the skin.

granulation tissue (*p. 202*)
A combination of fibrin, fibroblasts, and capillaries that forms during tissue repair following inflammation.

contraction (*p. 202*)
A pulling together of the edges of a wound during the healing process.

pruritis (proo-RĪ-tus) (*p. 206*)
An irritating itching sensation, common in skin conditions.

skin signs (*p. 206*)
Characteristic abnormalities in the skin surface that can assist in diagnosing skin conditions.

Additional Terms of Clinical Importance

SKIN CANCERS (*p. 189*): **basal cell carcinoma, squamous cell carcinoma**

TUMORS IN THE DERMIS (*p. 192*): **hemangioma**

BURNS AND GRAFTS (*p. 203*): **erythema, blister, split-thickness graft, full-thickness graft**

SYMPTOMS AND SIGNS (*p. 206*): **flat macule, wheal, papule, nodule, vesicle (blister), pustule, erosion (ulcer), crust, scales, fissure**

Study Outline

Integumentary Structure and Function (pp. 184–199).

Related Key Terms

1. The **integumentary system** consists of the **cutaneous membrane**, which includes the **epidermis** and **dermis**, and the **accessory structures**. Underneath lies the **subcutaneous layer**.

THE EPIDERMIS (pp. 184–190).

2. Cell divisions in the **stratum germinativum** replace more superficial cells.
3. As epidermal cells age they pass through the **stratum spinosum**, the **stratum granulosum**, the **stratum lucidum** (if thick skin), and the **stratum corneum**. In the process they accumulate large amounts of **keratin**. Ultimately the cells are shed or lost.
   — **keratohyalin • eleidin keratinized • cornified insensible perspiration xerosis**
4. **Thin skin** covers most of the body; heavily abraded body surfaces may be covered by **thick skin**.
5. **Epidermal ridges**, such as those on the palms and soles, improve our gripping ability and increase the skin's sensitivity.
   — **dermal papillae**
6. The color of the epidermis depends on two factors: blood supply and pigment composition and concentration. **Melanocytes** protect us from **ultraviolet radiation**.
   — **cyanosis • carotene melanin • albino melanocyte-stimulating hormone (MSH)**
7. **Langerhans cells** are part of the immune system; **Merkel cells** provide information about objects touching the skin.

THE DERMIS (pp. 190–192).

8. The **dermis** consists of the **papillary layer** and the deeper **reticular layer**.
   — **stretch marks • tretinoin lines of cleavage**
9. The papillary layer of the dermis contains blood vessels, lymphatics, and sensory nerves. This layer supports and nourishes the overlying epidermis. The reticular layer consists of a meshwork of collagen and elastic fibers oriented to resist tension in the skin.

THE SUBCUTANEOUS LAYER (pp. 192–193).

10. The subcutaneous layer or **hypodermis** stabilizes the skin's position against underlying organs and tissues.
    — **subcutaneous injection hypodermic needle**

ACCESSORY STRUCTURES (pp. 193–199).

11. Hairs originate in complex organs called **hair follicles**. Each hair has a **root** and a **shaft**. Hair production involves cell specialization to form a soft core or **medulla** surrounded by a **cortex**. The **cuticle** is a hard layer that coats the hair.
12. There are **vellus hairs** ("peach fuzz"), heavy **terminal hairs**, and **intermediate hairs** (of intermediate character) on our bodies.
13. The **arrector pili** muscles can elevate the hairs.
14. Our hairs grow and are shed according to the **hair growth cycle**. A single hair grows for 2–5 years, and is subsequently shed.
15. **Sebaceous glands** discharge the waxy **sebum** into hair follicles. **Sebaceous follicles** lack hairs but have large sebaceous glands.
16. **Apocrine sweat glands** produce an odorous secretion; the more numerous **merocrine**, or **eccrine**, **sweat glands** produce a watery secretion, known as *sensible perspiration*.
17. **Ceruminous glands** in the ear produce a waxy **cerumen**.
18. Nail production occurs at the **nail root**.

Local Control of Integumentary Function (pp. 199–205).

INFLAMMATION OF THE SKIN (p. 199)

1. The epidermis provides mechanical protection and keeps microorganisms outside of the body. Penetration of the epidermis triggers **inflammation**.

REGENERATION (pp. 199–205)

2. The skin can regenerate effectively even after considerable damage. The process includes formation of a **scab** and **granulation tissue**. **Contraction** then pulls the edges of the wound closer together.

Aging and the Integumentary System (p. 205)

1. Aging affects all of the components of the integumentary system.

Related Key Terms

papilla • matrix • soft keratin • hard keratin

root hair plexus • club hair

male pattern baldness

lumen • folliculitis furuncle • seborrheic dermatitis

myoepithelial cells • mammary glands

nail bed • cuticle (eponychium) • nail root

lunula • nail body • nail grooves • nail folds free edge • hyponychium skin grafts • keloid

Review Planner

		Level 1
1	Compare the structures and functions of the layers of the skin.	**2 6 9 15 16 17 23 27 28 30 31 34 35 36 41**
2	Discuss the functions of the skin's accessory structures.	**1 3 4 5 7 13 19 21 22 25 33 37 39**
3	Explain what accounts for individual and racial differences in skin, such as skin color.	**12 20 26 29 32**
4	Describe how the integumentary system helps to regulate body temperature.	**8 22**
5	Discuss the effects of ultraviolet radiation on the skin.	**14 26 29**
6	Describe the mechanisms that produce hair and determine hair texture and color.	**1 10 11**
7	Explain how the skin responds to injuries and repairs itself.	**18 24 27**
8	Summarize the effects of the aging process on the skin.	**11 38**

Level 2: **28 32 40 42 43**

Level 3: **44–46**

C M

Excessive Keratin Production
Dermatitis
Baldness and Hirsutism • Acne
Complications of Inflammation
A Classification of Wounds
Synthetic Skin

ABC NEWS

Ozone Damage and Skin Cancer
Second Skin for Burn Patients

Chapter Review

Review Questions

MULTIPLE CHOICE

1. Each individual hair has a root and a: a) vellus; b) dermis; c) shaft; d) sebaceous follicle.
2. The major components of the integumentary system are the accessory structures and the: a) cutaneous membrane; b) subcutaneous layer; c) epidermis; d) dermis.
3. The following are all accessory structures, *except*: a) nails; b) dermal papillae; c) hair; d) sweat glands.
4. Nail production occurs: a) at the free edge of the nail; b) in the cuticle; c) throughout the entire nail; d) at the nail root.
5. Intermediate hairs are: a) also called "peach fuzz"; b) also called "club hairs"; c) found on the human head, including eyebrows and eyelashes; d) found on the arms and legs.
6. Cornification does not occur: a) on the palms of the hands; b) in the armpit; c) over the surfaces of the eyes; d) on the scalp.
7. Sebaceous glands discharge a secretion called: a) lumen; b) sebum; c) matrix; d) keratin.
8. Which of the following normally occurs if body temperature rises above normal?: a) Circulation to the skin decreases; b) Sweat gland activity decreases; c) Evaporative cooling stops; d) Blood flow to the skin increases.

DIAGRAMS

9. Create a diagram that illustrates the two major components of the integumentary system, and the principal categories of those two components. Include brief statements that summarize the major functions of all of these entities.
10. Diagram the steps in the hair growth cycle.

TRUE/FALSE

(If a statement is false, your instructor may wish to have you correct it.)

11. **T/F:** White hair results from the presence of air bubbles within the hair shaft.
12. **T/F:** Blood with abundant oxygen takes on a bluish color.
13. **T/F:** Sebum is a breeding ground for bacteria.
14. **T/F:** The integumentary system requires exposure to ultraviolet radiation in order to synthesize vitamin D.

MATCHING QUESTIONS

(Match each statement with the most appropriate *term.)*

Statements:

15. The layer that is attached to the basement membrane and contains numerous stem cells.
16. The layer that is found only in areas of very thick skin.
17. The layers of flattened and dead cells that cover the epidermis.
18. The combination of fibrin clot, fibroblasts, and capillaries that forms at the site of an injury.
19. The secretions produced by the eccrine sweat glands.
20. A bluish coloration in the skin surface.
21. Sweat glands that produce a cloudy, viscous secretion.
22. Sweat glands that produce a clear watery secretion that cools the surface of the skin.
23. Epidermal cells attached to sensory neurons that provide information about objects touching the skin.
24. A process triggered by mast cells in the dermis that is an important defense against pathogens and injury.
25. An orange-yellow pigment that accumulates inside skin cells.
26. A brown, yellow-brown, or black pigment stored in cells of the stratum germinativum.

Terms:

a) Merkel cells
b) Carotene
c) Inflammation
d) Stratum lucidum
e) Stratum corneum
f) Cyanosis
g) Melanin
h) Granulation tissue
i) Stratum germinativum
j) Merocrine (eccrine) sweat glands
k) Apocrine sweat glands
l) Sensible perspiration

SHORT ANSWER/ESSAY

27. Why does the epidermis continually shed its cells?
28. Why does taking a long bath make the skin look "puckered"?
29. How does the ultraviolet radiation in sunlight affect the epidermis? Discuss both positive and negative effects.
30. What are the six principal functions of the integumentary system?
31. What is keratin? List three human anatomical structures that are made of keratin.
32. Explain the physiological reasons behind each of the following common expressions:
    a) "Turning white" with fear.
    b) A "hair-raising" experience.
33. What functions does hair serve in humans?
34. What causes lines of cleavage, and where are they located? What are their implications for surgery?
35. What are the characteristics of the hypodermis? Name four of its functions. Why is it a useful area for administering drugs?
36. Describe the life history of cells in the epidermis. Through what strata do they pass, and how long do these cells generally live?
37. What is the difference between a vellus hair and a terminal hair? In what areas of the body would you expect to find each type?

38. How does aging alter the integumentary system. How do these changes affect the physical condition of elderly people?
39. Compare the functions of the secretions produced by apocrine sweat glands and merocrine, or eccrine, sweat glands.
40. Every day after brushing your hair you notice that some hairs remain in the brush. Should you be concerned about this? Explain your answer.
41. What are the functions of epidermal ridges and dermal papillae?
42. Worn out by studying, you spend an entire day in bed and expend as little energy as possible. Would you still perspire? Explain your answer.
43. Many shampoo advertisements promise to give us "healthier" hair. Do you think this is a realistic claim? Explain your answer.

CRITICAL THINKING/APPLICATIONS

44. What are the advantages and disadvantages of transdermal medication compared to oral administration of a drug? For what types of health conditions do you think transdermal medication would be most useful?
45. Describe three common types of skin cancer. Which is the most serious? Why do you think skin cancers are the most common type of cancers?
46. A seven-year-old boy is admitted with third-degree burns of the following areas: left arm, front and back trunk, and both legs. Estimate the total surface area that has been burned. What are this patient's chances for recovery? Would your answer be different if the patient were a 68-year-old woman? Explain.

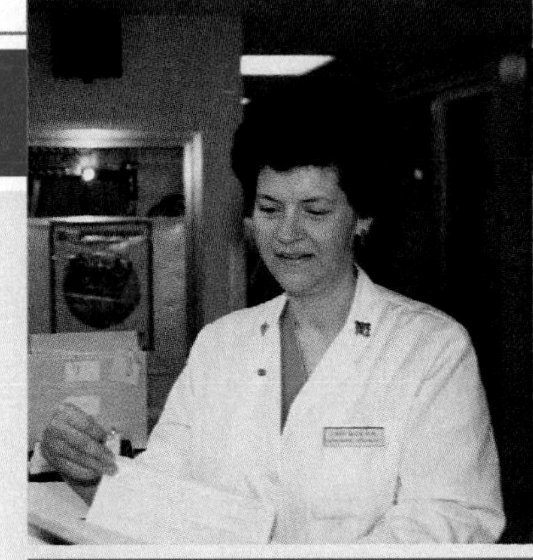

Career Close-up: Burn Nurse Specialist

"I loved every rotation I had during nursing school," recalls Linda K. Book, R.N., B.S.N. "Obstetrics, psychiatry–everything but the Burn Center. I dreaded dealing with burn patients, because their injuries seem so painful and I felt so inadequate to help them. But, when I applied for a job after graduation the only position available was in the Burn Center. I took it, on condition that I could transfer when another job came open. After six months they offered me the position I'd originally wanted–but I chose to stay in the Center."

Now, 24 years later, Linda is a Burn Nurse Specialist at a surburban medical center. The Burn Center where she works is the largest in the state, with a nine-bed intensive care unit that serves about 250 patients a year.

"Today the trend is toward holistic medicine–treating the entire person rather than just the physical injuries," Linda comments. "But in the Burn Center we've been practicing holistic medicine for years. A severe burn affects every system in the body. First, there's the damage to the integument, the largest organ system. Often there is respiratory and cardiac involvement; muscles, tendons, and bones may be injured too. And a burn takes a tremendous emotional toll on the patient. So you don't just treat the burn; you treat the whole situation."

The first three days after a burn are crucial, since the patient suffers massive fluid and electrolyte losses. After a severe burn the body's metabolism speeds up and the patient's temperature rises 1–2° F. "During this time, nutrition is our most important medication," Linda notes. "As a general rule, patients with major burns must double their caloric intake to avoid losing weight and muscle. Without adequate nutrition patients 'burn up' metabolically. A high protein diet also helps correct the immune deficiencies that occur after a burn and leave the patient more susceptible to infections. Of course, the actual burn wounds are common infection sites, but so are the lungs, bloodstream, and urinary tract–especially if breathing tubes, intravenous fluid lines, or bladder catheters are used."

Even after initial injuries have healed, burn survivors and their families face difficult adjustments, both physical and emotional. Crippling deformities appear as scar tissue proliferates and muscles and tendons contract. Years of painful reconstructive surgery loom ahead. With such a grim picture, why does Linda Book love her job?

"There's so much variety when you work in a Burn Center. You deal with multiple aspects of health care: trauma, psychiatry, even obstetrics—we've actually had deliveries here. You don't just deal with the same kind of problem over and over. And we really get to know the patients inside and out," Linda adds. "We meet their families, we follow them up over years, we help them reestablish their lives. In short, we become friends." The hospital sponsors a Burns Recovery Support Group that meets regularly to help patients deal with the aftermath of their injuries. Outpatient Services may follow up on a burn survivor for as long as 11 years, depending on the amount of reconstructive surgery required. Linda is closely involved in these services. She is also active in educating patients and the community about burn prevention. Linda sums it up simply: "I love my work because I feel I make a difference."

Have you oiled your joints lately?

Fortunately, unlike the Tin Man, we don't have to worry about this—our joints are normally self-lubricating. But take a moment to think about how important these joints are. What would happen to us if our bodies froze into position, as the Tin Man's at inconvenient times? We depend on our bones for support, but support without mobility would leave us little better than plants or statues. In this chapter we'll learn about the structure of bone, a very remarkable material—lightweight, adaptable, yet extremely strong. And we'll see how bones are linked together to give us our precious freedom of movement.

CHAPTER 7

Skeletal System: Osseous Tissue and Skeletal Structure

Chapter Objectives

After reading this chapter, you will be able to:

1. Describe the functions of the skeletal system.
2. Compare the structures and functions of compact and spongy bones.
3. Discuss the processes by which bones develop and grow and account for variations in their internal structure.
4. Describe the remodeling and homeostatic mechanisms of the skeleton.
5. Classify bones according to their shapes and give examples for each type.
6. Describe the different types of fractures and explain how fractures heal.
7. Discuss the effects of nutrition, hormones, exercise, and aging on bone development and the skeletal system.
8. Distinguish between different types of joints and link structural features to joint functions.
9. Describe the dynamic movements of the skeleton.

Introduction

This chapter begins our examination of the skeletal system. The skeleton has many functions, but the most obvious involves supporting the weight of the body. Bones work together with muscles to maintain body position and to produce controlled, precise movements. With the skeleton to pull against, contracting muscles can make us sit, stand, walk, or run. Without something to hold onto, contracting muscle fibers merely get shorter and fatter.

The skeletal system includes the bones of the skeleton and the cartilages, ligaments, and other connective tissues that stabilize or connect them. The functions of the skeletal system can be summarized as:

1. **Support:** The skeletal system provides structural support for the entire body. Individual bones or groups of bones provide a framework for the attachment of soft tissues and organs.
2. **Storage:** The calcium salts of bone represent a valuable mineral reserve that maintains normal

concentrations of calcium and phosphate ions in body fluids. In addition, fat cells in areas of *yellow marrow* store lipids that represent an important energy reserve.

3. **Blood cell production:** Red blood cells and other blood elements are produced within the *red marrow* that fills the internal cavities of many bones. The role of the bone marrow in blood cell formation will be discussed in later chapters dealing with the cardiovascular and lymphatic systems (Chapters 19–22).
4. **Protection:** Delicate tissues and organs are often surrounded by skeletal elements. The ribs protect the heart and lungs, the skull encloses the brain, the vertebrae shield the spinal cord, and the pelvis cradles delicate digestive and reproductive organs.
5. **Leverage:** The bones of the skeleton function as **levers**: they can change the magnitude and direction of the forces generated by skeletal muscles. The movements produced range from the delicate motion of a fingertip to powerful changes in the position of the entire body.

This chapter expands upon the discussion of bone presented in Chapter 5. You may wish to review the section on supporting connective tissues before proceeding. ∞ (p. 147) The following sections will focus on mechanisms responsible for skeletal growth, remodeling, and repair. We will also consider the ways bones interact at joints, or **articulations**.

As noted in the Systems Overview (Figure SO-4), the skeletal system can be divided into *axial* and *appendicular divisions.* ∞ (p. 170) Chapter 9 will consider the functional anatomy of the axial division, and Chapter 10 will complete the picture with a discussion of the appendicular division.

■ Structure of Bone

Bone, or **osseous tissue**, is one of the supporting connective tissues. Like other connective tissues, osseous tissue contains specialized cells, extracellular fibers, and a ground substance. In the supporting connective tissues, the fibers and ground substance interact to form a **matrix**. The matrix of cartilage is gelatinous—strength is provided by protein fibers and the perichondrium (Figure 5-16). ∞ (p. 148) The matrix of bone is solid. The distinctive stony character of bone results from the deposition of calcium salts in the matrix. Crystals of calcium phosphate, $Ca_3(PO_4)_2$, account for almost two-thirds of the weight of bone. The remaining third is dominated by collagen fibers and small amounts of other calcium salts, such as calcium carbonate. Osteocytes, other cell types, and proteoglycans contribute around 2 percent to the mass of a typical bone.

Calcium phosphate crystals are very strong, but inflexible. They can withstand compression, but the crystals are likely to shatter when exposed to bending, twisting, or sudden impacts. Collagen fibers are extremely tough, but quite flexible. They can easily tolerate stretching, twisting, and bending, but when compressed they simply bend out of the way. In bone, the collagen fibers provide an organic framework for the formation of mineral crystals. These crystals, a form of calcium phosphate called **hydroxyapatite**, form small plates that lie alongside the collagen fibers. The result is a protein-crystal combination with properties intermediate between those of collagen and those of pure mineral crystals.

HISTOLOGICAL ORGANIZATION

Figure 7-1 provides a detailed sectional view of the humerus, the bone of the upper arm. There are two types of bone visible in the section: *dense* (*compact*) *bone* and *spongy* (*cancellous*) *bone*. **Dense**, or **compact bone**, is relatively solid, whereas **spongy**, or **cancellous** (KAN-sel-us), **bone** resembles a network of bony struts separated by spaces that are normally filled with **bone marrow**. Both compact and spongy bone are present; compact bone forms the walls, and a layer of spongy bone surrounds the internal **marrow cavity**.

The general histology of bone was introduced in Chapter 5 (see Figure 5-18). ∞ (p. 151) Both compact and spongy bone contain bone cells, or **osteocytes** (OS-tē-ō-sīts; *osteon*, bone), in small pockets called **lacunae** (la-KOO-nē). Lacunae are found between narrow sheets of calcified matrix that are 4–12 μm thick. These layers of calcification are known as **lamellae** (lah-MEL-lē; *lamella*, thin plate). Small channels, called **canaliculi** (ka-na-LIK-ū-lē), radiate through the matrix, interconnecting lacunae and linking them to nearby blood vessels. The canaliculi contain cytoplasmic extensions of the osteocytes. The fluid that surrounds the osteocytes and their extensions provides a route for the diffusion of nutrients and waste products.

HISTOLOGICAL DIFFERENCES BETWEEN COMPACT AND SPONGY BONE

The basic functional unit of compact bone is the **osteon** (OS-tē-on), or *Haversian system*. Within an osteon the osteocytes are arranged in concentric layers around a **central canal**, or *Haversian canal*, that contains one or more blood vessels. (Clopton Havers was a seventeenth-century English anatomist who wrote a book on bone structure.) The lamellae are cylindrical, oriented parallel to the long axis of the central canal. Canaliculi radiating from the lacunae connect the osteocytes with one another and with the blood vessels of the Haversian canal. **Interstitial**

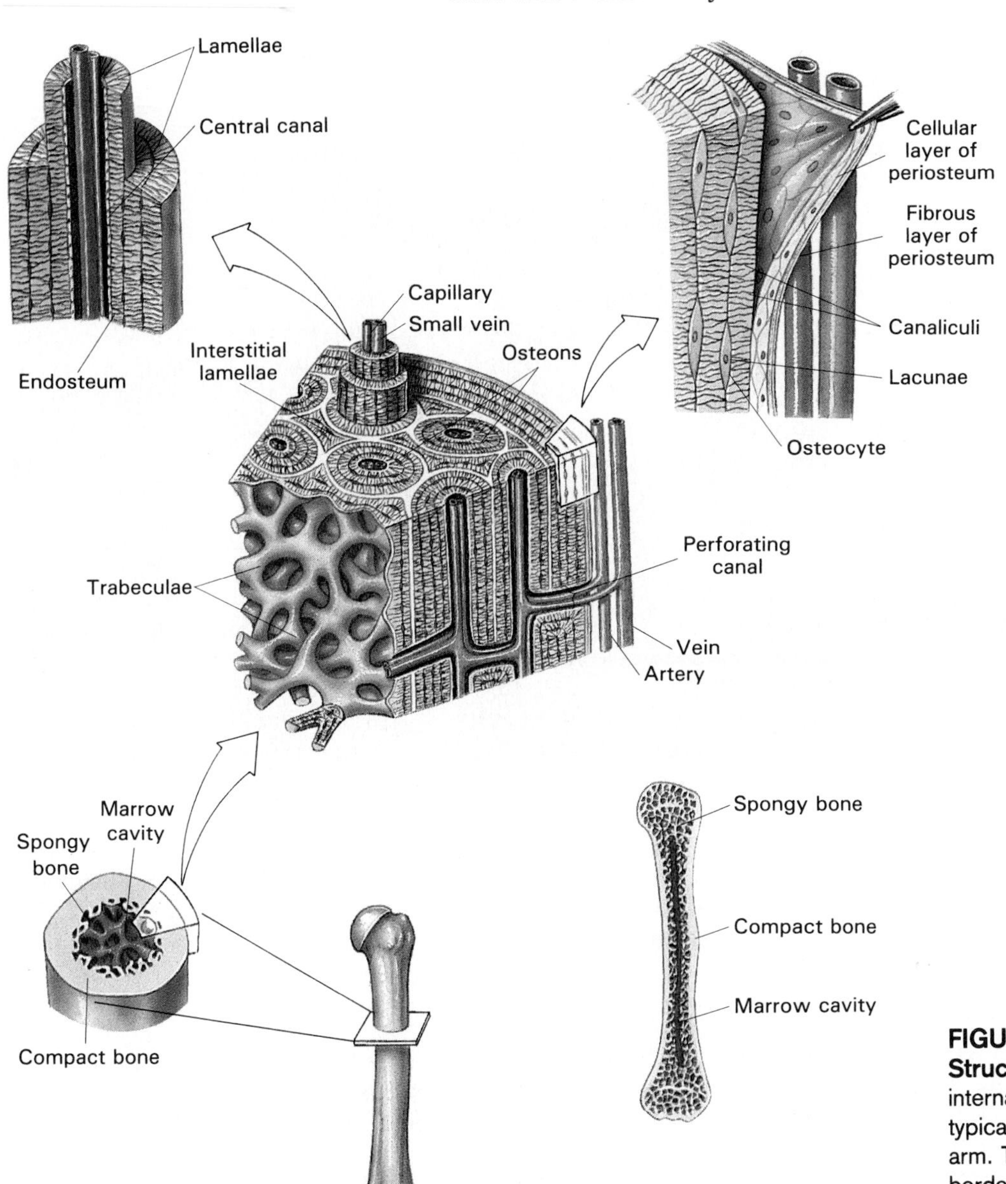

FIGURE 7-1
Structure of a Typical Bone. The internal and external appearance of a typical bone, the humerus of the upper arm. The bone has a marrow cavity bordered by spongy bone and enclosed within a sleeve of compact bone.

lamellae fill in the spaces between the osteons in compact bone. Because they have no blood vessels of their own, the osteocytes in these regions depend on diffusion from adjacent osteons.

With these structural features in mind, turn to the light micrographs of Figure 7-2a,b. These are complementary views of compact bone. Figure 7-2a was prepared after treating the bone with acids to dissolve the calcium salts. This procedure shows the size and distribution of osteocytes, blood vessels, and central canals. To prepare the specimen for Figure 7-2b, a small piece of bone was carefully sanded until it was thin enough for light to pass through it. The intact matrix appears white, and the lacunae, canaliculi, and central canals look black because they are filled with bone dust. Figure 7-2c adds the three-dimensional perspective provided by scanning electron microscopy.

Spongy bone has a quite different lamellar arrangement, and there are no osteons. The lamellae form struts or plates called **trabeculae** (tra-BEK-ū-lē). The thin trabeculae often branch, creating an open network. Canaliculi radiating from the lacunae end at the exposed surfaces of the trabeculae, and nutrients and wastes diffuse between the marrow spaces and osteocytes.

FUNCTIONAL DIFFERENCES BETWEEN COMPACT AND SPONGY BONE

A layer of compact bone covers bone surfaces everywhere except inside joint capsules, where articular cartilages protect opposing surfaces. Compact bone is usually found where stresses arrive from a limited range of directions. Osteons in compact bone are all

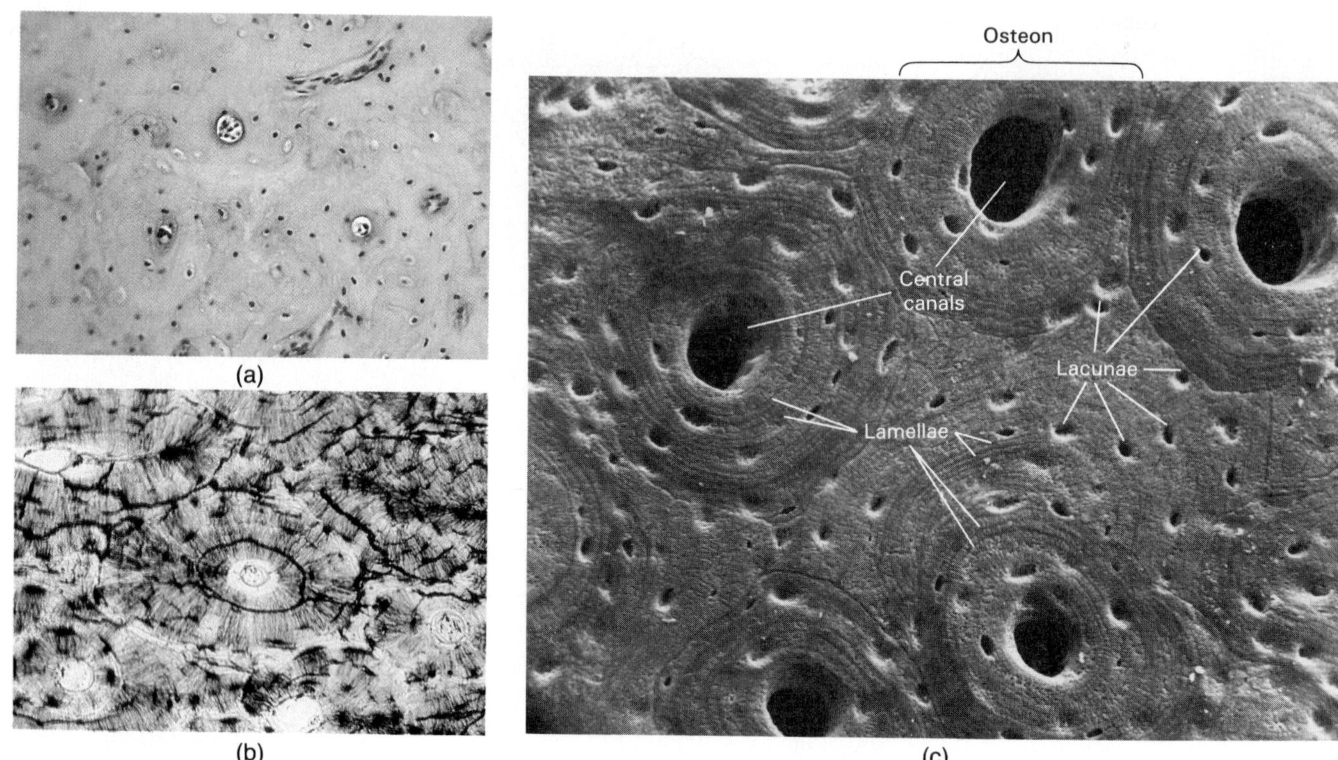

FIGURE 7-2
Compact Bone. (a) Light micrograph showing the structure of compact bone as it appears after the minerals have been dissolved to permit thin sectioning. (LM, × 200) (b) To prepare this specimen, a piece of compact bone was ground down until it became thin enough to allow light to penetrate. The dark lines are spaces filled with the bone dust produced during grinding. (LM, × 200) (c) A scanning electron micrograph of compact bone, showing several osteons. (SEM, × 270)

lined up the same way, and such bones are very strong when stressed along that axis. You might envision a single osteon as a drinking straw with very thick walls. When you attempt to push the ends of a straw together, it is quite strong; yet if you hold the ends and push from the side it breaks easily.

The arm and leg bones are good examples, because they are built to withstand forces applied at either end. When you stand on your hands or crawl along the floor, the bones of the arms distribute your weight to the hands. When standing, the leg bones conduct the body weight to your feet. Figure 7-3 shows the orientation of lamellae in the *femur*, the bone of the thigh. The compact bone of the **cortex** surrounds the marrow cavity, also known as the **medullary cavity** (*medulla*, innermost part). Stresses are normally applied along the axis of the shaft, or **diaphysis** (dī-A-fi-sis); an impact to the side of the shaft can lead to a broken leg. In the shaft, the osteons are parallel to the long axis of the shaft, so it does not bend when forces are applied to either end. But anyone dining on chicken wings can verify that long bones can easily be broken by snapping them across the long axis.

The hip joint consists of the head of the femur and a corresponding socket on the lateral surface of the hip bone. The femoral head projects medially, and body weight compresses the medial side of the diaphysis. Because the force is applied off-center, the bone has a tendency to bend into a lateral bow. The other side of the shaft, which resists this bending, is placed under a stretching load, or *tension*, as indicated in Figure 7-3b. Bending does occur, however, in disorders that reduce the amount of calcium salts in the skeleton, such as **rickets**. Because in rickets the bones are poorly mineralized, they are very flexible and are less able to resist compression and tension. Affected individuals develop a bowlegged appearance as the leg bones bend under the weight of the body.

Spongy bone is found where bones are not heavily stressed or where stresses arrive from many directions. It is present at the expanded ends of long bones, where they articulate with other skeltetal elements. These expanded regions are the heads, or **epiphyses** (ē-PIF-i-sēs), of the bone. Figure 7-3a shows the trabecular alignment in an epiphysis of the femur. The trabeculae are oriented along the stress lines, and there is extensive cross-bracing. The alignment can be seen clearly in Figure 7-3b,c. Note how the trabeculae are aligned along the force lines shown in part (a). At the hip and knee joints the spongy bone of the epiphyses transfers forces from pelvis to femur and from femur to *tibia*, the major bone of the lower leg.

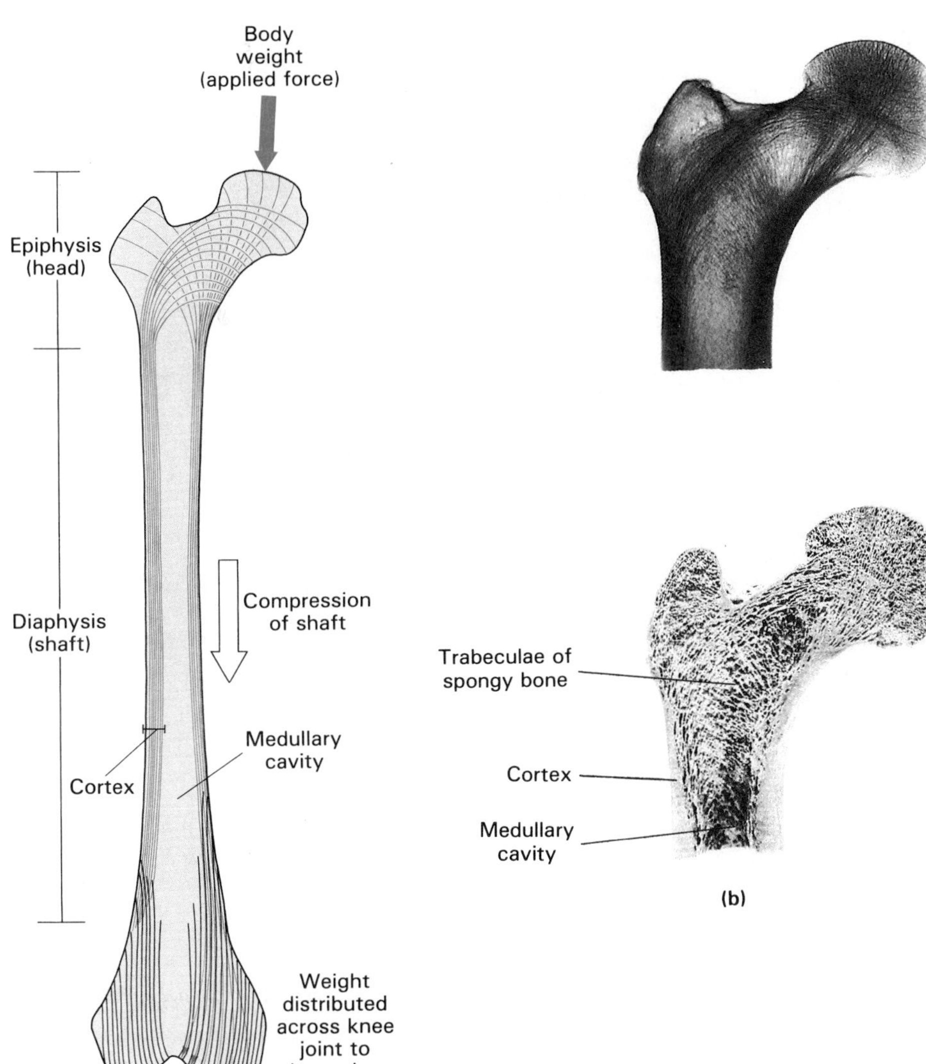

FIGURE 7-3
Lamellar Organization in a Long Bone. (a) The femur, or thigh bone, has a diaphysis (shaft) with walls of compact bone and epiphyses (heads) filled with spongy bone. The body weight is transferred to the femur at the hip joint. Because the hip joint is off-center relative to the axis of the shaft, the body weight is distributed along the bone so that the medial portion of the shaft is compressed and the lateral portion is stretched. (b) The upper photo is an X-ray showing the orientation of the trabeculae in the epiphysis. The lower photo shows the epiphysis after sectioning. Compare the orientation of trabeculae with the stress lines indicated in part (a).

In addition to being able to withstand stresses applied from many directions, spongy bone is much lighter than compact bone. This reduces the weight of the skeleton and makes it easier for muscles to move the bones. Finally, the trabecular framework protects the cells of the bone marrow, and areas of spongy bone, such as the epiphyses of the femur, are important sites of blood cell formation.

THE PERIOSTEUM AND ENDOSTEUM

The outer surface of a bone is covered by a **periosteum** that consists of a fibrous outer layer and a cellular inner layer. These layers were diagrammed in Figure 7-1. The periosteum isolates the bone from surrounding tissues, provides a route for circulatory and nervous supply, and actively participates in bone growth and repair. Near articulations, the periosteum becomes continuous with the collagen fibers of the **joint capsule**. The fibers of **tendons** intermingle with those of the periosteum, attaching skeletal muscles to the bones they move. Figure 7-4 diagrams these structural features.

Inside the bone, a cellular **endosteum** lines the marrow cavity. This layer, also shown in Figure 7-4, covers the trabeculae of spongy bone and lines the inner surfaces of the central canals. The endosteum is active during the growth of bone and whenever repair or remodeling is under way. The endosteal lining is not a complete epithelium, and the matrix is occasionally exposed. At these exposed sites cells called *osteoclasts* and *osteoblasts* can contact the mineralized surfaces.

Osteoclasts (OS-tē-ō-klasts; *clast*, break) are giant cells with 50 or more nuclei. Acids secreted by osteoclasts dissolve the bony matrix and release the stored minerals. This process, called **osteolysis** (os-tē-OL-i-sis), is important in the regulation of calcium and phosphate concentrations in body fluids.

Osteoblasts (OS-tē-ō-blasts; *blast*, precursor) are cuboidal cells that synthesize the organic components of the bone matrix. This material, called **osteoid** (OS-tē-oyd), later becomes mineralized through an

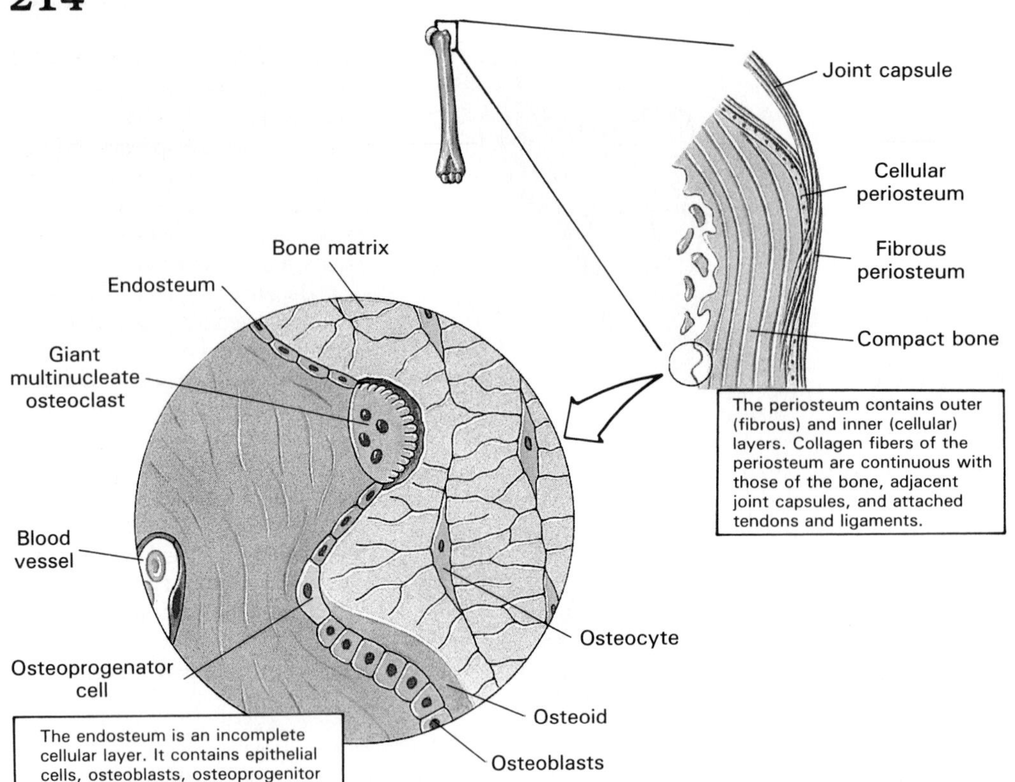

The periosteum contains outer (fibrous) and inner (cellular) layers. Collagen fibers of the periosteum are continuous with those of the bone, adjacent joint capsules, and attached tendons and ligaments.

The endosteum is an incomplete cellular layer. It contains epithelial cells, osteoblasts, osteoprogenitor cells, and osteoclasts.

FIGURE 7-4 The Periosteum and Endosteum

unknown mechanism. Osteoblasts are responsible for the production of new bone, a process called **osteogenesis** (os-tē-ō-JEN-e-sis; *gennan*, to produce). At any given moment, osteoclasts are removing matrix and osteoblasts are adding to it. A proper balance must be maintained; the control mechanisms involved will be considered later in this chapter.

The endosteum also contains small numbers of mesenchymal cells that can undergo mitosis, producing daughter cells that differentiate into osteoblasts. These **osteoprogenitor cells** (os-tē-ō-prō-JEN-i-tor; *progenitor*, ancestor) play an important role in fracture repair, a process discussed in a later section.

√ **How would the strength of a bone be affected if the ratio of collagen to calcium (hydroxyapatite) increased?**

√ **A sample of bone shows concentric layers surrounding a central canal. Is it from the shaft or the end of a long bone?**

√ **If the activity of osteoclasts exceeds the activity of osteoblasts in a bone, how will the mass of the bone be affected?**

Development and Growth

The growth of the skeleton determines the size and proportions of the body. The skeleton begins to form about 6 weeks after the egg is fertilized, when the embryo is approximately 12 mm long. (Before this time all supporting elements are cartilagenous.) During development, the bones undergo a tremendous increase in size. Bone growth continues through adolescence, and portions of the skeleton usually do not stop growing until age 18–25. The entire process is carefully regulated, and a breakdown in regulation ultimately affects all of the body systems. This section considers the physical process of osteogenesis (bone formation) and growth. The next section examines the maintenance and turnover of mineral reserves in the adult skeleton.

During development, mesenchyme or cartilage is replaced by bone. This process of replacing other tissues with bone is called **ossification**. There are important differences between ossification and **calcification**. Calcification refers to the deposition of calcium salts within a tissue. Many tissues will become calcified if the concentration of calcium salts in body fluids becomes abnormally high. The calcified tissues may be injured or even killed, but their other characteristics do not change. For example, the calcification of cartilage produces calcified cartilage, not bone. Ossification refers *specifically* to the formation of bone.

There are two major forms of ossification. In *intramembranous ossification* bone develops from mesenchyme or fibrous connective tissue. In *endochondral ossification* bone replaces an existing cartilage model.

INTRAMEMBRANOUS OSSIFICATION

Intramembranous (in-tra-MEM-bra-nus) **ossification** begins when osteoblasts differentiate within a connective tissue. This type of ossification normally occurs in the deeper layers of the dermis, and the bones that result are often called **dermal bones**. Examples of dermal bones include several bones of the skull, the lower jaw, and the collarbone (clavicle).

In the process of intramembranous ossification, osteoblasts first cluster together and start to secrete the organic components of the matrix. The resulting mixture of collagen fibers, proteoglycans, and glycoproteins, or osteoid, then becomes mineralized through the crystallization of calcium salts. The place where ossification first occurs is called an **ossification center**. As ossification proceeds, it traps some osteoblasts inside bony pockets; these cells differentiate into osteocytes.

Figure 7-5 provides additional details of intramembranous ossification and considers the three-dimensional changes that take place in the tissue. The developing bone grows outward from the ossification center in small struts, called **spicules**. Although osteoblasts are still being trapped in the expanding bone, mesenchymal cell divisions continue to produce additional osteoblasts. As a result, the rate of bone growth actually accelerates.

Bone growth is an active process, and osteoblasts require oxygen and a reliable supply of nutrients. Blood vessels that branch between the spicules meet these demands. Although initially the intramembranous bone resembles spongy bone, subsequent remodeling around the trapped blood vessels can produce compact bone.

ENDOCHONDRAL OSSIFICATION

Endochondral ossification (en-dō-KON-dral; *endo*, inside + *chondros*, cartilage) begins with the formation of a cartilaginous model. Limb bone development

FIGURE 7-5
A Three-Dimensional View of Intramembranous Ossification

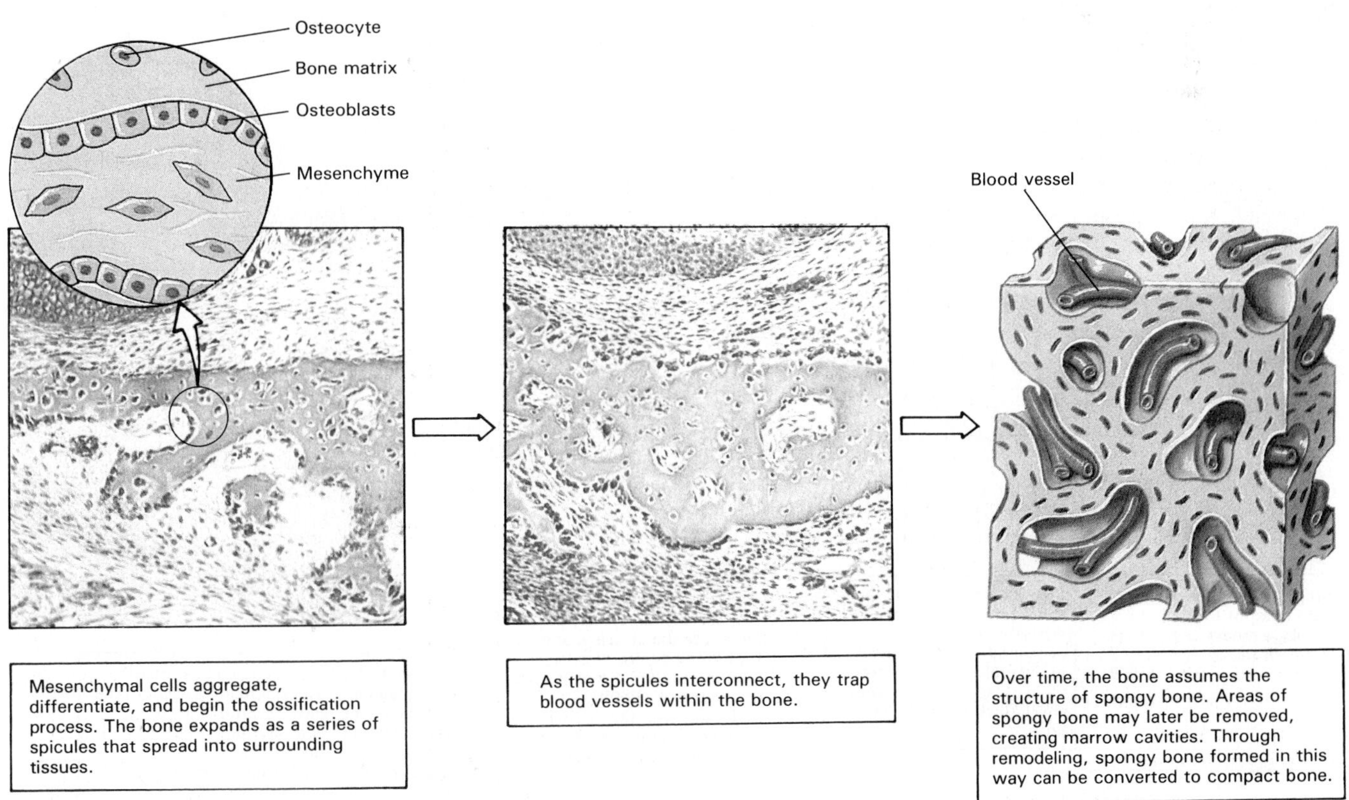
Osteocyte
Bone matrix
Osteoblasts
Mesenchyme
Blood vessel
Mesenchymal cells aggregate, differentiate, and begin the ossification process. The bone expands as a series of spicules that spread into surrounding tissues.
As the spicules interconnect, they trap blood vessels within the bone.
Over time, the bone assumes the structure of spongy bone. Areas of spongy bone may later be removed, creating marrow cavities. Through remodeling, spongy bone formed in this way can be converted to compact bone.

is a good example of this process. By the time an embryo is 6 weeks old, the proximal bone of the limb, either the humerus (upper arm) or femur (thigh), is present, but it is composed entirely of cartilage. This model continues to grow by expansion of the cartilage matrix (interstitial growth) and the production of new cartilage at the outer surface (appositional growth). These growth mechanisms were detailed in Chapter 5 and illustrated in Figure 5-17. (p. 150) Steps in the growth and ossification of a limb bone are diagrammed in Figure 7-6.

STEP 1: As the cartilage enlarges, chondrocytes near the center of the shaft increase greatly in size, and the surrounding matrix begins to calcify. Deprived of nutrients, these chondrocytes die and disintegrate.

STEP 2: Blood vessels now invade the perichondrium surrounding this portion of the shaft, and the cells of the inner, cellular layer closest to the cartilage develop into osteoblasts. The perichondrium has now been converted into a periosteum, and the inner **osteogenic layer** produces a layer of bone around the shaft of the cartilage.

STEP 3: Capillaries and osteoblasts migrate into the heart of the cartilage, invading the spaces left by the disintegrating chondrocytes. The calcified cartilaginous matrix breaks down, and osteoblasts replace it with spongy bone. Bone development proceeds from this **primary center of ossification** toward the ends of the cartilaginous model.

STEP 4: While the diameter is small, the entire diaphysis is filled with spongy bone, but as it enlarges osteoclasts erode the central portion and create a marrow cavity. Further growth involves two distinct processes: an enlargement in *diameter* and an increase in *length*.

Increasing the Diameter of a Developing Bone

The diameter of a bone enlarges through appositional growth at the outer surface. (This process was described in Chapter 5; see Figure 5-19.) (p. 154) As periosteal cells develop into osteoblasts and produce additional bony matrix, blood vessels and collagen fibers of the periosteum become incorporated into the bony structure. Examine the steps outlined in Figure 7-7. Blood vessels running parallel to the surface become the center of individual osteons. Other vessels, known as **perforating canals**, or the *canals of Volkmann*, deliver blood to more distant central canals and the marrow cavity of the interior. Wherever tendons or ligaments are woven into the periosteum, the collagen fibers become cemented into the superficial lamellae by the osteoblasts. This makes them

FIGURE 7-6
Initial Steps in Endochondral Ossification

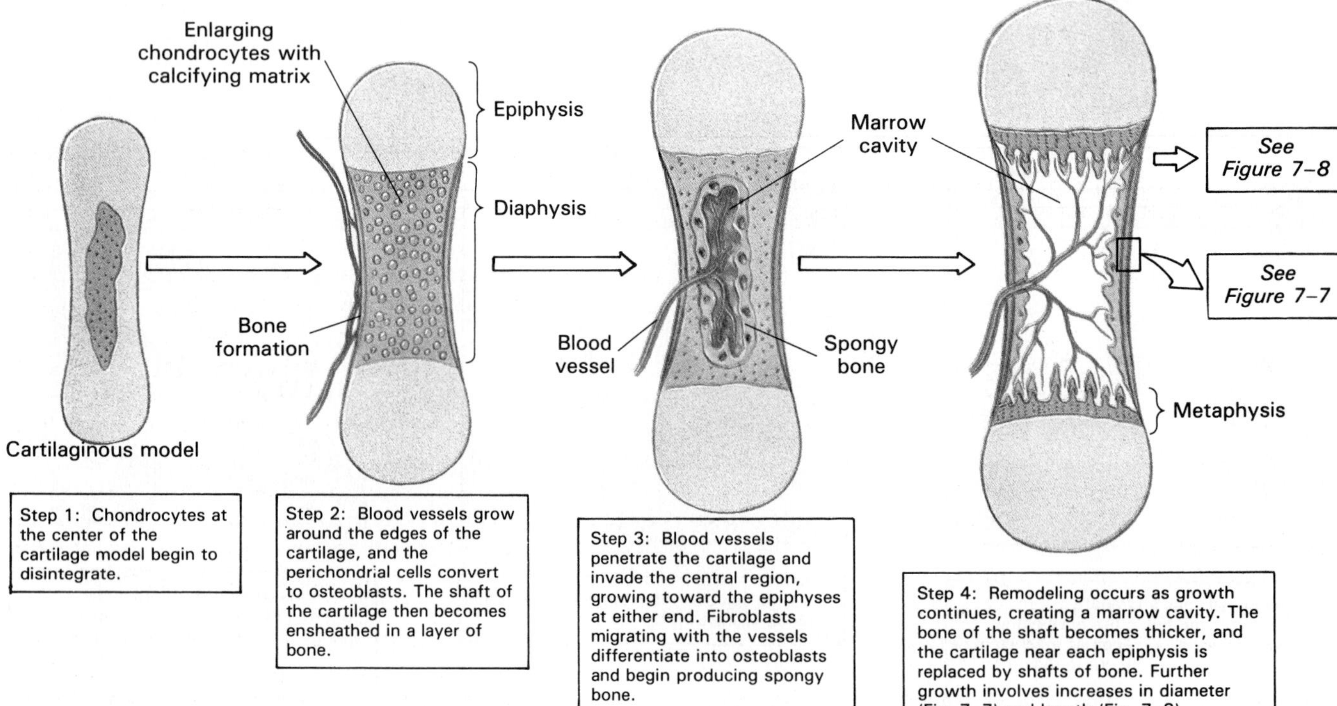
Enlarging chondrocytes with calcifying matrix
Epiphysis
Diaphysis
Bone formation
Cartilaginous model
Marrow cavity
Blood vessel
Spongy bone
See Figure 7–8
See Figure 7–7
Metaphysis
Step 1: Chondrocytes at the center of the cartilage model begin to disintegrate.
Step 2: Blood vessels grow around the edges of the cartilage, and the perichondrial cells convert to osteoblasts. The shaft of the cartilage then becomes ensheathed in a layer of bone.
Step 3: Blood vessels penetrate the cartilage and invade the central region, growing toward the epiphyses at either end. Fibroblasts migrating with the vessels differentiate into osteoblasts and begin producing spongy bone.
Step 4: Remodeling occurs as growth continues, creating a marrow cavity. The bone of the shaft becomes thicker, and the cartilage near each epiphysis is replaced by shafts of bone. Further growth involves increases in diameter (Fig. 7–7) and length (Fig. 7–8).

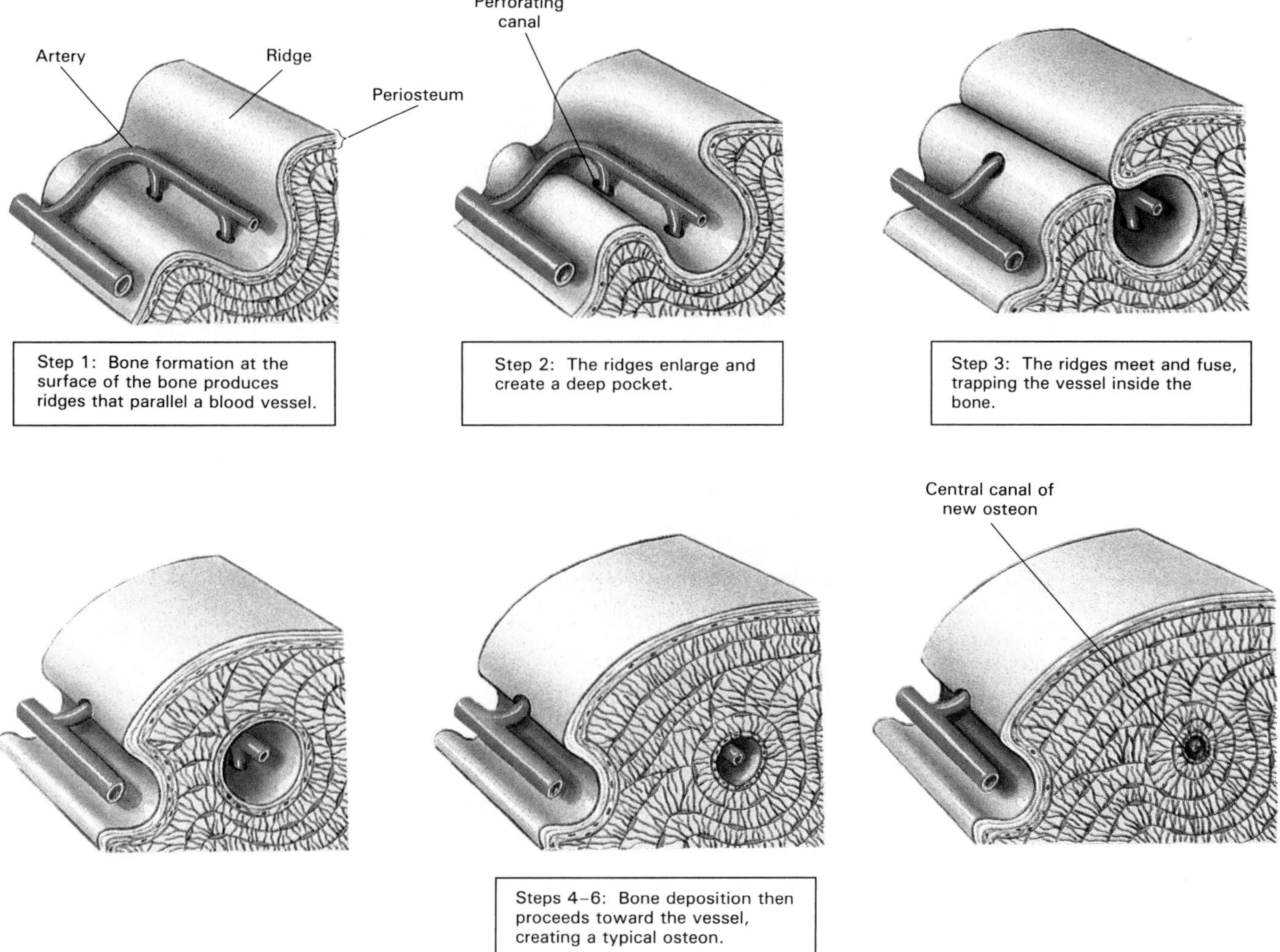

Artery
Ridge
Periosteum
Perforating canal
Step 1: Bone formation at the surface of the bone produces ridges that parallel a blood vessel.
Step 2: The ridges enlarge and create a deep pocket.
Step 3: The ridges meet and fuse, trapping the vessel inside the bone.
Central canal of new osteon
Steps 4–6: Bone deposition then proceeds toward the vessel, creating a typical osteon.

FIGURE 7-7
Appositional Bone Growth. Three-dimensional diagrams illustrating the mechanism responsible for increasing the diameter of a growing bone.

part of the general structure of the bone, providing a much stronger bond than would otherwise be possible. An extremely powerful pull on a tendon or ligament will usually break a bone rather than snap the collagen fibers at the bone surface.

Increasing the Length of a Developing Bone

Growth in bone length involves a very different mechanism. During the initial stages of osteogenesis, the osteoblasts move away from the center of ossification toward the epiphyses. But they do not manage to complete the ossification of the model immediately, because the epiphyseal cartilages continue to enlarge. The region where the cartilage is being replaced by bone lies at the junction between the diaphysis (shaft) and epiphyses (heads) of the bone. This region is known as the **metaphysis** (me-TA-fi-sis). On the shaft side of the metaphysis osteoblasts are continually invading the cartilage and converting it to bone. But on the epiphyseal side, new cartilage is produced at the same rate. As a result, the osteoblasts never quite catch up with the epiphysis, although the skeletal element continues to enlarge. This process, diagrammed in Figure 7-8, could be compared to the pacing of a jogger by a person on a bicycle. The jogger keeps advancing, but the bike stays a few feet ahead.

At around the time of birth, the center of each epiphysis begins to calcify. Capillaries and osteoblasts migrate into these areas, creating **secondary ossification centers**. Soon the epiphyses are filled with spongy bone. A thin cap of **articular cartilage** remains exposed to the joint cavity. This cartilage prevents damaging bone-to-bone contact within the joint. At the metaphysis a relatively narrow cartilaginous **epiphyseal plate** separates the epiphysis from the diaphysis. These developments are shown in Figure 7-8a.

Within the epiphyseal plate, the chondrocytes closest to the epiphysis continue to enlarge and divide, adding to the thickness of the plate. This forces the epiphysis farther from the shaft, as indicated

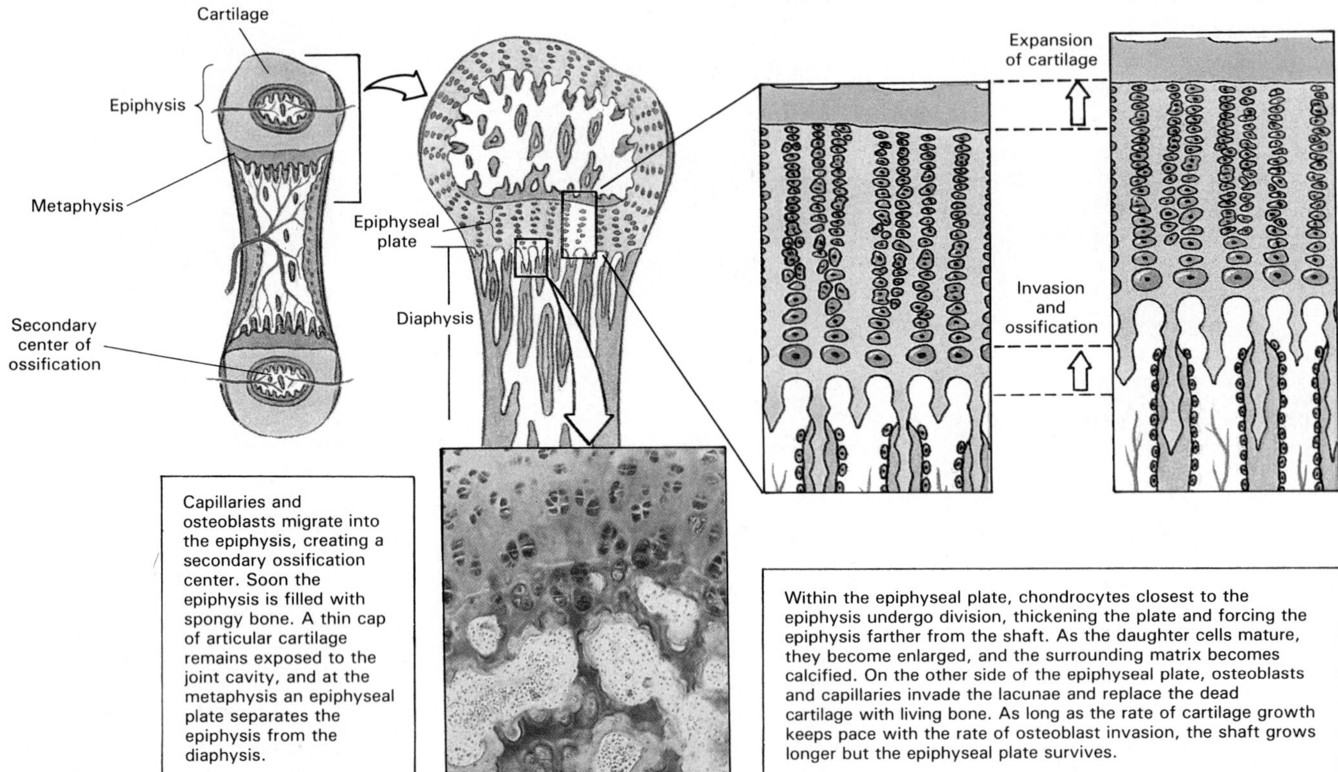
Cartilage
Epiphysis
Metaphysis
Secondary center of ossification
Epiphyseal plate
Diaphysis
Expansion of cartilage
Invasion and ossification
Capillaries and osteoblasts migrate into the epiphysis, creating a secondary ossification center. Soon the epiphysis is filled with spongy bone. A thin cap of articular cartilage remains exposed to the joint cavity, and at the metaphysis an epiphyseal plate separates the epiphysis from the diaphysis.
Within the epiphyseal plate, chondrocytes closest to the epiphysis undergo division, thickening the plate and forcing the epiphysis farther from the shaft. As the daughter cells mature, they become enlarged, and the surrounding matrix becomes calcified. On the other side of the epiphyseal plate, osteoblasts and capillaries invade the lacunae and replace the dead cartilage with living bone. As long as the rate of cartilage growth keeps pace with the rate of osteoblast invasion, the shaft grows longer but the epiphyseal plate survives.

FIGURE 7-8
Bone Growth at the Epiphyseal Plate. The light micrograph shows the interface between the degenerating cartilage and the advancing osteoblasts.

in Figure 7-8b. As the daughter cells mature, they become enlarged, and the surrounding matrix becomes calcified. On the other side of the epiphyseal plate, osteoblasts and capillaries continue to invade the lacunae and replace the dead cartilage with living bone. As long as the rate of cartilage growth keeps pace with the rate of osteoblast invasion, the shaft grows longer but the epiphyseal plate survives. Figure 7-8c is a light micrograph that shows the interface between the degenerating cartilage and the advancing osteoblasts. [**CM**: *Inherited Abnormalities in Skeletal Development*]

Hormonal Regulation of Bone Growth

Circulating hormones regulate the pattern of growth by changing the rates of osteoblast and chondrocyte activity. **Growth hormone**, produced by the pituitary gland, and **thyroxine**, from the thyroid gland, stimulate bone growth. In proper balance, these hormones maintain normal activity at the epiphyseal plates until roughly the time of puberty. Excessive or inadequate hormone production has a pronounced effect on activity at epiphyseal plates. **Gigantism** results from an overproduction of growth hormone before puberty; at age 22, one individual reached a height of 8 feet 11 inches and a weight of 475 pounds. The opposite extreme is **pituitary growth failure** (*pituitary dwarfism*), in which inadequate growth hormone production leads to reduced epiphyseal activity and abnormally short bones. This form of growth failure is becoming increasingly rare in the United States because children can be treated with human growth hormone prepared by the genetic engineering techniques described in Chapter 4. (p. 114)

When sex hormone production increases at puberty, bone growth accelerates dramatically, and osteoblasts begin to produce bone faster than the rate of epiphyseal cartilage expansion. In effect, the jogger speeds up, moving closer to the bicycle ahead. Over time, the epiphyseal plates narrow and eventually ossify, or "close." The overall effect is a period of sudden growth that ends as the individual reaches sexual and physical maturity. The location of the plate can still be detected as a distinct **epiphyseal line** that remains after epiphyseal growth has ended.

BONE GROWTH AND BODY PROPORTIONS

Growth at epiphyseal plates is characteristic of all bones that form through endochondral ossification. There are differences from bone to bone and individual to individual as to the timing of epiphyseal closure. The toes may complete their ossification by age 11, whereas portions of the pelvis or the wrist may continue to enlarge until age 25. On the average, the epiphyseal plates in the arms and legs will close by age 18 (women) or 20 (men). Differences in the stimulatory effects of male and female sex hormones account for the variation between the sexes and for related variations in body size and proportions. In *precocious puberty*, production of sex hormones escalates early, perhaps by age 8 or even earlier. This results in an abbreviated growth spurt, followed by premature closure of the epiphyseal plates. [**CM**: *Hyperostosis and Acromegaly*]

REQUIREMENTS FOR NORMAL BONE GROWTH

Normal osteogenesis cannot occur without a reliable source of minerals, especially calcium salts. During prenatal development these minerals are absorbed from the mother's bloodstream. The demands are great enough that the maternal skeleton often decreases in size and bone mass during pregnancy. From infancy to adulthood, the diet must provide adequate amounts of calcium and phosphate, and the individual must be able to absorb and transport these minerals to sites of bone formation.

Vitamin D plays an important role in normal calcium metabolism by stimulating the absorption and transport of calcium and phosphate ions. This vitamin, introduced in Chapter 6, can be obtained from dietary supplements or manufactured by epidermal cells exposed to ultraviolet radiation. (p. 190) Rickets, mentioned earlier in this chapter, is a condition marked by a softening and bending of bones. It occurs in growing children, usually as a result of vitamin D deficiency.

Vitamins **A** and **C** are also essential for normal bone growth and remodeling. For example, a deficiency of vitamin C will result in **scurvy**. One of the primary symptoms of this condition is a reduction in osteoblast activity that leads to weak and brittle bones. (Other symptoms will be considered in Chapter 25.) In addition to vitamins, hormones such as growth hormone, thyroid hormones, sex hormones, and those involved with calcium metabolism (described below) are essential to normal skeletal growth and development.

- **How could X-rays of the femur be used to determine whether a person had reached full height?**
- **In the Middle Ages, choirboys were sometimes castrated (had their testes removed) to prevent their voices from changing. How would this have affected their height?**
- **Why are pregnant women given calcium supplements and encouraged to drink milk even though their skeletons are fully formed?**

Remodeling and Homeostatic Mechanisms

Of the five major functions of the skeleton discussed earlier in this chapter, the first two—support and storage—depend on the dynamic nature of bone. In the adult, osteocytes in lacunae maintain the surrounding matrix, continually removing and replacing the surrounding calcium salts. But osteoclasts and osteoblasts also remain active, even after the epiphyseal plates have closed. Normally their activities are balanced: as one osteon forms through the activity of osteoblasts another is destroyed by osteoclasts. The turnover rate for bone is quite high, and up to 18 percent of the protein and mineral components are removed and replaced each year. Every part of every bone may not be affected, as there are regional and even local differences in the rate of turnover. For example, the spongy bone in the head of the femur may be replaced two or three times each year, whereas the compact bone along the shaft remains largely untouched.

Building and demolishing almost one-fifth of the skeleton each year represents a substantial commitment of energy. You are already familiar with the advantages of metabolic turnover for individual cells (Chapter 4). (p. 117) Similar factors make it advantageous for bones to undergo continual remodeling. That remodeling may involve a change in the shape or internal architecture of a bone, or a change in the total amount of minerals deposited in the skeleton.

CHANGES IN BONE SHAPE

Mineral turnover gives each bone the ability to adapt to new stresses. Osteoblast sensitivity to electrical events has been suggested as the mechanism that controls the internal organization and structure of bone. Whenever a bone is stressed, the mineral crystals generate minute electrical fields. Osteoblasts are apparently attracted to such fields, and once in the area they begin to produce bone. This appears to explain why the trabeculae of spongy bone follow lines of stress, and why the trabecular arrangement changes when the applied forces increase or decrease.

Similar factors affect the shapes of bony surfaces. For example, bumps and ridges on the surface of a bone mark the sites where tendons attach to the bone. If the muscle becomes more powerful, the corresponding ridge enlarges to withstand the increased force. Heavily stressed bones become thicker and stronger, whereas bones not subjected to ordinary stresses will become thin and brittle. Regular exercise is therefore important as a stimulus that maintains normal bone structure; champion weight lifters have massive bones with thick, prominent ridges.

Degenerative changes in the skeleton occur after relatively brief periods of inactivity. For example, using a crutch while wearing a cast takes the loading off the injured leg. After a few weeks, the unstressed leg will lose up to about a third of its bone mass. The bones rebuild equally quickly when normal loading resumes. However, the removal of calcium salts can be a potentially serious health hazard for astronauts remaining in a weightless environment and for bedridden or paralyzed patients who spend months or years without stressing the skeleton.

CHANGES IN THE MINERAL CONTENT OF THE SKELETON

The bones of the skeleton are more than just racks to hang muscles on. They are important mineral reservoirs, as indicated in Figure 7-9. Other minerals will be considered in later chapters; for the moment we will focus on the homeostatic regulation of calcium ion concentrations in body fluids. Calcium is the most abundant mineral in the human body. A typical human body contains 1–2 kg (2.2–4.4 lb) of calcium, with more than 98 percent of it deposited in the skeleton.

Calcium ion concentrations must be closely controlled to prevent damage to essential physiological systems. Even small variations from normal concentrations will have some effect on cellular operations. Larger changes can cause a clinical crisis. Nerve and muscle cells are particularly sensitive to changes in the concentration of calcium ions. If the calcium concentration in body fluids increases by 30 percent, neurons and muscle cells become relatively unresponsive. If calcium levels decrease by 35 percent, they become so excitable that convulsions may occur. A 50 percent reduction in calcium concentrations usually causes death. Such gross disturbances in calcium metabolism are relatively rare, for the calcium ion concentrations are so closely regulated that daily fluctuations of more than 10 percent are very unusual.

In **osteomalacia** (os-tē-ō-ma-LĀ-shē-ah; *malakia*, softness) the size of the skeletal elements remains the same but their mineral content decreases, softening the bones. In this condition the osteoblasts are working hard, but the matrix is not accumulating enough calcium salts. This can occur in adults or children whose diet contains inadequate levels of calcium or vitamin D.

FIGURE 7-9
An Analysis of Bone

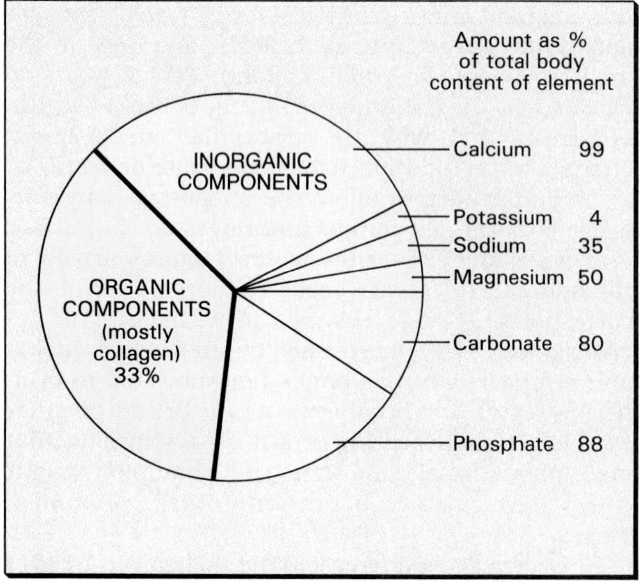

Hormones and the Regulation of Calcium Ion Concentrations

The calcium ion concentration in body fluids depends on activities in three organs: (1) the bones, (2) the intestinal tract, and (3) the kidneys. Interactions between these components are indicated in Figure 7-10. Factors that promote an elevation in calcium levels are indicated by red arrows; blue arrows indicate factors that depress calcium ion concentrations.

Calcium ion homeostasis is maintained by a negative feedback system involving a pair of hormones with opposing effects. These hormones coordinate the storage, absorption, and excretion of calcium ions. The target organs are the bones (storage), the digestive tract (absorption), and the kidneys (excretion). This regulatory process has been summarized in Figure 7-11.

When the calcium ion concentration of the blood rises above normal, special cells within the thyroid gland secrete the hormone **calcitonin** (kal-si-TŌ-nin). Calcitonin has three major functions:

- *inhibiting* osteoclast activity,
- *decreasing* the rate of intestinal absorption, and
- *increasing* the rate of calcium ion excretion.

Less calcium enters body fluids because osteoclasts leave the mineral matrix alone, and intestinal absorption declines. More calcium leaves body fluids, for osteoblasts continue to produce new bone matrix, and calcium ion excretion at the kidneys accelerates. The net result is a decline in calcium concentrations, restoring homeostasis.

If calcium concentrations fall below normal, cells of the **parathyroid gland** release **parathormone** into the circulatory system. Parathormone also has three major effects:

- *stimulating* osteoclast activity,
- *increasing* the rate of intestinal absorption, and
- *decreasing* the rate of calcium ion excretion.

More calcium enters body fluids, losses are restricted, and calcium concentrations increase to normal levels.

By providing a calcium reserve, the skeleton maintains calcium homeostasis in body fluids. This

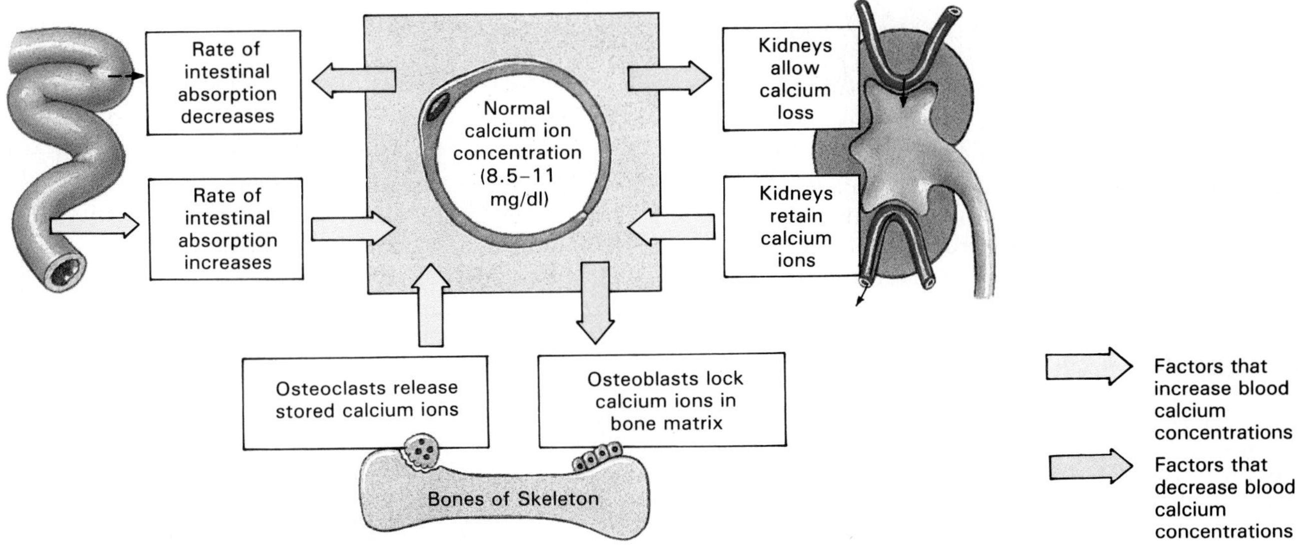

FIGURE 7-10
Factors that Alter the Concentration of Calcium Ions in Body Fluids

FIGURE 7-11
The Homeostatic Regulation of Calcium Ion Concentrations

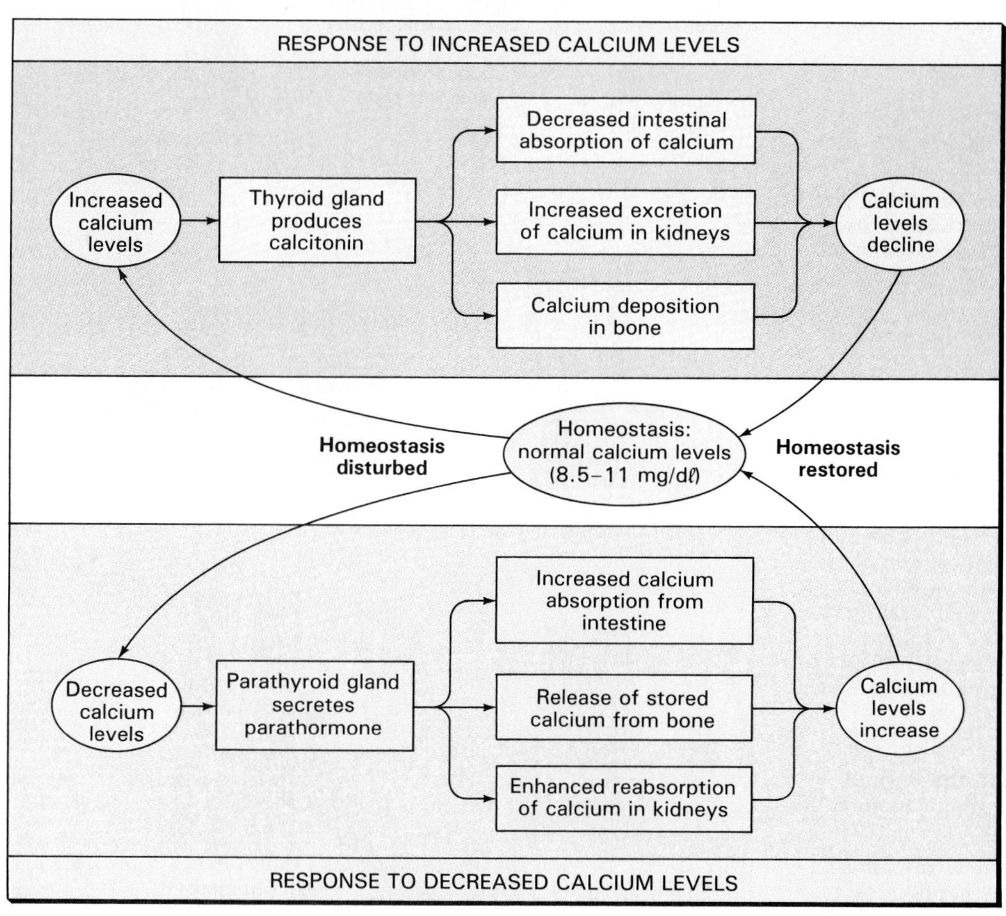

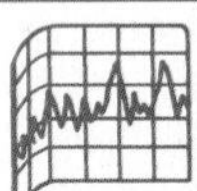

Diagnostics: Classification and Identification of Fractures

Fractures are classified according to their external appearance, the site of the fracture, and the nature of the crack or break in the bone. Important fracture types are indicated in Table 7-1, and several have been paired with representative X-rays. Many fractures fall in more than one category. For example, Colles' fracture is a transverse fracture, but depending on the injury it may also be a comminuted fracture that can be either open or closed.

TABLE 7-1 A Classification of Fractures

Fracture Type
Closed, or **simple,** fractures are completely internal; they do not involve a break in the skin
Open, or **compound,** fractures project through the skin; they are more dangerous because of the possibility of infection or uncontrolled bleeding
Comminuted fractures shatter the affected area into a multitude of bony fragments
Transverse fractures break a shaft bone across its long axis
In a **greenstick** fracture only one side of the shaft is broken, and the other is bent; this usually occurs in children whose long bones have yet to fully ossify
Spiral fractures, produced by twisting stresses, spread along the length of the bone
A **Colles'** fracture is a break in the distal portion of the radius, the slender bone of the forearm; it is often the result of reaching out to cushion a fall
A **Pott's** fracture occurs at the ankle and affects both bones of the lower leg
Compression fractures occur in vertebrae subjected to extreme stresses, as when landing on your seat after a fall
Epiphyseal fractures usually occur where the matrix is undergoing calcification and chondrocytes are dying. A clean transverse fracture along this line usually heals well. Fractures between the epiphysis and the epiphyseal plate can permanently halt further longitudinal growth unless carefully treated; often surgery is required
Nondisplaced fractures retain the normal alignment of the bone elements or fragments
Displaced fractures produce new and abnormal arrangements of bony elements

Description

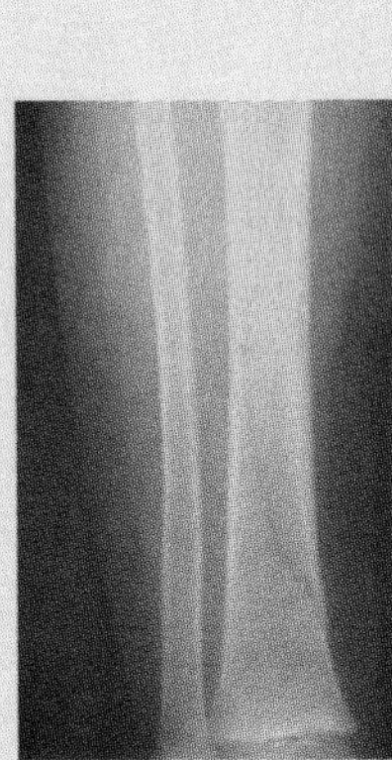

Radius spiral fracture

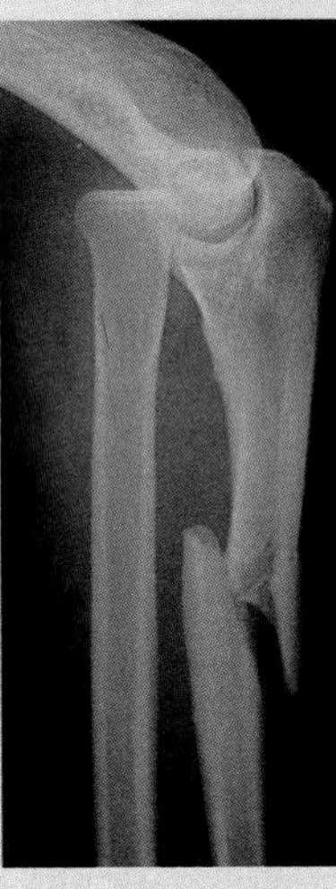

Dislocated radius and displaced ulnar fracture

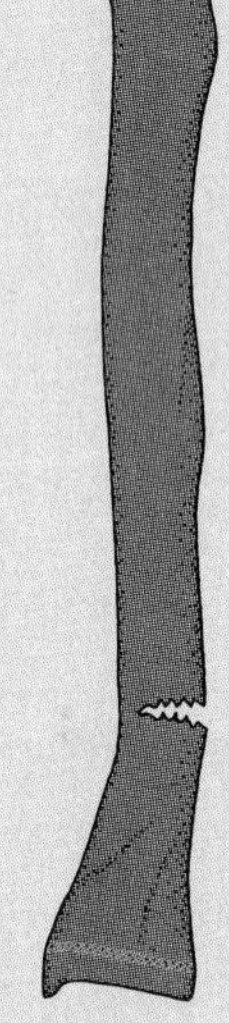

Greenstick fracture

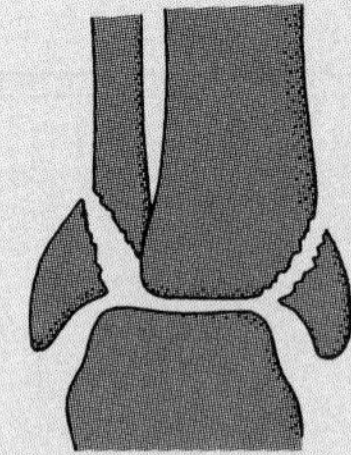

Pott's fracture–dislocation

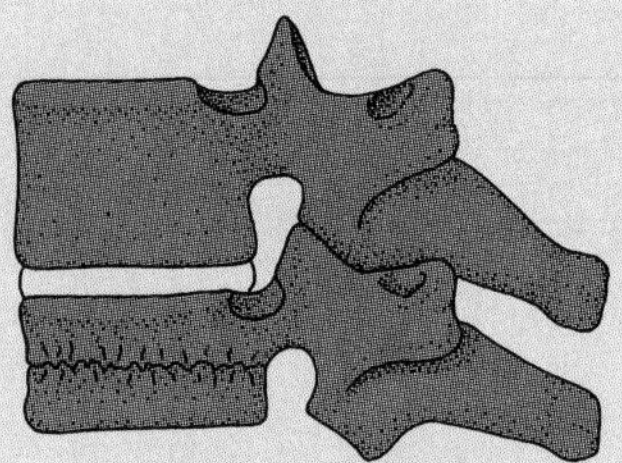

Compression fracture

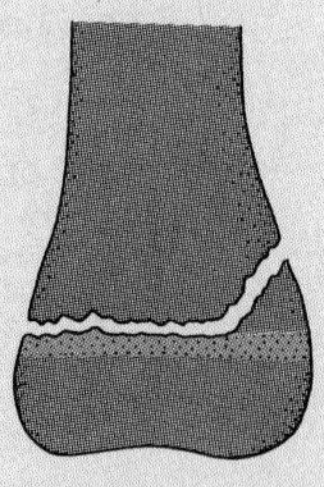

Epiphyseal plate fracture

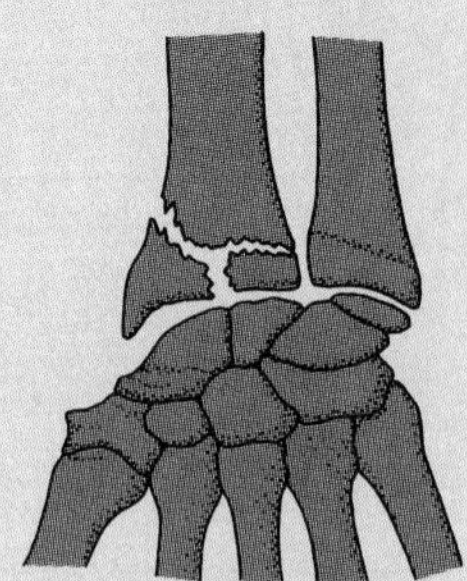

Colles' fracture

function can have a direct effect on the shape and strength of the bones in the skeleton. When large numbers of calcium ions are mobilized, the bones become weaker; when calcium salts are deposited, the bones become more massive.

INJURY AND REPAIR

Despite its mineral strength, bone cracks or even breaks if subjected to extreme loads, sudden impacts, or stresses from unusual directions. The damage produced constitutes a **fracture** (see Diagnostics: Classification and Identification of Fractures).

Healing of a fracture usually occurs even after severe damage, provided that the circulatory supply and the cellular components of the endosteum and periosteum survive. Stages in the repair process are diagrammed in Figure 7-12.

STEP 1: In even a small fracture, many blood vessels are broken and extensive bleeding occurs. A large clot, or **fracture hematoma**, soon closes off the injured vessels and leaves a fibrous meshwork in the damaged area. The disruption of circulation kills osteocytes around the fracture, increasing the size of the area affected. Dead bone soon extends along the shaft in either direction from the break.

STEP 2: In the adult the cells of the periosteum and endosteum are relatively inactive. When a fracture occurs, the cells of the intact endosteum and periosteum divide rapidly, and the daughter cells migrate into the fracture zone. An enlarged collar, or **external callus** (*callum*, hard skin), forms, encircling the bone at the level of the fracture. An extensive **internal callus** organizes within the marrow cavity and between the broken ends of the shaft. At the center of the external callus, the cells differentiate into chondrocytes and produce a block of cartilage. At the edges of each callus the cells differentiate into osteoblasts and begin creating a bridge of spongy bone that links the bone fragments on either side of the fracture.

STEP 3: As the repair continues, osteoblasts replace the central cartilage with spongy bone. When this conversion is complete the external and internal calluses form an extensive and continuous brace at the fracture site.

FIGURE 7-12
Repair of a Fracture

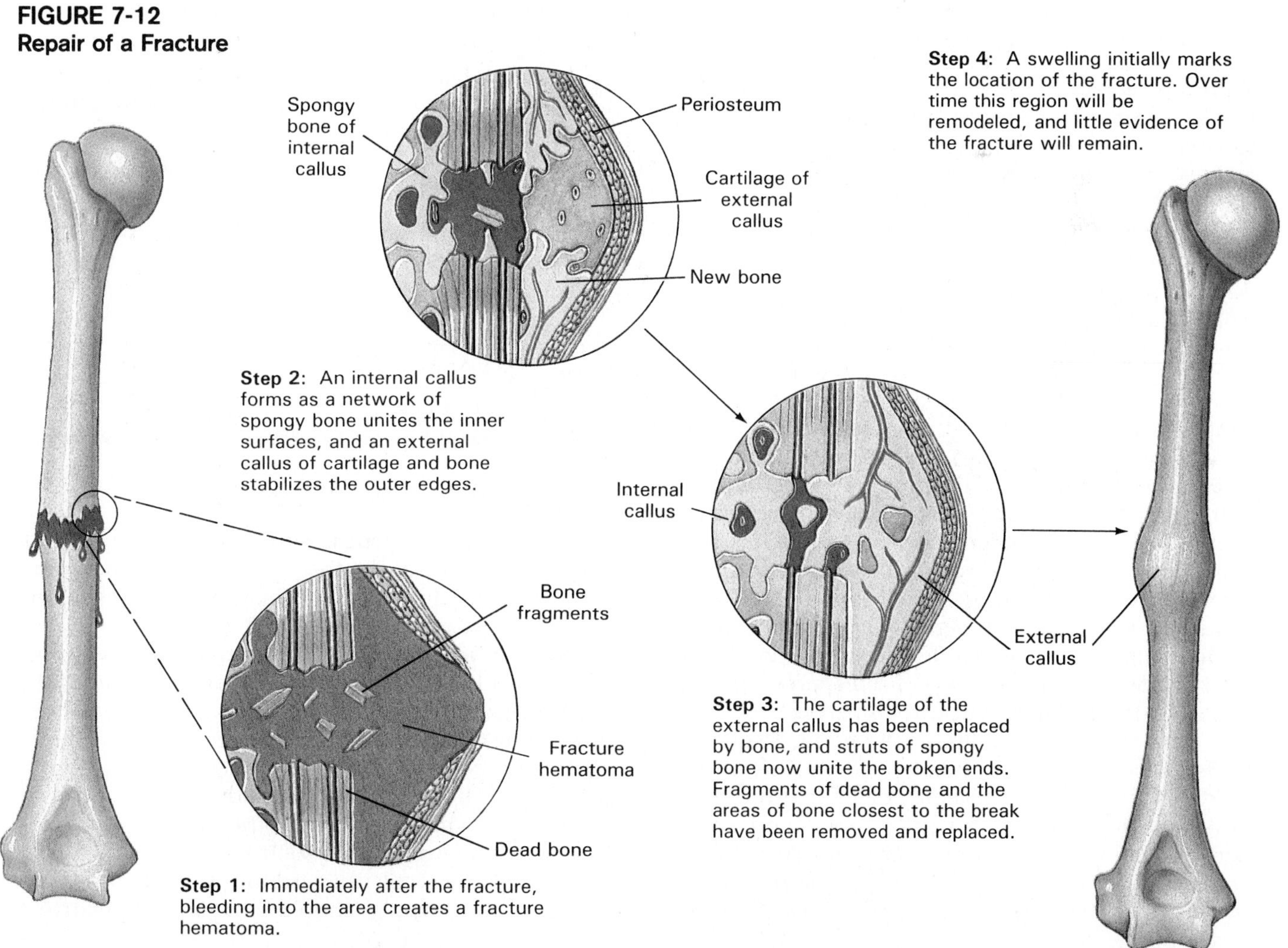

Spongy bone of internal callus
Periosteum
Cartilage of external callus
New bone
Step 4: A swelling initially marks the location of the fracture. Over time this region will be remodeled, and little evidence of the fracture will remain.
Step 2: An internal callus forms as a network of spongy bone unites the inner surfaces, and an external callus of cartilage and bone stabilizes the outer edges.
Internal callus
External callus
Bone fragments
Fracture hematoma
Dead bone
Step 1: Immediately after the fracture, bleeding into the area creates a fracture hematoma.
Step 3: The cartilage of the external callus has been replaced by bone, and struts of spongy bone now unite the broken ends. Fragments of dead bone and the areas of bone closest to the break have been removed and replaced.

STEP 4: Osteoclasts and osteoblasts then remodel the region for a period ranging from 4 months to well over a year. When the remodeling is complete, the fragments of dead bone and the trabecular bone of the calluses will be gone, and only living compact bone will remain. The repair may be "good as new," with no indications that a fracture ever occurred, but often the bone will be slightly thicker than normal at the fracture site.

Comparable events occur after more complex breaks or even after transplantation of a bone fragment. Roughly 200,000 people receive bone grafts each year in the United States to stimulate bone repair. The bone fragments are often taken from another part of the body, such as the hip. However, because the inserted bone is ultimately destroyed and replaced, dead and sterilized bone fragments from other donors or even other species of animals can be used to establish a framework for the repair process. [**CM**: *Stimulation of Bone Growth*]

AGING AND THE SKELETAL SYSTEM

The bones of the skeleton become thinner and relatively weaker as a normal part of the aging process. Inadequate ossification is called **osteopenia** (os-tē-ō-PĒ-nē-a; *penia*, lacking), and everyone becomes slightly osteopenic as they age. The reduction in bone mass occurs because between the ages of 30 and 40, osteoblast activity begins to decline while osteoclast activity continues at normal levels. Once the reduction begins, women lose roughly 8 percent of their skeletal mass every decade, whereas men's skeletons deteriorate at the slower rate of about 3 percent per decade. All parts of the skeleton are not equally affected. Epiphyses, vertebrae, and the jaws lose more than their fair share, resulting in fragile limbs, a reduction in height, and the loss of teeth.

Osteoporosis (os-tē-ō-por-Ō-sis; *porosus*, porous) is a condition that produces a reduction in bone mass sufficient to compromise normal function. The distinction between the "normal" osteopenia of aging and the clinical condition of osteoporosis is therefore a matter of degree. Current estimates indicate that 29 percent of women between the ages of 45 and 79 can be considered osteoporotic. The increase in incidence after menopause has been linked to decreases in the production of estrogens (female sex hormones). The incidence of osteoporosis in men of the same age is estimated at 18 percent.

The excessive fragility of the bones frequently leads to breakage, and subsequent healing is impaired. Vertebrae may collapse, distorting the vertebral articulations and putting pressure on spinal nerves. Therapies that boost estrogen levels, dietary changes to elevate calcium levels in the blood, and exercise that stresses bones and stimulates osteoblast activity appear to slow but not completely prevent the development of osteoporosis.

Osteoporosis can also develop as a secondary effect of many cancers. Cancers of the bone marrow, breast, or other tissues release a chemical known as **osteoclast-activating factor**. This compound increases both the number and activity of osteoclasts and produces a severe osteoporosis.

Infectious diseases that affect the skeletal system become more common in older individuals. In part this reflects the higher incidence of fractures, combined with slower healing and reduction in immune defenses noted in Chapter 5. (p. 162) **Osteomyelitis** (os-tē-ō-mī-e-LĪ-tis; *myelos*, marrow) is a painful infection of a bone most often caused by bacteria. This condition, most common in people over 50 years of age, can lead to dangerous systemic infections. A virus appears to be responsible for **Paget's disease**, also known as **osteitis deformans** (os-tē-Ī-tis de-FŌR-mans). This condition affects roughly 10 percent of the population over 70. Osteoclast activity accelerates, producing areas of acute osteoporosis, and osteoblasts produce abnormal matrix proteins. The result is a gradual deformation of the skeleton.

- √ **Would you expect to see any difference in the bones of a body builder before and after the addition of muscle mass? Why or why not?**
- √ **A person has a tumor of the parathyroid glands that causes oversecretion of PTH. What symptoms would you expect to result?**
- √ **It is rumored that as individuals age they gradually get shorter. Do you believe this? Why or why not?**

■ Anatomy of Skeletal Elements

CLASSIFICATION OF BONES

We can divide the bones of the body into six broad categories according to their individual shapes. You are already familiar with the basic appearance of the humerus, an example of a **long bone**. Long bones are also found in the forearm, lower leg, palms, soles, fingers, and toes. Figure 7-13 compares a long bone, the humerus (Figure 7-13a), with representatives of other bone types. **Short bones** (Figure 7-13b) are boxy in appearance. Their external surfaces are covered by compact bone, but the interior contains spongy bone. Short bones are found at the wrists and ankles. **Flat bones** (Figure 7-13c), such as the bones forming the roof of the skull, the sternum (breastbone), the ribs, and the scapula (shoulder blade), are relatively

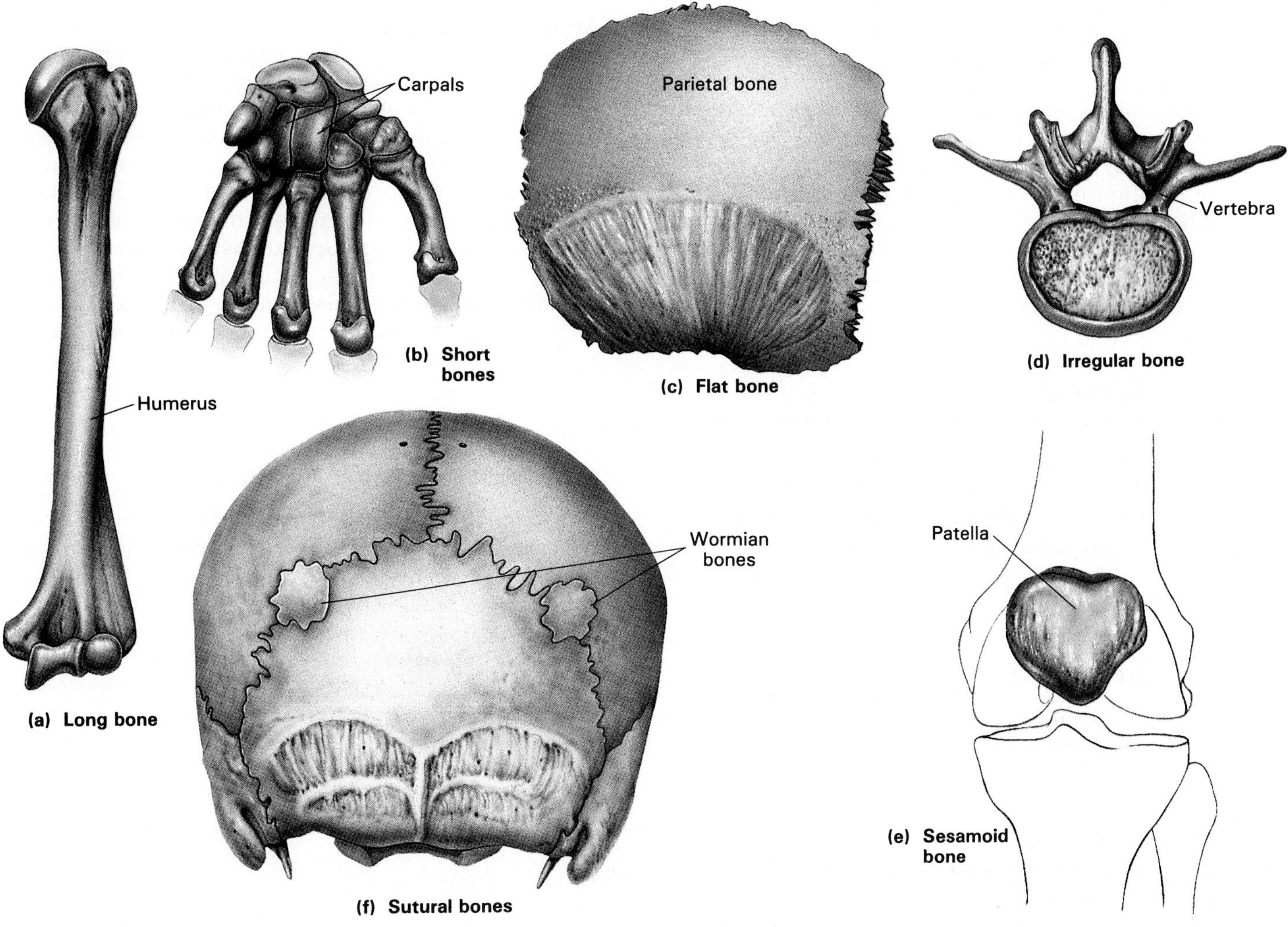

FIGURE 7-13
Shapes of Bones

thin. In structure they resemble a spongy bone sandwich; such bones are strong but relatively light. Flat bones provide protection for underlying soft tissues and offer an extensive surface area for the attachment of skeletal muscles.

Not all of the more than 206 bones of the human body fit into one of these three easily recognizable categories. **Irregular bones** (Figure 7-13d) have complex shapes with short, flat, notched, or ridged surfaces. Their internal structure is equally varied. The spinal vertebrae and several bones in the skull are examples of irregular bones. **Sesamoid bones** (Figure 7-13e) are usually small, round, and flat. They develop inside tendons and are most often encountered near joints at the knee, the hands, and the feet. Few individuals have sesamoid bones at every possible location, but everyone has sesamoid **patellae** (pa-TEL-lē), or kneecaps. **Sutural bones** (Figure 7-13f) are small, flat, oddly shaped bones found between the flat bones of the skull. There are individual variations in the number, shape, and position of the sutural bones. Sutural bones are also referred to as **Wormian bones**. Their borders are undeniably wild and wriggly, but, although this can jog your memory at exam time, it is *not* how the name originated. Instead, the name commemorates a Danish anatomist from the seventeenth century, Olaus Worm. [**CM**: *Heterotopic Bones*]

BONE MARKINGS

Each bone in the body has a distinctive shape and characteristic external and internal features. Elevations or projections form where tendons and ligaments attach and where adjacent bones articulate. Depressions and perforations indicate sites where blood vessels and nerves lie alongside or penetrate the bone. Detailed examination of these bone **markings** can yield an abundance of anatomical information. Anthropologists, criminologists, and pathologists can often determine the size, weight, sex, and general appearance of an individual on the basis of incomplete skeletal remains. (This topic will be discussed further in Chapter 9.)

We will ignore minor variations of individual bones and focus on prominent features that identify the bone. These markings are useful because they provide fixed landmarks that can help in determining the position of the soft tissue components of other systems. Specialized terms are used to describe the various bumps and dips. Before proceeding further it is important that you develop your skeletal vocabulary by studying Table 7-2.

Articulations

Joints, or **articulations**, exist wherever two bones meet. The function of each joint depends on its anatomy. Some joints permit extensive movement between the articulating elements, whereas others essentially interlock the bones to form a single, rigid structure.

CLASSIFICATION OF JOINTS

A convenient functional classification is based on the range of motion permitted. Three major categories of joints are recognized, as indicated in Table 7-3. An immovable joint is a **synarthrosis** (sin-ar-THRŌ-sis; *syn*, together + *arthros*, joint); a slightly movable joint is an **amphiarthrosis** (am-fē-ar-THRŌ-sis; *amphi*, on both sides); and a freely movable joint is a **diarthrosis** (dī-ar-THRŌ-sis; *dia*, through). Subdivisions within each major category indicate significant structural differences. Synarthrotic or amphiarthrotic joints are classified as fibrous or cartilaginous, and diarthrotic joints are subdivided according to the degree of movement permitted. An alternative classification scheme, based on the structure of joints (bony fusion, fibrous, cartilaginous, synovial) is presented in Table 7-4.

Immovable Joints (Synarthroses)

At a synarthrosis the bony edges are quite close together and may even interlock. A **suture** (*sutura*, a sewing together) is a synarthrotic joint between the bones of the skull. The edges of the bones are interlocked and bound together by dense connective tissue. Another synarthrosis binds each tooth to the surrounding bony socket. This fibrous connection is termed a **periodontal** (pe-rē-ō-DON-tal; *peri*, around + *odont*, tooth) **ligament**, and the articulation

TABLE 7-2 An Introduction to Skeletal Terminology

General Description	*Anatomical Term*	*Definition*	*Example*
Elevations and projections (general)	Process	Any projection or bump	See Figure 8-12
	Ramus	An extension of a bone making an angle to the rest of the structure	
Processes formed where tendons or ligaments attach	Trochanter	A large, rough projection	See Figures 9-13 and 9-16
	Tuberosity	A smaller rough projection	
	Tubercle	A small, rounded projection	
	Crest	A prominent ridge	
	Line	A low ridge	
Processes formed for articulation with adjacent bones	Head	The expanded articular end of an epiphysis, separated from the shaft by a narrower neck	See Figure 7-6
	Condyle	A smooth, rounded or articular process	
	Trochlea	A smooth, grooved articular process shaped like a pulley	
	Facet	A small, flat articular surface	
	Spine	A pointed process	
Depressions	Fossa	A shallow depression	See Figure 10-2
	Sulcus	A narrow groove	
Openings	Foramen	A rounded passageway for blood vessels and/ or nerves	See Figure 8-3
	Fissure	An elongate cleft	
	Meatus	A canal leading through the substance of a bone	
	Sinus or antrum	A chamber within a bone, normally filled with air	

TABLE 7-3 A Functional Classification of Articulations

Functional Category	*Structural Types*	*Description*	*Example*
SYNARTHROSIS (no movement)	**Fibrous**		
	Suture	Fibrous connections plus interdigitation	Between the bones of the skull
	Gomphosis	Fibrous connections plus insertion in alveolus	Between the teeth and jaws
	Cartilaginous		
	Synchondrosis	Interposition of cartilage plate	Epiphyseal plates
	Bony fusion		
	Synostosis	Conversion of other articular form to solid mass of bone	Portions of the skull
AMPHIARTHROSIS (little movement)	**Fibrous**		
	Syndesmosis	Ligamentous connection	Between the bones of the lower leg
	Cartilaginous		
	Symphysis	Connection by a fibrocartilage pad	Between right and left halves of pelvis; between adjacent vertebrae of spinal column
DIARTHROSIS (free movement)	**Synovial**	Complex joint bounded by joint capsule and containing synovial fluid	Numerous, subdivided by range of movement (see Figure 7–17)
	Monaxial	Permits movement in one plane	Elbow, ankle
	Biaxial	Permits movement in two planes	Ribs, wrist
	Triaxial	Permits movement in all three planes	Shoulder, hip

is a **gomphosis** (gom-FŌ-sis; *gomphosis*, a bolting together).

An epiphyseal plate also represents an articulation between two bones, although the two are part of the same skeletal element. Such a rigid, cartilaginous connection characterizes a **synchondrosis** (sin-kon-DRŌ-sis; *syn*, together + *chondros*, cartilage). Sometimes two separate bones actually fuse together, and the boundary between them disappears. This creates a **synostosis** (sin-os-TŌ-sis), surely the ultimate in immovable joints.

TABLE 7-4 A Structural Classification of Articulations

Structure	*Type*	*Functional Category*
Bony fusion	Synostosis	Synarthrosis
Fibrous joint	Suture	Synarthrosis
	Gomphosis	Synarthrosis
	Syndesmosis	Amphiarthrosis
Cartilaginous joint	Synchondrosis	Synarthrosis
	Symphysis	Amphiarthrosis
Synovial joint	Monaxial, biaxial, triaxial	All diarthroses

For examples see Table 7-3.

Slightly Movable Joints (Amphiarthroses)

An amphiarthrosis permits very limited movement, and the bones are usually farther apart than they are at a synarthrosis. The bones may be connected by collagen fibers or cartilage. At a **syndesmosis** (sin-dez-MŌ-sis; *desmo*, band or ligament) they are connected by a ligament. Examples include the articulations between the two bones of the lower leg, the tibia and fibula. At a **symphysis** the bones are separated by a broad disc of fibrocartilage. The articulations between the spinal vertebrae and the anterior connection between the two pelvic bones, or coxae, are examples of symphyses.

Freely Movable Joints (Diarthroses)

Diarthroses, or **synovial** (si-NŌ-vē-al) **joints**, permit a wide range of motion. The basic structure of a synovial joint was introduced in Chapter 5 during our discussion of synovial membranes. Figure 7-14a details the structure of a representative synovial joint.

Synovial joints are typically found at the ends of long bones, such as those of the arms and legs. Under normal conditions the bony surfaces do not contact one another, for they are covered with special **articular cartilages**. The joint is surrounded by a fibrous **joint capsule,** and the inner surfaces of the joint cavity are lined with a synovial membrane, described in Chapter 5. (p. 155) **Synovial fluid** diffuses across the synovial membrane and provides additional lubrication.

FIGURE 7-14
Structure of a Synovial Joint. (a) Diagrammatic view of a simple articulation. (b) A sectional view of the knee joint.

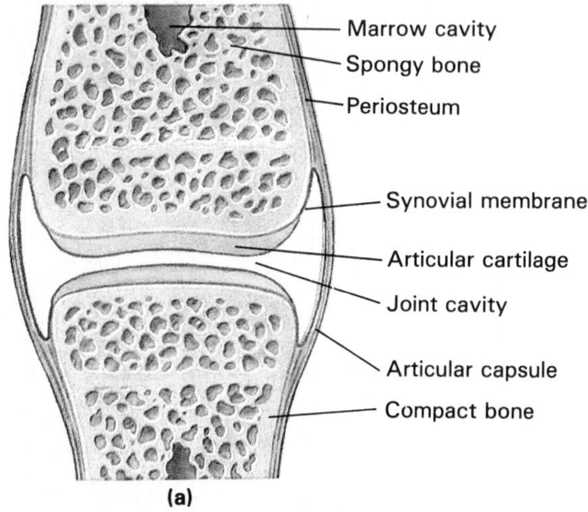

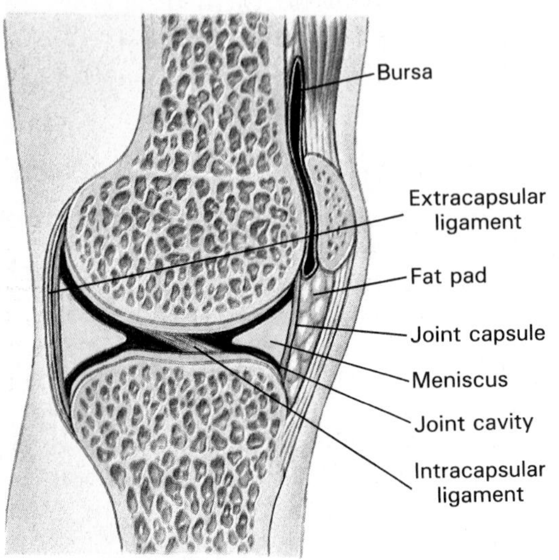

Articular cartilages differ from other hyaline cartilages, such as those of the nose or rib cage. There is no perichondrium, and the matrix is bathed by the synovial fluid of the joint cavity. The matrix also contains much more water than other cartilages. These cartilages act like sponges. When a portion of an articular cartilage is compressed, some of the water is squeezed out of the cartilage and into the space between the opposing surfaces. This thin layer of fluid reduces the friction between moving surfaces in a joint to around one-fifth of that between two pieces of ice. When the compression stops, fluid is sucked back into the cartilage.

In addition to providing lubrication, the synovial fluid nourishes the cartilage cells. The total quantity of fluid is less than 3 mℓ even in a large joint such as the knee, so the synovial fluid must be continually circulated to provide nutrients and a route for waste disposal. Movement of the joint keeps the fluid circulating, and the compression and expansion of the articular cartilages pump the fluid in and out of the matrix.

In complex joints such as the knee (Figure 7-14b), isolated fibrocartilage pads lie between the opposing articular surfaces. These pads may be of several types. **Menisci** (men-IS-kē; *meniscus*, crescent), or **articular discs**, may subdivide a synovial cavity, channel the flow of synovial fluid, or allow for variations in the shapes of the articular surfaces. **Fat pads** are often found around the edges of the joint, lightly covered by a layer of synovial membrane. Fat pads provide protection for the articular cartilages. They also act as packing material for the joint; when the bones move, the fat pads fill in the spaces created as the joint cavity changes shape.

The joint capsule that surrounds the entire joint is continuous with the periostea of the articulating bones. **Accessory ligaments** are localized thickenings of the capsule. **Extracapsular ligaments** can be seen on the outside of the capsule; **intracapsular ligaments** are found inside the capsule. Where a tendon or ligament rubs against other tissues, small pockets of synovial fluid form to reduce friction and act as a shock absorber. These **bursae** are characteristic of many synovial joints. They may also appear around tendon sheaths, beneath the skin covering a bone, or within other connective tissues exposed to friction or pressure.

A joint cannot be both highly mobile and very strong; the greater the range of motion at a joint, the weaker it becomes. A synarthrosis, the strongest type of joint, does not permit any movement. A very mobile diarthrosis, such as the shoulder joint, still has a limited range of movement, and exceeding that range damages the joint. When a **dislocation**, or **luxation** (luks-Ā-shun), occurs, the articulating surfaces

Clinical Comment: Rheumatism, Arthritis, and Synovial Function

Proper synovial function depends on healthy articular cartilages. When an articular cartilage has been damaged, the matrix begins to break down, and the exposed cartilage changes from a slick, smooth gliding surface to a rough feltwork of bristly collagen fibers. This feltwork drastically increases friction, damaging the cartilage further. Eventually the central area of the articular cartilage may completely disappear, exposing the underlying bone.

Fibroblasts are attracted to areas of friction, and they begin tying the opposing bones together with a network of collagen fibers. This network may later be converted to bone, locking the articulating elements into position. Such a bony fusion, called **ankylosis** (an-ke-LŌ-sis), eliminates the friction, but only by the drastic remedy of making movement impossible.

Degenerative changes may be caused by simply immobilizing a joint. When motion ceases, so does circulation of synovial fluid, and the cartilages begin to suffer. **Continuous passive motion (CPM)** of any injured joint appears to encourage the repair process by improving the circulation of synovial fluid. The movement is often performed by a physical therapist during the recovery process.

Rheumatism (ROO-ma-tizm) is a general term that indicates pain and stiffness affecting the skeletal system, the muscular system, or both. There are several major forms of rheumatism. **Arthritis** (ar-THRĪ-tis) includes all of the rheumatic diseases that affect synovial joints. Arthritis always involves damage to the articular cartilages, but the specific cause may vary. For example, arthritis can result from bacterial or viral infection, injury to the joint, metabolic problems, or severe physical stresses.

The diseases of arthritis are usually considered as either **degenerative** or **inflammatory** in nature. Degenerative diseases begin at the articular cartilages, and modification of the underlying bone and inflammation of the joint occur secondarily. Inflammatory diseases start with the inflammation of synovial tissues, and damage later spreads to the articular surfaces. We will consider a single example of each type.

Osteoarthritis (os-tē-ō-ar-THRĪ-tis), also known as **degenerative arthritis**, or **degenerative joint disease (DJD)**, usually affects older individuals. In the U.S. population 25 percent of women and 15 percent of men over 60 years of age show signs of this disease. The condition seems to result from cumulative wear and tear on the joint surfaces. Some individuals, how-

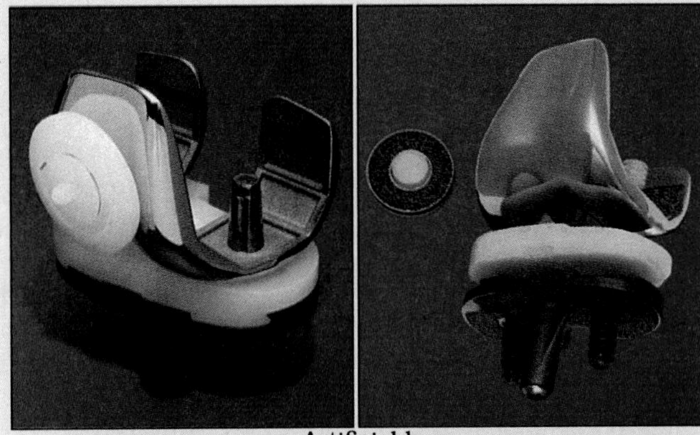

Artificial knee

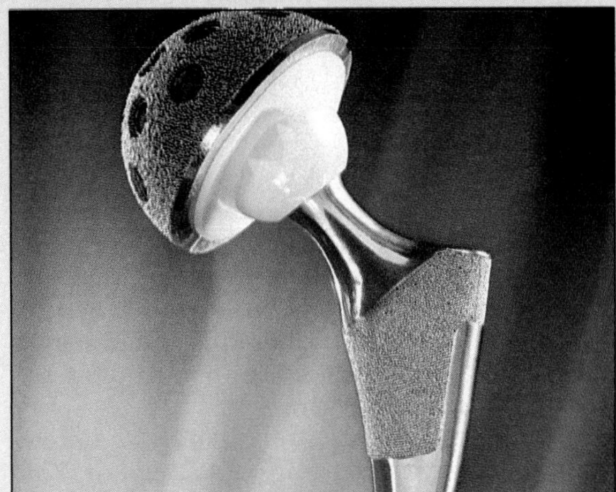

Artificial hip

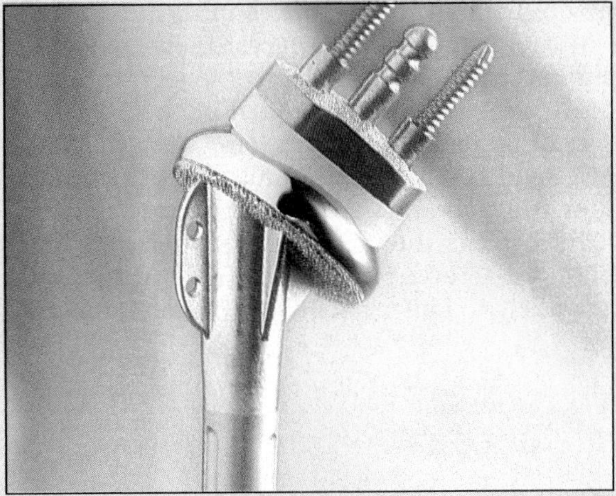

Artificial shoulder

FIGURE 7-15
Examples of Artificial Joints

ever, may have a genetic predisposition to develop osteoarthritis, for researchers have recently isolated a gene linked to the disease. This gene codes for an abnormal form of collagen that differs from the normal protein in only 1 of its 1000 amino acids.

Rheumatoid arthritis is an inflammatory condition that affects roughly 2.5 percent of the adult population. The cause is uncertain, although allergies, bacteria, viruses, and genetic factors have all been proposed. The synovial membrane becomes swollen and inflamed, a condition known as **synovitis** (sīn-ō-VĪ-tis). The cartilaginous matrix begins to break down, and the process accelerates as dying cartilage cells release lysosomal enzymes.

In their later stages, inflammatory and degenerative forms of arthritis produce an inflammation that spreads into the surrounding area. Ankylosis, common in the past when complete rest was routinely prescribed for arthritis patients, is rarely seen today. Regular exercise, physical therapy, and drugs that reduce inflammation, such as aspirin, can slow the progress of the disease. Surgical procedures can realign or redesign the affected joint, and in extreme cases involving the hip, knee, elbow, or shoulder the defective joint can be replaced by an artificial one. Joint replacement has the advantage of eliminating the pain and restoring full range of motion. Prosthetic (artificial) joints, such as those shown in Figure 7-15, are weaker than natural ones, but elderly people seldom stress them to their limits.

are forced out of position. This displacement can damage the articular cartilages, tear ligaments, or distort the joint capsule. Although the inside of a joint has no pain receptors, the nerves that monitor the capsule, ligaments, and tendons are quite sensitive, and dislocations are very painful. The damage accompanying a partial dislocation, or **subluxation** (sub-luks-Ā-shun), is less severe.

Dislocations are usually prevented by anatomical features that restrict the range of movement. For example, the accessory ligaments and the joint capsule provide stabilization and support, and the shapes of the articulating surfaces or the presence of other bones can prevent movement in certain directions. Movement can also be limited by the presence of muscles or fat pads around the joint. Finally, skeletal muscles can stabilize or limit movement at a joint by pulling on tendons. People who are "double-jointed" have joints that are weakly stabilized. Their joints permit a greater range of motion, but they are also more likely to suffer partial or complete dislocations.

ARTICULAR FORM AND FUNCTION

To *understand* human movement you must become aware of the relationship between form and function at each articulation. To *describe* human movement you need a frame of reference that permits accurate and precise communication. The synovial joints can be classified according to their anatomical and functional properties. To demonstrate the basis for that classification, we will describe the movements that can occur at a typical synovial joint, using a simplified model.

Planes of Motion

Take a pencil (or pen) as your model and stand it upright on the surface of a desk or table, as indicated in Figure 7-16a. The pencil represents a bone, and the desk is an articular surface. A little imagination and a lot of twisting, pushing, and pulling will demonstrate that there are only three ways to move the model. Considering them one at a time will provide a frame of reference for analyzing any complex movement.

Possible movement 1: The point can move

If you hold the pencil upright but do not secure the point, you can push the pencil across the surface. You could slide it forward or backward, or from one side to the other. This kind of motion is called **gliding** (Figure 7-16b). In a gliding movement the direction of motion can be described using two lines of reference. In this case one line, or **axis**, represents forward/backward motion, and the other movement left/right. For example, a specific move might be described as "forward and to the left" or even "backward 1 cm and to the right 2.5 cm."

Possible movement 2: The shaft can change its angle with the surface

With the tip held in position, you can still move the free (eraser) end forward and backward or from side to side. These movements differ from gliding because they change the angle between the shaft of the bone and the articular surface. Such movements are examples of **angular motion** (Figure 7-16c).

Any angular motion can be described with a pair of axes, but in one instance a special term is useful. Grasp the free end of the pencil and move it in any direction, so that the shaft is no longer vertical. Now move that end through a complete circle, and

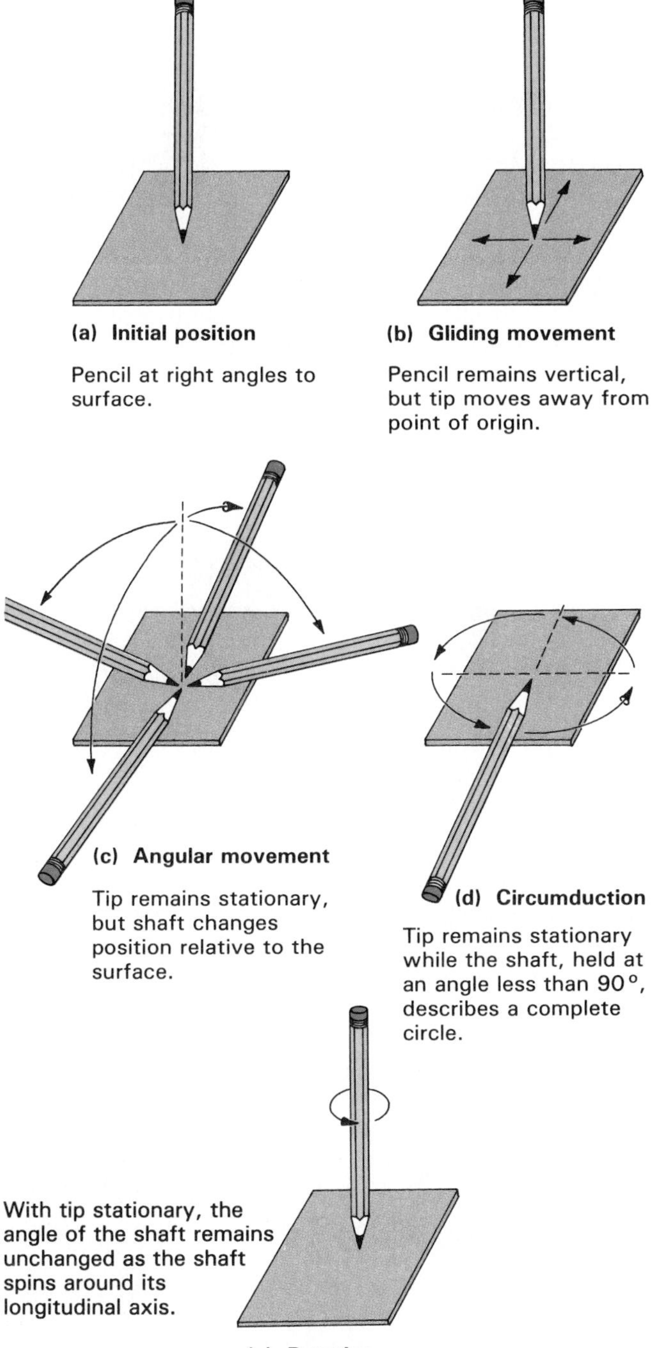

(a) Initial position
Pencil at right angles to surface.
(b) Gliding movement
Pencil remains vertical, but tip moves away from point of origin.
(c) Angular movement
Tip remains stationary, but shaft changes position relative to the surface.
(d) Circumduction
Tip remains stationary while the shaft, held at an angle less than 90°, describes a complete circle.
With tip stationary, the angle of the shaft remains unchanged as the shaft spins around its longitudinal axis.
(e) Rotation

FIGURE 7-16
A Simple Model of Articular Motion

try to describe the movement in terms of forward or backward and right or left. The description soon becomes rather complicated and difficult to follow. Anatomists avoid the problem by using a special term, **circumduction** (sir-kum-DUK-shun; *circum*, around), for the movement in Figure 7-16d.

Possible movement 3: The shaft can rotate

If you prevent movement of the base and keep the shaft vertical, you can still spin the shaft around. This movement is called **rotation** (Figure 7-16e). Several articulations will permit partial rotation, but none can rotate freely; such a movement would hopelessly tangle the blood vessels, nerves, and muscles that cross the joint.

Reviewing these possibilities, any gliding or angular motion can be described using the two axes we defined as "forward/backward" and "side/side." Rotation provides the third axis. Any specific joint may permit movement in one or more of these axes, and this forms the basis for a functional classification of diarthrotic joints. If an articulation permits movement that is limited to only one axis, it is called **monaxial** (mon-AKS-ē-al). If movement involving two axes can occur, the articulation is **biaxial** (bī-AKS-ē-al). The most mobile joints permit movement involving all three axes, so they are called **triaxial** (trī-AKS-ē-al). Each functional category can be subdivided on the basis of anatomical structure (Figure 7-17).

A Classification of Synovial Joints

1. **Monaxial joints:** The elbow and knee are called **hinge joints** (Figure 7-17a) because they permit angular movement in a single plane, like the opening and closing of a door. **Pivot joints** (Figure 7-17b) are also monaxial, but they permit only rotation. A pivot joint between the first two cervical vertebrae allows you to turn your head to either side.
2. **Biaxial joints: Gliding joints** (Figure 7-17c) have rather flattened faces, and the articular surfaces are able to slide across one another. Although rotation is theoretically possible at such a joint, ligaments usually prevent or restrict such movement. Gliding joints are found at the ends of the clavicles (collarbones), between vertebrae, and between the bones of the wrists and ankles.

   Because the articulating surfaces are flat, gliding involves little angular motion. In an **ellipsoidal joint** (Figure 7-17d) an oval articular face nestles within a depression on the opposing surface. With such an arrangement, angular motion occurs, but both gliding and rotation are prevented. In fact, angular movement is restricted to two directions, along or across the length of the oval. Ellipsoidal joints connect the fingers and toes with the bones of the palms and soles.

   Saddle joints (Figure 7-17e) have complex articular faces. Each one resembles a saddle because it is concave on one axis and convex on the other. Saddle joints are extremely mobile, allowing both gliding and angular motion. Moving the saddle joint at the base of your thumb is an excellent demonstration that also provides an excuse for twiddling your thumbs during a lecture.

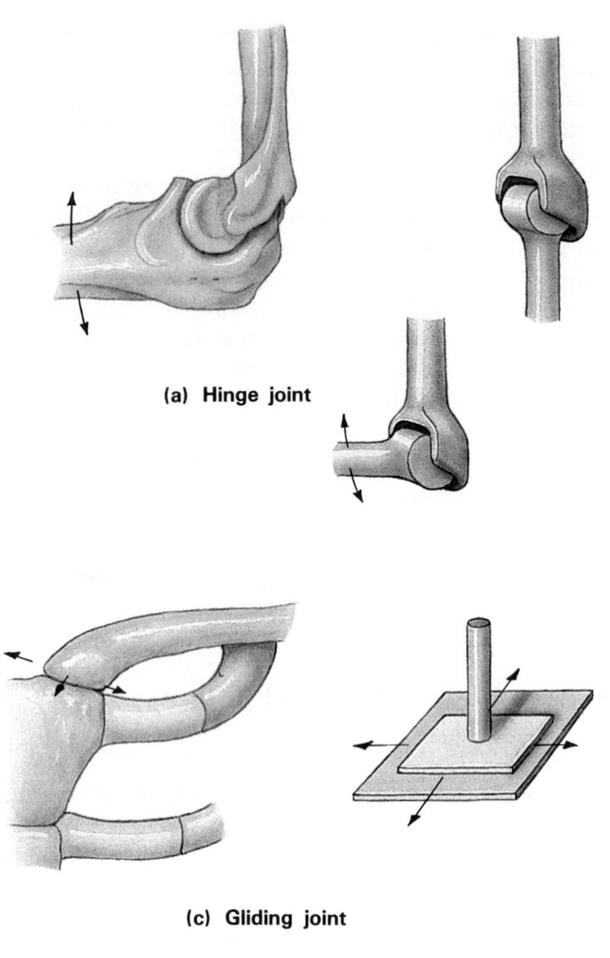

(a) Hinge joint

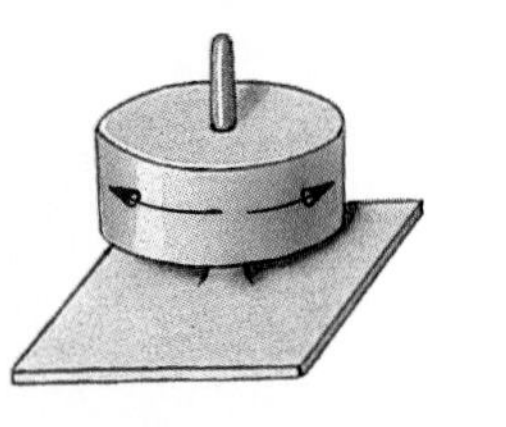

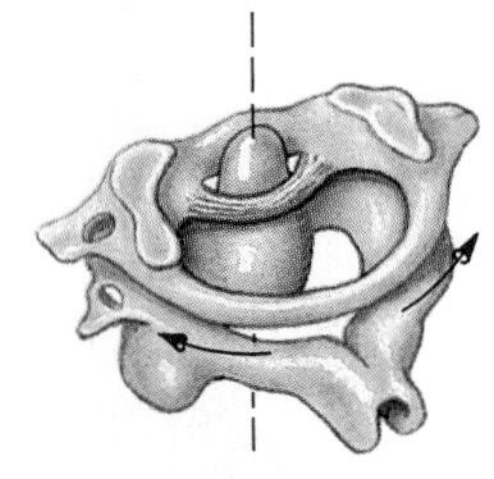

(b) Pivot joint

(c) Gliding joint

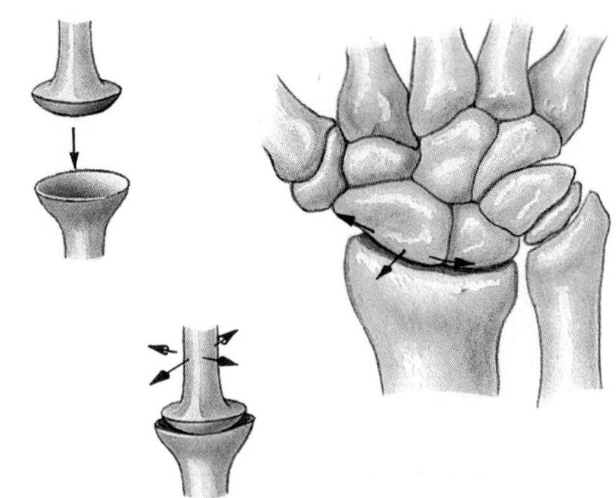

(d) Ellipsoidal joint

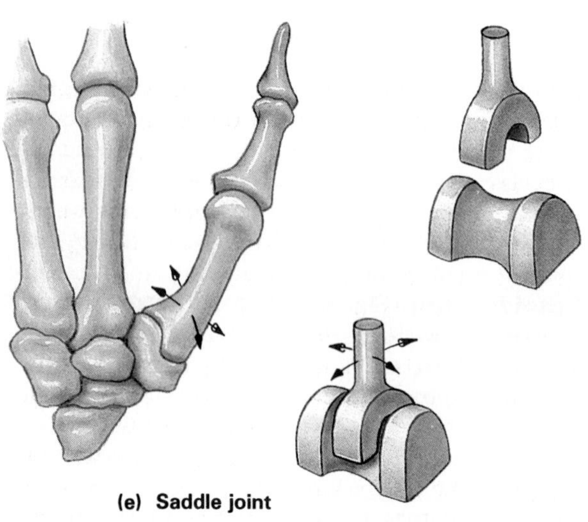

(e) Saddle joint

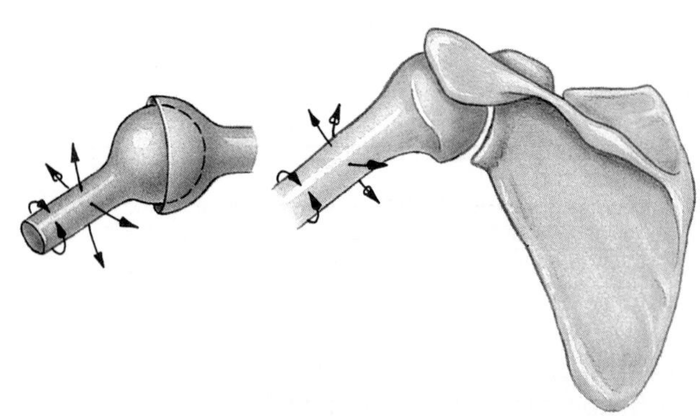

(f) Ball–and–socket joint

FIGURE 7-17
A Functional Classification of Synovial Joints

3. **Triaxial joints:** Triaxial articulations are known as **ball-and-socket joints** (Figure 7-17f), for the round head of one bone rests within a cup-shaped depression. All combinations of movements, including rotation, can be performed at the ball-and-socket joints of the shoulders and hips.

DESCRIBING DYNAMIC MOTION

When considering motion at triaxial joints, phrases such as "bends the leg" or "raises the arm" are not very informative. You need a reliable frame of reference and descriptive terms that have specific meanings. Let's start with a figure standing in the anatomi-

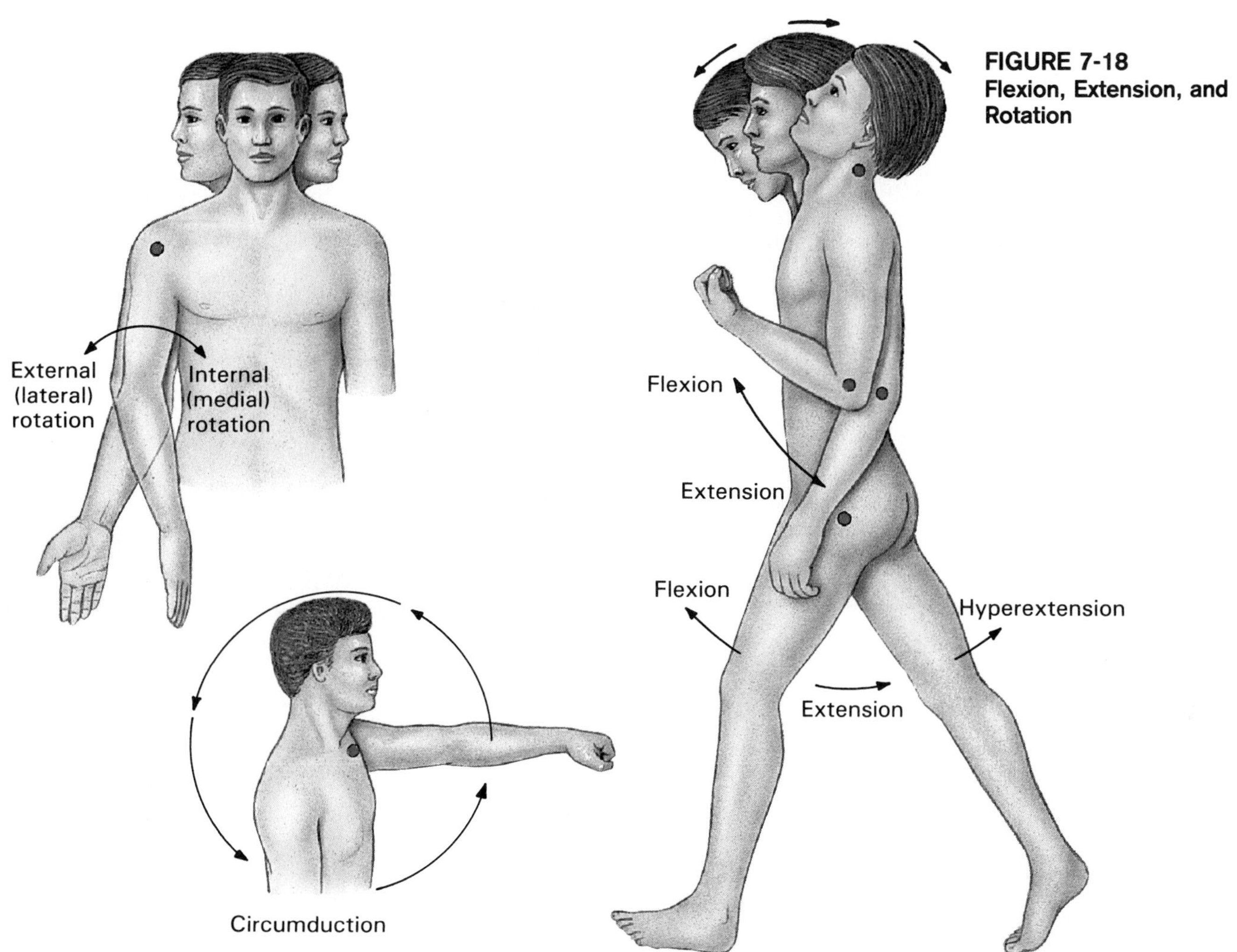

FIGURE 7-18
Flexion, Extension, and Rotation

cal position, and follow Figures 7-18 to 7-20. **Flexion** (FLEK-shun) can be defined as movement that *reduces the angle between articulating elements.* **Extension** *increases the angle between articulating elements.* These extremely useful terms can be applied to articulations throughout the body. When you bring your head toward your chest, you flex the head. When you bend down to touch your toes, you flex the spine. Extension reverses these movements. Examples of flexion and extension are presented in Figure 7-18.

Flexion at the shoulder or hip moves the limbs forward, whereas extension moves them back. Flexion of the wrist moves the palm forward, and extension moves it back. In each of these examples extension can be continued past the anatomical position, in which case **hyperextension** occurs. You can also hyperextend the head, a movement that allows you to gaze at the ceiling. Hyperextension of other joints is usually prevented by ligaments, bony processes, or soft tissues. Hyperextension of the hinge joints at the elbow and knee sometimes occurs in an injury, with painful and damaging results.

Rotational movements of the head can be adequately described in terms of left or right **rotation**. When analyzing movements of the limbs, if the anterior aspect of the limb rotates **inward**, toward the ventral surface of the body, you have **internal**, or **medial**, **rotation**. If it turns outward, you have **external**, or **lateral**, **rotation**. Moving your arm in a loop, as when drawing a large circle on a chalkboard, provides an example of **circumduction**. These rotational movements are also illustrated in Figure 7-18.

Swinging the distal end of the upper arm to the side, parallel to the frontal plane, represents an example of **abduction** (*ab*, from). Moving it medially constitutes **adduction** (*ad*, to). Abduction and adduction refer to motion relative to the midline of the body, so these terms always refer to movements of the appendicular skeleton. Figure 7-19 provides examples of abduction and adduction.

There are a number of special terms and interpretations that apply to specific articulations or unusual types of movement. Several are concerned with movements of the hands and feet, and these are also diagrammed in Figure 7-19. Adduction of the wrist moves the heel of the hand *toward* the body, whereas abduction moves it farther *away*. Comparable movements at the ankle are extremely difficult, but you can turn the sole inward to some degree. This motion

FIGURE 7-19
Other Movements of the Extremities

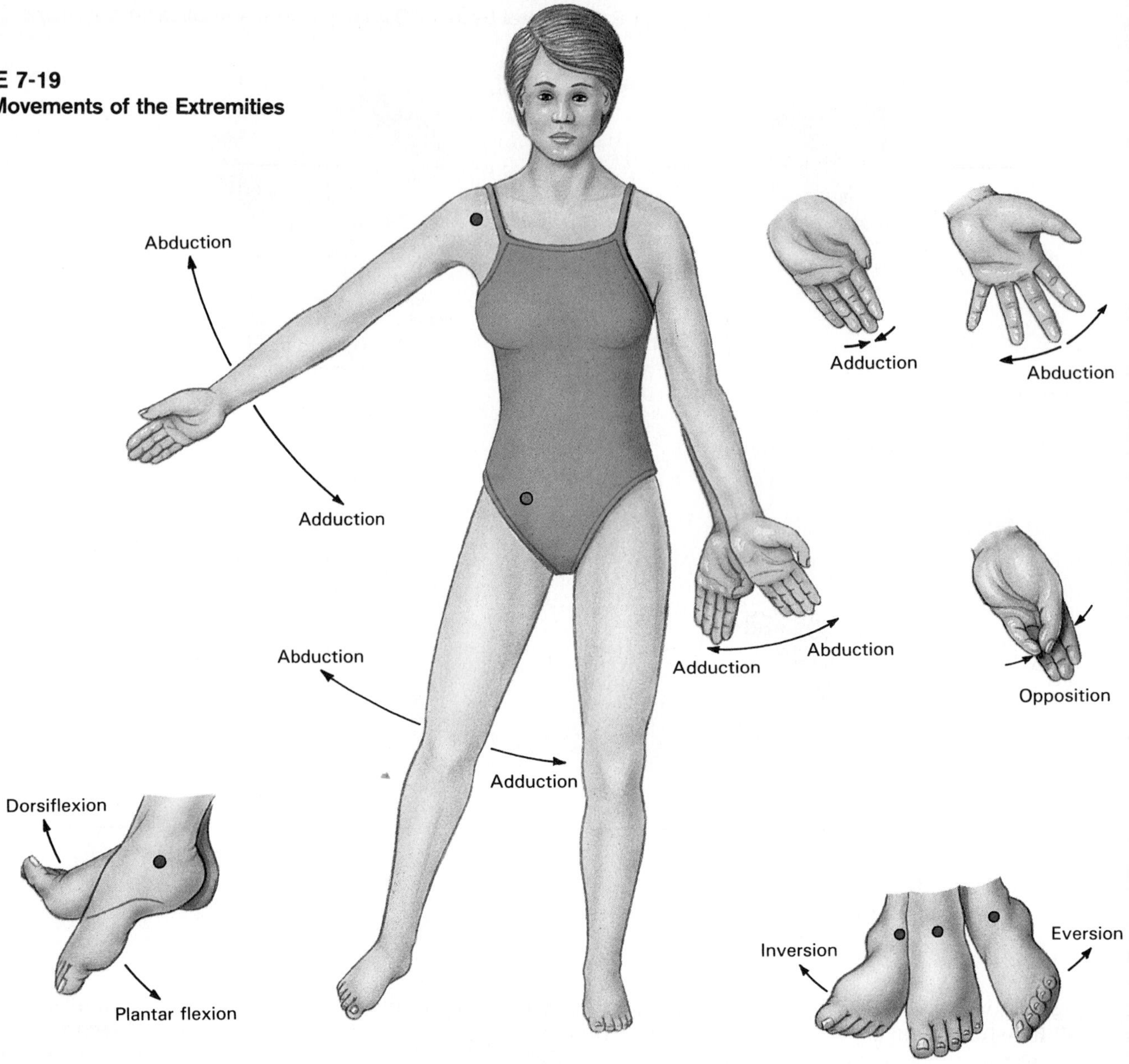

FIGURE 7-20
Additional Skeletal Movements

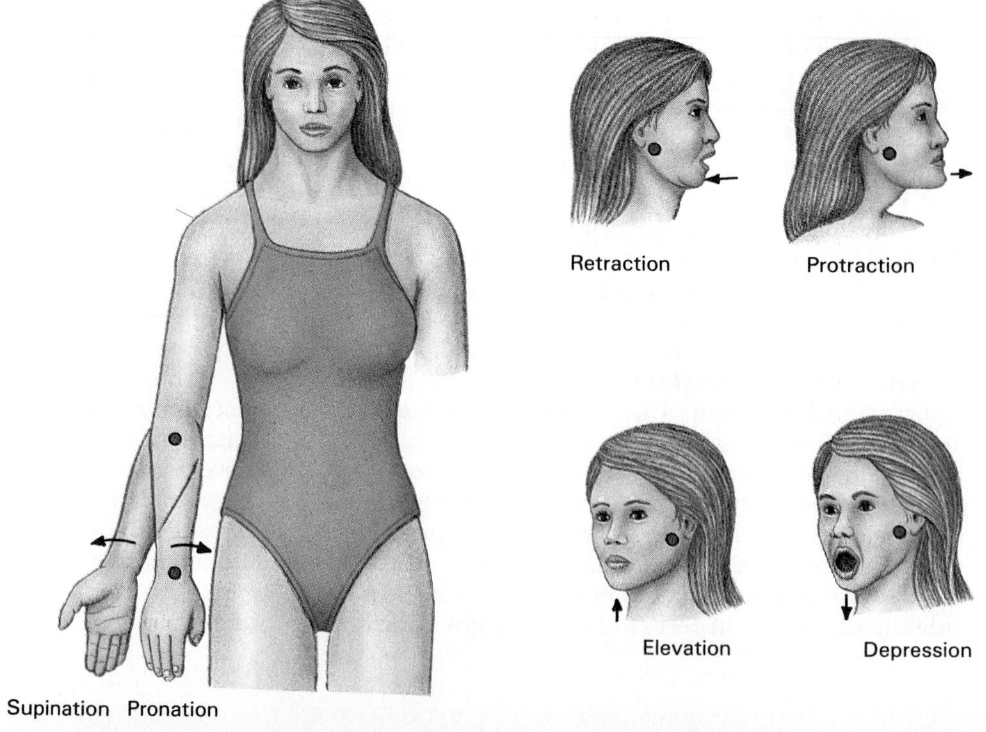

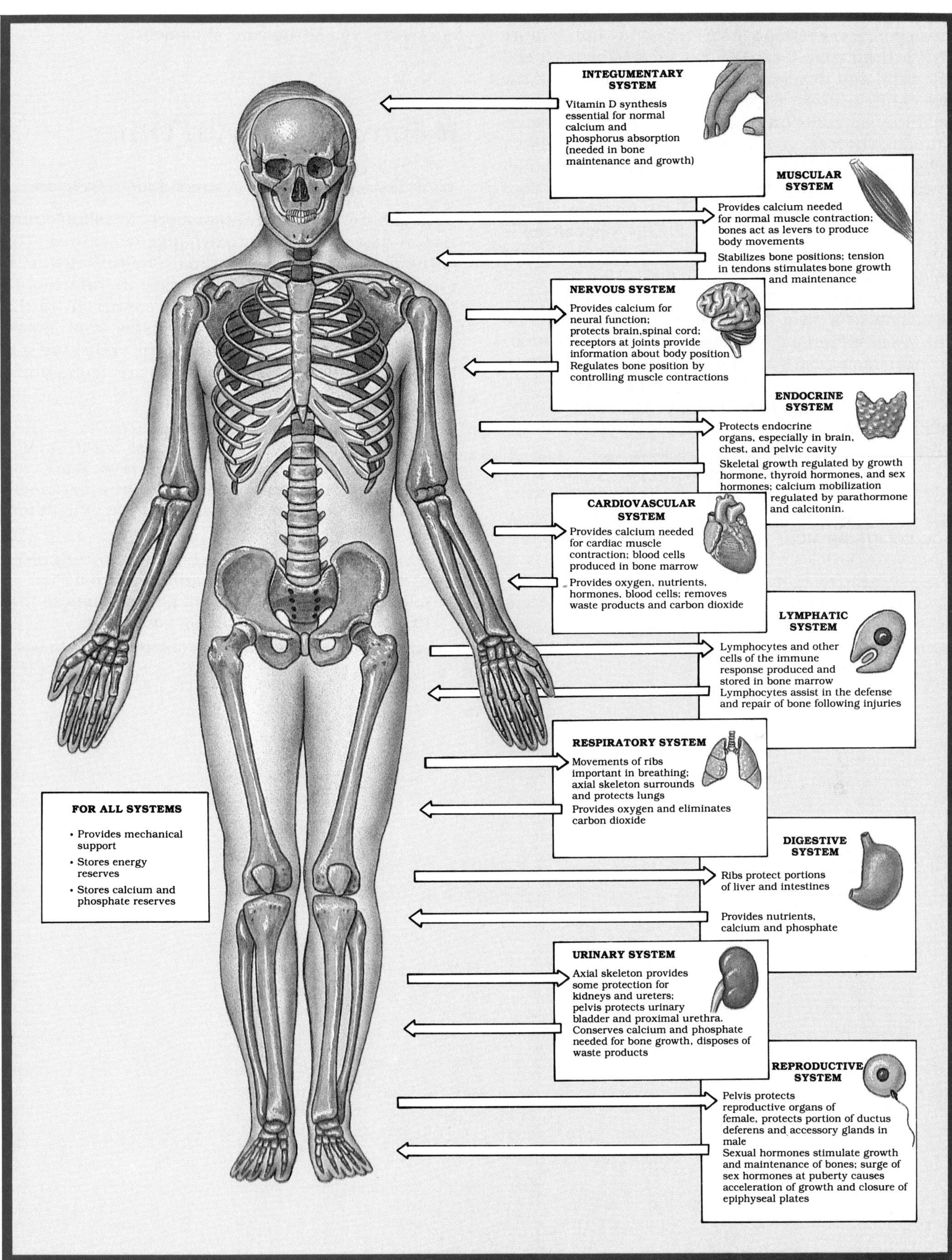

INTEGUMENTARY SYSTEM
Vitamin D synthesis essential for normal calcium and phosphorus absorption (needed in bone maintenance and growth)
MUSCULAR SYSTEM
Provides calcium needed for normal muscle contraction; bones act as levers to produce body movements
Stabilizes bone positions; tension in tendons stimulates bone growth and maintenance
NERVOUS SYSTEM
Provides calcium for neural function; protects brain,spinal cord; receptors at joints provide information about body position
Regulates bone position by controlling muscle contractions
ENDOCRINE SYSTEM
Protects endocrine organs, especially in brain, chest, and pelvic cavity
Skeletal growth regulated by growth hormone, thyroid hormones, and sex hormones; calcium mobilization regulated by parathormone and calcitonin.
CARDIOVASCULAR SYSTEM
Provides calcium needed for cardiac muscle contraction; blood cells produced in bone marrow
Provides oxygen, nutrients, hormones, blood cells; removes waste products and carbon dioxide
LYMPHATIC SYSTEM
Lymphocytes and other cells of the immune response produced and stored in bone marrow
Lymphocytes assist in the defense and repair of bone following injuries
RESPIRATORY SYSTEM
Movements of ribs important in breathing; axial skeleton surrounds and protects lungs
Provides oxygen and eliminates carbon dioxide
FOR ALL SYSTEMS
• Provides mechanical support
• Stores energy reserves
• Stores calcium and phosphate reserves
DIGESTIVE SYSTEM
Ribs protect portions of liver and intestines
Provides nutrients, calcium and phosphate
URINARY SYSTEM
Axial skeleton provides some protection for kidneys and ureters; pelvis protects urinary bladder and proximal urethra.
Conserves calcium and phosphate needed for bone growth, disposes of waste products
REPRODUCTIVE SYSTEM
Pelvis protects reproductive organs of female, protects portion of ductus deferens and accessory glands in male
Sexual hormones stimulate growth and maintenance of bones; surge of sex hormones at puberty causes acceleration of growth and closure of epiphyseal plates

FIGURE 7-21
Functional Relationships between the Skeletal System and Other Systems

is termed **inversion** (*in*, into + *vertere*, to turn), and its opposite is **eversion** (ē-VER-shun; *e*-, out) (Figure 7-19). Although you can flex, extend, and hyperextend the palm and fingers, comparable movements of the sole and toes are called **dorsiflexion** and **plantar** (*planta*, sole) **flexion** (Figure 7-19). For example, to stand on tiptoes, you must perform plantar flexion. When you spread your fingers or toes apart you abduct them, because they move *away* from a central digit (finger or toe). Bringing them back together constitutes adduction (Figure 7-19). Finally, **opposition** is the special movement of the thumb that enables it to grasp and hold an object (Figure 7-19).

The bones of the forearm—the radius and ulna—articulate at a pivot joint that permits you to rotate the forearm from palm-facing-front (supine position) to palm-facing-back (prone position). This motion, diagrammed in Figure 7-20, is called **pronation** (prō-NĀ-shun). The opposing movement, in which the palm is turned forward, is **supination** (sū-pi-NĀ-shun).

Protraction entails moving a part of the body forward in the horizontal plane. **Retraction** occurs when you move it back. You protract your jaw when you grasp your upper lip in your teeth, and you protract your clavicles when you cross your arms. Figure 7-20 illustrates examples of protraction and retraction. **Elevation** and **depression** (Figure 7-20) occur when a structure moves up and down. You depress your lower jaw when you open your mouth, and elevate it as you close it. Another familiar elevation occurs when you shrug your shoulders.

Integration with Other Systems

Although the bones may seem inert, you should now realize that they are quite dynamic structures. The entire skeletal system is intimately associated with other systems. For example, bones are attached to the muscular system, extensively connected to the cardiovascular and lymphatic systems, and largely under the physiological control of the endocrine system. These functional relationships are diagrammed in Figure 7-21.

√ **In a newborn infant the large bones of the skull are joined by fibrous connective tissue. What type of joint is this? These bones later grow, interlock, and form immovable joints. What type of joint are these?**

√ **Give the proper term for each of the following types of motion: (a) moving your arm away from the midline of the body, (b) turning your palms so that they face forward, (c) bending your elbow.**

CHAPTER REVIEW

Review of Selected Clinical Terms

rickets (*p. 212*)
A disorder that reduces the amount of calcium salts in the skeleton; often characterized by a "bowlegged" appearance because leg bones bend.

gigantism (*p. 218*)
A condition resulting from an overproduction of growth hormone before puberty.

pituitary growth failure (*p. 218*)
A type of dwarfism caused by inadequate growth hormone production.

scurvy (*p. 219*)
A condition involving weak, brittle bones; it results from a vitamin C deficiency.

osteomalacia (os-tē-ō-ma-LĀ-shē-ah) (*p. 220*)
A softening of bone due to a decrease in the mineral content.

fracture (*p. 223*)
A crack or break in a bone.

osteopenia (os-tē-ō-PĒ-nē-a) (*p. 224*)
Inadequate ossification, leading to thinner, weaker bones.

osteoporosis (os-tē-ō-por-Ō-sis) (*p. 224*)
A reduction in bone mass to a degree that compromises normal function.

osteomyelitis (os-tē-ō-mī-e-LĪ-tis) (*p. 224*)
A painful infection in a bone, usually caused by bacteria.

Paget's disease (osteitis deformans) (os-tē-Ī-tis de-FŌR-mans) (*p. 224*)
A condition characterized by gradual deformation of the skeleton.

ankylosis (an-ke-LŌ-sis) (*p. 229*)
A bony fusion between opposing bones that locks the articulating elements into position.

rheumatism (ROO-ma-tizm) (*p. 229*)
A general term indicating pain and stiffness affecting the skeletal and/or muscular systems.

arthritis (ar-THRĪ-tis) (*p. 229*)
A term including all of the rheumatic diseases affecting synovial joints.

osteoarthritis (os-tē-ō-ar-THRĪ-tis) (*p. 229*)
A form of arthritis which results from cumulative wear and tear on joint surfaces.

rheumatoid arthritis (*p. 230*)
A type of inflammatory arthritis.

synovitis (sīn-ō-VĪ-tis) (*p. 230*)
A condition in which the synovial membrane becomes inflamed; a symptom of rheumatoid arthritis.

Additional Terms of Clinical Importance

FRACTURE CLASSIFICATION AND IDENTIFICATION (*p. 222*): **closed fracture, open fracture, comminuted fracture, transverse fracture, greenstick fracture, spiral fracture, Colles' fracture, Pott's fracture, compression fracture, epiphyseal fracture, displaced fracture, nondisplaced fracture**

RHEUMATISM, ARTHRITIS, AND SYNOVIAL FUNCTION (*p. 229*): **continuous passive motion (CPM)**

Study Outline

Related Key Terms

Introduction (pp. 209–210)

1. The skeletal system includes the bones of the skeleton and the cartilages, ligaments, and other connective tissues that stabilize or interconnect bones. Its functions include: structural support, storage, blood cell production, protection, and leverage.

Related Key Terms: **articulations**

Structure of Bone (pp. 210–214)

1. **Osseous tissue** is a supporting connective tissue with a solid **matrix**.

Related Key Terms: **hydroxyapatite**

HISTOLOGICAL ORGANIZATION (p. 210)

2. There are two types of bone: **dense**, or **compact**, **bone** and **spongy**, or **cancellous**, **bone**. Both contain **osteocytes** in **lacunae**. Layers of calcified matrix are **lamellae**.

Related Key Terms: **marrow cavity** • **bone marrow**

HISTOLOGICAL DIFFERENCES BETWEEN COMPACT AND SPONGY BONE (pp. 210–211)

3. The basic functional unit of compact bone is the **osteon**, containing osteocytes arranged around a **central canal**.
4. Spongy bone contains **trabeculae**, often in an open network.

Related Key Terms: **interstitial lamellae**

FUNCTIONAL DIFFERENCES BETWEEN COMPACT AND SPONGY BONE (pp. 211–213)

5. Compact bone is found where stresses come from a limited range of directions; spongy bone is located where stresses are few or come from many different directions.

Related Key Terms: **cortex** • **medullary cavity** **diaphysis** • **rickets** **epiphyses**

Chapter Review

THE PERIOSTEUM AND ENDOSTEUM (pp. 213–214).

6. A bone is covered by a **periosteum** and lined with an **endosteum**. **Osteoclasts** dissolve the bony matrix through **osteolysis**. Other cells called **osteoblasts** synthesize the matrix in the process of **osteogenesis**.

Related Key Terms

joint capsule • **tendons**
osteoprogenitor cells

Development and Growth (pp. 214–219)

1. **Ossification** is the process of converting other tissues to bone; **calcification** is the process of depositing calcium salts within a tissue.

INTRAMEMBRANOUS OSSIFICATION (p. 215)

2. **Intramembranous ossification** begins when osteoblasts differentiate within connective tissue, and can produce spongy or compact bone.

dermal bones
ossification center • **spicules**

ENDOCHONDRAL OSSIFICATION (pp. 215–218)

3. **Endochondral ossification** begins by forming a cartilaginous model that is gradually replaced by bone.
4. **Growth hormone** and **thyroxine** stimulate bone growth.

osteogenic layer
primary center of ossification
perforating canals • **metaphysis**
secondary ossification centers
articular cartilage
epiphyseal plate

BONE GROWTH AND BODY PROPORTIONS (pp. 218–219)

5. There are differences between bone to bone and between individuals regarding the timing of epiphyseal closure.

gigantism
pituitary growth failure
epiphyseal line

REQUIREMENTS FOR NORMAL BONE GROWTH (p. 219)

6. Normal osteogenesis requires a reliable source of minerals, vitamins, and hormones.

vitamin D • **vitamin A**
vitamin C • **scurvy**

Remodeling and Homeostatic Mechanisms (pp. 219–224)

CHANGES IN BONE SHAPE (pp. 219–220)

1. Mineral turnover allows bone to adapt to new stresses.

CHANGES IN THE MINERAL CONTENT OF THE SKELETON (pp. 220–223)

2. Calcium is the most common mineral in the human body, with more than 98 percent of it located in the skeleton.
3. Two hormones, **calcitonin** and **parathormone**, regulate calcium ion homeostasis. Calcitonin leads to a decline in calcium concentrations, while parathormone increases calcium concentrations.

parathyroid gland

INJURY AND REPAIR (pp. 223–224)

4. A **fracture** is a crack or break in a bone. Repair of a fracture involves the formation of a **fracture hematoma**, an **external callus**, and an **internal callus**.

AGING AND THE SKELETAL SYSTEM (p. 224)

5. Effects of aging on the skeleton can include **osteopenia** and **osteoporosis**.

osteomyelitis
Paget's disease (osteitis deformans)

Anatomy of Skeletal Elements (pp. 224–226)

CLASSIFICATION OF BONES (pp. 224–225)

1. Categories of bones include: **long bones**, **short bones**, **flat bones**, **irregular bones**, **sesamoid bones**, and **sutural bones (Wormian bones)**.

patellae

BONE MARKINGS (pp. 225–226)

2. Bone **markings** can be used to identify specific bones within each category.

Articulations (pp. 226–236)

CLASSIFICATION OF JOINTS (pp. 226–230).

1. **Articulations** (joints) exist wherever two bones interact. Immovable joints are **synarthroses**, slightly movable joints are **amphiarthroses**, and those that are freely movable are called **diarthroses**.
2. Examples of synarthroses include a **suture**, a **gomphosis**, a **synchondrosis**, and a **synostosis**.

periodontal ligament

3. Examples of amphiarthroses are a **syndesmosis** and a **symphysis**.
4. The bony surfaces at diarthroses are covered by **articular cartilages**, lubricated by **synovial fluid**, and enclosed within a **joint capsule**. Other synovial structures can include **menisci**, or **articular discs**; **fat pads**; and **accessory ligaments**.

extracapsular ligaments
intracapsular ligaments • **bursae**

5. Extremes of motion may cause a **dislocation (luxation)** or a less severe **subluxation** (partial dislocation) of a joint. Such movements are usually prevented by the joint capsule, accessory ligaments,

the apposition of soft tissues, bony projections, tendons, and/or muscular contractions.

ARTICULAR FORM AND FUNCTION (pp. 230–232)

6. Possible movements can be classified as **gliding**, **angular motion**, **circumduction**, and **rotation**. Joints are called **monaxial**, **biaxial**, or **triaxial** depending on the degree of movement they allow.
7. Monaxial joints include **hinge joints** and **pivot joints**. Biaxial joints can be **gliding joints**, **ellipsoidal joints**, or **saddle joints**. Triaxial, or **ball-and-socket joints**, permit rotation.

DESCRIBING DYNAMIC MOTION (pp. 232–236)

8. Important terms that describe dynamic motion are **flexion**, **extension**, **hyperextension**, **rotation**, **circumduction**, **abduction**, and **adduction**.
9. Movements of the foot include **inversion**, **eversion**, **dorsiflexion**, and **plantar flexion**. **Opposition** is the thumb movement that enables us to grasp objects.
10. The bones in the forearm permit **pronation** and **supination**.
11. **Protraction** involves moving something forward; **retraction** involves moving it back. **Depression** and **elevation** occur when we move a structure down and up.

Related Key Terms

axis

internal (medial) rotation
external (lateral) rotation

Review Planner

		Level 1
1	Describe the functions of the skeletal system.	**3 9 16 35 36 45**
2	Compare the structures and functions of compact and spongy bones.	**1 2 8 11 12 44**
3	Discuss the processes by which bones develop and grow and account for variations in their internal structure.	**1 13 19 20 21 30 31 32 39 47 50**
4	Describe the remodeling and homeostatic mechanisms of the skeleton.	**10 17 20 30 32**
5	Classify bones according to their shapes and give examples for each type.	**6 16 37**
6	Describe the different types of fractures and explain how fractures heal.	**23 42 51**
7	Discuss the effects of nutrition, hormones, exercise, and aging on bone development and the skeletal system.	**17 23 39 40 41 43**
8	Distinguish between different types of joints and link structural features to joint functions.	**4 5 7 14 25 27 28 29 46**
9	Describe the dynamic movements of the skeleton.	**15 33**

Level 2	**34 38 48 49**
Level 3	**52 53 54 55**
C M	**Inherited Abnormalities in Skeletal Development • Hyperostosis Stimulation of Bone Growth Heterotopic Bones**
ABC NEWS	**Bone Loss and Women Athletes • Osteoporosis Treatment**

Chapter Review

Review Questions

MULTIPLE CHOICE

1. Bone cells are called: a) osteons; b) osteocytes; c) articulations; d) lamellae.
2. The basic functional units of compact bone are called: a) osteons; b) osteocytes; c) articulations; d) lamellae.
3. Bone markings can indicate sites where: a) tendons attach to the bone; b) blood vessels lie next to the bone; c) nerves penetrate the bone; d) all of the above.
4. You would expect to find synovial fluid in: a) a suture in the skull; b) a symphysis; c) bursae; d) none of the above.
5. An example of a biaxial joint is a: a) pivot joint; b) saddle joint; c) ball-and-socket joint; d) hinge joint.
6. The patella is an example of a(n): a) sesamoid bone; b) flat bone; c) irregular bone; d) short bone.
7. An example of a triaxial joint is a(n): a) pivot joint; b) saddle joint; c) ball-and-socket joint; d) ellipsoidal joint.
8. The struts or plates that often form an open network in spongy bone are called: a) canaliculi; b) trabeculae; c) epiphyseal plates; d) osteoblasts.
9. Osseous tissue is a type of: a) cardiac tissue; b) adipose tissue; c) muscle tissue; d) connective tissue.
10. Inadequate osteoblast activity or excessive osteoclast activity can cause: a) arthritis; b) osteomyelitis; c) osteopenia; d) calcification.
11. The canal around which the osteocytes in compact bone are arranged concentrically is the: a) osteon; b) periosteum; c) perforating canal; d) central canal.

DIAGRAMS

12. Draw a diagram that illustrates the structure of compact bone. Show an osteocyte, a lacuna, canaliculi, an osteon, central canal, interstitial lamellae, and blood vessels.
13. Draw two flow charts, one outlining the steps in intramembranous ossification and the other illustrating the steps in endochondral ossification.
14. Create a table that lists the three functional categories of articulations. For each one, list the joints in that category and an example of where these joints are found in the body.
15. Illustrate the following types of movements: a) flexion/extension/hyperextension; b) external (lateral) rotation/internal (medial) rotation; c) abduction/adduction; d) inversion/eversion; e) dorsiflexion/plantar flexion; f) pronation/supination; g) protraction/retraction; h) depression/elevation.

TRUE/FALSE

16. **T/F:** A long bone can best withstand stresses applied at either end.
17. **T/F:** Bone has a very limited ability to repair itself.
18. **T/F:** The greater the range of motion in a joint, the stronger it becomes.

MATCHING QUESTIONS

(*Match each statement with the* most appropriate *term.*)

Statements:

19. The process of bone formation.
20. Cells that remove bony matrix by secreting acids.
21. The dissolution of bony matrix.
22. A condition in which individuals often have a bowlegged appearance.
23. A crack or break in a bone.
24. The shaft of a bone.
25. Freely movable joints.
26. The broad end of a bone.
27. Slightly movable joints.
28. Immovable joints.
29. A dislocation in a joint.
30. Cells that synthesize the organic components of the bone matrix.
31. The junction between the broad end and the shaft of a bone.
32. The process of converting other tissues to bone.
33. Reducing the angle between articulating elements.

Terms:

a) Synarthroses
b) Diarthroses/synovial joints
c) Epiphysis
d) Osteoblasts
e) Amphiarthroses
f) Metaphysis
g) Fracture
h) Ossification
i) Rickets
j) Diaphysis
k) Osteogenesis
l) Flexion
m) Osteoclasts
n) Osteolysis
o) Luxation

SHORT ANSWER/ESSAY

34. Why does bone undergo continual remodeling?
35. Describe the functions of the skeletal system.
36. Discuss the role of the skeleton as a mineral reservoir. Which mineral is most important?
37. Describe the six categories of bones, and give an example of where each can be found in the body. On what basis are bones classified?
38. Explain how the structure of bone contributes to its distinctive character.
39. Why is vitamin D important to bone growth?
40. Discuss three conditions affecting the skeletal system that can result from the aging process.
41. What two hormones regulate the pattern of bone growth?
42. Place the steps involved in repairing a fracture into the correct sequence:
    a) Osteoblasts replace central cartilages with spongy bone.
    b) A fracture hematoma forms at the site of injury.
    c) An external callus encircles the bone.
    d) Osteoclasts and osteoblasts remodel the injured area.

43. What are calcitonin and parathormone? Compare their effects on osseous tissue.
44. Compare the structures and functions of compact and spongy bone.
45. Why is the skeletal system important for movement?
46. What do the joints between your thoracic vertebrae have in common with the joints that connect your fingers to the bones of your palms? How do they differ?
47. Compare the mechanisms that increase the diameter and the length of a developing bone.
48. You strap on a pair of ice skates for the first time in your life. While you're balancing on the skates your ankles frequently bend inward and outward, but you neither sprain nor dislocate your ankle joints. Why not?
49. Compare the functions of the periosteum and the endosteum. Where are they located?
50. Distinguish between intramembranous and endochondral ossification.
51. Pat fractures a bone. Although no bones protrude through her skin, an X-ray image reveals multiple bone fragments in the injured area. What type of fracture did she sustain? Is it open or closed?

CRITICAL THINKING/APPLICATIONS

52. A man comes to the emergency room with a fractured bone in his foot. He tells you that he hurt his foot while running a marathon; he had sprained that foot 3 weeks earlier and had been on crutches for 3 weeks. However, he is an experienced marathon runner and wonders why he injured the foot again when it had healed after the sprain. What would you tell him?
53. Mrs. Wiltfong, age 72, broke her arm when she stumbled and caught herself against a piece of furniture. While talking with her in the emergency room, you learn that she gets very little exercise and never drinks milk. For what condition of the skeletal system is she at risk? What risk factors does she have?
54. Why are victims of chronic starvation more likely to have fractures than well-fed individuals?
55. A skeleton is found encased in concrete during a bridge renovation. Weeks later, police know the murder victim's approximate age, weight, and fitness. How were these conclusions drawn?

Many societies throughout the world have chosen this sensible method of carrying heavy loads. Rather than bearing them in their arms—and possibly straining their backs—people place bundles on their heads. This Mali woman is carrying a load of grain. The weight is distributed along her axial skeleton, which forms the longitudinal axis of the body, while her hands are free for other tasks. In this chapter we will learn more about the axial skeleton, and find that it plays many roles besides carrying heavy loads.

CHAPTER 8

The Skeletal System: Axial Division

Chapter Objectives

After reading this chapter, you will be able to:

1 Name the components of the axial skeleton and their functions.
2 Identify the bones of the skull and explain the significance of the markings on individual bones.
3 Discuss the process by which the adult skull develops.
4 Explain the significance of the articulations between the thoracic vertebrae, the ribs, and the sternum.
5 Discuss the differences in structure and function of the various vertebrae.
6 Discuss the bones of the neck and trunk and their distinctive markings.

Introduction

Bone first evolved in fishes roughly 400 million years ago. It was dermal bone, developing in the skin, and it provided a mineral reserve and a protective armor plating for defense against predators. The internal skeletons of these fishes were composed primarily of cartilage, but in the evolutionary line that lead toward humans most of those cartilages were replaced by bones.

As you saw in Chapter 7, bones are remarkable structures. (p. 210) They are as strong or stronger than reinforced concrete, but considerably lighter. Better yet, they can be remodeled and reshaped to meet metabolic demands or to adapt to changing activity patterns. The basic features of the human skeleton have been shaped by evolution, but the detailed characteristics of each bone reflect the stresses placed upon it. As a result, the skeleton changes in the course of a lifetime. Examples discussed in the last chapter included the proportional changes at puberty and the gradual osteopenia of aging. This chapter will consider additional examples, such as the changes in the shape of the vertebral column during the transition from crawling to walking.

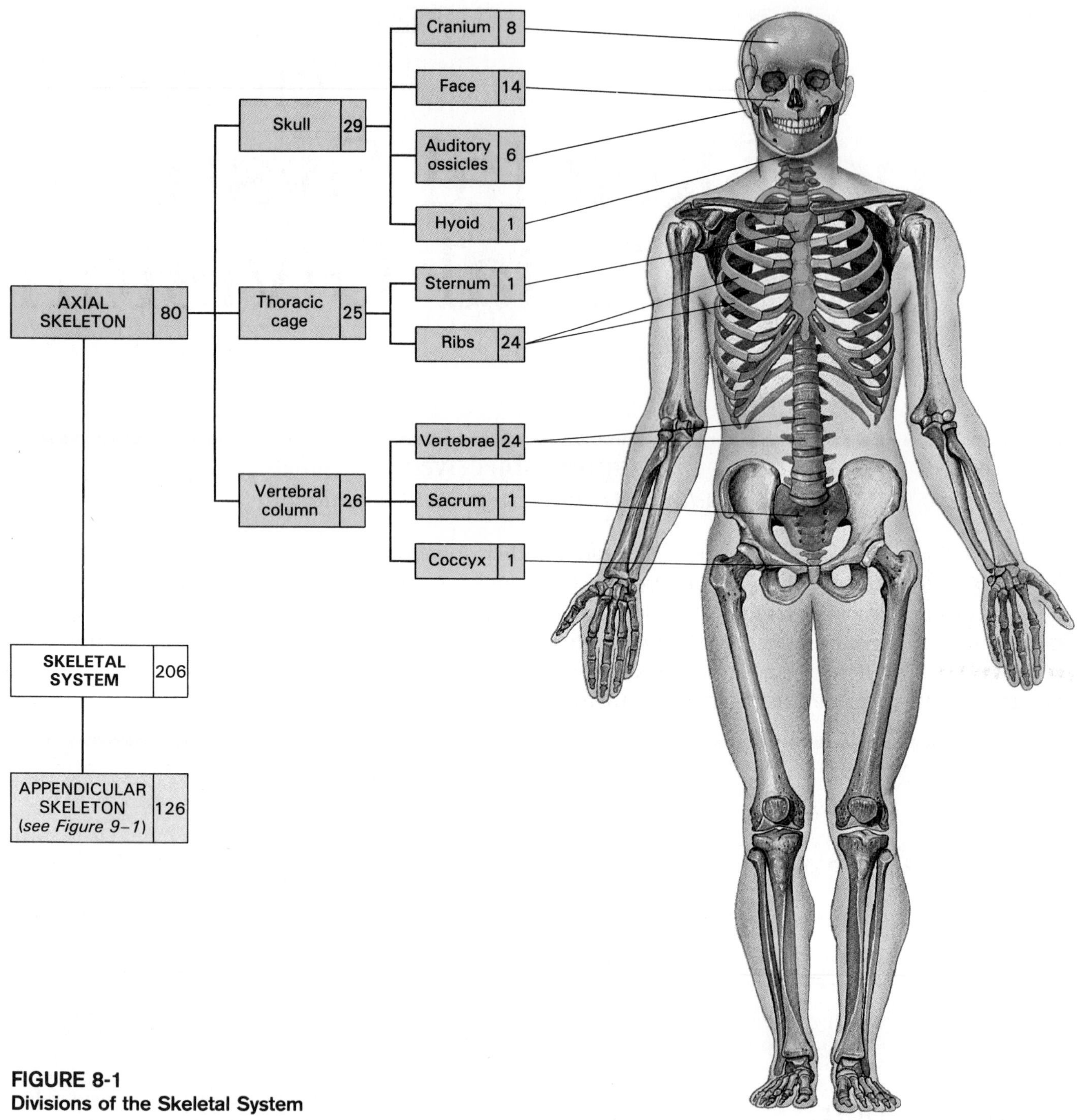

FIGURE 8-1
Divisions of the Skeletal System

The skeletal system can be divided into *axial* and *appendicular divisions* (Figure 8-1). The **axial skeleton** forms the longitudinal axis of the body. This division contains 80 bones, about 40 percent of the bones in the human body. That number can be further subdivided into the **skull** (22 bones), the **auditory ossicles** (6) and **hyoid bone** (1) associated with the skull, the **vertebral column** (26), and the **thoracic** (*rib*) **cage** composed of the **ribs** (24) and the **sternum** (1). All of these elements are shown in Figure 8-1, except for the auditory ossicles which are enclosed by the skull.

The axial skeleton creates a framework that supports and protects organ systems in the dorsal and ventral body cavities. In addition, it provides an extensive surface area for the attachment of muscles that (1) adjust the positions of the head, neck, and trunk, (2) perform respiratory movements, and (3) stabilize or position elements of the appendicular skeleton.

The **appendicular skeleton** includes the bones of the arms and legs and those of the **pectoral** and **pelvic girdles**, which attach the limbs to the trunk. All together there are 126 appendicular bones; 32 are associated with each arm and 31 with each leg.

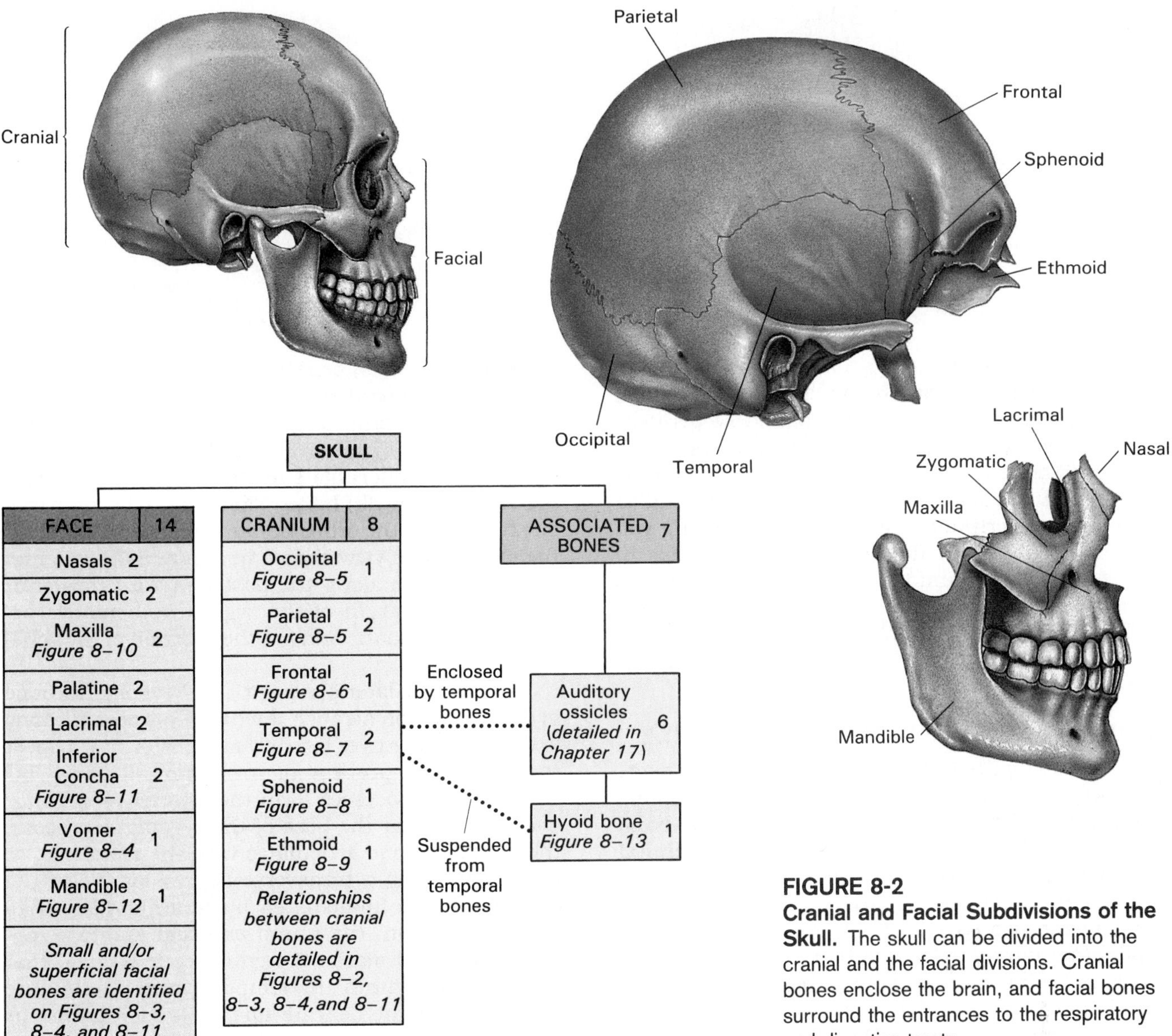

FIGURE 8-2
Cranial and Facial Subdivisions of the Skull. The skull can be divided into the cranial and the facial divisions. Cranial bones enclose the brain, and facial bones surround the entrances to the respiratory and digestive tracts.

This chapter considers the functional anatomy of the bones and articulations of the axial skeleton, beginning with those of the skull. The components of the appendicular skeleton will be examined in Chapter 9.

■ The Skull

The bones of the skull protect the brain and guard the entrances to the digestive and respiratory systems. The skull contains 22 bones: 8 form the **cranium**, or *brainbox*, and 14 are associated with the face. The major cranial and facial bones are shown in Figure 8-2.

The cranium encloses the **cranial cavity**, a fluid-filled chamber that cushions and supports the brain. Blood vessels, nerves, and the membranes that stabilize the position of the brain are attached to the inner surface of the cranium. Its outer surface provides an extensive area for the attachment of muscles that move the eyes, jaws, and head. A specialized articulation between the cranium and the first spinal vertebra stabilizes the positions of the brain and spinal cord while permitting a considerable range of head movements.

Facial bones form through intramembranous ossification after cranial development is already under way (see the Embryology Summary on pp. 252–253). If the cranium is the house where the brain resides, the facial complex is the front porch. Facial bones protect and support the entrances to the digestive and respiratory tracts. They also provide areas for the attachment of muscles that control facial expressions and assist in the manipulation of food. Both cranial and facial bones protect and support delicate sense organs involved with vision, hearing, balance, olfaction (smell), and gustation (taste).

We will begin with an examination of the adult skull (Figure 8-3). If you are reading this in lab, you should also look at a real skull so that you can learn to recognize key features without color coding. Figure 8-3 shows five views of the skull: (a) *posterior* (from the back), (b) *superior* (from the top), (c) *anterior* (from the front), (d) *lateral* (from the side), and (e)

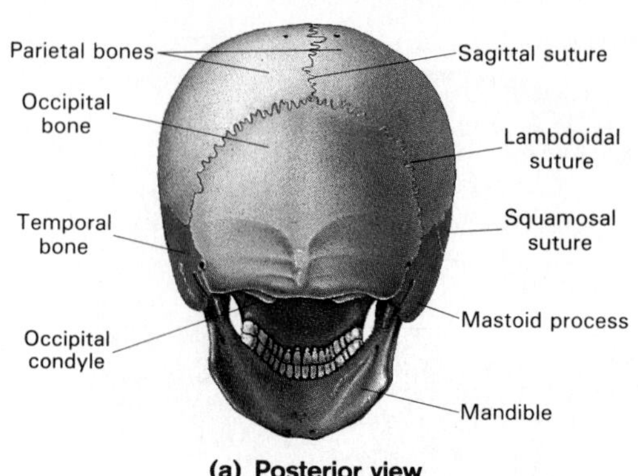

(a) Posterior view

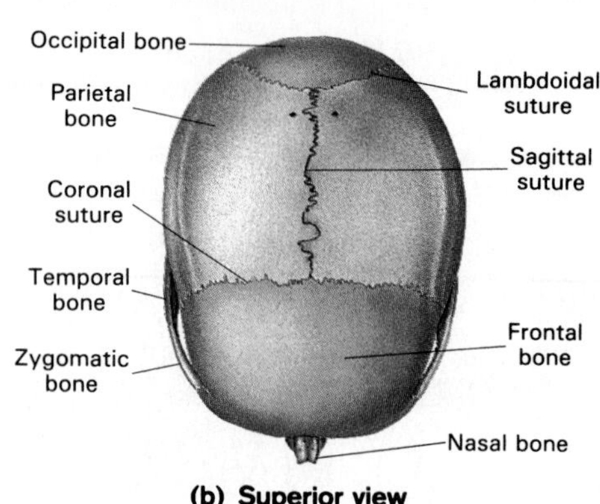

(b) Superior view

inferior (from below). The discussion that follows describes the bones of the skull as they appear during an examination of the surface from these perspectives. We will then consider sectional views of the skull, and then examine the individual bony elements in detail. This will be your first exposure to detailed anatomical descriptions, and before proceeding you might want to review the directional references included in Tables 1-1 and 1-2 and the anatomical terms introduced in Table 7-2. ∞ (pp. 13, 16, 226) (Specialized reference terms not used in earlier chapters will be shown in italics, with explanations in parentheses.)

SUPERFICIAL ANATOMY OF THE SKULL

Refer first to Figure 8-3a. The opening that connects the cranial cavity with the canal enclosed by the spinal column is the **foramen magnum**. The **occipital bone** surrounds the foramen magnum; at the **occipital condyles** the skull articulates with the vertebral column. Along its superior margin, the occipital bone contacts the two **parietal** (pa-RĪ-e-tal) **bones** at the **lambdoidal** (lam-DOYD-al) **suture** or *occipital-parietal suture*. One or more **sutural bones** may also be found along this line. The parietal bones articulate at the **sagittal suture**, which extends *rostrally* (toward the nose) until it intersects the **coronal suture** (Figure 8-3b). The coronal suture separates the two parietal bones from the single median **frontal bone**. The occipital, parietal, and frontal bones form the "skullcap," or **calvaria** (kal-VAR-ē-a).

Moving from Figure 8-3b to 8-3c, you can follow the frontal bone on to the face, where it forms the superior surface of each eye socket, or **orbit**. Midway between the orbits, the frontal bone articulates with a pair of **nasal bones** that extend to the superior border of the **external nares** (NA-rēz), or nasal openings, of the skull. The lateral surfaces of the nasal bones articulate with the **maxillary** (MAK-si-ler-ē) **bone** of either side. Each **maxilla** (mak-SIL-a) also contacts the frontal bone at the superior and medial border of the orbit.

Several bones are situated between the frontal and maxillary bones inside each orbit. Proceeding from medial to lateral, as you enter the orbit you encounter a small and delicate **lacrimal bone** (LAK-ri-mal; *lacrimae*, tears). Along its lateral margin the lacrimal contacts the **ethmoid**, which in turn articulates with the **sphenoid** (SFĒ-noid) **bone**. Several prominent *foramina* (passageways) and *fissures* (elongate openings) penetrate the sphenoid or lie between the sphenoid and maxillary bones. The sphenoid is actually quite large and important, but much of it is hidden by other bones. This can be demonstrated by considering its relationship with the **zygomatic** (zī-go-MA-tik) **bone**. The zygomatic completes our trip across the orbit, but as you continue around the skull, to the view shown in Figure 8-3d, you find that the suture along the *posteromedial* (posterior and medial) margin of the zygomatic bone lies between the zygomatic and another portion of the sphenoid.

Along its lateral margin, the zygomatic bone gives rise to the slender **temporal process**, a bony extension that curves laterally and posteriorly to meet the **zygomatic process** of the **temporal bone**. Together these processes form the **zygomatic arch**, or *cheekbone*. Near the base of the zygomatic process, the temporal bone articulates with the **mandible**, or lower jaw, along a transverse depression called the **condylar fossa**. Immediately posterior and lateral to that articulation, the round **external auditory meatus** (mē-Ā-tus) marks the entrance to a canal that extends deep within the temporal bone. In life, the canal ends at the delicate *tympanic membrane*, or *eardrum*. The prominent bulge just posterior and inferior to the meatus is the **mastoid process**.

The lateral face of the temporal bone superior to the meatus is broad and flat. This is the **squama**, or *squamous portion*, of the temporal bone. Along its curving border it articulates with the sphenoid, parietal, and occipital bones at the **sphenosquamosal**, **squamosal**, and **occipitomastoid sutures**.

Now examine Figure 8-3e, showing the inferior surface of the skull. You are already familiar with the lambdoidal suture, where the occipital bone articulates with the parietal bones. Along its lateral margins the occipital also articulates with the temporal bones. The occipitomastoid suture runs *anteroposteriorly* (from anterior to posterior) medial to the tip of the mastoid process. This articulation takes an abrupt turn medially at a point where a slender, spinous projection leaves the surface of the temporal bone. This landmark is the **styloid process**, and the turn points toward the **occipital condyles**. These are the articular surfaces that connect the skull with the first vertebra of the neck.

Anterior to the styloid process the occipital bone narrows considerably. At its anterior and lateral borders it articulates with the sphenoid bone. Farther

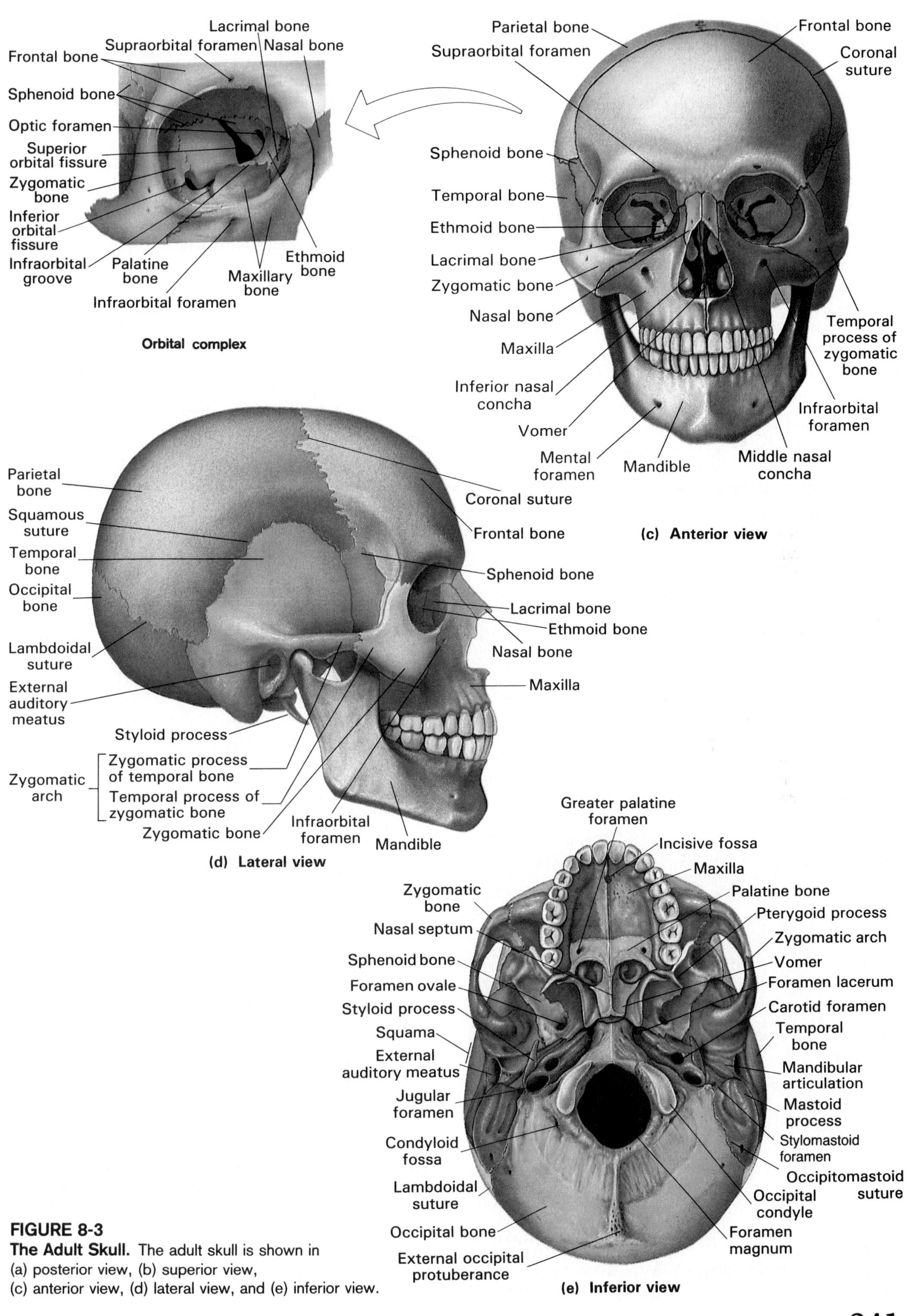

FIGURE 8-3
The Adult Skull. The adult skull is shown in (a) posterior view, (b) superior view, (c) anterior view, (d) lateral view, and (e) inferior view.

laterally, the temporal bones do likewise. Along its *anteromedial* (anterior and medial) border, the occipital contacts the **vomer**. The vomer supports a prominent partition that forms part of the **nasal septum**. The inferior margin of the vomer articulates with the paired **palatine bones** that form the posterior surface of the bony palate, or "roof of the mouth." The posterolateral margin of each palatine bone contacts the **pterygoid processes** (TER-i-goid; *pterygion*, wing), or **plates**, of the sphenoid. The lateral and rostral borders of the palatines articulate with the maxillae.

SECTIONAL ANATOMY OF THE SKULL

Figure 8-4 shows the skull as seen in horizontal, sagittal, and frontal section. You should try to integrate these sectional views with those of Figures 8-2 and

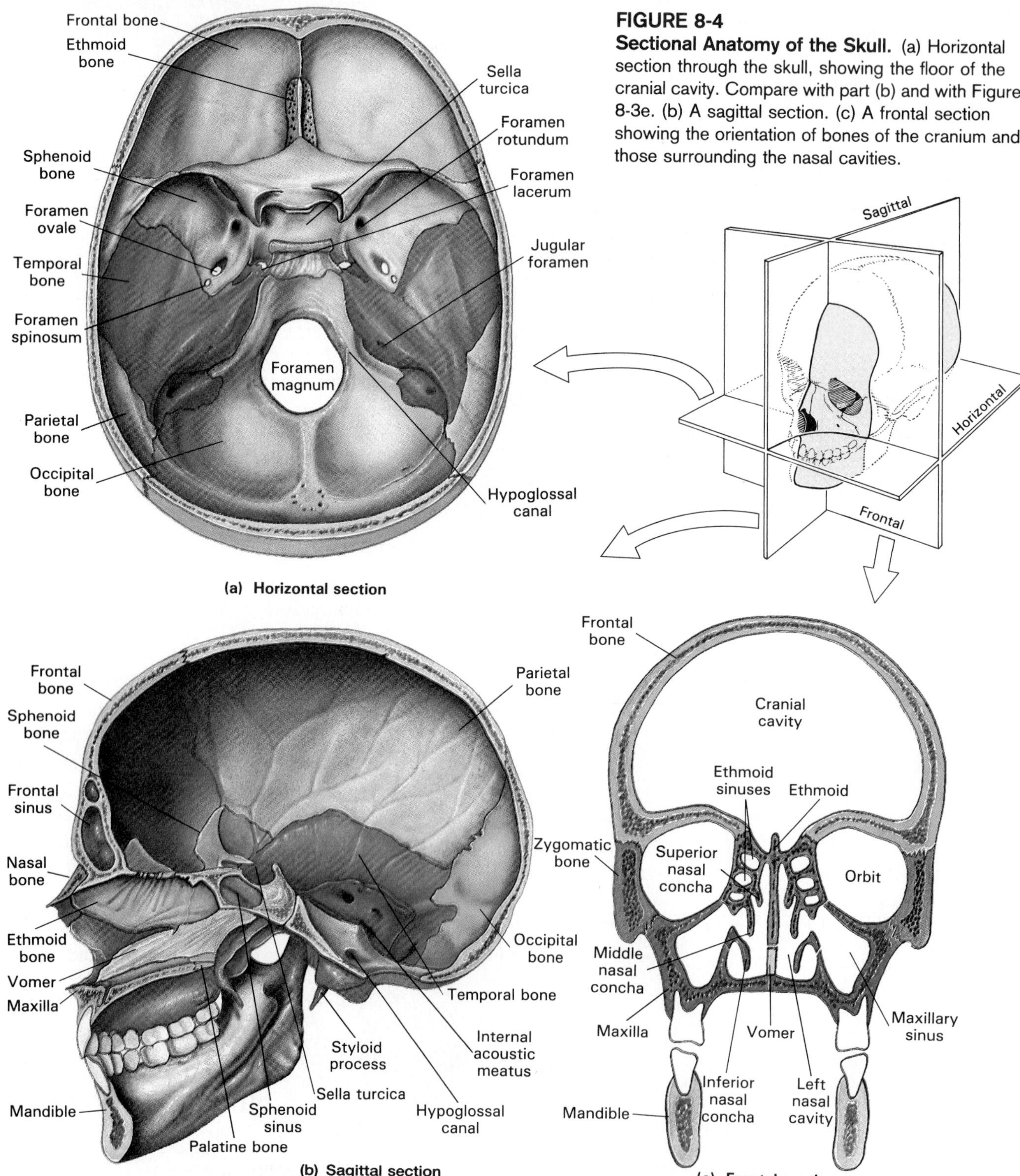

(a) Horizontal section

(b) Sagittal section

(c) Frontal section

FIGURE 8-4
Sectional Anatomy of the Skull. (a) Horizontal section through the skull, showing the floor of the cranial cavity. Compare with part (b) and with Figure 8-3e. (b) A sagittal section. (c) A frontal section showing the orientation of bones of the cranium and those surrounding the nasal cavities.

8-3. For example, Figure 8-4a shows the floor of the cranial cavity. Comparing this view with Figure 8-3e provides a much better perspective on the shape of the sphenoid, its *fossae* (depressions), and prominent foramina.

The sphenoid and frontal bones contain internal chambers, or *sinuses*, visible in Figure 8-4b. This figure illustrates the relationships between the bones of the palate. A comparison between this view and Figure 8-4c should enable you to visualize the internal anatomy of the nasal cavity. The vomer and ethmoid form the bony elements of the composite **nasal septum**, which continues into the fleshy portion of the nose as a sheet of hyaline cartilage. On either side of this partition lie the **nasal conchae** (KONG-kē; *concha*, shell). The superior conchae and medial conchae are part of the ethmoid; the inferior conchae are separate bones.

- **During baseball practice a ball hits Casey in the eye, fracturing the bones directly above and below the orbit. Which bones were broken?**
- **The mastoid and styloid processes are found on which of the skull bones?**
- **What bones articulate with the vomer?**

BONES OF THE CRANIUM

Now that you are acquainted with the bones of the adult skull, we can discuss the major bones in more detail. As we proceed, use Figures 8-3 and 8-4 to develop a three-dimensional perspective on the individual bones. Ridges and foramina that are detailed here mark the attachment of muscles or passage of nerves and blood vessels that will be discussed in later chapters. Each of the following figures includes information concerning important landmarks of the skull.

The Occipital Bone

The external surface of the occipital bone (Figure 8-5a) bears a number of prominent ridges. The **occipital crest** extends posteriorly from the foramen magnum, ending in a small bump called the **external occipital protuberance**. Two horizontal ridges intersect the crest, the **inferior** and **superior nuchal** (NOO-kal) **lines**. These landmarks indicate the attachment of muscles and ligaments that stabilize the joint and the occipital condyles and balance the weight of the head over the vertebrae of the neck. Lateral to each hypoglossal canal, the occipital forms part of the wall of the large **jugular foramen**. The *internal jugular*

FIGURE 8-5
The Occipital and Parietal Bones. (a) The occipital bone, inferior (external) view. (b) The occipital bone, superior (internal) view. (c) A lateral view of the right parietal bone.

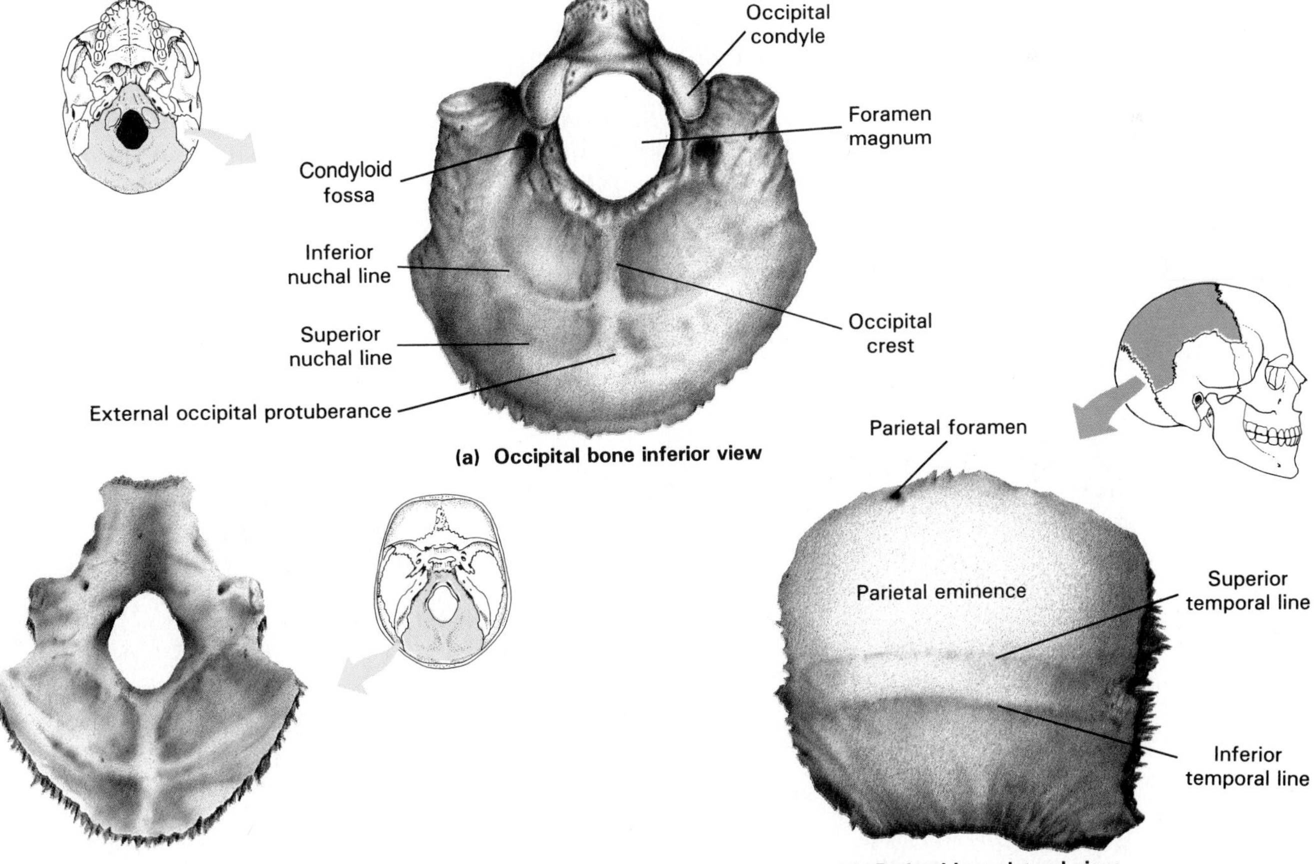

vein that passes through this foramen carries venous blood out of the cranial cavity. The **hypoglossal canals** begin on the lateral surfaces of the occipital, just superior to the condyles. The *hypoglossal nerves* that pass through these canals control the tongue muscles.

Inside the skull, the hypoglossal canals begin on the inner surface of the occipital bone near the foramen magnum (Figures 8-4a and 8-5b). Note that the internal surface of the occipital is concave. The prominent **internal occipital crest** is crossed by grooves that mark the path of prominent blood vessels. The curve closely follows the contours of the brain, and the crest marks the attachment site of membranes that stabilize the brain's position in the cranial cavity. The blood vessels are large veins that drain into the internal jugular vein.

The Parietal Bones

The lateral surface of each parietal bone (Figure 8-5c) bears a pair of low ridges, the **superior** and **inferior temporal lines**. These lines mark the attachment of the strong *temporalis muscle*, which closes the mouth. The smooth parietal surface above these lines forms the **parietal eminence**. The inner surface of the parietal, shown in Figure 8-4b, retains the impression of several veins and arteries that branch inside the cranium.

The Frontal Bone

During development the bones of the cranium form through the fusion of separate centers of ossification; this process will be described in a later section. At birth the fusions have not been completed, and there are two frontal bones connected at the **metopic suture**. Although the suture usually disappears by age 8 with the fusion of the bones, the adult skull may retain traces of the metopic suture. This suture, or what remains of it, runs down the center of the **frontal squama**, or *forehead* (Figure 8-6a). To either side, you will find a continuation of the superior temporal line already noted on the parietal surface.

The frontal squama ends at the **supraorbital margins** that support the eyebrows. Each margin is perforated by a single **supraorbital foramen**. The orbital surface is relatively smooth, but contains small openings for blood vessels and nerves heading to or from structures in the orbit. The shallow **lacrimal fossa** marks the location of the *lacrimal* (tear) *gland* that lubricates the surface of the eye.

The **frontal sinuses**, also illustrated in Figure 8-4b, are extremely variable in size and time of appearance. They usually appear after age 6, but some people never develop them at all. In addition to making bones lighter, sinuses produce mucus that helps keep the surfaces of the nasal cavities clean and moist. (Sinuses and sinus problems are discussed in the Clinical Comment: Septal Defects and Sinus Troubles on p. 249.)

The Temporal Bones

Most of the important landmarks on the temporal bone can be seen on its inferior surface (Figure 8-7a). Near the posterior portion of the occipitotemporal suture, the **mastoid process** provides an attachment site for muscles that rotate or extend the head. Near the base of this process, the **mastoid foramen** penetrates the temporal bone. Blood vessels travel through this passageway to reach the membranes surrounding the brain. Ligaments that support the hyoid bone attach to the sharp **styloid process** (STĪ-loyd; *stylos*, pillar). The **stylomastoid foramen** lies posterior to the base of the styloid process. The *facial*

FIGURE 8-6
The Frontal Bone. (a) Anterior surface. (b) Inferior surface.

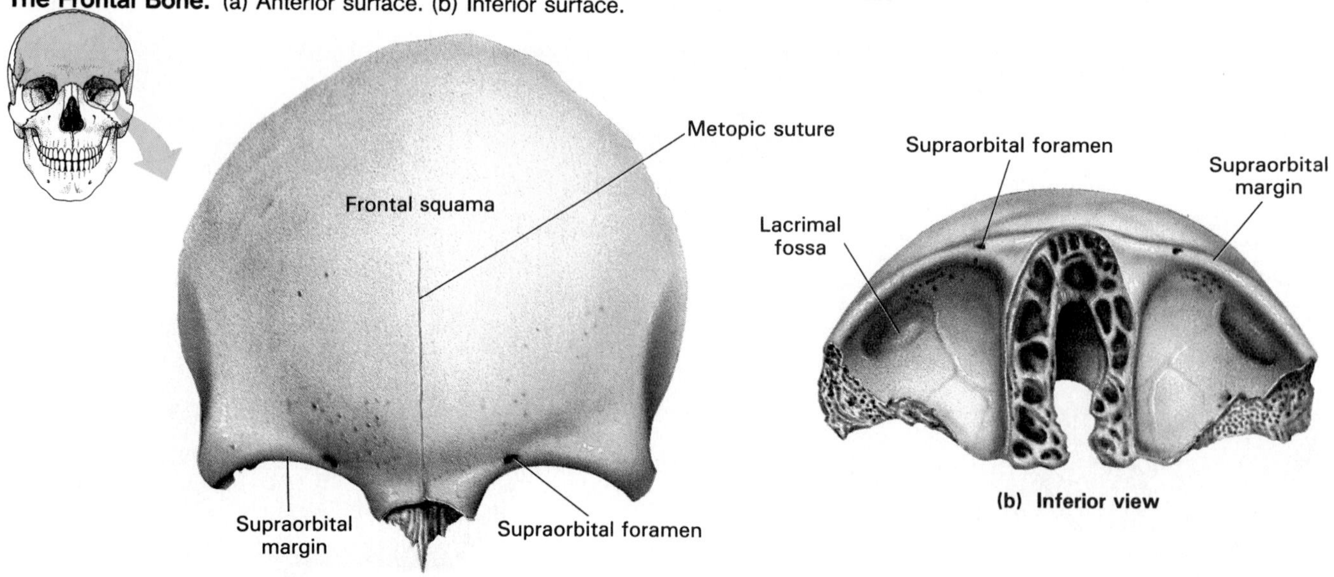

nerve, the nerve that passes through this foramen, controls the facial muscles.

Medially, each temporal completes the encirclement of the **jugular foramen** of that side, in cooperation with the adjacent portion of the occipital bone. Anterior and slightly medial to the jugular foramen, the round **carotid foramen** is the entrance to the *carotid canal*. The *carotid artery*, a major artery that delivers blood to the brain, enters the skull at this point. Anterior and medial to this opening, an elongate, jagged slit, the **foramen lacerum** (LA-se-rum; *lacerare*, to tear) extends between the occipital and

FIGURE 8-7
The Temporal Bone. (a) Inferior view of the right temporal bone, showing major anatomical landmarks. (b) Inferior view of the right temporal bone, shown as if transparent to indicate the location of internal structures associated with the middle and inner ear. (c) Lateral view of the right temporal bone. (d) Medial view of the right temporal bone.

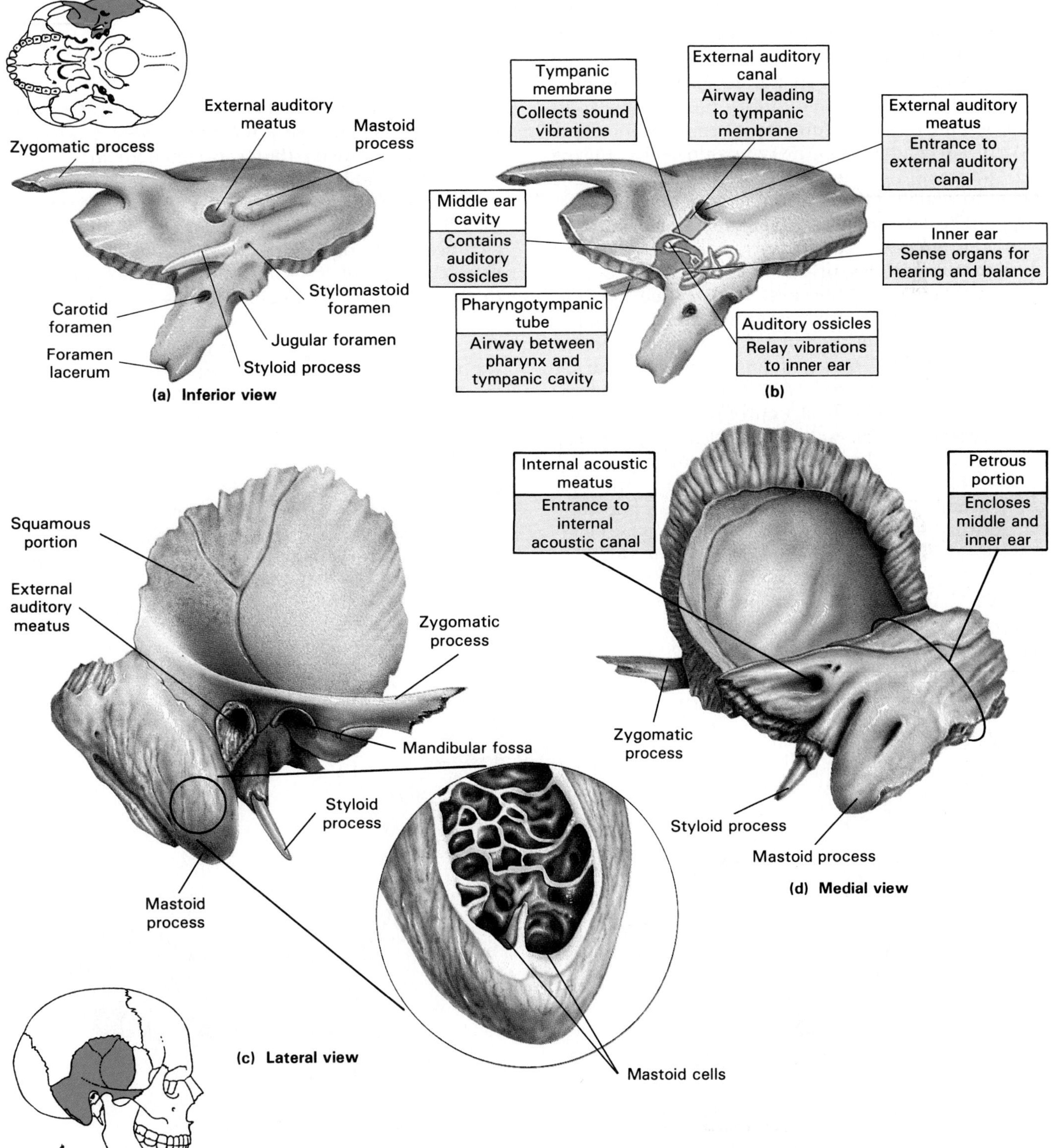

temporal bones. In life this region contains cartilage, but the tissue disintegrates during the preparation of a dried skull. Lateral to the foramen lacerum lie the **mandibular fossa** and the **articular tubercle**. These mark the location of the articulation between the temporal bone and the lower jaw.

Lateral and anterior to the carotid foramen, the temporal bone meets the sphenoid. At the base of the sphenoidal spine, a small canal begins that ends inside the mass of the temporal bone. This has long been known as the *Eustachian tube* (ū-STĀ-kē-an), commemorating an Italian anatomist of the sixteenth century, Bartolomeo Eustachio. His anatomical descriptions were exceedingly precise and detailed, although for some reason he failed to mention the passageway that now bears his name. The term **pharyngotympanic tube** or *auditory tube* will be used to designate this structure, for it begins in the pharynx (throat) and ends at the **tympanic cavity**. Figure 8-7b shows a transparent temporal bone, making it easier to visualize the location and orientation of the tympanic cavity and pharyngotympanic tube.

The tympanic cavity, or *middle ear*, contains the **auditory ossicles**, or *ear bones*. These tiny bones, three on each side of the skull, transfer sound vibrations from the **tympanic membrane** (*eardrum*) to the hearing receptors of the *inner ear*. (Details concerning the individual bones and their role in hearing will be discussed in Chapter 17.) The tympanic membrane is a soft, delicate structure that is not visible in a dried skull. The auditory ossicles and inner ear are completely enclosed by the temporal bone.

The middle ear and the pharyngotympanic tube are normally filled with air, and when you pop your ears during an airplane ride you are forcing air along this passage to adjust the pressure inside the tympanic cavity. The tympanic cavity also communicates with the **mastoid cells**, a network of air-filled spaces within the mastoid process. These spaces, sometimes called the *mastoid sinuses*, are shown in Figure 8-7c.

A medial view of the temporal bone (Figure 8-7d) shows additional features of interest. The thick **petrous portion**, or *pyramid*, of the temporal bone houses the receptor organs of the inner ear that provide sensations of hearing and balance. The **internal acoustic meatus** opens into the **internal acoustic canal** that carries blood vessels and nerves to the inner ear and the facial nerve to the stylomastoid foramen.

The Sphenoid Bone

The sphenoid bone acts as a bridge uniting the cranial and facial bones. A careful review of Figures 8-3 and 8-4 reveals that the sphenoid articulates with the frontal, occipital, ethmoid, and temporal bones of the cranium, and the palatine, zygomatic, maxilla, and vomer of the facial complex. It also acts as a brace, strengthening the sides of the skull. The general shape of the sphenoid has been compared to a giant butterfly or to a bat with its wings extended.

The wings can be seen most clearly on the superior surface, as illustrated in Figure 8-8a. Between the wings there is a central depression that marks the location of the pituitary gland. This gland, also called the *hypophysis*, sits below the brain, connected by a narrow stalk of neural tissue. The bony pocket that encloses the pituitary is called the **hypophyseal** (hī-po-FE-sē-al) **fossa**, or the **sella turcica** (TUR-si-ka). A literal translation of *sella turcica* is "Turkish saddle," which the depression supposedly resembles. If the Turk rode facing forward, his back would rest against the **posterior clinoid** (KLĪ-noid) **process** and he could reach forward to grip the **ante-**

FIGURE 8-8
The Sphenoid Bone. (a) Superior surface. (b) Anterior surface.

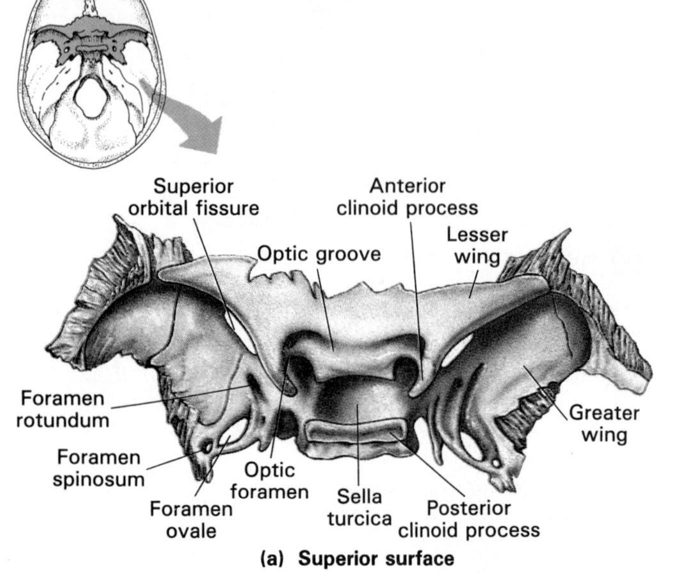

(a) Superior surface

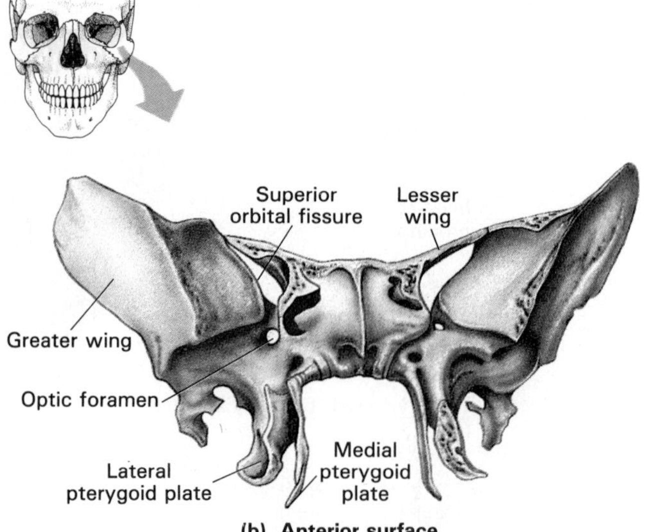

(b) Anterior surface

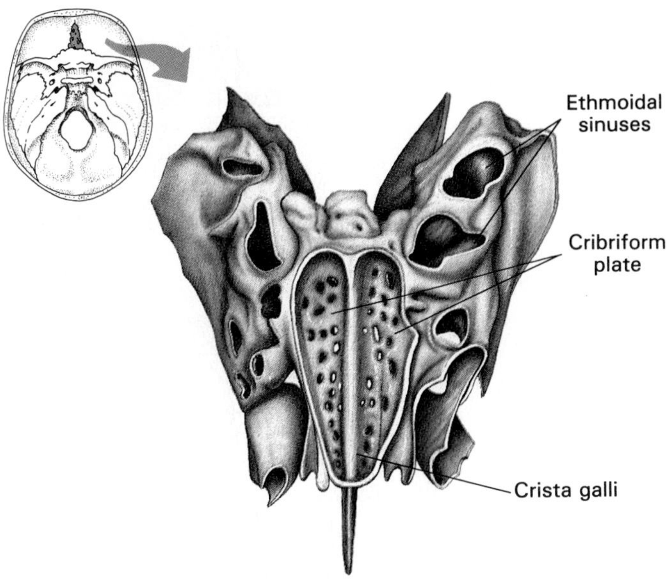

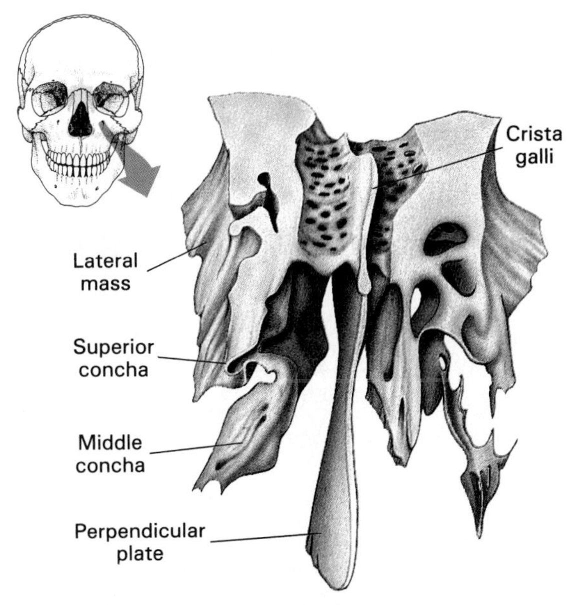

FIGURE 8-9
The Ethmoid Bone. (a) Superior surface. (b) Anterior surface.

rior clinoid processes on either side. The anterior clinoid processes continue rostrally as the **lesser wings** of the sphenoid.

The transverse groove that crosses the front of the saddle, above the level of the seat, is the **optic groove**. At either end of this groove is an **optic foramen**. The optic nerves that carry visual information from the eyes to the brain travel through these foramina. On either side of the sella turcica, the **superior orbital fissure**, the **foramen rotundum**, the **foramen ovale** (ō-VAH-la), and the **foramen spinosum** penetrate the sphenoid. These passages carry blood vessels and nerves to structures of the orbit, face, and jaws. The **greater wings** of the sphenoid extend laterally and rostrally from these foramina. The superior orbital fissures and optic foramina can also be seen in the anterior view of the sphenoid (Figure 8-8b).

The Ethmoid Bone

Figure 8-9 details the structure of the ethmoid bone. The superior surface (Figure 8-9a) contains a prominent ridge, the **crista galli**, or "cock's comb." Membranes that stabilize the position of the brain attach to this bony partition. The holes in the adjacent **cribriform plate** (*cribrum*, sieve) permit passage of sensory nerves from olfactory (smell) receptors in the nasal cavities to the brain.

Viewing the ethmoid from its anterior surface (Figure 8-9b), you can see the **lateral masses**, visible inside each orbit, and the **superior** and **medial conchae** that project into the nasal chamber. The lateral masses contain the **ethmoidal sinuses**; mucous secretions from these sinuses flush the surfaces of the nasal cavities. The projecting conchae break up the airflow, creating swirls and eddies. This mechanism slows air movement, but provides additional time for warming, humidification, and dust removal before the air reaches more delicate portions of the respiratory tract. The **perpendicular plate** forms part of the nasal septum, along with the vomer and a piece of hyaline cartilage. Olfactory (smell) receptors cover the upper portion of the plate and adjacent regions of the ethmoid.

- √ **The internal jugular veins are important blood vessels of the head. What bones do these blood vessels pass through?**
- √ **What bone contains the depression called the sella turcica? What is located in the depression?**
- √ **Which of the five senses would be affected if the cribriform plate of the ethmoid failed to form?**

BONES OF THE FACE

The bones of the face are somewhat easier to visualize than those of the cranium. Tiny facial bones and those whose primary features have already been noted are not considered in this section. However, before proceeding further, review the locations and appearance of the nasal and zygomatic bones (Figure 8-3c,d) and the palatines and vomer (Figures 8-3e and 8-4b).

(a) Lateral surface

(b) Sectional view

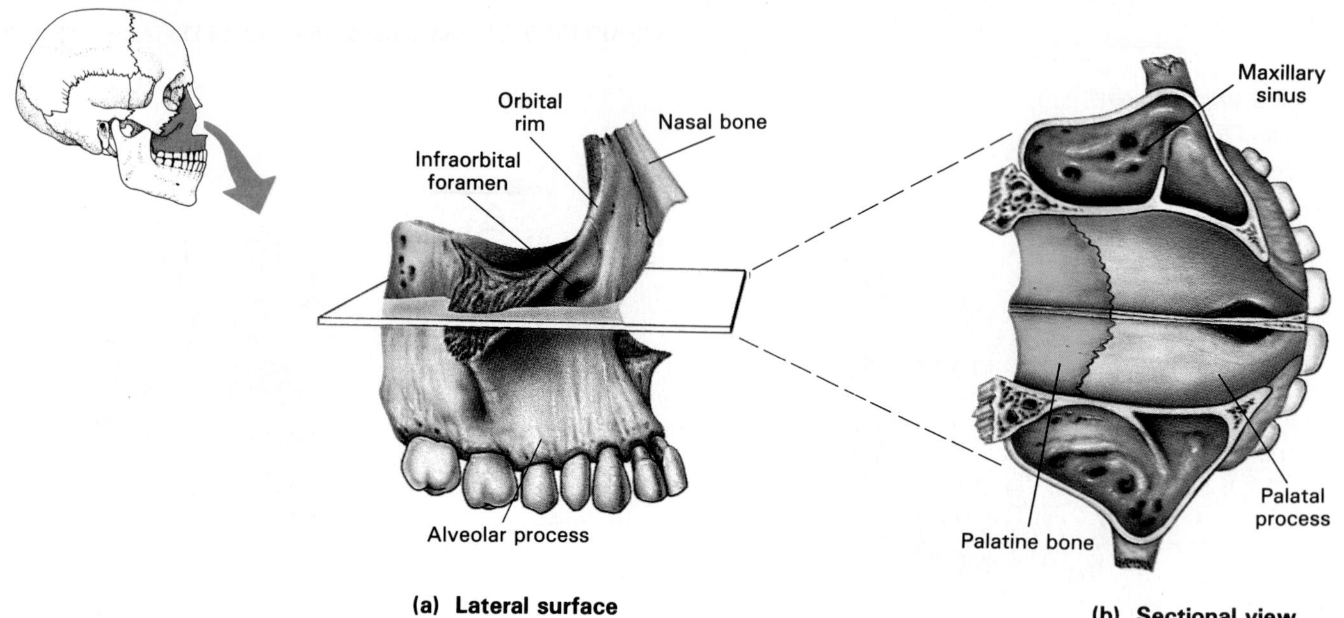

FIGURE 8-10
The Maxilla. (a) Anterior surface of right maxilla, showing superficial landmarks and the sectional plane for part (b). (b) Horizontal section through both maxillary bones, showing the size and orientation of the maxillary sinuses.

FIGURE 8-11
The Nasal Complex. (a) Sagittal section through the skull, with the nasal septum removed to show major features of the wall of the right nasal cavity. (b) A frontal section that indicates the positions of the paranasal sinuses.

(a) Sagittal section

(b) Frontal section

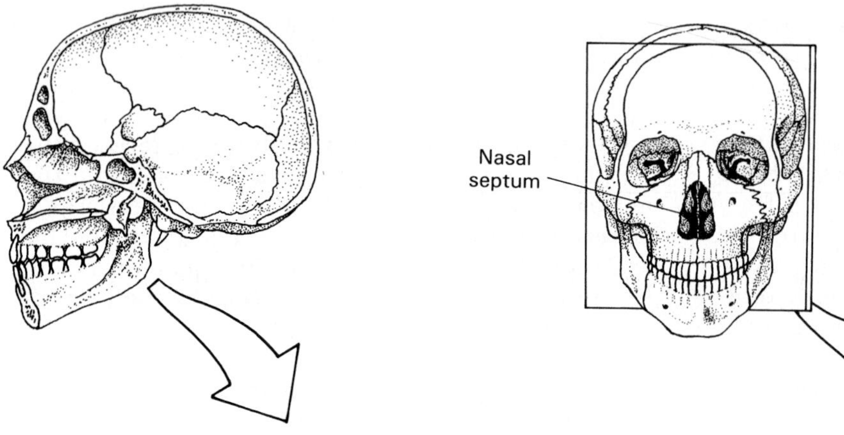

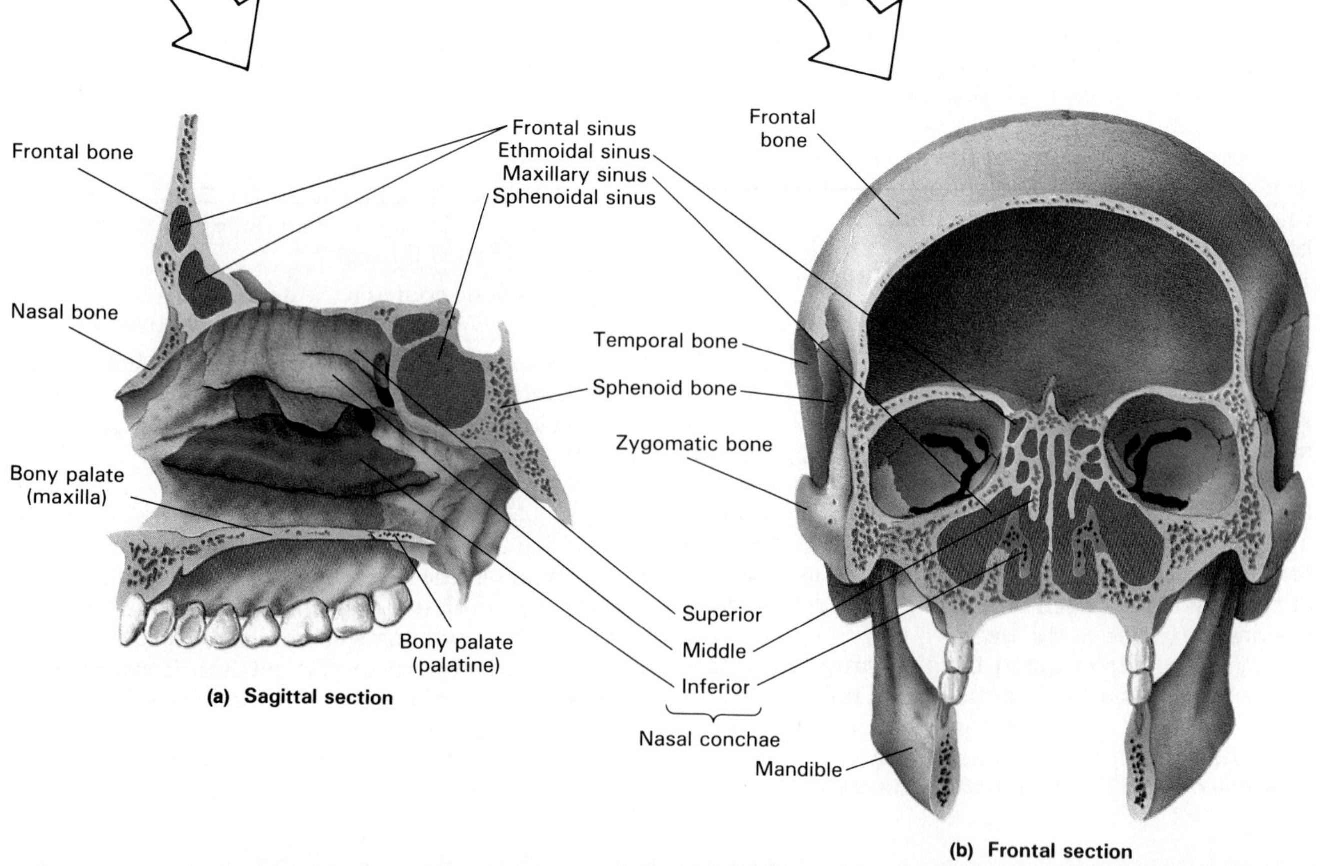

The Maxillary Bones

The maxillary bones articulate with all of the other facial elements except the mandible. Prominent superficial landmarks are indicated in Figure 8-10a. The **orbital rim** provides protection for the eye and other structures in the orbit. A large **infraorbital foramen** marks the entrance of a major sensory nerve from the face; it leaves the skull via the foramen rotundum of the sphenoid. An elongate **inferior orbital fissure** within each orbit lies between the maxilla and the sphenoid. The oral margins of the maxillae form the **alveolar processes** that harbor the upper teeth. As you will recall from Chapter 7, each tooth is locked in a socket, or *alveolus*, at a synarthrotic joint known as a *gomphosis*. (pp. 226–227)

In horizontal section (Figure 8-10b) you can see the large **maxillary sinuses** that lighten the portion of the maxillae above the embedded teeth. Infections of the gums or teeth can sometimes spread into the maxillary sinuses, increasing pain and making treatment more complicated. Infection of the maxillary sinus can in turn cause pain that is felt in the teeth, because of pressure exerted on sensory nerves. The sectional view also shows the extent of the **palatal processes** that form most of the roof of the mouth.

The attachment of the maxillary bones to the frontal, sphenoid, and ethmoid bones of the cranium completes the lateral walls of the *nasal cavities*. Together, the bones that enclose these airways and the associated *paranasal sinuses* are known as the **nasal complex**.

The Nasal Complex

The nasal complex includes the bones that form the superior and lateral walls of the nasal cavities and the sinuses that drain into them. The ethmoid and vomer form the **nasal septum** (*septum*, wall) that separates the nasal cavities. (Figures 8-4b and 8-11a).

In Figure 8-11b the right nasal cavity has been exposed by removing the nasal septum. As described earlier, portions of the maxillary and palatine bones form the floor of the nasal cavity. The maxillary, sphenoid, and ethmoid bones establish the wall; the **inferior nasal concha** projects into the cavity from an attachment on the lateral wall. Much of the anterior margin of the nasal cavity is formed by the soft tissues of the nose, but the bridge of the nose is supported by the maxillary and nasal bones. The frontal and ethmoid bones form the roof of the nasal cavity.

The frontal, sphenoid, ethmoid, and maxillary bones contain the **paranasal sinuses**. Figure 8-11b shows the location of the **frontal** and **sphenoidal sinuses**. The **ethmoidal** and **maxillary sinuses** are indicated in Figure 8-11c. Sinuses make the skull bones lighter and provide an extensive area of mucous epithelium. The mucous secretions are released into the nasal cavities, and the ciliated epithelium passes the mucus back toward the throat, where it is eventually swallowed. Incoming air is humidified and warmed as it flows across this carpet of mucus, and foreign particles, such as dust or bacteria, become trapped in the sticky mucus and swallowed. This mechanism helps protect more delicate portions of the respiratory tract.

Clinical Comment: Septal Defects and Sinus Troubles

The mucous membrane of the paranasal sinuses responds to environmental stress by accelerating the production of mucus. The mucus flushes irritants off the walls of the nasal cavities, just as water from a hose washes dirt from a car. A variety of stimuli produce this result, including sudden changes in temperature or humidity, irritating vapors, and bacterial or viral infections.

The flushing action often succeeds in removing a mild irritant. But a viral or bacterial infection produces an inflammation of the mucous membrane of the nasal cavity. As swelling occurs, the communicating passageways narrow. Mucus drainage slows, congestion increases, and the victim experiences headaches and a feeling of pressure within the facial bones. This condition of sinus inflammation and congestion is called **sinusitis**. The maxillary sinuses are often involved. Because gravity does little to assist mucus drainage from these sinuses, the effectiveness of the flushing action is reduced and pressure on the sinus walls tends to increase.

Temporary sinus problems may accompany allergies, or the exposure of the mucous epithelium to chemical irritants or invading microorganisms. Chronic sinusitis may occur as the result of a **deviated** (nasal) **septum**. In this condition the nasal septum has a bend in it, most often at the junction between the bony and cartilaginous portions of the septum. Septal deviation often blocks drainage of one or more sinuses, producing chronic bouts of infection and inflammation. Deviated septa can result from developmental abnormalities or injuries to the nose, and the condition can usually be corrected or improved by surgery.

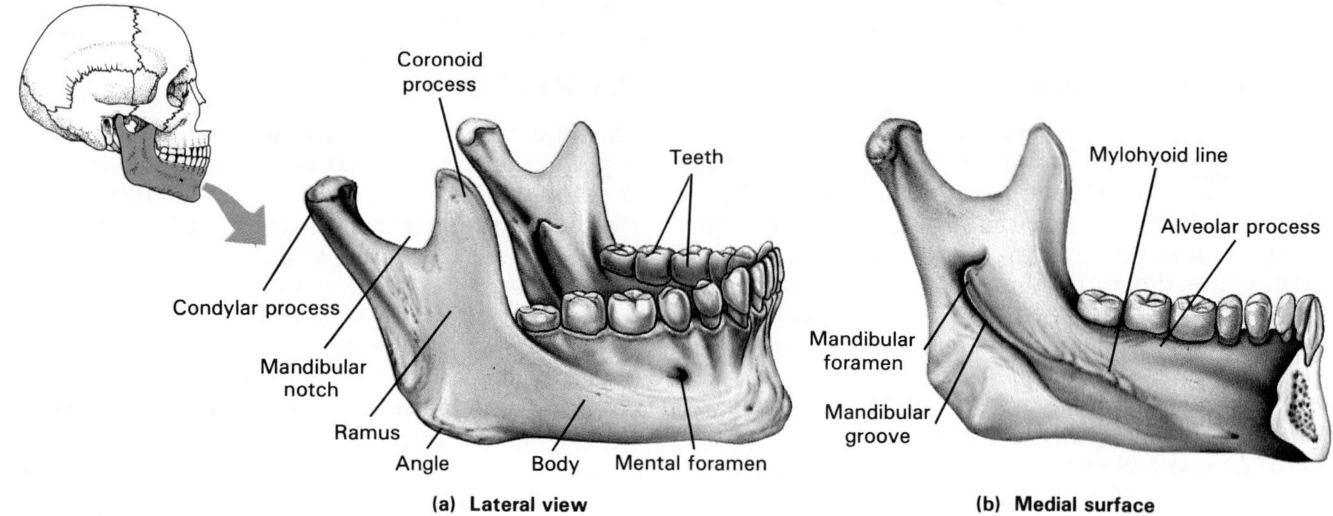

FIGURE 8-12
The Mandible. (a) Lateral view. (b) Medial view of the right side of the mandible.

The Mandible

The broad **mandible**, shown in Figure 8-12, is the entire lower jaw. This bone can be subdivided into the horizontal **body** and the ascending **rami** (branches). Each ramus meets the body at the mandibular **angle**. The **condylar processes** articulate with the mandibular fossae of the temporals (seen in Figure 8-7a). The articulation is quite mobile, as evidenced by the jaw movements during chewing. The disadvantage of such mobility is that the jaw can easily be dislocated by forceful protraction or lateral displacement.

At the **coronoid** (kō-RŌ-noid) **processes** the *temporalis* muscle that originates along the temporal lines inserts onto the mandible. This is one of the most powerful muscles involved in closing the mouth. Anteriorly, the **mental foramina** (*mentalis*, chin) penetrate the body on each side of the chin. Nerves pass through these foramina carrying sensory information from the lips and chin.

Medially (Figure 8-12b), an **alveolar process** covers the alveoli and the roots of the teeth in the lower jaw. The **mylohyoid line** extends horizontally below the alveolar process, marking the attachment site of a muscle (the *mylohyoid*) that supports the floor of the mouth. A depression inferior to this line, the **mandibular groove**, cradles one of the salivary glands. On the medial aspect of each ramus above the groove a prominent **mandibular foramen** leads into the **mandibular canal**. A nerve that uses this passage carries sensory information from the teeth and gums; dentists often anesthetize this nerve before working on the lower teeth.

The Hyoid Bone

The hyoid bone lies below the skull, suspended by the **stylohyoid ligaments** (see Figure 8-13). The **body** of the hyoid serves as a base for several muscles concerned with movements of the tongue and *larynx* (voicebox). Because muscles and ligaments form the only connections between the hyoid and other skeletal elements, the entire complex is quite mobile. The larger processes on the hyoid are the **greater cornu**, which help support the larynx and serve as the base for muscles that move the tongue. The **lesser cornu** are connected to the stylohyoid ligaments, and from these ligaments the hyoid and larynx hang beneath the skull like a child's swing below a tree limb.

Many superficial bumps and ridges are associated with the skeletal muscles described in Chapter 11; learning the names now will help you organize the material in that chapter. Table 8-1 summarizes information concerning the foramina and fissures introduced thus far. This table is intended as a reference that will be especially important in later chapters dealing with the nervous and cardiovascular systems.

FIGURE 8-13
The Hyoid Bone

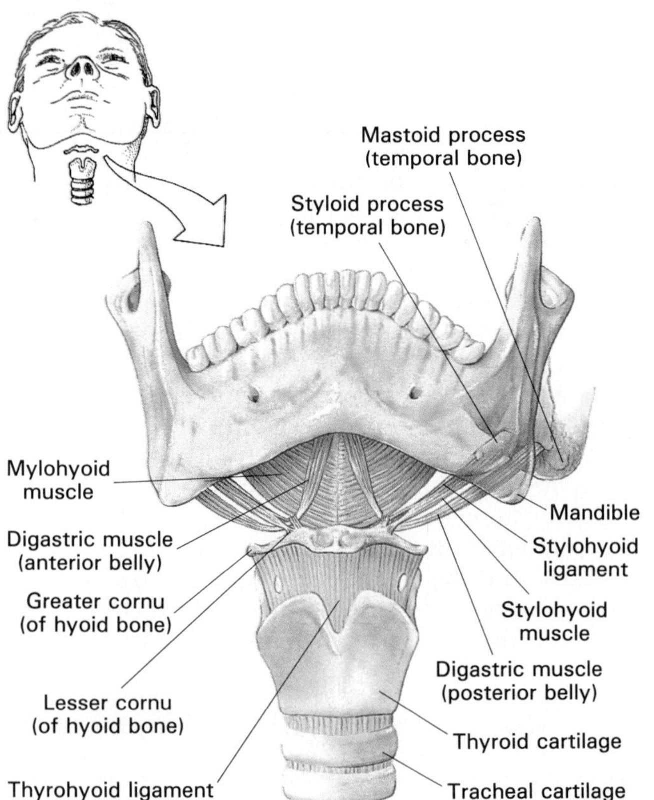

TABLE 8-1 A Key to the Foramina and Fissures of the Skull

Bone	Foramen/Fissure	Major Structures Using Passageway: Neural Tissue	Major Structures Using Passageway: Vessels and Other Structures
OCCIPITAL	Foramen magnum	Medulla (last portion of brain) and spinoaccessory nerve (XI) (Also supporting membranes around CNS)	Vertebral arteries
	Hypoglossal canal	Hypoglossal nerve (XII) provides motor control to muscles of the tongue.	
With temporal	Jugular foramen	Glossopharyngeal nerve (IX), vagus nerve (X), spinoaccessory nerve (XI). Nerve IX provides taste sensation; X is important for visceral functions; XI innervates important muscles of the back and neck.	Internal jugular vein, important vein returning blood from brain to heart.
FRONTAL	Supraorbital foramen	Supraorbital nerve, sensory branch of the ophthalmic nerve, innervating the eyebrow, eyelid, and frontal sinus.	Supraorbital artery delivers blood to same region.
LACRIMAL	Lacrimal foramen		Tear duct, drains into the nasal chamber.
TEMPORAL	Mastoid foramen		Vessels to membranes around CNS
	Stylomastoid foramen	Facial nerve (VII) provides motor control of facial muscles.	
	Carotid foramen		Internal carotid artery, major arterial supply to the brain.
	External auditory meatus and canal		Air conducts sound to eardrum.
	Internal acoustic meatus and canal	Acoustic nerve (VIII) goes to sense organs for hearing and balance. Facial nerve (VII) enters here, exits at stylomastoid foramen.	Internal acoustic artery to inner ear
SPHENOID	Optic foramen	Optic nerve (II) brings information from the eye to the brain.	Opthalmic artery brings blood into orbit.
	Superior orbital fissure	Oculomotor nerve (III), trochlear nerve (IV), ophthalmic branch of trigeminal nerve (V), abducens nerve (VI). Ophthalmic nerve provides sensory information about eye and orbit; other nerves control muscles that move the eye.	Ophthalmic vein returns blood from orbit.
	Foramen rotundum	Maxillary branch of trigeminal nerve (V) provides sensation to the face.	
	Foramen ovale	Mandibular branch of trigeminal nerve (V) controls the muscles that move the lower jaw and provides sensory information from that area.	
	Foramen spinosum		Vessels to membranes around CNS.
With temporal and occipital	Foramen lacerum		Internal carotid artery leaves carotid canal, enters cranium via foramen lacerum.
With maxilla	Inferior orbital fissure	Maxillary branch of trigeminal nerve (V). See *Foramen rotundum*.	
ETHMOID	Cribriform plate	Olfactory nerve (I) provides sense of smell.	
MAXILLA	Infraorbital foramen	Infraorbital nerve, maxillary branch of trigeminal nerve (V) from the inferior orbital fissure to face.	Infraorbital artery same distribution
MANDIBLE	Mental foramen	Mental nerve, sensory nerve branch of the mandibular nerve, provides sensation from the chin and lips.	Mental vessels to chin and lips
	Mandibular foramen	Inferior alveolar nerve, sensory branch of the mandibular nerve, provides sensation from the gums, teeth.	Inferior alveolar vessels, supply same region.

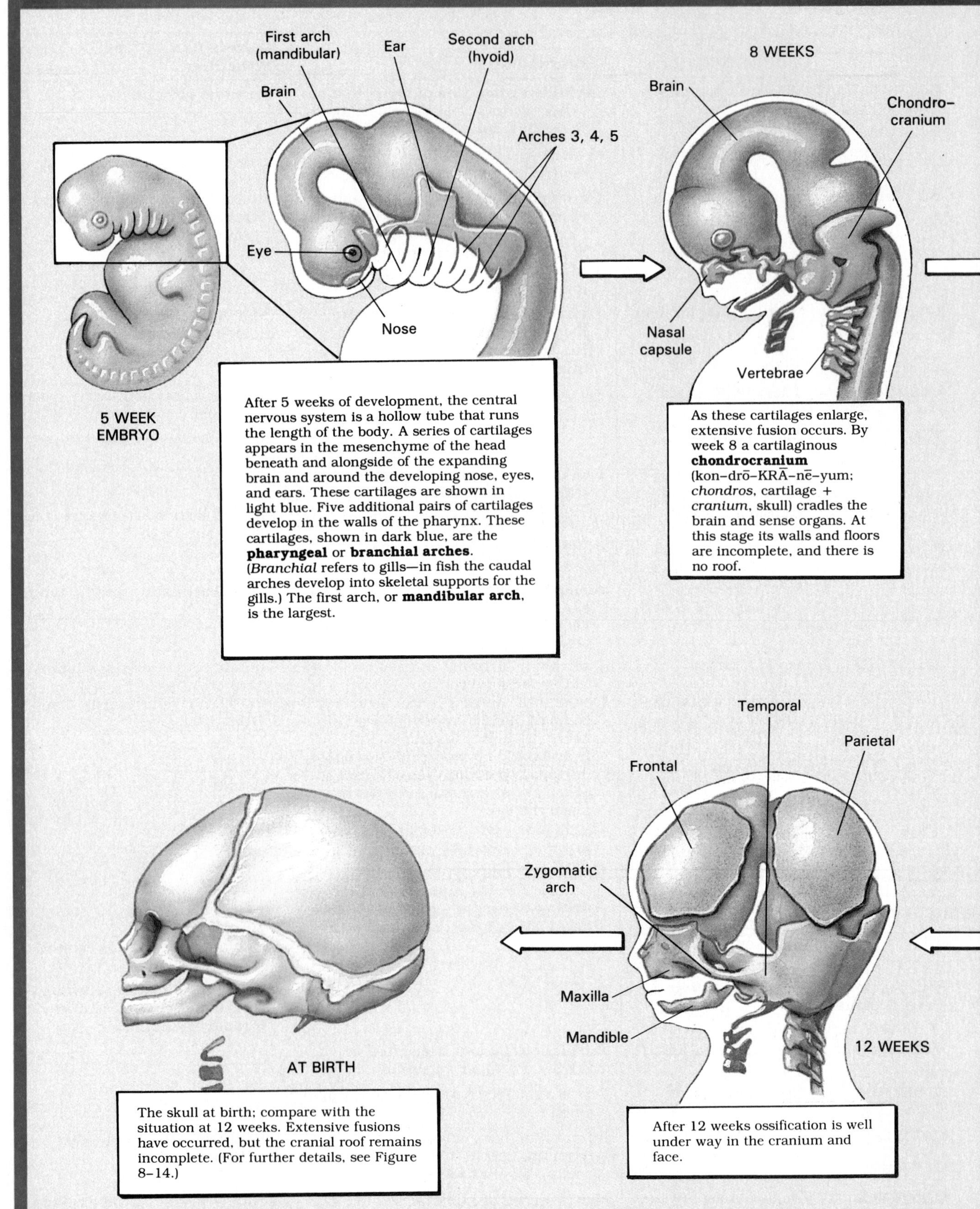

First arch
(mandibular)
Ear
Second arch
(hyoid)
Brain
Arches 3, 4, 5
Eye
Nose
5 WEEK
EMBRYO
After 5 weeks of development, the central nervous system is a hollow tube that runs the length of the body. A series of cartilages appears in the mesenchyme of the head beneath and alongside of the expanding brain and around the developing nose, eyes, and ears. These cartilages are shown in light blue. Five additional pairs of cartilages develop in the walls of the pharynx. These cartilages, shown in dark blue, are the **pharyngeal** or **branchial arches**. (*Branchial* refers to gills—in fish the caudal arches develop into skeletal supports for the gills.) The first arch, or **mandibular arch**, is the largest.
8 WEEKS
Brain
Chondro-
cranium
Nasal
capsule
Vertebrae
As these cartilages enlarge, extensive fusion occurs. By week 8 a cartilaginous **chondrocranium** (kon-drō-KRĀ-nē-yum; *chondros*, cartilage + *cranium*, skull) cradles the brain and sense organs. At this stage its walls and floors are incomplete, and there is no roof.
Temporal
Parietal
Frontal
Zygomatic
arch
Maxilla
Mandible
12 WEEKS
After 12 weeks ossification is well under way in the cranium and face.
AT BIRTH
The skull at birth; compare with the situation at 12 weeks. Extensive fusions have occurred, but the cranial roof remains incomplete. (For further details, see Figure 8–14.)

Embryology Summary: Development of the Skull

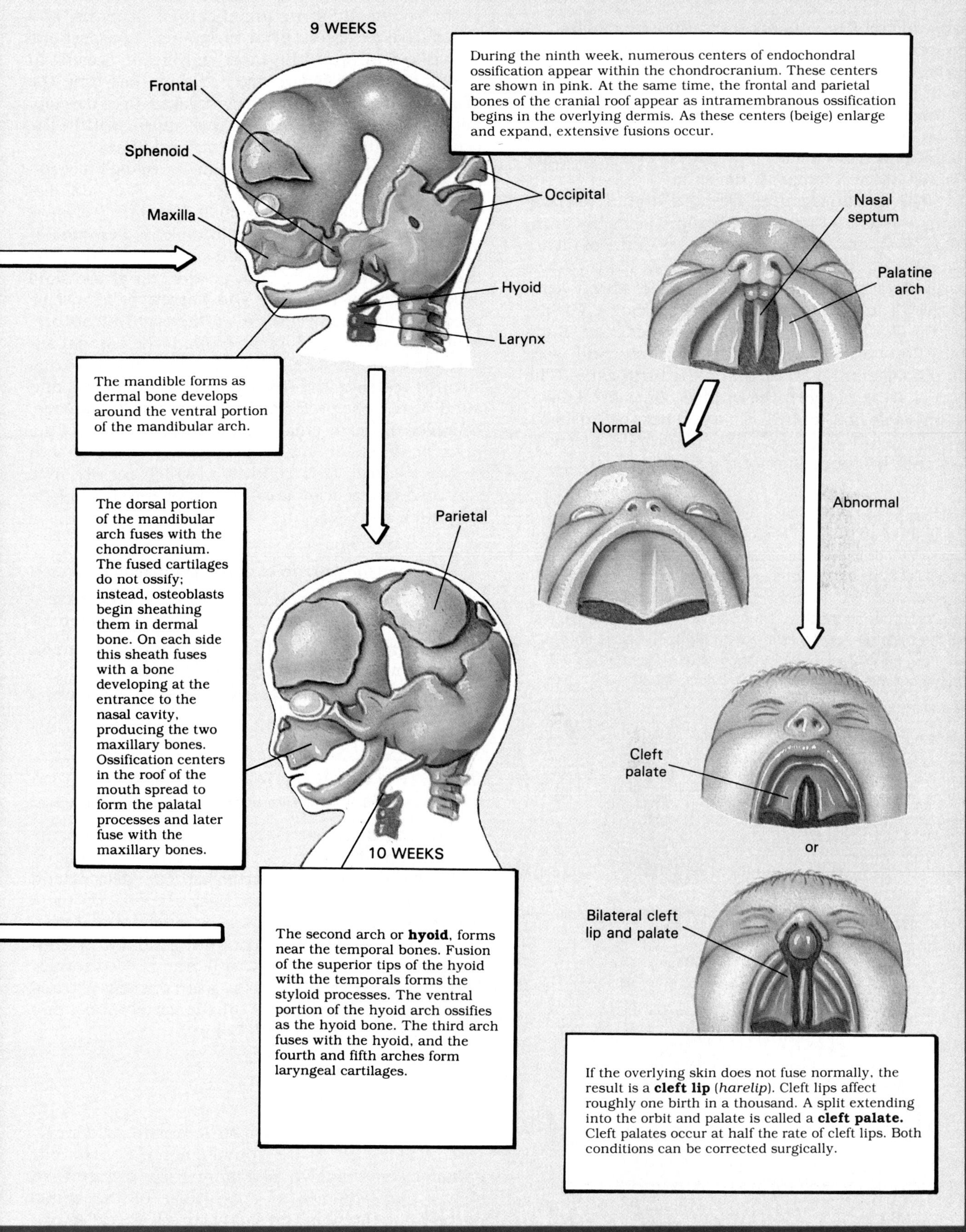

During the ninth week, numerous centers of endochondral ossification appear within the chondrocranium. These centers are shown in pink. At the same time, the frontal and parietal bones of the cranial roof appear as intramembranous ossification begins in the overlying dermis. As these centers (beige) enlarge and expand, extensive fusions occur.

The mandible forms as dermal bone develops around the ventral portion of the mandibular arch.

The dorsal portion of the mandibular arch fuses with the chondrocranium. The fused cartilages do not ossify; instead, osteoblasts begin sheathing them in dermal bone. On each side this sheath fuses with a bone developing at the entrance to the nasal cavity, producing the two maxillary bones. Ossification centers in the roof of the mouth spread to form the palatal processes and later fuse with the maxillary bones.

The second arch or **hyoid**, forms near the temporal bones. Fusion of the superior tips of the hyoid with the temporals forms the styloid processes. The ventral portion of the hyoid arch ossifies as the hyoid bone. The third arch fuses with the hyoid, and the fourth and fifth arches form laryngeal cartilages.

If the overlying skin does not fuse normally, the result is a **cleft lip** (*harelip*). Cleft lips affect roughly one birth in a thousand. A split extending into the orbit and palate is called a **cleft palate.** Cleft palates occur at half the rate of cleft lips. Both conditions can be corrected surgically.

THE SKULLS OF INFANTS AND CHILDREN

Many different centers of ossification are involved in the formation of the skull, but as development proceeds, fusion of the centers produces a smaller number of composite bones. For example, the sphenoid begins as 14 separate ossification centers. At birth fusion has not been completed, and there are two frontal bones, four occipital bones, and a number of sphenoid and temporal elements.

The skull organizes around the developing brain, and as the time of birth approaches the brain enlarges rapidly. Although the bones of the skull are also growing, they fail to keep pace, and at birth the cranial bones are connected by areas of fibrous connective tissue. These connections are quite flexible, and the skull can be distorted without damage. Such distortion normally occurs during delivery and eases the passage of the infant along the birth canal. The fibrous areas between the cranial bones are known as **fontanels** (fon-tah-NELS; sometimes spelled *fontanelles*). Figure 8-14 indicates the prominent fontanels and the appearance of the skull at birth.

FIGURE 8-14
The Skull of an Infant. The skull of an infant contains a greater number of individual bones than that of an adult. Many of the bones will eventually fuse; thus there will be fewer bones in the adult skull. The flat bones of the skull are separated by areas of fibrous connective tissue, allowing for cranial expansion and the distortion of the skull during birth. The large fibrous areas are called fontanels. By about age 4 these areas will disappear, and skull growth will be completed.

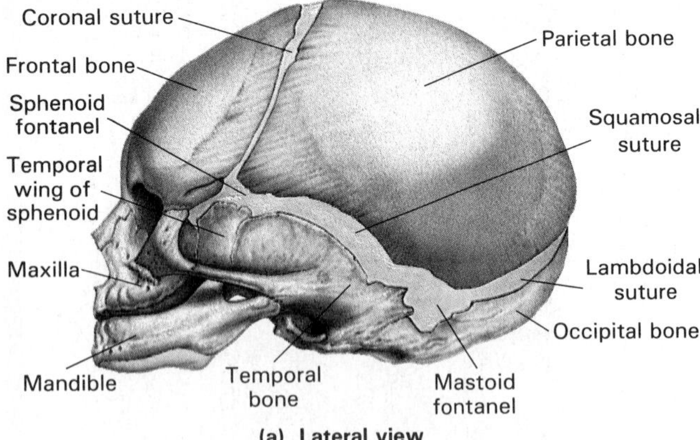

(a) Lateral view

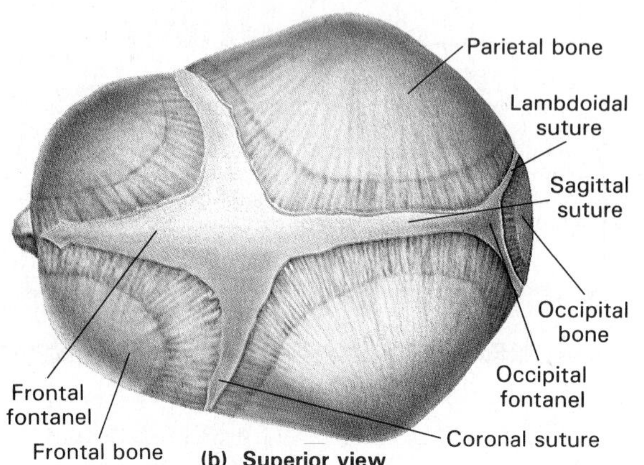

(b) Superior view

The skulls of infants and adults differ in terms of the shape and structure of cranial elements, and this difference accounts for variations in proportions as well as in size. The most significant growth in the skull occurs before age 5, for at that time the brain stops growing and the cranial sutures develop. Separate ossification centers that appear within the occipital fontanel produce the sutural bones.

The growth of the cranium is usually coordinated with the expansion of the brain. Unusual distortions of the skull result from the premature closure of one or more fontanel, a condition called **craniostenosis** (krā-nē-ō-sten-Ō-sis; *stenosis*, narrowing). As the brain continues to enlarge, the rest of the skull accommodates it. A long and narrow head will be produced by early closure of the sagittal suture, whereas a very broad skull results if the coronal suture forms prematurely. Closure of *all* of the cranial sutures restricts the development of the brain, and surgery must be performed to prevent brain damage. However, if brain enlargement stops because of genetic or developmental abnormalities, skull growth ceases as well. This condition, which results in a very undersized head, is called **microcephaly** (mī-krō-SEF-a-lē). [**CM**: *Phrenology*]

- √ **What are the functions of the paranasal sinuses?**
- √ **Why would a fracture of the coronoid process of the mandible make it difficult to close the mouth?**
- √ **What symptoms would you expect to see in a person suffering from a fractured hyoid bone?**

The Neck and Trunk

The rest of the axial skeleton is subdivided on the basis of vertebral structure, as indicated in Figure 8-15. The seven **cervical vertebrae** of the neck extend inferiorly as far as the trunk. Each of the 12 **thoracic vertebrae** articulates with one or more pairs of **ribs**. The five **lumbar vertebrae** continue caudally, the fifth articulating with the fused vertebrae of the **sacrum**. The small **coccyx** (KOK-siks) also consists of fused vertebrae. The total length of the vertebral column of an adult averages 71 cm (28 in.).

SPINAL CURVATURE

The vertebrae do not form a straight and rigid structure. A side view of the spinal column reveals four **spinal curves**, shown in Figure 8-15. Figure 8-16 shows the sequence of appearance of the spinal curves. The thoracic and sacral curves are called **primary curves** because they begin to appear late in fetal development. These are also called **accommodation curves** because they accommodate the thoracic

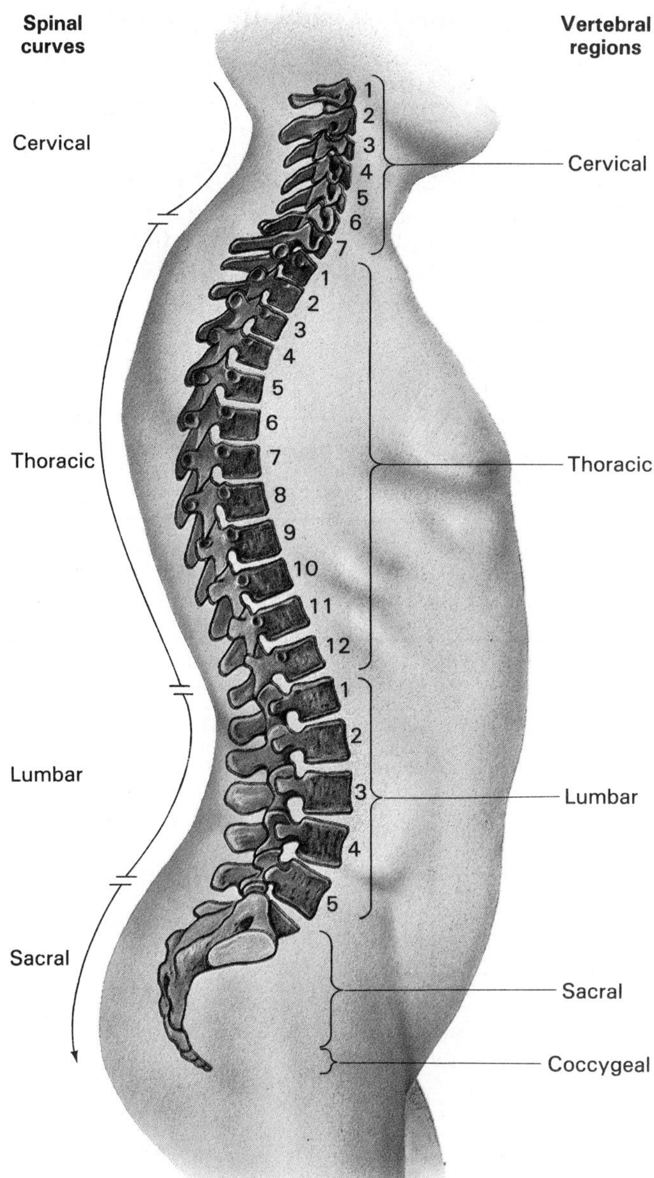

FIGURE 8-15
The Vertebral Column. The major divisions of the vertebral column, showing the four spinal curves.

FIGURE 8-16
The Development of Spinal Curvature. The secondary spinal curves, which balance the body weight over the legs, do not appear until the individual begins walking.

and abdominopelvic viscera. The lumbar and cervical curves, known as **secondary curves**, do not appear until several months after birth. These are also known as **compensation curves** because they help position the body weight over the legs.

When standing, the weight of the body must be transmitted through the spinal column to the pelvic girdle and ultimately to the legs. Yet most of the body weight lies in front of the spinal column. The various curves bring that weight in line with the body axis. Consider what people do automatically when they stand holding a heavy object. To avoid toppling forward, they exaggerate the lumbar curvature, bringing the weight closer to the body axis. This posture can lead to discomfort at the base of the spinal column. Similarly, women in the later stages of pregnancy often develop chronic back pain from the changes in lumbar curvature that adjust for the increasing weight of the fetus. No doubt you have seen pictures of African or South American people carrying heavy objects balanced on their heads. Such a practice increases the load on the spinal column, but because the weight is aligned with the axis of the spine, the spinal curves are not affected and strain is minimized.

Clinical Comment: Kyphosis, Lordosis, Scoliosis

There are several abnormal distortions of the normal spinal curvature. In **kyphosis** (kī-FŌ-sis), the normal thoracic curvature becomes exaggerated, producing a "roundback" appearance. This can be caused by (1) osteoporosis or a compression fracture affecting the anterior portions of vertebral centra, (2) chronic contractions in muscles that insert on the vertebrae, or (3) abnormal vertebral growth. In **lordosis** (lor-DŌ-sis), or "swayback," the abdomen and buttocks protrude because of an exaggerated lumbar curvature.

Scoliosis (skō-lē-Ō-sis) involves an abnormal lateral curvature. This curvature may result from developmental problems, such as incomplete vertebral formation, or from muscular paralysis affecting one side of the back. In four out of five cases it is impossible to determine the structural or functional cause of the abnormal spinal curvature. Scoliosis usually appears in adolescence, during periods of rapid growth. Treatment consists of a combination of exercises, braces, and sometimes surgical modifications of the affected vertebrae. Early detection greatly improves the chances for successful treatment.

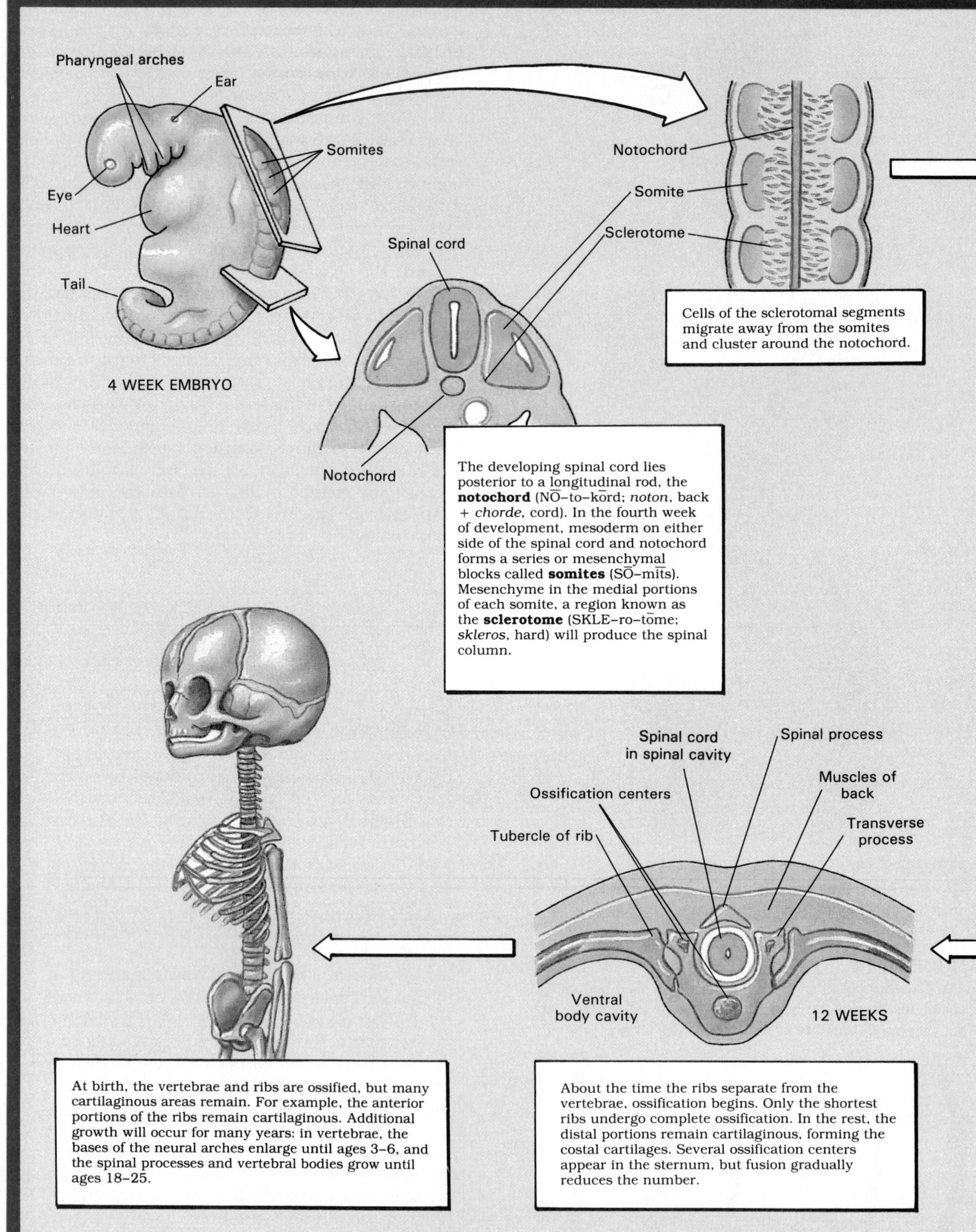
Pharyngeal arches
Ear
Somites
Eye
Heart
Tail
4 WEEK EMBRYO
Notochord
Somite
Sclerotome
Spinal cord
Notochord
Cells of the sclerotomal segments migrate away from the somites and cluster around the notochord.
The developing spinal cord lies posterior to a longitudinal rod, the **notochord** (NŌ–to–kōrd; *noton*, back + *chorde*, cord). In the fourth week of development, mesoderm on either side of the spinal cord and notochord forms a series or mesenchymal blocks called **somites** (SŌ–mīts). Mesenchyme in the medial portions of each somite, a region known as the **sclerotome** (SKLE–ro–tōme; *skleros*, hard) will produce the spinal column.
Spinal cord in spinal cavity
Spinal process
Muscles of back
Ossification centers
Transverse process
Tubercle of rib
Ventral body cavity
12 WEEKS
At birth, the vertebrae and ribs are ossified, but many cartilaginous areas remain. For example, the anterior portions of the ribs remain cartilaginous. Additional growth will occur for many years: in vertebrae, the bases of the neural arches enlarge until ages 3–6, and the spinal processes and vertebral bodies grow until ages 18–25.
About the time the ribs separate from the vertebrae, ossification begins. Only the shortest ribs undergo complete ossification. In the rest, the distal portions remain cartilaginous, forming the costal cartilages. Several ossification centers appear in the sternum, but fusion gradually reduces the number.

Embryology Summary: Development of the Spinal Column

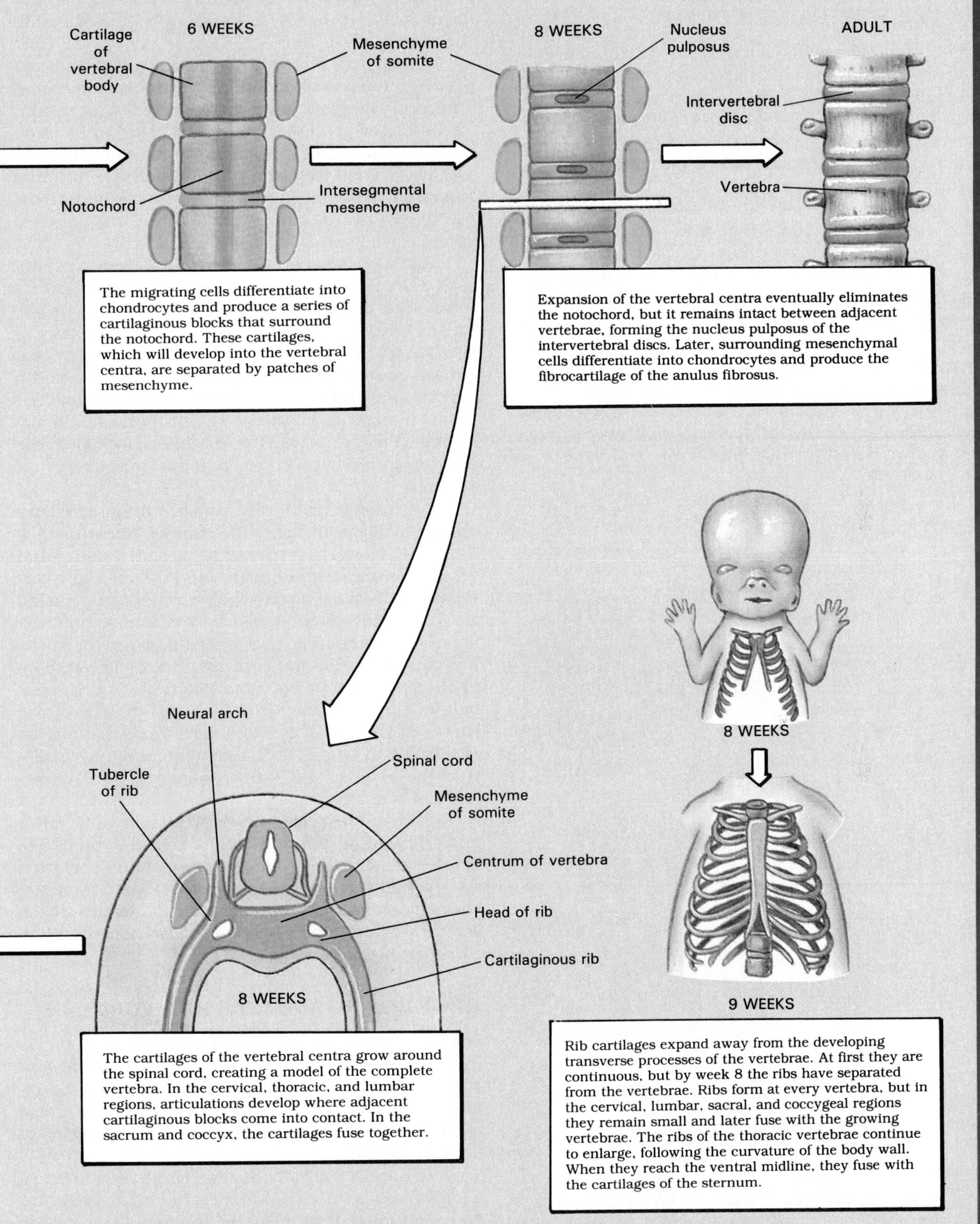

The migrating cells differentiate into chondrocytes and produce a series of cartilaginous blocks that surround the notochord. These cartilages, which will develop into the vertebral centra, are separated by patches of mesenchyme.

Expansion of the vertebral centra eventually eliminates the notochord, but it remains intact between adjacent vertebrae, forming the nucleus pulposus of the intervertebral discs. Later, surrounding mesenchymal cells differentiate into chondrocytes and produce the fibrocartilage of the anulus fibrosus.

The cartilages of the vertebral centra grow around the spinal cord, creating a model of the complete vertebra. In the cervical, thoracic, and lumbar regions, articulations develop where adjacent cartilaginous blocks come into contact. In the sacrum and coccyx, the cartilages fuse together.

Rib cartilages expand away from the developing transverse processes of the vertebrae. At first they are continuous, but by week 8 the ribs have separated from the vertebrae. Ribs form at every vertebra, but in the cervical, lumbar, sacral, and coccygeal regions they remain small and later fuse with the growing vertebrae. The ribs of the thoracic vertebrae continue to enlarge, following the curvature of the body wall. When they reach the ventral midline, they fuse with the cartilages of the sternum.

INTRODUCTORY ANATOMY OF THE VERTEBRAL COLUMN

Figure 8-17 shows a typical vertebra from several different perspectives. The mass of a vertebra is concentrated within the **body**, or **centrum** (plural, *centra*). The **pedicles** (PE-di-kls) arise along the *posterolateral* (posterior and lateral) margins of the body. The pedicles form the side walls of the **vertebral foramen**, or **vertebral canal**, that houses the spinal cord. The **laminae** (LA-mi-nē) extend *dorsomedially* (dorsally and medially) from the pedicles, uniting to form the **spinal** (or *spinous*) **process**. The laminae form the roof of the vertebral foramen, and together the pedicles and laminae constitute the **vertebral arch**, or **neural arch**. **Transverse processes** project laterally or dorsolaterally from the pedicles. They serve as sites for muscle attachment and may also articulate with the ribs.

FIGURE 8-17
Vertebral Anatomy. The anatomy of a typical vertebra and the structure of the intervertebral articulations. Note the position and structure of the intervertebral discs and the way the spinal column surrounds and protects the delicate neural tissue of the spinal cord.

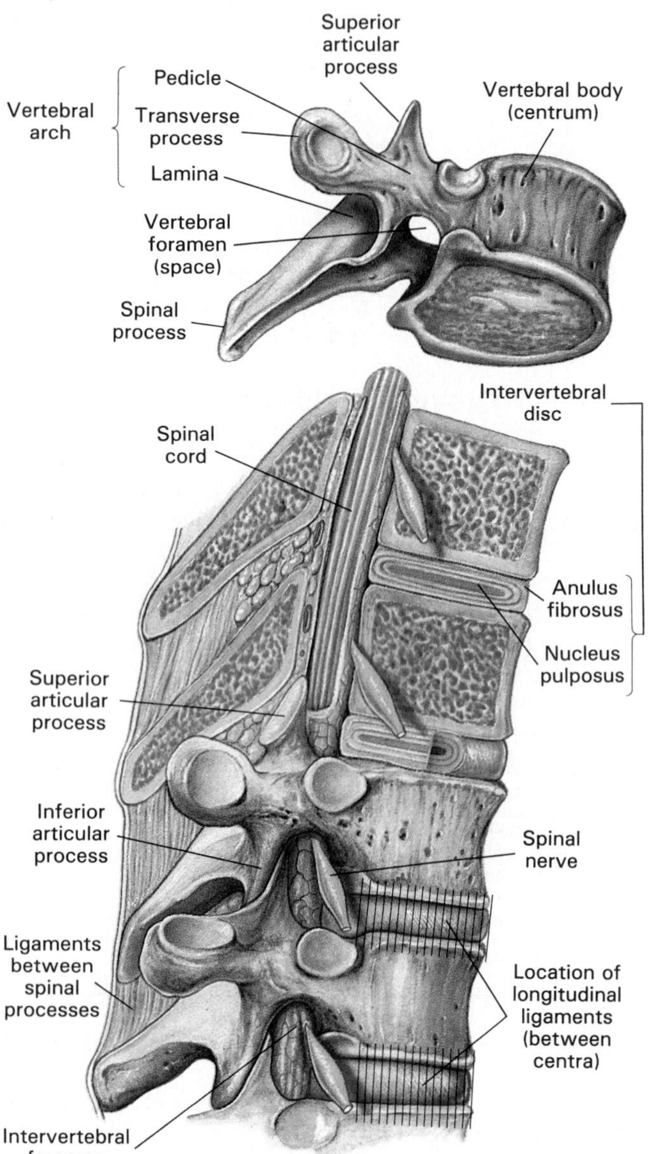

Successive vertebrae articulate at gliding joints between **superior** and **inferior articular processes** and at symphyseal joints between vertebral centra. The superior processes have their articular faces on their dorsal surfaces, whereas the inferior processes articulate along their ventral aspects. **Intervertebral foramina** between successive vertebrae permit the passage of nerves running to or from the enclosed spinal cord.

Little gliding occurs between adjacent vertebral centra because the articulations of symphyseal joints (discussed in Chapter 7) restrict relative movement. ∞ (p. 227) Figure 8-17 contains details of the articular structure. The periostea of adjacent centra are connected by longitudinal ligaments, and a cartilaginous **intervertebral disc** usually lies between them. Intervertebral discs are not found in the sacrum and coccyx, where the vertebrae have fused, nor are they found between the first and second cervical vertebrae.

An intervertebral disc consists of an extensive region of fibrocartilage, the **anulus fibrosus** (AN-ū-lus fī-BRŌ-sus), surrounding a soft, elastic, and gelatinous **nucleus pulposus** (pul-PŌ-sus). The intervertebral discs act as shock absorbers, compressing and distorting when stressed. This change prevents bone-to-bone contact that might damage the vertebrae or jolt the spinal cord and brain. These discs make a significant contribution to an individual's height; they account for roughly one-quarter of the length of the spinal column above the sacrum. Part of the loss in height that accompanies aging results from the decreasing size and resiliency of the intervertebral discs.

The symphyseal joints between vertebral centra and the gliding joints between articular processes severely restrict the possible range of motion between any two adjacent vertebrae. Nevertheless, there are so many such articulations along the length of the vertebral column that the spine as a whole can achieve considerable mobility.

REGIONAL STRUCTURE AND FUNCTION

Although each vertebra bears characteristic markings and articulations, you should focus on the general characteristics of each region and how the regional variations reflect differences in function. Table 8-2 compares typical vertebrae from each region of the spinal cord.

To specify an individual vertebrae, a numerical shorthand is used. A capital letter indicates the region and a number indicates the vertebra in question, starting with the vertebra closest to the skull. Thus C_3 refers to the third cervical vertebra, with C_1 in contact w th the skull; L_4 is the fourth lumbar vertebra, with L_1 in contact with the last thoracic vertebra (see Figure 8-15).

TABLE 8-2 Regional Differences in Vertebral Structure and Function

Type (Number)	Location	Centrum	Vertebral Foramen	Spinal Process	Transverse Process	Functions
Cervical vertebra (7)	Neck	Small, oval, curved faces	Large	Long, split, tip, points caudally	Has transverse foramen	Support skull, stabilize relative positions of brain and spinal cord, allow controlled head movement
Thoracic vertebra (12)	Chest	Medium, heart-shaped, flat faces, facets for rib articulations	Smaller	Long, slender, not split, points caudally	All but two have facets for rib articulations	Support weight of head, neck, organs of thoracic cavity, articulate other ribs to allow changes in volume of thoracic cage
Lumbar vertebrae (5)	Lower back	Massive, oval flat faces	Smallest	Blunt, broad, points posteriorly	Short, no articular facets or transverse foramina	Support weight of head, neck, organs of thoracic and abdominal cavities

Cervical Vertebrae

The seven cervical vertebrae extend from the head to the thorax. A typical cervical vertebra is illustrated in Table 8-2, with additional details provided in Figure 8-18. Notice that the body of the vertebra is relatively small compared with the size of the vertebral foramen. At this level the spinal cord still contains most of the nerves that connect the brain to the rest of the body. As you continue along the vertebral canal, the diameter of the spinal cord decreases, and so does the size of the neural arch. On the other hand, cervical vertebrae support only the weight of the head, so the vertebral bodies can be relatively small and light. As you continue caudally along the column, the centra gradually enlarge.

In a cervical vertebra, the superior face of the centrum is concave from side to side and also slopes in an anterior and caudal direction. The spinal process is relatively stumpy, usually shorter than the diameter of the vertebral foramen, and the tip of the process bears a prominent notch. Laterally, the transverse processes are fused to the **costal processes** that originate near the ventrolateral portion of the body. *Costal* refers to rib, and these processes represent the fused remnants of cervical ribs. The costal and transverse processes encircle prominent, round, **transverse foramina**. In life these passageways protect important blood vessels servicing the brain.

This description would be adequate to identify all but the first two cervical vertebrae. When cervical vertebrae C_3–C_7 articulate, their interlocking centra permit a relatively greater degree of flexibility than do those of other regions. The **atlas** (C_1) holds up the head, articulating with the occipital condyles of the skull. It is named after Atlas, a figure in Greek mythology who held up the world. The articulation between the occipital and the atlas is a pair of ellipsoidal joints that permit nodding (as when indicating "yes") but prevent twisting. The atlas in turn forms a pivot joint with the **axis** (C_2). This articulation permits rotation (as when shaking the head to indicate "no"). The organization of the atlas/axis joint can be seen in Figure 8-18c.

The two vertebrae can be recognized easily once their functions are understood. The atlas looks like a typical cervical vertebra that has lost its body. The body was not actually misplaced, for during development it fused to the centrum of the axis, forming the prominent **dens** (*denz*; tooth), or **odontoid** (ō-DON-toid; *odontos*, tooth) **process**. A transverse ligament anchors the dens to the inner surface of the atlas. Important muscles controlling the position of the head and neck attach to the especially robust spinal process of the axis.

The transition from one vertebral region to another is not abrupt, and the caudal vertebra of one region usually resembles the cranial vertebra of the next. The **vertebra prominens**, C_7, has a long, slender spinal process that ends in a broad tubercle that can be felt beneath the skin at the base of the neck. This vertebra, outlined in Figure 8-19, is the interface between the cervical curve, which arches forward, and the thoracic curve, which arches backward. The transverse processes are large, providing additional surface area for muscle attachment, and the transverse foramen is very small. The foramen may even be absent entirely, in which case the costal process forms a small but distinct rib.

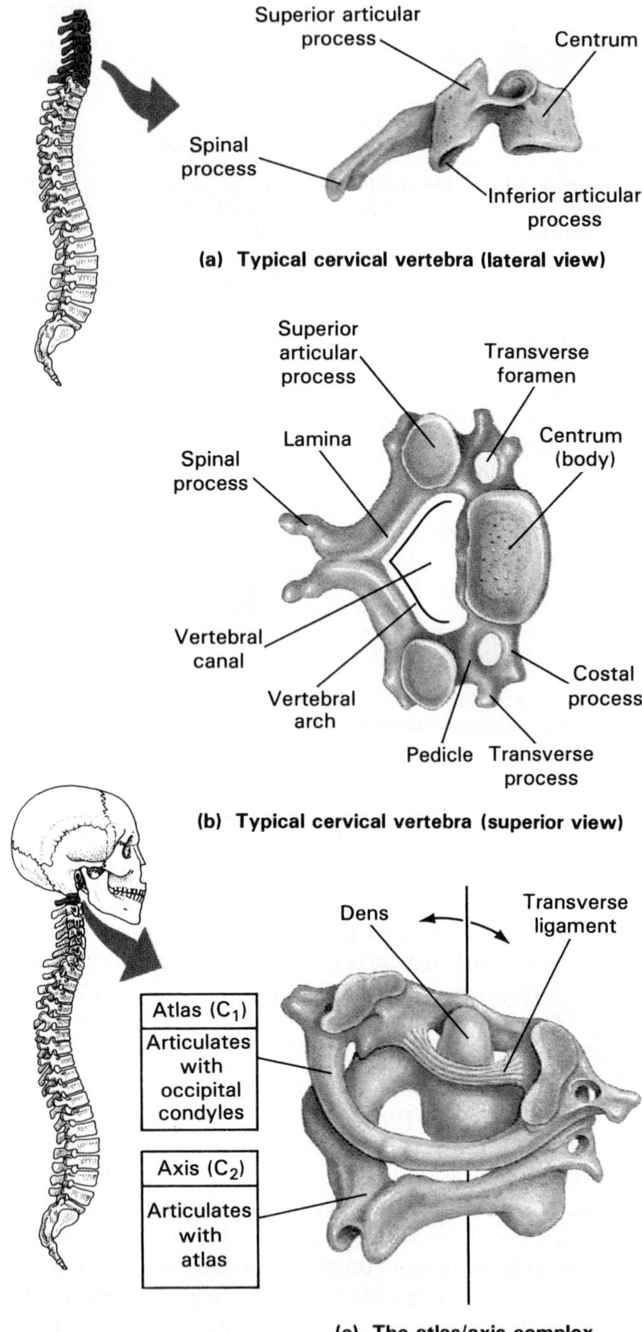

FIGURE 8-18
Cervical Vertebrae. (a) Lateral view of a typical cervical vertebra (C_3–C_6). (b) Superior view of the same vertebra. Note the characteristic features listed in Table 8-2. (c) The articulation between the atlas (C_1) and axis (C_2).

A large elastic ligament, the **ligamentum nuchae** (li-ga-MEN-tum NOO-kē; *nucha*, nape) begins at the vertebra prominens and extends cranially to an insertion along the external occipital crest. Along the way, it also attaches to the spinal processes of the other cervical vertebrae. When the head is upright, this ligament acts like the string on a bow, maintaining the cervical curvature without muscular effort. If the neck has been bent forward, the elasticity in this ligament helps return the head to an upright position.

The head is relatively massive, and it sits atop the cervical vertebrae like a soup bowl on the tip of a finger. With this arrangement, small muscles can produce significant effects by tipping the balance one way or another. But if the body suddenly changes position, as in a fall or during rapid acceleration (a jet taking off) or deceleration (a car crash), the balancing muscles are not strong enough to stabilize the head. A dangerous partial or complete dislocation of the cervical vertebrae can result, with injury to muscles and ligaments and potential injury to the spinal cord. The term **whiplash** is used to describe such an injury, because the movement of the head resembles the cracking of a whip.

A blow to the base of the skull can be particularly dangerous because of its potential effects on the atlas and axis. A dislocation here can force the dens into the base of the brain, with fatal results.

The Thorax

The skeleton of the chest, or **thorax**, consists of the thoracic vertebrae, the ribs, and the sternum. The ribs, or **costae**, and the sternum form the **rib cage** and establish the contours of the thoracic cavity. The rib cage protects the heart, lungs, and other internal organs and serves as a base for muscles involved with respiration.

Thoracic Vertebrae There are 12 thoracic vertebrae. A typical thoracic vertebra (Table 8-2 and Figure 8-19) has a distinctive heart-shaped body that is more massive than that of a cervical vertebra. The vertebral foramen is relatively smaller, and the long, slender spinal process projects posterocaudally. The spinal processes of T_{10}, T_{11}, and T_{12} increasingly resemble those of the lumbar series as the transition between the thoracic and lumbar curvatures approaches. Because of the weight carried by the lower thoracic and lumbar vertebrae, it is difficult to stabilize the transition between the thoracic and lumbar curves. As a result, after a hard fall compression fractures or compression/dislocation fractures most often involve the last thoracic and first two lumbar vertebrae.

Each thoracic vertebra articulates with ribs along the dorsolateral surfaces of the centrum. The location and structure of the articulations vary somewhat from vertebra to vertebra, as indicated in Figure 8-19b. Rib pairs 2 through 8 originate between adjacent vertebrae, so vertebrae T_2 to T_8 have **superior** and **inferior demifacets** on each side. The first rib originates at the body of T_1, so that vertebra has an articular facet and an inferior demifacet but no superior demifacet on each side. Vertebra T_9 has only a superior demifacet, whereas T_{10}, T_{11}, and T_{12} have a single articular facet on either side. The transverse processes of T_1 to T_{10} also contain facets for rib articulation.

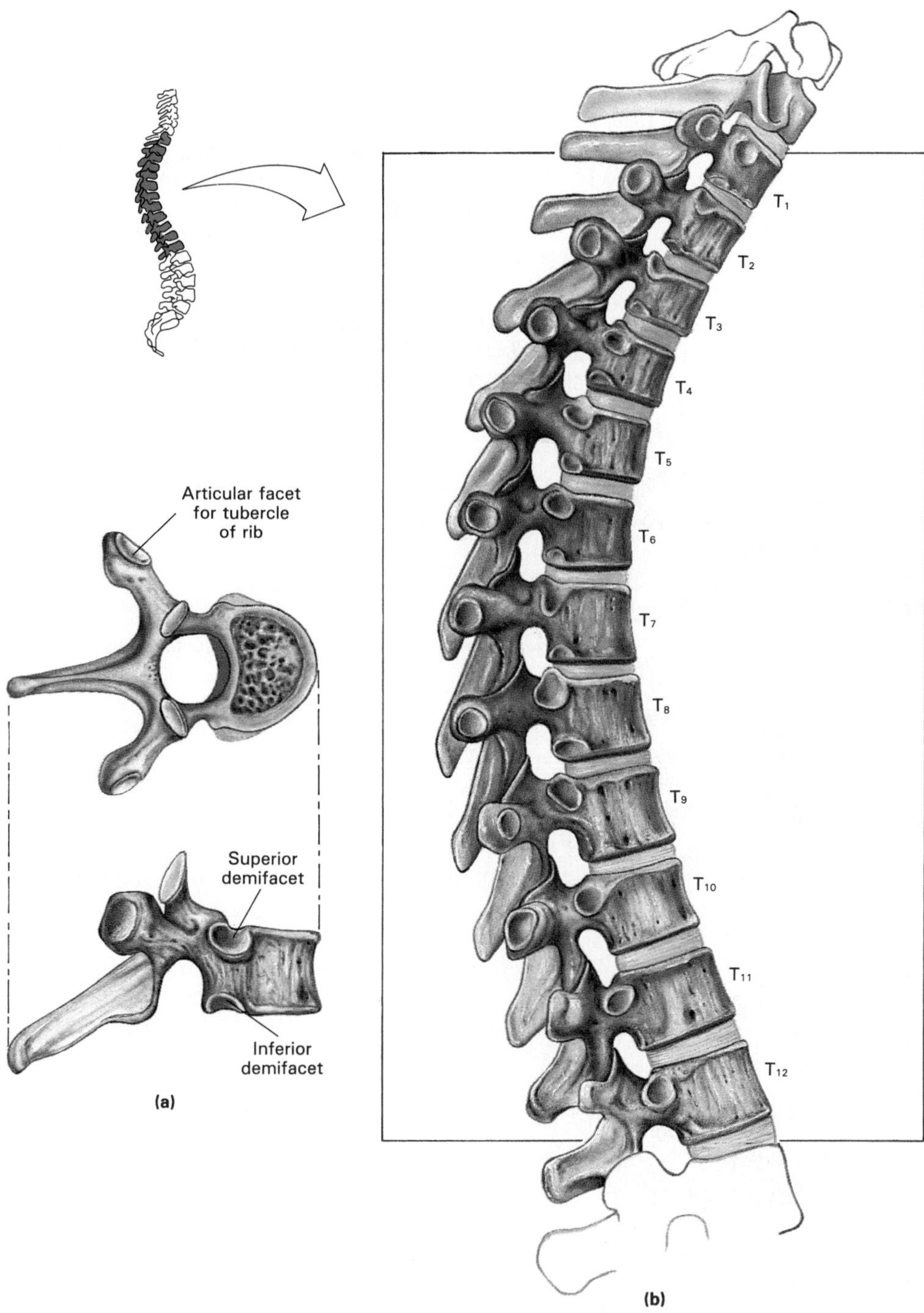

FIGURE 8-19
Thoracic Vertebrae. (a) Superior and lateral views of a typical thoracic vertebra (T_1–T_{10}). Note the characteristic features listed in Table 8-2. (b) Lateral view of the thoracic region of the spinal column. The last cervical vertebra (C_7, or the vertebra prominens) resembles the first thoracic vertebra. The last thoracic (T_{12}) resembles the first lumbar vertebra.

The Ribs and Sternum Ribs are elongate, flattened bones that originate on or between the thoracic vertebrae and end in the wall of the thoracic cavity. There are 12 pairs of ribs, detailed in Figure 8-20. The first seven pairs are called **true**, or **vertebrosternal**, **ribs**. They reach the anterior body wall and are connected to the sternum by separate cartilaginous extensions, the **costal cartilages**. Ribs 8–12 are called the **false ribs** because they do not attach directly to the sternum. The costal cartilages of ribs 8–10, the **vertebrochondral ribs**, fuse together. This fused cartilage merges with the costal cartilage of rib 7 before it reaches the sternum (see Figure 8-20a). The last two pairs of ribs are called **floating ribs** because they have no connection with the sternum.

Figure 8-20b shows the superior surface of a typical rib. The rib articulates with the vertebral column at the **head**, or **capitulum** (ka-PIT-ū-lum). When

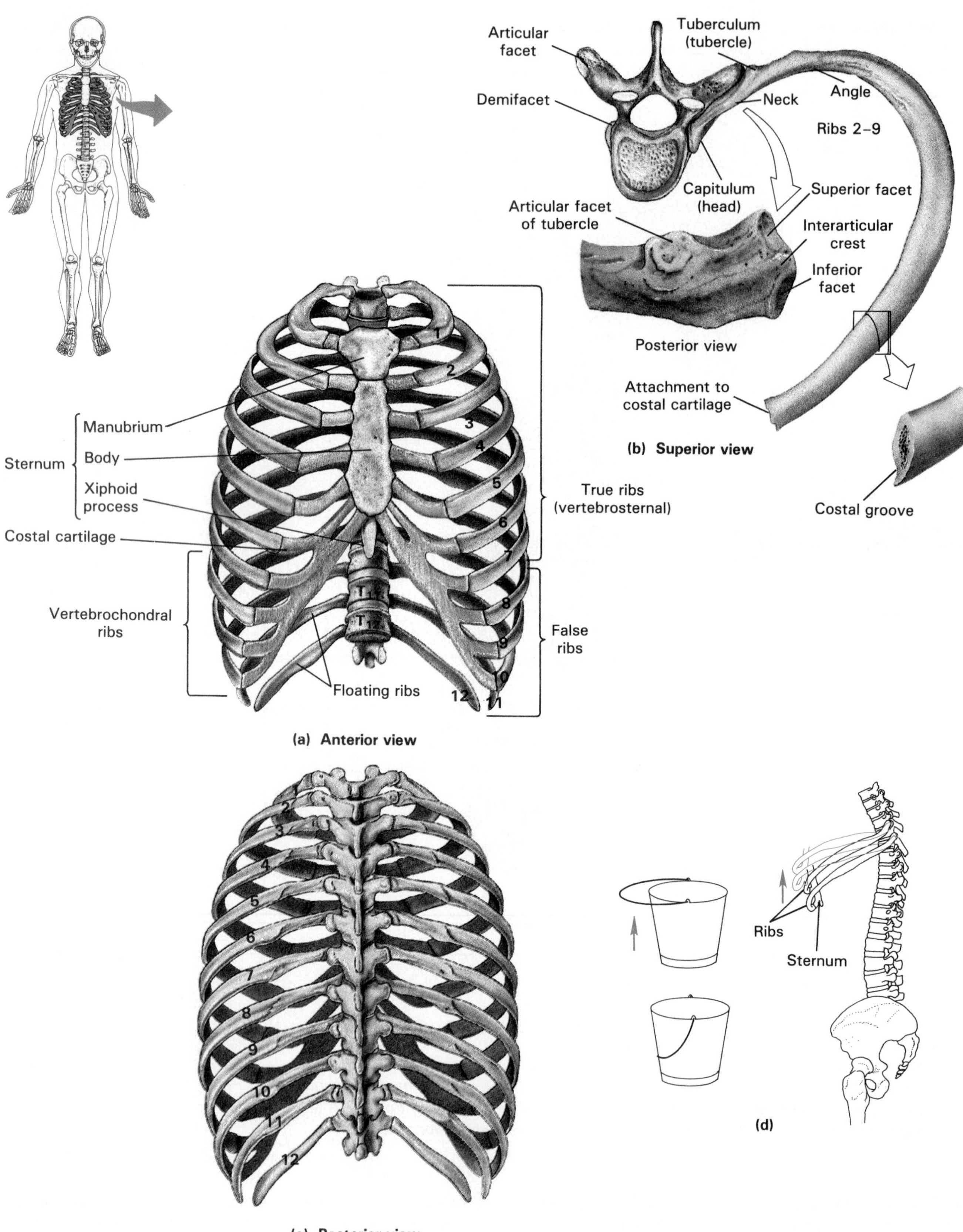

FIGURE 8-20

The Thoracic Cage. (a) Anterior view of the rib cage and sternum. (b) Details of rib structure and the articulations between the ribs and thoracic vertebrae. (c) Posterior view of the rib cage. (d) Effect of rib movement on the volume of the thoracic cavity. These movements are important in respiration.

articulating between adjacent vertebrae, the articular face is divided into **superior** and **inferior articular facets** by the **interarticular crest**. After a short **neck**, the **tubercle**, or **tuberculum** (tu-BER-kū-lum), projects dorsally. The inferior portion of the tubercle contains an articular facet that contacts the transverse process of the thoracic vertebra.

In the case of ribs that originate at demifacets, this articulation involves the more caudal member of the vertebral pair. From T_1 to T_{10}, ribs that originate at facets articulate with their respective vertebrae. The transverse processes of T_{11} and T_{12} do not contain articular facets, and neither do the corresponding tubercles. The difference in rib orientation can be seen in Figure 8-20c.

Returning to Figure 8-20b, the bend, or **angle**, of the rib indicates the site where the **body**, or **shaft**, begins to curve toward the sternum. The superficial surface is convex and provides an attachment site for muscles of the shoulder girdle and trunk. The superior and inferior surfaces connect to the **intercostal muscles**, which move the ribs. The internal surface is concave, and a prominent **costal groove** along its inferior border marks the path of nerves and blood vessels.

Now return to Figure 8-20a. The adult sternum has three components. The broad, triangular **manubrium** (ma-NŪ-brē-um) articulates with the clavicles and the cartilages of the first pair of ribs. The elongate **body** ends at the slender **xiphoid** (ZĪ-foid) **process**. Ossification of the sternum begins at six to ten different centers, and fusion is not completed until at least age 25. Before age 25, the sternal body consists of four separate bones. Their boundaries can be detected as a series of transverse lines crossing the adult sternum. The xiphoid process is usually the last of the sternal components to undergo ossification and fusion. Its connection to the body of the sternum can be broken by an impact or strong pressure, creating a spear of bone that can severely damage the liver. To reduce the chances of that happening, strong emphasis is placed on the proper positioning of the hand during cardiopulmonary resuscitation (CPR) training.

With their complex musculature, dual articulations at the vertebrae, and flexible connection to the sternum, the ribs are quite mobile. If you compare parts (a) and (c) of Figure 8-20 you will find that as the ribs curve away from the vertebral column they angle downward. Functionally, a typical rib acts as if it were the handle on a bucket, lying just below the horizontal plane (Figure 8-20d). Pushing it down forces it inward; pulling it up swings it outward. In addition, because of the curvature of the ribs the same movements change the position of the sternum. Depressing the ribs moves the sternum posteriorly, whereas elevation moves it anteriorly. As a result, movements of the ribs affect both the width and the depth of the thoracic cage, increasing or decreasing its volume accordingly. [**CM**: *The Thoracic Cage and Surgical Procedures*]

Sports and Fitness: Cracked Ribs

A hockey player is checked into the boards; a basketball player flies out of bounds after a loose ball, slamming into the first row of seats; a wide receiver is hit hard after catching a pass over the middle. Sudden impacts in the chest such as these are relatively common, and the ribs usually take the full force of the contact.

The ribs are composed of spongy bone with a thin outer covering of compact bone. They are firmly bound in connective tissues and are interconnected by layers of muscle. As a result, displaced fractures are uncommon, and rib injuries usually heal swiftly and effectively. In extreme injuries, a broken rib can be forced into the thoracic cavity, and can disrupt internal organization and damage organs. The entry of air into one of the pleural cavities, a condition known as a **pneumothorax** (nū-mō-THŌR-aks), may lead to a *collapsed lung*. Damage to a blood vessel or even the heart can cause bleeding into the thoracic cavity, a condition called a **hemothorax**. A hemothorax can also impair lung function because fluid accumulates and the lung is compressed.

Lumbar Vertebrae

The body of a typical lumbar vertebra (Table 8-2 and Figure 8-21a,b) is thicker than that of a thoracic vertebra, and the superior and inferior faces are oval rather than heart-shaped. There are no articular facets on either the body or the transverse processes, and the vertebral foramen is somewhat broader than that of a thoracic vertebra. The transverse processes are slender and project dorsolaterally; a short, stout spinal process extends dorsally. The relatively massive spinal process provides surface area for the attachment of the lower back muscles that reinforce or adjust the lumbar curvature.

The vertebral column is like a tree, rather slender and flexible at the top but increasingly massive as you approach the base. The cervical vertebrae are the smallest and most mobile, for they support only the skull. The lumbar vertebrae are the most massive and least mobile, for they support most of the body weight. As you increase the loading on the vertebrae, the intervertebral discs become increasingly important as shock absorbers. The lumbar discs, which are subjected to the most pressure, are the thickest of all. The lumbar articulations restrict the stresses on the discs by limiting vertebral motion.

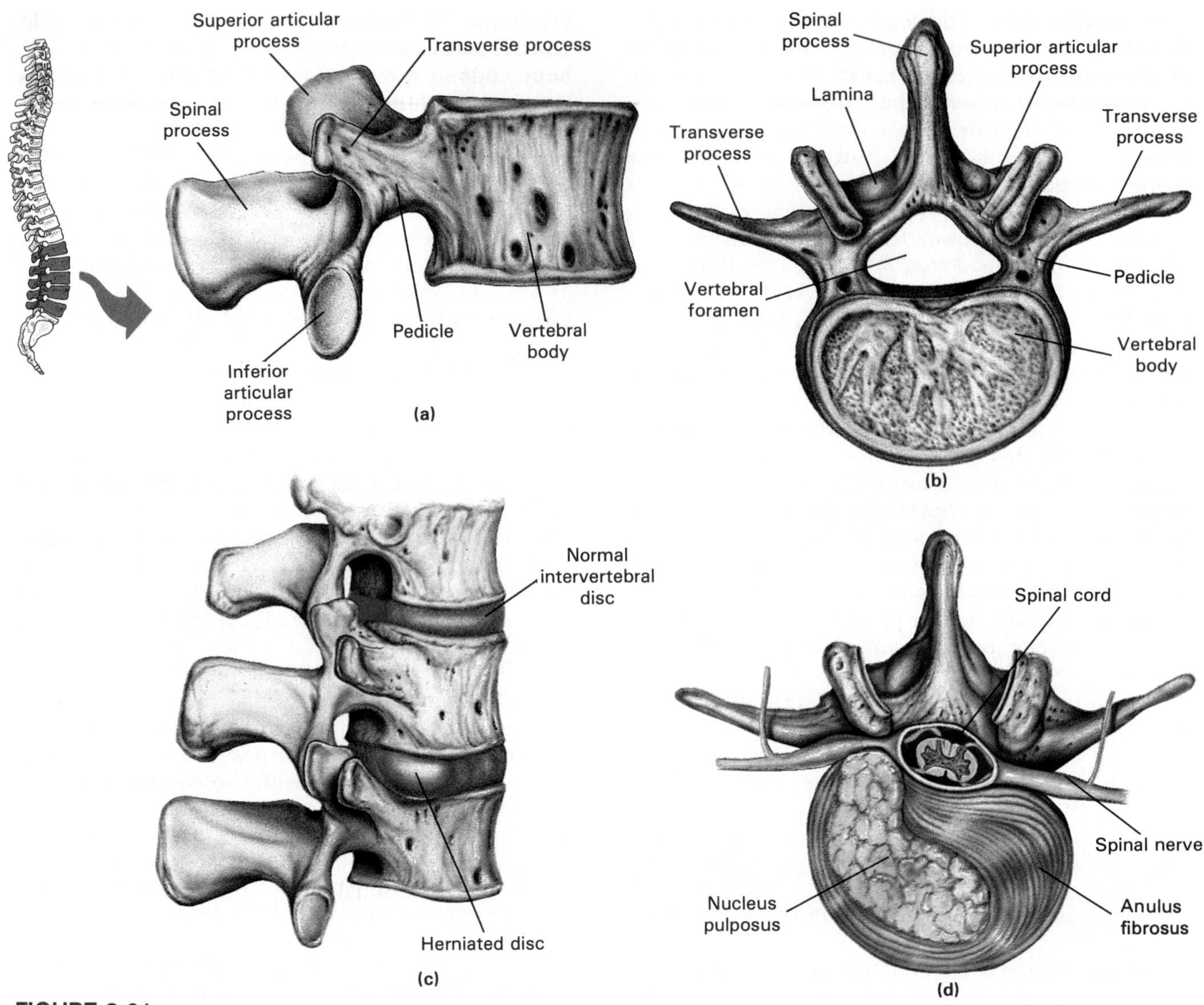

FIGURE 8-21
Lumbar Vertebrae. (a) Lateral view of a typical lumbar vertebra. (b) Superior view. (c) Lateral view of the lumbar region of the spinal column, showing a herniated intervertebral disc. (d) Sectional view through the herniated disc, showing the distortion of the nucleus pulposus and its effect on the spinal cord.

Shortly after physical maturity is reached, the nucleus pulposus within each disc begins to degenerate, and the "cushion" becomes less effective. Over the same period, the anulus fibrosus loses its elasticity. If the stresses are sufficient, the nucleus pulposus may break through the surrounding fibrocartilage and protrude beyond the intervertebral space, as indicated in Figure 8-21c,d. This condition, called a **herniated disc**, further reduces disc function. The term *slipped disc* is often used to describe this problem, although disc slippage does not actually occur.

The breakthrough usually occurs along the posterolateral surface of the disc, where no ligaments reinforce it. Sensory nerves are then distorted, producing pain, and the protruding mass can also compress the nerves passing through the intervertebral foramen. **Sciatica** (sī-AT-i-ka) is the painful result of compression of the roots of the sciatic nerve where they exit from between the lower lumbar vertebrae. The acute initial pain in the lower back is sometimes called **lumbago** (lum-BĀ-gō).

Most lumbar disc problems can be successfully treated with some combination of rest, back braces, analgesic (pain-killing) drugs, and physical therapy. Surgery to relieve the symptoms is required in only about 10 percent of cases involving lumbar disc herniation. To remove the offending disc, the nearest vertebral arch must be removed by shaving away the laminae. For this reason the procedure is known as a **laminectomy** (la-mi-NEK-to-mē).

The Sacrum and Coccyx

The **sacrum** consists of the fused elements of five sacral vertebrae. This structure affords protection for reproductive, digestive, and excretory organs, and attaches the axial skeleton to the appendicular skeleton by articulation with the pelvic girdle. The broad surface area of the sacrum provides an extensive area for the attachment of muscles, especially those responsible for leg movement. Figure 8-22 shows the posterior, lateral, and anterior surfaces of the sacrum.

Refer to Figure 8-22a. The sacrum is curved, with a convex dorsal surface. The narrow, caudal portion is the sacral **apex**, whereas the broad superior surface forms the **base**. The **articular processes** form synovial articulations with the last lumbar vertebra. The **sacral canal** begins between those processes and extends the length of the sacrum. Nerves and membranes that line the vertebral canal in the spinal cord continue into the sacral canal.

Before birth, five vertebrae fuse to form the sacrum, and their spinal processes form a series of elevations along the **median sacral crest**. The laminae of the fifth sacral vertebra fail to contact one another at the midline, and they form the **sacral cornua**. These ridges establish the margins of the **sacral hiatus** (hī-Ā-tus), the end of the vertebral canal. In life this opening is covered by connective tissues. On either side of the median sacral crest, the **sacral foramina** represent the intervertebral foramina, now enclosed by the fused sacral bones. A broad sacral **wing**, or **ala**, extends laterally from each **lateral sacral crest**. The median and lateral sacral crests provide surface area for the attachment of muscles of the lower back and hip. [**CM**: *Spina Bifida*]

Viewed from the side (Figure 8-22b) the sacral curvature is more apparent. Along its lateral border, a thickened, flattened area marks the **sacroiliac joint**, the site of articulation with the coxae (hip bones). These areas are called the **auricles** of the sacrum. The anterior surface of the sacrum is concave (Figure 8-22c), and prominent transverse lines mark the boundaries of the individual vertebrae. At the apex, a flattened area marks the site of articulation with the **coccyx**.

The coccyx provides an attachment site for a muscle that closes the anal opening. Only the first two coccygeal vertebrae have transverse processes and unfused neural arches. The prominent laminae of the first are known as the **coccygeal cornua**, and they curve to meet the cornua of the sacrum. The four coccygeal vertebrae do not complete their fusion until late in adulthood. In elderly individuals the coccyx may also fuse with the sacrum.

- √ **Joe suffered a hairline fracture at the base of the odontoid process. What bone is fractured and where would you find it?**
- √ **Improper administration of CPR (cardiopulmonary resuscitation) could result in a fracture of what bone?**
- √ **In adults, five large vertebrae fuse to form what single structure?**

Articulations of the Axial Skeleton

The articulations of the axial skeleton permit limited movement but they are very strong and are heavily reinforced with ligaments. Table 8-3 summarizes the information presented in this chapter concerning the joints between elements of the axial skeleton.

FIGURE 8-22
The Sacrum and Coccyx. (a) Posterior view. (b) Lateral view from the right side. (c) Anterior view.

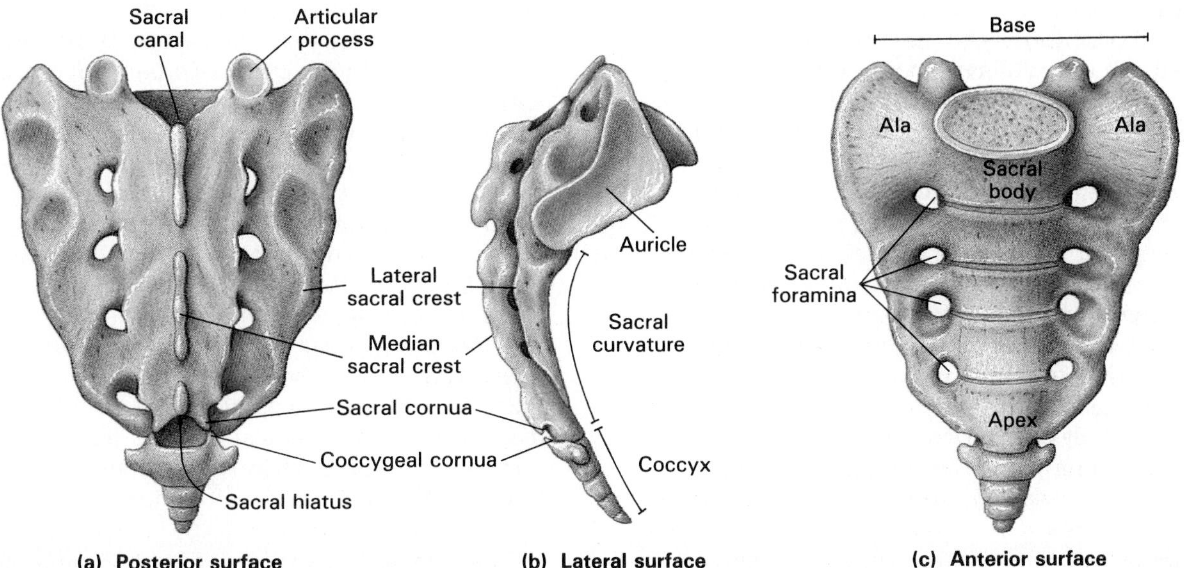

TABLE 8-3 Articulations of the Axial Skeleton

Element	*Joint*	*Type of Articulation*	*Movements*
Cranial and facial bones of skull	Various	Synarthroses (suture or synostosis)	None None
Maxilla/teeth **Mandible/teeth**		Synarthrosis (gomphosis)	None
Temporal bone/mandible	Temporomandibular	Hinge diarthrosis	Elevation and depression, lateral gliding
Occipital bone/atlas	Atlanto-occipital	Ellipsoidal diarthrosis	Flexion/extension
Atlas/axis	Atlanto-axial	Pivot diarthrosis	Rotation
Other vertebral elements	Intervertebral (between centra)	Amphiarthrosis (symphysis)	Slight movement
	Intervertebral (between articular processes)	Gliding diarthrosis	Slight rotation and flexion/extension
Thoracic vertebrae and ribs	Vertebrocostal	Gliding diarthrosis	Elevation/depression
Ribs and sternum	Sternocostal	Synarthrosis (synchondrosis)	No movement
Sternum and clavicle	Sternoclavicular	Gliding diarthrosis	Protraction/retraction, depression/elevation
L_5/sacrum	Between centrum and sacral body	Amphiarthrosis (symphysis)	Slight movement
	Between inferior articular processes of L_5 and articular processes of sacrum	Gliding diarthrosis	Slight flexion/extension
Sacrum/coxae	Sacroiliac	Gliding diarthrosis	Slight movement
Sacrum/coccyx	Sacrococcygeal	Gliding diarthrosis (may become fused)	Slight movement
Coccygeal bones		Synarthrosis (synostosis)	No movement

Health News: A Matter of Perspective

Several professions focus attention on the structure and function of the human skeleton. Each specialty has a different perspective, with its own techniques, traditions, and biases, and people seeking medical advice may find it difficult to choose the most appropriate specialist. Unfortunately, misinformation and misconceptions have polarized opinions regarding the relative merits of the various specialists.

An **orthopedist**, or **orthopedic surgeon**, is a physician (M.D.) who specializes in the structure and function of bones and articulations. Orthopedic surgery attempts to restore lost function or repair damage. An orthopedist might typically be involved in the reconstruction of a limb after a severe fracture, the replacement of a hip, knee surgery, and so forth. There are subspecialties that focus on particular regions such as the face, jaws, or hands. If an injury was not the primary cause of a particular problem, an orthopedist would be trained to recognize the underlying condition. As a physician, an orthopedist must complete 4 years of college, medical school, an internship, and a residency program, pass national boards, and be licensed by the state in which he or she practices medicine. All in all, this could add up to 12 or 13 years of training after high school.

Osteopaths (D.O.) and **osteopathic surgeons** have a slightly different perspective, but like orthopedists, osteopaths must complete undergraduate and medical school, serve an internship, and pass national and state examinations. Additional residency training may also be required. Osteopaths in general place a greater emphasis than orthopedists on preventive medicine, such as the maintenance of normal structural relationships, proper nutrition, exercise, and a healthy environment. When it comes to evaluating and treating bone or joint conditions, there is no substantive difference today between osteopaths and orthopedists.

A century ago, when osteopathy was founded, drug therapies for the treatment of diseases were unreliable, and osteopaths put their trust in rest, cleanliness, and proper nutrition.

The viewpoint of a **chiropractor** (D.C.) differs considerably from that of either orthopedists or osteopaths. In its original form, chiropractic proposed that all forms of disease resulted from abnormal functioning of the nervous system. This rather sweeping premise was modified as physiological principles became better understood. *Straight chiropractors* continue to place primary emphasis on vertebral alignment as the source of spinal nerve compression that causes peripheral symptoms. They use modern radiological techniques to evalulate vertebral position, but generally avoid the use of drugs. Treatment involves physical manipulation of the spinal column to bring the vertebral elements into proper alignment.

Chiropractors not classified as "straight" may have widely varying approaches and techniques. Many claim that spinal manipulation alone can cure conditions totally unrelated to the skeletal or nervous systems. There is no known mechanism that would lend plausibility to these claims, nor any supporting evidence.

Because there is so much variability among the practitioners, generalizations are difficult or impossible to make. However, chiropractors are licensed in every state. For example, to be licensed in Hawaii an applicant needs more than 2 years of undergraduate credits (60 semester hours), a diploma from one of 16 accredited 4-year chiropractic colleges, and completion of a national examination.

Podiatrists (D.P.M.), or *chiropodists*, are specialists in foot structure. They often treat sports injuries, design supports for feet affected by trauma or developmental problems, and perform limited surgical procedures. Podiatrists must be graduates of an accredited podiatry college and may be required to pass national and state examinations.

Where you take your aching bones is your choice, but the way the practitioner views your problem will obviously affect your treatment. Be wary of any individual who claims to have the answer to every question. A competent medical practitioner knows his or her personal and professional limitations and is willing to refer a patient to another specialist when appropriate.

CHAPTER REVIEW

Review of Selected Clinical Terms

sinusitis (*p. 249*)
Inflammation and congestion of the sinuses.

deviated nasal septum (*p. 249*)
A bent nasal septum that slows or prevents sinus drainage.

craniostenosis (**krā-nē-ō-sten-Ō-sis**) (*p. 254*)
Premature closure of one or more fontanels, which can lead to unusual distortions of the skull.

microcephaly (**mī-krō-SEF-a-lē**) (*p. 254*)
An undersized head resulting from genetic or developmental abnormalities.

kyphosis (**kī-FŌ-sis**) (*p. 255*)
Abnormal exaggeration of the thoracic curvature that produces a "roundback" appearance.

lordosis (**lōr-DŌ-sis**) (*p. 255*)
Abnormal lumbar curvature giving a "swayback" appearance.

scoliosis (**skō-lē-Ō-sis**) (*p. 255*)
Abnormal lateral curvature of the spine.

whiplash (*p. 260*)
An injury resulting from a sudden change in body position that can injure the cervical vertebrae.

pneumothorax (**nū-mō-THŌR-aks**) (*p. 263*)
The entry of air into a pleural cavity.

hemothorax (*p. 263*)
Bleeding into the thoracic cavity.

herniated disc (*p. 264*)
A condition in which the nucleus pulposus protrudes beyond the intervertebral space; often called a "slipped disc."

sciatica (**sī-AT-i-ka**) (*p. 264*)
Pain resulting from compression of the roots of the sciatic nerve.

lumbago (**lum-BĀ-gō**) (*p. 264*)
The acute initial pain in the lower back resulting from compression of the roots of the sciatic nerve.

laminectomy (**la-mi-NEK-to-mē**) (*p. 264*)
A surgical procedure done to relieve symptoms of lumbar disc problems by removing the involved disc.

MATTER OF PERSPECTIVE (*pp. 266–267*): **orthopedist (orthopedic surgeon), osteopath (osteopathic surgeon), chiropractor, podiatrist**

Study Outline

Related Key Terms

Introduction (pp. 237–239)

1. The skeletal system consists of the axial skeleton and the appendicular skeleton. The **axial skeleton** can be subdivided into the **skull**, the **auditory ossicles** and **hyoid**, the **vertebral column**, and the **thoracic cage** composed of the **ribs** and **sternum**.
2. The **appendicular skeleton** includes the arms, the legs, and the **pectoral** and **pelvic girdles**.

The Skull (pp. 239–254)

1. The **cranium** encloses the **cranial cavity**, a division of the dorsal body cavity.

SUPERFICIAL ANATOMY OF THE SKULL (pp. 240–242)

2. Prominent superficial landmarks on the skull include the **foramen magnum**; the **occipital bone**; the **parietal bones**; the **frontal bone**; the **lambdoidal**, **sagittal**, and **coronal sutures**; and **sutural bones**.
3. Prominent landmarks on the facial area include the **orbits**, **nasal bones**, **external nares**, and **maxillary bones** (each called a **maxilla**). Between the frontal and maxillary bones inside each orbit are the **lacrimal bone**, the **ethmoid bone**, the **sphenoid bone**, and the **zygomatic bone**.
4. The **temporal process** meets the **zygomatic process** of the **temporal bone** to form the **zygomatic arch**. Other features include the **mandible** and the **external auditory meatus**.
5. Important features on the inferior surface of the skull are the **styloid process**, the **occipital condyles**, the **nasal septum**, the **palatine bones**, and the **pterygoid processes** or **plates**. — **vomer**

SECTIONAL ANATOMY OF THE SKULL (pp. 242–243)

6. The sphenoid and frontal bones contain **sinuses**. In the nose lie the **nasal conchae**.

BONES OF THE CRANIUM (pp. 243–247)

7. The occipital bone surrounds the foramen magnum and articulates with the sphenoid, temporal, and parietal bones.
8. The parietal bones articulate with the occipital, the temporal, and the frontal bone. The frontal bone articulates with the parietals, the temporal, the sphenoid, the ethmoid, the nasals, the lacrimals, the zygomatics, and the maxillae.
9. The sphenoid articulates with all the cranial bones as well as the maxillary bones, the zygomatics, and the palatines. The maxillary bones articulate with every bone in the face except the mandible.
10. The **cribriform plate** contains perforations for olfactory nerves. The **perpendicular plate** forms part of the nasal septum. The **paranasal sinuses** contain air.
11. The hyoid bone, suspended by **stylohyoid ligaments**, consists of a **body**, the **greater cornu**, and the **lesser cornu**.

THE SKULLS OF INFANTS AND CHILDREN (p. 254)

12. Fibrous connections at **fontanels** permit the skulls of infants and children to continue growing. Premature closure of fontanels can lead to **craniostenosis** or **microcephaly**.

The Neck and Trunk (pp. 254–265)

1. There are 7 **cervical vertebrae**, 12 **thoracic vertebrae** (which articulate with **ribs**), and 5 **lumbar vertebrae** (which articulate with the **sacrum**). The **coccyx** consists of fused vertebrae.

SPINAL CURVATURE (pp. 254–255)

2. The spinal column has four **spinal curves**: the thoracic and sacral curves are called **primary**, or **accommodation**, **curves**; the lumbar and cervical curves are known as **secondary**, or **compensation**, **curves**.

INTRODUCTORY ANATOMY OF THE VERTEBRAL COLUMN (p. 258)

3. A typical vertebra has a **body**, or **centrum**, and a **vertebral** or **neural arch** and connects to other vertebrae at the **superior** and **inferior articular processes**. An **intervertebral disc** consists of an inner **nucleus pulposus** and an outer **annulus fibrosus**.

REGIONAL STRUCTURE AND FUNCTION (pp. 258–265)

4. Cervical vertebrae are distinguished by the shape of the centrum and in the presence of **costal processes** with **transverse foramina**.
5. The skeleton of the **thorax** consists of the thoracic vertebrae, the ribs, and the sternum. The ribs and sternum form the **rib cage**. Thoracic vertebrae have distinctive heart-shaped centra.
6. Ribs 1–7 are **true**, or **vertebrosternal**, **ribs**. Ribs 8–12 are called **false ribs**; they include the **vertebrochondral ribs** and two pairs of **floating ribs**. A typical rib has a **head**, or **capitulum**; **neck**; **tubercle**, or **tuberculum**; **angle**; and **body**, or **shaft**. An inferior **costal groove** marks the path of nerves and blood vessels.
7. The sternum consists of a **manubrium**, a **body**, and a **xiphoid process**.
8. The lumbar vertebrae are the most massive and least mobile; they are subjected to the greatest strains, and a **herniated disc** can occur.
9. The **sacrum** protects reproductive, digestive, and excretory organs. It has **auricles** and articulates with the **coccyx**.

Articulations of the Axial Skeleton (pp. 265–267)

1. The articulations of the axial skeleton permit limited movement, but are strong and heavily reinforced.

Related Key Terms

jugular foramen
hypoglossal canal

lacrimal fossa • **frontal sinuses**
mastoid process
mastoid foramen
styloid process
jugular foramen
carotid foramen
foramen lacerum
pharyngotympanic tube
tympanic cavity
auditory ossicles
tympanic membrane
internal acoustic meatus
internal acoustic canal

sella turcica • **hypophyseal fossa**
optic groove • **optic foramen**
superior orbital fissure
foramen rotundum
foramen ovale
foramen spinosum
superior/medial conchae
ethmoidal sinuses
inferior orbital fissure
maxillary sinuses
nasal complex • **nasal septum**
inferior nasal concha
frontal/sphenoidal sinuses
ethmoidal/maxillary sinuses
mandible
vertebral foramen
vertebral canal • **spinal process**
transverse processes
intervertebral foramina

atlas • **axis**
dens or odontoid process
vertebra prominens

costal cartilages • **false ribs**
costae • **intercostal muscles**

articular processes • **sacral canal**
sacral foramina • **sacroiliac joint**

Chapter Review

■ Review Planner

		Level 1	Level 2	
1	Name the components of the axial skeleton and their functions.	3 12 31 33 34 38 44	Level 2	29–38
2	Identify the bones of the skull and explain the significance of the markings on individual bones.	1 4 6 8 10 11 13 19–28 31 32 33 37 38 43	Level 3	46–50
3	Discuss the process by which the adult skull develops.	7 15 36 40		
4	Explain the significance of the articulations between the thoracic vertebrae, the ribs, and the sternum.	5 16 30 35	C M	**Phrenology** **The Thoracic Cage and Surgical Procedures** **Spina Bifida**
5	Discuss the differences in structure and function between the various vertebrae.	2 9 14 17 18 34 39 41 42 45		
6	Discuss the bones of the neck and trunk and their distinctive markings.	2 3 5 9 18 39		

■ Review Questions

MULTIPLE CHOICE

1. The foramen magnum is: a) the largest vertebra in the spinal column; b) the opening that connects the cranial cavity with the canal enclosed by the spinal column; c) the largest bone in the skull; d) the opening in which the sinuses are found.
2. The atlas is also known as: a) C_1; b) T_1; c) L_5; d) the coccyx.
3. Another term for "thorax" is: a) rib; b) vertebra; c) spinal curve; d) chest.
4. The eye socket is known as the: a) external nares; b) frontal bone; c) orbit; d) ethmoid.
5. The ribs originate on or between the ________ vertebrae. a) cervical; b) thoracic; c) lumbar; d) sacral.
6. The zygomatic arch is the: a) cheekbone; b) eyebrow ridge; c) forehead; d) outer ear.
7. The condition of having a very undersized head is called: a) lesser cornu; b) craniostenosis; c) microcephaly; d) sella turcica.
8. The mandible is the: a) nasal opening; b) lower jaw; c) upper jaw; d) forehead.
9. The sacrum is located between the: a) cervical and thoracic vertebrae; b) thoracic and lumbar vertebrae; c) lumbar vertebrae and coccyx; d) skull and cervical vertebrae.
10. The nasal openings of the skull are called: a) temporal processes; b) zygomatic arches; c) orbits; d) external nares.
11. If you struck your frontal squama against something, you would have a bruise on your: a) forehead; b) nose; c) chin; d) none of the above.

DIAGRAMS

12. Create a chart that illustrates the major subdivisions of the axial skeleton. Place the following on the chart under the appropriate subdivision: cranium, ribs, sternum, face, auditory ossicles, sacrum, hyoid, coccyx, vertebrae.
13. Draw a skull in profile (it doesn't have to be fancy). Label the following major cranial and facial bones: parietal, mandible, nasal, occipital, lacrimal, temporal, maxilla, frontal, sphenoid, zygomatic, ethmoid.
14. Create a table that lists the following information for each of the three types of vertebrae: total number; location; characteristics of centrum; vertebral canal; characteristics of spinal process; characteristics of transverse processes; functions.

TRUE/FALSE

15. **T/F:** Most significant growth in the bones of the skull occurs by the age of 5 years.
16. **T/F:** The volume of the thoracic cage remains constant since the ribs cannot move.
17. **T/F:** Lumbago is a painful condition that develops in the upper back.
18. **T/F:** The sacrum consists of fused vertebrae.

MATCHING QUESTIONS

(Match the locations with the letters of the appropriate terms. Note that some locations may be matched with more than one term.)

Locations:	Terms:
19. Sphenoid	a) Jugular foramen
20. Ethmoid	b) Stylomastoid foramen
21. Maxilla	c) Cribriform plate
22. Occipital	d) Sella turcica
23. Temporal	e) Optic foramen
24. Frontal	f) Greater cornu
25. Parietal	g) Nasal septum
26. Nasal complex	h) Condylar processes
27. Mandible	i) Superior conchae
28. Hyoid	j) Hypoglossal canals
	k) Mastoid process
	l) Alveolar processes
	m) Frontal squama
	n) Parietal eminence

SHORT ANSWER/ESSAY

29. Explain the importance of the axial skeleton.
30. Place in the correct sequence, moving top to bottom:
    a) Floating ribs
    b) Vertebrosternal ribs
    c) Vertebrochondral ribs
31. Where is the sphenoid, and what are its functions?
32. What are located at the ridges and foramina on the skull?
33. What are the functions of the cranium and facial bones? What subdivision of the axial skeleton do they form?
34. What are the functions of the sacrum and the coccyx?
35. What is meant by the terms "true ribs" and "false ribs"? Give the alternate term for each.
36. Is the metopic suture normally present? In whom, and where?
37. Where would you expect to find the pharyngotympanic tube?
38. Describe the relationship between the auditory ossicles and the tympanic membrane. To which of our senses do they contribute?
39. Place in the correct sequence, starting with the skull and moving caudally:
    a) Lumbar vertebrae
    b) Cervical vertebrae
    c) Coccyx
    d) Thoracic vertebrae
    e) Sacrum
40. Describe the process by which the skull develops to its adult form. How does it accommodate the growing brain?
41. Describe the structure and function of an intervertebral disc.
42. You are looking at two vertebrae. One is extremely small and light; the other is very heavy and massive. Identify the category to which each vertebra belongs, and explain your answer.
43. Name two landmarks on the skull that connect it to the spinal column.
44. What are the sinuses, and what are their functions?
45. Explain what is meant by C_3 and L_1.

CRITICAL THINKING/APPLICATIONS

46. Jan Cameron has been experiencing a great deal of pain. The doctor tells her she has a "pinched nerve" caused by a "slipped disc" in the C_5 area.
    a) What is a "disc," and what is its function?
    b) What is the location of her injury?
47. You're getting a cold and your nose is starting to "run." Describe what causes this reaction. Does it have a purpose?
48. A friend slips and falls on the ice. After several days of persistent pain in her side, she goes to the emergency room. She is told she cracked a rib, but is sent home without a cast. She is surprised by this, since she always thought that a cracked bone of any type requires immobilization in order to heal. What would you tell her?
49. Why is it important to keep the back straight when lifting a heavy object?
50. Mr. H. is driving on the highway and has to stop suddenly to avoid hitting a stalled car. Explain how this could affect his cervical vertebrae.

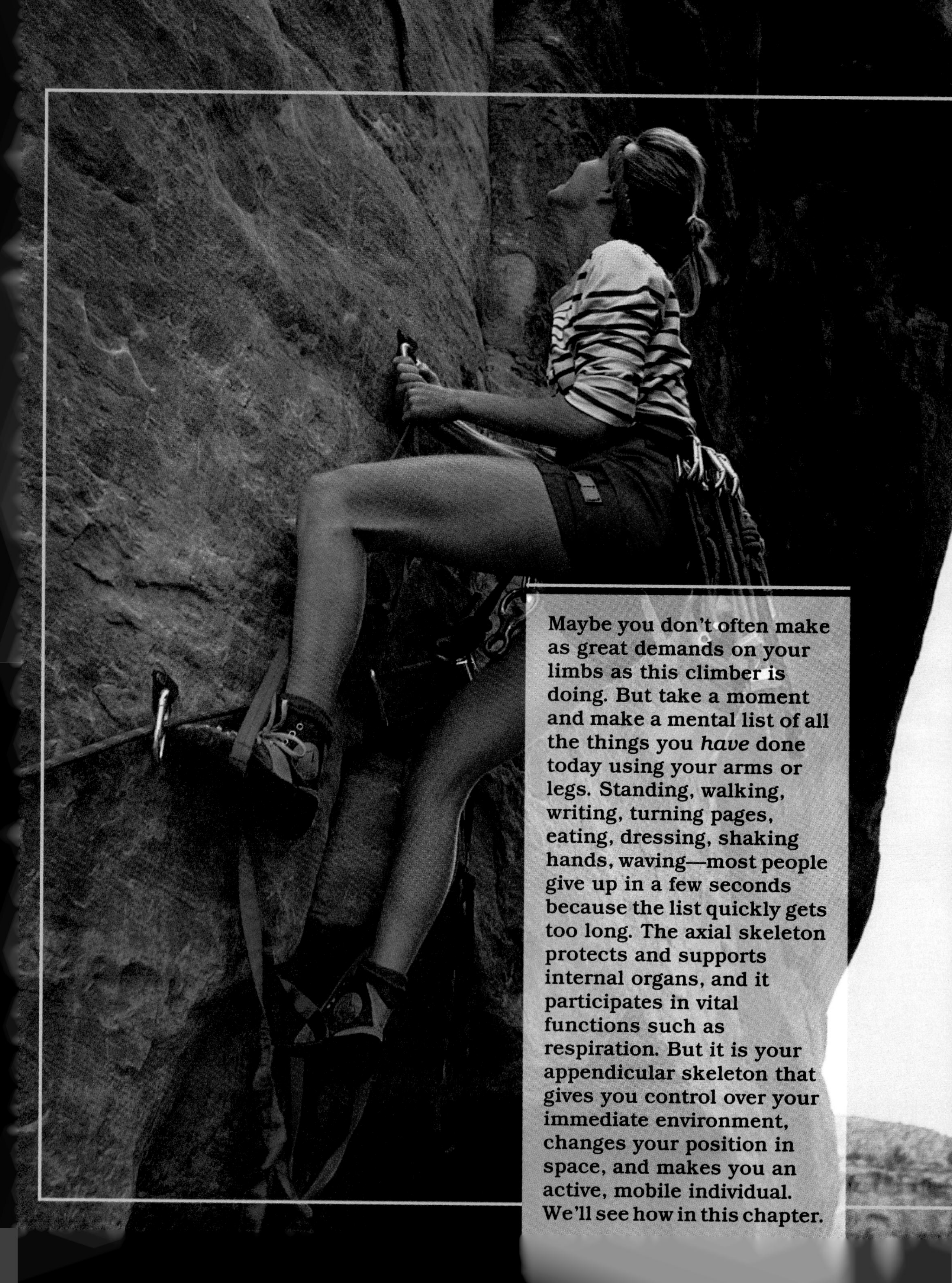

Maybe you don't often make as great demands on your limbs as this climber is doing. But take a moment and make a mental list of all the things you *have* done today using your arms or legs. Standing, walking, writing, turning pages, eating, dressing, shaking hands, waving—most people give up in a few seconds because the list quickly gets too long. The axial skeleton protects and supports internal organs, and it participates in vital functions such as respiration. But it is your appendicular skeleton that gives you control over your immediate environment, changes your position in space, and makes you an active, mobile individual. We'll see how in this chapter.

CHAPTER 9

The Skeletal System: Appendicular Division

Chapter Objectives

After reading this chapter, you will be able to:

1 Identify the components of the appendicular skeleton and their functions.

2 Relate the structural differences between the pectoral and pelvic girdles to the variations in their functional roles.

3 Identify the major structural features and markings of important bones of the appendicular skeleton.

4 Discuss the structures and functions of the joints in the appendicular skeleton.

5 Explain how study of the skeleton can reveal important information about an individual.

6 Describe the skeletal differences between males and females.

Introduction

The **appendicular skeleton** includes the bones of the arms and legs and the supporting elements that connect the limbs to the trunk. The basic components are indicated in Figure 9-1, but what the figure cannot capture is the dynamic way these components interact in the course of daily activities. The hands and arms reach, grab, pull, and twist; the legs swing, bend, and kick. Each articulation in the appendicular skeleton represents a unique compromise between strength and mobility. The stronger and more stable the joint, the more restricted the range of motion. The shoulder joint has an impressive range of motion, but it is relatively weak. It is the surrounding skeletal muscles and their tendons that perform the major stabilizing, positioning, and bracing functions. In contrast, the hip joint, which transfers the body weight to the femur, is massively reinforced by ligaments—muscles play a relatively small role in its stabilization. However, all the ligaments and other anatomical features that add strength and stability limit the range of movement; in effect, they get in the way.

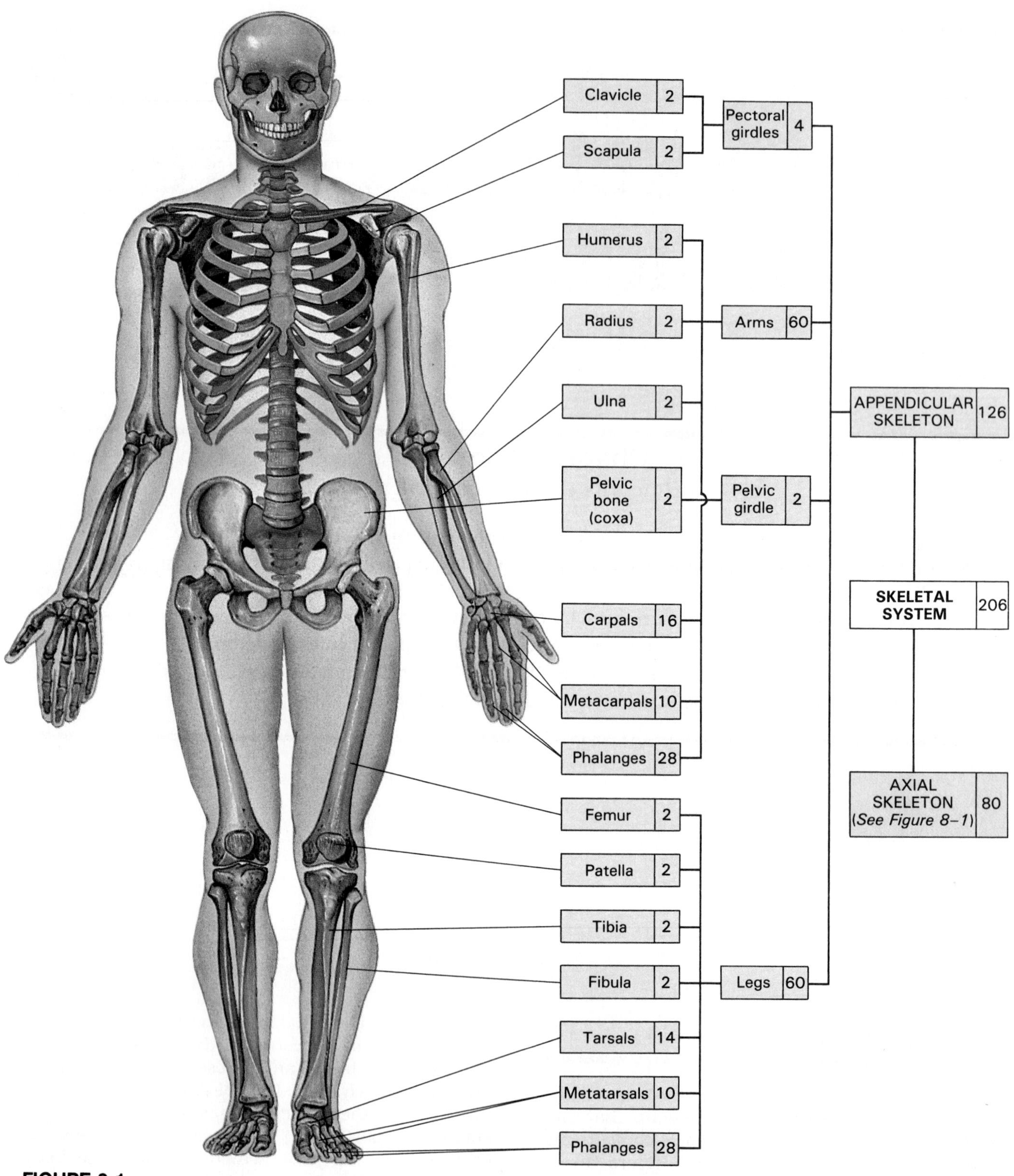

FIGURE 9-1
The Appendicular Skeleton

This chapter describes the bones of the appendicular skeleton. As in Chapter 8, the descriptions emphasize anatomical landmarks that either assist in performing one of the functions of the skeletal system or demonstrate functional interactions between this system and other systems.

The Pectoral Girdle and Arm

Many of the items in your list of activities undoubtedly involved precise and controlled movements of one or both arms. People who have lost the use of their legs find their mobility greatly restricted, but once adjusted they can lead relatively normal lives. Those who have lost their arms must face much greater changes in the quality of life.

Each arm articulates with the trunk at the **shoulder**, or **pectoral girdle**. The shoulder girdle consists of a broad, flat **scapula** (*shoulder blade*) and the short **clavicle** (*collarbone*). The clavicle articulates with the manubrium of the sternum—this is the *only* direct connection between the pectoral girdle and the axial skeleton. Skeletal muscles support and position the scapula, which has no bony or ligamentous bonds to the thoracic cage.

THE PECTORAL GIRDLE

Movements of the clavicle and scapula position the shoulder joint and provide a base for arm movement. Once the shoulder joint is in position, muscles that originate on the pectoral girdle help to move the arm. The surfaces of the scapula and clavicle are therefore extremely important as sites for muscle attachment. Where major muscles attach they leave their marks, creating bony ridges and flanges. Other superficial landmarks indicate the position of soft tissue structures, such as nerves or blood vessels, that monitor, control, and nourish these living bones and the muscles around them.

The Scapula

Figure 9-2 details the anatomy of the right scapula. The anterior aspect of the scapular **body** forms a broad triangle (Figure 9-2a). The three sides of that triangle are the **superior border**; the **vertebral**, or **medial**, **border**; and the **lateral**, or **axillary**, **border** (*axilla*, armpit). Muscles that position the scapula attach along these edges. The corners of the triangle are called the **superior angle**, the **inferior angle**, and the **lateral angle**. The region adjacent to the lateral angle is not flat, but thickened and rounded into a definite **neck** that supports the cup-shaped **glenoid**

FIGURE 9-2
The Scapula. (a) Anterior, (b) lateral, and (c) posterior views of the right scapula.

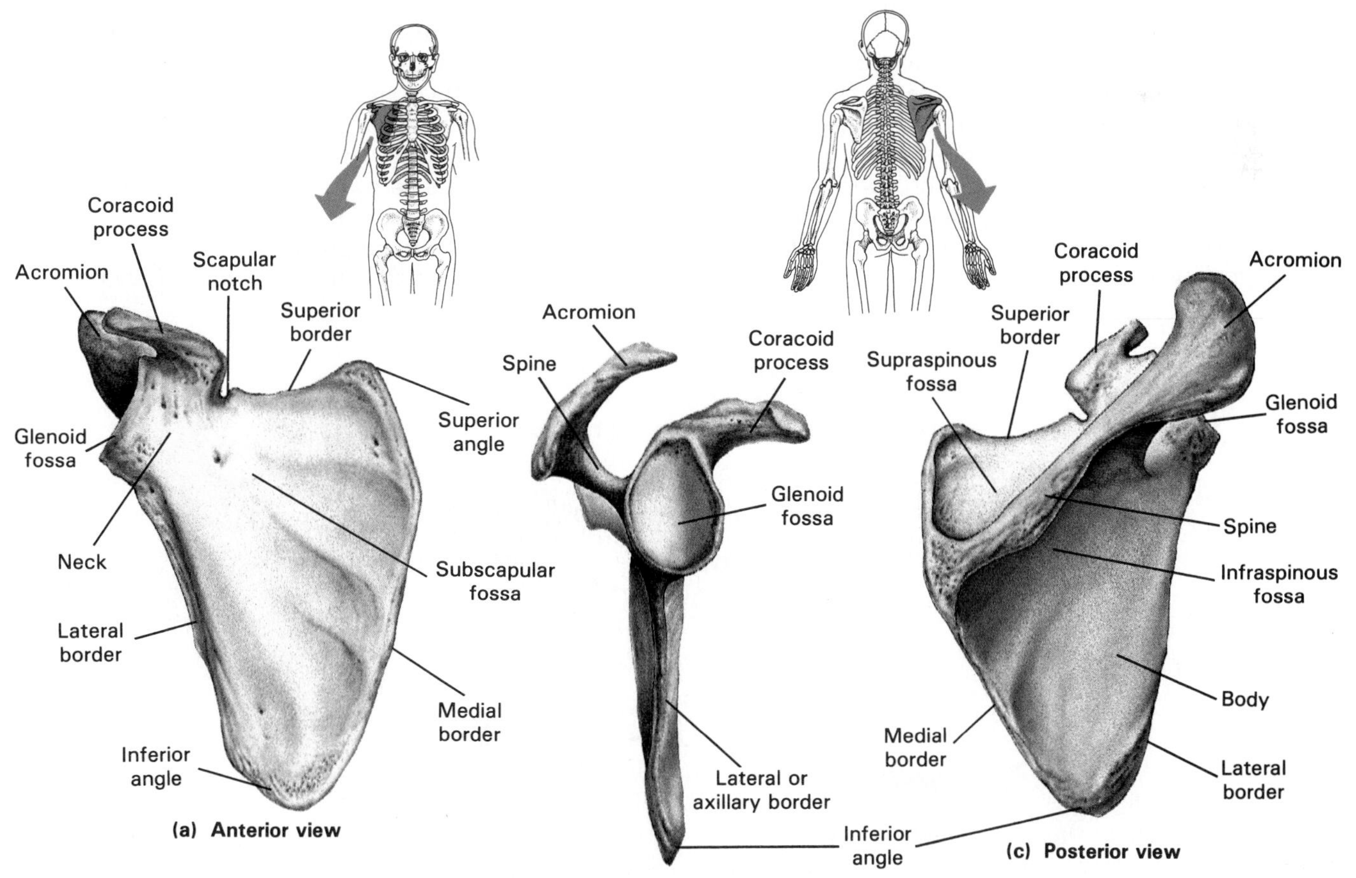

fossa. At the glenoid fossa, the scapula articulates with the proximal end of the **humerus**, the bone of the upper arm. This articulation is the **scapulohumeral joint**, or **shoulder joint**. The capsule of this joint originates at the scapular neck. The relatively smooth **subscapular fossa** forms most of the concave, ventral surface of the scapula. In life, this fossa is filled by the *subscapularis* muscle, which inserts on the humerus and rotates the arm.

Figure 9-2b shows a lateral view of the scapula. The articular head of the humerus is smooth, round, and several times the size of the glenoid fossa. Two large processes extend over the superior margin of the glenoid fossa, just above the head of the humerus. The smaller, anterior projection is the **coracoid** (kō-RA-koid) **process.** This Greek word means "like a crow"; the projection reminded early anatomists of a crow's beak. The **acromion** (a-KRŌ-mē-on) is the larger posterior process. If you run your fingers along the superior surface of the shoulder joint, you will feel this process. The acromion articulates with the clavicle at the **acromioclavicular joint**. Both the acromion and the coracoid process are attached to ligaments and tendons associated with the shoulder joint. (We will consider the functions of these tendons and ligaments when we discuss the shoulder joint later in this chapter.)

The acromion is continuous with the **scapular spine** (Figure 9-2c). This ridge crosses the scapular body before ending at the medial border. The scapular spine divides the convex dorsal surface of the body into two regions. The area superior to the spine constitutes the **supraspinous fossa**; the *supraspinatus* muscle attaches here. The region below the spine is the **infraspinous fossa**, home of the *infraspinatus* muscle. Distally, both the supraspinatus and infraspinatus muscles are attached to the upper arm. The entire dorsal surface of the scapula is marked by small ridges and lines that mark attachment sites for smaller muscles.

The Clavicle

The clavicle, shown in Figure 9-3, originates at the craniolateral border of the sternum. From this roughly triangular **sternal end** the clavicle curves laterally and dorsally until it articulates with the acromion of the scapula. The **acromial end** is larger than the sternal end. The smooth superior surface of the clavicle lies just beneath the skin. The rough inferior surface of the acromial end is marked by prominent lines and tubercles that indicate the attachment sites for muscles and ligaments.

You can easily explore the interaction between scapulae and clavicles. Place your fingers in the notch at the top of the sternum, and locate the clavicle on either side. When you move your shoulders you can feel the clavicles change their positions. Because the clavicles are so close to the skin, you can trace one laterally until it articulates with the acromion. If you are uncertain as to the precise location of the joint, shrug your shoulders a few times while feeling along the distal portion of the clavicle. The changing position of the clavicle relative to the acromion should highlight the spot. The movements are limited by the position of the clavicle, as indicated in Figure 9-4. Several inherited developmental abnormalities can cause a reduction or elimination of the clavicles. Individuals with such an abnormality have completely mobile scapulae, and their shoulders can be swung medially almost far enough to meet in front of the sternum.

Many wild and domesticated animals, including cats, dogs, and horses, either lack clavicles entirely or retain only small bony remnants within large tendons. Tracing their evolutionary lineages back 65 million years reveals that their ancestors possessed clavicles as substantial as those of human beings. Since clavicular injuries are so common (see the Sports and Fitness box), one might wonder why the clavicle was not reduced or eliminated in human evolution.

FIGURE 9-3
The Clavicle. (a) Anterior and (b) posterior views of the right clavicle.

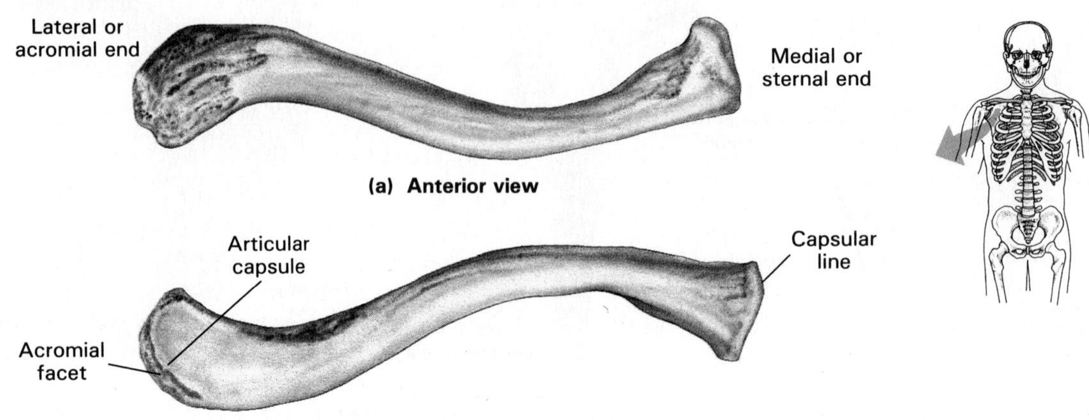

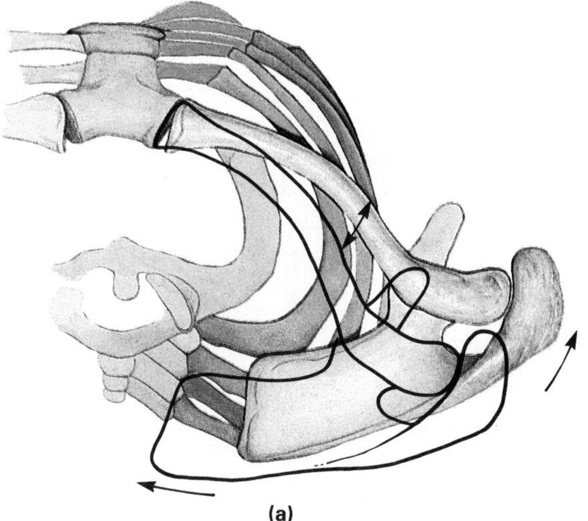

(a)

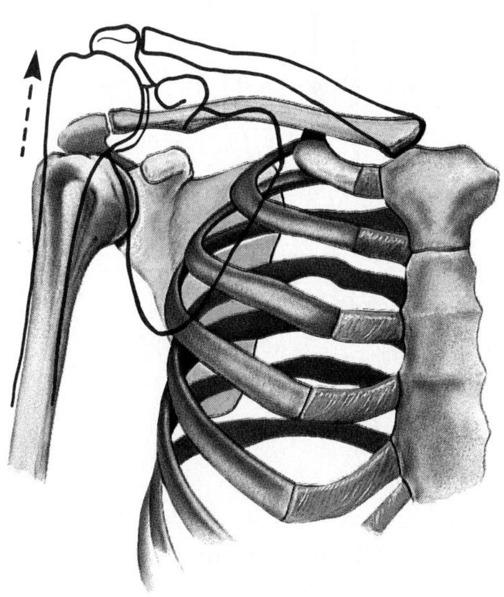

(b)

FIGURE 9-4
Mobility of the Pectoral Girdle. (a) Alterations in position that occur during protraction and retraction of the right shoulder. (b) Alterations in position that occur during elevation and depression of the right shoulder. In each instance note that the clavicle is responsible for limiting the range of motion.

The answer appears to lie in the basic lifestyles of our ancestors compared with those of other animals. Cats, for example, capture prey by running and leaping. Reducing the size of the clavicle frees the scapula, and with a mobile scapula they can reach farther and take longer strides. However, that mobility is not always an advantage. If a cat hangs by its front paws from a tree limb, it soon tires, because the muscles must bear the entire weight of the body. When the muscles fatigue, the choice is simple: drop off and hope for a soft landing, or damage tendons and ligaments.

Cats spend most of their time on the ground and relatively little time in trees. The situation was quite different in the evolutionary line that led to human beings; our distant ancestors spent most of their time in the trees. An animal with functional clavicles has a bony bridge between the shoulder and rib cage. When it hangs by its forelimbs, its shoulder muscles can be partially relaxed, because the clavicles transfer much of the body weight directly to the arms. Thus this anatomical feature was retained in the evolution of humans and other primates.

Sports and Fitness: Clavicular Injuries

Clavicular injuries may be all too familiar to readers who play contact sports. When a head-on charge leads to a collision, such as a block (in football) or check (in hockey), the shoulder usually lies in the impact zone. The clavicle provides the only fixed support for the pectoral girdle, and it cannot resist large forces. In a **shoulder separation**, the acromioclavicular joint undergoes partial or complete dislocation. This injury can result from a blow to the upper surface of the shoulder. The acromion is forcibly depressed, but the clavicle is held back by powerful muscles.

A fall onto an outstretched arm can produce enough compressive force to fracture the clavicle. Clavicular fractures are very common, both in sports and in household accidents, but the breaks usually heal rapidly without a cast.

THE ARM

The upper arm, or *brachium*, contains a single bone, the **humerus**. The humerus extends from the scapula to the elbow. At its proximal end, the round **head** of the humerus articulates with the scapula. At its distal end, it articulates with the bones of the forearm, or *antebrachium*, the **radius** and **ulna**.

The Humerus

Figure 9-5 illustrates the anatomy of the humerus, and you should refer to this illustration as our description proceeds. The prominent **greater tubercle** of the humerus is located near the head, on the lateral surface of the epiphysis (Figure 9-5a). It establishes the contour of the shoulder. You can verify its postion by feeling for a bump situated a few centimeters anterior and inferior to the tip of the acro-

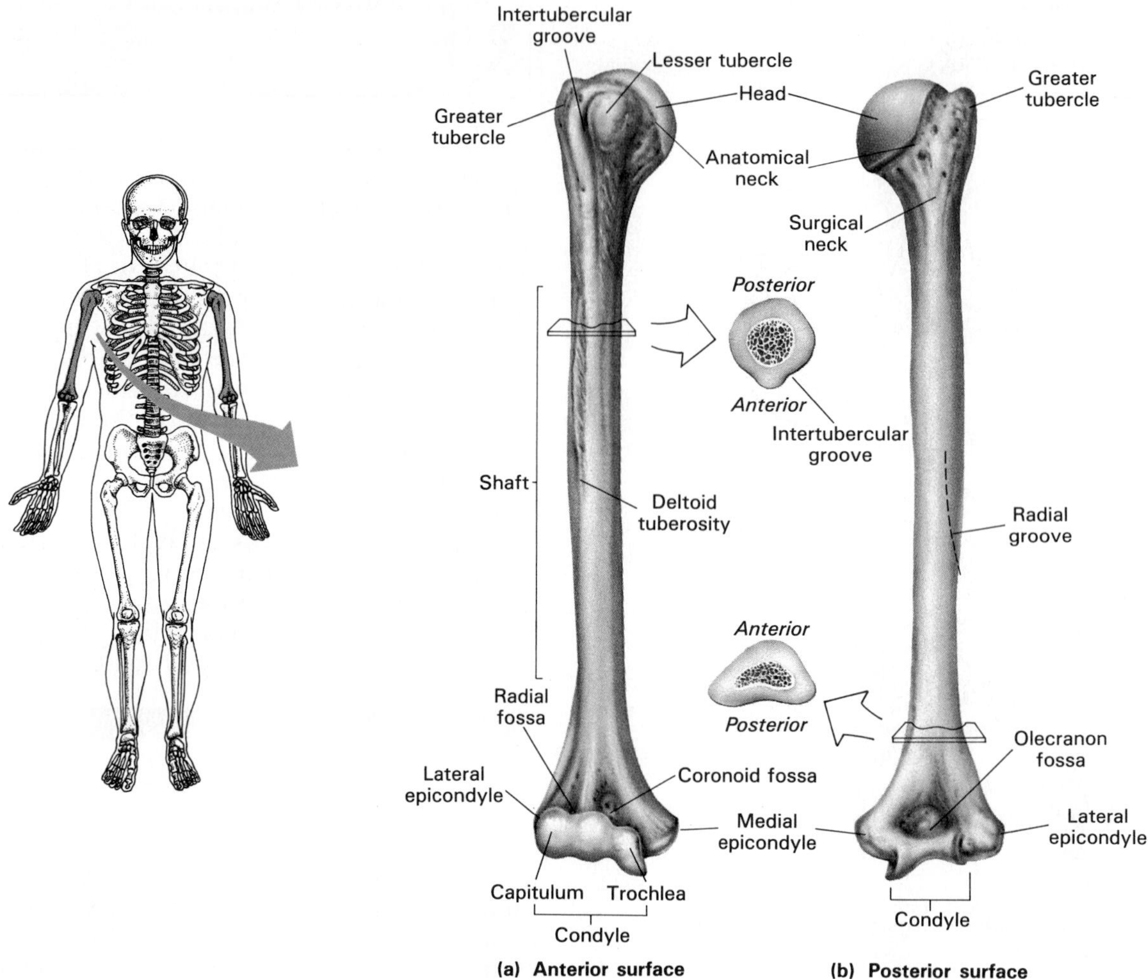

FIGURE 9-5
Bones of the Upper Arm. Major landmarks on (a) the anterior and (b) the posterior surface of the right humerus are shown.

mion. The **lesser tubercle** lies on the anterior and medial surface of the epiphysis, separated from the greater tubercle by the **intertubercular groove**, also known as the *intertubercular sulcus*. Both tubercles are important sites for muscle attachment; a large tendon runs along the groove. The **anatomical neck** marks the distal limits of the articular capsule. It lies between the tubercles and the articular surface of the head. Distal to the tubercles, the narrow **surgical neck** corresponds to the metaphysis of the growing bone. It earned its name by being a common fracture site.

The proximal **shaft** of the humerus is round in section. The elevated **deltoid tuberosity** runs along the lateral border of the shaft, extending more than halfway down its length. It is named after the *deltoid muscle* that attaches to it. On the anterior surface of the shaft, the intertubercular groove runs alongside the deltoid tuberosity.

On the posterior surface (Figure 9-5b), the deltoid tuberosity ends at the **radial groove**. This depression marks the path of the *radial nerve*, a large nerve that monitors sensory receptors on the back of the hand and controls large muscles that extend the forearm. Distal to the radial groove, the posterior surface is relatively flat. At the distal epiphysis, the humerus expands to either side, forming a broad triangle. **Medial** and **lateral epicondyles** project to either side, providing additional surface area for muscle attachment, and the articular **condyle** dominates the inferior surface of the humerus.

Returning to Figure 9-5a, a low ridge crosses the condyle, dividing it into two distinct regions. The **trochlea** is the large medial portion shaped like a spool or pulley. It extends from the base of the **coronoid fossa** (KOR-o-noyd; *corona*, crown) on the anterior surface to the **olecranon fossa** on the posterior surface. These depressions accept projections from the surface of the ulna as the elbow approaches full flexion or extension. Laterally, the rounded **capitulum** covers the anterior and inferior portions of the condyle. A shallow **radial fossa** proximal to the capitulum accommodates a small projection on the radius.

The Radius and Ulna

The **radius** and **ulna** are the bones of the forearm. In the anatomical position, the radius lies along the lateral margin of the forearm and extends toward the thumb. The ulna forms the medial support of the forearm. It articulates with the radius at a pair of pivot articulations, the **radioulnar joints**. The structure of these bones can be seen in Figure 9-6a,b.

The **olecranon** (ō-LEK-ra-non) **process** forms the superior and posterior portions of the proximal epiphysis of the ulna. The articular surface lies on its anterior surface, where the **trochlear notch** (or *semilunar notch*) accommodates the trochlea of the humerus at the **olecranal**, or **elbow, joint**. The olecranon process forms the superior lip of the notch, and the **coronoid process** provides a prominent inferior margin. During extreme extension, the olecranon process swings into the olecranon fossa on the posterior face the humerus. In extreme flexion, the coronoid process projects into the coronoid fossa on the anterior humeral surface. Lateral to the coronoid process, a smooth **radial notch** accommodates the head of the radius.

In section, the shaft of the ulna is triangular, with the smooth medial surface at the base of the triangle and the lateral margin at the apex. A fibrous sheet, the **interosseous membrane**, connects the lateral margin of the ulna to the radius as indicated in the section taken from Figure 9-6a. The ulnar shaft narrows distally before ending at a disc-shaped head whose posterior margin supports a short **styloid process** (*styloid*, long and pointed). A triangular **articular cartilage** attaches to the styloid process, isolating the ulnar head from the bones of the wrist. As a result, only the expansive distal portion of the radius participates in the wrist joint. Near the medial border of the articular surface the **ulnar notch** marks the distal articulation between the radius and ulna. The **styloid process** of the radius assists in the lateral stabilization of the wrist joint.

The shaft of the radius curves along its length, and near the elbow a prominent **radial tuberosity** marks the attachment site of a muscle that flexes the forearm. A narrow neck extends from the tuberosity to the head of the radius. The disc-shaped head articulates with the capitulum of the humerus and forms a pivot articulation with the ulna at the radial notch. As rotation at this site pronates the forearm, the distal ulnar notch of the radius rolls across the

FIGURE 9-6
Bones of the Forearm. The radius and ulna are shown in (a) posterior view and (b) anterior view. (c) Note the changes that occur during pronation.

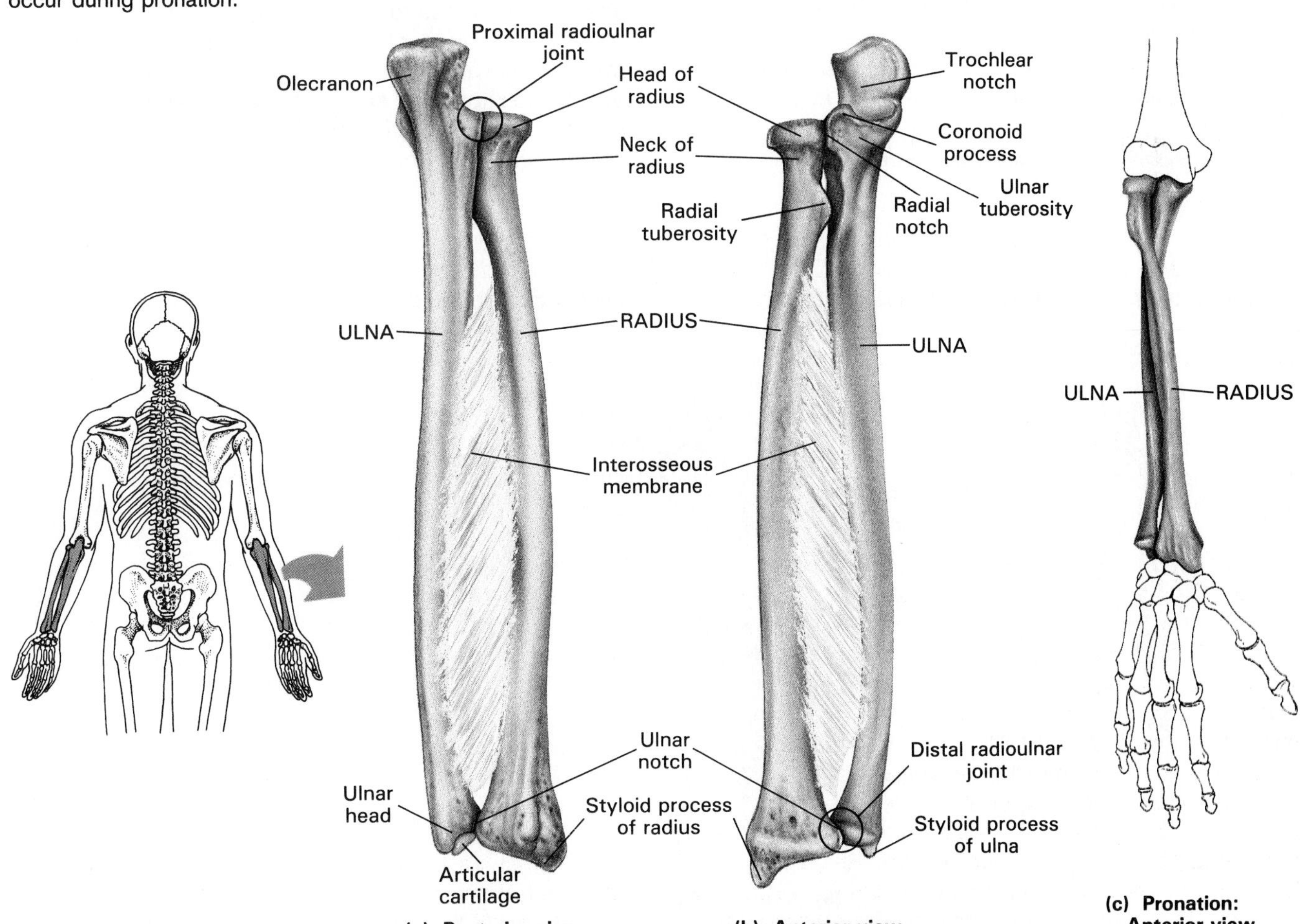

rounded surface of the ulnar head, as indicated in Figure 9-6c.

The Wrist and Hand

The eight bones of the wrist, or **carpus**, are shown in Figure 9-7. They form two rows, with four **proximal carpals—scaphoid**, **lunate**, **triquetal** (tri-QWE-tal), and **pisiform** (PI-si-form)—and four **distal carpals—trapezium**, **trapezoid**, **capitate**, and **hamate**. Surfaces that do not participate in articulations are roughened by the attachment of ligaments and the passage of tendons. A fibrous capsule, reinforced by broad ligaments, surrounds the wrist complex and stabilizes the positions of the individual carpals.

There are 14 phalangeal bones in each hand. Five **metacarpals** (met-a-KAR-pals) articulate with the distal carpals and form the palm of the hand. The metacarpals in turn articulate with the finger bones, or **phalanges** (fa-LAN-jēs). Four of the fingers contain three phalanges (proximal, middle, and distal) but the thumb, or **pollex** (POL-eks), has only two (proximal and distal).

√ **Why would a broken clavicle affect the mobility of the scapula?**

√ **The rounded projections on either side of the elbow are parts of what bone?**

√ **What bone of the forearm is lateral when the arm is pronated and medial when the arm is supinated?**

ARTICULATIONS OF THE ARM

The shoulder, elbow, and wrist are responsible for positioning the hand, which performs precise and controlled movements. The shoulder has great mobility, the elbow has great strength, and the wrist makes fine adjustments in the orientation of the palm and fingers.

Functional Anatomy of the Shoulder Joint

The shoulder joint permits the greatest range of motion of any joint in the body. Because it is also the most frequently dislocated joint, it provides an excellent demonstration of the principle that strength and stability must be sacrificed to obtain mobility.

Figure 9-8 details the structure of the shoulder joint. Figure 9-8a reviews the relationship between the bones involved in this joint, and Figure 9-8b indicates the size of the capsule and notes the location of several bursae at the shoulder. The relatively loose articular capsule extends from the scapular neck to the humerus, and this oversized capsule permits an extensive range of motion. As at other joints, bursae at the shoulder reduce friction where large muscles and tendons pass across the joint capsule. Bursae, introduced in Chapter 7, are chambers lined by synovial membrane and filled with synovial fluid. (p. 228) The enclosed fluid acts like an elastic cushion, absorbing shocks and reducing friction.

Bursae are characteristic of many synovial joints, but those of the shoulder are especially large and numerous. The **subacromial bursa** (Figure 9-8b) and the **subcoracoid bursa** (Figure 9-8c) prevent contact between the acromial and coracoid processes and

FIGURE 9-7
Bones of the Wrist and Hand. (a) Posterior view and (b) anterior view of the right hand.

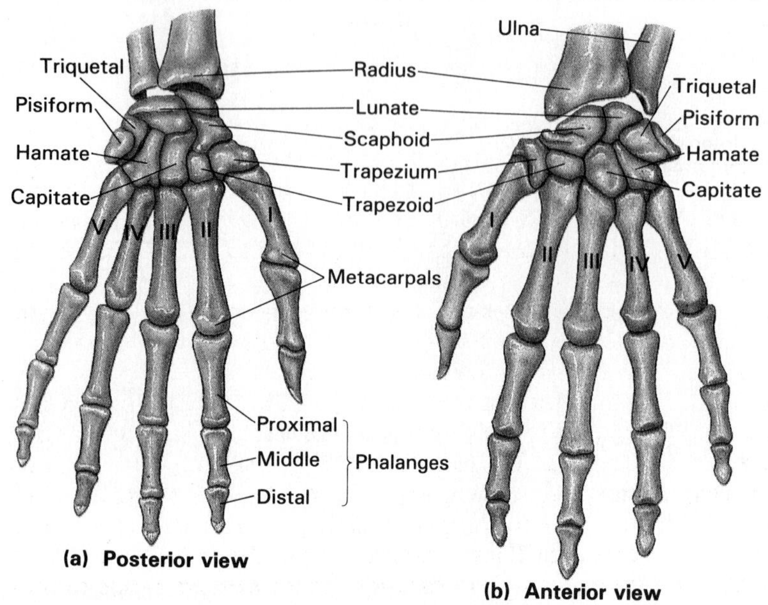

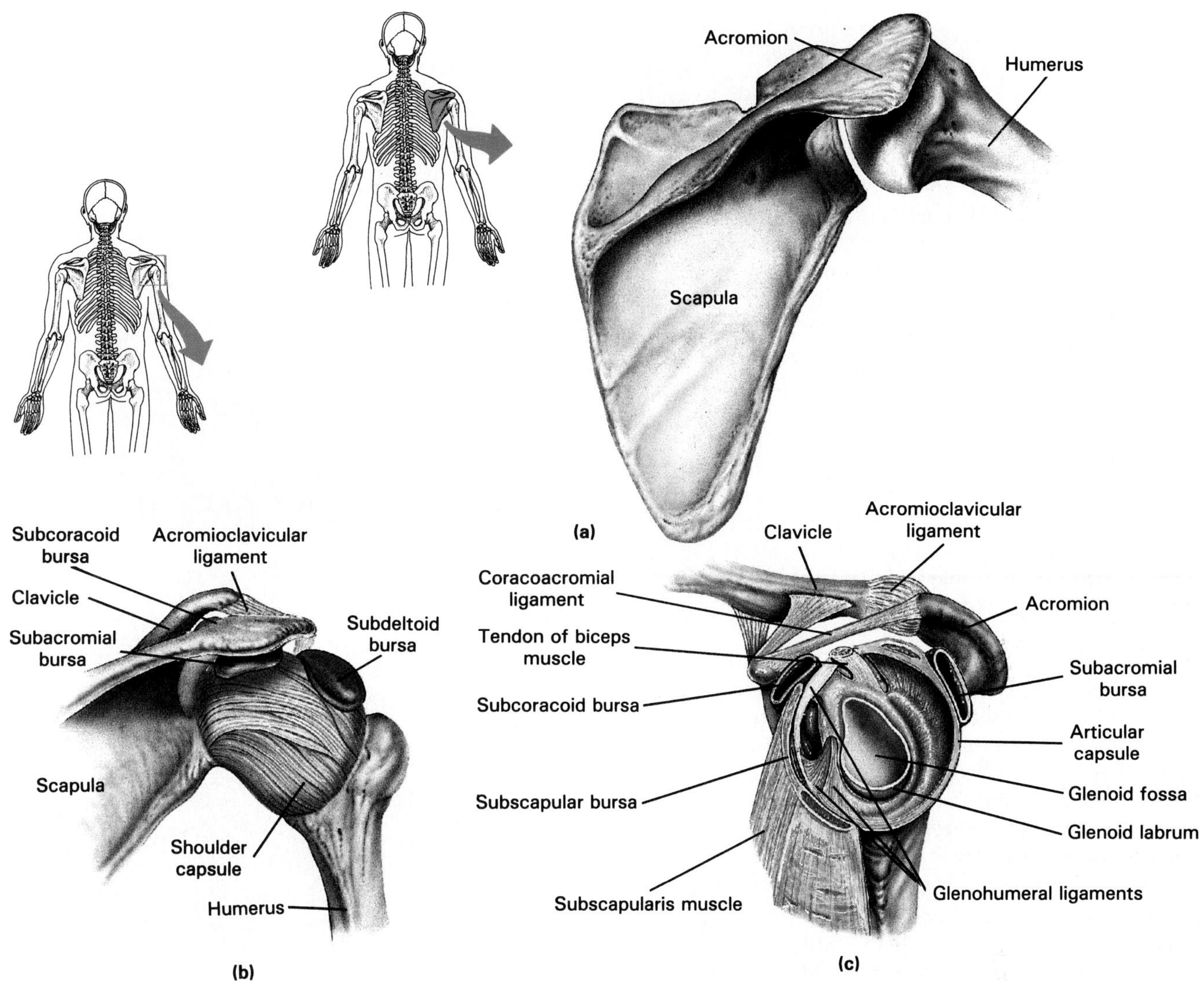

FIGURE 9-8
The Shoulder Joint. (a) Posterior view of the pectoral girdle. Note the relatively large size of the humeral head compared with the glenoid fossa. (b) Posterior view showing the orientation and extent of the articular capsule and two of the associated bursae. (c) A lateral view of the shoulder joint with the humerus removed. Notice that the acromion, clavicle, and associated ligaments form a bridge that reinforces the superior margin of the capsule. Bursae around the joint reduce friction between ligaments, tendons, and bony surfaces.

the capsule. The other bursae indicated in Figure 9–8b,c, the **subdeltoid bursa** and the **subscapular bursa**, lie between large muscles and the capsular wall. Inflammation of one or more of these bursae can restrict motion and produce the painful symptoms of **bursitis**, discussed in the Clinical Comment box.

In life, the surface of the glenoid fossa is covered by a cartilaginous **glenoid labrum** (*labrum*, lip or edge), and the glenoid/humeral articulation is not a snug fit. The capsular wall is relatively thin, but it thickens anteriorly in regions known as the **glenohumeral ligaments**. Because the capsular fibers are usually loose, these ligaments participate in joint stabilization only as the humerus approaches or exceeds the limits of normal motion. The large **coracohumeral ligament** originates at the base of the coracoid process and inserts on the head of the humerus. The **coracoacromial ligament** spans the gap between the coracoid process and the acromion, just above the capsule. This provides additional support to the superior surface of the capsule. In addition, a large **coracoclavicular ligament** ties the clavicle to the coracoid process and helps to limit the relative motion between the clavicle and scapula.

Muscles that move the humerus do more to stabilize the shoulder joint than all the ligaments and capsular fibers combined. Powerful muscles originating on the trunk, shoulder girdle, and humerus cover the anterior, superior, and posterior surfaces of the capsule. These muscles form the *rotator cuff* that swings the arm through its impressive range of mo-

Clinical Comment: Bursitis

In **bursitis**, bursae become inflamed, causing pain whenever the tendon or ligament moves. Inflammation can result from the friction associated with repetitive motion, pressure over the joint, irritation by chemical stimuli, infection, or trauma. Bursitis associated with repetitive motion often occurs at the shoulder; for example, golfers, pitchers, and tennis players may develop bursitis, usually at the subscapular bursa. The most common pressure-related bursitis is a **bunion**. Bunions form over the base of the big toe as a result of the friction and distortion of the joint caused by tight shoes, especially those with pointed toes. There is chronic inflammation of the region, and as the wall of the bursa thickens, fluid builds up in the surrounding tissues. The result is a firm, tender nodule.

There are special names for bursitis at other locations; the names of these conditions indicate the occupations most often associated with them. In "housemaid's knee," which accompanies prolonged kneeling, the affected bursa lies between the patella (*kneecap*) and the skin. "Weaver's bottom" is produced by pressure on the posterior and inferior tip of the pelvic girdle and can result from prolonged sitting on hard surfaces. Finally, "student's elbow" is a form of bursitis that can result from propping your head above a desk while struggling through your anatomy and physiology textbook.

Most of the symptoms of bursitis subside if the stimulus is removed, and a variety of anti-inflammatory drugs may help. In extreme cases the affected bursae can be surgically removed in a **bursectomy**.

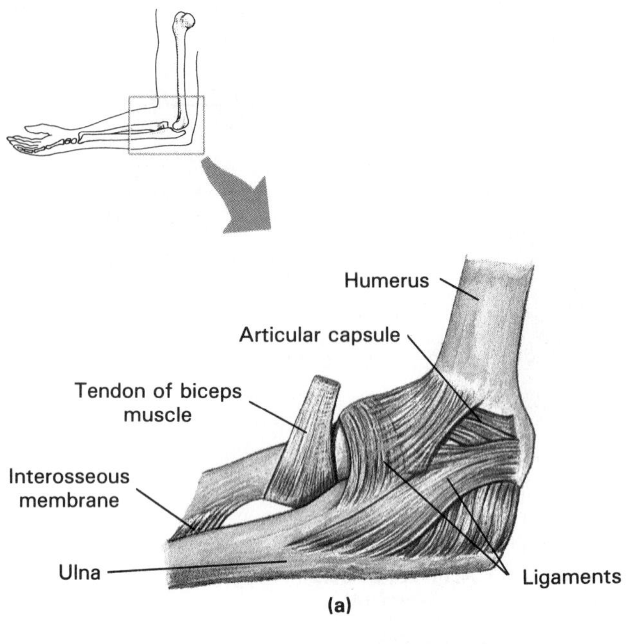

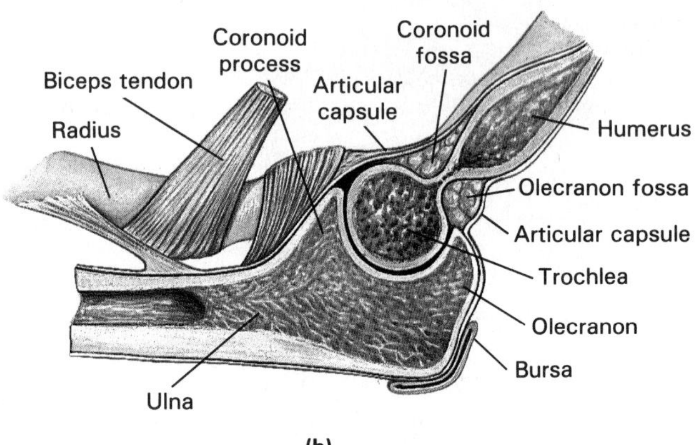

FIGURE 9-9
The Elbow Joint. (a) Medial aspect of the right elbow joint, showing the ligaments that stabilize the joint. (b) Medial aspect of the right elbow joint, showing the orientation of the articulation between the humerus and ulna.

tion. The inferior surface of the capsule is poorly reinforced, however. If you abduct your arm fully, you may be able to feel the head of the humerus as it presses against the inferior portion of the joint capsule. An impact or violent muscle contraction most often dislocates the arm at this site. Such a dislocation can tear the inferior capsular wall and the glenoid labrum. The healing process often leaves a weakness that increases the chances for future dislocations.

Functional Anatomy of the Elbow Joint

Figure 9-9 diagrams the structure of the elbow joint, also known as the **olecranal joint**. Muscles that extend the elbow attach to the rough surface of the olecranon process. These muscles are primarily under the control of the nerve that passes along the radial groove of the humerus. On the front of the arm, the large *biceps brachii* muscle pulls on the tendon that attaches to the **ulnar tuberosity**, producing flexion of the elbow.

The elbow joint is extremely stable because: (1) the bony surfaces of the humerus and ulna interlock; (2) the articular capsule is very thick; and (3) the capsule is reinforced by stout ligaments. Nevertheless, the joint can be damaged by severe impacts or unusual stresses. When a person falls on his hand with a partially flexed elbow, powerful contractions of the muscles that extend the elbow can break the ulna at the center of the trochlear notch. A less common but equally distressing injury occurs when the weight of the body pushes the humerus forward, through the articular capsule. Less violent stresses can produce dislocations or other injuries to the elbow if epiphyseal growth has not been completed. For example, parents in a hurry may drag a toddler along behind them, exerting an upward, twisting pull on the elbow joint that can result in a partial dislocation known as a "nursemaid's elbow."

The Wrist Joint and Hand

The structure of the wrist joint is shown in Figure 9-10. The **radiocarpal articulation** is an ellipsoidal joint, and the **intercarpal articulations** are gliding joints. Together these joints permit flexion, extension, abduction, and adduction of the wrist.

Distally, the carpals articulate with the metacarpals of the palm. The first metacarpal has a saddle-type articulation at the wrist, which increases the mobility of the thumb. All other carpal/metacarpal articulations are gliding joints. The articulations between metacarpals and their phalanges are ellipsoidal, permitting flexion/extension and adduction/abduction. The interphalangeal joints function as hinges, allowing only flexion and extension.

Table 9-1 summarizes the characteristics of the articulations considered in our discussion of the pectoral girdle and arm.

✓ **Would a tennis player or a jogger be more likely to develop inflammation of the subscapular bursa? Why?**

✓ **Mary falls on her hands with her elbows slightly flexed. After the fall, she can't move her left arm at the elbow. If a fracture exists, what bone is most likely broken?**

FIGURE 9-10
Articulations of the Wrist and Hand. (a) Anterior view of the right wrist, identifying the different types of articulations. (b) The wrist bones and (c) a sectional view through the wrist, showing the radiocarpal, intercarpal, and carpometacarpal joints.

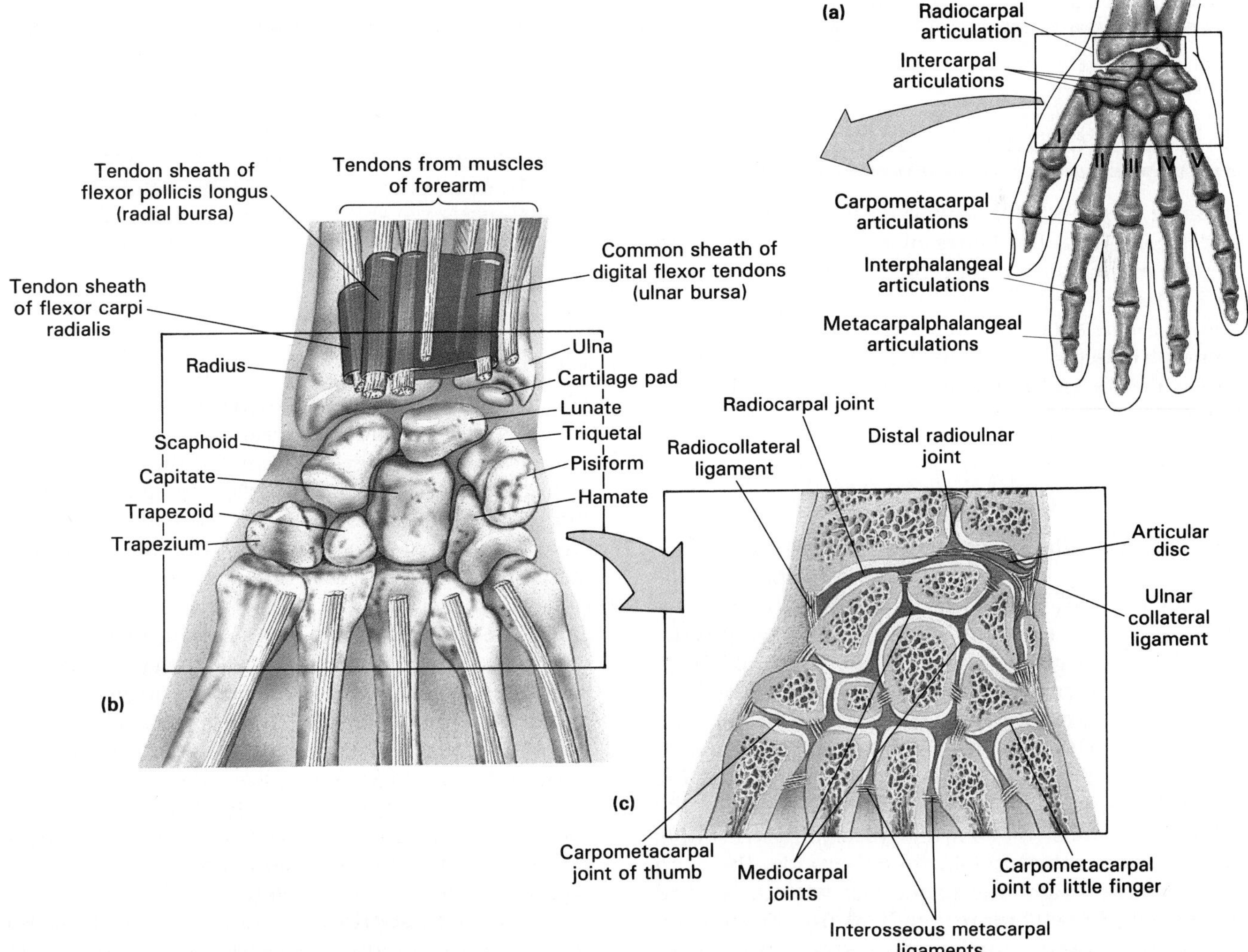

TABLE 9-1 Articulations of the Pectoral Girdle and Arm

Elements	*Joint*	*Type of Articulation*	*Movement*
Clavicle/sternum	Sternoclavicular	Gliding diarthrosis	Gliding
Scapula/clavicle	Acromioclavicular	Gliding diarthrosis	Gliding
Scapula/humerus	Scapulohumeral, or shoulder	Ball-and-socket diarthrosis	Flexion/extension, adduction/abduction, circumduction, rotation
Humerus/ulna and radius	Olecranal, or elbow	Hinge diarthrosis	Flexion/extension
Radius/ulna	Proximal radioulnar	Pivot diarthrosis	Pronation/supination
	Distal radioulnar	Pivot diarthrosis	Pronation/supination
Radius/carpals	Radiocarpal, or wrist	Ellipsoidal diarthrosis	Flexion/extension, adduction/abduction, circumduction
Carpal/carpal	Intercarpal	Gliding diarthrosis	Gliding
Carpal/metacarpal (first)	Carpometacarpal of thumb	Saddle diarthrosis	Flexion/extension, adduction/abduction, circumduction, opposition
Carpal/metacarpal (2–5)	Carpometacarpals	Gliding diarthrosis	Slight flexion/extension adduction/abduction
Metacarpals/phalanges	Metacarpophalangeal	Ellipsoidal diarthrosis	Flexion/extension, adduction/abduction, circumduction
Phalanges/phalanges	Interphalangeal	Hinge diarthrosis	Flexion/extension

The Pelvic Girdle and Leg

Because of the stresses involved in weight bearing and locomotion, the bones of the pelvic girdle and legs are more massive than those of the pectoral complex. The pelvic girdle is also much more firmly attached to the axial skeleton. Large tendons and ligaments produce relatively massive superficial marks, which can be quite useful in identifying individual bones.

THE PELVIC GIRDLE

The **pelvic girdle** articulates with the thigh bones of the legs. Dorsally, the two halves of the pelvic girdle contact the auricular surfaces of the sacrum. Ventrally, the pelvic elements are joined at a symphysis.

The pelvic girdle consists of two large hip bones, or **coxae**. Each coxa forms through the fusion of three bones, the **ilium** (IL-ē-um), the **ischium** (IS-kē-um), and the **pubis** (PŪ-bis). Dorsally the hip bones articulate with the auricular surfaces of the sacrum at the **sacroiliac joint**, a gliding diarthrosis. Ventrally the coxae are connected at a symphyseal joint, the *pubic symphysis*. At the hip joint to either side, the head of the **femur** (thighbone) articulates with the curved surface of the **acetabulum** (a-se-TAB-ū-lum; *acetabulum*, cup). Viewed from the side, the acetabulum lies inferior and anterior to the center of the coxa. The walls of the acetabulum enclose a space, the **acetabular fossa**, with a diameter of approximately 5 cm (2 in.). The coxal bones meet inside the acetabular fossa, as if it were a pie sliced into three pieces. The structure of the hip, or coxal, bones of the pelvic girdle is shown in Figure 9-11.

The Coxa

The **ilium**, the largest coxal bone, provides the superior slice that includes around two-fifths of the acetabular surface. Refer to Figure 9-11a, which shows a lateral view of the pelvic girdle and sacrum. Above the acetabulum, the ilium forms a broad, curved surface that provides an extensive area for the attachment of muscles, tendons, and ligaments. Prominent ridges, notches, and spines mark the attachment sites. The iliac expansion begins at the **anterior inferior iliac spine**, above the **inferior iliac notch**, and continues anteriorly to the **anterior superior iliac spine**. Curving posteriorly, the superior border supports the **iliac crest**, a ridge marking the attachments of both ligaments and muscles. The iliac crest ends at the **posterior superior iliac spine**. Below the spine, the ilial margin continues inferiorly to the rounded **posterior inferior iliac spine** that sits above the **greater sciatic** (sī-A-tik) **notch**.

Near the superior and posterior margin of the acetabulum, the ilium fuses with the **ischium**, which accounts for the posterior two-fifths of the acetabular

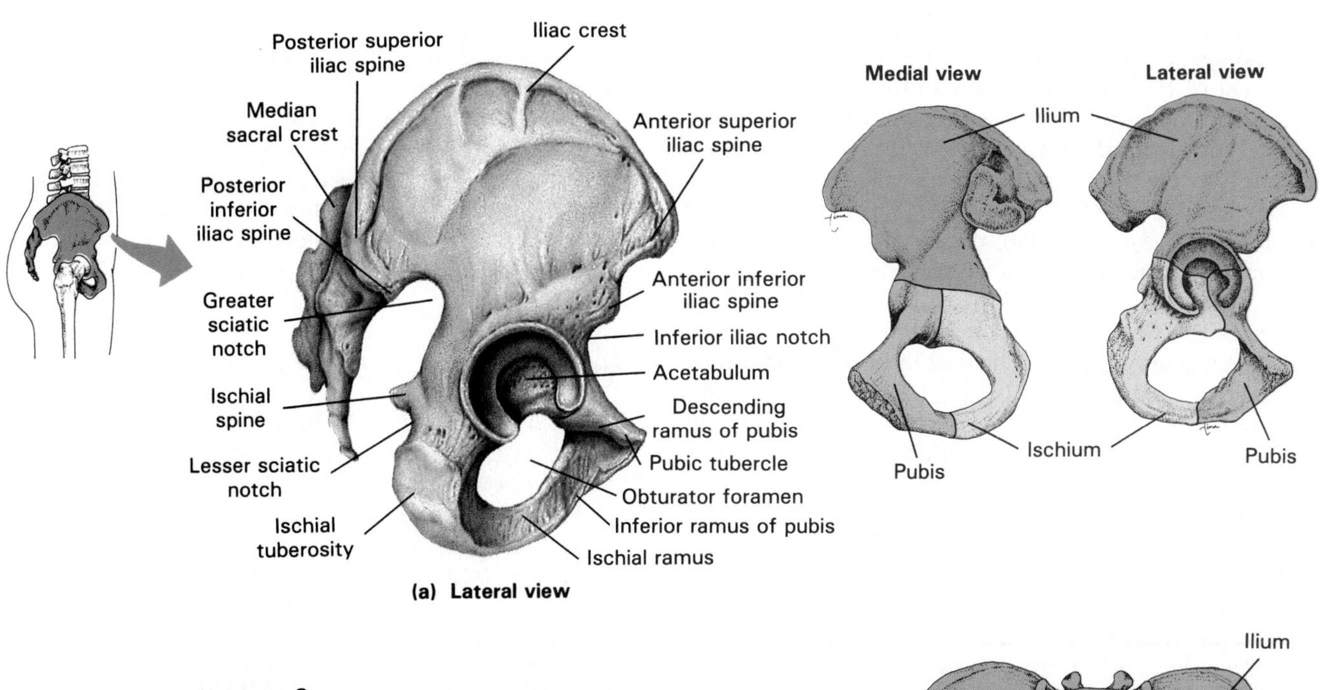

(a) Lateral view

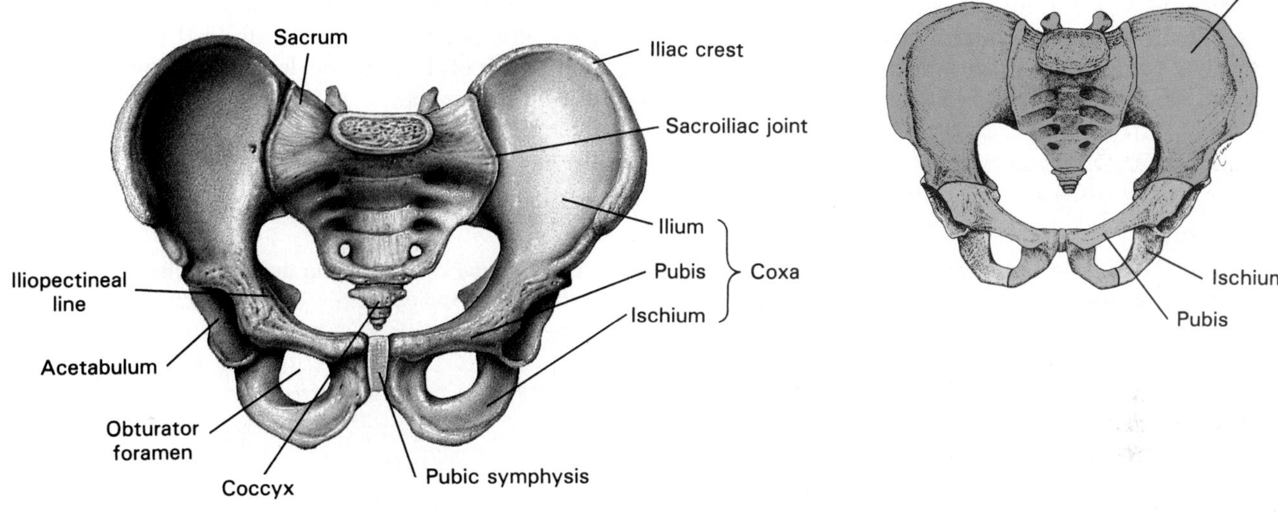

(b) Anterior view

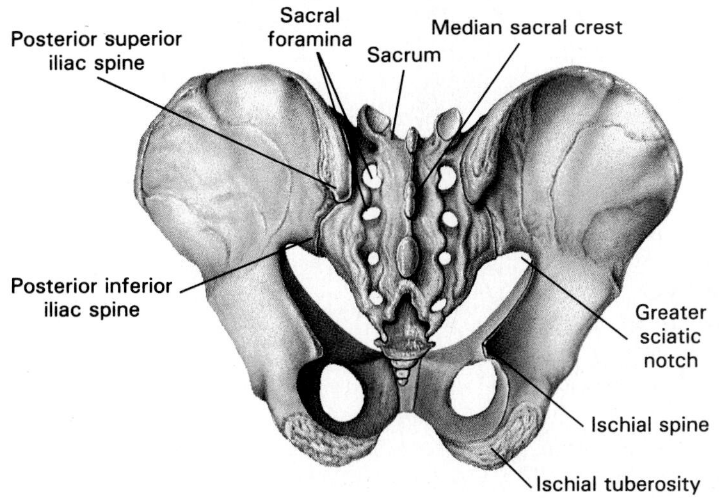

(c) Posterior view

FIGURE 9-11
The Pelvic Girdle. (a) Lateral view, (b) anterior view, and (c) posterior view.

surface. Posterior to the acetabulum, the prominent **ischial spine** projects above the **lesser sciatic notch**. The rest of the ischium forms a sturdy process that turns medially and inferiorly. Its posterolateral border bears the **ischial tuberosity**. The narrow **ramus** (branch) of the ischium continues toward its anterior fusion with the **pubis**.

At the point of fusion, the ramus of the ischium meets the **inferior ramus** of the pubis. Anteriorly, the inferior ramus ends at the **pubic tubercle**, where it meets the **descending ramus** that originates near the acetabular margin. The pubic and ischial rami encircle the **obturator** (OB-tū-rā-tor) **foramen**. In life, this space is closed by a sheet of collagen fibers whose inner and outer surfaces provide a firm base for the attachment of muscles and visceral structures. At the juncture of the pubic ramus with the body of the pelvis, the pubis contacts the ilium and ischium within the acetabular fossa.

Now examine Figure 9-11b. The anterior and medial surface of the pubis contains a roughened area that marks the articulation with the pubis of the opposite side. An amphiarthrotic articulation, the **pubic symphysis**, limits movement between the two pubic bones. A prominent ridge, the **iliopectineal** (il-ē-ō-pek-TIN-ē-al) **line**, extends from the area of the symphysis diagonally across the pubic and iliac surfaces, ending at the **articular surface** of the ilium. Ligaments arising at the **iliac tuberosity** stabilize the sacroiliac joint. Anterior to tuberosity and articulation, the broad **iliac fossa** provides additional surface area for leg muscle attachment.

Figure 9-11c gives a posterior view of the pelvic girdle. Note the positions of the sacrum and coccyx, and locate the major anatomical landmarks from parts (a) and (b) from this perspective.

The Pelvis

The **pelvis** consists of the coxae, the sacrum, and the coccyx. It is thus a composite structure that includes portions of both the appendicular and axial skeletons. An extensive network of ligaments connects the lateral borders of the sacrum with the iliac crest, the ischial tuberosity, the ischial spine, and the iliopectineal line. Other ligaments tie the ilia to the posterior lumbar vertebrae. These interconnections increase the structural stability of the pelvis.

The pelvis may be subdivided into the **greater**, or **false**, **pelvis** and the **lesser**, or **true**, **pelvis**. The boundaries of each are indicated in Figure 9-12. The false pelvis consists of the expanded, bladelike portions of each ilium above the iliopectineal line. The

FIGURE 9-12
The Pelvis. (a) Superior view, showing the pelvic brim and pelvic inlet. (b) Lateral view, showing the boundaries of the true (lesser) and false (greater) pelvis. (c) Inferior view, showing the limits of the pelvic outlet.

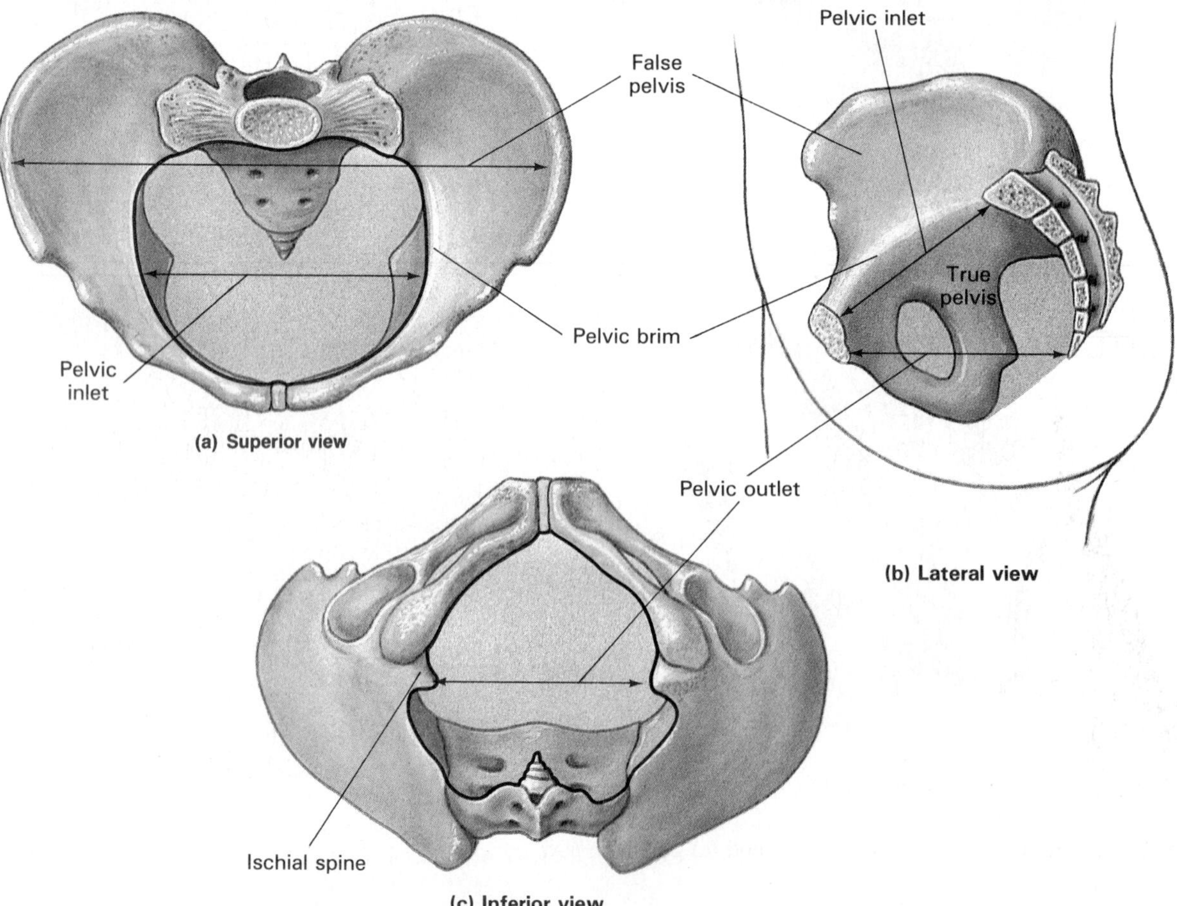

pubis, ischium, and the rest of the ilium form the lesser, or true, pelvis.

The true pelvis encloses the *pelvic cavity*, a subdivision of the peritoneal cavity (see Figure 1-13). (p. 19) In anterior view (Figure 9-12b) the upper limit of the true pelvis is a line that extends from either side of the base of the sacrum, along the iliopectineal lines to the superior margin of the pubic symphysis. The bony edge of the true pelvis is called the **pelvic brim**, and the enclosed space is the **pelvic inlet**.

The **pelvic outlet** is the opening bounded by the inferior edges of the pelvis, as indicated in Figure 9-12c. In life this region, the **perineum** (per-i-NE-um), extends between the coccyx, the ischial tuberosities, and the inferior border of the pubic symphysis. This forms the floor of the pelvic cavity and supports the enclosed organs.

THE LEG

The leg consists of the thighbone, or **femur**, the **tibia** and **fibula** of the lower leg, and the bones of the ankle and foot.

The Femur

The **femur**, or *thighbone*, is the longest and heaviest bone in the body. Distally, the femur articulates with the **tibia** of the lower leg at the knee joint. The rounded epiphysis, or head, of the femur, shown in Figure 9-13, articulates with the pelvis at the acetabulum. Distally, the neck joins the shaft at an angle of 125°–140°. The **greater trochanter** arises lateral to the juncture of the neck and shaft; the **lesser trochanter** originates along the crest near the medial surface of the femur. Both trochanters develop where large tendons attach to the femoral shaft. On the anterior surface of the femur, the raised **trochanteric** (tro-kan-TER-ik) **line** marks the distal limits of the articular capsule. This line continues around to the posterior surface, passing beneath the trochanters as the **trochanteric crest**.

The proximal femoral shaft is round in cross section. A prominent elevation, the **linea aspera** (*aspera*, rough), runs along the center of the posterior surface, marking the attachment site of powerful muscles that abduct the femur. Distally, the linea

FIGURE 9-13
Bones of the Thigh. Anatomical landmarks on the right femur are presented as seen from (a) the anterior and (b) the posterior surface.

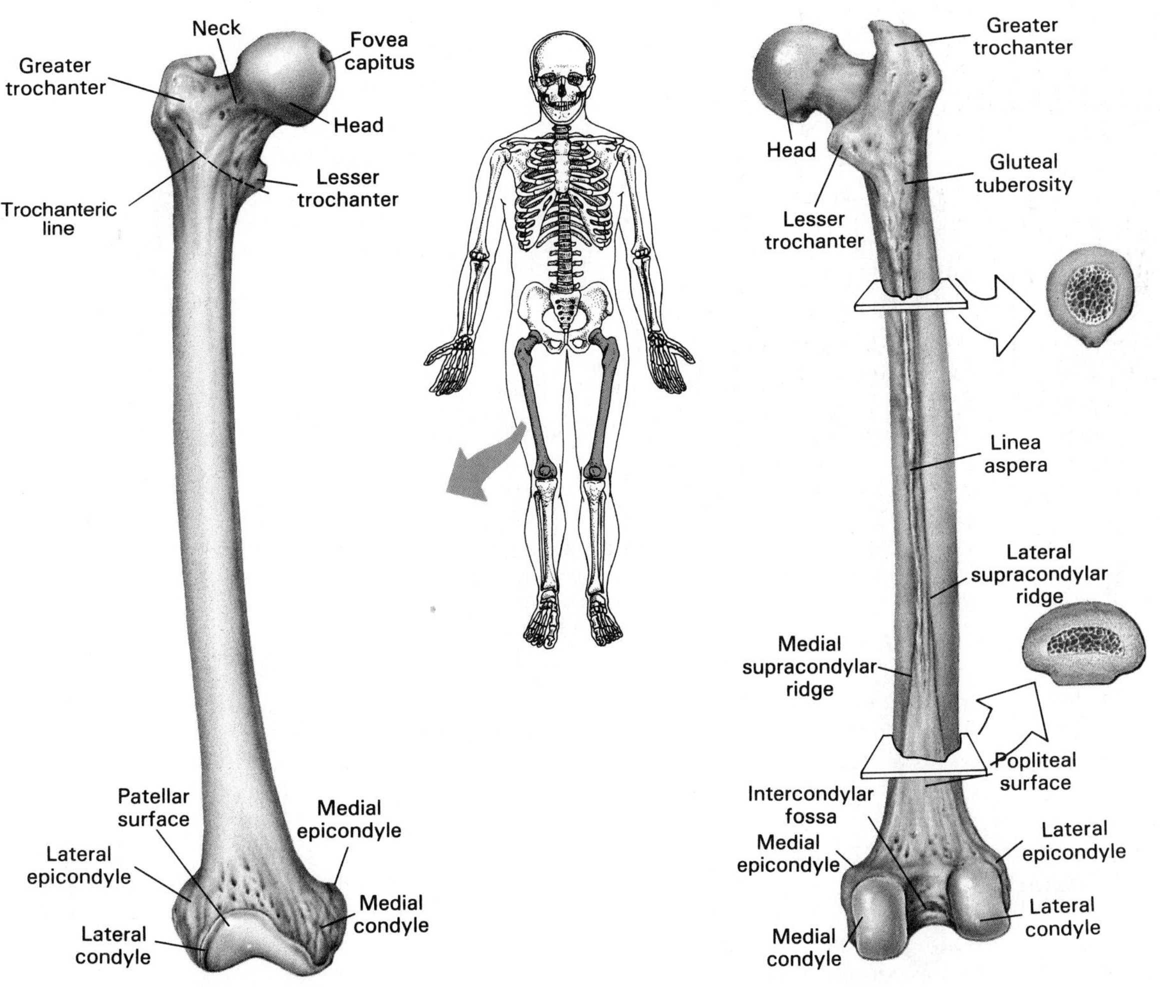

aspera divides distally to pass alongside a flattened area, the **popliteal surface** of the femur. Two large **articular condyles** can be seen at the distal and posterior border of the shaft. They are separated by a deep **intercondylar fossa**, and a pair of **epicondyles** project to either side of the round condylar surfaces.

The articular condyles continue across the inferior surface of the femur, but the intercondylar fossa does not extend completely around to the anterior surface. As a result, the smooth articular faces merge, producing an articular surface with elevated lateral borders. This is the **patellar surface** over which the **patella** (*kneecap*) glides.

The Tibia and Fibula

Figure 9-14 shows the structure of the *tibia* and *fibula*. The condyles of the femur articulate with the **tibial condyles** of the **tibia**, the large medial bone of the lower leg. The anterior surface of the tibia, or shin bone, can easily be investigated with your fingers, as it lies just beneath the skin. A ligament from the patella attaches to the **tibial tuberosity** just below the knee joint.

A projecting **anterior crest** extends almost the entire length of the anterior surface. As the distal end of the tibia is reached, the tibia broadens, and the medial border ends in a large process, the **medial malleolus** (ma-LĒ-o-lus; *malleolus*, hammer). The inferior surface of the tibia forms a hinge joint with the proximal bone of the ankle; the medial malleolus provides medial support for the ankle.

The slender **fibula** parallels the lateral border of the tibia. It does not help in transferring weight to the foot, for it is completely excluded from the knee joint. The fibular **head** articulates along the lateral margin of the tibia, inferior and slightly posterior to the lateral condyle. The medial border is bound to the tibia by the interosseous membrane along the fibular **interosseous crest**. The fibula does not participate in the knee joint, and it does not bear weight. However, it is an important surface for muscle attachment, and the distal **lateral malleolus** provides lateral stability to the ankle.

Figure 9-14b shows a posterior view of the tibia and fibula. A ridge, the **intercondylar eminence**, sep-

FIGURE 9-14
Bones of the Lower Leg. (a) Anterior view of the right tibia and fibula. (b) Posterior view. (c) Sectional view at the level indicated in (b).

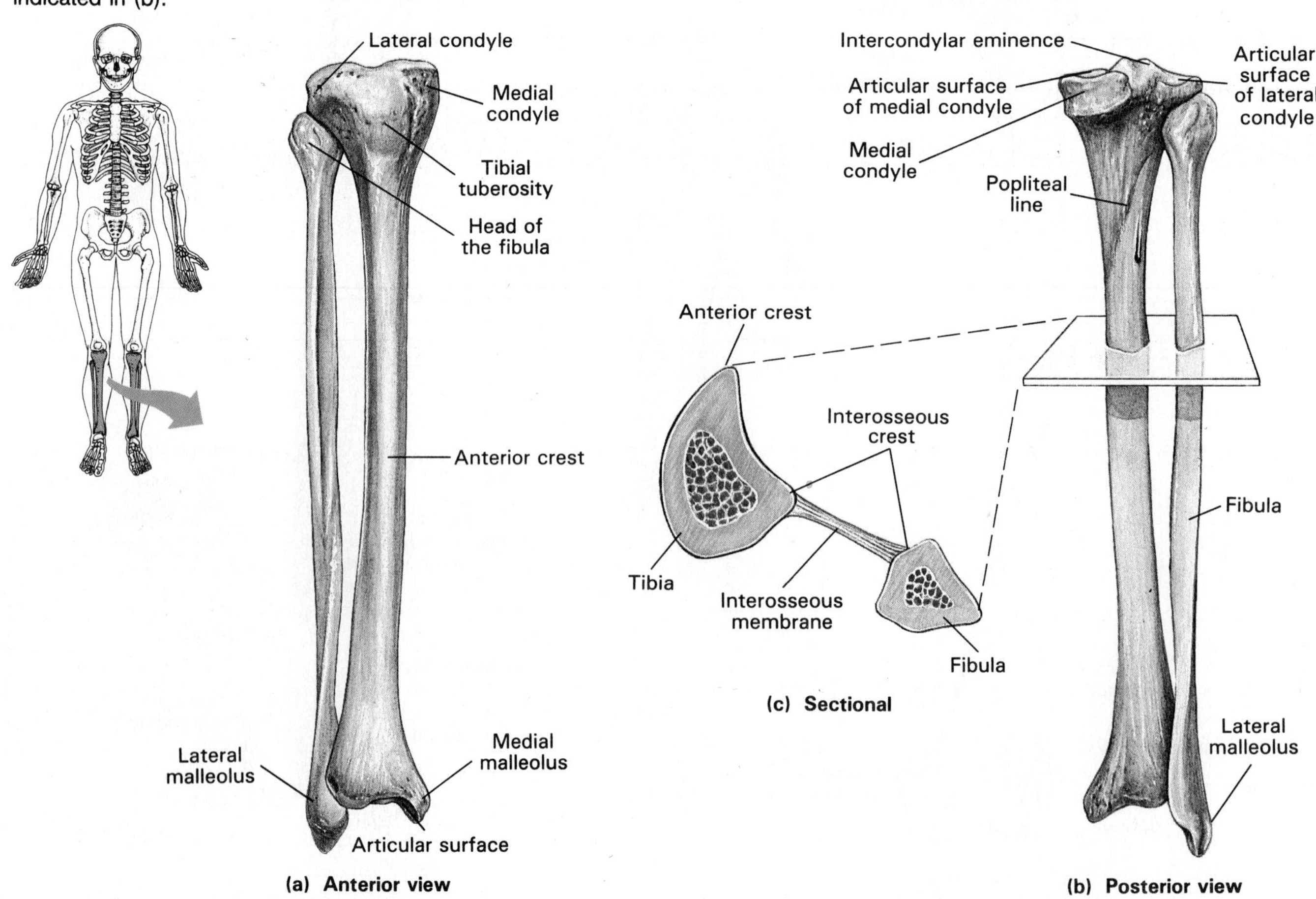

arates the medial and lateral condyles of the tibia. A sectional view through the shafts of the tibia and fibula (Figure 9-14c) shows the location of **interosseous crests** along their opposing margins. A fibrous **interosseous membrane** extends between the crests. This membrane helps stabilize the positions of these bones and provides additional surface area for muscle attachment.

The Ankle and Foot

The ankle, or **tarsus**, includes seven separate bones: the **calcaneus**, **talus**, **navicular**, **cuboid**, **medial cuneiform**, **intermediate cuneiform**, and **lateral cuneiform** (Figure 9-15a). Only the proximal tarsal, the **talus**, articulates with the tibia and fibula. The talus then passes the weight to the ground via other bones of the foot.

When standing normally, most of your weight is transmitted to the ground by the inferior surface of the talus to the large **calcaneus** (kal-KĀ-nē-us), or *heel bone.* The posterior projection of the calcaneus receives the composite **calcaneal tendon**, or *Achilles tendon*, of the calf muscles that raise the heel and depress the sole (plantar flexion). The rest of the body weight is passed via the cuboid and cuneiform to the **metatarsals** that support the sole of the

FIGURE 9-15
Bones of the Ankle and Foot. (a) Bones of the right foot as viewed from above. Note the orientation of the tarsals that convey the weight of the body to the heel and the plantar surfaces of the foot. (b) Medial view, showing the relative positions of the tarsals and the orientation of the transverse and longitudinal arches. (c) Lateral view of the right foot. Note the firm contact with the surface along the lateral border of the sole.

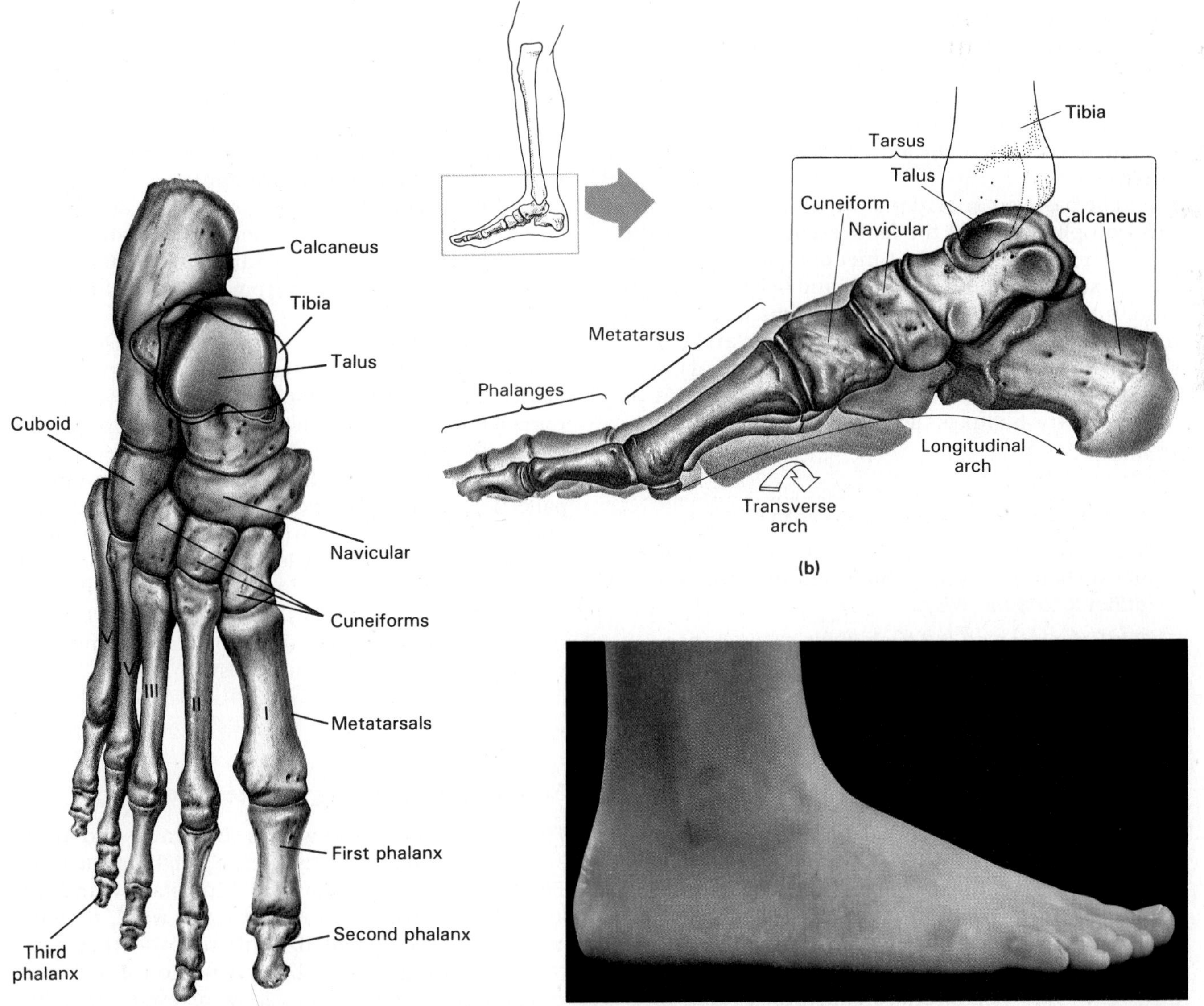

foot. The amount transferred forward depends on the position of the foot and the placement of body weight. During dorsiflexion of the foot, as when "digging in the heels," all of the body weight rests on the calcaneus. During plantar flexion and "standing on tiptoe," the talus and calcaneus transfer the weight to the metatarsals and phalanges through more anterior tarsal bones.

Weight transfer occurs along the **longitudinal arch** of the foot, outlined in Figure 9-15b. Strong ligaments and tendons maintain this arch by tying the calcaneus to the distal portions of the metatarsals. The lateral, calcaneal side of the foot carries most of the weight of the body while standing normally. This portion of the arch has much less curvature than the medial, talar portion, which in addition to its more pronounced curvature also possesses considerable elasticity. As a result, the medial, plantar surface remains elevated, and the muscles, nerves, and blood vessels that service the inferior surface of the foot are not squeezed between the metatarsals and the ground. The elasticity of the talar arch absorbs the shocks that accompany sudden changes in weight loading. For example, the stresses involved with leaping, landing, or dancing on the toes are cushioned by the elasticity of this portion of the arch. Because the degree of curvature changes from the medial to the lateral borders of the foot, a **transverse arch** also exists.

The basic organizational pattern at the metatarsals and phalanges of the foot resembles that of the hand. The most significant functional differences are: (1) ligaments and tendons above and below the **tarsometatarsal joints** restrict movement much more than the carpometacarpal joints at the wrist; and (2) although the first toe, or **hallux**, has two phalangeal bones, its mobility is limited because the joint at the base of the hallux is ellipsoidal rather than saddle-shaped.

√ **What three bones make up the coxa?**

√ **The fibula does not participate in the knee joint nor does it bear weight, but when fractured it is difficult to walk. Why?**

√ **While jumping off the back steps at his house, 10-year-old Joey lands on his right heel and breaks his foot. What foot bone is most likely broken?**

ARTICULATIONS OF THE LEG

The functional requirements of the leg joints are very different from those of the arms. These joints must transfer the body weight to the ground, and during movements such as running, jumping, or twisting the muscles develop forces considerably greater than the weight of the body. Joints of the hip, ankle, and foot are sturdier than those at corresponding locations in the arms, and their ranges of motion are less. The knee has a range of motion comparable to that of the elbow, but it is subjected to much greater forces and so is less stable.

Functional Anatomy of the Hip Joint

Figure 9-16 details the structure of the **hip joint**. Refer first to Figure 9-16a. A fibrocartilage pad covers the articular surface of the acetabulum and extends like a horseshoe along the sides of the **acetabular notch**. A fat pad covered by synovial membrane covers the central portion of the acetabulum, and ligaments crossing the acetabular notch complete the inferior border of the articular depression. The fat pad acts as a shock absorber, and the adipose tissue stretches and distorts without damage.

Compared with that of the shoulder, the articular capsule of the hip joint is extremely dense and strong. As indicated in Figure 9-16b,c it extends from the lateral and inferior surfaces of the pelvic girdle to the trochanteric line and crest of the femur, enclosing both the femoral head and neck. This arrangement helps keep the head from moving away from the acetabulum. Three broad ligaments reinforce the articular capsule: the **iliofemoral**, **pubofemoral**, and **ischiofemoral ligaments**. A fourth ligament, the **ligamentum teres** (*teres*, long and round) originates inside of the acetabulum and attaches to the center of the femoral head (Figure 9-16d). Additional stabilization comes from the bulk of the surrounding muscles.

The combination of an almost complete bony socket, a strong articular capsule, supporting ligaments, and muscular padding makes this an extremely stable joint. Fractures of the femoral neck or between the trochanters are actually more common than hip dislocations. See the Health News box on page 288. Although flexion, extension, adduction, abduction, and rotation are permitted, the total range of motion is considerably less than that of the shoulder. Hip flexion is the most important normal movement, and the primary limits are imposed by the surrounding muscles. Other directions of movement are restricted by ligaments and capsular fibers. [**CM:** *Hip Fractures and the Elderly*]

Functional Anatomy of the Knee Joint

The hip joint passes weight to the femur, and at the knee joint the femur transfers the weight to the tibia. The shoulder is mobile; the hip, stable; and the knee . . .? If you had to choose one word, it would probably be "complicated." Although the knee functions as a hinge joint, the articulation is far more complex than that of the elbow or even the ankle. The rounded femoral condyles roll across the top of

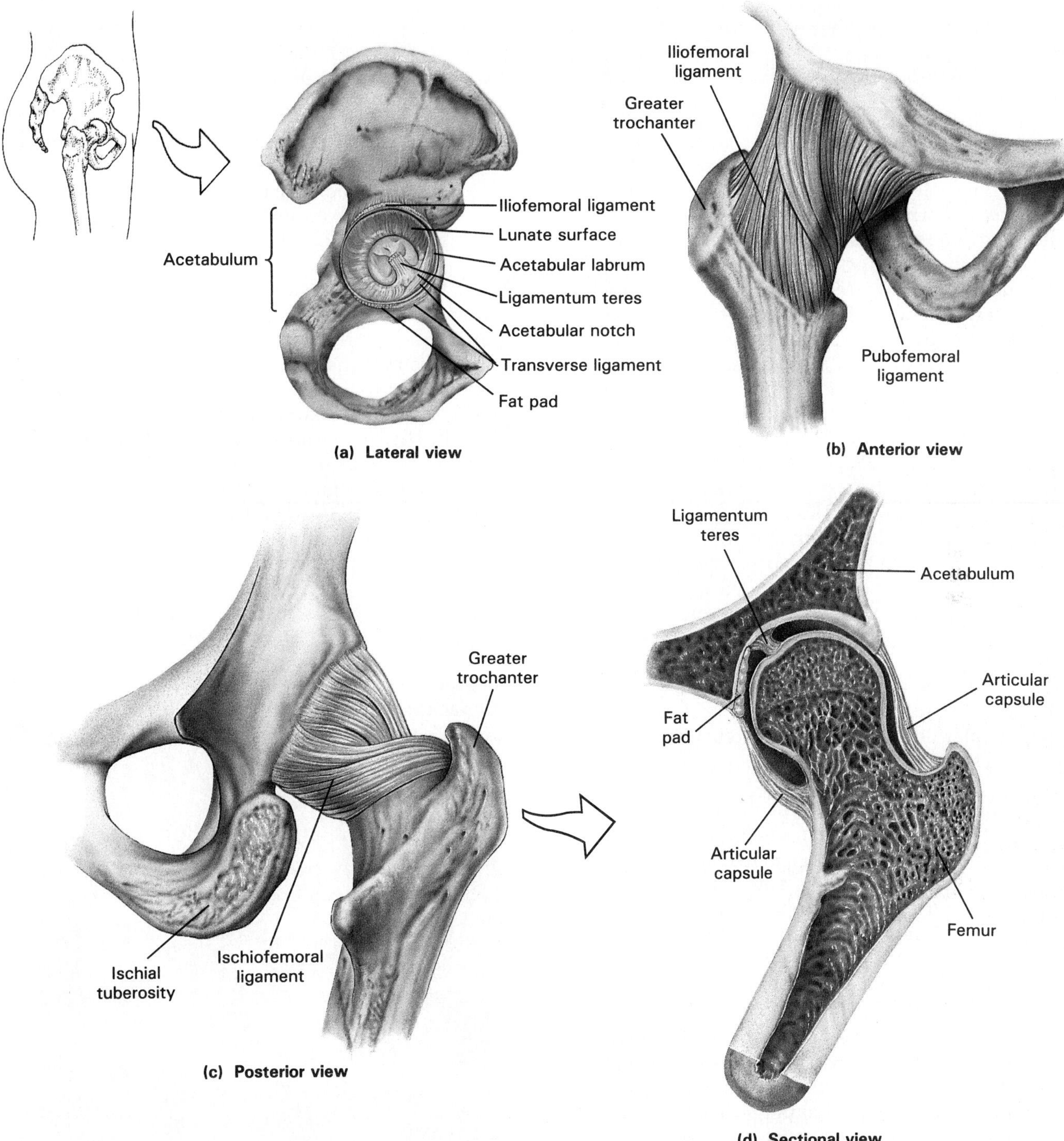

FIGURE 9-16
The Hip Joint. (a) Lateral view of the right hip joint with the femur removed. (b) Anterior view. This joint is extremely strong and stable, in part because of the massive capsule. (c) Posterior view, showing additional ligaments that add strength to the capsule. (d) Sectional view, showing the position and orientation of the ligamentum teres.

the tibia, so the points of contact are constantly changing. Important features of the knee joint are indicated in Figure 9-18.

Structurally the knee resembles three separate joints, two between the femur and tibia (medial to medial condyle and lateral to lateral condyle) and one between the patella and the femur. There is no single unified capsule, nor is there a common synovial cavity. A pair of fibrocartilage pads, the **medial** and **lateral menisci**, lie between the femoral and tibial surfaces. They act as cushions and conform to the shape of the articulating surfaces as the femur changes posi-

Health News: A Case Study: Bo Jackson

Everyone agrees that Bo Jackson was an athletic phenomenon. At age 28 he was playing professional football with the LA Raiders and professional baseball for the Kansas City Royals–and starring in both sports. He was also the centerpiece for successful advertising campaigns for sporting goods, using the "Bo knows" slogan. But on January 13, 1991, things changed–perhaps permanently–when Bo was tackled near the sidelines in an NFL playoff game. The combination of pressure and twisting applied by the tackler and the tremendous power of Bo's thigh muscles produced an unusual injury, a fracture-dislocation of the hip. Roughly 15 percent of the inferior acetabular fossa was broken away. The femur was not broken, but it was dislocated. Although he experienced severe pain, the immediate damage to the femur and hip was sufficiently limited that initial reports were optimistic about his return to professional sports.

The optimism began to fade when it became apparent that the complications of the injury were more damaging than the injury itself. The dislocation apparently tore blood vessels in the capsule and along the femoral neck, where the capsular fibers attach. Two problems gradually developed as a result:

1. The mineral deposits in the bone of the pelvis and femur are turned over very rapidly (see Chapter 7). The osteocytes have high energy demands, and a reduction in blood flow first injures and then kills them. When bone maintenance stops in the affected region, the matrix begins to break down. This process is called *avascular necrosis.*
2. The chondrocytes in the articular cartilages absorb nutrients from the synovial fluid, which circulates around the joint cavity as the bones change position. In this injury, the initial impact and damage to the articular cartilages was followed by joint immobility and poor circulation to the synovial membrane, depriving the cells of adequate nourishment. In other words, the chondrocytes were first damaged and then starved. The combination resulted in the gradual loss of the articular cartilages of the femur and acetabulum.

Will Bo Jackson return to competitive sports? At this point, his future remains uncertain. In an elderly individual, the likely outcome would be eventual insertion of an artificial hip (see the related discussion in the Clinical Manual for details). Because he is young and in good health, the bones may heal—but the cartilage damage may pose a more serious long-term problem. Cartilage has a very limited ability to repair itself, and articular cartilages are no exception. So even if the avascular necrosis reverses itself and the hip joint becomes strong enough to bear weight, it is questionable whether Bo will be able to use that joint without pain, and without risk of further injury.

FIGURE 9-17
Bo Jackson at Work Prior to His Injury

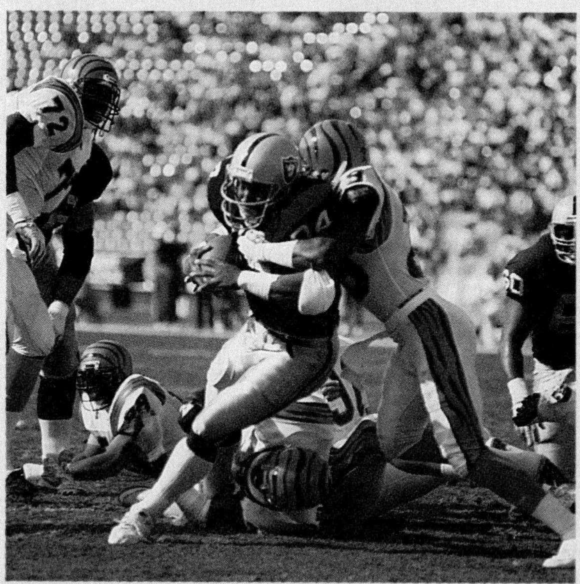

tion. Prominent **fat pads** provide padding around the margins of the joint and assist the bursae in reducing friction between the patella and other tissues.

Seven major ligaments stabilize this joint; as a result, a complete dislocation of the knee is an extremely rare event. The tendon from the muscles responsible for extending the knee passes over the anterior surface of the joint. The patella is embedded within this tendon, and the **patellar ligament** continues to its attachment on the anterior surface of the tibia. The patellar ligament provides support to the front of the knee joint. The **popliteal ligaments** extend between the femur and the heads of the tibia and fibula. These ligaments reinforce the back of the knee joint.

Inside the joint capsule, the **anterior cruciate** and **posterior cruciate** ligaments attach the intercondylar area of the tibia to the condyles of the femur. Anterior and posterior refer to their sites of origin on the tibia, and they cross one another as they proceed to their destinations on the femur. The term *cruciate* is derived from the Latin word *crucialis*, meaning a cross. These ligaments limit the anterior and posterior movement of the femur and maintain the alignment of the femoral and tibial condyles.

The **tibial collateral ligament** reinforces the medial surface of the knee joint, and the **fibular collateral ligament** reinforces the lateral surface. These ligaments tighten only at full extension, and in this position they act to stabilize the joint. In addition, the knee "locks" in the extended position. At full extension a slight lateral rotation of the tibia tightens the anterior cruciate ligament and jams the meniscus between the tibia and femur. This mechanism allows you to stand for prolonged periods without continually contracting the extensor muscles. Unlocking the joint requires muscular contractions that produce medial rotation of the lower leg or lateral rotation of the femur.

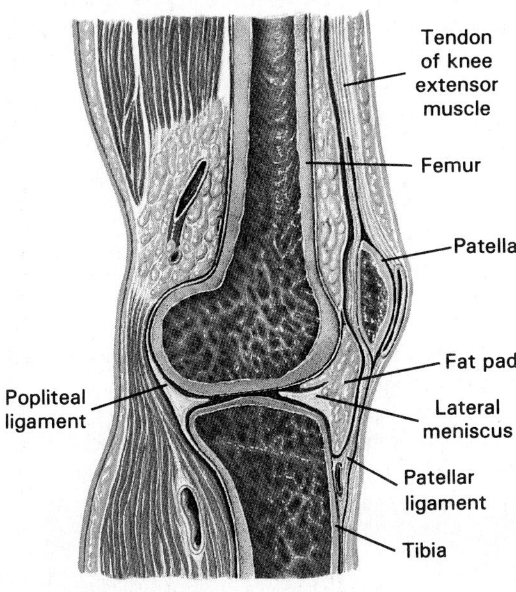

The Ankle and Joints of the Foot

Weight is transferred from the tibia to the superior surface of the talus. The tibiotalar joint is a classic hinge joint that permits limited dorsiflexion (sole elevated) and plantar flexion (sole depressed). An extensive array of ligaments interconnects the bones of the ankle, and the medial and lateral malleoli prevent sliding from side to side. These features are indicated on Figure 9-20.

The articulations between the tarsals, comparable to those between the carpals of the wrist, permit a limited gliding motion. The tarsometatarsal joints are also gliding diarthroses. Joints between the metatarsals and phalanges resemble those of the metacar-

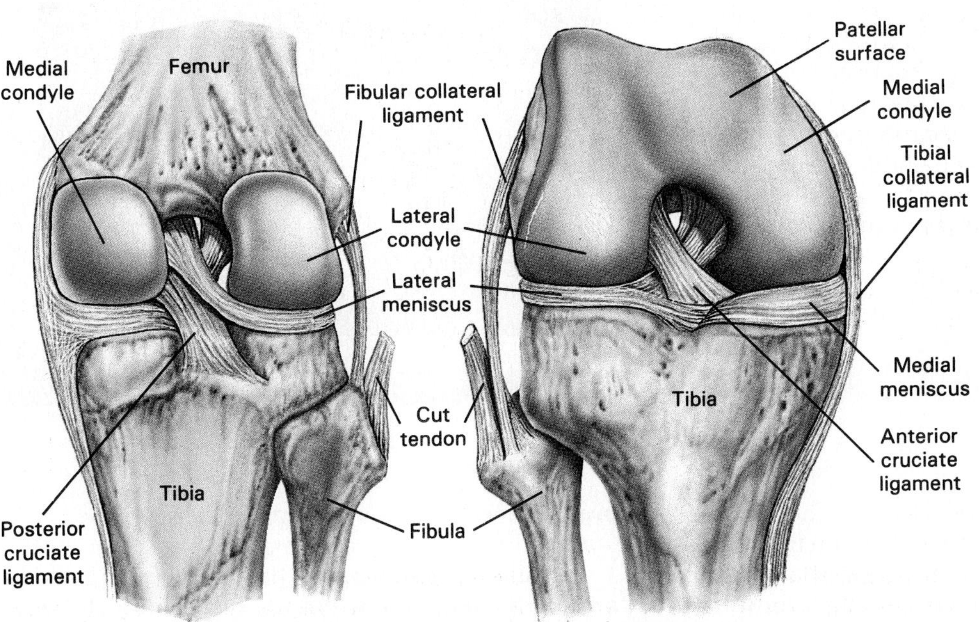

FIGURE 9-18
The Knee Joint. (a) The extended right knee as seen in section, showing major anatomical features. (b) Anterior view of the flexed knee. (c) Posterior view of the knee in full extension.

Sports and Fitness: Knee Injuries

Athletes place tremendous stresses on their knees. Ordinarily, the medial and lateral menisci move as the femoral position changes. Placing a lot of weight on the knee while it is partially flexed can trap a meniscus between the tibia and femur, resulting in a break or tear in the cartilage. In the most common injury, the lateral surface of the lower leg is driven medially, tearing the medial meniscus. In addition to being quite painful, the torn cartilage may restrict movement at the joint. It can also lead to chronic problems and the development of a "trick knee" that feels rather unstable. Sometimes the meniscus can be heard popping in and out of position when the knee is extended.

An **arthroscope** uses fiber optics to permit exploration of a joint without major surgery. Optical fibers are thin threads of glass or plastic that conduct light. The fibers can be bent around corners, so they can be introduced into a knee or other joint and moved around, enabling the physician to see what is going on inside the joint. If necessary, the apparatus can be modified to perform surgical modification of the joint at the same time. This procedure, called **arthroscopic surgery**, has greatly simplified the treatment of knee and other joint injuries. Figure 9-19 is an arthroscopic view of the interior of an injured knee, showing a damaged meniscus. Although small pieces of cartilage can be removed and the meniscus surgically trimmed, the only sure cure is **meniscectomy**, the removal of the affected cartilage. New tissue-culturing techniques may someday permit the replacement of the meniscus.

An arthroscope cannot show the physician soft tissue details outside the joint cavity, and repeated arthroscopy eventually leads to the formation of scar tissue and other joint problems. MRI (magnetic resonance imaging, Chapter 1) is a cost-effective and noninvasive method of viewing without injury and examining soft tissues around the joint. (p. 22)

Less common knee injuries involve tearing one or more stabilizing ligaments or damaging the patella. Torn ligaments are often difficult to correct surgically, and healing is slow. The patella can be injured in a number of ways. If the lower leg is immobilized (as it might be in a football pileup, for example) while you try to straighten your leg, the muscles that extend the knee are powerful enough to pull the patella apart. Impacts to the anterior surface of the knee may also shatter the kneecap. Treatment involves surgical removal of fragments and repair of the tendons and ligaments, followed by immobilization of the joint. Total knee replacements are rarely performed on young people, but are becoming increasingly common among elderly patients with severe arthritis; an artifical knee was shown in Figure 7-15. (p. 229)

FIGURE 9-19
Arthroscopy and the Knee. This photograph, taken through an arthroscope, shows the internal structure of a damaged knee. Injuries like this are relatively common in athletes engaged in contact sports, such as football.

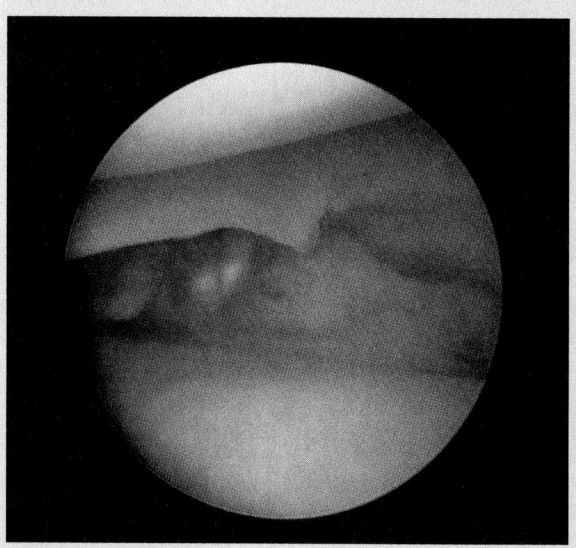

pals and phalanges of the hand. Because the articulation between the hallux and the first metatarsal is a hinge rather than a saddle joint, opposition cannot occur. A pair of sesamoid bones usually form in the tendons that cross the surface of this joint, and their presence restricts movement further. [**CM:** *Problems with the Ankle and Foot*]

Table 9-2 summarizes the characteristics of the articulations mentioned in the discussion of the pelvic girdle and legs.

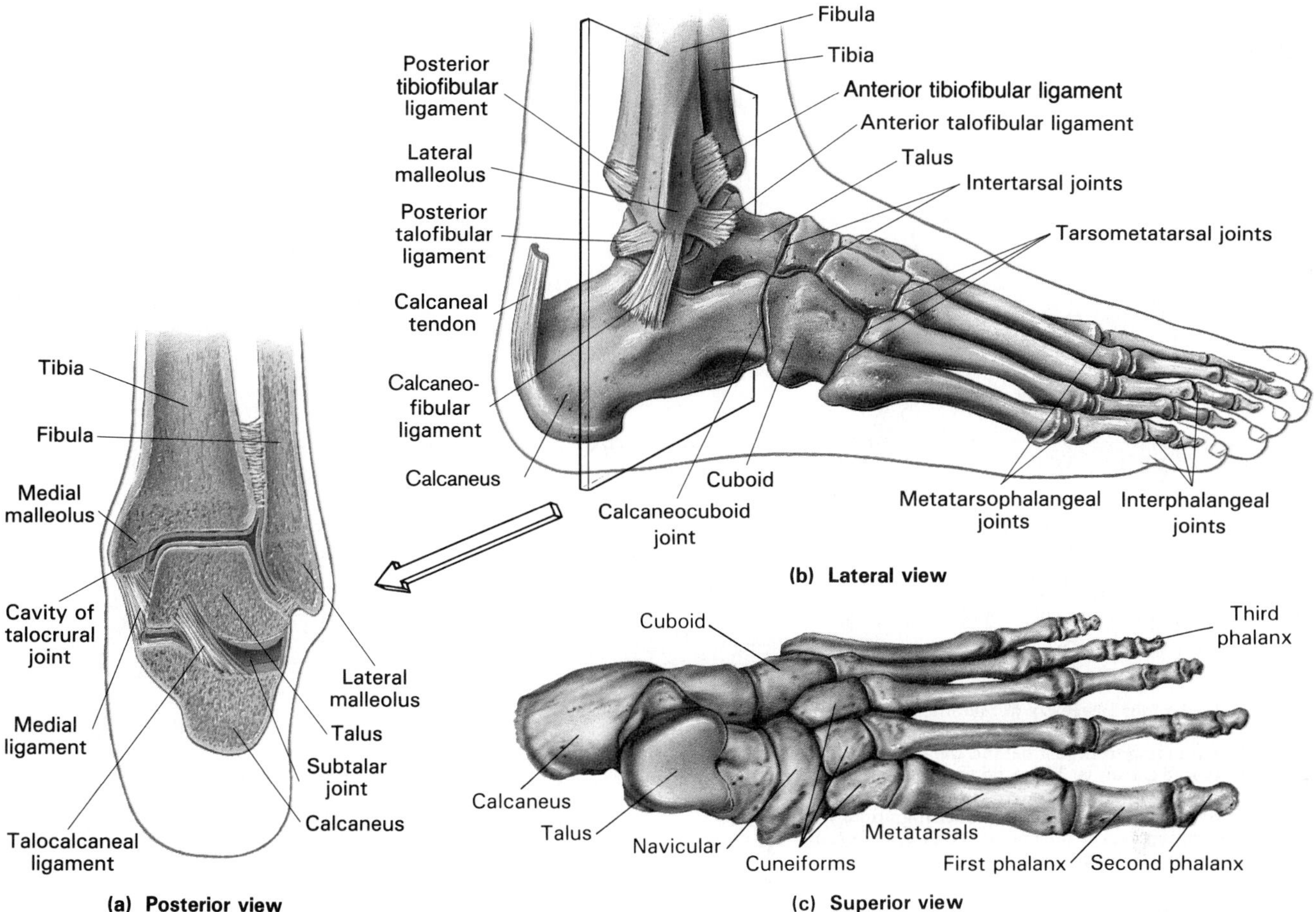

Figure 9-20
Joints of the Ankle and Foot. (a) Lateral view of the ankle joint, showing the ligamentous and bony props that help stabilize the joint. (b) Posterior view, with a transparent talus, showing the orientation of the tibiotalar joint and the placement of the medial and lateral malleoli. (c) Superior view of the foot, with other joints identified.

TABLE 9-2 Articulations of the Pelvic Girdle and Leg

Elements	*Joint*	*Type of Articulation*	*Movements*
Sacrum/ilium of coxa	Sacroiliac	Gliding diarthrosis	Gliding
Coxa/coxa	Symphysis pubis	Amphiarthrotic symphysis	Slight
Coxa/femur	Hip	Ball-and-socket diarthrosis	Flexion/extension, adduction/abduction, circumduction, rotation
Femur/tibia	Knee	Complex, functions as hinge	Flexion/extension, limited rotation
Tibia/fibula	Tibiofibular (proximal)	Gliding diarthrosis	Gliding
	Tibiofibular (distal)	Gliding diarthrosis and amphiarthrotic syndesmosis	Slight gliding
Tibia and fibula with talus	Ankle, or tibiotalar	Hinge diarthrosis	Dorsiflexion/plantar flexion
Tarsal/tarsal	Intertarsal	Gliding diarthrosis	Gliding
Tarsal/metatarsal	Tarsometatarsal	Gliding diarthrosis	Gliding
Metatarsal/phalanges	Metartarsophalangeal	Ellipsoidal diarthrosis	Dorsiflexion/plantar flexion, adduction/abduction
Phalanges/phalanges	Interphalangeal	Hinge diarthrosis	Dorsiflexion/plantar flexion

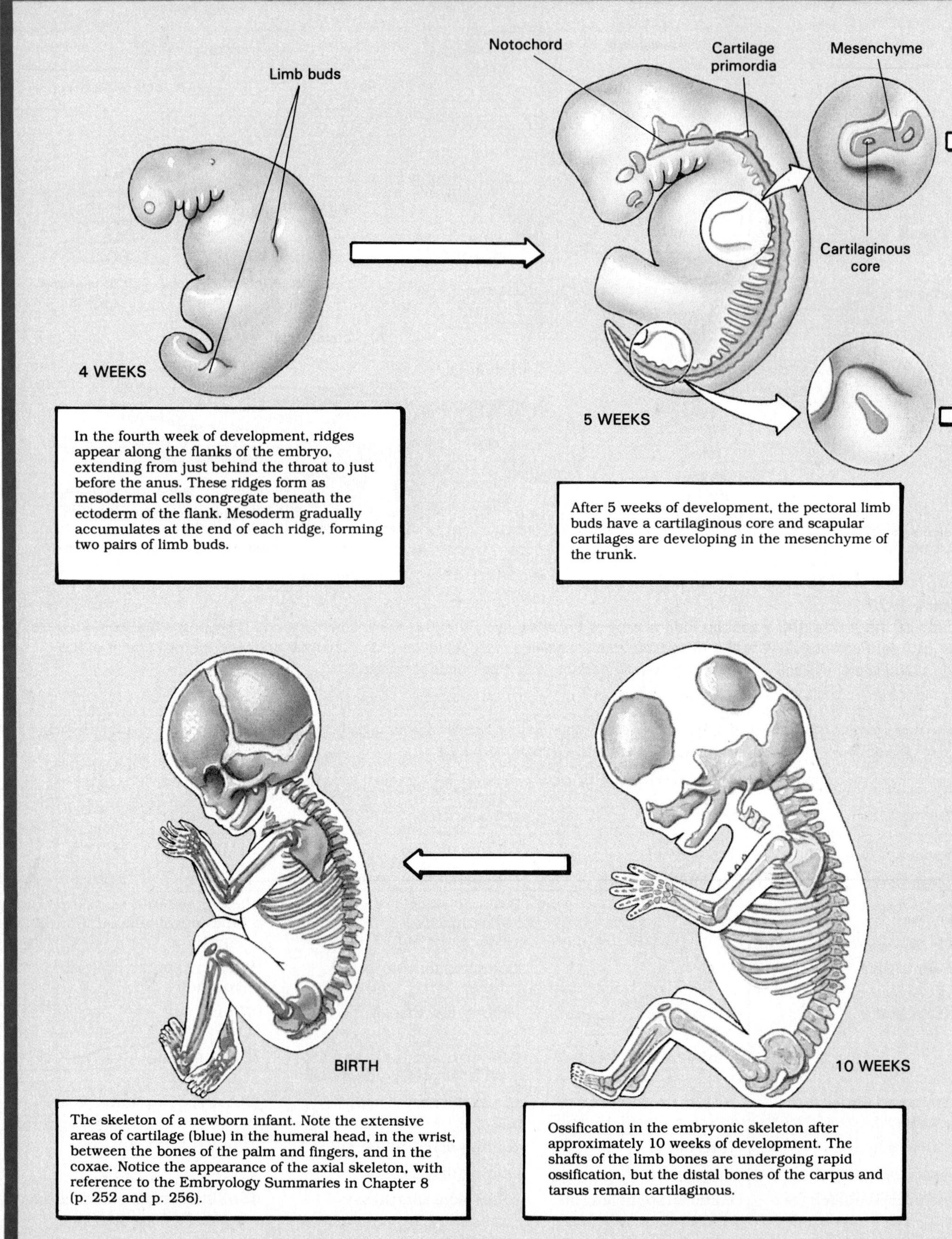

In the fourth week of development, ridges appear along the flanks of the embryo, extending from just behind the throat to just before the anus. These ridges form as mesodermal cells congregate beneath the ectoderm of the flank. Mesoderm gradually accumulates at the end of each ridge, forming two pairs of limb buds.

After 5 weeks of development, the pectoral limb buds have a cartilaginous core and scapular cartilages are developing in the mesenchyme of the trunk.

Ossification in the embryonic skeleton after approximately 10 weeks of development. The shafts of the limb bones are undergoing rapid ossification, but the distal bones of the carpus and tarsus remain cartilaginous.

The skeleton of a newborn infant. Note the extensive areas of cartilage (blue) in the humeral head, in the wrist, between the bones of the palm and fingers, and in the coxae. Notice the appearance of the axial skeleton, with reference to the Embryology Summaries in Chapter 8 (p. 252 and p. 256).

Embryology Summary: Development of the Appendicular Skeleton

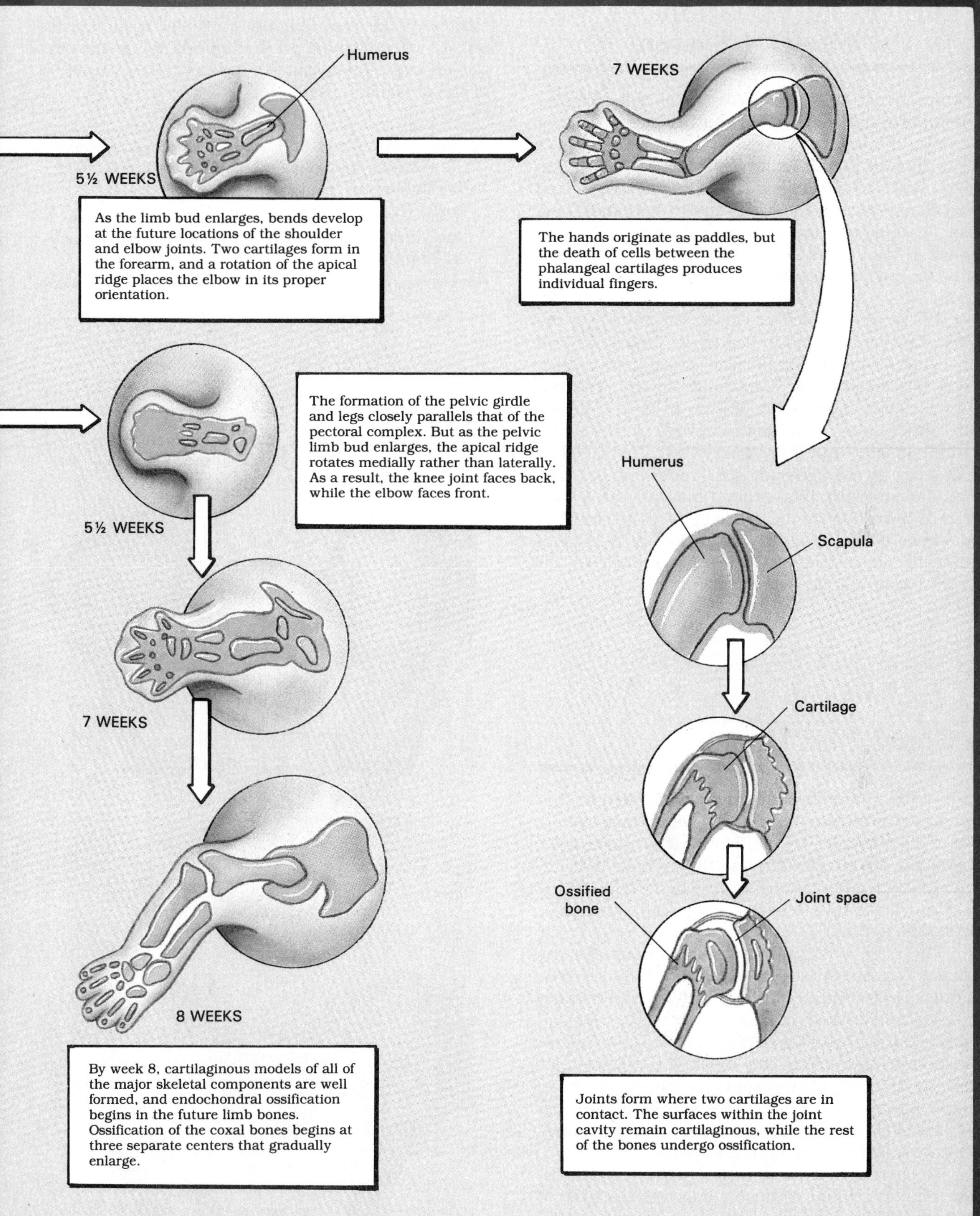

As the limb bud enlarges, bends develop at the future locations of the shoulder and elbow joints. Two cartilages form in the forearm, and a rotation of the apical ridge places the elbow in its proper orientation.

The hands originate as paddles, but the death of cells between the phalangeal cartilages produces individual fingers.

The formation of the pelvic girdle and legs closely parallels that of the pectoral complex. But as the pelvic limb bud enlarges, the apical ridge rotates medially rather than laterally. As a result, the knee joint faces back, while the elbow faces front.

By week 8, cartilaginous models of all of the major skeletal components are well formed, and endochondral ossification begins in the future limb bones. Ossification of the coxal bones begins at three separate centers that gradually enlarge.

Joints form where two cartilages are in contact. The surfaces within the joint cavity remain cartilaginous, while the rest of the bones undergo ossification.

■ Individual Variation in the Skeletal System

A comprehensive study of a human skeleton can reveal important information about the individual. For example, there are characteristic racial differences in portions of the skeleton, especially the skull and pelvis, and the development of various ridges and general bone mass can permit an estimation of muscular development and body weight. Details such as the condition of the teeth or the presence of healed fractures can provide information about the individual's medical history. Two important details, sex and age, can be determined or closely estimated on the basis of measurements indicated in Tables 9-3 and 9-4. Table 9-3 considers characteristic differences between the skeletons of males and females, but not every skeleton shows every feature in classic detail. Many differences, including markings on the skull, cranial capacity, and general skeletal features, reflect differences in average body size, muscle mass, and muscular strength. The general mass of the female pelvis is lower for similar reasons, but the shape of the female pelvis, diagrammed in Figure 9-21, is a genetically determined feature that represents an adaptation for childbearing.

■ Bones and Muscles

Many of the anatomical landmarks identified in this chapter are attachment sites of skeletal muscles, organs of the muscular system. The skeletal and muscular systems are structurally and functionally interdependent; their interactions are so extensive that they are often considered to be parts of a single *musculoskeletal system*.

There are direct physical connections, for the connective tissues that surround the individual muscle cells are continuous with those that establish the organic framework of an attached bone. Muscles and bones are also physiologically linked, because muscle contractions can occur only when the extracellular concentration of calcium remains within relatively narrow limits. With most of the body's calcium tied up in the skeleton, abnormalities that affect the bones can have a direct effect on the muscles.

The bones, in turn, are directly affected by muscular activity. When a muscle enlarges because of regular exercise, the increased forces exerted on the skeleton will make bones become stronger and more massive. The reverse holds true as well, for both muscles and bones diminish in size and strength after a period of inactivity.

Chapters 7, 8, and 9 have established the anatomical framework that the muscular system relies on. We will now build on that foundation as the next two chapters investigate the structure and function of the muscular system.

- √ **Where would you find the following ligaments: iliofemoral ligament, pubofemoral ligament, and ischiofemoral ligament?**
- √ **What symptoms would you expect to see in an individual who has damaged the menisci of the knee joint?**

FIGURE 9-21
Sexual Differences in the Male and Female Pelvis. Note the much sharper pubic angle in the pelvis of a male (a) than in that of a female (b).

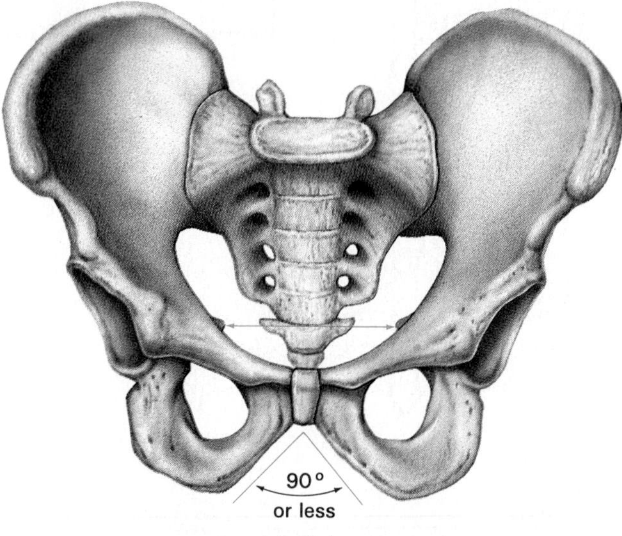

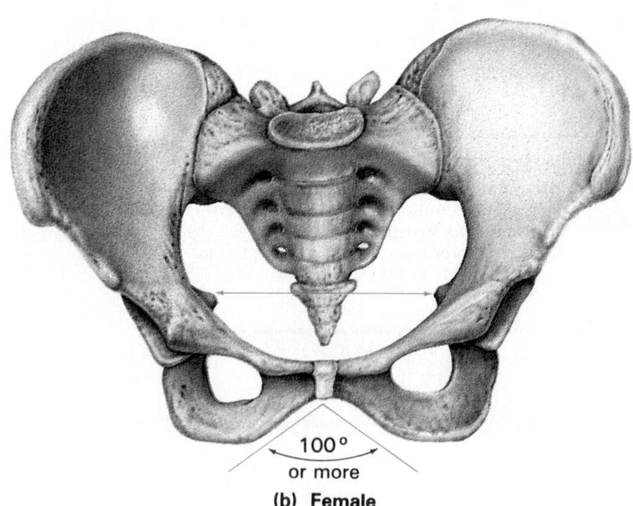

TABLE 9-3 Sexual Differences in the Human Skeleton

Region/Feature	Male	Female
SKULL		
General appearance	Heavier, rougher	Lighter, smoother
Forehead	Sloping	More vertical
Sinuses	Larger	Smaller
Cranium	About 10% larger (average)	About 10% smaller
Mandible	Larger, robust	Lighter, smaller
Teeth	Larger	Smaller
PELVIS		
General appearance	Narrow, robust, heavy, rough	Broad, light, smoother
Pelvic inlet	Heart-shaped	Oval to round
Iliac fossa	Relatively deep	Relatively shallow
Ilium	Extends farther above sacral articulation	More vertical; less extension above sacroiliac joint
Angle inferior to pubic symphysis	Under 90°	100° or more
Acetabulum	Directed laterally	Faces slightly anteriorly as well as laterally
Obturator foramen	Oval	Triangular
Ischial spine	Points medially	Points posteriorly
Sacrum	Long, narrow triangle with pronounced sacral curvature	Broad, short triangle with less curvature
Coccyx	Points anteriorly	Points inferiorly
OTHER SKELETAL ELEMENTS		
Bone weight	Heavier	Lighter
Bone markings	More prominent	Less prominent

Note: Refer to text for comments concerning the relationships between bone size, markings, and body size.

TABLE 9-4 Age-Related Changes in the Skeleton

Region/Structure	Event(s)	Age (Years)
GENERAL SKELETON		
Bony matrix	Reduction in mineral content	Standard "normal" values differ for males versus females between age 45 and 65; similar reductions occur in both sexes after age 65.
Markings	Reduction in size, roughness	Gradual reduction with increasing age and decreasing muscular strength and mass.
SKULL		
Fontanels	Closure	5+
Metopic suture	Fusion	2–8
Occipital bone	Fusion of ossification centers	1–4
Styloid process	Fusion with temporal bone	12–16
Hyoid bone	Complete ossification and fusion	25–30
Teeth	Loss of "baby teeth"; appearance of secondary dentition; eruption of posterior molars	Detailed in Chapter 24 (Digestive System).
Mandible	Loss of teeth; reduction in bone mass; change in angle at mandibular notch	Accelerates in later years (60+)
VERTEBRAE		
Curvature	Appearance of major curves	Described in Figure 8-16.
Intervertebral discs	Reduction in size, percentage contribution to height	Accelerates in later years (60+).
LONG BONES		
Epiphyseal plates	Fusion	Ranges vary according to specific bone under discussion, but general analysis permits determination of approximate age (3–7, 15–22, etc.).
PECTORAL AND PELVIC GIRDLES		
Epiphyses	Fusion	Overlapping ranges somewhat narrower than the above, including 14–16, 16–18, 22–25 years.

CHAPTER REVIEW

Review of Selected Clinical Terms

shoulder separation (*p. 273*)
The partial or complete dislocation of the acromioclavicular joint.

bursitis (*p. 278*)
Inflammation of the bursae.

bunion (*p. 278*)
The most common pressure-related bursitis; forms over the base of the big toe.

bursectomy (*p. 278*)
The surgical removal of inflamed bursae.

arthroscope (*p. 290*)
An instrument that uses fiber optics to explore a joint without major surgery.

arthroscopic surgery (*p. 290*)
The surgical modification of a joint using an arthroscope.

meniscectomy (*p. 290*)
The surgical removal of an injured meniscus.

Study Outline

Related Key Terms

Introduction (pp. 269–270)

1. The **appendicular division** includes the bones of the arms, legs, and supporting elements that connect the limbs to the trunk.

The Pectoral Girdle and Arm (pp. 271–280)

1. Each arm articulates with the trunk at the **shoulder,** or **pectoral, girdle**, which consists of the **scapula** and **clavicle**.

THE PECTORAL GIRDLE (pp. 271–273)

2. The clavicle and scapula position the shoulder joint, help move the arm, and provide a base for arm movement and muscle attachment.
3. The scapula articulates with the **humerus** at the **scapulohumeral joint**, or **shoulder joint**. Both the **coracoid process** and the **acromion** are attached to ligaments and tendons. The **scapular spine** crosses the scapular body.

   Related key terms: **neck** • **glenoid fossa**, **subscapular fossa**, **acromioclavicular joint**, **supraspinous fossa**, **infraspinous fossa**

THE ARM (pp. 273–276)

4. The capsule of the shoulder attaches to the **humerus** at the **anatomical neck**; the **greater tubercle** and **lesser tubercle** are important sites for muscle attachment. Other prominent landmarks include the **deltoid tuberosity**, the **radial groove**, the **medial** and **lateral epicondyles**, and the articular **condyle**.
5. Distally the humerus articulates with the **radius** and **ulna**. The medial **trochlea** extends from the **coronoid fossa** to the **olecranon fossa**.
6. The **radius** and **ulna** are the bones of the forearm. The olecranon fossa accommodates the **olecranon process** during extension of the **olecranal** or **elbow joint**. The coronoid and radial fossae accommodate the **coronoid process**.

   Related key terms: **radioulnar joints**, **interosseous membrane**, **styloid process**

7. The bones of the wrist, or **carpus**, form two rows of **carpals**. Four of the fingers contain three **phalanges**; the **pollex** has only two.

   Related key terms: **scaphoid** • **lunate**, **triquetal** • **pisiform**, **trapezium** • **trapezoid**, **capitate** • **hamate**, **metacarpals**, **olecranal joint**, **radiocarpal articulation**, **intercarpal articulations**

The Pelvic Girdle and Leg (pp. 280–291)

THE PELVIC GIRDLE (pp. 280–283)

1. The **pelvic girdle** articulates with the thighbones. It consists of two **coxae**; each coxa forms through the fusion of an ilium, an ischium, and a pubis. The head of the **femur** articulates with the **acetabulum**.
2. The largest coxal bone, the **ilium**, fuses with the **ischium**, which in turn fuses with the **pubis**. At the point of fusion, the ramus of the ischium meets the **inferior ramus** of the pubis. The **pubic symphysis** limits movement between the pubic bones. Ligaments stabilize the **sacroiliac joint**.

   Related key terms: **acetabular fossa** • **iliac crest**, **ischial tuberosity**, **obturator foramen**

3. The pelvis may be subdivided into the **greater**, or **false**, **pelvis** and the **lesser**, or **true**, **pelvis**.

Related Key Terms

iliac fossa • pelvic brim
pelvic inlet • pelvic outlet
perineum

THE LEG (pp. 283–286)

4. The **femur**, or thighbone, is the longest bone in the body. It articulates with the **tibia** at the **tibial condyles**. A ligament from the patella attaches at the **tibial tuberosity**.
5. Other tibial landmarks include the **anterior crest**, the **interosseous crest**, the **interosseous membrane**, which connects to the **fibula**, and the **medial malleolus**. The **fibular head** articulates with the tibia below the knee, and the **lateral malleolus** stabilizes the ankle.
6. The **tarsus**, or ankle, includes seven bones; only the **talus** articulates with the tibia and fibula. When standing normally, most of our weight is transferred to the **calcaneus**, and the rest is passed on to the **metatarsals**. Weight transfer occurs along the **longitudinal arch**; there is also a **transverse arch**.

navicular • cuboid
medial/intermediate/lateral cuneiforms • calcaneal tendon
tarsometatarsal joint

7. The basic organizational pattern of the metatarsals and phalanges of the foot resembles that of the hand, except for the hallux.
8. The femur articulates with the coxa at the **hip joint**, a ball-and-socket diarthrosis. The **patella**, or kneecap, is embedded within a tendon that supports the front of the joint.
9. The knee joint is a complicated hinge joint that actually consists of three joints.

Individual Variation in the Skeletal System (p. 294)

1. Studying a human skeleton can reveal important information such as race, medical history, weight, sex, body size, muscle mass, and age.

Bones and Muscles (pp. 294–295)

1. The interactions between the muscular and skeletal systems are so extensive that they are often considered to be an integrated "musculoskeletal system."

Review Planner

	Objective	Level 1
1	Identify the components of the appendicular skeleton and their functions.	**3 5 9 10 11–43**
2	Relate the structural differences between the pectoral and pelvic girdles to the variations in their functional roles.	**1 2 6 8 11 16 17 21 22 27**
3	Identify the major structural features and markings of important bones of the appendicular skeleton.	**3 9 14 15 20 22 24 26 27 28 29 31 34 36**
4	Discuss the structures and functions of the joints in the appendicular skeleton.	**6 8 19 23 25 33 39 40 45**
5	Explain how study of the skeleton can reveal important information about an individual.	**4**
6	Describe the skeletal differences between males and females.	**7**

Level 2	**37 41 42**
Level 3	**44–47**
C M	**Hip Fractures and the Elderly** **Problems with the Ankle and Foot**

Chapter Review

Review Questions

MULTIPLE CHOICE

1. The only direct connection between the pectoral girdle and the axial skeleton is the connection between the: a) scapula and sternum; b) clavicle and sternum; c) shoulder joint and bursae; d) cervical and thoracic vertebrae.
2. The pelvic girdle consists of two: a) coxae; b) ilia; c) ischia; d) rami.
3. The longest single bone in the body is the: a) humerus; b) scapular spine; c) tibia; d) femur.
4. Which of the following can be learned from studying a person's skeleton: a) age; b) gender; c) muscular strength; d) all of the above.

DIAGRAMS

5. Cover the labels in Figure 9-1 and identify the principal elements of the axial and appendicular divisions.
6. a) Create a table that lists the name of the joint found at each of the following locations of the pelvic girdle and leg, the type of articulation, and the movement(s) it permits: sacrum/ilium of coxa; coxa/coxa; coxa/femur; femur/tibia; tibia/fibula; tibia and fibula with talus; tarsal/tarsal; tarsal/metatarsal; metatarsal/phalanges; phalanges/phalanges.

   b) Create a table that lists the name of the joint found at each of the following locations of the pectoral girdle and arm, the type of articulation, and the movement(s) it permits: clavicle/sternum; scapula/clavicle; scapula/humerus; humerus/ulna and radius; radius/ulna; radius/carpals; carpal/carpal; carpal/metacarpal (first); carpal/metacarpal (2–5); metacarpals/phalanges; phalanges/phalanges.
7. Create a table that summarizes the major differences between the male and female skeleton for the skull and pelvis.

TRUE/FALSE

8. **T/F:** The shoulder joint permits the greatest range of motion of any joint in the body.
9. **T/F:** The term *phalanges* can be applied to both the fingers and the toes.
10. **T/F:** The axial and appendicular skeletons are totally independent systems that are not connected.

MATCHING QUESTIONS

(*Match each element with the appropriate location.*)

Elements:	Locations:
11. Ilium	a) Scapula
12. Metacarpals	b) Clavicle
13. Metatarsals	c) Humerus
14. Sternal end	d) Radius/ulna
15. Ulnar tuberosity	e) Wrist/hand
16. Pelvic inlet	f) Olecranal joint
17. Ischium	g) Coxa
18. Pollex	h) Pelvis
19. Medial menisci	i) Femur
20. Deltoid tuberosity	j) Tibia/fibula
21. Pubis	k) Ankle/foot
22. Acromion	l) Knee joint
23. Intercarpal articulations	
24. Calcaneal tendon	
25. Radioulnar joints	
26. Greater trochanter	
27. Scapular spine	
28. Medial malleolus	
29. Radial groove	
30. Patella	

SHORT ANSWER/ESSAY

Identify and give a function for each of the following:

31. Glenohumeral ligaments.
32. Longitudinal arch.
33. Bursae.
34. Obturator foramen.
35. Fibula.
36. Lateral malleolus.

37. The chapter notes that each joint in the appendicular skeleton represents a compromise between strength and mobility. Explain this statement.
38. What are the functions of the scapula and clavicle?
39. Discuss the structure of the knee joint. Why is a complete dislocation of the knee relatively rare?
40. Discuss the functional anatomy of the hip joint. Of which types of movement is it capable?
41. Compare the organizational pattern of the metatarsals/phalanges with that of the metacarpals/phalanges. Are there any differences? If so, why do you think these differences exist?
42. Discuss how the shoulder, elbow, and wrist work together to control movements of the hand.
43. Name three factors that stabilize the elbow joint.

CRITICAL THINKING APPLICATIONS

44. A high-school student comes to the emergency room, complaining of persistent pain beneath his right shoulder blade. In talking with him, you discover that he has been spending many hours trying to improve his pitching skills for his school's softball team. What do you think is causing the pain?
45. Bob injured his right knee during a basketball game when he jumped to retrieve the ball and landed off-balance on that leg. Since then he has pain and limited mobility of the knee joint. What type of injury do you think Bob sustained? What

techniques are available to explore the extent of the damage?

46. Two patients sustain hip fractures. In one case, a pin is inserted into the joint and the injury heals well. In the other, the fracture fails to heal. Identify the types of fractures that are probably involved. Why did the second patient's fracture not heal, and what steps can be taken to restore normal function?

47. What are bunions? Would you expect them to be more common in men or in women? Explain your answer.

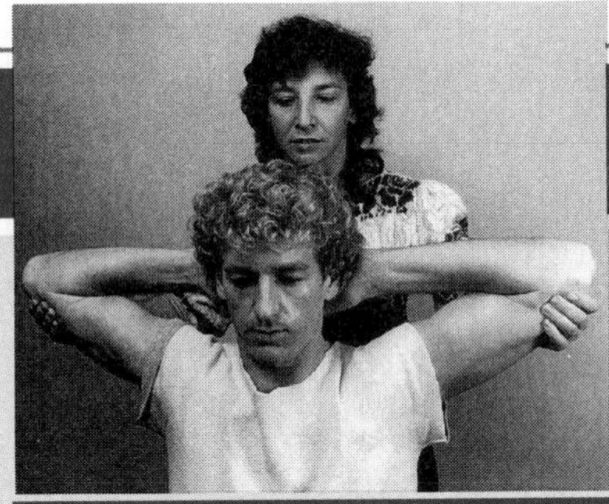

Career Close-up: Physical Therapist

Marika Molnar can watch every performance of the New York City Ballet free because she works for the company. But Marika is no ballerina. She is the company's physical therapist. "When I was studying dance at NYU, I saw dancers who were injured and couldn't continue their careers. They had no place to go. I said to myself, 'There must be some way to help people recover from these injuries. You can't just tell these people to stop moving.' So I applied to a physical therapy graduate program to study the functioning human body and learn how to use my knowledge to help dancers."

Now she treats and rehabilitates injured professionals. When a dancer comes to her, she examines the injured area to determine if she can treat the problem or whether it requires a doctor or hospital. Marika may, for example, put a sprained ankle into an air cast and then teach the dancer how to ice or tape the injury. She suggests exercises to strengthen the area and reduce swelling. "Most of the injuries I treat do not need surgery, but they do take time to heal. A dancer with a mild sprained ankle takes two weeks to get back to class, and two more weeks after that to get on a full performance schedule."

Marika uses her coursework every day. "I've had Anatomy and Physiology at every school I attended. Cadaver study was invaluable because there I saw how the body actually works. I learned to respect it and began thinking of it as an instrument that needs maintenance. We went layer by layer on several different cadavers to see what was underneath, and I learned that every body is different. The New York City Ballet Company has a lot of people, and that means a lot of body types. When I examine someone, I must know where everything is by the bony landmarks. I need to be good at palpitation–for example, feeling around the ankle area for a tendon or ligament injury. Motion palpitation can be very important–I might look for something like tendinitis by feeling for thickening in the tendon or listening for abnormal sounds during movement. It's almost like having X-ray vision. It's as if I can see under the skin."

There are other factors Marika must consider. "I have to watch the dancers to make sure they are not getting worn out. Fatigued performers are prone to injury because they cannot pay attention as closely. The nervous system controls movement, and it doesn't function as well when you are fatigued."

Marika's office at the theatre is staffed six hours a day during performance season. "The size of my therapy room at the dance company was literally the janitor's closet the first four years. But soon there were lines outside, so they built me an area in the locker room. Now I have a beautiful office in a brand new building. It just goes to show the trend in helping performing artists."

Marika is always on the alert. "This spring a principal dancer tore her anterior cruciate ligament in mid-performance. She felt something happen to her knee while she was in the air, so she landed on the other leg and hobbled offstage. Another dancer immediately stepped into her place, and the show went on as usual." But performances do not always run so smoothly. "Once I was watching another company perform when suddenly the curtain came down, a rare occurrence. I ran backstage to find that the principal had torn her Achilles tendon. That needs immediate attention, so she was taken to the hospital."

"Sometimes I watch ballets because I like them, and sometimes because I am watching a dancer who is returning after an injury. I love to watch dance. But it's even more enjoyable when I see the performance from backstage and know that I made a major contribution to a dancer's performance."

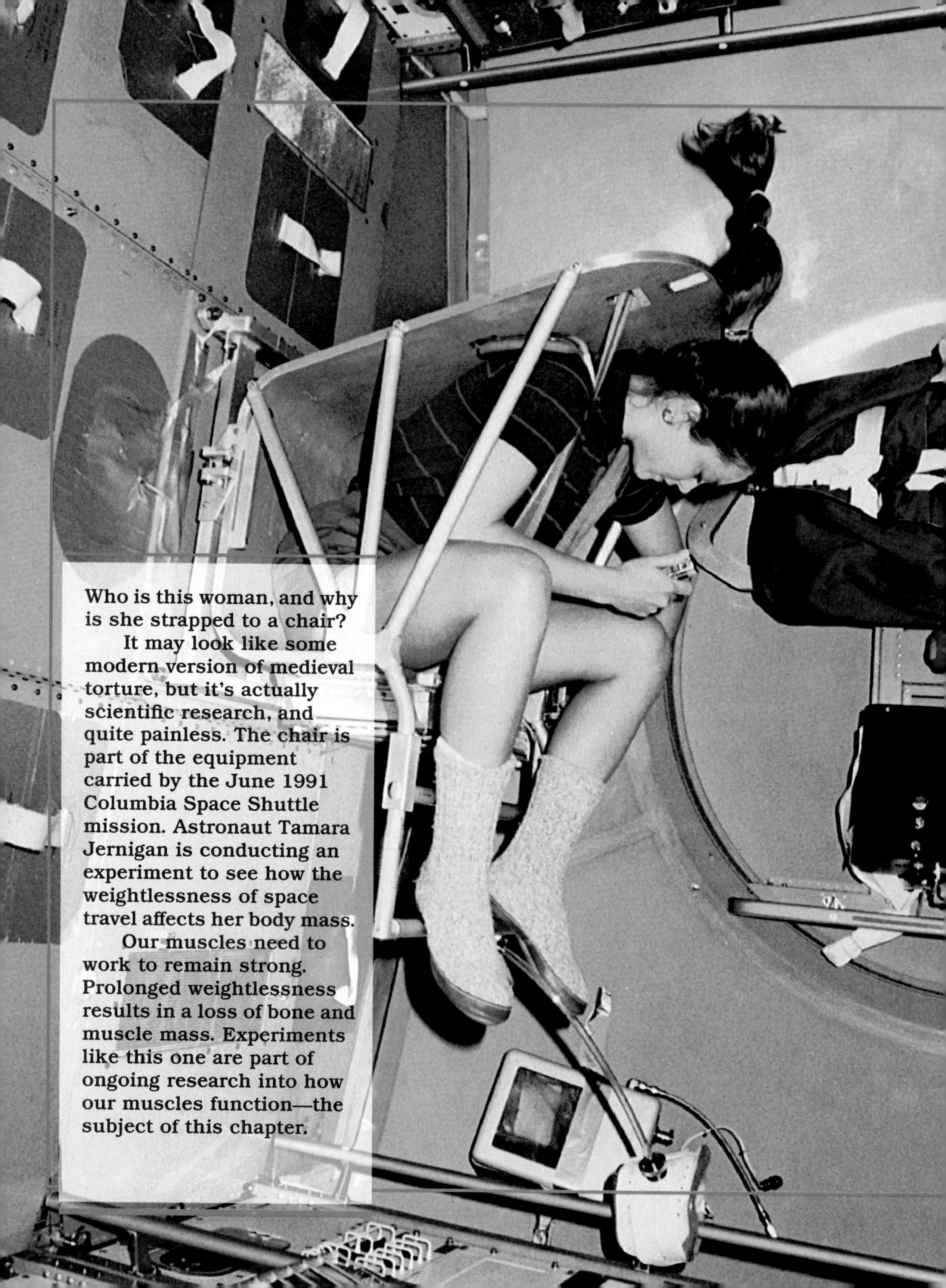

Who is this woman, and why is she strapped to a chair?

It may look like some modern version of medieval torture, but it's actually scientific research, and quite painless. The chair is part of the equipment carried by the June 1991 Columbia Space Shuttle mission. Astronaut Tamara Jernigan is conducting an experiment to see how the weightlessness of space travel affects her body mass.

Our muscles need to work to remain strong. Prolonged weightlessness results in a loss of bone and muscle mass. Experiments like this one are part of ongoing research into how our muscles function—the subject of this chapter.

CHAPTER 10

The Muscular System: Skeletal Muscle Tissue

Chapter Objectives

After reading this chapter, you will be able to:

1 Describe the characteristics and functions of muscle tissue.
2 Discuss the organization of muscle at the tissue level.
3 Identify the unique characteristics of skeletal muscle fibers.
4 Explain the process of muscular contraction and the mechanisms that control it.
5 Show how the events of a twitch are related to the development of muscle tension.
6 Compare the different types of muscle contractions.
7 Describe the mechanisms by which muscles obtain and use energy to power contractions.
8 Relate types of muscle fibers to muscular performance.
9 Distinguish between aerobic and anaerobic endurance, and explain their implications for muscular performance.
10 Describe the effects of exercise and aging on muscle.

Introduction

Think for a moment what life would be like without muscle tissue. Imagine being unable to sit, stand, walk, speak, or grasp objects in your environment. Imagine, too, how your internal functions would be affected. Blood would not circulate, because there would be no heartbeat to propel it through the vessels. The lungs could not rhythmically empty and fill, nor could food move through the digestive tract. In fact, there would be practically no movement through any of the internal passageways of the body.

This is not to say that all life depends on muscle tissue. There are large organisms that get by very nicely without it—we call them plants. But life as we live it would be impossible, for many of our physiological processes, and virtually all our dynamic interactions with the environment, involve muscle tissue.

Muscle tissue, one of the four primary tissue types, consists chiefly of muscle cells (usually called *muscle fibers*)—cells highly specialized for contraction. It also includes the connective tissue fibers that harness those contractions to perform useful work. There are three types of muscle tissue: **skeletal muscle**, **cardiac muscle**, and **smooth muscle**. Without these muscle tissues, introduced in Chapter 5, nothing in the body would move, and no body movement could occur. (p. 156) Skeletal muscles move the body by pulling on bones of the skeleton, making it possible for us to walk, dance, bite into an apple, or play the ukelele. Cardiac and smooth muscles push fluids and solids along internal passageways—blood through our arteries and veins and food through our digestive tracts.

This chapter considers the structure and function of skeletal muscle tissue, while Chapter 11 details the functional anatomy of the muscular system. Chapter 20 discusses the cardiac muscle tissue of the heart, and Chapter 24 examines the smooth muscle tissue along the digestive tract.

Functions of Skeletal Muscle

Skeletal muscle tissue, connective tissues, and neural tissue combine to form *skeletal muscles*. These contractile organs are directly or indirectly attached to the bones of the skeleton. Skeletal muscles perform the following functions.

1. **Produce skeletal movement**: Muscle contractions pull on tendons and move the bones of the skeleton. The effects range from simple motions such as extending the arm to the highly coordinated movements of swimming, skiing, or typing.
2. **Maintain posture and body position**: Tension in our muscles also maintains body posture—for example, holding the head in position when reading a book, or balancing the weight of the body above the feet when walking. Without constant muscular activity we could not sit upright without collapsing into a heap or stand without toppling over.
3. **Support soft tissues**: The abdominal wall and the floor of the pelvic cavity consist of layers of skeletal muscle. These muscles support the weight of visceral organs and shield internal tissues from injury.
4. **Guard entrances and exits**: The openings that lead into the digestive and urinary tracts are encircled by skeletal muscles. These muscles provide voluntary control over swallowing, defecation, and urination.
5. **Maintain body temperature**: Muscle contractions require energy, and whenever energy is used in the body, some of it is converted to heat. The heat lost by working muscles keeps our body temperature in the range required for normal functioning.

Anatomy of Skeletal Muscles

When naming structural features of muscles and their components, anatomists often used the Greek words *sarkos* (flesh) and *mys* (muscle). These word roots should be kept in mind as our discussion proceeds. We will discuss the gross anatomy of skeletal muscle first and then move on to the microstructure that makes contraction possible.

GROSS ANATOMY

Figure 10-1 illustrates the appearance and organization of a typical skeletal muscle. You can readily see that connective tissue plays a major role in its structural organization.

Connective Tissue Organization

Three layers of connective tissue are part of each muscle: an outer *epimysium*, a central *perimysium*, and an inner *endomysium*.

The entire muscle is surrounded by the **epimysium** (ep-i-MĪS-ē-um; *epi-*, on + *mys*, muscle), a dense layer of collagen fibers. The epimysium separates the muscle from surrounding tissues and organs. It is one component of the deep fascia, the dense connective tissue layer described in Chapter 5 (see Figure 5-22). (p. 159)

The connective tissue fibers of the **perimysium** (per-i-MĪS-ē-um; *peri-*, around) divide the skeletal muscle into a series of compartments, each containing a bundle of muscle fibers called a **fascicle** (FA-sik-ul; *fasciculus*, a bundle). In addition to collagen and elastic fibers, the perimysium contains blood vessels and nerves that maintain blood flow and provide innervation to the fascicles. Each fascicle receives branches of the blood vessels and nerves of the perimysium.

Within the fascicle, the delicate connective tissue of the **endomysium** (en-do-MĪS-ē-um; *endo-*, inside) surrounds the skeletal muscle fibers and ties adjacent muscle fibers together. Scattered **satellite cells** lie between the endomysium and the muscle fibers. These cells function in the repair of damaged muscle tissue.

The collagen fibers of the endomysium and perimysium are interwoven, and those of the perimysium blend into those of the epimysium. At each end of the muscle, the collagen fibers of the epimysium come together to form a bundle known as a **tendon**. Tendons, an example of dense regular connective tissue described in Chapter 5, attach skeletal muscles to bones. (p. 144) The tendon fibers are interwoven into the connective tissues of the bone, providing a firm attachment. As a result, any contraction of the muscle will exert a pull on its tendon and thereby on the attached bone. [**CM**: *Trichinosis*]

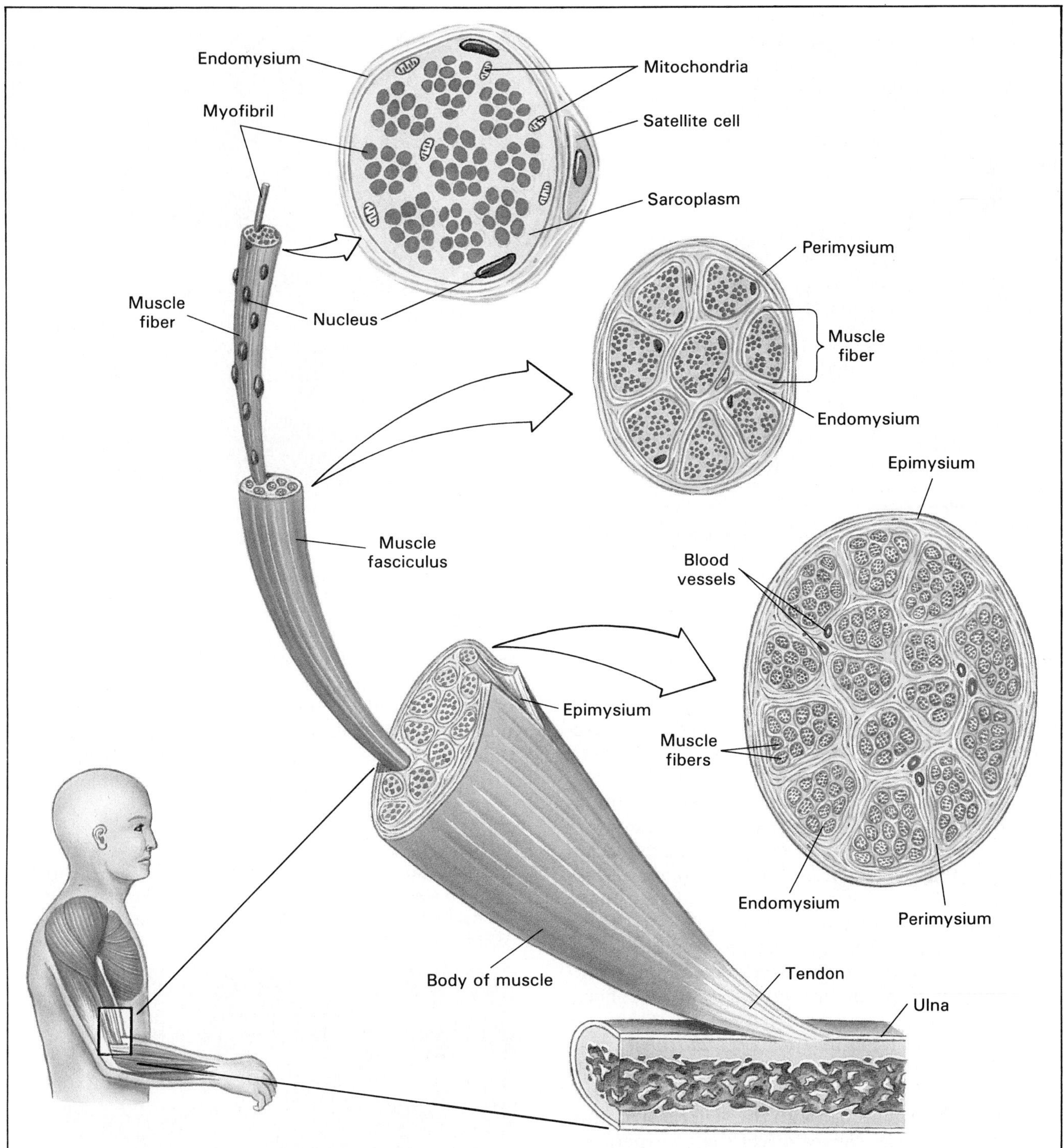

FIGURE 10-1
Organization of Skeletal Muscles. A skeletal muscle consists of fascicles (bundles of muscle fibers) enclosed by the epimysium. The bundles are separated by connective tissue fibers of the perimysium, and within each bundle the muscle fibers are surrounded by the endomysium. Each muscle fiber has many nuclei. The cytoplasm contains mitochondria and other organelles seen in this figure and in Figure 10-4.

Nerves and Blood Vessels

The connective tissues of the epimysium and perimysium contain the nerves and blood vessels that service the muscle fibers. Skeletal muscles are often called *voluntary muscles* because they contract when stimulated by motor nerves of the central nervous system. (p. 172) Axons, or *nerve fibers*, penetrate the epi-

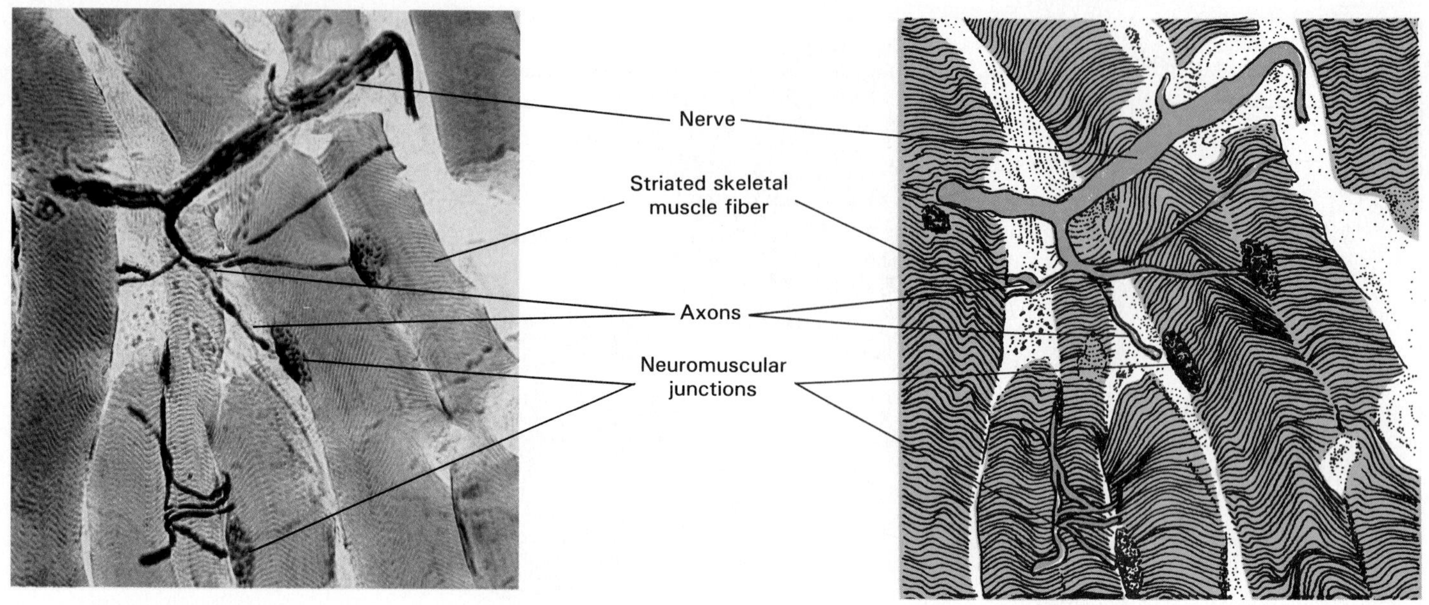

FIGURE 10-2
Skeletal Muscle Innervation. Several neuromuscular junctions are seen on the muscle fibers of this fascicle. (LM, × 160)

mysium, branch through the perimysium, and enter the endomysium to reach individual muscle fibers. Communication between nerve and muscle fiber occurs across a specialized junction, called the **neuromuscular junction**, or *myoneural junction*. Several of these junctions are shown in Figure 10-2.

Muscle contraction requires tremendous quantities of energy. An extensive vascular supply delivers the oxygen and nutrients needed for the production of ATP in active skeletal muscles. These blood vessels often enter the muscle in company with the nerve supply, and the vessels and nerves follow the same branching pattern through the perimysium. Once within the endomysium, the arteries supply an extensive capillary network around each muscle fiber. Capillaries surrounding a portion of a single muscle fiber can be seen in Figure 10-3. Because these capillaries are coiled, rather than straight, they are able to tolerate changes in the length of the muscle fiber.

You are now familiar with the general structure of a skeletal muscle. Nerves direct the contractions, connective tissues harness the forces generated, and vessels deliver oxygen and nutrients and remove wastes. Next we will examine the microscopic structure of a typical skeletal muscle fiber, and then proceed to relate that microstructure to the physiology of the contraction process.

MICROANATOMY OF SKELETAL MUSCLE FIBERS

Muscle fibers are quite different from the "typical" cell described in Chapter 3. One obvious difference is size, for a skeletal muscle fiber is an enormous cell. A fiber from a leg muscle could have a diameter of 100 μm and a length equal to that of the entire muscle (30–40 cm, or 10–16 in.). A second obvious difference is that each skeletal muscle fiber contains hundreds of nuclei just beneath the cell membrane.

FIGURE 10-3
Capillary Supply to Skeletal Muscle. Capillaries along the axis of a skeletal muscle fiber. Note the looping and curling of the capillaries that allow cycles of stretching and contraction. (SEM, x 151)

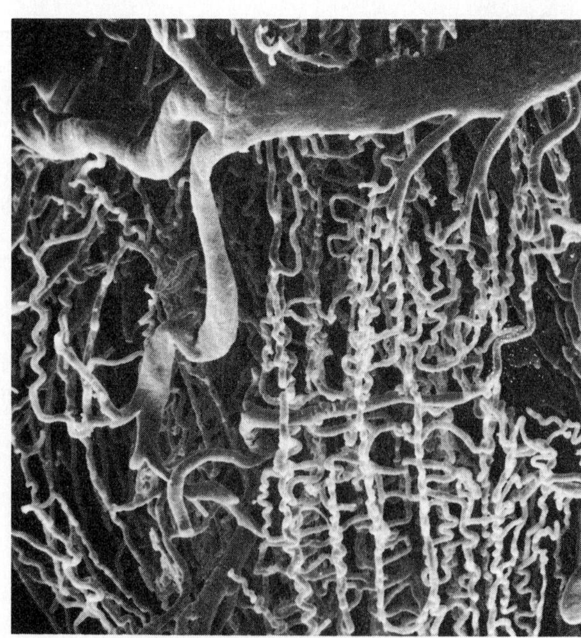

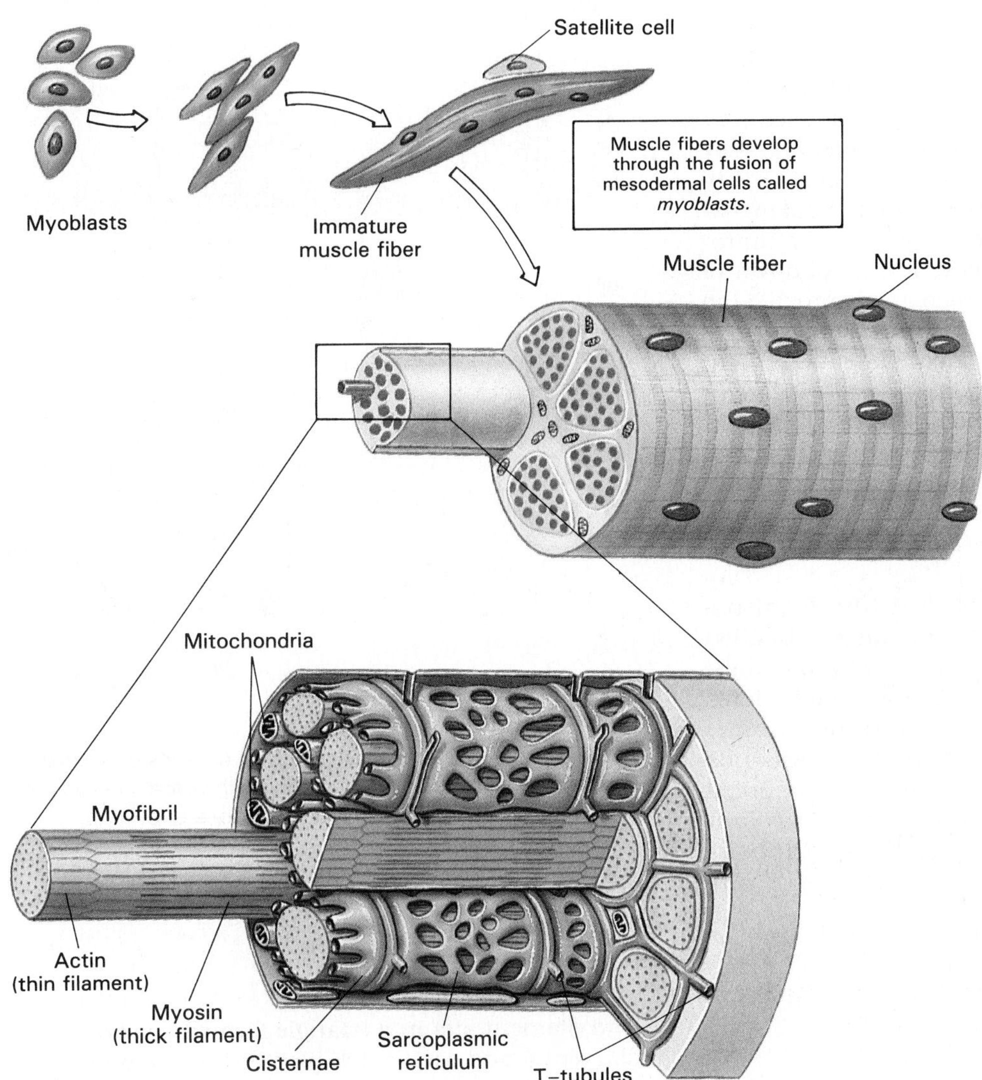

FIGURE 10-4 Organization of Muscle Tissue

This *multinucleate* condition also distinguishes cells of skeletal muscle from those of other muscle types. The genes contained in these nuclei direct the production of enzymes and structural proteins required for normal contraction, and the presence of multiple gene copies speeds up the process. This is particularly important in muscle cells, where metabolic turnover tends to be very rapid (see Chapter 4). (p. 117)

The distinctive features of size and multiple nuclei are related. During development, groups of embryonic cells called **myoblasts** fuse together to create individual skeletal muscle fibers, as indicated in Figure 10-4. Each nucleus in a skeletal muscle fiber reflects the contribution of a single myoblast. Some myoblasts do not fuse with developing muscle fibers. These unfused cells remain in adult skeletal muscle tissue as the satellite cells shown in Figure 10-2. After an injury, they may enlarge, divide, and fuse with damaged muscle fibers, thereby assisting in the regeneration of the tissue.

The cell membrane, or **sarcolemma** (sar-cō-LEM-a; *sarkos*, flesh + *lemma*, husk) of a muscle fiber surrounds the cytoplasm, or **sarcoplasm** (SAR-kō-plazm). Openings scattered across the surface of the sarcolemma lead into a network of narrow tubules that extend into the sarcoplasm. These **transverse tubules**, or **T tubules**, help distribute the command to contract throughout the muscle fiber.

Myofibrils and Myofilaments

Just inside the sarcolemma is a thin layer of sarcoplasm. This layer, which contains the cell nuclei, surrounds hundreds to thousands of **myofibrils**. Each myofibril is a cylindrical structure 1–2 μm in diameter and as long as the entire cell. Myofibrils can actively shorten; they are the structures responsible for muscle fiber contraction. Because the myofibrils are attached to the sarcolemma at each end of the cell, their contraction shortens the entire cell.

Surrounding each myofibril is a sleeve made up of membranes of the **sarcoplasmic reticulum (SR)**, a membrane complex similar to the endoplasmic reticulum of other cells. We will see in a subsequent section how this membrane network, which is closely associated with the transverse tubules, plays a key role in controlling the contraction of individual myofibrils. Scattered between the myofibrils are mitochondria and glycogen granules. The breakdown of glycogen and the activity of mitochondria provide the ATP needed to power muscular contractions.

Myofibrils are bundles of **myofilaments**, protein filaments consisting primarily of the proteins *actin* and *myosin*. The actin filaments are called **thin filaments** and the myosin filaments are called **thick filaments**; both were introduced in Chapter 3. (p. 66) Myofilaments are organized in repeating functional units called **sarcomeres** (SAR-kō-mērs; *sarkos*, flesh + *meros*, part).

The arrangement of thick and thin filaments within the sarcomere gives each sarcomere a banded appearance. All of the myofibrils are arranged parallel to the long axis of the cell, with their sarcomeres lying side by side. As a result, the entire muscle fiber has a banded appearance corresponding to the bands of the individual sarcomeres. In Chapter 5 we noted that skeletal muscle is also known as *striated voluntary muscle* because of these bands, or *striations* (stripes), on the muscle fibers. (p. 157)

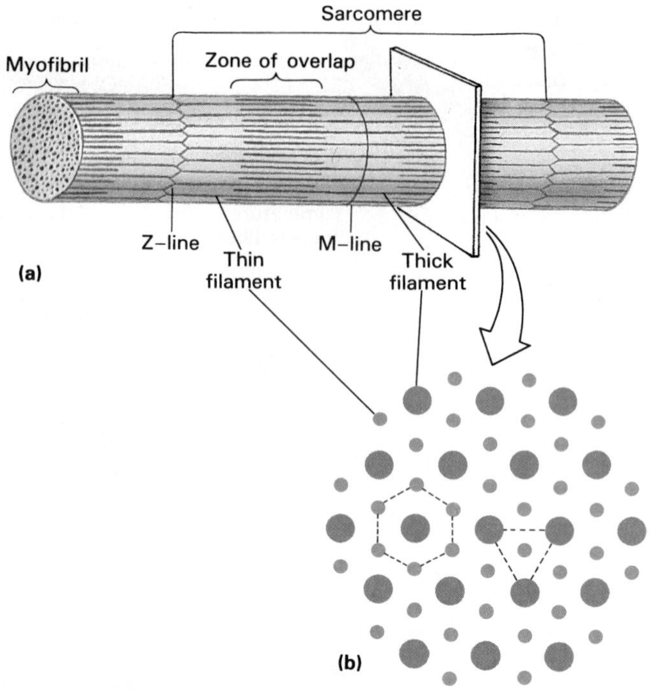

FIGURE 10-5
Sarcomere Structure. Each myofibril consists of a linear series of sarcomeres. (a) Organization of thick and thin filaments. (b) A cross-sectional view through the zone of overlap.

Sarcomere Organization

Each myofibril consists of a linear series of approximately 10,000 sarcomeres. *Sarcomeres are the smallest functional units of the muscle fiber—interactions between the thick and thin filaments of sarcomeres are responsible for muscle contraction.*

Figure 10-5a diagrams the structure of an individual sarcomere. Each sarcomere has a resting length of about 2.6 μm. The thick filaments lie in the center of the sarcomere, linked by filaments of the **M line**. Thin filaments at either end of the sarcomere, attached to interconnecting filaments that make up the **Z lines**, extend toward the M line. In the **zone of overlap** the thin filaments pass between the thick filaments.

The differences in the size and density of thick filaments and thin filaments account for the banded appearance of the sarcomere. The **A band** is the area containing thick filaments; it includes the M line, the **H band** (thick filaments only), and the zone of overlap (thick and thin filaments). Between the A band and the the Z line is the **I band**, which contains only thin filaments.

Figure 10-5b is a cross section through the zone of overlap, showing the relative sizes and arrangement of thick and thin filaments. In this region each thin filament sits in a triangle formed by three thick filaments, and each thick filament is surrounded by six thin filaments.

Thin and Thick Filaments

Each thin filament (Figure 10-6a) consists of a pair of protein strands wound together. These strands are chains of actin molecules. Thin filaments measure 5–6 nm in diameter and 1 μm in length. Thick filaments (Figure 10-6b) are 10–12 nm in diameter and 1.6 μm long. These protein filaments are composed of a helical array of myosin molecules. Each myosin molecule consists of a double myosin strand with an attached, elongate *tail* and a free globular *head*. The myosin molecules are oriented away from the center of the thick filament, with the heads projecting outward.

Transverse Tubules and the Sarcoplasmic Reticulum

Electrical events at the sarcolemmal surface trigger a contraction by altering the chemical environment around every sarcomere in the muscle fiber. The electrical "message" is distributed by the transverse tu-

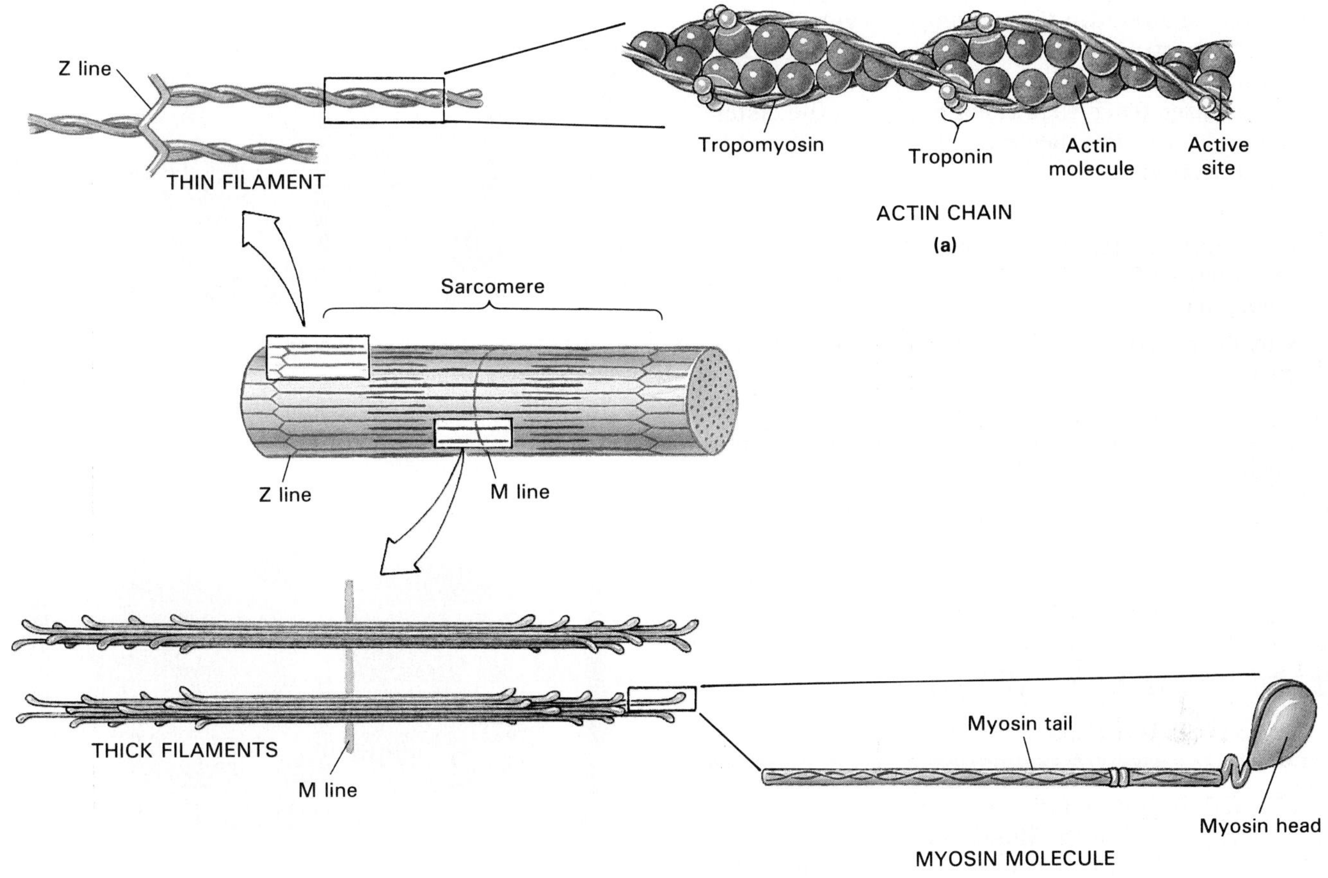

FIGURE 10-6
Thick and Thin Filaments

bules that extend deep into the sarcoplasm of the muscle fiber. As indicated in Figure 10-7, a transverse tubule begins at the sarcolemma and travels inward at right angles to the membrane surface. Along the way, branches from the transverse tubule encircle each of the individual sarcomeres at the boundary between the A band and the I band.

Where it encircles the sarcomere, the transverse tubule makes close contact with the membranes of the sarcoplasmic reticulum. The SR is most extensive around the thick filaments. At the zones of overlap between the thin and thick filaments the individual tubules of the reticulum enlarge, fuse, and form expanded chambers called **terminal cisternae**. A transverse tubule lies sandwiched between each pair of adjacent terminal cisternae, forming a complex known as a **triad**. Although the membranes of the triad are in close contact and tightly bound together, there is no direct connection between them, and their fluid contents remain separate and distinct.

In Chapter 3 we noted the existence of special ion pumps; these keep the intracellular concentration of calcium ion very low. (p. 84) Most cells pump the calcium ions across their cell membranes and into the extracellular fluid. Although skeletal muscle fibers do pump calcium ions out of the cell in this

FIGURE 10-7
Sarcoplasmic Reticulum and Transverse Tubules

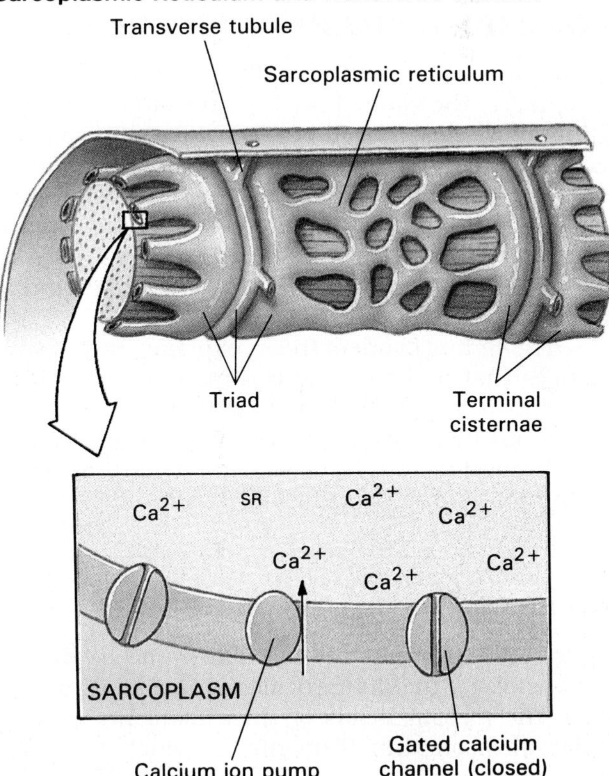

way, they also remove them from the cytosol by actively transporting them into the cisternae of the sarcoplasmic reticulum. A muscle contraction begins when the large reserves of calcium ions in the cisternae are released; the mechanism will be explored in the next section.

- ✓ **How would severing the tendon that was attached to a muscle affect the ability of the muscle to move a body part?**
- ✓ **Why does skeletal muscle appear striated when viewed with a microscope?**
- ✓ **Where would you expect to find the greatest concentration of calcium ions in resting skeletal muscle?**

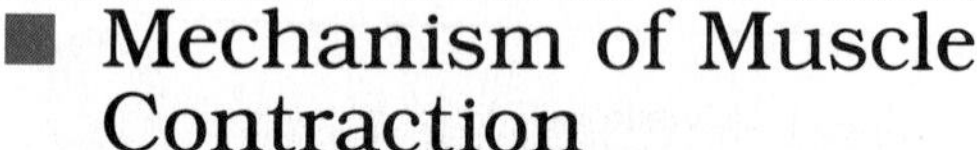

Mechanism of Muscle Contraction

A contracting muscle fiber exerts a pull, or **tension**, and shortens in length. Muscle fiber contraction results from interactions between the thick and thin filaments in each sarcomere. The explanation for the contraction process is called the *sliding filament theory*.

THE SLIDING FILAMENT THEORY

When a sarcomere contracts, the H band and I band get smaller, the zone of overlap gets larger, and the Z lines move closer together. The width of the A band remains constant throughout the contraction. These observations, summarized in Figure 10-8, make sense only if *the thin filaments are sliding toward the center of the sarcomere, alongside the thick filaments*. This interpretation is known as the **sliding filament theory**.

The sliding filament theory explains the physical changes that occur during contraction, but it does not explain why it begins, how it uses energy, or what stops the contraction. These questions lead us to a discussion of the molecular events that occur during the contraction process.

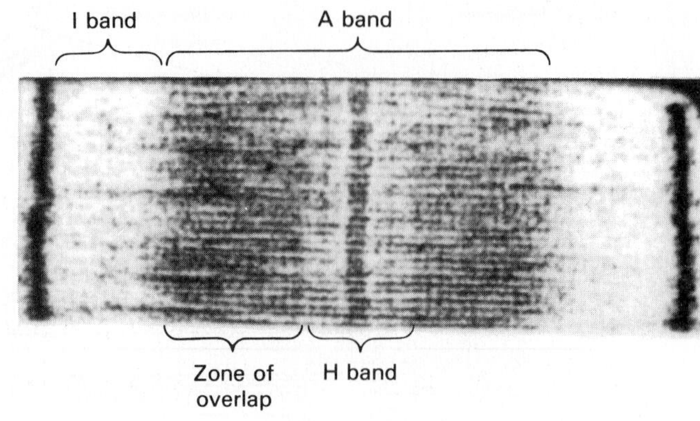

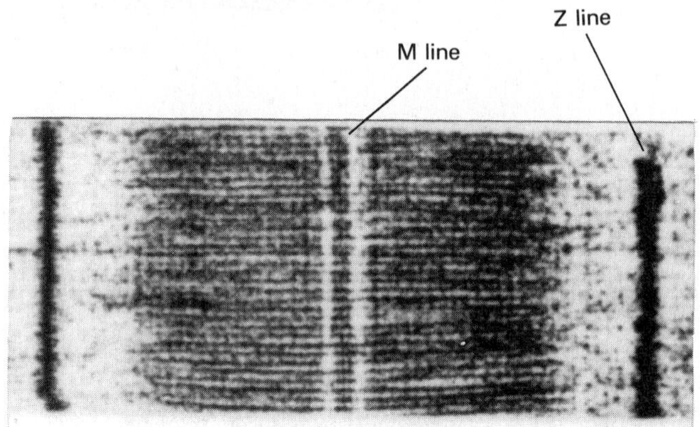

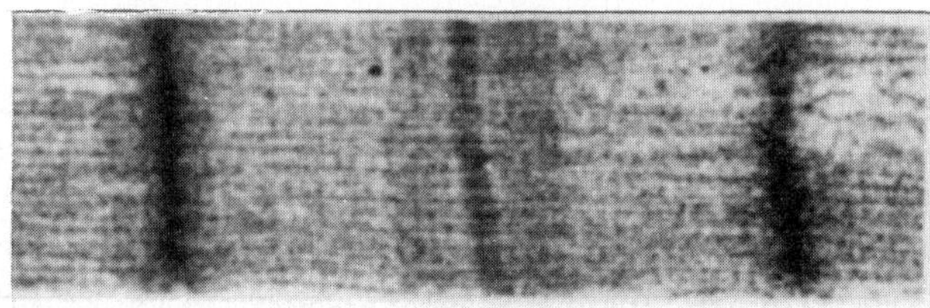

FIGURE 10-8
Changes in the Appearance of a Sarcomere during Contraction of a Skeletal Muscle Fiber. During a contraction the A band stays the same width, but the Z lines move closer together and the I band gets smaller.

The Sliding Mechanism

Figure 10-9 diagrams the mechanism believed to be responsible for the sliding of filaments. Sliding occurs when the myosin heads of thick filaments bind to **active sites** on thin filaments, in much the same way that a substrate molecule binds to the active site of an enzyme. Because they connect thick filaments and thin filaments, the myosin heads are also known as **cross-bridges**. When a cross-bridge binds to an active site, it pivots toward the M line, pulling the thin filament toward the center of the sarcomere. The cross-bridge then detaches and returns to its original position, ready to repeat the cycle of "attach, pivot, detach, and return."

Within the zone of overlap, each thick filament makes cross-bridge connections with six thin fila-

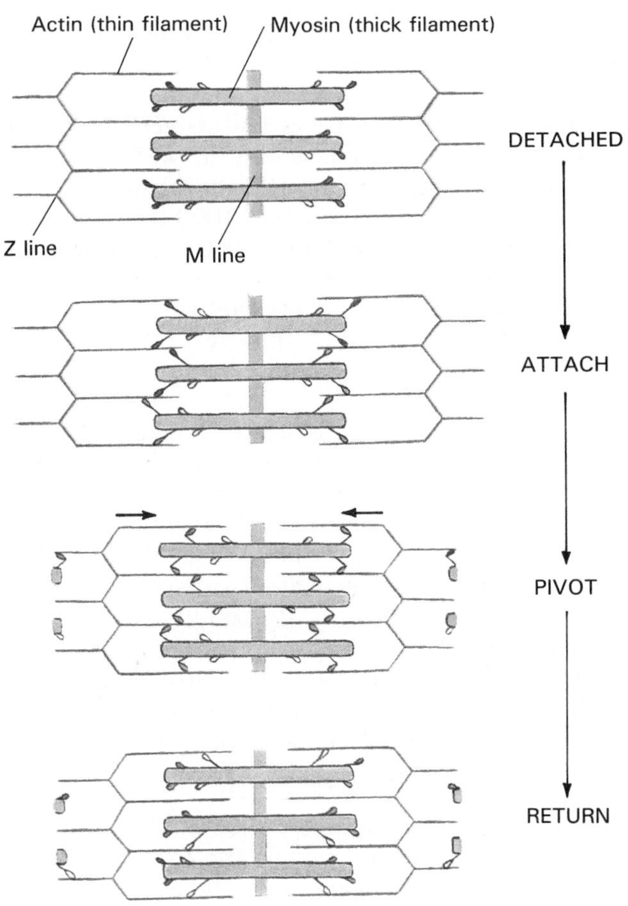

ments, and each thin filament binds to three thick filaments. At any given moment, some of the cross-bridges are free, some binding, and some pivoting. As a result there is always tension applied to the thin filaments, and sliding proceeds smoothly.

The Molecular Basis for Muscle Contraction

As noted above, contraction involves an interaction between the myosin heads and active sites along the thin filaments. At rest these interactions are prevented by two other proteins, **tropomyosin** (tro-po-MĪ-o-sin) and **troponin** (TRŌ-po-nin; *trope*, turning), that lie in the groove between the intertwined actin strands. The tropomyosin molecules form a long chain that covers the active sites, blocking actin-myosin interaction (Figure 10-10a). Troponin binds to both actin and tropomyosin, holding the tropomyosin strand in place. Until a change in the position of the troponin-tropomyosin complex exposes the active sites, no contraction can take place.

FIGURE 10-9
The Sliding Filament Theory. During a contraction, binding occurs between the myosin heads of the thick filaments and special sites on the thin (actin) filaments. The heads pivot, pulling the thin filaments toward the center of the sarcomere. Repeated cycles of attachment, pivoting, and release move the Z lines toward the ends of the thick filaments.

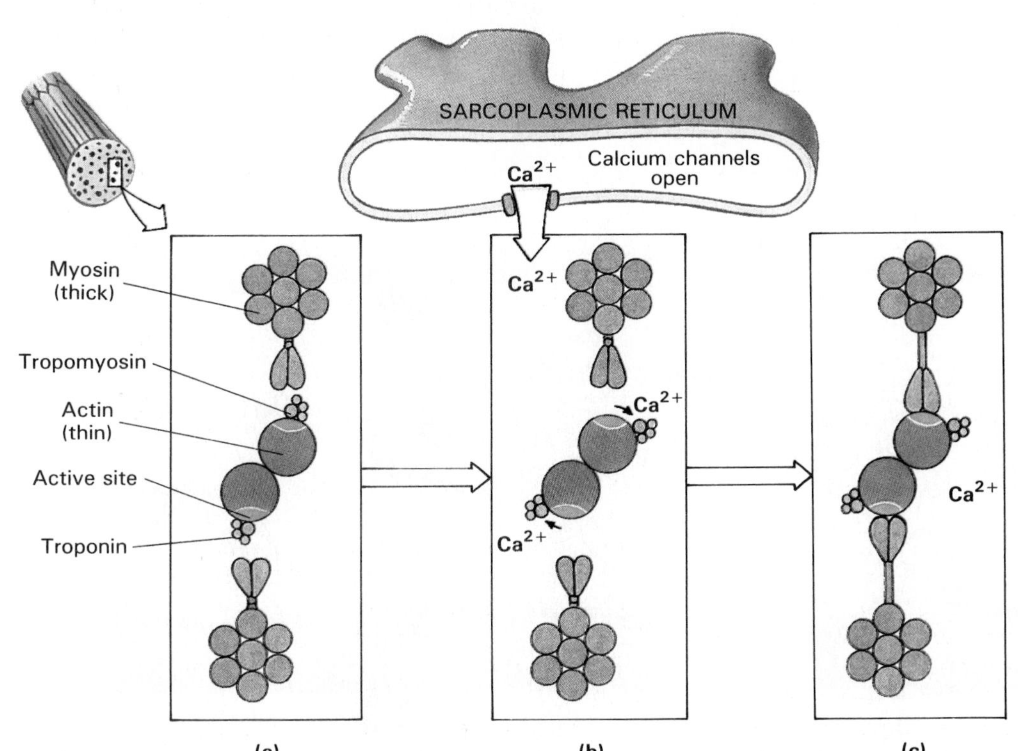

FIGURE 10-10
Molecular Events That Initiate Interactions of Thick and Thin Filaments. (a) In a resting sarcomere the tropomyosin strands lie over the active sites on the thin filaments, preventing cross-bridge interactions. (b) When calcium ions enter the sarcomere they bind to troponin, which rotates, moving the tropomyosin and exposing the active sites. (c) Cross-bridge binding then occurs, and the contraction begins.

Calcium Release Troponin is the lock that keeps the active sites inaccessible. The key that unlocks the active sites is calcium. Each troponin molecule, in addition to having binding sites for both tropomyosin and actin, also has a third binding site available for a calcium ion. When a calcium ion binds to troponin, it weakens the bond between troponin and actin. The troponin molecule then changes position, moving the tropomyosin molecule away from the active site (Figure 10-10b). Cross-bridge binding can occur, and a contraction begins.

The sarcoplasm around the sarcomeres of a resting muscle fiber contains very low concentrations of calcium ions (Ca^{2+}), somewhere around 10^{-7} M/ℓ. The calcium concentration inside the terminal cisternae, however, may be as much as 1000 times higher. These cisternae are the source of the calcium that triggers muscle contraction. Because the cisternae are situated at the zones of overlap, where the thick and thin filaments interact, the effect of calcium release on the sarcomere is almost instantaneous.

The Contraction Cycle Figure 10-11 details the molecular events that occur during the contraction process. In the resting sarcomere each cross-bridge is bound to a molecule of ADP and phosphate (PO_4^{3-}), the products released by the breakdown of a molecule of ATP. The cross-bridge acts as an **ATPase**, an enzyme that can break down ATP. In addition to binding the breakdown products, the cross-bridge stores the energy released by the rupture of the high-energy bond.

The contraction process involves five interlocking steps:

STEP 1: Active site exposure. Calcium binds to the troponin molecule, moving the troponin-tropomyosin complex away from the active sites as shown in Figure 10-10.

STEP 2: Cross-bridge attachment. When the active sites are exposed, myosin cross-bridges bind to them.

STEP 3: Pivoting. The attached myosin head pivots toward the center of the sarcomere. This step uses the energy stored in the myosin molecule.

FIGURE 10-11
Molecular Events of the Contraction Process

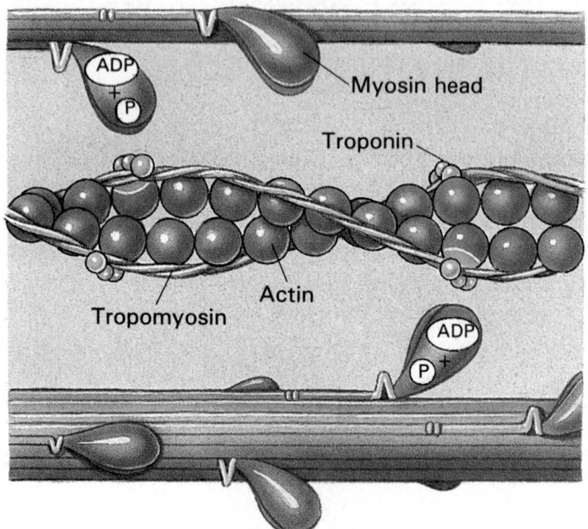

Resting sarcomere

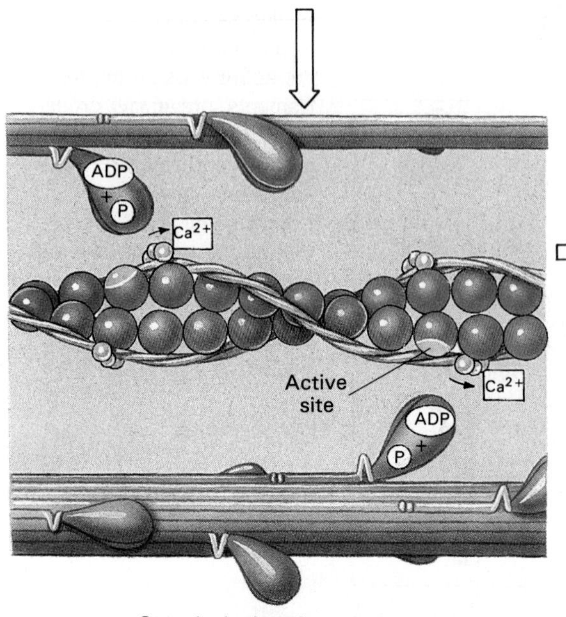

Step 1: Active site exposure

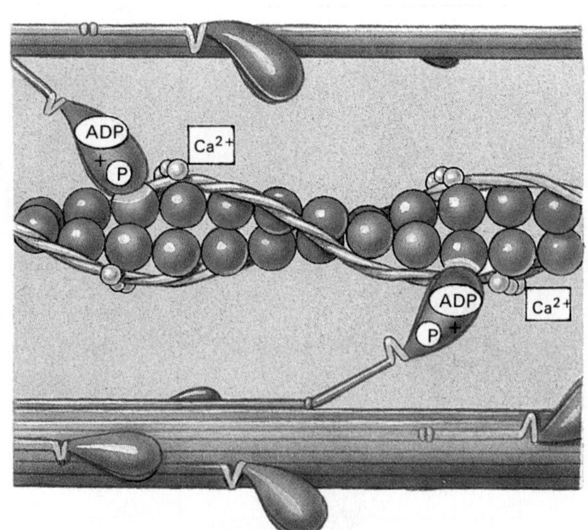

Step 2: Cross-bridge attachment

In the process, ADP and a phosphate group are released.

STEP 4: Cross-bridge detachment. The bond between the active site on the actin molecule and the myosin cross-bridge remains intact until the myosin head binds another ATP molecule.

STEP 5: Myosin activation. Activation occurs as the free myosin head splits ATP, retaining the ADP, the phosphate group, and energy released by the reaction. The entire cycle can now be repeated, beginning with Step 2, as long as the active sites on the actin molecule remain exposed.

This cycle is broken when calcium ion concentrations return to normal resting levels. When that occurs there are no longer enough calcium ions to "unlock" the troponin molecules. Tropomyosin once again covers the active sites, and contraction can no longer take place. Calcium ion concentrations are reduced by (1) active transport across the cell membrane into the extracellular fluid and (2) active transport into the sarcoplasmic reticulum. Of the two, transport into the sarcoplasmic reticulum is the most important.

NEURAL CONTROL OF MUSCLE FIBER CONTRACTION

Chapter 12 examines neural physiology in detail, so our discussion here will focus on the effects of neural stimulation of skeletal muscles. Before proceeding, however, you should take a moment to review the details of neuron structure that were presented in Chapter 5. (p. 157) The basic sequence of events involved in the neural control of skeletal muscle function can be summarized as follows:

- Chemicals released by the motor neuron at the neuromuscular junction alter the transmembrane potential of the sarcolemma. This change sweeps across the surface of the sarcolemma and into the transverse tubules.
- The change in the transmembrane potential of the T tubules triggers the release of calcium ions by the sarcoplasmic reticulum.

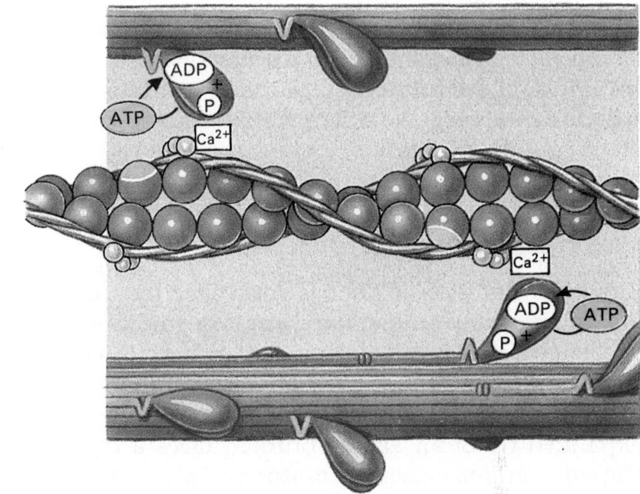

Step 5: Myosin reactivation

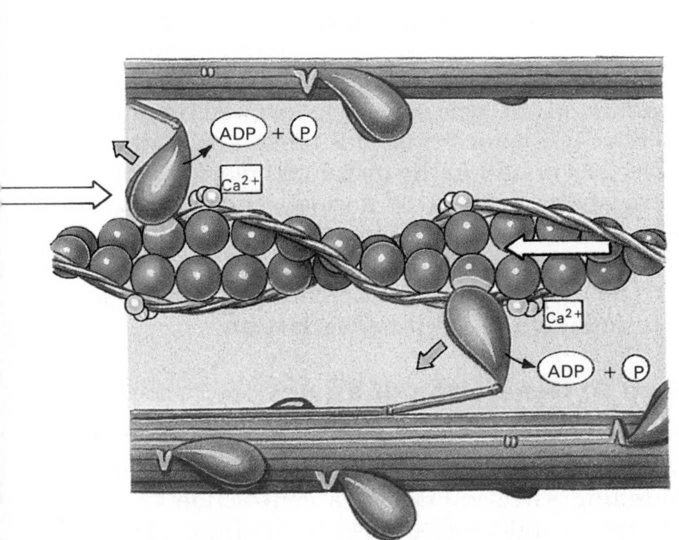

Step 3: Pivoting of myosin head

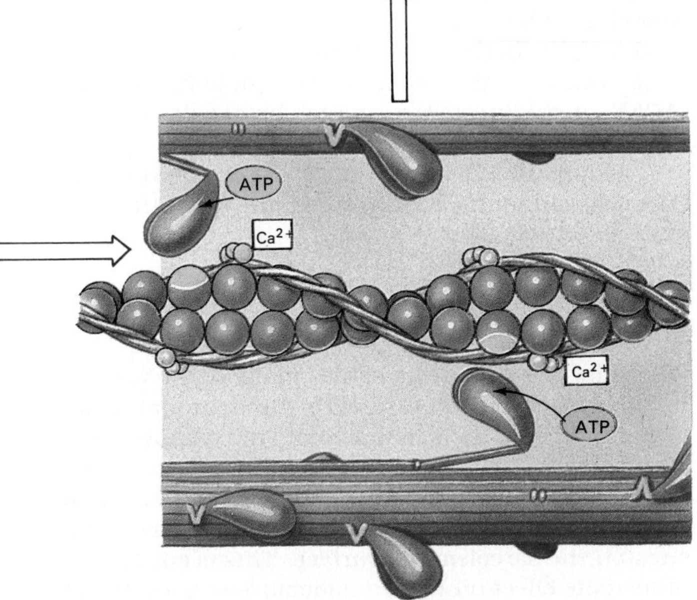

Step 4: Cross–bridge detachment

Clinical Comment: Rigor Mortis

When death occurs, circulation ceases and the skeletal muscles are deprived of nutrients and oxygen. Within a few hours, the skeletal muscle fibers have run out of ATP, and the sarcoplasmic reticulum becomes unable to remove calcium ions from the sarcoplasm. Calcium ions diffusing into the sarcoplasm from the extracellular fluid or leaking out of the sarcoplasmic reticulum then trigger a sustained contraction. Without ATP, the cross-bridges cannot detach from the active sites, and the muscle locks in the contracted position. All of the body's skeletal muscles are involved, and the individual becomes "stiff as a board." This physical state, called **rigor mortis**, lasts until the lysosomal enzymes released by autolysis break down the myofilaments 15–25 hours later.

This sequence of events, which forms the link between electrical activity in the sarcolemma and the initiation of a contraction, is called **excitation-contraction coupling**. We will now examine each of these steps in greater detail.

The Neuromuscular Junction

Each skeletal muscle fiber is controlled by a neuron at a single **neuromuscular junction** midway along its length. Several neuromuscular junctions were seen in Figure 10-2. Figure 10-12a summarizes key features of this structure. The **synaptic knob**, the expanded tip of an axonal branch, faces a region of the sarcolemma called the **motor end plate**. A narrow space, the **synaptic cleft**, separates the two. The cytoplasm of the synaptic knob contains mitochondria and vesicles filled with molecules of **acetylcholine** (as-ē-til-KŌ-len), usually abbreviated **ACh**. The opposing motor end plate contains receptors that will bind to ACh. Both the synaptic cleft and motor end plate contain the enzyme **acetylcholinesterase** (**AChE**, or *cholinesterase*), which breaks down molecules of ACh.

Figure 10-12b details the steps involved in the chemical communication between the synaptic knob and the muscle fiber.

Step 1: Release of Acetylcholine

When the nerve becomes active, many of the vesicles in the synaptic knob fuse with the neuronal membrane and dump their contents into the synaptic cleft. The synaptic cleft is very narrow (20–30 nm), and the ACh molecules rapidly diffuse across it. When they arrive at the motor end plate they bind to receptor sites on the sarcolemmal surface. This event has an immediate effect on the transmembrane potential at that location. [**CM**: *Botulism*]

Step 2: Depolarization of the Motor End Plate

Chapter 3 introduced the factors that create and maintain the transmembrane potential in all living cells. The transmembrane potential of a resting skeletal muscle fiber averages around –85 mV, with an excess of negative charges on the inside of the membrane. A state of equilibrium exists, with the sodium-potassium exchange pump keeping pace with the rate of sodium ion entry and potassium ion loss. (p. 85)

Any stimulus that disturbs the balance between sodium entry and potassium loss will alter the resting potential. In the case of a motor end plate, binding of ACh to the receptors causes an increase in the membrane permeability to sodium ions. Sodium ions now enter the cell faster than the sodium-potassium exchange pump can remove them. The inside of the membrane gains positive charges, and the transmembrane potential shifts toward 0 mV. The potential difference *de*creases, a change that is called a **depolarization** of the cell membrane. [**CM**: *Myasthenia Gravis*]

This depolarization lasts from 2–20 msec, depending on the type of muscle involved, because by then acetylcholinesterase has broken down the ACh molecules bound to the motor end plate. Membrane permeability returns to normal, and the sodium-potassium exchange pump soon brings the transmembrane potential back to –85 mV. The reestablishment of the normal membrane potential after depolarization is called **repolarization**.

Step 3: Generation of an Action Potential

The temporary depolarization produced by ACh binding is restricted to the motor end plate and the surrounding sarcolemma. For a contraction to begin, the events under way at this site must spread to every portion of the cell membrane. This can occur because

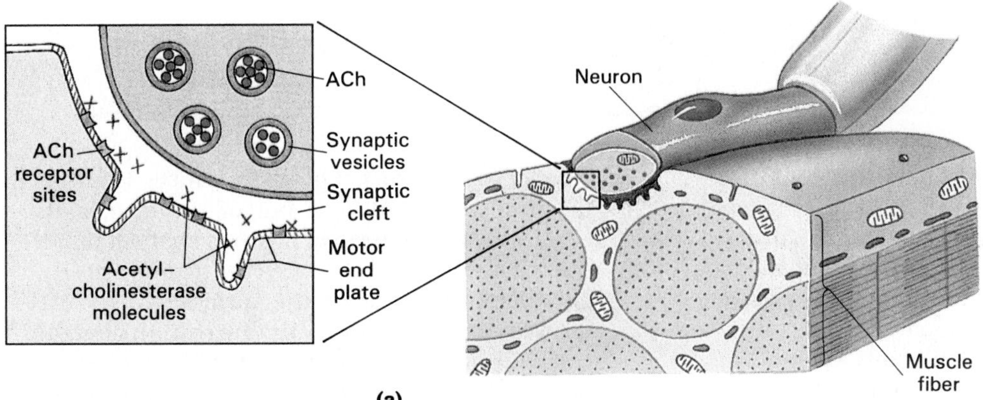

FIGURE 10-12 The Neuromuscular Junction. Steps in the chemical communication between the synaptic knob and motor end plate.

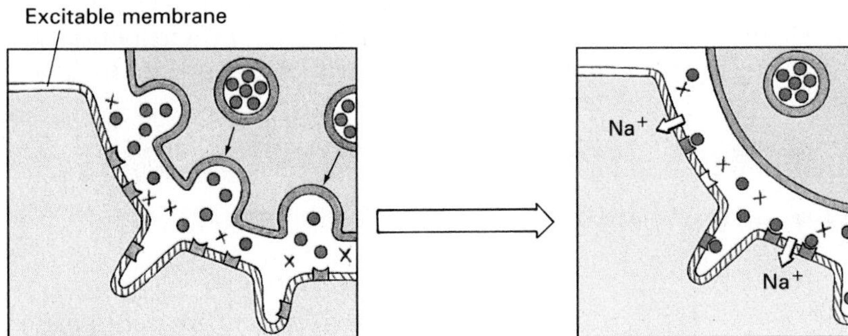

Step 1: Release of Acetylcholine
Vesicles in the synaptic knob fuse with the neuronal membrane and dump their contents into the synaptic cleft.

Step 2: Depolarization of the Motor End Plate
The binding of ACh to the receptors increases the membrane permeability to sodium ions. Sodium ions enter the cell at an increased rate, and the transmembrane potential shifts toward 0 mV.

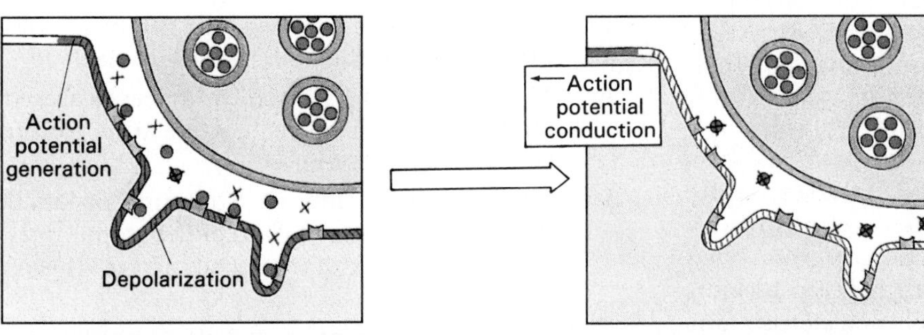

Step 3: Generation of Action Potential
When the sarcolemma around the motor end plate depolarizes to around −60 mV, there is a sudden, temporary change in the transmembrane potential. This electrical event is called an *action potential*.

Step 4: Conduction of Action Potential
The action potential is immediately conducted across the membrane surface, as a ripple moves away from a rock dropped into a pond. It reaches all parts of the sarcolemmal surface, and travels down each of the transverse tubules, triggering the release of calcium ions at the terminal cisternae.

(b)

the sarcolemma has the property of **excitability**: the ability to conduct an electrical impulse. Only muscle fibers and nerve fibers have excitable membranes. When the sarcolemma around the motor end plate depolarizes to around –60 mV there is a massive change in ion permeability. The result is a sudden, temporary depolarization that sweeps across the entire surface of the cell. This *conducted change* in the transmembrane potential is called an **action potential**. (The mechanism of action potential generation and conduction will be described in Chapter 12.)

Step 4: Conduction of an Action Potential

When an action potential appears in the sarcolemma at the motor end plate, it is immediately conducted across the membrane surface. To visualize this event, think of a ripple moving away from a rock dropped into a pond. The action potential spreads over all parts of the sarcolemmal surface and travels down each of the transverse tubules, quickly reaching all of the triads that encircle the sarcomeres of the muscle fiber.

Clinical Comment: Muscle Sensitivity to Electrolyte Concentrations

Skeletal and other muscle fibers are sensitive to changes in the concentration of calcium or potassium ions in their surroundings. Unusually high or low calcium or potassium concentrations in the interstitial fluids will affect muscle activity by altering the transmembrane potential and the activity of the sarcoplasmic reticulum. For example, a high extracellular potassium concentration, or **hyperkalemia** (hī-per-ka-LĒ-mē-a; *kalium*, potassium + *haima*, blood) reduces the difference between the potassium concentrations inside and outside the cell. The smaller concentration gradient causes fewer potassium ions to diffuse out of the cytoplasm, and the accumulation of positive charges inside the cell gradually depolarizes the membrane. Hyperkalemia can lead to the paralysis of skeletal muscles and death by cardiac arrest. In general, cardiac muscle is more sensitive to ionic disturbances than skeletal muscle. These clinical conditions will be discussed further in Chapter 20, and the regulation of calcium and potassium ion concentrations will be considered in Chapters 26 and 27.

Step 5: Release of Calcium Ions

As we noted earlier, nearly all of the calcium ions in the resting muscle fiber are held within the terminal cisternae of the sarcoplasmic reticulum. When an action potential arrives at the T tubule of a triad, it triggers a sudden, massive release of calcium ions from the adjacent terminal cisternae. For a brief moment, the membranes of the SR become unusually permeable to calcium ions. Within a millisecond calcium ions diffuse out of the SR, and the concentration of Ca^{2+} around the sarcomere reaches 100 times the resting levels.

As the calcium ion concentration rises, active sites are exposed on the thin filaments, cross-bridge interactions occur, and a contraction begins. Even though the sarcoplasmic reticulum regains its normal properties almost immediately, the calcium ions cannot be recaptured fast enough to prevent a contraction. Because all of the triads in the muscle fiber are affected, this contraction is a combined effort involving every sarcomere on every myofibril. [**CM:** *Duchenne's Muscular Dystrophy*]

MUSCLE CONTRACTION: A SUMMARY

The next section will consider the forces that are generated by muscle fiber contraction. Before proceeding, however, let's step back and summarize the entire sequence of events from neural activation through excitation-contraction coupling to the completion of a contraction. Figure 10-13 provides a visual summary.

1. At the neuromuscular junction, ACh released by the synaptic knob binds to receptors on the motor end plate. Binding produces a temporary depolarization that affects adjacent portions of the sarcolemma. (The effect is temporary because the ACh is rapidly broken down by acetylcholinesterase.)
2. When the sarcolemma is sufficiently depolarized, an action potential appears. It is conducted away from the motor end plate in all directions.
3. The action potential reaches the triads by conduction along the transverse tubules.
4. The sarcoplasmic reticulum releases stored calcium ions, increasing the calcium concentration at the sarcomeres.
5. Calcium ions bind to troponin, producing a change in the orientation of the troponin-tropomyosin complex that exposes active sites on the thin filaments.
6. Repeated cycles of cross-bridge binding, pivoting, and detachment occur, powered by the breakdown of ATP. These events produce filament sliding, and the muscle fiber shortens.
7. The sarcoplasmic reticulum reabsorbs calcium ions, and the concentration of calcium ions in the sarcoplasm declines.
8. When calcium ion concentrations approach normal resting levels, the troponin-tropomyosin complex returns to its normal position. This change covers the active sites and prevents further cross-bridge interaction. The contraction is now at an end.

√ **How would a drug that interferes with cross-bridge formation affect muscle contraction?**

√ **What would you expect to happen to a resting skeletal muscle if the sarcolemma suddenly became very permeable to calcium ions?**

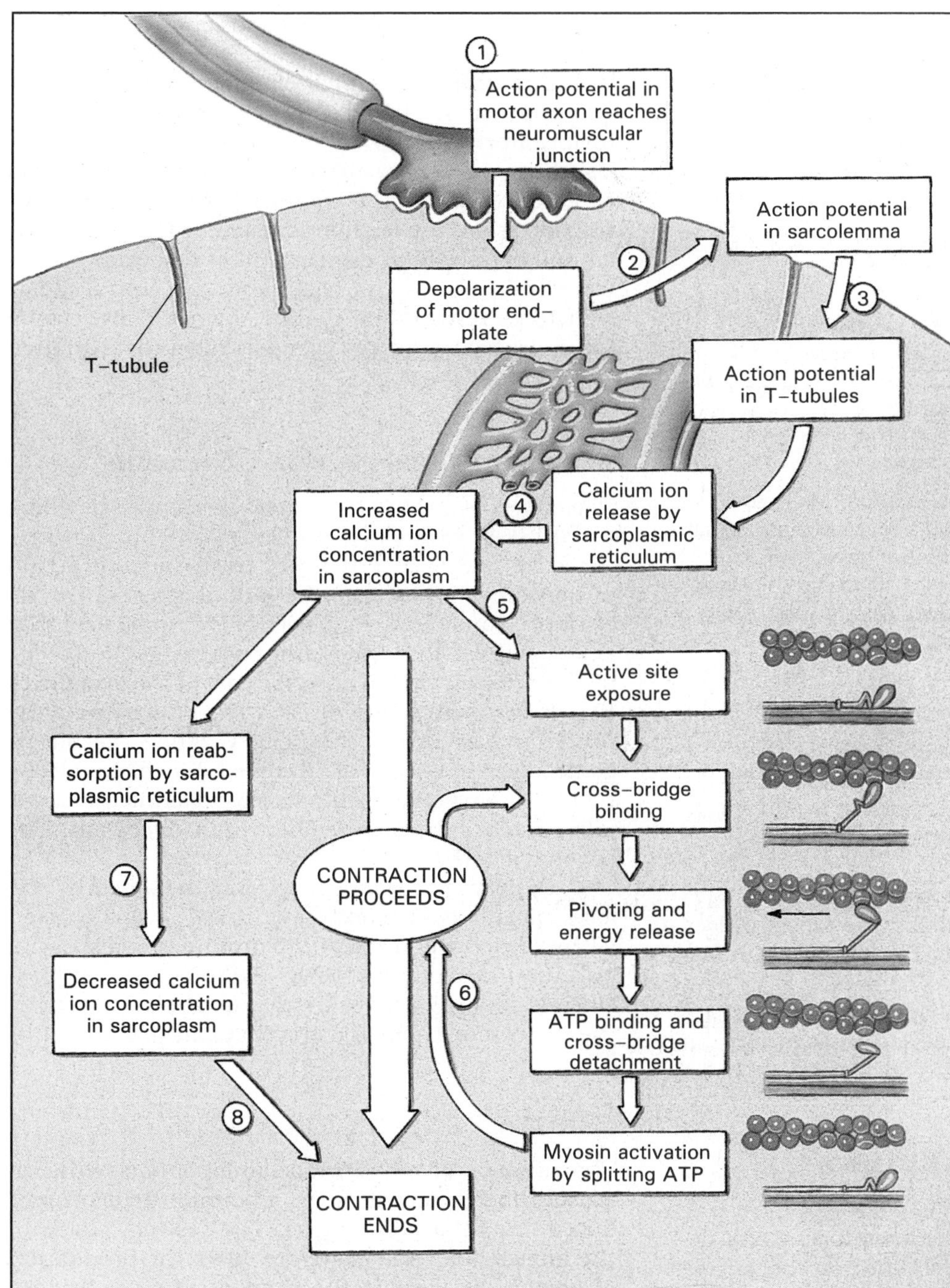

FIGURE 10-13 Summary of the Events in Excitation-Contraction Coupling and Muscle Contraction

√ **Predict what would happen to a muscle if the motor endplate failed to produce any acetylcholinesterase.**

The Twitch and Development of Tension

A **twitch** is a single stimulus-contraction-relaxation sequence in a muscle fiber. The duration of a single twitch is variable. Twitches in an eye muscle fiber can be as brief as 10 msec, but a twitch in a calf muscle fiber lasts around 100 msec. Figure 10-14 is a graph, or **myogram**, of the development of tension in a stimulated calf muscle fiber during a twitch.

STAGES OF A TWITCH

A single twitch can be divided into a *latent period*, a *contraction phase*, and a *relaxation phase*. The **latent period** begins at stimulation and typically lasts about 10 msec. Over this period the action potential sweeps across the sarcolemma, and calcium ions are

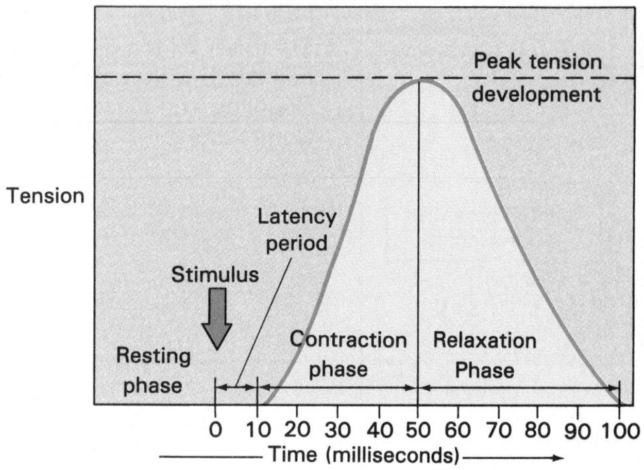

FIGURE 10-14
The Twitch and Development of Tension. Myogram showing details of the time course of a single isometric twitch contraction. Note the presence of a latent period, which corresponds to the time needed for the conduction of action potential and the subsequent release of calcium ions by the sarcoplasmic reticulum.

released by the sarcoplasmic reticulum. During the latent period the muscle fiber does not produce tension, because the contractile mechanism is not yet activated.

In the 40-msec **contraction phase**, tension rises to a peak. Throughout this period the cross-bridges are interacting with the active sites on the actin filaments.

The **relaxation phase** then continues for another 50 msec; during this period muscle tension falls to resting levels as the cross-bridges detach.

RELAXATION AND THE RETURN TO RESTING LENGTH

Our discussion of the contraction process has detailed how muscle fibers actively shorten. Because a muscle fiber enters the relaxation phase at its contracted length, you might wonder how it can return to its normal resting length. There is no active mechanism for muscle fiber elongation; contraction is active, but elongation is passive. After a contraction, a muscle fiber returns to its original length through (1) elastic forces and (2) the movement of other, opposing muscles.

Elastic Forces

In addition to myofibrils, a muscle fiber contains a variety of other intracellular structures—such as mitochondria, sarcoplasmic reticulum, and nuclei—and the entire package is surrounded by a sarcolemma. The muscle fiber is wrapped in the fibers of the endomysium, which transfer the contraction forces to the tendon.

These fibers and intracellular structures are flexible and elastic. When the muscle fiber contracts, the myofibrils pull on the sarcolemma, the extracellular fibers are stretched, and intracellular elements are compressed or otherwise distorted. When the contraction ends, the tension is relieved. The intracellular and extracellular elements then rebound; the extracellular fibers contract, and the intracellular structures regain their normal shapes. These elastic forces gradually return the muscle fiber to its original resting length.

Active or Passive Muscle Movements

Elastic forces gradually stretch relaxed muscle fibers toward their normal resting lengths. More rapid returns to resting length result from the contraction of opposing muscles or the pull of gravity. For an example, consider the muscles of the upper arm that flex or extend the elbow. Contraction of the *biceps brachii* muscle on the anterior part of the arm flexes the elbow; contraction of the *triceps* muscle on the posterior part extends the elbow. When the biceps contracts, the triceps is stretched. When the biceps relaxes, contraction of the triceps extends the elbow and stretches the muscle fibers of the biceps to their original length.

Opposing muscle groups may be aided by the force of gravity. For example, if the biceps relaxed with the elbow pointed at the ground, gravity would pull the forearm back down. Although gravity provides assistance, active muscles are needed to control the rate of movement and prevent damage to the joint.

LENGTH-TENSION RELATIONSHIPS

The number of cross-bridge interactions within a muscle fiber determines the amount of tension produced during a contraction. The resting length of the muscle fiber will therefore affect the production of tension by varying the number of cross-bridges within the zone of overlap. Figure 10-15 indicates the relationship between the length of a muscle fiber and the maximum tension it can develop.

Because a contraction involves all the sarcomeres along all the myofibrils, the tension produced by the intact muscle fiber can be related to the structure of individual sarcomeres. When the sarcomeres are as short as they can be, the thick filaments are jammed against the Z lines, and a contraction cannot occur (Figure 10-15a). Even if the sarcomeres are somewhat longer, the thin filaments that extend across the center of the sarcomere collide with or overlap the thin filaments of the opposite side (Figure 10-15b). This changes the three-dimensional arrangement of thick and thin filaments and interferes with the normal binding of cross-bridges to active

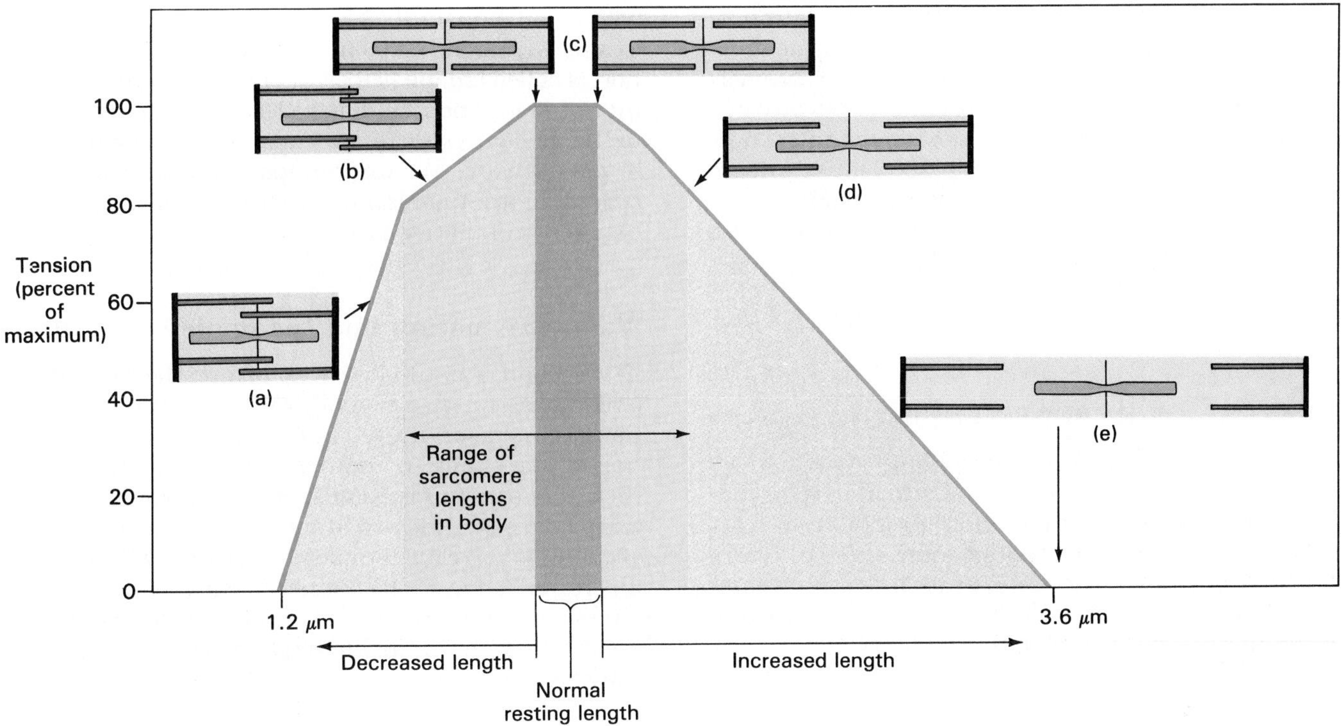

FIGURE 10-15
The Effect of Sarcomere Length on Tension. The tension produced on stimulation depends on the number of cross-bridges formed. When the sarcomeres are too short (a,b), contraction cannot occur because the thick filaments come in contact with the Z lines. If the sarcomeres are stretched too far, the zone of overlap is reduced (d) or disappears (e), and cross-bridge interactions are reduced or cannot occur. The tension produced varies over intermediate sarcomere lengths, reaching a maximum when the zone of overlap is large but the thin filaments do not extend across the center of the sarcomere (c).

sites. The result is a reduction in the amount of tension produced.

Within the optimal range of sarcomere lengths (Figure 10-15c), the maximum number of cross-bridges can form, and the tension produced is highest. Any further increase in sarcomere length reduces the tension produced by reducing the size of the zone of overlap and the number of potential cross-bridge interactions (Figure 10-15d). When the zone of overlap disappears altogether, thin and thick filaments cannot interact, and the muscle fiber cannot contract (Figure 10-15e).

In summary, muscle fibers can contract most forcefully when stimulated over a relatively narrow range of resting lengths. The normal range of sarcomere lengths in the body is indicated in Figure 10-15. The arrangement of skeletal muscles, connective tissues, and bones normally prevents extreme compression or excessive stretching that would correspond to Figure 10-15a or 10-15e. For example, straightening the elbow stretches the biceps brachii muscle, but stabilizing ligaments and the bony structure of the elbow stop this movement before the muscle fibers stretch too far. During normal movements our muscle fibers perform over a broad range of intermediate lengths; the tension produced varies, depending on the initial length of the muscle fiber. During an activity such as walking, in which muscles contract and relax in a cyclical fashion, muscle fibers are stretched to a length very close to "ideal" before they are stimulated to contract.

Muscle Mechanics

Now that you are acquainted with the dynamics of muscle contraction at the level of the individual muscle fiber we can take a step back and discuss the performance of skeletal muscles, the organs of the muscular system.

MOTOR UNITS AND MUSCLE CONTROL

A typical skeletal muscle contains thousands of muscle fibers. Although some motor neurons control a single muscle fiber, most control hundreds or thousands of muscle fibers. All of the muscle fibers controlled by a single motor neuron constitute a **motor**

unit. The size of a motor unit is an indication of how fine the control of movement can be. In the muscles of the eye, where precise control is extremely important, a motor neuron may control two or three muscle fibers. We have much less precise control over our leg muscles, where more than 2000 muscle fibers respond to the call of a single motor neuron.

A skeletal muscle contracts when its motor units are stimulated. The amount of tension produced depends on two factors: (1) the frequency of stimulation and (2) the number of motor units involved.

EFFECTS OF REPEATED STIMULATIONS

Figure 10-16 presents the myograms generated by stimulating an entire skeletal muscle at varying frequencies. A single stimulation produces a brief twitch, with the time course indicated in Figure 10-14. Twitches in a skeletal muscle do not accomplish anything useful; all normal activities involve more sustained, controlled contractions.

Treppe

If a muscle is stimulated a second time immediately after the relaxation phase has ended, the contraction that occurs will develop a slightly higher maximum tension. The increase in peak tension indicated in Figure 10-16a will continue over the first 30–50 stimulations. Thereafter the amount of tension produced will remain constant. Because the peak tension rises in stages, like the steps in a staircase, this phenomenon is called **treppe** (TREP-e), a German word meaning "stairs." The rise may reflect a gradual increase in the concentration of calcium ions in the cytoplasm, in part because the ion pumps in the sarcoplasmic reticulum are unable to recapture them in the time between stimulations.

Wave Summation and Incomplete Tetanus

If a second stimulus arrives before the relaxation phase has ended, a second, more powerful contraction occurs, as indicated in Figure 10-16b. The addition of one twitch to another in this way constitutes the **summation of twitches**, or simply **wave summation**. If you continue to stimulate the muscle, never allowing it to relax completely, tension will peak as illustrated in Figure 10-16c. A muscle producing peak tension during rapid cycles of contraction and relaxation is said to be in **incomplete tetanus** (*tetanos*, convulsive tension).

Complete Tetanus

Complete tetanus can be obtained by increasing the rate of stimulation until the relaxation phase is completely eliminated (Figure 10-16d). In complete tetanus the action potentials are arriving so fast that the sarcoplasmic reticulum does not have time to

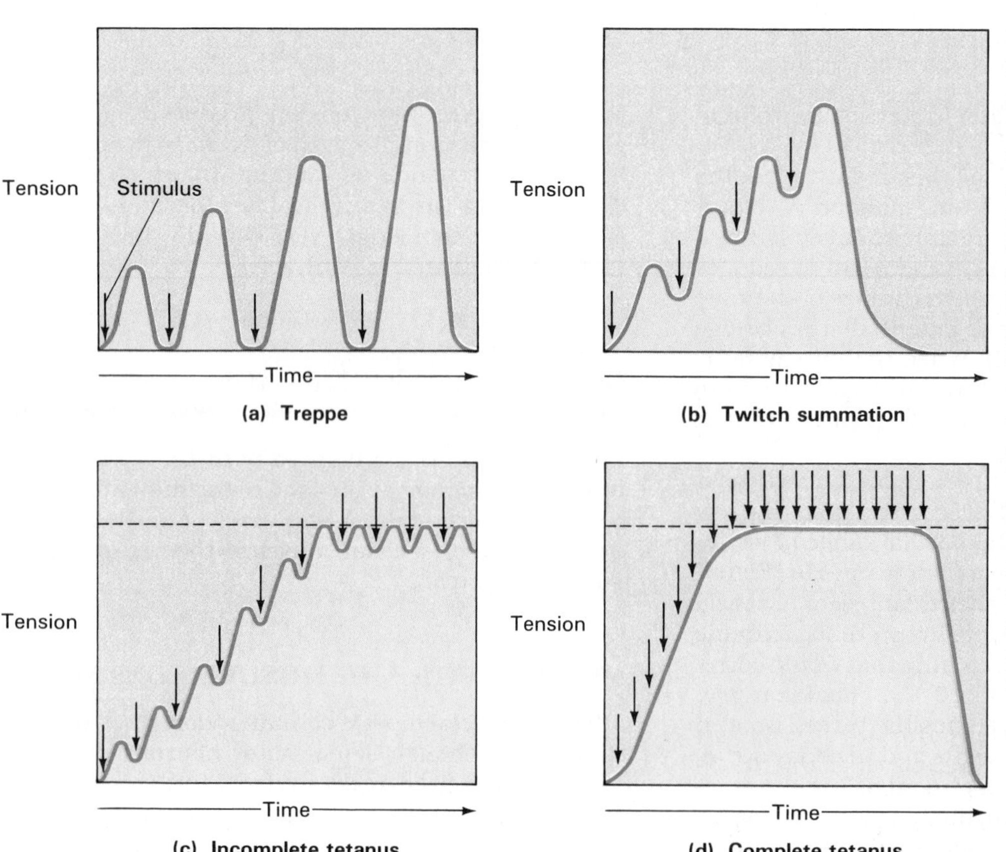

FIGURE 10-16 Effects of Repeated Stimulations. (a) Treppe occurs when additional stimuli arrive shortly after the completion of the relaxation phase of each twitch. (b) Wave summation occurs when successive stimuli arrive before relaxation has been completed. (c) If the rate of stimulation increases further, tension production will rise to a peak, and the periods of relaxation will be very brief. This condition is incomplete tetanus. (d) In complete tetanus the frequency of stimulation is so high that the relaxation phase has been completely eliminated. Tension rises and plateaus at maximal levels.

Clinical Comment: Tetanus

Children are often told to "watch out" for rusty nails. Parents are not worrying about the rust or the nail, but about infection with a very common bacterium, *Clostridium tetani*. This bacterium can cause **tetanus**, a disease that has no relationship to the normal response to neural stimulation.

The bacteria that cause tetanus, though found virtually everywhere, can thrive only in tissues that contain abnormally low amounts of oxygen. For this reason a deep puncture wound, such as that from a nail, carries a much greater risk than a shallow, open cut that bleeds freely.

When active, these bacteria release a powerful toxin that affects the central nervous system. Motor neurons are particularly sensitive to it, and their stimulation produces a sustained, powerful contraction of skeletal muscles throughout the body. This toxin also has a direct effect on skeletal muscle fibers, producing contraction even in the absence of neural commands.

After exposure, the incubation period (the time before symptoms develop) is usually less than 2 weeks. The most common complaints are headache, muscle stiffness, and difficulty in swallowing. Because it soon becomes difficult to open the mouth, this disease is also called "lockjaw." Widespread muscle spasms usually develop within 2–3 days of the initial symptoms, and they often continue for a week before subsiding. After 2–4 weeks, symptoms in surviving patients disappear with no after effects.

Although severe tetanus has a 40–60 percent mortality rate, immunization is effective in preventing the disease. There are approximately 500,000 cases of tetanus worldwide each year, but only about 100 of them occur in the United States, thanks to an effective immunization program. (Most readers will have had "tetanus shots" or "tetanus boosters" sometime within the last 10 years.) Severe symptoms in unimmunized patients can be prevented by early administration of an antitoxin, usually **human tetanus immune globulin**. Such treatment does not reduce symptoms that have already appeared, however, and there is no generally effective treatment for this disease.

reclaim the calcium ions. The high calcium ion concentration in the cytoplasm prolongs the state of contraction, making it continuous. Virtually all normal muscular contractions involve the complete tetanus of the participating motor units.

RECRUITMENT

We have a remarkable ability to control the amount of pull exerted by our skeletal muscles. During a normal contraction, tension rises smoothly, not jerkily, because activated muscle fibers are responding in complete tetanus (see Figure 10-16d). The total force exerted by the muscle as a whole depends on how many motor units are activated. By varying the number of motor neurons activated at any one time, the nervous system provides precise control over the pull of a muscle.

When a decision is made to perform a specific movement, specific groups of motor neurons within the central nervous system are stimulated. The stimulated neurons do not respond simultaneously, and over time the number of activated motor units gradually increases. As indicated in Figure 10-17, the muscle fibers of each motor unit are intermingled with those of other units. Because of this intermingling the direction of pull exerted on the tendon does not change as more motor units are activated, although the total amount of force steadily increases. The smooth but steady increase in muscular tension produced by increasing the number of active motor units is called **recruitment**, or **multiple motor unit summation**.

Peak tension production occurs when all of the motor units in the muscle are contracting in complete tetanus. However, such powerful contractions do not last long, because the individual muscle fibers soon use up their available energy reserves. During a sustained tetanic contraction motor units are activated on a rotating basis, so that some of them are resting and recovering while others are actively contracting.

ISOMETRIC AND ISOTONIC CONTRACTIONS

A maximally stimulated skeletal muscle 1 square inch in cross-sectional area can develop roughly 50 pounds of force. If we hang a 40-pound weight from that

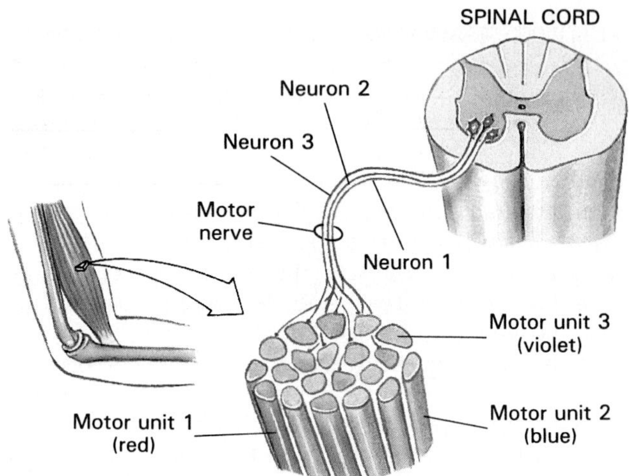

FIGURE 10-17
Arrangement of Motor Units in a Skeletal Muscle. Muscle fibers of different motor units are intermingled, so that the net distribution of force applied to the tendon remains constant even when individual muscle groups cycle between contraction and relaxation.

muscle and stimulate it, the muscle will contract (Figure 10-18a). Before sliding can occur, the cross-bridges must produce enough "pull," or **tension**, to overcome the **resistance**, in this case a 40-pound weight. Tension in the muscle builds until it exceeds the amount of resistance, and then the muscle shortens. As it shortens, the tension in the muscle remains constant, at a value that just exceeds the applied

FIGURE 10-18
Isotonic and Isometric Contractions. (a,b) A muscle attached to a weight is stimulated and develops enough tension to lift the weight. Tension remains constant for the duration of the contraction, although the length of the muscle changes. This is an example of isotonic contraction. (c,d) If the muscle is attached to a larger weight, on stimulation, tension will rise to a peak but the muscle will be unable to shorten. This is an example of isometric contraction.

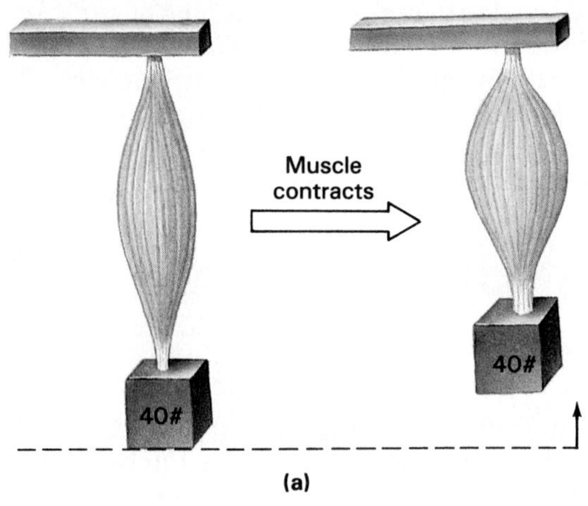

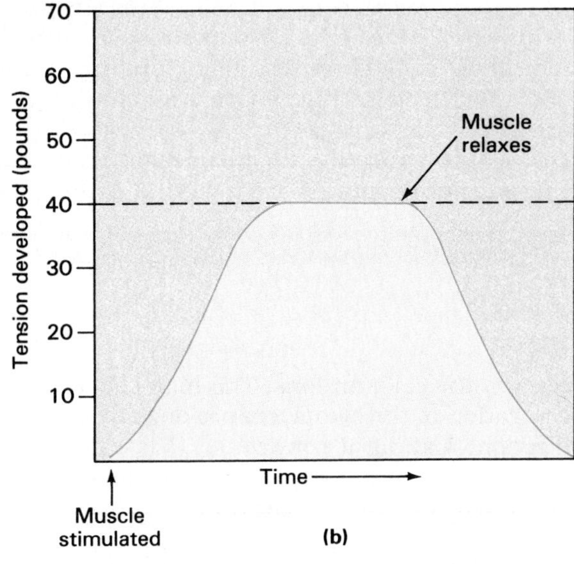

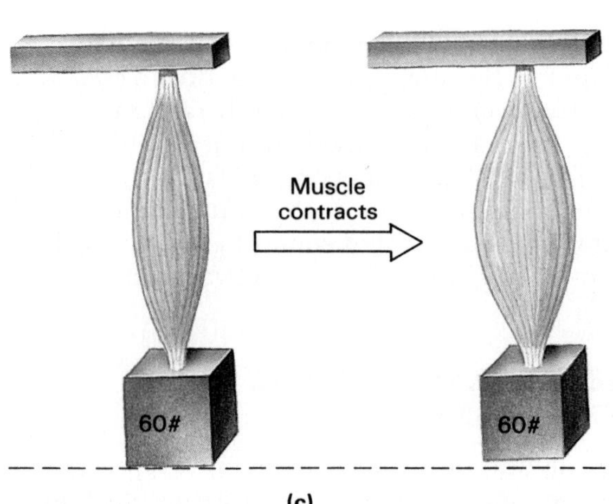

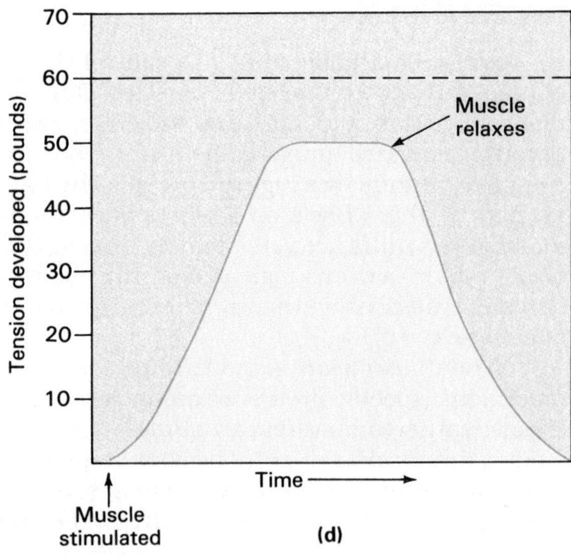

resistance (Figure 10-18b). This kind of contraction is called **isotonic** (*iso-*, equal + *tonos*, tension).

Now suppose we attach a 60-pound weight to the muscle and stimulate it, as in Figure 10-18c. This time the length of the muscle cannot change. Although cross-bridges form, and tension rises to peak values, the muscle cannot overcome the resistance of the weight (Figure 10-18d). This kind of contraction, in which tension rises but the length of the muscle remains constant, is called **isometric** (*metric*, measure).

Normal daily activities involve a combination of isotonic and isometric muscular contractions. As you sit reading this text, isometric contractions of postural muscles maintain your position. When you next turn a page, the movements of your arm, hand, and fingers will be produced by isotonic contractions.

MUSCLE TONE

Some of the motor units within any particular muscle are always active, even when the entire muscle is not contracting. Their contractions do not produce enough tension to cause movement, but they do tense and firm the muscle. This resting tension in a skeletal muscle is called **muscle tone**. A muscle with little muscle tone appears limp and flaccid, whereas one with moderate muscle tone is quite firm and solid. The identity of the stimulated motor units changes constantly, so that a constant tension in the attached tendon is maintained but individual muscle fibers can relax.

Resting muscle tone stabilizes the position of bones and joints. For example, in muscles involved with balance and posture, enough motor units are stimulated to produce the isometric tension needed to maintain body position. Muscle tone also helps prevent sudden, uncontrolled changes in the position of bones and joints. In addition to bracing, the elastic nature of muscles and tendons lets skeletal muscles act as shock absorbers that cushion the impact of a sudden bump or shock.

Specialized muscle cells, called **muscle spindles**, are monitored by sensory nerves that control the muscle tone in the surrounding muscle tissue. Reflexes triggered by activity in these sensory nerves play an important role in the reflex control of position and posture. The activity and sensitivity of muscle spindles increase with exercise training, and the background muscle tone that they create accelerates the recruitment process during a voluntary contraction. The stimulated muscle fibers also enlarge and become more powerful. This process, called *hypertrophy*, is discussed more fully later in this chapter.

At the other extreme, a skeletal muscle that is not stimulated by a motor neuron on a regular basis will lose muscle tone and mass. The muscle becomes limp, or **flaccid**, and the muscle fibers become smaller and weaker. This reduction in muscle size, tone, and power is called **atrophy**. Individuals paralyzed by spinal injuries or other damage to the nervous system will gradually lose muscle tone and volume in the areas affected. Even a temporary reduction in muscle use can lead to muscular atrophy; this may easily be seen by comparing limb muscles before and after a cast has been worn. Muscle atrophy is initially reversible, but dying muscle fibers are not replaced, and in extreme atrophy the functional losses are permanent. That is why physical therapy is so important in cases where patients are temporarily unable to move normally. [**CM**: *Polio*]

- √ **Why is it difficult to contract a muscle that has been overstretched?**
- √ **A motor unit from a skeletal muscle contains 1500 muscle fibers. Would this muscle be involved in fine, delicate movements or powerful, gross movements? Explain.**
- √ **Is it possible for a muscle to contract without shortening? Explain.**

■ Energetics of Muscular Activity

Muscle contraction requires large amounts of energy. A single muscle fiber may contain 15 billion thick filaments, each composed of 400 myosin molecules. When actively contracting each myosin cross-bridge breaks down around 100 ATP molecules per second. In other words, an active skeletal muscle fiber may require some 600 trillion molecules of ATP each second, and this does not include the energy needed to pump the calcium ions back into the sarcoplasmic reticulum. Moreover, even a small muscle is likely to consist of many thousands of fibers.

Resting skeletal muscle cells contain large energy reserves. Those reserves are in the form of (1) ATP, (2) other high-energy compounds, especially *creatine phosphate*, and (3) glycogen. Table 10-1 details the energy reserves of a representative skeletal muscle fiber. (Before proceeding, you may wish to review the material on catabolism and energy production in Chapter 4.) (p. 102)

ATP RESERVES

A resting muscle generates ATP through aerobic respiration performed by its mitochondria. The muscle fiber absorbs glucose and fatty acids from the bloodstream. These compounds are broken down by way of glycolysis, the tricarboxylic acid cycle, and the elec-

TABLE 10-1 Sources of Energy Stored in a Fast-Twitch Muscle Fiber

Energy Stored as	*Utilized through*	*Initial Quantity*	*Number of Twitches Supported by Each Energy Source Alone*	*Duration of Isometric Tetanic Contraction Supported by Each Energy Source Alone (in seconds)*
ATP	ATP → ADP + P	3 mM	10	2
CP	CP + ADP → ATP + C	20 mM	70	15
Glycogen	Glycolysis (anaerobic)	100 mM	670	130
	Aerobic respiration		12,000	2400 (40 min)

tron transport system, as described in Chapter 4, with a high energy yield in the form of ATP. (pp. 107–108)

CREATINE PHOSPHATE

At rest the cell produces more ATP than can be stored among the myofilaments. The excess energy is transferred to another high-energy compound, **creatine phosphate** (**CP**). This reaction can be summarized as:

ATP + creatine → ADP + creatine phosphate

During a contraction each cross-bridge breaks down ATP, producing ADP and a phosphate group. Creatine phosphate then releases its stored energy and converts ADP to ATP, through the reverse reaction:

ADP + creatine phosphate → ATP + creatine

The enzyme that facilitates this reaction is called **creatine phosphokinase** (**CPK** or **CK**). When muscle cells are damaged, CPK leaks across the cell membranes and into the circulation. Thus a high blood concentration of CPK usually indicates serious muscle damage.

At rest a skeletal muscle cell contains about six times as much creatine phosphate as ATP. But when a muscle fiber is contracting repeatedly, it takes only around 30 seconds to exhaust these energy reserves, and the cell must rely on other mechanisms to convert ADP to ATP. At such times the cell calls upon the energy contained in stored glycogen molecules.

GLYCOGEN AND PRODUCTION OF ATP

Glycogen, diagrammed in Figure 2-15a, is a polysaccharide chain of glucose molecules. Typical skeletal muscle fibers contain large glycogen reserves that may account for 1.5 percent of their total weight. These reserves form insoluble granules scattered between the myofibrils. When the muscle fiber begins to run short of ATP and CP, enzymes break the glycogen molecules apart, releasing glucose, which can be catabolized to provide ATP.

Two mechanisms are used to generate ATP from glucose. *Aerobic respiration*, introduced in Chapter 4 and briefly summarized here, takes place when the supply of oxygen is adequate for the energy needs of working muscles. (p. 103) *Anaerobic glycolysis*, introduced below, is resorted to when the energy demand cannot be met by aerobic respiration.

AEROBIC RESPIRATION

Figure 10-19 compares the metabolism of a resting skeletal muscle fiber with one working at minimal and one at peak levels. As we have said, resting skeletal muscle (Figure 10-19a) produces energy faster than it consumes it. It can thus afford to use part of the energy provided by its mitochondria to build ATP, CP, and glycogen reserves.

Even at low levels of activity, aerobic respiration can still provide most of the ATP needed to support muscle contractions. Glycogen molecules are split apart to produce glucose, which in turn is catabolized through the glycolysis pathway, and the mitochondria increase their rates of energy production via the TCA cycle and the electron transport system. In this reaction sequence, summarized in Figure 10-19b, glucose molecules are broken down in the sarcoplasm to produce a pair of pyruvic acid molecules and two molecules of ATP. In the process the coenzyme nicotinamide adenine dinucleotide (NAD) is converted to $NADH_2$. Mitochondria then break down the pyruvic acid molecules, yielding carbon dioxide, water, and 34 ATP molecules, and also reconvert the $NADH_2$ to NAD.

These mitochondrial activities are relatively efficient, but their rate of ATP generation is limited by the availability of oxygen. A sufficient supply of oxygen becomes a problem as the energy demands of the muscle fiber increase. Although oxygen consumption and energy production by mitochondria can increase

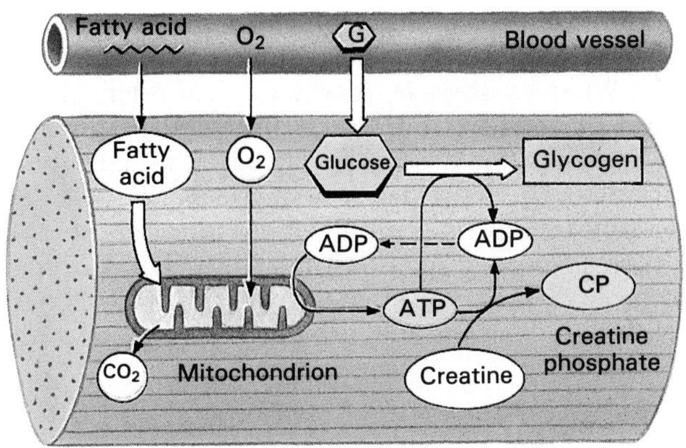

(a) Resting muscle: Fatty acids are catabolized; the ATP produced is used to build energy reserves of ATP, CP, and glycogen.

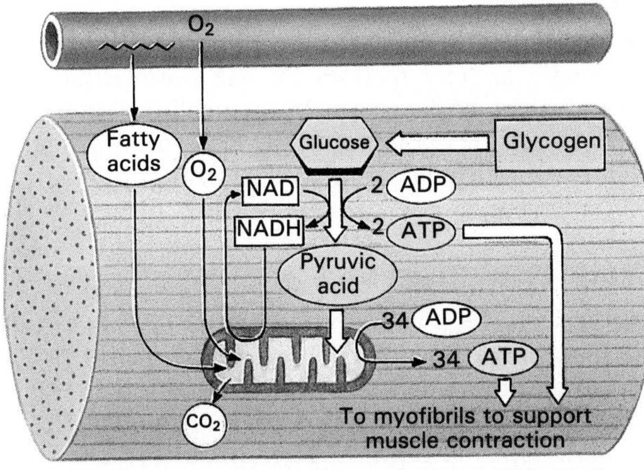

(b) Moderate activity: Glucose and fatty acids are catabolized; the ATP produced is used to power contraction.

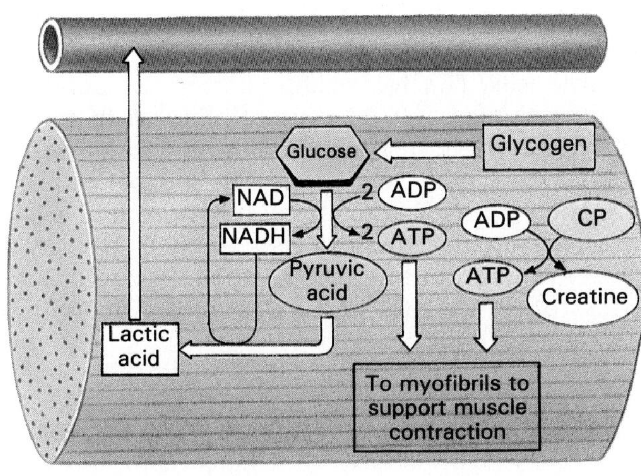

(c) Peak activity: Most ATP is produced through anaerobic glycolysis, with lactic acid as a by product.

FIGURE 10-19
Muscle Metabolism. (a) A resting muscle breaks down fatty acids via aerobic respiration to make ATP. Surplus ATP is used to build energy reserves of creatine phosphate and glycogen. (b) At modest levels of activity, mitochondria can still meet the energy demands of contraction by aerobic respiration, using fatty acids and glucose derived from glycogen as substrates. (c) At peak levels of activity the demand for ATP is very great. Mitochondria cannot get enough oxygen to meet the needs of the muscles through aerobic respiration. Most of the ATP is provided by anaerobic glycolysis, which produces lactic acid.

to 40 times resting levels, the energy demands of the muscle fiber may increase by 120 times. Thus at peak levels of exertion, mitochondrial activity provides only around one-third of the required ATP. The rest is produced through *anaerobic glycolysis* (*an*, not + *aer*, air + *bios*, life).

ANAEROBIC GLYCOLYSIS

Glycolysis was defined in Chapter 4 as the breakdown of glucose to pyruvic acid. (p. 102) This reaction sequence, summarized in Figure 10-19c, provides a net gain of 2 ATP and generates a molecule of $NADH_2$. In aerobic respiration the pyruvic acid is further catabolized in the TCA cycle, and mitochondria reconvert the $NADH_2$ to NAD. This last step is important, because NAD is a complex molecule needed in glycolysis, and the body cannot afford to continually make a fresh supply of it. Even without mitochondria, a cell could theoretically continue to perform glycolysis, producing pyruvic acid and generating 2 ATP per glucose molecule, if it had another way to convert $NADH_2$ to NAD. Fortunately, it does have another way.

In **anaerobic glycolysis**, glucose is broken down to pyruvic acid and the pyruvic acid is then converted to **lactic acid**, a related three-carbon molecule. In the course of this reaction, which requires the enzyme **lactate dehydrogenase**, or **LDH**, $NADH_2$ is converted back to NAD. Anaerobic glycolysis thus enables the cell to continue generating ATP when mitochondria cannot get enough oxygen for aerobic respiration. This pathway, however, has its drawbacks:

1. Lactic acid is an organic acid that can lower the intracellular pH. Buffers in the cytosol can resist pH shifts, but their defenses are limited. Eventually changes in pH will alter the functional characteristics of key enzymes.
2. Anaerobic glycolysis is relatively inefficient. Under anaerobic conditions *18* molecules of glucose must be converted to lactic acid molecules to obtain the same amount of energy produced by the aerobic catabolism of a single glucose molecule.

MUSCLE FATIGUE

A skeletal muscle fiber is said to be **fatigued** when it can no longer contract despite continued neural stimulation. Muscle fatigue may be caused by the exhaustion of energy reserves or the buildup of lactic acid. Muscle fatigue is cumulative, and the effects become more pronounced as more muscle fibers are affected. The result is a gradual reduction in the capabilities of the entire skeletal muscle.

If the muscle contractions use ATP at or below the maximum rate of mitochondrial ATP generation, the muscle fiber can function aerobically. Under these conditions fatigue will not occur until glycogen, lipid, and amino acid reserves are depleted. This type of fatigue affects the muscles of long-distance athletes, such as marathon runners, after hours of exertion.

When a muscle produces a sudden, intense burst of activity, the ATP is provided by anaerobic glycolysis. After a relatively short time (seconds to minutes), the rising lactic acid levels lower the tissue pH, and the muscle can no longer function normally. Athletes running sprints, such as the 100-yard dash, suffer from this type of muscle fatigue. We will return to the topics of fatigue, athletic training, and metabolic activity later in the chapter.

Sports and Fitness: Premature Muscle Fatigue and Cramps

Normal muscle function requires (1) substantial intracellular energy reserves, (2) a normal circulatory supply, and (3) a normal blood oxygen concentration. Anything that interferes with one or more of these factors will promote premature muscle fatigue. For example, reduced blood flow from tight clothing, a circulatory disorder, or loss of blood slows the delivery of oxygen and nutrients and accelerates the buildup of lactic acid. These factors combine to promote fatigue. Fatigue also accelerates in muscles that have low energy reserves because of starvation, illness, or metabolic disturbances, and in muscles that receive inadequate oxygen because of respiratory disorders such as pneumonia.

An extended and painful contraction, or **cramp**, may develop following severe exertion. During a cramp, the membrane of the muscle fiber conducts action potentials at abnormally high frequencies in the absence of neural stimulation. The problem appears to be caused by changes in the membrane's permeability. Such changes may result from alterations in ion concentrations of the tissue fluids. For example, profuse sweating can produce dehydration and a loss of sodium ions, and these events can cause muscle cramps. Local changes in pH can have the same effect. The role of lactic acid in production of cramps remains uncertain. Cramps may also occur independent of physical exertion, as a result of fatigue, chemical or physical damage to the sarcolemma, irritation of motor neurons, or various drugs.

THE RECOVERY PERIOD

The **recovery period** begins immediately after a period of muscle activity, and it lasts until the conditions inside the muscle have returned to normal, pre-exertion levels. It may take several hours for muscle fibers to recover from a brief period of intense activity. After severe maximal exertion, complete recovery may take a week. During the recovery period the muscle's metabolic activity focuses on the removal of lactic acid and the replacement of intracellular energy reserves, and the body as a whole loses the heat generated during intense muscular contraction.

Lactic Acid Recycling

The reaction that converts pyruvic acid to lactic acid is freely reversible. When pyruvic acid and $NADH_2$ concentrations rise, LDH generates lactic acid (Figure 10-20a). During the recovery period, when lactic acid concentration is high, LDH converts the lactic acid back to pyruvic acid. This pyruvic acid can then be used by mitochondria in the TCA cycle. The ATP produced is used to convert creatine to creatine phosphate and to rebuild muscle glycogen reserves. After conversion to pyruvic acid, lactic acid can also be used to synthesize glucose, which supplements the glucose absorbed from the blood (Figure 10–20b).

Following a period of exertion, blood lactate levels continue to rise as lactate ions diffuse out of the muscle cells. The liver readily absorbs the lactate ions and catabolizes 20–30 percent of them in the TCA cycle, thereby providing enough ATP to convert the rest of the lactate to glucose. The glucose molecules are then released into the circulation, and skeletal muscle cells use them to rebuild their glycogen reserves. This shuffling of lactate to the liver and glucose back to muscle cells is the **Cori cycle**, diagrammed in Figure 10-20c.

Oxygen Debt

During the recovery period, the body's oxygen demand goes up considerably. The **oxygen debt** created during exercise is the amount of oxygen used in the

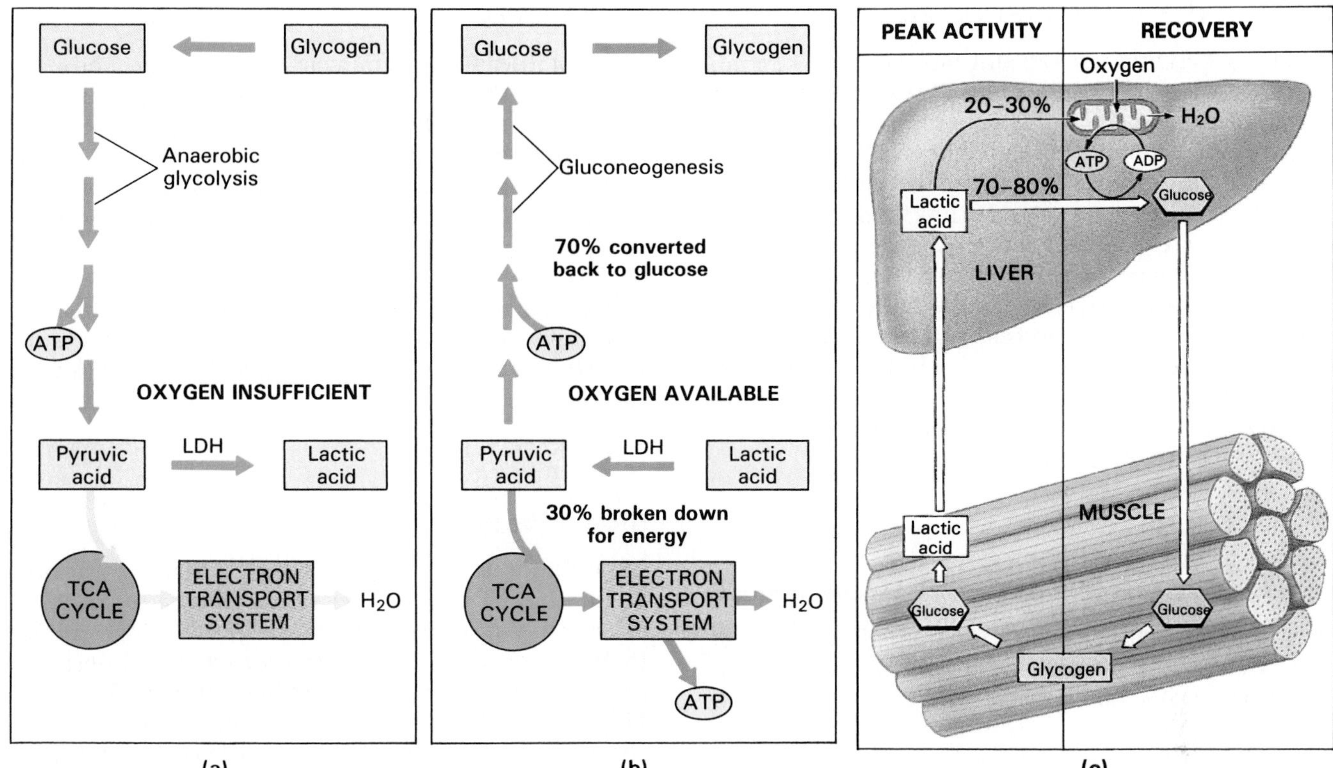

FIGURE 10-20
Lactic Acid Recycling and the Cori Cycle. (a) During strenuous activity skeletal muscles perform anaerobic glycolysis—they convert the pyruvate produced by the breakdown of glucose into lactic acid. (b) During the recovery period, when oxygen is plentiful, lactic acid is converted back to pyruvate. Some of the pyruvate is broken down aerobically, by the TCA cycle, to produce energy. The rest is converted back to glucose. (c) The work of lactic acid recycling is divided between the liver and the muscles in the Cori cycle. The liver absorbs the circulating lactic acid and produces glucose for discharge into the bloodstream. Muscle fibers then use the glucose to rebuild their glycogen reserves.

recovery period to restore normal pre-exertion conditions. Three major processes are involved: (1) oxygen for aerobic respiration is consumed by liver cells, which need to make a great deal of ATP to convert lactic acid to glucose; (2) oxygen for aerobic respiration is consumed by skeletal muscle fibers as they restore ATP, creatine phosphate, and glycogen concentrations to their former levels; and (3) the normal oxygen concentration in the blood and peripheral tissues is replenished. While the oxygen debt is being repaid, the breathing rate and depth are increased; as a result, you continue to breathe heavily long after you stop exercising.

Thermoregulatory Adjustments

Muscular activity generates substantial amounts of heat. When a catabolic reaction occurs, such as the breakdown of glycogen or the reactions of glycolysis, the muscle fiber captures only a portion of the released energy. The rest is radiated into the surrounding cytoplasm as heat. A resting muscle fiber relying on aerobic respiration captures about 42 percent of the energy released in catabolism. The other 58 percent warms the sarcoplasm, interstitial fluid, and circulating blood. This production of energy by muscles is an important mechanism for maintaining normal body temperature.

When muscles become active, their consumption of energy skyrockets. As anaerobic glycolysis becomes the primary method of ATP production, they become less efficient at capturing energy. At peak levels of exertion, only about 30 percent of the released energy is captured as ATP, and the remaining 70 percent warms the muscle and surrounding tissues. Body temperature soon begins to climb, and heat loss at the skin accelerates, through mechanisms described in Chapter 6. (p. 197)

The excess heat must be promptly carried away from the active muscle because skeletal muscles are very sensitive to high temperatures. If the temperature within a muscle rises more than a few degrees, a severe contraction, called **rigor**, will result. Even after the exercise period ends and **initial heat** production stops, energy use remains high as the oxygen debt is repaid. This produces **recovery heat** that keeps us perspiring for some time thereafter.

HORMONES AND MUSCLE METABOLISM

Metabolic activities in skeletal muscle fibers are adjusted by hormones of the endocrine system. *Growth hormone* from the pituitary gland and *testosterone* (male sex hormone) stimulate the synthesis of contractile proteins and the enlargement of skeletal muscles. *Thyroid hormones* elevate the rate of energy consumption by resting and active skeletal muscles. The increased heat production that follows thyroid stimulation is important in maintaining body temperature in cold climates. (This process will be discussed further in Chapter 25.) During a sudden crisis, hormones of the adrenal gland, notably *epinephrine* (adrenaline), stimulate muscle metabolism and increase the force of contraction.

Muscle Performance

Muscle performance can be considered in terms of sheer **power**, the maximum amount of tension produced by a particular muscle or muscle group, and **endurance**, the amount of time for which the individual can perform a particular activity. Two major factors determine the capabilities of a particular skeletal muscle: (1) the types of muscle fibers within the muscle and (2) physical conditioning or training.

TYPES OF SKELETAL MUSCLE FIBER

There are three major types of skeletal muscle fibers in the human body: *fast fibers*, *slow fibers*, and *intermediate fibers*.

Fast Fibers

Most of the skeletal muscle fibers in the body are called **fast fibers** because they can contract in 0.01 seconds or less following stimulation. Fast fibers are large in diameter; they contain densely packed myofibrils, large glycogen reserves, and relatively few mitochondria. The tension produced by a muscle fiber is directly proportional to the number of sarcomeres, so fast-fiber muscles produce powerful contractions. However, because these contractions use ATP in massive amounts, prolonged activity is primarily supported by anaerobic glycolysis, and fast fibers fatigue rapidly.

Slow Fibers

Slow fibers are only about half the diameter of fast fibers, and they take three times as long to contract after stimulation. Slow fibers are specialized to enable them to continue contracting for extended periods, long after a fast muscle would have become fatigued. The most important specializations improve mitochondrial performance. Slow muscle tissue contains a more extensive network of capillaries than is typical of fast muscle tissue, so oxygen supply is dramatically increased. In addition, slow muscle fibers contain the red pigment **myoglobin** (MĪ-ō-glō-bin). This globular protein, detailed in Figure 2-21c, is structurally related to hemoglobin, the oxygen-carrying pigment found in the blood. (p. 53) Myoglobin also binds oxygen molecules; thus resting slow muscle fibers contain substantial oxygen reserves that can be mobilized during a contraction.

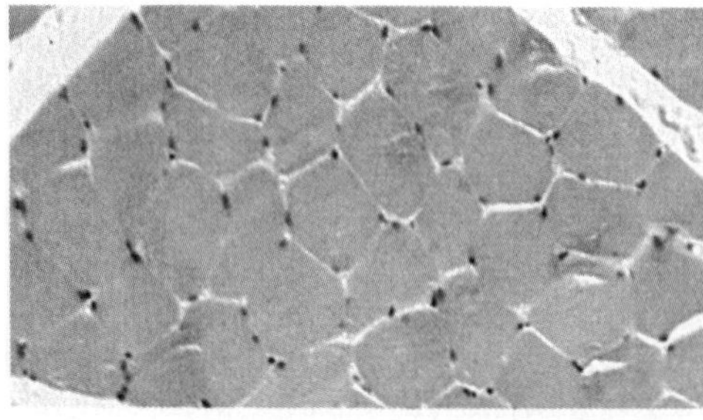

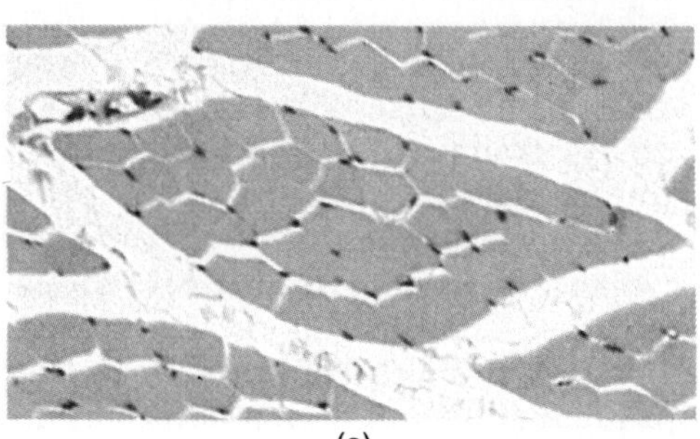

(a)

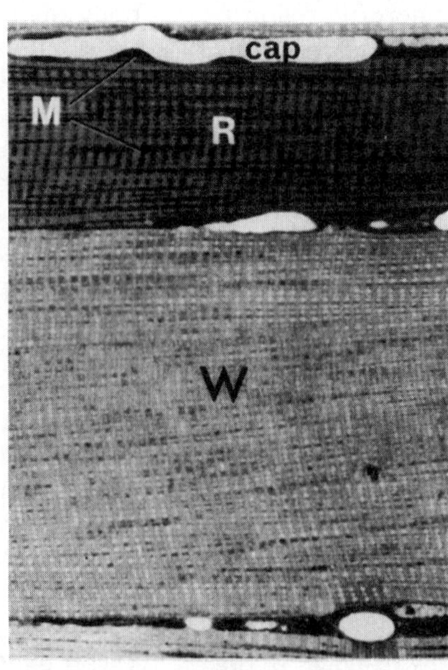

(b)

FIGURE 10-21
Fast and Slow Muscle Fibers. (a) Note the difference in the size of fast muscles, above, and slow muscles, below. (LM, x168) (b) The slender red muscle fiber (R) has more mitochondria (M) and a more extensive capillary supply (cap) than the white muscle fiber (W). (LM, x720)

TABLE 10-2 Properties of Skeletal Muscle Fiber Types

Property	*Slow*	*Intermediate*	*Fast*
Cross-sectional diameter	Small	Intermediate	Large
Tension	Low	Intermediate	High
Contraction speed	Slow	Fast	Fast
Fatigue resistance	High	Intermediate	Low
Color	Red	White	White
Myoglobin content	High	Low	Low
Capillary supply	Dense	Intermediate	Scarce
Mitochondria	Many	Intermediate	Few
Glycolytic enzyme concentration in sarcoplasm	Low	High	High
Substrates	Lipids, carbohydrates amino acids (aerobic)	Primarily carbohydrates (anaerobic)	Carbohydrates (anaerobic)
Ca^{2+} uptake by SR	Slow	Rapid	Rapid
Alternative names	Type I, S (slow), SO	Type II, FR (fast resistant)	Type II, FF (fast fatigue)

With oxygen reserves and a more efficient blood supply, the mitochondria of slow muscle fibers are able to contribute a greater amount of ATP while contractions are under way, making the cell less dependent on anaerobic metabolism than fast-fiber muscles are. Some of the mitochondrial energy production involves the breakdown of stored lipids rather than glycogen, so glycogen reserves are smaller than those of fast muscle cells. Slow muscles also contain a relatively larger number of mitochondria than do fast muscle fibers. Figure 10-21 compares the appearance of fast and slow skeletal muscle fibers.

Intermediate Fibers

Intermediate fibers have properties intermediate between those of fast fibers and slow fibers. The three types of muscle fiber are compared in Table 10-2. Histologically, intermediate fibers are very similar to fast fibers, although they have a greater resistance to fatigue.

Distribution of Muscle Fibers and Muscle Performance

The percentage of fast, intermediate, and slow muscle fibers in a particular skeletal muscle can be quite variable. Muscles dominated by fast fibers appear pale, and they are often called **white muscles**. Chicken breasts contain "white meat" because chickens use their wings only for brief intervals, as when fleeing from a predator, and the power for flight comes from fast fibers in their breast muscles. The extensive blood vessels, mitochondria, and myoglobin in slow muscle fibers give them a reddish color, and muscles dominated by slow fibers are therefore known as **red muscles**. Chickens walk around all day, and the movements are performed by the slow muscle fibers in the "dark meat" of their legs.

Most human muscles contain a mixture of fiber types, and so appear pink. However, there are no slow fibers in muscles of the eye and hand, where swift but brief contractions are required. Many back and calf muscles are dominated by slow fibers; these muscles contract almost continually to maintain an upright posture. The percentage of fast versus slow fibers in each muscle is genetically determined. The proportion of intermediate fibers changes with physical conditioning; if used repeatedly for endurance events, fast fibers can develop the appearance and functional capabilities of intermediate fibers.

Physical conditioning and training schedules enable athletes to improve both power and endurance. In practice, the training schedule varies depending on whether the activity is primarily supported by aerobic or anaerobic energy production.

ANAEROBIC ENDURANCE AND HYPERTROPHY

Anaerobic endurance is the ability to support sustained, powerful muscle contractions through anaerobic mechanisms. Examples of activities that require anaerobic endurance include a 50-yard dash or swim, a pole vault, or a weight-lifting competition. Such activities are performed by fast muscle fibers. When these muscles begin contracting at maximal levels, the energy for the first 10–15 seconds comes from the ATP and CP (creatine phosphate) reserves of the cytoplasm. As these reserves dwindle, glycogen breakdown and glycolysis provide additional energy. At peak levels of activity the ATP, CP, and glycogen reserves combined are only enough to sustain contractions for about a minute before fatigue occurs.

Athletes training to develop anaerobic endurance perform frequent, brief, intensive workouts. As a result of repeated, exhaustive stimulation, fast muscle fibers develop a larger number of mitochondria, a higher concentration of glycolytic enzymes, and larger glycogen reserves. These muscle fibers have more myofibrils, and each myofibril contains a larger number of thick and thin filaments. The net effect is an enlargement, or **hypertrophy**, of the stimulated muscle. Hypertrophy occurs in muscles that have been repeatedly stimulated to produce near-maximal tension; the intracellular changes that occur increase the amount of tension produced when these muscles contract. A champion weight lifter or body builder provides an excellent example of hypertrophied muscular development.

In general, the muscle mass of men exceeds that of women, because muscle cells are stimulated by the male sex hormone, **testosterone**. In recent years amateur and professional athletes have used this steroid hormone, or related drugs, to stimulate development of hypertrophied muscles. The steroids appear to have the desired effect only when taken by individuals already engaged in an intensive weight-training program, and who are on a high-protein, high-calorie diet. It should be noted that this combination of training and diet would *by itself* produce muscular hypertrophy, even without the use of steroids. The benefits of steroids thus do not outweigh the known risks, which include liver failure and infertility. (The use of steroids by athletes is discussed further in Chapter 18.)

AEROBIC ENDURANCE AND THE ANAEROBIC THRESHOLD

Aerobic endurance refers to the length of time for which a muscle can continue to contract while supported by mitochondrial activities. The contractions are fueled by aerobic energy production, and as carbohydrate reserves decline, the muscle cells can begin breaking down lipids and amino acids to obtain additional ATP. Because mitochondrial activities yield relatively large quantities of ATP, muscle contractions can continue for an extended period. Figure 10-22 compares the power/endurance curves for anaerobic (curve 1) and aerobic (curve 2) activities.

Initially many of the nutrients catabolized by the muscle fiber are obtained from reserves within the muscle cells themselves. Prolonged aerobic activity, however, must be supported by nutrients provided by the circulating blood. During exercise, blood vessels in the skeletal muscles dilate, and the increased blood flow brings oxygen and nutrients to the active muscle tissue. Warm-up periods are therefore important because they stimulate circulation in the muscles before the serious competition begins.

Training to improve aerobic endurance usually involves sustained low levels of muscular activity. Examples include jogging, distance swimming, and other exercises that do not require peak tension production. Improvements in aerobic endurance result from (1) altering the characteristics of muscle fibers and (2) improving the performance of the cardiovascular system.

The composition of fast and slow fibers in each muscle is genetically determined, and there are significant individual differences. These variations have an effect on aerobic endurance, for a person with more slow muscle fibers in a particular muscle will be better able to perform under aerobic conditions. However, skeletal muscle cells respond to changes in the pattern of neural stimulation. Fast muscle fibers trained for aerobic competition develop the characteristics of intermediate fibers, and this change improves aerobic endurance.

Cardiovascular activity affects muscular performance by delivering oxygen and nutrients to active muscles. Physical training alters cardiovascular function by accelerating blood flow, thus improving oxygen and nutrient availability. Factors involved in improving cardiovascular performance will be detailed in Chapter 21.

There are upper limits to the amount of oxygen that can be delivered by the circulatory system, even in trained athletes. So for maximum endurance, the energy demands of the muscles must be maintained at or below the **anaerobic threshold**, the boundary line between aerobic and anaerobic energy reliance. Energy demands above that threshold must be met

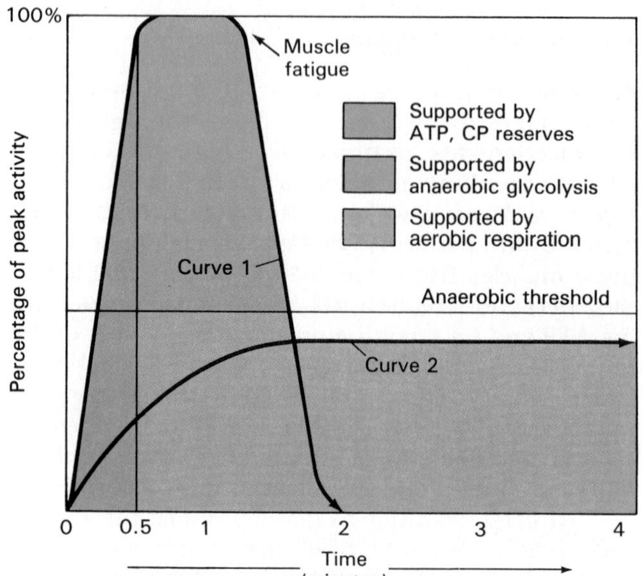

FIGURE 10-22
Muscular Performance and Endurance. Curve 1 indicates the performance of muscles operating at peak levels when they are forced to rely on anaerobic glycolysis for ATP production. Curve 2 describes muscular performance when ATP demands are kept below the anaerobic threshold. Muscular fatigue occurs after 1–2 minutes of peak anaerobic activity, but activity supported by aerobic respiration can continue for several hours.

through anaerobic glycolysis, which promotes fatigue. Many popular racquet sports, such as tennis or squash, are dominated by aerobic activities, but punctuated by brief periods with energy demands above the anaerobic threshold.

If activity demands are consistently maintained below the anaerobic threshold, fatigue will not occur until nutrient reserves are exhausted. Because glucose is a preferred energy source, aerobic athletes, such as marathon runners, often "load" or "bulk up" on carbohydrates on the day before an event.

Aging and the Muscular System

As the body ages, there is a general reduction in the size and power of all muscle tissues. The effects on the muscular system can be summarized as follows:

1. *Skeletal muscle fibers become smaller in diameter.* This reduction in size primarily reflects a decrease in the number of myofibrils. In addition, the muscle fibers contain smaller ATP, CP, and glycogen reserves, and less myoglobin. The overall effect is a reduction in muscle strength and endurance, and a tendency to fatigue rapidly. Because cardiovascular performance also decreases with age, blood flow to active muscles does not increase with exercise as rapidly as it does in younger people. These factors interact to produce decreases in anaerobic and aerobic performance of 30–50 percent at age 65.
2. *Skeletal muscles become smaller and less elastic.* Aging skeletal muscles develop increasing amounts of fibrous connective tissue, a process called **fibrosis**. Fibrosis makes the muscle less flexible, and the collagen fibers can restrict movement and circulation.
3. *Tolerance for exercise decreases.* A lower tolerance for exercise results in part from the tendency for rapid fatigue, and in part from the reduction in thermoregulatory ability described in Chapter 6. Because the elderly cannot eliminate the heat generated during muscular contraction as effectively as younger people, they are subject to overheating.
4. *Ability to recover from muscular injuries decreases.* The number of satellite cells steadily decreases with age, and the amount of fibrous tissue increases. As a result, when an injury occurs, repair capabilities are limited, and scar tissue formation is the usual result.

The rate of decline in muscular performance is the same in all individuals, regardless of their exercise patterns or lifestyle. Therefore to be in good shape late in life, one must be in *very* good shape early in life. Regular exercise helps control body weight, strengthens bones, and generally improves the quality of life at all ages. Extremely demanding exercise is not as important as regular exercise. In fact, extreme exercise in the elderly may lead to problems with tendons, bones, and joints. Although it has obvious effects on the quality of life, there is no clear evidence that exercise prolongs life expectancy. [**CM**: *Fibromyalgia and Chronic Fatigue Syndrome*]

Integration with Other Systems

To operate at maximum efficiency, the muscular system must be supported by many other systems. The changes that occur during exercise provide a good example of such interaction. As noted in earlier sections, active muscles consume oxygen and generate carbon dioxide and heat. Responses of other systems include:

1. *Cardiovascular system*: Dilation of blood vessels in the active muscles and the skin, and an increase in the heart rate. These adjustments accelerate oxygen delivery and carbon dioxide removal at the muscle and bring heat to the skin for radiation into the environment.
2. *Respiratory system*: Increased respiratory rate and depth of respiration. Air moves in and out of the lungs more quickly, keeping pace with the increased rate of blood flow through the lungs.
3. *Integumentary system*: Dilation of blood vessels and increased sweat gland secretion. This combination helps promote evaporation at the skin surface and removes the excess heat generated by muscular activity.
4. *Nervous and endocrine systems*: Direct the responses of other systems by controlling heart rate, respiratory rate, and sweat gland activity.

However, the muscular system has extensive interactions with other systems even at rest. Figure 10-23 summarizes the range of interactions between the muscular systems and other vital systems.

- √ **Why would a sprinter experience muscle fatigue before a marathon runner would?**
- √ **Which activity would be more likely to create an oxygen debt, swimming laps or lifting weights?**
- √ **What type of muscle fibers would you expect to predominate in the large leg muscles of someone who excels at endurance activities such as cycling or long-distance running?**

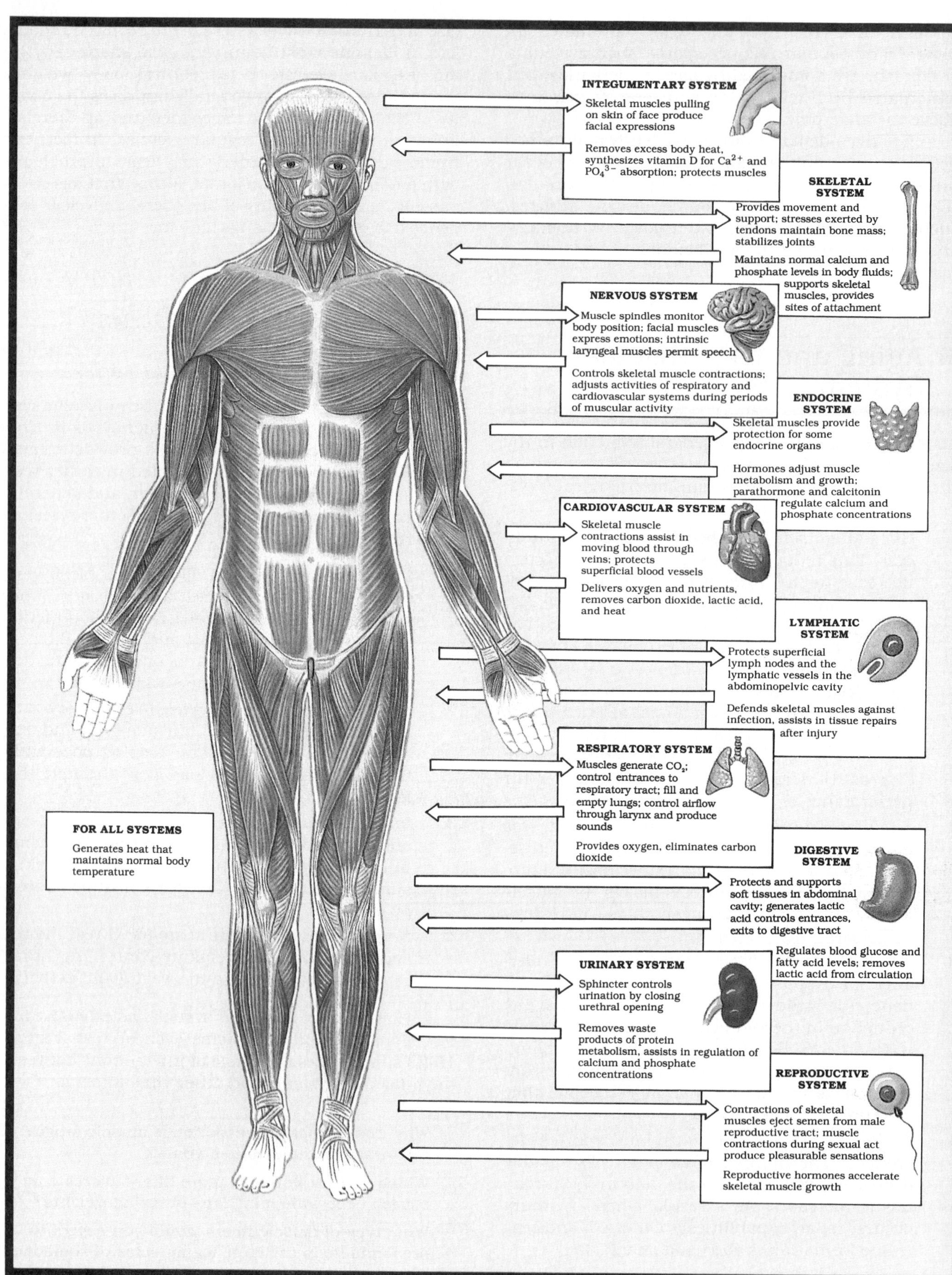

INTEGUMENTARY SYSTEM
Skeletal muscles pulling on skin of face produce facial expressions
Removes excess body heat, synthesizes vitamin D for Ca^{2+} and PO_4^{3-} absorption; protects muscles
SKELETAL SYSTEM
Provides movement and support; stresses exerted by tendons maintain bone mass; stabilizes joints
Maintains normal calcium and phosphate levels in body fluids; supports skeletal muscles, provides sites of attachment
NERVOUS SYSTEM
Muscle spindles monitor body position; facial muscles express emotions; intrinsic laryngeal muscles permit speech
Controls skeletal muscle contractions; adjusts activities of respiratory and cardiovascular systems during periods of muscular activity
ENDOCRINE SYSTEM
Skeletal muscles provide protection for some endocrine organs
Hormones adjust muscle metabolism and growth; parathormone and calcitonin regulate calcium and phosphate concentrations
CARDIOVASCULAR SYSTEM
Skeletal muscle contractions assist in moving blood through veins; protects superficial blood vessels
Delivers oxygen and nutrients, removes carbon dioxide, lactic acid, and heat
LYMPHATIC SYSTEM
Protects superficial lymph nodes and the lymphatic vessels in the abdominopelvic cavity
Defends skeletal muscles against infection, assists in tissue repairs after injury
RESPIRATORY SYSTEM
Muscles generate CO_2; control entrances to respiratory tract; fill and empty lungs; control airflow through larynx and produce sounds
Provides oxygen, eliminates carbon dioxide
FOR ALL SYSTEMS
Generates heat that maintains normal body temperature
DIGESTIVE SYSTEM
Protects and supports soft tissues in abdominal cavity; generates lactic acid controls entrances, exits to digestive tract
Regulates blood glucose and fatty acid levels; removes lactic acid from circulation
URINARY SYSTEM
Sphincter controls urination by closing urethral opening
Removes waste products of protein metabolism, assists in regulation of calcium and phosphate concentrations
REPRODUCTIVE SYSTEM
Contractions of skeletal muscles eject semen from male reproductive tract; muscle contractions during sexual act produce pleasurable sensations
Reproductive hormones accelerate skeletal muscle growth

FIGURE 10–23
Functional Relationships between the Muscular System and Other Systems

CHAPTER REVIEW

Review of Selected Clinical Terms

rigor mortis (*p. 308*)
A state following death during which muscles are locked in the contracted position, making the body extremely stiff.

hyperkalemia (hī-per-ka-LĒ-mē-a) (*p. 310*)
A high extracellular concentration of potassium.

tetanus (*p. 315*)
A disease caused by a bacterial toxin that causes sustained, powerful contractions of skeletal muscles.

cramp (*p. 320*)
An extended, painful muscular contraction.

fibrosis (*p. 325*)
A process in which increasing amounts of fibrous connective tissue develop, making muscles less flexible.

Additional Terms of Clinical Importance

TETANUS (*p. 315*) **human tetanus immune globulin**

Study Outline

Related Key Terms

Introduction (pp. 297–298)

1. There are three types of muscle tissue: **skeletal muscle**, **cardiac muscle**, and **smooth muscle**.

Functions of Skeletal Muscle (p. 298)

1. **Skeletal muscles** attach to bones directly or indirectly, and perform these functions: (1) produce skeletal movement; (2) maintain posture and body position; (3) support soft tissues; (4) guard entrances and exits; (5) maintain body temperature.

Anatomy of Skeletal Muscles (pp. 298–304)

GROSS ANATOMY (pp. 298–300)

1. Each muscle fiber is surrounded by an **epimysium**, a **perimysium**, and an **endomysium**. At the end of the muscle is a **tendon**.
2. Communication between a nerve and a muscle fiber occurs across the **neuromuscular junction**.

Related key terms: **fascicle** • **satellite cells**

MICROANATOMY OF SKELETAL MUSCLE FIBERS (pp. 300–304)

3. A muscle cell has a cell membrane, or **sarcolemma**, **sarcoplasm** (cytoplasm), and **sarcoplasmic reticulum (SR)**, similar to the endoplasmic reticulum of other cells. **Transverse tubules (T tubules)** and **myofibrils** aid in contraction. Filaments in a myofibril are organized into repeating functional units called **sarcomeres**.
4. Myofilaments consist of **thin filaments** (actin) and **thick filaments** (myosin).

Related key terms: **myoblasts** • **myofilaments**

M line • **Z line**
zone of overlap • **A band**
H band • **I band**
terminal cisternae • **triad**

Mechanism of Muscle Contraction (pp. 304–311)

THE SLIDING FILAMENT THEORY (pp. 304–307)

1. The **sliding filament theory** explains contraction, during which a muscle fiber exerts **tension** (a pull) and shortens.
2. Sliding involves a cycle of "attach, pivot, detach, and return." The five-step contraction process involves **active sites** on thin filaments and **cross-bridges** of the thick filaments. At rest the necessary interactions are prevented by two proteins, **tropomyosin** and **troponin**, on the thin filaments.

Related key terms: **ATPase**

NEURAL CONTROL OF MUSCLE FIBER CONTRACTION (pp. 307–310)

3. Neural control of muscle function involves a sequence of events called **excitation-contraction coupling**. This forms the link between electrical activity in the sarcolemma and the initiation of a contraction.
4. Each fiber is controlled by a neuron at a **neuromuscular junction**; the junction includes the **synaptic knob**, the **synaptic cleft**, and

Related Key Terms

the **motor end plate**. **Acetylcholine (ACh)** and **acetylcholinesterase (AChE)** play a role in the chemical communication between synaptic knob and muscle fiber.

5. The steps involved in contraction include: acetylcholine release; **depolarization** of the motor end plate followed by **repolarization** to a normal membrane potential; generation of an **action potential**, made possible by the **excitability** of the sarcolemma; action potential conduction; and calcium ion release.

The Twitch and Development of Tension (pp. 311–313)

STAGES OF A TWITCH (pp. 311–312)

1. A **twitch** (a single stimulus-contraction-relaxation sequence) consists of a **latent period**, a **contraction phase**, and a **relaxation phase**. — **myogram**

RELAXATION AND THE RETURN TO RESTING LENGTH (p. 312)

2. Contraction is active, but elongation of a muscle fiber is a passive process which can occur either through elastic forces or through the movement of other, opposing muscles.

LENGTH-TENSION RELATIONSHIPS (pp. 312–313)

3. The number of cross-bridge interactions within a muscle fiber determines the amount of tension produced during a contraction. The fibers can contract most forcefully when stimulated over a relatively narrow range of resting lengths.

Muscle Mechanics (pp. 313–317)

MOTOR UNITS AND MUSCLE CONTROL (pp. 313–314)

1. The number and size of a muscle's **motor units** indicate how precisely controlled its movements are.

EFFECTS OF REPEATED STIMULATIONS (pp. 314–315)

2. Repeated stimulation after the relaxation phase produces **treppe**, in which the peak tension rises in stages. Repeated stimulation before the relaxation phase ends may produce **summation of twitches (wave summation)**, in which one twitch is added to another, **incomplete tetanus** (in which tension will peak because the muscle is never allowed to relax completely), or **complete tetanus** (in which the relaxation phase is completely eliminated). Almost all normal muscular contractions involve the complete tetanus of motor units in the participating muscles. — **recruitment**; **multiple motor unit summation**

ISOMETRIC AND ISOTONIC CONTRACTIONS (pp. 315–317)

3. Normal activities usually include both **isotonic** contractions (in which the tension in a muscle remains constant, at a value that just exceeds the applied resistance) and **isometric** contractions (in which tension rises but the length of the muscle remains constant). — **tension** • **resistance**

MUSCLE TONE (p. 317)

4. Resting **muscle tone** stabilizes bones and joints. Inadequate stimulation causes muscles to become **flaccid** and undergo **atrophy**. — **muscle spindles**

Energetics of Muscular Activity (pp. 317–322)

ATP RESERVES (pp. 317–318)

1. Muscle contractions require large amounts of energy. A resting muscle generates ATP through aerobic respiration performed by its mitochondria.

CREATINE PHOSPHATE (p. 318)

2. **Creatine phosphate (CP)** can release stored energy to convert ADP to ATP. — **creatine phosphokinase (CPK, CK)**

GLYCOGEN AND PRODUCTION OF ATP (p. 318)

3. When a muscle fiber runs short of ATP and CP, enzymes can break down glycogen molecules to release glucose.

AEROBIC RESPIRATION (pp. 318–319)

4. Even at low levels of activity, aerobic respiration can provide most of the necessary ATP to support muscle contractions.

ANAEROBIC GLYCOLYSIS (p. 319)

5. **Anaerobic glycolysis** enables the cell to continue generating ATP when mitochondria cannot obtain enough oxygen for aerobic respiration. — **lactic acid**; **lactate dehydrogenase (LDH)**

MUSCLE FATIGUE (p. 320)

6. A **fatigued** muscle can no longer contract.

THE RECOVERY PERIOD (pp. 320–321)

7. The **recovery period** begins immediately after a period of muscle activity, and continues until conditions inside the muscle have returned to pre-exertion levels. The **oxygen debt** created during exercise is the amount of oxygen used in the **recovery period** to restore normal conditions. Excess heat can produce a severe contraction called **rigor**.

Related Key Terms

Cori cycle • **initial heat** • **recovery heat**

HORMONES AND MUSCLE METABOLISM (p. 322)

8. Hormones of the endocrine system adjust metabolic activities in skeletal muscle fibers.

Muscle Performance (pp. 322–325)

1. Muscle performance can be considered in terms of **power** (the maximum amount of tension produced by a particular muscle or muscle group) and **endurance** (the duration of muscular activity).

TYPES OF SKELETAL MUSCLE FIBER (pp. 322–323)

2. The three types of skeletal muscle fibers are **fast fibers**, **slow fibers**, and **intermediate fibers**. Muscles dominated by fast fibers are **white muscles**; those with more slow fibers are **red muscles**.

myoglobin

ANAEROBIC ENDURANCE AND HYPERTROPHY (pp. 323–324)

3. **Anaerobic endurance** is the ability to support sustained, powerful muscle contractions through anaerobic mechanisms. Training to develop anaerobic endurance can lead to **hypertrophy** (enlargement) of the stimulated muscles. **Testosterone** can also lead to hypertrophy.

AEROBIC ENDURANCE AND THE ANAEROBIC THRESHOLD (pp. 324–325)

4. **Aerobic endurance** is the time over which a muscle can continue to contract while supported by mitochondrial activities. For maximum endurance, energy demands should remain at or below the **anaerobic threshold**.

Aging and the Muscular System (p. 325)

1. The aging process reduces the size, elasticity, and power of all muscle tissues. Exercise tolerance and the ability to recover from muscular injuries both decrease.

fibrosis

Integration with Other Systems (pp. 325–326)

1. To operate at maximum efficiency, the muscular system must be supported by many other systems. Even at rest, it interacts extensively with other systems.

Review Planner

	Objective	Level 1
1	Describe the characteristics and functions of muscle tissue.	6 13
2	Discuss the organization of muscle at the tissue level.	3 12 13 20 27 30 33 34 43
3	Identify the unique characteristics of skeletal muscle fibers.	1 2 12 20 30 34 44 45 48
4	Explain the process of muscular contraction and the mechanisms that control it.	9 10 11 13 14 15 23 24 25 32 33 36 42 47 50

Level	Questions
Level 2	39 40 50 52 53
Level 3	54–58

5	Show how the events of a twitch are related to the development of muscle tension.	**7 14 19 49**
6	Compare the different types of muscle contractions.	**15 21 25 28 29 40 41 47 49**
7	Describe the mechanisms by which muscles obtain and use energy to power contractions.	**5 16 31 35 46 51**
8	Relate types of muscle fibers to muscular performance.	**3 22 31 35 37**
9	Distinguish between aerobic and anaerobic endurance, and explain their implications for muscular performance.	**16 17 22 26 31 37**
10	Describe the effects of exercise and aging on muscle.	**4 17 22**

C M

Trichinosis • Botulism
Myasthenia Gravis
Duchenne's Muscular Dystrophy
Polio • Fibromyalgia and Chronic Fatigue Syndrome

Experimental Muscular Dystrophy Treatment

Review Questions

MULTIPLE CHOICE

1. The dense layer of collagen fibers that surrounds the entire muscle fiber is called the: a) tendon; b) perimysium; c) epimysium; d) fascicle.
2. The connective tissue fibers of the _______ divide a skeletal muscle into a series of compartments: a) tendon; b) perimysium; c) epimysium; d) fascicle.
3. Red muscles are dominated by: a) fast fibers; b) slow fibers; c) intermediate fibers; d) none of the above.
4. Hypertrophy occurs when a stimulated muscle becomes: a) larger; b) smaller; c) faster; d) slower.
5. An ATPase is a substance that can _______ ATP: a) break down; b) stimulate the production of; c) generate.
6. The three types of muscle tissue are skeletal, cardiac, and _______ muscle: a) connective; b) epithelial; c) smooth; d) striated.
7. A twitch consists of a latent period, a _______ phase, and a relaxation phase: a) contraction; b) flaccid; c) rigor; d) pivoting.
8. Aging skeletal muscles tend to develop: a) treppe; b) hypertrophy; c) tetanus; d) fibrosis.
9. Communication between nerves and muscle fibers occurs at: a) neuromuscular junctions; b) fascicles; c) active sites; d) cross-bridges.
10. A muscle contraction involves an interaction between myosin heads and the: a) neuromuscular junctions; b) treppe; c) transverse tubules; d) active sites.
11. Cross-bridges connect: a) the synaptic cleft to the synaptic knob; b) acetylcholine to acetylcholinesterase; c) thin filaments to thick filaments; d) the transmembrane potential to the action potential.

DIAGRAMS

12. Draw a diagram illustrating the major components of a skeletal muscle fiber. Include the sarcolemma, a myofibril, the sarcoplasmic reticulum, a transverse tubule, and a sarcomere.
13. Illustrate the key components of a neuromuscular junction. Include the neuron and muscle fiber, the synaptic knob, the motor end plate, and the synaptic cleft.

TRUE/FALSE

(Your instructor may wish to have you correct the statement if it is false.)

14. **T/F:** Muscle fibers can contract most forcefully when stimulated over a relatively narrow range of resting lengths.
15. **T/F:** The more muscle fibers a single motor unit controls, the more precise its control becomes.
16. **T/F:** Anaerobic glycolysis is a more efficient process than aerobic respiration.
17. **T/F:** Warm-up exercises before an athletic competition are important because they stimulate circulation in the muscles.
18. **T/F:** Satellite cells become more numerous in aging muscle tissue.

MATCHING QUESTIONS

(Match each term with the most appropriate statement.)

Statements:

19. A single stimulus-contraction-relaxation sequence.
20. A network of narrow tubes in the sarcoplasm that encircle sarcomeres.
21. Resting tension in a skeletal muscle.
22. Large muscle fibers that produce powerful contractions but fatigue rapidly.

Terms:

a) Sarcoplasm
b) Transverse tubules
c) Excitability
d) Sarcolemma
e) Thin filaments
f) Thick filaments
g) Depolarization
h) Repolarization
i) Action potential
j) Twitch
k) Motor unit
l) Treppe

23. The reestablishment of normal membrane potential.
24. The ability to conduct an electrical impulse.
25. All of the muscle fibers controlled by a single motor neuron.
26. The amount of oxygen used in the recovery period to restore normal pre-exertion conditions.
27. Helical arrays of myosin molecules.
28. "Stairstep" increase in tension production following repeated stimulation of a muscle.
29. Smooth, steady increase in muscular tension produced by increasing the number of active motor units.
30. The cell membrane of a muscle fiber.
31. The time period over which a muscle can continue to contract while supported by mitochondrial activities.
32. A conducted change in the transmembrane potential.
33. Chains of actin molecules on which the active sites are located.
34. The cytoplasm of a muscle fiber.
35. A high-energy compound that is an important energy reserve for muscle cells.
36. A shift in the transmembrane potential from around –85 mV toward 0 mV.
37. Muscle fibers that are specialized to continue contracting for extended periods.

m) Recruitment
n) Muscle tone
o) Creatine phosphate
p) Oxygen debt
q) Fast fibers
r) Slow fibers
s) Aerobic endurance

SHORT ANSWER/ESSAY

38. What are the three types of tissue that combine to form skeletal muscles?
39. Discuss the significance of muscle tissue.
40. During a sustained contraction motor units are activated on a rotating basis. Why is this important? What happens if all of the motor units in a muscle are activated?
41. Distinguish between incomplete and complete tetanus. Which is generally involved in normal muscular contractions?
42. Place the following events in the correct sequence:
    a) Repeated cycles of cross-bridge binding occur.
    b) An action potential appears in the sarcolemma and is conducted away in all directions.
    c) The sarcoplasmic concentration of calcium ions declines.
    d) Calcium ion concentrations approach resting levels and the troponin-tropomyosin complex covers the active sites.
    e) The action potential travels along the transverse tubules to the triads.
    f) ACh binds to receptors on the motor end plate.
    g) The sarcoplasmic reticulum releases stored calcium ions, increasing the calcium concentration at the sarcomeres.
    h) Calcium ions bind to troponin molecules.
43. Distinguish between thick and thin filaments. What are their functions?
44. Where are satellite cells found, and what is their function?
45. How are skeletal muscle fibers different from the "typical" cells discussed earlier in the book?
46. Discuss what happens during the recovery period.
47. Why don't muscle contractions occur constantly?
48. What produces the "striped" appearance of skeletal muscle fibers?
49. How does a contracted muscle fiber return to its original length?
50. Discuss the steps involved in the neural control of skeletal muscle function.
51. Describe anaerobic glycolysis. What are its benefits and its drawbacks?
52. Discuss the steps involved in the process of a muscle contraction.
53. Would you expect that an athletic teenager will be in "better shape" physically at the age of 60, than a teenager who leads a sedentary lifestyle? Explain your answer.

CRITICAL THINKING/APPLICATIONS

54. Atracurium is a drug that blocks the binding of ACh. Give an example of a site where such binding normally occurs, and predict the physical effect of this drug.
55. One Saturday you do a lot of work in your yard. Later that night you are awakened by a sudden painful contraction in a leg muscle. a) What is this contraction called? b) What may have caused this reaction?
56. Mr. C., 57 years old, has noticed that his eyelids sag, sometimes obscuring his vision. He also has occasional difficulty with chewing and swallowing his food. Explain a possible cause for these symptoms and describe why they occur.
57. Describe what happens to muscles that are not "used" on a regular basis. What can be done to offset this effect?
58. In mystery stories, the time of a murder victim's death is often estimated by the flexibility of the body. Explain how this can be a clue to the "time of death."

Ah, a perfect dive! Think about the coordination needed to execute such a skilled maneuver: all of the body's muscles must work together, each exerting just the right amount of tension to achieve the desired result.

We all perform skilled movements every day. Maybe the Bolshoi Ballet hasn't asked to feature us in their next production, and the Olympic committee has somehow failed to ask us to try out for the team, but our muscles are star performers nevertheless. Even the simplest of our actions calls on dozens of the body's 700-odd skeletal muscles to do our bidding. You're so good at managing this team that most of the time you don't even need to think about it! In this chapter we'll meet most of the major muscles of the body and find out how they contribute to our ability to move and manipulate objects in the world around us.

CHAPTER 11

The Muscular System: Organization

Chapter Objectives

After reading this chapter, you will be able to:

1. Describe the arrangement of fascicles in the various types of muscles, and explain the resulting functional differences.
2. Describe the different classes of levers and how they make muscles more efficient.
3. Predict the actions of a muscle on the basis of its origin and insertion.
4. Explain how muscles interact to produce or oppose movements.
5. Use the name of a muscle to help identify and remember its location, appearance, and function.
6. Identify the principal axial muscles of the body, together with their origins and insertions.
7. Identify the principal appendicular muscles of the body, together with their origins and insertions.
8. Compare the major muscle groups of the arms and legs and relate their differences to their functional roles.

■ Introduction

The **muscular system** includes all of the skeletal muscles that can be controlled voluntarily. Most of the muscle tissue in the body is part of this system, and approximately 700 skeletal muscles have been identified. Some are attached to bony processes, others to broad sheets of connective tissue, but all are directly or indirectly associated with the skeletal system. Rather than attempt to survey all 700 skeletal muscles, we will focus on a relatively small but representative number of muscles, about 20 percent of the total. To ease the strain of memorization, these muscles have been organized into a much smaller number of anatomical and functional groups.

The general appearance of each muscle provides clues to its primary function. Muscles involved with locomotion and posture work across joints, producing skeletal movement. Those that support soft tissue form slings or sheets between relatively stable bony elements, whereas those that guard an entrance or exit completely encircle the opening.

At the level of the individual skeletal muscle, two factors interact to determine the effects of its contraction: the anatomical arrangement of the muscle fibers and the way the muscle attaches to the skeletal system. The observed performance of muscles in the body can be understood in terms of basic mechanical laws. The analysis of biological systems in mechanical terms is the study of **biomechanics**. This chapter examines the biomechanics and gross anatomy of the muscular system.

Biomechanics and Muscle Anatomy

Although most skeletal muscle fibers contract at comparable rates and shorten to the same degree, variations in microscopic and macroscopic organization can dramatically affect the power, range, and speed of movement produced when a muscle contracts.

ORGANIZATION OF SKELETAL MUSCLE FIBERS

As you will recall from Chapter 10, the muscle fibers within a skeletal muscle form bundles called *fascicles.* (p. 298) The muscle fibers in a single fascicle are always arranged side by side, but the fascicles in the skeletal muscle can be organized in various ways. These different patterns of organization give rise to four types of muscles: *parallel muscles, convergent muscles, pennate muscles,* and *circular muscles.*

Parallel Muscles

In a **parallel muscle** the fascicles are parallel to the long axis of the muscle. The functional characteristics of a parallel muscle resemble those of an individual muscle fiber. Consider the skeletal muscle shown in Figure 11-1a. It has a firm attachment and a tendon that extends from the free tip to a movable bone of the skeleton.

A skeletal muscle cell can contract effectively until it has shortened by roughly 30 percent. Because the muscle fibers are parallel to the long axis of the muscle, when they contract together the entire muscle shortens by the same amount. For example, if the skeletal muscle is 10 cm long, the end of the tendon will move 3 cm when the muscle contracts. The tension developed by the muscle during this contraction depends on the total number of myofibrils

FIGURE 11-1
Different Arrangements of Skeletal Muscle Fibers

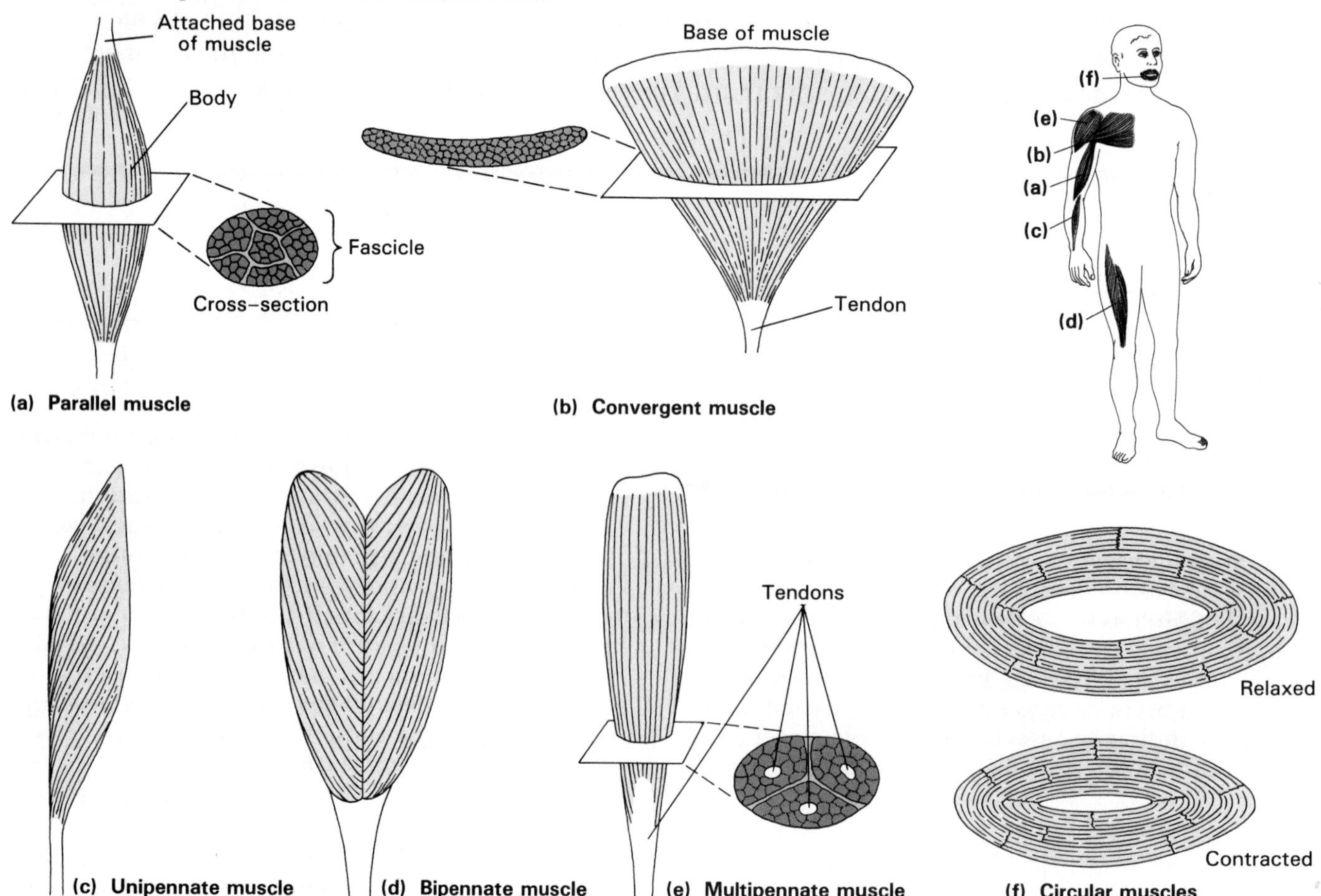

it contains. Because the myofibrils are distributed evenly through the sarcoplasm of each cell, the tension can be estimated on the basis of the cross-sectional area of the resting muscle. A parallel skeletal muscle 1 square inch in cross-sectional area can develop approximately 50 pounds of force.

Most of the skeletal muscles in the body are parallel muscles. Some are flat bands, with broad attachments at each end; others are plump and cylindrical, with tendons at one or both ends. In the latter case the muscle is spindle-shaped (Figure 11-1a), with a central **body**, also known as the *belly*, or *gaster* (GAS-ter; *gaster*, stomach). The **biceps brachii** muscle of the upper arm is an example of a parallel muscle with a central body. The bulge of the contracting biceps can be seen on the front of the upper arm when the elbow is flexed.

Convergent Muscles

In a **convergent muscle** the muscle fibers are based over a broad area, but all the fibers come together at a common attachment site. They may pull on a tendon, a tendinous sheet, or a slender band of collagen fibers known as a **raphe** (RAY-fē; seam). The muscle fibers often spread out, like a fan or a broad triangle, with a tendon at the tip (Figure 11-1b). The prominent chest muscles of the *pectoralis group* have this shape. Such a muscle has versatility, for the direction of pull can be changed by stimulating only one group of muscle cells at any one time. But when they all contract at once, they do not pull as hard on the tendon as a parallel muscle of the same size. The reason is that the muscle fibers on opposite sides of the tendon are pulling in different directions, rather than working together.

Another example of a convergent muscle is the *mylohyoid muscle* that tenses the floor of the mouth. This muscle has a median raphe that extends from the chin to the hyoid bone. The muscle fibers extend from the median raphe to the *mylohyoid line*, a landmark on the inner surface of the mandible (see Figure 8-12). The mylohyoid is illustrated in Figure 11-8.

Pennate Muscles

In a **pennate muscle** (*penna*, feather) the fascicles form a common angle with the tendon. Because the muscle cells pull at an angle, contracting pennate muscles do not move their tendons as far as parallel muscles do. But a pennate muscle will contain more muscle fibers than a parallel muscle of the same size. A muscle that has more muscle fibers also has more myofibrils and sarcomeres, and as a result the contraction of the pennate muscle generates more tension than a parallel muscle of the same size.

If all of the muscle cells are found on the same side of the tendon, the muscle is **unipennate** (Figure 11-1c). A long muscle that extends the fingers, the *extensor digitorum* (Figure 11-17), is an example of a unipennate muscle. More commonly, there are muscle fibers on both sides of the tendon. A prominent muscle of the thigh, the *rectus femoris* (Figure 11-21), is a **bipennate** muscle (Figure 11-1d) that helps to extend the knee. If the tendon branches within the muscle, the muscle is **multipennate** (Figure 11-1e). The triangular *deltoid* muscle that covers the superior surface of the shoulder joint (Figure 11-14) is an example of a multipennate muscle.

Circular Muscles

In a **circular muscle**, or **sphincter** (SFINK-ter), the fibers are concentrically arranged around an opening or recess (Figure 11-1f). When the muscle contracts the diameter of the opening decreases. Circular muscles guard entrances and exits of internal passageways such as the digestive and urinary tracts.

SKELETAL MUSCLE LENGTH-TENSION RELATIONSHIPS

As we saw in Chapter 10, skeletal muscle fibers operate most efficiently, producing maximum tension, over a relatively narrow range of sarcomere lengths. ∞ (p. 312) Stretched too far, the muscle loses power because the overlap between myofilaments is reduced. Excessive overlap also reduces the production of tension by disrupting the normal relationship between thick and thin filaments. (If this is unclear, examine Figure 10-15 before reading further.)

Normally other muscles, tendons, bones, and ligaments keep the degree of extension or compression of muscle fibers within tolerable limits. However, variations in resting sarcomere length still have an effect on the amount of tension that can be produced when the muscle is stimulated. For example, consider what happens when you try to lift a heavy bucket by flexing the elbow. This action is difficult if your forearm is extended, but it becomes easier as your elbow approaches a 90° angle. As you near full flexion, tension production declines, and movement becomes more difficult.

To see how sarcomere length affects tension, look at Figure 11-2. At full extension the tendon is almost parallel to the bones of the limb. Most of the pull exerted simply compresses the articular faces within the joint rather than producing movement. The joint does not suffer, however, because the sarcomeres are stretched and tension development is relatively low. The most effective use of force occurs when the limb is partially flexed, for when the tendon approaches a 90° angle all of the pull exerted by the muscle produces movement. If we then examined the muscle cells in the biceps brachii, the major muscle responsible for bending the elbow, we would find the sarcomeres at the optimal length for tension development. As the contraction nears its limits, tension falls once more. This decreased tension again protects

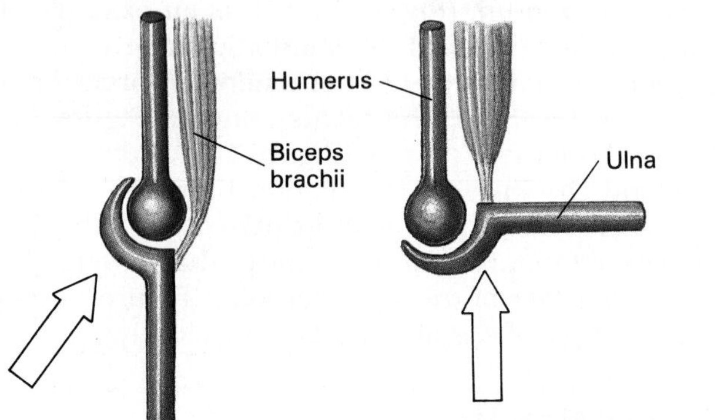

(a) Contraction at full extension produces compression at the articular surfaces, but little movement.

(b) As a right angle is reached, there is little compression on the joint, and all of the force exerted produces movement of the limb.

(c) As full flexion is approached, the direction of force tends to pull the articulation apart.

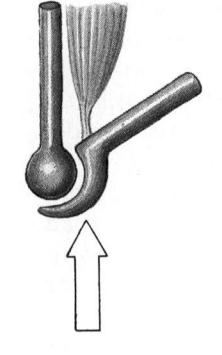

FIGURE 11-2
Movement and the Forces Acting at an Articulation. The forces acting at a simple hinge joint, such as the elbow, change as the attached muscles contract. This simplified diagram shows the direction of forces exerted by the biceps during a contraction.

the joint, for as the angle between the bony elements decreases, more and more of the force will be acting to pull the articulating elements apart.

When complex movements occur, muscles often work in groups rather than as individuals. Their cooperation helps to improve the efficiency of a particular movement. For example, large arm and leg muscles produce flexion or extension over an extended range of motion. Although these muscles cannot develop much tension at full extension, they are usually paired with smaller muscles that provide assistance until the larger muscle can perform at maximum efficiency. The sarcomeres of the smaller muscle are at optimal length at full extension, so when the two contract together, the power of the smaller muscle will be at a maximum while that of the larger muscle is at a minimum.

Figure 11-14b depicts an example: two muscles, the *latissimus dorsi* and the *teres major*, that pull the arm downward. With the arm pointed at the ceiling, the muscle fibers of the massive latissimus ("lats" to body builders) are at maximum stretch, and they are aligned parallel to the humerus. The situation resembles that in Figure 11-2a for the stretched biceps, and the latissimus muscle cannot develop much power in this position. However, the orientation of the teres major, which is based on the scapula, is closer to that of Figure 11-2b. It can therefore contract more efficiently, and its contraction assists the latissimus in starting a downward movement. The importance of this smaller "assistant" decreases as the movement proceeds.

LEVERS

Skeletal muscles do not work in isolation. When a muscle is attached to the skeleton, the nature and site of the connection will determine the power, speed, and range of the movement produced. These characteristics are interdependent, and the relationships can explain a great deal about the general organization of the muscular and skeletal systems.

The power, speed, or direction of movement produced by contraction of a muscle can be modified by attaching the muscle to a lever. A **lever** is a rigid structure—such as a board, a crowbar, or a bone—that moves on a fixed point, called the **fulcrum**. In the body, each bone is a lever and each joint a fulcrum. A child's teeter-totter, or seesaw, provides a more familiar example of lever action. Levers can change (1) the direction of an applied force, (2) the distance and speed of movement produced by a force, and (3) the strength of a force.

Classes of Levers

Three classes of levers are found in the human body. The seesaw is an example of a **first-class lever**: one in which the fulcrum lies between the applied force and the resistance (Figure 11-3a). There are not many examples of first-class levers in the body. One, involving the muscles that extend the neck, is shown in Figure 11-3a.

In a **second-class lever** (Figure 11-3b) the resistance is located between the applied force and the fulcrum. A familiar example of such a lever is a loaded wheelbarrow. The weight of the load is the resistance, and the upward lift on the handle is the applied force. Because in this arrangement the force is always farther from the fulcrum than the resistance is, a small force can balance a larger weight. In other words, the force is magnified. Notice, however, that when a force moves the handle, the resistance moves more

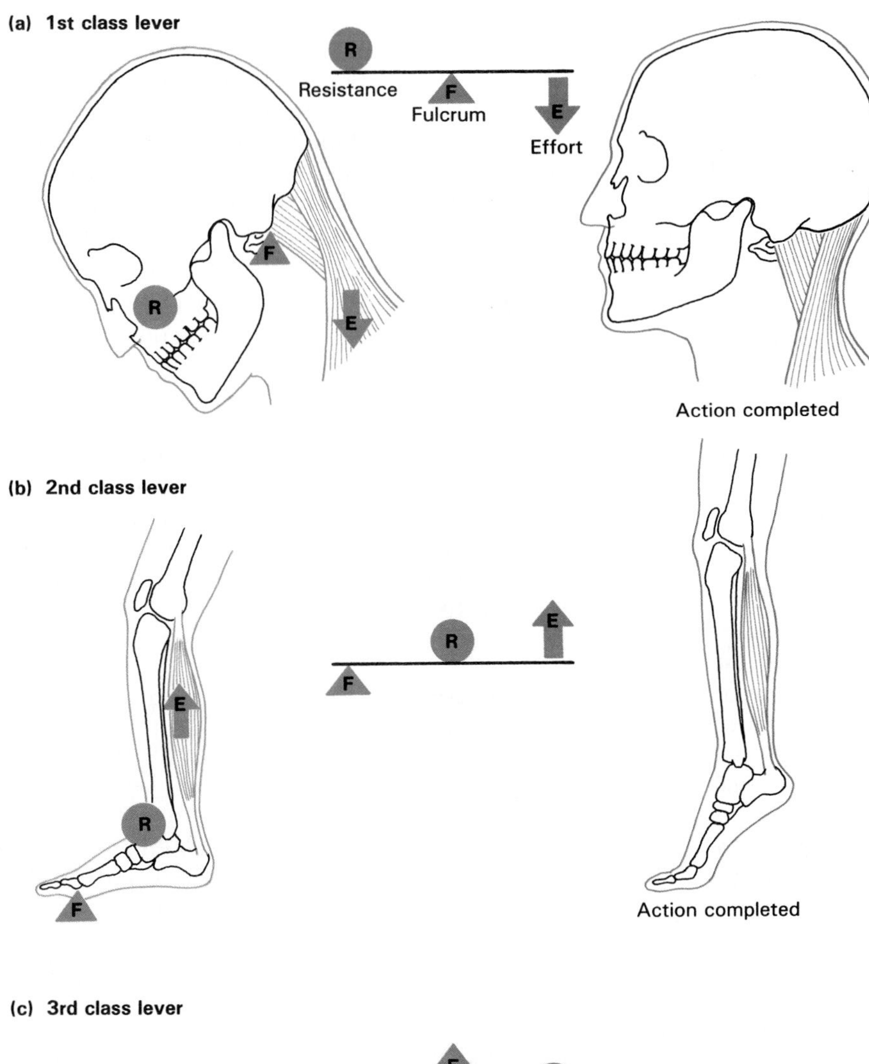

FIGURE 11-3
The Three Classes of Levers. (a) In a first-class lever the applied force and the resistance are on opposite sides of the fulcrum. First-class levers can change the amount of force transmitted to the resistance and alter the direction and speed of movement. (b) In a second-class lever the resistance lies between the applied force and the fulcrum. This arrangement magnifies force at the expense of distance and speed; the direction of movement remains unchanged. (c) In a third-class lever the force is applied between the resistance and the fulcrum. This arrangement increases speed and distance moved but requires a larger applied force. The example shows the force and distance relationships involved in the contraction of the biceps brachii muscle that flexes the forearm.

slowly and covers a shorter distance. There are few biological examples of second-class levers, but an important one is included in Figure 11-3b.

Third-class levers are the most common levers in the body. In this lever system, a force is applied between the resistance and the fulcrum (Figure 11-3c). The effect of this arrangement is just the reverse of that produced by a second-class lever: speed and distance traveled are increased at the expense of force. In the example illustrated (the biceps brachii that flexes the forearm), the resistance is six times farther away from the fulcrum than the applied force. The effective force is therefore reduced from 180 kg to 30 kg. However, the distance traveled and the speed of movement are *increased* by the same ratio: the resistance travels 45 cm while the insertion point moves only 7.5 cm.

Although every muscle does not operate as part of a lever system, the presence of levers provides speed and versatility far in excess of what we would predict on the basis of muscle physiology alone. Skeletal muscle cells resemble one another closely, and their abilities to contract and generate tension are quite similar. Consider a skeletal muscle that can contract in

500 msec and shorten 1 cm while exerting a 10–kg pull. Without using a lever, this muscle would be performing efficiently only when moving a 10–kg weight a distance of 1 cm. But by using a lever, the same muscle operating at the same efficiency could move 20 kg a distance of 0.5 cm, 5 kg a distance of 2 cm, or 1 kg a distance of 10 cm.

Muscle Terminology

This chapter focuses on the functional anatomy of skeletal muscles and muscle groups. Once again you will be faced with a number of new terms, and this section attempts to give you some assistance in understanding them.

ORIGINS, INSERTIONS, AND ACTIONS

Each muscle begins at an **origin**, ends at an **insertion**, and contracts to produce a specific **action**. In general, the origin remains stationary while the insertion moves. For example, the triceps inserts on the olecranon process and originates closer to the shoulder. Such determinations are made during normal movement. Part of the fun of studying the muscular system is that you can actually do the movements and think about the muscles involved, and laboratory discussions of the muscular system often resemble completely disorganized aerobics classes.

When the origins and insertions can not be determined easily on the basis of movement, other rules are used. If a muscle extends between a broad aponeurosis and a narrow tendon, the aponeurosis is the origin and the tendon is the insertion. If there are several tendons at one end and just one at the other, there are multiple origins and a single insertion. These simple rules cannot cover every situation, and knowing which end is the origin and which is the insertion is ultimately less important that knowing where the two ends attach and what the muscle does when it contracts.

Almost all skeletal muscles either originate or insert upon the skeleton. When a muscle moves a portion of the skeleton, that movement may involve *flexion, extension, adduction, abduction, protraction, retraction, elevation, depression, rotation, circumduction, pronation, supination, inversion,* or *eversion*. If your memory of Chapter 7 is dim, you may wish to review Figures 7-18, 7-19, and 7-20 at this time. (pp. 233-34)

Muscles can be grouped according to their **primary actions**.

- A **prime mover**, or **agonist**, is a muscle whose contraction is chiefly responsible for producing a particular movement, such as flexion of the forearm.
- When a **synergist** (*syn-*, together + *ergon*, work) contracts, it assists the prime mover in performing that action. Synergists may provide additional pull near the insertion, or stabilize the point of origin. Their importance in assisting a particular movement may change as the movement progresses; in many cases they are most useful at the start, when the prime mover is stretched and its power is relatively low.
- **Antagonists** are prime movers whose actions oppose that of the agonist under consideration. Thus agonists and antagonists are functional opposites—if one produces flexion, the other will have extension as its primary action. When an agonist contracts to produce a particular movement the coresponding antagonist will be stretched, but it will usually not relax completely. Instead, its tension will be adjusted to control the speed of the movement and ensure its smoothness.

NAMES OF SKELETAL MUSCLES

You will not need to learn every one of the nearly 700 muscles in the human body, but you will have to become familiar with the most important ones. Fortunately, anatomists of the past did not have memories any better than ours. Rather than writing the answers on their sleeves, they assigned names to the muscles that provided clues to their identification. If you can learn to recognize the clues, you will find it easier to remember the names and identify the muscles.

Some names refer to the orientation of the muscle fibers within a particular skeletal muscle. Often the terms have Greek or Latin roots. **Rectus** means "straight," and rectus muscles are parallel muscles whose fibers generally run along the long axis of the body. Because there are several rectus muscles, the name usually includes a second term that refers to a precise region of the body. For example, the *rectus abdominis* is found on the abdomen, and the *rectus femoris* on the thigh. Other directional indicators include **transversus** and **obliquus** for muscles whose fibers run across or at an oblique angle to the longitudinal axis of the body.

Table 11-1 includes a useful summary of terms that designate specific regions of the body. They are usually found as modifiers that help to identify individual muscles, as in the case of the rectus muscles noted previously. In a few cases, the muscle is such a prominent feature of the region that the regional name alone will identify it. Examples would include the *temporalis* of the head and the *brachialis* of the upper arm.

Other muscles were named after specific and unusual structural features. The *biceps* muscle has two tendons of origin (*bi-*, two + *caput*, head), the *triceps* has three, and the *quadriceps* four. Shape

TABLE 11-1 Muscle Terminology

Terms Indicating Direction Relative to the Longitudinal Axis of the Body	*Terms Indicating Actions*	*Terms Indicating Specific Regions of the Body*[a]	*Terms Indicating Structural Characteristics of the Muscle*
Lateralis = lateral	**General**	Abdominis (abdomen)	**Origin/Insertion**
Medialis/medius = medial	Abductor	Anconeus (elbow)	Biceps (two heads)
Obliquus = oblique	Adductor	Auricularis (auricle of ear)	Triceps (three heads)
Rectus = straight, parallel	Depressor	Brachialis (brachium)	Quadriceps (four heads)
Transversus = transverse	Extensor	Capitis (head)	**Shape**
	Flexor	Capri (wrist)	Deltoid (triangle)
	Levator	Cervicis (neck)	Orbicularis (circle)
	Pronator	Cleido/clavius (clavicle)	Pectinate (comblike)
	Rotator	Coccygeus (coccyx)	Piriformis (pear-shaped)
	Supinator	Costalis (ribs)	Platy- (flat)
	Tensor	Cutaneous (skin)	Pyramidal (pyramid)
	Specific	Femoris (femur)	Rhomboideus (rhomboid)
	Buccinator (trumpeter)	Genio- (chin)	Serratus (serrated)
	Risorius (laugher)	Glossal (tongue)	Splenius (bandage)
	Sartorius (like a tailor)	Hallucis (big toe)	-tendinosus (tendinous)
		Ilio- (ilium)	Trapezius (diamond)
		Inguinal (groin)	**Other Striking Features**
		Lumborum (lumbar region)	Alba (white)
		Nasalis (nose)	Anterior
		Nuchal (back of neck)	Brevis (short)
		Oculo- (eye)	Externus (superficial)
		Oris (mouth)	Extrinsic (outside)
		Palpebrae (eyelid)	Gracilis (slender)
		Pollicis (thumb)	Inferioris (inferior)
		Popliteus (behind knee)	Internus (deep, internal)
		Psoas (loin)	Intrinsic (inside)
		Radialis (radius)	Lateralis (lateral)
		Scapularis (scapula)	Latissimus (widest)
		Temporalis (temples)	Longis (long)
		Thoracis (thoracic region)	Longissimus (longest)
		Tibialis (tibia)	Magnus (large)
		Ulnaris (ulna)	Major (larger)
		Uro- (urinary)	Maximus (largest)
			Medialis (medial)
			Minimus (smallest)
			Minor (smaller)
			Posterior
			Profundus (deep)
			Superficialis (superficial)
			Superioris (superior)
			Teres (long and round)
			Vastus (great)

[a] For other regional terms, refer to Figure 1-7 on anatomical landmarks.

is sometimes an important clue to the name of a muscle. For example, the *trapezius* (tra-PĒ-zē-us), *deltoid*, *rhomboideus* (rom-BOI-dē-us), and *orbicularis* (or-bik-ū-LAR-is) refer to prominent muscles that look like a trapezoid, a triangle, a rhomboid, or a circle, respectively. Long muscles are called **longis** (long) or **longissimus** (longest), and **teres** muscles are both long and round. Short muscles are called **brevis**; large ones are called **magnus** (big), **major** (bigger), or **maximus** (biggest); and small ones are called **minor** (smaller) or **minimus** (smallest).

Muscles visible at the body surface are often called **externus** or **superficialis**, whereas those lying beneath are termed **internus** or **profundus**. Superficial muscles that position or stabilize an organ are called **extrinsic** muscles; those that operate within the organ are called **intrinsic** muscles.

Many names tell you the specific origin and insertion of each muscle. In such cases, the first part of the name indicates the origin and the second part the insertion. The *genioglossus*, for example, originates at the chin (*geneion*) and inserts in the tongue (*glossus*). Although the names may be long and occasionally difficult to pronounce, most can be figured out with the help of Table 11–1 and the anatomical terms introduced in Chapter 1. (pp. 11–16)

Names that indicate the primary function of the muscle are particularly useful. Many muscles are named *flexor*, *extensor*, *retractor*, and so on. These are such common actions that the names almost always include other clues concerning the appearance or location of the muscle. For example, the *extensor carpi radialis longus* is a long muscle found along the radial (lateral) border of the forearm. When it contracts its primary function is extension of the carpus (wrist). Most names are not this long, but they usually contain enough information that your memory isn't left without a clue.

A few muscles are named after the specific movements associated with special occupations or habits. The *sartorius* (sar-TŌ-rē-us) muscle is active when crossing the legs. Before sewing machines were invented, a tailor would sit on the floor cross-legged, and the name of the muscle was derived from *sartor*, the Latin word for "tailor." On the face, the *buccinator* (BUK-si-nā-tor) muscle compresses the cheeks, as when pursing the lips and blowing forcefully. *Buccinator* translates as "trumpet player." Finally, another facial muscle, the *risorius* (ri-SŌ-rē-us), was supposedly named after the mood expressed. However, the Latin term *risor* means "laugher," while a more appropriate description for the effect would be "grimace."

The separation of the skeletal system into axial and appendicular divisions provides a useful guideline for subdividing the muscular system as well.

- The **axial musculature** arises on the axial skeleton. It positions the head and spinal column and also moves the rib cage, assisting in the movements that make breathing possible. It does not play a role in movement or support of

TABLE 11-2 Muscles of facial expression

Region/Muscle	Origin	Insertion	Action	Innervation
Mouth				
Buccinator	Alveolar processes of maxillae and mandible	Blends into fibers of orbicularis oris	Compresses cheeks	Facial nerve (N VII)
Depressor labii	Mandible	Lower lip	Depresses lip	Same
Levator labii	Maxillae	Orbicularis oris	Raises upper lip	Same
Mentalis	Mandible	Skin of chin	Elevates and protrudes lower lip	Same
Orbicularis oris	Maxilla and mandible	Lips	Compresses, purses lips	Same
Risorius	Fascia surrounding parotid salivary gland	Angle of mouth	Draws corner of mouth to the side	Same
Zygomaticus	Zygomatic bone	Angle of mouth	Draws corner of mouth back and up	Same
Eye				
Corrugator supercilii	Frontal bone near nasal suture	Eyebrow	Pulls skin down and forward; wrinkles brow	Same
Levator palpebrae	Tendinous band around optic foramen	Upper eyelid	Raises upper lid	Oculomotor nerve (N III)[a]
Orbicularis oculi	Medial margin of orbit	Skin around eyelids	Closes eye	Facial nerve (N VII)
Nose				
Procerus	Skull	Skin, cartilages of nose	Moves nose, changes position, shape of nostrils	Same
Nasalis	Maxilla	Bridge, inferior corners and tip of nose	Compresses bridge, depresses tip, elevates corners	Same
Ear (extrinsic)				
Auricularis	Galea aponeurotica	External ear	Moves external ear	Same
Scalp				
Frontalis	Galea aponeurotica	Skin of eyebrow and bridge of nose	Raises eyebrows, wrinkles forehead	Same
Occipitalis	Superior nuchal line	Galea aponeurotica	Tenses, retracts scalp	Same
Neck				
Platysma	Upper thorax between cartilage of second rib and acromion of scapula	Mandible and skin of cheek	Tenses skin of neck, depresses mandible	Same

[a] This muscle originates in association with the extrinsic oculomotor muscles, so its innervation is unusual.

the pectoral or pelvic girdles or appendages. This category encompasses roughly 60 percent of the skeletal muscles in the body.

- The **appendicular musculature** stabilizes or moves components of the appendicular skeleton.

We will now examine representatives of these muscular divisions, paying attention to patterns of origin, insertion, and action. This discussion relies heavily on an understanding of skeletal anatomy, and you may find it helpful to review appropriate figures in Chapters 8 and 9 as you proceed. For convenience, the relevant figures in Chapters 8 and 9 are indicated in the figure captions in this chapter.

√ **What type of muscle would you expect to find guarding the opening between the stomach and the small intestine?**

√ **The joint between the occipital bone of the skull and the first cervical vertebra (atlas) is an example of what type of lever system?**

√ **What muscle would be the antagonist of the biceps brachii?**

√ **What does the name flexor carpi radialis longus tell you about this muscle?**

The Axial Musculature

The axial muscles fall into logical groups based on location, function, or both. The groups do not always have distinct anatomical boundaries. For example, a function such as the extension of the spine involves muscles along the entire length of the spinal column.

The first group includes the *muscles of the head and neck* that are not associated with the spinal column. These muscles include those that move the face, tongue, and larynx. They are therefore responsi-

Figure 11-4
Muscles of Facial Expression. *See also Figure 8-3.*

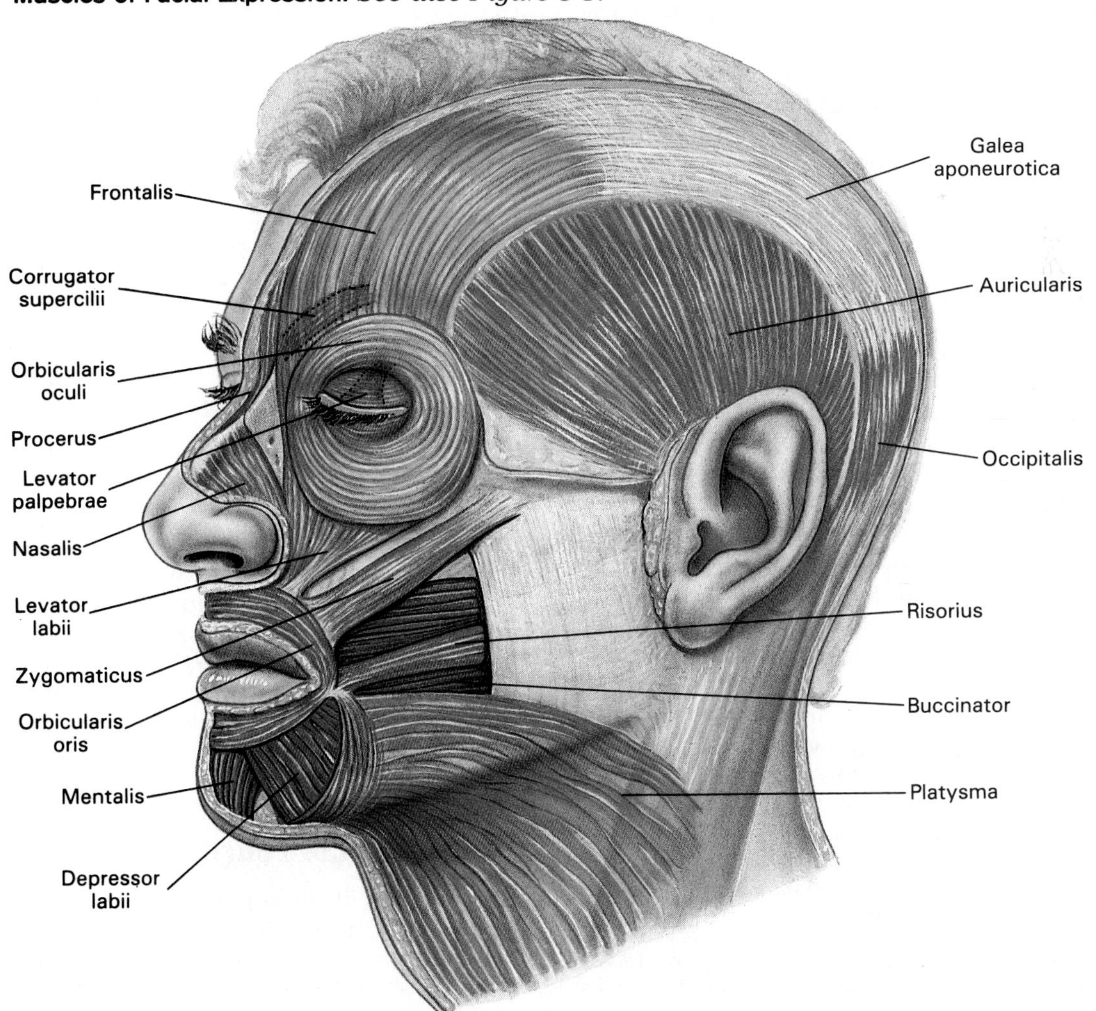

ble for verbal and nonverbal communication—laughing, talking, frowning, smiling, whistling. This group of muscles also performs feeding activities, such as sucking or chewing, as well as contractions of the eye muscles that help us look around for something to eat.

The second group, the *muscles of the spine*, includes numerous flexors and extensors of the head, neck, and spinal column.

The third group, the *oblique* and *rectus* muscles, form the muscular walls of the thoracic and abdominopelvic cavities between the first thoracic vertebra and the pelvis. In the thoracic area these muscles are partitioned by the ribs, but over the abdominal surface they form broad muscular sheets. Although they do not form a complete muscular wall, there are also oblique and rectus muscles in the neck.

The fourth group, the *muscles of the pelvic floor*, extend between the sacrum and pelvic girdle, forming the muscular *perineum* that closes the pelvic outlet.

The tables that follow summarize information concerning the origin, insertion, and action of each muscle. You will also find information concerning the innervation of the individual muscles. **Innervation** refers to the identity of the nerve that controls the muscle. The names of the nerves provide information concerning the distribution of the nerve or the site at which the nerve leaves the confines of the axial skeleton. For example, the *facial nerve* innervates the facial musculature, and the various *spinal nerves* exit via the intervertebral foramina. You will learn more about the individual nerves involved when we discuss the peripheral nervous system in Chapters 13 amd 14.

MUSCLES OF THE HEAD AND NECK

The muscles of the head and neck can be divided into several groups. The *muscles of facial expression*, the *muscles of mastication*, and the *muscles of the tongue* originate on the skull or hyoid bone. Muscles involved with sight and hearing are also based on the skull. The *extrinsic eye muscles*—those associated with movements of the eye—will be considered here. The intrinsic eye muscles, which control the diameter of the pupil and the shape of the lens, are discussed in Chapter 17, along with the muscles associated with the ear and hearing. In the neck, the *extrinsic muscles of the larynx* adjust the position of the hyoid bone and larynx. The intrinsic laryngeal muscles, including those of the vocal cords, and the pharyngeal musculature will be discussed in Chapters 23 (respiratory system) and 24 (digestive system).

Muscles of Facial Expression

The muscles of facial expression, which are shown in Figure 11-4 and detailed in Table 11-2 (page 334), originate on the surface of the skull. At their insertions the epimysial fibers are woven into those of the superficial fascia and the dermis of the skin; when they contract, the skin moves. These muscles are innervated by the seventh cranial nerve. The largest group of facial muscles is associated with the mouth. The **orbicularis oris** constricts the opening, while other muscles move the lips or the corners of the mouth. The **buccinator**, one of the muscles associated with the mouth, has two functions related to feeding (in addition to its importance to musicians). During chewing it cooperates with the masticatory muscles by moving food back across the teeth from the space inside the cheeks. In infants, the buccinator provides suction for suckling at the breast.

Smaller groups of muscles control movements of the eyebrows and eyelids, the scalp, the nose, and the external ear. The **epicranius** (ep-i-KRĀN-ē-us; *epi-*, on + *kranion*, skull) of the scalp consists of two muscles, the **frontalis** and the **occipitalis**, separated by a collagenous sheet, the **galea aponeurotica** (GĀ-lē-a ap-ō-nū-RŌT-i-ka; *galea*, a helmet, + aponeurosis). The **platysma** (pla-TIZ-ma; *platys*, flat) covers the ventral surface of the neck, extending from the base of the neck to the periosteum of the mandible and the fascia at the corners of the mouth.

Extrinsic Eye Muscles

Six **oculomotor** (ok-ū-lō-MŌ-ter) **muscles** originating on the surface of the orbit control the position of the eye. These muscles, shown in Figure 11-5 and detailed in Table 11-3, are the **inferior rectus**, **lateral rectus**, **medial rectus**, **superior rectus**, **inferior oblique**, and **superior oblique**. The oculomotor muscles are innervated by the third, fourth, and sixth cranial nerves.

Muscles of Mastication

The muscles of mastication, shown in Figure 11-6 and detailed in Table 11-4, open, close, protract, and retract the lower jaw and move it from side to side. The large **masseter** is the most powerful and important of the masticatory muscles. The **temporalis** assists in elevation of the mandible, whereas the **pterygoid** muscles used in various combinations can elevate or protract the mandible or slide it from side to side. These movements are important in maximizing the efficient use of the teeth while chewing foods of various consistencies.

Muscles of the Tongue

The muscles of the tongue have delightfully descriptive names, all ending in *glossus*, from the Greek word for "tongue." Once you can recall the structures referred to by *palato-*, *stylo-*, *genio-*, and *hyo-*, you

FIGURE 11-5
Oculomotor Muscles. *See also Figure 8-3.*

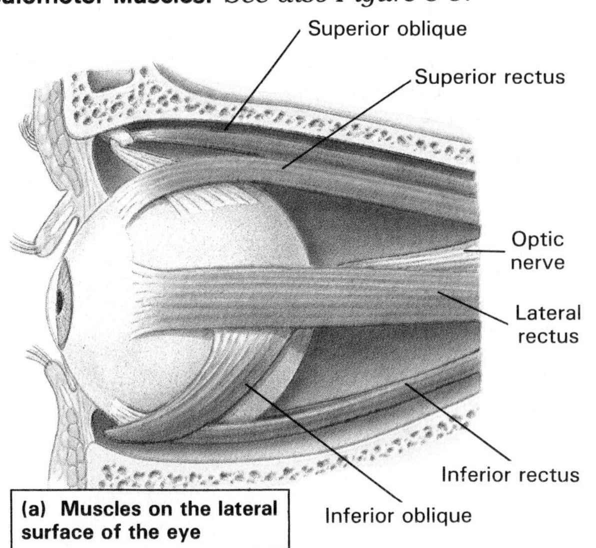

(a) Muscles on the lateral surface of the eye

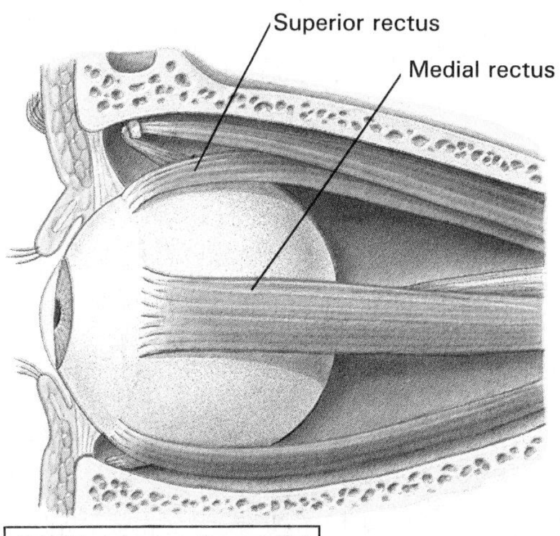

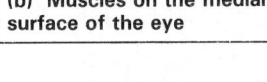

(b) Muscles on the medial surface of the eye

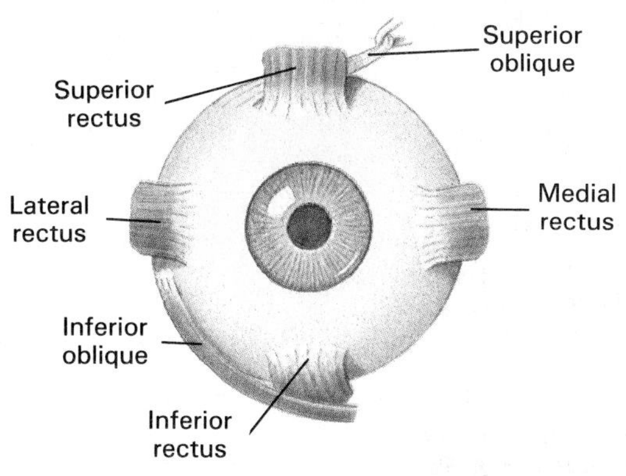

(c) An anterior view of the eye, showing the orientation of the oculomotor muscles

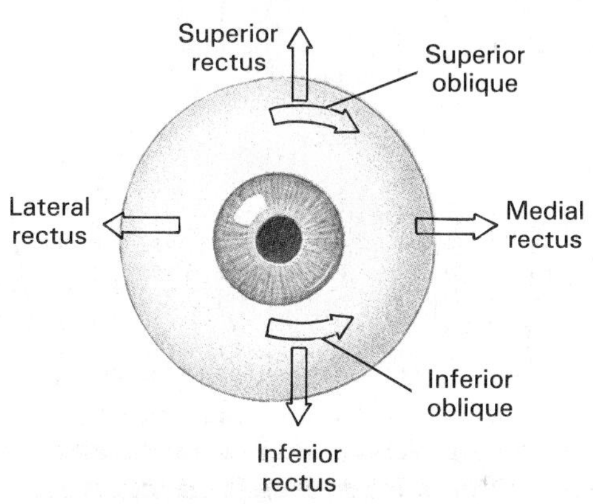

(d) An anterior view of the eye, showing the directions of eye movement produced by contractions of the individual muscles

TABLE 11-3 Extrinsic Oculomotor Muscles

Muscle	*Origin*	*Insertion*	*Action*	*Innervation*
Inferior rectus	Sphenoid around optic foramen	Inferior, medial surface of eyeball	Eye looks down	Oculomotor nerve (N III)
Lateral rectus	Same	Lateral surface of eyeball	Eye rotates laterally	Abducens nerve (N VI)
Medial rectus	Same	Medial surface of eyeball	Eye rotates medially	Oculomotor nerve (N III)
Superior rectus	Same	Superior, medial surface of eyeball	Eye looks up	Same
Inferior oblique	Maxilla at front of orbit	Inferior, lateral surface of eyeball	Eye rolls, looks up and to the side	Same
Superior oblique	Sphenoid around optic foramen	Superior, medial surface of eyeball	Eye rolls, looks down and to the side	Trochlear nerve (N IV)

FIGURE 11-6
Muscles of Mastication. *See also Figures 8-3 and 8-12.*

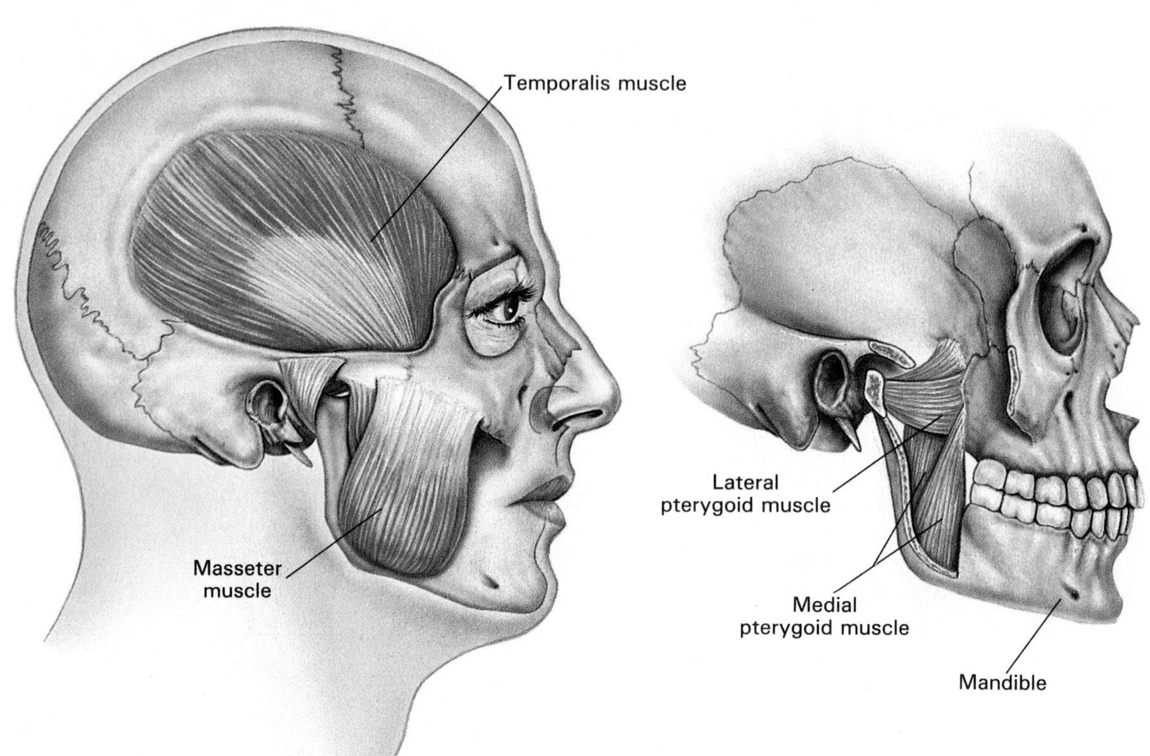

(a) The temporalis and masseter are prominent muscles on the lateral surface of the skull. The temporalis passes medial to the zygomatic arch to insert on the mandibular ramus. The masseter inserts on the angle and lateral surface of the mandible.

(b) The location and orientation of the pterygoid muscles can be seen after removing the overlying muscles, along with a portion of the mandible.

TABLE 11-4 Muscles of Mastication

Muscle	*Origin*	*Insertion*	*Action*	*Innervation*
Masseter	Zygomatic arch	Lateral surface of mandibular ramus	Elevates mandible	Trigeminal nerve (N V), mandibular branch
Temporalis	Along temporal lines of skull	Coronoid process of mandible	Elevates mandible	Same
Pterygoideus (medial and lateral)	Medial pterygoid plate and processes	Medial surface of mandibular ramus	Elevates, protracts, and/or moves mandible	Same

FIGURE 11-7
Muscles of the Tongue. *See also Figures 8-3 and 8-13.*

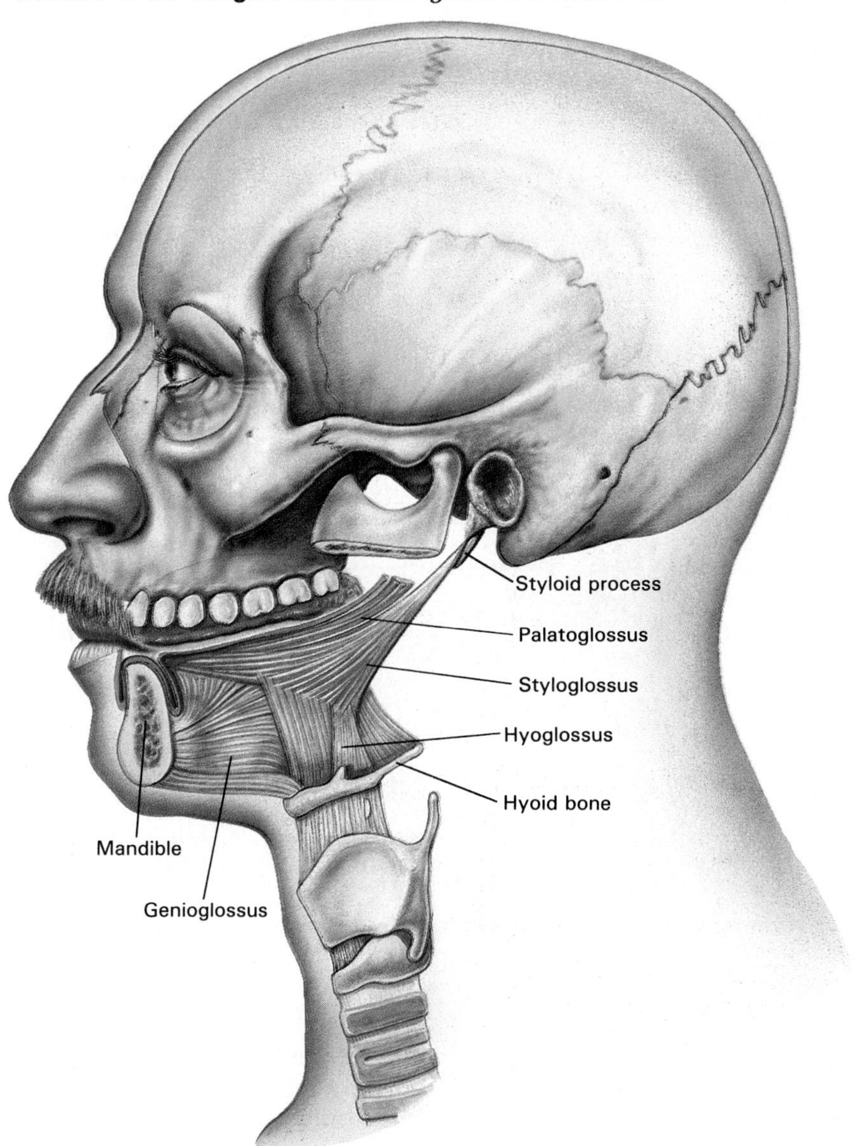

TABLE 11-5 Muscles of the Tongue

Muscle	*Origin*	*Insertion*	*Action*	*Innervation*
Genioglossus	Medial surface of mandible around chin	Body of tongue, hyoid bone	Depresses and protracts tongue	Hypoglossal nerve (N XII)
Hyoglossus	Body and greater cornu of hyoid bone	Sides of tongue	Depresses and retracts tongue	Same
Palatoglossus	Anterior surface of soft palate	Same	Elevates tongue, depresses fleshy palate	Same
Styloglossus	Styloid process of temporal bone	Via sides to the tip and base of tongue	Retracts tongue, elevates sides	Same

■ TABLE 11-6 **Extrinsic Muscles of the Larynx**

Muscle	*Origin*	*Insertion*	*Action*	*Innervation*
Digastricus	Hyoid bone	Two bellies: *posterior* to mastoid region of temporal; *anterior* to inferior surface of mandible at chin	Elevates larynx and/or depresses mandible	Hypoglossal nerve (N XII) to posterior belly; Trigeminal nerve (N V), mandibular branch, to anterior belly
Geniohyoid	Medial surface of mandible at chin	Hyoid bone	Same	Hypoglossal nerve (N XII)
Mylohyoid	Mylohyoid line of mandible	Medial connective tissue band (raphe) that runs to hyoid	Elevates floor of mouth and/or depresses mandible	Trigeminal nerve (N V), mandibular branch
Omohyoid	Central tendon attaches to clavicle and 1st rib	Two bellies: *anterior* attaches to hyoid bone; *posterior* to superior margin of scapula	Depresses larynx	Cervical spinal nerves
Sternohyoid	Clavicle	Hyoid	Same	Same
Sternothyroid	Dorsal surface of manubrium and 1st rib	Thyroid cartilage of larynx	Same	Same
Stylohyoid	Styloid process of temporal bone	Hyoid	Elevates larynx	Facial nerve (N VII)
Thyrohyoid	Thyroid cartilage of larynx	Hyoid	Elevates thyroid, depresses hyoid	Hypoglossal nerve (N XII)

shouldn't have much trouble with this group. The **palatoglossus** originates at the palate, the **styloglossus** at the styloid process, the **genioglossus** at the chin, and the **hyoglossus** at the hyoid bone (Figure 11-7). These muscles, used in various combinations, move the tongue in the delicate and complex patterns necessary for speech, and manipulate food within the mouth in preparation for swallowing. As indicated in Table 11-5, they are innervated by the *hypoglossal nerve*, a cranial nerve whose name indicates its function as well as its location.

Extrinsic Muscles of the Larynx

The extrinsic laryngeal muscles (Figure 11-8 and Table 11-6) control the position of the larynx, depress the mandible, tense the floor of the mouth, and provide a stable foundation for muscles of the tongue and pharynx. The **digastricus** has two bellies, as the name implies (*di-*, two + *gaster*, stomach). One belly extends from the chin to the hyoid, and the other continues from the hyoid to the mastoid portion of the temporal bone. This muscle opens the mouth by depressing the mandible. It overlies the broad, flat **mylohyoid**, which provides a muscular floor to the mouth. The **stylohyoid** forms a muscular connection between the hyoid apparatus and the styloid process of the skull. The other members of this group are straplike muscles that run between the sternum and the chin.

MUSCLES OF THE SPINE

The superficial muscles of the spine can be subdivided into **spinalis**, **longissimus** and **iliocostalis** divisions, as shown in Figure 11-9 on page 343. In the lower lumbar and sacral regions the distinction between the longissimus and iliocostalis muscles becomes indistinct, and they are sometimes known as the **sacrospinalis** muscles. When contracting together, these muscles extend the spinal column. When only

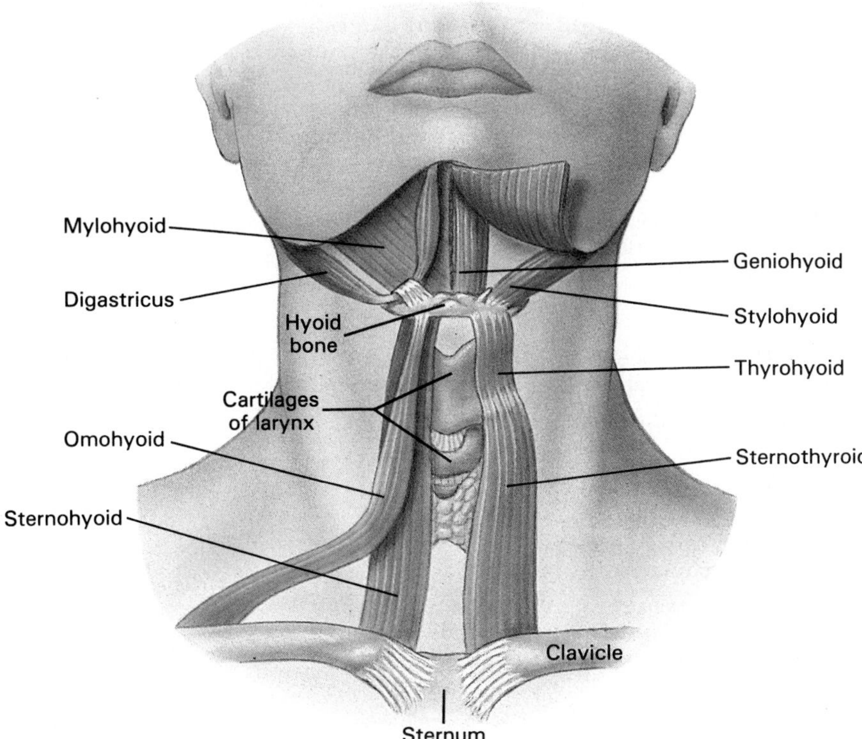

FIGURE 11-8
Extrinsic Muscles of the Larynx. *See also Figures 8-3 and 8-13.*

the muscles on one side contract, the spine is bent laterally.

Beneath the spinalis muscles, the **transversus muscles** interconnect and stabilize the vertebrae. These muscles are important in making fine adjustments in vertebral position, producing slight extension or rotation of the spinal column.

Among the muscles that move the spine there are many dorsal extensors but only a few ventral flexors. The spinal column does not need a massive series of attached flexor muscles because many of the large trunk muscles, although not directly associated with the spine, nevertheless flex the spine when they contract. As a result, relatively few muscles are found associated with the anterior surface of the spinal column. In the neck, the **longus capitis** and the **longus cervicis** rotate or flex the neck, depending on whether the muscles of one or both sides are contracting. In the lumbar region, the large **quadratus lumborum** muscles flex the spine and depress the ribs. Details concerning the individual muscles can be found in Table 11-7.

OBLIQUE AND RECTUS MUSCLES

The muscles of the oblique and rectus groups lie between the vertebral spines and the ventral midline. They are shown in Figure 11-10 and summarized in Table 11-8 on page 344. The oblique muscles can compress underlying structures or rotate the spinal column, depending on whether one or both sides are contracting. The rectus muscles are important flexors of the spinal column, acting in opposition to the varied extensors of the spine.

The oblique series includes the **scalenes** of the neck and the **intercostal** and **transversus** muscles of the thoracic region. In the thorax, the oblique muscles lie between the ribs, and the **external intercostals** cover the **internal intercostals**. Both muscles are important in respiratory movements of the ribs. A small **transversus thoracis** crosses the inner surface of the rib cage and is covered by the serous membrane that lines the pleural cavities. The sternum occupies the place where one might expect to find thoracic rectus muscles—no members of the rectus group are ex-

TABLE 11-7 Muscles of the Spine

Group/Muscle	Origin	Insertion	Action	Innervation
SUPERFICIAL SPINAL EXTENSORS				
Spinalis group				
Semispinalis capitis	Processes of lower cervical and upper thoracic vertebrae	Occipital bone, between nuchal lines	The two sides act together to extend head; either alone extends and tilts head to that side	Cervical spinal nerves
Splenius	Spinal processes and ligaments connecting upper cervical vertebrae	Mastoid process and occipital bone of skull	The two sides act together to extend head; either alone rotates and tilts head to that side	Same
Spinalis dorsi	Spinal processes of lower thoracic and upper lumbar vertebrae	Spinal processes of upper thoracic vertebrae	Extends spinal column	Thoracic spinal nerves
Longissimus group				
Longissimus capitis	Processes of lower cervical and upper thoracic vertebrae	Mastoid processes of temporal bone	The two sides act together to extend head; either alone rotates and tilts head to that side	Cervical and thoracic spinal nerves
Longissimus cervicis	Transverse processes of upper thoracic vertebrae	Transverse processes of middle and upper cervical vertebrae	Same	Same
Longissimus dorsi	Broad aponeurosis and at transverse processes of lower thoracic and upper lumbar vertebrae; joins iliocostalis to form "sacrospinalis"	Transverse processes of higher vertebrae and inferior surfaces of ribs	Extends and/or bends spine to the side	Same
Iliocostalis group				
Iliocostalis cervicis	Superior borders of vertebrosternal ribs near the angles	Transverse processes of middle and lower cervical vertebrae	Extends or bends neck, elevates ribs	Same
Iliocostalis thoracis	Superior borders of lower 7 ribs medial to the angles	Upper ribs and transverse process of last cervical vertebra	Stabilizes thoracic vertebrae in extension	Thoracic spinal nerves
Iliocostalis lumborum	Sacrospinal aponeurosis and iliac crest	Inferior surfaces of lower 7 ribs near their angles	Extends spine, depresses ribs	Lumbar spinal nerves
DEEP SPINAL EXTENSORS				
Transversus group	Between the dorsal processes of adjacent vertebrae		Stabilize, rotate, and/or extend the spine	Cervical, thoracic, and lumbar spinal nerves
SPINAL FLEXORS				
Longus capitis	Anterior processes of cervical vertebrae	Base of the occipital bone	The two sides act together to bend head forward; either alone rotates head to that side	Cervical spinal nerves
Longus cervicis	Anterior processes and bodies of cervical and upper thoracic vertebrae	Transverse processes of upper cervical vertebrae	Flexes and/or rotates neck; limits hyperextension	Same
Quadratus lumborum	Iliac crest	Last rib and transverse processes of lumbar vertebrae	Together they depress ribs, flex spine; one side alone flexes laterally	Thoracic and lumbar spinal nerves

FIGURE 11-9
Muscles of the Spine. *See also Figures 8-15, 8-17, and 8-21.*

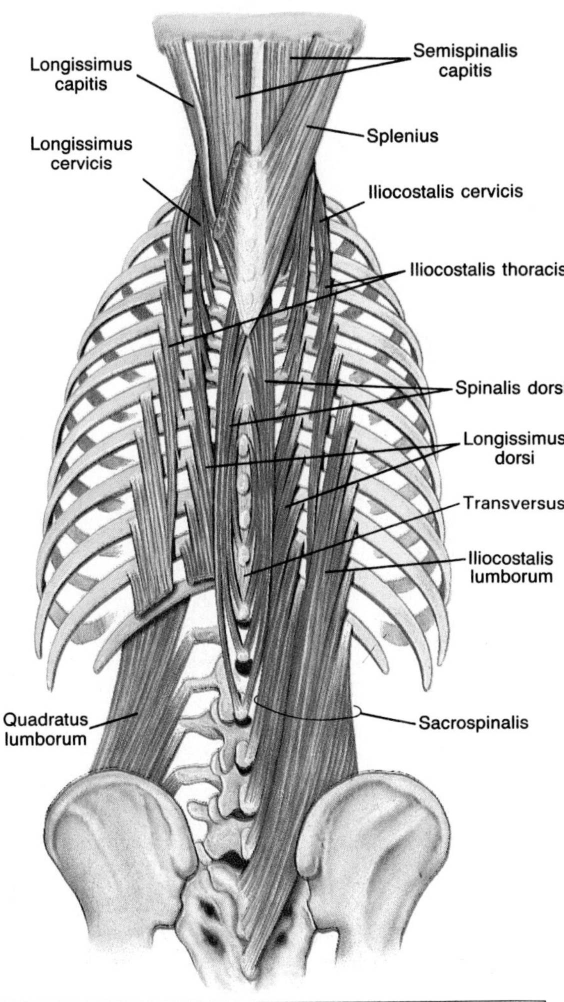

(a) Superficial muscles (right) and deep muscles (left) of the back.

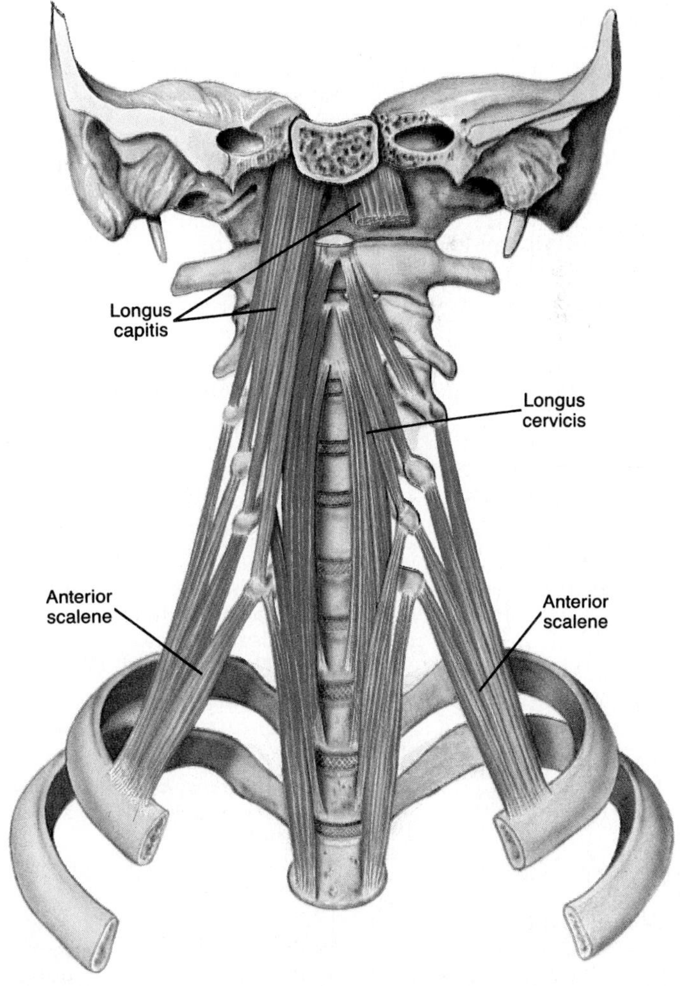

(b) Muscles arising from the anterior surfaces of the vertebrae.

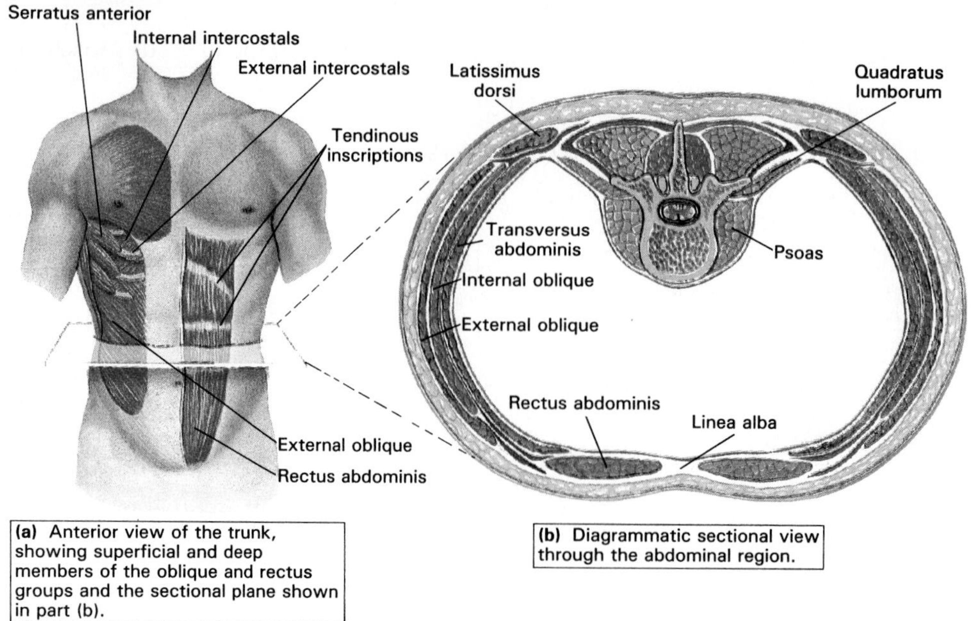

FIGURE 11-10
Oblique and Rectus Muscles. *See also Figures 8-13, 8-20, and 8-21.*

TABLE 11-8 Oblique and Rectus Muscles

Group/Muscle	*Origin*	*Insertion*	*Action*	*Innervation*
OBLIQUE GROUP				
Cervical region				
Scalene	Transverse and costal processes of cervical vertebrae	Superior surfaces of first two ribs	Elevates ribs, and/or flexes neck	Cervical spinal nerves
Thorac region				
External intercostals	Inferior border of each rib	Superior border of the next rib	Elevate ribs	Intercostal nerves (branches of thoracic spinal nerves)
Internal intercostals	Superior border of each rib	Inferior border of the previous rib	Depress ribs	Same
Transversus thoracis	Medial surface of sternum	Cartilages of ribs	Same	Same
Abdominal region				
External oblique	Lower eight ribs	Linea alba and iliac crest	Compresses abdomen, depresses ribs, flexes or bends spine	Intercostals and iliohypogastric nerves
Internal oblique	Lumbodorsal fascia and iliac crest	Lower ribs, xiphoid of sternum, and linea alba	Same	Same
Transversus abdominis	Cartilages of lower ribs, iliac crest, and lumbodorsal fascia	Linea alba and pubis	Compresses abdomen	Intercostals, iliohypogastric, and ilioinguinal nerves
RECTUS GROUP				
Cervical region	*See muscles in Table 11-6*			
Thoracic region				
Diaphragm	Xiphoid process, cartilages of ribs 4–10, and anterior surfaces of lumbar vertebrae	Central tendinous sheet	Contraction expands thoracic cavity, compresses abdominopelvic cavity	Phrenic nerves
Abdominal region				
Rectus abdominis	Inferior surfaces of costal cartilages (ribs 5–7) and xiphoid process of sternum	Superior surface of pubis around symphysis	Depresses ribs, flexes vertebral column	Thoracic spinal nerves (T_7–T_{12})

Sports and Fitness: Hernias

When the abdominal muscles contract forcefully, pressure in the abdominopelvic cavity can skyrocket, and those pressures are applied to internal organs. If the individual exhales at the same time, the pressure is relieved, because the diaphragm can move upward as the lungs collapse. But during vigorous isometric exercises or when lifting a weight while holding one's breath, pressure in the abdominopelvic cavity can rise to 1500 pounds per square inch, roughly 100 times normal pressures. Pressures this high can cause a variety of problems, among them the development of a hernia.

A **hernia** develops when an organ protrudes through an abnormal opening. There are many types of hernias; we will consider only *inguinal* (groin) *hernias* and *diaphragmatic hernias* here.

Late in the development of the male, the testes descend into the scrotum by passing through the abdominal wall at the **inguinal canals**. In the adult male, the spermatic ducts and associated blood vessels penetrate the abdominal musculature at the inguinal canals on their way to the abdominal reproductive organs. In an **inguinal hernia** (Figure 11-11) the inguinal canal enlarges, and abdominal contents such as a portion of the intestine (or more rarely the bladder) are forced into the inguinal canal. If the herniated structures become trapped or twisted within the inguinal sac, surgery may be required to prevent serious complications. Inguinal hernias are not always caused by unusually high abdominal pressures. Injuries to the abdomen, or inherited weakness or distensibility of the canal, may have the same effect.

The esophagus and major blood vessels pass through an opening in the muscular diaphragm. In a **diaphragmatic hernia**, also called a *hiatal hernia* (hī-Ā-tal; *hiatus*, a gap or opening), abdominal organs slide into the thoracic cavity, most often through the **esophageal hiatus**, the opening used by the esophagus. The severity of the condition will depend on the location and size of the herniated organ(s). Hiatal hernias are actually very common, and most go unnoticed. Radiologists see them in about 30 percent of patients examined with barium contrast techniques (Figure 1-14). (p. 20) When clinical complications develop, they usually occur because abdominal organs that have pushed into the thoracic cavity are exerting pressure on structures or organs there. As is the case with inguinal hernias, a diaphragmatic hernia may result from congenital factors or from an injury that weakens or tears the diaphragmatic muscle.

FIGURE 11-11
Inguinal Hernia

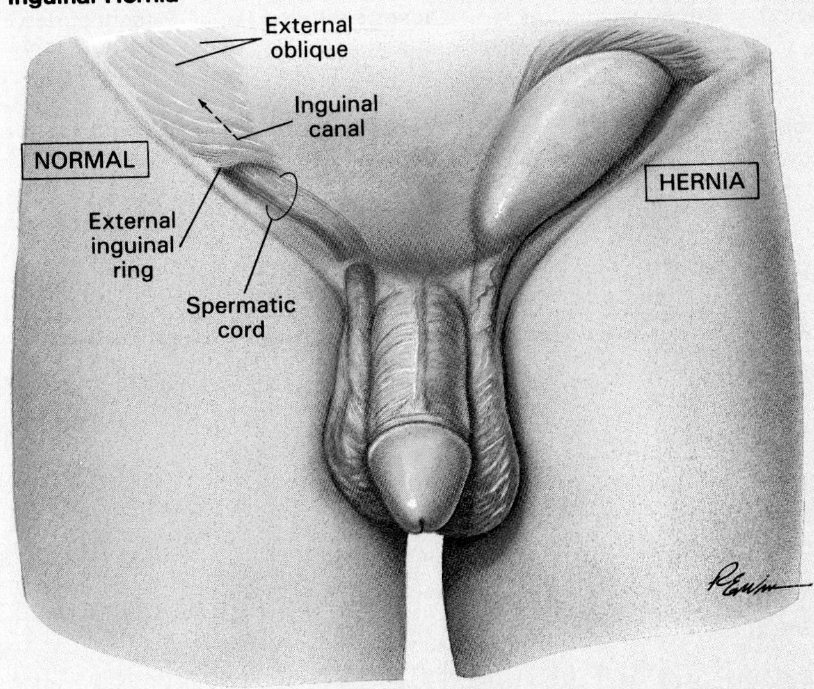

TABLE 11-9 Muscles of the Pelvic Floor

Group/Muscle	*Origin*	*Insertion*	*Action*	*Innervation*
UROGENITAL TRIANGLE				
Superficial muscles				
Bulbocavernosus:				
male	Collagen sheath at base of penis; fibers cross over urethra	Median raphe and central tendon of perineum	Compresses base, stiffens penis, ejects urine or semen	Pudendal nerve, perineal branch
female	Collagen sheath at base of clitoris; fibers run on either side of urethral and vaginal openings	Central tendon of perineum	Compresses and stiffens clitoris, narrows vaginal opening	Same
Ischiocavernosus	Inferior ramus and tuberosity of ischium	Symphysis pubis anterior to base of penis or clitoris	Compresses and stiffens penis or clitoris	Same
Superficial transverse perineus	Inferior ischial ramus	Central tendon of perineum	Stabilizes central tendon of perineum	Same
Deep muscles: urogenital diaphragm				
Deep transverse perineus	Inferior ischial ramus	Median raphe of urogenital diaphragm	Stabilizes central tendon of perineum	Pudendal nerve, perineal branch
Urethral sphincter:				
male	Ischial and pubic rami	To median raphe at base of penis; inner fibers encircle urethra	Closes urethra, compresses prostate and bulbourethral glands	Same
female	Ischial and pubic rami	To median raphe; inner fibers encircle urethra	Closes urethra, compresses vagina and greater vestibular glands	Same
ANAL TRIANGLE				
Pelvic diaphragm				
Coccygeus	Ischial spine	Lateral, inferior borders of the sacrum	Flexes coccyx and coccygeal vertebrae	Pudendal nerve
External anal sphincter	Via tendon to coccyx	Encircles anal opening	Closes anal opening	Pudendal nerve, hemorrhoidal branch
Levator ani:				
Iliococcygeus	Ischial spine, pubis	Coccyx	Tenses floor of pelvis, supports pelvic organs, flexes coccyx, elevates and retracts anus	Pudendal nerve
Pubococcygeus	Inner margins of pubis	Coccyx	Same	Same

FIGURE 11-12
Muscles of the Pelvic Floor. *See also Figures 9-11 and 9-12.*

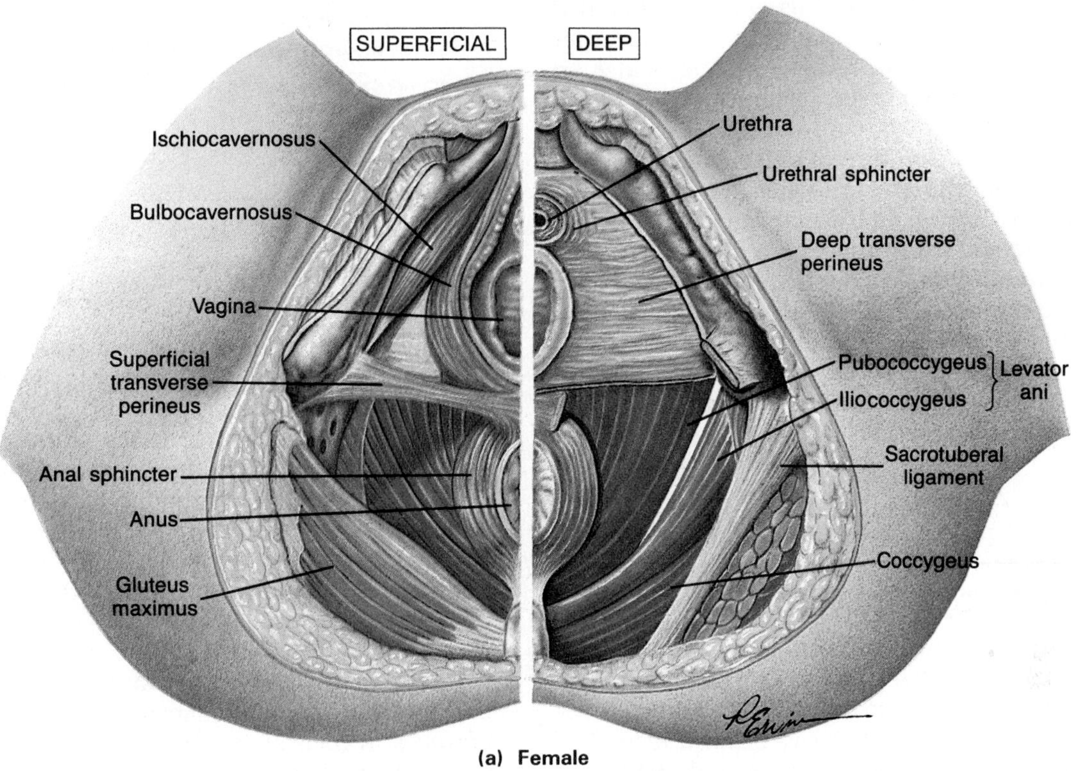

(a) Female

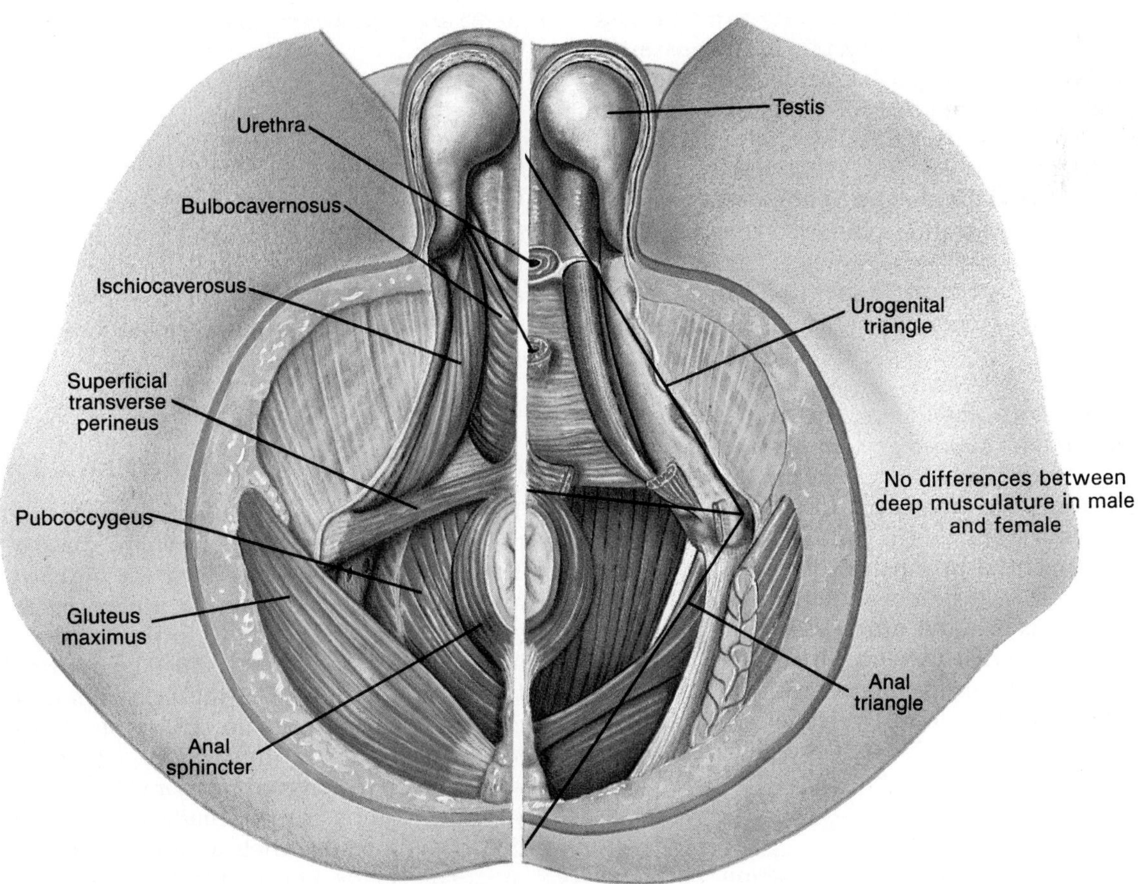

(b) Male

posed on the ventral surface of the spinal column.

The muscular **diaphragm** separates the thoracic and abdominopelvic cavities. This muscle is included here because it is developmentally linked to the other muscles of the chest wall. The diaphragm is also an important respiratory muscle. (The functions of the respiratory muscles will be considered in Chapter 23.)

The same basic pattern of musculature extends unbroken across the abdominopelvic surface. Here the muscles are known as the **external obliques**, the **internal obliques**, the **transversus abdominis**, and the **rectus abdominis**. The rectus abdominis begins at the xiphoid process and ends near the pubic symphysis. This muscle is longitudinally divided by a median collagenous partition, the **linea alba** (white line). The transverse **tendinous inscriptions** divide this muscle into segments.

MUSCLES OF THE PELVIC FLOOR

The boundaries of the **perineum** are established by the inferior margins of the pelvis. The muscular floor of the pelvic cavity is created by muscles that connect the sacrum and coccyx to the ischium and pubis. These are illustrated in Figure 11-12, and additional details are provided in Table 11-9.

If you draw a line between the ischial tuberosities you will divide the perineum into two triangles, an **anterior**, or **urogenital, triangle** and a **posterior**, or **anal, triangle**. The superficial muscles of the anterior triangle are the muscles of the external genitalia. They overlie deeper muscles that strengthen the pelvic floor and encircle the urethra. These muscles constitute the **urogenital diaphragm**, a muscular layer that extends between the pubic bones. (The term *diaphragm* refers to any muscular sheet that forms a wall. When used without a modifier, however, *diaphragm*, or *diaphragmatic muscle*, specifies the muscular partition that separates the abdominopelvic and thoracic cavities.)

Another and even more extensive muscular sheet, the **pelvic diaphragm**, forms the muscular foundation of the anal triangle. This layer extends anteriorly above the urogenital diaphragm as far as the pubic symphysis.

The urogenital and pelvic diaphragms do not completely close the pelvic outlet, for the urethra, vagina (in females), and anus pass through them and open to the environment. Muscular sphincters surround their openings and permit voluntary control of urination and defecation. Muscles, nerves, and blood vessels also pass through the pelvic outlet as they travel to or from the legs.

√ **If you were contracting and relaxing your masseter muscle, what would you probably be doing?**

√ **Damage to the external intercostal muscles would interfere with what important process?**

√ **If someone hits you in your rectus abdominis muscle, how would your body position change?**

■ The Appendicular Musculature

The appendicular musculature positions and directs the movements of the appendicular skeleton. There are two major groups: (1) the muscles of the shoulders and arms and (2) the muscles of the pelvic girdle and legs. There are few similarities between the two groups, because the functions and required ranges of motion are very different. In addition to increasing the mobility of the arm, the muscular connections between the pectoral girdle and the axial skeleton must act as shock absorbers. For example, people who are jogging can still perform delicate hand movements because the muscular connections between the axial and appendicular skeleton smooth out the bounces in their stride. In contrast, the pelvic girdle has evolved to transfer weight from the axial to the appendicular skeleton. A muscular connection would reduce the efficiency of the transfer, and the emphasis is on sheer power rather than extreme versatility.

MUSCLES OF THE SHOULDERS AND ARMS

Muscles associated with the shoulders and arms can be divided into four groups: (1) muscles that position the shoulder girdle, (2) muscles that move the upper arm, (3) muscles that move the forearm and wrist, and (4) muscles that move the palm and fingers.

Muscles That Position the Shoulder Girdle

Two large, superficial muscles cover the back and portions of the neck, reaching to the base of the skull. The **trapezius** originates along the middle of the neck and back, and inserts upon the clavicles and the scapular spines. The muscle forms a broad diamond, as illustrated in Figure 11-13. The **sternocleidomastoid** (ster-nō-klī-dō-MAS-toid) extends from the clavicles and the sternum to the mastoid region of the skull. As indicated in Table 11-10, these extensive muscles are innervated by more than one nerve, and specific regions can be made to contract independently. As a result, their actions are quite varied.

The removal of the trapezius reveals the **rhomboideus** muscles and the **levator scapulae**. These

FIGURE 11-13
Muscles of the Shoulder Girdle. *See also Figures 8-20, 9-4, and 9-8.*

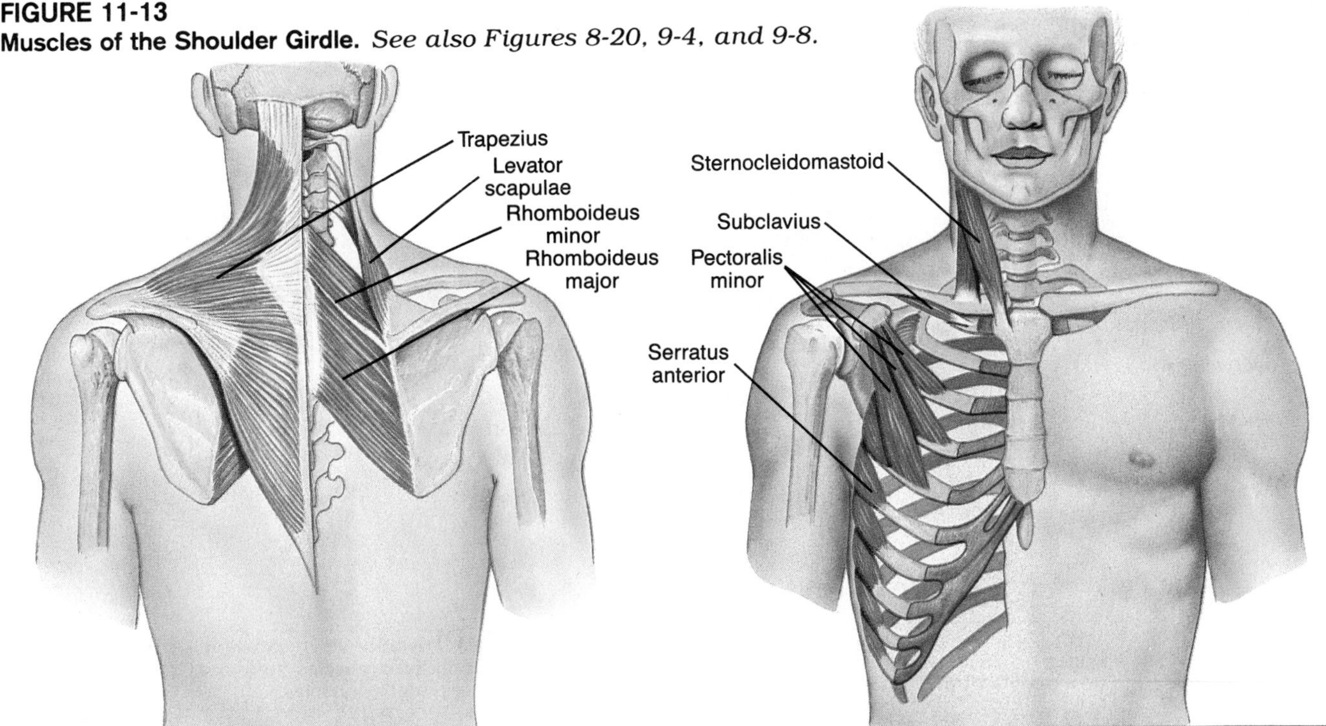

(a) Posterior view, showing superficial muscles (left) and deep muscles (right) of the shoulder girdle

(b) Deep muscles best viewed from the anterior surface

TABLE 11-10 Muscles That Move the Shoulder Girdle

Muscle	*Origin*	*Insertion*	*Action*	*Innervation*
Levator scapulae	Dorsal surfaces of first 4 cervical vertebrae	Vertebral border of scapula near superior angle	Elevates scapula	Dorsal scapular nerve
Pectoralis minor	Ventral surfaces of ribs 3–5	Coracoid process of scapula	Depresses and protracts shoulder; rotates scapula laterally; elevates ribs if scapula is stationary	Median pectoral nerve
Rhomboideus major	Spinal processes of upper thoracic vertebrae	Vertebral border of scapula from spine to inferior angle	Adducts and rotates scapula laterally	Dorsal scapular nerve
Rhomboideus minor	Spinal processes of vertebrae C_7–T_1	Vertebral border near spine	Same	Same
Serratus anterior	Ventral and superior margins of ribs 1–9	Ventral surface of vertebral border of scapula	Protracts shoulder, abducts and medially rotates scapula	Long thoracic nerve
Sternocleidomastoid	Superior margins of manubrium and clavicle	Mastoid region of skull	Together they flex the neck; alone one side bends head toward shoulder and turns face to opposite side	Spinoaccessory nerve (N XI) and cervical spinal nerves
Subclavius	First rib	Clavicle	Depresses and protracts shoulder	Subclavian nerve
Trapezius	Occipital bone, ligamentum nuchae, and spinal processes of thoracic vertebrae	Clavicle and scapula (acromion and scapular spine)	Depends on active region and state of other muscles; may elevate, adduct, depress, or rotate scapula and/or elevate clavicle; can also extend head and neck	Spinoaccessory nerve (N XI) and cervical spinal nerves

FIGURE 11-14
Muscles That Move the Upper Arm. *See also Figures 9-4, 9-5, and 9-8.*

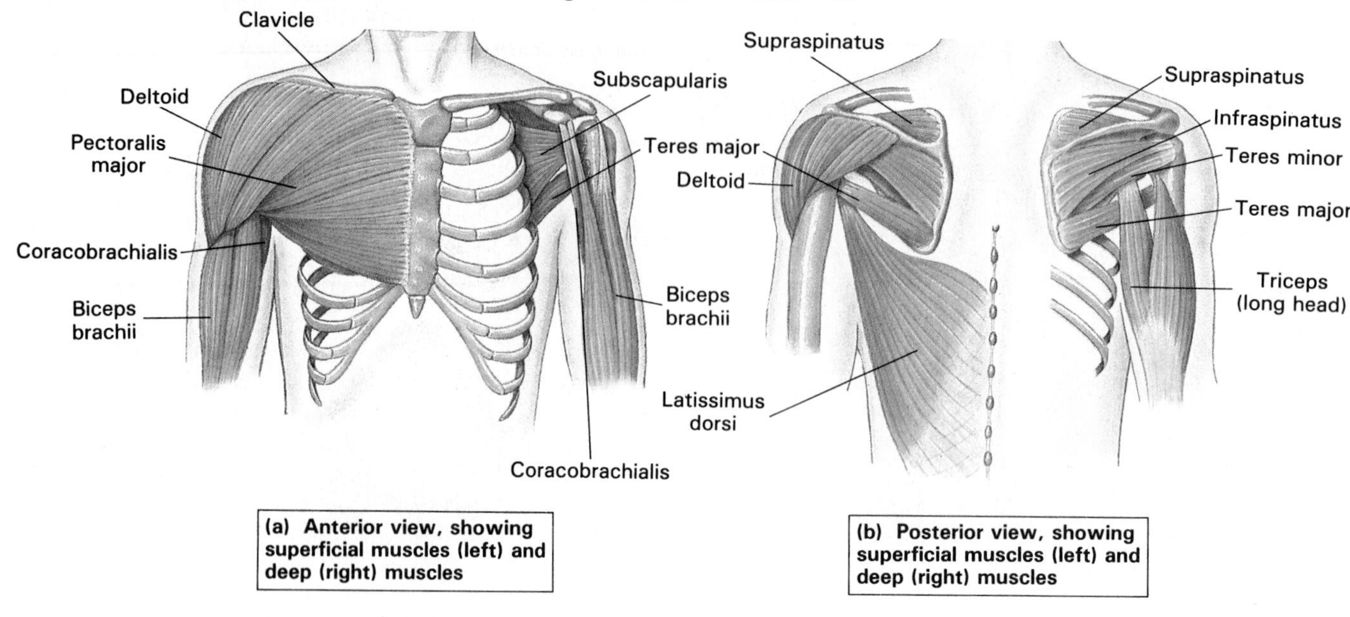

TABLE 11-11 Muscles That Move the Upper Arm

Muscle	*Origin*	*Insertion*	*Action*	*Innervation*
Coracobrachialis	Coracoid process	Medial margin of shaft of humerus	Adducts and flexes humerus	Musculocutaneous nerve
Deltoid	Clavicle and scapula (acromion and adjacent scapular spine)	Deltoid tuberosity of humerus	Abducts arm	Axillary nerve
Supraspinatus	Supraspinous fossa of scapula	Greater tubercle of humerus	Abducts arm	Suprascapular nerve
Infraspinatus	Infraspinous fossa of scapula	Greater tubercle of humerus	Lateral rotation of humerus	Suprascapular nerve
Subscapularis	Subscapular fossa of scapula	Lesser tubercle of humerus	Medial rotation of humerus	Subscapular nerve
Teres major	Inferior angle of scapula	Intertubercular groove of humerus	Adducts and medially rotates arm	Lower subscapular nerve
Teres minor	Axillary border of scapula	Greater tubercle of humerus	Lateral rotation of humerus	Axillary nerve
Triceps (long head)	*See Table 11-12*			
Latissimus dorsi	Spinal processes of lower thoracic vertebrae, ribs 8–12, the spines of lumbar vertebrae, and the lumbodorsal fascia	Lesser tubercle, intertubercular groove of humerus	Extends, adducts, and medially rotates humerus	Thoracodorsal nerve
Pectoralis major	Cartilages of ribs 2–6, body of sternum, and inferior, medial portion of clavicle	Greater tubercle of humerus	Flexes, adducts, and medially rotates humerus	Pectoral nerves

muscles are attached to the dorsal surfaces of the cervical and thoracic vertebrae, and they insert along the vertebral border of each scapula, between the superior and inferior angles. Contraction of the rhomboideus muscles performs adduction, pulling the scapula toward the center of the back. The levator scapulae, as the name implies, elevates the scapula, as when you shrug your shoulders.

On the chest, the **serratus anterior** originates along the anterior surfaces of several ribs. This fan-shaped muscle inserts along the anterior margin of the vertebral border of the scapula. When the serratus anterior contracts, it abducts the scapula and swings the shoulder forward.

Two other deep chest muscles arise along the ventral surfaces of the ribs. The **subclavius** (*sub*, below + *clavius*, clavicle) inserts upon the inferior border of the clavicle. When it contracts, it depresses and protracts the scapular end of the clavicle. Because ligaments connect this end to the shoulder joint and scapula, these structures move as well. The **pectoralis minor** attaches to the coracoid process of the scapula. Its contraction usually complements that of the subclavius.

Muscles That Move the Upper Arm

The muscles that move the upper arm (Figure 11-14 and Table 11-11) are easiest to remember when grouped by primary actions. The **deltoid** is the major abductor of the upper arm, but the **supraspinatus** assists at the start of this movement. The **subscapularis** and **teres major** rotate the arm medially, whereas the **infraspinatus** and the **teres minor** perform lateral rotation. All of these muscles originate on the scapula. The small **coracobrachialis** is the only muscle attached to the scapula that has flexion and adduction of the humerus as its primary actions.

The **pectoralis major** extends between the chest and the greater tubercle of the humerus; the **latissimus dorsi** extends between the thoracic vertebrae and the lesser tubercle. The pectoralis major flexes the upper arm, and the latissimus dorsi extends it. These two muscles can also work together to produce adduction and medial rotation of the humerus.

The muscles of this group provide substantial support for the loosely built scapulohumeral joint of the shoulder. The tendons of the supraspinatus, infraspinatus, subscapularis, and teres minor blend with and support the capsular fibers that enclose the shoulder joint. They are the muscles of the **rotator cuff**, a frequent site of sports injuries. As noted in Chapter 9, these muscles must stabilize the shoulder joint while controlling an extensive range of movement. ∞ (p. 277) Powerful, repetitive arm movements, such as pitching a fastball at 96 mph for nine innings, can place intolerable strains on the muscles of the rotator cuff, leading to muscle strains, bursitis, and other painful injuries. (Sports injuries are discussed further at the end of this chapter.)

Muscles That Move the Forearm and Wrist

Although most of the muscles that insert upon the forearm and wrist (Figure 11-15 and Table 11-12) originate on the humerus, there are two noteworthy exceptions. The **biceps brachii** and the long head of the **triceps**, which insert upon the bones of the forearm, originate on the scapula. Although their contractions can have a secondary effect on the shoulder, their primary actions are on the elbow. The triceps extends the forearm when, for example, we do push-ups. The biceps both flexes and supinates the forearm. With the forearm pronated (palm facing back) the biceps cannot function effectively. As a result, we are strongest when flexing the supinated forearm; the biceps brachii then makes a prominent bulge.

The **brachialis** and **brachioradialis** also flex the forearm, opposed by the **anconeus** and the triceps. The **flexor carpi ulnaris**, the **flexor carpi radialis**, and the **palmaris longus** are superficial muscles that work together to produce flexion of the wrist. Because they originate on opposite sides of the humerus, the flexor carpi radialis flexes and *ab*ducts while the flexor carpi ulnaris flexes and *ad*ducts. The **extensor carpi radialis** muscles and the **extensor carpi ulnaris** have a similar relationship; the former produces extension and abduction, the latter extension and adduction.

The **pronator teres** and the **supinator** arise on both the humerus and forearm. They rotate the radius without producing either flexion or extension of the elbow. The **pronator quadratus** arises on the ulna and assists the pronator teres in opposing the actions of the supinator or biceps. The muscles involved in pronation and supination can be seen in Figure 11-16. Note the changes in orientation that occur as the pronator teres and pronator quadratus contract. During pronation the tendon of the biceps rolls under the radius, and a bursa prevents abrasion against the tendon.

As you study the muscles included in Table 11-12, you may find it useful to know that a connection exists between the location of a muscle and its function. In general, the extensor muscles lie along the dorsal and lateral surfaces of the arm, whereas the flexors are found on the ventral and medial surfaces.

Muscles That Move the Palm and Fingers

The superficial and deep muscles of the forearm, detailed in Table 11-13 on page 354 are concerned with digital flexion and extension. These muscles stop before reaching the wrist, and only their tendons cross the articulation. As indicated in Figure 11-17, these are relatively large muscles, and keeping them clear

TABLE 11-12 Muscles That Move the Forearm and Wrist

Muscle	*Origin*	*Insertion*	*Action*	*Innervation*
PRIMARY ACTION AT THE ELBOW				
Flexors				
Biceps brachii	Short head from the coracoid process; long head from the supraglenoid tuberosity (both on the scapula)	Tuberosity of radius	Flexes and supinates forearm	Musculocutaneous nerve
Brachialis	Anterior, distal surface of humerus	Tuberosity of ulna	Flexes forearm	Same
Brachoradialis	Lateral epicondyle of humerus	Lateral aspect of styloid process of radius	Same	Radial nerve
Extensors				
Anconeus	Posterior surface of lateral epicondyle of humerus	Lateral margin of olecranon on ulna	Extends forearm, moves ulna laterally during pronation	Same
Triceps				
lateral head	Superior, lateral margin of humerus	Olecranon process of ulna	Extends forearm	Same
long head	Infraglenoid tuberosity of scapula	Same	Same	Same
medial head	Posterior margin of humerus inferior to radial groove	Same	Same	Same
PRONATORS/SUPINATORS				
Pronator				
quadratus	Medial surface of distal portion of ulna	Anterolateral surface of distal portion of radius	Pronates forearm	Median nerve
teres	Medial epicondyle of humerus and coronoid process of ulna	Distal lateral surface of radius	Same	Same
Supinator	Lateral condyle of humerus	Anterolateral surface of radius distal to the radial tuberosity	Supinates forearm	Radial nerve
PRIMARY ACTION AT THE WRIST				
Flexors				
Flexor carpi radialis	Medial epicondyle of humerus	Base of 2nd metacarpal	Flexes and abducts palm	Median nerve
Flexor carpi ulnaris	Medial epicondyle of humerus; adjacent medial surface of olecranon and anteromedial portion of ulna	Bases of metacarpals 3–5	Flexes and adducts palm	Ulnar nerve
Palmaris longus	Medial epicondyle of humerus	Palmar aponeurosis	Flexes palm	Median nerve
Extensors				
Extensor carpi radialis	Lateral epicondyle of humerus	Base of 2nd metacarpal	Extends and abducts palm	Same
Extensor carpi ulnaris	Lateral epicondyle of humerus; adjacent dorsal surface of ulna	Base of 5th metacarpal	Extends and adducts palm	Deep radial nerve

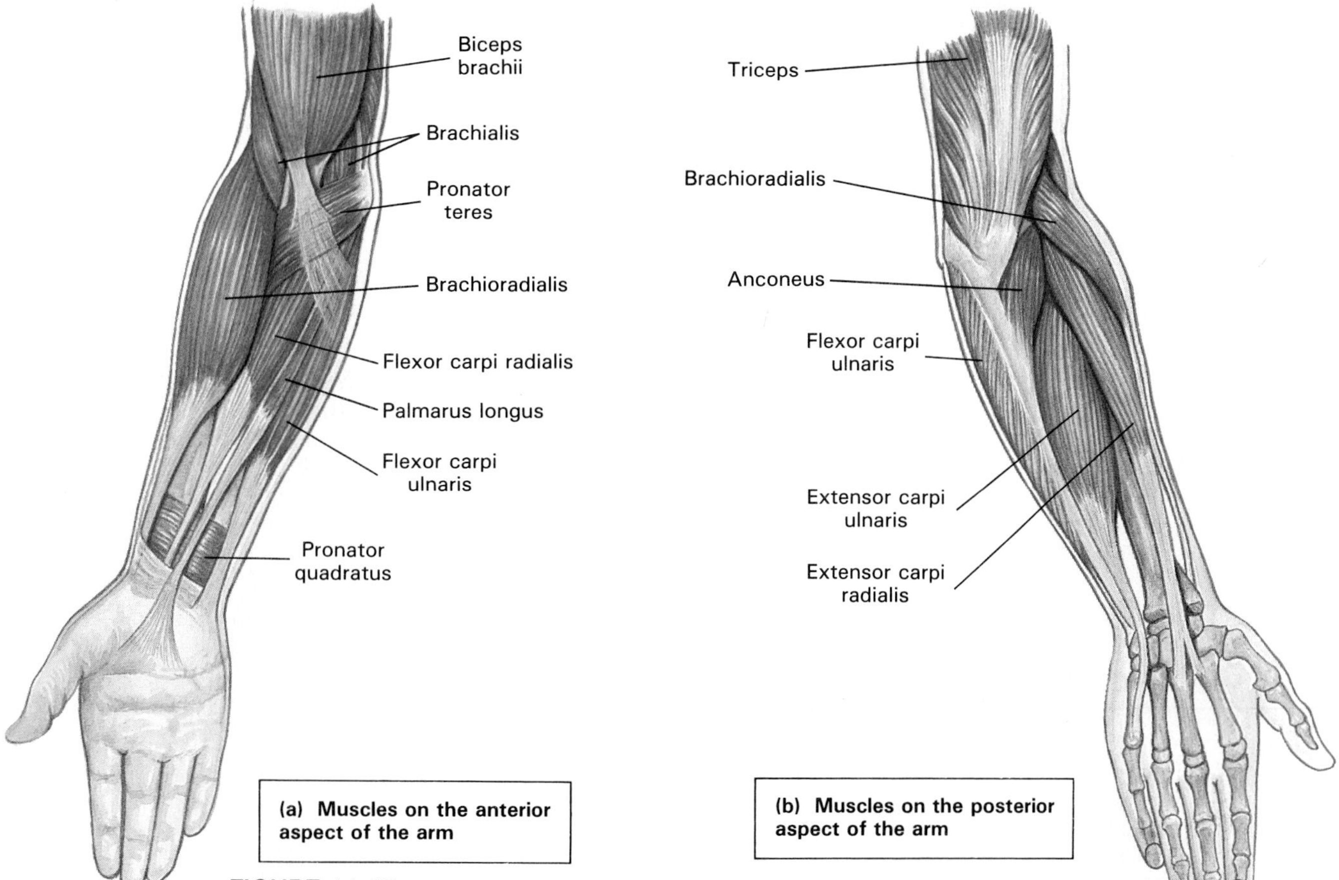

Biceps brachii
Brachialis
Pronator teres
Brachioradialis
Flexor carpi radialis
Palmarus longus
Flexor carpi ulnaris
Pronator quadratus
Triceps
Brachioradialis
Anconeus
Flexor carpi ulnaris
Extensor carpi ulnaris
Extensor carpi radialis

(a) Muscles on the anterior aspect of the arm

(b) Muscles on the posterior aspect of the arm

FIGURE 11-15
Muscles That Move the Forearm and Wrist. *See also Figures 9-5, 9-6, and 9-7.*

FIGURE 11-16
Muscles of Pronation and Supination. *See also Figure 9-6.*

Pronator teres
Radius
Ulna
Pronator quadratus
Radius
Supinator
Biceps
Ulna
Bursa

(a) Anterior view of the supine forearm

(b) Pronation

(c) Sectional view in pronation at the level of the biceps tendon insertion

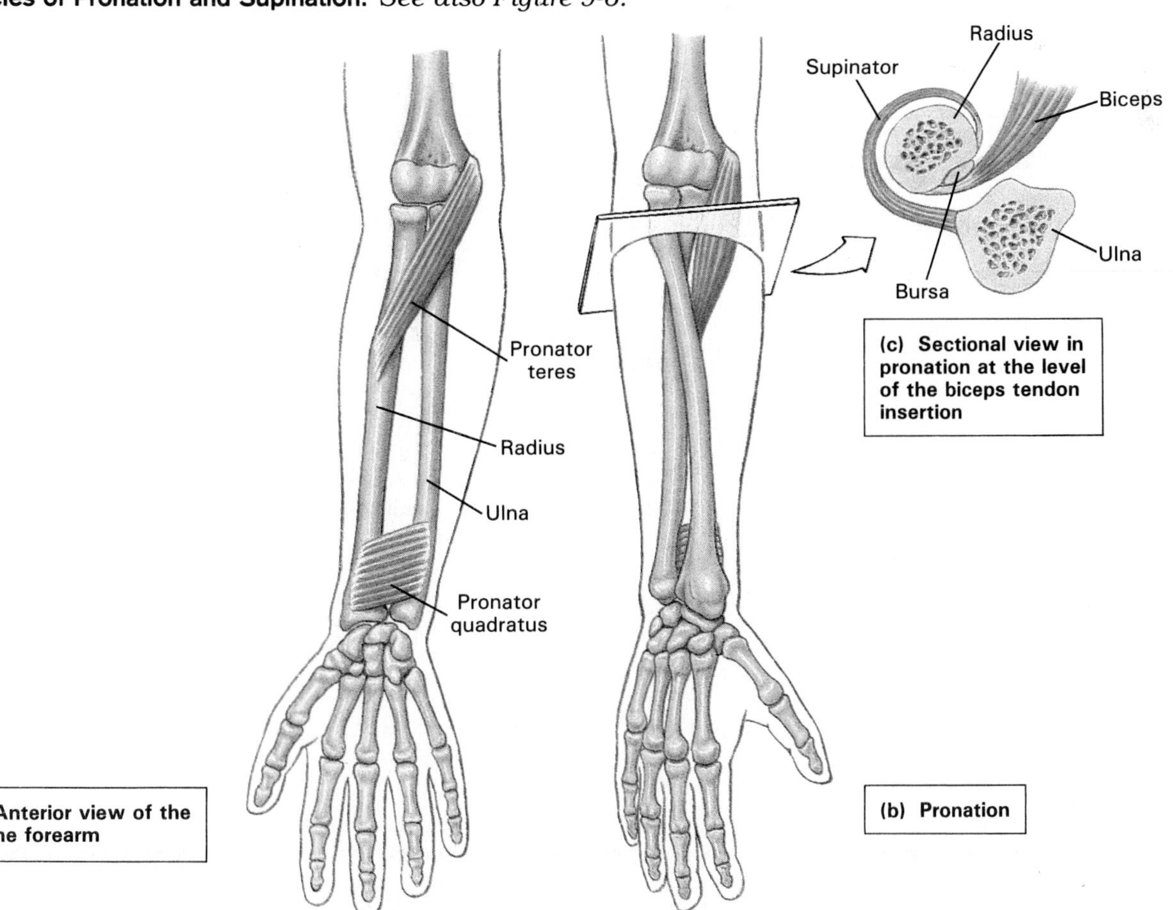

FIGURE 11-17
Muscles That Move the Palm and Fingers. *See also Figures 9-6 and 9-7.*

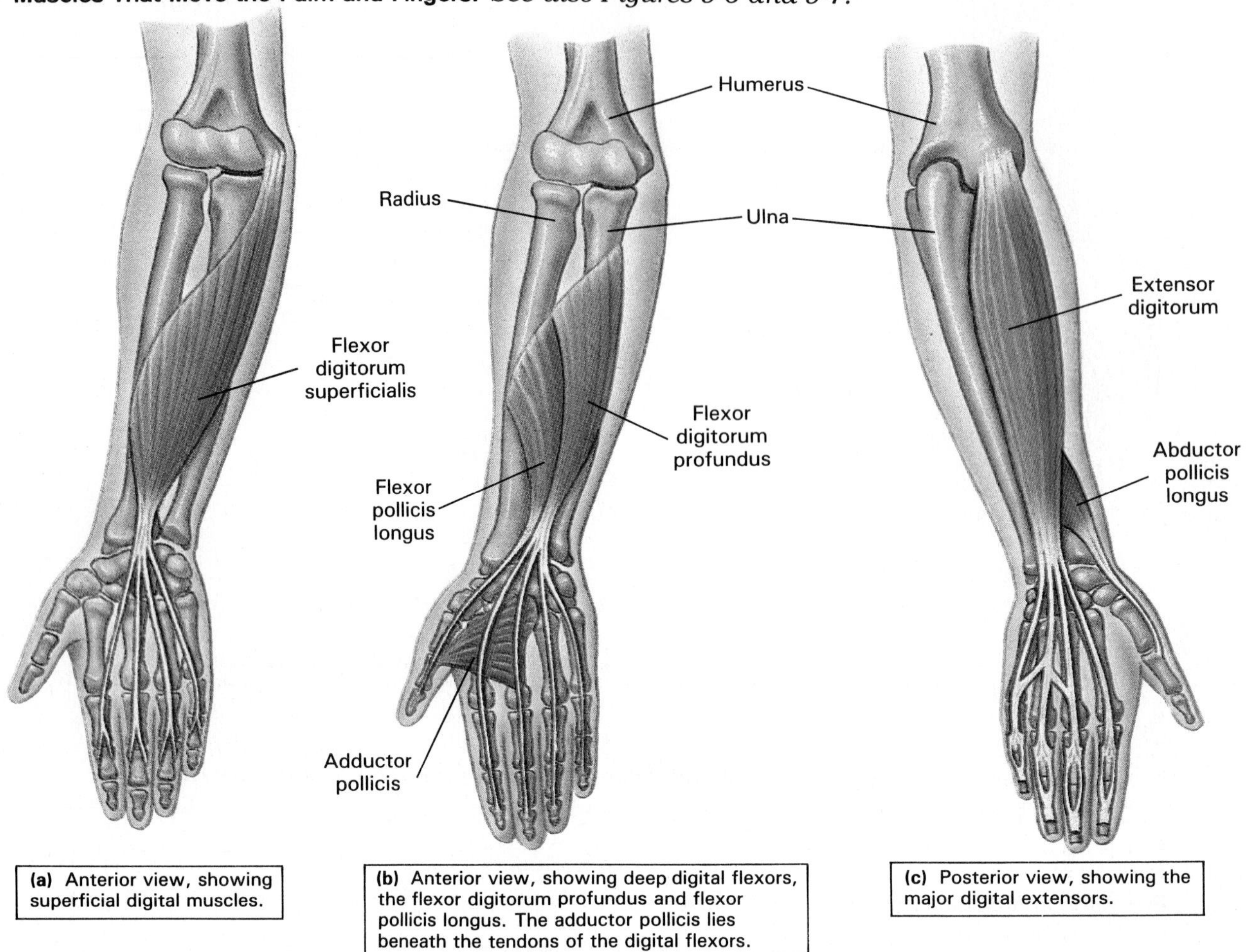

(a) Anterior view, showing superficial digital muscles.

(b) Anterior view, showing deep digital flexors, the flexor digitorum profundus and flexor pollicis longus. The adductor pollicis lies beneath the tendons of the digital flexors.

(c) Posterior view, showing the major digital extensors.

TABLE 11-13 Muscles That Move the Palm and Fingers

Muscle	*Origin*	*Insertion*	*Action*	*Innervation*
Adductor pollicis	**Metacarpals and carpals**	**Proximal phalanx of thumb**	**Adducts thumb**	**Ulnar nerve**
Abductor pollicis longus	**Proximal dorsal surfaces of ulna and radius**	**Lateral margin of 1st metacarpal**	**Abducts thumb**	**Deep radial nerve**
Extensor digitorum	**Lateral epicondyle of humerus**	**Dorsal surfaces of the phalanges**	**Extends fingers and palm**	**Same**
Flexor digitorum profundus	**Proximal anteromedial surface of ulna**	**Bases of distal phalanges**	**Flexes fingers, specifically 3rd phalanx on 2nd, and 2nd on 1st; flexes palm**	**Median nerve**
Flexor digitorum superficialis	**Medial epicondyle of humerus; adjacent anterior surfaces of ulna and radius**	**Midlateral surface of 2nd phalanx; connected by ligaments to others**	**Flexes fingers, specifically 2nd phalanx on 1st; flexes palm**	**Same**
Flexor pollicis longus	**Shaft of radius and interosseous membrane**	**Distal phalanx of thumb**	**Flexes thumb**	**Same**

Clinical Comment: Carpal Tunnel Syndrome

In **carpal tunnel syndrome**, inflammation of the sheath surrounding the flexor tendons of the palm leads to compression of the *median nerve*, a mixed (sensory and motor) nerve that innervates the palm. Symptoms include pain, especially on palmar flexion, a tingling sensation or numbness on the palm, and weakness in the abductor pollicis muscle. This condition is fairly common, and often strikes those engaged in repetitive hand movements, such as typing, working at a computer keyboard, or playing the piano. Treatment involves administration of anti-inflammatory drugs such as aspirin, and use of a splint to prevent wrist flexion and stabilize the region.

of the joints ensures maximum mobility at both the wrist and hand. The tendons that cross the dorsal and ventral surfaces of the wrist pass through **tendon sheaths**, elongate bursae that reduce friction. These sheaths can be seen in Figure 9-10. (p. 279) Inflammation of tendon sheaths can restrict movement and irritate the median nerve. This causes chronic pain, a condition known as *carpal tunnel syndrome.*

The muscles of the forearm provide strength and crude control of the palm and fingers. Fine control of the hand involves small muscles that originate on the carpals and metacarpals. No muscles originate on the phalanges, and only tendons extend across the distal joints of the fingers.

- √ **What muscle are you using when you shrug your shoulders?**
- √ **Sometimes baseball pitchers will suffer from rotator cuff injuries. What muscles are involved in this type of injury?**
- √ **Injury to the flexor carpi ulnaris would impair what two movements?**

MUSCLES OF THE LEGS

The pelvic girdle is tightly bound to the axial skeleton, and little relative movement is permitted. The few muscles that can influence the position of the pelvis were therefore considered in our discussion of the axial musculature. The leg muscles can be divided into three functional groups: (1) muscles that move the thigh, (2) muscles that move the lower leg, and (3) muscles that affect the ankles, feet, and toes.

Muscles That Move the Thigh

One method for organizing the diverse muscles that originate on the pelvis considers the orientation of each muscle around the hip joint. Figure 11-18 presents the basic concepts involved. Muscles originating on the surface of the pelvis and inserting on the leg will produce characteristic movements determined by their position relative to the acetabulum. These muscles are detailed in Table 11-14 and diagrammed in Figures 11-19 and 11-20.

Gluteal muscles cover the lateral surface of the ilium. The **gluteus maximus** is the largest and most posterior of the gluteal muscles. It originates along

FIGURE 11-18
The Attachment of Muscles to the Pelvic Girdle. *See also Figures 9-11 and 9-12.* The attachment of muscles that originate on the pelvis can be roughly predicted on the basis of their positions relative to the acetabulum. For each major muscle group the direction of the pull exerted on the femur is indicated by an arrow. Specific muscles are detailed in Figures 11-19 and 11-20.

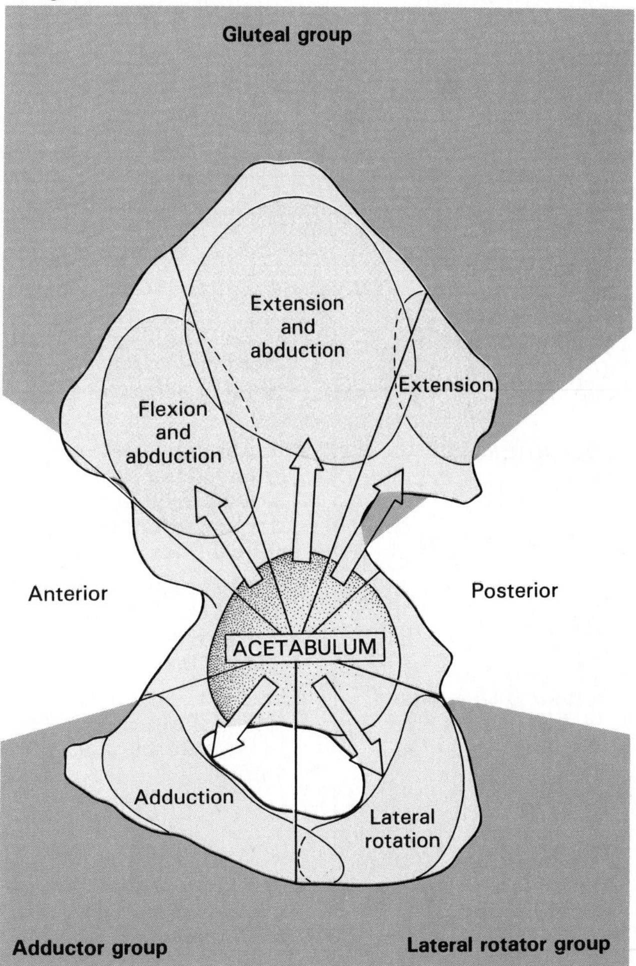

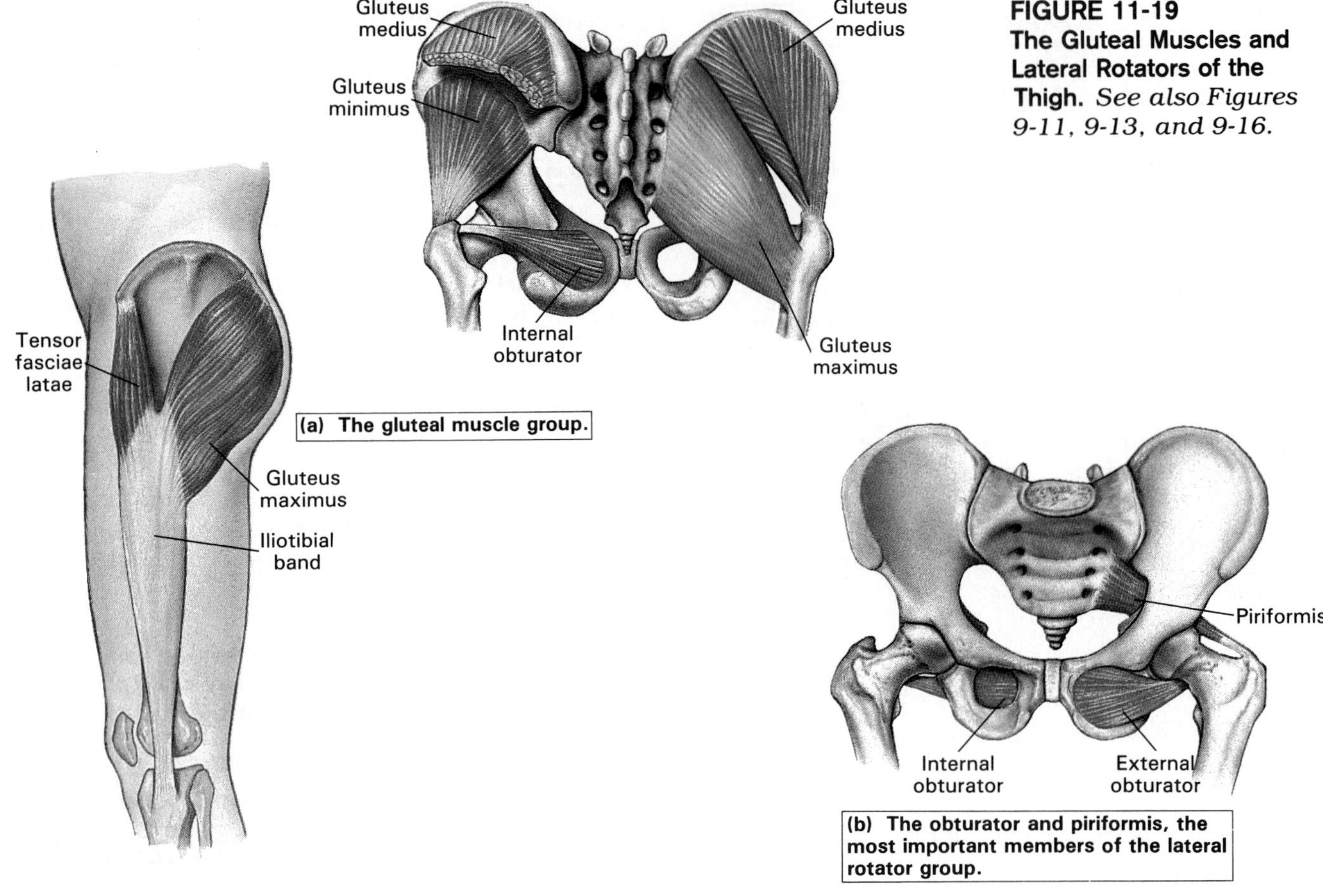

(a) The gluteal muscle group.

(b) The obturator and piriformis, the most important members of the lateral rotator group.

FIGURE 11-19 The Gluteal Muscles and Lateral Rotators of the Thigh. *See also Figures 9-11, 9-13, and 9-16.*

TABLE 11-14 Muscles That Move the Thigh

Group/Muscle	*Origin*	*Insertion*	*Action*	*Innervation*
Gluteal group				
Gluteus maximus	Iliac crest of ilium, sacrum, coccyx, and lumbodorsal fascia	Iliotibial tract and gluteal tuberosity of femur	Extends and laterally rotates thigh	Inferior gluteal nerve
Gluteus medius	Anterior iliac crest of ilium, lateral surface between superior and inferior gluteal lines	Greater trochanter of femur	Abducts and medially rotates thigh	Superior gluteal nerve
Gluteus minimus	Lateral surface of ilium between inferior and anterior gluteal lines	Greater trochanter of femur	Abducts and medially rotates thigh	Same
Tensor fasciae latae	Iliac crest and surface of ilium between anterior iliac spines	Iliotibial tract	Flexes, abducts, and medially rotates thigh; tenses fasciae latae, which laterally supports the knee	Same
Lateral Rotator Group				
Obturators (externus and internus)	Lateral and medial margins of obturator foramen	Trochanteric fossa of femur	Laterally rotates thigh	Nerve to obturator internus
Piriformis	Anterolateral surface of sacrum	Greater trochanter of femur	Laterally rotates and adducts thigh	Same

FIGURE 11-20
The Adductors and Hamstrings of the Thigh.
See also Figures 9-11, 9-13, and 9-16.

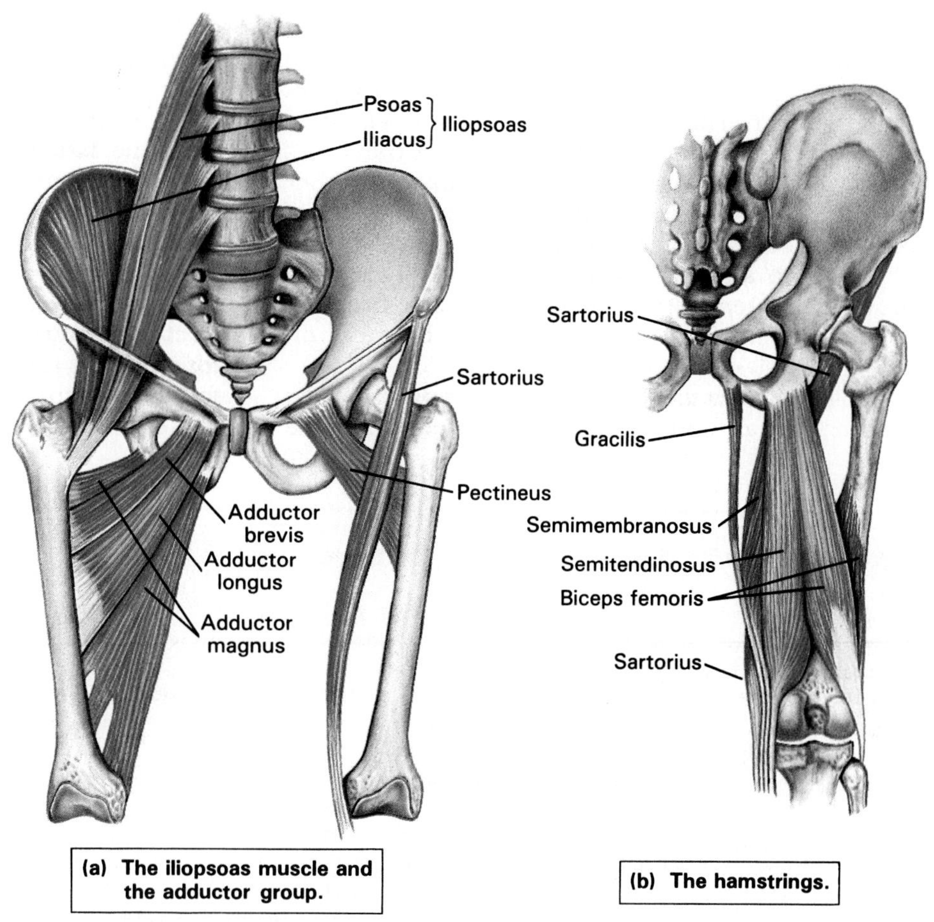

(a) The iliopsoas muscle and the adductor group.

(b) The hamstrings.

Group/Muscle	*Origin*	*Insertion*	*Action*	*Innervation*
Adductor Group				
Adductor brevis	Inferior ramus of pubis	Linea aspera of femur	Adducts thigh	Obturator nerve
Adductor longus	Inferior ramus of pubis anterior to brevis	Same	Adducts, flexes, and medially rotates thigh	Same
Adductor magnus	Inferior ramus of pubis posterior to brevis	Same	Adducts thigh; anterior portion flexes thigh, posterior portion extends thigh	Obturator and sciatic nerves
Pectineus	Superior surface of pubis	Pectineal line inferior to lesser trochanter of femur	Flexes and adducts thigh	Femoral nerve
Iliopsoas				
Iliacus	Iliac fossa of ilium	Femur distal to lesser trochanter; tendon fused with that of psoas	Flexes hip and/or lumbar spine	Same
Psoas	Anterior surfaces of vertebrae T_{12}–L_5	Femur distal to lesser trochanter in company with iliacus	Same	Same

the edge of the posterior superior iliac spine and the ligaments that bind the sacrum to the ilium. Acting alone, this massive muscle extends and laterally rotates the thigh. The gluteus maximus shares an insertion with the **tensor fasciae latae**, a muscle that originates on the anterior superior iliac spine. Together these muscles pull on the **iliotibial tract**, a band of collagen fibers that extends along the lateral surface of the thigh and inserts upon the tibia. This tract provides a lateral brace for the knee that becomes particularly important when a person balances on one leg.

The **gluteus medius** and **gluteus minimus** take their origins anterior to that of the gluteus maximus and insert upon the greater trochanter of the femur. The anterior and inferior gluteal lines on the lateral surface of the ilium mark the boundaries between these muscles.

The **lateral rotators** arise close to or inferior to the horizontal axis of the acetabulum. There are six muscles in all, of which the **piriformis** and the **obturator** muscles are the most important.

The **adductors** are located inferior to the acetabular surface. This muscle group includes the **adductor magnus**, the **adductor brevis**, the **adductor longus**, and the **pectineus**. All but the magnus are found both anterior and inferior to the joint, so they can contribute to flexion as well as adduction. The

TABLE 11-15 Muscles That Move the Lower Leg

Muscle	*Origin*	*Insertion*	*Action*	*Innervation*
Flexors of the Leg:				
Biceps femoris	Tuberosity of ischium and linea aspera of femur	Head of fibula, lateral condyle of tibia	Flexes leg, extends and adducts thigh	Sciatic nerve, peroneal branch
Semimembranosus	Tuberosity of ischium	Posterior surface of medial condyle of tibia	Flexes leg, extends, adducts, and medially rotates thigh	Sciatic nerve
Semitendinosus	Tuberosity of ischium	Proximal, posteromedial surface of tibia near insertion of gracilis	Same	Same
Gracilis	Interior rami of pubis and ischium	Anterior surface of tibia inferior to medial condyle	Flexes leg and adducts thigh	Obturator nerve
Sartorius	Anterior superior spine of ilium	Medial surface of tibia near tibial tuberosity	Flexes leg, flexes and laterally rotates thigh	Femoral nerve
Popliteus	Proximal shaft of tibia	Lateral condyle of femur	Medially rotates tibia (or laterally rotates femur)	Sciatic nerve, tibial branch
Extensors of the Leg:				
Rectus femoris	Anterior inferior spine and superior acetabular rim of ilium	Tibial tuberosity via patellar ligament	Extends leg, flexes thigh	Femoral nerve
Vastus intermedius	Anterolateral surface of femur along linea aspera (distal half)	Same	Extends leg	Same
Vastus lateralis	Anterior and inferior to greater trochanter of femur and along linea aspera (proximal half)	Same	Same	Same
Vastus medialis	Entire length of linea aspera of femur	Same	Same	Same

magnus can produce either adduction and flexion or adduction and extension, depending on the region stimulated. These muscles insert upon low ridges along the posterior surface of the femur, and as a result their contractions also produce lateral rotation.

The medial surface of the pelvis is dominated by a single pair of muscles. The large **psoas major** arises alongside the lower thoracic and lumbar vertebrae, and its insertion lies on the lesser trochanter of the femur. Before reaching this insertion, its tendon merges with that of the **iliacus**, which lies nestled within the iliac fossa. These two muscles are powerful flexors of the thigh, and are often referred to as the **iliopsoas** muscle.

Muscles That Move the Lower Leg

Muscles involved with movements at the knee, ankle, and foot (Figure 11-21 and Table 11-15) follow the same distribution pattern noted in the section dealing with the arm. Extensors are found along the anterior and lateral surfaces of the leg, and flexors lie along the posterior and medial surfaces. Figure 11-21 includes a sectional view of the thigh. Dense connective tissue partitions separate these groups, and each division, or *compartment*, receives its own blood supply and innervation. Although the flexors and adductors originate on the pelvic girdle, most of the extensors originate on the femoral surface.

FIGURE 11-21
The Quadriceps Group. *See also Figures 9-11, 9-13, 9-14, and 9-16.*

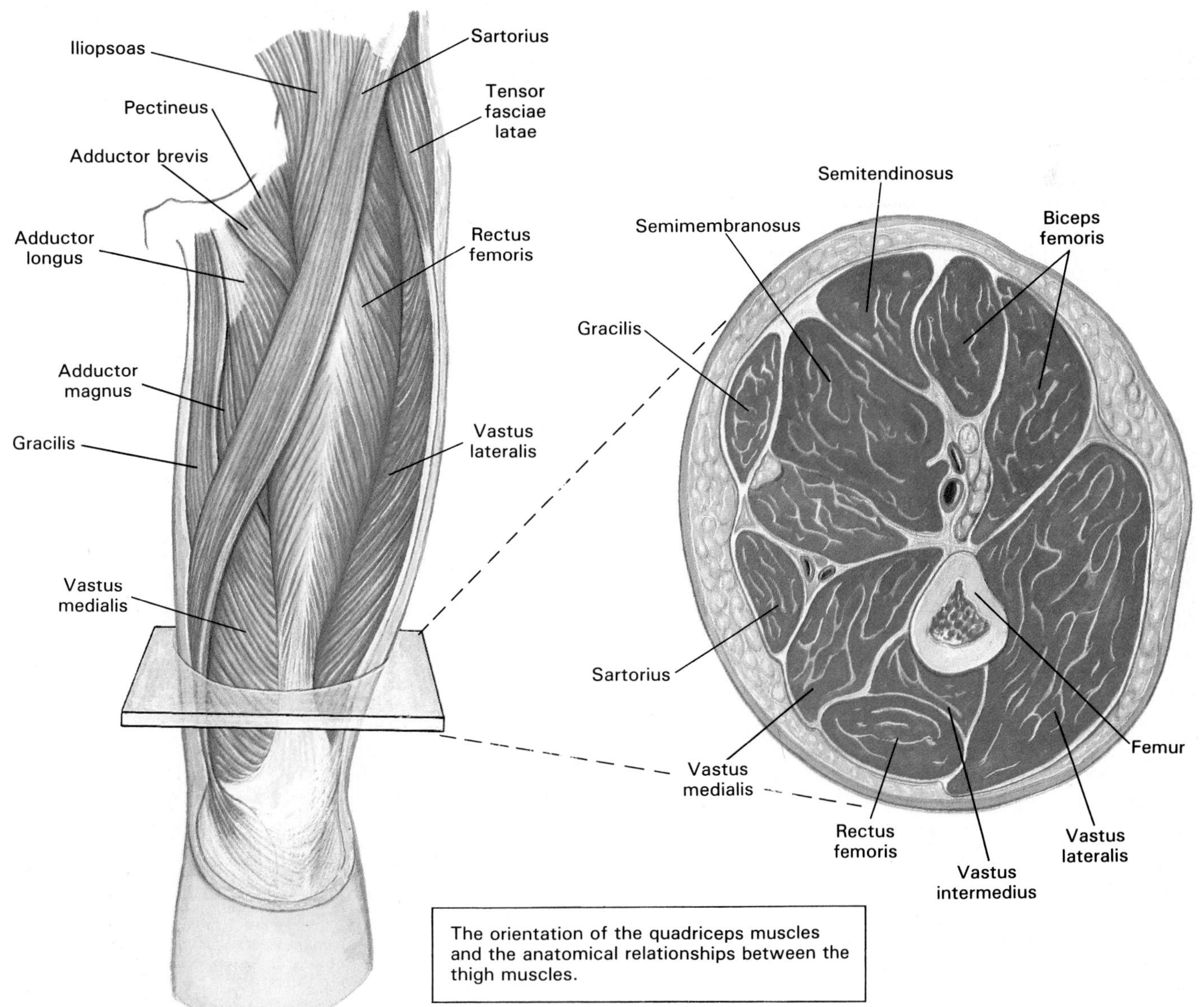

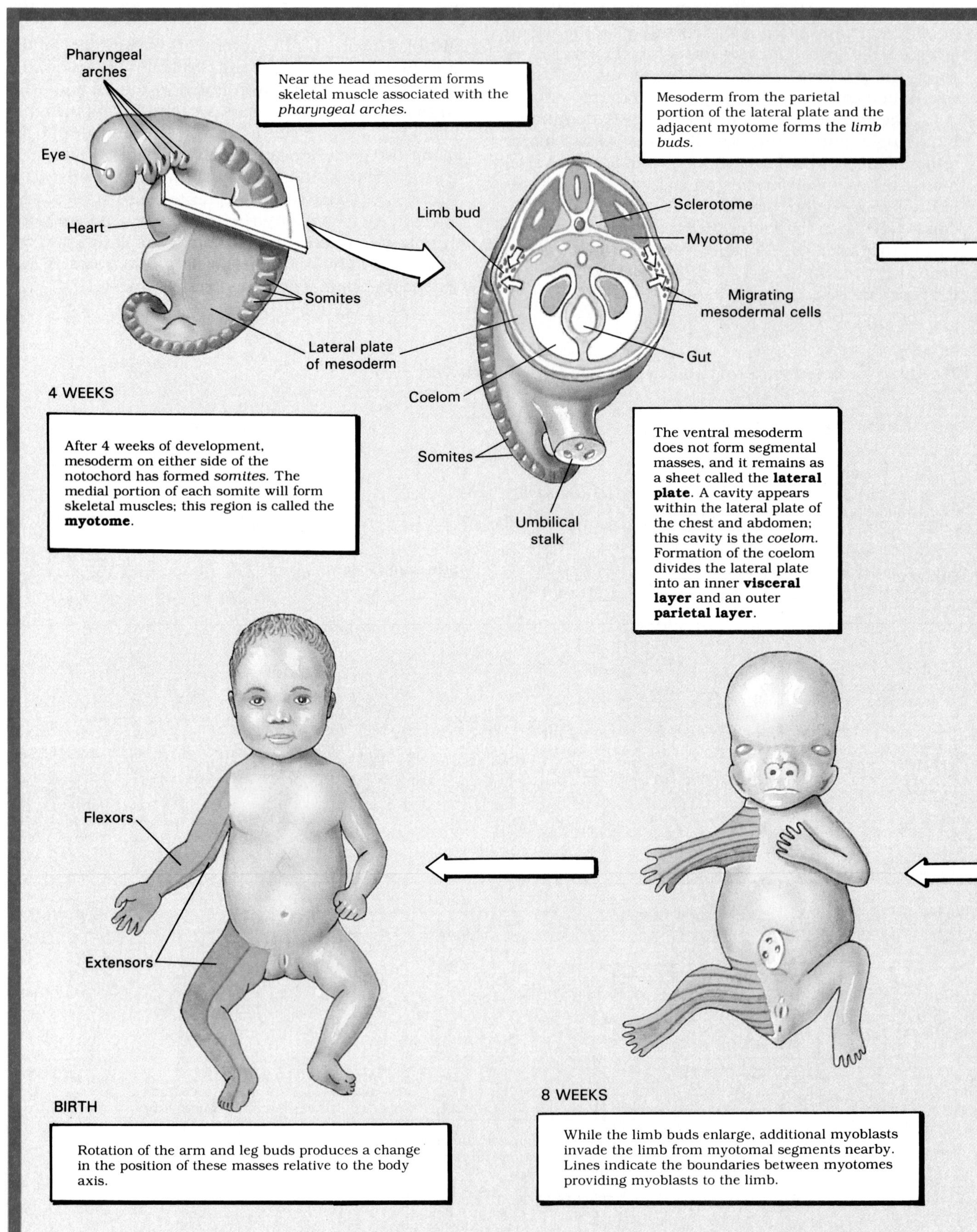

Pharyngeal arches
Eye
Heart
Somites
Lateral plate of mesoderm
4 WEEKS
Near the head mesoderm forms skeletal muscle associated with the *pharyngeal arches.*
Mesoderm from the parietal portion of the lateral plate and the adjacent myotome forms the *limb buds.*
Limb bud
Sclerotome
Myotome
Migrating mesodermal cells
Gut
Coelom
Somites
Umbilical stalk
After 4 weeks of development, mesoderm on either side of the notochord has formed *somites.* The medial portion of each somite will form skeletal muscles; this region is called the **myotome**.
The ventral mesoderm does not form segmental masses, and it remains as a sheet called the **lateral plate**. A cavity appears within the lateral plate of the chest and abdomen; this cavity is the *coelom.* Formation of the coelom divides the lateral plate into an inner **visceral layer** and an outer **parietal layer**.
Flexors
Extensors
BIRTH
Rotation of the arm and leg buds produces a change in the position of these masses relative to the body axis.
8 WEEKS
While the limb buds enlarge, additional myoblasts invade the limb from myotomal segments nearby. Lines indicate the boundaries between myotomes providing myoblasts to the limb.

Embryology Summary: Development of the Axial Musculature

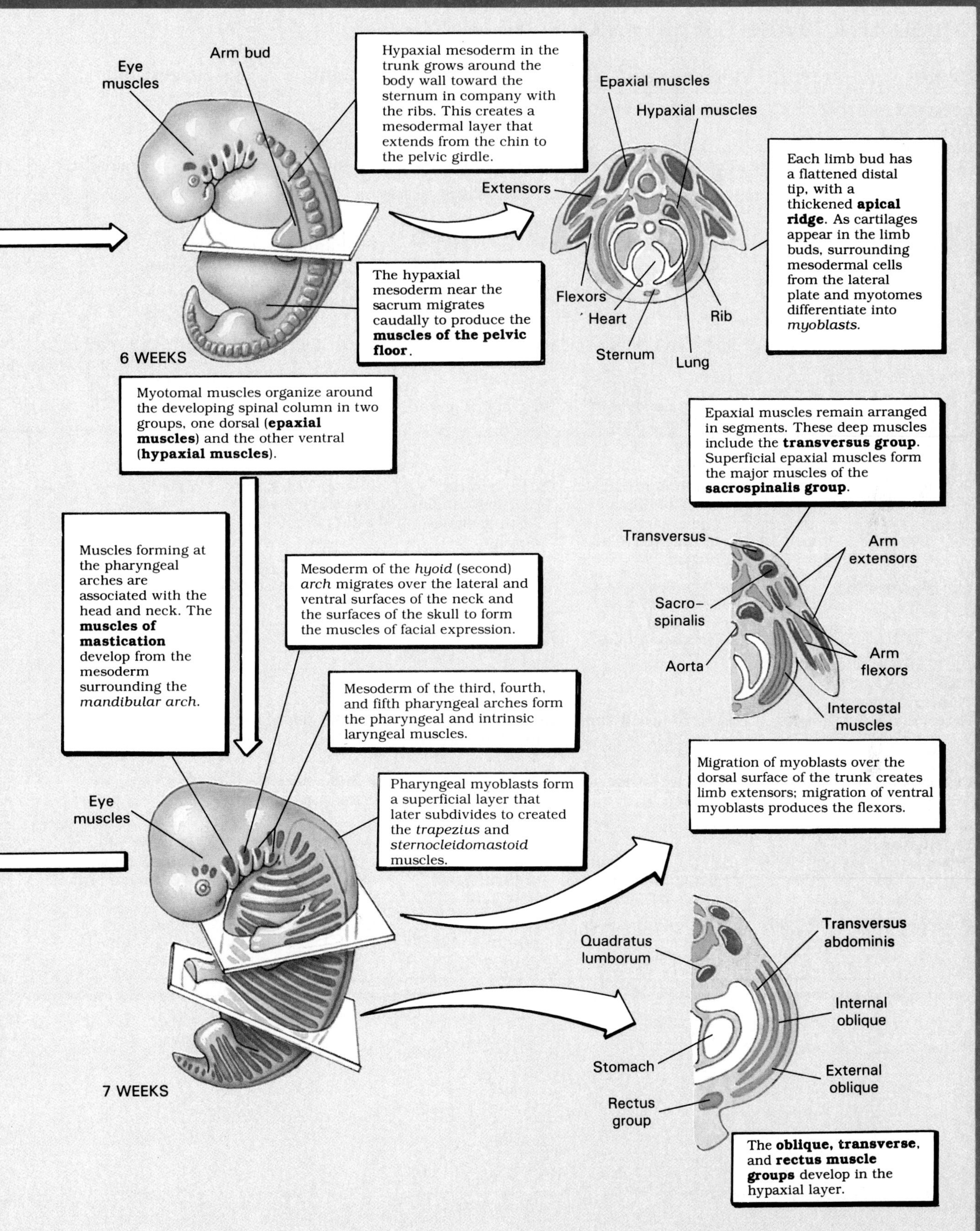

Eye muscles
Arm bud
Hypaxial mesoderm in the trunk grows around the body wall toward the sternum in company with the ribs. This creates a mesodermal layer that extends from the chin to the pelvic girdle.
Epaxial muscles
Hypaxial muscles
Extensors
Each limb bud has a flattened distal tip, with a thickened **apical ridge**. As cartilages appear in the limb buds, surrounding mesodermal cells from the lateral plate and myotomes differentiate into *myoblasts*.
The hypaxial mesoderm near the sacrum migrates caudally to produce the **muscles of the pelvic floor**.
Flexors
Heart
Rib
Sternum
Lung
6 WEEKS
Myotomal muscles organize around the developing spinal column in two groups, one dorsal (**epaxial muscles**) and the other ventral (**hypaxial muscles**).
Epaxial muscles remain arranged in segments. These deep muscles include the **transversus group**. Superficial epaxial muscles form the major muscles of the **sacrospinalis group**.
Transversus
Arm extensors
Muscles forming at the pharyngeal arches are associated with the head and neck. The **muscles of mastication** develop from the mesoderm surrounding the *mandibular arch*.
Mesoderm of the *hyoid* (second) *arch* migrates over the lateral and ventral surfaces of the neck and the surfaces of the skull to form the muscles of facial expression.
Sacro-spinalis
Aorta
Arm flexors
Mesoderm of the third, fourth, and fifth pharyngeal arches form the pharyngeal and intrinsic laryngeal muscles.
Intercostal muscles
Eye muscles
Pharyngeal myoblasts form a superficial layer that later subdivides to created the *trapezius* and *sternocleidomastoid* muscles.
Migration of myoblasts over the dorsal surface of the trunk creates limb extensors; migration of ventral myoblasts produces the flexors.
Quadratus lumborum
Transversus abdominis
Internal oblique
Stomach
External oblique
Rectus group
7 WEEKS
The **oblique, transverse,** and **rectus muscle groups** develop in the hypaxial layer.

TABLE 11-16 Muscles That Move the Ankle, Foot, and Toes

Muscle	*Origin*	*Insertion*	*Action*	*Innervation*
PRIMARY ACTION AT THE ANKLE				
Dorsiflexors				
Tibialis anterior	Lateral condyle and proximal shaft of tibia	Base of first metatarsal	Dorsiflexes foot	Sciatic nerve, deep peroneal branch
Plantar flexors				
Gastrocnemius	Above femoral condyles	Calcaneus via Achilles tendon	Plantar flexes, inverts, and adducts foot; flexes leg	Sciatic nerve, tibial branch
Peroneus brevis	Midlateral margin of fibula	Base of 5th metatarsal	Everts foot	Sciatic nerve, superficial peroneal branch
longus	Lateral condyle of tibia and head of fibula	Base of 1st metatarsal	Everts and plantar flexes foot; supports longitudinal arch	Same
Soleus	Head and proximal shaft of fibula, and adjacent posteromedial shaft of tibia	Calcaneus via Achilles tendon (with gastrocnemius)	Plantar flexes, inverts, and adducts foot	Sciatic nerve, tibial branch
Tibialis posterior	Interosseous membrane and adjacent shafts of tibia and fibula	Tarsals and metatarsals	Adducts and inverts foot	Same
PRIMARY ACTION AT THE TOES				
Plantar flexors				
Flexor digitorum longus	Posteromedial surface of tibia	Inferior surfaces of phalanges, toes 2–5	Plantar flexes toes 2–5	Same
Flexor hallucis longus	Posterior surface of fibula	Inferior surface, terminal phalanx of big toe	Plantar flexes big toe	Same
Dorsiflexors				
Extensor digitorum longus	Lateral condyle of tibia, anterior surface of fibula	Superior surfaces of phalanges, toes 2–5	Dorsiflexes toes 2–5	Sciatic nerve, deep peroneal branch
Extensor hallucis longus	Anterior surface of fibula	Superior surface, terminal phalanx of big toe	Dorsiflexes big toe	Same

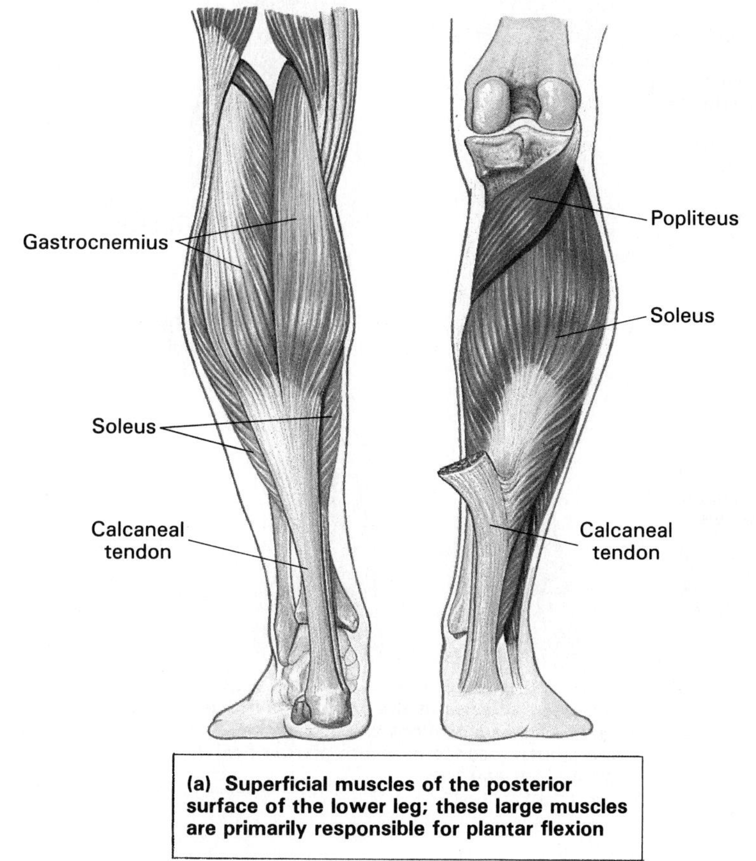

(a) Superficial muscles of the posterior surface of the lower leg; these large muscles are primarily responsible for plantar flexion

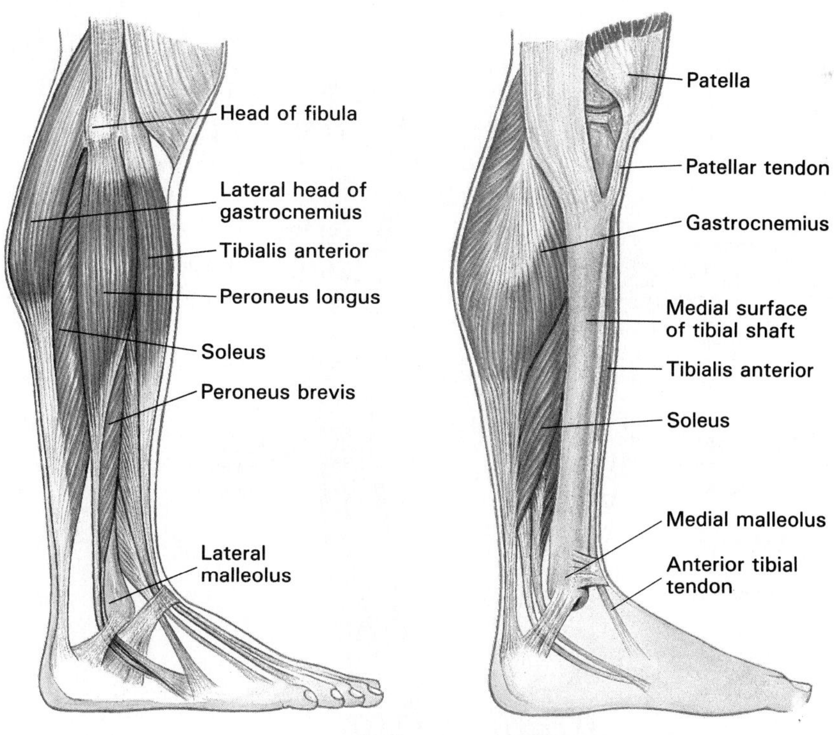

(b) Lateral and medial views

FIGURE 11-22
Muscles That Move the Ankle, Foot, and Toes. *See also Figures 9-13, 9-14, 9-15, and 9-18.*

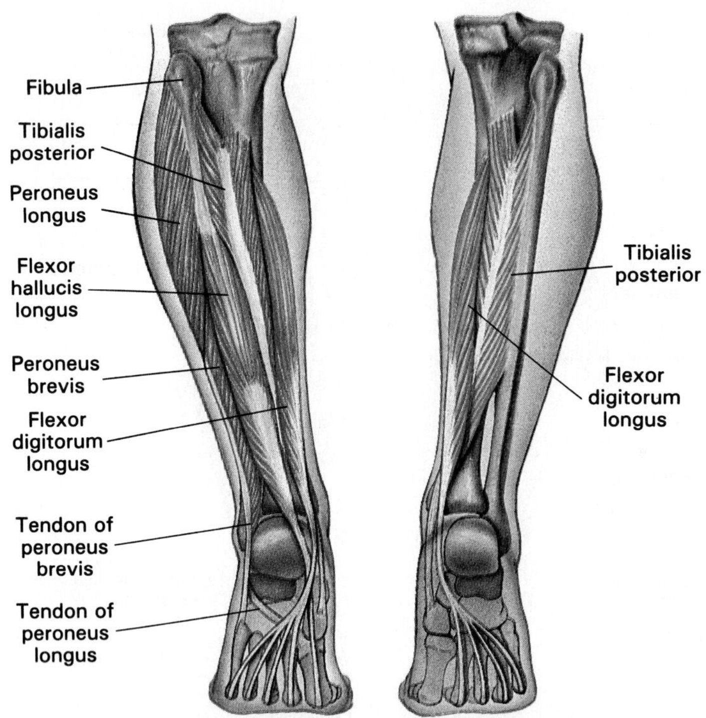

(c) Deep muscles of the posterior surface of the lower leg; these smaller muscles are primarily concerned with plantar flexion of the foot and toes

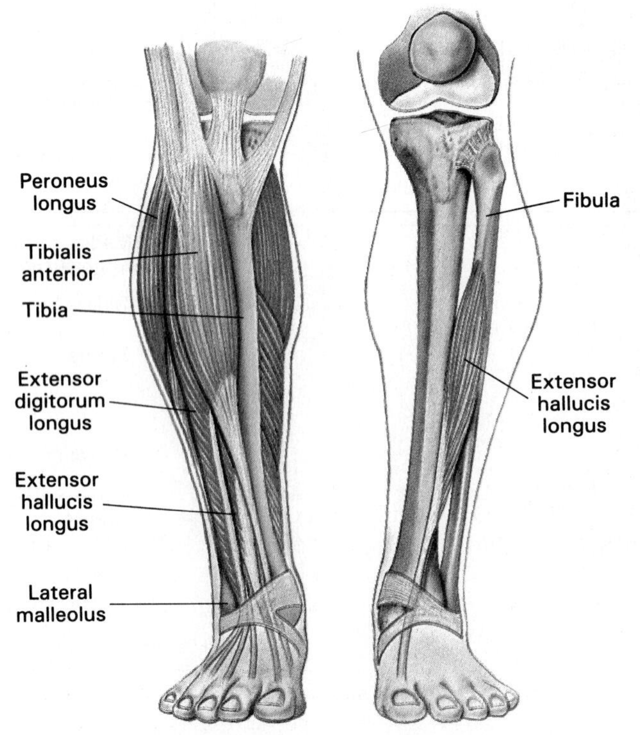

(d) These muscles on the anterior surface of the lower leg are primarily concerned with dorsiflexion of the foot and extension of toes

Clinical Comment: Intramuscular Injections

Drugs are frequently injected into tissues, rather than directly into the circulation. This method enables the physician to introduce a large amount of a drug at one time, yet have it enter the circulation gradually. In an **intramuscular (IM) injection** the drug is introduced into the mass of a large skeletal muscle. Uptake is usually faster and accompanied by less tissue irritation than when drugs are administered *intradermally* or *subcutaneously* (injected into the dermis or subcutaneous layers; see Figure 6-2). (p. 185). Up to 5 mℓ of fluid may be injected at one time, and multiple injections are possible.

The most common complications involve accidental injection into a blood vessel or piercing of a nerve. The sudden entry of massive quantities of a drug into the blood stream can have unpleasant or even fatal consequences, while damage to a nerve can cause motor paralysis or sensory loss. As a result, the site of injection must be selected with care. Bulky muscles that contain few large vessels or nerves make ideal targets, and the posterior, lateral, superior portion of the gluteus maximus is often selected. The deltoid muscle of the arm, about 1 inch distal to the acromion, is another popular site. Probably the most satisfactory from a technical point of view is the vastus lateralis of the thigh, for an injection into this thick muscle will not encounter vessels or nerves. This is the preferred site in infants and young children whose gluteal and deltoid muscles are relatively small.

The *flexors of the leg* include five muscles collectively known as the **hamstrings**: the **biceps femoris**, the **semimembranosus**, the **semitendinosus**, the **gracilis**, and the **sartorius**. These muscles arise along the edges of the pelvis and insert upon the tibia and fibula, and their contractions produce flexion of the knee. Because the biceps femoris, semimembranosus, and semitendinosus originate on the pelvic surface inferior and posterior to the acetabulum, their contractions also produce extension of the hip. The gracilis takes its origin along the anterior and inferior margin of the pubis, so its contraction leads to hip flexion and adduction. The sartorius is the only knee flexor that originates superior to the acetabulum, and its insertion lies along the medial aspect of the tibia. When it contracts, it flexes and laterally rotates the thigh, as when crossing the legs.

In Chapter 9 we noted that the knee joint can be locked at full extension by a slight lateral rotation of the tibia. (p. 289) The small **popliteus** (pop-LI-tē-us) muscle arises on the posterior tibial shaft and inserts on the femur near the lateral condyle. When flexion is initiated, this muscle contracts to produce a slight medial rotation of the tibia that unlocks the joint.

Collectively the *knee extensors* are known as the **quadriceps femoris**. The three **vastus** muscles originate along the body of the femur, and with the assistance of the **rectus femoris** they cradle the femur the way a bun surrounds a hot dog. All four muscles insert upon the patella and reach the tibial tuberosity by way of the patellar ligament. The rectus originates on the anterior inferior iliac spine, so in addition to producing extension of the knee it can assist in flexion of the thigh.

Muscles That Move the Ankle, Foot, and Toes

Muscles that move the ankle, foot, and toes are shown in Figure 11-22 and detailed in Table 11-16 on page 362. Most of the muscles that move the ankle produce the plantar flexion involved with walking and running movements. The large **gastrocnemius** (gas-trok-NĒ-mē-us; *gaster*, stomach + *kneme*, knee) of the calf is an important plantar flexor, but the slow muscle fibers of the underlying **soleus** are more powerful. The gastrocnemius arises from two heads located on the medial and lateral epicondyles of the femur just proximal to the knee. A sesamoid bone, the **fabella**, is usually found within the gastrocnemius. The gastrocnemius and soleus muscles share a common tendon, the **calcaneal**, or *Achilles*, **tendon**. A pair of deep **peroneus** muscles produce eversion as well as plantar flexion. Inversion is caused by contraction of the **tibialis** muscles; the large **tibialis anterior** opposes the gastrocnemius and dorsiflex the foot.

Important digital muscles originate on the surface of the tibia, the fibula, or both (Figure 11-22c,d). Smaller muscles of digital extension are found on the dorsal surfaces of the metatarsals. Others concerned with digital flexion originate at the anterior border of the calcaneus; their tone contributes to maintenance of the longitudinal arches of the foot.

√ **What leg movement would be impaired by injury to the obturator muscle?**

√ **You often hear of athletes suffering a "pulled hamstring." To what does this phrase refer?**

√ **How would you expect a torn achilles tendon to affect movement of the foot?**

Clinical Comment: Compartment Syndromes

In the arms and legs the interconnections between the superficial fascia, the deep fascia of the muscles, and the periostea of the appendicular skeleton are quite substantial. The muscles within a limb are effectively isolated in **compartments** formed by dense collagenous sheets, as shown in Figure 11-23. Blood vessels and nerves traveling to specific muscles within the limb enter and branch within the appropriate compartments.

When a crushing injury, severe contusion, or strain occurs, the blood vessels within one or more compartments may be damaged. These compartments then become swollen with blood and fluid leaked from damaged vessels. Because the connective tissue partitions are very strong, the accumulated fluid cannot escape, and pressures rise within the affected compartments. Eventually compartment pressures may become so high that they compress the regional blood vessels and eliminate the circulatory supply to the muscles and nerves of the compartment. This produces a condition of **ischemia** (is-KĒ-mē-a), or "blood starvation," known as the **compartment syndrome**.

Slicing into the compartment along its longitudinal axis or implanting a drain are emergency measures used to relieve the pressure. If such steps are not taken, the contents of the compartment will suffer severe damage. Nerves in the affected compartment will be destroyed after 2–4 hours of ischemia, although they can regenerate to some degree if the circulation is restored. After 6 hours or more, the muscle tissue will also be destroyed, and no regeneration will occur. The muscles will be replaced by scar tissue, and shortening of the connective tissue fibers may result in *contracture*.

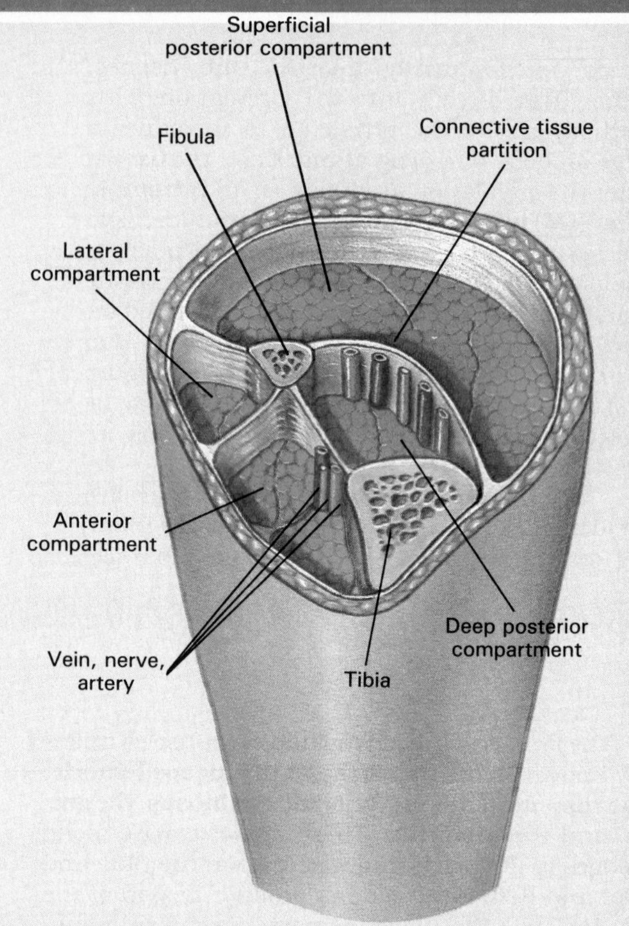

FIGURE 11-23
Musculoskeletal Compartments. A diagrammatic section through the lower leg, with the muscles removed to show the arrangement of the compartments. A section through the upper leg or arm would show a comparable arrangement of dense connective tissue partitions. The anterior and lateral compartments of the leg contain muscles of the extensor/dorsiflexor series, while the posterior compartments ensheath the flexor/plantar flexor muscles.

Sports and Fitness: Sports Injuries

Chapters 7-10 cited a number of health benefits attributed to regular exercise. For example, the application of stress to the skeleton encourages osteoblast activity and reduces the rate of bone loss with aging (see Chapter 7). (p. 224) Yet exercise carries a number of risks due to the stresses placed on muscles, joints, and connective tissues.

Many Americans participate in exercise programs and sports on a regular basis; more

FIGURE 11-24
Common Sports Injuries

Baseball finger: a tear of the tendon that extends the fingertip, caused by the impact of a baseball on the extended finger

Pitcher's arm: an inflammation at the origin of the flexor carpi muscles at the medial epicondyle of the humerus, from flexing of the wrist just prior to the release of the ball

Blocker's arm: the formation of heterotopic bone along the lateral surface of the forearm, due to repeated impact with an opponent

Kicker's ankle: the formation of bony spurs on the anterior surface of the tibia and proximal portion of the talus, due to repeated contact with the football

Charleyhorse: a painful tear and contusion of a muscle, often the entire quadriceps group (leg extensors)

Pulled groin: a strain affecting the adductor muscles of the thigh

Pulled hamstrings: a strain affecting the hamstring muscles (leg flexors)

Shin splints: an inflammation of the dorsiflexor muscles over the anterior surface of the lower leg

Tennis elbow: inflammation at the origin of the extensor carpi muscles at the lateral epicondyle of the humerus; these muscles are tensed while performing a backhand swing, and contact with the ball stresses the muscle attachments

Tennis leg: a partial tear at the lateral origin of the gastrocnemius muscle as a result of playing hard with the body weight transferred to the toes rather than the heels than 30 million Americans go jogging, and millions more participate in various amateur and professional sports. As a result, *sports injuries* are very common, and sports medicine has become an active area of professional and academic research interest.

Sports injuries affect amateurs and professionals alike. For example, in 1986 the approximately 1500 players in the National Football League reported 1500 injuries. At the amateur level, a 5-year study of college football players indicated that 73.5 percent experienced mild injuries, 21.5 percent moderate injuries, and 11.6 percent severe injuries in the course of their playing careers. Contact sports are not the only activities that show a significant injury rate; a study of 1650 joggers running at least 27 miles per week reported 1819 injuries in a single year.

Muscles and bones respond to increased use by enlarging and strengthening. Poorly conditioned individuals are therefore more likely to subject their bones and muscles to intolerable stresses. Training is also important in minimizing the use of antagonistic muscle groups and keeping joint movements within the intended ranges of motion. Planned warm-up exercises before athletic events stimulate circulation, improve muscular performance and control, and help prevent injuries to muscles, joints, and ligaments. Stretching exercises stimulate muscle circulation and help keep ligaments and joint capsules supple. Such conditioning extends the range of motion and prevents sprains and strains when sudden loads are applied.

Dietary planning can also be important in preventing injuries to muscles during endurance events, such as marathon running. Emphasis has often been placed on the importance of carbohydrates, leading to the practice of "carbohydrate loading" before a marathon. But muscles also utilize amino acids extensively while operating within aerobic limits, and an adequate diet must include both carbohydrates and proteins.

Improved playing conditions, equipment, and regulations also play a role in reducing the incidence of sports injuries. Jogging shoes, ankle or knee braces, helmets, and body padding are examples of equipment that can be effective. The substantial penalties now earned for "personal fouls" in contact sports have reduced the numbers of neck and knee injuries.

Several injuries common to those engaged in active sports may also affect nonathletes, although the primary causes may differ considerably. A partial listing of activity-related conditions would include the following:

Bone bruise: bleeding within the periosteum of a bone

Bursitis: inflammation of the bursae around one or more joints

Stress fractures: cracks or breaks in bones subjected to repeated stresses or trauma

Muscle cramps: prolonged, involuntary, and painful muscular contractions

Sprains: tears or breaks in ligaments or tendons

Strains: tears or breaks in muscles

Tendinitis: inflammation of the connective tissue surrounding a tendon

Many of these conditions have been discussed in previous chapters. Other injuries are characteristic of particular sports—examples are included in Figure 11-24.

Finally, many sports injuries would be prevented if people who engage in regular exercise would use common sense and recognize their personal limitations. It can be argued that some athletic events, such as the "ultramarathon," place such excessive stresses on the cardiovascular, muscular, respiratory, and urinary systems that they cannot be recommended, even for athletes in peak condition.

CHAPTER REVIEW

Review of Selected Clinical Terms

hernia (*p. 345*)
A condition involving an organ or body part that protrudes through an abnormal opening.

inguinal hernia (*p. 345*)
A condition in which the inguinal canal enlarges and abdominal contents are forced into the inguinal canal.

diaphragmatic hernia (hiatal hernia) (*p. 345*)
A hernia that occurs when abdominal organs slide into the thoracic cavity.

rotator cuff (*p. 351*)
The muscles that surround the shoulder joint; a frequent site of sports injuries.

carpal tunnel syndrome (*p. 355*)
An inflammation of the sheath surrounding the flexor tendons of the palm.

intramuscular (IM) injection (*p. 365*)
Administering a drug by injecting it into the mass of a large skeletal muscle.

ischemia (is-KĒ-mē-a) (*p. 366*)
A condition of "blood starvation" resulting from compression of regional blood vessels.

compartment syndrome (*p. 366*)
Ischemia resulting from accumulated blood and fluid trapped within a musculoskeletal compartment.

bone bruise (*p. 368*)
Bleeding within the periosteum of a bone.

bursitis (*p. 368*)
Inflammation of the bursae around one or more joints.

stress fractures (*p. 368*)
Cracks or breaks in bones subjected to repeated stresses or trauma.

muscle cramps (*p. 368*)
Prolonged, involuntary, painful muscular contractions.

sprains (*p. 368*)
Tears or breaks in ligaments or tendons.

strains (*p. 368*)
Tears or breaks in muscles.

tendinitis (*p. 368*)
Inflammation of the connective tissue surrounding a tendon.

Additional Terms of Clinical Importance

HERNIAS (*p. 345*): **inguinal canals, esophageal hiatus**

COMPARTMENT SYNDROMES (*p. 366*): **compartments**

Study Outline

Related Key Terms

Introduction (pp. 327–328)

1. The **muscular system** includes all the skeletal muscle tissue that can be controlled voluntarily. The analysis of biological systems in mechanical terms is the study of **biomechanics.**

Biomechanics and Muscle Anatomy (pp. 328–332)

ORGANIZATION OF SKELETAL MUSCLE FIBERS (pp. 328–329)

1. A muscle can be classified according to the arrangement of fibers and fascicles as a: **parallel muscle**, **convergent muscle**, **pennate muscle**, or a **circular muscle** or **sphincter**. A pennate muscle may be **unipennate**, **bipennate**, or **multipennate**. — *Related Key Terms:* **body • raphe**

SKELETAL MUSCLE LENGTH-TENSION RELATIONSHIPS (pp. 329–330)

2. Skeletal muscles operate most efficiently over a relatively narrow range of sarcomere lengths.

LEVERS (pp. 330–332)

3. A **lever** can change the direction, speed, and distance of muscle movements, and modify the force applied to them. — *Related Key Terms:* **fulcrum**
4. Levers may be classified as **first class**, **second class**, or **third class levers**; the last are the most common type of lever in the body.

Muscle Terminology (pp. 332–335)

ORIGINS, INSERTIONS, AND ACTIONS (p. 332)

1. Each muscle may be identified by its **origin**, **insertion**, and **primary action**. A muscle may be classified as a **prime mover** or **agonist**, a **synergist**, or an **antagonist**.

NAMES OF SKELETAL MUSCLES (pp. 332–335)

2. The names of muscles often provide clues to their location, orientation, or function (see Table 11-1).

Chapter Review

3. The **axial musculature** arises on the axial skeleton; it positions the head and spinal column and moves the ribcage. The **appendicular skeleton** stabilizes or moves components of the appendicular skeleton.

Related Key Terms

The Axial Musculature (pp. 335–348)

1. The axial muscles fall into logical groups based on location and/or function.
2. **Innervation** refers to the identity of the nerve that controls a muscle.

MUSCLES OF THE HEAD AND NECK (pp. 336–340)

3. Six **oculomotor muscles** control the position of the eye.

MUSCLES OF THE SPINE (pp. 340–341)

4. The superficial muscles of the spine can be classified into the **spinalis**, **longissimus**, and **iliocostalis** divisions. In the lower lumbar and sacral regions the longissimus and iliocostalis are sometimes called the **sacrospinalis** muscles.
5. Other muscles of the spine include the **transverse muscles**, the **longus capitis**, the **longus cervicis**, and the **quadratus lumborum** muscles.

orbicularis oris
buccinator • **epicranius**
frontalis • **occipitalis**
galea aponeurotica • **platysma**
inferior rectus • **lateral rectus**
medial rectus • **superior rectus**
inferior oblique
superior oblique • **masseter**
temporalis • **pterygoid**
palatoglossus • **styloglossus**
genioglossus • **hyoglossus**
digastricus • **mylohyoid**
stylohyoid

OBLIQUE AND RECTUS MUSCLES (pp. 341–348)

6. The oblique muscles include the **scalenes** and the **intercostal** and **transversus** muscles. The **external intercostals** and the **internal intercostals** are important in respiratory movements of the ribs. Also important to respiration is the **diaphragm**.

MUSCLES OF THE PELVIC FLOOR (p. 348)

7. The **perineum** can be divided into an **anterior** or **urogenital triangle** and a **posterior** or **anal triangle**. The pelvic floor consists of the **urogenital diaphragm** and the **pelvic diaphragm**.

transversus thoracis
external oblique
internal oblique
transversus abdominis
rectus abdominis • **linea alba**
tendinous inscriptions

The Appendicular Musculature (pp. 348–368)

MUSCLES OF THE SHOULDERS AND ARMS (pp. 348–355)

1. The **trapezius** and the **sternocleidomastoid** affect the position of the shoulder girdle, head, and neck. Other muscles inserting on the scapula include the **rhomboideus**, the **levator scapulae**, the **serratus anterior**, the **subclavius**, and the **pectoralis minor**.
2. The **deltoid** and the **supraspinatus** are important abductors. The **subscapularis** and the **teres major** rotate the arm medially; the **infraspinatus** and **teres minor** perform lateral rotation; and the **coracobrachialis** flexes and adducts the humerus.
3. The **pectoralis major** flexes the upper arm, while the **latissimus dorsi** extends it.

rotator cuff

4. The primary actions of the **biceps brachii** and the **triceps** (long head) affect the elbow. The **brachialis** and **brachioradialis** flex the forearm, opposed by the **anconeus**. The **flexor carpi ulnaris**, the **flexor carpi radialis**, and the **palmaris longus** cooperate to flex the wrist. They are opposed by the **extensor carpi radialis** and the **extensor carpi ulnaris**. The **pronator teres** and **pronator quadratus** pronate the forearm, opposed by the **supinator**.

tendon sheaths

MUSCLES OF THE LEGS (pp. 355–365)

5. **Gluteal muscles** cover the lateral surface of the ilium. The largest is the **gluteus maximus**, which shares an insertion with the **tensor fasciae latae**; together these muscles pull on the **iliotibial tract**.

gluteus medius
gluteus minimus

6. The **piriformis** and the **obturator muscles** are the most important **lateral rotators**. The **adductors** can produce a variety of movements.
7. The **psoas major** and the **iliacus** merge to form the **iliopsoas** muscle, a powerful flexor of the thigh.

adductor magnus
adductor brevis
adductor longus • **pectineus**

8. The **flexors of the leg**, or **hamstrings**, include the **biceps femoris**, **semimembranosus**, **semitendinosus**, **gracilis**, and **sartorius**. The **popliteus** unlocks the knee joint.
9. Collectively the **knee extensors** are known as the **quadriceps femoris**. This group includes the three **vastus** muscles and the **rectus femoris**.
10. The **gastrocnemius** and **soleus muscles** produce plantar flexion. A pair of **peroneus muscles** produce eversion as well as plantar flexion.

fabella • **calcaneal tendon**
tibialis • **tibialis anterior**

Review Planner

	Objective	Level 1	Level 2	Level 3
1	Describe the arrangement of fascicles in the various types of muscles, and explain the resulting functional differences.	**1 6 9 18**	**17 21**	**33–36**
2	Describe the different classes of levers and how they make muscles more efficient.	**3 23**		
3	Predict the actions of a muscle on the basis of its origin and insertion.	**19 31 32**		
4	Explain how muscles interact to produce or oppose movements.	**2 5 7 8 20**		
5	Use the name of a muscle to help identify and remember its location, appearance, and function.	**25 31 32**		
6	Identify the principal axial muscles of the body, together with their origins and insertions.	**10 13 15 16 31**		
7	Identify the principal appendicular muscles of the body, together with their origins and insertions.	**11 12 14 22 24 32**		
8	Compare the major muscle groups of the arms and legs and relate their differences to their functional roles.	**11 12 14 22 24 32**		

Review Questions

MULTIPLE CHOICE

1. Most of the skeletal muscles in the body are __________ muscles: a) parallel; b) convergent; c) pennate; d) circular.
2. The muscular system includes all of the skeletal muscle tissue that can be controlled: a) involuntarily; b) voluntarily; c) both voluntarily and involuntarily; d) none of the above.
3. If a bone is a lever, its fulcrum is a(n): a) axial muscle; b) appendicular muscle; c) joint; d) tendon.
4. Innervation refers to a particular __________ that controls a muscle: a) hormone; b) nucleus; c) catalyst; d) nerve.

MATCHING QUESTIONS

(*Match each term with the most appropriate statement.*)

Statements:

5. A muscle whose actions oppose those of an agonist.
6. A muscle with concentric fibers.
7. A muscle that assists an agonist in performing a particular action.
8. A muscle whose contraction produces a particular movement.

Terms:

a) Synergist
b) Prime mover
c) Antagonist
d) Sphincter

Chapter Review

SHORT ANSWER ESSAY

9. Describe the four general categories of muscles, and give an example of each.

Identify at least one muscle involved in each of the following movements:

10. Chewing a mouthful of pizza.
11. Shrugging the shoulders.
12. Throwing a textbook across the room.
13. "Pursing" the lips to whistle.
14. Dorsiflexing the foot.
15. Taking a deep breath.
16. Opening the mouth.
17. Define *diaphragm* and name three diaphragm muscles. Where are they found?
18. Why do skeletal muscles become less efficient when stretched too far or contracted too much?
19. Fill in the blanks:
    Each muscle ends at a(n) ________, contracts to produce a specific ________, and begins at a(n) ________.
20. Explain how muscles working in groups can improve the efficiency of movement.
21. Your tennis coach tells you to practice your serve and be sure to "follow through" after you hit the ball. What muscles are you using to perform this activity? How do they work together?
22. Identify the muscles that make up the hamstrings. What are their functions? Name an activity that involves them.
23. Why are levers important in the production of normal movement?
24. After a hard day you go home, kick off your shoes, and wiggle your toes. What muscles are you using?

Identify the following terms:

25. Rectus muscles.
26. Obliquus muscles.
27. Magnus/major/maximus muscles.
28. Brevis muscles.
29. Extrinsic muscles.
30. Internus or profundus muscles.

DIAGRAMS

31. Complete the following table:

Muscle	*Group*	*Origin*	*Insertion*	*Action*	*Innervation*
Masseter	Muscles of mastication	Zygomatic arch	Mandible	Elevates mandible	N V
	Rectus group	Xiphoid process and rib cartilages			Intercostal nerves
		Pterygoid processes			N V
Splenius			Occipital bone and mastoid process		
	Muscles of spine, deep flexors	Iliac crest	Twelfth rib		
		Mandible at chin	Hyoid bone		
Styloglossus					
External intercostals					
Transversus thoracis					

32. Complete the following table:

Muscle	*Origin*	*Insertion*	*Action*	*Innervation*
	Lumbodorsal fascia			
			Extends, and laterally rotates thigh	
		Scapular spine, acromion, clavicle		**N XI, and cervical nerves**
Deltoid				
	Lateral epicondyle		**Flexes and abducts wrist**	
		Greater tubercle of humerus **Pectineal line of femur**	**Abducts arm**	**Femoral nerve**
Levator scapulae				
	Subscapular fossa			
		Vertebral border of scapula from spine to inferior angle		
		Lateral to styloid process of radius		
Tibialis anterior				

CRITICAL THINKING/APPLICATIONS

33. While lifting his 8-year-old daughter one day, Jim experiences a sharp pain in his groin. He can feel a small bulge in his inguinal region. a) What is your diagnosis? b) Describe possible causes of this problem. c) Would you expect it to be more common in men or in women, or equally common in both? Explain your answer.
34. After a hard night of studying you awaken with a painful sensation that extends down the middle of your neck and back, fanning out in a diamond shape. What may have caused this pain, and which muscle is most likely affected? Where does it originate and insert?
35. In searching for a good aerobics class, you visit several health clubs and observe classes in progress. Describe the features that you would evaluate in choosing a safe and effective exercise class.
36. Discuss the benefits and drawbacks of intramuscular injection. What are the best sites for injection?

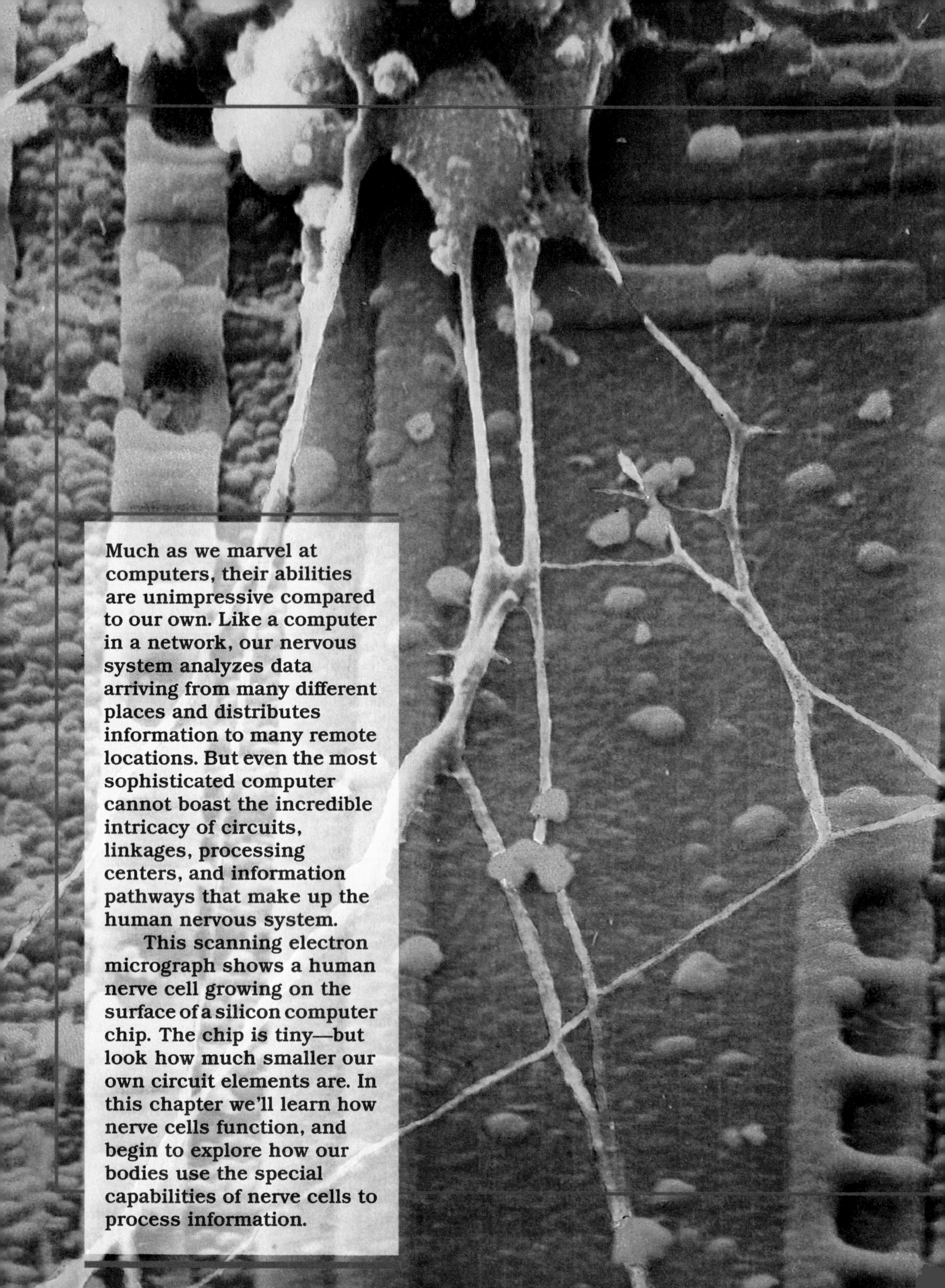

Much as we marvel at computers, their abilities are unimpressive compared to our own. Like a computer in a network, our nervous system analyzes data arriving from many different places and distributes information to many remote locations. But even the most sophisticated computer cannot boast the incredible intricacy of circuits, linkages, processing centers, and information pathways that make up the human nervous system.

This scanning electron micrograph shows a human nerve cell growing on the surface of a silicon computer chip. The chip is tiny—but look how much smaller our own circuit elements are. In this chapter we'll learn how nerve cells function, and begin to explore how our bodies use the special capabilities of nerve cells to process information.

CHAPTER 12

The Nervous System: Neural Tissue

Chapter Objectives

After reading this chapter, you will be able to:

1 Describe the anatomical organization and general functions of the nervous system.
2 Distinguish between neurons and neuroglia and compare their structures and functions.
3 Discuss the events that generate action potentials in the membranes of nerve cells.
4 Identify the factors that determine the frequency and speed of nerve impulse conduction.
5 Explain the mechanism of synaptic transmission and describe the types and effects of the most important neurotransmitters.
6 Discuss the interactions that make possible the processing of information in neural tissue.
7 Classify neurons into functional groups and discuss the interactions between these groups.

■ Introduction

In the next seven chapters our attention will shift to mechanisms that coordinate the activities of the body's organ systems. These activities must be tightly controlled and adjusted to meet changing environmental conditions. Two organ systems, the *nervous system* and the *endocrine system*, provide the necessary regulation. These systems share several structural and functional characteristics, and they usually act in a complementary fashion. The nervous system provides relatively swift but brief responses to stimuli, usually by temporarily modifying the activities of other organ systems. The modifications may appear almost immediately—in 1–2 msec—but the effects disappear soon after neural activity ceases.

The endocrine system adjusts the metabolic operations of other systems in response to changes in the availability of nutrients and the demand for energy. It also directs activities that continue for extended periods, such as growth and maturation, sexual development, pregnancy, or responses to chronic environmental stresses. In general, endocrine responses are slower to develop, but often last much longer than those of the nervous system.

This chapter introduces basic principles of neural function. Subsequent chapters explore increasing levels of complexity.

The Nervous System: An Overview

The **nervous system** includes all of the **neural tissue** in the body. Neural tissue, introduced in Chapter 5, carries information or instructions from one region of the body to another. (p. 157) The functions of the nervous system include:

1. Providing sensation of the internal and external environments;
2. Integrating sensory information;
3. Coordinating voluntary and involuntary activities; and
4. Regulating or controlling peripheral structures and systems.

Each of these functions will be considered in subsequent chapters of this unit.

There are two major *anatomical* subdivisions of the nervous system: the central nervous system and the peripheral nervous system. These divisions were introduced in the Systems Overview (see Figure S0-6). (p. 172) The **central nervous system (CNS)** consists of the *brain* and *spinal cord*. The CNS is responsible for integrating, processing, and coordinating sensory data and motor commands. It is also the seat of higher functions, such as intelligence, memory, learning, and emotion. The **peripheral nervous system (PNS)** includes all of the neural tissue outside the CNS. The PNS provides sensory information to the CNS and carries motor commands to peripheral tissues and systems.

Figure 12-1 shows the functional divisions of the nervous system. The **afferent division** brings sensory information to the CNS, and the **efferent division** carries motor commands to muscles and glands. The efferent division has *somatic* and *visceral* components. The **somatic nervous system (SNS)** provides voluntary control over skeletal muscle contractions. The *visceral motor system*, or **autonomic nervous system (ANS)**, provides automatic, involuntary regulation of smooth muscle, cardiac muscle, and glandular activity or secretions.

The CNS and PNS are not simply masses of neural tissue; they are complex organs that include blood vessels and the connective tissues that provide physical protection and mechanical support. Nevertheless, all of the varied and essential functions of the nervous system are performed by individual nerve cells that must be kept safe, secure, and fully functional. Our discussion of the nervous system therefore begins at the cellular level, with the detailed histology of neural tissue.

Cellular Organization in Neural Tissue

Neural tissue includes two distinct cell populations: nerve cells, or *neurons*, and supporting cells, or *neuroglia*. **Neurons** (*neuro*, nerve) are responsible for information transfer and processing in the nervous system. Neuron structure was briefly considered in

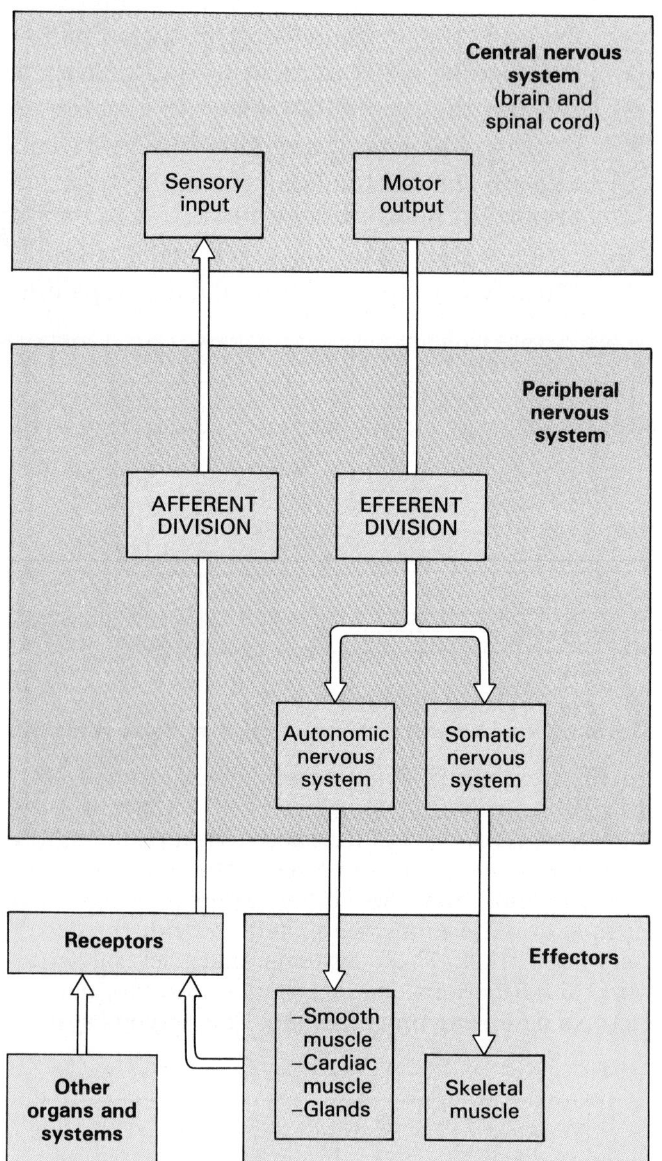

FIGURE 12–1
Overview of the Nervous System

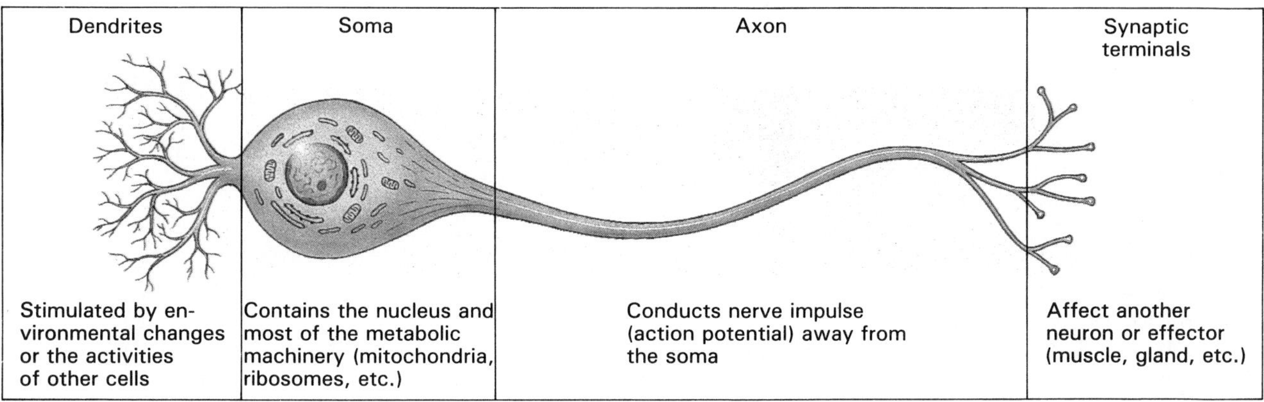

FIGURE 12–2
A Review of Neuron Structure

Chapter 5, and Figure 12-2 reviews pertinent details. (p. 157) A typical neuron has a cell body, or **soma**, several branching, sensitive **dendrites**, and an elongate **axon** that ends at one or more **synaptic terminals**. At each synaptic terminal the neuron communicates with other cells. The relatively large soma contains the organelles responsible for energy production and the synthesis of organic compounds, such as enzymes. This is only a general description; the proportions and shapes of neurons are actually quite variable.

Supporting cells, or **neuroglia** (noo-RŌG-lē-a; *glia*, glue), isolate the neurons, provide a supporting framework for the tissue, and act as phagocytes. Neuroglia, also called **glial cells**, far outnumber the neurons, and account for roughly half of the volume of the nervous system. Although neural tissue in both the central and peripheral divisions of the nervous system contains neurons and neuroglia, there are significant differences in the organization of the tissue in each instance, due primarily to the presence of distinctive glial cell populations.

NEURON STRUCTURE

The billions of neurons in the nervous system exhibit considerable diversity of form. Figure 12-3 depicts a generalized structure, but there are many variations on this basic anatomical theme (see also Figures 3-1 and 5-21). (pp. 60, 158)

- **Anaxonic** (an-ak-SON-ik) neurons are small, and there are no anatomical clues to distinguish dendrites from axons. Anaxonic neurons (Figure 12-3a) are found in the CNS and in special sense organs, but their functions are poorly understood.
- In a **unipolar** neuron the dendritic and axonal processes are continuous, and the soma lies off to one side (Figure 12-3b). Sensory neurons of the peripheral nervous system are usually unipolar.
- **Bipolar** neurons (Figure 12-3c) have one dendrite and one axon, with the soma between them. Bipolar neurons are relatively rare but important components of the CNS and special sense organs such as the eye and ear.
- **Multipolar neurons** have several dendrites and a single axon that may have one or more branches. Multipolar neurons are relatively common within the CNS. For example, all of the motor neurons that control skeletal muscles are multipolar. Figure 12-3d shows a typical multipolar neuron.

Structure of Multipolar Neurons

Physiologists have been able to study the activities of individual multipolar neurons, because they are relatively large. We will use them to examine typical neuron structure in greater detail (Figure 12-4). The soma contains a relatively large, round nucleus with a prominent nucleolus. The surrounding cytoplasm constitutes the **perikaryon** (per-i-KAR-ē-on; *karyon*, nucleus). The cytoskeleton of the perikaryon contains **neurofilaments** and **neurotubules**. Bundles of neurofilaments, called **neurofibrils**, extend into the dendrites and axon.

The perikaryon contains organelles that provide energy and perform synthetic activities. The numerous mitochondria, free and fixed ribosomes, and membranes of the rough endoplasmic reticulum (RER) give the perikaryon a coarse, grainy appearance. Mitochondria generate ATP to meet the high energy demands of an active neuron. The ribosomes and RER synthesize peptides and proteins. Aggregations of fixed and free ribosomes are conspicuously present. They are called *Nissl bodies* because they were first described by the German microscopist Franz Nissl. Nissl bodies account for the gray color of areas containing neuron cell bodies—the *gray matter* seen in dissection.

Most neurons lack an important organelle complex, the *centrosome*. In other cells, centrioles of the

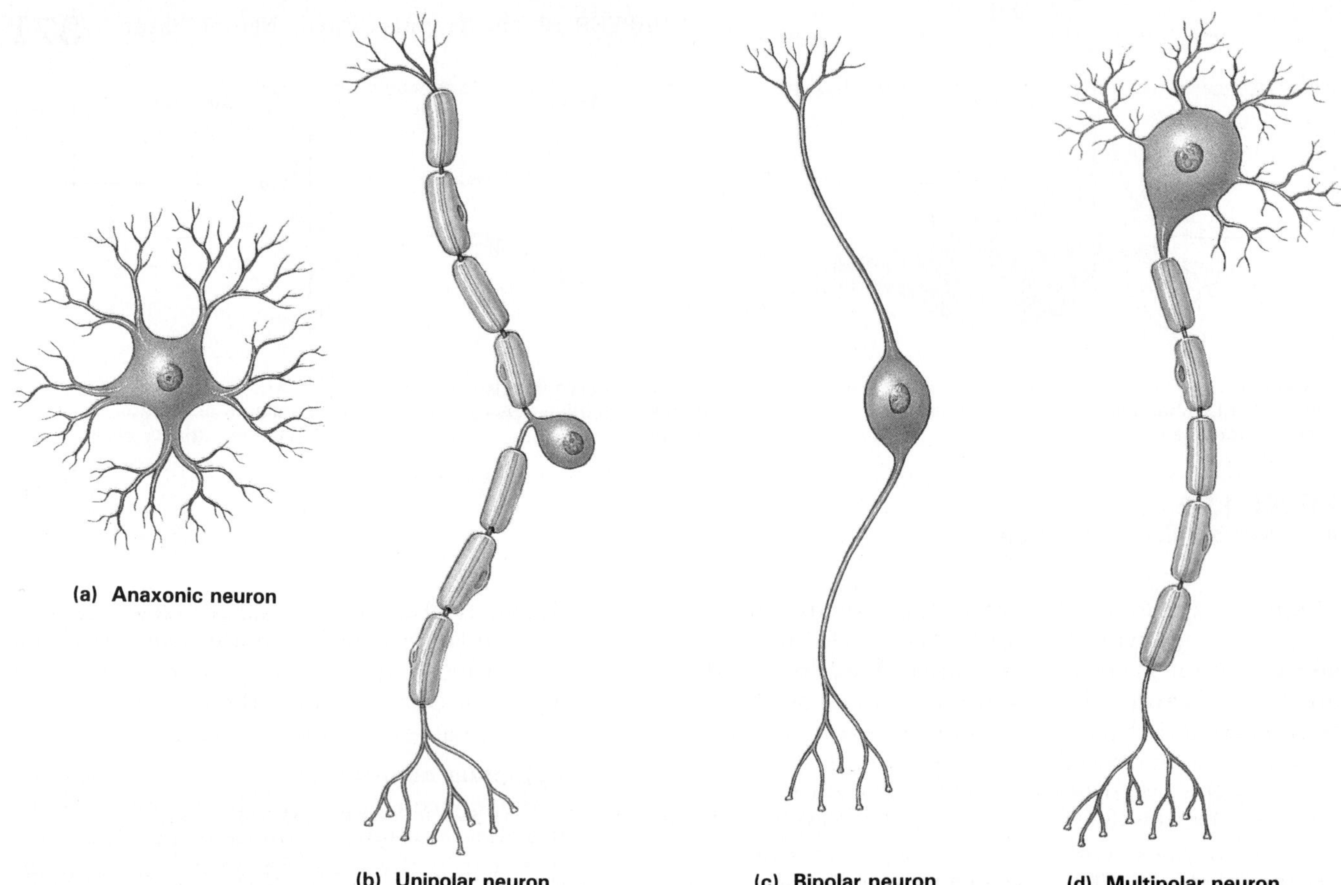

FIGURE 12–3
An Anatomical Classification of Neurons

centrosome form the spindle fibers that move chromosomes during cell division. Neurons usually lose their centrioles during differentiation and become incapable of undergoing mitosis. As a result, neurons lost to injury or disease cannot be replaced.

Projecting from the soma are a variable number of dendrites and a single large axon. The dendrites are highly branched, and the dendritic complex may account for 80–90 percent of the entire surface area of the neuron. The cell membrane of the dendrites and soma is sensitive to chemical, mechanical, or electrical stimulation, and the extensive dendritic branching greatly increases the size of the area monitored. Stimulation of any portion of this surface produces a localized change in the transmembrane potential that can lead to the generation of an action potential at the axon.

A specialized region, the **axon hillock**, connects the **initial segment** of the axon to the soma. The **axoplasm** (AK-so-plazm), or cytoplasm of the axon, contains neurofibrils, neurotubules, small vesicles, lysosomes, mitochondria, and various enzymes. An axon may branch along its length, producing branches called **collaterals**. The main trunk and the collaterals end in a series of fine extensions, called **telodendria** (te-lō-DEN-drē-a; *telo-*, end + *dendron*, tree).

Expanded **synaptic knobs** at the tips of the telodendria form synaptic terminals. Each terminal is part of a **synapse**, a site of intercellular communication. At a synapse the activity of one neuron affects the membrane characteristics of another cell. The communication most often involves the release of chemicals, called **neurotransmitters**, at the synaptic knob. This process will be detailed in a later section.

When one neuron communicates with another, the synapse may occur on a dendrite, on the soma, or along the length of the axon. A synapse may also permit communication between a neuron and another cell type, and these synapses are called **neuroeffector junctions**. There are two major classes of neuroeffector junctions: *neuromuscular junctions* and *neuroglandular junctions*. At a **neuromuscular junction** the neuron communicates with a muscle cell; Chapter 10 examined the structure of a neuromuscular junction. (p. 309) At a **neuroglandular junction** a neuron controls or regulates the activity of a secretory cell. Neuroeffector junctions involving other cell types are less common and will be described in later chapters.

Axoplasmic Transport

Each synaptic knob contains RER, ribosomes, and mitochondria needed to synthesize the neurotransmitter released at the synapse. Much synthetic activity also occurs at the soma: additional neurotransmit-

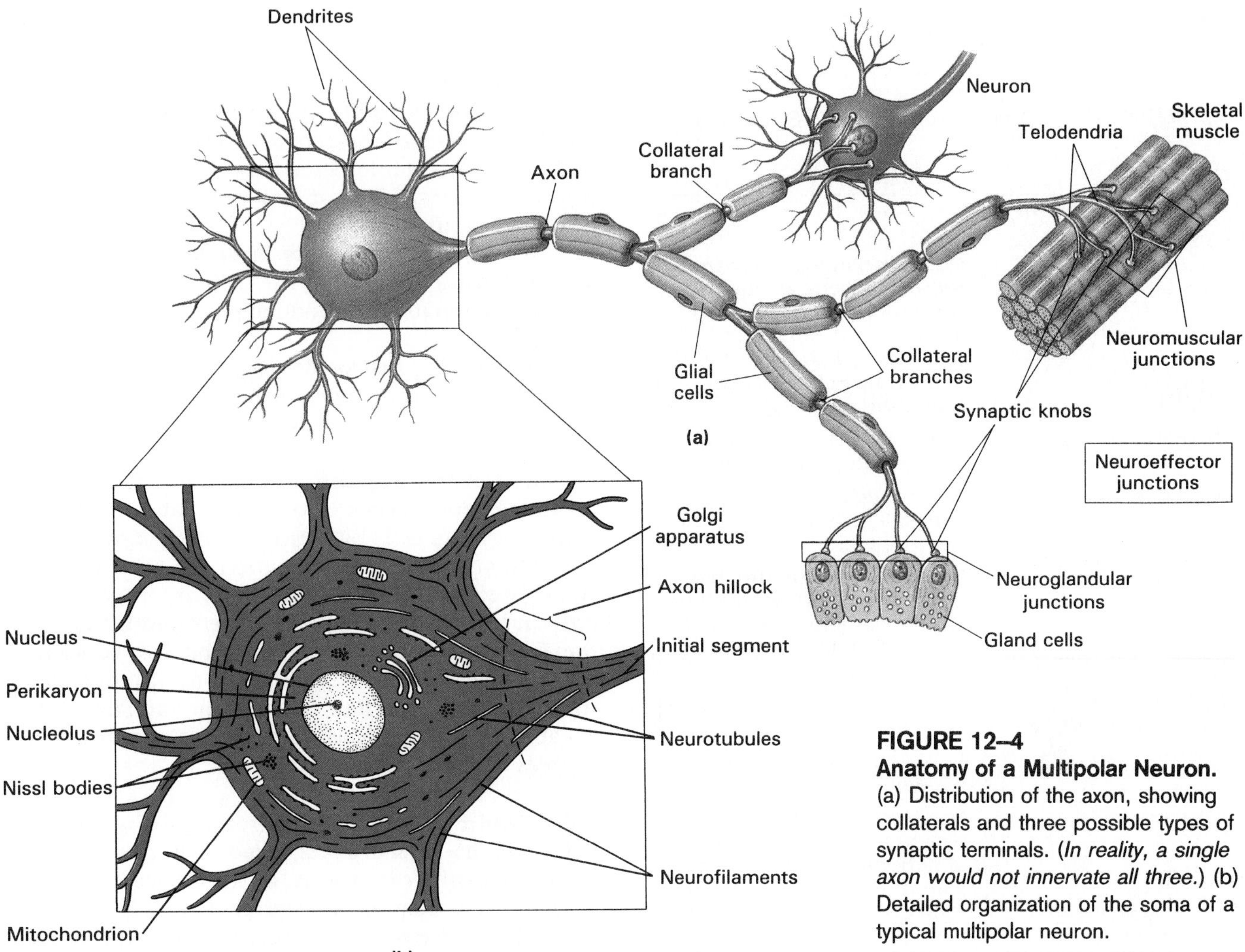

FIGURE 12–4
Anatomy of a Multipolar Neuron.
(a) Distribution of the axon, showing collaterals and three possible types of synaptic terminals. (*In reality, a single axon would not innervate all three.*) (b) Detailed organization of the soma of a typical multipolar neuron.

ter, enzymes, mitochondria, and lysosomes all form within the perikaryon. Many of these products are exported to the synaptic knobs along the length of the axon. This **axoplasmic transport** occurs along neurotubules, and the process requires ATP. Some materials travel in the "slow stream" at a few millimeters per day, but compounds important to synaptic function move along the axon in the "fast stream" at 5–10 mm per hour.

At the same time that some materials are traveling from the soma to the periphery of the neuron, other substances are being transported toward the soma. When conditions change at the synaptic knob, this **retrograde** (RET-rō-grād) **flow** soon causes a change in the environment of the soma. For example, retrograde flow may deliver chemicals or debris that indicate damage to the axon or synaptic knobs. This flow will trigger an immediate change in cellular operations by switching appropriate genes on or off. [**CM:** *Axoplasmic Transport and Disease*]

NEUROGLIA

The greatest variety of glial cells is found within the central nervous system. Although histological descriptions have been available for a century, the technical problems involved in isolating and manipulating individual glial cells have limited our understanding of their functions. Table 12-1 summarizes information concerning the major glial cell populations in the CNS and PNS.

Neuroglia of the Central Nervous System

There are four types of glial cells in the central nervous system: *astrocytes*, *oligodendrocytes*, *microglia*, and *ependymal cells*. These cell types can be distinguished on the basis of size, intracellular organization, and the presence of characteristic cytoplasmic processes.

Astrocytes **Astrocytes** (AS-trō-sīts; *astro-*, star + *cyte*, cell) are the largest and most numerous glial cells. Astrocytic processes contact the surfaces of nerve cell bodies and axons, in company with the processes of oligodendrocytes. Together they shield the neurons from direct contact with other neurons as well as from the surrounding interstitial fluid. Astrocytes, illustrated in Figure 12-5a,b, have a variety

TABLE 12-1 Glial Cells in the CNS and PNS

Cell type	*Functions*
CENTRAL NERVOUS SYSTEM	
Astrocytes	Maintain blood-brain barrier; provide structural support; regulate ion, nutrient, and dissolved gas concentrations; absorb and recycle neurotransmitters; form scar tissue after injury
Oligodendrocytes	Myelinate CNS axons; provide structural framework
Microglia	Remove cell debris, wastes, and pathogens by phagocytosis
Ependymal cells	Line ventricles (brain) and central canal (spinal cavity); assist in production, circulation, and monitoring of cerebrospinal fluid
PERIPHERAL NERVOUS SYSTEM	
Satellite cells	Surround nerve cell bodies in ganglia
Schwann cells	Surround all axons in PNS; responsible for myelination of peripheral axons; participate in repair process after injury

of functions, many of them poorly understood. These functions can be summarized as:

1. Maintaining the blood-brain barrier: Compounds dissolved in the circulating blood do not have free access to the interstitial fluid of the CNS. Neural tissue must be isolated from the general circulation because hormones or other chemicals normally present in the blood could have disruptive effects on neuron function. The endothelial cells lining CNS capillaries control the chemical exchange between the blood and interstitial fluid. These cells have very restricted permeability characteristics, and so create a **blood-brain barrier** that isolates the CNS from the general circulation. The slender cytoplasmic extensions of astrocytes end in expanded "feet" that wrap around capillaries. There are so many astrocytes, and so many astrocyte feet, that they form a complete cytoplasmic blanket around the capillaries, interrupted only where other glial cells contact the capillary walls. Chemicals secreted by astrocytes are somehow responsible for maintaining the special permeability characteristics of the endothelial cells. (The blood-brain barrier will be discussed further in Chapter 14.)

2. Creating a three-dimensional framework for the CNS: Astrocytes are packed with microfilaments that extend from foot to foot across the breadth of the cell. This reinforcement assists them in providing a structural framework for the neurons of the brain and spinal cord.

3. Performing repairs in damaged neural tissue: Following damage to the CNS, astrocytes make structural repairs by producing scar tissue at the injury site.

4. Guiding neuron development: Astrocytes in the embryonic brain appear to be involved in directing the growth and interconnection of developing neurons.

5. Controlling the interstitial environment: Although much needs to be learned about astrocyte physiology, there is evidence that astrocytes adjust the interstitial fluid composition by: (1) regulating the concentration of sodium ions, potassium ions, and carbon dioxide; (2) providing a rapid-transit system for the transport of nutrients, ions, and dissolved gases between capillaries and neurons; (3) controlling the volume of blood flow through the capillaries; and (4) absorbing and recycling some of the neurotransmitters released by active neurons.

Oligodendrocytes Like astrocytes, **oligodendrocytes** (o-li-gō-DEN-drō-sītz; *oligo*, few) possess slender cytoplasmic extensions, but their cell bodies are smaller, and they have fewer processes (Figure 12-5a,b). The processes usually contact the exposed surfaces of neurons, but the functions of processes ending at the soma have yet to be determined. Much more is known about the processes that end on the surfaces of axons.

Many axons in the CNS are completely sheathed in the processes of oligodendrocytes. This covering is so complete that when talking about an axon you have to be quite specific about which "outer membrane" you are referring to. The cell membrane of the axon is the **axolemma** (*lemma*, husk), but in this case the *real* outer boundary, or **neurilemma** (noo-ri-LEM-ma), is provided by the glial cell. Near the tip of each oligodendritic process, the cytoplasm becomes very thin, but the cell membrane expands to form an enormous membranous pad. This flattened pancake, diagrammed in Figure 12–5b, somehow gets wound around the axon, creating a multilayered sheath composed of 80 percent lipid (primarily phospholipids) and 20 percent protein. This membranous wrapping is called **myelin** (MĪ-e-lin), and the axon is said to be **myelinated**. Myelin improves the rate of action potential conduction along an axon; the mechanism will be described later in the chapter.

Many oligodendrocytes cooperate in the formation of a complete myelin sheath, and small gaps occur between adjacent wrappings. These gaps are called **nodes**, or the *nodes of Ranvier* (RAHN-vē-ā), and the relatively large areas wrapped in myelin are

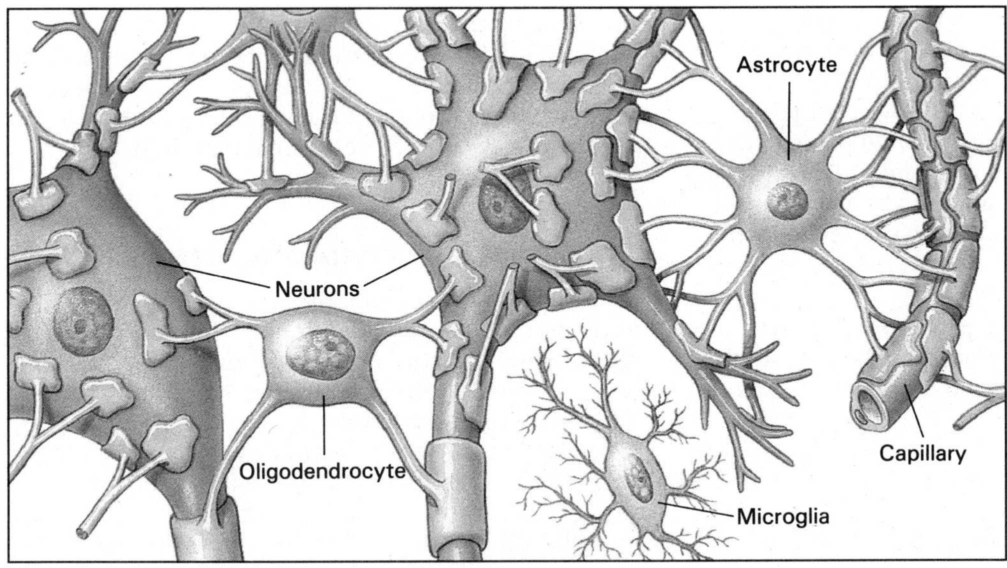

(a) Gray matter

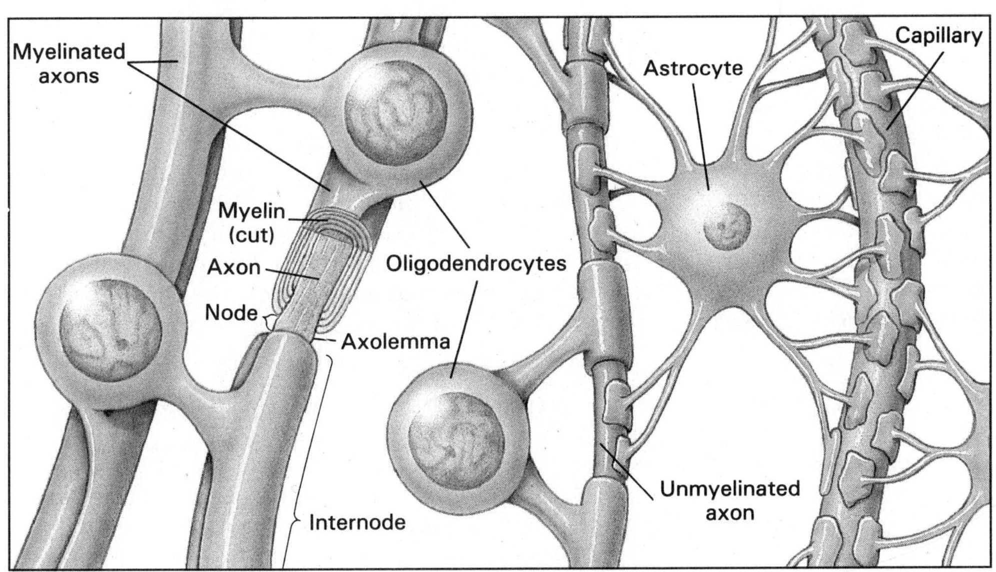

(b) White matter

FIGURE 12–5
Histology of Neural Tissue in the CNS. (a) A diagrammatic view of gray matter, showing relationships between major glial elements and nerve cell bodies. (b) A comparable view of white matter, showing relationships between glial cells and axons.

called **internodes** (*inter*, between). In dissection myelinated axons appear a glossy white, primarily due to the lipids present, and regions dominated by myelinated axons constitute the **white matter** of the CNS. In contrast, areas dominated by nerve cell bodies are called **gray matter** because of their dusky gray color. Not all axons in the CNS are myelinated, and **unmyelinated** axons may not be completely covered by glial cell processes.

In summary, oligodendrocytes play a role in structural organization by tying clusters of axons together and improve the functional performance of neurons by coating axons with myelin.

Microglia Roughly 5 percent of the CNS glial cells are **microglia** (mī-KRŌG-lē-ah). Microglia (Figure 12-5a) are smaller than the other glial elements, and their slender cytoplasmic processes have many fine branches. They are also capable of migrating through the surrounding neural tissue. Microglia do not develop in neural tissue (see the Embryology Summary, p. 377); they are phagocytic white blood cells that have migrated across capillary walls in the neural tissue of the CNS. Microglia act as a wandering police force and janitorial service, engulfing cellular debris, waste products, and pathogens. Under ordinary circumstances the few microglia present are able to perform the necessary cleanup operations. In times of infection or injury their numbers increase dramatically, as other phagocytic cells are attracted to the damaged area.

Ependymal cells The CNS consists of a mass of neural tissue organized into a hollow tube. The thickness of the walls and the diameter of the central space vary from one region to another. The narrow passageway within the spinal cord is called the *central canal*; the *ventricles* are expanded chambers found in portions of the brain. The ventricles and central canal are lined by a cellular layer called the **ependyma** (ep-EN-di-mah), and are filled with **cerebrospinal fluid (CSF)**. This fluid, which also surrounds the brain and spinal cord, provides a protective cushion and distributes dissolved gases, nutrients, wastes, and other materials.

Ependymal cells (Figure 12-6) are cuboidal to columnar in form, and their free surfaces are usually covered with cilia or microvilli. In a few parts of the brain, specialized ependymal cells participate in the secretion of the CSF. Other regions of the ependyma assist in the circulation of the CSF and probably perform additional functions as well. Unlike typical epithelial cells, ependymal cells have slender processes that branch extensively and make direct contact with glial cells in the surrounding neural tissue. Experimental evidence suggests that ependymal cells function as sensory receptors, responding to changes in the composition of the cerebrospinal fluid.

Neuroglia of the Peripheral Nervous System

Nerve cell bodies in the PNS are clustered together in masses called **ganglia** (singular *ganglion*). Axons are bundled together and wrapped in connective tissue, forming **peripheral nerves**, or simply *nerves*. Nerve cell bodies and axons in the PNS are completely insulated from their surroundings by the processes of glial cells. The two glial cell types involved are called *satellite cells* and *Schwann cells*.

Satellite cells, or *amphicytes* (AM-fi-sītz), surround the nerve cell bodies in peripheral ganglia, as indicated in Figure 12-7. **Schwann cells** produce a complete neurilemma around every peripheral axon, whether it is unmyelinated or myelinated. The physical relationships differ from those seen in the CNS. A Schwann cell may surround several different unmyelinated axons (Figure 12-8a), but it can myelinate only one segment of a single axon. The presumed mode of myelin formation in the PNS is summarized in Figure 12-8b, which should be compared with Figure 12-5b. Although the mechanism of myelination differs, myelinated axons in both the CNS and PNS have nodes and internodes, and the presence of my-

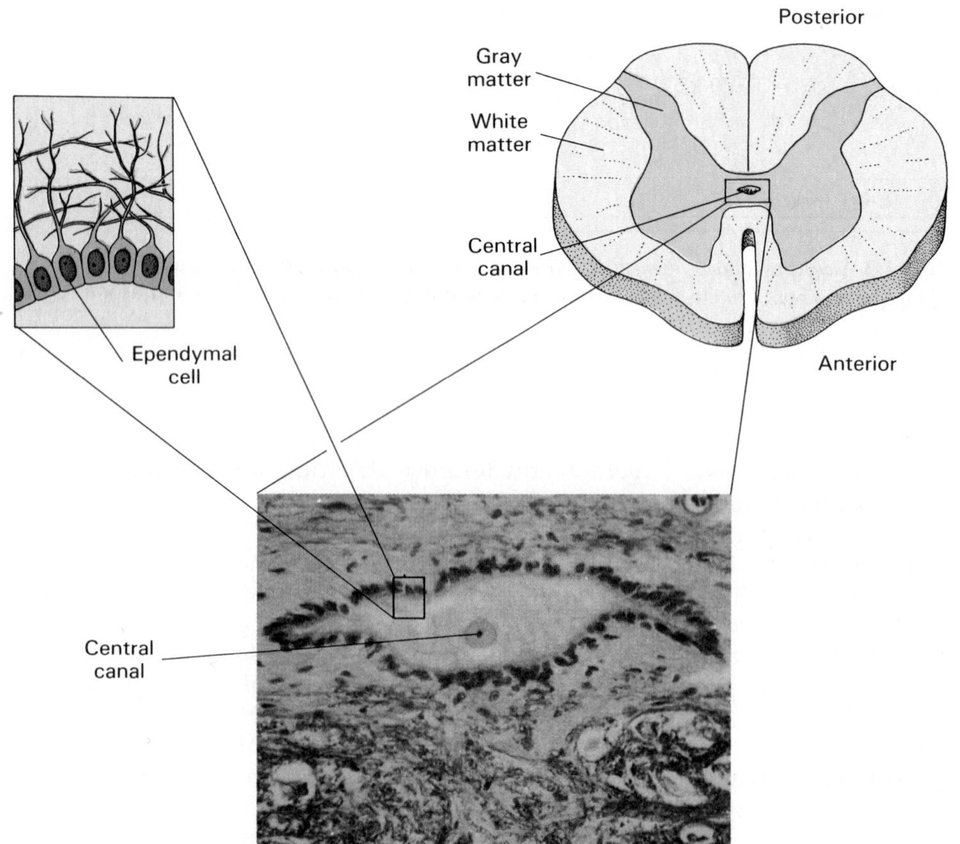

FIGURE 12–6
The Ependyma. Light micrograph, showing the ependymal lining of the central canal of the spinal cord (LM, x267). The diagrammatic view of ependymal organization is based on information provided by electron microscopy.

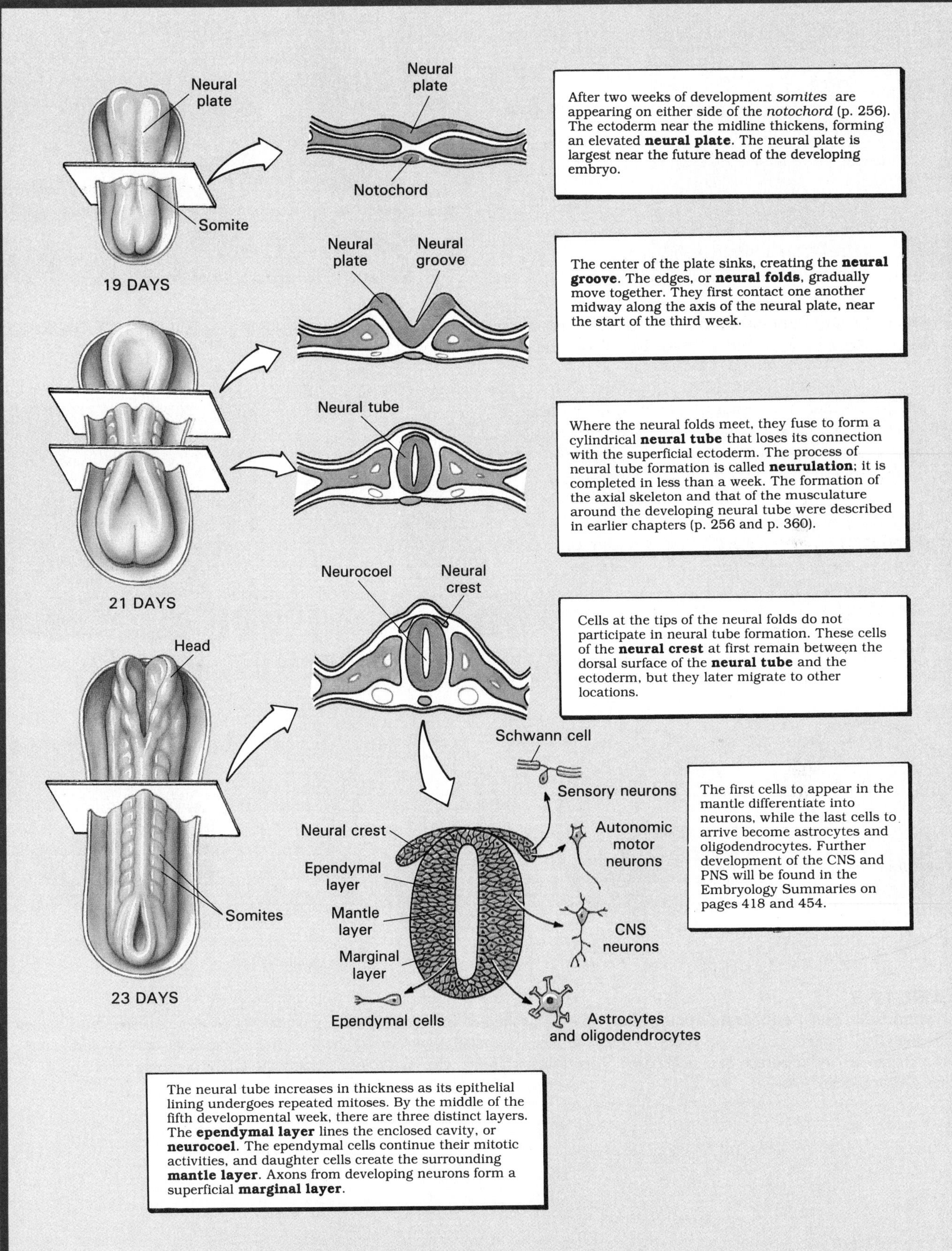

Neural plate
Somite
19 DAYS
Neural plate
Notochord
After two weeks of development *somites* are appearing on either side of the *notochord* (p. 256). The ectoderm near the midline thickens, forming an elevated **neural plate**. The neural plate is largest near the future head of the developing embryo.
Neural plate
Neural groove
The center of the plate sinks, creating the **neural groove**. The edges, or **neural folds**, gradually move together. They first contact one another midway along the axis of the neural plate, near the start of the third week.
Neural tube
Where the neural folds meet, they fuse to form a cylindrical **neural tube** that loses its connection with the superficial ectoderm. The process of neural tube formation is called **neurulation**; it is completed in less than a week. The formation of the axial skeleton and that of the musculature around the developing neural tube were described in earlier chapters (p. 256 and p. 360).
21 DAYS
Neurocoel
Neural crest
Cells at the tips of the neural folds do not participate in neural tube formation. These cells of the **neural crest** at first remain between the dorsal surface of the **neural tube** and the ectoderm, but they later migrate to other locations.
Head
Somites
23 DAYS
Schwann cell
Sensory neurons
Autonomic motor neurons
CNS neurons
Astrocytes and oligodendrocytes
Ependymal cells
Neural crest
Ependymal layer
Mantle layer
Marginal layer
The first cells to appear in the mantle differentiate into neurons, while the last cells to arrive become astrocytes and oligodendrocytes. Further development of the CNS and PNS will be found in the Embryology Summaries on pages 418 and 454.
The neural tube increases in thickness as its epithelial lining undergoes repeated mitoses. By the middle of the fifth developmental week, there are three distinct layers. The **ependymal layer** lines the enclosed cavity, or **neurocoel**. The ependymal cells continue their mitotic activities, and daughter cells create the surrounding **mantle layer**. Axons from developing neurons form a superficial **marginal layer**.

Embryology Summary: Introduction to the Development of the Nervous System

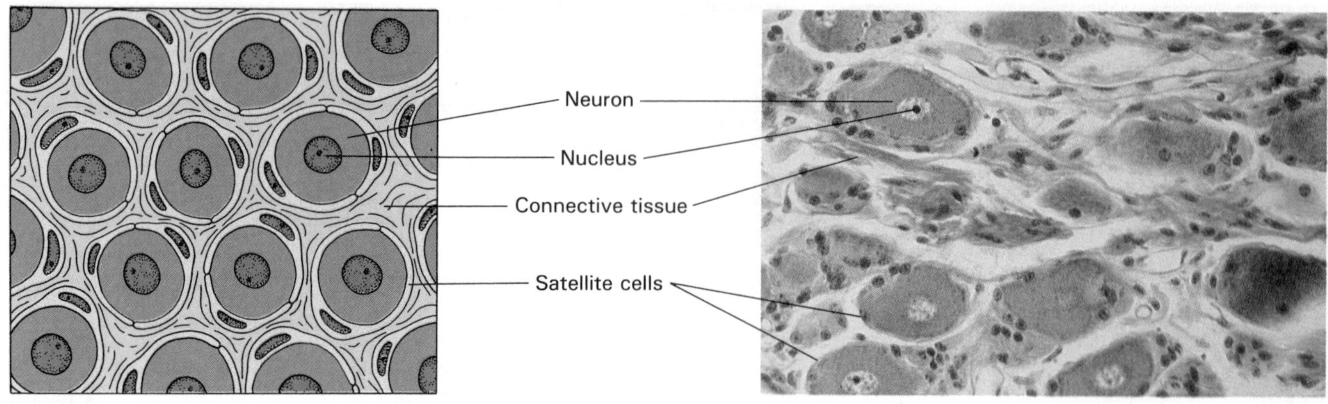

FIGURE 12–7
Satellite Cells and Peripheral Neurons. Satellite cells surround nerve cell bodies in peripheral ganglia (LM, x109). The somae are round in cross section because these are unipolar neurons (see Figure 12-3).

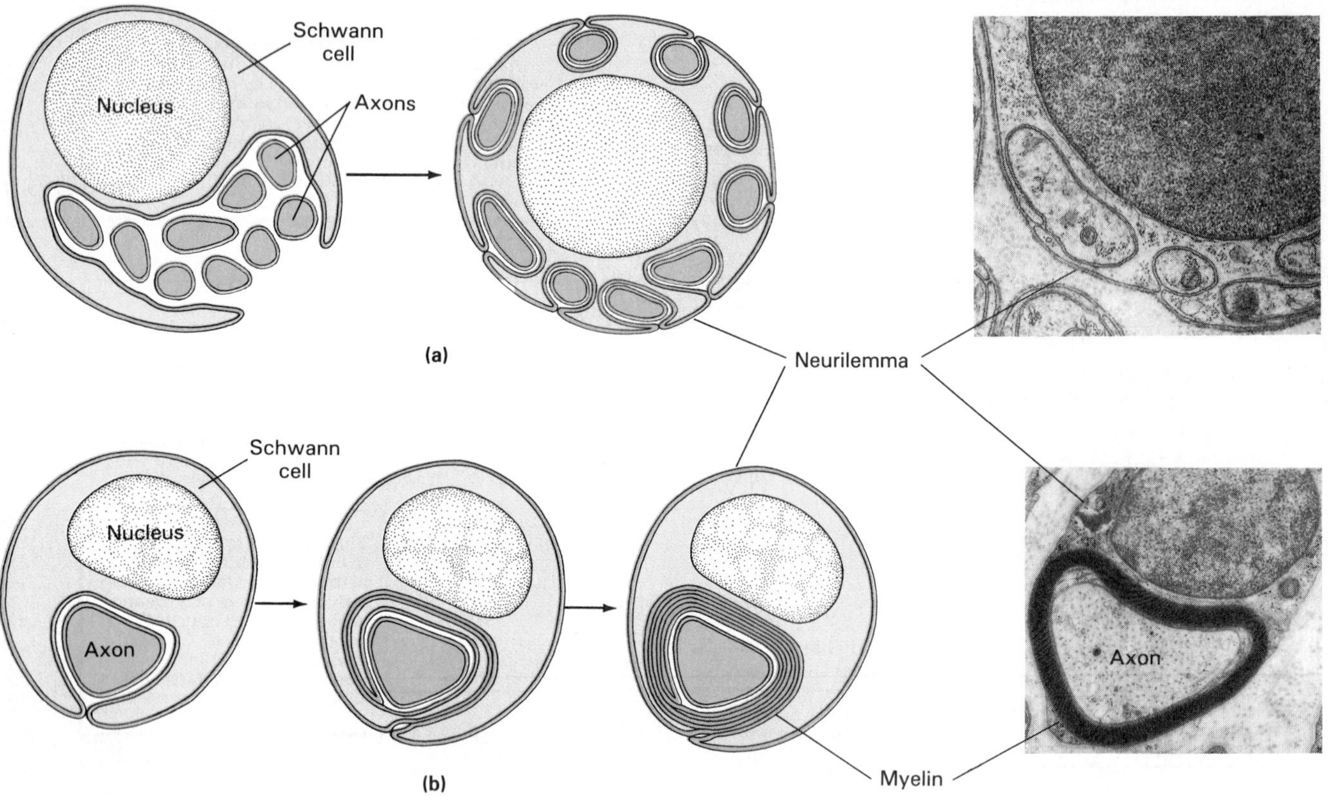

FIGURE 12–8
Schwann Cells and Peripheral Axons. (a) A single Schwann cell can encircle several unmyelinated axons. Unlike the situation inside the CNS (Figure 12-5b), every axon in the PNS has a complete neurilemmal sheath. (b) A single Schwann cell forms the myelin sheath around a portion of a single axon. This differs from the way myelin forms inside the CNS; compare with Figure 12-5b (TEM, x6638).

elin—however formed—increases the rate of action potential conduction.

- √ **What would damage to the afferent division of the nervous system interfere with?**
- √ **Examination of a tissue sample shows unipolar neurons. Are these more likely to be sensory neurons or motor neurons?**
- √ **What type of glial cell would you expect to find in large numbers in brain tissue from a person suffering from a CNS infection?**

Neurophysiology

We will now begin to explore the ways in which neurons, aided by glial cells, process information and communicate with one another and with peripheral effectors. Neurons and glial cells create a complex network that has a relatively stable three-dimensional structure. Yet the functions of the nervous system are dynamic and everchanging. For example, as you read these words your nervous system is performing complicated intellectual processing, while monitoring other systems and issuing appropriate voluntary and involuntary motor commands. In a few hours, at mealtime or while you are sleeping, the pattern of activity will be very different. The switch can be almost instantaneous because neural function does not involve substantial changes in physical structure, such as the sliding of myofilaments during muscle contraction. *All of the important steps occur at the surfaces of neurons, via changes in the transmembrane potential.* These are electrical events, which proceed at great speed. Understanding the mechanics of the nervous system thus requires a familiarity with basic electrical principles.

AN INTRODUCTION TO ELECTRICITY

Earlier chapters introduced several concepts relevant to this discussion, including *potential energy*, the energy of position, and *kinetic energy*, the energy of motion (Chapter 4), as well as the *transmembrane potential* (Chapters 3 and 10). ∞ (pp. 100, 87, 307) Electrical forces act between positive and negative charges. Opposite charges (+ and −) attract, whereas like charges (+ and + or − and −) repel one another. When a barrier prevents the movement of opposite charges toward one another, a *potential difference* exists. A potential difference is a form of potential energy that is measured in terms of *volts* or (for small potentials such as those found in cell membranes) *millivolts*. A **current** is a movement of charged objects, such as ions, are free to move in response to a potential difference. As movement occurs, potential energy is converted to kinetic energy that can be used to perform work. One example of this process, the generation of ATP through chemiosmosis, was described in Chapter 4. ∞ (p. 107)

The potential difference that exists across a membrane or other barrier is expressed as a **voltage** (V). The barrier has a certain amount of **resistance** (R) that restricts the movement of charges; the movement itself is a current (I). The relationship between voltage, resistance, and current can be summarized as:

$$I = V/R$$

$$\text{Current} = \text{Voltage/Resistance}$$

This simple equation, known as *Ohm's law*, makes two important points:

1. Current is *directly* proportional to voltage. Voltage is the potential difference, or *electrical gradient*, that drives current flow. If the resistance remains unchanged, the greater the voltage the higher the current.
2. Current is *inversely* proportional to resistance. If the resistance goes up, current decreases; if resistance declines, current increases.

ELECTRICAL FORCES AND THE TRANSMEMBRANE POTENTIAL

Ions cannot diffuse across the lipid portion of a membrane—they must pass through membrane channels. The number and type of channels therefore determine the resistance of the membrane. A synthetic membrane without any channels has an infinitely high resistance: no current flows through it, even at high voltages. A freely permeable membrane offers no resistance, and ion movements quickly eliminate any potential differences. Cell membranes are selectively permeable, and each ion moves through a specific membrane channel.

The intracellular and extracellular fluids differ in ion composition, so electrical forces are not the only forces that act across the cell membrane. Because the intracellular concentration of potassium ions is relatively high, potassium ions tend to move out of the cell. This movement is driven by a concentration gradient, or *chemical gradient*. Similarly, a chemical gradient for sodium ions tends to drive them into the cell. An electrical gradient exists as well, however, because the interior of the cell contains an excess of negative ions. This electrical gradient opposes the potassium chemical gradient and reinforces the sodium chemical gradient. The sum of all of the chemical and electrical forces active across the membrane is known as the **electrochemical gradient**.

Chapter 3 introduced the concept of *resting potential*, the transmembrane potential of an undisturbed cell. (p. 80) The resting potential remains stable because at this transmembrane potential an equilibrium exists between the electrochemical forces and the sodium-potassium exchange pump. Sodium ions enter and potassium ions leave, driven by their electrochemical gradients. The sodium-potassium exchange pump balances these gains and losses by exporting sodium and bringing in potassium. The resting potential of a typical neuron is −70 mV, as compared with the −85 mV typical of skeletal muscle fibers.

Any stimulus that (1) alters membrane permeability to sodium or potassium or (2) alters the activity of the exchange pump will disturb the resting potential. For the moment we will restrict the discussion to stimuli that alter the membrane permeability to sodium or potassium. Many different stimuli can have this effect, including mechanical pressure, changes in temperature, shifts in the extracellular ion concentrations, or exposure to chemicals. These stimuli alter membrane permeability by affecting specific membrane channels.

MEMBRANE CHANNELS

There are several types of sodium and potassium ion channels. **Passive channels** are always open, like the holes in a sieve or strainer. These "leaky" channels are important in establishing the normal resting potential of the cell. Cell membranes also contain **active channels**, often called **gated channels**. Gated channels are ion channels that open or close in response to specific stimuli. Each gated channel can be in one of three states: (1) closed but capable of opening, (2) open (**activated**), or (3) closed and incapable of opening (**inactivated**).

There are two major classes of gated channels:[1]

1. **Chemically regulated channels**: These channels open or close when they bind specific extracellular chemicals. The receptors that bind acetylcholine at the motor end plate (Chapter 10) are chemically regulated channels. (p. 308) In the presence of acetylcholine these channels open, allowing an influx of sodium ions that depolarizes a portion of the sarcolemma.
2. **Voltage-regulated channels**: Areas of excitable membrane also contain channels that open or close in response to changes in the transmembrane potential. For example, when the sarcolemma around the motor end plate depolarizes to around −60 mV, the opening of voltage-regulated sodium channels is the first step in the generation of an action potential.

[1] A third class of gated channels, *mechanically regulated channels*, open or close in response to physical distortion of the membrane surface. Such channels are particularly important in sensory receptors, such as the Merkel cells of the skin, that respond to touch, pressure, and vibration. These receptors will be discussed in detail in Chapter 17.

CHEMICALLY REGULATED CHANNELS AND GRADED POTENTIALS

A chemical that opens chemically regulated sodium channels will accelerate sodium entry into the cell. This accelerated influx of sodium shifts the transmembrane potential toward 0 mV, depolarizing the membrane. Opening a chemically gated potassium channel has the opposite effect: the rate of potassium outflow increases, and the interior of the cell loses additional positive ions. This loss of positive ions produces **hyperpolarization**, a shift in the resting potential further from 0 mV, perhaps to as much as −80 mV.

When such depolarization or hyperpolarization occurs on the membrane surface, the effects are confined to a relatively limited area. Consider Figure 12-9, in which a chemical stimulus is applied to the initial segment of a neuron. This figure diagrams the state of the membrane channels at the normal resting potential (−70 mV). Sodium channels outnumber potassium channels, but most of the sodium channels are gated, and the gates are closed. The entry of sodium ions through passive sodium channels is counterbalanced by the sodium-potassium exchange pump, which also reclaims potassium ions that diffuse through the passive potassium channels. The chemical opens chemically regulated sodium channels, and depolarization occurs. The sodium ions that cross the membrane spread along the inner surface, attracted by the excess of negative charges. This ion movement, called a **local current**, depolarizes the surrounding membrane. Because the sodium ions are spreading out, the degree of depolarization decreases with distance. The potential change that results is called a **graded potential**. Graded potentials have four basic characteristics:

1. The transmembrane potential is most affected at the site of stimulation, and the effect decreases with distance.
2. The effect spreads passively across the membrane surface, because of local currents.
3. The graded potential change may involve either depolarization, as in Figure 12-9, or hyperpolarization. The nature of the change is determined by the identity and properties of the activated membrane channel involved.
4. The stronger the stimulus, the greater the change in the transmembrane potential and the larger the area affected.

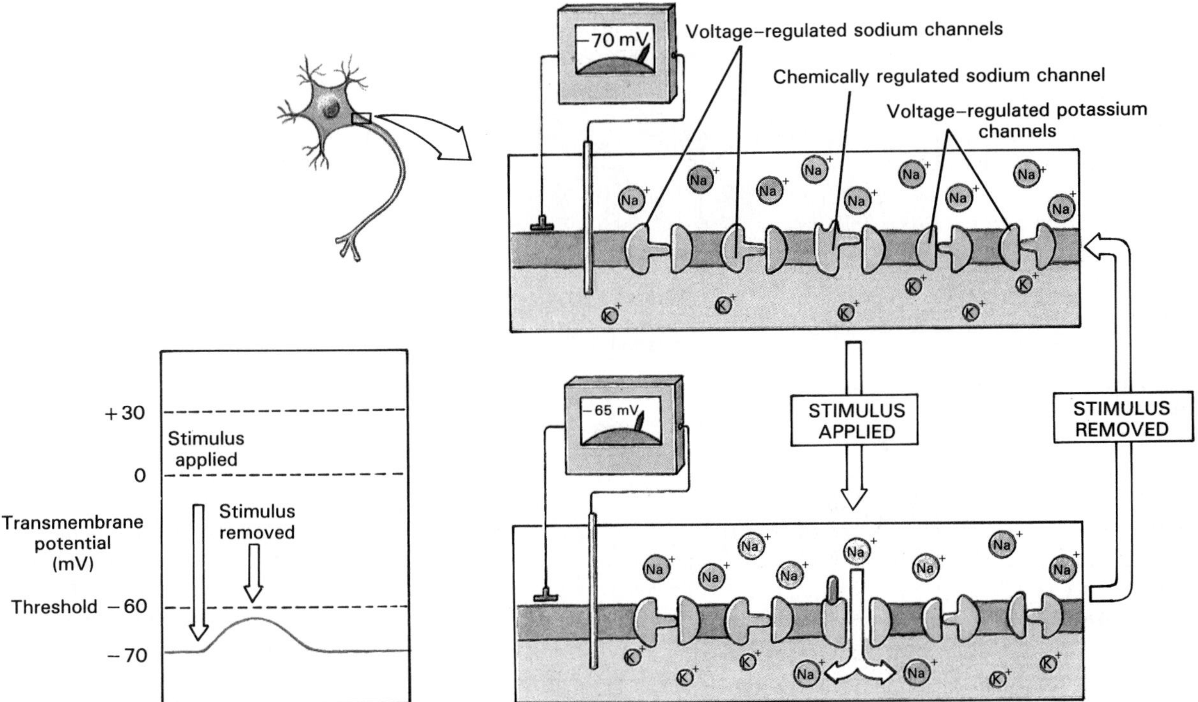

FIGURE 12–9
Graded Potentials An excitable membrane differs from other membranes because it contains many voltage-regulated sodium channels and a few voltage-regulated potassium channels. At the normal resting potential most of the gates are closed, and the sodium-potassium exchange pump keeps pace with the rates of sodium entry and potassium loss through passive membrane channels (not shown). When a chemical is applied to the soma of a neuron, chemically regulated sodium channels open, producing a graded depolarization. When the stimulus is removed, the channels close and the normal resting potential is restored.

Depolarization of a motor end plate exposed to acetylcholine is one example of a graded potential. Graded potentials occur in the membranes of many different cell types. *Action potentials* are restricted to areas of excitable membrane. These membranes have both chemically regulated and voltage-regulated ion channels.

VOLTAGE-REGULATED CHANNELS

Voltage-regulated channels open or close in response to changes in the local transmembrane potential. This type of gated channel is characteristic of muscle and nerve cells, the only two cell types that have *excitable membranes*. In a muscle fiber an excitable membrane covers the general sarcolemmal surface, including the transverse tubules. The motor end plate does not contain voltage-regulated channels, and only graded potentials can occur there. In a typical neuron, an excitable membrane covers the surface of the axon and its branches. The dendrites, soma, and presynaptic surfaces lack voltage-regulated channels, and these areas will support only graded potentials.

GENERATION OF AN ACTION POTENTIAL

The steps involved in the generation of an **action potential** are shown in Figure 12-10.

Depolarization to Threshold

Before an action potential can occur, there must be a graded depolarization of the cell membrane. This localized depolarization acts like pressure on the trigger of a gun. A slight pressure can be applied, and the gun will not fire. The gun fires only when a certain minimum pressure is applied to the trigger. Once the trigger pressure reaches this preset level, however, the firing pin drops and the gun discharges. At that point, it no longer matters whether the pressure was applied gradually or suddenly, or whether the shooter is a 4-year-old who can barely squeeze the trigger or a champion weight lifter who applies massive force. The bullet that leaves the gun always has the same speed and range, regardless of the forces that were applied to the trigger.

In the case of an axon, the graded potential is the pressure on the trigger and the action potential is the firing of the gun. An action potential will not appear unless the membrane depolarizes sufficiently, to a level known as the **threshold**. Threshold for an axon is usually between –60 mV and –55 mV. Thus a stimulus that lowers the resting membrane potential from −70 mV to only –65 mV or –62 mV will not produce an action potential. When such a stimulus is removed, the transmembrane potential simply returns to resting levels.

Any stimulus that brings the membrane to threshold will generate an identical action potential.

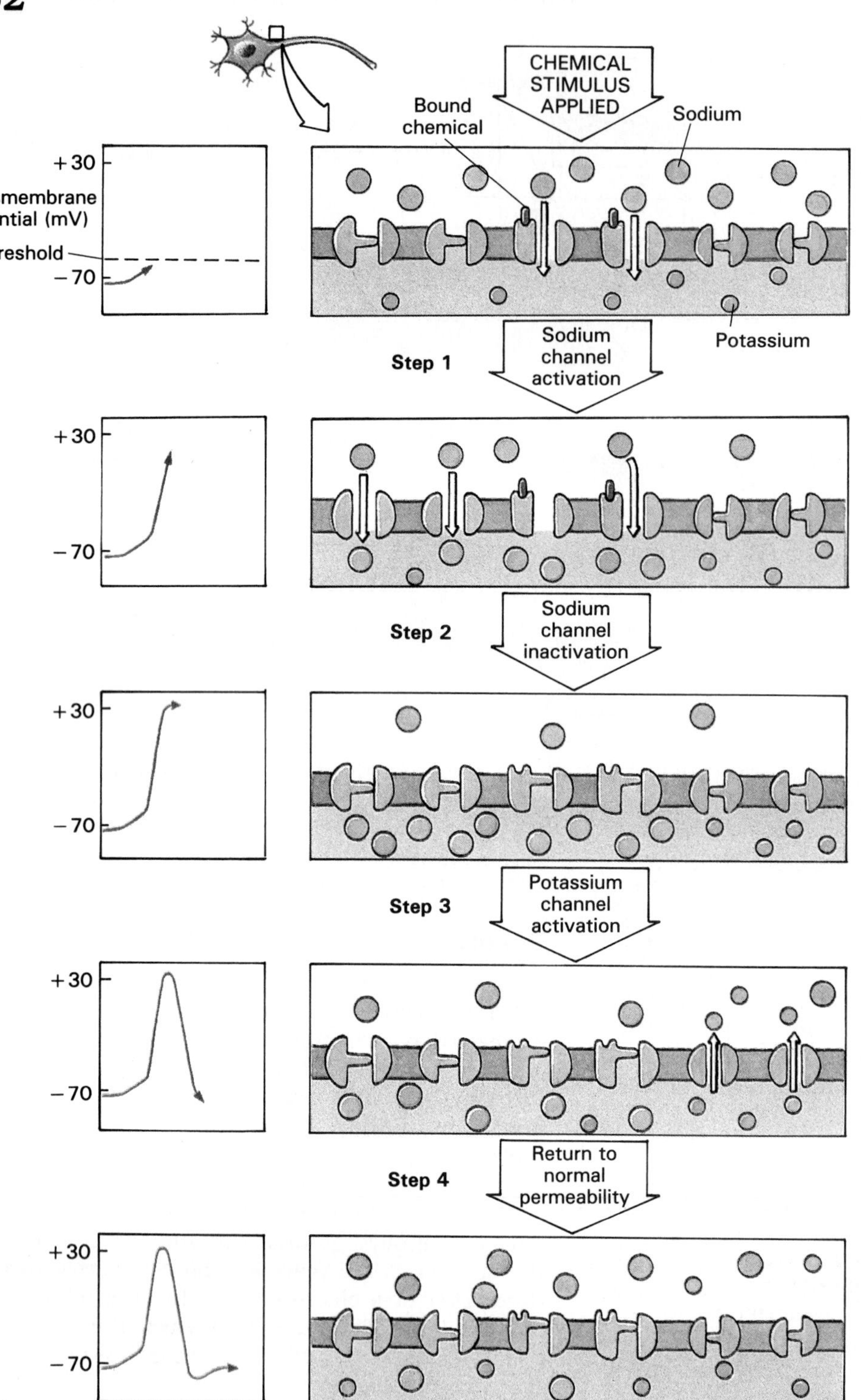

FIGURE 12–10 Generation of an Action Potential. If a graded depolarization brings the membrane potential to threshold, voltage-regulated sodium channels open (Step 1). An action potential begins as sodium ions flood into the cell, changing the transmembrane potential from −70 mV to +30 mV. At this point the sodium channels close again (Step 2) and potassium channels open (Step 3), allowing a movement of potassium ions out of the cell that restores the resting transmembrane potential. When this occurs, first the sodium channels and then the potassium channels return to their normal state (step 4). For clarity the passive sodium and potassium channels involved in the normal

That is, properties of the action potential are independent of the relative strength of the depolarizing stimulus. This is called the **all-or-none principle**, because a given stimulus either triggers a typical action potential or does not produce one at all. The all-or-none principle applies to all excitable membranes.

Step 1: Activation of Sodium Channels and Membrane Depolarization If depolarized to threshold, as in Figure 12-10, an action potential appears. The first step is the activation of voltage-regulated sodium channels. When these channels open, the membrane becomes extremely permeable to sodium ions. They rush into the cytoplasm, and the transmembrane potential at this site falls rapidly.

Step 2: Sodium Channel Inactivation In less than a millisecond the inner membrane surface contains more positive ions than negative ones, and the transmembrane potential has changed from –70 mV to +30 mV. At this voltage the sodium gates close, a step known as **sodium channel inactivation**.

Step 3: Potassium Channel Activation The change in membrane potential that causes the voltage-regulated sodium channels to close also causes the voltage-regulated potassium channels to open. Driven by their concentration gradient and repelled by the surplus of positive charges in the area, potassium ions rush out of the cell. This loss of positive charges shifts the transmembrane potential back toward resting levels, a process called *repolarization*.

Step 4: Return to Normal Permeability The voltage-regulated sodium channels remain inactivated until the membrane has repolarized to –70 mV. At this time they begin regaining their normal status (closed but capable of opening). The voltage-regulated potassium channels begin closing at the same time, but the process takes about a millisecond. Over that period potassium continues to move out of the cell at an accelerated rate, producing a brief hyperpolarization. As the voltage-regulated channels close, the transmembrane potential returns to normal resting levels. The membrane is now in its prestimulation condition, and the action potential is over.

Figure 12-11 shows the changes in membrane permeability that occur during an action potential. Table 12-2 summarizes the membrane events during an action potential, and Table 12-3 summarizes important differences between graded potentials and action potentials.

TABLE 12-2 Steps in the Generation of an Action Potential

1. A graded depolarization brings an area of excitable membrane to threshold.
2. The gates in voltage-regulated sodium channels open (sodium channel activation).
3. Sodium ions, driven by charge attraction and the concentration gradient, flood into the cell. The transmembrane potential goes from resting levels, −70 mV, to +30 mV.
4. The sodium channel gates close (sodium channel inactivation).
5. Voltage-regulated potassium channels open and potassium ions move out of the cell, driven by their concentration gradient and charge repulsion. Repolarization begins.
6. Sodium channels regain their normal properties when the resting potential is reached.
7. Because the gated potassium channels close slowly, a temporary hyperpolarization occurs. When all the gated potassium channels have closed, the membrane regains its normal resting potential.

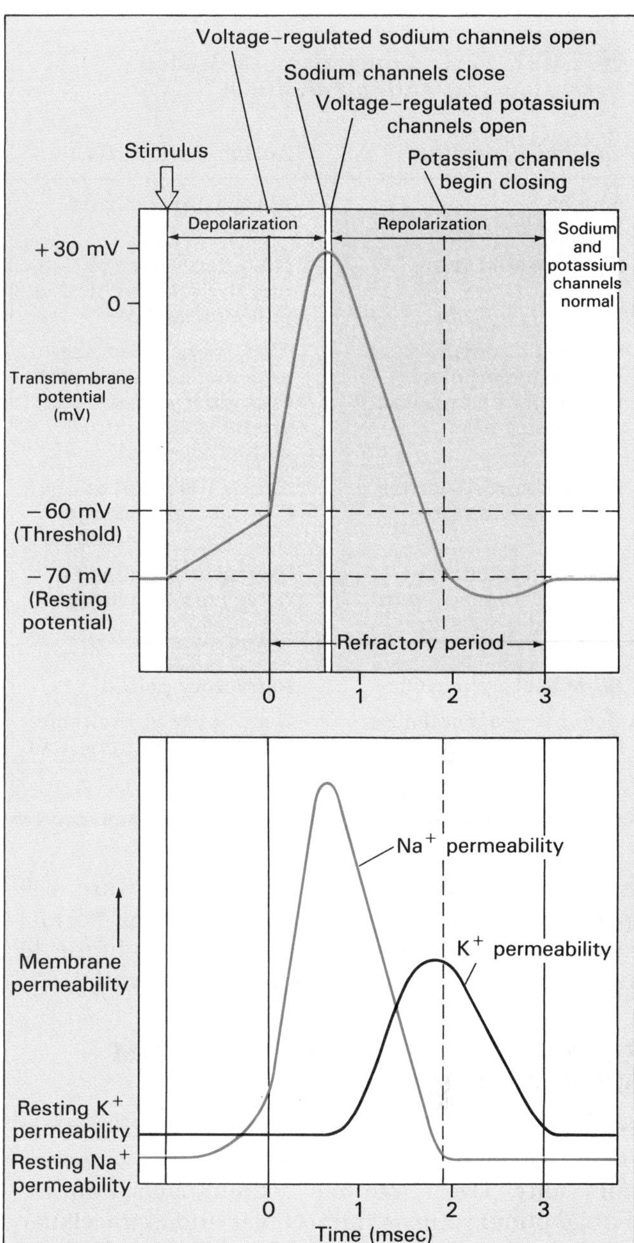

FIGURE 12–11
Permeability Changes and Transmembrane Potential Changes During an Action Potential. A sufficiently strong depolarizing stimulus will bring the membrane potential to threshold and trigger an action potential. The upper graph indicates the changes in the transmembrane potential that occur and the factors involved. The lower graph details the permeability changes in the membrane that result from the opening and closing of voltage-regulated ion channels.

REFRACTORY PERIODS

From the moment that the voltage-regulated sodium channels open at threshold until the inactivation period ends, the membrane cannot respond to further stimulation. This is the **absolute refractory period**. During the **relative refractory period** the sodium channels are in their normal resting condition but

TABLE 12-3 Comparison of Graded Potentials and Action Potentials

Graded Potentials	*Action Potentials*
May be depolarizing or hyperpolarizing	Always depolarizing
No threshold value	Must depolarize to threshold before action potential begins
Amount of depolarization or hyperpolarization depends on intensity of stimulus	All-or-none phenomenon; any stimulus that exceeds threshold will produce an identical action potential
Passive spread across membrane surface	Action potential at one site depolarizes adjacent sites to threshold
Effect on membrane potential decreases with distance from stimulation site	Propagated across entire membrane surface
No refractory period	Refractory period
Occur in most cell membranes	Occur only in excitable membranes of nerve and muscle cells

the voltage-regulated potassium channels are still open. Under these conditions a larger-than-normal depolarizing stimulus can bring the membrane to threshold and initiate a second action potential.

ROLE OF THE SODIUM-POTASSIUM EXCHANGE PUMP

Depolarization occurs through the influx of sodium ions, and repolarization involves the loss of potassium ions. Over time the sodium-potassium exchange pump returns intracellular and extracellular ion concentrations to prestimulation levels. But compared with the total number of ions inside and outside the cell, the number involved in a single action potential is relatively insignificant. Tens of thousands of action potentials can occur before the intracellular ion concentrations change enough to disrupt the entire mechanism. Thus the sodium-potassium exchange pump is not essential to any *single* action potential. Yet a maximally stimulated neuron may generate action potentials at a rate of 1000 per second. Thus activity of the exchange pump is needed to maintain ion concentrations within acceptable limits over time. If the pump is inactivated by a poison, or the cell runs out of ATP, the neuron soon loses its ability to function.

CONDUCTION OF AN ACTION POTENTIAL

An action potential involves a relatively small segment of the total membrane surface. But unlike graded potentials, which diminish rapidly with distance, action potentials affect the entire membrane surface. In effect, the permeability changes that occur during an action potential at one site start a chain reaction that ultimately affects the rest of the excitable membrane. The basic mechanism of action potential conduction is shown in Figure 12-12, which diagrams the structure of an unmyelinated axon.

For a brief moment at the peak of the action potential, the transmembrane potential becomes positive rather than negative (Step 1). At the site of the action potential, the inside of the membrane contains an excess of positive ions and the outside an excess of negative ions. Because opposite charges attract one another, ions immediately begin moving in the cytoplasm and in the extracellular fluid. This *local current* depolarizes adjacent portions of the membrane, and when threshold is reached, action potentials occur at these locations (Step 2).

The process continues like falling dominoes until the most distant portions of the cell membrane have been affected. Although an action potential develops at each individual location along the way, the effect is the same as if a single action potential traveled across the membrane surface. This form of action potential transmission is known as **continuous conduction**.

An action potential always proceeds away from the point of initial stimulation. It cannot go back in the direction it came from, because when it is occurring in an adjacent section of the membrane, the section where it originated is still in its refractory period. By the time that section recovers, the action potential is too far away to affect it any further. If there is to be a second action potential at the same site, a second stimulus must be applied.

CONDUCTION VELOCITY

The rate of action potential conduction varies, depending on the characteristics of the axon. The two most important factors are (1) the presence or absence of a myelin sheath and (2) the diameter of the axon.

Myelin and Saltatory Conduction

In continuous conduction an action potential appears to move across the membrane surface in a series of tiny steps (Figure 12-12a). Even though the events at any one location take only a millisecond, the sequence must be repeated at each step along the way. Continuous conduction occurs along unmyelinated axons, at speeds of around 1 m per second (2 mph).

In a myelinated fiber, the axon is wrapped in layers of myelin. This wrapping is complete except at the nodes, where adjacent glial cells contact one another. Between the nodes, the lipid content of the myelin blocks the flow of ions across the membrane,

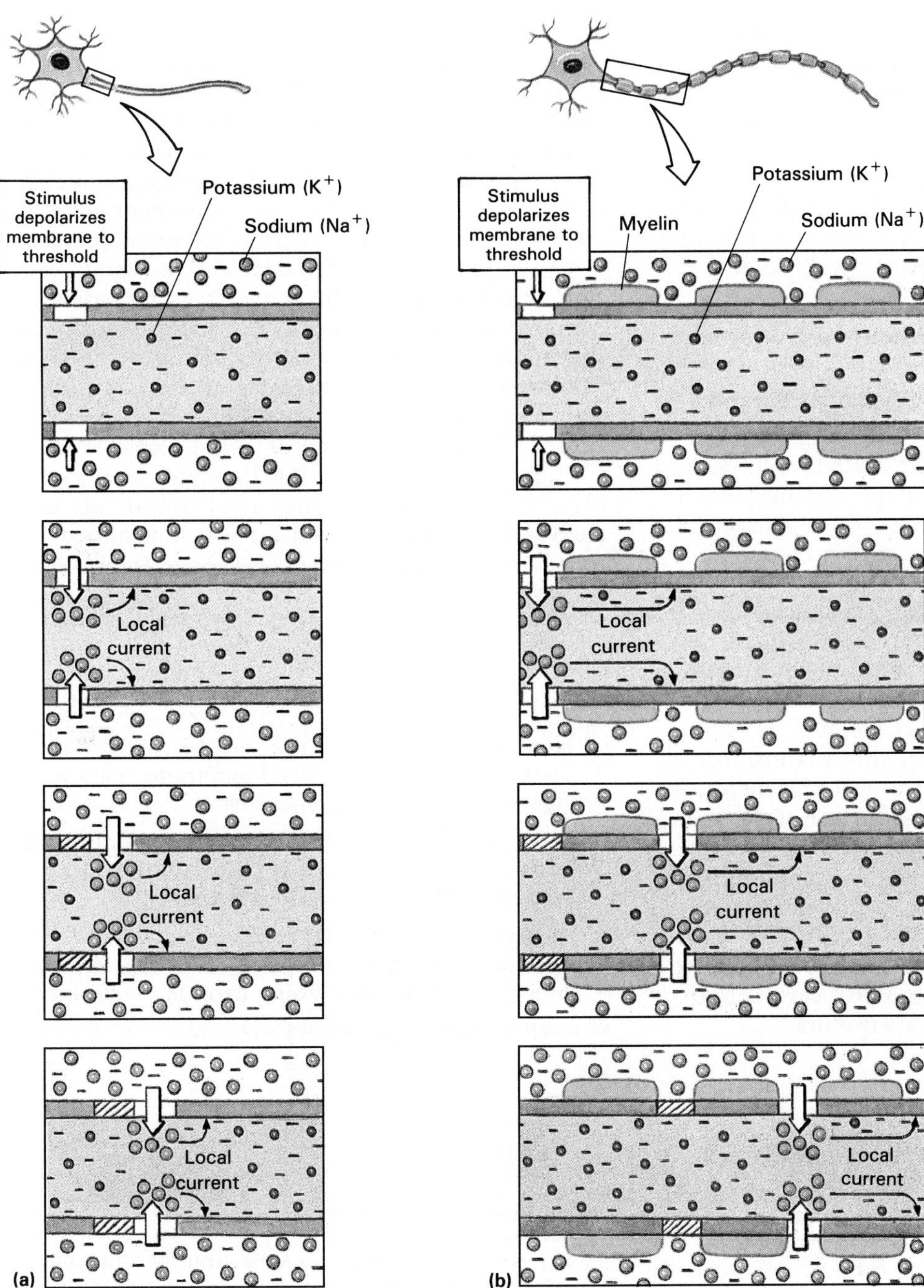

FIGURE 12–12 Action Potential Conduction over Unmyelinated and Myelinated Axons. (a) Continuous conduction along an unmyelinated axon. (b) Saltatory conduction along a myelinated axon.

and thus prevents continuous conduction of an action potential. Because ions can cross the membrane only at the nodes, *only a node can respond to a depolarizing stimulus.*

When an action potential appears at the initial segment, or at a node along the axon, the local current skips the internodes and depolarizes the closest node to threshold. Because the nodes may be 1–2 mm apart in a large myelinated axon, the action potential appears to leap from node to node, rather than moving along the axon in a series of tiny steps. This process is called **saltatory conduction**, taking its name from *saltare*, the Latin word for "leaping." Saltatory conduction, illustrated in Figure 12-12b, carries nerve impulses along an axon five to seven times faster than continuous conduction. [**CM:** *Neurotoxins in Seafood*]

Axon Diameter and Rate of Impulse Conduction

Axon diameter also has an effect on conduction velocity. To depolarize adjacent portions of the cell membrane, ions must move through the cytoplasm. The larger the diameter of the axon, the easier that movement becomes, and the more rapidly an action potential will be conducted. Axons are classified according to the relationships between diameter, myelination, and conduction speed, as detailed in Table 12-4.

- Type A fibers are the largest axons, with diameters ranging from 4 to 20 μm. These are myelinated axons that conduct action potentials at speeds of up to 140 m per second, the equivalent of over 300 mph.
- Type B fibers are smaller myelinated axons, with diameters of 2–4 μm. Their conduction speeds average around 18 m per second, or roughly 40 mph.
- Type C fibers are unmyelinated and under 2 μm in diameter. These axons conduct action potentials at the leisurely pace of 1 m per second, a mere 2 mph.

The relative importance of myelin can be seen by noting that in going from Type C to Type A fibers you find a tenfold increase in diameter, but the conduction speed increases by 150 times! [**CM:** *Growth and Myelination of the Nervous System*]

Type A fibers carry sensory information to the CNS concerning position, balance, and delicate touch and pressure sensations from the surface of the skin. The motor neurons that control skeletal muscles also send their commands over large, myelinated Type A axons. Type B fibers and Type C fibers bring less urgent information concerning temperature, pain, and general touch and pressure sensations. They also carry instructions to smooth muscles, cardiac muscle, glands, and other peripheral effectors.

When we need to tell a friend urgent news or receive an immediate response, we usually pick up the telephone. For general correspondence we usually send a letter first-class, which allows a swift but not immediate response. If we have to distribute an enormous volume of information to a huge mailing list, and there's no particular rush, bulk mail offers efficiency at a considerable savings. Instead of representing a compromise between time and money, information transfer within the nervous system reflects a compromise between time and space. Axons carry the information, and the larger the axon, the faster the rate of transmission. But if all sensory information were carried by Type A fibers, the peripheral nerves would be the size of garden hoses and the spinal cord would have the diameter of a garbage can. Instead, only around one-third of all axons carrying sensory information are myelinated, and most sensory data arrive over slender Type C fibers. [**CM:** *Demyelination Disorders*]

√ **How would a chemical that blocks the sodium channels in the nerve cell membrane affect a neuron's ability to depolarize?**

√ **What effect would decreasing the concentration of extracellular potassium ion have on the transmembrane potential of a neuron?**

√ **Two axons are tested for conduction velocities. One conducts action potentials at 50 meters/sec, the other at 1 meter/sec. Which axon is myelinated?**

TABLE 12-4 A Classification of Axons

Type	*Diameter*	*Myelination*	*Conduction Speed*	*Functions*
A	4–20 μm	Heavy	Up to 140 m/sec	*Sensory:* carries information concerning muscles and joints (proprioception), also fine touch and pressure; able to localize stimulus as arising in a particular area *Motor:* carries precise commands to skeletal muscles
B	2–4 μm	Light	Up to 18 m/sec	*Sensory:* carries information from some pain and temperature receptors; permits some localization of stimuli *Motor:* motor commands from CNS to ANS ganglia
C	Less than 2 μm	Unmyelinated	Around 1 m/sec	*Sensory:* crude touch, and pressure, pain, temperature; poor localization of sensations *Motor:* motor commands from ANS ganglia to smooth muscles, glands, etc.

Synaptic Communication

In the nervous system, information moves from one location to another in the form of action potentials. An action potential traveling along an axon is called a **nerve impulse**. At the end of an axon, a nerve impulse triggers events at a synapse that transfer the information to another neuron or effector cell. A synapse may be **chemical**, involving a neurotransmitter, or **electrical**, with direct physical contact between the cells. **Chemical synapses** are by far the most abundant, and there are several different types. Most interactions between neurons and all communications between neurons and peripheral effectors involve chemical synapses.

CHOLINERGIC SYNAPSES

We shall focus on chemical synapses releasing the neurotransmitter **acetylcholine**, or **ACh**. These are known as **cholinergic** synapses. The neuromuscular junction described in Chapter 10 is an example of a cholinergic synapse. (p. 309) Figure 12-13 shows the structure of a cholinergic synapse involving two neurons, rather than a neuron and a peripheral effector. Communication normally can occur in only one direction across a synapse, from the **presynaptic neuron** to the **postsynaptic neuron.** At a cholinergic synapse the **presynaptic membrane** and the **postsynaptic membrane** are separated by a synaptic cleft that averages 20 nm (0.02 μm) in width. Each synaptic knob contains mitochondria, vesicles, and areas of endoplasmic reticulum. Roughly half of the neurotransmitter molecules present in the terminal are dissolved in the cytoplasm, and the rest are packaged within synaptic vesicles. Every synaptic vesicle contains several thousand molecules of ACh, and there may be a million vesicles in a single synaptic knob.

Figure 12-14 diagrams the events that occur at a cholinergic synapse following the arrival of an action potential. These events can be summarized as follows:

STEP 1: The normal stimulus for neurotransmitter release is the depolarization of the presynaptic membrane following the arrival of an action potential. This depolarization involves not only an increase in sodium permeability, but a brief increase in the membrane permeability to calcium ions. Calcium ions (Ca^{2+}) then diffuse into the cytoplasm of the synaptic knob.

STEP 2: The increased calcium ion concentration activates enzymes that promote the release of ACh into the synaptic cleft. This release involves both the exocytosis of synaptic vesicles and diffusion through special membrane channels. The ACh diffuses across the synaptic cleft toward receptors on the postsynaptic membrane.

STEP 3: The release of ACh stops very soon because calcium ions are rapidly removed from the cytoplasm. In addition to being bound to special proteins and pumped out of the cell, they are absorbed by the endoplasmic reticulum and the remaining synaptic vesicles. At the postsynaptic membrane, the ACh receptors are chemically regulated channels. The primary response to ACh binding is an increased permeability to sodium ions, producing a depolarization that lasts about 20 msec. This response is a graded poten-

FIGURE 12–13
Structure of a Cholinergic Synapse

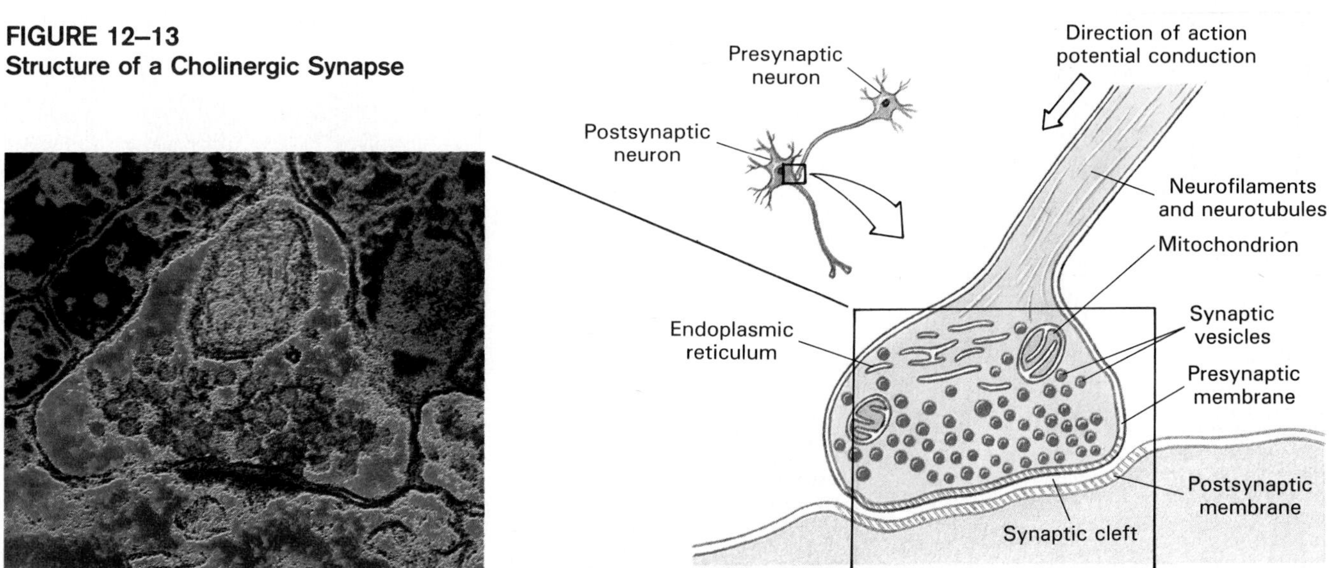

FIGURE 12–14
Function of a Cholinergic Synapse

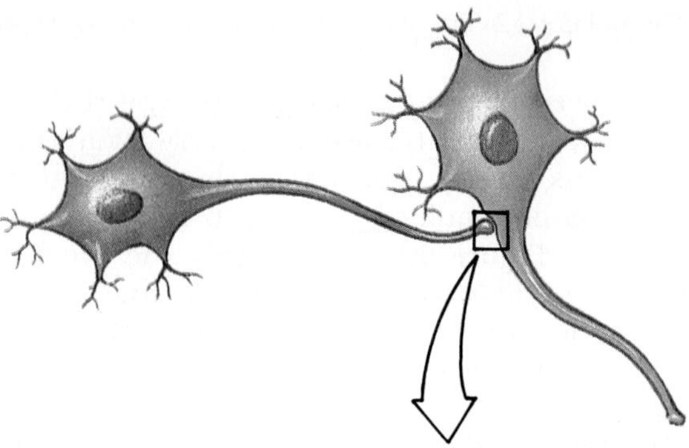

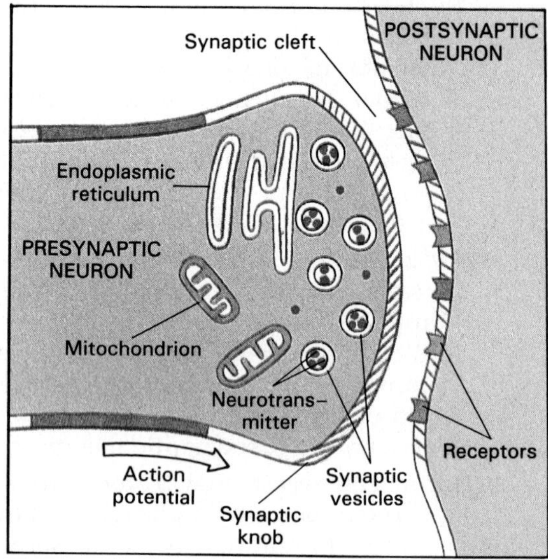

A narrow synaptic cleft separates presynaptic and postsynaptic neurons at a cholinergic synapse. The synaptic knob of the presynaptic neuron contains billions of molecules of the neurotransmitter acetylcholine—some dissolved in the cytoplasm, others held in synaptic vesicles.

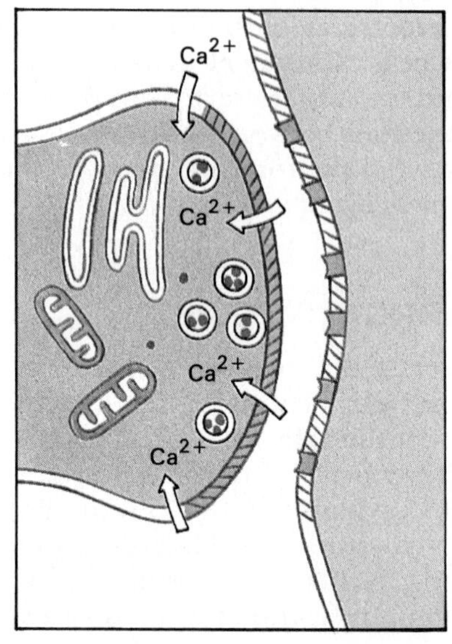

STEP 1: Arrival of an action potential at the synaptic knob depolarizes the presynaptic membrane, causing its permeability to calcium ions (Ca^{2+}) to increase briefly. In response, calcium ions diffuse into the cytoplasm of the synaptic knob.

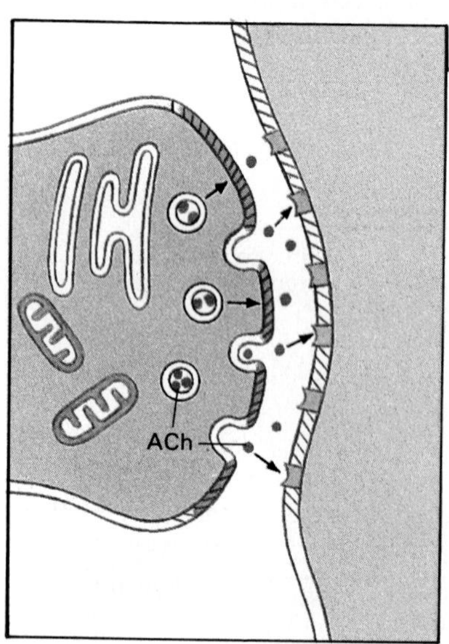

STEP 2: Calcium ions trigger the release of acetylcholine (ACh) into the synaptic cleft by exocytosis and diffusion. ACh molecules diffuse across the synaptic cleft toward receptors on the postsynaptic membrane.

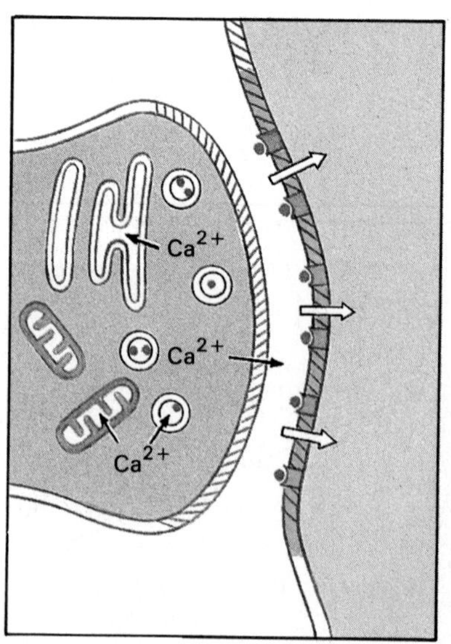

STEP 3: ACh binds to the postsynaptic receptors, which are chemically regulated channels. The usual response is an increased permeability to sodium ions, resulting in a depolarization lasting about 20 msec. ACh release at the synaptic knob ends as calcium ions are removed from the cytoplasm.

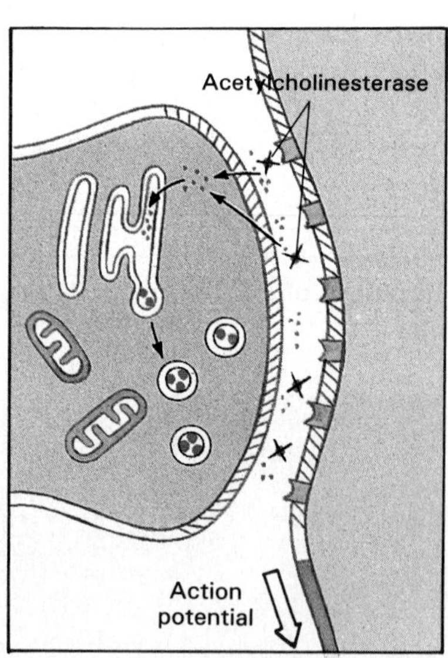

STEP 4: The effects on the postsynaptic membrane are temporary, because acetylcholinesterase breaks down ACh in the synaptic cleft and removes it from the postsynaptic membrane. The breakdown products are removed from the synaptic cleft by absorption at the synaptic knob.

tial: the more ACh released at the presynaptic membrane, the larger the depolarization. There is a 0.2–0.5 msec **synaptic delay** between the arrival of the stimulus at the synaptic knob and the effect on the postsynaptic membrane. Most of that delay reflects the time involved in calcium influx and neurotransmitter release; the synaptic cleft is relatively narrow, and diffusion of neurotransmitters across it takes only around 0.05 msec.

STEP 4: The effects on the postsynaptic membrane are temporary, rather than permanent, because the synaptic cleft and the postsynaptic membrane contain **acetylcholinesterase (AChE**, or *cholinesterase*), which hydrolyzes molecules of ACh into **acetate** and **choline**. This enzyme breaks down some of the released ACh before it even reaches the postsynaptic membrane. ACh molecules that do succeed in binding to receptor sites are usually dismantled within 20 msec of their arrival. The choline produced during the enzymatic breakdown of ACh is actively absorbed by the synaptic knob and eventually used to resynthesize ACh; the enzymes necessary for this process are present in the cytoplasm. Figure 12-15 shows the general pattern of ACh release and recycling at a cholinergic synapse.

FIGURE 12–15
ACh Release and Recycling at a Cholinergic Synapse. ACh at the synaptic knob is stored in the cytoplasm and in synaptic vesicles. Following its release into the synaptic cleft, ACh is broken down by hydrolysis into acetate and choline by the enzyme acetylcholinesterase. The choline is reabsorbed by the synaptic knob and used to synthesize ACh.

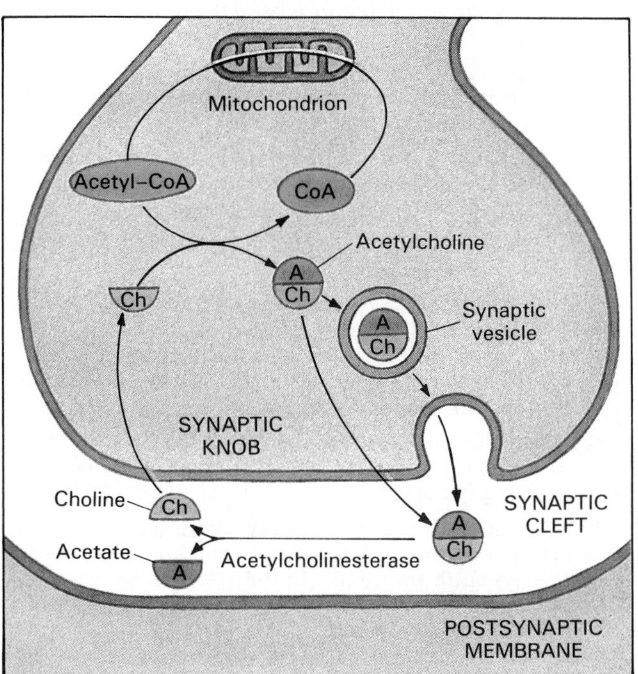

■ TABLE 12-5 Sequence of Events at a Typical Cholinergic Synapse

1. An arriving action potential depolarizes the synaptic knob and the presynaptic membrane.
2. Calcium ions enter the cytoplasm of the synaptic knob.
3. ACh release occurs through diffusion and exocytosis of neurotransmitter vesicles.
4. ACh diffuses across the synaptic cleft and binds to receptors on the postsynaptic membrane.
5. Chemically regulated sodium channels on the postsynaptic surface are activated, producing a graded depolarization.
6. ACh release ceases because calcium ions are removed from the cytoplasm of the synaptic knob.
7. The depolarization ends as ACh is broken down into acetate and choline by AChE.
8. The synaptic knob reabsorbs choline from the synaptic cleft and uses it to resynthesize ACh.

Because ACh molecules are recycled, the synaptic knob is not totally dependent on the ACh synthesized in the soma and delivered by axoplasmic transport. But under excessive stimulation the resynthesis and transport mechanisms may be unable to keep pace with the demand for neurotransmitter. **Synaptic fatigue** then occurs, and the synapse remains inactive until stores of ACh have been replenished.

Table 12-5 summarizes the events that occur at a cholinergic synapse.

ADRENERGIC SYNAPSES

Although acetylcholine is the neurotransmitter that has received the most attention, there are many other important chemical transmitters. **Norepinephrine** (nōr-ep-i-NEF-rin), or **NE**, is important in the brain and in portions of the autonomic nervous system. Norepinephrine is also called **noradrenaline**, and synapses releasing NE are described as **adrenergic**. An adrenergic synapse is diagrammed in Figure 12-16.

Norepinephrine usually has an excitatory, depolarizing effect on the postsynaptic membrane, but the mechanism is quite distinct from that of ACh. On the postsynaptic membrane surface, NE activates an enzyme, **adenyl cyclase**, that catalyzes the conversion of ATP to **cyclic-AMP (cAMP)** on the inner surface of the membrane. Cyclic-AMP then activates cytoplasmic enzymes that open ion channels and produce depolarization. The process thus involves two steps, with NE carrying the message from the presynaptic membrane and cAMP carrying the message into the cytoplasm. For this reason, cAMP is called a **second messenger**.

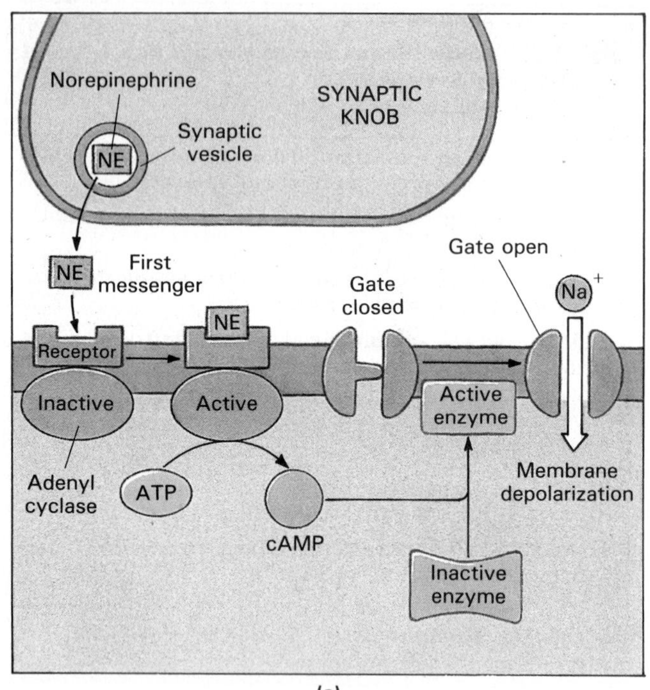

(a)

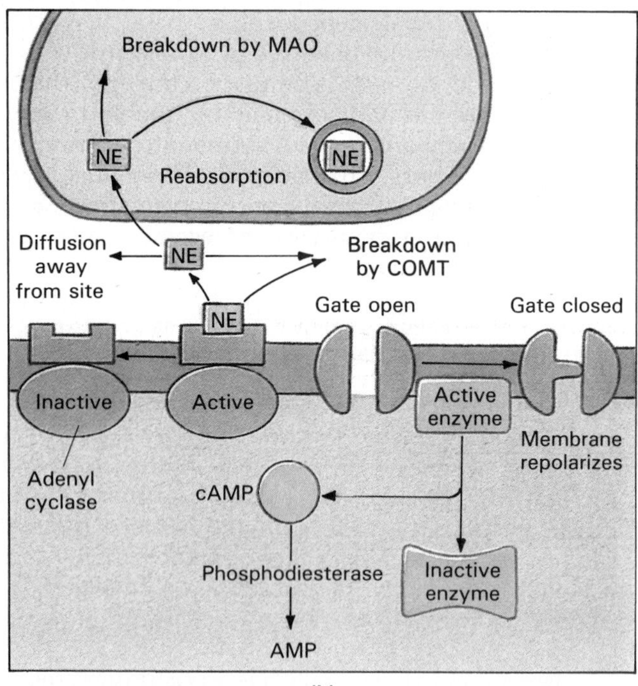

(b)

FIGURE 12–16
Events at an Adrenergic Synapse. (a) Norepinephrine released at the presynaptic membrane binds to a receptor on the postsynaptic surface. Binding activates the enzyme adenyl cyclase, which promotes the production of cyclic AMP. The cAMP then activates enzymes that open ion channels and cause depolarization of the postsynaptic cell membrane. (b) Cyclic-AMP is broken down by phosphodiesterase, and the cytoplasmic enzymes are inactivated. NE diffuses away from the site; it may be broken down by COMT or reabsorbed by the synaptic knob. Some of the absorbed NE will be broken down by MAO, but most will be recycled.

The effects on membrane potential are short-lived because another cytoplasmic enzyme, **phosphodiesterase**, quickly converts cyclic-AMP to AMP, an inactive molecule. In addition, NE is rapidly removed from the synapse. From 50 to 80 percent of the NE is reabsorbed and recycled by the synaptic knob; the rest diffuses out of the area or gets broken down by the enzymes **monoamine oxidase (MAO)** or **catechol-O-methyltransferase (COMT)**.

The sequence of events that take place at an adrenergic synapse are summarized in Table 12-6.

OTHER CHEMICAL SYNAPSES AND NEUROTRANSMITTERS

Dopamine (DŌ-pah-mēn), or **DOPA**, **gamma aminobutyric** (GAM-ma a-MĒ-nō-bū-TIR-ik) **acid**, also known as **GABA**, and **serotonin** (ser-ō-TŌ-nin) are CNS neurotransmitters whose effects are usually inhibitory (hyperpolarizing). There are also many other neurotransmitters whose functions are poorly understood. In a clear demonstration of the principle "the more you look the more you see," at least 50 neurotransmitters have now been identified, including certain amino acids, peptides, polypeptides, prostaglandins, and ATP. We know very little about the significance of these compounds, and next to nothing about the mechanisms involved in their synthesis and release. Table 12-7 lists major neurotransmitters

TABLE 12-6 Sequence of Events at an Adrenergic Synapse

1. An arriving action potential depolarizes the synaptic knob.
2. Calcium ions enter the cytoplasm, triggering release of norepinephrine (the first messenger) into the synaptic cleft.
3. NE diffuses across the cleft and binds to receptors on the postsynaptic membrane.
4. NE binding leads to activation of adenyl cyclase, which converts intracellular ATP to cyclic-AMP (the second messenger).
5. Cyclic-AMP activates an enzyme that opens chemically regulated ion gates, allowing an influx of sodium ions that depolarizes the membrane. Other activated enzymes may have diverse effects on cellular metabolism.
6. The effect on ion channels is limited by the action of phosphodiesterase, which converts cyclic-AMP to inactive AMP.
7. NE stimulation of the postsynaptic membrane decreases over time because of (a) NE absorption by the synaptic terminal, (b) diffusion away from the site, (c) destruction of NE by COMT (extracellular) or MAO (intracellular).

TABLE 12-7 Representative Neurotransmitters and Their Effects

Neurotransmitter	*Primary Distribution*	*Typical Effects*[1]	*Mechanism*
Acetylcholine	**Widespread throughout CNS and PNS**	**Usually excitatory (depolarizing)**	**Direct by effects on membrane channels**
AMINES			
Norepinephrine (NE)	**Many areas of the brain, some autonomic neurons in PNS**	**Usually excitatory**	**Indirect via second messengers**
Epinephrine	**Brain**	**Usually excitatory**	
Serotonin	**Brain (widespread)**	**Usually inhibitory**	
Dopamine (DOPA)	**Brain (specific areas)**	**Usually inhibitory**	
Histamine	**Brain (specific areas)**	**Usually inhibitory**	
AMINO ACIDS			
Gamma Aminobutyric Acid (GABA)	**Widespread throughout brain and spinal cord**	**Usually inhibitory (hyperpolarizing)**	
Glycine	**Spinal cord**	**Usually inhibitory**	**Direct and indirect**
Glutamate	**Brain and spinal cord**	**Usually excitatory**	
Aspartate	**Brain and spinal cord**	**Usually excitatory**	
NEUROPEPTIDES			
Substance P	**Spinal cord**	**Usually excitatory**	
Endorphins	**Brain and spinal cord**	**Usually inhibitory**	
OTHER SYNAPTIC CHEMICALS			
High-energy compounds (ATP, GTP)	**Brain**	**???**	
Hormones (insulin, glucagon, ACTH, etc.)	**Brain**	**???**	
Prostaglandins	**Brain**	**???**	

[1] Neurotransmitter effects vary because they depend on the nature of the postsynaptic receptor.

of the brain and spinal cord and their primary effects (when known).

A typical neuron synthesizes and releases a single neurotransmitter. Acetylcholine is released at synapses in the central and peripheral nervous systems, and in most cases ACh produces a depolarization in the postsynaptic membrane. But the ACh released at neuromuscular junctions in the heart produces a transient *hyperpolarization* of the membrane, moving the transmembrane potential farther from threshold. This difference highlights an important aspect of neurotransmitter function: *the effect on the postsynaptic membrane depends on the characteristics of the receptor, not on the nature of the neurotransmitter.*

NEUROMODULATORS

Chemical synapses are always active to some degree, and small numbers of neurotransmitter molecules are continually leaking through the presynaptic membranes. Other chemicals may also be released, some serving as **neuromodulators** (nu-rō-MOD-ū-lā-tōrz), compounds that influence the postsynaptic cell's response to the neurotransmitter. They are often **neuropeptides**, small peptide chains synthesized and released at synaptic terminals.

It can be very difficult to distinguish between neurotransmitters and neuromodulators. In general, neuromodulators: (1) have long-term effects that are relatively slow to appear, (2) trigger responses usually involving receptors and second messengers, (3) may affect the presynaptic membrane, the postsynaptic membrane, or both, and (4) can be released alone or in the company of a neurotransmitter. However, the same compound may function as a neurotransmitter in one site and as a neuromodulator in another. For this reason Table 12–7 does not draw distinctions between neurotransmitters and neuromodulators.

Functionally, the neurotransmitters and neuromodulators fall into two groups.

1. Neurotransmitters that have a direct effect on the membrane potential, by opening or closing chemically regulated channels. These neurotransmitters include acetylcholine and the amino acids *glutamate* and *aspartate*. In addition, there are chemically regulated channels

Clinical Comment: Endorphins

Neuropeptides called **endorphins** (en-DŌR-fins) are produced in the brain and spinal cord. The endorphins include several different **enkephalins** (en-KEF-a-lins), **beta-endorphin**, and **dynorphin** (di-NŌR-fin). Although their exact function is uncertain, their primary role is probably the relief of pain. These neuropeptides are structurally similar to morphine, and they are thought to bind to the same receptor sites. Pain relief occurs through inhibition of the release of Substance P at synapses that relay pain sensations. Dynorphin has far more powerful analgesic (pain-relieving) effects than either morphine or the endorphins.

sensitive to GABA and norepinephrine, although other membrane receptors for these neurotransmitters produce indirect effects.

2. Neurotransmitters that have an indirect effect on membrane potential, via *second messengers*. Membrane receptors for epinephrine, norepinephrine, dopamine, serotonin, histamine, GABA, and the various neuromodulators activate adenyl cyclase as depicted in Figure 12-16. This enzyme catalyzes the formation of cyclic-AMP, which opens membrane channels and activates intracellular enzymes.

ELECTRICAL SYNAPSES

Chemical synapses dominate the nervous system, and the responses of individual neurons are determined as different neurotransmitters arrive, neuromodulator concentrations change, and the transmembrane potential rises or falls. **Electrical synapses** lack this versatility. Electrical synapses are found between neurons in the CNS and PNS, but they are relatively rare. At an electrical synapse, the presynaptic and postsynaptic cell membranes are tightly bound together, and *gap junctions* (see Figures 3-29 and 5-2) permit the passage of ions between the cells. (pp. 90, 127) Because the two cells are linked in this way, changes in the transmembrane potential of one cell will produce local currents that affect the other cell as if the two shared a common cell membrane. This arrangement eliminates synaptic delay, and an electrical synapse efficiently carries nerve impulses from cell to cell. The transfer is direct and immediate, and the postsynaptic cell faithfully transmits the arriving impulses.

The situation at a chemical synapse is far more dynamic. The cells are not directly coupled, and interactions between excitatory and inhibitory neurotransmitters and neuromodulators make the response of the postsynaptic cell quite variable. That response reflects all of the stimuli, both excitatory and inhibitory, affecting the postsynaptic neuron at that moment. The postsynaptic cell is not a slave to the presynaptic neuron, and its activity reflects information processing at the cellular level.

Cellular Information Processing

Axons have excitable membranes that conduct action potentials over relatively great distances. At a synapse, the action potential triggers chemical or electrical events that affect another cell. When the communication involves two neurons, the synapse usually occurs at a dendrite or on the soma of the postsynaptic neuron. As noted earlier, these regions are not covered by an excitable membrane; they contain chemically regulated channels, but lack voltage-regulated channels. As a result, although the postsynaptic membrane at a chemical synapse may depolarize almost to 0 mV, action potentials do not appear.

Although the postsynaptic membrane cannot conduct action potentials, it can establish local currents that depolarize areas of excitable membrane to threshold. Consider Figure 12-18a (p. 395). The soma and dendrites are carpeted with synapses, many of which may be active at any given moment. Action potentials cannot occur in these inexcitable regions of the membrane. Nevertheless, stimulation at any one site produces a graded potential change that spreads across the membrane surface and onto the axon hillock (Figure 12-18b). In the figure, the degree of depolarization is indicated by the color intensity, as in earlier figures. If the axon hillock depolarizes to threshold, usually around –60 mV, an action potential forms in the initial segment and travels along the axon.

The inability of the soma and dendrites to support action potentials means that the soma and dendrites are unaffected by action potentials carried by the axon. Although most action potentials begin at the initial segment, one may be produced at a synapse somewhere along the axon. Such an action potential will be conducted toward the cell body as well as toward the telodendria. The action potential headed for the telodendria will depolarize the synaptic knob and cause the release of neurotransmitter. The one headed in the other direction, however, will not be conducted past the axon hillock. As a result, the soma and dendrites remain unaffected, preserving their sensitivity to depolarizing or hyperpolarizing stimuli.

The transmembrane potential at the axon hill-

Clinical Comment: Drugs and Synaptic Function

Many drugs interfere with key steps in the process of synaptic transmission. These drugs may either (1) interfere with transmitter synthesis, (2) alter the rate of transmitter release, (3) prevent transmitter inactivation, or (4) prevent transmitter binding to receptors. Table 12-8 lists several drugs that produce these effects at cholinergic synapses; a comparable array of drugs could be assembled that affect adrenergic synapses. The discussion that follows is limited

TABLE 12-8 Drugs Affecting Acetylcholine Activity at Synapses

Drug	*Mechanism*	*Effects*	*Remarks*
Hemicholinium	Blocks ACh synthesis	Produces symptoms of synaptic fatigue	
Botulinus toxin	Blocks ACh release directly	Paralyzes voluntary muscles	Produced by bacteria; responsible for serious incidents of food poisoning
Barbiturates	Decrease rate of ACh released	Muscular weakness, depression of CNS activity	Administered as sedatives and anesthetics
Procaine (Novocain)	Reduces membrane permeability to sodium	Prevents stimulation of sensory neurons	Used as a local anesthetic
Tetrodotoxin (also related compounds STX, CTX)	Blocks ACh release indirectly	Eliminates production of action potentials	Produced by some marine organisms during normal metabolic activity
Neostigmine	Prevents ACh inactivation by cholinesterase	Sustained contraction of skeletal muscles; other effects on cardiac and smooth muscles, glands, etc.	Used clinically to treat myasthenia gravis and to counteract overdoses of tubocurarine; related compound produced by Calabar bean
Insecticides (*malathion*, *parathion*, etc.); **nerve gases**	Prevents inactivation of ACh by cholinesterase	Same	Related compounds used in military nerve gases
d-tubocurarine	Prevents ACh binding to postsynaptic receptor sites	Paralysis of voluntary muscles	Curare produced by South American plant
Nicotine	Binds to ACh receptor sites	Low doses facilitate voluntary muscles; high doses cause paralysis	An active ingredient in cigarette smoke
Succinylcholine	Reduces sensitivity to ACh	Paralysis of voluntary muscles	Used to produce muscular relaxation during surgery
Atropine	Competes with ACh for binding sites on postsynaptic membrane	Voluntary muscle weakness, other effects on cardiac and smooth muscles	Produced by deadly nightshade plant

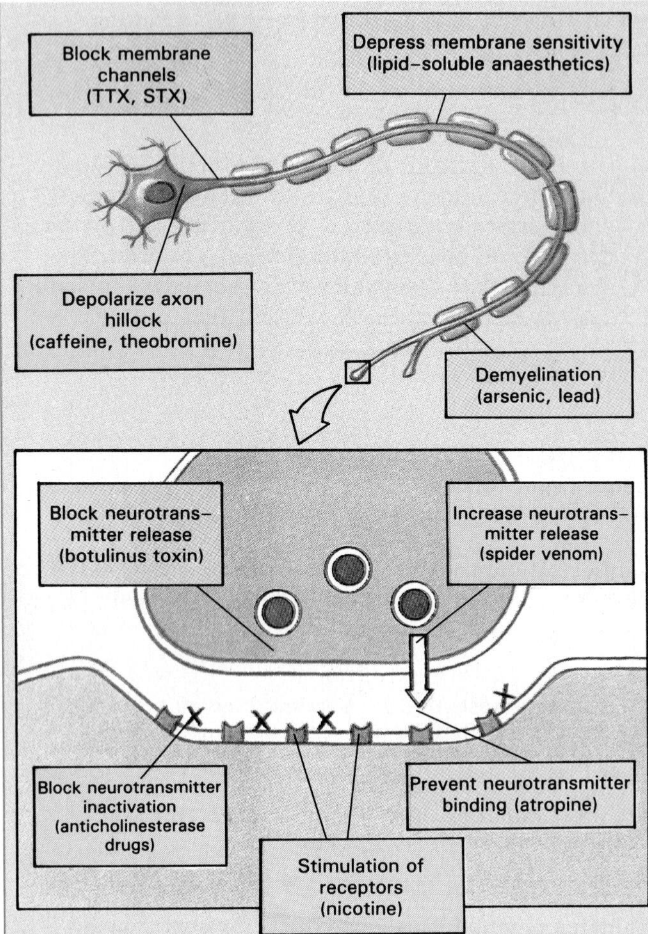

FIGURE 12–17
Mechanism of Drug Action at a Cholinergic Synapse. Factors that facilitate nerve function and make neurons more excitable are shown in red. Factors that inhibit or depress neural function are shown in blue.

to compounds that are clinically important; their mechanisms of activity are shown in Figure 12-17.

Botulinus toxin (CM, Chapter 10) was discussed in connection with the muscular system because the primary symptom of *botulism* is muscular paralysis. Botulinus toxin blocks the release of ACh at the presynaptic membrane of cholinergic neurons. The venom of the black widow spider has the opposite effect. It causes a massive release of ACh that produces intense muscular cramps and spasms.

Anticholinesterase drugs, sometimes called *cholinesterase inhibitors*, block the breakdown of ACh by acetylcholinesterase. The result is an exaggerated and prolonged stimulation of the postsynaptic membrane. At the neuromuscular junctions, this abnormal stimulation produces an extended and extreme state of contraction. Military nerve gases block cholinesterase activity for weeks, although few of those exposed are likely to live long enough to regain normal synaptic function. Most animals utilize ACh as a neurotransmitter, and anticholinesterase drugs, such as *malathion*, are in widespread use in pest-control projects.

Drugs such as **atropine** or **d-tubocurarine** prevent ACh from binding to the postsynaptic receptors. The latter compound is a derivative of *curare*, a plant extract used by certain South American tribes to paralyze their prey. Curare and related compounds induce paralysis by preventing stimulation of the neuromuscular junction by ACh. Atropine can also be administered intentionally to counteract the effects of anticholinesterase poisoning. Other compounds, including **nicotine**, an active ingredient in cigarette smoke, bind to the receptor sites and stimulate the postsynaptic membrane. There are no enzymes to remove these compounds, and the effects are relatively prolonged.

Health News: Coffee and Other Stimulants

The active ingredients of coffee, tea, and cocoa (caffeine, theobromine, and theophylline) lower the threshold at the axon hillock, making the neurons more sensitive to depolarizing stimuli. For this reason consuming large quantities of these beverages tends to facilitate neurons throughout the CNS. Under these conditions a relatively minor stimulus, such as the slamming of a nearby door, can have a sudden, pronounced effect. The phrase "coffee makes you jumpy" is not an exaggeration! Because nicotine stimulates ACh receptors and caffeine makes neurons more sensitive, the combination of cigarettes and coffee has a particularly strong stimulatory effect.

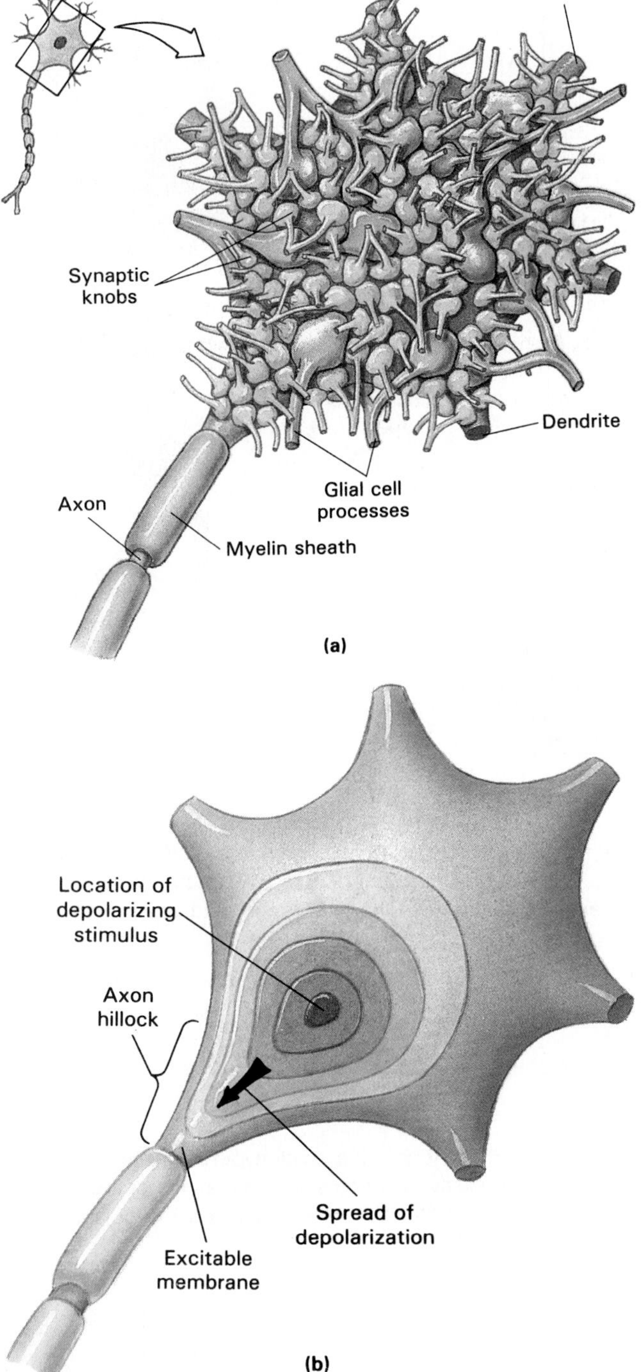

FIGURE 12–18
Synaptic Effects on the Transmembrane Potential of the Soma. (a) A single motor neuron may receive inputs across thousands of chemical synapses simultaneously. The soma is virtually blanketed with synaptic knobs. (b) When a single synapse becomes active, the arrival of neurotransmitter molecules leads to a graded change in the transmembrane potential at that site.

ock determines whether or not an action potential develops in the initial segment. Because the transmembrane potential at the axon hillock determines the rate of action potential generation, the axon hillock represents the actual site of information processing.

EXCITATORY POSTSYNAPTIC POTENTIALS

Literally thousands of synaptic knobs contact the soma of a single neuron. Some of those synapses will depolarize the postsynaptic membrane. A depolarization produced by the arrival of a neurotransmitter is called an **excitatory postsynaptic potential**, or **EPSP**. An EPSP results from the opening of chemically regulated ion channels that allow sodium ions into the cell.[2] Because this is a graded potential, an EPSP affects only the area immediately surrounding the synapse, as in Figure 12-18b.

A typical EPSP produces a depolarization of around 0.5 mV, much less than the 15–20 mV depolarization needed to bring the axon hillock to threshold. But individual EPSPs can combine through the process of **summation**. There are two forms of summation, *temporal summation* and *spatial summation*.

Temporal Summation

Temporal summation occurs at a single synapse, when a second EPSP arrives before the effects of the first have disappeared. A typical EPSP lasts about 20 msec, but under maximum stimulation one action potential may reach the synaptic knob each *millisecond*. Every time another batch of ACh molecules arrives at the postsynaptic membrane, additional receptor sites will be occupied, and the degree of depolarization will increase. In this way, a series of small steps can eventually bring the axon hillock to threshold, as illustrated in Figure 12-19a. Such addition of stimuli occurring in rapid succession is called **temporal summation** (*tempus*, time).

Spatial Summation

A typical motor neuron may be influenced by 6000 synaptic knobs, with 80–90 percent of them contacting the dendrites and most of the rest scattered across the soma. The activity of a single synapse produces a graded potential with localized effects. The effect on the axon hillock varies, depending on the strength of the stimulus and its distance from the hillock. But at any one moment, more than one synapse will be active, and the effects on the axon hillock are cumulative, as illustrated in Figure 12–23b. When the transmembrane potential at the axon hillock reaches threshold, an action potential will appear in the initial segment. Such addition of stimuli arriving at different locations of the nerve cell membrane is called **spatial summation**.

[2]These channels also let potassium ions out of the cell, but because sodium ions are driven by both electrical and chemical gradients the net effect is a slight depolarization.

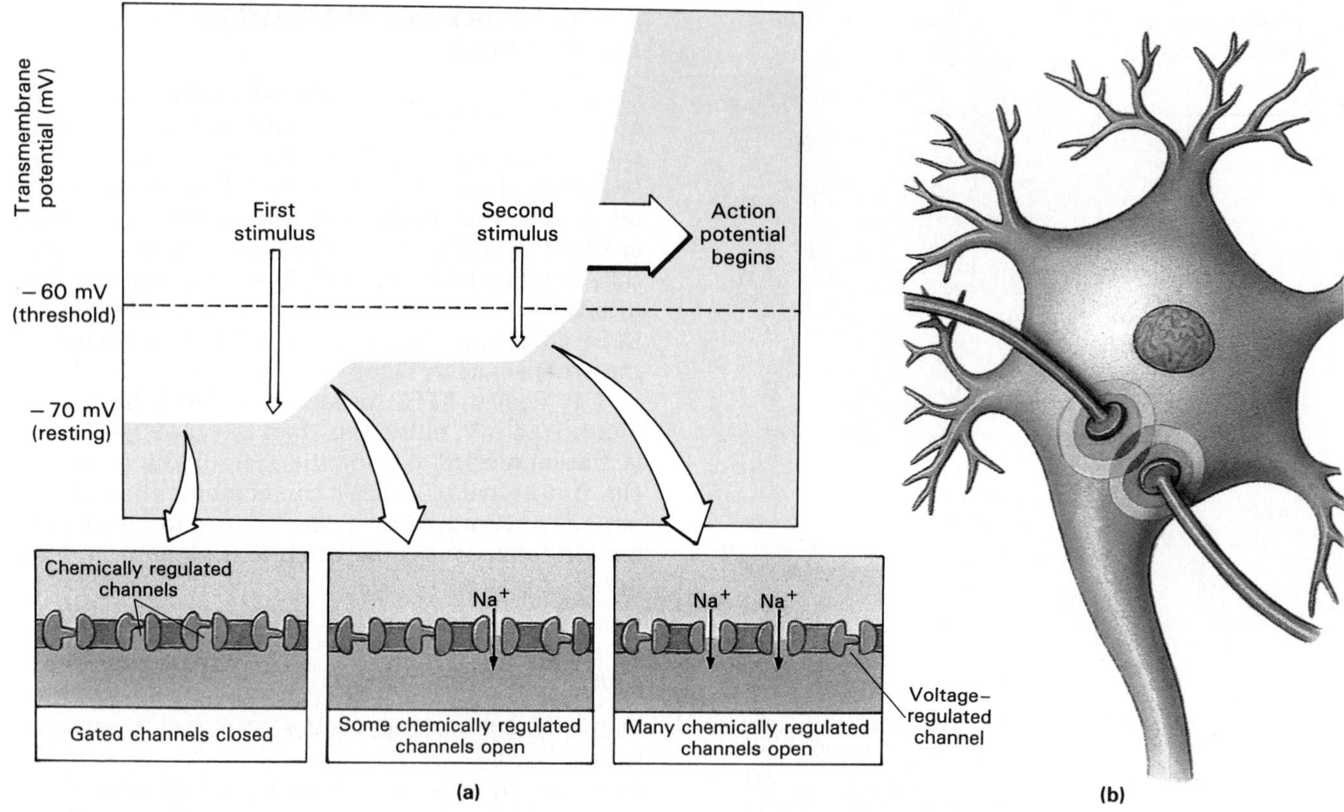

FIGURE 12–19
Temporal and Spatial Summation. (a) Temporal summation occurs on a membrane receiving two depolarizing stimuli separated in time. The effects of the second stimulus are added to those of the first. (b) Spatial summation occurs when sources of stimulation arrive simultaneously but at different locations. The effects of each spread across the membrane, and intervening areas feel the combined effects.

FIGURE 12–20
EPSP-IPSP Interactions. At time 1 a small depolarizing stimulus produces an EPSP. At time 2 a small hyperpolarizing stimulus produces an IPSP of comparable magnitude. If the two stimuli are applied simultaneously, as at time 3, summation occurs. Because the two are equal in size, the membrane potential remains at resting levels. If the EPSP were larger, a net depolarization would result; if the IPSP were larger, a net hyperpolarization would be seen.

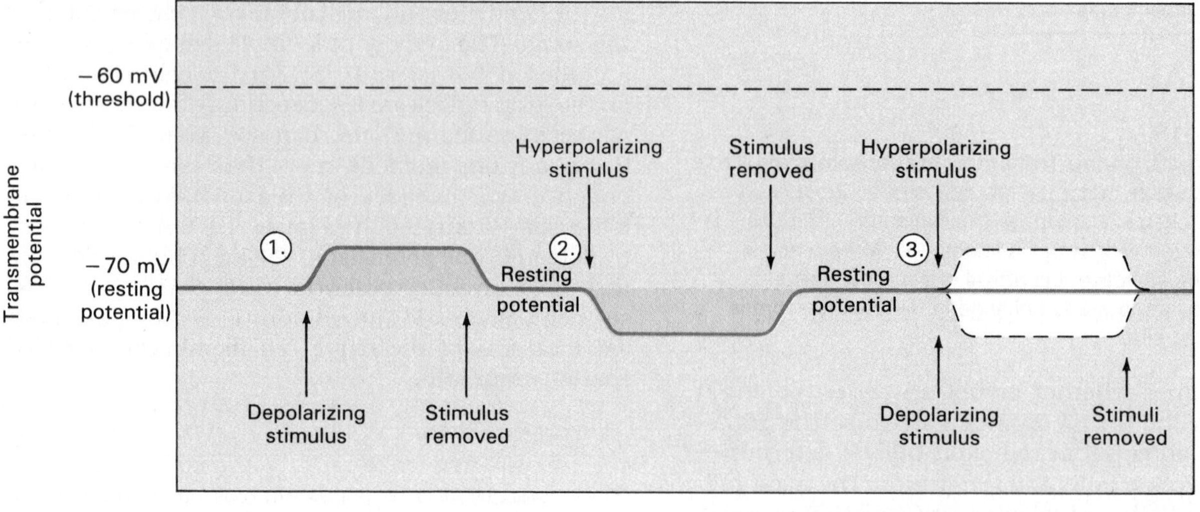

Spatial or temporal summation of EPSPs will not necessarily depolarize the axon hillock to threshold. But any shift in the transmembrane potential toward threshold makes the cell more sensitive to further stimulation. Such a neuron is said to be **facilitated**. The larger the degree of facilitation, the smaller the stimulus needed to trigger an action potential. In a highly facilitated neuron, even a small depolarizing stimulus will produce an action potential.

On the other hand, summation can keep the transmembrane potential of the axon hillock well above its threshold. Under these circumstances action potentials will zip down the axon one right after the other, separated only by the time it takes to produce an action potential and recover from it. The refractory period typically lasts about 0.5 msec, so an axon could theoretically conduct action potentials at a rate of 2500 per second; however, the highest recorded frequencies range between 500 and 1000 per second.

Clinical Comment: Ions and Neural Function

Changes in the ionic composition of the extracellular fluids have a direct effect on neural function. Fluctuations in sodium or potassium ion concentrations, caused by dehydration or kidney disease, may facilitate or depress neural activity by depolarizing or hyperpolarizing the cell membrane. Abnormally high or low extracellular calcium ion concentrations have a direct effect on synaptic function by reducing or exaggerating the amount of neurotransmitter released.

Changes in hydrogen ion concentration (pH) can have equally dramatic effects. The normal extracellular pH hovers around 7.4. If the pH rises, neurons are facilitated, and at a pH near 7.8 they begin to generate action potentials spontaneously, producing severe convulsions. If the pH declines, neurons are inhibited; at a pH around 7.0 the nervous system shuts down, and the individual becomes completely unresponsive. A variety of mechanisms exist to control the pH of the cerebrospinal fluid and other body fluids; these mechanisms will be discussed in later chapters.

INHIBITORY POSTSYNAPTIC POTENTIALS

Not all neurotransmitters have an excitatory (depolarizing) effect. An **inhibitory postsynaptic potential**, or **IPSP**, is a transient *hyperpolarization* of the postsynaptic membrane. An IPSP results from the opening of chemically regulated potassium or chloride channels. While the hyperpolarization continues, the neuron is **inhibited** because a larger-than-usual depolarizing stimulus must be provided to bring the membrane potential to threshold. For example, a stimulus sufficient to shift the transmembrane potential 10 mV, from –70 mV to –60 mV would normally produce an action potential. But if the transmembrane potential is reset at –85 mV by an IPSP, the same stimulus would depolarize it only to –75 mV—still well below threshold. Like EPSPs, IPSPs can summate spatially and temporally.

SUMMATION OF EPSPs AND IPSPs

Because EPSPs and IPSPs reflect the activation of different types of chemically regulated channels, the two processes can occur in the same portion of the cell membrane, and they have opposing effects on the transmembrane potential. This antagonism between IPSPs and EPSPs is an important mechanism for information processing at the cellular level. The activity of each neuron is determined by the transmembrane potential at the axon hillock, and at any given moment that potential will reflect the dynamic balance of EPSPs and IPSPs (Figure 12-20).

EPSP-IPSP interactions are the most important determinants of neural activity. Nevertheless, other factors can have significant impacts on the rate of action potential generation. For instance:

1. Neuromodulators or hormones can change the sensitivity of the postsynaptic membrane to excitatory or inhibitory neurotransmitters and promote facilitation or inhibition.
2. Inhibitory or excitatory synapses may occur at an **axoaxonal synapse** on the synaptic knob. Activity at this synapse modifies the rate of neurotransmitter release at the presynaptic membrane. In **presynaptic inhibition**, *GABA* release inhibits the opening of voltage-regulated calcium channels in the synaptic knob. This inhibition reduces the amount of neurotransmitter released when the action potential arrives at the presynaptic membrane, and so limits stimulation of the postsynaptic membrane. In **presynaptic facilitation** (Figure 12-21b), the calcium channels remain open for a longer period, increasing the amount of neurotransmitter released. Serotonin release at an axoaxonal synapse can cause presynaptic facilitation by prolonging the depolarization that follows the arrival of an action potential.

■ TABLE 12-9 Summary of Cellular Information Processing

1. The cell membrane of the dendrites and soma will not conduct action potentials. It contains chemically regulated ion channels that respond to the presence of neurotransmitters producing graded potentials (EPSPs or IPSPs).
2. Exposure to a specific neurotransmitter may produce an EPSP or an IPSP. These are graded potential changes whose size and extent depend on the concentration of neurotransmitter.
3. EPSPs and IPSPs can summate spatially or temporally; they have antagonistic effects on the transmembrane potential.
4. Action potentials begin at the initial segment of the axon. The rate of action potential generation is determined by the membrane potential at the adjacent axon hillock.
5. At any given moment the membrane potential at the axon hillock reflects the interaction between all of the EPSPs and IPSPs affecting the soma and dendrites.
6. Neuromodulators, hormones, and environmental factors can influence action potential generation by affecting the membrane potential or disturbing vital metabolic operations.

3. Environmental factors, such as pH, ion, or temperature changes, can alter the resting membrane potential or disrupt the metabolic operations that support action potential generation.

Table 12–9 summarizes important aspects of cellular information processing.

√ What effect would blocking the calcium channels at the presynaptic membrane of a cholinergic synapse have on communication at that synapse?

√ What would happen at an adrenergic synapse if the enzyme that converts ATP to cAMP at the postsynaptic terminal were blocked?

√ A neurotransmitter causes potassium channels to open but not sodium channels. What type of postsynaptic potential would this neurotransmitter produce?

■ Neurons and Metabolic Processes

The brain contributes just 2 percent to the body weight, but it accounts for 18 percent of resting energy consumption. Active neurons need ATP to support (1) the synthesis, release, and recycling of neurotransmitter molecules, (2) the movement of materials to and from the soma via axoplasmic flow, and (3) the recovery from action potentials. Each time an action potential occurs, sodium ions enter and potassium ions leave the cell, and ATP must be expended to maintain normal cytoplasmic ion concentrations. When impulses are generated at high frequencies, the energy demands are enormous.

Neurons normally derive ATP solely through aerobic glycolysis. Because their cytoplasm does not contain glycogen reserves, these cells are totally dependent on a continual and reliable supply of both oxygen *and* glucose from the blood. If the circulation is interrupted for just a few seconds, the nerve cell will be injured, and the longer the interruption, the more severe the injury. In a **stroke** the blood supply to the brain is interrupted by a circulatory blockage or other vascular problem. The degree of functional impairment after a stroke is determined by (1) the region deprived of circulation and (2) the duration of the circulatory interruption. The severity of brain damage after a stroke can be drastically reduced by reestablishing normal circulation; methods of accomplishing this are discussed in Chapter 21. [**CM:** *Tay-Sachs Disease*]

■ Neural Response to Injuries

A neuron responds to injury in a very limited, stereotyped fashion. Within the soma, the Nissl bodies disappear and the nucleus moves away from its centralized location. If the neuron recovers its functional abilities, it will return to a normal appearance. The key to recovery seems to be the events under way in the axon. Consider the response of a neuron to mechanical stresses, such as the pressure applied during a crushing injury. The pressure produces a local decrease in blood flow and oxygen availability, and the affected membrane becomes inexcitable. If the pressure is released after an hour or two, the nerve cell will recover within a few weeks. More severe or prolonged pressure will produce effects similar to those caused by cutting off the distal portion of the axon.

In the peripheral nervous system, the Schwann cells participate in the repair of damaged nerves. In the process known as **Wallerian degeneration**, illustrated in Figure 12-21, the axon distal to the injury site deteriorates, and macrophages migrate in to phagocytize the debris. The Schwann cells undergo mitosis and form a solid cellular cord that follows the path of the original axon. As the neuron recovers, its axon grows into the injury site, and the Schwann cells wrap around it.

If the axon continues to grow into the periphery alongside the appropriate cord of Schwann cells, it may eventually reestablish its normal synaptic contacts. If it stops growing, or wanders off in some new

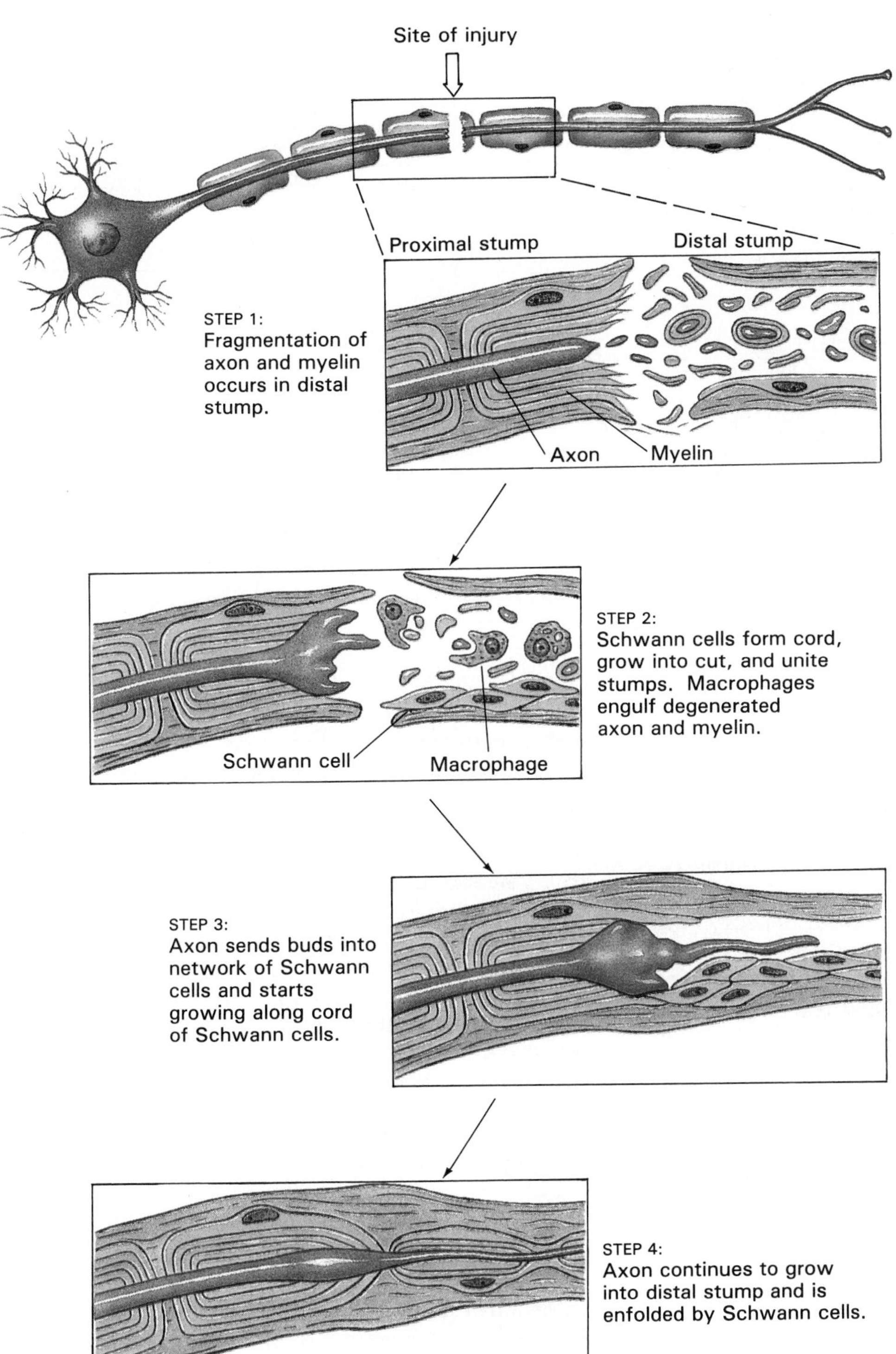
Site of injury
Proximal stump
Distal stump
STEP 1:
Fragmentation of axon and myelin occurs in distal stump.
Axon
Myelin
STEP 2:
Schwann cells form cord, grow into cut, and unite stumps. Macrophages engulf degenerated axon and myelin.
Schwann cell
Macrophage
STEP 3:
Axon sends buds into network of Schwann cells and starts growing along cord of Schwann cells.
STEP 4:
Axon continues to grow into distal stump and is enfolded by Schwann cells.

FIGURE 12–21
Nerve Regeneration Following Injury

direction, normal function will not return. The growing axon is most likely to arrive at its appropriate destination if the cut edges remain in contact. Additional information concerning neural repairs and surgical procedures will be presented in Chapter 13. A comparable series of events occurs inside the central nervous system, although the repairs take much longer to complete.

Functional Patterns of Neural Organization

Neurons can be functionally categorized into three groups: (1) sensory neurons, (2) motor neurons, and (3) interneurons, or association neurons. Their relationships are diagrammed in Figure 12-22.

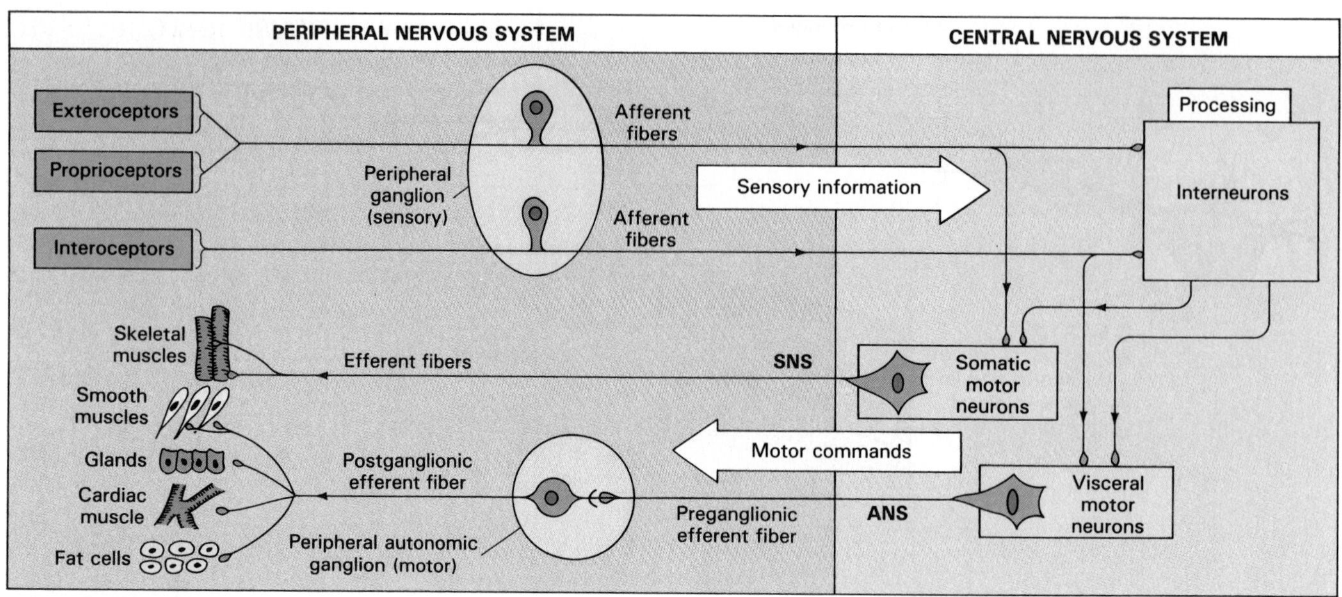

FIGURE 12–22
A Functional Classification of Neurons

SENSORY NEURONS

Sensory neurons form the *afferent division* (*ad*, to + *ferre*, to carry) of the PNS. Their function is to deliver information to the CNS. The cell bodies of sensory neurons are found in peripheral ganglia. These are unipolar neurons, and their processes, known as **afferent fibers**, extend between a sensory **receptor** and the spinal cord or brain. Sensory neurons collect information concerning the external or internal environment. The receptor may be a modified portion of a dendrite or a specialized cell that communicates with a sensory neuron across a chemical synapse. **Somatic sensory neurons** monitor the outside world and our position within it. **Exteroceptors** (*extero-*, outside) provide information about the external environment through the senses of sight, smell, hearing, and touch. **Proprioceptors** (prō-prē-ō-SEP-tōrz) monitor the position of skeletal muscles and joints. **Visceral sensory neurons**, or **interoceptors** (*intero-*, inside), monitor internal operations, including those of the digestive, respiratory, cardiovascular, urinary, and reproductive systems, and convey sensations of taste, deep pressure, and pain.

MOTOR NEURONS

Motor neurons of the *efferent division* (*ex*, from) carry instructions from the CNS to peripheral effectors. A motor neuron stimulates or modifies the activity of a peripheral tissue, organ, or organ system. Axons traveling away from the CNS are called **efferent fibers**. There are two major efferent divisions in the PNS. The *somatic nervous system* includes all of the **somatic motor neurons** that innervate skeletal muscles. The soma of a somatic motor neuron lies inside the CNS, and its axon extends into the periphery to end at neuromuscular junctions. We have voluntary control over the activity of the somatic nervous system.

The activities of the *autonomic nervous system* are primarily involuntary, or under "automatic" control. **Visceral motor neurons** innervate all peripheral effectors other than skeletal muscles. The axons of visceral motor neurons inside the CNS synapse on neurons in peripheral ganglia, and the ganglion cells control peripheral effectors. Axons extending from the CNS to a ganglion are called **preganglionic fibers**. Axons connecting the ganglion cells with the peripheral effectors are known as **postganglionic fibers**. This arrangement clearly distinguishes the autonomic (visceral motor) system from the somatic motor system.

INTERNEURONS

Interneurons, or **association neurons**, may be situated between sensory and motor neurons. Interneurons are located entirely within the brain and spinal cord, where they outnumber all other neurons combined. Interneurons are responsible for the analysis of sensory inputs and the coordination of motor outputs. The more complex the response to a given stimulus, the greater the number of interneurons involved.

Interneurons can be classified as **excitatory** or **inhibitory** on the basis of their effects on postsynaptic membranes. Excitatory interneurons produce a depolarization that may cause facilitation or stimulation; inhibitory interneurons produce an inhibitory hyperpolarization.

HIGHER LEVELS OF ORGANIZATION AND PROCESSING

There are around 10 million sensory neurons, 20 *billion* interneurons, and half a million motor neurons. The interneurons are organized into a smaller number of **neuronal pools**. A neuronal pool is a group of interconnected neurons with specific functions. Estimates concerning the actual number of pools vary, but range between a few hundred and a few thousand. Each pool has a limited number of input sources and output destinations, and the pool may contain excitatory and inhibitory neurons. The output of the entire pool may stimulate or depress the activity of other pools, or it may exert direct control over motor neurons or peripheral effectors.

The pattern of interaction between neurons provides clues to the functional characteristics of a neuronal pool. For example, a single excitatory interneuron may synapse on several others, as illustrated in Figure 12-23a. When this neuron is activated, some of the postsynaptic neurons will depolarize to threshold and generate action potentials of their own. These neurons lie within the **discharge zone** of the presynaptic neuron. Other postsynaptic neurons will not reach threshold, but they will undergo facilitation. These neurons form the **facilitated zone** of the presynaptic neuron. During a complex movement, motor neurons that control specific motor units are stimulated. Some of these motor neurons are brought to threshold, and their motor units contract at once. Other motor neurons are facilitated, and these will be recruited as the movement proceeds (see the discussion of recruitment in Chapter 10). (p. 315)

Divergence is the spread of information from one neuron to several neurons, as in Figure 12-23a, or from one pool to multiple pools. Divergence permits the broad distribution of a specific input. Considerable divergence occurs when sensory neurons bring information into the CNS, for the information is distributed to neuronal pools throughout the spinal cord and brain.

In **convergence** (Figure 12-23b), several neurons synapse on the same postsynaptic neuron. Several different patterns of activity in the presynaptic neurons can have the same effect on the postsynaptic neuron. Convergence permits the variable control of motor neurons by providing a mechanism for their voluntary and involuntary control. For example, the movements of your diaphragm and ribs are now being involuntarily controlled by respiratory centers in the brain. But the same motor neurons can also be controlled voluntarily, as when you take a deep breath

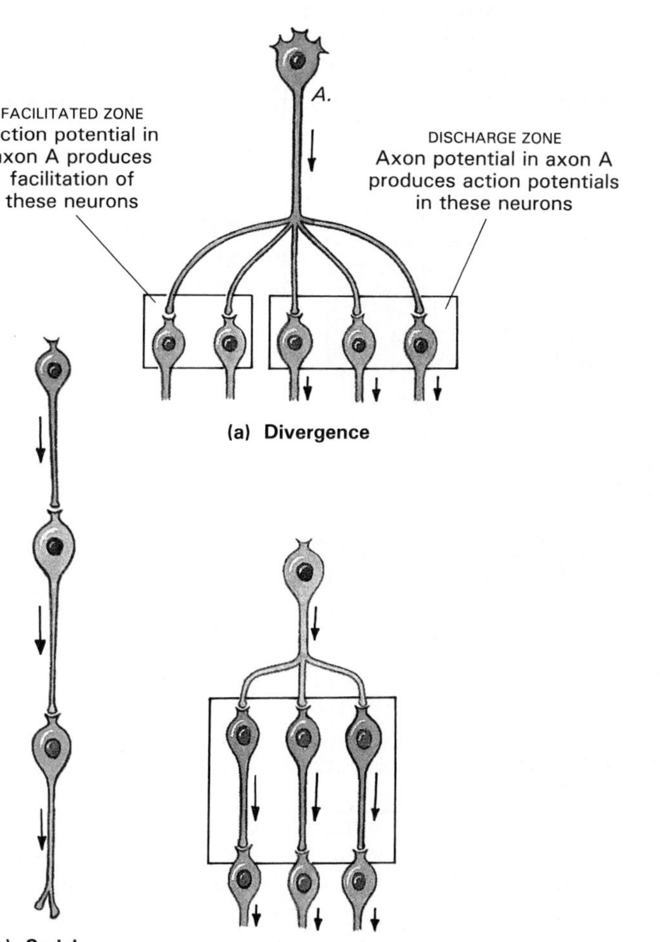

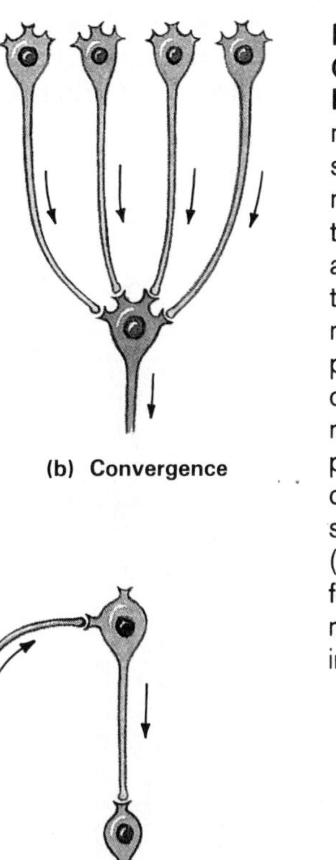

Figure 12–23 Organization of Neuronal Pools. (a) Divergence, a mechanism for spreading stimulation to multiple neurons or neuronal pools in the CNS. (b) Convergence, a mechanism providing input to a single neuron from multiple sources. (c) Serial processing, in which neurons or pools work in a sequential manner. (d) Parallel processing, in which neurons or pools process information simultaneously. (e) Reverberation, a feedback mechanism that may be excitatory or inhibitory.

and hold it. Two different neuronal pools are involved, each synapsing on the same motor neurons.

Neuronal pools may function in sequence, with one pool activating a second, which in turn activates a third, and so forth. This pattern is called **serial processing** (Figure 12-23c). Serial processing occurs as sensory information is relayed from one processing center to another in the brain.

Parallel processing (Figure 12-23d) occurs when several neuronal pools are processing the same information at one time. Thanks to parallel processing, many different responses occur simultaneously. For example, stepping on a tack stimulates sensory neurons that distribute the information to a number of neuronal pools. As a result of parallel processing you might withdraw your foot, shift your weight, move your arms, feel the pain, cry "ouch," and remember the last time you stepped on a tack, all at the same time.

Some neural circuits utilize positive feedback to produce **reverberation**. In this arrangement, collateral axons extend back toward the source of an impulse and further stimulate the presynaptic neurons. Once a reverberating circuit has been activated it will continue to function for an extended period, until synaptic fatigue or inhibitory stimuli break the cycle. As with convergence or divergence, reverberation can occur within a single neuronal pool, or it may involve a series of interconnected pools. A simple example of reverberation is diagrammed in Figure 12-23e; much more complicated examples of reverberation between neuronal pools in the brain may be involved in the maintenance of consciousness, muscular coordination, and normal breathing patterns. We will discuss these and other "wiring patterns" as we consider the organization of the spinal cord and brain in the following chapters.

Anatomical Organization of the Nervous System

The functions of the nervous system depend on the interactions between neurons in neuronal pools, and the most complex neural processing steps occur in the spinal cord and brain, the components of the central nervous system (CNS). The arriving sensory information and the outgoing motor commands are carried by the peripheral nervous system (PNS). The axons and somae of the CNS and PNS are not randomly scattered. Instead, they form masses or bundles with distinct anatomical boundaries. Major categories are detailed in Figure 12-24.

As previously mentioned, in the peripheral nervous system the nerve cell bodies are found in ganglia and the axons are bundled together in peripheral nerves. **Spinal nerves** communicate with the spinal cord, and **cranial nerves** are connected to the brain. Collections of nerve cell bodies in the CNS are termed **centers**. There are many types of centers, each with

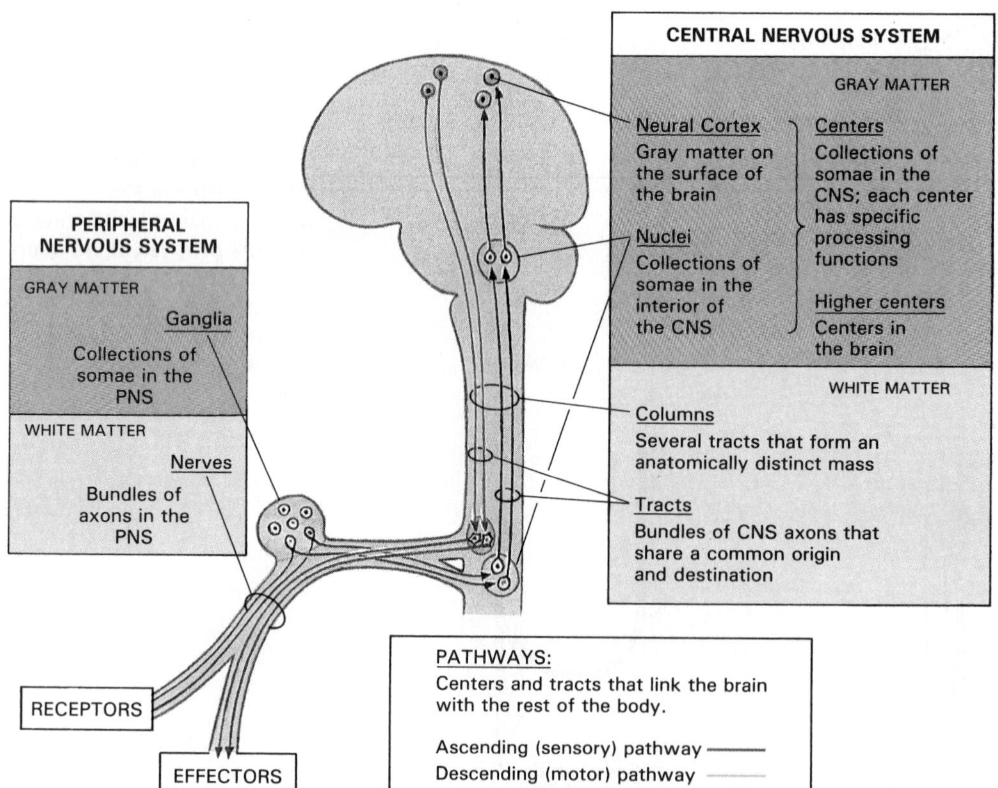

FIGURE 12–24
Anatomical Organization of the Nervous System

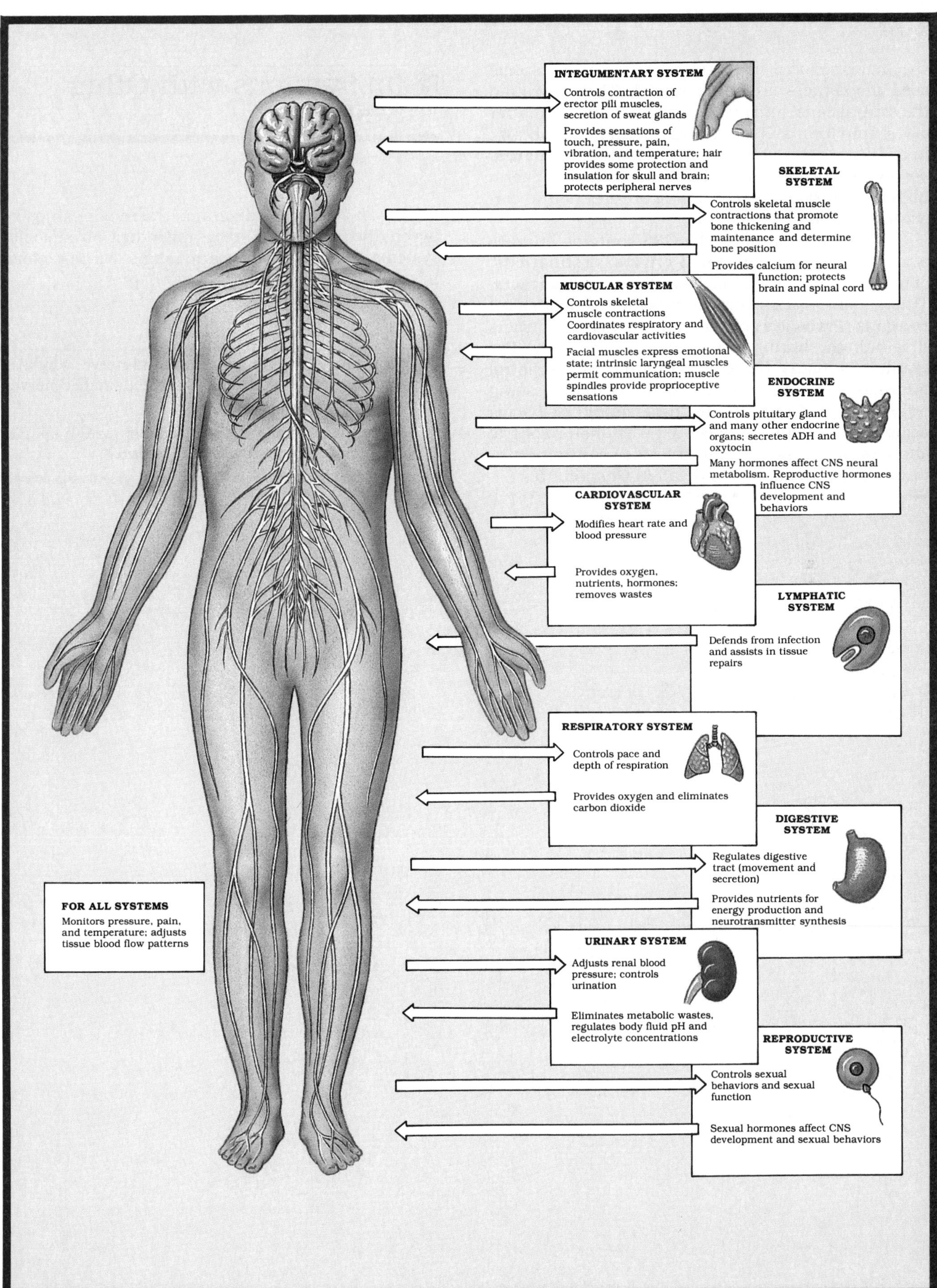

INTEGUMENTARY SYSTEM
Controls contraction of erector pili muscles, secretion of sweat glands
Provides sensations of touch, pressure, pain, vibration, and temperature; hair provides some protection and insulation for skull and brain; protects peripheral nerves
SKELETAL SYSTEM
Controls skeletal muscle contractions that promote bone thickening and maintenance and determine bone position
Provides calcium for neural function; protects brain and spinal cord
MUSCULAR SYSTEM
Controls skeletal muscle contractions
Coordinates respiratory and cardiovascular activities
Facial muscles express emotional state; intrinsic laryngeal muscles permit communication; muscle spindles provide proprioceptive sensations
ENDOCRINE SYSTEM
Controls pituitary gland and many other endocrine organs; secretes ADH and oxytocin
Many hormones affect CNS neural metabolism. Reproductive hormones influence CNS development and behaviors
CARDIOVASCULAR SYSTEM
Modifies heart rate and blood pressure
Provides oxygen, nutrients, hormones; removes wastes
LYMPHATIC SYSTEM
Defends from infection and assists in tissue repairs
RESPIRATORY SYSTEM
Controls pace and depth of respiration
Provides oxygen and eliminates carbon dioxide
DIGESTIVE SYSTEM
Regulates digestive tract (movement and secretion)
Provides nutrients for energy production and neurotransmitter synthesis
URINARY SYSTEM
Adjusts renal blood pressure; controls urination
Eliminates metabolic wastes, regulates body fluid pH and electrolyte concentrations
REPRODUCTIVE SYSTEM
Controls sexual behaviors and sexual function
Sexual hormones affect CNS development and sexual behaviors
FOR ALL SYSTEMS
Monitors pressure, pain, and temperature; adjusts tissue blood flow patterns

FIGURE 12–25
Functional Relationships between the Nervous System and Other Systems

specific functions. In the gray matter of the spinal cord the centers are called **nuclei**. They surround the central canal and are covered by the white matter of the spinal cord. Nuclei are also found in the brain; in addition, a thick blanket of gray matter, the **neural cortex**, covers portions of the brain surface. The term **higher centers** refers to nuclei or areas of neural cortex in the brain.

The white matter of the CNS contains bundles of axons that share common origins, destinations, and functions. These axonal bundles are called **tracts**. Tracts in the spinal cord form larger groups, called **columns**. **Pathways** consist of the centers and tracts that link the brain with the rest of the body. For example, **sensory (ascending) pathways** distribute information from peripheral receptors to processing centers in the brain; **motor (descending) pathways** begin at CNS centers concerned with motor control and end at the skeletal muscles they control. Pathways will be considered in detail in Chapter 15.

■ Interactions with Other Systems

Figure 12-25, p. 403 diagrams the relationships between the nervous system and other physiological systems. Many of these relationships will be explored in greater detail in subsequent chapters.

√ **In an injury involving a peripheral nerve, why is it important for the two ends of the damaged nerve to be closely aligned?**

√ **How would damage to interneurons in the spinal cord affect nervous system function?**

CHAPTER REVIEW

Review of Selected Clinical Terms

endorphins (en-DŌR-finz) (*p. 392*) Neuropeptides produced in the brain and spinal cord that appear to relieve pain.

anticholinesterase drug (*p. 394*) A drug which blocks the breakdown of ACh by acetylcholinesterase.

atropine (*p. 394*) A drug which prevents ACh from binding to the postsynaptic membrane.

d-tubocurarine (*p. 394*) A drug derived from curare which prevents ACh from binding to the postsynaptic membrane.

nicotine (*p. 394*) A compound which binds to receptor sites and stimulates the postsynaptic membrane; found in tobacco.

Additional Terms of Clinical Importance

ENDORPHINS (*p. 392*): **enkephalin, beta-endorphin, dynorphin**

Study Outline

Related Key Terms

Introduction (pp. 369–370)

1. Two organ systems, the nervous and endocrine systems, coordinate organ system activity. The nervous system provides swift but brief responses to stimuli; the endocrine system adjusts metabolic operations and directs long-term changes.

The Nervous System: An Overview (p. 370)

1. The **nervous system** includes all the **neural tissue** in the body. Its anatomical divisions include the **central nervous system (CNS)** (the brain and spinal cord) and the **peripheral nervous system (PNS)** (all of the neural tissue outside the CNS). Functionally it can be divided into an **afferent division**, which brings sensory information to the CNS and an **efferent division** (that carries motor commands to muscles and glands). The efferent division includes the **somatic nervous system (SNS)** (voluntary control over skeletal muscle contractions), and the **autonomic nervous system (ANS)** (automatic, involuntary regulation of smooth muscle, cardiac muscle, and glandular activity).

Cellular Organization in Neural Tissue (pp. 370–379)

1. There are two types of cells in neural tissue: **neurons,** which are responsible for information transfer and processing, and **neuroglia,** or **glial cells,** which provide a supporting framework and act as phagocytes.
2. A typical neuron has a **soma** (cell body), an **axon,** and several branching, sensitive **dendrites.**

Related Key Terms: **synaptic terminals**

NEURON STRUCTURE (pp. 371–373)

3. Neurons may be described as **anaxonic, unipolar, bipolar,** or **multipolar.**
4. Within a multipolar neuron, the **perikaryon** contains organelles, including **neurofilaments, neurotubules,** and **neurofibrils.** The **axon hillock** connects the **initial segment** of the axon to the soma, and the **axoplasm** contains numerous organelles.
5. **Collaterals** may branch from an axon, with **telodendria** branching from the axon's tip.
6. A **synapse** is a site of intercellular communication. A synapse where neurons communicate with other cell types is a **neuroeffector junction. Axoplasmic transport** carries substances along neurotubules, and **retrograde flow** returns materials toward the soma.

Related Key Terms: **synaptic knobs** **neurotransmitters** **neuromuscular junction** **neuroglandular junction**

NEUROGLIA (pp. 373–379)

7. There are four types of neuroglia: (1) **astrocytes** (largest and most numerous); (2) **oligodendrocytes,** which form the **neurilemma** and are responsible for the **myelination** of CNS axons; (3) **microglia**

Related Key Terms: **blood-brain barrier** **axolemma** • **myelin** • **nodes** **internodes** • **white matter** (phagocytic white blood cells); and (4) **ependymal cells** (with functions related to the **cerebrospinal fluid (CSF)**.

8. Nerve cell bodies in the PNS are clustered into **ganglia** (singular **ganglion**), and their axons form **peripheral nerves.**

Related Key Terms

gray matter • **unmyelinated**
ependyma
satellite cells • **Schwann cells**

Neurophysiology (pp. 379–386)

AN INTRODUCTION TO ELECTRICITY (p. 379)

1. Current is directly proportional to voltage, and inversely proportional to resistance.

voltage • **resistance** • **current**

ELECTRICAL FORCES AND THE TRANSMEMBRANE POTENTIAL (pp. 379–380)

2. The **electrochemical gradient** is the sum of all the chemical and electrical forces active across the membrane.

MEMBRANE CHANNELS (p. 380)

3. There are two types of gated channels: **chemically regulated channels** (which open or close when they bind specific chemicals) and **voltage-regulated channels** (which open or close in response to changes in the transmembrane potential).

passive channels
active channels
gated channels • **activated**
inactivated

CHEMICALLY REGULATED CHANNELS AND GRADED POTENTIALS (pp. 380–381)

4. **Depolarization** or **hyperpolarization** can lead to a **graded potential** (a change in potential that results when the degree of depolarization decreases with distance).

local current

VOLTAGE-REGULATED CHANNELS (p. 381)

5. Voltage-regulated channels open and close in response to changes in the local transmembrane potential.

GENERATION OF AN ACTION POTENTIAL (pp. 381–383)

6. An **action potential** appears when the membrane depolarizes to a level known as the **threshold.** The steps involved include: activation of sodium channels and membrane depolarization; sodium channel inactivation; potassium channel activation; and return to normal permeability.

all-or-none principle
sodium channel inactivation
absolute/relative refractory period

ROLE OF THE SODIUM-POTASSIUM EXCHANGE PUMP (p. 384)

7. The activity of the sodium-potassium exchange pump is necessary to maintain ion concentrations within acceptable limits over time.

CONDUCTION OF AN ACTION POTENTIAL (p. 384)

8. In **continuous conduction** an action potential spreads across the entire excitable membrane surface in a series of small steps.

CONDUCTION VELOCITY (pp. 384–386)

9. During **saltatory conduction** the action potential appears to leap from node to node, skipping the intervening membrane surface. Saltatory conduction carries nerve impulses five to seven times faster than continuous conduction.
10. Axons can be classified as **Type A, Type B,** or **Type C** on the basis of diameter, myelination, and conduction speed.

Synaptic Communication (pp. 387–392)

1. An action potential traveling along an axon is called a **nerve impulse.** A synapse may be **chemical** (involving a neurotransmitter) or **electrical** (with direct physical contact between cells). **Chemical synapses** are more common.

CHOLINERGIC SYNAPSES (pp. 387–389)

2. **Cholinergic** synapses release the neurotransmitter **acetylcholine (ACh).** Communication moves from the **presynaptic neuron** to the **postsynaptic neuron.** The **presynaptic membrane** and **postsynaptic membrane** are separated by a cleft, which causes a **synaptic delay.** If stores of ACh are exhausted, **synaptic fatigue** can occur.

acetylcholinesterase (AChE)
acetate • **choline**

ADRENERGIC SYNAPSES (pp. 389–390)

3. **Adrenergic** synapses release **norepinephrine (NE),** also called **noradrenaline.**

OTHER CHEMICAL SYNAPSES AND NEUROTRANSMITTERS (pp. 390–391)

4. Other neurotransmitters include **dopamine (DOPA), gamma aminobutyric acid (GABA),** and **serotonin.** Their effects on the postsynaptic membrane depend on the characteristics of the receptor, not on the nature of the neurotransmitter.

adenyl cyclase
cyclic-AMP (cAMP)
second messenger
phosphodiesterase
monoamine oxidase (MAO)
catechol-O-methyltransferase (COMT)

Related Key Terms

NEUROMODULATORS (pp. 391–392)

5. **Neuromodulators** influence the postsynaptic cell's response to neurotransmitters.

neuropeptides

ELECTRICAL SYNAPSES (p. 392)

6. **Electrical synapses** transmit arriving impulses faithfully, while the situation at chemical synapses can be more variable.

Cellular Information Processing (pp. 392–398)

1. The inability of the soma and dendrites to support action potentials has two important consequences: the soma and dendrites are unaffected by action potentials carried by the axon, and it is the transmembrane potential at the axon hillock that determines whether an action potential develops in the initial segment of the axon.

EXCITATORY POSTSYNAPTIC POTENTIALS (pp. 395–397)

2. A depolarization caused by a neurotransmitter is an **excitatory postsynaptic potential (EPSP).** Individual EPSPs can combine through **summation;** the two types of summation are **temporal summation** (which occurs at a single synapse when a second EPSP arrives before the effects of the first have disappeared) and **spatial summation** (which results from the cumulative effects of multiple synapses at various locations).

facilitated

INHIBITORY POSTSYNAPTIC POTENTIALS (p. 397)

3. Hyperpolarization of the postsynaptic membrane is an **inhibitory postsynaptic potential (IPSP).**

inhibited

SUMMATION OF EPSPs AND IPSPs (pp. 397–398)

4. EPSP/IPSP interactions are the most important determinants of neural activity.

axoaxonal synapse
presynaptic inhibition
presynaptic facilitation

Neurons and Metabolic Processes (p. 398)

1. Active neurons consume a great deal of ATP.

stroke

Neural Response to Injuries (pp. 398–399)

1. Neurons respond to injury in a limited, stereotyped fashion.

Wallerian degeneration

Functional Patterns of Neural Organization (pp. 399–402)

1. There are three functional categories of neurons: *sensory neurons*, *motor neurons*, and *interneurons* (association neurons).

SENSORY NEURONS (p. 400)

2. **Sensory neurons** form the afferent division of the PNS, and deliver information to the CNS.

afferent fibers • **receptor**
somatic sensory neurons
exteroceptors • **proprioceptors**
visceral sensory neurons (interoceptors)

MOTOR NEURONS (p. 400)

3. **Motor neurons** stimulate or modify the activity of a peripheral tissue, organ, or organ system.

efferent fibers
somatic motor neurons
visceral motor neurons
preganglionic fibers
postganglionic fibers

INTERNEURONS (p. 400)

4. **Interneurons (association neurons)** may be located between sensory and motor neurons; they analyze sensory inputs and coordinate motor outputs.

excitatory • **inhibitory**

HIGHER LEVELS OF ORGANIZATION AND PROCESSING (pp. 401–402)

5. The roughly 20 billion interneurons can be classified into **neuronal pools** (groups of interconnected neurons with specific functions). Neuron interaction can vary according to whether other neurons lie within the **discharge zone** or the **facilitated zone** of the presynaptic neuron.
6. **Divergence** is the spread of information from one neuron to several, or from one pool to several pools. In **convergence** several neurons synapse on the same postsynaptic neuron. Neuronal pools may also function in sequence **(serial processing)** or may process the same information at one time **(parallel processing).** In **reverberation** collateral axons establish a circuit that further stimulates presynaptic neurons.

Anatomical Organization of the Nervous System (pp. 402–404)

1. The functions of the nervous system as a whole depend on interactions between neurons in neuronal pools. **Spinal nerves** communicate with the spinal cord, and **cranial nerves** are connected to the brain.

centers • **nuclei**
higher centers
neural cortex

Chapter Review

2. **Sensory (ascending) pathways** carry information from peripheral receptors to the brain; **motor (descending) pathways** extend from CNS centers concerned with motor control to the associated skeletal muscles.

Related Key Terms
tracts • **columns** • **pathways**

Review Planner

	Level 1	Level 2
		31 33 34
1 Describe the anatomical organization and general functions of the nervous system.	**1 14 17 23 25 27 28 30**	
		Level 3: **46–48**
2 Distinguish between neurons and neuroglia and compare their functions and structures.	**2 12 21 29 37 42 43 44 45**	
3 Discuss the events that generate action potentials in the membranes of nerve cells.	**6 8 13 18 22 32 35 36 38 40**	 C M: **Axoplasmic Transport and Disease**; **Neurotoxins in Seafood**; **Growth and Myelination of the Nervous System**; **Demyelination Disorders**; **Tay-Sachs Disease**
4 Identify the factors that determine the frequency and speed of nerve impulse conduction.	**6 11 20 24 35 39 40 41 42 44**	
5 Explain the mechanism of synaptic transmission and describe the types and effects of the most important neurotransmitters.	**7 15 16 18 19 26**	 **Multiple Sclerosis**
6 Discuss the interactions that make possible the processing of information in neural tissue.	**9 10 14 17**	
7 Classify neurons into functional groups and discuss the interactions between these groups.	**3 4 5 14 17 23 28**	

Review Questions

MULTIPLE CHOICE

1. The autonomic nervous system would control which of the following actions? a) throwing a softball; b) your heart rate; c) thinking about the answer to this question; d) writing the answer.
2. Which of the following is *not* a component of a typical neuron? a) astrocyte; b) soma; c) dendrites; d) axon.
3. You notice a stop sign at the corner. What type of receptor provided that information? a) exteroceptor; b) proprioceptor; c) interoceptor; d) none of the above.
4. You're feeling hungry. What type of receptor provided this data? a) exteroceptor; b) proprioceptor; c) interoceptor; d) none of the above.
5. Interneurons are also known as: a) motor neurons; b) association neurons; c) sensory neurons; d) somatic sensory neurons.
6. An action potential will not appear until the membrane depolarizes to a level called the: a) synapse; b) axon hillock; c) transmembrane potential; d) threshold.
7. DOPA and ACh are examples of: a) synaptic knobs; b) glial cells; c) neurotransmitters; d) collaterals.
8. The sum of all the chemical and electrical forces active across a membrane is called a(n): a) neurotransmitter; b) hyperpolarization; c) synapse; d) electrochemical gradient.
9. The spread of information from one neuron to several neurons is called: a) convergence; b) divergence; c) serial processing; d) reverberation.
10. __________ occurs when several neuronal pools are processing the same information at the same time: a) parallel processing; b) serial processing; c) reverberation; d) facilitation.
11. You would expect that the most urgent sensory information would be carried to the CNS by: a) Type A fibers; b) Type B fibers; c) Type C fibers.

TRUE/FALSE

(If a statement is false, your instructor may wish to have you correct it.)

12. **T/F:** There are far more neurons in the body than neuroglia.
13. **T/F:** IPSPs and EPSPs can occur simultaneously on the same neuron.
14. **T/F:** Many ascending pathways distribute information from peripheral receptors to the brain.

MATCHING QUESTIONS

(Match each term with the most appropriate statement.)

Statements:

15. Synapses that release norepinephrine.
16. A synapse between a neuron and another type of cell.
17. Neurons that carry information from the PNS to the CNS.
18. A depolarization produced by the arrival of a neurotransmitter.
19. The region at which a neuron communicates with other cells.
20. A process by which an action potential jumps from node to node.
21. The process by which materials are transported from the soma to the synaptic knobs.
22. A transient hyperpolarization of the postsynaptic membrane.
23. Neurons that stimulate or modify the activity of a peripheral tissue, organ, or organ system.
24. An action potential traveling along an axon.
25. This division carries sensory information to the brain.
26. Synapses that release acetylcholine.
27. This division gives you voluntary control over the contractions of your leg muscles.
28. Neurons that are responsible for analyzing sensory input and coordinating motor output.
29. The process by which materials are transported to the soma from the periphery of a neuron.
30. The brain and spinal cord belong to this division.

Terms:

a) CNS
b) SNS
c) Afferent division
d) Synaptic terminal
e) Neuroeffector junction
f) Axoplasmic transport
g) Retrograde flow
h) Nerve impulse
i) Cholinergic
j) Adrenergic
k) EPSP
l) IPSP
m) Sensory neurons
n) Motor neurons
o) Interneurons
p) Saltatory conduction

SHORT ANSWER/ESSAY

31. What is the nervous system? Compare its functions with those of the endocrine system.
32. Place the following events in the correct sequence. What process is being described?:
    a) Potassium channels are activated.
    b) The membrane becomes extremely permeable to sodium ions.
    c) Voltage-regulated channels close.
    d) The inner membrane surface contains more positive ions than negative ions.
33. You touch a hot stove and quickly jerk your hand away. Describe which functional divisions of the nervous system were involved.
34. What two unusual features are shared by nerve cells and skeletal muscle cells? Why are calcium ions important to each?
35. A typical EPSP does not produce a depolarization strong enough to cause an action potential. Explain how EPSPs can lead to a nerve impulse.
36. What do we mean by the all-or-none principle?
37. What is the function of the blood-brain barrier? Which cells are responsible for maintaining this barrier, and how do they do it?
38. Fill in the blanks: Depolarization occurs through the influx of ________ ions, and repolarization involves the loss of ________ ions.
39. Which type of axon will conduct a nerve impulse the fastest, and which type is the slowest? Why?
40. What factors can affect the rate of action potential generation?
41. Compare the processes of continuous and saltatory conduction. Which is faster, and why?
42. What do we mean when we say that an axon is myelinated? What advantage does myelination provide?
43. Compare the functions of the four types of neuroglia in the CNS. Which type is the most numerous?
44. What is myelin, how is it produced, and why is it important?
45. Discuss the functions of the dendrites, the soma, and the synaptic knobs.

CRITICAL THINKING/APPLICATIONS

46. Four-year-old Tiffany lives with her parents in an old apartment building where the paint is peeling, and Tiffany has gotten into the habit of eating paint chips. Lately she has been showing signs of increased lethargy and fatigue and is having trouble walking. You know that some older buildings were painted with a lead-based paint. Discuss a possible cause of Tiffany's condition and why it could produce these symptoms.
47. One year-old Angela's mother is very much into health foods. She is determined to teach her daughter good nutritional habits from the beginning, and feeds Angela a strict diet that includes fruit juice, lean meat, grains, and skim milk. Could this affect the child's development? Explain.
48. You're studying for your Anatomy & Physiology midterm exam, and end up staying up all night and drinking lots of coffee to stay awake. In the morning you feel shaky and nauseated. Explain why.

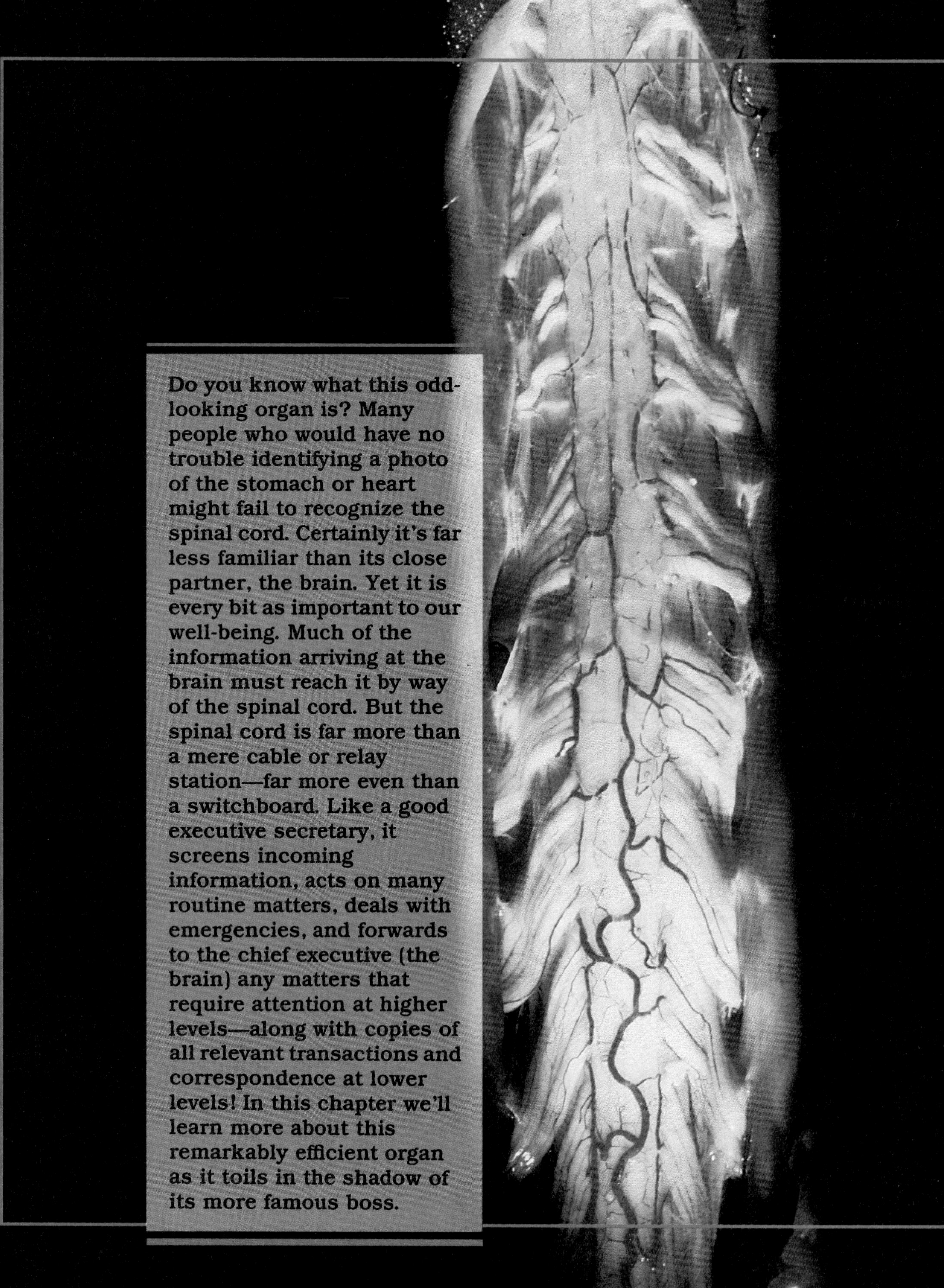

Do you know what this odd-looking organ is? Many people who would have no trouble identifying a photo of the stomach or heart might fail to recognize the spinal cord. Certainly it's far less familiar than its close partner, the brain. Yet it is every bit as important to our well-being. Much of the information arriving at the brain must reach it by way of the spinal cord. But the spinal cord is far more than a mere cable or relay station—far more even than a switchboard. Like a good executive secretary, it screens incoming information, acts on many routine matters, deals with emergencies, and forwards to the chief executive (the brain) any matters that require attention at higher levels—along with copies of all relevant transactions and correspondence at lower levels! In this chapter we'll learn more about this remarkably efficient organ as it toils in the shadow of its more famous boss.

CHAPTER 13

The Nervous System: The Spinal Cord and Spinal Nerves

Chapter Objectives

After reading this chapter, you will be able to:

1. Discuss the structure and functions of the spinal cord.
2. Explain the role of white matter and gray matter in processing and relaying sensory and motor information.
3. Describe the major components of a spinal nerve.
4. Relate spinal nerves to the regions that they innervate.
5. Describe the process of a neural reflex.
6. Classify the different types of reflexes.
7. Distinguish between the types of motor responses produced by various reflexes.
8. Explain how reflexes combine to produce complex behaviors.
9. Explain how higher centers control and modify reflex responses.

■ Introduction

Chapter 12 focused attention on the cellular level, where neural processing begins. At this level, chemical transmitters cross innumerable synapses, raising or lowering the transmembrane potential of countless individual neurons. The stimuli reaching each neuron are integrated to determine whether an action potential will be transmitted. This chapter begins to bridge the gap between cellular mechanisms and the higher-level neural functions that make life worth living. As you will see, all of our thoughts, feelings, memories, and behaviors begin as changes in the resting potentials of individual neurons.

The **central nervous system (CNS)** consists of the *spinal cord* and *brain*. Despite the fact that the two are intimately connected, the brain and spinal cord show significant degrees of functional independence. The spinal cord is far more than just a highway for information traveling to or from the brain. Although most sensory data will be relayed to the brain, the spinal cord also integrates and processes information on its own. This chapter describes the anatomy of the spinal cord and examines the integrative activities that occur in this portion of the CNS. The functional patterns encountered will help prepare us for a discussion of the more complex processing events that take place in the brain.

Gross Anatomy of the Spinal Cord

The adult spinal cord, detailed in Figure 13-1a,b, measures approximately 45 cm (18 in.) in length. The dorsal surface of the spinal cord bears a shallow longitudinal groove, the **posterior median sulcus**. A deep crease along the ventral surface forms the **anterior median fissure**. The diameter of the cord decreases in size as you proceed from cervical to sacral segments. The cervical spinal cord contains all of the ascending and descending tracts linking the spinal cord with the brain. The thoracic spinal cord includes all of the tracts involved with thoracic, lumbar, and sacral segments. The lumbar spinal cord carries tracts for the lumbar and sacral sections, while the sacral spinal cord, the narrowest of all, consists only of tracts that begin or end in that region.

The amount of gray matter is substantially increased in segments of the spinal cord concerned with the sensory and motor control of the limbs. These areas are expanded, forming **enlargements** of the spinal cord. The **cervical enlargement** supplies nerves to the shoulder girdle and arms; the **lumbar enlargement** provides innervation to structures of the pelvis and legs. Below the lumbar enlargement the spinal cord becomes tapered and conical; this region is known as the **conus medullaris**. A slender strand of fibrous tissue, the **filum terminale** ("terminal thread"), extends from the inferior tip of the conus medullaris. This filamentous extension continues along the length of the vertebral canal as far as the second sacral vertebra. There it provides longitudinal support to the spinal cord as a component of the *coccygeal ligament*.

Figure 13-1a provides a series of sectional views that demonstrate the variations in the relative mass of gray matter in the cervical, thoracic, lumbar, and sacral regions of the spinal cord. The entire spinal cord can be divided into 31 segments. Each segment is identified by a letter and number designation. For example, C_3, the segment in the uppermost section in Figure 13-1, is the third cervical segment.

Every spinal segment is associated with a pair of **dorsal root ganglia** that contains the cell bodies of sensory neurons. The **dorsal roots**, which contain the axons of these neurons, bring sensory information to the spinal cord. A pair of **ventral roots** contains the axons of somatic and visceral motor neurons that control peripheral effectors. On either side, the dorsal and ventral roots from each segment leave the vertebral column between adjacent vertebrae at the *intervertebral foramen*. The spinal nerves on either side form outside of the vertebral canal, where the ventral and dorsal roots unite, so that the dorsal root ganglion lies between the pedicles of succeeding vertebra. (You may wish to review the description of vertebral anatomy in Chapter 8.) ∞ (p. 258) Distal to each dorsal root ganglion the sensory and motor roots are bound together into a single **spinal nerve**. Spinal nerves are classified as **mixed nerves**, because they contain both afferent (sensory) and efferent (motor) fibers. [**CM:** *Shingles*]

The spinal cord continues to enlarge and elongate until an individual is approximately 4 years old. Up to that time enlargement of the spinal cord keeps pace with the growth of the vertebral column, as shown in Figure 13-1c. Throughout this period the short ventral and dorsal roots leave the vertebral canal through the adjacent intervertebral foramina to reach the dorsal root ganglia and spinal nerves.

After age 4 the vertebral column continues to grow, but the spinal cord does not. This vertebral growth carries the dorsal roots and spinal nerves farther and farther away from their original position relative to the spinal cord. As a result, the dorsal and ventral roots gradually elongate. The adult spinal cord extends only to the level of the first or second lumbar vertebra.

When seen in gross dissection, the filum terminale and the long ventral and dorsal roots caudal to the conus medullaris reminded early anatomists of a horse's tail. With this in mind the complex was called the **cauda equina** (KAW-da ek-WI-na; *cauda*, tail + *equus*, horse).

Spinal Meninges

The vertebral column and its surrounding ligaments, tendons, and muscles isolate the spinal cord from the external environment. The delicate neural tissues must also be defended against damaging contacts with the surrounding bony walls of the vertebral canal. A series of specialized membranes, the **spinal meninges** (men-IN-jēz), provide the necessary physical stability and shock absorption. Blood vessels branching within these layers also deliver oxygen and nutrients to the spinal cord. The structure of the spinal meninges can be seen in Figure 13-2a. There are three meningeal layers: the *dura mater*, the *arachnoid*, and the *pia mater*. At the foramen magnum of the skull, the spinal meninges are continuous with the **cranial meninges** that surround the brain. (The cranial meninges, which have the same three layers, are discussed in Chapter 14.) [**CM:** *Spinal Meningitis*]

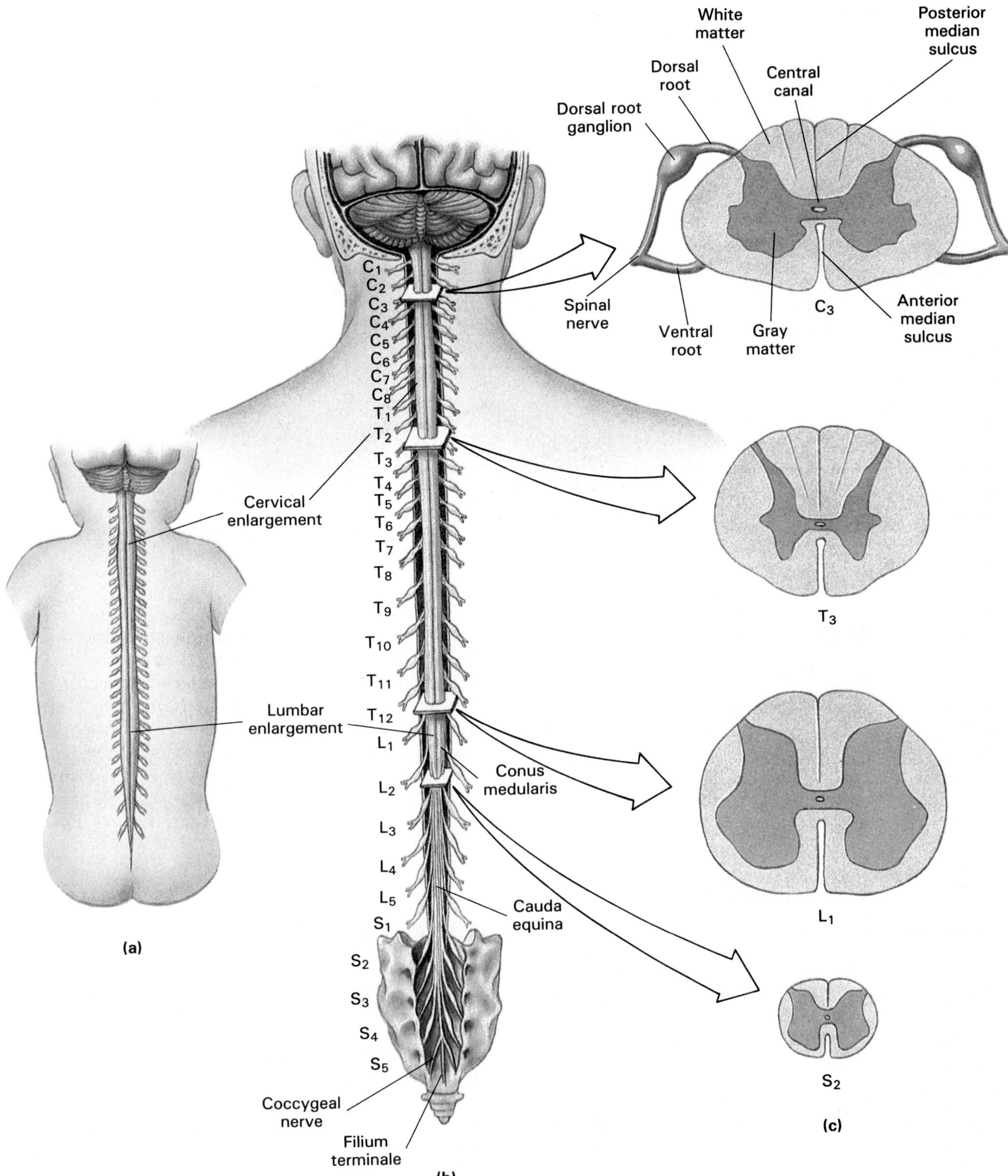

FIGURE 13-1
Gross Anatomy of the Spinal Cord. (a) Position and extent of the spinal cord in a 4-year-old child; compare with part (b). (b) Superficial anatomy and orientation of the adult spinal cord. (c) Cross sections through representative regions of the spinal cord, showing the arrangement of gray and white matter.

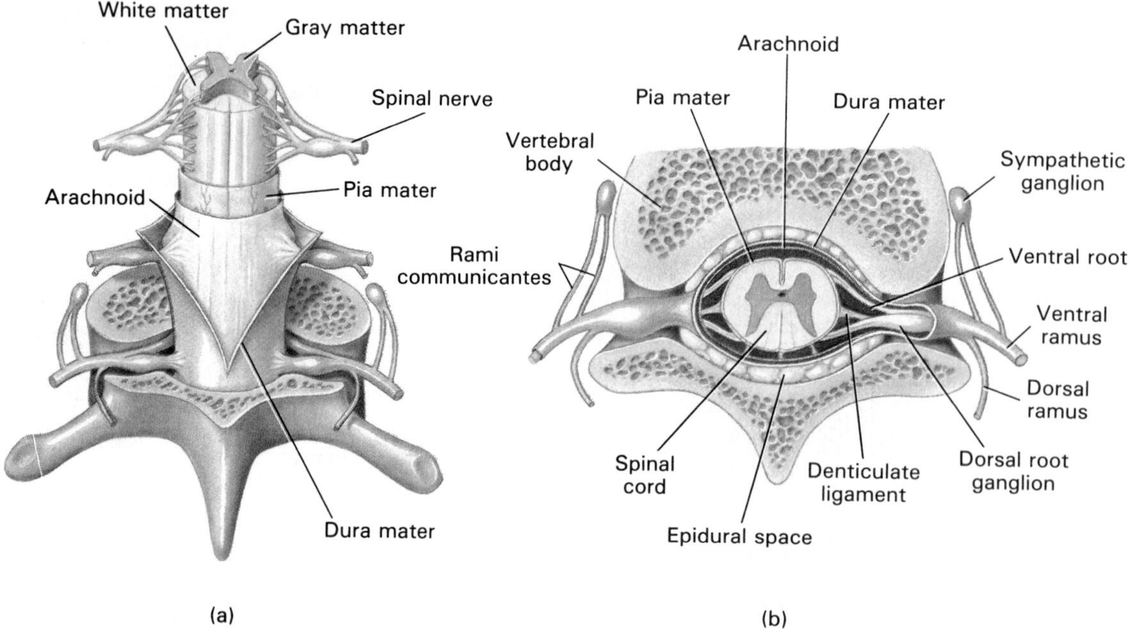

FIGURE 13-2
The Spinal Cord and Spinal Meninges. (a) Posterior view of the spinal cord, showing the meningeal layers, superficial landmarks, and distribution of gray and white matter. (b) Sectional view through the spinal cord and meninges, showing the peripheral distribution of the spinal nerves.

THE DURA MATER

The tough, fibrous **dura mater** (DŪ-ra MA-ter; *dura*, hard + *mater*, mother) forms the outermost covering of the spinal cord. The dense collagen fibers of the dura are oriented along the longitudinal axis of the cord. Between the dura mater and the walls of the vertebral canal lies the **epidural space**, which contains loose connective tissue, blood vessels, and adipose tissue.

The dura does not have extensive, firm connections to the surrounding vertebrae. Longitudinal stability is provided by localized attachment sites at either end of the vertebral canal. Cranially, the dura mater fuses with the periosteum of the occipital bone around the margins of the foramen magnum. Caudally the dura mater tapers from a sheath to a dense cord of collagen fibers that blend with components of the filum terminale. This **coccygeal ligament** continues along the sacral canal, ultimately blending into the periosteum of the coccyx. Lateral support is provided by the connective tissues within the epidural space. In addition, the dura mater extends between adjacent vertebrae at each intervertebral foramen, fusing with the connective tissues that surround the spinal nerves.

THE ARACHNOID

A narrow **subdural space** separates the inner surface of the dura mater from the second meningeal layer, the **arachnoid** (a-RAK-noyd; *arachne*, spider). The inner surface of the dura and the outer surface of the arachnoid are lined by simple squamous epithelia. The intervening space contains a small quantity of lymphatic fluid, which reduces friction between the opposing surfaces. Beneath the arachnoid epithelium lies the **subarachnoid space**, a delicate network of collagen and elastin fibers. The subarachnoid space contains a considerable quantity of **cerebrospinal fluid** that acts as a shock absorber as well as a diffusion medium for dissolved gases, nutrients, chemical messengers, and waste products. [**CM:** *Spinal Anesthesia*]

THE PIA MATER

The subarachnoid space bridges the gap between the arachnoid epithelium and the innermost meningeal layer, the **pia mater** (*pia*, delicate + *mater*, mother). The meshwork of elastin and collagen fibers of the pia are interwoven with those of the subarachnoid space. The blood vessels servicing the spinal cord are found here. Unlike more superficial meninges the pia mater is firmly bound to the underlying neural tissue. The surface of the spinal cord consists of a thin layer of astrocytes, and cytoplasmic extensions of these glial cells lock the collagen fibers of the pia mater in place.

The combination of pia mater and arachnoid is sometimes referred to as the *pia-arachnoid*. Along the length of the spinal cord, paired **denticulate ligaments** connect the pia-arachnoid to the dura mater. The denticulate ligaments originate along either side of the spinal cord, between the ventral and dorsal roots. They help stabilize the position of the spinal

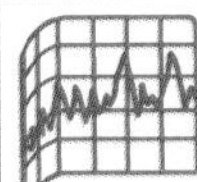

Diagnostics: Spinal Taps and Myelography

Tissue samples, or biopsies, are taken from many organs to assist in diagnosis. For example, when a liver or skin disorder is suspected, small plugs of tissue are removed and examined for signs of infection or cell damage, or used to identify the bacteria causing an infection. Unlike many other tissues, however, neural tissue consists largely of cells rather than extracellular fluids or fibers. Tissue samples are seldom removed for analysis because any extracted or damaged neurons will not be replaced. Instead, small volumes of cerebrospinal fluid (CSF) are extracted and analyzed. CSF is intimately associated with the neural tissue of the CNS, and pathogens, cell debris, or metabolic wastes in the CNS will therefore be detectable in the CSF.

The withdrawal of cerebrospinal fluid, known as a **spinal tap**, must be done with care to avoid injuring the spinal cord. The adult spinal cord extends only as far as the first or second lumbar vertebra. Between the second lumbar vertebra and the sacrum the meningeal layers remain intact, but they enclose only the relatively sturdy components of the cauda equina and a significant quantity of CSF. With the vertebral column flexed, a needle can be inserted between the lower lumbar vertebrae and into the subarachnoid spaces with minimal risk to the cauda equina. In this procedure, known as a **lumbar puncture** (Figure 13-3a), 3-9 ml of fluid are taken from the subarachnoid space. Spinal taps are performed when CNS infection is suspected, or when diagnosing severe back pain, headaches, disc problems, and some types of strokes.

Myelography involves the introduction of radiopaque dyes into the CSF of the subarachnoid space. Because the dyes are opaque to X-rays, the CSF appears white on an X-ray photograph, as in Figure 13-3b. Any tumors, inflammations, or adhesions that distort or divert CSF circulation will be shown in silhouette.

In the event of severe infection, inflammation, or leukemia (cancer of the white blood cells), antibiotics, steroids, or anticancer drugs can be injected into the subarachnoid space.

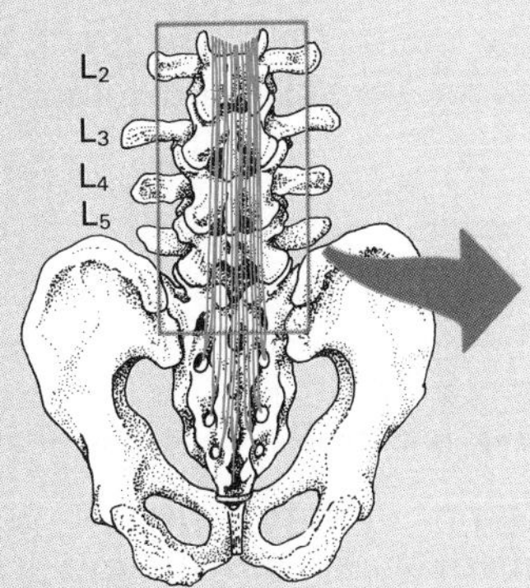

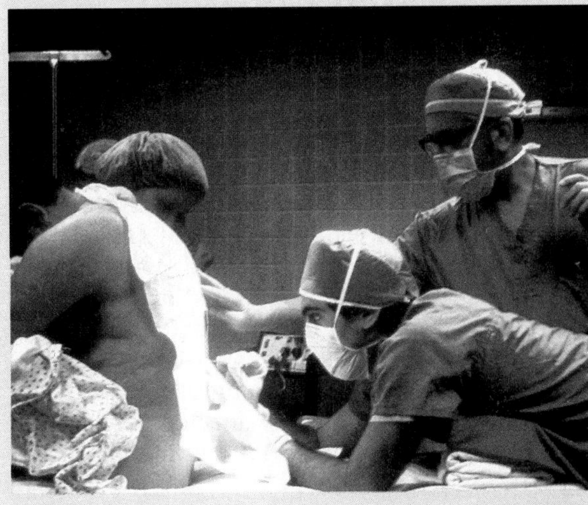

(a)

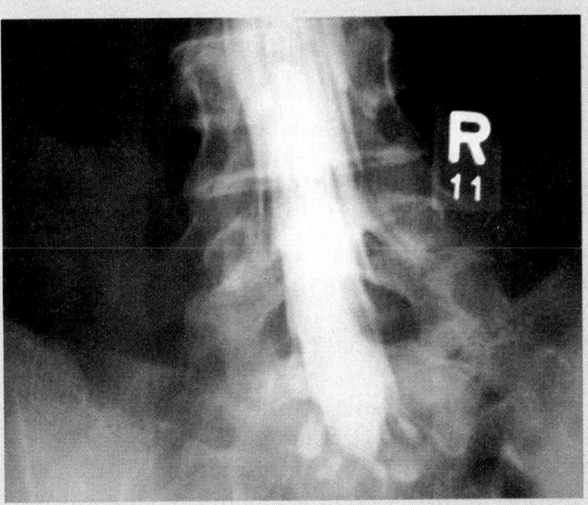

(b)

FIGURE 13-3
Spinal Taps and Myelography. (a) Position and procedure used in a typical (lumbar) spinal tap to obtain a sample of CSF. (b) A myelogram–an X-ray photograph of the spinal cord after introduction of a radiopaque dye into the CSF–showing the cauda equina in the lower lumbar region.

cord inside the vertebral canal by preventing side-to-side movement, while the connections to the skull, at the foramen magnum, and to the sacrum, via the coccygeal ligament, prevent superior-inferior movement.

The spinal meninges accompany the dorsal and ventral roots as they exit the vertebral cavity via the intervertebral foramina. As indicated in the sectional view of Figure 13-2b, the meningeal membranes are continuous with the connective tissues surrounding the spinal nerves and their peripheral branches.

√ **Damage to which root of a spinal nerve would interfere with motor function?**

√ **Where is the cerebrospinal fluid that surrounds the spinal cord located?**

■ Sectional Anatomy of the Spinal Cord

To understand the functional organization of the spinal cord, you must become familiar with its sectional organization. Consider Figure 13-4, which presents a "typical" section through the spinal cord. The **anterior median fissure** and the **posterior median sulcus** mark the division between left and right sides of the spinal cord. The peripherally situated *white matter* contains large numbers of myelinated and unmyelinated axons. The *gray matter*, dominated by the cell bodies of neurons and glial cells, surrounds the narrow **central canal** and forms a rough H, or butterfly shape. The projections of gray matter toward the outer surface of the spinal cord are called **horns**, as indicated in Figure 13-4a.

ORGANIZATION OF GRAY MATTER

The cell bodies of neurons in the gray matter of the spinal cord are organized into groups, called *nuclei*, with specific functions. **Sensory nuclei** receive and relay sensory information from peripheral receptors. **Motor nuclei** issue motor commands to peripheral effectors. Although sensory and motor nuclei appear rather small in transverse section, they may extend for a considerable distance along the length of the spinal cord. A frontal section along the length of the central canal of the spinal cord will separate the sensory (dorsal) nuclei from the motor (ventral) nuclei. The **posterior gray horns** contain somatic and visceral sensory nuclei, whereas the **anterior gray horns** are concerned with somatic motor control. The **lateral gray horns**, found in most segments, are prominent expansions of the posterior and lateral portions of the anterior horns. Nuclei containing visceral motor neurons are found in this area. The **gray commissures** (*commissura*, a joining together) above and below the central canal contain axons crossing from one side of the cord to the other before reaching a destination within the gray matter.

Figure 13-4a shows the relationship between the function of a particular nucleus (sensory or motor) and its relative position within the gray matter of the spinal cord. The nuclei within each gray horn are also highly organized. Figure 13-4b illustrates the distribution of somatic motor nuclei in the anterior gray horns of the cervical enlargement.

ORGANIZATION OF WHITE MATTER

The white matter can be divided into a half-dozen regions, or *columns*, as shown in Figure 13-4b. The **posterior white columns** extend between the posterior gray horns and the posterior median sulcus. The **anterior white columns** lie between the anterior gray horns and the anterior median fissure; they are interconnected by the **anterior white commissure**. The white matter between the anterior and posterior columns constitutes the **lateral white columns**.

Each column contains **tracts**, or **fasciculi**, whose axons share functional and structural characteristics. A specific tract conveys either sensory data or motor commands, and the axons are relatively uniform with respect to diameter, myelination, and conduction speed. All of the axons within a tract relay information in the same direction. Small tracts carry sensory or motor signals between segments of the spinal cord, and larger tracts connect the spinal cord with the brain. **Ascending tracts** carry sensory information up toward the brain, and **descending tracts** convey motor commands down into the spinal cord. Within each column the tracts show a regional organization comparable to that found in the nuclei of the gray matter; Figure 13-4b contains an example. The identities of the major CNS tracts will be discussed when we consider sensory and motor pathways in Chapter 15.

■ Spinal Nerves

There are 31 pairs of spinal nerves, each of which can be identified by its association with adjacent vertebrae. For example, you may speak of "cervical spinal nerves" or even "cervical nerves" when making a general reference to spinal nerves of the neck. When indicating specific spinal nerves, it is customary to give them a regional number, as indicated in Figure 13-1. The spinal nerves caudal to the first thoracic vertebra take their names from the vertebra immediately preceding them. Thus the spinal nerve T_1 emerges immediately caudal to vertebra T_1, spinal nerve T_2 follows vertebra T_2, and so forth.

Clinical Comment: Spinal Cord Injuries

Injuries affecting the spinal cord produce symptoms of sensory loss or motor paralysis that reflect the specific nuclei and tracts involved. At the outset, any severe injury to the spinal cord produces a period of sensory and motor paralysis termed **spinal shock**. The skeletal muscles become flaccid; neither somatic nor visceral reflexes function; and the brain no longer receives sensations of touch, pain, heat, or cold. The location and severity of the injury determine how long these symptoms persist and how completely the individual recovers.

Violent jolts, such as those associated with blows or gunshot wounds, may cause **spinal concussion** without visibly damaging the spinal cord. Spinal concussion produces a period of spinal shock, but the symptoms are only temporary and recovery may be complete in a matter of hours. More serious injuries, such as whiplash or falls, usually involve physical damage to the spinal cord. In a **spinal contusion** hemorrhages occur in the meninges, pressure rises in the cerebrospinal fluid, and the white matter of the spinal cord may degenerate at the site of injury. Gradual recovery over a period of weeks may leave some functional losses. Recovery from a **spinal laceration** by vertebral fragments or other foreign bodies will usually be far slower and less complete. **Spinal compression** occurs when the spinal cord becomes physically squeezed or distorted within the vertebral canal. In a **spinal transection** the spinal cord is completely severed. At present surgical procedures cannot repair a severed spinal cord, but experimental techniques may restore partial function (see Health News: Technology and Motor Paralysis on p. 432).

Spinal injuries often involve some combination of compression, laceration, contusion, and partial transection. Relieving pressure and stabilizing the affected area through surgery may prevent further damage and allow the injured spinal cord to recover as much as possible.

Extensive damage at the fourth or fifth cervical vertebra will eliminate sensation and motor control of the arms and legs. The extensive paralysis produced is called **quadriplegia**. If the damage extends from C_3 to C_5, the motor paralysis will include all of the major respiratory muscles, and the patient will usually need mechanical assistance in breathing. **Paraplegia**, the loss of motor control of the legs, may follow damage to the thoracic vertebrae. Injuries to the lower lumbar vertebrae may compress or distort the elements of the cauda equina, causing problems with peripheral nerve function.

FIGURE 13-4
Sectional Organization of the Spinal Cord. (a) The left half of this sectional view shows important anatomical landmarks; the right half indicates the functional organization of the gray matter in the anterior and posterior gray horns. (b) The left half shows the major regions of white matter. The right half indicates the anatomical organization of sensory tracts in the posterior white column. Note that both sensory and motor components of the spinal cord have a definite regional organization.

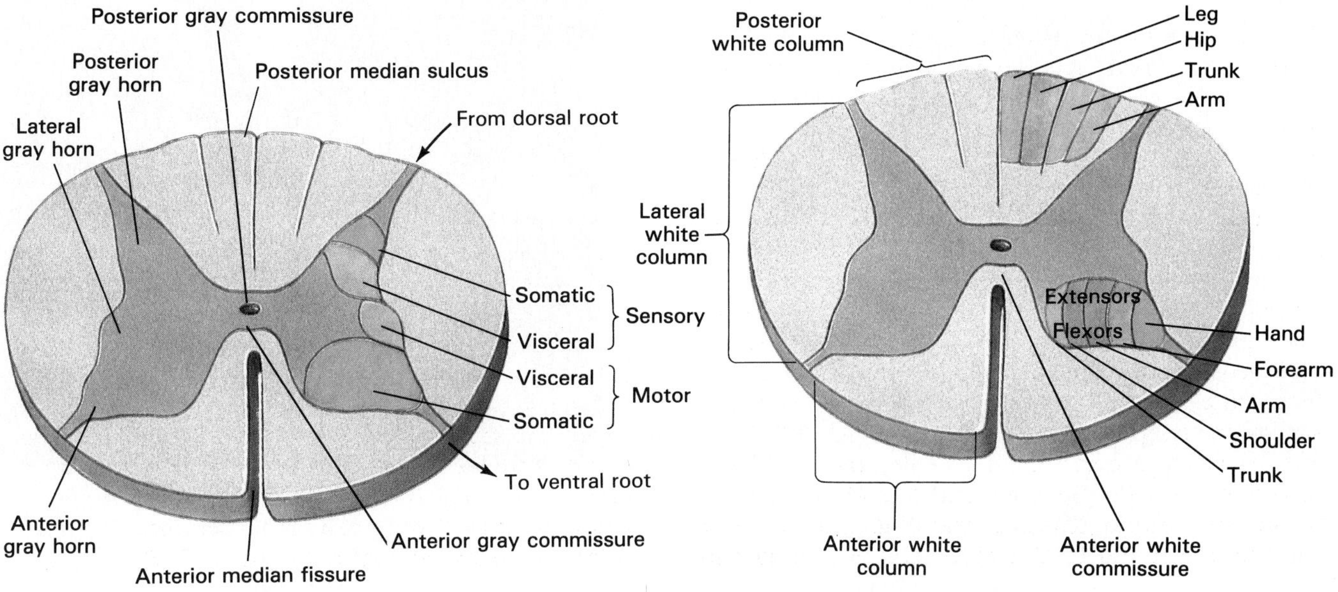

The arrangement differs in the cervical region because the first pair of spinal nerves, C_1, exits between the skull and the first cervical vertebra. For this reason, cervical nerves take their names from the vertebra immediately *following* them. In other words, cervical nerve C_2 *precedes* vertebra C_2, and the same system is used for the rest of the cervical series. The transition from one pattern of nomenclature to the other occurs between the last cervical and first thoracic vertebrae. Because the spinal nerve lying between them has been designated C_8, there are seven cervical vertebrae but *eight* cervical nerves.

A series of connective tissue layers surrounds each spinal nerve and continues along all of its peripheral branches. These layers, best seen in sectional view (Figure 13-5) are comparable to those associated with skeletal muscles (Figure 10-2). (p. 300) The outermost layer, or **epineurium**, consists of a dense network of collagen fibers. The fibers of the **perineurium** divide the nerve into a series of compartments that contain bundles of axons. Arteries and veins penetrate the epineurium and branch within the perineurium. The **endoneurium** consists of delicate connective tissue fibers that surround individual axons. Capillaries leaving the perineurium branch in the endoneurium and provide oxygen and nutrients to the axons and Schwann cells of the nerve. [**CM:** *Multiple Sclerosis*]

PERIPHERAL DISTRIBUTION OF SPINAL NERVES

Figure 13-6 diagrams the distribution of a typical spinal nerve. In the thoracic and lumbar regions, the first branch of each spinal nerve carries visceral motor fibers to a nearby **autonomic ganglion**. Because preganglionic axons are myelinated, this branch has a light color and is known as the **white ramus** ("branch"). The postganglionic fibers leaving the ganglion are unmyelinated, and those innervating glands and smooth muscles in the body wall or limbs form the **gray ramus** that rejoins the spinal nerve. The gray and white rami are collectively termed the **rami communicantes**, or "communicating

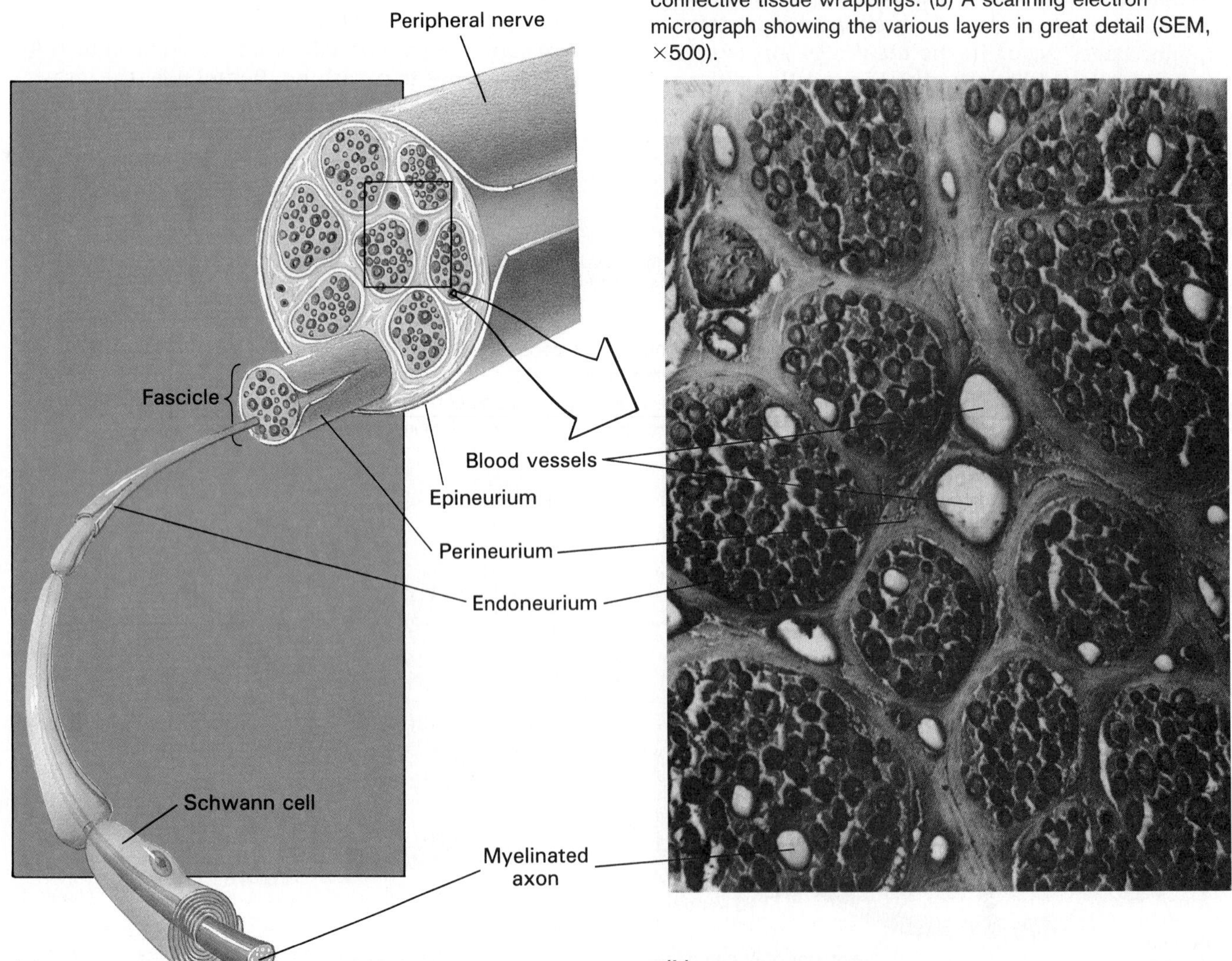

FIGURE 13-5
Peripheral Nerves. (a) A typical peripheral nerve and its connective tissue wrappings. (b) A scanning electron micrograph showing the various layers in great detail (SEM, ×500).

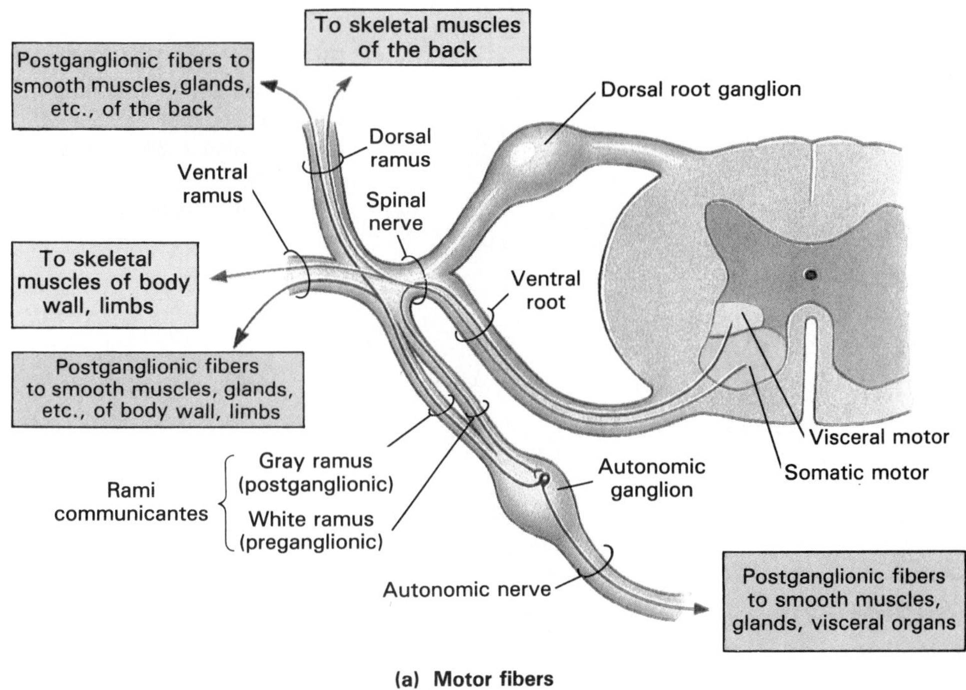

(a) Motor fibers

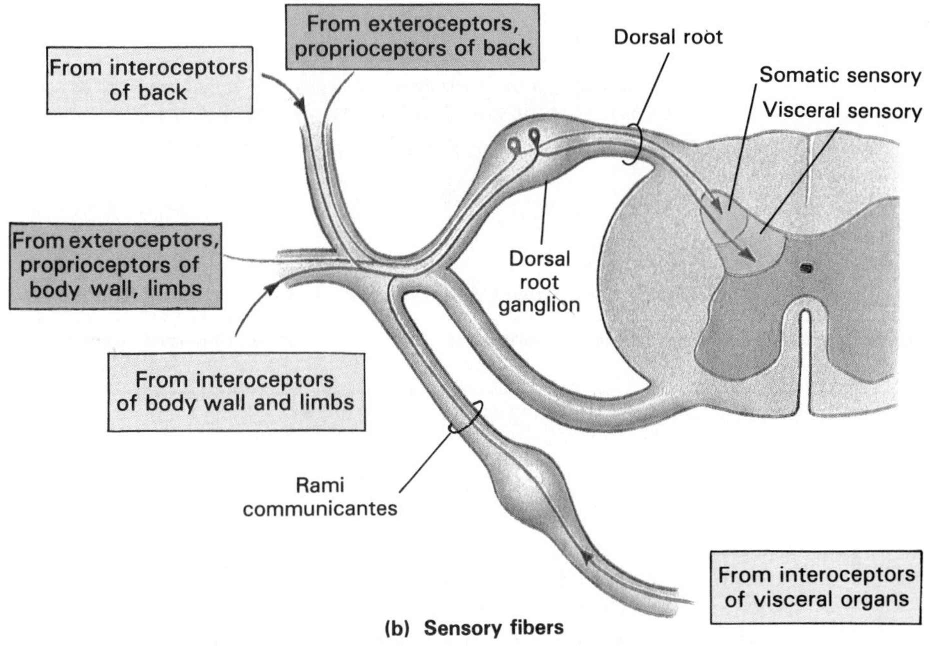

(b) Sensory fibers

FIGURE 13-6
Peripheral Distribution of Spinal Nerves.
(a) Diagrammatic view indicating the distribution of motor fibers in the major branches of a typical spinal nerve. (b) A comparable view, detailing the distribution of sensory fibers.

branches." Postganglionic fibers headed for internal organs do not rejoin the spinal nerve, but form a series of separate **autonomic nerves**.

The **dorsal ramus** of each spinal nerve provides sensory and motor innervation to the skin and muscles of the back. The relatively large **ventral ramus** supplies the ventrolateral body surface, structures in the body wall, and the limbs. The distribution of the dorsal and ventral rami provides a superb illustration of the segmental division of labor along the length of the spinal cord. Each pair of spinal nerves monitors a specific region of the body surface, an area known as a **dermatome**. Dermatomes, illustrated in Figure 13-7, are clinically important because damage to a spinal nerve or dorsal root ganglion will produce a characteristic loss of sensation in the skin.

NERVE PLEXUSES

The relatively simple distribution pattern illustrated in Figure 13-6 applies to spinal nerves T_2–T_{12}. But in segments controlling the skeletal musculature of the neck, arms, or legs the situation becomes more

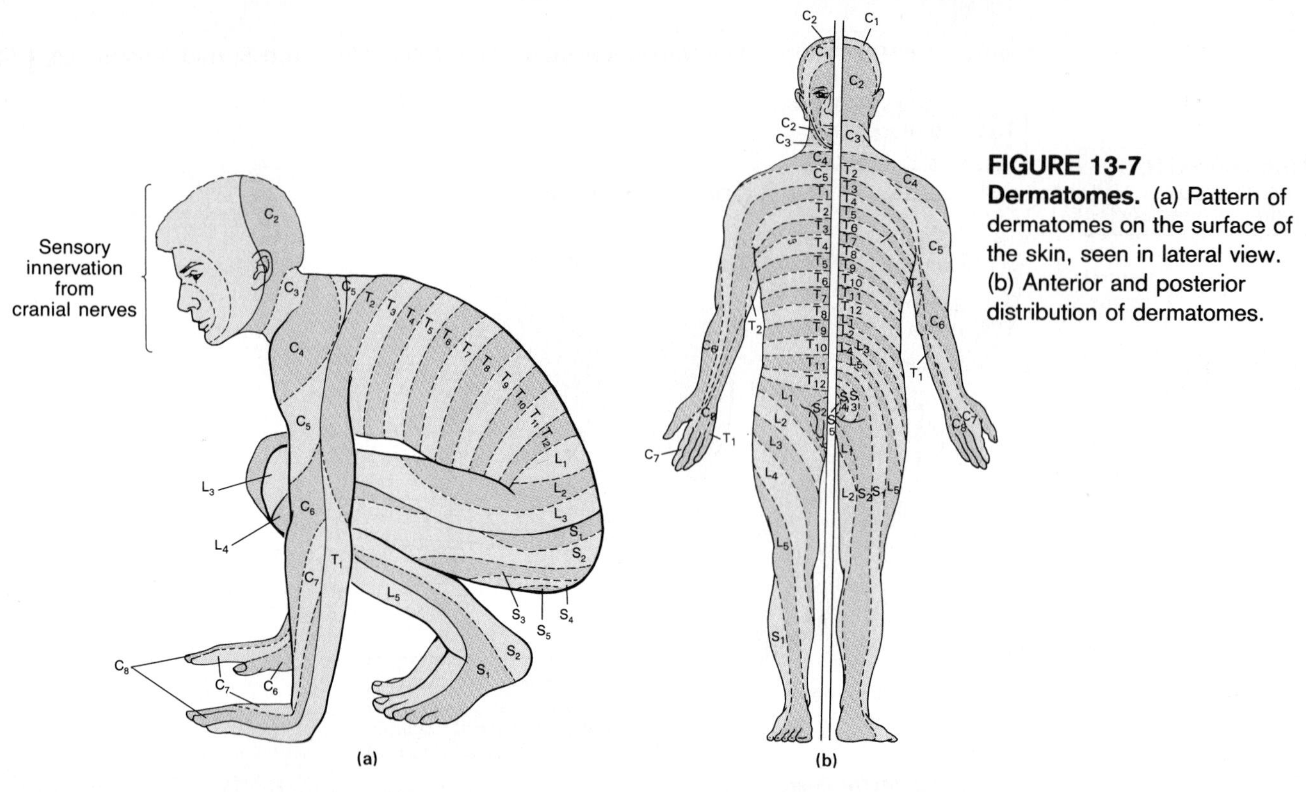

FIGURE 13-7
Dermatomes. (a) Pattern of dermatomes on the surface of the skin, seen in lateral view. (b) Anterior and posterior distribution of dermatomes.

TABLE 13-1 The Cervical Plexus

Spinal Segment	Nerves	Distribution
C_1–C_4	Ansa cervicalis (superior and inferior branches)	Five of the extrinsic laryngeal muscles (geniohyoid, thyrohyoid, sternothyroid, sternohyoid, omohyoid)
C_2–C_3	Lesser occipital, transverse cervical, supraclavicular, and greater auricular nerves	Skin of upper chest, shoulder, neck, and ear
C_3–C_5	Phrenic nerve	Diaphragm
C_1–C_5	Cervical nerves	Levator scapulae, trapezius, scalenes, sternocleidomastoid

TABLE 13-2 The Brachial Plexus

Spinal Segment	Nerves	Distribution
C_5, C_6	Axillary nerve	Deltoid and teres minor muscles Skin of shoulder
C_5–T_1	Radial nerve	Extensor muscles on the upper arm and forearm (triceps brachii, brachoradialis, extensor carpi radialis and ulnaris) Digital extensors and abductor pollicis Skin over the posterolateral surface of the arm
C_5–C_7	Musculocutaneous nerve	Flexor muscles on upper arm (biceps brachii, brachialis, coracobrachialis) Skin over lateral surface of forearm
C_6–T_1	Median nerve	Flexor muscles on forearm (flexor carpi radialis, palmaris longus) Pronators (p. quadratus and p. teres) Digital flexors Skin over lateral surface of hand
C_8, T_1	Ulnar nerve	Flexor muscle on forearm (flexor carpi ulnaris) Adductor pollicis and small digital muscles Skin over medial surface of hand

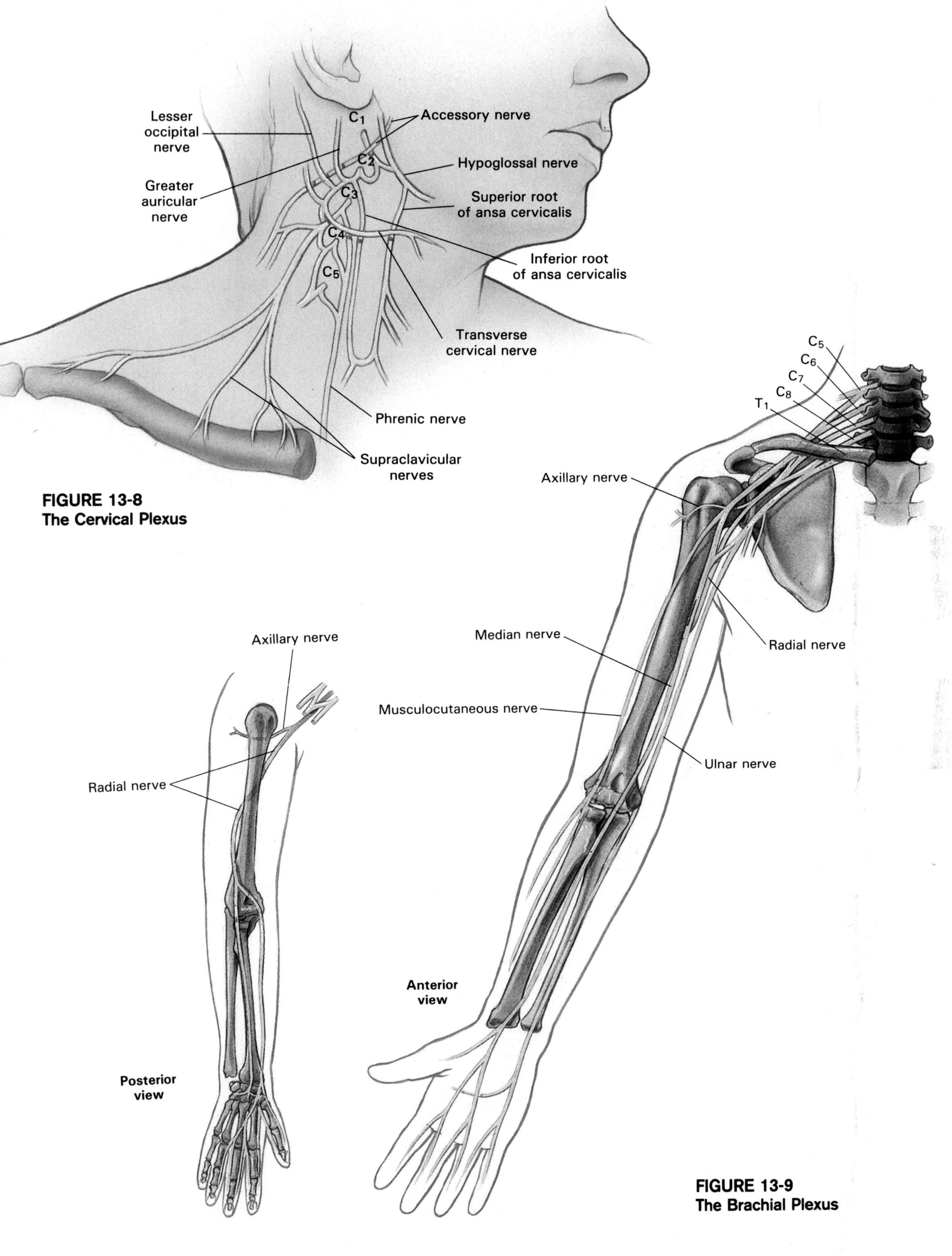

FIGURE 13-8
The Cervical Plexus

FIGURE 13-9
The Brachial Plexus

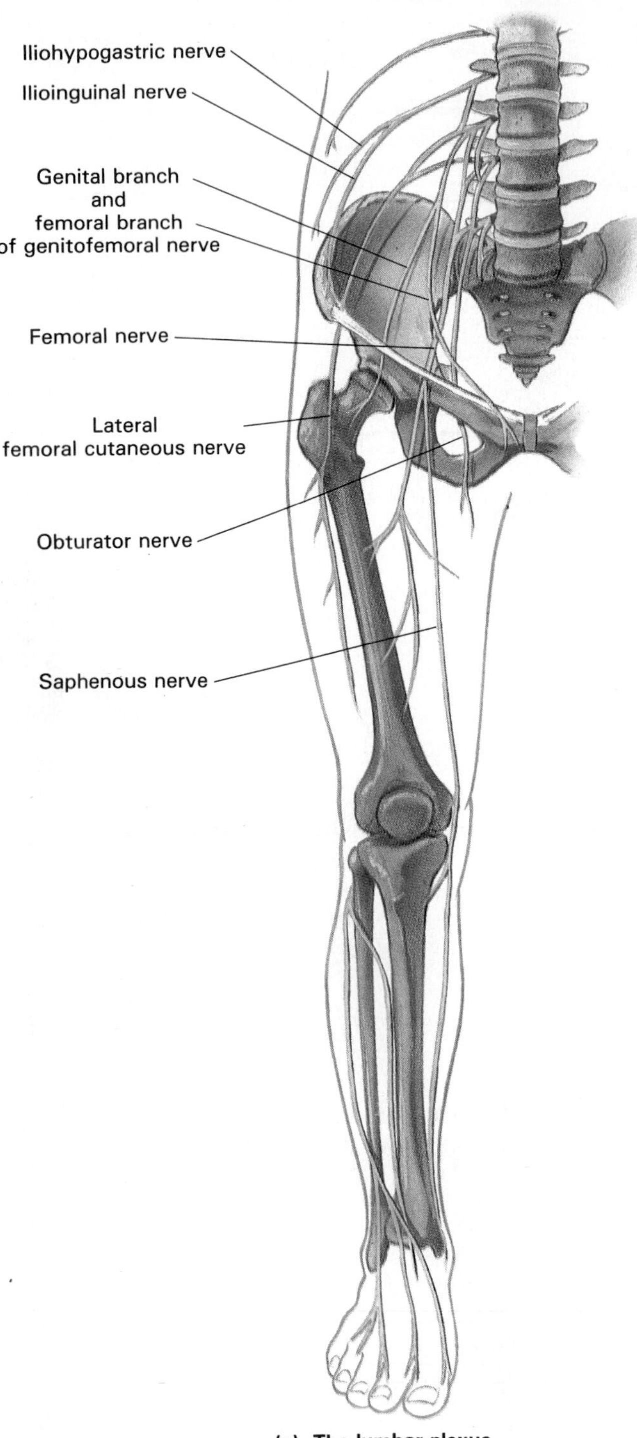

(a) The lumbar plexus

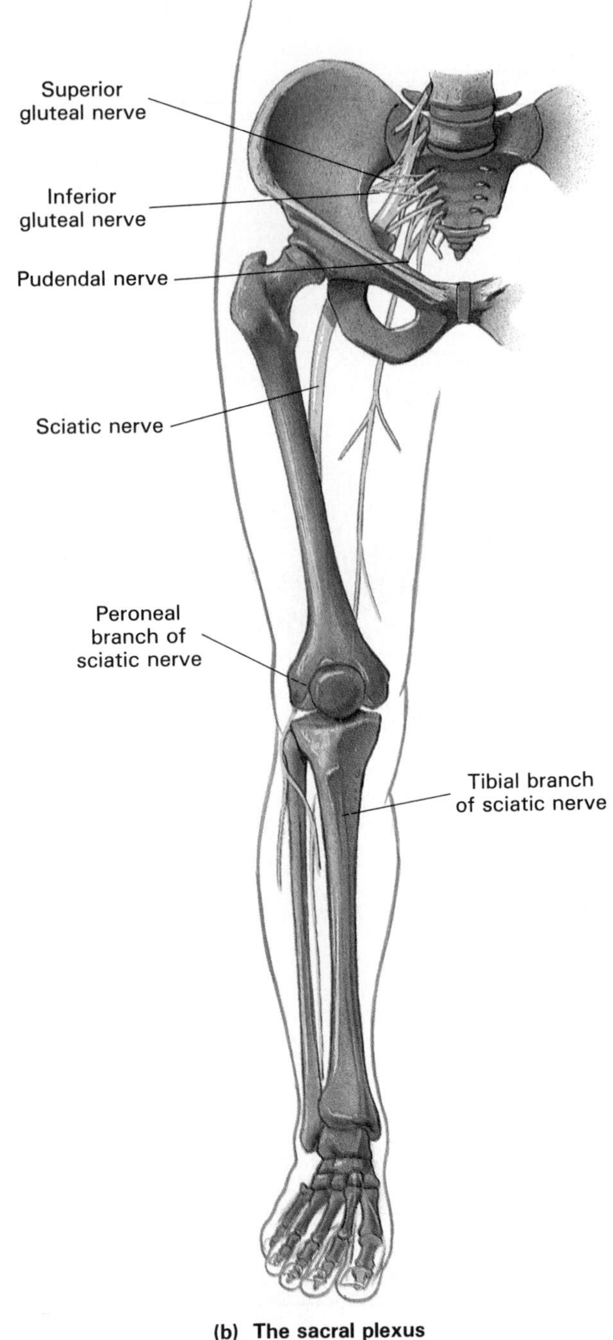

(b) The sacral plexus

FIGURE 13-10
The Lumbosacral Plexus

complicated. During development, skeletal muscles fuse with their neighbors to form larger muscles with compound origins. Although the anatomical distinctions may disappear, ventral rami from the associated spinal segments continue to provide innervation and motor control. As they converge, the ventral rami of adjacent spinal nerves blend their fibers to produce a series of compound nerve trunks. Such a complex, interwoven network of nerves is called a **nerve plexus** (PLEK-sus; a braid).

The **cervical plexus** (Figure 13-8 on p. 415 and Table 13-1 on p. 414) consists of the ventral rami of spinal nerves C_1–C_5. Its branches innervate the muscles of the neck and extend into the thoracic cavity to control the diaphragmatic muscles. The **brachial plexus** (Figure 13-9 on p. 415 and Table 13-2 on p. 414 innervates the shoulder girdle and arm, with contributions from nerves C_5 to T_1. The **lumbosacral plexus** supplies the pelvic girdle and leg. It can be further subdivided into a **lumbar plexus** (T_{12}–

TABLE 13-3 The Lumbosacral Plexus

Spinal Segment	*Nerves*	*Distribution*
THE LUMBAR PLEXUS		
T_{12}, L_1	Iliohypogastric nerve	Abdominal muscles (external and internal obliques, transversus abdominis) Skin over lower abdomen and buttocks
L_1	Ilioinguinal nerve	Abdominal muscles (with iliohypogastric) Skin over medial upper thigh and portions of external genitalia
L_1, L_2	Genitofemoral nerve	Skin over anteromedial surface of thigh and portions of external genitalia
L_2, L_3	Lateral femoral cutaneous nerve	Skin over anterior, lateral, and posterior surfaces of thigh
L_2–L_4	Femoral nerve	Anterior muscles of thigh (sartorius and quadriceps) Adductors of thigh (pectineus and iliopsoas) Skin over anteromedial surface of thigh, medial surface of leg and foot
L_2–L_4	Obturator nerve	Adductors of thigh (adductor magnus, brevis, longus) Gracilis muscle Skin over medial surface of thigh
L_2–L_4	Saphenous nerve	Skin over medial surface of leg
THE SACRAL PLEXUS		
L_4–S_2	Gluteal nerves:	
	Superior	Abductors of thigh (gluteus minimus, gluteus medius, and tensor fasciae latae)
	Inferior	Extensor of thigh (gluteus maximus)
L_4–S_3	Sciatic nerve:	Two of the hamstrings (semimembranosus, semitendinosus) Adductor magnus (with obturator nerve)
	Tibial branch	Flexors of leg and plantar flexors of foot (popliteus, gastrocnemius, soleus, tibialis posterior) Flexors (plantar flexors) of toes Skin over posterior surface of leg, plantar surface of foot
	Peroneal branch	Biceps femoris of hamstrings Peroneus (brevis and longus) and tibialis anterior Extensors (dorsiflexors) of toes Skin over anterior surface of leg and dorsal surface of foot
S_2–S_4	Pudendal nerve	Muscles of perineum Skin of external genitalia

L_4) and a **sacral plexus** (L_4–S_4). Figure 13-10 and Table 13-3 detail the branches of the lumbosacral plexus. [**CM:** *Palsies*]

Chapter 11 introduced the peripheral nerves that control the major appendicular muscle groups. (pp. 334-362) Tables 11-2 through 11-16 should provide a useful review and highlight the anatomical and functional relationships between nerves and muscles.

Although dermatomes can provide clues to the location of injuries along the spinal cord, the loss of sensation at the skin does not provide precise information concerning the site of injury, because the boundaries of dermatomes are not precise, clearly defined lines. More exact conclusions can be drawn from the loss of motor control, based on the origin and distribution of the peripheral nerves originating at nerve plexuses. [**CM:** *Leprosy*]

In the assessment of motor performance, a distinction is made between the conscious ability to control motor activities and the performance of automatic, involuntary motor responses. These latter, programmed, motor patterns, called *reflexes*, will now be examined in detail.

- **A patient suffering from polio has lost the use of his leg muscles. In what area of the spinal cord would you expect to locate the virally infected motor neurons in this individual?**
- **An anesthetic blocks the function of the dorsal rami of the cervical spinal nerves. What area of the body would be affected?**
- **Injury to which of the nerve plexuses would interfere with the ability to breathe?**

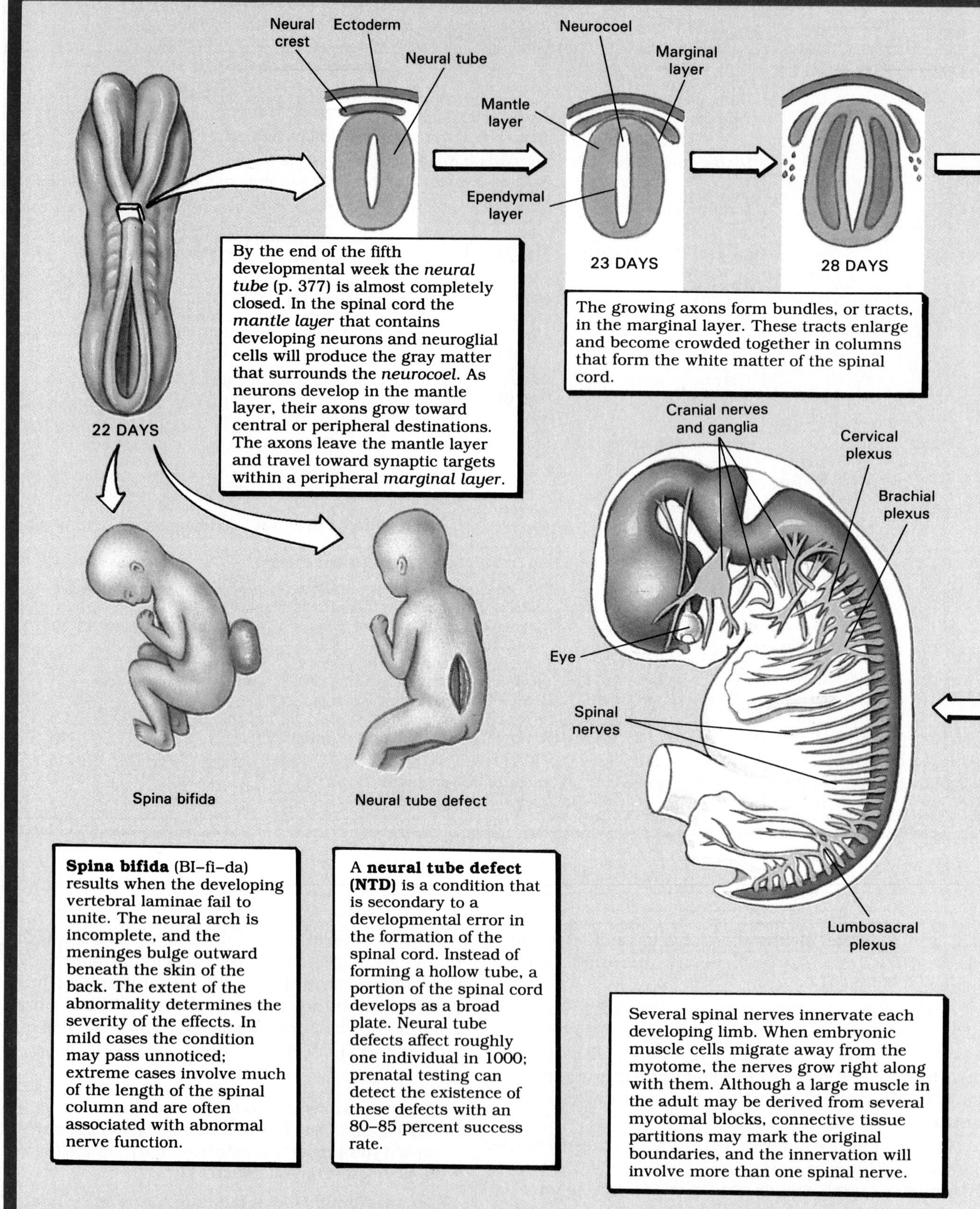

By the end of the fifth developmental week the *neural tube* (p. 377) is almost completely closed. In the spinal cord the *mantle layer* that contains developing neurons and neuroglial cells will produce the gray matter that surrounds the *neurocoel*. As neurons develop in the mantle layer, their axons grow toward central or peripheral destinations. The axons leave the mantle layer and travel toward synaptic targets within a peripheral *marginal layer*.

The growing axons form bundles, or tracts, in the marginal layer. These tracts enlarge and become crowded together in columns that form the white matter of the spinal cord.

Spina bifida (BI–fi–da) results when the developing vertebral laminae fail to unite. The neural arch is incomplete, and the meninges bulge outward beneath the skin of the back. The extent of the abnormality determines the severity of the effects. In mild cases the condition may pass unnoticed; extreme cases involve much of the length of the spinal column and are often associated with abnormal nerve function.

A **neural tube defect (NTD)** is a condition that is secondary to a developmental error in the formation of the spinal cord. Instead of forming a hollow tube, a portion of the spinal cord develops as a broad plate. Neural tube defects affect roughly one individual in 1000; prenatal testing can detect the existence of these defects with an 80–85 percent success rate.

Several spinal nerves innervate each developing limb. When embryonic muscle cells migrate away from the myotome, the nerves grow right along with them. Although a large muscle in the adult may be derived from several myotomal blocks, connective tissue partitions may mark the original boundaries, and the innervation will involve more than one spinal nerve.

Embryology Summary: Development of the Spinal Cord and Spinal Nerves

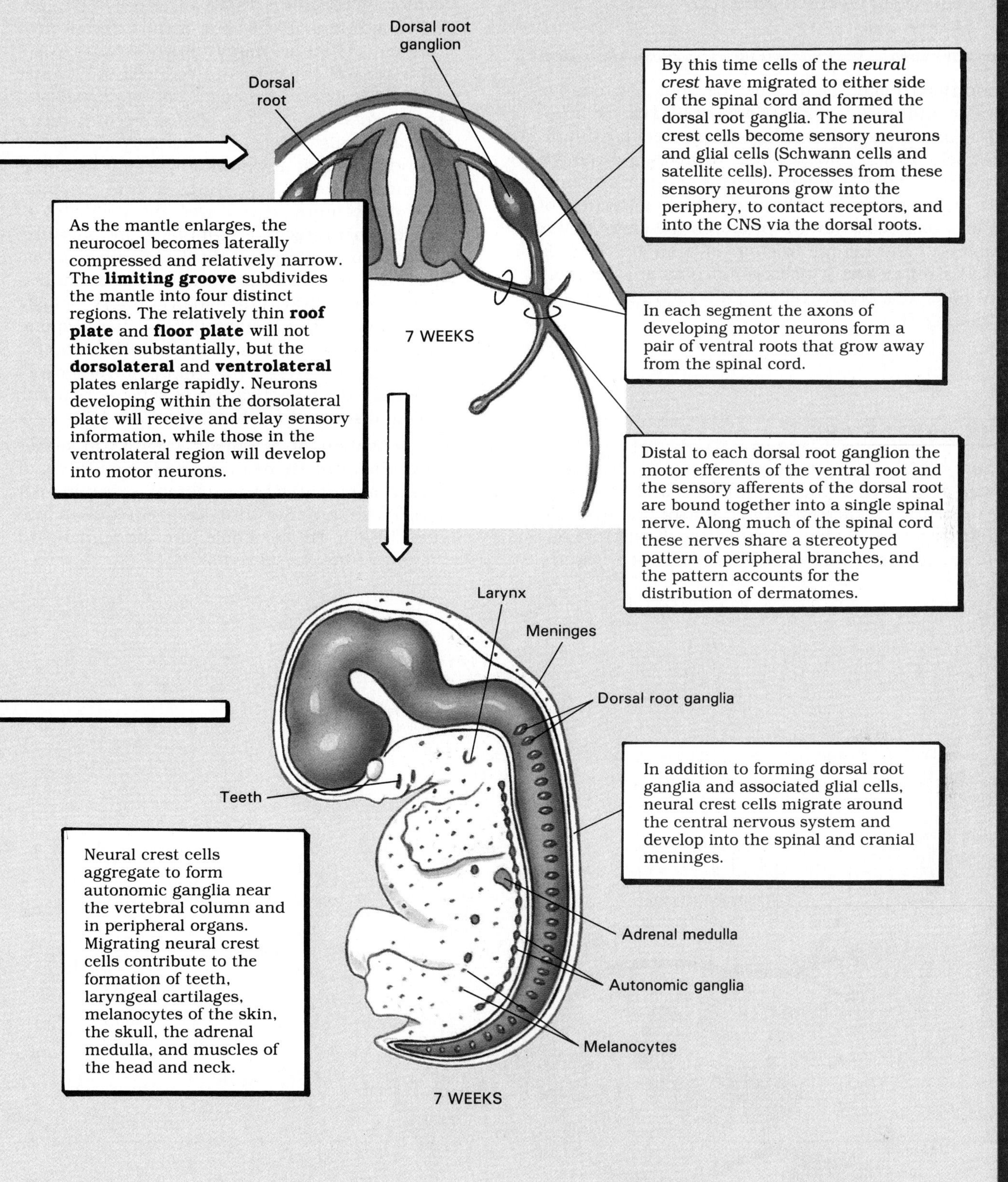

By this time cells of the *neural crest* have migrated to either side of the spinal cord and formed the dorsal root ganglia. The neural crest cells become sensory neurons and glial cells (Schwann cells and satellite cells). Processes from these sensory neurons grow into the periphery, to contact receptors, and into the CNS via the dorsal roots.

As the mantle enlarges, the neurocoel becomes laterally compressed and relatively narrow. The **limiting groove** subdivides the mantle into four distinct regions. The relatively thin **roof plate** and **floor plate** will not thicken substantially, but the **dorsolateral** and **ventrolateral** plates enlarge rapidly. Neurons developing within the dorsolateral plate will receive and relay sensory information, while those in the ventrolateral region will develop into motor neurons.

In each segment the axons of developing motor neurons form a pair of ventral roots that grow away from the spinal cord.

Distal to each dorsal root ganglion the motor efferents of the ventral root and the sensory afferents of the dorsal root are bound together into a single spinal nerve. Along much of the spinal cord these nerves share a stereotyped pattern of peripheral branches, and the pattern accounts for the distribution of dermatomes.

In addition to forming dorsal root ganglia and associated glial cells, neural crest cells migrate around the central nervous system and develop into the spinal and cranial meninges.

Neural crest cells aggregate to form autonomic ganglia near the vertebral column and in peripheral organs. Migrating neural crest cells contribute to the formation of teeth, laryngeal cartilages, melanocytes of the skin, the skull, the adrenal medulla, and muscles of the head and neck.

An Introduction to Reflexes

Conditions inside or outside the body can change rapidly and unexpectedly. **Neural reflexes** are automatic motor responses, triggered by specific stimuli, that help preserve homeostasis by making rapid adjustments in the function of organs or organ systems. The response shows little variability—activation of a particular reflex always produces the same motor response. The neural "wiring" of a single reflex is called a **reflex arc**. A reflex arc begins at a receptor and ends at a peripheral effector, such as a muscle or gland cell.

THE REFLEX ARC

Figure 13-11 diagrams the five steps involved in a neural reflex: (1) arrival of a stimulus and activation of a receptor, (2) activation of a sensory neuron, (3) information processing, (4) activation of a motor neuron, and (5) response by an effector (muscle or gland).

STEP 1: Arrival of a Stimulus and Activation of a Receptor A **receptor** is a specialized cell that monitors conditions in the body or the external environment. There are many types of sensory receptors; general categories introduced in Chapter 12 include *exteroceptors*, *interoceptors*, and *proprioceptors*. (p. 400) Each receptor has a characteristic range of sensitivity. Some respond to almost any stimulus. In our example, a painful stimulus activates a pain receptor. These receptors, themselves the dendrites of sensory neurons, are stimulated by pressure, temperature extremes, physical damage, or exposure to abnormal chemicals. Other receptors, such as those providing visual, auditory, or taste sensations, are specialized cells that respond only to a limited range of stimuli. (The specific mechanisms that link the stimulation of sensory receptors to the activation of sensory neurons will be discussed further in Chapter 17.)

STEP 2: Activation of a Sensory Neuron In either case, the ultimate result is the transmission of an action potential by an afferent (sensory) neuron. This neuron conducts nerve impulses into the CNS—in this example into the spinal cord via one of the dorsal roots.

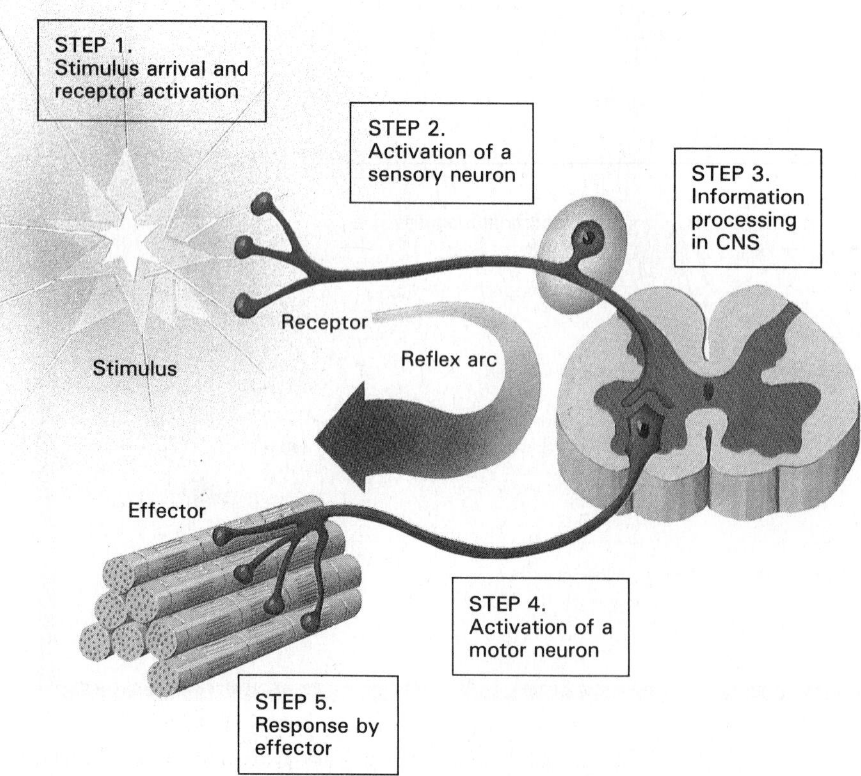

FIGURE 13-11
Components of a Reflex Arc

STEP 3: Information Processing Information processing begins when a neurotransmitter released by synaptic terminals of the sensory neuron reaches the postsynaptic membrane of a motor neuron or interneuron. In the simplest reflexes, such as the one diagrammed in Figure 13-11, this processing is performed by the motor neuron that controls peripheral effectors. (You may want to review the membrane events involved in cellular processing of nerve impulses, discussed in Chapter 12.) (p. 392–398) In more complex reflexes, several pools of interneurons are interposed between the sensory and motor neurons, and both serial and parallel processing occur. The goals of this information processing are the selection of an appropriate motor response and the activation of specific motor neurons.

STEP 4: Activation of a Motor Neuron Once stimulated to threshold, the axon of a motor neuron carries nerve impulses into the periphery, in this example over the ventral root of a spinal nerve.

STEP 5: Response of a Peripheral Effector Activation of the motor neuron then leads to a response by a peripheral effector, such as a skeletal muscle or gland. In general, this response is aimed at removing or counteracting the original stimulus. The reflex is thus an example of a *negative feedback control mechanism* (Chapter 1). (p. 8) By opposing potentially harmful changes in the internal or external environment, such reflexes play an important role in homeostatic maintenance.

CLASSIFICATION OF REFLEXES

Reflexes can be classified according to (1) their development, (2) the site where information processing occurs, (3) the nature of the resulting motor response, or (4) the complexity of the neural circuit involved. These categories, diagrammed in Figure 13-12, are not mutually exclusive; they represent different ways of describing a single reflex.

Development of Reflexes

Innate reflexes result from the connections that form between neurons during development. They usually appear in a predictable sequence, from the simplest reflex responses (withdrawal from pain) to more complex motor patterns (chewing, suckling, or tracking objects with the eyes). The basic motor patterns of an innate reflex are genetically programmed. Examples include the reflexive removal of a hand from a hot stovetop and blinking when the eyelashes are touched.

More complex, learned motor patterns are sometimes called **acquired reflexes**. An experienced driver steps on the brake when trouble appears on the road ahead; a professional skier makes rapid, automatic adjustments in body position while racing. These motor responses are rapid and automatic, but they were learned rather than preestablished. The distinction between innate and acquired reflexes is not absolute; some people can learn motor patterns more quickly than others, and the differences probably have a genetic basis.

Most reflexes, whether innate or acquired, can be modified over time or suppressed through con-

FIGURE 13-12
Methods of Classifying Reflexes

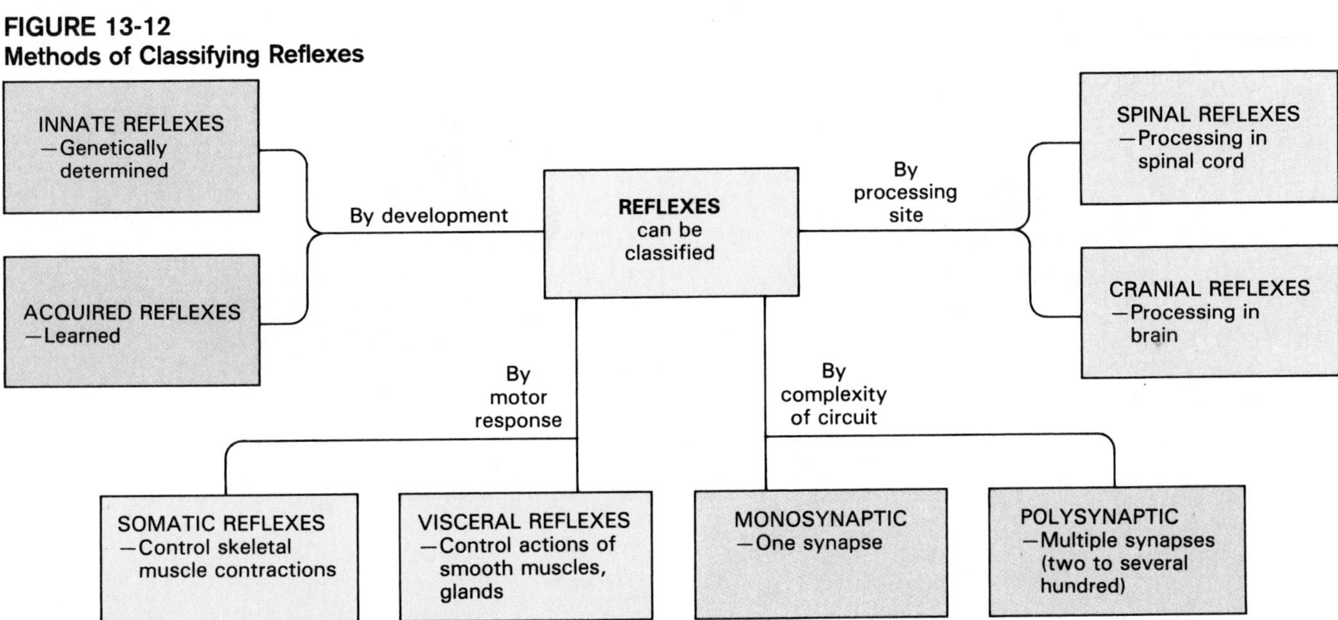

scious effort. For example, while walking a tightrope over the Grand Canyon you would probably ignore a bee sting, although under other circumstances it would probably cause you to abruptly withdraw the foot or hand involved.

Processing Sites

Reflexes processed in the brain are called **cranial reflexes**. Several examples of cranial reflexes will be found in Chapters 15 and 16. In a **spinal reflex** the important interconnections and processing events occur inside the spinal cord. These reflexes are discussed in detail in the next section.

Nature of the Motor Response

Somatic reflexes control activities of the muscular system, whereas **visceral**, or **autonomic**, **reflexes** control the activities of other systems. This chapter considers somatic reflexes in detail; visceral reflexes are examined in Chapter 15.

Complexity of the Neural Circuitry

In the simplest reflex arc a sensory neuron synapses directly on a motor neuron, which itself serves as the processing center. Such a reflex is termed a **monosynaptic reflex**. Transmission across a chemical synapse always involves a synaptic delay, but with only one synapse the delay between stimulus and response is minimized. Circuit 1 in Figure 13-13 is a typical monosynaptic reflex.

Many other reflexes have at least one interneuron placed between the sensory afferent and the motor efferent, as indicated in Circuit 2. These **polysynaptic reflexes** have a longer delay between stimulus and response, the length of the delay being proportional to the number of synapses involved. Polysynaptic reflexes can produce far more complicated responses because the interneurons can control several different muscle groups.

The next section describes representative monosynaptic and polysynaptic spinal reflexes. Many of the motor responses are extremely complicated; for example, stepping on a tack not only causes with-

FIGURE 13-13
Neural Organization and Simple Reflexes. (a) A monosynaptic reflex involves a peripheral sensory neuron and a central motor neuron. In this example, stimulation of the receptor will lead to a reflexive contraction in a skeletal muscle. (b) A polysynaptic reflex involves a sensory neuron, interneurons, and motor neurons. In this example the stimulation of the receptor leads to the coordinated contractions of two different skeletal muscles.

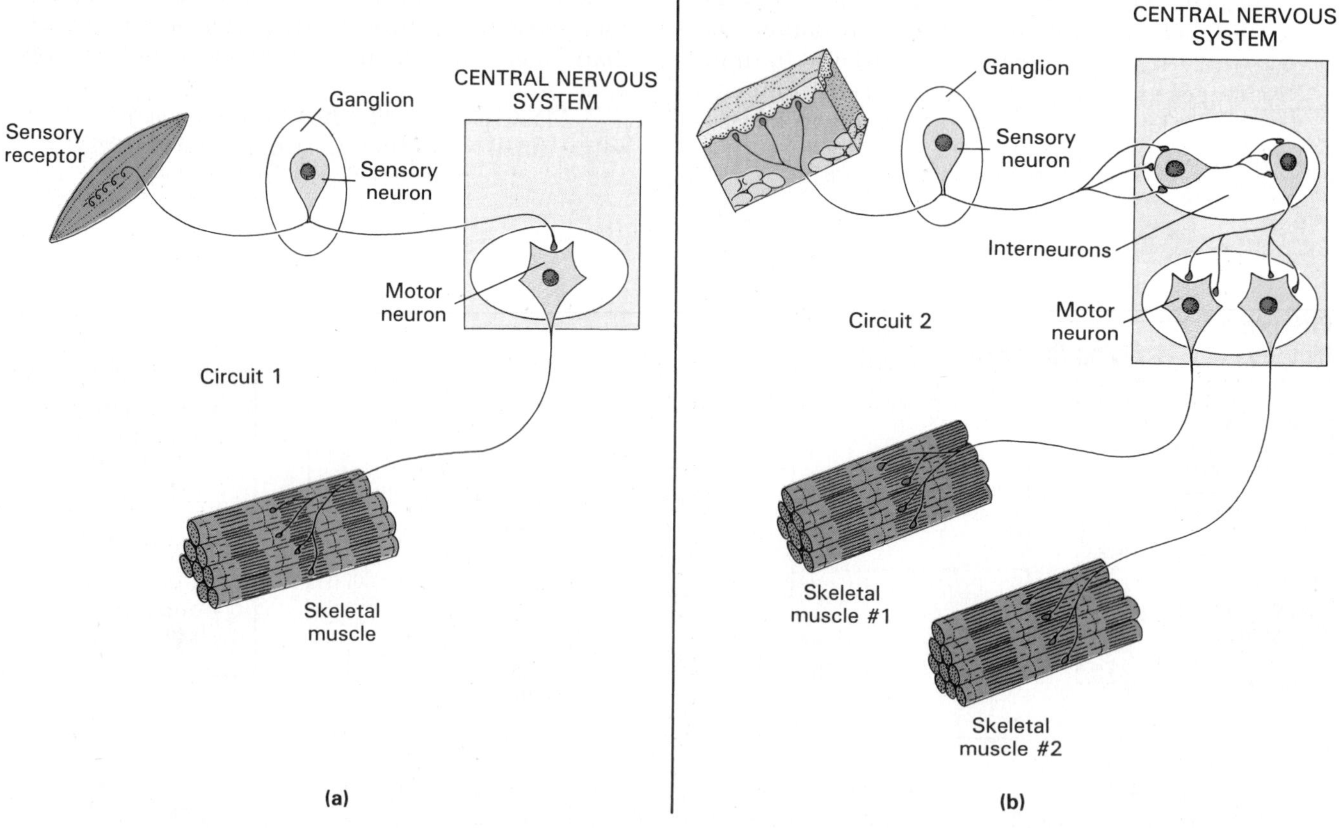

drawal of the foot, but triggers all of the muscular adjustments needed to prevent a fall. Such complicated responses result from the interactions between multiple interneuron pools.

Spinal Reflexes

The neurons of the spinal cord participate in a variety of reflex arcs. These **spinal reflexes** range in complexity from simple monosynaptic reflexes involving a single segment of the spinal cord to polysynaptic reflexes that involve motor output from many different segments. The most complicated spinal reflexes are called **intersegmental reflex arcs** because many segments interact to produce a coordinated motor response. These polysynaptic reflexes include excitatory and inhibitory synapses and produce highly variable motor patterns.

MONOSYNAPTIC REFLEXES

In a monosynaptic reflex there is little delay between sensory input and motor output. These reflexes control the most rapid, stereotyped motor responses of the nervous system to specific stimuli. The best known example of a monosynaptic reflex is the *stretch reflex*.

The Stretch Reflex

The **stretch reflex** provides automatic regulation of skeletal muscle length. This reflex is diagrammed in Figure 13-14a. The stimulus (increasing muscle length) activates a sensory neuron that triggers an immediate motor response (contraction of the stretched muscle) that counteracts the stimulus. Action potentials traveling toward and away from the spinal cord are conducted along large myelinated Type A fibers (Chapter 12), and the entire reflex is completed within 20 msec. (p. 386)

Muscle Spindles The sensory receptors in the stretch reflex are the **muscle spindles**. These receptors, which are scattered throughout skeletal muscles, were introduced in Chapter 10. (p. 317) As indicated in Figure 13-14b, each muscle spindle consists of a bundle of small, specialized skeletal muscle fibers, called **intrafusal muscle fibers**. The muscle

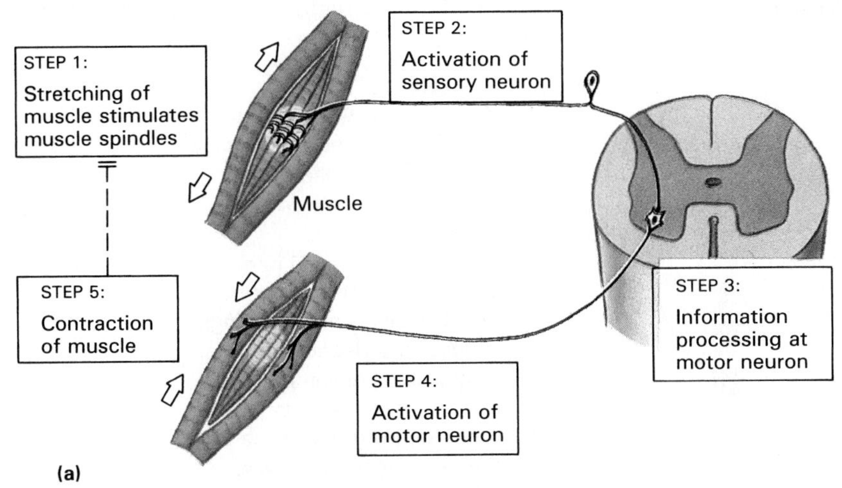

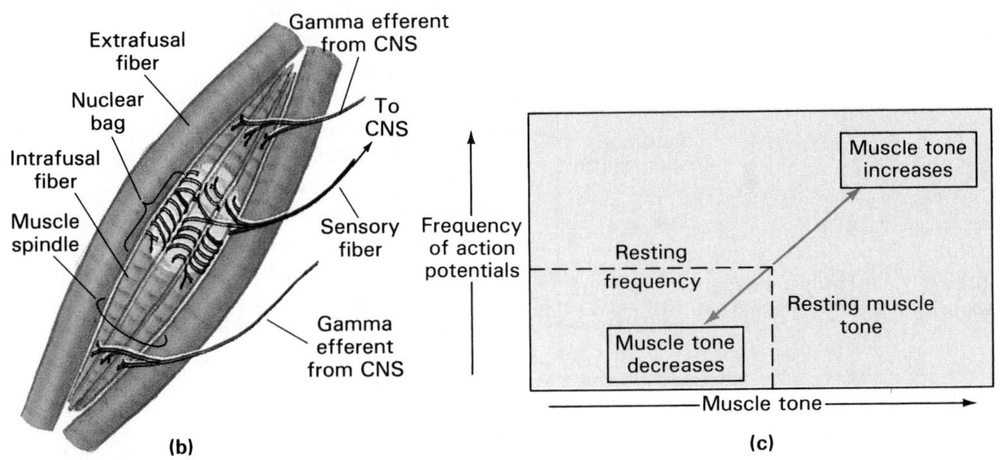

FIGURE 13-14 Components of the Stretch Reflex. (a) Diagram of the activities in a stretch reflex. (b) Structure of a muscle spindle. (c) Background activity in the sensory neuron maintains a stable resting muscle tone. An increase in the action potential rate elevates muscle tone; a decrease leads to muscle relaxation.

spindle is surrounded by larger **extrafusal muscle fibers** that are responsible for the resting muscle tone and, at greater levels of stimulation, for the contraction of the entire muscle.

Figure 13-14b details the structure of a single intrafusal fiber. *Myofibrils* in the intrafusal fiber are attached to a central area called the **nuclear bag**. The surface of the nuclear bag is monitored by the dendrites of a sensory neuron. This neuron is always active, conducting impulses to the CNS. Stretching the nuclear bag stimulates the sensory neuron, increasing the frequency of action potentials. Compressing the nuclear bag inhibits the sensory neuron, lowering the rate at which it transmits action potentials.

The axon of the sensory neuron synapses on CNS motor neurons that control the extrafusal muscle fibers *of the same muscle*. An increase in stimulation of the sensory neuron will cause a rise in muscle tension; a decrease in stimulation of the sensory neuron will cause a decline in muscle tension. This arrangement is the basis for the several functional roles of the stretch reflex.

- The stretch reflex provides automatic adjustment of muscle tone, increasing or decreasing it in response to information provided by the stretch receptors of the muscle spindles. Heightened stimulation of the sensory neurons in the spindles will increase muscle tone, whereas a decline in stimulation of these neurons will lead to muscle relaxation and a loss of tone (Figure 13-14c).
- When the muscle shortens passively, the sensory nerve endings at the muscle spindles are compressed, the rate of action potential production declines, and muscle tone decreases (Figure 13-15a).
- When the muscle elongates suddenly, the dendrites are stretched, the sensory neuron becomes very active, and muscle tone increases dramatically (Figure 13-15b). This reflexive contraction slows the rate of elongation and protects the muscle from extreme stretching. Figure 13-15c summarizes the effects of passive compression and stretching on muscle tone.
- During active muscle contraction (Figure 13-17), higher centers stimulate **gamma (γ) motor neurons** that innervate intrafusal muscle fibers. Impulses arriving over **gamma efferents** cause contraction of the myofibrils within intrafusal muscle fibers. These myofibrils pull on the nuclear bag and prevent the compression that occurs during passive shortening.

When the CNS issues somatic motor commands, both extrafusal and intrafusal fibers are stimulated.

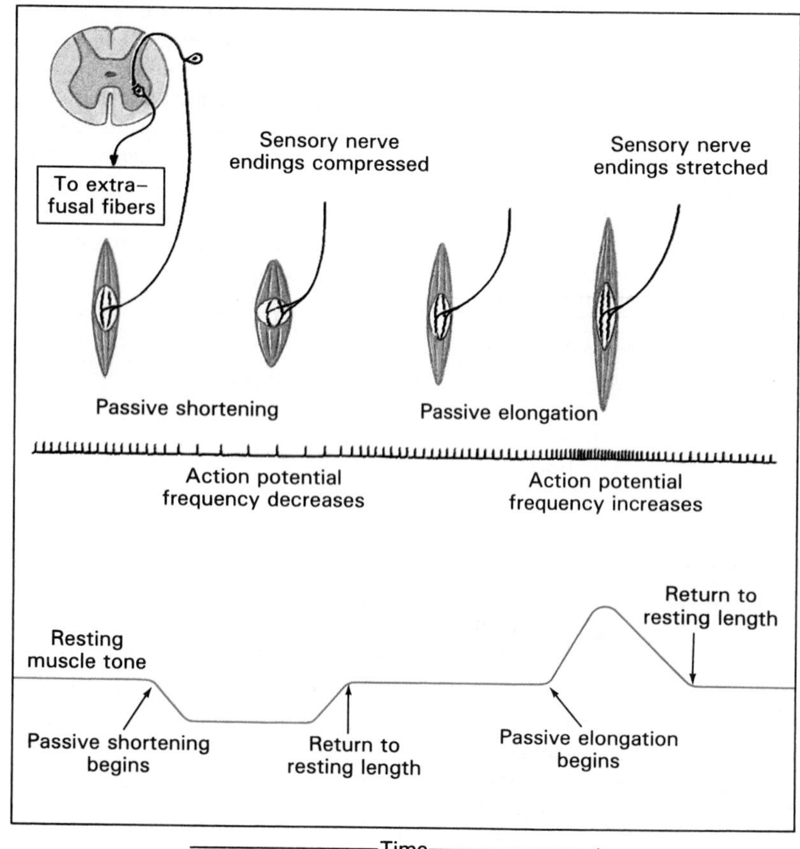

FIGURE 13-15
The Stretch Reflex and Passive Changes in Muscle Length. When the muscle shortens passively (left), the intrafusal fibers are compressed. This in turn compresses the nuclear bag and its sensory nerve endings, producing a decrease in the frequency of action potentials. The result is a drop in muscle tone, reducing resistance to the movement under way. When the muscle elongates passively (right), the stretching applied to the nuclear bag stimulates the sensory neuron, increasing the rate of action potential generation. This produces an increase in muscle tone that opposes additional stretching. Below, a graph of muscle tension as the muscle undergoes passive shortening and elongation. These changes result from the activity of sensory neurons monitoring intrafusal muscle fibers.

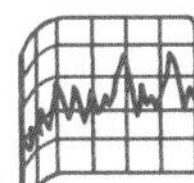

Diagnostics: Stretch Reflexes

Physicians use the sensitivity of the stretch reflex to test the general condition of the spinal cord, peripheral nerves, and muscles. For example, in the **knee jerk**, or **patellar**, **reflex**, a sharp rap on the patellar ligament stretches muscle spindles in the quadriceps (thigh) muscles (Figure 13-16a). With so brief a stimulus, the reflexive contraction occurs unopposed and produces a noticeable kick. If this contraction shortens the muscle spindles below their original resting lengths, the sensory nerve endings are compressed, the sensory neuron inhibited, and the leg drops back. If it swings back far enough to stretch the spindles a second time, there will be a small secondary kick (Figure 13-16b).

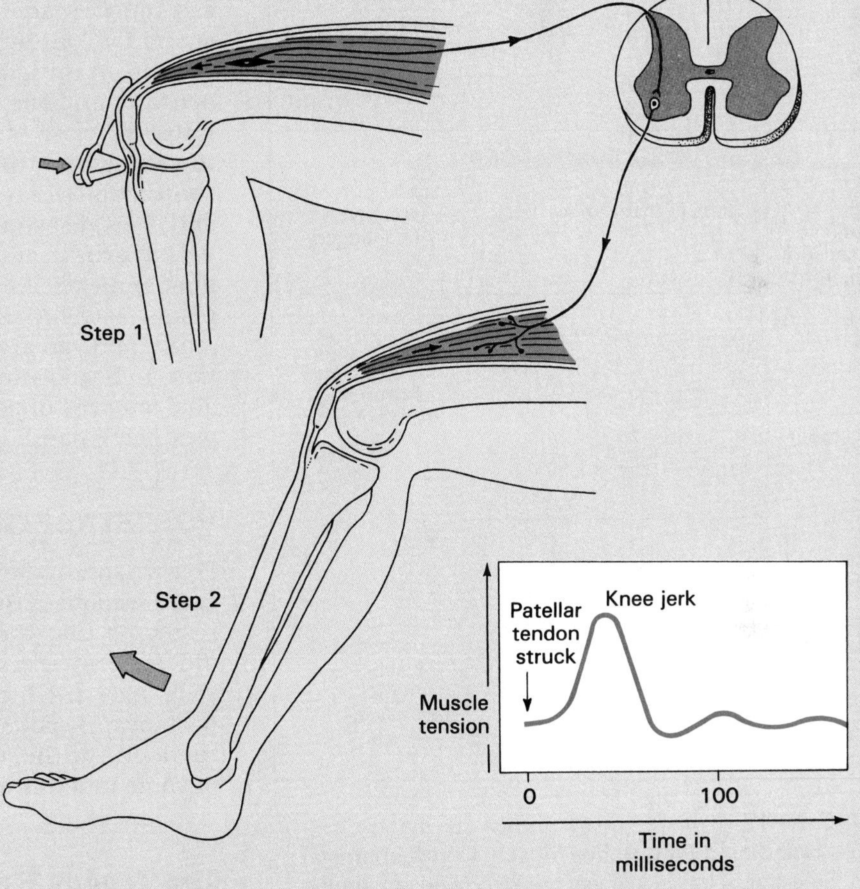

FIGURE 13-16
The Patellar Reflex. This stretch reflex is controlled by muscle spindles in the muscles that straighten the knee. In Step 1 a reflex hammer strikes the muscle tendon, stretching the spindle fibers. This results in a sudden increase in the activity of the sensory neurons. These neurons synapse on spinal motor neurons. In Step 2, the activation of extrafusal motor units produces an immediate increase in muscle tone and a reflexive kick. The changes in muscle tension can be graphically recorded. Note the small secondary kick that may occur if the leg rebounds past its resting position.

If there is no resistance, such as a weight in the hand, the extrafusal muscle fibers contract and the muscle shortens. As this occurs, the stimulation of the intrafusal muscle fibers prevents the reflexive decrease in muscle tone that would ordinarily result from compression of the nuclear bag (Figure 13-18a).

When there is resistance to the movement, such as a heavy weight in the hand, stimulation of gamma motor neurons triggers an automatic increase in the amount of force produced by the extrafusal muscle fibers. The mechanism is diagrammed in Figure 13-18b. Stimulation of the extrafusal muscle fibers does not immediately shorten the muscle, because tension must rise enough to overcome the resistance of the weight. But the impulses arriving over the gamma efferents stimulate the intrafusal fibers anyway. This stretches the nuclear bags, stimulating the sensory receptors *as if the entire muscle were being*

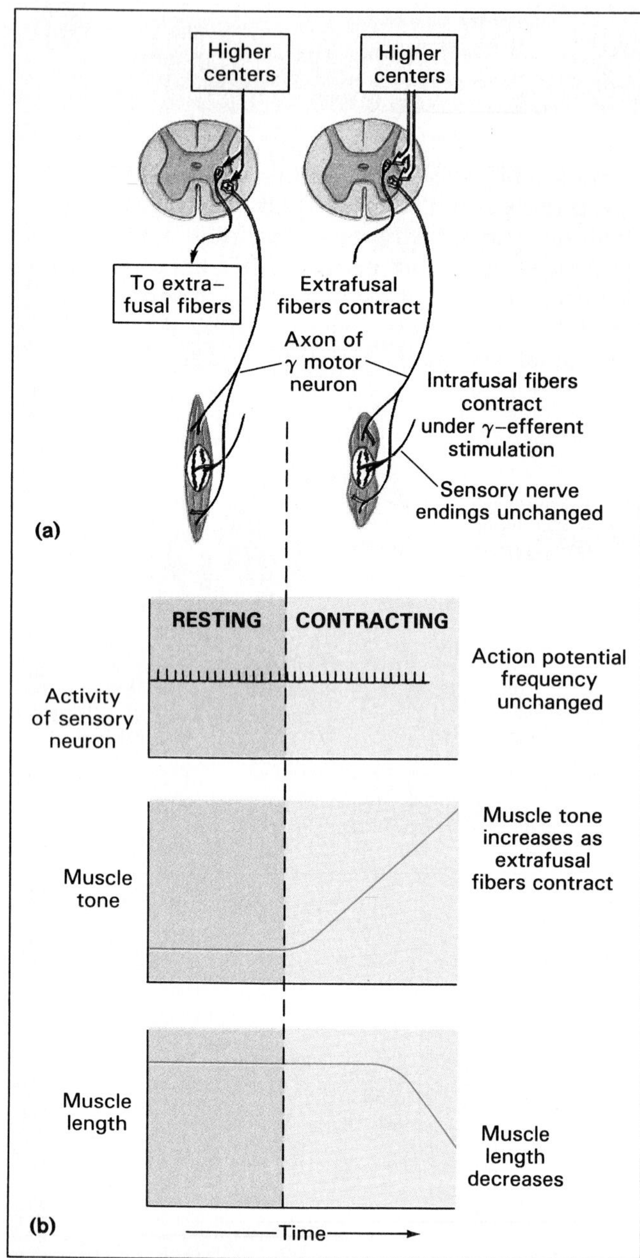

FIGURE 13-17
Muscle Spindles during Active Muscle Contraction. (a) When the extrafusal muscle fibers are stimulated, so are the gamma efferents that innervate the intrafusal fibers. Contraction of the intrafusal fibers pulls on the nuclear bag and prevents a reflexive reduction in muscle tone. (b) A graph of changes in muscle length and tension during an active contraction. Note that the muscle tone begins to rise immediately, due to stimulation of the gamma efferents, before the extrafusal fibers have developed enough tension to shorten.

stretched. The sensory neurons then direct an increase in muscle tone, and more motor units are activated. This process of *recruitment* continues until tension rises enough to overcome the resistance. (Recruitment, or *multiple motor unit summation,* was discussed in Chapter 10.) (p. 315) One indication of the significance of intrafusal muscle fibers in voluntary muscle contractions is that almost one-third of the motor fibers entering a skeletal muscle control the intrafusal fibers of muscle spindles, rather than extrafusal muscle fibers responsible for contraction.

The Stretch Reflex and Regulation of Posture The stretch reflex is an example of a **postural reflex**, a reflex that maintains normal upright posture. For example, standing involves a cooperative effort by many different muscle groups. Some of these muscles work in opposition to one another, exerting forces that keep the body's weight balanced over the feet. If the body leans forward, stretch receptors in the calf muscles are stimulated, and those muscles respond by contracting. This contraction returns the body to an upright position. If the muscles overcompensate, and the body begins to lean back, the calf muscles relax, but stretch receptors in muscles of the shins and thighs are stimulated. The muscular contractions that result reposition the body weight and avert disaster.

Postural muscles usually have a firm muscle tone and extremely sensitive stretch receptors. As a result, very fine adjustments are continually being made, and you are not aware of the cycles of contraction and relaxation that occur. Stretch reflexes are only one type of postural reflex; there are many complex polysynaptic postural reflexes.

POLYSYNAPTIC REFLEXES

Polysynaptic reflexes can produce far more complicated responses than can monosynaptic reflexes. One reason is that the interneurons can control several different muscle groups. Moreover, these interneurons may produce either excitatory or inhibitory postsnyaptic potentials (EPSPs or IPSPs) at CNS motor nuclei, so the response can involve the stimulation of some muscles and the inhibition of others.

The Tendon Reflex

The stretch reflex regulates the length of a skeletal muscle. The **tendon reflex** monitors the tension produced during a muscular contraction and prevents damage to the tendons by excessive stresses. The receptors for this reflex are nerve endings called **Golgi tendon organs** (Figure 13-19). These sensory receptors monitor tension in the collagen fibers of the tendon. When the collagen fibers stretch, the sensory neuron is stimulated. Within the spinal cord these neurons stimulate inhibitory interneurons that synapse upon motor neurons controlling the skeletal muscle(s) responsible for the tension. When extremely powerful forces are generated, this reflex can relax the entire muscle before the muscle or tendon tears apart.

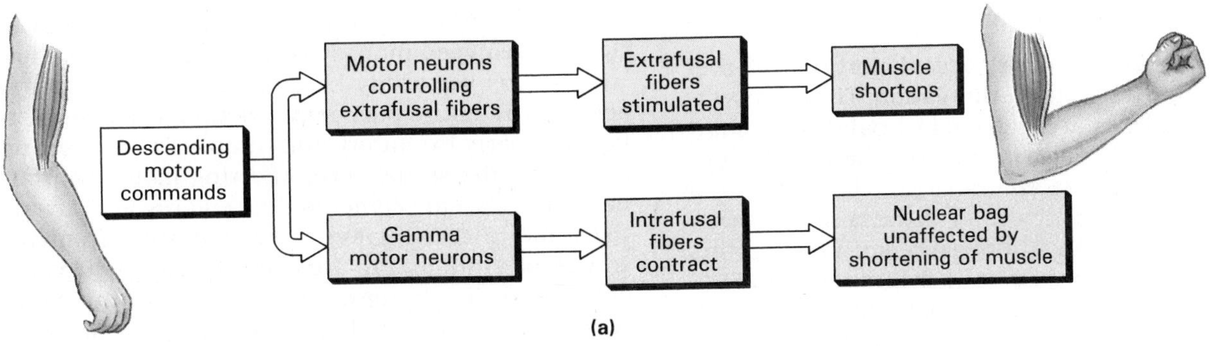

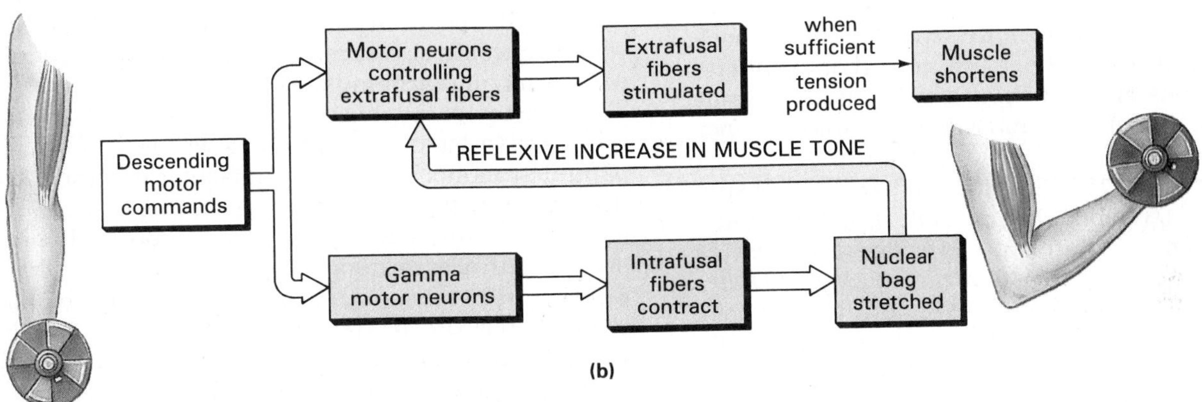

FIGURE 13-18
Muscle Spindle Function during Active Muscle Contraction. (a) Without resistance. (b) With resistance.

FIGURE 13-19
The Tendon Reflex. This reflex protects the connective tissues of tendons against excessive force produced during muscular contraction.

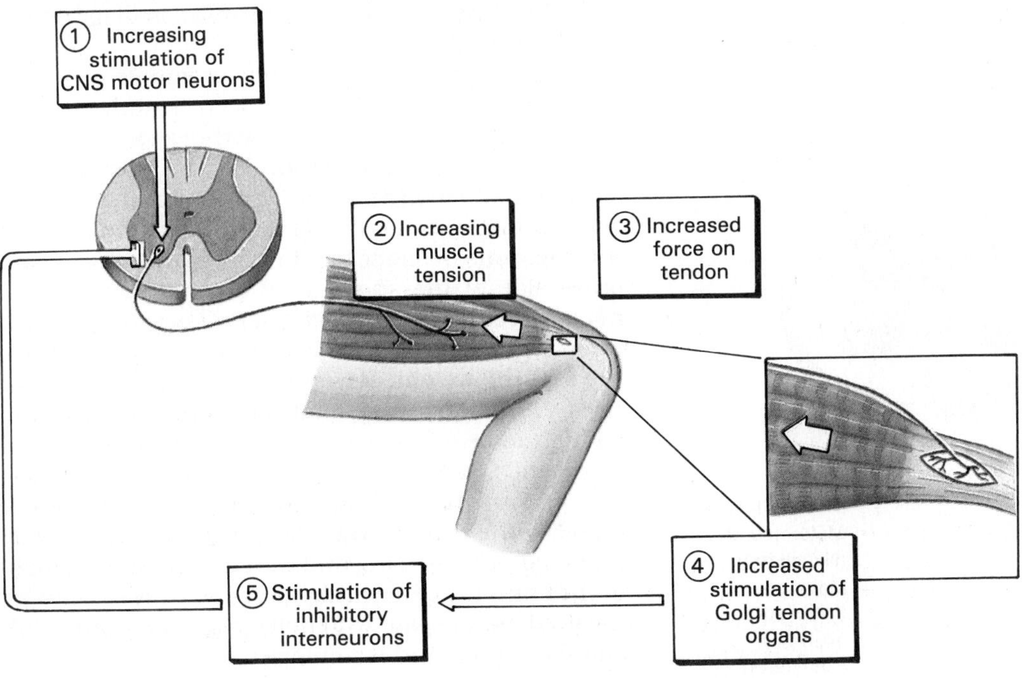

Withdrawal Reflexes

Withdrawal reflexes move affected portions of the body away from a source of stimulation. The strongest withdrawal reflexes are triggered by painful stimuli, but these reflexes are also initiated by the stimulation of touch or pressure receptors.

There are several types of withdrawal reflexes. The **flexor reflex** is a withdrawal reflex affecting the muscles of a limb; an example is diagrammed in Figure 13-20. As you will recall from Chapters 7 and 11, *flexion* is a reduction in the angle between two articulating bones, and the contractions of *flexor muscles* perform this movement. (pp. 233, 332–334) Stepping on a tack produces a dramatic flexor reflex in the affected leg. When the pain receptors in the foot are stimulated, the sensory neurons activate interneurons in the spinal cord that stimulate motor neurons in the anterior gray horns. The result is a contraction of flexor muscles that yanks the foot off the ground.

When a specific muscle contracts, opposing muscles are stretched. For example, the flexor muscles that bend the leg are opposed by *extensor muscles* that straighten it out. A potential conflict exists here: contraction of a flexor muscle should theoretically trigger a stretch reflex in the extensors that would cause them to contract, opposing the movement that is under way. Interneurons in the spinal cord prevent such competition through **reciprocal inhibition**. When one set of motor neurons is stimulated, those controlling antagonistic muscles are inhibited. The term *reciprocal* refers to the fact that the system works both ways. When the flexors contract, the extensors relax; when the extensors contract, the flexors relax.

FIGURE 13-20
The Withdrawal (Flexor) Reflex

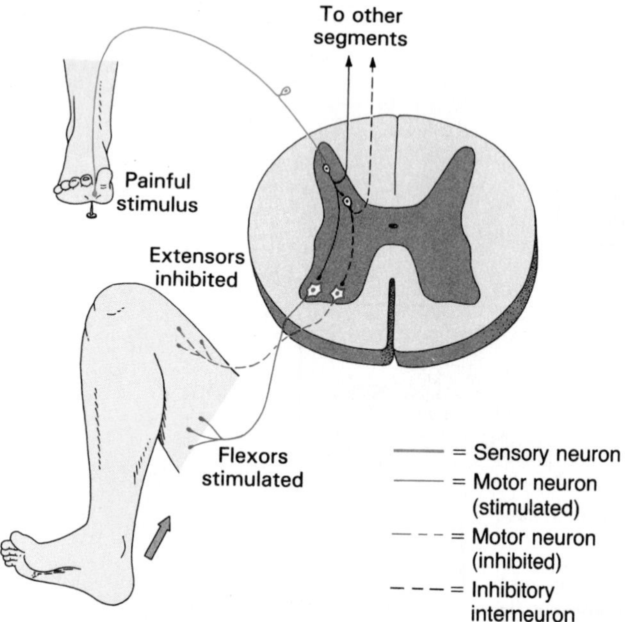

Flexor reflexes are very different from the relatively stereotyped monosynaptic reflexes. They show tremendous versatility because the sensory neuron activates an entire pool of interneurons. These interneurons carry excitatory and inhibitory impulses up and down the spinal cord, affecting motor neurons in many different segments. The end result is always the same: a coordinated movement away from the source of stimulation. But the distribution of the effects and the strength and character of the motor responses depend on the intensity and location of the stimulus. When you step on something sharp, mild discomfort might provoke a brief contraction in muscles of the ankle and foot. More powerful stimuli would produce coordinated muscular contractions affecting the position of the ankle, foot, and leg. Severe pain would also stimulate contractions of shoulder, trunk, and arm muscles.

The motor responses produced in a withdrawal reflex are variable in duration as well as in form. A painful stimulus will trigger muscular contractions that continue for several seconds. In this case, reverberating circuits among the interneuron pools produce a controlled, extended contraction rather than a momentary muscular spasm. In contrast, monosynaptic reflexes are relatively invariable and brief in duration; the entire knee jerk reflex is completed in roughly 20 msec.

Crossed Extensor Reflexes

The stretch, tendon, and withdrawal reflexes are **ipsilateral reflex arcs** (*ipsi*, same + *lateral*, side), where the sensory stimulus and the motor response occur on the same side of the body. The **crossed extensor reflex** is a **contralateral reflex arc** (*contra*, opposite); the motor response occurs on the side opposite the stimulus. The crossed extensor reflex is diagrammed in Figure 13-21.

The crossed extensor reflex complements the withdrawal reflex, and the two occur simultaneously. When you step on a thorn, the withdrawal reflex pulls your foot away from the ground as the crossed extensor reflex stiffens the other leg to support the body's weight. In the crossed extensor reflex, the axons of interneurons responding to the pain cross to the other side of the spinal cord and stimulate motor neurons controlling the extensor muscles of the uninjured leg. The opposite leg straightens to support the shifting weight. Reverberating circuits use positive feedback to ensure that the movement lasts long enough to be effective, despite the absence of motor commands from higher centers.

In summary, the crossed extensor is the mirror image of the withdrawal reflex. It can produce a variety of motor responses, and the location and intensity of the stimulus determine the specific muscle groups involved. Severe pain will elicit a response from trunk, shoulder, arm, and leg muscles.

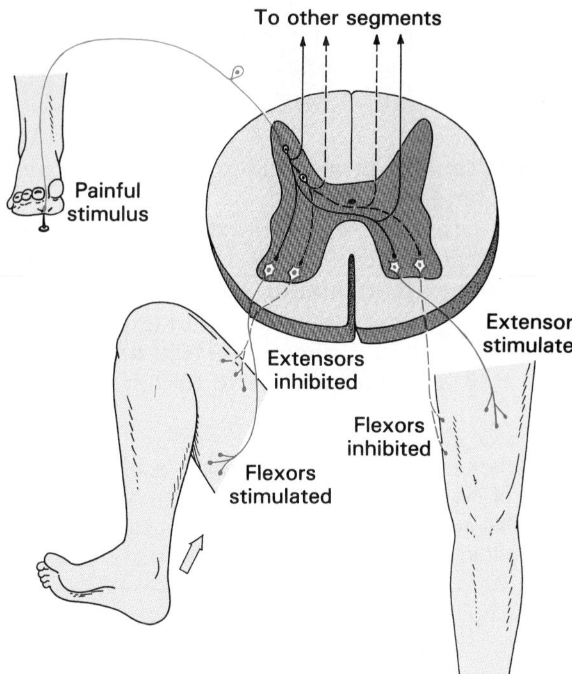

FIGURE 13-21
The Crossed Extensor Reflex

The withdrawal and crossed extensor reflexes are not the most complicated spinal reflexes. Postural and motor reflexes associated with standing, walking, and running are also spinal reflexes. But all polysynaptic reflexes share the same basic characteristics:

1. **They involve pools of interneurons.** Sensory information activates pools of interneurons. Processing occurs in those pools before motor neurons are affected. The output of the neuronal pools may have an excitatory or inhibitory effect, depending on the nature of the neurotransmitter released. The tendon reflex produces inhibition of motor neurons, whereas the flexor, withdrawal, and crossed extensor reflexes direct specific muscle contractions.
2. **They are intersegmental in distribution.** The interneuron pools extend across spinal segments, and they may activate muscle groups in many parts of the body.
3. **They involve reciprocal innervation.** Reciprocal innervation coordinates muscular contractions and reduces resistance to movement. In the flexor, withdrawal, and crossed extensor reflexes the contraction of one muscle group was associated with the inhibition of opposing muscles. This process requires two different populations of interneurons, some releasing excitatory neurotransmitters and others inhibitory neurotransmitters.
4. **They have reverberating circuits that prolong the reflexive motor response.** Positive feedback between interneurons that innervate motor neurons and the processing pool maintains the stimulation even after the initial stimulus has faded.
5. **Several reflexes may cooperate to produce a coordinated, controlled response.** In this process, antagonistic reflexes are inhibited. For example, reflexive or voluntary stimulation of a muscle is accompanied by the stimulation of gamma efferents that innervate intrafusal muscle fibers. As a result, muscle tone increases and assists the movement, rather than acting to oppose it.

Integration and Control of Spinal Reflexes

Reflex motor behaviors occur automatically, without instructions from higher centers. However, higher centers can have a profound effect on reflex performance. Processing centers in the brain can facilitate or inhibit reflex motor patterns based in the spinal cord. Descending tracts originating in the brain

FIGURE 13-22
The Babinski Reflex. (a) The positive Babinski reflex (*Babinski sign*) occurs in the absence of descending inhibition. It is normal in infants but pathological in adults. (b) The negative Babinski reflex, a curling of the toes, seen in normal adults.

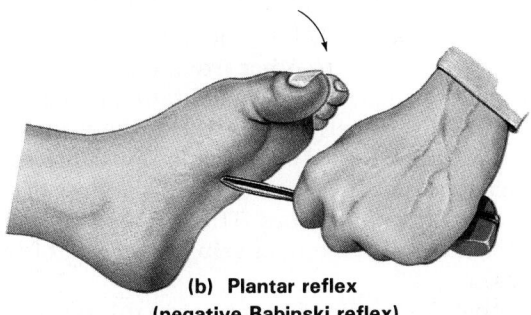

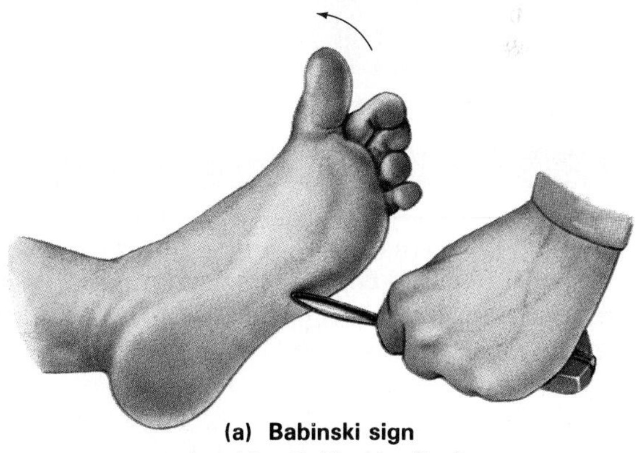

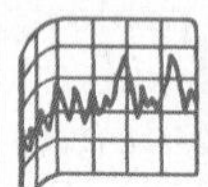

Diagnostics: Reflexes and Diagnostic Testing

Many reflexes can be assessed through careful observation and the use of simple tools. The procedures are easy to perform, and the results can provide valuable information about damage to the spinal cord or spinal nerves. By testing a series of spinal and cranial reflexes a physician can assess the function of sensory pathways and motor centers throughout the spinal cord and brain. (Cranial reflexes will be discussed further in Chapter 14.)

Neurologists test many different reflexes; only a few are so generally useful that physicians make them part of a standard physical examination. These reflexes are shown in Figure 13-23.

The *patellar* (Figure 13-16), *ankle jerk* (Figure 13-23a), *biceps* (Figure 13-23b), and *tri-*

FIGURE 13-23
Reflexes Useful in Clinical Testing. (a) The ankle jerk reflex, a plantar flexion of the foot following a blow to the calcaneal tendon. (b) The biceps reflex, produced by stretching the tendon of the biceps brachii at its insertion on the forearm. (c) The triceps reflex, produced by tapping on the tendon near its insertion on the olecranon process. (d) The abdominal reflex. A normal reflex occurs when descending facilatory tracts are intact; damage to these tracts causes the disappearance of this reflex.

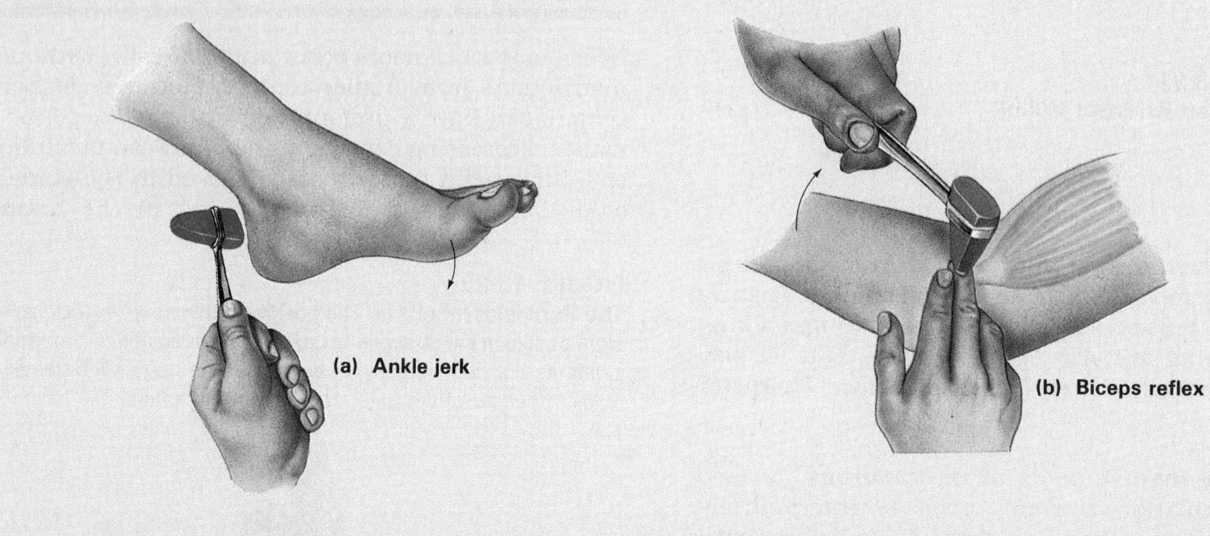

synapse on interneurons and motor neurons throughout the spinal cord. These synapses are continuously active, producing EPSPs or IPSPs at the postsynaptic membrane.

REINFORCEMENT AND INHIBITION

Synapses producing EPSPs make postsynaptic neurons more sensitive to other excitatory stimuli. The resulting level of generalized facilitation (Chapter 12) rises and falls, depending on the activity in higher centers. (p. 397) For example, an effort to pull apart clasped hands elevates the general state of facilitation along the spinal cord. This elevated facilitation leads to an enhancement, or **reinforcement**, of spinal reflexes.

Other descending fibers have an inhibitory effect on spinal reflexes. Stroking an infant's foot on the side of the sole produces a fanning of the toes known as the **Babinski sign**, or *positive Babinski reflex* (Figure 13-22a on page 429). This response disappears as descending inhibitory synapses develop. In the adult, the same stimulus produces a curling of the toes, called a **plantar reflex** or *negative Babinski reflex*, after about a 1–second delay (Figure 13-22b). If either the higher centers or the descending tracts are damaged, the Babinski sign will reappear. As a result, this reflex is often tested if CNS injury is suspected.

When higher centers issue motor commands, they can activate the complex motor patterns already programmed into the spinal cord. By making use of preexisting patterns, a relatively small number of descending fibers can control complex motor functions. For example, the motor patterns for walking, running, and jumping are primarily directed by neuronal *ceps reflexes* (Figure 13-23c) are stretch reflexes controlled by specific segments of the spinal cord. Testing these reflexes provides information about the corresponding spinal segments. For example, a normal patellar reflex, or knee jerk (Figure 13-16), indicates that spinal nerves and spinal segments L_2–L_4 are undamaged.

The Babinski sign is normally absent in the adult, due to descending spinal inhibition. The **abdominal reflex**, present in the normal adult, results from descending spinal facilitation. In this reflex, a light stroking of the skin produces a reflexive twitch in the abdominal muscles that moves the navel toward the stimulus (Figure 13-23d). This reflex disappears following damage to descending tracts.

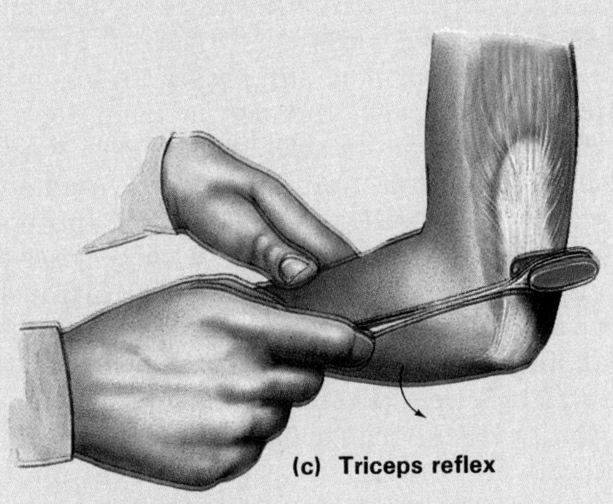

(c) Triceps reflex

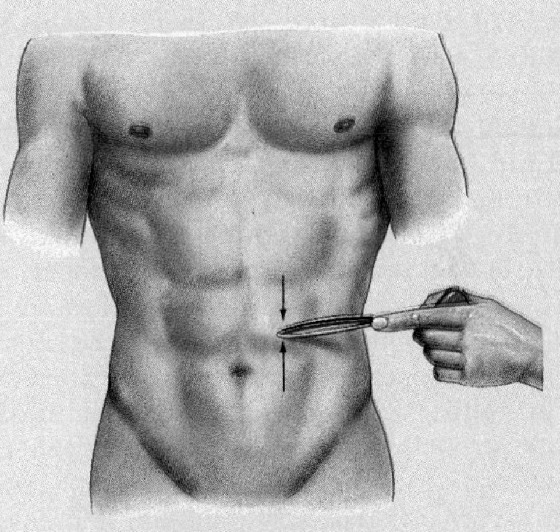

(d) Abdominal reflex

pools in the spinal cord. The descending pathways provide appropriate facilitation, inhibition, or "fine tuning" of the established patterns.

VOLUNTARY MOVEMENTS AND REFLEX MOTOR PATTERNS

When complicated voluntary movements are under way, spinal reflexes such as those we have described in this chapter assist with muscular coordination and control. For example, the descending tracts that stimulate the motor neurons responsible for bending the arm also (1) stimulate the gamma efferents to the intrafusal fibers (stretch reflex) and (2) stimulate the interneurons that inhibit antagonistic muscle groups (withdrawal and crossed extensor reflexes). As the muscle contraction proceeds, tendon reflexes adjust the tension produced.

Motor control therefore involves a series of interacting levels. At the lowest level are monosynaptic reflexes that are rapid but stereotyped and relatively inflexible. At the highest level are centers in the brain that can modulate or build upon reflexive motor patterns. [CM: *Abnormal Reflex Activity*]

- √ **One of the first somatic reflexes to develop is the suckling reflex. This is an example of what type of reflex?**
- √ **How would increased stimulation of the muscle spindles involved in the patellar (knee jerk) reflex by gamma motor neurons affect the speed of the reflex?**
- √ **After suffering an injury to his back, Tom exhibits a positive Babinski reflex. What does this imply about Tom's injury?**

Health News: Technology and Motor Paralysis

If a peripheral axon is damaged but not displaced, normal function may eventually return as the cut stump grows across the injury site away from the soma and along its former path. The mechanics of this process were detailed at the close of the last chapter. (p. 398)

In order for normal function to be restored, several things must happen. The severed ends must be relatively close together (1–2 mm), they must remain in proper alignment, and there must not be any physical obstacles between them, such as the dense fibers of scar tissue. These conditions can be created in the laboratory, using experimental animals and individual axons or small axonal bundles. But in accidental injuries to peripheral nerves the edges are likely to be jagged, intervening segments may be lost entirely, and elastic contraction in the surrounding connective tissues may pull the cut ends apart and misalign them.

Until recently the surgical response would involve trimming the injured nerve ends, neatly sewing them together, and hoping for the best. This procedure was often unsuccessful, in part because scalpels do not produce a smoothly cut surface and the thousands of broken axons would never be perfectly aligned. Moreover, nerve axons are not highly elastic, so if a large segment of the nerve was removed, crushed, or otherwise destroyed there would be no way to bring the intact ends close enough to permit any regeneration. In such instances a **nerve graft** could be inserted, using a section from some other peripheral nerve. The functional results were even less likely to be wholly satisfactory, for the growing axonal tips had to find their way across not one but two gaps, and the chances for successful alignment were proportionately smaller. Nevertheless, any return of function was certainly better than none at all!

Two very different research strategies are now being pursued. One focuses on the physical and biochemical control of nerve regeneration, while the other looks for a more "high-tech" solution. An example of a biological approach that focuses on physical factors is the use of a synthetic sleeve to guide nerve growth. The sleeve is a tube with an outer layer of silicone around an inner layer of cowhide collagen bound to the proteoglycans from shark cartilage. Using this sleeve as a guide, axons can grow across gaps as large as 20 mm (0.75 in.). The procedure has yet to be tried on humans, and functional restoration is incomplete because proper alignment does not always occur. Experiments on other mammals have successfully reunited the cut sections of large peripheral nerves.

A second biological line of investigation involves the biochemical control of nerve growth and regeneration. Neurons are influenced by a combination of growth promoters and growth inhibitors. Damaged myelin sheaths apparently release an inhibitory factor that slows the repair process. Researchers have made a monoclonal antibody, *IN–1*, that will inactivate the inhibitory factor released in the damaged spinal cords of rats. The treatment stimulates repairs, even in severed spinal cords.

In the meantime, other research teams are experimenting with the use of computers to stimulate specific muscles and muscle groups electrically. The technique is called *functional electrical stimulation*, or FES. This approach often involves implanting a network of wires beneath the skin with their tips in skeletal muscle tissue. The wires are connected to a computer that may be small enough to be worn at the waist. The wires deliver minute electrical stimuli to the muscles, depolarizing their membranes and causing contractions. With this equipment and lightweight braces, quadriplegics have walked several hundred yards and paraplegics several thousand.

Even more impressive results have been obtained using a network of wires woven into the fabric of close-fitting garments. This provides the necessary stimulation without the complications and maintenance problems that accompany implanted wires. A paraplegic woman in a set of electronic "hot pants" completed several miles of the 1985 Honolulu Marathon, and more recently a paraplegic woman walked down the aisle at her wedding. A commercial version of the system may be on the market in the near future.

Such technological solutions can provide only a degree of motor control without accompanying sensation. Everyone would prefer a biological procedure that would restore the functional integrity of the nervous system. For now, however, computer-assisted programs such as FES can improve the quality of life for thousands of paralyzed individuals.

CHAPTER REVIEW

Review of Selected Clinical Terms

spinal tap (*p. 409*)
A procedure in which fluid is extracted from the subarachnoid space through a needle inserted between the vertebrae.

lumbar puncture (*p. 409*)
A spinal tap performed between adjacent lumbar vertebrae.

myelography (*p. 409*)
A diagnostic procedure in which a radiopaque dye is introduced into the cerebrospinal fluid in order to obtain an X-ray of the spinal cord.

spinal shock (*p. 411*)
A period of sensory and motor paralysis following any severe injury to the spinal cord.

quadriplegia (*p. 411*)
Paralysis involving loss of sensation and motor control of the arms and legs.

paraplegia (*p. 411*)
Paralysis involving loss of motor control of the legs.

patellar reflex (*p. 425*)
The "knee jerk" reflex; often used to provide information about the related spinal segments.

Babinski sign (positive Babinski reflex) (*p. 430*)
A spinal reflex in infants, consisting of a fanning of the toes, produced by stroking the foot on the side of the sole; in adults, a sign of CNS injury.

plantar reflex (negative Babinski reflex) (*p. 430*)
A spinal reflex in adults, consisting of a curling of the toes, produced by stroking the foot on the side of the sole.

abdominal reflex (*p. 431*)
A reflexive twitch in abdominal muscles that moves the navel toward a stimulus; often used to provide information about descending tracts in the spinal cord.

Additional Terms of Clinical Importance

SPINAL CORD INJURIES (*p. 411*): **spinal concussion, spinal contusion, spinal laceration, spinal compression, spinal transection**

TECHNOLOGY AND MOTOR PARALYSIS (*p. 432*): **nerve graft**

Study Outline

Related Key Terms

Introduction (pp. 405–406)

1. The **central nervous system (CNS)** consists of the spinal cord and brain. In addition to relaying information to and from the brain, the spinal cord integrates and processes information on its own.

Gross Anatomy of the Spinal Cord (p. 406)

1. The adult spinal cord includes localized **enlargements** that provide innervation to the limbs. The spinal cord has 31 segments, each associated with a pair of **dorsal roots** and a pair of **ventral roots.**
2. The **filum terminale** (a strand of fibrous tissue) that originates at the **conus medullaris** ultimately becomes part of the coccygeal ligament.

Related key terms: **posterior median sulcus**, **anterior median fissure**, **cervical enlargement**, **lumbar enlargement**, **dorsal root ganglia**, **mixed nerves** • **spinal nerve**, **cauda equina**

Spinal Meninges (pp. 406–410)

1. The **spinal meninges** provide physical stability and shock absorption for neural tissues of the spinal cord; the **cranial meninges** surround the brain.

THE DURA MATER (p. 408)

2. The **dura mater** covers the spinal cord; caudally it tapers into the **coccygeal ligament.** The **epidural space** separates the dura mater from the walls of the vertebral canal.

THE ARACHNOID (p. 408)

3. Beneath the inner surface of the dura mater are the **subdural space,** the **arachnoid** (the second meningeal layer), and the **subarachnoid space.** The latter contains **cerebrospinal fluid,** which acts as a shock absorber and a diffusion medium for dissolved gases, nutrients, chemical messengers, and waste products.

THE PIA MATER (pp. 408–410)

4. The **pia mater,** a meshwork of elastin and collagen fibers, is the innermost meningeal layer. Unlike more superficial meninges, it is bound to the underlying neural tissue.

Related key terms: **denticulate ligaments**

Chapter Review

Sectional Anatomy of the Spinal Cord (p. 410)

Related Key Terms

1. The **white matter** contains myelinated and unmyelinated axons, while the **gray matter** contains cell bodies of neurons and glial cells. The projections of gray matter toward the outer surface of the spinal cord are called **horns.**

anterior median fissure
posterior median sulcus
central canal

ORGANIZATION OF GRAY MATTER (p. 410)

2. The **posterior gray horns** contain somatic and visceral sensory nuclei, while nuclei in the **anterior gray horns** are concerned with somatic motor control. The **lateral gray horns** contain visceral motor neurons. The **gray commissures** contain axons that cross from one side of the cord to the other.

ORGANIZATION OF WHITE MATTER (p. 410)

3. The white matter can be divided into six **columns,** each of which contains **tracts (fasciculi). Ascending tracts** relay information from the spinal cord to the brain, and **descending tracts** carry information from the brain to the spinal cord.

posterior white column
anterior white column
lateral white column
anterior white commissure

Spinal Nerves (pp. 410–417)

1. There are 31 pairs of spinal nerves. Each has an **epineurium** (outermost layer), **perineurium,** and **endoneurium** (innermost layer).

PERIPHERAL DISTRIBUTION OF SPINAL NERVES (pp. 412–413)

2. A typical spinal nerve has a **white ramus** (which contains myelinated axons), a **gray ramus** (containing unmyelinated fibers that innervate glands and smooth muscles in the body wall or limbs), a **dorsal ramus** (providing sensory and motor innervation to the skin and muscles of the back), and a **ventral ramus** (supplying the ventrolateral body surface, structures in the body wall, and the limbs). Each pair of nerves monitors a region of the body surface called a **dermatome.**

autonomic ganglion
rami communicantes
autonomic nerves

NERVE PLEXUSES (pp. 413–417)

3. A complex, interwoven network of nerves is called a **nerve plexus.** The three large plexuses are the **cervical plexus,** the **brachial plexus,** and the **lumbosacral plexus.** The latter can be divided into the **lumbar plexus** and the **sacral plexus.**

An Introduction to Reflexes (pp. 420–423)

1. A **neural reflex** is an automatic, involuntary motor response that helps preserve homeostasis by rapidly adjusting the functions of organs or organ systems.
2. A **reflex arc** is the neural "wiring" of a single reflex.

THE REFLEX ARC (pp. 420–421)

3. There are five steps involved in a neural reflex: (1) arrival of a stimulus and activation of a receptor; (2) activation of a sensory neuron; (3) information processing; (4) activation of a motor neuron; (5) response by an effector.
4. A **receptor** is a specialized cell that monitors conditions in the body or external environment. Each receptor has a characteristic range of sensitivity.

CLASSIFICATION OF REFLEXES (pp. 421–423)

5. **Innate reflexes** result from the connections that form between neurons during development. **Acquired reflexes** are learned, and often more complex.
6. Reflexes processed in the brain are **cranial reflexes.** In a **spinal reflex** the important interconnections and processing occur inside the spinal cord.
7. **Somatic reflexes** control skeletal muscles, and **visceral (autonomic) reflexes** control the activities of other systems.
8. A **monosynaptic reflex** is the simplest reflex arc, in which a sensory neuron synapses directly on a motor neuron which acts as the processing center. **Polysynaptic reflexes,** which have at least one interneuron placed between the sensory afferent and the motor efferent, have a longer delay between stimulus and response.

Spinal Reflexes (pp. 423–429)

1. Spinal reflexes range from simple monosynaptic reflexes to more complex polysynaptic reflexes. **Intersegmental reflex arcs,** in which many segments interact to produce a coordinated motor response, are the most complicated spinal reflexes.

MONOSYNAPTIC REFLEXES (pp. 423–426)

2. The **stretch reflex** is a monosynaptic reflex that automatically regulates skeletal muscle length and muscle tone. The sensory receptors involved are **muscle spindles.**
3. A **postural reflex** maintains normal upright posture.

POLYSYNAPTIC REFLEXES (pp. 426–429)

5. **Polysynaptic reflexes** can produce more complicated responses. Examples include the **tendon reflex** (which monitors the tension produced during muscular contractions and prevents damage to tendons), **withdrawal reflexes** (which move affected portions of the body away from a source of stimulation; the **flexor reflex** is a withdrawal reflex affecting the muscles of a limb), and the **crossed extensor reflex** (which complements the withdrawal reflex).
6. In an **ipsilateral reflex arc** the sensory stimulus and motor response occur on the same side of the body. In a **contralateral reflex arc** the motor response occurs on the side opposite the stimulus.
7. All polysynaptic reflexes share the following traits: (1) they involve pools of interneurons; (2) they are intersegmental in distribution; (3) they involve reciprocal innervation; (4) they have reverberating circuits that prolong the reflexive motor response; and (5) several reflexes may cooperate to produce a coordinated response.

Related Key Terms

intrafusal muscle fibers
extrafusal muscle fibers
nuclear bag
gamma motor neurons
gamma efferents
Golgi tendon organs
reciprocal inhibition

Integration and Control of Spinal Reflexes (pp. 429–431)

1. The brain can facilitate or inhibit reflex motor patterns based in the spinal cord.

REINFORCEMENT AND INHIBITION (pp. 430–431)

2. The enhancement of spinal reflexes is called **reinforcement.**

VOLUNTARY MOVEMENTS AND REFLEX MOTOR PATTERNS (p. 431)

3. Motor control involves a series of interacting levels; monosynaptic reflexes form the lowest level, while at the highest level are the centers in the brain that can modulate or build upon reflexive motor patterns.

Babinski sign
plantar reflex

Review Planner

	Objective	Level 1
1	Discuss the structure and functions of the spinal cord.	**6 8 10 12 14 15 18 21 25 36 42 43 47 49 50**
2	Explain the role of white matter and gray matter in processing and relaying sensory and motor information.	**10 14 25 42 43 50**
3	Describe the major components of a spinal nerve.	**18 24 27 37**
4	Relate spinal nerves to the regions that they innervate.	**11 22 28–35 45**
5	Describe the process of a neural reflex.	**16 17 20 23 39**
6	Classify the different types of reflexes.	**2 3 4 9 16 23 30 40 46**
7	Distinguish between the types of motor responses produced by various reflexes.	**2 3 5 17 20**

Level 2: **41 48 50**

Level 3: **51–54**

C M: **Shingles • Spinal Meningitis**
Spinal Anesthesia
Multiple Sclerosis
Palsies • Leprosy
Abnormal Reflex Activity

ABC NEWS: **Treatment For New Spinal Cord Injuries**

Chapter Review

8	Explain how reflexes combine to produce complex behaviors.	**5 9 23**
9	Explain how higher centers control and modify reflex responses.	**13**

Review Questions

MULTIPLE CHOICE

1. The spinal cord consists of ________ segments: a) 10; b) 23; c) 31; d) 45.
2. You touch a hot soldering iron and jerk your hand away. What type of reflex is this? a) innate; b) acquired; c) plantar; d) cranial.
3. In question 2, what type of motor response was involved? a) visceral reflex; b) autonomic reflex; c) somatic reflex; d) postural reflex.
4. The stretch reflex is an example of a: a) monosynaptic reflex; b) polysynaptic reflex.
5. The tendon reflex: a) regulates the length of a skeletal muscle; b) moves portions of the body away from a stimulus; c) maintains upright posture; d) adjusts the tension produced during a muscular contraction.
6. The spinal cord typically stops growing at approximately what age? a) birth; b) 4 years; c) puberty; d) adulthood.
7. A specialized cell that monitors conditions in the body or the external environment is called a: a) nuclear bag; b) dermatome; c) receptor; d) pia-arachnoid.
8. The projections of gray matter toward the outer surface of the spinal cord are called: a) horns; b) denticulate ligaments; c) arachnoids; d) white matter.
9. The flexor reflex is an example of a: a) tendon reflex; b) withdrawal reflex; c) stretch reflex; d) postural reflex.
10. The regions within the white matter of the spinal cord are called: a) spinal meninges; b) spinal reflexes; c) dermatomes; d) columns.

TRUE/FALSE

(If a statement is false, your instructor may wish to have you correct it.)

11. **T/F:** The diaphragm is innervated by a branch of the brachial plexus.
12. **T/F:** The only function of the spinal cord is to convey information to and from the brain.
13. **T/F:** In adulthood, the Babinski sign is inhibited by higher centers.

MATCHING QUESTIONS

(Match each term with the most appropriate statement.)

Statements:

14. Areas of gray matter in the spinal cord that contain somatic and visceral sensory nuclei.
15. The outermost meningeal covering of the spinal cord.
16. A reflex arc in which a sensory neuron synapses directly on a motor neuron.
17. A reflex arc in which the sensory stimulus and motor response occur on the same side of the body.
18. The innermost layer surrounding a spinal nerve.
19. A fluid that circulates within the subarachnoid space and acts as a shock absorber.
20. A reflex arc in which the motor response occurs on the side opposite the stimulus.
21. The innermost meningeal layer in contact with the spinal cord.
22. A specific region of the body surface that is monitored by a particular pair of spinal nerves.
23. A reflex arc in which at least one interneuron is placed between the sensory neuron and the motor neuron.
24. The outermost layer surrounding a spinal nerve, consisting of collagen fibers.
25. Areas of gray matter in the spinal cord that contain visceral motor nuclei.
26. A series of specialized membranes that surround the brain to stabilize it and absorb shocks.
27. The middle layer of connective tissue surrounding a spinal nerve that divides the nerve into compartments.

Terms:

a) Cranial meninges
b) Dura mater
c) Pia mater
d) Posterior gray horns
e) Lateral gray horns
f) CSF
g) Endoneurium
h) Perineurium
i) Epineurium
j) Dermatome
k) Polysynaptic reflex
l) Monosynaptic reflex
m) Ipsilateral reflex arc
n) Contralateral reflex arc

SHORT ANSWER/ESSAY

Which ramus would be involved if you felt something brushing against your:

28. Abdomen.
29. Left forearm.
30. Upper back.
31. Right ankle.

Which plexus has primary responsibility for innervating the following muscles?

32. Diaphragm.
33. Pelvic diaphragm.
34. Deltoid.
35. Gastrocnemius.

36. A nurse refers to an adult patient's spinal cord injuries as injuries to L_3 and C_3 segments. Where are these injuries located in relation to the vertebral column?
37. What are mixed nerves? Give an example.
38. Discuss five characteristics common to all polysynaptic reflexes.
39. Place these steps in the correct sequence. Identify the process that is being described:
    a) Activation of a motor neuron.
    b) Activation of a sensory neuron.
    c) Activation of a receptor.
    d) Response of a skeletal muscle.
    e) Processing of information.
40. What is a stretch reflex, and what purpose(s) does it serve? Identify the sensory receptors involved.
41. While driving one day, you slam on the brakes quickly when another car cuts in front of you. Identify the reflex involved in terms of its development, the nature of the resulting response, and neural circuits involved.
42. Differentiate between gray matter and white matter.
43. Fill in the blanks: ________ tracts carry sensory information to the brain, while ________ tracts carry motor commands to the spinal cord.
44. Where is cerebrospinal fluid found, and what are its functions?
45. Your Anatomy & Physiology instructor asks a question. You don't know the answer, so you shrug your shoulders. Identify the peripheral nerves that enabled you to do this.
46. You lean forward to look at something more closely and then straighten up. Identify the type of reflex that enables you to maintain your upright position.
47. Identify the spinal segments involved in the lumbar plexus and sacral plexus.
48. Ms. Shay slips and falls on her back. Discuss clues that can be used to assess the location of any injuries to her spinal cord.
49. Why do cervical nerves outnumber cervical vertebrae?
50. Discuss the organization of the white matter in the spinal cord. How is information conveyed?

CRITICAL THINKING/APPLICATIONS

51. Two young men are involved in an automobile accident. Six months later one has lost motor control of his legs, while the other has lost sensation and motor control of his arms and legs. What are the terms of these conditions, and what areas of the spinal cord are involved?
52. Mr. Rouse comes to the emergency room complaining of bilateral arm and shoulder weakness. What portion of the spinal cord may be affected?
53. Discuss diagnostic techniques that could be employed to isolate the cause of Mr. Rouse's problem. What conditions can be revealed with these techniques?
54. A patient is brought to the emergency room after a work-related accident. The physician who examines him mentions the terms "spinal concussion" and "spinal contusion." Distinguish between these terms. Which is more serious?

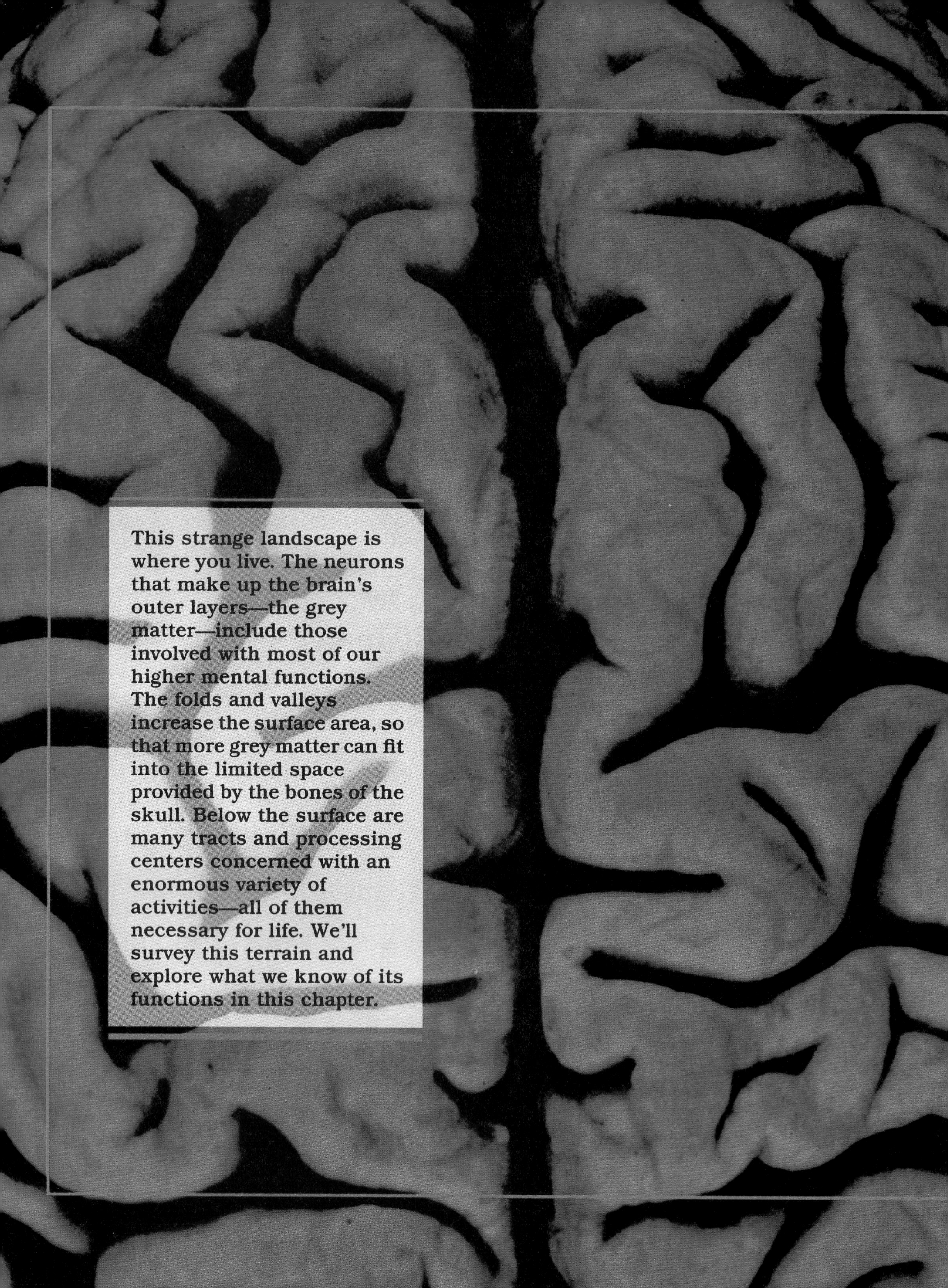
This strange landscape is where you live. The neurons that make up the brain's outer layers—the grey matter—include those involved with most of our higher mental functions. The folds and valleys increase the surface area, so that more grey matter can fit into the limited space provided by the bones of the skull. Below the surface are many tracts and processing centers concerned with an enormous variety of activities—all of them necessary for life. We'll survey this terrain and explore what we know of its functions in this chapter.

CHAPTER 14

The Nervous System: The Brain and Cranial Nerves

Chapter Objectives

After reading this chapter, you will be able to:

1. Name the major regions of the brain and describe their functions.
2. Identify important structures within each region and explain their functions.
3. Distinguish between motor, sensory, and association areas of the cerebral cortex.
4. Explain the role of "higher centers."
5. Explain how the brain is protected.
6. Discuss the circulation and functions of cerebrospinal fluid.
7. Identify the cranial nerves and relate each pair of cranial nerves to its principal functions.
8. Discuss important cranial reflexes.

Introduction

Previous chapters dealing with the nervous system have paved the way for our discussion of the brain. One cannot help but be a bit intimidated at the start of such a project. The brain contains tens of billions of neurons organized into hundreds of neuronal pools; it has a complex three-dimensional structure and performs a bewildering array of functions. Hardly anyone would deny that the brain is the most fascinating organ in the body. All of our dreams, passions, plans, and memories are the result of brain activity. If a heart, liver, lung, or kidney stops working, heroic measures, including organ transplantation, are taken to preserve the life of the individual. But if the brain permanently stops working, so that the person is classified as "brain dead," for all intents and purposes that individual ceases to exist.

This chapter begins an exploration of brain function that continues through the next three chapters. Our discussion is based on a simple concept: despite the complexity and versatility of the brain, the basic principles of neural activity and information processing introduced in Chapters 12 and 13 apply to the brain as well as the spinal cord.

The brain is far more complex than the spinal cord, and it can respond to stimuli with greater versatility. That versatility results from (1) the tremendous number of neurons and neuronal pools in the brain and (2) the complexity of the interconnections between these neurons and neuronal pools. The brain contains roughly 35 billion neurons, each of which may receive information across as many as 80,000 synapses at one time. The neurons are organized into pools with extensive interconnections. Excitatory and inhibitory interactions between these pools ensure that the response can vary to meet changing circumstances. But adaptability has a price. A response cannot be immediate, precise, and adaptable all at the same time. Adaptability means multiple processing steps, and every synapse adds to the delay between stimulus and response. As we saw in the previous chapter, one of the major functions of spinal reflexes is to provide an *immediate* response that can be fine-tuned or elaborated on by more versatile but slower processing centers in the brain.

We now begin a detailed examination of the brain. This chapter focuses attention on the major structures of the brain and their relationships with the cranial nerves.

■ Gross Anatomy and Organization

The adult human brain contains almost 98 percent of the neural tissue in the body. A "typical" brain weighs 1.4 kg (3 lb) and has a volume of 1200 cc (71 in.^{3}). There is considerable individual variation, and the brains of males average about 10 percent larger than those of females, owing to differences in average body size. No correlation exists between brain size and intelligence, and individuals with the smallest (750 cc) and largest (2100 cc) brains are functionally normal.

Despite the differences in size and weight, all human brains share characteristic anatomical landmarks. These landmarks are quite useful as reference points in discussions of brain function. As you study the descriptions that follow, keep in mind that early anatomists devoted considerable time and attention to the description of the brain, and almost every major structure has had a commemorative title at some time in the past. Although attempts are being made to discourage the creation and use of eponyms, many texts continue to rely on them. For your reference the most common eponyms associated with the brain are listed in a table preceding the Glossary.

MAJOR DIVISIONS AND PROMINENT LANDMARKS

There are six major regions in the adult brain: (1) the *cerebrum*, (2) the *diencephalon*, (3) the *mesencephalon*, or midbrain, (4) the *cerebellum*, (5) the *pons*, and (6) the *medulla oblongata*. Figure 14-1a,b introduces these regions and their general functions.

Figure 14-1a presents a superficial view of the adult brain (for divisions in the embryonic brain, see the Embryology Summary on pp. 454–455). Viewed from the superior surface, the **cerebrum** (SER-e-brum) can be divided into large, paired **cerebral hemispheres**. Conscious thought processes, sensations, intellectual functions, memory storage and retrieval, and complex motor patterns originate in the cerebrum. Immediately behind the cerebrum are the somewhat smaller hemispheres of the **cerebellum** (ser-e-BEL-um). The cerebellum adjusts voluntary and involuntary motor activities on the basis of sensory information and stored memories of previous movements.

Other regions of the brain can best be examined after removing the cerebrum and cerebellum. What remains (the diencephalon, mesencephalon, pons, and medulla oblongata) is called the **brain stem**. The brain stem contains a variety of important processing centers and also provides relay stations for information headed to or from the cerebrum or cerebellum.

Landmarks on the brain stem are shown in Figure 14-1b.

- The portion of the brain stem attached to the cerebrum is the **diencephalon** (di-en-SEF-a-lon; *dia*, through). The walls of the diencephalon form the **thalamus**, which contains relay and processing centers for sensory information. A narrow stalk connects the floor of the diencephalon, or **hypothalamus** (*hypo-*, below), to the **pituitary gland**. The hypothalamus contains centers involved with emotions, autonomic function, and hormone production. The pituitary gland is the primary link between the nervous and endocrine systems.
- Nuclei in the **midbrain**, or **mesencephalon** (mez-en-SEF-a-lon; *meso*, middle) process visual and auditory information and generate involuntary somatic motor responses. This region also contains centers involved with the maintenance of consciousness.
- The term **pons** refers to a bridge, and the pons of the brain connects the cerebellum to the brain stem. In addition to tracts and relay centers, this region of the brain also contains nuclei involved with somatic and visceral motor control.
- The spinal cord connects to the brain at the **medulla oblongata**, or simply the *medulla*. The superior portion of the medulla has a thin, membranous roof, but the caudal portion resembles

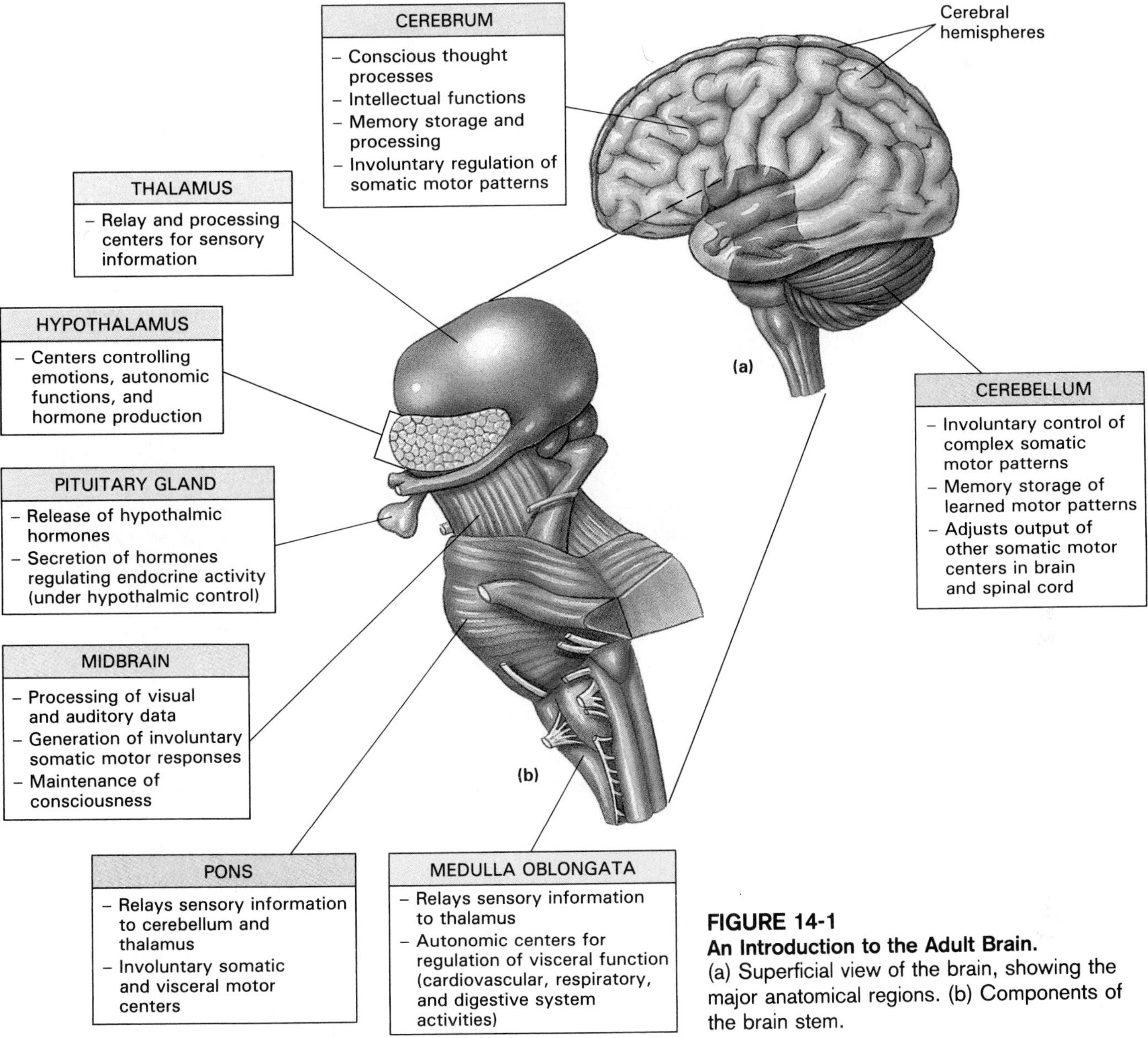

FIGURE 14-1
An Introduction to the Adult Brain.
(a) Superficial view of the brain, showing the major anatomical regions. (b) Components of the brain stem.

the spinal cord. The medulla relays sensory information to the thalamus and other brain stem centers; it also contains major centers concerned with the regulation of autonomic function, such as heart rate, blood pressure, respiration, and digestive activities.

VENTRICLES OF THE BRAIN

The brain is hollow, like the spinal cord, and has a central passageway filled with cerebrospinal fluid. Inside the cerebral hemispheres, diencephalon, pons, and medulla this passageway expands to form chambers called **ventricles** (VEN-tri-kls). Figure 14-2a shows the position of the ventricles in an intact brain, and Figure 14-2b-d includes sectional views indicating the orientations of the ventricles in the major brain regions.

Each cerebral hemisphere contains an enlarged ventricular chamber (Figure 14-2b). A thin partition, the **septum pellucidum**, separates this pair of **lateral ventricles**. There is no direct connection between the two lateral ventricles, but each communicates with the ventricle of the diencephalon through the **interventricular foramen** (*foramen of Munro*). Because there are two lateral ventricles, the diencephalic chamber is called the **third ventricle**.

Instead of a ventricle, the mesencephalon has a slender canal known as the **mesencephalic aqueduct** (*aqueduct of Sylvius* or *cerebral aqueduct*). This passageway, seen in section in Figure 14-2c, connects the third ventricle with the **fourth ventricle** of the pons and upper portion of the medulla oblon-

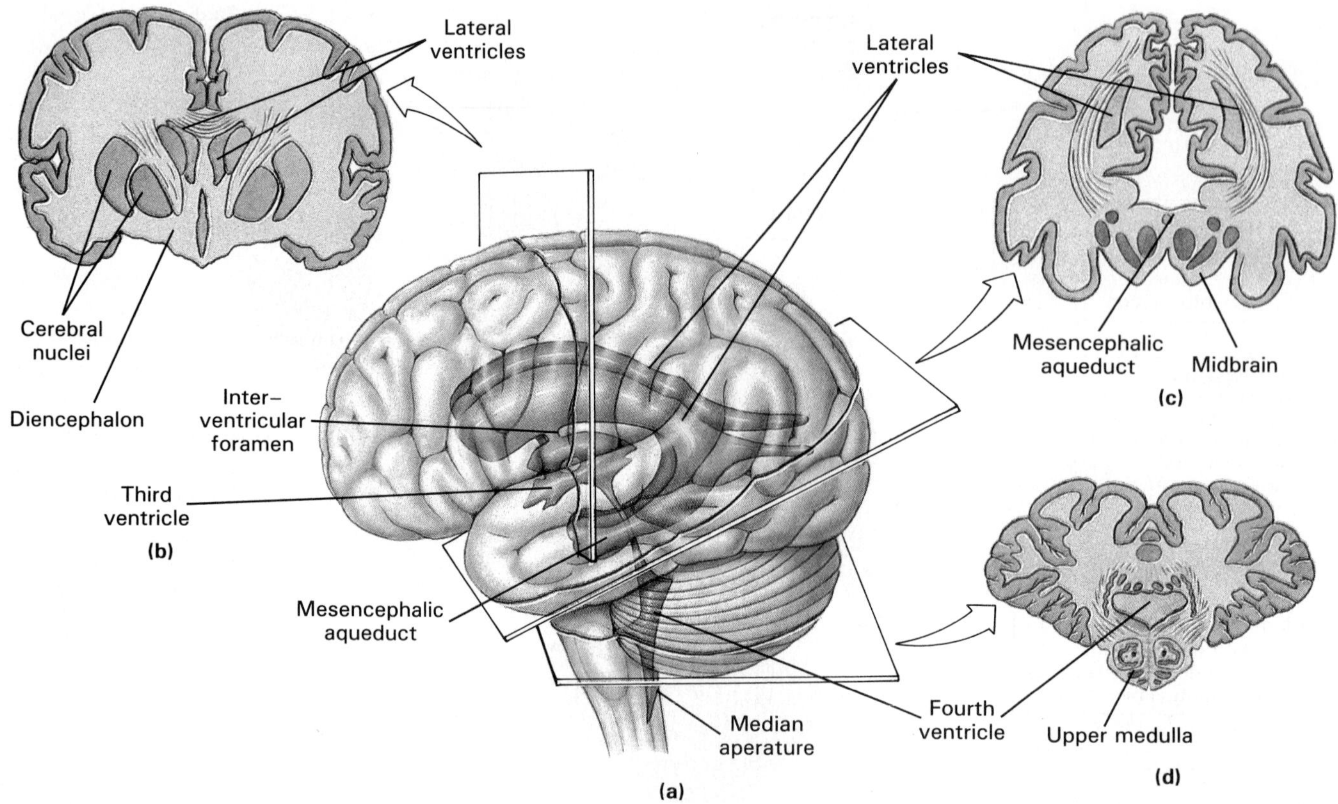

FIGURE 14-2
Ventricles of the Brain. (a) Orientation and extent of the ventricles as seen through a transparent brain. (b) The two lateral ventricles of the cerebral hemispheres and the third ventricle in the diencephalon. (c) The mesencephalic aqueduct connects the third and fourth ventricles. (d) The fourth ventricle extends between the cerebellum and pons and continues over the superior portion of the medulla.

gata (Figure 14-2d). In the caudal half of the medulla the fourth ventricle narrows and becomes continuous with the central canal of the spinal cord.

The ventricles are normally filled with cerebrospinal fluid. There is a continuous circulation of cerebrospinal fluid from the ventricles and central canal into the *subarachnoid space* of the meninges that surround the CNS. Details of the circulatory pattern and its clinical implications will be found in a later section.

DISTRIBUTION OF GRAY AND WHITE MATTER

Figure 14-2 also diagrams the general organization of gray and white matter inside the brain. In the brain stem the distribution of gray matter resembles that in the spinal cord. *Nuclei* are clustered around the ventricles, surrounded by *tracts* of white matter. However, the arrangement of tracts and nuclei is not as regular as that in the spinal cord. Tracts cross, merge, and branch, sometimes passing around or through the nuclei in their path. As a result, the nuclei in the brain stem lack the clear functional organization of nuclei in the spinal cord, with its dorsal sensory nuclei and ventral motor nuclei.

The brain also contains extensive areas of **neural cortex** (*cortex*, rind). Neural cortex is a layer of gray matter found on the surface of the cerebrum. A blanket of neural cortex covers the cerebral and cerebellar hemispheres, and tracts of white matter connect these cortical areas with other processing centers in the brain. As noted in Chapter 12, the brain and spinal cord have nuclei and *centers* where processing occurs. (p. 404) The term *higher centers* refers to nuclei, centers, and cortical areas of the cerebrum, cerebellum, and upper brain stem (diencephalon and mesencephalon). Output from these processing centers modifies the activities of nuclei and centers in the lower brain stem and spinal cord. The nuclei and cortical areas of the brain can receive sensory information and issue motor commands to peripheral effectors indirectly, via the spinal cord and spinal nerves, or directly, via the cranial nerves.

We will now examine each of the six regions of the brain, beginning with the cerebrum and moving toward the medulla oblongata.

■ The Cerebrum

The cerebrum is the largest region of the brain. Conscious thought processes and all intellectual functions originate in the cerebral hemispheres. Much of the cerebrum is involved in the processing of somatic sensory and motor information. Somatic sensory information relayed to the cerebrum reaches our conscious awareness, and cerebral neurons exert voluntary or involuntary control over somatic motor neurons. Most visceral sensory processing and visceral motor (autonomic) control, by contrast, occur at centers elsewhere in the brain, usually outside our conscious awareness.

The cerebrum includes gray matter and white matter. Gray matter is found in a superficial layer of neural cortex and in deeper *cerebral nuclei*. The *central white matter* lies beneath the neural cortex and around the cerebral nuclei.

THE CEREBRAL HEMISPHERES

Figure 14-3 is a superficial view of the cerebrum. A thick layer of neural cortex covers the paired **cerebral hemispheres** that form the superior surface of the cerebrum. The cortical surface forms a series of elevated ridges, or **gyri** (JĪ-re), separated by shallow depressions, called **sulci** (SUL-kē), or deeper grooves, called **fissures**. Gyri increase the surface area of the cerebral hemispheres and the number of neurons in the cortical regions. The total surface area of the cerebral hemispheres is roughly equivalent to 2.25 square meters (25 sq. ft) of flat surface, and that large an area can be packed into the skull only when folded, like a crumpled piece of paper. Complex analytical and integrative functions require large numbers of neurons; even with extensive folding of the cerebral hemispheres, the size of the cranium has had to enlarge greatly in the course of human evolution to accommodate the modern human cerebrum.

Boundaries between Cerebral Lobes

The two cerebral hemispheres are separated by a deep **longitudinal fissure**, and each hemisphere can be divided into **lobes** named after the overlying bones of the skull. There are individual differences in the appearance of the sulci and gyri of each brain, but the boundaries between lobes are reliable landmarks. A deep groove, the **central sulcus**, extends laterally from the longitudinal fissure. The area anterior to the central sulcus is the **frontal lobe**, and the **lateral sulcus** marks its inferior border. The cortex inferior to the lateral sulcus is the **temporal lobe**. Pushing this lobe to the side (as in Figure 14-3) exposes the **insula** (IN-su-la), an "island" of cortex that is otherwise invisible. The **parietal lobe** extends between the central sulcus and the **parieto-occipital sulcus**. What remains constitutes the **occipital lobe**.

Each lobe contains functional regions whose boundaries are less clearly defined. Some of these regions are concerned with sensory information and others with motor commands. Three points should be kept in mind as this discussion proceeds:

1. Each cerebral hemisphere receives sensory information and generates motor commands that concern the opposite side of the body. The left hemisphere controls the right side, and the right hemisphere controls the left side. This crossing over has no known functional significance.
2. Although the two hemispheres are very similar in appearance, there are significant functional differences between them. These differences primarily affect higher-order functions, the topic of Chapter 15, but some examples will be noted below.
3. The assignment of a specific function to a region of the cerebral cortex is at best imprecise. The boundaries are indistinct, with considerable overlap; any one region may have several different functions. In addition, some aspects of cortical function, such as consciousness, cannot easily be assigned to any single region. [**CM**: *Phrenology*]

Motor and Sensory Areas of the Cortex

Figure 14-3 and Table 14-1 detail the major motor and sensory regions of the cerebral cortex. The central sulcus separates the motor and sensory portions of the cortex. The **precentral gyrus** of the frontal lobe forms the anterior margin of the central sulcus. The surface of this gyrus is the **primary motor cortex**. Neurons of the primary motor cortex direct voluntary movements by controlling somatic motor neurons in the brain stem and spinal cord. These neurons are called **pyramidal cells**, and the pathway that provides voluntary motor control is known as the **pyramidal system**.

The **postcentral gyrus** of the parietal lobe forms the posterior margin of the central sulcus, and its surface contains the **primary sensory cortex**. Neurons in this region receive somatic sensory information from touch, pressure, pain, taste, and temperature receptors. We are consciously aware of these sensations because the brain stem nuclei relay sensory information to the primary sensory cortex. This is an example of *serial processing*, a functional arrangement introduced in Chapter 12. (p. 402) *Parallel processing* occurs at the same time, as collaterals deliver information to the cerebral nuclei and other centers. Because both serial and parallel processing occur, sensory information is monitored at both conscious and unconscious levels.

TABLE 14-1 The Cerebral Cortex

Region/Nucleus	*Functions*
Primary motor cortex	Voluntary control of skeletal muscles
Primary sensory cortex	Conscious perception of touch, pressure, vibration, pain, temperature, and taste
Visual cortex (occipital lobe); **auditory cortex and olfactory cortex** (temporal lobe)	Conscious perception of visual, auditory, and olfactory stimuli
Association areas	Integration and processing of sensory data; processing and initiation of motor activities

FIGURE 14-3
The Cerebral Hemispheres. Major anatomical landmarks on the surface of the left cerebral hemisphere. To expose the insula, the lateral sulcus has been opened. The insert shows the multiple layers of neurons that are characteristic of the cerebral cortex.

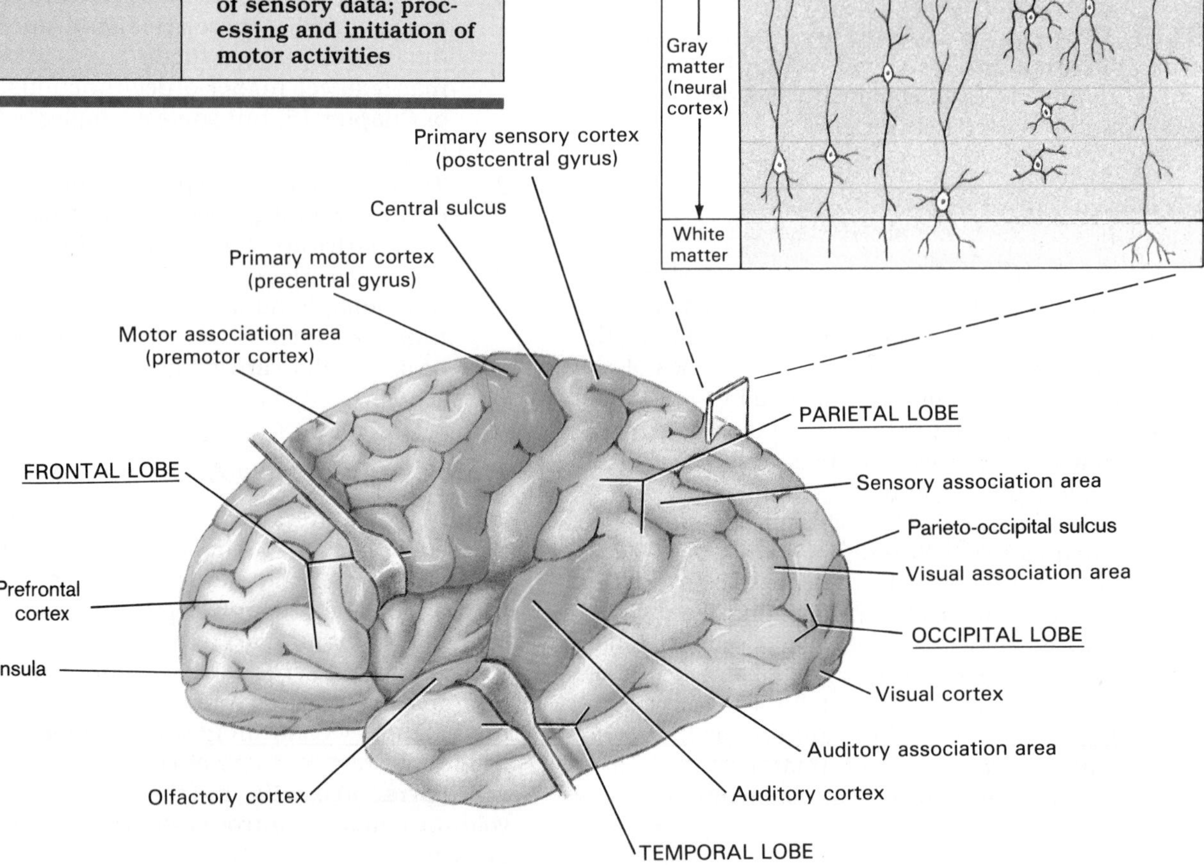

Sensory information concerning sensations of sight, sound, and smell arrive at other portions of the cerebral cortex. The **visual cortex** of the occipital lobe receives visual information, and the **auditory cortex** and **olfactory cortex** of the temporal lobe receive information concerned with hearing and smell respectively.

Association Areas

The sensory and motor regions of the cortex are connected to nearby **association areas** that interpret incoming data or coordinate a motor response. The **somatic motor association area**, or **premotor cortex**, is responsible for the coordination of learned motor responses. The functional distinctions between the sensory and motor association areas are most evident after localized brain damage has occurred. For example, an individual with a damaged **visual association area** sees letters quite clearly, but is unable to recognize or interpret them. This person scans the lines of a printed page and sees rows of clear symbols that convey no meaning. Someone with damage to the area of the premotor cortex concerned with coordination of eye movements can understand written letters and words, but cannot read because his or her eyes cannot track along the lines on a printed page.

Integrative Centers

"Higher-order" integrative centers receive information from many different association areas. These regions direct extremely complex motor activities and perform complicated analytical functions. For example, the **prefrontal cortex** of the frontal lobe integrates information from sensory association areas and performs abstract intellectual functions, such as predicting the consequences of possible responses.

These lobes and cortical areas are found on both cerebral hemispheres. Higher-order integrative centers concerned with complex processes, such as speech, writing, mathematical computation, or understanding spatial relationships, are restricted to the left or right hemisphere. These centers and their functions will be considered in Chapter 15.

THE CENTRAL WHITE MATTER

The **central white matter** contains three major groups of axons: (1) *association fibers*, tracts that interconnect areas of neural cortex within a single cerebral hemisphere; (2) *commissural fibers*, tracts that connect the two cerebral hemispheres; and (3) *projection fibers*, tracts that link the cerebrum with other regions of the brain and the spinal cord. These relationships are illustrated in Figure 14-4 and listed in Table 14-2.

- Association fibers interconnect portions of the cerebral cortex. The shortest association fibers are called **arcuate** (AR-ku-at) **fibers** because they curve in an arc to pass from one gyrus to another. The longer association fibers are organized into discrete bundles. The **longitudinal fasciculi** connect the frontal lobe to the other lobes of the same hemisphere.
- A dense band of **commissural fibers** (kom-MI-su-ral; *commissura*, a crossing over) permit communication between the two hemispheres. Prominent commissural bundles include the **corpus callosum** and the **anterior commissure**.
- **Projection fibers** link the cerebral cortex to the brain stem, cerebellum, and spinal cord. All must pass through the diencephalon, where the converging fibers heading to sensory areas of the cortex run past the diverging fibers leaving the motor cortex. In gross dissection the afferent fibers and efferent fibers look alike, and the entire array is known as the **internal capsule**.

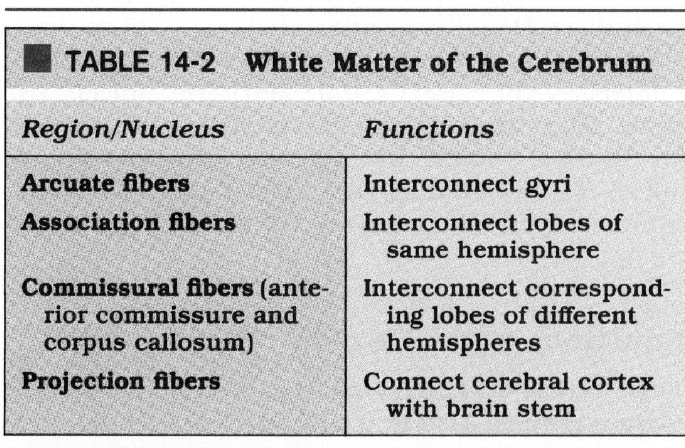

TABLE 14-2 White Matter of the Cerebrum

Region/Nucleus	*Functions*
Arcuate fibers	Interconnect gyri
Association fibers	Interconnect lobes of same hemisphere
Commissural fibers (anterior commissure and corpus callosum)	Interconnect corresponding lobes of different hemispheres
Projection fibers	Connect cerebral cortex with brain stem

THE CEREBRAL NUCLEI

The **cerebral nuclei** lie within each hemisphere beneath the floor of the lateral ventricle, as indicated in Figure 14-5. They sit within the central white

FIGURE 14-4
The Central White Matter. (a) Lateral aspect of the brain, showing arcuate fibers and longitudinal fasciculi. (b) Anterior view of the brain, showing orientation of the commissural and projection fibers.

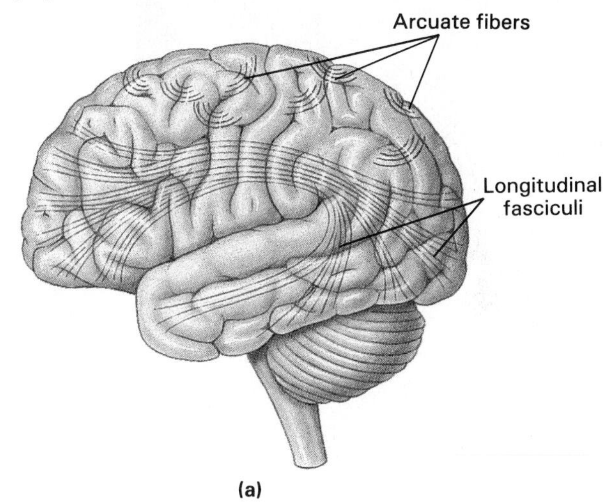

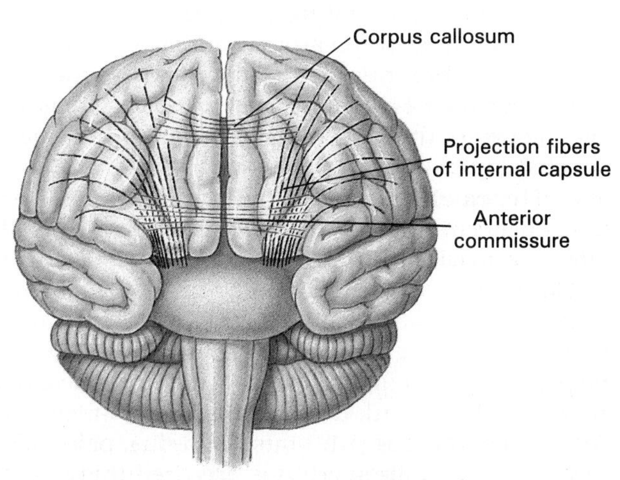

TABLE 14-3 The Cerebral Nuclei

	Region/Nucleus	*Functions*
	Amygdaloid body	Component of limbic system
Corpus striatum	**Caudate nucleus** **Lentiform nucleus: putamen and globus pallidus**	Involuntary adjustment and modification of voluntary motor commands
	Claustrum	Plays uncertain role in processing of visual information

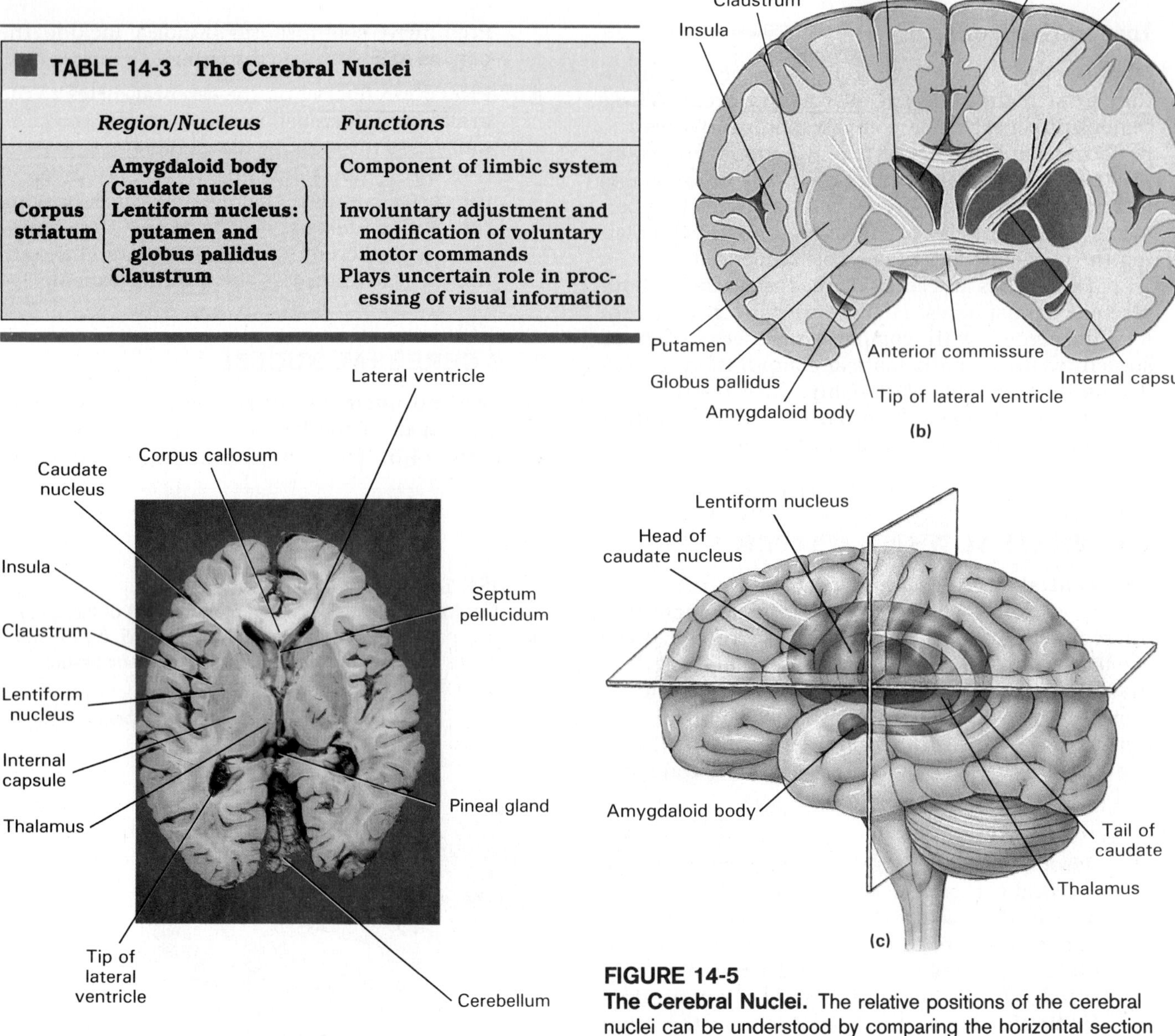

FIGURE 14-5
The Cerebral Nuclei. The relative positions of the cerebral nuclei can be understood by comparing the horizontal section (a) and the frontal section (b) with the three-dimensional representation (c).

matter, and the radiating projection and commissural fibers must travel around or between these nuclei. The cerebral nuclei are also known as the **basal nuclei**, or *basal ganglia*. The latter term has persisted despite the fact that ganglia are otherwise restricted to the peripheral nervous system.

The **caudate nucleus** has a massive head and a slender, curving tail that follows the curve of the lateral ventricle. At the tip of the tail there is a separate nucleus, the **amygdaloid body** (ah-MIG-da-loyd; *amygdale*, almond). Three masses of gray matter lie between the bulging surface of the insula and the lateral wall of the diencephalon. These are the **claustrum** (KLAWS-trum), the **putamen** (pū-TĀ-men), and the **globus pallidus** (GLŌ-bus PAL-i-dus; pale globe).

Several additional terms are used to designate specific anatomical or functional subdivisions of the cerebral nuclei. The putamen and globus pallidus are often considered as subdivisions of a larger **lentiform** (lens-shaped) **nucleus**, for when exposed on gross dissection they form a rather compact, rounded mass. The term **corpus striatum** (striated body) encompasses the caudate *and* lentiform nuclei. The name refers to the striated (striped) appearance of the internal capsule as it passes between the two nuclei. Table 14-3 diagrams these relationships and summarizes the functions of the cerebral nuclei.

Functions of the Cerebral Nuclei

The cerebral nuclei are important components of the **extrapyramidal system**, a pathway that controls muscle tone and coordinates learned movement patterns and other somatic motor activities. These nuclei are processing centers whose functions blur the distinctions between "conscious" and "unconscious" motor control. Under normal conditions these nuclei do not *initiate* a particular movement. But once a movement is under way the cerebral nuclei provide the general pattern and rhythm. For example, when walking, the caudate nucleus and putamen control the cycles of arm and leg movements that occur between the time the decision is made to "start walking" and the time the "stop" order is given.

The globus pallidus controls and adjusts muscle tone and sets body position in preparation for a voluntary movement. For example, when you decide to pick up a pencil, the globus pallidus positions the shoulder and stabilizes the arm as you consciously reach and grasp.

The functions of other cerebral nuclei are poorly understood. The claustrum appears to be involved in the unconscious processing of visual information. The amygdaloid body is an important component of the *limbic system.*

THE LIMBIC SYSTEM

The **limbic system** (LIM-bik; *limbus,* a border) includes nuclei and tracts along the border between the cerebrum and diencephalon. The functions of the limbic system involve emotional states and related behavioral drives. This system also provides a link between the conscious, intellectual functions of the cerebral cortex and the unconscious and autonomic functions of the brain stem. Table 14-4 and Figure 14-6 summarize the organization and functions of the limbic system.

The **amygdaloid body** at the "tail end" of the caudate nucleus has already been described as a cerebral contribution to the limbic system. It appears to act as an interface between the limbic system, the cerebrum, and various sensory systems. The location of other important cerebral components can be seen in Figure 14-6. The gyri that curve along the corpus callosum and onto the medial surface of the temporal lobe constitute the **limbic lobe** of the cerebral hemisphere. The **cingulate gyrus** (SIN-gū-lāt; *cingulum,* a girdle or belt) sits above the corpus callosum. The **parahippocampal** (pa-ra-hip-ō-KAM-pal) **gyrus** conceals an underlying nucleus, the **hippocampus,** which lies beneath the floor of the lateral ventricle. Early anatomists thought this nucleus resembled a sea horse (*hippocampus*); it appears to be important in learning and the storage of long-term memories.

The **fornix** (FŌR-niks) is a tract of white matter that connects the hippocampus with the hypothalamus. From the hippocampus the fornix curves medially to meet its counterpart from the opposing hemisphere and proceeds anteriorly, below the corpus callosum, before arching downward to the hypothalamus. Many of the fibers end in the **mamillary bodies** (MAM-i-lar-ē; *mamilla,* a breast), prominent nuclei in the floor of the hypothalamus. The mamillary bodies contain motor nuclei that control reflex movements associated with eating, such as chewing, licking, and swallowing.

FIGURE 14-6
The Limbic System. A sagittal section through the cerebrum, showing the cortical areas associated with the limbic system. The parahippocampal gyrus is shown as if transparent so that deeper limbic components can be seen.

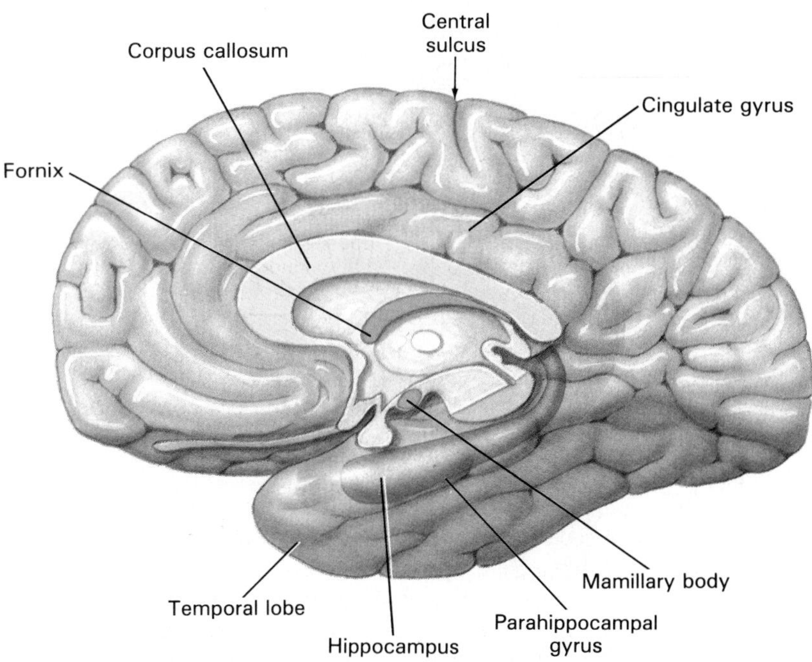

TABLE 14-4 The Limbic System

FUNCTION:
Processing of memories, creation of emotional states, drives, and associated behaviors

CEREBRAL COMPONENTS:
Cortical areas: limbic lobe (cingulate gyrus and parahippocampal gyrus)

Nuclei: hippocampus, amygdaloid body

Tracts: fornix

DIENCEPHALIC COMPONENTS:
Thalamus: anterior nuclear group

Hypothalamus: centers concerned with emotions, appetites (thirst, hunger), and related behaviors (see Table 14-6)

OTHER COMPONENTS:
Reticular formation: network of interconnected nuclei through brain stem

Several other nuclei in the walls (thalamus) and floor (hypothalamus) of the diencephalon are components of the limbic system. The **anterior nucleus** relays visceral sensations from the hypothalamus to the cerebral components of the system. The boundaries between the hypothalamic nuclei are often poorly defined, but experimental stimulation has outlined a number of important hypothalamic centers responsible for the emotions of rage, fear, pain, sexual arousal, and pleasure.

Stimulation of the hypothalamus can also produce heightened alertness and a generalized excitement. This response is caused by widespread stimulation of the **reticular formation**, an interconnected network of nuclei that extends the entire length of the brain stem. Conversely, stimulation of adjacent portions of the hypothalamus or thalamus will depress reticular activity, resulting in generalized lethargy or actual sleep.

- **What would happen if an interventricular foramen became blocked?**
- **Mary suffers a head injury that damages her primary motor cortex. Where is this area located?**
- **What symptoms would you expect to observe in an individual who has damage to the extrapyramidal system?**

The Diencephalon

The diencephalon provides the switching and relay centers required to integrate the conscious and unconscious sensory and motor pathways. Figure 14-7 shows the position of the diencephalon, along with other landmarks on the brain stem.

The diencephalic roof, or **epithalamus**, contains a vascular network involved in the production of cerebrospinal fluid. This network, the *choroid plexus*, will be considered later in the chapter. The epithalamus also contains the **pineal gland**, an endocrine structure that secretes the hormone **melatonin**. Melatonin indirectly suppresses the production of melanin in the skin. It accomplishes this by inhibiting the release of a pituitary hormone that stimulates melanocyte activity. (Details of melatonin secretion and pineal regulation will be found in Chapter 18.)

The epithalamus does not contain neural tissue; all of the neural tissue in the diencephalon is concentrated in the *thalamus* (walls) and *hypothalamus* (floor). Ascending sensory information from the spinal cord and cranial nerves (other than the olfactory) is processed in the thalamus. The hypothalamus contains centers involved with emotions and visceral processes that affect the cerebrum as well as other components of the brain stem. It also controls a variety of autonomic functions and forms the link between the nervous and endocrine systems.

FIGURE 14-7
The Brain Stem. (a) Lateral view of the brain stem, seen from the left side. (b) Posterior view of the brain stem.

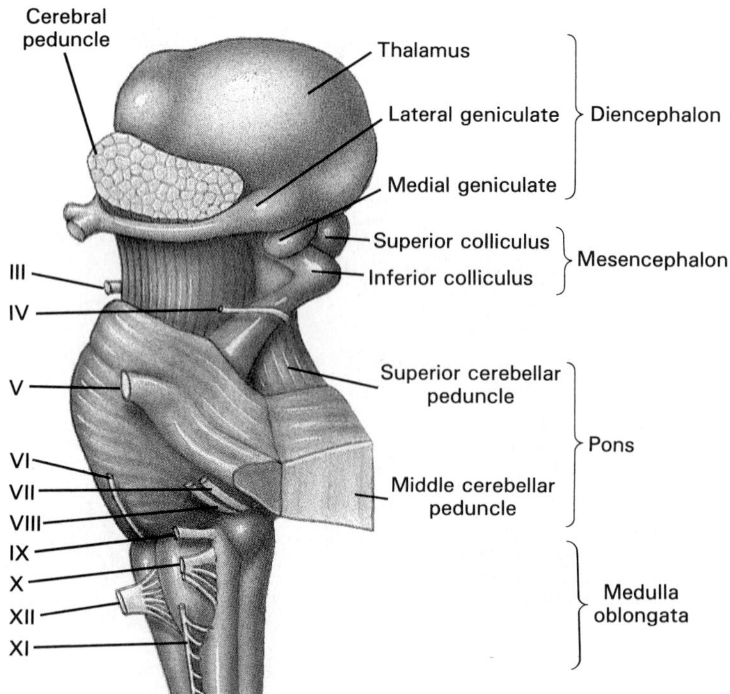

(a) Lateral view

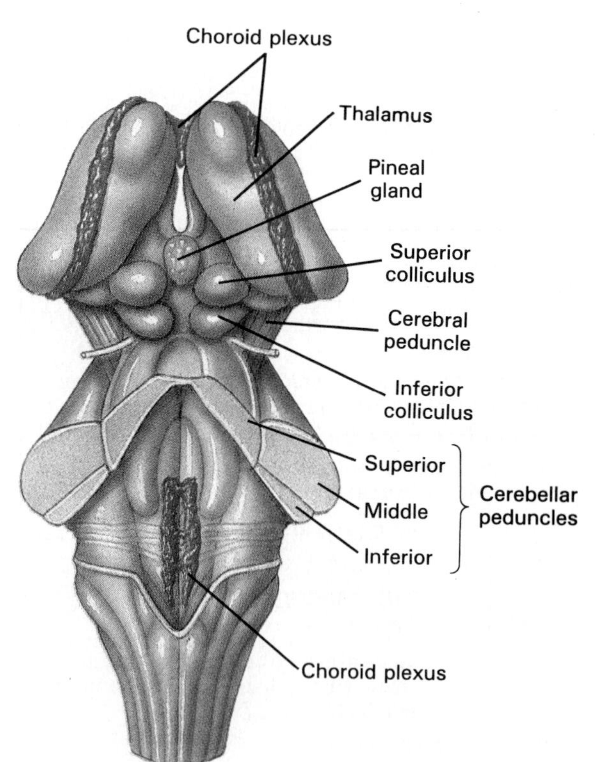

(b) Posterior view

THE THALAMUS

The thalamus is the final relay point for ascending sensory information that will be projected to the primary sensory cortex. It acts as a filter, passing on only a small portion of the arriving sensory information. The thalamus also coordinates the activities of the pyramidal and extrapyramidal systems, discussed earlier. The principal thalamic structures are indicated in Figure 14-8.

The four major groups of thalamic nuclei, detailed in Table 14-5, are (1) the *anterior group*, (2) the *medial group*, (3) the *ventral group*, and (4) the *posterior group*.

1. **The Anterior Group:** The **anterior nuclei** are part of the limbic system, discussed previously.
2. **The Medial Group:** The **medial nuclei** of the thalamus provide a conscious awareness of emotional states by connecting the emotional centers in the hypothalamus with the frontal lobes of the cerebrum. These nuclei also integrate sensory information arriving at other portions of the thalamus for relay to the frontal lobes.
3. **The Ventral Group:** The **ventral nuclei** relay information from the cerebral nuclei and cerebellum, specifically to the primary motor cortex and the motor association area of the frontal lobe. They also relay sensory data concerning touch, pressure, pain, temperature, proprioception, and taste to the primary sensory cortex of the parietal lobe.
4. **The Posterior Group:** The **posterior nuclei** include the *pulvinar* and the *geniculates*. The **pulvinar** integrates sensory information for projection to the association areas of the cerebral cortex. The **lateral geniculates** (je-NIK-ū-lāts; *genicula*, a little knee) receive visual information from the eyes, brought by the **optic tracts**. Efferent fibers project to the visual cortex and descend to the mesencephalon. The **medial geniculates** relay auditory information to the auditory cortex from the specialized receptors of the inner ear.

TABLE 14-5 The Thalamus

Structure/Nuclei	*Functions*
Anterior group	Part of the limbic system
Medial group	Integrates sensory information and other data arriving at the thalamus for projection to the frontal lobes of the cerebral hemispheres
Ventral group	Projects sensory information to the primary sensory cortex of the cerebral hemisphere; relays information from cerebellum and cerebral nuclei to motor areas of cerebral cortex
Posterior group: Pulvinar	Integrates sensory information for projection to association areas of cerebral cortex
Lateral geniculates	Project visual information to the visual cortex of occipital lobe
Medial geniculates	Project auditory information to auditory cortex of temporal lobe

THE HYPOTHALAMUS

The structure of the hypothalamus is detailed in Figure 14-9. This portion of the diencephalon contains

FIGURE 14-8
The Thalamus. (Above) Lateral view of the brain, showing the positions of the major thalamic structures. Functional areas of cerebral cortex are also indicated. (Below) Enlarged view of the thalamic nuclei of the left side. The color of each nucleus or group of nuclei matches the color of the associated cortical region.

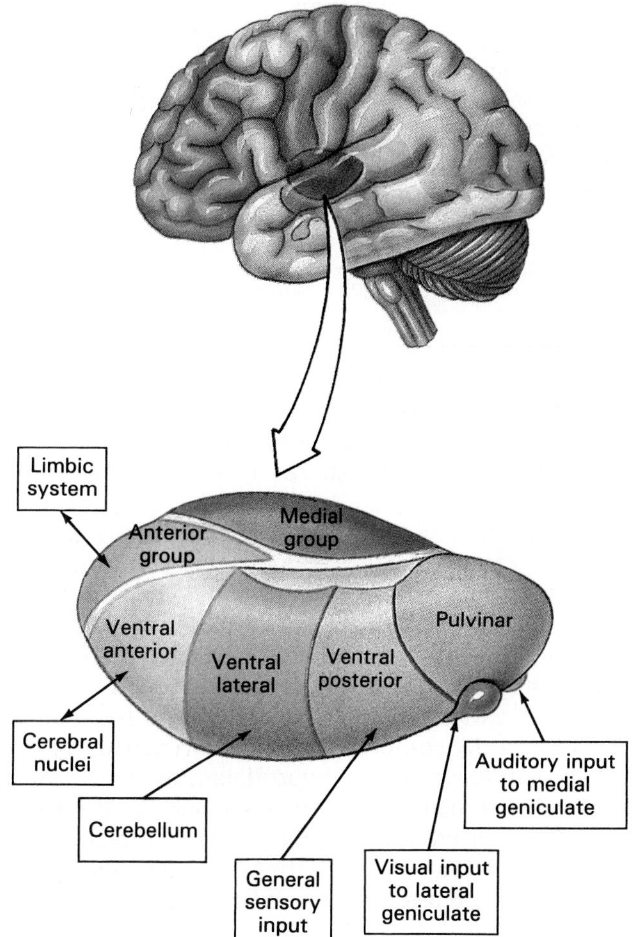

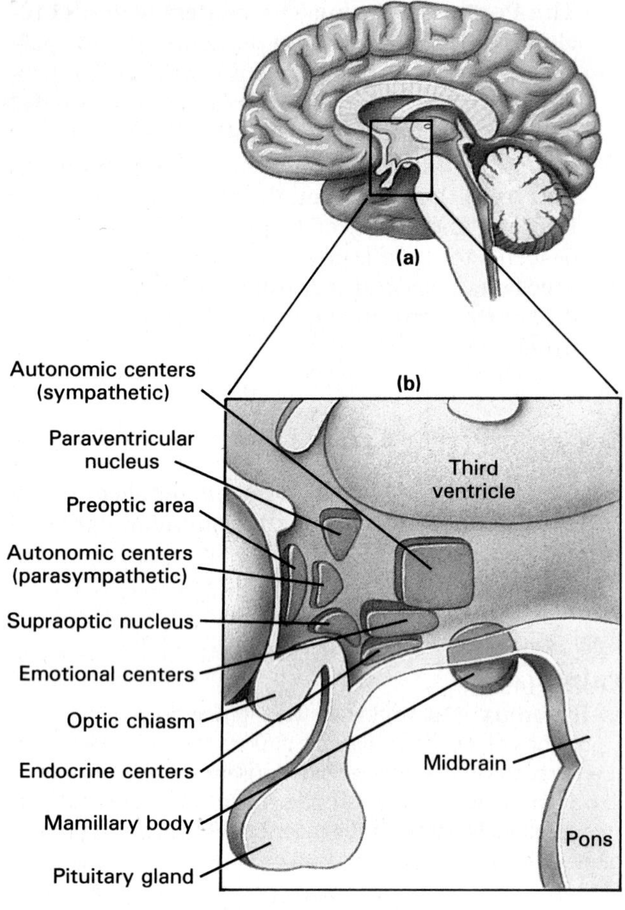

FIGURE 14-9
The Hypothalamus. (a) Sagittal section through the brain, showing the location of the hypothalamus and its relationship to other brain stem components. (b) Enlarged view of the hypothalamus, showing the locations of major nuclei and centers. Functions for these centers are included in Table 14-6.

TABLE 14-6 Components and Functions of the Hypothalamus

Region/Nucleus	*Functions*
Hypothalamus in general	Controls autonomic functions; sets appetitive drives (thirst, hunger, sexual desire) and behaviors; sets emotional states (with limbic system); integrates with endocrine system (see Chapter 18)
Supraoptic nucleus	Secretes antidiuretic hormone, restricting water loss at the kidneys
Paraventricular nucleus	Secretes oxytocin, stimulating smooth muscle contractions in uterus and mammary glands
Preoptic area	Regulates body temperature via control of autonomic centers in the medulla
Tuberal area	Produces inhibitory and releasing hormones that target endocrine cells of the anterior pituitary
Autonomic centers	Control heart rate and blood pressure via regulation of autonomic centers in the medulla
Mamillary bodies	Control feeding reflexes (licking, swallowing, etc.)

a variety of important control and integrative centers in addition to those associated with the limbic system. Table 14-6 lists the major nuclei and centers of the hypothalamus, together with their chief functions.

Hypothalamic centers are continually stimulated by collaterals delivering sensory information from the peripheral nervous system. The hypothalamus receives collateral input from other centers in the cerebrum and brain stem as well. Hypothalamic neurons detect and respond to changes in the surrounding extracellular fluids. They also respond to changes in the composition of the circulating blood, for the capillaries in the hypothalamus are far more permeable than those found elsewhere in the CNS.

Axons leaving hypothalamic nuclei form tracts that run to other centers in the brain stem and spinal cord. Hypothalamic output can have one of the following functions:

1. Control of involuntary somatic motor activities. The mamillary bodies, which direct motor patterns involved with eating, were considered as part of the limbic system. By stimulation of appropriate centers in other portions of the brain, hypothalamic nuclei direct somatic motor patterns associated with the emotions of rage, pleasure, pain, and sexual arousal. For example, the changes in facial expression that accompany rage and the basic movements associated with sexual activity are controlled by hypothalamic centers.

2. Control of autonomic function. The hypothalamus contains nuclei that adjust the activities of autonomic centers in other portions of the brain stem and spinal cord. These hypothalamic centers coordinate the activities of autonomic centers concerned with regulating heart rate, blood pressure, respiration, digestive function, and the elimination of fecal and urinary wastes.

3. Coordination of activities of the nervous and endocrine systems. Most endocrine organs are under direct or indirect hypothalamic control. Much of the regulatory control is exerted through inhibition or stimulation of the anterior portion of the pituitary gland, or **hypophysis** (hī-POF-i-sis; *hypo-*, below + *phyein*, to grow). **Releasing hormones** and **inhibiting hormones** secreted by nuclei in the **tuberal area** of the hypothalamus promote or inhibit secretion of hormones by the anterior pituitary gland.

4. Secretion of hormones. In addition to producing releasing and inhibiting factors, the hypothalamus secretes two hormones, *ADH* and *oxytocin*. These hormones are produced by the **anterior nuclei** of the hypothalamus and are transported along axons for release into the circulation in the posterior pituitary gland. **Antidiuretic hormone** (an-tī-dī-ū-RET-ik; *anti-*, opposite + *diourein*, to urinate), or **ADH**, produced by the **supraoptic nucleus**, restricts water loss at the kidneys. The **paraventricular nucleus** secretes **oxytocin** (oks-i-TŌ-sin), a hormone that stimulates smooth muscle contractions in the uterus and mammary glands. The uterine contractions are essential to normal childbirth, and the smooth muscle contractions at the mammary glands eject the milk in response to suckling at the breast.

5. Production of emotions and behavioral drives. Activity in specific hypothalamic centers triggers emotions of rage, aggression, and pleasure. Other centers produce sensations that lead to changes in voluntary or involuntary behavior patterns. For example, when body temperature declines, you may take voluntary steps to warm yourself, perhaps by putting on an extra sweater, because you "feel cold." That feeling represents a vague, poorly localized, but very effective input from the hypothalamus. These unfocused "impressions" originating in the hypothalamus are called **drives**.

Hunger and thirst are two other important examples of hypothalamic drives. Stimulation of the **feeding center** of the hypothalamus produces intense hunger. This craving for food, normally triggered by a decline in blood glucose level, can be eliminated by stimulation of the **satiety center**. Activity in the **thirst center** produces the conscious urge to take a drink. Hypothalamic neurons in this center detect changes in the osmotic concentration of the blood. When the concentration rises, the thirst center is stimulated. The thirst center is also stimulated by ADH. Stimulation of the thirst center triggers a behavioral response (drinking) that complements the physiological response to this hormone (water conservation by the kidneys). As in the case of temperature regulation, the conscious sensations are only part of the hypothalamic response. For example, the thirst center also orders the release of ADH by neurons in the supraoptic nucleus.

6. Coordination between voluntary and autonomic functions. Hypothalamic activity can produce the conscious emotions of rage and pleasure. Conversely, cerebral activity can trigger changes in autonomic function by stimulating hypothalamic centers. When you think about doing something dangerous or facing a stressful situation, your heart rate and respiratory rate go up and your body prepares for an emergency. These autonomic adjustments are made because cerebral activities are monitored by the hypothalamus. Because of these integrative activities, chronic psychological stresses can cause clinical abnormalities in a variety of organ systems. These **psychosomatic** conditions have a psychological basis, but the physiological symptoms are quite real, and usually reflect inadequate or excessive levels of autonomic activity.

7. Regulation of body temperature. The **preoptic area** of the hypothalamus controls the physiological responses to changes in body temperature. In doing so, it coordinates the activities of other CNS centers and regulates other physiological systems. If body temperature falls, the preoptic area sends instructions to the *vasomotor center* in the medulla, an autonomic center that controls blood flow by regulating the diameter of peripheral blood vessels. In response, the vasomotor center decreases the blood supply to the skin, reducing the rate of heat loss. Other hypothalamic efferents target nuclei that increase muscle tone, perhaps to the point that shivering begins. These represent short-term solutions to the problem; more effective responses require a readjustment in the rate of heat production by peripheral tissues. The endocrine system accomplishes this by releasing the appropriate hormones, and these secretory activities are controlled by the same hypothalamic center.

The Mesencephalon or Midbrain

The external anatomy of the mesencephalon can be seen in Figure 14-7, and the major nuclei are detailed in Figure 14-10 and Table 14-7. The roof, or **tectum**, of the mesencephalon contains two pairs of sensory nuclei known collectively as the **corpora quadrigemina** (KŌR-pō-ra quad-ri-JEM-i-na). These nuclei, the *superior* and *inferior colliculi*, are relay stations concerned with the processing of visual and auditory sensations. The **superior colliculus** (kol-IK-ū-lus; *colliculus*, a small hill) receives visual inputs from the lateral geniculate of the thalamus. The **inferior colliculus** receives auditory data from nuclei in the medulla, some of which may be forwarded to the medial geniculate.

The mesencephalon also contains the headquarters of the reticular formation. Specific patterns of stimulation in this region can produce a variety of involuntary motor responses. On each side the reticular formation contains a pair of nuclei, the **red nucleus** and the **substantia nigra**. The red nucleus is

TABLE 14-7 Components and Functions of the Mesencephalon (Midbrain)

Subdivision	*Region/Nucleus*	*Functions*
GRAY MATTER Tectum (roof)	Superior colliculus	Integrates visual information with other sensory inputs; initiates involuntary motor responses
	Inferior colliculus	Relays auditory information to medial geniculate
Walls and floor	Red nucleus	Involuntary control of muscle tone and posture
	Substantia nigra	Regulates activity in the cerebral nuclei
	Reticular formation (headquarters)	Automatic processing of incoming sensations and outgoing motor commands; can initiate involuntary motor responses to stimuli
	Other nuclei/centers	Nuclei associated with two cranial nerves (N III, IV)
White Matter	Cerebral peduncles	Connect primary motor cortex with motor neurons in brain and spinal cord; carry ascending sensory information to thalamus

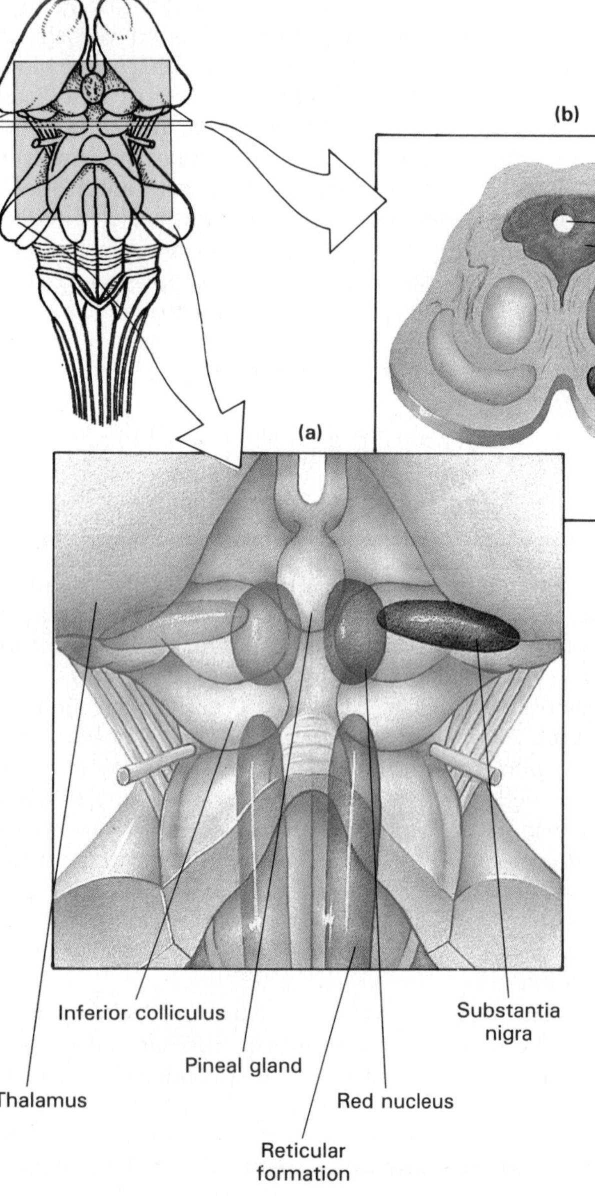

FIGURE 14-10
The Mesencephalon. (a) Posterior view, showing the position of important nuclei. (b) Sectional view, taken at the region indicated in (a).

extremely well provided with blood vessels, giving it a strong red coloration. This nucleus integrates information from the cerebrum and cerebellum and issues involuntary motor commands concerned with the maintenance of muscle tone and posture. The **substantia nigra** (NĪ-grah; black) lies lateral to the red nucleus. The gray matter in this region contains darkly pigmented cells, giving it a black color. The substantia nigra plays an important role in regulating the motor output of the cerebral nuclei.

The nerve fiber bundles on the ventrolateral surfaces of the mesencephalon (see Figures 14-7a and 14-10b) are the **cerebral peduncles** ("little feet"). They contain ascending fibers headed for thalamic nuclei

Clinical Comment: The Cerebral Nuclei, Substantia Nigra, and Parkinson's Disease

The cerebral nuclei contain two discrete populations of neurons. One group stimulates motor neurons by releasing ACh, and the other inhibits motor neurons by the release of GABA. Under normal conditions the excitatory neurons remain inactive, and the descending tracts are primarily responsible for inhibiting motor neuron activity. If the descending tracts are severed in an accident, the loss of inhibitory control leads to a generalized state of muscular contraction known as **decerebrate rigidity**.

The excitatory neurons are quiet because they are continually exposed to the inhibitory effects of the neurotransmitter dopamine. This compound is manufactured by neurons in the substantia nigra and carried by axoplasmic flow to synapses in the cerebral nuclei. If the ascending tract or the dopamine-producing neurons are damaged, this inhibition is lost, and the excitatory neurons become increasingly active. This increased activity produces the motor symptoms of **Parkinson's disease**, or *paralysis agitans*.

Parkinson's disease is characterized by a pronounced increase in muscle tone. Voluntary movements become hesitant and jerky, a condition called **spasticity**, for a movement cannot occur until one muscle group manages to overpower its antagonists. Individuals with Parkinson's disease show spasticity during voluntary movement and a continual **tremor** when at rest. A tremor represents a tug of war between antagonistic muscle groups that produces a background shaking of the limbs, in this case at a frequency of 4–6 cycles per second. Individuals with Parkinson's disease also have difficulty starting voluntary movements. Even changing one's facial expression requires intense concentration, and the individual acquires a blank, static expression. Finally, the positioning and preparatory adjustments normally performed automatically no longer occur. Every aspect of each movement must be voluntarily controlled, and the extra effort requires intense concentration that may prove tiring and extremely frustrating. In the late stages of this condition, other CNS effects, such as depression and hallucinations, often appear.

Providing the cerebral nuclei with dopamine can significantly reduce the symptoms for two-thirds of Parkinson's patients, but intravenous dopamine injection is not effective because the molecule cannot cross the blood-brain barrier. The most common procedure involves the oral administration of the drug L-DOPA (levodopa), a related compound that crosses the capillaries and is then converted to dopamine. Unfortunately, it appears that with repeated treatment the capillaries become less permeable to L-DOPA, and so the required dosage increases. The effectiveness of L-DOPA can be increased by giving it in combination with other drugs, such as *Amantadine* or *Bromocriptine*. Amantadine accelerates dopamine release at synaptic terminals and Bromocriptine, a dopamine agonist, stimulates dopamine receptors on postsynaptic membranes.

Surgery to control Parkinson's symptoms focuses on the destruction of large areas within the cerebral nuclei or thalamus to control the motor symptoms of tremor and rigidity. The high rate of success for drug therapy has greatly reduced the number of surgical procedures. Recent attempts to transplant tissues producing dopamine or related compounds into the cerebral nuclei have met with limited success. Variable results have been obtained with the transplantation of tissue from the adrenal gland; the transplantation of fetal tissue into adult brains has been more successful.

Individuals with Parkinson's disease are usually elderly. However, since 1983 an increasing number of young people have developed this condition. In that year a drug appeared on the streets rumored to be "synthetic heroin." In addition to the compound that produced the "high" sought by users, the drug contained several contaminants, including a complex molecule with the abbreviated name **MPTP**. This accidental byproduct of the synthetic process destroys neurons of the substantia nigra, eliminating the manufacture and transport of dopamine to the cerebral nuclei. As a result of exposure to this drug, approximately 200 young, healthy adults have developed symptoms of severe Parkinson's disease. Why MPTP targets these particular neurons, and not all of the CNS neurons that produce dopamine, remains a mystery.

and the descending fibers of the pyramidal system that carry voluntary motor commands from the primary motor cortex of each cerebral hemisphere.

■ The Cerebellum

The cerebellum, detailed in Figure 14-11 and Table 14-8, is an automatic processing center. It has two primary functions:

1. The cerebellum *oversees the postural muscles of the body*, making rapid adjustments to maintain balance and equilibrium. These alterations in muscle tone and position are made by modifying the activity of the red nucleus.
2. The cerebellum *programs and tunes voluntary and involuntary movements*. These functions are performed indirectly, by regulating activity along both pyramidal and extrapyramidal motor pathways at the cerebral cortex, cerebral nuclei, and motor centers in the brain stem.

The cerebellum has a complex, highly convoluted surface composed of neural cortex. The folds, or **folia** (FŌ-lē-ah), of the surface are less prominent than the gyri of the cerebral hemispheres. The **anterior** and **posterior lobes** are separated by a transverse fissure, as indicated in Figure 14-11a. Along the mid-

■ TABLE 14-8 Components of the Cerebellum

Subdivision	*Region/Nucleus*	*Functions*
Gray matter	Cerebellar cortex	Involuntary coordination and control of ongoing movements of body parts
	Cerebellar nuclei	As above
White matter	Arbor vitae	Connects cerebellar cortex and nuclei with cerebellar peduncles

FIGURE 14-11
The Cerebellum. (a) Posterior and superior surface of the cerebellum, showing major anatomical landmarks and regions. (b) Sectional view of the cerebellum, showing the arrangement of gray matter and white matter. A single Purkinje cell is seen in the photograph; these large neurons are found in the cerebellar cortex.

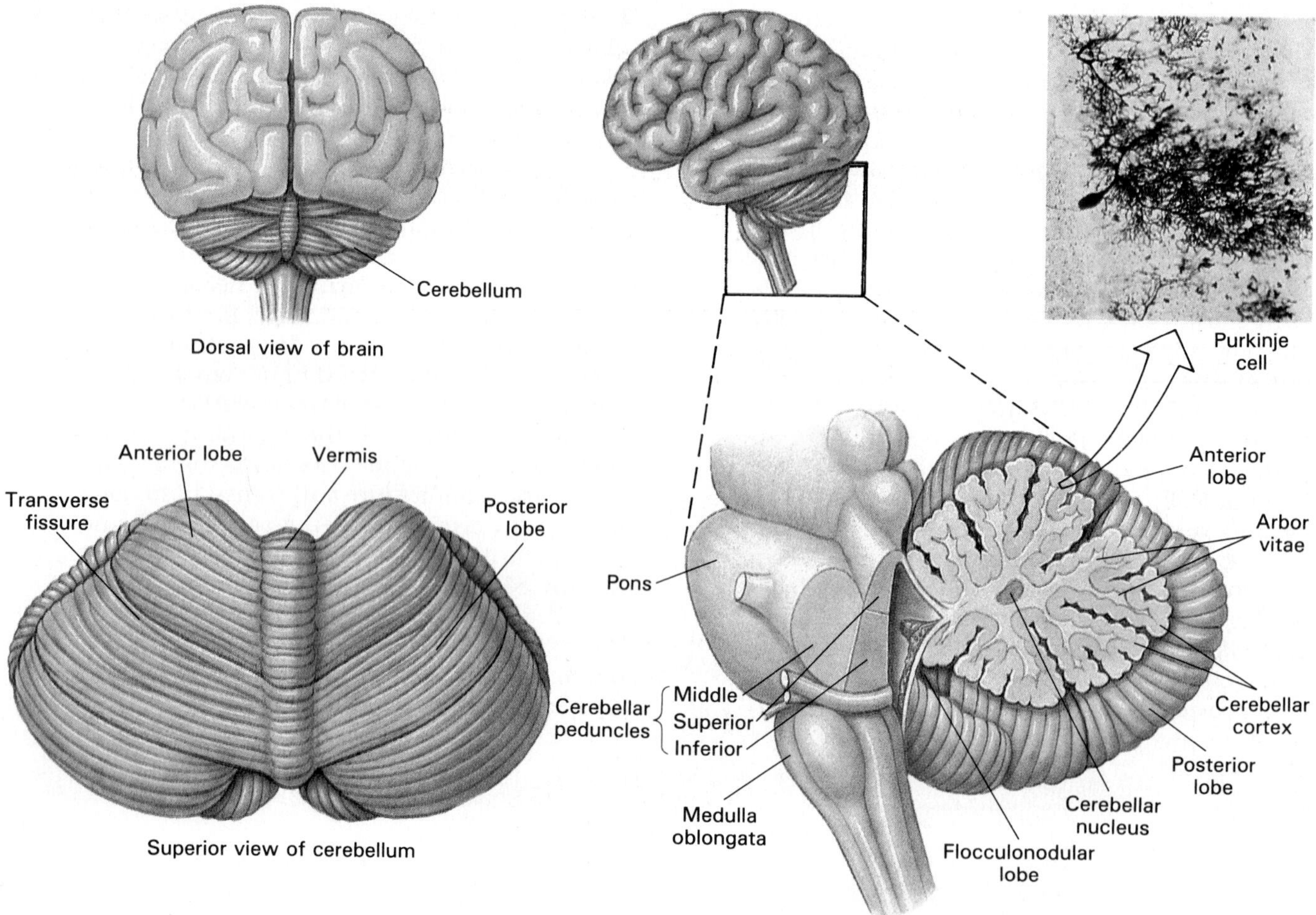

line a narrow band of cortex known as the **vermis** (VER-mis; worm) separates the **cerebellar hemispheres** of the posterior lobe. The slender **flocculonodular** (flok-ū-lō-NOD-ū-lar) **lobe** lies between the roof of the fourth ventricle and the cerebellar hemispheres and vermis. These relationships are diagrammed in Figure 14-11b.

The cerebellar cortex contains huge, highly branched **Purkinje** (pur-KIN-jē), or **basket, cells**. Internally, the white matter of the cerebellum forms a branching array that in sectional view resembles a tree. Anatomists call it the **arbor vitae**, or "tree of life." The cerebellum receives proprioceptive information (position sense) from the spinal cord, and monitors all proprioceptive, visual, tactile, balance, and auditory sensations received by the brain. Information concerning the voluntary and involuntary motor commands issued by higher centers reaches the cerebellum via tracts from the cerebral cortex and thalamus. These synapse within **cerebellar nuclei** before projecting to the cerebellar cortex. Axons carrying sensory information from any of the above sources will pass through the deeper layers of the cerebellar cortex to terminate near the cortical surface. There they synapse with the dendritic processes of the Purkinje cells that perform the final processing steps. Tracts containing the axons of Purkinje cells then relay motor commands to nuclei within the cerebrum and brain stem. [**CM**: *Cerebellar Dysfunction*]

The cerebellum stores memories of learned movement patterns. When tennis players or golfers, for example, spend hours practicing, they are firmly establishing these cerebellar motor patterns. The effortless serve or swing proceeds without conscious direction. In fact, thinking about one's performance is almost certain to disrupt the pattern, for conscious motor activities contradict or override the programmed commands issued by the cerebellum.

The Pons

The external appearance of the pons can be seen in Figure 14-7; important nuclei are included in Figure 14-12 and Table 14-9. The pons contains:

TABLE 14-9 Components of the Pons

Subdivisions	*Region/Nucleus*	*Functions*
Gray matter	**Respiratory centers**	**Modify output of respiratory centers in the medulla**
	Other nuclei/centers	**Nuclei associated with four cranial nerves**
White matter	**Cerebellar peduncles**	
	Superior	**Link the cerebellum with mesencephalon, diencephalon, and cerebrum**
	Middle	**Contain transverse fibers and carry communications between the cerebellum and pons**
	Inferior	**Link the cerebellum with the medulla and spinal cord**
	Transverse fibers	**Interconnect cerebellar hemispheres**

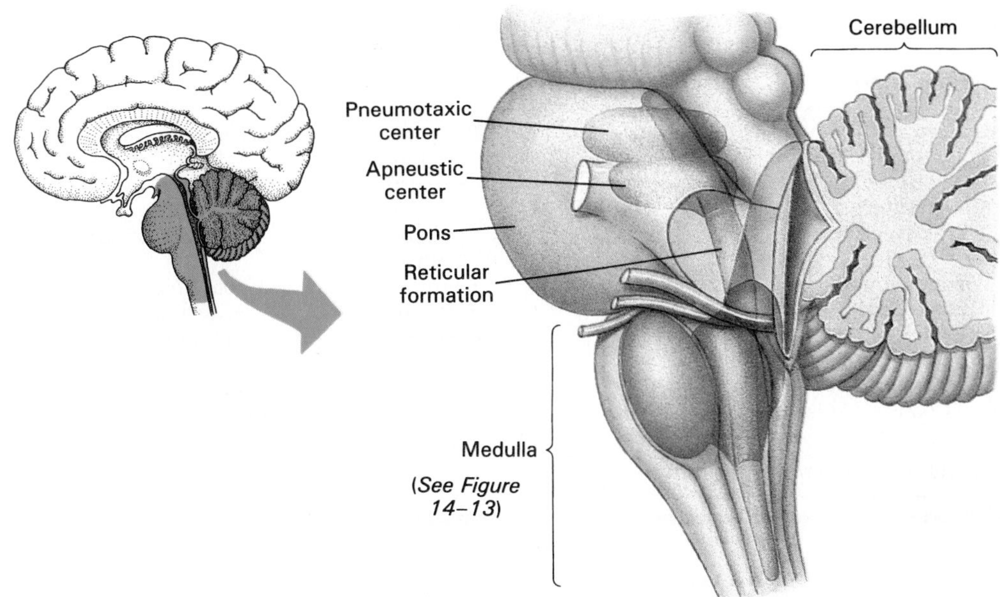

FIGURE 14-12 The Pons

1. Sensory and motor nuclei for four of the cranial nerves. The cranial nerves associated with the pons innervate the jaw muscles, the anterior surface of the face, one of the extrinsic eye muscles (the lateral rectus), and the sense organs of the inner ear.

2. Nuclei concerned with the involuntary control of respiration. The reticular formation in this region contains two respiratory centers, the *apneustic center* and the *pneumotaxic center*. These centers modify the activity of the *respiratory rhythmicity center* in the medulla oblongata.

3. Tracts that link the cerebellum with the brain stem, cerebrum, and spinal cord. These tracts form prominent superficial landmarks known as the *superior, middle,* and *inferior cerebellar peduncles*. The **superior cerebellar penduncles** link the cerebellum with nuclei in the midbrain, diencephalon, and cerebrum. The **middle cerebellar penduncles** are connected to a broad band of fibers that cross the ventral surface of the pons at right angles to the axis of the brain stem. These **transverse fibers** permit communication between the cerebellar hemispheres of opposite sides. The middle cerebellar peduncles also connect the cerebellar hemispheres with sensory and motor nuclei in the pons. The **inferior cerebellar peduncles** permit communication between the cerebellum and nuclei in the medulla oblongata and carry ascending and descending cerebellar tracts from the spinal cord.

4. Ascending and descending tracts. These longitudinal tracts interconnect other portions of the CNS.

■ The Medulla Oblongata (Myelencephalon)

The medulla oblongata, often simply called the *medulla*, is connected to the spinal cord. Figure 14-7 details the external surface of the medulla; Figure 14-13 and Table 14-10 provide additional information concerning its functional components. In sectional view the caudal portion of the medulla resembles the spinal cord in having a rounded shape and a narrow central canal. Closer to the pons, the central canal becomes enlarged and continuous with the fourth ventricle.

The medulla physically connects the brain with the spinal cord, and many of its functions are directly related to this fact. For example, all communication between the brain and spinal cord involves tracts that ascend or descend through the medulla. In many instances these tracts synapse in the medulla in sensory or motor nuclei that act as relay stations and processing centers.

In addition to these nuclei, the medulla contains sensory and motor nuclei associated with five of the cranial nerves. These cranial nerves innervate muscles of the pharynx, neck, and back, as well as visceral organs of the thoracic and peritoneal cavities.

The medulla also contains a variety of nuclei involved in the regulation of vital autonomic functions. Only a few representative nuclei will be described here.

Some medullary nuclei relay sensory information to higher centers. For example, the **nucleus gracilis** and the **nucleus cuneatus** pass somatic sen-

■ TABLE 14-10 Components and Functions of the Medulla Oblongata

Subdivision	*Region/Nucleus*	*Functions*
Gray matter	**Nucleus gracilis** **Nucleus cuneatus**	**Relay somatic sensory information to the thalamus**
	Olivary nuclei	**Relay information to the cerebellum**
	Reflex centers:	
	Cardiac centers	**Regulate heart rate and force of contraction**
	Vasomotor center	**Regulates distribution of blood flow**
	Respiratory rhythmicity center	**Sets the pace of respiratory movements**
	Other nuclei/centers	**Sensory and motor nuclei of five cranial nerves**
		Nuclei relaying ascending sensory information from the spinal cord to higher centers
White matter	**Ascending and descending tracts**	**Link the brain with the spinal cord**

sory information to the thalamus, and the **olivary nuclei** relay information from the spinal cord, the cerebral cortex, and brain stem to the cerebellar cortex. The bulk of these nuclei create the **olives**, prominent bulges along the ventrolateral surface of the medulla. The medulla also contains the sensory and motor nuclei associated with five of the cranial nerves.

The reticular formation in the medulla contains nuclei and centers responsible for the regulation of vital autonomic functions. These **reflex centers** receive inputs from cranial nerves, the cerebral cortex, and the brain stem, and their output controls or adjusts the activities of one or more peripheral systems. The **cardiovascular centers** adjust heart rate, the strength of cardiac contractions, and the flow of blood through peripheral tissues. On functional grounds the cardiovascular centers may be subdivided into **cardiac** (*kardia*, heart) and **vasomotor** (*vas*, canal) centers, but their anatomical boundaries are difficult to determine. The **respiratory rhythmicity center** sets the basic pace for respiratory movements, and its activity is regulated by inputs from the apneustic and pneumotaxic centers of the pons.

- √ **Damage to the lateral geniculate nuclei of the thalamus would interfere with the functions of which of the senses?**
- √ **What area of the diencephalon would be stimulated by changes in body temperature?**
- √ **In what part of the brain would you find a worm (vermis) and a tree (arbor vitae)?**

FIGURE 14-13
The Medulla

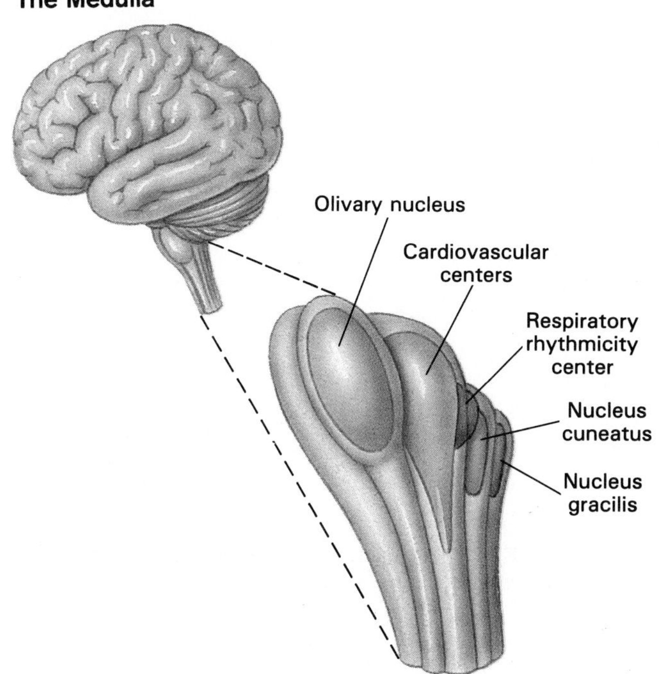

- √ **The medulla is one of the smallest sections of the brain, yet damage there can cause death, whereas similar damage in the cerebrum might go unnoticed. Why?**

Protection and Support of the Brain

Despite its complex functions, the brain is relatively small and extremely delicate. It has a very high metabolic rate and requires abundant nutrients and a constant supply of oxygen. At the same time, it must be isolated from a variety of compounds in the blood that could interfere with its complex operations. These special needs necessitate a variety of special adaptations for the protection and support of the brain.

THE CRANIAL MENINGES

The brain lies nestled within the cranium of the skull, and there is an obvious correspondence between the shape of the brain and that of the cranial cavity. The relationships are indicated in Figure 14-14. The brain must be protected from a variety of environmental hazards, including jolts, impacts, and extremes of temperature. The intervertebral discs of the vertebral column and the muscles and vertebrae of the neck reduce the shocks generated during locomotion. Mechanical protection is afforded by the bones of the skull, but the massive cranial bones also pose a threat. The brain is like a person driving a 5000-pound automobile. If the car hits a tree, the car protects the driver from contact with the tree, but serious injury will occur unless a seat belt or airbag also protects the driver from contact with the car. [**CM**: *Cranial Trauma*]

Within the cranial cavity, the *cranial meninges* that surround the brain provide this protection, acting as shock absorbers that prevent contact with surrounding bones. The membrane layers that make up the cranial meninges—the *dura mater, arachnoid*, and *pia mater*—are continuous with those of the spinal cord (Chapter 13). (p. 406) However, the cranial meninges have distinctive anatomical and functional characteristics that differ from those of the spinal meninges. Within the skull, the dura mater consists of two fibrous layers, the outermost fused to the periosteum of the skull. The inner and outer layers are separated by a slender gap that contains tissue fluids and blood vessels, including the large veins known as **dural sinuses**.

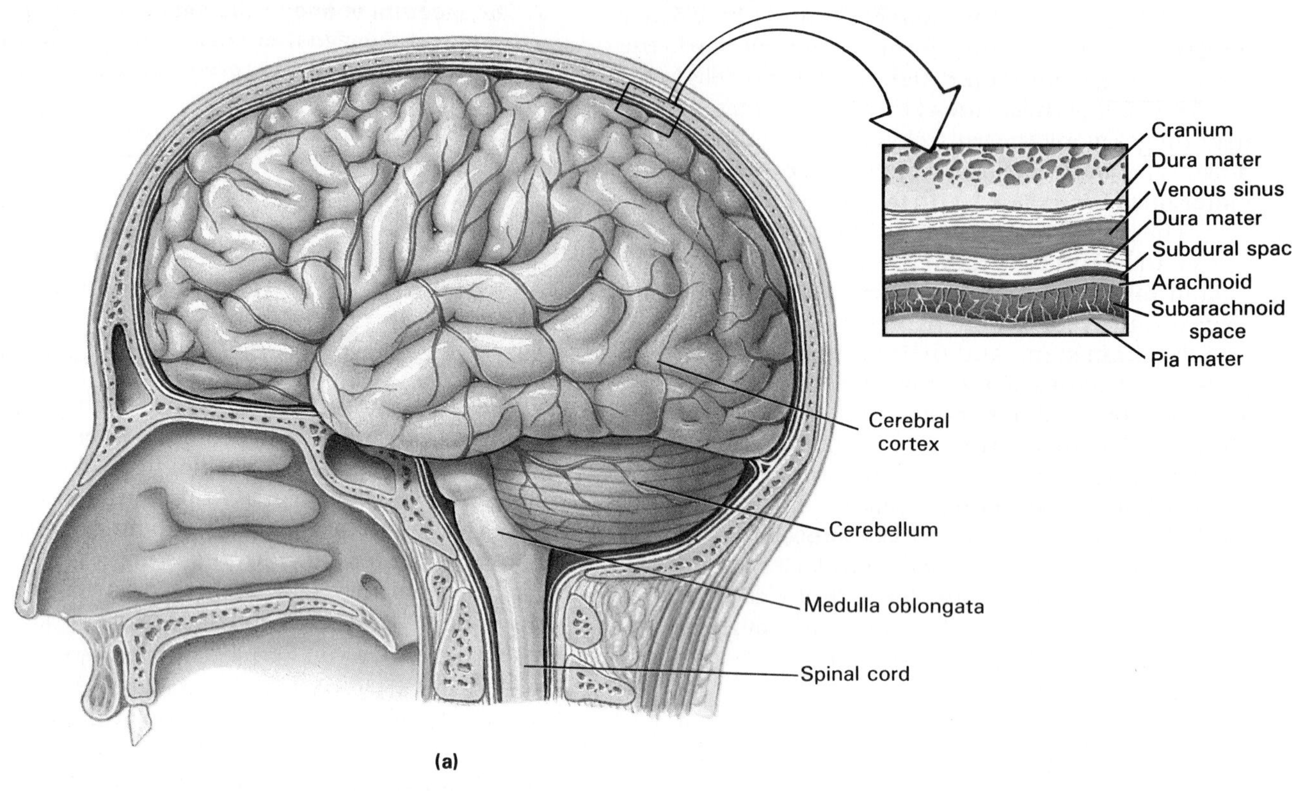

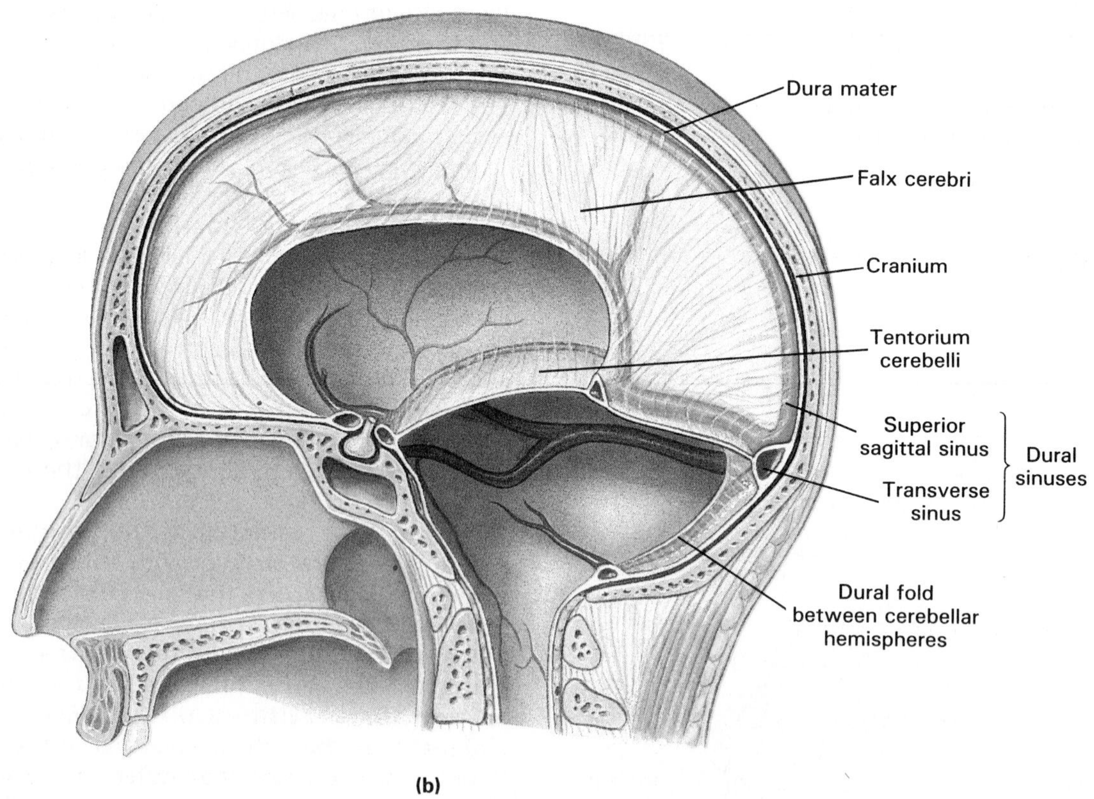

FIGURE 14-14
Relationship between the Brain, Cranium, and Meninges. (a) Lateral view of the brain, showing its position in the cranium and the organization of the meningeal coverings. (b) A view of the cranial cavity with the brain removed, showing the orientation and extent of the falx cerebri and tentorium cerebelli.

Clinical Comment: Epidural and Subdural Hemorrhages

A severe head injury may damage meningeal vessels and cause bleeding into the epidural or subdural spaces. The most dangerous cases of epidural bleeding, or **epidural hemorrhage**, involve an arterial break. The arterial blood pressure usually forces considerable quantities of blood into the epidural space, distorting the underlying soft tissues of the brain. The individual loses consciousness from minutes to hours after the injury, and death follows in untreated cases.

An epidural hemorrhage involving a damaged vein does not produce massive symptoms immediately, and the individual may become unconscious from several hours to several days or even weeks after the original incident. Consequently, the problem may not be noticed until the nervous tissue has been severely damaged by distortion, compression, and secondary hemorrhaging. Epidural hemorrhages are not common, occurring in fewer than 1 percent of head injuries. This is rather fortunate, for the mortality rate is 100 percent in untreated cases and over 50 percent even after removal of the blood pool and closure of the damaged vessels.

In a **subdural hemorrhage** the blood accumulates between the dura and the arachnoid. Subdural hemorrhages are roughly twice as common as epidural hemorrhages. The most common source of blood is a small vein or one of the dural sinuses. Because the blood pressure is somewhat lower, the extent and effects of the condition may be quite variable.

At several locations the innermost layer of the dura mater extends deep into the cranial cavity, providing additional stabilization and support to the brain. The **falx cerebri** (falks ser-E-brē; *falx*, curving, or sickle-shaped) is a fold of dura mater that projects between the cerebral hemispheres in the midsagittal plane. Its posterior margin fuses with a second partition that crosses the cranial cavity at right angles to the falx cerebri. This dural sheet, the **tentorium cerebelli** (ten-TŌR-ē-um ser-e-BEL-ē; *tentorium*, a covering), separates and protects the cerebellar hemispheres from those of the cerebrum. A third, smaller projection of the inner dural membrane extends from the midsagittal line into the space between the cerebellar hemispheres. These structures can be seen in Figure 14-14b,c.

A slender subdural space separates the opposing epithelia of the dura mater and the arachnoid. The arachnoid covers the brain, providing a smooth surface that does not follow the underlying neural contours. The subarachnoid space contains a delicate meshwork of collagen and elastin fibers, through which cerebrospinal fluid percolates. The pia mater closely adheres to the surface of the brain, anchored by the processes of astrocytes. The pia mater is highly vascular, and large vessels branch over the surface of the brain, supplying the superficial areas of neural cortex. This extensive circulatory supply is extremely important, for the brain has a very high rate of metabolism; at rest the 1.4–kg brain uses as much oxygen as 28 kg of skeletal muscle.

THE BLOOD-BRAIN BARRIER

As noted in Chapter 12, neural tissue in the CNS is isolated from the general circulation by the *blood-brain barrier.* (p. 374) This barrier exists because the capillaries of the CNS are impermeable to many compounds. In general, only lipid-soluble compounds can diffuse into the interstitial fluid of the brain and spinal cord. Water-soluble compounds cannot cross the endothelial lining without the assistance of active or passive carriers. Many different transport mechanisms are involved. For example, there are separate transport systems for glucose, large amino acids, and glycine (the smallest of the amino acids).

The transport process is both selective and directional. Large amino acids and glucose are transported out of the blood, whereas glycine is transported into the blood. There are functional reasons for directional transport. The neurons have a constant need for glucose that must be met regardless of the relative concentrations in the blood and interstitial fluid. However, glycine is a neurotransmitter, and its concentration in neural tissue must be kept relatively low, much lower than that in the circulating blood. Nevertheless, most of the transport mechanisms of the blood-brain barrier involve facilitated diffusion (Chapter 3) and so occur down concentration gradients. (p. 83)

The blood-brain barrier remains intact throughout the CNS, with two notable exceptions:

1. In portions of the hypothalamus the capillary endothelium is extremely permeable. This permeability exposes the hypothalamic nuclei in the anterior and tuberal regions to circulating hormones, a key step that permits the hypothalamic monitoring of endocrine function. It also permits the diffusion of releasing factors into the circulation for delivery to the anterior pituitary gland.
2. In the membranous roof of the diencephalon and medulla, the pia mater supports extensive capillary networks that project into the ventri-

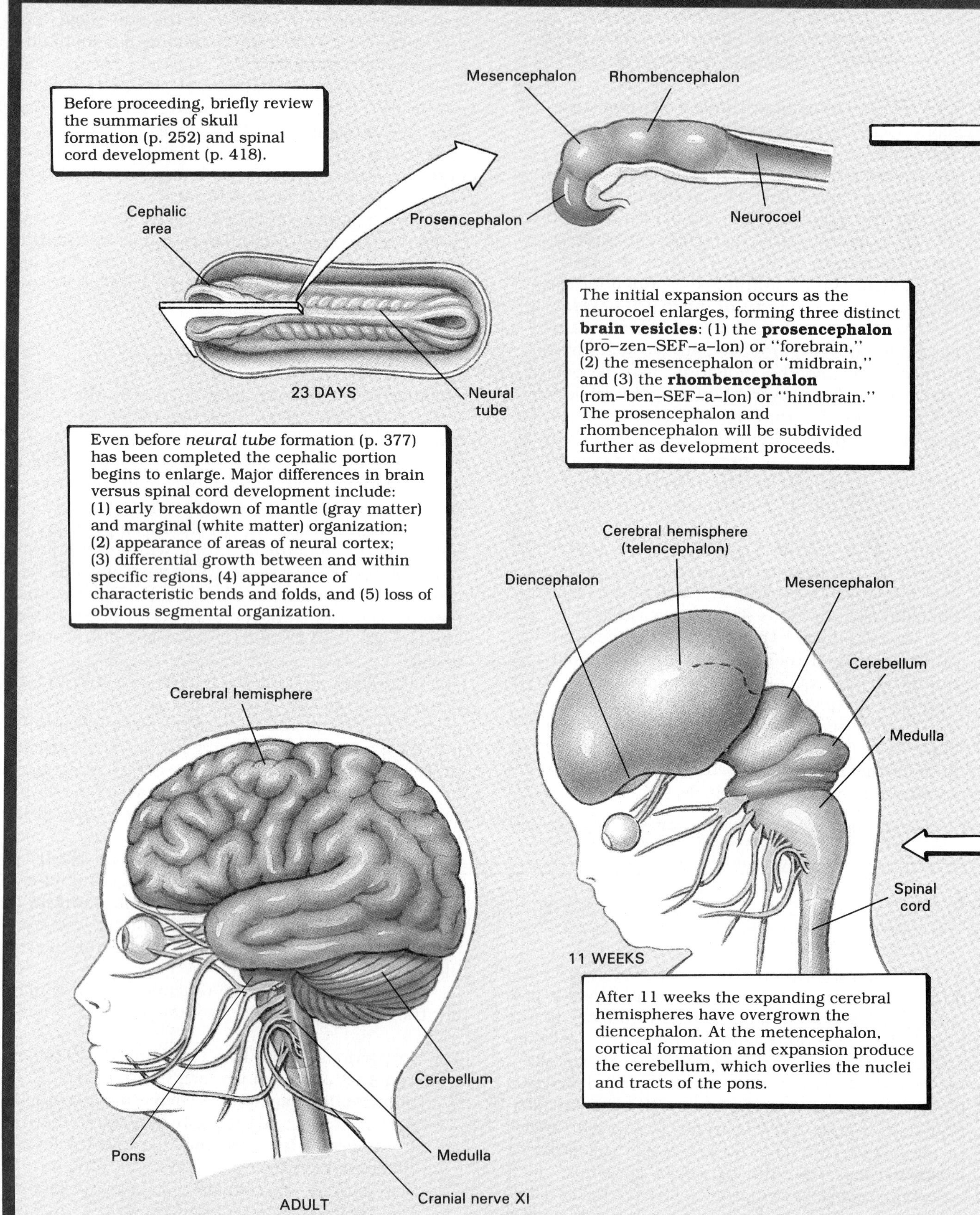

Before proceeding, briefly review the summaries of skull formation (p. 252) and spinal cord development (p. 418).
Mesencephalon
Rhombencephalon
Prosencephalon
Neurocoel
Cephalic area
23 DAYS
Neural tube
The initial expansion occurs as the neurocoel enlarges, forming three distinct **brain vesicles**: (1) the **prosencephalon** (prō–zen–SEF–a–lon) or "forebrain," (2) the mesencephalon or "midbrain," and (3) the **rhombencephalon** (rom–ben–SEF–a–lon) or "hindbrain." The prosencephalon and rhombencephalon will be subdivided further as development proceeds.
Even before *neural tube* formation (p. 377) has been completed the cephalic portion begins to enlarge. Major differences in brain versus spinal cord development include: (1) early breakdown of mantle (gray matter) and marginal (white matter) organization; (2) appearance of areas of neural cortex; (3) differential growth between and within specific regions, (4) appearance of characteristic bends and folds, and (5) loss of obvious segmental organization.
Cerebral hemisphere (telencephalon)
Diencephalon
Mesencephalon
Cerebellum
Medulla
Spinal cord
11 WEEKS
After 11 weeks the expanding cerebral hemispheres have overgrown the diencephalon. At the metencephalon, cortical formation and expansion produce the cerebellum, which overlies the nuclei and tracts of the pons.
Cerebral hemisphere
Cerebellum
Pons
Medulla
Cranial nerve XI
ADULT

Embryology Summary: Development of the Brain and Cranial Nerves

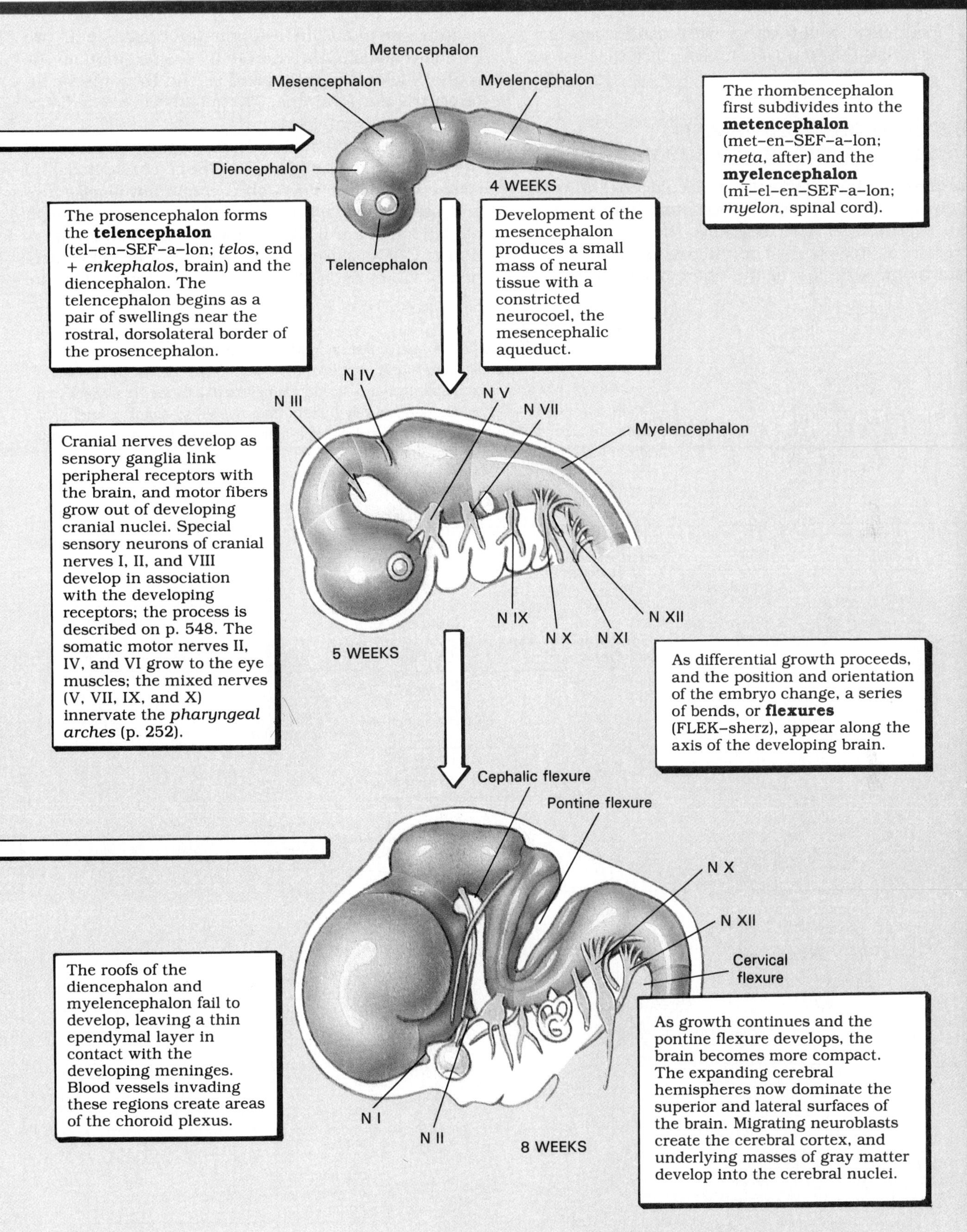
Metencephalon
Mesencephalon
Myelencephalon
Diencephalon
Telencephalon
4 WEEKS
The rhombencephalon first subdivides into the **metencephalon** (met-en-SEF-a-lon; *meta*, after) and the **myelencephalon** (mī-el-en-SEF-a-lon; *myelon*, spinal cord).
The prosencephalon forms the **telencephalon** (tel-en-SEF-a-lon; *telos*, end + *enkephalos*, brain) and the diencephalon. The telencephalon begins as a pair of swellings near the rostral, dorsolateral border of the prosencephalon.
Development of the mesencephalon produces a small mass of neural tissue with a constricted neurocoel, the mesencephalic aqueduct.
N IV
N III
N V
N VII
Myelencephalon
Cranial nerves develop as sensory ganglia link peripheral receptors with the brain, and motor fibers grow out of developing cranial nuclei. Special sensory neurons of cranial nerves I, II, and VIII develop in association with the developing receptors; the process is described on p. 548. The somatic motor nerves II, IV, and VI grow to the eye muscles; the mixed nerves (V, VII, IX, and X) innervate the *pharyngeal arches* (p. 252).
N IX
N X
N XI
N XII
5 WEEKS
As differential growth proceeds, and the position and orientation of the embryo change, a series of bends, or **flexures** (FLEK-sherz), appear along the axis of the developing brain.
Cephalic flexure
Pontine flexure
N X
N XII
Cervical flexure
The roofs of the diencephalon and myelencephalon fail to develop, leaving a thin ependymal layer in contact with the developing meninges. Blood vessels invading these regions create areas of the choroid plexus.
N I
N II
8 WEEKS
As growth continues and the pontine flexure develops, the brain becomes more compact. The expanding cerebral hemispheres now dominate the superior and lateral surfaces of the brain. Migrating neuroblasts create the cerebral cortex, and underlying masses of gray matter develop into the cerebral nuclei.

cles of the brain. These capillaries are unusually permeable, and together with modified ependymal cells they form a complex called the *choroid plexus*.

THE CHOROID PLEXUS AND CEREBROSPINAL FLUID FORMATION

The **choroid plexus** (*choroid*, a vascular coat, *plexus*, a network) is the site of cerebrospinal fluid production. Figure 14-15a,b diagrams the location and structure of areas of the choroid plexus. In the lower brain stem, a region of the choroid plexus in the roof of the fourth ventricle projects between the cerebellum and pons. In the anterior brain stem, two extensive folds of the choroid plexus originate in the roof of the diencephalon and extend through the interventricular foramina. These folds cover the floors of the lateral ventricles.

Large ependymal cells cover the capillaries of the choroid plexus and contact the cerebrospinal fluid of the ventricles. Through a combination of active and passive transport these cells secrete cerebrospinal fluid at a rate of about 1200 mℓ/day. The composition of CSF is closely regulated. Even the concentration of vitamins, such as the C and B vitamins, are

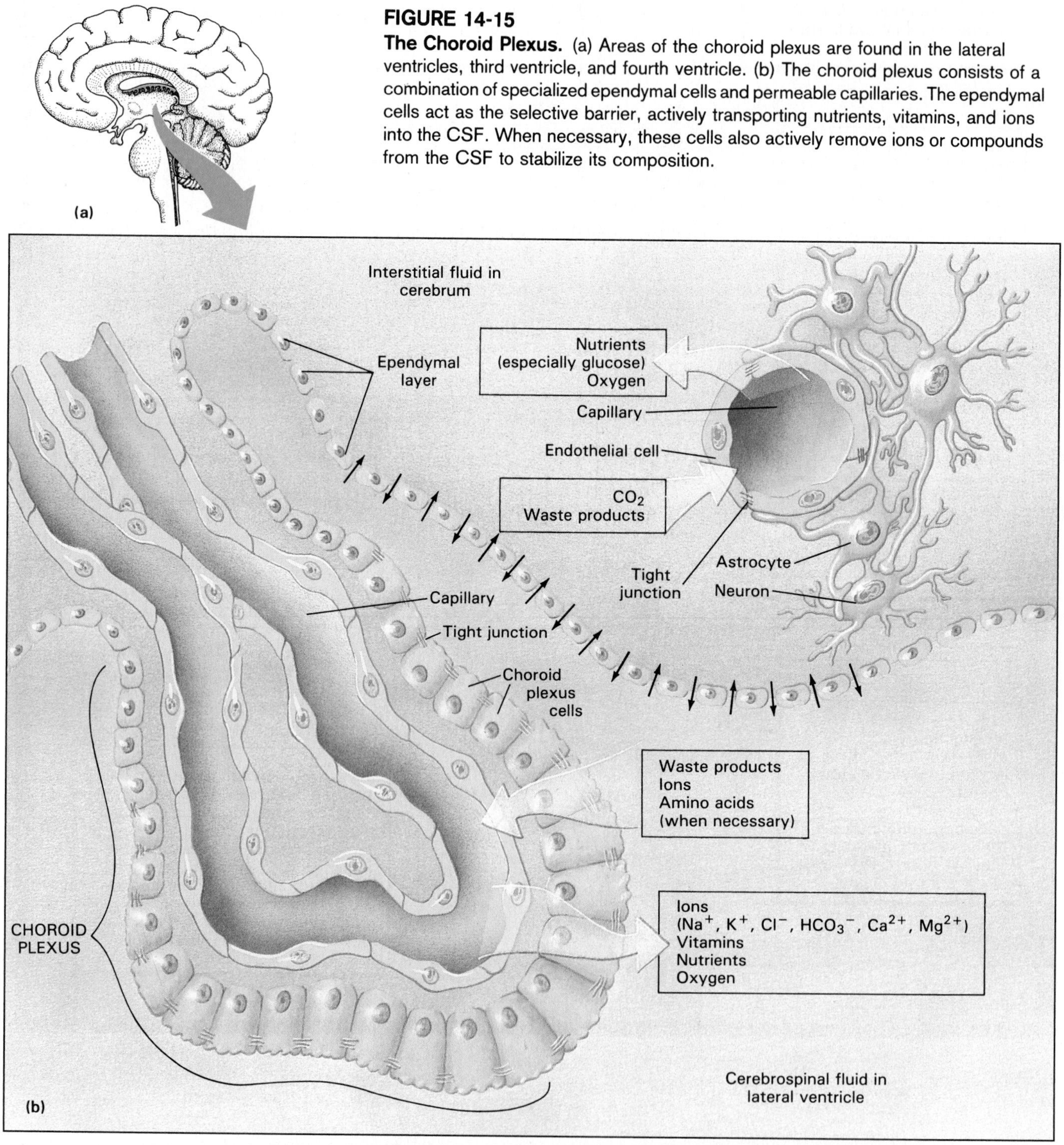

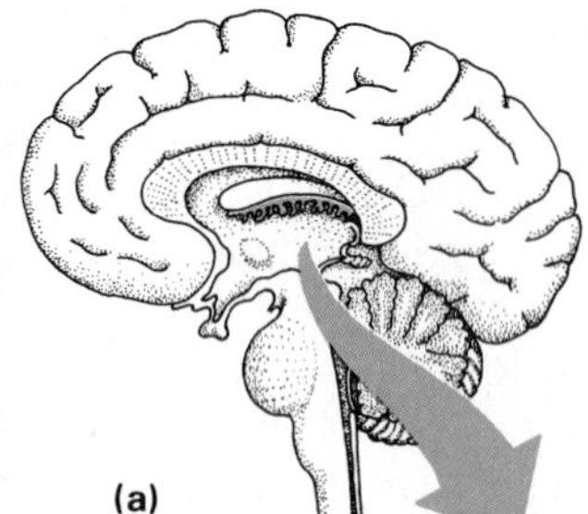

FIGURE 14-15
The Choroid Plexus. (a) Areas of the choroid plexus are found in the lateral ventricles, third ventricle, and fourth ventricle. (b) The choroid plexus consists of a combination of specialized ependymal cells and permeable capillaries. The ependymal cells act as the selective barrier, actively transporting nutrients, vitamins, and ions into the CSF. When necessary, these cells also actively remove ions or compounds from the CSF to stabilize its composition.

held within narrow limits. In contrast to the blood-brain barrier, the choroid plexus relies on active transport mechanisms that consume ATP. The regulation involves transport in both directions, and the choroid plexus removes waste products from the CSF and fine-tunes its composition over time. The differences in composition between cerebrospinal fluid and blood plasma (blood with the cellular elements removed) are quite pronounced; a detailed analysis can be found in Appendix V.

CEREBROSPINAL FLUID AND ITS CIRCULATION

Cerebrospinal fluid completely surrounds and bathes the exposed surfaces of the central nervous system. It has several vital functions:

1. It *provides cushioning* for delicate neural structures, like the air bag in a car.
2. It *provides support*, because the brain essentially floats in the cerebrospinal fluid. A human brain weighs about 1400 g in air, but only about 50 g when supported by the cerebrospinal fluid.
3. It *transports nutrients, chemical messengers, and waste products*. Except at the choroid plexus, the ependymal lining is freely permeable, and the CSF is in constant chemical communication with the interstitial fluid of the CNS.

Because free exchange occurs between the interstitial fluid and CSF, changes in CNS function may produce changes in the composition of the CSF. As noted in Chapter 13, a spinal tap can provide useful clinical information concerning CNS injury, infection, or disease. (p. 409) The total volume of CSF averages around 150 ml. Although the choroid plexus secretes that volume every few hours, the rate of removal normally keeps pace with the rate of production. If it does not, a variety of clinical problems may appear (see Clinical Comment: Hydrocephalus).

Figure 14-16 diagrams the circulation of cerebrospinal fluid. CSF forms at the choroid plexus and circulates between the ventricles, passes along the central canal, and reaches the subarachnoid space via three holes in the roof of the fourth ventricle. These passageways consist of the paired **lateral apertures** and a single **median aperture**.

Once inside the subarachnoid space, the CSF circulates around the spinal cord and cauda equina and across the surfaces of the brain. Along the falx cerebri, slender extensions of the arachnoid cross the subdural space and penetrate the inner layer of the dura mater. These **arachnoid villi** project into the **superior sagittal sinus**, a large vein situated within the falx cerebri. Diffusion across the arachnoid villi returns excess cerebrospinal fluid to the venous circulation.

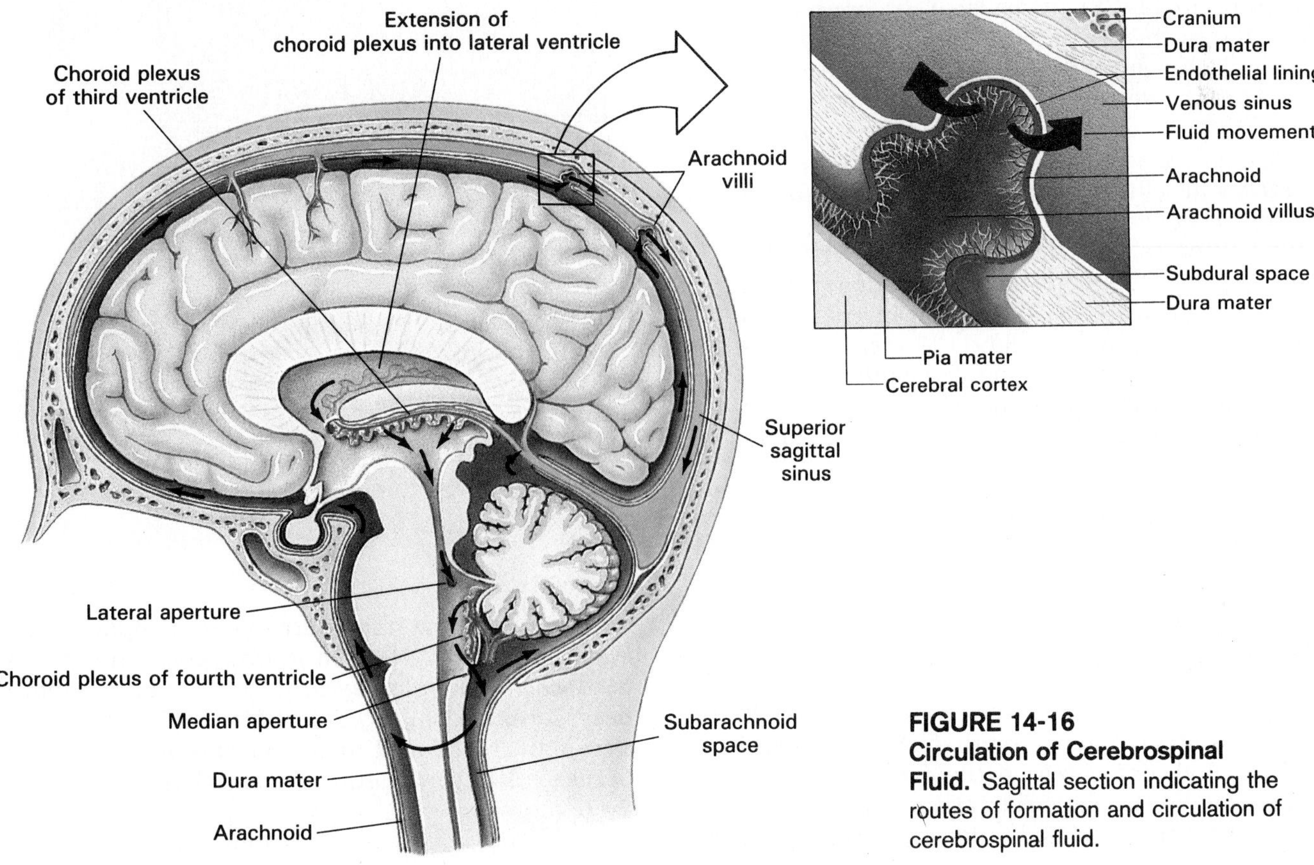

FIGURE 14-16 Circulation of Cerebrospinal Fluid. Sagittal section indicating the routes of formation and circulation of cerebrospinal fluid.

Clinical Comment: Hydrocephalus

The adult brain is surrounded by the inflexible bones of the cranium. The cranial cavity contains two fluids, blood and cerebrospinal fluid, and the relatively firm tissues of the brain. Because the total volume cannot change, when the volume of blood or CSF increases, the volume of the brain must decrease. In a subdural or epidural hemorrhage the fluid volume increases as blood collects within the cranial cavity. The rising **intracranial pressure** compresses the brain, leading to neural dysfunction that often ends in unconsciousness and death.

Any alteration in the rate of cerebrospinal fluid production will usually be matched by an increase in the rate of removal at the arachnoid villi. If this equilibrium is disturbed, clinical problems appear as the intracranial pressure changes. The volume of cerebrospinal fluid will increase if the rate of formation accelerates or the rate of removal decreases. In either event the increased fluid volume leads to compression and distortion of the brain. Increased rates of formation may accompany head injuries, but the most common problems arise from masses, such as tumors or abscesses, or from congenital abnormalities. These conditions have the same effect: they restrict the normal circulation and absorption of CSF. Because CSF production continues, the ventricles gradually expand, distorting the surrounding neural tissues and causing the deterioration of brain function.

Infants are especially sensitive to alterations in intracranial pressure, because the arachnoid villi do not appear until roughly 3 years of age. As in an adult, if intracranial pressure becomes abnormally high, the ventricles will expand. But in an infant the cranial sutures have yet to fuse, and the skull can enlarge to accommodate the extra fluid volume. This produces an enormously expanded skull, a condition called **hydrocephalus** or "water on the brain." Infant hydrocephalus (Figure 14-17) often results from blockage of the mesencephalic aqueduct due to a congenital defect or infection. Untreated infants often suffer some degree of mental retardation. Successful treatment usually involves the installation of a **shunt**, a bypass that drains the excess cerebrospinal fluid and reduces the intracranial pressure. The shunt can often be removed once the arachnoid villi begin functioning.

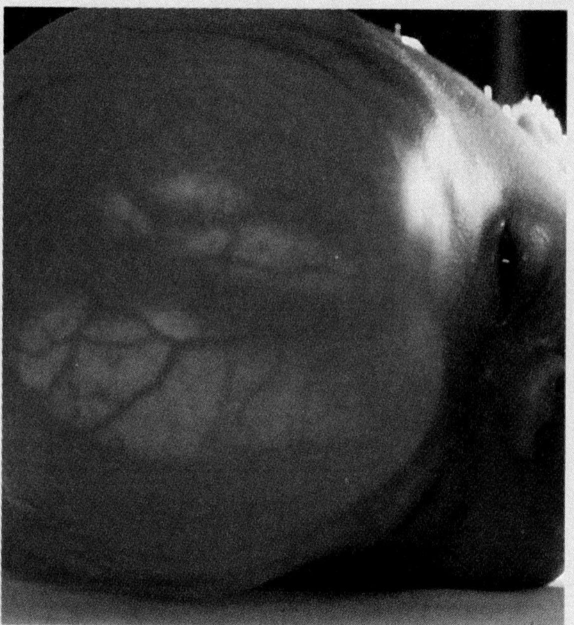

FIGURE 14-17
Hydrocephalus. This infant suffers from hydrocephalus, a condition usually caused by impaired circulation and removal of cerebrospinal fluid. CSF buildup leads to distortion of the brain and enlargement of the cranium.

■ The Cranial Nerves

The cranial nerves are components of the peripheral nervous system that connect to the brain rather than to the spinal cord. Twelve pairs of cranial nerves can be seen along the ventrolateral surface of the brain, each with a name related to its appearance or function. In order, the names are *olfactory, optic, oculomotor, trochlear, trigeminal, abducens, facial, vestibulocochlear, glossopharyngeal, vagus, spinal accessory*, and *hypoglossal*. Many of these names were first introduced in the discussion of foramina of the skull (review Figures 8-3 and 8-4). (pp. 241, 242)

Cranial nerves are also numbered according to

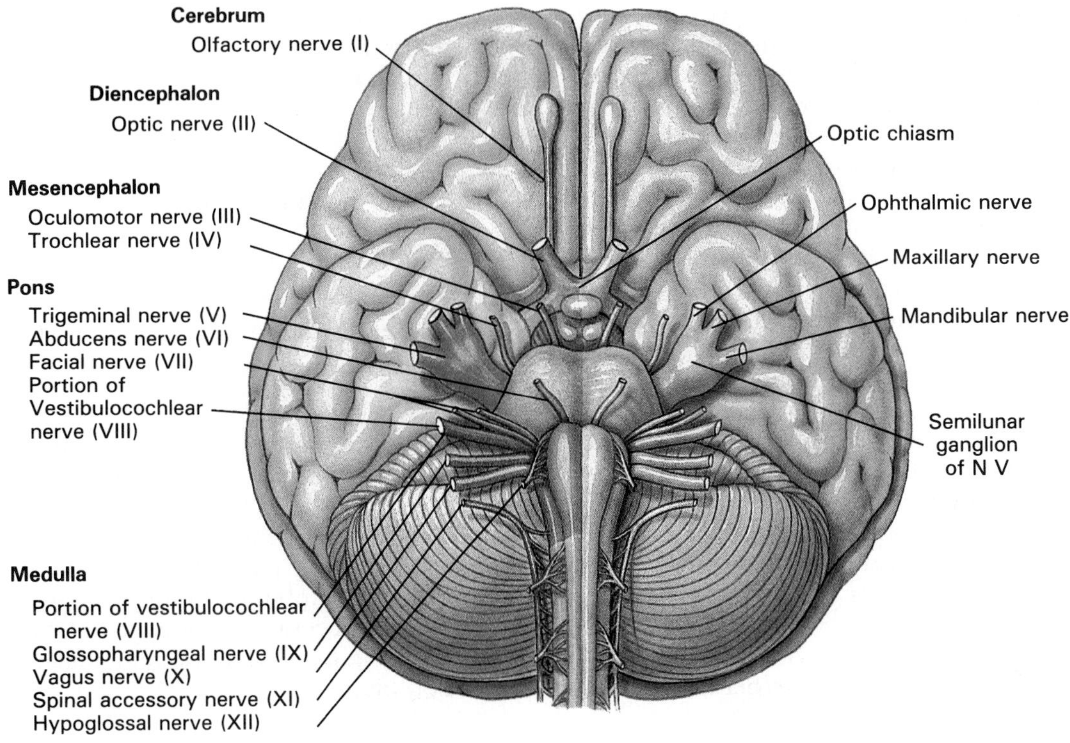

FIGURE 14-18
Origins of the Cranial Nerves. On the left half of this inferior view of the brain, colors highlight the relationships between the cranial nerves and specfic regions of the brain. On the right the brain surface is shown as it appears on gross dissection.

their position along the longitudinal axis of the brain. Roman numerals are usually used, either alone or with the prefix N or CN. For example, the abbreviations I, N I, and CN I refer to the first cranial nerve (the olfactory nerve). We will use the abbreviation N to avoid possible confusion with the cervical spinal nerves, designated by the letter C.

Each cranial nerve attaches to the brain near the associated sensory or motor nuclei. The sensory nuclei act as switching centers, with the postsynaptic neurons relaying the information to other nuclei or to processing centers within the cerebral or cerebellar cortex. In a similar fashion the motor nuclei receive convergent inputs from higher centers or from other nuclei along the brain stem.

The discussion that follows focuses on the *primary functions* of each cranial nerve. For convenience, each nerve will be classified as primarily sensory, motor, or mixed (sensory and motor). As is the case elsewhere in the PNS, a nerve containing thousands of motor fibers to a skeletal muscle will always carry a few sensory fibers from muscle spindles and tendon organs within that muscle. These sensory fibers will be assumed to be present, but ignored in the classification of the nerve. Similarly, regardless of their other functional roles, several cranial nerves (N III, VII, IX, and X) distribute autonomic fibers to peripheral ganglia, just as spinal nerves deliver them to ganglia along the spinal cord. These autonomic fibers will be discussed further in Chapter 16.

Figure 14-18 indicates the sites where the cranial nerves attach to the brain, and you should refer to this figure as the individual nerves are described. You may also find it useful to review Figures 8-3 and 8-4 to refresh your memory of the locations of the cranial bones and foramina. (pp. 241, 242)

The Olfactory Nerve (N I)

Primary function: Special sensory, smell
Origin: Receptors of olfactory epithelium
Passes through: Cribriform plate of ethmoid
Destination: Olfactory bulbs

FIGURE 14-19
The Olfactory Nerve

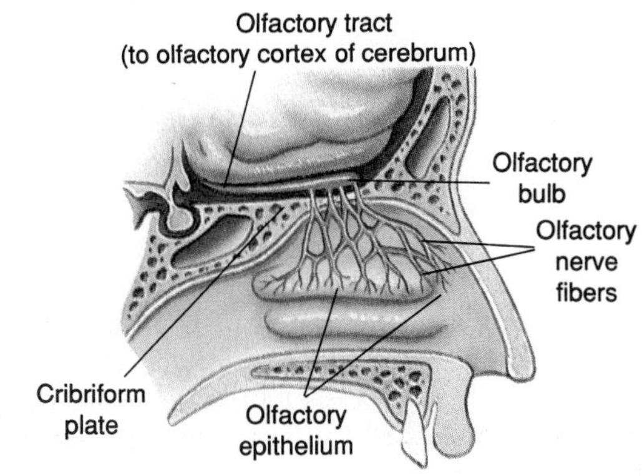

The first pair of cranial nerves carries special sensory information responsible for the sense of smell. As indicated in Figure 14-19, the receptors are specialized neurons in the epithelium covering the surfaces of the superior nasal conchae and nasal septum. Axons from these sensory neurons collect to form 20 or more bundles that penetrate the cribriform plate of the ethmoid. Almost at once they enter the **olfactory bulbs**, neural masses that lie on either side of the crista galli. The olfactory afferents synapse within the olfactory bulbs, and the axons of the postsynaptic neurons proceed to the cerebrum along the slender **olfactory tracts**.

The olfactory nerves are not as anatomically distinct as the peripheral nerves discussed in Chapter 13. Because they looked like a normal nerve, anatomists a hundred years ago called the *olfactory tracts* the first cranial nerve. Later studies demonstrated that the olfactory tracts and bulbs are actually part of the cerebrum, so the texts had to be revised. By then the numbering system was already firmly established, and anatomists were left with a forest of tiny olfactory nerve bundles lumped together as N I.

The olfactory nerves are the only cranial nerves attached to the cerebrum. The rest originate or terminate within nuclei of the brain stem.

The Optic Nerve (N II)

Primary function: Special sensory, vision
Origin: Retina of eye
Passes through: Optic foramen of sphenoid
Destination: Diencephalon via optic chiasm

The **optic nerves** (N II) carry visual information from special sensory ganglia in the eyes. These nerves, shown in Figure 14-20, pass through the optic foramina of the sphenoid before reaching the **optic chiasm** ("crossing") at the ventral and anterior margin of the diencephalon. Some reorganization of the axon bundles occurs at the chiasm before they continue to the thalamus (lateral geniculates) and mesencephalon (superior colliculus).

FIGURE 14-20
The Optic Nerve

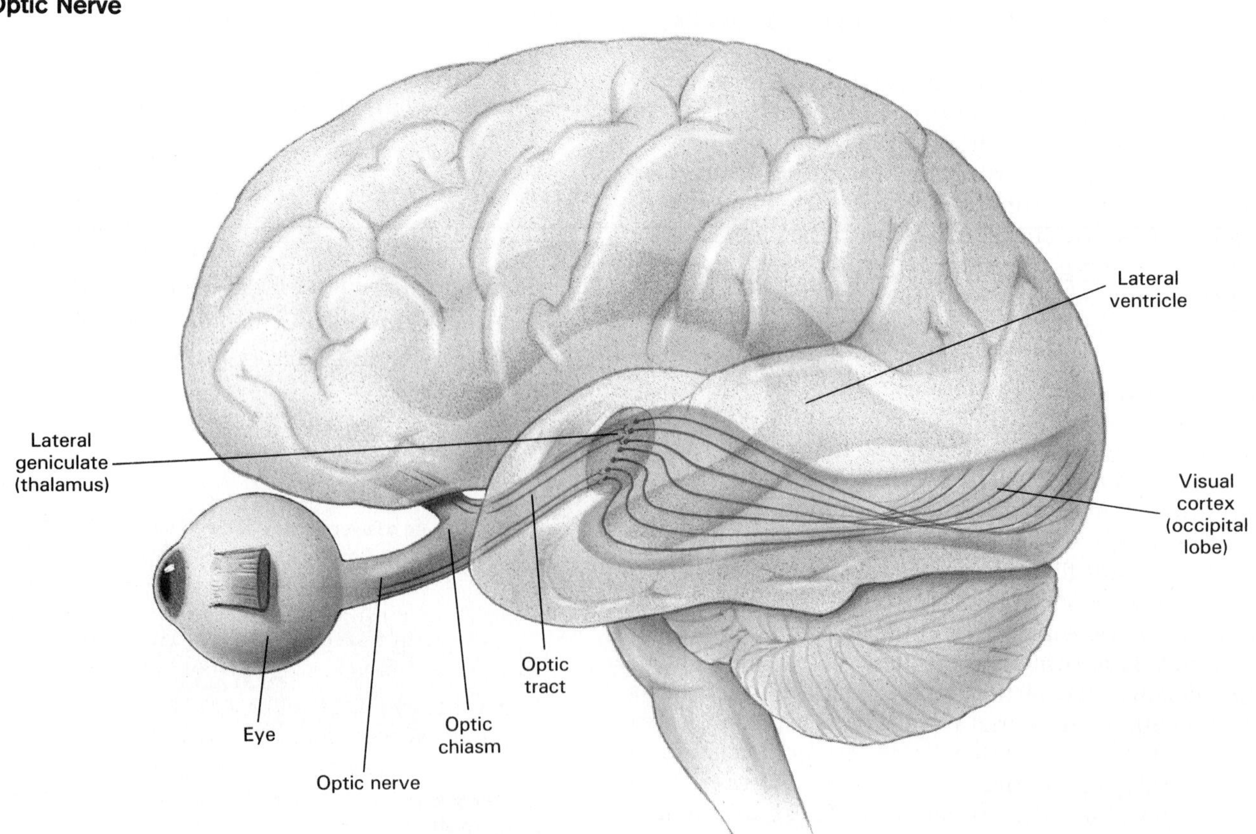

The Oculomotor Nerve (N III)

Primary function: Motor, eye movements

Origin: Mesencephalon

Passes through: Superior orbital fissure of sphenoid

Destination: Inferior, medial, and superior rectus, inferior oblique, and levator palpebrae muscles

The mesencephalon contains the motor nuclei controlling the third and fourth cranial nerves. These nuclei and their nerves are shown in Figure 14-21.

The **oculomotor nerve** (N III) is the primary source of innervation for the extrinsic oculomotor muscles, the muscles that move the eyeball (see Figure 11-5). (p. 337) This cranial nerve emerges from the ventral surface of the mesencephalon and pierces the orbit wall at the superior orbital fissure. It controls four of the six oculomotor muscles and also services the levator palpebrae, the muscle that raises the upper eyelid (see Figure 11-4). (p. 335)

The oculomotor nerve also delivers preganglionic autonomic (*parasympathetic*) fibers to neurons of the **ciliary ganglion**. The ganglionic neurons control intrinsic eye muscles.

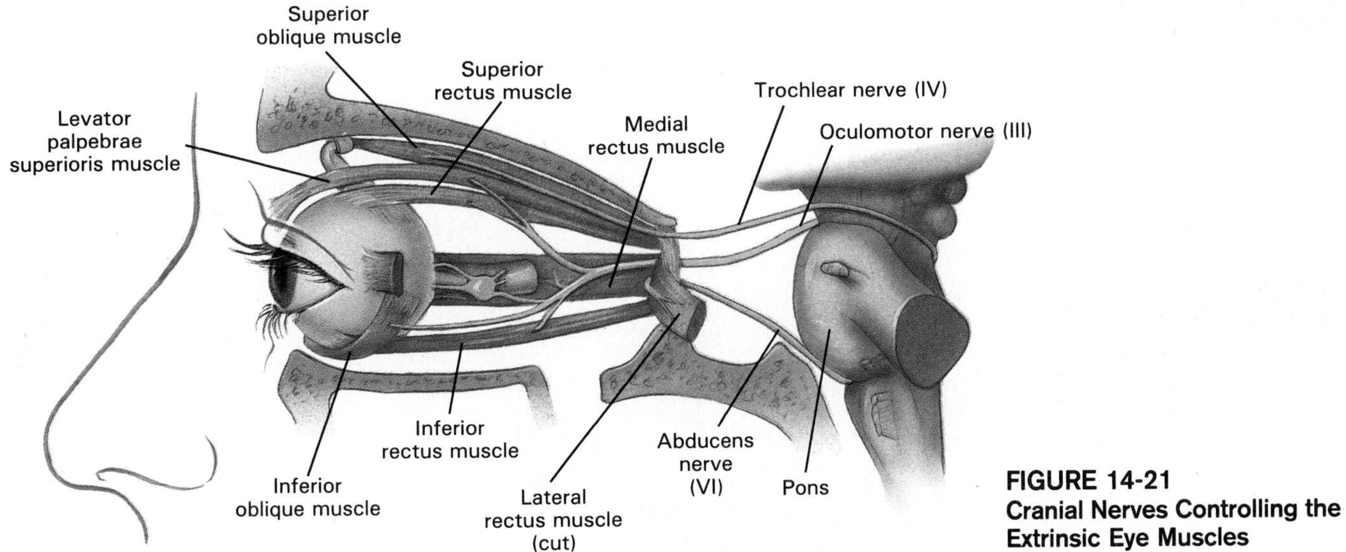

FIGURE 14-21
Cranial Nerves Controlling the Extrinsic Eye Muscles

The Trochlear Nerve (N IV)

Primary function: Motor, eye movements

Origin: Mesencephalon

Passes through: Superior orbital fissure of sphenoid

Destination: Superior oblique muscle

The **trochlear** (TRŌK-lē-ar) **nerve**, smallest of the cranial nerves, innervates the superior oblique muscle of the eye. The origin and distribution of this nerve are shown in Figure 14-21. The motor nucleus lies in the ventrolateral portion of the mesencephalon, but the fibers emerge from the surface of the tectum to enter the orbit at the superior orbital fissure. The name "trochlear nerve" should remind you that the innervated muscle passes through a ligamentous pulley (= trochlea) on its way to an insertion on the surface of the eye.

The Trigeminal Nerve (N V)

Primary function: Mixed (sensory and motor); ophthalmic and maxillary branches sensory, mandibular branch mixed

Origin:

Ophthalmic branch (sensory): orbital structures, nasal cavity, skin of forehead, upper eyelid, eyebrow, nose (part)

Maxillary branch (sensory): lower eyelid, upper lip, gums, and teeth; cheek; nose, palate, and pharynx (part)

Mandibular branch (mixed): sensory from lower gums, teeth, and lips; palate and tongue (part); motor from motor nuclei of pons

Passes through: Ophthalmic branch via superior orbital fissure, maxillary branch via foramen rotundum, mandibular branch via foramen ovale

Destination: Ophthalmic and maxillary branches to sensory nuclei in pons; mandibular branch innervates muscles of mastication

The pons contains the nuclei associated with three cranial nerves (N V, VI, and VII) and contributes to the control of a fourth (N VIII). The trochlear nerve is easily overlooked, but the **trigeminal** (trī-JEM-i-nal) **nerve** is hard to miss, for it is the largest of the cranial nerves. This mixed nerve, detailed in Figure 14-22, provides sensory information from the head and face and motor control over the muscles of mastication. In structure this nerve resembles an enlarged spinal nerve. Sensory (dorsal) and motor (ventral) roots originate on the lateral surface of the pons. The sensory branch is larger, and the enormous **semilunar ganglion** contains the cell bodies of the sensory neurons. As the name implies, the *tri*geminal

FIGURE 14-22
The Trigeminal Nerve

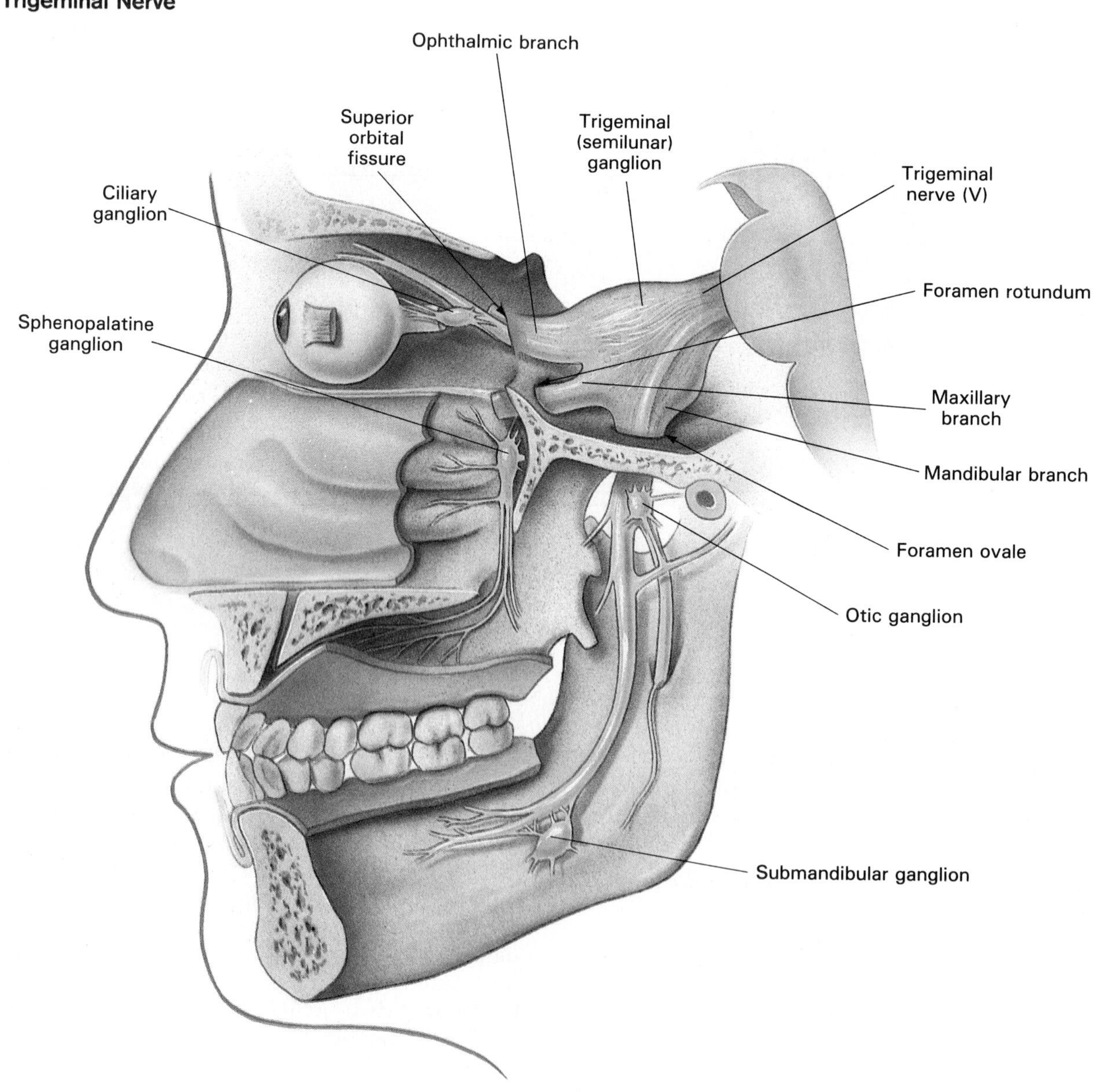

has three major branches; the relatively small motor root contributes to only one of the three.

The first branch, or **ophthalmic branch**, of the trigeminal is purely sensory. It enters the orbit via the superior orbital fissure before branching further. This nerve innervates orbital structures, the nasal cavity and sinuses, and the skin of the forehead, eyebrows, eyelids, and nose.

The second branch, or **maxillary branch**, leaves the cranium at the foramen rotundum, entering the floor of the orbit through the inferior orbital fissure and reaching the face via the infraorbital foramen. It then radiates to supply the lower eyelid, upper lip, cheek, and nose. Deeper sensory structures of the upper gums and teeth, the palate, and portions of the pharynx are also innervated by the maxillary nerve. Preganglionic autonomic (parasympathetic) fibers travel in this branch to reach the **sphenopalatine ganglion**. Motor neurons in this ganglion innervate the lacrimal (tear) gland of the eye.

The third branch, or **mandibular branch**, is the largest, and includes all of the fibers contained within the motor root. This branch exits the cranium through the foramen ovale. The somatic motor components of the mandibular nerve innervate the muscles of mastication. Autonomic (parasympathetic) fibers synapse in the **submandibular ganglion** whose neurons innervate salivary glands in the floor of the mouth. The sensory fibers carry proprioceptive information from those muscles and monitor (1) the skin of the temples; (2) the lateral surfaces, gums, and teeth of the mandible; (3) the salivary glands; and (4) the anterior portions of the tongue. [**CM**: *Tic Douloureux*]

The Abducens Nerve (N VI)

Primary function: Motor, eye movements

Origin: Pons

Passes through: Superior orbital fissure of sphenoid

Destination: Lateral rectus muscle

The **abducens** (ab-DU-senz) **nerve** innervates the lateral rectus, the sixth of the extrinsic oculomotor muscles. The nerve emerges from the inferior surface of the brain at the border between the pons and the medulla. It reaches the orbit through the superior orbital fissure in company with the oculomotor and trochlear nerves, as indicated in Figure 14-21.

The Facial Nerve (N VII)

Primary function: Mixed (sensory and motor)

Origin: Sensory from taste receptors on anterior two-thirds of tongue; motor from motor nuclei of pons

Passes through: Internal acoustic meatus of temporal, along internal acoustic canal and facial canal to reach stylomastoid foramen

Destination: Sensory to sensory nuclei of pons; motor to muscles of facial expression

The **facial nerve**, like the trigeminal, is a mixed nerve. As indicated in Figure 14-23, the sensory neurons reside within the **geniculate ganglion**, and the motor nuclei lie within the pons. The sensory and motor roots emerge from the side of the pons and enter the internal acoustic meatus of the temporal bone. The nerve then passes through the facial canal to reach the face via the stylomastoid foramen. The motor fibers control the superficial muscles of the scalp and face and deep muscles near the ear. The sensory neurons monitor proprioceptors in the facial muscles, provide deep pressure sensations over the face, and receive taste information from receptors along the anterior two-thirds of the tongue. [**CM**: *Bell's Palsy*]

FIGURE 14-23
The Facial Nerve

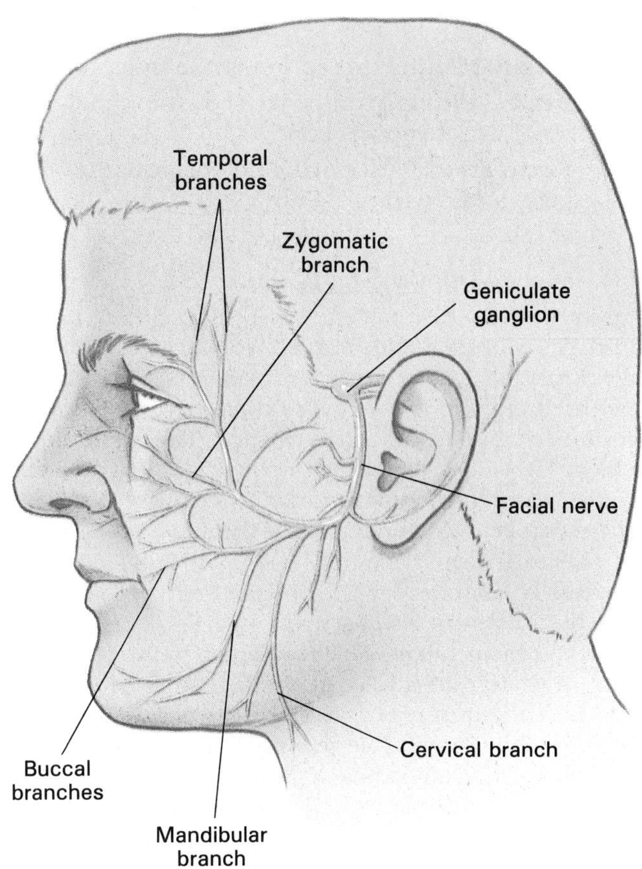

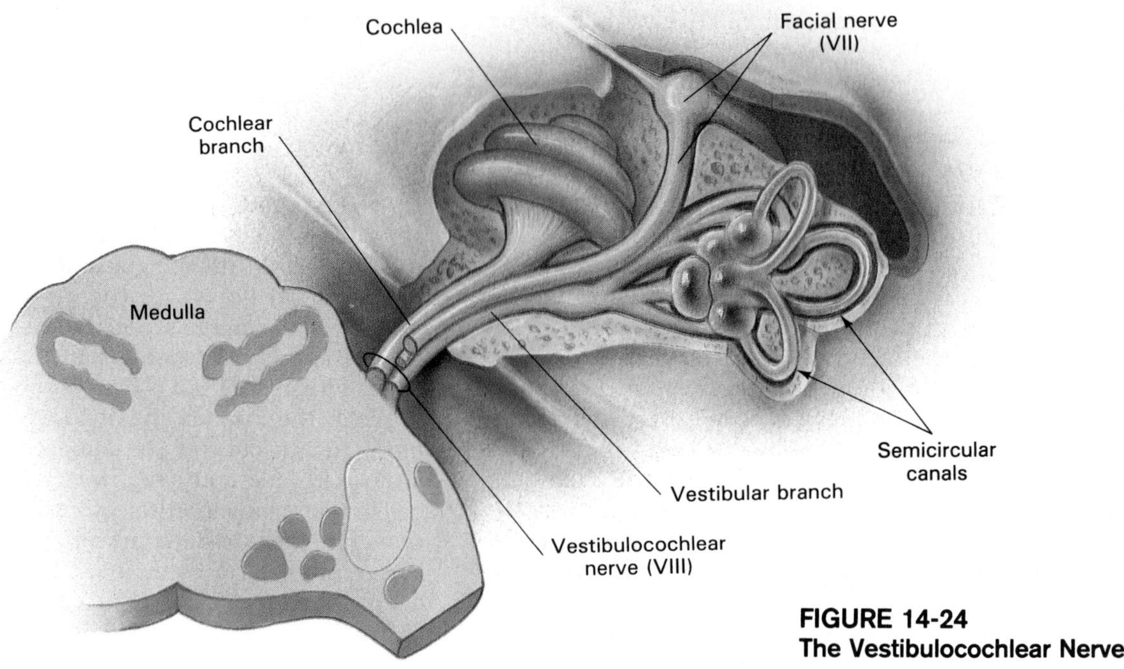

FIGURE 14-24
The Vestibulocochlear Nerve

The Vestibulocochlear Nerve (N VIII)

Primary function: Sensory (special), balance (vestibular branch) and hearing (cochlear branch)

Origin: Monitors receptors of the inner ear

Passes through: Internal acoustic canal and meatus of temporal bone

Destination: Vestibular and cochlear nuclei of pons and medulla

The **vestibulocochlear nerve** (Figure 14-24) is also known as the *acoustic nerve*, the *auditory nerve*, and the *statoacoustic nerve*. We will use the term vestibulocochlear because it makes it easy to remember the names of its two major branches: the *vestibular branch* and the *cochlear branch*. The vestibulocochlear nerve lies posterior to the origin of the facial nerve, straddling the boundary between the pons and the medulla. This nerve reaches the sensory receptors of the inner ear by entering the internal acoustic meatus in company with the facial nerve. There are two distinct bundles of sensory fibers within the vestibulocochlear nerve. The **vestibular nerve** (*vestibulum*, a cavity) originates at the receptors of the *vestibule*, the portion of the inner ear concerned with balance sensations. The sensory neurons are located within an adjacent sensory ganglion, and their axons target the **vestibular nuclei** of the medulla. These afferents convey information concerning position, movement, and balance. The **cochlear nerve** (KOK-lē-ar; *cochlea*, snail shell) monitors the receptors providing the sense of hearing. The nerve cells are located within a peripheral ganglion, and their axons synapse within the **cochlear nuclei** of the medulla. Axons leaving the vestibular and cochlear nuclei relay the sensory information to other centers or initiate reflexive motor responses. The sensory mechanisms, pathways, and motor responses to vestibular and cochlear stimulation will be discussed in detail in Chapter 17.

The Glossopharyngeal Nerve (N IX)

Primary function: Mixed (sensory and motor)

Origin: Sensory from posterior one-third of the tongue, part of the pharynx and palate, the carotid arteries of the neck; motor from motor nuclei of medulla

Passes through: Jugular foramen of inferior surface of temporal bone to foramen lacerum between occipital and temporal bones (superior surface)

Destination: Sensory to sensory nuclei of medulla; motor to pharyngeal muscles involved in swallowing

In addition to the vestibular nucleus of N VIII, the medulla contains the sensory and motor nuclei for the ninth, tenth, eleventh, and twelfth cranial nerves. The **glossopharyngeal nerve** (glos-ō-fah-RIN-je-al; *glossum*, tongue) innervates the tongue and pharynx (Figure 14-25). The glossopharyngeal nerve

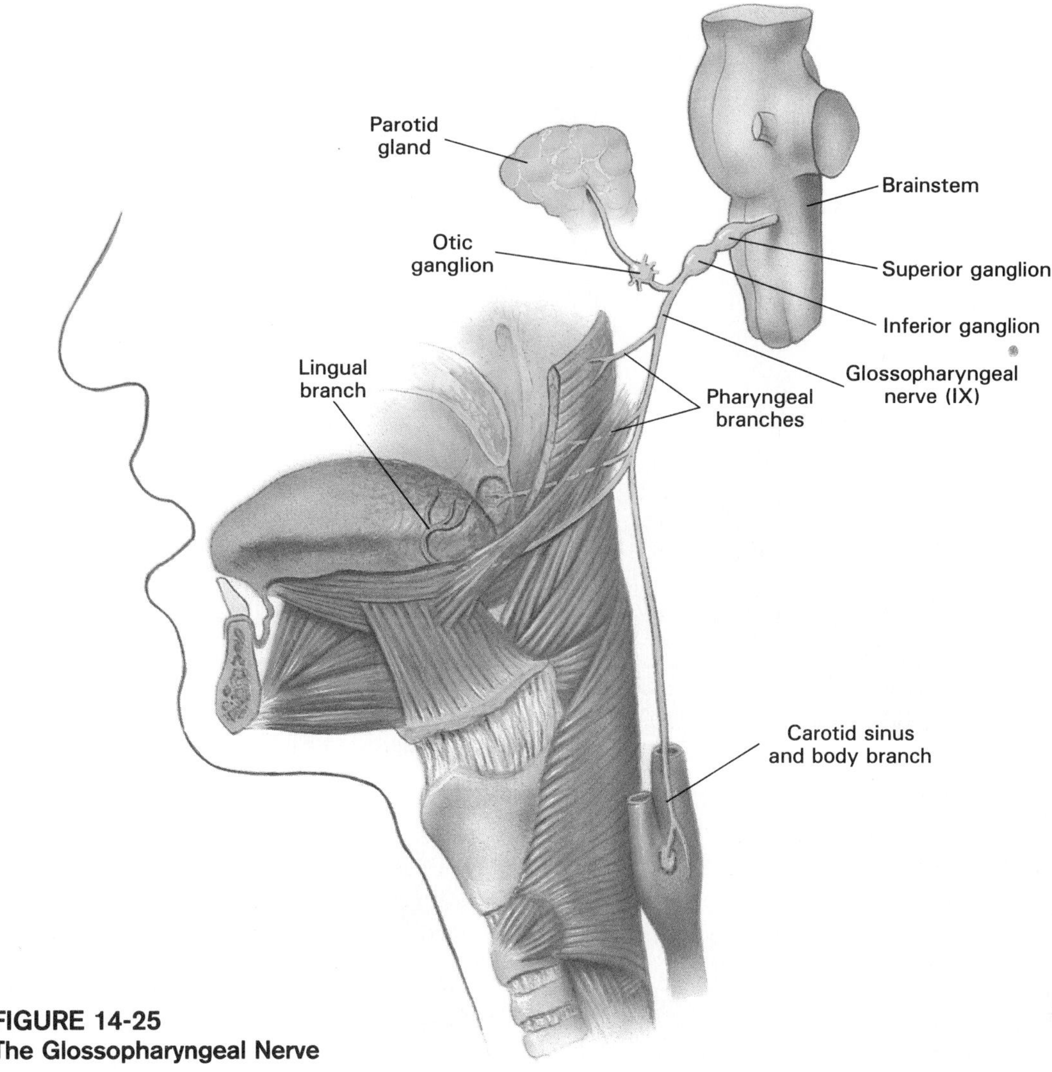

FIGURE 14-25
The Glossopharyngeal Nerve

exits the cranium via the jugular foramen in company with N X and N XI. It is a mixed nerve, but sensory fibers are most abundant; the sensory neurons are located in the **petrosal ganglion**. The efferent fibers control the pharyngeal muscles involved in swallowing.

The sensory neurons are located within a pair of small ganglia near the base of the nerve. The afferents carry general sensory information from the lining of the pharynx and the soft palate to a nucleus in the medulla. The glossopharyngeal nerve also provides taste sensations from the posterior third of the tongue and has special receptors monitoring the blood pressure and dissolved gas concentrations within major blood vessels.

The Vagus Nerve (N X)

Primary function: Mixed (sensory and motor)

Origin: Sensory from pharynx (part); pinna and external auditory meatus and canal, diaphragm, visceral organs in thoracic and abdominopelvic cavities, motor from motor nuclei in medulla

Passes through: Jugular foramen between occipital and temporal bones

Destination: Sensory fibers to sensory nuclei and autonomic centers of medulla; motor fibers to palatal and pharyngeal muscles, to smooth muscles of the digestive, respiratory, and cardiovascular systems, to cardiac muscle of the heart

FIGURE 14-26
The Vagus Nerve

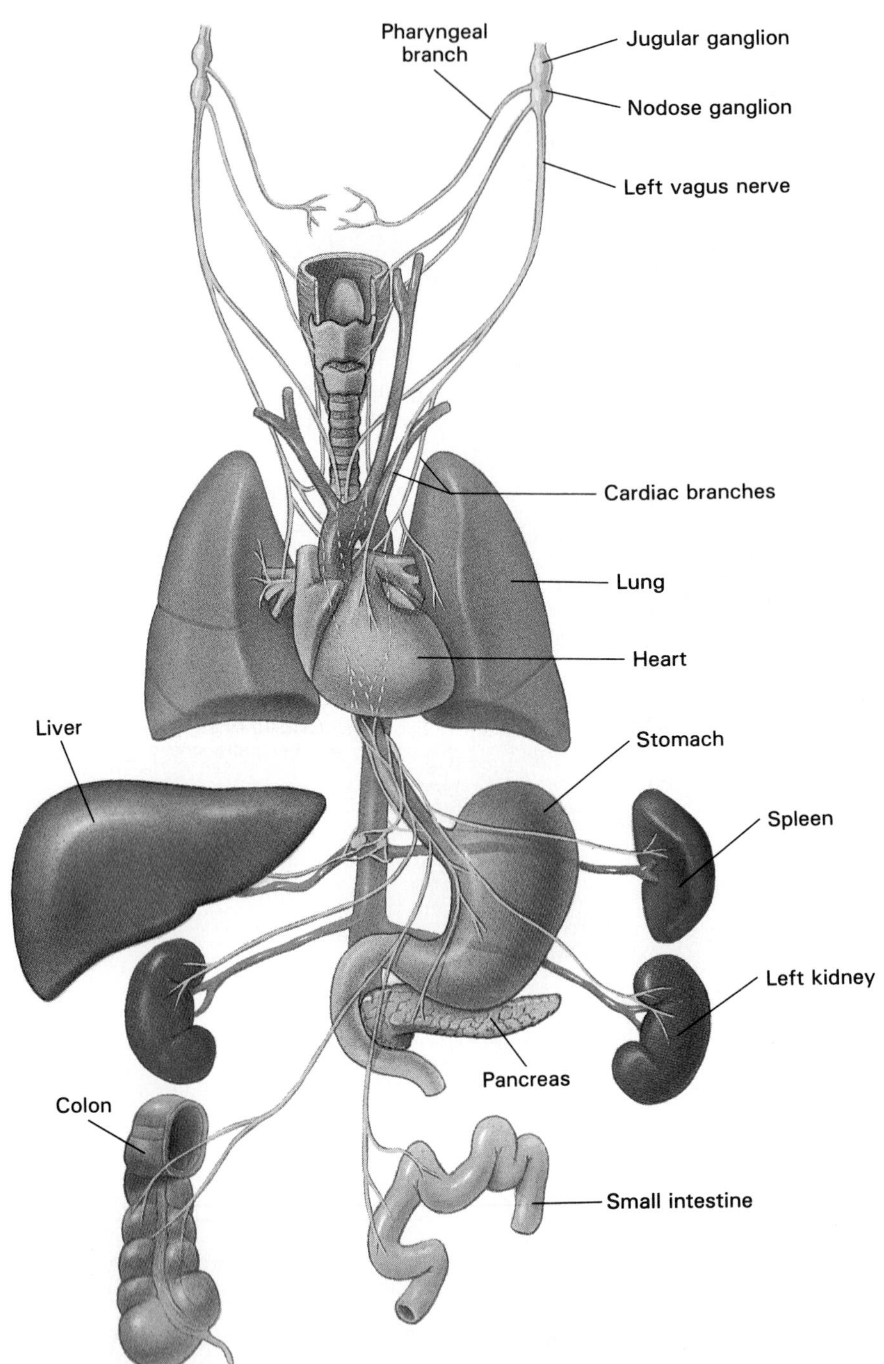

The **vagus** (VĀ-gus) **nerve** arises immediately posterior to the glossopharyngeal. Many small rootlets contribute to its formation, and developmental studies indicate that this nerve probably represents the fusion of several smaller cranial nerves during our evolution. As its name suggests (*vagus*, wandering) the vagus nerve branches and radiates extensively; Figure 14-26 indicates only the general pattern of distribution.

Sensory neurons are located within the **jugular** and **nodose ganglia** (NŌ-dos; knot). The vagus provides somatic sensory information concerning the external auditory meatus, a portion of the ear, and the diaphragm, and special sensory information from pharyngeal taste receptors. But the majority of the vagal afferents provide visceral sensory information from receptors along the esophagus, respiratory tract, and abdominal viscera as distant as the terminal segments of the large intestine. Vagal afferents are vital to the autonomic control of visceral function, but because the information often fails to reach the cerebral cortex we are not consciously aware of the sensations they provide.

The motor components of the vagus are equally diverse. Vagal efferents control several skeletal muscles of the soft palate and pharynx. They also carry preganglionic autonomic fibers that affect the heart and control smooth muscles and glands within the areas monitored by vagal sensory afferents, including the stomach, intestines, and gallbladder.

The Spinal Accessory Nerve (N XI)

Primary function: Motor

Origin: Motor nuclei of spinal cord and medulla

Passes through: Jugular foramen between occipital and temporal bones

Destination: Medullary branch innervates voluntary muscles of palate, pharynx, and larynx; spinal branch controls sternocleidomastoid and trapezius muscles

The **spinal accessory nerve**, sometimes called simply the *accessory nerve*, differs from other cranial nerves in that some of its motor fibers originate in the lateral gray horns of the first five cervical vertebrae (Figure 14-27a). These fibers enter the cranium through the foramen magnum, unite with the motor fibers arising at a nucleus in the medulla, and leave the cranium through the jugular foramen. Outside the skull, the **medullary branch** innervates the voluntary swallowing muscles of the soft palate and pharynx and the intrinsic laryngeal muscles that control the vocal cords. The **spinal branch** controls the sternocleidomastoid and trapezius muscles associated with the pectoral girdle (see Table 11-10). (p. 349)

The Hypoglossal Nerve (N XII)

Primary function: Motor, tongue movements

Origin: Motor nuclei of medulla

Passes through: Hypoglossal canal of occipital bone

Destination: Muscles of the tongue

The **hypoglossal** (hī-pō-GLOS-al) **nerve** leaves the cranium through the hypoglossal canal of the occipital bone (Figure 14-27b). It then follows a curving path to reach the muscles of the tongue. This nerve provides voluntary motor control over movements of the tongue.

A SUMMARY OF CRANIAL NERVE BRANCHES AND FUNCTIONS

Few people are able to remember the names, numbers, and functions of the cranial nerves without a struggle. To aid memorization, mnemonic devices may prove useful. The most famous and oft-repeated is *On Old Olympus' Towering Top A Finn And Ger-*

FIGURE 14-27
The Spinal Accessory and Hypoglossal Nerves

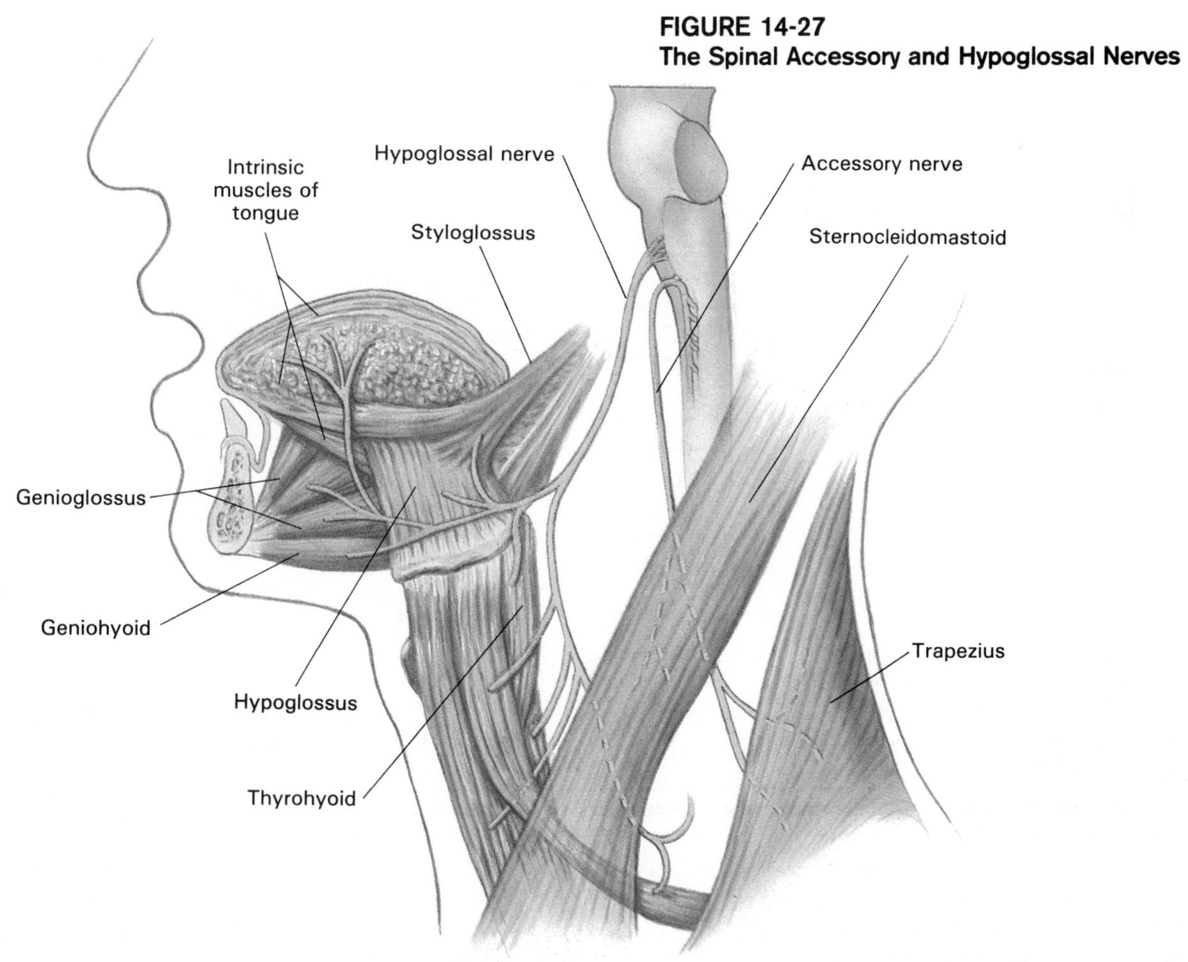

TABLE 14-11 A Functional Classification of Cranial Nerves

Classification	*Primary Functions*	*Cranial Nerves*
Special sensory	Carry information from specialized exteroceptive organs to the brain	I, II, VIII (olfactory, optic, and vestibulocochlear)
Motor	Control the extrinsic oculomotor muscles	III, IV, VI (oculomotor, trochlear, and abducens)
	Provides voluntary motor control over large superficial muscles of the back	XI (spinal accessory)
	Controls the tongue muscles	XII (hypoglossal)
Mixed	Carry sensory information and voluntary/involuntary motor commands	V, VII, IX, X (trigeminal, facial, glossopharyngeal, and vagus)

man Viewed Some Hops.[1] A more modern one, *Oh, Once One Takes The Anatomy Final, Very Good Vacations Seem Heavenly*, may be a bit easier to remember. After learning the names, recognizing the basic distribution and function of a specific nerve becomes easier if you reorganize the list. Three basic categories are presented in Table 14-11. This categorization minimizes the diversity of functions present, but subdividing the "deadly dozen" does highlight the general patterns.

[1] The *and* refers to the acoustic nerve.

Cranial Reflexes

Cranial reflexes are reflex arcs that involve the sensory and motor fibers of cranial nerves. Many examples of cranial reflexes are discussed in later chapters. For example, cranial somatic reflexes involving special sensory stimuli are discussed in Chapter 17, and visceral reflexes are detailed in Chapters 15, 21, 23, and 25. Therefore, this section will simply provide an overview and general introduction.

TABLE 14-12 Cranial Reflexes

Reflex	*Stimulus*	*Afferents*	*Central Synapse*	*Efferents*	*Response*
Corneal reflex	Contact with corneal surface	N V (trigeminal)	Motor nucleus for N VII (facial)	N VII	Blinking of eye
Tympanic reflex	Loud noise	N VIII (vestibulocochlear)	Inferior colliculus (midbrain)	N VII	Reduced movement of auditory ossicles
Auditory reflexes	Loud noise	N VIII	Motor nuclei of brain stem and spinal cord	N III, IV, VI, VII, X, cervical nerves	Movements triggered by sudden sounds
Vestibulo-ocular reflexes	Rotation of head	N VIII	Motor nuclei controlling eye muscles	N III, IV, VI	Opposite movement of eyes to stabilize field of vision
Direct light reflex	Light striking photoreceptors	N II (optic)	Superior colliculus (midbrain)	N III (oculomotor)	Constriction of ipsilateral pupil
Consensual light reflex	Light striking photoreceptors	N II	Superior colliculus	N III	Constriction of contralateral pupil

Table 14-12 lists representative examples of cranial reflexes and their functions. These reflexes are clinically important because they provide a quick and easy method for checking the condition of cranial nerves and specific nuclei and tracts in the brain.

Cranial somatic reflexes are seldom more complex than the somatic reflexes of the spinal cord. This table includes four somatic reflexes: the *corneal reflex*, the *tympanic reflex*, the *auditory reflexes*, and the *vestibulo-ocular reflexes*.

1. The **corneal reflex** produces rapid blinking of the affected eye following contact to the *cornea*, the transparent region on the anterior surface of the eyeball. The blinking action helps to sweep away irritants and protect the delicate corneal surface. This reflex is often used to test the sensory function of the trigeminal nerve.

2. The **tympanic reflex** protects the auditory (hearing) receptors of the inner ear from loud noises. Distortion of the eardrum, or *tympanum*, stimulates sensory neurons of the trigeminal nerve. The afferent fibers synapse in the inferior colliculus of the midbrain, and the postsynaptic neurons control motor neurons of the facial nerve. These motor neurons direct the contraction of small skeletal muscles that tense the auditory ossicles and limit the amount of sound energy relayed to the inner ear. (The relationship between the tympanum, auditory ossicles, and inner ear was described in Chapter 8.) ∞ (p. 246)

3. **Auditory reflexes** trigger somatic motor responses to sounds, especially loud or sudden noises. The pattern of motor response varies depending on the volume of the sound and the direction to the source. For example, a sudden sound behind and to the right may produce a rotation of the eyes, head, and neck to that side. The more complex auditory responses may include motor commands traveling over spinal nerves as well as cranial nerves.

4. The **vestibulo-ocular reflexes** are also highly variable. The stimulus is a rotational movement of the head; the response is an eye movement in the opposite direction. For a demonstration of this reflex, sit on a stool and stare at an object in front of you. Now have someone rotate the stool to the right. Your eyes will automatically swing to the left, keeping the object in view.

It also lists two very simple examples of cranial reflexes that control visceral motor activity: the light reflexes involving cranial nerves II and III. These reflexes control the amount of light entering the eye. When a bright light shines into one eye, the pupil in that eye constricts, reducing the amount of light reaching the visual receptors. This is the **direct light reflex**. The reflex arc proceeds from the visual receptors to neurons in the retina, along afferent fibers in the optic nerve to processing centers in the midbrain, then back along efferent fibers to the pupillary constrictors. The pupil of the other eye constricts at the same time, a response called the **consensual light reflex**. These reflexes are often used to check the integrity of the optic nerves and processing centers in the midbrain.

The brain contains many reflex centers that control visceral motor activity. Many of these reflex centers are in the medulla, and they can direct very complex visceral motor responses to stimuli. Examples discussed in later chapters include the vomiting, coughing, sneezing, gagging, and swallowing reflexes and reflexes involved in the control of respiratory, digestive, and cardiovascular function.

Chapters 12–14 introduced (1) the organization of neural tissue, (2) the spinal cord and spinal nerves, and (3) the brain and cranial nerves. The next three chapters examine the integrated functioning of the nervous system. Chapter 15 focuses on the processing and integration of somatic sensory information, and the neural control of voluntary and involuntary somatic motor activities. Chapter 15 also considers higher-order functions, such as memory and learning, that can affect the processing of sensory information and the patterns of somatic motor response. Chapter 16 explores the autonomic nervous system and the regulation of visceral motor function.

- √ **How would decreased diffusion across the arachnoid villi affect the volume of cerebrospinal fluid in the ventricles?**
- √ **Damage to which of the cranial nerves can result in death?**
- √ **John is experiencing problems in moving his tongue. His doctor tells him it is due to pressure on a cranial nerve. Which cranial nerve is involved?**
- √ **What symptoms would you associate with damage to the abducens nerve (N VI)?**

CHAPTER REVIEW

Review of Selected Clinical Terms

psychosomatic (*p. 445*)
A term describing physiological symptoms which have a psychological basis.

decerebrate rigidity (*p. 447*)
A generalized state of muscular contraction resulting from loss of CNS inhibitory control.

Parkinson's disease (paralysis agitans) (*p. 447*)
A condition characterized by a pronounced increase in muscle tone, resulting from uninhibited excitatory neurons.

spasticity (*p. 447*)
A condition characterized by hesitant, jerky voluntary movements and increased muscle tone.

tremor (*p. 447*)
A background shaking of the limbs resulting from a "tug of war" between antagonistic muscle groups.

epidural hemorrhage (*p. 453*)
A condition involving bleeding into the epidural spaces.

subdural hemorrhage (*p. 453*)
A condition in which blood accumulates between the dura and the arachnoid.

hydrocephalus (*p. 458*)
Also known as "water on the brain"; a condition in which the skull expands to accommodate extra fluid.

Additional Terms of Clinical Importance

THE CEREBRAL NUCLEI, SUBSTANTIA NIGRA, AND PARKINSON'S DISEASE (*p. 447*): **MPTP**

HYDROCEPHALUS (*p. 458*)
intracranial pressure

Study Outline

Related Key Terms

Introduction (pp. 433–434)

1. The brain is far more complex than the spinal cord; its complexity makes it adaptable but slower in response than spinal reflexes.

Gross Anatomy and Organization (pp. 434–436)

MAJOR DIVISIONS AND PROMINENT LANDMARKS (pp. 434–435)

1. There are six regions in the adult brain: cerebrum, diencephalon, mesencephalon (midbrain), cerebellum, pons, and medulla oblongata.
2. Conscious thought, intellectual functions, memory, and complex involuntary motor patterns originate in the **cerebrum**. The **cerebellum** adjusts voluntary and involuntary motor activities based on sensory data and stored memories. — **cerebral hemispheres**, **brain stem**
3. The walls of the **diencephalon** form the **thalamus** that contains relay and processing centers for sensory data. The **hypothalamus** contains centers involved with emotions, autonomic function, and hormone production.
4. The **midbrain (mesencephalon)** processes visual and auditory information and generates involuntary somatic motor responses. The **pons** connects the cerebellum to the brain stem, and is involved with somatic and visceral motor control. The spinal cord connects to the brain at the **medulla oblongata**, which relays sensory information and regulates autonomic functions.

VENTRICLES OF THE BRAIN (pp. 435–436)

5. The central passageway of the brain expands to form chambers called **ventricles**. Cerebrospinal fluid continually circulates from the ventricles and central canal of the spinal cord into the subarachnoid space of the meninges that surround the CNS. — **septum pellucidum**, **lateral ventricles**, **interventricular foramen**, **mesencephalic aqueduct**, **fourth ventricle**

DISTRIBUTION OF GRAY AND WHITE MATTER (p. 436).

6. The brain contains extensive areas of **neural cortex**, a layer of gray matter on the surfaces of the cerebrum and cerebellum.

The Cerebrum (pp. 437–442)

THE CEREBRAL HEMISPHERES (pp. 437–439)

1. The cortical surface contains **gyri** (elevated ridges) separated by **sulci** (shallow depressions) or deeper grooves (**fissures**). The **longitudinal fissure** separates the two **cerebral hemispheres**. The **cen-** — **lobes** • **lateral sulcus**, **insula** • **parieto-occipital sulcus**

Related Key Terms

tral sulcus marks the boundary between the **frontal lobe** and the **parietal lobe**. Other sulci form the boundaries of the **temporal lobe** and the **occipital lobe**.

2. Each cerebral hemisphere receives sensory information and generates motor commands that concern the opposite side of the body. There are significant functional differences between the two hemispheres, and the assignment of specific functions to a region of cerebral cortex is imprecise.
3. The **primary motor cortex** of the **precentral gyrus** directs voluntary movements. The **primary sensory cortex** of the **postcentral gyrus** receives somatic sensory information from touch, pressure, pain, taste, and temperature receptors.

pyramidal cells
pyramidal system • **visual cortex**
auditory cortex • **olfactory cortex**

4. **Association areas**, such as the **visual association area** and **somatic motor association area (premotor cortex)**, control our ability to understand sensory information and coordinate a motor response. "Higher-order" integrative centers receive information from many different association areas and direct complex motor activities and analytical functions.

prefrontal cortex

THE CENTRAL WHITE MATTER (p. 439)

5. The **central white matter** contains three major groups of axons: (1) **association fibers** (tracts that interconnect areas of neural cortex within a single cerebral hemisphere); (2) **commissural fibers** (tracts connecting the two cerebral hemispheres); and (3) **projection fibers** (tracts that link the cerebrum with other regions of the brain and spinal cord).

arcuate fibers
longitudinal fasciculi
corpus callosum
anterior commissure
internal capsule

THE CEREBRAL NUCLEI (pp. 439–441)

6. The **cerebral nuclei (basal nuclei)** within the central white matter include the **caudate nucleus**, **amygdaloid body**, **claustrum**, and **putamen**. The cerebral nuclei are part of the **extrapyramidal system**, which controls muscle tone and coordinates learned movement patterns and other somatic motor activities.

lentiform nucleus
corpus striatum

THE LIMBIC SYSTEM (pp. 441–442)

7. The **limbic system** includes the **amygdaloid body**, **cingulate gyrus**, **parahippocampal gyrus**, **hippocampus**, and the **fornix**. The **mamillary bodies** control reflex movements associated with eating. The functions of the limbic system involve emotional states and related behavioral drives.
8. The **anterior nucleus** relays visceral sensations, and stimulating the **reticular formation** produces heightened awareness and a generalized excitement.

limbic lobe

The Diencephalon (pp. 442–445)

1. The diencephalon provides the switching and relay centers necessary to integrate the conscious and unconscious sensory and motor pathways. The diencephalic roof (**epithalamus**) contains the **pineal gland** and a vascular network that produces cerebrospinal fluid.

melatonin

THE THALAMUS (p. 443)

2. The thalamus is the final relay point for ascending sensory information and coordinates the pyramidal and extrapyramidal systems.

anterior nuclei • **medial nuclei**
ventral nuclei • **posterior nuclei**
pulvinar • **lateral geniculates**
optic tracts • **medial geniculates**

THE HYPOTHALAMUS (pp. 443–445)

3. The hypothalamus contains important control and integrative centers. It can: (1) control involuntary somatic motor activities; (2) control autonomic function; (3) coordinate activities of the nervous and endocrine systems; (4) secrete hormones; (5) produce emotions and behavioral drives; (6) coordinate voluntary and autonomic functions; (7) regulate body temperature.

hypophysis • **releasing hormones**
inhibiting hormones
tuberal area • **anterior nuclei**
antidiuretic hormone (ADH)
supraoptic nucleus
paraventricular nucleus
oxytocin • **drives**
feeding center
satiety center • **thirst center**
psychosomatic • **preoptic area**

The Mesencephalon or Midbrain (pp. 445–448)

1. The **tectum** (roof) of the mesencephalon contains two pairs of sensory nuclei, the **corpora quadrigemina**. On each side the **superior colliculus** receives visual inputs from the thalamus and the **inferior colliculus** receives auditory data from the medulla. The **red nucleus** integrates information from the cerebrum and issues involuntary motor commands related to muscle tone and posture. The **substantia nigra** regulates the motor output of the cerebral nuclei. The **cerebral peduncles** contain ascending fibers headed for thalamic nuclei, and descending fibers of the pyramidal system that carry voluntary motor commands from the primary motor cortex of each cerebral hemisphere.

Chapter Review

The Cerebellum (pp. 448–449)

1. The cerebellum oversees the body's postural muscles and programs and tunes voluntary and involuntary movements. The **cerebellar hemispheres** consist of neural cortex formed into folds or **folia**. The surface can be divided into the **anterior** and **posterior lobes**, the **vermis**, and the **flocculonodular lobe**.

Related Key Terms

Purkinje (basket) cells
arbor vitae • cerebellar nuclei

The Pons (pp. 449–450)

1. The pons contains: (1) sensory and motor nuclei for four cranial nerves; (2) nuclei concerned with involuntary control of respiration; (3) tracts linking the cerebellum with the brain stem, cerebrum, and spinal cord; and (4) ascending and descending tracts.

superior cerebellar peduncles
middle cerebellar peduncles
inferior cerebellar peduncles
transverse fibers

The Medulla Oblongata (Myelencephalon) (pp. 450–451)

1. The medulla oblongata (medulla) connects the brain to the spinal cord. It contains **olivary nuclei**, which relay information from the spinal cord, the cerebral cortex, and brain stem to the cerebellar cortex. Its **reflex centers**, including the **cardiovascular centers** and the **respiratory rhythmicity center**, control or adjust the activities of peripheral systems.

nucleus gracilis
nucleus cuneatus • olives
cardiac centers
vasomotor center

Protection and Support of the Brain (pp. 451–458)

THE CRANIAL MENINGES (pp. 451–453)

1. Special adaptations are necessary to protect and support the delicate brain. The **falx cerebri** (a fold of dura mater) and the **tentorium cerebelli** (a dural sheet) help stabilize and support the brain.

dural sinuses

THE BLOOD-BRAIN BARRIER (pp. 453–456)

2. The blood-brain barrier isolates neural tissue from the general circulation.
3. The blood-brain barrier remains intact throughout the CNS except in portions of the hypothalamus and at the choroid plexus in the membranous roof of the diencephalon and medulla.

THE CHOROID PLEXUS AND CEREBROSPINAL FLUID FORMATION (pp. 456–457)

4. The **choroid plexus** is the site of cerebrospinal fluid production.

CEREBROSPINAL FLUID AND ITS CIRCULATION (p. 457)

5. Cerebrospinal fluid: (1) cushions delicate neural structures, (2) supports the brain, and (3) transports nutrients, chemical messengers, and waste products. Cerebrospinal fluid reaches the subarachnoid space via the **lateral apertures** and a **median aperture**. Diffusion across the **arachnoid villi** into the **superior sagittal sinus** returns CSF to the venous circulation.

The Cranial Nerves (pp. 458–468)

1. There are 12 pairs of cranial nerves.
2. The **olfactory nerve** (N I) carries sensory information responsible for the sense of smell. The olfactory afferents synapse within the **olfactory bulbs**. — **olfactory tracts**
3. The **optic nerve** (N II) carries visual information from special sensory receptors in the eyes. — **optic chiasm**
4. The **oculomotor nerve** (N III) is the primary source of innervation for the extrinsic oculomotor muscles that move the eyeball. — **ciliary ganglion**
5. The **trochlear nerve** (N IV), the smallest cranial nerve, innervates the superior oblique muscle of the eye.
6. The **trigeminal nerve** (N V), the largest cranial nerve, is a mixed nerve with **ophthalmic**, **maxillary**, and **mandibular branches**. — **semilunar ganglion**, **sphenopalatine ganglion**, **submandibular ganglion**
7. The **abducens nerve** (N VI) innervates the sixth extrinsic oculomotor muscle, the lateral rectus.
8. The **facial nerve** (N VII) is a mixed nerve that controls muscles of the scalp and face. It provides pressure sensations over the face and receives taste information from the tongue. — **geniculate ganglion**
9. The **vestibulocochlear nerve** (N VIII) contains the **vestibular nerve,** which monitors sensations of balance, position, and movement, and the **cochlear nerve**, which monitors hearing receptors. — **vestibular nuclei**, **cochlear nuclei**
10. The **glossopharyngeal nerve** (N IX) is a mixed nerve that innervates the tongue and pharynx and controls the action of swallowing. — **petrosal ganglion**
11. The **vagus nerve** (N X) is a mixed nerve that is vital to the autonomic control of visceral function and has a variety of motor components. — **nodose ganglia • jugular ganglia**

Related Key Terms

12. The **spinal accessory nerve** (N XI) has a **medullary branch**, which innervates voluntary swallowing muscles of the soft palate and pharynx, and a **spinal branch**, which controls muscles associated with the pectoral girdle.
13. The **hypoglossal nerve** (N XII) provides voluntary motor control over tongue movements.

Cranial Reflexes (pp. 468–469)

1. **Cranial reflexes** are reflex arcs that involve the sensory and motor fibers of cranial nerves. Important somatic cranial reflexes include the: **corneal reflex** (rapid blinking of the affected eye following contact to the cornea); **tympanic reflex** (protects auditory receptors of the inner ear from loud noises); **auditory reflexes** (trigger somatic motor responses to sounds); and **vestibulo-ocular reflexes** (eye movement in the opposite direction when the head is rotated). The **light reflexes** control the amount of light entering the eye.

direct light reflex
consensual light reflex

Review Planner

	Objective	Level 1
1	Name the major regions of the brain and describe their functions.	**8 12–24 28 36 38 43**
2	Identify important structures within each region and explain their functions.	**1 2 3 6 8 12–24 25–35 39 42**
3	Distinguish between motor, sensory, and association areas of the cerebral cortex.	**8 22 23 34 45**
4	Explain the role of "higher centers."	**33**
5	Explain how the brain is protected.	**7 10**
6	Discuss the circulation and functions of cerebrospinal fluid.	**7 11**
7	Identify the cranial nerves and relate each pair of cranial nerves to its principal functions.	**9 40 41 44**
8	Discuss important cranial reflexes.	**4 5 40**

Level 2	**37**
Level 3	**46–48**
C M	**Cerebellar Dysfunction** **Cranial Trauma • Tic Douloureux** **Bell's Palsy**
C M	**Phrenology**

Review Questions

MULTIPLE CHOICE

1. Stimulating the reticular formation will cause: a) pain; b) heightened alertness and a generalized excitement; c) generalized lethargy; d) sleep.
2. "Drives" such as hunger and thirst originate in the: a) cerebrospinal fluid; b) thalamus; c) cerebral nuclei; d) hypothalamus.
3. At the cerebral hemispheres, the central passageway of the brain expands to form chambers called: a) ventricles; b) auricles; c) internal capsules; d) corpora quadrigemina.
4. The cranial reflex that protects the auditory receptors of the inner ear from loud noises is the: a) corneal reflex; b) auditory reflex; c) tympanic reflex; d) vestibulo-ocular reflex.
5. The cranial reflex that causes you to blink after a light touch to your cornea is the: a) corneal reflex; b) light reflex; c) vestibulo-ocular reflex; d) consensual light reflex.
6. The layer of gray matter found on the surface of the brain is called: a) white matter; b) neural cortex; c) cerebral nuclei; d) cerebral peduncles.
7. Which of the following is *not* a function of the

CSF? a) it cushions neural structures; b) it transports nutrients; c) it relieves intracranial pressure; d) it supports the brain.

DIAGRAMS

8. Draw an outline of a cerebrum and label the following: temporal lobe, parietal lobe, frontal lobe, occipital lobe, visual cortex, auditory cortex, primary motor cortex, and primary sensory cortex.
9. Complete the following table:

Classification	*Primary Functions*	*Cranial Nerves*
Special sensory		
Motor	Control extrinsic oculomotor muscles	
		XII (hypoglossal)
Mixed		

TRUE/FALSE

(If a statement is false, your instructor may wish to have you correct it.)

10. **T/F:** The cranial meninges consist of three membrane layers continuous with those of the spinal cord.
11. **T/F:** Cerebrospinal fluid returns to the venous circulation at the choroid plexus.

MATCHING QUESTIONS

(Match these terms with the region of the brain in which they are found. Use the following key for the regions):

Terms:

12. Arbor vitae
13. Olives
14. Thalamus
15. Transverse fibers
16. Flocculonodular lobe
17. Red nucleus
18. Temporal lobe
19. Amygdaloid body
20. Substantia nigra
21. Hypothalamus
22. Visual association area
23. Primary motor cortex
24. Reflex centers

Regions:

a) Cerebrum
b) Diencephalon
c) Mesencephalon
d) Cerebellum
e) Pons
f) Medulla oblongata

(Match the following structures of the brain with their functions):

Functions:

25. The pathway that provides voluntary motor control.
26. Sets the basic pace for respiratory movements.
27. Coordinates the activities of the nervous and endocrine systems.
28. Oversees the body's postural muscles, programs and tunes movements.
29. Integrates information from the cerebrum and cerebellum and issues involuntary motor commands related to maintaining muscle tone and posture.
30. The pathway that controls involuntary somatic motor activity.
31. Adjust heart rate, strength of cardiac contractions, and peripheral blood flow.
32. Links the conscious, intellectual functions of the cerebral cortex with the unconscious, autonomic functions of the brain stem.
33. A "higher-order" center that integrates information from sensory association areas and performs abstract intellectual functions.
34. Neurons in this area receive somatic sensory information from touch, pressure, pain, taste, and temperature receptors.
35. Contain motor nuclei that control reflex movements associated with eating.

Structures:

a) Mamillary bodies
b) Pyramidal system
c) Prefrontal cortex
d) Extrapyramidal system
e) Limbic system
f) Hypothalamus
g) Red nucleus
h) Cerebellum
i) Primary sensory cortex
j) Cardiovascular centers
k) Respiratory rhythmicity center

SHORT ANSWER/ESSAY

36. What is the brain stem? List its functions.
37. What area of the brain is responsible for psychosomatic conditions? Are they "real"? Explain.
38. As you answer these questions, you find that you are trying to remember information and use it to "reason out" responses. What region of the brain is responsible for this type of activity?
39. Distinguish between the thalamus, epithalamus, and hypothalamus. Where are they found, and what are their functions?
40. While shopping at a mall, you watch the antics of a young child and follow her with your eyes as she runs into a store. What are the sensory and motor cranial nerves involved in your actions? Include the nerve numbers and origins.
41. You are chewing popcorn and enjoying the salty taste. What are the cranial nerves involved? Include the nerve numbers and origins.
42. You are dining at an outdoor cafe, and as the sun sets you put on a sweater because you start to feel chilly. In what area of the brain does this sensation originate? What are such feelings called?
43. Fill in the blanks: The superior surface of the cerebrum forms the large, paired ________ ________.
44. What cranial nerve(s) are responsible for monitoring sensations of balance? Are they sensory, motor, or mixed, and where do they originate?
45. Distinguish between association areas and integrative centers of the neural cortex. For what types of activities are they responsible?

CRITICAL THINKING/APPLICATIONS

46. Mrs. Gammons, 73 years old, has been finding it increasingly difficult to control her movements, and even when sitting still she experiences a slight continual tremor. Identify her condition and discuss the parts of the brain that are involved.
47. An unconscious teenager is brought to the emergency room. His parents tell you he was playing football and his head struck the ground. Initially he appeared unhurt, but 2 hours later he became lethargic and lost consciousness. Identify a problem that could explain the delayed response.
48. Having memorized a piano piece, you find that as long as you don't think about it consciously you have no trouble playing it. When you think about where to place your fingers next, however, you can't continue. Explain what is happening and the region of the brain involved.

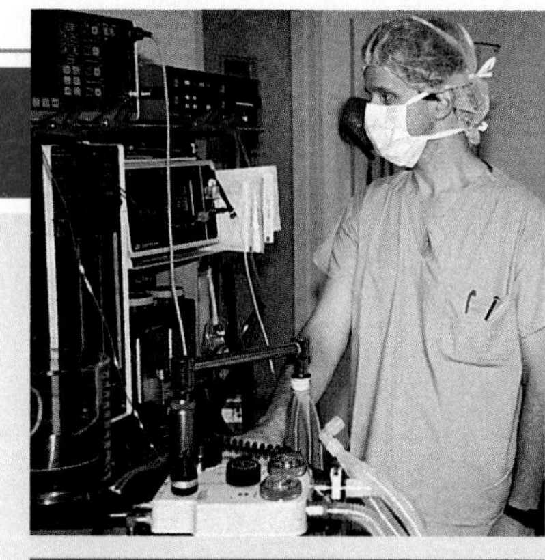

Career Close-up: Certified Registered Nurse Anesthetist

"When you give anesthesia," says Steve LoGrasso, "you're essentially administering a series of poisons. Anesthetic agents are neuro-blocking drugs that keep patients from registering the pain of surgery. My goal is to create a stress-free response to surgery."

Steve is a CRNA (Certified Registered Nurse Anesthetist). He is responsible for choosing the appropriate anesthetic drugs for each patient, administering them throughout a surgical procedure, monitoring the patient's vital signs, and (post-operatively) waking up the patient smoothly. Behind this basic job description, however, lurk many variables.

"Every time I anesthetize a patient, it's a new journey," says Steve. Many factors can affect how people respond to drugs: their usual alcohol or drug intake, preexisting physical conditions, emotional states. Generally, heavier and taller people need more of a drug than smaller, thinner types, but not always: "I've seen big guys who black out immediately," notes Steve, "and frail elderly women who require large doses to become drowsy."

Typically Steve first meets a patient on the way to the operating room. "You learn to read people quickly, to figure out who's frightened and needs comforting. Some people want reassurance; others appreciate a little humor to make the situation less tense. If I think I can kid with them, I say 'It's OK; I just read up on this last night.' "

Once in the operating room, Steve attaches monitors to the patient and proceeds quickly: "I try to give some sedation as soon as possible. People get anxious when they hear the surgical instruments being laid out." Ninety percent of the procedures at which he assists involve general anesthesia, in which the patient is put to sleep. In most of these cases Steve administers oxygen and anesthetic gases through an endotracheal tube that he inserts into the patient's airway. Throughout the surgery Steve keeps an eye on both the patient and the monitors that measure vital signs such as heart rate, blood pressure, body temperature, and circulating oxygen and carbon dioxide levels. He also assesses blood loss, urine output, hydration, and how much longer the surgery will last.

"It's like landing a plane," he explains—"you start by reducing your altitude gradually. Similarly, toward the end of a procedure you start tapering off the anesthetic." Awakening after surgery can be traumatic; if the anesthetist doesn't time the removal of the endotracheal tube correctly, the patient could start to choke as the cough reflex becomes active again.

Other anesthetic techniques include spinals and epidurals, in which the patient is awake but insensitive to pain. Steve must remain acutely aware of the pros and cons of each approach. "In a spinal, you block nerve impulses by injecting anesthetic into the cerebrospinal fluid of the subarachnoid space," he notes. "This produces a rapid-acting, profound block of sensory and motor impulses; you can maintain high levels of sensory blockade without actually administering very much drug. For an epidural, you also give an injection at the L_2-L_3 or L_3-L_4 level, but you insert a catheter into the epidural space without penetrating the subarachnoid membrane. Thus, the medication diffuses across the membrane rather than entering the space directly. You can fine-tune an epidural more than a spinal; even though the pain is blocked, patients can still move their legs. That's why epidurals are often used in labor and delivery."

After earning his RN and BSN, Steve worked for two years in a Surgical Intensive Care Unit (an entrance requirement for many CRNA training programs). He completed a two-year, full-time Nurse Anesthesia Program and passed a national certifying exam before joining the CRNA staff of a large suburban hospital.

"People have two great fears: pain and the unknown," he says, "and surgery involves both. My biggest enjoyment is making patients feel good—alleviating pain and reducing anxiety. When there's no change in vital signs during a procedure, I've done my job right."

All of the body's organ systems are vital, and a serious malfunction in any one of them will cause death. Yet from a personal standpoint, the nervous system is surely the most important of all systems. The nervous system creates our unique personalities and establishes our consciousness. Dreaming, planning, wondering, working, playing, learning, creating . . . all of these are among the functions of the nervous system. They are not easy to localize on an anatomical diagram, and perhaps they never will be. But they are what determine the quality of our lives. In this chapter we will explore some of these "higher" functions, as well as those that enable us to control the movements of our bodies.

CHAPTER 15

The Nervous System: Pathways, Processing, and Higher-Order Functions

Chapter Objectives

After reading this chapter, you will be able to:

1 Identify the principal sensory and motor pathways.
2 Compare the processes and functions of the pyramidal and extrapyramidal systems.
3 Explain how we can distinguish between sensations that originate in different areas of the body.
4 Describe the levels of information processing involved in motor control.
5 Discuss how the brain integrates sensory information and coordinates responses.
6 Explain the importance of hemispheric specialization.
7 Explain how memories are created, stored, and recalled.
8 Discuss levels of consciousness and sleep.
9 Summarize the effects of aging on the nervous system.

■ Introduction

There is a saying that "big cities never sleep." If you visit New York or Los Angeles at 3:00 A.M., there are shops open, deliveries being made, people on the street, and traffic moving briskly. The central nervous system (CNS) is much more complex than any city, and far busier. There is a continuous flow of information between the brain, spinal cord, and peripheral nerves. At any given moment, millions of sensory neurons are delivering information to processing centers in the CNS, and millions of motor neurons are controlling or adjusting the activities of peripheral effectors. This process continues 24 hours a day, whether we are awake or asleep. You may sleep soundly, but your nervous system does not; many brain stem centers are active throughout our lives, performing vital autonomic functions.

The communication between the CNS, the PNS (peripheral nervous system), and peripheral organs and systems occurs over pathways, nerve tracts, and nuclei that relay sensory and motor information between the periphery and higher centers of the brain. There are usually several successive levels of processing along the sensory and motor pathways. For example, ascending sensory information may be analyzed at the nucleus gracilis or cuneatus in the medulla, in nuclei of the reticular formation, and in thalamic nuclei—all before it reaches the cerebral cortex. These processing steps may block, reduce, or heighten our conscious and unconscious awareness of the stimulus. For example, while the cerebral cortex considers a voluntary response, collateral processing within the brain stem and cerebellum may already be initiating an involuntary one.

Many subtle forms of interaction, feedback, and regulation link higher centers with the various components of the brain stem. Only a few are understood in any detail. The following sections focus on basic patterns and principles in the sensory, motor, and higher-order functions of the brain.

Sensory and Motor Pathways

Our attention will focus on pathways that utilize the major ascending (sensory) and descending (motor) tracts of the spinal cord. As we proceed with our discussion, it may help you to remember that:

- The tract names indicate the origins and destinations of the axons. If the name of a tract begins with *spino-* it must *start* in the spinal cord and

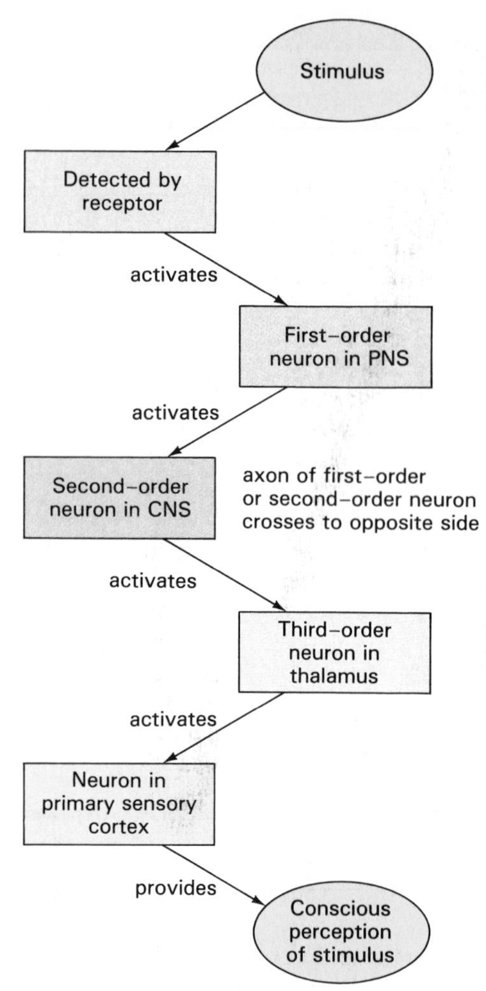

FIGURE 15-1
Components of a Somatic Sensory Pathway

TABLE 15-1 Principal Ascending (Sensory) Tracts in the Spinal Cord, and a Summary of the Functional

Tract	*Origin*	*Destination*[a]
THE POSTERIOR COLUMN PATHWAY:		
Fasciculus gracilis	Proprioceptors and fine touch, pressure, and vibration receptors of lower body	Nucleus gracilis (medulla)
Fasciculus cuneatus	Proprioceptors and fine touch, pressure, and vibration receptors of upper body	Nucleus cuneatus (medulla)
THE SPINOCEREBELLAR PATHWAY:		
Posterior spinocerebellar tract	Interneurons relaying information from proprioceptors	Cerebellum
Anterior spinocerebellar tract	Interneurons relaying information from proprioceptors	Cerebellum
THE SPINOTHALAMIC PATHWAY:		
Lateral spinothalamic tract	Interneurons relaying information from pain and temperature receptors	Thalamus (ventral nuclei)
Anterior spinothalamic tract	Interneurons relaying information from crude touch and pressure receptors	Thalamus (ventral nuclei)

[a] Location of first synapse.

end in the brain, and it therefore carries sensory information. If the name of a tract ends in *-spinal* its axons must *start* in the higher centers and *end* in the spinal cord, bearing motor commands.

- The rest of the tract name indicates the associated nucleus or cortical area of the brain. You should already be familiar with the primary functions of these nuclei and areas from Chapter 14.

SENSORY PATHWAYS

Sensory receptors monitor conditions in the body or the external environment. When stimulated, a receptor passes information to the central nervous system. This information, called a **sensation**, arrives in the form of action potentials in an afferent (sensory) fiber. Chapter 17 will consider the origins of sensations and the pathways involved in relaying the information to conscious and unconscious processing centers in the CNS. These processing steps result in motor responses that preserve homeostasis. Most of the processing occurs in centers along the sensory pathways in the spinal cord or brain stem; only about 1 percent of the information provided by afferent fibers reaches the cerebral cortex and our conscious awareness. For example, we usually do not feel the clothes we wear or hear the hum of the engine in our car.

This section discusses pathways that carry somatic sensory information to the primary sensory cortex of the cerebral hemispheres. Chapter 16 describes pathways involved in the distribution of visceral sensory information, and Chapter 17 considers the central processing of special sensory data.

Figure 15-1 presents a diagrammatic view of the components of a somatic sensory pathway. The sensory neuron that delivers the sensations to the CNS is often called a **first-order neuron**. Inside the CNS the axon of the first-order neuron synapses on an interneuron known as a **second-order neuron**. The second-order neuron, which may be located in the spinal cord or brain stem, in turn synapses on a **third-order neuron** in the thalamus. Somewhere along its length, the axon of either the first-order or second-order neuron crosses over to the opposite side of the body. As a result, the right side of the thalamus receives sensory information from the left side of the body. The axons of the third-order neurons synapse on neurons of the primary sensory cortex of the cerebral hemisphere. Because these axons do not cross over, the right cerebral hemisphere receives sensory information from the left side, and vice versa.

Table 15-1 describes three major somatic sensory pathways, or **somatosensory pathways**: the *posterior column pathway*, the *spinothalamic pathway*, and the *spinocerebellar pathway*. Figure 15-2 indicates their relative positions in the spinal cord.

The Posterior Column Pathway

In the **posterior column pathway**, highly localized ("fine") touch, pressure, vibration, and proprioceptive (position) sensations arrive along the dorsal roots of spinal nerves. The axons ascend within the **fasciculus gracilis** and the **fasciculus cuneatus**, synaps-

Roles of the Associated Nuclei in the Brain

Sensations	*Comments*
Position, fine touch, pressure, vibration	Ascends on same side as stimulus
Position, fine touch, pressure, vibration	Ascends on same side as stimulus
Proprioception	Ascends on same side as stimulus
Proprioception	Crosses to ascend on side opposite stimulus
Pain and temperature	Crosses to ascend on side opposite stimulus
Crude touch and pressure	Crosses to ascend on side opposite stimulus

FIGURE 15-2
Ascending Tracts in the Spinal Cord. A cross-sectional view indicating the locations of the major ascending sensory tracts in the spinal cord. For information about these tracts, see Table 15-1. Descending motor tracts are shown in outline; these tracts are identified in Figure 15-5.

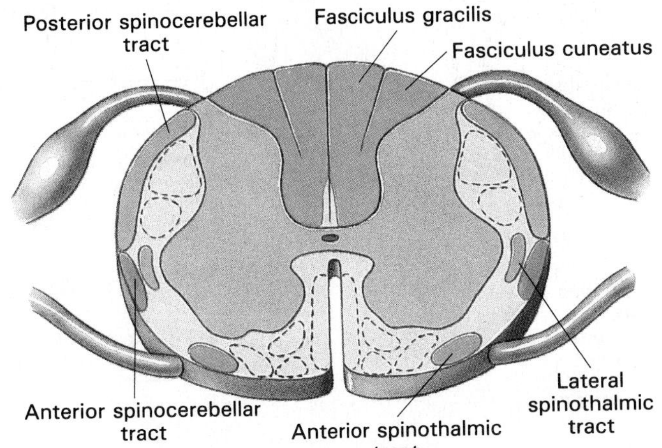

ing at the nucleus gracilis and the nucleus cuneatus, two medullary nuclei introduced in Chapter 14. (p. 450) The postsynaptic neurons then relay the information to the thalamus along a tract called the **medial lemniscus** (*lemniskos*, ribbon).

As the axons leave the nuclei to enter the medial lemniscus, they cross over, or **decussate**, to the opposite side of the brain stem. As it travels toward the thalamus, the medial lemniscus incorporates sensory information collected by cranial nerves V, VII, IX, and X. The posterior column pathway is diagrammed in Figure 15-3a. This figure details the pathway carrying sensory information from the left side of the body to the primary sensory cortex of the right cerebral hemisphere. (The pathway for sensations originating on the right side, which ends at the left cerebral hemisphere, is not shown.)

As you may recall from Figure 13-4, ascending sensory information maintains a strict regional organization along the pathway from center to center. (p. 411) In the thalamus, data arriving over the posterior column pathway are integrated, sorted according to the region of the body involved, and projected to the primary sensory cortex. The sensations arrive organized so that sensory information from the toes arrives at one end of the primary sensory cortex, and information from the head arrives at the other.

When sensations originating in one region of the body are projected to the corresponding area of the primary sensory cortex, the individual can state the nature of the stimulus and its location. If that area of the sensory cortex is electrically stimulated, the individual reports the same perceptions. By electrically stimulating the cortical surface, experimenters have been able to create a map of the primary sensory cortex. The results are shown in Figure 15-3a. This sensory map is called a **sensory homuncu-**

FIGURE 15-3

The Posterior Column and Spinothalamic Pathways. (a) The posterior column pathway delivers fine touch, vibration, and proprioception information to the primary sensory cortex of the cerebral hemisphere on the opposite side of the body. The cross-over occurs in the medulla, after a synapse in the nucleus gracilis or nucleus cuneatus. (For clarity, this figure shows only the pathway for sensations originating on the left side of the body.) (b) The spinothalamic pathways carry sensations of pain and temperature (lateral spinothalamic tract) and crude touch and pressure (anterior spinothalamic tract) to the sensory cortex on the opposite side. The cross-over occurs in the spinal cord, at the level of entry. (Only one tract is detailed on each side, although each side has both tracts.)

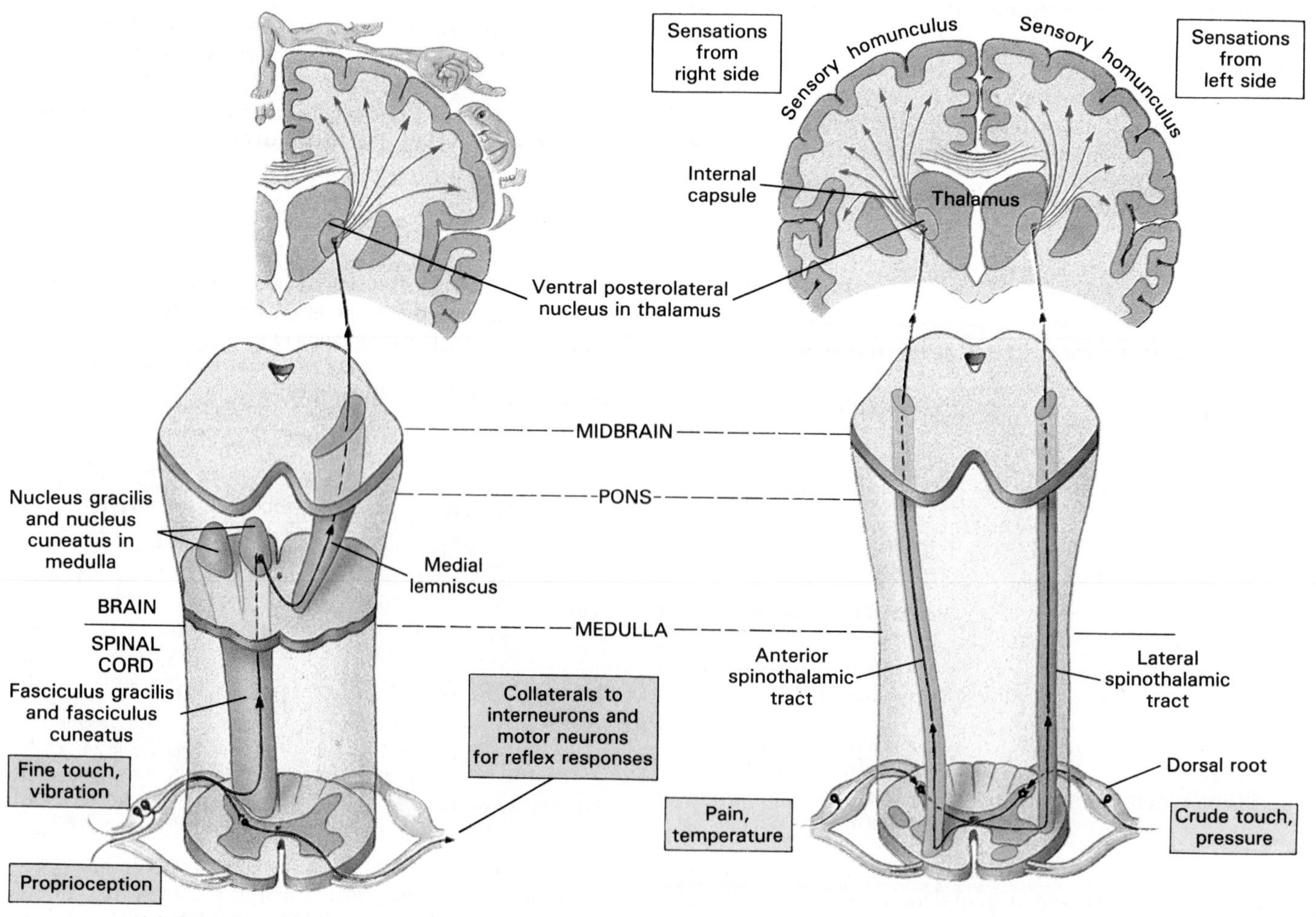

lus ("little man"). The proportions of the homunculus are obviously very different from those of the individual. For example, the face is huge and distorted, with enormous lips and tongue, whereas the back is relatively tiny. These distortions occur because the area of sensory cortex devoted to a particular region is proportional not to its absolute size, but rather to the *number of sensory receptors* it contains. In other words, it takes many more cortical neurons to process sensory information arriving from the tongue, which has tens of thousands of taste and touch receptors, than it does to analyze sensations originating on the back, where touch receptors are few and far between.

The organized distribution of sensory information on the sensory cortex enables us to determine what specific portion of the body has been affected by a stimulus. If a sensation arrives at the wrong part of the sensory cortex, we will reach improper conclusions about the source of the stimulus. For example, the pain of a heart attack is often felt in the left arm; this mechanism of *referred pain* will be discussed in Chapter 17. Another example is the phenomenon of *phantom limb pain*, in which someone continues to experience pain in a limb after it has been lost through accident or disease due to activity along the sensory pathway that monitored the intact limb.

Our perception of a given sensation as pain, rather than as temperature or touch, depends on processing in the thalamus. If the cerebral cortex were damaged or the projection fibers cut we could still detect a light touch, but we would be unable to determine its source.

The Spinothalamic Pathway

The **spinothalamic pathway**, diagrammed in Figure 15-3b, begins as poorly localized ("crude") sensations of touch, pressure, pain, and temperature enter the spinal cord and synapse within the posterior gray horns. Interneurons send the information to the opposite side, to ascend within the **anterior** and **lateral spinothalamic tracts**. These tracts and the medial lemniscus converge on the **ventral posterolateral nuclei** of the thalamus. Projection fibers then carry the information to the primary sensory cortex.

To make it easier to follow, Figure 15-3b shows the distribution route for pain and temperature sensations from the right side of the body and crude touch and pressure from the left side. However, remember that receptors monitor pain, temperature, touch, and pressure on each side of the body, and each side has anterior and lateral spinothalamic tracts.

The Spinocerebellar Pathway

The **spinocerebellar pathway** includes the **posterior** and **anterior spinocerebellar tracts**. These axons carry proprioceptive information concerning the position of muscles, tendons, and joints to the cerebellum for processing. Additional details will be considered when we examine cerebellar function later in the chapter.

MOTOR PATHWAYS

The central nervous system issues motor commands in response to information provided by sensory systems. These commands are distributed by the somatic nervous system (SNS) and the autonomic nervous system (ANS). The SNS issues somatic motor commands that direct the contractions of skeletal muscles. The autonomic nervous system, sometimes known as the visceral motor system, controls visceral effectors, such as smooth and cardiac muscles, glands, and fat cells. The output of the somatic nervous system is under voluntary control, but autonomic activity is regulated outside of our conscious awareness.

The motor neurons of the SNS and ANS are organized in different ways. Figure 15-4 provides a summary diagram. Somatic motor pathways (Figure 15-4a) always involve at least two motor neurons: an **upper motor neuron**, whose soma lies in a CNS processing center, and a **lower motor neuron**, located in a motor nucleus of the brain stem or spinal cord. Activity in the upper motor neuron can facilitate or inhibit the lower motor neuron, but only the axon of the lower motor neuron extends outside of the CNS and contacts skeletal muscle fibers. Destruction or damage to a lower motor neuron produces a flaccid paralysis of the innervated motor unit. Damage to an upper motor neuron may produce muscle rigidity, flaccidity, or uncoordinated contractions, depending on the identity of the upper motor neuron. For example, cutting descending motor tracts (the axons of upper motor neurons) causes the reappearance of the Babinski reflex and the disappearance of the abdominal reflex, two spinal reflexes dependent on lower motor neuron function. (You may wish to review the discussion of spinal reflexes in Chapter 13.) (pp. 423–430)

Two neurons are also involved in the autonomic nervous system, but one of them is always located in the periphery. As indicated in Figure 15-4b, autonomic motor control requires a **preganglionic neuron** in the CNS and a **ganglionic neuron** in a peripheral ganglion. This chapter considers the motor pathways of the SNS; those of the ANS will be detailed in Chapter 16.

Voluntary and involuntary somatic motor commands issued by the brain reach peripheral targets by traveling over two integrated motor pathways, the *pyramidal system* and the *extrapyramidal system*. These systems were originally thought to be completely separate, with the pyramidal system providing voluntary control and the extrapyramidal system regulating involuntary motor activity. In fact these systems are so extensively integrated that it is very difficult to distinguish between them on either anatomical or functional grounds. (Examples will be

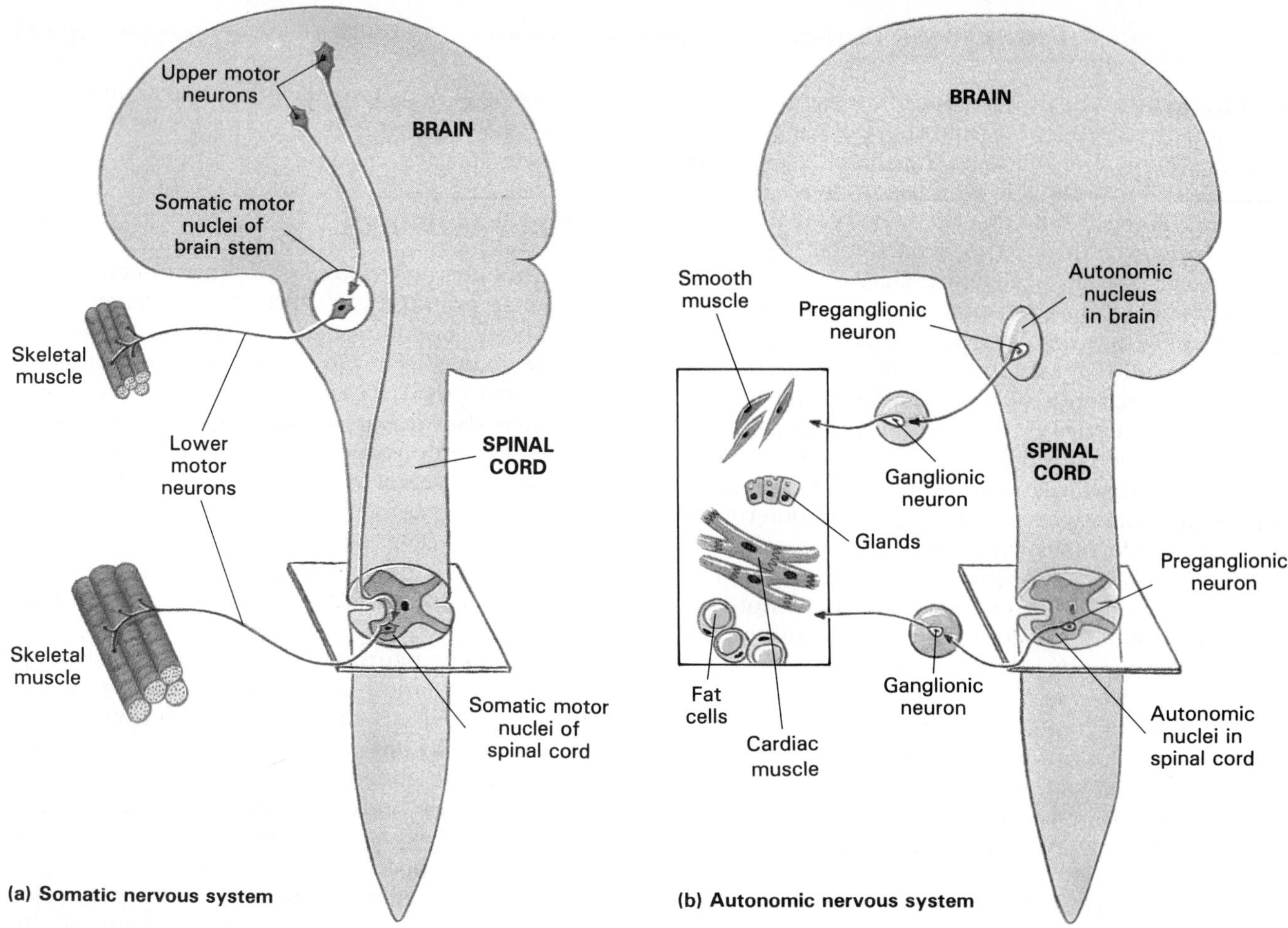

FIGURE 15-4
The Somatic and Autonomic Nervous Systems. (a) In the somatic nervous system, an upper motor neuron in a CNS processing center controls a lower motor neuron in the brain or spinal cord. The axon of the lower motor neuron has direct control over skeletal muscle fibers in the periphery. Stimulation of the lower motor neuron always has an excitatory effect on these muscle fibers. (b) In the autonomic nervous system the axon of a first-order motor neuron in the CNS controls a second-order neuron in a peripheral ganglion. The stimulation of the second-order neuron may lead to excitation or inhibition of the visceral effector.

TABLE 15-2 Principal Descending (Motor) Tracts in the Spinal Cord, and a Summary of the Functional

Tract	*Origin*	*Destination*[a]	*Actions*
PYRAMIDAL TRACTS			
Corticobulbar tracts	Primary motor cortex (cerebral hemispheres)	Motor neurons of cranial nerve nuclei in brain stem	Voluntary motor control of skeletal muscles
Lateral corticospinal tract	Same	Motor neurons of anterior gray horns of spinal cord	Same
Anterior corticospinal tract	Same	Same	Same
EXTRAPYRAMIDAL TRACTS			
Rubrospinal tract	Red nucleus (midbrain)	Motor neurons of anterior gray horns	Involuntary regulation of posture and muscle tone
Reticulospinal tract	Reticular formation (network of nuclei in brain stem)	Somatic and visceral motor neurons of anterior and lateral gray horns	Involuntary regulation of reflex activity and autonomic functions
Vestibulospinal tract	Vestibular nucleus (near rostral border of medulla)	Motor neurons of anterior gray horns	Involuntary regulation of balance and muscle tone
Tectospinal tract	Tectum (midbrain)	Motor neurons of anterior gray horns of cervical spinal cord	Involuntary regulation of eye, head, neck, and arm position in response to visual and auditory stimuli

[a] Location of first synapse.

noted as the discussion proceeds.) Figure 15-5 indicates the positions of the major pyramidal and extrapyramidal tracts in the spinal cord, and Table 15-2 summarizes information concerning these systems.

The Pyramidal System

Voluntary control of skeletal muscles is provided by the **pyramidal system**. One way to observe the activities of this system is through direct electrical stimulation of the primary motor cortex in the precentral gyrus. Such stimulation will produce specific peripheral muscular contractions. As in the case of the sensory cortex, there is a fine point-to-point correspondence between the motor cortex and specific regions of the body. The cortical areas can be mapped out in diagrammatic form, creating a **motor homunculus**. A motor homunculus is seen in Figure 15-6, which shows the pyramidal system components involved in controlling the right side of the body.

The proportions of the motor homunculus are also quite different from those of the actual body, because the motor area devoted to a specific region of the cortex is proportional to the number of motor units involved. As a result, the homunculus provides an indication of the degree of fine motor control available. For example, the hands, face, and tongue are very large, and the trunk relatively small. The grossly oversized hands show how important they are to our daily lives, and how many different muscles and motor units are involved in writing, grasping, manipulating, and so on. If you compare the motor homunculus with the sensory homunculus, detailed in Figure 15-3, you will see many similarities. For example, the areas devoted to the hands and mouth are disproportionately large in both. There are differences, however, because some highly sensitive regions (such as the sole of the foot) contain few motor units, whereas others (such as the facial and eye muscles) have an extraordinarily large number of motor units.

The pyramidal system owes its name to the unusual shapes of the neurons of the primary motor cortex. These neurons are called **pyramidal cells** because their cell bodies are shaped like miniature pyramids. A pyramidal cell has its base facing the cerebral surface and the tip, which is attached to the axon, pointing toward the internal capsule. The axons of the pyramidal cells extend into the brain stem and spinal cord to synapse on somatic motor neurons. The **corticobulbar tracts** (kōr-ti-kō-BUL-bar; *bulbar*, brain stem) terminate at the nuclei of cranial nerves, whereas the **corticospinal tracts** synapse on motor neurons in the anterior gray horns of the spinal cord. The corticobulbar and corticospinal tracts enter the internal capsule, descend, and emerge briefly on either side of the mesencephalon as the cerebral peduncles. [**CM**: *Cerebral Palsy*]

The corticobulbar tracts branch and synapse in the brain stem, at the motor nuclei of cranial nerves. The corticospinal tracts are visible along the ventral surface of the medulla as a pair of thick bands, the **pyramids**. Along the length of the pyramids roughly 85 percent of the axons decussate, crossing the midline to enter the descending **lateral corticospinal tracts** on the opposite side of the spinal cord. The remaining 15 percent continue uncrossed along the spinal cord as the **anterior corticospinal tracts**, but these fibers too will cross over within the anterior

Roles of the Associated Nuclei in the Brain
Comments
Crosses to opposite side within brain stem
Crosses to opposite side before entering spinal cord
Descends uncrossed but crosses to opposite side before synapsing
Crosses to opposite side before entering spinal cord
Descends without crossing to opposite side
Descends without crossing to opposite side
Crosses to opposite side before entering spinal cord

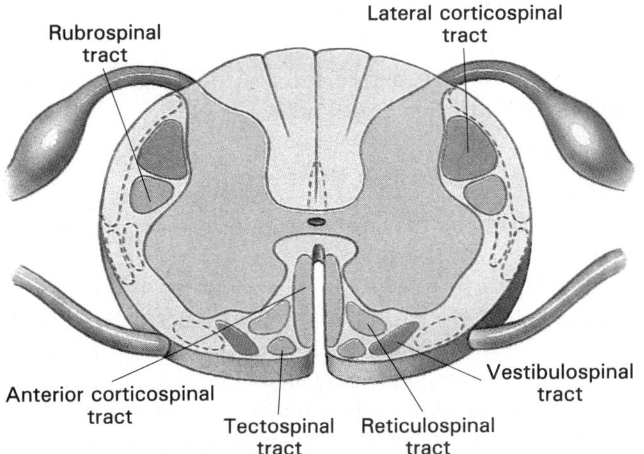

FIGURE 15-5
Descending Motor Pathways in the Spinal Cord. A cross-sectional view indicating the locations of the major descending motor tracts that contain the axons of upper motor neurons. Sensory tracts, detailed in Figure 15-2, are outlined.

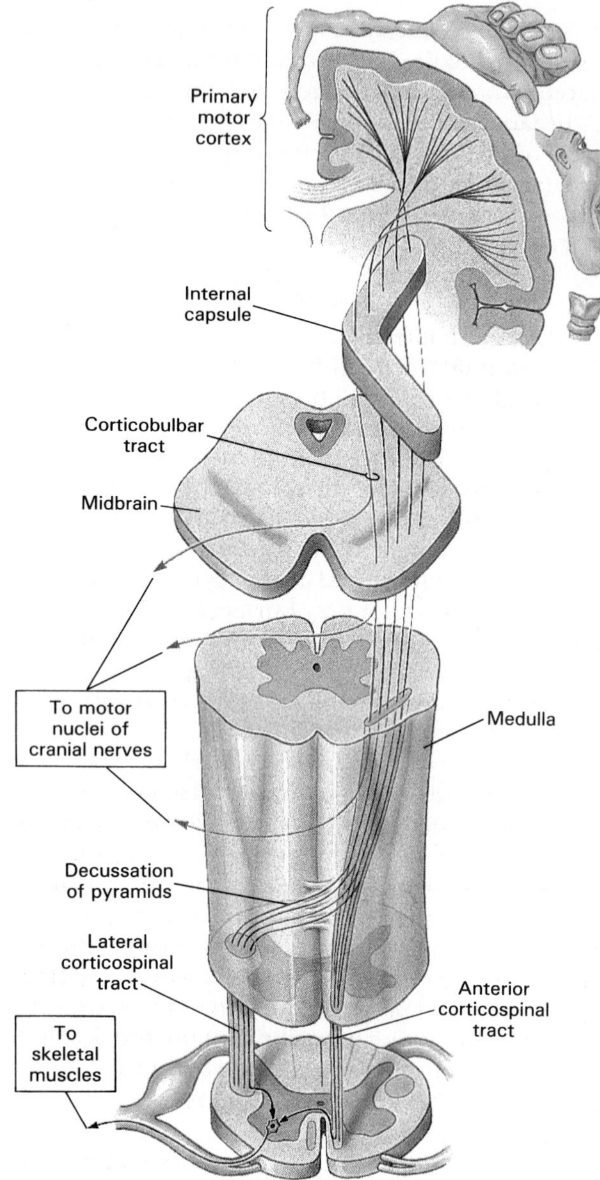

FIGURE 15-6
The Pyramidal System. The pyramidal system originates at the primary motor cortex. Axons of the pyramidal cells descend in the internal capsule. The corticobulbar tracts end at the motor nuclei of cranial nerves. Most of the remaining pyramidal fibers cross over in the medulla before descending into the spinal cord.

white commissure before synapsing on motor neurons in the anterior gray horns.

The Extrapyramidal System

The axons of the pyramidal cells of the motor cortex descend to synapse on motor neurons in the spinal cord and brain. Because there are no intervening synapses, the pyramidal system provides a rapid and direct mechanism for the control of skeletal muscles. Several other centers may issue motor commands as a result of processing performed at an unconscious, involuntary level. These centers and their associated tracts constitute the **extrapyramidal system**, or **EPS**.

Figure 15-7 indicates the major components of the extrapyramidal system, which controls motor neurons in the brain stem and spinal cord. The effects produced by EPS activation vary from direct stimulation through a mild facilitation to an outright inhibition. By altering the sensitivity of motor neurons controlling muscles or groups of muscles, the extrapyramidal system can modify or even override the commands arriving over the pyramidal tracts. Although complicated motor activities involving the entire body can be directed by extrapyramidal centers, the movements produced are relatively fixed and invariable.

Processing centers of the EPS include the vestibular nucleus, the superior colliculus, the red nucleus, the reticular formation, and the cerebral nuclei. Their outputs may target the motor nuclei of cranial nerves or descend into the spinal cord as the **vestibulospinal**, **tectospinal**, **rubrospinal**, and **reticulospinal tracts**. Figure 15-5 indicates the location of these tracts in the spinal cord.

The Vestibular Nuclei and the Vestibulospinal Tracts The vestibular nuclei receive information, via the eighth cranial nerve, from receptors in the inner ear that monitor position and movement. They respond to changes in the orientation of the head, sending motor commands that alter the muscle tone, extension, and/or position of the neck, eyes, head, and limbs. The primary goal is the maintenance of posture and balance. The descending fibers within the spinal cord constitute the **vestibulospinal tracts**.

The Tectum and the Tectospinal Tracts Commands carried by the **tectospinal tracts** change the position of the eyes, head, neck, and arms in response to bright lights, sudden movements, or loud noises. These tracts originate in the roof (tectum) of the midbrain, in the superior and inferior colliculi, and cross to the opposite side immediately. The motor commands carried by the tectospinal tracts are triggered by visual and auditory stimuli; the colliculi also process inputs from the cerebrum, the cerebellum, the reticular formation, and other nuclei in the brain stem.

The Red Nucleus and the Rubrospinal Tracts The red nucleus receives extensive inputs from cerebral nuclei, the cerebellum, and the reticular formation. The **rubrospinal tracts** (*ruber*, red) carry the motor responses to spinal motor neurons. Direct stimulation of the red nucleus increases muscle tone and may produce either flexion or extension of the axial skeleton, depending on the area affected.

The Reticular Formation and the Reticulospinal Tracts The reticular formation receives collateral inputs from almost every ascending and descending

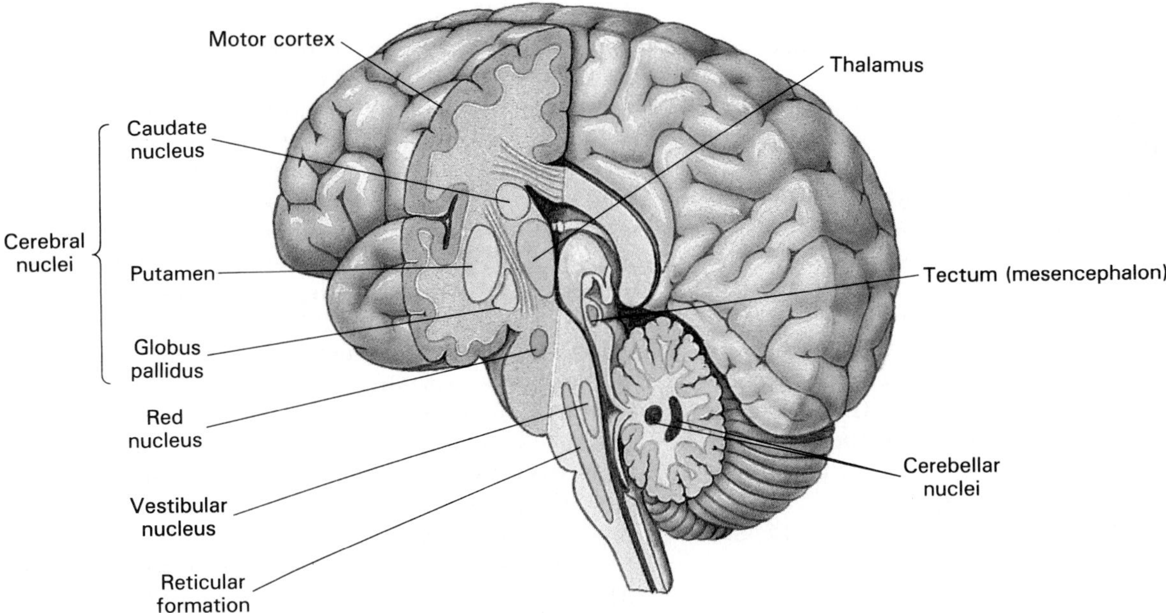

FIGURE 15-7
Components of the Extrapyramidal System. Cutaway view showing the location of major components of the extrapyramidal system.

pathway, as well as via extensive interconnections with the cerebrum, the cerebellum, and nuclei within the brain stem. The motor commands carried by the **reticulospinal tracts** vary depending on the region stimulated. Thus, stimulation of one region produces involuntary eye movements, whereas stimulation of a different area affects the respiratory muscles.

The Cerebral Nuclei The cerebral nuclei, introduced in Chapter 14 (see Figure 14-5 and Table 14-3), are the most important and complex components of the extrapyramidal system. (p. 440) These nuclei are processing centers whose functions blur the distinctions between "conscious" and "unconscious" motor control. In effect, they provide the background patterns of movement involved in the performance of voluntary motor activities. These nuclei do not exert direct control over CNS motor neurons; instead, they adjust the motor commands issued in other processing centers.

The cerebral nuclei use three major pathways to accomplish this:

1. One group of axons synapses with thalamic neurons, which then send their axons to the primary motor cortex. This arrangement creates a feedback loop that changes the sensitivity of the pyramidal cells and alters the pattern of instructions carried by the corticospinal tracts.
2. The second group innervates the red nucleus and alters the activity in the rubrospinal tracts.
3. The third group travels through the thalamus to reach centers in the reticular formation, where they adjust the output of the reticulospinal tracts.

Complex mechanisms control the activities of the cerebral nuclei. Two distinct populations of neurons exist: one that stimulates motor neurons by releasing acetylcholine (ACh) and another that inhibits motor neurons by the release of gamma aminobutyric acid (GABA). Under normal conditions the excitatory neurons are kept inactive, and the descending tracts are primarily responsible for inhibiting motor neuron activity. If the descending tracts are severed in an accident, the loss of inhibitory control leads to a generalized state of muscular contraction known as *decerebrate rigidity* (see Chapter 14). (p. 447)

Because their background activity results in inhibition, the cerebral nuclei do not *initiate* specific movements. However, once a movement is under way the cerebral nuclei provide a background pattern and rhythm.

Consider what happens at the start of a brief stroll. You decide to start walking, and off you go, without consciously thinking about the details. But the basic rhythm and the motor patterns involved in moving your legs, shifting your weight, swinging your arms, and so forth are directed by the cerebral nuclei. When you start the movement, branches from the pyramidal tracts synapse in the cerebral nuclei. The caudate nucleus and putamen respond by issuing the characteristic stereotyped motor commands that continue until the decision is made to stop the stroll.

Injuries to the motor cortex eliminate the ability to exert fine control over motor units, but gross movements may still be produced by the cerebral nuclei using the reticulospinal or rubrospinal tracts. An individual whose primary motor cortex has been destroyed retains the ability to walk, maintain balance, and perform other voluntary and involuntary movements. The movements lack precision, and in general are awkward and poorly controlled.

The Cerebellum

The pyramidal and extrapyramidal systems could not function effectively if their outputs were not continually readjusted as the movements evolved. The complex processing and integration of proprioceptive data from peripheral structures, visual information from the eyes, and equilibrium-related sensations from the inner ear are not performed by either the cerebral cortex or the cerebral nuclei. Instead, these correlations occur within the cerebellum, and the cerebellar output regulates the activity along both pyramidal and extrapyramidal motor pathways.

The cerebellum, discussed in Chapter 14 (see Figure 14-11 and Table 14-8), has as its primary functions (1) overseeing the postural muscles of the body and (2) adjusting voluntary and involuntary motor patterns. (p. 448) The cerebellum receives proprioceptive information from the spinal cord along the spinocerebellar tracts. Additional sensory information arrives from the reticular formation and the vestibular nuclei. Some of the afferents pass through the arbor vitae to reach the superficial layers of the cerebellar cortex, while others synapse in cerebellar nuclei.

The integrative activities performed by neurons in the cerebellar cortex and cerebellar nuclei are essential to the precise control of voluntary and involuntary movements. For example, consider the motor problems involved with a seemingly simple movement, such as touching the tip of the nose with the eyes closed. That movement begins simply enough, as the primary motor cortex sends voluntary instructions to the motor neurons involved, and the cerebral nuclei inhibit the opposing muscle groups. But as the movement proceeds, the relative degree of stimulation and inhibition changes. At precisely the right moment, the opposing muscles are actively stimulated to ensure that the movement ends with a light touch. Throughout the movement the cerebellum performs the necessary adjustments, balancing the activities of muscle groups and modulating the degree of stimulation versus inhibition.

The relationships between the cerebellum and other motor centers are diagrammed in Figure 15-8. Both pyramidal and extrapyramidal centers send information to the cerebellum when motor commands are issued. As the movement proceeds, the cerebellum monitors proprioceptive (position) and vestibular (balance) information and adjusts the activities of the voluntary and involuntary motor centers. In general, any voluntary movement begins with the activation of far more motor units than are actually required or even desirable. The cerebellum pro-

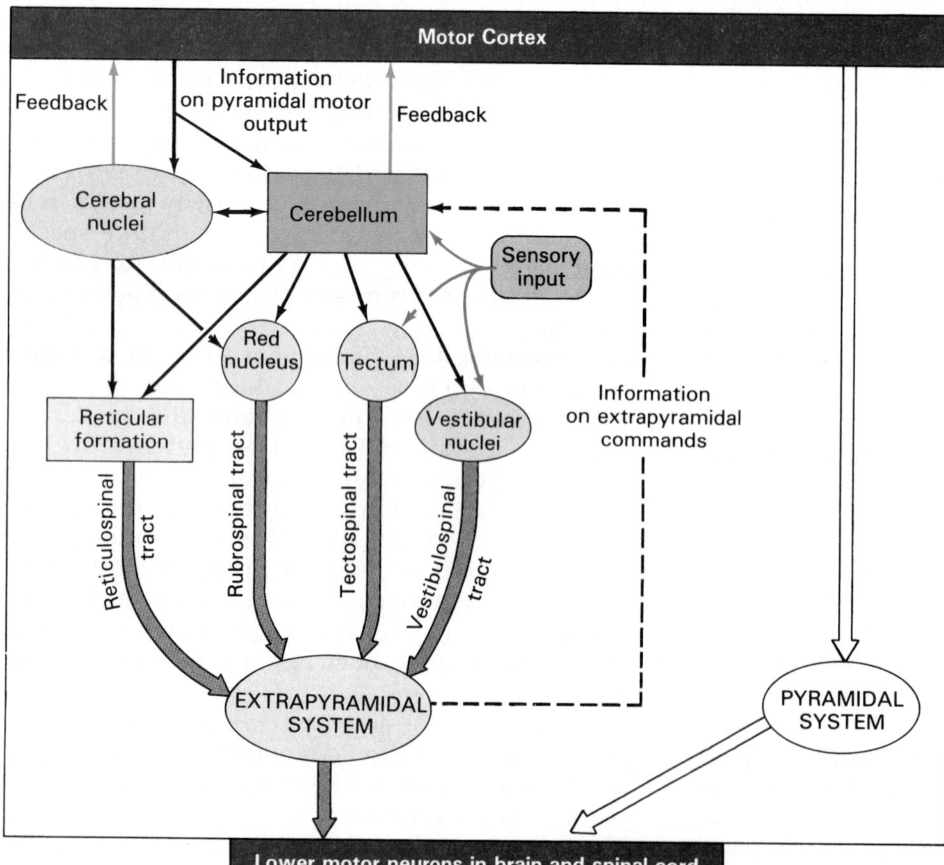

FIGURE 15-8
The Extrapyramidal System. A diagram showing important interconnections and tracts of the pyramidal and extrapyramidal systems. Note the central role played by the cerebellum.

vides the necessary inhibition, reducing the number of motor commands to an efficient minimum. As the movement proceeds, the pattern and degree of inhibition change, producing the desired result.

To direct rapid and precise changes in position and movement the cerebellum must not only "know" what muscles are needed now, but also what muscles are going to be needed next. The cerebellum has to make predictions about how long the complete movement will take and what distance will be covered. Without the ability to make such predictions, smooth movements would be impossible, for the information arriving in the cerebellum at any given moment concerns the state of affairs in the periphery a few milliseconds earlier. Imagine driving a car while looking only through the rear window and you will understand the problem.

These predictive abilities must be learned by repetition and trial and error. Developing the necessary patterns of motion and response takes time. Many of the basic patterns are established early in life; examples would include the fine balancing adjustments made while running. Motor patterns involved in extremely complex behaviors require considerable effort and practice. For example, the cerebellum controls the relaxed, smooth, and automatic movements of acrobats, dancers, golfers, tennis players, and sushi chefs.

Clinical Comment: Anencephaly

Although it may sound strange, physicians usually take a newborn infant into a dark room and shine a light against the skull. They are checking for **anencephaly** (an-en-SEF-a-lē), a rare condition in which the brain fails to develop at levels above the mesencephalon or lower diencephalon. In such instances the cranial vault is empty and translucent enough to transmit light.

Unless the condition is discovered right away, the parents may take the infant home, totally unaware of the problem. All the normal behavior patterns expected of a newborn are present, including suckling, stretching, yawning, crying, kicking, sticking fingers in the mouth, and tracking movements with the eyes. However, death will occur naturally over a period of days to months.

This tragic condition provides a striking demonstration of the role of the brain stem in controlling complex involuntary motor patterns. During normal development these patterns become incorporated into variable and versatile behaviors as control and analytical centers appear in the cerebral cortex.

Levels of Processing and Motor Control

Ascending sensory information is relayed from one nucleus or center to another in a series of steps. For example, somatic sensory information from the spinal cord goes from a nucleus in the medulla to a nucleus in the thalamus before it reaches the primary sensory cortex. Information processing occurs at each step along the way, with results that may block, reduce, or heighten conscious and unconscious awareness of the stimulus.

These processing steps are important, but they take time. Every synapse means another synaptic delay. Between the conduction time and the synaptic delays, several milliseconds can pass before the primary sensory cortex receives information from a peripheral receptor. Additional time will pass before the primary motor cortex orders a voluntary motor response.

This delay is not dangerous because involuntary motor commands are issued by relay stations in the spinal cord and brain stem. While the conscious mind is still processing the information, neural reflexes provide an immediate response that can later be "fine-tuned." For example, when someone touches a hot stove top, the response (withdrawing the hand) occurs before the individual is aware of the injury. Voluntary motor responses, such as shaking the hand, stepping back, and crying out, occur somewhat later. In this case the initial reflexive response, directed by neurons in the spinal cord, was supplemented by a voluntary response controlled by the cerebral cortex. The spinal reflex provided a rapid, automatic, preprogrammed response that preserved homeostasis. The cortical response was more complex, but it required more time to prepare and execute.

Nuclei in the brain stem are also involved in a variety of complex reflexes. Some of these nuclei receive sensory information and generate appropriate motor responses. These motor responses may involve direct control over motor neurons or the regulation of reflex centers in other parts of the brain. This pattern can be seen in Figure 15-9a, which diagrams the levels of somatic motor control.

Motor neurons controlled by simple reflexes are at the bottom level. These reflexes are directed by nuclei in the spinal cord and brain stem. Higher levels perform more elaborate processing; as one moves from the medulla to the cerebral cortex the motor patterns become increasingly complex and variable. For example, the respiratory rhythmicity center of the medulla sets a basic breathing rate. Centers in

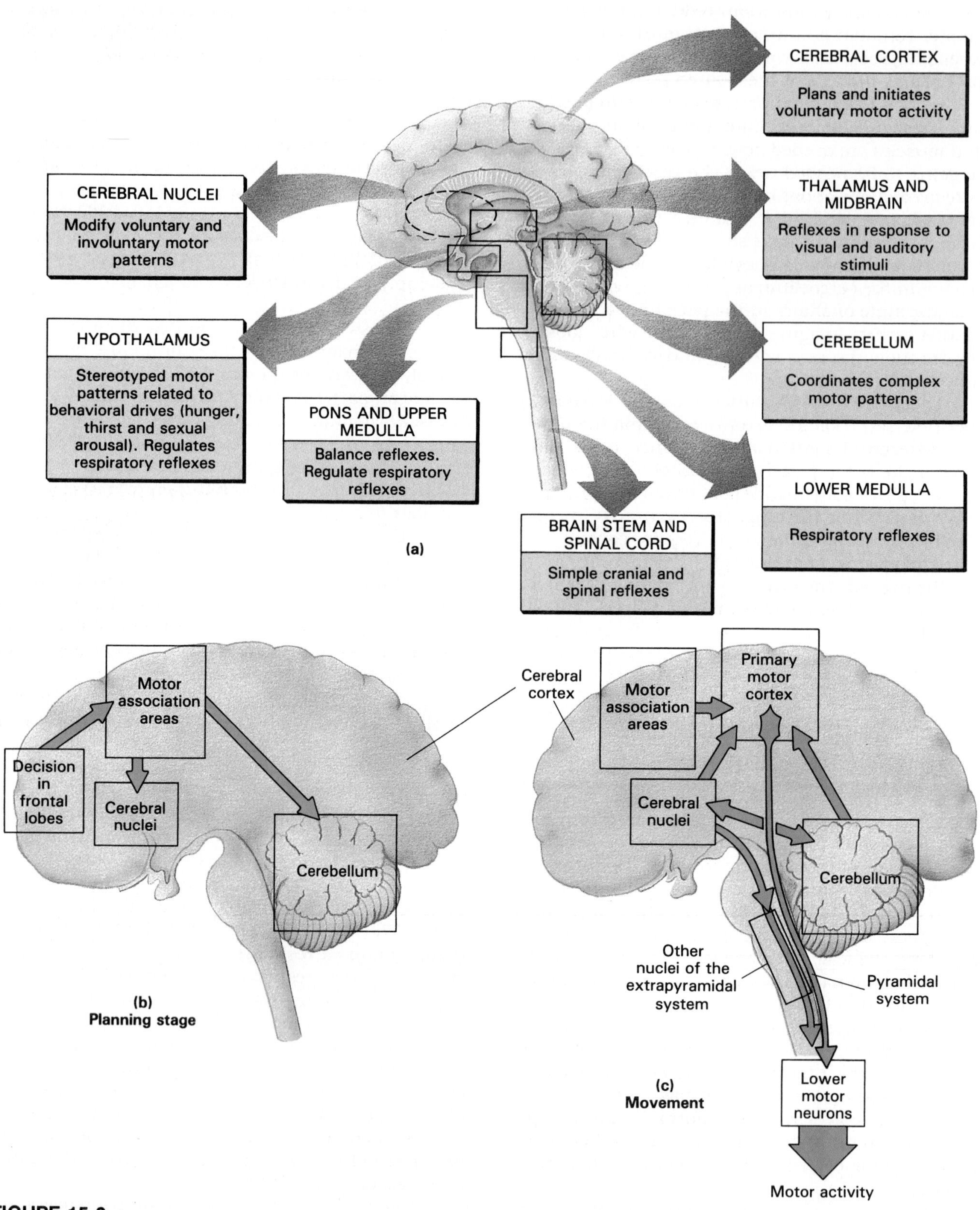

FIGURE 15-9

Levels of Somatic Motor Control. (a) Somatic motor control involves a series of levels, with simple spinal and cranial reflexes at the bottom and complex voluntary motor patterns at the top. (b) The planning stage: When a conscious decision is made to perform a specific movement, information is relayed from the frontal lobes to motor association areas. These areas in turn relay the information to the cerebellum and cerebral nuclei. (c) Movement: As the movement begins, the motor association areas send instructions to the primary motor cortex. Feedback from the cerebral nuclei and cerebellum modifies those commands, and extrapyramidal motor output directs involuntary adjustments in position and muscle tone.

the pons adjust that rate in response to commands received from the hypothalamus (involuntary) or cerebral cortex (voluntary).

The cerebral nuclei, cerebellum, midbrain, and hypothalamus control the most complicated involuntary motor patterns. Examples include motor patterns associated with eating and reproduction (hypothalamus), walking and body positioning (cerebral nuclei), learned movement patterns (cerebellum), and movements in response to sudden visual or auditory stimuli (midbrain).

At the highest level are the complex, variable, and voluntary motor patterns dictated by the cerebral cortex. Motor commands may be given to specific motor neurons directly, or indirectly by altering the activity of a reflex control center. Figure 15-9b,c provides a simple diagram of the steps involved in the planning and execution of a voluntary movement.

During development the control levels appear in sequence, beginning with the spinal reflexes. More complex reflexes develop as the neurons grow and interconnect. The process proceeds relatively slowly, as billions of neurons establish trillions of synaptic connections. At birth neither the cerebral nor the cerebellar cortex is fully functional, and their abilities take years to mature. A number of anatomical factors, noted in earlier chapters, contribute to this maturation:

1. Cortical neurons continue to increase in number until at least age 1;
2. The brain grows in size and complexity until at least age 4;
3. Myelination of CNS neurons continues at least until puberty.

As these events occur, cortical neurons continue to establish new interconnections that will have a long-term effect on the functional capabilities of the individual.

- **As a result of pressure on her spinal cord, Jill cannot feel touch or pressure on her legs. What spinal tract is being compressed?**
- **What is the anatomical reason for the left side of the brain controlling motor function on the right side of the body?**
- **An injury to the superior portion of the motor cortex would affect what part of the body?**

Monitoring Brain Activity: The Electroencephalogram

Sensory and motor pathways have relatively distinct anatomical boundaries, and the associated cortical areas are well defined. Higher-order functions, however, are much harder to assign to specific tracts and regions. Such functions as memory, learning, intellectual processes, and consciousness involve complex interactions between many components of the brain, especially portions of the cerebral hemispheres. These are among the most intriguing aspects of brain activity, but they have also proven to

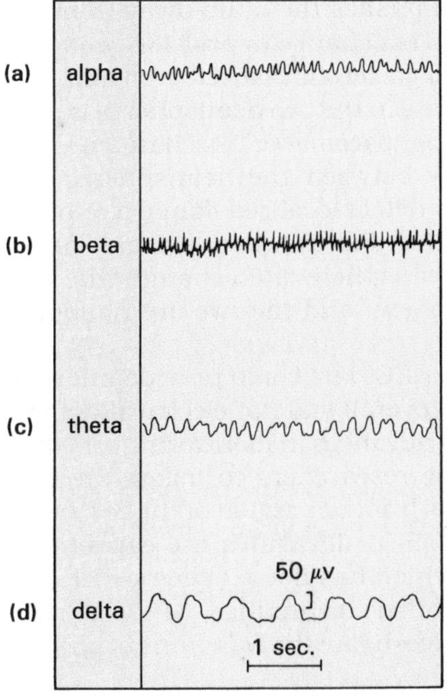

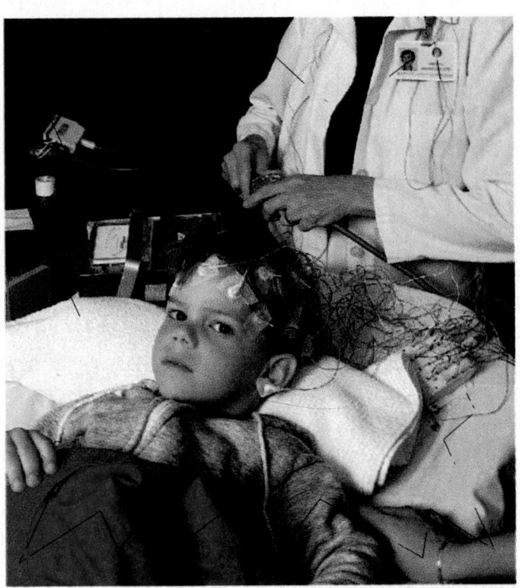

(e)

FIGURE 15-10
Brain Waves. (a) Alpha waves are characteristic of normal resting adults. (b) Beta waves typically accompany intense concentration. (c) Theta waves are seen in children and in frustrated adults. (d) Delta waves occur during deep sleep and in certain pathological states. (e) A patient wired for EEG monitoring.

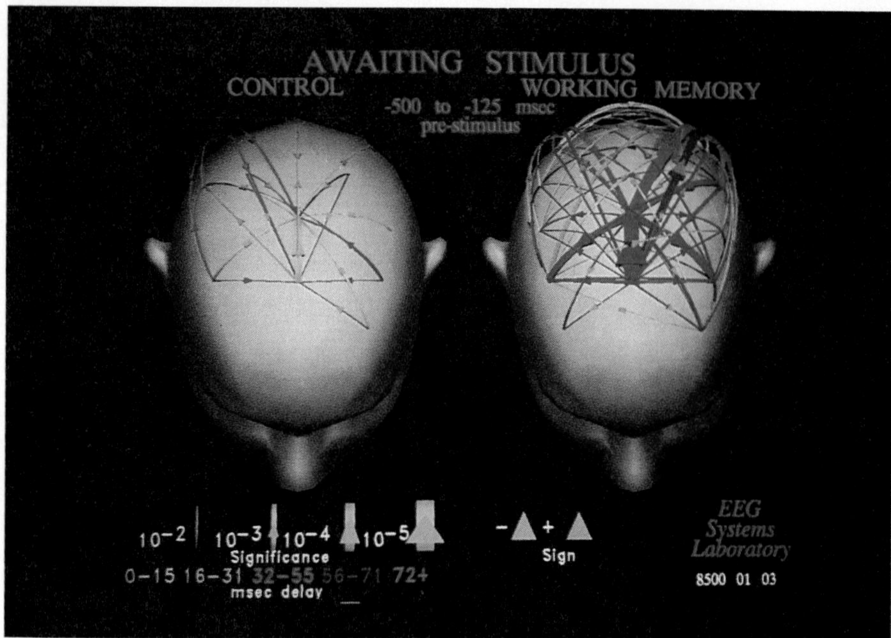

FIGURE 15-11
EEG Traces of Cerebral Activity Involved in Memory. To create these images, EEG data was recorded at 33 points on the skull and processed by computer at 8-millisecond intervals to create a dynamic record of cerebral activity. The image at left is a composite of data from five subjects during an instant when they were waiting for a number to be flashed on a blank screen. At right, the same subjects were looking at the screen while trying to remember two previously displayed numbers. The degree of similarity in activity at two points is indicated by the thickness of the arrow joining them. Different colors are used to indicate the time delay between related peaks of activity. (This technique was developed by Alan Gevins at the EEG Systems Laboratory in San Francisco.)

be the most difficult to localize anatomically and investigate experimentally. Experiments with other animals do not always provide reliable information about cerebral processing in humans, and opportunities to perform experiments on human subjects are understandably limited.

The primary sensory cortex and the primary motor cortex have been mapped by direct stimulation in patients undergoing brain surgery. The functions of other regions of the cerebrum can be inferred from behavioral changes that follow localized injuries or strokes, although in many cases the precise nature of the damage can be determined only at autopsy. However, there are noninvasive methods to follow cortical activity. One of the most common methods involves monitoring the electrical activity of the brain.

Neural function depends on electrical events within the cell membrane. The brain contains billions of nerve cells, and their activity generates an electrical field that can be measured by placing electrodes on the brain or the outer surface of the skull. The electrical activity changes constantly, as nuclei and cortical areas are stimulated or quiet down. An **electroencephalogram (EEG)** is a printed record of the electrical activity over time. The **brain waves** observed provide an indication of the ongoing level of activity, and the patterns can be correlated with the individual's state of attention or level of consciousness.

Typical brain waves are shown in Figure 15-10 on p. 483. **Alpha waves** are found in the brains of normal resting adults when their eyes are closed. Alpha waves disappear during sleep, but they also vanish when the individual begins to concentrate on some specific task. At that time the alpha waves are replaced by higher-frequency **beta waves**. Beta waves are typical of individuals under stress or in states of psychological tension. **Theta waves** are seen in the brains of children and in intensely frustrated adults. They may also indicate the presence of a brain disorder, such as a tumor. **Delta waves** are very large amplitude, low-frequency waves. These waves are seen in the brains of infants (in whom cortical development is still incomplete), during deep sleep at all ages, and when a tumor, vascular blockage, or inflammation has damaged portions of the brain.

Electroencephalograms can sometimes provide useful diagnostic information. Because they monitor electrical events that can only be detected over small distances, conditions that change the shape of the brain can alter the EEG. For example, in a subdural hemorrhage (Chapter 14), blood accumulates in the subdural space and pushes the brain away from the skull. (p. 453) This change in position produces a change in the EEG pattern measured over that site.

Electrical activity in the two hemispheres is usually synchronized; the "pacemaker" mechanism is not known. Asynchrony between the hemispheres can therefore be used to detect localized damage or other cerebral abnormalities. For example, a tumor or injury affecting one hemisphere often changes the pattern in that hemisphere, and the two are no longer aligned. [**CM**: *Seizures and Epilepsy*]

Until recently, EEG data could provide information only about the overall state of electrical activity in the brain. Recent advances in monitoring and data analysis now enable researchers to follow the flow of electrical impulses from one region of the cerebrum to another. Figure 15-11 illustrates the capabilities of this technique, which has not yet been widely utilized. At present, other techniques, such as PET scans, are used to investigate the functions of specific regions. [**CM**: *PET Scans of the Brain*]

Higher-Order Functions

Higher-order functions have the following characteristics:

1. They are performed by the cerebral cortex.
2. They involve complex interactions between areas of the cortex and between the cerebral cortex and other areas of the brain.
3. They involve both conscious and unconscious information processing.
4. They are not part of the programmed "wiring" of the brain; therefore, the functions are subject to modification and adjustment over time.

Our discussion of higher-order function begins by identifying the cortical areas involved and considering functional differences between the left and right cerebral hemispheres. We will then describe the mechanisms of memory and learning and detail the neural interactions responsible for consciousness, sleep, and arousal. The final section provides an overview of brain chemistry and its effects on behavior and personality.

CORTICAL INTEGRATION OF INFORMATION

The sensory, motor, and association areas of the cerebral hemispheres were introduced in Chapter 14. (pp. 437–438) The primary sensory cortex receives touch, pressure, vibration, pain, and temperature information relayed from the thalamus; these are examples of **general senses**. The thalamus also relays information concerning the **special senses** of sight, smell, taste, and hearing to appropriate cortical areas. Figure 15-12a reviews the major cortical regions of the left cerebral hemisphere introduced in Chapter 14.

Scanning data, EEG traces, and clinical observation have shown that several cortical areas act as higher-order integrative centers for complex sensory stimuli and motor responses. These centers, indicated in Figure 15-12b, include the *general interpretive area*, the *speech center*, and the *prefrontal cortex*.

FIGURE 15-12
Functional Areas of the Cerebral Cortex. Centers responsible for higher-order functions are not found in both hemispheres. The left hemisphere usually contains the general interpretive area and the speech center. Specializations of the right cerebral hemisphere are shown in Figure 15-14.

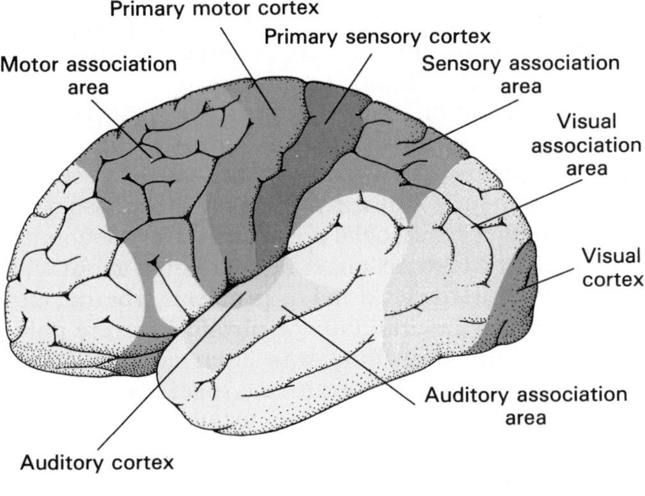

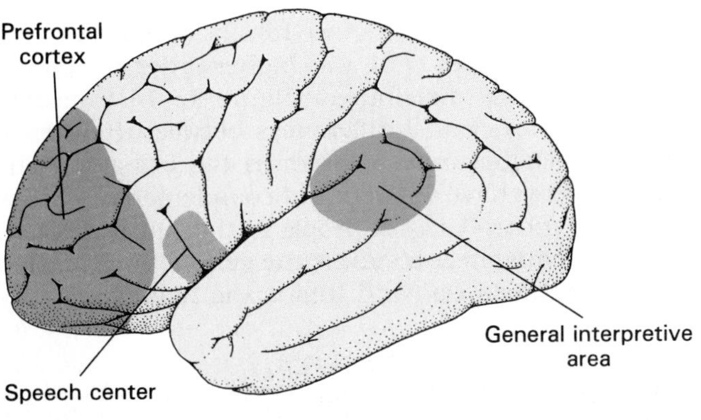

The General Interpretive Area

The **general interpretive area** (*Wernicke's area*) receives information from all the sensory association areas. This analytical center is present in only one hemisphere, usually the left. Damage to the general interpretive area affects the ability to interpret what is read or heard, even though the words are understood as individual entities. For example, an individual might understand the meaning of the words "sit" and "here" but be totally bewildered by the request "sit here."

Aphasia (*a*-, without + *phasia*, speech) is a disorder affecting the ability to speak or read. Extreme, or **global aphasia**, results from extensive damage to the general interpretive area or to the associated sensory tracts. Affected individuals are totally unable to speak, to read, or to understand or interpret the speech of others. Global aphasia often accompanies a severe stroke or tumor that affects a large area of cortex including the speech and language areas. Recovery is possible when the condition results from edema or hemorrhage, but the process often takes months or even years.

Major motor aphasia may develop in some individuals after a brief period of global aphasia. This condition is extremely frustrating for the patient, because the individual can understand language and knows how to respond, but lacks the motor control necessary to produce the right combinations of sounds.

Lesser degrees of aphasia often follow minor strokes. There is no initial period of global aphasia, and the individuals can understand spoken and written words. The problems encountered with speaking or writing gradually fade, and many individuals with minor aphasia recover completely.

Dyslexia (*lexis*, diction) is a disorder affecting the comprehension and use of words. **Developmental dyslexia** affects children; there are estimates that up to 15 percent of children in the United States suffer from some degree of dyslexia. These children have difficulty reading and writing, although their other intellectual functions may be normal or above normal. Their writing looks uneven and disorganized; letters are often reversed or written in the wrong order (Figure 15-13). Recent evidence suggests that at least some forms of dyslexia result from problems in processing and sorting visual information. For example, several investigators have reported significant improvement in the performance of dyslexic children using special glasses or blue-tinted overlays. [**CM**: *Huntington's Disease*]

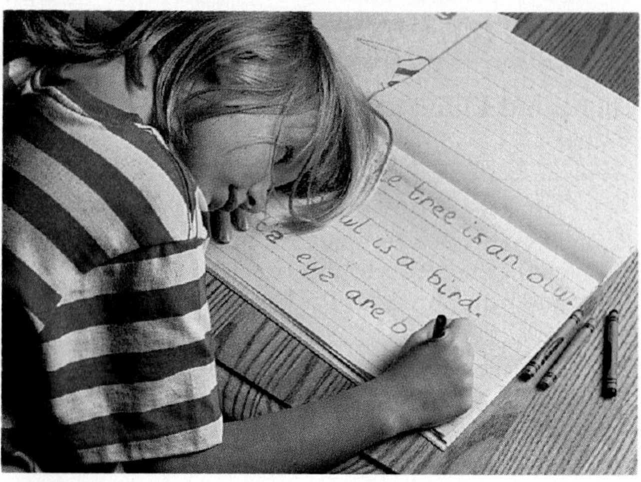

FIGURE 15-13
Dyslexia. In dyslexia an individual has difficulty with the comprehension and use of words. Children with developmental dyslexia have problems learning to read and write; letters are often written backwards and in the wrong order.

The Speech Center

Efferents from the general interpretive area target the **speech center** (*Broca's area*). This center lies along the edge of the premotor cortex in the same hemisphere as the general interpretive area. The speech center regulates the patterns of breathing and vocalization needed for normal speech. The corresponding regions on the opposite hemisphere are not "inactive," but their functions are less well defined. Damage to the speech center can manifest itself in various ways. Some individuals have difficulty speaking, although they know exactly what words to use; others talk constantly but use all the wrong words.

The Prefrontal Cortex

The **prefrontal cortex** of the frontal lobe coordinates information from the secondary and special association areas of the entire cortex. In doing so it performs such abstract intellectual functions as predicting the future consequences of events or actions. Damage to the prefrontal cortex leads to difficulties in estimating the temporal relationships between events; questions such as "How long ago did this happen?" or "What happened first?" become difficult to answer.

The prefrontal cortex has extensive connections with other cortical areas and with other portions of the brain, such as the limbic system. Feelings of frustration, tension, and anxiety are generated at the prefrontal cortex as it interprets ongoing events and makes predictions about future situations or consequences. If the connections between the prefrontal cortex and other brain regions are severed, the tensions, frustrations, and anxieties are removed. Earlier in this century this rather drastic procedure, called a **prefrontal lobotomy**, was used to "cure" a variety of mental illnesses, especially those associated with violent or antisocial behavior.[1] After a lobotomy, the patient would no longer be concerned about what had previously been a major problem, whether psychological (hallucinations) or physical (severe pain). However, the individual was often equally unconcerned about tact, decorum, and toilet training. Now that drugs have been developed to target specific pathways and regions of the CNS, lobotomies are no longer used to change behavior.

HEMISPHERIC SPECIALIZATION

The regions seen in Figure 15-12a are present on both hemispheres, but the higher-order functions are not equally distributed. Figure 15-14 indicates the major functional differences between the hemispheres. Higher-order centers in the left and right hemispheres have different but complementary functions. In the majority of people in the United States, the left hemisphere contains the general interpretive and speech centers, and this is the hemisphere re-

[1] In the movie *One Flew over the Cuckoo's Nest* Jack Nicholson played the part of a mental patient "cured" by a prefrontal lobotomy.

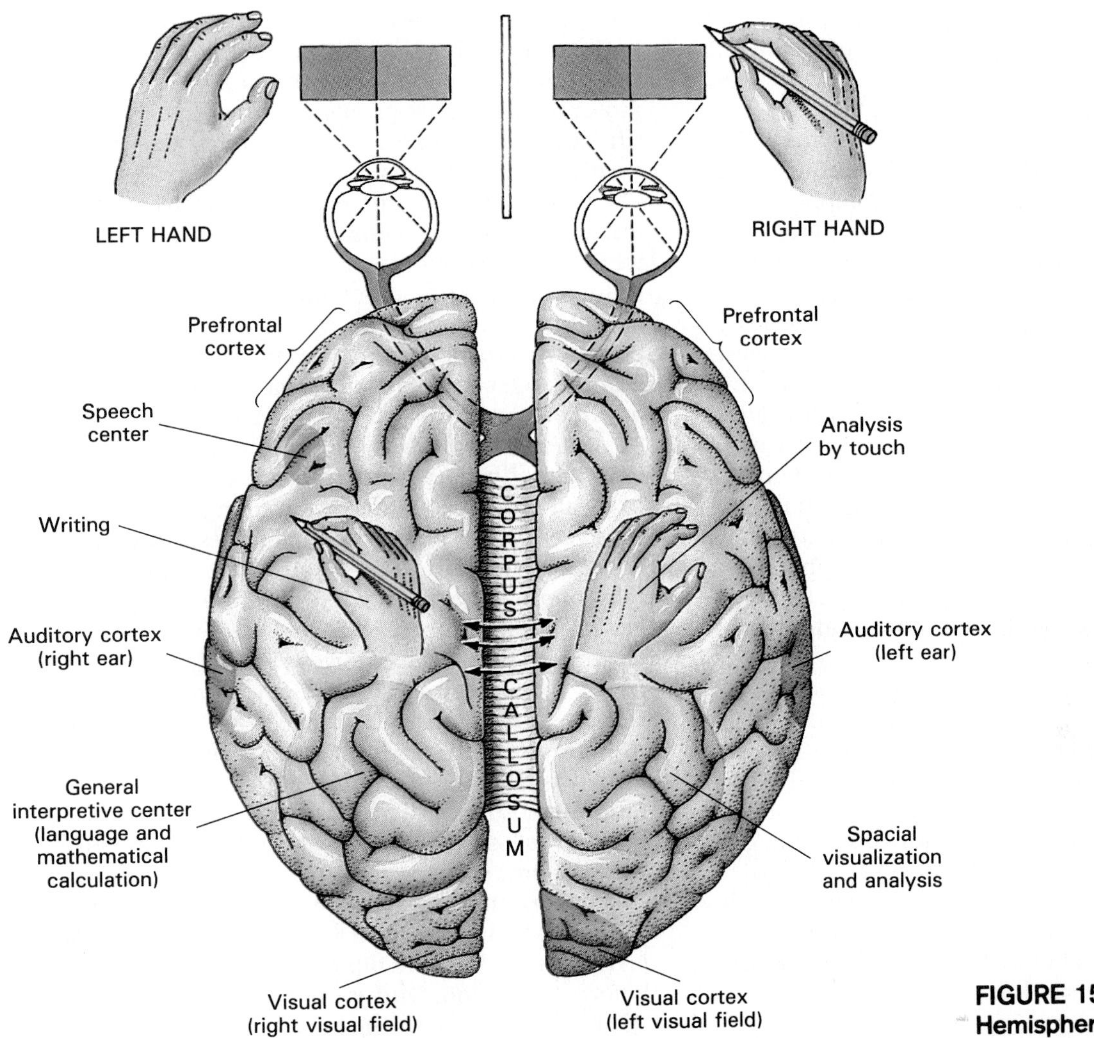

FIGURE 15-14
Hemispheric Specialization. Functional differences between the left and right cerebral hemispheres.

sponsible for language-based skills. For example, reading, writing, and speaking are dependent on processing done in the left cerebral hemisphere. This hemisphere is also important in performing analytical tasks, such as mathematical calculations and logical decision making. Although the left hemisphere was formerly called the *dominant hemisphere*, a more appropriate term is the **categorical hemisphere**, because the right hemisphere has many important functions.

The right cerebral hemisphere analyzes sensory information and relates the body to the sensory environment. Interpretive centers in this hemisphere permit the identification of familiar objects by touch, smell, taste, or feel. Because it is concerned with spatial relationships and analyses, the term **representational hemisphere** is used to refer to the right hemisphere.

The designation of a hemisphere as categorical, as opposed to representational, depends on the location of the major functional centers, especially the general interpretive and speech centers. Because the left hemisphere is categorical in the majority of both left-handed and right-handed individuals, there is probably a genetic basis for the distribution of functions. An estimated 90 percent of the population have an enlarged left hemisphere at birth. (The remaining 10 percent may have hemispheres of equal size or an enlarged right hemisphere.)

The functional significance (if any) of having the categorical hemisphere on the left side is unknown. Interestingly, there may be a link between handedness and sensory/spatial abilities. An unusually high percentage of musicians and artists are left-handed; the complex motor activities performed by these individuals are directed by the primary motor cortex and association areas on the right (representational) hemisphere.

Hemispheric specialization does not mean that the two hemispheres are independent, merely that certain centers have evolved to process information gathered by the system as a whole. The intercommunication occurs over commissural fibers, especially those of the corpus callosum. The corpus callosum alone contains over 200 million axons, carrying an estimated 4 billion impulses per second!

This communication across the corpus callosum permits the integration of sensory information and motor commands. Yet the two hemispheres are significantly different in terms of their ongoing processing activities. Otherwise untreatable seizures may sometimes be treated by severing the corpus collosum. This surgery produces symptoms of **disconnection syndrome**. In this condition the two hemispheres function independently, each remaining "unaware" of stimuli or motor commands involving their counterpart. The result is a number of rather interesting changes in the individual's abilities. For example, objects touched by the left hand can be recognized but not verbally identified, because the sensory information arrives at the right hemisphere and the speech center is on the left. The object can be verbally identified if felt with the right hand, but the person will not be able to say whether or not it is the same object previously touched with the left hand. This problem with cross-referencing sensory information applies to all incoming sensations.

Two years after a surgical sectioning of the corpus callosum, the most striking behavioral abnormalities have disappeared, and the individual may test normally. In addition, individuals born without a functional corpus callosum do not show obvious sensory or motor deficits. In some way the CNS adapts to the situation, probably by increasing the amount of information transferred across the anterior commissure.

MEMORY

What was the topic of the last sentence you read? What do your parents look like? What is your Social Security number? When did Columbus discover America? What does a red traffic light mean? What does a hot dog taste like? Answering these questions involves accessing *memories*, stored bits of information gathered through prior experience. Memories that can be voluntarily retrieved and verbally expressed are called **declarative memories**.

Declarative memories can be classified according to the nature of the information stored. **Fact memories** are specific bits of information, such as the color of a stop sign or the smell of perfume. **Skill memories** are learned motor behaviors. For example, you can probably remember how to light a match or open a screw-top jar. With repetition, skill memories become incorporated at the unconscious level. They are then called *reflexive memories*.

Reflexive memories are not voluntarily accessible. Examples of reflexive memories would include salivating at the smell of your favorite food or the complex motor patterns involved in skiing, playing the violin, and similar activities. Memories related to programmed behaviors, such as eating, are stored in appropriate portions of the brain stem. Complex motor memories are probably stored in the cerebellum (the importance of cerebellar memory was discussed earlier in this chapter).

Two classes of memories appear to exist. **Short-term**, or **primary, memories** do not last long, but while they persist the information can be recalled immediately. Primary memories contain small bits of information, such as a person's name or a telephone number. **Long-term memories** remain for much longer periods, in some cases for an entire lifetime. There are two types of long-term memories. **Secondary memories** are long-term memories that fade with time and may require considerable effort to recall. **Tertiary memories** are long-term memories that seem to be part of consciousness, such as your name or the contours of your own body. Proposed relationships between these memory classes are diagrammed in Figure 15-15.

Each class of memories probably involves a different physiological mechanism. Short-term memories are thought to be maintained by small reverberating circuits between regions of the cerebral cortex or between the cortex and thalamus. As the neurons fatigue, or some other stimulus arrives, the original memory is lost. Sudden shocks or reductions in brain metabolism, such as might be produced by anesthesia or hypothermia, can "wipe the slate clean" and cancel short-term memories. For this reason accident victims are often unable to remember the moments before a collision, and surgery patients often forget the ride to the operating room.

Repeating a phone number or other bit of information reinforces the original short-term memory and helps ensure its conversion to a long-term memory. That conversion is called **memory consolidation**.

Brain Regions Involved in Memory Consolidation and Access

Two components of the limbic system, the amygdaloid body and the hippocampus, are essential to memory consolidation. Damage to either of these areas will interfere with normal memory consolidation. Damage to the hippocampus leads to an immediate loss of short-term memory, although long-term memories remain intact and accessible. Tracts leading from the amygdaloid body to the hypothalamus may link memories to specific emotions. A cerebral nucleus near the diencephalon, the **nucleus basalis**, plays an uncertain role in memory storage and retrieval. Tracts connect this nucleus with the hippocampus, amygdaloid body, and all areas of the cerebral cortex. Damage to this nucleus is associated with changes in emotional states, memory, and intellectual function (see the discussion of Alzheimer's disease later in this chapter.)

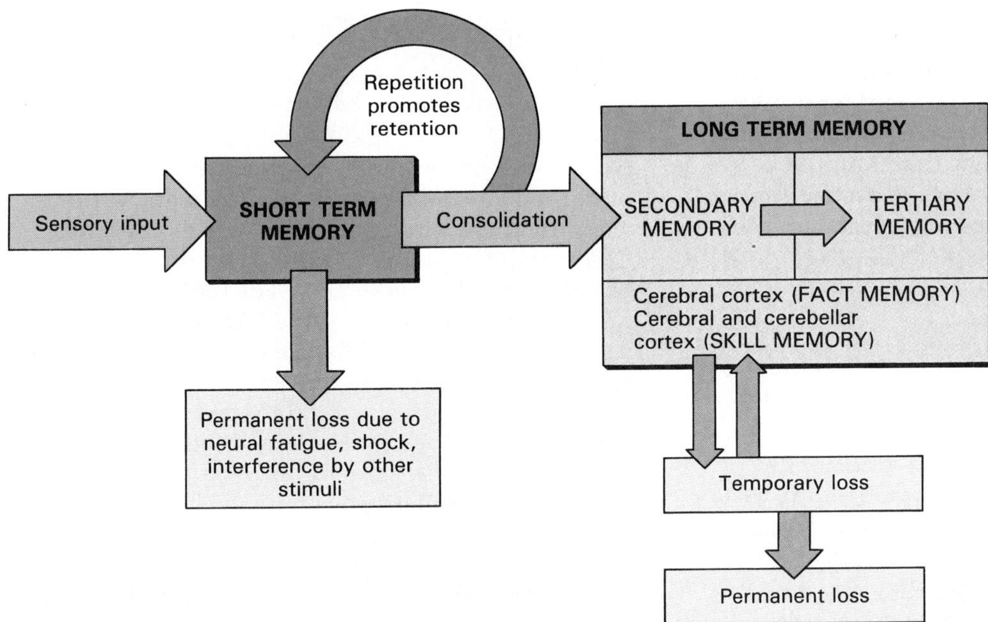

FIGURE 15-15
Memory Storage. Steps in the storage of memories and the conversion from short-term to long-term memories.

Long-term memories are stored in the cerebral cortex. Conscious motor and sensory memories are referred to the appropriate association areas. For example, visual memories are stored in the visual association area, and memories of voluntary motor activity in the premotor cortex. Special portions of the occipital and temporal lobes retain the memories of faces, voices, and words. In at least some cases, a specific memory probably reflects the activity of a single neuron. For example, in one portion of the temporal lobe an individual neuron responds to the sound of one word and ignores others. A specific neuron may also be activated by the proper combination of sensory stimuli associated with a particular individual, such as your grandmother. As a result these neurons are called "grandmother cells." On a still larger scale, stimulation of other areas of the temporal lobe produce a generalized sense of familiarity. This feeling, known as *deja vu*, makes you feel that you've heard, seen, or done something before, even though you know (objectively) that this is the first time.

Cellular Mechanisms of Memory Formation and Storage

At the cellular level, memory consolidation involves anatomical or physiological changes, or both. For legal, ethical, and logistical reasons, it is not possible to perform very much research on these mechanisms with human subjects. Research on other animals, often those with relatively simple nervous systems, has indicated that the following mechanisms may be involved:

1. Increased neurotransmitter release: A synapse that is frequently active increases the amount of neurotransmitter that it stores, and it releases more on each stimulation. The greater the amount of neurotransmitter released, the greater the effect on the postsynaptic neuron. In the establishment of a neural circuit, this process occurs at each synapse along the way.

2. Facilitation at synapses: When a neural circuit is repeatedly activated, the synaptic terminals begin releasing neurotransmitter in small quantities on a continual basis. The neurotransmitter binds to receptors on the postsynaptic membrane, producing a graded depolarization that brings the membrane closer to threshold. This facilitation affects all of the neurons in the circuit. The circuit then acts like a gun with a "hair trigger" that will fire at the slightest provocation.

3. Formation of additional synaptic connections: There is evidence that when one neuron repeatedly communicates with another, branches grow outward from the axon near the synaptic terminal. These branches, known as "sprouts," grow toward the postsynaptic neuron and establish additional synaptic connections. As a result, activity in the presynaptic neuron will have a greater effect on the transmembrane potential of the postsynaptic neuron.

The facilitated circuit that results from these processes represents a **memory engram**. Efficient conversion of a short-term memory into a memory engram takes time, usually at least an hour and often longer. Whether or not that conversion occurs depends on several factors, including the nature, intensity, and frequency of the original stimulus. Very strong, repeated, or exceedingly pleasant (or unpleasant) events are excellent candidates for conversion to long-term memories. Drugs that stimulate the CNS, such as caffeine and nicotine, seem to enhance memory consolidation; membrane effects of these drugs were discussed in Chapter 12. (p. 394) [**CM**: *Amnesia*]

√ **Towards the end of her A & P test, Tina begins to feel a great deal of stress. What type of brain wave pattern would you expect her to exhibit?**

√ **After suffering a head injury in an automobile accident, David has difficulty comprehending what he hears or reads. This might indicate damage to what portion of the brain?**

√ **As you recall facts while taking your A & P test, what type of memory are you using?**

CONSCIOUSNESS

A conscious individual is alert and attentive; an unconscious individual is not. The difference is obvious, but there are many gradations of both the conscious and unconscious states. For example, a normal conscious person can be almost asleep, wide awake, or

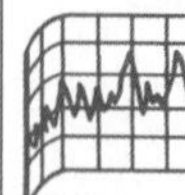

Diagnostics: Altered States

A fine line sometimes separates normal from abnormal states of awareness, and many variations in these states are clinically significant. The state of **delirium** involves wild oscillations in the level of consciousness. The individual has little or no grasp of reality and often experiences hallucinations. When capable of communication, the person seems restless, confused, and unable to deal with situations and events. Delirium often develops (1) in individuals with very high fevers, (2) in patients experiencing withdrawal from addictive drugs, such as alcohol, (3) after taking hallucinogenic drugs, such as LSD, (4) as a result of brain tumors affecting the temporal lobes, and (5) in the late stages of some infectious diseases, including syphilis and AIDS.

The term **dementia** implies a more stable, chronic state characterized by deficits in memory, spatial orientation, language, or personality. *Senile dementia*, a form of dementia that may develop in the elderly, is discussed in a later section.

TABLE 15-3 States of Awareness

Level or State	*Description*
CONSCIOUS STATES	
Delirium	Disorientation, restlessness, confusion, hallucinations, agitation, alternating with other conscious states
Dementia	Difficulties with spatial orientation, memory, language, changes in personality
Confusion	Reduced awareness, easily distracted, easily startled by sensory stimuli, alternates between drowsiness and excitability; resembles minor form of delirium state
Normal consciousness	Aware of self and external environment, well-oriented, responsive
Somnolence	Extreme drowsiness, but will respond normally to stimuli
Chronic vegetative state	Conscious but unresponsive, no evidence of cortical function
UNCONSCIOUS STATES	
Asleep	Can be aroused by normal stimuli (light touch, sound, etc.)
Stupor	Can be aroused by extreme and/or repeated stimuli
Coma	Cannot be aroused and does not respond to stimuli (coma states can be further subdivided according to the effect on reflex responses to stimuli)

"high strung" and jumpy; a sleeping person can be dozing lightly or so deeply asleep that he cannot easily be awakened.

Normal individuals cycle between the alert, conscious state and the asleep state each day. This section considers the maintenance of normal consciousness and various aspects of the awake and asleep states. For reference purposes, Table 15-3 indicates the entire range of conscious and unconscious states, ranging from *delirium* through *coma*. It is important to realize that these states are external indications of the level of ongoing CNS activity. When CNS function becomes abnormal, the state of consciousness can be affected. As a result, clinicians are quick to note any abnormalities in the state of consciousness in their patients. (See Diagnostics: Altered States.)

The Reticular Activating System

The state of consciousness experienced by an individual is determined by complex interactions between the brain stem and the cerebral cortex. One of the most important brain stem components is a poorly defined network in the reticular formation known as the **reticular activating system** (**RAS**). This network, diagrammed in Figure 15-16, extends from the hypothalamus to the medulla. The output of the RAS projects throughout the cerebral cortex. When the RAS is inactive, so is the cerebral cortex; stimulation of the RAS produces a widespread activation of the cerebral cortex.

The mesencephalic portion of the RAS appears to be the center of the system, and stimulation of this area produces the most pronounced and long-lasting effects on the cerebral cortex. Stimulating other portions of the RAS seems to have an effect only insofar as it changes the activity of this region. The greater the stimulation to the mesencephalic region of the RAS, the more alert and attentive the individual will be to incoming sensory information. The thalamic portion of the RAS may also play an important role in focusing attention on specific mental processes.

After many hours of activity, the reticular formation becomes less responsive to stimulation. The individual becomes less alert and more lethargic. The precise mechanism remains unknown, but neural fatigue probably plays a relatively minor role in the reduction of RAS activity. It seems much more likely that the regulation of awake-asleep cycles involves an interplay between brain stem nuclei. One group of nuclei stimulates the RAS and maintains the awake, alert state. These nuclei produce norepinephrine. The other group depresses RAS activity and promotes deep sleep. These nuclei produce serotonin. These "dueling nuclei" are located in the cerebrum, hypothalamus, pons, and medulla.

FIGURE 15-16
The Reticular Activating System. The mesencephalic headquarters of the reticular formation receives collateral inputs from a variety of sensory pathways. Stimulation of this region produces arousal and heightened states of attentiveness.

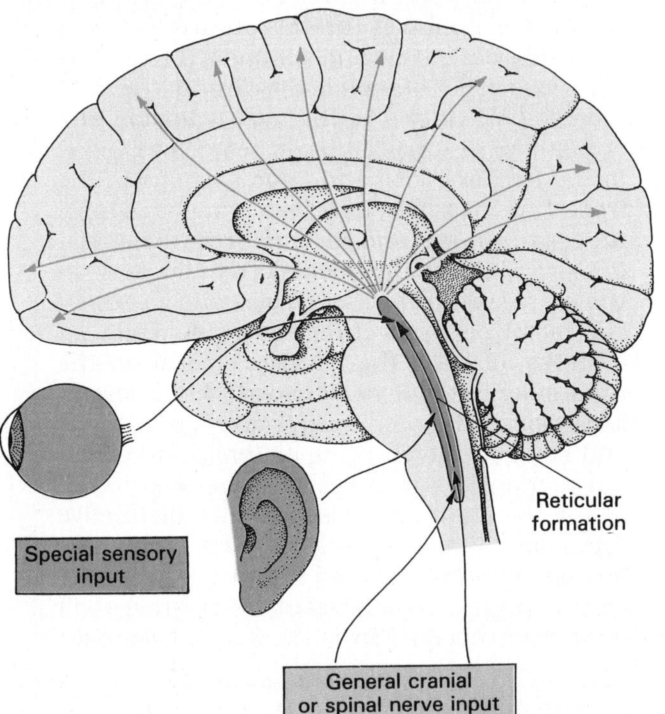

Sleep

When the activity of the cerebral cortex becomes low enough, the individual becomes unconscious. *Unconscious* can imply a number of different conditions, ranging from the deep, unresponsive state induced by anesthesia prior to major surgery to the light drifting "nod" that occasionally plagues students reading anatomy and physiology textbooks. An individual is considered to be asleep when he or she is unconscious but can still be awakened by normal sensory stimuli. There are several different levels of sleep, each typified by characteristic patterns of brain wave activity. These levels are illustrated in Figure 15-17.

In **deep sleep**, also called **slow wave** or **non-REM sleep**, the entire body relaxes, and activity at the cerebral cortex is at a minimum. Heart rate, blood pressure, respiratory rate, and energy utilization decline by up to 30 percent. During **rapid eye movement** (**REM**) **sleep** active dreaming occurs, accompanied by alterations in blood pressure and respiratory rates. Although the EEG shows traces resembling those of the awake state, the individual becomes even less receptive to outside stimuli, and muscle tone decreases markedly. Intense inhibition of skeletal mus-

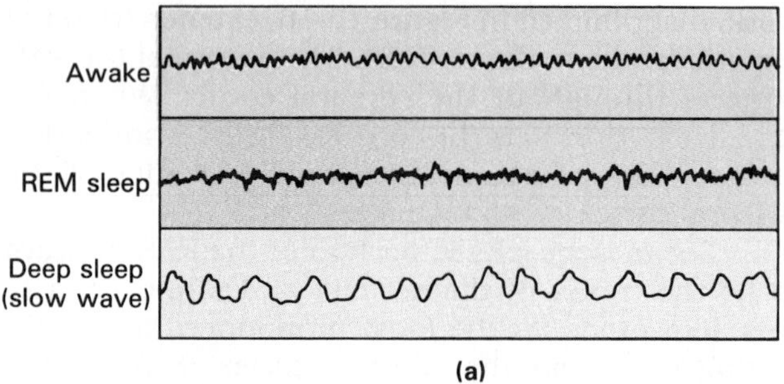

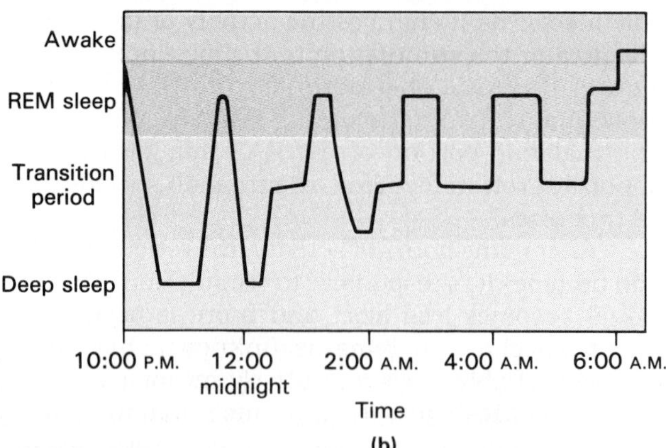

FIGURE 15-17
Levels of Sleep. (a) EEG from the awake, REM, and slow wave (deep) sleep states. The EEG pattern during REM sleep resembles the alpha waves typical of awake adults. (b) Typical pattern of oscillation between sleep stages during a single night's sleep for a healthy young adult.

cles probably prevents us from physically producing the responses we envision during the dream sequence. The oculomotor muscles escape this inhibitory influence, and the eyes move rapidly as the imaginary events unfold.

Periods of REM and deep sleep alternate throughout the night, beginning with a period of deep sleep that lasts around an hour and a half. REM periods initially average about 5 minutes in length, but they gradually increase to about 20 minutes over an 8-hour night. Although each night we probably spend less than 2 hours in dreamland, there are many sources of variation. For example, children devote more time to REM sleep than do adults, and extremely tired individuals have very short and infrequent REM periods.

Sleep produces only minor alterations in the physiological activities of other organs and systems, and none that appear to be essential to normal function. Its significance must lie in its impact on the central nervous system, but the physiological or biochemical basis remains to be determined. It is known that extended periods without sleep will lead to a variety of disturbances in mental function. Slowed reaction times, irritability, and even bizarre behavioral changes may appear. [**CM**: *Sleep Disorders*]

Arousal

Arousal appears to be one of the functions of the reticular formation. The reticular formation is especially well suited for providing "watchdog" services, for it has extensive interconnections with the sensory, motor, and integrative nuclei and pathways all along the brain stem. In short, your sleep may be ended by any stimulus sufficient to activate the reticular formation and RAS.

Arousal occurs rapidly, and the effects of a single stimulation of the RAS last less than a minute. Thereafter consciousness can be maintained by positive feedback, since activity in the cerebral cortex, cerebral nuclei, sensory and motor pathways, and so on will continue to stimulate the RAS.

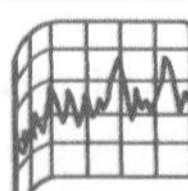

Diagnostics: Cerebrovascular Diseases

Cerebrovascular diseases are circulatory disorders that interfere with the normal circulatory supply to the brain. The particular distribution of the vessel involved will determine the symptoms, and the degree of oxygen or nutrient starvation will determine their severity. A stroke, or **cerebrovascular accident (CVA)**, occurs when the blood supply to a portion of the brain is shut off. Affected neurons begin to die in a matter of minutes.

The symptoms of a stroke provide an indication of the vessel and region of the brain involved. For example, the carotid artery enters the skull via the carotid foramen (Figure 21-12b). One major branch of the carotid, the *middle cerebral artery*, is the most common site of a stroke. Superficial branches deliver blood to the temporal lobe and large portions of the frontal and parietal lobes; deep branches supply the cerebral nuclei and portions of the thalamus. If a stroke blocks the middle cerebral artery on the left side of the brain, aphasia and a sensory and motor paralysis of the right side result. In a stroke affecting the middle cerebral on the right side, the individual experiences a loss of sensation and motor control over the left side and has difficulty drawing or interpreting spatial relationships. Strokes affecting vessels supplying the brain stem also produce distinctive symptoms; those affecting the lower brain stem are often fatal. (Further information on the causes, diagnosis, and treatment of strokes will be found in Chapter 21 of the Clinical Manual.)

BRAIN CHEMISTRY AND BEHAVIOR

The general distribution of neurotransmitters in the central nervous system was discussed in Chapter 12. (p. 390) Of the roughly 50 known neurotransmitters, the most important and widespread are acetylcholine (ACh), norepinephrine (NE), gamma aminobutyric acid (GABA), dopamine (DOPA), serotonin, histamine, Substance P, and the enkephalins and endorphins. ACh and GABA are widely distributed throughout the brain and spinal cord. The distribution of neurons releasing other compounds varies from region to region in the brain. However, tracts originating at each nucleus distribute these neurotransmitters throughout the CNS. The release of one of these chemicals can have dramatic effects on neural processing. One example, the release of dopamine at the cerebral nuclei by neurons of the substantia nigra, was discussed in Chapter 14. (p. 447) A second example, the interplay between serotonin and norepinephrine that appears to be involved in the regulation of asleep-awake cycles, was discussed earlier in this chapter.

It has become clear that changes in the normal balance between these neurotransmitters can produce profound alterations in brain function. In many cases the relationship has been discovered by determining the mechanism repsonsible for the effects of administered drugs. Three examples demonstrate the pattern that is emerging:

1. An extensive network of tracts delivers serotonin to nuclei and higher centers throughout the brain. Compounds that enhance the effects of serotonin produce hallucinations; *LSD (lysergic acid diethylamide)* is a powerful hallucinogenic drug that activates serotonin receptors in the brain stem, hypothalamus, limbic system, and spinal cord. Compounds that inhibit serotonin production or block its action cause severe depression and anxiety. The most effective antidepressive drug now in use, *fluoxetine (Prozac®)*, slows the removal of serotonin at synapses, causing an increase in serotonin concentrations at the postsynaptic membrane.
2. Norepinephrine is another important neurotransmitter with pathways throughout the brain. Drugs that stimulate norepinephrine release cause exhilaration, and those that depress NE release cause depression. One inherited form of depression has been linked to a defective enzyme involved in norepinephrine synthesis.
3. Pathways distributing dopamine are less numerous but still functionally important. Inadequate dopamine production causes the motor problems of Parkinson's disease. As noted in Chapter 14, this condition can result from exposure to the illicit drug MPTP. (p. 447) It is thought that excessive production of dopamine is somehow linked to **schizophrenia**, a psychological disorder marked by pronounced disturbances of mood, thought patterns, and behavior. Amphetamine, or "speed," stimulates dopamine secretion, and in large doses it can produce symptoms resembling those of schizophrenia. (It also affects other neurotransmitter systems, producing a variety of changes in CNS function.)

For additional information concerning the effects and classification of CNS-active drugs see Health News: Pharmacology and Drug Abuse.

PERSONALITY AND OTHER MYSTERIES

The basic pathways and components of the CNS have been traced, and memory acquisition, learning, and associated mental processes can now be manipulated and explored in animal experiments. But the truly remarkable human characteristics of self-awareness and personality remain phenomena without discrete anatomical foundations. That is not to say that an anatomical basis does not exist, just that self-awareness and personality are more characteristic of the brain as an integrated system than a function of any one specific nucleus or region.

As a result, the origins of human consciousness and personality may remain a mystery for the immediate future. But our knowledge of the anatomy and biochemistry of specific tracts, nuclei, and regions will continue to provide useful clinical information. In recent years significant advances have been made by correlating anatomical or physiological deficits with observed behavioral disorders. In some cases, such as the use of L-DOPA to treat Parkinson's disease, this procedure has led to new and effective forms of treatment.

Aging and the Nervous System

The aging process affects all bodily systems, and the nervous system is no exception. Anatomical and physiological changes begin shortly after maturity (probably by age 30) and accumulate over time. Although an estimated 85 percent of the elderly (above age 65) lead relatively normal lives, there are noticeable changes in mental performance and CNS functioning.

Health News: Pharmacology and Drug Abuse

The drug industry certainly qualifies as "big business." Yearly sales approach $10 billion, and each year well over $1 billion are spent on advertising campaigns. If you watch television, you are presented with a dazzling array of advertisements for alcoholic beverages, cold medicines, diet pills, headache remedies, anxiety relievers, and sleep promoters. But although roughly one-third of the drugs sold affect the CNS, few consumers have any understanding of how these drugs exert their effects. Despite an overwhelming interest in medicinal and "recreational" (often illicit) drugs, Americans seldom encounter accurate information about the mechanism of action or the associated hazards. Unfounded and inaccurate rumors are therefore quite common; the concept that "natural" drugs are safer or more effective than "artificial" (synthetic) drugs is an example of the dangerous but popular misconceptions that have appeared in recent years.

The major categories of drugs affecting CNS function are presented in Table 15-4. Considerable overlap exists between these categories. For example, any drug that causes heavy sedation will also depress the perception of pain and alter mood at the same time. Well-known prescription, nonprescription, and illicit drugs are included in this table where appropriate. Some of the nonprescription drugs, most of the prescription drugs, and all of the illegal CNS-active drugs are prone to abuse because tolerance and addiction occur.

Sedatives lower the general level of CNS activity, reduce anxiety, and have a calming effect. **Hypnotic** drugs further depress activity, producing drowsiness and promoting sleep. All levels of CNS activity are reduced, so the effects can range from a mild relaxation to a general anesthesia, depending on the drug and the dosage administered. Sedatives and hypnotic drugs are prescribed more often than any others—almost half of the CNS-active drugs sold are sedatives or hypnotics.

Analgesics, the second most popular group, provide relief from pain. Some, like aspirin, act in the periphery by reducing the source of the painful stimulation. (Aspirin slows the release of prostaglandins that promote inflammation and stimulate pain receptors.) Others, especially the drugs structurally related to opium, target the CNS processing of pain sensations. The compounds bind to receptors on the surfaces of CNS neurons, notably in the cerebral nuclei, the limbic system, thalamus, hypothalamus, midbrain, medulla, and spinal cord. The drugs traditionally administered, such as morphine, mimic the activity and structure of endorphins already present in the CNS (see Clinical Comment: Endorphins in Chapter 12). ∞ (p. 392)

Psychotropics are used to produce changes in mood and emotional state. Certain forms of **depression** have been shown to result from inadequate production of norepinephrine, or its structural relatives, in key nuclei of the brain. Overexcitement, or **mania**, has been correlated with an overproduction of these compounds. Enhancing or blocking the production of these transmitters can often provide the mental stabilization needed to alleviate conditions such as these.

Antipsychotics reduce the hallucinations and behavioral or emotional extremes that characterize the severe mental disorders known as **psychoses**, but leave other mental functions relatively intact. *Phenothiazines*, notably *chlorpromazine* (*Thorazine*), were initially used for control of nausea and vomiting and to promote sedation and relaxation. However, they ultimately proved to be more useful in controlling psychotic behavior.

The term *tranquilizer* was first used for the early antipsychotic drugs, such as *Reserpine*. Later, the category was expanded to in-

TABLE 15-4 A Simple Classification of

Category (Common Name)	*Actions*
Sedatives and hypnotics (downers)	Depress CNS activity; may promote sleep, reduce anxiety, create calm
Analgesics (pain killers)	Relieve pain at source or along CNS pathways
Psychotropics (mood changers)	Alter CNS function and change mental state and/or mood
Anticonvulsants	Inhibit spread of cortical stimulation
Stimulants (uppers)	Facilitate CNS activity

clude "minor tranquilizers," such as *Valium*, whose effects more closely resembled those of sedatives. Eventually it became apparent that "tranquilization" was more useful as a descriptive term than as a classification for specific drugs. For example, a tranquilizing effect can be produced by sedatives (alcohol, *Valium*), hypnotics (barbiturates), and antipsychotics.

The first **anticonvulsants**, used to control seizures, were sedatives such as phenobarbital. Over time, however, patients required larger and larger doses to achieve the same level of effect. This phenomenon, called **tolerance**, frequently appears when any CNS-active drug is administered. *Dilantin* and related drugs have powerful anticonvulsant effects without producing tolerance.

Stimulants facilitate activity in the central nervous system. Few are used clinically, as they may trigger convulsions or hallucinations. Several of the ingredients found in coffee and tea are CNS stimulants, caffeine being the most familiar example. **Amphetamines** prompt the release of NE, stimulating the respiratory and cardiovascular control centers and elevating muscle tone to the point that tremors often begin. Amphetamines also stimulate dopamine and serotonin release in the CNS and facilitate cranial and spinal reflexes.

Drug abuse has a very hazy definition, for it implies that the individual voluntarily utilizes a drug in some inappropriate manner. Any drug use that violates medical advice, prevalent social mores, or common law can be included within this definition. **Drug addiction** refers to an overwhelming dependence and compulsion to use a specific drug, despite the medical or legal risks involved. If use is stopped or prevented, the individual will often suffer symptoms of **withdrawal**. The physiological foundations for drug dependence and addiction vary, depending on the individual compound considered, and in general the mechanisms are poorly understood.

The dangers inherent in recreational drug use have been repeatedly overlooked or ignored. Until recently, the abuse of common addictive drugs such as alcohol or nicotine was considered relatively normal, or at least excusable. The risks of alcohol abuse can hardly be overstated; it has been estimated that 40 percent of young men have problems with alcohol consumption, and 7 percent of the adult population show signs of alcohol abuse or addiction. Long-term abuse increases the risks of diabetes, liver and kidney disorders, cancer, cardiovascular disease, and digestive system malfunctions. In addition, CNS disturbances often lead to accidents, due to poor motor control, and violent behaviors, due to interference with normal emotional balance and analytical function.

Drugs that Affect the CNS

Examples		
Prescription	*Nonprescription*	*Comments*
Barbiturates (phenobarbital, Nembutal, amytal); benzodiazepines (Librium, Valium)	Sominex, Ny-tol, Benadryl, alcohol	Those used to depress seizures may be considered as "anticonvulsants"; prescription forms addictive.
Opiates (morphine, Demerol, codeine)	Aspirin, Tylenol, Ibuprofen	Heroin and cocaine are members of the addictive group of CNS-active drugs.
Antipsychotics (chlorpromazine); antidepressants (imipramine); antianxiety (Librium, Valium); mood stabilization (lithium)	Caffeine, alcohol show psychotropic effects	Many of these are also addictive.
Dilantin; also may use sedative-hypnotics		
Xanthines (caffeine); amphetamines	Diet pills, Sudafed, Actifed, caffeine	Prescription and nonprescription forms addictive.

AGE-RELATED ANATOMICAL AND FUNCTIONAL CHANGES

Age-related anatomical changes in the nervous system that are commonly seen include:

1. **A reduction in brain size and weight.** This reduction results primarily from a decrease in the volume of the cerebral cortex. The brains of elderly individuals have narrower gyri and wider sulci than those of young persons, and the subarachnoid space is enlarged.

2. **A reduction in the number of neurons.** Brain shrinkage has been linked to a loss of cortical neurons, although evidence exists that neuronal loss does not occur (at least to the same degree) in brain stem nuclei.

3. **A decrease in blood flow to the brain.** With age, fatty deposits gradually accumulate in the walls of blood vessels. Like a kink in a garden hose or a clog in a drain, these deposits reduce the rate of blood flow through arteries. (This process, called *arteriosclerosis*, affects arteries throughout the body; it is discussed further in Chapter 21.) The reduction in blood flow does not cause a cerebral crisis, but it does increase the chances that the individual will suffer a stroke.

4. **Changes in synaptic organization of the brain.** The number of dendritic branchings and interconnections appears to decrease. Synaptic connections are lost, and the rate of neurotransmitter production declines.

5. **Intracellular and extracellular changes in CNS neurons.** Many neurons in the brain begin accumulating abnormal intracellular deposits, including *lipofuscin*, *plaques*, and *neurofibrillary tangles*. **Lipofuscin** is a granular pigment that has no known function. **Plaques** are accumulations of an unusual fibrillar protein, **amyloid**, surrounded by abnormal dendrites and axons. **Neurofibrillary tangles** are masses of neurofibrils that form dense mats inside the soma. The significance of these cellular and extracellular abnormalities remains to be determined. There is evidence that they appear in all aging brains (see below), but when present in excess they seem to be associated with clinical abnormalities.

These anatomical changes are linked to a series of functional alterations. In general, neural processing becomes less efficient. For example, memory consolidation often becomes more difficult, and secondary memories, especially those of the recent past, become harder to access. The sensory systems of the elderly, notably hearing, balance, vision, smell, and taste, become less acute. Light must be brighter, sounds louder, and smells stronger before they are perceived. Reaction times are slowed, and reflexes—even some withdrawal reflexes—become weaker or even disappear. There is a decrease in the precision of motor control, and it takes longer to perform a given motor pattern than it did 20 years earlier.

For roughly 85 percent of the elderly population, these changes do not interfere with their abilities to function in society. But for as yet unknown reasons, many elderly individuals become incapacitated by progressive CNS changes. By far the most common such incapacitating condition is *Alzheimer's disease*.

ALZHEIMER'S DISEASE

Alzheimer's disease is a progressive disorder characterized by the loss of higher cerebral functions. It is the most common cause of **senile dementia**, or "senility." The first symptoms usually appear at 50–60 years of age, although the disease occasionally affects younger individuals. Alzheimer's disease has widespread impact on the elderly; an estimated 2 million people in the United States, including roughly 15 percent of those over 65, suffer from some form of the condition, and it causes approximately 100,000 deaths each year.

In its characteristic form, Alzheimer's disease produces a gradual deterioration of mental organization. The afflicted individual loses memories, verbal and reading skills, and emotional control. Initial symptoms are subtle—moodiness, irritability, depression, and a general lack of energy. These symptoms are often ignored, overlooked, or dismissed. Elderly relations are viewed as "eccentric" or "irrascible," and humored whenever possible. As the condition progresses, however, it becomes more difficult to ignore or accommodate. The victim has difficulty making decisions, even minor ones. Mistakes—sometimes dangerous ones—are made, either through bad judgment or simple forgetfulness. For example, the person might decide to make dinner, light the gas burner, place a pot on the stove top, and go into the living room. Two hours later, the pot, still on the stove, melts into a shapeless blob and starts a fire that destroys the house.

The memory losses continue, and the problems become more severe. The affected person may forget relatives, her home address, or how to use the telephone. The memory loss often starts with an inability to store long-term memories, followed by the loss of recently stored memories, and eventually the loss of basic long-term memories, such as the sound of the victim's own name. The loss of memory affects both intellectual and motor abilities, and a patient with severe Alzheimer's disease has difficulty in performing even the simplest motor tasks. Although by this time victims are relatively unconcerned about their mental state or motor abilities, the condition can have devastating emotional effects on the immediate family.

Individuals with Alzheimer's disease show a pronounced decrease in the number of cortical neurons, especially in the frontal and temporal lobes. It now appears that this loss is correlated with inadequate ACh production in the nucleus basalis of the cerebrum. Axons leaving this region project throughout the cerebral cortex, and when ACh production declines cortical function deteriorates.

Most cases of Alzheimer's disease are associated with unusually large concentrations of plaques and neurofibrillary tangles in the nucleus basalis, hippocampus, and parahippocampal gyrus. In addition, an abnormal protein, called **Alzheimer's disease-associated protein** (ADAP) appears in brain regions, such as the hippocampus, specifically associated with memory processing. Because this protein also appears in small quantities in the cerebrospinal fluid of many Alzheimer's patients, a blood screening test is now being developed to detect the condition before mental deterioration becomes pronounced.

Although the link remains uncertain, the areas containing concentrations of plaques, tangles, and ADAP are the same regions involved with memory, emotions, and intellectual function. It remains to be determined whether the plaques, tangles, and ADAP deposits *cause* Alzheimer's or whether they are secondary signs of ongoing metabolic alterations that have an environmental, hereditary, or infectious basis. There appears to be some link between Alzheimer's disease and Down syndrome. The majority of individuals with Down syndrome develop Alzheimer's disease relatively early in life, although the connection between the two conditions has yet to be determined.

This chapter began with somatic sensory and motor pathways and ended with a discussion of higher-order function. The next chapter examines the autonomic nervous system, an integrative system that normally operates outside our conscious awareness but nevertheless exerts essential control over all other physiological systems.

- √ **What would happen to a sleeping individual if his or her recticular activating system (RAS) were suddenly stimulated?**
- √ **What would you expect to be the effect of a drug that substantially increases the amount of serotonin released in the brain?**
- √ **One of the problems associated with aging is difficulty in recalling things and even a total loss of memory. What are some possible reasons for these changes?**

CHAPTER REVIEW

Review of Selected Clinical Terms

electroencephalogram (EEG) (*p. 484*)
A printed record of the brain's electrical activity over time.

aphasia (*p. 485*)
A disorder affecting the ability to speak or read.

dyslexia (*p. 486*)
A disorder affecting the comprehension and use of words.

delirium (*p. 490*)
A conscious state involving confusion and wild oscillations in the level of consciousness.

dementia (*p. 490*)
A chronic state of consciousness characterized by deficits in memory, spatial orientation, language, or personality.

cerebrovascular diseases (*p. 492*)
Circulatory disorders that interfere with the normal circulatory supply to the brain.

cerebrovascular accident (CVA) (*p. 492*)
Also known as "stroke"; occurs when the blood supply to a portion of the brain is shut off.

schizophrenia (*p. 493*)
A psychological disorder marked by pronounced disturbances of mood, thought patterns, and behavior.

Alzheimer's disease (*p. 496*)
A progressive disorder marked by the loss of higher cerebral functions.

Additional Terms of Clinical Importance

ANENCEPHALY (*p. 481*): **anencephaly**

THE ELECTROENCEPHALOGRAM (*p. 484*): **brain waves, alpha waves, beta waves, theta waves, delta waves**

PHARMACOLOGY AND DRUG ABUSE (*p. 494*): **sedatives, hypnotic drugs, analgesics, psychotropics, depression, mania, antipsychotics, psychoses, anticonvulsants, tolerance, stimulants, amphetamines, drug abuse, drug addiction, withdrawal**

AGE-RELATED ANATOMICAL AND FUNCTIONAL CHANGES (*p. 496*): **lipofuscin, plaques, amyloid, neurofibrillary tangles**

ALZHEIMER'S DISEASE (*p. 496*): **senile dementia, Alzheimer's disease-associated protein**

Study Outline

Introduction (pp. 471–472)

Related Key Terms

1. There is continual communication between the brain, spinal cord, and peripheral nerves.

Sensory and Motor Pathways (pp. 472–481)

SENSORY PATHWAYS (pp. 473–475)

1. A **sensation** arrives in the form of an action potential in an afferent fiber. The **posterior column pathway** carries fine touch, pressure, and proprioceptive sensations. The axons ascend within the **fasciculus gracilis** and **fasciculus cuneatus**, to be relayed to the thalamus via the **medial lemniscus**. As the axons enter the medial lemniscus, they **decussate** (cross over) to the opposite side of the brain stem.

   Related key terms: **first-order neuron**, **second-order neuron**, **third-order neuron**, **somatosensory pathways**, **sensory homunculus**

2. The **spinothalamic pathway** carries poorly localized sensations of touch, pressure, pain, and temperature. These decussate in the spinal cord and ascend via the **anterior** and **lateral spinothalamic tracts** to the **ventral posterolateral nuclei** of the thalamus.
3. The **spinocerebellar pathway**, including the **posterior** and **anterior spinocerebellar tracts**, carries sensations to the cerebellum concerning the position of muscles, tendons, and joints.

MOTOR PATHWAYS (pp. 475–481)

4. Somatic motor pathways always involve an **upper motor neuron** (whose soma lies in a CNS processing center) and a **lower motor neuron** (located in a motor nucleus of the brain stem or spinal cord). Autonomic motor control requires a **preganglionic neuron** (in the CNS) and a **ganglionic neuron** (in a peripheral ganglion).
5. The neurons of the primary motor cortex are **pyramidal cells**; the **pyramidal system** provides voluntary skeletal muscle control. The **corticobulbar tracts** terminate at the cranial nerves, while the **corticospinal tracts** synapse on motor neurons in the anterior gray horns of the spinal cord. The corticospinal tracts are visible along the medulla as a pair of thick bands, the **pyramids**, where

   Related key terms: **motor homunculus**

most of the axons decussate to enter the descending **lateral corticospinal tracts** or the **anterior corticospinal tracts**. The pyramidal system provides a rapid, direct mechanism for controlling skeletal muscles.

6. The **extrapyramidal system (EPS)** consists of several other centers that may issue motor commands as a result of processing performed at an unconscious, involuntary level. Its outputs may descend via the **vestibulospinal**, **tectospinal**, **rubrospinal**, or **reticulospinal tracts**.
7. The **vestibulospinal tracts** carry information related to maintaining balance and posture. Commands carried by the **tectospinal tracts** change the position of the eyes, head, neck, and arms in response to bright lights, sudden movements, or loud noises. The **rubrospinal tracts** carry motor responses to spinal motor neurons. Motor commands carried by the **reticulospinal tracts** vary according to the region stimulated.
8. The **cerebral nuclei** (the most important and complex components of the extrapyramidal system) use three major pathways to adjust the motor commands issued in other processing centers: (1) One group of axons synapses with thalamic neurons and creates a feedback loop that changes the sensitivity of pyramidal cells and alters the instructions carried by the corticospinal tracts; (2) Another group innervates the red nucleus and alters activity in the rubrospinal tracts; (3) The third group travels through the thalamus to reach centers in the reticular formation, where they adjust the output of the reticulospinal tracts.
9. The **cerebellum** regulates the activity along both pyramidal and extrapyramidal motor pathways. The integrative activities performed by neurons in the cerebellar cortex and cerebellar nuclei are essential for precise control of voluntary and involuntary movements.

Levels of Processing and Motor Control (pp. 481–483)

1. Ascending sensory information is relayed in a series of steps and is processed at each step with results that may block, reduce, or heighten response to the stimulus. Every synapse means another synaptic delay.

Monitoring Brain Activity: The Electroencephalogram (pp. 483–484)

1. Higher-order functions involve complex interactions between many components of the brain.
2. An **electroencephalogram (EEG)** is a printed record of **brain waves**. **Alpha waves** are found in normal adults under resting conditions, but are replaced by **beta waves** during times of concentration. **Theta waves** appear in the brains of children and in stressed adults and may indicate a brain disorder. **Delta waves** appear in the brains of infants and in the brains of adults during deep sleep or in cases of brain damage.

Higher-Order Functions (pp. 484–493)

1. Higher-order functions have the following characteristics: (1) They are performed by the cerebral cortex; (2) They involve complex interactions between areas of the cortex and between the cerebral cortex and other areas of the brain; (3) They involve both conscious and unconscious information processing; (4) They are subject to modification and adjustment over time.

CORTICAL INTEGRATION OF INFORMATION (pp. 485–486)

2. The primary sensory cortex receives information concerning the **general senses** (touch, pressure, vibration, pain, and temperature) relayed from the thalamus. The thalamus also relays information concerning the **special senses** of sight, smell, taste, and hearing to appropriate cortical areas.
3. The **general interpretive area** receives information from all the sensory association areas. It is present in only one hemisphere, usually the left.
4. The **speech center** regulates the patterns of breathing and vocalization needed for normal speech.

Related Key Terms

aphasia • **global aphasia**
major motor aphasia
dyslexia
developmental dyslexia

Chapter Review

5. The **prefrontal cortex** coordinates information from the secondary and special association areas of the entire cortex, and performs abstract intellectual functions.

Related Key Terms
prefrontal lobotomy

HEMISPHERIC SPECIALIZATION (pp. 486–488)

6. The left hemisphere is usually the **categorical hemisphere**, which contains the general interpretive and speech centers and is responsible for language-based skills. The other hemisphere, which is concerned with spatial relationships and analyses, is the **representational hemisphere**.
7. In the **disconnection syndrome** the two hemispheres function independently, each remaining "unaware" of stimuli or motor commands involving the other.

MEMORY (pp. 488–490)

8. **Declarative memories** (memories that can be voluntarily retrieved and explicitly stated) can be classified into **short-term**, or **primary, memories** (short-lived, but while they persist the information can be recalled immediately) and **long-term memories** (which remain for much longer periods). Long-term memories in turn can be subdivided into **secondary memories** (which fade with time and may be difficult to recall) and **tertiary memories** (which seem to be part of consciousness). Conversion from short-term to long-term memory is **memory consolidation**. The amygdaloid body and the hippocampus are essential to memory consolidation.

nucleus basalis

9. Memories can also be classified according to the nature of the stored information: **fact memories** are specific bits of information, while **skill memories** are learned motor behaviors. With repetition, skill memories are incorporated at the unconscious level and become **reflexive memories**, which are not voluntarily accessible.
10. Cellular mechanisms which seem to be involved in memory formation and storage include: (1) increased neurotransmitter release; (2) facilitation at synapses; and (3) formation of additional synaptic connections.

memory engram

CONSCIOUSNESS (pp. 490–492)

11. There are many gradations of conscious and unconscious states. One's state of consciousness is determined by interactions between the brain stem and cerebral cortex; one of the most important brain stem components is a network in the reticular formation called the **reticular activating system (RAS)**.
12. In **deep sleep** (**slow wave** or **non-REM sleep**) the body relaxes and cerebral cortex activity is low. During **rapid eye movement (REM) sleep** active dreaming occurs. Periods of REM and deep sleep alternate throughout the night.
13. **Arousal** follows activation of the reticular formation.

BRAIN CHEMISTRY AND BEHAVIOR (p. 493)

14. Changes in the normal balance between neurotransmitters can profoundly alter brain function. The most important and widespread neurotransmitters are acetylcholine (ACh), norepinephrine (NE), gamma-aminobutyric acid (GABA), dopamine (DOPA), serotonin, histamine, substance P, and the enkephalins and endorphins.

schizophrenia

PERSONALITY AND OTHER MYSTERIES (p. 493)

15. The human characteristics of self-awareness and personality remain a mystery; they appear to be characteristic of the brain as an integrated system, rather than a function of any specific component.

Aging and the Nervous System (pp. 493–497)

AGE-RELATED ANATOMICAL AND FUNCTIONAL CHANGES (p. 496)

1. Age-related changes in the nervous system include: (1) reduction in brain size and weight; (2) reduction in number of neurons; (3) decrease in blood flow to the brain; (4) changes in synaptic organization of the brain; and (5) intracellular and extracellular changes in CNS neurons.

lipofuscin • plaques • amyloid neurofibrillary tangles

ALZHEIMER'S DISEASE (pp. 496–497)

2. **Alzheimer's disease** is a progressive disorder characterized by the loss of higher cerebral functions; it is the most common cause of **senile dementia**.

Alzheimer's disease-associated protein

Review Planner

		Level 1
1	Identify the principal sensory and motor pathways.	**1 9 10 11 12 20 27 34 35–39**
2	Compare the processes and functions of the pyramidal and extrapyramidal systems.	**1 10 11 26 29**
3	Explain why we can distinguish between sensations that originate in different areas of the body.	**9 10 11 12 30 31 32**
4	Describe the levels of information processing involved in motor control.	**2 5 16 19 21 29**
5	Discuss how the brain integrates sensory information and coordinates responses.	**2 3 4 5 12 13 14 16 17 19 21 23**
6	Explain the importance of hemispheric specialization.	**13 24**
7	Explain how memories are created, stored, and recalled.	**8 15 18 25**
8	Discuss levels of consciousness and sleep.	**7 22**
9	Summarize the effects of aging on the nervous system.	**6 28**

Level 2	**21 24 29 32 33**
Level 3	**40–43**
C M	**Cerebral Palsy** **Seizures and Epilepsy** **Huntington's Disease • Amnesia** **Sleep Disorders**
C M	**PET Scans of the Brain**
	In the Mind's Eye • Lost in Time **A New Life for Andrew** **Brain Attack**

Review Questions

MULTIPLE CHOICE

1. The pyramidal and extrapyramidal systems carry: a) sensory commands; b) autonomic motor commands; c) somatic motor commands.
2. The region of the CNS that processes and integrates proprioceptive data and regulates activity along both pyramidal and extrapyramidal pathways is the: a) medulla; b) cerebellum; c) spinal cord; d) pons.
3. An electroencephalogram depicts: a) electrical activity in the brain; b) Metabolic activity in the brain; c) changes in blood flow within the brain; d) the types of memories stored at that moment in the brain.
4. A PET scan can depict: a) electrical activity in the brain; b) abnormal brain waves; c) changes in blood flow within the brain; d) the types of memories stored at that moment in the brain.
5. Higher-order functions are performed by the: a) ventral posterolateral nuclei; b) cerebral cortex; c) cerebellum; d) special senses.
6. Which of the following is *not* associated with age-related changes in the nervous system? a) amyloid plaques; b) memory engrams; c) neurofibrillary tangles; d) lipofuscin.
7. Arousal from sleep occurs as the result of: a) stimulation of the RAS; b) beta waves; c) reactions to REM sleep; d) disrupted communication between cerebral hemispheres.
8. Memories that seem to be part of consciousness, such as one's own name, are called: a) primary memories; b) secondary memories; c) tertiary memories; d) memory engrams.
9. Damage to the spinothalamic pathway would mean that: a) you could not detect highly localized proprioceptive sensations; b) you could not detect poorly localized sensations of touch; c) sensations would not travel within the fasciculus gracilis.

Chapter Review

TABLES

10. Complete the following table:

Tracts	*Origin*	*Destination*	*Sensation/Actions*	*Comments*
Fasciculus gracilis				**Ascends on same side as stimulus**
	Proprioceptors and fine touch, pressure, vibration receptors of upper body	**Nucleus cuneatus (medulla)**		
	Interneurons relaying information from proprioceptors	**Cerebellum**		**Ascends on same side as stimulus**
Anterior spinocerebellar tract				
	Interneurons relaying information from pain and temperature receptors	**Thalamus (ventral nuclei)**		**Crosses to ascend on side opposite stimulus**
Anterior spinothalamic tract		**Thalamus (ventral nuclei)**		**Crosses to ascend on side opposite stimulus**

11. Complete the following table:

	Origin	*Destination*	*Sensation/Actions*	*Comments*
	Primary motor cortex (cerebral hemispheres)	**Motor neurons of anterior gray horns**		**Crosses to opposite side before entering spinal cord**
Anterior corticospinal tract				
	Red nucleus (midbrain)	**Motor neurons of anterior gray horns**		**Crosses to opposite side before entering spinal cord**
Reticulospinal tract		**Somatic and visceral motor neurons of anterior and lateral gray horns**		**Descends without crossing to opposite side**
	Vestibular nucleus (near rostral border of medulla)			**Descends without crossing to opposite side**
		Motor neurons of anterior gray horns of cervical spinal cord		

TRUE/FALSE

(If the statement is false, your instructor may wish to have you correct it.)

12. **T/F:** A sensation originates as an action potential in an efferent fiber.

13. **T/F:** The two cerebral hemispheres normally function independently of each other.

MATCHING QUESTIONS

(Match the terms with the most appropriate statement.)

Statements:

14. Brain waves found in normal adults who are

Terms:

a) Decussation
b) Alpha waves
c) Beta waves

resting with their eyes closed.

15. Memories that are not voluntarily accessible.

16. A condition in which each cerebral hemisphere is "unaware" of stimuli or motor commands involving the other hemisphere.

17. Brain waves that indicate concentration on a particular task.

18. Memories that can be voluntarily retrieved.

19. The process during which axons cross over to the opposite side of the brain stem.

d) Disconnection syndrome
e) Declarative memories
f) Reflexive memories

SHORT ANSWER/ESSAY

20. Fill in the blanks: If the name of a tract begins with *spino-*, the tract __________ in the spinal cord and __________ in the brain. If the tract name ends with *-spinal*, the tract __________ in the brain and __________ in the spinal cord.
21. You're sitting in the library studying. Suddenly several books fall to the floor with a loud crash. In response, you move your eyes and head toward the source of the sound. Along what tracts did the motor commands travel that caused you to do this, and where do the tracts originate?
22. Distinguish between the physiological effects of deep sleep and REM sleep. During which stage does dreaming occur?
23. An eighth-grader has trouble reading and writing. Although he tests normally on intelligence tests, he repeatedly reverses the shapes of letters when he writes and often reverses patterns of letters within words. What is this condition called?
24. Discuss hemispheric specialization. Which hemisphere is generally responsible for performing analytical and language-based tasks?
25. Final exams are approaching. All through the semester, you've diligently studied each chapter of your textbook; however, being pressed for time, you have to study the material in the last chapter in a hurry. A year later, you find that you have trouble remembering the material in the last chapter, while the information in the rest of the book is much easier to remember. What types of memories are involved?
26. If a person's primary motor cortex is damaged, how will this affect his ability to move? What part of the body will direct his movements?
27. Describe the pathway that would be traveled by sensations of vibration.
28. Explain why elderly people are generally at greater risk for a stroke.
29. Discuss the role of the cerebral nuclei in the extrapyramidal system. How do they affect movement?
30. You are reclining with your eyes closed. You feel your cat bite the little toe on your left foot. A moment later, seeking variety, the cat bites the little finger on your right hand. Explain why, even with your eyes closed, you can tell where the cat bit you.
31. Even with your eyes closed, you can tell when your arm is raised or when your knee joint is flexed. What pathway carries these sensations, and in what area of the brain are they processed?
32. What do we learn from a sensory homunculus and a motor homunculus? Would the sensory and motor depictions of the same part of the body resemble each other?
33. Discuss the neural tissue abnormalities often found in patients with Alzheimer's disease.
34. Along what pathway would you expect sensations of temperature to travel?

Name the pathway or system of which these tracts are part:

35. Anterior spinocerebellar tracts.
36. Corticobulbar tracts.
37. Medial lemniscus.
38. Lateral spinothalamic tracts.
39. Rubrospinal tracts.

CRITICAL THINKING/APPLICATIONS

40. Rebecca has been waking up lately with sore jaw muscles. A routine visit to the dentist discloses that the surfaces of several molars are unusually worn. What behavior might she be subject to? When do such behaviors occur?
41. Mrs. H., 78 years old, has experienced a sudden feeling of numbness in her right arm and right leg and finds it more difficult to move them than before. Relatives notice that she also has some difficulty in speaking and reading. What do you think caused these symptoms? What part(s) of the CNS are affected?
42. Mr. R. has noticed for some time that he has occasional brief losses of consciousness that come and go without warning; afterward he cannot remember anything that occurred during that time. No one else seems to notice anything, although his wife once remarked that his facial muscles twitched during one such episode. What is your diagnosis? What treatments are available for this condition?
43. One patient suffers a CVA in the left hemisphere; another suffers a CVA in the right hemisphere. What functions could be affected in each case?

If you think your life on earth is difficult, how would you like to be an astronaut? Survival in a frigid, airless, weightless environment poses special problems—problems not encountered on the surface of our hospitable planet. In such an environment, even the most mundane body functions are impossible. How could your body continue to breathe, maintain a stable temperature, keep its blood circulating? In short, how would you remain poised on that knife-edge known as homeostasis?

In this photo we see how NASA answered these questions for its astronauts. This astronaut is encased in a Manned Maneuvering Unit (MMU), a mobile life-support system that monitors his breathing, heart rate, oxygen and carbon dioxide levels, and body temperature. While not exactly elegant, the MMU allows the astronaut's body to continue performing all those vital, unconscious functions that we normally take for granted. In this chapter we will learn about the autonomic nervous system, which supervises those functions—on earth or off it.

CHAPTER 16

The Nervous System: Autonomic Division

Chapter Objectives

After reading this chapter, you will be able to:

1 Compare the autonomic nervous system with the other divisions of the nervous system.
2 Explain the functions and structures of the sympathetic and parasympathetic divisions.
3 Discuss the mechanisms of neurotransmitter release in the autonomic nervous system.
4 Compare the effects of autonomic neurotransmitters on target organs and tissues.
5 Discuss the relationship between the sympathetic and parasympathetic divisions and explain the implications of dual innervation.
6 Describe the levels of integration and control of the autonomic system.
7 Explain the importance of autonomic tone.

■ Introduction

Your conscious thoughts, plans, and actions represent only a tiny fraction of the activities of your nervous system. Events that occur at the conscious level are merely the tip of the iceberg. In practical terms, your conscious thoughts and the somatic motor system that operates under voluntary control have little to do with your immediate or long-term survival. Of course, the somatic motor system can be important in moving you out of the way of a speeding bus or pulling a hand from a hot stove—but it was your conscious movements that put you in jeopardy in the first place. If all consciousness were eliminated, vital physiological processes would continue virtually unchanged; a night's sleep is not a life-threatening event. Longer, deeper states of unconsciousness are not necessarily more dangerous, as long as nourishment is provided in some way. People who have suffered severe brain injuries can survive in a coma for decades.

Survival under these conditions is possible because routine homeostatic adjustments in physiological systems are made by the autonomic nervous system (ANS). It is the ANS that coordinates cardiovascular, respiratory, digestive, excretory, and reproductive functions. In doing so, the ANS adjusts internal water, electrolyte, nutrient, and dissolved gas concentrations in body fluids—all without instructions or interference from the conscious mind.

This chapter examines the anatomical and physiological divisions of the autonomic nervous system. However, before exploring the details we will take a moment to develop a perspective on the system as a whole.

An Overview of the ANS

It will be useful to compare the organization of the ANS with that of the somatic nervous system (SNS), which was considered in earlier chapters. We will focus on (1) the neural interactions that direct motor output and (2) the subdivisions of the ANS, based on structural and functional patterns of peripheral innervation.

NEURAL INTERACTIONS THAT DIRECT MOTOR OUTPUT

As we saw in Chapter 15, the lower motor neurons of the SNS exert direct control over skeletal muscles. (p. 475) In the autonomic system, by contrast, there is always a synapse interposed between the CNS and the peripheral effector. Visceral motor neurons in the CNS, known as **first-order** or **preganglionic neurons**, send their axons to synapse on **second-order** or **ganglionic neurons**, located in **autonomic ganglia** outside the CNS. Considerable divergence occurs—a single myelinated preganglionic fiber may activate over two dozen ganglionic neurons. Axons leaving the autonomic ganglia are smaller, unmyelinated **postganglionic fibers**. (Note that the axon of a *ganglionic neuron* is called a *postganglionic fiber* because it carries impulses away from the ganglion.) Postganglionic fibers innervate peripheral organs, such as cardiac muscle, smooth muscles, glands, and adipose tissues. This organizational scheme is diagrammed in Figure 16-1.

FIGURE 16-1
An Overview of the Autonomic Nervous System

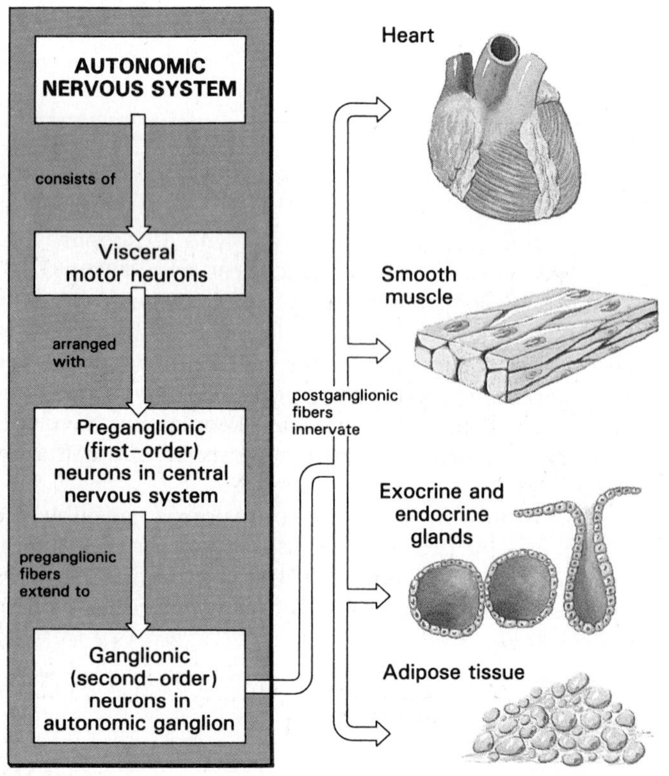

FIGURE 16-2
Components of the ANS

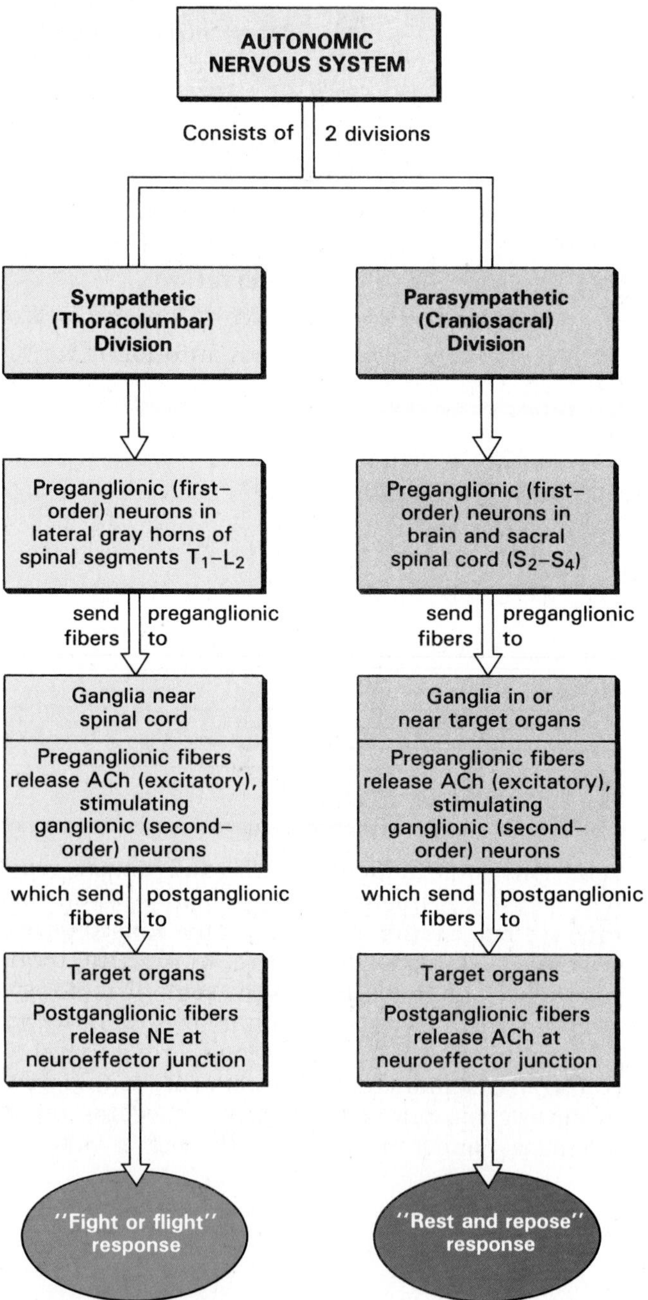

SUBDIVISIONS OF THE ANS

The SNS can be roughly divided into the pyramidal system (which exerts voluntary somatic motor control) and the extrapyramidal system (which exerts involuntary somatic motor control and adjusts pyramidal commands). The ANS also contains two subdivisions, detailed in Figure 16-2.

- Preganglionic fibers from the thoracic and lumbar spinal segments synapse in ganglia near the spinal cord. These axons and ganglia are part of the **thoracolumbar** (thō-ra-kō-LUM-bar), or **sympathetic, division** of the ANS. The sympathetic division is often called the "fight or flight" system because it usually stimulates tissue metabolism, increases alertness, and generally prepares the body to deal with emergencies.
- Preganglionic fibers originating in the brain and the sacral spinal segments synapse on neurons of **intramural ganglia** (*murus*, wall), located inside the tissues of visceral organs. These components are part of the **craniosacral** (krā-nē-ō-SĀK-ral), or **parasympathetic, division** of the ANS. The parasympathetic division is often known as the "rest and repose" system because it conserves energy and promotes sedentary activities, such as digestion.

The sympathetic and parasympathetic divisions affect target organs through the controlled release of specific neurotransmitters by the postganglionic fibers. Whether the result is a stimulation or inhibition of activity depends on the response of the membrane receptor to the presence of the neurotransmitter. Three general patterns, detailed in later sections, are worth noting at this time:

1. All preganglionic autonomic fibers are *cholinergic*: they release acetylcholine (ACh) at their synaptic terminals. The effects are always excitatory.
2. Postganglionic parasympathetic fibers are also cholinergic, but the effects may be excitatory or inhibitory, depending on the nature of the receptor.
3. Most postganglionic sympathetic terminals are *adrenergic*: they release norepinephrine (NE). The effects are usually excitatory.

Each of these divisions has a characteristic anatomical and functional organization. Our detailed examination of the ANS begins with a description of the sympathetic and parasympathetic divisions; we will then consider the ways these autonomic divisions interact to maintain homeostasis.

■ The Sympathetic Division

The sympathetic division, diagrammed in Figure 16-3, consists of:

FIGURE 16-3
Organization of the Sympathetic Division of the ANS

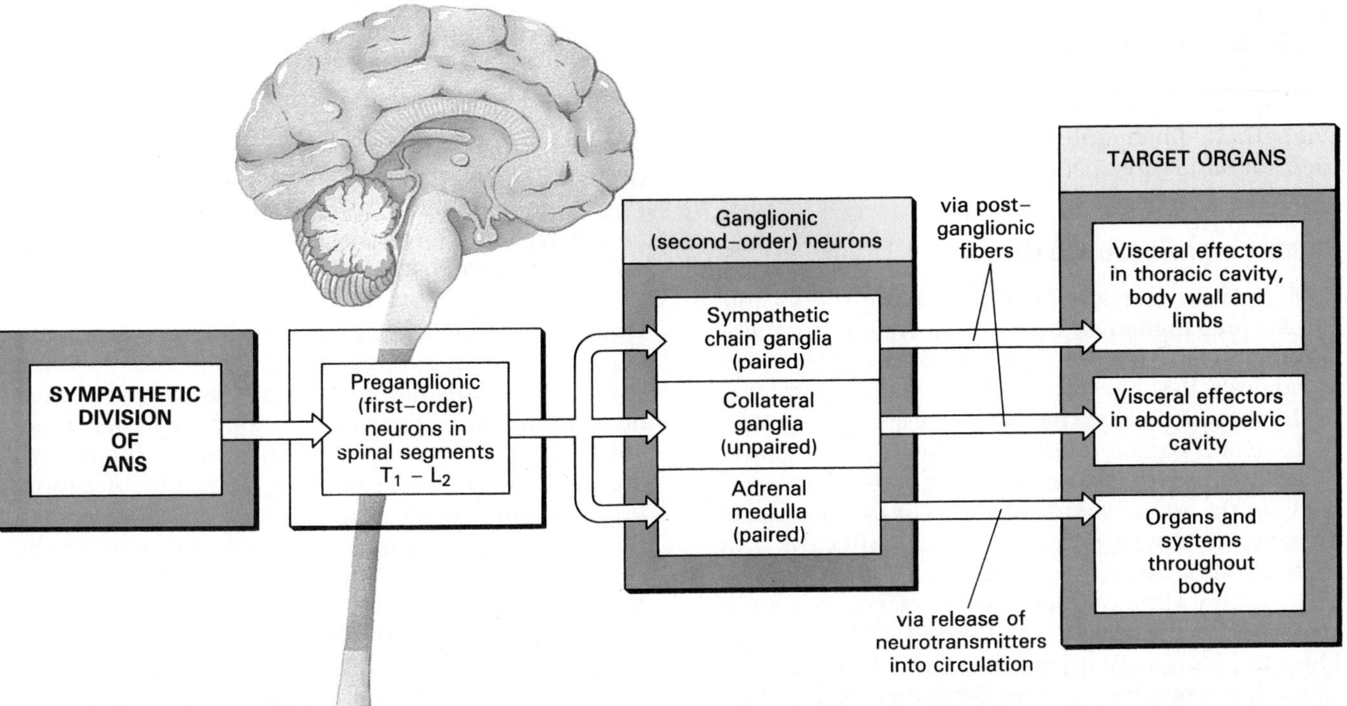

1. *Preganglionic (first-order) neurons located between segments T_1 and L_2 of the spinal cord.* These neurons are situated in the lateral gray horns, and their axons enter the ventral roots of these segments.
2. *Ganglionic (second-order) neurons located in ganglia near the vertebral column.* There are two different types of sympathetic ganglia:
   - **Sympathetic chain ganglia**, also called *paravertebral ganglia*, lie on either side of the vertebral column. These ganglia contain second-order neurons that control effectors in the body wall and inside the thoracic cavity.
   - **Collateral ganglia**, also known as *prevertebral ganglia*, are found anterior to the vertebral centra. Collateral ganglia contain second-order neurons that innervate tissues and organs in the abdominopelvic cavity.
3. *Specialized second-order neurons in the interior of the adrenal gland.* The interior of each adrenal gland, a region known as the *adrenal medulla*, contains a modified sympathetic ganglion. The ganglionic neurons of the medulla have very short axons, and when stimulated they release neurotransmitters into the general circulation.

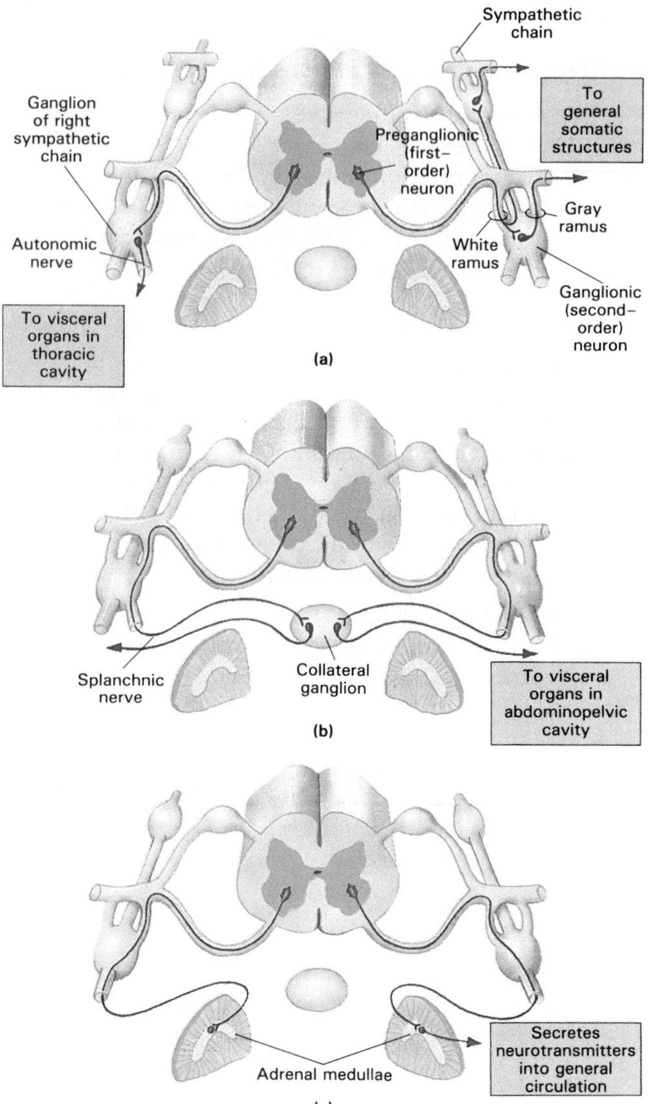

FIGURE 16-4
Sympathetic Pathways. Preganglionic fibers leave the spinal cord in the ventral roots of spinal nerves. They synapse on ganglionic neurons in sympathetic chain ganglia (a), in collateral ganglia (b), or in the adrenal medulla (c).

PATTERNS OF SYMPATHETIC INNERVATION

From spinal segments T_1 to L_2, sympathetic preganglionic fibers join the ventral root of each spinal nerve. Chapter 13 detailed the organization of a typical spinal nerve (see Figure 13-6). (p. 413) Figure 16-4 follows the distribution of sympathetic preganglionic fibers to the sympathetic chain ganglia, collateral ganglia, and the adrenal medulla. Figure 16-5 provides a more detailed look at the distribution of sympathetic fibers and the paths they take to reach their target organs.

The Sympathetic Chain

Each spinal nerve gives rise to a *white ramus* that carries preganglionic fibers to a nearby sympathetic chain ganglion. If a preganglionic fiber carries motor commands that target structures in the body wall or the thoracic cavity, the synapse occurs in a sympathetic chain ganglion (Figure 16-4a). Postganglionic fibers that control visceral effectors in the body wall, such as the sweat glands of the skin or the smooth muscles in superficial blood vessels, enter the gray ramus and return to the spinal nerve for subsequent distribution. However, spinal nerves do not innervate structures in the ventral body cavities. Postganglionic fibers targeting structures in the thoracic cavity, such as the heart and lungs, form **autonomic nerves** that proceed directly to their visceral destination. (Although Figure 16-4 shows autonomic nerves on the right side and spinal nerve distribution on the left, bear in mind that in reality *both* innervation patterns are found on *each* side of the body.)

Organization of the Sympathetic Chain A preganglionic fiber may synapse within the closest sympathetic chain ganglion, but collaterals may also reach neurons in other ganglia. Extensive divergence occurs, with one preganglionic fiber synapsing on two dozen or more second-order (ganglionic) neurons. Preganglionic fibers running between the sympathetic chain ganglia interconnect them, forming the elongate *sympathetic chain* that resembles a string of pop-beads.

Each ganglion in the sympathetic chain innervates a particular body segment or group of segments. *Every spinal nerve has a gray ramus that carries sympathetic postganglionic fibers.* Roughly 8 percent of the axons in each spinal nerve are sympathetic postganglionic fibers. As a result, the dorsal and ventral rami of the spinal nerves provide extensive sympathetic innervation to structures in the body wall and limbs. (For a review of the distribution of spinal nerves, see Figures 13-8, 13-9, and 13-10.) (pp. 415 and 416)

Anatomy of the Sympathetic Chain With these principles in mind, consider the additional details shown in Figure 16-5. Although first-order sympathetic neurons are limited to segments T_1 to L_2, sympathetic ganglia provide postganglionic fibers to every spinal nerve and, via autonomic nerves, to the regions serviced by cranial nerves III, VII, IX, and X. In all, there are 3 cervical, 11 thoracic, 4 lumbar, and 4 sacral sympathetic ganglia. Note that (1) *only the thoracic and upper lumbar ganglia receive preganglionic fibers via white rami,* and (2) *the cervical, lower lumbar, and sacral ganglia receive preganglionic innervation via collateral fibers.* Thus damage to the ventral roots of thoracic spinal nerves can eliminate sympathetic motor function on the affected side of the head, neck, and trunk. In contrast, damage to the ventral roots of cervical spinal nerves will produce voluntary muscle paralysis on the affected side, but leave sympathetic function intact.

Functions of the Sympathetic Chain Postganglionic fibers leaving the sympathetic chain reach their peripheral targets via the spinal nerves and autonomic nerves of the chest. A partial listing of the effects produced by sympathetic postganglionic fibers in spinal nerves includes:

- stimulation of secretion by sweat glands
- stimulation of arrector pili muscles (producing "goosebumps")

FIGURE 16-5
The Distribution of Sympathetic Postganglionic Fibers

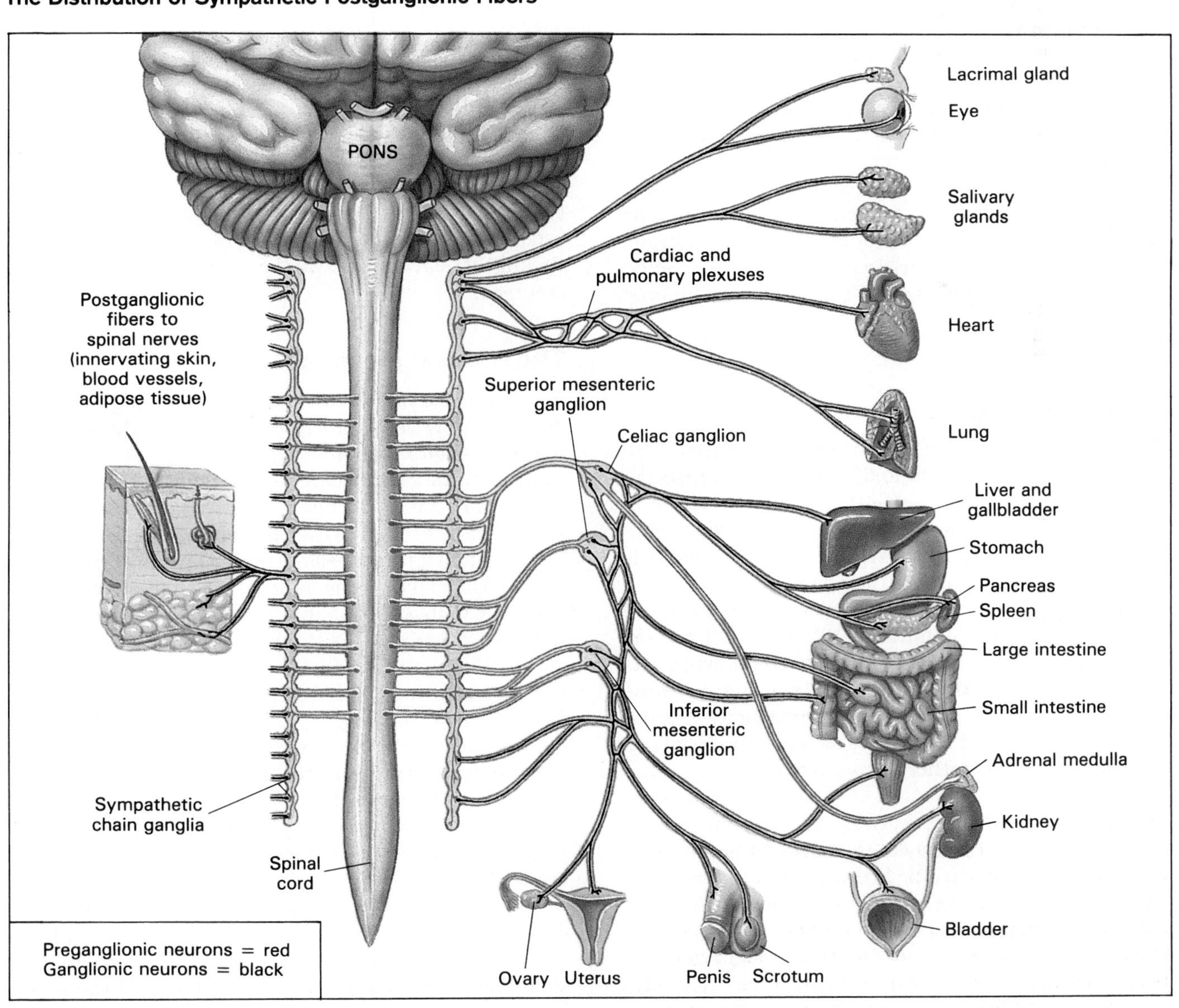

- constriction of cutaneous blood vessels and reduction in circulation to the skin, as well as to most other organs in the body wall
- acceleration of blood flow to skeletal muscles
- stimulation of energy production and use by skeletal muscle tissue
- release of stored lipids from subcutaneous adipose tissue
- dilation of the pupils to allow more light into the eyes and focusing of the eyes for viewing distant objects

The postganglionic fibers that enter the thoracic cavity in autonomic nerves also have a number of important functions, including:

- accelerating the heart rate
- increasing the force of cardiac contractions
- dilating the respiratory passageways

These changes help prepare the individual for a crisis that will require sudden, intensive physical activity. The heart works harder, moving blood faster. The muscles receive more blood, and their utilization of stored and absorbed nutrients accelerates. Lipids, a potential energy source, are released. The lungs deliver more oxygen and prepare to eliminate the carbon dioxide that will be produced by contracting muscles. Sweat glands become active, anticipating the need to lose the heat that these muscles will generate. The eyes look for approaching dangers. The mechanisms responsible for these effects will be considered in a later section.

Collateral Ganglia

The abdominopelvic viscera receive sympathetic innervation via preganglionic fibers that pass through the sympathetic chain without synapsing. These fibers originate at first-order neurons in the lower thoracic and upper lumbar segments. They synapse within separate *collateral ganglia* (Figure 16-4b). Preganglionic fibers that innervate the collateral ganglia form the **splanchnic** (SPLANK-nik) **nerves** in the dorsal wall of the abdominal cavity. Splanchnic nerves from both sides of the body converge on these ganglia; although there are two sympathetic chains, one on each side of the vertebral column, the collateral ganglia are single, rather than paired.

Anatomy of the Collateral Ganglia The splanchnic nerves innervate three collateral ganglia that are diagrammed in Figure 16-5. Preganglionic fibers from the seven lower thoracic segments end at the **celiac** (SĒ-lē-ak) **ganglion** and the **superior mesenteric ganglion**. The celiac ganglion innervates the stomach, liver, pancreas, and spleen; the superior mesenteric ganglion sends postganglionic fibers to the small intestine and the initial segments of the large intestine. Preganglionic fibers from the lumbar segments form splanchnic nerves that end at the **inferior mesenteric ganglion**. This ganglion provides sympathetic innervation to the terminal portions of the large intestine, the kidney and bladder, and the sex organs.

Functions of the Collateral Ganglia Postganglionic fibers leaving the collateral ganglia extend throughout the abdominopelvic cavity, innervating a variety of visceral tissues and organs. A summary of their effects includes:

- constricting small arteries and reducing the flow of blood to visceral organs
- decreasing the activity of digestive glands and organs, including the stomach, intestines, pancreas, and gallbladder
- stimulating the release of glucose from glycogen reserves in the liver
- stimulating the release of lipids from adipose tissue
- relaxing the smooth muscle in the wall of the urinary bladder
- controlling some aspects of sexual function, such as ejaculation in the male

The general pattern here is (1) the reduction of blood flow and energy use by visceral organs that are not important to short-term survival, such as the digestive tract, and (2) the release of stored energy reserves.

The Adrenal Medulla

Preganglionic fibers entering the adrenal gland proceed to its center, to the region called the **adrenal medulla** (Figure 16-4c). They synapse on modified neurons that perform an endocrine function. As you will recall from Chapter 5, endocrine cells produce secretions (*hormones*) that are released into the circulation to affect target cells in distant parts of the body. ∞ (p. 126) The cells of the adrenal medulla are neurons with short axons that end on an extensive network of capillaries in the center of the adrenal gland. When stimulated, these cells release the neurotransmitters norepinephrine (NE) and epinephrine (E) into surrounding capillaries. This arrangement provides a graphic example of the difficulties encountered when attempting to differentiate between the nervous and endocrine systems. The cells are obviously neurons, and they are found in a modified sympathetic ganglion, but their functions are definitely endocrine!

Once inside the circulation, the neurotransmitters are carried throughout the body. As they diffuse into peripheral tissues, they cause changes in the metabolic activities of many different cells. In general, these effects resemble those produced by the stimulation of sympathetic postganglionic fibers. However, (1) cells not innervated by sympathetic postganglionic fibers are affected as well, and (2) the effects last significantly longer than those produced by direct sympathetic innervation.

CENTRAL EFFECTS OF SYMPATHETIC STIMULATION

The sympathetic division can change tissue and organ activities by releasing NE at peripheral synapses and distributing norepinephrine and epinephrine throughout the body in the bloodstream. The motor fibers that target specific effectors, such as smooth muscle fibers in blood vessels of the skin, can be activated in reflexes that do not involve other peripheral effectors. For example, vessel diameters may be decreased slightly to elevate blood pressure, but the individual will not begin sweating and will not stop digesting dinner. In a crisis, however, the entire division responds. This event is called **sympathetic activation**.

Sympathetic activation is controlled by sympathetic centers in the hypothalamus. The effects are not limited to peripheral tissues, and sympathetic activation also alters CNS activity. For example, when sympathetic activation occurs the individual experiences:

- increased alertness, via stimulation of the reticular activating system, causing the individual to feel "on edge"
- a feeling of energy and euphoria, often associated with a disregard for danger and a temporary insensitivity to painful stimuli
- increased activity in the cardiovascular and respiratory centers of the pons and medulla, leading to elevations in blood pressure, heart rate, breathing rate, and depth of respiration
- a general elevation in muscle tone through stimulation of the extrapyramidal system, so that the person *looks* tense, and may even begin to shiver

These changes, coupled with the peripheral changes already noted, complete the preparations necessary for the individual to cope with stressful and potentially dangerous situations.

We have examined the distribution of sympathetic impulses and the general effects of sympathetic activation. We will now consider the cellular basis for these effects on peripheral organs.

SYMPATHETIC ACTIVATION AND NEUROTRANSMITTER RELEASE

Figure 16-6 diagrams the basic sequence of events that follows sympathetic activation. On stimulation, sympathetic preganglionic fibers release ACh at synapses with ganglionic neurons. As you will recall

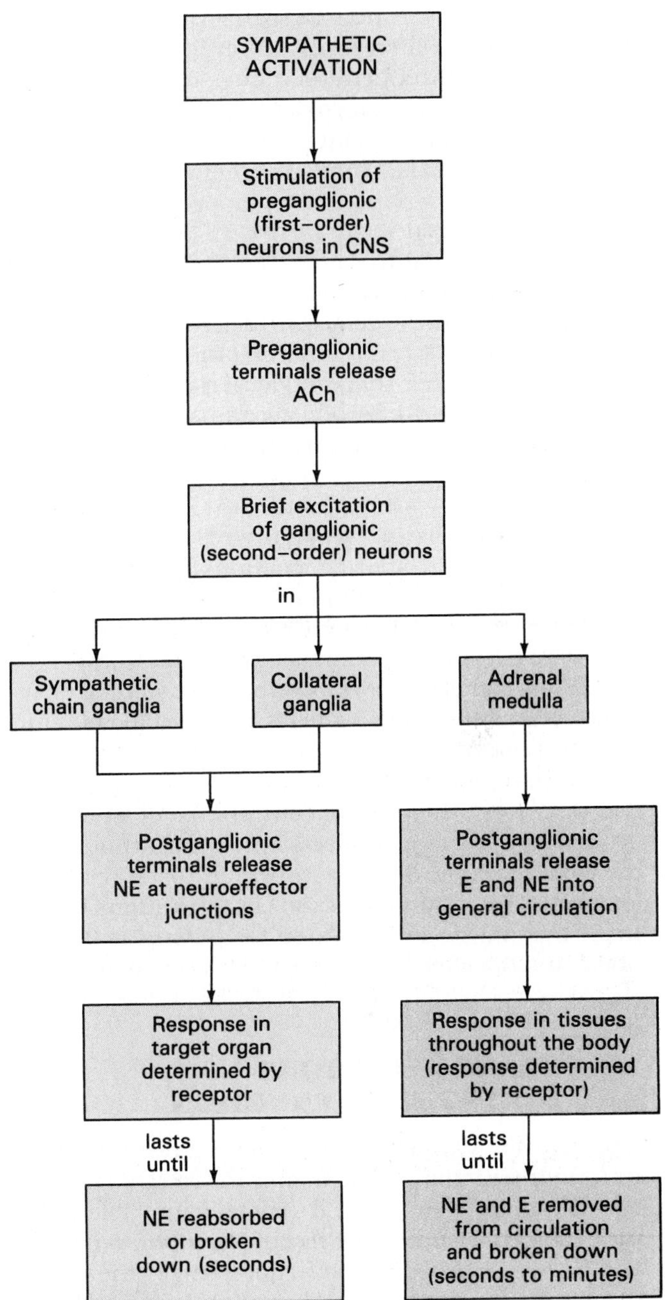

FIGURE 16-6
The Sequence of Events Following Sympathetic Activation

from Chapter 12, synapses using ACh as a transmitter are called *cholinergic*. (p. 387) The effect of stimulation on these ganglionic neurons is always excitatory.

Typical sympathetic postganglionic fibers release norepinephrine at their neuroeffector junctions; these are *adrenergic* synapses. The synaptic terminals do not resemble the neuroeffector junctions of the SNS. Instead, the terminal axon branches form a network or chain of enlarged terminal knobs that either contact the target cells or end in the adjacent connective tissues. Neurotransmitter release follows the general pattern described for adrenergic synapses in Chapter 12. After entering the interstitial fluid, the norepinephrine affects the postsynaptic membrane for several seconds. Over that period the terminal knobs gradually reabsorb from 50 to 80 percent of the NE for subsequent recycling. The rest diffuses out of the area or is broken down by enzymes such as *monoamine oxidase* (MAO) and *catechol-O-methyltransferase* (COMT). (For a more detailed description of an adrenergic synapse, review appropriate sections of Chapter 12, especially Figure 12-16). (p. 390)

ACh release at the adrenal gland stimulates the secretory activities of the adrenal medulla. The modified neurons in the adrenal medulla have short axons that release epinephrine and norepinephrine into the bloodstream, rather than into local tissues. Epinephrine, also called *adrenaline*, accounts for 75–80 percent of the secretory output; the rest is norepinephrine.

In summary, stimulation of the sympathetic division has two distinctive results: (1) the release of norepinephrine at specific locations and (2) the secretion of epinephrine (and modest amounts of norepinephrine) into the general circulation. The norepinephrine released by terminal knobs affects its target for a few seconds before resorption, diffusion, or enzymatic degradation inactivates it. But the bloodstream does not contain enzymes that break down epinephrine or norepinephrine, and most peripheral tissues contain relatively low concentrations of such enzymes. As a result, the effects of adrenal stimulation are widespread, and they continue for a relatively long time. For example, tissue concentrations of epinephrine may remain elevated for as long as 30 seconds, and the effects may persist for several minutes.

MEMBRANE RECEPTORS AND SYMPATHETIC FUNCTION

The effects of sympathetic stimulation result primarily from interactions with membrane receptors sensitive to norepinephrine and epinephrine. There are two classes of sympathetic receptors, *alpha receptors* and *beta receptors*. In general, norepinephrine stimulates alpha receptors more than it does beta receptors; epinephrine stimulates both classes of receptors. To produce their observed effects, these receptors activate or inactivate intracellular enzymes. Depending on the nature of the enzyme involved, the result can be stimulatory or inhibitory.

Alpha Receptors

There are two types of **alpha (α) receptors**, called $\boldsymbol{\alpha_1}$ (alpha-1) and $\boldsymbol{\alpha_2}$ (alpha-2). Stimulation of the most common alpha receptors, α_1, on the target cell, triggers a depolarization that has an excitatory effect. The mechanism, diagrammed in Figure 16-7, involves activation of an enzyme that causes the release of calcium ions inside the cell. The surfaces of smooth muscle and gland cells contain abundant α_1 receptors. Stimulation of these receptors on smooth muscle cells during sympathetic activation triggers the constriction of peripheral blood vessels and the closure of sphincters along the digestive tract. In an emergency these responses help to elevate blood pressure, reduce blood flow to digestive organs, and reduce the oxygen demands of organs that cannot help the individual fight or flee.

In addition to elevating intracellular calcium concentrations, stimulation of α_1 receptors can speed up cellular metabolism by activating enzymes that regulate important metabolic pathways. For example, stimulation of alpha receptors in skeletal muscle tissue accelerates the breakdown of glycogen and enhances energy production during a crisis.

The second type of alpha receptor, α_2, has a much more limited distribution. Stimulation of these receptors leads to a decrease in the intracellular concentration of cyclic-AMP, which usually has an inhibitory effect. (As noted in Chapter 12, cyclic-AMP is an important second messenger that triggers alterations in cellular activity.) (p. 389) The effects of α_2 stimulation are diagrammed in Figure 16-7b.

Alpha-2 receptors are found on the *presynaptic* surfaces at autonomic neuroeffector junctions. In the sympathetic division, this arrangement provides a negative feedback mechanism that controls the amount of NE released at the synapse. A synaptic terminal releases NE until concentrations within the synapse rise enough to stimulate the presynaptic α_2 receptors. This stimulation has an inhibitory effect, and neurotransmitter release stops. (Presynaptic inhibition was introduced in Chapter 12.) (p. 397)

Because α_2 receptors are also present at *parasympathetic* neuroeffector junctions, *release of E or NE inhibits parasympathetic activity*. This fact probably accounts for the widespread inhibitory effects of sympathetic stimulation on the digestive system noted earlier in the chapter.

Beta Receptors

There are also two classes of **beta (β) receptors**, $\boldsymbol{\beta_1}$ (beta-1) and $\boldsymbol{\beta_2}$ (beta-2). Both types of beta receptors act by increasing intracellular concentrations of cyclic-AMP (Figure 16-8). The observed effects vary depending on the nature of the enzymes stimulated by this second messenger. Beta-1 receptors are equally sensitive to epinephrine and norepinephrine, and their stimulation has an excitatory effect. These receptors are found in the heart, and when stimulated they produce an increase in heart rate and force of contraction. These responses accelerate blood flow to active tissues and help maintain adequate blood pressure and flow in the event of a severe injury and blood loss.

Beta-2 receptors are found in the smooth muscles surrounding blood vessels in the heart, skeletal muscles, and lungs and in smooth muscles in the walls of the respiratory passageways. Sympathetic stimulation of the β_2 receptors produces a generalized

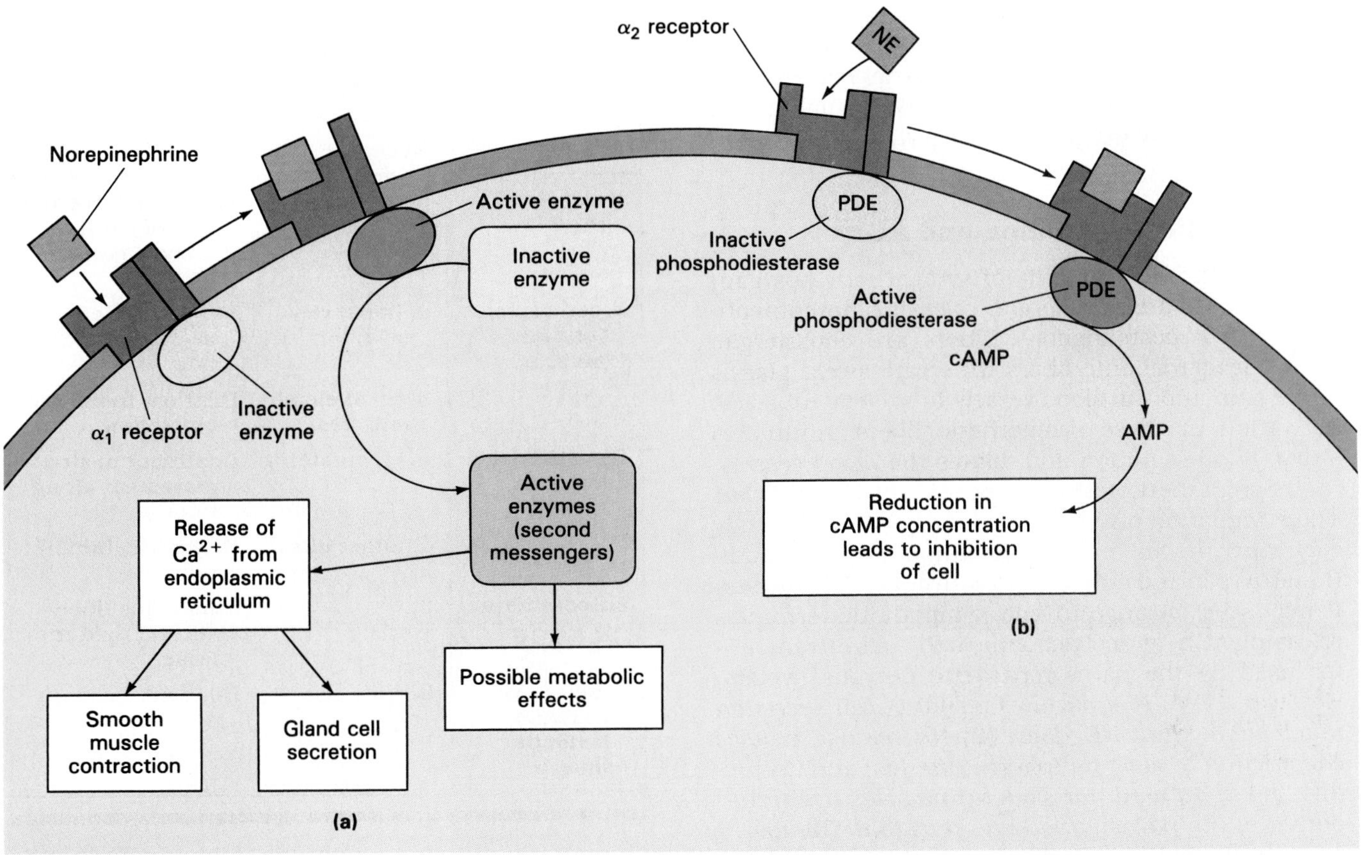

α_2 receptor
NE
Norepinephrine
Active enzyme
PDE
Inactive enzyme
Inactive phosphodiesterase
Active phosphodiesterase
PDE
cAMP
Inactive enzyme
α_1 receptor
AMP
Active enzymes (second messengers)
Reduction in cAMP concentration leads to inhibition of cell
Release of Ca^{2+} from endoplasmic reticulum
(b)
Possible metabolic effects
Smooth muscle contraction
Gland cell secretion
(a)

FIGURE 16-7
Alpha Receptor Function. (a) Binding of NE or epinephrine to an α_1 receptor starts a chain of enzymatic events leading to the release of calcium ions into the cytoplasm. The increased calcium concentration triggers contraction of a smooth muscle cell or secretion by a gland cell. (b) Binding of norepinephrine or epinephrine to an α_2 receptor leads to a reduction in the concentration of cyclic-AMP in the cytoplasm. The result is inhibition of the target cell.

FIGURE 16-8
Beta Receptor Function

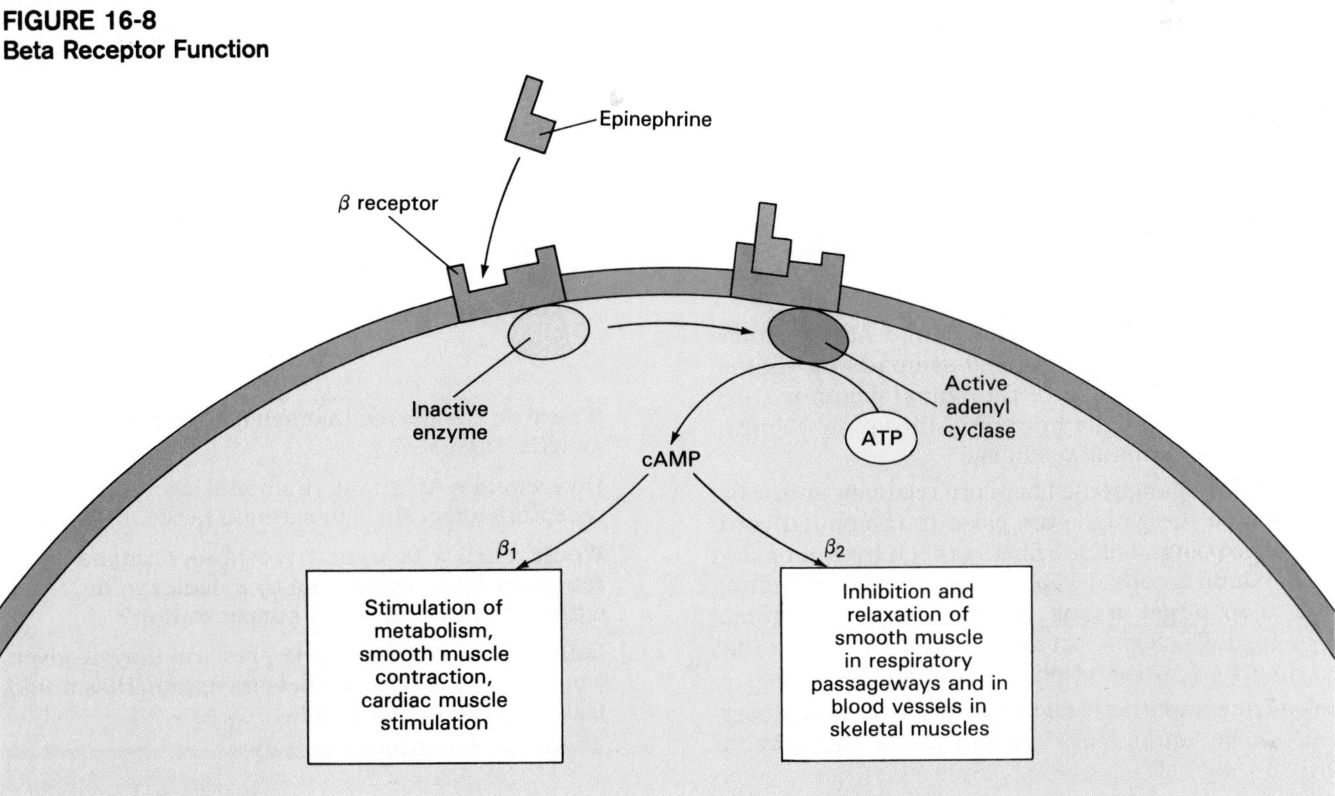

inhibition and relaxation of smooth muscle fibers. The result is a dilation of the blood vessels where these receptors are present and an increase in the diameter of the respiratory passageways, making it easier to breathe quickly and deeply.

Sympathetic Stimulation and ACh

Although the vast majority of sympathetic postganglionic fibers are adrenergic, releasing norepinephrine, a few postganglionic fibers are cholinergic. These postganglionic fibers innervate sweat glands of the skin and the blood vessels to skeletal muscles. Activation of these sympathetic fibers stimulates sweat gland secretion and dilates the blood vessels.

Sympathetic activation occurs during exercise. The stimulation of sweat glands helps keep the body cool, and the increased circulation provides additional oxygen and nutrients to active skeletal muscles. It may seem strange to find sympathetic terminals releasing ACh, as acetylcholine is the neurotransmitter used by the parasympathetic nervous system. However, (1) *ACh stimulates sweat gland secretion much more than NE does*; (2) *NE release causes constriction of most peripheral arteries*; and (3) *neither the body wall nor skeletal muscles are innervated by the parasympathetic division.* The distribution of cholinergic fibers via the sympathetic division provides a method of regulating sweat gland secretion and selectively controlling blood flow to skeletal muscles while reducing the flow to other tissues in the body wall.

TABLE 16-1 Examples of Alpha and Beta Receptor Distribution

Organ	*Receptor*	*Response*
Heart (cardiac muscle)	β_1	Increased force and rate of cardiac muscle contraction
Blood vessels (smooth muscle)	α_1 (most vessels)	Vasoconstriction, decreased blood flow
	β_2 (in skeletal muscles)	Dilation, increased blood flow
Digestive tract (smooth muscle)	α_1 (sphincters)	Contraction, stops movement along tract
	α_2 (other sites)	Relaxation inhibition
Adipose tissue	β_1	Stimulates lipolysis and lipid release
Respiratory passages (smooth muscle)	β_2	Dilation

RECEPTOR DISTRIBUTION AND EFFECTOR RESPONSES

Table 16-1 includes the distribution of alpha and beta receptors in major organs. This table also indicates the effects of sympathetic stimulation.

A SUMMARY OF THE SYMPATHETIC DIVISION

1. The sympathetic division of the ANS includes two segmentally arranged sympathetic chains, one on each side of the spinal column, three collateral ganglia in front of the spinal column, and two adrenal medullae.
2. The preganglionic fibers are relatively short, because the ganglia are close to the spinal cord. The postganglionic fibers are relatively long and extend a considerable distance before reaching their target organs. (In the case of the adrenal medulla, very short axons end at capillaries that carry their secretions throughout the body.)
3. The sympathetic division shows extensive divergence, and a single preganglionic fiber may innervate as many as 32 second-order neurons in different ganglia. As a result, a single sympathetic motor neuron inside the CNS can control a variety of peripheral effectors and produce a complex and coordinated response.
4. All preganglionic neurons release ACh at their synapses with ganglionic neurons. Most of the postganglionic fibers release norepinephrine, but a few release ACh.
5. The effector response depends on the nature of the enzymes activated when NE or E binds to alpha or beta receptors. A single effector cell may have more than one type of receptor on its surface, and the interaction between the different receptors determines the results of sympathetic stimulation.

- √ **Where do the nerves that synapse in the collateral ganglia originate?**
- √ **How would a drug that stimulates acetylcholine receptors affect the sympathetic nervous system?**
- √ **Would a cell with alpha-1 receptors or alpha-2 receptors be most affected by a decrease in extracellular calcium ion concentration?**
- √ **Individuals with high blood pressure may be given a medication that blocks beta receptors. How would this help their condition?**

The Parasympathetic Division

The parasympathetic division of the ANS includes:

1. *Preganglionic (first-order) neurons in the brain stem and in sacral segments of the spinal cord.* In the brain, the mesencephalon, pons, and medulla contain autonomic nuclei. In the sacral segments of the spinal cord, the autonomic nuclei lie in the lateral gray horns of spinal segments S_2–S_4.
2. *Ganglionic (second-order) neurons in peripheral ganglia located within or adjacent to the target organs.* The preganglionic fibers of the parasympathetic division do not diverge as extensively as do those of the sympathetic division. A typical preganglionic fiber synapses on 6–8 ganglionic neurons. In contrast to the pattern in the sympathetic division, these second-order neurons are all located in the same ganglion, and their postganglionic fibers influence the same target organ. As a result, *the effects of parasympathetic stimulation are more specific and localized than those of the sympathetic division.*

ORGANIZATION OF THE PARASYMPATHETIC DIVISION

Figure 16-9 diagrams the pattern of parasympathetic innervation. Preganglionic fibers leaving the brain travel within cranial nerves III (oculomotor), VII (facial), IX (glossopharyngeal), and X (vagus). Parasympathetic fibers in the oculomotor, facial, and glossopharyngeal nerves are concerned with the control of visceral structures in the head. These fibers synapse in the **ciliary**, **sphenopalatine**, **submandibular**, and **otic ganglia.** (These ganglia were shown in Figure 14-22.) (p. 462) Short postganglionic fibers then continue to their peripheral targets. The vagus nerve provides preganglionic parasympathetic innervation to structures in the thoracic and abdominopelvic cavity as distant as the last segments of the large intestine. This one nerve provides roughly 75 percent of all parasympathetic outflow.

The sacral parasympathetic outflow leaves the sacral segments of the spinal cord but does not join the ventral roots of the spinal nerves. Instead, the preganglionic fibers form distinct **pelvic nerves** that innervate intramural ganglia in the kidney and bladder, the terminal portions of the large intestine, and the sex organs.

FIGURE 16-9
Organization of the Parasympathetic Division of the ANS

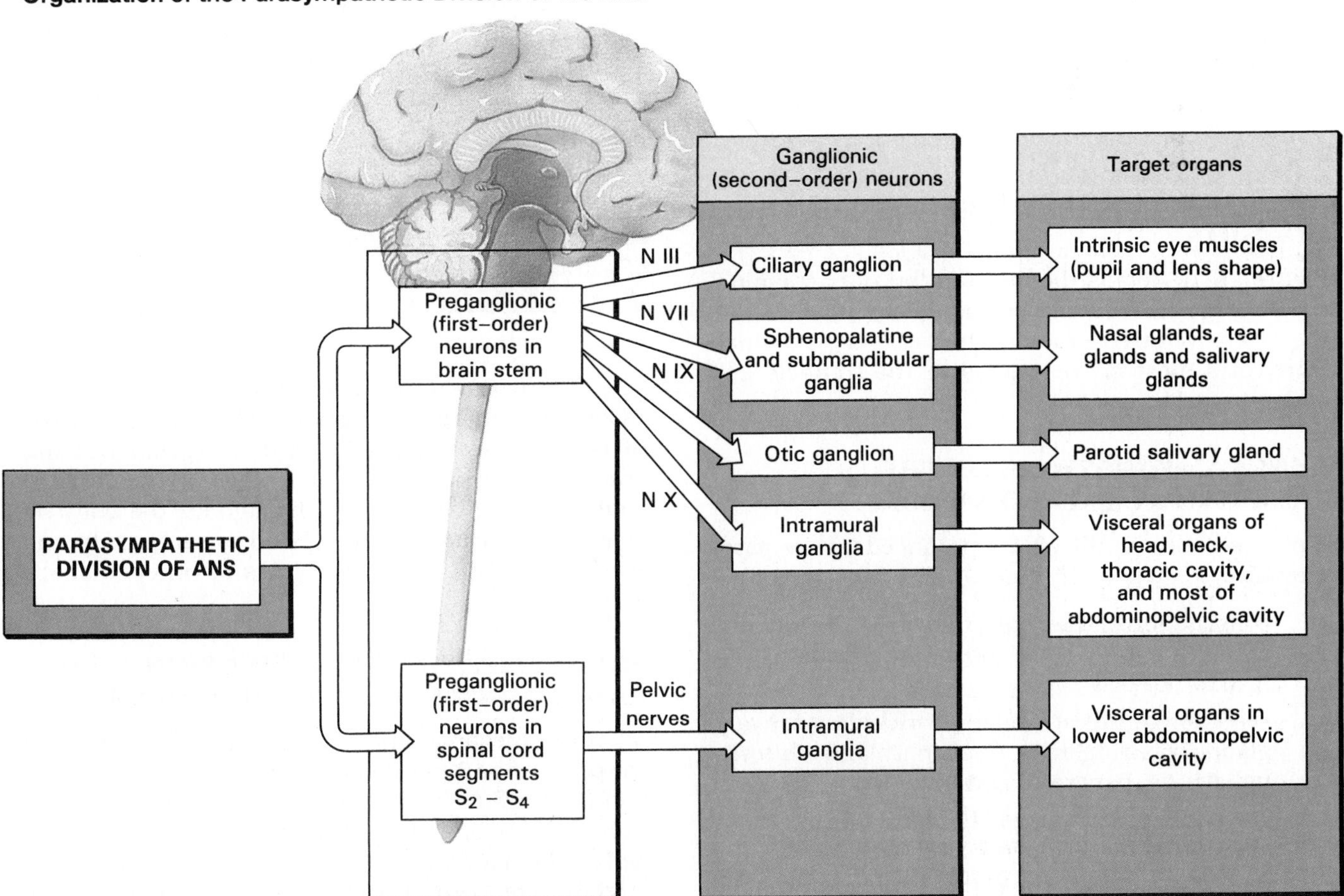

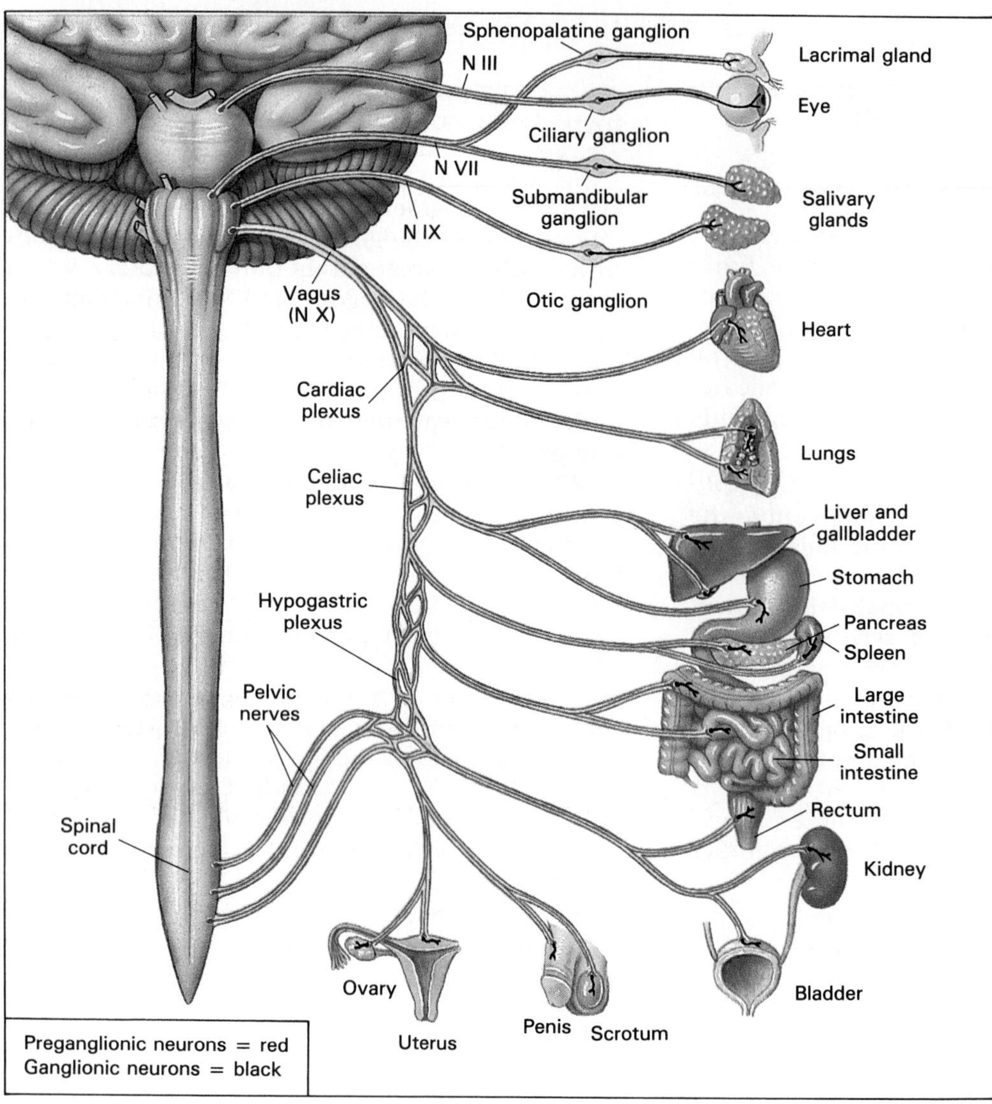

FIGURE 16-10
Distribution of the Parasympathetic Output

ANATOMY OF THE PARASYMPATHETIC DIVISION

Figure 16-10 details the distribution of parasympathetic preganglionic and postganglionic fibers. Note the extensive distribution of the vagus nerve, which innervates most of the organs in the thoracic and abdominopelvic cavities.

GENERAL FUNCTIONS OF THE PARASYMPATHETIC DIVISION

A partial listing of the effects produced by the parasympathetic division includes:

- constriction of the pupils to restrict the amount of light entering the eyes, and focusing on nearby objects
- secretion by digestive glands, including salivary glands, gastric glands, duodenal glands, intestinal glands, pancreas, and liver
- secretion of hormones that promote nutrient absorption by peripheral cells
- increased smooth muscle activity along the digestive tract
- stimulation and coordination of defecation
- contraction of the urinary bladder during urination
- constriction of the respiratory passageways
- reduction in heart rate and force of contraction

These functions center on relaxation, food processing, and energy absorption. The parasympathetic division has been called the *anabolic system* because stimulation leads to a general increase in the nutrient content of the blood. Cells throughout the body respond to this increase by absorbing nutrients and using them to support growth and other anabolic activities.

PARASYMPATHETIC ACTIVATION AND NEUROTRANSMITTER RELEASE

All of the preganglionic and postganglionic fibers in the parasympathetic division release ACh at synapses and neuroeffector junctions. The neuroeffector junctions are small, with narrow synaptic clefts, as typified by the neuromuscular junctions of the muscular system. (The neuromuscular junction was described in Chapter 10, and cholinergic synapses were detailed in Chapter 12—see Figures 10-12, 12-13, and 12-15.) (pp. 309, 387, 389) The effects of stimulation are short-lived, for most of the ACh released is inactivated by acetylcholinesterase within the synapse. The choline produced is resorbed by the synaptic knob. Any ACh diffusing into the surrounding tissues will be attacked by another enzyme, **tissue cholinesterase**. As a result, the effects of parasympathetic stimulation are quite localized and last a few seconds at most.

RECEPTOR DISTRIBUTION AND EFFECTOR RESPONSES

Although all the synapses (neuron to neuron) and neuroeffector junctions (neuron to effector) of the parasympathetic division use the same transmitter, acetylcholine, two different types of ACh receptors are found on the postsynaptic membranes.

- **Nicotinic** (nik-ō-TIN-ik) **receptors** *are found on the surfaces of ganglion cells of both the parasympathetic and sympathetic divisions, as well as at neuromuscular junctions of the SNS.* The receptor molecules are chemically regulated sodium ion channels, and exposure to ACh always causes excitation of the ganglionic (second-order) neuron or muscle fiber.
- **Muscarinic** (mus-kar-IN-ik) **receptors** produce a longer-lasting effect than do nicotinic receptors. *Muscarinic receptors are found at cholinergic neuroeffector junctions in the parasympathetic division, as well as at the few cholinergic neuroeffector junctions in the sympathetic division.* Muscarinic receptors activate enzymes that may open potassium ion channels, producing hyperpolarization and inhibition, or close potassium channels, producing depolarization and excitation.

Examples of inhibitory muscarinic receptors are found at cholinergic neuroeffector junctions:

- at sphincter muscles regulating passage along the digestive tract, where parasympathetic stimulation causes relaxation and promotes movement of materials
- on the surfaces of cardiac muscle fibers, where parasympathetic stimulation leads to a decrease in the rate and force of contraction
- at smooth muscle in the walls of small arteries along the digestive tract (parasympathetic innervation) and in skeletal muscle (sympathetic innervation), where ACh exposure causes vessel dilation.

Examples of excitatory muscarinic receptors are found at cholinergic neuroeffector junctions:

- in the pancreas, salivary glands, and other secretory glands of the digestive tract, where stimulation accelerates secretion
- on smooth muscles surrounding the respiratory passageways, leading to muscle contraction and airway obstruction
- in the walls of the urinary bladder, causing bladder contraction.

Table 16-2 summarizes details concerning the adrenergic and cholinergic receptors in the ANS.

TABLE 16-2 Adrenergic and Cholinergic Receptors in the ANS

Receptor	*Location*	*Response*	*Mechanism*
ADRENERGIC RECEPTORS			
α_1	Widespread, found in most tissues; not in heart	Excitation, stimulation of metabolism	Activation of enzymes, release of intracellular calcium ions
α_2	Sympathetic and parasympathetic neuroeffector junctions	Inhibition of neurotransmitter release	Reduction in cAMP concentrations
β_1	Heart, kidneys, liver, adipose tissue	Stimulation, increased energy consumption	Enzyme activation
β_2	Smooth muscle in vessels of heart and skeletal muscle, intestinal muscle, lungs, and bronchi	Inhibition, relaxation	Enzyme activation
CHOLINERGIC RECEPTORS			
Nicotinic	All autonomic synapses between preganglionic and ganglionic neurons; also neuromuscular junctions of the SNS	Stimulation, excitation	Opening of chemically regulated Na^+ channels
Muscarinic	All parasympathetic neuroeffector junctions; cholinergic sympathetic neuroeffector junctions	Variable	Enzyme activation causing changes in membrane permeability to K^+

The names *nicotinic* and *muscarinic* indicate chemical compounds that stimulate these receptor sites. The toxin nicotine can be obtained from a variety of sources, including tobacco leaves; the poison muscarine can be extracted from toadstools. These compounds have discrete actions, targeting either the autonomic ganglia and skeletal neuromuscular junctions (nicotine) or the parasympathetic neuroeffector junctions (muscarine). They produce dangerously exaggerated, uncontrolled responses that parallel those produced by normal receptor stimulation. For example, nicotine poisoning occurs if as little as 50 mg of the compound is ingested or absorbed through the skin. The symptoms reflect widespread autonomic activation—vomiting, diarrhea, high blood pressure, rapid heart rate, sweating, and profuse salivation. Because the neuromuscular junctions of the SNS are stimulated, convulsions occur. In severe cases, stimulation of nicotinic receptors inside the CNS may lead to coma and death within minutes.

A SUMMARY OF THE PARASYMPATHETIC DIVISION

1. The parasympathetic division includes visceral motor nuclei associated with four cranial nerves (III, VII, IX, and X) and in sacral segments S_2–S_4.
2. The second-order neurons are situated in intramural ganglia or in ganglia closely associated with the target organs.
3. The parasympathetic division innervates areas serviced by the cranial nerves and organs in the thoracic and abdominopelvic cavities.
4. All parasympathetic neurons are cholinergic. Ganglionic neurons have nicotinic receptors that are excited by ACh. Muscarinic receptors present at neuroeffector junctions may produce either excitation or inhibition, depending on the nature of the enzymes activated when ACh binds to the receptor.
5. The effects of parasympathetic stimulation are usually brief in duration and restricted to specific organs and sites.

Anatomical and Functional Relationships between the Sympathetic and Parasympathetic Divisions

The sympathetic division has widespread impact, reaching visceral and somatic structures throughout the body. The parasympathetic division innervates only visceral structures serviced by the cranial nerves or lying within the abdominopelvic cavity. Although some organs are innervated by one division or the other, most vital organs receive **dual innervation**—that is, they receive instructions from both autonomic divisions. Where dual innervation exists, the two divisions often have opposing effects. Sympathetic-parasympathetic opposition can be seen along the digestive tract, at the heart, in the lungs, and elsewhere. There are a few exceptions to this general rule, where the two control specific functions within a complex organ or system. Secretory control of the salivary glands or the sexual functions of the male reproductive tract are examples.

ANATOMY OF DUAL INNERVATION

Figure 16-11 highlights the basic pattern of dual innervation. Major organizational patterns include:

- In the head, parasympathetic postganglionic fibers from the ciliary, sphenopalatine, submandibular, and otic ganglia accompany the cranial nerves to their peripheral destinations. The sympathetic innervation reaches the same structures by traveling directly from the superior ganglia of the sympathetic chain.
- In the thoracic and abdominopelvic cavities, the sympathetic postganglionic fibers mingle with parasympathetic preganglionic fibers at a series of anatomical "crossroads." Considerable branching occurs, forming a nerve network known as a *plexus*.[1]
- Autonomic fibers entering the thoracic cavity intersect at the **cardiac plexus** and the **pulmonary plexus**. These plexuses contain sympathetic and parasympathetic fibers bound for the heart and lungs, respectively, as well as the parasympathetic ganglia whose output affects these organs.
- Parasympathetic preganglionic fibers of the vagus nerve entering the abdominopelvic cavity join the **celiac plexus**, also known as the *solar plexus*. The celiac plexus innervates viscera down to the initial segments of the large intestine. The **hypogastric plexus** contains the parasympathetic outflow of the pelvic nerves and the sympathetic output of the inferior mesenteric ganglion. These plexuses are illustrated in Figure 16-11.

Table 16-3 provides an organizational comparison of the sympathetic and parasympathetic divisions of the ANS. The distinctions have physiological

[1]Chapter 13 (Figures 13-8 through 13-12) discussed the plexuses created by intersecting sensory and somatic motor nerves traveling to and from the limbs.

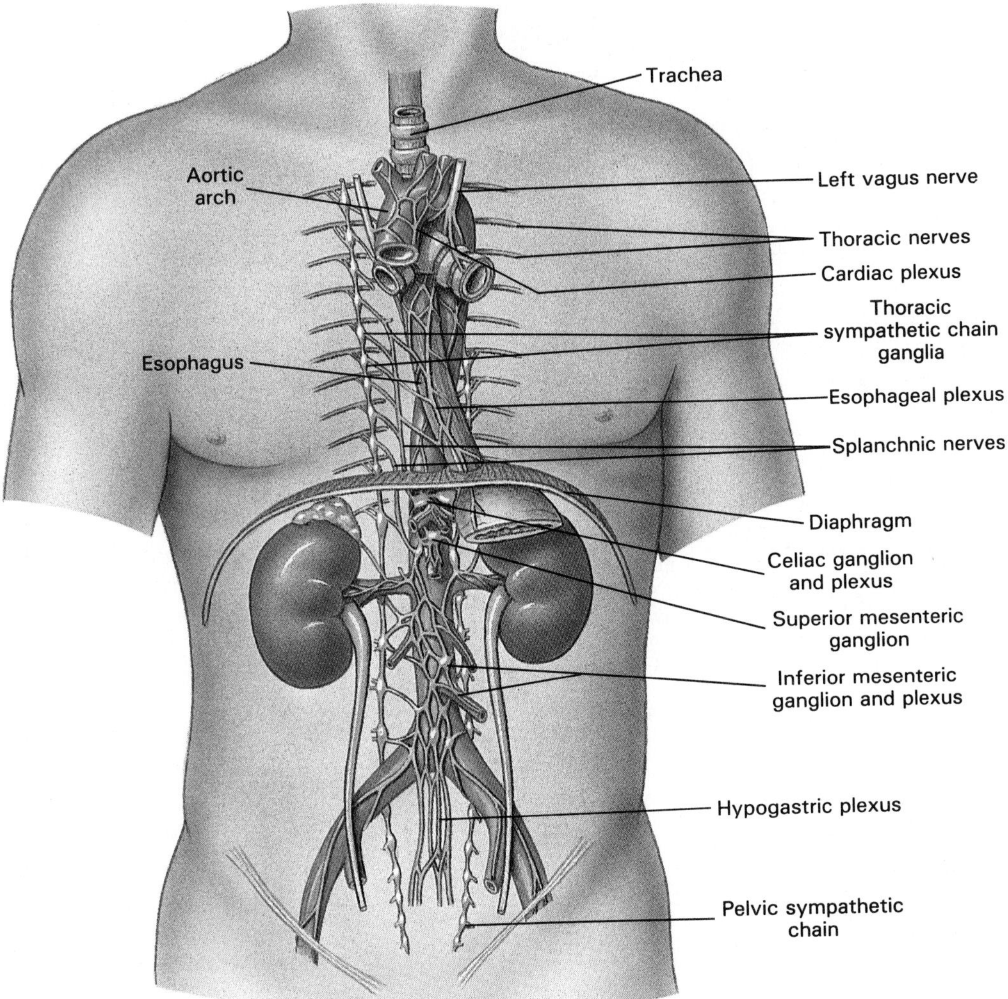

FIGURE 16-11
The Peripheral Autonomic Plexuses

TABLE 16-3 A Comparison of the Sympathetic and Parasympathetic Divisions of the ANS

Characteristic	*Sympathetic Division*	*Parasympathetic Division*
Location of CNS visceral motor neuron	Lateral gray horns, spinal segments T_1–L_2	Brain stem and spinal segments S_2–S_4
Location of PNS ganglia	Near spinal column	Typically intramural
Preganglionic fibers		
Length	Relatively short	Relatively long
Neurotransmitter released	Acetylcholine	Acetylcholine
Postganglionic fibers		
Length	Relatively long	Relatively short
Neurotransmitter released	Usually norepinephrine	Always acetylcholine
Neuroeffector junction	Enlarged terminal knobs that release transmitter near target cells	Neuroeffector junctions releasing transmitter to special receptor surface
Degree of divergence from CNS to ganglion cells	Approximately 1:32	Approximately 1:6
General function	Stimulate metabolism, increase alertness, prepare for emergency "fight or flight"	Promote relaxation, nutrient uptake, energy storage, "rest and repose"

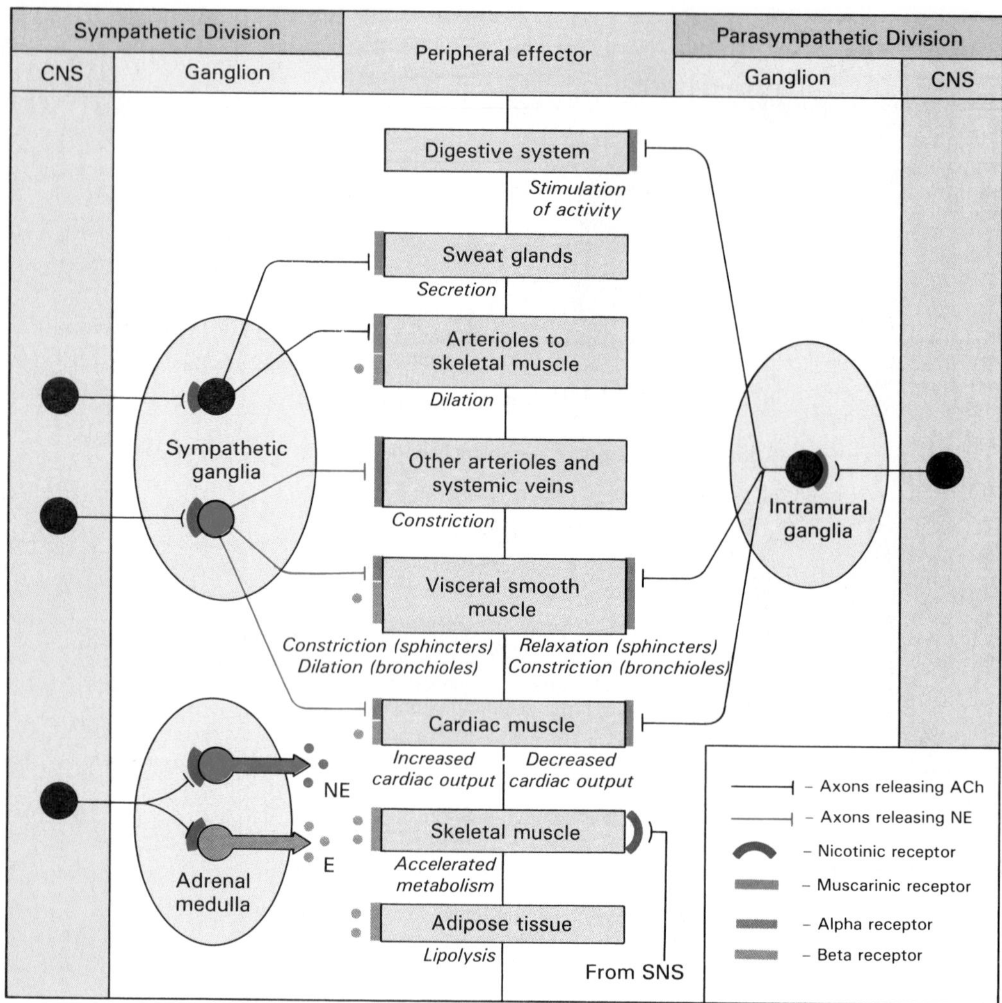

FIGURE 16-12
A Schematic Diagram of Sympathetic and Parasympathetic Interactions

and functional correlates. Figure 16-12 diagrams the distribution of sympathetic and parasympathetic fibers and indicates the neurotransmitters involved. This figure also provides an overview of the effects of sympathetic versus parasympathetic stimulation. Table 16-4 provides a more detailed comparison.

AUTONOMIC TONE

Even in the absence of stimuli, autonomic motor neurons show a resting level of spontaneous activity. The level of activation determines the **autonomic tone** of the individual. Autonomic tone is an important aspect of ANS function. *If a nerve is absolutely silent under normal conditions, then all it can do is increase its activity on demand. But if the nerve maintains a background level of activity, it may either increase or decrease its activity, providing a range of control options.*

Autonomic tone is significant where dual innervation occurs and the ANS divisions have opposing effects. It is even more important in situations where dual innervation does not occur. We will consider one example of each arrangement, to demonstrate how autonomic tone affects ANS function.

Autonomic Tone in Dual Innervation

The heart is an example of an organ that receives dual innervation and the two autonomic divisions have opposing effects. ACh released by the parasympathetic division causes a reduction in heart rate, whereas NE released by the sympathetic division accelerates the heart rate. Because autonomic tone exists, small amounts of both these neurotransmitters are released on a continual basis. By means of small adjustments in parasympathetic versus sympathetic stimulation, the heart rate can be controlled very precisely. In a crisis, stimulation of the sympathetic innervation and inhibition of the parasympathetic innervation accelerate the heart rate to the maximum extent possible.

Autonomic Tone in the Absence of Dual Innervation

Several organs are innervated by one division only. For example, most structures in the body wall, such as blood vessels, glands, adipose tissue, and skeletal muscles, do not receive parasympathetic innervation. On the other hand, the constrictor muscles of the pupil, the tear glands, and nasal glands are not innervated by the sympathetic division.

The sympathetic control of blood vessel diameter

TABLE 16-4 A Functional Comparison of the Sympathetic and Parasympathetic Divisions of the ANS

Structure	*Sympathetic Receptor Type*	*Sympathetic Innervation Effect*	*Parasympathetic Innervation Effect (All muscarinic receptors)*
EYE	α_1	Dilates pupil, accommodation for distance vision	Constricts pupil, accommodation for near vision
SALIVARY GLANDS	α_1, β_1	Serous secretion stimulated	Watery secretion stimulated
SWEAT GLANDS	α_1	Increased secretion[a]	None (not innervated)
TEAR GLANDS		None (not innervated)	Secretion
CARDIOVASCULAR SYSTEM			
Blood vessels			None (not innervated)
To integument	α_1	Vasoconstriction	
To skeletal muscles	β_2	Vasodilation[a]	
To digestive viscera	α_1	Vasoconstriction	
Veins	α_1, β_1	Constriction	
Heart			
Rate	β_1	Increases	Decreases
Force of contraction	β_1	Increases	Decreases[b]
Blood pressure	β_1	Increases	Decreases[b]
ADRENAL GLAND		Medulla secretes epinephrine, norepinephrine	None
RESPIRATORY SYSTEM			
Diameter of passageways	β_2	Increases	Decreases
Respiratory rate		Increases	Decreases
DIGESTIVE SYSTEM			
Sphincters	α_1	Constricts	Dilates
General level of activity	α_2, β_2	Decreases	Increases
Secretory glands	α_2	Inhibited	Stimulated
Liver	α_1, β_2	Glycogen breakdown	Glycogen synthesis
ADIPOSE TISSUE	β_1	Lipolysis	
URINARY SYSTEM			
Kidneys	β_2	Decreases urine production	Increases urine production
Bladder	α_1, β_2	Constricts sphincter, relaxes bladder	Tenses bladder, relaxes sphincter to eliminate urine
MALE REPRODUCTIVE SYSTEM	α_1	Increases glandular secretion and ejaculation	Erection
FEMALE REPRODUCTIVE SYSTEM	α_1	Increased glandular secretion; contraction of pregnant uterus	Variable (depending on hormones present)
	β_2	Relaxation of nonpregnant uterus	

[a] Sympathetic terminals release acetylcholine instead of norepinephrine at this site.
[b] Effects on force and blood pressure may vary depending on the diameter of peripheral vessels.

provides an example of how autonomic tone allows fine adjustment of peripheral activities when the target organ is not innervated by both ANS divisions. Sympathetic postganglionic fibers innervate the smooth muscle cells in the walls of peripheral vessels. The background sympathetic tone keeps these muscles partially contracted, so that the vessels are ordinarily at roughly half of their maximum diameter. Because the normal diameter is maintained by sympathetic tone, increasing or decreasing sympathetic stimulation provides precise control of vessel diameter over its entire range. When the vessel dilates, blood flow to the region increases; when the diameter decreases, blood flow is reduced. [**CM**: *Autonomic Tone and Hypersensitivity*]

AUTONOMIC PATTERNS OF ACTIVITY

One way to recall the different functions of the sympathetic and parasympathetic divisions is to imagine a situation dominated by one system or the other. The parasympathetic division is responsible for the state of "rest and repose" that follows a big dinner. The body relaxes, energy demands are minimal, and both heart rate and blood pressure are relatively low. Meanwhile, the organs of the digestive tract are highly stimulated. Salivary glands and other secretory glands are active, the stomach is contracting, and smooth muscle contractions move materials along the digestive tract. This movement promotes defecation; at the same time, smooth muscle contractions along the urinary tract promote urination. The overall pattern is: (1) decreased metabolic rate, (2) decreased heart rate and blood pressure, (3) salivary and digestive gland secretion, (4) increased digestive tract activity, and (5) urination and defecation.

The functions of the sympathetic nervous system can be remembered by envisioning a walk down a dark alley. Strange noises in the darkness ahead will almost certainly activate the sympathetic "fight or flight" response. The result is an increase in the level of alertness and a rise in the metabolic rate to as much as twice resting levels. Metabolic fuels are dumped into the circulation as the liver releases glucose and fat cells release fatty acids. Digestive and urinary activities are suspended temporarily, and blood flow to the skeletal muscles increases. In addition, sympathetic activation leads to dilation of the respiratory passageways and an increased respiratory rate. Both heart rate and blood pressure increase, moving blood around the body more quickly. Finally, heat production increases as energy consumption rises, and sweat gland activity promotes cooling. The general pattern can be summarized as follows: (1) mental alertness, (2) increased metabolic rate, (3) suspension of digestive and urinary tract function, (4) activation of energy reserves, (5) increased respiratory rate and efficiency, (6) increased heart rate and blood pressure, and (7) activation of sweat glands.

INTEGRATION AND CONTROL OF AUTONOMIC FUNCTIONS

Figure 15-9 diagrammed the levels of motor control in the nervous system, with particular attention to the somatic nervous system (SNS). ∞ (p. 482) The lowest level consisted of the lower motor neurons involved in the cranial and spinal reflex arcs. The highest level consisted of the pyramidal motor neurons of the primary motor cortex, operating with the assistance of extrapyramidal and cerebellar nuclei.

The ANS is also organized into a series of interacting levels. At the bottom are the lower motor neurons that participate in cranial and spinal *visceral reflexes*. These motor neurons are located in the spinal cord and lower brain stem.

Visceral reflexes provide automatic motor responses that can be modified, facilitated, or inhibited by higher centers, especially those of the hypothalamus. For example, shining a light in the eye triggers a visceral reflex that constricts the pupils of both eyes (the consensual light reflex, described in Chapter 14). ∞ (p. 469) In total darkness, the pupils dilate. But the motor nuclei directing pupillary constriction or dilation are also controlled by hypothalamic centers concerned with emotional states. When you are queasy or nauseated, your pupils constrict; when you are sexually aroused, your pupils dilate.

Visceral Reflexes

Visceral reflexes are the simplest functional units in the autonomic nervous system. As indicated in Figure 16-13, each visceral reflex arc consists of a receptor, a sensory nerve, a processing center (interneuron or motor neuron), and two visceral motor neurons, preganglionic (first-order) and ganglionic (second-order). Sensory nerves deliver information to the CNS along spinal nerves, cranial nerves, and the autonomic nerves that innervate peripheral effectors. Whether the afferent fibers synapse directly on first-order motor neurons or on CNS interneurons, all visceral reflexes are polysynaptic.

Many autonomic reflexes will be considered in later chapters, and only a brief summary will be presented here and in Table 16-5. Examples of parasympathetic reflexes include:

1. *Reflexes that coordinate the digestive activities, including the movement of materials along the digestive tract and the secretion of*

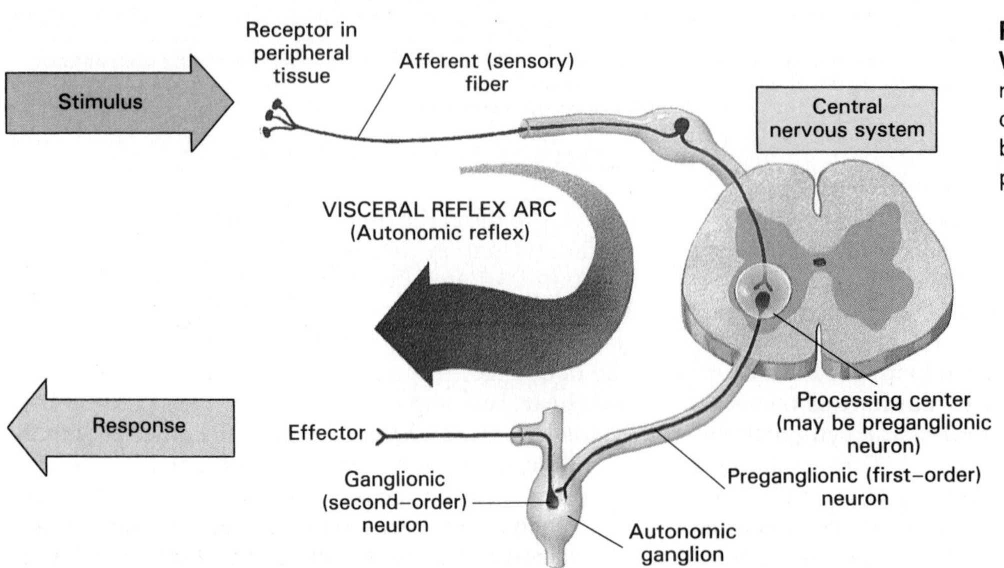

FIGURE 16-13
Visceral Reflexes. Visceral reflexes have the same basic components as somatic reflexes, but all visceral reflexes are polysynaptic.

TABLE 16-5 Representative Visceral Reflexes

Reflex	*Stimulus*	*Response*	*Comments*
PARASYMPATHETIC REFLEXES			
Gastric and intestinal reflexes	Pressure and physical contact with materials	Smooth muscle contractions that propel materials and mix with secretions	Via vagus nerve
Defecation	Distention of rectum	Relaxation of internal anal sphincter	Requires voluntary relaxation of external sphincter
Urination	Distention of urinary bladder	Contraction of bladder walls	Urine passage requires voluntary relaxation of external sphincter
Light and consensual light reflexes	Bright light shining in eye(s)	Constriction of pupils of both eyes	
Swallowing reflex	Movement of material into upper pharynx	Smooth muscle contractions moving material to stomach	Coordinated by medullary swallowing center
Vomiting reflex	Irritation of digestive tract lining	Reversal of normal smooth muscle action to eject contents	Coordinated by medullary vomiting center
Coughing reflex	Irritation of respiratory tract lining	Sudden explosive ejection of air	Coordinated by medullary coughing center
Cardioinhibitory reflex	Sudden rise in blood pressure in carotid artery	Reduction in heart rate and force of contraction	Coordinated in cardiac center in medulla oblongata
Sexual arousal	Erotic stimuli (visual or tactile)	Increased glandular secretions, sensitivity	
SYMPATHETIC REFLEXES			
Cardioacceleratory reflex	Sudden decline in blood pressure in carotid artery	Increase in heart rate and force of contraction	Coordinated in cardiac center in medulla oblongata
Vasomotor reflexes	Changes in blood pressure in major arteries	Changes in diameter of peripheral vessels to maintain normal range of blood pressures	Coordinated in vasomotor center in medulla oblongata
Pupillary reflex	Low light level reaching visual receptors	Dilation of pupil	
Ejaculation (in male)	Erotic stimuli (tactile)	Skeletal muscle contractions ejecting semen	

digestive glands and organs. The reflex responses are directed primarily by the vagus nerve (N X).

2. *The reflexive processes of defecation and urination*. These processes involve a reflex motor stage, directed by the pelvic nerves, and a voluntary stage, controlled by the somatic motor system.
3. *The light and consensual light reflexes*, discussed in Chapter 14. (pp. 468–469)
4. *The reflex responses of swallowing, coughing, gagging, sneezing, and vomiting*. These reflexes are controlled by parasympathetic centers in the medulla, with motor commands issued over the glossopharyngeal and vagus nerves (N IX and X).
5. *Baroreceptor reflexes*. Several different baroreceptor reflexes detect changes in blood pressure and direct compensatory responses: increases or decreases in heart rate and contractile force, and constriction or dilation of blood vessels. The control centers are located in one of the cardiac centers described in Chapter 14. (p. 451) The parasympathetic component of these reflexes reduces heart rate and lowers blood pressure. The output is distributed by the vagus nerve.
6. *The reflex pattern of sexual arousal in response to erotic stimuli*. Erotic sights, sounds, smells, and touch sensations can trigger sexual arousal, effected primarily via parasympathetic motor commands traveling over the pelvic nerves.

Sympathetic reflexes include:

1. *Baroreceptor reflexes*: The sympathetic component of these reflexes increases heart rate and force of contraction. Output is carried over the sympathetic cardiac nerves; the response is assisted by the release of NE and E at the adrenal medulla.

2. *Vasomotor reflexes*: Vasomotor reflexes regulate the diameter of peripheral blood vessels and control regional blood flow. Receptors involved with these reflexes monitor blood pressure in the walls of major arteries, such as the carotid artery that carries blood to the brain. The vasomotor center of the medulla oblongata, described in Chapter 14, is the processing center for this reflex. (p. 451) The sympathetic motor commands are distributed to vessels of the skin and body wall by spinal nerves and to organs of the thoracic and abdominopelvic cavities by sympathetic nerves from the sympathetic chain and collateral ganglia.
3. *Pupillary dilation reflex*: This reflex, which dilates the pupils and allows more light into the eyes, is important in increasing visual sensitivity in poor light conditions. It is the functional opposite of the light reflex, although dual innervation is not involved. (Sympathetic stimulation causes contraction of a dilator muscle, whereas parasympathetic stimulation leads to contraction of a constrictor muscle.)
4. *Sexual reflexes, including ejaculation in the male*: These reflexes are triggered by various erotic stimuli.

This is only a partial listing, and many examples of autonomic reflexes involved in respiration, cardiovascular function, and other visceral activities will be examined in subsequent chapters. The parasympathetic division participates in reflexes affecting individual organs and systems. This reflects the relatively specific and restricted pattern of innervation: little divergence in general, and no multi-organ divergence. In contrast, there are fewer sympathetic reflexes. This division is typically activated as a whole, in part because of the degree of divergence and in part because the release of hormones by the adrenal medulla produces widespread peripheral effects.

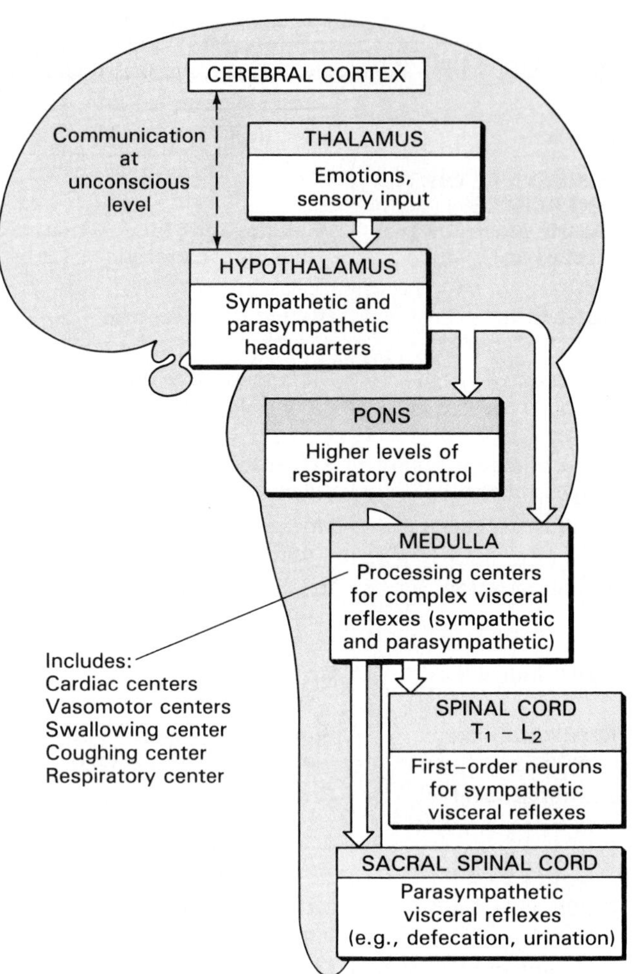

FIGURE 16-14
Levels of Autonomic Control

Higher Levels of Autonomic Control

The levels of autonomic tone in the sympathetic and parasympathetic nervous systems are controlled by centers in the brain stem concerned with specific visceral functions. Figure 16-14 diagrams the relationships involved. In addition to the cardiovascular and respiratory centers in the medulla oblongata, there are other centers and nuclei involved with respiration, digestive secretions, peristalsis, and urinary function. These, in turn, are subject to regulation by the hypothalamus. In general, centers in the posterior and lateral hypothalamus are concerned with the coordination and regulation of sympathetic function, and portions of the anterior and medial hypothalamus control the parasympathetic division.

The term *autonomic* was originally applied to the visceral motor system because it was thought that the regulatory centers functioned without regard for other CNS activities. This view has been drastically revised in light of subsequent research. Because the hypothalamus interacts with all other portions of the brain, activity in the limbic system (memories, emotional states), thalamus (sensory information), or cerebral cortex (conscious thought processes) can have dramatic effects on autonomic function. When you become angry, your heart rate accelerates, your blood pressure rises, your respiratory rate increases; when you remember your last big dinner your stomach "growls" and your mouth waters.

Although conscious thought processes have an effect on the autonomic nervous system, the normal individual does not perceive this because visceral sensory information does not reach the cerebral cortex. Even when a conscious mental process triggers a physiological shift, such as a change in blood pressure, sweat gland activity, skin temperature, or muscle tone, the information never arrives at the sensory cortex. **Biofeedback** is an attempt to bridge this gap. In this technique, a person's physiological processes are monitored and a visual or auditory signal is used to alert the subject when a particular change takes place. These signals let the individual know when ongoing conscious thought processes have triggered a desirable change in autonomic function. For example, when biofeedback is used to regulate blood pres-

Health News: Pharmacology and the Autonomic Nervous System

In 1960, the 5-year survival rate for patients surviving their first heart attack was very low, primarily because it was difficult and sometimes impossible to control high blood pressure. Thirty years later, the situation has changed radically; with treatment, the survivors of heart attacks often lead relatively normal lives. The major change that occurred over the past three decades was the development of drugs and procedures that can selectively target organ systems and receptors in the autonomic nervous system.

Many clinical conditions are characterized by symptoms either caused or aggravated by autonomic activities. The drugs administered to counteract or reduce such symptoms are called **mimetic** if they mimic the activity of one of the normal autonomic transmitters. Drugs that reduce the effects of autonomic stimulation by keeping the neurotransmitter from affecting the postsynaptic membranes are known as **blocking agents**. Table 16-6 relates important mimetic drugs and blocking agents to specific autonomic activities.

Mimetic drugs have a number of advantages over the neurotransmitters whose actions they simulate. For example, the sympathetic transmitters norepinephrine and epinephrine must be administered into the bloodstream by injection or infusion, for they do not survive absorption across the digestive tract and passage through the liver. Moreover, although these compounds have widespread effects, the duration of those effects is limited. These drawbacks can be avoided with drugs that mimic the effects of sympathetic stimulation. Such **sympathomimetic drugs** may survive oral administration, produce longer-lasting effects, or have more specific actions. For example, they may be applied topically to reduce hemorrhaging, by spray to reduce nasal congestion, by inhalation to dilate the respiratory passageways of asthmatics, or in drops to dilate the pupils before an eye examination. They are injected to elevate blood pressure and improve cardiac performance following severe blood loss or heart muscle failure. These drugs are used to treat a variety of disorders, and only representative examples will be considered here.

Phenylephrine stimulates alpha receptors, causing a constriction of peripheral vessels and elevating blood pressure. It is sometimes prescribed in cases of low blood pressure. Two other sympathomimetic drugs, *ephedrine* and *pseudoephedrine*, stimulate beta receptors. These drugs form the basis for several cold remedies (*Actifed®*, *Sudafed®*) that reduce nasal congestion and open respiratory passages through their effects on β_2 receptors. Undesirable side effects, such as jitteriness, anxiety, sleeplessness, and high blood pressure, result from the facilitation of CNS pathways and stimulation of peripheral alpha and β_1 receptors.

Sympathomimetic drugs that selectively target β_1 receptors, such as *dobutamine*, are especially valuable in increasing heart rate and blood pressure; they are often prescribed to improve the performance of a failing heart. Recently, sympathomimetic drugs targeting β_2 receptors, such as *albuteral* or *terbutaline*, have been developed to treat the bronchial constriction that accompanies an asthma attack.

Sympathetic blocking agents bind to the receptor sites and prevent a normal response to the presence of neurotransmitters or sympathomimetic drugs. The **alpha-blockers** eliminate the peripheral vasoconstriction that accompanies sympathetic stimulation. The alpha-blockers include *prazosin*, which selectively targets α_1 receptors and is used to reduce high blood pressure. For treating chronic high blood pressure and other forms of cardiovascular disease, **beta-blockers** are effective and clinically useful. In general, beta-blockers decrease heart rate and force of contraction, reducing the strain on the heart and simultaneously lowering peripheral blood pressure. *Propranolol* and *metoprolol* are two of the most popular beta-blockers currently on the market. Propranolol affects both β_1 and β_2 receptors, so patients receiving high doses may experience difficulties in breathing as their respiratory passageways constrict. Metoprolol targets β_1 receptors almost exclusively, leaving the respiratory smooth muscles relatively unaffected.

Parasympathomimetic drugs may also be used to increase the activity along the digestive tract and encourage defecation and urination. *Physostigmine* and *neostigmine* are important parasympathomimetic drugs that work by blocking the action of acetylcholinesterase. Because this enzyme is rendered inoperative, levels of ACh within the synapses climb, and parasympathetic activity is enhanced. Chapter 12 noted a blocking agent called *d-tubocurarine* that blocks neuromuscular transmission. ∞ (p. 394) The administration of physostigmine or neostigmine can counteract the paralytic effects of this drug.

Parasympathetic blocking agents such as *atropine* target the muscarinic receptors at the neuroeffector junctions. These drugs have diverse effects, but they are often used to control the diarrhea and cramps associated with various forms of food poisoning. The drug *Lomotil®*, known as the "traveler's friend," can provide temporary relief from diarrhea for the duration of a plane flight home. Among its other effects, atropine causes an elevation of the heart rate, due to a loss of parasympathetic tone. *Scopolamine* has similar effects on peripheral tissues, but has greater influence on the CNS. Its most useful effects are promoting drowsiness, reducing nausea, and relieving anxiety. As a result, scopolamine is often given when preparing a patient for surgery, prior to the administration of the anesthetic agent. (Scopolamine is also adminstered transdermally to control nausea, as discussed in Chapter 6.) (p. 187)

TABLE 16-6 Drugs and the ANS

Drug	*Mechanism*	*Action*	*Clinical Uses*
SYMPATHOMIMETIC			
Phenylephrine (Neosynephrine)	Stimulates α_1 receptors	Elevates blood pressure, stimulates smooth muscle	As a nasal decongestant and to elevate low blood pressure
Clonidine	Stimulates α_2 receptors	Lowers blood pressure	Treatment of high blood pressure
Isoproterenol	Stimulates β receptors	Stimulates heart rate, dilates respiratory passages	Treatment of respiratory disorders and as a cardiac stimulant during cardiac resuscitation
Albuteral, terbutaline	Stimulates β_2 receptors	Dilates respiratory passages	Treatment of asthma, severe allergies, and other respiratory disorders
Ephedrine	Stimulates NE release at neuroeffector junctions	Similar to effects of epinephrine	As a nasal decongestant, and to elevate blood pressure or dilate respiratory passageways
PARASYMPATHOMIMETIC			
Physostigmine, neostigmine	Blocks action of acetylcholinesterase	Increases ACh concentrations at parasympathetic neuroeffector junctions	Stimulates digestive tract and smooth muscles of urinary bladder
Pilocarpine	Stimulates muscarinic receptors	Similar to effects of ACh	Applied topically to cornea of eye to cause pupillary contraction
SYMPATHETIC BLOCKING AGENTS			
Prazosin (Minipress)	Blocks α_1 receptors	Lowers blood pressure	Treatment of high blood pressure
Propranolol (Inderal)	Blocks β, and β_2 receptors	Reduces metabolic activity in cardiac muscle but may constrict respiratory passageways	Treatment of high blood pressure: used to reduce heart rate and force of contraction in heart disease
Metoprolol	Blocks β_1 receptors	Reduces metabolic activity in cardiac muscle	Similar to those of Inderal but less effect on respiratory muscles
PARASYMPATHETIC BLOCKING AGENTS			
Atropine, related drugs	Blocks muscarinic receptors	Inhibits parasympathetic activity	Treating diarrhea; dilating pupils; raising heart rate; blocking secretions of digestive and respiratory tracts prior to surgery; used to treat accidental exposure to anticholinesterase drugs, such as pesticides or military nerve gases

sure, a light or tone informs the subject when blood pressure drops. With practice, some people can learn to recreate the proper mood or thought pattern that will lower their blood pressure.

Biofeedback techniques have been used to promote conscious control of blood pressure, heart rate, circulatory pattern, skin temperature, brain waves, and so forth. By reducing stress, lowering blood pressure, and improving circulation these techniques can reduce the severity of clinical symptoms and lower the risks for serious complications, such as heart attacks or strokes, in patients with high blood pressure. Unfortunately, not everyone can learn to influence autonomic functions, and the combination of variable results and expensive equipment make biofeedback unsuitable for widespread application.

- √ **What effect would stimulation of the muscarinic receptors in cardiac muscle have on the heart?**
- √ **What effect would loss of sympathetic tone have on blood flow to a tissue?**
- √ **What physiological changes would you expect to observe in a patient who is about to have a root canal done and who is quite anxious about the procedure?**

CHAPTER REVIEW

Review of Selected Clinical Terms

sympathomimetic drugs (*p. 519*) Drugs that mimic the effects of sympathetic stimulation.

sympathetic blocking agents (*p. 519*) Drugs that bind to receptor sites, preventing a normal response to neurotransmitters or sympathomimetic drugs.

alpha-blockers (*p. 519*) Drugs that eliminate the peripheral vasoconstriction that accompanies sympathetic stimulation.

beta-blockers (*p. 519*) Drugs that decrease heart rate and force of contraction, lowering peripheral blood pressure.

parasympathomimetic drugs (*p. 519*) Drugs that mimic parasympathetic stimulation and increase the activity along the digestive tract.

parasympathetic blocking agents (*p. 519*) Drugs that target the muscarinic receptors at neuroeffector junctions.

Additional Terms of Clinical Importance

PHARMACOLOGY AND THE AUTONOMIC NERVOUS SYSTEM (*p. 519*): **mimetic, blocking agents**

Study Outline

Related Key Terms

Introduction (pp. 499–500)

1. The autonomic nervous system (ANS) coordinates cardiovascular, respiratory, digestive, excretory, and reproductive functions.

An Overview of the ANS (pp. 500–501)

NEURAL INTERACTIONS THAT DIRECT MOTOR OUTPUT (p. 500)

1. **First-order (preganglionic) neurons** in the CNS send axons to synapse on **second-order (ganglionic) neurons** in **autonomic ganglia** outside the CNS. — **postganglionic fibers**

SUBDIVISIONS OF THE ANS (p. 501)

2. Visceral efferents from the thoracic and lumbar segments form the **thoracolumbar**, or **sympathetic**, **division** ("fight or flight" system) of the ANS. Visceral efferents leaving the brain and sacral segments form the **craniosacral**, or **parasympathetic**, **division** ("rest and repose" system). — **intramural ganglia**

The Sympathetic Division (pp. 501–508)

1. The sympathetic division consists of preganglionic (first-order) neurons (between segments T_1 and L_2), ganglionic (second-order) neurons (in ganglia near the vertebral column), and specialized second-order neurons (inside the adrenal gland). There are two types of sympathetic ganglia: **sympathetic chain ganglia** (paravertebral ganglia) and **collateral ganglia** (prevertebral ganglia).

PATTERNS OF SYMPATHETIC INNERVATION (pp. 502–505)

2. Postganglionic fibers targeting thoracic cavity structures form **autonomic nerves** that go directly to their visceral destination. Preganglionic fibers running between the sympathetic chain ganglia interconnect them to form an elongated sympathetic chain.
3. The abdominopelvic viscera receive sympathetic innervation via preganglionic fibers that synapse within collateral ganglia. The preganglionic fibers that innervate the collateral ganglia form the **splanchnic nerves**.
4. The **celiac ganglion** innervates the stomach, liver, pancreas, and spleen; the **superior mesenteric ganglion** innervates the small intestine and initial segments of the large intestine; and the **inferior mesenteric ganglion** innervates the kidney, bladder, sex organs, and terminal portions of the large intestine.
5. Preganglionic fibers entering the adrenal gland synapse within the **adrenal medulla**.

CENTRAL EFFECTS OF SYMPATHETIC STIMULATION (p. 505)

6. In a crisis, the entire division responds, an event called **sympathetic activation**. Its effects include: increased alertness, a feeling

of energy and euphoria, increased cardiovascular and respiratory activity, and general elevation in muscle tone.

SYMPATHETIC ACTIVATION AND NEUROTRANSMITTER RELEASE (pp. 505–506)

7. Stimulation of the sympathetic division has two distinctive results: the release of norepinephrine at specific locations, and secretion of epinephrine and norepinephrine into the general circulation.

MEMBRANE RECEPTORS AND SYMPATHETIC FUNCTION (pp. 506–508)

8. There are two types of sympathetic receptors: **alpha receptors** (which respond to norepinephrine or epinephrine by depolarizing the membrane) and **beta receptors** (which are particularly sensitive to epinephrine).
9. Most postganglionic fibers are adrenergic, but a few are cholinergic. Postganglionic fibers innervating sweat glands of the skin and blood vessels to skeletal muscles release ACh.

alpha-1 (α_1) receptors
alpha-2 (α_2) receptors
beta-1 (β_1) receptors
beta-2 (β_2) receptors

A SUMMARYOF THE SYMPATHETIC DIVISION (p. 508)

10. Preganglionic sympathetic fibers are relatively short. Except for those of the adrenal medulla, postganglionic fibers are quite long. Extensive divergence typically occurs, with a single preganglionic fiber synapsing with many ganglionic neurons in different ganglia. A single effector cell may have more than one type of receptor, so that the result of sympathetic stimulation depends on interaction between the receptors.

The Parasympathetic Division (pp. 509–512)

1. The parasympathetic division includes preganglionic (first-order) neurons in the brain stem and sacral segments of the spinal cord, and ganglionic (second-order) neurons in peripheral ganglia located within or next to target organs.

ORGANIZATION OF THE PARASYMPATHETIC DIVISION (p. 510)

2. Preganglionic fibers leaving the sacral segments form **pelvic nerves**.

ciliary ganglion
sphenopalatine ganglion
submandibular ganglion
otic ganglion

GENERAL FUNCTIONS OF THE PARASYMPATHETIC DIVISION (p. 510)

3. The effects produced by the parasympathetic division center on relaxation, food processing, and energy absorption.

PARASYMPATHETIC ACTIVATION AND NEUROTRANSMITTER RELEASE (pp. 510–511)

4. All the parasympathetic preganglionic and postganglionic fibers release ACh at synapses and neuroeffector junctions. The effects are short-lived due to the actions of **tissue cholinesterase**.

RECEPTOR DISTRIBUTION AND EFFECTOR RESPONSES (pp. 511–512)

5. Two different ACh receptors are found in postsynaptic membranes. Stimulation of **muscarinic receptors** produces a longer-lasting effect than does stimulation of **nicotinic receptors**.

A SUMMARY OF THE PARASYMPATHETIC DIVISION (p. 512)

6. The parasympathetic division innervates areas serviced by cranial nerves and organs in the thoracic and abdominopelvic cavities. All preganglionic and postganglionic parasympathetic neurons are cholinergic, and the effects of stimulation are usually brief and restricted to specific sites.

Anatomical and Functional Relationships between the Sympathetic and Parasympathetic Divisions (pp. 512–521)

1. The sympathetic division has widespread impact, reaching visceral and somatic structures throughout the body.
2. The parasympathetic division only innervates visceral structures serviced by cranial nerves or lying within the abdominopelvic cavity. Organs with **dual innervation** receive instructions from both divisions.

ANATOMY OF DUAL INNERVATION (pp. 512–514)

3. In body cavities the parasympathetic and sympathetic nerves intermingle to form a series of characteristic nerve plexuses (nerve networks), which include the **cardiac**, **pulmonary**, **celiac**, and **hypogastric plexuses**.

AUTONOMIC TONE (pp. 514–515)

4. Even when stimuli are absent, autonomic motor neurons show a resting level of activation, the **autonomic tone**.

AUTONOMIC PATTERNS OF ACTIVITY (pp. 515–516)

5. The overall pattern of parasympathetic functions is: decreased metabolic rate, decreased heart rate and blood pressure, salivary and

digestive gland secretion, increased digestive tract activity, and urination and defecation.

6. The general pattern of sympathetic functions is: mental alertness, increased metabolic rate, suspended digestive and urinary tract function, activation of energy reserves, increased respiratory rate and efficiency, increased heart rate and blood pressure, and activated sweat glands.

INTEGRATION AND CONTROL OF AUTONOMIC FUNCTIONS (pp. 516–521)

7. **Visceral reflexes** are the simplest functional units in the ANS. Some parasympathetic reflexes are: (1) reflexes that coordinate digestive activities; (2) defecation and urination; (3) light and consensual light reflexes; (4) swallowing, coughing, gagging, sneezing, and vomiting; (5) baroreceptor reflexes; and (6) sexual arousal in response to erotic stimuli. Some sympathetic reflexes include: (1) baroreceptor reflexes; (2) vasomotor reflexes; (3) pupillary dilation reflex; and (4) sexual reflexes.
8. In general, centers in the posterior and lateral hypothalamus are concerned with the coordination and regulation of sympathetic function, and portions of the anterior and medial hypothalamus control the parasympathetic division. **Biofeedback** techniques can sometimes help to shift autonomic functions in a desired direction.

Review Planner

	Objective	Level 1
1	Compare the autonomic nervous system with the other divisions of the nervous system.	**1 27**
2	Explain the functions and structures of the sympathetic and parasympathetic divisions.	**1 2 8 10 11 12–26**
3	Discuss the mechanisms of neurotransmitter release in the autonomic nervous system.	**3 11**
4	Compare the effects of autonomic neurotransmitters on target organs and tissues.	**4 5 11 14 15 29**
5	Discuss the relationship between the sympathetic and parasympathetic divisions and explain the implications of dual innervation.	**5 7 30**
6	Describe the levels of integration and control of the autonomic system.	**1 20 32**
7	Explain the importance of autonomic tone.	**6 9 31**

Level 2	**28 33**
Level 3	**34–36**
CM	**Autonomic Tone and Hypersensitivity**

Review Questions

MULTIPLE CHOICE

1. The simplest functional units of the autonomic nervous system are: a) motor neurons; b) first-order neurons; c) visceral reflexes; d) neurotransmitters.

2. Which of the following is *not* a result of the functions of the sympathetic nervous system?: a) increased metabolic rate; b) faster breathing; c) increased hunger; d) increased blood pressure.

3. How many neurotransmitters are used in the parasympathetic nervous system?: a) one; b) two; c) three; d) the number depends on the particular stimulus involved.
4. Stimulation of ________ receptors can constrict peripheral blood vessels and close sphincters along the digestive tract: a) alpha-1; b) alpha-2; c) beta-1; d) beta-2.
5. The effects of the parasympathetic division most often ________ those of the sympathetic division. a) reinforce; b) substitute for; c) oppose; d) are unrelated to.
6. Autonomic tone exists because: a) ANS neurons from both divisions often innervate the same organs; b) ANS neurons rarely innervate the same organs as SNS neurons; c) ANS neurons are inactive unless stimulated by higher centers; d) ANS neurons are always active to some degree.
7. Organs with dual innervation receive input from both: a) alpha and beta receptors; b) sympathetic and parasympathetic divisions; c) autonomic and visceral nervous systems; d) muscarinic and nicotinic receptors.
8. You would expect the reflexive processes of defecation and urination to be controlled by the: a) sympathetic division; b) parasympathetic division; c) both the sympathetic and parasympathetic divisions.
9. Autonomic tone is *most* important when: a) an organ is innervated by only one division of the ANS; b) an organ is innervated by both divisions of the ANS; c) an organ is innervated only by the SNS; d) an organ is innervated only by preganglionic neurons.
10. You would expect a reflex that dilates the pupils, allowing more light to enter the eyes, to be controlled by the: a) sympathetic division; b) parasympathetic division; c) both the sympathetic and parasympathetic divisions.

DIAGRAMS

11. Complete the following table:

Characteristic	*Sympathetic Division*	*Parasympathetic Division*
Location of CNS visceral motor neuron		
Location of PNS ganglia		
Preganglionic fibers		
Length		
Neurotransmitter released		
Postganglionic fibers		
Length		
Neurotransmitter released		
Neuroeffector junction		
Degree of divergence from CNS to ganglion cells		
General function		

MATCHING QUESTIONS

(Identify each term or statement with a P for parasympathetic or an S for sympathetic.)

12. Paravertebral ganglia.
13. Pelvic nerves.
14. Stimulation exerts relatively long-lasting effects.
15. Stimulation exerts relatively specific, localized effects.
16. Alpha receptors.
17. Muscarinic receptors.
18. Innervates areas serviced by the cranial nerves and organs in the thoracic and abdominopelvic cavities.
19. Innervates visceral and somatic structures throughout the body.
20. Second-order neurons in the adrenal medulla.
21. Splanchnic nerves.
22. Celiac ganglion.
23. Nicotinic receptors.
24. Beta receptors.
25. The presence of both cholinergic and adrenergic synapses.
26. The presence of cholinergic synapses only.

SHORT ANSWER/ESSAY

27. Is the term "autonomic nervous system" accurate? Why or why not?
28. You're studying for an important exam. Which do you think would be more sensible: to study before or after dinner? Explain your answer.
29. Tissue concentrations of these neurotransmitters can remain elevated for as long as 30 seconds after they are released. Identify the neurotransmitters and explain why this occurs.
30. Why is it beneficial that dual innervation often exerts opposing effects? Give an example.
31. Explain the importance of autonomic tone.
32. Can activity in the cerebral cortex alter autonomic function? Explain your answer.
33. How does the ANS stimulate or inhibit the activities of organs? What determines which type of effect it has?

CRITICAL THINKING/APPLICATIONS

34. You're waiting to go on stage to make a speech before a large audience. You feel your heart begin to pound and you're breathing faster. Explain what is happening, and identify the autonomic division, neurotransmitter(s), and receptor(s) involved.
35. Explain why mental or emotional stress could affect one's physical health.
36. Mr. M., 65 years old, is recovering from his first heart attack. His physician has prescribed a beta-blocker as part of his ongoing treatment plan. Explain why beta-blockers could be beneficial to this patient.

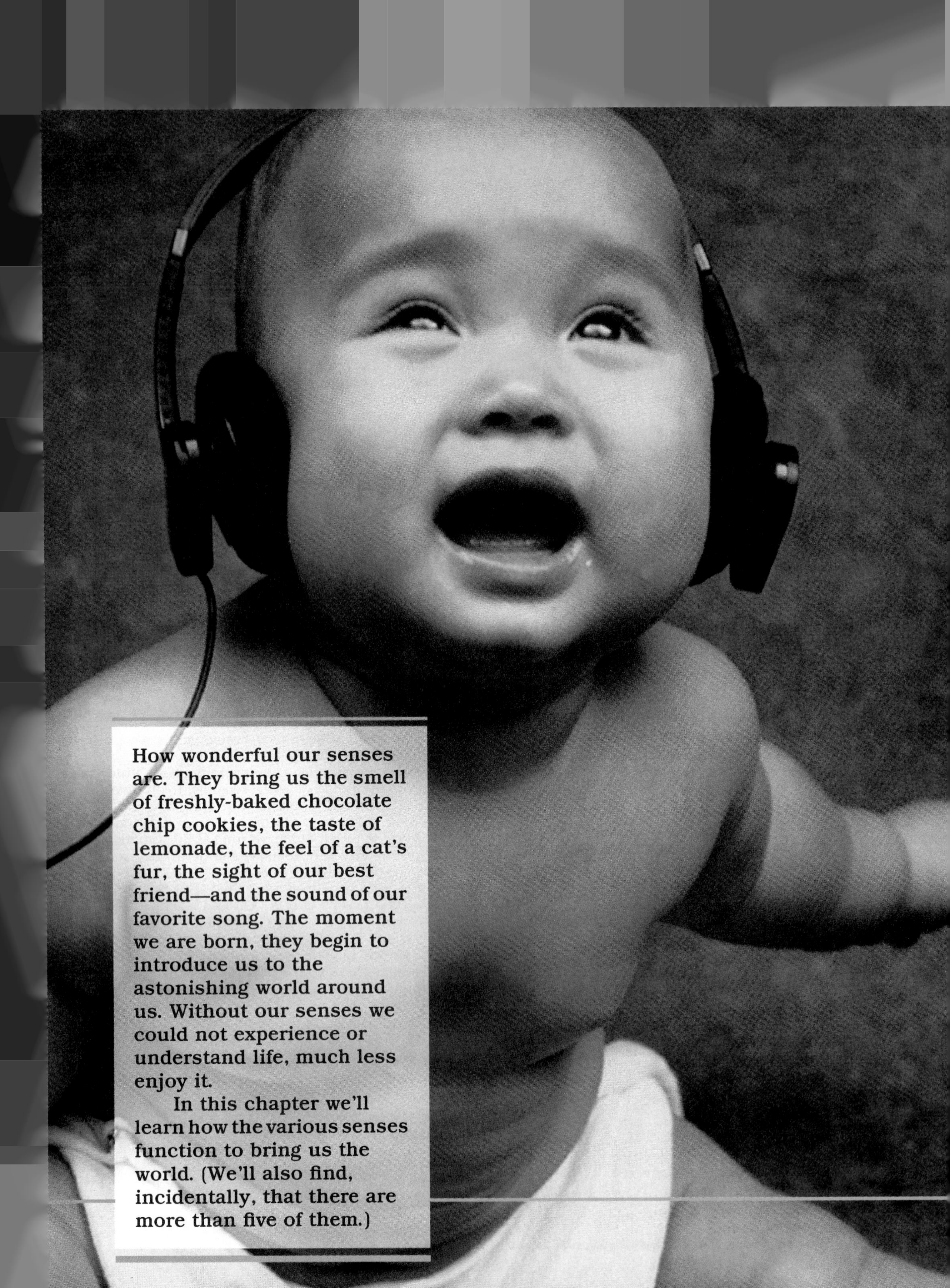

How wonderful our senses are. They bring us the smell of freshly-baked chocolate chip cookies, the taste of lemonade, the feel of a cat's fur, the sight of our best friend—and the sound of our favorite song. The moment we are born, they begin to introduce us to the astonishing world around us. Without our senses we could not experience or understand life, much less enjoy it.

In this chapter we'll learn how the various senses function to bring us the world. (We'll also find, incidentally, that there are more than five of them.)

CHAPTER 17

Sensory Function

Chapter Objectives

After reading this chapter, you will be able to:

1. Distinguish between the general and special senses.
2. Identify the receptors for the general senses and describe how they function.
3. Explain why receptors respond to specific stimuli and how the organization of a receptor affects its sensitivity.
4. Discuss the receptors and processes involved in the senses of smell, taste, and equilibrium.
5. Describe the parts of the ear and their roles in the process of hearing.
6. Identify the parts of the eye and their functions.
7. Explain how we are able to see objects and distinguish colors and depth.
8. Discuss how the central nervous system processes information related to all the senses.

■ Introduction

Every cell membrane is a receptor that responds to changes in the extracellular environment. For example, the appearance of epinephrine triggers widespread changes in the metabolic activities of peripheral tissues. These adjustments are quite different from those produced by a different stimulus, such as a drop in temperature or a change in sodium or potassium ion concentrations. Due to small differences in the structure of their cell membranes, an environmental stimulus significant to one cell may leave another totally unaffected. The body contains billions of cells, each with specific sensitivities. What excites a kidney cell may be ignored by an osteocyte and "turn off" a muscle cell.

The cells of the body are organized into tissues, organs, and organ systems with well-defined functions. The nervous system provides centralized control and coordination, directing relatively swift responses to peripheral stimuli. Sensory receptors represent the interface between the nervous system and the internal and external environments. A

sensory receptor is a specialized cell that monitors conditions in the body or the external environment. When stimulated, a receptor passes information to the central nervous system. This information, called a **sensation**, arrives in the form of action potentials in an afferent (sensory) fiber. Sensations processed at the unconscious and conscious levels result in motor responses that preserve homeostasis.

The term **general senses** refers to sensations of temperature, pain, touch, pressure, vibration, and proprioception (body position). General sensory receptors are distributed throughout the body, and they are relatively simple in structure. Some of the information they send to the CNS reaches the cerebral cortex and our conscious awareness. These sensations arrive at the primary sensory cortex, or *somatosensory cortex*, via pathways described in Chapters 14 and 15. (pp. 437, 472–475)

The **special senses** are smell (**olfaction**), taste (**gustation**), balance (**equilibrium**), hearing, and vision. These sensations are provided by specialized receptor cells that are structurally more complex than those of the general senses. These receptors are found in sensory areas protected by surrounding tissues; in the nose, eye, and ear they occupy a central position within complex **sense organs**. The information provided by these receptors is distributed to specific areas of the cerebral cortex (the auditory cortex, the visual cortex, and so forth) and to centers throughout the brain stem.

Because the CNS receives all of this information indirectly, our knowledge of the environment is limited to those characteristics that stimulate our peripheral receptors. Colors invisible to us guide insects to flowers; sounds we cannot hear and smells we cannot detect provide dogs, cats, and dolphins with important information about their surroundings. What we *do* perceive varies considerably with the state of our nervous system and the other activities that it may be conducting. For example, when sympathetic activation stimulates the reticular activating system, we experience heightened awareness of sensory information and hear sounds that would normally escape our notice. Yet when concentrating on a difficult problem, we may remain unaware of even relatively loud and unfamiliar noises. Finally, it should be realized that our perception of any stimulus is a neural event and not necessarily an environmental reality. In phantom limb pain (Chapter 15) a person "feels" pain in a missing limb; during an epileptic seizure the individual often experiences sights, sounds, or smells that originate in the cerebral cortex, not in the outside world. (p. 475) [**CM:** *Epilepsy*]

This chapter begins by examining receptor function and basic concepts in sensory processing. We will then apply this information to each of the general and special senses.

Receptor Physiology and Sensory Coding

Each receptor has a characteristic sensitivity. For example, a touch receptor is very sensitive to pressure but relatively insensitive to chemical stimuli; a taste receptor is sensitive to dissolved chemicals, but insensitive to pressure. This concept is called **receptor specificity**. Specificity may result from the structure of the receptor cell itself or from the presence of accessory cells or structures that shield it from other stimuli. The simplest receptors are the dendrites of sensory neurons. The dendritic processes, called **free nerve endings**, are not protected by accessory structures. They can be stimulated by many different stimuli (Figure 17-1a). For example, free nerve endings that provide the sensation of pain may respond to chemical stimulation, pressure, temperature changes, or physical damage. Complex receptors, such as the visual receptors of the eye, are protected by accessory cells and layers of connective tissue. These cells are seldom exposed to any stimulus other than light, and so provide very specific information.

The area monitored by a single receptor cell is its **receptive field** (Figure 17-1b). Whenever a stimulus arrives in the receptive field, the CNS receives the information "stimulus arriving at receptor X." The larger the receptive field, the poorer our ability to localize a stimulus. For example, a touch receptor on the general body surface may have a receptive field 7 cm (2.5 in.) in diameter. As a result, a light touch can be described only as affecting a general area, not an exact spot. On the tongue, where the receptive fields are less than a millimeter in diameter, we can be very precise about the location of a stimulus.

An arriving stimulus can take many different forms; it may be a physical force, such as pressure, a dissolved chemical, a sound, or a beam of light. Regardless of the nature of the stimulus, however, sensory information must be sent to the CNS in the form of action potentials, which are electrical events. The process of translating a stimulus into an action potential is called **transduction**.

THE TRANSDUCTION PROCESS

Transduction can be divided into three steps:

Step 1: An arriving stimulus alters the permeability of the receptor membrane.

The nature of the interaction between stimulus and receptor determines the receptor's specificity. For example, a receptor membrane containing chemically

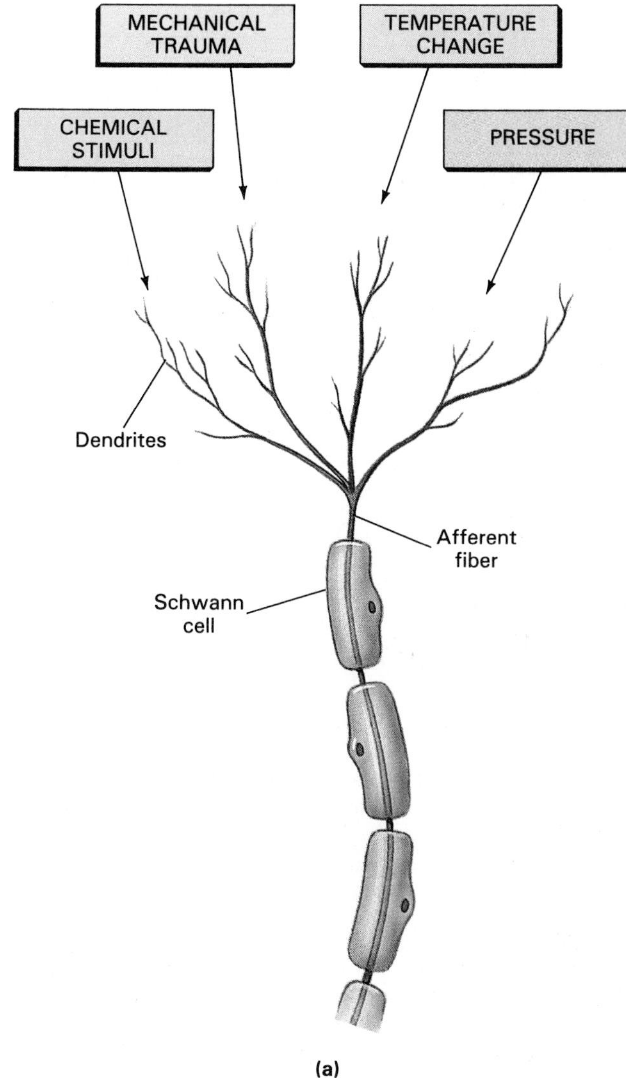

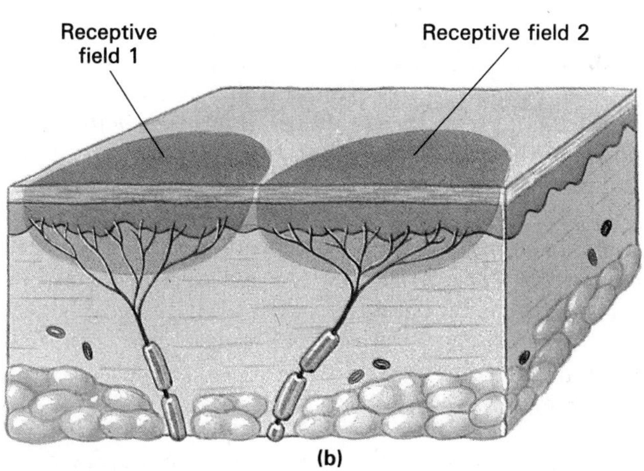

FIGURE 17-1
Receptors and Receptive Fields. (a) A free nerve ending consists of sensory dendrites that may be stimulated by a variety of stimuli. (b) Each receptor monitors a specific area known as the receptive field.

regulated ion channels will respond to the presence of specific dissolved molecules. Regardless of the nature of the interaction, however, the result is always the same: ion channels in the receptor membrane open or close, and the transmembrane potential changes. The change in the transmembrane potential that accompanies receptor stimulation is called a **receptor potential**. The receptor potential is usually a depolarization. It is a graded potential change: the stronger the stimulus the larger the receptor potential.

Step 2: The receptor potential produces a generator potential.

Receptor membranes cannot conduct action potentials. To send information to the CNS, the receptor potential must affect the excitable membrane of a sensory neuron. A membrane depolarization that leads to an action potential in an excitable membrane is called a **generator potential**.

The receptors for the general senses are the dendrites of sensory neurons. In these receptors, therefore, the receptor potential and the generator potential occur in the same cell. As indicated in Figure 17-2a, the dendrites are connected to the initial segment of the axon. An action potential will appear when this segment depolarizes to threshold. In some cases a single, large receptor potential may be sufficient; in other cases, temporal or spatial summation may be required.

Sensations of taste, hearing, equilibrium, and vision are provided by specialized receptor cells that communicate with sensory neurons by way of chemical synapses. The receptor cells develop graded receptor potentials in response to stimulation, and the change in membrane potential alters the rate of neurotransmitter release at the synapse. Figure 17-2b details receptor activity for receptor cells providing taste and hearing sensations. When a receptor potential appears, these cells release neurotransmitters that depolarize the membrane of the postsynaptic sensory neuron. The larger the receptor potential, the more transmitter molecules are released.

Step 3: An action potential travels to the CNS over an afferent fiber.

The arriving information is then processed and interpreted by the CNS at the conscious and unconscious levels.

INTERPRETATION OF SENSORY INFORMATION

When sensory information arrives at the CNS it is routed according to the location and nature of the stimulus. Previous chapters have emphasized the fact that axons in the CNS are organized in bundles with

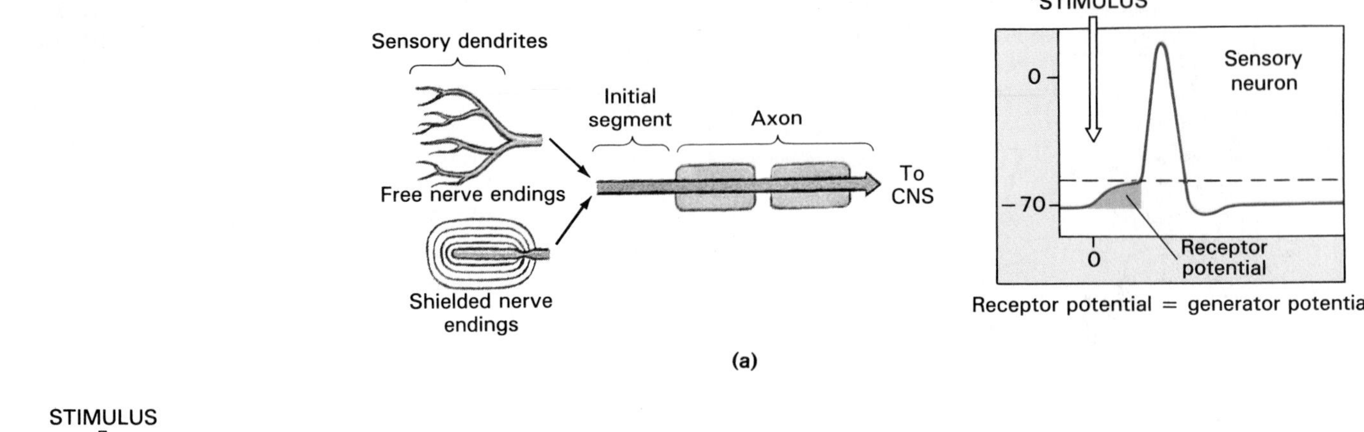

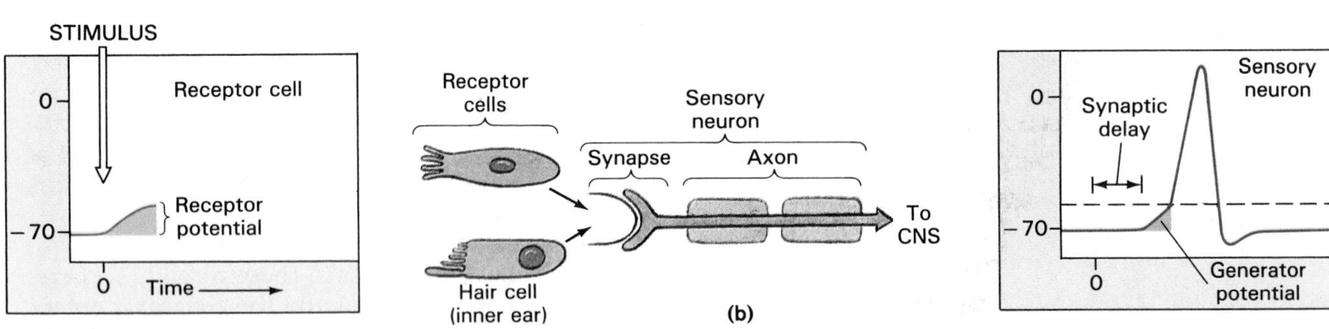

FIGURE 17-2
Receptor and Generator Potentials. (a) When the sensory neuron acts as the receptor, a stimulus depolarizes the dendrites and brings the initial segment of the axon to threshold. The receptor and the neuron are the same cell, and the receptor potential acts as a generator potential. (b) In the special senses of taste, equilibrium, hearing, and vision the receptor cells are specialized cells that communicate with neurons across chemical synapses. The receptor cell shows a graded response to stimulation. In this example stimulation produces a depolarization that accelerates neurotransmitter release; the neurotransmitter produces a generator potential in the postsynaptic membrane.

specific origins and destinations. Along sensory pathways, axons relay information from point A (the receptor) to point B (a sensory neuron at a specific site in the cerebral cortex). For example, touch, pressure, pain, temperature, and taste sensations arrive at the primary sensory cortex; visual, auditory, and olfactory information reach the visual, auditory, and olfactory regions of the cortex. The neural link between receptor and cortical neuron is called a **labeled line.** Each labeled line carries information concerning a specific sensation (touch, pressure, vision, and so forth) from receptors in a specific part of the body.

Because the CNS interprets sensory information entirely on the basis of the line over which it arrives, it cannot tell the difference between a "true" sensation and a "false" one generated somewhere along the line. For example, when rubbing the eyes, one often "sees" flashes of light. Although the stimulus is mechanical, rather than visual, any activity along the optic nerve is projected to the visual cortex and experienced as a visual perception.

The identity of the active labeled line indicates the location and nature of the stimulus. *All other characteristics of the stimulus are conveyed by the pattern of action potentials in the afferent fibers.* This **sensory coding** provides information about the strength, duration, variation, and movement of the stimulus.

Some sensory neurons, called **tonic receptors**, are always active. The frequency with which they generate action potentials indicates the background level of stimulation. When the stimulus increases or decreases, their rate of action potential generation changes accordingly. Other receptors are normally inactive, but become active for a short time whenever there is a change in the conditions they are monitoring. These **phasic receptors** provide information on the intensity and rate of change of a stimulus by altering the frequency and rate of change of the action potentials they generate. Receptors that combine phasic and tonic coding convey extremely complicated sensory information.

CENTRAL PROCESSING AND ADAPTATION

Adaptation is a reduction in sensitivity in the presence of a constant stimulus. **Peripheral**, or **sensory**, **adaptation** occurs when the receptors or sensory neurons alter their levels of activity. The receptor responds strongly at first, but thereafter the activity along the afferent fiber gradually declines, in part due to synaptic fatigue. This response is characteristic of phasic receptors, which are also called **fast-adapting receptors**. Tonic receptors show little peripheral adaptation, and so are called **slow-adapting receptors**.

Adaptation also occurs inside the CNS along the sensory pathways. For example, a few seconds after exposure to a new smell, conscious awareness of the stimulus virtually disappears, although the sensory neurons are still quite active. This process is known as **central adaptation**. Central adaptation usually involves the inhibition of nuclei along a sensory pathway.

Peripheral adaptation reduces the amount of information reaching the CNS. Central adaptation at the unconscious level further restricts the amount of detail arriving at the cerebral cortex. Most of the incoming sensory information is processed in centers along the spinal cord or brain stem, potentially triggering involuntary reflexes, such as the withdrawal reflex. Only about 1 percent of the information provided by afferent fibers reaches the cerebral cortex and our conscious awareness.

However, the output from higher centers can increase receptor sensitivity or facilitate transmission along a sensory pathway. The reticular activating system in the midbrain helps focus attention, and thus heightens or reduces awareness of arriving sensations. This adjustment of sensitivity can occur under conscious or unconscious direction. When we "listen carefully" our sensitivity and awareness of auditory stimuli increase. The reverse occurs when we enter a noisy factory or walk along a crowded city street, as we automatically "tune out" the high level of background noise.

SENSORY LIMITATIONS

Our sensory receptors provide a detailed picture of our bodies and our surroundings. Yet that picture is incomplete for several reasons.

First, humans do not have receptors for every possible stimulus. If a stimulus does not affect one or more receptors, as far as the CNS is concerned it does not exist. For example, humans cannot detect infrared radiation, although many snakes use infrared receptors to home in on the body heat of their prey.

Second, our receptors have characteristic ranges of sensitivity. We can neither hear high frequency sounds that dolphins respond to nor detect smells that excite a bloodhound.

Third, the transduction process converts a real stimulus into a neural event that must be interpreted by the CNS. Our perception of a particular stimulus is an interpretation, not an environmental reality. Abnormal receptor function or inappropriate stimulation can produce sensations that have no basis in fact. One example, seeing "stars" after pressure is applied to the eye, was already noted. Perception can also be altered by facilitation or inhibition along the sensory pathways. For example, the release of endorphins along pain pathways can inhibit transmission of painful sensations. As a result, painful stimuli can be "ignored" because the impulses do not reach the primary sensory cortex. Facilitation of sensory pathways can produce heightened sensitivity to specific stimuli. It can also lead to unusual perceptions that do not match environmental realities. For example, descending facilitation of thermoreceptors occurs when someone anticipates touching a hot stove. On contact the facilitation may become so intense that the person may "feel" intense heat even if the surface of the stove is freezing cold.

This discussion has introduced basic concepts of receptor function and sensory processing. We will next consider how these concepts apply to the general senses.

The General Senses

Receptors for the general senses are scattered throughout the body. They are relatively simple in structure. These receptors are classified according to the nature of the stimulus that excites them:

- **Nociceptors** (nō-sē-SEP-tōrz; *noceo*, hurt) respond to a variety of stimuli usually associated with tissue damage. Receptor activation causes the sensation of pain.
- **Thermoreceptors** respond to changes in temperature.
- **Mechanoreceptors** are stimulated or inhibited by physical distortion, contact, or pressure on their cell membranes. There are many different types of mechanoreceptors.
- **Chemoreceptors** monitor the chemical composition of body fluids and respond to the presence of specific molecules.

Each class of receptors has distinct structural and functional characteristics.

NOCICEPTION

Pain receptors, or nociceptors, are especially common in the superficial portions of the skin, in joint capsules, within the periostea of bones, and around the walls of blood vessels. There are few nociceptors in other deep tissues or in most visceral organs. Pain receptors are free nerve endings with large receptive fields. As a consequence it is often difficult to determine the exact origin of a painful sensation.

There are several different populations of nociceptors. Some are most sensitive to extremes of temperature, others to mechanical damage, and a third group to dissolved chemicals. However, very strong stimuli will excite all three receptor types. Stimulation causes depolarization, and when the initial segment of the axon reaches threshold, an action potential heads toward the CNS.

Two types of axons carry painful sensations. Myelinated Type A fibers carry sensations of **fast**, or **prickling, pain**. An injection or a deep cut produces this type of pain. These sensations reach the CNS very quickly, where they often trigger somatic reflexes. They are also relayed to the primary sensory cortex, and so receive conscious attention. The arriving information usually permits localization of the stimulus to an area several inches in diameter.

Slower Type C fibers carry sensations of **slow**, or **burning and aching, pain**. These sensations cause a generalized activation of the reticular formation and thalamus. The individual becomes aware of the pain, but has only a general idea of the area affected.

When deep or visceral structures are involved the location can be very difficult to determine. Pain sensations from visceral organs are often perceived as originating in more superficial regions innervated by the same spinal nerves. The precise mechanism responsible for this **referred pain** remains to be determined, but several clinical examples are given in Figure 17-3. Cardiac pain, for example, is often perceived as originating in the upper chest and left arm.

Pain receptors exhibit a tonic response to stimulation. Significant peripheral adaptation does not occur: the receptors continue to respond as long as the painful stimulus remains. Painful sensations cease only after tissue damage has ended. However, central adaptation may reduce *perception* of the pain while the pain receptors are still stimulated. This effect involves the inhibition of centers in the thalamus, reticular formation, lower brain stem, and spinal cord. Endorphin release has been proposed as an important mechanism for the inhibition of pain centers in the brain. These centers use the neuropeptide *substance P* as a neurotransmitter. Endorphins bind to the presynaptic membrane and prevent the release of substance P. [**CM**: *The Control of Pain*]

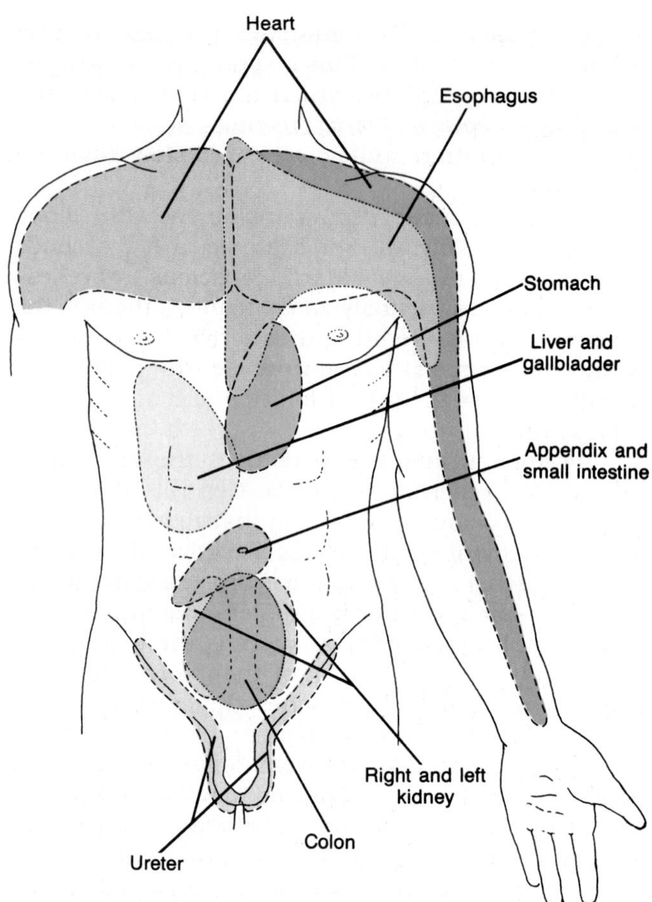

FIGURE 17-3
Referred Pain. Pain sensations originating in visceral organs are often perceived as involving specific regions of the body surface innervated by the same spinal nerves.

THERMORECEPTION

Temperature receptors are scattered immediately beneath the surface of the skin. They are also found in skeletal muscles, in the liver, and in the hypothalamus. There seem to be several distinct populations of thermoreceptors, and their abundance varies from place to place across the body surface. In general, cold receptors are three or four times more numerous than warm receptors. The receptors are free nerve endings, and there are no known structural differences between warm and cold thermoreceptors.

Thermal sensations are conducted along the same pathways that carry pain sensations. They are sent to the reticular formation, the thalamus, and (to a lesser extent) the primary sensory cortex. Thermoreceptors are phasic receptors. They are very active when the temperature is changing, but they quickly adapt to a stable temperature. When we enter an air-conditioned classroom on a hot summer day or a toasty lecture hall on a brisk fall evening, the temperature seems unpleasant at first, but the discomfort fades as adaptation occurs.

MECHANORECEPTION

Mechanoreceptors are sensitive to stimuli that distort their cell membranes. These membranes contain **mechanically regulated ion channels** whose gates open or close in response to stretching, compression, twisting, or other distortions of the membrane. There are three classes of mechanoreceptors:

1. **Tactile receptors** provide sensations of touch, pressure, and vibration.

2. **Baroreceptors** (bar-ō-rē-SEP-tōrz; *baro-*, pressure) detect pressure changes in the walls of blood vessels and in portions of the digestive, reproductive, and urinary tracts.
3. **Proprioceptors** monitor the positions of joints and muscles. The proprioceptors are the most structurally and functionally complex of the general sensory receptors.

Tactile Receptors

The distinctions between sensations of touch, pressure, and vibration are hazy, for a touch also represents a pressure, and a vibration consists of an oscillating touch/pressure stimulus. **Fine touch and pressure receptors** provide detailed information about a source of stimulation, including its exact location, shape, size, texture, and movement. These receptors are extremely sensitive and have relatively narrow receptive fields. **Crude touch and pressure receptors** provide poor localization and little additional information about the stimulus.

Tactile receptors range in complexity from free nerve endings to specialized sensory complexes with accessory cells and supporting structures. Figure 17-4 shows six different types of tactile receptors in the skin. Free nerve endings between adjacent epidermal cells are sensitive to touch and pressure; there are no apparent structural differences between these receptors and the free nerve endings that provide temperature or pain sensations. These are the only sensory receptors found on the corneal surface of the eye, but in other portions of the body surface more specialized tactile receptors are probably more important.

Free nerve endings associated with hair follicles provide more specific information. Wherever hairs are found, the free nerve endings of the **root hair plexus** monitor distortions and movements across the body surface. A network of dendritic branches surrounds the dermal base of each hair follicle and penetrates its epithelial lining. When the hair is displaced, the movement of the follicle distorts the sensory dendrites and produces action potentials. These receptors adapt rapidly, so they are best at detecting initial contact and subsequent movements. For example, most people feel their clothing only when they move or when they consciously focus attention on their skin.

Two other receptors provide fine touch and pressure information. **Merkel's** (MER-kelz) **discs** are tonically active, extremely sensitive, and have narrow receptive fields. The dendritic processes of a single myelinated afferent fiber make close contact with unusually large epithelial cells in the stratum germinativum of the skin. (These *Merkel cells* were described in Chapter 6.) (p. 190) **Meissner's** (MĪS-nerz) **corpuscles** are also found where tactile sensitivities are extremely well developed. They are especially common at the eyelids, lips, fingertips, nipples, and external genitalia. The dendrites are highly coiled and interwoven, and they are surrounded by modified Schwann cells. A fibrous capsule surrounds the entire complex and anchors it within the dermis. Meissner's corpuscles are fairly large structures, measuring roughly 100 μm in length and 50 μm in width. They detect light touch, movement, and vibration; they adapt to stimulation within a second after contact.

Pacinian (pa-SIN-ē-an) **corpuscles** are considerably larger encapsulated receptors. The dendritic process lies within a series of concentric cellular layers, and the entire collection may reach 4 mm in length and 1 mm in diameter. The accessory structure shields the dendrite from virtually every source of stimulation other than direct pressure. When pressure is first applied, distortion of the dendrite stimulates the receptor. This response is phasic because the capsule soon stretches and relieves the pressure. As a result, Pacinian corpuscles are most sensitive to pulsing or vibrating stimuli. These receptors are scattered throughout the integument, notably in the fingers, breasts, and external genitalia. They are also encountered in the superficial and deep fasciae, in joint capsules, in mesenteries, and in the wall of the urinary bladder.

FIGURE 17-4
Tactile Receptors in the Skin. The location and general appearance of six important tactile receptors.

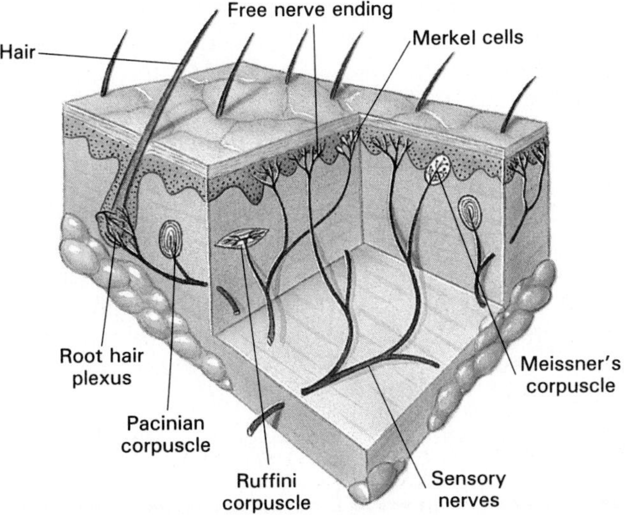

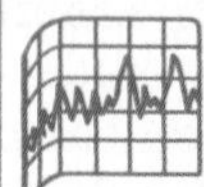

Diagnostics: Assessment of Tactile Sensitivities

Because tactile sensitivities can be affected by various pathological conditions or damage to neural pathways, mapping tactile responses can sometimes aid clinical assessment. Sensory losses with clear regional boundaries indicate trauma to spinal nerves. For example, sensory loss along a dermatomal boundary (described in Chapter 13) can permit a reasonably precise determination of the affected spinal nerve(s). More widespread sensory loss may indicate damage to ascending tracts in the spinal cord. **Anesthesia** implies a total loss of sensation; **hypesthesia**, a reduction in sensitivity; and **paresthesia**, abnormal sensations, such as the pins-and-needles sensation when an arm or leg "falls asleep" due to pressure on a peripheral nerve. (Several types of *pressure palsies* producing temporary paresthesia are discussed in the Clinical Manual.) [**CM**: *Palsies*]

Regional sensitivity to light touch can be checked by gentle contact with a fingertip or a slender wisp of cotton. The **two-point discrimination test** provides a more detailed sensory map for tactile receptors. Two fine points of a drawing compass, bent paper clip, or other object, are applied to the skin surface simultaneously. The subject then describes the contact. When the points fall within a single receptive field, the individual will report only one point of contact. A normal individual loses two-point discrimination at 1 mm (0.04 in.) on the surface of the tongue, at 2–3 mm (0.08–0.12 in.) on the lips, at 3–5 mm (0.12–0.20 in.) on the backs of the hands and feet, and at 4–7 cm (1.6–2.75 in.) over the general body surface.

Vibration receptors are tested by applying the base of a tuning fork to the skin. Damage to an individual spinal nerve produces insensitivity to vibration along the paths of the related sensory nerves. If the sensory loss results from spinal cord damage, the injury site can often be located by walking the tuning fork down the spinal column, resting its base on the vertebral spines.

Ruffini (ru-FĒ-nē) **corpuscles** are also sensitive to pressure and distortion of the skin, but they are tonically active and show little if any adaptation. The capsule surrounds a core of collagen fibers that are continuous with those of the surrounding dermis. Inside the capsule a network of dendrites is intertwined with the collagen fibers. Any tension or distortion of the dermis tugs or twists the capsular fibers, stretching or compressing the attached dendrites and altering the activity in the myelinated afferent fiber.

Table 17-1 summarizes the functions and characteristics of the six tactile receptors discussed. The central distribution of tactile sensations, via the posterior column and spinothalamic pathways, was considered in Chapter 15. (pp. 473–475) Tactile sensitivities may be altered by peripheral infection, disease processes, and damage to sensory afferents or central pathways. Important clinical tests are used to evaluate tactile sensitivity (see Diagnostics: Assessment of Tactile Sensitivities).

TABLE 17-1 Touch and Pressure Receptors

Sensation	*Receptor*	*Responds to*	*Coding*	*Localization*	*Receptive Field*
Fine touch	Free nerve ending	Light contact with skin	Tonic	Excellent	Small
	Merkel's disc	Same	Phasic/tonic	Excellent	Very small
	Root hair plexus	Initial contact with hair shaft	Phasic	Excellent	Small
Pressure and vibration	Meissner's corpuscle	Initial contact and low-frequency vibrations	Phasic	Excellent	Small
	Pacinian corpuscle	Initial contact (deep) and high-frequency vibrations	Phasic	Good	Variable
Deep pressure	Ruffini corpuscle	Stretching and distortion of the dermis	Tonic	Good	Variable

Baroreceptors

Baroreceptors monitor changes in pressure. The receptor itself consists of free nerve endings that branch within the elastic tissues in the wall of a distensible organ, such as a blood vessel or a portion of the respiratory, digestive, or urinary tracts. When the pressure changes, the elastic walls of these tracts contract or expand. This movement distorts the dendritic branches and alters the rate of action potential generation. Baroreceptors respond immediately to a change in pressure, but they adapt rapidly and the output along the afferent fibers gradually returns to "normal."

Figure 17-5 provides examples of important baroreceptor functions. Baroreceptors monitor blood pressure in the walls of major vessels, including the carotid artery (at the *carotid sinus*) and the aorta (at the *aortic sinus*). The information plays a major role in regulating cardiac function and adjusting blood flow to vital tissues. Baroreceptors in the lungs monitor the degree of lung expansion. This information is relayed to the respiratory rhythmicity center, which sets the pace of respiration. Baroreceptors in the digestive and urinary tracts trigger a variety of visceral reflexes, including those of urination and defecation. These baroreceptor reflexes will be detailed in chapters considering specific physiological systems.

Proprioceptors

Proprioceptors monitor the position of joints, the tension in tendons and ligaments, and the state of muscular contraction. In general, proprioceptors do not adapt to constant stimulation. Two representative examples were described in earlier chapters: *tendon organs*, which monitor the strain on a tendon, and *muscle spindles*, which monitor the length of a skeletal muscle.

In a Golgi tendon organ, dendrites branch repeatedly and wind around the densely packed collagen fibers in a tendon. Tension or distortion of the tendon stimulates the receptor, and excessive stimulation triggers a reduction in the strength of muscle contractions. This *tendon reflex* was discussed in Chapter 13. Chapters 10 and 13 described muscle spindles and considered their role in regulating skeletal muscle tone and maintaining normal posture and balance. (pp. 317, 423–426)

CHEMORECEPTION

Every neuron functions as a chemoreceptor when it responds to an excitatory or inhibitory neurotransmitter. The free nerve endings of nociceptors may be stimulated by a variety of chemicals including ions, acids, histamine, various enzymes, and prostaglandins. More specialized chemoreceptive neurons can detect small changes in the concentration of specific chemicals or compounds. The special senses of taste (gustation) and smell (olfaction) are provided by the most sensitive chemoreceptors in the body. Specific details concerning these senses will be presented in a later section. In general, chemoreceptors respond only to water-soluble and lipid-soluble substances that are dissolved in the surrounding fluid. The receptors show a pronounced peripheral adaptation over a period of seconds, and central adaptation may also occur.

Excluding the special senses of taste and smell, there are no well-defined chemosensory pathways in the brain or spinal cord. The locations of important chemosensory receptors are indicated in Figure 17-6. Neurons within the respiratory centers of the

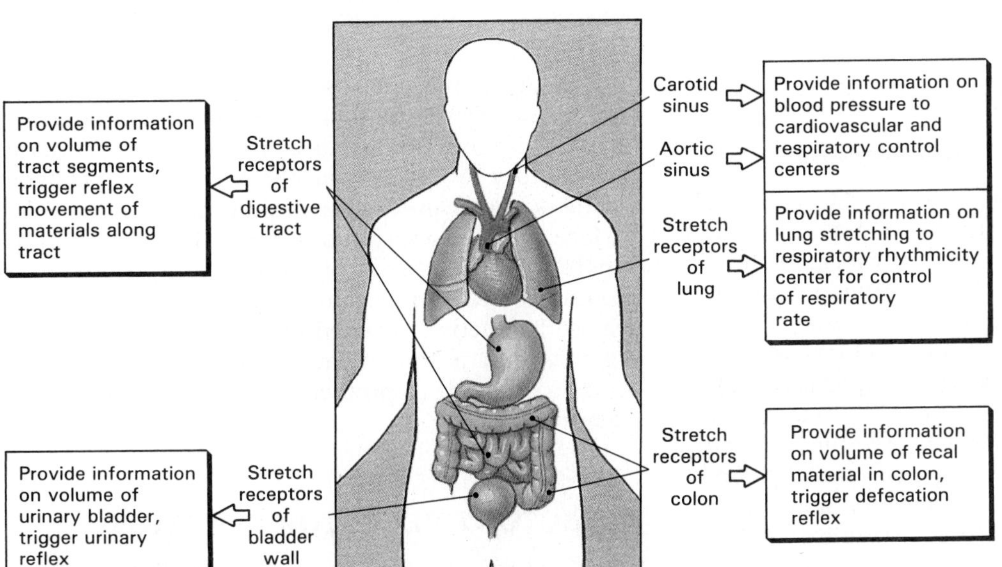

FIGURE 17-5 Baroreceptors and the Regulation of Autonomic Functions. Baroreceptors provide information essential to the regulation of autonomic activities, including respiration, digestion, urination, and defecation.

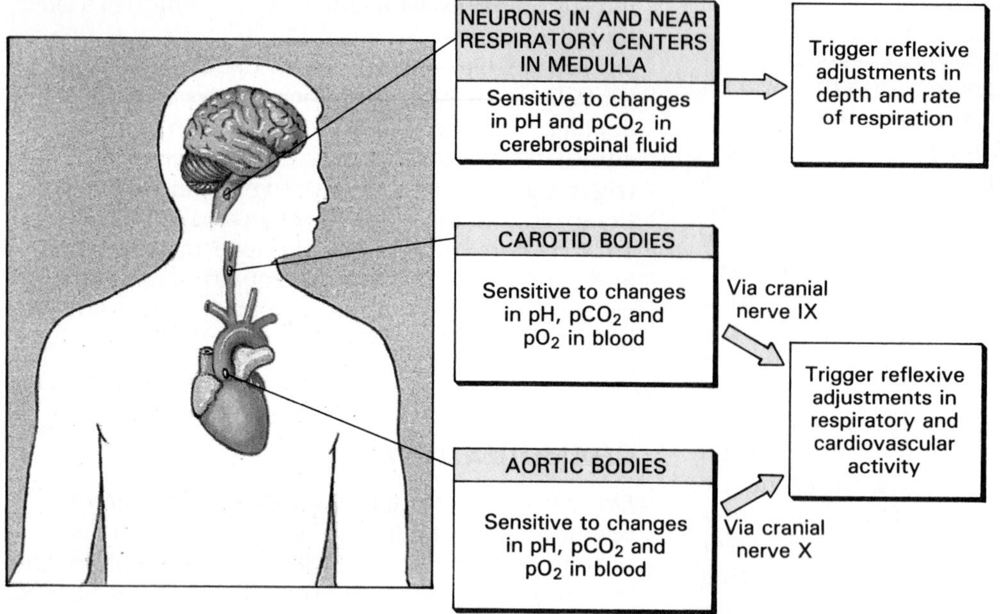

FIGURE 17-6
Chemoreceptors. Chemoreceptors are found inside the CNS, on the ventrolateral surfaces of the medulla, and in the aortic and carotid bodies. These receptors are involved in the autonomic regulation of respiratory and cardiovascular function.

brain respond to the concentration of hydrogen ions (pH) and carbon dioxide molecules in the cerebrospinal fluid. Other receptors in the periphery monitor the oxygen concentration of arterial blood. These chemoreceptive neurons are found within the **carotid bodies**, near the origin of the internal carotid arteries on each side of the neck, and in the **aortic bodies** between the major branches of the aortic arch. The afferent fibers leaving the carotid and aortic bodies reach the respiratory centers by traveling along the ninth (glossopharyngeal) and tenth (vagus) cranial nerves. These chemoreceptors play an important role in the reflexive control of respiration and cardiovascular function.

■ Olfaction

The paired **olfactory organs** are located on either side of the nasal septum in the upper portion of the nasal cavity just inferior to the cribriform plate of the ethmoid. (For orientation, you may wish to review Figure 8-11.) (p. 248) In gross dissection the olfactory organs have a yellow-brown coloration, due to the presence of pigments in the epithelial cells and the overlying mucus. Figure 17-7 details the structure and location of these sensory organs. Each olfactory organ contains mucous glands and an **olfactory epithelium**. The epithelium contains supporting cells and the **olfactory receptors**, neurons sensitive to chemicals dissolved in the overlying mucus. In addition to olfactory receptors, the olfactory epithelium contains columnar **supporting cells** and small **basal (stem) cells**. Beneath the basement membrane, large **olfactory glands** produce a pigmented mucus that covers the epithelium.

The olfactory epithelium covers the roof of each nasal cavity and extends across portions of the nasal septum and the superior conchae. When air is drawn in through the nose, turbulence generates eddies that bring airborne compounds to the olfactory organs. A normal, relaxed inspiration provides a small sample of the inspired air (around 2 percent) to the olfactory organs. Sniffing repeatedly creates more extensive eddies, increasing the flow of air across the sensitive epithelium and intensifying the stimulation of the receptors.

Once the compounds have reached the olfactory organs, water-soluble and lipid-soluble materials must diffuse into the mucus before they can stimulate the olfactory receptors. The olfactory glands produce a continual stream of mucus that passes across the surface of the olfactory organ, preventing the buildup of potentially dangerous or overpowering stimuli and keeping the area moist and free from dust or other debris. The densely packed microvilli of the supporting cells may assist in mucus propulsion and also act as a cushion to protect the cilia of the olfactory receptors.

OLFACTORY RECEPTOR FUNCTION

The olfactory receptors are highly modified neurons. The apical portion of each receptor forms a prominent

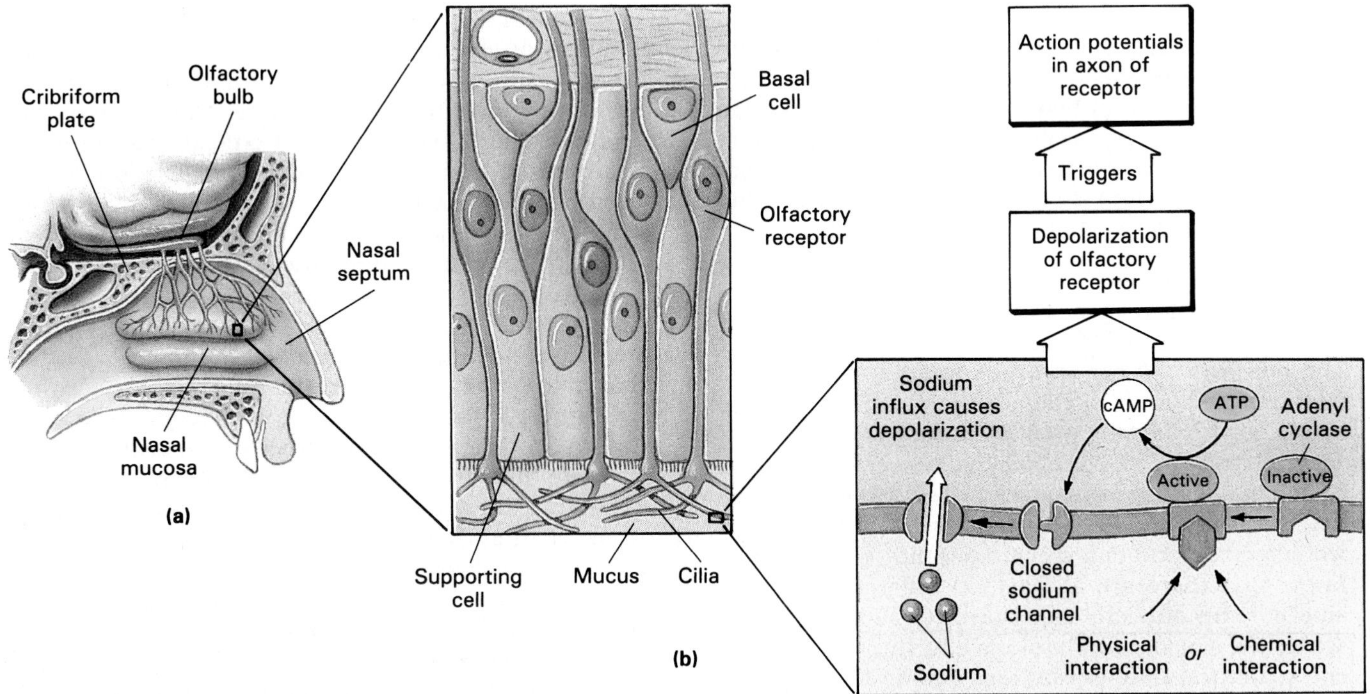

FIGURE 17-7
The Olfactory Organs. (a) The structure of the olfactory organ on the right side of the nasal septum. (b) An olfactory receptor is a modified neuron with multiple cilia extending from its free surface. Odorant particles may trigger receptor activity by binding to receptors specific to molecular structure (the chemical theory) or three-dimensional shape (the physical theory).

knob that projects above the epithelial surface. That projection provides a base for up to 20 cilia that extend into the surrounding mucus for distances of up to 250 μm. These elongate cilia are in constant motion, assisting in the movement of mucus and exposing their considerable surface area to the dissolved chemical compounds. All together there are somewhere between 10 and 20 million olfactory receptors, packed into an area of roughly 5 square centimeters. Taking the exposed ciliary surfaces into account, the actual sensory area probably approaches that of the entire body surface. Nevertheless, our olfactory sensitivities cannot compare to those of other vertebrates such as dogs, cats, or fishes. A German shepherd sniffing for smuggled drugs or explosives has an olfactory receptor surface 72 times greater than that of the nearby Customs Inspector.

The first step in olfactory reception occurs on the surface of an olfactory cilium, but how the dissolved chemicals interact with the receptor surface remains to be determined. There may be specific chemical receptors incorporated into the membrane surface (the *chemical theory*), or the membrane receptors may respond to certain molecular shapes (the *physical theory*). These two possibilities are diagrammed in Figure 17-7b; either one could produce changes in the permeability characteristics of the receptor membrane and give rise to action potentials.

When binding occurs, it leads to the activation of adenyl cyclase, the enzyme that converts ATP to cyclic-AMP. The cyclic-AMP then opens sodium channels in the membrane, which as a result begins to depolarize. If sufficient depolarization occurs, an action potential appears in the axon, and the information is relayed to the central nervous system.

OLFACTORY PATHWAYS

The olfactory system is very sensitive. As few as four molecules of an odorous substance can activate an olfactory receptor. However, the activation of an afferent fiber does not guarantee a conscious awareness of the stimulus. Considerable convergence occurs along the olfactory pathway, and inhibition at the intervening synapses can prevent the sensations from reaching the cerebral cortex. Axons leaving the olfactory epithelium collect into 20 or more bundles that penetrate the cribriform plate of the ethmoid to reach the olfactory bulbs of the cerebrum. Roughly 26,000 afferent fibers synapse on each second-order neuron, and there are complex interconnections between the neurons. Efferent fibers from nuclei elsewhere in the brain also contact the neurons of the olfactory bulb and provide a mechanism for central modification of olfactory sensitivities.

Axons leaving the olfactory bulb travel along the olfactory tract (N I) to reach the olfactory cortex, the hypothalamus, and portions of the limbic system. Olfactory stimuli are the only type of sensory information that reaches the cerebral cortex without first synapsing in the thalamus. The extensive limbic and hypothalamic connections help to explain the profound emotional and behavioral responses that can be produced by certain smells. The perfume industry understands the practical implications of these connections, expending considerable effort to develop odors that trigger sexual responses.

OLFACTORY DISCRIMINATION

The olfactory system can make subtle distinctions between thousands of chemical stimuli. Attempts to break down olfactory perceptions into a handful of "primary smells" have not been successful. At one time it was thought that any smell could be described with as few as seven descriptive terms, and that each would correspond to a specific receptor type. We now know that there are at least 50 different "primary smells," and our language does not allow us to describe the sensory impressions effectively. No apparent structural differences exist among the olfactory cells, but the epithelium as a whole contains receptor populations with distinctly different sensitivities. The CNS interprets the smell on the basis of the particular pattern of receptor activity.

AGING AND OLFACTORY SENSITIVITY

The olfactory receptor population shows considerable turnover, with new receptor cells being produced by division and differentiation of basal cells in the epithelium. This is the only known example of neuronal replacement in the adult human. Despite this process, the total number of receptors declines with age, and elderly individuals have difficulty detecting odors in low concentrations. This decline in the number of receptors accounts for Grandmother's tendency to apply perfume in excessive quantities and explains why Grandfather's aftershave seems so strong — they must apply more to be able to smell it themselves.

Gustation

Gustation, or tasting, provides information required to make a conscious decision about whether or not to actually swallow a foreign object. This is the last chance for voluntary decision making, though involuntary visceral reflexes, such as vomiting, may later reject the item.

The **gustatory** (GUS-ta-tōr-ē), or **taste**, **receptors** are distributed over the surface of the tongue and adjacent portions of the pharynx and larynx. The receptors are clustered in individual **taste buds** such as those illustrated in Figure 17-8. Each taste bud contains around 40 slender sensory receptors, known as **gustatory cells**, and a number of supporting cells.

Tasting occurs before significant digestive action has occurred, while the tongue, lips, cheeks, and teeth hold and manipulate the material and facilitate its solution in the salivary secretions. Sensory cells would be damaged by direct exposure to relatively unprocessed oral contents, but the taste buds are recessed into the surrounding epithelium. Each gustatory cell extends slender microvilli, sometimes called *taste hairs*, into the surrounding fluids through the narrow **taste pore**.

At birth there are taste buds scattered across the dorsal surface and margins of the tongue, in the pharyngeal walls, and over the superior surfaces of the larynx. By adulthood the taste receptors on the pharynx and larynx have declined in importance, and the taste buds of the tongue are the major gustatory receptors. These taste organs are particularly well protected, for they lie along the sides of epithelial projections, called **papillae** (pa-PIL-lē). There are three different types of papillae on the human tongue: **filiform** (*filum*, thread), **fungiform** (*fungus*, mushroom), and **circumvallate** (sir-kum-VAL-āt; *circum-*, around + *vallum*, wall). There are regional differences in the distribution of the papillae, as indicated in Figure 17-8. The largest number of taste buds are associated with the circumvallate papillae, which form a V near the posterior margin of the tongue. Despite this relatively protected position, it's still a hard life, and a typical gustatory cell survives for only about 10 days before being replaced by the division of nearby epithelial cells.

GUSTATORY RECEPTOR FUNCTION

The mechanism behind gustatory reception seems to parallel that of olfaction. Dissolved chemicals contacting the taste hairs provide the stimulus that produces a change in the transmembrane potential of the taste cell. The dendrites of the sensory afferents are tightly wrapped by folds of the receptor cell membrane. It is not clear how the two communicate, but stimulation of the gustatory cell leads to action potentials in the afferent fiber.

There are four **primary taste sensations**: sweet, salt, sour, and bitter. Each taste bud shows a particular sensitivity to one of these tastes, and a sensory map of the tongue indicates that there are regional differences in primary sensitivity. The threshold for receptor stimulation varies for each of the primary taste sensations, and the taste receptors respond most readily to unpleasant rather than attractive stimuli. We are almost a thousand times more sensitive to acids, which give a sour taste, than to either sweet or salty chemicals. But we are a hundred times more sensitive to bitter compounds than we are to acids! This pattern of sensitivity has survival value, for acids can damage the mucous membranes of the mouth and pharynx, and many potent biological toxins produce an extremely bitter taste.

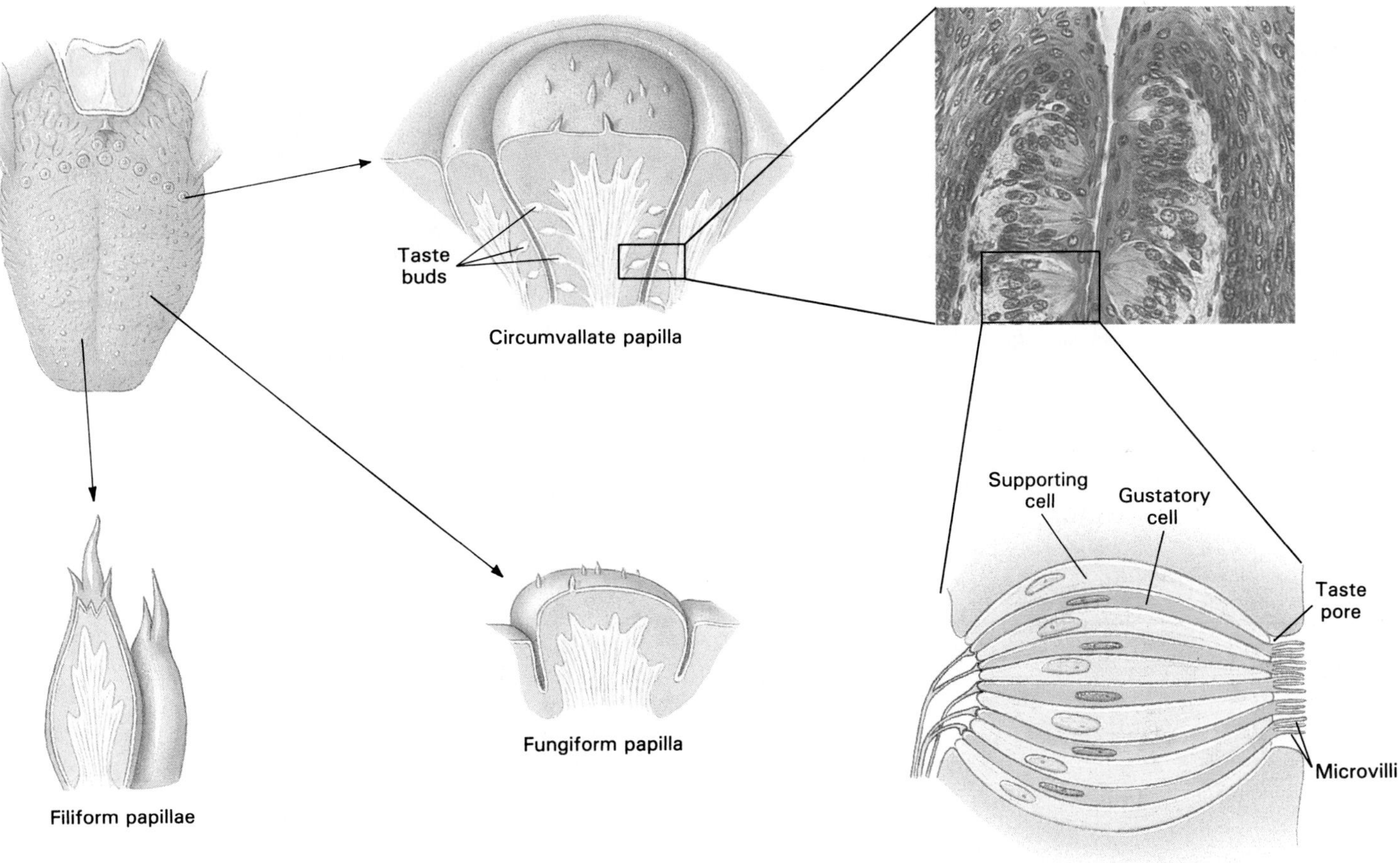

FIGURE 17-8
Gustatory Reception. Gustatory receptors are found in taste buds that form pockets in the epithelium of the tongue.

GUSTATORY PATHWAYS

Taste buds are monitored by the seventh (facial), ninth (glossopharyngeal), and tenth (vagus) cranial nerves, as indicated in Figure 17-9. The sensory afferents synapse within the **nucleus solitarius** of the medulla, and the axons of the postsynaptic neurons enter the medial lemniscus. After another synapse in the thalamus, the information is projected to the appropriate portions of the primary sensory cortex.

In assembling a conscious perception of taste, the information received from the taste buds is correlated with other sensory data. Our perception of the general texture of the food, together with the taste-related sensations of "peppery" or "burning," result from the stimulation of general sensory afferents contained in the trigeminal nerve (N V). In addition, information from the olfactory receptors plays an overwhelming role in taste perception. We are several thousand times more sensitive to "tastes" when our olfactory organs are fully functional. When a person has a cold and airborne molecules cannot reach the olfactory receptors meals are likely to taste rather dull and unappealing. This reduction in taste sensation will occur even though the taste buds may be responding normally.

INDIVIDUAL DIFFERENCES IN GUSTATORY SENSITIVITY

There are significant individual differences in taste sensitivity. Many of these conditions are inherited. The best-known example involves sensitivity to the compound phenylthiourea, also known as phenylthiocarbamide, or **PTC**. In the Caucasian population roughly 70 percent can taste this substance; the rest are unable to detect it.

FIGURE 17-9
Gustatory Pathways. Three different cranial nerves (VII, IX, and X) carry gustatory information to the primary sensory cortex of the cerebrum.

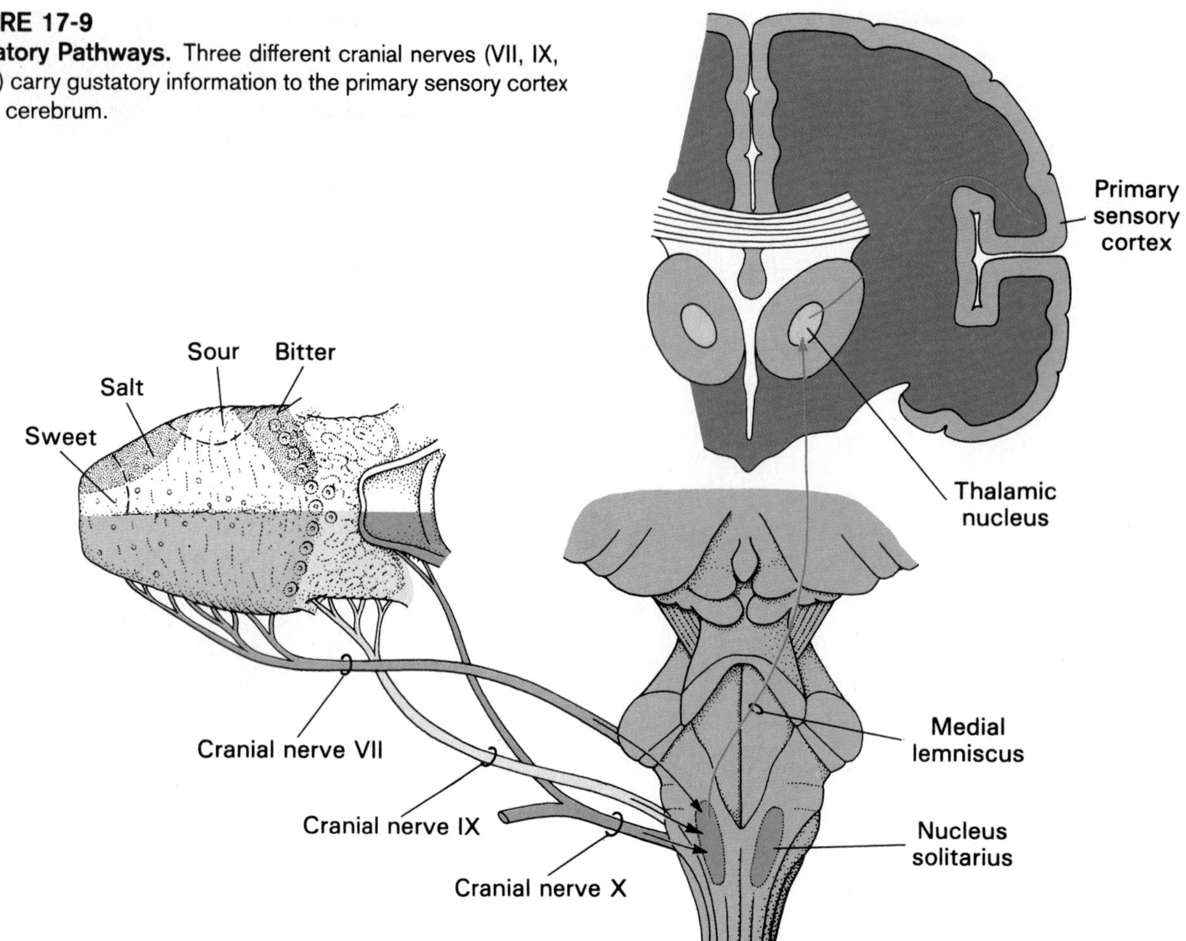

AGING AND GUSTATORY SENSITIVITY

Our tasting abilities change with age. We begin life with over 10,000 taste buds, but the number starts declining dramatically by age 50. The sensory loss becomes especially significant because aging individuals also experience a decline in the population of olfactory receptors. As a result, many elderly people find that their food tastes bland and unappetizing, whereas children often find spicy foods too strongly seasoned.

- **When you first entered the A & P lab for dissection, you are very aware of the odor, but by the end of the lab period, the smell doesn't seem to be nearly as strong. Why?**
- **When the nociceptors in your hand are stimulated, what sensation do you perceive?**
- **What would happen to an individual if the information from proprioceptors in the legs was blocked from reaching the CNS?**
- **If you completely dry the surface of the tongue, then place salt or sugar crystals on it, they can't be tasted. Why?**

Equilibrium and Hearing

The senses of equilibrium and hearing are provided by the receptors of the **inner ear**, a collection of fluid-filled tubes and chambers. Figure 17-10 depicts the position and structure of the inner ear, also known as the **membranous labyrinth** (*labyrinthos*, a network of canals). Its chambers and canals contain a fluid, the **endolymph** (EN-dō-limf), with unusual electrolyte concentrations.

The **bony labyrinth** is a region of dense bone that surrounds and protects the membranous labyrinth. Its inner contours closely follow the contours of the membranous labyrinth, while its outer walls are fused with the surrounding temporal bone. Between the bony and membranous labyrinths flows the **perilymph** (PER-i-limf), a liquid whose properties closely resemble those of cerebrospinal fluid.

The bony labyrinth can be subdivided into the **vestibule** (VES-ti-būl), the **semicircular canals**, and the **cochlea** (KOK-lē-a; *cochlea*, snail shell). Receptors in the vestibule and semicircular canals provide the sense of equilibrium; those in the cochlea provide the sense of hearing. The entire inner ear, including the cochlea, is buried in the temporal bone

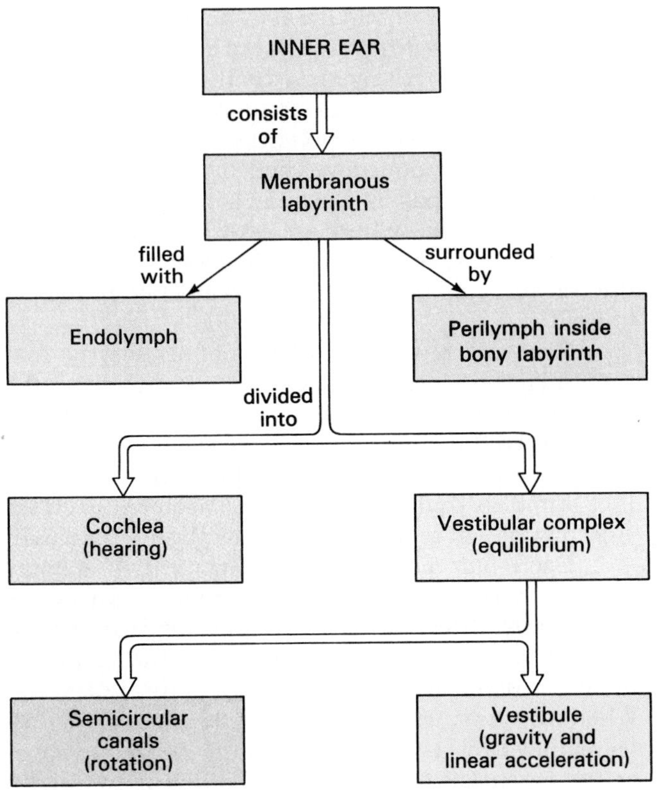

FIGURE 17-10
Functional Relationships of the Inner Ear

of the skull. (The location is shown in Figure 8-7.) (p. 245) The structures and airspaces of the **external ear** and **middle ear** function in the capture and transmission of sound to the cochlea.

ANATOMY AND RECEPTOR FUNCTION

The External and Middle Ear

The **external ear** (Figure 17–11a) includes the cartilaginous **pinna**, which surrounds the **external auditory meatus**, the entrance to the **external auditory canal**. Special **ceruminous glands** along the canal secrete a waxy material, and many small, outwardly projecting hairs help to prevent the entrance of foreign objects or insects. The pinna affords additional protection and helps to provide directional sensitivity to the ear by blocking or encouraging the passage of sound to the tympanic membrane. The external auditory canal ends at the **tympanic membrane**, the dividing line between the external ear and the middle ear.

Although the tympanic membrane, or eardrum, normally prevents direct contact between the external environment and the middle ear, this compartment is not totally isolated. It communicates with the nasopharynx via the slender **pharyngotympanic** (*Eustachian*, or *auditory*) **tube**, and with the mastoid sinuses via a number of small and variable connections. The

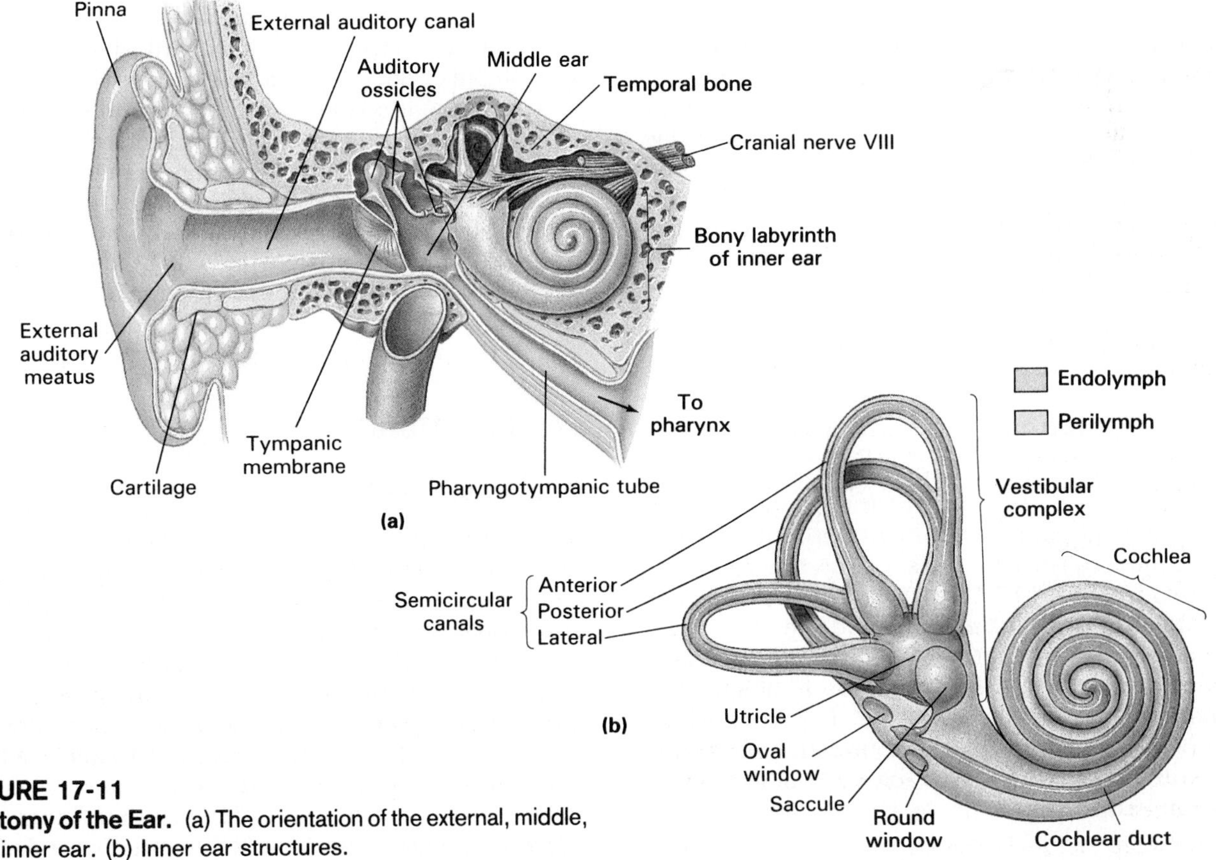

FIGURE 17-11
Anatomy of the Ear. (a) The orientation of the external, middle, and inner ear. (b) Inner ear structures.

middle ear encloses and protects the tiny **auditory ossicles** that connect the tympanic membrane with the receptor complex of the inner ear. [**CM**: *Otitis Media and Mastoiditis*]

Structure of the Inner Ear

The inner ear, detailed in Figure 17-11b, includes the vestibule, semicircular canals, and cochlea. The combination of vestibule and semicircular canals is called the *vestibular complex*, because the fluid-filled chambers of the vestibule are broadly continuous with those of the semicircular canals. The vestibule includes a pair of membranous sacs, the **saccule** (SAK-ūl) and the **utricle** (Ū-tre-kl), or the *sacculus* and *utriculus*. Receptors in the saccule and utricle provide sensations of gravity and linear acceleration. Those in the semicircular canals are stimulated by rotation of the head. Together, these perceptions of gravity, linear acceleration, and rotation provide us with our sense of equilibrium, or balance. The physiological and anatomical basis for these sensitivities will be considered below.

The cochlea contains a slender, elongate portion of the membranous labyrinth known as the **cochlear duct**. The cochlear duct sits sandwiched between a pair of perilymph-filled chambers, and the entire complex is coiled around a central bony hub. In sectional view the spiral arrangement resembles that of a snail shell, *cochlea* in Latin.

The outer walls of the perilymphatic chambers consist of dense bone everywhere except at two small areas near the base of the cochlear spiral. The **round window** is a thin, membranous partition that separates the perilymph of the cochlear chambers from the air spaces of the middle ear. The **oval window** has a similar structure, but it is firmly bound to the auditory ossicles of the middle ear. When a sound vibrates the tympanic membrane the movements are conducted over the ossicular chain to the surface of the oval window. This process leads to the stimulation of sensory receptors along the cochlear duct, and we "hear" the sound.

Receptor Function in the Inner Ear

The sensory versatility of the inner ear provides a striking example of the way similar receptor cells can monitor very different environmental stimuli. The basic receptors, called **hair cells**, of the inner ear are illustrated in Figure 17-12a. These receptor cells are surrounded by **supporting cells** and are monitored by sensory afferents. Often there are also synapses with CNS efferents that may function in adaptation or sensitization of the receptor cell. The free surface of the receptor supports 80–100 unusual microvilli, called **stereocilia**, that may reach 100 μm in length. Each hair cell in the vestibule also contains a **kinocilium**, a single large cilium.

Cells of the intestinal tract, the respiratory tract, and many other epithelial surfaces have cilia or microvilli. These cells actively move their cilia or microvilli, and those movements are accompanied by changes in the transmembrane potential. The hair cells of the inner ear do not move their cilia and microvilli; instead, they sit passively and let a mechanical stimulus do the work. When an external force pushes against the cilia or microvilli, the movement distorts the cell surface and changes the transmembrane potential.

Each cell has a specific axis of sensitivity. In a vestibular hair cell, the axis is determined by the position of the kinocilium. Displacement of stereocilia toward the kinocilium depolarizes the cell, and a movement away from the kinocilium hyperpolarizes it. A stimulus that produces a displacement to either side has no effect on the transmembrane potential.

Each hair cell communicates with a sensory neuron by continually releasing small quantities of neurotransmitter. When the hair cell hyperpolarizes, less neurotransmitter is released, and the frequency of action potentials in the sensory neuron declines. When the hair cell depolarizes, the rate of neurotransmitter release accelerates, and the frequency of action potentials increases. This change in frequency is proportional to the strength of the stimulus. Moreover, the cell will respond only to movement along one axis, as explained above. Thus the output of a single hair cell provides information on *strength* and *direction of movement* at the same time.

Hair cells provide information concerning the direction and strength of mechanical stimuli. The stimuli involved, however, are quite varied: gravity or acceleration in the vestibule, rotation in the semicircular canals, and sound in the cochlea. The sensitivities of the hair cells differ because in each of these regions there are different accessory structures that determine the source of stimulation. The importance of these accessory structures will become apparent as we consider the semicircular canals, utricle, and saccule.

EQUILIBRIUM

The Semicircular Canals: Rotational Motion

Receptors in the semicircular canals respond to rotational movements. Figure 17-12b depicts the vestibular complex in greater detail. The **anterior**, **posterior**, and **lateral** semicircular canals are continuous with the utricle. Each semicircular canal contains a swollen region, the **ampulla**, which contains the sensory receptors. Hair cells attached to the wall of the ampulla form a raised structure known as a **crista** (Figure 17-12c). The kinocilia and stereocilia of the hair cells are embedded in a gelatinous structure, the **cupula** (KŪ-pū-la). Because the cupula has a density

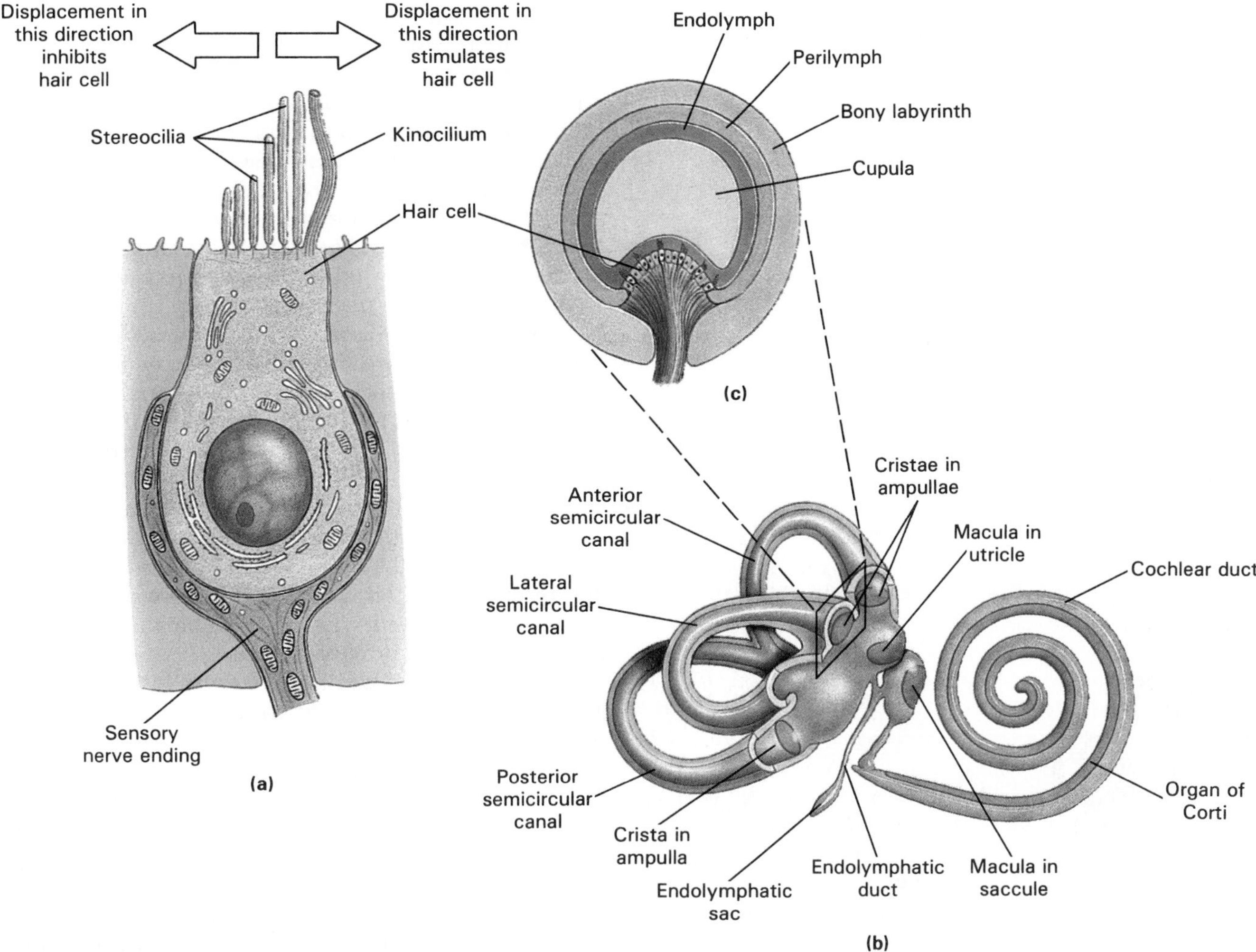

FIGURE 17-12
Hair Cells of the Inner Ear. (a) Structure of a typical hair cell, showing details revealed by electron microscopy. An individual hair cell has a kinocilium and numerous stereocilia. Each cell contacts a sensory neuron. Bending the stereocilia toward the kinocilium depolarizes the cell and stimulates the sensory neuron. Displacement in the opposite direction inhibits the sensory neuron. (b) The semicircular canals are part of the vestibule. This figure also indicates the distribution of hair cells in the vestibule and cochlea. (c) A section through the ampulla of a semicircular canal. Displacement of the cupula stimulates the hair cells.

very close to that of the surrounding endolymph, it essentially "floats" above the receptor surface, nearly filling the ampulla. When the head rotates in the plane of the canal, movement of the endolymph along the canal axis pushes the cupula and distorts the receptor processes. Fluid movement in one direction stimulates the hair cells, and movement in the opposite direction inhibits them. When the endolymph stops moving, the elastic nature of the cupula makes it "bounce back" to its normal position.

Each semicircular canal responds to one of three possible rotational movements, as indicated in Figure 17-13. The sensitivity of a particular canal is determined by its orientation. To distort the cupula and stimulate the receptors, endolymph must flow along the axis of the canal. Any movement other than a rotation in that plane will fail to register. A horizontal rotation, as in shaking the head "no," stimulates the hair cells of the lateral semicircular canal. Nodding "yes" excites the anterior canal, while tilting the head from side to side activates the receptors in the posterior canal. Of course, very few everyday movements consist of rotations in a single plane. In effect, however, the three planes monitored by the semicircular canals correspond to the three dimensions in the world around us. Even the most complex movement can be analyzed in terms of motion in three rotational planes.

Automatic eye movements occur in response to sensations of motion (whether real or illusory) under the direction of the superior colliculus. These movements, discussed in Chapter 14, attempt to keep the gaze focused on a specific point in space. (p. 469) When you spin around, your eyes fix on one point for a moment, then jump ahead to another, in a series of short, jerky movements. These eye movements sometimes appear after damage to the brain stem or inner ear, a condition termed **nystagmus**.

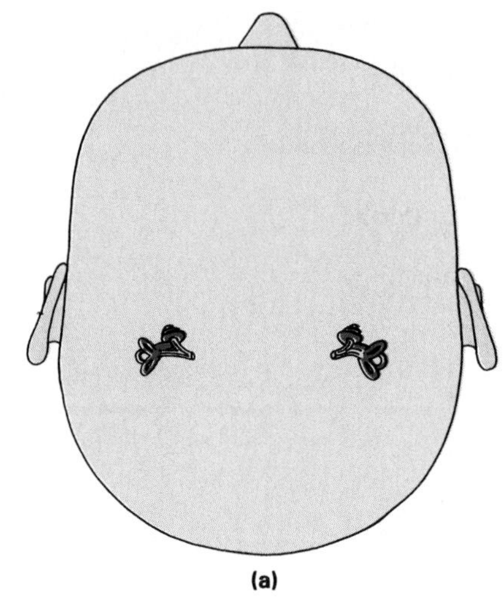

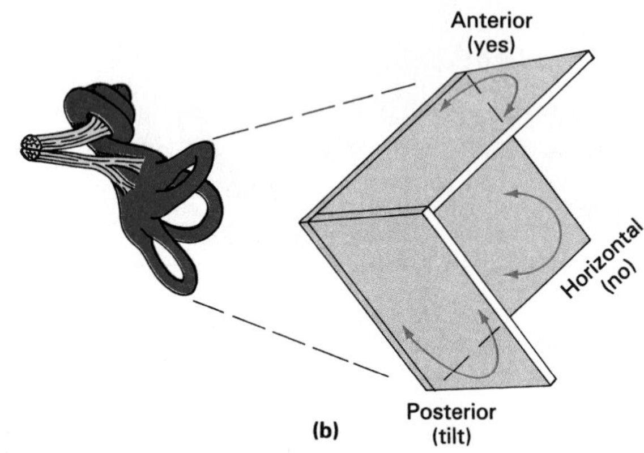

FIGURE 17-13
Function of the Semicircular Canals. (a) A superior view of the head, showing the location of the semicircular canals. (b) Planes of sensitivity for the semicircular canals.

The Vestibule (Utricle and Saccule): Gravity and Linear Acceleration

Receptors in the utricle and saccule respond to gravity and linear acceleration. The two chambers, detailed in Figure 17-14a, are connected by a slender passageway continuous with the narrow **endolymphatic duct**. The duct terminates in a blind pouch, the **endolymphatic sac**, that projects into the dura of the temporal bone. Portions of the cochlear duct continually secrete endolymph, and at the endolymphatic sac excess fluids return to the general circulation.

The hair cells of the utricle and saccule are clustered in oval **maculae** (MAK-ū-lē; *macula*, spot). As in the ampullae, the hair cell processes are embedded in a gelatinous mass, but the macular receptors lie under a thin layer containing densely packed mineral crystals. These **otoconia** (ō-tō-KŌ-nē-a; *oto-*, ear + *conia*, dust), sometimes called **otoliths**, can be seen in Figure 17-14b. When the head is in the normal, upright position the otoconia sit atop the maculae. Their weight presses down on the macular surfaces, pushing the sensory hairs down rather than to one side or another. When the head is tilted, the pull of gravity on the otoconia shifts the mass to the side. This shift distorts the sensory hairs, and the change in receptor activity tells the CNS that the head is no longer level (Figure 17-14c).

Otoconia have considerable inertia, and they are connected to the rest of the body only by the sensory processes of the macular cells. So whenever the rest of the body makes a sudden movement, the otoconia lag behind. When an elevator starts its downward plunge, we are immediately aware of it because the otoconia no longer push so forcefully against the surface of the receptor cells. Once they catch up, we are no longer aware of any movement until the elevator brakes to a halt. As the body slows down, the otoconia press harder against the hair cells, and we "feel" the force of gravity increase.

A similar mechanism accounts for our perception of linear acceleration in a car that speeds up suddenly. The otoconia lag behind, distorting the sensory hairs and changing the activity in the sensory neurons. A comparable otoconial movement occurs when the head is tilted back, so that gravity pulls the otoconia at an angle with respect to the sensory hairs. The two stimuli are not identical, and under normal circumstances the nervous system avoids confusion by noting the subtle variations and integrating these sensations with visual information.

Central Processing of Vestibular Sensations

Hair cells of the vestibule and semicircular canal are monitored by sensory neurons located in adjacent **vestibular ganglia**. Sensory fibers from each ganglion form the **vestibular branch** of the vestibulocochlear nerve, N VIII. These fibers synapse on neurons within the **vestibular nuclei** at the boundary between the pons and medulla. The two vestibular nuclei (1) integrate the sensory information arriving from each side of the head, (2) relay information to the cerebellum, (3) relay information to the cerebral cortex, providing a conscious sense of position and movement, and (4) send commands to motor nuclei in the brain stem and spinal cord. These reflexive motor commands are distributed to the motor nuclei for cranial nerves involved with eye, head, and neck movements (N III, IV, VI, and XI). Descending instructions along the **vestibulospinal tracts** of the spinal cord adjust peripheral muscle tone to complement the reflexive movements of the head or neck. These pathways are diagrammed in Figure 17-15 on page 542. [**CM**: *Vertigo, Dizziness, and Motion Sickness*]

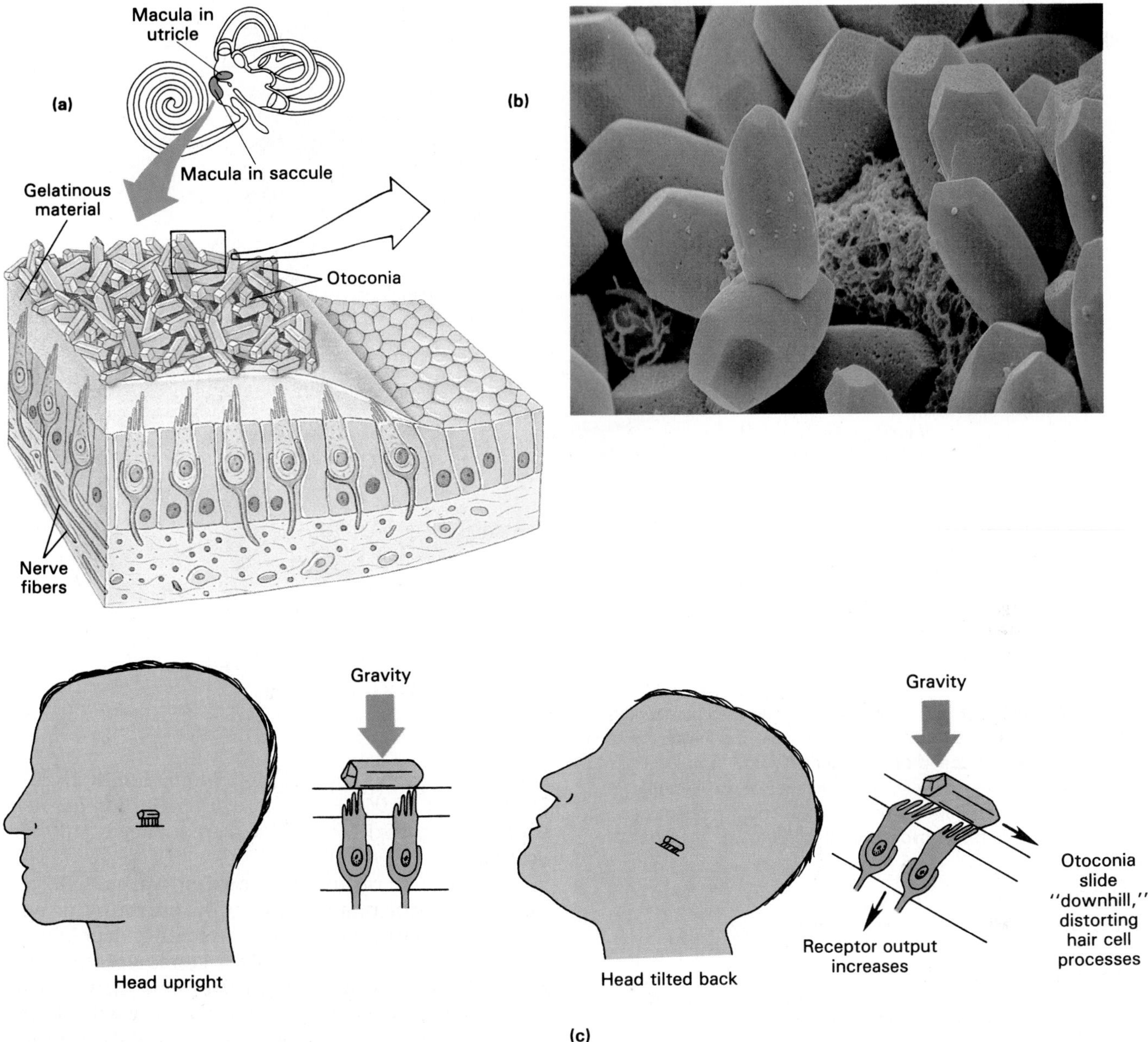

FIGURE 17-14
The Maculae of the Vestibule. (a) Location of the maculae within the utricle and saccule. (b) Detailed structure of a sensory macula and a scanning electron micrograph showing the crystalline structure of an otoconia (otolith). (c) Diagrammatic view of changes in otoconial position during tilting of the head.

THE COCHLEA AND HEARING

The bony cochlea encloses the cochlear duct, whose receptors provide the sensation of hearing. Hearing is the detection of sound, which consists of pressure waves conducted through air or water. Sound waves are conducted through the external ear and amplified by the middle ear before they are analyzed by cochlear receptors. This discussion will consider: (1) the physics of sound; (2) sound collection, amplification, and conduction; (3) receptor function in the cochlea; and (4) central processing of auditory information.

The Physics of Sound

Sound travels through the air in a series of waves, as indicated in Figure 17-16. Each wave consists of a region where the air molecules are crowded together and an adjacent zone where they are relatively far apart. These waves travel through the air at approximately 1235 km/h (768 mph). Physicists use the term **cycles** rather than waves, and the number of cycles per second (cps), or **Hertz (Hz)**, represents the **frequency** of the sound. What we perceive as the **pitch** of a sound is our sensory response to its frequency.

FIGURE 17-15
Pathways for Equilibrium Sensations

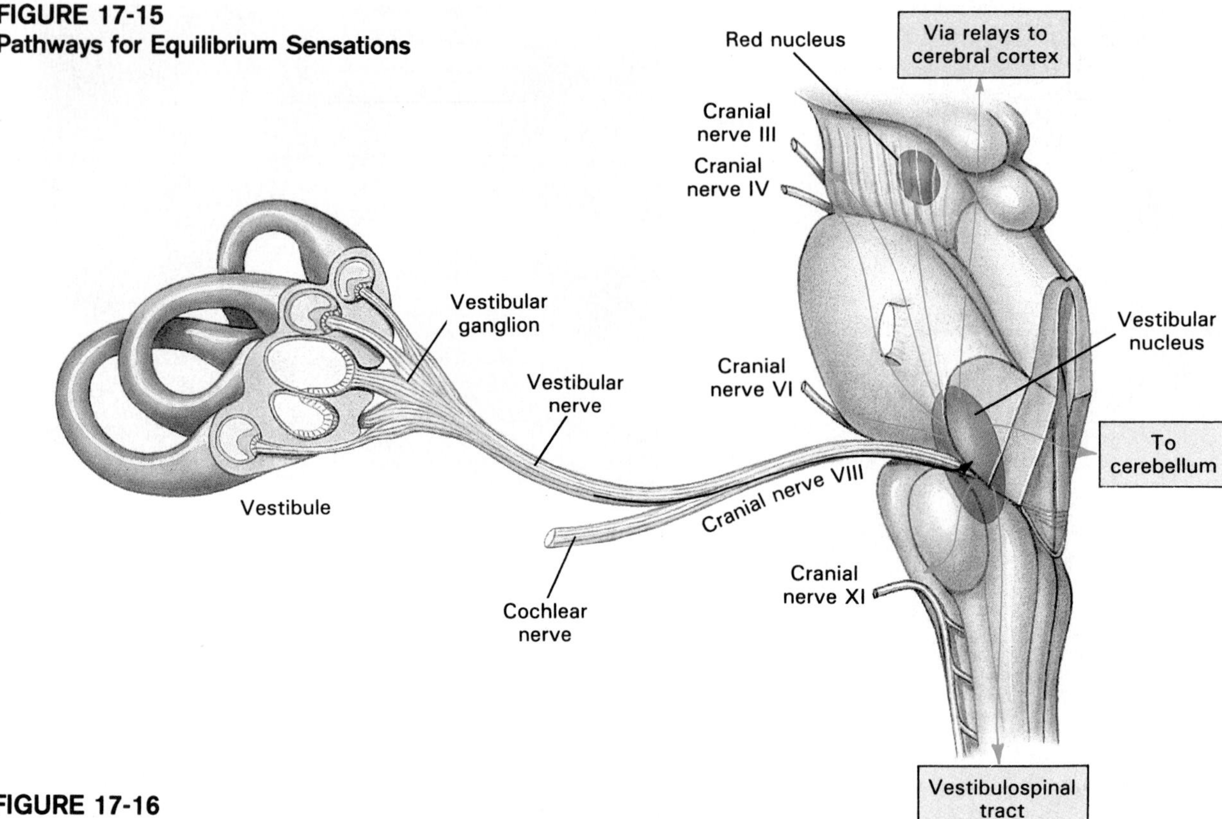

FIGURE 17-16
Sound Waves in Air. (a) Sound waves generated by a tuning fork travel through the air as pressure waves. The waves move through the air at a speed of approximately 343 m/s. (b) The frequency of the sound wave is the number of wavelengths that pass a fixed reference point each second. Frequencies are reported in terms of cycles per second (cps), or Hertz (Hz).

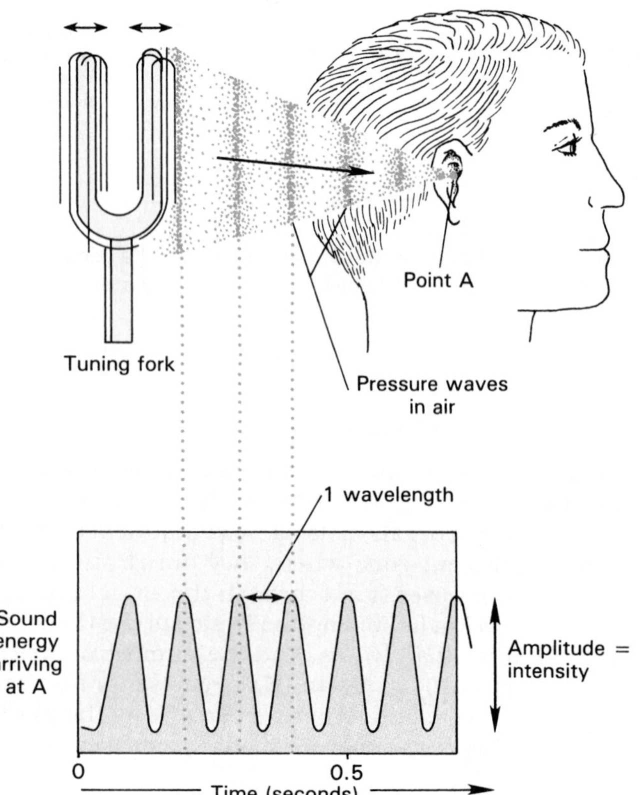

A *high-frequency* sound (high pitch) might have a frequency of 15,000 Hz or more; a very *low-frequency* sound (low pitch) could have a frequency of 100 Hz or less.

It takes energy to produce sound waves. A vibrating guitar string pushes against the surrounding air, producing sound waves whose frequency depends on the frequency of vibration. At a stereo speaker, the speaker surface vibrates with the same results. The sound waves then carry that energy away with them. The energy content, or power, of a sound determines its **intensity**, or loudness, which is reported in **decibels** (DES-i-bels). Table 17-2 ranks the decibel levels of familiar sounds.

When sound waves strike an object, their energy can be felt as a physical pressure. Given the right combination of frequencies and intensities, the object will begin to vibrate at the same frequency as the sound, a phenomenon called *resonance*. The greater the sound intensity, the greater the amount of movement produced.

Sound Collection, Amplification, and Conduction

Structures involved in sound collection and conduction are diagrammed in Figure 17-17a. Sound waves enter the external auditory meatus and travel along the external auditory canal toward the tympanic membrane. The orientation of the canal helps provide

TABLE 17-2 Sound Intensities

Typical Level (Decibels)	*Example*	*Dangerous Time Exposure*
0	Lowest audible sound	
30	Quiet library, soft whisper	
40	Quiet office, living room, bedroom away from traffic	
50	Light traffic at a distance, refrigerator, gentle breeze	
60	Air conditioner at 20 feet, conversation, sewing machine in operation	
70	Busy traffic, office calculator, noisy restaurant	Some damage if continuous
80	Subway, heavy city traffic, alarm clock at 2 feet, factory noise	Over 8 hours
90	Truck traffic, noisy home appliances, shop tools, lawnmower	Less than 8 hours
100	Chain saw, boiler shop, pneumatic drill	2 hours
120	"Heavy metal" rock concert, sandblasting, thunderclap nearby	Immediate danger
140	Gunshot, jet plane	Immediate danger
180	Rocket launching pad	Hearing loss inevitable

FIGURE 17-17
The Middle Ear. (a) The structures of the external and middle ear. (b) The auditory ossicles. (c) Movement of the tympanic membrane affects the auditory ossicles and relays vibrations to the oval window of the inner ear.

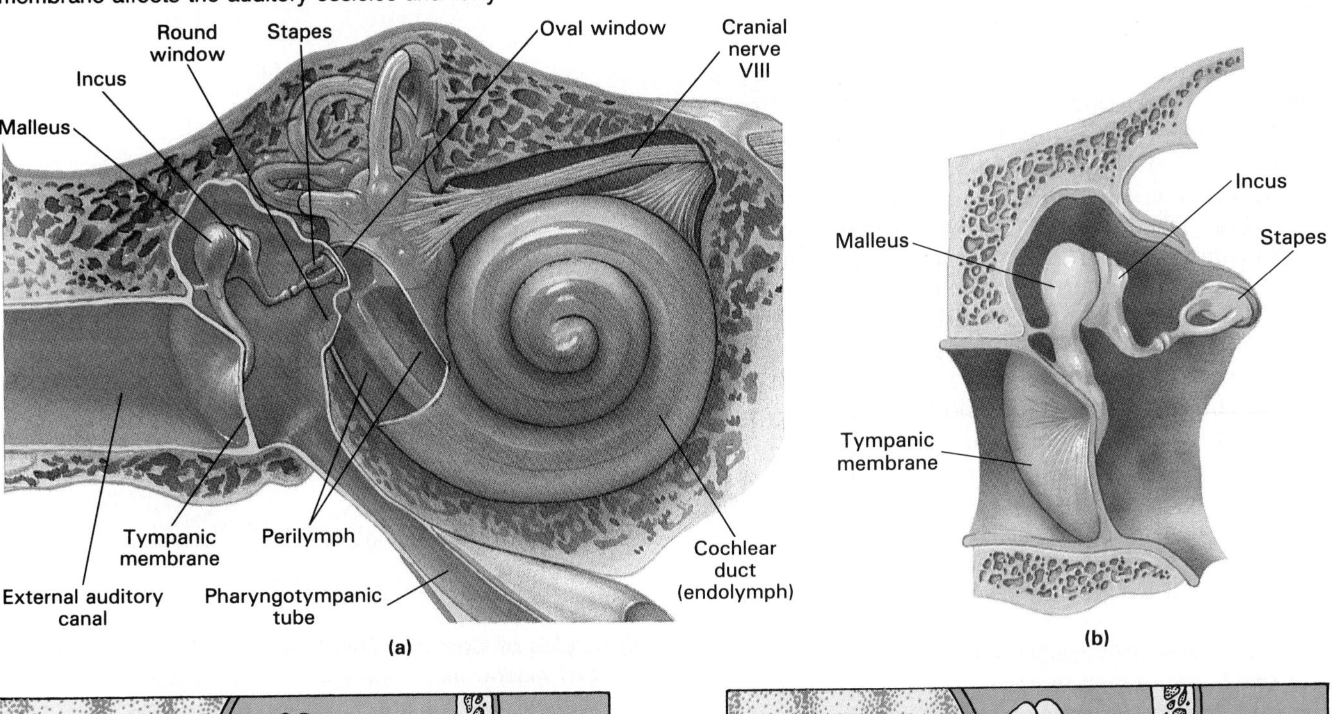

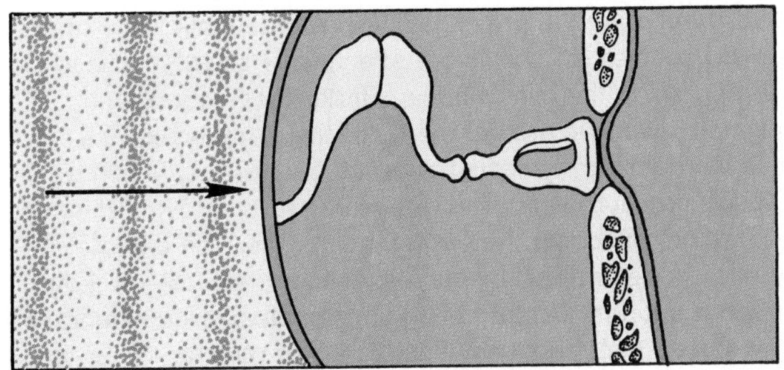
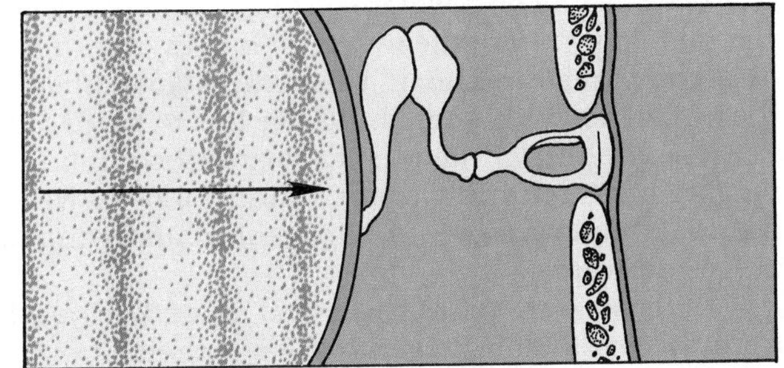

some directional sensitivity. Sound waves approaching the side of the head have reasonably direct access to the tympanic membrane on that side, whereas those arriving from another direction must bend around corners or pass through the pinna or other body tissues. Each of these routes will reduce the energy content of the sound and make it less likely to affect the tympanic membrane.

The tympanic membrane provides the surface for sound collection, and it vibrates in response to sound waves with frequencies between approximately 20 and 20,000 Hz. The vibrational frequency depends on the frequency of the sound, and the amount of movement depends on its intensity. The higher the frequency, the faster the rate of vibration; the greater the intensity, the greater the distance the membrane moves as it vibrates.

The tympanic membrane converts the sound waves into mechanical movements. Three tiny bones, the **auditory ossicles** of the middle ear, amplify the movements of the tympanic membrane and conduct the vibrations to the inner ear. The mechanism is indicated in Figure 17-17b,c. The tympanic membrane is bound to the **malleus**, and the **stapes** is attached to the membranous oval window of the cochlea. The **incus** lies between the malleus and the stapes. Because of the way these ossicles are connected, an in-out movement of the tympanic membrane will produce a rocking motion of the stapes. In the process, the amount of movement increases markedly. From our earlier discussion of lever action and muscle power, you will recall that the same applied force can move a heavy object a short distance or a light object a long distance. Because the tympanic membrane is 22 times larger and heavier than the oval window, a 1-μm movement of the tympanic membrane produces a 22-μm deflection of the oval window.

Movement at the oval window applies pressure to the perilymph of the **vestibular duct** (*scala vestibuli*). As indicated in Figure 17-18, the vestibular duct is connected to the **tympanic duct** (*scala tympani*) at the tip of the cochlear spiral. The tympanic duct also has a membranous region, the round window. The cochlear duct (*scala media*) lies between the vestibular duct and the tympanic duct. (This figure ignores the anatomical details and focuses attention on the functional relationships between these chambers.)

Liquids are virtually incompressible. If you have ever tried sitting on a waterbed, you know that when you push down *here* the waterbed bulges over *there.* In a comparable manner, when the stapes pushes the oval window in, something else has to bulge out. Because the rest of the cochlea is sheathed in bone, bulging can occur only at the round window. When the oval window moves inward, the round window bulges outward.

A sustained pressure would force perilymph from the vestibular duct into the tympanic duct, as indicated in Figure 17-18a. But instead of applying

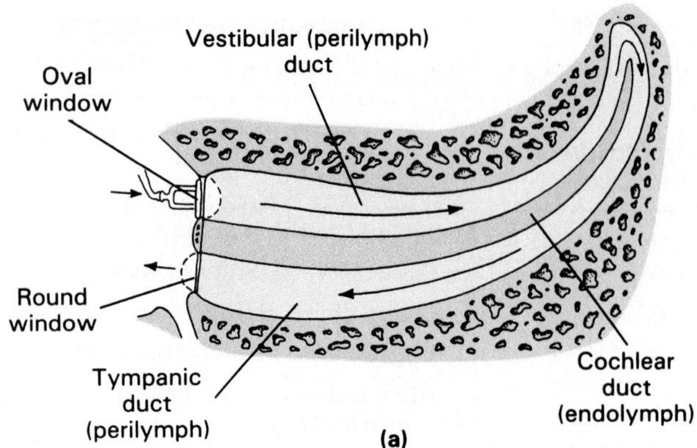

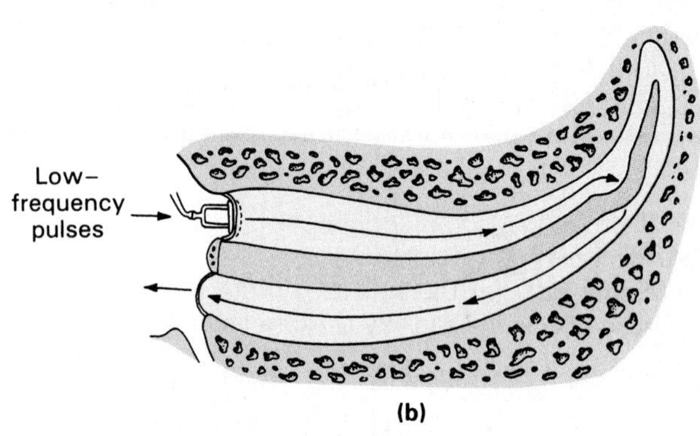

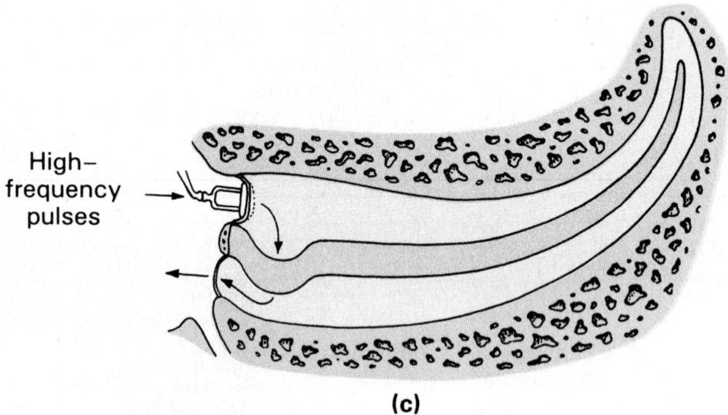

FIGURE 17-18
Principles of Cochlear Function. In the inner ear, when the oval window moves inward, the round window moves outward, relieving the pressure. This diagrammatic view of the uncoiled cochlea (a) shows the effects of a slow, sustained push applied to the oval window. Under these conditions perilymph will move around the tip of the spiral where the vestibular duct and tympanic duct are continuous. When a pulsating pressure is applied, pressure waves move through the cochlea, distorting the basilar membrane. The site of distortion is determined by the frequency of stimulation. Low-frequency sounds affect distal portions of the basilar membrane (b), whereas high-frequency sounds affect regions closer to the round window (c).

a sustained pressure, the stapes vibrates against the oval window, generating a series of brief pressure pulses. The louder the sound, the more forceful the movement, and the higher the frequency, the shorter the duration of the individual pulse. Except at extremely low frequencies (below 20 Hz) there is no fluid movement from the vestibular duct to the tympanic duct. Instead, the pressure waves take a shortcut through the cochlear duct, producing a local distortion (Figure 17-18b,c).

The precise location of the distortion varies, according to the frequency of the pressure waves. The basilar membrane near the round window is relatively stiff and narrow, whereas the distal tip is relatively broad and loose. High-frequency pulses affect the basilar membrane near the oval window; the stiff membrane responds like a tightly strung guitar string. The lower the frequency of the sound, the farther away from the oval window the distortion will be. The actual amount of movement at a given location will depend on the amount of force applied to the oval window. Hair cells in the cochlear duct can therefore provide information on the frequency and intensity of a particular sound by reporting *where* and *how much* distortion occurs.

Receptor Function in the Cochlea

Figure 17-19a shows the anatomical organization of the cochlea, which coils around a central spire containing the cell bodies of sensory neurons. The hair cells of the cochlear duct are found in the **organ of Corti**. This sensory layer sits above the **basilar membrane** that separates the cochlear duct from the tympanic duct.

The hair cells of the organ of Corti are arranged in a series of longitudinal rows with their stereocilia contacting the overlying **tectorial membrane** (tek-TŌR-ē-al; *tectum*, roof). This membrane is firmly attached to the inner wall of the cochlear duct. When a portion of the basilar membrane bounces up and down, the stereocilia of the hair cells are distorted (Figure 17-19b), producing graded receptor potentials. The greater the amount of distortion, the larger the receptor potential; the larger the receptor potential, the greater the stimulation of the sensory neuron with which the receptor synapses.

A single row of hair cells will be stimulated by a faint sound; as the intensity of the sound increases, more and more rows become involved. The organ of Corti therefore provides information to the CNS in

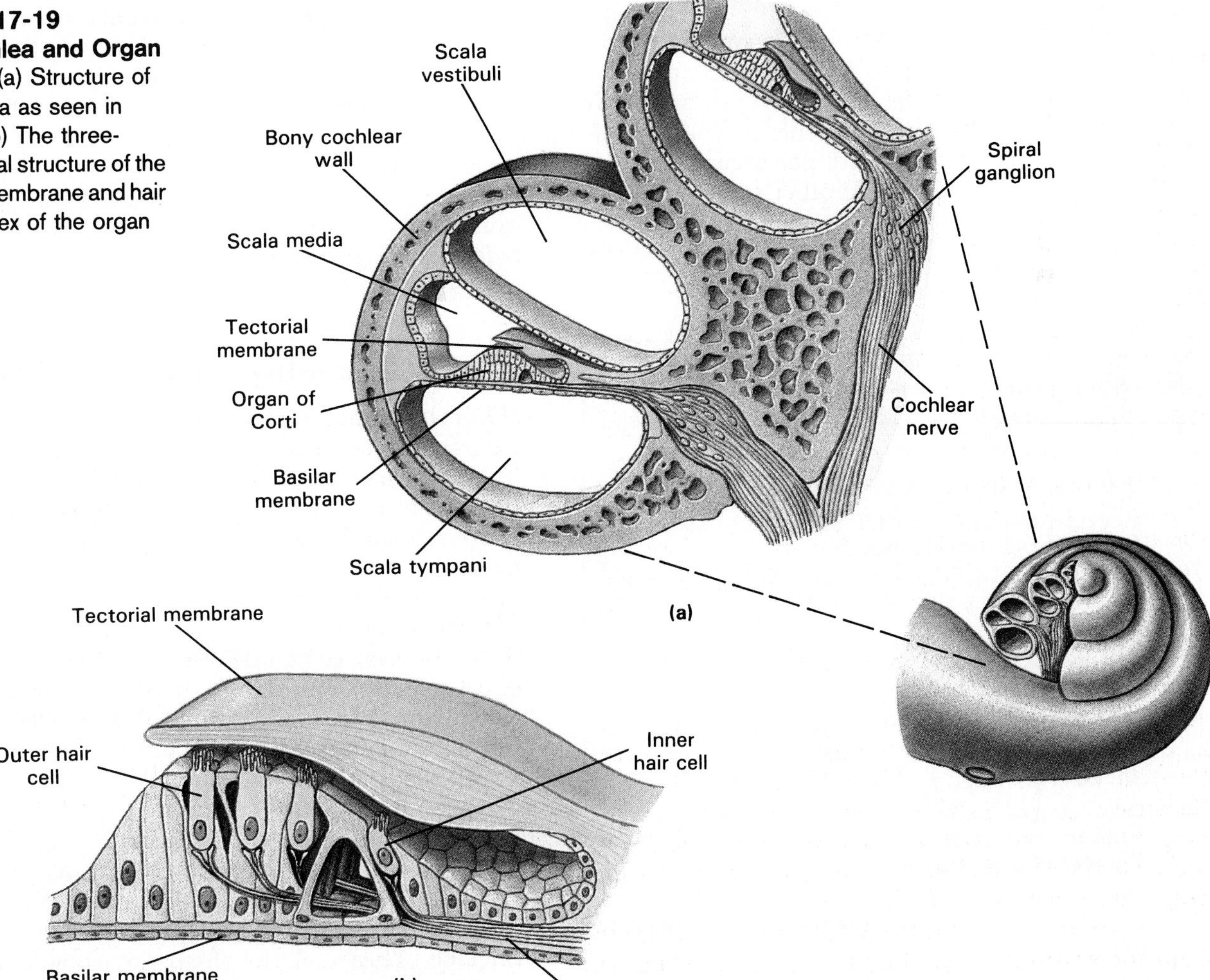

FIGURE 17-19
The Cochlea and Organ of Corti. (a) Structure of the cochlea as seen in section. (b) The three-dimensional structure of the tectorial membrane and hair cell complex of the organ of Corti.

two different ways. The frequency is indicated by the region of stimulation. The intensity is indicated by the rate at which action potentials are sent along sensory neurons monitoring a single row of hair cells (faint sounds) or multiple rows (loud sounds).

Very high intensity sounds produce hearing losses by actually breaking the stereocilia off the surfaces of the hair cells. Two muscles, the **tensor tympani** (TEN-sor tim-PAN-ē) and the **stapedius** (stā-PĒ-dē-us), reduce the degree of amplification when they contract. The arrival of a loud sound triggers their contraction through a cranial reflex that activates these muscles in less than a tenth of a second (see Table 14-12). ∞ (p. 468) Unfortunately, this is not necessarily fast enough, and the damage done prior to their contraction may be both serious and painful.

Table 17-3 summarizes the steps involved in translating a sound wave into an auditory sensation.

Auditory Sensitivity

Our hearing abilities are remarkable, though it is difficult to assess the absolute sensitivity of the system. From the softest audible sound to the loudest tolerable blast represents a trillionfold increase in power. Theoretically, if we were to remove the stapes, the receptor mechanism is so sensitive that we could "hear" the sound of air molecules bouncing off the oval window, responding to displacements as small as one-tenth the diameter of a hydrogen atom. We never utilize the full potential of this system because body movements and our internal organs produce squeaks, groans, thumps, and other sounds that are "tuned out" by central and peripheral adaptation. When other environmental noises fade away, the level of adaptation drops and the system becomes increasingly sensitive. If we relax in a quiet room, our heartbeat seems to get louder and louder as the auditory system adjusts to the level of background noise.

Young children have the greatest hearing range; they can detect sounds ranging from a 20-Hz buzz to a 20,000-Hz whine. With age damage due to loud noises or other injuries accumulates; the eardrum gets less flexible, the articulations between the ossicles stiffen, and the round window may begin to ossify. As a result, older individuals show some degree of hearing loss.

Hearing Loss There are probably over 6 million people in the United States alone who have at least a partial hearing deficit. **Conductive deafness** results from conditions in the middle ear that block the normal transfer of vibration from the tympanic membrane to the oval window. A plugged external auditory canal from accumulated wax or trapped water may cause a temporary hearing loss. Scarring or perforation of the tympanic membrane and immobilization of one or more of the auditory ossicles are more serious examples of conduction deafness.

In **nerve deafness** the problem lies within the cochlea or somewhere along the auditory pathway. The vibrations are reaching the oval window, but the receptors either cannot respond or their response cannot reach its central destinations. Toxic drugs entering the endolymph may kill the receptors, and bacterial infections may damage the hair cells or affect the cochlear nerve. Hair cells can also be damaged by exposure to aminoglycoside antibiotics, such as neomycin or gentamicin; this potential side effect must be balanced against the severity of infection before these drugs are prescribed. [**CM**: *Testing and Treating Hearing Deficits*]

TABLE 17-3 Steps in the Production of an Auditory Sensation

1. Sound waves arrive at the tympanic membrane.
2. Movement of the tympanic membrane causes displacement of the malleus.
3. Movement of the malleus causes movement of the incus and stapes.
4. Movement of the stapes against the oval window establishes pressure waves in the perilymph of the vestibular duct.
5. The pressure wave distorts the basilar membrane on its way to the round window of the tympanic duct.
6. Distortion of the basilar membrane forces the hair cells of the organ of Corti toward or away from the tectorial membrane.
7. This movement leads to displacement of stereocilia and stimulation of sensory neurons of the cochlear nerve.

Central Processing of Auditory Information

Hair cell stimulation leads to the activation of sensory neurons that carry auditory information to the brain. The neuronal somae form the **spiral ganglion** in the bony hub of the cochlea. Their afferent fibers form the **cochlear branch** of the vestibulcochlear nerve (N VIII). These axons enter the medulla and synapse at the **cochlear nucleus**. From here the information crosses to the opposite side of the brain and ascends to the inferior colliculus of the midbrain. This processing center coordinates a number of responses to acoustic stimuli, including auditory reflexes involving skeletal muscles of the head, face, and trunk. These reflexes automatically change the position of the head in response to a sudden loud noise.

Before reaching the cerebral cortex and our conscious awareness, ascending auditory sensations synapse in the thalamus. Projection fibers then deliver the information to the auditory cortex of the temporal lobe. These relationships are shown in Figure 17-20.

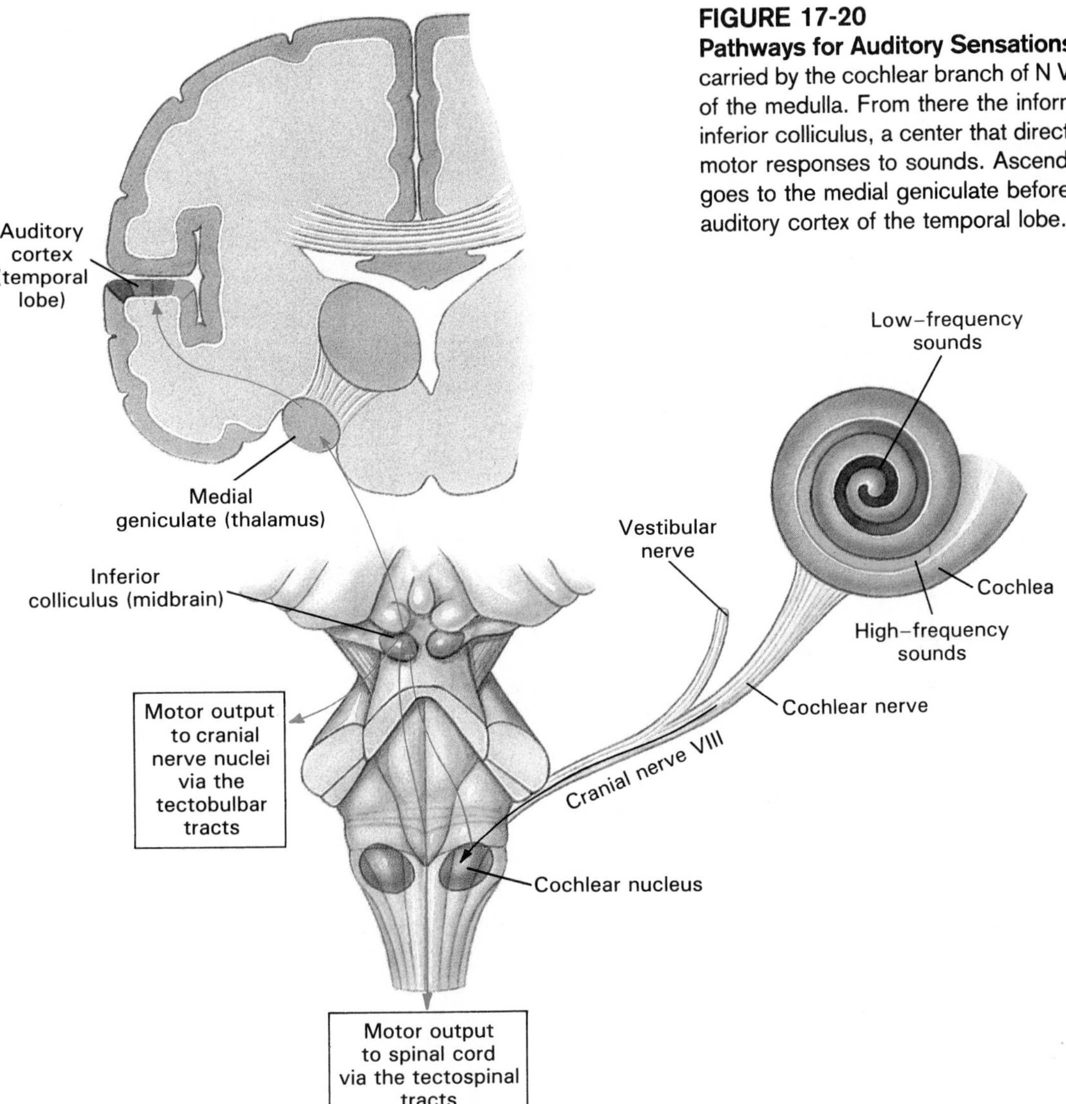

FIGURE 17-20
Pathways for Auditory Sensations. Auditory sensations are carried by the cochlear branch of N VIII to the cochlear nucleus of the medulla. From there the information is relayed to the inferior colliculus, a center that directs a variety of unconscious motor responses to sounds. Ascending acoustic information goes to the medial geniculate before being forwarded to the auditory cortex of the temporal lobe.

The auditory cortex contains a map of the organ of Corti. High-frequency sounds activate one portion of the cortex and low-frequency sounds affect another. If the auditory cortex is damaged, the individual will respond to sounds and have normal acoustic reflexes, but sound interpretation and pattern recognition will be difficult or impossible. Damage to the adjacent association area leaves the ability to detect the tones and patterns, but produces an inability to comprehend their meaning.

An intact auditory cortex is also important in determining the relative direction from which a particular sound originated. With the assistance of other nuclei along the auditory pathway, the cortical neurons compare the time at which a sound arrives at one ear with the time at which it arrives at the other. For example, if a sound arrives at the right ear before it reaches the left, the source must be closer to the right ear. Other clues, such as the relative intensity of the sound at each ear, may supply additional information and permit more precise localization. Behavioral responses, such as turning the head, provide comparative data that help locate the source. Because these processing, integration, and behavioral tests take time, it is difficult to determine the source of a very brief, very soft sound.

- √ **Why does blockage of the pharyngotympanic tube produce an earache?**
- √ **If the round window were not able to bulge out with increased pressure in the perilymph, how would sound perception be affected?**
- √ **How would loss of stereocilia from the hair cells of the organ to Corti affect hearing?**

■ The Eyes and Vision

Unlike many other animals, humans rely more on vision than on any other sense. Our visual receptors

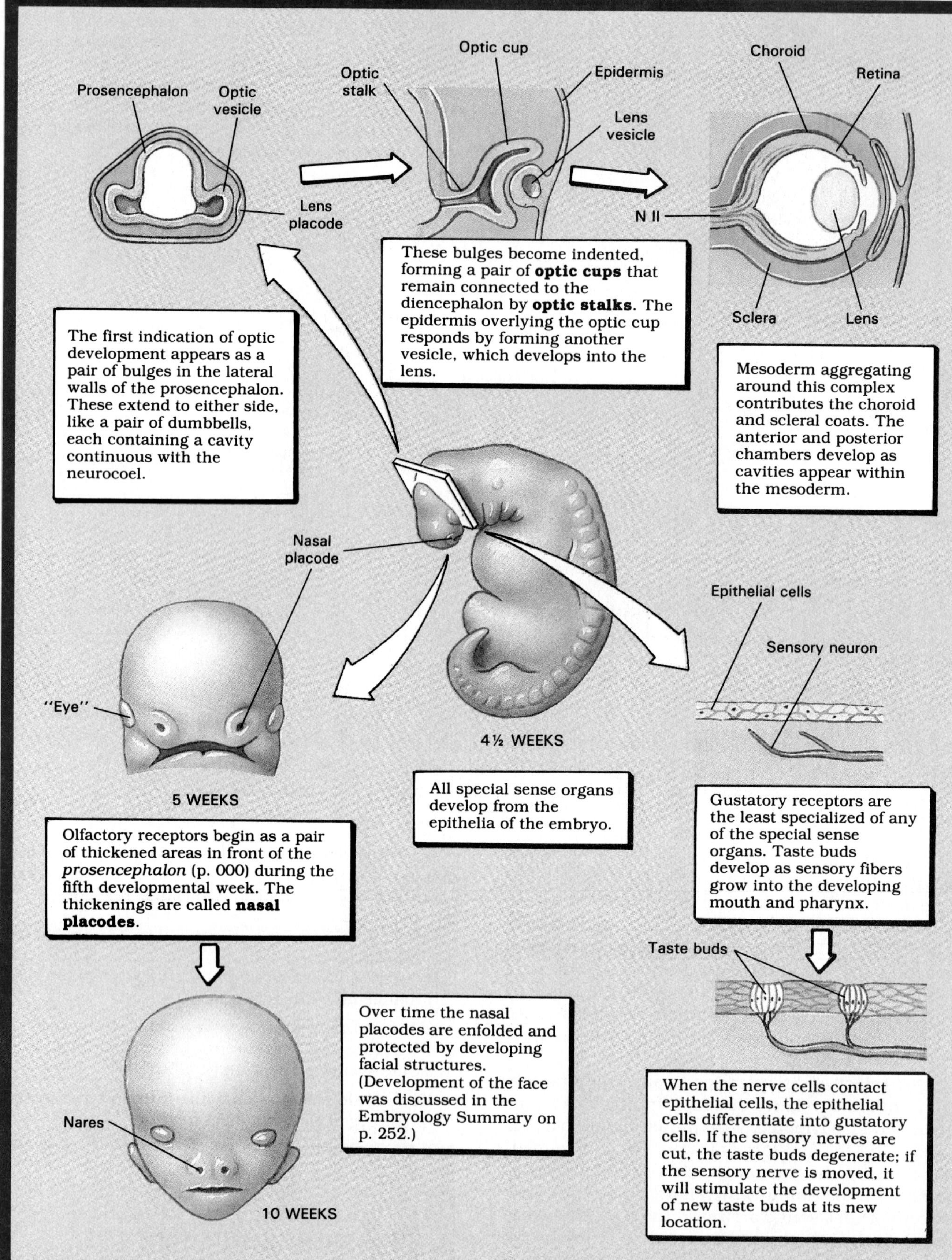

Prosencephalon
Optic vesicle
Lens placode
Optic stalk
Optic cup
Epidermis
Lens vesicle
Choroid
Retina
N II
Sclera
Lens
These bulges become indented, forming a pair of **optic cups** that remain connected to the diencephalon by **optic stalks**. The epidermis overlying the optic cup responds by forming another vesicle, which develops into the lens.
The first indication of optic development appears as a pair of bulges in the lateral walls of the prosencephalon. These extend to either side, like a pair of dumbbells, each containing a cavity continuous with the neurocoel.
Mesoderm aggregating around this complex contributes the choroid and scleral coats. The anterior and posterior chambers develop as cavities appear within the mesoderm.
Nasal placode
Epithelial cells
Sensory neuron
"Eye"
4½ WEEKS
5 WEEKS
All special sense organs develop from the epithelia of the embryo.
Olfactory receptors begin as a pair of thickened areas in front of the *prosencephalon* (p. 000) during the fifth developmental week. The thickenings are called **nasal placodes**.
Gustatory receptors are the least specialized of any of the special sense organs. Taste buds develop as sensory fibers grow into the developing mouth and pharynx.
Taste buds
Over time the nasal placodes are enfolded and protected by developing facial structures. (Development of the face was discussed in the Embryology Summary on p. 252.)
Nares
10 WEEKS
When the nerve cells contact epithelial cells, the epithelial cells differentiate into gustatory cells. If the sensory nerves are cut, the taste buds degenerate; if the sensory nerve is moved, it will stimulate the development of new taste buds at its new location.

Embryology Summary: Development of Special Sense Organs

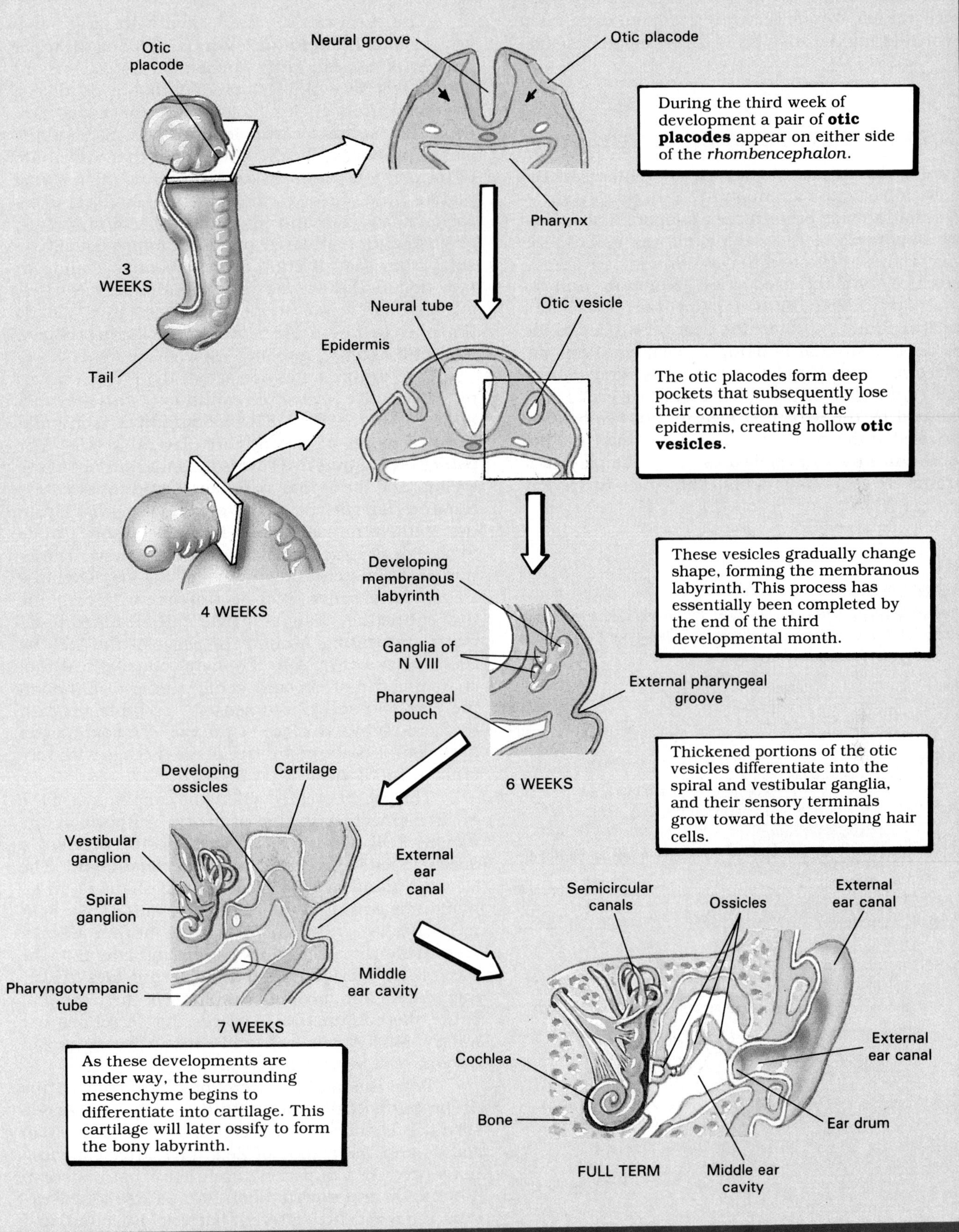
Otic placode
Neural groove
Otic placode
During the third week of development a pair of **otic placodes** appear on either side of the *rhombencephalon*.
Pharynx
3 WEEKS
Tail
Neural tube
Otic vesicle
Epidermis
The otic placodes form deep pockets that subsequently lose their connection with the epidermis, creating hollow **otic vesicles**.
4 WEEKS
Developing membranous labyrinth
These vesicles gradually change shape, forming the membranous labyrinth. This process has essentially been completed by the end of the third developmental month.
Ganglia of N VIII
External pharyngeal groove
Pharyngeal pouch
6 WEEKS
Thickened portions of the otic vesicles differentiate into the spiral and vestibular ganglia, and their sensory terminals grow toward the developing hair cells.
Developing ossicles
Cartilage
Vestibular ganglion
Spiral ganglion
External ear canal
Pharyngotympanic tube
Middle ear cavity
7 WEEKS
As these developments are under way, the surrounding mesenchyme begins to differentiate into cartilage. This cartilage will later ossify to form the bony labyrinth.
Semicircular canals
Ossicles
External ear canal
External ear canal
Cochlea
Bone
Ear drum
FULL TERM
Middle ear cavity

are contained in elaborate structures, the eyes, which enable us not only to detect light but to create detailed visual images. We will therefore begin our discussion by considering the anatomy of these complex organs.

ACCESSORY STRUCTURES OF THE EYE

Figure 17-21 details the superficial anatomy of the eye, with particular attention to the *accessory structures* that provide protection or support. The **accessory structures** of the eye include the eyelids, the superficial epithelium of the eye, and the structures associated with the production, secretion, and removal of tears. The eyelids, or **palpebrae** (pal-PĒ-brē), can close firmly to protect the delicate surface of the eye. Their continual blinking movements keep the surface lubricated and free from dust and debris. The free margins of the upper and lower eyelids are separated by the **palpebral fissure**, but the two are connected at the **medial canthus** (KAN-thus) and the **lateral canthus**. The **eyelashes** along the palpebral margins are very robust hairs. These help to prevent foreign particles and insects from reaching the surface of the eye.

FIGURE 17-21
Accessory Structures of the Eye. (a) Gross and superficial anatomy of the accessory structures. (b) Details of the organization of the lacrimal apparatus.

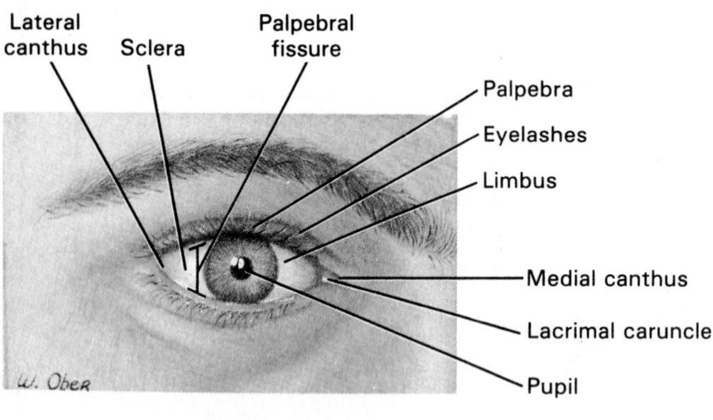

(a)

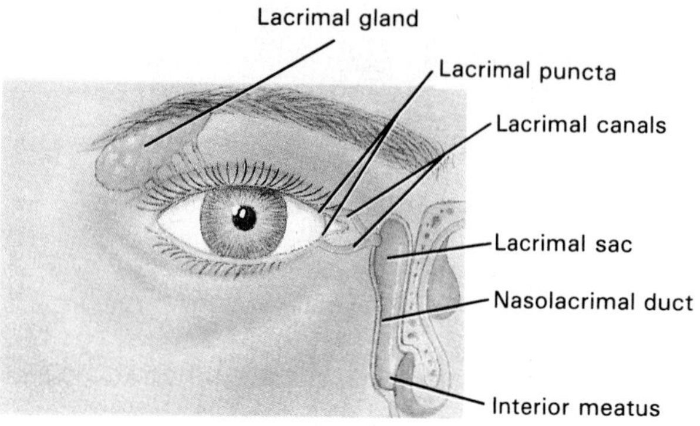

(b)

The eyelashes are associated with large sebaceous glands, the *glands of Zeis* (zīs). Along the inner margin of the lid, large **Meibomian** (mī-BŌ-mē-an) **glands** secrete a lipid-rich product that helps to keep the eyelids from sticking together. At the medial canthus the **lacrimal caruncle** (KAR-unk-ul) contains glands producing the thick secretions that contribute to the gritty deposits occasionally found after a good night's sleep. These various glands are subject to occasional invasion and infection by bacteria. A cyst, or **chalazion** (kah-LA-zē-on; small lump) usually results from the infection of a Meibomian gland. An infection in a sebaceous gland of one of the eyelashes, a Meibomian gland, or one of the many sweat glands that open to the surface between the follicles produces a painful localized swelling known as a **sty**.

The visible outer surface of the eye is covered by a distinctive epithelium continuous with the inner lining of the eyelids. This **conjunctiva** (kon-junk-TĪ-va) consists of two or three layers of cells. The **palpebral conjunctiva** covers the inner surface of the eyelids, and the **bulbar**, or **ocular**, **conjunctiva** covers the anterior surface of the eye. Goblet cells within the epithelium assist the various accessory glands in providing a superficial lubricant that prevents friction and moistens the surface of the eye. Over the transparent **cornea** (KŌR-nē-a) of the eye the conjunctival epithelium changes to a delicate stratified squamous epithelium. Around the edges of the lids the conjunctiva converts to the more robust epithelium characteristic of exposed bodily surfaces. Although there are no specialized sensory receptors beneath the conjunctival surface of the eye, the conjunctiva itself contains abundant free nerve endings with very broad sensitivities. [**CM**: *Conjunctivitis*]

The pocket created where the conjunctiva of the eyelid connects with that of the eye is known as the **fornix** (FŌR-niks). The superior fornix receives a dozen or more ducts from the **lacrimal gland**. The lacrimal gland has the approximate shape of an almond, measuring roughly 12 × 20 mm (0.5 × 0.75 in.). It nestles within a depression in the frontal bone, just inside the orbit and superior and lateral to the eyeball. The lacrimal gland normally provides the key ingredients and most of the volume of the tears that bathe the conjunctival surfaces. Its secretions are watery, slightly alkaline, and contain **lysozyme**, an enzyme that attacks bacteria.

Once the lacrimal secretions have reached the ocular surface, they mix with the products of accessory glands and the oily secretions of the Meibomian glands and the glands of Zeis. The latter contributions produce a superficial "oil slick" that assists in lubrication and slows evaporation. Tears reduce friction, remove debris, prevent bacterial infection, and provide nutrients and oxygen to portions of the conjunctival epithelium.

Each time the eyelids flicker across the surface

of the eye, microvilli on the surface of the palpebral conjunctiva provide a fresh supply of tears with a gentle sweeping action. There is little space between the eyelids and the orbital surface, so the lacrimal gland need only produce tears at a rate of around 1 mℓ/day to fulfill its functional obligations. The tears arrive via 10–12 ducts that open into the lateral portion of the superior fornix. They circulate across the ocular surface and accumulate at the medial canthus, in an area known as the **lacus lacrimalis**, or "lake of tears."

Two small pores, the **lacrimal puncta**, drain the lacrymal lake, emptying into the **lacrimal canals** that run along grooves in the surface of the lacrimal bone. These passageways terminate at the **lacrimal sac**, and from there the **nasolacrimal duct** carries the tears to the inferior meatus of the nose. Blockage of the lacrimal puncta or oversecretion of the lacrimal glands can produce "watery eyes" that are constantly overflowing. Dry eyes due to inadequate tear production are far more common. Lubricating "artificial tears" in the form of eye drops are the usual answer, but more serious cases may be treated by surgically closing the lacrimal puncta.

ANATOMY OF THE EYE

The specialized photoreceptors of the visual system form a single layer within the eyeball. Each roughly spherical eyeball has a diameter of nearly 2.5 cm (1 in.), and weighs around 8 g (0.28 oz). The eyeball sits within the orbit, in the company of the extrinsic oculomotor muscles, the lacrimal gland, and the various cranial nerves and blood vessels that service the eye and adjacent portions of the orbit and face. Any orbital space not committed to other structures contains a mass of **orbital fat** that provides padding and insulation.

The anatomical organization of the eye can most easily be understood by considering a sectional view, as presented in Figure 17-22. There are three distinct layers: an outer *fibrous tunic*, a medial *vascular tunic*, and an inner *neural tunic*.

The Fibrous Tunic

The **fibrous tunic**, the outermost layer covering the eye, consists of the *sclera* and the *cornea*. The fibrous tunic: (1) provides mechanical support and some de-

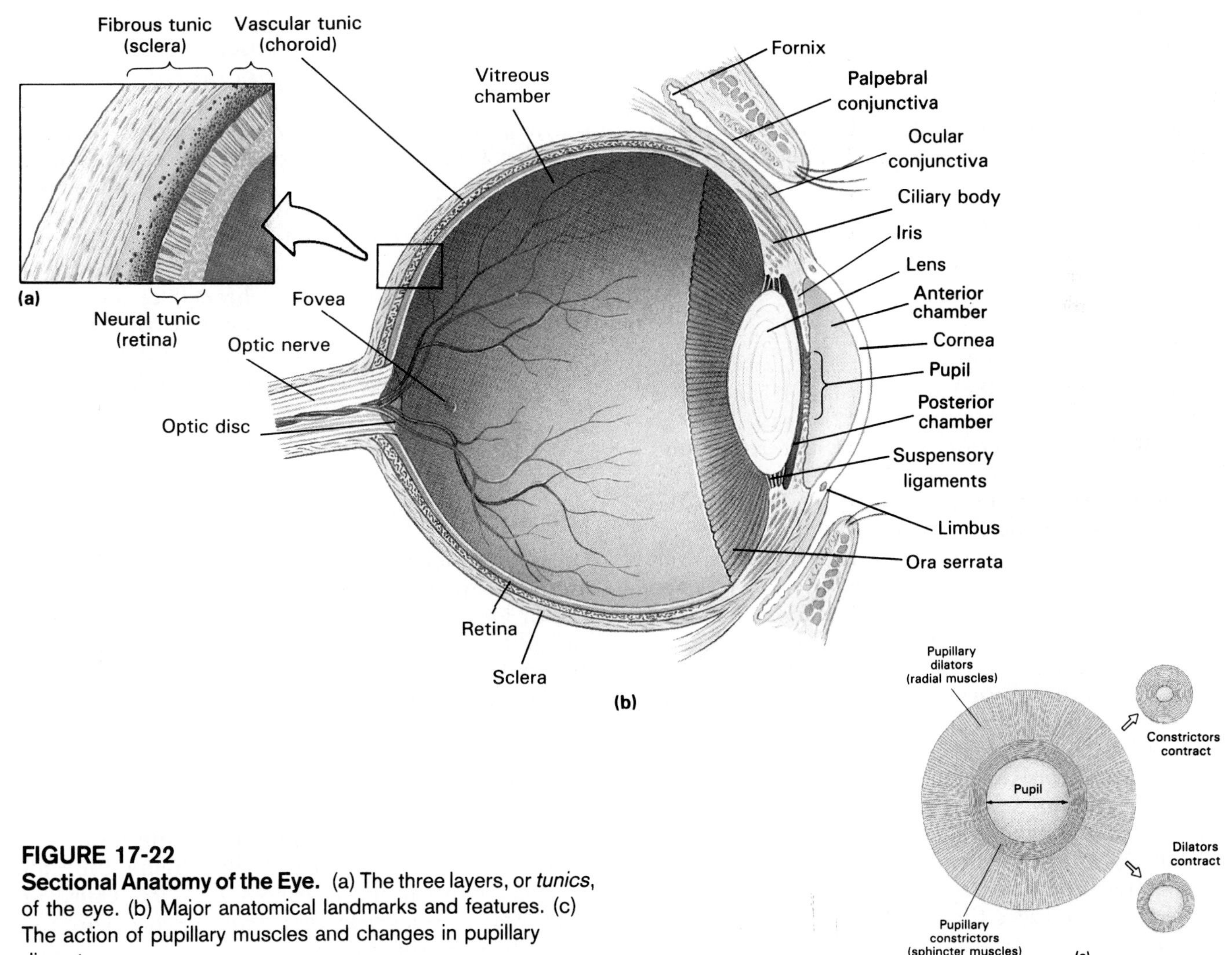

FIGURE 17-22
Sectional Anatomy of the Eye. (a) The three layers, or *tunics*, of the eye. (b) Major anatomical landmarks and features. (c) The action of pupillary muscles and changes in pupillary diameter.

gree of physical protection, (2) serves as an attachment site for the extrinsic eye muscles, and (3) assists in the focusing process.

Most of the ocular surface is covered by the **sclera** (SKLER-a). The sclera consists of a dense fibrous connective tissue containing both collagen and elastin fibers. This layer is thickest at the back of the eye, near the exit of the optic nerve, and thinnest over the anterior surface. The six extrinsic oculomotor muscles insert upon the sclera, blending their collagenous fibers with those of the outer tunic.

The connective tissue covering the sclera contains small blood vessels and nerves, and those communicating with the internal tissues of the eye must penetrate the sclera. The network of small capillaries that lies beneath the ocular conjunctiva usually does not carry enough blood to lend an obvious color to the sclera, and the collagen fibers are visible as the "white of the eye."

The transparent cornea is structurally continuous with the sclera; the **limbus** is the border between the two. The dense collagenous fibers of the cornea are organized into a series of layers that do not interfere with the passage of light. There are no blood vessels between the cornea and the overlying conjunctiva, and the superficial epithelial cells must obtain oxygen and nutrients from the tears that flow across their free surfaces. When light travels from the air into the relatively dense cornea, the light path is bent. This effect is an important first step in the focusing process, described in a later section. [**CM**: *Corneal Transplants*]

The Vascular Tunic

The **vascular tunic**, or **uvea**, contains numerous blood vessels, lymphatics, and all of the intrinsic oculomotor muscles. The functions of this layer include: (1) providing a route for blood vessels and lymphatics that supply tissues of the eye, (2) regulating the amount of light entering the eye, (3) secreting and reabsorbing the *aqueous humor* that circulates within the eye, and (4) controlling the shape of the lens, an essential part of the focusing process.

The vascular tunic includes three distinct structures, the **iris**, the **ciliary body**, and the **choroid** (see Figure 17-22). The iris can be seen through the transparent corneal surface. The iris contains blood vessels, pigment cells, and two layers of smooth muscle fibers. When these muscles contract they change the diameter of the central opening, or **pupil**, of the iris. One group of smooth muscle fibers forms a series of concentric circles around the pupil (Figure 17-23c). When these **pupillary constrictor muscles** contract, the diameter of the pupil decreases. A second group of smooth muscles extends radially away from the edge of the pupil. Contraction of these **pupillary dilator muscles** enlarges the pupil. Both muscle groups are controlled by the autonomic nervous system.

Pupillary reflexes are triggered by sudden changes in light intensity. Exposure to bright light produces a rapid reflexive decrease in pupillary diameter, under parasympathetic stimulation. A sudden reduction in light levels produces a much slower pupillary dilation, under the control of the sympathetic division. Pupillary diameter ranges from 1.5–2 mm in bright light to 7–8 mm in the dark. Illumination changes affecting either eye will cause reflexive adjustments in both pupils (the consensual light reflex, described in Chapters 14 and 16). (pp. 469, 517)

The space between the iris and the cornea constitutes the **anterior chamber** of the eye (Figure 17-23a). The body of the iris consists of a connective tissue that contains abundant pigment cells, and other pigment cells are found in the epithelium covering the posterior surface of the iris. The thickness of the iris and number and distribution of pigment cells determine its apparent color. When there are many pigment cells in the body of the iris, the eye appears brown. When there are very few, light passes through the iris and is reflected off the inner surface of the pigmented epithelium, giving the iris a bluish coloration.

At the periphery the iris attaches to the anterior portion of the ciliary body. The bulk of the ciliary body consists of the **ciliary muscle**, a muscular ring that projects into the interior of the eye. The ciliary body begins at the junction between the cornea and sclera and extends posteriorly to the **ora serrata** (Ō-ra ser-RĀT-a; serrated mouth). The epithelium is thrown into numerous folds, called **ciliary processes**. The **suspensory ligaments** of the lens attach to these processes. These fibers hold the lens posterior to the iris and centered on the pupil. As a result, any light passing through the pupil and headed for the photoreceptors will have to pass through the lens. The space between the suspensory ligament and the iris represents the **posterior chamber** of the eye. Like the anterior chamber, this region contains a fluid, the **aqueous humor**.

The Aqueous Humor

There are many blood vessels in the connective tissues in and around the ciliary processes. Aqueous humor forms as the interstitial fluids pass between the epithelial cells of the ciliary processes and enter the posterior chamber. The epithelial cells appear to regulate its composition, which resembles that of cerebrospinal fluid. The aqueous humor circulates, as diagrammed in Figure 17-23b, so that in addition to forming a fluid cushion it provides an important route for nutrient and waste transport. Aqueous humor returns to the circulation in the anterior chamber near the edge of the iris. After diffusing through the local epithelium, it passes into the **canal of Schlemm**, a passageway that communicates with the veins servicing the eye. [**CM**: *Glaucoma*]

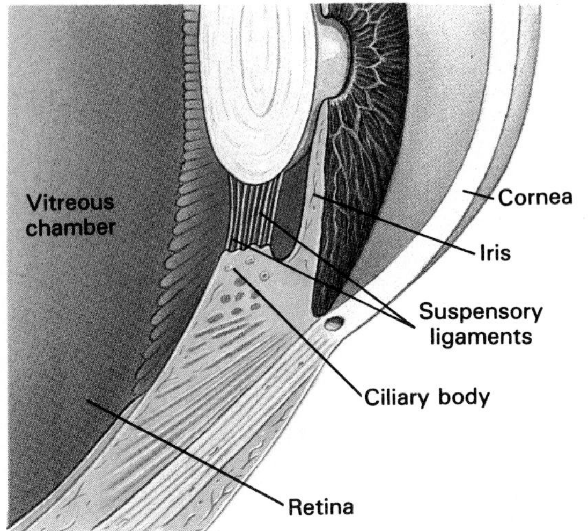

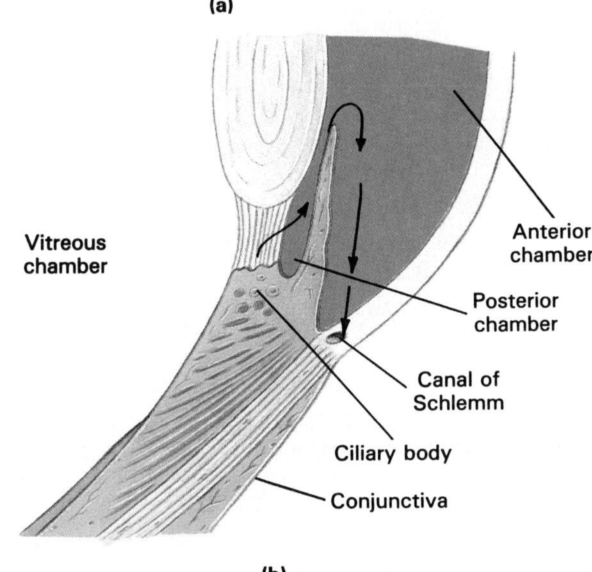

FIGURE 17-23
The Circulation of Aqueous Humor. (a) The lens is suspended between the vitreous chamber and the posterior chamber. Its position is maintained by the suspensory ligaments that attach the lens to the ciliary body. (b) Aqueous humor secreted at the ciliary body circulates through the posterior and anterior chambers before being reabsorbed via the canal of Schlemm.

The Vitreous Chamber

The **lens** lies behind the cornea, held in place by the suspensory ligaments that originate on the ciliary body of the choroid. The lens and its attached suspensory ligaments form the anterior boundary of the **vitreous chamber**. This chamber contains the **vitreous body**, a gelatinous mass sometimes called the "vitreous humor." The vitreous body helps to stabilize the shape of the eye and gives physical support to the retina. Specialized cells embedded within the vitreous body produce the collagen fibers and proteoglycans, especially hyaluronic acid, that account for its consistency. Aqueous humor produced in the posterior chamber freely diffuses through the vitreous body and across the retinal surface.

The Neural Tunic

The neural tunic consists of two distinct layers, an outer **pigment layer** and an inner **retina** that contains the visual receptors and associated neurons. The pigment layer (1) absorbs light after it passes through the retina and (2) has important biochemical interactions with the photoreceptor layer of the retina. The retina contains (1) the photoreceptors that respond to light, (2) supporting cells and neurons that perform preliminary processing and integration of visual information, and (3) blood vessels supplying tissues lining the vitreous chamber.

The retina and pigment layers are normally very close together, but not tightly interconnected. The pigmented layer is continuous with the pigmented epithelium of the ciliary body and iris. The retina extends anteriorly only as far as the ora serrata. The retina thus forms a cup that establishes the posterior and lateral boundaries of the vitreous chamber.

THE LENS AND IMAGE FORMATION

The primary function of the lens is to focus the visual image on the retinal receptors. It accomplishes this by changing its shape.

Structure of the Lens

The lens consists of concentric layers of cells that are precisely organized. Cells deep inside the lens are inert; they have lost most of their organelles, and the cytoplasm is filled with regularly oriented microfilaments. Cells at the surface remain metabolically active and maintain the structure of the lens.

A dense fibrous capsule covers the entire lens. Many of the capsular fibers are elastic, and unless an outside force is applied, they will contract and make the lens spherical. Around the edges of the lens, the capsular fibers intermingle with those of the suspensory ligaments. Tension in the suspensory ligaments overpowers the elastic capsule and gives the lens the shape of a flattened oval.

The transparency of the lens depends on a precise combination of structural and biochemical characteristics. When that balance becomes disturbed the lens loses its transparency, and the abnormal lens is known as a **cataract**. Cataracts may result from drug reactions, injuries, or radiation, but **senile cataracts** are the most common form. As aging proceeds, the lens becomes less elastic, takes on a yellowish hue, and eventually begins to lose its transparency. As the lens becomes opaque, or "cloudy," the individual needs brighter and brighter reading lights, and visual clarity begins to fade. When the lens becomes completely opaque, the person will be functionally blind, even though the retinal receptors are alive and well. Modern surgical procedures involve removing the lens, either intact or in pieces, after shattering it with high-frequency sound. The missing lens can be replaced by an artificial substitute, and vision can then be fine-tuned with glasses or contact lenses.

The Lens and Accommodation

There are approximately 130 million photoreceptors in the retina, each monitoring a specific location. A visual image results from the processing of information provided by the entire receptor population. The eye has often been compared to a camera, and like a camera the arriving image must be in focus if it is to provide any useful information. "In focus" means that the rays of light arriving fróm the object strike the sensitive surface of the film precisely ordered so as to form a miniature image of the original. If they are not perfectly focused, the image will be blurry. Focusing normally occurs in two steps, as light passes through the cornea and lens.

Light is bent, or **refracted**, when it passes between media of differing densities. You can see refraction clearly by sticking a pencil into a glass of water. Because refraction occurs as the light passes into the air from the much denser water, the pencil shaft appears to bend sharply at the air-water interface. In the human eye, the greatest amount of refraction occurs when light passes from the air into the cornea. Corneal tissues and the aqueous humor have similar

Sports and Fitness: Underwater Vision

The cornea loses its refractive powers when immersed in water, for light then passes from one fluid into another of nearly identical density. Without assistance from the cornea, the lens cannot refract the incoming light enough to produce a focused image. As a result, things look blurry and indistinct when eyes are opened under water. Vision can be improved by wearing a mask or goggles, which place a pocket of air in front of the cornea. Refraction then occurs at the front of the mask (water to glass to air) before the light reaches the cornea. Although the cornea now functions normally, and vision is clear, the mask refraction magnifies the image by about 25 percent. Objects therefore seem much larger and closer when viewed under water.

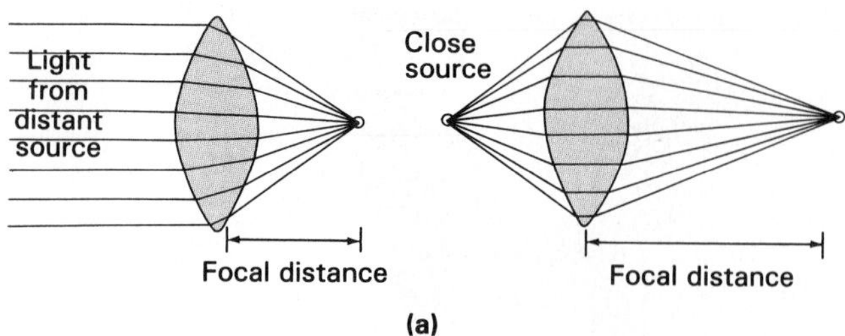

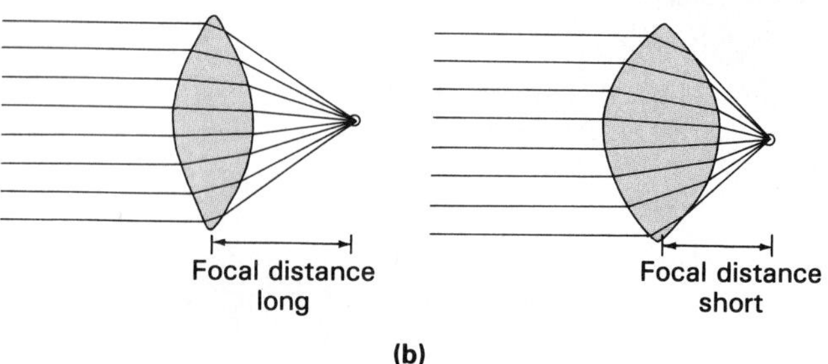

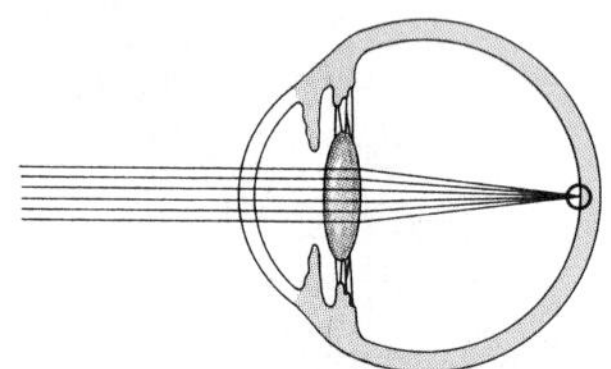

Ciliary muscle relaxed, lens flattened for distant vision

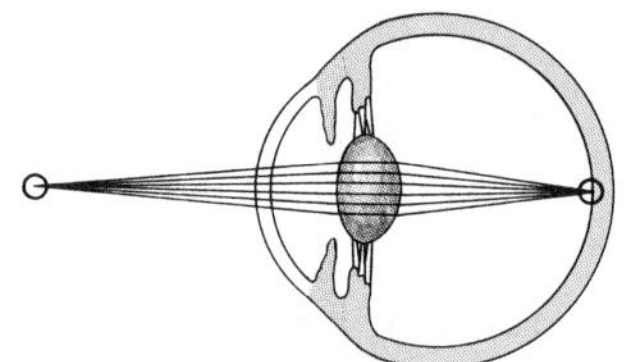

Ciliary muscle contracted, lens rounded for close vision
(c)

FIGURE 17-24
Visual Accommodation. (a) A lens refracts light toward a specific point. The distance from the center of the lens to that point is the focal distance of the lens. Light from a distant source arrives with all of the light waves traveling parallel to one another. Light from a nearby source will still be diverging when it strikes the lens. Note the difference in focal distance after refraction. (b) The rounder the lens, the shorter the focal distance. (c) For the eye to form a sharp image, the focal distance must equal the distance between the center of the lens and the retinal receptors, which cannot be varied. The lens compensates for variations in the distance between the eye and the object in view by changing its shape, a process called accommodation. The lens changes shape when tension changes in the suspensory ligaments. When the ciliary muscle relaxes, the ligaments pull against the margins of the lens and flatten it. When the ciliary muscle contracts, the suspensory ligaments allow the lens to round up.

densities, so little refraction occurs at their interface, but significant additional refraction takes place when the light enters the relatively dense lens.

The lens provides the additional refraction needed to focus the light rays from an object toward a specific **focal point**. The distance between the center of the lens and the focal point is the **focal distance** (Figure 17-24). The focal distance is determined by:

1. *The distance of the object from the lens*: The focal distance increases as an object moves closer to a lens, as indicated in Figure 17-24a.
2. *The shape of the lens*: The rounder the lens, the more refraction occurs, so a very round lens has a shorter focal distance than a flatter one (Figure 17-24b).

During **accommodation** the lens becomes rounder to focus the image of a nearby object on the retina. The mechanism of accommodation is diagrammed in Figure 17-24c. The lens is held in place by the suspensory ligaments that originate at the ciliary body. Smooth muscle fibers in the ciliary body act like the sphincter muscles of the iris. When the ciliary muscle relaxes, the suspensory ligaments pull at the circumference of the lens, making it relatively flat. When the ciliary muscles contract, the ciliary body moves toward the lens. This movement reduces the tension in the suspensory ligaments, and the elastic capsule pulls the lens into a more spherical shape. The rounder shape increases the refractive power of the lens, enabling it to bring light from nearby objects to a focus on the retina.

As indicated in Figure 17-24c, the greatest amount of refraction is required to view objects that are very close to the lens. In the normal eye, when the ciliary muscles are relaxed and the lens is flattened, a distant image will be focused on the retinal surface (Figure 17-25a). This condition is called **em-**

FIGURE 17-25
Visual Abnormalities. In normal vision (a), the lens focuses the visual image on the retina. Common problems with the accommodation mechanism involve an inability to lengthen the focal distance enough to focus the image of a distant object on the retina (myopia, b) and an inability to shorten the focal distance adequately for near objects (hyperopia, c). (d) Corneal astigmatism will cause asymmetrical blurring or darkening of these lines.

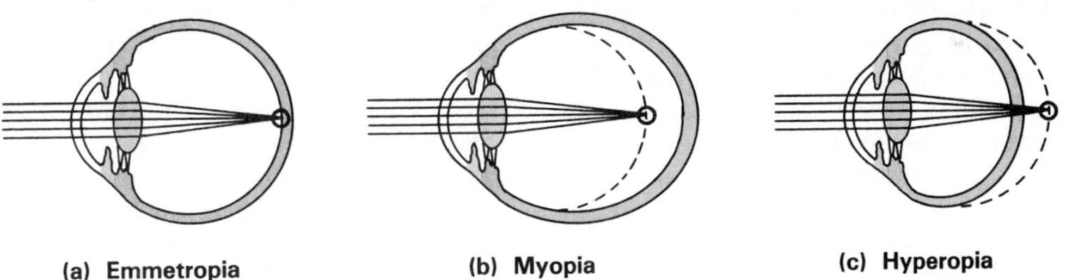

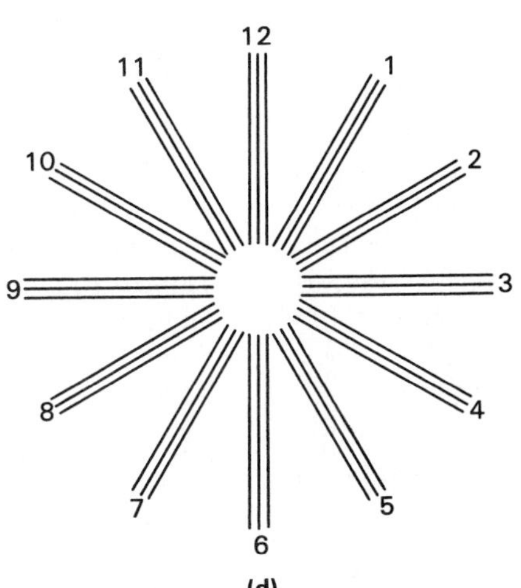

Clinical Comment: Problems with the Accommodation Mechanism

Figure 17-25b,c diagrams two common problems with the accommodation mechanism. If the eyeball is too deep, the image of a distant object will form in front of the retina, and the retinal picture will be blurry and out of focus. Vision at close range will be normal, because the lens will be able to round up as needed to focus the image on the retina. As a result, such individuals are said to be "nearsighted." Their condition is more formally termed **myopia** (*myein*, to shut + *ops*, eye).

If the eyeball is too shallow, **hyperopia** results. The ciliary muscles must contract to focus even a distant object on the retina, and at close range the lens cannot provide enough refraction. These individuals are said to be "farsighted" because they can see distant objects most clearly. Older individuals become farsighted as their lenses lose elasticity; this form of hyperopia is called **presbyopia** (*presbys*, old man).

Myopia and hyperopia can usually be corrected with glasses or contact lenses. These supplemental lenses provide the precise amount of refraction needed to compensate for abnormalities in the shape of the cornea, stiffness of the lens, or depth of the eyeball. Variable success has been achieved by surgically reshaping the cornea to alter its refractive powers. This procedure, called **radial keratotomy**, remains controversial. Although roughly two-thirds of patients are satisfied with the results, corneal healing takes several years.

A more recent and possibly more effective procedure is **photorefractive keratectomy** (**PRK**), in which a computer-guided laser shapes the cornea to exact specifications. Tissue is removed only to a depth of 10–20 μm—no more than about 10 percent of the cornea's thickness. The entire procedure can be done in less than a minute and costs about $2000 per eye. Clinical trials are now in progress and will probably continue for at least another year.

If the cornea and lens are not smoothly curved and symmetrical the visual image will be distorted to some degree. This **astigmatism** can also be corrected by glasses or special contact lenses. Figure 17-25d contains a simple test for astigmatism; minor astigmatism is relatively common.

metropia (*emmetro-*, proper). The inner limit of clear vision is determined by the degree of elasticity in the lens. Children can usually focus on something 2–3 inches from the eye, but over time the lens tends to become stiffer and less responsive. An adult can usually focus on objects 6–8 inches away, and as aging proceeds this distance gradually increases.

The lens does not actually go from being flat as a board to round as a soccer ball. There are limits to the elastic abilities of the lens, and most lenses have minor imperfections or irregularities in shape. For any given lens shape the variation from "ideal" will usually be greatest around the periphery of the lens. Light passing through these regions would fail to refract properly, disrupting the clarity of the visual image. Such problems are reduced during normal accommodation by the simultaneous adjustment of the pupillary diameter. Constriction or dilation of the pupil therefore not only limits the amount of light entering the eye, but also controls the path of the light so that it reaches portions of the lens that have the necessary refractive abilities.

Visual Acuity

In rating the clarity of vision, or **visual acuity**, a person whose vision is rated 20/20 is seeing details at a distance of 20 feet as clearly as a "normal" individual would. Vision noted as 20/15 is better than average, for at 20 feet the person is able to see details that would be clear to a normal eye only at a distance of 15 feet. Conversely, a person with 20/30 vision must be 20 feet from an object to discern details that a person with normal vision could make out at a distance of 30 feet.

When visual acuity falls below 20/200, even with the help of glasses or contact lenses, the individual is considered to be legally blind. There are probably fewer than 400,000 legally blind people in the United States; more than half are over 65 years of age. Common causes of blindness include diabetes mellitus, cataracts, glaucoma, corneal scarring, retinal detachment, accidental injuries, and hereditary factors that are as yet poorly understood.

LIGHT AND PHOTORECEPTION

Retinal Organization

In sectional view, the retina contains several layers of cells, detailed in Figure 17-26. The outermost layer, closest to the choroid, consists of pigment cells. The next layer contains the visual receptors that communicate with an adjacent layer of **bipolar cells**. **Horizontal cells** at this level form a network that inhibits or facilitates communication between visual receptors and bipolar cells. Bipolar cells synapse within the layer of **ganglion cells** that faces the vitreous

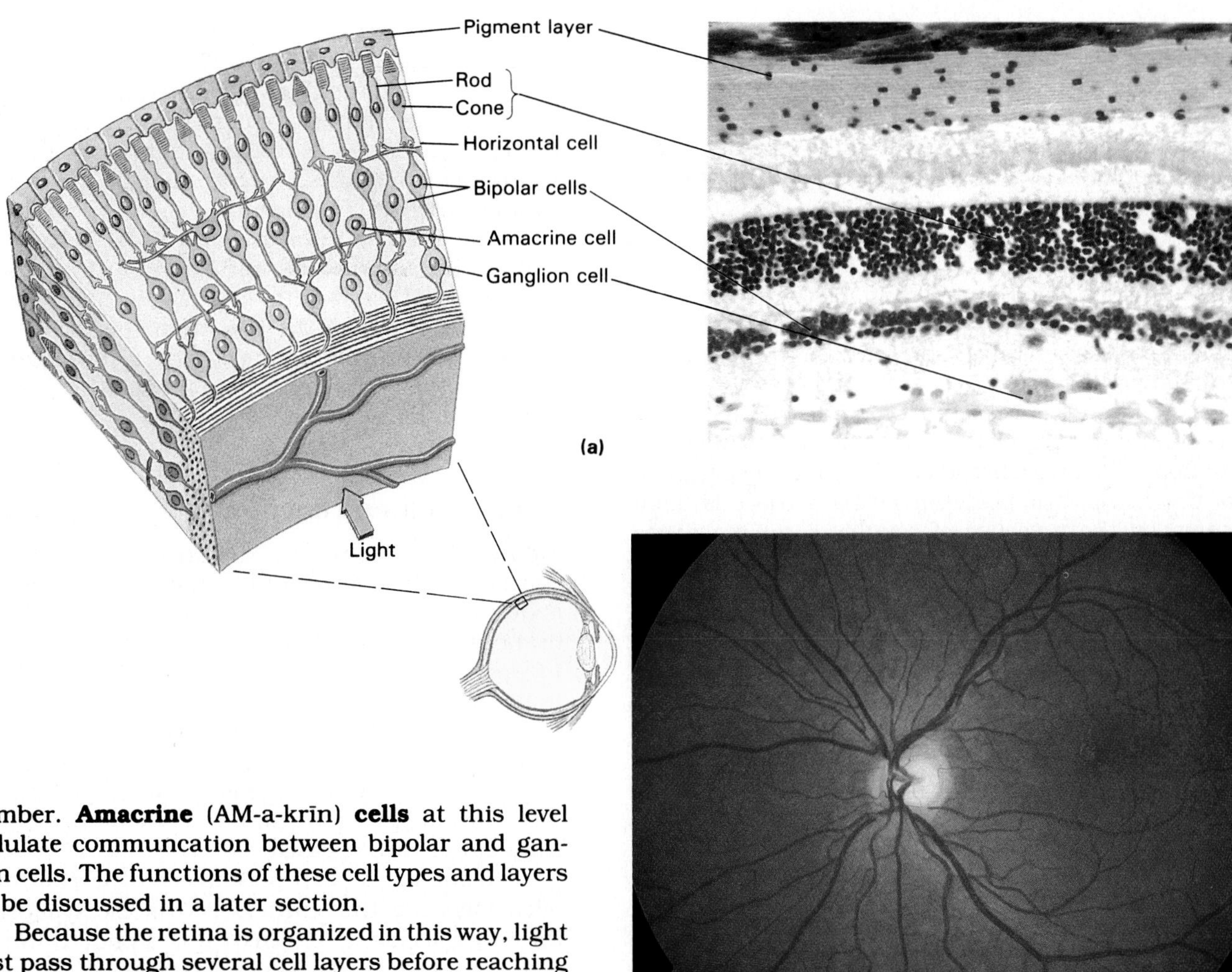

FIGURE 17-26
Retinal Organization. (a) Cellular organization of the retina. Note that the photoreceptors are located closest to the choroid rather than near the vitreous chamber. (b) A photograph taken through the pupil of the eye, showing the retinal blood vessels.

chamber. **Amacrine** (AM-a-krīn) **cells** at this level modulate communcation between bipolar and ganglion cells. The functions of these cell types and layers will be discussed in a later section.

Because the retina is organized in this way, light must pass through several cell layers before reaching the visual receptors. When a physician examines eyes using a bright light the patient often sees a complicated "road map." The lines are the shadows of blood vessels crossing the retinal surface. In Figure 17-26b, a photograph of the retinal surface that was taken through the cornea and lens, the retinal vessels are quite apparent.

We will now trace the chain of events that begins with the arrival of light at a photoreceptor and ends with the perception of a visual image.

Light and Energy

There are many different forms of radiant energy. Energy is radiated in waves that have a characteristic **frequency** (cycles per second) and **wavelength** (distance between wave peaks). The entire spectrum of electromagnetic radiation ranges from extremely long-wavelength radio waves to extremely short-wavelength gamma rays. Our eyes are sensitive to wavelengths of 400–700 nm, the spectrum of **visible light**. This spectrum, seen in a rainbow, can be remembered by the acronym "ROY G. BIV" (red, orange, yellow, green, blue, indigo, violet).

A **photon** is a single energy packet of visible light. Photons of red light carry the least energy, and those from the violet portion of the spectrum contain the most. An ordinary light bulb or the sun emits photons of all wavelengths. These photons stimulate both rods and cones, producing a perception of "white" light. Our eyes also detect the photons bouncing off objects around us. If all of the photons bounce off, the object appears white; if they are all absorbed, and nothing bounces off, the object appears black. The object will appear to have a particular color if it reflects photons from a portion of the visible spectrum.

Rods and Cones

The visual receptors of the retina are called **photoreceptors** because they detect photons. There are two types of photoreceptors, **rods** and **cones**. Rods respond to the arrival of almost any photon, regardless of its energy content. Because rods are relatively unselective, they do not discriminate between different colors of light. They are very sensitive, however, and rods enable us to see in dimly lit rooms, at twilight, or in pale moonlight.

There are three types of cones: **blue cones**, **green cones**, and **red cones**. Each type has a different range of sensitivity. Their stimulation in various combinations accounts for the perception of colors. Because cones are less sensitive than rods, they function only in relatively bright light.

Rods and cones are not evenly distributed across the outer surface of the retina. Approximately 125 million rods form a broad band around the periphery of the retina. The posterior retinal surface is dominated by the presence of roughly 6 million cones. These are especially densely packed within the **fovea** (FŌ-vē-a; shallow depression) or **macula lutea** (LOO-tē-a; yellow spot), which lacks rods entirely. The fovea is the site of sharpest vision; when you look directly at an object its image falls upon this portion of the retina. Figure 17-27 diagrams the pattern of rod and cone distribution across the midline of the retina and relates these patterns to visual acuity.

FIGURE 17-27
Retinal Distribution of Rods and Cones. The two color bands across the retina illustrate the relative concentrations of rods and cones in this area. Note the correlation between the distribution of cones and the sharpness of vision (visual acuity), as indicated by the graph.

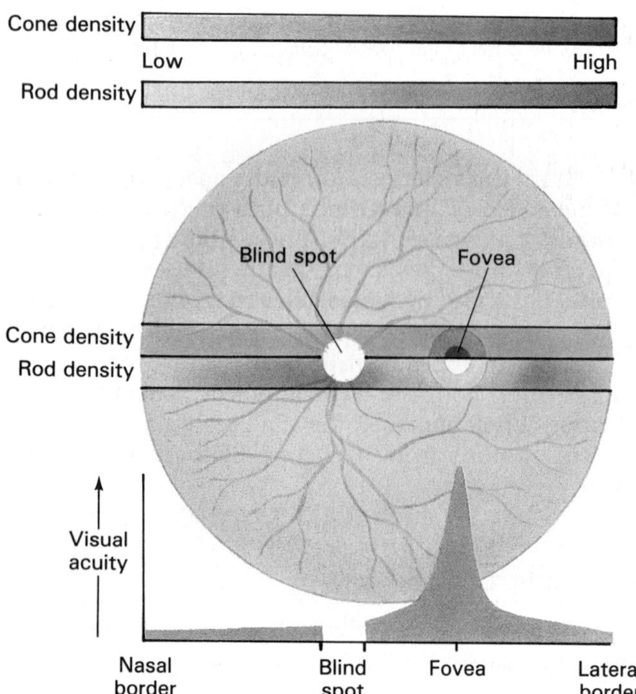

You are probably already aware of the visual consequences of this distribution. When you look directly at something, you are focusing the image on the fovea, the center of color vision. This gives you a very good image, as long as there is enough light to stimulate the cones. In very dim light, cones simply cannot function. For example, when you try to stare at a dim star, you are unable to see it. But if you look a little to one side, rather than directly at the star, you will see it quite clearly. Shifting your gaze moves the image of the star from the fovea, where it does not provide enough light to stimulate the cones, to the periphery, where it can affect the more sensitive rods.

Photoreceptor Function

Figure 17-28a compares the structure of rods and cones. The elongate **outer segment** of a photoreceptor contains hundreds to thousands of flattened membranous plates, or **discs**. In a rod each disc is an independent entity, but in the cones the discs are infoldings of the cell membrane. A narrow connecting stalk attaches the outer segment to the **inner segment**, a region that contains all of the usual cellular organelles. The inner segment makes synaptic contact with other cells, and it is here that neurotransmitters are released.

In the dark, each photoreceptor is continually releasing neurotransmitters across the synapses at the inner segment. The arrival of a photon initiates a chain of events that alters the transmembrane potential and changes the rate of neurotransmitter release.

Neurotransmitter release occurs at a resting transmembrane potential of approximately –40 mV. The resting potential is relatively low compared with that of a typical neuron or skeletal muscle fiber because sodium channels in the outer segment are open (Figure 17-28b). The open channels allow a continual leakage of positively charged sodium ions into the cell, making the interior less negative with respect to the extracellular fluid than it would otherwise be. These extra ions are pumped out of the cytosol by active transport. When the outer segment absorbs a photon, the sodium channels close, but the pumps keep working. Sodium concentration within the cell falls, and the transmembrane potential rises above resting levels. This hyperpolarization causes the rate of neurotransmitter release to decline (Figure 17-28c).

Visual Pigments

Light absorption, the key step in this sequence of events, requires special organic compounds called **visual pigments**. These compounds, found in the outer segments of the photoreceptor cells, are derivatives of the compound **rhodopsin** (rō-DOP-sin). Rho-

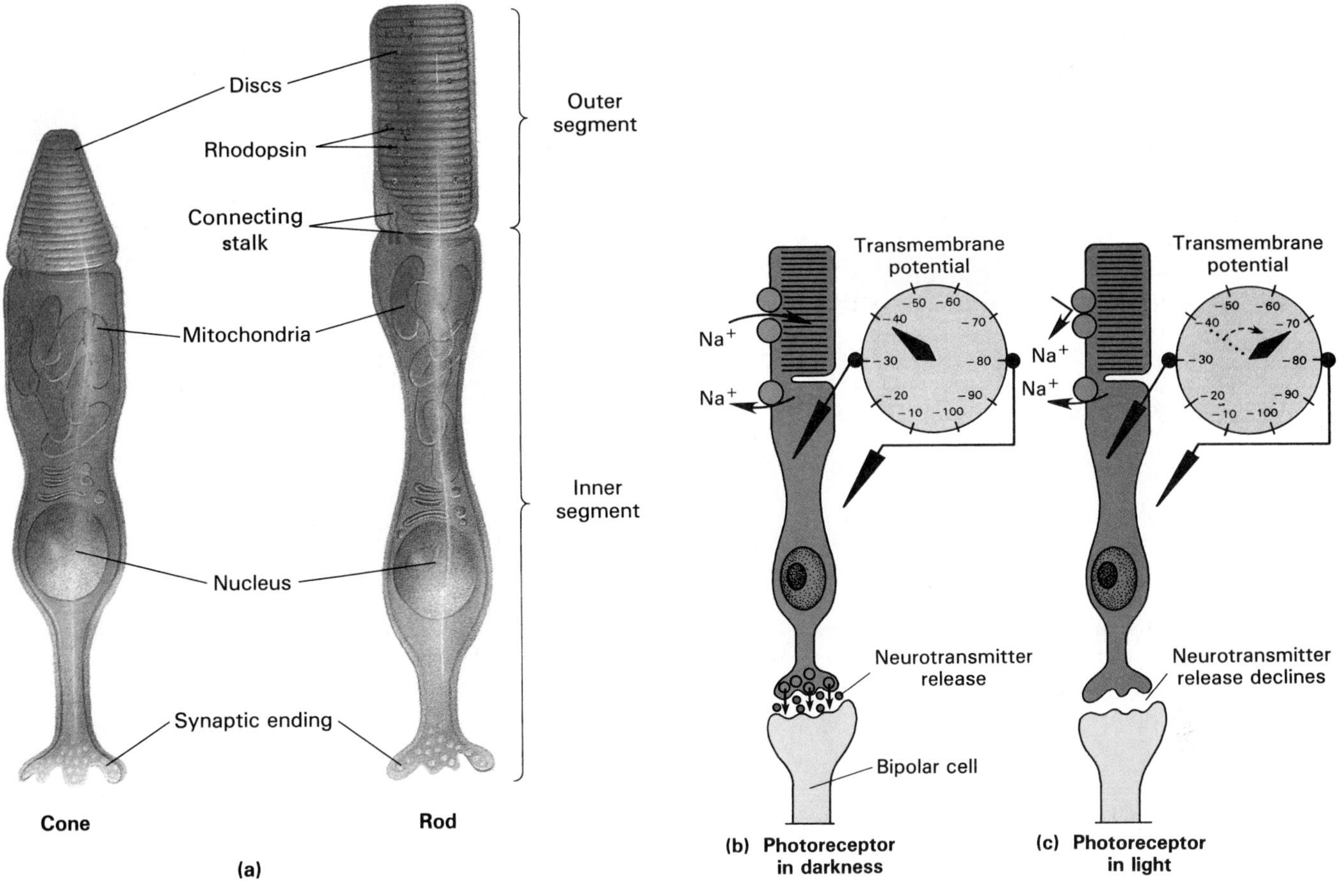

FIGURE 17-28
Retinal Photoreceptors. (a) A three-dimensional view of a rod and a cone, based on data from electron microscopy studies. (b) Photoreceptor in darkness: Sodium channels are open, and the membrane is depolarized to –40 mV. (c) Photoreceptor in light: Photon absorption initiates a chain of events leading to closure of the sodium channels and hyperpolarization of the membrane.

FIGURE 17-29
Rhodopsin. A rhodopsin molecule consists of the pigment retinal bound to the protein opsin.

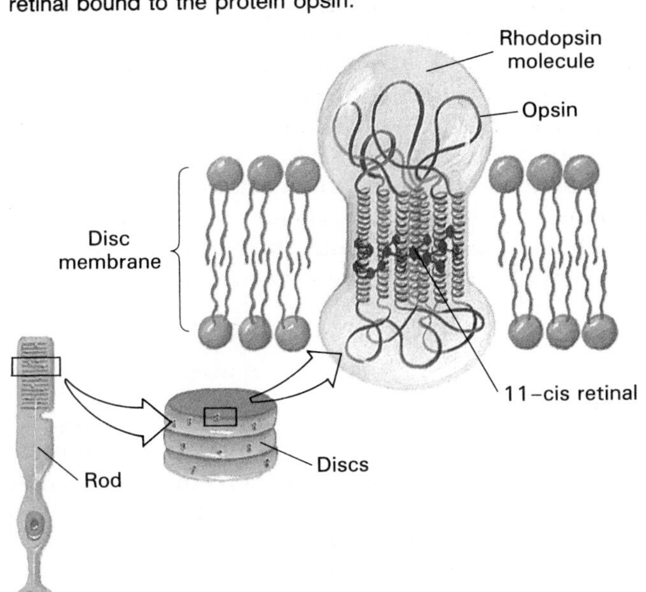

dopsin consists of a protein, **opsin**, bound to a pigment, **retinal** (RET-i-nal), synthesized from **vitamin A**. Figure 17-29 diagrams the structure of a rhodopsin molecule. The pigment retinal is identical in both rods and cones. However, each of four different receptor types (one rod, three cones) contains a different form of opsin. We will now follow the steps in rhodopsin-based photoreception as it occurs in rods.

There are two possible configurations for the bound retinal molecule. It is normally in the **11-cis** form; on absorbing light, it adopts the more linear **11-trans** form. This change in shape triggers a chain of enzymatic steps, diagrammed in Figure 17-30.

Step 1: Opsin activation occurs. Opsin functions as a receptor protein embedded in the membrane. When retinal changes from the 11-cis to the 11-trans form, the opsin is "turned on," gaining the ability to act as an enzyme.

Step 2: Opsin activates a second enzyme, called **transducin**, or **G-protein**.

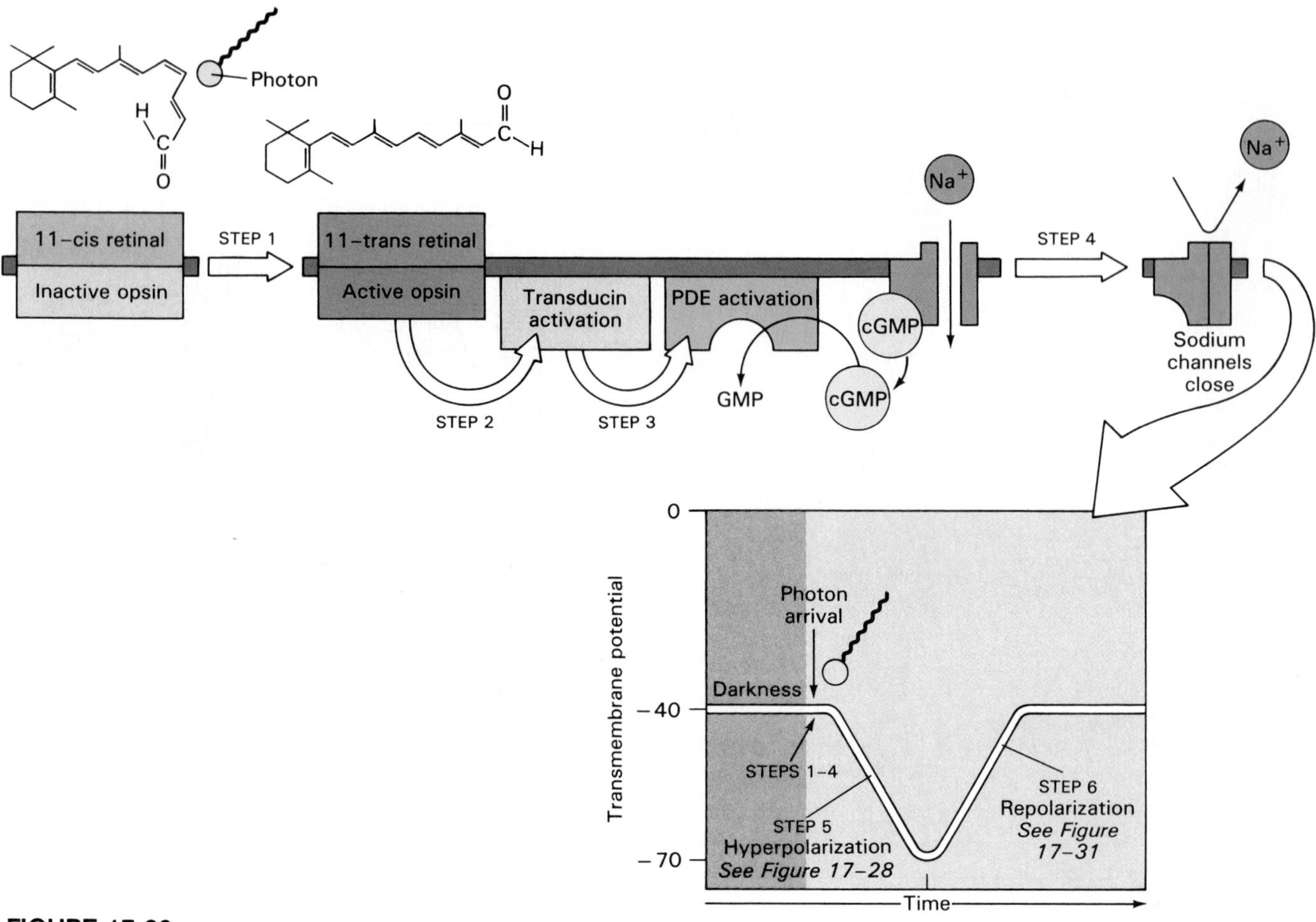

FIGURE 17-30
Photoreception. Absorption of a photon converts 11-cis retinal to 11-trans retinal and activates opsin (Step 1). In Step 2, opsin activates transducin (G-protein), and this initiates Step 3, the activation of PDE (phosphodiesterase). Phosphodiesterase then breaks down cytoplasmic cGMP, without which the sodium channels close (Step 4). This closure leads to Step 5, the hyperpolarization of the photoreceptor membrane. This hyperpolarization causes a reduction in the rate of neurotransmitter release by the inner segment.

Step 3: Transducin activates **phosphodiesterase (PDE)**. This enzyme breaks down **cyclic-GMP (cGMP)**, a derivative of the high-energy compound GTP (guanosine triphosphate).

Step 4: The membrane sodium channels close. These chemically regulated channels must bind cGMP to remain open; when phosphodiesterase breaks down the cGMP, the channels close.

Step 5: The membrane hyperpolarizes, and the rate of neurotransmitter release declines. Although the sodium channels are now closed, active transport continues to remove sodium ions from the cytoplasm, and the transmembrane potential drops. This decline is the signal that light has struck the retina at that location.

Shortly after the conformational change occurs, the rhodopsin molecule begins to "bleach," or break down into retinal and opsin (Figure 17-31a). The retinal must be enzymatically converted to the 11-cis form before it can recombine with opsin. This conversion requires energy in the form of ATP (adenosine triphosphate), and it takes time. Bleaching contributes to the lingering visual impression after a flash bulb goes off. After an intense exposure to light, a photoreceptor cannot respond to further stimulation until its rhodopsin molecules have been regenerated. As a result, a "ghost" image remains on the retina. Staring at the checkerboard image in Figure 17-31b will also produce a ghost image that results in part from pigment bleaching. Bleaching is seldom noticeable under ordinary circumstances because the eyes are constantly making small, involuntary changes in position that move the image across the retinal surface. [**CM**: *Night Blindness*]

Color Vision

In a normal individual the cone population consists of 16 percent blue cones, 10 percent green cones, and 74 percent red cones. Although their sensitivities overlap, each type is most sensitive to a specific portion of the visual spectrum (see Figure 17-32a). Color

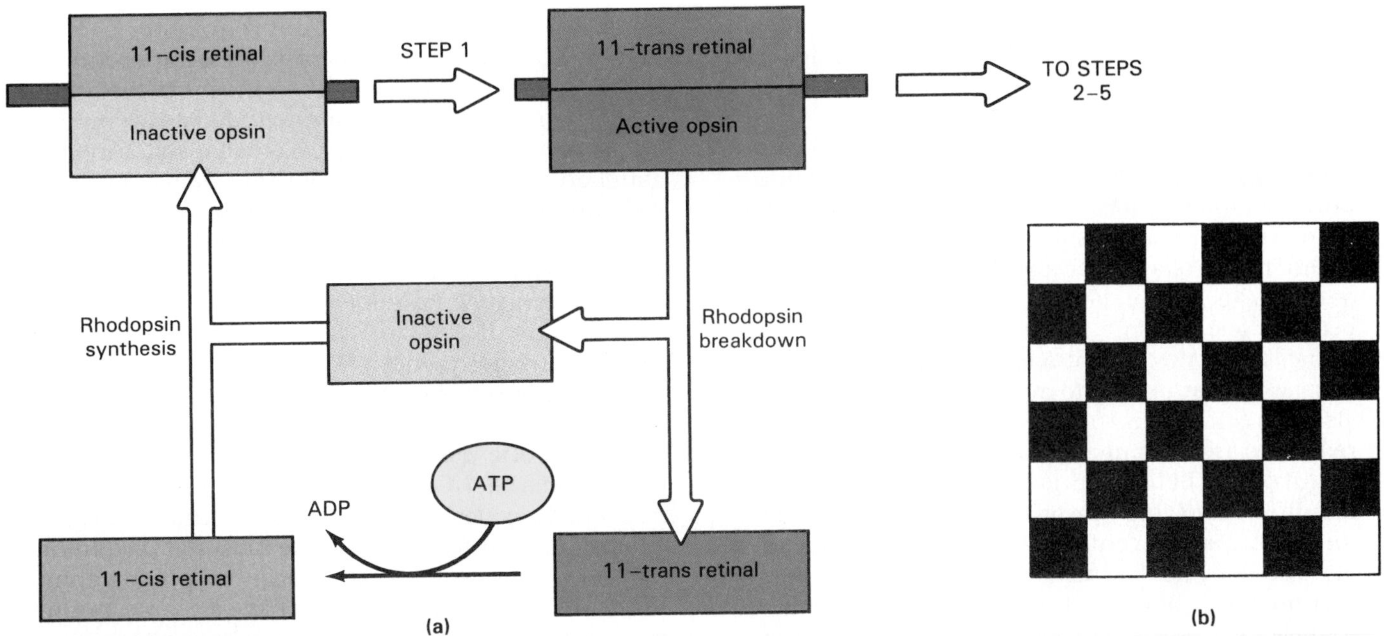

FIGURE 17-31
Bleaching of Visual Pigments. (a) Immediately after a photon is absorbed and opsin is activated, rhodopsin breaks down into 11-trans retinal and opsin. Reconversion to 11-cis retinal requires ATP; after conversion, the rhodopsin molecule reforms. (b) Staring at this checkerboard will bleach visual pigments from retinal areas monitoring white patches. If you shift your gaze to a blank wall, the corresponding bleaching pattern will be evident.

discrimination occurs through the integration of information arriving from all types of cones. Our perception of colors results from different patterns of response. For example, the perception of yellow results from a combination of inputs from green cones (highly stimulated), red cones (stimulated), and blue cones (relatively unaffected). If all three cone populations are stimulated, we perceive the color as white;

FIGURE 17-32
Cones and Color Vision. (a) A graph comparing the absorptive characteristics of blue, green, and red cones with those of typical rods. Notice that the rod sensitivities overlap those of the cones, and the various cone types have overlapping sensitivity curves. (b) Part of a standard test for color vision. Lack of one or more populations of cones will produce an inability to distinguish the patterned images.

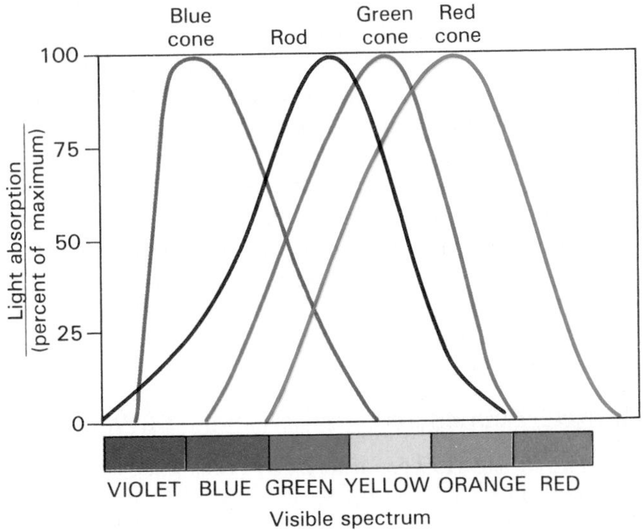

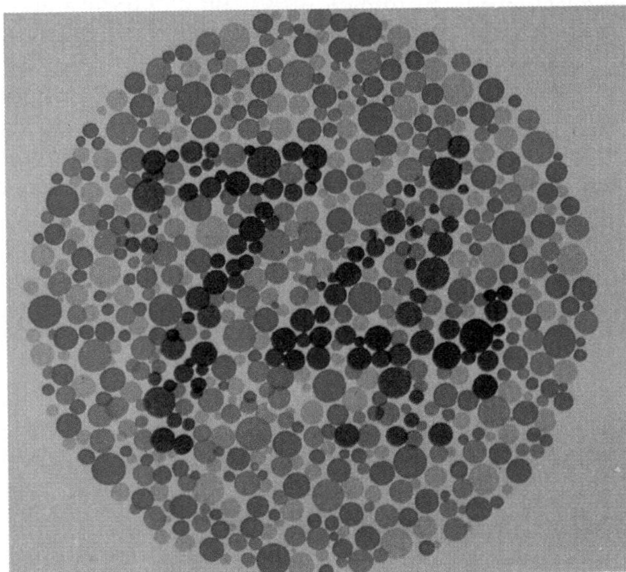

we also perceive white if only rods are stimulated. For this reason everything appears to be black and white when we enter dimly lit surroundings or walk by starlight.

Color Blindness Persons unable to distinguish certain colors have a form of **color blindness**. The standard tests for color vision involve picking numbers or letters out of a complex and colorful picture, such as those in Figure 17-32b. Color blindness occurs because one or more classes of cones are nonfunctional. The cones may be absent entirely, or they may be present but unable to manufacture the necessary visual pigments. In the most common condition the red cones are missing, and the individual cannot distinguish red light from green light. Inherited color blindness involving one or two cone pigments is not unusual. Ten percent of all men show some color blindness, while the incidence among women is only around 0.67 percent. Total color blindness is extremely rare; only 1 person in 300,000 fails to manufacture any cone pigments at all.

Adaptation to Changing Light Levels

The sensitivity of the visual system varies, depending on the intensity of illumination. After 30 minutes or more in the dark, almost all visual pigments will be fully receptive to stimulation. This is the **dark-adapted** state. When dark-adapted, the visual system is extremely sensitive. For example, a single rod will hyperpolarize in response to a single photon of light. Even more remarkable, if as few as seven rods absorb photons at one time, the individual will see a flash of light.

When the lights come on, at first they seem almost unbearably bright, but over the next few minutes sensitivity decreases as bleaching occurs. Eventually the rate of visual pigment breakdown is balanced by the rate of reformation. This is the **light-adapted** state. In moving from the depths of a cave to the full light of midday the receptor sensitivity decreases by about 25,000 times.

A variety of central responses further adjust light sensitivity. Constriction of the pupil, via the pupillary constrictor reflex, reduces the amount of light entering the eye to 1/30th the maximum dark-adapted levels. Since dilating the pupil fully can produce a 30-fold increase in the amount of light entering the eye, and facilitating some of the synapses along the visual pathway can perhaps triple its sensitivity, the entire system may increase its efficiency by a factor of over 1 million.

PROCESSING OF VISUAL INFORMATION

Retinal Processing

Each photoreceptor monitors a specific receptive field. A visual image results from the processing of information provided by the entire receptor population. A significant amount of processing occurs in the retina before the information is sent to the brain.

The retina contains approximately 130 million photoreceptors, 6 million bipolar cells, and 1 million ganglion cells. The amount of convergence differs in rods and cones. As many as a thousand rods may pass information via their bipolar cells to a single ganglion cell. Cones, by contrast, typically show very little convergence; in the fovea the ratio of cone cells to ganglion cells is 1 to 1. This difference has direct functional consequences. When a ganglion cell monitoring a group of rods becomes active, that message indicates "light arriving in this general area." The activation of a ganglion cell that monitors a single cone means "light arriving at this precise location." As a result, cones not only work best under daylight conditions, but also provide the most detailed information about the visual image. In photographic terms, pictures formed by rods have a coarse, grainy appearance that blurs the details, whereas those produced by cones are extremely sharp and clear.

Horizontal cells adjust the sensitivities of bipolar cells to stimulation by photoreceptors; amacrine cells perform the same function at synapses between bipolar and ganglion cells. The net effects are to coordinate the activities of multiple ganglion cells and to heighten contrast and improve the visual image. Whether it receives information from a single cone or a multitude of rods, each ganglion cell monitors a specific portion of the visual image. The sensory field is usually circular, and ganglion cells typically respond differently to stimuli arriving in the center and at the edges of their receptive fields. Some ganglion cells are excited by light arriving in the center of their sensory field (**on-center cells**). Others are inhibited by light in the central zone but stimulated by peripheral illumination (**off-center cells**). The relationships are diagrammed in Figure 17-33.

Axons from the entire population of ganglion cells converge on the **optic disc**, penetrate the wall of the eye, and proceed toward the diencephalon as the optic nerve (N II). There are no photoreceptors or other retinal components at the optic disc. Because light striking this area goes unnoticed, it is also known as the **blind spot**. You do not "notice" a blank spot in the visual field because the same involuntary eye movements that prevent excessive bleaching of visual pigments move the image across the blind spot. This movement allows the brain to fill in the perceptual gap. A simple experiment, shown in Figure 17-34, will demonstrate the presence and location of the blind spot within the visual field. [**CM**: *Scotomas and Floaters*]

Central Processing

The two optic nerves, one from each eye, reach the diencephalon at the optic chiasm (Figure 17-35). From this point approximately one-half of the fibers proceed toward the lateral geniculate of the same side of the brain, while the other half cross over to reach

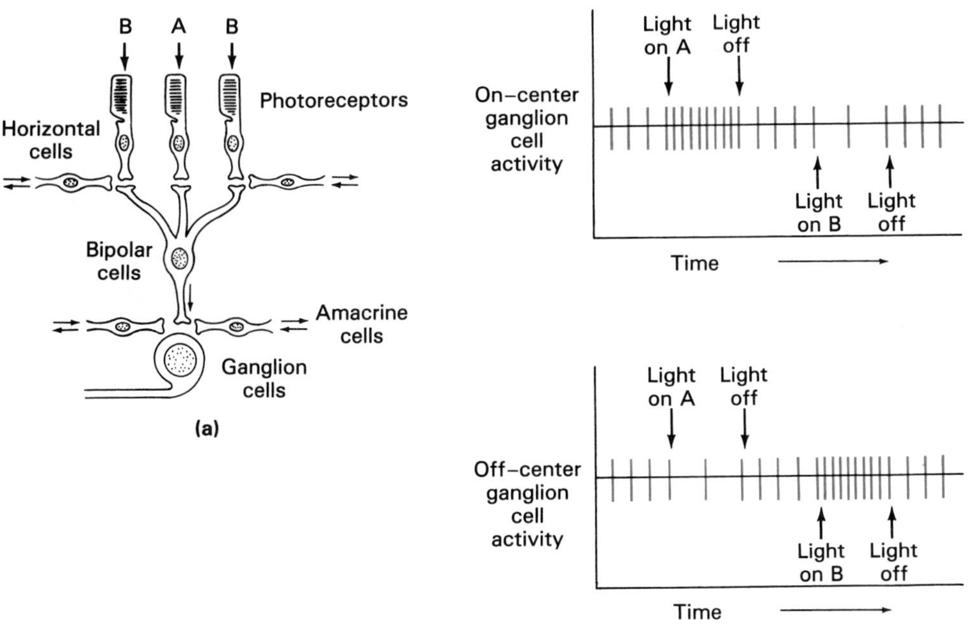

FIGURE 17-33 Convergence and Ganglion Cell Function. (a) Ganglion cells monitor a well-defined portion of the visual field. Horizontal and amacrine cells modulate the communications between receptor cells, bipolar cells, and ganglion cells. (b) Some ganglion cells respond strongly to light arriving at the center of their receptive fields (on-center neurons). Illumination of the edges of the receptive field results in an inhibition of their activity. Off-center neurons show an opposite response to illumination.

the lateral geniculate of the opposite side. Visual information from the left half of each retina arrives at the lateral geniculate of the left side; information from the right half of each retina goes to the right side. The lateral geniculates act as switching and processing centers that relay visual information to reflex centers in the brain stem as well as to the cerebral cortex. For example, the pupillary reflexes and reflexes that control eye movement are triggered by information relayed by the lateral geniculates.

Cortical Integration

The sensation of vision arises from the integration of information arriving at the visual cortex of the

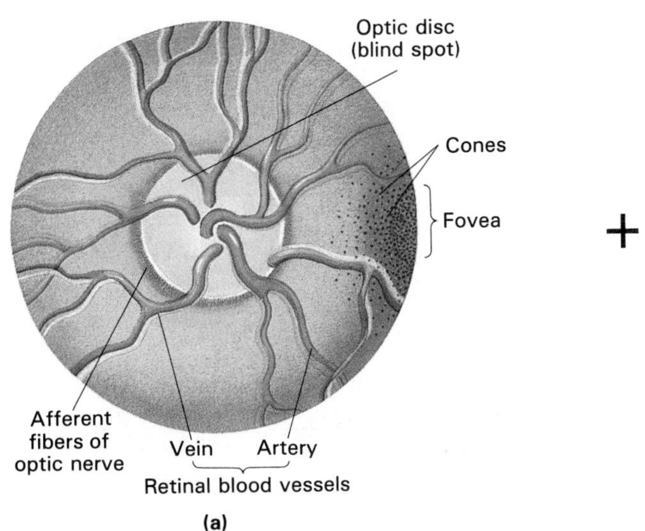

+ ●

(b)

FIGURE 17-34 The Optic Disc. (a) There are no photoreceptors at the origin of the optic nerve; this area is the optic disc, or "blind spot." (b) Stare at the cross with your right eye and vary the distance between your eye and the figure; the spot will seem to disappear when its image strikes the optic disc. For your left eye, watch the dot.

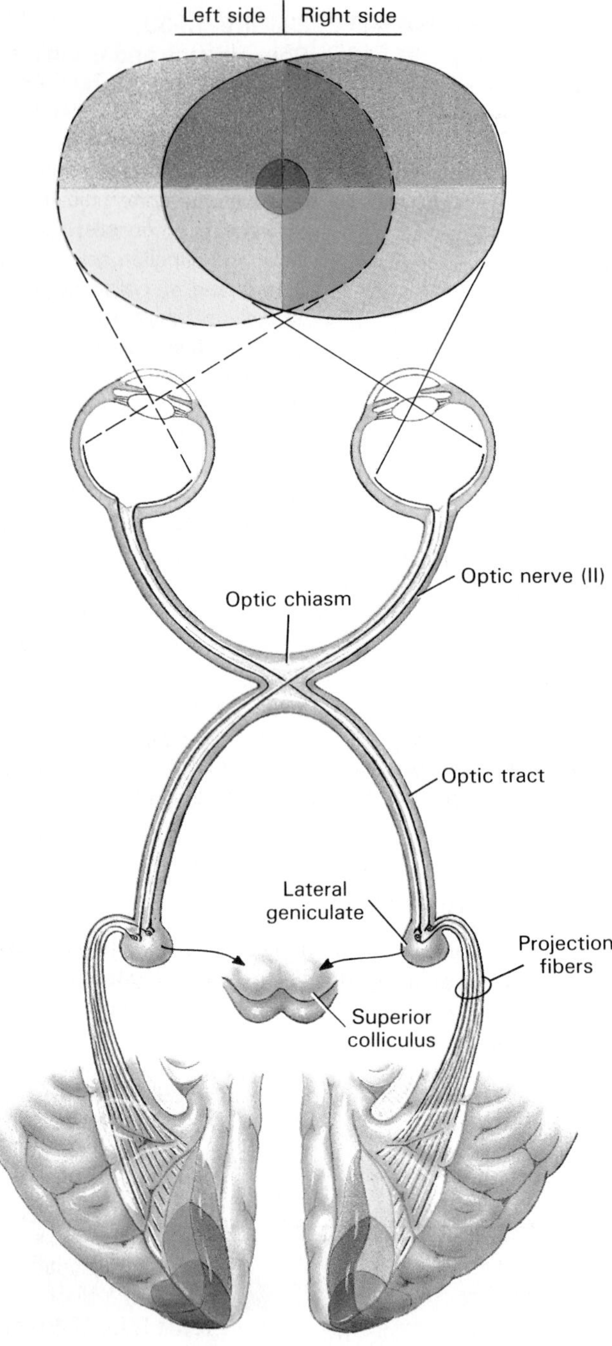

FIGURE 17-35
The Visual Pathways. At the optic chiasm, a partial cross-over of nerve fibers occurs. As a result, each hemisphere receives visual information from the lateral half of the visual field of the eye on that side and from the medial half of the visual field of the eye on the opposite side. Visual association areas integrate this information to develop a composite picture of the entire visual field.

cerebrum. The visual cortex contains a sensory map of the entire field of vision. As in the case of the primary sensory cortex, the map does not faithfully duplicate the relative areas within the sensory field. For example, the area assigned to the fovea covers about 35 times the surface it would cover if the map were proportionally accurate.

There is extensive convergence within the visual system, first in the retina and later in the visual cortex. As indicated in Figure 17-36, information from the retinal ganglion cells converges on neurons of the visual cortex. Cortical neurons called **simple cells** respond best when a stimulus affects a row of ganglion cells. A simple cell integrates information and enhances contrast and image detail. **Complex cells** respond best to moving linear stimuli, probably by monitoring the activities of several simple cells. More complicated interactions between complex cells provide sensitivity to shapes and borders affecting large areas of the retina.

Image Inversion

The visual image arrives at the retina upside down, as Figure 17-37 demonstrates. The brain makes allowance for this, and we are not consciously aware that we are seeing everything inverted. Interesting experiments have been done, giving volunteers a pair of special glasses that invert the image and produce a retinal image with a "proper" top to bottom orienta-

FIGURE 17-36
Convergence and Processing at the Visual Cortex

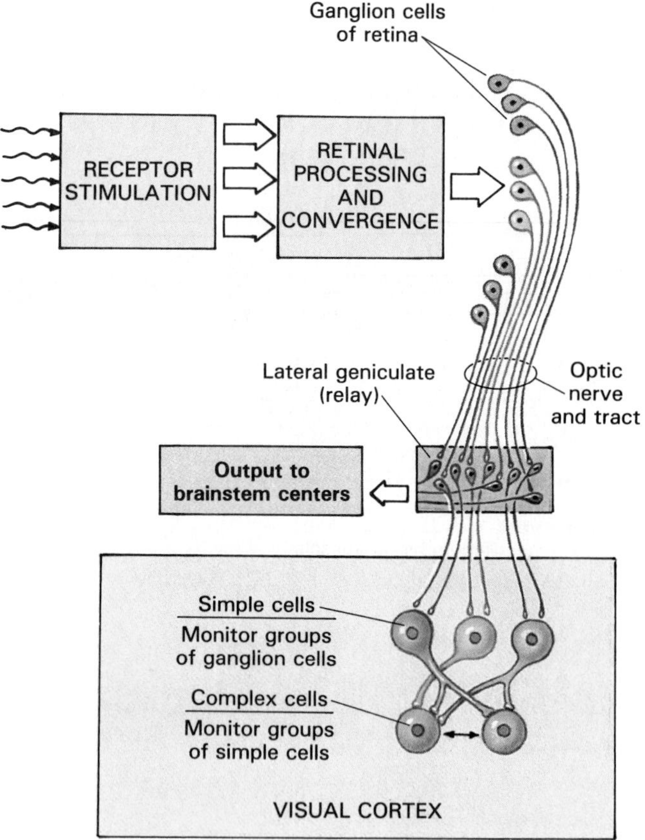

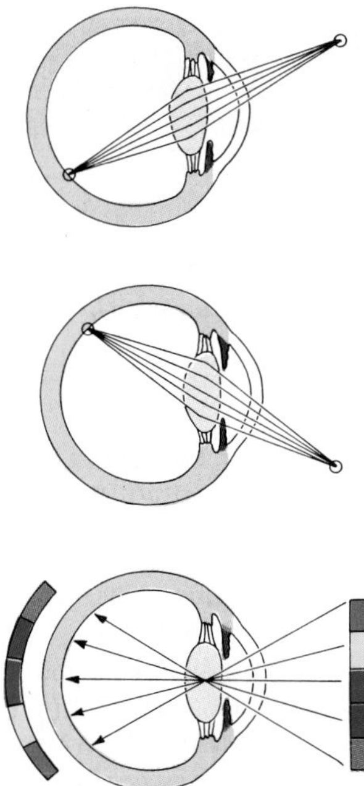

FIGURE 17-37
Image Projection on the Retina. Light arriving at an angle will be focused on the retina at a point directly opposite the center of the lens. A complex object, such as a colored pole, can be considered to be creating a series of individual images on the retinal surface. When these are mapped out they form an upside-down image.

tion. At first the volunteers find this quite disturbing, and the initial days are extremely difficult. Yet the brain somehow makes the necessary adjustments, and the individual's motor skills and orientation gradually improve, even to the point of being able to ride a bicycle. Subsequent removal of the glasses produces a comparable but shorter period of sensory confusion.

Depth Perception

If you close your right eye, then open it and close your left, you will notice that the picture changes somewhat. Each eye receives a slightly different image, because their foveas are 2–3 inches apart. For one thing, each eye provides some visual information that is unavailable to the other because your nose and eye socket block its view of the opposite side. The partial cross-over that occurs at the optic chiasm ensures that the visual cortex receives a *composite* picture of the entire visual field. But other, subtler differences between the images provided by the right and left eyes are not totally ignored. Each eye also sees the world from a slightly different angle. The association and integrative areas of the cortex compare these views and use them to provide us with depth perception.

The Brain Stem and Visual Processing

Many centers in the brain stem receive visual information, either from the lateral geniculates or via collaterals from the optic tracts. Collaterals that bypass the lateral geniculates synapse in the superior colliculus or hypothalamus. The superior colliculus of the midbrain issues motor commands controlling unconscious eye, head, or neck movements in response to visual stimuli. Visual inputs to the **suprachiasmatic** (sū-pra-kī-az-MA-tic) **nucleus** of the hypothalamus affect the function of other brain stem nuclei. This nucleus establishes a daily pattern of visceral activity that is tied to the day-night cycle. This **circadian rhythm** (*circa*, about + *dies*, day) affects metabolic rate, endocrine function, blood pressure, digestive activities, the awake-asleep cycle, and other physiological and behavioral processes discussed in Chapters 14–16. [**CM**: *Analyzing Sensory Disorders*]

- **What layer of the eye would be the first to be affected by inadequate tear production?**
- **When the lens is very round, are you looking at an object that is close to you or distant from you?**
- **If a person was born without cones in her eyes, would she still be able to see? Explain.**
- **How could a diet that is deficient in vitamin A affect vision?**

CHAPTER REVIEW

Review of Selected Clinical Terms

referred pain (*p. 528*)
Pain sensations from visceral organs, often perceived as originating in more superficial areas innervated by the same spinal nerves.

anesthesia (*p. 530*)
A total loss of sensation.

hypesthesia (*p. 530*)
A reduction in sensitivity.

paresthesia (*p. 530*)
Abnormal sensations.

nystagmus (*p. 540*)
Short, jerky eye movements that sometimes appear after damage to the brain stem or inner ear.

conductive deafness (*p. 546*)
Deafness resulting from conditions in the middle ear that block the transfer of vibrations from the tympanic membrane to the oval window.

nerve deafness (*p. 546*)
Deafness resulting from problems within the cochlea or along the auditory pathway.

cataract (*p. 553*)
An abnormal lens that has lost its transparency.

myopia (*p. 556*)
A condition in which vision at close range is normal but distant objects appear blurry; "nearsightedness."

hyperopia (*p. 556*)
A condition in which nearby objects are blurry but distant objects are clear; "farsightedness."

presbyopia (*p. 556*)
A type of hyperopia that develops with age as lenses become less elastic.

color blindness (*p. 562*)
A condition in which people are unable to distinguish certain colors.

Additional Terms of Clinical Importance

ASSESSMENT OF TACTILE SENSITIVITIES (*p. 530*): **two-point discrimination test**

PROBLEMS WITH THE ACCOMMODATION MECHANISM (*p. 556*):
radial keratotomy, photorefractive keratectomy (PRK)

Study Outline

Introduction (pp. 523–524)

1. The **general senses** are temperature, pain, touch, pressure, vibration, and proprioception; receptors for these sensations are distributed throughout the body. Receptors for the **special senses** (smell, taste, sight, balance, and hearing) are located in specialized areas or in **sense organs**. A **sensory receptor** is a specialized cell that, when stimulated, sends a **sensation** to the CNS.

Receptor Physiology and Sensory Coding (pp. 524–527)

1. **Receptor specificity** allows each receptor to respond to particular stimuli. The simplest receptors are **free nerve endings**; the area monitored by a single receptor cell is the **receptive field**. Transduction is the translation of a stimulus into an action potential.

THE TRANSDUCTION PROCESS (pp. 524–525)

2. Transduction involves the development of a **receptor potential** and a **generator potential**. The resulting action potential travels to the CNS over an afferent fiber.

INTERPRETATION OF SENSORY INFORMATION (pp. 525–526)

3. **Tonic receptors** are always sending signals to the CNS; **phasic receptors** become active only when the conditions that they monitor change.

Related Key Terms: **labeled line** • **sensory coding**

CENTRAL PROCESSING AND ADAPTATION (pp. 526–527)

4. **Adaptation** (a reduction in sensitivity in the presence of a constant stimulus) may involve changes in receptor sensitivity (**peripheral**, or **sensory**, **adaptation**) or inhibition along the sensory pathways (**central adaptation**). **Fast-adapting** receptors are phasic; **slow-adapting receptors** are tonic.

SENSORY LIMITATIONS (p. 527)

5. The information provided by our sensory receptors is incomplete because: (1) We do not have receptors for every stimulus; (2) Our receptors have limited ranges of sensitivity; (3) A stimulus produces a neural event that must be interpreted by the CNS.

Chapter Review

The General Senses (pp. 527–532)

Related Key Terms

NOCICEPTION (pp. 527–528)

1. **Nociceptors** respond to a variety of stimuli usually associated with tissue damage. There are two types of these painful sensations: **fast** (*prickling*) **pain** and **slow** (*burning and aching*) **pain**.

referred pain

THERMORECEPTION (p. 528)

2. **Thermoreceptors** respond to changes in temperature.

MECHANORECEPTION (pp. 528–531)

3. **Mechanoreceptors** respond to physical distortion, contact, or pressure on their cell membranes; **tactile receptors** to touch, pressure, and vibration; **baroreceptors** to pressure changes in the walls of blood vessels and the digestive, reproductive, and urinary tracts; **proprioceptors** monitor positions of joints and muscles.

mechanically regulated ion channels

4. **Fine touch and pressure receptors** provide detailed information about a source of stimulation; **crude touch and pressure receptors** are poorly localized. Important tactile receptors include the **root hair plexus**, **Merkel's discs**, **Meissner's corpuscles**, **Pacinian corpuscles**, and **Ruffini corpuscles**.
5. Baroreceptors monitor changes in pressure; they respond immediately but adapt rapidly. Baroreceptors in the walls of major arteries and veins respond to changes in blood pressure. Receptors along the digestive tract help coordinate reflex activities of digestion.
6. Proprioceptors monitor the position of joints, tension in tendons and ligaments, and the state of muscular contraction. Proprioceptors include tendon organs and muscle spindles.

CHEMORECEPTION (pp. 531–532)

7. In general, **chemoreceptors** respond to water-soluble and lipid-soluble substances that are dissolved in the surrounding fluid. They monitor the chemical composition of body fluids.

carotid bodies • **aortic bodies**

Olfaction (pp. 532–534)

1. The **olfactory organs** contain the **olfactory epithelium** with **olfactory receptors** (neurons sensitive to chemicals dissolved in the overlying mucus), **supporting cells**, and **basal** (**stem**) **cells**. Their surfaces are coated with the secretions of the **olfactory glands**.

OLFACTORY RECEPTOR FUNCTION (pp. 532–533)

2. The olfactory receptors are modified neurons. Our olfactory sensitivities are much lower than those of many other vertebrates.

OLFACTORY PATHWAYS (pp. 533–534)

3. The olfactory system is very sensitive; its extensive limbic and hypothalamic connections help explain the emotional and behavioral responses that can be produced by certain smells.

OLFACTORY DISCRIMINATION (p. 534)

4. The olfactory system can make subtle distinctions between thousands of chemical stimuli; the CNS interprets the smell based on the particular pattern of receptor activity.

AGING AND OLFACTORY SENSITIVITY (p. 534)

5. The olfactory receptor population shows considerable turnover and is the only known example of neuronal replacement in the adult human. However, the total number of receptors declines with age.

Gustation (pp. 534–536)

1. **Gustatory** (taste) **receptors** are clustered in **taste buds**, each of which contains **gustatory cells**, which extend taste hairs through a narrow **taste pore**.
2. Taste buds are associated with epithelial projections (**papillae**).

filiform papillae
fungiform papillae
circumvallate papillae

GUSTATORY RECEPTOR FUNCTION (p. 534)

3. The **primary taste sensations** are sweet, salt, sour, and bitter.

GUSTATORY PATHWAYS (p. 535)

4. The taste buds are monitored by cranial nerves that synapse within the **nucleus solitarius**.

INDIVIDUAL DIFFERENCES IN GUSTATORY SENSITIVITY (p. 535)

5. There are significant individual differences in taste sensitivity, some of which are inherited.

PTC

AGING AND GUSTATORY SENSITIVITY (p. 536)

6. The number of taste buds declines with age.

Chapter Review

Equilibrium and Hearing (pp. 536–547)

Related Key Terms

1. The senses of equilibrium and hearing are provided by the receptors of the **inner ear** (also known as the **membranous labyrinth**). Its chambers and canals contain the fluid **endolymph**. The **bony labyrinth** surrounds and protects the membranous labyrinth. The bony labyrinth can be subdivided into the **vestibule** and **semicircular canals** (receptors in the vestibule and semicircular canals provide the sense of equilibrium) and the **cochlea** (these receptors provide the sense of hearing). The structures and airspaces of the **external ear** and **middle ear** help capture and transmit sound to the cochlea.

   perilymph

ANATOMY AND RECEPTOR FUNCTION (pp. 537–538)

2. The external ear includes the **pinna** that surrounds the **external auditory meatus**, the entrance to the **external auditory canal** that ends at the **tympanic membrane** ("eardrum"). The tympanic membrane communicates with the nasopharynx via the **pharyngotympanic tube**. The middle ear encloses and protects the **auditory ossicles** that connect the tympanic membrane with the receptor complex of the inner ear.

   ceruminous glands

3. The vestibule includes a pair of membranous sacs, the **saccule** and **utricle**, whose receptors provide sensations of gravity and linear acceleration. The cochlea contains the **cochlear duct**, an elongated portion of the membranous labyrinth.

   round window • **oval window**

4. The basic receptors of the inner ear are **hair cells** whose surfaces support **stereocilia**. Hair cells provide information about the direction and strength of varied mechanical stimuli.

   supporting cells • **kinocilium**

EQUILIBRIUM (pp. 538–540)

5. The **anterior**, **posterior**, and **lateral** semicircular canals are continuous with the utricle. Each contains an **ampulla** with sensory receptors. Here the cilia contact a gelatinous **cupula**.

   crista • **nystagmus**

6. The saccule and utricle are connected by a passageway continuous with the **endolymphatic duct**, which terminates in the **endolymphatic sac**. In the saccule and utricle hair cells cluster within **maculae** where their cilia contact **otoconia** (densely packed mineral crystals). When the head tilts, the mass of otoconia shifts, and the resulting distortion in the sensory hairs signals the CNS.

   otoliths

7. The vestibular receptors activate sensory neurons of the **vestibular ganglia**. The axons form the **vestibular branch** of the vestibulocochlear nerve (N VIII), synapsing within the **vestibular nuclei**.

   vestibulospinal tracts

THE COCHLEA AND HEARING (pp. 541–547)

8. The energy content of a sound determines its **intensity**, measured in **decibels**. Sound waves travel toward the tympanic membrane, which vibrates; the **auditory ossicles** conduct the vibrations to the inner ear. Movement at the oval window applies pressure to the perilymph of the **vestibular duct**.

   cycles • **Hertz (Hz)** • **frequency** • **pitch** • **malleus** • **incus** • **stapes** • **tympanic duct**

9. Pressure waves distort the **basilar membrane** and push the hair cells of the **organ of Corti** against the **tectorial membrane**. The **tensor tympani** and **stapedius** muscles contract to reduce the amount of motion when very loud sounds arrive.
10. **Conductive deafness** results from conditions in the middle ear that block sound transmission to the oval window. **Nerve deafness** reflects problems in the cochlea or along central pathways.
11. The sensory neurons are located in the **spiral ganglion** of the cochlea. Their afferent fibers form the **cochlear branch** of the vestibulocochlear nerve (N VIII), synapsing at the **cochlear nucleus**.

The Eyes and Vision (pp. 547–565)

ACCESSORY STRUCTURES OF THE EYE (pp. 550–551)

1. The **accessory structures** of the eye include the **palpebrae** (eyelids), which are separated by the **palpebral fissure**. The **eyelashes** line the palpebral margins. Along the inner margin of the lid are **Meibomian glands**, which secrete a lipid-rich product. Glands at the **lacrimal caruncle** produce other secretions.

   medial canthus • **lateral canthus** • **sty** • **chalazion**

2. An epithelium called the **conjunctiva** covers most of the exposed surface of the eye; the **bulbar**, or **ocular**, conjunctiva covers the

anterior surface of the eye and the **palpebral conjunctiva** lines the inner surface of the eyelids. The **cornea** is transparent.

3. The secretions of the **lacrimal gland** bathe the conjunctiva; these secretions are slightly alkaline and contain **lysozymes** (enzymes that attack bacteria). Tears collect in the **lacus lacrimalis**. The tears reach the inferior meatus of the nose after passing through the **lacrimal puncta**, the **lacrimal canals**, the **lacrimal sac**, and the **nasolacrimal duct.**

Related Key Terms: **fornix**

ANATOMY OF THE EYE (pp. 551–553)

4. The eye has three layers: an outer **fibrous tunic**, a vascular tunic, and an inner neural tunic. Most of the ocular surface is covered by the **sclera** (a dense fibrous connective tissue); the **limbus** is the border between the sclera and the cornea.

Related Key Terms: **orbital fat**

5. The **vascular tunic**, or **uvea**, includes the **iris**, the **ciliary body**, and the **choroid**. The iris forms the boundary between the **anterior** and **posterior chambers**. The ciliary body contains the **ciliary muscle** and the **ciliary processes**, which attach to the **suspensory ligaments** of the lens.

Related Key Terms: **pupil** • **pupillary constrictor muscles** • **pupillary dilator muscles** • **ora serrata**

6. The fluid **aqueous humor** circulates within the eye and reenters the circulation after diffusing through the walls of the anterior chamber and into the **canal of Schlemm**.
7. The **lens**, held in place by the suspensory ligaments, lies behind the cornea and forms the anterior boundary of the **vitreous chamber**. This chamber contains the **vitreous body**, a gelatinous mass that helps stabilize the shape of the eye and support the retina.
8. The **neural tunic** consists of an outer **pigment layer** and an inner **retina**; the latter contains visual receptors and associated neurons.

THE LENS AND IMAGE FORMATION (pp. 553–556)

9. The lens focuses a visual image on the retinal receptors; a lens that has lost its transparency is a **cataract**. Light is **refracted** (bent) when it passes through the cornea and lens. During **accommodation** the shape of the lens changes to focus an image on the retina. Normal **visual acuity** is rated 20/20.

Related Key Terms: **senile cataracts** • **focal point** • **focal distance** • **emmetropia**

LIGHT AND PHOTORECEPTION (pp. 556–562)

10. The direct line to the CNS proceeds from the photoreceptors to **bipolar cells**, then to **ganglion cells**, and to the brain via the optic nerve. **Horizontal cells** and **amacrine cells** modify the signals passed between other retinal components.
11. Light is radiated in waves with a characteristic **frequency** and **wavelength**. A **photon** is a single energy packet of **visible light**. There are two types of **photoreceptors** (visual receptors of the retina): **rods** and **cones**. Rods respond to almost any photon, regardless of its energy content; cones have characteristic ranges of sensitivity. Many cones are densely packed within the **fovea** (**macula lutea**), the site of sharpest vision.

Related Key Terms: **blue cones** • **green cones** • **red cones**

12. Each photoreceptor contains an **outer segment** with membranous **discs**. A narrow stalk attaches the outer to the **inner segment**. Light absorption occurs in the **visual pigments**, which are derivatives of **rhodopsin** (**opsin** plus a pigment, **retinal**, that is synthesized from **vitamin A**). The retinal molecule can be in either the normal **11-cis** or (after absorbing light) the **11-trans** form.

Related Key Terms: **transducin (G-protein)** • **phosphodiesterase (PDE)** • **cyclic-GMP (cGMP)**

13. **Color blindness** is the inability to detect certain colors.
14. In the **dark-adapted** state, almost all visual pigments will be fully receptive to stimulation. The **light-adapted** state is characterized by constriction of the pupil and bleaching of the visual pigments.

PROCESSING OF VISUAL INFORMATION (pp. 562–565)

15. Each photoreceptor monitors a specific receptive field. Cortical neurons called **simple cells** integrate information and enhance contrast and image detail. **Complex cells** respond best to moving linear stimuli, probably by monitoring the activities of several simple cells. The visual image arrives at the retina upside down, but the brain adjusts for this.

Related Key Terms: **on-center cells** • **off-center cells** • **optic disc** • **blind spot**

16. Visual inputs to the **suprachiasmatic nucleus** affect the function of other brain stem nuclei. This nucleus establishes a visceral **circadian rhythm** that is tied to the day/night cycle and affects other metabolic processes.

Chapter Review

Review Planner

	Objective	Level 1
1	Distinguish between the general and special senses.	6 9 24
2	Identify the receptors for the general senses and describe how they function.	1 2 8 16 19 22 34 40 47 49 51
3	Explain why receptors respond to specific stimuli and how the organization of a receptor affects its sensitivity.	1 2 8 14 33
4	Discuss the receptors and processes involved in the senses of smell, taste, and equilibrium.	6 7 10 14 28 38 39 44 45 48 50
5	Describe the parts of the ear and their roles in the process of hearing.	3 4 11 17 20 41 46
6	Identify the parts of the eye and their functions.	12 13 18 21 23 25 26 27 42 43
7	Explain how we are able to see objects and distinguish colors and depth.	21 23 35 36
8	Discuss how the central nervous system processes information related to all the senses.	10 11 15 37

Level 2: **29 30 31 32 33 37**

Level 3: **52–55**

CM: **The Control of Pain**
Otitis Media and Mastoiditis
Vertigo, Dizziness, and Motion Sickness • Conjunctivitis
Corneal Transplants
Glaucoma • Night Blindness
Scotomas and Floaters

CM: **Testing and Treating Hearing Deficits**
Analyzing Sensory Disorders

ABC News: **Enough to Make You Sick**

Review Questions

MULTIPLE CHOICE

1. The area monitored by a single receptor cell is the: a) receptor potential; b) central adaptation; c) receptive field; d) proprioceptor.
2. Sensory receptors that are normally inactive but respond briefly to a changed level of stimulation are called: a) tonic receptors; b) phasic receptors; c) ionic receptors; d) adaptable receptors.
3. A high-frequency sound would have a high: a) decibel level; b) pitch; c) wavelength.
4. Decibels are used to measure: a) intensity; b) frequency; c) wavelength; d) pitch.
5. Which of the following would stimulate the olfactory epithelium?: a) a red stop sign; b) a song played on the radio; c) being thrown off balance; d) entering a rose garden.
6. Which of the following is *false*? Receptors for special senses differ from those for general senses in that they are: a) faster to respond; b) better protected; c) more complex.
7. The primary taste sensations are: a) sweet, salt, sour, and hot; b) sweet, salt, bitter, and chocolate; c) sweet, bitter, sour, and salt; d) sweet, acid, alkaline, and salt.
8. Someone who is suffering a heart attack may feel pain radiating down the left arm. This is known as: a) prickling pain; b) peripheral adaptation; c) transduction; d) referred pain.
9. The special senses include smell, taste, hearing, vision, and: a) equilibrium; b) proprioception; c) pain; d) touch.
10. Damage to the nucleus solitarius would: a) produce partial blindness; b) eliminate the sense of smell; c) affect hearing and balance sensations; d) change the taste of food.
11. Damage to the cochlear nucleus would: a) produce partial blindness; b) eliminate the sense of smell; c) affect hearing; d) change the taste of food.
12. The palpebrae are also known as the: a) eyelashes; b) eardrums; c) eyelids; d) tear ducts.

DIAGRAMS

13. Draw a diagram of an eyeball. Label the lens, cornea, iris, retina, sclera, optic nerve, and pupil.

MATCHING QUESTIONS

(Match each term with the most appropriate statement.)

Statements:

14. Openings through which a gustatory cell extends its taste hairs into the surrounding fluids.
15. Short, jerky eye movements that can appear after damage to the brain stem or inner ear.
16. Tactile receptors that are scattered throughout the integument.
17. The structure that vibrates in response to sounds.
18. The area where tears accumulate.
19. Sensory receptors that monitor the chemical composition of body fluids and respond to the presence of specific molecules.
20. Mineral crystals that press against the sensory hairs of the utricle and saccule and provide clues to the orientation of the head.
21. The area of sharpest vision in the retina.
22. Sensory receptors that monitor blood pressure in the walls of major vessels.
23. Focuses a visual image on the retinal receptors.

Terms:

a) Pacinian corpuscles
b) Baroreceptors
c) Chemoreceptors
d) Taste pore
e) Tympanic membrane
f) Nystagmus
g) Otoconia
h) Lacus lacrimalis
i) Lens
j) Fovea

SHORT ANSWER/ESSAY

24. Distinguish between the general senses and the special senses. How do their receptors differ?
25. What happens when the pupillary constrictor muscles contract? How does this compare with the result when the pupillary dilator muscles contract? What are the functions of these muscles?
26. How do the cornea and lens affect light that enters the eye? What role does this play in vision?
27. What are the functions of tears? What gland provides the main ingredients for tears?
28. To which taste sensations are we most sensitive? Why?
29. In a vision test, what is meant by a rating of "20/20"? Is a rating of 20/30 better or worse?
30. Several rock'n'roll musicians have admitted during interviews that they have become partially deaf. Why do you think this happened?
31. Jane is looking through a telescope at a dim star. Her astronomy instructor tells her that if she does not look directly at the star she will be able to see it better. Is this true, and if so why?
32. Mrs. N., 72 years old, had a small gas leak in her apartment. The leak was not detected until her neighbor smelled the gas and alerted the landlord. Why didn't Mrs. N. detect the leak?
33. Define "transduction," and describe the steps involved.
34. What do baroreceptors, tendon organs, Ruffini corpuscles, and muscle spindles have in common?
35. When 6-year-old Matthew is asked to pick out the red block from a pile of toys, he randomly hands you a gray block, a green block, or a red one. What could be the reason? Explain what causes this condition.
36. Mr. Brennan is told to wear a patch over his right eye for a week during recovery from a serious corneal abrasion. He finds that while wearing the patch he has trouble grasping objects because he reaches too far. Explain why this occurs.
37. A friend comments that she hears a faucet dripping. Previously, you had not noticed the sound; after her remark, you find that you continue to hear it. Describe what has happened.
38. What are the functions of the mucus glands in our olfactory organs?

Identify the sense that is associated with the following:

39. Saccule.
40. Merkel's discs.
41. Cochlea.
42. Meibomian glands.
43. Choroid.
44. Olfactory glands.
45. Taste buds.
46. Round window.

Identify the type of receptor that would respond to these stimuli:

47. Cutting your finger with a knife
48. Tilting back too far in your chair and feeling you are off-balance.
49. Entering a highly air-conditioned classroom.
50. Smelling your favorite food.
51. Something brushing against the hair on your forearm.

CRITICAL THINKING/APPLICATIONS

52. A child complains that he can't read what his teacher writes on the chalkboard during class, and that after several hours his head hurts. What is your diagnosis, and what causes this condition?
53. Mr. Romero, 62, has trouble hearing people during conversations, and his family persuades him to have his hearing tested. How can the physician determine whether Mr. Romero's problem results from nerve deafness or conductive deafness? How can these conditions be treated?
54. Five-year-old Bill tells his mother that he sees "spots" when he looks at the sky or focuses "up close" on something. They don't interfere with his vision and he isn't sure how long he's had them. Is this something to be concerned about? What would you recommend?
55. Mr. Willis visits his doctor because he is having trouble with his vision. The physician performs a series of tests and determines the following: Stimuli are reaching Mr. Willis's sensory receptors; the receptors respond normally; and the pathways are intact and normal. Classify Mr. Willis's sensory dysfunction, and give an example of a condition that could cause this type of dysfunction.

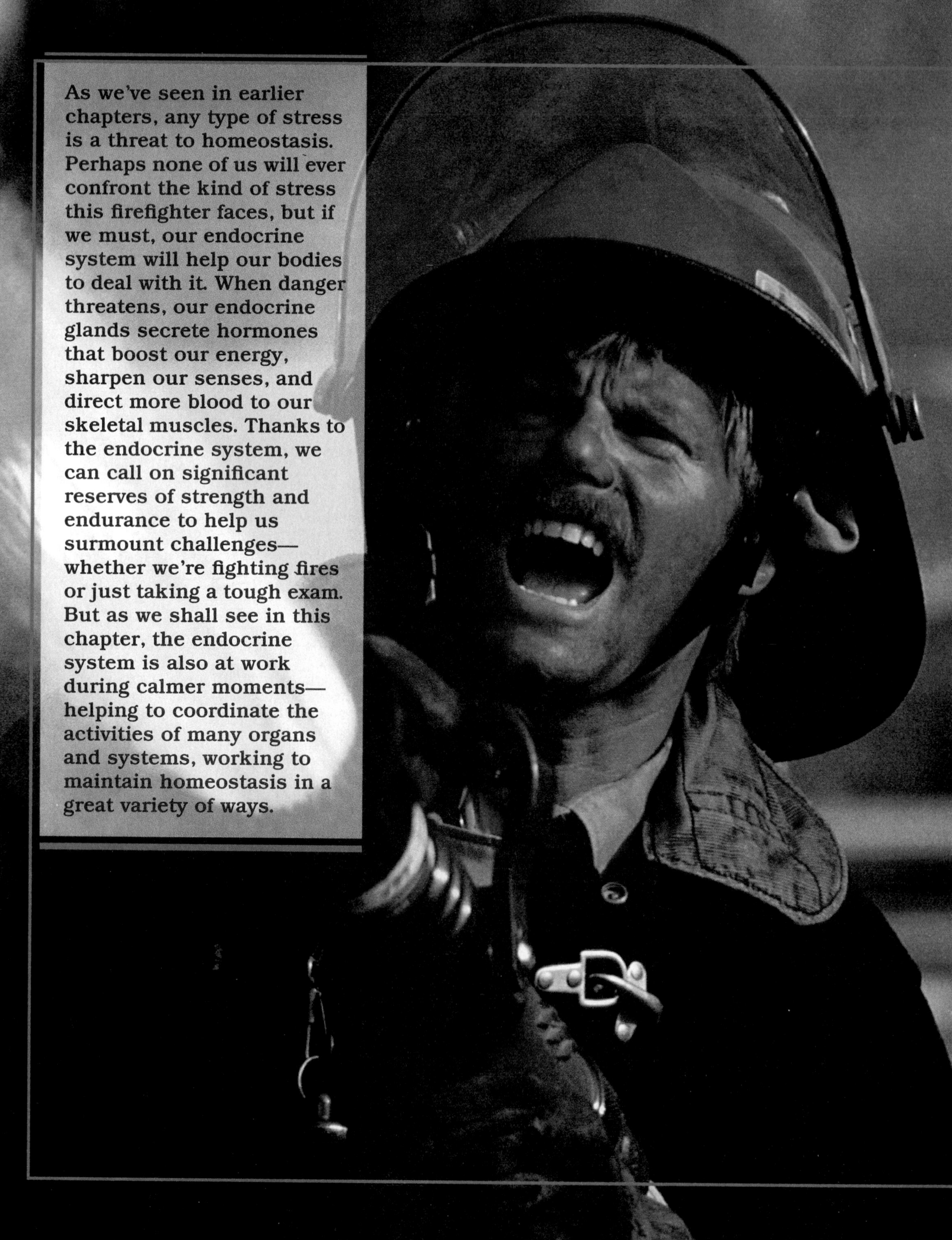

As we've seen in earlier chapters, any type of stress is a threat to homeostasis. Perhaps none of us will ever confront the kind of stress this firefighter faces, but if we must, our endocrine system will help our bodies to deal with it. When danger threatens, our endocrine glands secrete hormones that boost our energy, sharpen our senses, and direct more blood to our skeletal muscles. Thanks to the endocrine system, we can call on significant reserves of strength and endurance to help us surmount challenges—whether we're fighting fires or just taking a tough exam. But as we shall see in this chapter, the endocrine system is also at work during calmer moments—helping to coordinate the activities of many organs and systems, working to maintain homeostasis in a great variety of ways.

CHAPTER 18

The Endocrine System

Chapter Objectives

After reading this chapter, you will be able to:

1 Compare the endocrine and nervous systems.
2 Compare the cellular components of the endocrine system with those of other tissues and systems.
3 Identify the endocrine organs and the hormones they produce.
4 Explain how hormones exert their effects, and identify the effects of various important hormones.
5 Relate the structure of hormones to their functions.
6 Discuss the results of abnormal hormone production.
7 Explain how hormones interact to produce coordinated physiological responses.
8 Describe how endocrine organs are controlled.
9 Identify five hormones that are especially important to normal growth.
10 Explain how the endocrine system responds to stress.

Introduction

Homeostatic regulation involves coordinating the activities of organs and systems throughout the body. The cells of the nervous and endocrine systems work together to monitor and adjust the physiological activities under way at any given moment. Their activities are closely coordinated, and their effects are usually complementary.

In general, the nervous system performs crisis management by producing rapid, short-term, and very specific responses to environmental stimuli. Neurons communicate with one another and with effectors by releasing chemical neurotransmitters. The actions of these transmitters are highly localized, limited to specific target cells. The effects are immediate, but they last only as long as the neural output continues, usually a matter of seconds.

Endocrine cells, by contrast, release chemicals into the bloodstream for distribution throughout the body. These chemicals, called **hormones**, alter the metabolic activities of many different tissues and organs simultaneously. The hormonal effects may not be immediately apparent, but when they do appear they often persist for days, even in the absence of further hormonal stimulation. This response pattern makes the endocrine system particularly effective in regulating ongoing metabolic processes. One example, the endocrine regulation of calcium ion concentrations, was detailed in Chapter 7. (p. 220) Gradual changes in the pattern of hormone production orchestrate complex, long-term processes such as embryological development, growth, sexual maturation, and reproduction.

When viewed from this general perspective, the differences between the nervous and endocrine systems are relatively clear. But when considered in detail, the two systems are virtually impossible to separate on either functional or anatomical grounds. For example, many compounds used as hormones by the endocrine system, such as dopamine, norepinephrine, and epinephrine, also function as neurotransmitters inside the CNS.

Anatomical boundaries are equally difficult to determine. The adrenal medulla is a modified sympathetic ganglion whose neurons secrete neurotransmitters (primarily epinephrine) into the circulation. As discussed in Chapter 16, epinephrine produces swift peripheral responses that help prepare the body for a crisis. (p. 504) In this case we have an endocrine gland that functions as a component of the nervous system and releases a hormone that produces an immediate effect on peripheral systems.

The hypothalamus is an even more dramatic example of the hazy boundary between these two systems. This portion of the brain (1) secretes two hormones (ADH and oxytocin) that target peripheral effectors, (2) controls the secretory output of the adrenal medulla, and (3) releases hormones that control the pituitary gland. In essence, the hypothalamus is the interface between the neural and endocrine systems, and a later section will discuss this role in detail.

This chapter introduces the components and functions of the endocrine system and examines the interactions between the nervous and endocrine systems. Subsequent chapters will consider specific endocrine organs, hormones, and functions in greater detail.

An Overview of the Endocrine System

The endocrine system includes all of the endocrine cells and tissues of the body. As noted in Chapter 5, endocrine cells are glandular secretory cells that release their secretions internally, rather than onto an epithelial surface. (p. 126) The components of the endocrine system are introduced in Figure 18-1. This figure also lists the major hormones produced at each endocrine organ.

STRUCTURE OF HORMONES

Hormones can be divided into three different groups on the basis of chemical structure: *amino acid derivatives*, *peptide hormones*, and *steroids*. Examples of each group are included in Figure 18-2.

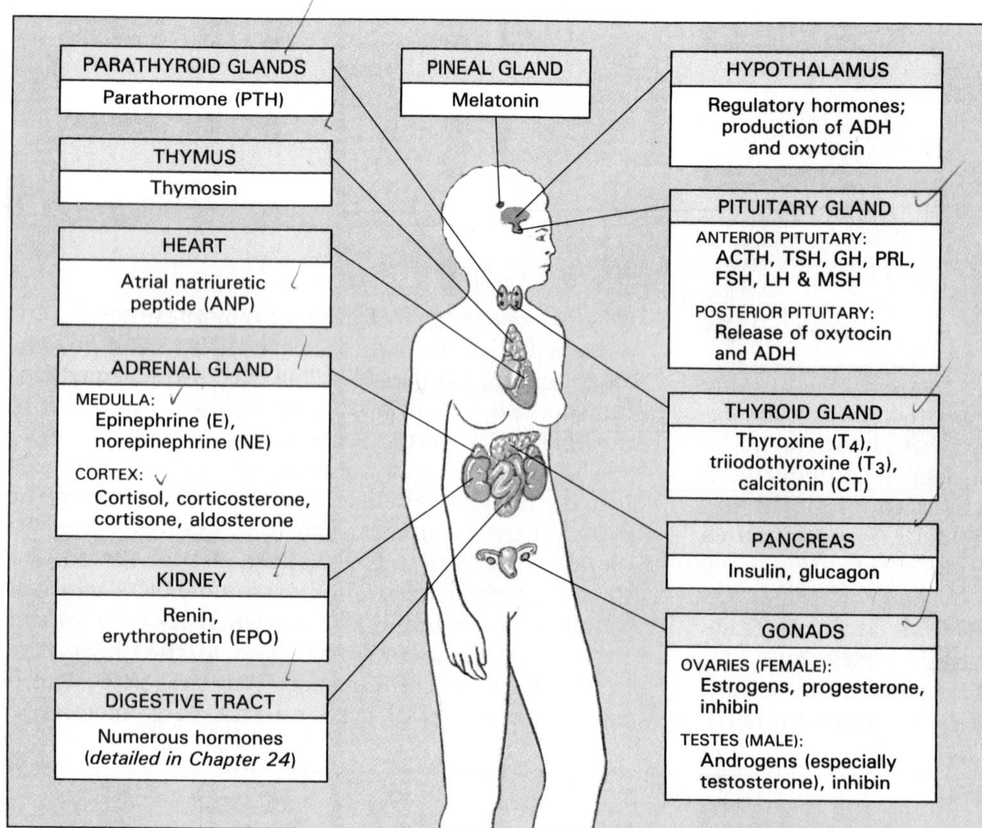

FIGURE 18-1
The Endocrine System

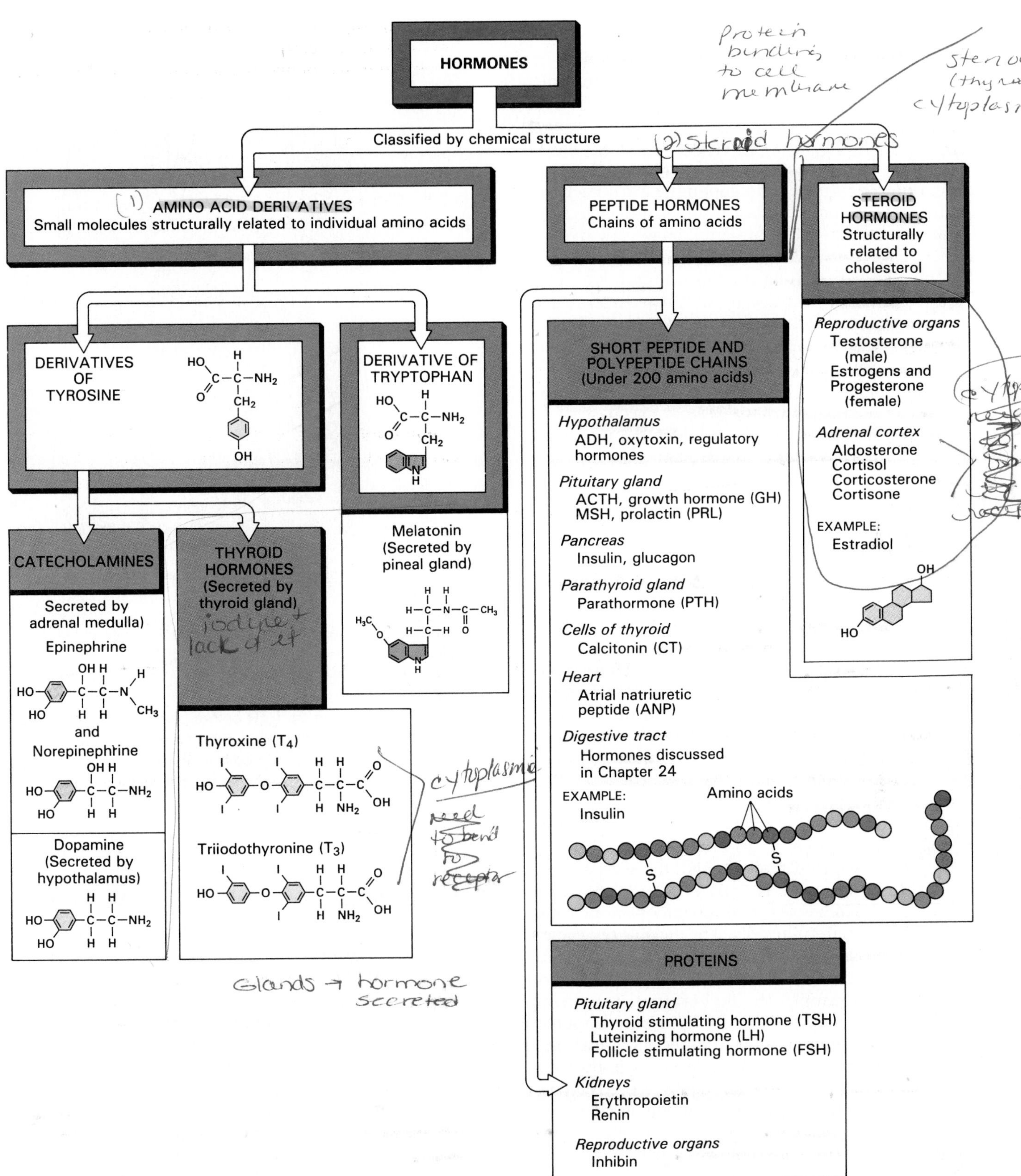

FIGURE 18-2
A Structural Classification of Hormones

Amino Acid Derivatives

Some hormones are relatively small molecules that are structurally similar to amino acids. (Amino acids, the building blocks of proteins, were introduced in Chapter 2.) (p. 51) This group includes *epinephrine*, *norepinephrine*, *dopamine*, the *thyroid hormones*, and the pineal hormone *melatonin*. Epinephrine, norepinephrine, and dopamine are structurally similar; these compounds are sometimes called **catecholamines** (kat-e-KŌL-am-ēnz). As we saw in Chapter 16, epinephrine and norepinephrine are secreted by the adrenal medulla during sympathetic activation. Dopamine, secreted by the hypothalamus, assists in the regulation of pituitary gland function. **Thyroid hormones** are released by the thyroid gland, and **melatonin** is produced by the pineal gland. As indicated in Figure 18-2, catecholamines and thyroid hormones are synthesized from molecules of the amino acid **tyrosine** (TĪ-rō-sēn). Melatonin is manufactured from molecules of the amino acid *tryptophan*.

Peptide Hormones

Peptide hormones consist of chains of amino acids. These molecules range from short amino acid chains, such as ADH or oxytocin (9 amino acids apiece), to polypeptides such as *growth hormone* (191 amino acids) and *prolactin* (198 amino acids). This category also includes proteins, such as *thyroid-stimulating hormone*, that contain thousands of amino acids. The majority of hormones fall within this group, including all of the hormones secreted by the hypothalamus, pituitary gland, heart, kidneys, thymus, digestive tract, and pancreas.

Figure 18-3 diagrams stages in the synthesis of peptide hormones. In general, the process follows the pattern of protein synthesis and secretion discussed in Chapter 4. (pp. 110–113) Synthesis begins at ribosomes of the rough endoplasmic reticulum (RER) of glandular cells. The peptide chains produced at the RER are called **prehormones** because they are structurally distinct from the final hormone product. For example, the prehormone may contain extra amino acids, or it may be missing important carbohydrate components. Further modification in the endoplasmic reticulum converts the prehormone to a **prohormone**, an inactive form of the hormone. Final conversion to an active hormone usually occurs either as the prohormone passes through the Golgi apparatus or within a secretory vesicle.

Steroid Hormones

Steroid hormones are lipids structurally similar to cholesterol (Chapter 2). (p. 49) These hormones are released by the reproductive organs and the adrenal glands.

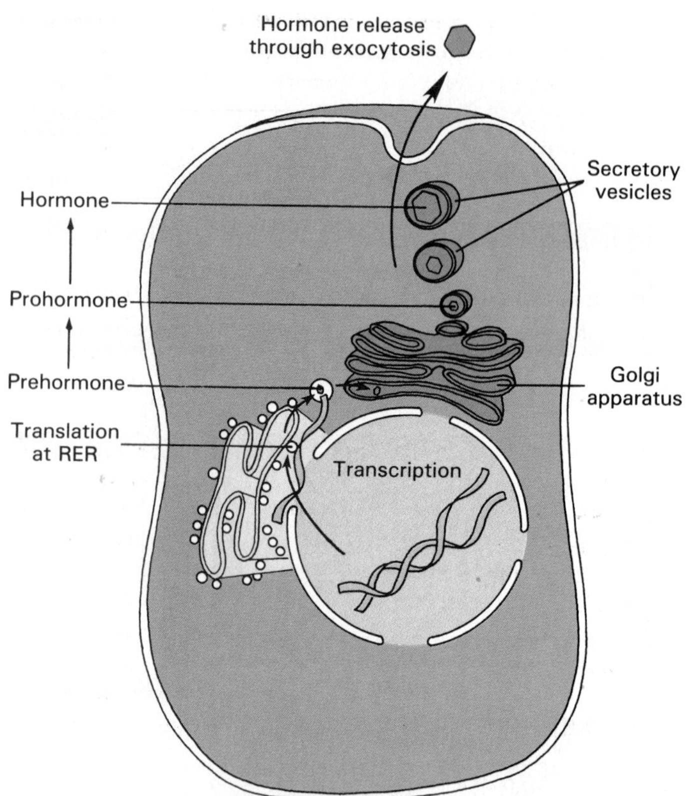

FIGURE 18-3
Peptide Hormone Production. Activated genes in the nucleus are transcribed, producing mRNA that travels to the rough endoplasmic reticulum. Translation at the RER then produces molecules of a prehormone. These molecules undergo further modification in the RER before being exported in transport vesicles as molecules of prohormone. In the Golgi apparatus the final structural modifications are made, producing a hormone that will be released at the cell surface through exocytosis.

MECHANISMS OF HORMONAL ACTION

As we learned in Chapter 4, all cellular activities and metabolic pathways are controlled by enzymes. (pp. 98–101) Hormones alter cellular operations by changing the *types*, *activities*, *quantities*, or *properties* of important enzymes. Through one or more of these mechanisms, a hormone modifies the activities of its **target cells**, peripheral cells sensitive to its presence.

The first step in the process involves an interaction between the hormone and a specific *receptor complex*. The presence or absence of the necessary receptor determines each cell's hormonal sensitivities. Binding of a hormone to its receptor starts a biochemical chain of events that changes the pattern of enzymatic activity within the cell.

The receptor complex may be found either on the cell membrane or in the cytoplasm. The catecholamines and peptide hormones target receptors on the cell membrane. The thyroid hormones and steroids interact with receptors inside the cell. We will now consider the basic mechanisms involved, using a few specific examples. (Further discussion and additional examples will be found in later sections dealing with individual hormones and their effects.)

Hormones and the Cell Membrane

The receptors for catecholamines (epinephrine, norepinephrine, dopamine) and peptide hormones are found in the cell membranes of their respective target cells. These hormones do not exert their effects directly. For example, a hormone does not enter a cell and begin building a protein or catalyzing a specific reaction. The hormone acts as a **first messenger** that causes the appearance of a **second messenger** in the cytoplasm. The second messenger may function as an enzyme activator, inhibitor, or cofactor to change the direction and rate of cytoplasmic reactions. Several different compounds can function as second messengers, but the most important are cyclic-AMP and calcium ions.

The link between the first messenger and the second messenger is a **G protein**, an enzyme attached to the membrane receptor. There are several different types of G proteins; transducin, a G protein involved in photoreception, was discussed in Chapter 17. (pp. 559–560) Binding of a hormone to its receptor activates the associated G protein. Figure 18-4 diagrams the possible events set in motion by activation of a G protein, each of which leads to the appearance of a second messenger in the cytoplasm:

1. Activation of adenyl cyclase. The activation of adenyl cyclase was described in Chapter 12 in the discussion of epinephrine's effects on β_1 receptors. (p. 389) When activated, this enzyme converts ATP to a ring-shaped molecule, cyclic-AMP (cAMP). Cyclic-AMP then functions as the second messenger, opening ion channels or activating key enzymes in the cytoplasm. Many other hormones that we will encounter later in this chapter, including calcitonin, parathormone, ADH, ACTH, FSH, LH, TSH, and glucagon, produce their effects by increasing intracellular concentrations of cyclic-AMP.

2. Release or entry of calcium ions. An activated G protein can trigger the opening of calcium ion channels in the membrane or the release of calcium ions (Ca^{2+}) from intracellular stores. The cal-

FIGURE 18-4 Mechanisms of Peptide Hormone Activity. Peptide hormones act by binding to membrane receptors and activating G proteins. Four possible results are shown, each leading to the release of a second messenger in the cytoplasm. G protein activation may lead to (1) activation of adenyl cyclase and formation of cyclic-AMP, (2) opening of calcium ion channels and calcium release into the cytoplasm, (3) activation of phosphodiesterase and a reduction in intracellular cAMP concentrations, or (4) activation of specific cytoplasmic enzymes.

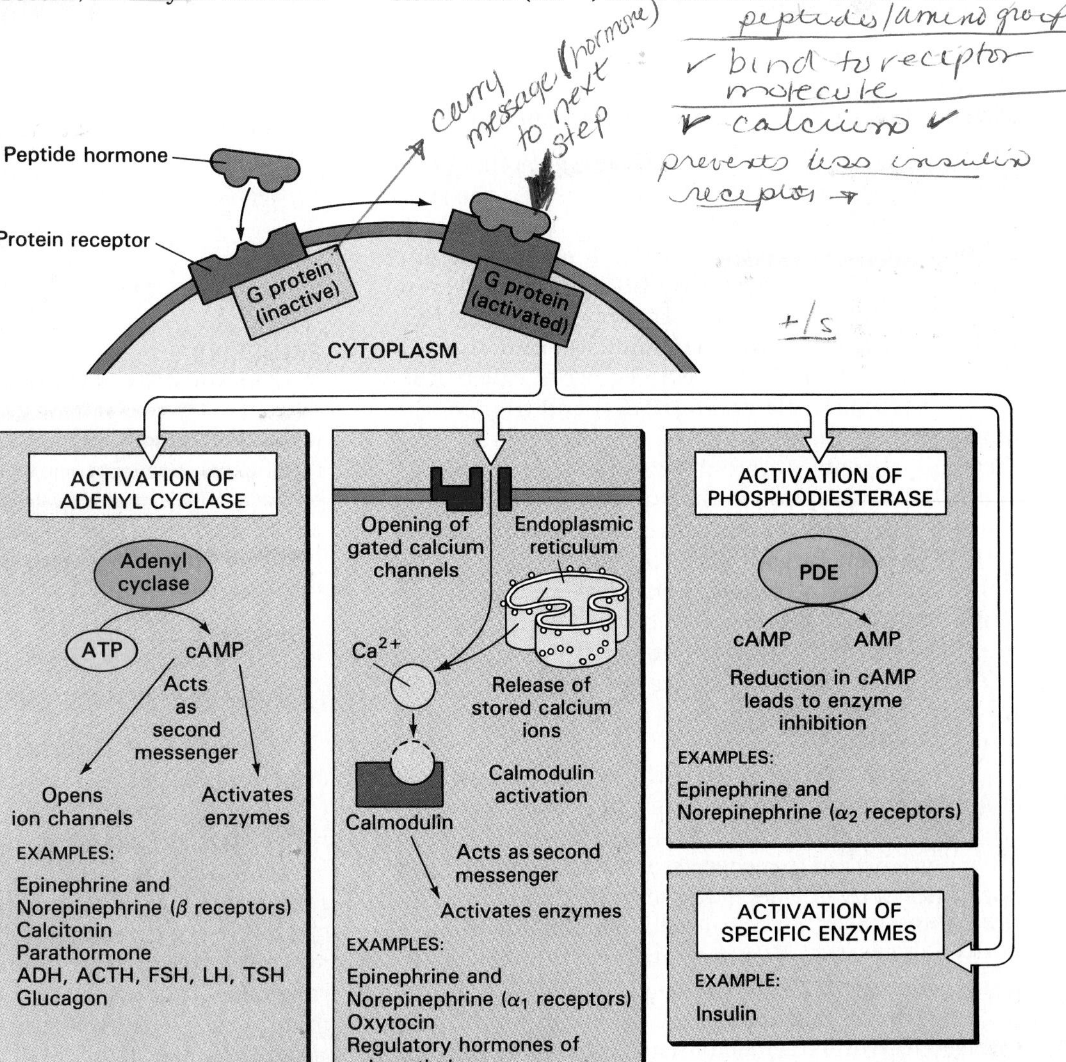

cium ions then serve as the second messenger, usually in conjunction with an intracellular protein called **calmodulin**. Once it has bound calcium ions, calmodulin can activate specific cytoplasmic enzymes. This chain of events is responsible for the stimulatory effects that follow activation of α_1 receptors by epinephrine or norepinephrine. Calmodulin also acts as an intermediary for oxytocin and several regulatory hormones secreted by the hypothalamus.

3. Activation of phosphodiesterase. Phosphodiesterase is an enzyme that converts cAMP to AMP. Its activation therefore results in a decreased concentration of cAMP in the cytoplasm. This decrease usually has an inhibitory effect, due to inactivation of enzymes dependent on cAMP. This mechanism accounts for the inhibitory effects of norepinephrine and epinephrine via α_2 receptors.

4. Activation of other intracellular enzymes. G proteins may also activate intracellular enzymes that have specific functions. For example, *insulin* is a pancreatic hormone involved with the regulation of blood glucose concentrations. When insulin binds to its receptor, a G protein activates various enzymes that accelerate glucose transport across the cell membrane.

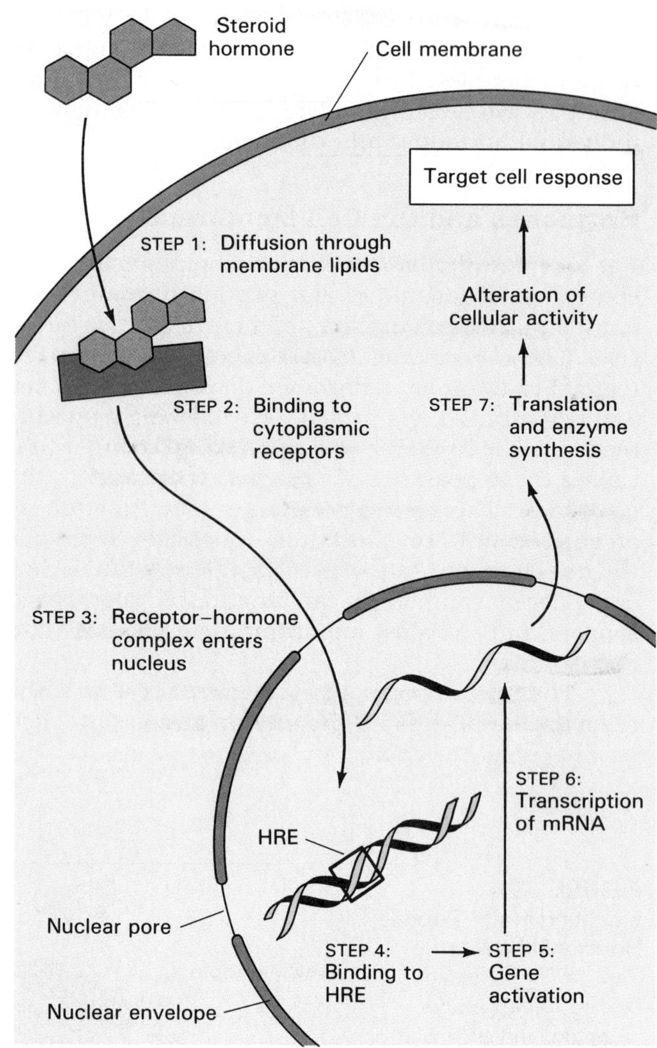

FIGURE 18-5
Mechanism of Steroid Hormone Activity. Steroid hormones diffuse through the membrane lipids and bind to cytoplasmic receptors that then enter the nucleus. Once inside the nucleus, the hormone-receptor complex binds to hormone-responsive elements (HREs) that activate genes. This leads to the production of mRNA molecules coding for the production of specific enzymes. These enzymes alter the metabolic activities of the cell and produce a cellular response.

Hormones and Cytoplasmic Receptors

The thyroid hormones and steroid hormones cross the cell membrane and bind to receptors in the cytoplasm or nucleus. Steroid hormones diffuse rapidly through the lipid portion of the cell membrane and enter the cytosol, where they attach themselves to specific receptor molecules (Figure 18-5). The hormone-receptor complex then diffuses into the nucleus, where it binds to DNA segments called **hormone-responsive elements** (**HRE**s) and triggers the activation or inactivation of specific genes. This mechanism enables steroid hormones to alter the rate of mRNA transcription in the nucleus, thereby changing the pattern of enzyme synthesis and metabolic activity in their target cells. (You may wish to review the discussions of protein synthesis and the enzymatic control of cellular activity in Chapter 4.) For example, the male sex hormone *testosterone* stimulates the production of enzymes and proteins in skeletal muscle fibers, causing an increase in muscle size and power.

Thyroid hormones cross the cell membrane and enter the nucleus without binding to cytoplasmic receptors. In effect, they bypass Step 2 in Figure 18-5. Once bound to receptors in the nucleus, thyroid hormones activate specific genes and alter transcription. This alteration affects the metabolic activities of the cell by changing the nature or number of enzyme molecules in the cytoplasm.

Figure 18-6 summarizes the information provided concerning the mechanisms of hormone action.

CONTROL OF ENDOCRINE ACTIVITY

There are many functional parallels between the organization of the nervous and endocrine systems. Chapter 13 considered the basic operation of neural reflex arcs, the simplest organizational units in the nervous system. (p. 420) The most direct arrangement was a monosynaptic reflex, such as the stretch reflex diagrammed in Figure 18-7a. Polysynaptic reflexes provide more complex and variable responses to stimuli, and higher centers that integrate multiple inputs can facilitate or inhibit these reflexes as needed.

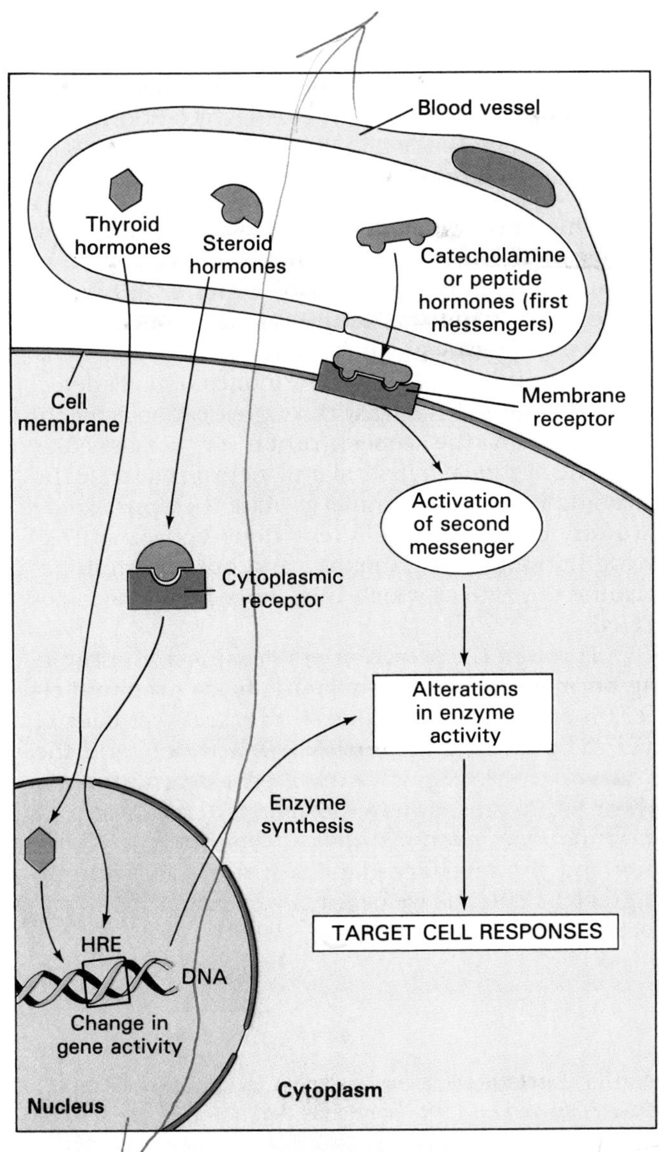

FIGURE 18-6
A Summary of the Mechanisms of Hormone Action. Catecholamines and peptide hormones act via second messengers released when the hormones bind to receptors at the membrane surface. Steroid hormones bind to receptors in the cytoplasm that then enter the nucleus. Thyroid hormones proceed directly to the nucleus to reach hormonal receptors.

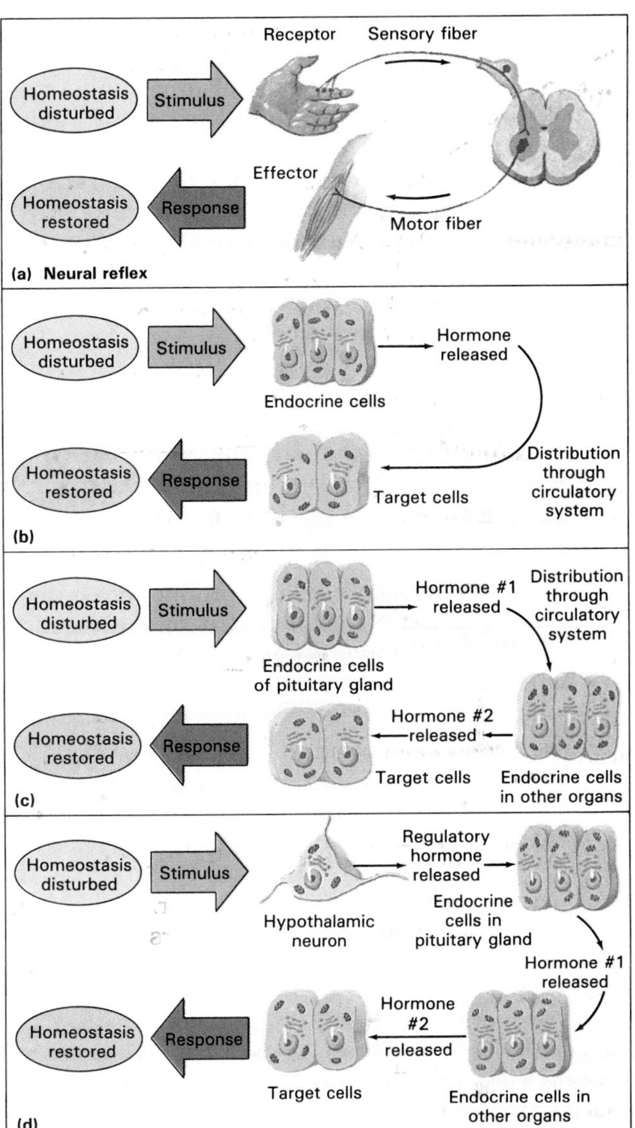

FIGURE 18-7
Patterns of Endocrine Activity and Control. (a) A neural reflex, for comparison with endocrine control mechanisms. (b) A simple endocrine reflex. The endocrine cell responds to a change in the composition of the extracellular fluid, and the hormone released stimulates a target cell to respond in a way that restores homeostasis. (c) A more complex endocrine reflex. Cells of the pituitary gland respond to a stimulus by releasing a hormone that triggers hormone secretion at another endocrine organ. This hormone stimulates an appropriate response by peripheral target cells. (d) The most complex endocrine responses are directed by the hypothalamus. In this case the stimulus may be a chemical change in the blood or some change in CNS activity. In response, the hypothalamus produces a regulatory hormone that controls the hormonal output of the pituitary gland.

Endocrine Reflexes

Endocrine reflexes are the functional counterparts of neural reflexes. A simple example, introduced in Chapter 7 and diagrammed in Figure 18-7b, is the control of calcium ion concentrations by parathormone or calcitonin. The endocrine cells in such a reflex are responding directly to changes in the composition of the extracellular fluid. The hormones they secrete trigger a response that restores homeostatic conditions. Reflexes of this type represent the primary control mechanism for endocrine cells in the heart, pancreas, and digestive tract.

More complex endocrine reflexes involve intermediary steps. The endocrine cells of the thyroid, adrenal gland, and reproductive organs produce hormones with widespread and diverse effects. These hormones are released only when chemical instructions arrive from the pituitary gland, in the form of circulating hormones. The pituitary therefore acts as a "master gland" that integrates chemical information and directs the responses of other endocrine organs. This mechanism is diagrammed in Figure 18-7c.

The most complex endocrine responses involve the hypothalamus. As indicated in Figure 18-7d, the hypothalamus represents the highest level of endocrine control.

The Hypothalamus and Endocrine Regulation

Coordinating centers in the hypothalamus regulate the activities of the nervous and endocrine systems via three different mechanisms, diagrammed in Figure 18-8:

1. The hypothalamus contains autonomic centers that exert direct neural control over the endocrine cells of the adrenal medulla. When the sympathetic division is activated, the adrenal medulla releases hormones into the bloodstream.
2. The hypothalamus itself acts as an endocrine organ, releasing hormones into the circulation at the posterior pituitary. Two of these, ADH and oxytocin, were introduced in Chapter 14. (p. 445)
3. The hypothalamus secretes **regulatory hormones**, or *regulatory factors*, special hormones that regulate the activities of endocrine cells in the pituitary gland.

There are two classes of regulatory hormones. A **releasing hormone (RH)** stimulates production of one or more hormones at the anterior pituitary, whereas an **inhibiting hormone (IH)** prevents the synthesis and secretion of hormones. The concentration of circulating pituitary hormones usually determines the rate of RH release through a feedback mechanism. When the concentration of a particular hormone declines, hypothalamic neurons secrete the releasing factor needed to stimulate the appropriate pituitary cells. Complex interactions between circulating hormones, metabolites, and other stimuli determine the rate at which inhibitory hormone is secreted.

Through the secretion of releasing and inhibiting hormones, the hypothalamus has a profound effect on endocrine functions. It exerts direct control over the production of pituitary hormones, and the pituitary in turn regulates the production of thyroid, adrenal, and reproductive hormones. The structural and functional relationships between the hypothalamus and the pituitary gland will now be examined in greater detail, as we begin our examination of endocrine tissues and organs.

FIGURE 18-8
Three Types of Hypothalamic Control over Endocrine Organs. (1) The hypothalamus exerts direct neural control over the secretory activity of the adrenal medulla. (2) Hypothalamic neurons secrete ADH and oxytocin, hormones that produce specific responses in peripheral target organs. (3) Hypothalamic neurons release regulatory hormones that control the secretory activity of the pituitary gland.

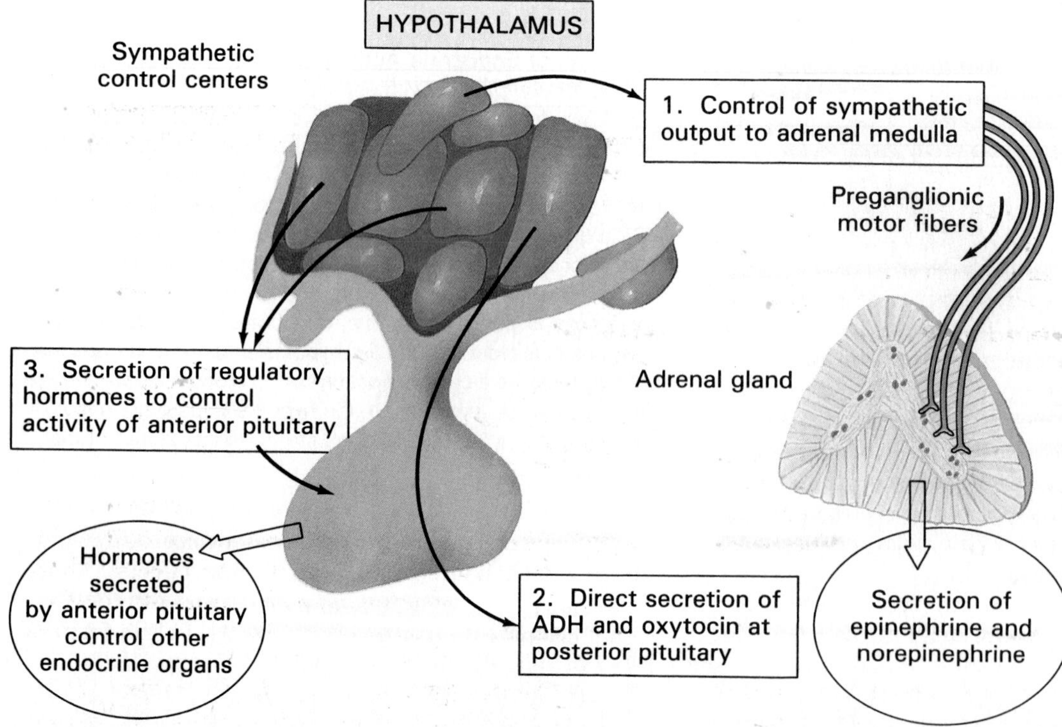

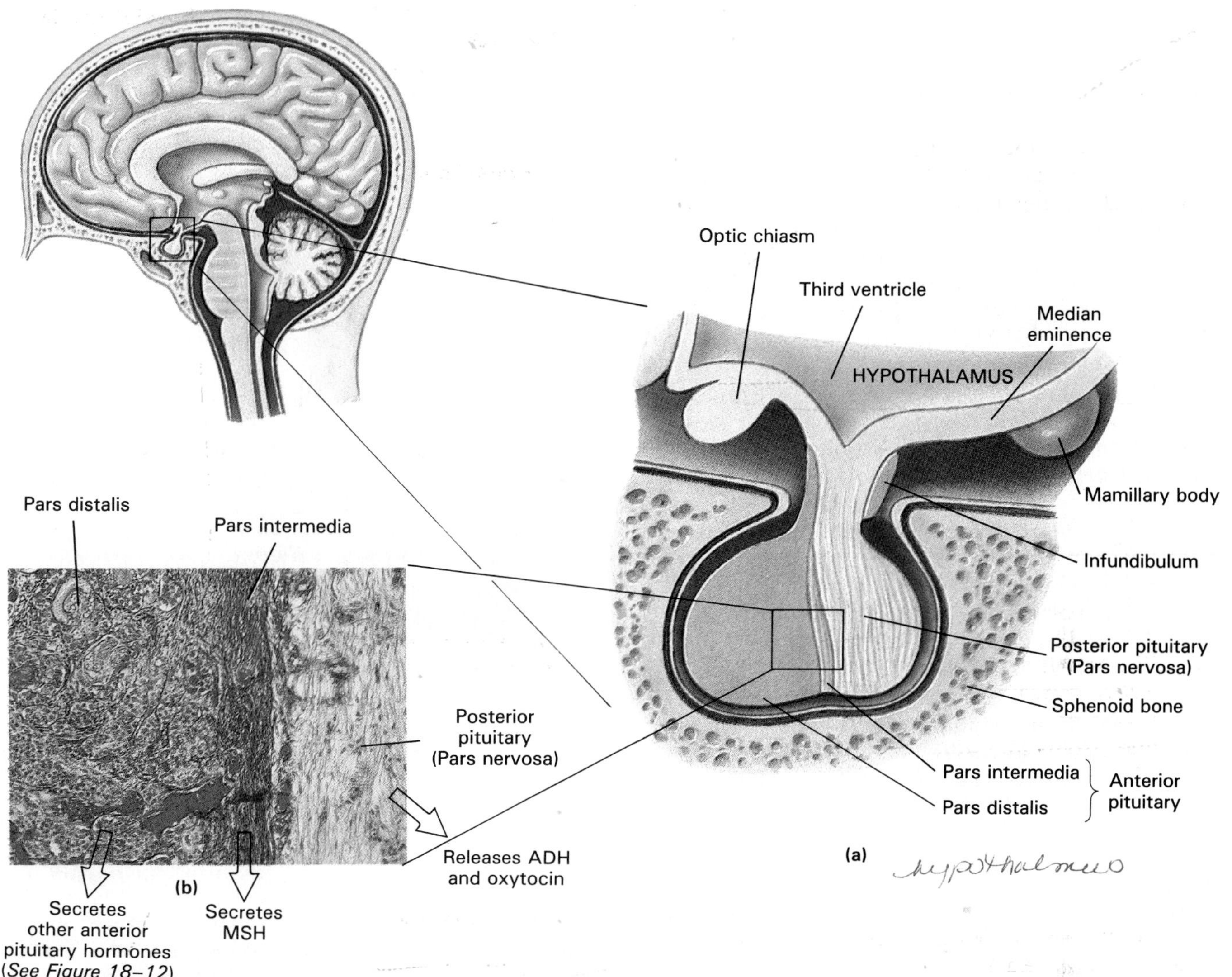

FIGURE 18-9
Anatomy and Orientation of the Pituitary Gland (LM, ×64)

■ The Pituitary Gland

Figure 18-9 details the anatomical organization of the pituitary gland, or **hypophysis** (hī-POF-i-sis). This small, oval gland lies nestled within a depression in the sphenoid bone, the sella turcica (see Figure 8-8). It hangs beneath the hypothalamus, connected by a slender stalk, the **infundibulum** (in-fun-DIB-ū-lum; funnel) (see Figure 14-11). The pituitary gland can be divided into posterior and anterior divisions on anatomical and developmental grounds (see the Embryology Summary on p. 490). Nine important hormones are released by the pituitary gland. All of these are peptide hormones that bind to membrane receptors and use cyclic-AMP as a second messenger.

THE POSTERIOR PITUITARY

The **posterior pituitary** is also called the **neurohypophysis** (Noo-rō-hī-POF-i-sis) or *pars nervosa* (nervous part), because it contains the axons of hypothalamic neurons. Neurons of the **supraoptic** and **paraventricular nuclei** manufacture ADH and oxytocin, respectively. These secretions move via axoplasmic transport along the infundibulum to the basement membranes of capillaries in the posterior pituitary.

Antidiuretic hormone (ADH) is released in response to a variety of stimuli, most notably a rise in the concentration of electrolytes in the blood or a fall in blood volume or pressure. The primary function of ADH is to decrease the amount of water lost at the kidneys. With losses minimized, any water absorbed from the digestive tract will be retained, reducing the concentration of electrolytes. In high concentrations ADH also causes the constriction of peripheral blood vessels, which helps to elevate blood pressure. The production of ADH is inhibited by alcohol, which explains the increased fluid excretion that follows the consumption of alcoholic beverages.

In women, **oxytocin** (*oxy-*, quick + *tokos*, childbirth) stimulates smooth muscle cells in the uterus

Clinical Comment: Diabetes Insipidus

There are several different forms of **diabetes**, all characterized by excessive urine production (**polyuria**). Although diabetes can be caused by physical damage to the kidneys, most forms are the result of endocrine abnormalities. The two most important forms are *diabetes insipidus* and *diabetes mellitus*. (Diabetes mellitus is described on p. 594.)

Diabetes insipidus develops when the posterior pituitary no longer releases adequate amounts of ADH. Water conservation at the kidneys is impaired, and excessive amounts of water are lost in the urine. As a result, an individual with diabetes insipidus is constantly thirsty, but the fluids consumed are not retained by the body. Mild cases may not require treatment, as long as fluid and electrolyte intake keep pace with urinary losses. In severe diabetes insipidus the fluid losses can reach 10 liters per day, and a fatal dehydration will occur unless treatment is provided. One innovative treatment method involves administering a synthetic form of ADH, *desmopressin acetate* (DDAVP) in a nasal spray. The drug enters the bloodstream after diffusing through the nasal epithelium.

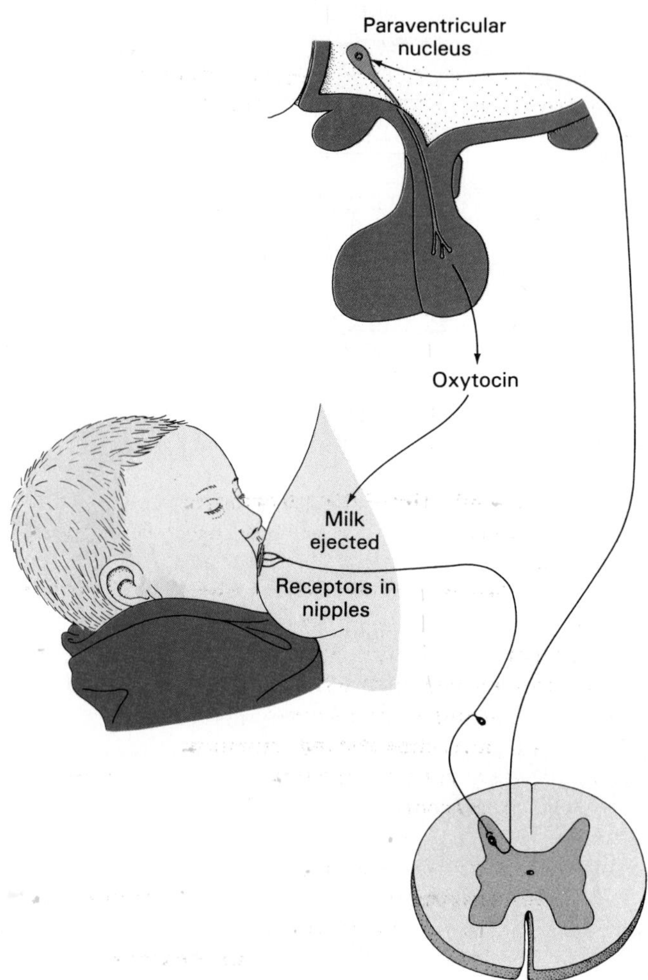

FIGURE 18-10
The Milk Let-Down Reflex. The suckling of an infant at the breast stimulates receptors whose afferent fibers cause the release of oxytocin at the posterior pituitary. Oxytocin arriving in the circulation then stimulates the contraction of smooth muscle cells in the mammary glands, causing discharge of milk into collecting ducts that lead to the nipple.

and contractile (myeloepithelial) cells surrounding the secretory cells of the mammary glands. Until the last stages of pregnancy the uterine muscles are insensitive to oxytocin, but sensitivity becomes more pronounced as the time of delivery approaches. The stimulation of uterine muscles by oxytocin plays a role in the maintenance and completion of normal labor and childbirth. (The mechanism will be detailed in Chapter 29.) After delivery, oxytocin stimulates the contraction of contractile cells around the mammary glands. The infant suckling at the breast provides the necessary stimulus, and afferent fibers innervating the nipples relay the information to the hypothalamus. Oxytocin is then released, and the contractile cells respond by squeezing milk from the secretory pockets into large collecting chambers.

The "milk let-down" reflex is diagrammed in Figure 18-10. The hypothalamus plays a key role in this reflex. As noted in Chapter 14, this portion of the brain receives inputs from most other centers in the brain stem and cerebrum. ∞ (pp. 443–445) These inputs can inhibit or facilitate the milk let-down reflex. For example, anxiety, stress, and other factors can prevent the flow of milk, even when the mammary glands are fully functional. By contrast, nursing mothers can become conditioned to associate a baby's crying with suckling. These women may begin milk let-down as soon as they hear a baby cry.

Oxytocin has no proven function in the human male. However, evidence is accumulating that in other mammals oxytocin in both sexes may stimulate affectionate emotions, sexual attraction, and satisfaction. These responses presumably result from effects of oxytocin on the CNS.

THE ANTERIOR PITUITARY

The **anterior pituitary**, or **adenohypophysis** (ad-e-nō-hī-POF-i-sis), contains a variety of endocrine cell types. It can be subdivided into a large **pars distalis** (dis-TAL-is; distal part) and a slender **pars intermedia** (in-ter-MĒ-dē-a; intermediate part). The pars intermedia forms a narrow band adjacent to the neurohypophysis. Most of the endocrine cells of the anterior pituitary are found in the pars distalis.

Hormones of the Anterior Pituitary

We will restrict our discussion to seven hormones whose functions and control mechanisms are reasonably well understood. All but one of these hormones are produced by the pars distalis, and four of them regulate the production of hormones by other endocrine glands. The names of these hormones indicate their activities, but the phrases are often so long that abbreviations are used instead.

1. **Thyroid-stimulating hormone** (**TSH**) targets the thyroid gland and triggers the release of thyroid hormones.

2. **Adrenocorticotrophic hormone** (**ACTH**) stimulates the release of steroid hormones by the adrenal gland. ACTH specifically targets cells producing hormones called **glucocorticoids** (gloo-kō-KŌR-ti-koids) that affect glucose metabolism.

3. **Follicle-stimulating hormone** (**FSH**) promotes egg development in women and stimulates the secretion of **estrogen**, the female sex hormone, by ovarian cells. In men, FSH production supports sperm production in the testes.

4. **Luteinizing** (LOO-tē-in-ī-zing) **hormone** (**LH**) induces ovulation in women and promotes the ovarian secretion of a hormone, **progesterone**, that prepares the body for possible pregnancy. In men the same hormone is often called **interstitial cell-stimulating hormone** (**ICSH**) because it stimulates the production of male sex hormones, called **androgens** (AN-drō-jenz; *andros*, man), by the interstitial cells of the testes.

FSH and LH are called **gonadotrophins** (gō-nad-ō-TRŌ-finz; *trophikos*, nourishing) because they regulate the activities of the male and female sex organs (gonads).

5. **Prolactin** (prō-LAK-tin; *pro-*, before + *lac*, milk), or **PRL**, stimulates the development of the mammary glands and the production of milk. Although PRL exerts the dominant effect on the glandular cells, normal development of the mammary glands is regulated by the interaction of a number of hormones. Prolactin, estrogens, progesterone, glucocorticoids, pancreatic hormones, and hormones produced by the placenta cooperate in preparing the mammary glands for secretion, and milk let-down occurs only in response to oxytocin release at the posterior pituitary.

6. **Growth hormone** (**GH**), also called *human growth hormone* (HGH) or **somatotrophin** (*soma*, body), stimulates cell growth and replication by accelerating the rate of protein synthesis. Although virtually every tissue responds to some degree, skeletal muscle cells and chondrocytes (cartilage cells) are particularly sensitive to levels of growth hormone.

The stimulation of growth by GH occurs indirectly. Liver cells respond to the presence of growth hormone by releasing hormones called **somatomedins** that bind to receptor sites on a variety of cell membranes. In skeletal muscle cells, cartilage cells, and other target cells somatomedins increase the rate of amino acid uptake. The extra amino acids are used to synthesize proteins. These effects develop almost immediately after GH release occurs, and they are particularly important when the blood contains high concentrations of glucose and amino acids.

Other effects appear hours later, as glucose and amino acid concentrations fall to normal or below normal levels. Under these conditions growth hormone mobilizes energy reserves stored in adipose tissue and liver cells. Adipocytes (fat cells) break down stored triglycerides (fats) and release fatty acids into the blood. As circulating lipid concentrations rise, other tissues begin breaking down fatty acids instead of glucose to meet their energy demands. Liver cells respond by reducing their glycogen reserves and releasing glucose into the circulation. Because most peripheral cells are metabolizing lipids, rather than glucose, blood glucose concentrations begin climbing.

These changes in glucose utilization involve an interaction between growth hormone and somatomedins in peripheral tissues. The combination may suppress peripheral glucose use so effectively that the blood glucose concentration rises to above normal levels. The elevation of blood glucose levels by growth hormone has been called a **diabetogenic effect**, because *diabetes mellitus*, considered later in the chapter, is characterized by abnormally high blood glucose concentrations.

Growth hormone stimulates muscular and skeletal development, and if it is administered before the epiphyseal plates have closed it will cause an increase in height, weight, and muscle mass. Children unable to produce adequate concentrations of growth hormone suffer from *pituitary growth failure*, sometimes called *pituitary dwarfism*, a condition introduced in Chapter 7. (p. 218) The steady growth and maturation that normally precede and accompany puberty do not occur in these individuals, who have short stature, slow epiphyseal growth, and larger than normal adipose tissue reserves.

Normal growth patterns can be restored by the administration of growth hormone. Prior to the advent of gene splicing and recombinant-DNA techniques (Chapter 4), GH had to be carefully extracted and purified from the pituitaries of cadavers at considerable expense. (p. 116) It is now possible to use genetically manipulated bacteria to produce pure growth hormone in commercial quantities. In addition to treating pituitary growth failure, growth hormone supplements may slow or even reverse the losses in bone and muscle mass that accompany aging.

Health News: Growth Hormone—Too Much of a Good Thing?

The current availability of purified human growth hormone has unfortunately led to its use under medically questionable circumstances. For example, many physicians are approached by parents seeking GH for short but otherwise normal children. These parents view short stature as a handicap that merits treatment. There are several problems with this approach. First, the hormonal regulation of growth is quite complex (see p. 598) and subject to considerable individual variation. It is perfectly normal for a child to grow in spurts, and being relatively short (compared with other children) at age 4 does not mean the individual will be shorter than others in his age group at age 18. Second, it is important to realize that growth hormone does not have just a single effect. It affects many different tissues and has widespread metabolic impacts. For example, children exposed to growth hormone may grow faster, but their body fat content declines drastically. This decline is associated with metabolic changes in many organs. The range and significance of these metabolic side effects are now the subject of long-term studies. Until the details are understood, it would seem unwise to use this–or any other–hormone unless a clear medical need exists.

Despite the negative publicity that has surrounded the practice, a significant number (estimated at up to 30 percent) of amateur and professional athletes continue to use hormones to improve their performance. Although synthetic forms of testosterone (male sex hormone) are used most often, young athletes may combine testosterone and growth hormone. The risks of GH use remain uncertain, but the risks associated with testosterone use are quite clear; they will be considered later in the chapter (see pp. 600–601).

7. **Melanocyte-stimulating hormone (MSH)** is the only hormone released by the pars intermedia. As the name indicates, MSH stimulates the melanocytes of the skin, increasing their production of melanin, a brown-black pigment.

HYPOTHALAMIC CONTROL OF THE ANTERIOR PITUITARY

The production of hormones in the anterior pituitary is controlled by the hypothalamus through the secretion of specific regulatory factors. At the **median eminence**, a swelling near the attachment of the infundibulum, neurons release regulatory factors into the surrounding interstitial fluids. Their secretions enter the circulation quite easily because the capillaries in this region are more permeable than those found elsewhere in the CNS. The lining of these capillaries has a "swiss cheese" appearance, and there are open spaces between adjacent endothelial cells. Wherever a "hole" exists, only the endothelial basement membrane separates the blood from the surrounding interstitial fluid. Such vessels are described as **fenestrated** (FEN-es-trat-ed; *fenestra*, window), and they are found only where relatively large molecules enter or leave the circulatory system.

The Hypophyseal Portal System

Before leaving the hypothalamus, the capillary network unites to form a series of larger vessels that spiral around the infundibulum to reach the pituitary gland. Once within the anterior pituitary, these vessels form a second capillary network that branches among the endocrine cells. This circulatory arrangement, illustrated in Figure 18-11, is an anatomical oddity. An artery usually conducts blood from the heart to a capillary network, and a typical vein carries blood from a capillary network back to the heart. The vessels between the median eminence and the anterior pituitary, by contrast, carry blood from one capillary network to another.

Blood vessels that link two capillary networks are called **portal vessels**, and the entire complex is termed a **portal system**. Portal systems provide an efficient means of chemical communication by ensuring that all of the blood entering the portal vessels will reach the intended target cells before returning to the general circulation. The communication is strictly one-way, however, because any chemicals released by the cells "downstream" must do a complete tour of the circulatory system before reaching the capillaries at the start of the portal system. Portal vessels are named after their destinations, so this particular network of vessels represents the **hypophyseal portal system**.

Hypothalamic Control Mechanisms

An endocrine cell in the anterior pituitary may be controlled by releasing hormones, inhibiting hormones, or some combination of the two.

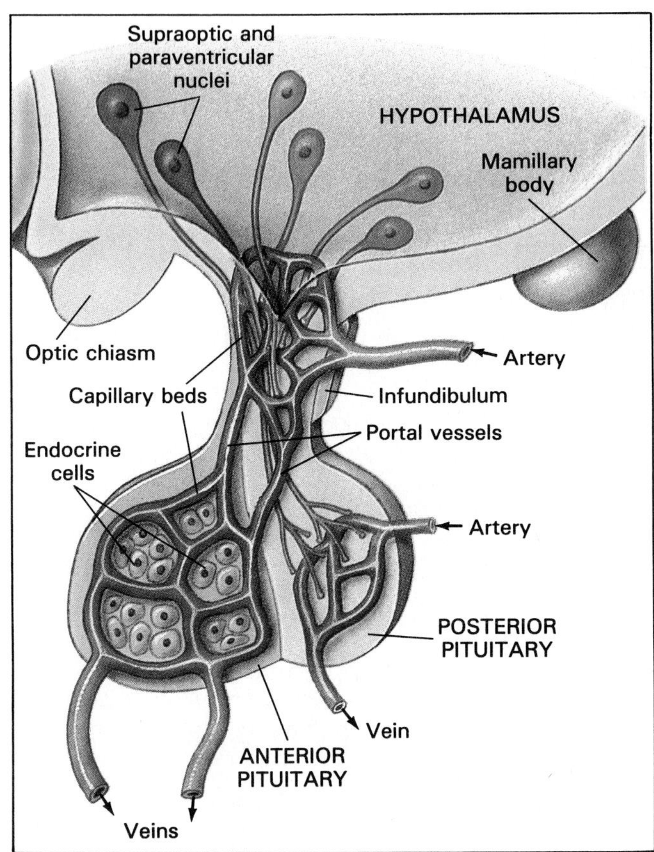

FIGURE 18-11
The Hypophyseal Portal System

Control by Releasing Hormones The production of TSH, ACTH, and the gonadotrophic hormones FSH and LH occurs in response to the presence of specific releasing hormones. Thyroid hormone-releasing hormone (**TRH**) stimulates the secretion of TSH, and corticotrophin-releasing hormone (**CRH**) causes the release of ACTH. FSH and LH secretion occur in response to a single gonadotrophin-releasing hormone, **GnRH**.

In each of these instances the secretion of hypothalamic releasing hormone is regulated by a feedback mechanism. The controlling factor is the concentration of hormones from the corresponding peripheral endocrine organ. For example, CRH causes the anterior pituitary to release ACTH, which in turn stimulates the secretion of steroid hormones by the adrenal gland. As the concentration of these steroid hormones increases, CRH production is inhibited. Secretion of ACTH falls accordingly, and the production of adrenal hormones is reduced. In addition, the adrenal steroids have a direct inhibitory effect on the pituitary endocrine cells that produce ACTH. *In general, a pituitary endocrine cell that is stimulated by a releasing hormone is usually inhibited by the peripheral hormone it controls.* These relationships are diagrammed in Figure 18-12.

Control by Inhibiting Hormones Prolactin and MSH production are controlled by specific inhibiting factors. Prolactin secretion is suppressed by hypothalamic centers releasing prolactin-inhibiting hormone, **PIH**. This compound is actually an important neurotransmitter, *dopamine*, but in this case the neurons release it into the portal circulation rather than across a synapse. Although there is no prolactin-releasing hormone, the presence of TRH, estrogens, or progesterone in the blood can stimulate prolactin secretion to some degree. MSH production is inhibited by melanocyte-stimulating hormone-inhibiting hormone, **MSH-IH**.

Regulation by Releasing and Inhibiting Hormones The production of growth hormone is regulated by the appearance of growth hormone-releasing hormone (**GH-RH**) and growth hormone-inhibiting hormone (**GH-IH**) in the hypophyseal circulation. The hypothalamic nuclei secreting these regulatory factors respond to a variety of stimuli. In addition to monitoring the concentration of circulating growth hormone, these hypothalamic neurons are influenced by other hormones, neurotransmitters, and the blood concentrations of glucose, amino acids, and lipids. The anterior pituitary cells that produce growth hormone also vary their secretory activities in response to circulating adrenal, thyroid, or reproductive hormones.

Primary and Secondary Effects of Releasing Hormones A single regulatory hormone may have related secondary effects, and these multiple effects may help to integrate endocrine responses to stimuli. For example, the release of a single inhibiting hormone, GH-IH, both suspends growth and reduces the peripheral demand for energy. The *primary effect* on growth reflects lower GH concentrations. The *secondary effect*, the reduction in peripheral energy demand, results from an inhibition of TSH production and a decrease in thyroid gland activity.

Table 18-1 summarizes important information concerning the hormonal products of the pituitary gland; review this table carefully before considering the structure and function of other endocrine organs.

- ✓ **How would blocking the activity of phosphodiesterase affect a cell that responds to hormonal stimulation by the cAMP second messenger system?**
- ✓ **How would dehydration affect the level of ADH released by the neurohypophysis?**
- ✓ **A blood sample shows elevated levels of somatomedins. What pituitary hormone would you expect to be elevated as well?**
- ✓ **What effect would elevated levels of cortisol, a hormone from the adrenal cortex, have on the level of ACTH?**

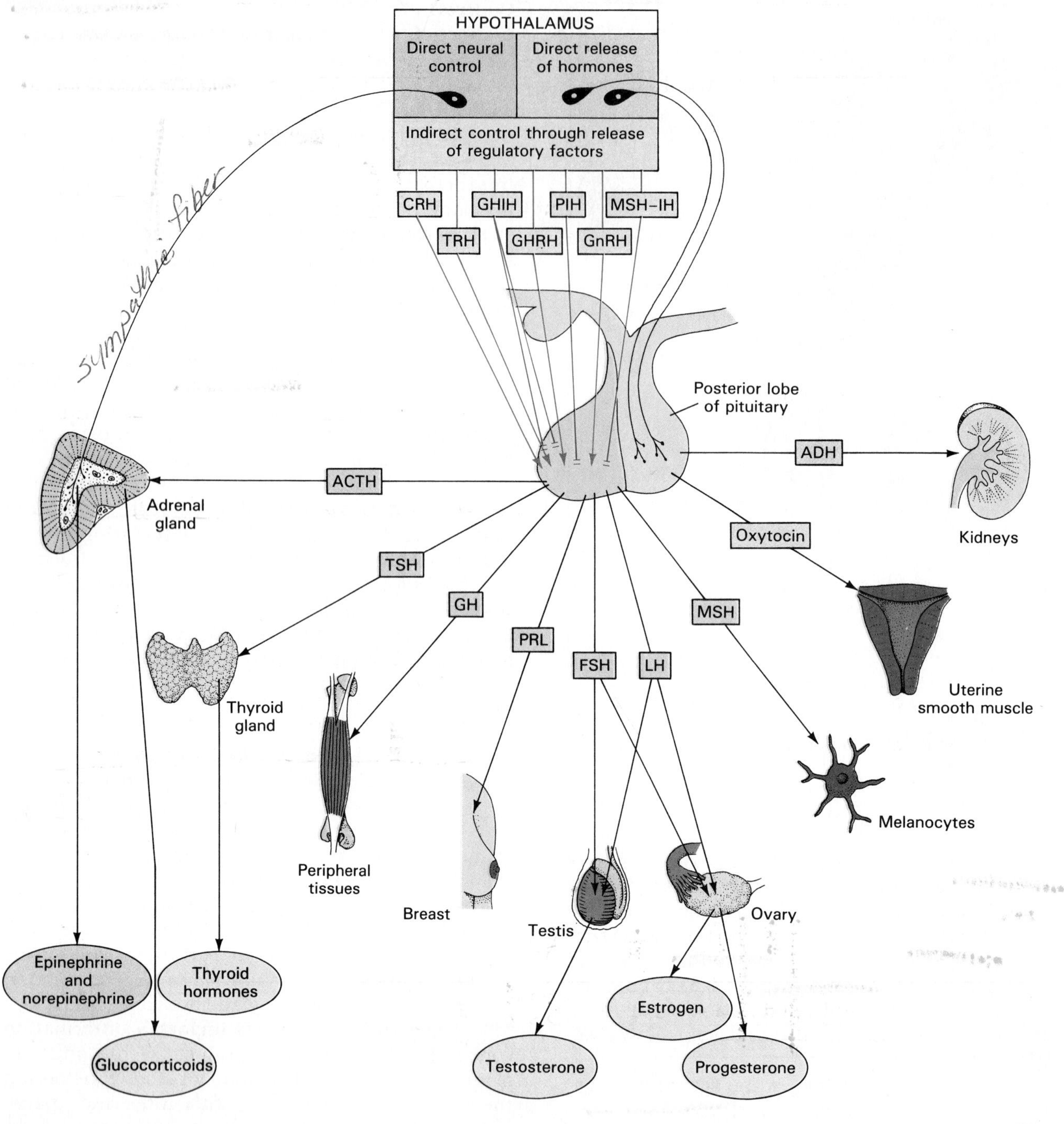

CRH Corticotrophin–releasing hormone
GHIH Growth hormone–inhibiting hormone
PIH Prolactin–inhibiting hormone
MSHIH Melanocyte–stimulating hormone inhibiting hormone
GnRH Gonadotrophin–releasing hormone
GHRH Growth hormone–releasing hormone
TRH Thyrotrophin–releasing hormone

ACTH Adrenocorticotrophic hormone
TSH Thyroid–stimulating hormone
GH Growth hormone
PRL Prolactin
FSH Follicle–stimulating hormone
LH Luteinizing hormone
MSH Melanocyte–stimulating hormone
ADH Antidiuretic hormone

FIGURE 18-12
Pituitary Hormones and Their Targets

TABLE 18-1 The Pituitary Hormones

Region/Area	*Hormones*	*Hypothalamic Control Mechanism*	*Targets*	*Hormonal Effects*
ANTERIOR PITUITARY (ADENOHYPOPHYSIS)				
Pars distalis	Thyroid-stimulating hormone (TSH)	Thyrotrophin-releasing hormone (TRH)	Thyroid gland	Secretion of thyroid hormones
	Adrenocorticotrophic hormone (ACTH)	Corticotrophin-releasing hormone (CRH)	Adrenal cortex, zona fasciculata	Glucocorticoid secretion
	Gonadotrophic hormones:	Gonadotrophin-releasing hormone (GnRF)		
	Follicle-stimulating hormone (FSH)		Follicle cells of ovaries in female	Estrogen secretion, ovum development
			Germinative cells of testes in male	Continued mitoses and sperm formation
	Luteinizing hormone (LH) or Interstitial cell stimulating hormone (ICSH)		Follicle cells of ovaries in female	Ovulation, formation of corpus luteum, progesterone secretion
			Interstitial cells of testis in male	Testosterone secretion
	Prolactin (PRL)	Prolactin-inhibiting hormone (PIH)	Mammary glands	Production of milk
	Growth hormone (GH)	Growth hormone-releasing hormone (GH-RH), Growth hormone-inhibiting hormone (GH-IH)	All cells	Growth, protein synthesis, lipid mobilization, and catabolism
Pars intermedia	Melanocyte-stimulating hormone (MSH)	Melanocyte-stimulating hormone-inhibiting hormone (MSH-IH)	Melanocytes	Increased melanin synthesis
POSTERIOR PITUITARY (NEUROHYPOPHYSIS PARS NERVOSA)				
	Antidiuretic hormone (ADH)	Transported over axons from supraoptic nucleus	Kidneys	Reabsorption of water; elevation of blood volume and pressure
	Oxytocin	Transported over axons from paraventricular nucleus	Uterus, mammary glands	Labor contractions, milk ejection

The Thyroid Gland

The thyroid gland curves across the anterior surface of the trachea just below the **thyroid** ("shield-shaped") **cartilage** that dominates the anterior surface of the larynx. The two **lobes** of the thyroid gland are united by a slender connection, the **isthmus** (IS-mus). The size of the gland is quite variable, depending on heredity, environment, and nutritional factors, but a weight of about 34 g (1.2 oz) could be considered average. The thyroid gland has a deep red coloration because of the large number of blood vessels servicing the glandular cells.

THYROID FOLLICLES AND THYROID HORMONES

The thyroid gland contains large numbers of **thyroid follicles**. Several follicles are shown in sectional view in Figure 18-13, but in three dimensions they resemble tennis balls. The glandular lining of a follicle consists of a simple cuboidal epithelium. A network of capillaries surrounds each follicle, delivering nutrients and regulatory hormones to the glandular cells and accepting their secretory products and metabolic wastes.

Thyroid follicles release several hormones into the circulation. All are structural derivatives of the amino acid **tyrosine**, to which three or four iodine

FIGURE 18-13
The Thyroid Gland. (a) Location and anatomy of the thyroid. (b) Histological organization of the thyroid (LM, ×108). (c) Histological details of the thyroid gland.

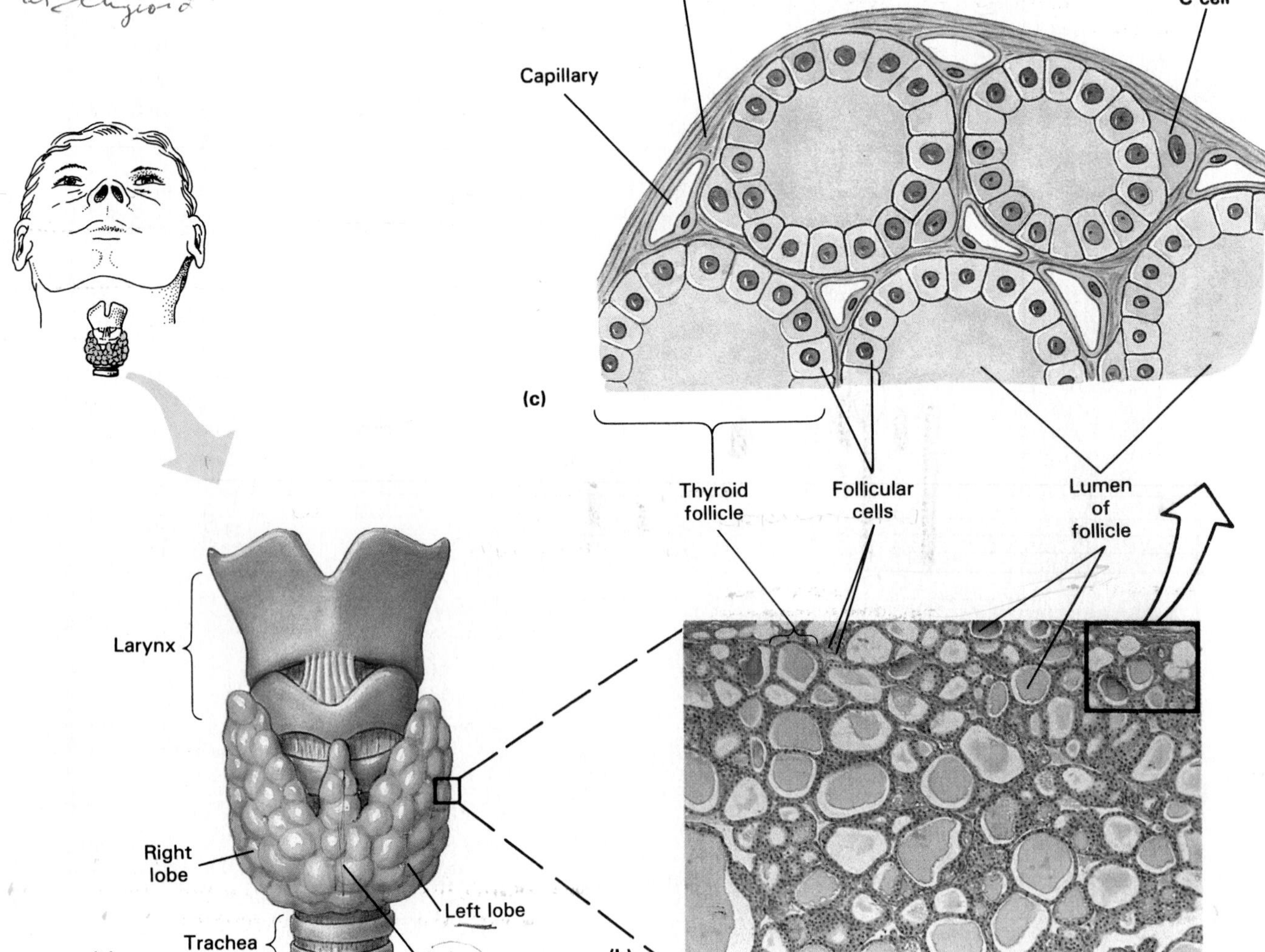

atoms have been attached. The hormone **thyroxine** (thī-ROKS-ēn), or **TX**, contains four atoms of iodine; it is also known as *tetraiodothyronine* (tet-ra-ī-ō-dō-THĪ-rō-nēn), or $\mathbf{T_4}$. Thyroxine accounts for roughly 90 percent of all thyroid secretions. **Triiodothyronine**, or $\mathbf{T_3}$, a related molecule containing three iodine atoms, is secreted in small amounts.

These hormones affect almost every cell in the body because they diffuse through the cytoplasm to reach receptor sites in the nucleus. One effect is accelerated production of the membrane proteins of the sodium-potassium exchange pump. As noted in Chapter 3, these proteins are enzymes that convert ATP to ADP, and their activity consumes large amounts of energy. (p. 85) Thyroid hormones also activate genes coding for the synthesis of enzymes involved in glycolysis and energy production. The overall effect is an increase in cellular rates of metabolism and oxygen consumption.

This **calorigenic effect** of thyroid hormones enables us to adapt to cold temperatures. When the metabolic rate increases, more heat is generated, replacing the heat lost to a chilly environment. In growing children, thyroid hormones are also essential to normal development of the skeletal, muscular, and nervous systems. Table 18-2 summarizes the effects of thyroid hormones on major organs and systems. [**CM:** *Thyroid Gland Disorders*]

The two thyroid hormones have complementary effects. T_3 produces a strong and immediate increase in cellular metabolism, but the effects are short-lived. T_4 gradually diffuses into peripheral tissues, and its slow entry produces a gradual, long-term response. Once in the cytoplasm, molecules of T_4 are converted to T_3, and the metabolic effects of the two hormones are identical.

TABLE 18-2 Effects of Thyroid Hormones on Peripheral Tissues

1. Elevate oxygen consumption and rate of energy consumption; usually cause a rise in body temperature
2. Increase heart rate and force of contraction; usually cause a rise in blood pressure
3. Increase sensitivity to sympathetic stimulation
4. Maintain normal sensitivity of respiratory centers to changes in oxygen and carbon dioxide concentrations
5. Stimulate formation of red blood cells and enhance oxygen delivery
6. Stimulate activity of other endocrine tissues
7. Accelerate turnover of minerals in bone

REGULATION OF THYROID HORMONE RELEASE

A thyroid hormone is a very small molecule, barely larger than its parent amino acid, and it can easily diffuse across the follicular cell membrane. This ability poses an interesting regulatory problem, because whenever the molecules appear in the cytoplasm they begin diffusing out of the cell. Yet the demand for thyroid hormones changes constantly, and the thyroid follicles must maintain a substantial reserve of thyroid hormones.

Thyroid hormone production, storage, and release are diagrammed in Figure 18-14. Follicular cells synthesize a globular protein called **thyroglobulin** (thī-rō-GLOB-ū-lin) and release it into the lumen (central chamber) of the follicle. The follicular contents represent a **colloid**, a viscous fluid containing large quantities of suspended proteins. Each thyroglobulin molecule contains enough tyrosine to produce several molecules of either T_3 or T_4, but at the time of secretion the iodine has yet to be attached. As they are synthesizing and secreting thyroglobulins, the endocrine cells are also obtaining iodine molecules from the surrounding interstitial fluids. The iodine is transported across the cell and released into the follicle, where it binds to the tyrosine molecules of thyroglobulin. Eventually each molecule of thyroglobulin contains several potentially active thyroid hormones. But because they are attached to a large protein, rather than floating around on their own, the thyroid hormones cannot escape by diffusing through the follicular walls.

Hormonal secretion occurs when endocytosis brings a small packet of thyroglobulin molecules into the cytoplasm of a follicular cell. Lysosomal enzymes then break down the thyroglobulin complex, releasing thyroid hormones into the surrounding cytoplasm. Within moments these hormones diffuse out of the cell and into the local circulation. Roughly three-fourths of the thyroid hormone molecules entering the circulation become attached to special proteins, called **thyroid-binding globulins**, or **TBG**. The hormone molecules that remain unbound diffuse into peripheral tissues. This produces an immediate effect on cellular processes. The bound molecules are gradually released over the next week or more, prolonging the effects of thyroid stimulation.

Thyroid follicles are continually manufacturing, secreting, absorbing, and breaking down thyroglobulin molecules. As a result, thyroid hormones are released at a low background level. When stimulated, the follicle cells speed up their rates of thyroglobulin synthesis and breakdown. The most important factor controlling the rate of thyroid hormone release is the concentration of TSH in the circulating blood; Figure 18-14c diagrams the control mechanism involved.

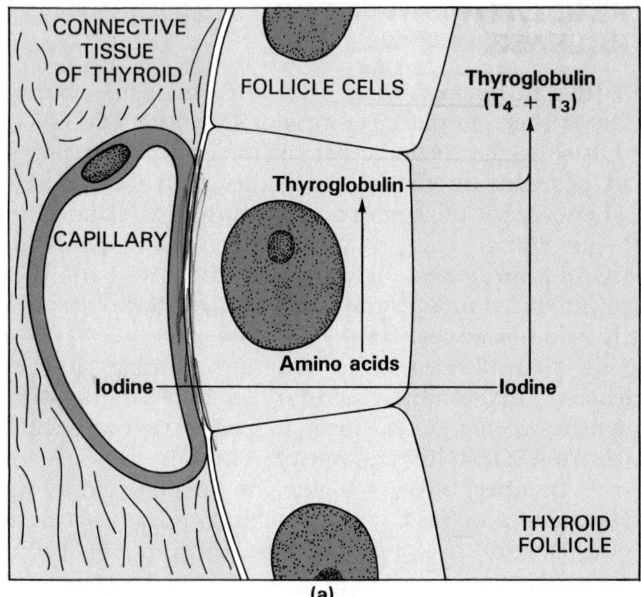

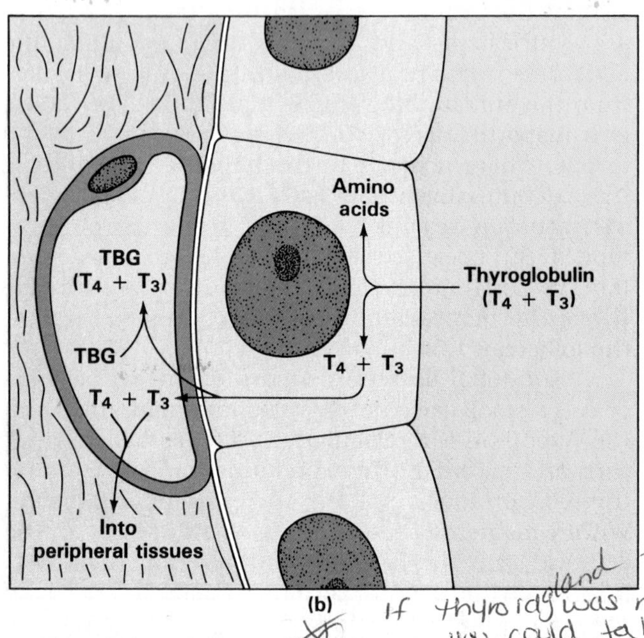

THE C CELLS OF THE THYROID GLAND: CALCITONIN

A second population of endocrine cells lies sandwiched between the cuboidal follicle cells and their basement membrane. These cells are larger than those of the follicular epithelium, and they do not stain as clearly. These are the **C cells** that produce the hormone **calcitonin** (**CT**). Calcitonin assists in the regulation of calcium ion concentrations in body fluids. The functions of this hormone were discussed in Chapter 7. (p. 220) In summary, calcitonin stimulates osteoblasts, inhibits osteoclasts, reduces intestinal absorption of calcium, and stimulates calcium ion excretion at the kidneys. The overall effect is to decrease the amount of calcium dissolved in body fluids. The C cells do not need any hormonal instructions to tell them when to secrete calcitonin, for they are continually monitoring the local calcium ion concentrations. When those concentrations rise, the C cells respond by secreting calcitonin into the interstitial fluids. This lowers the calcium ion concentrations, eliminating the stimulus and "turning off" the C cells.

Several chapters have dealt with the importance of calcium ions in controlling muscle and nerve cell activities. Calcium ion concentrations also affect the sodium permeabilities of excitable membranes. At high calcium ion concentrations, sodium permeability decreases and membranes become less responsive. Such problems are prevented by the secretion of calcitonin under appropriate conditions. But lower than normal concentrations are equally dangerous, because sodium permeabilities then increase and cells become extremely excitable. To prevent convulsions or muscular spasms, calcium levels must not be allowed to fall too far, and this half of the regulatory story involves the parathyroid glands and **parathormone** (**PTH**).

FIGURE 18-14
The Functions of the Thyroid Follicles. (a) Synthesis and storage of thyroid hormones. Follicle cells manufacture thyroglobulin molecules that are stored in the lumen of the follicle. Iodine then combines with the thyroglobulin to produce T_3 and T_4. (b) Secretion and transport of thyroid hormones. Follicular cells reabsorb the thyroglobulin molecules and break them down. The released molecules of T_3 and T_4 then diffuse out of the cells and into the circulation. (c) Regulation of thyroid activity. The control mechanism shown follows the pattern described in Figure 18-7d.

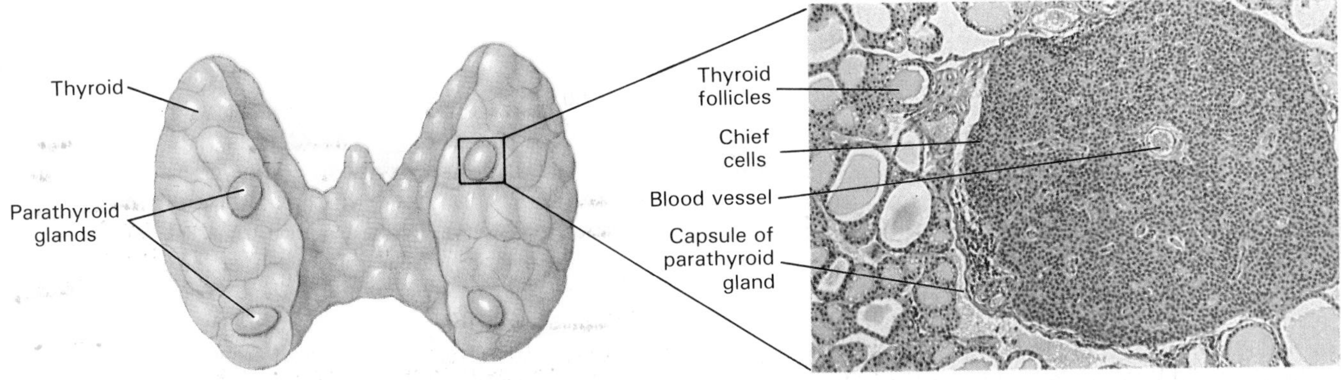

FIGURE 18-15
The Parathyroid Glands (LM, ×88)

The Parathyroid Glands

Two pairs of parathyroid glands are found embedded in the posterior surfaces of the thyroid gland. The gland cells are separated by the dense capsular fibers of the thyroid. All together the four parathyroid glands weigh a mere 1.6 g. The histological appearance of a parathyroid gland is shown in Figure 18-15. There are at least two different cell populations found in the parathyroid. The **chief cells** produce parathormone; the functions of the other cell types are unknown.

Like the C cells of the thyroid, the chief cells monitor the circulating concentration of calcium ions. When the calcium concentration falls below normal, the chief cells secrete parathormone. As discussed in Chapter 7, parathormone stimulates osteoclasts, inhibits osteoblasts, promotes intestinal absorption of calcium, and reduces urinary excretion of calcium ions until blood concentrations return to normal. (p. 220) Together the C cells and chief cells can regulate calcium ion concentrations within relatively narrow limits.

↑ Ca.
stimulates osteoclasts
promotes intestinal abs. of Ca
✓ Targets bones, kidneys
intestines

The Thymus

The thymus is embedded in a mass of connective tissue inside the thoracic cavity, usually just behind the sternum. In a newborn infant the thymus is relatively enormous, often extending from the base of the neck to the superior border of the heart. Although its relative size decreases as the child grows, the thymus continues to enlarge slowly, reaching its maximum size just before puberty, at a weight of around 40 g. After puberty it gradually diminishes in size; by age 50 the thymus may weigh less than 12 g. [**CM:** *Disorders of Parathyroid Function*]

The thymus produces several hormones, but only one has a known function. **Thymosin** (thī-MŌ-sin) plays a key role in the development and maintenance of normal immunological defenses. It has been suggested that the gradual decrease in the size and secretory abilities of the thymus may make the elderly more susceptible to disease. The histological organization of the thymus and the functions of thymosin will be further considered in Chapter 22.

Information concerning the hormones of the thyroid, parathyroid, and thymus glands is summarized in Table 18-3.

TABLE 18-3 Hormones of the Thyroid, Parathyroids, and Thymus

Gland/Cells	*Hormones*	*Targets*	*Effects*
THYROID			
Follicular epithelium	Thyroxine (TX,T_4), triiodothyronine (T_3)	Most cells	Increases energy utilization, oxygen consumption, growth, and development
C cells	Calcitonin (CT)	Bone, kidneys, intestines	Decreases calcium ion concentrations in body fluids
PARATHYROIDS			
Chief cells	Parathormone (PTH)	Bone, kidneys, intestines	Increases calcium ion concentrations in body fluids
THYMUS	Thymosin	Lymphocytes	Maturation and functional competence of immune system

The Adrenal Gland

A single **adrenal gland** sits along the superior border of each kidney. For this reason the adrenals are also called the **suprarenal glands** (sū-pra-RĒ-nal; *supra-*, above + *renes*, kidneys). Each adrenal gland forms a curving, flattened cap nestled between the kidney, the diaphragm, and the major arteries and veins running along the dorsal wall of the abdominopelvic cavity. The adrenal glands project into the abdominopelvic cavity, but they remain separated from it by the peritoneal lining. Each gland has a dense fibrous **capsule** whose fibers are tied into those of the adipose tissue mass surrounding the adrenals and kidneys.

A typical adrenal gland weighs about 7.5 g, but adrenal size can vary greatly as secretory demands change. The adrenal gland can be divided into two parts, a superficial **adrenal cortex** and an inner **adrenal medulla**. These regions can be seen in the sectional view provided in Figure 18-16.

THE ADRENAL CORTEX

The adrenal cortex has a grayish yellow coloration due to the presence of stored lipids, especially cholesterol and various fatty acids. The adrenal cortex produces more than two dozen different steroid hormones, collectively called **adrenocortical steroids**, or simply **corticosteroids**. These hormones are vital; if the adrenal glands are destroyed or removed, corticosteroids must be administered or the individual will not survive. The corticosteroids exert their effects by determining which genes are transcribed in the nuclei of their target cells, and at what rates. The resulting changes in the nature and concentration of enzymes in the cytoplasm can shift the direction of metabolic operations.

The cortex can be subdivided into three *zones*, as indicated in Figure 18-16. Each zone synthesizes different steroids in response to specific regulatory mechanisms.

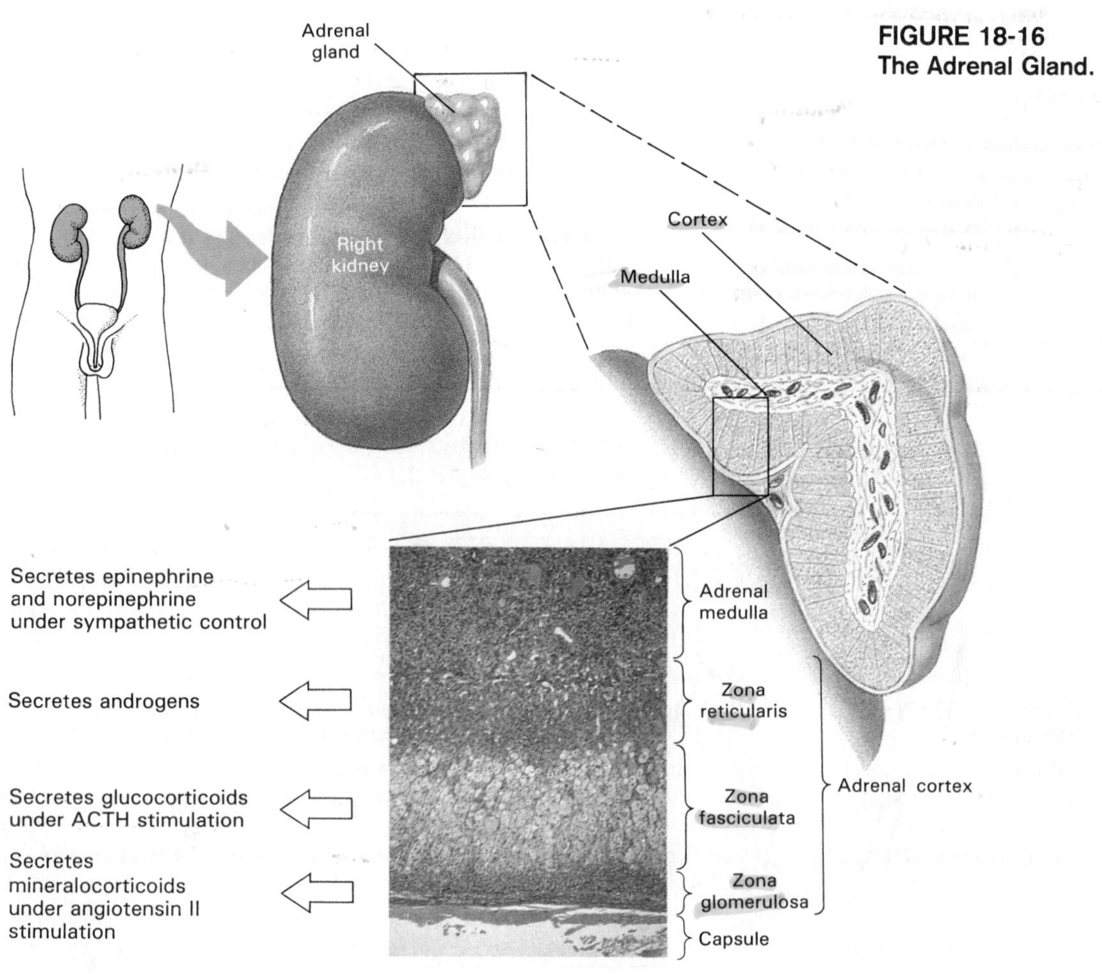

FIGURE 18-16
The Adrenal Gland.

The Zona Reticularis

The **zona reticularis** (re-ti-kū-LAR-is; *reticulum*, network) forms a narrow band bordering the adrenal medulla. In total the zona reticularis accounts for only around 7 percent of the total volume of the adrenal cortex. The endocrine cells of the zona reticularis form a folded, branching network, and fenestrated blood vessels wind between the cells. The zona reticularis normally secretes small amounts of male sex hormones, or *androgens*. Androgens are produced in large quantities by the male testes, and the significance of the small adrenal contribution—in both males *and* females—remains uncertain. ACTH stimulates the zona reticularis to a slight degree, but the effects are usually insignificant.

The Zona Fasciculata

The **zona fasciculata** (fa-sik-ū-LA-ta; *fasciculus*, little bundle) begins at the outer border of the zona reticularis and extends toward the adrenal capsule. It contributes about 78 percent of the cortical volume. The endocrine cells form cords that radiate like a sunburst from the zona reticularis. The individual cords are stacks of cells, and adjacent cords are separated by flattened vessels with fenestrated walls. Each endocrine cell is packed with lipid droplets that give the cytoplasm a pale, foamy appearance.

Steroid production in the zona fasciculata is stimulated by ACTH from the anterior pituitary. This zone produces steroid hormones collectively known as **glucocorticoids (GC)** because of their effects on glucose metabolism. **Cortisol** (KŌR-ti-sol; also called *hydrocortisone*), **corticosterone** (kor-ti-KOS-te-rōn), and **cortisone** are the three most important glucocorticoids. These hormones accelerate the rates of glucose synthesis and glycogen formation, especially within the liver. Adipose tissue responds by releasing fatty acids into the blood, and other tissues begin to break down fatty acids instead of glucose. This process has been called the **glucose-sparing effect** of glucocorticoids.

"Steroid creams" are often used to control irritating allergic rashes, such as those produced by poison ivy, and injections of glucocorticoids may be used to control more severe allergic reactions. The **anti-inflammatory** activities of these steroids result from their effects on white blood cells and other components of the immune system. Glucocorticoids slow the migration of phagocytic cells into an injury site, and phagocytic cells already in the area become less active. In addition, mast cells exposed to these steroids are less likely to release the chemicals that promote inflammation. As a result, swelling and further irritation are dramatically reduced. On the negative side, wound healing slows to a crawl, and the weakening of the region's defenses makes it an easy target for infecting organisms. For this reason topical steroids are used to treat superficial rashes, but are never applied to open wounds.

The Zona Glomerulosa

The **zona glomerulosa** (glo-mer-ū-LŌ-sa) accounts for about 15 percent of the cortical volume. This zone extends from the radiating cords of the zona fasciculata to the overlying capsule. A *glomerulus* is a little ball or knot, and here the endocrine cells form densely packed clusters. The cells are smaller than those of the zona fasciculata, and they contain fewer lipid droplets.

The zona glomerulosa produces **mineralocorticoids (MC)**, steroid hormones that affect the electrolyte composition of bodily fluids. **Aldosterone** is the principal mineralocorticoid produced by the human adrenal cortex. Aldosterone targets kidney cells that regulate the ionic composition of the urine. It causes the retention of sodium ions and water, reducing fluid losses in the urine. Aldosterone also reduces sodium and water losses at the sweat glands, salivary glands, and along the digestive tract.

Aldosterone secretion occurs when the zona glomerulosa is stimulated by the hormone **angiotensin II** (*angeion*, vessel + *teinein*, to stretch). The appearance of this hormone in the circulation involves a series of steps that begins with the secretion of an enzyme, **renin**, by kidney cells. In addition to its other effects, angiotensin II stimulates the production of aldosterone and thereby assists in the restoration of normal blood volume and pressure. (Renin and angiotensin will be discussed further in a later section.) [CM: *Disorders of the Adrenal Cortex*]

Information concerning the hormones of the adrenal cortex is summarized in Table 18-4.

THE ADRENAL MEDULLA

The boundary between the adrenal cortex and medulla does not form a straight line, and the supporting connective tissues and blood vessels are extensively interconnected. The adrenal medulla has a reddish brown coloration due in part to the many blood vessels in this area. It contains large, rounded cells similar to those found in other sympathetic ganglia, and these cells are innervated by preganglionic sympathetic fibers. (The nature of the innervation of the adrenal medulla and its relationship to the sympathetic division of the ANS was discussed in Chapter 16.) (p. 504)

The adrenal medulla contains two populations of secretory cells, one producing epinephrine (adrenaline) and the other norepinephrine (noradrenaline). There is evidence that the two cell types are distributed in different areas of the medulla and that their secretory activities can be independently con-

■ **TABLE 18-4 The Adrenal Hormones**

Region/Zone	*Hormone*	*Targets*	*Effects*
CORTEX			
Zona glomerulosa	Mineralocorticoids (MC), primarily aldosterone	Kidneys	Increases reabsorption of sodium ions and water from the urine
Zona fasciculata	Glucocorticoids (GC): cortisol (hydrocortisone), corticosterone, cortisone	Most cells	Releases amino acids from skeletal muscles, lipids from adipose tissues; promotes liver glycogen and glucose formation; promotes peripheral utilization of lipids (glucose-sparing); anti-inflammatory effects
Zona reticularis	Androgens		Uncertain significance under normal conditions
MEDULLA	Epinephrine (adrenaline, E), norepinephrine (noradrenaline, NE)	Most cells	Increased cardiac activity, blood pressure, glycogen breakdown, blood glucose; release of lipids by adipose tissue (Chapter 16)

trolled. The catecholamine secretions of the medullary cells are packaged in vesicles that form dense clusters just inside the cell membranes. These hormones are released slowly through exocytosis, but sympathetic stimulation accelerates the rate of discharge dramatically.

Epinephrine makes up 75–80 percent of the secretions from the medulla, the rest being norepinephrine. The peripheral effects of these hormones, which result from interaction with alpha and beta receptors on cell membranes, were detailed in Chapter 16. (pp. 506–508) Stimulation of α_1 and β_1 receptors, the most common types, accelerates cellular energy utilization and the mobilization of energy reserves. For example, α_1 receptors are found on skeletal muscle fibers, in adipose tissues, and in the liver. Catecholamine secretion triggers a mobilization of glycogen reserves in skeletal muscles and accelerates the breakdown of glucose to provide ATP. This combination increases muscular power and endurance. In adipose tissue, stored fats are broken down to fatty acids, and in the liver, glycogen molecules are converted to glucose. The fatty acids and glucose are then released into the circulation for utilization by peripheral tissues. At the heart, stimulation of β_1 receptors triggers an increase in the rate and force of cardiac muscle contraction.

The metabolic changes that follow catecholamine release are at their peak 30 seconds after adrenal stimulation, and they linger for several minutes thereafter. As a result, the effects produced by stimulation of the adrenal medulla outlast the other signs of sympathetic activation. [**CM:** *Disorders of the Adrenal Medulla*]

What symptoms would you expect to see in an individual whose diet lacks iodine?

When a person's thyroid gland is removed, signs of decreased thyroid hormone concentration do not appear until about one week later. Why?

Removal of the parathyroid glands would result in a decrease in the blood of what important mineral?

What effect would elevated cortisol levels have on the level of glucose in the blood?

■ Endocrine Functions of the Kidneys and Heart

Endocrine cells in the kidneys and heart produce hormones important for the regulation of blood pressure and blood volume. Kidney cells sensitive to reductions in blood flow respond to a decline in blood volume and/or blood pressure by releasing an enzyme and a hormone. The enzyme, **renin**, initiates a chain of events diagrammed in Figure 18-17a. The liver normally secretes inactive molecules of the globular protein **angiotensinogen**. Renin acts on angiotensinogen, converting it to **angiotensin I**. In the vessels of the lungs this compound is further modified to angiotensin II by a **converting enzyme**. Angiotensin II stimulates the adrenal production of aldosterone, causing water retention at the kidneys and preventing further reductions in blood volume. (Additional details concerning the actions of this hormone will be found in Chapters 21 and 26.)

The kidney hormone, **erythropoietin** (e-rith-rō-poy-Ē-tin; *erythros*, red + *poiesis*, making), or **EPO**, stimulates the production of red blood cells by the bone marrow. The increase in the number of erythrocytes elevates blood volume to some degree and, because these cells transport oxygen, the increase in their number improves oxygen delivery to peripheral tissues.

The glandular cells of the heart respond when the blood pressure and/or volume become excessive,

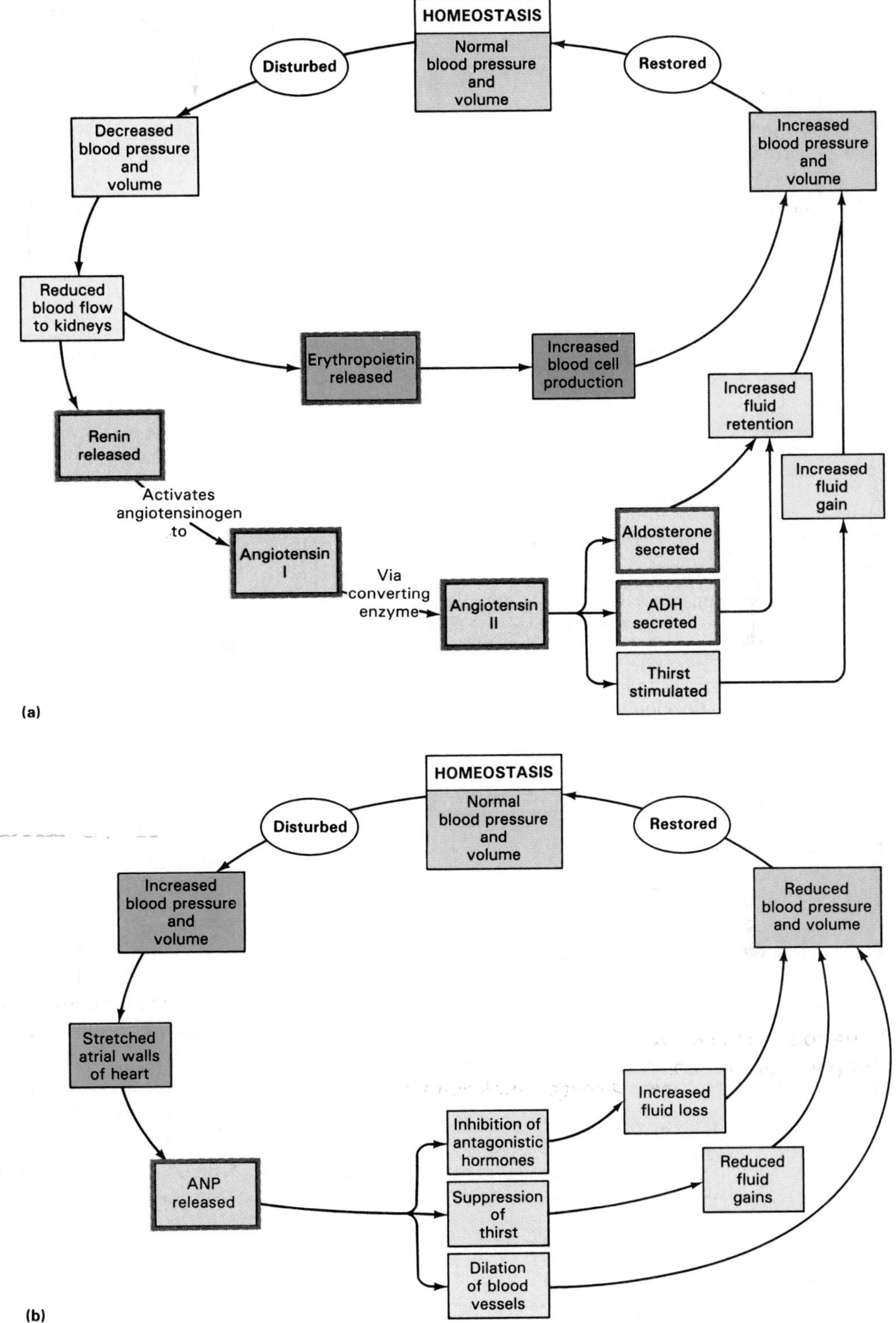

FIGURE 18-17
The Endocrine Functions of the Kidneys and Heart. (a) The renin-angiotensin system assists in the elevation of blood volume and blood pressure. (b) These actions are opposed by those of atrial natriuretic peptide.

as diagrammed in Figure 18-17b. Under these conditions, specialized muscle cells in the heart produce a hormone, **atrial natriuretic peptide** (**ANP**) (nā-trē-ū-RET-ik; *natrium*, sodium + *ouresis*, making water). This hormone has four major effects:

1. It promotes the loss of sodium ions and water at the kidneys;
2. It depresses thirst;
3. It prevents the secretion of water-conserving hormones, such as ADH, aldosterone, and catecholamines; and
4. It lowers blood pressure by dilating peripheral blood vessels and inhibiting the secretion of catecholamines.

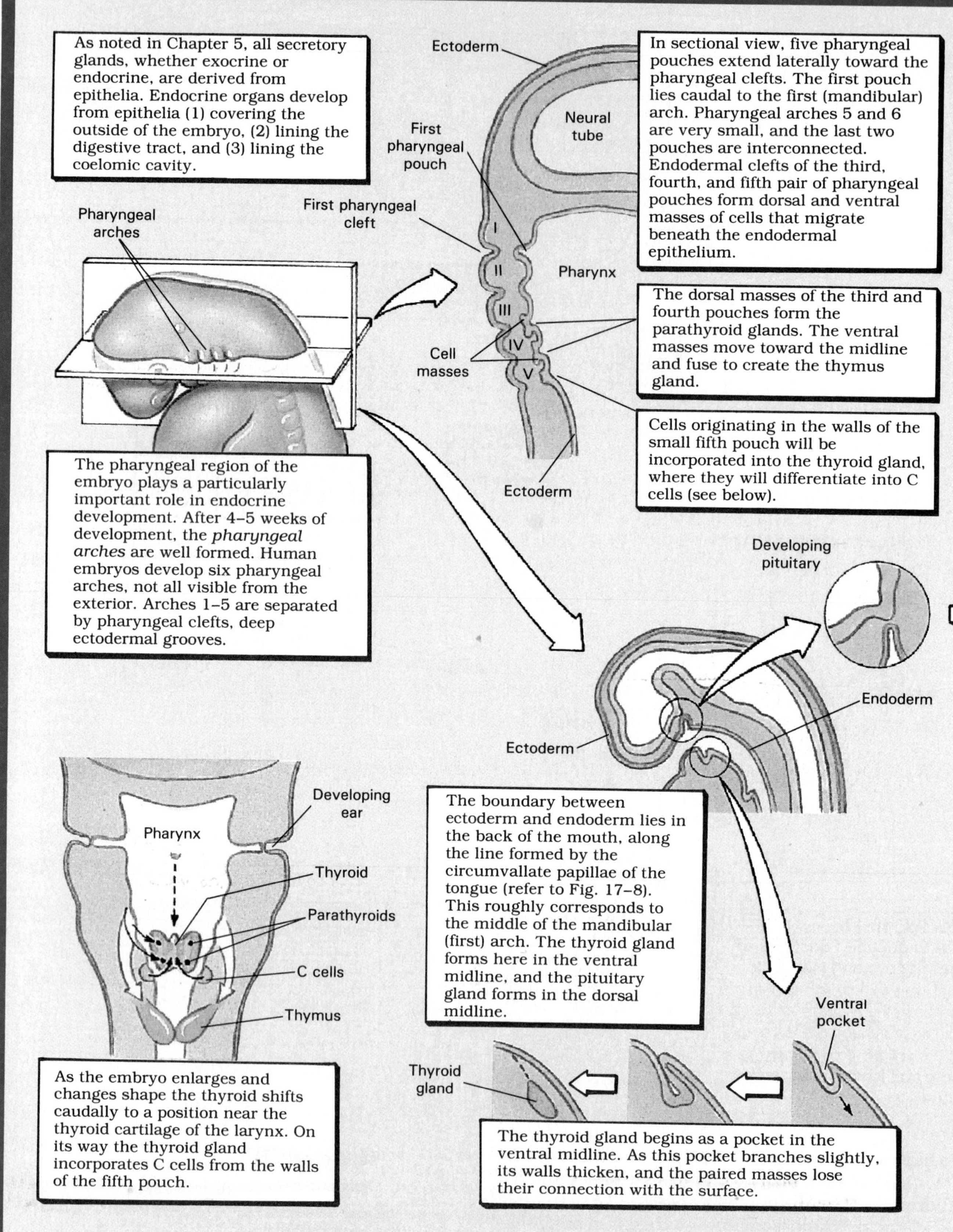

As noted in Chapter 5, all secretory glands, whether exocrine or endocrine, are derived from epithelia. Endocrine organs develop from epithelia (1) covering the outside of the embryo, (2) lining the digestive tract, and (3) lining the coelomic cavity.
Ectoderm
Neural tube
First pharyngeal pouch
First pharyngeal cleft
Pharyngeal arches
In sectional view, five pharyngeal pouches extend laterally toward the pharyngeal clefts. The first pouch lies caudal to the first (mandibular) arch. Pharyngeal arches 5 and 6 are very small, and the last two pouches are interconnected. Endodermal clefts of the third, fourth, and fifth pair of pharyngeal pouches form dorsal and ventral masses of cells that migrate beneath the endodermal epithelium.
I
II
III
IV
V
Pharynx
Cell masses
The dorsal masses of the third and fourth pouches form the parathyroid glands. The ventral masses move toward the midline and fuse to create the thymus gland.
Cells originating in the walls of the small fifth pouch will be incorporated into the thyroid gland, where they will differentiate into C cells (see below).
Ectoderm
The pharyngeal region of the embryo plays a particularly important role in endocrine development. After 4–5 weeks of development, the *pharyngeal arches* are well formed. Human embryos develop six pharyngeal arches, not all visible from the exterior. Arches 1–5 are separated by pharyngeal clefts, deep ectodermal grooves.
Developing pituitary
Endoderm
Ectoderm
Developing ear
Pharynx
Thyroid
Parathyroids
C cells
Thymus
The boundary between ectoderm and endoderm lies in the back of the mouth, along the line formed by the circumvallate papillae of the tongue (refer to Fig. 17–8). This roughly corresponds to the middle of the mandibular (first) arch. The thyroid gland forms here in the ventral midline, and the pituitary gland forms in the dorsal midline.
Ventral pocket
Thyroid gland
As the embryo enlarges and changes shape the thyroid shifts caudally to a position near the thyroid cartilage of the larynx. On its way the thyroid gland incorporates C cells from the walls of the fifth pouch.
The thyroid gland begins as a pocket in the ventral midline. As this pocket branches slightly, its walls thicken, and the paired masses lose their connection with the surface.

Embryology Summary: Development of the Endocrine System

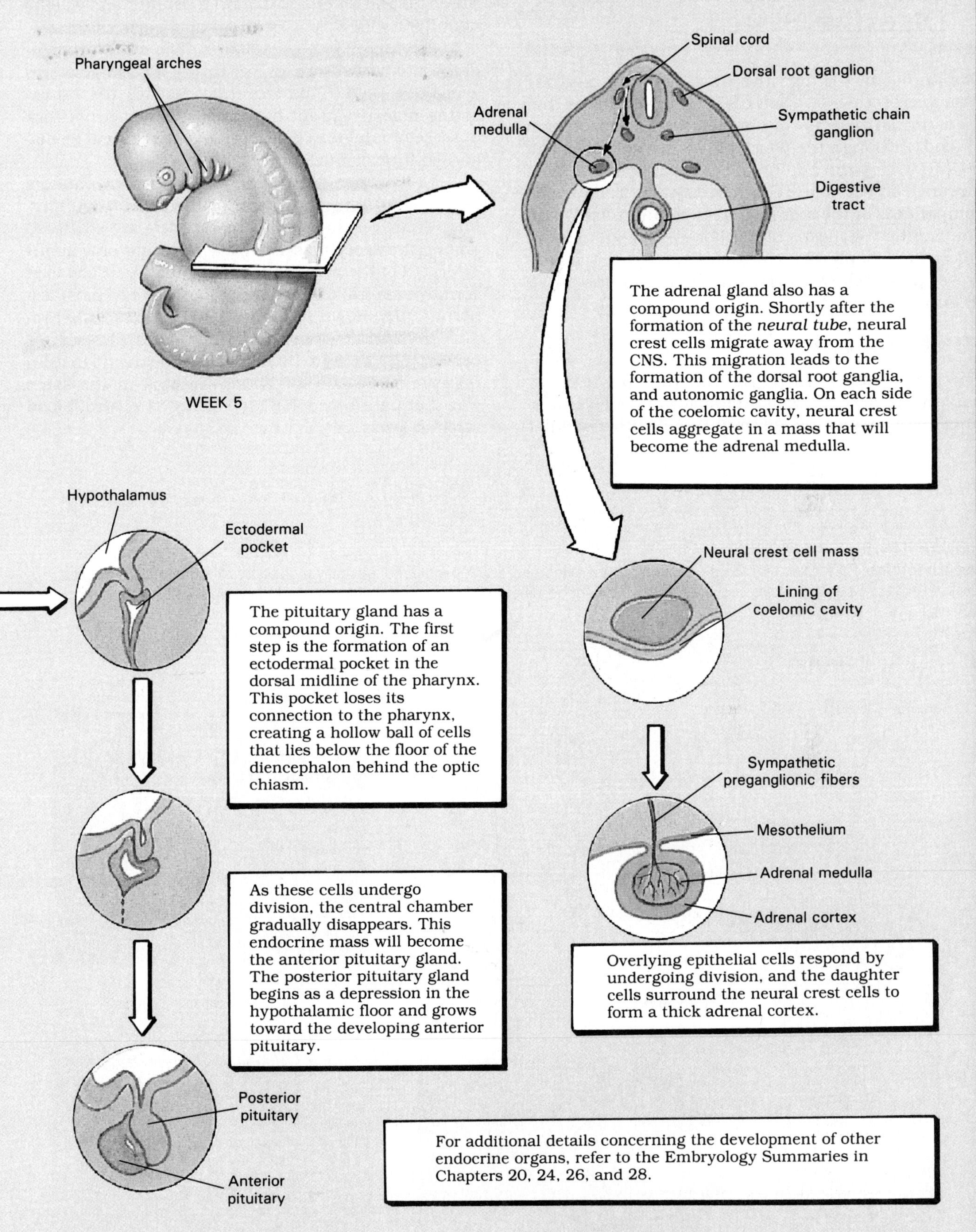
Pharyngeal arches
WEEK 5
Spinal cord
Dorsal root ganglion
Adrenal medulla
Sympathetic chain ganglion
Digestive tract
The adrenal gland also has a compound origin. Shortly after the formation of the *neural tube*, neural crest cells migrate away from the CNS. This migration leads to the formation of the dorsal root ganglia, and autonomic ganglia. On each side of the coelomic cavity, neural crest cells aggregate in a mass that will become the adrenal medulla.
Hypothalamus
Ectodermal pocket
The pituitary gland has a compound origin. The first step is the formation of an ectodermal pocket in the dorsal midline of the pharynx. This pocket loses its connection to the pharynx, creating a hollow ball of cells that lies below the floor of the diencephalon behind the optic chiasm.
Neural crest cell mass
Lining of coelomic cavity
Sympathetic preganglionic fibers
Mesothelium
Adrenal medulla
Adrenal cortex
As these cells undergo division, the central chamber gradually disappears. This endocrine mass will become the anterior pituitary gland. The posterior pituitary gland begins as a depression in the hypothalamic floor and grows toward the developing anterior pituitary.
Overlying epithelial cells respond by undergoing division, and the daughter cells surround the neural crest cells to form a thick adrenal cortex.
Posterior pituitary
Anterior pituitary
For additional details concerning the development of other endocrine organs, refer to the Embryology Summaries in Chapters 20, 24, 26, and 28.

Endocrine Tissues of the Digestive System

The linings of the digestive tract, the liver, and the pancreas produce a variety of exocrine secretions that are essential to the normal breakdown and absorption of food. Although the pace of digestive activities can be affected by the autonomic nervous system, most digestive processes are controlled locally. The various components of the digestive tract communicate with one another by means of hormones that will be considered in Chapter 24. One digestive organ, the pancreas, produces two hormones with widespread effects.

THE PANCREAS

The **pancreas** lies within the abdominopelvic cavity near the border between the stomach and small intestine. It is a slender, usually pink organ with a nodular (lumpy) consistency. The adult pancreas ranges between 20 and 25 cm (8–10 in.) in length, and weighs about 80 g (2.8 oz). Both exocrine and endocrine cells are found in the pancreas. The **exocrine pancreas** produces large quantities of an alkaline, enzyme-rich fluid. This secretion reaches the lumen of the digestive tract by traveling along a network of secretory ducts. The exocrine pancreas will be discussed further in Chapter 24.

Cells of the **endocrine pancreas** form clusters known as **pancreatic islets**, or the *islets of Langerhans* (LAN-ger-hanz). Pancreatic islets are scattered among the exocrine cells and account for only about 1 percent of the pancreatic cell population. Each islet contains several different cell types; the two most important are shown in Figure 18-18. **Alpha cells** produce the hormone **glucagon** (GLOO-ka-gon), and **beta cells** secrete **insulin** (IN-su-lin). Glucagon and insulin regulate blood glucose concentrations in the same way that parathormone and calcitonin control blood calcium levels.

FIGURE 18-18
The Endocrine Pancreas. (a) Gross anatomy of the pancreas. (b) A pancreatic islet surrounded by exocrine-secreting cells. (LM, ×285) (c) Special staining techniques can be used to differentiate between alpha cells (left) and beta cells (right) in a pancreatic islet. (LM, ×140)

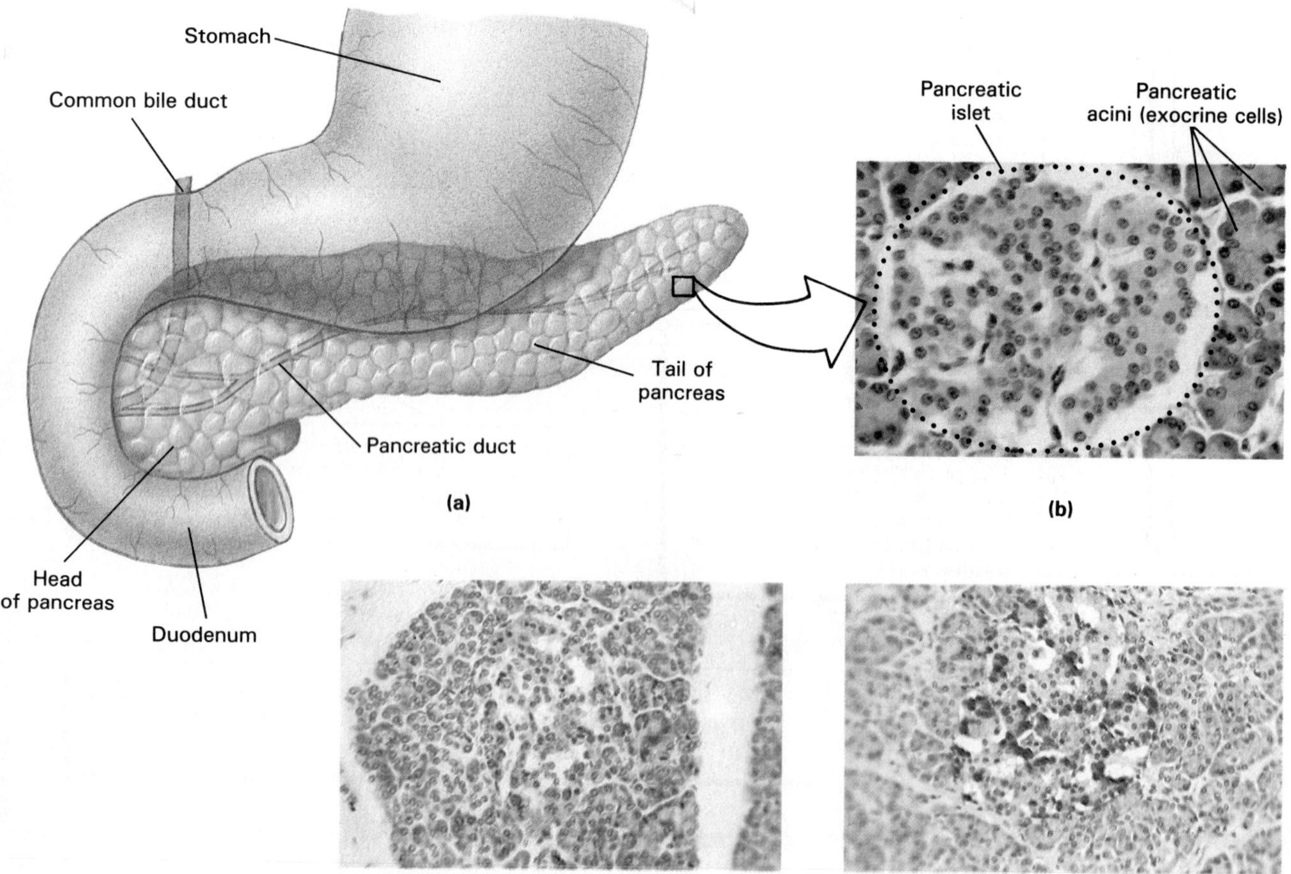

Regulation of Blood Glucose Concentrations

Figure 18-19 diagrams the mechanism of hormonal regulation of blood glucose levels. When blood glucose levels rise, beta cells release insulin, and this hormone stimulates the transport of glucose across cell membranes. The sudden increase in the amount of glucose in the cytoplasm then changes the metabolic activities in peripheral cells. When glucose is abundant, cells use it as an energy source, and they stop breaking down amino acids and lipids. For the duration of insulin stimulation, even skeletal muscles use glucose for an energy source, rather than fatty acids.

The ATP generated by the breakdown of glucose molecules is used to build proteins and to increase energy reserves. Most cells increase their rates of protein synthesis in response to insulin, and a secondary effect of insulin is an increase in the rate of amino acid transport across cell membranes. Insulin stimulates adipocytes to increase their rates of triglyceride

FIGURE 18-19
Regulation of Blood Glucose Concentrations

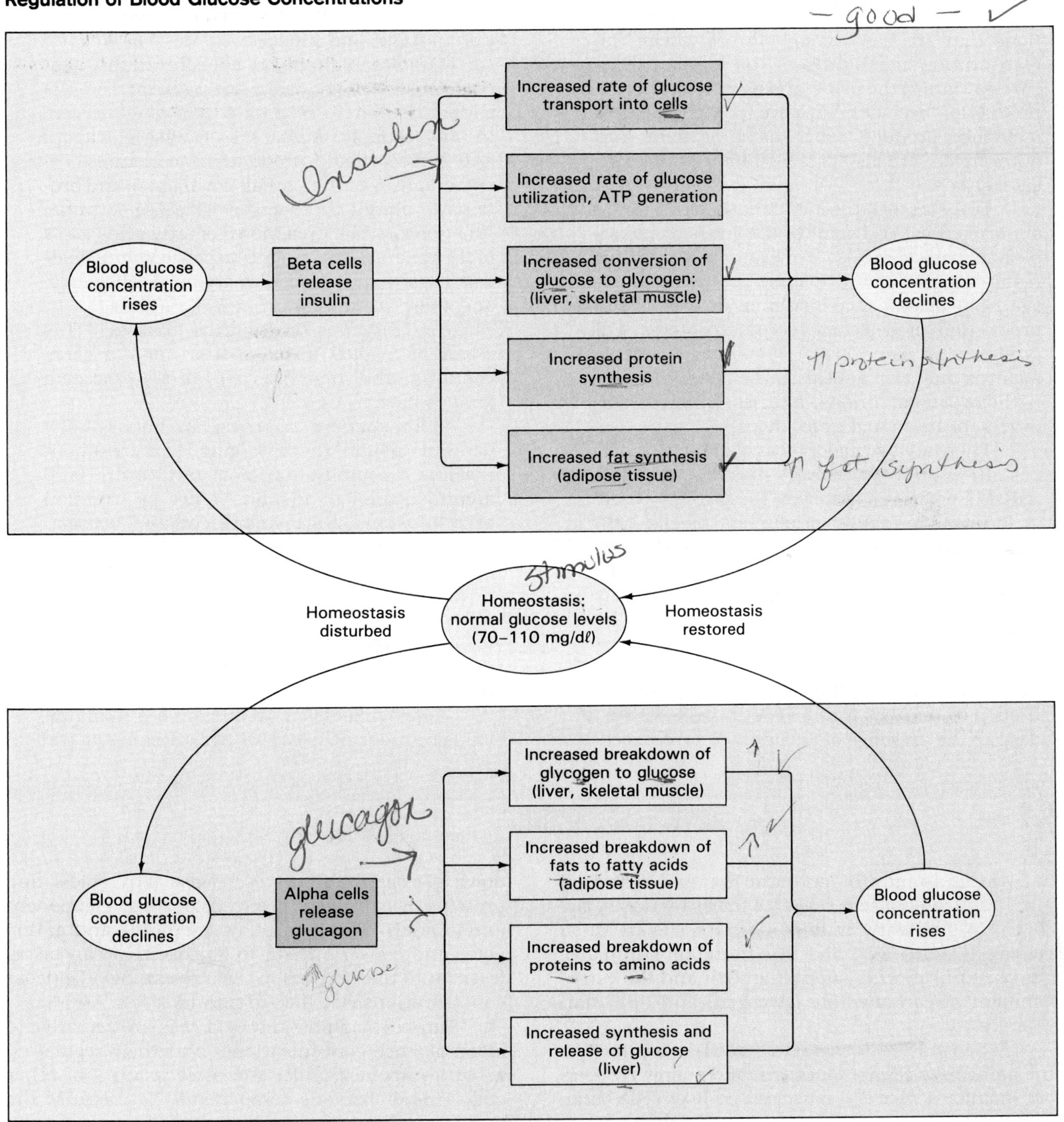

Clinical Comment: Diabetes Mellitus

Blood glucose levels are usually very closely regulated by insulin and glucagon. Whether glucose is absorbed across the digestive tract or manufactured and released by the liver, very little leaves the body intact once it has entered the circulation. Glucose does enter the urine, but the cells lining the excretory passageways of the kidneys usually reabsorb virtually all of it.

Diabetes mellitus (mel-LI-tus; *mellitum*, honey) is characterized by glucose concentrations that are high enough to overwhelm the reabsorption capabilities of the kidneys. Glucose appears in the urine (**glycosuria**), and urine production becomes excessive (polyuria). Other metabolic products, such as fatty acids and other lipids, are also present in abnormal concentrations.

Diabetes mellitus may be caused by genetic abnormalities, and some of the genes responsible have been identified. For example, genes that result in inadequate insulin production, the synthesis of abnormal insulin molecules, or the production of defective receptor proteins will produce the same basic symptoms. Diabetes mellitus may also appear as the result of other pathological conditions, injuries, immune disorders, or hormonal imbalances.

There are two major types of diabetes mellitus. In **insulin-dependent diabetes mellitus** (**IDDM**), or **Type I diabetes**, the primary cause is inadequate insulin production by the beta cells of the pancreatic islets. As noted in Chapter 3, glucose normally enters peripheral cells through facilitated diffusion. (p. 83) This transport mechanism does not function in the absence of insulin, and when insulin concentrations decline, cells can no longer absorb glucose from their surroundings. Under these conditions peripheral tissues remain glucose-starved despite the presence of adequate or even excessive amounts of glucose in the circulation. After a meal rich in glucose, blood concentrations may become so elevated that the kidney cells cannot reclaim all of the glucose molecules entering the urine. The high urinary concentration of glucose limits the ability of the kidney to conserve water, so the individual urinates frequently and may become dehydrated. The chronic dehydration leads to disturbances of neural function (blurred vision, tingling sensations, disorientation, fatigue) and muscle weakness.

Despite high blood concentrations, glucose cannot enter endocrine tissues, and the endocrine system responds as if glucose were in short supply. Alpha cells release glucagon, and glucocorticoid production accelerates. Peripheral tissues then break down lipids and proteins to obtain the energy needed to continue functioning. The breakdown of fatty acids leads to the generation of molecules called **ketone bodies**. These small molecules are metabolic acids, and their accumulation in large numbers can cause a dangerous reduction in blood pH. This condition, called **ketoacidosis**, often triggers vomiting, and in severe cases it can precede a fatal coma.

If the patient survives, an impossibility without insulin therapy, long-term treatment involves a combination of dietary control and administration of insulin, either by injection or by infusion using an **insulin pump**. The treatment is complicated by the fact that tissue glucose demands cycle up and down, depending on physical activity, emotional state, stress, and other factors that are hard to assess or predict. It is therefore difficult to maintain stable and normal blood glucose levels over long periods of time.

Type I diabetes most often appears in individuals under 40 years of age. Because it fre-

(fat) synthesis and storage. In the liver and in skeletal muscles, insulin also accelerates the formation of glycogen. In summary, insulin secretion occurs when glucose is abundant, and this hormone stimulates glucose utilization to support growth and the establishment of carbohydrate (glycogen) and lipid (fat) reserves.

When glucose concentrations fall below normal, the alpha cells release glucagon, and energy reserves are mobilized. Skeletal muscles and liver cells break down glycogen, adipocytes release fatty acids, and proteins are broken down into their component amino acids. The liver takes the lipids and amino acids and converts them to glucose that can be released into the circulation. As a result, blood glucose concentrations rise toward normal levels.

Pancreatic alpha and beta cells are sensitive to blood glucose concentrations, and their regulatory activities are not under the direct control of other endocrine or nervous components. Yet because the quently appears in childhood, it has been called **juvenile-onset diabetes**. Most people with this type of diabetes (roughly 80 percent) have circulating antibodies that target the surfaces of beta cells. The disease may therefore be an example of an *autoimmune disorder*, a condition that results when the immune system attacks normal body cells. (Possible mechanisms responsible for autoimmune disorders will be discussed in Chapter 22.) Consequently, attempts have been made to prevent the appearance of Type I diabetes with *azathioprine* (*Imuran*), a drug that suppresses the immune system. This procedure is somewhat dangerous, however, because compromising immune function increases the risk of acquiring serious infections or developing cancer.

Non-insulin-dependent diabetes mellitus (NIDDM), or **Type II diabetes**, typically affects obese individuals over 40 years of age. Because of the age factor this condition is also called **maturity-onset diabetes**. Insulin levels are normal or elevated, but peripheral tissues no longer respond normally. Treatment consists of weight loss and dietary restrictions that may elevate insulin production and tissue response.

These forms of diabetes mellitus may affect 1 percent of the U.S. population. Maturity-onset diabetes is roughly three times as common as insulin-dependent diabetes. The standard testing procedures check the primary diabetic signs: high fasting blood glucose concentrations, the appearance of glucose in the urine, and an inability to reduce elevated glucose levels. The latter capability is examined by a *glucose tolerance test*. Blood concentrations are monitored after a fasting subject consumes roughly 75 g (2.6 oz) of glucose. In a normal individual the glucose will enter the circulation, insulin production will rise, and peripheral tissues will absorb the glucose so rapidly that blood concentrations will remain relatively normal. Without adequate insulin production, glucose concentrations skyrocket to more than twice normal levels.

Probably because glucose levels cannot be stabilized adequately, even with treatment, patients with diabetes mellitus often develop chronic medical problems. In general, these problems are related to circulatory system abnormalities. The most common examples include the following:

1. Vascular changes at the retina, including proliferation of capillaries and hemorrhaging, often cause disturbances of vision. This condition is called **diabetic retinopathy**.
2. Changes occur in the clarity of the lens, producing cataracts.
3. Small hemorrhages and inflammation at the kidneys cause degenerative changes that can lead to kidney failure. This condition is called **diabetic nephropathy**.
4. A variety of neural problems appear, including nerve palsies, paresthesias, and autonomic dysfunction. These disorders, collectively termed **diabetic neuropathy**, are probably related to disturbances in the blood supply to neural tissues.
5. Degenerative changes in cardiac circulation can lead to early heart attacks. For a given age group, heart attacks are 3–5 times more likely in diabetic individuals.
6. Other peripheral changes in the vascular system can disrupt normal circulation to the extremities. For example, a reduction in blood flow to the feet can lead to tissue death, ulceration, infection, and loss of toes or a major portion of one or both feet.

islet cells are extremely sensitive to variations in blood glucose levels, any hormone that affects blood glucose concentrations will indirectly affect the production of insulin and glucagon.

Other Pancreatic Hormones

The islets have at least four other cell populations that have been given letter designations. One group of islet cells (*D cells*) produces **somatostatin**, a hormone identical to the hypothalamic regulatory hormone GH-IH. This hormone targets the digestive system rather than the pituitary gland. It inhibits insulin and glucagon production by other islet cells and slows the rates of absorption and secretion along the digestive tract. Although these actions can be demonstrated experimentally, as yet no consensus exists concerning the functional significance of any islet cells other than those classified as alpha or beta cells. Information concerning the major pancreatic hormones is included in Table 18-5.

■ **TABLE 18-5 Hormones of the Pancreas**

Structure/Cells	*Hormones*	*Primary Targets*	*Effects*
PANCREATIC ISLETS			
Alpha cells	**Glucagon**	**Liver, adipose tissues**	**Mobilization of lipid reserves, glucose synthesis and glycogen breakdown in liver, elevation of blood glucose concentrations**
Beta cells	**Insulin**	**All cells**	**Facilitates uptake of glucose by cells, stimulates lipid and glycogen formation and storage**

■ Endocrine Tissues of the Reproductive System

The endocrine tissues of the reproductive system are primarily restricted to the reproductive organs. In the male, the **interstitial cells** of the testis produce the male hormones known as androgens. **Testosterone** (tes-TOS-ter-ōn) is the most important androgen. This hormone promotes the production of functional sperm, maintains the secretory glands of the male reproductive tract, and determines secondary sexual characteristics such as the distribution of facial hair and body fat. Testosterone also affects metabolic operations throughout the body, notably stimulating protein synthesis and muscle growth, and produces aggressive behavioral responses. During embryonic development, the production of testosterone affects the development of CNS structures, including hypothalamic nuclei that will later affect sexual behaviors.

Cells directly associated with the formation of functional sperm secrete an additional hormone, called **inhibin**. Inhibin production, which occurs under FSH stimulation, depresses the secretion of GnRH and of FSH at the anterior pituitary. Throughout adult life, these two hormones interact to maintain sperm production at normal levels.

In the ovaries, eggs develop in specialized structures called **follicles**, under stimulation by FSH. Follicle cells surrounding the eggs produce **estrogens** (ES-trō-jenz). These steroid hormones support the maturation of the eggs and stimulate the growth of the uterine lining. Under FSH stimulation, ovarian cells outside the follicles secrete inhibin, which suppresses FSH release through a feedback mechanism comparable to that in males. After ovulation has occurred, under LH stimulation, the follicular cells reorganize into a **corpus luteum** (LOO-tē-um; "yellow body"). The luteal cells then begin to release a mixture of estrogens and **progesterone** (prō-JES-ter-ōn). Progesterone accelerates the movement of the fertilized egg along the oviducts and prepares the uterus for the arrival of the developing embryo. Progesterone also causes an enlargement of the mammary glands.

The production of androgens, estrogens, and progesterone is controlled by regulatory hormones

■ **TABLE 18-6 Hormones of the Reproductive System**

Structure/Cells	*Hormones*	*Primary Targets*	*Effects*
TESTES			
Interstitial cells	**Testosterone**	**Most cells**	**Supports functional maturation of sperm, protein synthesis in skeletal muscles, male secondary sexual characteristics, and associated behaviors**
	Inhibin	**Anterior pituitary**	**Inhibits secretion of FSH**
OVARIES			
Follicular cells	**Estrogens**	**Most cells**	**Supports follicle maturation, female secondary sexual characteristics, and associated behaviors**
	Inhibin	**Anterior pituitary**	**Inhibits secretion of FSH**
Corpus luteum	**Progesterone**	**Uterus, mammary glands**	**Prepares uterus for implantation; prepares mammary glands for secretory functions**

released by the anterior pituitary gland. During a pregnancy the placenta itself functions as an endocrine organ, working together with the ovaries and the pituitary gland to promote normal fetal development and delivery. (The anatomical and endocrinological aspects of reproduction and fetal development will be detailed in Chapters 28 and 29.)

Information concerning the reproductive hormones can be found in Table 18-6.

■ The Pineal Gland

The pineal gland lies in the roof of the thalamus. It contains neurons, glial cells, and special secretory cells called **pinealocytes** (pi-NĒ-al-ō-sīts). These cells synthesize the hormone **melatonin** (mel-a-TŌ-nin) from molecules of the neurotransmitter serotonin. Collaterals from the visual pathways enter the pineal gland and affect the rate of melatonin production. Melatonin production is lowest during daylight hours and highest in the dark of night. This hormone has a direct effect on the melanocytes of the skin, where it inhibits the production of melanin. Melatonin also affects two hypothalamic nuclei:

1. Melatonin stimulates the production of MSH-IH and blocks the release of melanocyte-stimulating hormone (MSH) at the anterior pituitary gland. This indirect action nicely complements its direct effects on melanocyte activity.
2. Melatonin also slows the maturation of sperm, eggs, and reproductive organs by reducing the rate of gonadotrophin-releasing hormone (GnRH) secretion. The significance of this effect in humans remains uncertain. Because of the cyclical nature of pineal activity this organ may also be involved with the establishment or maintenance of basic *circadian rhythms*, daily changes in physiological processes that follow a regular pattern.

How would an increase in the amount of atrial natriuretic peptide affect the concentration of sodium ion in the urine?

Why does a person with Type I diabetes urinate frequently and have a pronounced thirst?

What affect would increased levels of glucagon have on the amount of glycogen stored in the liver?

Increased amounts of light would inhibit the production of which hormone?

NEWS

Health News: Light and Behavior

Exposure to sunlight can do more than stimulate a tan or promote the formation of vitamin D. There is evidence that daily light-dark cycles have widespread effects on the central nervous system, with melatonin playing a key role. Several studies have indicated that residents of temperate and higher latitudes in the Northern Hemisphere undergo seasonal changes in mood and activity patterns. These people feel most energetic from June through September, whereas the period of December through March finds them with relatively low spirits. (The situation in the Southern Hemisphere, where the winter and summer seasons are reversed, is just the opposite.) The degree of seasonal variation differs from individual to individual; some people are affected so severely that they seek medical attention. The observed symptoms have recently been termed **seasonal affective disorder**, or **SAD**. Individuals with SAD experience depression and lethargy, and find it difficult to concentrate. Often they sleep for long periods, perhaps 10 hours or more per day. They also frequently go on eating binges and have a craving for carbohydrates.

Melatonin secretion appears to be regulated by sunlight exposure, not simply by light exposure. Normal interior lights are apparently not strong enough or do not release the right mixture of light wavelengths to depress melatonin production. Because many people spend very little time outdoors, melatonin production increases in the winter, and the depression, lethargy, and concentration problems appear to be linked to elevated melatonin levels in the blood. Comparable symptoms can be produced in a normal experimental subject by an injection of melatonin.

The cyclical eating binges may have a different origin. One of the serotonin pathways in the brain stem controls the craving for carbohydrates. For example, in some individuals drugs that increase serotonin production reduce interest in carbohydrates and promote weight loss. Whether or not increased melatonin production causes decreased serotonin release has yet to be determined.

Many SAD patients may be successfully treated by exposure to sun lamps that produce full-spectrum light. Experiments are under way to define exactly how intense the light must be and determine the minimal effective time of exposure.

Patterns of Hormonal Interaction

Although hormones are usually studied individually, the extracellular fluids contain a mixture of hormones whose concentrations change daily and even hourly. When a cell receives instructions from two different hormones at the same time, there are several possible results:

1. The two hormones may have opposing, or **antagonistic**, effects, as in the case of parathormone and calcitonin, or insulin and glucagon. The net result will depend on the balance between the two hormones. However, the presence of an antagonistic hormone means that the observed effects will be smaller than those produced by either hormone acting unopposed.
2. The two hormones may have additive effects. Sometimes the net result is not only greater than the effect that each would produce acting alone, but is actually greater than the *sum* of their individual effects. An example of such a **synergistic** (sin-er-JIS-tik; *synairesis*, a drawing together) effect is the glucose-sparing actions of growth hormone and glucocorticoids.
3. One hormone can have a **permissive** effect on another. In such cases the first hormone is needed for the second to produce its effect. For example, epinephrine has no apparent effect on energy consumption unless thyroid hormones are also present in normal concentrations.
4. Finally, hormones may produce different but complementary results in specific tissues and organs. These **integrative** effects are important in coordinating the activities of diverse physiological systems.

This section will present three examples of the ways hormones interact to produce complex, well-coordinated results. The patterns introduced will provide the background needed to understand the more detailed discussions found in chapters dealing with cardiovascular function, metabolism, excretion, and reproduction.

HORMONES AND GROWTH

Normal growth requires the cooperation of several endocrine organs. Five different hormones are especially important, although many others have secondary effects on growth rates and patterns.

1. **Growth hormone**: Growth hormone assists in the maintenance of normal blood glucose concentrations and the mobilization of lipid reserves stored in adipose tissues. It is not the primary hormone involved, however, and an adult with a growth hormone deficiency but normal levels of thyroxine, insulin, and glucocorticoids will have no physiological problems. The effects of GH on protein synthesis and cellular growth are most apparent in children, where GH supports muscular and skeletal development. Undersecretion or oversecretion of GH can lead to dwarfism or gigantism, disorders considered in Chapter 7. (p. 218)
2. **Thyroid hormones**: Normal growth also requires appropriate levels of thyroid hormones. If these hormones are absent for the first year after birth, the nervous system fails to develop normally, producing mental retardation. If thyroxine concentrations decline later in life, but before puberty, normal skeletal development does not continue.
3. **Insulin**: Growing cells need adequate supplies of energy and nutrients. Without insulin, produced by the pancreas, the passage of glucose and amino acids across cell membranes will be drastically reduced or eliminated.
4. **Parathormone**: Parathormone promotes the absorption of calcium salts for subsequent deposition in bone.
5. **Gonadal hormones**: The activity of osteoblasts in key locations and the growth of specific cell populations are affected by the presence or absence of sexual hormones. The differential growth induced by these hormones changes skeletal proportions and triggers the development of secondary sexual characteristics.

The circulating concentrations of these hormones are regulated independently, and every time the hormonal mixture changes, metabolic operations are modified to some degree. The alterations vary in duration and intensity, producing unique individual growth patterns.

HORMONES AND STRESS

Any threat to homeostasis represents a form of **stress**. In our lives there are many different sources of stress, which can be categorized according to the nature of the threat. They include physical stresses such as illness or injury, emotional stresses produced by depression or anxiety, environmental stresses such as extreme heat or cold, and metabolic stresses ranging from dietary inadequacies to outright starvation. Despite the variety of potential stresses, the human body responds to each one with the same basic pattern of hormonal and physiological adjustments. These responses constitute the **general adaptation syndrome**, or **GAS**. The GAS can be divided into three phases: the *alarm phase*, the *resistance phase*, and the *exhaustion phase*.

The Alarm Phase

The **alarm phase** represents a swift, immediate response to the stress, under the direction of the sympathetic division of the autonomic nervous system. During this phase energy reserves are mobilized, mainly in the form of glucose, and the body prepares for any physical activities needed to eliminate or escape from the source of the stress. Epinephrine (adrenaline) is the dominant hormone of the alarm phase, and its secretion accompanies the sympathetic activation that produces the "fight or flight" response discussed in Chapter 16. (p. 501) The characteristics of the alarm phase include increased mental alertness, increased energy consumption by skeletal muscles and other peripheral tissues, the mobilization of energy reserves (glycogen and lipids), an increase in blood flow to skeletal muscles, a decrease in blood flow to the skin, kidneys, and digestive organs, a drastic reduction in digestion and urine production, increased sweat gland secretion, and increases in blood pressure, heart rate, and respiratory rate.

Although the effects of epinephrine are most apparent during the alarm phase, other hormones play supporting roles. The secretion of ADH further reduces fluid losses at the kidneys and conserves blood volume. In addition, when sympathetic activation reduces the circulatory supply to the kidneys, it also triggers the release of renin and stimulates the secretion of aldosterone. The reduction of water losses by circulatory changes, ADH production, and aldosterone secretion may be very important if the stress(es) result in a loss of blood.

The Resistance Phase

In many instances the temporary adjustments of the alarm phase are sufficient to remove or overcome the stressful stimulus. But some stresses, such as starvation, acute illness, or severe anxiety, can persist for hours, days, or weeks. If a stress lasts longer than a few hours, the individual will enter the **resistance phase** of the GAS. Glucocorticoids are the dominant hormones of the resistance phase, but many other hormones are involved. Throughout the resistance phase energy demands remain higher than normal, due to the background production of glucocorticoids, epinephrine, growth hormone, and thyroid hormones.

Neural tissue has a high demand for energy, and neurons must have a reliable supply of glucose; if blood glucose concentrations fall too far, neural function deteriorates. Glycogen reserves are adequate to maintain normal glucose concentrations during the alarm phase, but after several hours these reserves are nearly exhausted. The endocrine secretions of the resistance phase are coordinated to achieve three integrated results:

1. **The mobilization of lipid and protein reserves**: The hypothalamus produces GH-RH and CRH, stimulating the release of growth hormone and, via ACTH, the secretion of glucocorticoids. Adipose tissues respond to glucocorticoids by releasing stored fatty acids; skeletal muscles respond by breaking down proteins and releasing amino acids into the circulation.
2. **The elevation and stabilization of blood glucose concentrations**: As blood glucose levels decline, glucagon and glucocorticoids stimulate the liver to manufacture glucose from the fatty acids and amino acids provided by adipose tissue and skeletal muscles. The glucose molecules are then released into the circulation, and blood glucose concentrations return to normal levels.
3. **The conservation of glucose for neural tissues**: Glucocorticoids and growth hormone from the anterior pituitary stimulate lipid metabolism in peripheral tissues. These glucose-sparing effects maintain normal blood glucose levels even after long periods of starvation. However, neural tissues do not alter their metabolic activities, and they continue to use glucose as an energy source.

The Exhaustion Phase

Once these adjustments have been made, the lipid reserves of the body are sufficient to maintain the resistance phase for a period of weeks or even months. Thus the metabolic changes that characterize the resistance phase also provide the ability to endure lengthy periods of starvation. But the resistance phase cannot be sustained indefinitely. During starvation the situation becomes critical as lipid reserves are exhausted, and more and more of the cellular energy demands must be met by breaking down structural proteins. Even if nutrient intake continues at normal levels, however, the individual will eventually "crash." Although the metabolic effects of glucocorticoids are essential to normal resistance, their anti-inflammatory action slows wound healing and increases susceptibility to infection. The continued conservation of fluids under the influence of ADH and aldosterone stresses the cardiovascular system by producing elevated blood volumes and higher than normal blood pressures. Alternatively, the adrenal cortex may simply be unable to continue producing glucocorticoids, and this quickly eliminates the ability to maintain acceptable blood glucose concentrations. Poor nutrition, emotional or physical trauma, chronic illness, or damage to key organs such as the heart, liver, or kidneys will significantly hasten the onset of the exhaustion phase.

Whatever the mechanism, sooner or later homeostatic regulation breaks down, and the malfunctioning of nerve and muscle cells marks the arrival of the exhaustion phase. Unless corrective actions are taken almost immediately, the failure of one or more organ systems will prove fatal. Although a single cause, such as heart failure, may be listed as the cause of death, the underlying problem is the inability to support the endocrine and metabolic adjustments of the resistance phase. The three phases of the GAS are summarized in Figure 18-20.

HORMONES AND BEHAVIOR

As we have seen, many endocrine functions are regulated by the hypothalamus, and hypothalamic neurons monitor the levels of many circulating hormones. Other portions of the central nervous system are also quite sensitive to hormonal stimulation.

The clearest demonstrations of the effects of specific hormones involve individuals whose endocrine glands are oversecreting or undersecreting. Normal changes in circulating hormone levels can also cause behavioral changes. Chapter 7 noted that one of the triggers for closure of the epiphyses is the increase in sexual hormone production at the time of puberty. (p. 218) In precocious (premature) puberty, sex-hormones are produced at an inappropriate time, perhaps as early as 5 or 6 years of age. The affected children not only begin to develop adult secondary sexual characteristics, but also undergo significant behavioral changes. The "nice little kid" disappears, and the child becomes aggressive and assertive. These behavioral alterations represent the effects of sexual hormones on CNS function. Thus behaviors that in normal teenagers are usually attributed to environmental stimuli, such as peer pressure, actually have some physiological basis as well. In the adult, changes in the mixture of hormones reaching the CNS can

FIGURE 18-20
The General Adaptation Syndrome

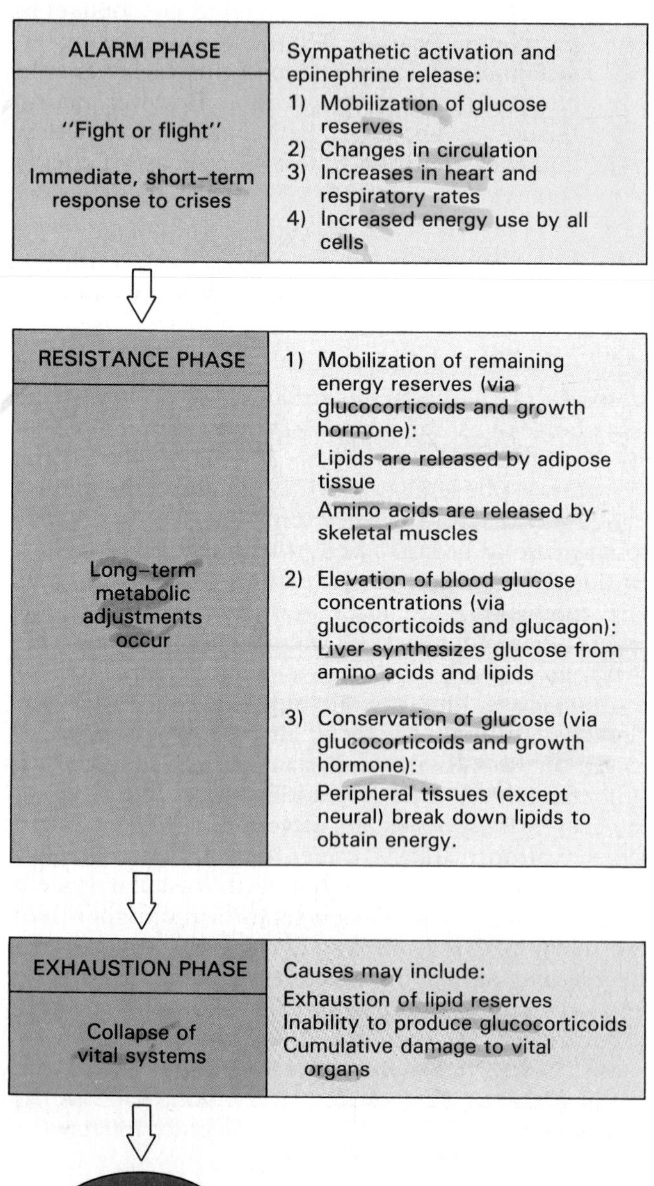

Sports and Fitness: Androgen Abuse

One of the first endocrinological "mass experiments" occurred early in World War II, when the German government administered testosterone to Nazi SS officers in an attempt to make them more aggressive. (There is no evidence that the experiment succeeded.) Nowadays, such practices are universally deplored in all civilized societies, and medical research involving humans is generally subject to tight ethical constraints and meticulous scientific scrutiny. Yet a clandestine, unscientific, and potentially quite dangerous program of "experimentation" with hormones is today being pursued by athletes in many countries. Despite being banned by the International Olympic Committee, the United States Olympic Committee, the NCAA, and the NFL, and condemned by the American Medical Association and the American College of Sports Medicine, the use of androgens, or "anabolic steroids," has become popular with many athletes, both amateur and professional.

The goal of steroid use is to increase muscle mass, endurance, and "competitive spirit." It has been suggested that as many as 30 percent of college and professional athletes and 10–20 percent of male high school athletes may be using anabolic steroids (with or without growth hormone) to improve their performance. Among body builders, the proportion using steroids in this country may be as high as 80 percent. Black market sales of these products probably exceed $100 million annually. The compounds are administered orally or by injection—typically in doses 10–1000 times higher than those normally prescribed in medical treatment.

One supposed justification for this practice has been the unfounded opinion that compounds manufactured in the body are not only safe, but "good for you." In reality the administration of natural *or* synthetic androgens in abnormal amounts carries unacceptable health risks. Androgens affect many tissues in a variety of ways. Known complications include premature epiphyseal closure; various liver dysfunctions (including jaundice and hepatic tumors); prostate enlargement and urinary tract obstruction; and testicular atrophy and lowered sperm count, leading to infertility. A link to heart attacks, impaired cardiac function, and strokes has also been suggested. Moreover, the normal regulation of androgen production involves a feedback mechanism comparable to that described for adrenal steroids earlier in this chapter. GnRH stimulates the production of ICSH (LH), and ICSH stimulates the secretion of testosterone and other androgens by the interstitial cells of the testes. Circulating androgens, in turn, inhibit the production of both GnRH and ICSH. Thus when synthetic androgens are administered in high doses, they can (1) suppress the normal production of testosterone, and (2) depress the manufacture of GnRH by the hypothalamus. *This suppression of GnRH release may be permanent.* The use of androgenic "bulking agents" by female body builders may not only add muscle mass, but alter muscular proportions and secondary sexual characteristics. For example, women taking steroids can develop irregular menstrual periods and changes in body hair distribution (including baldness). Finally, androgen abuse may cause a generalized depression of the immune system.

Ironically, although anabolic steroids may improve performance over the short term, it is not clear that chronic steroid use is more effective than a rigorous training program. Steroids can accelerate growth and muscle mass development, but the athlete could reach the same levels of performance using more traditional (and safer) methods.

have significant effects on intellectual capabilities, memory, learning, and emotional states.

HORMONES AND AGING

The endocrine system shows relatively few functional changes with age. The most dramatic exception is the decline in the concentration of reproductive hormones. Effects of these hormonal changes on the skeletal system were noted in Chapter 7; further discussion will be found in Chapter 29. (p. 224)

Blood and tissue concentrations of many other hormones, including TSH, thyroid hormones, ADH, PTH, prolactin, and glucocorticoids, remain unchanged. But, although circulating hormone levels may remain within normal limits, some endocrine tissues become less responsive to stimulation. For example, in elderly individuals less GH and insulin is secreted after a carbohydrate-rich meal or in a glucose tolerance test.

Finally, it should be noted that age-related changes in other tissues affect their abilities to respond to hormonal stimulation. As a result, peripheral tissues may become less responsive to some hormones. This loss of sensitivity has been documented for glucocorticoids and ADH.

ENDOCRINE DISORDERS

Endocrine disorders fall into two basic categories: they reflect either abnormal hormone production or abnormal cellular sensitivity. The symptoms are interesting because they highlight the significance of normally "silent" hormonal contributions. An abbreviated summary is presented in Table 18-7.

Interactions with Other Systems

The relationships between the endocrine system and other systems are summarized in Figure 18-21.

- **Insulin lowers the level of glucose in the blood and glucagon causes glucose levels to rise. What is this type of hormonal interaction called?**
- **The lack of which hormones would inhibit skeletal formation?**
- **Why do levels of GH-RH and CRH rise during the resistance phase of the general adaptation syndrome (GAS)?**

TABLE 18-7 Clinical Implications of Endocrine Malfunctions

Hormone	*Underproduction Syndrome*	*Principal Symptoms*	*Overproduction Syndrome*	*Principal Symptoms*
Growth hormone (GH)	Pituitary growth failure (p. 577)	Retarded growth, abnormal fat distribution, low blood glucose hours after a meal	Gigantism, acromegaly (CM)	Excessive growth
Antidiuretic hormone (ADH)	Diabetes insipidus (p. 576)	Polyuria	SIADH (Syndrome of inappropriate ADH secretion (CM)	Increased body weight and water content
Thyroxine (TX, T_4)	Myxedema, cretinism (CM)	Low metabolic rate, body temperature; impaired physical and mental development	Graves' disease (CM)	High metabolic rate, body temperature
Parathyroid hormone (PTH)	Hypoparathyroidism (CM)	Muscular weakness, neurological problems, formation of dense bones, tetany due to low blood calcium concentrations	Hyperparathyroidism (CM)	Neurological, mental, muscular problems due to high blood calcium concentrations; weak and brittle bones
Insulin	Diabetes mellitus (Type I) (p. 594)	High blood glucose, impaired glucose utilization, dependence on lipids for energy, glycosuria	Excess insulin production or administration	Low blood glucose levels, possibly causing coma
Mineralocorticoids (MC)	Hypoaldosteronism (CM)	Polyuria, low blood volume, high blood potassium concentrations	Aldosteronism (CM)	Increased body weight due to water retention; low blood potassium concentration
Glucocorticoids (GC)	Addison's disease (CM)	Inability to tolerate stress, mobilize energy reserves, maintain normal blood glucose concentrations	Cushing's disease (CM)	Excessive breakdown of tissue proteins and lipid reserves; impaired glucose metabolism
Epinephrine (E), norepinephrine (NE)	None identified		Pheochromocytoma (CM)	High metabolic rate, body temperature, and heart rate; elevated blood glucose levels; other symptoms comparable to those of excessive autonomic stimulation
Estrogens (female)	Hypogonadism	Sterility, lack of secondary sexual characteristics	Androgenital syndrome (CM)	Overproduction of androgens by z. reticularis of adrenal leads to masculinization
	Menopause	Cessation of ovulation	Precocious puberty (p. 600)	Early production of developing follicles and estrogen secretion
Androgens (male)	Hypogonadism, eunuchoidism	Sterility, lack of secondary sexual characteristics	Gynecomastia (CM)	Abnormal production of estrogens, sometimes due to adrenal or interstitial cell tumors, leads to breast enlargement
			Precocious puberty (p. 600)	Early production of androgens leading to premature physical development and behavioral changes

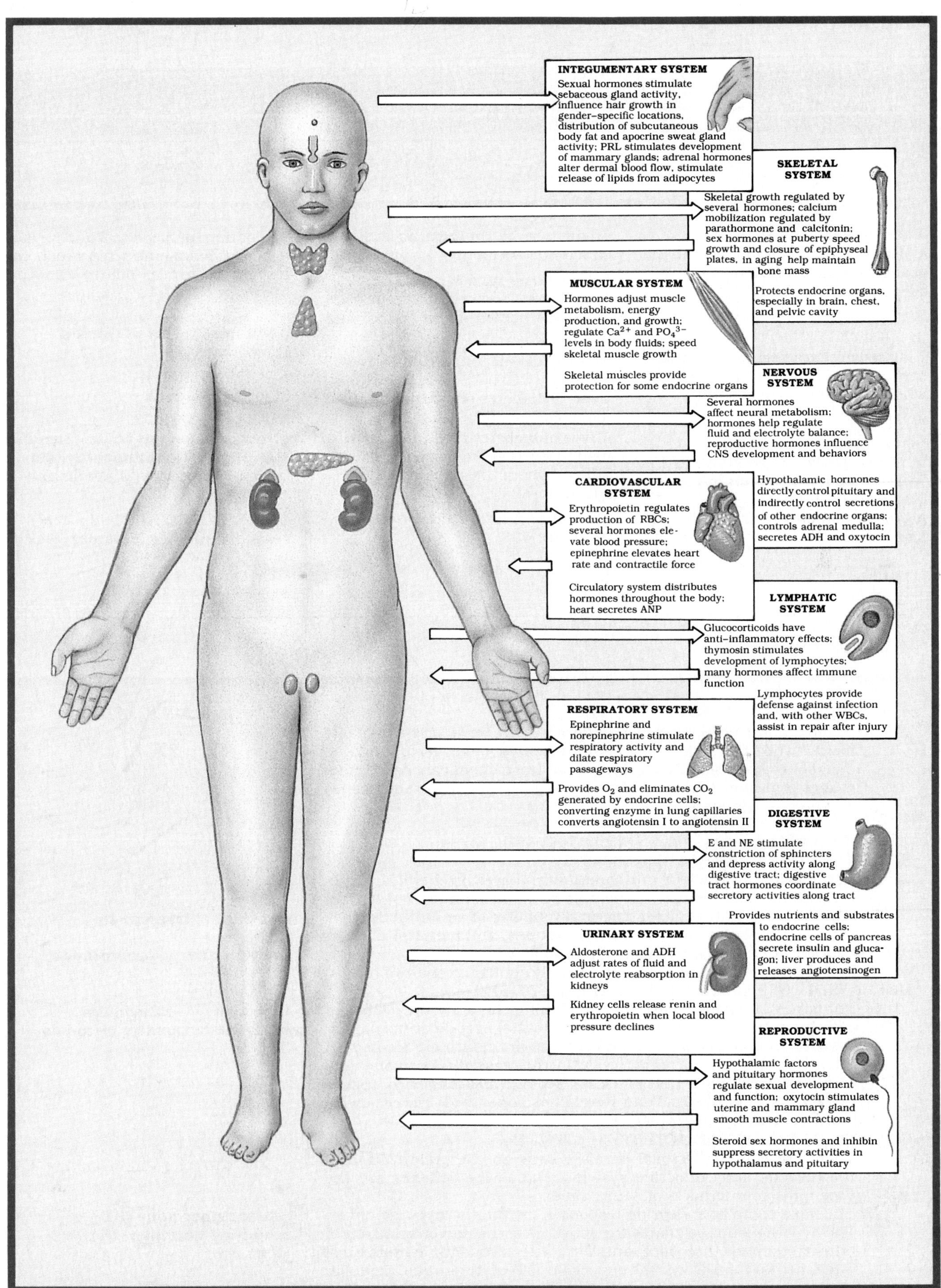

INTEGUMENTARY SYSTEM
Sexual hormones stimulate sebaceous gland activity, influence hair growth in gender-specific locations, distribution of subcutaneous body fat and apocrine sweat gland activity; PRL stimulates development of mammary glands; adrenal hormones alter dermal blood flow, stimulate release of lipids from adipocytes
SKELETAL SYSTEM
Skeletal growth regulated by several hormones; calcium mobilization regulated by parathormone and calcitonin; sex hormones at puberty speed growth and closure of epiphyseal plates, in aging help maintain bone mass
Protects endocrine organs, especially in brain, chest, and pelvic cavity
MUSCULAR SYSTEM
Hormones adjust muscle metabolism, energy production, and growth; regulate Ca^{2+} and PO_4^{3-} levels in body fluids; speed skeletal muscle growth
Skeletal muscles provide protection for some endocrine organs
NERVOUS SYSTEM
Several hormones affect neural metabolism; hormones help regulate fluid and electrolyte balance; reproductive hormones influence CNS development and behaviors
Hypothalamic hormones directly control pituitary and indirectly control secretions of other endocrine organs; controls adrenal medulla; secretes ADH and oxytocin
CARDIOVASCULAR SYSTEM
Erythropoietin regulates production of RBCs; several hormones elevate blood pressure; epinephrine elevates heart rate and contractile force
Circulatory system distributes hormones throughout the body; heart secretes ANP
LYMPHATIC SYSTEM
Glucocorticoids have anti-inflammatory effects; thymosin stimulates development of lymphocytes; many hormones affect immune function
Lymphocytes provide defense against infection and, with other WBCs, assist in repair after injury
RESPIRATORY SYSTEM
Epinephrine and norepinephrine stimulate respiratory activity and dilate respiratory passageways
Provides O_2 and eliminates CO_2 generated by endocrine cells; converting enzyme in lung capillaries converts angiotensin I to angiotensin II
DIGESTIVE SYSTEM
E and NE stimulate constriction of sphincters and depress activity along digestive tract; digestive tract hormones coordinate secretory activities along tract
Provides nutrients and substrates to endocrine cells; endocrine cells of pancreas secrete insulin and glucagon; liver produces and releases angiotensinogen
URINARY SYSTEM
Aldosterone and ADH adjust rates of fluid and electrolyte reabsorption in kidneys
Kidney cells release renin and erythropoietin when local blood pressure declines
REPRODUCTIVE SYSTEM
Hypothalamic factors and pituitary hormones regulate sexual development and function; oxytocin stimulates uterine and mammary gland smooth muscle contractions
Steroid sex hormones and inhibin suppress secretory activities in hypothalamus and pituitary

FIGURE 18-21
Functional Relationships between the Endocrine System and Other Systems

CHAPTER REVIEW

Review of Selected Clinical Terms

diabetes insipidus (*p. 576*)
A disorder that develops when the posterior pituitary no longer releases adequate amounts of ADH.

diabetes mellitus (mel-LI-tus) (*p. 594*)
A disorder characterized by glucose concentrations high enough to overwhelm the kidneys' reabsorption capabilities.

insulin-dependent diabetes mellitus (IDDM); (also known as **Type I diabetes** or **juvenile-onset diabetes**) (*p. 594*)
A type of diabetes mellitus; the primary cause is inadequate insulin production by the beta cells of the pancreatic islets.

ketoacidosis (*p. 594*)
A condition in which large numbers of ketone bodies in the blood lead to a dangerously low blood pH.

non-insulin-dependent diabetes mellitus (NIDDM); (also known as **Type II diabetes** or **maturity-onset diabetes**) (*p. 595*)
A type of diabetes mellitus in which insulin levels are normal or elevated, but peripheral tissues no longer respond normally.

general adaptation syndrome (GAS) (*p. 598*)
The pattern of hormonal and physiological adjustments with which the human body responds to all forms of stress.

Additional Terms of Clinical Importance

DIABETES INSIPIDUS (*p. 576*): **diabetes, polyuria**

DIABETES MELLITUS (*p. 594*): **glycosuria, ketone bodies, insulin pump, diabetic retinopathy, diabetic nephropathy, diabetic neuropathy**

LIGHT AND BEHAVIOR (*p. 597*) **seasonal affective disorder (SAD)**

Study Outline

Introduction (pp. 567–568)

Related Key Terms

1. In general, the nervous system performs short-term "crisis management," while the endocrine system regulates longer-term, ongoing metabolic processes. Endocrine cells release chemicals called **hormones** that alter the metabolic activities of many different tissues and organs simultaneously.

An Overview of the Endocrine System (pp. 568–574)

STRUCTURE OF HORMONES (pp. 568–570)

1. Hormones can be divided into three groups based on chemical structure: amino acid derivatives, peptide hormones, and steroids.
2. Amino acid derivatives are structurally similar to amino acids; they include **catecholamines, thyroid hormones,** and **melatonin.** — **tyrosine • tryptophan**
3. **Peptide hormones** are chains of amino acids. — **prehormone • prohormone**
4. **Steroid hormones** are lipids structurally similar to cholesterol.

MECHANISMS OF HORMONAL ACTION (pp. 570–572)

5. Hormones exert their effects by modifying the activities of **target cells** (peripheral cells that are sensitive to that particular hormone). Receptors for catecholamine and peptide hormones are located on the cell membranes of target cells; in this case the hormone acts as a **first messenger** that causes a **second messenger** to appear in the cytoplasm. Thyroid and steroid hormones cross the cell membrane and bind to receptors in the cytoplasm or nucleus. — **G protein • calmodulin hormone-responsive elements (HREs)**

CONTROL OF ENDOCRINE ACTIVITY (pp. 572–574)

1. There are many functional parallels between the organization of the nervous and endocrine systems. **Endocrine reflexes** are the functional counterparts of neural reflexes.
2. The most complex endocrine responses involve the hypothalamus. The hypothalamus regulates the activities of the nervous and endocrine systems via three mechanisms: (1) Its autonomic centers exert direct neural control over the endocrine cells of the adrenal medulla; (2) It acts as an endocrine organ itself by releasing hormones into the circulation; (3) It secretes **regulatory hormones** that control the activities of endocrine cells in the pituitary gland. — **releasing hormone (RH) inhibiting hormone (IH)**

The Pituitary Gland (pp. 575–581)

Related Key Terms

1. The gland releases nine important peptide hormones; all bind to membrane receptors and use cyclic AMP as a second messenger.

THE POSTERIOR PITUITARY (pp. 575–576).

2. The **posterior pituitary (neurohypophysis)** contains the axons of hypothalamic neurons. Neurons of the **supraoptic** and **paraventricular nuclei** manufacture antidiuretic hormone (ADH) and oxytocin, respectively. ADH decreases the amount of water lost at the kidneys. In women, **oxytocin** stimulates smooth muscle cells in the uterus and contractile cells in the mammary glands.

THE ANTERIOR PITUITARY (pp. 576–578)

3. The **anterior pituitary (adenohypophysis)** can be subdivided into the large **pars distalis** and the slender **pars intermedia.** Important hormones released by the pars distalis include: (1) **thyroid-stimulating hormone (TSH)** (triggers the release of thyroid hormones); (2) **adrenocorticotrophic hormone (ACTH)** (stimulates the release of **glucocorticoids** by the adrenal gland; (3) **follicle-stimulating hormone (FSH)** (stimulates **estrogen** secretion and egg development in women and sperm production in men); (4) **luteinizing hormone/interstitial cell-stimulating hormone (LH/ICSH)** (causes ovulation and **progesterone** production in women and **androgen** production in men); (5) **prolactin (PRL)** (stimulates the development of the mammary glands and the production of milk); and (6) **growth hormone (GH or somatotropin)** (stimulates cells' growth and replication by triggering the release of **somatomedins** from liver cells).

   hypophysis • infundibulum
   gonadotrophins
   diabetogenic effect

4. **Melanocyte-stimulating hormone (MSH),** released by the pars intermedia, stimulates melanocytes to produce melanin.

HYPOTHALAMIC CONTROL OF THE ANTERIOR PITUITARY (pp. 578–581)

5. At the **median eminence,** neurons release regulatory factors in to the surrounding interstitial fluids. Their secretions enter the circulation easily since the capillaries in this area are **fenestrated.**
6. The **hypophyseal portal system** ensures that all of the blood entering the **portal vessels** will reach the intended target cells before returning to the general circulation.

   portal system

7. Thyroid-hormone-releasing hormone **(TRH)** promotes the secretion of TSH; corticotrophin-releasing hormone **(CRH)** causes the secretion of ACTH; gonadotrophin-releasing hormone **(GnRH)** promotes the secretion of FSH and LH.
8. Inhibiting factors control the release of prolactin (inhibited by prolactin-inhibiting factor or **PIH**) and MSH (controlled by melanocyte-stimulating hormone-inhibiting hormone or **MSH-IH**).
9. The production of growth hormone is regulated by growth hormone-releasing hormone **(GR-RH)** and growth hormone-inhibiting hormone **(GH-IH).** A single regulatory hormone may have related secondary effects.

The Thyroid Gland (pp. 582–584)

1. The thyroid gland lies near the **thyroid cartilage** of the larynx, and consists of two **lobes** connected by a narrow **isthmus.**

THYROID FOLLICLES AND THYROID HORMONES (pp. 582–583)

2. The thyroid gland contains numerous **thyroid follicles.** Thyroid follicles release several hormones, including **thyroxine (TX** or **T_4)** and **triiodothyronine (T_3).**
3. Thyroid hormones exert a **calorigenic effect** which enables us to adapt to cold temperatures.

REGULATION OF THYROID HORMONE RELEASE (p. 583)

4. The follicle cells manufacture **thyroglobulin** and store it as a **colloid** (a viscous fluid containing suspended proteins). The cells also transport iodine from the extracellular fluids into the follicular chamber, where they complex with the thyroglobulin molecules.
5. Most of the thyroid hormones entering the bloodstream are attached to special **thyroid-binding globulins (TBG);** the unbound hormones affect peripheral tissues immediately, while the binding globulins gradually release their hormones over a week or more.

THE C CELLS OF THE THYROID GLAND: CALCITONIN (p. 584)

6. The **C cells** of the follicles produce **calcitonin (CT),** which helps to regulate calcium ion concentrations in body fluids.

Chapter Review

The Parathyroid Glands (p. 585)

Related Key Terms

1. Four parathyroid glands are embedded in the posterior surface of the thyroid gland. The **chief cells** of the parathyroid produce **parathormone (PTH)** in response to lower than normal concentrations of calcium ions. These and the C cells of the thyroid gland maintain calcium ion levels within relatively narrow limits.

The Thymus (p. 585)

1. The thymus produces several hormones, but only one has a known function: **thymosin,** which plays a role in developing and maintaining normal immunological defenses.

The Adrenal Gland (pp. 586–588)

1. A single **adrenal (suprarenal) gland** lies along the superior border of each kidney. Each gland is surrounded by a fibrous **capsule.** The adrenal gland can be subdivided into the superficial **adrenal cortex** and the inner **adrenal medulla.**

THE ADRENAL CORTEX (pp. 586–587)

2. The adrenal cortex manufactures steroid hormones called **adrenocortical steroids (corticosteroids.)** The cortex can be subdivided into three areas: (1) the **zona reticularis,** which produces androgens of uncertain significance; (2) the **zona fasciculata,** which produces **glucocorticoids (GC),** notably **cortisol, corticosterone,** and **cortisone** (all of which exert a **glucose-sparing effect** on peripheral tissues); and (3) the **zona glomerulosa,** which releases **mineralocorticoids (MC),** principally **aldosterone,** which restrict sodium and water losses at the kidneys, sweat glands, digestive tract, and salivary glands. The zona glomerulosa responds to the presence of **angiotensin II,** which appears following the secretion of the enzyme **renin** by kidney cells exposed to a decline in blood volume and/or blood pressure.

anti-inflammatory effects

THE ADRENAL MEDULLA (pp. 587–588)

3. The adrenal medulla produces epinephrine (75–80 percent) and norepinephrine (20–24 percent.)

Endocrine Functions of the Kidneys and Heart (pp. 588–589)

1. Endocrine cells in the kidneys and heart produce hormones important for the regulation of blood volume and blood pressure. **Renin** converts **angiotensinogen** to **angiotensin I.** In the lungs this compound is converted to **angiotensin II**, the hormone that stimulates the adrenal production of aldosterone.
2. The kidney hormone, **erythropoietin (EPO),** stimulates red blood cell production by the bone marrow.
3. Specialized muscle cells in the heart produce **atrial natriuretic peptide (ANP)** when blood pressure or blood volume become excessive.

converting enzyme

Endocrine Tissues of the Digestive System (pp. 592–595)

1. The lining of the digestive tract, the liver, and the pancreas produce exocrine secretions that are essential to the normal breakdown and absorption of food.

THE PANCREAS (pp. 592–595)

2. The **pancreas** contains both exocrine and endocrine cells. The **exocrine pancreas** secretes an enzyme-rich fluid which travels to the lumen of the digestive tract. Cells of the **endocrine pancreas** form clusters called **pancreatic islets** (*islets of Langerhans*), containing **alpha cells** (which produce the hormone **glucagon**) and **beta cells** (which secrete **insulin).** Insulin lowers blood glucose by increasing the rate of glucose uptake and utilization; glucagon raises blood glucose by increasing the rates of glycogen breakdown and glucose manufacture in the liver.

somatostatin

Endocrine Tissues of the Reproductive System (pp. 596–597)

1. The **interstitial cells** of the male testis produce androgens. **Testosterone** is the most important male sexual hormone.
2. In women, eggs develop in **follicles;** follicle cells surrounding the eggs produce **estrogens.** After ovulation, the cells reorganize into a **corpus luteum** that releases a mixture of estrogens and **progesterone.** If pregnancy occurs, the placenta functions as an endocrine organ.

inhibin

Related Key Terms

The Pineal Gland (p. 597)

1. The pineal gland contains **pinealocytes** that synthesize **melatonin.** Melatonin inhibits the production of melanin; stimulates the production of MSH-IH and blocks the release of melanocyte-stimulating hormone (MSH); and slows the maturation of sperm, eggs, and reproductive organs by reducing the rate of GnRH secretion.

Patterns of Hormonal Interaction (pp. 598–601)

1. The endocrine system functions as an integrated unit, and hormones often interact. These interactions may exert several effects: (1) **antagonistic** (opposing) effects; (2) **synergistic** (additive) effects; (3) **permissive** effects, in which one hormone is necessary for another to produce its effect; or (4) **integrative** effects, in which hormones produce different but complementary results.

HORMONES AND GROWTH (p. 598)

2. Normal growth requires the cooperation of several endocrine organs. Five hormones are especially important: growth hormone, thyroid hormones, insulin, parathormone, and gonadal hormones.

HORMONES AND STRESS (pp. 598–600)

3. Any threat to homeostasis represents a **stress.** Our bodies respond to all types of stress with the **general adaptation syndrome (GAS).** The GAS can be divided into three phases: (1) the **alarm phase** (an immediate, "fight or flight" response to the stress, under the direction of the sympathetic division of the autonomic nervous system; (2) the **resistance phase** (glucocorticoids are the dominant hormones of this phase, which mobilizes lipid and protein reserves; elevates and stabilizes blood glucose concentrations; and conserves glucose for neural tissues); and finally, (3) the **exhaustion phase** (the eventual breakdown of homeostatic regulation and failure of one or more organ systems.)

HORMONES AND BEHAVIOR (p. 600)

4. Many hormones affect the functional state of the nervous system, producing changes in mood, emotional states, and various behaviors.

HORMONES AND AGING (p. 601)

5. The endocrine system shows relatively few functional changes with advanced age. The most dramatic endocrine change is the decline in the concentration of reproductive hormones.

ENDOCRINE DISORDERS (p. 601)

6. Endocrine disorders fall into two categories: they reflect either abnormal hormone production or abnormal cellular sensitivity.

Chapter Review

Review Planner

	Objective	Level 1
1	Compare the endocrine and nervous systems.	**8 12**
2	Compare the cellular components of the endocrine system with those of other tissues and systems.	**3 5 10 13 15 26**
3	Identify the endocrine organs and the hormones they produce.	**1 3 4 5 6 10 15**
4	Explain how hormones exert their effects, and identify the effects of various important hormones.	**2 4 5 9 14 16 18 19 24 27 28 31–43**
5	Relate the structure of hormones to their functions.	**7 17**
6	Discuss the results of abnormal hormone production.	**4 5 6**
7	Explain how hormones interact to produce coordinated physiological responses.	**14 16 18 19 21 24**
8	Describe how endocrine organs are controlled.	**16 30**
9	Identify five hormones that are especially important to normal growth.	**11 25**
10	Explain how the endocrine system responds to stress.	**9 22**

Level 2	**20 23 29 30**
Level 3	**44–49**
CM	**Thyroid Gland Disorders Disorders of Parathyroid Function Disorders of the Adrenal Cortex Disorders of the Adrenal Medulla**
ABCNEWS	**Use of Growth Hormones in Children • Steroids**

Review Questions

MULTIPLE CHOICE

1. The hypophysis is the: a) thyroid gland; b) hypothalamus; c) pituitary gland; d) adrenal gland.
2. The hormone which stimulates uterine muscles to contract is: a) oxytocin; b) ADH; c) MSH; d) PTH.
3. Chief cells are found in the ________ and produce ________: a) parathyroid/calcitonin; b) adrenal gland/parathormone; c) pituitary gland/ ACTH; d) parathyroid/parathormone.
4. Removing the adrenal glands would: a) be lethal in the absence of hormone therapy; b) have no effects; c) eliminate the production of renin; d) promote the resistance phase of the GAS.
5. Damage to the beta cells of the pancreatic islets would: a) reduce the production of glucagon; b) decrease the production of insulin; c) reduce the level of glucose concentrations in the blood; d) none of the above.
6. Removing the thyroid gland would affect the production of: a) TRH; b) TSH; c) T_4; d) thymosin.
7. Which of the following is *not* one of the catecholamines?: a) dopamine; b) norepinephrine; c) epinephrine; d) acetylcholine.
8. Two anatomical structures closely related to both the endocrine and nervous systems are the: a) hypophysis and adrenal medulla; b) adrenal cortex and adrenal medulla; c) hypophysis and pancreas; d) adrenal cortex and pancreas.
9. The physiological responses of the GAS alarm phase are related to those produced by activating the: a) parasympathetic nervous system; b) sympathetic nervous system; c) somatic nervous system.
10. The C cells produce: a) calcitonin; b) calcium; c) chief cells; d) parathormone.
11. A hormone that does *not* play a major role in growth is: a) parathormone; b) thyroid hormone; c) oxytocin; d) testosterone.
12. Which of the following is *false*? The endocrine system differs from the nervous system in that its effects are usually: a) more powerful; b) longer lasting; c) less localized; d) slower to appear.

MATCHING QUESTIONS

(Match each term with the most appropriate statement.)

Statements:

13. Blood vessels that link two capillary networks.
14. A hormone whose effect depends on activation of an intracellular intermediary.
15. Within the ovaries, eggs develop in these specialized structures.
16. A hypothalamic hormone that stimulates production of hormones at the anterior pituitary.
17. Most hormones fall into this group, which consists of chains of amino acids.
18. Steroid hormones that affect glucose metabolism.
19. A hypothalamic hormone that prevents the synthesis and secretion of hormones at the anterior pituitary.

Terms:

a) Peptide hormones
b) First messenger
c) Inhibiting hormone
d) Releasing hormone
e) Portal vessels
f) Follicles
g) Glucocorticoids

SHORT ANSWER/ESSAY

20. How do the nervous and endocrine systems cooperate to maintain homeostasis?
21. Discuss four ways in which hormones can interact.
22. Identify the phases of the general adaptation syndrome that are characterized by the following responses. If the phases are out of sequence, put them in the correct sequence:
    a) Lipid and protein reserves are mobilized; blood glucose concentrations rise and stabilize; glucose is conserved for neural tissues.
    b) Increased mental alertness and energy consumption by skeletal muscles; increased heart rate, blood pressure, and respiratory rate; decreased blood flow to the kidneys and digestive organs.
    c) Malfunctioning of nerve and muscle cells; adrenal cortex may stop producing glucocorticoids.
23. You apply a steroid cream to a bothersome case of poison ivy, and the itching decreases. Explain why this happens. Should you also rub a steroid cream on a cut if it becomes infected?
24. Fill in the blanks: Glucagon and ________ regulate ________ concentrations in the blood, in the same way that ________ and parathormone regulate ________ concentrations in the blood.
25. Describe how inadequate levels of the following could affect a child's growth:
    a) Growth hormone
    b) Thyroid hormone
    c) Parathormone
    d) Gonadal hormones
    e) Insulin
26. What are fenestrated vessels? What is their function?
27. Explain the link between insulin and the body's energy reserves.
28. What is the calorigenic effect, and how is it helpful? What hormones are responsible?
29. How does aging affect the endocrine system?
30. What role does the hypothalamus play in the endocrine system? What are the mechanisms involved in this role?

Identify the hormones that are responsible for the following events:

31. Induces ovulation in women; promotes ovarian secretion of progesterone.
32. Causes the retention of sodium ions and water, thus reducing fluid losses in the urine.
33. Stimulates the mammary glands to produce milk.
34. Stimulates the bone marrow to produce red blood cells.
35. Stimulates osteoblasts, inhibits osteoclasts, and stimulates the kidneys to secrete calcium ions.
36. Released by the pars intermedia; causes darkening of the skin.
37. Triggers the release of thyroid hormones.
38. Blocks the release of MSH and stimulates the production of MSH-IH.
39. Plays a key role in developing and maintaining normal immunological responses.
40. Triggers the secretion of FSH and LH.
41. Promotes the production of functional sperm, maintains the secretory glands of the male reproductive tract, and determines secondary sexual characteristics.
42. Decreases the amount of water lost at the kidneys; in high concentrations can cause the constriction of peripheral blood vessels.
43. Stimulates cell growth and replication by accelerating the rate of protein synthesis.

CRITICAL THINKING/APPLICATIONS

44. Rebecca M. has just been diagnosed as having diabetes mellitus. Her physician tells her to be sure to have her eyes examined regularly, and to check her feet regularly. Explain the reasons for this advice.
45. Sixteen-year-old John is a promising athlete who is below the average height for his age. He wants to play football professionally, but is convinced that he needs to be taller and stronger in order to accomplish his dream. He persuades his parents to visit a doctor and inquire about GH treatments. Should the physician provide these treatments? Explain your answer.
46. Mr. H. has been experiencing severe, sharp pain in his lower back. Other symptoms include depression and frequent nausea. Recently he fractured a bone in his foot when he slipped while walking from his car to the front door. What is your diagnosis? Which gland is involved? Explain your answer.
47. Roger M. has been suffering from extreme thirst; he drinks numerous glasses of water every day and urinates a great deal. Name two disorders that could produce these symptoms. What test could a clinician perform to determine which disorder is present?
48. Some cities of the northwestern United States that tend to have many cloudy days have been shown to have suicide rates that are higher than average. Describe a possible reason for this phenomenon. What hormone is involved?
49. Suppose you wanted to be a professional athlete, and improving your athletic performance was an important goal for you. Would you consider using anabolic steroids? Explain your answer.

Do you think that this diver is swimming through a hostile, alien environment? Well, yes and no. While we cannot function in a watery environment, lots of organisms can. More to the point, we ourselves aren't nearly as dry as we tend to think. Almost every physiological process that keeps us alive is performed in a watery environment: the interior of our cells, our tissue fluids—and our bloodstream. It has been said that, while our remote ancestors left the ocean many millions of years ago, we still carry our ocean around inside us. In this chapter we'll learn about one of the major components of this internal sea—the blood that keeps all of our cells supplied with the oxygen and nutrients that smaller, simpler creatures can obtain directly from the waters around them.

CHAPTER 19

The Cardiovascular System: The Blood

Chapter Objectives

After reading this chapter, you will be able to:

1 Describe the important components and major functions of blood.
2 Discuss the composition and functions of plasma.
3 List the characteristics and functions of red blood cells.
4 Identify the locations where the components of blood are produced and discuss the factors that regulate their production.
5 Explain what determines a person's blood type and why blood types are important.
6 Categorize the various white blood cells on the basis of their structures and functions.
7 Describe how white blood cells fight infection.
8 Explain the mechanisms that control blood loss after an injury.
9 Describe the life cycles of blood cells.

Introduction

The living body is in constant chemical communication with its external environment. Nutrients are absorbed through the lining of the digestive tract, gases move across the delicate epithelium of the lungs, and wastes are excreted in the feces and urine. These chemical exchanges occur at specialized sites in the body. Yet because of all parts of the body are linked by a transport network, the *cardiovascular system*, they may affect every cell, tissue, and organ in a matter of moments.

Diffusion across exposed surfaces can meet the demands of small embryos. By the time the embryo has reached a few millimeters in length, however, developing tissues will consume oxygen and nutrients and generate waste products faster than they can be provided or removed by simple diffusion. If growth is to continue, the cardiovascular system must provide a mechanism for the rapid transport of nutrients, waste products, and cells within the body. The heart begins beating by the end of the third week of embryonic life, when most other organ systems have barely started their development. Once this internal distribution network is functioning, the embryo can make more efficient use of the nutrients obtained from the maternal bloodstream, and the embryo doubles its size in the next week.

The cardiovascular system can be compared to the cooling system of a car. The basic components include a circulating fluid (blood), a pump (the heart), and an assortment of conducting pipes (the circulatory system). Although the cardiovascular system is far more complicated and versatile, both mechanical and biological systems can malfunction due to similar causes: fluid losses, pump failures, or damaged pipes. The first three chapters of this unit examine these components individually. Chapter 19 discusses the nature of the circulating blood, Chapter 20 considers the structure and function of the heart, and Chapter 21 examines the circulatory network and the integrated functioning of the cardiovascular system.

The cardiovascular system is intimately bound up with the *lymphatic system*. The two systems are physically integrated–vessels of the lymphatic system empty their contents into the bloodstream, and cell populations of the lymphatic system travel through the circulatory system to reach all parts of the body. These cells, along with the associated tissues and organs, are part of the body's system of defense against the many threats to which it is constantly exposed. Dangerous organisms, foreign substances, and chemical poisons are common in the environment, and their entry into living tissues can threaten homeostasis and cause illness or even death. Chapter 22 details the components of the lymphatic system and considers the array of defenses that the body deploys to protect itself from internal and external hazards.

■ Functions and Composition of the Blood

This chapter examines the structure and functions of blood, a specialized connective tissue introduced in Chapter 5. ∞ (p. 146) The circulating blood provides nutrients, oxygen, chemical instructions, and a mechanism for waste removal to each of the roughly 75 trillion individual cells in the human body. The blood also transports specialized cells that defend peripheral tissues from infection and disease. These services are absolutely essential, and a region completely deprived of circulation may die in a matter of minutes. The functions of the blood include the following:

1. Blood *transports dissolved gases*, bringing oxygen from the lungs to the tissues and carrying carbon dioxide from the tissues to the lungs.
2. Blood *distributes nutrients* absorbed at the digestive tract or released from storage in adipose tissue or the liver.
3. Blood *transports metabolic wastes* from peripheral tissues to sites of excretion, such as the kidneys.
4. Blood *delivers enzymes and hormones* to specific target tissues.
5. Blood *regulates the pH and electrolyte composition of interstitial fluids* throughout the body. An extensive array of buffers enables the bloodstream to deal with the acids generated by active tissues, such as the lactic acid produced by skeletal muscles.
6. Blood *restricts fluid losses* through damaged vessels or at other injury sites. The **clotting reaction** seals the breaks in the vessel walls, preventing changes in blood volume that could seriously affect blood pressure and cardiovascular function.
7. Blood *defends the body against toxins and pathogens*. It transports *white blood cells*, specialized cells that migrate into peripheral tissues to fight infections or remove debris, and delivers *antibodies*, special proteins that attack invading organisms or foreign compounds. The blood also receives toxins produced by infection, physical damage, or metabolic activity and delivers them to the liver and kidneys, where they can be inactivated or excreted.
8. Blood *helps regulate body temperature* by absorbing and redistributing heat. Blood is nearly 50 percent water, and as we saw in Chapter 2, water has an extraordinarily high heat capacity. ∞ (p. 38) In a car, the heat generated by the engine is absorbed by cooling water, pumped to the radiator, and released into the environment. In the body, active skeletal muscles and other tissues generate heat, and the bloodstream carries it away. When body temperature is already high, that heat is lost across the surface of the skin. When body temperature is too low, warm blood is directed to the most temperature-sensitive organs.

Blood is normally confined to the circulatory system, and it has a characteristic and unique composition (Figure 19-1).

- **Plasma** (PLAZ-mah), the ground substance of blood, has a density only slightly greater than that of water. It contains dissolved proteins rather than the network of insoluble fibers found in loose connective tissue or cartilage.
- **Formed elements** are blood cells and cell fragments that are suspended in the plasma. These elements are present in great abundance and are highly specialized. **Red blood cells (RBCs)** transport oxygen and carbon dioxide. The less numerous **white blood cells (WBCs)** are components of the immune system. **Platelets** are small

packets of cytoplasm that contain enzymes and factors important to the blood clotting process.

Together the plasma and formed elements constitute **whole blood**. These components may be separated, or **fractionated**, for analytical or clinical purposes.

Interactions between the dissolved proteins and the surrounding water molecules make plasma relatively sticky, cohesive, and resistant to flow. These characteristics determine the **viscosity** of a solution. Solutions are usually compared with pure water, which has a viscosity of 1.0. Gasoline has a very low viscosity (0.4); solutions such as antifreeze (20) or molasses (300+) have relatively high viscosities. Plasma has a viscosity of 1.5, 50 percent more than pure water. The viscosity of whole blood is much greater–about 5.0–because of interactions between water molecules and the formed elements.

Blood volume in liters can be estimated quite easily, by calculating 7 percent of the body weight in kilograms. For example, a 75-kg individual would have a blood volume of approximately 5.25 liters of blood ($75 \times 0.07 = 5.25$). More precise determinations can be obtained using special dyes or radioisotopes, such as phosphorus-32. There are 5–6 liters of whole blood in the cardiovascular system of an adult man, and 4–5 liters in an adult woman. Clinicians use the terms **hypovolemic** (hī-pō-vo-LĒ-mik), **normovolemic** (nōr-mō-vo-LĒ-mik), and **hypervolemic** (hī-per-vo-LĒ-mik) to refer to low, normal, or excessive blood volumes, respectively. These conditions produce characteristic symptoms because variations in blood volume affect other components of the cardiovascular system. For example, an abnormally large blood volume can place a severe stress on the heart, which must push the extra fluid around the circulatory system.

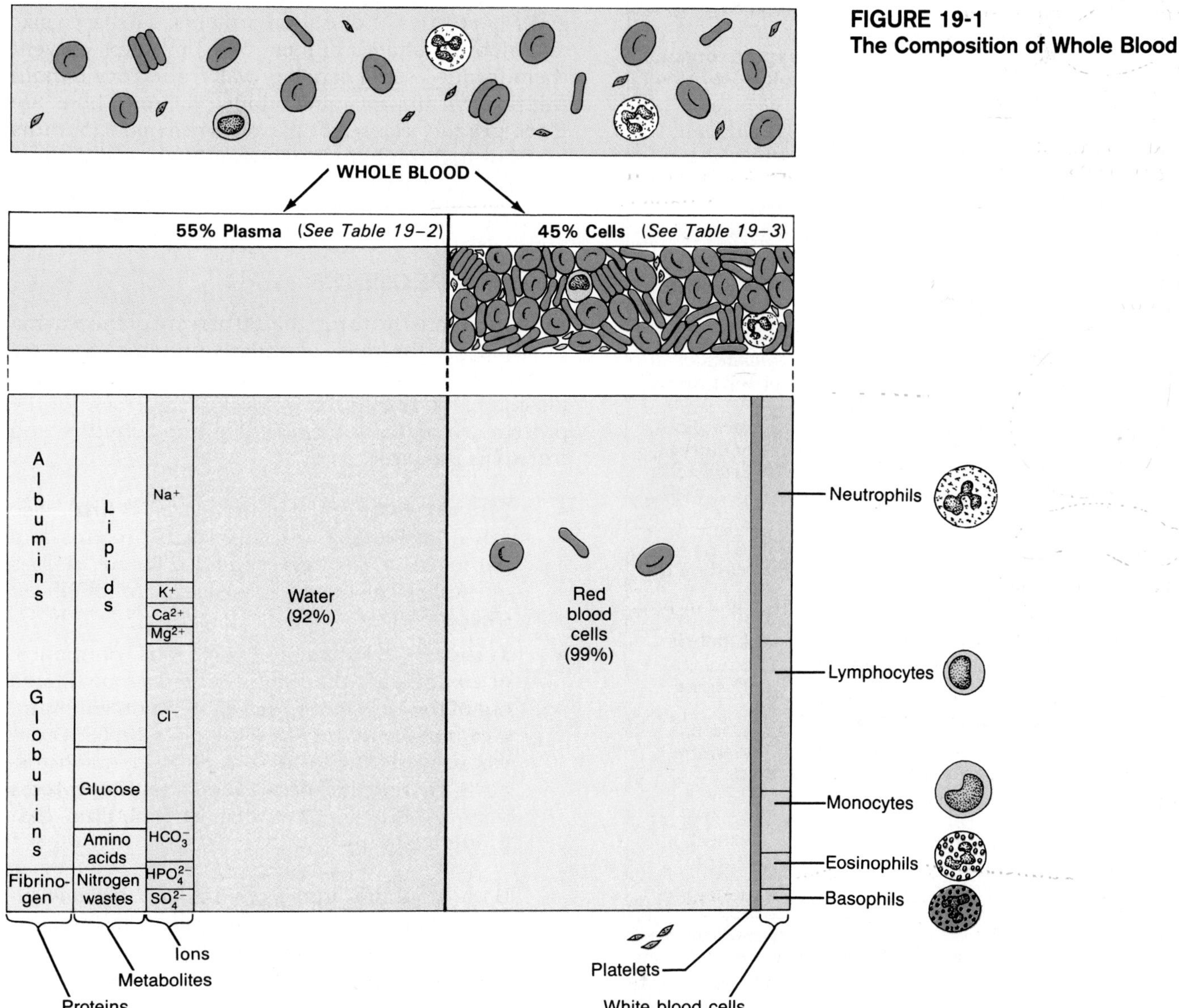

FIGURE 19-1
The Composition of Whole Blood

Plasma

Figure 19-1 and Table 19-1 summarize information concerning the composition of whole blood. Plasma contributes approximately 55 percent of the volume of whole blood, and water accounts for 92 percent of the plasma volume. Table 19-2 contains a more detailed analysis of plasma, interstitial fluid, and cytoplasm. The concentrations reported are average values, and the actual concentrations vary depending on the area sampled and the ongoing activity within that particular region.

TABLE 19-1 Composition of Whole Blood

Component	*Significance*
PLASMA	
Water	Dissolves and transports organic and inorganic molecules, blood cells, and heat
Electrolytes	Normal extracellular fluid ion composition essential for vital cellular activities
Nutrients	Used for energy production, growth, and maintenance of cells
Organic wastes	Travel to sites of breakdown or excretion
Proteins	
Albumins	Major contributor to osmotic concentration of plasma; transport lipids
Globulins	Transport ions, hormones, lipids
Fibrinogen	Essential component of clotting system; can be converted to insoluble fibrin
FORMED ELEMENTS	
Red blood cells	Transport oxygen and carbon dioxide
White blood cells	
Neutrophils	Phagocytic cells; engulf debris and pathogens
Eosinophils	Phagocytic cells; engulf items coated in antibodies
Basophils	Stimulate inflammation in tissues by releasing histamine
Monocytes	Phagocytic cells; engulf debris and pathogens
Lymphocytes	Immune defense against specific pathogens, toxins, or foreign proteins
Platelets	Participate in clotting response

Note: Detailed information about individual components can be found in Tables 19-2 and 19-3.

DIFFERENCES BETWEEN PLASMA AND INTERSTITIAL FLUID

In many respects, the composition of the plasma resembles that of tissue fluid. The concentrations of the major plasma ions, for example, are similar to those of interstitial fluid and differ markedly from those found inside living cells. The chief differences between plasma and tissue fluid involve the concentrations of dissolved gases and proteins.

Because the dissolved oxygen concentration of plasma is higher than that of interstitial fluid, oxygen diffuses out of the bloodstream and into peripheral tissues. Carbon dioxide concentrations in the tissues are much higher than those in plasma, so CO_2 diffuses in the opposite direction. Diffusion cannot eliminate these concentration differences, however, because (1) the plasma is in constant motion, carrying carbon dioxide away and delivering more oxygen, and (2) the tissue cells continue to absorb oxygen and release additional carbon dioxide.

A second significant difference between plasma and interstitial fluid is that the plasma contains great quantities of dissolved proteins. The large size and globular shapes of most blood proteins prevent them from crossing capillary walls, and they remain trapped within the circulatory system. There are three primary classes of plasma proteins, *albumins* (al-BŪ-mins), *globulins* (GLOB-ū-lins), and *fibrinogen* (fī-BRIN-ō-jen).

PLASMA PROTEINS

Albumins constitute roughly 60 percent of the plasma proteins. As the most abundant proteins, they are major contributors to the osmotic pressure of the plasma. **Globulins**, accounting for 33 percent of the protein population, include *immunoglobulins* and *transport proteins*.

- **Immunoglobulins** (i-mū-nō-GLOB-ū-linz), also called **antibodies**, attack foreign proteins and pathogens. There are several different classes of immunoglobulins, discussed in detail in Chapter 22.
- **Transport proteins** bind small ions, hormones, or compounds that might otherwise be filtered out of the blood at the kidneys. Important examples include thyroid-binding globulin (TBG, introduced in Chapter 18), which binds and transports thyroid hormones, and **metalloproteins** (met-al-ō-PRŌ-tēnz), transport globulins that bind metal ions. (p. 583)

Both albumins and globulins can become attached to lipids, such as triglycerides, fatty acids, or cholesterol, that are not water-soluble. The protein-lipid combination readily dissolves in plasma, and

TABLE 19-2 The Chemical Composition of Body Fluids

	Plasma[a] *(Average)*	*Interstitial Fluid (Average)*	*Cytoplasm (Skeletal Muscle)*
Water (%)	92	94	66
pH	7.4	7.4	7.0
Dissolved gases (pressure, in mm Hg)			
Oxygen	90 (arterial)	35	20
Carbon dioxide	39 (arterial)	46	50
Electrolytes (mEq/ℓ)			
Na^+ (sodium)	145	145	10
K^+ (potassium)	4	4	160
Mg^{2+} (magnesium)	2	2	26
Ca^{2+} (calcium)	5	5	2
Cl^- (chloride)	102	114	3
HCO_3^- (bicarbonate)	20	31	10
Metabolites (mg/dℓ)			
Carbohydrates	90	90	0–20
Lipids (g/dℓ)	0.6	0.6	2–95
Amino acids	61	30	200
Organic wastes	39		
Dissolved proteins (g/dℓ)			
Total proteins	7.6	1.8	16
Albumins	4.6		
Globulins	2.7		
Fibrinogen	0.3		

[a] Normal ranges for plasma composition are included in Appendix V.

in this way the circulatory system transports insoluble lipids to peripheral tissues. Globulins involved in lipid transport are called **lipoproteins** (lī-pō-PRŌ-tēnz).

The third type of protein, **fibrinogen**, functions in the clotting reaction. Under certain conditions fibrinogen molecules interact, forming large, insoluble strands of **fibrin** (FĪ-brin). These fibers provide the basic framework for a blood clot. If steps are not taken to prevent clotting, the conversion of fibrinogen to fibrin will occur in a sample of plasma. This conversion removes the clotting proteins, leaving a fluid known as **serum**.

The liver synthesizes and releases more than 90 percent of the plasma proteins. This includes all of the albumin and fibrinogen and most of the globulins in plasma. However, the immunoglobulins and protein hormones are synthesized elsewhere. Immunoglobulins (antibodies) are produced by *plasma cells*, cells of the immune system that were introduced in Chapter 5. (p. 139) Protein hormones, including thyroid-stimulating hormone (TSH), follicle-stimulating hormone (FSH), luteinizing hormone (LH), and prolactin (PRL), are produced by endocrine organs. Because the liver is the primary source of plasma proteins, liver disorders can alter the composition and functional properties of the blood. For example, some forms of liver disease can lead to uncontrolled bleeding, caused by inadequate synthesis of fibrinogen and other plasma proteins involved in the clotting response.

Formed Elements

The major cellular components of blood, red blood cells and white blood cells, were introduced in Chapter 5 and earlier in this chapter. (pp. 146–147) There are two major classes of white blood cells, *granular* (with granules) and *agranular* (without granules). In addition, blood contains the noncellular formed elements called *platelets*, small packets of cytoplasm that function in the clotting response. Table 19-3 summarizes information concerning the formed elements of the blood. The sections that follow will consider each of these components individually.

RED BLOOD CELLS

Red blood cells, or **erythrocytes** (e-RITH-rō-sītz; *erythros*, red) account for slightly less than half of the

TABLE 19-3 A Review of the Formed Elements of the Blood

Cell	*Abundance (per µℓ)*	*Characteristics*	*Functions*	*Remarks*
RED BLOOD CELLS	5.2 million (range: 4.4–6.0 million)	Flattened, circular, no nucleus, mitochondria, or ribosomes; red color due to presence of hemoglobin molecules	Transport oxygen from lungs to tissues, and carbon dioxide from tissues to lungs	120-day life expectancy; amino acids and iron recycled; produced in bone marrow
WHITE BLOOD CELLS	7000 (range: 6000–9000)			
Granulocytes				
Neutrophils	4150 (range: 1800–7300) differential count: 57%	Round, nucleus resembles a series of beads; cytoplasm contains large, pale inclusions	Phagocytic; engulf pathogens or debris in tissues	Survive minutes to days, depending on activity; produced in bone marrow
Eosinophils	165 (range: 0–700) differential count: 2.4%	Round, nucleus usually in two lobes; cytoplasm contains large granules that are usually stained bright red	Phagocytic: engulf anything in tissues that is labeled with antibodies	Produced in bone marrow
Basophils	44 (range: 0–150) differential count: 0.6%	Round, nucleus usually cannot be seen due to dense, blue granules in cytoplasm	Enter damaged tissues and release histamine and other chemicals	Assist mast cells of tissues in producing inflammation; produced in bone marrow
Agranulocytes				
Monocytes	456 (range: 200–950) differential count: 6.5%	Very large, lima bean-shaped nucleus, abundant pale cytoplasm	Enter tissues to become free macrophages; engulf pathogens or debris	Primarily produced in bone marrow
Lymphocytes	2185 (range: 1500–4000) differential count: 31%	Slightly larger than RBC, round nucleus, very little cytoplasm	Cells of lymphatic system, providing defense against specific pathogens or toxins	T cells attack directly; B cells form plasma cells that attack with antibodies; produced in bone marrow and lymphatic tissues
Platelets	350,000 (range: 150,000–500,000)	Cytoplasmic fragments, contain enzymes and proenzymes; no nucleus	Hemostasis: clump together and stick to vessel wall (platelet phase); activate intrinsic pathway of coagulation phase	Produced by megakaryocytes in bone marrow

[a] Differential count: percentage of circulating white blood cells.

total blood volume. The **hematocrit** (hē-MA-tō-krit) value indicates the percentage of whole blood occupied by cellular elements. The normal hematocrit in adult men averages 46 (range: 40–54); the average for adult women is 42 (range: 37–47). Because whole blood contains roughly 1000 red blood cells for each white blood cell, the hematocrit closely approximates the volume of erythrocytes. As a result, hematocrit values are often reported as the **volume of packed red cells** (**VPRC**) or simply the **packed cell volume** (**PCV**).

The number of erythrocytes in the blood of a normal individual staggers the imagination. A standard blood test checks the number per **microliter** ($\mu\ell$) of whole blood. One microliter, or cubic millimeter (mm^3), of whole blood from a man contains roughly 5.4 million erythrocytes; a microliter of blood from a woman contains about 4.8 million erythrocytes. There are approximately 260 million red blood cells in a single drop of whole blood, and 25 trillion (2.5×10^{13}) RBCs in the blood of an average adult. The number of erythrocytes thus approximates that of all other cells combined!

Clinical Comment: Anemia and Polycythemia

Anemia (a-NĒ-mē-ah) exists when the oxygen-carrying capacity of the blood is reduced, diminishing the delivery of oxygen to peripheral tissues. Such a reduction causes a variety of symptoms, including premature muscle fatigue, weakness, lethargy, and a general lack of energy. Anemia may exist because the hematocrit is abnormally low or because the amount of hemoglobin in the RBCs is reduced. Standard laboratory tests can be used to differentiate between the various forms of anemia on the basis of the number, size, shape, and hemoglobin content of red blood cells; these procedures are detailed in the Diagnostics box on page 616.

An elevated hematocrit with a normal blood volume constitutes **polycythemia** (po-lē-sī-THĒ-mē-ah). There are several different types of polycythemia. *Erythrocytosis* (e-rith-rō-sī-TŌ-sis), a polycythemia affecting only red blood cells, will be considered later in the chapter. **Polycythemia vera** ("true polycythemia") results from an increase in the numbers of all blood cells. The hematocrit may reach 80–90, at which point the tissues become oxygen-starved because red blood cells are blocking the smaller vessels. This condition seldom strikes young people; most cases involve patients age 60–80. There are several treatment options, but none cures the condition. The cause of polycythemia vera is unknown, although there is some evidence that the condition is linked to radiation exposure.

Structure of RBCs

Erythrocytes transport oxygen and carbon dioxide within the bloodstream. They are among the most specialized cells of the body, and their specializations are apparent when RBCs are compared with "ordinary" cells. Figure 19-2 indicates the significant differences seen using light and electron microscopy. Each red blood cell has a thin central region and a thick outer margin. A typical erythrocyte has a diameter of 7.7 μm and a maximum thickness of roughly 2.6 μm, although the center narrows to about 0.8 μm.

This unusual shape gives each RBC a relatively large surface area, allowing rapid diffusion between the cytoplasm and surrounding plasma. The total surface area of the red blood cells in the blood of a typical adult is roughly 3800 square meters, 2000 times the total surface area of the body. The flattened shape also enables them to form stacks, like dinner plates. These stacks, called **rouleaux** (roo-LŌ; "little rolls"), form and dissociate repeatedly without affecting the cells involved. An entire rouleau can pass along a blood vessel little larger than the diameter of a single erythrocyte, whereas individual cells would bump the walls, bang together, and form log jams that could block the circulatory passageway. Finally, the slender profile of an erythrocyte gives the cell considerable flexibility, and erythrocytes can bend and flex with apparent ease. By changing their shape, individual red blood cells can squeeze through capillaries as narrow as 4 μm.

RBC Longevity and Circulation

During their differentiation and maturation, red blood cells lose most of their organelles; circulating RBCs lack mitochondria, ribosomes, and nuclei. Without mitochondria, these cells can obtain energy only through anaerobic glycolysis, and they rely on glucose obtained from the surrounding plasma. This characteristic makes RBCs relatively inefficient from an energy standpoint, but it ensures that absorbed oxygen will be carried to peripheral tissues, not "stolen" by mitochondria in the cell. Without a nucleus or ribosomes, protein synthesis cannot occur, so as cellular enzymes and structural proteins age and deteriorate, they cannot be replaced.

An erythrocyte is exposed to severe stresses. A single circuit of the circulatory system usually takes

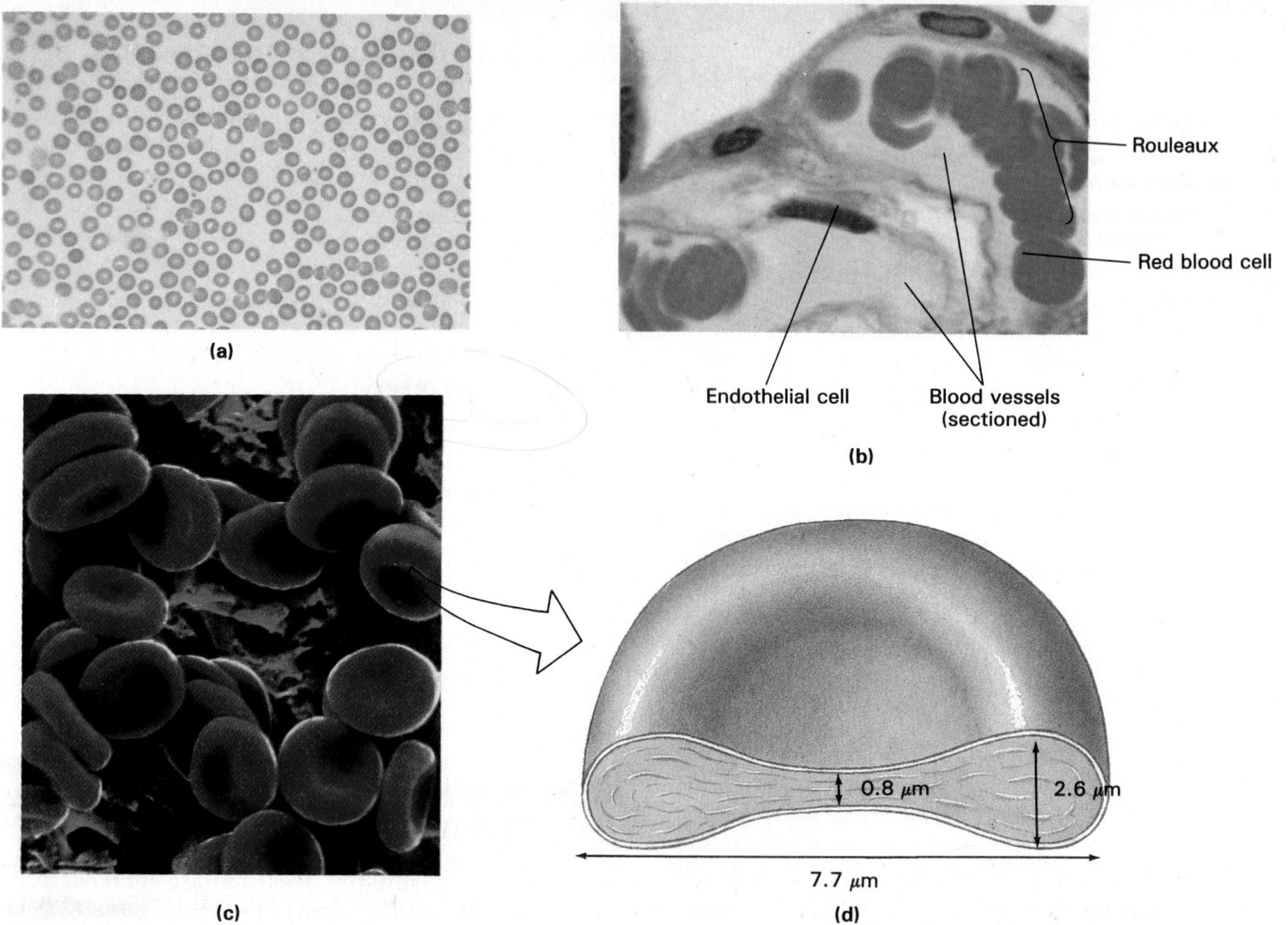

FIGURE 19-2
Anatomy of Red Blood Cells. (a) When viewed in a standard blood smear, red blood cells appear as two-dimensional objects because they are flattened against the surface of the slide. (LM, × 318) (b) When traveling through relatively narrow capillaries, erythrocytes may stack like dinner plates, forming rouleaux. (LM, × 2000) (c) A scanning electron micrograph of red blood cells reveals their three-dimensional structure quite clearly. (× 1250) (d) A sectional view of a red blood cell.

less than 30 seconds. In that time a red blood cell gets pumped out of the heart and forced along vessels where it bounces off the walls and collides with other red cells. It stacks in rouleaux, contorts and squeezes through tiny capillaries, and then joins its comrades in a headlong rush back to the heart for another round. With all this wear and tear and no repair mechanisms, a typical red blood cell has a relatively short lifespan. After traveling about 700 miles in 120 days, it either breaks down or is destroyed by phagocytic cells. The continual elimination of red blood cells is usually unnoticed because new erythrocytes are entering the circulation at a comparable rate. About 1 percent of the circulating erythrocytes are replaced each day, and in the process approximately 3 million new erythrocytes enter the circulation *each second*!

RBCs and Hemoglobin

In effect, a developing erythrocyte loses any intracellular component not directly associated with its primary function—the transport of respiratory gases. A mature red blood cell consists of a cell membrane surrounding a compact mass of transport proteins. The cytoplasm contains water (66 percent) and proteins (about 33 percent). Molecules of **hemoglobin** (HĒ-mō-glō-bin) (**Hb**) account for over 95 percent of the erythrocyte's proteins and give the cell its red color. Hemoglobin is responsible for the cell's ability to transport oxygen and carbon dioxide. Most of the other intracellular proteins are enzymes involved with anaerobic energy production and the maintenance of the hemoglobin molecules.

The Hemoglobin Molecule and Gas Transport Four globular protein subunits combine to form a single molecule of hemoglobin. Each of the subunits contains a single molecule of **heme**, a complex structure shown in Figure 19-3. Heme is an example of a **porphyrin**, a special organic compound usually associated with metal ions. Each heme unit holds an iron ion in such a way that it can interact with an oxygen molecule. The iron-oxygen interaction is very weak, and the two can be easily separated without damage to either the hemoglobin or the oxygen molecule.

There are approximately 280 million molecules of hemoglobin in each red blood cell, and because a hemoglobin molecule contains four heme units, each erythrocyte can potentially carry more than a billion molecules of oxygen. The actual amount of oxygen bound in each erythrocyte depends on the conditions in the surrounding plasma. When oxygen is abundant in the plasma, the hemoglobin molecules gain oxygen

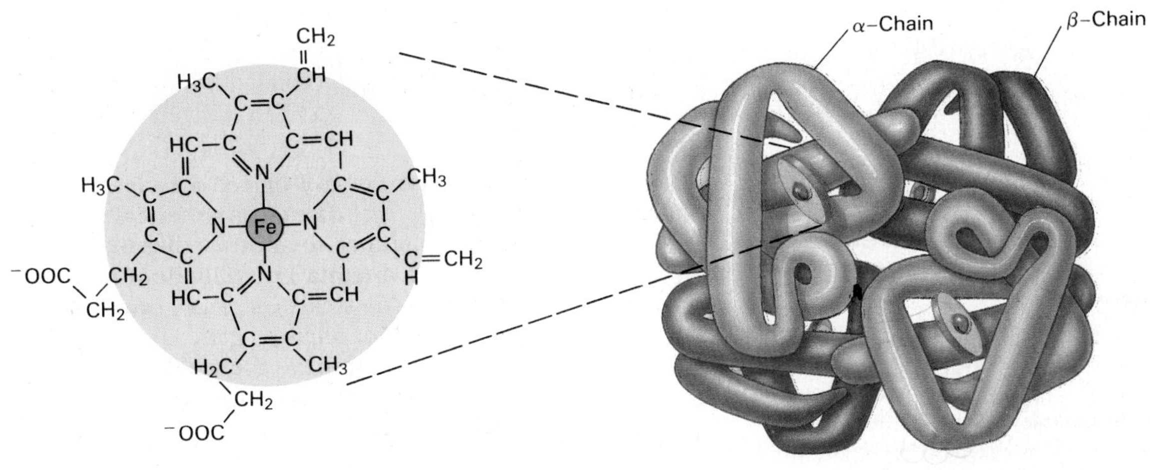

FIGURE 19-3
The Structure of Hemoglobin. Hemoglobin consists of four globular protein subunits. Each subunit contains a single molecule of heme, a porphyrin ring surrounding a single ion of iron.

until all of the heme molecules are occupied. As the plasma oxygen levels decline, the hemoglobin molecules release their oxygen reserves.

When plasma oxygen concentrations are falling, plasma carbon dioxide levels are usually rising. Under these conditions the globin portion of each hemoglobin molecule begins to bind carbon dioxide molecules (a process discussed in more detail in a later section). The binding of carbon dioxide to the globin portion of the molecule is just as reversible as the binding of oxygen to heme, and the CO_2 will be released when plasma carbon dioxide concentrations decline.

As red blood cells circulate, they are exposed to varying combinations of oxygen and carbon dioxide concentrations. At the lungs, diffusion brings oxygen into the plasma and removes carbon dioxide. The hemoglobin molecules in red blood cells respond by absorbing oxygen and releasing carbon dioxide. In the peripheral tissues, the situation is reversed because active cells are consuming oxygen and producing carbon dioxide. As blood flows through these areas oxygen diffuses out of the plasma, and carbon dioxide diffuses in. Under these conditions hemoglobin releases its stored oxygen and binds carbon dioxide.

Hemoglobin Conservation and Recycling Hemoglobin works efficiently only when packaged in red blood cells, in company with essential enzymes. If a damaged or aged erythrocyte ruptures, the hemoglobin does not remain in solution for very long. It breaks down, and the individual subunits are small enough to pass through the filtration mechanism at the kidneys and be lost in the urine. When abnormally large numbers of erythrocytes are breaking down in the circulation, the urine may develop a reddish or brown coloration. This condition, called **hematuria** (hēm-a-TŪ-rē-ah) or **hemoglobinuria**, is usually prevented by specialized mechanisms that reclaim and recycle hemoglobin. Only about 10 percent of the red cells survive long enough to rupture, or **hemolyze** (HĒ-mō-liz), within the circulation. Phagocytic cells of the liver, spleen, and bone marrow monitor the condition of circulating erythrocytes, and they usually recognize and engulf erythrocytes *before* they hemolyze. These phagocytes also remove hemoglobin and red blood cell fragments from the circulation. (Phagocytosis and phagocytic cells were introduced in Chapter 3; additional details will be found in Chapter 22.) (p. 86)

Once a red blood cell has been engulfed and broken down by a phagocytic cell, each component of the hemoglobin molecule has a different fate (Figure 19-5):

1. The globular proteins are disassembled into their component amino acids. These amino acids are either metabolized by the cell or released into the circulation for use by other cells.
2. The heme is stripped of its iron and converted to **bilirubin** (bil-ē-ROO-bin), a porphyrin derivative. The phagocytic cells then release the bilirubin into the circulation. Liver cells absorb the bilirubin and use it to synthesize a complex molecule called **conjugated bilirubin**, or *bilirubin diglucuronide*. Most of the conjugated bilirubin is excreted in the bile. A small amount reenters the circulation; under normal circumstances it will be removed at the kidneys and eliminated in the urine. Urinary concentrations of conjugated bilirubin are very low, and unconjugated bilirubin does not normally enter the urine at all. [**CM**: *Bilirubin Tests and Jaundice*]
3. Iron extracted from the heme molecules may be stored within the phagocytic cell or released

Clinical Comment: Sickle Cell Anemia

Sickle cell anemia affects roughly 0.14 percent of the black population in the United States. This condition results from a change in the amino acid sequence in one of the globin chains of the hemoglobin molecule. When the blood contains an abundance of oxygen, the hemoglobin molecules and the erythrocytes that carry them appear normal. But when the defective hemoglobin gives up enough of its stored oxygen, the complex of hemoglobin molecules within the red blood cells changes shape, and the cells become stiff and markedly curved (Figure 19-4).

This "sickling" does not affect the oxygen-carrying capabilities of the erythrocytes, but it causes them to become more fragile and easily damaged. Moreover, an RBC that has folded to squeeze into a narrow capillary may lose its oxygen, change shape, and become stuck. A circulatory blockage results, and the nearby tissues become oxygen-starved. Symptoms of this condition include pain and damage to a variety of organs and systems, depending on the location of the obstructions. In addition, the trapped red blood cells eventually die and break down, producing a characteristic anemia. Transfusions of normal blood can temporarily prevent additional complications, and there are experimental drugs that can control or reduce sickling.

Interestingly, this disease is genetically associated with resistance to another disease. To develop sickle cell anemia an individual must have two copies of the sickle cell gene, one from each parent. If only one sickling gene is present, the individual is said to have the **sickling trait**. In such cases most of the hemoglobin is of the normal form, and the erythrocytes function normally. But the presence of the abnormal hemoglobin somehow gives the individual the ability to resist the parasitic infections that produce the symptoms of **malaria**. The malaria parasites enter the bloodstream when an individual is bitten by an infected mosquito. In a normal person the microorganisms then invade and reproduce within the erythrocytes. But when they enter an erythrocyte from a person with the sickling trait, the cell responds by sickling. Either the sickling itself kills the parasite, or the sickling attracts the attention of a phagocyte. In either event the individual remains unaffected by the disease, while normal individuals sicken and often die. Roughly 8 percent of the U.S. black population carries a single gene for sickle cell anemia; in parts of Africa where malaria is common the percentage is 3–4 times higher.

FIGURE 19-4
Sickling in Red Blood Cells. (a) When fully oxygenated, the cells of an individual with the sickling trait appear relatively normal. (SEM, × 15,554) (b) At lower oxygen concentrations the RBCs change shape, becoming relatively rigid and sharply curved.

(a)

(b)

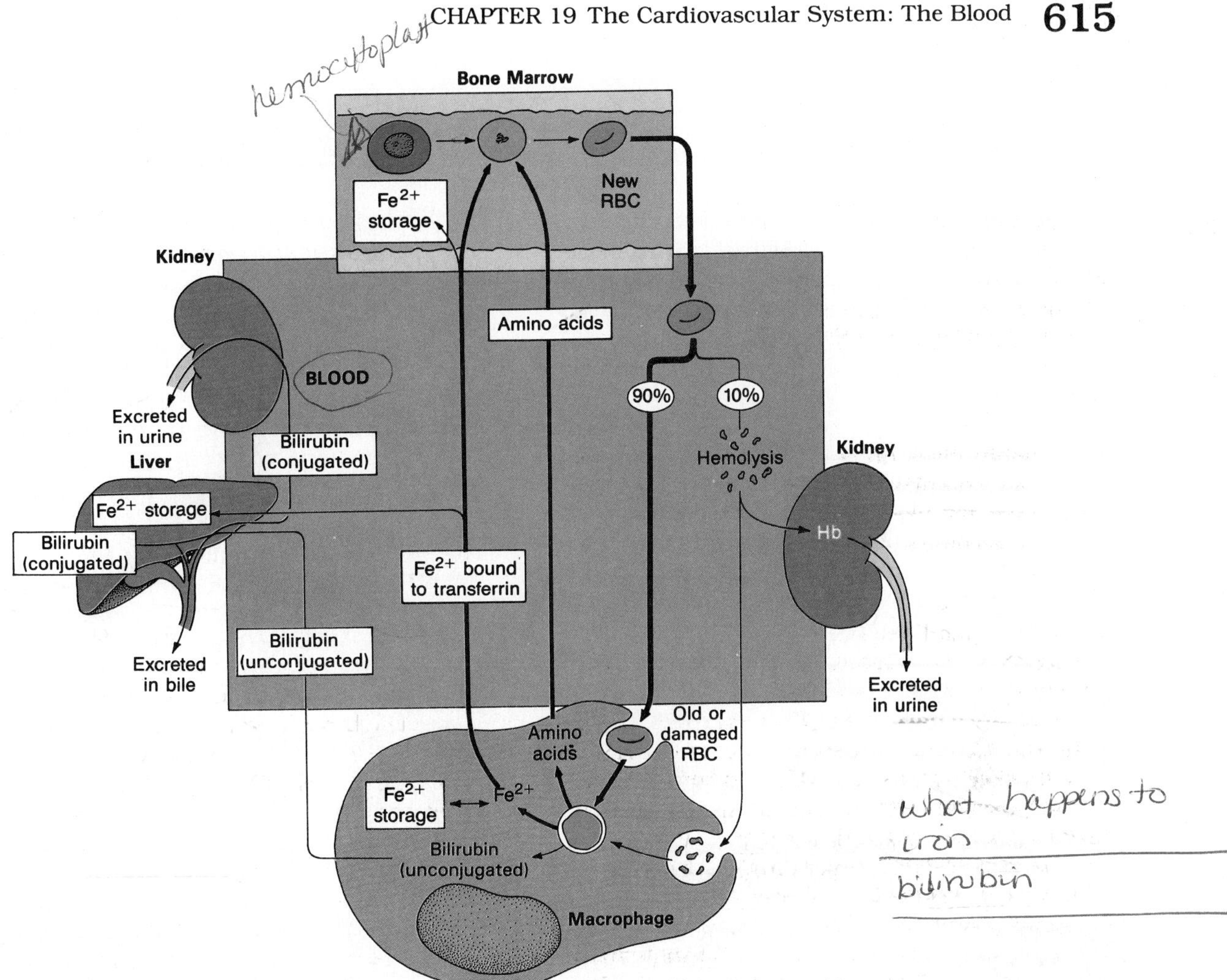

FIGURE 19-5
Red Blood Cell Turnover. The normal pathways for recycling amino acids and iron from aging or damaged red blood cells. The amino acids are absorbed, especially by developing cells in the bone marrow. The iron is stored in many different sites. The porphyrins of the heme units are converted to bilirubin, absorbed by the liver, and subsequently excreted in the bile or urine.

into the bloodstream, where it binds to a plasma protein, **transferrin** (tranz-FER-in). Red blood cells developing in the bone marrow absorb the amino acids and transferrins from the circulation and use them to synthesize new hemoglobin molecules. Excess transferrins are removed by the liver and bone marrow, and the iron is stored in two special protein-iron complexes, **ferritin** (FER-i-tin) and **hemosiderin** (hē-mō-SID-e-rin).

In summary, most of the components of an individual erythrocyte are recycled following hemolysis or phagocytosis. The entire system is remarkably efficient; although roughly 26 mg of iron are incorporated into hemoglobin molecules each day, a dietary supply of 1–2 mg will keep pace with the incidental losses that occur at the kidney and the digestive tract.

Abnormalities in iron uptake, metabolism, or excretion can cause serious clinical problems. Women are especially dependent on a normal dietary supply of iron, because their iron reserves are smaller than those of men. The body of a normal man contains around 3.5 g of iron in the ionic form Fe^{2+}. Of that amount, 2.5 g are bound to the hemoglobin of circulating red blood cells, and the rest is stored in the liver and bone marrow. In women, the total body iron content averages 2.4 g, with roughly 1.9 g incorporated into red blood cells. Thus a woman's iron reserves consist of only 0.5 g, half that of a typical man. If dietary supplies of iron are inadequate, hemoglobin production slows down, and symptoms of *iron deficiency anemia* appear. (This condition is discussed in the Clinical Manual for Chapter 25.) [**CM**: *Iron Deficiencies and Excesses*]

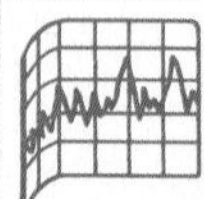

Diagnostics: Blood Tests and RBCs

Blood tests provide information about the general health of an individual, usually with a minimum of trouble and expense. Several common blood tests focus attention on red blood cells, the most common and abundant cellular elements. This section describes several common examples.

- **Reticulocyte count.** *Reticulocytes* are immature red blood cells that are still synthesizing hemoglobin. Most reticulocytes remain in the bone marrow until they complete their maturation, but some enter the circulation. Reticulocytes normally account for around 0.8 percent of the erythrocyte population. Values above 1.5 percent or below 0.5 percent indicate that something is wrong with the rates of RBC survival or maturation.
- **Hematocrit (Hct).** As mentioned previously, the hematocrit value is the percentage of whole blood occupied by cells. Normal adult hematocrits average 46 for men and 42 for women, with ranges of 40–54 for men and 37–47 for women. Because the numbers of RBCs far exceed those of WBCs, the hematocrit is normally used as a monitor for circulating erythrocytes. Although an abnormal hematocrit indicates that a problem exists, additional tests are needed to make a more definitive diagnosis. These procedures examine the size, age, abundance, and hemoglobin content of the erythrocytes.
- **Hemoglobin concentration (Hb).** This test determines the amount of hemoglobin in the blood, expressed in grams per 100 mℓ (g/dℓ). Normal values range from 14 to 18 g/dℓ in males and 12 to 16 g/dℓ in females.

  The differences in hemoglobin concentration reflect the differences in hematocrit. For both sexes, a single RBC contains 27–33 picograms (pg) of hemoglobin. RBCs containing normal amounts of hemoglobin are termed **normochromic**, while **hyperchromic** and **hypochromic** indicate higher or lower than normal hemoglobin content, respectively.
- **RBC count.** The number of RBCs per microliter of blood; a normal value is approximately 5.2 million.

  Calculations based on the hematocrit, hemoglobin content, and RBC count can be used to develop a better picture of the condition of the RBCs. Values often reported in blood screens include:
- **Mean corpuscular volume (MCV).** The average volume of an individual red blood cell, in cubic micrometers. Calculated by dividing the volume of red cells per microliter by the RBC count, using the formula:

$$\text{MCV} = \frac{\text{Hct}}{\text{RBC count (in millions)}} \times 10$$

  Normal values range from 82.2 to 100.6. Using the values given above, the mean corpuscular volume would be:

$$\text{MCV} = \frac{46}{5.2} \times 10 = 88.5\ \mu\text{m}^3$$

  Cells of normal size are **normocytic**, while larger or smaller than normal RBCs are called **macrocytic** or **microcytic**, respectively.
- **Mean corpuscular hemoglobin concentration (MCHC).** The amount of hemoglobin within a single RBC, expressed in

✓ **How would a decrease in the amount of globulin proteins in the blood affect the state of a person's health?**

✓ **How would an individual's hematocrit change after suffering a hemorrhage?**

✓ **How would the level of bilirubin in the blood be affected by diseases that cause damage to the liver?**

✓ **What factors are common to all forms of anemia?**

Blood Types

An individual's **blood type** is determined by the presence or absence of specific components in the erythrocyte cell membranes. The cell membrane of a typical red blood cell contains a number of **agglutinogens** (a-glu-TIN-o-jenz) exposed to the plasma. Agglutinogens are integral glycoproteins or glycolipids whose characteristics are genetically determined. There are at least 50 different kinds of agglutinogens on the surfaces of RBCs. Three of particular importance have been designated agglutinogens **A**, **B**, and **D** (**Rh**).

TABLE 19-4 RBC Tests and Anemias

Anemia type	*Hct*	*Hb*	*Reticulocyte count*	*MCV*	*MCHC*
Hemorrhagic	low	low	normal	normal	normal
Aplastic	low	low	very low	normal	normal
Iron deficiency	low	low	normal	low	low
Pernicious	low	low	very low	high	high

picograms. Normal values range from 27 to 34 pg. Calculated as:

MCHC = Hb/RBC count (in millions) × 10

As an example, Table 19-4 shows how this information can be used to distinguish among four major types of anemia.

1. **Hemorrhagic anemia** results from severe bleeding. Erythrocytes are of normal size, each contains a normal amount of hemoglobin, and reticulocytes are present in normal concentrations, at least initially. Blood tests would therefore show a low hematocrit and low hemoglobin, but the MCV, MCHC, and reticulocyte counts would be normal.
2. In **aplastic** (ā-PLAS-tik) **anemia** the bone marrow fails to produce new red blood cells. The 1986 nuclear accident in Chernobyl (U.S.S.R.) caused a number of cases of aplastic anemia. The condition is usually fatal unless surviving stem cells repopulate the marrow or a marrow transplant is performed. In aplastic anemia the circulating red blood cells are normal in all respects, but because new RBCs are not being produced the reticulocyte count is extremely low.

- In **iron deficiency anemia**, normal hemoglobin synthesis cannot occur because iron reserves are inadequate. Developing red blood cells cannot synthesize functional hemoglobin, and as a result they are unusually small. A blood test therefore shows a low hematocrit, low hemoglobin content, low MCV, and low MCHC, but a normal reticulocyte count.
- Finally, in **pernicious** (per-NISH-us) **anemia** normal red blood cell maturation ceases because of an inadequate supply of vitamin B_{12}. Erythrocyte production declines, and the red blood cells are abnormally large and may develop a variety of bizarre shapes. Blood tests from a person with pernicious anemia indicate a low hematocrit with a very high MCV and a low reticulocyte count.

The red blood cells of a particular individual may have any combination of these agglutinogens on their surfaces. For example, **Type A** blood has agglutinogen A, **Type B** has agglutinogen B, **Type AB** has both, and **Type O** has neither. These blood types are not evenly distributed through the population. Type O is the most common, followed in decreasing order of occurrence by Type A, Type B, and Type AB (statistics are included in Figure 19-6). These are average values for the U.S. population; there are variations due to racial and ethnic differences.

The presence of the Rh agglutinogen, sometimes called the *Rh factor*, is indicated by the terms **Rh-positive** (present) or **Rh-negative** (absent). In recording the complete blood type, the term *Rh* is usually omitted, and the data are reported as O-negative, A-positive, and so forth.

Agglutinins and Cross-Reactions You are probably already aware that your blood type must be checked before you give or receive blood. The surface agglutinogens on your own red blood cells are ignored by your immune system. (This ability to recognize normal body cells will be examined in Chapter 22.) How-

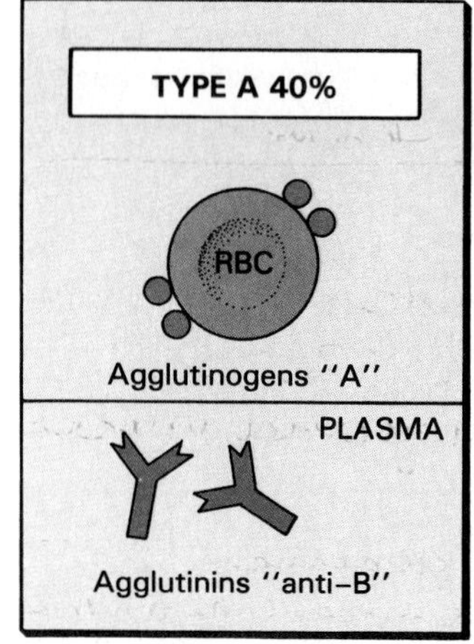

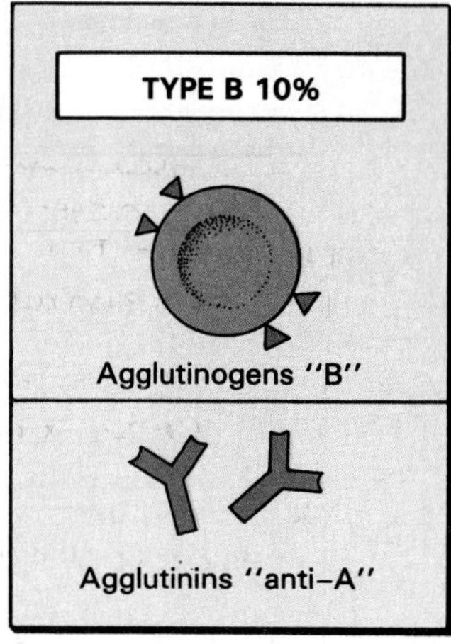

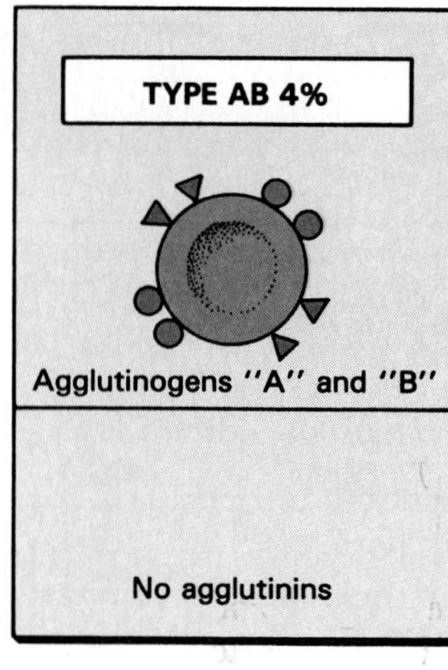

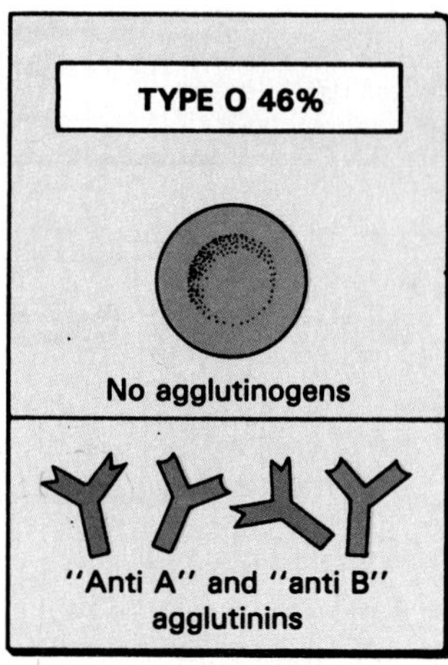

FIGURE 19-6
Blood Typing. The blood type depends on the presence of agglutinogens on RBC surfaces. The plasma contains agglutinins, antibodies that will react with foreign agglutinogens. Note the relative frequencies of each blood type in the U.S. population.

ever, your plasma contains immunoglobulins called **agglutinins** (a-GLOO-ti-ninz) that will attack "foreign" agglutinogens. For example, if you have Type A blood, your plasma holds circulating anti-B agglutinins that will attack Type B erythrocytes. If you are Type B, your plasma contains anti-A agglutinins. The red blood cells of an individual with Type O blood lack agglutinogens altogether, so the plasma of such an individual contains both anti-A and anti-B agglutinins. At the other extreme, Type AB individuals lack agglutinins sensitive to either A or B agglutinogens.

In contrast to the situation with agglutinogens A and B, the plasma of an Rh-negative individual does not normally contain anti-Rh agglutinins. These agglutinins are present only if the individual has been **sensitized** by previous exposure to Rh-positive erythrocytes. Such exposure may occur accidentally, during a transfusion, but it may also accompany a seemingly normal pregnancy involving an Rh-negative mother and an Rh-positive fetus (see Clinical Comment: Hemolytic Disease of the Newborn).

As in the distribution of A and B agglutinogens, there are racial and regional differences in Rh type. For example, the fraction of Rh-negative individuals in the Caucasian population is roughly 15 percent. That figure is 5 percent for the black American population, 1 percent for Asians, and essentially 0 for the black African population.

When an agglutinin meets its specific agglutinogen, a **cross-reaction** occurs. Initially the red blood cells clump together, a process called **agglutination** (a-glu-ti-NĀ-shun), and they may also **hemolyze**, or rupture (Figure 19-7). Clumps and fragments of red blood cells under attack form drifting masses that

FIGURE 19-7
Cross-reactions. A cross-reaction occurs when agglutinins encounter their target agglutinogens. The result is extensive clumping of the affected RBCs, followed by hemolysis.

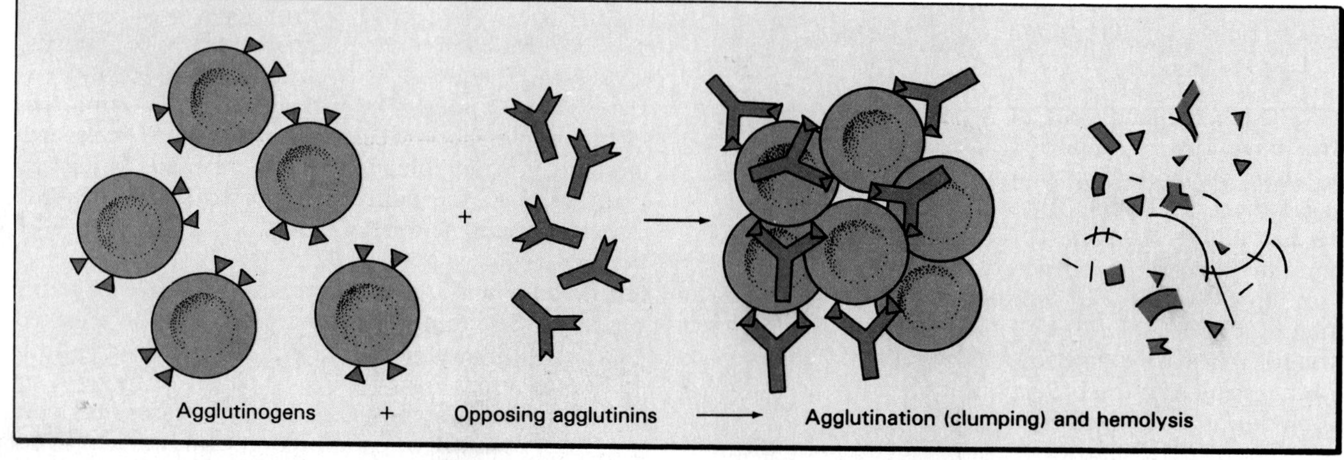

can plug small vessels in the kidneys, lungs, heart, or brain, damaging or destroying dependent tissues. To avoid such reactions, care must be taken to ensure that the blood types of the donor and the recipient are **compatible**. In practice, this involves choosing a donor whose blood cells will not undergo cross-reaction with the plasma of the recipient. Unless large volumes of whole blood or plasma are transferred, cross-reactions between the donor's plasma and the recipient's blood cells fail to produce significant agglutination because the donated plasma gets diluted through mixing with the plasma of the recipient.

Testing for Compatibility Testing for compatibility normally involves two steps: a determination of blood type and a cross-match test. At least 50 agglutinogens have been identified on red cell surfaces, but the standard test for blood type categorizes a blood sample on the basis of the three most likely to produce dangerous cross-reactions. The test involves taking drops of blood and mixing them with solutions containing anti-A, anti-B, *and* anti-Rh (anti-D) agglutinins. Any cross-reactions are then recorded. For example, if the red blood cells clump together when exposed to anti-A *and* anti-B, the individual has Type AB blood. If no reactions occur, the person must be Type O. The presence or absence of the Rh agglutinogen is also noted, and the individual is classified as Rh-positive or Rh-negative on that basis. Type O-positive is the most common blood type. The red blood cells of these individuals do not have agglutinogens A and B, but they do have agglutinogen D.

Standard blood typing can be completed in a matter of minutes, and Type O blood can be safely administered in an emergency. For example, a patient with a severe gunshot wound may require 5 *liters* or more of blood before the damage can be repaired. Under these circumstances it may not be possible to do more than collect blood, make certain that it is Type O, and adminster it to the victim. Because their blood cells are unlikely to produce severe cross-reactions in a recipient, Type O individuals (especially O^-) are sometimes called *universal donors*. However, it should be realized that with at least 48 other possible agglutinogens on the cell surface, cross-reactions can occur, even to Type O blood. Whenever time and facilities permit, further testing is performed to ensure complete compatibility.

Cross-match testing involves exposing the donor's red blood cells to a sample of the recipient's plasma under controlled conditions. This procedure reveals the presence of significant cross-reactions involving other aggluntinogens and agglutinins.

WHITE BLOOD CELLS

White blood cells, or **leukocytes** (LOO-kō-sītz; *leukos*, white), are scattered throughout peripheral tissues. These cells help defend the body against invasion by pathogens and remove toxins, wastes, and abnormal or damaged cells. There are two major classes of white blood cells. The **granular leukocytes**, or **granulocytes** (GRAN-ū-lō-sītz), have large granular inclusions in their cytoplasm. The **agranular leukocytes**, or **agranulocytes**, have several distinctive features, among them the absence of comparable cytoplasmic granules. Typical granular and agranular leukocytes are shown in Figure 19-9 on page 622. Most of the white blood cells in the body are found in peripheral tissues—the circulating leukocytes represent only a small fraction of the total population.

A typical microliter of blood contains 6000–9000 leukocytes. By examining a stained blood smear a **differential count** of the white blood cell population can be determined. The values obtained indicate the number of each type of cell encountered in a sample of 100 white blood cells. The normal range for each cell type is indicated in Table 19-3. The term **leukopenia** (loo-kō-PĒ-nē-ah; *penia*, poverty) indicates inadequate numbers of white blood cells. **Leukocytosis** (loo-kō-sī-TŌ-sis) refers to excessive numbers. The endings *-penia* and *-osis* can also be used to indicate low or high numbers of specific types of white blood cells. For example, *lymphopenia* means too few lymphocytes, whereas in *lymphocytosis* their numbers are unusually high.

Unlike erythrocytes, leukocytes do not circulate for extended periods. The bloodstream provides rapid transportation to areas of invasion or injury, and as they travel along the miles of capillaries the circulating white blood cells monitor the composition of the plasma, looking for the chemical signs of trouble brewing in the adjacent interstitial fluids. They are attracted to specific chemical stimuli, and this **chemotaxis** (kē-mō-TAK-sis) draws them to invading pathogens, damaged tissues, and other white blood cells. A leukocyte can migrate across the endothelial lining of a capillary by squeezing between adjacent endothelial cells, a process known as **diapedesis** (dī-a-pe-DĒ-sis).

Granular Leukocytes

Granular leukocytes are subdivided into **neutrophils**, **eosinophils**, and **basophils** on the basis of their staining characteristics. Neutrophils and eosinophils are important phagocytic cells that participate in the immune response. To avoid confusion with the larger phagocytes found in the blood and peripheral tissues, neutrophils and eosinophils are sometimes called *microphages*.

Neutrophils Fifty to seventy percent of the circulating white blood cells are **neutrophils** (NOO-trō-fils). This name was selected because their granules are chemically neutral and thus difficult to stain with either acid or basic dyes. A mature neutrophil (Figure

Clinical Comment: Hemolytic Disease of the Newborn

Genes controlling the presence or absence of any agglutinogen in the erythrocyte membrane are provided by both parents, so a child may have a blood type different from that of either parent. During pregnancy, when fetal and maternal circulatory systems are closely intertwined, the mother's agglutinins may cross the placental barrier, attacking and destroying fetal red blood cells. The resulting condition is called **hemolytic disease of the newborn, or HDN**.

There are many forms of HDN, some so mild as to remain undetected, but those involving the Rh agglutinogen are potentially quite dangerous. Because HDN results from the passage of *maternal* agglutinins across the placental barrier, an Rh-positive mother (who lacks anti-Rh agglutinins) can carry an Rh-negative fetus without difficulty. Potential problems appear when an Rh-negative woman carries an Rh-positive fetus. Sensitization usually occurs at delivery, when bleeding occurs at the placenta and uterus. This event can trigger the maternal production of anti-Rh agglutinins. Within 6 months of delivery, roughly 20 percent of Rh-negative mothers who carried Rh-positive children have become sensitized.

Because the anti-Rh agglutinins are not produced in significant amounts until after delivery, this first infant will not be affected. But if a second pregnancy occurs involving an Rh-positive fetus, the mother will respond by producing massive amounts of anti-Rh agglutinins. These agglutinins then enter the fetal circulation, destroying fetal red blood cells and producing a dangerous anemia. The fetal demand for blood cells increases, and they begin leaving the bone marrow and entering the circulation before completing their development. Because these immature RBCs are called **erythroblasts**, the condition is known as **erythroblastosis fetalis** (e-rith-rō-blas-TŌ-sis fē-TAL-is). The entire sequence of events is summarized in Figure 19-8. Without treatment, the fetus will probably die before delivery or shortly thereafter.

A newborn with severe HDN is anemic, and the high concentration of circulating bilirubin produces jaundice. Because the maternal agglutinins will remain active for 1–2 months after delivery, the infant may need to have its entire blood volume replaced. This blood replacement removes most of the maternal agglutinins as well as the affected erythrocytes, reducing the complications and the chance that the infant will die. When there is a danger that the fetus may not survive to full term, premature delivery may be induced after 7–8 months of development. In a severe case affecting a fetus at an earlier stage, one or more transfusions can be given while the fetus continues to develop within the uterus.

To avoid the problem completely, the maternal production of agglutinins is prevented by the administration of anti-Rh agglutinins (available under the name *RhoGam*) during and following delivery. These "foreign" agglutinins destroy any fetal red blood cells that cross the placental barrier. Thus there are no exposed agglutinogens to stimulate the maternal immune system, sensitization does not occur, and anti-Rh agglutinins are not produced. This relatively simple procedure could almost entirely prevent HDN mortality caused by Rh incompatibilities.

19-9a) has a very dense, contorted nucleus that may be condensed into a series of lobes resembling beads on a chain. This attribute has given these cells another name, **polymorphonuclear leukocytes** (pol-ē-mōr-fō-NŪ-klē-ar; *poly*, many + *morphe*, form), or **PMNs**. "Polymorphs," or "polys," are roughly 12 μm in diameter, and their cytoplasm is packed with pale granules containing lysosomal enzymes and bactericidal (bacteria-killing) compounds.

Neutrophils are highly mobile, and they are usually the first of the white blood cells to arrive at an injury site. They are very active phagocytes, specializing in attacking and digesting bacteria. Neutrophils usually have a short lifespan, surviving for about 12 hours. When actively engulfing debris or pathogens, their time may be even more limited. After ingesting a couple of dozen bacteria a neutrophil dies, but its breakdown releases chemicals that attract other neutrophils to the site.

Eosinophils **Eosinophils** (ē-ō-SIN-ō-fils) were so named because their granules stained darkly with a red dye, eosin. They will also stain with other acid dyes, so the name **acidophil** (a-SID-ō-fils) is also used. Eosinophils (Figure 19-9b) usually represent 2–4 percent of the circulating white blood cells. These cells

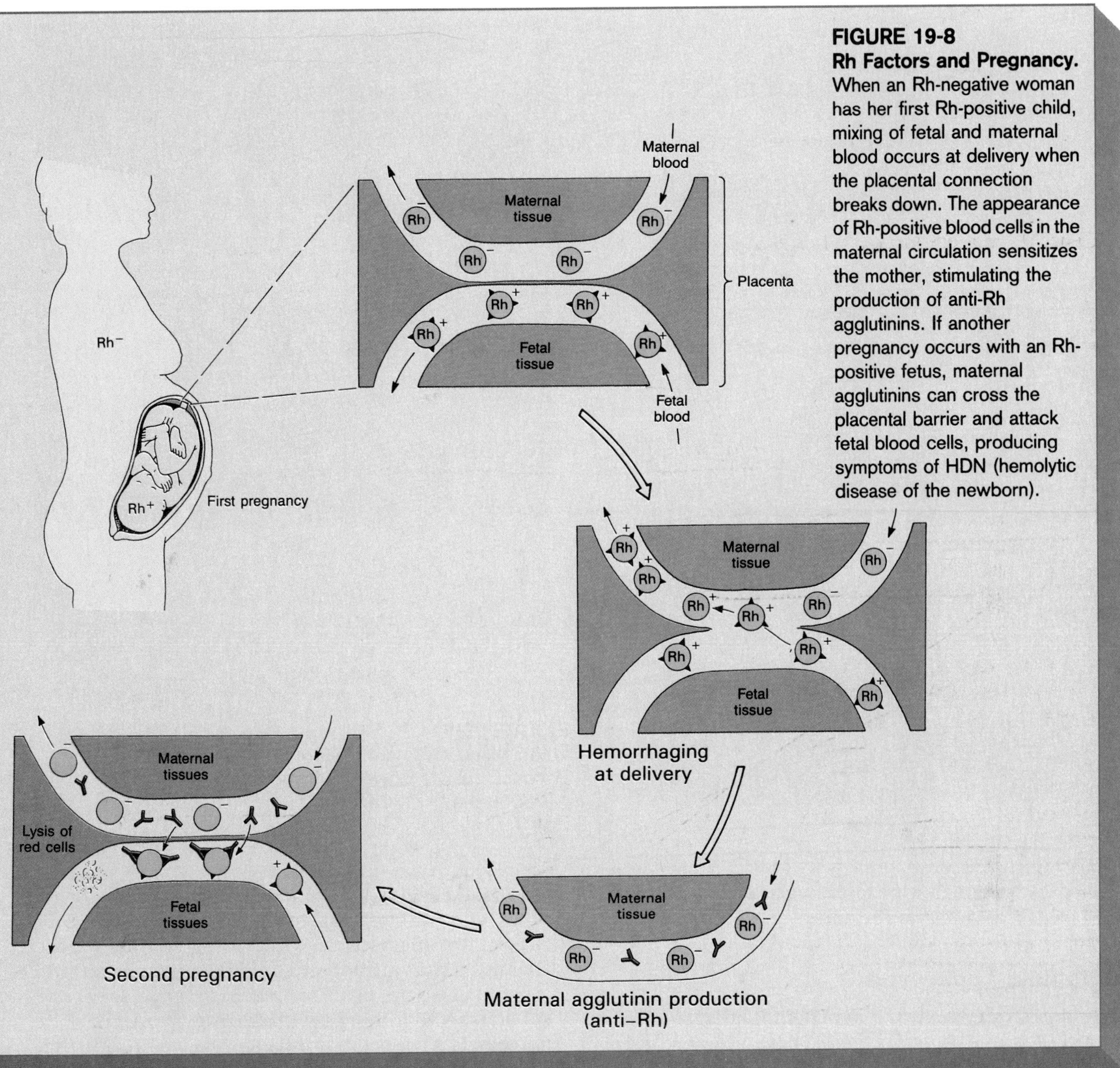

FIGURE 19-8
Rh Factors and Pregnancy. When an Rh-negative woman has her first Rh-positive child, mixing of fetal and maternal blood occurs at delivery when the placental connection breaks down. The appearance of Rh-positive blood cells in the maternal circulation sensitizes the mother, stimulating the production of anti-Rh agglutinins. If another pregnancy occurs with an Rh-positive fetus, maternal agglutinins can cross the placental barrier and attack fetal blood cells, producing symptoms of HDN (hemolytic disease of the newborn).

are similar in size to neutrophils, but the combination of deep red granules and a bilobed (two-lobed) nucleus makes an eosinophil easy to identify. Eosinophils are phagocytic cells, but they generally ignore bacteria and cellular debris. Instead they are attracted to foreign compounds that have reacted with circulating antibodies. Eosinophil numbers increase dramatically during an allergic reaction or a parasitic infection.

Basophils **Basophils** (BĀ-sō-fils) have numerous granules that stain darkly with basic dyes, and in a standard blood smear the inclusions are a deep purple/blue color (Figure 19-9c). They are smaller than the other granulocytes, ranging between 8 and 10 μm in diameter. They are also relatively rare, accounting for less than 1 percent of the leukocyte population. Basophils migrate to sites of injury and cross the capillary endothelium to accumulate within the damaged tissues, where they discharge their granules into the interstitial fluids. The granules contain histamine, and its release exaggerates the inflammation response at the injury site. Similar compounds are produced by the mast cells of damaged connective tissues, but mast cells and basophils are distinct populations with separate origins. Other chemicals re-

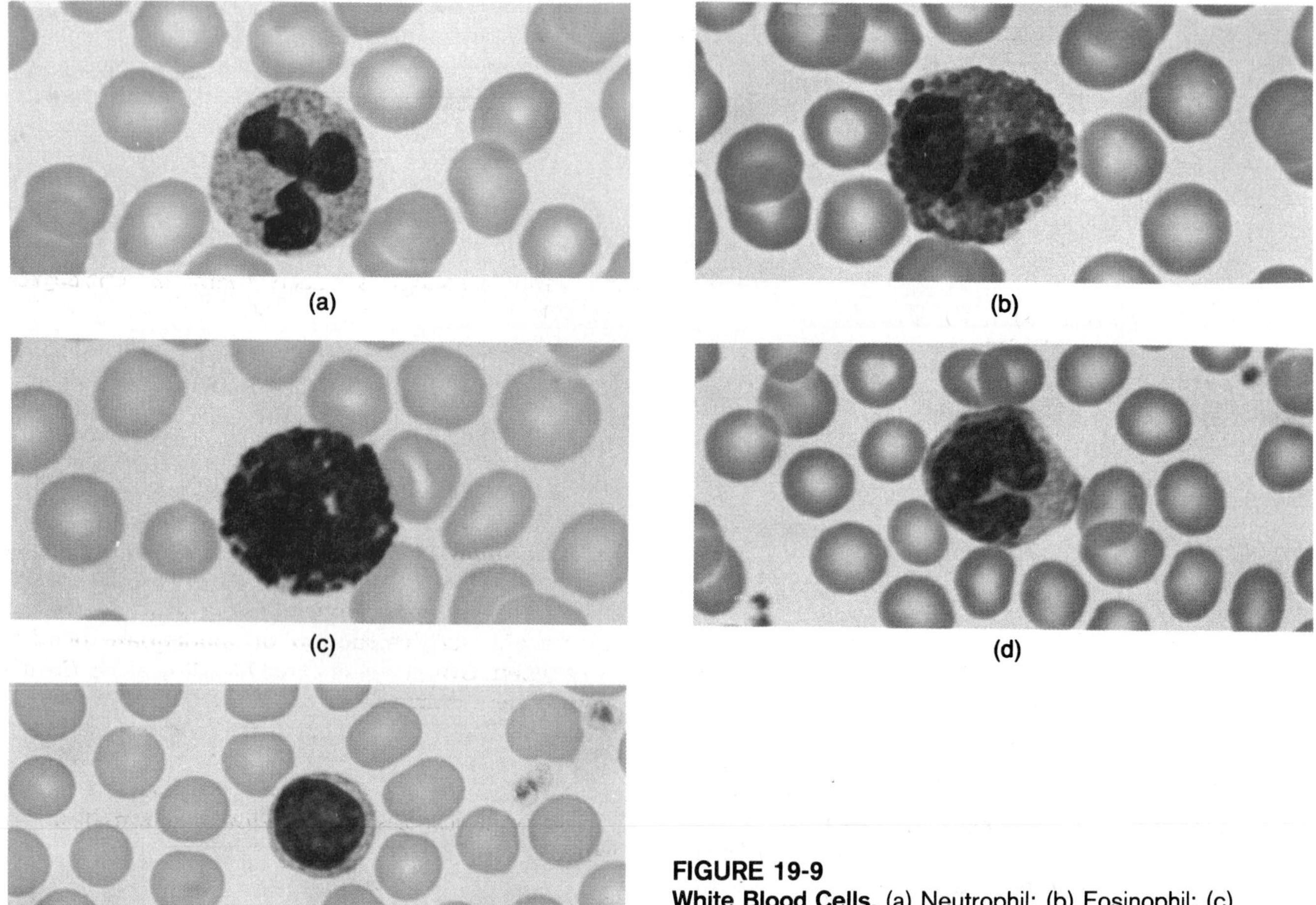

FIGURE 19-9
White Blood Cells. (a) Neutrophil; (b) Eosinophil; (c) Basophil; (d) Monocyte; (e) Lymphocyte. (LMs; a-c, × 1500); d-e, × 1325)

leased by stimulated basophils attract eosinophils and other basophils to the area.

Agranular Leukocytes

The blood contains two types of agranular leukocytes: *monocytes* and *lymphocytes*. These cells are seen in Figure 19-9d,e.

Monocytes **Monocytes** (MON-ō-sītz) may exceed 15 μm in diameter, nearly twice that of a typical erythrocyte. When flattened in a blood smear they look even larger, so monocytes are relatively easy to identify. The nucleus is large and often has the shape of an oval or kidney bean. Monocytes normally account for 2–8 percent of the circulating leukocytes. An individual monocyte uses the bloodstream as a highway, remaining in circulation for just a few days before entering peripheral tissues. Monocytes outside the bloodstream are usually called *free macrophages*, to distinguish them from the immobile *fixed macrophages* found in many connective tissues. (Tissue macrophages were described in Chapter 5.) (p. 139) Free macrophages are highly mobile cells, and they usually arrive at an injury site almost immediately after the incident. They are enthusiastic phagocytes, often attempting to engulf items as large or larger than themselves. While doing so they release chemicals that attract and stimulate monocytes and other phagocytic cells. Free macrophages also secrete substances that lure fibroblasts into the region. The fibroblasts then begin producing the scar tissue that will wall off the injured area. When free macrophages encounter a foreign object too large for a single cell to engulf, several of them may fuse together to create a single **phagocytic giant cell** big enough to do the job.

Lymphocytes Typical **lymphocytes** (LIM-fō-sītz) are roughly the same size as red blood cells, and they lack the abundant granules seen in the cytoplasm of the granulocytes. In fact, in a blood smear of lymphocytes you usually see very little cytoplasm, just a thin halo around a relatively large nucleus. Lymphocytes account for 20–30 percent of the leukocyte population of the blood. Blood lymphocytes represent a minute segment of the entire lymphocyte population, for lymphocytes are the primary cells of the **lymphatic system**, a network of special vessels and organs distinct from those of the circulatory system.

Lymphocytes differ from the other white blood cells discussed so far, all of which are part of the body's system of *nonspecific defenses*. These defenses respond to a variety of stimuli, but always in the same way–they do not discriminate between one threat and another. Lymphocytes, by contrast, are the cell's responsible for *specific immunity*: the ability of the body to mount a counterattack against invading pathogens or foreign proteins *on an individual basis*.

Lymphocytes respond to such threats in two ways. Lymphocytes called **T cells** enter peripheral tissues and attack foreign cells directly. Another group of lymphocytes, the **B cells**, differentiate into plasma cells that secrete antibodies that attack alien cells or proteins in distant portions of the body. T cells and B cells cannot be distinguished with the light microscope. (The lymphatic system and immunity are discussed in Chapter 22.) [**CM**: *The Leukemias*]

PLATELETS

Bone marrow contains a number of very unusual cells, called **megakaryocytes** (meg-a-KĀR-ē-ō-sitz; *megas*, big + *karyon*, nucleus + *-cyte*, cell). As the name suggests, these are enormous cells (up to 160 μm in diameter) with large nuclei. The dense nucleus of a megakaryocyte has a distinctive doughnut shape, and the surrounding cytoplasm contains Golgi apparatus, ribosomes, and mitochondria in abundance. The cell membrane communicates with an extensive membrane network that radiates throughout the peripheral cytoplasm, as illustrated in Figure 19-10.

Megakaryocytes are very active cells that continually manufacture proteins, enzymes, and membranes. But instead of growing ever larger, these cells shed small membrane-enclosed packets of cytoplasm that enter the circulation. These **platelets** (PLĀT-lets) are flattened discs, round when viewed from above and spindle-shaped in section. They average about 4 μm in diameter and are roughly 1 μm thick. Platelets were once thought to be cells that had lost their nuclei, and histologists called them **thrombocytes** (THROM-bō-sītz; *thrombos*, clot). The term is still in use, although *platelet* is more suitable because these are enzyme packets, not individual cells.

Platelets are continually replaced, and an individual platelet circulates for 10–12 days before being removed by phagocytes. There are 150,000–500,000 platelets in each microliter of circulating blood; 350,000/μℓ represents the average concentration. An abnormally low platelet count (80,000/μℓ or less) is known as **thrombocytopenia** (throm-bō-sī-tō-PĒ-nē-ah). Thrombocytopenia usually indicates excessive platelet destruction or inadequate platelet production. Symptoms include bleeding along the digestive tract, within the skin, and occasionally inside the CNS. Platelet counts in **thrombocytosis** (throm-bō-sī-TŌ-sis) may exceed 1,000,000/μℓ. Thrombocytosis usually results from accelerated platelet formation in response to infection, inflammation, or cancer.

The functions of platelets include:

1. *Transport of chemicals important to the clotting process*. By releasing enzymes and other factors at the appropriate times, platelets help initiate and control the clotting process.

FIGURE 19-10
Megakaryocytes and Platelet Formation. Megakaryocytes stand out in bone marrow sections because of their relatively enormous size and the unusual shape of their nuclei. These cells are continually shedding chunks of cytoplasm that enter the circulation as platelets. (LM, × 360)

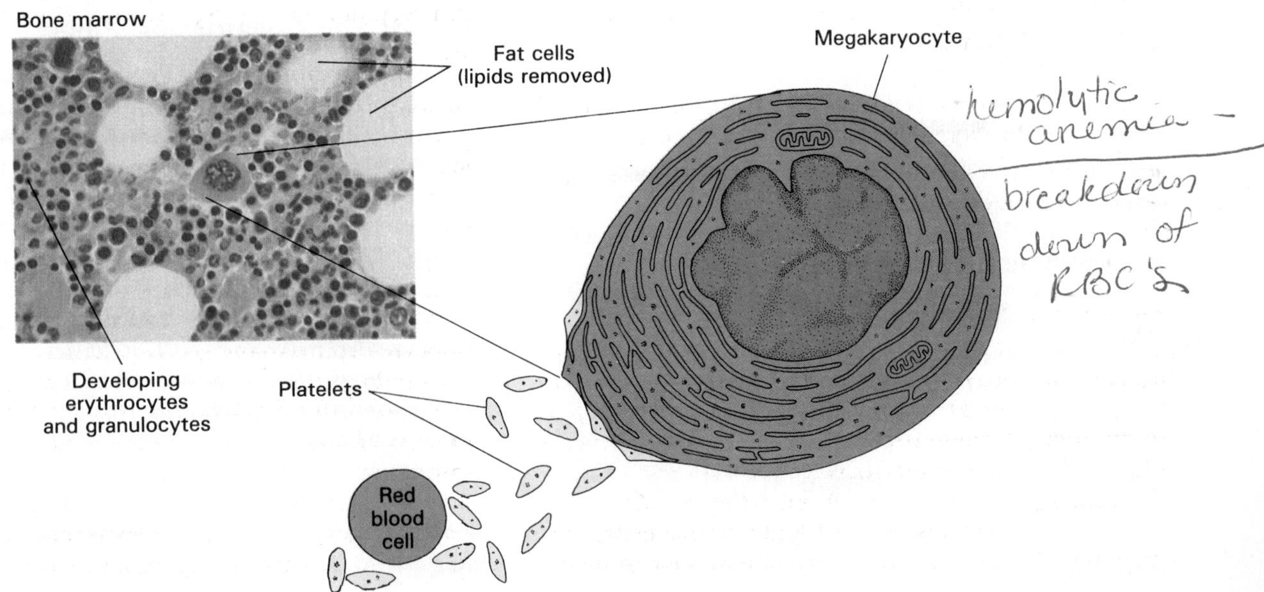

2. *Formation of a temporary patch in the walls of damaged blood vessels.* Platelets clump together at an injury site, forming a *platelet plug* that can slow the rate of blood loss while clotting occurs.
3. *Active contraction after clot formation has occurred.* Platelets contain filaments of actin and myosin. After a blood clot has formed, the contraction of platelets shrinks the clot and reduces the size of the break in the vessel wall.

Platelets are one participant in a vascular *clotting system* that also includes plasma proteins and the cells and tissues of the circulatory network.

- Why can't a person with type A blood receive blood from a person with type B blood?
- What type of white blood cell would you expect to find in the greatest numbers in an infected cut?
- What cell type would you expect to find in elevated numbers in a person producing large amounts of circulating antibodies to combat a virus?
- A sample of bone marrow has fewer than normal numbers of megakaryocytes. What body process would you expect to be impaired as a result?

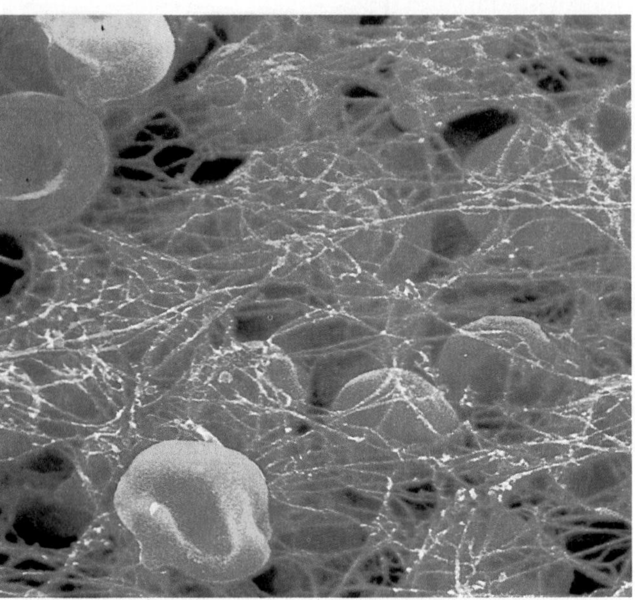

FIGURE 19-11
Structure of a Blood Clot. A scanning electron micrograph showing the network of fibrin that forms the framework of a clot. (SEM, × 4375) Red blood cells trapped in those fibers add to the mass of the blood clot and give it a red color. Platelets that stick to the fibrin strands gradually contract, shrinking the clot and tightly packing the RBCs. Figure 19-12 details the sequence of events that produces a blood clot.

Hemostasis

The process of **hemostasis** (*haima*, blood + *stasis*, halt) prevents the loss of blood through the walls of damaged vessels. In doing so it not only restricts blood loss but establishes a framework for tissue repairs. Although hemostasis can be analyzed as a series of steps, it is more like a chain reaction. In essence, each step modifies the events already under way; as a result, it is easier to say when a particular step begins than when it ends.

Step 1: The Vascular Phase

Cutting the wall of a blood vessel triggers a contraction in the smooth muscle fibers in the vessel wall. This contraction produces a local **vascular spasm** that decreases the diameter of the vessel at the site of injury. Such a constriction can slow or even stop the loss of blood through the vessel wall. The period of local vasoconstriction, called the **vascular phase** of the clotting reaction, lasts about 30 minutes.

Step 2: The Platelet Phase

During the vascular phase there are changes in the membrane characteristics of the endothelial cells at the injury site. Their membranes become "sticky," and in small capillaries endothelial cells may stick together and block the opening completely. In larger vessels, platelets begin to attach to exposed endothelial surfaces. This attachment marks the start of the **platelet phase** of the clotting process. As more and more platelets arrive they stick to one another as well. This reaction forms a mass that may plug the break in the vascular lining if the damage is not severe or the vessel is relatively small.

Step 3: The Coagulation Phase

The vascular and platelet phases begin within a few seconds after the injury. The **coagulation** (cō-ag-ū-LĀ-shun) **phase** does not start until 30 seconds to several minutes later. Blood clotting, or **coagulation**, involves a complex sequence of steps leading to the conversion of circulating fibrinogen into the insoluble protein **fibrin**. As the fibrin network grows, blood cells and additional platelets are trapped within the fibrous tangle, forming a **blood clot** that effectively seals off the damaged portion of the vessel. Figure 19-11 shows the structure of a blood clot viewed with a scanning electron microscope, and key stages of the coagulation phase are summarized in Figure 19-12.

Clotting Factors Normal coagulation cannot occur unless the plasma contains the necessary **clotting**

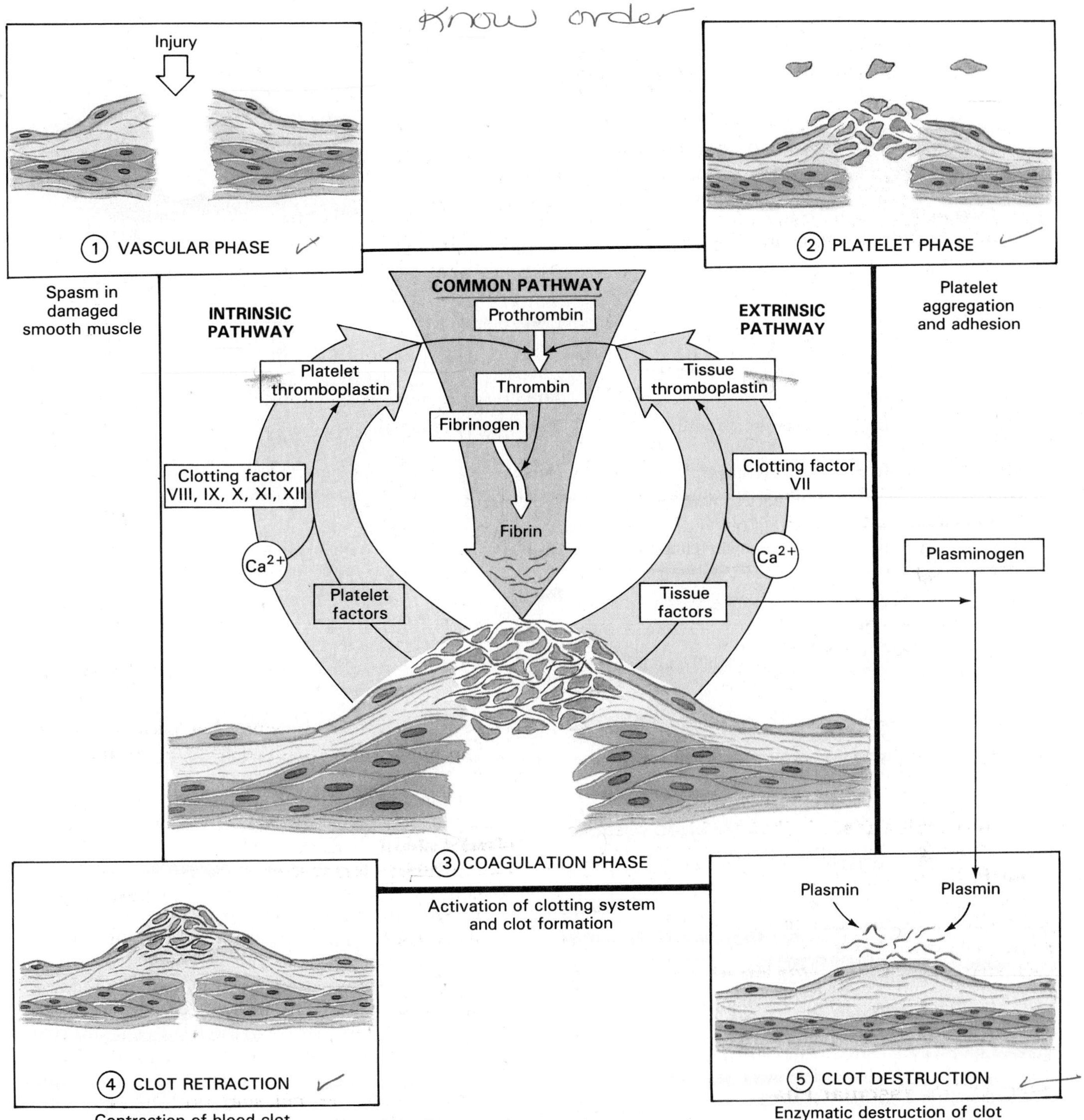

FIGURE 19-12
The Clotting Response. A flow chart summarizing the events in the clotting response (additional details are discussed in the text).

factors. Important clotting factors include calcium ions and 11 different proteins, called **procoagulants**. Under the proper circumstances these procoagulants can be converted to active enzymes that direct essential reactions in the clotting response. As a result, procoagulants are also called **proenzymes**.

For reference, specific clotting factors are identified in Table 19-5. Many are identified by Roman numerals; calcium ions, for example, are also known as clotting factor IV. All but two of the proenzymes (Factors III and VIII) are synthesized and released by the liver. During the coagulation phase, proenzymes interact, with the conversion of one proenzyme often creating an enzyme that activates a second proenzyme, and so on, in a chain reaction, or *cascade*. Figure 19-12 provides an overview of the sequence. [**CM**: *Disseminated Intravascular Coagulation*]

TABLE 19-5 Procoagulants

Factor	Structure	Name	Source	Concentration in Plasma (μg/mℓ)	Pathway
I	Protein	Fibrinogen	Liver	2500–3500	Common
II	Protein	Prothrombin	Liver, requires vitamin K	100	Common
III	Lipoprotein	Thromboplastin	Damaged tissue, activated platelets	0	Extrinsic and intrinsic
IV	Ion	Calcium ions	Bone, gut (see Ch. 7)	100	Entire process
V	Protein	Proaccelerin	Liver, platelets	10	Extrinsic and intrinsic
VI		(no longer used)			
VII	Protein	Proconvertin	Liver, requires vitamin K	0.5	Extrinsic
VIII	Protein	Antihemophilic factor (AHF)	Platelets, endothelial cells	15	Intrinsic
IX	Protein	Christmas factor	Liver, requires vitamin K	3	Intrinsic
X	Protein	Stuart-Prower factor	Liver, requires vitamin K	10	Extrinsic and intrinsic
XI	Protein	Plasma thromboplastin antecedent (PTA)	Liver	under 5	Intrinsic
XII	Protein	Hageman factor	Liver	under 5	Intrinsic; also activates plasmin
XIII	Protein	Fibrin-stabilizing factor (FSF)	Liver	20	Stabilizes fibrin, slows fibrinolysis

The Extrinsic, Intrinsic, and Common Pathways The initial steps in the coagulation process may involve clotting factors of the *extrinsic pathway* or the *intrinsic pathway*. The **extrinsic pathway** begins with the release of **tissue factor** by damaged endothelial cells or peripheral tissues. The greater the damage, the more tissue factor is released and the faster clotting occurs. Tissue factor then combines with calcium ions and one of the clotting factors to form an enzyme called **tissue thromboplastin**.

The **intrinsic pathway** begins with the activation of a proenzyme exposed to collagen fibers at the injury site. This pathway proceeds with the assistance of **platelet thromboplastic factor** from aggregating platelets. Platelets also release a variety of other factors that accelerate the reactions of the intrinsic pathway. After a series of linked reactions, several enzymes, calcium ions, and platelet thromboplastic factor then combine to form a complex called **platelet thromboplastin**.

Once thromboplastin from either the extrinsic or intrinsic pathways appears in the plasma, the **common pathway** begins. The first step involves the activation of a clotting factor responsible for the conversion of the proenzyme **prothrombin** into the enzyme **thrombin** (THROM-bin). Thrombin then completes the coagulation process by converting fibrinogen to fibrin.

The time it takes to complete the clotting process and establish a fibrin network varies. When a blood vessel is damaged, both the extrinsic and intrinsic pathways respond, and clotting usually occurs in about 15 seconds. The immediacy of the response reflects the speed of the extrinsic pathway, for if only the intrinsic pathway is activated, clotting takes 2–3 minutes. Once it begins to form, the clot gradually enlarges, and despite its leisurely pace the intrinsic pathway ultimately produces a larger, more effective clot than the extrinsic pathway alone. Thus the extrinsic pathway produces a small amount of thrombin very quickly, whereas the intrinsic pathway produces greater amounts of thrombin somewhat later. In effect, the extrinsic pathway applies a quick patch that the intrinsic pathway reinforces. [**CM**: *Hemophilia*]

Step 4: Clot Retraction

Once the fibrin meshwork has appeared, platelets and red blood cells stick to the fibrin strands. The platelets then contract, and the entire clot begins to undergo **retraction**, or **syneresis** (si-NER-e-sis; "a drawing together"). Retraction pulls the torn edges closer together and stabilizes the injury site. This process reduces the size of the damaged area, making it easier for the fibroblasts, smooth muscle cells, and endothelial cells in the area to carry out the necessary repairs.

Step 5: Fibrinolysis

As the repairs proceed, the clot gradually dissolves. This process, called **fibrinolysis** (fī-brin-OL-i-sis), begins with the activation of the proenzyme **plasminogen** (plaz-MIN-ō-jen) by another enzyme, **tissue plasminogen activator**, or **t-PA**, released by damaged tissues. Activation of plasminogen produces the enzyme **plasmin** (PLAZ-min), which begins digesting the fibrin strands and eroding the foundation of the clot.

Abnormalities in the Clotting Response

The coagulation process involves a complex chain of events, and a disorder that affects any individual clotting factor may disrupt the entire process. As a result, there are many different clinical conditions involving the clotting system. Calcium ions and a single vitamin, **vitamin K**, have an effect on almost every aspect of the clotting process. All three pathways (intrinsic, extrinsic, and common) require the presence of calcium ions, and disorders affecting calcium metabolism can have a direct effect on the coagulation process. Adequate amounts of vitamin K must be present in the diet for the liver to be able to synthesize four of the clotting factors, including prothrombin. A vitamin K deficiency therefore leads to the breakdown of the common pathway, inactivating the clotting system.

An overactive clotting system can also cause serious problems. The endothelial lining of blood vessels is very smooth and slick, but there are occasional bends and ripples where vessels branch or where adjacent epithelial cells interlock. If the clotting system becomes oversensitive, any of these irregular surfaces may break platelets apart or cause platelet aggregation. Each of these possibilities can have unpleasant consequences. When platelets break apart, clot formation begins in the circulation. These blood clots do not stick to the wall of the vessel but continue to drift around until plasmin digests them or they become stuck in a small blood vessel. A drifting blood clot is an example of an **embolus** (EM-bo-lus; *embolos*, plug). When an embolus becomes stuck in a blood vessel, it blocks circulation to the area downstream, killing the affected tissues. This condition is called an **embolism**. Embolus formation has been a recurring problem for individuals receiving artificial heart valves because clots form on the artificial surfaces. When these clots break free, they form emboli that may get stuck in capillaries in the brain, causing strokes. If an embolus forms in the venous system it will probably become lodged in one of the capillaries of the lungs, causing a condition known as a *pulmonary embolism*.

A **thrombus** (*thrombos*, clot) begins to form when platelets stick to the wall of an intact blood vessel. Often the platelets are attracted to areas called **plaques**, where endothelial and smooth muscle cells are storing large quantities of lipids. The blood clot gradually enlarges, projecting into the lumen of the vessel and reducing its diameter. Eventually the vessel may be completely blocked, or a large chunk of the clot may break off, creating an equally dangerous embolus.

Clinicians may attempt to prevent unwanted clotting by administering drugs that depress the clotting response or dissolve clots already present. Important anticoagulant drugs include **heparin** (HEP-a-rin), which inactivates thrombin, and **coumadin** (COO-ma-din), which depresses the synthesis of several clotting factors. To treat specific conditions, including pulmonary or venous embolisms, clots may be removed by the enzymes **streptokinase** (strep-tō-KĪ-nāce) or **urokinase** (ū-rō-KĪ-nāce). These enzymes, carefully introduced into the region of the clot, convert plasminogen to plasmin, but undesirable side effects limit their use. Because it accelerates the removal of blood clots, t-PA can also be employed. This enzyme is now commercially produced using recombinant DNA techniques. Initial clinical trials suggested that it has fewer side effects than either streptokinase or urokinase, but subsequent studies have called these findings into question. Current evidence indicates that there is little reason to choose t-PA over streptokinase, especially in view of its cost (about $4000, compared with about $400 for streptokinase). [CM: *Testing the Clotting System*]

Hemopoiesis

The process of blood cell formation is called **hemopoiesis** (hēm-o-poi-Ē-sis), or *hematopoiesis*. Blood cells appear in the circulation during the third week of embryonic development. These cells divide repeatedly, increasing their numbers. As other organ systems appear, some of the embryonic blood cells move out of the circulation and into the liver, spleen, thymus, and bone marrow. These embryonic cells differentiate into **stem cells**, which produce blood cells by their divisions. As the skeleton enlarges, the bone marrow becomes increasingly important; in the adult it is the primary site of blood cell formation. Figure 19-13 indicates the changing focus of hemopoietic activity during embryonic development.

Stem cells called **hemocytoblasts** produce all of the blood cells, but the process occurs in a series of steps. Hemocytoblast divisions produce at least four different types of stem cells with relatively restricted futures. These cells, called **progenitor cells**, remain

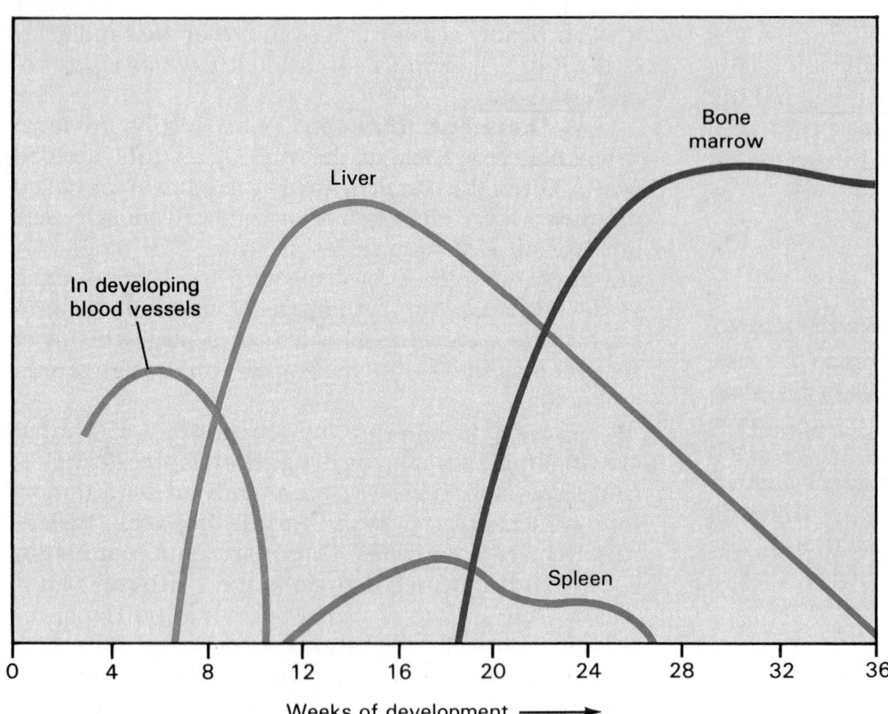

FIGURE 19-13
Blood Cell Production. During development the primary site of blood cell production changes.

capable of division, but their daughter cells will differentiate only into specific types of blood cells. For example, one type of progenitor cell produces daughter cells that mature into red blood cells; another gives rise to certain types of white blood cells. Figure 19-14 summarizes important details concerning the formation of the various cellular components of the blood.

ERYTHROPOIESIS

Erythropoiesis (e-rith-rō-poi-Ē-sis) refers specifically to the formation of erythrocytes. The bone marrow, or **myeloid tissue** (MĪ-e-loyd; *myelos*, marrow), is the primary site of erythropoiesis in the adult. The active **red marrow** is found in portions of the vertebrae, sternum, ribs, skull, scapulae, pelvis, and proximal limb bones. Other marrow areas contain a fatty tissue, known as **yellow marrow**. Under extreme stimulation, as following a severe and sustained blood loss, areas of yellow marrow can be converted to red marrow, increasing the rate of red blood cell formation. For erythropoiesis to proceed normally, the myeloid tissues must receive adequate supplies of amino acids, iron, and **vitamin B_{12}**, a vitamin obtained from dairy products and meat.

Regulation of Erythropoiesis

Erythropoiesis is closely regulated by circulating hormones. One important hormone, *erythropoietin*, was introduced in Chapter 18. (p. 588) Erythropoietin, also called **EPO** or **erythropoiesis-stimulating hormome**, appears in the plasma when peripheral tissues, especially the kidneys, are exposed to low oxygen concentrations (Figure 19-15a). For example, EPO release occurs during anemia, and when blood flow to the kidneys declines as a result of a thrombus or hemorrhage. Once in the circulation, EPO travels to areas of red marrow, where it stimulates stem cells and developing erythrocytes. Erythropoietin has two major effects: (1) it stimulates increased rates of mitotic divisions in erythroblasts and in the stem cells that produce erythroblasts; and (2) it speeds up the maturation of red blood cells, primarily by accelerating the rate of hemoglobin synthesis. Under maximum EPO stimulation the bone marrow can increase the rate of red blood cell formation tenfold, to around 30 million per second.

Stages in RBC Maturation

During its maturation, a red blood cell passes through a series of stages. **Hematologists** (hē-ma-TOL-o-jists), specialists in blood formation and function, have given specific names to key stages, as indicated in Figure 19-14. **Erythroblasts** are very immature red blood cells that are actively synthesizing hemoglobin. **Reticulocytes** (re-TIK-ū-lō-sītz) are taking the last step in the maturation process. After shedding their nuclei, these cells will develop the appearance of mature erythrocytes and enter the circulation. Some reticulocytes enter the circulation before they have completed their maturation; these immature red cells normally account for about 0.8 percent of the erythrocyte population.

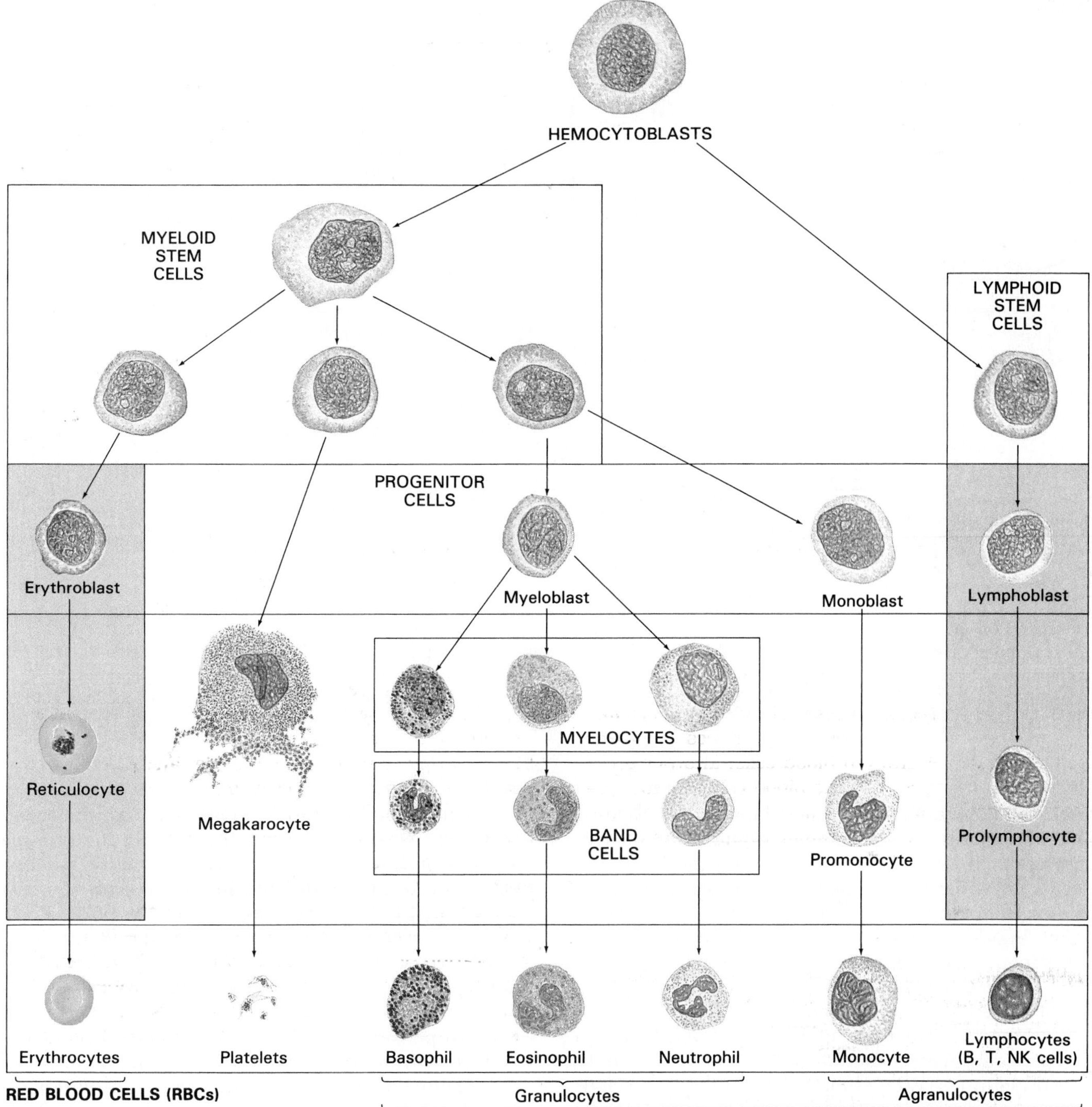

FIGURE 19-14
The Origins and Differentiation of Blood Cells. Hemocytoblast divisions give rise to myeloid and lymphoid stem cells. Myeloid stem cells with more restricted fates produce progenitor cells which divide to produce the various classes of blood cells.

WHITE BLOOD CELL FORMATION

Stem cells responsible for the production of white blood cells originate in the bone marrow. Granulocytes complete their development in myeloid tissue; monocytes begin their differentiation in the marrow, enter the circulation, and complete their development when they become free macrophages in peripheral tissues. Stem cells responsible for the production of lymphocytes, a process called **lymphopoiesis**, also originate in the bone marrow, but many of these subsequently migrate to peripheral **lymphoid tissues**, including the thymus, spleen, and lymph nodes. As a result, lymphocytes are produced in these organs as well as in the bone marrow.

Regulation of Lymphopoiesis

Factors that regulate lymphocyte maturation are as yet incompletely understood. Prior to maturity, thymosin and perhaps other hormones of the thymus gland promote the differentiation and maintenance

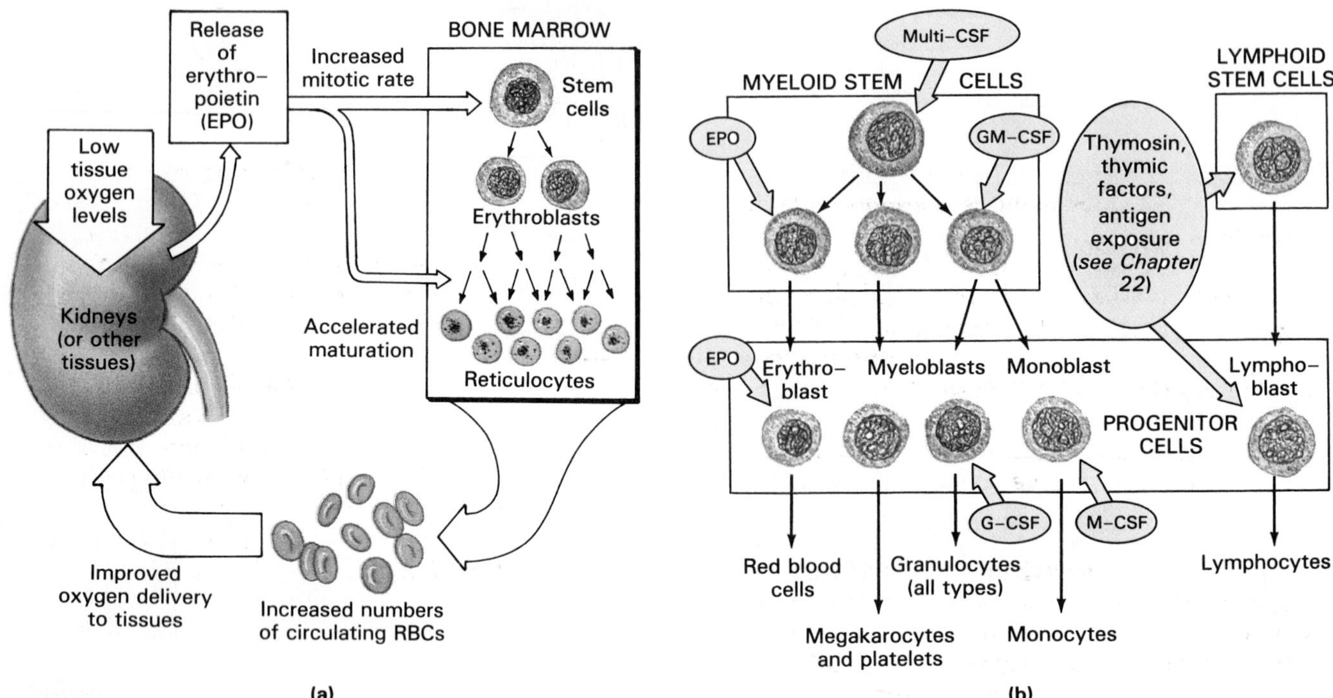

FIGURE 19-15
The Control of Hemopoiesis. (a) Tissues deprived of oxygen release erythropoietin (EPO). This hormone stimulates division of stem cells and erythroblasts and accelerates the maturation of RBCs. More red cells then enter the circulation, improving delivery of oxygen to the tissues. (b) The factors regulating hemopoiesis. The four colony-stimulating factors have specific stem or progenitor cell targets.

of T cell populations. The importance of the thymus gland in adulthood, especially in aging, remains controversial. In the adult, production of B and T lymphocytes is primarily regulated by exposure to antigens (foreign proteins, cells, or toxins). When foreign antigens appear, lymphocyte production escalates; the control mechanisms will be described in Chapter 22.

Several hormones are involved in the regulation of other white blood cell populations. Figure 19-15b diagrams the functions of these hormones, called **colony-stimulating factors (CSF)**. Four CSFs have been identified, each stimulating white blood cell or white *and* red blood cell formation. The abbreviations for each factor indicate their targets:

- **M-CSF** stimulates activity in the monocyte/macrophage line;
- **G-CSF** stimulates production of granulocytes;[1]
- **GM-CSF** stimulates production of both granulocytes and monocytes;
- **multi-CSF** accelerates production of granulocytes, monocytes, megakaryocytes, and erythrocytes.

[1] A genetically engineered form of G-CSF, sold under the name of *filgrastim* (*Neupogen*), is now used to stimulate production of neutrophils in patients undergoing cancer chemotherapy.

Chemical communication between lymphocytes and other leukocytes assists in the coordination of the immune response. For example, active macrophages release chemicals that make lymphocytes more sensitive to antigens and accelerate the development of specific immunity. In turn, active lymphocytes release multi-CSF and GM-CSF, reinforcing nonspecific defenses. (These interactions will be explored further in Chapter 22.)

To perform its vital functions, blood must be kept in motion; if the circulatory supply is cut off, dependent tissues may be destroyed in a matter of minutes. Individual RBCs complete two trips around the circulatory system each minute. The circulation of blood begins in the third week of embryonic development and continues throughout life. The next chapter examines the structure and function of the heart, the pump that maintains this vital blood flow.

About half of our vitamin K is produced by bacteria in the intestine. How would you expect extended use of broad-spectrum antibiotics to affect blood clotting?

How would an increase in the level of oxygen supplied to the kidneys affect the level of erythropoietin in the blood? ↓ EPO secretion

Clinical Comment: Erythrocytosis and Blood Doping

In **erythrocytosis** (e-rith-rō-sī-TŌ-sis) the blood contains abnormally large numbers of red blood cells. Erythrocytosis usually results from the massive release of erythropoietin by tissues deprived of oxygen. People moving to high altitudes usually experience erythrocytosis following their arrival, because the air contains less oxygen than it does at sea level. The increased number of red blood cells compensates for the fact that individually each RBC is carrying less oxygen than it would at sea level. Mountaineers and those living at altitudes of 10,000–12,000 feet may have hematocrits as high as 65.

Individuals whose hearts or lungs are functioning inadequately may also develop erythrocytosis. For example, this condition is often seen in heart failure and emphysema, two conditions discussed in later chapters. Whether the blood fails to circulate efficiently or the lungs do not deliver enough oxygen to the blood, peripheral tissues remain oxygen-poor despite the rising hematocrit. Having a higher concentration of red blood cells increases the oxygen-carrying capacity of the blood, but it also makes the blood thicker and harder to push around the circulatory system. This increases the workload on the heart, making a bad situation even worse.

The practice of **blood doping** has become widespread among competitive athletes involved with endurance sports such as cycling. The procedure entails removing whole blood from the athlete in the weeks before an event. The packed red cells are separated from the plasma and stored. By the time of the race the competitor's bone marrow will have replaced the lost blood. Immediately before the event the packed red cells are reinfused, increasing the hematocrit. The objective is to elevate the oxygen-carrying capacity of the blood, and so increase endurance. The consequence is that the athlete's heart is placed under a tremendous strain. The long-term effects are unknown, but the practice obviously carries a significant risk; it has recently been banned in amateur sports.

Because it is now being synthesized using recombinant DNA techniques, EPO can be obtained by individuals attempting to circumvent these rules. Although use of this drug will be difficult for officials to detect, individuals using it are taking the same risks as those who practice blood doping. Recently the widow of a European cyclist who died suddenly and unexpectedly publically blamed her husband's death on EPO abuse.

Health News: Transfusions and Synthetic Blood

In a **transfusion**, blood components are provided to an individual whose blood volume has been reduced or whose blood is deficient in some way. The blood is obtained under sterile conditions, treated to prevent clotting and stabilize the red blood cells, and refrigerated. Transfusions of whole blood are most often used to restore blood volume after massive hemorrhaging has occurred. In an **exchange transfusion** the entire blood volume of an individual is drained off and simultaneously replaced with whole blood from another source.

Chilled whole blood remains usable for around 3.5 weeks. For longer storage the blood must be fractionated. The red blood cells are separated from the plasma, treated with a special antifreeze solution, and frozen. The plasma can then be stored chilled, frozen, or freeze-dried. This procedure permits long-term storage of rare blood types that might not otherwise be available for emergency use.

Fractionated blood has many uses. **Packed red blood cells (PRBC)**, with most of the plasma removed, are preferred for cases of anemia, in which the volume of blood may be close to normal but its oxygen-carrying capabilities have been reduced. Plasma may be administered when massive fluid losses are occurring, such as following severe burns. Plasma samples can be further fractionated to yield albumins, immunoglobulins, clotting factors, and plasma enzymes that can be administered separately. White blood cells and platelets can also be collected, sorted, and stored for subsequent transfusion.

Some 12 million pints of blood are used each year in the United States alone, and the demand for blood or blood components often exceeds the supply. Moreover, there has been increasing concern over the danger of transfusion recipients becoming infected with hepatitis virus or HIV (the virus that causes AIDS) from contaminated blood. The result has been a number of changes in transfusion practices over recent years. In general, fewer units of blood are now administered. There has also been an increase in **autologous transfusion**, in which blood is removed from a patient (or potential patient), stored, and later transfused back into the original donor when needed—for example, following a surgical procedure. Genetically engineered EPO can be used as an additional measure to help the patient's body restore its full complement of red cells more quickly. Moreover, new technology permits the reuse of blood "lost" during surgery. The blood is collected and filtered; the platelets are removed, and the remainder of the blood is reinfused into the patient.

Despite these economies and improvements, shortages of blood and anxieties over the safety of existing stockpiles persist. In addition, some people may be unable or unwilling to accept transfusions for medical or religious reasons. Thus there has been widespread interest in a number of recent attempts to develop synthetic blood components.

Genetic engineering procedures are now being used to synthesize one of the subunits of normal human hemoglobin, which can then be introduced into the circulation to increase oxygen transport and total blood volume. An alternative strategy involves removing hemoglobin molecules from red cells and attaching them to inert carrier molecules that will prevent their loss at the kidneys. Still a third approach draws on natural but nonhuman sources: the FDA recently approved testing of a blood substitute called *Hemopure* that contains purified hemoglobin from cattle. Since the hemoglobin is removed from erythrocytes and infused without plasma, cross-reactions are not expected to occur.

Plasma expanders can be used to increase blood volume temporarily, over a period of hours, while preliminary blood typing is under way. These solutions contain large carbohydrate molecules, rather than dissolved proteins, to maintain proper osmolarity. (The emergency use of the carbohydrate *dextran* in sodium chloride solutions was discussed in Chapter 3.) ∞ (p. 82) Although these carbohydrates are not metabolized, they are gradually removed from circulation by phagocytes, and the blood volume steadily declines. Plasma expanders are easily stored, and their sterile preparation ensures that there are no problems with viral or bacterial contamination. Although they provide a temporary solution to hypovolemia (low blood volume), plasma expanders fail to increase the amount of oxygen delivered to peripheral tissues.

Whole blood substitutes are highly experimental solutions still undergoing clinical evaluation. In addition to the osmotic agents found in plasma expanders, these solutions contain small clusters of synthetic molecules built of carbon and fluorine atoms. The mixtures, known as **perfluorochemical (PFC) emulsions**, can carry roughly 70 percent of the oxygen of whole blood. Animals have been kept alive after an exchange transfusion that completely replaced their circulating blood with a PFC emulsion.

PFC solutions have the same advantages of other plasma expanders, plus the added benefits of transporting oxygen. Because there are no RBCs involved, the PFC emulsions can carry oxygen to regions whose capillaries have been partially blocked by fatty deposits or blood clots. Unfortunately, PFCs do not absorb oxygen as effectively as normal blood. To ensure that they deliver adequate oxygen to peripheral tissues, the individual must breathe air rich in oxygen, usually through an oxygen mask. In addition, phagocytes appear to engulf the PFC clusters. These problems have limited the use of PFC emulsions on humans. However, one PFC solution, *Fluosol*, has been used to enhance oxygen delivery to cardiac muscle during heart surgery.

Another approach involves the manufacture of miniature erythrocytes by enclosing small bundles of hemoglobin in a lipid membrane. These **neohematocytes** (nē-ō-hēm-A-tō-sītz) are spherical, with a diameter of under 1 μm, and they can easily pass through narrow or partially blocked vessels. The major problem with this technique is that phagocytes treat neohematocytes like fragments of normal erythrocytes, so they remain in circulation only for about 5 hours.

CHAPTER REVIEW

Review of Selected Clinical Terms

normovolemic (nor-mo-vo-LĒ-mik) (*p. 607*)
The condition of having normal blood volume.

hematocrit (he-MA-to-krit) (*p. 611*)
The value that indicates the percentage of whole blood occupied by cellular elements.

anemia (a-NĒ-mē-ah) (*p. 611*)
A condition in which the oxygen-carrying capacity of the blood is reduced, due to low hematocrit or low blood hemoglobin concentrations.

polycythemia (po-lē-sī-THĒ-mē-ah) (*p. 611*)
A blood condition showing an elevated hematocrit with a normal blood volume.

normochromic (*p. 616*)
The condition in which red blood cells contain normal amounts of hemoglobin.

normocytic (*p. 616*)
A term referring to cells of normal size.

hemolytic disease of the newborn (HDN) (*p. 620*)
A condition in which fetal red blood cells have been destroyed by maternal agglutinins.

embolism (*p. 627*)
A condition in which a drifting blood clot becomes stuck in a blood vessel, blocking circulation to the area downstream.

thrombus (*p. 627*)
A blood clot.

plaque (*p. 627*)
An abnormal area within a blood vessel where large quantities of lipids accumulate.

transfusion (*p. 631*)
A procedure in which blood components are given to someone whose blood volume has been reduced or whose blood is defective.

packed red cells (*p. 631*)
Red blood cells from which most of the plasma has been removed.

Additional Terms of Clinical Importance

ANEMIA AND POLYCYTHEMIA (*p. 611*): **polycythemia vera**

SICKLE CELL ANEMIA (*p. 614*): **sickle cell anemia, sickling trait**

BLOOD TESTS AND RBCs (*p. 616*): **reticulocyte count, hematocrit (Hct), hemoglobin concentration, hyperchromic, hypochromic, RBC count, mean corpuscular volume (MCV), macrocytic, microcytic, mean corpuscular hemoglobin (MCHC), hemorrhagic anemia, aplastic anemia, iron deficiency anemia, pernicious anemia**

HEMOLYTIC DISEASE OF THE NEWBORN (*p. 620*): **erythroblasts, erythroblastosis fetalis**

ERYTHROCYTOSIS AND BLOOD DOPING (*p. 631*): **erythrocytosis, blood doping**

TRANSFUSIONS AND SYNTHETIC BLOOD (*p. 631*): **exchange transfusion, autologous transfusion, plasma expanders, whole blood substitutes, perfluorochemical (PFC) emulsions, neohematocytes**

Study Outline

Related Key Terms

Introduction (pp. 605–606)

1. The cardiovascular system provides a mechanism for the rapid transport of nutrients, waste products, and cells within the body.

Functions and Composition of the Blood (pp. 606–607)

1. Blood is a specialized connective tissue. Its functions include: (1) transporting dissolve gases; (2) distributing nutrients; (3) transporting metabolic wastes; (4) delivering enzymes and hormones; (5) regulating the pH and electrolyte composition of interstitial fluids; (6) restricting fluid losses through damaged vessels or injuries via the **clotting reaction**; (7) defending the body against toxins and pathogens; and (8) regulating body temperature by absorbing and redistributing heat.
2. Blood contains **plasma**, **red blood cells (RBCs)**, **white blood cells (WBCs)**, and **platelets**. The plasma and **formed elements** constitute **whole blood**, which can be **fractionated** for analytical or clinical purposes.

viscosity • hypovolemic
normovolemic • hypervolemic

Plasma (pp. 608–609)

1. Plasma accounts for about 55 percent of the volume of blood; roughly 92 percent of plasma is water.

DIFFERENCES BETWEEN PLASMA AND INTERSTITIAL FLUID (p. 608)

2. Plasma differs from interstitial fluid because it has a higher dissolved oxygen concentration and large numbers of dissolved pro-

teins. There are three classes of plasma proteins: *albumins*, *globulins*, and *fibrinogen*.

PLASMA PROTEINS (pp. 608–609)

3. **Albumins** constitute about 60 percent of plasma proteins. **Globulins** constitute roughly 33 percent of plasma proteins; they include **immunoglobulins (antibodies)**, which attack foreign proteins and pathogens, and **transport proteins**, which bind ions, hormones, and other compounds. **Fibrinogen** molecules function in the clotting reaction by interacting to form **fibrin**; removing fibrinogen from plasma leaves a fluid called **serum**.

Related Key Terms: **metalloproteins** • **lipoproteins**

Formed Elements (pp. 609–624)

RED BLOOD CELLS (pp. 609–619)

1. Red blood cells **(erythrocytes)** account for slightly less than half the blood volume. The **hematocrit** value indicates the percentage of whole blood occupied by cellular elements.
2. RBCs transport oxygen and carbon dioxide within the bloodstream. They are highly specialized cells with large surface-to-volume ratios. RBCs lack mitochondria, ribosomes, and nuclei. This leaves them unable to perform normal maintenance operations, so they usually degenerate after about 120 days in the circulation.

Related Key Terms: **volume of packed red cells (VPRC)** **packed cell volume (PCV)** **microliter** **rouleaux**

3. Molecules of **hemoglobin (Hb)** account for over 95 percent of the RBCs proteins. Hemoglobin is a globular protein formed from four subunits. Each subunit contains a single molecule of **heme** (a **porphyrin**), and can reversibly bind an oxygen molecule. Damaged or dead RBCs are recycled by phagocytes.

Related Key Terms: **hematuria** • **hemoglobinuria** **hemolysis** • **bilirubin** **conjugated bilirubin** **transferrin** • **ferritin** **hemosiderin**

4. One's **blood type** is determined by the presence or absence of specific **agglutinogens** in the RBC cell membranes: agglutinogens **A**, **B**, and **D (Rh)**. **Agglutinins** within the plasma will react with RBCs bearing different agglutinogens. Anti-Rh agglutinins are only synthesized after an Rh-negative individual becomes **sensitized** to the Rh agglutinogen.

Related Key Terms: **type A** • **type B** • **type AB** **type O** • **Rh-positive** **Rh-negative**

5. To avoid dangerous reactions, the blood types of a donor and recipient must be **compatible**. **Cross-match** testing helps insure compatibility.

Related Key Terms: **cross-reaction** • **agglutination**

WHITE BLOOD CELLS (pp. 619–623)

6. White blood cells **(leukocytes)** defend the body against pathogens and remove toxins, wastes, and abnormal or damaged cells.
7. Leukocytes show **chemotaxis** (attraction to specific chemicals) and **diapedesis** (the ability to move through vessel walls).
8. Granular leukocytes are subdivided into **neutrophils**, **eosinophils**, and **basophils**. 50 to 70 percent of circulating WBCs are neutrophils, which are highly mobile phagocytes. The much less common eosinophils are phagocytes that are attracted to foreign compounds that have reacted with circulating antibodies. The relatively rare basophils migrate to damaged tissues and release histamines, aiding the inflammation response.

Related Key Terms: **granular leukocytes (granulocytes)** **agranular leukocytes (agranulocytes)** **differential count** **leukopenia** • **leukocytosis** **polymorphonuclear leukocytes (PMNs)**

9. Agranular leukocytes are subdivided into **monocytes** and **lymphocytes**. Monocytes migrating into peripheral tissues become free macrophages. Lymphocytes, the primary cells of the **lymphatic system**, include **T cells** (which enter peripheral tissues and attack foreign cells directly) and **B cells** (which produce antibodies).

Related Key Terms: **phagocytic giant cell**

PLATELETS (pp. 623–624)

10. **Megakaryoctyes** in the bone marrow release packets of cytoplasm **(platelets)** into the circulating blood. The functions of platelets include: 1) transporting chemicals important to the clotting process; 2) forming a temporary patch in the walls of damaged blood vessels; and 3) contracting after a clot has formed in order to reduce the size of the break in the vessel wall.

Related Key Terms: **thrombocytes** **thrombocytopenia** **thrombocytosis**

Hemostasis (pp. 624–627)

1. The five-step process of **hemostasis** prevents the loss of blood through the walls of damaged vessels. The initial step, the **vascular phase**, is a period of local vasoconstriction resulting from a **vascular spasm** at the injury site. The **platelet phase** follows as platelets stick to damaged surfaces. The **coagulation phase** occurs as factors released by endothelial cells or peripheral tissues **(extrinsic pathway)** and platelets **(intrinsic pathway)** interact with **clotting factors** to form a **blood clot**. During the next step the entire clot begins to undergo **retraction (syneresis)**, pulling the torn edges closer together. During the final step of **fibrinolysis**, the clot gradually dis-

Related Key Terms: **fibrin** • **procoagulants** **proenzymes** • **tissue factor** **platelet thromboplastic factor** **platelet thromboplastin** **common pathway** • **prothrombin** **thrombin**

solves through the action of **plasmin**, the activated form of circulating **plasminogen**.
2. Calcium ions and **vitamin K** affect almost every aspect of the clotting process. Excessive or inappropriate clotting can lead to the formation of an **embolus** (a drifting blood clot) or a **thrombus** (a clot attached to the wall of an intact vessel). Anticoagulant drugs include **heparin** and **coumadin**. Clots may be removed by activating plasminogen with the enzymes **streptokinase**, **urokinase**, or **tissue plasminogen activator (t-PA)**.

Related Key Terms

embolism • **plaques**

Hemopoiesis (pp. 627–630)

1. **Hemopoiesis** is the process of blood cell formation. Circulating **stem cells** called **hemocytoblasts** divide to form all of the blood cells.

progenitor cells

ERYTHROPOIESIS (p. 628)

2. **Erythropoiesis**, the formation of erythrocytes, occurs mainly within the **myeloid tissue** (bone marrow) in adults. RBC formation increases under **erythropoiesis-stimulating hormone (erythropoietin, EPO)** stimulation, which occurs when peripheral tissues are exposed to low oxygen concentrations. Stages in RBC development include **erythroblasts** and **reticulocytes**.

red marrow • **yellow marrow**
vitamin B_{12} • **hematologist**

WHITE BLOOD CELL FORMATION (pp. 629–630)

1. Granuloctyes and monocytes are produced by stem cells in the bone marrow. Stem cells responsible for **lymphopoiesis** (production of lymphocytes) also originate in the bone marrow, but many migrate to peripheral **lymphoid tissues**.
2. Factors that regulate lymphocyte maturation are not completely understood. Several **colony-stimulating factors (CSF)** are involved in regulating other WBC populations and coordinating RBC and WBC production.

M-CSF • **G-CSF** • **GM-CSF**
multi-CSF

Review Planner

	Objective	Level 1
1	Describe the important components and major functions of blood.	**1 2 5 10 11 12 26 27**
2	Discuss the composition and functions of plasma.	**1 7 19 31 41**
3	List the characteristics and functions of red blood cells.	**14 25 39 42**
4	Identify the locations where the components of blood are produced and discuss the factors that regulate their production.	**7 18**
5	Explain what determines a person's blood type and why blood types are important.	**3 8 16 30 32**
6	Categorize the various white blood cells on the basis of their structures and functions.	**5 6 15 21**
7	Describe how white blood cells fight infection.	**5 6 15 21 24 40**
8	Explain the mechanisms that control blood loss after an injury.	**9 10 17 22 33 34**
9	Describe the life cycles of blood cells.	**18 20 23**

Level 2: **28 29 36–40**

Level 3: **43–48**

CM: **Bilirubin Tests and Jaundice**
Testing the Clotting System

CM: **The Leukemias** • **Disseminated Intravascular Coagulation**
Hemophilia

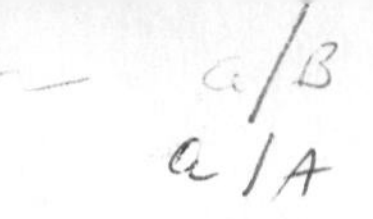

Chapter Review

Review Questions

MULTIPLE CHOICE

1. The ground substance of blood consists of: a) leukocytes; b) erythrocytes; c) plasma; d) platelets.
2. Which of the following is *not* a function of blood? a) delivering enzymes and hormones; b) processing information; c) defending the body against toxins; d) regulating body temperature.
3. The most common blood type in the U.S. population is type: a) A; b) B; c) AB; d) O.
4. A WBC count of 4000 per microliter indicates: a) leukocytosis; b) anemia; c) polycythemia; d) leukopenia.
5. Damage to basophils would affect the ability of blood to: a) phagocytize foreign compounds that have reacted with circulating antibodies; b) attack bacteria; c) exaggerate the inflammation response; d) transport oxygen.
6. Damage to lymphocytes would affect: a) specific immunity; b) blood clotting; c) the inflammation response; d) oxygen and CO_2 transport.
7. More than 90 percent of the plasma proteins are synthesized by the: a) heart; b) plasma cells; c) thymus gland; d) liver.
8. In a blood test for compatibility, the RBCs clump when exposed to anti-A agglutinins, but not when exposed to anti-B. This blood type is: a) A; b) B; c) AB; d) O.
9. Which of the following could result from thrombocytopenia?: a) excessive bleeding from a cut; b) increased bacterial infections; c) lowered populations of T cells; d) none of the above.
10. All of the following are involved in hemostasis *except*: a) platelets; b) fibrin; c) albumin; d) clotting factors.
11. The term "formed elements" refers to: a) blood cells and platelets; b) erythrocytes and plasma; c) plasma; d) crucial ions found in the interstitial fluid.
12. You would expect a person who is hypovolemic to have: a) excessive blood volume; b) low blood volume; c) normal blood volume.
13. The hematocrit value measures: a) the percentage of whole blood occupied by cellular elements; b) blood volume; c) the percentage of platelets per microliter; d) the coagulation time of a sample of blood.

MATCHING QUESTIONS

(*Match each term with the most appropriate statement.*)

Statements:

14. This component is responsible for an RBC's ability to transport dissolved gases.
15. Leukocytes that are responsible for specific immunity.
16. Combinations of these membrane proteins determine blood type.
17. A drifting blood clot in the circulation.
18. The process of red blood cell formation.
19. The most abundant component of plasma.
20. Hormones that help regulate WBC populations.
21. Granular leukocytes that attack bacteria.
22. The process of blood clotting.
23. The process of blood cell formation.
24. Immunoglobulins that will attack "foreign" membrane proteins in red blood cells.
25. "Stacks" of RBCs.
26. Plasma and formed elements.
27. Blood cells and cell fragments.

Terms:

a) Formed elements
b) Whole blood
c) Water
d) Rouleaux
e) Hemoglobin
f) Agglutinogens
g) Agglutinins
h) Neutrophils
i) Lymphocytes
j) Coagulation
k) Embolus
l) Hemopoiesis
m) Erythropoiesis
n) Colony-stimulating factors

SHORT ANSWER/ESSAY

28. Why would tissues that are completely deprived of circulation die quickly?
29. Explain how blood maintains homeostasis by making it possible for other systems in the body to perform their functions.
30. One of the blood types could be called the "universal recipient." Which one, and why?
31. Fill in the blanks: Plasma has a viscosity of ______, compared with a viscosity of 1.0 for ______.
32. Why is it important to test for blood compatibility before giving a transfusion?
33. If its calcium ions are removed, will whole blood coagulate? Explain your answer.
34. Place the following steps in the correct sequence. What process do these steps make up, and what is its function?
    a) Platelets bind to exposed endothelial surfaces.
    b) A fibrin network grows and forms a blood clot.
    c) Local vasoconstriction decreases the diameter of an injured blood vessel.
    d) Plasmin digests the fibrin strands and erodes the foundation of the blood clot.
    e) Retraction pulls the edges of the injured area closer together.
35. Give another term for each of the following:
    a) Erythrocyte.
    b) Leukocyte.
    c) Hematocrit.
36. Estimate your own blood volume.
37. Advertisements for some products claim that "iron-poor blood" poses a health risk. Would you agree? Why or why not?

38. Explain how blood maintains homeostasis of the oxygen and CO_2 concentrations in the body.
39. Explain the significance of the lack of organelles in an RBC. Is this an advantage or a disadvantage for the cell? Explain your answer.
40. What is meant by nonspecific and specific defenses? Name a cell that is important in each type of immunity.
41. Distinguish between the function of albumins, globulins, and fibrinogen.
42. Relate the shape of erythrocytes to their functions.

CRITICAL THINKING/APPLICATIONS

43. Why are sickled red cells harmful?
44. Mr. G. is admitted to the hospital. A blood test reveals a conjugated direct bilirubin concentration of 32 mg/dℓ. Explain this finding and identify Mr. G.'s condition.
45. A blood test of a patient shows a low reticulocyte count. Which types of anemia are unlikely considering this finding? What can be done to narrow down the remaining possibilities?
46. Blood tests show that a pregnant Rh-positive woman is carrying an Rh-negative fetus. What health risks can result? Why?
47. Suppose that an Rh-negative woman is carrying an Rh-positive fetus. What health risks can result, and why? How would your answer differ depending on whether this was her first child?
48. Mr. Rodriguez is admitted for hip surgery. Postoperatively he is given *heparin*, an anticoagulant, in small doses. Why?

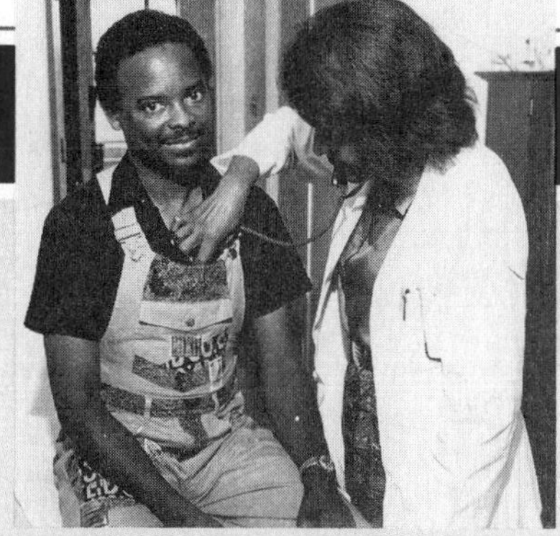

Career Close-up: Certified Medical Nurse Practitioner

Iver Gandy often finds herself trying to help people who refuse to help themselves. Gandy, a nurse practitioner specializing in sickle cell conditions, teaches her patients how to maintain a nutritious diet and a healthy lifestyle that will avert sickle cell crisis. Sometimes, they just don't listen.

"There was a young woman who came to the hospital who had sickle cell anemia and was pregnant," Gandy recalls. "She was in really tough shape because she had developed cholecystitis—her gallbaldder was inflamed, she had stones, and she had gone into sickle cell crisis. We put a cholecystotomy tube into her gallbladder to drain the bile and put her on a low-fat diet. But she kept sneaking junk food! And then she'd have more gallbladder attacks. We couldn't follow her around all day to stop her from eating the wrong foods. So we finally had to perform a cholecystectomy. After her gallbladder was removed, she had a normal delivery and a healthy baby."

Gandy is used to monitoring the lives of her sickle cell patients. It's an important aspect of her demanding job. She runs a sickle cell clinic attended weekly by about eight patients at St. Louis Regional Medical Center, St. Louis, MO. "When I began work, I was in general medicine, after which I worked for a diabetic clinic, and then as a renal nurse practitioner," Gandy says. "But sickle cell was a hematological subspecialty that always fascinated me. It's an inherited genetic disorder that's caused by a defect in the red blood cells. And 95 percent of the people who have sickle cell disease are black."

There's no cure for sickle cell conditions. It is possible to control the disease, however, and to decrease the severity and number of sickle cell crisis. During these crises—which can be brought on by anything from stress to dehydration—a patient's blood thickens and the capillaries are blocked by sickled cells. Often, a sickle cell patient will have to be hospitalized because of severe joint pain resulting from these crises. Gandy's clinic serves two functions: It teaches sickle cell patients how to use behavior modification, diet, and medication to control their disease, and it charts their condition on an ongoing basis.

"We try to teach our patients how to keep themselves nutritionally balanced and also to drink up to two gallons of fluid every day," Gandy says. "Sickle cell disease causes a kidney defect that results in our patients' inability to concentrate their urine, so they need to drink more than normal people. They need to help their kidneys out by giving them an ongoing supply of non-alcoholic, nutritious beverages." Gandy has gone as far as to call difficult patients every six hours to make sure that they're drinking on schedule.

Ongoing research keeps changing the way that medical professionals deal with sickle cell conditions. Gandy works hard to keep up. She reads the journals, attends national conferences, and takes courses, such as a recent class in sickle cell educator counseling. "We have a psychologist who works with our group," Gandy says. "But theirs can be such a tough life—many of our patients can't hold full-time jobs. And some of the kids have seen older siblings die of the disease and they're scared."

"Having sickle cell anemia is a lifetime concern," says Gandy. "But we are making great strides in the field. . . . It used to be a pediatric condition—most sickle cell patients never lived past age 20. Now we have patients in their 60s or 70s because our research has resulted in better care and medication."

You wouldn't think that any organ that contracts 70 to 80 times each minute would need exercise, would you? But regular exercise, such as bicycling, can significantly improve the efficiency of your heart, which translates into years of better health for you.

With every pump of the pedals, this biker is pushing his body closer to its limits. In this chapter we will see how the heart makes its vital contribution to homeostasis and how it continues to perform, with never an interruption, in the face of widely varying physical demands.

CHAPTER 20

The Cardiovascular System: The Heart

Chapter Objectives

After reading this chapter, you will be able to:

1. Relate the anatomy of the heart to its functions.
2. Identify important differences between cardiac and skeletal muscle.
3. Trace the path of blood flow through the heart.
4. Identify the major arteries and veins of the pulmonary and systemic circuits.
5. Describe the physiological mechanism of cardiac contraction.
6. Describe how cardiac performance is measured.
7. Explain what can be learned by listening to the heart or analyzing an ECG.
8. Describe the mechanisms that regulate heart rate and cardiac output.

■ Introduction

Every living cell relies on the surrounding interstitial fluid for oxygen, nutrients, and waste disposal. Conditions in the interstitial fluid are kept stable through continuous exchange between the peripheral tissues and the circulating blood. But the blood can help to maintain homeostasis only as long as it stays in motion. If blood remains stationary, its oxygen and nutrient supplies are quickly exhausted, its capacity to absorb wastes is soon saturated, and neither hormones nor white blood cells can reach their intended targets. Thus all of the functions of the cardiovascular system ultimately depend on the heart. This muscular organ beats approximately 100,000 times each day, forcing blood through the vessels of the circulatory system.

For a practical demonstration of the heart's pumping abilities, go into the kitchen and open the faucet. To deliver an amount of water equal to the volume of blood pumped by the heart in an average lifetime, that faucet would have to be left on for at least *45 years*. Equally remarkable, the volume of blood pumped by the heart can be varied over a wide range, from 5 to 30 liters per minute. The performance of the heart is closely monitored and finely regulated to ensure that conditions in the peripheral tissues remain within normal limits.

This chapter begins by examining the structural features that enable the heart to perform so reliably. We will then consider the physiological mechanisms that regulate cardiac activity to meet changing circumstances.

An Overview of the Circulatory System

The circulatory system consists of a network of blood vessels that carry blood between the heart and peripheral tissues. It can be subdivided into a **pulmonary circuit** that carries blood to and from the exchange surfaces of the lungs and a **systemic circuit** that transports blood to and from the rest of the body. Each circuit begins and ends at the heart. **Arteries**, or *efferent vessels*, carry blood away from the heart; **veins**, or *afferent vessels*, return blood to the heart. **Capillaries** are small, thin-walled vessels between the smallest arteries and veins.

As indicated in Figure 20-1, blood travels through these circuits in sequence. For example, blood returning to the heart in the systemic veins must complete the pulmonary circuit before reentering the systemic arteries. The heart contains four muscular chambers, two associated with each circuit. The **right atrium** (Ā-trē-um; chamber) receives blood from the systemic circuit, and the **right ventricle** (VEN-tri-k'l; little belly) discharges it into the pulmonary circuit. The **left atrium** collects blood from the pulmonary circuit, and the **left ventricle** ejects it into the systemic circuit. When the heart beats, the two ventricles contract at the same time and eject equal volumes of blood.

Structure of the Heart

POSITION AND SUPERFICIAL ANATOMY

Despite its impressive workload, the heart is a small organ, roughly the size of a clenched fist. The position of the heart in the thoracic cavity was detailed in Chapter 1, and you may wish to review Figure 1-13 before proceeding. (p. 19) The heart lies near the center of the thoracic cavity, enclosed by the connective tissues of the mediastinum. The mediastinum,

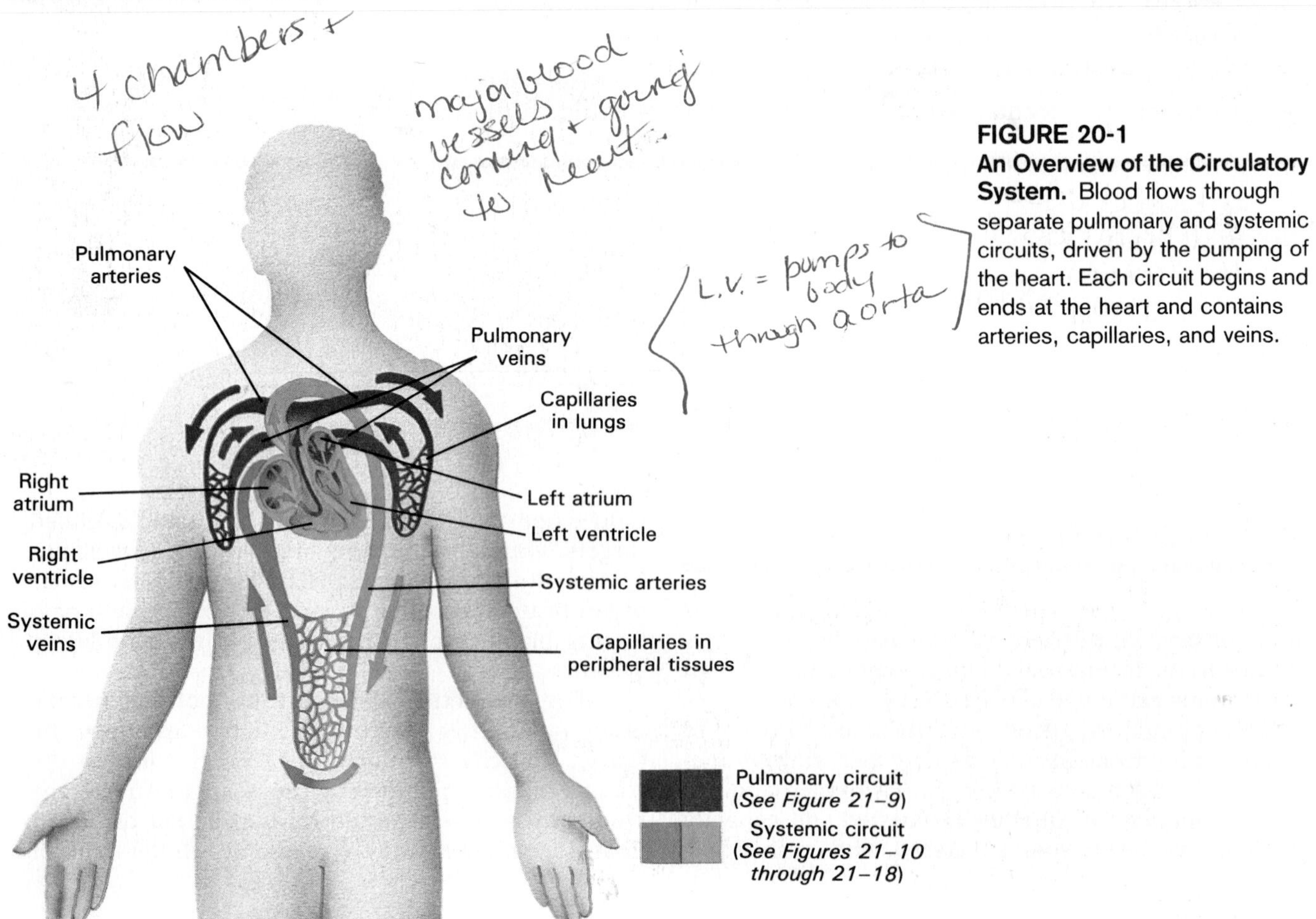

FIGURE 20-1
An Overview of the Circulatory System. Blood flows through separate pulmonary and systemic circuits, driven by the pumping of the heart. Each circuit begins and ends at the heart and contains arteries, capillaries, and veins.

which also contains the thymus, esophagus, and trachea, divides the thoracic cavity into two pleural cavities. Figure 20-2a shows an anterior view of the open chest cavity.

The heart lies near the anterior chest wall, directly behind the sternum. Figure 20-2b is a sectional view that illustrates the position of the heart relative to other structures in the mediastinum. The heart is surrounded by the **pericardial** (per-i-KAR-dē-al) **cavity**, one of the three ventral body cavities introduced in Chapter 1. (p. 17) This cavity is lined by a serous membrane called the **pericardium**. (Se-

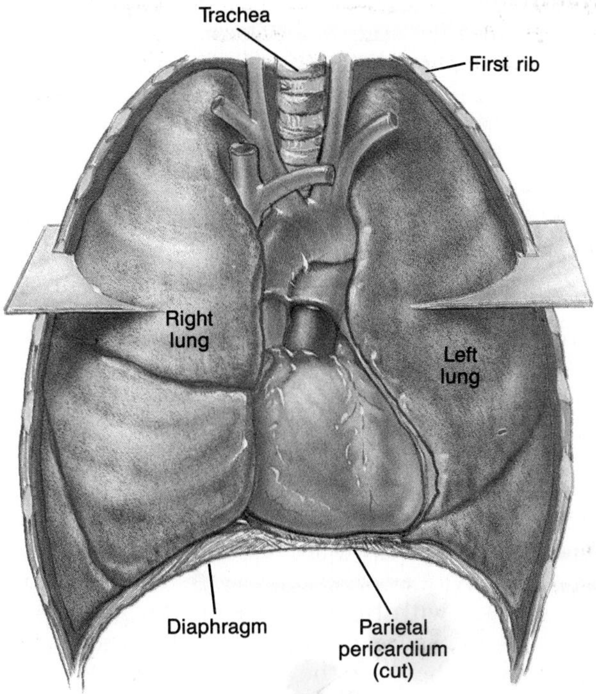

FIGURE 20-2
Location of the Heart in the Thoracic Cavity. The heart is situated within the anterior portion of the mediastium, immediately posterior to the sternum. (a) Anterior view of the open chest cavity, showing the position of the heart and major vessels relative to the lungs. The sectional plane indicates the orientation of (b). (b) Sectional view showing the position of the heart in the mediastinum and the location of other mediastinal organs. (c) Relationships between the heart and the pericardial cavity.

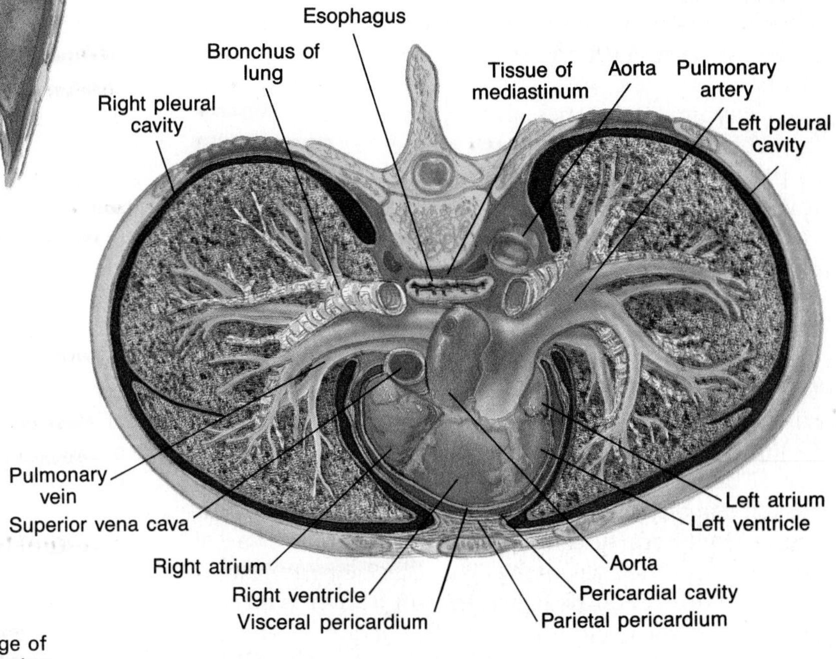

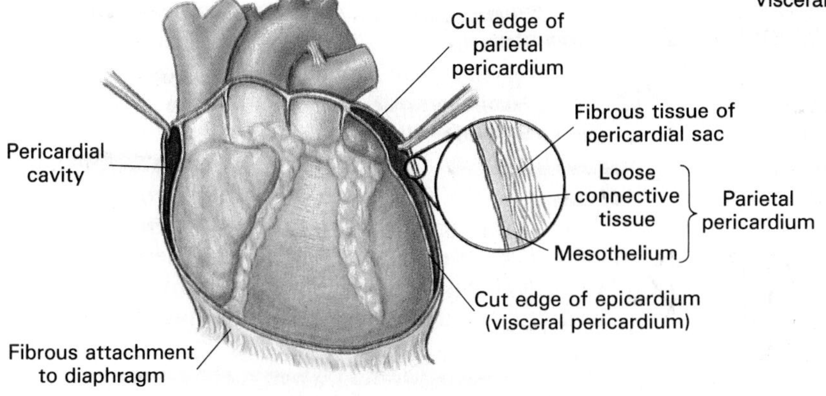

rous membranes, introduced in Chapter 5, consist of a layer of loose connective tissue covered by a moist and delicate mesothelium. ∞ (p. 154) To visualize the relationship between the heart and the pericardial cavity, imagine pushing your fist toward the center of a large balloon (Figure 20-2c). The balloon represents the pericardium, and your fist the heart. Your wrist, where the balloon folds back upon itself, corresponds to the **base** of the heart. At the base, the heart is connected to the major veins and arteries of the circulatory system. The smooth, moist pericardial lining prevents friction as the heart beats, and the connective tissues binding the base to the mediastinum limit distortion of the major vessels during a contraction.

The pericardium can be subdivided into the *visceral pericardium* and the *parietal pericardium*. The **visceral pericardium**, or **epicardium**, covers the outer surface of the heart; the **parietal pericardium** lines the inner surface of the **pericardial sac** that surrounds the heart. The pericardial sac is reinforced by a dense network of collagen fibers that stabilizes the positions of the pericardium, heart, and associated vessels in the mediastinum. The slender gap between the opposing parietal and visceral surfaces is the pericardial cavity. This space normally contains a small quantity of pericardial fluid that acts as a lubricant, reducing friction between the opposing layers.

A typical heart measures approximately 12.5 cm (5 in.) from the attached base to the free tip, or **apex** (Ā-peks). As the result of changes in orientation during development, the heart does not lie directly on the body's midline. (See the Embryology Summary on p. 647.) The base sits behind the sternum at the level of the third costal cartilage, centered about 1.2 cm (0.5 in.) to the left side. The apex reaches the fifth intercostal space approximately 7.5 cm (3 in.) to the left of the midline. Although this orientation can be seen in Figure 20-2a, it is shown more clearly in Figure 20-3a.

The heart also sits at an angle to the longitudinal axis of the body. The base occupies the **superior border** of the heart, and includes the bases of the major vessels and the two atria. The right atrium forms the **right border** of the heart; the **left border** is formed by the left ventricle and a small portion of the left atrium. The left border extends to the apex, where it meets the **inferior border**. The wall of the right ventricle forms most of the inferior border. The heart is also rotated slightly toward the left, so that the **anterior**, or **sternocostal** (ster-nō-KOS-tal) **surface** (Figure 20-3b) primarily consists of the right atrium and right ventricle. The posterior wall of the left ventricle forms much of the sloping **diaphragmatic surface** (Figure 20-3c) between the base and the apex of the heart. You should locate these various borders and surfaces in Figure 20-3 before proceeding further.

Several external features distinguish the atria from the ventricles. The atria have relatively thin muscular walls, and they are highly distensible. When not filled with blood, the outer portion of the atrium deflates and becomes a rather lumpy and wrinkled flap. This expandable extension of the atrium is called an **auricle** (AW-ri-k'l; *auris*, ear) because it reminded early anatomists of the external ear. A deep groove, the **coronary sulcus**, marks the border between the atria and the ventricles. Another depression, the **interventricular sulcus**, marks the boundary line between the left and right ventricles. The connective tissue of the epicardium at the coronary and interventricular sulci usually contains substantial amounts of fat, and in fresh or preserved hearts this fat must be stripped away to expose the underlying grooves. These sulci also contain the major arteries and veins that supply blood to the cardiac muscle tissue of the heart.

SECTIONAL ANATOMY AND ORGANIZATION

Figure 20-4a illustrates the four internal chambers of the heart. The two atria are separated by the **interatrial septum** (*septum*, wall), and the **interventricular septum** divides the two ventricles. Each atrium communicates with the ventricle on the same side. Folds of tissue, called **valves**, extend from the sides of the openings between the atria and ventricles. The valves are arranged in such a way as to ensure a one-way flow of blood from the atria into the ventricles, preventing any backflow. (Valve function will be detailed in a later section.)

The right atrium receives blood from the systemic circuit via two large veins, the **superior vena cava** (VĒ-na CĀ-va) and **inferior vena cava**. The superior vena cava delivers blood from the head, neck, arms, and chest. The inferior vena cava carries blood returning from the rest of the trunk, the viscera, and the legs. Prominent muscular ridges, the **pectinate muscles** (*pectin*, comb), or *musculi pectinati*, run along the inner surface of the auricle and across the adjacent anterior atrial wall.

Blood travels from the right atrium into the right ventricle through a broad opening bounded by three flaps of fibrous tissue. These flaps, or **cusps**, are part of the **right atrioventricular (AV) valve**, also known as the **tricuspid valve** (trī-KUS-pid; *tri*, three). Each cusp is braced by tendinous fibers, the **chordae tendineae** (KOR-dē TEN-di-nē; "tendinous cords"), that are connected to special **papillary muscles** (PAP-i-ler-ē) on the inner surface of the right ventricle. The internal surface of the ventricle also contains a series of deep grooves and folds, the **trabeculae carnae** (tra-BEK-ū-lē CAR-nē; *carneus*, fleshy).

Blood leaving the right ventricle flows into the large **pulmonary trunk** that marks the start of the

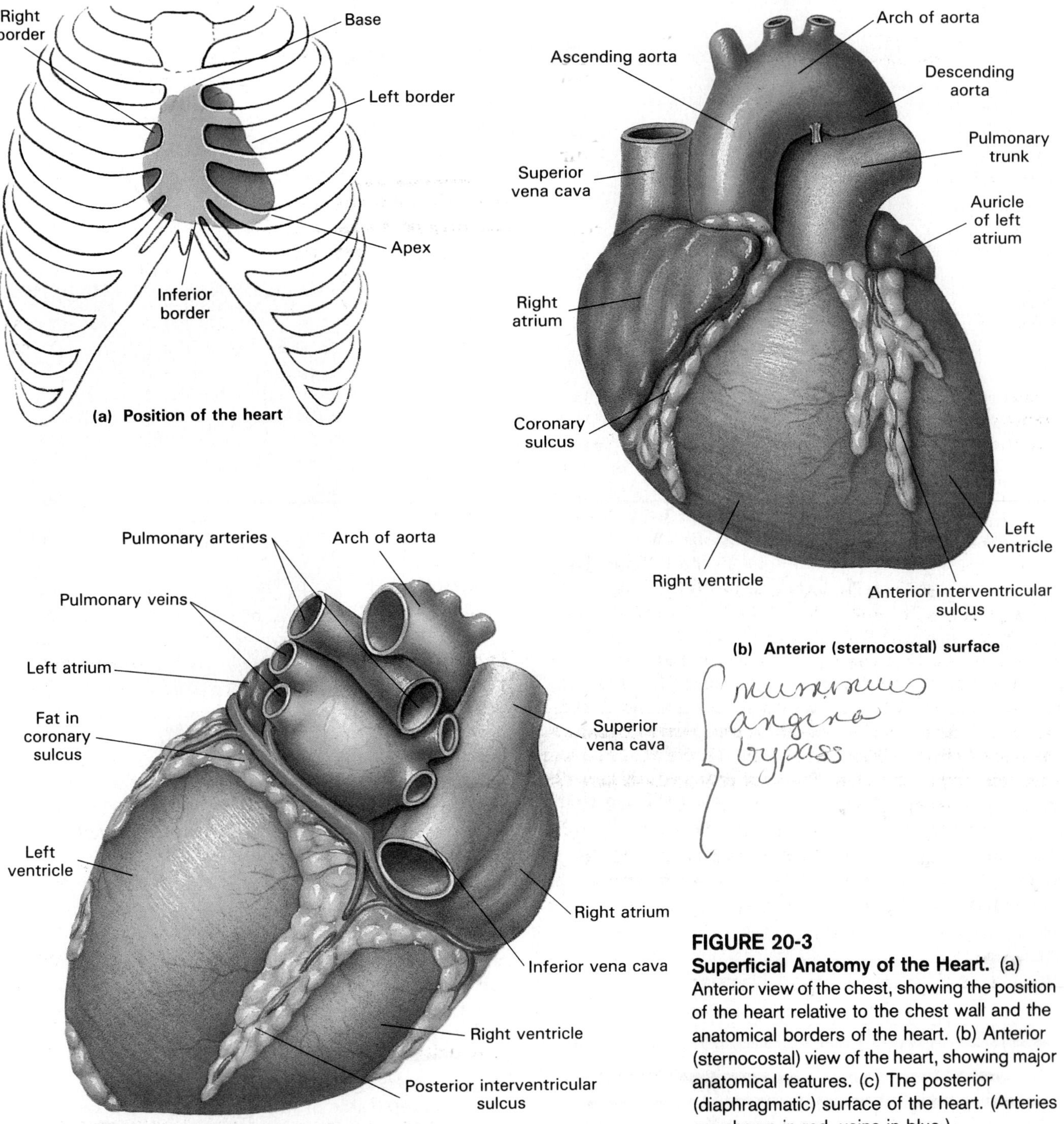

FIGURE 20-3
Superficial Anatomy of the Heart. (a) Anterior view of the chest, showing the position of the heart relative to the chest wall and the anatomical borders of the heart. (b) Anterior (sternocostal) view of the heart, showing major anatomical features. (c) The posterior (diaphragmatic) surface of the heart. (Arteries are shown in red, veins in blue.)

pulmonary circuit. The **pulmonary semilunar valve** guards the entrance to this efferent trunk. Once within the pulmonary trunk, blood flows into the **left** and **right pulmonary arteries**. These vessels branch repeatedly within the lungs, supplying capillaries where gas exchange occurs. From the respiratory capillaries blood collects into the **left** and **right pulmonary veins** that empty into the left atrium.

Like the right atrium, the left atrium has an auricle and a valve, the **left atrioventricular (AV) valve**, or **bicuspid** (bī-KUS-pid) **valve**. As the name *bicuspid* implies, the left AV valve contains a pair of cusps rather than a trio. Clinicians often use the term **mitral** (MĪ-tral) when referring to this valve. Regardless of the term selected, this "valve of many names" permits the flow of blood from the left atrium into the left ventricle.

The internal organization of the left ventricle resembles that of the right ventricle. The trabeculae carnae are prominent, and a pair of papillary muscles (Figure 20-4b) brace the chordae tendineae that insert on the bicuspid valve. Blood leaving the left ventricle passes through the **aortic semilunar valve** and into the systemic circuit via the **ascending aorta**.

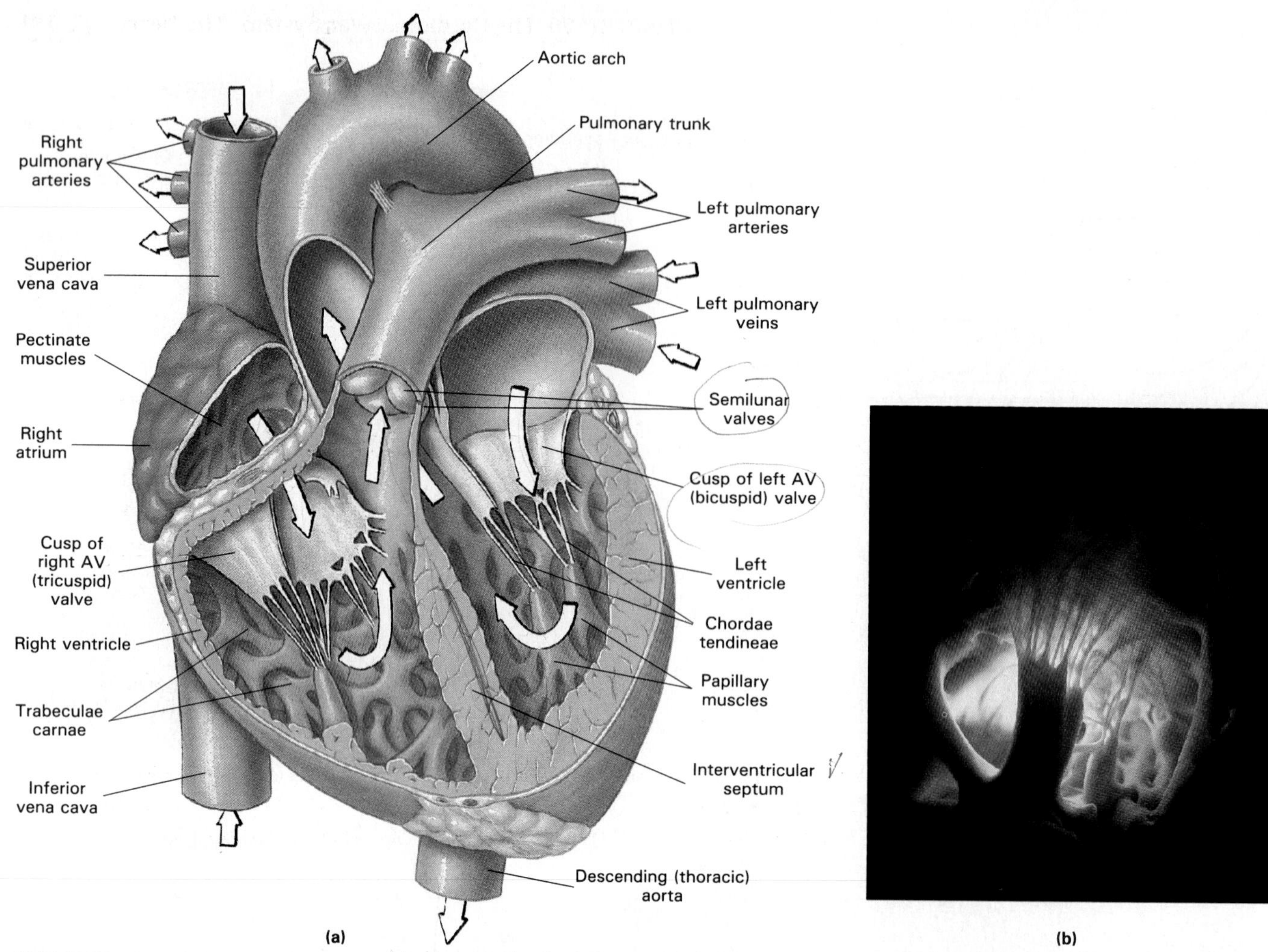

FIGURE 20-4
Sectional Anatomy of the Heart. (a) A diagrammatic frontal section through the heart, showing major landmarks and the path of blood flow through the atria and ventricles. (b) Photograph of papillary muscles and chordae tendineae supporting the tricuspid (right AV) valve. The picture was taken inside the right ventricle, looking toward a light shining from the right atrium.

Functional Anatomy of the Heart

The function of an atrium is to collect blood returning to the heart and deliver it to the attached ventricle. The functional demands placed on the right and left atria are very similar, and the two chambers look almost identical. But the demands placed on the right and left ventricles are very different, and there are characteristic anatomical differences between the two.

These differences can best be examined in a three-dimensional view such as that provided by Figure 20-5. The lungs are close to the heart, and the pulmonary arteries and veins are relatively short and wide. Thus the right ventricle normally does not need to push very hard to propel blood through the pulmonary circuit. The wall of the right ventricle is relatively thin, and in sectional view it resembles a pouch attached to the massive wall of the left ventricle. When the right ventricle contracts it acts like a bellows pump, squeezing the blood against the mass of the left ventricle. This mechanism moves blood very efficiently with minimal effort, but it develops relatively low pressures (normally 15–28 mm Hg).

A comparable pumping arrangement would not be suitable for the left ventricle, because six to seven times as much force must be exerted to push blood around the systemic circuit. The left ventricle has an extremely thick muscular wall, and it is round in cross section. When this ventricle contracts, two things happen: the distance between the base and apex decreases, and the diameter of the ventricular chamber decreases. If you imagine the effects of simultaneously squeezing and rolling up the end of a toothpaste tube you will get the idea. The forces generated are quite powerful, more than enough to open the semilunar valve and eject blood into the ascending aorta. As the powerful left ventricle contracts, it also bulges into the right ventricular cavity. This improves

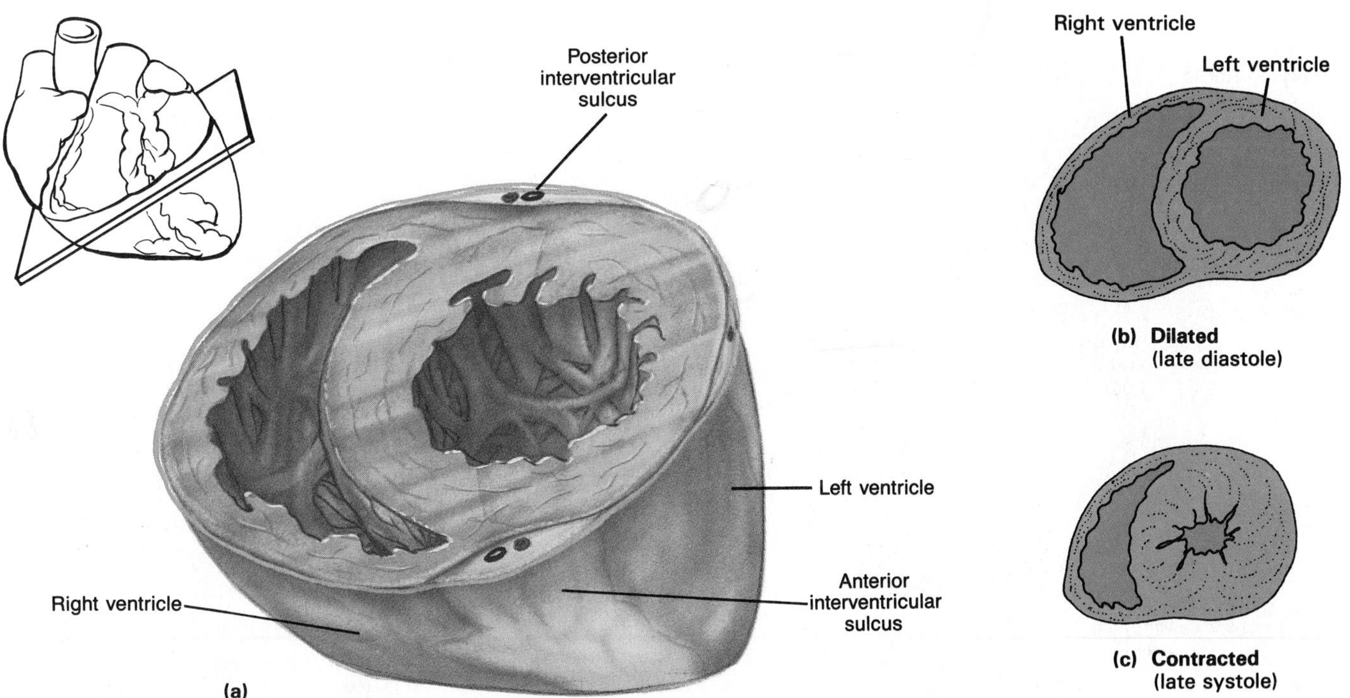

FIGURE 20-5
Structural Differences between the Left and Right Ventricles. (a) Detailed sectional view through the heart, showing the relative thicknesses of the two ventricles. Note the pouchlike shape of the right ventricle and the mass of the left ventricle. (b) Diagrammatic view of the relaxed ventricles just before a contraction, when they are filled with blood. (c) Comparable view of the contracted ventricles.

the efficiency of the right ventricle's efforts. Individuals whose right ventricular musculature has been severely damaged may continue to survive because of the extra push provided by the contraction of the left ventricle.

CORONARY CIRCULATION

The heart works continuously, and cardiac muscle fibers require reliable supplies of oxygen and nutrients. An average heart weighs around 300 g (11 oz), a mere 0.43 percent of the total body weight. Yet it accounts for more than 9 percent of the oxygen demand, and at rest almost 4 percent of the blood ejected by the left ventricle enters the **coronary circulation** that supplies blood to the muscles of the heart. During maximum exertion the oxygen demand rises considerably, and the blood flow to the heart may increase to nine times resting levels.

The coronary circulation involves an extensive network of vessels. Figure 20-6a shows the characteristic vasculature in exquisite detail. This cast was made by injecting the vessels of an intact heart with a fluid latex that filled the circulatory passageways before solidifying. The cardiac tissue was then dissolved with acids, leaving a rubberized cast of the cardiac circulatory network.

Major coronary vessels are identified in Figure 20-6b,c. The **coronary arteries** originate at the base of the ascending aorta. Blood pressure here is the highest found anywhere in the systemic circuit, and this pressure ensures a continuous, forceful blood flow to meet the demands of active cardiac muscle tissue. The **right coronary artery** follows the coronary sulcus around the heart. It usually gives rise to two branches. The **marginal branch** extends along the right border, and the **posterior interventricular branch** runs toward the apex within the posterior interventricular sulcus. The **left coronary artery** also has two major branches. The **circumflex branch** curves to the left around the coronary sulcus, eventually meeting and fusing with small branches of the right coronary artery. The **anterior interventricular branch**, or **left anterior descending artery**, swings around the pulmonary trunk and runs along the anterior surface within the interventricular sulcus. This branch supplies small tributaries continuous with those of the right coronary artery. Such interconnections between arteries are called **anastomoses** (a-nas-to-MŌ-sēs; *anastomosis*, outlet). Because the arteries are interconnected in this way, the blood supply to the cardiac muscle remains relatively constant, regardless of pressure fluctuations within the left and right coronary arteries. The **great** and **middle cardiac veins** carry blood away from the coronary capillaries. They drain into the **coronary sinus**, a large, thin-walled vein that lies in the posterior portion of the coronary sulcus. The coronary sinus communicates with the right atrium near the base of the inferior vena cava.

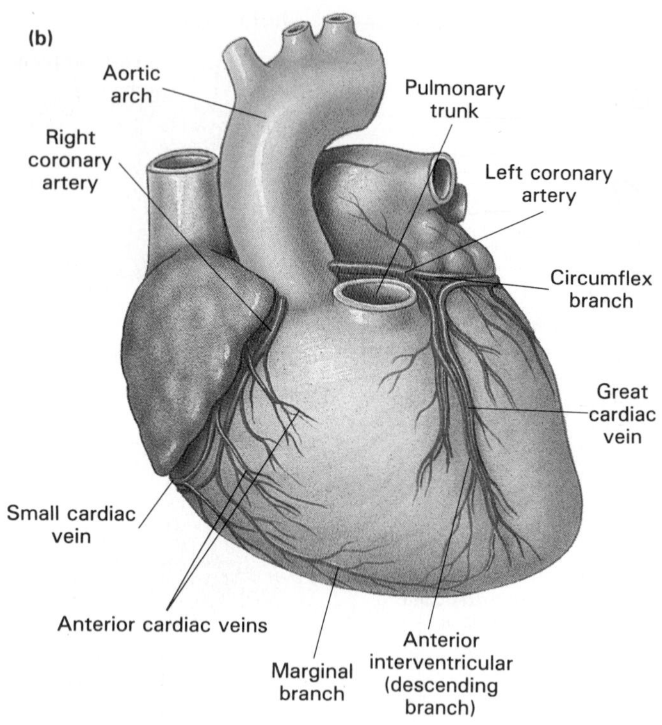

FIGURE 20-6
Coronary Circulation. (a) A cast of the coronary vessels, showing the complexity and extent of the coronary circulation. (b) Coronary vessels supplying the anterior surface of the heart. (c) Coronary vessels supplying the posterior surface of the heart.

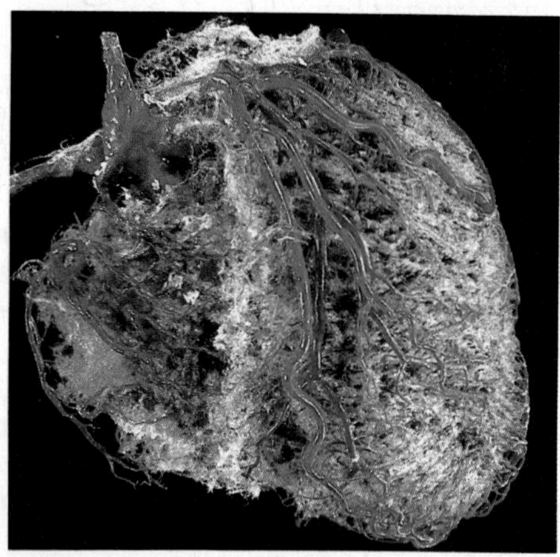

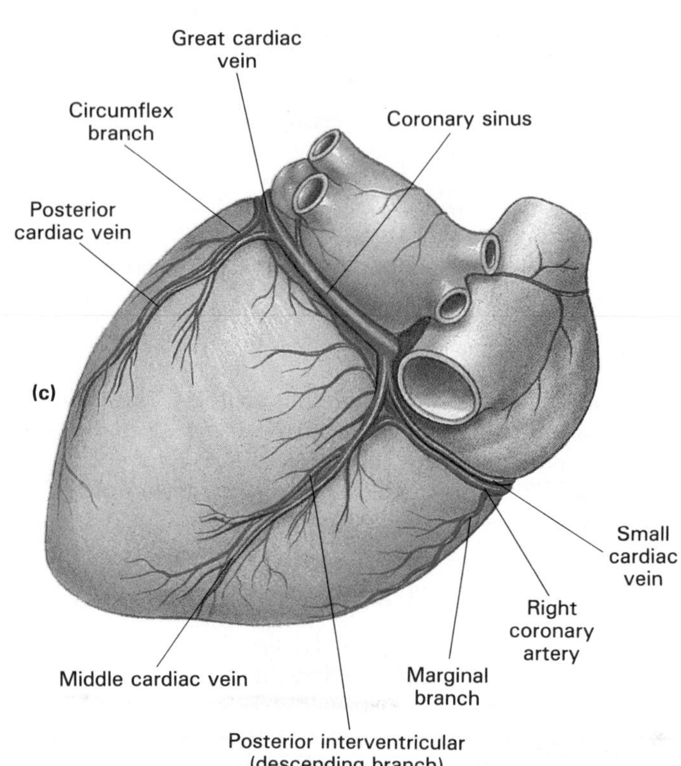

TISSUES OF THE HEART WALL

A section through the wall of the heart reveals several distinct layers: the *epicardium* (visceral pericardium), the *myocardium*, and the *endocardium*. These layers are diagrammed in Figure 20-7a. The epicardium covers the outer surface of the heart. This serous membrane consists of an exposed mesothelium and an underlying layer of loose connective tissue. The **myocardium**, or muscular wall of the heart, contains cardiac muscle tissue and associated connective tissues, blood vessels, and nerves. The myocardium consists of concentric layers of cardiac muscle tissue. These layers wrap around the atria and spiral around the ventricles (Figure 20-7b). The inner surfaces of the heart, including the valves, are covered by a squamous epithelium, the **endocardium** (en-dō-KAR-dē-um; *endo-*, inside), that is continuous with the endothelium of the attached blood vessels. [**CM**: *Infection and Inflammation of the Heart*]

The Fibrous Skeleton

The connective tissues of the heart include large numbers of collagen and elastin fibers. Each cardiac muscle cell is wrapped in a strong but elastic sheath, and adjacent cells are tied together by fibrous cross-

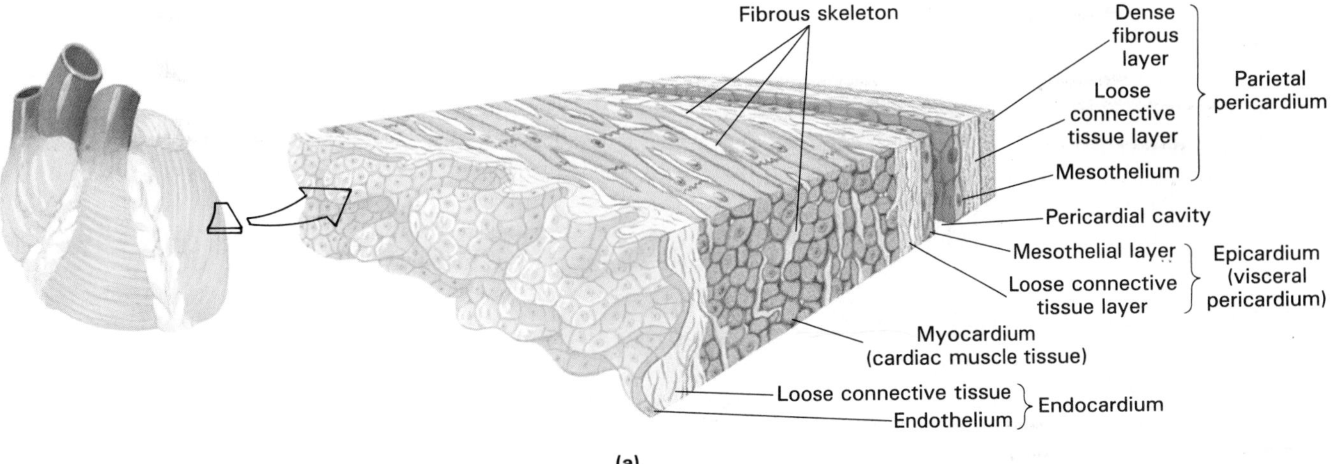

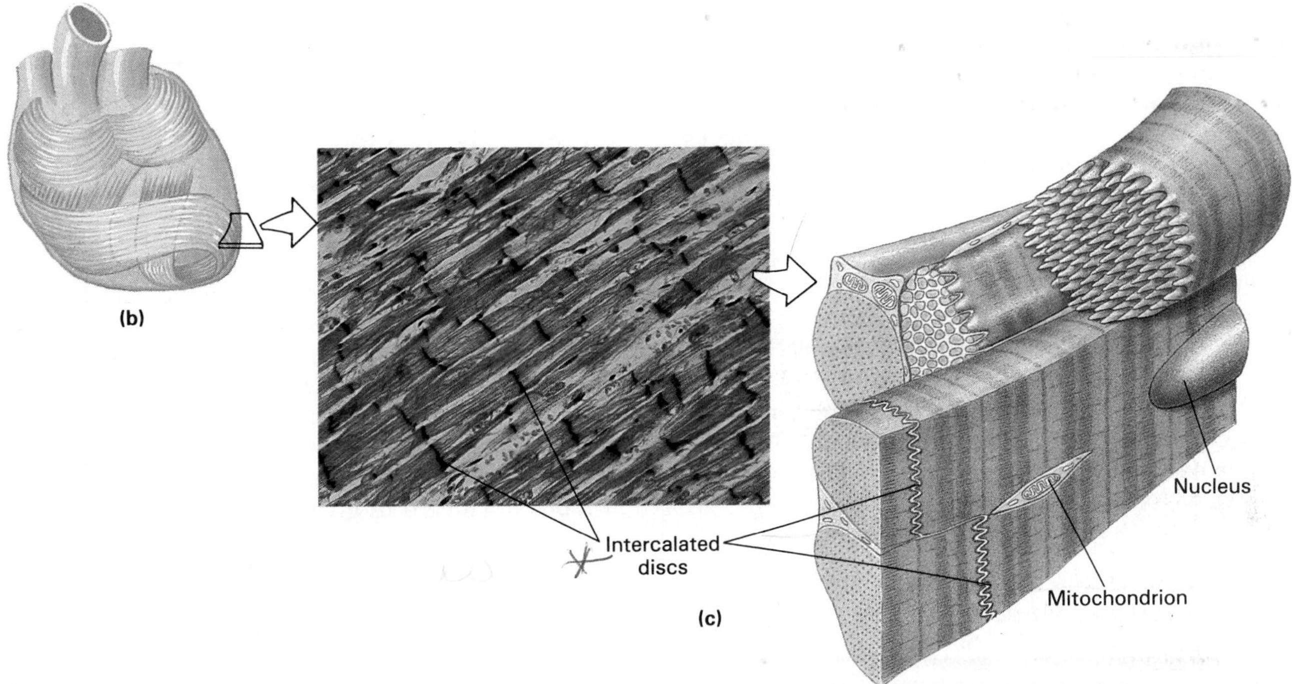

FIGURE 20-7
Organization of Muscle Tissue in the Heart. (a) A diagrammatic section through the heart wall showing the relative positions of the epicardium, myocardium, and endocardium. (b) Cardiac muscle tissue in the heart forms concentric layers that wrap around the atria and spiral within the walls of the ventricles. (c) Sectional and diagrammatic views of cardiac muscle tissue. Histological characteristics of cardiac muscle fibers compared with those of skeletal muscle fibers include: (1) small size, (2) single, centrally placed nucleus, (3) branching interconnections between fibers, and (4) presence of intercalated discs.

links, or "struts." These fibers are in turn interwoven with more extensive sheets of fibrous tissue that separate concentric layers of cardiac muscle fibers and encircle each of the heart valves. This internal connective tissue network is called the **fibrous skeleton** of the heart.

The fibrous skeleton has several important functions:

1. It stabilizes the positions of the muscle fibers and valves in the heart;
2. It provides physical support for the cardiac muscle fibers and for the blood vessels and nerves in the myocardium;
3. It helps distribute the forces of contraction;
4. It adds strength and prevents overexpansion of the heart;

5. It helps to maintain the normal shape of the heart;
6. It provides elasticity that helps return the heart to its original shape after each contraction; and
7. It physically isolates the muscle fibers of the atria from those in the ventricles (the importance of this isolation in normal cardiac function will be detailed below).

Structure of Cardiac Muscle Fiber

Cardiac muscle tissue was introduced in Chapter 5, and its properties were briefly compared with those of other muscle types. Cardiac muscle fibers, or **cardiocytes**, are relatively small, averaging 10–20 μm in diameter and 50–100 μm in length. Each cell has a single, centrally placed nucleus. Typical cardiac muscle fibers are detailed in Figure 20-7c.

As in skeletal muscle fibers, each cardiac muscle fiber contains organized myofibrils, and the presence of many aligned sarcomeres gives it a striated appearance. Contraction involves the shortening of individual sarcomeres, but there are significant differences between skeletal and cardiac muscle fibers with regard to metabolism, membrane characteristics, T tubule structure, and the sarcoplasmic reticulum. (These differences will be detailed in later sections of this chapter.)

Cardiac muscle fibers are almost totally dependent on aerobic respiration to obtain the energy needed to continue contracting. The sarcoplasm of a cardiac muscle fiber thus contains large numbers of mitochondria and abundant reserves of myoglobin (to store oxygen). Energy reserves are maintained in the form of glycogen and lipid inclusions.

Intercalated Discs and Communication between Cardiac Muscle Fibers

Each cardiac muscle fiber contacts several others at specialized sites known as **intercalated** (in-TER-ka-lā-ted) **discs**. At an intercalated disc:

1. The cell membranes of two cardiac muscle fibers are extensively intertwined and bound together by tight junctions. This connection helps to stabilize their relative positions and maintain the three-dimensional structure of the tissue.
2. Myofibrils in the two interlocking muscle fibers are firmly anchored to the membrane at the intercalated disc. Because their myofibrils are essentially locked together, the two muscle fibers can "pull together" with maximum efficiency.
3. Cardiac muscle fibers at an intercalated disc are also connected by gap junctions, allowing ions and small molecules to move from one cell to another. This creates a direct electrical connection between the two muscle fibers, and an action potential can travel across an intercalated disk, moving quickly from one cardiac muscle fiber to another.

Because the cardiac muscle fibers are mechanically, chemically, and electrically connected to one another, the entire tissue resembles a single, enormous muscle fiber. For this reason cardiac muscle has been called a *functional syncytium* (sin-SIT-ē-um; a fused mass of cells).

Damage to the semilunar valves on the right side of the heart would interfere with blood flow to what vessel?

How could you distinguish a sample of cardiac muscle tissue from a sample of skeletal muscle tissue?

Cardiac Physiology

The functional properties of the heart are a direct reflection of the unusual histological characteristics of cardiac muscle tissue. Our discussion of cardiac physiology therefore begins at the cellular level, with an examination of the basic mechanisms of cardiac muscle contraction. We will then proceed to consider how the heart performs as a pump. The last section provides an introduction to the heart as an integrated component of the cardiovascular system, a topic that will be considered in greater detail in Chapter 21.

CARDIAC CONTRACTIONS

We will begin by examining the contraction process in an individual cardiac muscle fiber and then consider the coordination of cellular activities throughout the myocardium. This discussion assumes that you are familiar with the mechanics of skeletal muscle contraction detailed in Chapter 10. (pp. 304–311) (You may wish to review the relevant material before going on.)

The Stimulus for Contraction

As in skeletal muscle, the first step in triggering a contraction is the appearance of an action potential in the sarcolemma. However, the pattern of ion movement in a cardiac muscle fiber action potential differs from that in a skeletal muscle fiber. Figure 20-8 on the next page compares the action potential of a skeletal muscle fiber (a) with that of a cardiac muscle fiber (b). Instead of repolarizing immediately, the transmembrane potential of the cardiac muscle fiber remains near 0 mV for around 200 msec (0.2 sec) before returning to resting levels. Until the membrane repolarizes, it cannot respond to further stimulation, so the refractory period of a cardiac muscle fiber membrane is relatively long. From start to finish, the complete depolarization-repolarization process lasts around 300 msec. This is some 30 times the duration of an action potential in a skeletal muscle sarcolemma. With a single twitch lasting 300 msec, a normal cardiac muscle fiber can increase its rate of contraction to only about 200 per minute, even under maximum stimulation.

An action potential triggers a single contraction, indicated by the dashed lines in Figure 20-9b. Because the muscle fiber has entered the relaxation phase before the refractory period ends, rapid stimulation of a cardiac muscle fiber produces a series of isolated contractions rather than the tetanic contractions seen in skeletal muscles. This distinction is very important, because a tetanic contraction in the heart would prove fatal. (A heart permanently locked in the contracted state cannot pump blood, and the interruption in circulation to the brain will normally cause death in a matter of minutes.)

When an action potential passes across the surface of a cardiac muscle fiber, it activates the contractile machinery inside the cell. This process, called *excitation-contraction coupling*, also differs from that seen in skeletal muscle fibers.

FIGURE 20-8
The Mechanics of Cardiac Muscle Fiber Contraction.
(a) In a skeletal muscle fiber, the refractory period is relatively brief, averaging about 10 msec. Because the membrane has repolarized before the contraction phase ends, repeated stimulation can produce a tetanic contraction. (b) In a cardiac muscle fiber, the refractory period is prolonged, and the muscle fiber enters the relaxation phase before repolarization has been completed. As a result, repeated stimulation cannot produce a tetanic contraction.

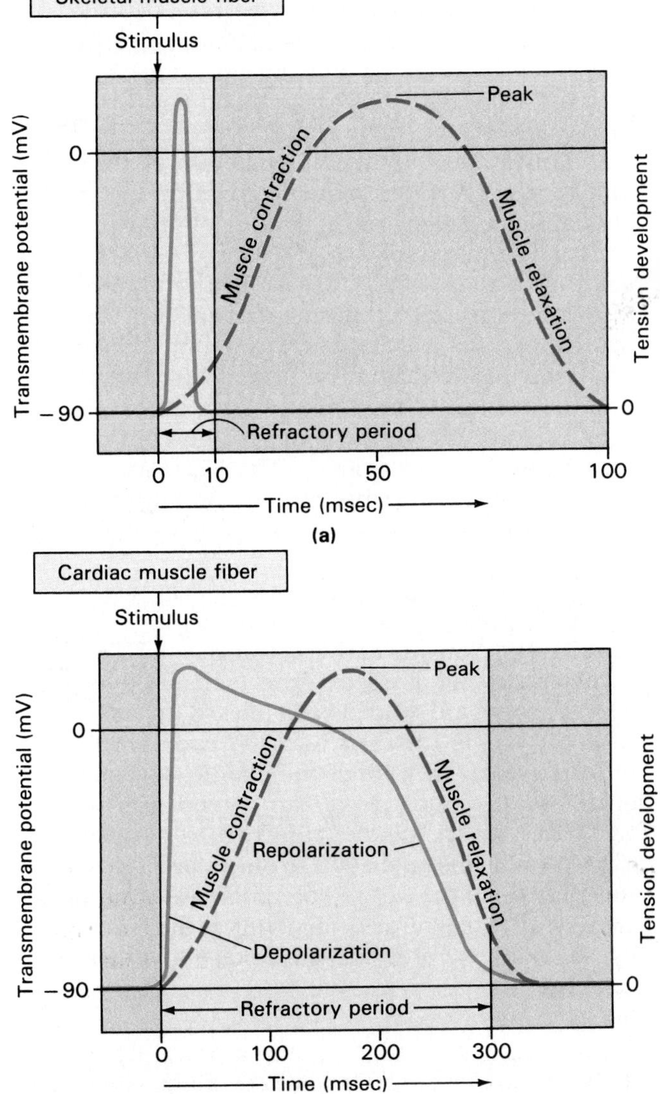

Excitation-Contraction Coupling

In the resting cardiac muscle fiber, calcium ions are absorbed by the sarcoplasmic reticulum (SR) and actively transported out of the cell. As a result, sarcoplasmic calcium concentrations are very low. The event that ultimately triggers a contraction in a cardiac muscle fiber is the appearance of free calcium ions in the sarcoplasm.

Cardiac muscle fibers are much smaller than skeletal muscle fibers, and their internal organization differs from that of skeletal muscle fibers. For example, the T tubules are short and broad, and there are no triads. However, expanded portions of the SR contact the cell membrane or the enlarged T tubules, and when an action potential sweeps across the surface of the cells, calcium ions are released by the SR. In addition, passage of an action potential opens voltage-regulated calcium ion channels, and calcium ions flood into the muscle fiber. The opening of these channels and the continued movement of calcium ions into the sarcoplasm account for the prolonged plateau typical of an action potential in a cardiac muscle fiber. Roughly one-fifth of the calcium ions involved in a contraction cross the cell membrane in this way. More important, the entry of calcium ions through membrane channels in some way promotes the release of calcium reserves in the SR. Extracellular calcium thus contributes to sarcomere activation directly and indirectly. Cardiac muscle fibers are therefore very sensitive to changes in the calcium ion concentration of the extracellular fluid. (This sensitivity also explains the effectiveness of calcium-channel blockers in treating angina, a condition detailed in the Clinical Comment on p. 644.) [**CM:** *Cardiac Contractions and the Composition of the Extracellular Fluid*]

Clinical Comment: Coronary Artery Disease and Myocardial Infarction

Coronary artery disease The term **coronary artery disease** (**CAD**) refers to degenerative changes in the coronary circulation. The effects produced range from the virtually unnoticeable to the immediately fatal. Cardiac muscle fibers need a constant supply of oxygen and nutrients, and any reduction in coronary circulation produces a corresponding reduction in cardiac performance. Such reduced circulatory supply, known as **coronary ischemia** (is-KĒ-mē-a), usually results from partial or complete blockage of the coronary arteries. The usual cause is the formation of a plaque in the wall of a coronary vessel. (The bulk of a plaque consists of lipid deposits, especially cholesterol; other aspects of plaque structure will be detailed in Chapter 21.) The plaque deposit, or an associated thrombus (Chapter 19), then narrows the passageway and reduces blood flow. (pp. 626–627) Spasms in the smooth muscles of the vessel wall can further decrease blood flow or even stop it altogether. A variety of imaging procedures can be used to visualize coronary circulation. Figure 20-9 compares a PET scan of the normal heart (a) with a heart that has severely restricted coronary circulation (b).

CAD is progressive, and in the early stages the individual may be unaware of the problem. One of its first symptoms is often **angina pectoris** (an-JĪ-na PEK-tor-is; *angina*, pain spasm + *pectoris*, of the chest). In the most common form of angina, temporary insufficiency and ischemia develop when the workload of the heart increases. Although the individual may feel comfortable at rest, any unusual exertion or emotional stress can produce a sensation of pressure, chest constriction, and pain that may radiate from the sternal area to the arms, back, and neck.

Angina can often be controlled by a combination of drug treatment and changes in lifestyle. Lifestyle changes to combat angina include (1) limiting activities known to trigger angina attacks, such as strenuous exercise, and avoiding stressful situations; (2) stopping smoking; and (3) modifying the diet to lower fat consumption so that plaque deposition and growth will be slowed. Among the medications useful for controlling angina, beta-blockers such as *propranolol* or *metoprolol* are often prescribed because these drugs prevent the stimulation of heart muscle by the sympathetic nervous system. When these blockers are present, the existing parasympathetic tone causes a reduction in heart rate. (The effects of beta-receptor stimulation and autonomic tone were considered in Chapter 16. (pp. 506–508, 514–515))

The drug *nitroglycerin* (nī-trō-GLIS-er-in) is usually administered transdermally, by means of a patch worn on the skin or orally, by means of a small tablet placed under the tongue. Nitroglycerin is useful because it is a powerful vasodilator (a compound that dilates blood vessels). When peripheral vessels dilate, the heart does not have to work as hard to force blood around the systemic circuit; when the coronary vessels dilate, blood flow to the heart muscle improves. This combination eases the strain on the heart, reduces its oxygen demand, and relieves the symptoms of angina. Another vasodilator, the hormone atrial natriuretic peptide, is now being produced by genetic engineering techniques. This compound, discussed in Chapter 18, can increase coronary blood flow by up to 25 percent. (p. 589)

Other drugs, known as **calcium channel blockers**, may be administered during acute angina attacks. Following an action potential, calcium ions moving across the cell membrane trigger the cardiac contraction. Calcium channel blockers restrict the movement of calcium ions into the cardiac muscle cells, weakening the contractions and lowering the demand for oxygen and nutrients. These drugs also affect the smooth muscles in the walls of the coronary vessels, and as the smooth muscles relax, the vessels dilate and cardiac circulation improves. The combined effects relieve the angina symptoms.

Angina can also be treated surgically. A single, soft plaque may be removed with the aid of a slender, elongate **catheter** (KATH-e-ter). The catheter, a small-diameter tube, is inserted into a large artery and guided back toward the heart. Ultimately it is directed into the coronary arteries to reach the site of the plaque. A variety of different surgical tools can be slid into the catheter, and the plaque can then be attacked with laser beams or chewed to pieces by a miniature version of the Roto-Rooter machine. Debris created during plaque destruction is sucked up by the catheter, preventing blockage of smaller vessels.

(a)

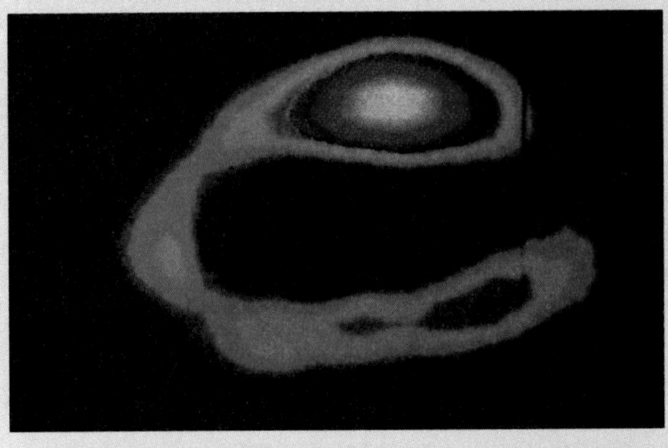
(b)

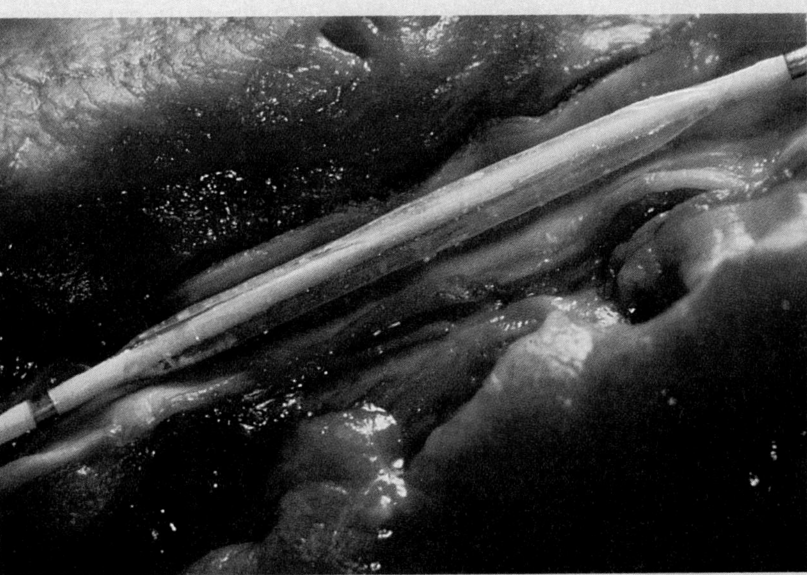
(c)

FIGURE 20-9
Coronary Circulation and Clinical Testing. (a) A PET scan of the heart, taken after introduction of an imaging dye containing the radioisotope technetium-99. The brighter the color, the greater the blood flow through the tissue. The wall of this heart has an extensive circulatory supply. (b) A PET scan of a damaged heart produced using the same procedure. Most of the heart is deprived of circulation. (c) Balloon angioplasty can sometimes be used to remove a circulatory blockage. The catheter is guided through the coronary arteries to the site of blockage and inflated to press the soft plaque against the vessel wall.

In **balloon angioplasty** (AN-jē-ō-plas-tē; *angeion*, vessel) the catheter tip contains an inflatable balloon. Once in position, the balloon is inflated, pressing the plaque against the vessel walls (Figure 20-7c). This procedure works best in small (under 10 mm), soft plaques. Several factors make this procedure highly attractive: (1) the mortality rate during surgery is only around 1 percent, (2) the success rate is over 90 percent, and (3) it can be performed on an outpatient basis. Although in about 20 percent of patients the plaque deposit returns to its original size within 6 months, the procedure can be repeated. Unfortunately only around 10 percent of severe angina patients have isolated problems suitable for balloon angioplasty.

Coronary bypass surgery involves taking a small section from either a small artery (often the *internal thoracic artery*) or a peripheral vein and using it to create a detour around the obstructed portion of a coronary artery. As many as four coronary arteries can be rerouted this way during a single operation. The procedures are named according to the number of vessels repaired, so one speaks of single, double, triple, or quadruple coronary bypass operations. When performed before significant heart damage has occurred, the mortality rate during surgery is relatively low (1–2 percent). Under these conditions the procedure completely eliminates the angina symptoms in 70 percent of the cases and provides partial relief to an another 20 percent.

Bypass surgery was initially hailed as the best treatment for angina, and patients who might benefit from other therapies often elected to undergo coronary bypass surgery. Although it does offer certain advantages, recent studies have shown that for mild angina, coronary bypass surgery does not yield significantly better results than drug therapy. Current recommendations are that coronary bypass surgery should be reserved for cases of severe angina that do not respond to other treatment.

Myocardial infarction In a **myocardial** (mī-ō-KAR-dē-al) **infarction** (**MI**), or *heart attack*, the coronary circulation becomes blocked and the cardiac muscle cells die from lack of oxygen. The affected tissue then degenerates, creating a nonfunctional area known as an *infarct*. Heart attacks most often result from severe coronary artery disease. The consequences depend on the site and nature of the circulatory blockage. If it occurs near the base of one of the coronary arteries the damage will be widespread, and the heart will probably stop beating. If the blockage involves one of the smaller arterial branches, the individual may survive the immediate crisis, but there are many potential complications, all unpleasant. As scar tissue forms in the damaged area, the heartbeat may become irregular, and other vessels can become constricted, creating additional circulatory problems.

Myocardial infarctions are most often associated with fixed blockages, such as those seen in CAD. When the crisis develops due to thrombus (clot) formation at a plaque, the condition is called **coronary thrombosis**. A vessel already narrowed by plaque formation may also become blocked by a sudden spasm in the smooth muscles of the vascular wall. The individual then experiences intense pain, similar to that of an angina attack but persisting even at rest.

The cytoplasm of a damaged cardiac muscle fiber differs from that of a normal muscle cell. As the supply of oxygen decreases, the cells become more dependent on anaerobic glycolysis to meet their energy needs. (The balance between aerobic and anaerobic ATP generation in muscle fibers was discussed in Chapter 10.) ∞ (pp. 318–321) Over time the cytoplasm accumulates large numbers of enzymes involved with anaerobic metabolism.

As their cell membranes deteriorate, dying cardiac muscle fibers release these enzymes into the surrounding intercellular fluids. The appearance of anaerobic metabolic enzymes in the circulation thus indicates that an infarct has occurred. The enzymes tested for in a diagnostic blood test include **lactate dehydrogenase** (**LDH**), **serum glutamic oxaloacetic transaminase** (**SGOT,** also called *aspartate aminotransferase*), **creatine phosphokinase** (**CPK**, or **CK**), and a special form of **creatine phosphokinase** found only in cardiac muscle (**MB-CK**).

Roughly 25 percent of MI patients die before obtaining medical assistance, and 65 percent of MI deaths among those under age 50 occur within an hour after the initial infarct. The goals of treatment are to limit the size of the infarct and prevent additional complications by preventing irregular contractions, improving circulation with vasodilators, providing additional oxygen, reducing the cardiac workload, and, if possible, eliminating the cause of the circulatory blockage. Anticoagulants may help prevent the formation of additional thrombi, and clot-dissolving enzymes may reduce the extent of the damage if they are administered within 6 hours after the MI has occurred.

There are roughly 1.3 million MIs in the United States each year, and half of the victims die within a year of the incident. A number of factors have been identified that appear to increase the risk of a heart attack. They include smoking, high blood pressure, high blood cholesterol levels, diabetes, and obesity. There are also hereditary factors that may predispose an individual to coronary artery disease. The presence of two risk factors more than doubles the risk, so eliminating as many risk factors as possible will improve one's chances of preventing or surviving a heart attack. For example, changes in eating habits to limit dietary cholesterol, exercise to lower weight, and seeking treatment for high blood pressure are relatively easy steps in the right direction. The benefits are considerable; it has been estimated that reduction of coronary risk factors could prevent 150,000 deaths each year in the United States alone.

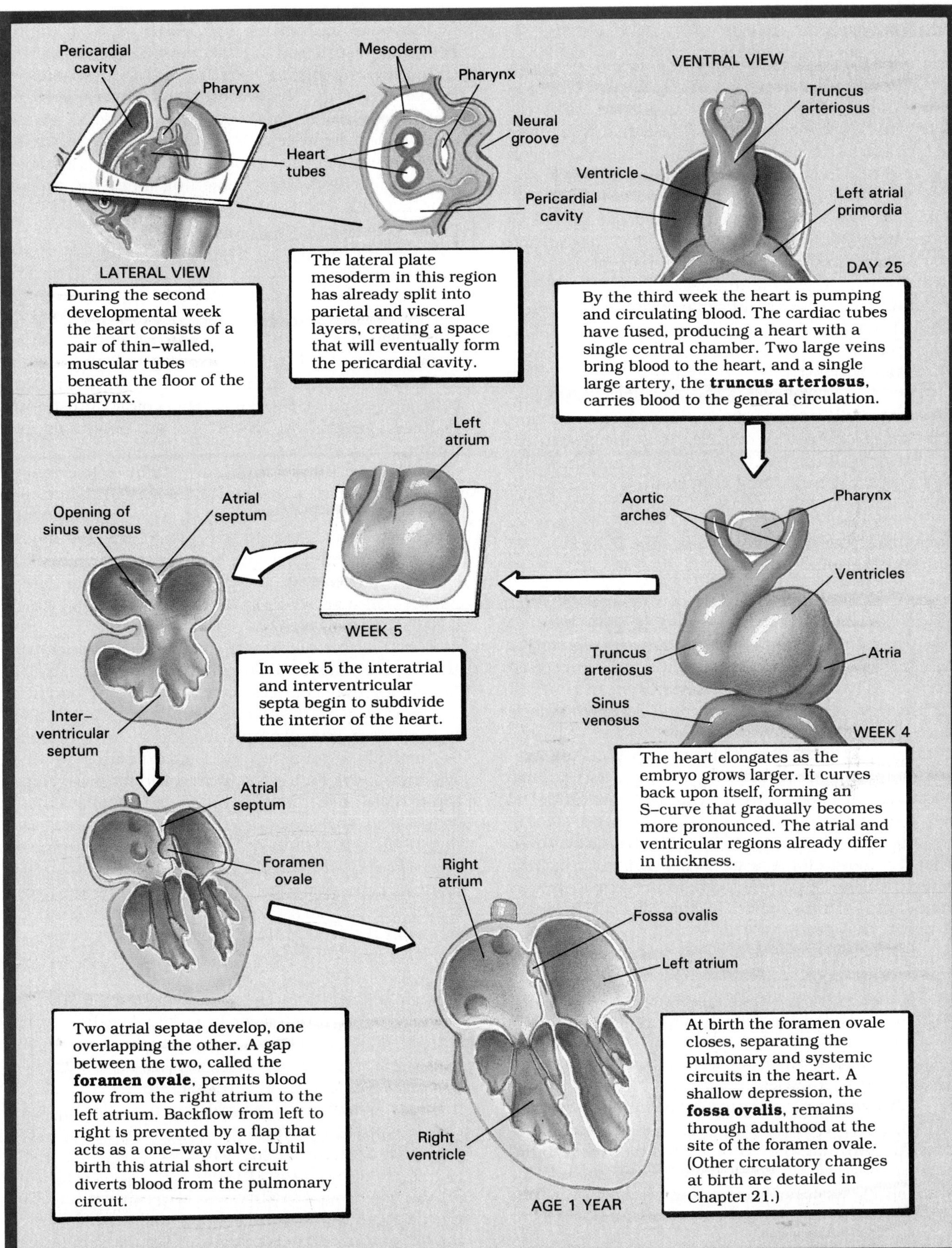

During the second developmental week the heart consists of a pair of thin-walled, muscular tubes beneath the floor of the pharynx.

The lateral plate mesoderm in this region has already split into parietal and visceral layers, creating a space that will eventually form the pericardial cavity.

By the third week the heart is pumping and circulating blood. The cardiac tubes have fused, producing a heart with a single central chamber. Two large veins bring blood to the heart, and a single large artery, the **truncus arteriosus**, carries blood to the general circulation.

The heart elongates as the embryo grows larger. It curves back upon itself, forming an S-curve that gradually becomes more pronounced. The atrial and ventricular regions already differ in thickness.

In week 5 the interatrial and interventricular septa begin to subdivide the interior of the heart.

Two atrial septae develop, one overlapping the other. A gap between the two, called the **foramen ovale**, permits blood flow from the right atrium to the left atrium. Backflow from left to right is prevented by a flap that acts as a one-way valve. Until birth this atrial short circuit diverts blood from the pulmonary circuit.

At birth the foramen ovale closes, separating the pulmonary and systemic circuits in the heart. A shallow depression, the **fossa ovalis**, remains through adulthood at the site of the foramen ovale. (Other circulatory changes at birth are detailed in Chapter 21.)

Embryology Summary: Development of the Heart

Energetics and Cardiac Muscle Fiber Contractions

Once a contraction has started, the necessary energy is obtained by the mitochondrial breakdown of fatty acids (stored as lipid droplets) and glucose (stored as glycogen). When necessary, cardiac muscle can also metabolize lactic acid generated by skeletal muscle and other tissues. These are aerobic reactions, and they can occur only when oxygen is readily available. (You may wish to review the discussion of aerobic metabolism in Chapter 4. (p. 103) In addition to obtaining oxygen from the coronary circulation, cardiac muscle fibers maintain sizable reserves of oxygen bound to the heme units of myoglobin molecules. (The structure of this globular protein was detailed in Chapter 2, and its function in slow skeletal muscle fibers was discussed in Chapter 10. (pp. 53, 322) The presence of myoglobin gives the myocardium its deep red color. Under normal circumstances the combination of circulatory supplies plus myoglobin reserves is enough to meet the oxygen demands of the heart, even when working at maximum capacity.

Coordination of Cardiac Muscle Fiber Contractions

At each intercalated disc, one cardiac muscle fiber is mechanically and electrically connected to another. As a result, the contraction of any one fiber will trigger the contraction of several others, and the contraction will spread throughout the myocardium. In contrast to the situation in skeletal muscle, cardiac muscle tissue contracts on its own, in the absence of neural or hormonal stimulation. This property is called *automaticity*, or *autorhythmicity*. (Automaticity is also characteristic of some types of smooth muscle tissue discussed in Chapter 24.) Neural or hormonal stimuli can alter the basic rhythm of contraction, but even a heart removed for a heart transplant will continue to beat unless steps are taken to prevent it. (Heart transplants are discussed in the Clinical Manual.) [**CM**: *The Cardiomyopathies*]

Each contraction follows a precise sequence: the atria contract first, followed by the ventricles. If the contractions follow another sequence, the normal pattern of blood flow is disturbed. For example, if all of the chambers contract at the same time, the closing of the AV valves prevents blood flow between the atria and ventricles.

Cardiac contractions are coordinated by specialized cardiac muscle fibers incapable of undergoing powerful contractions. There are two distinct populations of these cells. **Nodal cells** are responsible for establishing the rate of cardiac contraction, and **conducting fibers** distribute the contractile stimulus to the general myocardium.

Nodal Cells Nodal cells are unusual because their cell membranes depolarize spontaneously. When threshold is reached, an action potential develops in the nodal cell membrane. Each time an action potential occurs and the membrane repolarizes, the cycle begins again, generating action potentials at regular intervals. Nodal cells are electrically coupled to one another, to conducting fibers, and to normal cardiac muscle fibers. As a result, when an action potential appears in a nodal cell, it sweeps through the conducting system, reaching all of the cardiac muscle tissue and causing a contraction. In this way nodal cells determine the heart rate.

Not all nodal cells depolarize at the same rate, and the normal rate of contraction is established by the nodal cells that reach threshold first. These **pacemaker cells** are found in the **cardiac pacemaker**, or **sinoatrial** (sī-nō-Ā-trē-al) **node** (**SA node**). The SA node is embedded in the posterior wall of the right atrium, near the entrance of the superior vena cava. Pacemaker cells depolarize rapidly and spontaneously, generating 70–80 action potentials per minute.

The spontaneous depolarization of pacemaker cells results from a gradual reduction in membrane permeability to potassium ions. This reduction upsets the balance between the rates of potassium loss and sodium entry; because the cell now gains Na^+ faster than it loses K^+, depolarization occurs. Figure 20-10 diagrams this gradual depolarization, which is called a **prepotential**.

In Figure 20-10a the nodal cell is generating action potentials at a rate of around 60 per minute. Any factor that changes either the resting potential or the rate of spontaneous depolarization will alter that rhythm. For example, consider the way nodal cell activity is adjusted by the autonomic nervous system. Acetylcholine (ACh) released by parasympathetic motor neurons hyperpolarizes nodal cell membranes and slows the rate of spontaneous depolarization (Figure 20-10b). These actions lower the heart rate. Norepinephrine (NE) released by sympathetic neurons depolarizes the nodal membranes and accelerates the depolarization rate. The result is an increase in the heart rate (Figure 20-10c).

Conducting Fibers The stimulus for a contraction is generated at the SA node, but it must be distributed so that: (1) the atria contract together, before the ventricles, and (2) the ventricles contract together, in a wave that begins at the apex and spreads toward the base. When the ventricles contract in this way, blood is pushed toward the base of the heart, into the aortic and pulmonary trunks.

If the contractile stimulus simply passed from cell to cell, this sequence would be disrupted. The right atrium would contract first, followed closely by the left atrium. Action potentials would then spread from the atrial myocardium to the ventricular myocardium, producing a wave of contraction that would begin near the base and spread toward the apex.

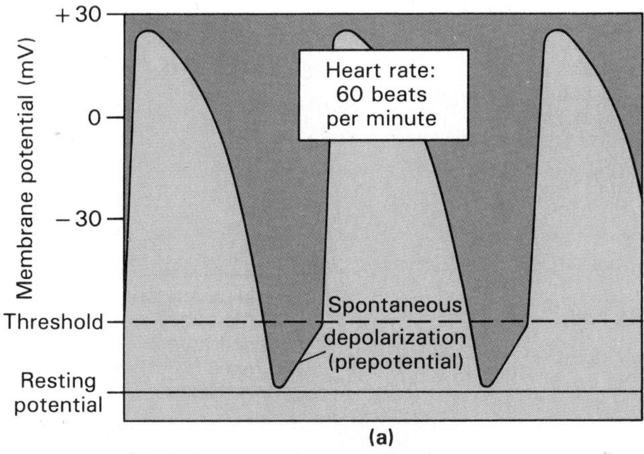

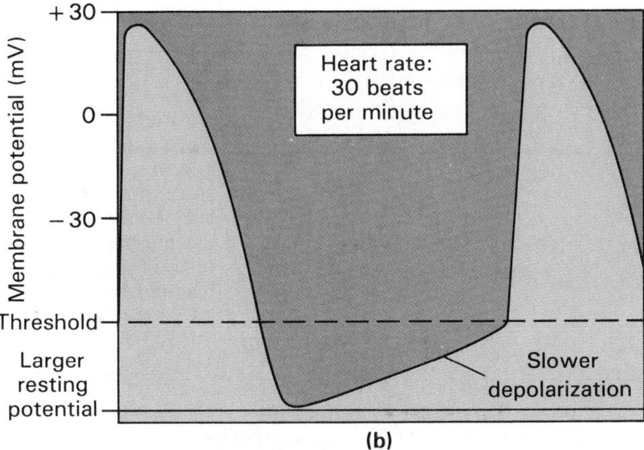

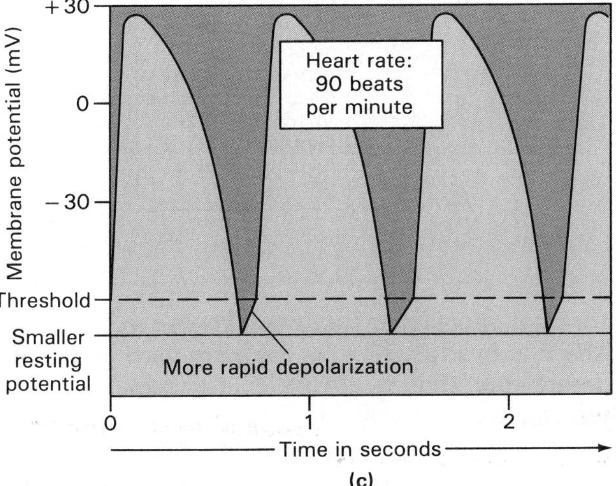

FIGURE 20-10
Pacemaker Function. (a) Pacemaker cells have membrane potentials slightly lower than those of other cardiac muscle fibers. Their cell membranes undergo spontaneous depolarization to threshold, producing action potentials at a frequency determined by (1) the resting membrane potential and (2) the rate of spontaneous depolarization.
(b) Parasympathetic stimulation releases ACh, which hyperpolarizes pacemaker cells and reduces the rate of spontaneous depolarization. This slows the heart rate.
(c) Sympathetic stimulation releases NE, which depolarizes pacemaker cells and accelerates the rate of spontaneous depolarization. This increases the heart rate.

The relationship between nodal cells and conducting fibers prevents this abnormal contraction sequence. Three basic principles are involved:

1. The membranes of conducting fibers distribute action potentials much more rapidly than do typical cardiac muscle fiber membranes. Conducting fibers create a rapid transit system that delivers the contractile stimulus to distant portions of the heart before the stimulus can arrive by relay from cell to cell.
2. The fibrous skeleton of the heart separates the atrial and ventricular myocardia. This separation blocks impulse conduction between cardiac muscle fibers in the atria and those in the ventricles. Conducting fibers thus form the only electrical connection between them.
3. The rate of conduction varies from one part of the conducting system to another. This variation in rate helps coordinate the contraction process. For example, a delay in conduction between the atria and ventricles allows time for atrial contraction and blood movement before ventricular contraction begins.

The conducting network of the heart is illustrated in Figure 20-11. The SA nodal cells are electrically connected to those of the larger **atrioventricular** (ā-trē-ō-ven-TRIK-ū-lar) **node** (**AV node**) via conducting fibers in the atrial walls. Although these nodal cells also depolarize spontaneously, they generate only 40–60 action potentials per minute. Under normal circumstances, before an AV cell depolarizes to threshold spontaneously, it is stimulated by an action potential generated by the SA node. However, if the AV node no longer receives action potentials from the SA node, it will become the pacemaker for the heart. (Further details will be found in the discussion of cardiac abnormalities in a later section.)

The AV node sits within the floor of the right atrium near the opening of the coronary sinus. From here the action potentials travel to the **AV bundle**, also known as the *bundle of His* (hiss). This rather massive bundle of conducting fibers travels along the interventricular septum a short distance before dividing into **bundle branches** that radiate across the inner surfaces of the right and left ventricles. At this point specialized **Purkinje** (pur-KIN-jē) **cells** (*Purkinje fibers*) convey the impulses to the contractile cells of the ventricular myocardium.

Pacemaker cells in the SA node usually generate action potentials at rates of between 60 and 100 per minute. It takes roughly 66 msec for an action potential to travel from the SA node to the AV node over the conducting pathways. Along the way, the conducting fibers pass the contractile stimulus to cardiac muscle fibers of the right and left atria. The action potential then spreads across the atrial surfaces

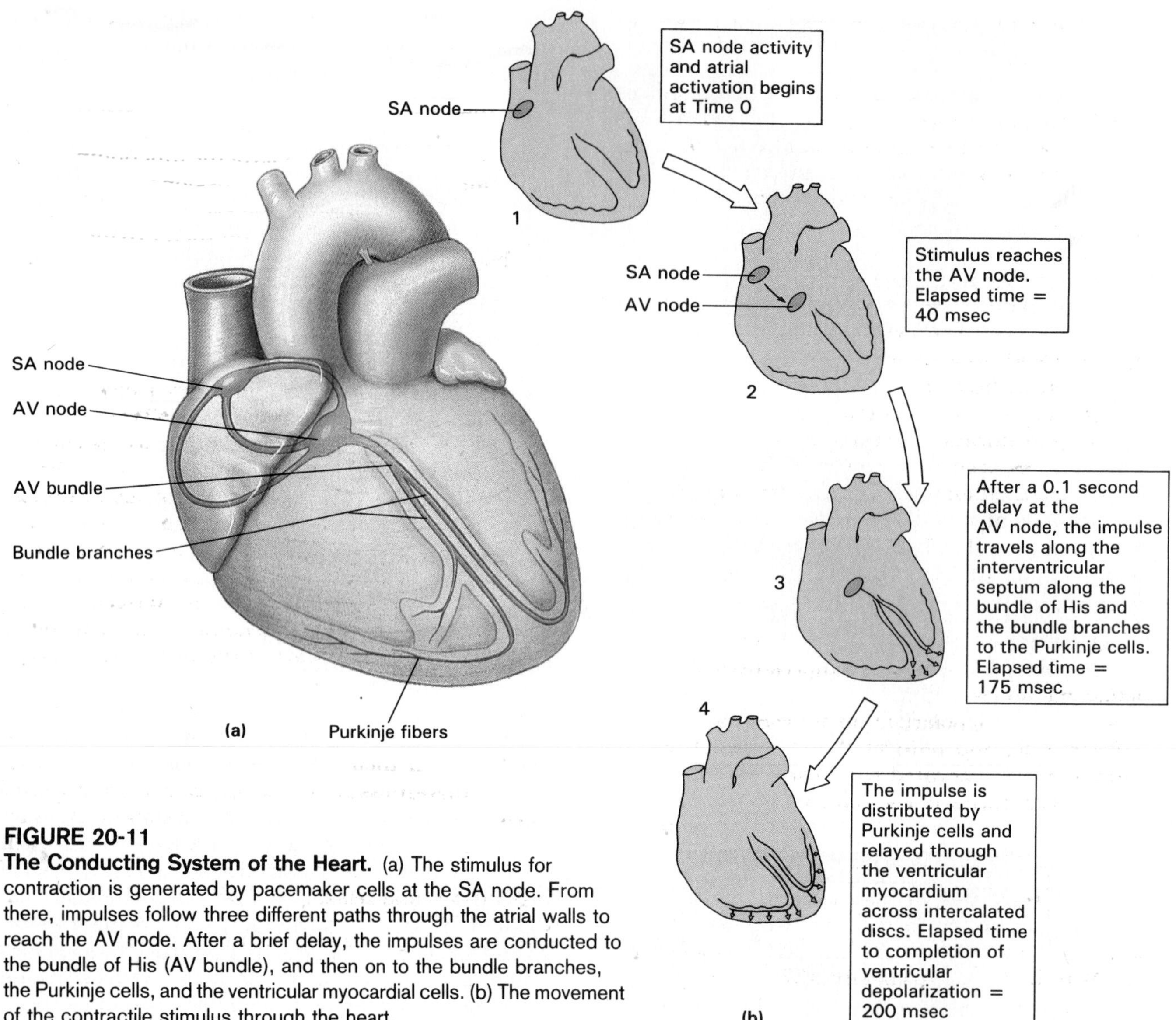

SA node
SA node activity and atrial activation begins at Time 0
1
SA node
AV node
Stimulus reaches the AV node. Elapsed time = 40 msec
2
After a 0.1 second delay at the AV node, the impulse travels along the interventricular septum along the bundle of His and the bundle branches to the Purkinje cells. Elapsed time = 175 msec
3
4
The impulse is distributed by Purkinje cells and relayed through the ventricular myocardium across intercalated discs. Elapsed time to completion of ventricular depolarization = 200 msec
(b)
SA node
AV node
AV bundle
Bundle branches
(a)
Purkinje fibers

FIGURE 20-11
The Conducting System of the Heart. (a) The stimulus for contraction is generated by pacemaker cells at the SA node. From there, impulses follow three different paths through the atrial walls to reach the AV node. After a brief delay, the impulses are conducted to the bundle of His (AV bundle), and then on to the bundle branches, the Purkinje cells, and the ventricular myocardial cells. (b) The movement of the contractile stimulus through the heart.

through cell-to-cell contact. The stimulus affects only the atria, because the fibrous skeleton electrically isolates the atria from the ventricles everywhere except at the AV bundle.

At the AV node the impulse slows down, and another 50 msec passes before it reaches the AV bundle. This delay is important, because the atria must be contracting, and blood movement must be under way, before the ventricles are stimulated. Once the impulse enters the AV bundle it flashes down the septum, along the bundle branches, and into the ventricular myocardium via Purkinje cells. These large-diameter fibers conduct action potentials very rapidly, as fast as small myelinated axons. Within about 150 msec the signal to begin a contraction has reached all of the ventricular cardiac muscle fibers.

A number of clinical problems are the result of abnormal pacemaker function. **Bradycardia** (brād-e-KAR-dē-a; *bradys*, slow) is the term used to indicate a heart rate that is slower than normal, whereas **tachycardia** (tak-e-KAR-dē-a; *tachys*, swift) indicates a faster than normal heart rate. These are relative terms, and in clinical practice the definition varies depending on the normal resting heart rate of the individual. [**CM**: *Problems with Pacemaker Function*]

Extracellular Fluid and Cardiac Contractions

The pace established by the SA node depends on (1) the resting membrane potential of the nodal cells and (2) the rate of spontaneous depolarization. Any factor that changes the resting potential or the rate of depolarization will therefore have a direct effect on the heart rate, and the force of contraction may be changed as well. Examples of important factors include:

1. Ion concentrations in the extracellular fluid: Abnormal extracellular *potassium ion* concentrations alter the resting potential at the SA node and change the heart rate. For example, when the extracellular concentration of potassium declines, the rate of potassium diffusion out of the muscle fiber increases. (Recall from Chapter 3 that an increase in a concentration gradient will accelerate the rate of diffusion down the gradient. (pp. 78–79) As a result, a decrease in extracellular potassium concentration leads to membrane hyperpolarization and a reduction in heart rate. Changes in extracellular *calcium ion* concentrations primarily affect the strength and duration of cardiac contractions. For example, a rise in calcium ion concentrations leads to increased excitability and prolonged contractions. (Additional details concerning the effects of changes in potassium and calcium ion concentrations can be found in the Clinical Manual.)

2. Changes in body temperature: Temperature changes affect metabolic operations throughout the body. For example, lowering temperatures depresses cellular metabolism. At the heart, this depression slows the rate of depolarization at the SA node, lowers the heart rate, and reduces the strength of cardiac contractions. An elevated body temperature accelerates the heart rate and the contractile force, one reason why your heart seems to be racing and pounding whenever you have a fever.

3. Autonomic activity: Autonomic effects on heart rate primarily reflect the responses of nodal cells, particularly those of the SA node, to acetylcholine (ACh) and norepinephrine (NE). The effects of these neurotransmitters on nodal cells have already been discussed. Epinephrine released by the adrenal medulla during sympathetic activation has a similar effect on heart rate, as do stimulants such as caffeine. In excessive amounts these compounds can make the myocardium so excitable that abnormal contractions occur. Epinephrine and norepinephrine also increase the strength of cardiac contractions, by a mechanism detailed in a later section.

How do you think a decrease in extracellular calcium ion concentration would affect the strength of a cardiac contraction?

If the cells of the SA node were not functioning, what affect would this have on heart rate?

Why is it important for the impulses from the atria to be delayed at the AV node before passing into the ventricles?

The Heart as a Pump

Now that we have described the histological and physiological basis for coordinated contractions, we can proceed with an analysis of the functioning heart as an integrated unit. Our discussion will begin with a detailed examination of the events that take place during a typical heartbeat, and then expand to consider how cardiac performance can be varied to meet the changing needs of peripheral tissues.

THE CARDIAC CYCLE

The period between the start of one heartbeat and the beginning of the next is a single **cardiac cycle**. The cardiac cycle therefore includes alternate periods of contraction and relaxation. For any one chamber in the heart, the cardiac cycle can be divided into two phases. During contraction, or **systole** (SIS-to-lē), the chamber pushes blood into an adjacent chamber or into an arterial trunk. Systole is followed by the second phase, one of relaxation, or **diastole** (dī-AS-to-lē). During diastole the chamber fills with blood and prepares for the start of the next cardiac cycle.

The function of any pump is to develop pressure and move a particular volume of fluid in a specific direction at an acceptable speed. Unless obstacles are placed in their path, fluids will move from an area of high pressure to one of relatively lower pressure. When you open a faucet, a constant pressure provided by mechanical pumps forces fluids through the pipes. By contrast, the heart works in cycles of contraction and relaxation, and the pressure within each chamber alternately rises and falls. Valves between adjacent chambers help to ensure that blood flows in the desired direction, but the mere presence of valves is not enough to accomplish this.

Blood will flow from one chamber to another only if the pressure in the first chamber exceeds that in the second. This basic principle governs the movement of blood between atria and ventricles, between ventricles and arterial trunks, and between the major veins and the atria. The correct pressure relationships are dependent on the careful timing of contractions. For example, blood movement could not occur in the desired direction if an atrium and its attached ventricle contracted at precisely the same moment. The elaborate pacemaking and conduction systems normally provide the required spacing between atrial and ventricular systole.

It is possible to monitor alterations in pressure within the chambers of a functioning heart. Figure 20-12 summarizes the information recorded from the left atrium and ventricle during a normal cardiac cycle. The cycle begins with atrial systole. As the atria contract, the elevated atrial pressures force blood into the ventricles. At this point the ventricles are already

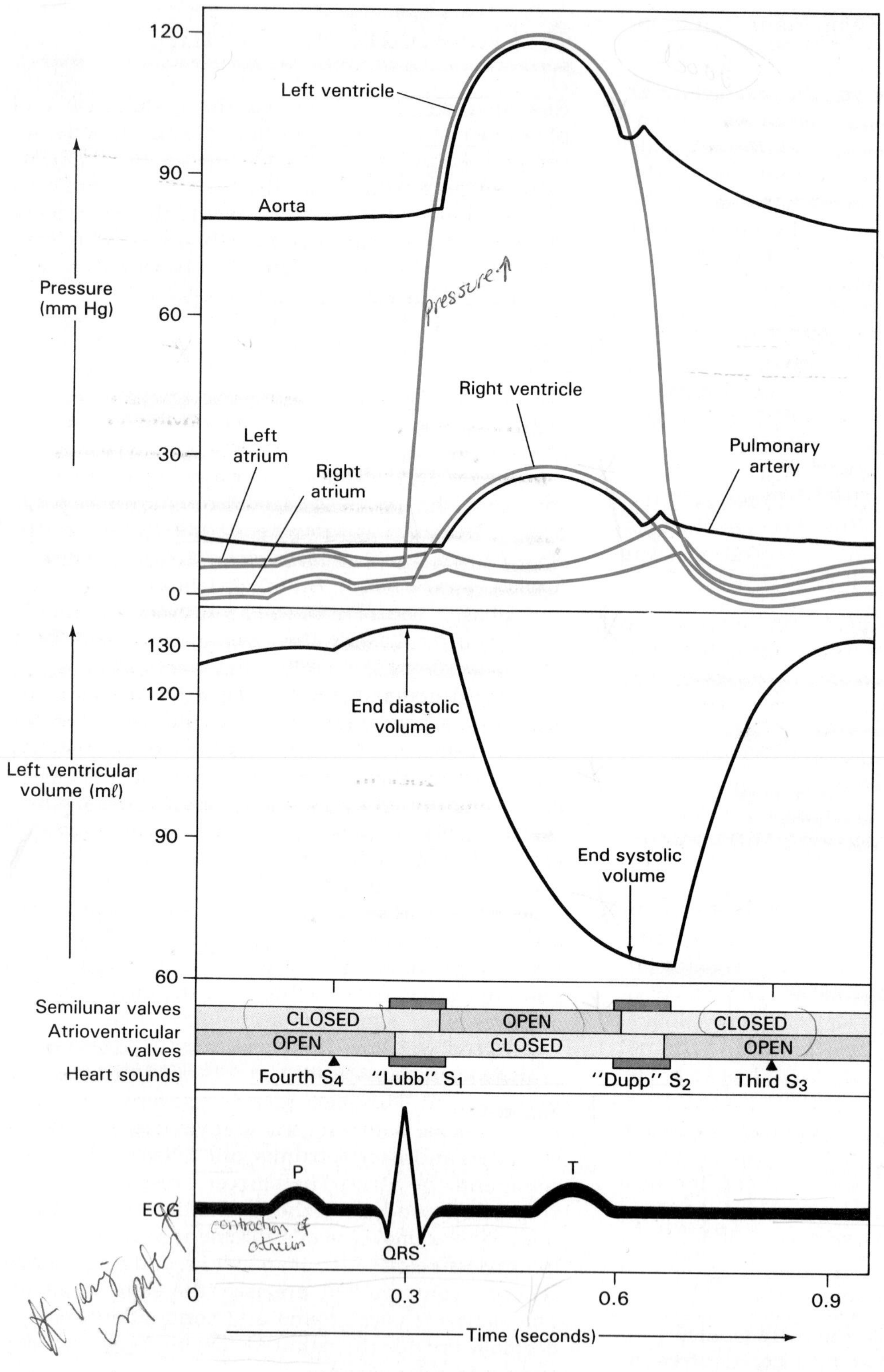

FIGURE 20-12 Details of a Cardiac Cycle. An integrated view of pressure and volume changes in the heart and great vessels, the ECG, heart sounds, and the status of coronary valves during a single cardiac cycle.

filled to around 70 percent of capacity, and atrial systole essentially "tops them off" by providing the additional 30 percent. As atrial systole ends, ventricular systole begins. As the internal pressures rise above those in the atrium, the AV valves swing shut. But blood does not begin moving into the arterial trunks until ventricular pressures exceed the arterial pressures. At this point the blood pushes open the semilunar valves and flows into the aortic and pulmonary trunks. This blood flow continues for the duration of ventricular systole.

When ventricular diastole begins, ventricular pressures decline rapidly. Once the ventricular pressures have fallen below those of the efferent vessels, blood starts to flow back toward the ventricles. This movement closes the semilunar valves. Ventricular pressures continue to decline, soon falling below those of the atria. The AV valves then open, and blood

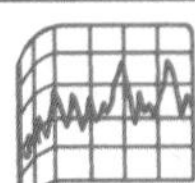

Diagnostics: Monitoring the Living Heart

Many different techniques can be used to examine the structure and performance of the living heart. No single diagnostic procedure can provide the complete picture, so the tests used will vary depending on the suspected nature of the problem. A standard chest X-ray will show the basic size, shape, and orientation of the heart. Additional details require more specialized procedures to enhance the clarity of the images.

Coronary arteriography (ar-tē-rē-OG-ra-fē) is often used to look for restrictions in the coronary circulation. In this procedure a catheter is inserted into a major artery in the arm or leg and then maneuvered back along the arterial passageways until its tip arrives at the heart. A radiopaque dye can then be released at the openings of the coronary arteries and its distribution followed in a series of high-speed X-rays. For direct analyses of cardiac performance and the collection of blood samples, a catheter may be introduced into the heart itself. The instrument can enter via the aorta, as already described, or from the venous system by way of the inferior vena cava.

Because the heart is constantly moving, ordinary computerized tomography (CT) and magnetic resonance imaging (MRI) scans create blurred images. With special instruments and computers, images generated at high speed can be used to develop three-dimensional still or moving pictures of the heart as it beats (Figure 20-13a-c). These procedures produce dramatic images, but the cost and complexity of the equipment have so far limited their use to major research institutions. Although PET scans can be used to diagnose disorders of coronary circulation (see Figure 20-7), cost factors are also responsible for limiting the clinical use of this technology.

Ultrasound analysis, called **echocardiography** (ek-ō-kar-dē-OG-ra-fē), provides images that lack the clarity of CT or MRI scans, but the equipment is relatively inexpensive and portable (Figure 20-13d). Recent advances in data processing have made the images suitable for following details of cardiac contractions, and echocardiography is now an important diagnostic tool. Another relatively widespread approach involves preparation of a **coronary angiogram** (Figure 20-13e). In this procedure, radiopaque dyes are injected into the coronary circulation, permitting X-ray analysis of coronary blood flow.

FIGURE 20-13
Monitoring the Living Heart. (a) Three-dimensional CT scan of a frontal section through the heart. (b) Three-dimensional CT scan of the anterior surface of the heart. (c) MRI scans showing changes in the heart during ventricular diastole (left) and systole (right). (d) Echocardiographic image of a cardiac section, with a matching diagram. (e) Coronary angiogram showing the cardiac circulation.

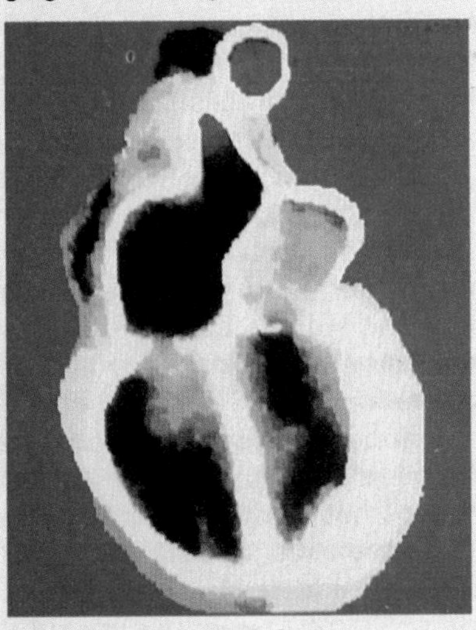

(a)

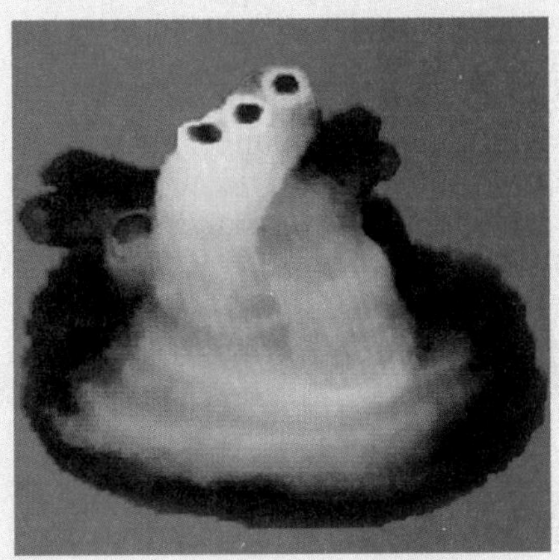

(b)

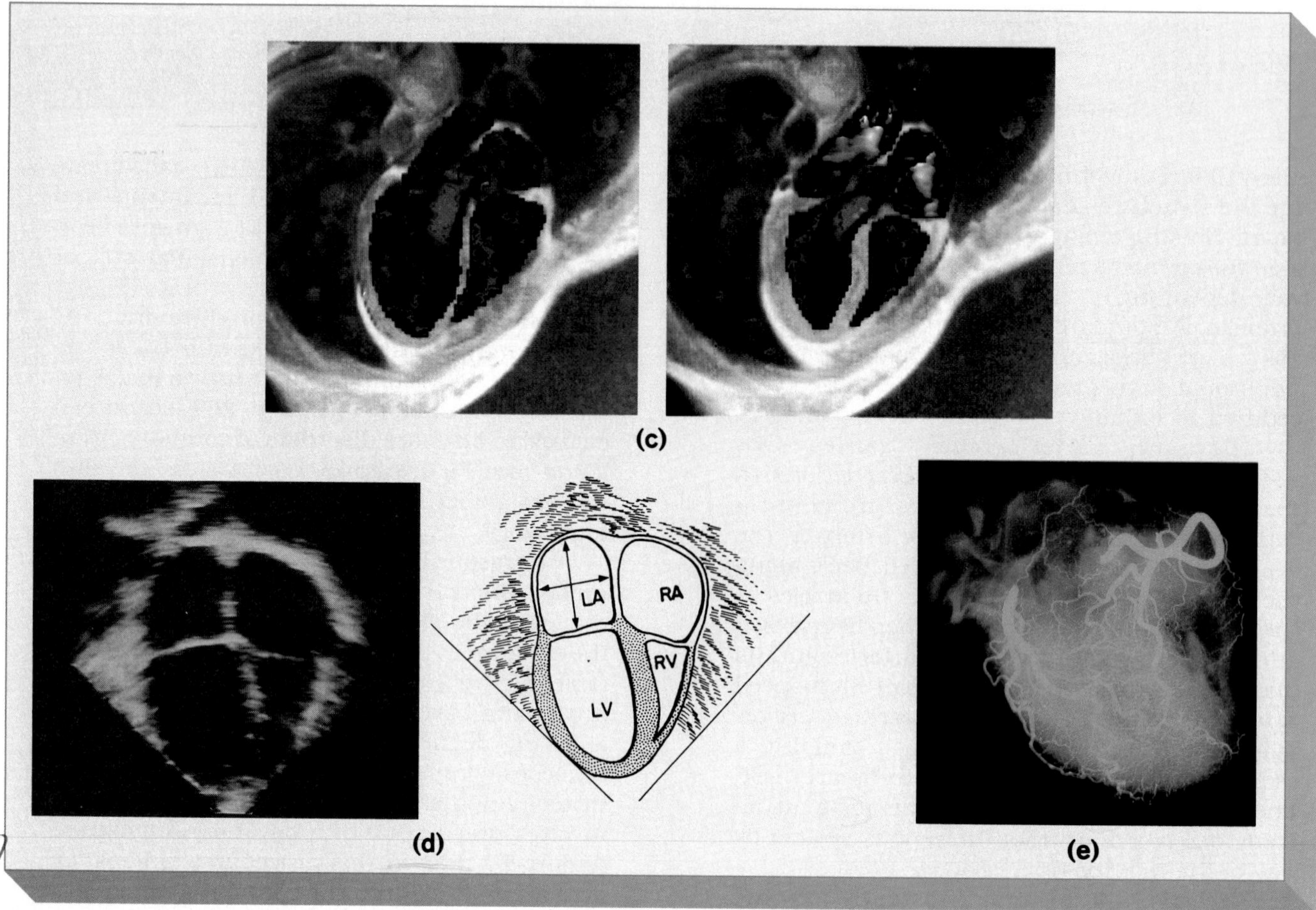

(c)

(d)

(e)

flows from the atria into the ventricles. Both the atria and the ventricles are now in diastole, but the atria, which have been in diastole for a longer period, are already partially filled with blood. The ventricles are almost empty, and ventricular pressures are very low. As the elasticity of the fibrous skeleton helps reexpand the relaxed ventricles, ventricular pressures drop further. These relatively low pressures pull blood from the major veins through the relaxed atria. By the time atrial systole marks the start of another cardiac cycle, the ventricles are nearly three-quarters full. The relatively minor contribution atrial systole makes to ventricular volume explains why individuals can survive quite normally when their atria have been so severely damaged that they can no longer function. By contrast, damage to one or both ventricles can leave the heart unable to maintain adequate cardiac output. A condition of **heart failure** then exists. Important aspects of heart failure are discussed in Chapters 20 and 21 of the Clinical Manual. [**CM**: *Valvular Heart Disease, The Cardiomyopathies, Congestive Heart Failure*]

STRUCTURE AND FUNCTION OF VALVES

Figure 20-14 presents a diagrammatic view of the structure and function of typical atrioventricular and semilunar valves. The chordae tendineae and papillary muscles play an important role in the normal function of the AV valves. When a ventricle is filling with blood, the muscles are relaxed and the AV valve offers no resistance to the flow of blood from atrium to ventricle. When the ventricle begins to contract, blood moving back toward the atrium swings the cusps together, closing the valve.

The normal timing of valve operation is shown in Figure 20-13. During ventricular systole, tension in the papillary muscles and chordae tendineae braces the cusps and keeps them from swinging into the atrium. This action prevents the backflow, or **regurgitation**, of blood into the atrium each time the ventricle contracts. The semilunar valves do not require muscular braces because the arterial walls do not contract, and the relative positions of the cusps are stable. When these valves close, the three symmetrical cusps support one another (Figure 20-14b).

Minor abnormalities in valve shape are relatively common. For example, an estimated 10 percent of normal individuals age 14–30 have some degree of **mitral valve prolapse**. In this condition the mitral valve cusps do not close properly. The problem may involve abnormally long (or short) chordae tendineae or malfunctioning papillary muscles. Because the valve does not work perfectly, some regurgitation occurs during left ventricular systole. The surges, swirls, and eddies that occur during regurgitation create a rushing, gurgling sound known as a **heart murmur**. Most of these individuals are completely

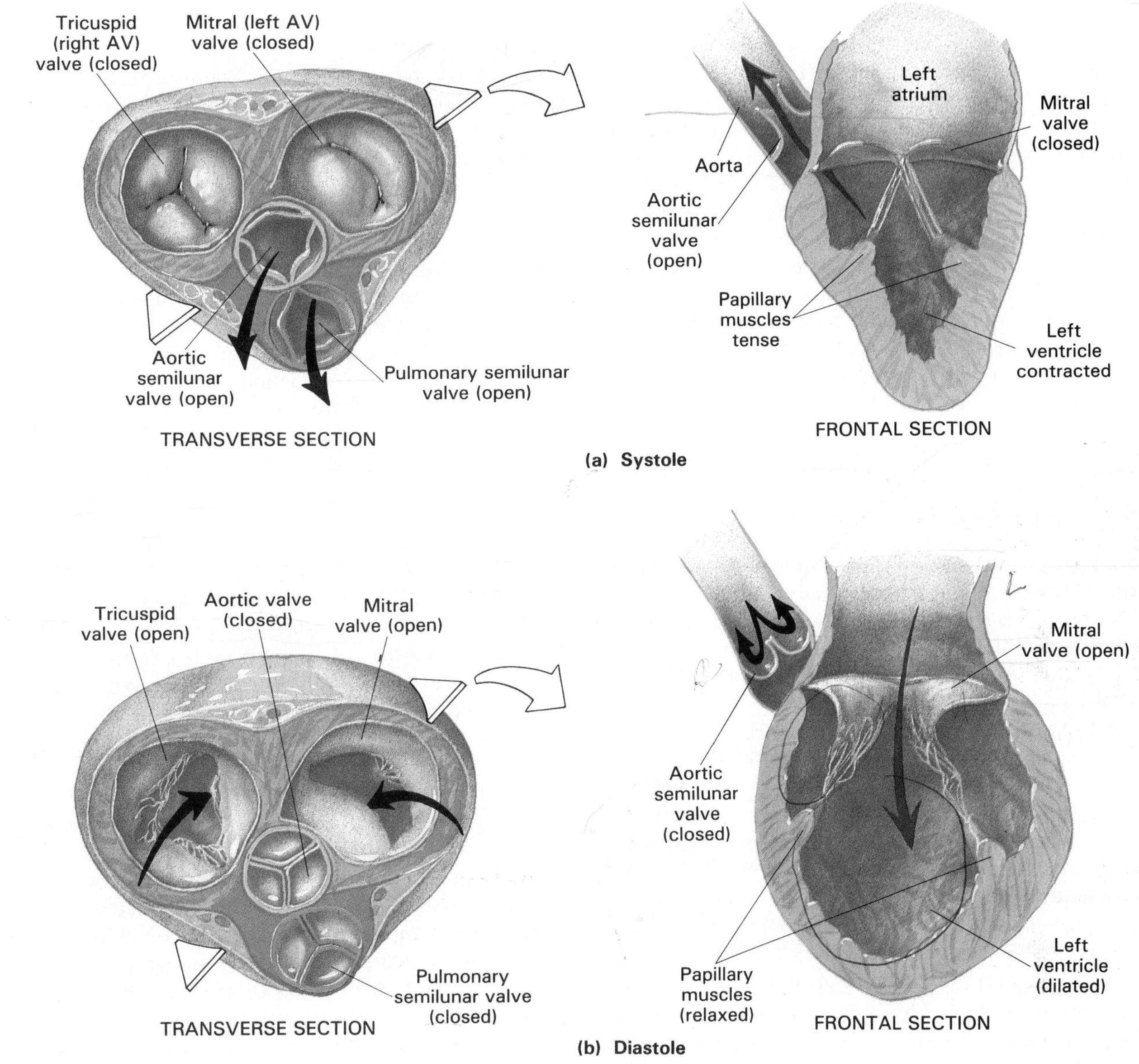

Tricuspid (right AV) valve (closed)
Mitral (left AV) valve (closed)
Aortic semilunar valve (open)
Pulmonary semilunar valve (open)
TRANSVERSE SECTION

Left atrium
Mitral valve (closed)
Aorta
Aortic semilunar valve (open)
Papillary muscles tense
Left ventricle contracted
FRONTAL SECTION

(a) Systole

Tricuspid valve (open)
Aortic valve (closed)
Mitral valve (open)
Pulmonary semilunar valve (closed)
TRANSVERSE SECTION

Mitral valve (open)
Aortic semilunar valve (closed)
Papillary muscles (relaxed)
Left ventricle (dilated)
FRONTAL SECTION

(b) Diastole

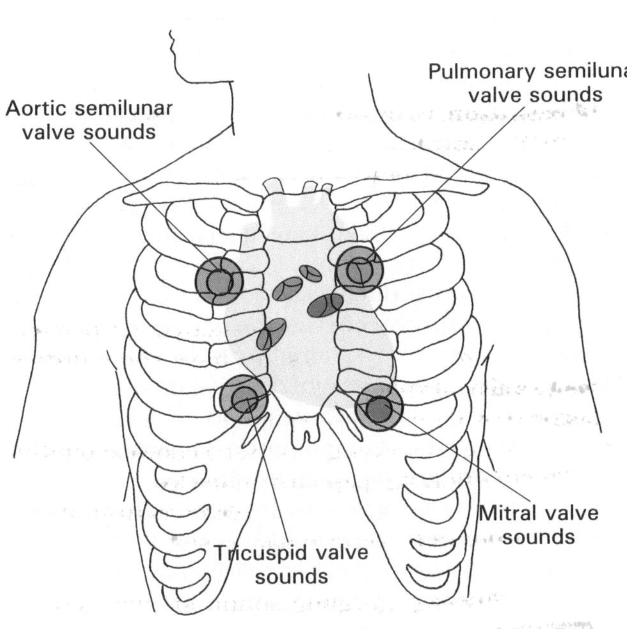

(c) Heart sounds

FIGURE 20-14
Valves of the Heart. (a) The appearance of the cardiac valves during ventricular systole, when the AV valves are closed and the semilunar valves are open. In the frontal section, note the bracing of the bicuspid valve by the chordae tendineae and papillary muscles. (b) Valve position during ventricular diastole, when the AV valves are open and the semilunar valves are closed. Note that the chordae tendineae are slack and the papillary muscles relaxed. (c) Placement of a stethoscope to listen to the sounds generated by the action of individual valves.

asymptomatic, and they live normal lives unaware of any circulatory malfunction.

More serious valvular abnormalities can interfere with cardiac function, and the timing and intensity of the related heart murmurs can provide useful diagnostic information. Physicians use an instrument called a **stethoscope** (STETH-o-scōp) to listen to normal and abnormal heart sounds. The placement of the stethoscope varies, depending on the valve under examination (Figure 20-14c). Valve sounds must pass through the pericardium, surrounding tissues, and the chest wall, and some tissues muffle sounds more than others. As a result, the stethoscope placement does not always correspond to the position of the valve under review. [**CM**: *Valvular Heart Disease*]

HEART SOUNDS

Figure 20-12 indicates the timing of the various heart sounds, including the familiar "lubb-dupp" that accompanies each heartbeat. This technique, called *auscultation* (Chapter 1), is a simple, inexpensive, and effective method of cardiac diagnosis. (p. 15)

When you listen to your own heart, you usually hear the first and second heart sounds. These sounds accompany the action of the heart valves. The first heart sound, known as "lubb," lasts a little longer than the second. It marks the start of ventricular systole, and the sound is produced as the AV valves close and the semilunars open. The second heart sound, "dupp," occurs at the beginning of ventricular diastole, when the semilunar valves close.

Third and fourth heart sounds may be audible as well, but even with a stethoscope they are usually very faint. These sounds are associated with systolic and diastolic blood flow, rather than valve action, and they are seldom detectable in normal adults.

More sophisticated procedures can be used to obtain a detailed assessment of cardiac function. One of the most important techniques involves the analysis of the electrical signals generated by the heart during the cardiac cycle.

THE ELECTROCARDIOGRAM (ECG)

The electrical events occurring in the heart are powerful enough that they can be detected by electrodes on the body surface. A recording of these electrical activities constitutes an **electrocardiogram** (ē-lek-trō-KAR-dē-ō-gram), also called an **ECG** or **EKG**. During each cardiac cycle a wave of depolarization radiates through the atria, reaches the AV node, travels down the septum to the apex, turns, and spreads through the ventricular myocardium toward the base. This electrical activity can be monitored from the body surface.

By comparing the information obtained from electrodes placed at different locations you can check the performance of specific nodal, conducting, and contractile components. For example, when a portion of the heart has been damaged by an infarct, cardiac muscle cells are severely damaged or killed. They no longer conduct action potentials, so an ECG will reveal an abnormal pattern of electrical conduction.

The appearance of the ECG tracing varies, depending on the placement of the monitoring electrodes, or leads. Figure 20-15 shows the important features of an electrocardiogram as analyzed with the leads in one of the standard configurations. The small **P wave** accompanies the depolarization of the atria. The **QRS complex** appears as the ventricles depolarize. This is a relatively strong electrical signal because the mass of the ventricular muscle is much larger than that of the atria. The smaller **T wave** indicates ventricular repolarization. You do not see a deflection corresponding to atrial repolarization because it occurs while the ventricles are depolarizing, and the electrical events are masked by the QRS complex.

Analyzing an ECG involves measuring the size of the voltage changes and determining the durations and temporal relationships of the various components. Attention usually focuses on the amount of depolarization occurring during the P wave and the QRS complex. For example, a smaller than normal electrical signal may mean that the mass of the heart muscle has decreased, and excessively strong depolarizations may mean that the heart muscle has become enlarged. The size and shape of the T wave may also be affected by any condition that slows ventricular repolarization. For example, starvation and low cardiac energy reserves, coronary ischemia, or abnormal ion concentrations will reduce the size of the T wave.

Measurements of the time between waves are also taken. The values are reported as segments or intervals. The terms used are indicated in Figure 20-15, and you will find that the names do not always seem to fit. For example, the P-R interval extends from the start of atrial depolarization to the start of the QRS complex (ventricular depolarization), rather than to R. This is because in abnormal ECGs the peak can be difficult to determine. Extension of the P-R interval to more than 0.2 seconds can indicate damage to the conducting system or AV node. Measuring the Q-T interval checks the time required for the ventricles to undergo a single depolarization-repolarization cycle. It also roughly corresponds to the duration of a ventricular contraction. The Q-T interval can be extended by conduction problems, coronary ischemia, or myocardial damage.

Despite the variety of sophisticated equipment available to assess or visualize cardiac function, in the vast majority of cases the electrocardiogram pro-

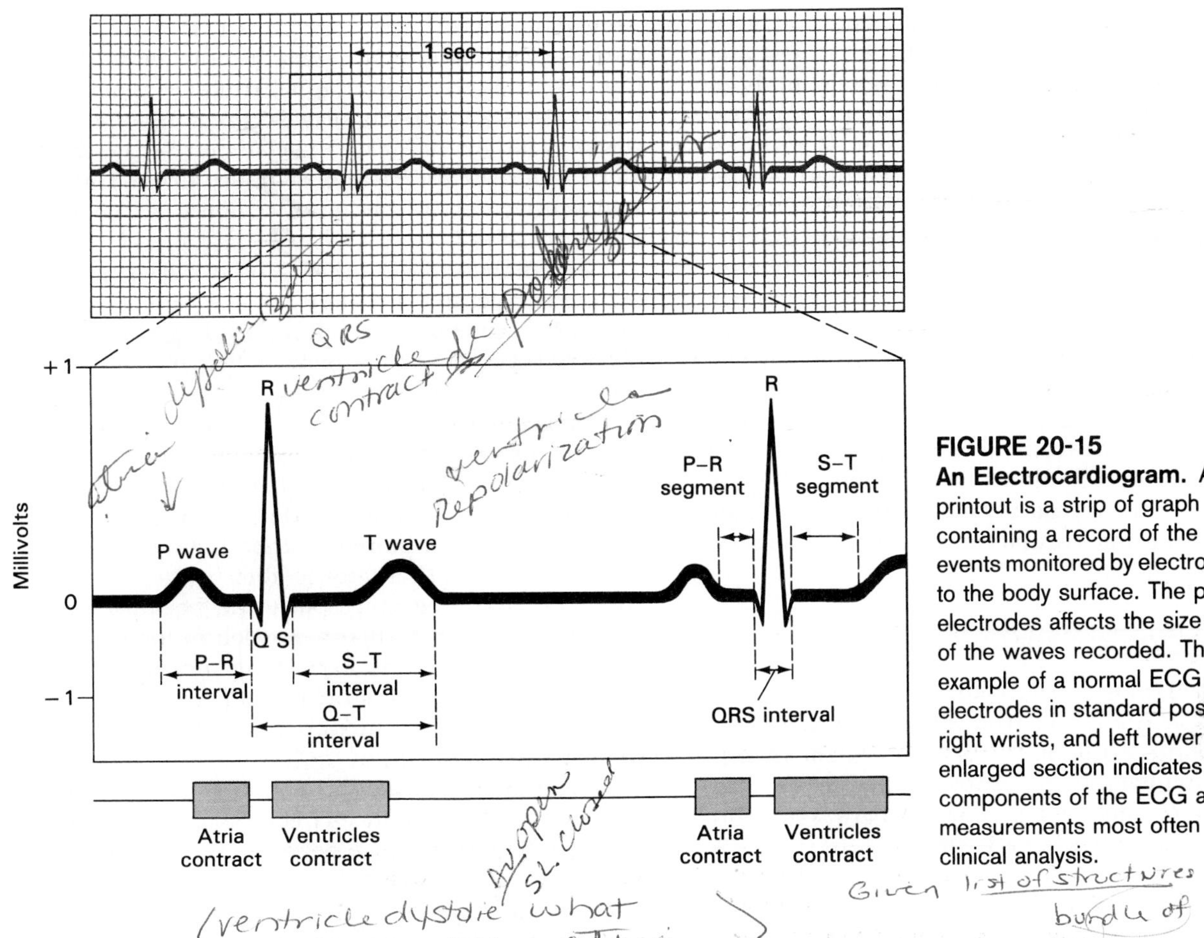

FIGURE 20-15
An Electrocardiogram. An ECG printout is a strip of graph paper containing a record of the electrical events monitored by electrodes attached to the body surface. The placement of electrodes affects the size and shape of the waves recorded. This is an example of a normal ECG using three electrodes in standard position (left and right wrists, and left lower leg). The enlarged section indicates the major components of the ECG and the measurements most often taken during clinical analysis.

vides the most important diagnostic information. ECG analysis is especially useful in detecting and diagnosing **cardiac arrhythmias** (a-RITH-mē-as), abnormal patterns of cardiac activity. Momentary arrhythmias are not inherently dangerous, and about 5 percent of the normal population experiences a few abnormal heartbeats each day. Clinical problems appear when the arrhythmias reduce the pumping efficiency of the heart. Serious arrhythmias may indicate damage to the myocardial musculature, injuries to the pacemakers or conduction pathways, exposure to drugs, or variations in the electrolyte composition of the extracellular fluids. For a discussion of the most common types of arrhythmias detected with the ECG see the Clinical Manual. [**CM**: *Interpreting Abnormal ECGs*]

- **When pressure in the left ventricle is rising is the heart pumping blood? Explain.**
- **What prevents the AV valves from opening back into the atria?**
- **Why is it a problem if the heart beats too fast?**

Cardiodynamics

The term **cardiodynamics** refers to the movements and forces generated during cardiac contractions. Each time the heart beats, the two ventricles eject equal amounts of blood. The amount ejected by a ventricle during a single beat is the **stroke volume** (**SV**). The stroke volume may vary from beat to beat, and physicians are often more interested in the **cardiac output** (**CO**), the amount of blood pumped each minute. The cardiac output provides a useful indication of ventricular efficiency over time. It can be calculated by taking the average stroke volume and multiplying it by the heart rate (HR). This relationship can be summarized as:

$$\underset{\text{(m}\ell\text{/min)}}{\overset{\text{CO}}{\text{cardiac output}}} = \underset{\text{(m}\ell\text{)}}{\overset{\text{SV}}{\text{stroke volume}}} \times \underset{\text{(bpm)}}{\overset{\text{HR}}{\text{heart rate}}}$$

For example, if the average stroke volume is 80 mℓ and the heart rate is 70 beats per minute (bpm), the cardiac output will be:

$$80\text{ m}\ell \times 70/\text{min} = 5600\text{ m}\ell/\text{min} = 5.6\text{ liters per minute}$$

Cardiac output is highly variable; a normal heart can increase its rate of contraction by 2.5 times, and its stroke volume can almost triple. Increasing both the heart rate and stroke volume together can raise the cardiac output by 600–700 percent. The difference between resting and maximal cardiac output is the **cardiac reserve**.

Cardiac output is precisely regulated so that peripheral tissues receive an adequate circulatory supply under a variety of conditions. For convenience the regulation of heart rate and stroke volume can be considered separately, although alterations in cardiac output usually reflect changes in both aspects of cardiac function. Figure 20-16 identifies the major factors involved in the control of cardiac output. The rest of this chapter examines these factors; Chapter 21 will describe the regulatory mechanisms responsible for maintaining cardiac output within homeostatic limits.

THE CONTROL OF HEART RATE

The basic heart rate is established by the pacemaker cells of the SA node, but this intrinsic rate can be modified by the autonomic nervous system (ANS). Anatomical details were presented in Chapter 16, and only a brief summary will be provided here. The sympathetic and parasympathetic divisions of the ANS provide innervation to the heart via the cardiac plexus. Postganglionic sympathetic neurons are found in the cervical and upper thoracic ganglia. The vagus carries parasympathetic preganglionic fibers to small ganglia in the cardiac plexus. Both ANS divisions innervate the SA and AV nodes as well the atrial and ventricular cardiac muscle fibers, as indicated in Figure 20-17.

The effects of NE and ACh on nodal tissues were detailed earlier in this chapter. In the general myocardium, NE binds to β_1 receptors, leading to activation of adenyl cyclase and an increase in intracellular concentrations of cyclic-AMP (cAMP). This second messenger activates enzymes with two major effects: (1) calcium ion channels open, allowing more calcium ions into the cell, and (2) calcium transport at the SR accelerates, so the additional calcium entering the sarcoplasm is stored, rather than pumped out of the cell. When this muscle fiber is stimulated, the contraction will be more powerful, because the SR will release larger numbers of calcium ions. In summary, NE release leads to both an increase in heart rate through its effects on the SA node and an increased force of contraction through its effects on the general myocardium.

The primary effect of ACh is at the membrane surface, where it produces hyperpolarization and inhibition. The result is a decrease in both the heart

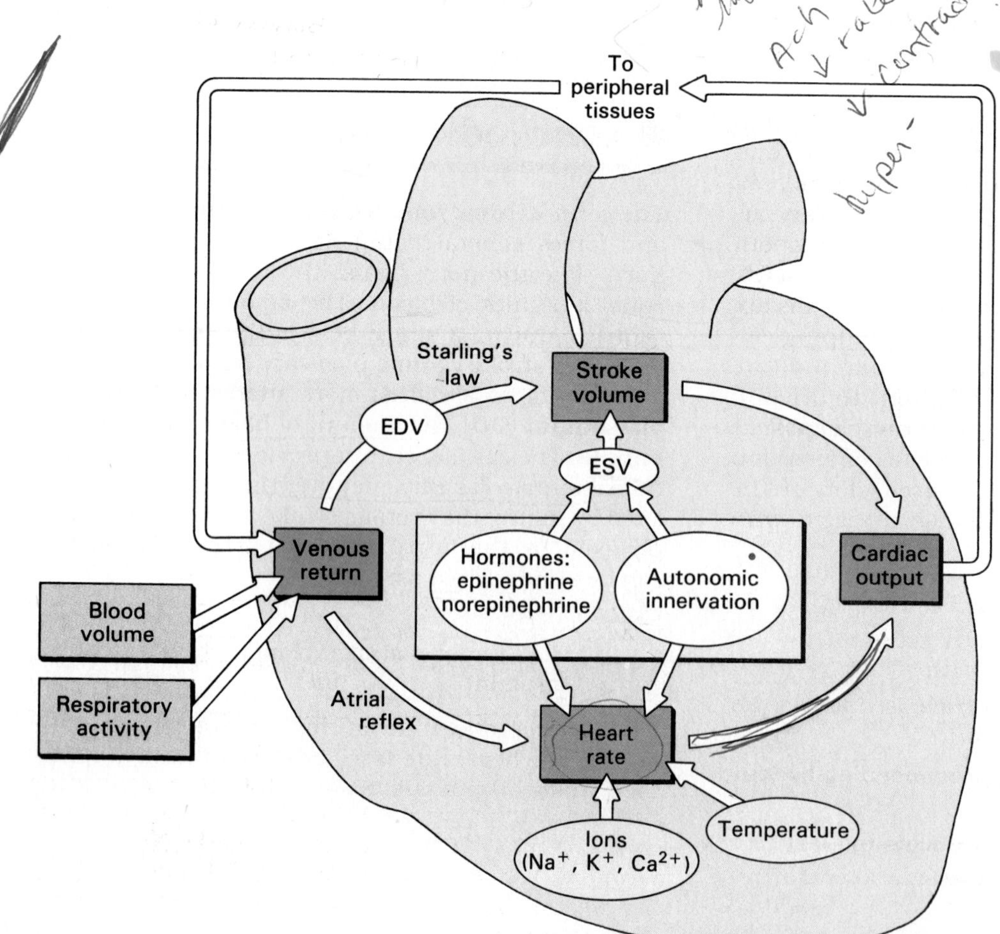

FIGURE 20-16
Factors that Influence Cardiac Output.

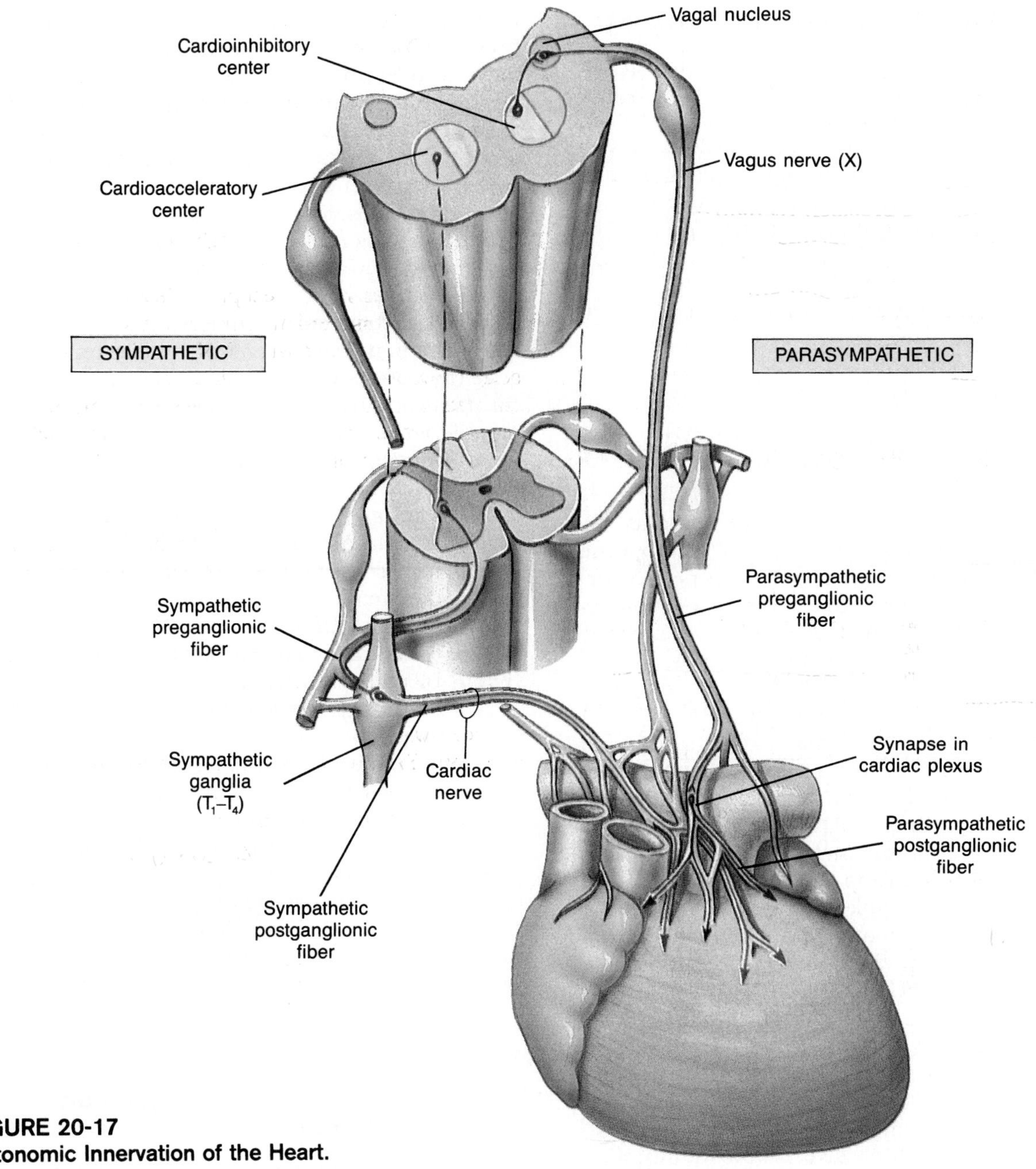

FIGURE 20-17
Autonomic Innervation of the Heart.

rate and force of cardiac contractions. Because parasympathetic innervation of the ventricles is relatively limited, the atria show the greatest changes in contractile force.

As in other organs with dual innervation, there is a resting autonomic tone. Both autonomic divisions are normally active at a steady background level, releasing ACh and NE both at the nodes and into the myocardium. Thus, cutting the vagus nerves increases the heart rate, and sympathetic blocking agents slow the heart rate. Through dual innervation and adjustments in autonomic tone, the ANS can make very delicate adjustments in cardiovascular function to meet the demands of other systems.

The **atrial reflex** (*Bainbridge reflex*) involves a combination of intrinsic and extrinsic adjustments in heart rate that affect cardiac output. This reflex is triggered by an increase in the flow of venous blood to the heart. When the walls of the right atrium are stretched, the cells of the SA node depolarize faster, and the heart rate accelerates. In addition, the stimulation of stretch receptors in the atrial walls triggers a reflexive increase in heart rate due to increased sympathetic activity.

THE REGULATION OF STROKE VOLUME

When a ventricle contracts, it does not pump out every drop of blood it contains. At the start of a typical cardiac cycle the left ventricle contains about 130 mℓ of blood. This constitutes the **end-diastolic volume**, or **EDV**. After ventricular systole has been completed, the ventricle still contains about 50 mℓ of blood. This is the **end-systolic volume**, or **ESV**. The difference between the EDV (130 mℓ) and the ESV (50 mℓ) represents the stroke volume (130 − 50 = 80 mℓ). Changing either the EDV or the ESV can have a significant effect on the stroke volume, and thus on cardiac output.

The EDV and Intrinsic Regulation of Stroke Volume

Two factors interact to determine the end-diastolic volume: *filling time* and *venous return*. **Filling time** is the duration of ventricular diastole, the period during which blood can flow into the ventricles. **Venous return** is the rate at which blood enters the ventricles over the same period. Filling time depends entirely on the heart rate. The faster the heart rate, the shorter the available filling time. Venous return changes in response to alterations in cardiac output, peripheral circulation, or other factors that affect the rate of blood flow through the venae cavae. These factors will be explored in Chapter 21.

Intrinsic regulation includes changes in cardiac output that result from factors affecting cardiac muscle directly. For example, changes in extracellular ion concentrations or changes in body temperature alter the heart rate by their direct effects on nodal cells and cardiac muscle fibers. Intrinsic regulation also includes the responses of the heart to changes in filling time or venous return that affect the end-diastolic volume. These regulatory activities are performed at the level of the individual muscle cell.

Cardiac muscle contraction is an active process, but relaxation is entirely passive. Unlike the situation in skeletal muscle, in cardiac muscle tissue there are no antagonistic muscle groups to extend the cardiac muscle fibers after each contraction. The necessary stretching force is provided by the blood pouring into the heart, aided by the elasticity of the fibrous skeleton. As a result, there is a direct relationship between the amount of blood entering the heart and the amount of blood ejected during the next contraction.

As in skeletal muscle tissue, the degree of contraction and the amount of force developed depend on the initial length of the sarcomeres. (You may wish to review Figure 11-15 at this time.) In a resting individual the force provided by the venous return stretches the myocardium only slightly. At the start of the next systolic contraction the sarcomeres are still relatively short, so the cardiac muscle fibers contract only a short distance and develop little power. If the venous return suddenly increases, more blood flows into the heart and the myocardium stretches further. As the sarcomeres approach optimal lengths the muscle fibers are able to contract more powerfully and over a greater distance, pumping out more blood.[1] This general rule of "more in = more out" was first proposed by Starling, a physiologist, so the principle is often called **Starling's law of the heart.**

Starling's law of the heart can be demonstrated most effectively in animal experiments, using hearts deprived of their normal innervation. In the human heart, intrinsic regulation functions primarily to balance the output of the two ventricles. When the heart contracts and blood leaves the right ventricle, a comparable volume of blood arrives at the left atrium, to be ejected by the left ventricle at the next contraction.

On inhalation, the expansion of the thoracic cavity pulls air into the lungs. At the same time, blood is pulled into the vena cava and right atrium from the smaller veins of the abdominal cavity and lower body. On exhalation, the thoracic cavity decreases in size. This forces air out of the lungs and pushes venous blood into the right atrium. This mechanism is called the **thoracoabdominal** (thō-ra-kō-ab-DOM-i-nal) **pump.**

The venous return changes as this accessory pump operates. On inhalation, venous blood accumulates in the vena cava, and the rate of venous return increases. On exhalation, the rate of venous return decreases suddenly. These changes in venous return have a direct effect on cardiac output. On inhalation, the increased venous return stretches the atrial walls and elevates the stroke volume (Starling's law) and heart rate (atrial reflex). [**CM**: The Thoracoabdominal Pump, Cardiac Output, and Sinus Arrhythmias]

The ESV and Autonomic Regulation of Stroke Volume

Autonomic activity affects stroke volume primarily by changing the end-systolic volume (ESV). Sympathetic activation causes the release of norepinephrine by postganglionic fibers and the secretion of norepinephrine and epinephrine by the adrenal medulla. In addition to their effects on heart rate, these compounds stimulate cardiac muscle cell metabolism and increase the force and degree of contraction. The heart then contracts more completely, ejecting more blood and decreasing the ESV.

As the heart rate rises, the available filling time becomes shorter and shorter. At moderate levels of sympathetic stimulation the end-diastolic volume remains relatively normal, due to the increased rate

[1] Stretching past optimal length does not normally occur because further expansion is prevented by the fibrous skeleton and the pericardium.

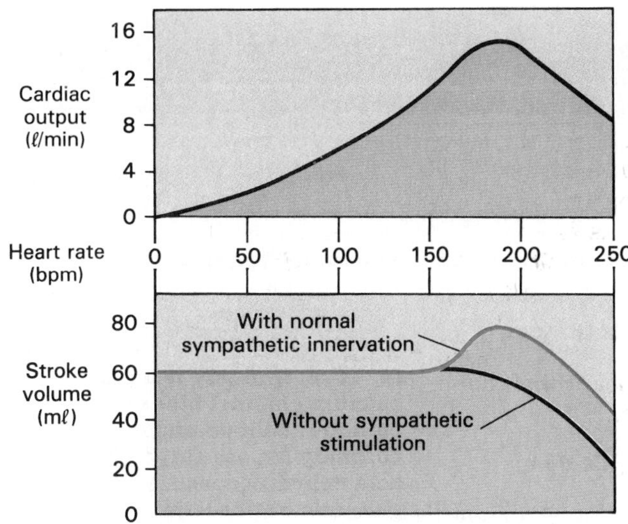

FIGURE 20-18
The Effect of Heart Rate on Stroke Volume and Cardiac Output

of venous return. But as the heart rate continues to climb, the reduced filling time affects the end-diastolic volume. Stroke volume peaks at a heart rate of about 180 bpm. At higher heart rates the stroke volume is reduced because the rate of venous return cannot make up for the reduction in filling time. The effects of these changes on cardiac output are diagrammed in Figure 20-18.

Parasympathetic stimulation reduces the force of contraction, lowering the stroke volume and elevating the ESV. Parasympathetic stimulation also slows the heart rate, further reducing the cardiac output.

COORDINATION OF AUTONOMIC ACTIVITY

The cardiac centers of the medulla contain the autonomic headquarters for cardiac control. (These centers were discussed in Chapter 14.) (p. 451) Stimulation of the **cardioacceleratory center** activates the necessary sympathetic neurons; the nearby **cardioinhibitory center** governs the activities of the parasympathetic neurons. The cardiac centers receive inputs from higher centers, especially from the parasympathetic and sympathetic headquarters in the hypothalamus. Information concerning the status of the cardiovascular system arrives via visceral sensory fibers in the vagus and in sympathetic nerves of the cardiac plexus.

The cardiac centers also monitor baroreceptors and chemoreceptors innervated by the glossopharyngeal and vagus nerves (N IX and X). On the basis of this information, the cardiac centers adjust the cardiac performance to maintain adequate circulation to vital organs, such as the brain. These centers respond to changes in blood pressure and in the arterial concentrations of dissolved oxygen and carbon dioxide. For example, a decline in blood pressure or oxygen concentrations, or an increase in carbon dioxide levels, usually indicates that the heart must work harder to meet the demands of peripheral tissues. The cardiac centers then call for an increase in the cardiac output. [**CM**: *The Cardiomyopathies*]

THE HEART AND THE CARDIOVASCULAR SYSTEM

The goal of cardiovascular regulation is the maintenance of adequate tissue blood flow. The heart cannot accomplish this by itself, and it does not work in isolation. For example, when blood pressure changes, cardiovascular centers not only adjust the heart rate, but alter the diameters of peripheral blood vessels. These adjustments work together to keep blood pressure within normal limits and to maintain circulation to vital tissues and organs. Chapter 21 begins by examining the structure of the circulatory network. It then completes our discussion of the cardiovascular system by considering its responses to changing activity patterns and circulatory emergencies.

- **What effect would stimulating the acetylcholine receptors of the heart have on cardiac output?**
- **What effect would a decreased venous return have on the stroke volume?**
- **How would increased sympathetic stimulation of the heart affect the end systolic volume?**

CHAPTER REVIEW

Review of Selected Clinical Terms

coronary artery disease (CAD) (*p. 644*)
Degenerative changes in the coronary circulation.

angina pectoris (an-JĪ-na PEK-tor-is) (*p. 644*)
A condition in which exertion or stress can produce severe chest pain, resulting from temporary insufficiency and ischemia when the heart's workload increases.

myocardial (mī-ō-KAR-dē-al) infarction (MI) (*p. 646*)
A condition in which the coronary circulation becomes blocked and the cardiac muscle cells die from oxygen starvation; also called a heart attack.

coronary thrombosis (*p. 646*)
A blockage due to the formation of a clot (thrombus) at a plaque in a coronary artery.

bradycardia (brād-e-KAR-dē-a) (*p. 650*)
A heart rate that is slower than normal.

tachycardia (tak-e-KAR-dē-a) (*p. 650*)
A heart rate that is faster than normal.

electrocardiogram (ē-lek-trō-KAR-dē-ō-gram) (ECG or EKG) (*p. 656*)
A recording of the electrical activities of the heart over time.

cardiac arrhythmias (a-RITH-mē-as) (*p. 657*)
Abnormal patterns of cardiac contraction.

Additional Terms of Clinical Importance

CORONARY ARTERY DISEASE AND MYOCARDIAL INFARCTION (*p. 644*): **coronary ischemia, calcium channel blockers, catheter, balloon angioplasty, coronary bypass surgery, lactic acid dehydrogenase (LDH), serum glutamic oxaloacetic transaminase (SGOT), creatine phosphokinase (CPK or CK), MB-CK**

MONITORING THE LIVING HEART (*p. 653*): **coronary arteriography, echocardiography, coronary angiogram**

THE ELECTROCARDIOGRAM (ECG) (*p. 656*): **P wave, QRS complex, T wave.**

Study Outline

Related Key Terms

An Overview of the Circulatory System (p. 634)

1. The circulatory system can be subdivided into the **pulmonary circuit** (which carries blood to and from the lungs) and the **systemic circuit** (which transports blood to and from the rest of the body). **Arteries** carry blood away from the heart; **veins** return blood to the heart. **Capillaries** are tiny vessels between the smallest arteries and veins.

Structure of the Heart (pp. 634–637)

1. The heart has four chambers: the **right atrium** and **right ventricle**, and the **left atrium** and **left ventricle**. The heart is surrounded by the **pericardial cavity** (lined by the **pericardium**); the **visceral pericardium (epicardium)** covers the heart's outer surface, and the **parietal pericardium** lines the inner surface of the **pericardial sac** that surrounds the heart.

Related Key Terms: **base • apex • auricle coronary sulcus interventricular sulcus**

SECTIONAL ANATOMY AND ORGANIZATION (pp. 636–637)

2. The atria are separated by the **interatrial septum**, and the ventricles are divided by the **interventricular septum**. The right atrium receives blood from the systemic circuit via two large veins, the **superior vena cava** and **inferior vena cava**. The atrial walls contain prominent muscular ridges, the **pectinate muscles**.
3. Blood flows from the right atrium into the right ventricle via the **right atrioventricular (AV) valve (tricuspid valve)**. This opening is bounded by three **cusps** of fibrous tissue braced by the tendinous **chordae tendineae** that are connected to **papillary muscles**.

Related Key Terms: **trabeculae carnae**

4. Blood leaving the right ventricle enters the **pulmonary trunk** after passing through the **pulmonary semilunar valve**. The pulmonary trunk divides to form the **left** and **right pulmonary arteries**. The **left** and **right pulmonary veins** return blood to the left atrium. Blood leaving the left atrium flows into the left ventricle via the **left atrioventricular (AV) valve (bicuspid valve** or **mitral valve)**. Blood leaving the left ventricle passes through the **aortic semilunar valve** and into the systemic circuit via the **ascending aorta**.

Chapter Review

Functional Anatomy of the Heart (pp. 638–642)

Related Key Terms

1. Anatomical differences between the ventricles reflect the functional demands placed on them. The wall of the right ventricle is relatively thin, while the left ventricle has a massive muscular wall.

CORONARY CIRCULATION (p. 639)

2. The **coronary circulation** meets the high oxygen and nutrient demands of cardiac muscle cells. The **coronary arteries** originate at the base of the ascending aorta. Interconnections between arteries called **anastomoses** ensure a constant blood supply. The **great** and **middle cardiac veins** carry blood from the coronary capillaries to the **coronary sinus**.

anterior (sternocostal) surface
diaphragmatic surface
right coronary artery
marginal branch
posterior interventricular branch
left coronary artery
circumflex branch
anterior interventricular branch
left anterior descending artery

TISSUES OF THE HEART WALL (pp. 640–642)

3. The bulk of the heart consists of the muscular **myocardium**. The **endocardium** lines the inner surfaces of the heart. The **fibrous skeleton** supports the heart's contractile cells and valves.
4. **Cardiac muscle fibers (cardiocytes)** are interconnected by **intercalated discs** that convey the force of contraction from cell to cell and conduct action potentials.

Cardiac Physiology (pp. 642–651)

CARDIAC CONTRACTIONS (pp. 642–651)

1. Cardiac muscle fibers have a long refractory period, so rapid stimulation produces isolated rather than tetanic contractions.
2. Cardiac muscle fibers obtain energy by aerobic reactions that break down glucose and fatty acids. **Nodal cells** establish the rate of cardiac contraction, and **conducting fibers** distribute the contractile stimulus to the general myocardium. Unlike skeletal muscle, cardiac muscle contracts without neural or hormonal stimulation. **Pacemaker cells** found in the **cardiac pacemaker (sinoatrial [SA] node)** normally establish the rate of contraction. From the SA node the stimulus travels to the **atrioventricular (AV) node**, then to the **AV bundle** which divides into **bundle branches**. From here **Purkinje cells** convey the impulses to the ventricular myocardium.

prepotential • **bradycardia**
tachycardia

The Heart as a Pump (pp. 651–657)

THE CARDIAC CYCLE (pp. 651–654)

1. The **cardiac cycle** consists of **systole** (contraction), followed by **diastole** (relaxation).

STRUCTURE AND FUNCTION OF VALVES (pp. 654–656)

2. Valves normally permit blood flow in only one direction, preventing the **regurgitation** (backflow) of blood.

mitral valve prolapse
heart murmur • **stethoscope**
heart failure

HEART SOUNDS (p. 656)

3. The closure of valves and rushing of blood through the heart causes characteristic heart sounds that can be heard during auscultation.

THE ELECTROCARDIOGRAM (ECG) (pp. 656–657)

4. A recording of electrical activities in the heart is an **electrocardiogram (ECG** or **EKG)**. Important landmarks of an ECG include the **P wave** (atrial depolarization), **QRS complex** (ventricular depolarization), and **T wave** (ventricular repolarization).

cardiac arrhythmias

Cardiodynamics (pp. 657–661)

1. The amount of blood ejected by a ventricle during a single beat is the **stroke volume (SV)**; the amount of blood pumped each minute is the **cardiac output (CO)**. The difference between resting and maximal cardiac output is the **cardiac reserve**.

CONTROL OF HEART RATE (pp. 658–659)

2. The basic heart rate is established by the pacemaker cells, but it can be modified by the ANS. The **atrial reflex** accelerates the heart rate when the walls of the right atrium are stretched.

REGULATION OF STROKE VOLUME (pp. 660–661)

3. The stroke volume is the difference between the **end-diastolic volume (EDV)** and the **end-systolic volume (ESV)**. The **filling time** and **venous return** interact to determine the end-diastolic volume. Normally, the greater the EDV, the more powerful the succeeding contraction (**Starling's law of the heart**).
4. Sympathetic activity produces more powerful contractions that reduce the ESV, while parasympathetic stimulation slows the heart rate, reduces the contractile strength, and raises the ESV.

intrinsic regulation

Chapter Review

COORDINATION OF AUTONOMIC ACTIVITY (p. 661)

5. The **cardioacceleratory center** in the medulla activates sympathetic neurons; the **cardioinhibitory center** governs the activities of the parasympathetic neurons. The cardiac centers receive inputs from higher centers, and from receptors monitoring blood pressure and the concentrations of dissolved gases.

Review Planner

	Objective	Level 1
1	Relate the anatomy of the heart to its functions.	**1 2 8 33 36 39 43 44 45**
2	Identify important differences between cardiac and skeletal muscle.	**4 21 29 34 38**
3	Trace the path of blood flow through the heart.	**5 6 14 15 22 24 40 41 42**
4	Identify the major arteries and veins of the pulmonary and systemic circuits.	**10 15 24 27**
5	Describe the physiological mechanism of cardiac contractions.	**4 9 11 12 13 18 19 25 29 33 37**
6	Describe how cardiac performance is measured.	**3 11 20 23 25**
7	Explain what can be learned by listening to the heart or analyzing an ECG.	**3 11 13 20 30**
8	Describe the mechanisms that regulate or influence heart rate and cardiac output.	**4 7 12 16 17 35**

Level 2	**26 28 31 32 45**
Level 3	**46–50**
C M	**Infection and Inflammation of the Heart • Cardiac Contractions and the Composition of the Extracellular Fluid • Problems with Pacemaker Function Valvular Heart Disease • The Thoracoabdominal Pump, Cardiac Output, and Sinus Arrhythmias The Cardiomyopathies**
C M	**Interpreting Abnormal ECGs**

Review Questions

MULTIPLE CHOICE

1. The heart is surrounded by the: a) dorsal cavity; b) pericardial cavity; c) peritoneal cavity; d) pleural cavity.
2. The sternocostal surface of the heart is the: a) posterior surface; b) dorsal surface; c) anterior surface; d) rostral surface.
3. A heart murmur may indicate: a) an atrial reflex; b) a valvular defect; c) a functional syncytium; d) none of the above.
4. The rate of cardiac contraction is established by the: a) pacemaker cells; b) conducting fibers; c) intercalated discs; d) Purkinje cells.
5. Blood flows from the left atrium into the left ventricle via the: a) bicuspid valve; b) tricuspid valve; c) papillary muscles; d) ascending aorta.
6. If the pulmonary semilunar valve were damaged, it would have an immediate effect on: a) blood flow from the left atrium to the left ventricle; b) blood flow into the right atrium; c) blood flow from the right atrium into the right ventricle; d) blood flow from the right ventricle to the pulmonary trunk.
7. The autonomic centers that regulate cardiac function are found in the: a) spinal cord; b) cerebrum; c) cerebellum; d) medulla.
8. The layer of the heart wall that contains cardiac muscle tissue is the: a) endocardium; b) epicardium; c) myocardium.
9. The period between the start of one heartbeat and the beginning of the next heartbeat is the: a) systole; b) cardiac cycle; c) diastole; d) T wave.

DIAGRAMS

10. Create a diagram that illustrates the sequence involved in the circulatory system. Include the pulmonary and systemic circuits, the left and right atria, the left and right ventricles, systemic arteries and veins, pulmonary arteries and veins, and the capillaries in lungs and peripheral tissues.

11. Diagram a normal ECG tracing. Label the P wave, QRS complex, and T wave, and briefly describe what occurs in the heart at each point.

TRUE/FALSE

(If the statement is false, your instructor may wish to have you correct it.)

12. **T/F:** Cardiac muscle tissue requires neural stimulation in order to contract.
13. **T/F:** In normal adults, the first heart sound heard during auscultation is usually longer than the second heart sound.
14. **T/F:** At the peak of ventricular systole, pressure is higher in the aorta than in the ventricle.

MATCHING QUESTIONS

(Match each term with the most appropriate statement.)

Statements:

15. The circulatory path that carries blood to and from the lungs. C
16. Neurotransmitter that decreases the heart rate and force of contraction. K
17. Neurotransmitter that increases the heart rate and force of contraction. J
18. The contraction phase of the cardiac cycle. F
19. Nodal cells. E
20. A recording of the electrical events occurring in the heart. H
21. Specialized sites where one cardiac muscle fiber contacts others. D
22. Small, thin-walled vessels that connect small arteries and veins. A
23. The difference between resting and maximal cardiac output. I
24. The circulatory path that carries blood to and from the body, excluding the lungs. B
25. The relaxation phase of the cardiac cycle. G

Terms:

a) Capillaries
b) Systemic circuit
c) Pulmonary circuit
d) Intercalated discs
e) Pacemaker cells
f) Systole
g) Diastole
h) Electrocardiogram
i) Cardiac reserve
j) Norepinephrine
k) Acetylcholine

SHORT ANSWER/ESSAY

26. Explain why the heart is crucial to maintaining homeostasis.
27. Fill in the blanks: ______ carry blood away from the heart; ______ return blood to the heart.
28. Which is more likely to be life-threatening: damage to the left or to the right ventricle? Why?
29. What cell junctions are present at an intercalated disc? Explain why these junctions are important.
30. Explain what each of the following ECG readings could indicate:
    a) A P-R interval of 0.5 seconds.
    b) Extended Q-T interval.
31. Explain the importance of pressure relationships to the cardiovascular system. How do contractions maintain these pressure relationships?
32. What is cardiac output? Calculate your own cardiac output, using a resting pulse rate and assuming an average stroke volume of 80 mℓ.
33. What are conducting fibers? Why are they important to the heart's function?
34. How does the refractory period of a cardiac muscle fiber membrane compare with that of a skeletal muscle fiber? Why is this significant?
35. Explain Starling's law of the heart.
36. Define anastomoses, and explain their significance to the body.
37. Fill in the blanks: Normally fluids move from an area of ______ pressure to one of ______ pressure.
38. Identify two important ways in which cardiac muscle fibers differ from skeletal muscle fibers, and explain the significance of each difference in terms of the heart's functions.
39. What is the function of the pericardial fluid found within the pericardial cavity?
40. An elderly man develops a thrombus in the inferior vena cava. How would this affect his circulation?
41. Fill in the blanks: The ______ ______ receives blood from the systemic circuit; the ______ ______ ejects blood into the systemic circuit.
42. Fill in the blanks: The ______ ______ collects blood from the pulmonary circuit; the ______ ______ ejects blood into the pulmonary circuit.
43. What is the fibrous skeleton of the heart, and why is it important?
44. Distinguish between the functions of the atria and the ventricles.
45. Relate the anatomy and pumping mechanisms of the left and right ventricles to their functions.

CRITICAL THINKING/APPLICATIONS

46. What is CAD? Explain why it can be so dangerous.
47. Compare the actions of drugs that are commonly used to treat angina. How do these drugs control the symptoms? Do they work in the same way?
48. Distinguish between hypercalcemia and hyperkalemia.
49. A patient's ECG tracing shows the following: two P waves followed by a normal QRS complex. What condition is indicated and is it serious?
50. Distinguish between atrial fibrillation and ventricular fibrillation. What is more likely to be dangerous, and why?

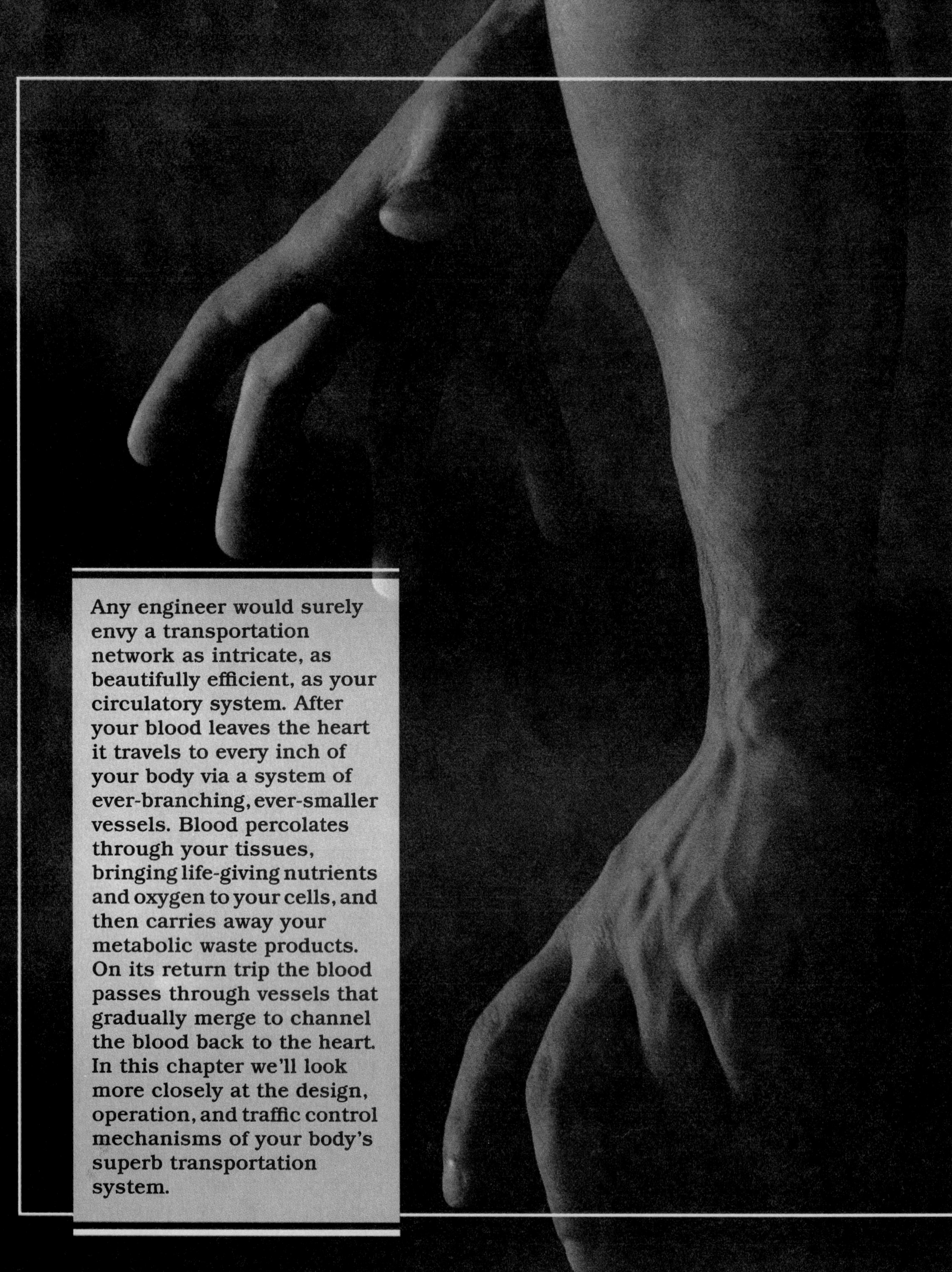

Any engineer would surely envy a transportation network as intricate, as beautifully efficient, as your circulatory system. After your blood leaves the heart it travels to every inch of your body via a system of ever-branching, ever-smaller vessels. Blood percolates through your tissues, bringing life-giving nutrients and oxygen to your cells, and then carries away your metabolic waste products. On its return trip the blood passes through vessels that gradually merge to channel the blood back to the heart. In this chapter we'll look more closely at the design, operation, and traffic control mechanisms of your body's superb transportation system.

CHAPTER 21

The Cardiovascular System: Vessels and Circulation

Chapter Objectives

After reading this chapter, you will be able to:

1 Relate the structural characteristics of blood vessels to their functions.
2 Describe how and where dissolved materials enter and leave the circulatory system.
3 Explain the mechanisms that regulate blood flow through arteries, capillaries, and veins.
4 Identify major blood vessels and the areas they serve.
5 Describe how blood pressure is measured.
6 Explain the factors that influence blood pressure.
7 Describe how central and local controls interact to regulate blood flow and pressure in tissues.
8 Explain how the activities of the cardiac, vasomotor, and respiratory centers are coordinated.
9 Explain how the circulatory system responds to the demands of exercise and blood loss.
10 Describe the age-related changes that occur in the cardiovascular system.

Introduction

Blood leaves the heart in the pulmonary and aortic trunks, each with a diameter of around 2.5 cm (1 in.). These vessels branch repeatedly, forming the major arteries that distribute blood to body organs. Within these organs further branching occurs, creating several hundred million tiny arteries that provide blood to more than 10 billion capillaries barely the diameter of a single red blood cell. These capillaries form extensive, branching networks; if all of the capillaries in the body were placed end to end they would easily span the United States, with a combined length of over 5000 miles.

All chemical and gaseous exchange between the blood and interstitial fluid takes place across capillary walls. Tissue cells rely on capillary diffusion to obtain nutrients and oxygen and to remove metabolic wastes, such as carbon dioxide and urea. The diffusion distances involved are very small—few living cells lie farther than 125 μm (0.005 in.) from a capillary.

That proximity is important, for as we noted in Chapter 3, diffusion can occur very quickly over short distances. (p. 79)

This chapter begins with a description of the histological and anatomical organization of arteries, veins, and capillaries. The rest of the chapter considers the regulation of cardiovascular function. The preservation of homeostasis involves interaction among all of the components of the circulatory system, as well as between the cardiovascular system and other organ systems.

Functional Anatomy of the Circulatory System

Arteries and veins form an internal distribution system, with the heart providing the necessary propulsion. The pulmonary and systemic circuits begin at the heart, as large-diameter arterial trunks receive blood passing through the semilunar valves. The vessels connected to these trunks branch repeatedly. As the branching proceeds the arteries gradually decrease in size until they become **arterioles** (ar-TĒ-rē-ōlz). Arterioles are the smallest vessels of the arterial system; from the arterioles blood enters the capillary networks that service local tissues.

Blood flowing out of the capillary complex first enters small **venules** (VEN-ūlz). Venules are the smallest vessels of the venous system, averaging 20 μm in diameter. These slender vessels subsequently merge with their neighbors to form small veins. Blood then passes through medium-sized and large veins before reaching the venae cavae (in the systemic circuit) or pulmonary veins (in the pulmonary circuit).

HISTOLOGICAL ORGANIZATION

Figure 21-1 introduces the histological characteristics of typical arteries and veins. Arteries and veins have three distinct layers within their walls. The **tunica interna** (in-TER-na), or *tunica intima*, includes the endothelial lining of the vessel and an underlying layer of connective tissue dominated by elastic fibers. The **tunica media** is the middle layer that contains concentric sheets of smooth muscle tissue in a framework of collagen and elastic fibers. The surrounding **tunica externa** (eks-TER-na), or *tunica adventitia*, forms a connective tissue sheath. The collagen fibers of the tunica externa may blend into those of the adjacent tissues, stabilizing and anchoring the blood vessel.

These multiple layers give arteries and veins considerable strength. The muscular and elastic components also permit controlled alterations in diameter as blood pressure or blood volume changes. However, the walls of arteries and veins are too thick to allow diffusion between the bloodstream and surrounding tissues, or even between the blood and the tissues of the vessel itself. For this reason the walls of large vessels themselves contain small arteries and veins that supply the smooth muscle fibers and fibroblasts of the tunica media and tunica externa. These blood vessels are called the **vasa vasorum** ("vessels of vessels"). [**CM:** *Aneurysms*]

Arteries and veins often lie side by side in a narrow band of connective tissue, as in Figure 21-1. Several important characteristic structural differences between arteries and veins are apparent in this figure:

1. In general, the walls of arteries are thicker than those of veins. The tunica media of an artery contains more smooth muscle and elastic fibers

FIGURE 21-1
A Comparison of Typical Arteries and Veins. (LM, x44)

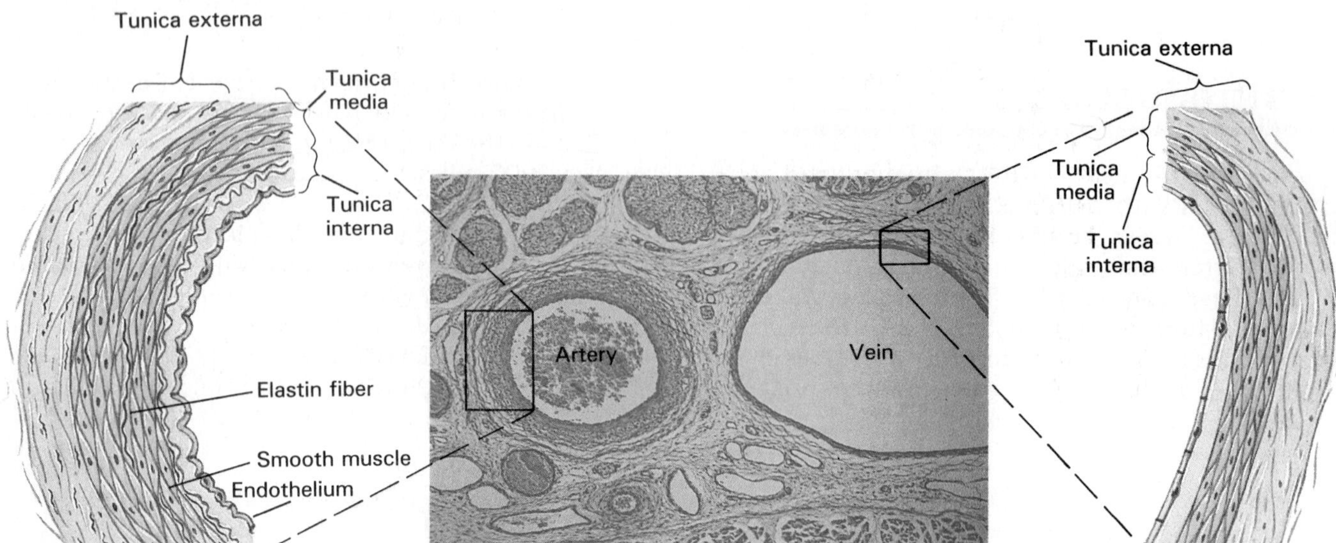

than does that of a vein. These contractile and elastic components resist the pressure generated by the heart as it forces blood into the circuit.

2. The endothelial lining of an artery cannot contract, so it is thrown into folds that give arterial sections a pleated appearance. The lining of a vein looks like a typical endothelial layer. (Endothelial cells were described in Chapter 5.) (p. 132)

3. In a sectional view, a cut artery looks smaller than the corresponding vein. When blood pressure no longer pushes against the smooth muscle and elastic tissue in their walls, arteries constrict. Veins, which have little elastic tissue in their walls, constrict very little. Because the walls of arteries are relatively thick and strong, they retain their circular shape in section. Cut veins tend to collapse, and in section they often look flattened or grossly distorted.

ARTERIES

In traveling from the heart to peripheral capillaries, blood passes through *elastic arteries*, *muscular arteries*, and *arterioles*. **Elastic arteries** are large vessels with diameters of up to 2.5 cm (1 in.). The pulmonary and aortic trunks and their major arterial branches are examples of elastic arteries. The walls of elastic arteries are not very thick, but they are extremely resilient. The tunica media of these vessels contains a high proportion of elastic fibers and relatively few smooth muscle cells. As a result, elastic arteries are able to tolerate the pressure shock produced each time ventricular systole occurs and blood leaves the heart. When pressures rise suddenly, elastic arteries stretch rather than break. During ventricular dias-

FIGURE 21-2
Histological Structure of the Circulatory System

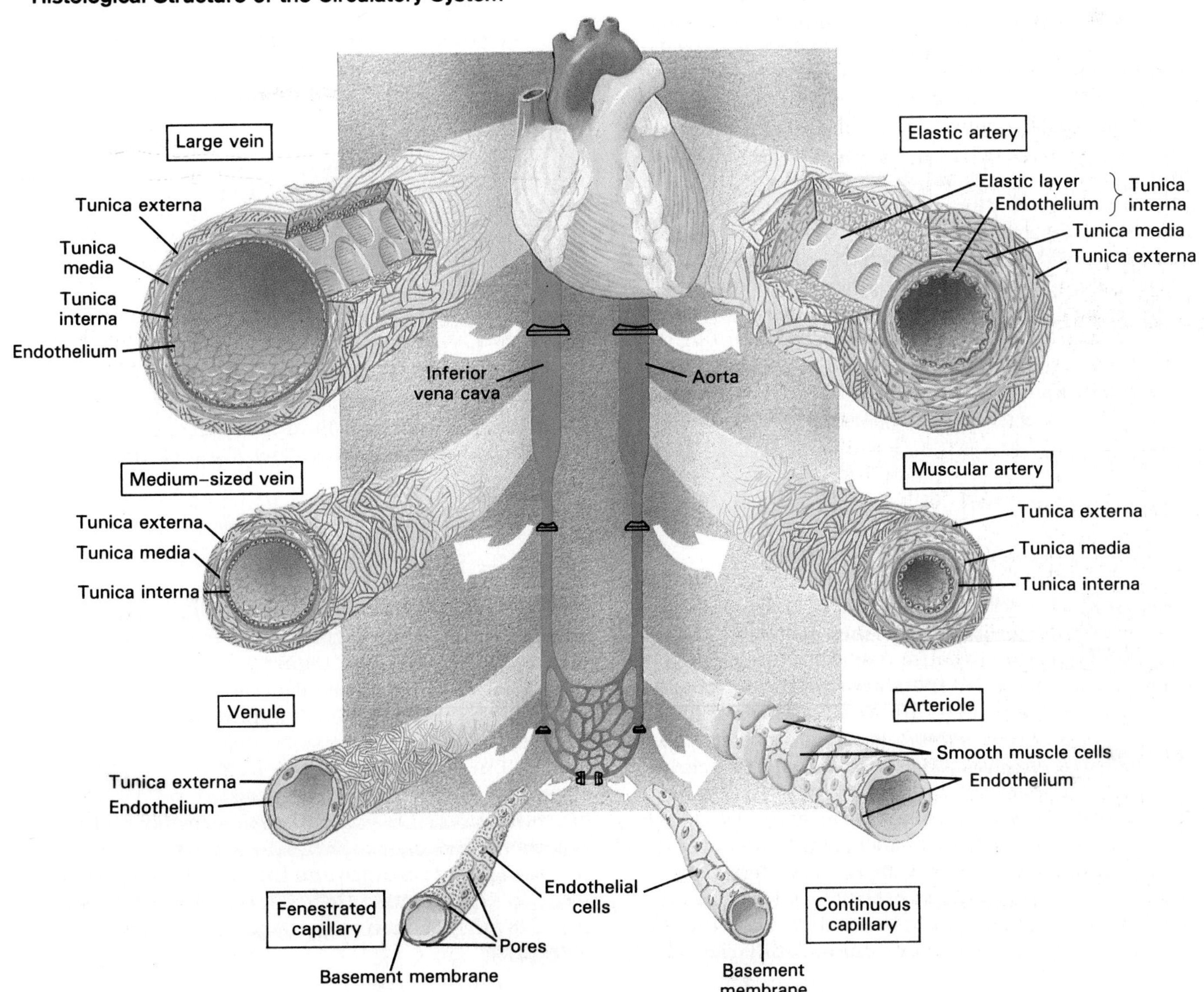

tole, blood pressure within the arterial system falls, and the elastic fibers recoil to their original dimensions. Their expansion cushions the sudden rise in pressure during ventricular systole, and their contraction slows the decline in pressure during ventricular diastole. (This process will be examined more closely in a later section.)

Medium-sized arteries, also known as **muscular arteries** or *distribution arteries*, distribute blood to peripheral organs. A typical muscular artery has a diameter of approximately 0.4 cm (0.15 in.). The carotid artery of the neck is a muscular artery. The thick tunica media in a muscular artery contains a greater amount of smooth muscle and fewer elastic fibers than are found in an elastic artery.

Arterioles are much smaller than medium-sized arteries, with an average diameter of about 30 μm. The tunica media of an arteriole of this size consists of an incomplete layer of smooth muscle fibers. The smooth muscle fibers within the walls of the muscular arteries and arterioles enable these vessels to change their diameter, altering the blood pressure and changing the rate of flow through the dependent tissues. These variations occur under local, autonomic, or endocrine stimulation. For example, most arterioles dilate when tissue oxygen levels decline or under the influence of atrial natriuretic peptide (ANP) (Chapter 20). (p. 644) Arterioles in most tissues constrict under sympathetic stimulation, although those in skeletal and cardiac muscle respond by dilating.

Histological distinctions between elastic arteries, muscular arteries, and arterioles can be seen in the "typical" sections presented in Figure 21-2 on page 665, but it should be remembered that these arteries are interconnected, and vessel characteristics change gradually as you travel away from the heart. For example, the largest of the muscular arteries contain a considerable amount of elastin, while the smallest resemble heavily muscled arterioles.

CAPILLARIES

Capillaries are the only blood vessels whose walls permit exchange between the blood and the surrounding interstitial fluids. Because the walls are relatively thin, the diffusion distances are small and exchange can occur quickly. In addition, blood flows through capillaries relatively slowly, allowing sufficient time for diffusion or active transport of materials across the capillary walls.

A typical capillary consists of an endothelial tube inside a delicate basement membrane. The average diameter of a capillary is a mere 8 μm, very close to that of a single red blood cell. In most regions the endothelium forms a complete lining, with the endothelial cells of these **continuous capillaries** connected by tight junctions. **Fenestrated capillaries** (FEN-es-trā-ted; *fenestra*, window), are capillaries that have an incomplete endothelial lining. These two capillary types are shown in Figure 21-3.

The walls of a fenestrated capillary have a "Swiss cheese" appearance, and the gaps are big enough for large molecules, including proteins, to enter or leave the circulation. Examples of fenestrated capillaries noted in earlier chapters include the choroid plexus of the brain (Chapter 14) and the blood vessels in the median eminence of the hypothalamus (Chapter 18). (pp. 456, 578) Fenestrated capillaries are also found at filtration sites in the kidneys.

Sinusoids (SĪ-nus-oidz) are specialized fenestrated capillaries found in the liver, bone marrow, and the adrenal glands. They form flattened, irregular passageways, so blood flow through these tissues occurs relatively slowly. The slow blood flow maximizes the time available for absorption and secretion across the sinusoidal walls. The basement membrane that surrounds the endothelium of a sinusoid is usually very thin, and this facilitates the exchange of very large molecules.

Movement of Materials across Capillary Walls

Diffusion through gaps between adjacent endothelial cells represents the primary route for substances entering or leaving a continuous capillary. Lipid-soluble materials and dissolved gases can also diffuse through the endothelial cells, and some larger compounds may be actively transported and secreted. Molecules the size of small proteins may occasionally cross the endothelia of fenestrated capillaries, where the gaps permit more extensive exchange with the interstitial fluids. In a few cases endothelial cells may also perform bulk transport, actively moving vesicles containing solutes and fluid between the blood and the interstitial environment (see Figure 3-25). (p. 85)

Capillary Beds

Capillaries do not function as individual units, but as part of an interconnected network called a **capillary plexus**, or **capillary bed**. Figure 21-4a presents a simplified version of the organizational arrangement in a capillary bed. Upon reaching its target area a single arteriole usually gives rise to dozens of capillaries that will in turn collect into several venules. The entrance to each capillary is guarded by a band of smooth muscle, the **precapillary sphincter**. Contraction of the smooth muscle fibers narrows the diameter of the capillary entrance and thereby reduces the flow of blood. Relaxation of the sphincter dilates the opening, allowing blood to enter the capillary at an accelerated rate.

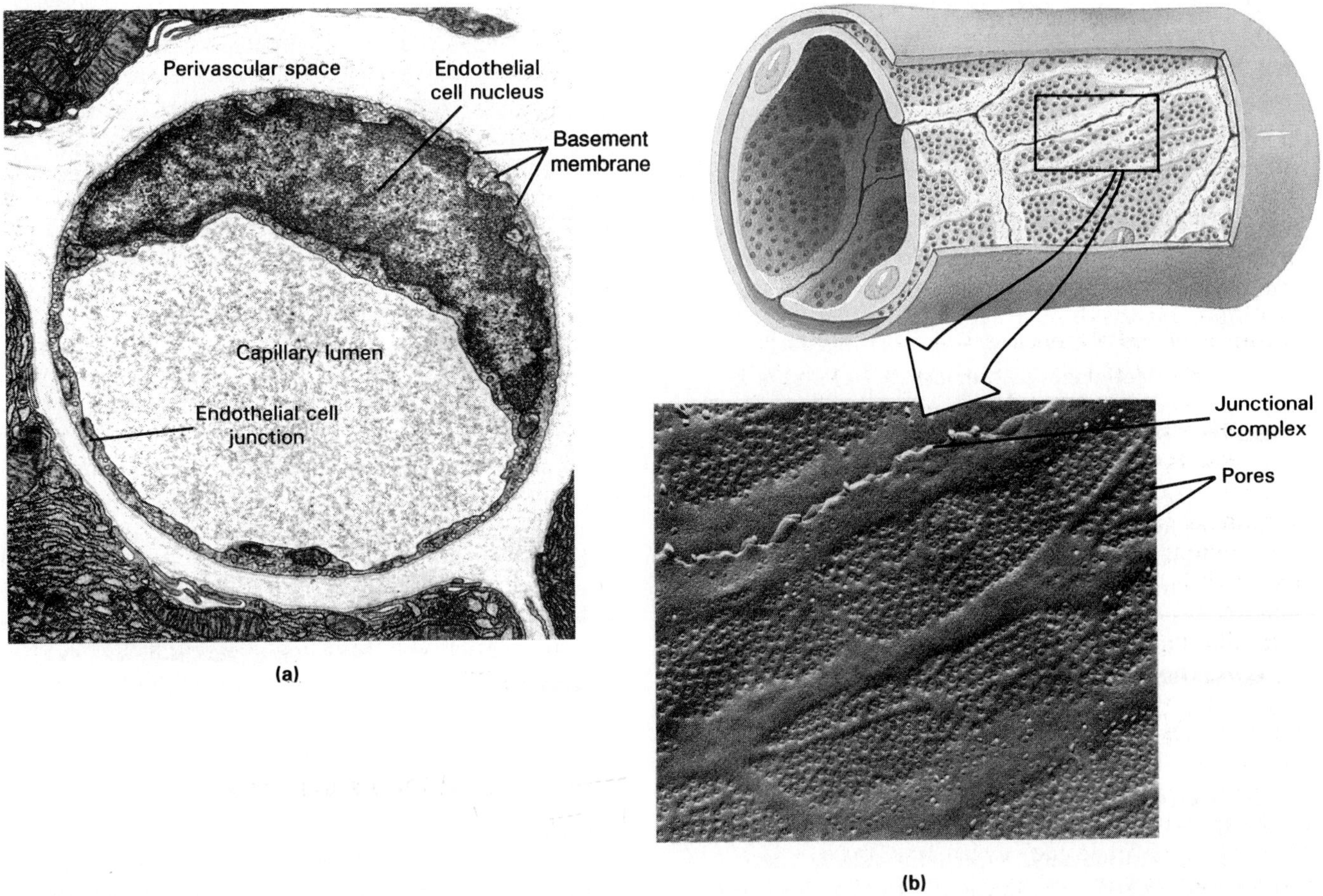

Perivascular space
Endothelial cell nucleus
Basement membrane
Capillary lumen
Endothelial cell junction
(a)

Junctional complex
Pores
(b)

FIGURE 21-3
Structure of Capillaries. (a) TEM of cross section through a continuous capillary. A single endothelial cell forms a complete lining around this portion of the capillary. (b) Longitudinal section (above) and SEM (below) showing the wall of a fenestrated capillary. The pores are gaps in the endothelial lining that permit the passage of large volumes of fluid and solutes. (SEM, x8875)

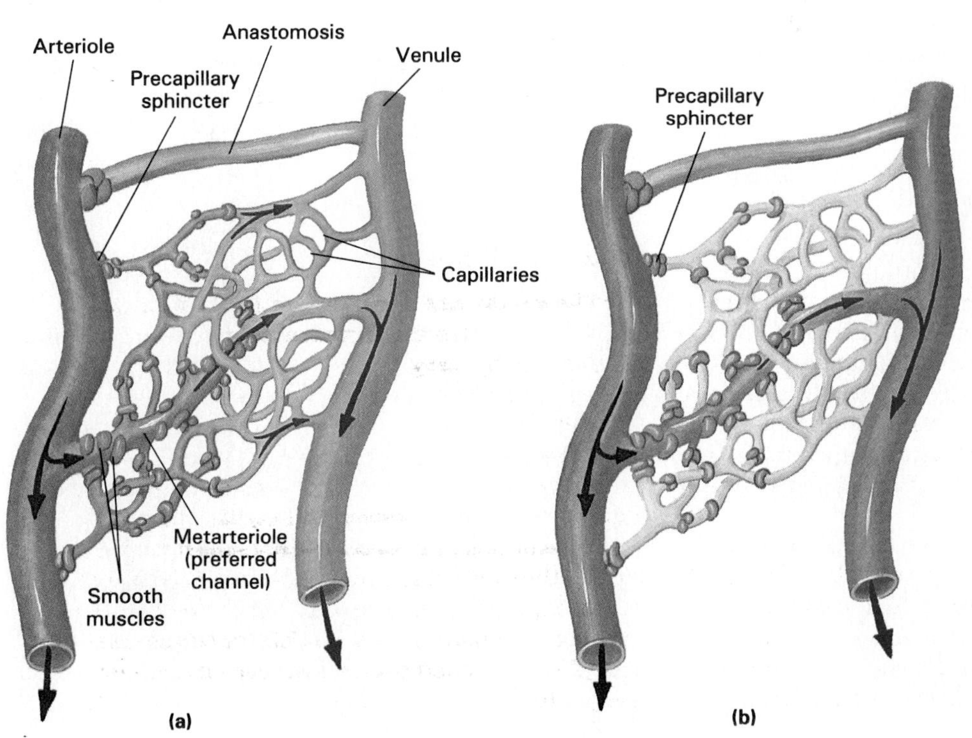

FIGURE 21-4
Organization of a Capillary Bed. Basic features of a typical capillary bed. The two panels show two possible patterns of blood flow through the capillary network as vasomotion occurs. The pattern changes continually in response to regional alterations in tissue oxygen demand.

Clinical Comment: Arteriosclerosis

Arteriosclerosis (ar-tē-rē-ō-skle-RŌ-sis) is a thickening and toughening of arterial walls. Although this may not sound like a life-threatening problem, complications related to arteriosclerosis account for roughly one-half of all deaths in the United States. There are many different forms of arteriosclerosis; one example, coronary artery disease (CAD), was discussed in Chapter 20. (p. 644)

Atherosclerosis (ath-er-ō-skle-RŌ-sis) is a type of arteriosclerosis characterized by changes in the endothelial lining. Evidence now indicates that this condition begins when smooth muscle fibers near the tunica interna begin dividing repeatedly. Monocytes then invade the area, migrating between the endothelial cells, and both monocytes and smooth muscle fibers begin phagocytizing circulating lipids, primarily cholesterol. Even the endothelial cells begin to accumulate lipids. As the cells become engorged, gaps soon appear in the endothelial lining, and platelets begin sticking to the exposed collagen fibers. The result is a **plaque**, a fatty mass of tissue that projects into the lumen of the vessel.

A typical plaque can be seen in Figure 21-5. Elderly individuals, especially elderly men, are most likely to develop atherosclerotic plaques. There is evidence that estrogens (female sex hormones) may slow plaque formation; this may account for the lower incidence of coronary artery disease, myocardial infarctions (MIs), and strokes in women. Young women are at relatively low risk for CAD as compared with young men. After menopause, when estrogen production declines, the risk of CAD in women increases markedly. In addition to advanced age and male sex, other important risk factors include high blood cholesterol levels, high blood pressure, and cigarette smoking. For example, roughly 20 percent of middle-aged men have all three of these risk factors; these individuals are four times more likely to experience an MI or cardiac arrest than other men in their age group. Although fewer women develop this condition, elderly women smokers with high blood cholesterol and high blood pressure are at much greater risk than other women. Other factors that may promote development of atherosclerosis in both men and women include diabetes mellitus, obesity, and stress.

Potential treatments for atherosclerotic plaques, such as catheterization and balloon angioplasty, were discussed in Chapter 20. (p. 644) However, the best approach is to try to avoid atherosclerosis by eliminating or reducing risk factors. Suggestions include: (1) reducing the amount of dietary cholesterol and saturated fats, by restricting consumption of fatty meats (such as beef, lamb, and pork), egg yolks, and cream; (2) giving up smoking; (3) checking your blood pressure, and taking steps to lower it if necessary; (4) having your blood cholesterol levels checked at annual physical examinations; (5) controlling your weight; and (6) exercising regularly.

FIGURE 21-5
A Plaque Blocking a Peripheral Artery. (LM, x36)

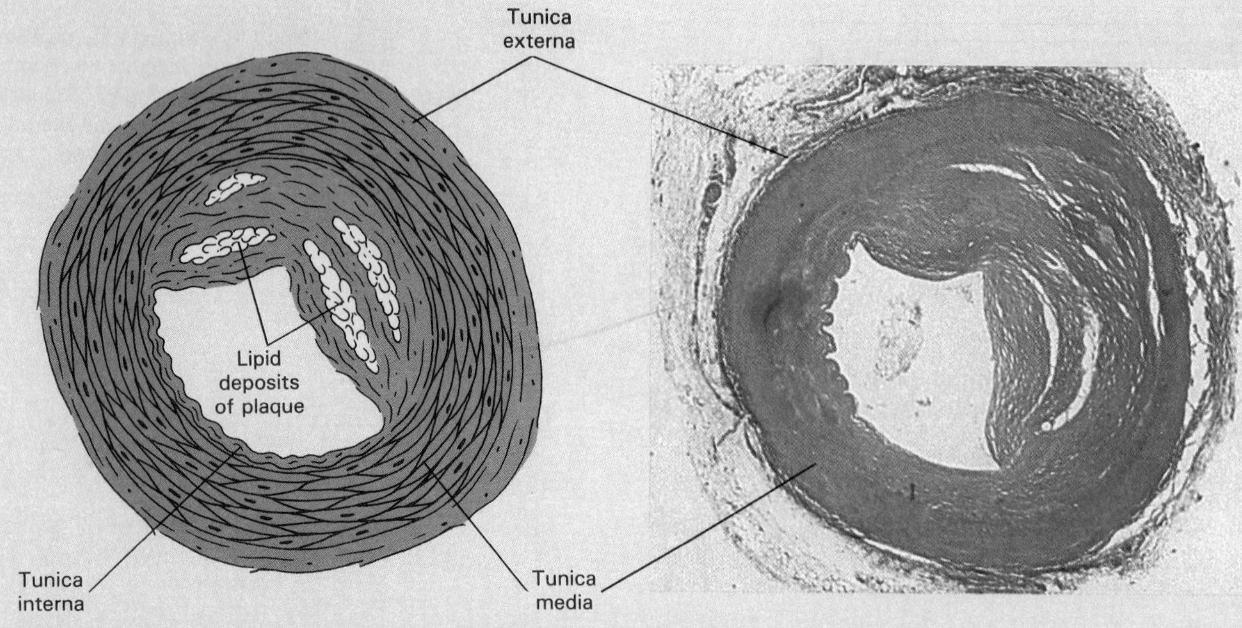

Autoregulation of Capillary Blood Flow

Blood usually flows from the arterioles to the venules at a constant rate, but the blood flow within a single capillary can be quite variable. Each precapillary sphincter goes through cycles of activity, alternately contracting and relaxing perhaps a dozen times each minute. As a result of this cyclical change in vessel diameter, a process called **vasomotion**, the blood flow within any one capillary occurs in a series of pulses rather than as a steady and constant stream. The net effect is that blood may reach the venules by one route now, and by a quite different route later, as shown in Figure 21-4b. This process is regulated at the tissue level, as smooth muscle fibers respond to localized changes in the composition of the interstitial fluid. The regulation of blood flow at the tissue level is called **autoregulation**.

Central Regulation of Capillary Blood Flow

There are also mechanisms to modify the circulatory supply to the entire capillary complex. In many instances the capillary networks within an area may be serviced by more than one artery. The arteries, called **collaterals**, enter the region and fuse, rather than ending in a forest of arterioles. The interconnection constitutes an **arterial anastomosis**. Such an arrangement guarantees a reliable blood supply to the tissues, for if one arterial supply should become blocked, another will supply blood to the capillary bed. **Arteriovenous** (ar-tē-rē-ō-VĒ-nus) **anastomoses** are direct connections between arterioles and venules. Smooth muscles in the walls of these vessels can contract or relax to regulate the amount of blood reaching the capillary bed. For example, when the arteriovenous anastomoses are dilated, blood will bypass the capillary bed and flow directly into the venous circulation. Collateral blood flow is regulated primarily by sympathetic innervation, under the control of cardiovascular centers in the medulla.

Within the capillary bed, **central**, or **preferred, channels** provide another relatively direct means of arteriole-venule communication. The arteriolar segment of the channel contains smooth muscles capable of altering its diameter, and this region is often called a **metarteriole** (met-ar-TĒ-rē-ōl). The rest of the central channel resembles a typical capillary.

VEINS

Venules vary widely in size and character, but an average venule has an internal diameter of roughly 20 μm. The smallest venules resemble expanded capillaries, and venules smaller than 50 μm lack a tunica media altogether. **Medium-sized veins** range from 2 to 9 mm in diameter. In these veins the tunica media contains several smooth muscle layers, and there are longitudinal bundles of elastic and collagen fibers in a relatively thick tunica externa. **Large veins** include the venae cavae and their tributaries within the abdominopelvic and thoracic cavities. The slender tunica media is surrounded by a thick tunica externa composed of a mixture of elastic and collagenous fibers. Typical veins are illustrated in Figures 21-1 and 21-2.

Veins and Venous Pressures

The arterial system is a high-pressure system, for it takes almost all of the force developed by the heart to push blood through the network of arteries and across miles of capillaries. Blood pressure within a peripheral venule is only about 10 percent of that in the ascending aorta, and once in the venous system the pressures continue to fall. As a result, the relatively thin walls of veins are adequate to withstand venous pressures.

The blood pressure in venules and medium-sized veins is actually so low that it cannot oppose the force of gravity without assistance. Valves in these veins act like the valves in the heart, preventing the backflow of blood. As long as the valves function normally, any movement that distorts or compresses a vein will push blood toward the heart (Figure 21-6).

FIGURE 21-6
Function of Valves in the Venous System. Valves in the walls of medium-sized veins prevent the backflow of blood. Venous compression caused by the contraction of adjacent skeletal muscles assists in maintaining venous blood flow. Changes in body position and the thoracoabdominal pump may provide additional assistance.

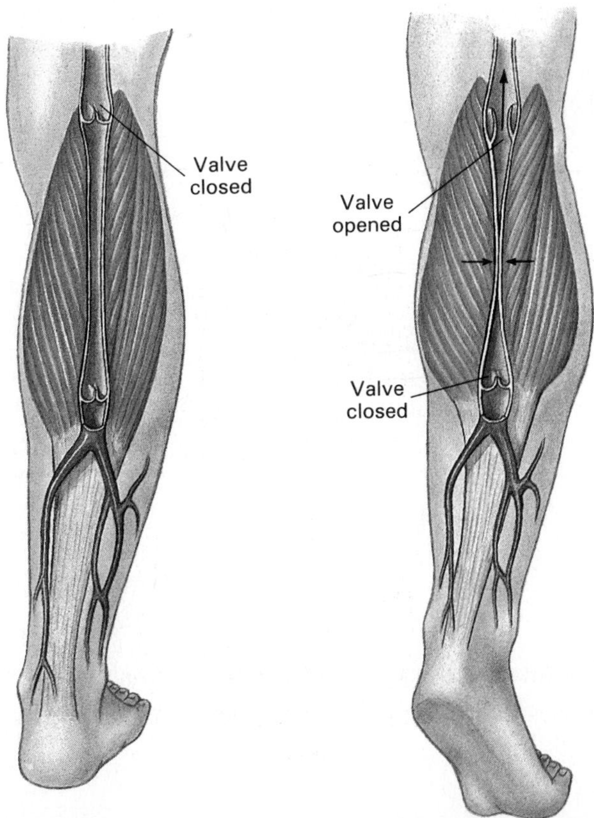

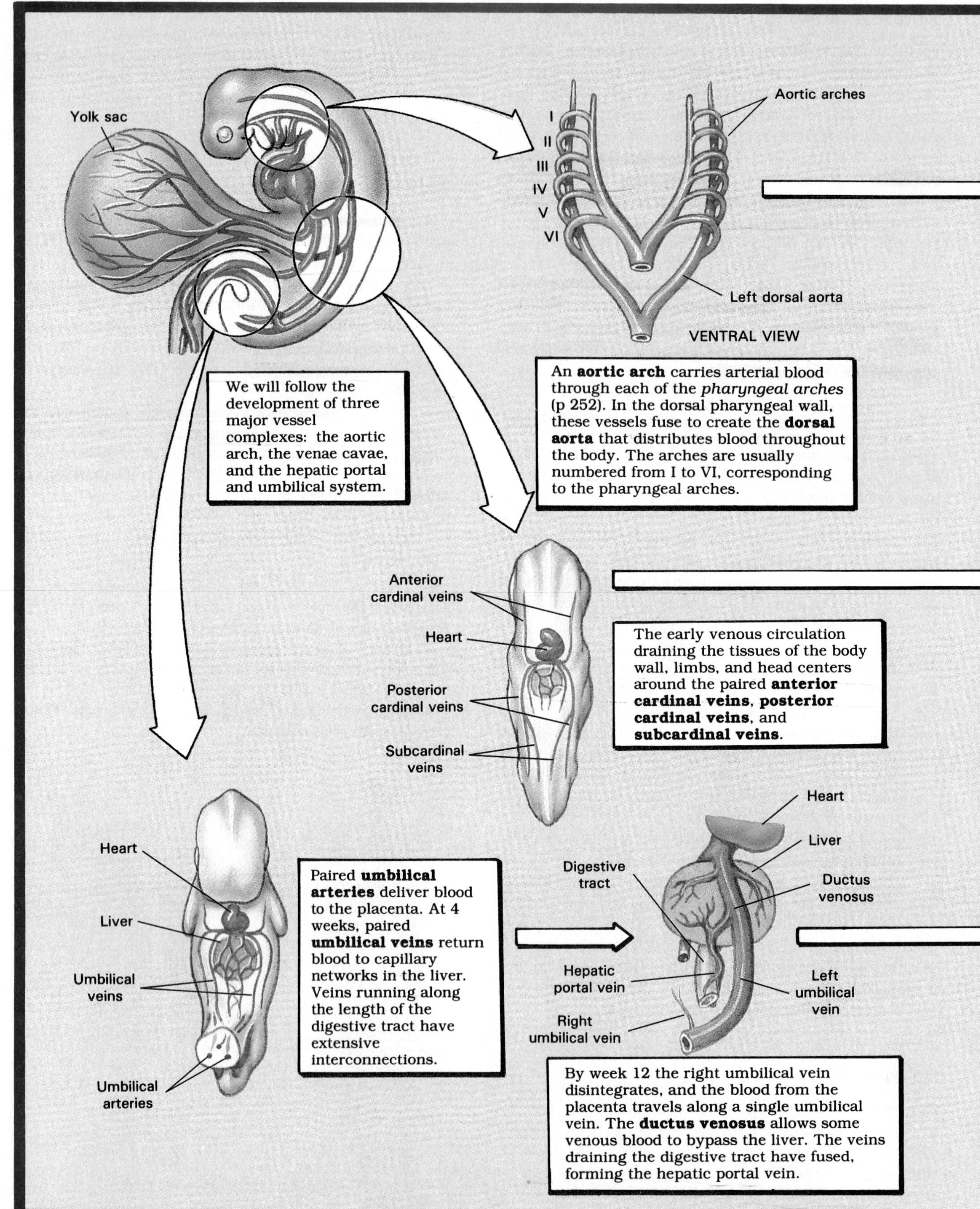
Yolk sac
Aortic arches
I
II
III
IV
V
VI
Left dorsal aorta
VENTRAL VIEW
We will follow the development of three major vessel complexes: the aortic arch, the venae cavae, and the hepatic portal and umbilical system.
An **aortic arch** carries arterial blood through each of the *pharyngeal arches* (p 252). In the dorsal pharyngeal wall, these vessels fuse to create the **dorsal aorta** that distributes blood throughout the body. The arches are usually numbered from I to VI, corresponding to the pharyngeal arches.
Anterior cardinal veins
Heart
Posterior cardinal veins
Subcardinal veins
The early venous circulation draining the tissues of the body wall, limbs, and head centers around the paired **anterior cardinal veins**, **posterior cardinal veins**, and **subcardinal veins**.
Heart
Liver
Umbilical veins
Umbilical arteries
Paired **umbilical arteries** deliver blood to the placenta. At 4 weeks, paired **umbilical veins** return blood to capillary networks in the liver. Veins running along the length of the digestive tract have extensive interconnections.
Heart
Liver
Digestive tract
Ductus venosus
Hepatic portal vein
Right umbilical vein
Left umbilical vein
By week 12 the right umbilical vein disintegrates, and the blood from the placenta travels along a single umbilical vein. The **ductus venosus** allows some venous blood to bypass the liver. The veins draining the digestive tract have fused, forming the hepatic portal vein.

Embryology Summary: Development of the Circulatory System

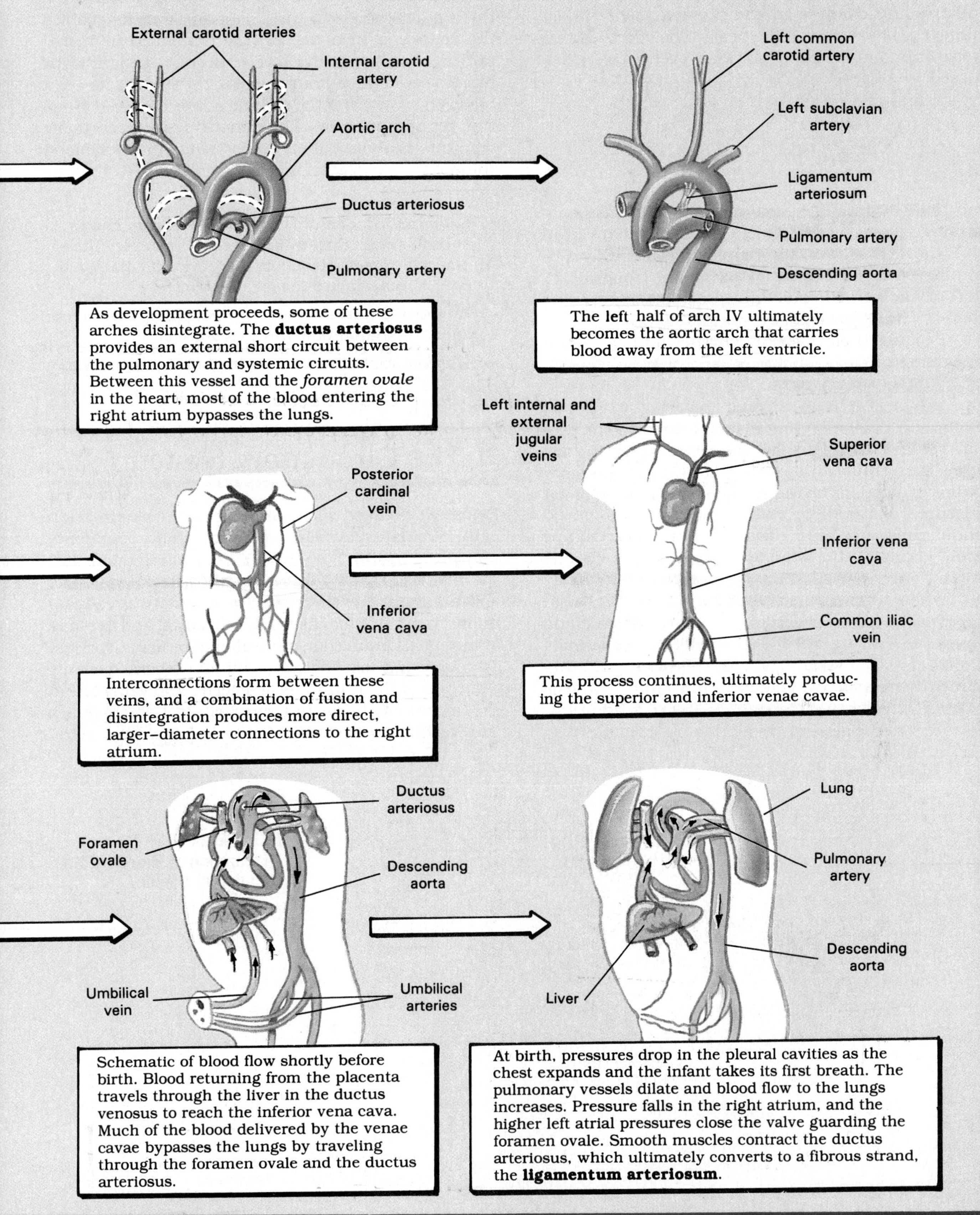

As development proceeds, some of these arches disintegrate. The **ductus arteriosus** provides an external short circuit between the pulmonary and systemic circuits. Between this vessel and the *foramen ovale* in the heart, most of the blood entering the right atrium bypasses the lungs.

The left half of arch IV ultimately becomes the aortic arch that carries blood away from the left ventricle.

Interconnections form between these veins, and a combination of fusion and disintegration produces more direct, larger-diameter connections to the right atrium.

This process continues, ultimately producing the superior and inferior venae cavae.

Schematic of blood flow shortly before birth. Blood returning from the placenta travels through the liver in the ductus venosus to reach the inferior vena cava. Much of the blood delivered by the venae cavae bypasses the lungs by traveling through the foramen ovale and the ductus arteriosus.

At birth, pressures drop in the pleural cavities as the chest expands and the infant takes its first breath. The pulmonary vessels dilate and blood flow to the lungs increases. Pressure falls in the right atrium, and the higher left atrial pressures close the valve guarding the foramen ovale. Smooth muscles contract the ductus arteriosus, which ultimately converts to a fibrous strand, the **ligamentum arteriosum**.

Large veins such as the vena cava do not have valves, but pressure changes in the thoracic cavity assist in moving blood toward the heart. The effects of this *thoracoabdominal pump* on venous return were discussed in Chapter 20. (p. 660) [**CM:** *Problems with Valve Function*]

DISTRIBUTION OF BLOOD

The total blood volume is unevenly distributed among arteries, veins, and capillaries. As indicated in Figure 21-7, the heart, arteries, and capillaries normally contain 30–35 percent of the blood volume (roughly 1.5 liters of whole blood), and the venous system contains the rest (65–70 percent, or around 3.5 liters).

Because their walls are thinner and contain a lower proportion of smooth muscle, veins are much more elastic than arteries. For a given rise in pressure, a typical vein will stretch about eight times as much as a corresponding artery. Added to the fact that veins are larger than their corresponding arteries, this ability to stretch accounts for the large volume of blood in the venous system at normal blood pressures. If the blood volume rises or falls, so do venous pressures. The elastic walls then stretch or recoil, changing the volume of blood in the venous system. The veins can thus act as a **blood reservoir**.

When serious hemorrhaging occurs, the vasomotor center of the medulla stimulates sympathetic nerves innervating smooth muscle fibers in the walls of medium-sized veins. The veins contract, and this **venoconstriction** (vē-nō-kon-STRIK-shun) further reduces the volume of the venous system. In addition, blood enters the general circulation from venous networks in the liver, bone marrow, and skin. Reducing the amount of blood in the venous system enables the volume within the arterial system to be maintained at near-normal levels, despite a significant blood loss. The venous system therefore acts as a blood reservoir, and the change in volume constitutes the **venous reserve**. The venous reserve normally amounts to about 1 liter, 20 percent of the total blood volume.

- **Examination of a cross section of tissue shows several small, thin-walled vessels with very little smooth muscle tissue in the tunica media. What type of vessels are these?**
- **How would relaxation of the precapillary sphincters affect the blood flow through a tissue?**
- **Why are valves found in veins but not in arteries?**

The Circulatory System

Before we proceed with a more detailed examination of the circulatory system, a few preliminary comments may prove useful. It should hearten you to know that the name of a blood vessel usually gives clues to its appearance or distribution. Since you are already familiar with the major skeletal, muscular, and nervous anatomical landmarks, there should be few surprises. The following text, figures, and tables consider only a portion of the hundreds of arteries and veins found in the human body, but most clinical emergencies involve the relatively large vessels considered here.

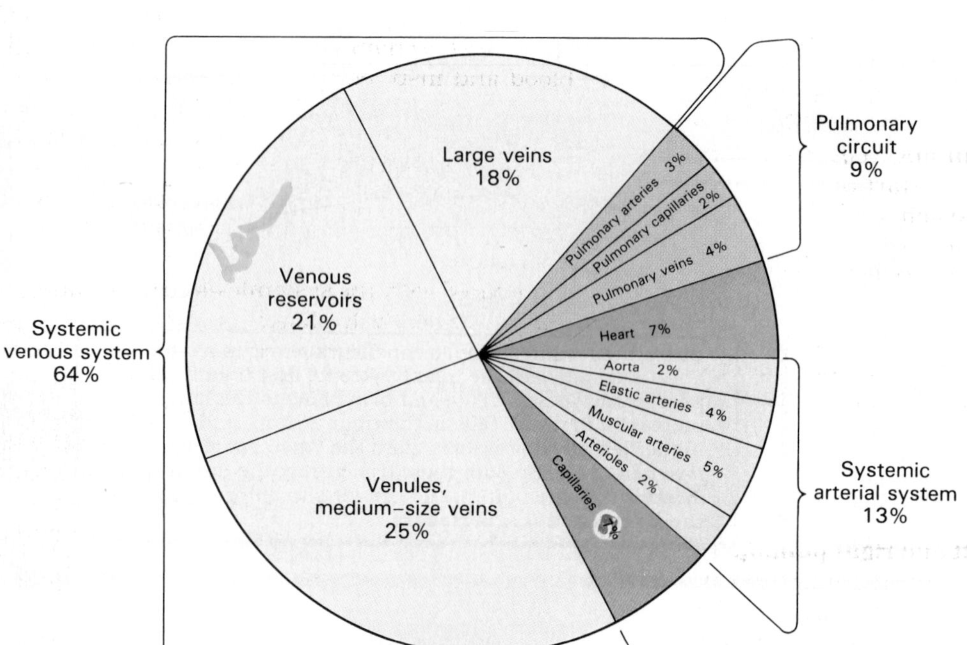

FIGURE 21-7
Distribution of Blood in the Circulatory System

Three important functional patterns will emerge from the tables and figures that follow:

1. The peripheral distribution of arteries and veins on the left and right sides are usually identical except near the heart, where large vessels connect to the atria or ventricles.
2. A single vessel may have several different names as it passes specific anatomical boundaries, making accurate anatomical descriptions possible when the vessel extends far into the periphery.
3. Arteries and veins often make anastomotic connections that reduce the impact of a temporary or even permanent occlusion (blockage) of a single vessel.

A final precautionary note concerns the memorization of this material. Concentrate on understanding the anatomical details concerning each vessel, with particular attention to its distribution. Although the accompanying figures will help you visualize the information, keep in mind that these are stylized representations of the circulatory system. Real vessels are not color-coded with red arteries and blue veins (although there are differences in their general coloration related to the degree of oxygenation of their hemoglobin). Moreover, there can be considerable variation from one individual to another in the sites of origin and termination of a particular vessel.

Figure 21-8 presents a stylized summary of the primary circulatory routes within the pulmonary and systemic circuits. We will now examine each of these circuits in detail.

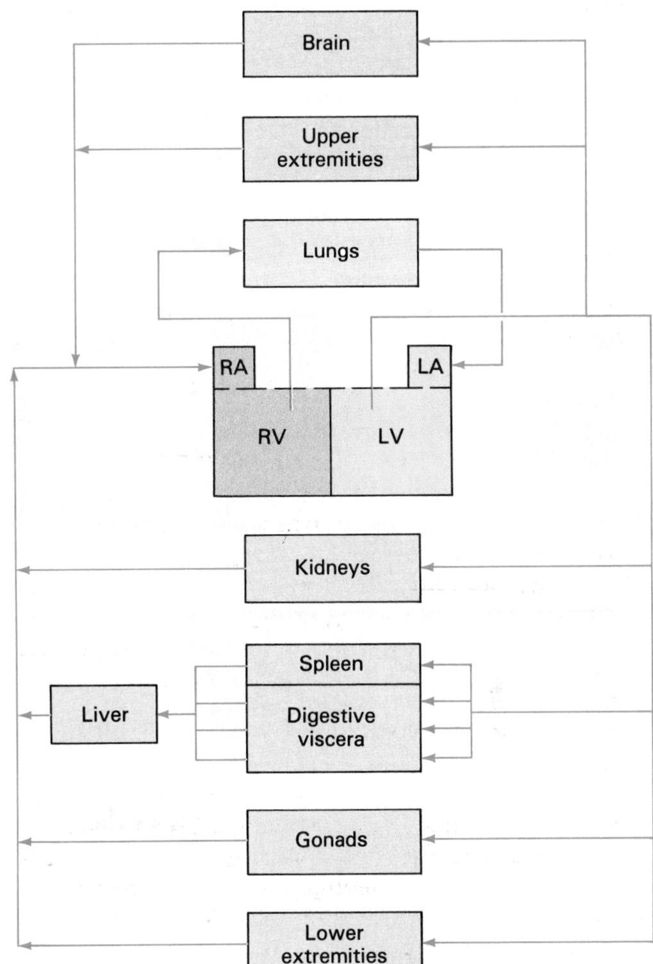

FIGURE 21-8
An Overview of the Pattern of Circulation

THE PULMONARY CIRCULATION

Blood entering the right atrium has just returned from a trip to peripheral capillary beds, where oxygen was released and carbon dioxide absorbed. After traveling through the right atrium and ventricle, blood enters the pulmonary trunk, the start of the pulmonary circuit. In this circuit, oxygen stores will be replenished, carbon dioxide excreted, and the "renewed" blood returned to the heart for distribution in the systemic circuit.

Figure 21-9 details the anatomy of the pulmonary circuit. The arteries of the pulmonary circuit differ from those of the systemic circuit in that they carry deoxygenated blood. (For this reason, color-coded diagrams usually show the pulmonary arteries in blue, the same color as systemic veins.) As the pulmonary trunk curves over the superior border of the heart it gives rise to the **left** and **right pulmonary arteries**. These large arteries enter the lungs before branching repeatedly, giving rise to smaller and smaller arteries. The smallest branches, the pulmonary arterioles, provide blood to capillary networks that surround small air pockets, or **alveoli** (al-VĒ-ō-lī; *alveolus*, sac). The walls of alveoli are thin enough for gas exchange to occur between the capillary blood and inspired air. (This exchange process will be detailed in Chapter 23.) As it leaves the alveolar capillaries, oxygenated blood enters venules that in turn unite to form larger vessels carrying blood to the **pulmonary veins**. These four veins, two from each lung, empty into the left atrium, completing the pulmonary circuit.

Compared with the systemic circuit, the pulmonary circuit is relatively short. The heart and lungs are only a few centimeters apart. To force blood around the pulmonary circuit, the right ventricle need develop a pressure of only about 15 mm Hg.[1] By comparison, the left ventricle must develop 6–8 times as much pressure to maintain systemic blood flow. These pressure differences account for the differences in ventricular structure detailed in Chapter 20. ∞ (pp. 638–639)

[1] mm Hg = millimeters of mercury, a standardized unit of pressure.

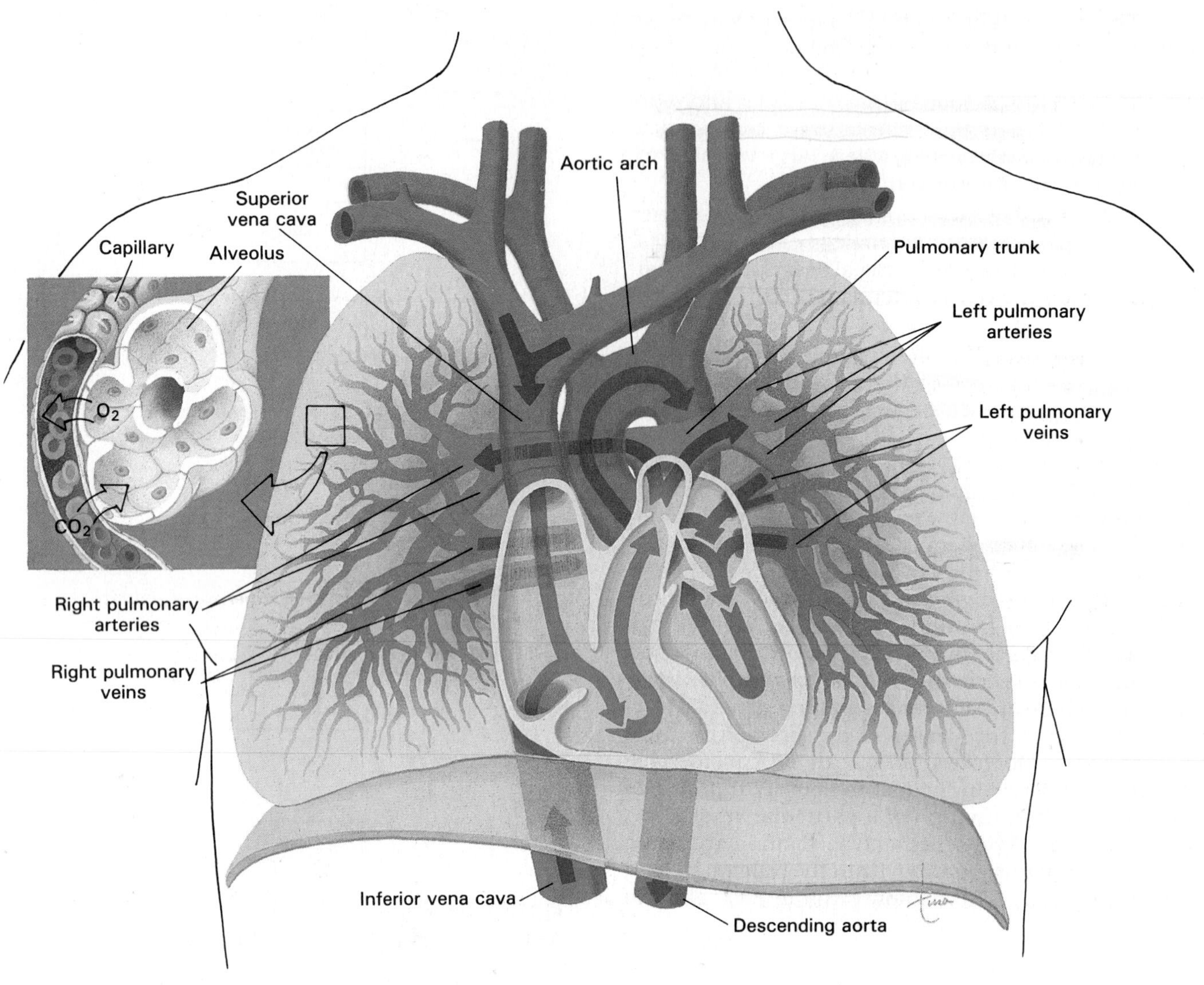

FIGURE 21-9
The Pulmonary Circuit

THE SYSTEMIC CIRCULATION

The systemic circulation supplies the capillary beds in all other parts of the body. This circuit, which at any given moment contains about 91 percent of the total blood volume, begins at the left ventricle and ends at the right atrium.

Systemic Arteries

Figure 21-10 is an overview of the arterial system. This figure indicates the relative locations of major systemic arteries. The detailed distribution of these vessels and their branches will be found in Figures 21-11 to 21-13.

The **ascending aorta** begins at the aortic semilunar valve of the left ventricle. The two coronary arteries originate near the base of the ascending aorta; their distribution was illustrated in Figure 20-6. The **aortic arch** curves across the superior surface of the heart, connecting the ascending aorta with the caudally directed descending aorta.

Arteries Originating at the Aortic Arch Three elastic arteries originate along the aortic arch (Figure 21-11). These arteries, the **brachiocephalic** (brāk-ē-ō-se-FAL-ik), the **left common carotid**, and the **left subclavian** (sub-CLĀ-vē-an), deliver blood to the head, neck, shoulders, and arms. The brachiocephalic artery, also called the **innominate artery** (i-NOM-i-nāt; unnamed), ascends for a short distance before branching to form the **right common carotid artery** and the **right subclavian artery**. There is only one brachiocephalic artery, and the left common carotid and left subclavian arteries arise separately from the aortic arch. However, in terms of their peripheral distribution the vessels on the left side are mirror images of those on the right side. Because the descrip-

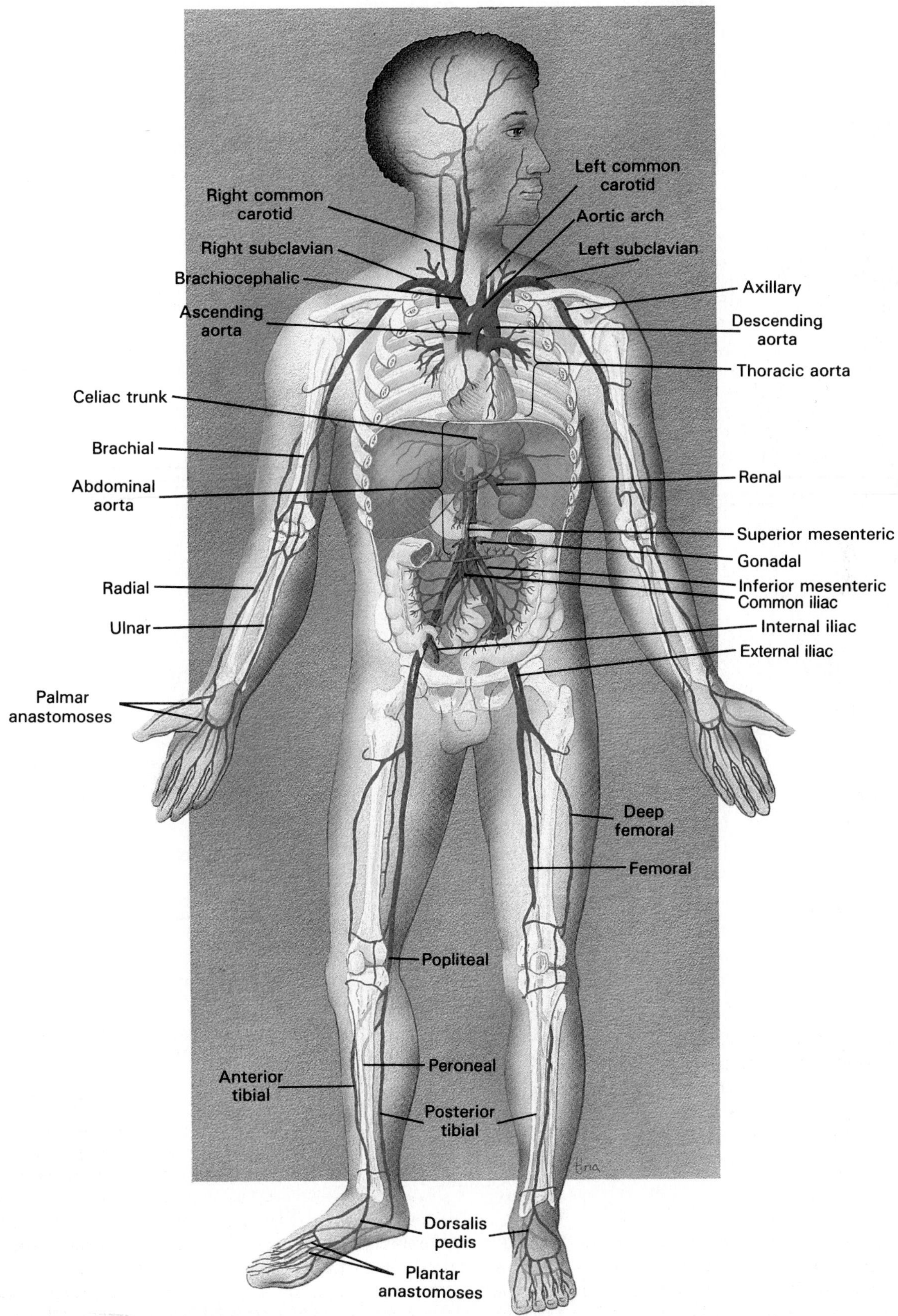
Right common carotid
Left common carotid
Aortic arch
Right subclavian
Left subclavian
Brachiocephalic
Axillary
Ascending aorta
Descending aorta
Thoracic aorta
Celiac trunk
Brachial
Renal
Abdominal aorta
Superior mesenteric
Gonadal
Inferior mesenteric
Common iliac
Radial
Ulnar
Internal iliac
External iliac
Palmar anastomoses
Deep femoral
Femoral
Popliteal
Peroneal
Anterior tibial
Posterior tibial
Dorsalis pedis
Plantar anastomoses

FIGURE 21-10
An Overview of the Arterial System

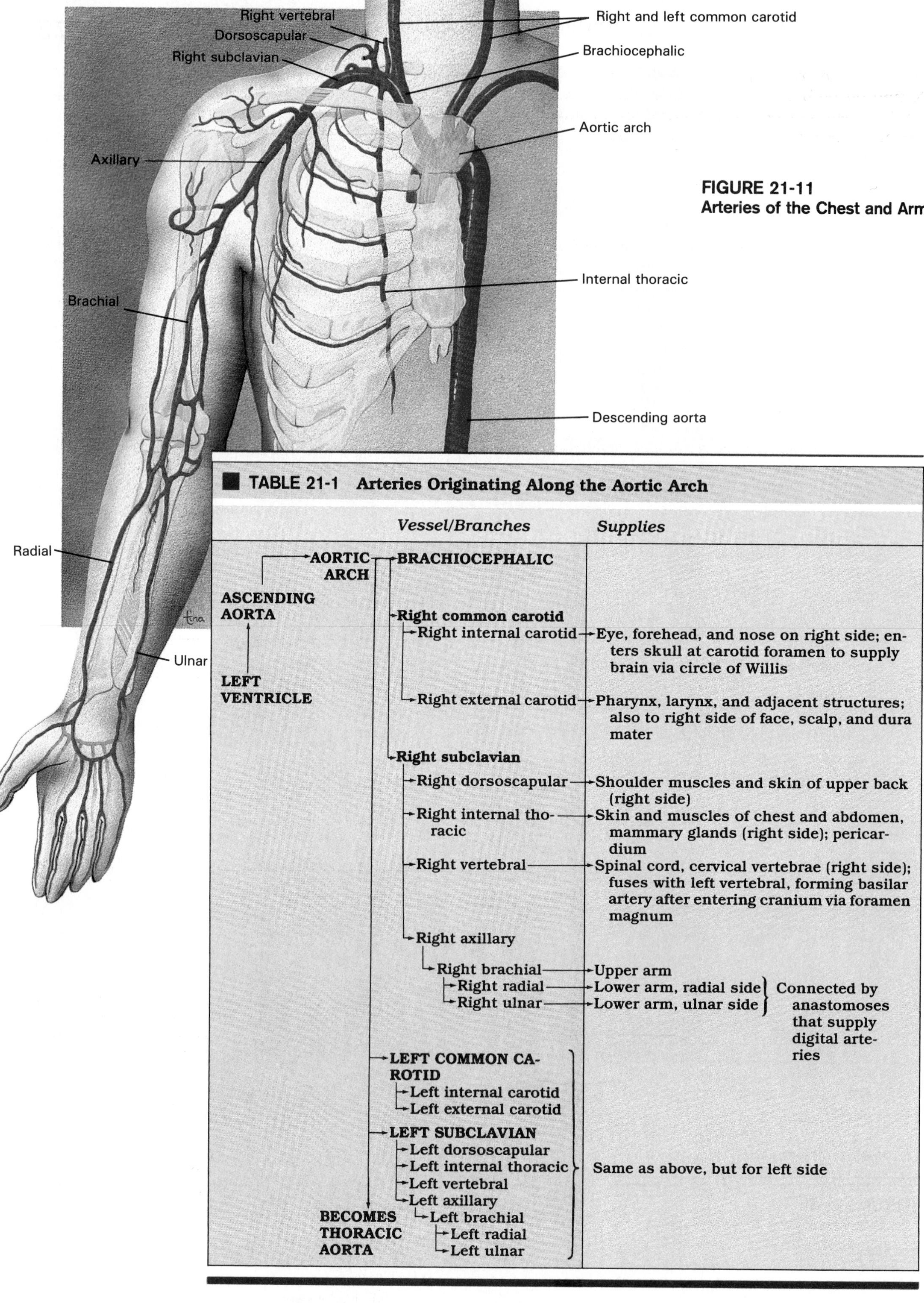

FIGURE 21-11
Arteries of the Chest and Arm

■ TABLE 21-1 **Arteries Originating Along the Aortic Arch**

Vessel/Branches	Supplies
LEFT VENTRICLE → ASCENDING AORTA → AORTIC ARCH → **BRACHIOCEPHALIC**	
→ **Right common carotid**	
→ Right internal carotid	Eye, forehead, and nose on right side; enters skull at carotid foramen to supply brain via circle of Willis
→ Right external carotid	Pharynx, larynx, and adjacent structures; also to right side of face, scalp, and dura mater
→ **Right subclavian**	
→ Right dorsoscapular	Shoulder muscles and skin of upper back (right side)
→ Right internal thoracic	Skin and muscles of chest and abdomen, mammary glands (right side); pericardium
→ Right vertebral	Spinal cord, cervical vertebrae (right side); fuses with left vertebral, forming basilar artery after entering cranium via foramen magnum
→ Right axillary	
→ Right brachial	Upper arm
→ Right radial	Lower arm, radial side } Connected by anastomoses that supply digital arteries
→ Right ulnar	Lower arm, ulnar side } Connected by anastomoses that supply digital arteries
→ **LEFT COMMON CAROTID**	Same as above, but for left side
→ Left internal carotid	Same as above, but for left side
→ Left external carotid	Same as above, but for left side
→ **LEFT SUBCLAVIAN**	Same as above, but for left side
→ Left dorsoscapular	Same as above, but for left side
→ Left internal thoracic	Same as above, but for left side
→ Left vertebral	Same as above, but for left side
→ Left axillary	Same as above, but for left side
→ Left brachial	Same as above, but for left side
→ Left radial	Same as above, but for left side
→ Left ulnar	Same as above, but for left side
BECOMES THORACIC AORTA	

tions that follow focus attention on major branches found on both sides of the body, the terms "right" or "left" will not be used. Figures 21-11 and 21-12 illustrate the major branches of these arteries, and additional details are included in Table 21-1.

The Subclavian Arteries The subclavian arteries supply blood to the arms, chest wall, shoulders, back, and central nervous system. Three major branches arise before a subclavian artery leaves the thoracic cavity: (1) a **dorsoscapular artery**, which provides blood to muscles and other tissues of the shoulder and upper back; (2) an **internal thoracic artery**, supplying the pericardium and anterior wall of the chest; and (3) a **vertebral artery**, which provides blood to the brain and spinal cord.

After passing the first rib, the subclavian gets a new name, the **axillary artery**. The axillary artery crosses the axilla (armpit) to enter the arm, where its name changes again, becoming the **brachial artery**. The brachial artery provides blood to the upper arm before branching to create the **radial artery** and **ulnar artery** of the forearm. These arteries are connected by anastomoses at the palm, and the **digital arteries** originate from these arterial connections.

The Carotid Artery and the Blood Supply to the Brain Figure 21-12a follows the arterial supply to the brain. The common carotids ascend deep in the tissues of the neck. A carotid artery can usually be located by pressing gently along either side of the trachea until a strong pulse is felt. Each common carotid artery divides into an external carotid and an internal carotid artery at an expanded chamber, the **carotid sinus**. (This sinus, which contains baroreceptors involved in cardiovascular regulation, was introduced in Chapter 17.) (p. 531) The external carotids supply blood to the pharynx, larynx, and face. The internal carotids enter the skull through the carotid foramina of the temporal bones, delivering blood to the brain. (The carotid foramina of the skull are shown in Figures 8-3 and 8-4.) (pp. 241–242)

Earlier chapters emphasized that the brain is extremely sensitive to changes in its circulatory supply. An interruption of circulation for several seconds will produce unconsciousness, and after 4 minutes

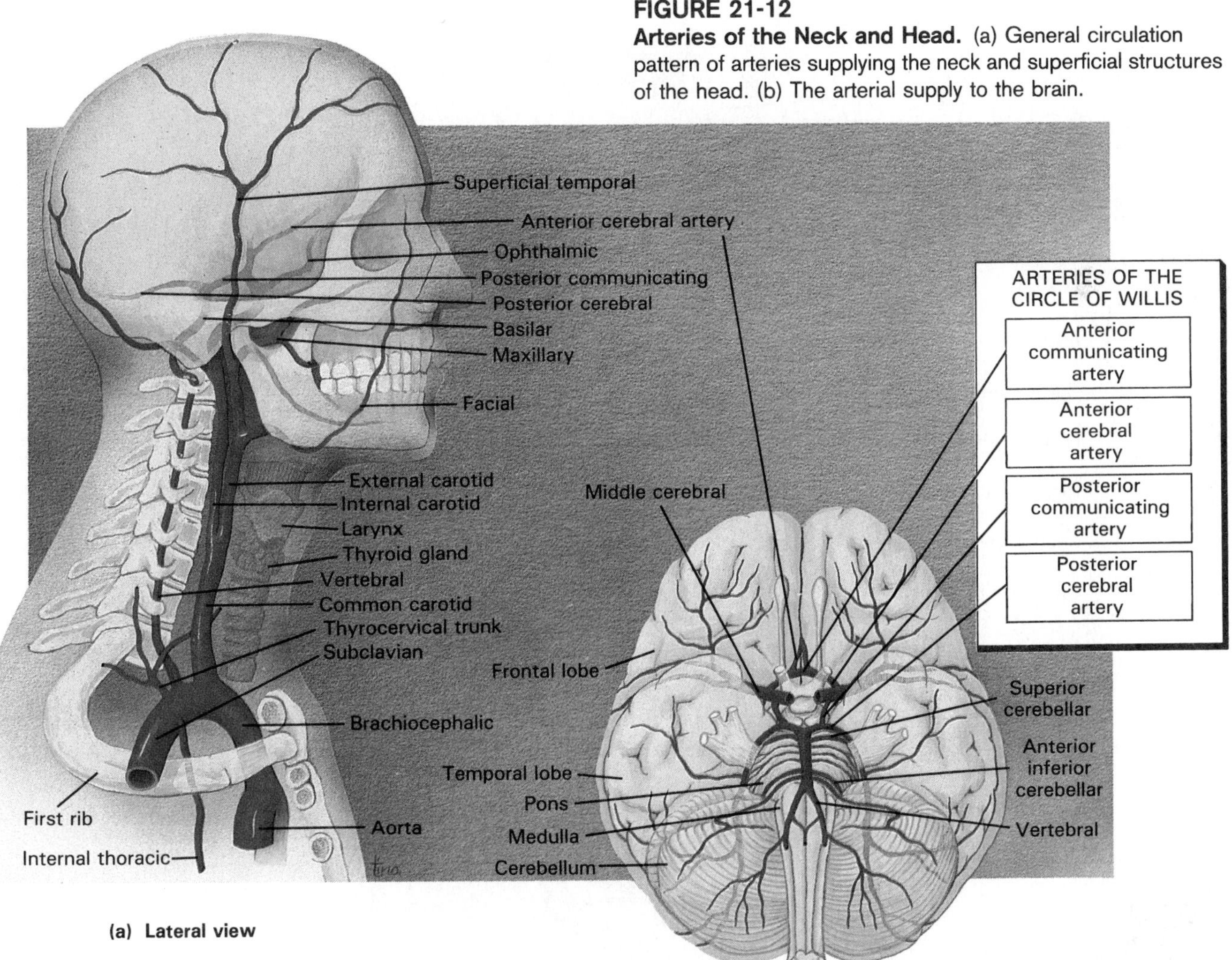

FIGURE 21-12
Arteries of the Neck and Head. (a) General circulation pattern of arteries supplying the neck and superficial structures of the head. (b) The arterial supply to the brain.

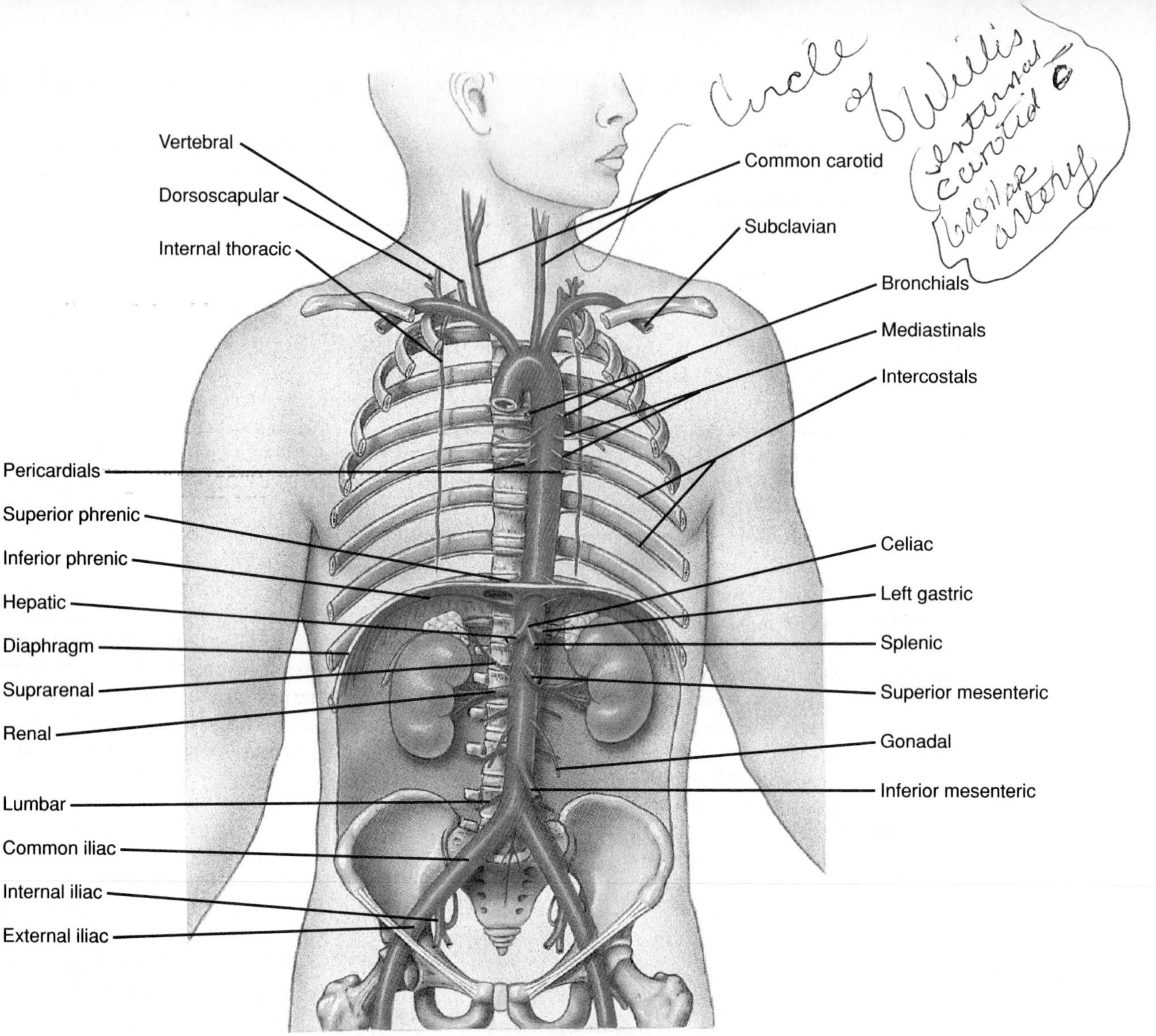

Vertebral
Dorsoscapular
Internal thoracic
Common carotid
Subclavian
Bronchials
Mediastinals
Intercostals
Pericardials
Superior phrenic
Inferior phrenic
Hepatic
Diaphragm
Suprarenal
Renal
Lumbar
Common iliac
Internal iliac
External iliac
Celiac
Left gastric
Splenic
Superior mesenteric
Gonadal
Inferior mesenteric
Circle of Willis
internal carotid
basilar artery

(a)

Vena cava
Aorta
Liver
Celiac
Hepatic
Right gastric
Superior mesenteric
Gastroduodenal
Ascending colon
Small intestine
Internal iliac
External iliac
Stomach
Left gastric
Splenic
Spleen
Gastroepiploic
Pancreas
Inferior mesenteric
Common iliac
Descending colon
Rectum

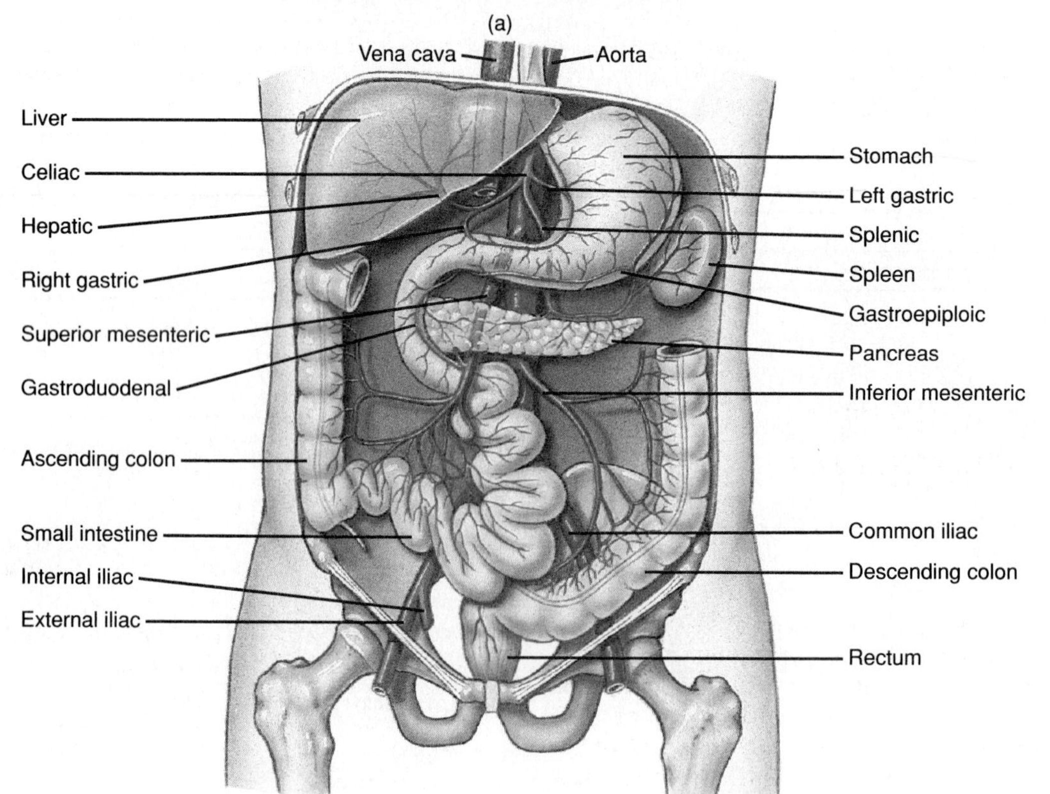

(b)

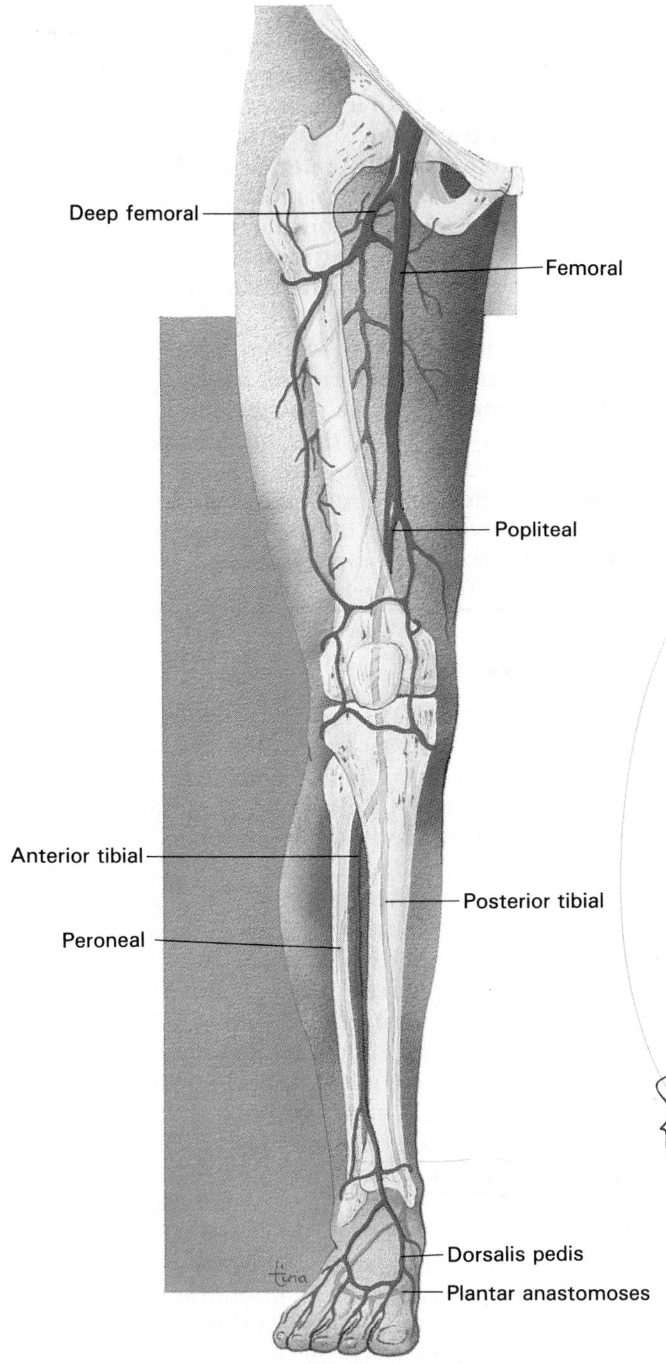

FIGURE 21-13
Major Arteries of the Trunk and Leg

there may be some permanent neural damage. Such circulatory crises are rare, because blood reaches the brain through the vertebral arteries as well as by way of the internal carotids. The vertebral arteries ascend within the transverse foramina of the cervical vertebrae (shown in Figure 8-18), penetrating the skull at the foramen magnum. (p. 260) Inside the cranium the two vertebral arteries fuse to form a large **basilar artery** that continues along the ventral surface of the brain. These vessels are indicated in Figure 21-12b.

Normally, the internal carotids supply the arteries of the anterior half of the cerebrum, and the rest of the brain receives blood from the vertebral arteries. But this circulatory pattern can easily change, because the internal carotids and the basilar artery are interconnected in a ring-shaped anastomosis, the **circle of Willis**, that encircles the infundibulum of the pituitary gland. With this arrangement, the brain can receive blood from either the carotids or the vertebrals, and the chances for a serious interruption of circulation are reduced. [**CM:** *Treatment of Cerebrovascular Disease*]

The Descending Aorta The diaphragm divides the descending aorta into a superior **thoracic aorta** and an inferior **abdominal aorta**. The thoracic aorta travels within the mediastinum, providing blood to the intercostal arteries that carry blood to the spinal cord and the body wall. This artery also gives rise to small arteries that end in capillary beds in the esophagus, pericardium, and other mediastinal structures. Near the diaphragm a pair of **superior phrenic** (FREN-ik) **arteries** deliver blood to the muscular diaphragm that separates the thoracic and abdominopelvic cavities. The branches of the thoracic aorta are detailed in Table 21-2 and Figure 21-13a.

The abdominal aorta (Table 21-3 and Figure 21-13b) descends posterior to the peritoneal cavity, often surrounded by a cushion of adipose tissue. This vessel delivers blood to elastic arteries that distribute blood to visceral organs of the digestive system and to the kidneys and adrenal glands. The **celiac** (SĒL-ē-ak) **artery**, **superior mesenteric** (mez-en-TER-ik) **artery**, and **inferior mesenteric artery** arise on the anterior surface of the abdominal aorta and branch in the connective tissues of the mesenteries. These

TABLE 21-2 Branches of the Thoracic Aorta

Vessel	*Supplies*
Intercostals[a,b]	Spinal cord, back muscles, body wall, and skin
Esophageals[a,b]	Esophagus
Bronchials[a,b]	Conducting passageways of respiratory tract (not involved with gas exchange)
Mediastinals[a,b]	Mediastinal structures
Pericardials	Pericardium
Superior phrenics[a]	Diaphragm

[a] Left and right
[b] Numerous

TABLE 21-3 Branches of the Abdominal Aorta

Vessel/Branches	*Supplies*
Inferior phrenics[a]	Diaphragm
Celiac	
Left gastric	Stomach
Hepatic	Liver
Splenic	Spleen, pancreas, stomach
Superior mesenteric	Pancreas, small intestine, and first two-thirds of large intestine
Inferior mesenteric	Last third of large intestine, colon, and rectum
Gonadal[a] (testicular or ovarian)	Gonads (testes or ovaries)
Suprarenal[a]	Adrenal glands
Renal[a]	Kidneys
Lumbar[b]	Spinal cord and abdominal wall
Common iliac[a]	Pelvis and legs
Internal iliac	Pelvic muscles, skin, urinary and reproductive organs of pelvic cavity
External iliac	
Deep femoral[a]	Hip joint, femoral head, femoral shaft, deep muscles of thigh
Femoral	Thigh
Popliteal	Lower leg and foot
Anterior tibial	Connected by anastomoses of dorsalis pedis, dorsal arch, and plantar arch that service distal portion of foot and toes
Posterior tibial	
Peroneal	

[a] Left and right
[b] Numerous

three vessels provide blood to all of the digestive organs in the abdominopelvic cavity:

- the celiac delivers blood to the liver, stomach, and spleen
- the superior mesenteric supplies the pancreas, small intestine, and most of the large intestine
- the inferior mesenteric delivers blood to the last portion of the large intestine and rectum.

Paired **gonadal** (gō-NAD-al) **arteries** originate between the superior and inferior mesenteric arteries; in males they are called *testicular arteries*, in females, *ovarian arteries*. The peripheral distribution of gonadal vessels (both arteries and veins) differs in males and females; the differences will be detailed in Chapter 28.

The **suprarenal arteries** and **renal arteries** arise along the posterolateral surface of the abdominal aorta and travel behind the peritoneal lining to reach the adrenal glands and kidneys. Small **lumbar arteries** begin on the posterior surface of the aorta and supply the spinal cord and the abdominal wall.

Near the level of vertebra L_4 the abdominal aorta divides to form a pair of muscular arteries. These **common iliac** (IL-ē-ak) **arteries** carry blood to the pelvis and legs. As it travels along the inner surface of the ilium, each common iliac divides to form an **internal iliac artery** that supplies smaller arteries of the pelvis and an **external iliac artery** that enters the leg.

Once in the leg, the external iliac artery branches, forming the **femoral artery** and the **deep femoral artery**. When it reaches the lower leg, the femoral artery becomes the **popliteal artery**, which almost immediately branches to form the **anterior tibial**, **posterior tibial**, and **peroneal** arteries. These arteries are connected by two anastomoses, one on the top of the foot (the **dorsalis pedis**) and one the bottom (the **plantar arch**).

Systemic Veins

Figure 21-14 illustrates the distribution of the major vessels of the venous system. A comparison with Figure 21-10 reveals that complementary arteries and veins often run side by side, and in many cases they have comparable names. For example, the axillary arteries run alongside the axillary veins. In addition, arteries and veins often travel in the company of peripheral nerves that have the same names and innervate the same structures. (This anatomical relationship becomes apparent when comparing Figures 13-9, 13-10, 21-10, and 21-14.) (pp. 415–416)

One significant difference between the arterial and venous systems concerns the distribution of major veins in the neck, arms, and legs. Arteries in these areas are not found at the body surface; instead, they are deep beneath the skin, protected by bones and surrounding soft tissues. In contrast, there are usually two sets of peripheral veins, one superficial and the other deep. The superficial veins are so close to the surface that they can be seen quite easily. This makes them easy targets for obtaining blood samples, and most blood tests are performed on venous blood collected from the superficial veins of the arm (usually where they cross the elbow).

This dual-venous drainage plays an important role in the control of body temperature. When body temperature becomes abnormally low, the arterial blood supply to the skin is reduced, and the superficial veins are bypassed. Blood entering the limbs then returns to the trunk in the deep veins. When overheating occurs, the blood supply to the skin increases, and the superficial veins dilate. This is one reason why superficial veins in the arms and legs become prominent during periods of heavy exercise, or when sitting in a sauna, hot tub, or steam bath.

The branching pattern of peripheral veins is much more variable than that of arteries. Arterial

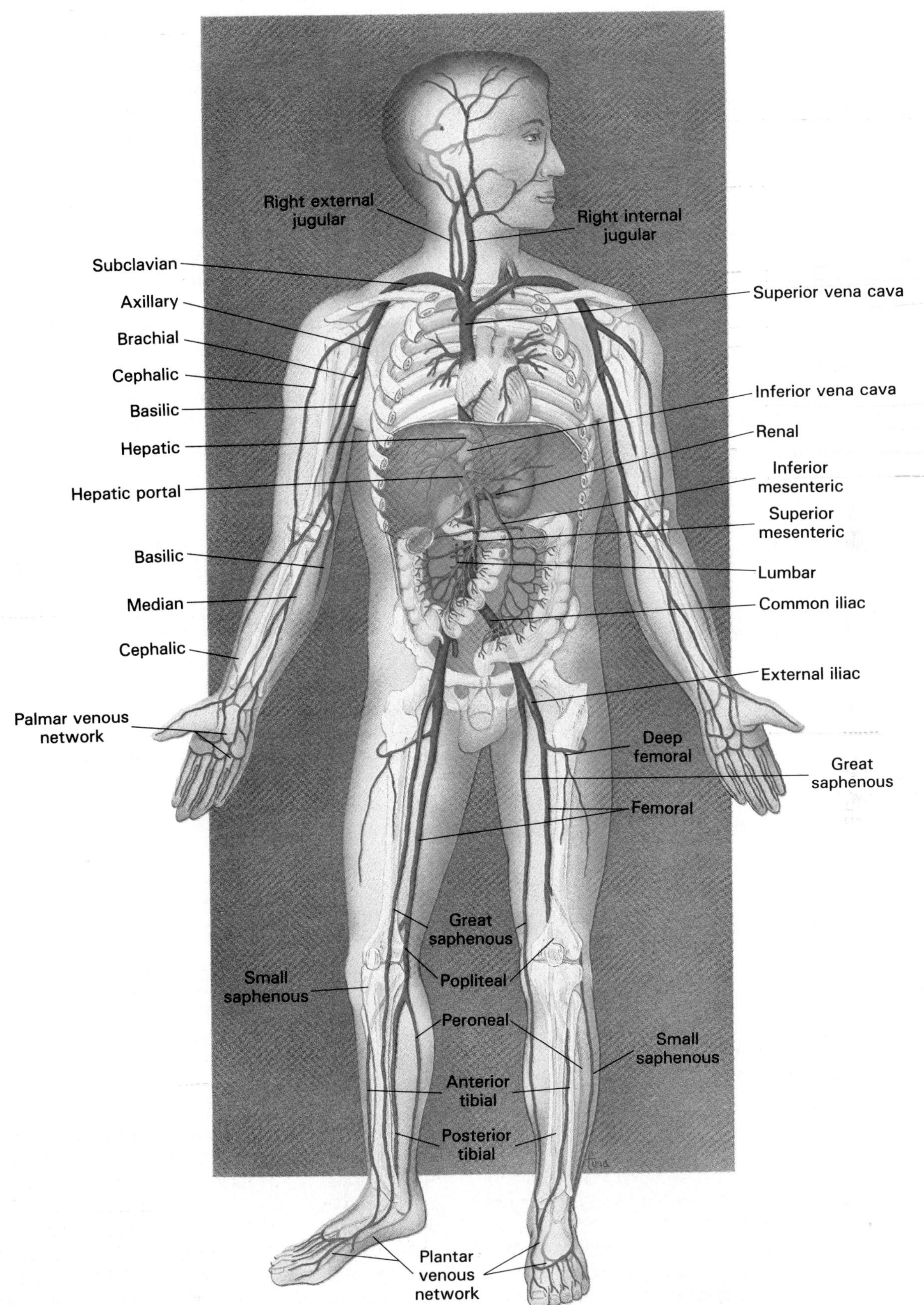
Right external jugular
Right internal jugular
Subclavian
Superior vena cava
Axillary
Brachial
Cephalic
Basilic
Inferior vena cava
Hepatic
Renal
Hepatic portal
Inferior mesenteric
Superior mesenteric
Basilic
Lumbar
Median
Common iliac
Cephalic
External iliac
Palmar venous network
Deep femoral
Great saphenous
Femoral
Great saphenous
Popliteal
Small saphenous
Peroneal
Small saphenous
Anterior tibial
Posterior tibial
Plantar venous network

FIGURE 21-14
An Overview of the Venous System

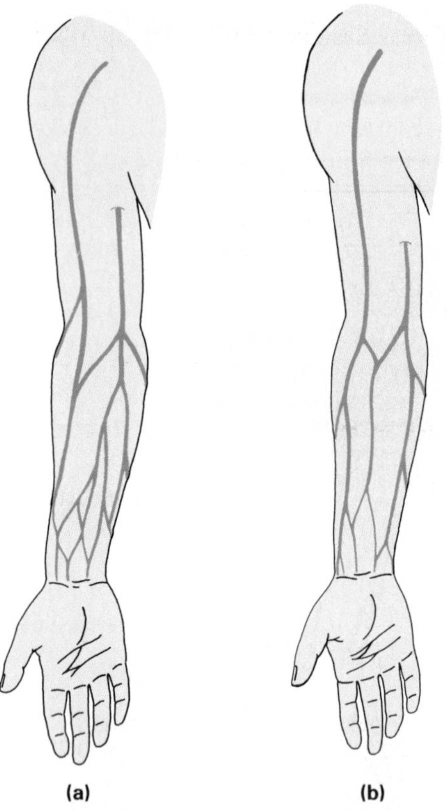

FIGURE 21-15
Variations in Venous Distribution. (a) The typical distribution of superficial veins on the anterior surface of the arm. (b) A relatively common alternative venous arrangement.

pathways are usually direct, because developing arteries grow toward active tissues. By the time blood reaches the venous system, pressures are low, and routing variations make little functional difference.

During development, large veins form through the fusion of many smaller, interconnected veins. The final result is extremely variable. Consider the superficial veins of the arm diagrammed in Figure 21-15a. This is the most common arrangement, but the venous pattern in Figure 21-15b may be found instead, and it is just as effective in returning blood toward the heart. Anyone attempting to draw a blood sample from the arm must be aware of the different venous possibilities, because needle placement varies from individual to individual.

The discussion that follows is based on the most common arrangement of veins. We will begin with the tributaries of the superior vena cava and then proceed to the veins draining into the inferior vena cava.

The Superior Vena Cava The **superior vena cava** (SVC) receives blood from the head, neck, chest, shoulders, and arms.

Venous Return from the Head and Neck Small veins in the neural tissue of the brain empty into a network of thin-walled channels, the **dural sinuses**. (The relationships between these vessels and the

FIGURE 21-16
Major Veins of the Head and Neck. (a) Veins draining superficial and deep portions of the head and neck. (b) Veins draining the cranium. For the relationship of these veins to meningeal layers, see Figure 14–14.

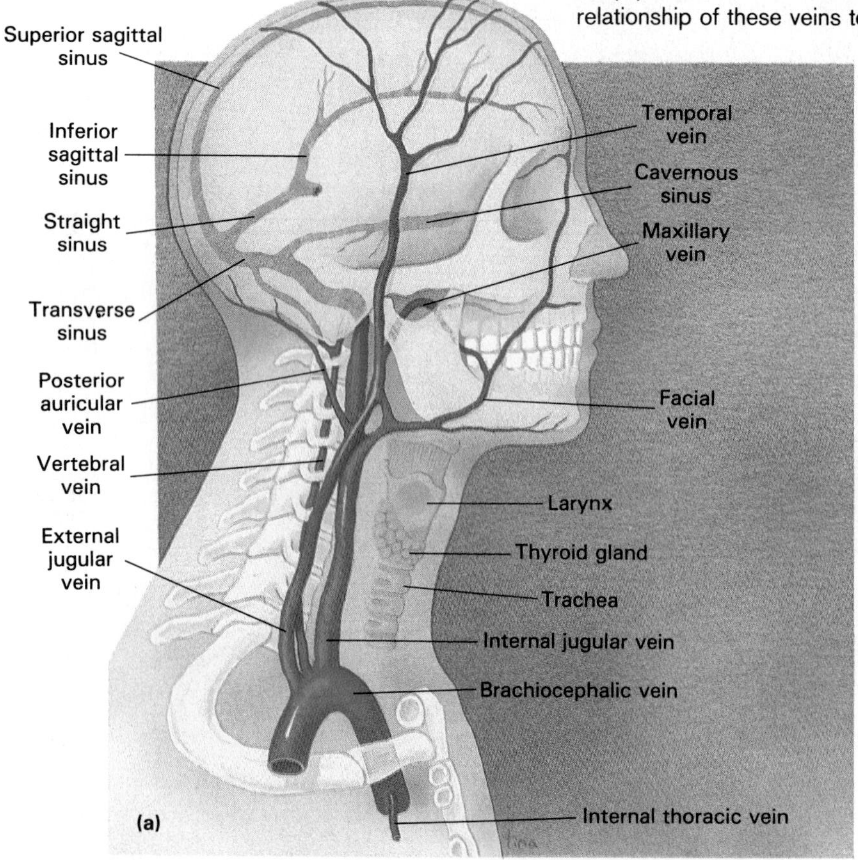

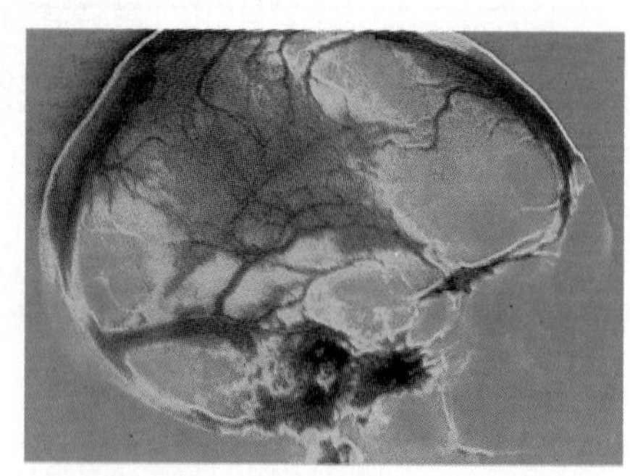

meninges of the brain were detailed in Figure 14-14.) (p. 452) The largest and most important sinus, the **superior sagittal sinus**, is found in the falx cerebri. Most of the blood leaving the brain passes through one of the dural sinuses and enters one of the internal jugular veins that penetrate the jugular foramen and descend in the deep tissues of the neck. (The path of the dural veins and the locations of the jugular foramina can be reviewed in Figures 8-3 to 8-5.) (p. 241–243) The more superficial **external jugular veins** collect blood from the overlying structures of the head and neck. These veins travel just beneath the skin, and a *jugular venous pulse* (*JVP*) can sometimes be seen at the base of the neck. **Vertebral veins** drain the cervical spinal cord and the posterior surface of the skull. These vessels, diagrammed in Figure 21-16, descend within the transverse foramina of the cervical vertebrae, in company with the vertebral arteries.

Venous Return from the Limbs and Chest The major veins of the upper body are indicated in Figure 21-17a, and information concerning the venous tributaries of the superior vena cava can be found

FIGURE 21-17
The Venous Drainage of the Abdomen and Chest

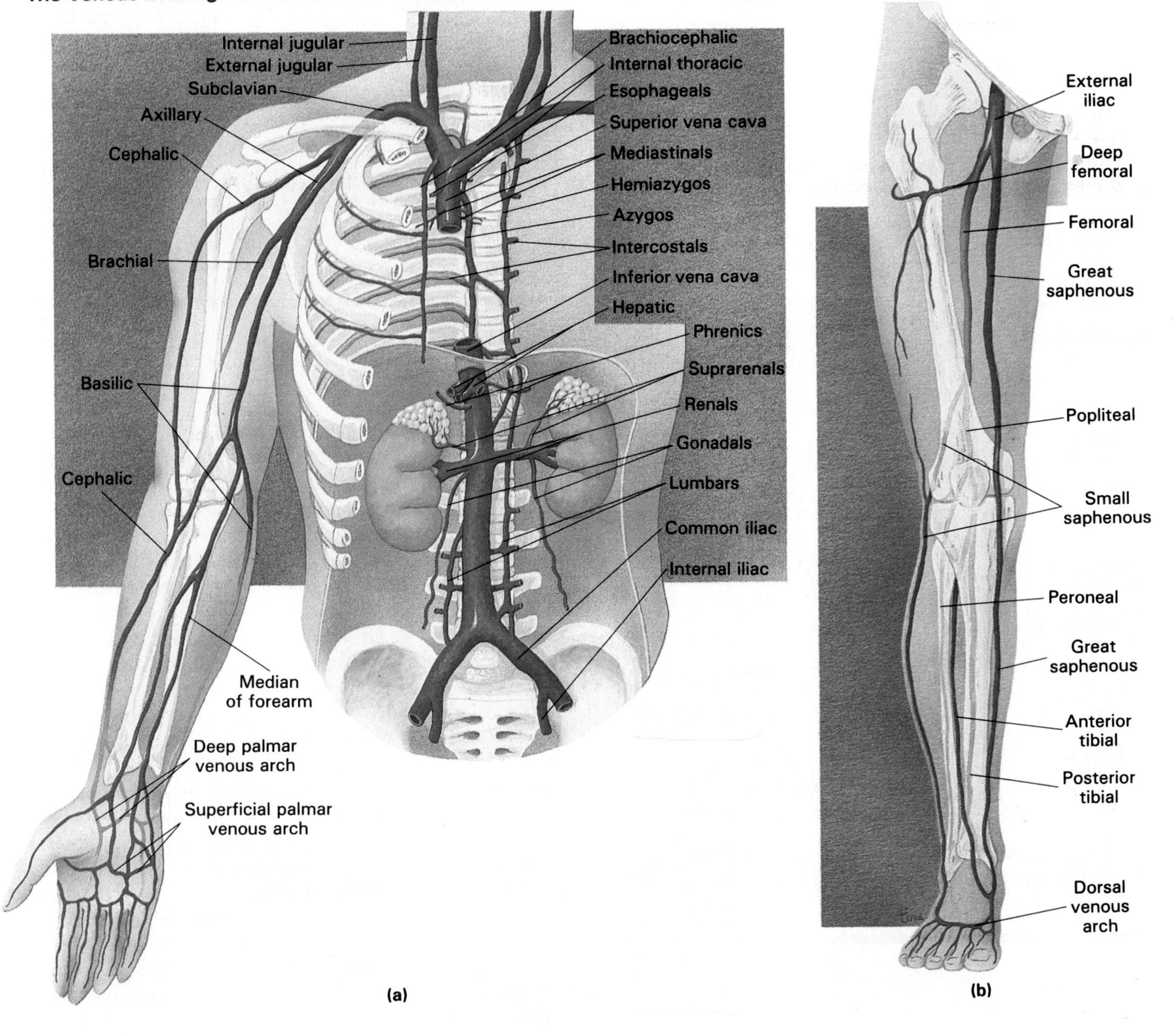

Internal jugular
External jugular
Subclavian
Axillary
Cephalic
Brachial
Basilic
Cephalic
Median of forearm
Deep palmar venous arch
Superficial palmar venous arch
Brachiocephalic
Internal thoracic
Esophageals
Superior vena cava
Mediastinals
Hemiazygos
Azygos
Intercostals
Inferior vena cava
Hepatic
Phrenics
Suprarenals
Renals
Gonadals
Lumbars
Common iliac
Internal iliac

(a)

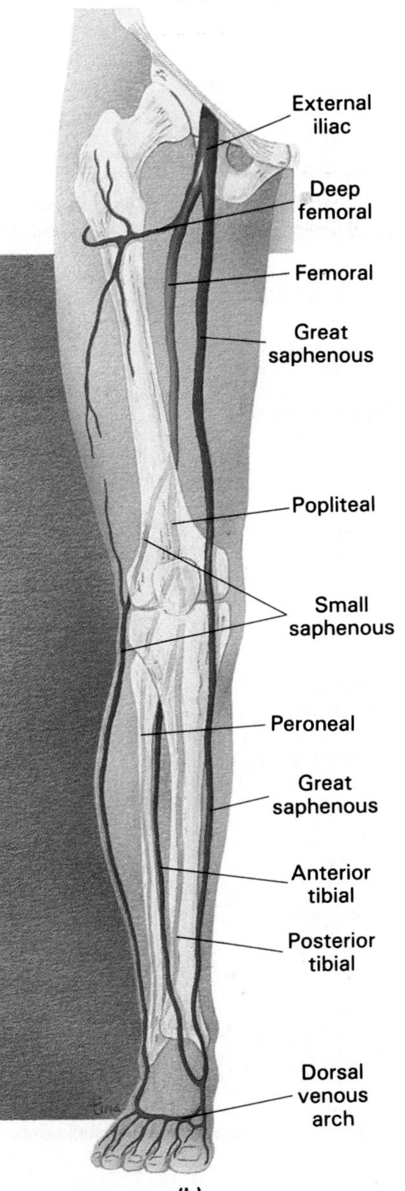

(b)

in Table 21-4. A venous network in the palms collects blood from the digital veins. As illustrated in Figure 21-15, the superficial veins of the forearm are variable. These vessels drain into the **cephalic vein** and the **basilic vein**. The deeper veins of the forearm consist of a **radial vein** and an **ulnar vein**. After crossing the elbow, these veins fuse to form the **brachial vein**. As the brachial vein continues toward the trunk, it receives blood from the cephalic and basilic veins before entering the axilla as the **axillary vein**.

The axillary vein then continues into the trunk, and at the level of the first rib it becomes the **subclavian vein**. After traveling a short distance inside the thoracic cavity, the subclavian meets and merges with the external and internal jugular veins of that side. This fusion creates the large **brachiocephalic vein**, also known as the **innominate vein**. Near the heart, the two brachiocephalic veins (one from each side of the body) combine to create the superior vena cava, or SVC.

The superior vena cava receives blood from the thoracic body wall via the **azygos** (AZ-i-gos) **veins** before arriving at the right atrium. *Azygos* means "unpaired," and the name was originally given to the vein on the right side. More thorough studies of the thoracic venous distribution revealed an almost identical vessel on the left side. Rather than have right and left unpaired veins, anatomists named the vessel on the left the **hemiazygos** (half-unpaired) **vein**.

TABLE 21-4 Venous Distribution of the Superior Vena Cava

Vessel/Branches	*Drains*
BRACHIOCEPHALIC[a]	
Internal jugular[a]	Cranium, face, and neck common carotid artery
Vertebral[a]	Cranium, spinal cord, vertebrae
External jugular[a]	Neck, face, salivary glands, scalp
Subclavian[a]	
Axillary	
Superficial	
Cephalic	Lateral surface of arm
Basilic	Posterior surfaces of arm and hand
	Interconnected by median cubital vein and median antebrachial network
Deep	
Brachial	
Radial	Radial side of forearm
Ulnar	Ulnar side of forearm
	Venous network of wrist and hand
AZYGOS[b]	Anastomotic connections to vein on left side, known as the hemiazygos
Intercostals[a,b]	Vertebrae and body wall
Esophageals[b]	Esophagus
Mediastinals[b]	Mediastinum

[a] Left and right
[b] Numerous

TABLE 21-5 Venous Distribution of the Inferior Vena Cava

Vessel/Branches	*Drains*
PHRENIC[a]	Diaphragm
HEPATIC	Liver
SUPRARENAL[a]	Adrenal glands
RENAL[a]	Kidneys
GONADAL[a] (testicular or ovarian)	Gonads (testes or ovaries)
LUMBAR[b]	Spinal cord, body wall
COMMON ILIAC[a]	Pelvis and legs
Internal iliac	Pelvic muscles, skin, urinary and reproductive organs of pelvic cavity
External iliac	
Femoral	Thigh
Great saphenous	Upper and lower leg and foot
Popliteal	Lower leg and foot
Superficial	
Small saphenous	
Deep	
Anterior tibial	
Posterior tibial	Extensive anastomoses interconnect veins of the ankle and foot
Peroneal	
Plantar	

[a] Left and right
[b] Numerous

The Inferior Vena Cava The **inferior vena cava** (IVC) collects most of the venous blood from organs below the diaphragm (a small amount reaches the superior vena cava via the azygos and hemiazygos veins). Veins of the abdomen and legs are illustrated in Figure 21-17b and detailed in Table 21-5. Blood leaving the capillaries in the sole of each foot collects into a network of **plantar veins**. The plantar network provides blood to the **anterior tibial vein**, the **posterior tibial vein**, and the **peroneal vein**, the deep veins of the lower leg. A **dorsal venous arch** drains blood from capillaries on the superior surface of the foot. This arch is drained by two superficial veins, the **great saphenous vein** (sa-FĒ-nus; *saphenes*, prominent) and the **small saphenous vein**. There are exten-

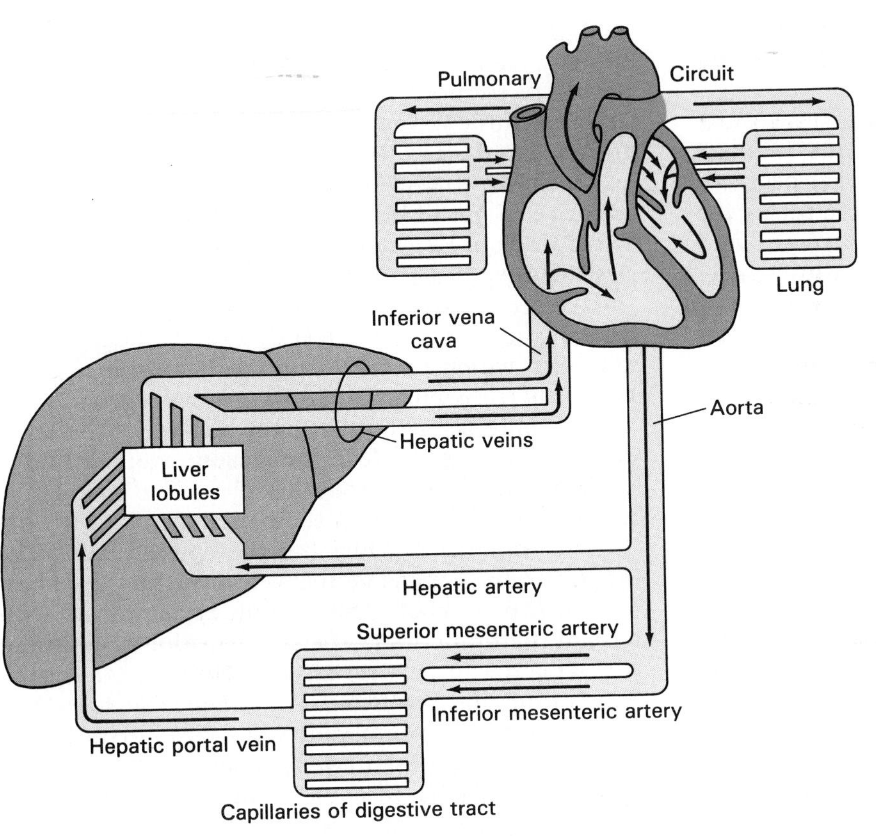

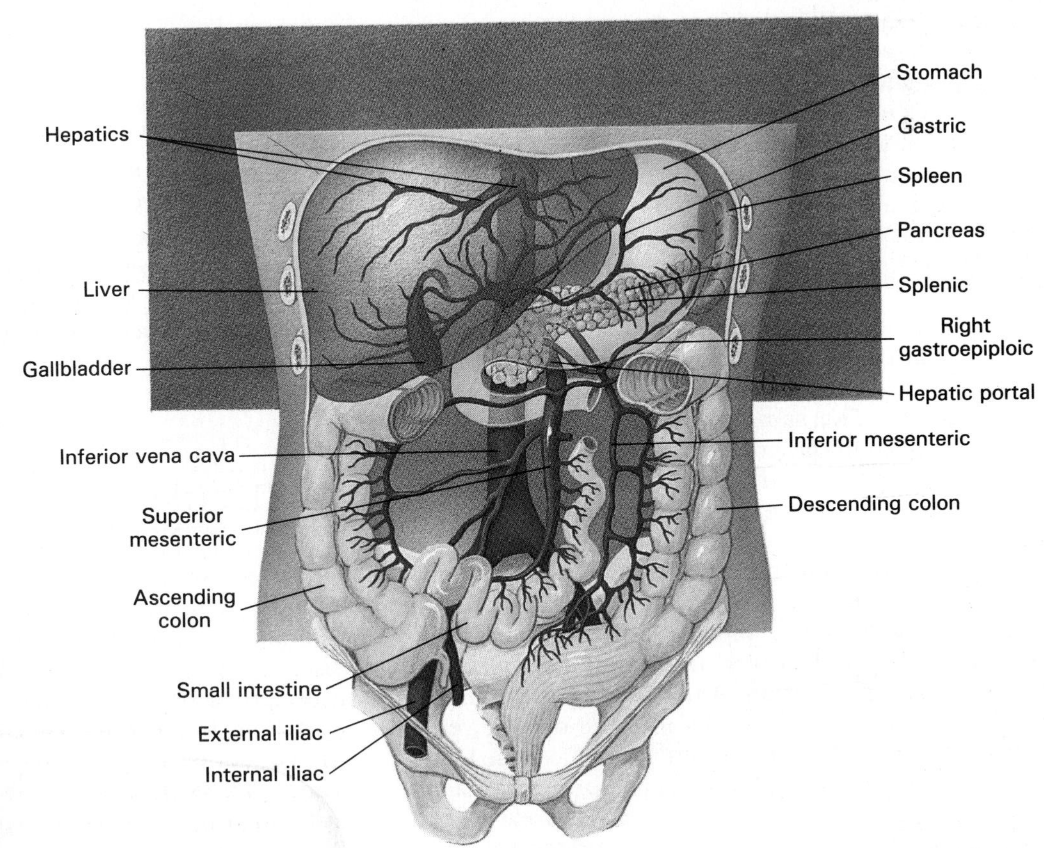

FIGURE 21-18
The Hepatic Portal System. (a) Generalized flow chart for circulation of blood to the liver via the hepatic artery and the hepatic portal vein. (b) Vessels of the hepatic portal system.

sive interconnections between the plantar arch and the dorsal arch, and the path of blood flow can easily shift from superficial to deep veins.

At the knee, the tibial and peroneal veins unite to form the **popliteal vein**, which also drains the small saphenous. When it reaches the femur, it becomes the **femoral vein**. Immediately before penetrating the abdominal wall, the femoral, great saphenous, and **deep femoral** unite. The large vein that results penetrates the body wall as the **external iliac vein**. As it travels across the inner surface of the ilium, the external iliac fuses with the **internal iliac vein**, which drains the pelvic organs. The resulting **common iliac vein** then meets its counterpart from the opposite side to form the inferior vena cava (IVC).

Like the aorta, the inferior vena cava lies posterior to the abdominopelvic cavity. As it ascends to the heart, the inferior vena cava collects blood from several **lumbar veins**; these veins may also empty into the common iliac or the hemiazygos. In addition, the IVC receives blood from the **gonadal**, **renal**, **suprarenal**, **phrenic**, and **hepatic veins** before reaching the right atrium.

The Hepatic Portal System You may have noticed that the list of veins did not include any names that refer to digestive organs other than the liver. Instead of traveling directly to the inferior vena cava, blood leaving the capillaries supplied by the celiac, superior, and inferior mesenteric arteries flows into the **hepatic portal system**. Blood flowing in the hepatic portal system is quite different from that in other systemic veins, because the hepatic portal vessels contain substances absorbed by the digestive tract. For example, levels of blood glucose, amino acids, fatty acids, and vitamins in the hepatic portal vein often exceed those found anywhere else in the circulatory system.

A portal system carries blood from one capillary bed to another; Chapter 18 discussed the anatomy and function of the hypophyseal portal system. (p. 578) The hypophyseal portal system carries regulatory hormones directly to their targets in the anterior pituitary, and small quantities are effective because they are not diluted by mixing in the entire bloodstream. The same principle of increased efficiency applies to the hepatic portal system. The liver regulates the concentrations of nutrients, such as glucose or amino acids, in the circulating blood. When digestion is under way, the digestive tract absorbs high concentrations of nutrients, along with various wastes and an occasional toxin. The hepatic portal system delivers these compounds directly to the liver, where liver cells absorb them for storage, metabolic conversion, or excretion. After passing through the liver sinusoids, blood collects into the hepatic veins that empty into the inferior vena cava. Because blood goes to the liver first, the composition of the blood in the general circulation remains relatively stable, regardless of the digestive activities under way.

Blood delivered to the liver by the hepatic portal system often contains high concentrations of absorbed nutrients, but it is venous blood that contains relatively little oxygen. The liver cells obtain oxygen from arterial blood from the hepatic artery, a branch of the celiac trunk. Liver cells are thus exposed to a mixture of portal blood and arterial blood, as indicated in Figure 21-18a on p. 685 (histological details are included in Chapter 24).

Figure 21-18b details the anatomy of the hepatic portal system. The system ends with the **hepatic portal vein**, which empties into the liver sinusoids. It begins in the capillaries of the digestive organs. Blood from capillaries along the lower portion of the large intestine enters the **inferior mesenteric vein**. As it nears the liver, veins from the spleen, the lateral border of the stomach, and the pancreas fuse with the inferior mesenteric, forming the **splenic vein**. The **superior mesenteric vein** also drains the lateral border of the stomach, through an anastomosis with one of the branches of the splenic vein. In addition, the superior mesenteric collects blood from the entire small intestine and two-thirds of the large intestine.

The hepatic portal vein forms through the fusion of the superior mesenteric and splenic veins. Of the two, the superior mesenteric normally contributes the greater volume of blood and most of the nutrients. As it proceeds toward the liver, the hepatic portal receives blood from the **gastric veins**, which drain the medial border of the stomach, and the **cystic vein** from the gallbladder.

- ✓ **Blockage of which branch of the aortic arch would interfere with the blood flow to the left arm?**
- ✓ **Why would compression of the common carotid artery cause a person to lose consciousness?**
- ✓ **Grace is in an automobile accident and ruptures her celiac artery. What organs would be affected most directly by this injury?**

■ Circulatory Physiology

We have now completed our discussions of the blood (Chapter 19), the heart (Chapter 20), and vessels of the circulatory system. The rest of this chapter will put these pieces together and consider the integrated functioning of the cardiovascular system.

It is important to recognize at the start that the goal of cardiovascular regulation is the maintenance of adequate blood flow through peripheral tissues and organs. Under normal circumstances blood flow is equal to cardiac output. When cardiac output

BF = CO

goes up, so does capillary blood flow; when cardiac output declines, blood flow is reduced.

Chapter 20 introduced the factors involved in cardiac output regulation, and the process will be examined more closely later in this chapter. ∞ (p. 657) In the meantime, we will continue to use the term *blood flow* rather than cardiac output to focus attention on events at the capillary level.

Two factors, *pressure* and *resistance*, affect flow rates. The relationships between these factors and blood flow must be understood before we can proceed to a discussion of circulatory function.

FLOW AND PRESSURE

Fluids, including blood, will flow from an area of higher pressure toward an area of relatively lower pressure. The flow rate is directly proportional to the pressure difference; the greater the difference in pressure, the faster the flow. Using symbols, this relationship can be expressed as:

F	~	P
flow	is proportional to	the difference in pressure

In the systemic circuit, P is equal to the **circulatory pressure**, the pressure difference between the base of the ascending aorta and the entrance to the right atrium. (Circulatory pressures, which average around 100 mm Hg, are examined more closely in a later section.) Pressures in the venous system are quite low, and the relatively high circulatory pressure is needed primarily to force blood through the arterial system. When referring to arterial pressure, the term **blood pressure** (**BP**) will be used to distinguish it from the total circulatory pressure.

Capillary blood flow is directly proportional to blood pressure, and blood pressure is closely regulated by a combination of neural and hormonal mechanisms. Blood pressure must be kept relatively high because several forces exist that oppose blood flow.

FLOW AND RESISTANCE

A resistance is a force that opposes movement. The resistance of the circulatory system opposes the movement of blood: the greater the resistance, the slower the blood flow. In effect, overcoming the resistance uses some of the energy that would otherwise produce movement. Using symbols, this relationship can be expressed as:

F	~ $\frac{1}{R}$
flow	is inversely proportional to resistance

For circulation to occur, the circulatory pressure must be greater than the **total peripheral resistance**, the resistance of the entire circulatory system. Because the resistance of the venous system is very low, for reasons detailed below, attention focuses on the **peripheral resistance** (**PR**), the resistance of the arterial system. For blood to flow into peripheral capillaries, blood pressure must be greater than the peripheral resistance; the higher the peripheral resistance, the lower the rate of blood flow.

Neural and hormonal control mechanisms regulate blood pressure, keeping it relatively stable. Adjustments in the peripheral resistance of vessels supplying specific organs allow the rate of blood flow to be precisely controlled. For example, Chapters 10 and 16 discussed the increase in blood flow to skeletal muscles during exercise. That increase occurs because there is a drop in the peripheral resistance of those vessels. The relationship can be described as:

$$F \sim \frac{P}{R} \quad or \quad F \sim \frac{BP}{PR}$$

The equation on the left combines the two equations presented previously: blood flow is *directly* proportional to pressure and *inversely* proportional to resistance. The equation on the right translates this general statement into terms that describe blood flow through peripheral capillaries. In the capillaries, P is the blood pressure (BP) and R is the peripheral resistance (PR).

Sources of peripheral resistance include *vascular resistance*, *viscosity*, and *turbulence*. Of these, only vascular resistance can be adjusted by the nervous or endocrine system to regulate blood flow. Viscosity and turbulence, which contribute to peripheral resistance, are normally constant. However, changes in viscosity or turbulence can occur as a result of pathological conditions discussed in this chapter and elsewhere in the text.

Vascular Resistance

Vascular resistance is the resistance of the blood vessels, and it is the largest component of peripheral resistance. The most important factor in vascular resistance is friction between the blood and the vessel walls. The amount of friction depends on the diameter of the vessel and its length. A decrease in diameter or an increase in length will produce an increase in resistance. However, differences in diameter have much more significant effects than differences in length. If there are two vessels of equal diameter, one twice as long as the other, the longer vessel will offer twice as much resistance to blood flow. But with two vessels of equal length, one twice the diameter of the other, the smaller one will offer 16 times as much resistance to blood flow.[2]

[2] The relationship is usually expressed in terms of the vessel radius, r, using the formula $R \sim 1/r^4$.

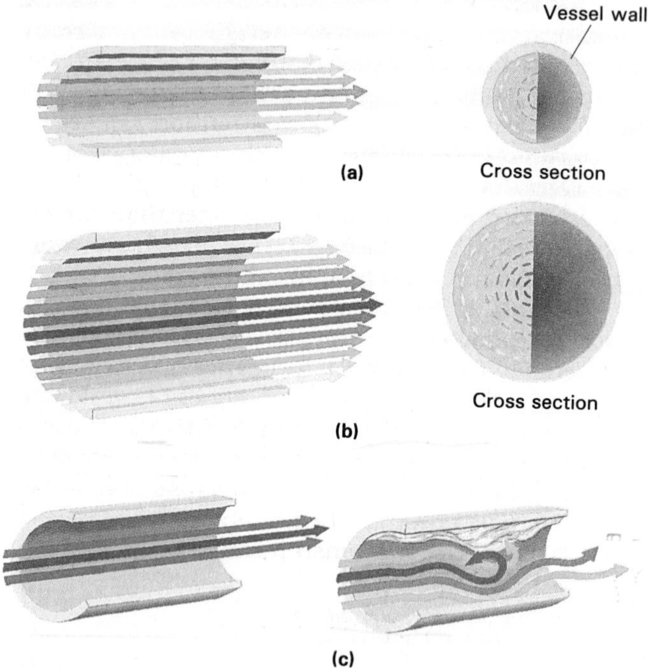

FIGURE 21-19
Blood Flow and Vessel Structure. Friction occurs between the vessel wall and the blood that contacts it, slowing the rate of blood flow near the wall. Friction then occurs between this slow-moving blood and blood farther from the wall. Frictional forces are lowest at the center of the vessel. (a) When the diameter is small, friction affects all of the blood flowing through the vessel. Resistance is therefore high, and the rate of blood flow is relatively low. (b) When the diameter is large, friction with the walls has negligible effect on blood flowing in the center. Resistance is therefore low, and the rate of blood flow is relatively high. (c) Irregularities or blockages in the path of blood flow cause the generation of eddies resembling whirlpools or rapids in a stream. This turbulence slows blood movement and increases resistance.

Consider Figure 21-19, which diagrams the blood flow through two vessels with different diameters. In the longitudinal sections, the relative lengths of the arrows indicate the flow rates measured at different distances from the walls. In the cross sections, the darker the color, the faster the flow rate. At the vessel wall, where friction is highest, blood moves very slowly. In a small vessel, all of the blood flowing past will be affected by friction with the vessel walls (Figure 21-19a). As a result, resistance is high, and the flow rate is low. The greater the distance between the wall and the center of the vessel, the faster the flow rate. If the vessel is large enough, the blood at the center slips by with minimal friction (Figure 21-19b). The resistance is therefore much lower than in the smaller vessel, and blood flows much more quickly.

Large arteries such as the aorta, brachiocephalic artery, or carotid artery, contribute little to the peripheral resistance. Most of the peripheral resistance occurs in the arterioles and capillaries. As noted earlier in the chapter, arterioles are extremely muscular; an arteriole 30 μm in diameter is wrapped in a 20 μm thick layer of smooth muscle. Local, autonomic, and hormonal stimuli that stimulate or inhibit the arteriolar smooth muscle tissue can adjust the diameters of these vessels. Because a small change in diameter will produce a very large change in resistance, as described above, this mechanism permits extensive control of peripheral resistance, and thereby of blood flow.

The diameters of small arterioles and capillaries are very small; many capillaries are smaller in diameter than a single red blood cell. If the blood pressure drops too far, it cannot overcome their resistance, and blood flow stops. When this occurs, the capillaries collapse like deflating balloons. The minimum pressure required to maintain blood flow and keep the capillary networks open is the **critical closing pressure**. A number of homeostatic mechanisms operate to ensure that systemic pressures remain within normal limits, and these mechanisms will be detailed later in the chapter.

Viscosity

Viscosity, introduced in Chapter 19, is resistance to flow caused by interactions among molecules and suspended materials in a fluid. (p. 607) Fluids of low viscosity, such as water (viscosity 1.0), flow at low pressures, whereas thick, syrupy fluids such as molasses (viscosity 300) flow only under relatively high pressures. Whole blood has a viscosity about five times that of water, because of the presence of plasma proteins and blood cells. Under normal conditions the viscosity of the blood remains stable, but anemia, polycythemia, or other disorders that affect the hematocrit (Chapter 19) also change blood viscosity and alter peripheral resistance. (p. 611)

Turbulence

Blood flow through a vessel is usually smooth. Blood flows slowly near the walls, and the speed gradually increases as one approaches the center of the vessel (Figure 21-19b, c). High flow rates, irregular surfaces, or sudden changes in vessel diameter upset this smooth flow, creating eddies and swirls as indicated in Figure 21-19c. This phenomenon, called **turbulence**, slows the rate of flow and increases resistance.

Turbulence normally occurs when blood flows between the chambers of the heart and from the heart into the aortic and pulmonary trunks. In addition to increasing resistance, this turbulence generates a sound that can often be heard through a stethoscope. (For example, turbulence in aortic stenosis produces a *heart murmur.* [**CM:** *Valvular Heart Disease*]) Turbulence also occurs in the centers of large arteries, such as the aorta, when cardiac output and arterial flow rates are very high.

Turbulence seldom occurs in smaller vessels unless their walls are damaged. For example, scar tissue formation at an injury site or the development of an atherosclerotic plaque will create abnormal turbulence and restrict blood flow (see the Clinical Comment on page 668). (Because of the sound, or *bruit*, produced by this turbulence, the presence of large plaques can often be detected with a stethoscope.)

A SUMMARY OF RELATIONSHIPS BETWEEN BLOOD FLOW, PRESSURE, AND RESISTANCE

1. Blood flow occurs in response to a pressure difference between the aorta and the entrance to the right atrium. This pressure gradient, called the circulatory pressure, is generated by the left ventricle.
2. Blood pressure (BP), the pressure in the arterial system, is responsible for pushing blood into peripheral capillary beds.
3. To maintain blood flow, blood pressure must be greater than the peripheral resistance (PR), the resistance of the arterial system.
4. The relationship between blood pressure, peripheral resistance, and blood flow can be summarized as:

$$F \sim \frac{BP}{PR}$$

5. The most important determinant of peripheral resistance is the diameter of arterioles, which can change in response to local, neural, or hormonal stimuli.
6. Although vessel length and blood viscosity contribute to peripheral resistance, under normal conditions these factors can be ignored because (a) the effects of vessel diameter are much greater than those of vessel length, and (b) blood viscosity changes only under pathological conditions.
7. Turbulence contributes to the resistance to flow through the heart and along large-diameter arteries, such as the aorta.

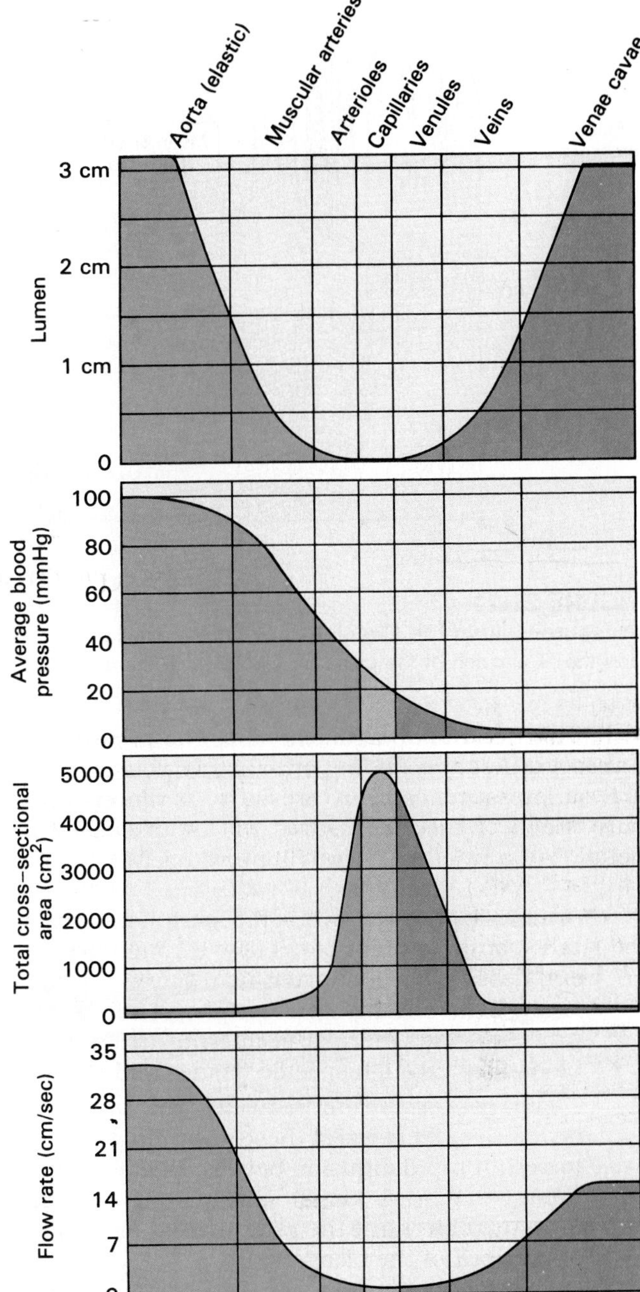

FIGURE 21-20
Pressure, Diameter, and Blood Flow Relationships in the Systemic Circuit

CIRCULATORY PRESSURES

With these relationships in mind, consider Figure 21-20, which indicates the pressures, vessel diameters, and flow rates in the systemic circuit. As blood travels from the aorta toward the capillaries, the resistance increases and blood pressure falls. Because flow rate varies directly with pressure, the rate of blood flow declines as the arterial pressure drops. By the time blood enters the capillaries, it is moving very slowly, and under low pressure. Venous pressure continues to decline, but as the vessels increase in diameter, the resistance decreases rapidly. This decreased resistance leads to an increase in flow rates.

Figure 21-21 details the blood pressure throughout the circulatory system. Systemic pressures are highest in the aorta, peaking at around 120 mm Hg, and they reach a minimum at the entrance to the right atrium. You will notice that pressures in the pulmonary circuit are much lower than those of the systemic circuit. This primarily reflects the fact that the pulmonary vessels are much shorter and more elastic, so they provide less resistance to blood flow.

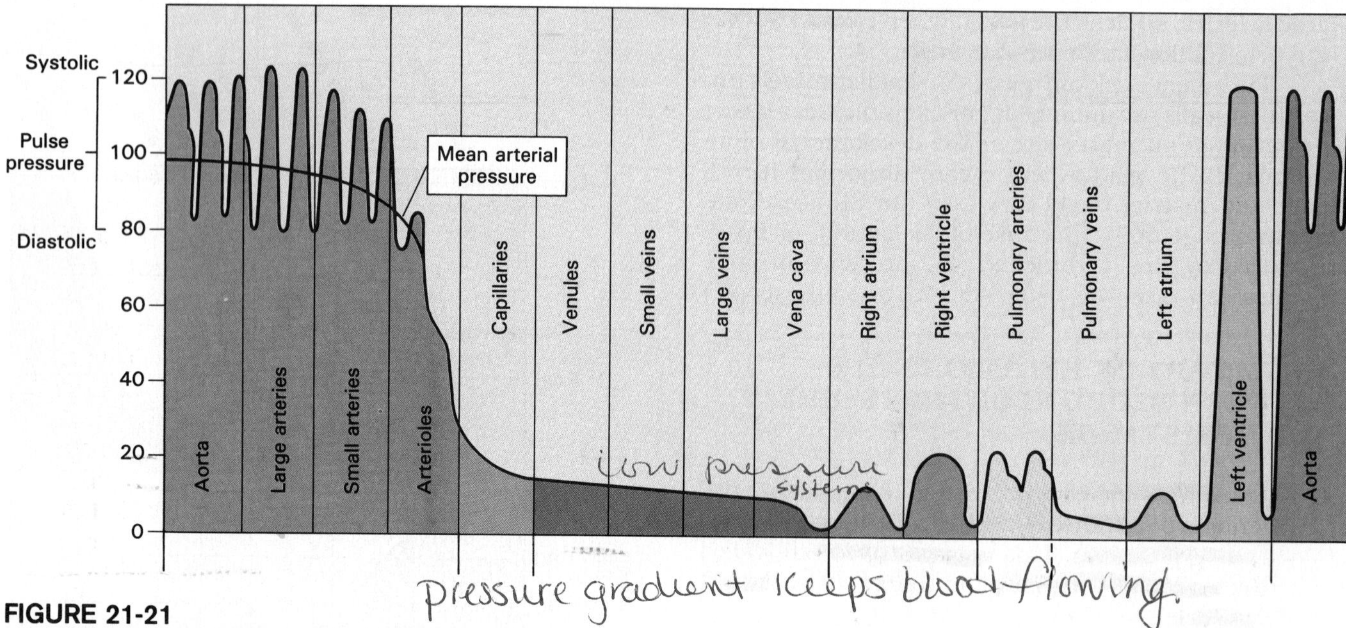

FIGURE 21-21
Pressures within the Circulatory System. Notice the general reduction in circulatory pressures within the systemic circuit and the elimination of the pulse pressure within the arterioles.

The functional significance of the pressure in a vessel differs, depending on whether you consider arterial pressure, capillary pressure, or venous pressure. Each of these pressures will be discussed in detail, but a brief overview will prove useful:

1. Arterial pressures overcome peripheral resistance and maintain blood flow through peripheral tissues. Because arterial pressures are relatively high, changes in peripheral resistance provide a rapid means of altering the rates of blood flow through specific tissues and organs.
2. Capillary pressures must be kept low to protect delicate capillary walls; these walls must be thin to permit rapid diffusion between the blood and interstitial fluid. Small changes in capillary pressure determine the rates of fluid movement into or out of the bloodstream.
3. Venous pressure, although low, determines the venous return, and this has a direct impact on cardiac output and peripheral blood flow.

Arterial Blood Pressure

The graph in Figure 21-21 indicates that blood pressure in large and small arteries is not stable. Blood pressure rises during ventricular systole and falls during ventricular diastole. **Systolic pressure** is the peak blood pressure measured during ventricular systole, and **diastolic pressure** is the minimum blood pressure at the end of ventricular diastole. The difference between the systolic and diastolic pressures is the **pulse pressure**. When reporting a single value for blood pressure, the **mean arterial pressure (MAP)** is used. The mean arterial pressure is calculated by adding one-third of the pulse pressure to the diastolic pressure. For example, if the systolic pressure is 120 mm Hg and the diastolic pressure is 90 mm Hg, the mean arterial pressure is 100 mm Hg $(90 + [(120 - 90)/3] = 90 + 10 = 100)$.

Blood pressure not only forces blood through the circulatory system, it also pushes outward against the walls of the containing vessels, just as air pushes against the walls of an inflated balloon. As a result, blood pressure can be measured indirectly, by determining how forcefully the blood presses against the vascular walls. This is usually done with the aid of an inflatable collar placed around a muscular artery, such as the brachial artery of the upper arm. (Practical details are provided in the accompanying Diagnostics box.) When recording the blood pressure, systolic and diastolic pressures are usually separated by a slashmark, as in "120/80" or "110/75."

The mean arterial pressure and the pulse pressure become smaller as the distance from the heart increases. The average pressure declines because of friction between the blood and the vessel walls. The pulse pressure fades because arteries are elastic tubes rather than solid pipes. As systolic pressure climbs, the arterial walls stretch, just as an extra puff of air expands a partially inflated balloon. This expansion allows the arterial system to accommodate some of the blood provided by ventricular systole. When diastole begins and pressures fall, the arteries recoil. Because the aortic semilunar valve prevents the return of blood to the heart, the arterial recoil pushes blood toward the capillaries. In essence, the arteries absorb part of the energy provided by ventricular systole and give it back during ventricular diastole. This phenomenon is called **elastic rebound**.

Elastic rebound reduces the size of pulse pressures. The effect is cumulative, with each segment

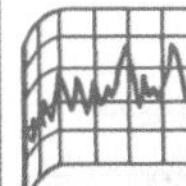

Diagnostics: Checking the Pulse and Blood Pressure

The pulse can be felt within any of the large or medium-sized arteries. The usual procedure involves using the fingertips to squeeze an artery against a relatively solid mass, preferably a bone. When the vessel is compressed, the pulse is felt as a pressure against the fingertips.

Figure 21-22a indicates the locations used to check the pulse. The inside of the wrist is often used, because the radial artery can easily be pressed against the distal portion of the radius. Other accessible arteries include the temporal, facial, carotid, brachial, femoral, and popliteal arteries. These points can be important in other ways. Firm pressure exerted at these locations, called **pressure points**, can reduce or eliminate arterial bleeding distal to the site.

The instrument used to determine blood pressure is called a **sphygmomanometer** (sfig-mō-ma-NOM-e-ter; *sphygmos*, pulse + *manometer*, device for measuring pressure), shown in Figure 21-22b. An inflatable cuff is placed around the upper arm in such a position that its inflation squeezes the brachial artery. A stethoscope is placed over the artery distal to the cuff, and the cuff is then inflated. A tube connects the cuff to a glass chamber containing liquid mercury, and as the pressure in the cuff rises it pushes the mercury up into a vertical column. A scale along the column permits the determination of the cuff pressure in millimeters of mercury (mm Hg). Inflation continues until cuff pressure is roughly 30 mm Hg above the pressure sufficient to completely collapse the brachial artery, stop the flow of blood, and eliminate the sound of the pulse.

The investigator then slowly lets the air out of the cuff. When the pressure in the cuff falls below systolic pressure, blood can again enter the artery. At first, blood enters only at

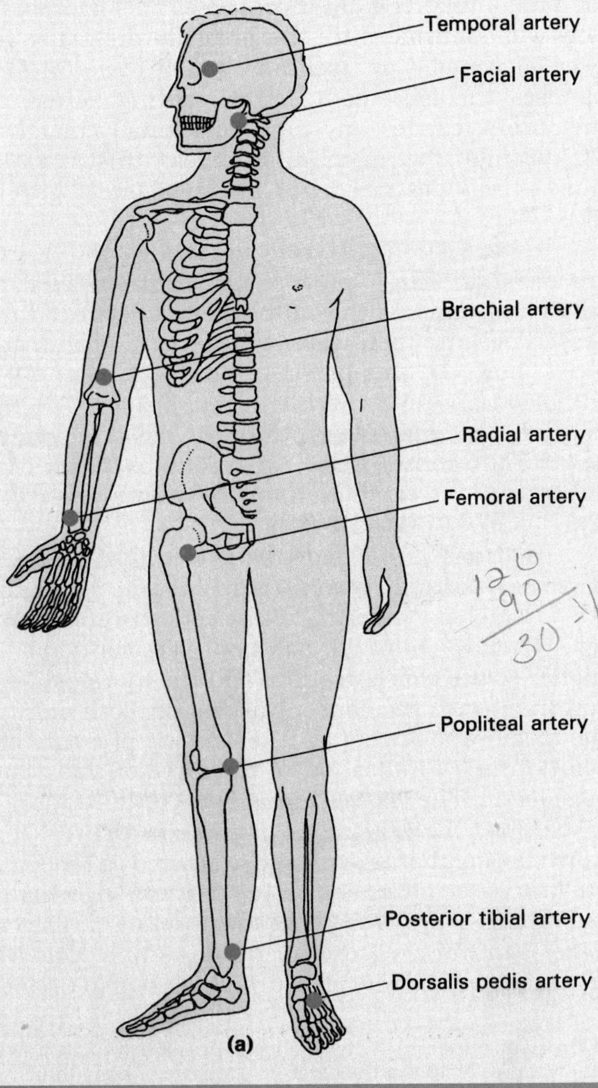

Temporal artery
Facial artery
Brachial artery
Radial artery
Femoral artery
Popliteal artery
Posterior tibial artery
Dorsalis pedis artery
(a)

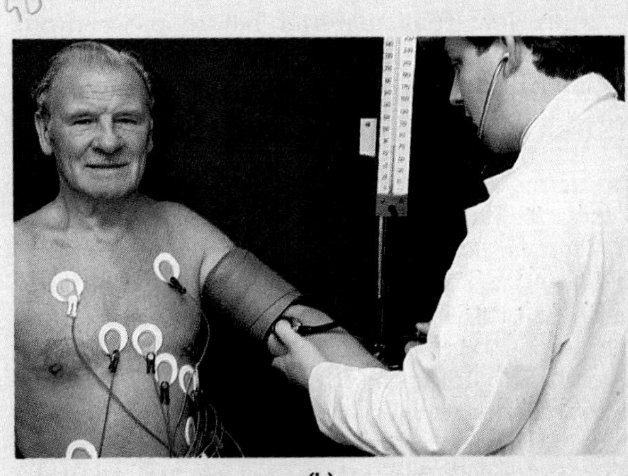

(b)

FIGURE 21-22
Checking the Pulse and Blood Pressure. (a) Pressure points used to check the presence and strength of the pulse. (b) Use of a sphygmomanometer to check arterial blood pressure. The chest electrodes and wires are electrocardiogram leads (Chapter 20); they are not involved in blood pressure measurement.

peak systolic pressures, and the stethoscope picks up the sound of blood pulsing through the artery. As the pressure falls further, the sound changes because the vessel is remaining open for longer and longer periods. When the cuff pressure falls below diastolic pressure, blood flow becomes continuous and the sound of the pulse becomes muffled, or disappears completely. Thus the pressure at which the pulse appears corresponds to the peak systolic pressure; when the pulse fades the pressure has reached diastolic levels. The distinctive sounds heard during this test are called **sounds of Korotkoff** (sometimes spelled *Korotkov* or *Korotkow*). They are produced by turbulence as blood flows past the constricted portion of the artery.

Blood pressure changes on a moment-to-moment basis, and the pressure measured in a doctor's office may not be an accurate indication of the daily average. For this reason physicians sometimes recommend *ambulatory pressure monitoring*. This procedure involves attaching an inexpensive monitor to a pressure cuff. The monitor provides a continual printout of the patient's blood pressure on a 24-hour basis. Typically, pressure oscillates from a low at around 2:00 A.M. to peak values at intervals during daylight hours.

reducing the magnitude of the pressure change experienced by its downstream neighbors. Along the arterioles the average pressures fall rapidly, and the pulse pressures disappear. By the time blood reaches a precapillary sphincter, there are no pressure oscillations, and the blood pressure remains steady at about 30 mm Hg. Along the length of a typical capillary, blood pressure gradually falls from about 30 mm Hg to roughly 18 mm Hg, the pressure at the start of the venous system. [**CM:** *Hypertension and Hypotension*]

Capillary Pressures and Capillary Dynamics

The blood pressure within a capillary pushes against the capillary walls, just as it does in the arteries. But unlike the situation in other portions of the circulatory system, capillary walls are quite permeable to small ions, nutrients, organic wastes, dissolved gases, and water.

Capillary Exchange Chapter 3 described the forces that move water and solutes across membranes. (Only a brief overview will be provided here, so you may wish to review that material before proceeding.) (p. 78) Solute molecules tend to diffuse across the capillary lining, driven by their individual concentration gradients. Water-soluble materials, including ions and small organic molecules such as glucose, amino acids, or urea, diffuse through small spaces between adjacent endothelial cells. Larger water-soluble molecules, such as plasma proteins, cannot normally leave the bloodstream. Lipid-soluble materials, including steroids, fatty acids, and dissolved gases, diffuse across the endothelial lining, passing through the membrane lipids. In a few specialized locations, such as the choroid plexus of the brain (Chapter 14), endothelial cells actively transport fluids and solutes within endocytic vesicles (shown in Figure 3-25). (pp. 456, 85)

Water molecules will move when driven by either *hydrostatic pressure* (**HP**) or *osmotic pressure* (**OP**). Hydrostatic pressure is a physical force that pushes water molecules from an area of high pressure to an area of relatively lower pressure. This hydrostatic flow will continue until the pressure difference has been eliminated or until an equally powerful force opposes the movement. Water moving across the membrane carries any molecules small enough to fit through the membrane pores; this filtration process was described in Chapter 3 (see Figure 3-22). (p. 83)

Hydrostatic pressure generated by the heart—the blood pressure—pushes against the walls of arteries and veins, but filtration cannot occur in these vessels because their walls are too thick. At the capillaries, however, this pressure can force water out of the blood and into the interstitial spaces. The tendency for water to move out of the blood is greatest at the start of a capillary, where the blood pressure is highest, and declines along the length of the capillary as blood pressure falls.

Osmosis is the movement of water molecules across a selectively permeable membrane separating two solutions of differing solute concentrations. Water will move into the solution that contains the higher solute concentration, and the movement will continue until the concentrations on both sides of the membrane are equal. The osmotic pressure of a solution is an indication of the force of that water movement. The higher the solute concentration of a solution, the greater its osmotic pressure.[3] It is worth noting that hydrostatic pressure is a force that pushes water molecules *out of* a solution, whereas osmotic pressure pulls water *into* a solution. The two forces can work together or in opposition. One way of measuring osmotic pressure, described in Chapter

[3] Clinicians often use the term *oncotic pressure* (*onkos*, a swelling) when referring to the osmotic pressure of body fluids.

3, involves the application of an opposing hydrostatic pressure.

Chapter 19 noted that plasma and interstitial fluids differ in composition. (p. 608) The osmotic pressure of the blood (OP_b) exceeds that of the interstitial fluid (OP_i), primarily because the plasma contains significant quantities of dissolved proteins. As a result, whenever plasma and interstitial fluid are separated by a permeable barrier water will tend to move from the interstitial fluid into the plasma. The concentration difference is enough to generate an osmotic pressure of about 25 mm Hg.

The Dynamic Center The two strongest forces affecting fluid movement across the capillary walls are HP_b and OP_b, and as we have just seen, they oppose one another. The blood hydrostatic pressure tends to push water out of the capillary, while the osmotic pressure tends to pull water back in. Figure 21-23 illustrates the balance of forces along the length of a capillary. The purple arrows represent hydrostatic pressures, and the blue arrows are osmotic pressures. *The direction of water movement is determined by the balance between these two opposing pressures.*

Hydrostatic pressure at the start of a capillary is strong enough to push water into the interstitial fluid. As we proceed along the capillary, however, two things happen:

1. The hydrostatic pressure declines; and
2. The blood becomes increasingly concentrated as the departing water leaves the plasma proteins behind. In short, the osmotic pressure increases as the hydrostatic pressure decreases.

At a point known as the **dynamic center**, the hydrostatic and osmotic forces are equal, and there is no net water movement into or out of the capillary. Distal to the dynamic center, the blood osmotic pressure exceeds the hydrostatic pressure, and there is a net water movement into the capillary. As the end of the capillary is approached, this water movement accelerates as the hydrostatic pressure falls to 18 mm Hg.

If the dynamic center lay midway along the length of the capillary, just as much water would move into the capillary over the venous half as flowed out in the arterial half, so the blood volume would remain constant. In reality, the dynamic center normally lies slightly past the midpoint, and the arterioles deliver more fluid to the capillaries than the venules carry away. The water and dissolved materials that leave the plasma flow through the tissues and eventually enter a network of **lymphatic vessels** that drain into the venous system. This continual fluid movement has several important functions:

1. It accelerates the distribution of nutrients, hormones, and dissolved gases throughout the tissue;
2. It assists in the transport of insoluble lipids and tissue proteins that cannot enter the circulation by crossing the capillary lining; and
3. It has a flushing action that helps speed the removal of hormones and that also carries bacterial toxins and other chemical stimuli to cells of the immune system.

FIGURE 21-23
Capillary Dynamics and Fluid Movements. The green arrows represent blood hydrostatic pressure and the blue arrows represent blood osmotic pressure. The net fluid movement is determined by the difference between these pressures. At the dynamic center, the two are equal.

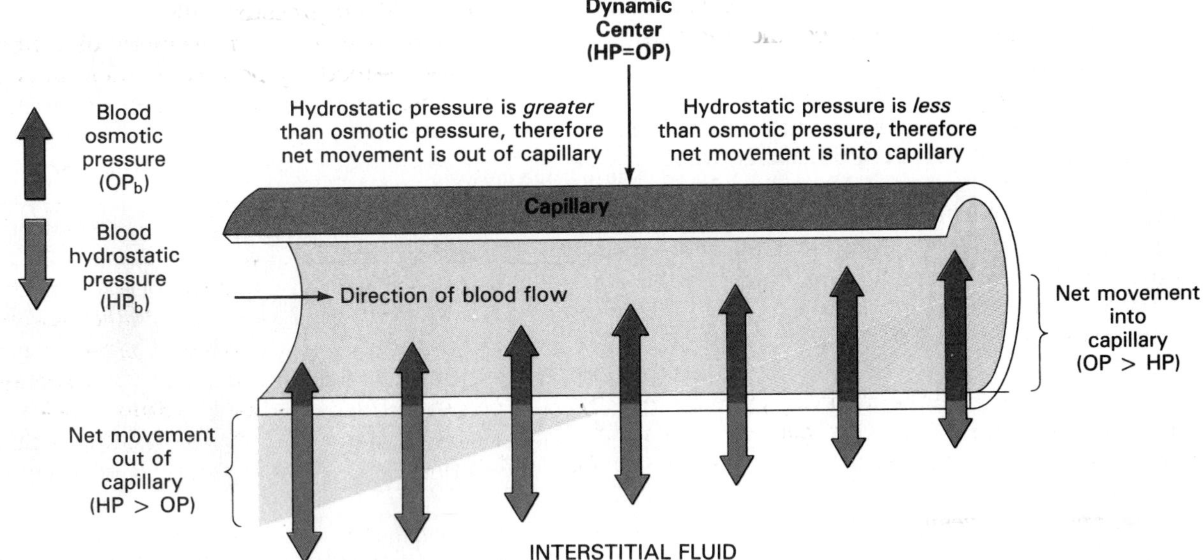

Capillary Dynamics and Blood Volume Normally, the dynamic center lies closer to the venous end of the capillary than to the arterial end, and there is a net movement of water into the interstitial spaces. Any condition that affects hydrostatic or osmotic pressures in the blood or tissues will shift the dynamic center. The effects can then be predicted on the basis of an understanding of capillary dynamics.

Consider what happens at the capillary level to an accident victim who has lost blood through hemorrhaging or a person lost in the desert, feeling the effects of dehydration. In both cases, the decrease in blood volume causes a drop in blood pressure. In the second case, the loss in blood volume is accompanied by a rise in the blood osmotic pressure, because as water is lost the blood becomes more concentrated. In either case the dynamic center, where hydrostatic and osmotic forces are equal, shifts toward the arterial end of the capillary (Figure 21-24a). Now instead of a net movement out of the blood, there is a net movement of water into the bloodstream. The blood volume increases at the expense of interstitial fluids, a process called a **recall of fluids**. During minor bleeding or gradual dehydration, fluid recall can reduce the effects and slow the appearance of clinical symptoms. It cannot completely restore homeostasis after severe bleeding or in acute dehydration, but it acts to complement the more complex neural and endocrine responses considered later in the chapter.

Edema (e-DĒ-ma) is an abnormal accumulation of interstitial fluid. There are many different causes of edema, and specific examples will be encountered in later chapters. The underlying problem in all types of edema is a disturbance in the normal balance between hydrostatic and osmotic forces at the capillary level. For example, a localized edema often occurs around a bruise. The fluid shift occurs because damaged capillaries at the injury site allow plasma proteins into the interstitial fluid. This leakage decreases the osmotic pressure of the blood and elevates that of the tissues. The reduction in OP_b shifts the dynamic center farther toward the venous end of the

FIGURE 21-24
Effects of Changes in Capillary Blood Pressure or Blood Osmolarity. (a) If the blood presssure declines, the dynamic center will shift closer to the arterial end of the capillary. More water will then move into the capillary. This net transfer of water from the interstitial fluids to the blood is called a recall of fluids. (b) If the pressure rises or blood osmotic pressure declines, the dynamic center shifts toward the venous end of the capillary. Water then moves into the interstitial fluids, potentially producing edema.

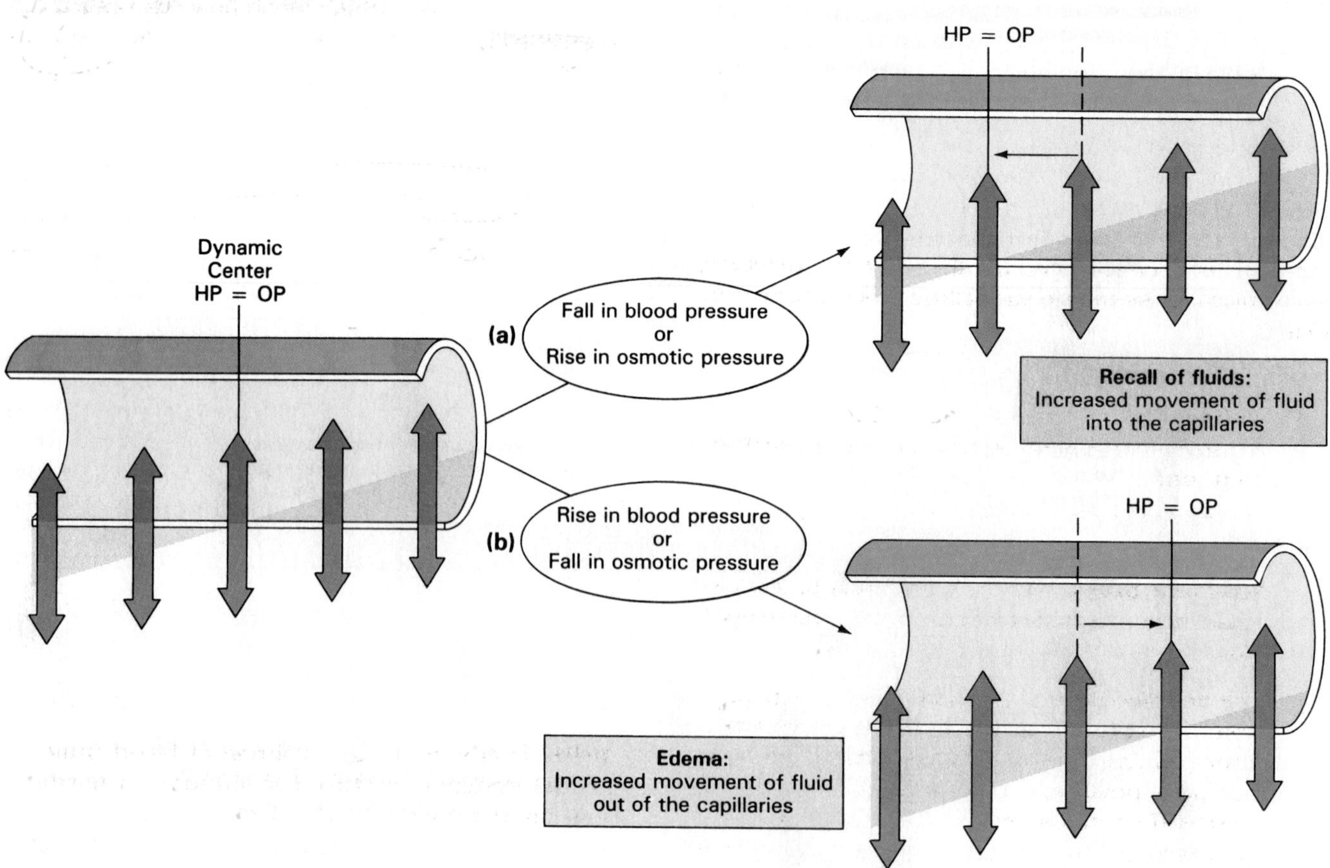

capillary (Figure 21-24b). More water then moves into the tissue, and edema results.

In acute starvation, the liver cannot synthesize enough proteins to maintain normal concentrations of protein in the blood. The osmotic pressure of the blood declines relative to that of the tissues, and fluids begin moving from the blood into peripheral tissues throughout the body. In children, fluid accumulates in the abdominopelvic cavity, producing the swollen bellies typical of starvation victims.

In the U.S. population, serious cases of edema most often result from an increase in the blood pressure in the arterial system, the venous system, or both. This often occurs during *congestive heart failure*, a condition discussed in the Clinical Manual. [CM: *Heart Failure*]

Venous Pressure

Although pressure at the start of the venous system is only about one-tenth that at the start of the arterial system, the blood must still travel through a vascular network as complex as the arterial system before returning to the heart. However, although venous pressures are low, the veins offer little resistance, and once in the venous system pressure declines very slowly. As blood travels through the venous system toward the heart, the veins become larger, resistance drops further, and the flow rate increases. Pressures at the entrance to the right atrium fluctuate, but they average around 2 mm Hg. This means that the driving force pushing blood through the venous system is roughly 16 mm Hg (18 mm Hg in the venules − 2 mm Hg in the venae cavae = 16 mm Hg).

When you are lying down, this pressure is sufficient, but when you are standing the venous blood returning from the body below the heart must overcome gravity as it ascends within the inferior vena cava. Three factors cooperate to assist the relatively low venous pressures in propelling blood toward the heart:

1. **Valves**: The valves in small and medium-sized veins ensure that blood flow occurs in one direction only. Valves are particularly important when considered in conjunction with factor 2 below.
2. **Muscular pumps**: The contractions of skeletal muscles near a vein compress it, helping to push blood toward the heart (see Figure 21-6).
3. **The thoracoabdominal pump**: During inspiration, decreased pressure in the thoracic cavity draws air into the lungs. This drop in pressure also pulls blood into the venae cavae and atria, increasing venous return. On exhalation, the increased pressure that forces air out of the lungs compresses the venae cavae, reducing venous return. (These fluctuations in venous return were described in Chapter 20.) (p. 660)

During exercise all three factors cooperate to elevate venous return and push cardiac output to maximal levels. When an individual stands at attention, with knees locked and leg muscles immobile, these assist mechanisms are impaired. The reduction in venous return then leads to a fall in cardiac output. This in turn reduces the blood supply to the brain, sometimes enough to cause **fainting**, a temporary loss of consciousness. The person then collapses, and in the horizontal position both venous return and cardiac output return to normal.

In a normal individual, where would you expect the blood pressure to be the greatest, the aorta or the inferior vena cava? Explain.

While standing in the hot sun, Sally begins to feel light-headed and faints. Explain.

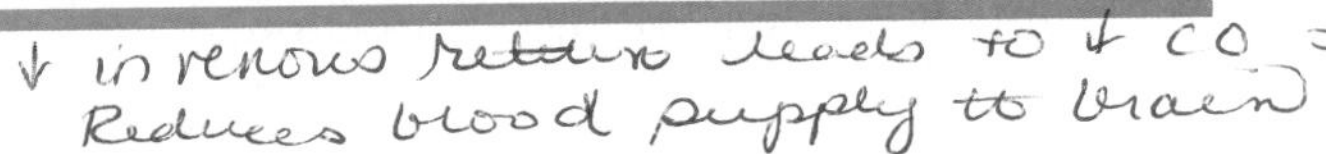

Cardiovascular Regulation

Homeostatic mechanisms regulate cardiovascular activity to ensure that tissue blood flow, also called tissue **perfusion**, meets the demand for oxygen and nutrients. The three variable factors are cardiac output, discussed in Chapter 20, peripheral resistance, and blood pressure. (pp. 657–661) As we have seen, the relationship between flow, blood pressure, and peripheral resistance can be expressed as $F \sim BP/PR$. Since blood flow and cardiac output (CO) are normally the same, we can rewrite this relationship as

$$CO \sim \frac{BP}{PR} \quad or \quad BP \sim CO \times PR.$$

We will begin this discussion by considering these components individually, concentrating on factors that can produce changes in each one under normal or abnormal conditions. We will then proceed to examine how adjustments in one or more of these components maintain homeostasis under different conditions.

VARIATIONS IN CARDIAC OUTPUT

Cardiac output (CO), the volume of blood pumped into the systemic circuit each minute, is normally equal to peripheral blood flow; for every 80 mℓ pumped out of the left ventricle, 80 mℓ moves into the right atrium. Increasing cardiac output increases the rate of peripheral blood flow, and reducing cardiac output slows this rate. The cardiac output changes in response to local, autonomic, and hormonal factors.

1. **Local factors**: Increases or decreases in venous return have a direct effect on cardiac output via intrinsic regulation, or Starling's law of the heart (Chapter 20). (p. 660) Changes in the ionic concentration of the interstitial fluid or damage to the cardiac muscle tissue may also limit or reduce the pumping abilities of the heart.
2. **Autonomic control**: The central regulation of cardiac output primarily involves the activities of the autonomic nervous system. Sympathetic stimulation increases cardiac output, and parasympathetic stimulation decreases it.
3. **Hormonal regulation**: The adrenal hormones epinephrine and norepinephrine elevate cardiac output. This response normally occurs during widespread sympathetic activation.

VARIATIONS IN PERIPHERAL RESISTANCE

Changes in peripheral resistance can occur under local control, in response to commands issued by autonomic centers in the CNS, or in response to circulating hormones.

1. **Local factors**: The metarterioles and precapillary sphincters respond automatically to alterations in the local environment. Vasomotion, the rhythmic alterations in blood flow within a capillary bed, is controlled by precapillary sphincters that respond to minor changes in the oxygen, pH, and carbon dioxide levels in the immediate area. When oxygen is abundant, the smooth muscles contract and slow down the flow of blood. Oxygen levels decline in active tissues as the cells absorb O_2 for use in aerobic respiration. As tissue oxygen supplies dwindle, carbon dioxide levels rise, and the pH falls. This combination causes the relaxation of smooth muscle fibers in the precapillary sphincter, and blood flow increases. The appearance of specific chemicals in the interstitial fluids can produce the same effect. For example, in the inflammation response vasodilation occurs at an injury site because histamine, bacterial toxins, and prostaglandins cause a relaxation of precapillary sphincters.
2. **Autonomic control**: The diameters of the arterioles are controlled primarily by neural mechanisms involving the vasomotor center of the medulla (see Figure 14-13). (p. 451) Inhibition of the vasomotor center leads to **vasodilation**, a dilation of arterioles that reduces peripheral resistance. Stimulation of the vasomotor center causes **vasoconstriction** (the constriction of peripheral arterioles), which increases peripheral resistance. Very strong stimulation of the vasomotor center also produces *venoconstriction*, (constriction of peripheral veins) elevating venous pressures and mobilizing the venous reserve.
3. **Hormonal regulation**: Epinephrine, norepinephrine, antidiuretic hormone (ADH), and angiotensin II are hormones that cause an extensive peripheral vasoconstriction that elevates peripheral resistance. Atrial natriuretic peptide (ANP) causes vasodilation, which lowers peripheral resistance.

Peripheral resistance also varies with changes in blood viscosity or turbulence. Changes in viscosity are relatively rare, but they can be important in patients with anemia, polycythemia, or liver disorders that affect plasma protein synthesis. Turbulence in large vessels is also seldom an important factor except in pathological conditions, such as valvular heart disease or atherosclerosis (see the Clinical Comment on page 668). [**CM:** *Valvular Heart Disease*]

ALTERATIONS IN BLOOD PRESSURE

Cardiac output and peripheral resistance are controlled by local, neural, or hormonal mechanisms. Blood pressure changes in response to alterations in either of these values, in accordance with the relationship BP $\sim$ CO $\times$ PR. The entire range of factors influencing blood pressure is diagrammed in Figure 21-25. Important relationships can be summarized as follows:

1. *Blood pressure varies directly with cardiac output.* Increases or decreases in cardiac output caused by changes in heart rate, stroke volume, or venous return cause corresponding changes in blood pressure.
2. *Blood pressure varies directly with peripheral resistance.* When peripheral resistance rises or falls due to vasoconstriction or vasodilation, there are corresponding elevations or declines in blood pressure.
3. *Blood pressure varies directly with blood volume.* Even a small increase or decrease in blood volume has an effect on blood pressure. Changes in blood volume affect venous return, and this in turn alters cardiac output. For example, hemorrhaging causes a drop in blood volume. This drop reduces venous return, causing a decline in both cardiac output and blood pressure. An infusion of whole blood or a plasma expander (Chapter 19) will elevate venous return and help restore normal cardiac output and blood pressure.

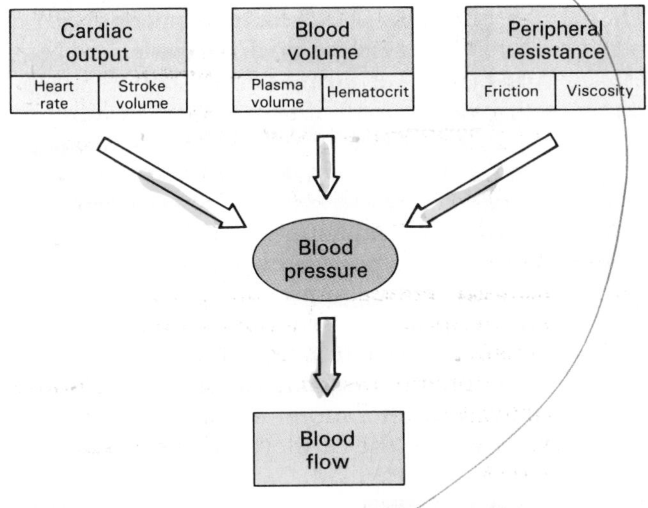

FIGURE 21-25
Major Factors that Influence Blood Pressure and Blood Flow

As indicated in Figure 21-25, blood pressure occupies a central position in cardiovascular regulation. Homeostatic mechanisms monitor blood pressure, the driving force for tissue perfusion. To keep blood pressure within normal limits, a combination of neural and endocrine responses adjusts cardiac output, peripheral resistance, and blood volume. We will now examine cardiovascular control under normal and abnormal conditions.

Homeostasis and Cardiovascular Function

Most cells are relatively close to capillaries. When a group of cells becomes active, the circulation to that region must increase to deliver the necessary oxygen and nutrients and to carry away the waste products and carbon dioxide that they generate. The goal of cardiovascular regulation is to ensure that these blood flow changes occur (1) at an appropriate time, (2) in the right area, and (3) without drastically altering blood pressure and blood flow to any vital organs.

Discussions earlier in the chapter listed local, autonomic, and endocrine factors that influence cardiac output and peripheral resistance. The same three factors are involved in the coordinated regulation of cardiovascular function (Figure 21-26):

1. *Local factors* change the pattern of blood flow within capillary beds in response to chemical changes in the interstitial fluids. At any given moment some of the precapillary sphincters are open and others closed. As long as the proportion remains constant the total blood flow through the tissues will remain constant. When widespread changes occur, they affect arterial blood pressure and oxygen and carbon dioxide levels in the blood and cerebrospinal fluid.

FIGURE 21-26
Normal Sequence of Events in Blood Pressure Regulation. Normal blood pressure regulation involves local, autonomic, and endocrine responses.

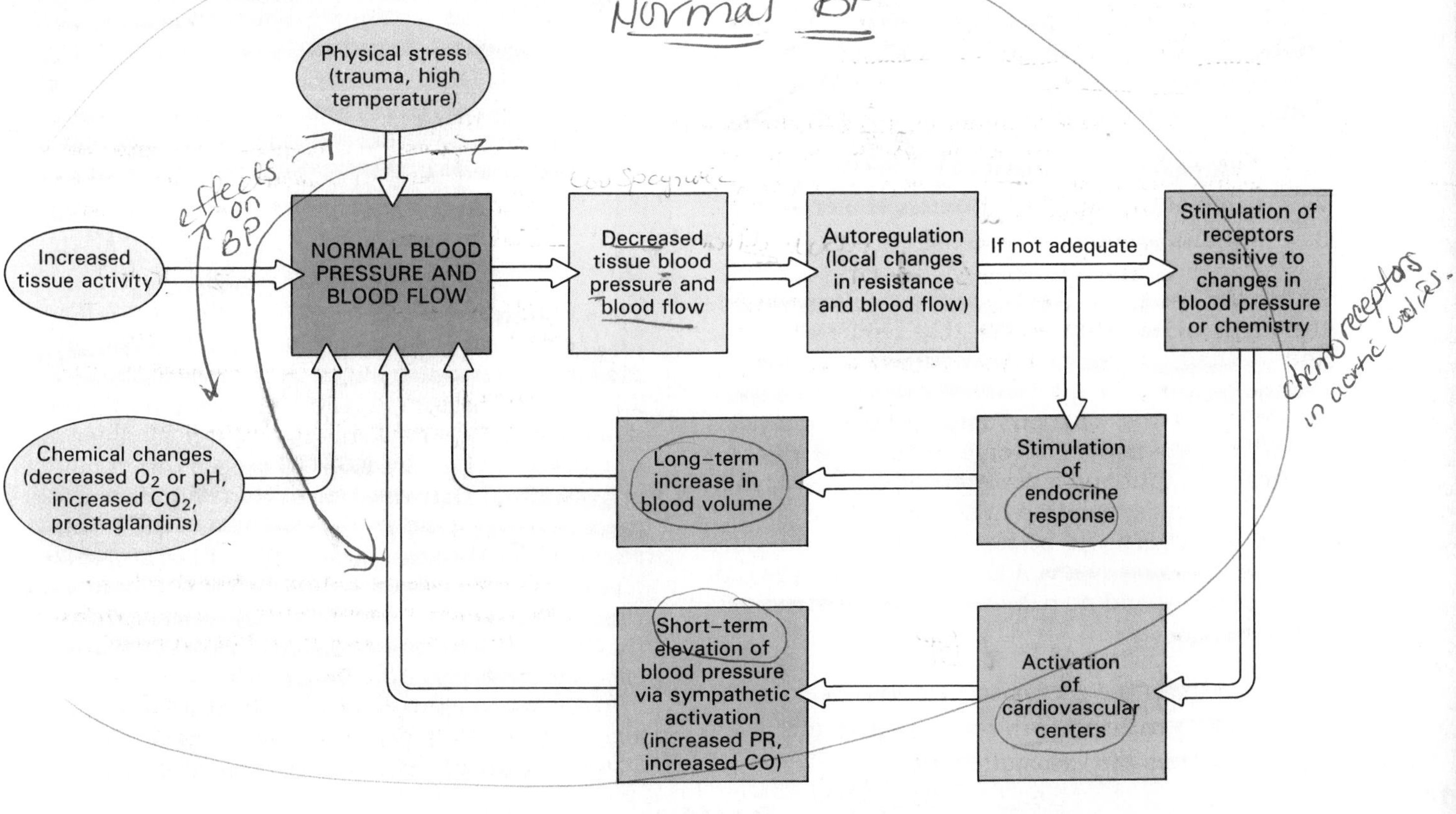

2. *Central mechanisms* respond to these changes in arterial pressure or changes in blood gas levels. The cardiovascular centers (cardiac and vasomotor) of the autonomic nervous system (ANS) respond first, adjusting cardiac output and peripheral resistance.
3. The *endocrine system* releases hormones that enhance short-term adjustments and direct long-term changes in cardiovascular performance.

Short-term responses adjust cardiac output and peripheral resistance to stabilize blood pressure and tissue blood flow. Long-term adjustments involve alterations in blood volume that affect cardiac output and the transport of oxygen and carbon dioxide to and from peripheral tissues.

NEURAL MECHANISMS AND THE SHORT-TERM REGULATION OF BLOOD PRESSURE

The nervous system provides rapid, immediate responses to changes in blood pressure and dissolved gas concentrations in the circulating blood. Two classes of reflexes are recognized: *baroreceptor reflexes* (sensitive to pressure) and *chemoreceptor reflexes* (sensitive to dissolved gases).

Baroreceptor Reflexes

Figure 21-27a presents the basic organization of the **baroreceptor reflexes**, autonomic reflexes that adjust cardiac output and peripheral resistance to maintain normal arterial pressures. When blood pressure climbs, the increased output from the baroreceptors

1. inhibits the cardioacceleratory center,
2. stimulates the cardioinhibitory center, and
3. inhibits the vasomotor center.

Under the command of the cardioinhibitory center, the vagus nerve releases ACh that reduces the rate and strength of the cardiac contractions, lowering cardiac output. The inhibition of the vasomotor center leads to the dilation of peripheral arterioles throughout the body. The combination of a reduction in cardiac output and a decrease in peripheral resistance then leads to a drop in blood pressure.

This pattern will be reversed if blood pressure becomes abnormally low. A fall in blood pressure produces a corresponding reduction in baroreceptor output, and this

1. stimulates the cardioacceleratory center,
2. inhibits the cardioinhibitory center, and
3. stimulates the vasomotor center.

The cardioacceleratory center stimulates sympathetic neurons innervating the SA node, AV node, and general myocardium. This stimulation produces an increase in heart rate and stroke volume, leading to an immediate increase in cardiac output. Vasomotor activity, also carried by sympathetic motor neurons, produces a rapid vasoconstriction, increasing peripheral resistance. These adjustments, increased cardiac output and increased peripheral resistance, work together to elevate blood pressure.

The receptor principles involved in baroreceptor function were detailed in Chapter 17. (p. 531) Three major baroreceptor populations enable the system to respond to alterations in blood pressure at key locations within the cardiovascular system.

1. *Aortic baroreceptors* monitor blood pressure within the ascending aorta. The aortic reflex adjusts blood pressure in response to changes in pressure at this location.
2. *Carotid sinus baroreceptors* respond to changes in blood pressure at the carotid sinus. Because pressure changes at this location affect the blood flow to the brain, the carotid sinus reflex is both extremely sensitive and quite important.
3. *Atrial baroreceptors* monitor the blood pressure at the end of the systemic circuit: at the venae cavae and the right atrium. The responses produced by the atrial reflex differ from those of the aortic or carotid reflexes. Under normal circumstances the heart pumps blood into the aorta at the same rate that it is arriving at the right atrium. When blood pressure rises at the atrium it means that a circulatory traffic jam exists, with blood arriving at the heart faster than it is being pumped out. The atrial baroreceptors solve the problem by stimulating the cardioacceleratory center, increasing cardiac output until the backlog of venous blood is removed and atrial pressure returns to normal. [**CM:** *Baroreceptor Accommodation*]

Chemoreceptor Reflexes

The **chemoreceptor reflexes**, diagrammed in Figure 21-27b, respond to changes in the oxygen or carbon dioxide levels in the blood and cerebrospinal fluid. (Chemoreceptor function and the locations of important chemoreceptors were discussed in Chapter 17.) (pp. 531–532) A fall in the oxygen content of the arterial blood detected by the **aortic bodies** or the **carotid bodies** leads to a stimulation of the cardioacceleratory and vasomotor centers. This elevates arterial pressure and increases blood flow through peripheral tissues. A rise in carbon dioxide levels in the blood or cerebrospinal fluid will have the same effects on cardiovascular function.

Chemoreceptor output also affects the respiratory centers. As a result, a rise in blood flow and blood pressure is associated with an elevated respira-

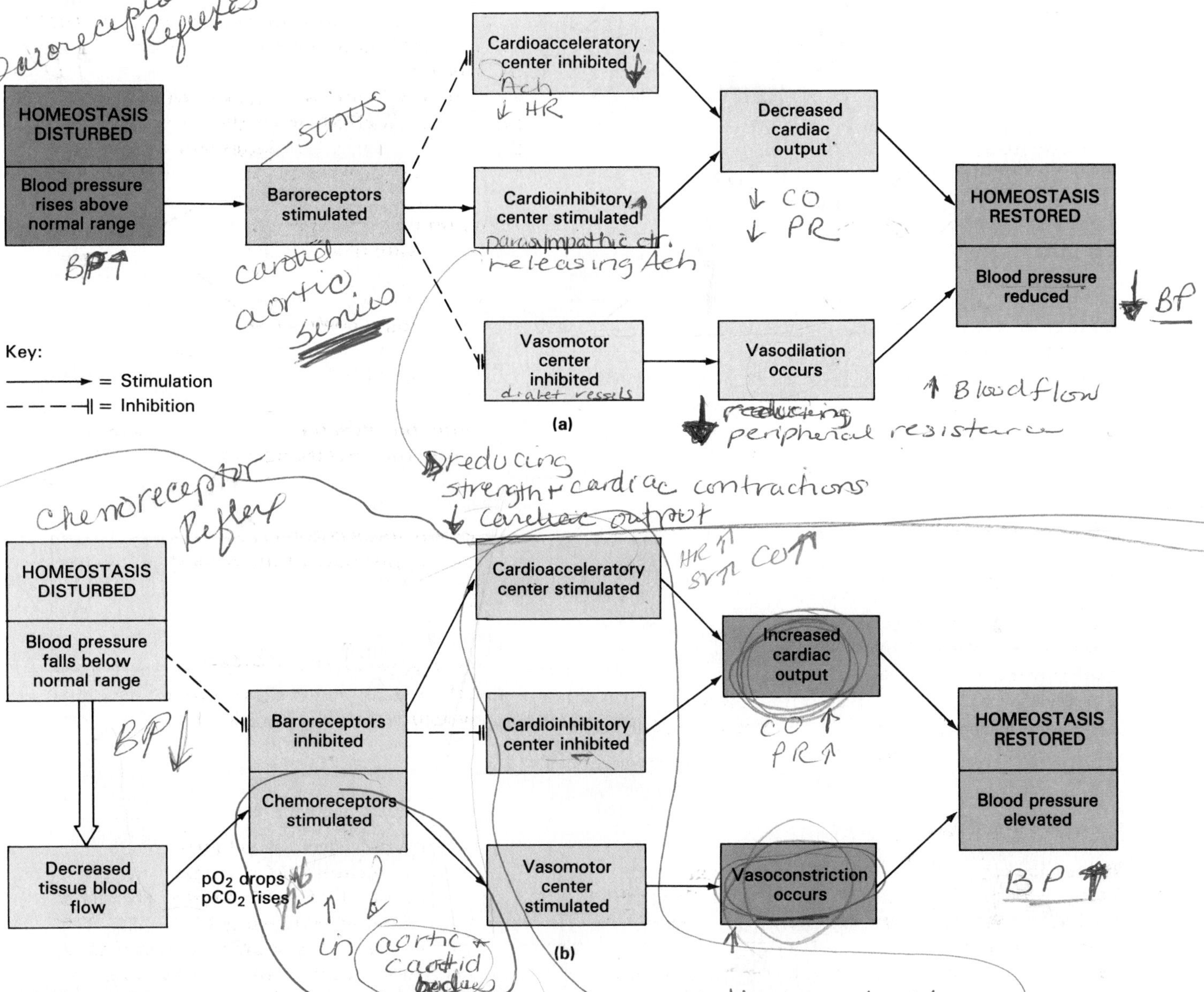

HOMEOSTASIS DISTURBED
Blood pressure rises above normal range
Baroreceptors stimulated
Cardioacceleratory center inhibited
Cardioinhibitory center stimulated
Vasomotor center inhibited
Decreased cardiac output
Vasodilation occurs
HOMEOSTASIS RESTORED
Blood pressure reduced

Key:
→ = Stimulation
– – –‖ = Inhibition

(a)

HOMEOSTASIS DISTURBED
Blood pressure falls below normal range
Decreased tissue blood flow
pO_2 drops
pCO_2 rises
Baroreceptors inhibited
Chemoreceptors stimulated
Cardioacceleratory center stimulated
Cardioinhibitory center inhibited
Vasomotor center stimulated
Increased cardiac output
Vasoconstriction occurs
HOMEOSTASIS RESTORED
Blood pressure elevated

(b)

FIGURE 21-27
Reflexes that Assist in the Regulation of Blood Pressure. (a) A diagrammatic flow chart for the baroreceptor reflexes. (b) Integrated activity of the chemoreceptor and baroreceptor reflexes. The chemoreceptor reflexes can have these effects even in the absence of baroreceptor stimulation and a fall in blood pressure.

tory rate. Coordination of cardiovascular and respiratory activity is vital, because accelerating tissue blood flow is useful only if the circulating blood contains adequate oxygen. In addition, a rise in the respiratory rate accelerates venous return through the action of the thoracoabdominal pump. (Other aspects of chemoreceptor activity and respiratory control will be considered in Chapter 23.)

Influence of the ANS and Higher Brain Centers

The cardiac and vasomotor centers may also be influenced by the activities of other areas of the brain. Stimulation by sympathetic neurons of the ANS, assisted by the release of epinephrine and norepinephrine by the adrenal medulla, acts on these centers to increase cardiac output and cause vasoconstriction. Parasympathetic stimulation affects the cardioinhibitory center, producing reduction in cardiac output. It does not directly affect the vasomotor center, but vasodilation occurs as sympathetic activity declines.

The activities of higher brain centers can also affect blood pressure. Our thought processes or emotional states can produce significant changes in blood pressure by influencing cardiac output and vasomotor tone. For example, strong emotions of anxiety, fear, or rage are accompanied by an elevation in blood pressure, caused by cardiac stimulation and vasoconstriction.

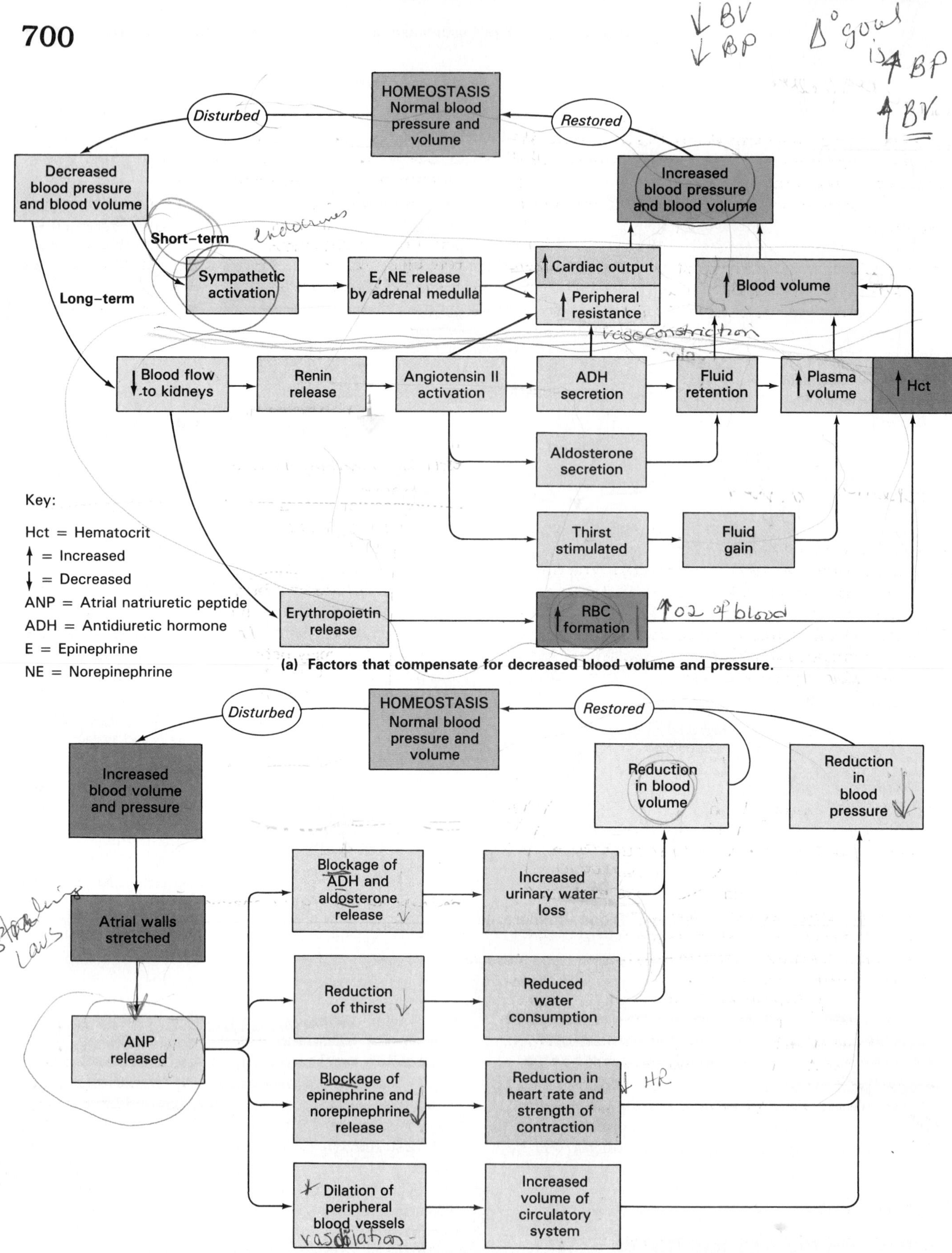

FIGURE 21-28
Homeostatic Regulation of Blood Pressure and Blood Volume. (a) Factors that compensate for decreased blood volume and pressure. (b) Factors that compensate for increased blood volume and pressure.

HORMONES AND CARDIOVASCULAR REGULATION

The endocrine system provides both short-term and long-term regulation of cardiovascular performance. Epinephrine and norepinephrine from the adrenal medulla stimulate cardiac output and peripheral vasoconstriction. Other hormones important in regulating cardiovascular function include (1) antidiuretic hormone (ADH), (2) angiotensin II, (3) erythropoietin (EPO), and (4) atrial natriuretic peptide (ANP). (These hormones and their functions were described in Chapter 18. (pp. 575, 588–589)) Although ADH and angiotensin II affect blood pressure, all four hormones are concerned primarily with the long-term regulation of blood volume, as diagrammed in Figure 21-28.

Antidiuretic Hormone

ADH is released at the posterior pituitary in response to a decrease in blood pressure or an increase in the osmotic concentration of the plasma. The immediate result is a peripheral vasoconstriction that elevates blood pressure. This hormone also stimulates the conservation of water at the kidneys, thus preventing a reduction in blood volume that would further reduce blood pressure.

Angiotensin II

Angiotensin II appears in the blood following the release of renin by specialized kidney cells stimulated by a fall in blood pressure. Renin starts a chain reaction that ultimately converts an inactive protein, angiotensinogen, to the hormone angiotensin II. Angiotensin II causes an extremely powerful vasoconstriction that elevates blood pressure almost at once. It also stimulates the secretion of ADH by the pituitary and aldosterone by the adrenal cortex. Aldosterone stimulates the reabsorption of sodium ions and water from the urine; its effects are therefore complementary to those of ADH. In addition, angiotensin II stimulates thirst, and the presence of ADH and aldosterone ensures that the additional water consumed will be retained, elevating the plasma volume.

Erythropoietin

Erythropoietin (EPO) is released at the kidneys if the blood pressure declines or if the oxygen content of the blood becomes abnormally low. This hormone stimulates red blood cell production, elevating the blood volume and improving the oxygen-carrying capacity of the blood.

Atrial Natriuretic Peptide

ANP is produced by specialized cardiac muscle fibers in the atrial walls when they are stretched by excessive venous return. ANP reduces blood volume and blood pressure by (1) increasing water losses at the kidneys; (2) reducing thirst; (3) blocking the release of ADH, aldosterone, and catecholamines; and (4) stimulating peripheral vasodilation. As blood volume and blood pressure decline, the stress on the atrial walls is removed, and ANP production ceases.

Patterns of Cardiovascular Response

In this and the previous two chapters we have considered the blood, the heart, and the circulatory system as individual entities. Yet in our day-to-day lives the cardiovascular system operates as an integrated complex. The interactions are quite fascinating, and of considerable importance when physical or physiological conditions are changing rapidly. Two common stresses, exercise and blood loss, provide examples of the adaptability of this system and its ability to maintain homeostasis. The homeostatic responses involve an interplay between the cardiovascular system (CVS) and other systems, and the central mechanisms are aided by automatic adjustments at the tissue level. We will also consider the physiological mechanisms involved in shock and heart failure, two important cardiovascular disorders.

EXERCISE AND THE CARDIOVASCULAR SYSTEM

At rest the cardiac output averages around 6 liters per minute. That value changes dramatically during exercise. In addition, the pattern of blood distribution

TABLE 21-6 Distribution of Blood During Exercise

Tissue	*Blood Flow (mℓ/min)* *Rest*	*Light Exercise*	*Strenuous Exercise*
Muscle	1200	4500	12,500
Heart	250	350	750
Brain	750	750	750
Skin	500	1500	1900
Kidney	1100	900	600
Abdominal viscera	1400	1100	600
Miscellaneous	600	400	400
Total cardiac output	5800	9500	17,500

changes markedly, as detailed in Table 21-6. As exercise begins, a number of interrelated changes occur.

1. *Extensive vasodilation occurs* as the rate of skeletal muscle oxygen consumption increases. Peripheral resistance drops, blood flow through the capillaries increases, and blood enters the venous system at an accelerated rate.
2. *The venous return increases*, as skeletal muscle contractions squeeze blood along the peripheral veins and an increased breathing rate pulls blood into the venae cavae, via the thoracoabdominal pump.
3. *There is a rise in cardiac output* caused by the increase in venous return. This increase occurs in direct response to ventricular stretching (Starling's law) and in a reflexive response to atrial stretching (the atrial reflex). The increased cardiac output keeps pace with the elevated demand, and arterial pressures are maintained despite the drop in peripheral resistance.

This regulation by venous feedback produces a gradual increase in cardiac output to about double resting levels. Over this range, typical of light exercise, the pattern of blood distribution remains relatively unchanged. However, there are minor increases in the blood flow to skeletal muscle, cardiac muscle, and the skin. The increased flow to the muscles reflects the dilation of precapillary sphincters in response to local factors; the increased blood flow to the skin occurs in response to the rise in body temperature.

At higher levels of exertion, other physiological adjustments occur as the sympathetic nervous system stimulates the cardiac and vasomotor centers. Cardiac output increases, and major changes in the peripheral distribution of blood increase the flow to active skeletal muscles. Under massive sympathetic stimulation the cardioacceleratory center can increase cardiac output to levels as high as 20–25 liters per minute. Even at these rates, the increased circulatory demands of the skeletal muscles can be met only if the vasomotor center severely restricts the blood flow to "nonessential" organs such as those of the viscera. When exercising at maximal levels the blood essentially races between the skeletal muscles, the lungs, and the heart. Only the blood supply to the brain remains unaffected.

Exercise, Cardiovascular Fitness, and Health

Cardiovascular performance improves significantly with training. Table 21-7 compares the cardiac performance of athletes with that of nonathletes. The first noteworthy point is that athletes have larger hearts: the average stroke volume is also greater than that of a nonathlete. These are important functional changes. Cardiac output is equal to the stroke volume times the heart rate, so for the same cardiac output, an individual with a larger stroke volume will have a slower heart rate. A professional athlete at rest can maintain normal blood flow to peripheral tissues at a heart rate as low as 50 bpm (beats per minute), and when necessary the cardiac output can increase tremendously. The heart rate can more than triple and more blood can be ejected with each beat. This combination increases the cardiac reserve to maximal levels.

Exercise and Cardiovascular Disease

Regular exercise has several beneficial effects. Even a moderate exercise routine (jogging 5 miles per week, for example) can lower total blood cholesterol levels. This is one of the major risk factors for atherosclerosis, leading to cardiovascular disease and strokes. In addition, a healthy lifestyle with regular exercise, a balanced diet, weight control, and no smoking, reduces stress, lowers blood pressure, and slows plaque formation.

Large-scale statistical studies indicate that regular moderate exercise may cut the incidence of heart attacks almost in half. However, at present only an estimated 8 percent of adults in the United States exercise at recommended levels.

Exercise is also beneficial in accelerating recovery after a heart attack. Regular light to moderate exercise, such as walking, jogging, or bicycling, coupled with a low-fat diet and low-stress lifestyle, not

TABLE 21-7 Effects of Training on Cardiovascular Performance

Subject	*Heart Weight (g)*	*Stroke Volume (mℓ)*	*Heart Rate (bpm)*	*Cardiac Output (ℓ/min)*	*Blood Pressure (systolic/diastolic)*
Nonathlete (rest)	300	60	83	5.0	120/80
Nonathlete (maximum)		104	192	19.9	187/75
Trained athlete (rest)	500	100	53	5.3	120/80
Trained athlete (maximum)		167	182	30.4	200/see note

Note: Diastolic pressures during maximal activity have not been accurately measured.
Data sources: Asmussen and Nielsen, Acta Physiologica Scandinavia, 27:217 (1952) and Reindell et al., Schweiz. Zeitschr. f. Sportmed. 1:97 (1953).

only reduces symptoms, such as angina, but also improves both mood and the overall quality of life. However, exercise does not remove the underlying medical problem, and atherosclerotic plaques do not disappear or grow smaller with exercise.

There is no evidence that *intense* athletic training lowers the incidence of cardiovascular disease. On the contrary, the strains placed on all physiological systems, including the cardiovascular system, during an ultramarathon or other athletic extreme can be severe. Before entering such competitions, a thorough physical is recommended. Individuals with congenital aneurysms, cardiomyopathy, or cardiovascular disease risk fatal circulatory problems, such as an arrhythmia or MI, during severe exercise. Even normal individuals can develop acute physiological disorders, such as kidney failure, after extreme exercise. The effects of exercise on other systems will be detailed in later chapters.

CARDIOVASCULAR RESPONSE TO HEMORRHAGING

In Chapter 19 we considered the local circulatory reaction to a break in the wall of a blood vessel. (p. 624) When the clotting response fails to prevent a significant blood loss, the entire cardiovascular system begins making adjustments to maintain blood pressure and restore blood volume. Those adjustments, diagrammed in Figure 21-29, are triggered by the fall in arterial blood pressure. The immediate problem is the maintenance of adequate blood pressure and peripheral blood flow. The long-term problem is the restoration of normal blood volume.

Elevation of Blood Pressure

Short-term responses appear almost as soon as the pressures start to decline, when the carotid and aortic reflexes increase cardiac output and cause peripheral vasoconstriction. With the blood volume reduced, cardiac output is maintained by increasing the heart rate, often to rates of 180–200 bpm. Sympathetic activation provides assistance by increasing vasomotor tone, constricting the arterioles, and elevating blood pressure. At the same time, the muscular arteries and veins decrease in diameter, and the venous reserve is called upon.

Mobilizing the Venous Reserve When you donate blood at a blood bank, the amount collected is usually 500 mℓ, roughly 10 percent of the total blood volume. Such a loss initially causes a drop in cardiac output, but the venoconstriction demanded by the vasomotor center quickly improves venous return and restores

FIGURE 21-29
Cardiovascular Responses to Hemorrhaging and Blood Loss

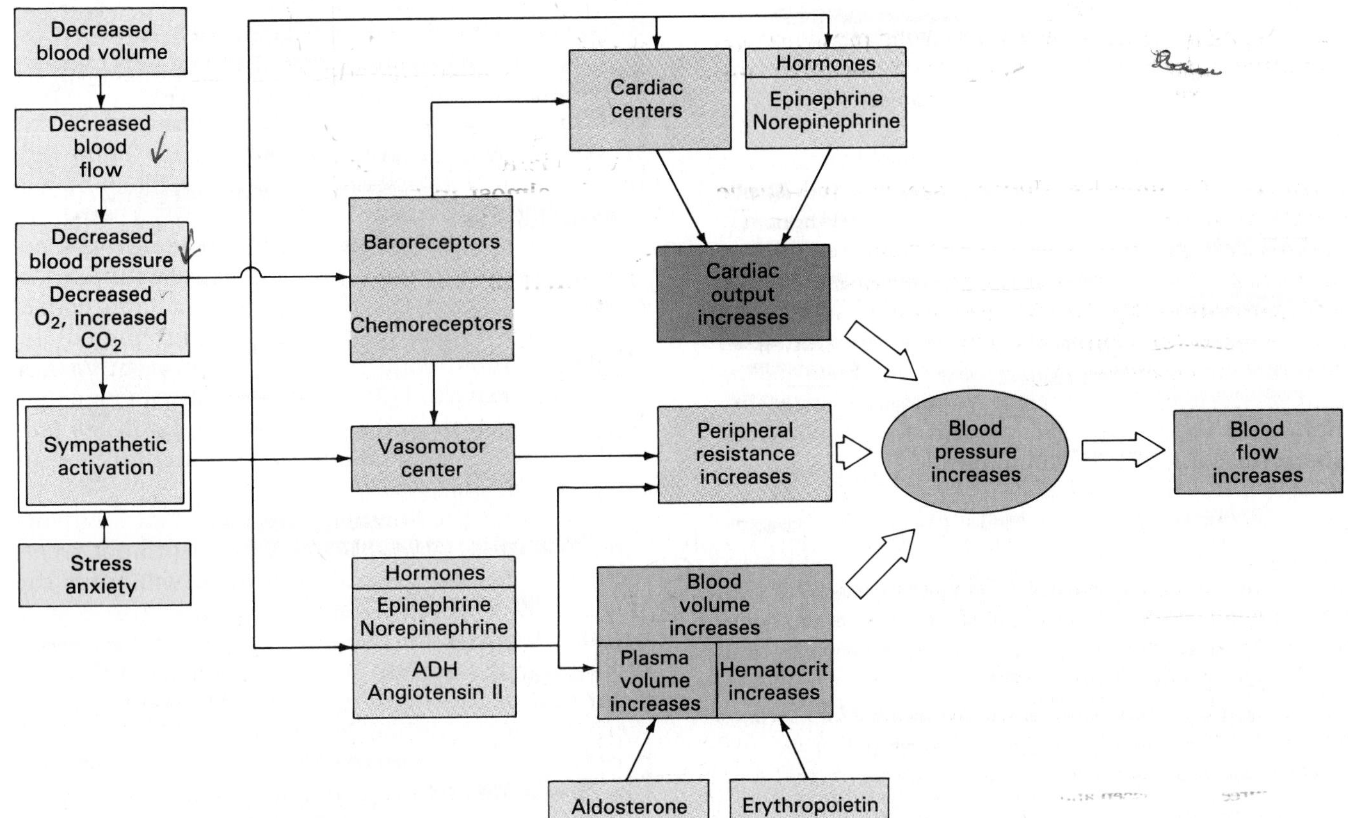

cardiac output to normal levels. This venous compensation can restore normal arterial pressures and peripheral blood flow after losses that amount to 15–20 percent of the total blood volume.

Hormonal Responses Sympathetic activation also causes the secretion of epinephrine and norepinephrine by the adrenal medulla. At the same time, ADH is released by the posterior pituitary, and at the kidneys the fall in blood pressure causes the release of renin, initiating the activation of angiotensin II. The catecholamines increase cardiac output, and in combination with ADH and angiotensin II they cause a powerful vasoconstriction that elevates blood pressure and improves peripheral blood flow.

Restoration of Blood Volume

Following a serious hemorrhage, several days may pass before the blood volume returns to normal. When the short-term responses are unable to maintain normal cardiac output and blood pressure, the decline in capillary blood pressure triggers the recall of fluids from the interstitial spaces. Over this period ADH and aldosterone promote fluid retention and reabsorption at the kidneys, preventing further reductions in blood volume. Thirst increases, and additional water is obtained by absorption across the digestive tract. This intake of fluid elevates plasma volume and ultimately replaces the interstitial fluids "borrowed" at the capillaries. Erythropoietin targets the bone marrow, stimulating the maturation of red blood cells, which increase the blood volume and improve oxygen delivery to peripheral tissues.

SHOCK

Shock is an acute circulatory crisis marked by low blood pressure (hypotension) and inadequate peripheral blood flow. Severe and potentially fatal symptoms develop as vital tissues become starved for oxygen and nutrients. Common causes of shock are (1) a fall in cardiac output after hemorrhaging or other fluid losses, (2) damage to the heart, (3) external pressure on the heart, or (4) extensive peripheral vasodilation. [**CM:** *Heart Failure*] All forms of shock share the same basic symptoms:

1. Hypotension, with systolic pressures below 90 mm Hg.
2. Pale, cool, and moist ("clammy") skin. The skin is pale and cool because of peripheral vasoconstriction; the moisture reflects sympathetic activation of the sweat glands.
3. Frequent confusion and disorientation, due to a fall in blood pressure at the brain.
4. Rapid, weak pulse.
5. Cessation of urination, because the reduced blood to the kidneys slows or stops urine production.
6. Acidosis, due to lactic acid generation in oxygen-deprived tissues.

We will focus on the cause, symptoms, and treatment of circulatory shock. Those interested in other types of shock should consult the Clinical Manual. [**CM:** *Other Types of Shock*]

Circulatory Shock

A severe reduction in blood volume produces symptoms of **circulatory shock**. The cardiovascular system can cope with fluid losses of up to about 30 percent of the total blood volume before symptoms of circulatory shock appear. The cause can be hemorrhaging or fluid losses to the environment, as in dehydration or after severe burns. Circulatory shock is often divided into *compensated* and *progressive* stages. During the **compensated stage** homeostatic adjustments can cope with the situation; the responses detailed in the previous section are part of the compensation process.

When blood volume declines by more than 35 percent, the individual enters the **progressive stage** of circulatory shock. Homeostatic mechanisms are unable to cope with the situation, and a vicious cycle begins: low blood pressure and low venous return lead to decreased cardiac output and myocardial damage, causing a further reduction in cardiac output. This sequence is diagrammed in Figure 21-30a. When the mean arterial blood pressure falls to about 50 mm Hg, carotid sinus baroreceptors trigger a massive activation of sympathetic vasoconstrictors. Cerebral vessels are not affected, and blood pressure in the carotid arteries remains relatively high (70 mm Hg). However, the maintenance of adequate blood flow to the brain occurs at the expense of other tissues. In this reflex, called the **central ischemic response**, peripheral circulation is reduced to an absolute minimum. To prevent the development of fatal consequences, treatment must be prompt and immediate. Treatment must concentrate on (1) preventing further fluid losses and (2) giving a transfusion of whole blood, plasma expanders, or blood substitutes (see Health News: Transfusions and Synthetic Blood in Chapter 19). (p. 631)

Together the neural and hormonal mechanisms can slow the progression of shock. If prompt treatment is not obtained, the individual will enter the stage of **irreversible shock**, diagrammed in Figure 21-30b. At this point, conditions in the heart, liver, kidneys, and CNS are rapidly deteriorating to the point that death will occur, even with medical treatment. When conditions deteriorate sufficiently in the periphery, smooth muscles will be unable to maintain the vasoconstriction demanded by the central ischemic response. The result is a widespread peripheral

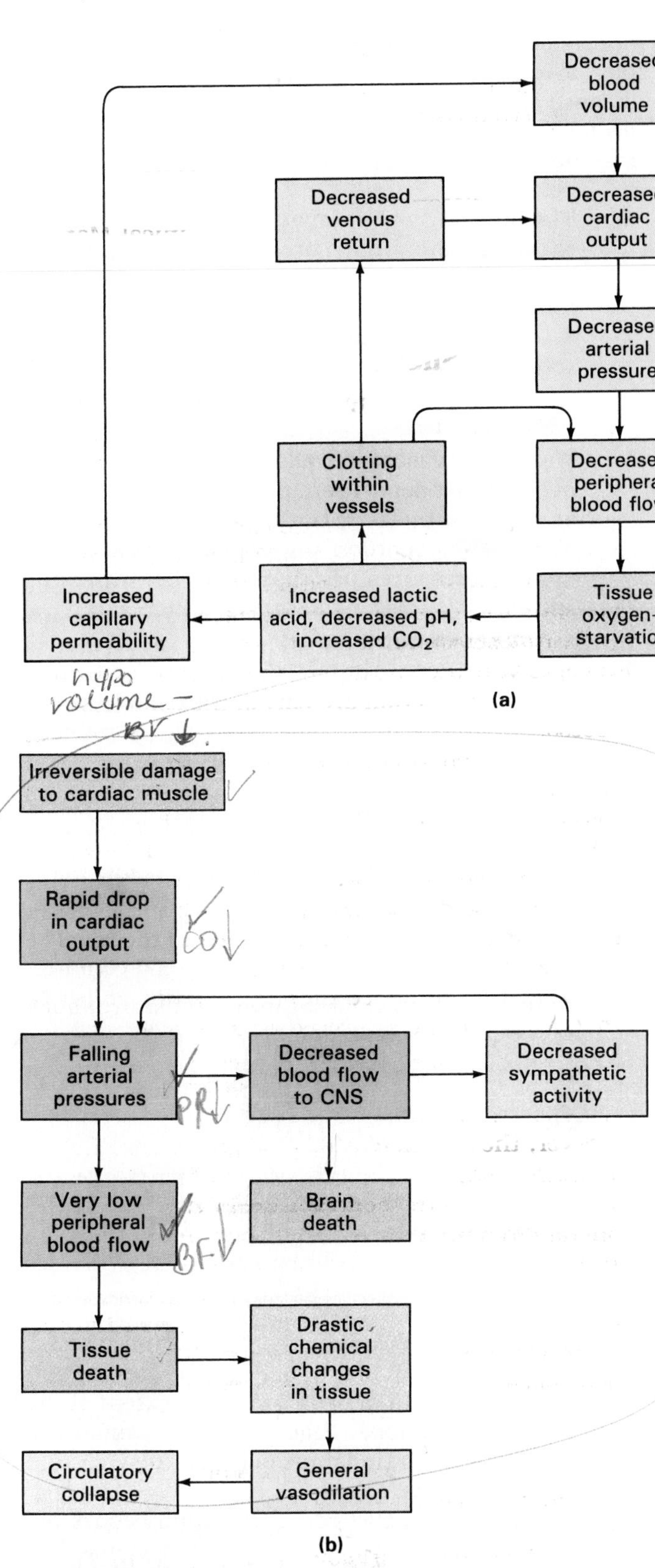

FIGURE 21-30
Shock. (a) Progressive shock, characterized by a gradual decline in systemic blood pressure, tissue blood flow, and cardiac output. (b) The stage of irreversible shock involves a series of integrated chain reactions leading to a rapid decline in cardiac output, a dramatic and irreversible fall in blood pressure, circulatory collapse, and eventual brain death.

vasodilation and an immediate and fatal decline in blood pressure. This event is called **circulatory collapse**. Following circulatory collapse, blood pressure in many tissues falls below the critical closing pressure of capillaries. Circulation in these areas then stops completely, and the cells die.

Aging and the Cardiovascular System

The capabilities of the cardiovascular system gradually decline with age. The major changes will be listed and discussed in the same sequence as the cardiovascular chapters: blood, heart, and vessels.

AGE-RELATED CHANGES IN THE BLOOD

1. The hematocrit often decreases in elderly individuals, lowering the oxygen-carrying capacity of the blood. It is not clear whether this decrease results from changes in the rate of bone marrow activity, from changes in diet (such as vitamin deficiency), or from a sedentary lifestyle.
2. Peripheral veins may become constricted or blocked by formation of a thrombus (stationary blood clot). This condition, called **venous thrombosis**, affects an estimated 25 percent of individuals over age 50. Thrombus formation occurs in blood when it remains stationary for an extended period; a sedentary lifestyle and poor venous circulation combine to produce venous thrombosis. The thrombi most often appear in

the veins of the legs, where gravity slows the return of blood toward the heart. There is a significant chance that a thrombus can become detached, pass through the heart, and become wedged in a small artery, most often in the lungs, causing a *pulmonary embolism*. This condition, which blocks circulation and elevates pulmonary blood pressures, is detailed in Chapter 23.

3. Even in the absence of venous thrombosis, blood often tends to pool in the veins of the lower legs because the valves are not working effectively. The elevated hydrostatic pressures at the venous side of the capillaries leads to edema in the feet and ankles. Many people prevent this by wearing elastic "support hose" that compress the tissues of the lower leg. The compression squeezes the tissues, forcing fluid back into the bloodstream.

THE AGING HEART

1. There is a reduction in the maximum cardiac output, primarily due to a decline in the maximum heart rate. The maximum heart rate of a person over age 65 is around 170 bpm, versus 195 bpm for a young individual.
2. The activities of the nodal and conducting fibers begin to change. One sign of these changes is the decrease in maximal heart rate. In addition, arrhythmias become more common. An estimated 30 percent of individuals over age 65 have ECG abnormalities.
3. The fibrous skeleton changes, becoming less elastic. With age, connective tissues throughout the body lose their resilience. In the heart, the amount of fibrous tissue increases; most of the fibers are composed of collagen, rather than elastin. Because the heart is less elastic, it takes more time to fill the ventricular chambers. This places another limit on cardiac output.
4. Progressive atherosclerosis leads to a restriction of coronary circulation. This often results in symptoms of *angina pectoris*, a condition described in Chapter 20. (p. 644)
5. Damage to ventricular cardiac muscle fibers is also progressive, and injured or dead cells are replaced by scar tissue. This substitution reduces the strength and thickness of the heart wall. Because scar tissue is not elastic, the sudden rise in pressure during ventricular systole may ultimately break through the heart wall.

AGING AND THE CIRCULATORY NETWORK

Most of the age-related changes in the circulatory system are related to arteriosclerosis, a condition described earlier in the chapter. Limited arteriosclerosis occurs as a normal consequence of aging. For example, arteries lose their elasticity, the amount of smooth muscle that they contain decreases, and they become stiff and relatively inflexible. These changes may have a number of secondary effects:

1. The inelastic walls of major arteries are less able to tolerate the sudden pressure increases during ventricular systole. Over time this weakening may result in formation of a swelling, or *aneurysm*, along the aortic or other large arterial trunk. These weakened areas may eventually rupture, causing a stroke, infarct, or massive blood loss, depending on the vessel involved.
2. Weakened vascular walls can become sites of calcium salt deposition, producing arteries that are not just inelastic but actually rigid. Without elasticity to cushion the pressure oscillations, systolic pressures peak much higher, exposing the walls to greater stresses and increasing the risks of a stroke or infarct.
3. Atherosclerotic plaques act as sites for thrombus formation, increasing the risk of a local circulatory blockage or formation of an embolus that can cause a stroke or heart attack.

- √ **Why does blood pressure increase during exercise?**
- √ **How would applying a small pressure below the carotid bifurcation affect your heat rate?**
- √ **What affect would vasoconstriction of the renal artery have on blood pressure and blood volume?**
- √ **Why does a person suffering from circulatory shock have a rapid and weak pulse?**

■ Integration with Other Systems

The cardiovascular system is both anatomically and functionally linked to all other systems. The section on vessel distribution demonstrated the extent of the anatomical connections. Figure 21-31 summarizes the physiological relationships between the CVS and other organ systems.

The most extensive communication occurs between the cardiovascular and lymphatic systems. Not only are the two systems physically interconnected, but cell populations of the lymphatic system use the cardiovascular system as a highway to move from one part of the body to another. The next chapter examines the lymphatic system in detail and considers the role of the lymphatic system in the immune response.

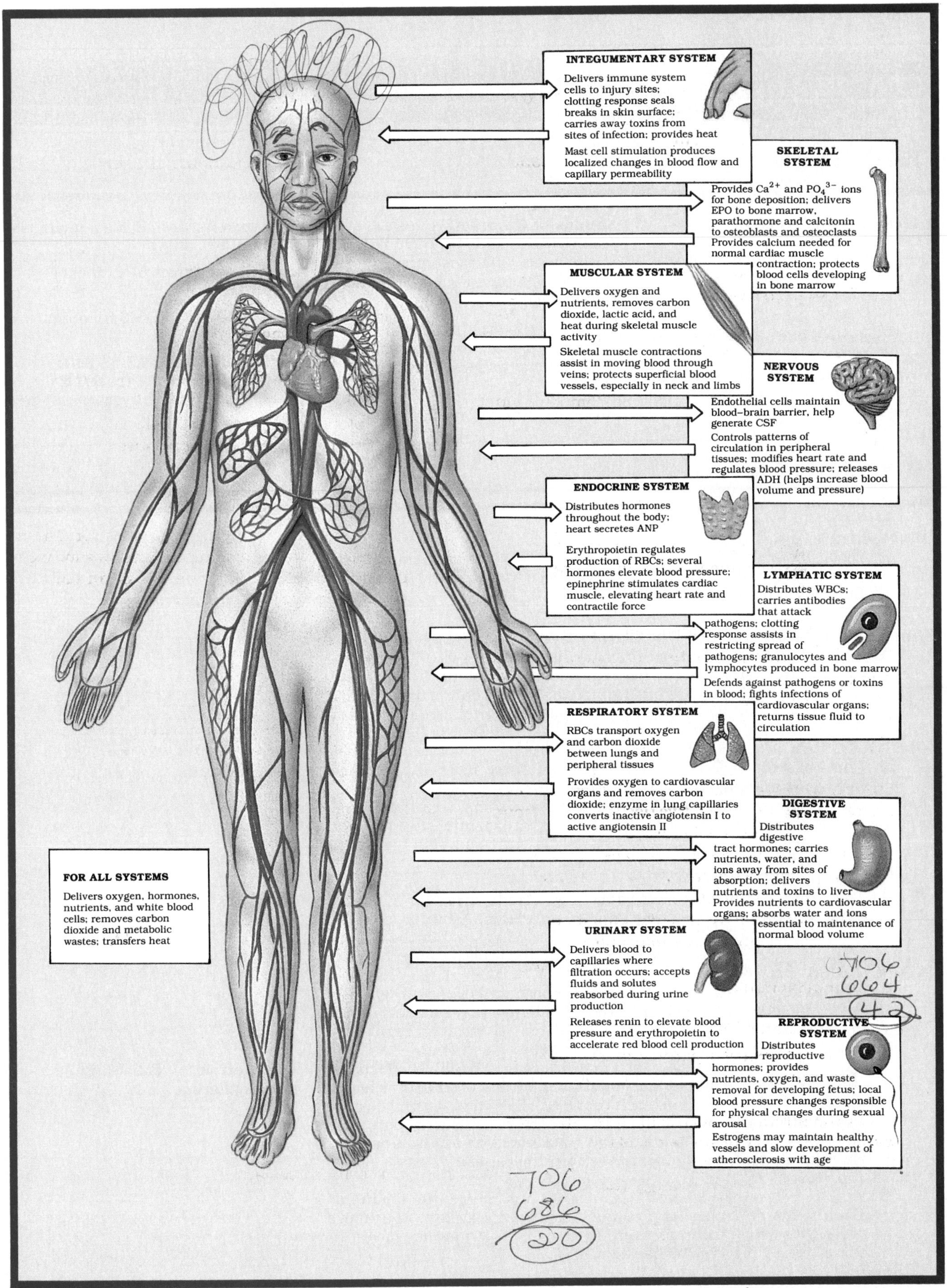

INTEGUMENTARY SYSTEM

Delivers immune system cells to injury sites; clotting response seals breaks in skin surface; carries away toxins from sites of infection; provides heat

Mast cell stimulation produces localized changes in blood flow and capillary permeability

SKELETAL SYSTEM

Provides Ca^{2+} and PO_4^{3-} ions for bone deposition; delivers EPO to bone marrow, parathormone and calcitonin to osteoblasts and osteoclasts

Provides calcium needed for normal cardiac muscle contraction; protects blood cells developing in bone marrow

MUSCULAR SYSTEM

Delivers oxygen and nutrients, removes carbon dioxide, lactic acid, and heat during skeletal muscle activity

Skeletal muscle contractions assist in moving blood through veins; protects superficial blood vessels, especially in neck and limbs

NERVOUS SYSTEM

Endothelial cells maintain blood–brain barrier, help generate CSF

Controls patterns of circulation in peripheral tissues; modifies heart rate and regulates blood pressure; releases ADH (helps increase blood volume and pressure)

ENDOCRINE SYSTEM

Distributes hormones throughout the body; heart secretes ANP

Erythropoietin regulates production of RBCs; several hormones elevate blood pressure; epinephrine stimulates cardiac muscle, elevating heart rate and contractile force

LYMPHATIC SYSTEM

Distributes WBCs; carries antibodies that attack pathogens; clotting response assists in restricting spread of pathogens; granulocytes and lymphocytes produced in bone marrow

Defends against pathogens or toxins in blood; fights infections of cardiovascular organs; returns tissue fluid to circulation

RESPIRATORY SYSTEM

RBCs transport oxygen and carbon dioxide between lungs and peripheral tissues

Provides oxygen to cardiovascular organs and removes carbon dioxide; enzyme in lung capillaries converts inactive angiotensin I to active angiotensin II

DIGESTIVE SYSTEM

Distributes digestive tract hormones; carries nutrients, water, and ions away from sites of absorption; delivers nutrients and toxins to liver

Provides nutrients to cardiovascular organs; absorbs water and ions essential to maintenance of normal blood volume

FOR ALL SYSTEMS

Delivers oxygen, hormones, nutrients, and white blood cells; removes carbon dioxide and metabolic wastes; transfers heat

URINARY SYSTEM

Delivers blood to capillaries where filtration occurs; accepts fluids and solutes reabsorbed during urine production

Releases renin to elevate blood pressure and erythropoietin to accelerate red blood cell production

REPRODUCTIVE SYSTEM

Distributes reproductive hormones; provides nutrients, oxygen, and waste removal for developing fetus; local blood pressure changes responsible for physical changes during sexual arousal

Estrogens may maintain healthy vessels and slow development of atherosclerosis with age

FIGURE 21-31
Functional Relationships Between the Cardiovascular System and Other Systems

CHAPTER REVIEW

Review of Selected Clinical Terms

arteriosclerosis (ar-tē-rē-ō-skle-RŌ-sis) (*p. 668*)
A thickening and toughening of arterial walls.

atherosclerosis (ath-er-ō-skle-RŌ-sis) (*p. 668*)
A type of arteriosclerosis characterized by changes in the endothelial lining.

sounds of Korotkoff (*p. 692*)
Distinctive sounds heard while measuring a person's blood pressure using a sphygmomanometer; produced by turbulence.

edema (e-DĒ-ma) (*p. 694*)
An abnormal accumulation of fluid in peripheral tissues.

shock (*p. 704*)
An acute circulatory crisis marked by hypotension and inadequate peripheral blood flow.

Additional Terms of Clinical Importance

ARTERIOSCLEROSIS (*p. 668*): **plaque**

CHECKING THE PULSE AND BLOOD PRESSURE (*p. 691*): **pressure points, sphygmomanometer**

Study Outline

Related Key Terms

Introduction (pp. 663–664)

1. Blood flows through a network of arteries, veins, and capillaries. All chemical and gaseous exchange between the blood and interstitial fluid takes place across capillary walls.

Functional Anatomy of the Circulatory System (pp. 664–672)

1. Arteries and veins form an internal distribution system, propelled by the heart. Arteries branch repeatedly, decreasing in size until they become **arterioles**; from the arterioles blood enters the capillary networks. Blood flowing from the capillaries enters small **venules** before entering larger veins.

HISTOLOGICAL ORGANIZATION (pp. 664–665)

2. The walls of arteries and veins contain three layers: the **tunica interna**, **tunica media**, and outermost **tunica externa**. In general, the walls of arteries are thicker than those of veins. The endothelial lining of an artery cannot contract, so it is thrown into folds. Arteries constrict when blood pressure does not distend them; veins constrict very little.
   Related Key Terms: **vasa vasorum**

ARTERIES (pp. 665–666)

3. The arterial system includes the large **elastic arteries**, medium-sized **muscular arteries**, and smaller **arterioles**. As we proceed toward the capillaries the number of vessels increases, but the diameter of the individual vessels decreases and the walls become thinner.

CAPILLARIES (pp. 666–669)

4. Capillaries are the only blood vessels whose walls permit exchange between blood and interstitial fluid. Capillaries may be **continuous** or **fenestrated**. **Sinusoids** are specialized fenestrated capillaries found in certain tissues that allow very slow blood flow.
5. Capillaries form interconnected networks called **capillary plexuses (capillary beds)**. A **precapillary sphincter** (a band of smooth muscle) adjusts the blood flow into each capillary. Blood flow within a capillary changes as **vasomotion** occurs. The entire capillary plexus may be bypassed by blood flow through **arteriovenous anastomoses** or via **central (preferred) channels** within the capillary plexus.
   Related Key Terms: **metarteriole • collaterals autoregulation**

VEINS (pp. 669–672)

6. Venules collect blood from the capillaries and merge into **medium-sized veins** and then **large veins**. The arterial system is a high-pressure system; blood pressure in veins is much lower. Valves in these vessels prevent the backflow of blood.

DISTRIBUTION OF BLOOD (p. 672)

7. Peripheral **venoconstriction** helps maintain adequate blood volume in the arterial system after a hemorrhage. The **venous reserve** normally accounts for about 20 percent of the total blood volume.
   Related Key Terms: **blood reservoir**

The Circulatory System (pp. 672–686)

Related Key Terms

THE PULMONARY CIRCULATION (p. 673)

1. The pulmonary circuit includes the pulmonary trunk, the **left** and **right pulmonary arteries**, and the **pulmonary veins** that empty into the left atrium. (See Figure 21–9.)

alveoli

THE SYSTEMIC CIRCULATION (pp. 674–686)

2. The **ascending aorta** gives rise to the coronary circulation. The **aortic arch** communicates with the **descending aorta.** (See Figures 21–10 to 21–13 and Tables 21–1 to 21–3.)
3. Arteries in the neck, arms, and legs are deep beneath the skin; in contrast, there are usually two sets of peripheral veins, one superficial and one deep. This dual-venous drainage is important for controlling body temperature.
4. The **superior vena cava** receives blood from the head, neck, chest, shoulders, and arms. (See Figures 21–14 and 21–16 and Table 21–4.) The **inferior vena cava** collects most of the venous blood from organs below the diaphragm. (See Figures 21–17 and 21–5.)

hemiazygos vein

Circulatory Physiology (pp. 686–695)

FLOW AND PRESSURE (p. 687)

1. Flow is proportional to the difference in pressure; blood will flow from an area of higher pressure to one of relatively lower pressure.

blood pressure (BP)

FLOW AND RESISTANCE (pp. 687–689)

2. For circulation to occur, the **circulatory pressure** must be greater than the **total peripheral resistance** (the resistance of the entire circulatory system). For blood to flow into peripheral capillaries, blood pressure must be greater than the **peripheral resistance (PR)** (the resistance of the arterial system). Neural and hormonal control mechanisms regulate blood pressure.

vascular resistance
turbulence
critical closing pressure

A SUMMARY OF RELATIONSHIPS BETWEEN BLOOD FLOW, PRESSURE, AND RESISTANCE (p. 689)

3. The most important determinant of peripheral resistance is the diameter of arterioles.

CIRCULATORY PRESSURES (pp. 689–695)

4. The high arterial pressures overcome peripheral resistance and maintain blood flow through peripheral tissues. Capillary pressures are normally low; small changes in capillary pressure determine the rate of fluid movement into or out of the bloodstream. Venous pressure, normally low, determines venous return and affects cardiac output and peripheral blood flow.

systolic pressure
diastolic pressure
pulse pressure
mean arterial pressure (MAP)
elastic rebound

5. At the capillaries, solute molecules diffuse across the capillary lining, water-soluble materials diffuse through small spaces between endothelial cells. Water will move when driven by either hydrostatic or osmotic pressure. The direction of water movement is determined by the balance between these two opposing pressures. Any change in hydrostatic or osmotic pressures in the blood or tissues will shift the **dynamic center** (the point where the two forces are equal).

lymphatic vessels
recall of fluids
edema

6. **Valves**, **muscular pumps**, and the **thoracoabdominal pump** help the relatively low venous pressures propel blood toward the heart.

fainting

Cardiovascular Regulation (pp. 695–697)

1. Homeostatic mechanisms ensure that tissue blood flow **(perfusion)** delivers adequate oxygen and nutrients.

VARIATIONS IN CARDIAC OUTPUT (pp. 695–696)

2. The cardiac output, normally equal to peripheral blood flow, changes in response to local factors, autonomic control, and hormonal regulation.

VARIATIONS IN PERIPHERAL RESISTANCE (p. 696)

3. Peripheral resistance can also be affected by local factors, autonomic control, and hormonal regulation. Changes in blood viscosity or turbulence can also change peripheral resistance.

vasodilation
vasoconstriction

ALTERATIONS IN BLOOD PRESSURE (pp. 696–697)

4. Blood pressure varies directly with cardiac output, peripheral resistance, and blood volume.

Homeostasis and Cardiovascular Function (pp. 697–701)

1. Local, autonomic, and endocrine factors influence the coordinated regulation of cardiovascular function. Local factors change the pat-

Related Key Terms

tern of blood flow within capillary beds in response to chemical changes in interstitial fluids. Central mechanisms respond to changes in arterial pressure or blood gas levels. Hormones can assist in short-term adjustments (changes in CO and peripheral resistance) and long-term adjustments (changes in blood volume that affect CO and gas transport.

NEURAL MECHANISMS AND THE SHORT-TERM REGULATION OF BLOOD PRESSURE (pp. 698–700)

2. **Baroreceptor reflexes** are autonomic reflexes that adjust CO and peripheral resistance to maintain normal arterial pressures. Baroreceptor populations include the aortic, carotid sinus, and atrial baroreceptors.
3. **Chemoreceptor reflexes** respond to changes in the oxygen or carbon dioxide levels in the blood and cerebrospinal fluid. Sympathetic activation leads to stimulation of the cardioacceleratory and vasomotor centers; parasympathetic activation stimulates the cardioinhibitory center. Epinephrine and norepinephrine stimulate cardiac output and peripheral vasoconstriction.

aortic bodies (carotid bodies)

HORMONES AND CARDIOVASCULAR REGULATION (p. 701)

4. ADH and angiotensin II also promote peripheral vasoconstriction in addition to their other functions. ADH and aldosterone promote water and electrolytes retention and stimulate thirst. EPO stimulates red blood cell production. ANP encourages fluid loss, reduces blood pressure, inhibits thirst, and lowers peripheral resistance.

Patterns of Cardiovascular Response (pp. 701–705)

EXERCISE AND THE CARDIOVASCULAR SYSTEM (pp. 701–703)

1. During exercise, blood flow to skeletal muscles increases at the expense of circulation to nonessential organs, and cardiac output rises. Cardiovascular performance improves with training. Athletes have larger stroke volumes, slower resting heart rates, and increased cardiac reserves.

CARDIOVASCULAR RESPONSES TO HEMORRHAGING (pp. 703–704)

2. Blood loss causes an increase in cardiac output, mobilization of venous reserves, peripheral vasoconstriction, and the liberation of hormones that promote fluid retention and the manufacture of erythrocytes.

SHOCK (pp. 704–705)

3. **Shock** is an acute circulatory crisis marked by hypotension and inadequate peripheral blood flow. A severe drop in blood volume produces symptoms of **circulatory shock**.

compensated stage
progressive stage
central ischemic response
irreversible shock
circulatory collapse

Aging and the Cardiovascular System (pp. 705–706)

AGE-RELATED CHANGES IN THE BLOOD (pp. 705–706)

1. Age-related changes can include: (1) decreased hematocrit; (2) constriction or blockage of peripheral veins by a thrombus (stationary blood clot); (3) pooling of blood in the veins of the lower legs because valves are not working effectively.

venous thrombosis

THE AGING HEART (p. 706)

2. Age-related changes include: (1) a reduction in the maximum cardiac output; (2) changes in the activities of the nodal and conducting fibers; (3) a reduction in the elasticity of the fibrous skeleton; (4) a progressive atherosclerosis that can restrict coronary circulation; (5) replacement of damaged cardiac muscle fibers by scar tissue.

AGING AND THE CIRCULATORY NETWORK (p. 706)

3. Age-related changes are often related to arteriosclerosis and include: (1) inelastic walls of arteries are less tolerant of sudden pressure increases, which may lead to an *aneurysm*; (2) calcium salts can deposit on weakened vascular walls, increasing the risk of a stroke or infarct; (3) thrombi can form at atherosclerotic plaques.

Review Planner

	Objective	Level 1
1	Relate the structural characteristics of blood vessels to their functions.	1 2 3 22 25 26 27 32 33 34 37 38 41 45
2	Describe how and where dissolved materials enter and leave the circulatory system.	16 24 38
3	Explain the mechanisms that regulate blood flow through arteries, capillaries, and veins.	5 6 11 14 15 18 21 37
4	Identify major blood vessels and the areas they serve.	7 8 9 10 12
5	Describe how blood pressure is measured.	20
6	Explain the factors that influence blood pressure.	6 14 15 21 40 43 44 47
7	Describe how central and local controls interact to regulate blood flow and pressure in tissues.	5 6 19 21 23 29 31 39
8	Explain how the circulatory system responds to the demands of exercise and blood loss.	4 17 30 35 36 46
9	Describe the age-related changes that occur in the cardiovascular system.	13 48

Level 2: **28 39 42 48**

54–58

C M

Aneurysms
Problems with Venous Valve Function
Treatment of Cerebrovascular Disease
Hypertension and Hypotension
Heart Failure
Baroreceptor Accommodation
Other Types of Shock

Brain Attack
Heart Disease Risk Factors
Hypertension and Black Americans

Review Questions

MULTIPLE CHOICE

1. The walls of arteries and veins have ______ layer(s): a) one; b) two; c) three; d) the number varies according to the function of the vessel.
2. A particular blood vessel has thick walls that contain much elastic tissue. This vessel is a(n): a) artery; b) vein; c) arteriole; d) capillary.
3. A blood vessel has a small diameter with delicate walls that permit the diffusion of lipid-soluble materials and dissolved gases. This vessel is a(n): a) artery; b) vein; c) arteriole; d) capillary.
4. Symptoms of circulatory shock normally appear following a loss exceeding ______ percent of the total blood volume: a) 10; b) 30; c) 50; d) the percentage varies depending on whether the blood loss is caused by hemorrhage or dehydration.
5. Carotid sinus baroreceptors help to: a) maintain blood flow to the brain; b) maintain blood flow from the brain; c) measure blood flow into the right atrium; d) monitor blood pressure within the ascending aorta.
6. Which of the following does *not* assist venous pressures to propel blood through veins?: a) valves; b) thoracoabdominal pump; c) muscular pumps; d) vasodilation.
7. Which of the following would *not* be affected immediately by a blockage in the subclavian arteries?: a) left arm; b) right index finger; c) the central nervous system; d) left little toe.
8. The circle of Willis helps ensure continuous blood supply to the: a) aorta; b) liver; c) carotid arteries; d) brain.

DIAGRAMS

9. Cover the labels in Figure 21-10 and identify the blood vessels of the arterial system.
10. Cover the labels in Figure 21-14 and identify the blood vessels of the venous system.

Chapter Review

TRUE/FALSE

(If a statement is false, your instructor may wish to have you correct it.)

11. **T/F:** Blood flows at a steady rate through capillaries.
12. **T/F:** The peripheral distribution of arteries and veins on the left and right sides of the body is identical.
13. **T/F:** Some degree of arteriosclerosis is a normal result of aging.

MATCHING QUESTIONS

(Match each term with the most appropriate statement.)

Statements:

14. The resistance of the arterial system.
15. The resistance of the blood vessels.
16. The point at which hydrostatic and osmotic forces within a capillary are equal, so that there is no net water movement in or out.
17. An acute circulatory crisis marked by hypotension and inadequate peripheral blood flow.
18. The changes in pressure in the thoracic cavity that help to pull blood into the venae cavae and then force it into the right atrium.
19. The control of blood flow at the tissue level.
20. The difference between the systolic and diastolic pressures.
21. Eddies in blood flow that can slow the flow rate and increase resistance.
22. An interconnection formed by arteries that enter a region and fuse.
23. An abnormal accumulation of fluid in peripheral tissues.
24. Specialized fenestrated capillaries that form irregular passageways, leading to relatively slow blood flow.
25. Blood vessel that carries blood from the heart to peripheral areas.
26. Blood vessel that returns blood to the heart.
27. A direct connection between an arteriole and a venule.

Terms:

a) Artery
b) Vein
c) Sinusoids
d) Autoregulation
e) Arterial anastomosis
f) Arteriovenous anastomosis
g) Pulse pressure
h) Peripheral resistance
i) Vascular resistance
j) Turbulence
k) Dynamic center
l) Edema
m) Thoracoabdominal pump
n) Shock

SHORT ANSWER/ESSAY

28. Explain the importance of blood vessels and circulation to homeostasis.
29. You're late for your A & P class, so you sprint from the parking lot to the classroom and arrive just in time. Describe the physiological adjustments your circulatory system just made.
30. An accident victim displays the following symptoms: hypotension; pale, cool, moist skin; confusion and disorientation. Identify her condition and explain why these symptoms occur. If you took her pulse, what would you find?
31. Describe the mechanisms that ensure a steady blood supply to peripheral tissues.
32. Place the following in sequence (largest diameter to smallest): muscular arteries, elastic arteries, capillaries, arterioles.
33. Place the following in sequence (largest diameter to smallest): venules, large veins, medium-sized veins, capillaries.
34. Relate any anatomical differences between arteries and veins to their functions.
35. You've been asked to present a talk on the cardiovascular effects of regular exercise. Summarize the major points you would make.
36. 23-year-old Carla has just given birth to her first child, and has suffered a major hemorrhage. Predict the physiological changes that will occur as her cardiovascular system attempts to restore homeostasis.
37. What is a precapillary sphincter? Explain its significance.
38. What is the hepatic portal system? Explain its importance in maintaining homeostasis.
39. Discuss the mechanisms that regulate cardiovascular activity in order to maintain homeostasis.
40. Fill in the blanks: Two factors, __________ and __________, affect the rate of blood flow.
41. What do venules and arterioles have in common? How are they different?
42. Why is it important for the brain to have a continuous supply of blood? Explain how this supply is maintained.
43. In the following examples, which blood vessel would offer more resistance to blood flow? Why? In which case would you find a greater difference in the vascular resistance?
    a) Two vessels have the same diameter, but one is twice as long as the other.
    b) Two vessels have the same length, but one is only half the diameter of the other.
44. Explain the significance of the formula $F \sim BP/PR$.
45. What are the vasa vasorum, and why are they important?
46. What is the venous reserve? Explain its role in maintaining homeostasis.
47. Where would you expect to find a higher blood pressure: in the aorta or the vena cava? Explain your answer.
48. Discuss the effects of aging on the cardiovascular system.

Identify the first part(s) of the body that would not *receive blood if a blockage occurred in the following:*

49. The pulmonary arteries; pulmonary veins
50. The left brachial vein
51. The right dorsoscapular branch of the right subclavian artery
52. The right common iliac artery
53. The celiac artery

CRITICAL THINKING/APPLICATIONS

54. Two patients have aneurysms. In one, the aneurysm is located in the subclavian vein; in the other, it is in the right carotid artery. Which condition is more likely to be life-threatening? Why?
55. Mr. Thomas, 55, has smoked cigarettes for many years and is 40 percent over normal weight for his height. During a routine physical exam, his blood pressure measures 160/95, and he is told he has atherosclerosis. Define "atherosclerosis." What treatment plan would you recommend?
56. Which have a higher incidence of aneurysms: veins or arteries? Explain why.
57. "Support hose" are tight-fitting elastic stockings or pantyhose (nylon stockings) that are often advertised as beneficial to people who suffer from varicose veins. Explain what causes varicose veins. Do you think this claim is valid? Why or why not?
58. A nurse practitioner tells Mrs. Bottega that her "blood pressure" is 150/90. Explain what these numbers represent, and calculate Mrs. Bottega's MAP. Should she seek treatment?

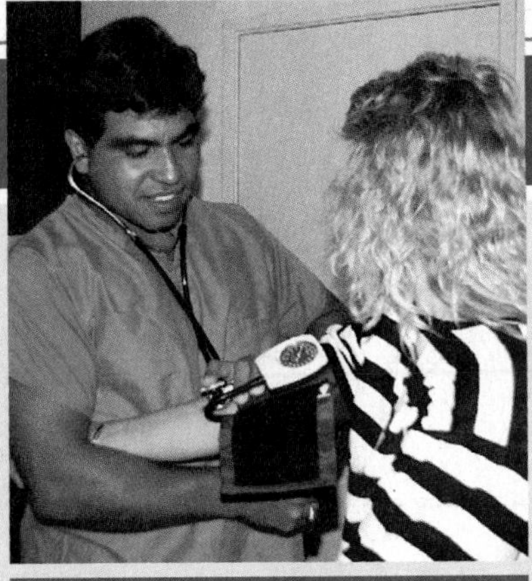

Career Close-up: Cardiovascular Technologist

"My name is Peter Calderon. My mother is from El Salvador; my dad is from Puerto Rico. They got divorced when I was 11. My mother had her hands full raising four kids by herself. There was no money for education. All of my relatives are uneducated; I'm the first one in my family to get even a high school diploma. When my father was young he was a gang member. Someone like me going to college and then doing postgraduate work is really amazing."

"I was lucky because I was involved in sports. The only reason I first went to school was because of football. I was able to get a football scholarship, so I could go to college and be just like the middle-class kids–which was my dream. I was a physical education major first–I thought I wanted to teach. Then I decided I'd rather go into sports medicine."

"Things were going well until my third year at San Diego State University, when I injured my ankle. That ended my football career! But I had always been interested in the body and in fitness, so I decided to stay with it. I always had jobs, I got financial aid, and I finished college."

"Then I did a one-year internship in cardiovascular rehabilitation at a local hospital. I worked with patients who had cardiovascular disease or surgery. I talked with them, learned what they had gone through, educated them about diet and exercise. It was wonderful . . . but I still wanted to do more. So I went back to school. I did postgraduate work at Grossmont College in San Diego, to become a cardiovascular technologist."

"I remember really looking forward to the Anatomy and Physiology class. I got an A in the first semester, and became a teaching assistant. That was a wonderful experience! I was able to show the other students how and why different body parts work. Blood and internal organs really don't bother me, because it's just our bodies . . . just us. In cardiovascular surgery, for example, you can actually watch the body working, and it's tremendously exciting."

"I took another internship—this time in invasive cardiovascular medicine. And then I became a special procedures technologist in the cardiac catheterization laboratory."

"The cardiovascular technologist works directly with the cardiologist in diagnosis and treatment of cardiovascular disease. The goal of diagnostic and therapeutic cardiology is to find a plan of treatment to avoid or delay bypass surgery. Cardiac catheterization is a valuable tool for us. It allows us to make a direct examination of the cardiovascular system by means of a fiber-optic device. The catheter is inserted into the femoral artery and pushed upward until it reaches the heart. Often, this procedure enables us to make a final determination as to whether the patient will need open heart surgery."

In the catheterization laboratory, I'm responsible for measuring hemodynamic pressures and blood saturations of the four chambers of the heart; monitoring the patient's electrocardiogram; manipulating the coronary catheters; and scrubbing in to assist the cardiologist with injections of contrast dye into the coronary arteries. I also set up and test the catheters before use."

"I work as part of a team, with a nurse, X-ray technologist, and another cardiovascular technician outside the room recording pressures and watching the EKG. We're all crosstrained in these areas so we rotate. That way we don't get burned out, and can fill in for one another if necessary."

"Of course, I love working with the patients. They come in very nervous and afraid; they don't know what's going on. I tell them what will be done, and then I ask if they have any questions. These patients are very cooperative. After the procedure, the ladies will often thank me with kisses, and the men will give me hugs and handshakes."

"No matter what I do in my career–and now I am a physician's assistant as well–I utilize what I learned in school and apply it to everyday work situations. I'm going back to school now to become an echocardiogram technologist too."

"I love my work–I love medicine, and I love people. I worked hard to get my life where it is, and I love it."

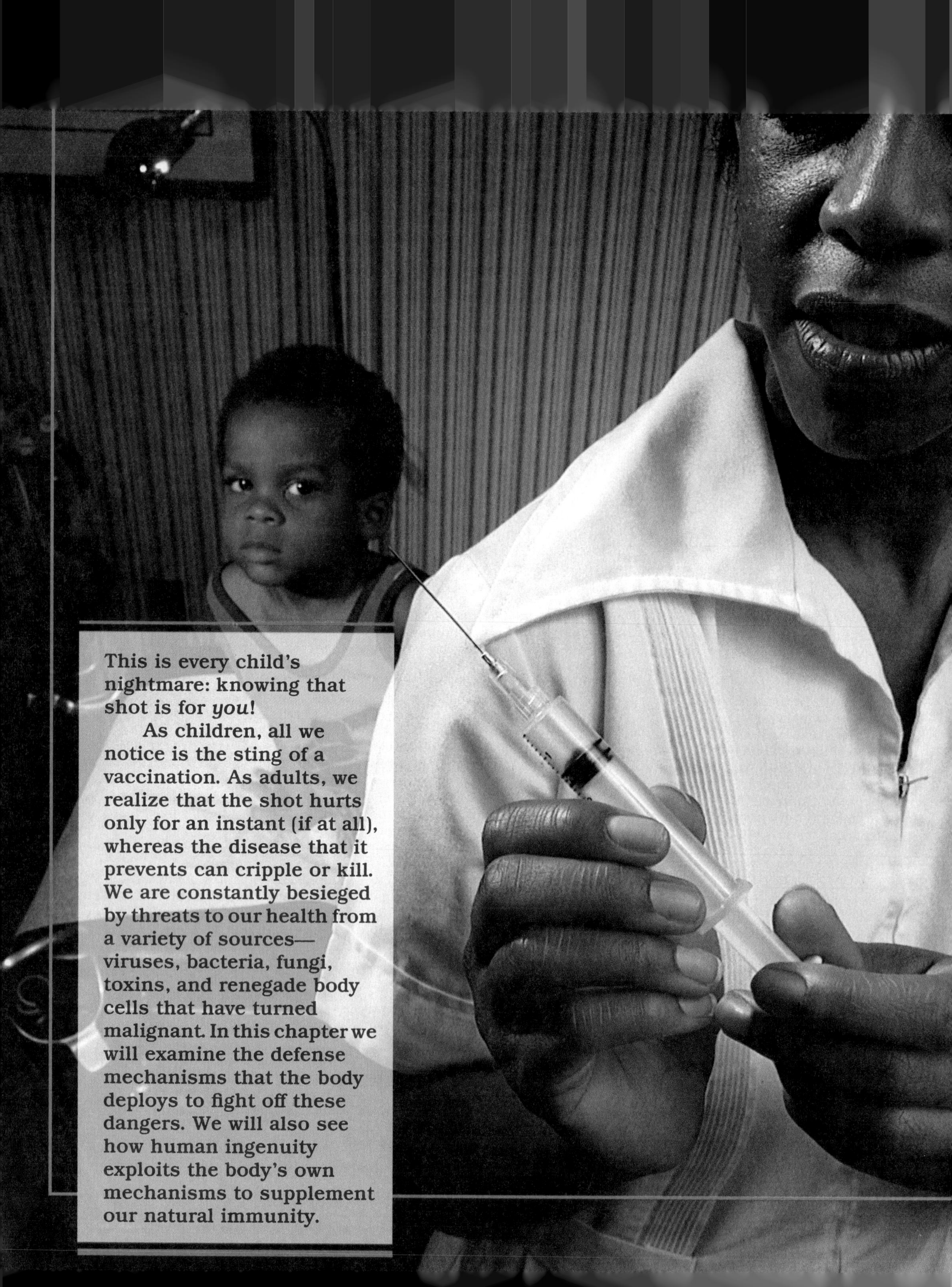

This is every child's nightmare: knowing that shot is for *you*!

As children, all we notice is the sting of a vaccination. As adults, we realize that the shot hurts only for an instant (if at all), whereas the disease that it prevents can cripple or kill. We are constantly besieged by threats to our health from a variety of sources—viruses, bacteria, fungi, toxins, and renegade body cells that have turned malignant. In this chapter we will examine the defense mechanisms that the body deploys to fight off these dangers. We will also see how human ingenuity exploits the body's own mechanisms to supplement our natural immunity.

CHAPTER 22

The Lymphatic System and Immunity

Chapter Objectives

After reading this chapter, you will be able to:

1 Identify the major components of the lymphatic system and explain their functions.
2 Discuss the importance of lymphocytes and describe where they are found in the body.
3 List the body's nonspecific defenses and explain how each functions.
4 Describe the characteristics of the body's specific defenses.
5 Compare the different types of immunity and their origins.
6 Explain how lymphocytes are activated and how they distinguish between foreign cells and cells of the body.
7 Discuss the roles of the various types of lymphocytes in an immune response.
8 Describe the structure of antibody molecules and explain how they function.
9 Relate allergic reactions and autoimmune disorders to immune mechanisms.
10 Specify the factors that can enhance or reduce one's resistance to disease.
11 Describe the changes in the immune system that occur with aging.

■ Introduction

The world is not always kind to the human body. Accidental bumps, cuts, and scrapes, chemical and thermal burns, extreme cold, and ultraviolet radiation are just a few of the hazards in the physical environment. Making matters worse, an assortment of viruses, bacteria, fungi, and parasites thrive in the environment. Many of these organisms are perfectly capable of not only surviving, but thriving inside our bodies—and potentially causing us great harm in the process. Staying alive and healthy involves a massive, combined effort involving many different organs and systems. In this ongoing struggle, the **lymphatic system** plays a central role. Lymphocytes, cells introduced in Chapters 5 and 19, are the dominant cells of the lymphatic system. (pp. 139, 622) They respond to the presence of

- invading organisms, such as bacteria or viruses;
- abnormal body cells, such as virus-infected cells or cancer cells;
- foreign proteins, such as the toxins released by some bacteria.

They attempt to eliminate these threats or render them harmless by a combination of physical and chemical attack.

The body also has other, less specialized defenses. For example, the cutaneous membrane of the skin is an effective physical barrier that both provides protection and prevents bacterial access to underlying tissues.

This chapter begins by examining the organization of the lymphatic system. We will then consider how the lymphatic system interacts with cells and tissues of other systems to defend the body against infection and disease.

Organization of the Lymphatic System

The lymphatic system includes:

1. A network of **lymphatic vessels** that begin in peripheral tissues and end by fusion with the venous system;
2. A fluid, called **lymph**, resembling plasma but containing a much lower concentration of proteins; and
3. **Lymphatic organs** connected to the lymphatic vessels and containing large numbers of lymphocytes.

Figure 22-1 provides a general overview of the components of this system.

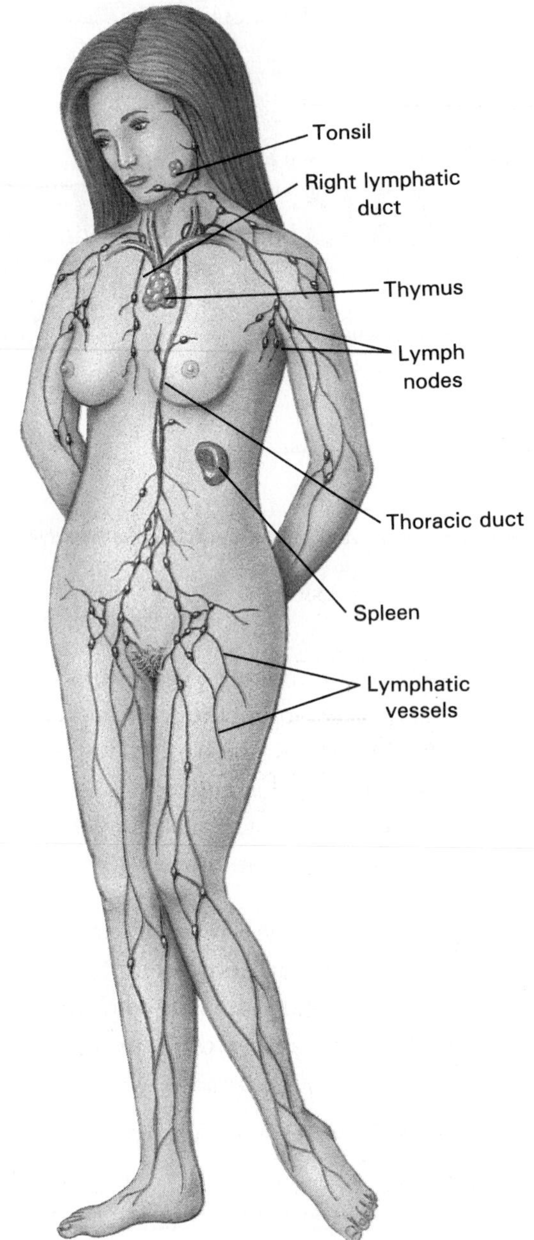

FIGURE 22-1
Components of the Lymphatic System

FUNCTIONS OF THE LYMPHATIC SYSTEM

The primary function of the lymphatic system is *the production, maintenance, and distribution of lymphocytes*. Lymphocytes are essential to the normal defense of the body, and most of this chapter will be devoted to investigating their activities. These cells are produced and stored within lymphatic organs, such as the spleen, thymus, and bone marrow.

Additional functions include *the maintenance of normal blood volume* and *the elimination of local variations in the composition of the interstitial fluid*. Capillaries normally deliver more fluid to peripheral tissues than they carry away (a process discussed in the last chapter). This fluid enters the smallest lymphatic vessels and is eventually returned to the circulatory system. The volume of flow is considerable—roughly 24 liters per day—and a break in a major lymphatic vessel can cause a rapid and potentially fatal decline in blood volume.

There is thus a continual movement of fluid from the capillaries, through the tissues, and back to the bloodstream via the lymphatic vessels. This movement helps to eliminate local differences in composition of tissue fluids and provides an alternative route for the distribution of hormones, nutrients, and waste products. For example, lipids absorbed by the digestive tract often fail to enter the circulation at the capillary level. However, they still reach the bloodstream via passage along lymphatic vessels (a process explored further in Chapter 24).

Additional details concerning each of these functions will be presented in our discussion of the individual system components.

LYMPHATIC VESSELS

Lymphatic vessels, often called **lymphatics**, carry lymph from the peripheral tissues to the venous system. Lymphatics begin as narrow passageways

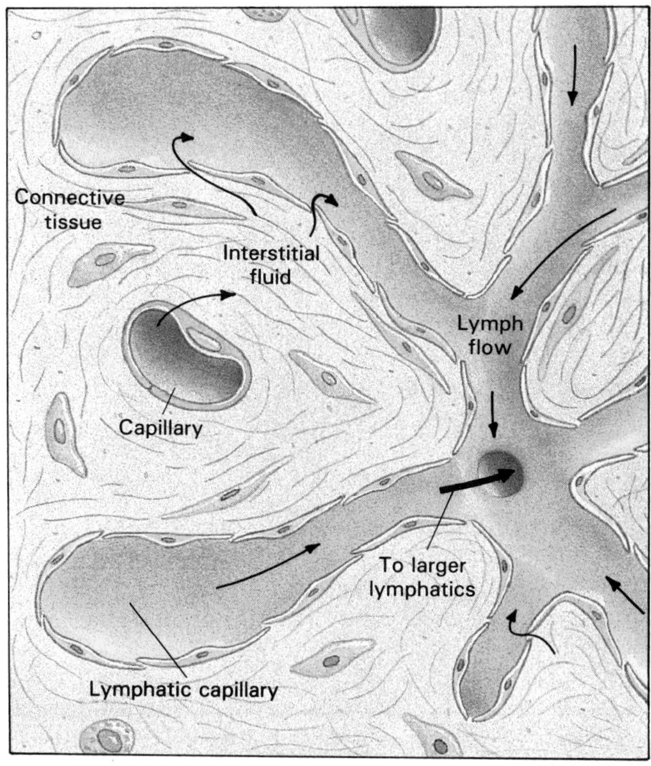

FIGURE 22-2
Lymphatic Capillaries. Lymphatic capillaries are blind pockets in areas of loose connective tissue. Interstitial fluid enters these pockets by passing between adjacent endothelial cells. From the lymphatic capillaries, this fluid, now called lymph, moves into larger lymphatic vessels.

branching through the tissues. These **lymphatic capillaries**, also known as *terminal lymphatics*, are lined by endothelial cells, but there is no underlying basement membrane, and there are no well-organized layers to separate the lymphatic from the surrounding tissue. In sectional view lymphatics are larger in diameter than capillaries, and they often have a flattened or irregular outline. As indicated in Figure 22-2, adjacent endothelial cells are not tightly bound together, but they do overlap. The region of overlap acts as a one-way valve, permitting fluid entry into the lymphatic but preventing its return to the intercellular spaces.

Lymphatic capillaries form a branching network in most tissues, but they are absent in areas not supplied by the circulatory system, such as the cornea of the eye. In addition, there are no lymphatics in the central nervous system.

From the lymphatic capillaries, lymph flows into larger lymphatic vessels that lead toward the trunk. The walls of these lymphatics contain layers comparable to those of veins, and like veins the larger lymphatics contain valves. The valves are quite close together, and at each valve the lymphatic vessel bulges noticeably, giving large lymphatics a beaded appearance. A typical valve can be seen in Figure 22-3a. Pressures within the lymphatic system are minimal, and the

FIGURE 22-3
Lymphatic Vessels and Valves. (a) Lymphatic valves resemble those of the venous system. Each valve consists of a pair of flaps that permit fluid movement in only one direction. (LM, × 51) (b) A diagrammatic view of loose connective tissue, showing small blood vessels and a lymphatic. The cross-sectional view emphasizes the structural differences between them.

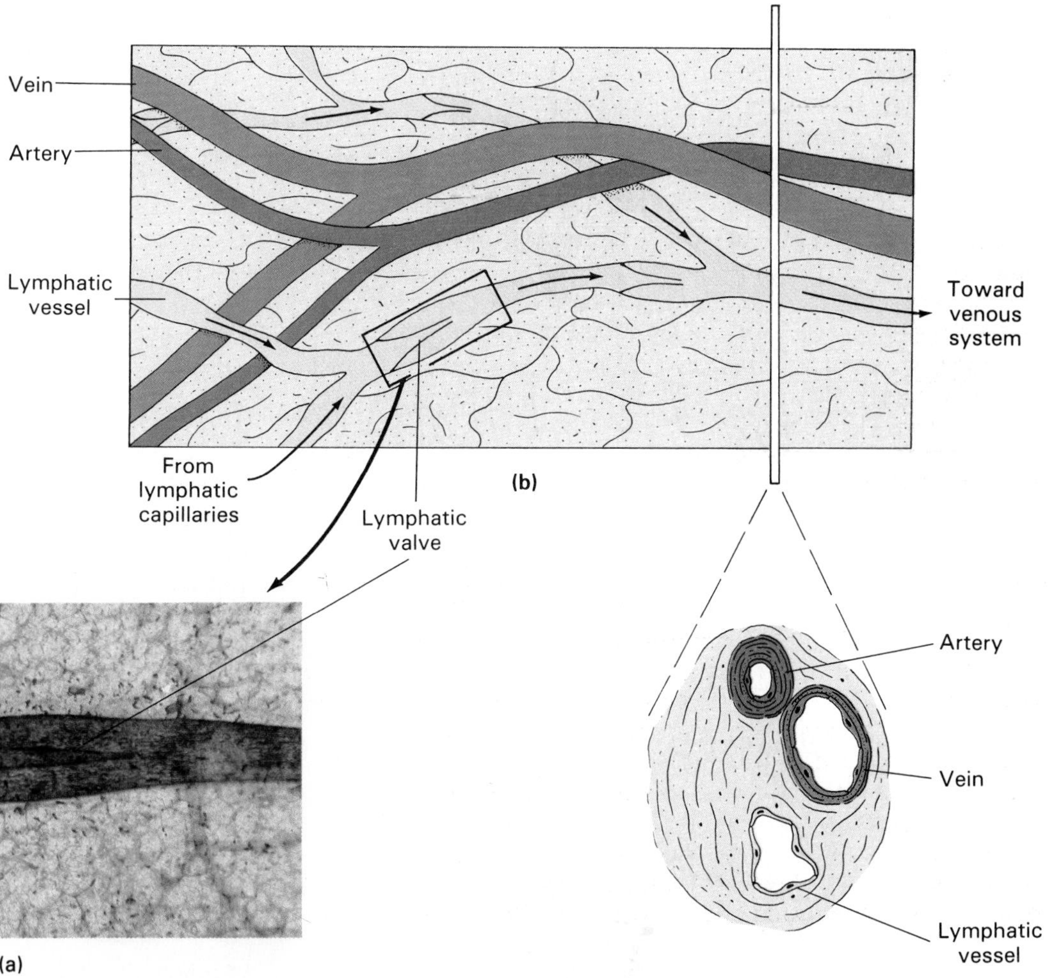

Vein
Artery
Lymphatic vessel
From lymphatic capillaries
Lymphatic valve
Toward venous system
(b)
(a)
Artery
Vein
Lymphatic vessel

valves are essential to maintaining normal lymph flow.

The lymphatic vessels are often found in association with blood vessels, as shown in Figure 22-3b. Note the differences in relative size, general appearance, and branching pattern that distinguish the lymphatics from arteries and veins. There are also characteristic color differences that are apparent on examining living tissues. Arteries are usually a bright red, veins a dark red, and lymphatics a pale golden color.

The lymphatic vessels ultimately empty into two large collecting ducts. The **thoracic duct** begins inferior to the diaphragm, as lymphatics from the lower abdomen, pelvis, and legs converge. These lymphatics deliver lymph to an expanded chamber, the **cisterna chyli**. As the thoracic duct ascends from the cisterna chyli, carrying all of the lymph from the body below the diaphragm, it also collects lymph from the left half of the head, neck, and chest. This large lymphatic duct empties into the venous system near the junction between the left internal jugular vein and the left subclavian vein. The smaller **right lymphatic duct**, which ends at the comparable location on the right side, delivers lymph from the right side of the body above the diaphragm. The distribution of the larger lymphatics and collecting ducts can be seen in Figure 22-4. [**CM**: *Lymphedema*]

LYMPHOCYTES

Lymphocytes were introduced in Chapter 19 because they account for roughly 25 percent of the circulating white blood cell population. (p. 622) However, circulating lymphocytes are only a small fraction of the total lymphocyte population. The body contains around 10^{12} lymphocytes, with a combined weight of over a kilogram.

Types of Lymphocytes

There are three different classes of lymphocytes in the blood: **T cells** (**t**hymus-dependent), **B cells** (**b**one marrow-derived), and **NK cells** (**n**atural **k**iller). Each type has distinctive biochemical and functional characteristics.

T Cells Approximately 80 percent of circulating lymphocytes are classified as T cells. **Cytotoxic T cells** attack foreign cells or body cells infected by viruses. Their attack often involves direct contact. These lymphocytes are responsible for providing **cellular immunity**. **Helper T cells** and **suppressor T cells** assist

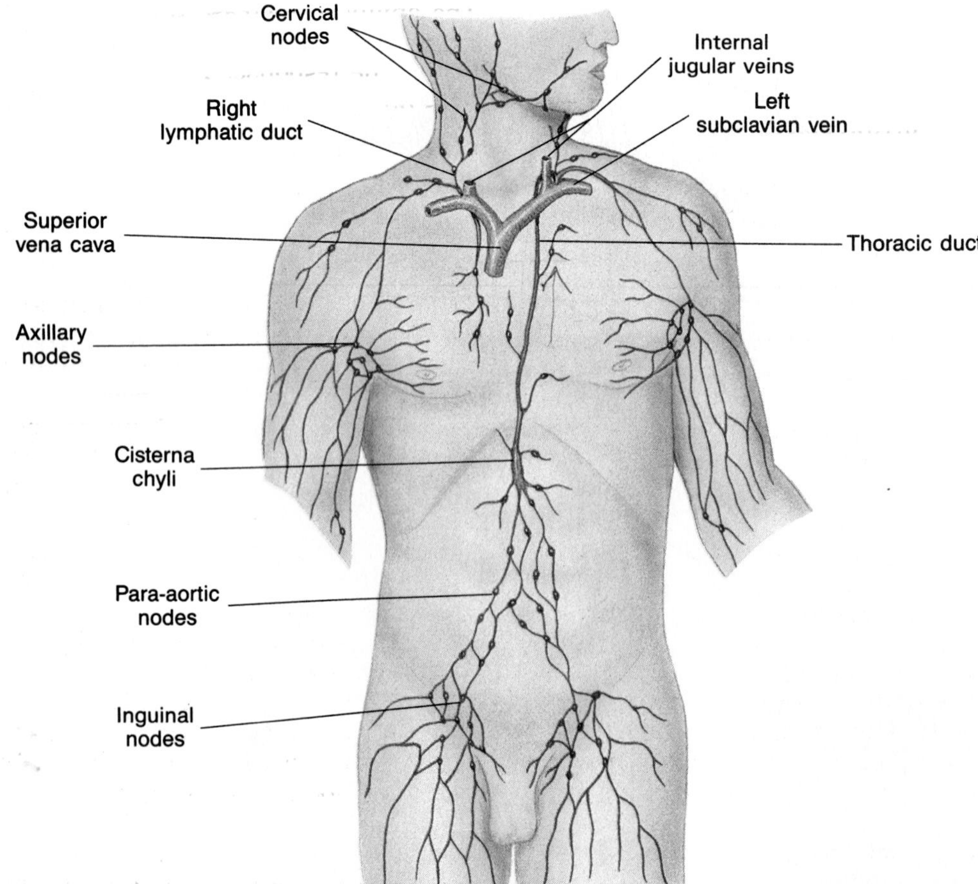

FIGURE 22-4
Lymphatic Ducts. The thoracic duct carries lymph originating in tissues inferior to the diaphragm and from the left side of the upper body. The right lymphatic duct drains the right half of the body superior to the diaphragm.

in the regulation and coordination of the immune response; for this reason they are also called **regulatory T cells**.

B Cells B cells account for 10–15 percent of circulating lymphocytes. When stimulated, B cells can differentiate into **plasma cells**. Plasma cells, introduced in Chapter 5, are responsible for the production and secretion of *antibodies*. (p. 139) These proteins react with specific chemical targets, called **antigens**. Antigens are usually pathogens, parts or products of pathogens, or other foreign compounds. Typical antigens are proteins, but some lipids, polysaccharides, and nucleic acids can also stimulate antibody production. When an antigen-antibody complex forms, it starts a chain of events leading to the destruction of the target compound or organism. Antibodies in body fluids are known as **immunoglobulins**. Because the blood is the primary distribution route for immunoglobulins, B cells are said to be responsible for **humoral** ("liquid") **immunity**.

NK Cells The remaining 5–10 percent of circulating lymphocytes are NK cells, also known as **large granular lymphocytes**. These lymphocytes will attack foreign cells, normal cells infected with viruses, and cancer cells that appear in normal tissues. Their continual policing of peripheral tissues has been termed **immunological surveillance**.

Lifespan and Circulation of Lymphocytes

Lymphocytes in the blood, bone marrow, spleen, thymus, and peripheral lymphatic tissues are not evenly distributed. The ratio of B cells to T cells varies depending on the tissue or organ considered. For example, B cells are seldom found in the thymus, and in the blood T cells outnumber B cells by a ratio of 8:1. However, this ratio changes to 1:1 in the spleen and 1:3 in the bone marrow.

The lymphocytes within these organs are visitors, not residents. Lymphocytes move throughout the body; they wander through a tissue and then enter a blood vessel or lymphatic for transport to another site. T cells, which must have close contact with their targets, move relatively quickly. For example, a wandering T cell may spend about 30 minutes in the blood, 5–6 hours in the spleen, and 15–20 hours in a lymph node. B cells, which are responsible for antibody production, move more slowly; a typical B cell spends around 30 hours in a lymph node before moving to another location.

In general, lymphocytes have relatively long lifespans. Roughly 80 percent survive for 4 years, and some last 20 years or more. Throughout life, normal lymphocyte populations are maintained through lymphopoiesis in the bone marrow and lymphatic tissues.

Lymphopoiesis

Chapter 19 discussed *hemopoiesis*, the formation of the cellular elements of blood. (pp. 627–630) Erythropoiesis (red blood cell formation) is normally confined to the bone marrow, but **lymphopoiesis** (lymphocyte production) involves the bone marrow, thymus, and peripheral lymphatic tissues. The relationships are diagrammed in Figure 22-5.

Hemocytoblasts in the bone marrow produce lymphocytic stem cells with two distinct fates. One group remains in the bone marrow and generates NK cells and B cells that migrate into peripheral tissues. The B cells take up residence in lymph nodes, the spleen, and other lymphatic tissues. The second group of stem cells migrates to the thymus. Under the influence of the hormone **thymosin**, and perhaps other thymic secretions, these cells divide repeatedly, producing T cells that subsequently migrate to the spleen, other lymphatic organs, and the bone marrow.

The bone marrow and thymus are essential for the differentiation of normal lymphocyte populations. After differentiation has been completed, each B cell and T cell gains the ability to respond to the presence of a specific antigen. NK cells will respond to the presence of abnormal cells.

As lymphocytes migrate through peripheral tissues, they retain the ability to divide. These divisions produce daughter cells of the same type; for example, a dividing B cell produces other B cells, not T cells or NK cells. The ability to increase the number of lymphocytes of a specific type is important to the success of the immune response. The mechanisms that regulate this proliferation will be considered in a later section.

LYMPHATIC TISSUES

Lymphatic tissues are connective tissues dominated by lymphocytes. In a **lymphatic nodule** the lymphocytes are densely packed in an area of loose connective tissue (Figure 22-6). Typical nodules average around a millimeter in diameter, but the boundaries are not distinct because no fibrous capsule surrounds them. They often have a pale, central zone, called a **germinal center**; the germinal center in Figure 22-6a appears relatively dark because the lymphocytes have been stained with a deep blue dye. Lymphocytes in the germinal center are actively dividing.

Lymphatic nodules are found beneath the epithelia lining the respiratory, digestive, and urinary tracts. Large nodules in the walls of the pharynx are called **tonsils**. There are usually five tonsils: a single **pharyngeal**, or **adenoid**, **tonsil**, a pair of **palatine tonsils**, and a pair of **lingual tonsils**. Their relative positions are indicated in Figure 22-6b. Clusters of lymphatic nodules beneath the epithelial lining of the intestine are known as **Peyer's patches**; Figure

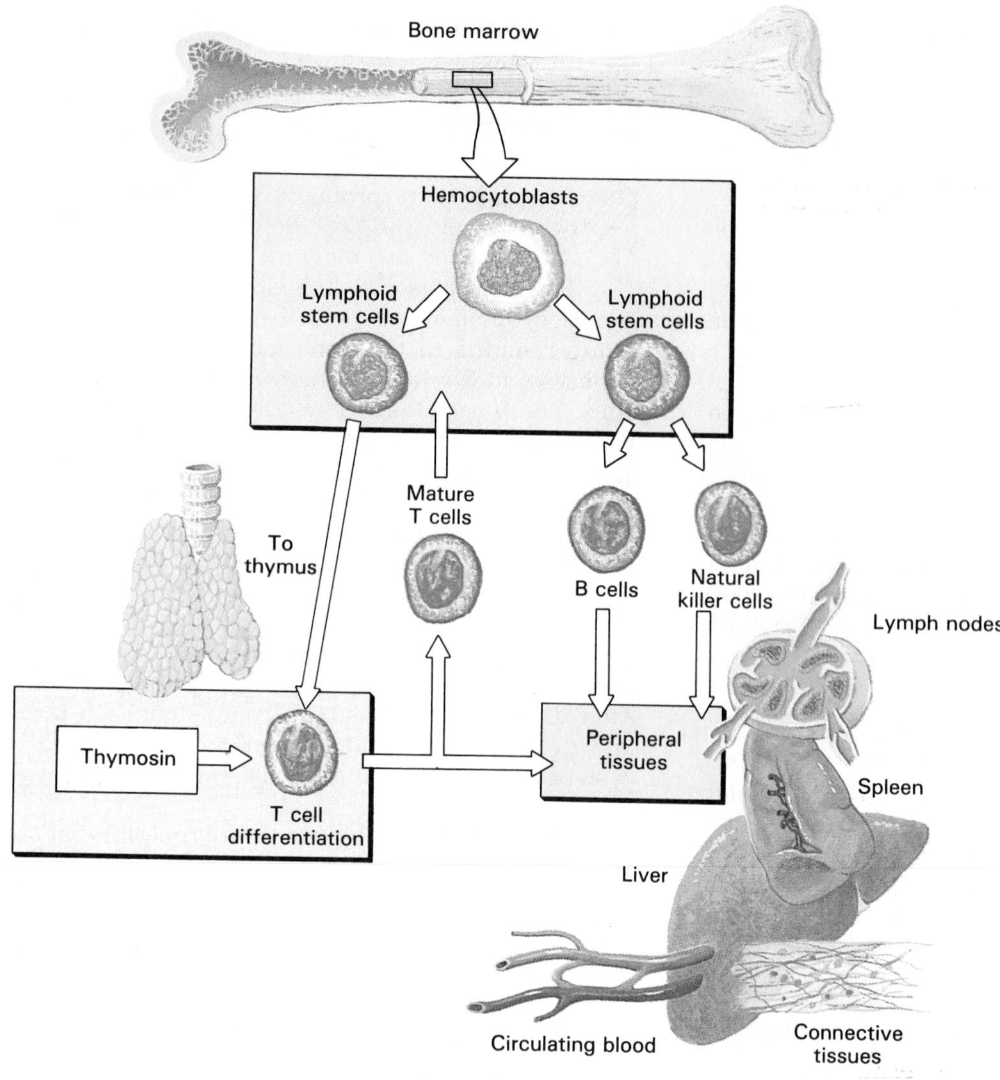

FIGURE 22-5
Derivation and Distribution of Lymphocytes. Hemocytoblast divisions produce lymphocytic stem cells with two different fates. One group remains in the bone marrow, producing daughter cells that mature into B cells and NK cells. The second group migrates to the thymus, where subsequent divisions produce daughter cells that mature into T cells. All three lymphocyte types circulate throughout the body in the bloodstream, leaving the circulation to take temporary residence in peripheral tissues.

FIGURE 22-6
Lymphatic Nodules. (a) Appearance of a typical nodule as seen with the light microscope. Note the relatively pale (pinkish) germinal center, where lymphocyte cell divisions occur. (LM, ×51) (b) The positions of the tonsils.

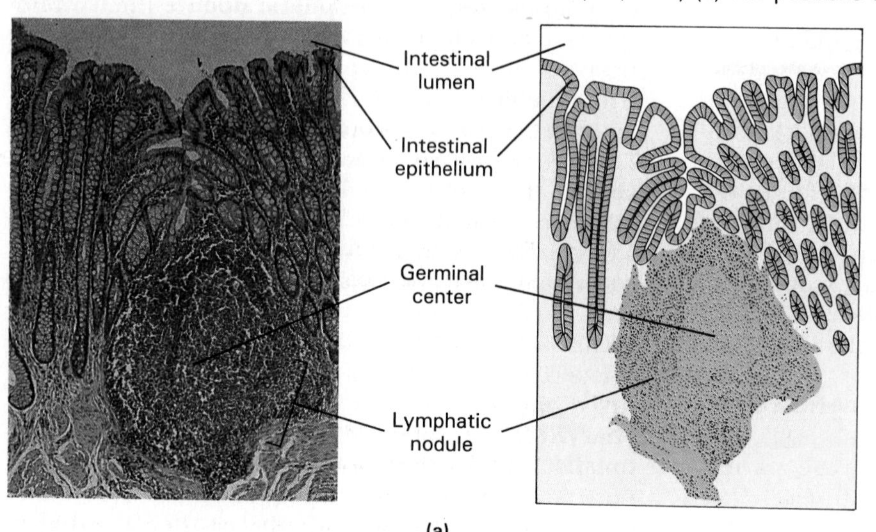

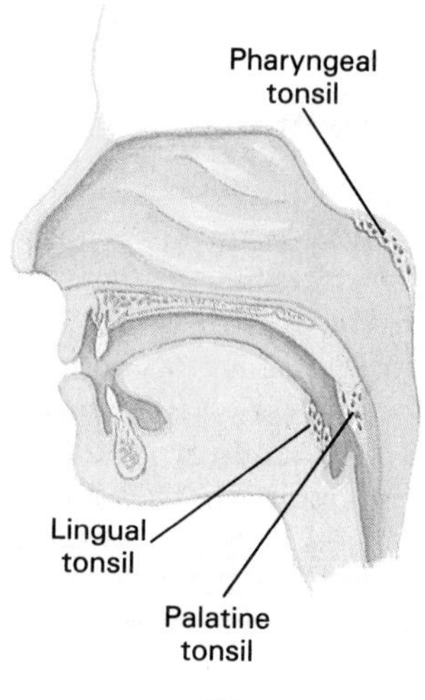

22-6a is an example. In addition, the walls of the appendix, a blind pouch that originates near the junction between the small and large intestines, contain a mass of fused lymphatic nodules. The extensive array of lymphatic tissue associated with the digestive system has been called **gut-associated lymphatic tissue (GALT)**. [**CM**: *Infected Lymphatic Nodules*]

LYMPHATIC ORGANS

Lymphatic organs have a clear internal organization, and they are separated from surrounding tissues by a fibrous capsule. Important lymphatic organs include the **lymph nodes**, the **thymus**, and the **spleen**.

Lymph Nodes

Lymph nodes are small, oval lymphatic organs ranging in diameter from 1 to 25 mm (up to around 1 in.). The entire organ is covered by a dense fibrous capsule, and the shape of a typical lymph node resembles that of a lima bean. At the indentation, or **hilus**, blood vessels, nerves, and lymphatic vessels penetrate the capsule to reach the interior of the lymph node. Inward extensions of the capsule, called **trabeculae** (tra-BEK-ū-lē), form a series of incomplete partitions, and a network of reticular fibers provides a stable three-dimensional organization, diagrammed in Figure 22-7. Lymph enters on the side opposite the hilus. It then flows through the node, departing in the lymphatics at the hilus.

Lymph traveling through a node passes through the outer **cortex** and the inner **medulla**. In the cortex the lymphatic tissues are organized into masses resembling lymphatic nodules, each with a pale-staining germinal center. The entire cortical zone is dominated by T cells. The medulla of the lymph node contains B cells arranged into elongate masses known as **medullary cords**. Removal or destruction of the thymus will produce a reduction in the size of the cortex, at least in part due to the elimination of thymosin production. The medullary cords are unaffected by the loss of the thymus, because B cells are not dependent on thymic secretions.

Within a lymph node lymph flows through a network of sinuses, broad channels with incomplete walls. Fixed macrophages in the walls of the sinuses engulf debris or pathogens in the lymph as it flows past. The lymph moves in one direction only, because of the presence of valves in the lymphatics connected to the node.

Lymph nodes monitor the contents of the lymph as it proceeds toward the venous system. Small clusters of lymph nodes are usually encountered near articulations. The largest nodes are found where peripheral lymphatics connect with the trunk, in regions such as the groin, the axillae (armpits), and at the base of the neck. These nodes are often called *lymph glands*, and "swollen glands" usually indicate inflammation or infection of peripheral structures. Dense collections of lymph nodes also exist within the mesenteries of the gut, near the trachea and passageways leading to the lungs, and in association with the thoracic duct. The overall pattern of node distribution can be seen in Figures 22-1 and 22-4.

A minor injury often produces a slight enlargement of the nodes along the lymphatics draining the region. The enlargement usually results from an increase in the number of lymphocytes and phagocytes in the node, in response to a minor, localized infection. Chronic or excessive enlargement of lymph nodes constitutes **lymphadenopathy** (lim-fad-e-NOP-a-thē). This may occur in response to bacterial or viral infections, endocrine disorders, or cancer.

Lymphatics are found in all portions of the body except the central nervous system, and the lymphatic capillaries offer little resistance to the passage of cancer cells. As a result, metastasizing cancer cells often spread along the lymphatics. Under these circumstances the lymph nodes serve as way stations for migrating cancer cells. Thus an analysis of lymph nodes can provide information on the spread of the cancer cells, and such information has a direct influence on the selection of appropriate therapies. One example, the staging of breast cancer by the degree of nodal involvement, will be detailed in Chapter 29. *Lymphomas*, one group of cancers originating within the lymphatic system, are discussed in the Clinical Manual. [**CM**: *Lymphomas*]

Lymphatic tissues and organs are distributed in areas particularly susceptible to injury or invasion. If we wanted to protect a house against intrusion

FIGURE 22-7
Structure of a Lymph Node

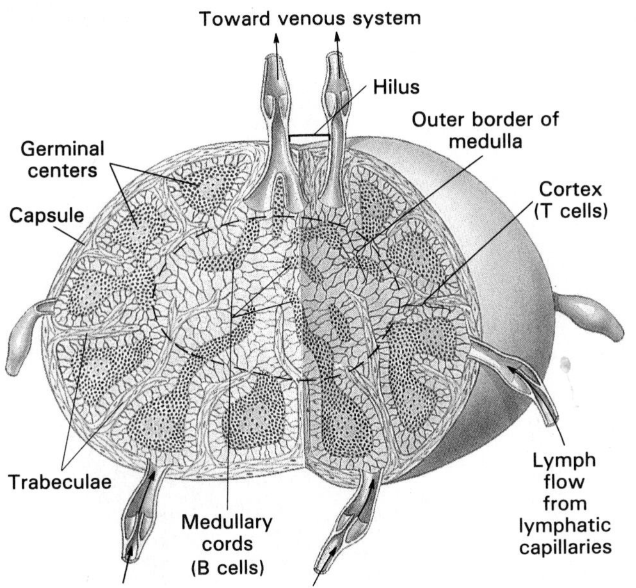

we might guard all entrances and exits, place traps by the windows, and perhaps even keep a pit bull indoors. The distribution of lymphatic tissues and lymph nodes follows this pattern. The cervical, axillary, and inguinal (groin) lymph nodes prevent entry from the head, neck, arms, and legs. The tonsils and respiratory nodules and nodes guard the entrances to the digestive and respiratory tracts. The abdominal lymph nodes monitor lymph arriving from the reproductive and urinary tracts, and the lymphatic tissues and lymph nodes of the intestines and mesenteries check the lymph originating at the digestive tract.

The Thymus

The thymus lies behind the sternum, in the anterior mediastinum. It has a pinkish coloration and a grainy consistency. The thymus reaches its greatest size (relative to body size) in the first year or two after birth and its maximum absolute size during puberty, when it weighs between 30 and 40 g. Thereafter the thymus gradually decreases in size and becomes increasingly fibrous, a process called **involution**.

Figure 22-8 presents the appearance and organization of the thymus. The capsule that covers the thymus divides it into two **thymic lobes** (Figure 22-8a,b). Fibrous partitions, or **septae**, from the capsule divide the lobes into **lobules** averaging 2 mm in width (Figure 22-8b,c). Each lobule consists of a densely packed outer **cortex** and a paler central **me-**

FIGURE 22-8
The Thymus. (a) Appearance and position of the thymus on gross dissection; note its relationship to other organs in the chest. (b) Anatomical landmarks on the thymus. (c) A low-power light micrograph of the thymus. Note the fibrous septae that divide the thymic tissue into lobes resembling interconnected lymphatic nodules. (d) At higher magnification the unusual structure of Hassall's corpuscles can be examined. The small cells in view are lymphocytes in various stages of development. (LM, ×336)

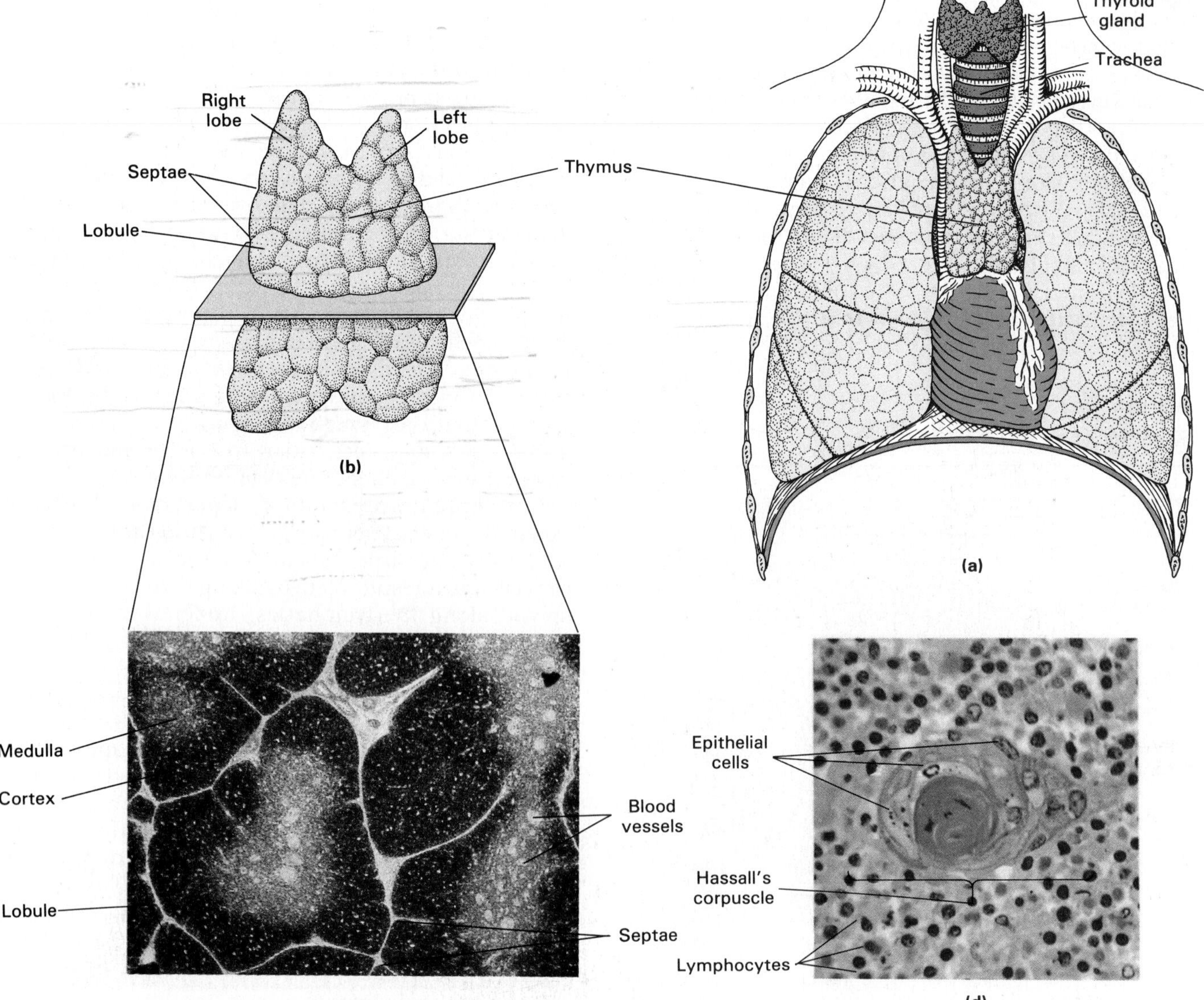

dulla. Lymphocytes in the cortex undergo mitosis, and as the T cells mature they migrate into the medulla, eventually entering one of the specialized blood vessels in that region.

Scattered among the lymphocytes are pale **epithelial cells** responsible for the production of thymic hormones. In the medullary region the epithelial cells cluster together in concentric layers, forming distinctive structures known as **Hassall's corpuscles** (Figure 22-8d). Despite their imposing appearance, the function of Hassall's corpuscles remains unknown.

The Spleen

Location and Structure of the Spleen The adult spleen contains the largest collection of lymphatic tissue in the body. It has a length of around 12 cm (5 in.) and an average weight approaching 160 g. The spleen lies along the curving lateral border of the stomach, extending between the ninth and eleventh ribs on the left side. Its shape primarily reflects its association with the structures around it. As indicated in Figure 22-9a, it sits wedged between the

FIGURE 22-9
The Spleen. (a) The shape of the spleen roughly conforms to those of adjacent organs. This transverse section through the trunk shows the typical position of the spleen within the abdominopelvic cavity. (b) External appearance of the intact spleen, showing major anatomical landmarks. This view should be compared with that of (a). (c) Histological appearance of the spleen. Areas of white pulp are dominated by lymphocytes; they appear blue because the nuclei of lymphocytes stain very darkly. Areas of red pulp contain a preponderance of red blood cells. (LM, ×41)

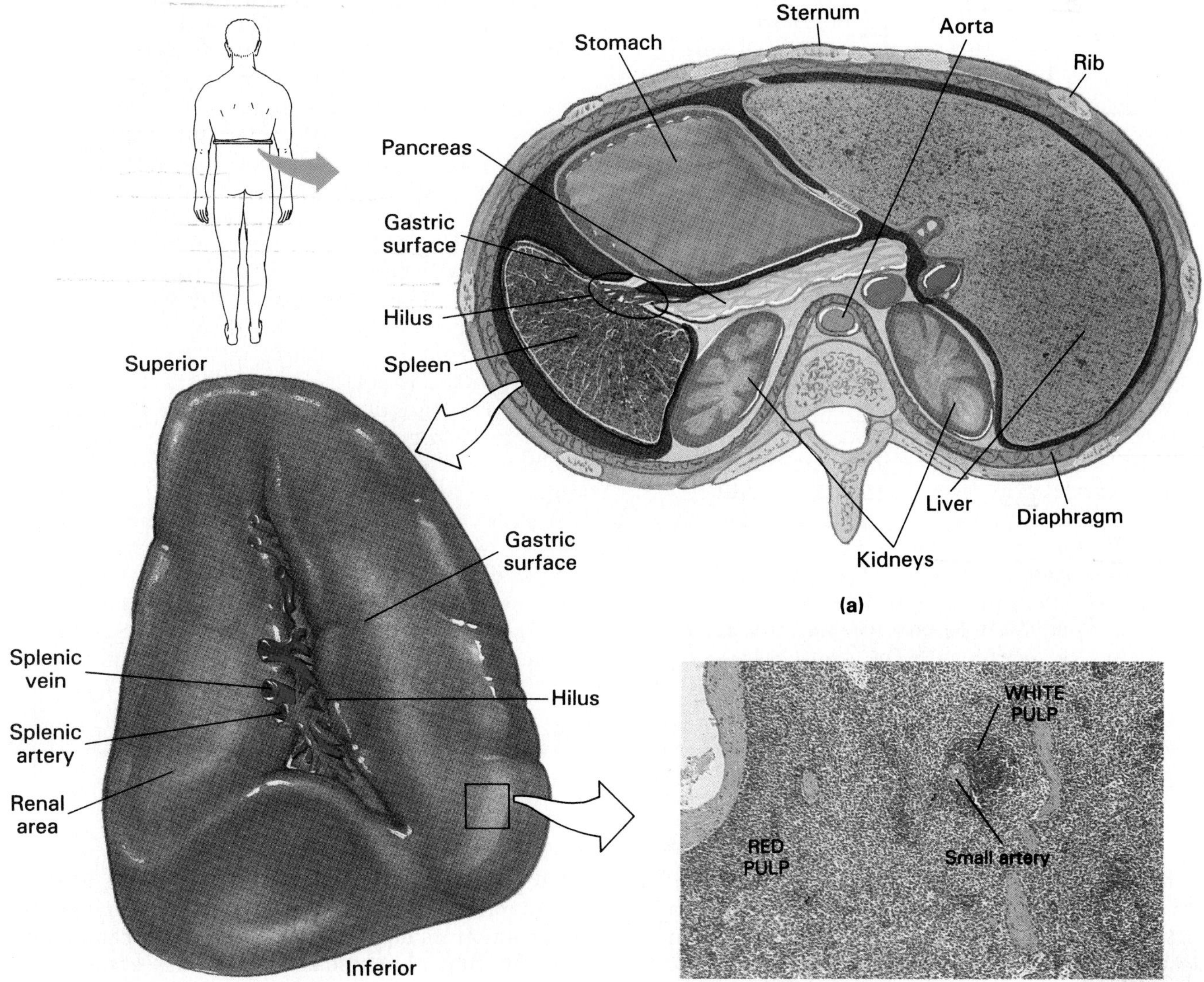

Clinical Comment: The Nature of the Enemy

Figure 3-2 presented the structure of a "typical" cell. (p. 61) The cellular organization depicted in that figure and described in Chapter 3 is that of **eukaryotic cells** (Ū-kar-yot-ic; *eu*, true + *karyon*, nucleus). The defining characteristic of these cells is the presence of a nucleus. All eukaryotic cells have similar membranes, organelles, and methods of cell division. All multicellular animals and plants (as well as some single-celled organisms) are composed of eukaryotic cells.

The eukaryotic plan of organization is not the only one found in the living world, however. There are organisms that do not consist of eukaryotic cells. These organisms are of great interest to us because they include most of the pathogens that can cause human diseases.

Bacteria are **prokaryotic cells**. These organisms do not have nuclei or any other membranous organelles. They do not have a cytoskeleton, and their cell membranes are surrounded by a semirigid cell wall made of carbohydrate and protein. Figure 22-10a shows the structure of a representative bacterium.

Bacteria are usually less than 2 μm in diameter. Many bacteria are quite harmless, and many more—including some that live within our bodies—are actually beneficial to us in a variety of ways. Other bacteria are dangerous pathogens that will destroy body tissues if given the opportunity. These bacteria absorb nutrients from their surroundings, and they release enzymes that damage cells and tissues. A few pathogenic bacteria also release toxic chemicals. Bacterial infections are responsible for many serious diseases, including tetanus, cholera, rheumatic fever, scarlet fever, pneumonia, meningitis, syphilis, gonorrhea, typhus, plague, tuberculosis, typhoid fever, and leprosy. These and other bacterial infections are considered in various chapters throughout the text.

Another type of pathogen conforms neither to the prokaryotic nor the eukaryotic organizational plan. These tiny pathogens, called **viruses**, are not cells at all. When free in the environment they do not show any of the characteristics of living organisms. They are classified as **infectious agents** because they can enter cells (either prokaryotic or eukaryotic) and replicate themselves.

Viruses consist of a core of nucleic acid (DNA or RNA) surrounded by a protein coat. (Some varieties have a membranous outer covering as well.) When a virus infects a cell, its nucleic acid enters the nucleus of the host and takes over the cell's metabolic machinery. These viral genes instruct the host cell to synthesize many copies of the virus's nucleic acid and proteins. These components then assemble themselves into new viruses that either bud out from the cell or cause it to rupture. Even if the host cell is not destroyed outright by these events, normal cell function is usually disrupted. Some viruses, however, can lie dormant within infected cells for long periods of time before initiating this process of replication.

The structure of a representative virus is shown in Figure 22-10b. Important viral diseases include influenza (flu), yellow fever, some leukemias, AIDS, hepatitis, polio, measles, mumps, rabies, and the common cold.

Bacteria and viruses are the best known human pathogens, but there are eukaryotic pathogens as well. Examples of the most important types are included in Figure 22-10c. **Protozoa** are unicellular eukaryotic organisms that are abundant in soil and water. They are responsible for a variety of serious human diseases, including *amoebic dysentery* and *malaria* (Chapter 19). (p. 614) **Fungi** (singular *fungus*) are eukaryotic organisms that absorb organic materials from the remains of dead cells. Mushrooms are familiar examples of very large fungi. In a fungal infection, a microscopic fungus spreads through living tissues, killing cells and absorbing nutrients. Several relatively common skin conditions (athlete's foot) and a few more serious diseases (histo

stomach, the left kidney, and the muscular diaphragm. The diaphragmatic surface is smooth and convex, but the visceral surface contains indentations that record the shapes of the stomach and kidney. The spleen has a soft consistency, and the shape changes as the stomach fills and empties. In dogs, cats, and many other mammals, the spleen contains large amounts of smooth muscle, and it can actively change shape. In these animals splenic contractions eject large quantities of blood into the circulation,

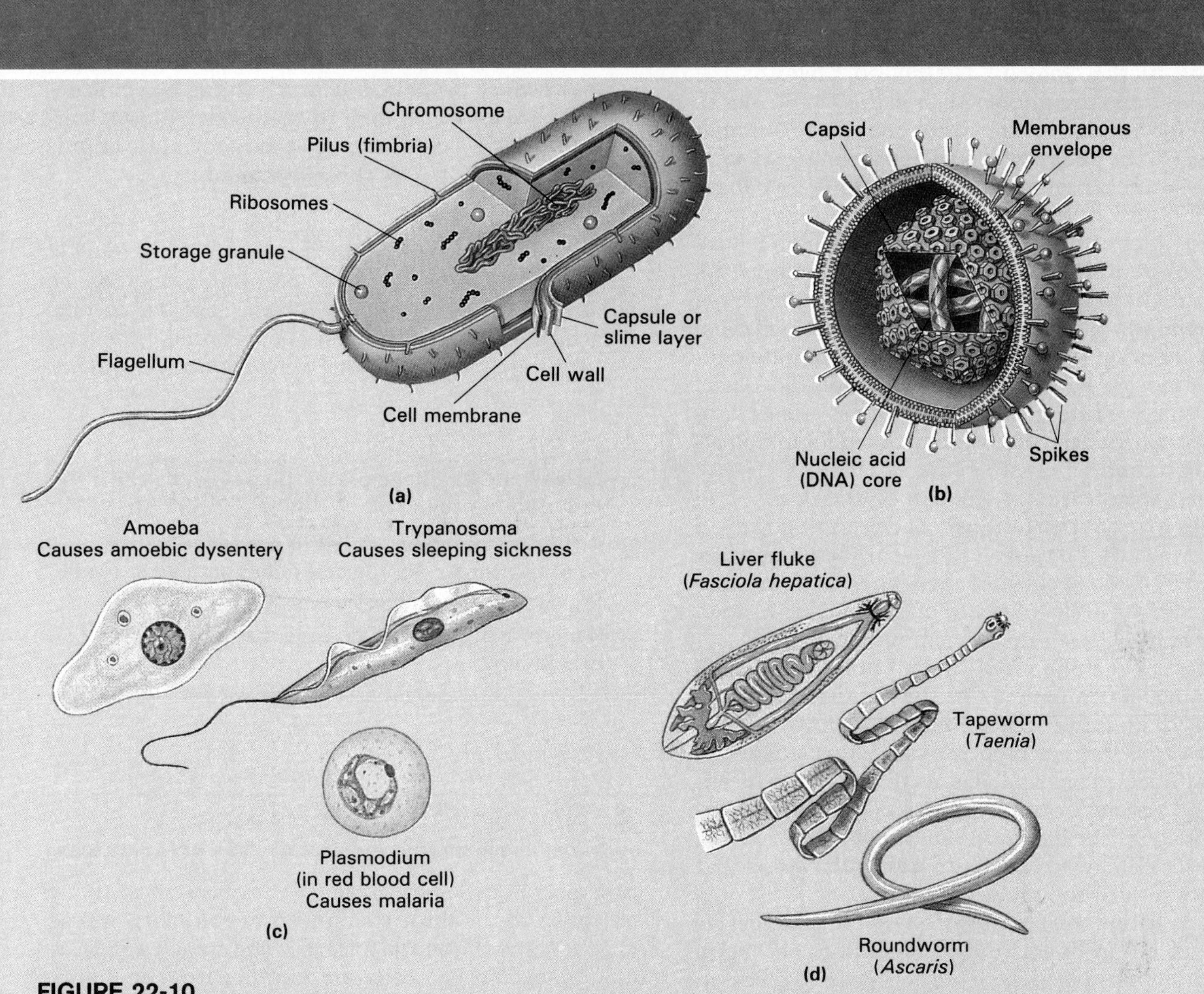

FIGURE 22-10
Pathogenic Organisms. (a) A bacterium, with prokaryotic characteristics indicated. Compare with Figure 3-3 (a typical eukaryotic cell). (b) A typical virus. Each virus has an inner chamber containing nucleic acid, surrounded by a protein *capsid* or by an inner capsid and an outer membranous *envelope*. The herpes virus is an enveloped DNA virus; it causes chickenpox, shingles, and herpes. (c) Protozoan pathogens. Protozoa are eukaryotic, single-celled organisms, common in soil and water. (d) Multicellular parasites. Several different groups of organisms are human parasites; many have complex life histories.

plasmosis) are the result of fungal infections.

Larger multicellular organisms (Figure 22-10d), generally referred to as *parasites*, can also invade the human body and cause diseases. These organisms, which range from microscopic flatworms to tapeworms a meter or more in length, usually cause weakness and discomfort, but do not *by themselves* kill their host. However, complications resulting from the parasitic infection, such as chronic bleeding or secondary infections by bacterial or viral pathogens, can ultimately prove fatal.

and this can be important in coping with a sudden blood loss. The human spleen is not muscular, and it neither contracts nor stores large quantities of blood.

The entire spleen is surrounded by a fibrous capsule. Splenic blood vessels and lymphatics communicate with the spleen on its visceral surface. The **splenic artery**, **splenic vein**, and the lymphatics draining the spleen pass through the **hilus**, a groove marking the border between the gastric and renal depressions. On gross dissection the spleen has a deep red color due to the blood it contains.

Function of the Spleen The spleen performs the same functions for the blood that the lymph nodes perform for lymph. Splenic functions can be summarized as: (1) the phagocytosis of abnormal blood components and (2) the initiation of immune responses by B cells and T cells. The correlations between structure and function are very apparent in the histological organization of the spleen. The cellular components constitute the **pulp** of the spleen (Figure 22-9c). Areas of **red pulp** contain large quantities of red blood cells, whereas areas of **white pulp** resemble lymphatic nodules. (Although concentrations of lymphatic tissue have a pasty white color seen in gross dissection, these areas turn deep blue when stained for histological examination.)

The splenic artery enters at the hilus and branches to produce a number of arteries that radiate outward toward the capsule. These **trabecular arteries** branch extensively, and their finer branches are surrounded by areas of white pulp. Capillaries then discharge the blood into the red pulp.

The cell population of the red pulp includes all of the normal components of the circulating blood, plus fixed and free macrophages. The structural framework of the red pulp consists of a network of reticular fibers. The blood passes through this meshwork and enters large sinusoids, also lined by fixed macrophages. The sinusoids empty into small veins, and these ultimately collect into **trabecular veins** that continue toward the hilus.

This circulatory pattern gives the phagocytes of the spleen an opportunity to identify and engulf any elderly, damaged, or infected cells present in the circulating blood. Lymphocytes are scattered throughout the red pulp, and the blood percolates around the areas of white pulp. Thus any microorganisms or abnormal plasma components will come to the attention of the lymphatic system as well. [**CM**: *Disorders of the Spleen*]

THE LYMPHATIC SYSTEM AND BODY DEFENSES

The human body has multiple defense mechanisms, but these can be sorted into two general categories:

1. **Nonspecific defenses** do not discriminate between one threat and another. These defenses, which are present at birth, include *physical barriers, phagocytic cells, immunological surveillance, interferon, complement, inflammation,* and *fever*. They provide the body with a defensive capability known as *nonspecific resistance* or *nonspecific immunity*.
2. **Specific defenses** provide protection against threats on an individual basis. For example, a specific defense may protect against infection by one type of bacteria, but ignore other bacteria and viruses. Many specific defenses develop after birth, as a result of accidental or deliberate exposure to environmental hazards. Specific defenses are dependent upon the activities of lymphocytes. Together, they produce a state of protection known as **specific immunity**.

Nonspecific and specific defenses are complementary, and both must function normally to provide adequate resistance to infection and disease. As we shall see in the following sections, all of the specific defense mechanisms and most of the nonspecific ones are intimately bound up with the lymphatic system.

How would blockage of the thoracic duct affect the circulation of lymph?

If the thymus gland failed to produce thymosin, what particular population of lymphocytes would be affected?

Why do lymph nodes enlarge during some infections?

■ Nonspecific Defenses

Nonspecific defenses prevent the approach of, deny entrance to, or limit the spread of microorganisms or other environmental hazards. The major categories of nonspecific defenses are summarized in Figure 22-11.

PHYSICAL BARRIERS

To cause trouble, a foreign compound or pathogen must enter the body tissues, and that means crossing an epithelium. The epithelial covering of the skin, described in Chapter 6, has multiple layers, a keratin coating, and a network of desmosomes that lock adjacent cells together. (p. 186) These create a very effective barrier that protects underlying tissues. Even along the more delicate internal passageways of the respiratory or digestive tracts, the epithelial cells are tied together by tight junctions and supported by a dense and fibrous basement membrane. In addition, most epithelia are protected by specialized accessory structures and secretions.

The exterior surface of the body has several layers of defense. The hairs found in most areas provide some protection against mechanical abrasion (especially on the scalp), and they often prevent hazardous materials or insects from contacting the skin's sur-

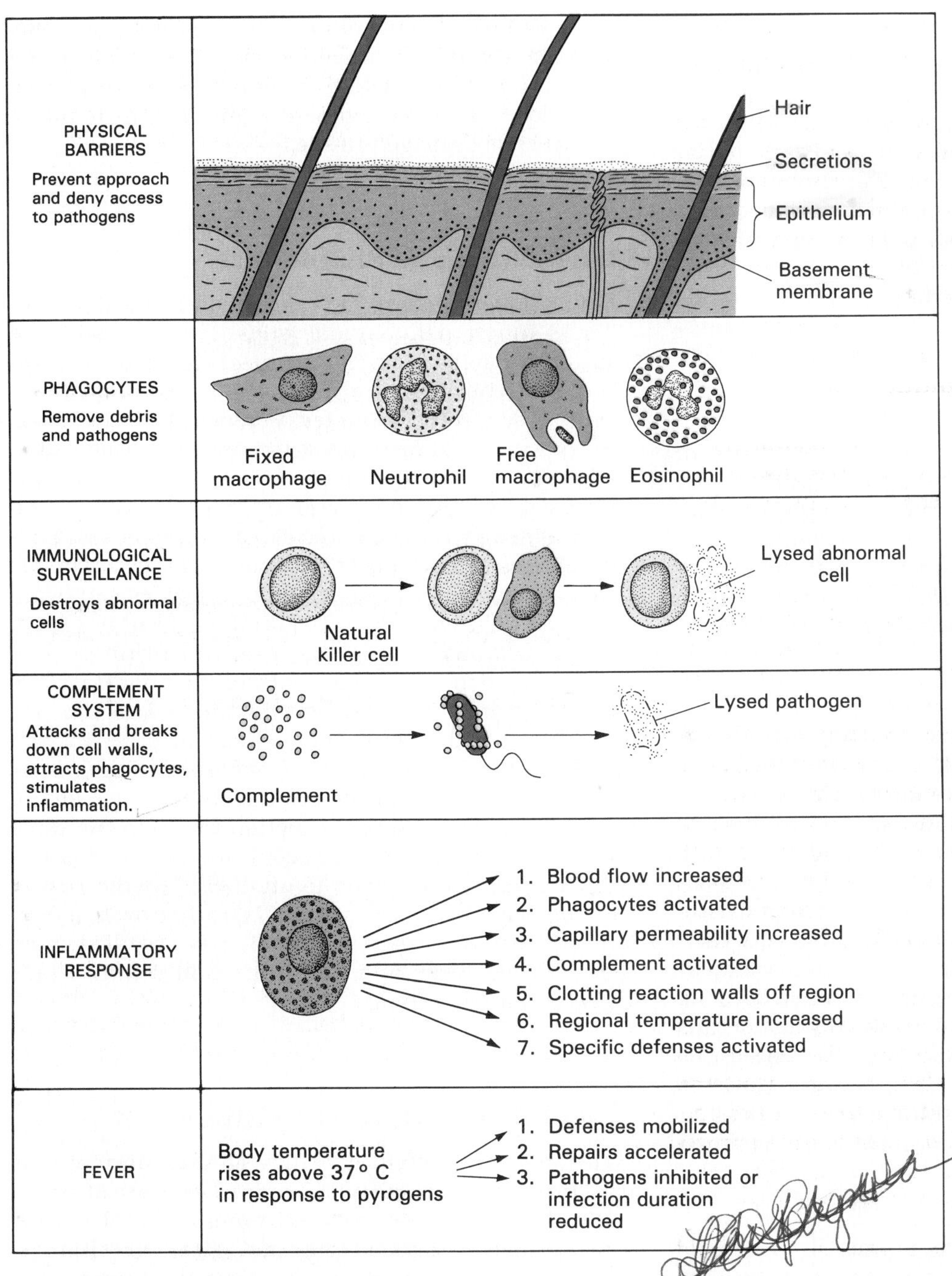

PHYSICAL BARRIERS
Prevent approach and deny access to pathogens
Hair
Secretions
Epithelium
Basement membrane
PHAGOCYTES
Remove debris and pathogens
Fixed macrophage
Neutrophil
Free macrophage
Eosinophil
IMMUNOLOGICAL SURVEILLANCE
Destroys abnormal cells
Natural killer cell
Lysed abnormal cell
COMPLEMENT SYSTEM
Attacks and breaks down cell walls, attracts phagocytes, stimulates inflammation.
Complement
Lysed pathogen
INFLAMMATORY RESPONSE
1. Blood flow increased
2. Phagocytes activated
3. Capillary permeability increased
4. Complement activated
5. Clotting reaction walls off region
6. Regional temperature increased
7. Specific defenses activated
FEVER
Body temperature rises above 37° C in response to pyrogens
1. Defenses mobilized
2. Repairs accelerated
3. Pathogens inhibited or infection duration reduced

FIGURE 22-11
Nonspecific Defenses

face. The epidermal surface also receives the secretions of sebaceous and sweat glands. These secretions flush the surface, washing away microorganisms and chemical agents. The secretions also contain bactericidal chemicals, destructive enzymes (lysozymes), and antibodies.

The epithelia lining the digestive, respiratory, urinary, and reproductive tracts are more delicate, but they are equally well defended. Mucus bathes most surfaces of the digestive tract, and the stomach contains a powerful acid that can destroy many potential pathogens. Mucus moves across the lining of the respiratory tract, urine flushes the urinary passageways, and glandular secretions do the same for the reproductive tract. Special enzymes, antibodies, and (most often) an acidic pH add to the effectiveness of these secretions.

PHAGOCYTIC CELLS

Phagocytes perform janitorial and police services in peripheral tissues, removing cellular debris and responding to invasion by foreign compounds or pathogenic organisms. These cells represent the "first line" of cellular defense against pathogenic invasion, and phagocytes often attack and remove the microorganisms before cells of the lymphatic system become aware of the incident. Two general classes of phagocytic cells are found in the human body: **microphages** and **macrophages**.

Microphages

Microphages are the neutrophils and eosinophils normally found in the circulating blood. These phagocytic cells leave the bloodstream and enter peripheral tissues subjected to injury or infection. As noted in Chapter 19, neutrophils are abundant, mobile, and quick to phagocytize cellular debris or invading bacteria. (pp. 619–620) Eosinophils, less abundant, target foreign compounds or pathogens that have been coated with antibodies.

Macrophages

The body also contains several different types of macrophages—large, actively phagocytic cells derived from the monocytes of the blood. Although there are no purely phagocytic organs or tissues, almost every tissue in the body shelters resident or visiting macrophages. This relatively diffuse collection of phagocytic cells has been called the **monocyte-macrophage system**.

Fixed Macrophages **Fixed macrophages** are permanent residents of specific tissues and organs. These cells are normally incapable of movement, so the objects of their phagocytic attention must diffuse or otherwise move through the surrounding tissue until they are within range. Fixed macrophages are scattered among connective tissues, usually in close association with collagen or reticular fibers, and their presence has been noted in the papillary and reticular layers of the dermis, in the pia-arachnoid layer of the meninges, and in bone marrow. In some organs the fixed macrophages have special names: **microglia** are macrophages inside the CNS, **Kupffer cells** are found in and around the liver sinusoids, and **Langerhans cells** reside within the epithelia of the skin and digestive tract.

Free Macrophages **Free macrophages**, or *mobile macrophages*, travel throughout the body, arriving at an injury site by migration through adjacent tissues or by recruitment from the circulating blood. When confronted with unusually large but highly attractive objects, several free macrophages may fuse their cell membranes to become **phagocytic** (or *foreign body*) **giant cells**. In the skeletal system, the fusion of macrophages produces the osteoclasts that dissolve and recycle the mineral content of bone. Some tissues contain free macrophages with distinctive characteristics. For example, the exchange surfaces of the lungs are patrolled by **alveolar macrophages**, also known as *phagocytic dust cells*.

The boundaries between fixed and free macrophages are not etched in stone. Both classes are derived from the monocytes of the blood; the primary difference is that fixed macrophages are always found within a given tissue, but the free macrophages come and go. During an infection this distinction often breaks down completely, for the fixed macrophages may lose their attachments and begin roaming around the damaged tissue.

Orientation and Phagocytosis

Mobile macrophages and microphages share a number of functional characteristics. They can all move through capillary walls by squeezing between adjacent endothelial cells, a process known as **diapedesis**. They may also be attracted or repelled by chemicals in the surrounding fluids, a phenomenon called **chemotaxis**. They are particularly sensitive to chemicals released by other body cells or by pathogens. When stimulated by these chemicals, macrophages and microphages head toward their source. If they are inside the bloodstream, they respond by diapedesis, moving into the adjacent interstitial fluid. Last but not least, they are all very active phagocytes. During phagocytosis, detailed in Figure 3-27, extracellular materials are packaged into a vesicle and digested when the vesicle fuses with one or more lysosomes. (p. 87) The residue is subsequently discharged as the vesicle wall becomes part of the cell membrane once again.

All phagocytic cells function in much the same way, although the items selected for phagocytosis may differ from one cell type to another. The lifespan of an actively phagocytic cell can be rather brief; a neutrophil usually expires before it has engulfed more than 25 bacteria, and in an infection it may attack that many in an hour.

IMMUNOLOGICAL SURVEILLANCE

The immune system generally ignores normal cells in the body's tissues, but abnormal cells are attacked and destroyed. The constant monitoring of normal tissues has been called **immunological surveillance**. This surveillance primarily involves the lymphocytes known as NK (natural killer) cells. These cells are sensitive to the presence of abnormal cell membranes; they are involved in the destruction of cancer cells and cells infected with viruses. The recognition mechanism differs from that involved in either cellular or humoral immunity. T cells and B cells provide defenses against specific threat, and their activation requires a relatively complex and time-consuming sequence of events. In contrast, NK cells respond immediately on encountering abnormal cells.

Upon contact with such a cell, the NK cell reacts in a predictable way. First, the Golgi apparatus moves around the nucleus until the maturing face points directly toward the abnormal cell. This process might be compared to the rotation of a tank turret to point the cannon toward the enemy. A flood of secretory

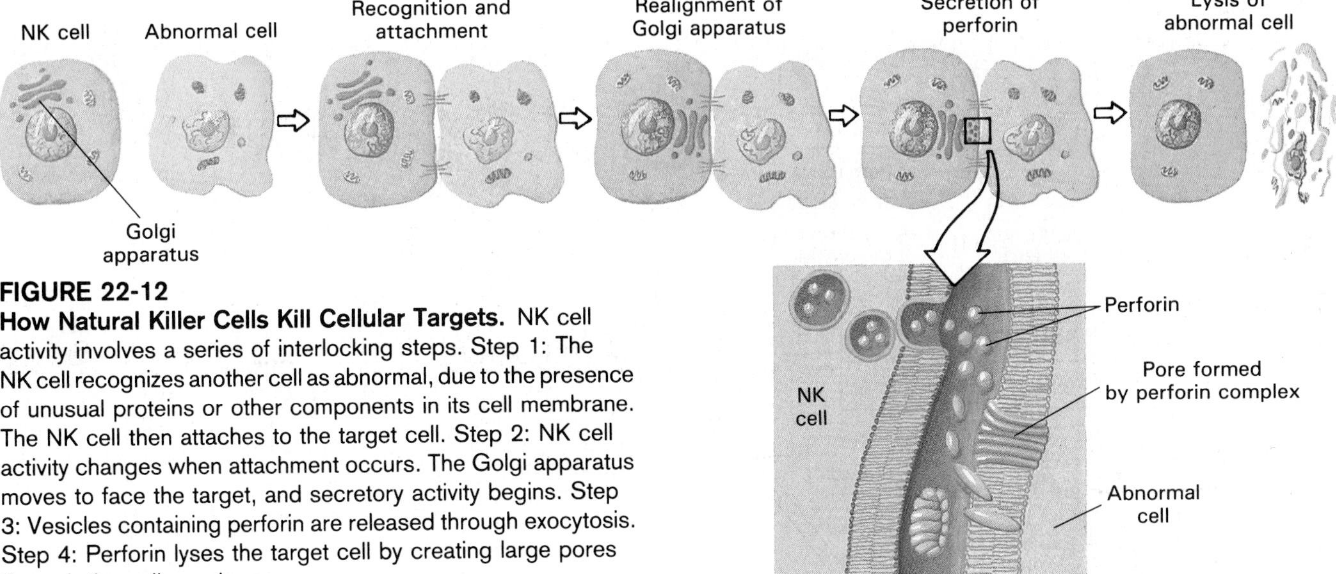

FIGURE 22-12
How Natural Killer Cells Kill Cellular Targets. NK cell activity involves a series of interlocking steps. Step 1: The NK cell recognizes another cell as abnormal, due to the presence of unusual proteins or other components in its cell membrane. The NK cell then attaches to the target cell. Step 2: NK cell activity changes when attachment occurs. The Golgi apparatus moves to face the target, and secretory activity begins. Step 3: Vesicles containing perforin are released through exocytosis. Step 4: Perforin lyses the target cell by creating large pores through the cell membrane.

vesicles then travel through the cytoplasm, to be released at the cell surface through exocytosis. These secretory vesicles contain proteins called **perforins**. Upon their release, perforins diffuse across the gap between the NK cell and its target. On reaching the opposing cell membrane, perforin molecules interact with one another and with the membrane.

The result, shown in Figure 22-12, is the creation of a network of pores in the cell membrane. These pores are large enough to permit the free passage of ions, proteins, and other intracellular materials. The target cell can no longer maintain internal homeostasis, and it quickly disintegrates. It is not clear why perforin does not affect the membrane of the NK cell itself; it has been proposed that the NK cell membranes contain a second protein, called *protectin*, that binds and inactivates perforin.

Cancer cells probably appear throughout life, but their cell membranes usually contain unusual proteins, called **tumor-specific antigens**, that NK cells recognize as abnormal. The affected cells are then destroyed, preserving tissue integrity. Unfortunately, some cancer cells avoid detection, perhaps because they lack tumor-specific antigens or these antigens are covered in some way. This process of avoiding detection is called **immunological escape**. Once immunological escape has occurred, cancer cells can multiply and spread without interference by NK cells.

NK cells are also important in fighting viral infections. Viruses reproduce inside living cells, beyond the reach of lymphocytes. However, infected cells incorporate viral antigens into their cell membranes. NK cells recognize these infected cells as abnormal. By destroying virally infected cells, NK cells can slow or prevent the spread of a viral infection.

INTERFERONS

Interferons are small proteins released by cells infected with viruses. When an interferon molecule reaches the membrane of a normal cell, it binds to surface receptors and, via second messengers, triggers the production of **antiviral proteins** in the cytoplasm. Antiviral proteins do not interfere with the entry of viruses, but they interfere with viral replication inside the cell. In addition to their role in slowing the spread of viral infections, interferons stimulate the activities of macrophages and NK cells.

There are at least three different interferons: α-interferon, produced by several different leukocytes; β-interferon, secreted by fibroblasts; and γ-interferon, secreted by T cells and NK cells. Interferons are examples of **cytokines** (SĪ-tō-kīns), chemical messengers released by tissue cells to coordinate local activities. In effect, cytokines are the hormones of the immune system; they are released to alter the activities of cells and tissues throughout the body. Other cytokines include *monokines*, released by cells of the monocyte-macrophage system, and *lymphokines*, produced by lymphocytes. Monokines and lymphokines are essential to the coordination and regulation of specific defenses, and further details concerning their function will be discussed below.

COMPLEMENT

The plasma contains 11 special **complement proteins** that form the **complement system**. These proteins interact with one another in chain reactions comparable to those of the clotting system. The basic sequence of events is summarized in Figure 22-13. There are two routes of complement activation, the *classic pathway* and the *alternative pathway*.

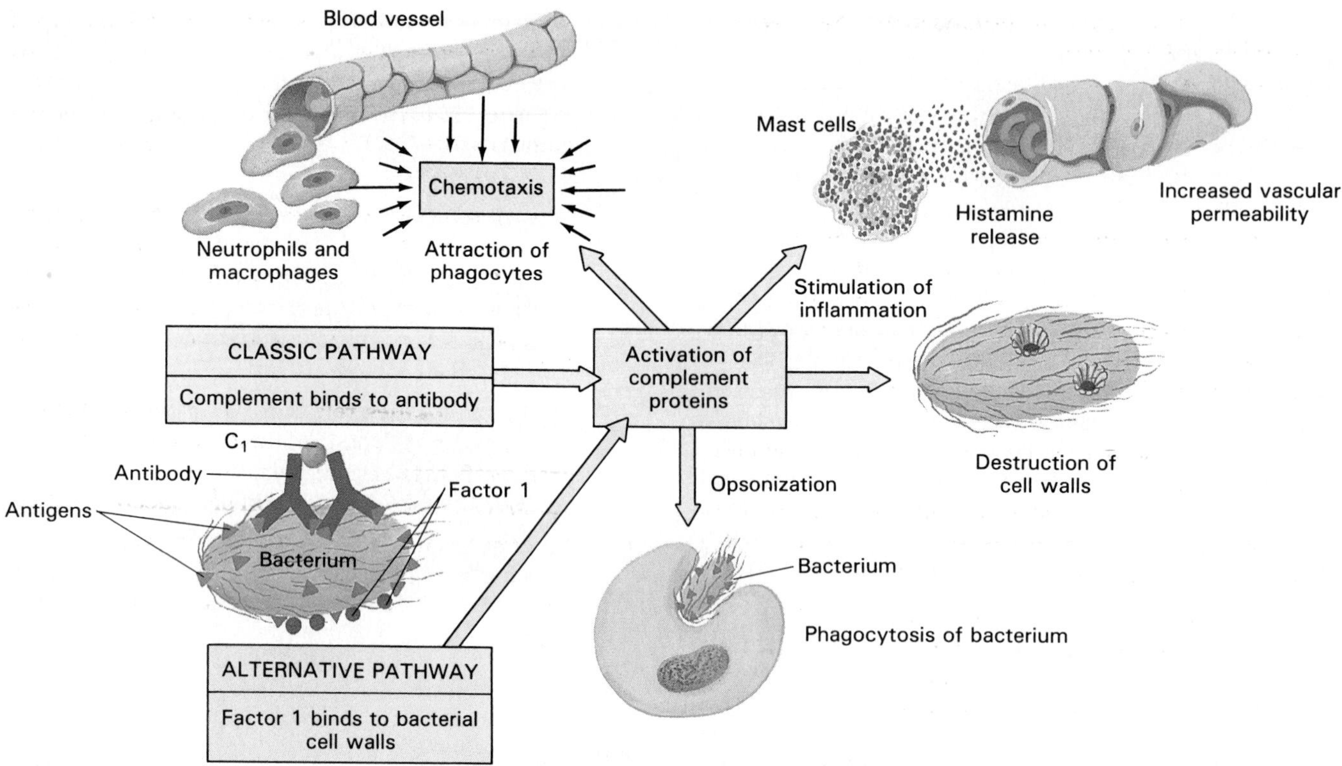

FIGURE 22-13
Complement Activation. Complement activation can be initiated by the classic pathway (complement binding to an antibody molecule) or the alternative pathway (complement binding to bacterial cell walls). Either one triggers a chain reaction between complement proteins in the blood. Complement interactions (1) attract phagocytes, (2) stimulate phagocytic activity, (3) promote inflammation, and (4) puncture cell membranes by creating pores comparable to those produced by perforin.

The most rapid and effective activation of the complement system occurs via the **classic pathway**. The process begins when one of the complement proteins (C_1) binds to an antibody molecule already attached to its specific antigen. (Antigen-antibody interactions will be described in a later section.) The bound complement protein then acts as an enzyme, catalyzing a series of reactions between other complement proteins. Known effects of complement activation include:

1. *Destruction of target cell membranes*: Five of the interacting complement proteins bind to the cell membrane, creating pores comparable to those produced by exposure to perforin. These pores have the same effect, and the target cell (prokaryotic or eukaryotic) is soon destroyed.
2. *Stimulation of inflammation*: Activated complement proteins enhance the release of histamine by mast cells and basophils. These substances increase the degree of local inflammation and accelerate circulation to the region.
3. *Attraction of phagocytes*: Activated complement proteins attract neutrophils and macrophages to the area, improving the chances that phagocytic cells will be able to cope with the injury or infection.
4. *Enhancement of phagocytosis*: A coating of complement proteins and antibodies both attracts phagocytes and apparently makes them easier to engulf. Because the antibodies involved are called **opsonins**, this effect is called **opsonization**.

A less effective, slower activation of the complement system occurs in the absence of antibody molecules. This **alternative pathway** starts when one of the complement proteins (*factor I*) binds to complex carbohydrates found in bacterial cell walls but not in the membranes of eukaryotic cells. This response can be sustained and enhanced by a complement factor called **properdin**.

INFLAMMATION

After crossing an epithelium and its basement membrane, a hazardous agent usually enters an area of loose connective tissue. Loose connective tissue, described in Chapter 5, contains a mixture of cell types, including both fixed and free macrophages. (p.

140) The appearance of foreign proteins or pathogens in this tissue triggers a complex process called *inflammation*, or the *inflammatory response*. Basic features of the inflammatory response were described in Chapter 5. (p. 159) Figure 22-14 contains a more detailed flow chart for the events that occur during inflammation.

Mast cells play a pivotal role in this process. When stimulated by mechanical stress or chemical changes in the local environment, these cells release histamine, serotonin, and heparin into the interstitial fluid. This release initiates a coordinated nonspecific response to connective tissue injury.

1. The histamine released increases capillary permeability and accelerates blood flow through the area. The increased circulation brings more cellular defenders to the site and carries away toxins and debris.
2. Clotting factors and complement proteins leave the bloodstream and enter the injured or infected area. Clotting does not occur at the actual site of injury because of the presence of heparin. However, a clot soon forms around the area, and this helps isolate the region and slow the spread of the chemical or pathogen. Meanwhile, complement activation breaks down cell walls and attracts phagocytes.
3. Debris or bacteria are attacked by fixed and free macrophages and by neutrophils drawn to the area by chemotaxis. Active macrophages secrete cytokines that attract other macrophages and microphages to the site. Neutrophils still within the circulation attack pathogens or toxins entering the bloodstream, and eosinophils may become involved if the foreign materials are coated with antibodies.
4. The increased blood flow elevates the local temperature, increasing the rate of enzymatic reactions and accelerating the activity of phagocytes. The rise in temperatures may also denature foreign proteins or vital enzymes of invading microorganisms.
5. Other monokines released by active phagocytes stimulate fibroblasts to begin barricading the area with scar tissue, reinforcing the clot and further slowing the invasion of adjacent tissues.
6. Foreign proteins, toxins, microorganisms, and active phagocytes in the area of inflammation stimulate the lymphatic system and activate the body's specific defenses.

FEVER

A **fever** is the maintenance of a body temperature greater than 37.2° C (99° F). Chapter 15 noted the presence of a temperature-regulating center in the preoptic area of the hypothalamus. This region acts as the body's thermostat. (The thermostatic control of body temperature was discussed in Chapter 1, and

FIGURE 22-14
The Inflammation Process

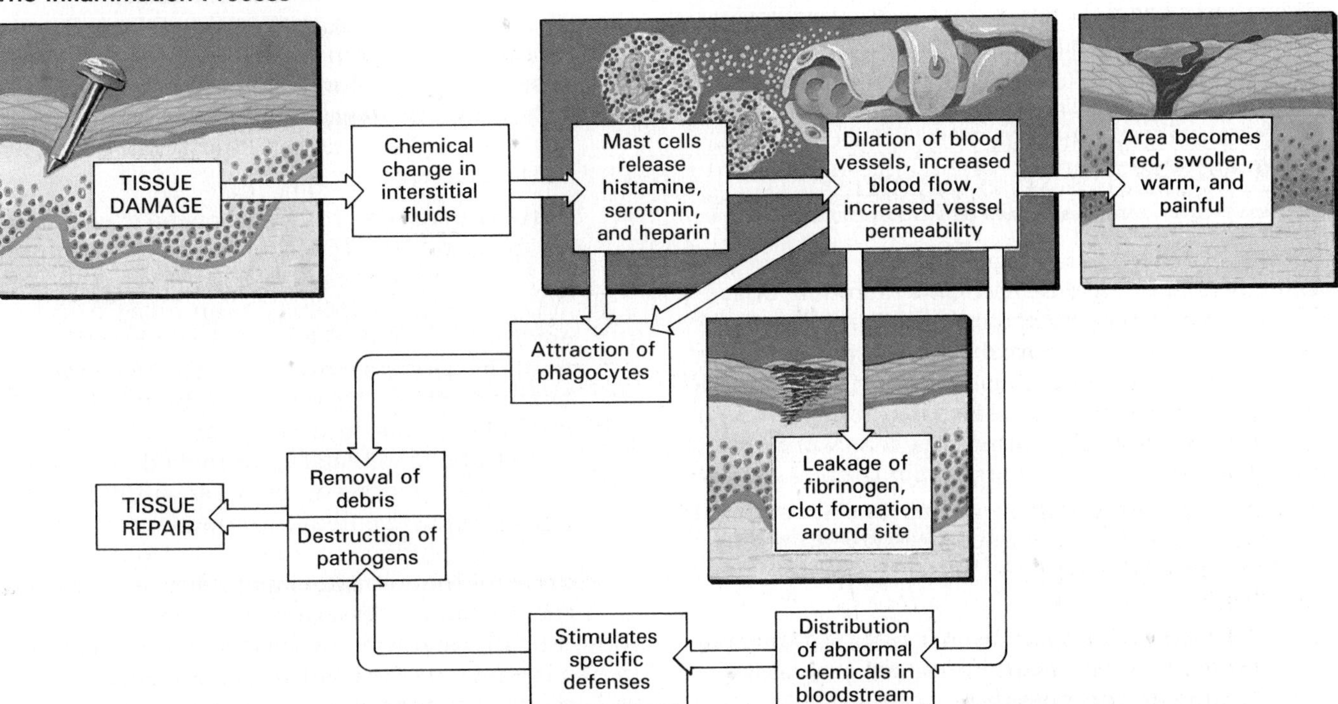

the preoptic area was described in Chapter 14.) (pp. 6–8, 445) Circulating proteins called **pyrogens** (PĪ-rō-jenz) can reset the thermostat and cause a rise in body temperature. A variety of stimuli, including pathogens, bacterial toxins, and antigen-antibody complexes, either act as pyrogens themselves or stimulate the release of pyrogens by macrophages. The pyrogen released by active macrophages is a monokine called **endogenous pyrogen**, or **interleukin-1** (in-ter-LOO-kin), abbreviated **Il-1**.

Within limits, a fever may be beneficial. High body temperatures may inhibit some viruses and bacteria, or it may speed up their reproductive rates so that the disease runs its course more rapidly. Metabolic processes accelerate as well. For each 1° C rise in temperature the metabolic rate jumps by 10 percent. This increased metabolic rate may help mobilize tissue defenses and speed the repair process. It does create an additional demand for nutrients, so the old adage about "feeding a fever" is well founded. High fevers (over 40° C, or 104° F) have negative effects on many different physiological systems. For example, CNS problems, such as nausea, disorientation, hallucinations, or convulsions, may develop. (Additional information on the origins and treatment of fevers will be found in Chapter 25.)

- **What types of cells would be affected by a decrease in the monocyte-forming cells in the bone marrow?**
- **A rise in the level of interferon in the body would suggest what kind of infection?**
- **What effects do pyrogens have in the body?**

Specific Defenses: The Immune Response

Specific defenses are provided by the lymphocytes of the lymphatic system. Our understanding of specific immunity has greatly improved in the past decade, and a comprehensive discussion would involve hundreds of pages and thousands of details. The discussion that follows emphasizes important patterns and introduces general principles that will provide a foundation for future courses in microbiology and immunology.

We will begin by considering four general characteristics of specific defenses:

1. *Specificity*: A specific defense is activated by an antigen, and the response targets that particular antigen and not others.
2. *Versatility*: In the course of a normal lifetime, an individual encounters tens of thousands of antigens. The immune system can differentiate between them, producing appropriate and specific responses to each.
3. *Memory*: During the initial response to the presence of an antigen, lymphocytes proliferate. Two kinds of cells are produced: some that attack the invader and others that remain inactive unless they are exposed to the same antigen at a later date. These inactive **memory cells** enable the immune system to "remember" antigens that it has previously enountered and launch a faster, stronger counterattack if one of them ever appears again.
4. *Tolerance*: **Tolerance** is said to exist when the immune system does not respond to a particular antigen. For example, all cells and tissues in the body contain antigens, but they normally fail to stimulate an immune response. Tolerance usually continues only as long as the individual is exposed to the antigen on a regular basis.

INNATE VERSUS ACQUIRED IMMUNITY

The functional relationship between specific and nonspecific immunity is diagrammed in Figure 22-15, which also indicates the various types of specific immunity. Specific immunity may be *innate* or *acquired*. There are two general types of acquired immunity, *active* and *passive*.

- **Innate immunity** is genetically determined; it is present at birth, and has no relation to previous exposure to the pathogen involved. For example, people are not subject to the same diseases as goldfish. Innate immunity breaks down only when the entire immune system malfunctions, as in AIDS (see the Clinical Comment later in the chapter).
- **Active immunity** appears following exposure to an antigen, as a consequence of the immune response. Active immunity may develop as a result of natural exposure to an antigen in the environment (**naturally acquired immunity**) or from deliberate exposure to an antigen (**active immunization**). Naturally acquired immunity normally begins to develop after birth, and it is continually adjusted as the individual encounters "new" pathogens or other antigens. The purpose of active immunization is to stimulate antibody production under controlled conditions, so that the individual will be able to overcome natural exposure to the pathogen at some time in the future. (Immunization is described more fully in the Clinical Comment: Principles of Immunization.)

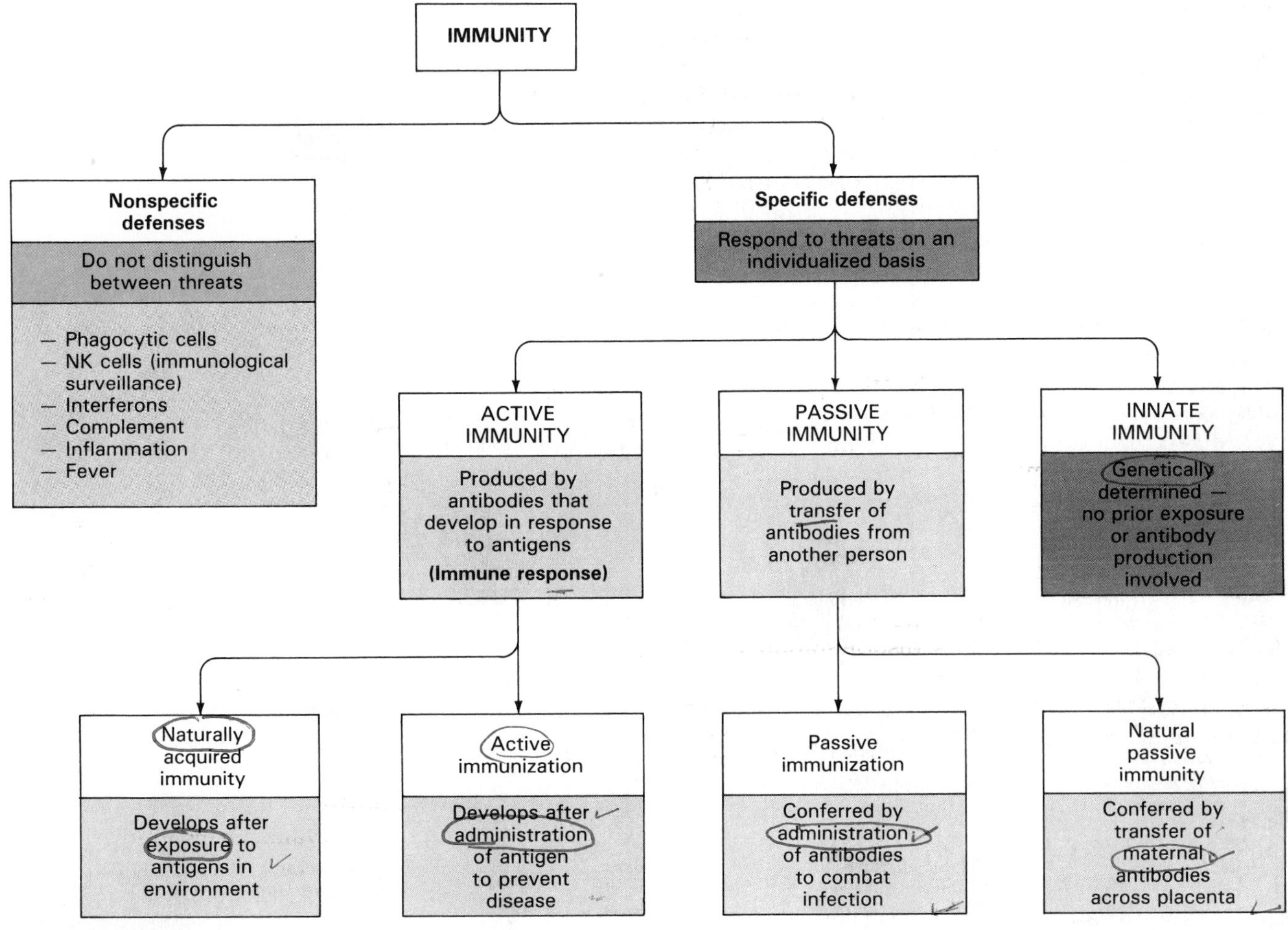

FIGURE 22-15
Types of Immunity

- **Passive immunity** is produced by the transfer of antibodies from another individual. **Natural passive immunity** results when antibodies produced by the mother cross the placental barrier to provide protection against embryonic or fetal infections. In **passive immunization** antibodies are administered to fight infection or prevent disease.

AN OVERVIEW OF THE IMMUNE RESPONSE

The goal of the **immune response** is the destruction or inactivation of pathogens, abnormal cells, and foreign molecules such as toxins. The body has two different ways to do this:

1. Direct attack by activated T cells (*cellular immunity*);
2. Attack by circulating antibodies released by *plasma cells* that are derived from activated B cells (*humoral immunity*).

Figure 22-16 provides an overview of the immune response. The process begins with the appearance of an antigen. Very quickly, specific T cells and B cells *preprogrammed to respond to that particular antigen* are activated. These lymphocytes respond because their cell membranes contain receptors capable of binding that antigen, a process known as **antigen recognition**.

There are millions of different antigens in the environment that may pose a potential threat to health. In the lifetime of any single individual, only a relatively small fraction of these antigens will enter body tissues. Your immune system, however, has no way of anticipating which antigens it will actually encounter. Its protective strategy is to prepare for *any* antigen that might appear. During development, differentiation of cells in the lymphatic system produces an enormous number of lymphocytes with varied antigen sensitivities. Among the trillion or so lymphocytes in the human body are millions of different

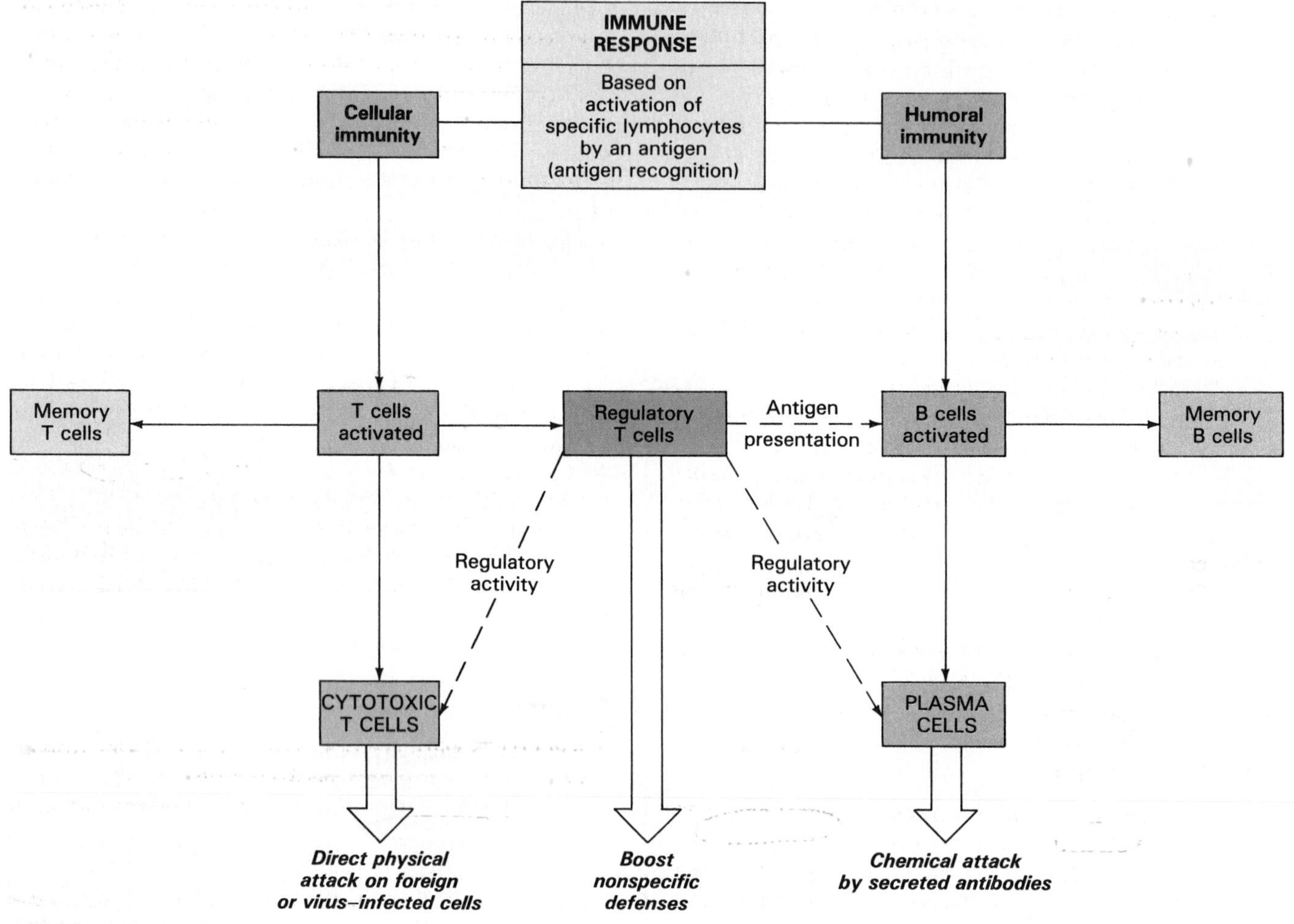

FIGURE 22-16
An Overview of the Immune Response

lymphocyte populations. Each population consists of several thousand cells that bear a specific receptor in their membranes, and are therefore prepared to recognize a specific antigen. When activated, these cells begin to divide and proliferate. *Thus whenever an antigen is encountered, more lymphocytes are produced that are sensitive to that particular antigen.*

The response of the immune system to the presence of an antigen includes both short-term and long-term components. The short-term response eliminates the antigen through a combination of physical and chemical attack. The long-term response involves the production of memory cells that will strike quickly if the same antigen appears at a later date. Because a single exposure leads to the production of large numbers of memory cells, the reappearance of that antigen will trigger a much more powerful, swift, and effective immune response.

The precise nature of the immune response varies depending on whether the activated lymphocyte is a T cell or a B cell.

- T cell activation leads to the formation of *cytotoxic T cells* and *memory T cells* that provide cellular immunity. Other activated T cells become *regulatory T cells* (helper T cells and suppressor T cells) that play a central role in coordinating cellular and humoral immune responses and integrating specific and nonspecific defense mechanisms.
- B cell activation leads to production of *plasma cells* and *memory B cells* that provide humoral immunity.

We will now consider these T cell and B cell functions in greater detail.

T CELLS AND CELLULAR IMMUNITY

Foreign compounds or pathogens entering a tissue often fail to stimulate an immediate immune response. In the case of foreign cells (including bacteria), viruses, and foreign proteins, the antigens are first engulfed by phagocytic cells. These phagocytic cells, called **accessory cells**, are most often fixed macrophages in connective tissues. However, any phagocytes of the monocyte-macrophage group, including free macrophages, Langerhans cells (skin), and Kupffer cells (liver), can perform this function.

Macrophages break down or alter the foreign antigens, creating antigenic fragments capable of stimulating lymphocyte activity. These *processed antigens* accumulate in the interstitial fluids of the region and appear on the surface of the macrophage, bound to HLA proteins. This process is called **antigen presentation** because T cells that contact the altered macrophage membrane can become activated, initiating an immune response.

T cells can also be activated by contact with antigens on the surface of almost any cell in the body. This is especially important in the defense against viral infection. Viruses entering body tissues enter their target cells and begin replicating. In this process, viral proteins appear in the cell membranes of infected cells.

This brings us to an interesting question: How can lymphocytes detect an abnormal antigen molecule among all of the proteins and glycoproteins normally found in cell membranes?

Antigen Recognition

All of the cells in the body contain integral glycoproteins whose structure is genetically determined. The genes directing the synthesis of these glycoproteins are found along one portion of chromosome 6, in a region called the **major histocompatibility complex (MHC)**. These membrane glycoproteins are called **human leukocyte antigens**, or **HLAs**. Their amino acid sequences and shapes differ from individual to individual. Two major classes of HLA proteins are recognized. **Class I** molecules are found in the membranes of all nucleated cells; **Class II** molecules are restricted to macrophages and lymphocytes. Although lymphocytes are continually exposed to these HLA proteins, an immune response does not normally occur.[1]

An immune response will develop, however, if antigenic material becomes attached to an HLA molecule. Each HLA molecule has a distinct three-dimensional shape with a relatively narrow central groove. An antigen that fits into this groove can be held in position by hydrogen bonding with adjacent portions of the HLA molecule. This binding directly or indirectly leads to lymphocyte activation. As diagrammed in Figure 22-17, antigens bound to Class I HLA molecules stimulate the production of cytotoxic T cells and suppressor T cells. Helper T cells are activated by contact with antigens bound to Class II HLA molecules.

Cytotoxic T Cells

Cytotoxic T cells are responsible for **cellular immunity**, also known as **cell-mediated immunity**. As indicated in Figure 22-17, these T cells are often activated

[1] It has been suggested that during development lymphocytes stimulated by "self" HLAs are killed by their own antibodies.

FIGURE 22-17
Interactions between Macrophages, T Cells, and B Cells

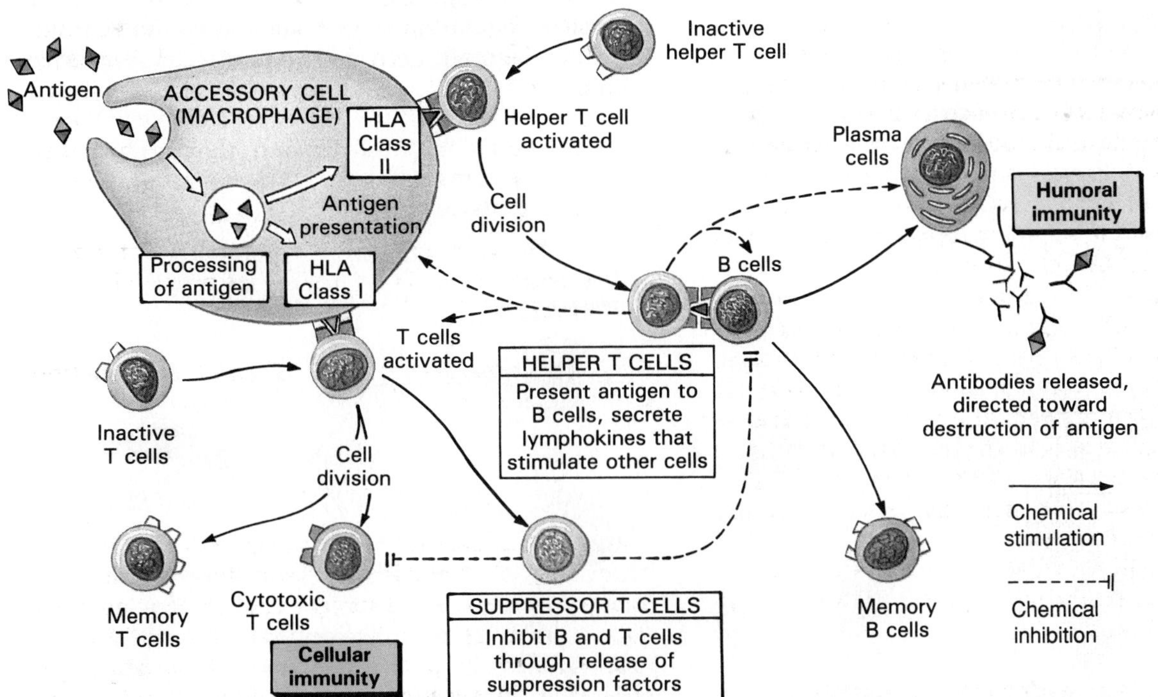

by contact with macrophage membranes containing processed antigens; they may also be stimulated by exposure to viral antigens in the membranes of infected cells. When activated, these T cells undergo a series of divisions that generate **cytotoxic T cells** and **memory T cells**. The cytotoxic T cells, also called **killer T cells**, track down and attack the bacteria, fungi, protozoa, or foreign tissues that contain the antigen. The cytotoxic T cell may contact and rupture the antigenic cell membrane through the release of perforin, or destroy its target at close range by secreting a poisonous **lymphotoxin** (lim-fō-TOK-sin). These mechanisms are diagrammed in Figure 22-18.

Cytotoxic T cells do not respond immediately following the appearance of an antigen in the body for several reasons:

1. The antigen must be engulfed, processed, and presented on the membranes of accessory cells;
2. T cells sensitive to that specific antigen must contact those membranes;
3. Several cycles of cell division must occur before the daughter cells differentiate into cytotoxic T cells and memory T cells;
4. The cytotoxic T cells must migrate to the focus of infection. As a result, it usually takes at least 2 days for the concentration of cytotoxic T cells to reach effective levels at the site of an injury or infection.

The memory T cells remain "in reserve." If the same antigen appears a second time, these cells will differentiate into cytotoxic T cells immediately, producing a more rapid and effective cellular response.

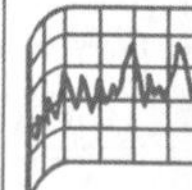

Diagnostics: Skin Tests

Skin tests can be used to determine whether or not an individual has been exposed to a particular antigen. The antigen is administered by shallow injection, and the site is inspected after a period of minutes to days. Most skin tests check for delayed hypersensitivity by injecting antigens taken from bacteria, fungi, viruses, or parasites. If the individual has previously been exposed to the antigen, in 2–4 days the injection site will become red, swollen, and invaded by macrophages, T cells, and neutrophils. These signs are considered a positive test. A positive skin test indicates previous exposure, but does not necessarily imply the presence of the disease. Many readers will have had skin tests that checked for sensitivity to the bacterium causing tuberculosis or the virus responsible for mumps. Many states require a tuberculosis test, called a *tuberculin skin test*, when someone applies for a job in the food service industry. If the test is positive, the individual has been exposed to the disease and developed antibodies. Further tests must be performed to determine whether an infection is under way. (Anyone with tuberculosis may accidentally transmit the bacteria when preparing or serving food.)

Regulatory T Cells

Regulatory T cells include *helper T cells* and *suppressor T cells*. **Helper T cells** release a variety of lymphokines that play a key role in the coordination of specific and nonspecific defenses, as well as in the chemical regulation of cellular and humoral immunity. For example, activated helper T cells secrete lymphokines that:

- Stimulate the T cell divisions that produce mem-

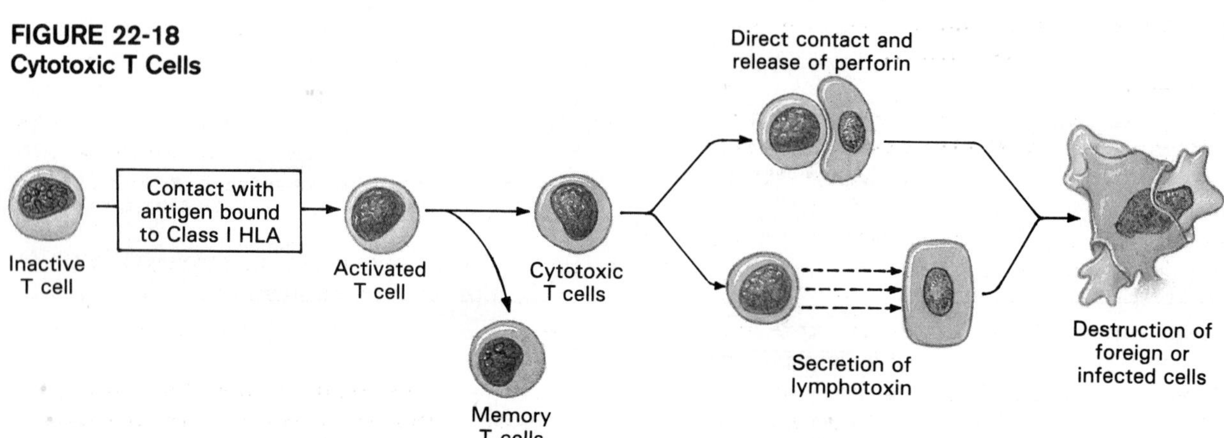

FIGURE 22-18
Cytotoxic T Cells

ory T cells and accelerate maturation of cytotoxic T cells;

- Enhance nonspecific defenses by (1) attracting macrophages to the area, (2) preventing their departure, and (3) stimulating their phagocytic activity and effectiveness;
- Attract and stimulate the activity of NK cells, providing another mechanism for the destruction of abnormal cells and pathogens;
- Promote B cell division, plasma cell maturation, and antibody production.

In addition, activated helper T cells participate in the direct activation of B cells sensitive to the antigen involved.

Suppressor T cells depress the responses of other T cells and B cells. They produce these effects by secreting *suppression factors*, inhibitory lymphokines of unknown structure. This suppression does not occur immediately, because suppressor T cell activation takes much longer than the activation of other types of T cells. As a result, suppressor T cells act *after* the initial immune response. In effect, these cells "put on the brakes" and limit the degree of immune system activation from a single stimulus.

B CELLS AND HUMORAL IMMUNITY

B cells are responsible for launching a chemical attack on antigens through the production of appropriate antibodies. Before we can discuss the details of this process, we need to take a closer look at the chemical structure of antigens, and how antibodies can provide an effective defense against them.

Antigen Structure

Antigens come in a variety of different sizes, ranging from a single polypeptide to an entire bacterium. Antibodies do not target the antigen as a whole, but instead focus on certain portions of its exposed surface. These regions are called **antigenic determinant sites** (Figure 22-19a). A **complete antigen** is an antigen that can trigger an immune response. To be a complete antigen a molecule must have at least two antigenic determinant sites. **Partial antigens**, or **haptens**, have only one antigenic site. Such molecules do not ordinarily stimulate the production of antibodies. Haptens include a variety of small compounds, among them several antibiotics including *penicillin*. Haptens may, however, become attached to carrier molecules, forming combinations with two or more antigenic determinant sites that can function as complete antigens. Most environmental antigens have multiple antigenic determinant sites; entire microorganisms may have thousands.

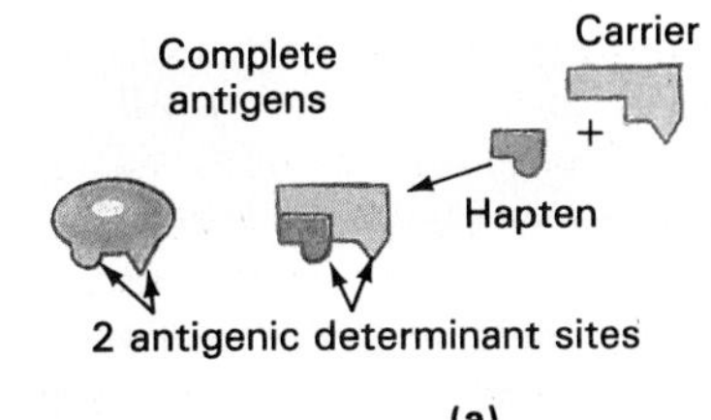

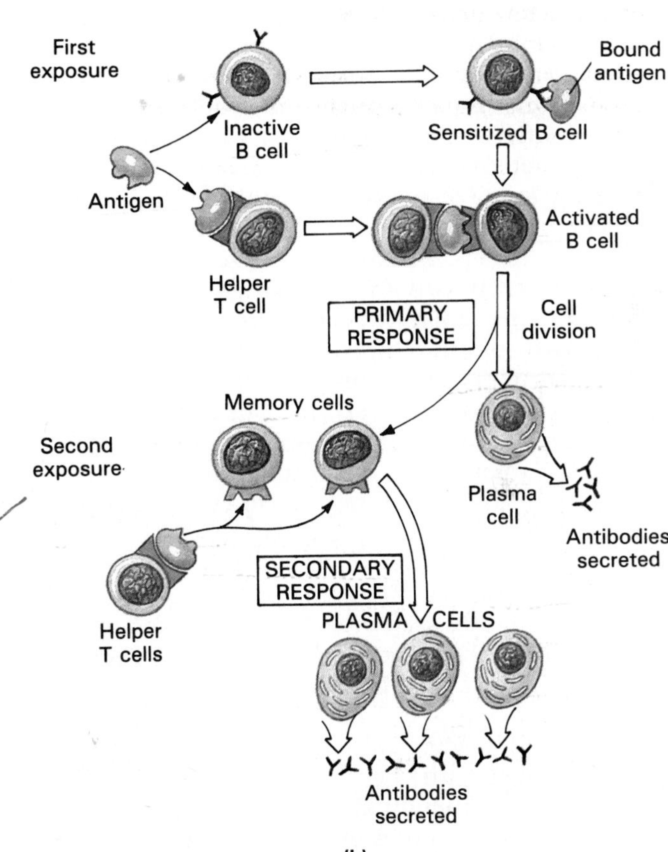

FIGURE 22-19
Humoral Immunity. (a) Formation of a complete antigen from a hapten and a carrier molecule. (b) Activation of B cells and the humoral immune response.

B Cell Activation

Each B cell carries some of its antibody molecules on the cell surface, where they can bind soluble antigens. Although antigen exposure sensitizes the cell and prepares it for activation, antigen exposure alone usually fails to activate B cells. *B cell activation primarily occurs through the activities of helper T cells.* Two different mechanisms may be involved:

1. Helper T cells may present antigenic materials to appropriate B cells and stimulate their activation through direct contact, as indicated in Figure 22-17.
2. Helper T cells also secrete lymphokines that promote the activation of B cells already exposed to appropriate antigens.

Health News: AIDS

Acquired immune deficiency syndrome, or **AIDS**, is caused by a virus known as **human immunodeficiency virus** (**HIV**), which selectively infects helper T cells (Figure 22-20). This by itself impairs the immune response, because these cells play a central role in coordinating cellular and humoral responses to antigens. To make matters worse, suppressor T cells are relatively unaffected by the virus, and over time the excess of suppressing factors "turns off" the normal immune response. Circulating antibody levels decline, cellular immunity is reduced, and the body is left without defenses against a wide variety of microbial invaders.

With immune function so reduced, ordinarily harmless pathogens can initiate lethal infections, known as *opportunistic infections*. AIDS patients are especially prone to lung infections and pneumonia, often caused by *Pneumocystis carinii* or other fungi, and to a wide variety of bacterial, viral, and protozoan diseases. Because immune surveillance is also depressed, the risk of cancer increases. One of the most common cancers seen in AIDS patients, though very rare in normal individuals, is **Kaposi's sarcoma**, characterized by rapid cell divisions in endothelial cells of cutaneous blood vessels.

Infection with HIV occurs through intimate contact with the body fluids of infected individuals. Although all body fluids carry the virus, the major routes of transmission involve contact with blood, semen, or vaginal secretions. Most AIDS patients become infected through sexual contact with an HIV-infected person (who may *not* necessarily be suffering from the clinical symptoms of AIDS). The next largest group of patients consists of intravenous drug users who shared contaminated needles. A relatively small number of individuals have become infected with the virus after receiving a transfusion of contaminated blood or blood products. Finally, an increasing number of infants are born with the disease, having acquired it in the womb from infected mothers.

The best defense against AIDS consists of avoiding sexual contact with infected individuals. *All* forms of sexual intercourse carry the potential risk of viral transmission. The use of synthetic condoms greatly reduces the chance of infection (although it does not completely eliminate it). Condoms that are not made of synthetic materials are effective in preventing pregnancy but do not block the passage of viruses.

Clinical symptoms of AIDS may not appear for 5–10 years or more following infection, and when they do appear they are often mild, consisting of lymphadenopathy and chronic but nonfatal infections. This condition has been called **AIDS-related complex**, or **ARC**. After a variable period of time, however, ARC gives way to full-blown AIDS. So far as is known, AIDS is invariably fatal, and it appears likely that in the absence of a major clinical advance all those who carry the virus will eventually die of the disease. Deaths in the United States have already climbed above 125,000, and estimates of the number of infected individuals range from 1 to 2 million. The numbers worldwide are even more frightening: the World Health Organization estimates that 10 million people may be infected.

Despite intensive efforts, a vaccine has yet to be developed that will provide immunity from HIV infection. However, the survival rate for AIDS patients has been steadily increasing because new drugs are avilable that slow the progress of the disease, and improved antibiotic therapies help combat secondary infections. This combination is extending the life span of patients while the search for more effective treatment continues. (For more details, see the Clinical Manual.) [**CM**: AIDS]

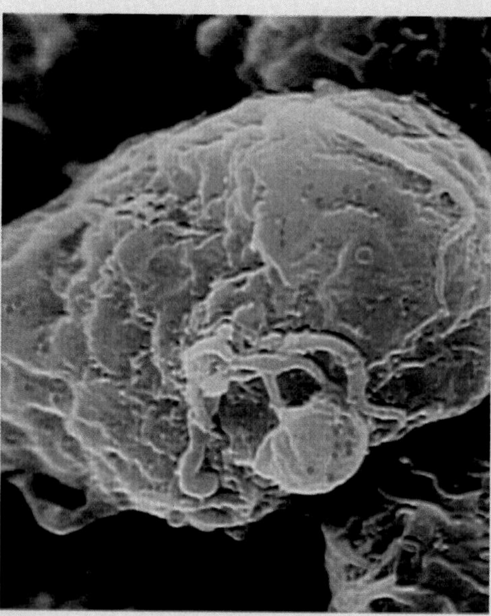

FIGURE 22-20
The AIDS Virus. A scanning electron micrograph of a lymphocyte infected with HIV. The blue dots scattered across the surface of the cell are individual viruses.

In addition to secreting hormones that stimulate B cell activation, helper T cells release other lymphokines that promote B cell division, plasma cell maturation, and antibody production.

Mechanism of Humoral Immunity

Figure 22-19b diagrams the events that occur when a B cell is activated by a complete antigen. The activated B cell may either (1) differentiate into a **plasma cell** immediately or (2) divide several times, producing daughter cells that differentiate into plasma cells and **memory B cells**. Helper T cells stimulate both of these processes by secreting lymphokines that promote B cell division and accelerate plasma cell maturation.

The antibodies produced by the plasma cells are the agents of the **primary response** to antigen exposure. Because the antigen must activate the appropriate B cells, and the B cells must then respond by differentiating into plasma cells, the primary response does not appear immediately. As indicated in Figure 22-21, there is a gradual, sustained rise in the concentration of circulating antibodies, and the antibody activity, or **antibody titer** ("standard"), in the plasma does not peak until 1–2 weeks after the initial exposure. Thereafter the antibody concentration declines, assuming that the individual is no longer exposed to the antigen. This reduction in antibody production occurs because (1) plasma cells have very high metabolic rates and survive only for a short time, and (2) further production of plasma cells is inhibited by suppression factors released by suppressor T cells. However, suppressor T cell activity does not begin immediately after antigen exposure, and under normal conditions helper cells outnumber suppressors by more than 3:1. As a result, many B cells are activated before suppressor T cell activity has a noticeable effect.

Two different types of antibodies are involved in the primary response. Molecules of *Immunoglobulin-M*, or *IgM*, are the first to appear in the circulation. IgM is secreted by the plasma cells that form immediately after B cell activation. These lymphocytes do not pause to produce memory cells. Levels of *Immunoglobulin-G*, or *IgG*, rise more slowly, because the stimulated lymphocytes undergo repeated cell divisions and generate large numbers of memory cells as well as plasma cells. In effect, IgM provides an immediate but limited defense that fights the infection until massive quantities of IgG can be produced. (Differences in the structure of these antibodies are discussed in a later section.)

Immunological Memory: The Secondary Response

Memory B cells do not differentiate into plasma cells unless they are exposed to the same antigen a second time. If and when that exposure occurs, these cells respond immediately, dividing and differentiating into plasma cells that secrete antibodies in massive quantities. This represents the **secondary response**, or **anamnestic response** (an-am-NES-tik; *anamnesis*, a memory), to antigen exposure.

As indicated in Figure 22-21, the secondary response produces an immediate rise in antibody concentrations to levels many times higher than those of the primary response. This response is so much faster and stronger than the primary response because the numerous memory cells constitute a kind of cellular rapid deployment force. They are primed to fight one particular enemy and can be mobilized and prepared for action on short notice. The secondary response appears even if the second exposure occurs years after the first, for memory cells are long-lived, potentially surviving for 20 years or more.

FIGURE 22-21
The Primary and Secondary Immune Responses. The primary response takes several weeks to develop peak antibody titers, and antibody concentrations do not remain elevated. The secondary response is characterized by a very rapid increase in antibody titer, to levels much higher than those of the primary response. Antibody activity remains elevated for an extended period following the second exposure to the antigen.

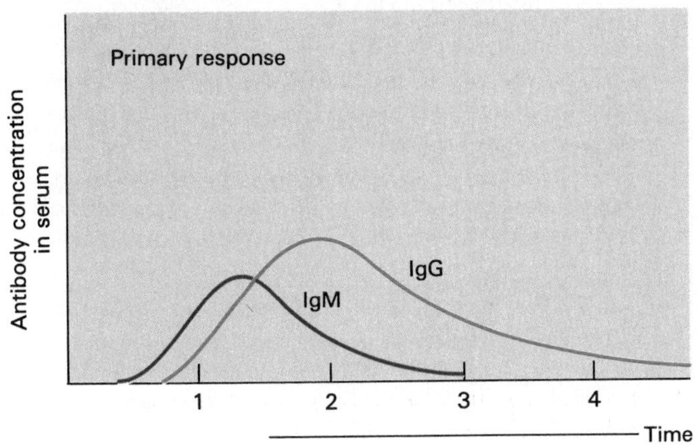

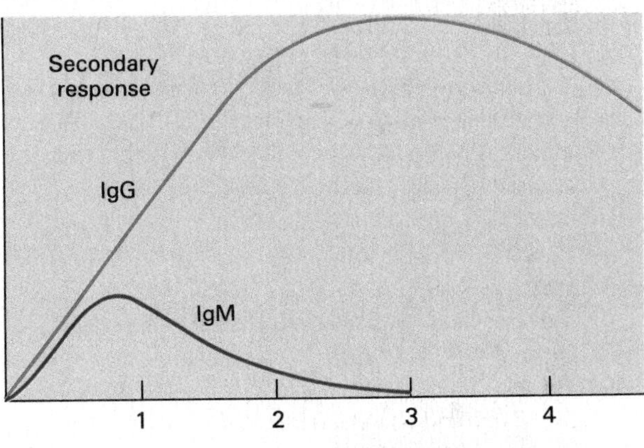

Time (weeks)

Clinical Comment: Principles of Immunization

The primary response to an antigen is not as effective a defense as the secondary response. The primary response develops slowly, and the antibodies are not produced in massive quantities. As a result the primary response may not prevent an infection, even though the secondary response would be more than adequate. From a practical standpoint this means that a person who survives the first infection will easily overcome a subsequent invasion by the same pathogen. Often the invading organisms will be destroyed before they have produced any symptoms of the disease.

This represents the basic principle behind the use of immunization to prevent disease. In **active immmunization** a primary response to a particular pathogen is intentionally stimulated before an individual encounters the pathogen in the environment. The result is lasting immunity against that pathogen. Immunization is accomplished by adminstering a **vaccine** (vak-SĒN), a preparation of antigens derived from a specific pathogen. The vaccine may be given orally or via intramuscular or subcutaneous injection. Most vaccines consist of the pathogenic organism, in whole or in part, living or dead. In some cases a vaccine contains one of the metabolic products of the pathogen.

Before live bacteria or viruses are administered they are weakened, or **attenuated** (a-TEN-ū-ā-ted), to lessen or eliminate the chance of a serious infection developing from exposure to the vaccine. The rubella, mumps, measles, smallpox, yellow fever, and oral polio vaccines are examples of vaccines using live attenuated viruses. Despite attenuation, the administration of live microorganisms may produce mild symptoms comparable to those of the actual disease. However, the risks of serious illness developing as a result of vaccination are very small compared with the risks posed by pathogen exposure *without* prior vaccination.

Inactivated or "killed" vaccines consist of bacterial cell walls or viral protein coats. These vaccines have the advantage that they cannot produce even mild symptoms of the disease. Unfortunately, inactivated vaccines do not stimulate as strong an immune response and so do not confer as long-lasting immunity as do live-organism vaccines. In the years following exposure, the antibody titer declines and the system eventually fails to produce an adequate secondary response. As a result, the immune system must be "reminded" of the antigen periodically by the administration of *boosters*. Influenza, typhoid, cholera, typhus, plague, and injected polio vaccines use inactivated viruses or bacteria. In some cases fragments of the bacterial or viral walls, or their toxic products, can be used to produce a vaccine. The tetanus, diphtheria, and hepatitis B vaccines are good examples. Data concerning attenuated and inactivated vaccines are presented in Table 22-1.

Gene-splicing techniques can now be used to incorporate antigenic compounds from pathogens into the cell walls of harmless bacteria. When exposed to these bacteria, the immune system responds by producing antibodies and memory B cells that are equally effective against the engineered bacterium and the pathogen.

Passive immunization is usually selected if the individual has already been exposed to a dangerous pathogen or toxin, so that there is not enough time for active immunization to take effect. In passive immunization the patient receives a dose of antibodies that will attack the pathogen and overcome the infection, even without the help of the host's own immune system. Passive immunization provides only short-term resistance to infection, for the antibodies are gradually removed from circulation and are not replaced.

The antibodies provided during passive immunization have traditionally been acquired by collecting and combining antibodies from the sera of many other individuals. This *pooled sera* is used to obtain large quantities of antibodies, but the procedure is very expensive, and improper treatment of the sera carries the risk of accidental transmission of an infectious agent, such as the hepatitis or AIDS virus. Antibodies can also be obtained from the blood of a domesticated animal (usually a horse) exposed to the same antigen. Unfortunately, recipients may suffer allergic reactions to horse serum proteins.

At present, antibody preparations are avail-

TABLE 22-1 Immunizations Currently Available

Immunization Target	*Type of Immunity Provided*	*Vaccine Type*	*Remarks*
VIRUSES			
Poliovirus	Active	Live, attenuated	Oral
	Active	Killed	Boosters every 2–3 years
Rubella	Active	Live, attenuated	
	Passive	Human antibodies (pooled)	
Mumps	Active	Live, attenuated	
Measles (rubeola)	Active	Live, attenuated	
Hepatitis A	Passive	Human antibodies (pooled)	
Hepatitis B	Active	Killed	May need periodic boosters
	Passive	Human antibodies (pooled)	
Smallpox	Active	Live, related virus	Boosters every 3–5 years (no longer required as disease appears to have been eliminated)
Yellow fever	Active	Live, attenuated	Boosters every 10 years
Herpes zoster	Passive	Human antibodies (pooled)	
Hemophilus influenza B (HIB)	Active	Killed	May need periodic boosters
Rabies	Passive	Human antibodies (pooled)	
	Passive	Horse antibodies	
BACTERIA			
Typhoid	Active	Killed	Boosters every 3 years
Tuberculosis	Active	Live, attenuated	
Plague	Active	Killed	Boosters every 2 years
Cholera	Active	Killed	Boosters every 6 months
Tetanus	Active	Toxins only	Boosters every 5–10 years
	Passive	Human antibodies (pooled)	
Diphtheria	Active	Toxins only	Boosters every 10 years
	Passive	Horse antibodies	
Streptococcal pneumonia	Active	Bacteria wall components	
Botulism	Passive	Horse antibodies	
Rickettsia: Typhus	Active	Killed	Boosters yearly
OTHER TOXINS			
Snake bite	Passive	Horse antibodies	
Spider bite	Passive	Horse antibodies	
Venomous fish spine	Passive	Horse antibodies	

able to treat hepatitis A, hepatitis B, diphtheria, tetanus, rabies, measles, rubella, botulism, and the venoms of certain fishes, snakes, and spiders. Gene-splicing technology can also be used to reproduce pure antibody preparations free from antigenic or viral contaminants, and this should eventually eliminate the need for pooled or foreign plasma.

It should also be noted that passive immunity occurs naturally during fetal development, because maternal IgG antibodies can cross the placental barriers and enter the fetal circulation.

Structure and Function of Antibodies

Figure 22-22 presents different views of a single antibody molecule. The molecule consists of two parallel pairs of polypeptide chains. Each chain contains fixed and variable segments. The fixed segments resemble those of every other antibody molecule of that particular class (the various classes of antibodies are described in the following section). The specificity of the antibody molecule depends on the structure of the variable segments. As we saw in Chapter 2, even a relatively short polypeptide chain can be made with an enormous number of different amino acid sequences. (p. 51) This fact accounts for the great diversity of antibody structure; different plasma cells can produce an estimated 27 million different antibodies.

When an antibody molecule binds to its proper antigen, an **antibody-antigen complex** is formed. The specificity of that binding depends initially on the three-dimensional "fit" between the variable segments of the antibody molecule and the corresponding antigenic determinant sites. Once the two molecules are in position, hydrogen bonding and other weak chemical forces lock them together. As bonding occurs, the antibody molecule changes from a T to a Y shape. This conformational change exposes regions of the molecule that play a role in several of the mechanisms leading to destruction of the antigen.

The binding of antibody to antigen is the first step in a number of processes, each a different way of eliminating the threat posed by the antigen.

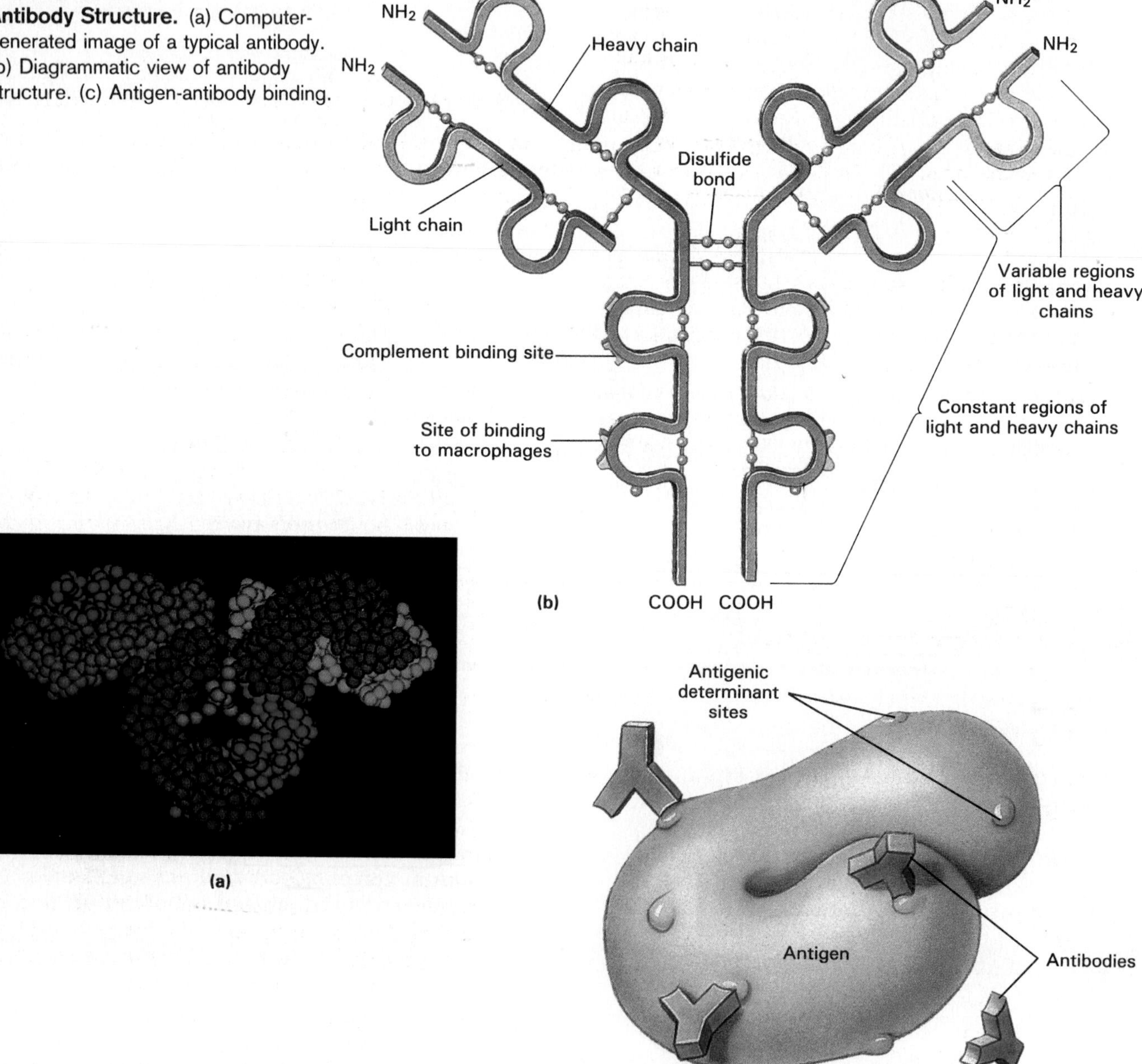

FIGURE 22-22
Antibody Structure. (a) Computer-generated image of a typical antibody. (b) Diagrammatic view of antibody structure. (c) Antigen-antibody binding.

1. *Neutralization*: Both viruses and bacterial toxins have specific sites that must bind to target regions on body cells in order to enter or injure them. Antibodies may bind to those sites, making the virus or toxin incapable of attaching itself to a cell. This mechanism is known as **neutralization**.
2. *Agglutination and precipitation*: Most antigens, as we have seen, possess many antigenic determinants to which antibody molecules can bind. But notice in Figure 22-22 that each antibody molecule has two binding sites and can therefore attach itself simultaneously to two antigens. Thus when large numbers of antibodies encounter their antigenic targets, they can interact to create a three-dimensional structure known as an **immune complex** (Figure 22-23). When the antigen is a soluble molecule, such as a toxin, this process may create complexes too large to remain in solution. The formation of insoluble immune complexes is called **precipitation**. When the antigenic target is on the surface of a cell (such as a bacterium) or a virus, the formation of large complexes is called **agglutination**. The clumping of red blood cells that occurs when incompatible blood types are mixed (Chapter 19) is an example of agglutination. (p. 618)
3. *Activation of complement*: Portions of the antibody molecules exposed by the change in shape bind complement proteins. The bound complement molecules then activate the complement system, destroying the antigen.
4. *Attraction of phagocytes*: Antigens covered with antibodies attract eosinophils, neutrophils, and macrophages—cells that phagocytize pathogens and destroy foreign or abnormal cell membranes.

FIGURE 22-23
Formation of Antigen-Antibody Complexes

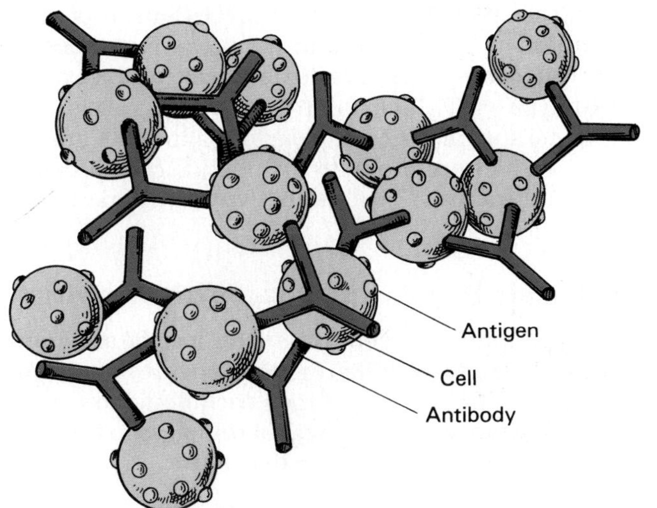

5. *Opsonization*: A coating of antibodies and complement proteins increases the effectiveness of phagocytosis. Microorganisms such as bacteria have slick cell membranes or capsules, and a phagocyte must be able to hang onto its prey before it can engulf it. Phagocytes can bind more easily to antibodies and complement proteins on the surface of a pathogen than they can to the bare surface itself.
6. *Stimulation of inflammation*: Antibodies may promote inflammation through the stimulation of basophils and mast cells.
7. *Prevention of bacterial and viral adhesion*: Antibodies dissolved in saliva, mucus, and perspiration coat epithelia and provide an additional layer of defense. A covering of antibodies makes it difficult for pathogens to attach to body surfaces and penetrate their defenses.

A Structural Classification of Antibodies

There are five different classes of antibodies, or **immunoglobulins** (Ig), in body fluids: *IgG*, *IgE*, *IgD*, *IgM*, and *IgA*.

1. **Immunoglobulin G**, or **IgG**, is the largest class of antibodies. There are several different types of IgG, which together account for 80 percent of all immunoglobulins. The IgG antibodies are responsible for resistance against many viruses, bacteria, and bacterial toxins.
2. **IgE** attaches to the exposed surfaces of basophils and mast cells. When a suitable antigen appears and binds to the IgE molecules, the cell is stimulated to release histamine and other chemicals that accelerate inflammation in the immediate area.
3. **IgD** is found on the surfaces of B cells, where it may help to bind antigen molecules. IgG, IgE, and IgD consist of individual molecules, as shown in Figure 22-22b.
4. **IgM** is the first antibody type secreted after the arrival of an antigen. The concentration of IgM then declines as IgG production accelerates. Although plasma cells secrete individual IgM molecules, they circulate as a five-antibody starburst (Figure 22-24a). This configuration makes them particularly effective in forming immune complexes. IgM antibodies are responsible for the agglutination of cross-matched blood, and so are used to determine an individual's blood type (Chapter 19). (p. 616) They may also attack bacteria that are insensitive to IgG.

IgA is found primarily in glandular secretions such as mucus, tears, and saliva. These antibodies attack pathogens *before* they gain access to internal tissues. IgA antibodies circulate in the

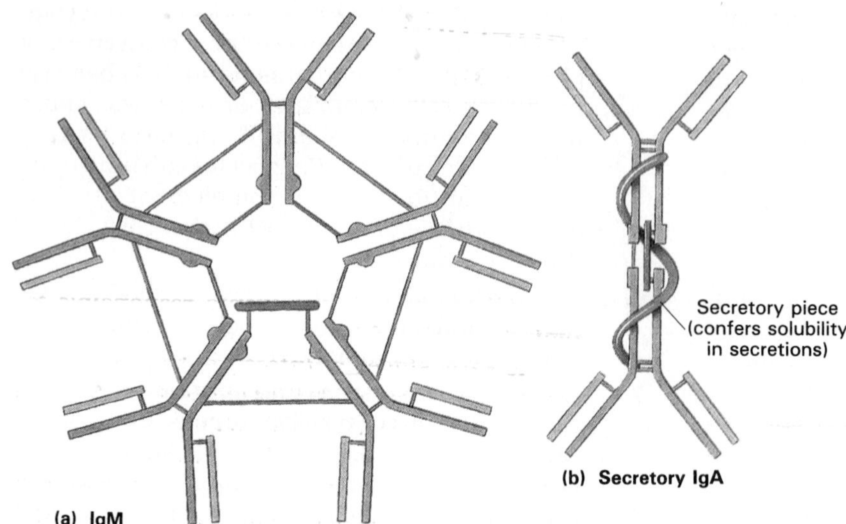

FIGURE 22-24
Classes of Antibodies. (a) IgM molecules circulate in groups of five. (b) IgA molecules are secreted in linked pairs. (IgG, IgE, and IgD circulate as individual antibody molecules, as shown in Figure 22-22b.)

blood as individual molecules or in pairs (Figure 22-24b). Epithelial cells absorb them from the blood and attach a *secretory piece* before secreting them onto the epithelial surface.

HORMONES OF THE IMMUNE SYSTEM

The specific and nonspecific defenses of the body are coordinated by physical interaction and by the release of chemical messengers. Examples of physical interaction include antigen presentation by activated macrophages and helper T cells. Examples of chemical messengers include the *lymphokines* secreted by lymphocytes and the *monokines* released by active macrophages.

Table 22-2 contains an abbreviated summary of the monokines and lymphokines identified to date. Four subgroups merit special attention: (1) *interleukins*, (2) *interferons*, (3) *tumor necrosis factors*, and (4) chemicals that regulate phagocytic activities.

Interleukins

Interleukins are probably the most diverse and important chemical messengers in the immune system. At least 10 different types of interleukins have been identified (Table 22-2). Interleukins have the following major functions:

1. *Increasing T cell sensitivity to antigens exposed on macrophage membranes.* This heightened sensitivity accelerates the production of cytotoxic and regulatory T cells.
2. *Stimulating B cell activity, plasma cell formation, and antibody production.* These events promote the production of antibodies and the development of humoral immunity.
3. *Enhancing nonspecific defenses.* Known effects of interleukin production include: (a) the stimulation of inflammation, (b) the formation of scar tissue by fibroblasts, (c) the elevation of body temperature via the preoptic nucleus of the CNS, (d) the stimulation of mast cell formation, and (e) the promotion of ACTH secretion by the anterior pituitary.

[**CM**: *Lyme Disease*]

Interferons

Interferons (in-ter-FĒR-ons) make the synthesizing cell and its neighbors resistant to viral infection, thereby slowing the spread of the virus. These compounds may have other beneficial effects in addition to their antiviral activity. For example, **alpha interferons** and **gamma interferons** attract and stimulate NK cells. Gamma interferons also stimulate the activities of macrophages, making them more effective at killing bacterial or fungal pathogens.

Tumor Necrosis Factors

Tumor necrosis factors (**TNF**s) slow tumor growth and kill sensitive tumor cells. Activated macrophages secrete one type of TNF and carry the molecules in their cell membranes. Cytotoxic T cells produce a different type of TNF. In addition to their effects on tumor cells, tumor necrosis factors stimulate granulocyte production, promote eosinophil activity, cause fever, and increase T cell sensitivity to interleukins.

Chemicals Regulating Phagocytic Activities

Several lymphokines coordinate the specific and nonspecific defenses by adjusting the activities of phagocytic cells. These lymphokines include factors that attract free macrophages and microphages to the area and prevent their premature departure.

TABLE 22-2 Chemical Mediators of the Immune Response

Compound	*Functions*
MONOKINES	
Interleukin-1 (Il-1)	Stimulates T cells to produce Il-2; promotes inflammation; causes fever
Leukotrienes	Stimulate regional inflammation
Tumor necrosis factor (TNF)	Kills tumor cells, slows tumor growth, stimulates activities of T cells and eosinophils
LYMPHOKINES	
Interleukin-2 (Il-2)	Stimulates growth and activation of other T cells and NK cells
Interleukin-3 (Il-3)	Stimulates production of mast cells
Il-4 (B cell differentiating factor), Il-5 (B cell growth factor), Il-6, Il-7 (B cell stimulating factors)	Promote differentiation and growth of B cells, and stimulate plasma cell formation and antibody production
Lymphotoxins	Kill cells, damage tissue, promote inflammation
Interferons	Activate other cells to prevent viral entry; inhibit viral replication
Alpha interferons	Attract and stimulate NK cells
Gamma interferon	Promotes activity of NK cells and macrophages
Transfer factor	Sensitizes other T cells to same antigen
Growth inhibitory factor (GIF)	Reduces/inhibits replication of target cells
Hemopoiesis-stimulating factor	Promotes blood cell production in bone marrow
Monocyte-chemotactic factor (MCF)	Attracts monocytes, activates them to macrophages
Macrophage-inhibitory factor (MIF)	Prevents macrophage migration from the area
Macrophage-activating factor (MAF)	Makes macrophages more active and aggressive
Microphage-chemotactic factor	Attracts microphages from the blood

- **A decrease in the number of cytotoxic T cells would affect what type of immunity?**
- **How would a lack of helper T cells affect the humoral immune response?**

Patterns of Immune Response

The basic chemical and cellular interactions that follow the appearance of a foreign antigen have now been detailed. Figure 22-25 presents a broader, integrated view of the immune response and its relationship to nonspecific defenses. Figure 22-26 provides different perspectives on the events responsible for overcoming a bacterial infection.

In the early stages of infection, before antigen processing has occurred, neutrophils and NK cells migrate into the threatened area and destroy bacterial cells. Over time, inflammation, lymphokines, and monokines draw increasing numbers of phagocytes to the region. Cytotoxic T cells appear as arriving T cells are activated by antigen presentation. Last of all, the population of plasma cells rises as activated B cells differentiate. This rise is followed by a gradual, sustained increase in the level of circulating antibodies.

The basic sequence of events is very similar when a viral infection occurs. However, the initial steps are different, because cytotoxic T cells and NK cells can be activated by contact with infected cells. Figure 22-27a,b contrasts the steps involved in defense against bacteria with the defense against infection by viruses.

IMMUNOLOGICAL COMPETENCE

The ability to demonstrate an immune response upon exposure to an antigen is called **immunological competence**. Cellular immunity can be demonstrated as early as the third month of fetal development, but active humoral immunity does not appear until somewhat later. Sometime after the fourth developmental month, the fetus may produce IgM antibodies if exposed to specific pathogens. However, fetal antibody production is uncommon, because the developing fetus has passive immunity due to the transfer of IgG antibodies from the maternal bloodstream. These are the only antibodies that cross the placenta, and they include the antibodies responsible for the clinical problems that accompany fetal-maternal Rh incompatibility. (Because the agglutinins anti-A and anti-B are IgM antibodies, and IgM cannot cross the placenta, problems with maternal-fetal incompatibilities involving the ABO blood groups rarely occur.)

Delivery eliminates the maternal supply of IgG, and although the mother provides IgA antibodies in the breast milk, the infant gradually loses its passive immunity. The amount of maternal IgG in the infant's bloodstream declines rapidly over the first 2 months after birth. During this period the infant becomes vulnerable to infection by bacteria or viruses that

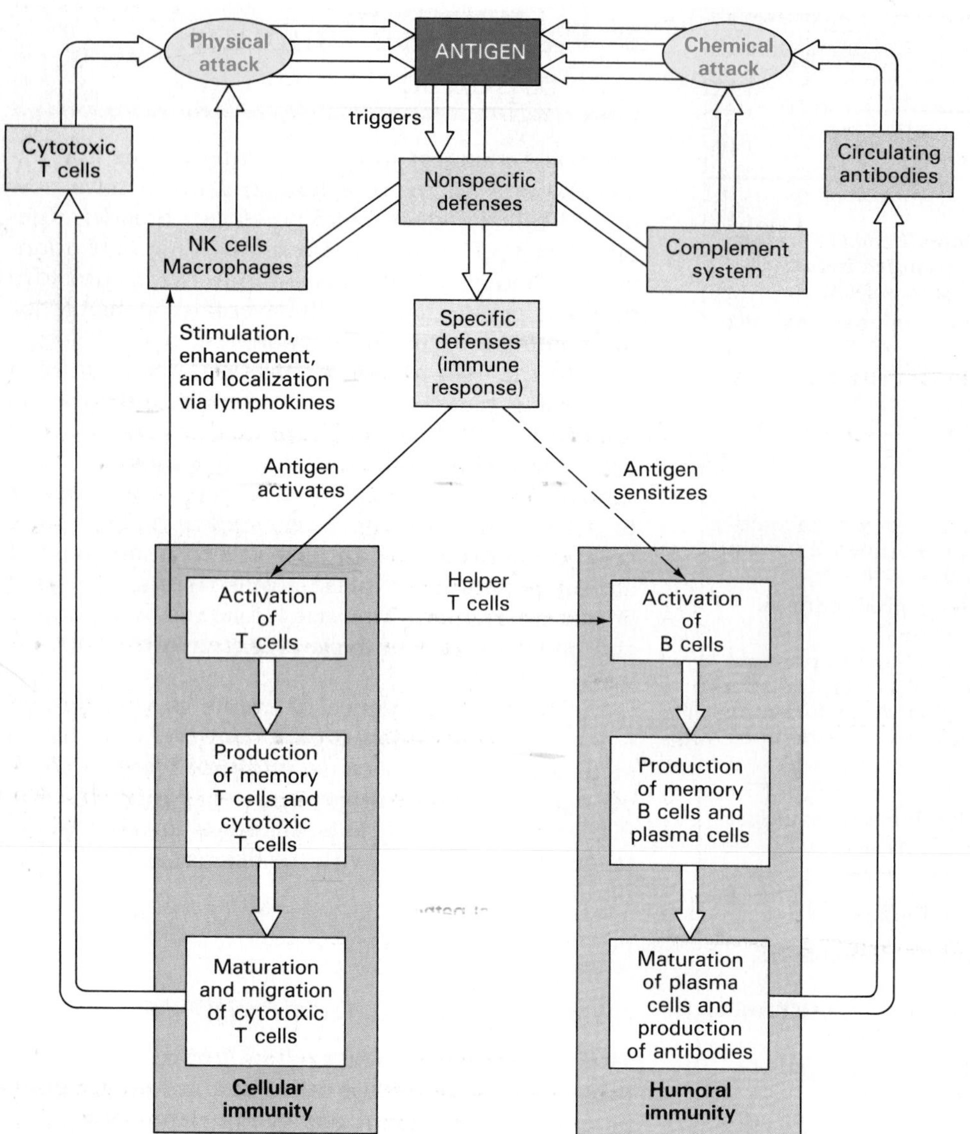

FIGURE 22-25
An Integrated Summary of the Immune Response

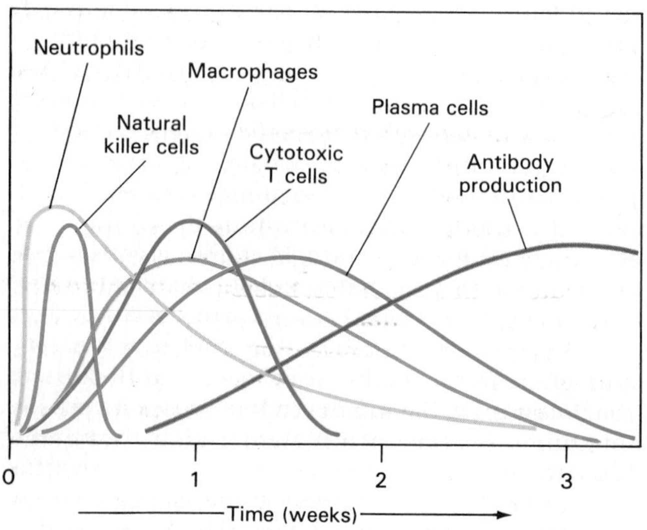

FIGURE 22-26
The Time Course of the Body's Response to Infection. This figure outlines the basic sequence of events that begins with the appearance of pathogens in peripheral tissues.

were previously overcome by maternal antibodies. The immune system also begins to respond to other environmental antigens and to the vaccinations administered by pediatricians. It has been estimated that children encounter a "new" antigen every 6 weeks from birth to age 12. (This fact explains why parents, exposed to the same antigens as children, usually remain healthy while *their* children develop runny noses and colds.) Over this period the concentration of circulating antibodies gradually rises toward normal adult levels, and the populations of memory B cells and T cells continue to increase. [**CM**: *Fetal Infections*]

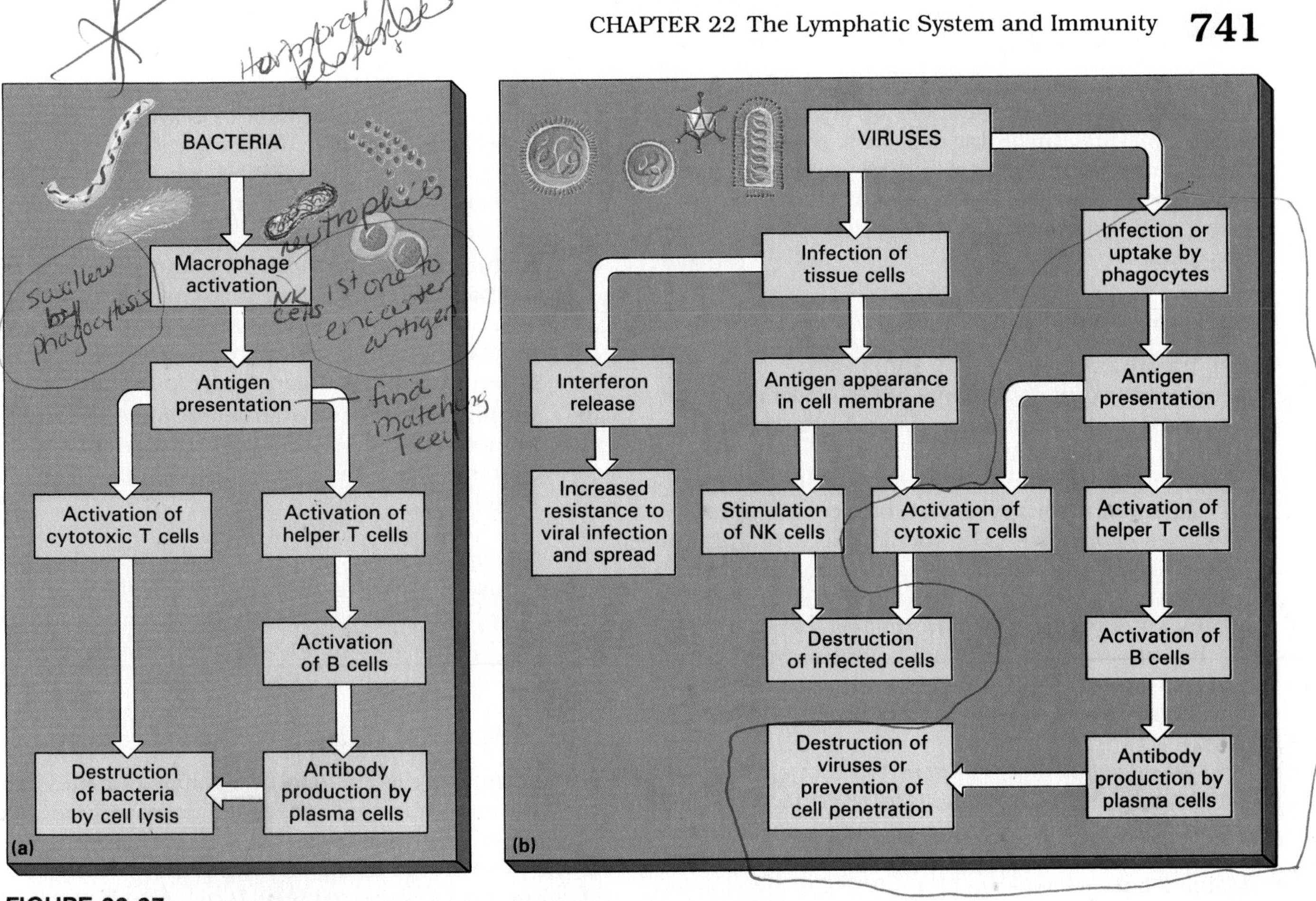

FIGURE 22-27
Defenses against Bacterial and Viral Pathogens. (a) Defenses against bacterial pathogens are usually initiated by active macrophages. (b) Defenses against viruses are usually activated following infection of normal cells.

IMMUNE DISORDERS

Because the immune response is so complex, there are many opportunities for things to go wrong. A great variety of clinical conditions may result from disorders of immune function. In an **immunodeficiency disease** either the immune system fails to develop normally or the immune response is blocked in some way. An important example of such diseases is *acquired immunodeficiency syndrome (AIDS)*. (See the Clinical Comment: AIDS on p. 732.)

Autoimmune Disorders

Autoimmune disorders develop when the immune response mistakenly targets normal body cells and tissues. Many examples of autoimmune disorders are noted in this and earlier chapters.

The immune system usually recognizes and ignores the antigens normally found in the body. The recognition system can malfunction, however, and when it does the activated B cells begin to manufacture antibodies against other cells and tissues. The trigger may be a reduction in suppressor T cell activity, excessive stimulation of helper T cells, tissue damage that releases large quantities of antigenic fragments, haptens bound to compounds normally ignored, viral or bacterial toxins, or some combination of factors.

The symptoms produced depend on the identity of the antigen attacked by these misguided antibodies, called **autoantibodies**. Several conditions described in earlier chapters are autoimmune disorders. For example, the inflammation of *thyroiditis* results from the release of autoantibodies against thyroglobulin. [**CM**: *Thyroid Gland Disorders*] *Rheumatoid arthritis* (Chapter 7) occurs when autoantibodies form immune complexes within connective tissues, especially around the joints. (p. 230) Many other autoimmune disorders appear to be cases of mistaken identity. For example, proteins associated with the measles, Epstein-Barr, influenza, and other viruses contain amino acid sequences that are similar to those of myelin proteins. As a result, antibodies that target these viruses may also attack myelin sheaths. This mechanism accounts for the neurologic complications that sometimes follow a vaccination or viral infection.

The risk of such complications increases if an individual has an unusual type of HLA proteins. At least 50 clinical conditions have been linked to specific variations in HLA structure. Many of these are disorders considered elsewhere in the text or in the Clinical Manual, including psoriasis, rheumatoid arthritis, myasthenia gravis, multiple sclerosis, narcolepsy, Type I diabetes, Graves' disease, Addison's disease, pernicious anemia, systemic lupus, and chronic hepatitis. [**CM**: *Systemic Lupus Erythematosus*]

Allergies

Immunodeficiency and autoimmune disorders are relatively rare conditions—clear evidence of the effectiveness of the immune system's control mechanisms. A far more common, and usually far less dangerous, class of immune disorders are the *allergies*.

Allergies are inappropriate or excessive immune responses to antigens. The sudden increase in cellular activity or antibody titers can have a number of unpleasant side effects. For example, neutrophils or cytotoxic T cells may destroy normal cells while attacking the antigen, or the antibody-antigen complex may trigger a massive inflammatory response. Antigens that trigger allergic reactions are often called **allergens**.

There are several different types of allergies. A complete classification recognizes four categories: *immediate hypersensitivity* (Type I), *cytotoxic reactions* (Type II), *immune complex disorders* (Type III), and *delayed hypersensitivity* (Type IV). The cross-reactions that occur following the transfusion of an incompatible blood type (Chapter 19) is an example of Type II hypersensitivity. (p. 618) Immediate hypersensitivity is considered in the Clinical Comment on page 744. For a discussion of other allergy types, consult the Clinical Manual. [**CM**: *Immune Complex Disorders and Delayed Hypersensitivity*]

STRESS AND THE IMMUNE RESPONSE

One of the normal effects of interleukin-1 secretion is the stimulation of ACTH production by the anterior pituitary. This in turn leads to the secretion of glucocorticoids by the adrenal cortex. (The functions of ACTH and glucocorticoids were described in Chapter 18. (p. 577)) The anti-inflammatory effects of the glucocorticoids may help control the size of the immune response. However, the long-term secretion of glucocorticoids, as in the resistance phase of the general adaptation syndrome (Chapter 18), can inhibit the immune response and lower resistance to disease. (p. 599) Several of the tissue responses to glucocorticoids alter the effectiveness of specific and nonspecific defenses. Examples include the following:

1. **Depression of the inflammatory response**: Glucocorticoids inhibit mast cells and make capillaries less permeable. Inflammation is therefore less likely, and when it does occur the capillary effects slow the entry of fibrinogen, complement, and cellular defenders.
2. **Reduction in the activities and numbers of phagocytes in peripheral tissues**: This reduction further impairs nonspecific defense mechanisms and interferes with the processing and presentation of antigens to lymphocytes.
3. **Inhibition of interleukin secretion**: A reduction in interleukin production depresses the lymphocytic response, even to antigens bound to HLA proteins.

The mechanisms responsible for these changes are still under investigation, but it should be noted that immune system depression due to chronic stress may represent a serious threat to health.

AGE AND THE IMMUNE RESPONSE

With advancing age the immune system becomes less effective at combatting disease. T cells become less responsive to antigens; as a result, fewer cytotoxic T cells respond to an infection. Because the number of helper T cells is also reduced, B cells are less responsive, and antibody levels do not rise as quickly after antigen exposure. The net result is an increased susceptibility to viral and bacterial infection. For this reason, vaccinations for acute viral diseases, such as the flu (influenza), are strongly recommended for elderly individuals. The increased incidence of cancer in the elderly reflects the fact that immune surveillance declines, and tumor cells are not eliminated as effectively.

- **What kind of immunity protects the developing fetus and how is it produced?**
- **How can increased stress decrease the effectiveness of the immune response?**

Integration with Other Systems

Figure 22-28 summarizes the interactions between the lymphatic system and other physiological systems.

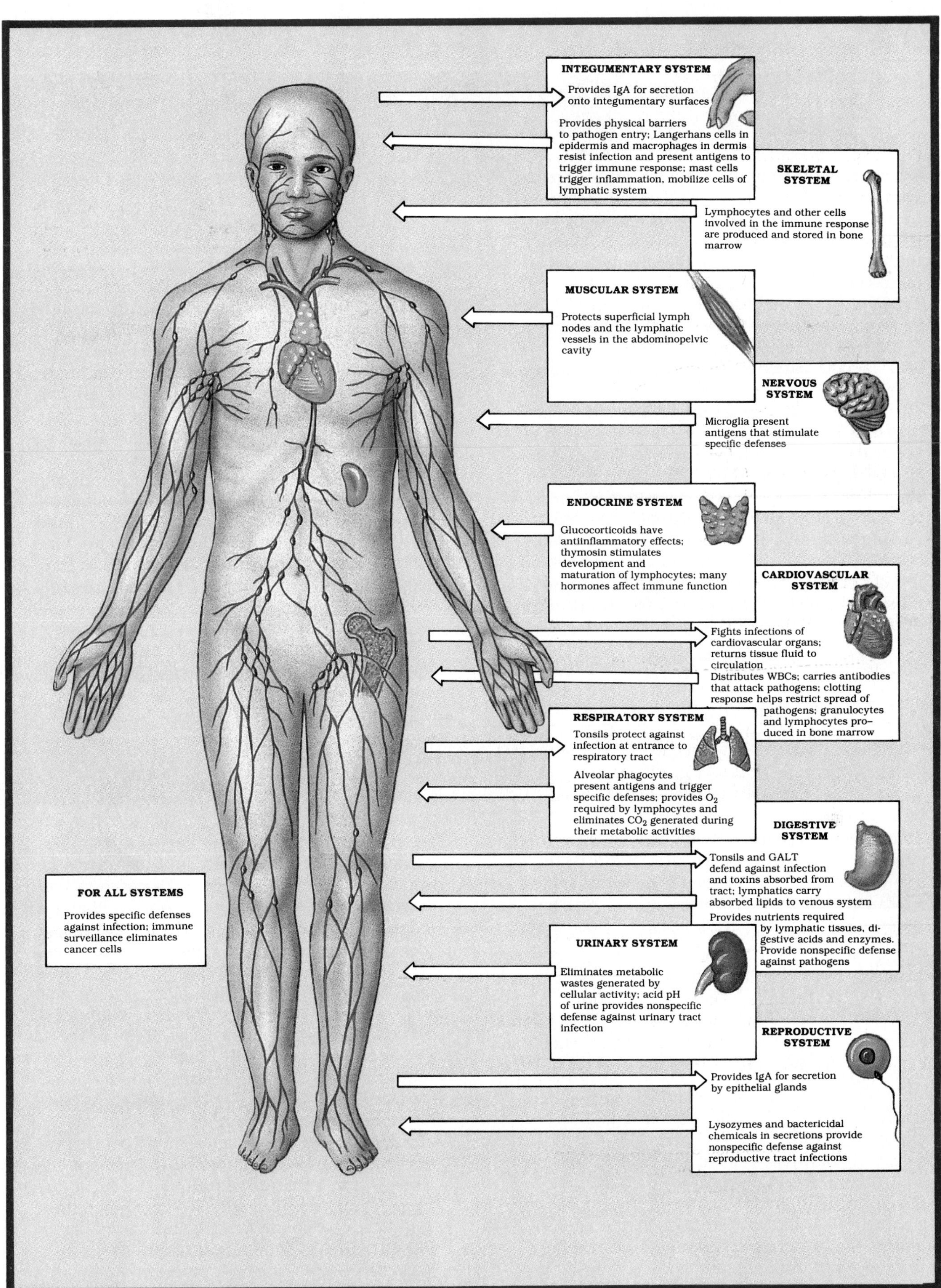

INTEGUMENTARY SYSTEM
Provides IgA for secretion onto integumentary surfaces
Provides physical barriers to pathogen entry; Langerhans cells in epidermis and macrophages in dermis resist infection and present antigens to trigger immune response; mast cells trigger inflammation, mobilize cells of lymphatic system
SKELETAL SYSTEM
Lymphocytes and other cells involved in the immune response are produced and stored in bone marrow
MUSCULAR SYSTEM
Protects superficial lymph nodes and the lymphatic vessels in the abdominopelvic cavity
NERVOUS SYSTEM
Microglia present antigens that stimulate specific defenses
ENDOCRINE SYSTEM
Glucocorticoids have antiinflammatory effects; thymosin stimulates development and maturation of lymphocytes; many hormones affect immune function
CARDIOVASCULAR SYSTEM
Fights infections of cardiovascular organs; returns tissue fluid to circulation
Distributes WBCs; carries antibodies that attack pathogens; clotting response helps restrict spread of pathogens; granulocytes and lymphocytes produced in bone marrow
RESPIRATORY SYSTEM
Tonsils protect against infection at entrance to respiratory tract
Alveolar phagocytes present antigens and trigger specific defenses; provides O_2 required by lymphocytes and eliminates CO_2 generated during their metabolic activities
DIGESTIVE SYSTEM
Tonsils and GALT defend against infection and toxins absorbed from tract; lymphatics carry absorbed lipids to venous system
Provides nutrients required by lymphatic tissues, digestive acids and enzymes. Provide nonspecific defense against pathogens
URINARY SYSTEM
Eliminates metabolic wastes generated by cellular activity; acid pH of urine provides nonspecific defense against urinary tract infection
REPRODUCTIVE SYSTEM
Provides IgA for secretion by epithelial glands
Lysozymes and bactericidal chemicals in secretions provide nonspecific defense against reproductive tract infections
FOR ALL SYSTEMS
Provides specific defenses against infection; immune surveillance eliminates cancer cells

FIGURE 22-28
Functional Relationships between the Lymphatic System and Other Systems

Clinical Comment: Transplantations and Graft Rejection

Organ transplantation is an important treatment option for patients with severe disorders of the kidneys, liver, heart, lungs, or pancreas. In 1989 surgeons transplanted approximately 8800 kidneys, 900 hearts, 500 livers, and 150 pancreases. These numbers are limited by the availability of suitable organ donors; during the same year there were 8000–10,000 patients who could have benefited from a heart transplant, and 16,000 patients were awaiting a kidney transplant.

Other than their primary illness, the major problem facing these patients before surgery is the availability of a suitable donor. Once surgery has been performed, the major problem is **graft rejection**. T cells sensitive to antigens bound to Class I HLA molecules can be activated by contact with cells other than macrophages and by the HLA proteins found on the surfaces of foreign cells. In the process of graft rejection T cells are activated by contact with the tissues of the donated organ. The cytotoxic T cells that develop then attack and destroy the foreign cells.

Graft rejection will not occur if the HLA molecules of two individuals are identical, and for this reason grafts made between identical twins are usually successful. The greater the difference in HLA structure, the greater the likelihood that graft rejection will occur. The process of *tissue typing* assesses the degree of similarity between the HLA complexes of two different individuals. For this process, lymphocytes are collected and examined, because they can easily be obtained from the blood and bear both Class I and Class II HLA molecules.

It has long been apparent that significant improvements in transplant success could be made by reducing the sensitivity of the immune system. Until recently the drugs used to produce this **immunosuppression** did not selectively target the immune system. For example, **prednisone** (PRED-ni-sōn), a corticosteroid, was used for its anti-inflammatory effect despite its effect

Clinical Comment: Immediate Hypersensitivity

Immediate hypersensitivity begins with the process of **sensitization**. Sensitization is the initial exposure to an allergen that leads to the production of large quantities of IgE. The tendency to produce IgE antibodies in response to specific allergens may be genetically determined. In drug reactions, such as penicillin allergies, IgE antibodies are produced to a hapten (partial antigen) bound to a larger molecule inside the body.

Following sensitization, the antibodies become attached to the cell membranes of basophils and mast cells throughout the body. When exposed to the same allergen at a later date, the bound antibodies stimulate these cells to release histamine, heparin, and other chemicals into the surrounding tissues. If allergen exposure occurs at the body surface, the response may be restricted to that area. If the allergen enters the systemic circulation, the response may be more dramatic and occasionally lethal.

In **anaphylaxis** (a-na-fi-LAK-sis; *ana-*, again + *phylaxis*, protection) a circulating allergen affects mast cells throughout the body (Figure 22-29). The entire range of symptoms can develop within a matter of minutes. Changes in capillary permeabilities produce swelling and edema in the dermis, and raised welts, or hives, appear on the surface of the skin. Smooth muscles along the respiratory passageways contract, and the narrowed passages make breathing extremely difficult. In severe cases of anaphylaxis an extensive peripheral vasodilation occurs, producing a fall in blood pressure that may lead to a circulatory collapse. This response has been called **anaphylactic shock**.

Many of the symptoms of immediate hypersensitivity can be prevented by the prompt administration of **antihistamines** (an-ti-HIS-ta-mēnz), drugs that block the action of histamine. *Benadryl* is a popular antihistamine available without a prescription. Treatment of severe anaphylaxis involves antihistamine corticosteroid, epinephrine and injections.

on glucose metabolism (Chapter 18). ∞ (p. 587) Two other drugs, **cyclophosphamide** (sī-klo-FOS-fa-mīd) and **azathioprine** (a-za-THĪ-ō-prēn) are more powerful, but they have greater associated risks. These drugs reduce the rates of cellular growth and replication throughout the body. Hematopoiesis slows dramatically, and undesirable side effects may develop in the reproductive, nervous, and integumentary systems.

An understanding of the chemical communication between helper and suppressor T cells, macrophages, and B cells has now led to the development of drugs with more selective effects. For example, **cyclosporin A (CsA)**, a compound derived from a fungus, was the most important immunosuppressive drug developed in the 1980s. The results were dramatic. In the early 1980s, before the use of cyclosporin, the 5-year survival rate for liver transplants was below 20 percent. In the early 1990s the survival rate is approximately 80 percent, and new drugs now undergoing clinical trials may further improve survival. One of the most promising drugs, tested under the name FK 506, specifically inhibits lymphocytes.

An obvious problem posed by the use of any immunosuppressive drug is that the individual becomes more susceptible to viral and bacterial infections. Immunosuppression may continue for a year or more after a transplant is performed, and the patient may not recover full immune function for months after treatment has been discontinued. Over this period the individual must be treated with injections of IgG from pooled sera (*gamma globulin*) and, when necessary, antibiotics. A more subtle risk of immunosuppression is the reduction in immune surveillance by NK cells. Transplant patients are 100 times more likely to develop cancer than others in their age group. Lymphoma-type cancers are the most common; these cancers appear to be linked to post-transplant infection with the Epstein-Barr virus.

FIGURE 22-29
The Mechanism Responsible for Anaphylaxis

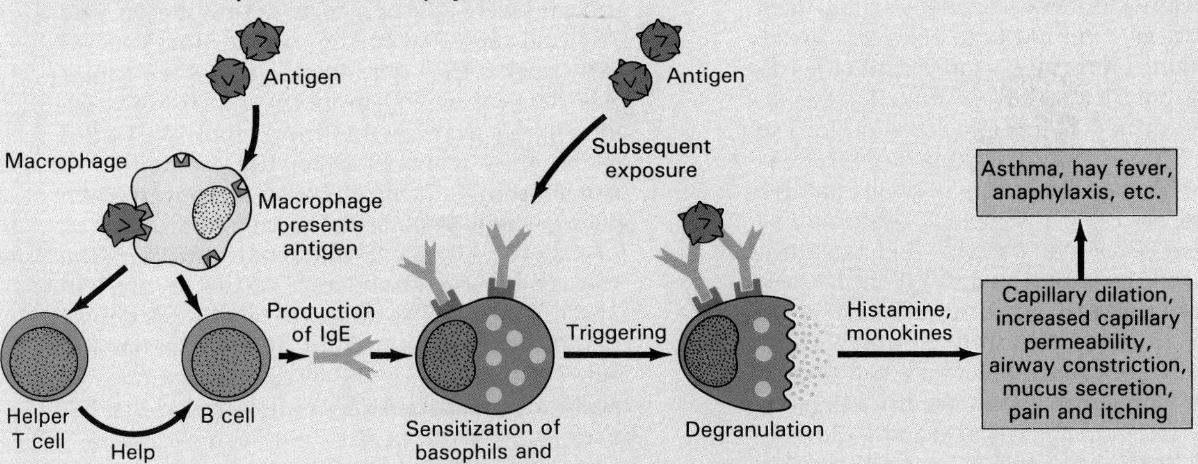

Health News: Technology, Immunity, and Disease

Our understanding of disease mechanisms has been profoundly influenced by recent advances in genetic engineering and protein analysis. The information and technical capabilities developed during the 1980s have already started to affect clinical procedures. This trend is sure to continue, and several lines of research are particularly exciting because of their broad potential application. These projects involve a mixture of genetic engineering, computer analysis, and protein biochemistry.

It is possible to identify an individual B cell responsible for producing a specific antibody. That B cell can then be isolated and fused with a cancer cell. This produces a **hybridoma** (hī-bri-DŌ-ma), a cancer cell that produces large quantities of a single antibody. Hybridomas undergo rapid mitotic divisions, like other cancer cells, so culturing that original hybridoma cell soon produces an entire population, or **clone**, of genetically identical cells. This particular clone will produce large quantities of a **monoclonal** (mo-nō-KLŌ-nal) **antibody**.

Monoclonal antibodies are useful because they are free from impurities. One important use for this technology has been the development of antibody tests for the clinical analysis of body fluids. Labeled antibodies can be used to detect small quantities of specific antigens in a sample of plasma or other body fluids. For example, a popular home pregnancy test relies on monoclonal antibodies that detect small amounts of a placental hormone, *human chorionic gonadotrophin* (*HCG*), in the urine of a pregnant woman. Other monoclonal antibodies are used in standard blood screening tests for venereal diseases and urine tests for ovulation.

Monoclonal antibodies can also be used to provide passive immunity to disease. Passive immunizations using monoclonal antibodies do not cause any of the unpleasant side effects associated with antibodies from pooled sera, for the product will not contain plasma proteins, viruses, or other contaminants. The antibodies can be made to order by exposing a population of B cells to a particular antigen, then isolating any B cells that produce the desired antibodies.

Genetic engineering techniques can also be used to promote immunity in other ways. One interesting approach involves gene-splicing techniques. The genes coding for an antigenic protein of a viral or bacterial pathogen are identified, isolated, and inserted into a harmless bacterium that can be cultured in the laboratory. The clone that eventually develops will produce large quantities of pure antigen that can then be used to stimulate a primary immune response. Vaccines against malaria and hepatitis were developed in this manner, and a similar strategy may be used to design an AIDS vaccine.

A more controversial experimental technique involves taking a pathogenic organism and adding or removing genes to make it harmless. The modified pathogen can then be used to produce active immunity without the risk of severe illness. Fears that the engineered organism could mutate or regain its pathogenic habits have so far limited the use of this approach, even in animal trials.

Hybridomas that manufacture other products of the immune system can also be produced. Interferons are not effective against all viruses, but interferon nasal sprays appear to provide resistance against the viruses responsible for the common cold. (Unfortunately these sprays can cause nasal bleeding and other unpleasant side effects, and they have not been approved for sale to the public.) Interferons can also control certain forms of virally induced cancers. Interleukins may prove useful in increasing the intensity of the immune response.

As our understanding of the immune system grows, complex therapies involving combinations of lymphokines, interleukins, and monoclonal antibodies are appearing. For example, in a recent clinical trial, cytotoxic T cells were removed from patients with malignant melanoma, a particularly dangerous cancer discussed in Chapter 6. (p. 189) These lymphocytes were able to recognize tumor cells, and they had migrated to the tumor. However, for some reason they were unable to kill the tumor cells. The extracted lymphocytes were cultured, and viruses were used to insert multiple copies of the genes responsible for the production of tumor necrosis factor. The patients were then given periodic infusions of these "supercharged" T cells. To enhance T cell activity further, the researchers also administered doses of interleukin-2. Although the outcome of this particular therapy remains uncertain, it is clear that the ability to manipulate the immune response will revolutionize the treatment of many serious diseases.

CHAPTER REVIEW

Review of Selected Clinical Terms

lymphadenopathy (lim-fad-e-NOP-a-thē) (*p. 715*)
Chronic or excessive enlargement of lymph nodes.

acquired immune deficiency syndrome (AIDS) (*p. 732*)
A disorder that develops, following HIV infection, characterized by reduced circulating antibody levels and depressed cellular immunity.

AIDS-related complex (ARC) (*p. 732*)
Early symptoms of HIV infection, consisting of lymphadenopathy and chronic infections.

active immunization (*p. 734*)
An immunization that intentionally stimulates a primary response to a pathogen before it is encountered in the environment.

vaccine (vak-SĒN) (*p. 734*)
A preparation of antigens derived from a specific pathogen; administered during immunization.

passive immunization (*p. 734*)
An immunization in which the patient receives a dose of antibodies that will attack a pathogen immediately.

immunodeficiency disease (*p. 741*)
A disease in which either the immune system fails to develop normally or the immune response is somehow blocked.

autoimmune disorder (*p. 741*)
A disorder that develops when the immune response mistakenly targets normal body cells and tissues.

allergy (*p. 742*)
An inappropriate or excessive immune response to antigens.

immunosuppression (*p. 744*)
A reduction in the sensitivity of the immune system.

anaphylaxis (a-na-fi-LAK-sis) (*p. 744*)
A type of allergy in which a circulating allergen affects mast cells throughout the body, producing numerous symptoms very quickly.

anaphylactic shock (*p. 744*)
A drop in blood pressure that may lead to circulatory collapse, resulting from severe cases of anaphylaxis.

hybridoma (hī-bri-DŌ-ma) (*p. 746*)
A cell formed by fusion of a cancer cell with a plasma cell; used to produce large amounts of a single antibody.

monoclonal (mo-nō-KLŌ-nal) antibody (*p. 746*)
An antibody produced in large quantities by a population of genetically identical cells.

Additional Terms of Clinical Importance

AIDS (*p. 732*): **human immunodeficiency virus (HIV), Karposi's sarcoma**

PRINCIPLES OF IMMUNIZATION (*p. 734*): **attenuated, inactivated**

TRANSPLANTATIONS AND GRAFT REJECTION (*p. 744*): **graft rejection, prednisone, cyclophosphamide, azathioprine, cyclosporin A (CsA)**

ALLERGIES (*p. 742*): **allergen**

IMMEDIATE HYPERSENSITIVITY (*p. 744*): **sensitization, antihistamine**

TECHNOLOGY, IMMUNITY, AND DISEASE (*p. 746*): **clone**

Study Outline

Related Key Terms

Introduction (pp. 709–710)

1. The cells, tissues, and organs of the **lymphatic system** play a central role in the body's defenses.

Organization of the Lymphatic System (pp. 710–720)

1. The lymphatic system includes a network of **lymphatic vessels** that carry **lymph** (a fluid similar to plasma but with a lower concentration of proteins). A series of **lymphatic organs** are connected to the lymphatic vessels.

FUNCTIONS OF THE LYMPHATIC SYSTEM (p. 710)

2. The lymphatic system produces, maintains, and distributes lymphocytes (cells that attack invading organisms, abnormal cells, and foreign proteins). The system also helps maintain blood volume and eliminate local variations in the composition of the interstitial fluid.

LYMPHATIC VESSELS (pp. 710–712)

3. Lymph flows along a network of **lymphatics** that originate in the **lymphatic capillaries** (*terminal lymphatics*). The lymphatic vessels empty into the **thoracic duct** and the **right lymphatic duct**. — Related key terms: **cisterna chyli**

Chapter Review

Related Key Terms

LYMPHOCYTES (pp. 712–713)

4. There are three different classes of lymphocytes: **T cells** (**t**hymus-dependent), **B cells** (**b**one marrow-derived), and **NK cells** (**n**atural **k**iller).
5. **Cytotoxic T cells** attack foreign cells or body cells infected by viruses; they provide **cellular immunity**. **Regulatory T cells** regulate and coordinate the immune response.
6. B cells can differentiate into **plasma cells**, which produce and secrete *antibodies* that react with specific chemical targets, or **antigens**. Antibodies in body fluids are called **immunoglobulins**. B cells are responsible for **humoral immunity**.
7. NK cells (also called **large granular lymphocytes**) attack foreign cells, normal cells infected with viruses, and cancer cells. They provide **immunological surveillance**.
8. Lymphocytes continually migrate in and out of the blood through the lymphatic tissues and organs. **Lymphopoiesis (lymphocyte production)** involves the **bone marrow**, **thymus**, and **peripheral lymphatic tissues**.

   thymosin

LYMPHATIC TISSUES (pp. 713–715)

9. **Lymphatic tissues** are connective tissues dominated by lymphocytes. In a **lymphatic nodule** the lymphocytes are densely packed in an area of loose connective tissue.

   pharyngeal (adenoid) tonsil
   palatine tonsils • **lingual tonsils**

LYMPHATIC ORGANS (pp. 715–720)

10. Important **lymphatic organs** include the **lymph nodes**, the **thymus**, and the **spleen**. Lymphatic tissues and organs are distributed in areas especially vulnerable to injury or invasion.
11. **Lymph nodes** are encapsulated masses of lymphatic tissue. The **cortex** is dominated by T cells; the inner **medulla** contains B cells arranged into **medullary cords**.

    trabeculae

12. The thymus lies behind the sternum, in the anterior mediastinum. **Epithelial cells** scattered among the lymphocytes produce thymic hormones.

    Hassall's corpuscles

13. The adult spleen contains the largest mass of lymphatic tissue in the body. The cellular components form the **pulp** of the spleen. **Red pulp** contains large numbers of red blood cells, and areas of **white pulp** resemble lymphatic nodules.

    trabecular arteries
    trabecular veins

THE LYMPHATIC SYSTEM AND BODY DEFENSES (p. 720)

14. The lymphatic system is a major component of the body's defenses. These fall into two categories: (1) **nonspecific defenses** that do not discriminate between one threat and another; and (2) **specific defenses** that protect against threats on an individual basis.

Nonspecific Defenses (pp. 720–726)

PHYSICAL BARRIERS (pp. 720–721)

1. Physical barriers include hair, epithelia, and various secretions of the integumentary and digestive systems.

PHAGOCYTIC CELLS (pp. 721–722)

2. There are two types of phagocytic cells: **microphages** and **macrophages** (cells of the **monocyte-macrophage system**). Microphages are the neutrophils and eosinophils in circulating blood.
3. Phagocytes move between cells through **diapedesis**, and they show **chemotaxis** (sensitivity and orientation to chemical stimuli).

   microglia • **Kupffer cells**
   Langerhans cells
   phagocytic giant cells
   alveolar macrophages

IMMUNOLOGICAL SURVEILLANCE (pp. 722–723)

4. **Immunological surveillance** involves constant monitoring of normal tissues by NK cells sensitive to abnormal antigens on the surfaces of otherwise normal cells. Cancer cells with **tumor-specific antigens** on their surfaces are killed.

   perforins

INTERFERONS (p. 723)

5. **Interferons**, small proteins released by cells infected with viruses, trigger the production of **antiviral proteins** that interfere with viral replication inside the cell. Interferons are **cytokines**, chemical messengers released by tissue cells to coordinate local activities.

COMPLEMENT (pp. 723–724)

6. At least 11 **complement proteins** make up the **complement system**. They interact with each other in chain reactions to destroy target cell membranes, stimulate inflammation, attract phagocytes, and/or enhance phagocytosis.

   classic pathway • **opsonins**
   opsonization
   alternative pathway
   properdin

Related Key Terms

INFLAMMATION (pp. 724–725)

7. Inflammation represents a coordinated nonspecific response to tissue injury.

FEVER (pp. 725–726)

8. A **fever** (body temperature greater than 37.2° C or 99° F) can inhibit pathogens and accelerate metabolic processes.

pyrogens
endogenous pyrogen
(interleukin-1 or Il-1)

Specific Defenses: The Immune Response (pp. 726–739)

1. Lymphocytes provide specific immunity, which has four general characteristics: specificity, versatility, memory, and tolerance. **Memory cells** enable the immune system to "remember" previous target antigens. **Tolerance** refers to the ability of the immune system to ignore some antigens, such as those of body cells.

INNATE VERSUS ACQUIRED IMMUNITY (pp. 726–727)

2. Specific immunity may involve **innate immunity** (genetically determined and present at birth) or **acquired immunity**. The two types of acquired immunity are **active immunity** (which appears following exposure to an antigen) and **passive immunity** (produced by the transfer of antibodies from another source).

naturally acquired immunity
active immunization
natural passive immunity
passive immunization

AN OVERVIEW OF THE IMMUNE RESPONSE (pp. 727–728)

3. The goal of the **immune response** is the destruction or inactivation of pathogens, abnormal cells, and foreign molecules.
4. Lymphocytes sensitive to a particular antigen have receptors in their membranes that can bind that particular antigen, a process known as **antigen recognition**. Binding activates the cell, causing it to divide and produce more cells with the same antigen sensitivity. The response to antigens includes both a short-term component (cytotoxic T cells and antibody-producing plasma cells) and a long-term component (memory T and B cells).

T CELLS AND CELLULAR IMMUNITY (pp. 728–731)

5. Foreign antigens must usually be processed by **accessory cells** (most often macrophages) and incorporated into their membranes (**antigen presentation**) before they can activate lymphocytes. T cells can also be activated by antigens displayed on the surface of most body cells—for example, viral antigens in the membranes of virus-infected cells.
6. All body cells have membrane glycoproteins called **human leukocyte antigens (HLAs)**. Lymphocytes are not activated by HLAs alone but will respond to an antigen bound to an HLA molecule.

Class I HLA • **Class II HLA**

7. **Cell-mediated immunity** (**cellular immunity**) results from the activation of T cells by antigens bound to Class I HLAs. When activated, these T cells divide to generate *cytotoxic T cells* and **memory T cells** that remain on reserve to guard against future such attacks.

lymphotoxin

8. **Helper T cells** secrete lymphokines that help coordinate specific and nonspecific defenses and regulate cellular and humoral immunity. **Suppressor T cells** depress the responses of other T and B cells.

B CELLS AND HUMORAL IMMUNITY (pp. 731–738)

9. Antigens range from a single polypeptide to an entire bacterium. Antibodies focus on specific **antigen determinant sites**.
10. B cell activation occurs primarily with help from helper T cells. An activated B cell may differentiate into a **plasma cell** or produce daughter cells that differentiate into plasma cells and **memory B cells**. The antibodies produced by plasma cells are the agents of the **primary response**. The maximum **antibody titer** appears during the **secondary** (**anamnestic**) **response** to antigen exposure.
11. When antibody molecules bind to an antigen they form an **antibody-antigen complex**. Effects that appear after binding include **neutralization; precipitation** (formation of an insoluble **immune complex**) and **agglutination** (formation of large complexes); **opsonization**; stimuation of **inflammation**; and prevention of bacterial/viral adhesion.
12. There are five classes of antibodies in body fluids: (1) **immunoglobulin G** (**IgG**), responsible for resistance against many viruses, bacteria, and bacterial toxins; (2) **IgE**, which releases chemicals that accelerate local inflammation; (3) **IgD**, found on the surfaces of B cells; (4) **IgM**, the first antibody type secreted after an antigen arrives; and (5) **IgA**, found in glandular secretions.

Chapter Review

	Related Key Terms
HORMONES OF THE IMMUNE SYSTEM (pp. 738–739)	
13. **Interleukins** increase T cell sensitivity to antigens exposed on macrophage membranes; stimulate B cell activity, plasma cell formation, and antibody production; and enhance nonspecific defenses.	
14. **Interferons** slow the spread of a virus by making the synthesizing cell and its neighbors resistant to viral infections.	**alpha interferons** **gamma interferons**
15. **Tumor necrosis factors** slow tumor growth and kill tumor cells.	
16. Several lymphokines adjust the activities of phagocytic cells in order to coordinate specific and nonspecific defenses.	

Patterns of Immune Response (pp. 739–742)

IMMUNOLOGICAL COMPETENCE (pp. 739–740)

1. **Immunological competence** is the ability to demonstrate an immune response upon exposure to an antigen. The developing fetus receives passive immunity from the maternal bloodstream. After delivery the infant begins developing acquired immunity following exposure to environment antigens.

IMMUNE DISORDERS (pp. 741–742)

2. In an **immunodeficiency disease** either the immune system does not develop normally or the immune response is somehow blocked. **Autoimmune disorders** develop when the immune response mistakenly targets normal body cells and tissues. — *Related key terms:* **autoantibodies**
3. **Allergies** are inappropriate or excessive immune responses to **allergens** (antigens that trigger allergic reactions). There are four types of allergies: *immediate hypersensitivity* (Type I), *cytotoxic reactions* (Type II), *immune complex disorders* (Type III), and *delayed hypersensitivity* (Type IV).

STRESS AND THE IMMUNE RESPONSE (p. 742)

4. Interleukin-1 released by active macrophages triggers the release of ACTH by the anterior pituitary. Glucocorticoids produced by the adrenal cortex moderate the immune response, but their long-term secretion can lower resistance to disease.

AGE AND THE IMMUNE RESPONSE (p. 742)

5. With aging, the immune system becomes less effective at combatting disease.

Review Planner

	Objective	Level 1
1	Identify the major components of the lymphatic system and explain their functions.	**1 11 20 26 30 33 40**
2	Discuss the importance of lymphocytes and describe where they are found in the body.	**2 6 13 14 24**
3	List the body's nonspecific defenses and explain how each functions.	**3 4 6 7 8 19 28 37 48**
4	Describe the characteristics of the body's specific defenses.	**3 4 9 31 32**
5	Compare the different types of immunity and their origins.	**4 5 10 15 23**
6	Explain how lymphocytes are activated and how they distinguish between foreign cells and cells of the body.	**9 16 36**
7	Discuss the roles of the various types of lymphocytes in an immune response.	**5 12 15 23 42 43 44 45 47 49 50**

Level 2: **25 27 34 35**

Level 3: **51–56**

CM: **Lymphedema • Infected Nodules**
Lymphomas
Disorders of the Spleen • AIDS
Lyme Disease • Fetal Infections
SCID
Systemic Lupus Erythematosus
Immune Complex Disorders and Delayed Hypersensitivity

ABC News: **Bone Marrow Transplants**
Multiple Sclerosis

8	Describe the structure of antibody molecules and explain how they function.	**5 17 22**
9	Relate allergic reactions and autoimmune disorders to immune mechanisms.	**18 29 41**
10	Specify the factors that can enhance or reduce one's resistance to disease.	**5 16 39**
11	Describe the changes in the immune system that occur with aging.	**38**

Review Questions

MULTIPLE CHOICE

1. A blockage of the thoracic duct would affect lymph flow from all of the following *except*: a) the left half of the head; b) the right half of the head; c) the left leg; d) the pelvis.
2. Approximately 80 percent of circulating lymphocytes are: a) B cells; b) T cells; c) NK cells; d) none of the above.
3. A cat scratches your finger, and the area around the scratch becomes red, swollen, and sore. This is an example of a: a) specific defense; b) nonspecific defense.
4. You had chickenpox as a child. Since then it has become your task to nurse younger brothers and sisters through their bouts with chickenpox. You owe this to a: a) specific defense; b) nonspecific defense.
5. In the example above, what mechanism protects you from developing chickenpox 10 years after your first exposure?: a) primary response; b) secondary response.
6. Decreased production of NK cells would impair: a) immune surveillance; b) the complement system; c) humoral immunity; d) the monocyte-macrophage system.
7. If a cell is infected by a virus, it can defend the body by releasing: a) perforins; b) tumor-specific antigens; c) eukaryotic cells; d) interferons.
8. Which of the following is *not* a direct result of complement activation?: a) immunological escape; b) destruction of target cell walls; c) enhanced inflammatory response; d) opsonization.
9. All of the cells in your body contain antigens, but normally these antigens do not stimulate an immune response. This is an example of: a) versatility; b) specificity; c) tolerance; d) immunological surveillance.
10. The fact that you usually do not catch the disease when your pet cat is ill is an example of: a) natural immunity; b) acquired immunity.

DIAGRAMS

11. Cover the labels in Figure 22-4 and identify the major lymphatic ducts and nodes as well as the veins illustrated.

TRUE/FALSE

(If a statement is false, your instructor may wish to have you correct it.)

12. **T/F:** The ratio of B cells to T cells is fairly constant throughout the body.
13. **T/F:** Fixed macrophages can become free macrophages under certain condtions.

MATCHING QUESTIONS

(Match each term with the most appropriate statement.)

Statements:

14. A phagocytic cell that lives in a specific organ.
15. Immunity resulting from the presence of circulating antibodies produced by plasma cells.
16. The ability to demonstrate an immune response upon exposure to an antigen.
17. The process in which target cells are coated with antibodies.
18. Inappropriate or excessive immune responses to antigens.
19. Chemical messengers released by tissue cells to coordinate local activities.
20. Large lymphatic nodules in the walls of the pharynx.
21. The body's ability to resist pathogens.
22. Antibodies found in body fluids.
23. Resistance to disease due to sensitized T cells that destroy antigen-bearing cells.
24. A phagocytic cell that travels through the body.

Terms:

a) Immunoglobulins
b) Tonsils
c) Immunity
d) Free macrophage
e) Fixed macrophage
f) Cytokines
g) Opsonization
h) Humoral immunity
i) Cell-mediated immunity
j) Immunological competence
k) Allergies

Chapter Review

SHORT ANSWER/ESSAY

25. Distinguish between the lymphatic system and the immune system.
26. Fill in the blank: The spleen performs the same functions for the ________ that the lymph nodes perform for lymph.
27. Explain the relationship between the lymphatic system and the circulatory system.
28. Cancer cells probably appear throughout life. Why don't more people develop cancer?
29. What do rheumatoid arthritis, psoriasis, and Graves' disease have in common?
30. A nursing instructor mentions that lymphopoiesis is severely reduced in a certain patient. Explain this term. In what part(s) of the body does this occur?
31. Fill in the blanks: In humoral immunity, activated ________ cells divide into daughter cells that differentiate into ________ cells and ________ ________ cells.
32. Fill in the blanks: In cell-mediated immunity, ________ cells divide to generate ________ ________ cells and ________ ________ cells.
33. What is GALT?
34. What do specific and nonspecific defenses have in common? How do they differ?
35. You inhale a virus; it enters one of your cells and begins replicating. Describe three mechanisms in your body that protect you from any ill effects.
36. Why do our bodies normally not develop an immune response against our own tissues?
37. Discuss the major categories of nonspecific defenses in the body.
38. Why are vaccinations against influenza more important for the elderly than for young people?
39. As any parent can attest, young children seem to "get sick" quite often. Explain why.
40. Describe the appearance of the large lymphatics that lead toward the trunk of the body. Relate their appearance to their function(s).
41. A friend is stung by a wasp and almost immediately becomes unable to speak, has difficulty breathing, and loses consciousness. Explain these events. How can these symptoms be treated?

Give a function for each of the following:

42. Cytotoxic T cells
43. Regulatory T cells
44. Plasma cells
45. Large granular lymphocytes
46. Thymus epithelial cells
47. NK cells
48. Pyrogens
49. Interleukins
50. Tumor necrosis factors

CRITICAL THINKING/APPLICATIONS

51. A patient is diagnosed as having a blockage of the right lymphatic duct. What condition could result? What symptoms would appear, and in what area(s) of the body?
52. An anesthesia technician is advised that she should be vaccinated against hepatitis B, which is caused by a virus. She is given one injection and told to come back for a second injection in a month and a third injection after six months. Explain the reason for these instructions.
53. A 10-year-old suffers from frequent sore throats. His mother asks the doctor if he should undergo a tonsillectomy. Would you recommend a tonsillectomy for this patient? Explain your answer.
54. An elderly man develops peritonitis following a ruptured appendix. Explain why this occurred.
55. What do bacteria and viruses have in common? How do they differ?
56. A pregnant woman develops rubella; it is a mild case, and she soon recovers. However, her baby is born with microcephaly and cataracts. Explain what happened. Could this have been prevented, and if so how?

Career Closeup: Doctor of Internal Medicine

Brian Kirchner had no intention of going to college when he was in high school. "I was interested in wood shop and making clay pots. I was going to be an industrial arts major. But I went to a community college for two years and then I became ambitious. I went for a degree in applied science and decided that I wanted to go to medical school.

"Since I was afraid that I if I didn't get into med school I wouldn't have a job, my 'insurance' was to become certified as a medical lab technician. At least I had a trade and I could earn my tuition money. The job not only supported me, but also gave good experience in drawing blood and understanding the kinds of testing doctors do routinely. By the time I reached med school, I was already an expert at things other students were learning for the first time. Besides, I got exposure to a health-related field so I could decide whether I really loved medicine."

Brian endured a lot to reach his goal. "Every day I hitch-hiked to college, even in the snow and rain." After six years of undergraduate study and experience as a lab technician, Brian made it into med school. "Some of my professors who used to pick me up hitch-hiking are now my patients. It's funny. At one time they taught me, and now I am using what they taught me to take care of them."

As a doctor of internal medicine, Brian treats patients for minor problems, such as sore throats and rashes, as well as more complicated conditions, such as heart disease and diabetes. "There are any number of things that can go wrong with the body. Today, for example, I treated patients suffering from shingles, diarrhea, high blood pressure, and dizziness. But I don't just treat people for physical conditions. Some people come in thinking they have a medical problem, but it's really connected to their mental state. They may be suffering from stress, grief, manic depression, or anxiety which will actually give them headaches, stomach pains, or a sore neck. After first talking to them and calming them, I might refer them to a psychologist or social worker or prescribe an anti-anxiety agent or antidepressant drug. Here in the real world I am treating life's stresses and circumstances."

Brian spends most of the day in his office, but before hours and at noon he sees patients at three area hospitals. "Sometimes specialists such as neurologists, ophthalmologists, or cardiologists need help from an internist because they are fine-tuned to deal exclusively with their own subjects whereas I am trained to take care of the whole body. An ophthalmologist might be doing eye surgery on a diabetic patient, so I would be called in to ensure that the patient is fit for surgery." Brian uses specialized doctors in his own practice as well. "I try to take care of someone as best I can. Most problems are uncomplicated enough for me to handle alone, but if need be, I will call in a specialist. For example, if someone has a heart problem, I will recommend consulting a cardiologist."

Brian finds his job challenging. "Every day is different—new to the point of being unpredictable. It really is a learning process. Once you get into private practice, you realize there is a lot you don't know. It takes time to get used to the people in the area, to working with the hospitals and the pharmacies. Often times it's intimidating, although now I'm feeling more comfortable."

Despite the struggles and challenges, Brian knows that being an internist is right for him. "I always wanted this, and now I'm here. It's nice to be my own boss and help people who are really in need and who appreciate me. It's fulfilling getting respect because I've helped someone get well."

A constant supply of oxygen is vital to our cells, as is a reliable method for removing the carbon dioxide they generate. The lungs provide the huge surface area needed for the exchange of gases, and breathing provides an elegantly simple mechanism to bring about the exchange. But what if there is not enough air to provide the oxygen? This F-14 jet fighter pilot must rely on technological help to supply oxygen so that this respiratory system can continue doing its job. We'll learn more about what that job entails in this chapter.

CHAPTER 23
The Respiratory System

Chapter Objectives

After reading this chapter you will be able to:

1 Describe the primary functions of the respiratory system.
2 Relate these functions to the anatomical and histological specializations of the tissues and organs in the system.
3 Explain how the delicate respiratory exchange surfaces are protected.
4 Describe the physical principles governing the movement of air into the lungs and the diffusion of gases into and out of the blood.
5 Describe how oxygen and carbon dioxide are transported in the blood.
6 Identify the reflexes that regulate respiration.
7 Explain how respiratory activities change to keep pace with metabolic needs.
8 Explain how the centers of respiratory control interact.
9 Describe the changes that occur in the respiratory system at birth and with aging.

■ Introduction

Living cells need energy for maintenance, growth, defense, and replication. Most eukaryotic cells obtain that energy through aerobic respiration (Chapter 4), a process that requires oxygen and produces carbon dioxide. ∞ (p. 107) To continue functioning, therefore, cells must have a way to obtain oxygen and eliminate carbon dioxide. Many aquatic organisms meet their respiratory demands by diffusion across the surface of the skin, or in specialized structures, such as the gills of a fish. Such arrangements are poorly suited for life on land because the exchange surfaces must be very thin and relatively delicate to encourage rapid diffusion. In air, the exposed membranes would collapse, evaporation and dehydration would reduce blood volume, and the delicate surfaces would be vulnerable to attack by pathogenic organisms. Our respiratory exchange surfaces are just as delicate as those of an aquatic organism, but they are confined to the inside of the lungs, in a warm, damp, and protected environment. Under these conditions diffusion can occur between the air and the blood. The cardiovascular system provides the link between the interstitial fluids and the exchange surfaces of the lungs. The circulating blood carries oxygen from the lungs to peripheral tissues; it also accepts and transports the carbon dioxide generated by these tissues, delivering it to the lungs.

Diffusion at the lungs occurs across the walls of capillaries, where the vessels pass very close to the enclosed air. Gas exchange occurs across the walls of delicate pockets, or **alveoli** (al-VĒ-ō-lī), in the lungs. The distance between the capillary wall and the air in an alveolus is usually less than 1 μm, and sometimes as small as 0.1 μm. To meet the metabolic requirements of our peripheral tissues, the exchange surfaces of the lungs must be very large, equal to around 140 square meters—roughly 80 times the total surface area of the body.

Our discussion of the respiratory system begins by following the air as it travels toward the alveoli. We will then consider the mechanics of breathing and the physiology of respiration.

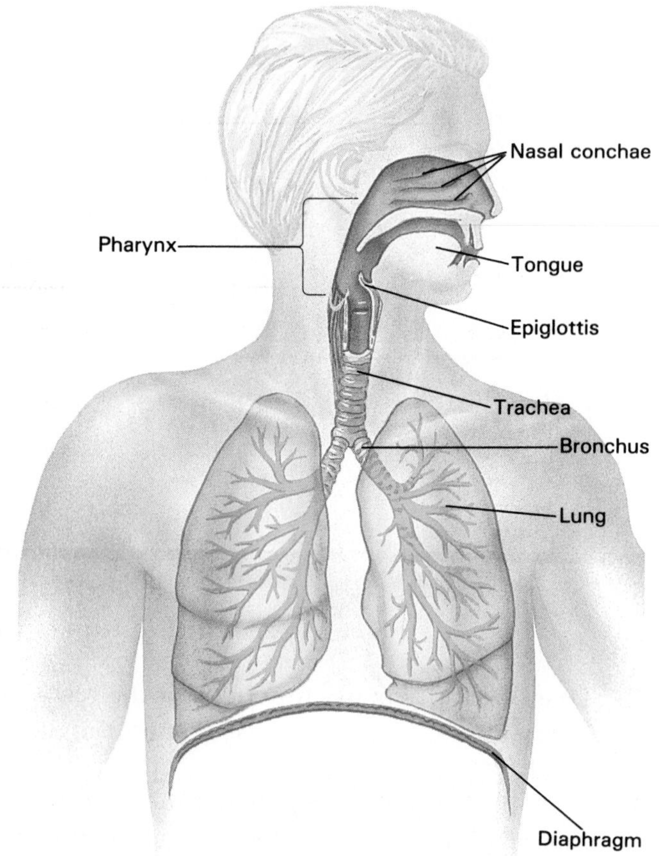

FIGURE 23-1
Components of the Respiratory System

Functional Anatomy and Organization

The **respiratory system** includes the nose, nasal cavity and sinuses, the pharynx, the larynx (voice box), the trachea (windpipe), and smaller conducting passageways leading to the exchange surfaces of the lungs. These components are illustrated in Figure 23-1.

FUNCTIONS OF THE RESPIRATORY SYSTEM

The primary functions of the respiratory system can be summarized as follows:

1. To provide an extensive area for gas exchange between air and circulating blood;
2. To move air to and from the exchange surfaces;
3. To protect respiratory surfaces from dehydration, temperature changes, or other environmental variations;
4. To defend the respiratory system and other tissues from pathogenic invasion;
5. To permit communication through speaking, singing, groaning, or the production of other sounds;
6. In addition, the respiratory system participates in the regulation of blood volume and pressure and the control of body fluid pH. These functions are considered in other chapters. Chapters 18 and 21 described the conversion of angiotensin I to angiotensin II in capillaries of the lungs, and the effects of angiotensin II on the cardiovascular, endocrine, nervous, and urinary systems. (pp. 588, 701) The influence of respiratory activity on pH will be detailed in Chapter 27.

THE UPPER AND LOWER RESPIRATORY TRACT

The **respiratory tract** consists of the conducting passageways that carry air to and from the alveoli. Air enters and leaves the respiratory tract through the action of the *respiratory muscles*, such as the diaphragm and the intercostals. The conducting passageways delivering air to the lungs constitute the **upper respiratory tract**. These passageways filter, warm, and humidify the air, protecting the more delicate conduction and exchange surfaces of the **lower respiratory tract**, or lungs, from debris, pathogens, and environmental extremes. Filtering, warming, and humidification of the inspired air begin at the entrance to the respiratory system and continue throughout the rest of the conducting system. By the time the air reaches the alveoli, most foreign particles and pathogens have been removed, and the humidity and temperature are within acceptable limits. To understand how these adjustments are made, and why they are important, you must become familiar with the functional anatomy of the entire system.

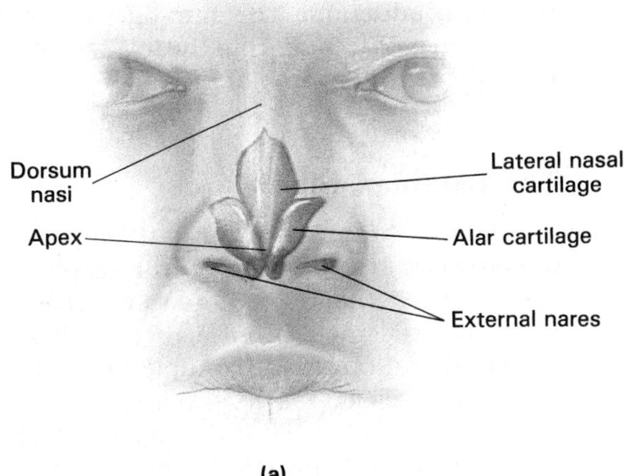

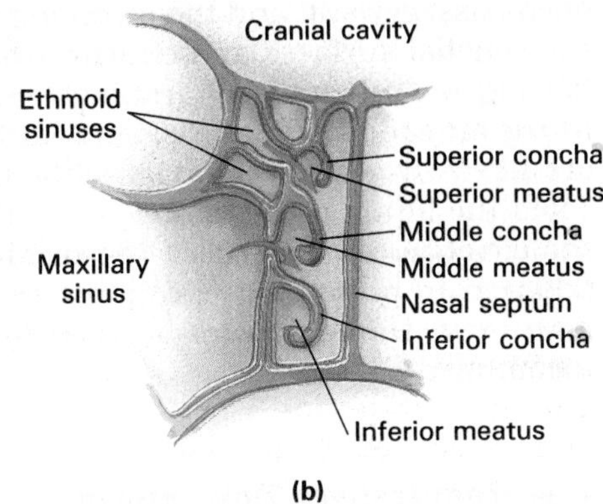

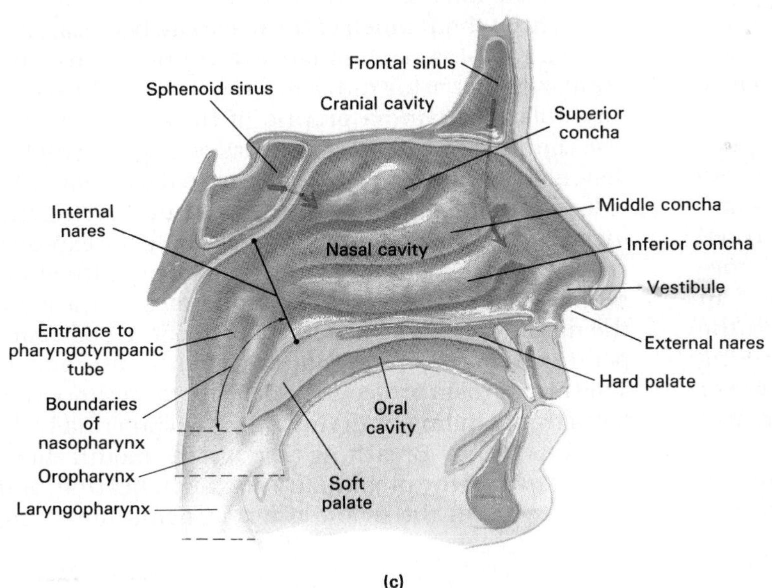

FIGURE 23-2
The Nose, Nasal Cavity, and Pharynx. (a) The nasal cartilages and external landmarks on the nose. (b) The meatuses and the positions of the entrances to the maxillary and ethmoidal sinuses. (c) The nasal cavity and pharynx as seen in sagittal section, with the nasal septum removed. The openings draining the frontal and sphenoidal sinuses are indicated by green arrows.

THE NOSE

External and internal features of the nose are indicated in Figure 23-2. Cartilages and bones associated with the nose were introduced in Chapter 8 and detailed in Figures 8-4 and 8-11; you should review these figures before proceeding. (pp. 242, 248) Air normally enters the respiratory system via the paired **external nares** (nostrils) that communicate with the **nasal cavity**. The **vestibule** (VES-ti-būl) is the portion of the nasal cavity contained within the flexible tissues of the external nose. The epithelium of the vestibule contains coarse hairs that extend across the external nares. Large airborne particles such as sand, sawdust, insects, or the eggs of various parasites are trapped in these hairs and prevented from entering the nasal cavity.

The maxillary, nasal, frontal, ethmoid, and sphenoid bones form the lateral and superior walls of the nasal cavity. The *nasal septum* divides the cavity into left and right sides, and a flexible cartilage in the anterior portion of the septum supports the bridge, or **dorsum nasi** (DOR-sum NĀ-zī), and **apex** of the nose. The bony posterior septum includes portions of the vomer and the ethmoid. A bony **hard palate**, formed by the palatine and maxillary bones, separates the oral and nasal cavities. A fleshy **soft palate** extends behind the hard palate, marking the boundary line between the superior **nasopharynx** (nā-zō-FĀR-inks) and the rest of the pharynx. The connections between the nasal cavity and the nasopharynx represent the **internal nares** (NĀ-rēz).

Superior, *middle*, and *inferior nasal conchae* project toward the septum from the lateral walls of the nasal cavity. To pass from the vestibule to the internal nares, air tends to flow between adjacent conchae, through the **superior**, **middle**, or **inferior meatuses** (mē-Ā-tus-es; *meatus*, passage) shown in Figure 23-2b. These are narrow grooves rather than open passageways, and the incoming air bounces off the conchal surfaces and churns around like water flowing over rapids. This turbulence serves a purpose: as the air eddies and swirls, small airborne particles are likely to come in contact with the mucus that coats the lining of the nasal cavity. For this reason the conchae are also called the **turbinate** bones. In addition to promoting filtration, the turbulence allows extra time for warming and humidifying the incoming air.

The Respiratory Epithelium

The respiratory epithelium, shown in Figure 23-3, consists of a pseudostratified, ciliated, columnar epithelium with numerous goblet cells. The goblet cells and mucous glands beneath the epithelium produce the mucus that bathes the exposed surfaces of the nasal cavity. Cilia sweep that mucus and any trapped debris or microorganisms toward the pharynx, where they will be swallowed and exposed to the acids and enzymes of the stomach. The surfaces are also flushed by mucus produced in the *paranasal sinuses* (*frontal, sphenoid, ethmoid,* and *maxillary*, detailed in Figure 8-11) and by the tears that flow through the nasolacrimal ducts. (p. 248) This filtering mechanism removes virtually all particles larger than around 10 μm from the inspired air. Smaller particles may be trapped by the mucus of the nasopharynx or secretions of the pharynx before proceeding farther along the conducting system. Exposure to unpleasant stimuli, such as noxious vapors, large quantities of dust and debris, allergens, or pathogens, usually causes a rapid increase in the rate of mucus production. The familiar symptoms of the "common cold" result from the invasion of the respiratory epithelium by one of many viruses.

The Lamina Propria

The **lamina propria** (LA-mi-na PRO-prē-a) is a layer of connective tissue between the respiratory epithelium and the underlying bones or cartilages. The epithelium and lamina propria are interdependent, and the combination is an example of a *mucous membrane*, or **mucosa** (mū-KŌ-sa), introduced in Chapter 5. (p. 152) The lamina propria usually contains scattered smooth muscle fibers; in some regions it forms relatively thick bands encircling or spiralling around the lumen.

Throughout much of the nasal cavity the lamina propria contains an abundance of arteries, veins, and capillaries that bring nutrients and water to the secretory cells. The lamina propria of the nasal conchae also contains an extensive network of large and highly distensible veins. This extensive vascularization provides a mechanism for warming and humidifying the incoming air. As cool, dry air passes over the exposed surfaces of the nasal cavity, the warm epithelium radiates heat and the mucus evaporates. Air leaving the nasal cavity has been heated almost to body temperature, and it is nearly saturated with water vapor. This mechanism protects the delicate respiratory surfaces from chilling or drying out—two potentially disastrous events. Breathing through the mouth eliminates much of the preliminary filtration, heating, and humidifying of the inspired air. Patients breathing

FIGURE 23-3
The Respiratory Epithelium. (a) Sketch and light micrograph showing the sectional appearance of the respiratory epithelium. (*LM*, ×810) (b) A surface view of the epithelium, as seen with the scanning electron microscope. The cilia of the epithelial cells form a dense layer that resembles a shag carpet. The movement of these cilia propels mucus across the epithelial surface. (*SEM*, ×1432)

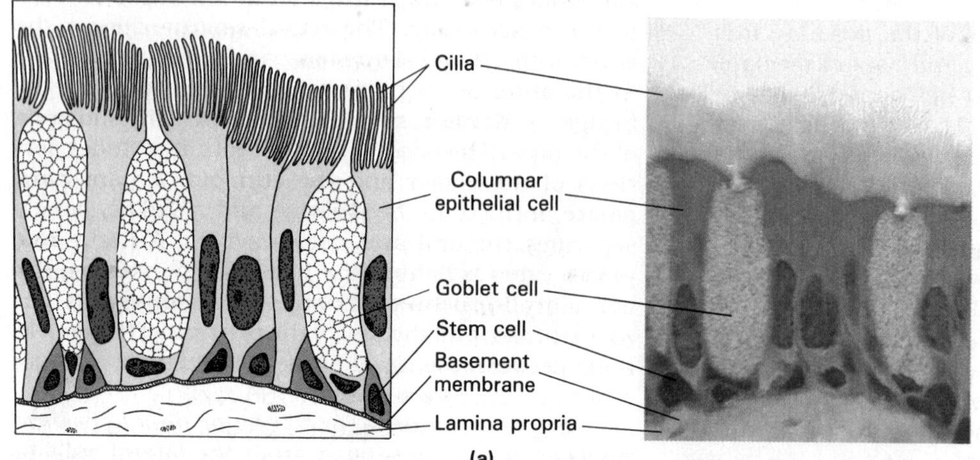

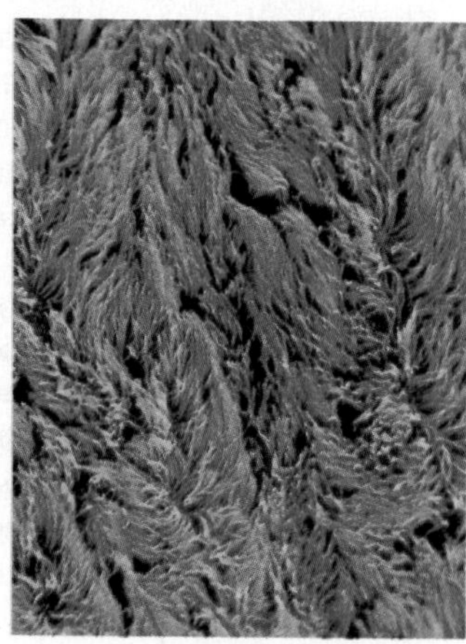

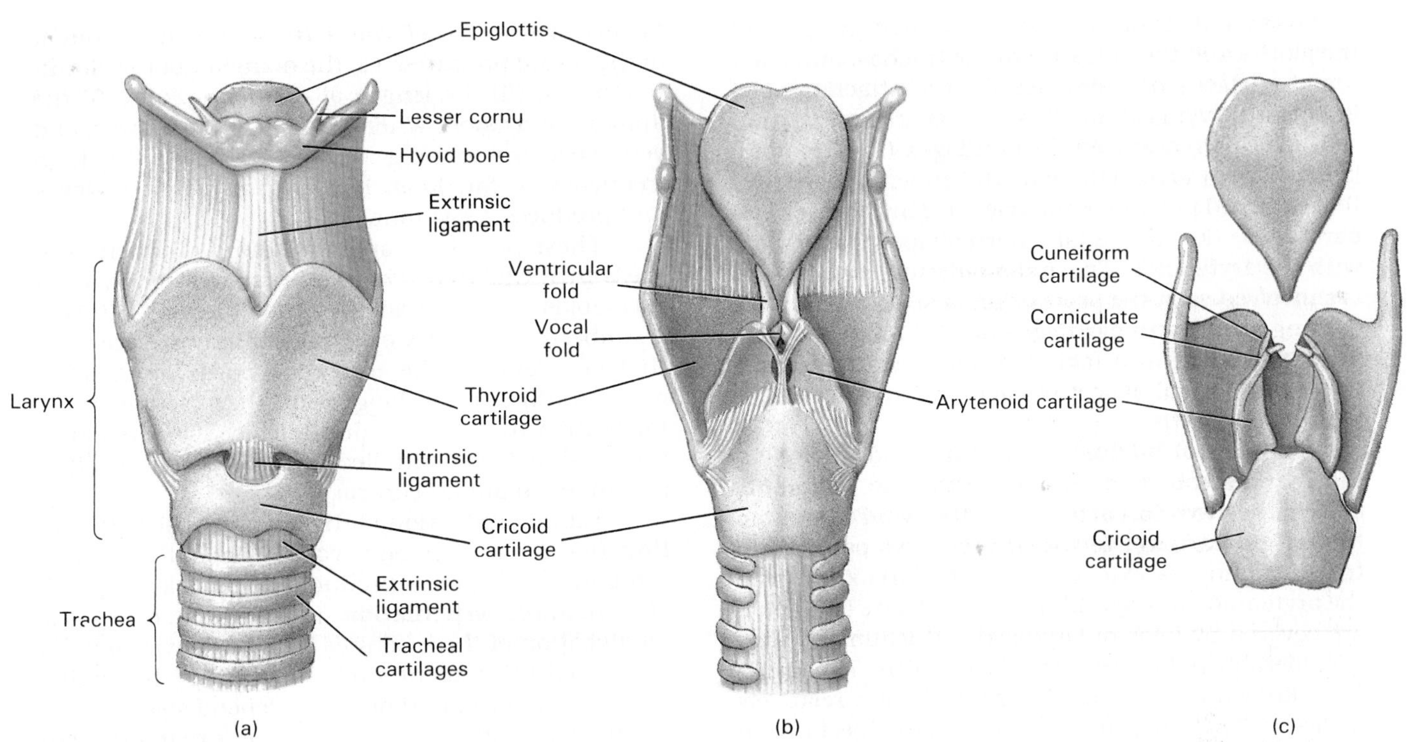

FIGURE 23-4
Anatomy of the Larynx. (a) Anterior view of the intact larynx. (b) Posterior view of the intact larynx. (c) Posterior view showing the relationships between the individual laryngeal cartilages.

on a respirator, which utilizes a tube to provide air directly into the trachea, must receive air that has been externally filtered and humidified or risk alveolar damage.

Several of the symptoms of the common cold result from changes in the permeabilities of these vessels. Fluid then moves into the lamina propria, and as it becomes swollen the flow of air is restricted, creating the sensation of nasal congestion. In addition, the excess fluid diffuses across the nasal epithelium, diluting the mucus secretions and producing a characteristic "runny nose." [**CM:** *Nosebleeds*]

THE PHARYNX

The **pharynx** is a chamber shared by the digestive and respiratory systems. It extends between the internal nares and the entrances to the larynx and esophagus. The curving superior and posterior walls are closely bound to the axial skeleton, but the lateral walls are quite flexible and muscular. The three subdivisions of the pharynx are indicated in Figure 23-2c. The soft palate separates the nasopharynx from the **oropharynx** (*oris*, mouth). The nasopharynx, lined by a typical respiratory epithelium, contains the entrances to the pharyngotympanic tubes and the pharyngeal (adenoid) tonsil. The oropharynx extends between the soft palate and the base of the tongue at the level of the hyoid bone. The palatine tonsils lie in the lateral walls of the oropharynx. The narrow **laryngopharynx** (lā-rin-gō-FĀR-inks) includes that portion of the pharynx lying between the hyoid and entrance to the esophagus. Materials entering the digestive tract pass through the oropharynx and laryngopharynx. These regions are lined by a stratified squamous epithelium that can resist mechanical abrasion, chemical attack, and pathogenic invasion.

THE LARYNX

Inspired (inhaled) air leaves the pharynx by passing through a narrow opening, the **glottis** (GLOT-is). The **larynx** (LAR-inks) surrounds and protects the glottis; important laryngeal landmarks are indicated in Figure 23-4. The larynx contains nine cartilages stabilized by ligaments and/or skeletal muscles. The elongate **epiglottis** (ep-i-GLOT-is) projects above the glottis. This structure, composed of elastic cartilage, has a flexible attachment to the anterior and superior borders of the hyoid and larynx. During swallowing the epiglottis folds back over the glottis, preventing the entry of liquids or solid food into the respiratory passageways. The curving **thyroid cartilage** (shield-shaped) forms much of the anterior and lateral surfaces of the larynx. A prominent ridge on the anterior surface of this cartilage forms the "Adam's apple." The thyroid sits atop the **cricoid cartilage** (KRĪ-koid; ring-shaped), and the posterior portion of the cricoid is greatly expanded, providing support in the absence of the thyroid cartilage. These two cartilages protect the glottis and the entrance to the trachea, and their broad surfaces provide sites for the attachment of important laryngeal muscles and ligaments.

The paired **arytenoid cartilages** (ar-i-TĒ-noid; ladle-shaped) articulate with the superior border of this enlarged portion of the cricoid. Tiny **corniculate cartilages** (kor-NIK-ū-lāt; horn-shaped) articulate with the arytenoids. The arytenoids and corniculates are involved with the production of sound. Elongate, curving **cuneiform cartilages** (kū-NĒ-i-form; wedge-shaped) lie within a fold of tissue that extends between the lateral aspect of each arytenoid cartilage and the epiglottis.

A series of **intrinsic ligaments** binds articulating cartilages to one another. **Extrinsic ligaments** attach the thyroid cartilage to the hyoid bone and the cricoid cartilage to the trachea. Two pairs of intrinsic ligaments extend across the larynx between the arytenoid and thyroid cartilages. These ligaments are covered by folds of laryngeal epithelium, so they considerably reduce the size of the glottis. The upper pair, known as the **ventricular folds**, are relatively inelastic. They help to prevent foreign objects from entering the glottis, and they protect the more delicate pair of **vocal folds**. The highly elastic vocal folds are involved with the production of sounds. For this reason they are known as the **true vocal cords.** Because the ventricular folds play no part in sound production, they are often called the **false vocal cords.** The orientation of these folds can be seen in Figure 23-5.

The Vocal Folds and Sound Production

Air passing through the glottis vibrates the vocal folds and produces sound waves. The factors that control the frequency of the sound produced can best be understood by comparing the vocal folds to the strings on a musical instrument like a guitar or violin. Three factors interact to determine the frequency, or pitch, of the sound produced: (1) the diameter of the vibrating string, (2) the length of the string, and (3) the amount of tension. When a string is thin, short, and very tense, the sound generated will have a very high frequency. A fat, long, loose string vibrates slowly and produces a low tone.

These principles apply to sound production in the larynx. Children of both sexes have slender, short vocal folds, and their voices tend to be high-pitched. At puberty the larynx of a male enlarges more than that of a female. The true vocal cords of an adult male are thicker and longer and they produce lower tones than those of an adult female. The tension in the vocal folds is controlled by skeletal muscles that are under voluntary control.

Sound production at the larynx is called **phonation** (fō-NĀ-shun; *phone*, voice). Phonation is one component of speech production, but clear speech also requires **articulation** (ar-tik-ū-LĀ-shun), the modification of these sounds by other structures. In a stringed instrument such as a guitar the quality of the sound produced does not depend solely on the nature of the vibrating string. The entire instrument becomes involved as the walls vibrate and the composite sound echoes within the hollow body. A similar amplification and resonance occur within the pharynx, the oral cavity, the nasal cavity, and the paranasal sinuses. The combination determines the particular and distinctive sound of each person's voice. The final production of distinct words further depends on voluntary movements of the tongue, lips, and cheeks.

The Laryngeal Musculature

The larynx is associated with two different groups of muscles, the *intrinsic laryngeal muscles* and the *extrinsic laryngeal muscles*. The **intrinsic laryngeal muscles** have two principal functions. One group regulates tension in the vocal folds, while a second set

FIGURE 23-5
The Vocal Cords. The glottis is shown in the open position (left) and closed position (center and right).

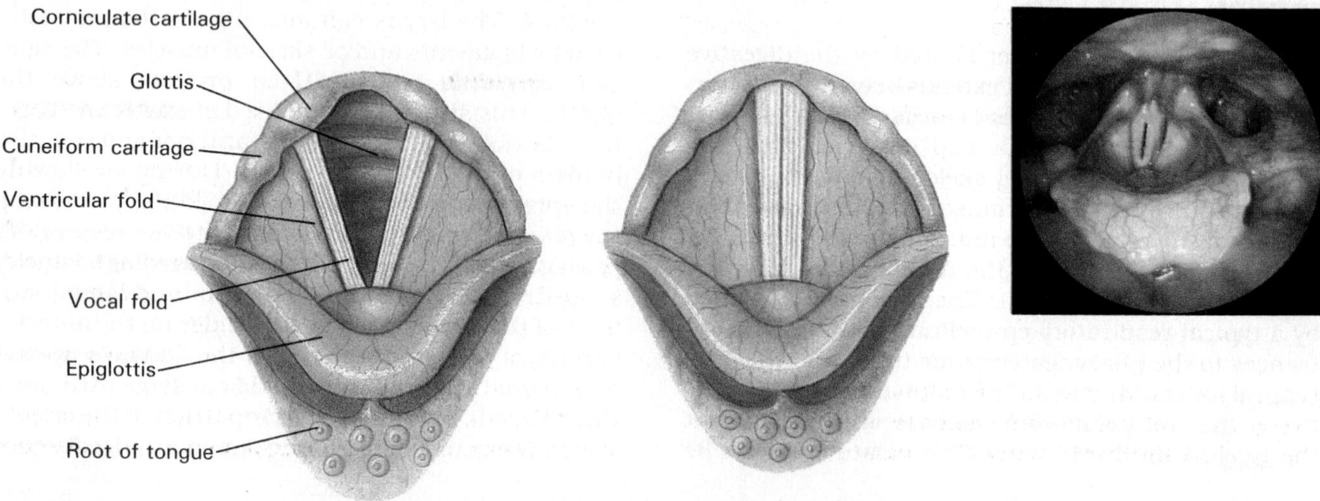

opens and closes the glottis. Those involved with the vocal folds insert upon the arytenoids and corniculates. By changing the positions of these cartilages, these muscles alter the tension in the vocal folds and adjust the frequency of the sound produced. The **extrinsic laryngeal musculature** positions and stabilizes the larynx. (Important extrinsic laryngeal muscles were described in Chapter 11 and detailed in Figure 11-8.) (p. 341)

During swallowing, both extrinsic and intrinsic muscles cooperate to prevent food or drink from entering the glottis. Prior to swallowing, the material is crushed and chewed into a pasty mass known as a **bolus**. Extrinsic muscles then elevate the larynx, bending the epiglottis over the entrance to the glottis, so that the bolus can glide across the epiglottis, rather than tumbling into the larynx. While this movement is under way, intrinsic muscles close the glottis. Any particles or liquids that do manage to contact the surfaces of the ventricular or vocal folds will trigger a fit of coughing that usually prevents them from entering the glottis. These laryngeal movements are summarized in Figure 23-6; the specific muscles involved in swallowing will be described in Chapter 24.

The larynx functions as a sentry for the passageways leading to the lungs, and anything that gets past the laryngeal entrance may block a respiratory passage or damage alveolar surfaces. Coughing and **laryngeal spasms** are reflexes that restrict the entrance of potentially dangerous materials. Irritation of the larynx often triggers the **cough reflex**. In a cough the glottis is forcibly closed while the lungs are still relatively full. The abdominal and internal intercostal muscles then contract suddenly, creating pressures that will blast air out of the respiratory passageways when the glottis reopens. Air leaving the larynx may travel at 100 mph, carrying mucus, foreign particles, and irritating gases out of the respiratory system. The cough reflex can also be triggered by irritation of respiratory passageways below the larynx. Laryngeal spasms result from the entry of chemical irritants, foreign objects, or fluids into the area around the glottis. This reflex temporarily closes the airway, but a very strong stimulus, such as a toxic gas, may close the glottis so powerfully that the individual may lose consciousness or even die before taking another breath. Fine chicken or fish bones that pierce the laryngeal walls may also stimulate laryngeal spasms and restrict the airway as swelling occurs. [**CM:** *Disorders of the Larynx*]

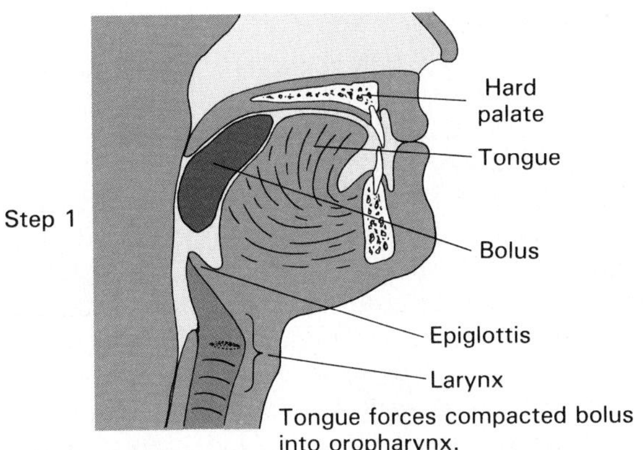

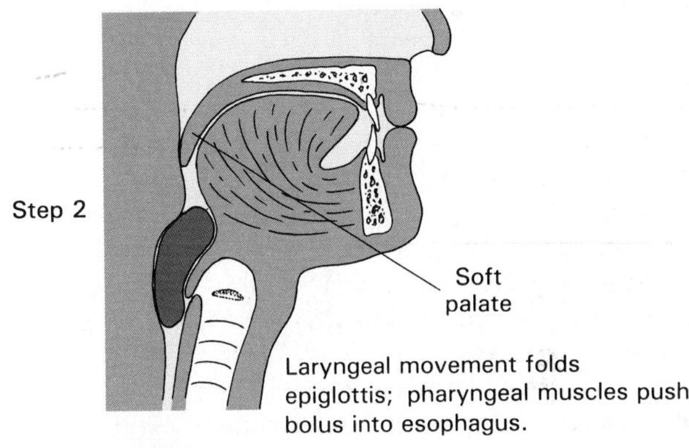

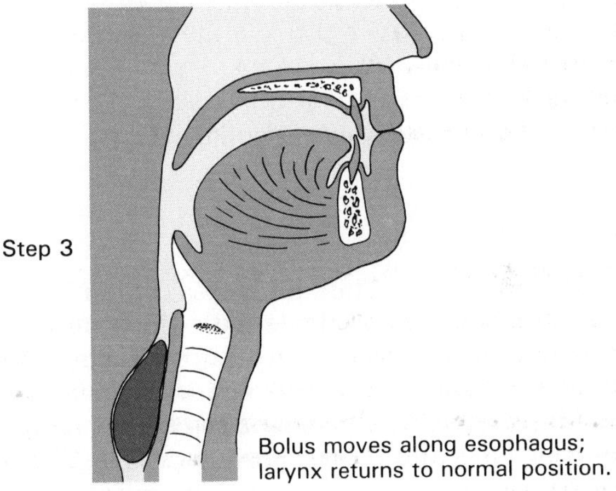

Tongue forces compacted bolus into oropharynx.

Step 3
Bolus moves along esophagus; larynx returns to normal position.

FIGURE 23-6
Movements of the Epiglottis During Swallowing. During swallowing the elevation of the larynx folds the epiglottis over the glottis, steering materials into the esophagus.

THE TRACHEA

The epithelium of the larynx is continuous with that of the **trachea** (TRĀ-kē-a), or "windpipe." The trachea extends between the level of the sixth cervical vertebra, where it attaches to the cricoid cartilage of the larynx, and that of the fifth thoracic vertebra, where it branches to form a pair of *primary bronchi*. The trachea is a tough, flexible tube with a diameter of around 2.5 cm (1 in.) and a length of approximately 11 cm (4.25 in.). Details of tracheal structure can be seen in Figure 23-7.

The mucosa of the trachea consists of a typical respiratory epithelium overlying a lamina propria that contains a network of elastin fibers. A thick layer of connective tissue, the **submucosa** (sub-mū-KŌ-sa),

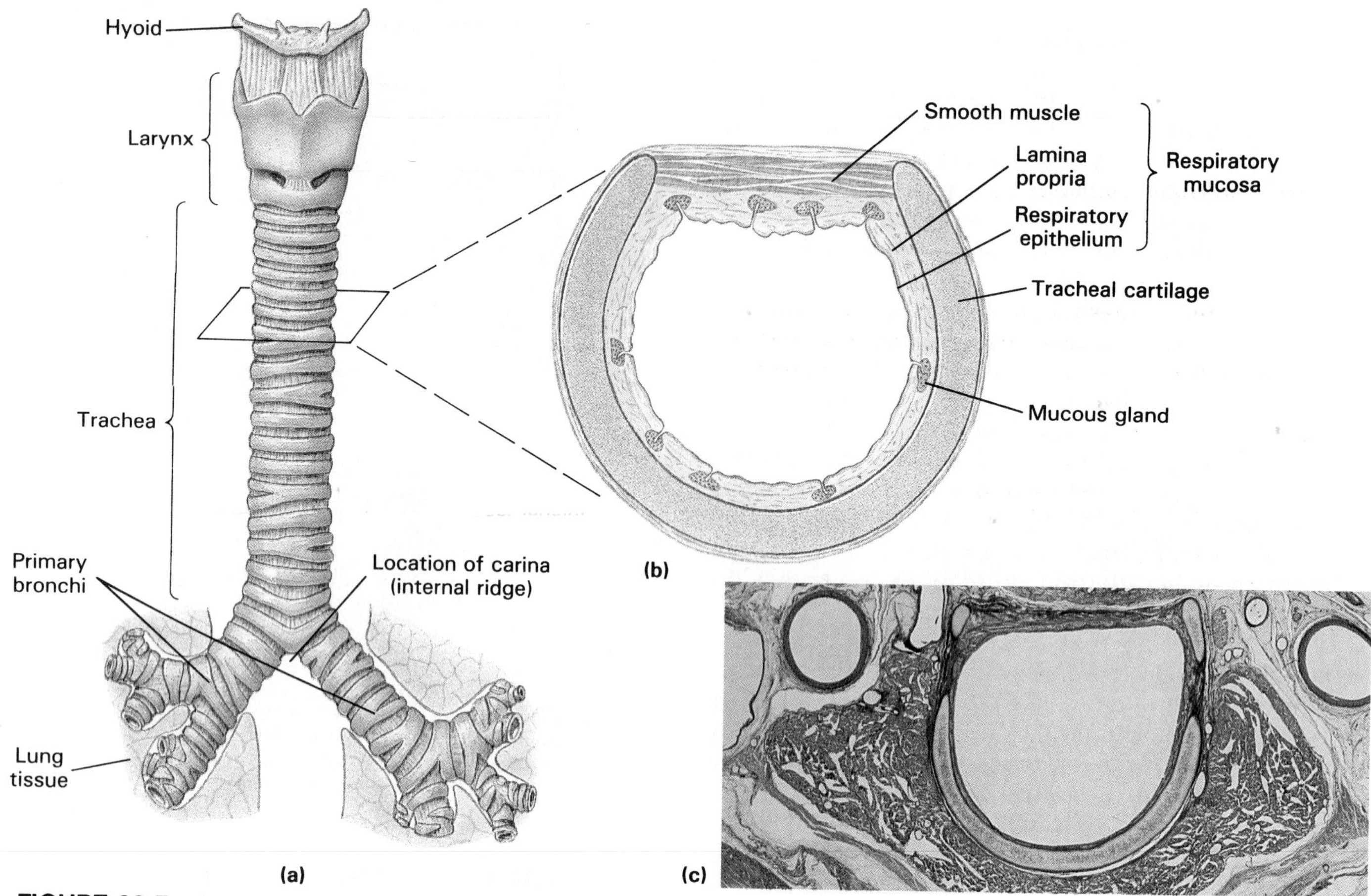

FIGURE 23-7
Anatomy of the Trachea. (a) Anterior view on dissection, showing the plane of section for (b). (b, c) Cross-sectional views of the trachea.

surrounds the mucosa. The submucosa contains mucous glands that communicate with the epithelial surface through a number of secretory ducts. The submucosa of the trachea also contains about 20 **tracheal cartilages**. These C-shaped cartilages stiffen the tracheal walls and protect the airway. They also prevent its collapse or overexpansion as pressures change in the respiratory system.

The open portions of the C-shaped tracheal cartilages face posteriorly, toward the esophagus. An elastic ligament and a band of smooth muscle, the **trachealis**, extend between the ends of the C. The ligament helps prevent overexpansion, and contraction of the trachealis alters the diameter of the tracheal lumen.

Because the cartilages do not continue around the trachea, the posterior tracheal wall can easily distort during swallowing, permitting the passage of large masses of food. The normal diameter of the trachea changes from moment to moment, primarily under the control of the sympathetic division of the ANS. Sympathetic stimulation increases the diameter of the trachea and makes it easier to move large volumes of air along the respiratory passageways. [**CM:** *Tracheal Blockage*]

THE PRIMARY BRONCHI

The trachea branches within the mediastinum, giving rise to the **right** and **left primary bronchi** (BRONG-kī). A ridge, the **carina** (ka-RĪ-na), marks the line of separation between the two bronchi. The histological organization of the primary bronchi resembles that of the trachea, complete with cartilaginous Cs. The right primary bronchus has a larger diameter than the left, and it descends toward the lung at a steeper angle. For this reason objects that manage to enter the trachea usually enter the right bronchus rather than the left.

Each primary bronchus travels to a groove along the medial surface of its lung before branching further. This groove, the **hilus** (HĪ-lus), also provides access for the pulmonary vessels and nerves. The entire array is firmly anchored in a meshwork of dense connective tissue. This complex, known as the **root** of the lung (Figures 23-8 and 23-9), attaches it to the mediastinum and fixes the positions of the major nerves, vessels, and lymphatics. The roots of the lungs are located at the level of vertebra T_5 (right) and T_6 (left).

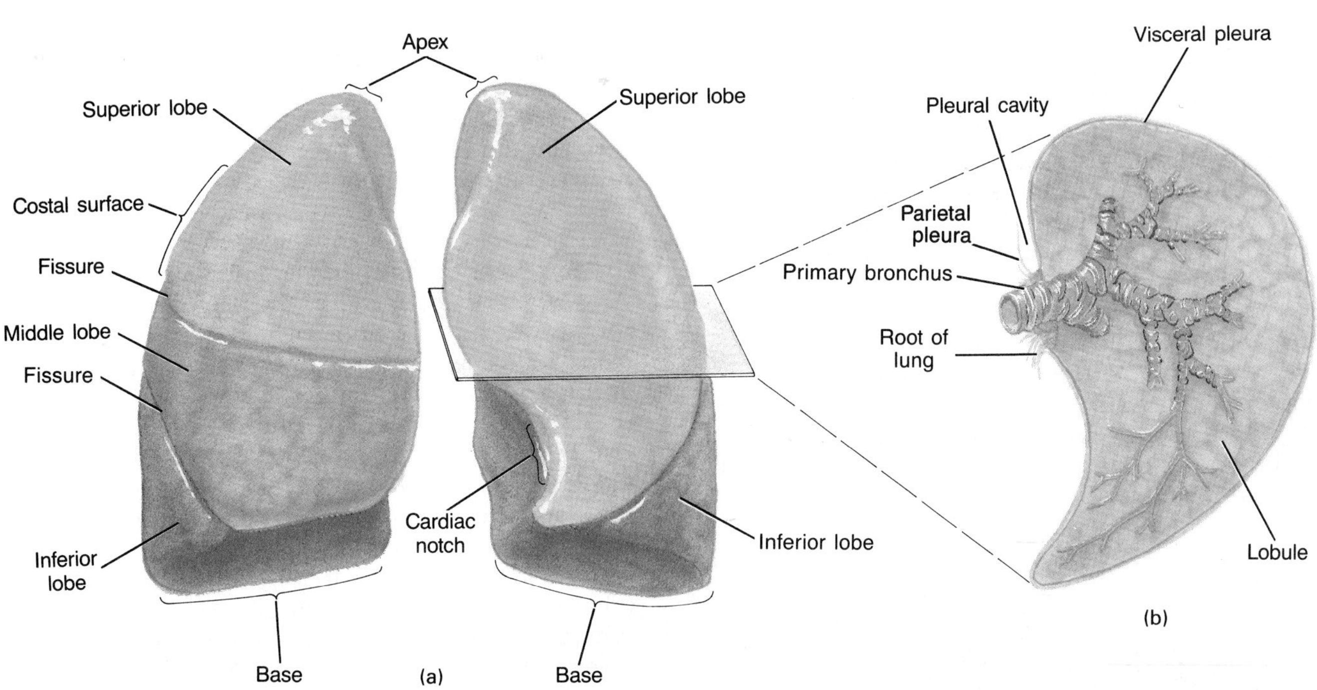

FIGURE 23-8
Gross Anatomy and Position of the Lungs. (a) Orientation of the lungs in the open chest. (b) A sectional view at the indicated level, showing the connective tissue distribution within the lung.

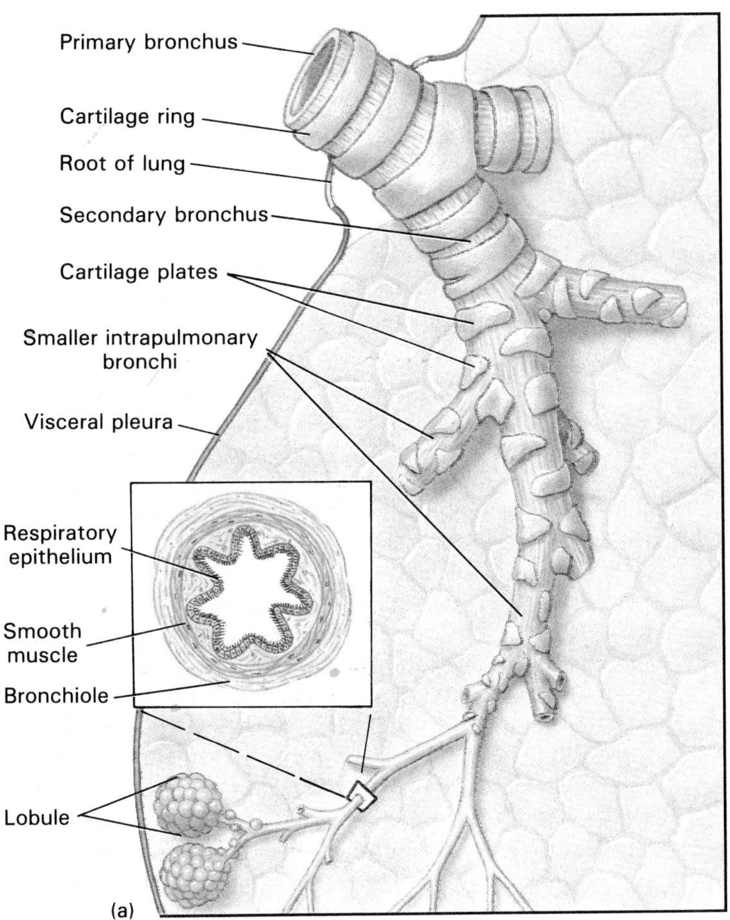

FIGURE 23-9
Organizational Pattern of the Lung. (a) Details of lung organization as seen on gross dissection and light microscopy. Note the characteristic and extensive branching that occurs along the airways. For clarity, the degree of branching has been reduced; an airway branches approximately 23 times before reaching the level of a lobule. (b) The pulmonary vessels branch in company with the bronchi and bronchioles, giving the lungs an extensive circulatory supply. In this preparation, yellow latex was injected into the trachea, filling all of the airways. The heart was also injected; the right side of the heart and pulmonary arteries are colored blue, and the pulmonary veins and left atrium and ventricle are red. This view highlights the structural integration between the cardiovascular and respiratory systems.

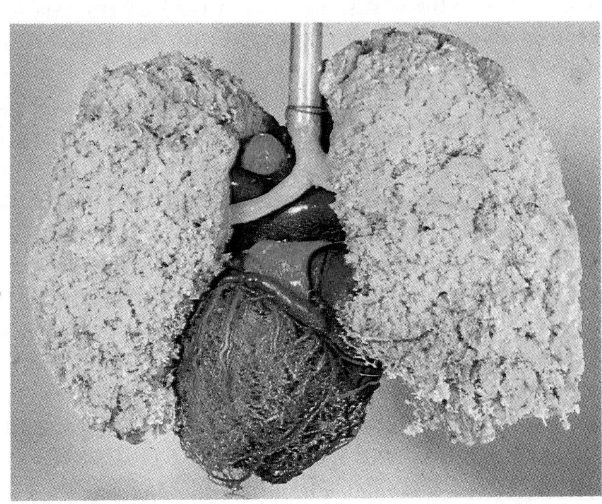

THE LUNGS

The superficial anatomy of the lungs can be seen in Figure 23-8a. The lungs have distinct **lobes** separated by deep fissures. The right lung has three lobes (**superior**, **middle**, and **inferior**) and the left lung has two (**superior** and **inferior**). The bluntly rounded **apex** of each lung extends into the base of the neck above the first rib, and the concave **base** rests on the superior surface of the diaphragm. The curving **costal surface** follows the inner contours of the rib cage. The **mediastinal surface** containing the hilus has a more irregular shape. The mediastinal surface of the left lung bears a concavity, the **cardiac notch**, that conforms to the shape of the pericardium, and both lungs bear grooves that mark the passage of vessels traveling to and from the heart.

The connective tissues of the root extend into the substance, or **parenchyma** (pār-ENG-kī-ma), of each lung, as indicated in Figure 23-8b. These fibrous partitions, or **trabeculae**, contain elastic fibers, smooth muscles, and lymphatics. They branch repeatedly, dividing the lobes into smaller and smaller compartments, and the branches of the conducting passageways, pulmonary vessels, and nerves of the lungs follow these trabeculae to reach their peripheral destinations. The terminal partitions, or **septa** divide the lung into **lobules** (LOB-ūlz), each serviced by tributaries of the pulmonary arteries, pulmonary veins, and respiratory passageways. The connective tissues of the septa are in turn continuous with those of the visceral pleura.

Because most of the actual volume of each lung consists of air-filled passageways and alveoli, the lung has a light and spongy consistency. Elastic fibers within the trabeculae, the septa, and the pleurae make the lung highly elastic and capable of tolerating great changes in volume. [**CM:** *Examining the Living Lung*]

BRONCHI AND BRONCHIOLES

As it enters the lung, each primary bronchus gives rise to secondary bronchi that enter the lobes of that lung. The secondary bronchi then divide repeatedly as they approach the respiratory lobules. Figure 23-9 shows the histological organization of bronchi and bronchioles. These branches within the lung, called **intrapulmonary bronchi**, differ in structure from the **extrapulmonary bronchi** (primary bronchi). Bands of smooth muscle completely encircle the lumen in the innermost portion of the submucosa. Tension in these muscles often throws the mucosa into a series of folds, and excessive stimulation can almost completely block the passageways. As noted in Chapter 16, sympathetic activation leads to a relaxation of these muscles and dilation of the respiratory passageways, especially the bronchioles. (p. 506) Histamine, released during inflammation, causes bronchiolar constriction.

Outside of the smooth muscle layer, the submucosa contains irregular blocks and bands of cartilage rather than cartilaginous Cs. The cartilages of the secondary bronchi are quite massive, but as we proceed farther along the branches of the bronchial "tree" the cartilages become smaller and smaller.

When the diameter of the passageway has narrowed to around 1 mm, cartilages disappear completely. This narrow passage represents a **bronchiole**, and each bronchiole enters a single lobule. Within the lobule the bronchiole branches further, giving rise to several terminal bronchioles. Terminal bronchioles have a diameter of less than 0.5 mm, and they lack the respiratory epithelium found along the larger respiratory passageways. The epithelial cells are cuboidal, cilia are rare, and there are no goblet cells or underlying mucous glands. Each terminal bronchiole gives rise to several respiratory bronchioles. These passages, the finest branches of the bronchial tree, deliver air to the respiratory surfaces of the lungs.

ALVEOLAR DUCTS AND ALVEOLI

Figure 23-10a shows the structure of the finest branches of the respiratory tree. Respiratory bronchioles open out into expansive chambers, called **alveolar ducts**. The blind pockets that open onto these chambers are the alveoli where gas exchange occurs. Each lung contains approximately 150 million alveoli, also known as "alveolar sacs," and their abundance gives the lung an open, spongy appearance. [**CM:** *Pneumonia*]

Pulmonary Circulation

The respiratory exchange surfaces are extensively connected to the circulatory system via the blood vessels of the *pulmonary circuit* (Figure 21-9). (You may find it useful to review the description of the circulation of the lungs at this time.) The pulmonary arteries enter the lungs at the hilus and branch to follow the bronchi to reach the lobules. Each lobule receives an arteriole and a venule, and a network of capillaries surrounds each alveolus directly beneath the *respiratory membrane* that lines the air spaces. After passing through the pulmonary venules, venous blood enters the pulmonary veins that deliver it to the left atrium.

Other portions of the respiratory tract receive blood from the external carotid arteries (nasal passages and larynx), the *thyrocervical arteries* (branches of the subclavian arteries that supply the inferior larynx and trachea), and the *bronchial arteries*. The capillaries supplied by the bronchial arteries provide oxygen and nutrients to the conducting

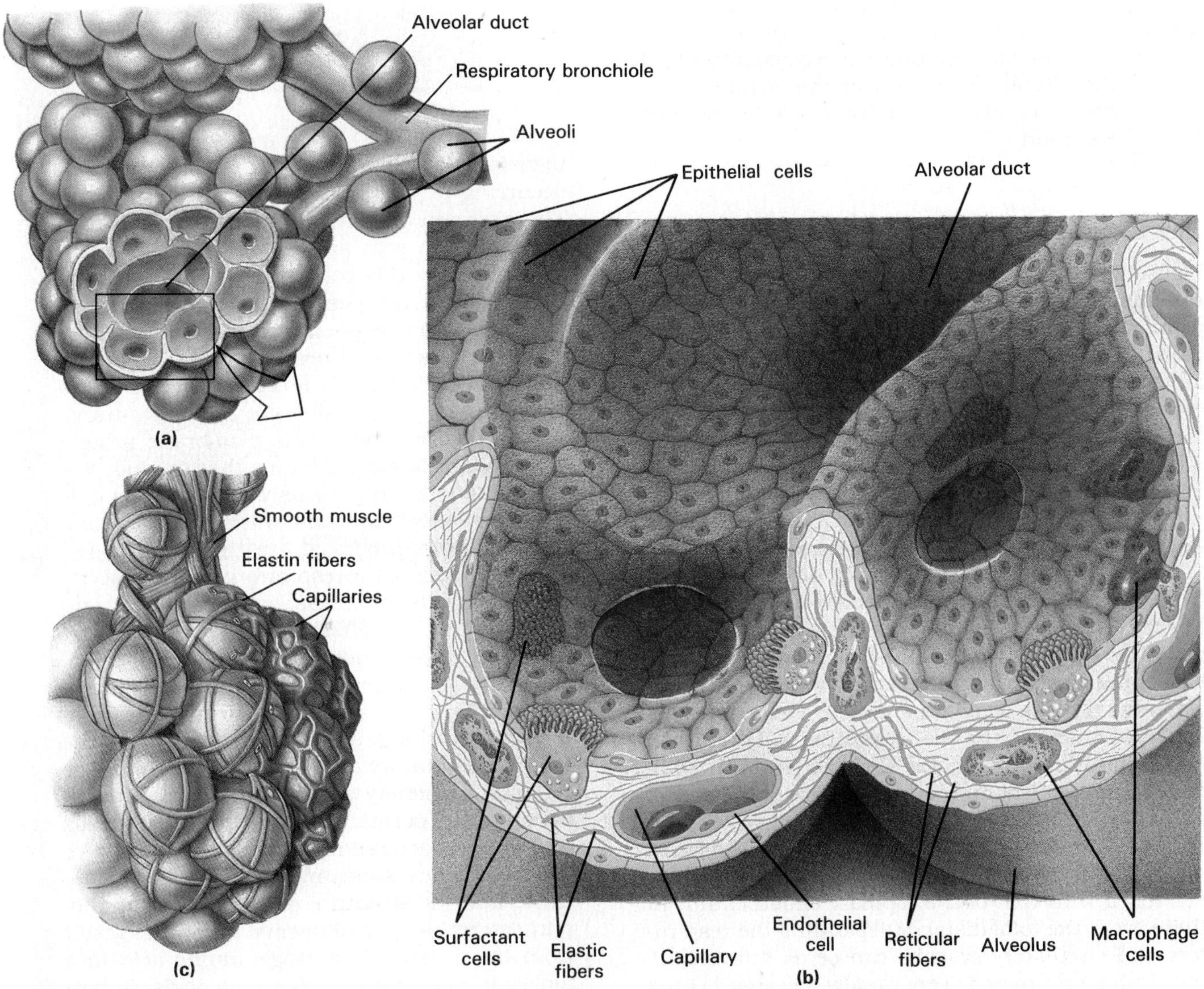

Alveolar duct
Respiratory bronchiole
Alveoli
(a)
Epithelial cells
Alveolar duct
Surfactant cells
Elastic fibers
Capillary
Endothelial cell
Reticular fibers
Alveolus
Macrophage cells
(b)
Smooth muscle
Elastin fibers
Capillaries
(c)

FIGURE 23-10
Alveolar Organization. (a) Basic structure of a lobule, cut to reveal the arrangement between the alveolar ducts and alveoli. (b) Diagrammatic view of alveolar structure and the respiratory membrane. (c) Connective tissue layers and vascular supply to the alveoli. A network of capillaries surrounds each alveolus. These capillaries are surrounded by elastic fibers. Respiratory bronchioles also contain wrappings of smooth muscle that can vary the diameter of these airways.

Clinical Comment: Pulmonary Embolism

The lungs are the only organs that receive the entire cardiac output. Blood pressure in the pulmonary circuit is usually relatively low, and pulmonary vessels can easily become blocked by blood clots, fat masses, or air bubbles in the pulmonary arteries. Blockage of a vessel stops blood flow to all of the alveoli serviced by the obstructed vessel. This condition is called a **pulmonary embolism**. Two conditions described in earlier chapters, *arteriosclerosis* and *venous thrombosis*, can promote development of a pulmonary embolism. If the blockage remains in place for several hours, the alveoli will permanently collapse. If the blockage occurs in a major pulmonary vessel, rather than a minor tributary, pulmonary resistance increases. This places extra strain on the right ventricle, which may be unable to maintain cardiac output. Congestive heart failure may be the result. [**CM:** *Heart Failure*] passageways of the lungs. The venous blood flows into the pulmonary veins, bypassing the rest of the systemic circuit and diluting the oxygenated blood leaving the alveoli. (The effect of this arrangement on the oxygen content of the blood will be detailed in a later section.)

The Respiratory Membrane

Figure 23-10b shows the alveolar lining, or **respiratory membrane**. The respiratory membrane consists primarily of a simple squamous epithelium. **Septal cells**, or **surfactant** (sur-FAK-tant) **cells**, scattered among the squamous cells produce an oily secretion containing a mixture of phospholipids. This **surfactant** coats the alveolar epithelium, reducing surface tension that would otherwise make the alveoli collapse. Roaming **alveolar macrophages** (*dust cells*) patrol the epithelium, phagocytizing any particles fine enough to reach the alveolar surfaces.

Each alveolus is surrounded by an extensive network of capillaries, as indicated in Figure 23-10c. There are no reinforcing layers of smooth muscle or cartilage. However, the lamina propria of each alveolus contains a thin layer of elastic fibers that is interwoven with those of adjacent alveoli. This elastic tissue helps maintain the relative positions of the alveoli and respiratory bronchioles. Contraction of these fibers during expiration reduces the size of the alveoli and assists in the process of expiration.

Gas exchange occurs in areas where the basement membrane underlying the alveolar epithelium has fused with that of the capillary endothelium. In these areas the total distance separating the respiratory and circulatory systems can be as little as 0.1 μm. Diffusion proceeds very rapidly, because (1) the distance is small and (2) the gases are lipid-soluble. The membranes of the epithelial and endothelial cells thus do not pose a barrier to the movement of oxygen and carbon dioxide between the blood and alveolar airspaces. [CM: *Tuberculosis*]

The Respiratory Defense System

The delicate surfaces of the respiratory system can be severely damaged if the inspired air becomes contaminated with debris or pathogens. The entire array of protective mechanisms is called the **respiratory defense system**. As already noted, the hairs, cilia, and mucus traps of the nasal cavity remove particles larger than about 10 μm. Particles as small as 5 μm are usually trapped by the mucus lining the larynx, trachea, and bronchi. All of the cilia in these regions beat toward the pharynx, forming a **mucus escalator** that carries the collected particles toward an acid bath in the stomach.

Particles 1–5 μm in diameter usually collect in the terminal or respiratory bronchioles, outside of the boundaries of the mucus escalator. There they are engulfed by alveolar macrophages that also remove smaller particles stuck in the alveolar surfac-

Clinical Comment: Cystic Fibrosis

Cystic fibrosis (CF) is the most common lethal inherited disease in the Caucasian population, occurring at a frequency of 1 birth in 1600. There are approximately 30,000 Americans with CF, and each year an additional 2000 babies are born with this condition. Individuals with CF seldom survive past age 30; death is usually the result of a massive bacterial infection of the lungs and associated heart failure.

The underlying problem involves a membrane protein responsible for the active transport of chloride ions. This membrane protein is abundant in exocrine cells that produce watery secretions. In persons with CF, these cells cannot transport salts and water effectively, and the secretions produced are thick and gooey. Mucous glands of the respiratory tract and secretory cells of the pancreas, salivary glands, and digestive tract are affected.

The most serious symptoms appear because the respiratory defense system cannot transport such dense mucus. The mucus escalator stops working, and mucus plugs block the smaller respiratory passageways. This blockage reduces pulmonary ventilation, and the inactivation of the normal respiratory defenses leads to frequent bacterial infections.

The gene responsible for CF has been identified, and the structure of the membrane protein determined. Its abnormal properties result from the absence of a single amino acid in a molecule containing 1480 amino acids. In normal individuals this molecule is produced in the endoplasmic reticulum, exported to the Golgi apparatus, and then moved to the cell surface in a vesicle (see Figure 3-25). (p. 85) In CF patients, the protein remains in the endoplasmic reticulum and never reaches the surface of the cell.

Now that the structure of the gene is understood, research continues with the goal of correcting the defect by the insertion of normal genes. (The process of gene therapy was introduced in Chapter 4 in the discussion of genetic engineering and the treatment of ADA deficiency.) (p. 114) In late 1990, researchers succeeded in using gene therapy to correct the defective chloride pump in cultured epithelial cells.

tant. Objects smaller than about 0.5 μm usually remain suspended in the air. [**CM:** *Overloading the Respiratory Defenses*]

THE PLEURAL CAVITIES

The thoracic cavity has the shape of a broad cone. Its walls are the rib cage, and the muscular diaphragm forms the floor. The mediastinum divides the thoracic cavity into two pleural cavities, as indicated in Figure 23-11. (The structure of the mediastinum and the relationships between the ventral body cavities were discussed in Chapter 1 and detailed in Figure 1-13.) (p. 19) Each lung occupies a single pleural cavity, lined by a serous membrane, or **pleura** (PLOO-ra). The **parietal pleura** covers the inner surface of the body wall and extends over the diaphragm and mediastinum. The **visceral pleura** covers the outer surfaces of the lungs, extending into the fissures between the lobes. A small amount of fluid crosses the pleural lining, giving it a moist, slippery coating that reduces friction between the parietal and visceral surfaces.

The pleural cavity actually represents a potential space rather than an open chamber, for the parietal and visceral layers are usually in close contact. A thin layer of fluid covering these mesothelial surfaces provides lubrication, preventing friction and irritation of the pleura. [**CM:** *Thoracentesis*]

RESPIRATORY CHANGES AT BIRTH

There are several important differences between the respiratory systems of a fetus and a newborn infant. Prior to delivery, pulmonary arterial resistance is high, because the pulmonary vessels are collapsed. The rib cage is compressed, and the lungs and conducting passageways contain only small amounts of fluid and no air. At birth the newborn infant takes a truly heroic first breath through powerful contractions of the diaphragmatic and external intercostal muscles. The inspired air enters the passageways with enough force to push the contained fluids out of the way and inflate the entire bronchial tree and most of the alveolar complexes. The same drop in pressure that pulls air into the lungs pulls blood into the pulmonary circulation; the changes in blood flow that occur lead to the closure of the *foramen ovale*, an interatrial connection, and the *ductus arteriosus*, the fetal connection between the pulmonary trunk and the aorta. (These events were detailed in the Embryology Summaries in Chapters 20 and 21.) (pp. 647, 670)

The exhalation that follows fails to completely empty the lungs, for the rib cage does not return to its former, fully compressed state. Cartilages and connective tissues keep the conducting passageways open, and the surfactant covering the alveolar surfaces prevents their collapse. Subsequent breaths complete the inflation of the alveoli. These physical changes are sometimes used by pathologists to determine whether a newborn infant died prior to delivery or shortly thereafter. Prior to the first breath, the lungs are completely filled with fluid, and they will sink if placed in water. After the first breath, even the collapsed lungs contain enough air to keep them afloat. [**CM:** *Neonatal Respiratory Distress Syndrome (NRDS)*]

- √ **When the tension in the vocal cords increases, what happens to the pitch of the voice?**
- √ **Why are the cartilages that reinforce the trachea C-shaped instead of complete circles?**
- √ **What would happen to the alveoli if surfactant were not produced?**
- √ **Why do chronic smokers develop a hacking "smoker's" cough?**

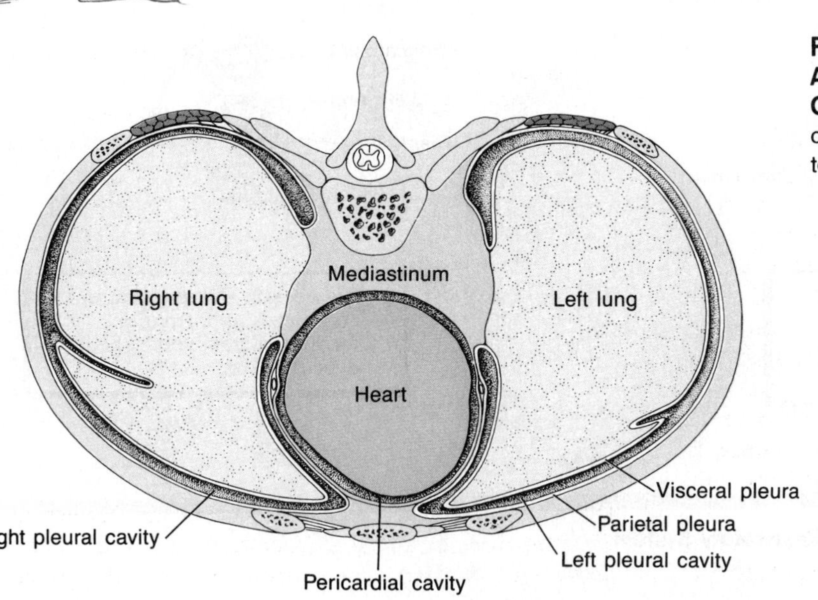

FIGURE 23-11
Anatomical Relationships in the Thoracic Cavity. Those interested in the appearance of other structures at this level should refer to Figures 1-13 and 20-2b.

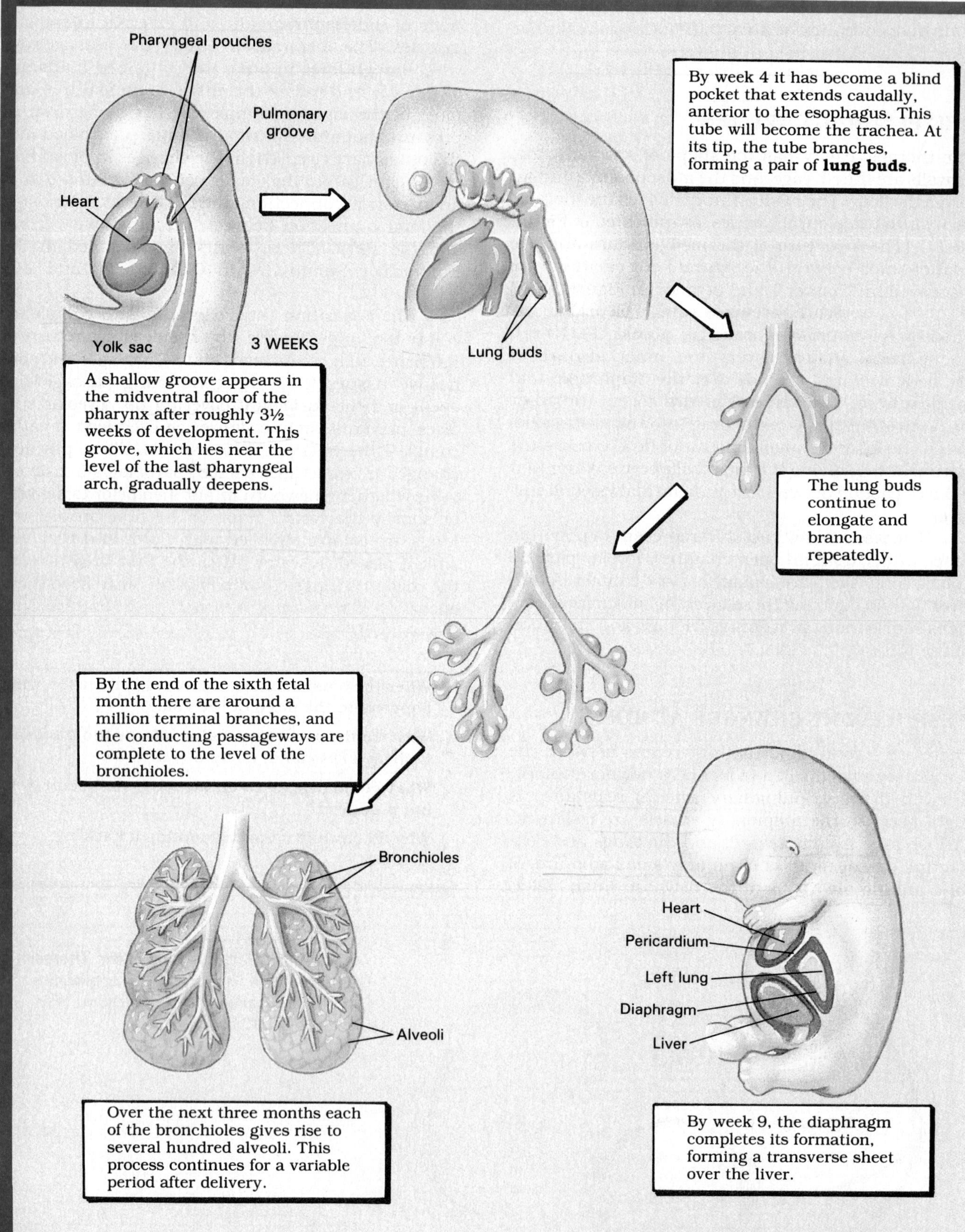

Pharyngeal pouches
Pulmonary groove
Heart
Yolk sac
3 WEEKS
A shallow groove appears in the midventral floor of the pharynx after roughly 3½ weeks of development. This groove, which lies near the level of the last pharyngeal arch, gradually deepens.
Lung buds
By week 4 it has become a blind pocket that extends caudally, anterior to the esophagus. This tube will become the trachea. At its tip, the tube branches, forming a pair of **lung buds**.
The lung buds continue to elongate and branch repeatedly.
By the end of the sixth fetal month there are around a million terminal branches, and the conducting passageways are complete to the level of the bronchioles.
Bronchioles
Alveoli
Over the next three months each of the bronchioles gives rise to several hundred alveoli. This process continues for a variable period after delivery.
Heart
Pericardium
Left lung
Diaphragm
Liver
By week 9, the diaphragm completes its formation, forming a transverse sheet over the liver.

Embryology Summary: Development of the Respiratory System

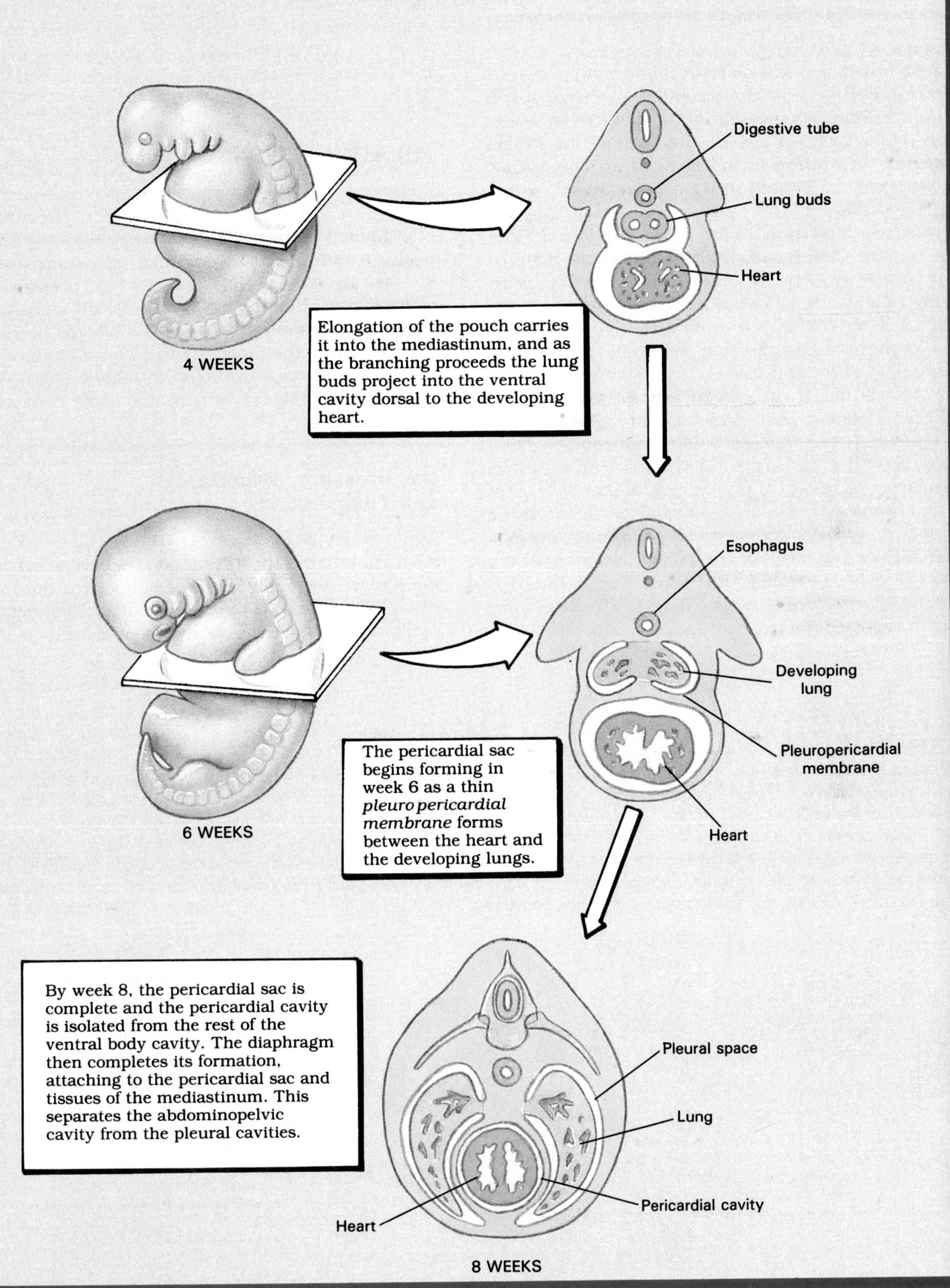
4 WEEKS
Digestive tube
Lung buds
Heart
Elongation of the pouch carries it into the mediastinum, and as the branching proceeds the lung buds project into the ventral cavity dorsal to the developing heart.
6 WEEKS
Esophagus
Developing lung
Pleuropericardial membrane
Heart
The pericardial sac begins forming in week 6 as a thin *pleuro pericardial membrane* forms between the heart and the developing lungs.
By week 8, the pericardial sac is complete and the pericardial cavity is isolated from the rest of the ventral body cavity. The diaphragm then completes its formation, attaching to the pericardial sac and tissues of the mediastinum. This separates the abdominopelvic cavity from the pleural cavities.
Pleural space
Lung
Pericardial cavity
Heart
8 WEEKS

Respiratory Physiology

Respiratory physiology focuses on a series of integrated processes: *pulmonary ventilation, external respiration, internal respiration,* and *cellular respiration.* **Pulmonary ventilation**, or breathing, refers to the movement of air into and out of the lungs. **External respiration** includes the diffusion of gases between the alveoli and the circulating blood. **Internal respiration** is the exchange of dissolved gases between the blood and the interstitial fluids in peripheral tissues. **Cellular respiration**, or *aerobic respiration,* is the absorption and utilization of oxygen by living cells, via biochemical pathways that generate ATP, carbon dioxide, and water.

Abnormalities affecting any single process will ultimately affect the gas concentrations of the interstitial fluids. If the oxygen content declines, the affected tissues will suffer from **hypoxia** (hī-POKS-ē-a). Hypoxia places severe limits on the metabolic activities of peripheral tissues. For example, the effects of *coronary ischemia* (Chapter 20) result from chronic hypoxia of cardiac muscle fibers. If the supply of oxygen gets cut off completely, **anoxia** (a-NOKS-ē-a) results. Anoxia kills cells very quickly. The effects of *cerebrovascular accidents* (Chapter 15) and *myocardial infarctions* (Chapter 20) are the result of localized anoxia. (pp. 492, 646)

This section will examine the processes of pulmonary ventilation, external respiration, and internal respiration. Cellular respiration, the major mechanism for ATP production in our cells, was detailed in Chapter 4. (p. 107)

PULMONARY VENTILATION

Pulmonary ventilation is the physical movement of air into and out of the respiratory tract. The function of pulmonary ventilation is to maintain adequate *alveolar ventilation,* the movement of air into and out of the alveoli. Alveolar ventilation prevents the buildup of carbon dioxide in the alveoli and ensures a continual supply of oxygen that keeps pace with absorption by the bloodstream. To understand this mechanical process we need to take a look at basic physical principles governing the movement of air.

Gas Pressure, Volume, and Temperature

The primary differences between liquids and gases such as air reflect the interactions between individual molecules. Although the molecules in a liquid are in constant motion, they are so close together that weak interactions such as hydrogen bonding can oc-

FIGURE 23-12
Pressure, Temperature, and Gas Volume. (a) Gas molecules within a container bounce off the walls and off one another, traveling the distance indicated in a given period of time. If the volume decreases, each molecule will travel the same distance, but will strike the walls more frequently. This increases the pressure exerted by the gas. Alternatively, if the volume of the container increases, the gas molecules will strike the walls less often, lowering the pressure in the container. (b) Decreasing the temperature of a gas takes energy away from the gas molecules, so they move more slowly and strike the walls less often. This lowers the pressure in the container. Elevating the temperature provides more energy, and the gas molecules move faster, have more frequent collisions, and increase pressure in the container.

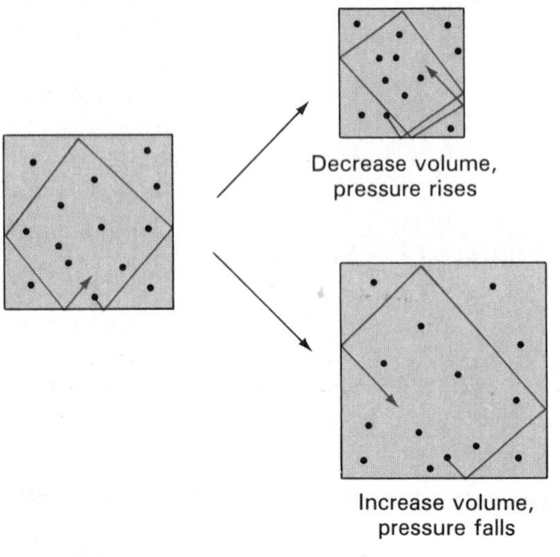

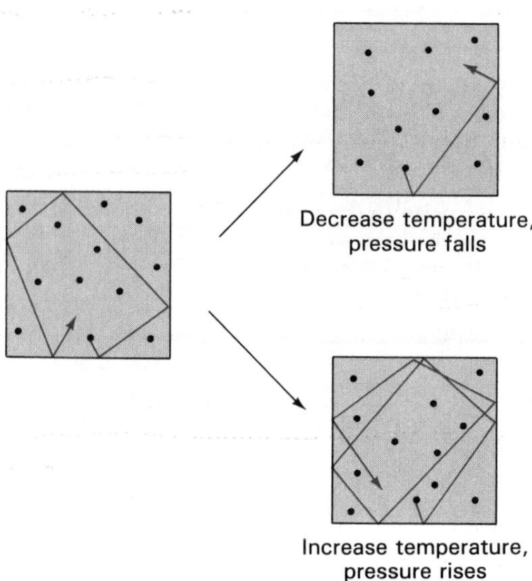

cur between the atoms of one molecule and those of another. As a result liquids are relatively dense. But because the electrons of adjacent atoms tend to repel those of their neighbors, liquids tend to resist compression. If you squeeze a balloon filled with water, it will distort into a different shape, but the volume of the new shape will be the same as that of the original.

In a gas the molecules are bouncing around as independent entities. At normal atmospheric pressures the molecules are fairly far apart, and the density of air is rather low. The forces acting between the molecules are minimal, so an applied pressure can push them closer together. For example, suppose that we have a sealed, elastic container of air at atmospheric pressure, such as the one depicted in Figure 23-12a. The pressure exerted by a gas results from the collision of gas molecules with objects in their environment. The greater the number of collisions the higher the pressure. If all the gas molecules are traveling at a constant speed, decreasing the volume of their container will result in more frequent collisions, and thus elevate the pressure. So an inverse relationship exists between pressure and volume: as the volume decreases, the pressure rises.

We could also change the number of collisions by making the individual molecules move at a different speed. If they move faster, they will hit the walls more often, and the pressure will go up. A slower pace will mean that there will be fewer collisions, and the pressure will drop. The speed of movement varies directly with the temperature, so higher temperatures mean higher pressures, and lower temperatures mean lower pressures (Figure 23-12b).

A change in the volume, pressure, or temperature of a quantity of gas will therefore affect the other parameters as well. For example, if we double the pressure on the outside of an elastic container filled with a gas, its volume will be reduced by one-half. If we reduce the pressure by one-half, the original volume will double. This relationship, first noted by Robert Boyle in the 1600s, is called **Boyle's law** of gases.

Air travelers are probably already aware of the practical consequences of Boyle's law. Many readers will be familiar with the flurry of yawning and gulping that occurs following take-off and before landing. As the plane gains altitude, air pressure decreases and the air in the middle ear expands. If the pharyngotympanic tubes are blocked by mucus, the tympanic membrane will be forced outward, creating a sensation of "fullness" in the ears. Yawning, swallowing, or gulping can often reopen the connecting passageways and relieve the pressure. Pinching the nose to force additional air into the middle ear not only does not help, but can actually make matters worse. That maneuver would be appropriate only when the plane loses altitude and the volume of air in the middle ear decreases. [**CM:** *Boyle's Law and Air Overexpansion Syndrome*]

Pressure and Airflow

Although we are seldom reminded of the fact, the body is under considerable pressure due to the weight of the earth's atmosphere. This **atmospheric pressure (atm)** pushes against everything around us. A column of air 1 inch square and the height of the atmosphere would weigh about 15 pounds, so 1 atm is equivalent to 15 pounds per square inch (**psi**). During our discussion of blood pressure, pressures were reported in millimeters of mercury (mm Hg); in these terms, atmospheric pressure is approximately 760 mm Hg.

Air under pressure tends to flow from an area of higher pressure to an area of relatively lower pressure. This tendency for directed airflow plus the pressure:volume relationships of Boyle's law provide the basis for pulmonary ventilation. A single respiratory cycle consists of an inhalation, or inspiration, and an exhalation, or expiration. These movements generate pressure differences that push air into or out of the lungs. The lungs work rather like the bellows pump shown in Figure 23-13. When the bellows expands, its volume increases and the pressure inside (P_i) falls below atmospheric levels (pressure outside, or P_o). Air rushes into the bellows until the volume stops increasing and the internal pressure is the same as that outside. When the bellows contracts, the volume decreases and the internal pressure rises. This forces the air back out of the bellows.

The volume of the lungs depends on the volume of the pleural cavities. When the diaphragm and/or external intercostal muscles contract, the volume of the thoracic cavity increases. The size of the mediastinum remains relatively constant, and the pleural cavities expand to fill the additional space. This expansion lowers the pressure inside the pleural cavities, and the lungs expand. Their expansion, in turn, lowers the pressure inside the lungs, and air enters along the respiratory passageways. At the end of inspiration the pleural cavities and lungs stop expanding, and air movement ceases when the pressure inside the lungs equals that of the surrounding atmosphere. The opposite chain of events occurs during exhalation. The pleural cavities decrease in volume, pressures rise, and the lungs are compressed. Because pressure in the lungs now exceeds that of the atmosphere, air moves out of the respiratory system. At the end of expiration air movement again ceases when the pressure difference has been eliminated.

The **intrapulmonary** (in-tra-PUL-mo-ner-ē) **pressure**, or **intra-alveolar** (in-tra-al-VĒ-ō-lar) **pressure**, is the pressure measured inside of the respiratory tract. The relationship between the intrapulmonary pressure and atmospheric pressure determines the direction of airflow. During quiet, relaxed breathing the pressure differences are relatively small, oscillating between −1 mm Hg (intrapulmonary 1 mm Hg below atmospheric) during inspiration to +1 mm Hg (intrapulmonary 1 mm Hg above atmospheric)

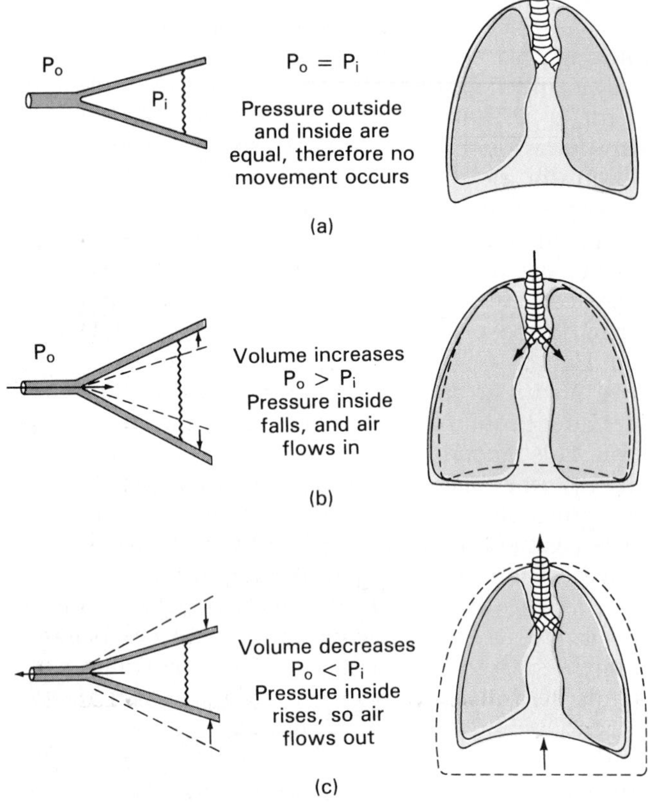
P_o
P_i
$P_o = P_i$
Pressure outside and inside are equal, therefore no movement occurs
(a)
P_o
Volume increases
$P_o > P_i$
Pressure inside falls, and air flows in
(b)
Volume decreases
$P_o < P_i$
Pressure inside rises, so air flows out
(c)

FIGURE 23-13
Pressure and Volume Relationships in the Lungs. (a) At the start of inspiration the pressure within the lungs is identical to that of the surrounding atmosphere. No net movement of air occurs under these conditions. (b) As inspiration commences, the volume of the thoracic cavity increases, so the pressure in the lungs decreases. Air then moves into the lungs. (c) During expiration the volume of the thoracic cavity decreases. Pressure inside the thoracic cavity then exceeds that of the surrounding atmosphere, forcing air out of the lungs.

during expiration. When a trained athlete breathes at maximum capacity the pressure differentials can become much larger, reaching −80 mm Hg during inspiration and +100 mm Hg during expiration.

The **intrapleural pressure** is the pressure measured in the slender space between the parietal and visceral pleurae. The parietal and visceral pleurae are separated by only a thin film of pleural fluid, and although the two membranes can slide across one another, they are held together by that fluid film. You encounter the same principle whenever you set a wet glass on a smooth surface. You can slide the glass quite easily, but when you try to lift it you encounter considerable resistance. As you pull the glass away from the tabletop you create a powerful suction, and the only way to defeat it is to tilt the glass so that air is pulled between the glass and the table, breaking the fluid bond.

A comparable fluid bond exists between the parietal pleura and the visceral pleura covering the lungs. After the first breath at birth, the lungs never return to their fully collapsed state. The elastic fibers in the walls of the lungs tend to contract and pull the lungs away from the pleural wall, but the elastic forces cannot break the fluid bond. The pull exerted does lower the intrapleural pressure, so that it averages about −4 mm Hg and reaches − 18 mm Hg during a powerful inspiration.

The graphs in Figure 23-14 follow the intrapleural and intrapulmonary pressures during a single quiet respiratory cycle and relate these changes to the volume of air that enters and leaves the lungs. Inspiration begins with the fall of intrapleural pressure that accompanies expansion of the thoracic cavity. The intrapleural pressure gradually falls to approximately −6 mm Hg. Over this period the intrapulmonary pressure declines to just under −1 mm Hg; it then begins to rise as air continues to flow into the lungs. When expiration begins, intrapleural and intrapulmonary pressures rise rapidly, forcing air out of the lungs. The cyclical changes in the intrapleural pressure are responsible for the *thoracoabdominal pump* that assists the venous return to the heart (discussed in Chapter 20). (p. 660)

An injury to the chest wall that penetrates the parietal pleura or damages the alveoli and the visceral pleura can allow air into the pleural cavity. This **pneumothorax** (nū-mō-THŌ-raks) breaks the fluid bond between the pleurae and allows the elastic fibers to contract. The result is called a collapsed lung, or **atelectasis** (at-e-LEK-ta-sis; *ateles*, imperfect + *ektasis*, expansion). Treatment for a collapsed lung involves removing as much of the air as possible before sealing the opening. This procedure lowers the intrapleural pressure and reinflates the lung.

Respiratory Muscles

The skeletal muscles involved in respiratory movements were introduced in Chapter 11. (p. 348) Of these the most important respiratory muscles are the diaphragm and the external and internal intercostals. Contraction of the diaphragm increases the volume of the thoracic cavity by tensing and flattening its floor, and this increase draws air into the lungs. The external intercostals may assist in inspiration by elevating the ribs, because this swings the ribs forward, increasing the width of the thoracic cage along its anterior-posterior axis. The internal intercostals depress the ribs and reduce the width of the thoracic cavity, thereby contributing to expiration. These muscles and their actions are diagrammed in Figure 23-15.

The **accessory muscles** become active when the depth and frequency of respiration must be increased markedly. The sternocleidomastoid, the serratus anterior, and the scalenes assist the external intercostals in elevating the ribs and performing inspiration. The abdominal obliques and rectus muscles assist the internal intercostals in expiration by compressing

FIGURE 23-14
Pressure Changes During Inspiration and Expiration. These graphs follow changes in the intrapleural and intrapulmonary pressures during a single respiratory cycle, and relate these changes to the tidal volume.

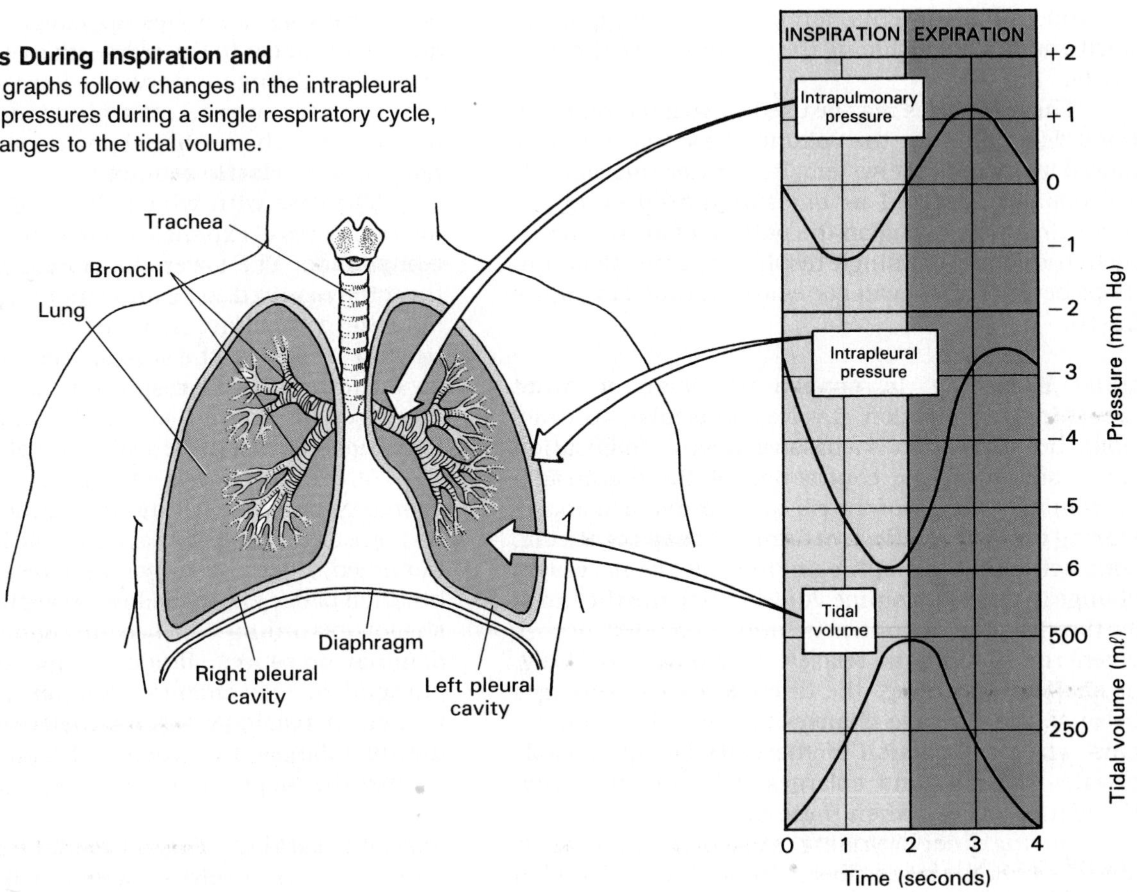

Trachea
Bronchi
Lung
Diaphragm
Right pleural cavity
Left pleural cavity
INSPIRATION
EXPIRATION
Intrapulmonary pressure
Intrapleural pressure
Tidal volume
+2
+1
0
−1
−2
−3
−4
−5
−6
Pressure (mm Hg)
500
250
Tidal volume (mℓ)
0 1 2 3 4
Time (seconds)

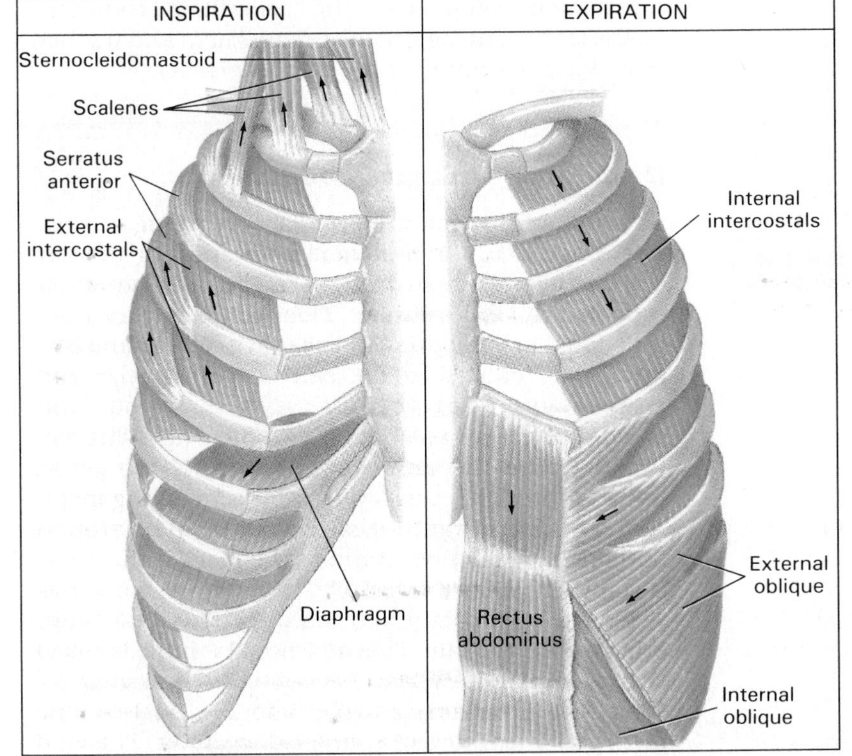

FIGURE 23-15
Respiratory Muscles

the abdominal contents, forcing the diaphragm upward and further reducing the volume of the thoracic cavity.

These muscles may be used in various combinations depending on the volume of air that must be moved in or out of the system. Respiratory movements are usually classified as *quiet breathing* or *forced breathing*, depending on the pattern of muscle activity in the course of a single respiratory cycle. Relationships between these categories are indicated in Figure 23-16.

Quiet Breathing In **eupnea** (ŪP-nē-a), or **quiet breathing**, inspiration involves muscular contractions, but expiration is a passive process. Inspiration can result from the contraction of the diaphragm or from movements of the rib cage, as detailed above. During **diaphragmatic breathing**, or **deep breathing**, contraction of the diaphragm provides the necessary change in thoracic volume. Air is drawn into the lungs as the diaphragm contracts, and exhalation occurs when the diaphragm relaxes. In **costal breathing**, or **shallow breathing**, the thoracic volume changes because the rib cage changes shape. Inhalation occurs when contraction of the external intercostals elevates the ribs and enlarges the thoracic cavity. Exhalation occurs when these muscles relax.

During quiet breathing, expansion of the lungs stretches their elastic fibers. In addition, elevation of the rib cage stretches opposing skeletal muscles and elastic fibers in the connective tissues of the body wall. When the inspiratory muscles relax, these elastic components contract, returning the diaphragm and/or rib cage to their original positions. This phenomenon is called **elastic rebound**.

The ease with which the lungs and chest wall tolerate cycles of expansion and contraction is termed **compliance**. The lower the compliance, the greater the force required to fill and empty the lungs. Factors reducing compliance include the formation of inelastic scar tissue in the lungs, the collapse of alveoli on expiration (as in respiratory distress syndrome, discussed previously), and arthritis affecting the articulations between the ribs and spinal column.

At minimal levels of activity, eupnea involves primarily deep breathing with little costal motion. As increased volumes of air are required, inspiratory movements become larger and result from a combination of deep and shallow breathing. Even at rest, shallow breathing can become important when abdominal pressures, fluids, or masses restrict diaphragmatic movements. For example, pregnant women increasingly rely on costal breathing as the uterus enlarges and pushes the abdominal viscera against the diaphragm.

Forced Breathing **Forced breathing**, or **hyperpnea** (hī-perp-NĒ-a), involves active inspiratory and expiratory movements. Forced breathing calls upon the accessory muscles to assist with inspiration, and expiration involves contraction of the internal intercostals. When breathing at absolute maximum levels, the abdominal muscles are used in exhalation. Their contraction compresses the abdominal contents, pushing them up against the diaphragm and further reducing the volume of the thoracic cavity.

FIGURE 23-16
A Classification of Respiratory Activity

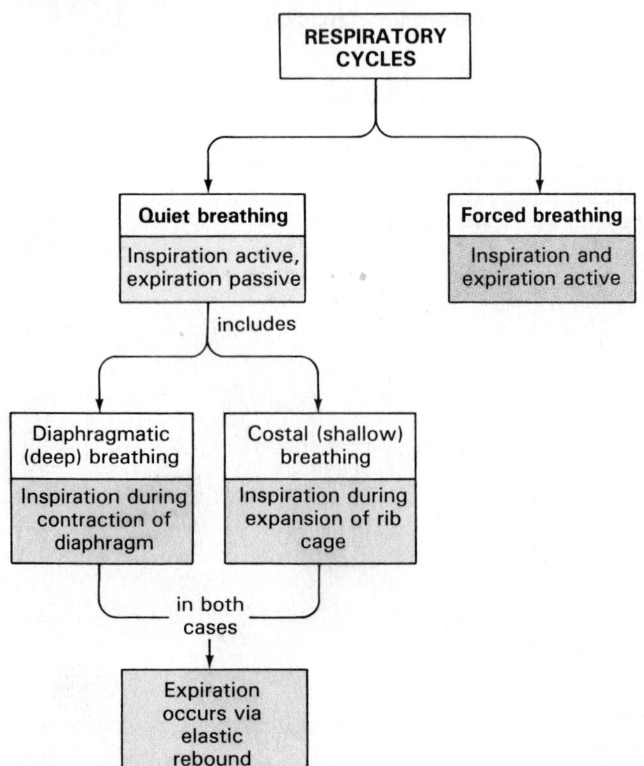

Respiratory Volumes and Rates

As you read this you are probably breathing quietly. The amount of air that moves in and out of your respiratory system during a normal quiet respiratory cycle is the **tidal volume**. Tidal volumes vary from individual to individual, but they average around 500 mℓ. If you paused at the end of a tidal cycle and exhaled until no more air could be forced out, you could probably expel about 1200 mℓ of air. This volume is called the **expiratory reserve**. If you pause midway through a tidal cycle, after completing inspiration, your respiratory system would contain around 500 mℓ more air than it did at the start of the cycle. If you then inhaled forcefully, expanding your lungs to maximum capacity, you could probably pull in another 3600 mℓ of air. This additional volume is called the **inspiratory reserve**. Adding the inspiratory reserve, the expiratory reserve, and the tidal volume gives you a measure of your **vital capacity**. The vital capacity is the maximum amount of air that can be

Clinical Comment: Artificial Respiration and CPR

Artificial respiration provides air to an individual whose cardiovascular system continues to function normally, but whose respiratory muscles are no longer operating. In **mouth-to-mouth resuscitation** a rescuer provides ventilation by exhaling into the mouth or mouth and nose of the victim. After each breath contact is broken to permit passive exhalation by the victim. Air provided in this way contains adequate oxygen to meet the needs of the victim.

There are also mechanical **resuscitators** (re-SUS-i-tā-tors), or *ventilators*, that deliver air under appropriate pressures. Their use often requires inserting a tube into the trachea and attaching a cyclic pump that alternately pushes air in and pulls air out. In more complex *tank respirators*, the victim's body is placed in a machine called an "iron lung." Pressure inside this sealed chamber can be regulated, and the victim's head remains exposed to air at regular atmospheric pressures. When pressure inside the chamber falls, the chest expands and air is drawn into the lungs. A subsequent period of increased pressure pushes the air out. Both of these techniques require a constant monitoring of the applied pressures to prevent damage to the lungs.

Cardiopulmonary resuscitation, or **CPR**, restores circulation and ventilation to an individual whose heart has stopped beating. Compression applied to the rib cage over the sternum reduces the volume of the thoracic cavity, squeezing the heart and propelling blood into the aorta and pulmonary trunk. When the pressure is removed the thorax expands, and blood moves into the great veins. Cycles of compression are interspersed with cycles of mouth-to-mouth breathing that maintain pulmonary ventilation.

This practice has been credited with saving thousands of lives each year. Basic CPR techniques can be mastered in about 8 hours of intensive training, using special equipment. Yearly recertification courses must be taken, for the skills fade with time, and CPR techniques cannot be practiced on a living person without causing severe injuries. Training is available at minimal cost through charitable organizations such as the Red Cross and the American Heart Association.

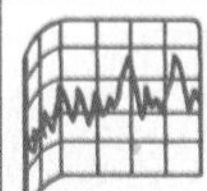

Diagnostics: Pulmonary Function Tests

Pulmonary function tests monitor several aspects of respiratory function. A **spirometer** (spī-ROM-e-ter) measures parameters such as vital capacity, expiratory reserve, and inspiratory reserve. A **pneumotachometer** provides additional information by determining the rate of air movement. Although these tests are relatively simple to perform, they have considerable diagnostic significance. For example, in *asthma* (see the Clinical Comment: Asthma), the constricted airways tend to close before an exhalation is completed. As a result, both the vital capacity and the expiratory reserve volume are diminished, and the narrow respiratory passageways reduce the flow rate as well. Air whistling through the constricted airways produces the characteristic "wheezing" that accompanies an asthmatic attack.

Conditions that restrict the maximum distensibility of the lungs have the same effect on vital capacity because they lower the inspiratory reserve. However, because the airways are not affected, the expiratory reserve and expiratory flow rates are relatively normal.

Clinical Comment: Asthma

Asthma (AZ-ma) affects an estimated 3–6 percent of the U.S. population. There are several different forms of asthma, but all are characterized by unusually sensitive and irritable conducting passageways. In many cases the trigger appears to be an immediate hypersensitivity reaction to an allergen in the inspired air. Drug reactions, air pollution, chronic respiratory infections, exercise, and/or emotional stress can also cause an asthmatic attack in sensitive individuals.

The most obvious and potentially dangerous symptoms include (1) the constriction of smooth muscles all along the bronchial tree, (2) edema and swelling of the walls of the respiratory passageways, and (3) accelerated production of mucus. The combination makes breathing very difficult. Exhalation is affected more than inhalation; the narrowed passageways often collapse before exhalation is completed. (This difficulty in completing normal exhalation can be important in the diagnosis of asthma, as discussed in Diagnostics: Pulmonary Function Tests.) Although mucus production increases, mucus transport slows, and fluids accumulate along the passageways. Coughing and wheezing then develop.

Severe asthmatic attacks reduce the functional capabilities of the respiratory system, and the peripheral tissues become oxygen-starved. This condition can prove fatal, and asthma fatalities have been increasing in recent years. The annual death rate from asthma in the United States is approximately 4 deaths per million population (for ages 5–34). Mortality among asthmatic blacks is twice that of whites.

Treatment of asthma involves dilation of the respiratory passageways by the administration of **bronchodilators** (brong-kō-dī-LĀ-torz). Important bronchodilators include *theophylline*, *epinephrine*, and other beta-adrenergic drugs. Although the strongest beta-adrenergic drugs are very useful in a crisis, they are effective only for relatively brief periods, and the individual must be closely monitored because of the potential effects on cardiovascular function.

FIGURE 23-17
Respiratory Performance and Volume Relationships

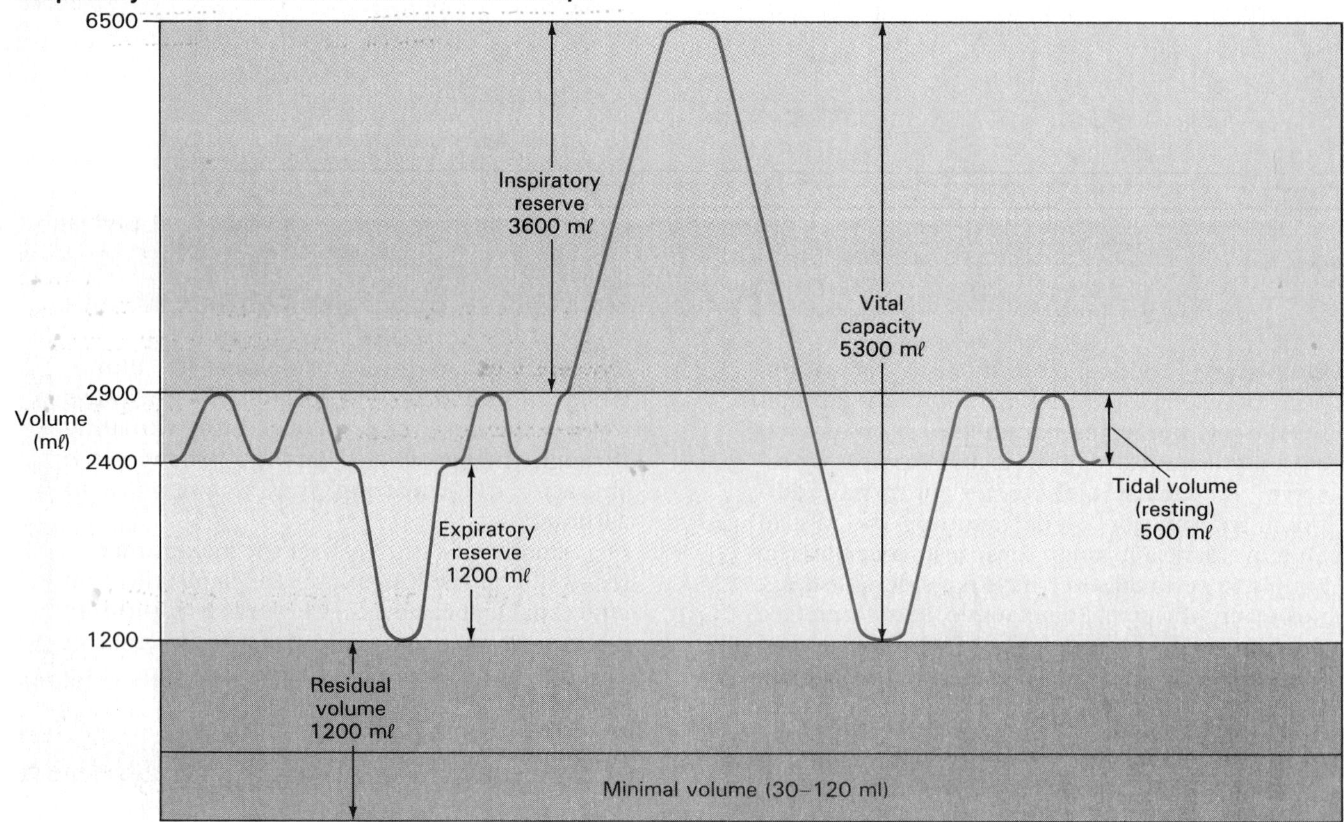

moved into and out of the respiratory system in a single respiratory cycle.

Roughly 1200 mℓ of air remain in the respiratory passageways and alveoli, even after exhausting the expiratory reserve. Most of this **residual volume** exists because the lungs are held against the thoracic wall, preventing their elastic fibers from contracting further. Opening the chest cavity, as in a pneumothorax, breaks the fluid bond, and the lungs collapse. This collapse reduces the amount of air in the respiratory system to the **minimal volume**. Some air remains in the lungs, even at minimal volume, because the surfactant coating the alveolar surfaces prevents their collapse. These volume relationships are graphically depicted in Figure 23-17.

The number of breaths per minute determines the **respiratory rate**. The normal adult respiratory rate at rest ranges from 12 to 18 breaths per minute. Children breath more rapidly, at rates of around 18–20 breaths per minute. The amount of air moved each minute can be calculated by multiplying the respiratory rate (f) by the tidal volume (V_t). This value is called the **respiratory minute volume**, $\dot{V}_e$. In other words:

$$\dot{V}_e \text{ (volume of air moved each minute)} = f \text{ (breaths per minute)} \times V_t \text{ (tidal volume)}$$

The respiratory minute volume averages around 6 liters per minute (12 × 500 mℓ). At peak respiratory rates of 40–50 breaths per minute and maximum cycles of inspiration and expiration, the respiratory minute volume can approach 200 liters (about 55 gal) per minute.

- ✓ **Mark breaks a rib that punctures the chest wall on his left side. What would you expect to happen to his left lung as a result?**
- ✓ **In emphysema, alveoli are replaced by large air spaces and elastic tissue by fibrous connective tissue. How would these changes affect the compliance of the lungs?**
- ✓ **In pneumonia, fluid accumulates in the alveoli of the lungs. How would this affect vital capacity?**

EXTERNAL RESPIRATION

Pulmonary ventilation ensures that the alveoli are supplied with oxygen, and it removes the carbon dioxide arriving from the bloodstream. Gas exchange between the alveolar air and the blood is the process of *external respiration*. To understand the events that occur during external respiration we must take a closer look at the process of diffusion. (You may wish to return to Chapter 3 for a quick review before proceeding.) ∞ (p. 78)

Diffusion between Liquids and Gases

Differences in pressure move gas molecules from one place to another. Pressure differences also affect the movement of gas molecules into and out of solution. **Solubility** (sol-ū-BIL-i-tē) is the ease with which a gas molecule enters a solution. Very little pressure is required to force large numbers of very soluble gas molecules into solution, whereas very high pressures are required to dissolve a comparable amount of a relatively insoluble gas. The relationship between pressure, solubility, and the number of gas molecules in solution is known as **Henry's law**.

When a gas under pressure contacts a liquid, the pressure tends to force gas molecules into solution. At a given pressure the number of dissolved gas molecules rises until an equilibrium state becomes established. At equilibrium, gas molecules are diffusing out of the liquid as fast as they are arriving, and the total number of gas molecules in solution remains constant. A change in pressure will temporarily disturb that equilibrium. A rise in pressure will force additional molecules into solution, and a fall in pressure will result in the movement of gas molecules out of the solution. These relationships are diagrammed in Figure 23-18. [**CM:** *Decompression Sickness*]

Mixed Gases and Dalton's Law

The air we breathe is not a single gas, but a mixture of gases. Nitrogen molecules (N_2) are the most abundant, accounting for about 78.6 percent of the atmospheric gas molecules. Oxygen molecules (O_2), the second most abundant, constitute roughly 20.8 percent of the atmospheric population. Most of the remaining 0.5 percent consists of water molecules, with carbon dioxide (CO_2) contributing a mere 0.04 percent to the atmospheric total.

Atmospheric pressure, 760 mm Hg, represents the combined effects of collisions involving each type of molecule. At any given moment, 78.6 percent of those collisions will involve nitrogen molecules, 20.8 percent oxygen molecules, and so on. Thus each of the gases contributes to the total pressure in proportion to its relative abundance. This principle is known as **Dalton's law**.

The pressure contributed by a single gas is the **partial pressure** of that gas, abbreviated by the prefix *p*. All of the partial pressures added together equal the total pressure exerted by the gas mixture. In the case of the atmosphere this relationship can be summarized as:

$$pN_2 + pO_2 + pH_2O + pCO_2 = 760 \text{ mm Hg}$$

Because we know the individual percentages, the partial pressure for each gas can be calculated easily. For example, the partial pressure of oxygen, pO_2, is 20.8 percent of 760 mm Hg, or roughly 159 mm Hg. The partial pressures for other atmospheric

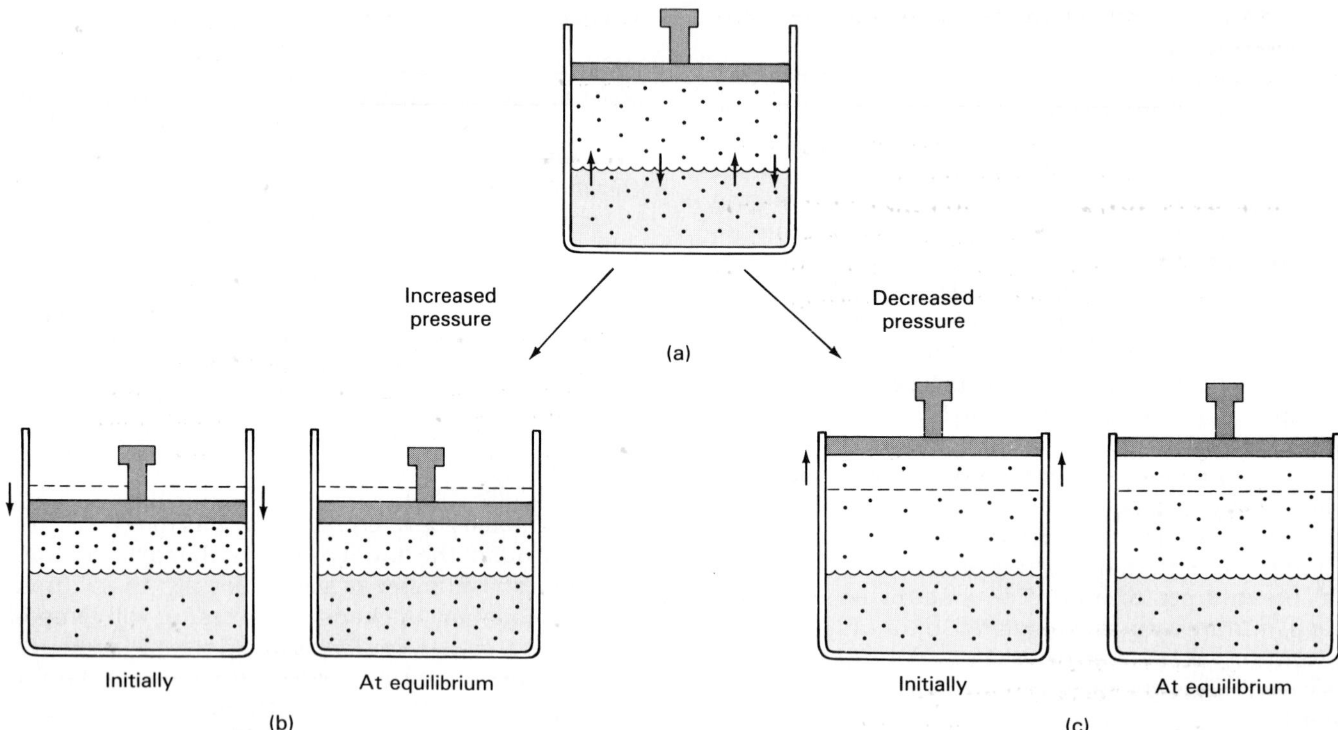

FIGURE 23-18
Henry's Law and the Relationship between Solubility and Pressure. (a) A solution containing dissolved gas molecules at equilibrium with air under a given pressure. (b) Increasing the pressure drives additional gas molecules into solution until a new equilibrium becomes established. (c) When the pressure decreases, some of the dissolved gas molecules leave the solution until equilibrium is restored.

gases are included in Table 23-1. These calculations are important because the partial pressure of an individual gas determines its rate of diffusion between the alveolar air and the bloodstream. For example, the partial pressure of oxygen determines how much oxygen enters solution, but it has no effect on the rate of nitrogen or carbon dioxide diffusion. This is clearly a case of "every gas for itself!"

Alveolar Ventilation

The respiratory minute volume ($\dot{V}_e$) measures pulmonary ventilation and provides an indication of how much air is moving in and out of the respiratory tract. Only a portion of the inspired air reaches the alveolar exchange surfaces. A typical tidal inspiration pulls around 500 mℓ of air into the respiratory sys-

TABLE 23-1 Partial Pressures and Normal Gas Concentrations in Air and Body Fluids

Source of Sample	*Nitrogen* (N_2)	*Oxygen* (O_2)	*Carbon Dioxide* (CO_2)	*Water Vapor* (H_2O)
Inspired air (dry)	597 (78.6%)	159 (20.8%)	0.3 (0.04%)	3.7 (0.5%)
Alveolar air (saturated)	573 (75.4%)	100 (13.2%)	40 (5.2%)	47 (6.2%)
Expired air (saturated)	569 (74.8%)	116 (15.3%)	28 (3.7%)	47 (6.2%)
Arterial blood (systemic)	573	95	40	
Peripheral tissues	573	40	45	
Venous blood (systemic)	573	40	45	

Note: Values are expressed in millimeters of mercury (mm Hg). Percentages indicate contribution to total atmospheric pressure at sea level (760 mm Hg).

tem. The first 350 mℓ inspired travels along the conducting passageways and enters the alveolar spaces. The last 150 mℓ of inspired air never gets farther than the conducting passageways and does not participate in gas exchange with the blood. The volume of air in the conducting passages is known as the **anatomic dead space**, or V_d. **Alveolar ventilation**, or $\dot{V}_e$, is the amount of air reaching the alveoli each minute. It can be calculated by subtracting the dead space from the tidal volume, using the formula:

$$\dot{V}_e = f \times (V_t - V_d)$$

At rest, alveolar ventilation rates are approximately 4.2 liters per minute (12 × 350 mℓ). However, the gas arriving in the alveoli is significantly different from that of the surrounding atmosphere.

Alveolar versus Atmospheric Air

As soon as air enters the respiratory tract its characteristics begin to change. In passing through the nasal cavity the air becomes warmer, and the amount of water vapor increases. Humidification and filtration continue as the air travels through the pharynx, trachea, and bronchial passageways. On reaching the alveoli, the incoming air mixes with air that remained in the alveoli after the previous respiratory cycle. The alveolar mixture thus contains more carbon dioxide and less oxygen than does atmospheric air. The last 150 mℓ of inspired air never gets farther than the conducting passageways and remains in the anatomic dead space of the lungs. During the subsequent expiration, the departing alveolar air mixes with air in the dead space to produce yet another mixture that differs from both atmospheric and alveolar samples.

The differences in composition between atmospheric (inspired) and alveolar air can be seen in Table 23-1. This table also indicates the partial pressures found in blood samples from the peripheral (systemic) arteries and veins. The blood delivered by the pulmonary arteries has a higher pCO_2 and a lower pO_2 than does alveolar air. Diffusion between the alveolar mixture and the pulmonary capillaries thus elevates the pO_2 of the blood while lowering its pCO_2. By the time the blood enters the pulmonary venules it has reached equilibrium with the alveolar air, so it departs the alveoli with a pO_2 of about 100 mm Hg and a pCO_2 of roughly 40 mm Hg. Diffusion between the blood in the pulmonary capillaries and the alveolar air occurs very rapidly. In a resting individual a red blood cell moves through a pulmonary capillary in about 0.75 seconds, and during exercise the passage takes less than 0.3 seconds.

As it enters the pulmonary veins this oxygenated blood mixes with blood that flowed through capillaries around conducting passageways rather than alveoli. Because gas exchange occurs only at alveoli, the blood leaving the conducting passageways carries relatively little oxygen. The partial pressure of oxygen in the pulmonary veins therefore drops to about 95 mm Hg. At this partial pressure, each 100 mℓ of blood contains about 20 mℓ of oxygen. No further changes in gas concentration occur until the blood reaches the peripheral capillaries, where exchange with the surrounding interstitial fluids takes place. [CM: *Bronchitis, Emphysema, and COPD*]

INTERNAL RESPIRATION

Normal interstitial fluid has a pO_2 of 40 mm Hg and a pCO_2 of 45 mm Hg. As a result, oxygen diffuses out of the capillaries and carbon dioxide diffuses in until the capillary partial pressures are the same as those in the adjacent tissues. At a normal tissue pO_2, the blood entering the venous system still contains around 15 mℓ of oxygen per 100 mℓ, having lost only

FIGURE 23-19
An Overview of Respiration and Respiratory Processes

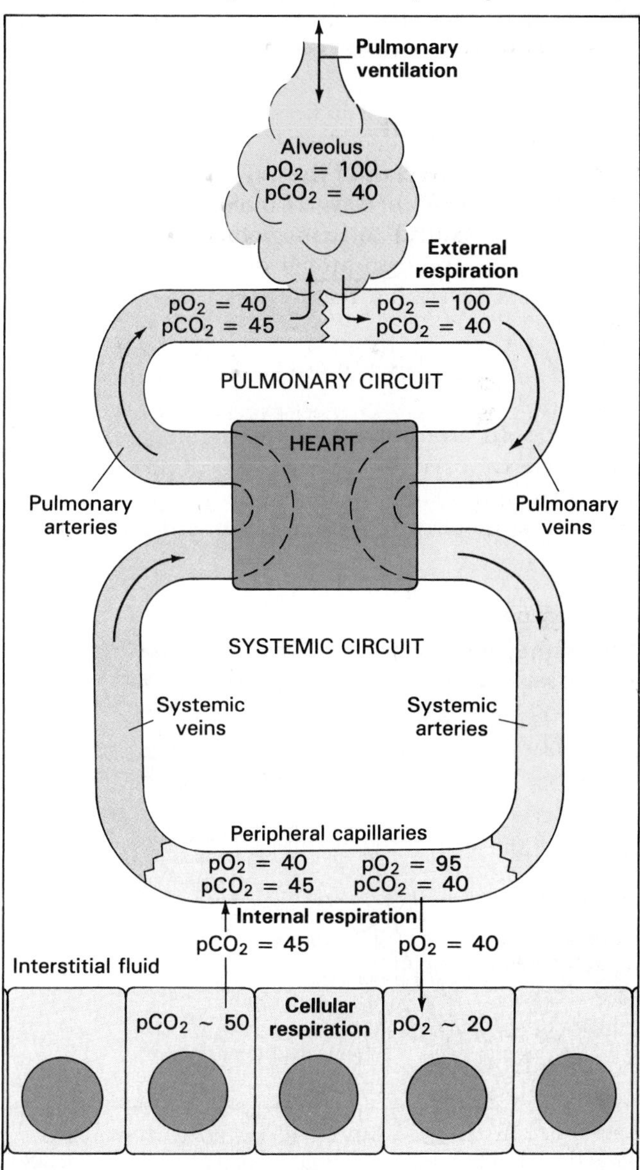

25 percent of its total oxygen content. That amount will be restored when the venous blood reaches the alveolar capillaries, at the same time that the excess CO_2 is lost. The interplay between cellular, internal, and external respiration is diagrammed in Figure 23-19. [**CM:** *Blood Gas Analysis*]

Gas Pickup and Delivery

Oxygen and carbon dioxide have limited solubilities in blood plasma. This limitation poses certain functional problems, for peripheral tissues need more oxygen and generate more carbon dioxide than the plasma can absorb and transport. The extra oxygen and carbon dioxide diffuse into the red blood cells, where the gas molecules are either tied up or used to manufacture soluble compounds. The important thing about these reactions is that they are *temporary* and *completely reversible*. When plasma oxygen or carbon dioxide concentrations are high, the excess molecules are removed by the red blood cells; when the plasma concentrations are falling, the red blood cells release their stored reserves.

Oxygen Transport

Only around 3 percent of the oxygen content of arterial blood consists of oxygen molecules in solution. The rest are bound to hemoglobin (Hb) molecules, specifically to the iron atoms in the center of heme units. This reversible reaction can be summarized as:

$$Hb + O_2 \rightleftharpoons HbO_2$$

Hemoglobin Saturation When hemoglobin molecules are exposed to oxygen, an equilibrium soon develops. If all of the hemoglobin were converted to HbO_2, the blood would be 100 percent oxygenated, or **saturated**, with oxygen. Complete saturation does not occur until the pO_2 reaches excessively high levels, around 250 mm Hg. At lower oxygen concentrations the blood is only partially saturated. At normal alveolar pressures (around 100 mm Hg) the **percent saturation** is still very high. Hemoglobin has a strong attraction for oxygen; at the oxygen concentration of alveolar air, 97.5 percent of the hemoglobin is in the form of HbO_2.

When oxygenated blood arrives in the peripheral capillaries, the pO_2 declines rapidly as a result of gas exchange with the interstitial fluid. This decline disturbs the equilibrium by removing one of the key reactants. As the plasma concentration of oxygen falls, hemoglobin gives up its oxygen. Plotting the percent saturation of hemoglobin against the pO_2 of the blood yields the **saturation curve** shown in Figure 23-20.

The shape of this curve reflects the particular properties of the hemoglobin molecule—specifically, its **oxygen affinity**, a term that refers to how readily the hemoglobin molecule binds oxygen. Because of the way in which the subunits of hemoglobin interact, the oxygen affinity of the molecule increases whenever

FIGURE 23-20
The Hemoglobin Saturation Curve. This curve indicates the normal saturation characteristics of hemoglobin and the effects of altering the pO_2, pH, temperature, and pCO_2.

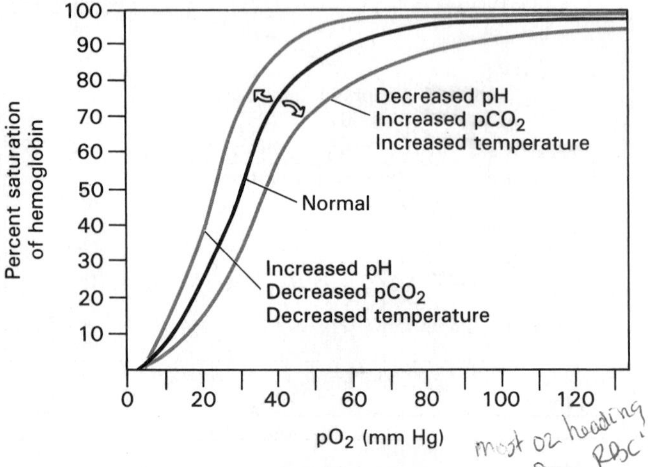

Sports and Fitness: Adaptation to High Altitudes

Atmospheric pressure declines with increasing altitude, and so do the partial pressures of the component gases, including oxygen. People living in Denver or Mexico City function normally with alveolar oxygen pressures in the 80–90 mm Hg range. At higher elevations, the alveolar pO_2 continues to decline. At 11,000 feet, an altitude familiar to many hikers and skiers, the alveolar pO_2 falls to around 60 mm Hg.

Despite the low alveolar pO_2, millions of people live and work at altitudes this high or higher. Important physiological adjustments include an increased respiratory rate, an increased heart rate, and an elevated hematocrit. Thus even though the hemoglobin is not fully saturated, there is more of it in circulation, and the round trip between the lungs and the peripheral tissues takes less time. These responses represent an excellent example of the functional interplay between the respiratory and cardiovascular systems. However, most of these adaptations take time (days to weeks) to appear. As a result, athletes planning to compete in events held at high altitude (such as the Olympics in Mexico City) must begin training well in advance.

an oxygen molecule is bound to one of the heme units. That is, binding of the first oxygen molecule makes it easier for Hb to bind a second; binding a second oxygen makes it easier to bind a third, and so forth. As a result, the saturation curve has a steep initial slope, where the percent saturation changes rapidly, followed by a prolonged plateau. [**CM:** *Carbon Monoxide Poisoning*]

Over the steep initial slope, a very small decrease in the plasma pO_2 will result in a large change in the amount of oxygen bound to or released from HbO_2. This ensures rapid oxygen uptake at the lungs, and efficient delivery to peripheral tissues. Because the curve rises quickly, hemoglobin will be more than 90 percent saturated if exposed to an alveolar pO_2 above 60 mm Hg. This means that near-normal oxygen transport can continue despite relatively large decreases in the oxygen content of alveolar air. Without this ability people would be unable to function at elevations much above sea level!

In the periphery, inactive tissues have little demand for oxygen, so the local pO_2 may rise to 50 mm Hg. Under these conditions hemoglobin will not release much oxygen. Active tissues consume oxygen at an accelerated rate, potentially lowering the tissue pO_2 to 20 mm Hg. This *automatically* increases the amount of oxygen released by hemoglobin molecules passing through local capillaries. You will notice that at a pO_2 of 40 mm Hg, an average tissue value, hemoglobin still retains 75 percent of its oxygen reserves. As a result the blood contains a relatively large oxygen reserve that can be called on in an emergency.

The Bohr Effect Active tissues also generate acids that lower the pH of the interstitial fluids. When extra hydrogen ions appear in the blood, the slope of the hemoglobin saturation curve changes. As indicated in Figure 23-20, a shift in pH from 7.4 (normal) to 7.2 (slightly more acidic) makes the hemoglobin molecules release their oxygen reserves more readily. Thus at a tissue pO_2 of 40 mm Hg, hemoglobin molecules will release 15 percent more oxygen at a pH of 7.2 than they would at a pH of 7.4. This effect of pH upon the hemoglobin saturation curve is called the **Bohr effect**.

Carbon dioxide is the primary compound responsible for the Bohr effect. When CO_2 diffuses into the blood, an enzyme called **carbonic anhydrase** (an-HĪ-drāz) catalyzes its reaction with water molecules:

$$CO_2 + H_2O \xrightleftharpoons{\text{carbonic anhydrase}} H_2CO_3$$

The product, H_2CO_3, is called carbonic acid because it dissociates into a bicarbonate ion (HCO_3^-) and a hydrogen ion, H^+. That free hydrogen ion then "bumps" the oxygen from one of the heme units. Other acids, such as the lactic acid generated by oxygen-starved tissues, will have the same effect.

Temperature Effects Temperature changes also affect the slope of the hemoglobin saturation curve, as indicated in Figure 23-20. As the temperature rises, the curve shifts to the right, and hemoglobin releases more oxygen.

All three of these factors (pO_2, the Bohr effect, and temperature) are important during periods of maximal exertion. When a skeletal muscle works hard, the regional temperature rises, the local pO_2 declines, the pCO_2 climbs, and lactic acid rushes into the interstitial fluids. The combination makes the hemoglobin entering the area release much larger amounts of oxygen. Without this automatic adjustment, tissue pO_2s would plummet, and the exertion would come to a premature halt.

DPG and HbO_2 When a low plasma pO_2 continues for extended periods, red blood cells generate larger than normal quantities of **2,3-diphosphoglycerate** (dī-fos-fō-GLIS-e-rāt), or **DPG**. DPG interacts with hemoglobin and affects its affinity for oxygen. When DPG levels are elevated, hemoglobin will release about 10 percent more oxygen at a given pO_2 than it would otherwise. DPG levels are also important in determining the length of time that fresh whole blood can be held by a blood bank. (Transfusions and blood storage were discussed in Chapter 19.) (p. 631) In stored whole blood DPG levels gradually decline, and in the presence of abnormally low DPG concentrations the hemoglobin of the red blood cells becomes more firmly bound to the available oxygen. This change makes the blood useless for transfusion purposes, because the red blood cells will no longer release oxygen to peripheral tissues, even at a low pO_2.

Fetal Hemoglobin A developing fetus obtains its oxygen from the maternal bloodstream, but the blood arriving at the placenta has a relatively low pO_2, ranging from 35 to 50 mm Hg. Survival is possible under these conditions only because fetal hemoglobin differs in structure from adult hemoglobin. The differences affect the shape of the saturation curve. Figure

FIGURE 23-21
A Functional Comparison of Fetal and Adult Hemoglobin

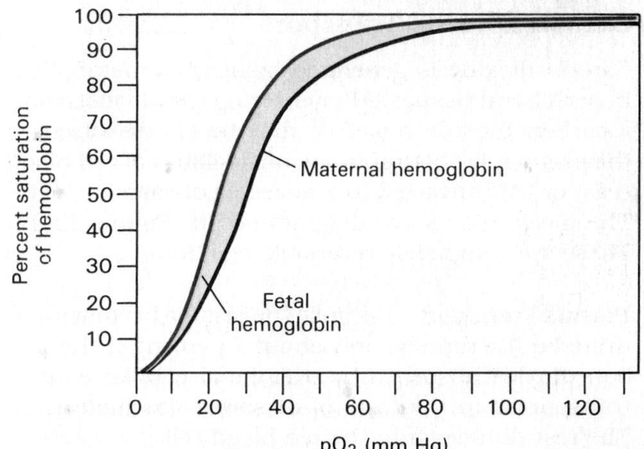

23-21 compares the saturation curves for adult and fetal hemoglobins.

Maternal blood arriving at the placenta has a pO_2 of 40 mm Hg, with roughly 75 percent hemoglobin saturation. The fetal blood arriving at the placenta has a pO_2 close to 20 mm Hg. Oxygen therefore diffuses into the fetal circulation until the pO_2s equilibrate at 30 mm Hg. The amount of oxygen transferred is increased because fetal hemoglobin has a much stronger affinity for oxygen. Fetal hemoglobin arrives at the placenta 58 percent saturated. By the time the exchange is over, the maternal hemoglobin will be less than 60 percent saturated, but the fetal saturation will be over 80 percent.

When the fetal red blood cells arrive in the peripheral tissues, the steeper slope of the saturation curve for fetal hemoglobin means that it will release a large amount of oxygen in response to a very small change in pO_2. The difference becomes apparent when you compare the uptake and delivery of oxygen in the maternal and fetal circulations. Maternal hemoglobin, with its gentler slope, leaves the lungs at a pO_2 of 100, with 97.5 percent saturation, and returns with a pO_2 of 40 and a 75 percent saturation. Thus it releases about 22 percent of its oxygen reserves over a 60-mm Hg drop in pO_2. Fetal hemoglobin leaves the placenta at a pO_2 of 30 mm Hg with an 80 percent saturation and returns with a pO_2 of 20 mm Hg and 58 percent saturation. It has also released 22 percent of its oxygen reserves, but over a drop of just 10 mm Hg.

As noted earlier, an adult living for an extended period with a maximum of 80 percent hemoglobin saturation would show several physiological adjustments. (See Sports and Fitness: Adaptation to High Altitudes on page 772.) The fetus exhibits comparable adjustments. In essence, a fetus resembles a mountain climber, because it has a significantly higher hematocrit than an adult at sea level, and 50 percent more hemoglobin per unit of blood. The fetal heart rate is also considerably faster, so the circulation time is reduced.

Carbon Dioxide Transport

Carbon dioxide is generated by aerobic metabolism in peripheral tissues. After entering the bloodstream, a carbon dioxide molecule may be (1) dissolved in the plasma, (2) bound to the hemoglobin of red blood cells, or (3) converted to a molecule of carbonic acid. The mechanisms are diagrammed in Figure 23-22. These are completely reversible reactions.

Plasma Transport Because the plasma becomes saturated quite rapidly, only about 7 percent of the carbon dioxide absorbed by peripheral capillaries gets transported in the form of dissolved gas molecules. The rest diffuses into the red blood cells.

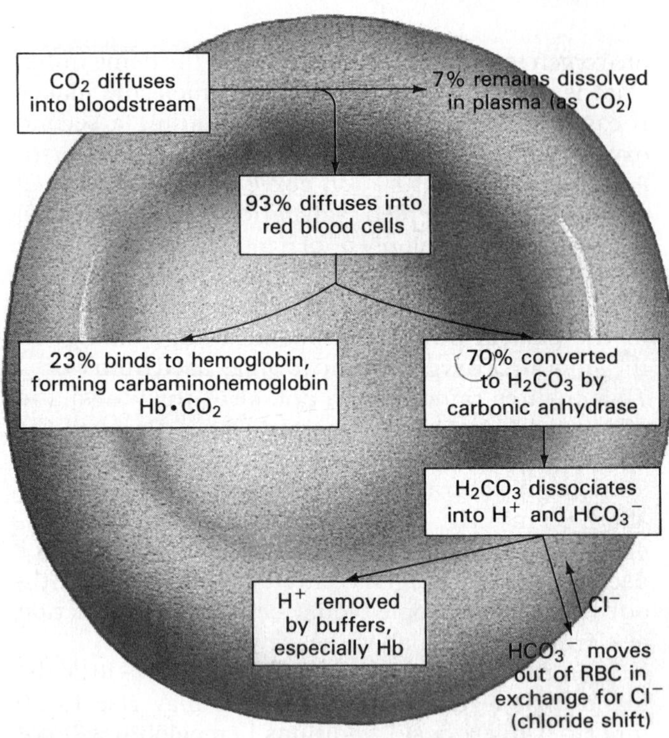

FIGURE 23-22
Carbon Dioxide Transport in the Blood

Hemoglobin Binding Once within the red blood cells, some of the carbon dioxide molecules are bound to exposed amino groups (NH_2) along the globin portions of hemoglobin molecules. The result is called **carbaminohemoglobin** (kar-bam-ē-nō-hē-mō-GLŌ-bin). This is a reversible reaction that can be summarized as:

$$CO_2 + HbNH_2 \rightleftharpoons HbNHCOOH \text{ (carbaminohemoglobin)}$$

At the pCO_2 of the peripheral tissues, about 23 percent of the carbon dioxide will be found in the form of carbaminohemoglobin. Upon arrival at the pulmonary capillaries, the plasma pCO_2 declines, and the stored carbon dioxide is released.

Carbonic Acid Formation The rest of the carbon dioxide molecules, roughly 70 percent of the total, are converted to carbonic acid through the activity of the enzyme carbonic anhydrase. The carbonic acid molecules do not remain intact, however; almost immediately each of these molecules breaks down into a hydrogen ion and a bicarbonate ion. The entire sequence can be summarized as:

$$CO_2 + H_2O \xrightleftharpoons{\text{carbonic anhydrase}} H_2CO_3 \rightleftharpoons H^+ + HCO_3^-$$

The reactions occur very rapidly and are completely reversible. Because most of the carbonic acid formed immediately dissociates into bicarbonate and hydrogen ions, we can ignore the intermediary step and summarize the reaction as:

$$CO_2 + H_2O \xrightleftharpoons{\text{carbonic anhydrase}} H^+ + HCO_3^-$$

In peripheral capillaries this reaction proceeds vigorously, tying up large numbers of carbon dioxide molecules. The reaction is driven from left to right because carbon dioxide continues to arrive, diffusing out of the interstitial fluids, and the hydrogen ions and bicarbonate ions are continually being removed. Most of the hydrogen ions get tied up by hemoglobin molecules. Bicarbonate ions diffuse into the surrounding plasma, where they associate with sodium ions to form **sodium bicarbonate** ($NaHCO_3$). Each time a bicarbonate anion diffuses across the cell membrane, a negatively charged chloride ion (Cl^-) diffuses into the red blood cell. When blood enters the peripheral tissues, the rising carbon dioxide concentration causes a rush of chloride ions into the red blood cells, an event known as the **chloride shift**. The chloride shift is an example of **countertransport**, a passive process involving membrane proteins. During countertransport, two ions move across a membrane in opposite directions. As in facilitated diffusion (Chapter 3, Figure 3-23), this process does not require ATP. (p. 84)

When venous blood reaches the alveoli, carbon dioxide diffuses out of the plasma and the pCO_2 declines. Because all of the carbon dioxide transport mechanisms are reversible, as carbon dioxide diffuses out of the red blood cells the reactions shown in Figure 23-22 proceed in the opposite direction. Hydrogen ions leave the hemoglobin molecules, bicarbonate ions diffuse into the cytoplasm of the red blood cells, and chloride ions return to the plasma.

Figure 23-23 summarizes the information presented concerning the transportation of carbon dioxide and oxygen. This is obviously a dynamic system, capable of varying its responses to meet changing circumstances. Some of the responses are automatic and result from the basic chemistry of the transport mechanisms. Other responses require coordinated adjustments in the activities of the cardiovascular and respiratory systems. We will now consider these levels of control and regulation.

Why does it take more energy to breathe on a hot humid day than on a cool dry day?

During exercise, hemoglobin releases more oxygen to the active skeletal muscles than it does when the muscles are at rest. Why?

How would an obstruction of the airways affect the body's pH?

Control of Respiration

Peripheral cells are continually absorbing oxygen from the interstitial fluids and generating carbon dioxide. Under normal conditions the cellular rates of absorption and generation are matched by the capil-

FIGURE 23-23
A Summary of the Primary Gas Transport Mechanisms

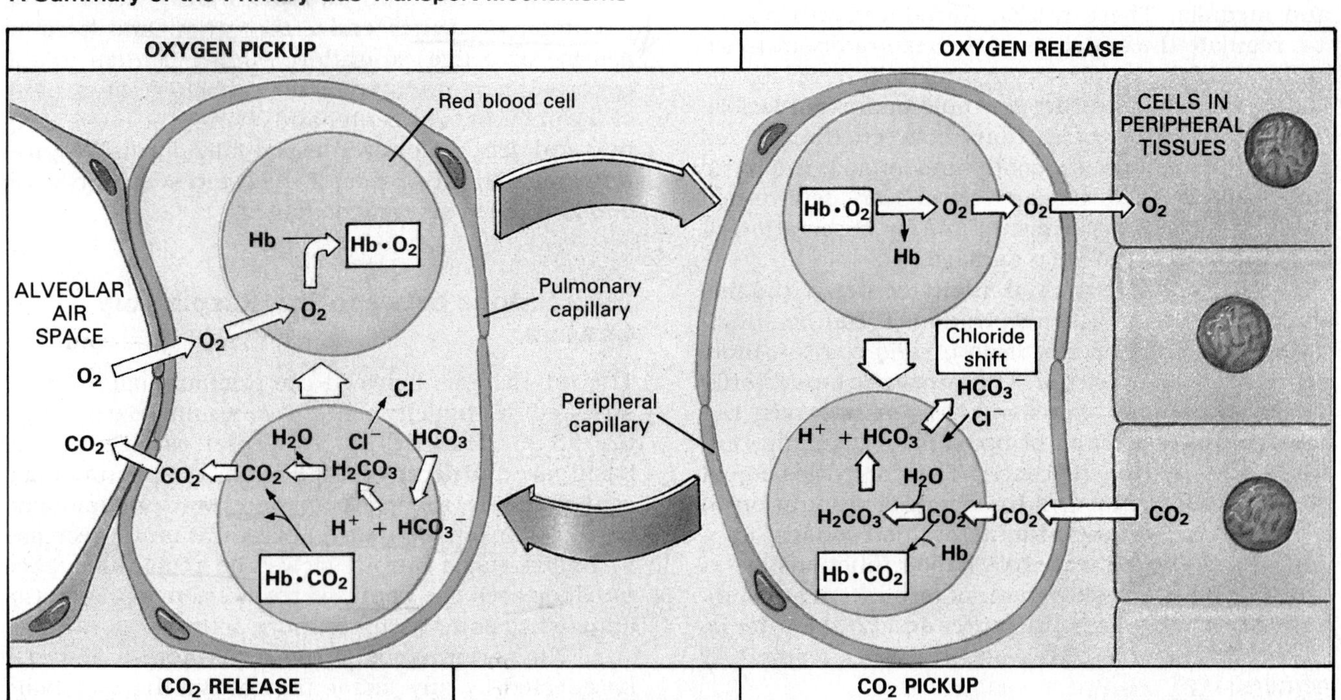

lary rates of delivery and removal. These rates are identical to those of oxygen absorption and carbon dioxide excretion at the lungs. If these rates become unbalanced, homeostatic mechanisms intervene to restore equilibrium.

The biochemical reactions involved in oxygen and carbon dioxide transport permit some degree of autoregulation, enough to deal with small, localized changes in peripheral oxygen demand. If a peripheral tissue becomes more active, the interstitial pO_2 falls and the pCO_2 rises. This increases the difference between the partial pressures in the tissues and in the arriving blood, so more oxygen is delivered and more carbon dioxide carried away. Because the venous blood from that area will mix with the blood from other, less active regions before reaching the pulmonary arteries, there will be little change in the composition of the blood arriving at the alveoli. The arterial pO_2 and pCO_2 remain unchanged, despite the minor alteration in the peripheral demand.

Autoregulation compensates for relatively small oscillations affecting individual tissues or even organs, but it cannot cope with large-scale or extended changes in peripheral oxygen demand. These homeostatic adjustments involve the integration of cardiovascular and respiratory responses, so that the respiratory rate and the cardiac output increase in a coordinated fashion. The regulatory centers integrating these responses are located in the pons and medulla.

THE RESPIRATORY CENTERS OF THE BRAIN

The **respiratory centers** include three pairs of loosely organized nuclei in the reticular formation of the pons and medulla. These nuclei, introduced in Chapter 14, regulate the activities of the respiratory muscles and control the respiratory minute volume by adjusting the frequency and depth of pulmonary ventilation. ∞ (p. 451) The interactions between the various respiratory centers are poorly understood; the basic functions of each nucleus have been determined through animal experiments and by observation of patients with brain stem damage.

The **respiratory rhythmicity center** of the medulla sets the pace for respiration. It can be subdivided into a dorsal *inspiratory center* and a more ventral *expiratory center*. The **apneustic** (ap-NŪ-stik) **center** of the lower pons and the **pneumotaxic** (nū-mō-TAKS-ik) **center** of the upper pons adjust the output of the rhythmicity center. Their activities adjust the respiratory rate and the depth of respiration in response to sensory stimuli or instructions from higher centers. We will explore their functions by examining (1) the basic pattern of respiration as established by the rhythmicity center and (2) the interactions between this center and the other respiratory centers.

Activities of the Respiratory Rhythmicity Center

During quiet respiration the inspiratory center gradually increases stimulation of the inspiratory muscles for 2 seconds and then becomes silent for the next 3 seconds. During its silence, the respiratory muscles relax, and passive exhalation occurs. The inspiratory center will maintain this basic rhythm even in the absence of sensory or regulatory stimuli. The expiratory center remains inactive during quiet respiration. It functions only during forced breathing, when the inspiratory movements are exaggerated. Under these conditions the increased activity of the inspiratory center in some way activates the expiratory center, and expiration becomes an active process. The relationships between the inspiratory and expiratory centers during quiet and forced ventilation are diagrammed in Figure 23-24.

The Apneustic Center

The apneustic center continually provides stimulation to the inspiratory center. Under normal conditions, this center is inhibited by input from the pneumotaxic center. If that inhibition is removed, following a stroke or other damage to the brain stem, the individual inhales to maximum capacity and maintains that state for 10–20 seconds at a time. Intervening exhalations are brief, and little pulmonary ventilation occurs.

The Pneumotaxic Center

The pneumotaxic center modifies the activities of the apneustic and rhythmicity centers. By inhibiting the apneustic center and the inspiratory center, the pneumotaxic center ends inspiration and permits passive or active expiration. Higher centers in the hypothalamus and cerebrum can alter the activity of the pneumotaxic center and change the respiratory rate and depth. However, essentially normal respiratory cycles continue even if the brain stem above the pons has been severely damaged.

Interactions between the Respiratory Centers

The interactions between the pneumotaxic, apneustic, and rhythmicity centers are summarized in Figure 23-25. The rhythmicity center establishes the basic pace and depth of respiration. The pneumotaxic center modifies that pace; an increase in pneumotaxic output quickens the pace of respiration. A decrease in pneumotaxic output slows the respiratory pace but increases the depth of respiration, because the apneustic center becomes more active.

The performance of the respiratory center can be affected by any factor that alters the metabolic

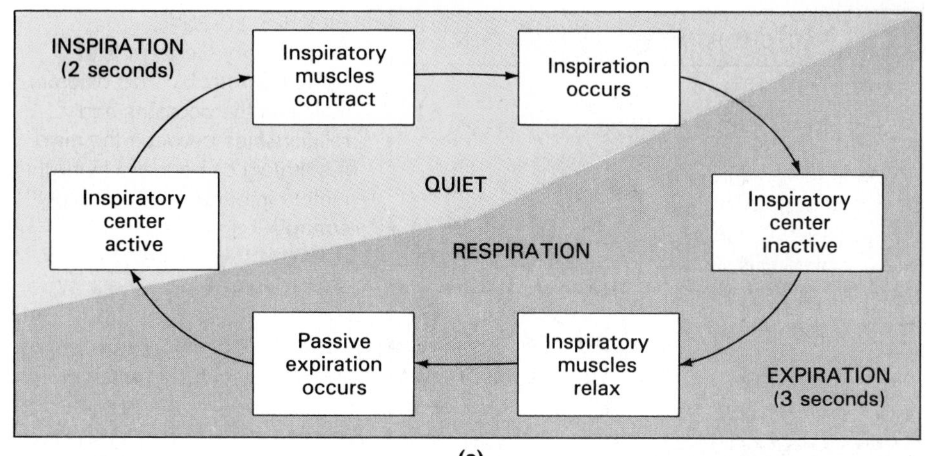

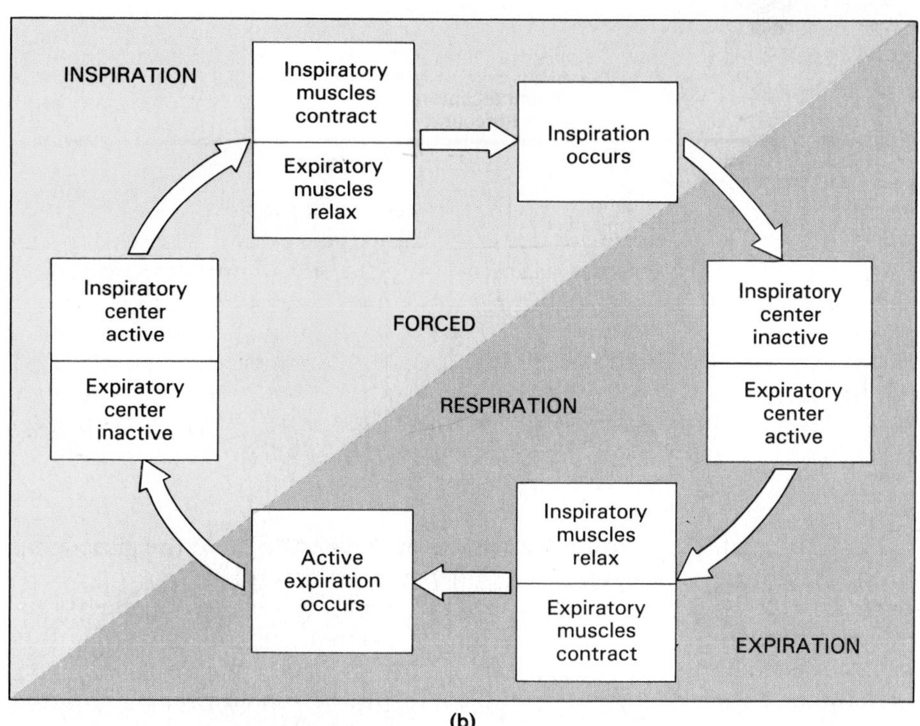

FIGURE 23-24
Basic Regulatory Patterns.
(a) Quiet respiration. (b) Forced respiration.

or chemical activities of neural tissues. Elevated body temperatures will accelerate respiration, while decreased temperatures will produce a respiratory slowdown. Central nervous system stimulants, such as amphetamines or even caffeine, will increase the respiratory rate. These actions can be opposed by CNS depressants, such as barbiturates or opiates. Respiratory activities are also strongly influenced by reflexes triggered by mechanical or chemical stimuli.

REFLEX CONTROL OF RESPIRATION

Normal breathing occurs automatically, without conscious control. Three different reflexes are involved in the regulation of respiration: (1) *mechanoreceptor reflexes* that respond to changes in the volume of the lungs or to changes in arterial blood pressure, (2) *chemoreceptor reflexes* that respond to changes in the pO_2 and pCO_2 of the blood and cerebrospinal fluid, and (3) *protective reflexes* that respond to physical injury or irritation of the respiratory tract.

Mechanoreceptor Reflexes

Several populations of baroreceptors are important in the regulation of respiratory function. (These receptors and their functions were described in Chapter 17.) (p. 531)

The Inflation and Deflation Reflexes The **inflation reflex** prevents overexpansion of the lungs during forced breathing. These stretch receptors are stimulated by expansion of the lungs. Sensory fibers leaving the stretch receptors reach the inspiratory and expiratory centers over the vagus nerve. As the volume of the lungs increases, the inspiratory center is gradu-

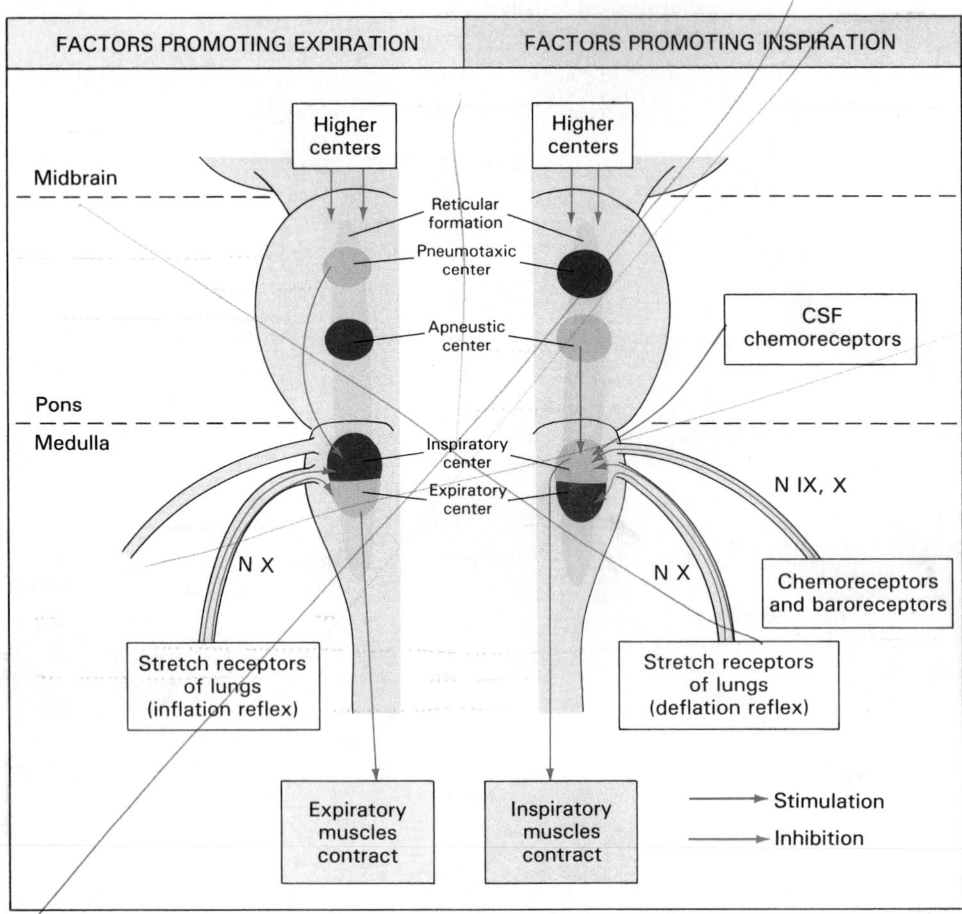

FIGURE 23-25
Respiratory Centers and Reflex Controls. This diagram indicates the positions and relationships between the major respiratory centers and highlights factors important to respiratory control.

ally inhibited and the expiratory center stimulated. Thus inspiration stops as the lungs near maximum volume, and active expiration then begins.

The **deflation reflex** inhibits the expiratory center and stimulates the inspiratory center when the lungs are collapsing. The smaller the volume of the lungs, the greater the degree of inhibition, until expiration stops and inspiration begins. This reflex normally functions only during forced expiration, when both the inspiratory and expiratory centers are active.

Although neither the inflation nor the deflation reflex is involved in normal quiet breathing, both are important in regulating the forced ventilations that accompany strenuous exercise. Together the inflation and deflation reflexes are known as the **Hering-Breuer reflexes**, after the physiologists who described them in 1865.

Baroreceptor Reflexes and Respiration The effects of the carotid and aortic baroreceptors on systemic blood pressure were described in Chapter 21. (p. 698) The output from these baroreceptors affects the respiratory centers as well as the cardiac and vasomotor centers. When blood pressure falls, the respiratory rate increases; when blood pressure rises the respiratory rate declines. This adjustment results from stimulation or inhibition of the inspiratory and expiratory centers by sensory fibers in the glossopharyngeal (IX) and vagus (X) nerves.

Curiously, stretching the anal sphincter stimulates the respiratory center and increases the rate of respiration. Although this reflex is occasionally used to stimulate respiration in an emergency situation, it is not clear what pathways are involved.

Chemoreceptor Reflexes

The respiratory centers are strongly influenced by chemoreceptor inputs from the ninth and tenth cranial nerves and from receptors monitoring cerebrospinal fluid (CSF) composition. The glossopharyngeal nerve (N IX) carries chemoreceptive information from the carotid bodies, adjacent to the carotid sinus. The vagus nerve (N X) monitors chemoreceptors in the aortic bodies, near the aortic arch. These receptors are sensitive to the pO_2 and pCO_2 in arterial blood. Chemoreceptors located in the ventrolateral floor of the medulla respond only to the pCO_2 of the cerebrospinal fluid. These medullary chemoreceptors are actually sensitive to the pH of the CSF. When the pCO_2 rises, the hydrogen ions released by the dissociation of carbonic acid make the CSF more acidic, and this change stimulates the chemoreceptors.

Under normal conditions the pO_2 has very little effect on the respiratory centers, and it is carbon dioxide that sets the respiratory pace. Chemoreceptors monitoring carbon dioxide levels are actually sensitive to the hydrogen ions liberated by the dissociation of carbonic acid. As a result, other conditions altering the pH of body fluids can affect respiratory performance. For example, the rise in lactic acid levels after exercise causes a drop in pH that helps stimulate respiratory activity.

Hypercapnia The term **hypercapnia** (hī-per-KAP-nē-a) refers to an increase in the pCO_2 of arterial blood. Figure 23-26a diagrams the central response to hypercapnia, which is triggered by stimulation of chemoreceptors in the carotid and aortic bodies. Carbon dioxide crosses the blood-brain barrier quite rapidly, so a rise in arterial pCO_2 almost immediately elevates CSF carbon dioxide levels and stimulates the chemoreceptive neurons of the medulla. These receptors stimulate the respiratory center to produce **hyperventilation** (hī-per-ven-ti-LĀ-shun), an increase in the rate and depth of respiration. Breathing becomes more rapid, and a greater amount of air moves in and out of the lungs with each breath. Because more air moves in and out of the alveoli each minute, alveolar concentrations of carbon dioxide decline, accelerating the diffusion of carbon dioxide from the alveolar capillaries.

If the arterial pCO_2 declines below normal levels (Figure 23-26b), chemoreceptor activity declines and the respiratory rate falls. This **hypoventilation** continues until the carbon dioxide partial pressure returns to normal.

Respiratory Drive and pCO_2 Carbon dioxide levels have a very powerful effect on respiratory activity. Increasing the pCO_2 from 40 mm Hg to 50 mm Hg more than quadruples the rate of pulmonary ventilation. By comparison, the effects of pO_2 alterations are relatively minor. Dropping the arterial pO_2 from 100 to 60 mm Hg has no significant effect on respiration, and a decline to 40 mm Hg, the normal pO_2 of peripheral tissues, increases pulmonary ventilation to only 50–70 percent above resting levels.

Carbon dioxide rather than oxygen has the greatest effect on the respiratory centers because under normal conditions the arterial pO_2 does not decline enough to activate the oxygen receptors. But when arterial pO_2 does fall, the two receptor populations cooperate. Carbon dioxide is generated during oxygen consumption, so when oxygen concentrations are falling rapidly carbon dioxide levels are usually increasing. This cooperation breaks down only under unusual circumstances. [**CM:** *Chemoreceptor Accommodation and Opposition*]

Chemoreceptor reflexes are extremely powerful respiratory stimulators, and they cannot be consciously suppressed. For example, you can hold your breath before diving into a swimming pool and thereby prevent the inhalation of water. But you cannot actually commit suicide by holding your breath "till you turn blue." Once the pCO_2 rises to critical levels, you will be forced to take a breath. [**CM:** *Shallow Water Blackout*]

Protective Reflexes

Protective reflexes operate on exposure to toxic vapors, chemical irritants, or mechanical stimulation of the respiratory tract (examples were noted earlier in the chapter). In general these reflexes involve a temporary period of **apnea** (AP-nē-a) when respiration is suspended. They are usually followed by a forceful

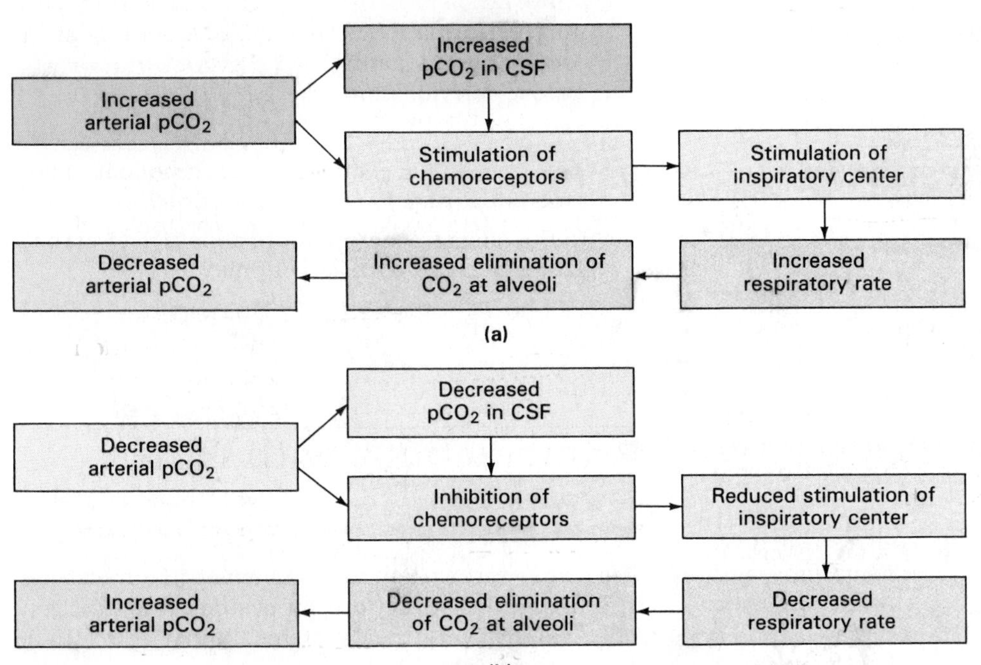

FIGURE 23-26 Chemoreceptor Response to Changes in pCO_2. (a) A rise in arterial pCO_2 stimulates chemoreceptors that accelerate breathing cycles at the inspiratory center. This increases the respiratory rate, encourages CO_2 loss at the lungs, and lowers arterial pCO_2. (b) A decline in arterial pCO_2 inhibits these chemoreceptors. In the absence of stimulation the rate of respiration decreases, slowing the rate of CO_2 loss and elevating arterial pCO_2.

expulsion of air intended to remove the offending stimulus. Sneezing and coughing are two excellent examples of protective reflexes. Sudden pain or immersion in cold water can also produce a temporary apnea.

As you are probably well aware, conscious and unconscious thought processes can have a direct effect on respiratory activities. But, although we can temporarily alter our respiratory performance, we cannot override the important respiratory reflexes.

HIGHER CENTERS AND RESPIRATORY PERFORMANCE

Higher centers influence respiration via inputs to the pneumotaxic center and by their direct influence on respiratory muscles. The higher centers involved are found in the cerebrum, especially the cerebral cortex, and in the hypothalamus. For example, respiratory activities can be modified by stimulation of association areas near the primary motor cortex and following stimulation of the speech (Broca's) center. These areas are important in the control of voluntary respiration and in the control of sound production and speech. Respiration can also be affected by stimulation of portions of the limbic cortex. Stimulation of the preoptic nucleus or the centers involved with rage, feeding, or sexual arousal changes the pattern of respiration.

The respiratory muscles receive instructions from higher centers primarily via extrapyramidal pathways. In addition, the respiratory centers are embedded in the reticular formation, and almost every sensory and motor nucleus has some connection with this complex. As a result, emotional and autonomic activities often affect the pace and depth of respiration. [**CM:** *Sudden Infant Death Syndrome*]

■ Aging and the Respiratory System

Many factors interact to reduce the efficiency of the respiratory system in elderly individuals. Three examples are particularly noteworthy:

1. With increasing age elastic tissue deteriorates throughout the body. This deterioration reduces the compliance of the lungs, lowering the vital capacity.
2. Movements of the chest cage are restricted by arthritic changes in the rib articulations and by decreased flexibility at the costal cartilages. In combination with the changes noted in (1), the stiffening and reduction in chest movement effectively limit the respiratory minute volume. This restriction contributes to the reduction in exercise performance and capabilities with increasing age.
3. Some degree of emphysema is normally found in individuals age 50–70. However, the extent varies widely depending on the lifetime exposure to cigarette smoke and other respiratory irritants. Figure 23-27 compares the respiratory performance of individuals who have never smoked with individuals who have smoked for varying periods of time. The message here is quite clear: although some decrease in respiratory performance is inevitable, stopping smoking (or never starting) can prevent serious respiratory deterioration.

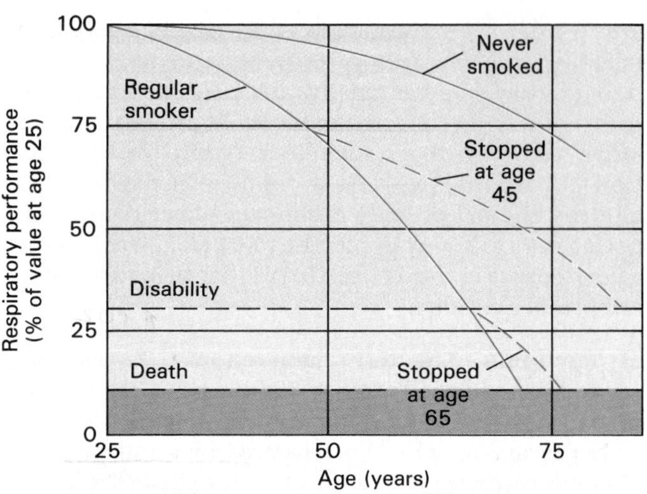

FIGURE 23-27
Aging and the Decline in Respiratory Performance. Graph comparing the relative respiratory performance of (1) individuals who have never smoked, (2) individuals who quit smoking at age 45, (3) individuals who quit smoking at age 65, and (4) lifelong smokers.

√ **What effect would exciting the pneumotaxic center in the brain stem have on respiration?**

√ **Are the chemoreceptors more sensitive to carbon dioxide levels or to oxygen levels?**

■ Interactions with Other Systems

The respiratory system has extensive anatomical connections to the cardiovascular system. It is functionally linked to all other systems, and Figure 23-28 details these interrelationships.

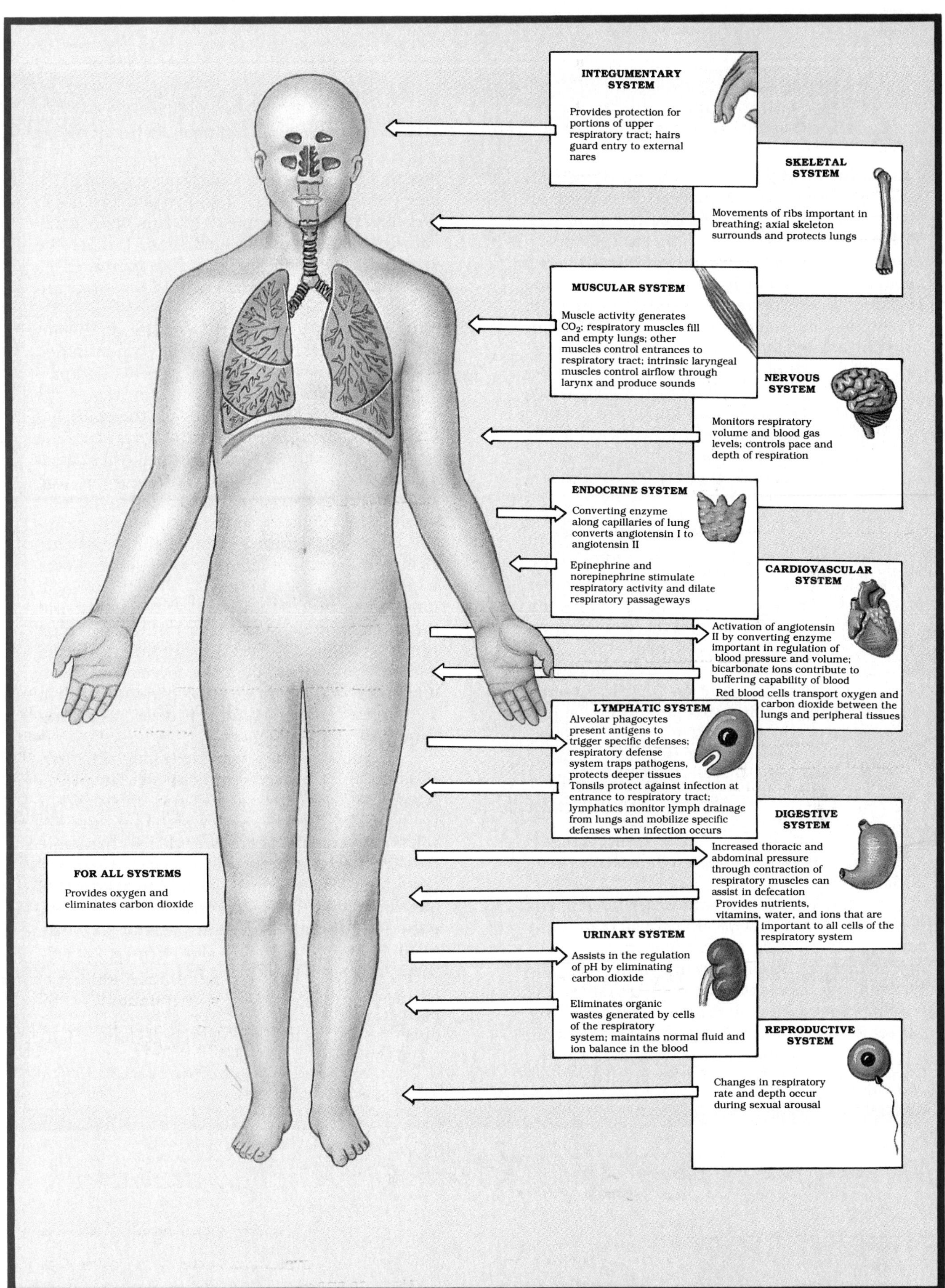
INTEGUMENTARY SYSTEM
Provides protection for portions of upper respiratory tract; hairs guard entry to external nares
SKELETAL SYSTEM
Movements of ribs important in breathing; axial skeleton surrounds and protects lungs
MUSCULAR SYSTEM
Muscle activity generates CO_2; respiratory muscles fill and empty lungs; other muscles control entrances to respiratory tract; intrinsic laryngeal muscles control airflow through larynx and produce sounds
NERVOUS SYSTEM
Monitors respiratory volume and blood gas levels; controls pace and depth of respiration
ENDOCRINE SYSTEM
Converting enzyme along capillaries of lung converts angiotensin I to angiotensin II
Epinephrine and norepinephrine stimulate respiratory activity and dilate respiratory passageways
CARDIOVASCULAR SYSTEM
Activation of angiotensin II by converting enzyme important in regulation of blood pressure and volume; bicarbonate ions contribute to buffering capability of blood
Red blood cells transport oxygen and carbon dioxide between the lungs and peripheral tissues
LYMPHATIC SYSTEM
Alveolar phagocytes present antigens to trigger specific defenses; respiratory defense system traps pathogens, protects deeper tissues
Tonsils protect against infection at entrance to respiratory tract; lymphatics monitor lymph drainage from lungs and mobilize specific defenses when infection occurs
DIGESTIVE SYSTEM
Increased thoracic and abdominal pressure through contraction of respiratory muscles can assist in defecation
Provides nutrients, vitamins, water, and ions that are important to all cells of the respiratory system
URINARY SYSTEM
Assists in the regulation of pH by eliminating carbon dioxide
Eliminates organic wastes generated by cells of the respiratory system; maintains normal fluid and ion balance in the blood
REPRODUCTIVE SYSTEM
Changes in respiratory rate and depth occur during sexual arousal
FOR ALL SYSTEMS
Provides oxygen and eliminates carbon dioxide

FIGURE 23-28
Functional Relationships between the Respiratory System and Other Systems

Health News: Lung Cancer, Smoking, and Diet

Lung cancer, or *pleuropulmonary neoplasm*, is an aggressive class of malignancies originating in the bronchial passageways or alveoli. These cancers affect the epithelial cells lining conducting passageways, mucous glands, or alveoli. Symptoms usually do not appear until the condition has progressed to the point that the tumor masses are restricting airflow or compressing adjacent mediastinal structures. Chest pain, shortness of breath, a cough or wheeze, and weight loss are common symptoms. Treatment programs vary depending on the cellular organization of the tumor and whether or not metastasis (cancer cell migration) has occurred, but surgery, radiation exposure, or chemotherapy may be involved.

Deaths from lung cancer were rare at the turn of the century, but there were 29,000 in 1956, 105,000 in 1978, and 142,100 in 1989. These figures continue to rise, with the number of diagnosed cases doubling every 15 years. Each year 22 percent of new cancers detected are lung cancers, and in 1989 101,000 men and 54,000 women were diagnosed with this condition. Lung cancers now account for 35 percent of all cancer deaths, making this condition the primary cause of cancer death in the U.S. population. Despite advances in the treatment of other forms of cancer, the survival statistics for lung cancer have not changed significantly. Even with early detection the 5-year survival rates are only 30 percent (men) to 50 percent (women), and most lung cancer patients die within a year of diagnosis.

Detailed statistical and experimental evidence has shown that *85–90 percent of all lung cancers are the direct result of cigarette smoking*. Claims to the contrary are simply unjustified and insupportable. The data are far too extensive to detail here, but the incidence of lung cancer for nonsmokers is 3.4 per 100,000 population, while the incidence for smokers ranges from 59.3 per 100,000 for those burning between a half-pack and a pack per day, to 217.3 per 100,000 for those smoking one to two packs per day. Prior to around 1970, this disease primarily affected middle-aged men, but as the number of women smokers has increased so has the number of women dying from lung cancer.

Smoking changes the quality of the inspired air, making it drier and contaminated with several carcinogenic compounds and particulate matter. The combination overloads the respiratory defenses and damages the epithelial cells throughout the respiratory system. The histological changes that follow were described in Chapter 5. (p. 160) Whether or not lung cancer develops appears to be related to the total cumulative exposure to the carcinogenic stimuli. The more cigarettes smoked, the greater the risk, whether those cigarettes are smoked over a period of weeks or years. The histological changes induced by smoking are reversible, and a normal epithelium will return if the stimulus is removed. At the same time the statistical risks decline to significantly lower levels. Ten years after quitting, a former smoker stands only a 10 percent greater chance of developing lung cancer than a nonsmoker.

The fact that cigarette smoking often causes cancer is not surprising in view of the toxic chemicals contained in the smoke. What is surprising is that more smokers do not develop lung cancer. There is evidence that some smokers have a genetic predisposition for developing one form of lung cancer. Dietary factors may also play a role in preventing lung cancer, although the details are controversial. In terms of their influence on the risk of lung cancer, there is general agreement that (1) vitamins C and A have no effect, (2) beta carotene, an orange pigment, and other vegetable components reduce the risk, and (3) a high-cholesterol, high-fat diet increases the risk.

CHAPTER REVIEW

Review of Selected Clinical Terms

pulmonary embolism (*p. 757*)
A condition in which a blocked blood vessel stops blood flow to a portion of the lungs.

cystic fibrosis (CF) (*p. 758*)
A relatively common lethal inherited disease in which mucus secretions become too thick to be transported easily.

hypoxia (hi-POKS-ē-a) (*p. 762*)
A condition of reduced tissue pO_2.

anoxia (a-NOKS-ē-a) (*p. 762*)
A condition that results when the tissue oxygen supply is cut off completely.

pneumothorax (nū-mō-THŌ-raks) (*p. 764*)
The entry of air into the pleural cavity .

atelectasis (at-e-LEK-ta-sis) (*p. 764*)
A collapsed lung.

cardiopulmonary resuscitation (CPR) (*p. 767*)
Applying cycles of compression to the rib cage and mouth-to-mouth breathing to maintain circulatory and respiratory function.

asthma (AZ-ma) (*p. 768*)
A condition characterized by unusually sensitive, irritable conducting passageways.

bronchodilators (brong-kō-di-LĀ-torz) (*p. 768*)
Drugs that dilate respiratory passageways.

lung cancer (pleuropulmonary neoplasm) (*p. 782*)
A class of aggressive malignancies originating in the bronchial passageways or alveoli.

Additional Terms of Clinical Importance

ARTIFICIAL RESPIRATION AND CPR (*p. 767*): **artificial respiration, mouth-to-mouth resuscitation, resuscitators**

PULMONARY FUNCTION TESTS (*p. 767*): **spirometer, pneumotachometer**

Study Outline

Related Key Terms

Introduction (pp. 747–748)

1. To continue functioning, most eukaryotic cells must obtain oxygen and eliminate carbon dioxide. In humans, gas exchange occurs at the **alveoli** of the lungs.

Functional Anatomy and Organization (pp. 748–759)

1. The **respiratory system** includes the nose, nasal cavity and sinuses, pharynx, larynx, trachea, and conducting passageways leading to the exchange surfaces of the lungs.

FUNCTIONS OF THE RESPIRATORY SYSTEM (p. 748)

2. The functions of the respiratory system include: (1) providing an area for gas exchange between air and circulating blood; (2) moving air to and from exchange surfaces; (3) protecting respiratory surfaces; (4) defending the respiratory system and other tissues from pathogens; (5) permitting vocal communication; and (6) helping to regulate blood volume and pressure, and body fluid pH.

THE UPPER AND LOWER RESPIRATORY TRACT (p. 748)

3. The **respiratory tract** consists of the conducting passageways that carry air to and from the alveoli. Those of the **upper respiratory tract** filter and humidify the incoming air. The **lower respiratory tract** includes delicate conduction passages and the alveolar exchange surfaces.

THE NOSE (pp. 749–751)

4. Air normally enters the respiratory system via the **external nares** that open into the **nasal cavity**. The **vestibule** (entrance) is guarded by hairs that screen out large particles.
5. The **hard palate** separates the oral and nasal cavities. The **soft palate** separates the superior **nasopharynx** from the rest of the pharynx. The connections between the nasal cavity and nasopharynx represent the **internal nares**. Incoming air flows through the **superior**, **middle**, or **inferior meatuses** (narrow grooves) and bounces off the conchal surfaces.

   Related key terms: **dorsum nasi • apex turbinate bones**
6. The respiratory epithelium produces mucus that traps incoming particles. Underneath is the **lamina propria** (a layer of connective tissue); the combined respiratory epithelium and lamina propria form a **mucosa** (mucous membrane).

Chapter Review

THE PHARYNX (p. 751)

Related Key Terms

7. The **pharynx** is a chamber shared by the digestive and respiratory systems. The **oropharynx** is continuous with the oral cavity; the **laryngopharynx** includes the narrow zone between the hyoid and the entrance to the esophagus.

THE LARYNX (pp. 751–753)

8. Inspired air passes through the **glottis** en route to the lungs; the **larynx** surrounds and protects the glottis. The **epiglottis** projects into the pharynx.
   - **true vocal cords** • **false vocal cords** • **thyroid cartilage** • **cricoid cartilage** • **arytenoid cartilage** • **corniculate cartilages** • **cuneiform cartilages** • **intrinsic ligaments** • **extrinsic ligaments**
9. Two pairs of folds span the glottal opening: the relatively inelastic **ventricular folds** and the more delicate **vocal folds**. Air passing through the glottis vibrates the vocal folds and produces sound.
   - **phonation** • **articulation**
10. The **intrinsic laryngeal muscles** regulate tension in the vocal folds, and open and close the glottis. The **extrinsic laryngeal musculature** position and stabilize the larynx. During swallowing, both sets of muscles help to prevent particles from entering the glottis.
    - **bolus** • **cough reflex** • **laryngeal spasms**

THE TRACHEA (pp. 753–754)

11. The **trachea** ("windpipe") extends from the sixth cervical vertebra to the fifth thoracic vertebra. The **submucosa** contains C-shaped **tracheal cartilages** that stiffen the tracheal walls and protect the airway. The posterior tracheal wall can distort to permit large masses of food to pass.
    - **trachealis**

THE PRIMARY BRONCHI (p. 754)

12. The trachea branches within the mediastinum to form the **right** and **left primary bronchi**. Each bronchus enters a lung at the **hilus** (a groove). The **root** of the lung is a connective tissue mass including the bronchus, pulmonary vessels, and nerves.
    - **carina**

THE LUNGS (p. 756)

13. The **lobes** of the lungs are separated by fissures; the right lung has three lobes and the left lung has two. The connective tissues of the root extend into the **parenchyma** of the lung as a series of **trabeculae** (partitions). These branch to form **septa** that divide the lung into **lobules.**
    - **superior lobe** • **middle lobe** • **inferior lobe** • **apex** • **base** • **costal surface** • **mediastinal surface** • **cardiac notch**

BRONCHI AND BRONCHIOLES (p. 756)

14. **Intrapulmonary bronchi** (branches within the lung) are surrounded by bands of smooth muscle.
    - **extrapulmonary bronchi**

ALVEOLAR DUCTS AND ALVEOLI (pp. 756–759)

15. The respiratory **bronchioles** open into **alveolar ducts**; many alveoli are interconnected at each duct. The respiratory exchange surfaces are extensively connected to the circulatory system via the vessels of the pulmonary circuit.
16. The **respiratory membrane** (alveolar lining) consists of a simple squamous epithelium; **septal cells (surfactant cells)** scattered in it produce an oily secretion that keep the alveoli from collapsing. **Alveolar macrophages** patrol the epithelium and engulf foreign particles.
17. The **respiratory defense system** includes the **mucus escalator** (which washes particles toward the stomach), alveolar macrophages, hairs, and cilia.

THE PLEURAL CAVITIES (p. 759)

18. Each lung occupies a single pleural cavity lined by a **pleura** (serous membrane).
    - **parietal pleura** • **visceral pleura**

RESPIRATORY CHANGES AT BIRTH (p. 759)

19. Before delivery the fetal lungs are fluid-filled and collapsed.

Respiratory Physiology (pp. 762–775)

1. Respiratory physiology focuses on a series of integrated processes: **pulmonary ventilation** (movement of air into and out of the lungs); **external respiration** (gas exchange between the alveoli and circulating blood); **internal respiration** (gas exchange between the blood and peripheral interstitial fluids); and **cellular respiration** (absorption and use of oxygen by living cells, via biochemical pathways that generate water, ATP, and carbon dioxide). If oxygen content declines, the affected tissues will suffer from **hypoxia**; if the oxygen supply is completely shut off, **anoxia** and tissue death result.

PULMONARY VENTILATION (pp. 762–769)

2. As pressure on a gas decreases, its volume expands; as pressure increases, volume contracts; this inverse relationship is known

as **Boyle's law**. Increased temperature elevates the volume of a gas, decreased temperature reduces it.

3. The relationship between **intrapulmonary pressure** (the pressure inside the respiratory tract) and **atmospheric pressure (atm)** determines the direction of air flow. The **intrapleural pressure** is the pressure in the space between the parietal and visceral pleurae.
4. The **vital capacity** includes the **tidal volume** plus the **expiratory reserve** and the **inspiratory reserve**. The air left in the lungs at the end of maximum expiration is the **residual volume**.

EXTERNAL RESPIRATION (pp. 769–771)

5. The relationship between pressure, solubility, and the number of gas molecules in solution is called **Henry's law**.
6. In a mixed gas the individual gases exert a pressure proportional to their abundance in the mixture **(Dalton's law).** The pressure contributed by a single gas in its **partial pressure.**
7. **Alveolar ventilation** is the amount of air reaching the alveoli each minute. Alveolar and atmospheric air differ in their composition.

INTERNAL RESPIRATION (pp. 771–775)

8. Blood entering peripheral capillaries delivers oxygen and absorbs carbon dioxide. The transport of oxygen and carbon dioxide in the blood involves reactions that are completely reversible.
9. Over the range of oxygen pressures normally present in the body, a small change in plasma pO_2 will mean a large change in the amount of oxygen bound or released. At alveolar pO_2 hemoglobin is almost fully saturated; at the pO_2 of peripheral tissues it retains a substantial oxygen reserve. When low plasma pO_2 continues for extended periods, red blood cells generate more **2,3-diphosphoglycerate (DPG)**, which reduces hemoglobin's affinity for oxygen.
10. Fetal hemoglobin has a stronger affinity for oxygen than does adult hemoglobin, which means that oxygen can be removed from the maternal bloodstream even though the maternal hemoglobin is only 70 percent saturated when it arrives at the placenta.
11. Aerobic metabolism in peripheral tissues generates carbon dioxide. Roughly 7 percent of the CO_2 transported in the blood is dissolved in the plasma; another 23 percent is bound as **carbaminohemoglobin**; the rest is converted to carbonic acid, which dissociates into a hydrogen ion and a bicarbonate ion.

Related Key Terms

psi • pneumothorax
atelectasis
accessory muscles
eupnea (quiet breathing)
diaphragmatic breathing (deep breathing)
costal breathing (shallow breathing)
elastic rebound • compliance
forced breathing (hyperpnea)
saturated

minimal volume
respiratory rate
respiratory minute volume

solubility

anatomic dead space

percent saturation
saturation curve
oxygen affinity • Bohr effect
carbonic anhydrase

sodium bicarbonate
chloride shift
countertransport

Control of Respiration (pp. 775–780)

1. Autoregulation compensates for small oscillations affecting individual tissues and organs; large-scale or extended changes require the integration of cardiovascular and respiratory responses.

THE RESPIRATORY CENTERS OF THE BRAIN (pp. 776–777)

2. The **respiratory centers** include three pairs of nuclei in the reticular formation of the pons and medulla. The **respiratory rhythmicity center** sets the pace for respiration. The **apneustic center** causes strong, sustained inspiratory movements, and the **pneumotaxic center** inhibits the apneustic center and the inspiratory center in the medulla.

REFLEX CONTROL OF RESPIRATION (pp. 777–780)

3. The **inflation reflex** prevents overexpansion of the lungs during forced breathing; the **deflation reflex** stimulates inspiration when the lungs are collapsing. Chemoreceptor reflexes respond to changes in the pO_2 and pCO_2 of the blood and cerebrospinal fluid.

HIGHER CENTERS AND RESPIRATORY PERFORMANCE (p. 780)

4. Conscious and unconscious thought processes can affect respiration by affecting the respiratory centers.

Hering-Breuer reflexes
hypercapnia • hyperventilation
hypoventilation • apnea

Aging and the Respiratory System (p. 780)

1. The respiratory system is generally less efficient in the elderly because: (1) elastic tissue deteriorates, lowering the vital capacity of the lungs; (2) movements of the chest cage are restricted by arthritic changes and decreased flexibility of costal cartilages; and (3) some degree of emphysema is normal in the elderly.

Chapter Review

Review Planner

	Objective	Level 1
1	Describe the primary functions of the respiratory system.	20 21 26 27 33 34 35 44
2	Relate these functions to the anatomical and histological specializations of the tissues and organs in the system.	1 2 15 22 23 24 31 32 54 55
3	Explain how the delicate respiratory exchange surfaces are protected.	22 23 24 29 48 49
4	Describe the physical principles governing the movement of air into the lungs and the diffusion of gases into and out of the blood.	4 5 6 7 17 26 32 33 35 37 41 42
5	Describe how oxygen and carbon dioxide are transported in the blood.	8 9 10
6	Identify the reflexes that regulate respiration.	11 13 14 19 39 43
7	Explain how respiratory activities change to keep pace with metabolic needs.	9 11 13 16 19 40
8	Explain how the centers of respiratory control interact.	12 14 19 25 28 30
9	Describe the changes that occur in the respiratory system at birth and with aging.	3 18 45 50 53

Level 2: 36 38 44 46 47 51 52

Level 3: 56–58

CM: Nosebleeds; Disorders of the Larynx; Tracheal Blockage; Pneumonia • Tuberculosis; Overloading the Respiratory Defenses; Neonatal Respiratory Distress Syndrome (NRDS). Decompression Sickness; Bronchitis, Emphysema, and COPD; Carbon Monoxide Poisoning; Chemoreceptor Accommodation and Opposition; Sudden Infant Death Syndrome (SIDS)

CM: Examining the Living Lung; Thoracentesis; Blood Gas Analysis

CM: Boyle's Law and Air Overexpansion Syndrome; Shallow Water Blackout

ABC NEWS: Learning about Asthma and its Treatment • Asthma's Deadly Comeback • Helping Premature Infants Breathe

Review Questions

MULTIPLE CHOICE

1. Someone in the building is baking chocolate chip cookies, and you sniff the aroma. The air enters through your: a) vestibule; b) alveoli; c) external nares; d) inferior meatus.
2. "Windpipe" is a common term for the: a) larynx; b) pharynx; c) epiglottis; d) trachea.
3. Before birth, the lungs contain: a) small amounts of fluid; b) large volumes of fluid; c) only carbon dioxide; d) small volumes of air.
4. If the accessory muscles help during inspiration, and active contraction of the internal intercostals permits expiration, this is called: a) eupnea; b) diaphragmatic breathing; c) intercostal breathing; d) hyperpnea.
5. The maximum amount of air that you can move in and out of your respiratory system in a single respiratory cycle is the: a) tidal volume; b) vital capacity; c) residual volume; d) inspiratory reserve.
6. What best describes the state of someone's respiratory system after a pneumothorax and atelectasis have occurred? a) the person's residual volume of air still remains; b) the person's minimal volume of air still remains; c) a vacuum exists within the lungs; d) none of the above.
7. The term "anatomic dead space" refers to: a) the amount of air reaching the alveoli each minute; b) an infant's inactive respiratory system before delivery; c) the volume of air in conducting pas-

sages; d) the way your head feels after a final exam.

8. Most of the oxygen in arterial blood is carried by: a) nitrogen molecules; b) CO_2 molecules; c) hemoglobin molecules; d) CO molecules.

9. Which of the following does *not* occur when DPG levels in blood decline? a) hemoglobin becomes more firmly bound to the available oxygen; b) stored whole blood becomes useless for transfusions; c) hemoglobin releases more oxygen.

10. Most of the CO_2 generated by aerobic metabolism in peripheral tissues: a) is bound to the hemoglobin of red blood cells; b) is dissolved in the plasma; c) is converted to carbonic acid; d) is converted to carbon monoxide.

11. Which of the following does *not* usually occur in response to hypercapnia? a) levels of CO_2 in the CSF rise; b) hyperventilation; c) the respiratory rate falls; d) chemoreceptive neurons in the medulla are stimulated.

12. Damage to your pneumotaxic center would: a) cause you to breathe faster; b) cause you to inhale more deeply; c) cause you to exhale deeply; d) inhibit the apneustic center.

13. A temporary period when respiration is suspended is known as: a) hypercapnia; b) the Bohr effect; c) atelectasis; d) apnea.

14. All of the following provide chemoreceptor inputs to the respiratory centers, *except* the: a) cerebrospinal fluid; b) olfactory nerve; c) glossopharyngeal nerve; d) vagus nerve.

DIAGRAMS

15. Cover the labels in Figure 23-1 and identify all the components of the respiratory system.

16. Draw two flow charts or diagrams, one illustrating the steps involved in quiet respiration, and the other illustrating the steps in forced respiration.

TRUE/FALSE

(Your instructor may wish to have you correct the statement if it is false.)

17. **T/F:** A forceful exhalation expels all of the air from one's respiratory system.

18. **T/F:** A certain amount of emphysema occurs normally with aging.

19. **T/F:** Under normal conditions the pO_2 establishes the respiratory pace.

MATCHING QUESTIONS

(Match each term with the most appropriate statement.)

Statements:

20. The process of sound production in the larynx.
21. The absorption and utilization of oxygen by living cells via biochemical pathways that generate water, ATP, and CO_2.
22. The serous membrane that covers the outer surfaces of the lungs.
23. The serous membrane that lines the inner surface of the pleural cavity.
24. An oily secretion that coats alveolar epithelia.
25. This center ends inspiration and permits passive or active expiration.
26. The movement of air in and out of the lungs.
27. Highly elastic structures that are involved with the production of sounds.
28. The center that establishes the pace of respiration.
29. The coordinated beating of cilia that helps to move trapped particles toward the pharynx.
30. This center continually stimulates the inspiratory center.
31. A pair of ligaments that help protect the "true vocal cords."
32. Delicate pockets in the lungs that serve as areas of gas exchange.
33. The exchange of dissolved gases between the blood and the interstitial fluids in peripheral tissues.
34. The modification of sounds produced in the larynx by other structures.
35. The diffusion of gases between the alveoli and the circulating blood.

Terms:

a) Alveoli
b) Phonation
c) Articulation
d) Vocal folds
e) Ventricular folds
f) Surfactant
g) Mucus escalator
h) Parietal pleura
i) Visceral pleura
j) Pulmonary ventilation
k) External respiration
l) Internal respiration
m) Cellular respiration
n) Respiratory rhythmicity center
o) Pneumotaxic center
p) Apneustic center

SHORT ANSWER/ESSAY

36. Discuss the mechanisms that control respiration to maintain homeostasis. Distinguish between those that involve autoregulation and those that involve higher centers.

37. Assume that P_a represents atmospheric pressures, and P_i represents pressures within the lungs.
    a) Describe the relationship between P_a and P_i if air is flowing into the lungs.
    b) Which would you expect to be greater, P_a or P_i, if air is leaving the lungs?
    c) Which would be greater if air is not flowing in either direction?

38. Your Anatomy & Physiology instructor asks a question; you answer it correctly. Explain how your respiratory system produced the necessary variety of sounds.

39. Fill in the blanks: The Hering-Breuer reflexes can be subdivided into the __________ and __________ reflexes.

40. A runner sprints the 100-yard dash and sets a new record. Describe what was occurring in her skeletal muscles during this time.

41. A friend's tidal volume measures 520 mℓ, and he is breathing approximately 45 times per minute. Calculate his respiratory minute volume in liters.

Chapter Review

42. Fill in the blanks: If temperature increases, pressure ________. If volume increases, pressure ________.
43. Fill in the blanks: If blood pressure rises, the respiratory rate ________; if blood pressure falls, the respiratory rate ________.
44. Discuss the primary functions of the respiratory system.
45. How does the respiratory system of a fetus change after delivery?
46. Your instructor announces a pop quiz, and you heave a great sigh. Describe how the characteristics of this air change as it enters your respiratory system and travels to your alveoli.
47. Distinguish between the upper and lower respiratory tracts. Which would you expect to be less susceptible to pathogens? Explain your answer.
48. It's a cold, dry winter day; the high school soccer coach advises her students to breathe through their noses, rather than their mouths. Explain why.
49. Explain how the respiratory defense system functions to protect your lungs against particles and viruses.
50. Describe the process by which a developing fetus obtains oxygen.
51. A patient's intrapulmonary and intrapleural pressures are falling, while his tidal volume is rising. Identify the stage of his respiratory cycle. What is occurring in the patient's circulatory system at this point?
52. While munching hamburgers with friends, you cough for several seconds because a crumb of food "went down the wrong way." Explain what happened and why you coughed.
53. Discuss changes that occur in the respiratory system with aging.
54. The chapter notes that the lungs normally have a light, spongy consistency. Explain why this is true. What is the significance of this structural characteristic?
55. Distinguish between the larynx and pharynx.

CRITICAL THINKING/APPLICATIONS

56. Five-year-old Billy suffers from recurrent bacterial pneumonias and chronic abdominal pain. He is underweight for his age and his skin has a dry, papery texture. Analysis of his perspiration reveals unusually high levels of sodium, potassium, and chloride ion concentrations. What condition might Billy have? How are the symptoms related?
57. Mrs. J. delivers her first child at 26 weeks. The physician notices that the infant's inhalations are long and forced, and the infant seems to tire rapidly. What do you think the physician will recommend? Explain why.
58. Chris has to attend an important meeting out of town, despite her bad cold. During the flight she feels pain in her sinuses and ears, and after the plane lands she finds that sounds seem muffled and indistinct. Explain what could have caused Chris's discomfort.

Career Close-up: Dental Hygienist

"It is amazing how the mouth is affected by all the rest of the body." To Heidi Lien, her job as a dental hygienist involves much more than just cleaning a patient's teeth. "I consider the mouth an open door through which bacteria can enter and cause infection. Because of this connection between the mouth and the immune system, it is important for people to get their teeth cleaned."

Heidi does what is called a prophylaxis: cleaning of all the diseased areas in the mouth and removal of plaque and calculus. "Calculus is hardened plaque, and it acts like a sliver sticking into the gums. This causes an infection and then an immune response; it's a vicious circle with the immune system." She is always considering the mouth's relation to the rest of the body. "I must take extra precautions for patients with diabetes and with smokers; their immune systems are not as effective and they don't heal easily. I must give antibiotics and other medications and a certain type of anesthetic to people with heart problems so that the medications I give do not interact with or oppose the other medications they might be taking. When I give an injection, I must know about every muscle, vein, and nerve in the face so that I properly numb the tooth and do not cause damage." Heidi views the mouth as an indicator of the body's other activities. "The mouth is so fascinating in health and disease because all the body's problems are manifested in the mouth. Even stress can show up in a person's mouth."

Today, Heidi still applies what she learned in her Anatomy and Physiology course. "Being able to understand the body is helpful not only on the job, but in everyday life as well. I can explain when my son asks questions like, 'Why does my elbow bend this way?', or I can help my husband better understand his knee problem."

In her job, Heidi values communicating the knowledge she has about the body to her patients. She thinks it is important that her patients realize how a clean mouth is better for the whole body. "I get a kick from seeing patients who are nervous and uneducated about their mouth become comfortable in the chair and then interested in their oral health. It's exciting to have a nervous child walk away liking the dentist instead of fearing the dentist, as I did when I was a child. The other day I had three tense and worried kids in the office. The oldest was terrified, and she asked me, 'Are you going to give me a shot?' I told her, 'It's only sleepy juice.' She asked me if I could do it instead of the dentist. She hardly knew what happened, because she asked if I had done it yet, and I had."

To Heidi, being close to her patients makes the job enjoyable. "They feel like an extended family. Some of them even visited me in the hospital after my son was born. As a dental hygienist, you need to get personal because it helps the patients feel more comfortable. The mouth is a tender, private place, and it can be embarrassing and difficult for people to allow me to work there."

N RDS

It's no secret that kids grow fast and use a lot of energy. But regardless of your age, whether your body is active or at rest, it never stops using energy—and at least some of its components are always growing, being repaired, or being replaced.

Both the energy and the raw materials for growth and maintenance are supplied by the digestive system, which converts what we eat into a form that the body can absorb and cells can use. Without processing by the digestive system, the food we put into our mouths would be of no more use to us than a lump of coal in the gas tank of a car. But of course, without food from which to extract nutrients our digestive systems would be of no use to us. So it's important to eat—even if the vegetables have a tendency to be left on the plate, or to end up on the table! In this chapter we'll learn more about the digestive system and what it does.

CHAPTER 24
The Digestive System

Chapter Objectives

After reading this chapter, you will be able to:

1. Identify the components of the digestive tract and discuss the functions of each.
2. Describe the histological characteristics of each segment of the digestive tract in relation to its function.
3. Discuss how food is processed in the mouth and describe the key events of the swallowing process.
4. Describe the characteristics of smooth muscle and explain how ingested materials are propelled through the digestive tract.
5. Identify the enzymes produced by the digestive system and what they accomplish.
6. Summarize the stages involved in the regulation of gastric function.
7. Discuss the homeostatic mechanisms that regulate the activities of the digestive system.
8. Discuss the roles played by the accessory organs in digestion.
9. Explain how various compounds in food are broken down and absorbed by the body.
10. Describe the changes in the digestive system that occur with aging.

■ Introduction

Few people give any serious thought to the digestive system unless it malfunctions. Yet we spend hours of conscious effort filling and emptying it. References to this system are part of our everyday language. Take a moment to think of all of the times people have used an expression relating to the digestive system in conversations with you. They may "have a gut feeling," "want to chew on" something, or find your opinions "hard to swallow." (Other, less polite remarks may be heard almost as often.) When something does go wrong with the digestive system, even something minor, most people will seek treatment immediately. For this reason television advertisements promote toothpaste and mouthwash, diet supplements, antacids, and laxatives on an hourly basis.

Chapter 23 discussed the respiratory system, which delivers the oxygen needed to "burn" metabolic fuels. The digestive system provides the fuel and performs many other chemical exchanges. It consists of a muscular tube, the **digestive tract**, and various **accessory organs**, including the salivary glands, liver, and pancreas.

Digestive functions can be considered as a series of integrated steps:

1. *Ingestion* occurs when materials enter the digestive tract via the mouth. Ingestion is an active

process, and it involves preliminary sorting and decision making.

2. *Mechanical processing* may or may not be required, depending on the nature of the ingested material. Liquids can be swallowed immediately, but solids must usually be processed first. Squashing with the tongue and tearing and crushing with the teeth are examples of mechanical processing that occurs before ingestion. Swirling, mixing, and churning motions of the digestive tract provide mechanical processing after ingestion.
3. *Digestion* refers to the chemical breakdown of food into small organic fragments suitable for absorption by the digestive epithelium.
4. Digestion usually involves the action of acids, enzymes, and buffers produced by active *secretion*. Some of these secretions are produced by the lining of the digestive tract, and the rest are provided by accessory organs. Accessory organs secrete mixtures of buffers and enzymes into ducts that empty into the digestive tract.
5. *Absorption*, the movement of organic substrates, electrolytes, vitamins, and water across the digestive epithelium and into the interstitial fluid of the digestive tract. Absorption occurs when digestion has proceeded to a suitable extent. Simple compounds such as glucose can be absorbed almost immediately, but complex proteins, carbohydrates, or lipids must first be broken down by digestive enzymes.
6. *Compaction* involves the progressive dehydration of the indigestible residue and the organic wastes contained in the various glandular secretions.
7. The resulting fecal material will be discharged into the environment, through the process of **defecation** (def-e-KĀ-shun), one form of *excretion*.

The lining of the digestive tract also plays a defensive role by protecting surrounding tissues against (1) the corrosive effects of digestive acids and enzymes, (2) mechanical stresses, such as abrasion,

FIGURE 24-1
Components of the Digestive System. This figure introduces the accessory organs and major regions of the digestive tract, together with their primary functions.

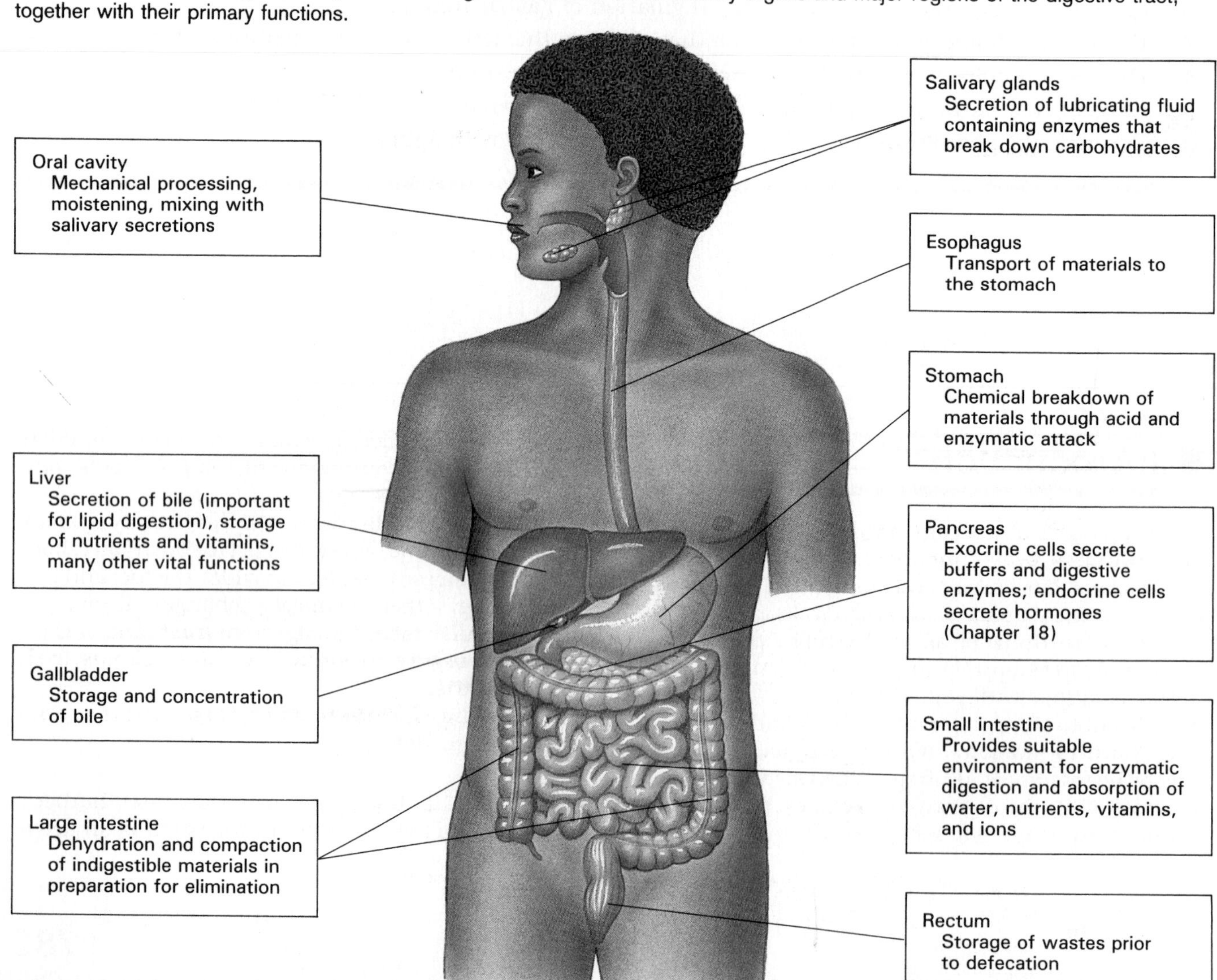

Oral cavity
Mechanical processing, moistening, mixing with salivary secretions
Liver
Secretion of bile (important for lipid digestion), storage of nutrients and vitamins, many other vital functions
Gallbladder
Storage and concentration of bile
Large intestine
Dehydration and compaction of indigestible materials in preparation for elimination
Salivary glands
Secretion of lubricating fluid containing enzymes that break down carbohydrates
Esophagus
Transport of materials to the stomach
Stomach
Chemical breakdown of materials through acid and enzymatic attack
Pancreas
Exocrine cells secrete buffers and digestive enzymes; endocrine cells secrete hormones (Chapter 18)
Small intestine
Provides suitable environment for enzymatic digestion and absorption of water, nutrients, vitamins, and ions
Rectum
Storage of wastes prior to defecation

and (3) pathogens that are either swallowed with food or residing inside the digestive tract. The digestive epithelium and its secretions provide a nonspecific defense against these bacteria, and bacteria reaching the underlying tissues are attacked by macrophages and other cells of the immune system.

■ An Overview of the Structure and Function of the Digestive Tract

Major anatomical subdivisions of the digestive tract are introduced in Figure 24-1. Although these subdivisions have overlapping functions, each region has certain areas of specialization and shows distinctive histological features that reflect those specializations.

HISTOLOGICAL ORGANIZATION

The digestive tract is more than a simple tube, and Figure 24-2 presents an idealized sectional view of the tract. The major layers, described in detail below, include (1) the *epithelium*, (2) the *lamina propria*, (3) the *muscularis mucosae*, (4) the *submucosa*, (5) the *muscularis externa*, and (6) the *adventitia*. Several of these terms were introduced in the discussion of the respiratory passageways.

Digestive Epithelium

The exposed epithelial surface is called a **mucous epithelium** because it is bathed in glandular secretions. Depending on the location, the epithelium may be stratified or simple, but it is usually thrown into a series of folds or pleats that allow the lumen to expand without stressing the lining. The oral cavity, the pharynx (inferior to the nasopharynx), the esophagus, and the anus are subjected to considerable mechanical abrasion, and each has a stratified squamous epithelial lining. The rest of the digestive tract is lined by a simple columnar epithelium, often containing secretory cells of various types.

The complex three-dimensional organization of this epithelium dramatically increases the surface area available for absorption. Ducts opening onto the epithelial surfaces carry the secretions of gland cells located in the surrounding layers of the digestive tract or within accessory organs.

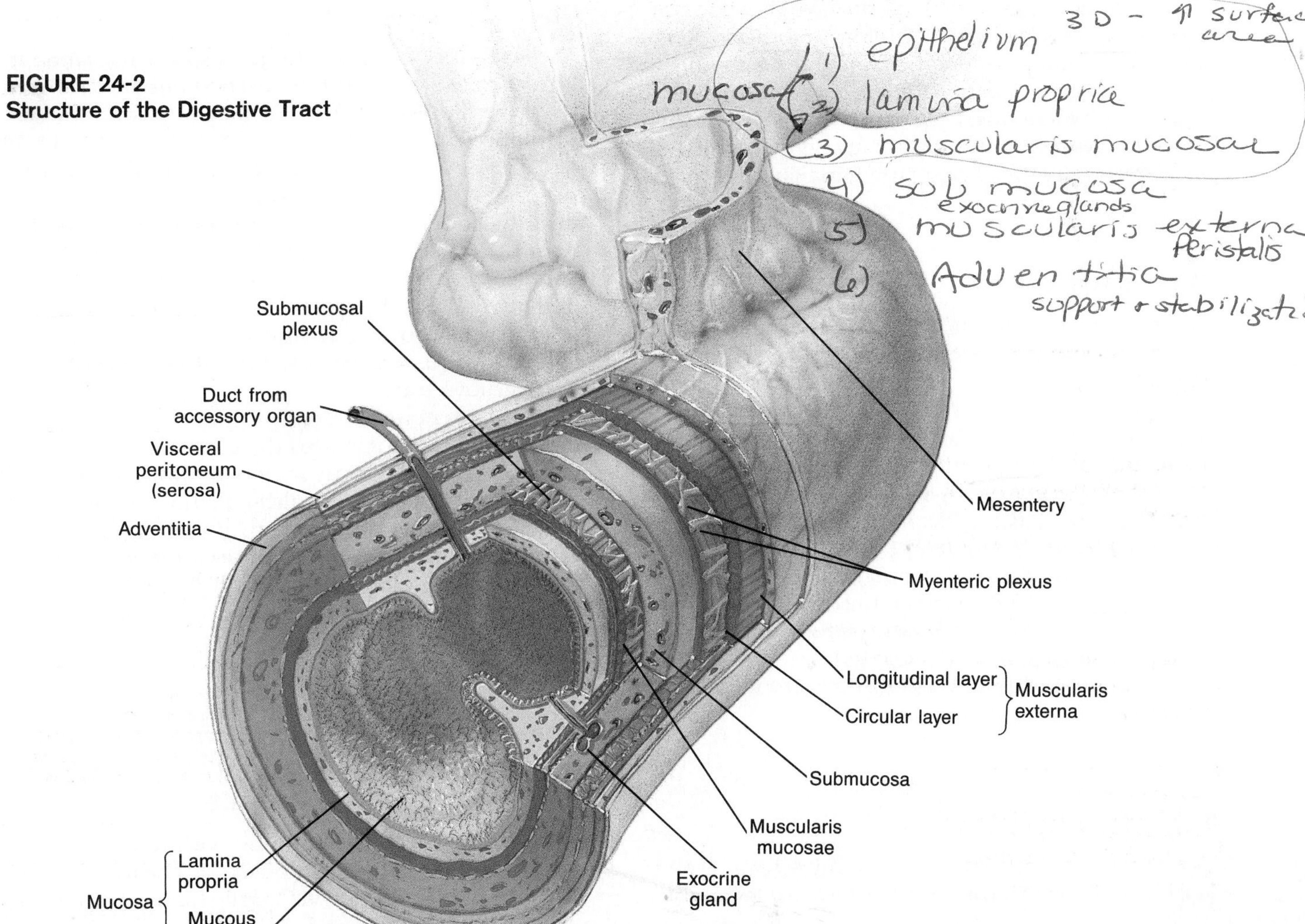

FIGURE 24-2
Structure of the Digestive Tract

The Lamina Propria

The digestive epithelium covers a layer of loose connective tissue called the **lamina propria**. The lamina propria contains blood vessels, sensory nerve endings, lymphatic vessels, smooth muscle fibers, and scattered areas of lymphatic tissue.

Together the epithelium and the lamina propria form the **mucosa**, or **mucous membrane**, of the digestive tract. In most regions of the digestive tract the outer portion of the mucosa contains a narrow band of smooth muscle and elastic fibers, the **muscularis** (mus-kū-LAR-is) **mucosae**. The smooth muscle fibers in the muscularis mucosae are arranged in two concentric layers. The inner layer encircles the internal passageway (the *circular muscle*), and the outer layer contains muscle fibers oriented along the long axis of the tract (the *longitudinal layer*). Contraction of these layers alters the shape of the lumen and moves the epithelial pleats and folds.

The Submucosa

The **submucosa** is a second layer of loose connective tissue that surrounds the muscularis mucosae. Large blood vessels and lymphatics are found in this layer, and in some regions the submucosa also contains exocrine glands that secrete buffers and enzymes into the lumen. Along its outer margin, the submucosa contains a network of nerve fibers and scattered nerve cells. This **submucosal plexus** (*plexus of Meissner*) contains sensory nerve cells, parasympathetic ganglia, and sympathetic postganglionic fibers.

Muscularis Externa

The submucosal plexus lies along the inner border of the **muscularis externa**, a collection of smooth muscle fibers arranged in an inner circular layer and an outer longitudinal one. Contractions of these layers in various combinations agitate or propel materials along the digestive tract. These movements are coordinated primarily by neurons of the **myenteric plexus** (mī-en-TER-ik; *mys*, muscle + *enteron*, intestine), or *plexus of Auerbach*. This network of parasympathetic ganglia and sympathetic postganglionic fibers lies sandwiched between the circular and longitudinal muscle layers. Parasympathetic stimulation increases muscular tone and activity, and sympathetic stimulation promotes muscular inhibition and relaxation.

The Adventitia

The muscularis externa is surrounded by a layer of loose connective tissue that is called the **adventitia** (ad-ven-TISH-a). In the oral cavity, esophagus, and rectum the adventitia contains a dense network of collagen fibers that firmly attach the tract to adjacent structures. Inside the peritoneal cavity, the adventitia of the digestive tract is covered by a mesothelium, forming a *serous membrane* known as the **serosa**.

Clinical Comment: Peritonitis

Inflammation of the peritoneal membrane produces symptoms of **peritonitis** (per-i-to-NĪ-tis), a painful condition that interferes with the normal functioning of the affected organs. Physical damage, chemical irritation, or bacterial invasion of the peritoneum can lead to severe and even fatal cases of peritonitis. Peritonitis due to bacterial infection is a potential complication of any surgery that involves opening the peritoneal cavity.

The Peritoneum

The serosa, also known as the *visceral peritoneum*, is continuous with the *parietal peritoneum* that lines the inner surfaces of the body wall. (These layers were introduced in Chapter 5.) (p. 154) Portions of the digestive tract are suspended within the peritoneal cavity by sheets of serous membrane that connect the parietal peritoneum with the visceral peritoneum. These **mesenteries** (MEZ-en-ter-ēz) are double sheets of peritoneal membrane, as indicated in Figure 24-2. The loose connective tissue between the epithelia provides an access route for the passage of the blood vessels, nerves, and lymphatics servicing the digestive tract.

The peritoneal lining continually produces peritoneal fluid that lubricates the opposing parietal and visceral surfaces. About 7 liters of fluid are secreted and reabsorbed each day, although the volume within the cavity at any one moment is very small. Under unusual conditions, the volume can increase markedly, reducing blood volume and distorting visceral organs. One example, *ascites*, will be considered later in the chapter.

SMOOTH MUSCLE AND THE MOVEMENT OF DIGESTIVE MATERIALS

Smooth muscle tissue is found within almost every organ, forming sheets, bundles, or sheaths around other tissues. As you will recall from Chapter 21,

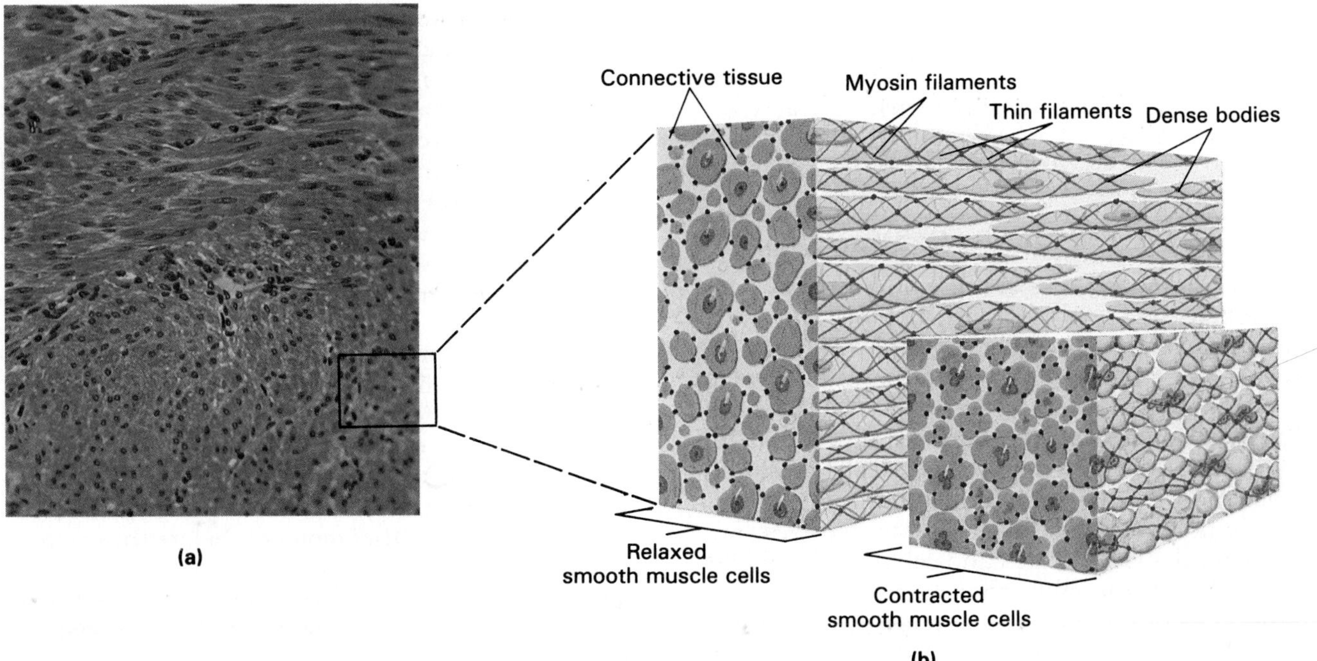

FIGURE 24-3
Smooth Muscle Tissue. (a) Many visceral organs contain several layers of smooth muscle tissue. As a result, a single sectional view shows these fibers in longitudinal and transverse section. (LM, ×243) (b) A diagrammatic section through smooth muscle tissue, showing the three-dimensional orientation of the muscle fibers. Note the changes in shape that occur during the contraction of a smooth muscle fiber.

smooth muscles encircling vessels of the circulatory system provide control over the peripheral distribution of blood. (p. 669) In the digestive system, extensive layers of smooth muscle in the musuclaris mucosae and muscularis externa surround the lumen of the digestive tract. These layers play an essential role in mechanical processing and in moving materials along the tract.

Smooth muscle fibers are surrounded by connective tissue, but the collagen fibers never unite to form tendons or aponeuroses as they do in skeletal muscles. Smooth muscle cells range from 5 to 10 μm in diameter and from 30 to 200 μm in length. There is a single nucleus, centrally located within each spindle-shaped fiber. Figure 24-3a shows typical smooth muscle fibers as seen in light microscopy, and Figure 24-3b a diagrammatic section through smooth muscle tissue.

Smooth Muscle Contraction

The mechanics of muscle contraction have already been described for skeletal muscle (Chapter 10) and cardiac muscle (Chapter 20). (p. 304, 642) However, there are important differences between the contraction of smooth muscle fibers and other muscle fiber types. Although actin and myosin filaments are present in smooth muscle, oriented along the long axis of the cell, there are no sarcomeres. As a result, smooth muscle fibers do not have striations. The thin filaments are attached to **dense bodies** that are firmly attached to the inner surface of the sarcolemma. Dense bodies anchor the thin filaments so that when sliding occurs between thin and thick filaments, the cell shortens. Because dense bodies are not arranged in straight lines, when a contraction occurs the muscle fiber twists like a corkscrew.

Excitation-Contraction Coupling

The trigger for contraction is the appearance of free calcium ions in the cytoplasm, but neither troponin nor tropomyosin is involved. Most of these calcium ions enter the cell from the extracellular fluid. Once in the cytoplasm, they interact with *calmodulin*, a binding protein introduced in Chapter 18. (p. 572) Calmodulin then activates an enzyme, **myosin light chain kinase,** that breaks down ATP and initiates the contraction.

Length-Tension Relationships in Smooth Muscle

Because the filaments are not rigidly organized, there is no direct relationship between tension development and resting length in smooth muscle. A stretched smooth muscle soon adapts to its new length and retains the ability to contract on demand. This ability to function over a wide range of lengths is called **plasticity**. Smooth muscle can contract over a range of lengths four times greater than that of skeletal muscle. This ability is especially important for digestive organs that undergo great changes in volume, such as the stomach. Despite the lack of sarcomere organization, smooth muscle contractions can be just as powerful as those of skeletal muscles.

Control of Smooth Muscle Contractions

Many smooth muscle cells are not innervated by motor neurons, and the neurons that do innervate smooth muscles are not under voluntary control. The nature of the connection with the nervous system provides a means of categorizing smooth muscle cells. **Multiunit smooth muscle fibers** are innervated in motor units comparable to those of skeletal muscles, but each muscle cell may be connected to several motor neurons. As in skeletal or cardiac muscles, an action potential is generated and conducted over the sarcolemma, but the contractions are more leisurely. Multiunit smooth muscle fibers are not typical of the digestive tract. They are found in the iris of the eye, where they regulate the diameter of the pupil, and along portions of the male reproductive tract.

In contrast, many **visceral smooth muscle fibers** lack a direct contact with any motor neuron. Visceral smooth muscle is found in the walls of the digestive tract, the gallbladder, the urinary bladder, and many other internal organs.

In visceral smooth muscle tissue the cells are arranged in sheets or layers, and adjacent muscle cells are electrically connected by gap junctions. Because they are connected in this way, the contraction stimulus passes from fiber to fiber with little delay. When one visceral smooth muscle fiber contracts, the contraction spreads in a wave that travels throughout the tissue.

The initial stimulus may be the activation of a motor neuron that contacts one of the muscle cells in the region. But smooth muscle fibers will contract or relax in response to chemicals, hormones, local concentrations of oxygen or carbon dioxide, or physical factors such as extreme stretching or irritation. (The effects of these factors on the smooth muscle fibers in capillary beds were described in Chapter 21.) ∞ (p. 696)

Many visceral smooth muscle networks show rhythmic cycles of activity in the absence of neural stimulation. This is especially true of the smooth muscle fibers of the muscularis layers of the digestive tract, where **pacesetter cells** undergo spontaneous depolarization and trigger contraction of entire muscular sheets, such as the layers of the muscularis mucosae and muscularis externa.

Peristalsis

The muscularis externa propels materials from one portion of the digestive tract to another through the contractions of **peristalsis** (per-i-STAL-sis). Peristalsis consists of waves of muscular contractions that move along the length of the digestive tract. The distance involved may vary, depending on the circumstances and the region involved. During a peristaltic movement, the circular muscles first contract behind the digestive contents. Longitudinal muscles contract next, shortening adjacent segments. A wave of contraction in the circular muscles then forces the materials in the desired direction, as diagrammed in Figure 24-4.

Regions of the small intestine also undergo **segmentation** movements that churn and fragment the digestive materials. During segmentation, diagrammed in Figure 24-5, a section of intestine resembles a chain of sausages due to the contractions of circular muscles in the muscularis externa. A given pattern exists for a brief moment, before those circular muscles relax and others contract, subdividing the sausages. Over time this action results in a thorough mixing of the contents with intestinal secretions.

Segmentation and peristalsis may be triggered

FIGURE 24-4
Peristalsis. Peristalsis propels materials along the length of the digestive tract.

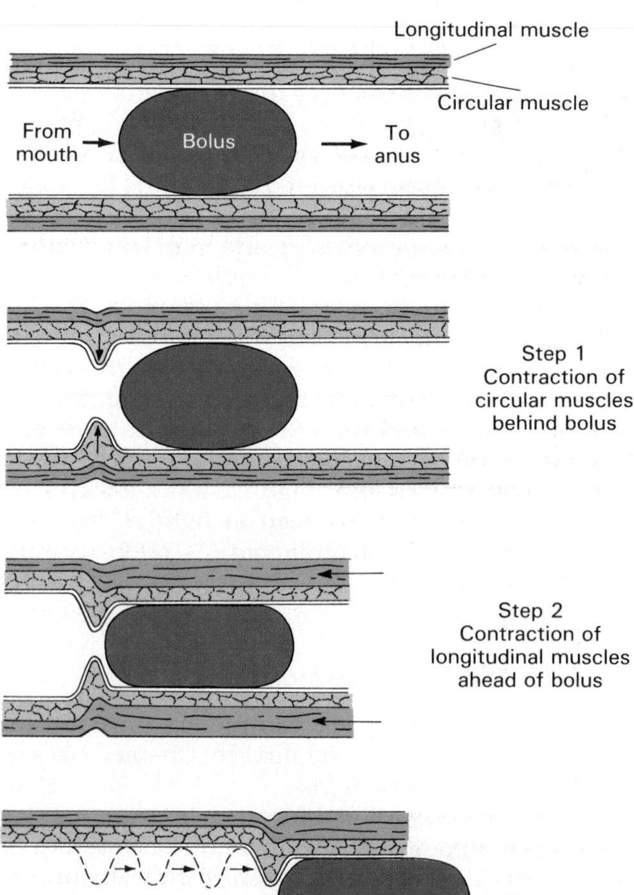

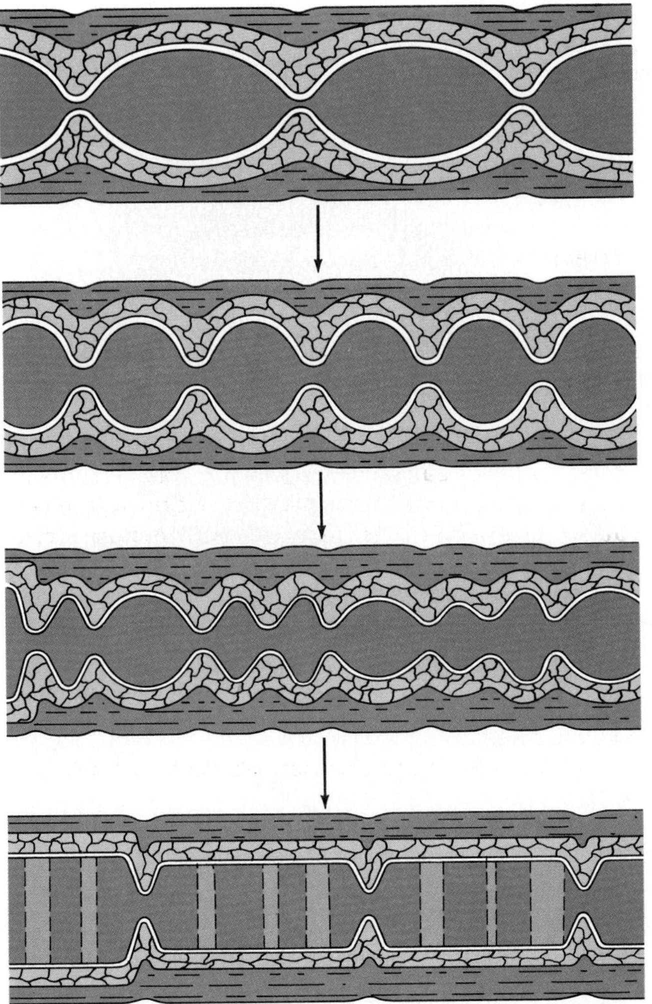

FIGURE 24-5
Segmentation Movements. Segmentation movements churn and mix the contents of the digestive tract but do not produce net movement in a particular direction.

by pacesetter cells or in response to chemical or physical stimulation. Peristaltic waves can also be initiated by afferent and efferent fibers within the glossopharyngeal, vagus, or pelvic nerves, but local peristaltic movements limited to a few centimeters of the stimulus point are triggered by sensory receptors in the walls of the digestive tract. These afferents synapse within the myenteric plexus to produce localized **myenteric reflexes**.

The Digestive Tract

Our discussion of the detailed mechanics of the digestive system begins with the functional anatomy of the digestive tract, and then proceeds to a discussion of the accessory organs. The final section of the chapter describes the chemical processes of digestion and begins a discussion of metabolism that continues in Chapter 25.

THE ORAL CAVITY

The functions of the oral cavity may be summarized as: (1) **analysis** of material before swallowing; (2) **mechanical processing** through the actions of the teeth, tongue, and palatal surfaces; (3) **lubrication** by mixing with mucus and salivary secretions; and (4) **digestion** of carbohydrates by salivary enzymes. Before continuing with the discussion of the anatomy of the oral cavity, you may wish to review the anatomy of this portion of the skull (Chapter 8) and the associated muscles (Chapter 11) and their innervation (Chapter 14). (pp. 249, 336, 462)

Anatomy of the Oral Cavity

Figure 24-6 shows the boundaries of the oral cavity, also known as the **buccal** (BUK-al) **cavity**. The **cheeks** form the lateral walls of this chamber; anteriorly the cheeks are continuous with the lips, or **labia** (LĀ-bē-a). The **vestibule**, a subdivision of the oral cavity, includes the space between the cheeks or lips and the teeth. A pink ridge, the gums, or **gingivae** (JIN-ji-vē), surrounds the bases of the teeth. The gums cover the tooth-bearing, or alveolar, surfaces of the upper and lower jaws. (The alveolar processes of the maxilla and mandible were described in Chapter 8—see Figures 8-10 and 8-12.) (pp. 248, 250)

Medially, the hard and soft palates provide a roof for the oral cavity, while the tongue dominates its floor. Inferior to the tongue the floor receives additional support from the *mylohyoid muscle* (Figure 11-8). (p. 341) The free anterior portion of the tongue is connected to the underlying epithelium by a thin fold of mucous membrane, the **lingual frenulum** (FREN-ū-lum; *frenulum*, a small bridle).

Three pairs of salivary glands secrete into the oral cavity. On each side the **parotid** (pa-ROT-id) **duct** draining a large **parotid gland** empties into the vestibule at the level of the second upper molar. The **sublingual** (sub-LING-gwal) **glands** secrete into the many **sublingual ducts** that open on either side of the lingual frenulum. **Submandibular ducts** that drain the **submandibular glands** open into the mouth behind the teeth on either side of the lingual frenulum.

The posterior margin of the soft palate supports the dangling **uvula** (Ū-vū-la) and two pairs of muscular **pharyngeal arches**. On either side a palatine tonsil lies between an anterior **palatoglossal** (pal-a-tō-GLOS-al) **arch** and a posterior **palatopharyngeal** (pal-a-tō-făr-IN-jē-al) **arch**. A curving line that connects the palatoglossal arches and uvula forms the boundaries of the **fauces** (FAW-sēz), the passageway between the oral cavity and the pharynx.

The Tongue

A muscular tongue manipulates materials inside the mouth and may occasionally be used to bring things

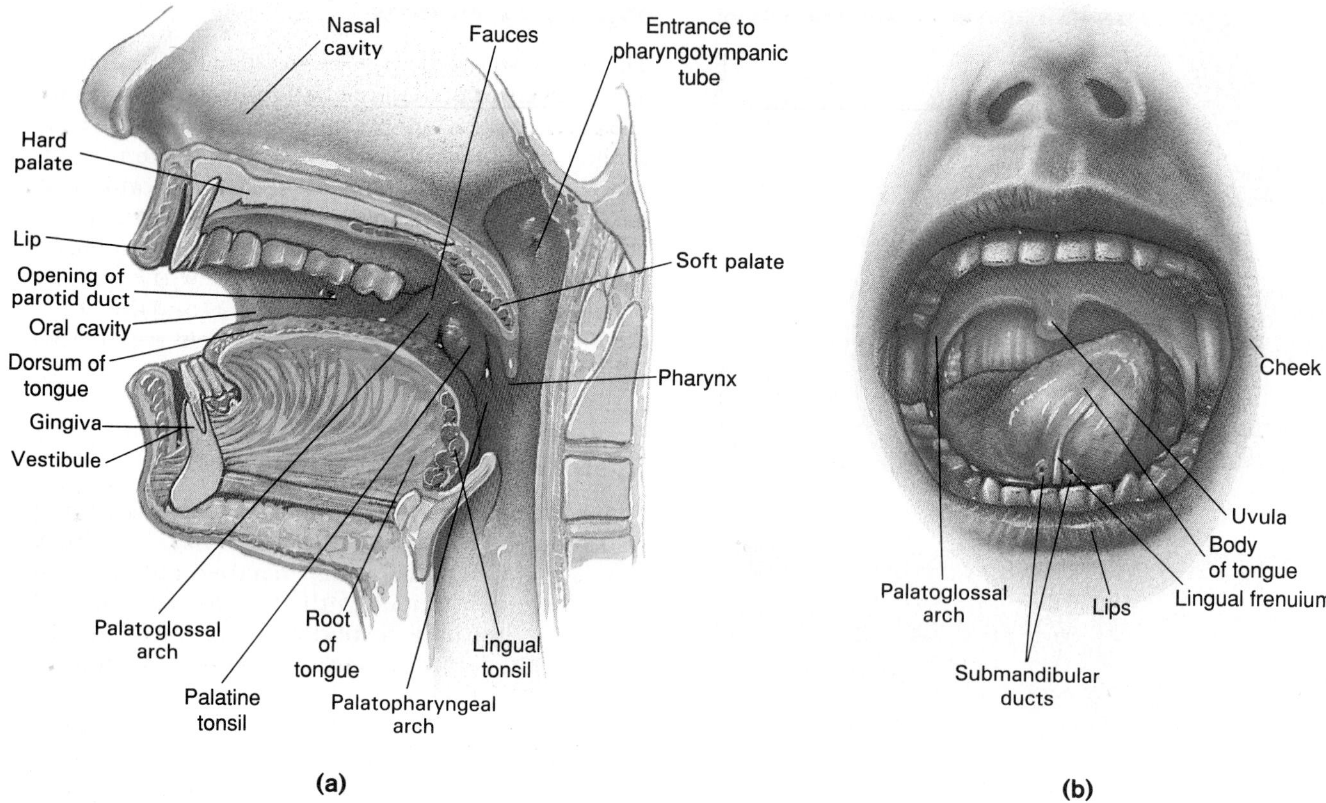

FIGURE 24-6
The Oral Cavity. (a) The oral cavity as seen in sagittal section. (b) An anterior view of the oral cavity, as seen through the open mouth.

(such as ice cream) into the oral cavity. The primary functions of the tongue are: (1) mechanical processing by compression, abrasion, and distortion; (2) manipulation to assist in chewing and prepare the material for swallowing; and (3) sensory analysis by touch, temperature, and taste receptors. Anatomical landmarks on the tongue are included in Figure 24-6; further details can be found in Figure 17-8. (p. 535)

The tongue can be divided into an anterior **body**, or *oral portion*, and a posterior **root**, or *pharyngeal portion*. The superior surface, or **dorsum**, of the body contains a forest of fine papillae. (Filiform, fungiform, and circumvallate papillae were described in Chapter 17 during a discussion of the sensory functions of the tongue.) (p. 534) A V-shaped line of circumvallate papillae roughly indicates the boundary between the body and the root of the tongue.

The thickened epithelium covering each papilla provides excellent traction and assists in the movement of materials by the tongue. A "hairy tongue" can result when keratinized layers fail to detach from the papillae. Severe anemia may cause the loss of papillae and the development of a smooth, shiny "bald tongue."

The inferior surface of the body of the tongue has a more delicate epithelial covering. Many small mucous glands discharge onto this surface. Along the inferior midline the *lingual frenulum* connects the body of the tongue to the mucosa of the oral floor, preventing extreme movements of the tongue. If the frenulum is too restrictive, the individual cannot eat or speak normally. When properly diagnosed, this condition, called **ankyloglossia** (ang-ki-lō-GLOS-ē-a), can be corrected surgically.

There are **intrinsic** and **extrinsic** tongue muscles, all under the control of the hypoglossal nerve (N XII). The extrinsic muscles, discussed in Chapter 11, include the *hyoglossus*, *styloglossus*, *genioglossus*, and *palatoglossus* muscles. (p. 336) They control gross, powerful movements of the tongue. The less massive intrinsic muscles alter the shape of the tongue and assist the extrinsic muscles during precise movements.

The root of the tongue actually lies in the pharynx, but for convenience we will discuss it now rather than later. There are no papillae on the surface of the root, but there are scattered taste buds. A pair of prominent lateral swellings mark the locations of the lingual tonsils.

The Teeth

The movements of the tongue are important in passing food across the surfaces of the teeth. Teeth perform chewing, or **mastication** (mas-ti-KĀ-shun), of materials in the mouth. Mastication breaks down tough connective tissues and plant fibers, and helps

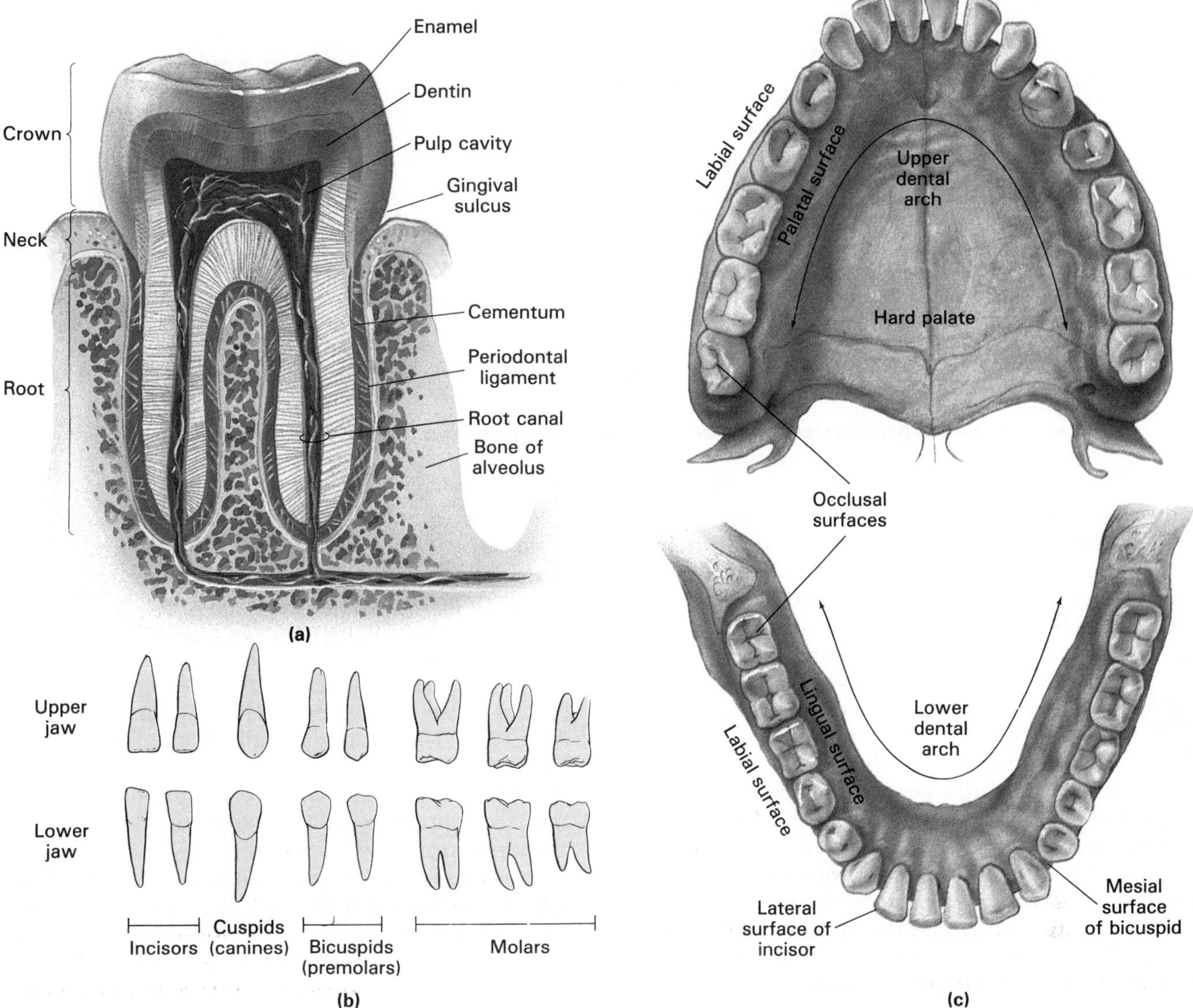

FIGURE 24-7
Teeth. (a) Diagrammatic section through a typical adult tooth. (b) The adult teeth. (c) Dental reference terms and the normal orientation of adult teeth.

saturate the materials with salivary lubricants and enzymes.

A typical adult tooth is diagrammed in Figure 24-7a. The bulk of each tooth consists of a mineralized matrix similar to that of bone. This material, called **dentin** (DEN-tin), differs from bone in that it does not contain living cells. Instead, cytoplasmic processes extend into the dentin from cells in the central **pulp cavity**. The pulp cavity receives blood vessels and nerves via a narrow **root canal** at the base, or **root**, of the tooth. The root sits within a bony socket, or *alveolus*. Collagen fibers of the periodontal ligament extend from the dentin of the root to the alveolar bone, creating a strong articulation known as a *gomphosis*. (This articulation type was introduced in Chapter 7.) (p. 227) A layer of **cementum** (se-MEN-tum) covers the dentin of the root, providing protection and firmly anchoring the **periodontal ligament**. Cementum also resembles bone, but it is softer, and remodeling does not occur following its deposition.

The **neck** of the tooth marks the boundary between the root and the **crown**. The dentin of the crown is covered by a layer of **enamel**. Enamel contains calcium phosphate in a crystalline form, and is the hardest biologically manufactured substance. Adequate amounts of calcium, phosphates, and vitamin D during childhood are essential if the enamel coating is to be complete and resistant to decay. Fluoride treatments or fluoridation of the water over the same period also assists, probably by increasing the density and hardness of the enamel layer.

TABLE 24-1 Tooth Eruption and Replacement

Teeth	Age at Eruption of Primary Teeth (in months)		Age at Eruption of Permanent Teeth (in years)	
	Lower	Upper	Lower	Upper
Central incisors	6	7.5	6–7	7–8
Lateral incisors	7	9	7–8	8–9
Cuspids (canines)	16	18	9–10	11–12
Primary first molar	12	14		
First bicuspids (premolars)			10–12	10–11
Primary second molar	20	24		
Second bicuspids (premolars)			11–12	10–12
First molars			6–7	6–7
Second molars			11–13	12–13
Third molars (wisdom teeth)			17–21	17–21

Epithelial cells of the **gingival** (JIN-ji-val) **sulcus** form tight attachments to the tooth above the neck, preventing bacterial access to the lamina propria or the relatively soft cementum of the root.

Adult teeth are shown in Figure 24-7b,c. There are several different types of teeth with specific functions. **Incisors** (in-SĪ-zerz), blade-shaped teeth found at the front of the mouth, are useful for clipping or cutting, as when nipping off the tip of a carrot stick. The **cuspids** (KUS-pidz), or *canines*, are conical with a sharp ridgeline and a pointed tip. They are used for tearing or slashing. A tough piece of celery might be weakened by the clipping action of the incisors, but then moved to one side to take advantage of the shearing action provided by the cuspids. Incisors and cuspids each have a single root. **Bicuspids** (bī-KUS-pidz), or *premolars*, have one or two roots, and **molars** have three. Premolars and molars have flattened crowns with prominent ridges. They are used for crushing, mashing, and grinding. A tough nut or sparerib will usually be shifted to the premolars and molars for successful crunching.

Dental Succession During development, two sets of teeth begin to form. The first to appear are the **deciduous teeth** (de-SID-ū-us; deciduus, falling off), also known as *primary teeth*, *milk teeth*, or *baby teeth*. There are usually 20 deciduous teeth, five on each side of the upper and lower jaws. These teeth will later be replaced by the adult **secondary dentition**, or *permanent dentition*. The larger adult jaws can accommodate more than 20 permanent teeth, and three additional teeth appear on each side of the upper and lower jaws as the individual ages. These teeth extend the length of the tooth rows posteriorly and bring the permanent tooth count to 32.

On each side of the upper or lower jaw the primary dentition consists of two incisors, one cuspid, and a pair of deciduous molars. These are gradually replaced by the *permanent dentition*. Table 24-1 presents the sequence of eruption of the primary dentition and the approximate ages at their replacement. In this process the periodontal ligaments and roots of the primary teeth are eroded away, until they fall out or are pushed aside by the emergence, or **eruption**, of the secondary teeth. The adult premolars take the place of the deciduous molars, and the definitive adult molars extend the tooth row as the jaw enlarges. The last molar, or *wisdom tooth*, may not erupt before age 21, if it appears at all. Wisdom teeth often develop in inappropriate positions, and they may be unable to erupt properly. Stuck, or impacted, teeth often degenerate, leading to the formation of abscesses that may be further complicated by bacterial invasion. Even if these teeth do erupt, they may be very difficult to brush, and gum infections are common. Many people elect to avoid the problem completely by having their wisdom teeth surgically removed in their late teens or early twenties.

A Dental Frame of Reference The upper and lower rows of teeth each form a curving dental arch. Relative positions along the arch are indicated by the use of special terms. **Labial** or **buccal**, refers to the outer margin of the dental arch, where exposed surfaces face the lips or cheeks. **Palatal** (upper) or **lingual** (lower) refers to the inner surface of the dental arch. **Mesial** (MĒ-zē-al) or **lateral** refers to the opposing surfaces between the teeth in a single dental arch. Mesial surfaces face forward or medially, and lateral surfaces face backward or laterally. For example, the mesial surface of each canine faces the lateral surface of the second incisor. The **occlusal surfaces** (o-KLU-sal; *occlusio*, closed) of the teeth face their counterparts on the opposing dental arch. The occlusal surfaces perform the actual clipping, tearing, crushing, and grinding actions of the teeth. [**CM:** *Dental Problems*]

Mastication

The *muscles of mastication* (Figure 11-6) close the jaws and slide or rock the lower jaw from side to side. (p. 338) During mastication the material is forced back and forth between the vestibule and the rest of the oral cavity, crossing and recrossing the occlusal surfaces. This movement results in part from the action of the masticatory muscles, but control would be impossible without the aid of the buccal, labial, and lingual muscles. Once the material has been shredded or torn to a satisfactory consistency, moistened with salivary secretions, and cleared by the taste receptors, the tongue begins compacting the debris into a small mass, or **bolus**, that can be swallowed relatively easily.

THE PHARYNX

The divisions and epithelial lining of the pharynx were discussed in Chapter 23. (p. 751) Beneath the lamina propria lies a dense layer of elastic fibers, usually bound to underlying skeletal muscles. The **pharyngeal constrictors** provide the impetus for bolus movement. The **palatopharyngeus** (pal-a-tō-fār-IN-jē-us) and **stylopharyngeus** (stī-lō-fār-IN-jē-us) elevate the larynx, and the **palatal muscles** raise the soft palate and adjacent portions of the pharyngeal wall. The latter muscles also pull open the entrance to the pharyngotympanic tube. As a result, swallowing repeatedly can help one adjust to pressure changes when flying or diving. These muscles are illustrated in Figure 24-8, and additional information can be found in Table 24-2.

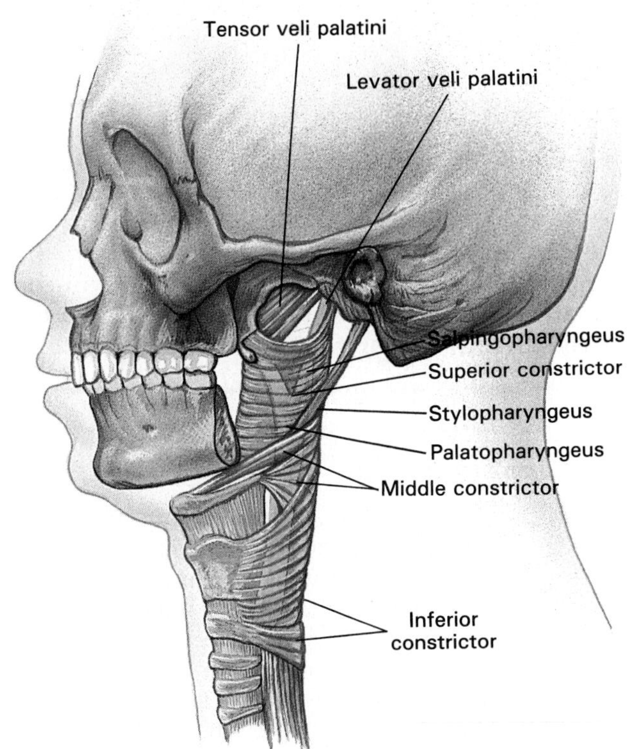

FIGURE 24-8
The Pharyngeal Musculature. (For information concerning the individual muscles, see Table 24-2.)

TABLE 24-2 The Pharyngeal Muscles

Muscle	*Origin*	*Insertion*	*Action*	*Innervation*
PHARYNGEAL CONSTRICTORS			Constrict pharynx to propel bolus into esophagus	Branches of pharyngeal plexus (N IX & X)
Superior constrictor	Pterygoid process of sphenoid, medial surfaces of mandible	Median raphe attached to occipital bone		
Middle constrictor	Cornu of hyoid	Median raphe		
Inferior constrictor	Cricoid and thyroid cartilages of larynx	Median raphe		
LARYNGEAL ELEVATORS[a]			Elevate larynx	Branches of pharyngeal plexus (N IX & X)
Palatopharyngeus	Soft palate	Thyroid cartilage		
Salpingopharyngeus	Cartilage around the inferior portion of the Eustachian tube	Thyroid cartilage		
Stylopharyngeus	Styloid process of temporal	Thyroid cartilage		
PALATAL MUSCLES			Elevate soft palate	
Levator veli palatini	Petrous portion of temporal and tissues around the Eustachian tube	Soft palate		N XI
Tensor veli palatini	Spine of sphenoid, tissues around the Eustachian tube	Soft palate		N V

[a] Assisted by the thyrohyoid, geniohyoid, stylohyoid, and hyoglossal muscles discussed in Chapter 11.

THE ESOPHAGUS

The esophagus lies posterior to the trachea in the neck. It passes along the dorsal wall of the mediastinum in the thoracic cavity and enters the peritoneal cavity through an opening in the diaphragm, the **esophageal hiatus** (hī-Ā-tus), before emptying into the stomach. The esophagus has a length of approximately 25 cm (1 ft) and a diameter of about 2 cm (0.75 in.).

The Esophageal Wall

Except during swallowing, normal muscular tone keeps the lumen closed, and the mucosa and submucosa are thrown into large folds. The muscularis mucosae consists of an irregular layer of smooth muscles between the mucous esophageal glands of the submucosa and the connective tissues of the lamina propria. The muscularis externa has inner circular and outer longitudinal layers. In the upper third of the esophagus these layers contain skeletal muscle fibers; in the middle third there is a mixture of skeletal and smooth muscle tissue; along the lower third only smooth muscles are found.

Distension of the esophagus caused by the passage of a bolus exerts pressure on surrounding structures. In the neck and upper thorax, the trachea accommodates by allowing compression of its posterior surface, the only region not reinforced by tracheal cartilages (see Figure 23-7). (p. 754)

Swallowing

The muscularis externa propels materials from one portion of the digestive tract to another through peristalsis, detailed earlier in the chapter. Swallowing, or **deglutition** (dēg-loo-TISH-un), can be divided into *buccal*, *pharyngeal*, and *esophageal phases*. Key aspects of each stage are diagrammed in Figure 24-9.

The buccal phase begins with the compression of the bolus against the hard palate. As indicated in Figure 24-9a,b the subsequent retraction of the tongue then forces the bolus into the pharynx and assists in the elevation of the soft palate, isolating the nasopharynx.

The **pharyngeal phase** begins as the bolus comes in contact with the palatal arches or the posterior pharyngeal wall or both (Figure 24-9c,d). Elevation of the larynx and folding of the epiglottis direct the bolus past the closed glottis, and in less than a second the pharyngeal muscles have propelled the bolus into the esophagus.

The **esophageal phase** of swallowing (Figure 24-9e–g) starts with the opening of the **upper esophageal**

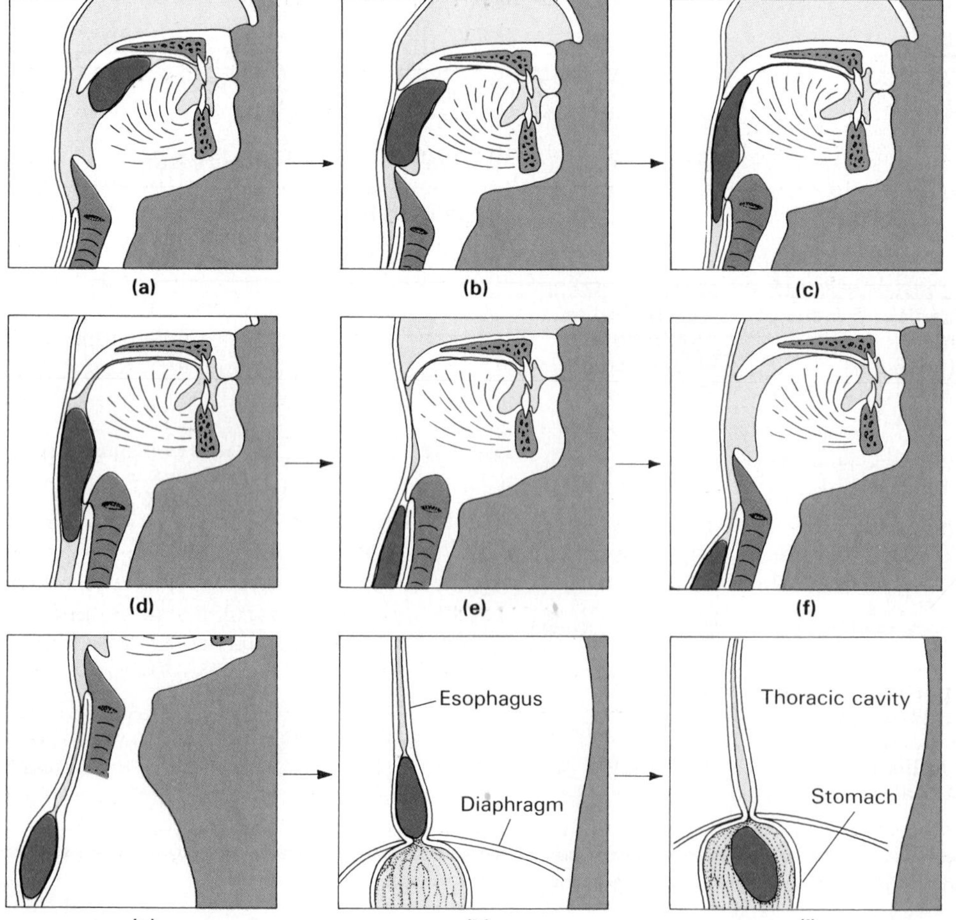

FIGURE 24-9
The Swallowing Process. This sequence, based on a series of X-rays, shows the stages of swallowing and the movement of materials from the mouth to the stomach. (See also Figure 23-6.)

sphincter. After passing through the open sphincter, the bolus is pushed down the length of the esophagus by a **primary peristaltic wave**. The approach of the bolus triggers the opening of the **lower esophageal sphincter**, and the bolus then continues into the stomach. For a typical bolus the entire trip takes about 9 seconds to complete. Fluids may make the journey in a few seconds, arriving ahead of the peristaltic contractions with the assistance of gravity. A relatively dry or poorly lubricated bolus travels much more slowly, and a series of **secondary peristaltic waves** may be required to push it all the way to the stomach. A completely dry bolus cannot be swallowed at all, for friction with the walls of the esophagus will make peristalsis ineffective. (For this reason it is not possible to swallow an entire slice of processed white bread without a drink.)

The Swallowing Reflex The oral cavity is the last chance for conscious decision making. Although the pharyngeal muscles are skeletal muscles, we have very limited voluntary control over them. Once a bolus has entered the laryngopharynx, swallowing will continue whether we are ready for it or not!

The **swallowing reflex** begins as tactile receptors on the palatal arches and uvula are stimulated by the passage of the bolus. The information is relayed to the **swallowing center** of the medulla over the trigeminal and glossopharyngeal nerves. Motor commands originating at this center then target the pharyngeal and esophageal musculature, producing a coordinated and stereotyped pattern of muscle contraction. The motor commands are distributed by cranial nerves V, IX, X, and XII. During the time it takes for the bolus to travel through the pharynx and into the esophagus, the respiratory centers are inhibited and breathing ceases. (Movements of the larynx during swallowing were detailed in Chapter 23). ∞ (p. 751)

Along the esophagus, primary peristaltic contractions are coordinated by afferent and efferent fibers within the glossopharyngeal and vagus nerves, but secondary peristaltic waves are local reflexes triggered by stimulation of sensory receptors in the esophageal walls. These receptors relay information via the submucosal and myenteric plexuses, producing peristaltic contractions in the absence of CNS instructions. Comparable myenteric reflexes typify the entire digestive tract between the lower esophagus and rectum. [**CM:** *Achalasia and Esophagitis*]

- **What effect would a drug that blocks parasympathetic stimulation of the digestive tract have on peristalsis?**
- **Where would you find the fauces?**
- **What is occurring when the soft palate and larynx elevate and the glottis closes?**

THE STOMACH

The stomach has three primary functions: (1) the bulk storage of ingested matter, (2) the mechanical breakdown of resistant materials, and (3) the disruption of chemical bonds through the action of acids and enzymes. The agitation of ingested materials with the gastric juices secreted by the glands of the stomach produces a viscous, soupy mixture called **chyme** (kīm).

Anatomy of the Stomach

Figure 24-10a indicates the principal anatomical landmarks of the stomach. This muscular organ has the shape of an expanded J, with a short **lesser curvature** and a long **greater curvature**. The esophagus connects to the medial aspect of the stomach at the **cardia** (KAR-dē-a), the boundary between the greater and lesser curvatures. The bulge of the greater curvature superior to the esophageal junction is the **fundus** (FUN-dus), and the large area between the fundus and the curve of the J is the gastric **body**. The curve of the J, the **pylorus** (pī-LŌR-us), connects with the proximal portion of the small intestine. A muscular **pyloric sphincter** regulates the flow of chyme between the stomach and small intestine.

The dimensions of the stomach are extremely variable. The "average" lesser curvature has a length of approximately 10 cm (4 in.), and the greater curvature measures around 40 cm (16 in.). When empty, the stomach resembles a muscular tube with a narrow and constricted lumen. When full, it can expand to contain 1–1.5 liters.

This extreme degree of distensibility requires a number of anatomical and histological specializations. On gross dissection the lining of the relaxed stomach contains a number of prominent ridges and folds. These **rugae** (ROO-gē; wrinkles) become less apparent as the stomach expands, and at maximum capacity they have all but disappeared.

The smooth muscle and mucosa of the stomach are so exceedingly elastic that huge meals can be tolerated. Usually the stretch receptors in the gastric walls provide enough information to the appetite center in the hypothalamus that the urge to eat disappears before the stomach expands to a potentially dangerous extent. Under unusual or extreme circumstances, the stomach wall can rupture and release the gastric contents into the peritoneal cavity, an event with a 63 percent mortality rate, even after surgery. In view of the current tendencies toward overindulgence in the westernized world, the fact that only 71 cases of gastric rupture have ever been reported stands as a testimonial to the tolerant nature of the gastric mucosa!

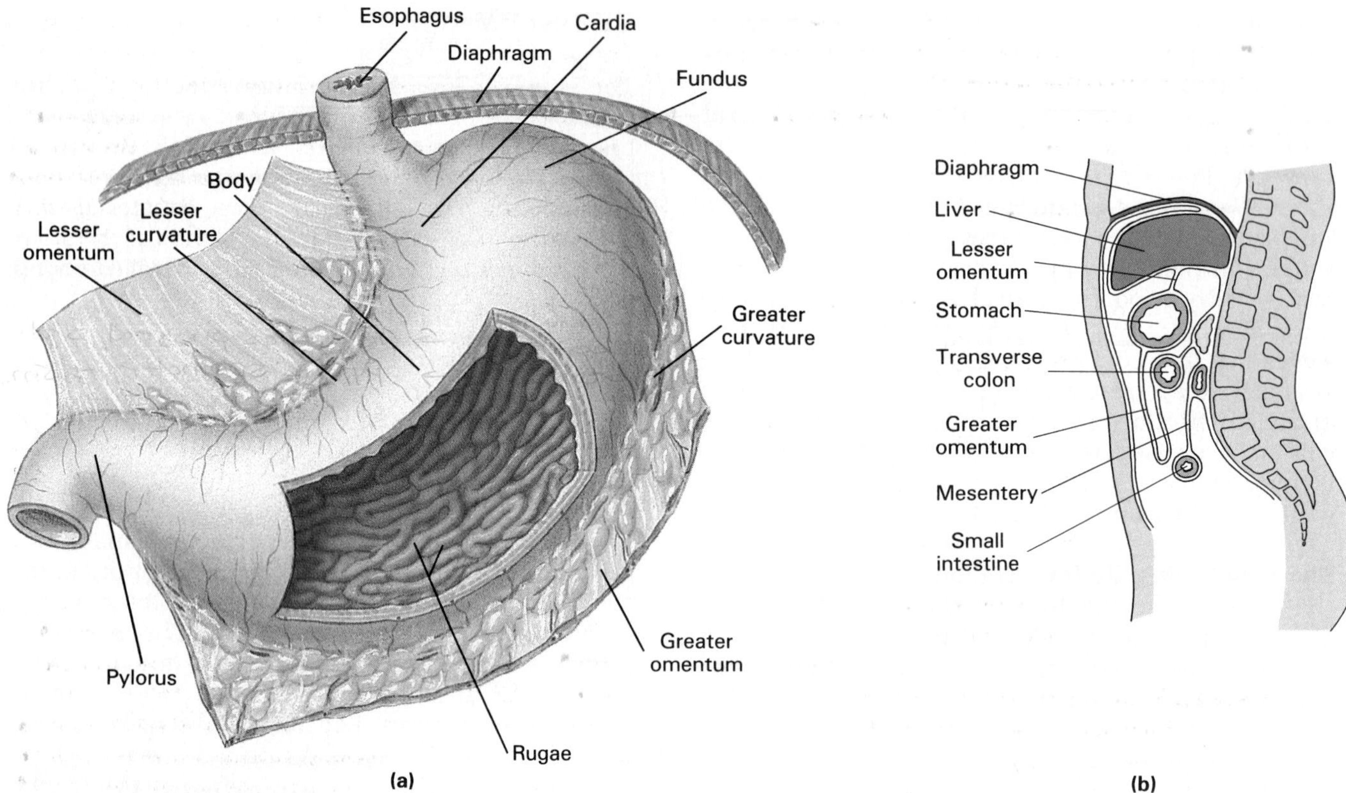

FIGURE 24-10
Gross Anatomy of the Stomach. (a) Anterior view of the stomach showing superficial landmarks and the stomach's position in the peritoneal cavity. (b) Diagrammatic sagittal section showing the relationships between the major mesenteries, the stomach, and other digestive organs.

Gastric Mesenteries

The visceral peritoneum covering the outer surface of the stomach is continuous with a pair of prominent mesenteries. The dorsal mesentery is greatly expanded, forming an enormous pouch known as the **greater omentum** (ō-MEN-tum; fat skin). This fatty sheet hangs like an apron over the abdominal viscera (see Figure 24-10b). Adipose tissue in the greater omentum conforms to the shapes of the surrounding organs, providing padding and protection across the anterior and lateral surfaces of the abdomen. The **lesser omentum** is a much smaller pocket in the ventral mesentery between the stomach and liver.

The Gastric Wall

Several different views of the gastric wall are included in Figure 24-11. The mucous membrane and submucosa are folded to produce the rugae visible on gross dissection. The superficial epithelium consists of a dense layer of simple columnar cells filled with mucous secretions, and the exposed gastric surface is carpeted in a viscous mucus. **Gastric pits**, pockets in the epithelium shown in Figure 24-11b,c, are also lined with mucous cells. Those at the base, or *neck*, of each pit are actively dividing. As the daughter cells mature, they are displaced toward the gastric lumen and eventually shed into the chyme.

The continual replacement of epithelial cells provides an additional defense against the gastric juices. If the acids and enzymes penetrate the mucous layers, the damaged cells will be detached and replaced almost immediately, and the epithelium will remain intact. A typical mucous epithelial cell survives for only about 3 days before being lost into the lumen. Exposure to strong alcohol or the chemicals found in certain foods will accelerate the destruction of these cells.

Each gastric pit communicates with **gastric glands** that extend deep into the underlying lamina propria. The gastric glands, seen in Figure 24-11c,d, are dominated by two types of secretory cells: **parietal cells** and **chief cells**. Together they secrete about 1500 mℓ of **gastric juice** each day.

Parietal Cells Parietal cells are especially common along the proximal portions of each gastric gland. These cells secrete *intrinsic factor* and *hydrochloric acid* (HCl). **Intrinsic factor** facilitates the absorption

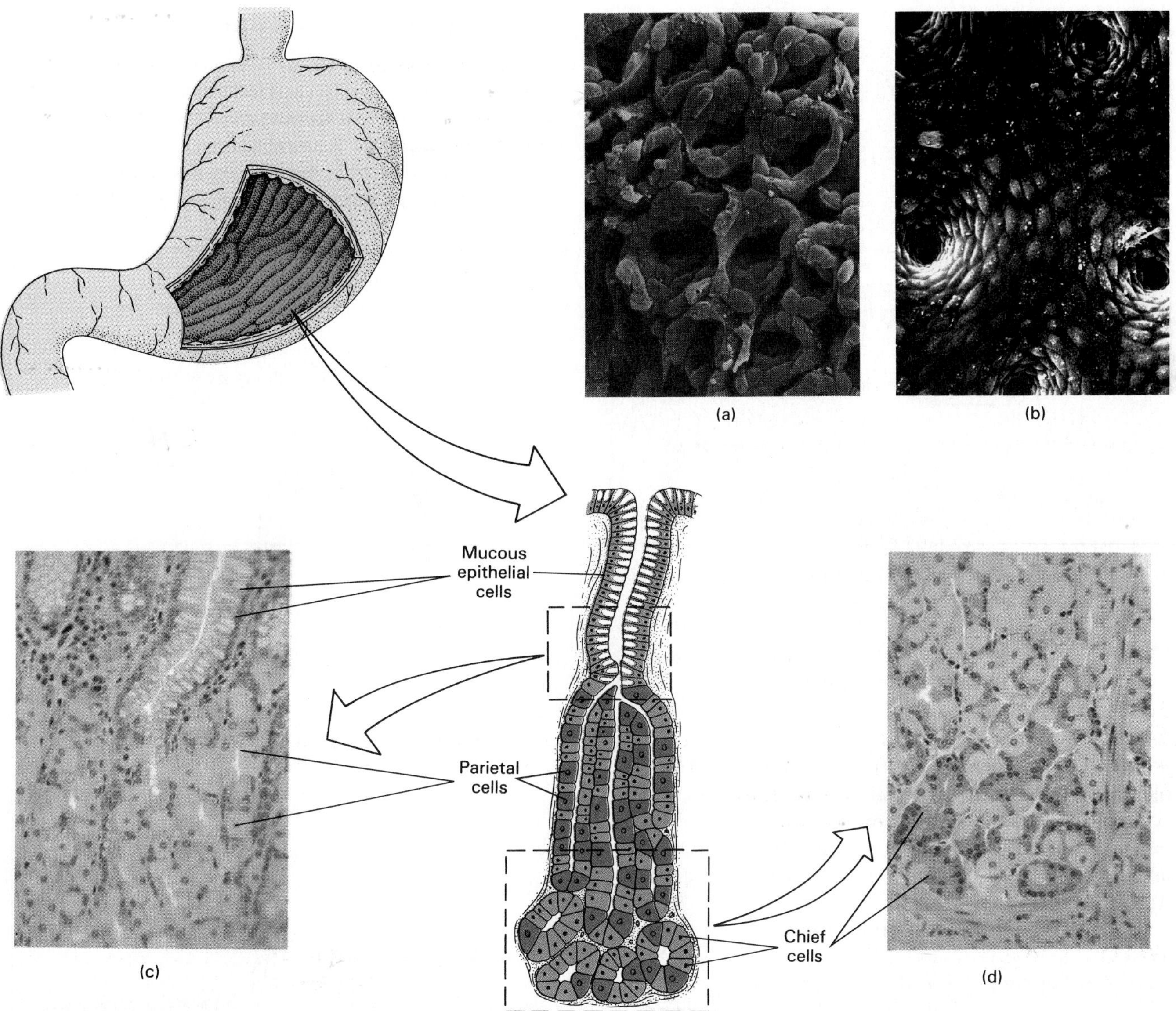

FIGURE 24-11
The Stomach Lining. (a) SEM of the empty stomach. (SEM, ×403) (b) SEM of the full stomach at the same magnification. Note the distension of the lining and the circular openings of the gastric pits. (SEM, ×405) (c,d) Micrographs showing cell organization in the superficial (c) and deep (d) portions of a gastric gland. (LMs, ×297)

of **vitamin B_{12}** across the intestinal lining. Hydrochloric acid lowers the pH of the gastric juice, kills microorganisms, breaks down cell walls and connective tissues in food, and activates the secretions of the chief cells. Hydrochloric acid secretion, diagrammed in Figure 24-12, begins as the enzyme *carbonic anhydrase* converts carbon dioxide and water to carbonic acid, which promptly dissociates into a hydrogen ion and a bicarbonate ion. (Carbonic anhydrase was introduced in Chapter 23, in the discussion of carbon dioxide transport.) ∞ (p. 773) As the carbonic acid dissociates the parietal cell actively pumps a hydrogen ion into the gastric pit. The bicarbonate ion diffuses into the lamina propria, in exchange for a chloride ion, and when the gastric glands are working extra hard, enough bicarbonate ions enter the circulation to significantly increase the pH of the blood. This sudden rush of bicarbonate ions into the circulation has been called the **alkaline tide**. Meanwhile, within the parietal cell, chloride ions are actively secreted into the gastric lumen.

With this arrangement, the pH inside the cell remains unchanged, but the concentration of hydrochloric acid in the gastric lumen continues to rise until the pH reaches 1.5–2.0. Such a low pH is essential for normal digestive activities, for pepsin, a proteolytic enzyme, works best in a strongly acid environment.

Chief Cells Chief cells are most abundant near the base of a gastric gland. These cells secrete an inactive proenzyme, **pepsinogen** (pep-SIN-ō-jen), and acids in

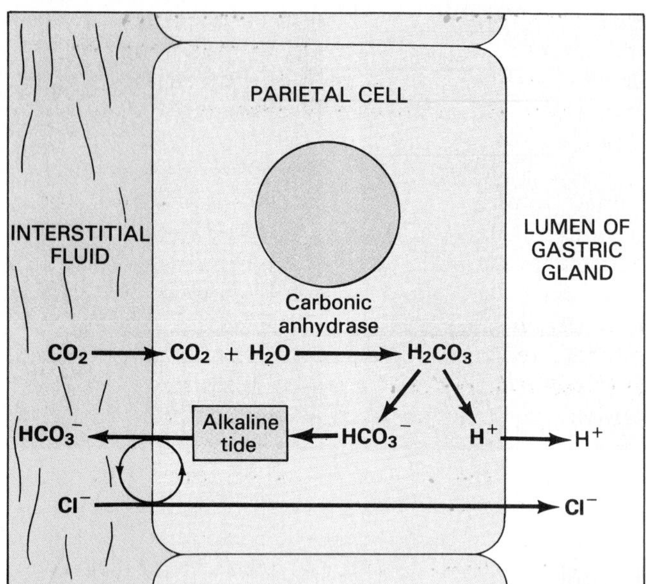

FIGURE 24-12
Ion Transport and Gastric Acid Secretion. An active parietal cell generates hydrogen ions through the dissociation of carbonic acid. Bicarbonate ions enter the interstitial fluid in exchange for chloride ions, and hydrogen and chloride ions diffuse into the lumen of the gastric gland.

the gastric lumen convert this to the active enzyme **pepsin**. Pepsin is an example of a **proteinase**, an enzyme that breaks down proteins. The stomachs of newborn infants also produce additional enzymes important for the digestion of milk, **rennin** and **gastric lipase**. Rennin coagulates milk proteins and gastric lipase initiates the digestion of milk fats.

Enteroendocrine Cells of the Stomach Scattered among the parietal and chief cells are **enteroendocrine** (en-ter-ō-EN-dō-krin) **cells** that secrete a variety of chemical compounds into their immediate surroundings. At least six different secretory products have been identified, but only one has a known function. One group of gastric enteroendocrine cells produces a polypeptide hormone, **gastrin** (GAS-trin), and releases it into the circulation of the lamina propria and perhaps into the gastric lumen as well. Gastrin stimulates the secretion of both parietal and chief cells; it is released when food enters the stomach.

Muscularis Layers The muscularis mucosae contains an extra layer of smooth muscle fibers, in addition to the normal circular and longitudinal layers, and many of these extend toward the lumen between the gastric glands. A third layer of smooth muscle is also found in the muscularis externa. In addition to the longitudinal and circular layers there is an inner, oblique layer. As noted earlier, the external surfaces of the stomach are covered by a layer of the visceral peritoneum. [**CM:** *Gastritis and Peptic Ulcers*]

Regulation of Gastric Function

The production of acid and enzymes by the gastric mucosa can be directly controlled by the central nervous system and indirectly regulated by local hormonal mechanisms. Several stages can be identified, although considerable overlap exists between them. These stages are summarized in Figure 24-13.

FIGURE 24-13
The Phases of Gastric Secretion

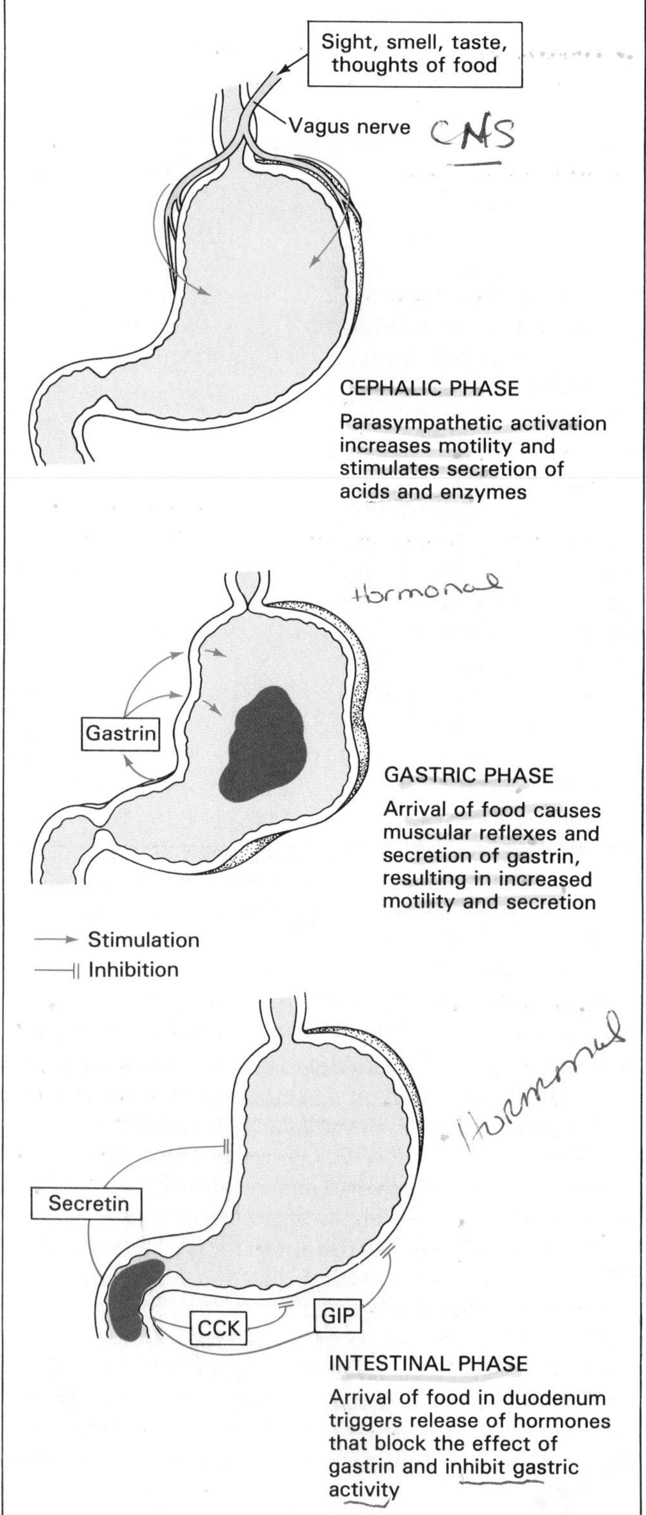

The Cephalic Phase The CNS regulation involves the vagus nerve. The sight or thought of food stimulates the vagus and initiates the **cephalic phase** of gastric secretion. The cephalic phase prepares the stomach to receive ingested materials. Postganglionic parasympathetic fibers innervate parietal cells, chief cells, and mucous cells of the stomach. In response to stimulation, the production of gastric juice accelerates, reaching rates of around 500 mℓ/h. This phase usually lasts for a relatively brief period before the *gastric phase* commences.

The Gastric Phase The **gastric phase** begins with the arrival of food in the stomach. Stimulation of stretch receptors in the stomach wall and chemoreceptors in the mucosa then trigger the release of gastrin into the circulation. Proteins, alcohol in small doses, and caffeine are potent stimulators of gastric secretion because they target the mucosal chemoreceptors.

Both parietal and chief cells respond to the presence of gastrin by accelerating their secretory activities. The effect on the parietal cells is the most pronounced, and the pH of the gastric contents declines accordingly. This phase may continue for several hours while the ingested materials are processed by the acids and enzymes.

During the gastric phase the muscularis externa is stimulated, and gastric contractions begin. As the stomach fills, only the material in contact with the gastric epithelium is exposed to digestive acids and enzymes. As a result, the central mass remains relatively unaffected, leaving the salivary enzymes free to continue the digestion of carbohydrates. These enzymes are still active 1–2 hours after a meal, only gradually becoming inoperative as the ingested mass becomes fragmented and the pH falls below 4.5.

Although gastric contractions begin almost at once, the initial contractions are merely weak pulsations in the gastric walls. These mixing waves occur several times per minute, and they gradually increase in intensity. After a half-hour or more the contractions are strong enough to swirl and churn the gastric contents, mixing the ingested materials with the gastric secretions to form chyme. As digestion proceeds, the contractions begin sweeping down the length of the stomach, and each time the pylorus contracts a small quantity of chyme squirts through the pyloric sphincter. Figure 1-14. includes an X-ray of an active stomach. (p. 20) This image shows the irregular shape of the stomach at a single instant; an X-ray taken moments later would reveal quite different contours.

In general, the greater the initial distension of the stomach, and the more gastrin released, the sooner the chyme will reach the small intestine. For example, a heavy meal containing meat (proteins), wine (alcohol), and after-dinner coffee (caffeine) will begin to leave the stomach after a relatively brief stay.

The Intestinal Phase The **intestinal phase** of gastric secretion begins when chyme starts to enter the small intestine. The purpose of the intestinal phase is to control the rate of gastric emptying. This ensures that the secretory, digestive, and absorptive functions of the small intestine can proceed at reasonable efficiency. Most of the regulatory controls are inhibitory, providing a brake for gastric activities. If the proximal portion of the small intestine becomes too full, too acid, unduly irritated by the chyme, or filled with partially digested proteins, carbohydrates, or fats, inhibitory reflexes will depress gastric activity. These **enterogastric reflexes** give the intestine extra time to deal with the problem and extend the digestive and processing time in the stomach.

The inhibition involves local, central, and endocrine mechanisms: (1) Inhibitory feedback to the submucosal and myenteric plexuses slows the contractions in the stomach walls. (2) Collaterals from these sensory fibers inhibit the parasympathetic nervous system and stimulate sympathetic innervation. The combination significantly reduces gastric activity. (3) In addition, the arrival of chyme stimulates enteroendocrine cells of the intestinal mucosa to release a variety of hormones.

Although a specific hormone responsible for turning off the secretory glands in the stomach has yet to be identified, at least two hormones released by the intestine inhibit gastric secretion to some degree. The release of these hormones, **secretin** (sē-KRĒ-tin) and **cholecystokinin** (kō-lē-sis-to-KĪ-nin), stimulates secretion by the pancreas and liver; the depression of gastric activity appears to be a secondary but complementary effect. [**CM:** *Stomach Cancer*]

In the stomach food becomes saturated with gastric juices and exposed to the digestive effects of a strong acid and a proteolytic enzyme, pepsin. Most of the important digestive processes are completed in the small intestine, where the products of digestion are absorbed. The small intestine produces only a few of the enzymes needed to break down the complex materials found in the diet. Most of the enzymes and buffers are contributed by the liver and pancreas, which will be discussed in a later section.

duo means 12

THE SMALL INTESTINE

The small intestine is about 6 meters (20 ft) long and has a diameter ranging from 4 cm at the stomach to about 2.5 cm at the junction with the large intestine. It has three subdivisions: the *duodenum*, the *jejunum*, and the *ileum*, illustrated in Figure 24-14a. The **duodenum** (dū-A-dē-num) is the 25 cm (1 ft) closest to the stomach. This portion of the small intestine receives chyme from the stomach and exocrine secretions from the pancreas and liver. A rather abrupt bend marks the boundary between the duodenum and the **jejunum** (je-JOO-num). Following the

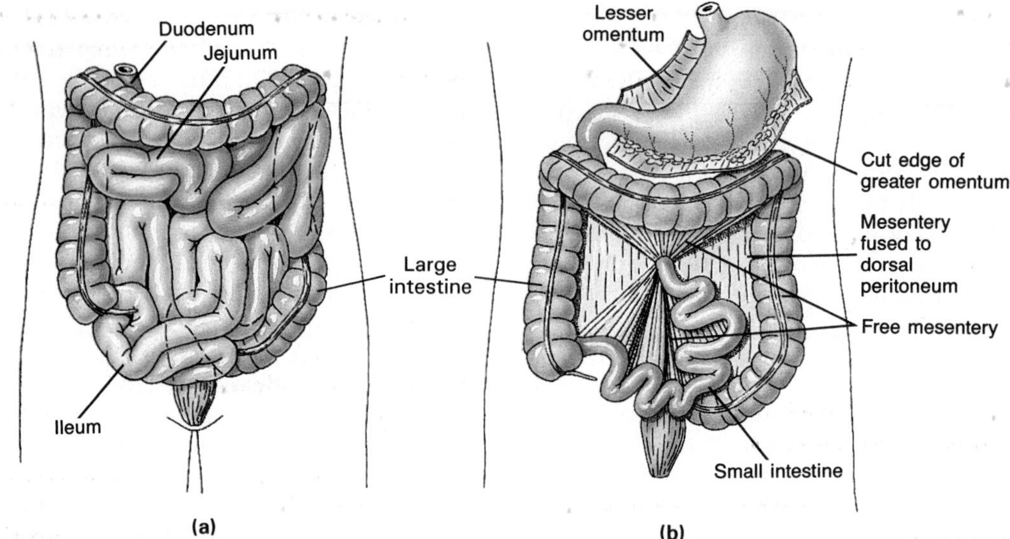

FIGURE 24-14
Gross Anatomy of the Small Intestine. (a) The small intestine includes the duodenum, jejunum, and ileum. The jejunum is the longest segment. (b) Most of the small intestine is supported by a free mesentery that fans out from a point of attachment to the dorsal body wall. The length of the small intestine has been reduced to emphasize the structures that restrict its movement, such as the stomach, the large intestine, and associated mesenteries. Its anterior and lateral movements are further restricted by the abdominal wall, and the pelvic girdle and pelvic diaphragm prevent inferior movement (these features are diagrammed in Figure 24-10b).

jejunum for another 2.5 meters (8 ft) leads us to the third segment, the **ileum** (IL-ē-um). The small intestine fits in the relatively small peritoneal cavity because it is well packed, and the position of each of the segments is stabilized by mesenteries attached to the dorsal body wall. These features can be seen in Figure 24-14b.

The Intestinal Wall

The intestinal mucosa bears a series of transverse folds called **plicae** (PLĪ-sē), shown in Figure 24-15a. Unlike the rugae in the stomach, each *plica* (PLĪ-ka) is a permanent feature of the intestinal lining. Roughly 800 plicae are found along the length of the small intestine, and their presence greatly increases the surface area available for absorption.

Figure 24-15b presents a more detailed view of the intestinal wall. Each plica bears a series of fingerlike projections, the **intestinal villi**. These villi are covered by a simple columnar epithelium that is carpeted with microvilli. If the small intestine were a simple tube with smooth walls, it would have a total absorptive area of around 3300 square centimeters, or roughly 3.6 square feet. Instead, the epithelium contains plicae, each branched plica supports a forest of villi, and each villus is covered by epithelial cells whose exposed surfaces are smothered in microvilli. This arrangement increases the total area for absorption to approximately 2 million square centimeters, or more than 2200 square feet!

Lacteals The blood vessels supplying the digestive tract were detailed in Figure 21-18. (p. 685) The lamina propria of each villus contains an extensive network of capillaries, seen in Figure 24-15c, that transports respiratory gases and carries absorbed materials to the hepatic portal circulation. Figure 24-15d diagrams the internal organization of a single villus. In addition to capillaries and nerve endings, each villus contains a terminal lymphatic called a **lacteal** (LAK-tē-al; *lacteus*, milky). This name refers to the pale, cloudy appearance of the lymph in these channels. The lacteals transport materials that fail to enter the local capillaries because they are unable to cross the capillary walls. For example, absorbed fatty acids are assembled into protein-lipid packages that are too large to diffuse into the bloodstream. These packets, called *chylomicrons*, reach the circulation by passage through the lymphatic system. (Chylomicrons, whose presence accounts for the milky appearance of fluid in the lacteals, will be discussed further in a later section.)

Intestinal Crypts Between the columnar epithelial cells, goblet cells eject mucus onto the intestinal surfaces. At the bases of the villi are found the entrances to the **intestinal crypts** (kripts). These pockets extend deep into the underlying lamina propria. Several dif-

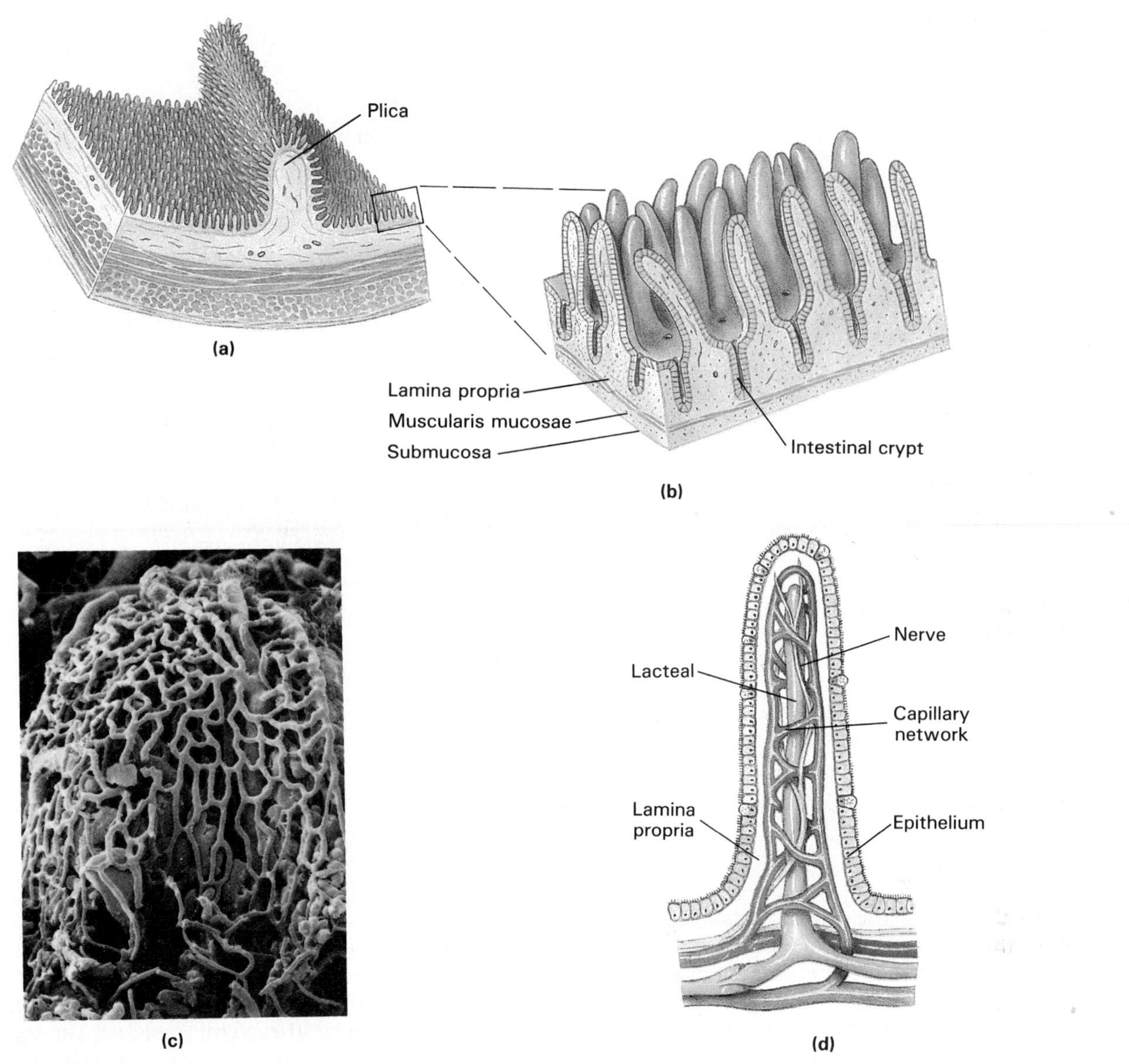

FIGURE 24-15
The Intestinal Wall. (a) Characteristic features of the intestinal lining. (b) The organization of villi and the intestinal crypts. (c) Scanning electron micrograph, showing the capillary supply of a villus. In this preparation the epithelium was removed after the capillaries were filled with latex rubber. (SEM, ×180) (d) Diagrammatic view of a single villus, showing the capillary and lymphatic supply.

ferent cell populations are found lining the intestinal crypts. Near the base of each crypt, stem cell divisions continually produce new generations of columnar and goblet cells. These new cells are continually displaced toward the intestinal surface, and within a few days they will have reached the tip of a villus. The continual shedding, or **exfoliation** (eks-fō-lē-Ā-shun; *exfolatio*, falling off), of intestinal cells renews the epithelial surface and adds intracellular enzymes to the intestinal contents. **Enterokinase** is one important enzyme that reaches the intestinal lumen in this way. Enterokinase does not directly participate in digestion, but it activates proenzymes secreted by the pancreas.

Intestinal crypts also contain enteroendocrine cells responsible for the production of several intestinal hormones, including cholecystokinin and secretin.

Regional Specializations

The regions of the small intestine have histological peculiarities that directly relate to their primary functions. The representative sections presented in Figure 24-16 should be consulted as you read further.

The Duodenum One reason that the stomach performs very little absorption is that the gastric con-

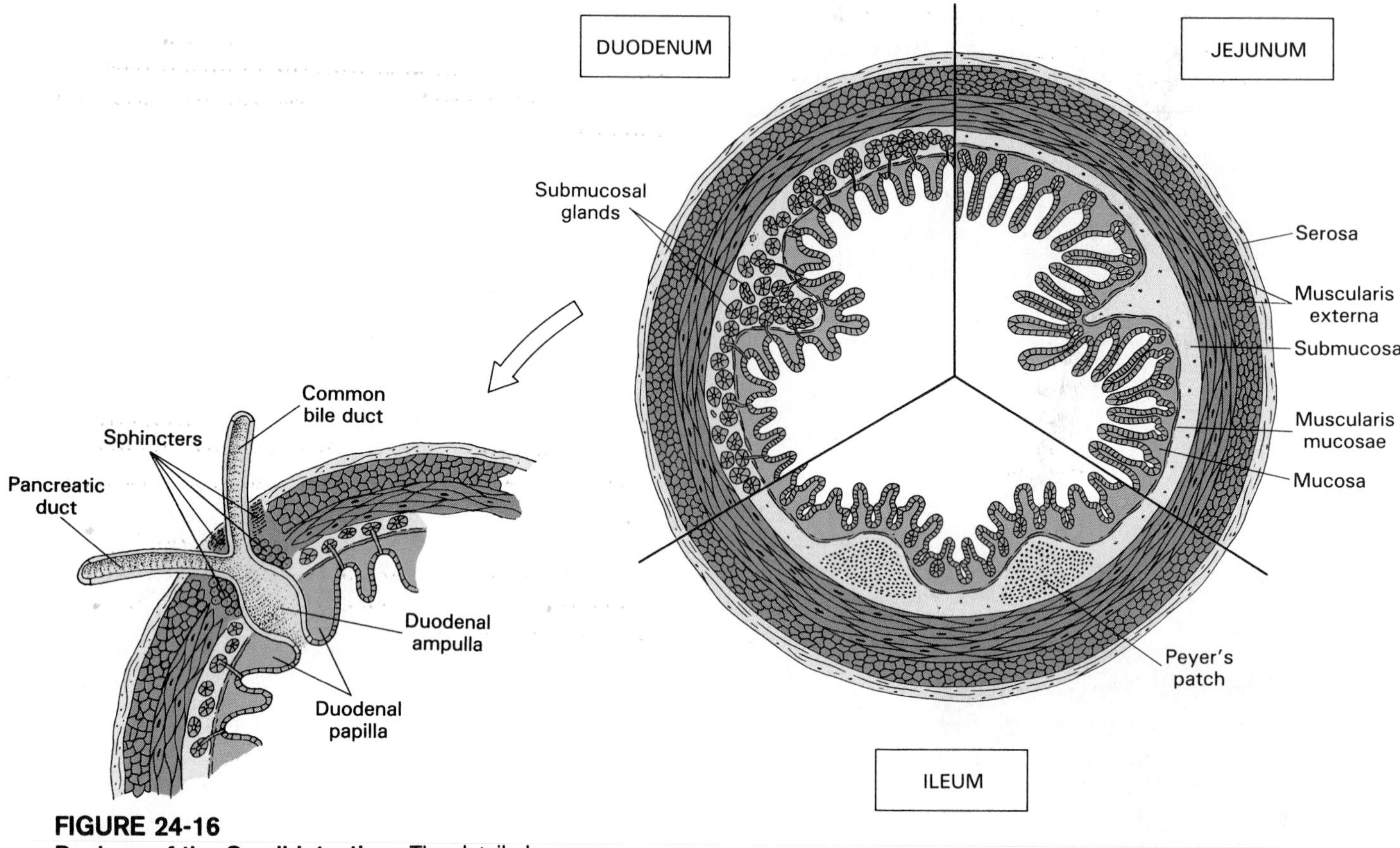

FIGURE 24-16
Regions of the Small Intestine. The detail shows a cross-section through the ampulla.

tents are so acidic that the entire gastric epithelium must be devoted to mucus production. The duodenum receives chyme from the stomach, and it must resist the gastric acids and enzymes until buffers can bring them under control. Therefore it should not surprise you to find that the most obvious histological characteristic of the duodenum is the presence of abundant mucus glands both within the epithelium and beneath it. The **submucosal glands**, also known as *Brunner's glands*, aid the duodenal crypts with the production of copious quantities of mucus. In addition to providing protection to the epithelium, the mucus contains buffers that elevate the pH.

Roughly halfway along its length, the duodenum receives additional assistance as buffers arrive from the pancreas. Within the duodenal wall, the common bile duct and pancreatic duct come together and form a muscular chamber, the **duodenal ampulla** (am-POOL-a). This chamber opens to the intestinal tract at a small mound called the **duodenal papilla** (*papilla of Vater*).

Submucosal glands are most abundant in the proximal portion of the duodenum, and their numbers decrease as we approach the jejunum. On the other hand, there are no plicae in the region adjacent to the stomach, and the villi are quite short. As we travel away from the stomach along the duodenum, the plicae and villi become more prominent, and they achieve their maximum size at the start of the jejunum. Once again, this pattern has functional implications. By the time the chyme reaches the jejunum, the pH has risen from 1–2 to 7–8, and the intestinal epithelium no longer needs extra mucous protection. Over the same period the chyme has been diluted by mixture with intestinal, pancreatic, and hepatic secretions. A variety of enzymes are present and digestion is well under way. Absorption can occur quite effectively under these relatively gentle and controlled conditions, and the exposed surface area increases accordingly.

The Jejunum and Ileum Plicae and villi remain prominent over the proximal half of the jejunum, and most intestinal absorption occurs here. As we approach the ileum the plicae and villi become smaller, and they continue to diminish in size along the length of the ileum. Under normal conditions essentially all of the absorption performed by the small intestine has occurred before the chyme reaches the terminal portions of the ileum. This region lacks plicae altogether, and the scattered villi are stumpy and conical. [**CM:** *Drastic Weight-Loss Techniques*]

The contents of the large intestine differ from those of the small intestine, in part due to bacterial activity. Bacteria are normal and even desirable inhabitants of the large intestine, and the surrounding mucosa nourishes these bacteria while confining them to the intestinal lumen. The epithelial barriers and underlying cells of the immune system protect the small intestine from bacteria migrating from the large intestine. In the ileum the scattered lymphatic nodules found in other segments of the small intestine are replaced by large masses of lymphatic tissue. These lymphatic centers, called *Peyer's patches*, are aggregations of lymphatic nodules that may be several inches in length and an inch in width. There are usually 20–30 of them, and they are most abundant near the entrance to the large intestine. Peyer's patches were considered in Chapter 22 in the discussion of lymphatic tissues (see Figure 22-6). (p. 714) [**CM:** *Giardiasis*]

Intestinal Movements

The adaptations for increasing absorptive area are certainly the most striking aspect of intestinal histology. Absorptive effectiveness is further enhanced by the fact that all of the elements involved are movable. The microvilli can be moved by their supporting microfilaments, the individual villi by scattered smooth muscle fibers, groups of villi by the muscularis mucosae, the plicae by the muscularis mucosae and the muscularis externa, and the lumenal contents by various combinations of the above. In short, the environment around each epithelial cell continually changes. This maximizes the opportunities for absorption, and roughly 80 percent of all absorption takes place in the small intestine, with the rest divided between the stomach and the large intestine.

As absorption occurs, weak peristaltic contractions slowly move the chyme along the length of the small intestine. These contractions are myenteric reflexes not under CNS control, and the effects are limited to within a few centimeters of the site of the original stimulus. More elaborate reflexes coordinate activities along the entire length of the small intestine. Two important examples are the *gastroenteric reflex* and the *gastroileal reflex*.

The Gastroenteric and Gastroileal Reflexes Distension of the stomach initiates the **gastroenteric** (gas-trō-en-TER-ik) **reflex**. This reflex produces an immediate increase in the rates of glandular secretion and peristaltic activity in all segments. The increased peristalsis distributes the chyme along the length of the small intestine and empties the duodenum. This action pushes additional chyme into the terminal segments of the ileum, but the **gastroileal** (gas-trō-IL-ē-al) **reflex** makes room for it.

The gastroileal reflex represents a combination of the gastroenteric reflex, neurally mediated, and a response to circulating levels of a digestive hormone, gastrin. The entry of food into the stomach triggers the release of gastrin, which relaxes the ileocecal sphincter. Because the sphincter is relaxed, the increased ileal peristalsis pushes chyme into the large intestine. On average it takes about 5 hours for chyme to pass from the duodenum to the end of the ileum, so the first of the materials to enter the duodenum after breakfast may leave the small intestine at lunch. [**CM:** *Vomiting and Intestinal Evacuation*]

The Control of Intestinal Movement Movements of the small intestine are controlled primarily by neural reflexes involving the submucosal and myenteric plexuses. Many of these are local reflexes, involving only neurons in the intestinal walls. Stimulation of the parasympathetic system increases the sensitivity of these reflexes and accelerates peristalsis and segmentation movements. In addition, some of the smooth muscle cells contract periodically, even without stimulation, establishing a basic contractile rhythm that then spreads from cell to cell. [**CM:** *Gastroenteritis*]

Intestinal Secretions

Roughly 1.8 liters of watery **intestinal juice** enter the intestinal lumen each day. Much of this fluid arrives through osmosis, as water flows out of the mucosa and into the relatively concentrated chyme. The rest is provided by intestinal glands stimulated by activation of touch and stretch receptors in the intestinal walls. Intestinal juice moistens the chyme, assists in buffering acids, and dissolves both the digestive enzymes and the products of digestion.

Regulating Secretory Activities Hormonal and CNS controls are important in regulating the secretory output of the larger digestive and accessory glands. These regulatory mechanisms are focused on the duodenum, for it is there that the acids must be neutralized and the appropriate enzymes added. The submucosal glands protect the duodenal epithelium from gastric acids and enzymes. Although they increase their secretory activities in response to local reflexes, parasympathetic (vagal) stimulation also has this effect. As a result the duodenal glands begin secreting during the cephalic phase of gastric secretion, long before the chyme reaches the pyloric sphincter. Sympathetic stimulation inhibits their activation, leaving the duodenal lining relatively unprepared for the arrival of the acid chyme. This probably accounts for the fact that duodenal ulcers can be caused by chronic stress or other factors that promote sympathetic activation.

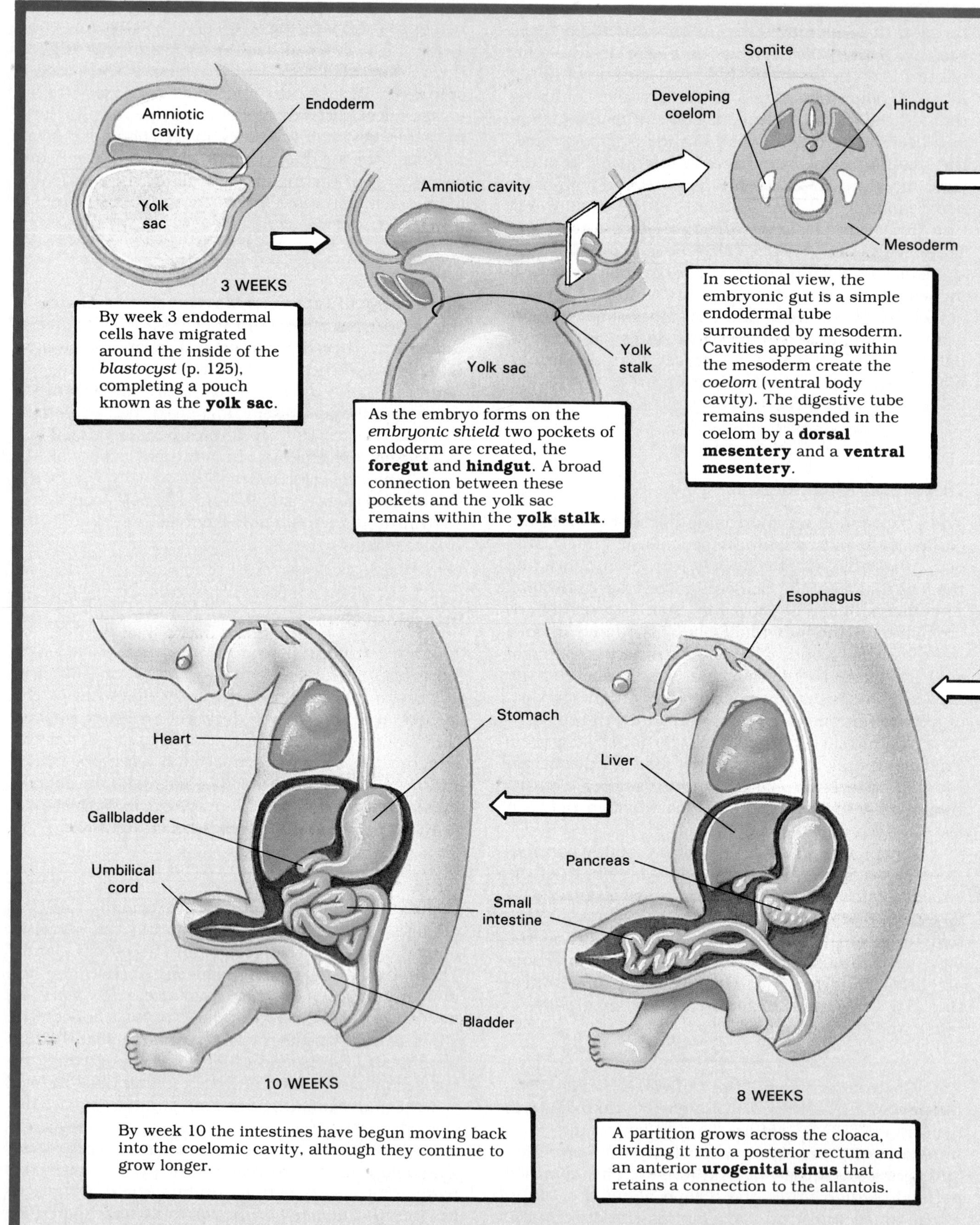
Amniotic cavity
Endoderm
Yolk sac
3 WEEKS
By week 3 endodermal cells have migrated around the inside of the *blastocyst* (p. 125), completing a pouch known as the **yolk sac**.
Amniotic cavity
Yolk sac
Yolk stalk
As the embryo forms on the *embryonic shield* two pockets of endoderm are created, the **foregut** and **hindgut**. A broad connection between these pockets and the yolk sac remains within the **yolk stalk**.
Somite
Developing coelom
Hindgut
Mesoderm
In sectional view, the embryonic gut is a simple endodermal tube surrounded by mesoderm. Cavities appearing within the mesoderm create the *coelom* (ventral body cavity). The digestive tube remains suspended in the coelom by a **dorsal mesentery** and a **ventral mesentery**.
Esophagus
Heart
Stomach
Liver
Gallbladder
Umbilical cord
Pancreas
Small intestine
Bladder
10 WEEKS
8 WEEKS
By week 10 the intestines have begun moving back into the coelomic cavity, although they continue to grow longer.
A partition grows across the cloaca, dividing it into a posterior rectum and an anterior **urogenital sinus** that retains a connection to the allantois.

Embryology Summary: Development of the Digestive System

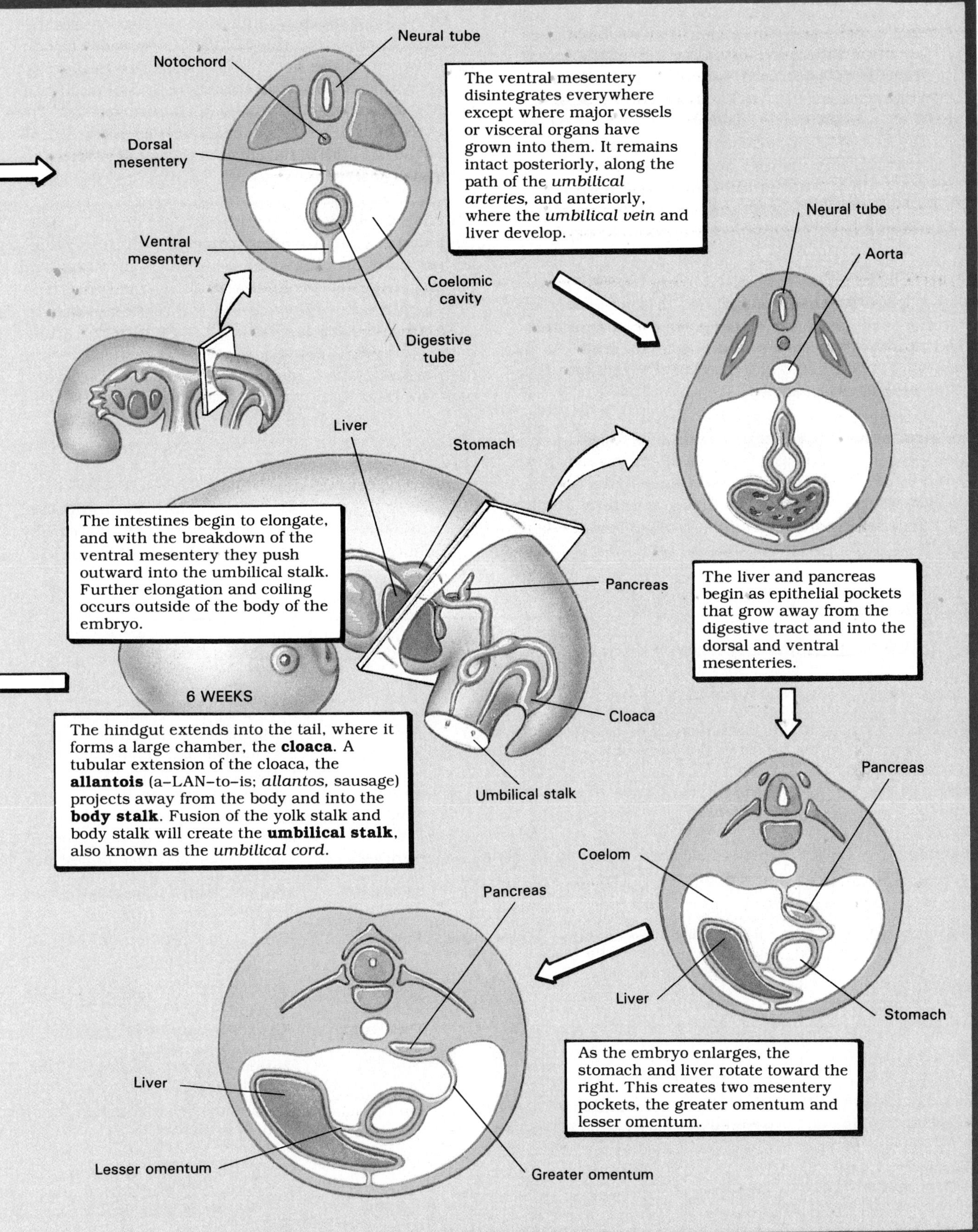

Neural tube
Notochord
Dorsal mesentery
Ventral mesentery
Coelomic cavity
Digestive tube
The ventral mesentery disintegrates everywhere except where major vessels or visceral organs have grown into them. It remains intact posteriorly, along the path of the *umbilical arteries*, and anteriorly, where the *umbilical vein* and liver develop.
Neural tube
Aorta
Liver
Stomach
The intestines begin to elongate, and with the breakdown of the ventral mesentery they push outward into the umbilical stalk. Further elongation and coiling occurs outside of the body of the embryo.
Pancreas
The liver and pancreas begin as epithelial pockets that grow away from the digestive tract and into the dorsal and ventral mesenteries.
6 WEEKS
Cloaca
The hindgut extends into the tail, where it forms a large chamber, the **cloaca**. A tubular extension of the cloaca, the **allantois** (a–LAN–to–is; *allantos*, sausage) projects away from the body and into the **body stalk**. Fusion of the yolk stalk and body stalk will create the **umbilical stalk**, also known as the *umbilical cord*.
Umbilical stalk
Pancreas
Coelom
Pancreas
Liver
Stomach
As the embryo enlarges, the stomach and liver rotate toward the right. This creates two mesentery pockets, the greater omentum and lesser omentum.
Liver
Lesser omentum
Greater omentum

Intestinal Hormones

Duodenal enteroendocrine cells produce hormones that coordinate the secretory activities of the stomach, duodenum, liver, and pancreas. **Enterocrinin**, a hormone released when acid chyme enters the small intestine, stimulates the duodenal glands. Several other important hormones have primary and secondary effects that are distinct but complementary. The three most important hormones involved in the regulation of intestinal activity are *secretin*, *cholecystokinin*, and *glucose-dependent insulinotropic peptide*.

Secretin **Secretin** is released when acids appear within the duodenal lumen. The primary effect of secretin is to cause an increase in the secretion of water and buffers by the pancreas and liver. As a secondary effect secretin further stimulates the duodenal submucosal glands.

Cholecystokinin **Cholecystokinin**, or **CCK**, provides a second important example of complementary hormonal effects. CCK is secreted when chyme arrives in the duodenum, especially when it contains lipids and partially digested proteins. This hormone also targets both the pancreas and the liver. In the pancreas, CCK accelerates the production and secretion of all types of digestive enzymes. It also causes the passage of bile from the gallbladder into the duodenum. In short, the net effects of CCK are to increase the secretion of pancreatic enzymes and to push pancreatic secretions and bile into the duodenum. The presence of either secretin or CCK in high concentrations has the additional effect of reducing gastric motility and secretory rates.

Glucose-dependent Insulinotropic Peptide **Glucose-dependent insulinotropic** (in-su-lin-ō-TRŌP-ik) **peptide**, or **GIP**, release occurs when fats and partially digested proteins enter the small intestine. This peptide hormone causes the release of insulin from the pancreatic islets. At high concentrations it also inhibits gastric activity, and for this reason the hormone was originally named *gastric inhibitory peptide*. The acronym GIP was retained, despite a revision of opinion as to its primary function.

Other Intestinal Hormones Several other hormones are produced in relatively small quantities. For example, large amounts of relatively undigested proteins also stimulate the release of gastrin by duodenal cells. **Vasoactive intestinal peptide**, or **VIP**,

TABLE 24-3 Important Gastrointestinal Hormones and Their Primary Effects

Hormone	*Stimulus*	*Origin*	*Target*	*Effects*
Gastrin	Vagal stimulation or arrival of food in the stomach	Stomach	Stomach	Stimulates production of acids and enzymes, increases motility
Enterocrinin	Arrival of acid chyme in the duodenum	Duodenum	Duodenal (Brunner's) glands	Stimulates alkaline mucus production
Secretin	Arrival of acid chyme in the duodenum	Duodenum	Pancreas	Stimulates alkaline buffer production
			Stomach	Inhibits gastric secretion and motility
Cholecystokinin (CCK)	Arrival of acid chyme containing lipids and partially digested proteins	Duodenum	Pancreas	Stimulates production of pancreatic enzymes
			Gallbladder	Stimulates contraction of gallbladder
			Duodenum	Causes relaxation of sphincter at base of bile duct
			Stomach	Inhibits gastric secretion and motion
Glucose-dependent insulinotropic hormone (GIP)	Arrival of chyme containing large quantities of glucose	Duodenum	Pancreas	Stimulates release of insulin by pancreatic islets
Vasoactive intestinal peptide (VIP)	Arrival of chyme in the duodenum	Duodenum	Duodenal glands, stomach	Stimulates buffer secretion, inhibits acid production, dilates intestinal capillaries
Gastrin	Arrival of chyme containing large quantities of undigested proteins	Duodenum	Stomach	Stimulates gastric secretion and motion (as above)

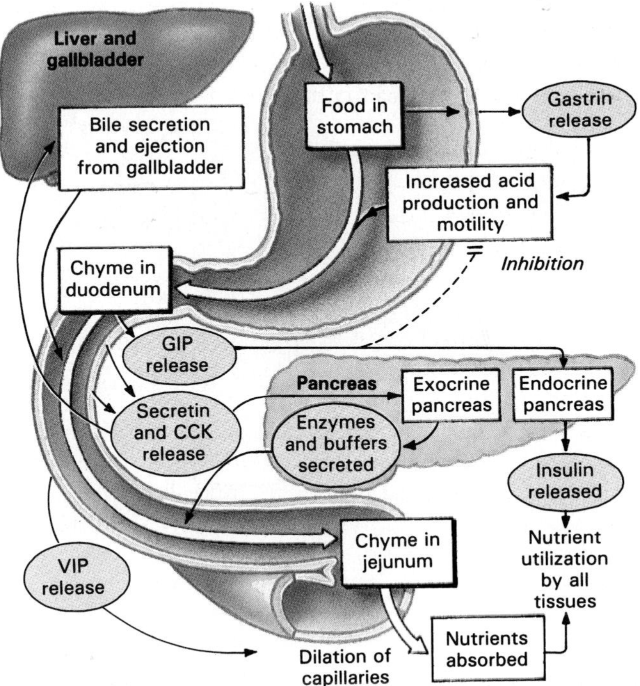

FIGURE 24-17
The Activities of Major Digestive Tract Hormones. This diagram follows the primary actions of gastrin, GIP, secretin, CCK, and VIP.

stimulates the secretion of intestinal glands, dilates regional capillaries, and inhibits acid production in the stomach. For many years physiologists suspected the presence of a hormone, *enterogastrone*, responsible for inhibiting gastric activity. It now appears likely that this inhibition reflects the combined actions of GIP and VIP. Data concerning all of the major gastrointestinal hormones are summarized in Table 24-3. Functional interactions between gastrin, secretin, CCK, GIP, and VIP are diagrammed in Figure 24-17.

The Diversity of Intestinal Hormones The gastrointestinal tract is the largest endocrine organ in the body, but the hormonal secretions are poorly understood. It has proven very difficult to determine the primary effects of these hormones, largely because all are peptide hormones with similar chemical structures. Careful analyses have led to a marked increase in the number of identified intestinal hormones, but their specific functions have yet to be sorted out to everyone's satisfaction.

√ **How would a large meal affect the pH of the blood that leaves the stomach?**

√ **When a person suffers from chronic ulcers in the stomach, the branches of the vagus nerve that serve the stomach are sometimes severed. Why?**

√ **How is the small intestine adapted for the absorption of nutrients?**

√ **How would a meal that is high in fat affect the level of cholecystokinin (CCK) in the blood?**

THE LARGE INTESTINE

The principal functions of the large intestine include: (1) the resorption of water and compaction of feces, (2) the absorption of important vitamins liberated by bacterial action, and (3) the storing of fecal material prior to defecation.

The large intestine, often called the **large bowel,** has an average length of approximately 1.5 meters (5 ft) and a width of 7.5 cm (3 in.). It can be divided into two major regions: the *colon*, the largest portion of the large intestine, and the *rectum*. The rectum is the last 15 cm (6 in.) of the large intestine and the end of the digestive tract.

The Colon

The most striking external feature of the colon is the pouches, or **haustrae** (HAWS-trē), that permit considerable distension and elongation. Cutting into the intestinal lumen reveals that the creases between the haustrae affect the mucosal lining as well, producing a series of internal creases. Three longitudinal bands of muscle, the **taenia coli** (TĒ-nē-a KŌ-lī), are visible on the outer surfaces of the colon just beneath the serosa. Muscle tone within these bands produces the haustrae.

The rest of the colon can be divided into regions using gross anatomical landmarks, as indicated in Figure 24-18. Chyme first enters an expanded chamber, called a **cecum** (SĒ-kum). A muscular sphincter, the **ileocecal** (il-ē-ō-SĒ-kal) **valve**, guards the connection between the ileum and the cecum. The cecum usually has the shape of a rounded sac, and the slender **appendix** attaches to the cecum along its posteromedial surface. The average appendix has a length of almost 9 cm (3.5 in.), and its walls are dominated by lymphatic tissue. It is not firmly attached to the surrounding mesenteries, and it often wriggles and twists as its muscular walls contract. Inflammation of the appendix produces the symptoms of *appendicitis*. [**CM:** *Infected Nodules*]

The **ascending colon** begins at the ileocecal valve. It ascends along the right side of the peritoneal cavity until reaching the inferior margin of the liver. The **right colic flexure**, or *hepatic flexure*, marks the transition to the **transverse colon**. The transverse colon continues toward the left side, passing below the stomach and following the curve of the body wall. Near the spleen it turns caudally at the **left colic flexure**, or *splenic flexure*. The **descending colon**

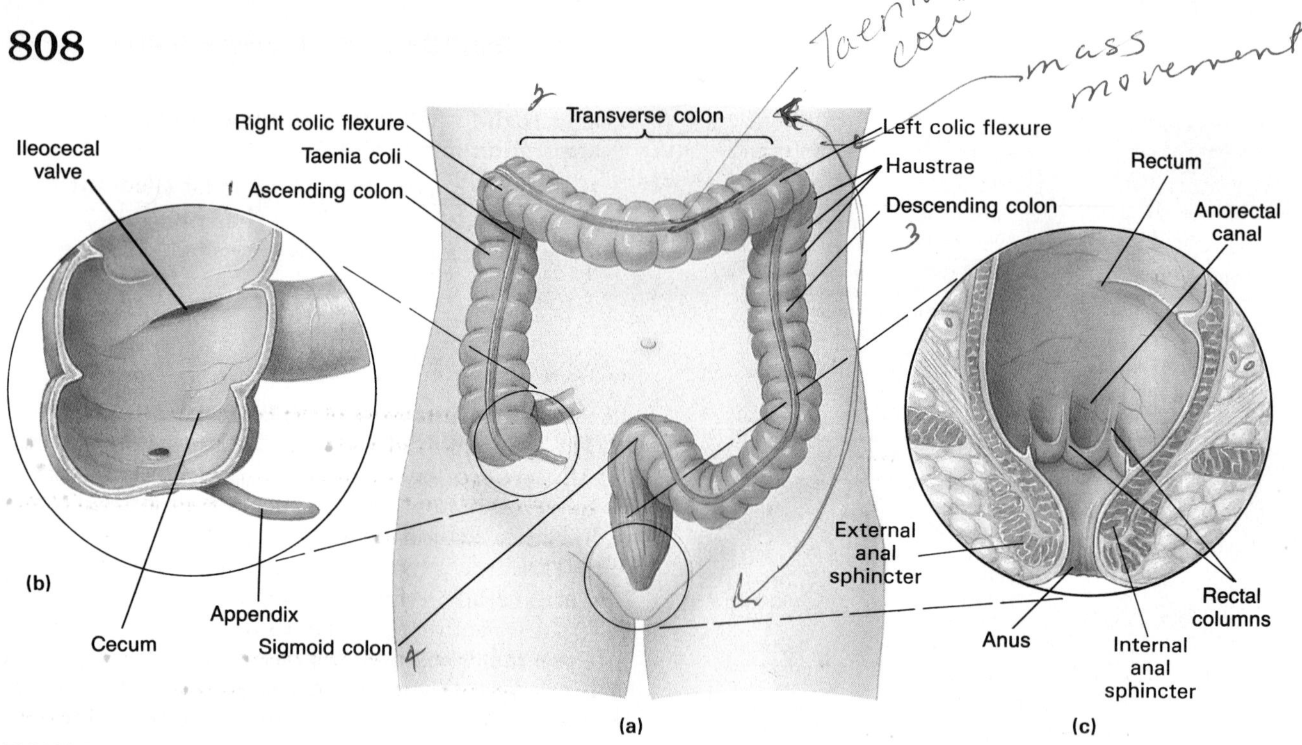

FIGURE 24-18
The Large Intestine. (a) Gross anatomy and regions of the large intestine. (b) Detailed anatomy of the cecum and appendix. (c) Detailed anatomy of the rectum and anus.

then continues along the left side until reaching the iliac fossa. There it curves and recurves as the **sigmoid colon** (SIG-moid; *sigmoides*, the Greek letter S), which empties into the rectum.

The Wall of the Colon Although the diameter of the colon is roughly three times that of the small intestine, the wall is much thinner. To develop a three-dimensional image of the mucosa, consider the effect of driving a miniature lawnmower over the surface of the small intestine. This would remove all of the villi that project above the mucosal surface, but would leave the crypts untouched. Now compare that mental image with Figure 24-19. The epithelium contains many more goblet cells than does the mucosa or the small intestine, but the columnar epithelial cells still have microvilli on their exposed surfaces. The crypts, dominated by goblet cells, are known as **intestinal glands**, and they are longer than those of the small intestine. Secretion of the colonic glands occurs as local stimuli trigger reflexes involving the local nerve plexuses. Large lymphatic nodules are common in the underlying lamina propria and submucosa.

The muscularis externa differs from that of other intestinal regions because the longitudinal layer has condensed to form the bands of the taenia coli. However, the mixing and propulsive contractions of the colon resemble those of the small intestine. [**CM:** *Diverticulitis and Colitis*]

FIGURE 24-19
The Lining of the Large Intestine. (LM, ×51)

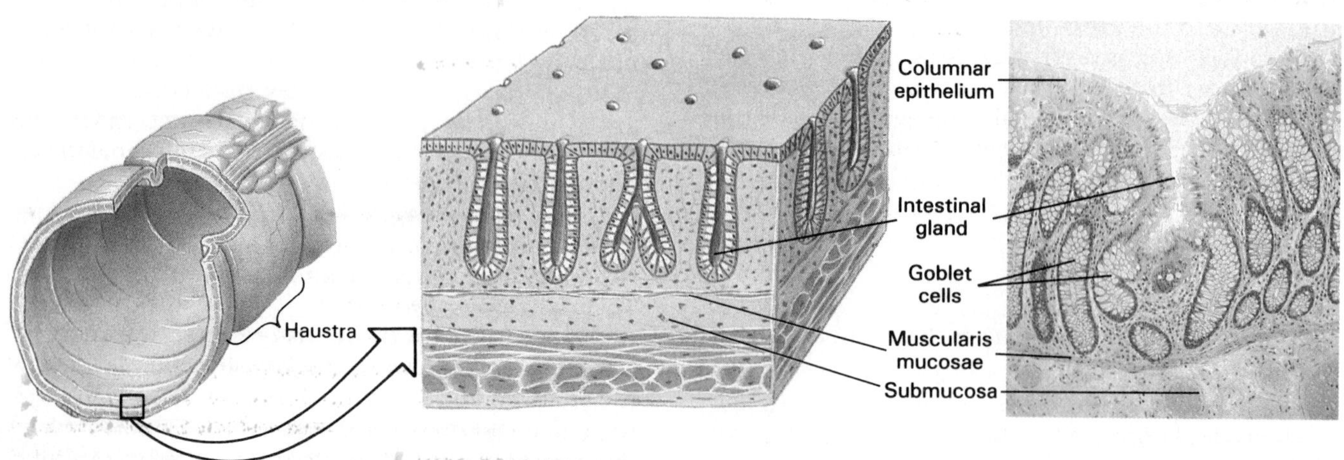

Colonic Movements Approximately 1500 mℓ of chyme enter the large intestine each day. Movement from the cecum to the transverse colon occurs very slowly, allowing hours for the thick chyme to be converted into a sludgy paste. Movement from the transverse colon through the rest of the large intestine results from powerful peristaltic contractions, called **mass movements**, that occur a few times each day. The normal stimulus is distension of the stomach and duodenum, and the commands are relayed over the intestinal nerve plexuses. The contractions force fecal materials into the rectum and produce the urge to defecate. Under unusual conditions these peristaltic contractions may be produced by distension of the colon or by colonic irritation and inflammation.

Although roughly 1500 mℓ of chyme arrive in the colon each day, only about 200 mℓ of feces are ejected. The remarkable efficiency of digestion can best be appreciated by considering the average composition of fecal wastes: 75 percent water, 5 percent bacteria, and the rest a mixture of indigestible materials, small quantities of inorganic matter, and the remains of epithelial cells.

The Rectum

The **rectum** (REK-tum) forms the last 15 cm (6 in.) of the digestive tract. The epithelium lining the proximal two-thirds of the rectum resembles that of the sigmoid colon, but the intestinal glands become progressively shorter as the exterior opening, or **anus**, approaches. Rectal organization is diagrammed in Figure 24-18c.

The last portion of the rectum, the **anorectal** (ā-nō-REK-tal) **canal**, contains small longitudinal folds, the **rectal columns**. The distal margins of the rectal columns are joined by transverse folds that mark the boundary between the columnar epithelium of the rectum and a stratified squamous epithelium similar to that found in the oral cavity. Very close to the anus the epidermis becomes keratinized and identical to the surface of the skin.

A network of veins in the lamina propria and submucosa of the anorectal canal occasionally becomes distended, producing *hemorrhoids*. [CM: *Problems with Valve Function*] The circular muscle layer of the muscularis externa in this region forms the **internal anal sphincter**. The **external anal sphincter** guards the exit of the anorectal canal. This sphincter, which consists of skeletal muscle fibers, is under voluntary control.

Defecation

The rectal chamber is usually empty except when one of those powerful peristaltic contractions forces fecal materials out of the sigmoid colon. Distension of the rectal wall then triggers the defecation reflex.

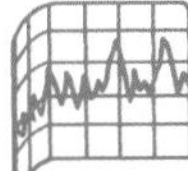

Diagnostics: Colonic Inspection and Cancer

Colon cancers are relatively common. There are approximately 120,000 cases diagnosed in the United States each year, and in 1987 there were an estimated 60,000 deaths from colon and rectal cancers. The mortality rate for these cancers remains high, and the best defense appears to be early detection and prompt treatment. Several procedures are especially useful in detecting potential problems in the colon and rectum.

The standard screening test involves checking the feces for blood. This is a simple procedure that can easily be performed on a stool (fecal) sample in the course of a routine physical. Individuals often ignore blood in the fecal products because they attribute it to "harmless" hemorrhoids; this should always be professionally verified.

If blood is detected in the feces, X-ray techniques are often used to take a closer look. In the usual procedure a large quantity of a liquid barium solution is introduced by enema. Because this solution is radiopaque, the X-rays will reveal any intestinal masses, such as a large tumor, blockages, or structural abnormalities.

Often the most precise surveys can be obtained with the aid of a flexible **colonoscope** (ko-LON-o-skōp). This instrument permits direct visual inspection of the lining of the large intestine. Variations on this procedure permit the collection of tissue samples. The colonoscope can also be used to remove small mucosal tumors, or **polyps** (POL-ips), that grow from the intestinal wall. This procedure avoids the potential complications of traditional surgery.

The defecation reflex (Figure 24-20) involves two positive feedback loops:

1. Stretch receptors in the rectal walls order a series of peristaltic contractions in the colon and rectum, moving feces toward the anus.
2. The sacral parasympathetic system, also activated by the stretch receptors, stimulates peristalsis via motor commands distributed by the pelvic nerves.

Movement of feces through the anorectal canal requires relaxation of the internal anal sphincter, but when this occurs the external sphincter automatically clamps shut. Thus the actual release of feces

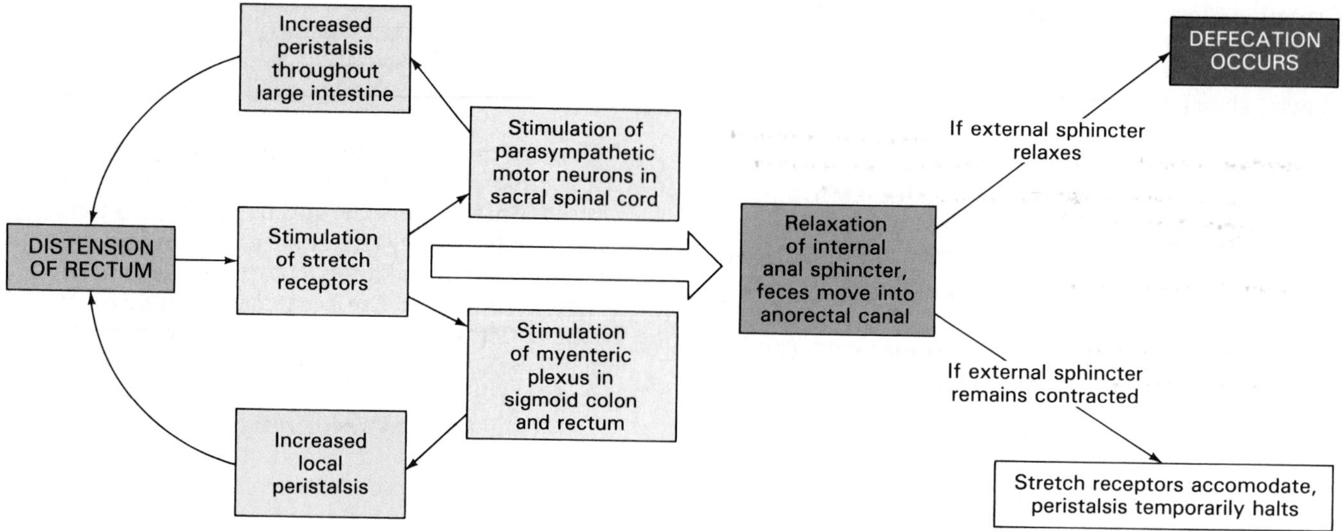

FIGURE 24-20
The Defecation Reflex

requires conscious effort to open the external sphincter voluntarily. If the commands do not arrive, the peristaltic contractions cease until additional rectal expansion triggers the defecation reflex a second time. In addition to opening the external sphincter, consciously directed activities such as tensing the abdominal muscles or making expiratory movements while closing the glottis elevate intra-abdominal pressures and help to force fecal materials out of the rectum. [**CM:** *Diarrhea and Constipation*]

The Accessory Organs

The accessory organs of the digestive system produce enzymes and buffers that are essential to normal digestive function. These organs also have vital secondary functions. The major accessory organs are the *salivary glands*, the *liver*, the *gallbladder*, and the *pancreas*.

THE SALIVARY GLANDS

Figure 24-21 details the locations and relative sizes of the salivary glands. The large parotid glands lie inferior to the zygomatic arch beneath the skin covering the lateral surface of the face. Each gland has an irregular shape, extending from the mastoid process of the temporal bone across the outer surface of the masseter muscle. The parotid duct runs anteriorly to empty into the vestibule at the level of the second upper molar. The sublingual glands are located beneath the mucous membrane of the floor of the mouth. Numerous sublingual ducts open along either side of the lingual frenulum. The submandibular glands are found in the floor of the mouth along the inner surfaces of the mandible (in the *mandibular groove*, shown in Figure 8-12). (p. 250) The submandibular ducts open into the mouth behind the teeth on either side of the lingual frenulum. (The openings of the salivary ducts are shown in Figure 24-6.)

These salivary glands produce 1.0–1.5 liters of saliva each day, with a composition of 99.4 percent water, plus an assortment of ions, buffers, waste products, metabolites, and enzymes. At mealtimes the production of large quantities of saliva lubricates the mouth and dissolves chemicals that stimulate the taste buds. A continual background level of secretion flushes the oral surfaces, and salivary immunoglobulins (IgA) and lysozymes help to control populations of oral bacteria. A reduction or elimination of salivary secretions, caused by radiation exposure, emotional distress, or other factors, triggers a bacterial population explosion in the oral cavity. This proliferation rapidly leads to recurring infections and the progressive erosion of the teeth and gums. [**CM:** *Mumps*]

Salivary Secretions

Each of the salivary glands has a distinctive cellular organization and produces saliva with slightly different properties. The parotid glands produce a thick, serous secretion containing large amounts of **salivary amylase**, also known as **alpha-amylase**. Salivary amylase begins breaking down complex carbohydrates such as starches or glycogen into smaller molecules. Saliva originating in the submandibular and sublingual salivary glands contains fewer enzymes, but larger numbers of glycoproteins called **mucins** (MŪ-sins). Mucins are responsible for the slick, lubricating properties of the mucous secretions of goblet cells and glands throughout the body. The histological ap-

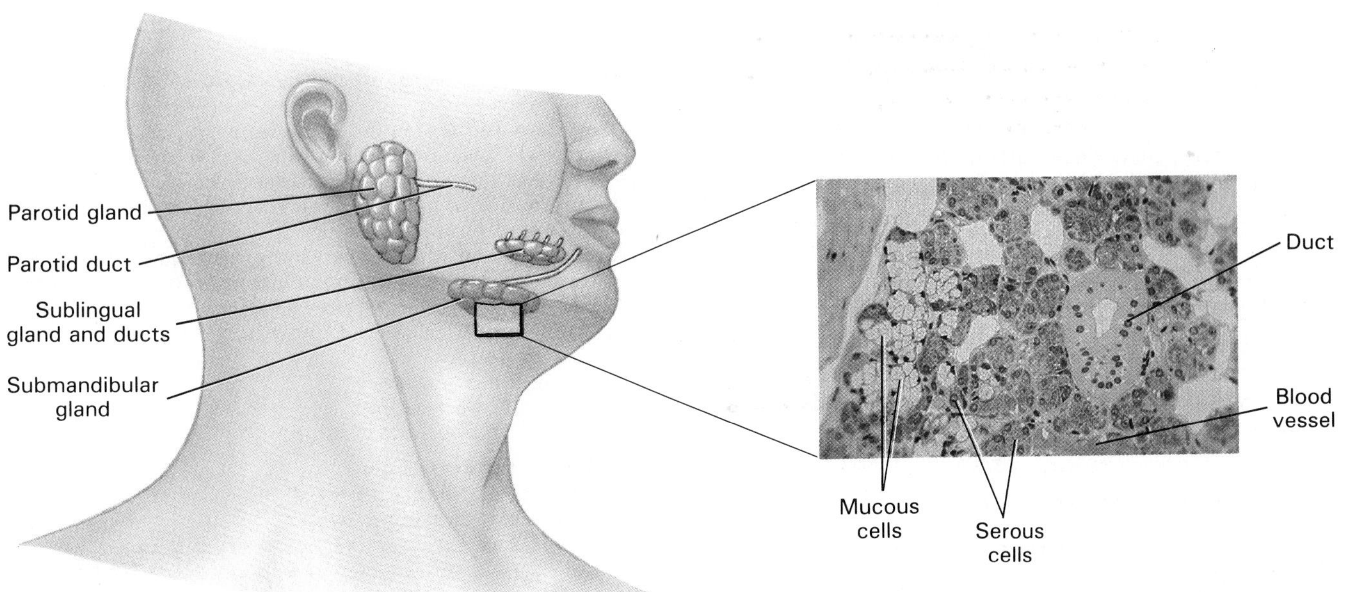

FIGURE 24-21
The Salivary Glands. Lateral view, showing the relative positions of the salivary glands and ducts on the right side of the head. For the positions of the ducts inside the oral cavity, see Figure 24-6. The submandibular gland secretes a mixture of mucins, produced by mucous cells, and enzymes, produced by serous cells. (LM, ×168)

pearance of the submandibular gland can be seen in Figure 24-21. This gland contains a mixture of serous and mucous secretory cells.

The saliva present in the mouth is a mixture of glandular secretions whose composition varies depending on the stimulus provided. At rest roughly 70 percent of the saliva entering the mouth originates within the submandibular glands, 25 percent arrives from the parotid glands, and 5 percent comes from the sublingual glands. At mealtimes, all three salivary glands elevate their rates of secretion, and salivary production may reach 7 mℓ per minute, with about 50 percent provided by the parotid glands. The increased secretory rates are accompanied by a gradual increase in pH, shifting from slightly acidic (pH 6.7) to slightly basic (pH 7.5).

Regulation of Salivary Gland Activity

Salivary secretions are usually controlled by the autonomic nervous system. Each gland receives parasympathetic and sympathetic innervation. The parasympathetic outflow originates within the **salivatory** (SAL-i-va-tō-rē) **nuclei** of the medulla and is relayed via the submandibular and otic ganglia. Any object placed within the mouth can trigger a salivary reflex by stimulating receptors monitored by the trigeminal nerve or by stimulating taste buds innervated by N VII, IX, or X. Parasympathetic stimulation accelerates secretion by all of the salivary glands, resulting in the production of large amounts of watery saliva containing abundant enzymes. The role of the sympathetic innervation remains uncertain; some evidence exists that it provokes the secretion of small amounts of very thick saliva.

The salivatory nuclei are also influenced by other brain stem nuclei as well as by the activities of higher centers. For example, chewing with an empty mouth, the smell of food, or even thinking about food will initiate an increase in salivary secretion rates. The presence of irritating stimuli in the esophagus, stomach, or intestines will also accelerate saliva production, as will the sensation of nausea. In functional terms, increased saliva production in response to unpleasant stimuli helps to reduce the magnitude of the stimulus, by dilution, a rinsing action, or by buffering strong acids or bases.

THE LIVER

The liver is the largest visceral organ, accounting for roughly 2.5 percent of the total body weight. This large, firm, reddish brown organ provides essential metabolic and synthetic services that fall into three basic categories: *metabolic regulation*, *hematological regulation*, and *bile production*.

Metabolic Regulation

The liver represents the central focus for metabolic regulation in the body. All blood leaving the absorptive areas of the digestive tract flows through the liver before reaching the general circulation. This arrangement gives the liver cells the opportunity to extract absorbed nutrients or toxins from the blood before it reaches the general systemic circulation. That reentry occurs via the hepatic veins that open into the inferior vena cava. This path also provides ample opportunity for the liver cells to monitor the circulating levels of metabolites and adjust them as necessary. Excesses are removed and stored, and deficiencies are corrected by mobilizing stored reserves or activating appropriate synthetic activities. Circulating toxins and metabolic waste products are also removed, for subsequent inactivation or excretion or both. Finally, fat-soluble vitamins (A, D, K, and E) are absorbed and stored.

Hematological Regulation

The liver is the largest blood reservoir in the body. In addition to the blood arriving over the hepatic portal vein, the liver receives about 25 percent of the cardiac output. As blood passes by, phagocytic cells in the liver remove aged or damaged red blood cells, debris, and pathogens from the circulation. Equally important, liver cells synthesize the plasma proteins that determine the osmotic concentration of the blood, transport nutrients, and establish the clotting and complement systems.

Synthesis and Secretion of Bile

Bile contains water, ions, conjugated bilirubin, and an assortment of lipids. The water and ions assist in the dilution and buffering of acids in chyme as it enters the small intestine. As noted in Chapter 19, bilirubin created during the recycling of heme from red blood cells is absorbed by the liver and excreted in the bile. Bacterial action in the intestinal tract converts the conjugated bilirubin to **urobilinogen**, a waste product that adds a yellow-brown color to the fecal materials. A small amount of urobilinogen diffuses through the intestinal lining and enters the bloodstream, to be removed and resecreted in the bile or excreted at the kidneys.

The lipids in bile are primarily derivatives of cholesterol. These steroids are collectively known as the **bile salts**. Bile salts are required for the normal digestion and absorption of fats. More than 90 percent of the bile salts entering the small intestine are reabsorbed and returned to the liver in the hepatic portal vein, to be secreted again at a later date. This recycling is known as the **enterohepatic circulation of bile**.

Other Liver Functions

To date, over 200 different functions have been assigned to the liver. (Table 24-4 contains a partial listing.) As a result, any condition that severely damages the liver represents a serious threat to life. The liver has the ability to partially regenerate after injury, but often the normal vascular pattern does not return, and liver function does not fully recover. The extensive vascularity also complicates the treatment of severe liver trauma, for the bleeding can be difficult to control and the texture of the liver makes normal suturing ineffective. Thus liver transplants are becoming increasingly common since the development of immunosuppressive drugs. Approximately 550 liver transplants were performed in the United States in 1989, with an 80 percent survival rate. (For further discussion of transplant technology, review Chapter 22.) (p. 744)

TABLE 24-4 Major Functions of the Liver

Digestive and Metabolic Functions
Synthesis and secretion of bile
Storage of glycogen and lipid reserves
Maintenance of normal blood glucose, amino acid, and fatty acid concentrations
Synthesis and interconversion of nutrient types (e.g., transamination of amino acids, or conversion of carbohydrates to lipids)
Synthesis and release of cholesterol bound to transport proteins
Inactivation of toxins
Storage of iron reserves
Storage of fat-soluble vitamins
Other Major Functions
Synthesis of plasma proteins
Synthesis of clotting factors
Synthesis of the inactive hormone, angiotensinogen
Phagocytosis of damaged red blood cells (by Kupffer cells)
Blood storage (major contributor to venous reserve)
Absorption and breakdown of circulating hormones (insulin, epinephrine, insulin) and immunoglobulins
Absorption and inactivation of lipid-soluble drugs

Anatomy of the Liver

Figure 24-22a,b indicates the position and anatomical landmarks of the liver. The liver is wrapped in a fibrous capsule covered by the visceral peritoneum. On the anterior surface a tough connective tissue fold, the **falciform** (FAL-si-fōrm) **ligament**, marks the division between the **left lobe** and **right lobe** of the liver. A thickening in the posterior margin of the falciform ligament is the **round ligament**, a fibrous band that marks the path of the fetal umbilical vein. (See the discussion of the development of the circulatory system in Chapter 21.) (p. 670)

The shape of the liver conforms to its surroundings. The **anterior surface**, or *parietal surface*, follows the smooth curve of the body wall (Figure 24-22b). The **posterior surface**, or *visceral surface*, (Figure 24-22c), bears the impressions of the stomach, small intestine, right kidney, and large intestine. The impression left by the inferior vena cava marks the

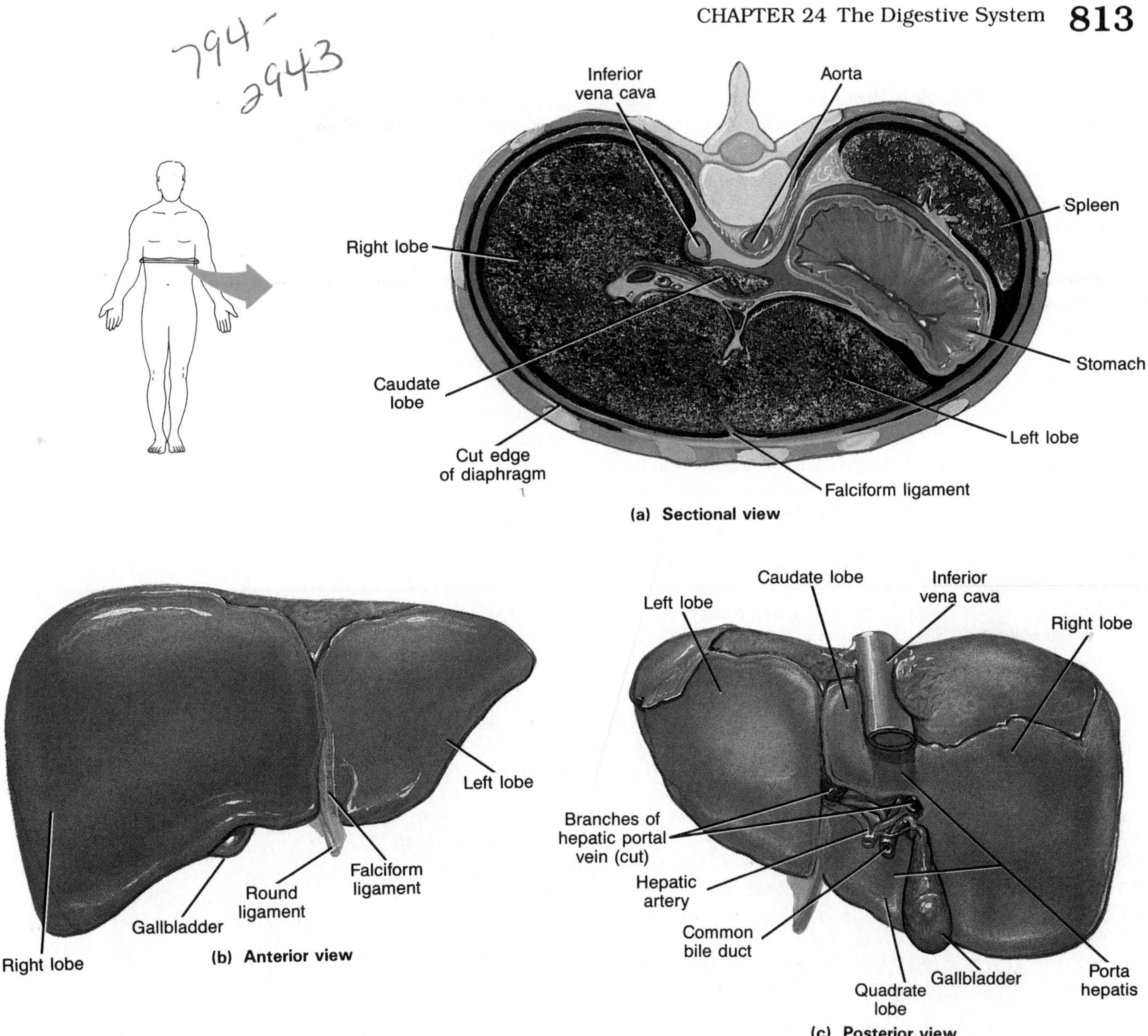

FIGURE 24-22
Anatomy of the Liver. (a) Sectional view through the upper abdomen, showing the position of the liver relative to other visceral organs. (b) Anatomical landmarks on the anterior surface of the liver. (c) The posterior surface of the liver.

division between the left lobe and the small **caudate** (KAW-dāt) **lobe**. Below the caudate lies the **quadrate lobe** sandwiched between the left lobe and the gallbladder. The gallbladder is a muscular sac that stores and concentrates bile prior to its excretion into the small intestine. A **cystic duct** leads from the gallbladder toward the **hilus** of the liver, at the center of the posterior surface. On the way it meets the **hepatic duct**, which collects bile from the liver lobes. Another tough fold of mesentery fuses with the liver capsule at the hilus. This fold encloses two vessels delivering blood to the liver (the hepatic portal vein and hepatic artery). It also contains the **common bile duct** created by the fusion of the hepatic and cystic ducts. The bile duct, hepatic portal vein, and hepatic artery are bound together by connective tissue. The entire complex is called the **porta hepatis**, which translates as the "doorway to the liver."

The shapes of the individual lobes are highly variable. As long as the blood vessels and ducts are unconstricted the differences do not affect the functional capabilities of the liver. An interesting anatomical consequence of the Victorian era was the appearance of "corset livers." Fashionable women were expected to wear corsets, complicated undergarments that resembled a cross between a girdle and an instrument of torture. The goal was to take a normal female form and, using a system of ropes and lashings, produce an "ideal" figure with a 20-inch waistline. Women who led what passed for normal lives at that time were found on autopsy to have livers marked by transverse fissures so deep that they nearly created superior and inferior lobes on either side. [CM: *Hepatitis*]

Hepatic Circulation The circulation to the liver was detailed in Chapter 21 (For a quick review refer to Figures 21-13 and 21-18). (pp. 678, 685) The hepatic artery, a branch of the celiac trunk, brings arterial blood to the liver. Roughly one-third of the normal hepatic blood flow arrives via the hepatic artery, and the rest is provided by the hepatic portal vein that forms through the fusion of the inferior mesenteric, superior mesenteric, and splenic veins. All of the blood leaving the absorptive areas of the digestive tract flows through the liver before reaching the general circulation. That reentry occurs via the hepatic veins that open into the inferior vena cava.

Histological Organization The basic functional unit of the liver is the **liver lobule**. There are approximately 100,000 lobules in the liver, and the orientation of adjacent lobules usually varies to some degree. As a result, a single histological section seldom reveals all of the details of lobular organization. Important features revealed by light and electron microscopic examination are diagrammed in Figure 24-23.

Liver cells, or **hepatocytes** (he-PAT-ō-sīts), within a lobule are arranged into a series of irregular plates like the spokes of a wheel. This arrangement can be seen in the lobule on the left side of Figure 24-23a and in the accompanying photomicrograph. The plates are only one-cell thick, and the exposed hepatocyte surfaces are covered with short microvilli.

The lobule on the right side of Figure 24-23a diagrams the circulatory supply to the lobule. Sinusoids between adjacent plates empty into the central

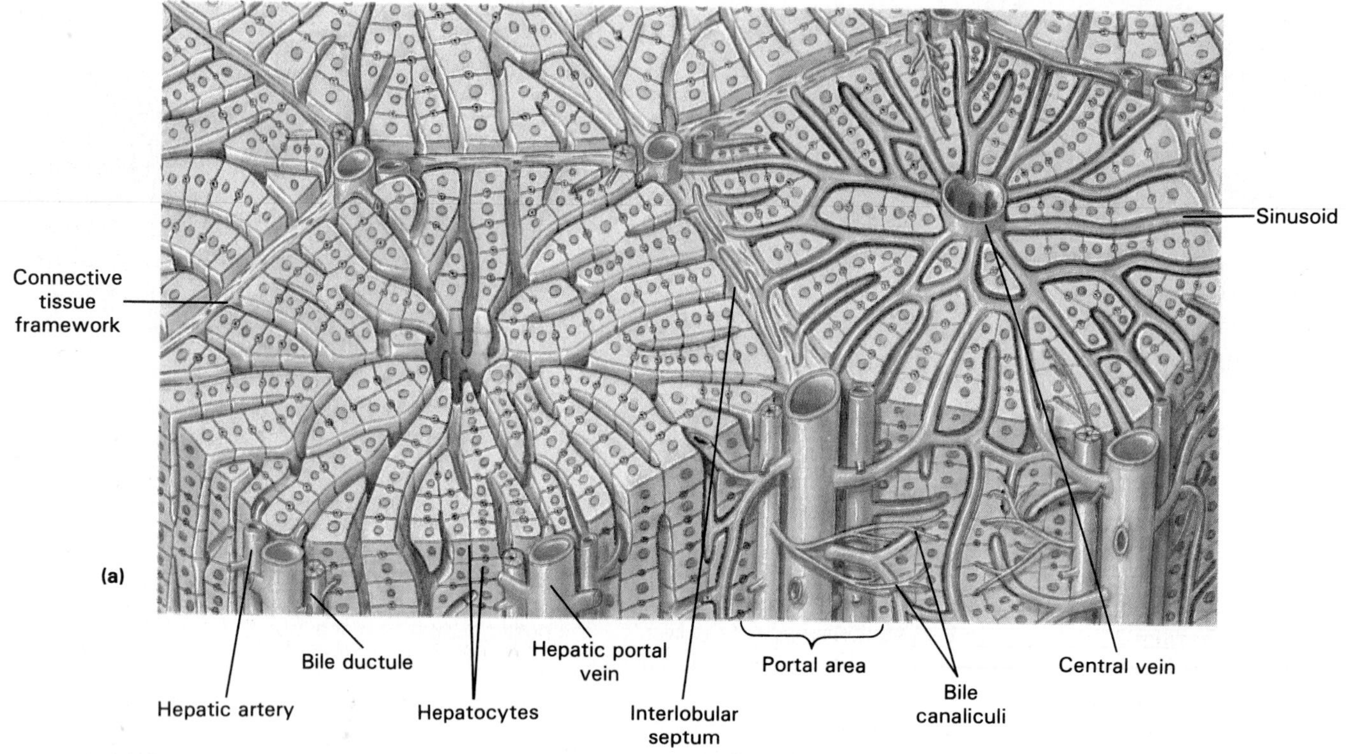

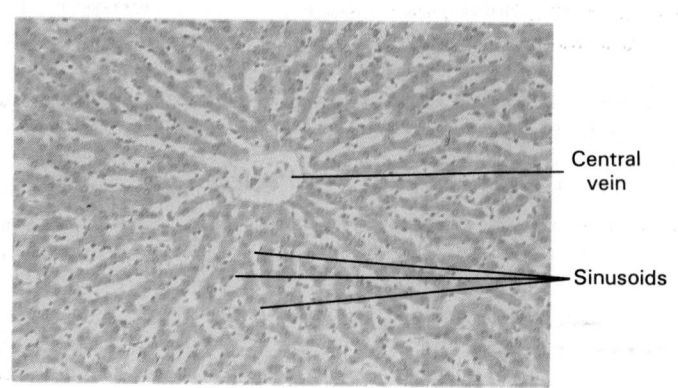

FIGURE 24-23
Liver Histology.
(a) Diagrammatic view of lobular organization. (b) Light micrograph showing a typical liver lobule (LM, ×53)

Clinical Comment: Portal Hypertension

Pressures in the hepatic portal system are usually low, averaging 10 mm Hg or less. This pressure can change markedly if blood flow through the liver becomes restricted. A thrombus in the hepatic or portal veins or degenerative changes in the structure of the liver can restrict blood flow and cause a gradual increase in portal pressures, a condition termed **portal hypertension**. As pressures rise, small peripheral veins and capillaries in the portal system become distended and are likely to rupture, and intestinal bleeding becomes a problem. At pressures 50–100 percent above normal, massive edema occurs across the capillary walls, creating a potentially fatal decline in blood volume.

In cases of portal hypertension resulting from liver damage, the rise in venous pressure is often accompanied by a leakage of fluid into the peritoneal cavity across the serosal surfaces of the liver and viscera. This condition, introduced in Chapter 5, is called **ascites** (a-SĪ-tēz). (p. 155) In addition to causing higher than normal portal pressures, the damaged liver cannot maintain plasma protein concentrations, so the osmotic pressure of the blood decreases and fluid losses accelerate. Ascites can also result from the generalized increase in venous pressures that accompanies congestive heart failure.

vein. (You may want to review the discussion of sinusoids in Chapter 21.) (p. 666) The walls of the sinusoids contain large openings that allow materials to pass out of the circulation and into the spaces surrounding the hepatocytes. In addition to typical endothelial cells, the sinusoidal lining includes a large number of phagocytic **Kupffer** (KOOP-fer) **cells**. Kupffer cells are part of the monocyte-macrophage system, and they engulf pathogens, cell debris, and damaged blood cells.

Blood enters the sinusoids from small branches of the portal vein and hepatic artery. A typical lobule has a hexagonal shape in cross section. There are six **portal areas**, one at each corner. A portal area contains: (1) a branch of the portal vein, (2) a branch of the hepatic artery, and (3) a tributary of the bile duct. Branches from the arteries and veins deliver blood to the sinusoids of adjacent lobules. As blood flows through the sinusoids, hepatocytes absorb and secrete materials into the bloodstream across their exposed surfaces. Blood then leaves the sinusoids and enters the **central vein** of the lobule. The central veins ultimately merge to form the hepatic veins that empty into the inferior vena cava. [**CM:** *Cirrhosis*]

Bile is secreted into a network of narrow channels between the opposing membranes of adjacent liver cells. These passageways, which resemble the interfacial canals between epithelial cells (Figure 5-2), are called **bile canaliculi**. (p. 127) The bile canaliculi extend outward, away from the central vein. These eventually connect with fine **bile ductules** (DUK-tūlz) that carry bile to the nearest portal area.

The small bile ducts from the various lobes unite to form the **hepatic duct**. At the porta hepatis, the hepatic duct and the cystic duct combine to create the common bile duct, which passes within the mesentery linking the liver to the stomach, turns, and penetrates the wall of the small intestine, where it merges with the **pancreatic duct** from the pancreas. As indicated in Figure 24-15, a muscular sphincter (the *sphincter of Oddi*) of variable size surrounds the lumen of the common bile duct and ampulla. Contraction of this sphincter seals off the passageway and prevents bile from entering the small intestine.

Liver cells produce roughly 1 liter of bile each day, but the sphincter remains closed except at mealtimes. Over the interim, when bile cannot flow along the common bile duct, it can instead enter the cystic duct for storage within the expandable *gallbladder*. [**CM:** *Analysis of Liver Structure and Function*]

THE GALLBLADDER

The gallbladder (Figure 24-24) is a muscular organ with the shape of a pear. It has two major functions, *bile storage* and *bile modification*. When filled to capacity, the gallbladder contains 40–70 mℓ of bile. It is lined by a columnar epithelium that forms complex folds in the empty gallbladder, but that flattens when the gallbladder is filled with bile. As bile remains in the gallbladder its composition gradually changes. Water is absorbed, and the bile salts and other components of bile become increasingly concentrated. The degree of concentration can be seen in Table 24-5, which compares the bile secreted by the liver with that ejected by the gallbladder. [**CM:** *Problems with Bile Storage and Secretion*]

The Physiological Role of Bile The arrival of the intestinal hormone cholecystokinin, or CCK, in the circulating blood provides the stimulus for the excretion of bile. Cholecystokinin relaxes the biliary sphincter, and contractions of the walls of the gallbladder then push bile into the small intestine. This hormone is released whenever chyme enters the intestine, but the amount secreted increases if the chyme contains large amounts of fat.

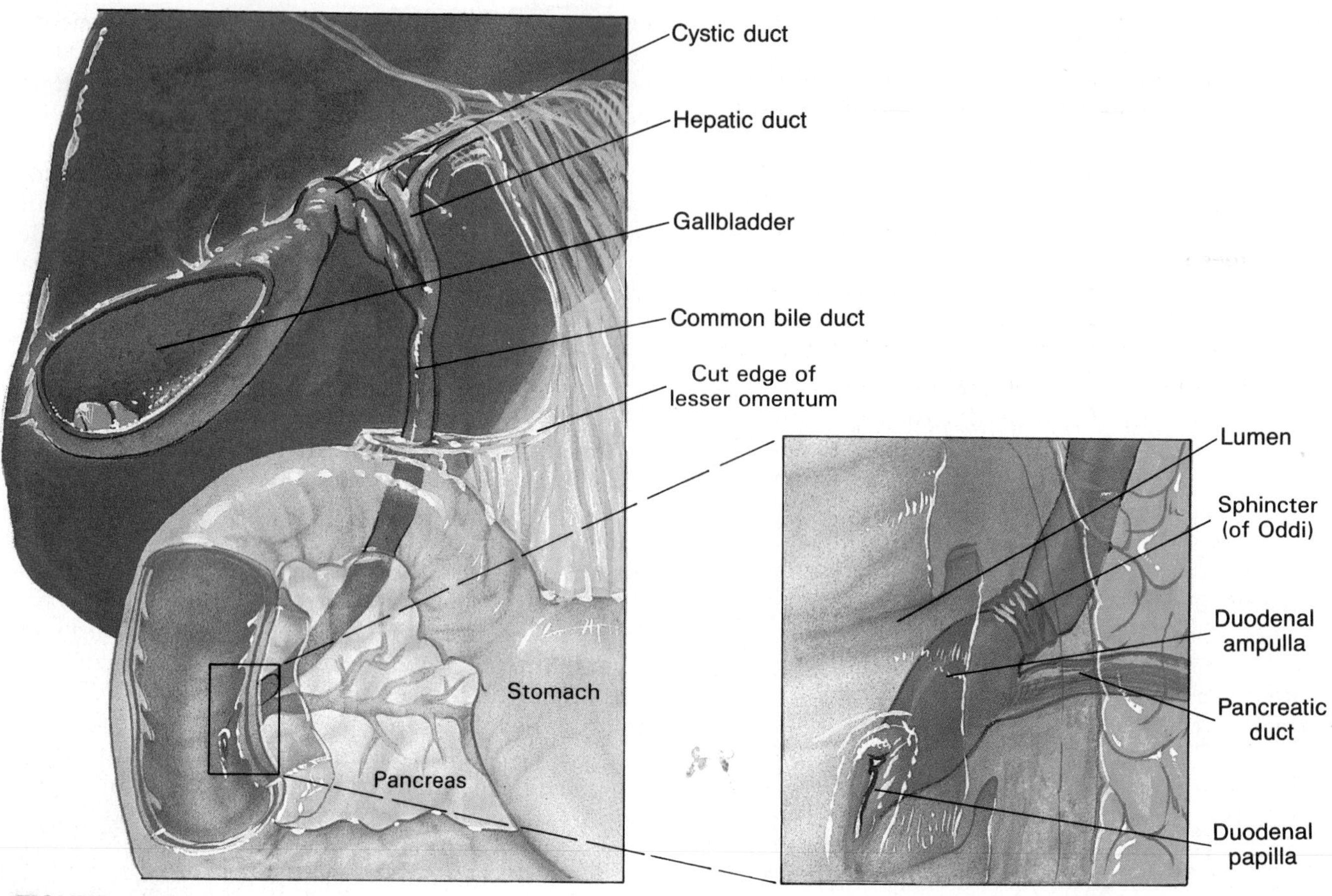

FIGURE 24-24
The Gallbladder. A view of the inferior surface of the liver, showing the position of the gallbladder and ducts that transport bile from the liver to the gallbladder and duodenum. A portion of the lesser omentum has been cut away to make it easier to see the relationship between the common bile duct, the hepatic duct, and the cystic duct.

TABLE 24-5 The Modification of Bile by the Gallbladder

Characteristic	*Bile Secreted by the Liver*	*Bile Ejected by the Gallbladder*
pH	7.15	7.3
Specific gravity	1.008	1.026
Water content (%)	97.5	86
Bile salts (g/dℓ)	0.9	9
Bilirubin (g/dℓ)	0.06	0.4
Cholesterol (g/dℓ)	0.1	0.3–0.9
Electrolytes (mEq/ℓ)		
Na^+	148	209
Ca^{2+}	4	25
Cl^-	110	30

See Appendix V for sources.

Most dietary lipids are not water-soluble. Mechanical processing along the digestive tract creates large drops containing a variety of different lipids. These drops are much too massive to be attacked by digestive enzymes. Bile salts break the drops apart, a process called **emulsification** (ē-mul-si-fi-KĀ-shun). Emulsification creates tiny *emulsion droplets* with a superficial coating of bile salts. The formation of tiny droplets increases the surface area available for enzymatic attack. In addition, the layer of bile salts facilitates interaction between the lipids and lipid-digesting enzymes provided by the pancreas. (We will return to the mechanism of lipid digestion in a later section.)

√ **A narrowing (stenosis) of the ileocaecal valve would interfere with the movement of chyme between what two organs?**

√ **A blockage of the ducts of the parotid glands would interfere with the digestion of which nutrient in the mouth?**

√ **How would a decrease in the amount of bile salts in bile affect the digestion and absorption of fat?**

THE PANCREAS

The pancreas has two distinct functions, one endocrine and the other exocrine. Endocrine cells of the pancreatic islets secrete insulin and glucagon into the bloodstream. These hormones and their actions were described in Chapter 18. (p. 592) The exocrine cells secrete a mixture of water, ions, and digestive enzymes into the small intestine. The fluids and ions assist in diluting and buffering the acids in the chyme. Pancreatic enzymes do most of the digestive work in the small intestine, breaking down ingested materials into small molecules suitable for absorption.

Pancreatic enzymes are broadly classified according to their intended targets. **Lipases** (LĪ-pā-zez) attack lipids, **carbohydrases** (kar-bō-HĪ-drā-zez) digest sugars and starches, and **proteolytic** (prō-tē-ō-LIT-ik) **enzymes** break proteins apart. Proteolytic enzymes include **proteinases** and **peptidases**; proteinases break apart large protein complexes, whereas peptidases break small peptide chains into individual amino acids.

Gross Anatomy of the Pancreas

The pancreas lies behind the stomach, extending laterally from the initial segment of the small intestine toward the spleen. It is an elongate, pinkish grey organ with a length of approximately 15 cm (6 in.) and a weight of around 80 g (3 oz). The broad **head** of the pancreas lies within the loop formed by the small intestine as it leaves the pylorus. The slender **body** extends transversely toward the spleen, and the **tail** is short and bluntly rounded. These landmarks are shown in Figure 24-25a.

The surface of the pancreas has a knobbly texture, and the tissue is soft and easily torn. The thin capsule permits a view of the individual lobules and the associated blood vessels and excretory ducts. Although a small **accessory duct** may be seen within the pancreatic head, only the relatively large **pancreatic duct** penetrates the wall of the duodenum in association with the common bile duct. Branches of the splenic, superior mesenteric, and hepatic arteries deliver blood to the pancreas (see Figure 21-18). (p. 685)

Histological Organization

Partitions of connective tissue divide the pancreatic tissue into distinct lobules, as seen in Figure 24-25b. The blood vessels and tributaries of the pancreatic ducts are found within these septa. Within each lobule, the ducts branch repeatedly before ending in blind pockets, the **pancreatic acini** (AS-i-nī), lined by a simple cuboidal epithelium. Pancreatic islets are scattered between the acini, but these endocrine tissues account for only around 1 percent of the cellular population of the pancreas. [**CM:** *Pancreatitis*]

Control of Pancreatic Secretion

The pancreatic acini produce a watery **pancreatic juice**. Secretion of pancreatic juice primarily occurs in response to hormonal instructions from the duodenum. When acid chyme arrives in the small intestine, secretin is released. This hormone triggers the pancreatic production of an alkaline fluid with a pH of 7.5–8.8. Among its other components, this secretion contains buffers, primarily sodium bicarbonate, that help bring the pH of the chyme under control. A different intestinal hormone, cholecystokinin, controls the production and secretion of pancreatic enzymes. The specific enzymes involved are **alpha-amylase**, almost identical to that produced by the salivary glands, **pancreatic lipase**, **nucleases** that break down nucleic acids, and four proteolytic enzymes.

Proteinases account for around 70 percent of the total pancreatic enzyme production. These enzymes are **trypsin** (TRIP-sin), **chymotrypsin** (kī-mō-TRIP-sin), **carboxypeptidase** (kar-bok-sē-PEP-ti-dāz), and **elastase** (ē-LAS-tāz). Each enzyme attacks peptide bonds with slightly different characteristics, but together they succeed in shattering complex proteins into a mixture of dipeptides, tripeptides, and amino acids. These enzymes are quite powerful, and the pancreatic cells protect themselves by secreting them as inactive **proenzymes**.[1] The proenzymes involved are **trypsinogen**, **chymotrypsinogen**, **procarboxypeptidase**, and **proelastase**. Enterokinase, an enzyme introduced in the discussion of the small intestine, triggers the conversion of trypsinogen to trypsin, and trypsin activates the other proenzymes. This conversion occurs in the duodenum.

The pancreatic duct fuses with the bile duct before emptying into the small intestine. The end of the pancreatic duct is not guarded by a sphincter, and when pancreatic secretion accelerates, so does the flow of pancreatic juice into the small intestine.

Digestion and Absorption

A typical meal contains a mixture of carbohydrates, proteins, and lipids, water, electrolytes, and vitamins. The digestive system handles each of these components differently. Large organic molecules must be broken down through digestion before absorption can occur. Water, electrolytes, and vitamins can be

[1] Examples of proenzymes discussed in earlier chapters include angiotensinogen, plasminogen, and many of the clotting factors and enzymes of the complement system.

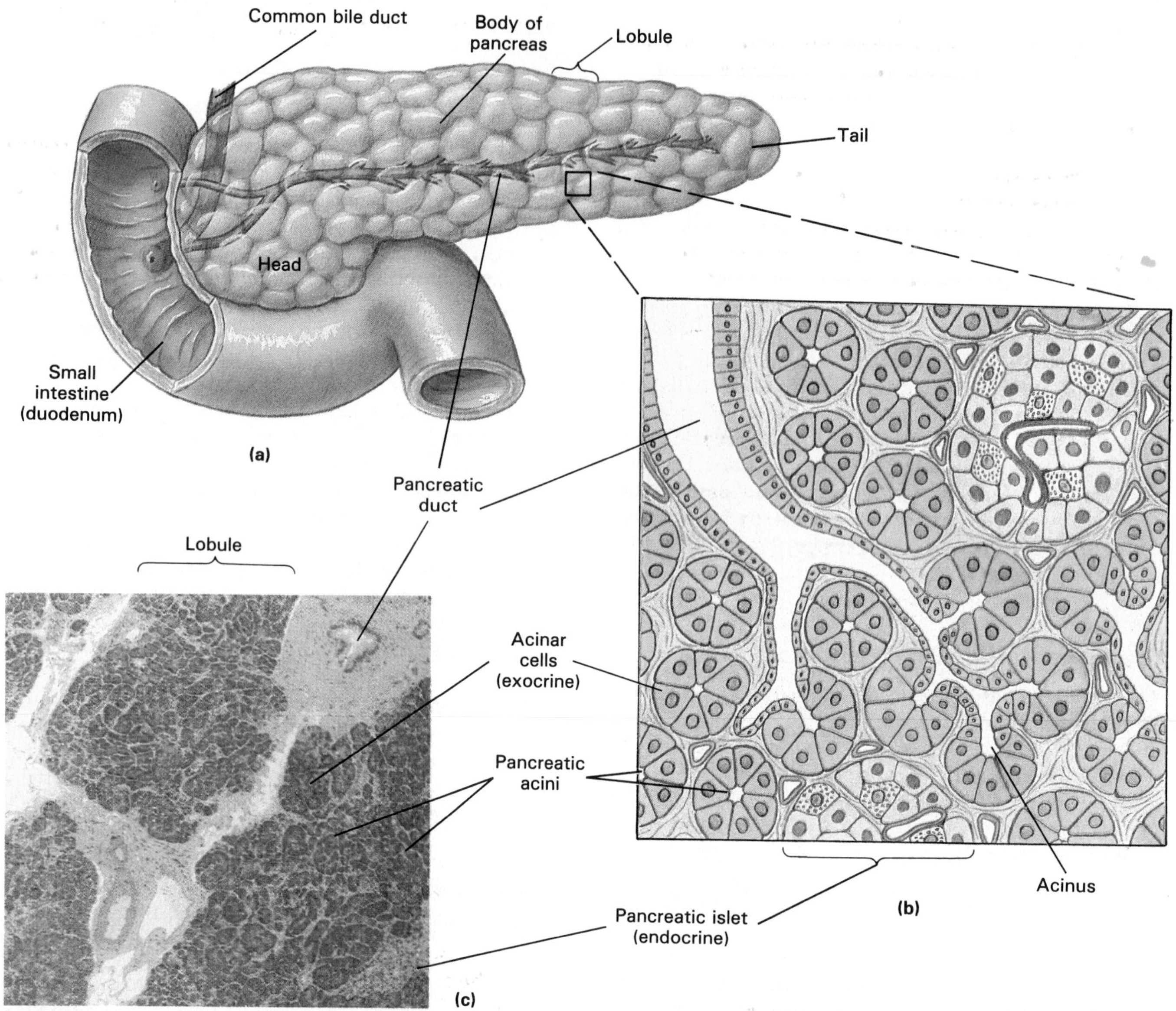

FIGURE 24-25
The Pancreas. (a) Gross anatomy of the pancreas. The head of the pancreas is tucked into a curve of the duodenum that begins at the pylorus of the stomach. (b) Diagrammatic view of the cellular organization of the pancreas, showing exocrine and endocrine regions. (c) A low-power light micrograph of the pancreas. This image shows the connective tissue septae that separate pancreatic lobules, the pancreatic duct, and exocrine and endocrine tissues. (LM, ×52) For other views of pancreatic structures, see Figure 18-18.

absorbed without preliminary processing, but special transport mechanisms are often involved.

THE PROCESSING AND ABSORPTION OF NUTRIENTS

Food contains large organic molecules, many of them insoluble. The digestive system first breaks down the physical structure of the ingested material and then proceeds to disassemble the component molecules into smaller fragments. This disassembly eliminates any antigenic properties, making the fragments suitable for absorption. The molecules released into the bloodstream will be used by the body to generate ATP and to synthesize its own carbohydrates, proteins, and lipids. This section will focus on the mechanics of digestion and absorption, while the fate of the compounds in the body will be considered in Chapter 25.

Ingested materials are usually complex chains of simpler molecules. In a typical dietary carbohydrate the basic molecules are simple sugars. In a protein, the building blocks are amino acids, and in lipids they are usually fatty acids. Digestive enzymes break the bonds between the component molecules in a

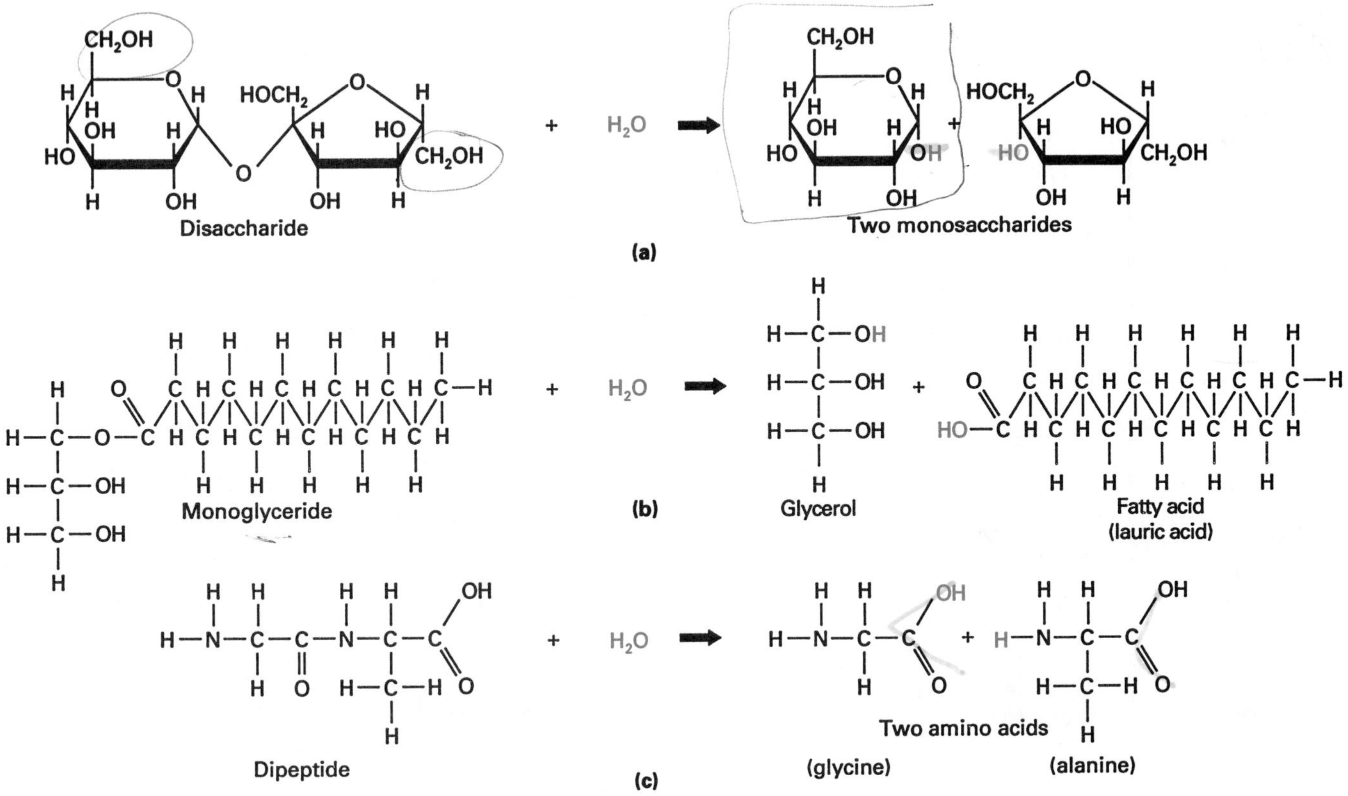

Disaccharide

H_2O

Two monosaccharides

(a)

Monoglyceride

Glycerol

Fatty acid (lauric acid)

(b)

Dipeptide

Two amino acids

(glycine)

(alanine)

(c)

FIGURE 24-26
Chemical Hydrolysis and Digestion. (a) Hydrolysis of a disaccharide into simple sugars. (b) Hydrolysis of a monoglyceride into glycerol and a fatty acid. (c) Hydrolysis of a depeptide into a pair of amino acids.

process called *hydrolysis*. (The hydrolysis of carbohydrates, lipids, and proteins was detailed in Chapter 2.) ∞ (p. 43) When considering the hydrolysis of carbohydrates or lipids, the reaction sequence can usually be described as:

$$X—O—Y + H_2O \rightarrow X—OH + HO—Y$$

In this reaction, two component molecules (X and Y) are bound together by their connection to a shared oxygen atom. In a disaccharide the two molecules would be simple sugars; in a monoglyceride one would be a fatty acid and the other a molecule of glycerol. Examples have been diagrammed in Figure 24-26a,b. In each case, hydrolysis splits the connection, creating two separate molecules bearing hydroxyl (OH^-) groups.

As indicated in Figure 24-26c, a peptide bond between two amino acids does not involve a shared oxygen atom. Such reactions can be summarized:

$$X—Y + H_2O \rightarrow X—OH + H—Y$$

The major classes of digestive enzymes differ in respect to their specific targets. Carbohydrases break the bonds between sugars, proteinases split the linkages between amino acids, and lipases separate the fatty acids from glycerides. Specific enzymes within each class may be even more selective, breaking bonds involving specific molecular participants. For example, a carbohydrase might ignore all bonds except those connecting two glucose molecules.

Figure 24-27 summarizes information concerning the digestive fates of carbohydrates, lipids, and proteins. Table 24-6 on page 821 reviews the major digestive enzymes and their functions. We will now take a closer look at the digestion and absorption of carbohydrates, lipids, and proteins.

Carbohydrate Digestion and Absorption

Carbohydrate digestion begins in the mouth during mastication, through the action of alpha-amylase from the salivary glands. Amylase breaks down complex carbohydrates into smaller fragments, producing a mixture primarily composed of disaccharides (two sugars) and trisaccharides (three sugars). Salivary amylase continues to digest the starches and glycogen in the meal for an hour or two before the stomach acids render it inactive. In the duodenum, the remaining complex carbohydrates are broken down through the action of alpha-amylase produced by the pancreas. Any disaccharides or trisaccharides already present in the food are completely ignored by the alpha-amylase produced by the salivary glands and pancreas, and additional hydrolysis does not occur until these molecules contact the intestinal mucosa.

Epithelial Processing and Absorption Prior to absorption, disaccharides and trisaccharides are fragmented into simple sugars by enzymes found on the surfaces of the intestinal microvilli. **Maltase** splits bonds between the two glucose molecules of the disaccharide **maltose**. **Sucrase** breaks the disaccharide **sucrose** into glucose and fructose, another six-carbon sugar. **Lactase** releases a molecule of glucose and

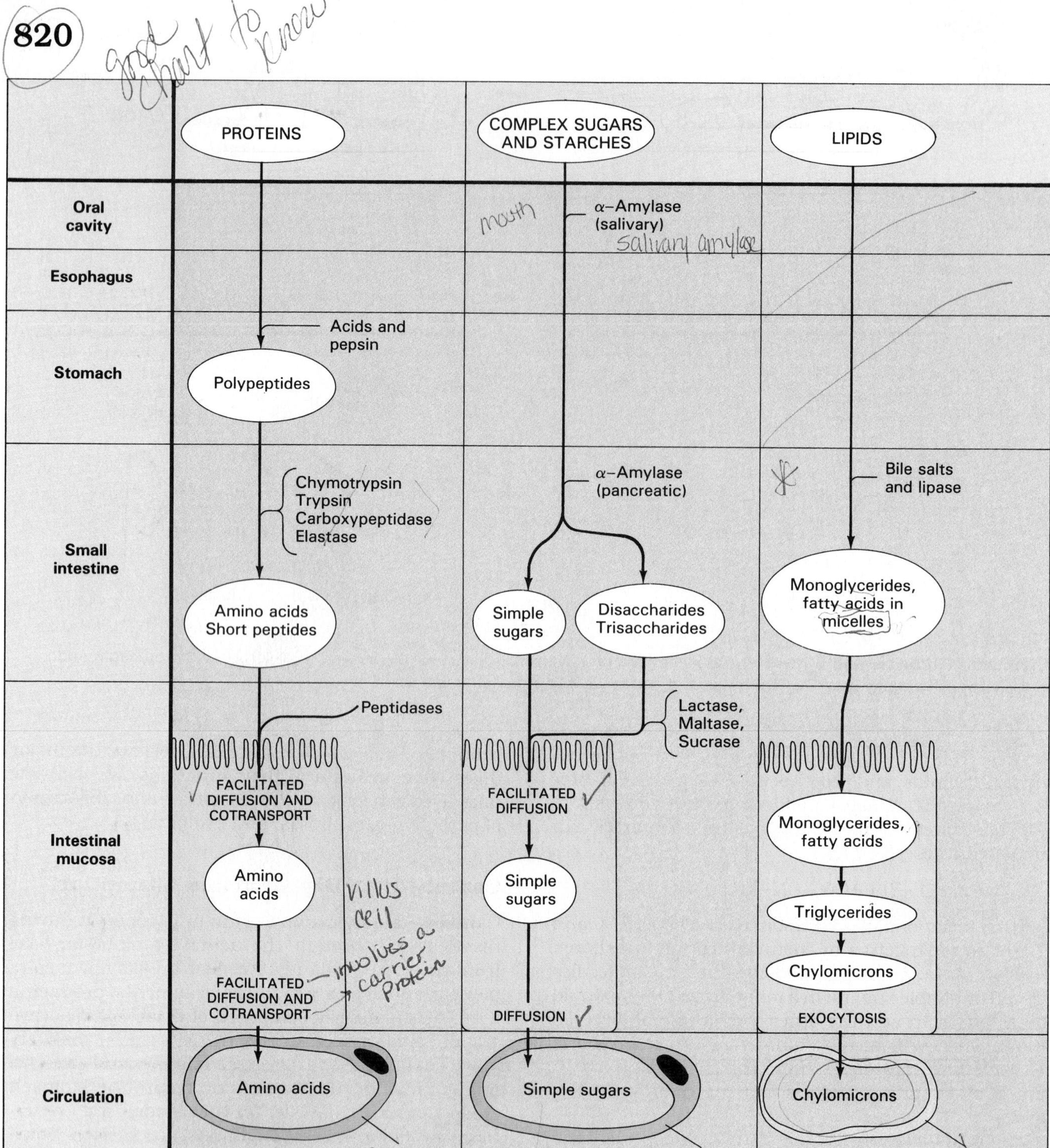

FIGURE 24-27
A Summary of the Chemical Events in Digestion

one of galactose from the hydrolysis of the disaccharide **lactose**.

The intestinal epithelium then absorbs the simple sugars through *facilitated diffusion* and **cotransport** mechanisms. Facilitated diffusion, which involves a carrier protein, was detailed in Chapter 3 (see Figure 3-23). (p. 84) Cotransport also involves a carrier protein, but there are three major differences between cotransport and facilitated diffusion:

1. Cotransport moves more than one molecule or ion through the membrane at the same time. The transported materials move in the same direction, which is down the concentration gra-

■ TABLE 24-6 Digestive Enzymes and Their Functions

Enzyme	*Source*	*Optimal pH*	*Target and Action*	*Products*	*Remarks*
Alpha-amylase	Salivary glands, pancreas	6.7–7.5	Breaks bonds between carbohydrate molecules	Disaccharides and trisaccharides	
Pepsin	Chief cells of stomach	1.5–2.0	Breaks bonds between amino acids in proteins	Short-chain polypeptides	Secreted as proenzyme, pepsinogen; activated by H^+ in stomach acid
Rennin			Coagulates milk proteins		Secreted only by stomachs of infants
Trypsin	Pancreas	7–8	Proteins, polypeptides	Short-chain peptides	Secreted as proenzyme, trypsinogen; activates other pancreatic proteinases
Chymotrypsin	Pancreas	7–8	Proteins, polypeptides	Short-chain peptides	Secreted as proenzyme, chymotrypsinogen
Carboxypeptidase	Pancreas	7–8	Proteins, polypeptides	Short-chain peptides and amino acids	Secreted as proenzyme, procarboxypeptidase
Elastase	Pancreas	7–8	Elastin	Short-chain peptides	Secreted as proenzyme, proelastase
Lipase	Pancreas	7–8	Triglycerides	Fatty acids and monoglycerides	Bile salts must be present
Nuclease	Pancreas	7–8	Nucleic acids	Nitrogen bases and simple sugars	Includes ribonuclease for RNA and deoxyribonuclease for DNA
Enterokinase	Mucosal cells of small intestine	7–8	Trypsinogen (proenzyme)	Trypsin	Reaches lumen through disintegration of shed epithelial cells
Maltase, sucrase, lactase	Mucosal cells of small intestine	7–8	Maltose, sucrose, lactose	Monosaccharides	Found in membrane surface of microvilli
Peptidase	Mucosal cells of small intestine	7–8	Dipeptides, tripeptides	Amino acids	Found in membrane surface of microvilli

dient for at least one of the transported substances.

2. Cotransport *by itself* does not require ATP. However, the cell often expends ATP to preserve homeostasis when cotransport is underway (see below).
3. Cotransport can occur despite an opposing concentration gradient for one of the transported substances. For example, cells lining the small intestine will continue to absorb glucose when glucose concentrations inside the cells are much higher than they are in the intestinal contents.

Figure 24-28a diagrams the cotransport system responsible for the uptake of glucose across the lining of the intestine. Extracellular sodium ion concentrations are relatively high, and membrane channels normally allow very little sodium to diffuse into the cell. However, sodium ions can reach the cytoplasm with the aid of a carrier protein. This is a passive process that resembles facilitated diffusion. The important distinction is that a glucose molecule must be attached to the carrier protein before it will carry the sodium ion to the interior of the cell. Once at the inner surface of the membrane, *both* the sodium ion *and* the glucose molecule are released.

Glucose cotransport is an example of **sodium-linked cotransport**. Comparable cotransport mechanisms exist for other simple sugars and for some amino acids. Although these mechanisms deliver

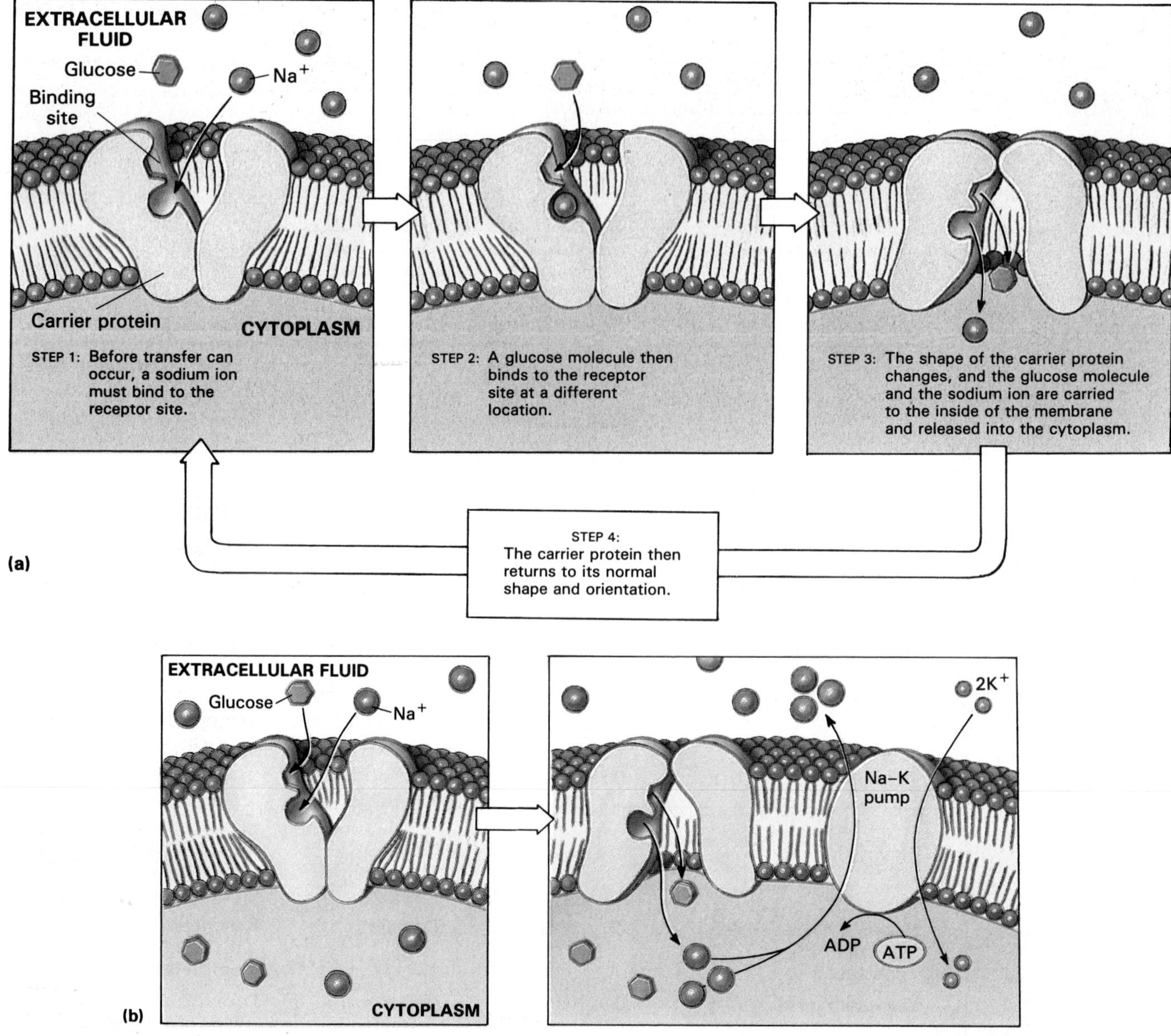
EXTRACELLULAR FLUID
Glucose
Na+
Binding site
Carrier protein
CYTOPLASM
STEP 1: Before transfer can occur, a sodium ion must bind to the receptor site.
STEP 2: A glucose molecule then binds to the receptor site at a different location.
STEP 3: The shape of the carrier protein changes, and the glucose molecule and the sodium ion are carried to the inside of the membrane and released into the cytoplasm.
STEP 4: The carrier protein then returns to its normal shape and orientation.
(a)
EXTRACELLULAR FLUID
Glucose
Na+
CYTOPLASM
(b)
2K+
Na–K pump
ADP
ATP

FIGURE 24-28
Sodium-Linked Cotransport of Glucose. (a) The steps involved with cotransport resemble those of facilitated diffusion, but the carrier protein has two receptor sites rather than one. Transport will occur only if both sites are occupied. (b) ATP is not required for cotransport, but the cell has to expend energy to remove the sodium ions. It "costs" 1 ATP for every three glucose molecules obtained through sodium-linked cotransport.

valuable nutrients to the cytoplasm, they also bring sodium ions. Those sodium ions must be ejected by the sodium-potassium exchange pump. As a result, it costs the cell one ATP for every three glucose or amino acid molecules transported into the cytoplasm. This relationship is diagrammed in Figure 24-28b.

Epithelial Transport The simple sugars entering the cell diffuse through the cytoplasm and across the basement membrane to enter the interstitial fluid. Further distribution occurs by diffusion into the intestinal capillaries for delivery to the hepatic portal vein.

The Fate of Indigestible Carbohydrates Indigestible carbohydrates, such as cellulose, are ignored by the intestinal enzymes and continue unaltered toward the large intestine. These complex polysaccharides provide a reliable nutrient source for colonic bacteria. Ordinarily the amount of suitable material available to the bacteria is relatively limited, and their metabolic activities are responsible for the small quantities of intestinal gas, or **flatus**, found in the large intestine. Much of the gas produced is absorbed through the intestinal walls, and the rest is passed with the fecal wastes. Consumption of meals containing large numbers of indigestible proteins or carbohydrates provides a nutritional bonanza for these bacteria. This stimulates a proportionate increase in gas production, leading to colonic distension, cramps, and the frequent discharge of intestinal gases. Beans contain a high concentration of indigestible poly-

Clinical Comment: Lactose Intolerance

Lactose is the primary carbohydrate in milk, so the enzyme lactase provides essential services throughout infancy and early childhood. The intestinal mucosa often stops producing lactase by adolescence, in which case the individual becomes **lactose intolerant**. This condition poses more of a problem than would be expected if the only important outcome were the inability to make use of this particular disaccharide. The clinical symptoms result because undigested lactose provides a particularly stimulating energy source for the bacterial inhabitants of the colon. Increased gas generation, cramps, and diarrhea often bring extreme discomfort and distress after the individual drinks a single glass of milk or eats small amounts of any other dairy product.

There appears to be a genetic basis for lactose intolerance, and in certain populations lactase production continues throughout adulthood. Only around 15 percent of Caucasians develop lactose intolerance, while estimates ranging from 80 to 90 percent have been suggested for the black and Oriental populations. These differences have an obvious effect on dietary preferences within these groups. They must also be kept in mind when planning relief efforts, for shipping powdered milk to starvation areas in Africa may actually make matters worse if supplies are distributed to adults rather than children.

saccharides, and this explains why a meal of "franks and beans" often triggers a marked increase in intestinal gas production.

Lipid Digestion and Absorption

Chapter 2 introduced the structure of *triglycerides*, the most abundant dietary lipids (see Figure 2-17). (p. 47) A triglyceride molecule consists of three fatty acids attached to a single molecule of glycerol. Triglycerides and other dietary fats are relatively unaffected by conditions in the stomach and enter the duodenum in the form of large lipid drops. Bile salts emulsify these drops into tiny emulsion droplets and provide an access route for pancreatic lipase. These enzymes convert the triglycerides into a mixture of fatty acids and monoglycerides.

As fatty acids and monoglycerides are released, they interact with bile salts in the surrounding chyme to form small lipid-bile salt complexes called **micelles** (mī-SELZ). A micelle is only about 2.5 nm (0.0025 μm) in diameter. When a micelle contacts the intestinal epithelium, the lipids diffuse across the cell membrane and enter the cytoplasm.

The intestinal cells use these lipids to manufacture new triglycerides that are coated with proteins, creating a complex known as a chylomicron (kī-lō-MĪ-kron). The chylomicrons are then secreted into the interstitial fluids. Most chylomicrons are unable to cross the basement membranes, and diffuse into the intestinal capillaries. They can easily enter the intestinal lacteals, which lack basement membranes and have large gaps between adjacent endothelial cells. From the lacteals they proceed along the lymphatics, through the thoracic duct, and finally enter the circulation at the left subclavian vein.

Protein Digestion and Absorption

Proteins have very complex structures, and several different techniques are used to disassemble them. The mechanical processing in the oral cavity increases the surface area exposed to gastric juices following ingestion. Placing the bolus into a strongly acid environment kills most pathogenic microorganisms, breaks down cell walls, and provides the proper environment for the efforts of pepsin, the proteolytic enzyme secreted by chief cells of the stomach. Pepsin does not complete the process, but it does reduce the relatively huge proteins of the chyme into smaller polypeptide fragments.

After the acid bath has ended and the pH has risen in the duodenum, pancreatic enzymes come into play. Working together, each with its own specificities, trypsin, chymotrypsin, elastase, and carboxypeptidase complete the disassembly of the fragments into a mixture of short peptide chains and individual amino acids. Peptidases on the surfaces of the microvilli complete the process by breaking the peptide chains into their component amino acids, and the amino acids are then absorbed by facilitated diffusion and cotransport mechanisms. Meanwhile, facilitated diffusion and cotransport at the inner surface of the cell dumps amino acids into the interstitial fluids. Once within the interstitial fluids, most of the amino acids diffuse into intestinal capillaries.

WATER AND ELECTROLYTE ABSORPTION

Each day 2–2.5 liters of water enter the digestive tract in the form of food or drink. The salivary, gastric, intestinal, and accessory gland secretions provide another 6–7 liters. Out of that total, only about 150 mℓ is lost in the fecal wastes. Yet our cells are totally unable to actively absorb or secrete water, and all of the water movements occur passively, following osmotic gradients.

As you will recall from previous chapters, when two fluids are separated by a membrane, water will

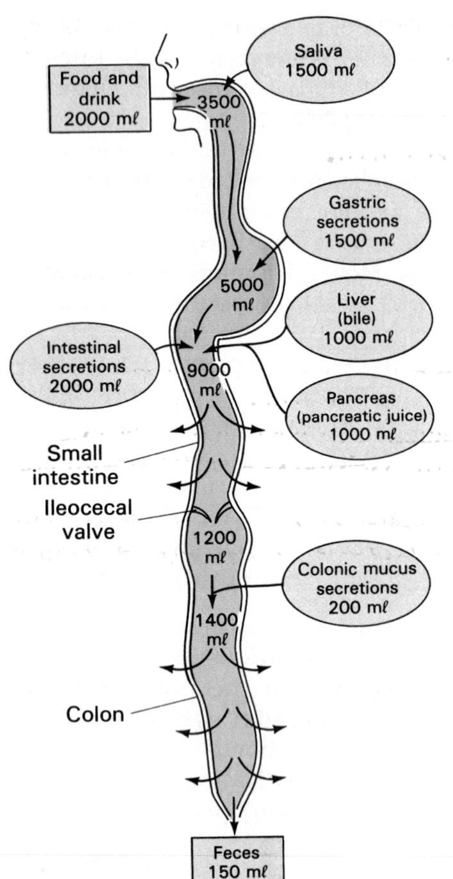

FIGURE 24-29
Fluid Absorption Along the Digestive Tract

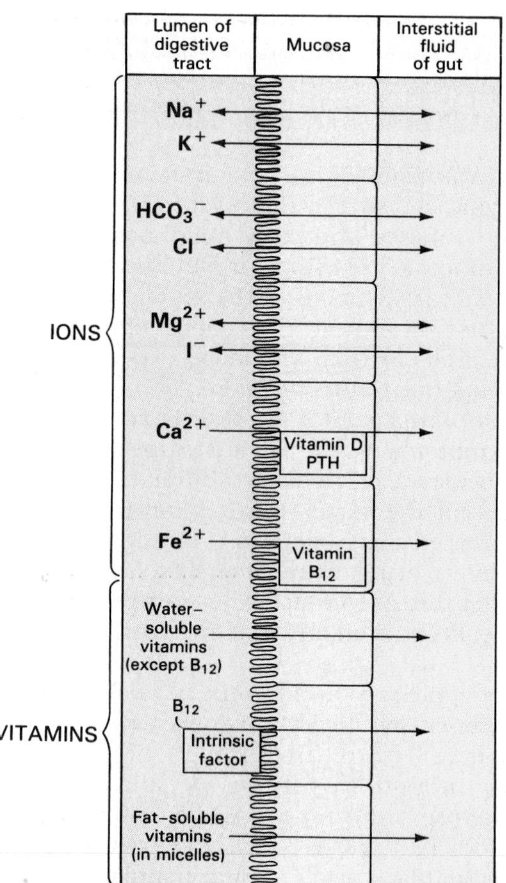

FIGURE 24-30
Ion and Vitamin Absorption by the Digestive Tract

tend to flow into the solution containing the largest number of dissolved materials. Osmotic movements are relatively rapid, so the interstitial fluids and those in the intestinal lumen always have the same concentration of dissolved particles. Yet the epithelial cells are continually absorbing dissolved nutrients and ions from the intestinal contents. As this lowers the concentration within the intestinal lumen, water moves into the surrounding tissues to maintain osmotic equilibrium. These movements are summarized in Figure 24-29.

For osmotic purposes, it really does not matter which ions are being transported. But for metabolic purposes it matters a great deal, and each ion is handled individually. Active or passive movements of sodium and chloride ions are most often responsible for moving water into or out of the lumen. Other important ions that are absorbed or secreted in smaller quantities include calcium, potassium, magnesium, iodine, bicarbonate, and iron. Calcium absorption occurs under hormonal control, requiring the presence of parathormone and vitamin D, as described in Chapter 7. (p. 220) Regulatory mechanisms governing the absorption or excretion of the other ions are poorly understood.

ABSORPTION OF VITAMINS

Vitamins, introduced in Chapter 4, are organic compounds required in very small quantities. (p. 98) The nine **water-soluble vitamins** function primarily as cofactors in enzymatic reactions. All but one, **vitamin B_{12}**, are easily absorbed by the digestive epithelium. Vitamin B_{12} cannot be absorbed by the intestinal mucosa unless it has been bound to *intrinsic factor*, a protein secreted by the parietal cells of the stomach. The bacteria residing in the intestinal tract are an important source for other water-soluble vitamins as they feast on the digestive residues.

Fat-soluble vitamins in the diet enter the duodenum in fat droplets, mixed with dietary lipids. They remain in association with those lipids when micelles form. The fat-soluble vitamins are then absorbed from the micelles in company with the products of lipid digestion. One fat-soluble vitamin, vitamin K, is also produced by bacterial action and absorbed in the colon. (This vitamin was introduced in Chapter 19 in the discussion of blood clotting.) (p. 627)

The information contained in Figure 24-30 summarizes the digestive fates of electrolytes and vitamins. Before continuing, review this figure carefully to ensure that you are reasonably familiar with the outlined mechanisms and events. The next chapter examines the interplay between the respiratory and digestive systems as it considers the metabolism of the entire body. [**CM:** *Malabsorption Syndromes*]

Aging and the Digestive System

Essentially normal digestion and absorption occur in elderly individuals. However, there are many changes in the digestive system that parallel age-related changes already described for other systems.

1. *The rate of epithelial stem cell division declines.* The digestive epithelium becomes more susceptible to damage by abrasion, acids, or enzymes. Peptic ulcers therefore become more likely. In the mouth, esophagus, and anus the stratified epithelium becomes thinner and more fragile.
2. *Smooth muscle tone decreases.* General motility decreases, and peristaltic contractions are weaker. This change slows the rate of chyme movement and promotes constipation. Sagging of the walls of haustrae in the colon can produce symptoms of *diverticulitis.* Straining to eliminate compacted fecal materials can stress the less-resilient walls of blood vessels, producing hemorrhoids. Problems are not restricted to the lower digestive tract. For example, weakening of muscular sphincters can lead to esophageal reflux and frequent bouts of "heartburn."
3. *The effects of cumulative damage become apparent.* A familiar example would be the gradual loss of teeth due to caries or gingivitis. Cumulative damage can involve internal organs as well. Toxins such as alcohol and other injurious chemicals that are absorbed by the digestive tract are transported to the liver for processing. The liver cells are not immune to these compounds, and chronic exposure can lead to *cirrhosis* or other types of liver disease.
4. *Cancer rates increase.* As noted in Chapter 5, cancers are most common in organs where stem cells divide to maintain epithelial cell populations. (p. 160) Rates of colon cancer and stomach cancer rise in the elderly; oral and pharyngeal cancers are particularly common in elderly smokers.
5. *Changes in other systems have direct or indirect effects on the digestive system.* For example, the reduction in bone mass and calcium content in the skeleton is associated with erosion of the tooth sockets and eventual tooth loss. The decline in olfactory and gustatory sensitivity with age can lead to dietary changes that affect the entire body.

√ **In cystic fibrosis, the pancreatic duct can become blocked with thickened secretions. This condition would interfere with the proper digestion of which group(s) of nutrients?**

√ **What component of a meal would increase the number of chylomicrons in the lacteals?**

√ **The absorption of which vitamin would be impaired by removal of the stomach?**

Integration with Other Systems

The digestive system is functionally linked to all other systems, and it has extensive anatomical connections to the nervous, cardiovascular, endocrine, and lymphatic systems. Figure 24-31 summarizes the physiological relationships between the digestive system and other organ systems.

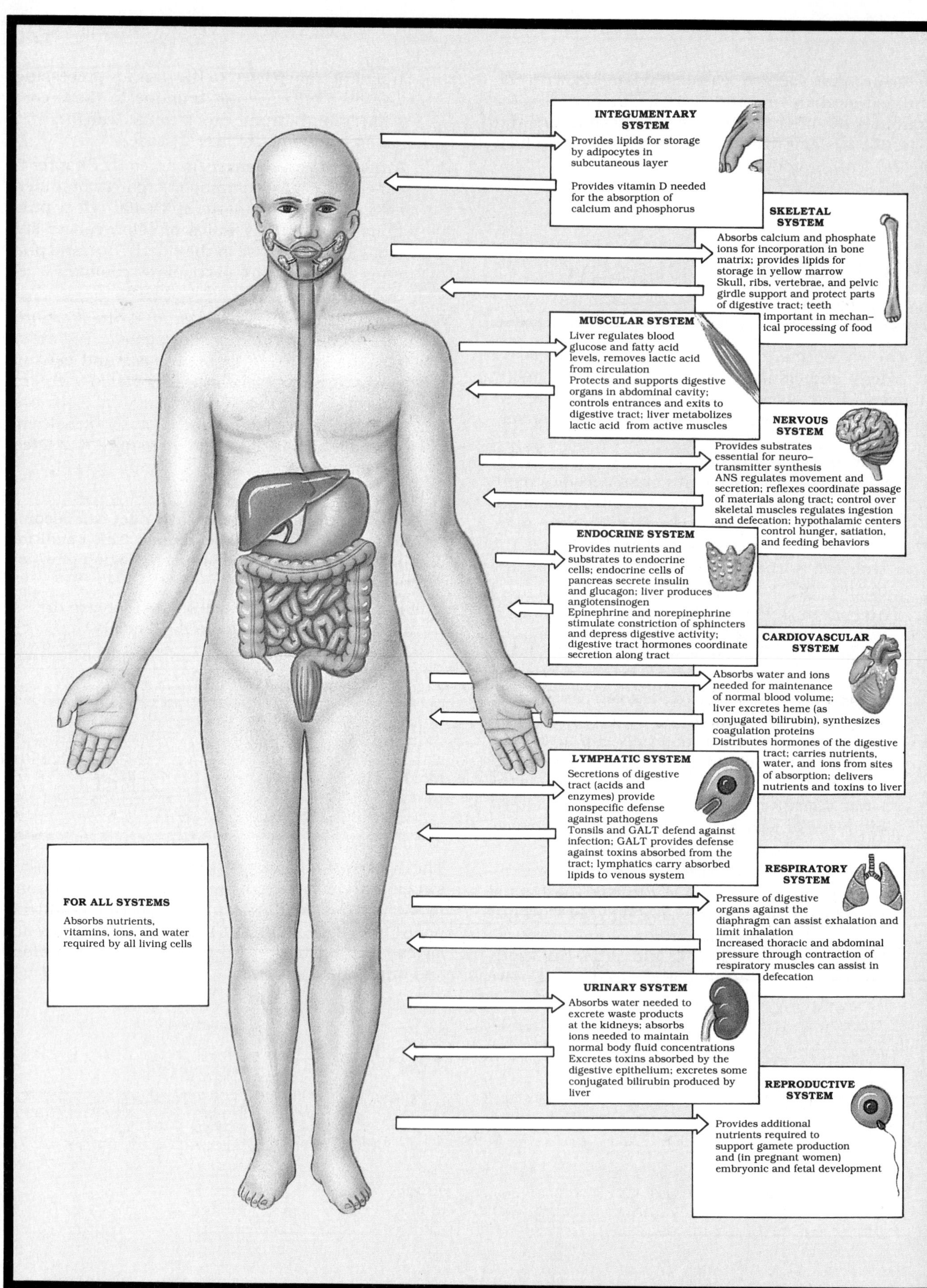
INTEGUMENTARY SYSTEM
Provides lipids for storage by adipocytes in subcutaneous layer
Provides vitamin D needed for the absorption of calcium and phosphorus
SKELETAL SYSTEM
Absorbs calcium and phosphate ions for incorporation in bone matrix; provides lipids for storage in yellow marrow
Skull, ribs, vertebrae, and pelvic girdle support and protect parts of digestive tract; teeth important in mechanical processing of food
MUSCULAR SYSTEM
Liver regulates blood glucose and fatty acid levels, removes lactic acid from circulation
Protects and supports digestive organs in abdominal cavity; controls entrances and exits to digestive tract; liver metabolizes lactic acid from active muscles
NERVOUS SYSTEM
Provides substrates essential for neurotransmitter synthesis
ANS regulates movement and secretion; reflexes coordinate passage of materials along tract; control over skeletal muscles regulates ingestion and defecation; hypothalamic centers control hunger, satiation, and feeding behaviors
ENDOCRINE SYSTEM
Provides nutrients and substrates to endocrine cells; endocrine cells of pancreas secrete insulin and glucagon; liver produces angiotensinogen
Epinephrine and norepinephrine stimulate constriction of sphincters and depress digestive activity; digestive tract hormones coordinate secretion along tract
CARDIOVASCULAR SYSTEM
Absorbs water and ions needed for maintenance of normal blood volume; liver excretes heme (as conjugated bilirubin), synthesizes coagulation proteins
Distributes hormones of the digestive tract; carries nutrients, water, and ions from sites of absorption; delivers nutrients and toxins to liver
LYMPHATIC SYSTEM
Secretions of digestive tract (acids and enzymes) provide nonspecific defense against pathogens
Tonsils and GALT defend against infection; GALT provides defense against toxins absorbed from the tract; lymphatics carry absorbed lipids to venous system
FOR ALL SYSTEMS
Absorbs nutrients, vitamins, ions, and water required by all living cells
RESPIRATORY SYSTEM
Pressure of digestive organs against the diaphragm can assist exhalation and limit inhalation
Increased thoracic and abdominal pressure through contraction of respiratory muscles can assist in defecation
URINARY SYSTEM
Absorbs water needed to excrete waste products at the kidneys; absorbs ions needed to maintain normal body fluid concentrations
Excretes toxins absorbed by the digestive epithelium; excretes some conjugated bilirubin produced by liver
REPRODUCTIVE SYSTEM
Provides additional nutrients required to support gamete production and (in pregnant women) embryonic and fetal development

FIGURE 24-31
Functional Relationships between the Digestive System and Other Systems

CHAPTER REVIEW

Review of Selected Clinical Terms

peritonitis (per-i-to-NĪ-tis) (*p. 786*) Inflammation of the peritoneal membrane.

polyps (POL-ips) (*p. 809*) Small mucosal tumors that grow from the intestinal wall.

portal hypertension (*p. 815*) High blood pressures in the hepatic portal system.

ascites (a-SĪ-tēz) (*p. 815*) Fluid leakage into the peritoneal cavity across the serosal surfaces of the liver and viscera.

Additional Terms of Clinical Importance

COLONIC INSPECTION AND CANCER (*p. 809*): **colonoscope**

LACTOSE INTOLERANCE (*p. 823*): **lactose intolerant**

Study Outline

Related Key Terms

Introduction (pp. 783–785)

1. The digestive system consists of the muscular **digestive tract** and various **accessory organs.**
2. Digestive functions include: ingestion, mechanical processing, digestion, secretion, absorption, compaction, and **defecation.**

An Overview of the Structure and Function of the Digestive Tract (pp. 785–789)

HISTOLOGICAL ORGANIZATION (pp. 785–786)

1. The digestive tract is lined by a **mucous epithelium** moistened by glandular secretions of the epithelial and accessory organs.
2. The **lamina propria** and epithelium form the **mucosa (mucous membrane)** of the digestive tract. Proceeding outward one enters the **submucosa**, the **muscularis externa**, and a layer of loose connective tissue called the **adventitia**. Within the peritoneal cavity this layer is covered by a serous membrane called the **serosa**. — **muscularis mucosae**, **submucosal plexus**, **myenteric plexus**
3. Double sheets of peritoneal membrane called **mesenteries** suspend the digestive tract.

SMOOTH MUSCLE AND THE MOVEMENT OF DIGESTIVE MATERIALS (pp. 786–789)

4. Smooth muscle lacks sarcomeres and the resulting striations. The thin filaments are anchored to **dense bodies**. Many smooth muscle cells are not innervated by motor neurons, and the neurons that do innervate smooth muscles are not under voluntary control. — **plasticity**, **myosin light chain kinase**, **multiunit smooth muscle fibers**, **visceral smooth muscle fibers**, **pacesetter cells**
5. The muscularis externa propels materials through the digestive tract through the contractions of **peristalsis. Segmentation** movements in areas of the small intestine churn digestive materials. — **myenteric reflexes**

The Digestive Tract (pp. 789–810)

THE ORAL CAVITY (pp. 789–793)

1. The functions of the oral cavity include: (1) **analysis** of potential foods; (2) **mechanical processing** using the teeth, tongue, and palatal surfaces: (3) **lubrication** by mixing with mucus and salivary secretions; and (4) **digestion** by salivary enzymes. — **buccal cavity** • **labia**, **vestibule** • **gingivae**, **lingual frenulum**
2. The **parotid**, **sublingual**, and **submandibular glands** discharge their secretions into the oral cavity. — **uvula** • **pharyngeal arches**, **palatoglossal arch**, **palatopharyngeal arch** • **fauces**, **paratid duct** • **sublingual ducts**, **submandibular ducts**
3. **Intrinsic** and **extrinsic** tongue muscles are controlled by the hypoglossal nerve. — **dorsum** • **ankyloglossia**
4. **Dentin** forms the basic structure of a tooth. The **crown** is coated with **enamel**, and the **root** with **cementum**. The **periodontal ligament** anchors the tooth in an alveolar socket. **Mastication** (chewing) occurs through the contact of the opposing **occlusal surfaces** of the teeth. — **pulp cavity** • **root canal**, **gingival sulcus**

THE PHARYNX (p. 793)

5. Propulsion of the **bolus** results from the contractions of the **pharyngeal constrictors** and the **palatal muscles**. — **palatopharyngeus**, **stylopharyngeus**

THE ESOPHAGUS (pp. 794–795)

6. **Deglutition** (swallowing) begins with the compaction of a bolus — **esophageal hiatus**

and its movement into the pharynx, followed by the elevation of the larynx, reflection of the epiglottis, and closure of the glottis. After opening of the **upper esophageal sphincter**, peristalsis moves the bolus down the esophagus to the **lower esophageal sphincter**.

Related Key Terms
pharyngeal phase
esophageal phase
peristaltic wave
swallowing reflex
swallowing center

THE STOMACH (pp. 795–799)

7. The stomach has three major functions: (1) bulk storage of ingested matter; (2) mechanical breakdown of resistant materials: and (3) disruption of chemical bonds using acids and enzymes. The **pyloric sphincter** guards the exit from the stomach. In a relaxed state the stomach lining contains numerous **rugae** (ridges and folds).

chyme • cardia • fundus
body • pylorus
greater omentum
lesser omentum

8. Within the **gastric glands**, **parietal cells** secrete **intrinsic factor** and hydrochloric acid. **Chief cells** secrete **pepsinogen**, which acids in the gastric lumen convert to the enzyme **pepsin**. **Enteroendocrine cells** of the stomach secrete several compounds, notably the hormone **gastrin**.

gastric juice • gastric pits
alkaline tide • proteinase
rennin

9. Gastric secretion includes: (1) the **cephalic phase**, which prepares the stomach to receive ingested materials; (2) the **gastric phase**, which begins with the arrival of food in the stomach; and (3) the **intestinal** phase, which controls the rate of gastric emptying.

enterogastric reflexes • secretin
cholecystokinin (CCK)

THE SMALL INTESTINE (pp. 799–807)

10. The small intestine includes the **duodenum**, the **jejunum**, and the **ileum**. The intestinal mucosa bears transverse folds called **plicae** and small projections, called **intestinal villi**, that increase the surface area for absorption. Each villus contains a terminal lymphatic called a **lacteal**. Pockets called **intestinal crypts** house enteroendocrine, goblet, and stem cells.

exfoliation • enterokinase
submucosal glands
duodenal ampulla
duodenal papilla
gastroenteric reflex

11. **Intestinal juice** moistens the chyme, helps to buffer acids, and dissolves digestive enzymes and the products of digestion.

enterocrinin
glucose-dependent insulinotropic peptide (GIP)
vasoactive intestinal peptide (VIP)

THE LARGE INTESTINE (pp. 807–810)

12. The main functions of the large intestine are to: (1) reabsorb water and compact the feces; (2) absorb vitamins liberated by bacteria; and (3) store fecal material prior to defecation.
13. The colon has a larger diameter and a thinner wall than the small intestine. It bears **haustrae** (pouches) and the **taenia coli** (longitudinal bands of muscle).
14. The rectum terminates in the **anorectal canal** leading to the **anus**. Muscular sphincters control the passage of fecal material to the anus. Distension of the rectal wall triggers the defecation reflex.

cecum • ileocecal valve
colic flexure • appendix
ascending colon
transverse colon
descending colon
sigmoid colon
rectum • mass movements

rectal columns
internal anal sphincter
external anal sphincter

The Accessory Organs (pp. 810–817)

THE SALIVARY GLANDS (pp. 810–811)

1. Saliva lubricates the mouth, dissolves chemicals, flushes the oral surfaces, and helps control bacteria. Salivation is usually controlled by the ANS, including the **salivatory nuclei** of the medulla.

salivary amylase (alpha-amylase)
mucins

THE LIVER (pp. 811–815)

2. The liver performs metabolic and hematological regulation, and produces **bile**. The bile ducts from each lobule unite to form the **hepatic duct**, which meets the **cystic duct** to form the **common bile duct** that empties into the duodenum.

urobilinogen • bile salts
enterohepatic circulation of bile
falciform ligament
round ligament
caudate lobe • hilus
porta hepatis

3. The **liver lobule** is the organ's basic functional unit. **Bile canaliculi** carry bile to the **bile ductules**, which lead to portal areas.

hepatocytes • Kupffer cells
central vein

THE GALLBLADDER (pp. 815–816)

4. The gallbladder stores and concentrates bile. During **emulsification** bile salts break apart large drops of lipids and make them accessible to **lipases** secreted by the pancreas.

THE PANCREAS (p. 817)

5. The pancreas has two functions: endocrine (secreting insulin and glucagon into the blood) and exocrine (secreting water, ions, and digestive enzymes into the small intestine). Pancreatic enzymes include lipases, **carbohydrases**, and **proteolytic enzymes**.

proteinases • peptidases

6. The **pancreatic duct** penetrates the wall of the duodenum. Within each lobule, ducts branch repeatedly before ending in the **pancreatic acini** (blind pockets).

accessory duct • pancreatic juice
pancreatic lipase • nuclease
trypsin • chymotrypsin
carboxypeptidase • elastase
proenzymes • trypsinogen
chymotrypsinogen
procarboxypeptidase
proclastase

Digestion and Absorption (pp. 817–824)

THE PROCESSING AND ABSORPTION OF NUTRIENTS (pp. 818–823)

1. The digestive system breaks down the physical structure of the ingested material and then disassembles the component molecules into smaller fragments through hydrolysis.

2. Amylase breaks down complex carbohydrates into disaccharides and trisaccharides. These are broken down into monosaccharides by enzymes at the epithelial surface and absorbed by the intestinal epithelium through facilitated diffusion and **cotransport**.
3. Triglycerides are emulsified into large lipid drops. The resulting fatty acids and monoglycerides interact with bile salts to form **micelles**, from which they diffuse across the intestinal epithelium.
4. Protein digestion involves the gastric enzyme pepsin and the various pancreatic proteases. Peptidases liberate amino acids that are absorbed and exported to the interstitial fluids.

WATER AND ELECTROLYTE ABSORPTION (pp. 823–824)

5. About 2.0–2.5 liters of water are ingested each day, and digestive secretions provide 6–7 liters. Nearly all is reabsorbed by osmosis.

ABSORPTION OF VITAMINS (pp. 824–825)

6. The nine **water-soluble vitamins** are important as cofactors in enzymatic reactions. **Fat-soluble vitamins** are enclosed within fat droplets and are absorbed with the products of lipid digestion.

Related Key Terms

maltase • maltose • sucrase
sucrose • lactase • lactose
sodium-linked cotransport
flatus

vitamin B_{12}

Review Planner

	Objective	Level 1
1	Identify the components of the digestive tract and discuss the functions of each.	2 4 11 12 13 14 15 16 28 34 43
2	Describe the histological characteristics of each segment of the digestive tract in relation to its function.	13 14 20 25 38 39 41 42 44 46
3	Discuss how food is processed in the mouth and describe the key events of the swallowing process.	1 10 19 31 32 35
4	Describe the characteristics of smooth muscle and explain how ingested materials are propelled through the digestive tracts	13 17 22 24 25 27 40 41 42 43
5	Identify the enzymes produced by the digestive system and what they accomplish.	3 7 19 21 45
6	Summarize the stages involved in the regulation of gastric function.	17 18 23 29 43 47
7	Discuss the homeostatic mechanisms that regulate the activities of the digestive system.	12 18 23 47
8	Discuss the roles played by the accessory organs in digestion.	2 3 5 7 11 15 34
9	Explain how various compounds in food are broken and absorbed by the body.	5 6 8 9 14 30 33
10	Describe the changes in the digestive system that occur with aging.	31

Level 2	**26 27 36 37**
Level 3	**48–52**
C M	**Dental Problems** **Achalasia and Esophagitis** **Gastritis and Peptic Ulcers** **Stomach Cancer • Giardiasis** **Vomiting and Intestinal Evacuation** **Gastroenteritis** **Diverticulitis and Colitis** **Diarrhea and Constipation** **Mumps • Hepatitis • Cirrhosis** **Problems with Bile Storage and Secretion • Pancreatitis** **Malabsorption Syndromes**
C M	**Analysis of Liver Structure and Function**
NEWS C M	**Drastic Weight-Loss Techniques**

Chapter Review

Review Questions

MULTIPLE CHOICE

1. You would expect "gingivitis" to be an inflammation of the: a) palatine tonsils; b) gums; c) tongue; d) root canal.
2. If the parotid duct became blocked, this would affect your ability to produce: a) bile; b) intrinsic factor; c) saliva; d) elastase.
3. Which of these is *not* produced by the pancreas? a) glucagon; b) lipase; c) trypsin; d) pepsin.
4. Which of the following is *not* a primary function of the large intestine? a) secreting hormones that regulate intestinal activity; b) reabsorbing water; c) absorbing important vitamins; d) storing fecal material.
5. Bile: a) is manufactured by the pancreas; b) flows into the stomach to aid digestion during especially large meals; c) is diluted in the gallbladder; d) assists in the digestion of fats.
6. Emulsification would help you to digest; a) an apple; b) a soft drink; c) butter; d) a carrot.
7. Pancreatic juice: a) contains strong acids; b) contains trypsin, chymotrypsin, carboxypeptidase, and elastase; c) enters the duodenum at the pyloric sphincter; d) is secreted in response to the hormone enterocrinin.
8. The digestive system absorbs ingested water by: a) sodium-linked cotransport; b) facilitated diffusion; c) osmosis; d) countertransport.
9. If the production of intrinsic factor decreased, this would affect the absorption of: a) extrinsic factor; b) vitamin B_{12}; c) vitamin K; d) vasoactive intestinal peptide and cholecystokinin (CCK).
10. You would expect the ________ surfaces of the teeth to show the most wear: a) labial; b) mesial; c) palatal; d) occlusal.

DIAGRAMS

11. Cover the labels in Figure 24-1. Identify the accessory organs and the major regions of the digestive tract, and list their primary functions.
12. List 7 to 10 ways in which the digestive system interacts functionally with other body systems.

TRUE/FALSE

(Your instructor may wish to have you correct the statement if it is false.)

13. **T/F:** The muscularis externa is responsible for peristalsis.
14. **T/F:** Most of the important absorptive activities occur in the large intestine.
15. **T/F:** Obstruction of the cystic duct will prevent the flow of bile.

MATCHING QUESTIONS

(Match each term with the most appropriate statement.)

Statements

16. Another term for the oral cavity.
17. A viscous mixture of food and gastric juices.
18. A hormone that increases the secretion of pancreatic enzymes and pushes pancreatic secretions and bile into the duodenum.
19. A salivary enzyme that breaks down complex carbohydrates.
20. A layer of loose connective tissue that surrounds the digestive tract.
21. An enzyme found in the stomachs of newborn infants that coagulates milk proteins.
22. Muscular contractions that propel ingested materials from the transverse colon to the rectum.
23. A peptide hormone, produced in the duodenum that stimulates the release of insulin.
24. Waves of muscular contractions that move along the length of the digestive tract.
25. The ability of a muscle fiber to contract over a wide range of lengths.

Terms:

a) Plasticity
b) Buccal cavity
c) GIP
d) Peristalsis
e) Chyme
f) Rennin
g) CCK
h) Mass movements
i) Alpha-amylase
j) Adventitia

SHORT ANSWER/ESSAY

26. Discuss the functions of the digestive system.
27. Compare smooth muscle to the other types of muscle. Explain how the presence of smooth muscle helps the digestive system to "do its job."
28. What are the primary distinctions between the subdivisions of the small intestine? Relate these characteristics to their functions.
29. Describe the phases of gastric secretion.
30. Discuss the ways in which the digestive system absorbs water, electrolytes, and vitamins.
31. Describe the changes that occur in the digestive system with aging. What disorders can result?
32. You bite off a mouthful of peanut brittle and chew it. Describe the role(s) played by each type of tooth in this action.
33. Compare the way(s) in which fats, carbohydrates, and proteins are broken down and absorbed by the digestive system.
34. Summarize the functions of the four major accessory organs.
35. Describe the stages in the process of deglutition.
36. Identify the names and numbers of the cranial nerve(s) that would allow you to detect the following sensations:
    a) A scratch in the lining of your buccal cavity.
    b) The taste of peppermint ice cream.
37. You scold a child and he sticks out his tongue at

you. Identify the muscle(s) and nerve(s) which made his response possible.

38. Explain the function(s) of plicae and intestinal villi. Where are they found and why are they important to the digestive system?
39. Fill in the blank: The hardest biologically manufactured substance is ________; it protects the ________.
40. What do peristalsis and segmentation have in common? How do they differ?

Identify and give a function for each of the following:

41. Muscularis mucosa
42. Pacesetter cells
43. Pyloric sphincter
44. Parietal cells
45. Gastrin
46. Submucosal glands (Brunner's glands)
47. Secretin

CRITICAL THINKING/APPLICATIONS

48. Why is uncontrolled diarrhea dangerous?
49. Explain why chronic emotional stress could cause duodenal ulcers.
50. Patients who have been diagnosed with liver cancer generally have a high mortality. Explain why.

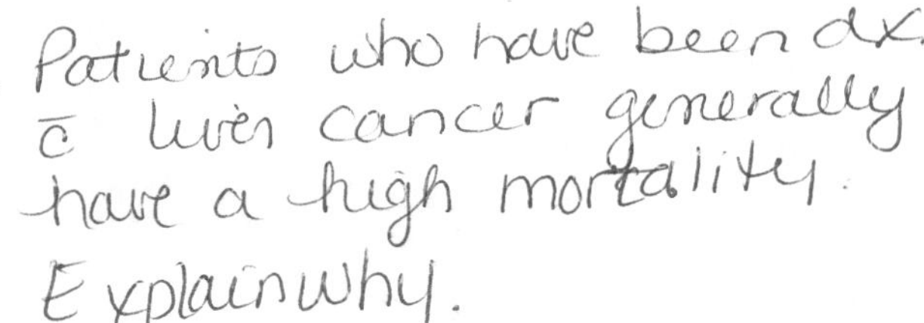
52. Patients who have been dx
c̄ liver cancer generally
have a high mortality.
Explain why.

Career Close-up: Sales Representative

For several years Nancy Wright has enjoyed a successful career with a large international pharmaceuticals firm. For two years she was the top sales representative in her division; one year she led the country in sales. In fact, she created so much new business that when she was promoted, the company hired two reps to replace her.

What does this have to do with Anatomy and Physiology? you may ask. A lot, actually, since Nancy credits much of her business success to her clinical background. She is a CETN (Certified Enterostomal Therapy Nurse), a nurse who specializes in the care and teaching of patients with surgically-made abdominal stomas (openings). "My nursing experience gives me credibility," she says. "I can identify with the patients and staff. I know what their needs are and I work with them to meet those needs."

When she first became a registered nurse, Nancy had no idea she would become a CETN. Her first contact with a stoma patient happened by chance. "I found a note on my desk one morning, asking me to meet a new patient," she recalls. "She was a 21-year-old graduate student with recurrent pelvic cancer; a year before she'd had a rectal tumor and the cancer had spread. It was a case of the patient teaching the clinician. Over a six-week period I helped her write a will, say goodbye to her fiancee, and arrange to return home to die. I realized I knew nothing about stoma patients, what their needs were, how to help them."

Nancy entered an 8-week Enterostomal Therapist training program offered by Houston's M.D. Anderson Hospital and Tumor Institute. After graduating and passing a national ceritfying exam, she joined the staff of a large Veterans Administration (VA) medical center. At the VA she directed patient care for all patients with abdominal stomas, as well as those with draining wounds, pressure ulcers, or other types of tissue breakdown.

"Many of our patients had stomas because of cancer," she notes. "Others had them because of diseases such as inflammatory bowel disease or diverticulitis. Some were there because of traumas such as spinal cord injuries, gunshot wounds, or motor vehicle accidents." In addition to her clinical responsibilities, Nancy worked with nursing students, conducted workshops, and taught at a university.

"I enjoyed being an Enterostomal Therapist because there's variety, yet continuity," she says. "You see patients move from the crisis stage to become outpatients who lead normal lives again. You get to know people and their families over the long term."

Her involvement with patients usually began before they underwent the stoma surgery. "I met with them and their families, helped them understand what was involved, answered questions. Post-operatively, I made sure the stoma was healthy and followed up on their progress. I taught them self-care and how to lead independent lives again."

Nancy feels her nursing experience was excellent training for a sales position. "There's a lot of similarity between the two roles," she notes. "In working with patients, I developed a rapport, educated them about the ostomy products they would use, and handled questions and objections. I worked with them on a continuing basis to use the products and improve their quality of life. As a sales rep I found myself taking a similar approach."

Her advice to anyone considering a career in health care?: "Allow yourself to be touched by opportunities and respond to them," she says. "I wouldn't be doing this now if it hadn't been for that first patient."

Our bodies are not isolated from the world around us—indeed, they could not survive without constant interchange. To live we must take in air, food, and water—as we see this person doing. But this thermogram, made with infrared (heat) radiation, reveals another transaction with the environment that we probably don't think much about. Our bodies must maintain a constant temperature of about 37° C. This isn't always an easy task—it requires balancing the amount of heat generated by our metabolic processes with the amount lost to a (usually) cooler environment. We can't afford to overheat—a fever that raises our body temperature a mere 5° C can cause death. But we can't afford to lose too much heat, either—that's why we build houses and wear clothing. In this chapter we'll focus on the water and nutrients that we obtain from our environment, and how they are used by the body. We'll also return to the subject of heat—how it is generated, conserved, and lost.

CHAPTER 25
Metabolism and Energetics

Chapter Objectives

After reading this chapter, you will be able to:

1 Compare the mechanisms of carbohydrate, lipid, protein, and nucleic acid metabolism.
2 Explain how the metabolic activities of tissues, organs, and organ systems are coordinated.
3 Describe the energy reserves of an average person.
4 Describe how we utilize our energy reserves to maintain homeostasis.
5 Explain what constitutes a balanced diet, and why it is important.
6 Discuss the functions of vitamins, minerals, and other important nutrients.
7 Discuss the homeostatic mechanisms that maintain a constant body temperature.

Introduction

Living cells are chemical factories that break down organic molecules to obtain energy, usually in the form of ATP. As we saw in Chapter 4, the major energy-producing activities usually occur inside mitochondria. (p. 101) To carry out these processes, cells in the human body must have reliable sources of water, organic substrates, vitamins, ions, and oxygen. Oxygen is absorbed at the alveoli of the lungs, the other items are obtained by the digestive tract, and the cardiovascular system ensures prompt distribution to individual cells throughout the body.

The energy released inside the cell supports metabolic turnover, growth, cell division, contraction, secretion, and a variety of other special functions that vary from cell to cell and tissue to tissue. Because each tissue type contains different populations of cells, the energy and nutrient requirements of any two tissues, such as loose connective tissue and cardiac muscle, are quite different. When cells, tissues, and organs vary their patterns of activity, the metabolic needs of the body change. There are short-term and long-term changes. For example, nutrient requirements vary from moment to moment (resting versus active), hour to hour (asleep versus awake), and year to year (growing child versus adult).

Differentiation creates tissues and organs that are specialized to store excess nutrients; the storage of lipids in adipose tissue is one familiar example. When nutrients are abundant, energy reserves are built up. These reserves, usually glycogen deposits or lipid inclusions, can then be called upon whenever the digestive tract cannot provide the right quantity or quality of nutrients. The endocrine system, with the assistance of the nervous system, adjusts and coordinates the metabolic activities of the body's tissues and controls the storage and mobilization of nutrient reserves.

Our diet provides carbohydrates, lipids, proteins, water, vitamins, and minerals that are absorbed through the lining of the digestive tract. The mechanisms involved were detailed in Chapter 24, and it may help to review Figures 24-27, 24-29, and 24-30 before proceeding. (pp. 820, 824) This chapter considers the fate of those absorbed nutrients once inside the body.

Cellular Metabolism

Figure 25-1 provides an overview of the ways cells use organic nutrients absorbed from the extracellular fluid. Simple sugars, amino acids, and lipids cross the cell membrane to join nutrients already in the cytoplasm. All of the anabolic and catabolic pathways in the cell utilize this collection of organic substrates, known as a *nutrient pool.*

In general, a cell with excess carbohydrates, lipids, and amino acids will break down carbohydrates to obtain energy. Lipids are a second choice, and amino acids are conserved. At the same time, the cell is replacing membranes, organelles, enzymes, and structural proteins. These anabolic activities require more amino acids than lipids, and relatively few carbohydrates.

CARBOHYDRATE METABOLISM

Chapter 4 discussed *glycolysis*, the major pathway of carbohydrate catabolism. Glycolysis and *aerobic respiration* provide most of the ATP used by typical cells. Figures 4-7 through 4-10 summarize important steps in carbohydrate catabolism, and you should take the time to review them before proceeding. (pp. 103, 104, 106, 108)

Because some of the steps in glycolysis are not reversible, carbohydrate synthesis involves a different set of regulatory enzymes, and carbohydrate breakdown and synthesis are independently regulated. The synthesis of glucose is called **gluconeogenesis** (gloo-kō-nē-ō-JEN-e-sis). Pyruvic acid or other three-carbon molecules can be used as starting materials. As indicated in Figure 25-2, this reaction sequence enables a cell to create glucose molecules from other carbohydrates, glycerol, or some amino acids. However, acetyl-CoA cannot be used to make glucose because the decarboxylation step between pyruvic acid and acetyl-CoA cannot be reversed. Fatty acids and many amino acids cannot be converted to glucose because their catabolic pathways, described in later sections, produce acetyl-CoA.

Glucose molecules created by gluconeogenesis can be used to manufacture other simple sugars, complex carbohydrates, proteoglycans, or nucleic acids. In the liver and in skeletal muscle, glucose molecules are stored as glycogen. The process of glycogen formation is known as **glycogenesis**. Glycogen is an important energy reserve that can be broken down when the cell cannot obtain enough glucose from the extracellular fluid. Although glycogen molecules are large, glycogen reserves take up very little space because they form compact, insoluble granules.

LIPID METABOLISM

Lipid molecules, like carbohydrates, contain carbon, hydrogen, and oxygen, but these atoms are present in different proportions. Because triglycerides are the most abundant lipid in the body, our discussion will

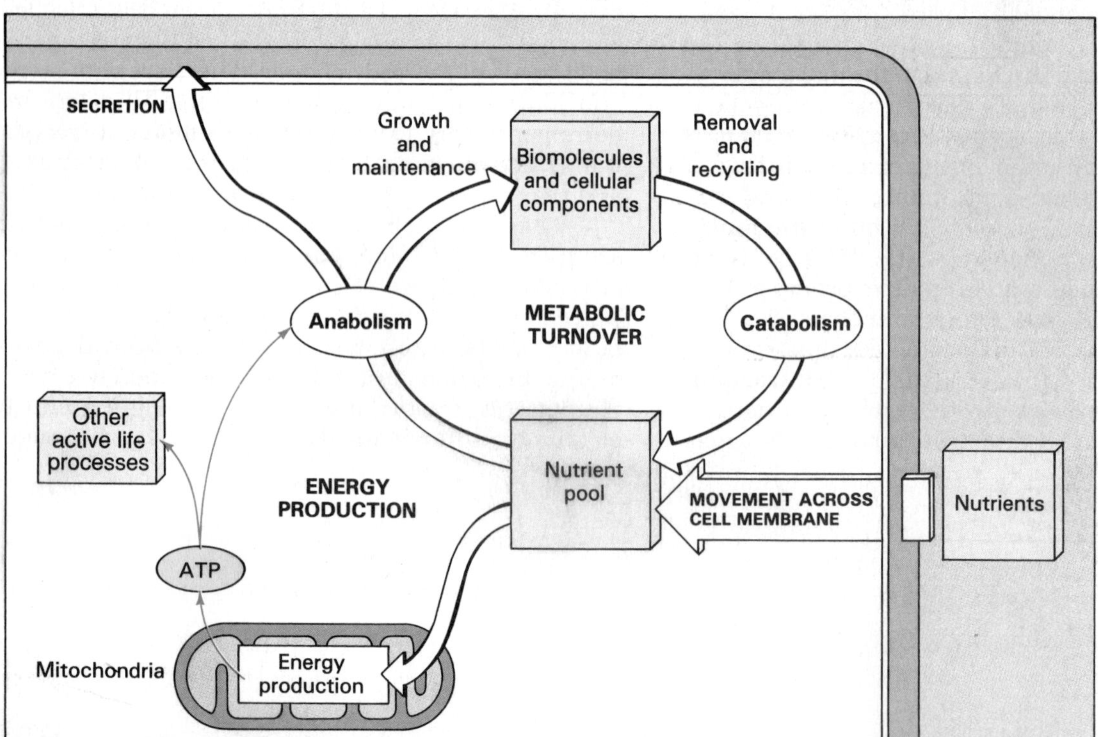

FIGURE 25-1 Nutrient Cycling. Cells absorb nutrients from the interstitial fluid and add them to the nutrient pool already present in the cytoplasm. These nutrients are used to support anabolic processes and the catabolic steps that provide ATP. (Compare Figure 4-15.)

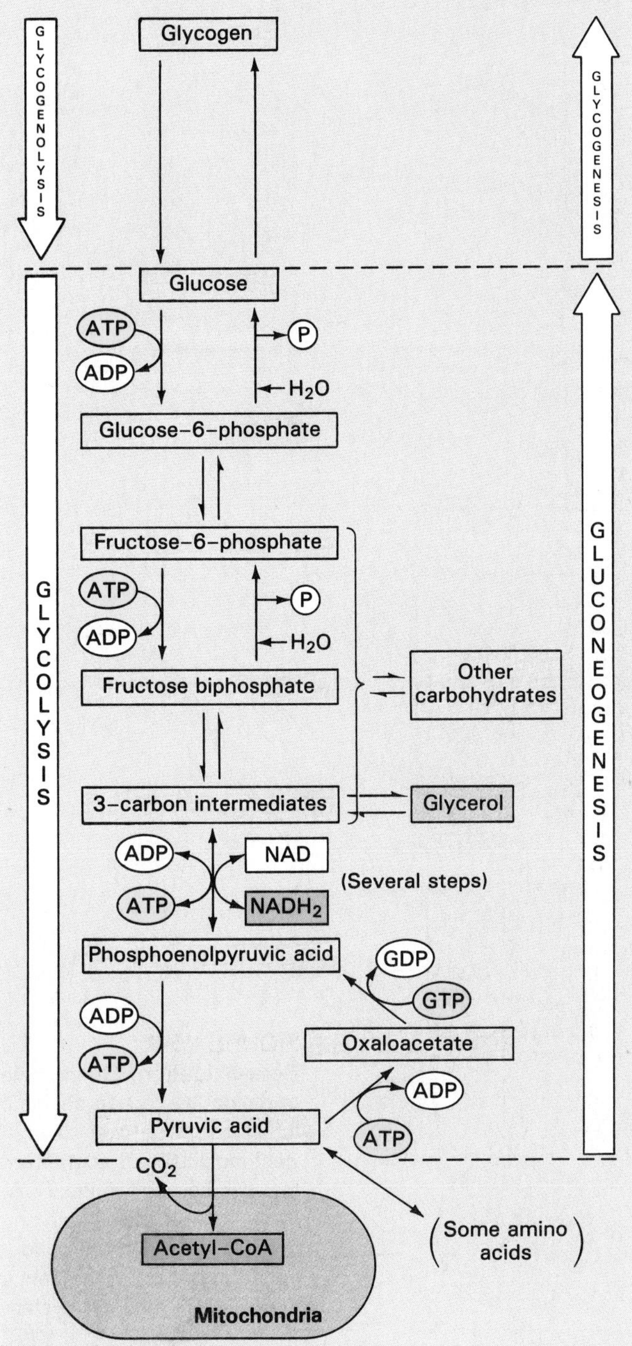

FIGURE 25-2
Carbohydrate Synthesis. This flow chart diagrams the major pathways for glycolysis and gluconeogenesis. Many of the reactions are freely reversible, but separate regulatory enzymes control key steps. Some amino acids, other carbohydrates, and glycerol can be converted to glucose. Notice that the enzymatic reaction that converts pyruvic acid to acetyl-CoA cannot be reversed.

focus on pathways for triglyceride breakdown and synthesis.

Lipid Catabolism

During lipid catabolism, or **lipolysis**, lipids are broken down into pieces that can be converted to pyruvic acid or channeled directly into the TCA cycle.

A triglyceride is first split into its component parts through hydrolysis. This step yields one molecule of glycerol and three of fatty acids. Glycerol enters the TCA cycle after enzymes in the cytosol convert it to pyruvic acid. The catabolism of fatty acids involves a completely different set of enzymes.

Beta-oxidation Fatty acid molecules are broken down into two-carbon fragments by means of a sequence of reactions known as **beta-oxidation**. This process occurs inside mitochondria, so the carbon chains can enter the TCA cycle immediately. Figure 25-3 details the process of beta-oxidation. The first step converts ATP to AMP and attaches coenzyme A. To reconvert AMP to ATP the cell must attach two high-energy phosphate groups, the equivalent of reconverting 2 ADP to ATP. Subsequent steps require coenzyme A but not ATP. Each of these steps generates molecules of acetyl-CoA, $FADH_2$, and $NADH_2$, and leaves a shorter carbon chain bound to coenzyme A.

This reaction sequence provides substantial energy benefits: 12 ATP from the processing of acetyl-CoA in the TCA cycle, plus 5 ATP from the $NADH_2$ and $FADH_2$. The cell gains 144 ATP from the breakdown of an 18-carbon fatty acid molecule. This is almost 1.5 times the energy obtained by the aerobic respiration of three 6-carbon glucose molecules. The catabolism of other lipids follows similar patterns, usually ending with the formation of acetyl-CoA.

Lipids and Energy Production

Lipids are important as an energy reserve because they can provide large amounts of ATP. They can be stored in compact droplets in the cytosol, because they are insoluble. However, this mode of storage makes it difficult for water-soluble enzymes to reach them. Lipid reserves are therefore more difficult to mobilize than carbohydrate reserves. In addition, most lipids are processed inside mitochondria, and mitochondrial activity is limited by the availability of oxygen. The net result is that lipids cannot provide large amounts of ATP in a short amount of time. However, cells with modest energy demands can shift over to lipid-based energy production when glucose supplies are limited. Skeletal muscle fibers normally cycle between lipid and carbohydrate metabolism. At rest, when energy demands are low, they break down fatty acids. When active, and energy demands are both large and immediate, skeletal muscle fibers shift over to metabolizing glucose.

Lipid Synthesis

The synthesis of lipids is known as **lipogenesis** (li-pō-JEN-e-sis). Figure 25-4 shows the major pathways of lipogenesis. Glycerol synthesis reverses the steps of glycolysis; the synthesis of most other lipids begins

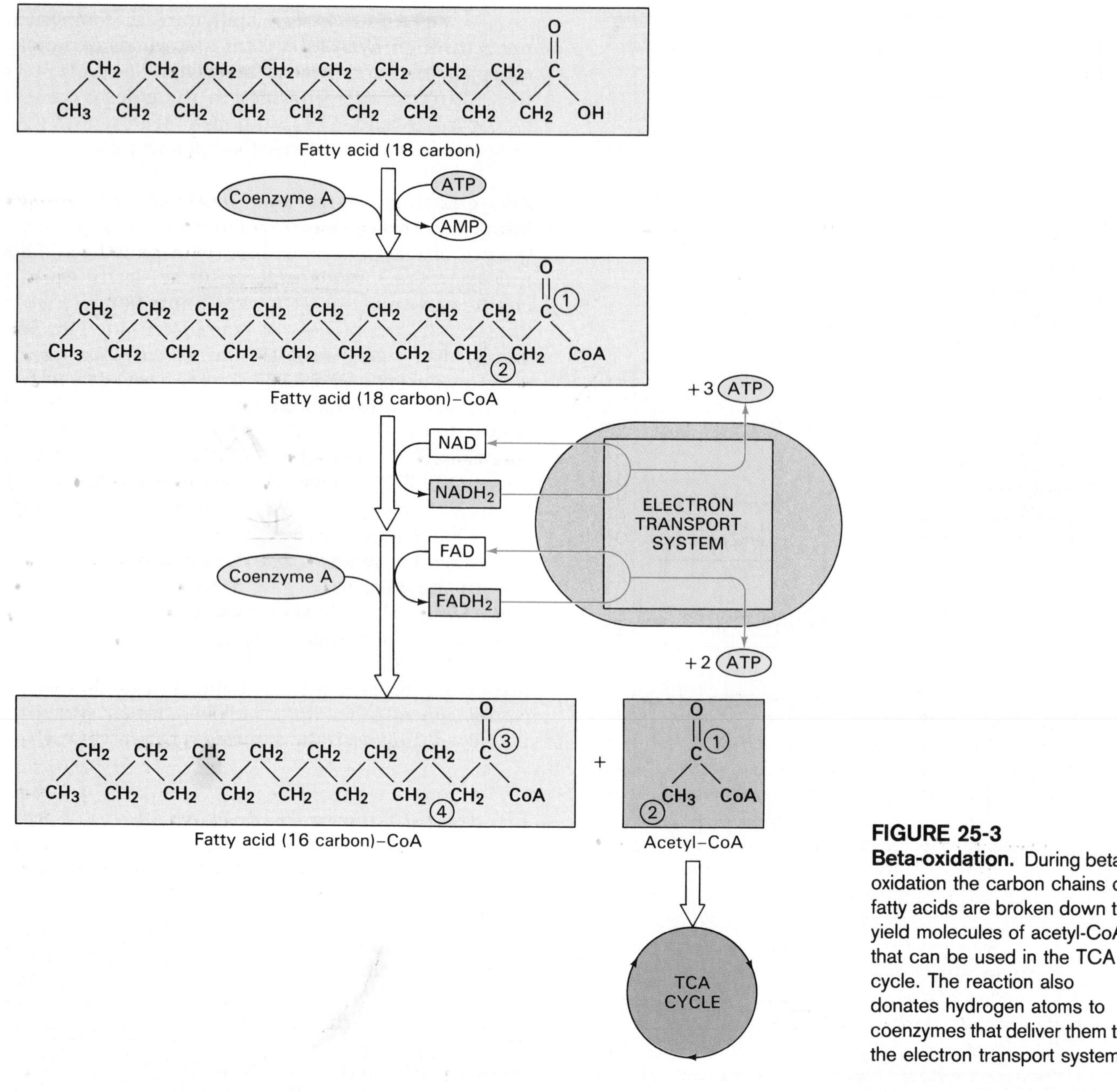

FIGURE 25-3
Beta-oxidation. During beta-oxidation the carbon chains of fatty acids are broken down to yield molecules of acetyl-CoA that can be used in the TCA cycle. The reaction also donates hydrogen atoms to coenzymes that deliver them to the electron transport system.

with acetyl-CoA. Lipogenesis can use almost any organic substrate because lipids, amino acids, and carbohydrates can be converted to acetyl-CoA.

Fatty acid synthesis involves a reaction sequence quite distinct from that of beta-oxidation. Our cells cannot *build* every fatty acid they can break down. **Linoleic acid**, an 18-carbon, unsaturated fatty acid, cannot be synthesized at all. A diet poor in linoleic acid slows growth and alters the appearance of the skin. **Arachidonic** and **linolenic acids** are other long-chain unsaturated fatty acids that the human body cannot synthesize. These three **essential fatty acids** must be included in the diet. They are needed to synthesize prostaglandins and phospholipids for cell membranes.

Lipid Transport and Distribution

Lipids circulate through the bloodstream as *lipoproteins* and *free fatty acids*. **Lipoproteins** are lipid-protein complexes that contain large insoluble glycerides and cholesterol with a superficial coating dominated by phospholipids and proteins. The proteins and phospholipids make the entire complex soluble, and the proteins play a role in the regulation of lipid absorption by cells. (The mechanism will be detailed below.)

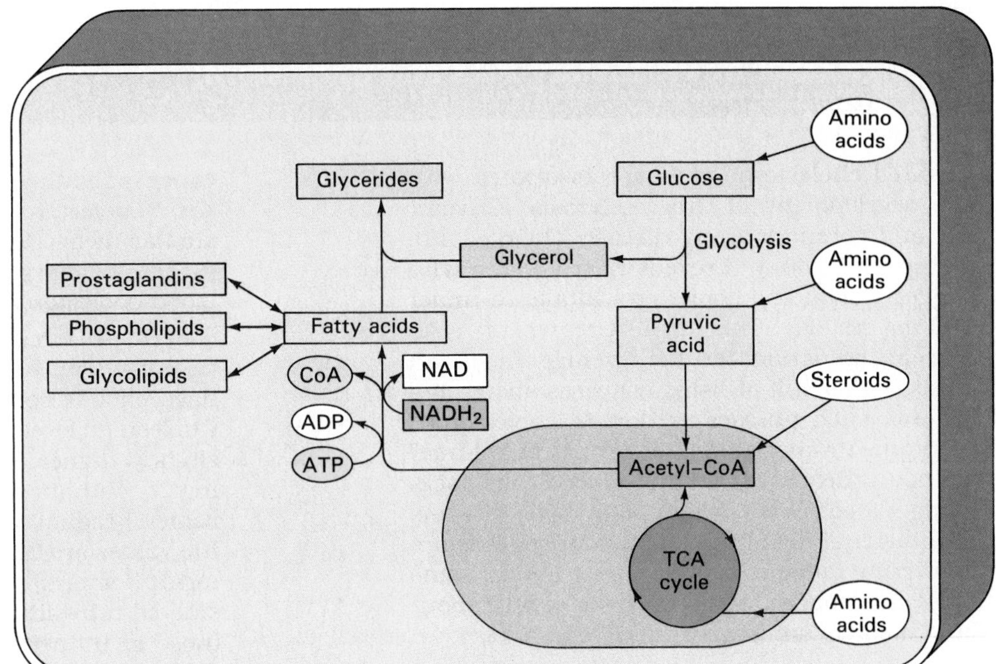

FIGURE 25-4
Lipid Synthesis. Pathways of lipid synthesis begin with acetyl-CoA. Molecules of acetyl-CoA can be strung together in the cytosol, yielding fatty acids. Those fatty acids can be used to synthesize glycerides or other lipid molecules. Lipids can be synthesized from amino acids or carbohydrates via acetyl-CoA.

Lipoproteins Lipoproteins are usually classified according to size and the relative proportions of lipid versus protein. The relationships between the different classes of lipoproteins will be examined after we introduce the major groups recognized at present:

1. **Chylomicrons**: Chylomicrons are the largest lipoproteins, ranging in diameter from 0.03 to 0.5 μm. They are produced by intestinal epithelial cells, as described in Chapter 24. (p. 800) Roughly 95 percent of the weight of a chylomicron consists of triglycerides.

   Chylomicrons carry absorbed lipids from the intestinal tract to the circulation. The other lipoproteins shuttle lipids between various tissues in the body, such as between the liver and adipose tissue. The liver is the primary source for all other types of lipoproteins.
2. **Very low-density lipoproteins (VLDLs)**: These lipid masses contain triglycerides manufactured by the liver plus small amounts of phospholipids and cholesterol. The primary function of VLDLs is the transport of these triglycerides to peripheral tissues. VLDLs range in diameter from 25 to 75 nm (0.025-0.075 μm).
3. **Intermediate-density lipoproteins (IDLs)** are intermediate in size and lipid composition between VLDLs and LDLs. They contain smaller amounts of triglycerides and relatively more phospholipids and cholesterol.
4. **Low-density lipoproteins (LDLs)**: These lipoproteins contain cholesterol, lesser amounts of phospholipids, and very few triglycerides. These lipoproteins, which are around 25 nm in diameter, deliver cholesterol to peripheral tissues.
5. **High-density lipoproteins (HDLs)**: High-density lipoproteins, around 10 nm in diameter, have roughly equal amounts of lipid and protein. The lipids are primarily cholesterol and phospholipids. The primary function of HDLs is transporting excess cholesterol from peripheral tissues back to the liver.

Figure 25-5 diagrams the probable relationships between these lipoproteins. Chylomicrons produced in the intestinal tract reach the venous circulation by entering lymphatic capillaries and traveling through the thoracic duct. Although chylomicrons are too large to diffuse across capillary walls, the endothelial lining of capillaries in adipose tissue, skeletal muscle, cardiac muscle, and the liver contains an enzyme, **lipoprotein lipase**, that breaks down complex lipids. When lipid complexes contact these endothelial walls, enzymatic activity releases fatty acids and monoglycerides that can diffuse across the endothelium and into the interstitial fluid.

Lipoprotein lipase also releases the lipid contents of other types of lipoproteins. The liver synthesizes VLDLs for discharge into the circulation. On arrival in peripheral capillaries, lipoprotein lipase removes many of the triglycerides, leaving an IDL. On return to the liver, additional triglycerides are removed, and the protein content is altered. This process creates an LDL that returns to peripheral tissues to deliver cholesterol.

These LDLs leave the circulation through capillary pores or cross the endothelium through the

Health News: Dietary Fats and Cholesterol

Elevated cholesterol levels are associated with the development of atherosclerosis (Chapter 21) and coronary artery disease (Chapter 20). (pp. 668, 644) Current nutritional advice suggests reducing cholesterol intake to under 300 mg per day. This amount represents a 40 percent reduction for the average American adult. As a result of rising concerns about cholesterol, such phrases as "low in cholesterol," "contains no cholesterol," and "cholesterol-free" are now widely used in the advertising and packaging of foods. What do they really mean in terms of individual health and diet planning? Before answering that question we must consider some basic information about cholesterol and about lipid metabolism in general:

1. Cholesterol has many vital functions in the human body. It serves as a waterproofing for the epidermis, a lipid component of all cell membranes, a key constituent of bile, and the precursor of several steroid hormones and one vitamin (vitamin D). Because cholesterol is so important, dietary restrictions should have the goal of keeping cholesterol levels within acceptable limits. The goal is *not* the elimination of cholesterol in the diet or the circulating blood.
2. Dietary restriction can modify cholesterol levels in the blood, but the diet is not the only source for circulating cholesterol. The human body can manufacture cholesterol from the acetyl-CoA obtained through glycolysis or the beta-oxidation of other lipids. If the diet contains an abundance of saturated fats, serum cholesterol levels will rise because excess lipids are broken down to acetyl-CoA that can be used to make cholesterol. This means that when trying to lower serum cholesterol by dietary control, other lipids must be restricted as well.
3. If the dietary supply of cholesterol is reduced, the body synthesizes more to maintain "acceptable" concentrations in the blood. The acceptable level depends on the genetic programming of the individual. Because individuals differ in genetic makeup, their cholesterol levels can vary even on similar diets. However, in virtually all instances dietary restrictions can lower blood cholesterol substantially.
4. Cholesterol levels vary with age and physical condition. At age 19, 3 out of 4 males have cholesterol levels below 170 mg/dℓ. Cholesterol levels in females of this age are slightly higher, typically at or below 175 mg/dℓ. With increasing age the cholesterol values gradually climb, and over age 70 the values are 230 mg/dℓ (males) and 250 mg/dℓ (females). Cholesterol levels are considered unhealthy if they are higher than those of 90 percent of the population in that age group. For males this value ranges from 185 mg/dℓ at age 19 to 250 mg/dℓ at age 70. For females the comparable values are 190 mg/dℓ and 275 mg/dℓ.

To determine whether you need to do anything about your cholesterol level without performing calculations regarding age, weight, and sex, just remember three simple rules:

- Individuals of any age with total cholesterol values below 200 mg/dℓ probably do not need to change their lifestyles unless they have a family history of coronary artery disease and atherosclerosis.
- Those with cholesterol levels between 200 and 239 mg/dℓ should modify their diets, lose weight (if overweight), and have annual checkups.
- Cholesterol levels over 240 mg/dℓ warrant drastic changes in dietary lipid consumption, perhaps coupled with drug treatment. Drug therapies are always recommended in cases where the serum cholesterol level exceeds 350 mg/dℓ. Examples of drugs used to lower cholesterol levels include *cholestyramine*, *colestipol*, and a newly approved drug, *lovastatin*.

mechanism shown in Figure 3-25. (p. 85) Once in peripheral tissues, the LDLs are absorbed by means of *receptor-mediated endocytosis* (see Figure 3-26). (p. 86) The endocytic vesicles formed fuse with lysosomes, and the lysosomal enzymes break down the contents. This step releases amino acids and cholesterol that diffuse into the cytoplasm. The cholesterol is then available for use by the cell in

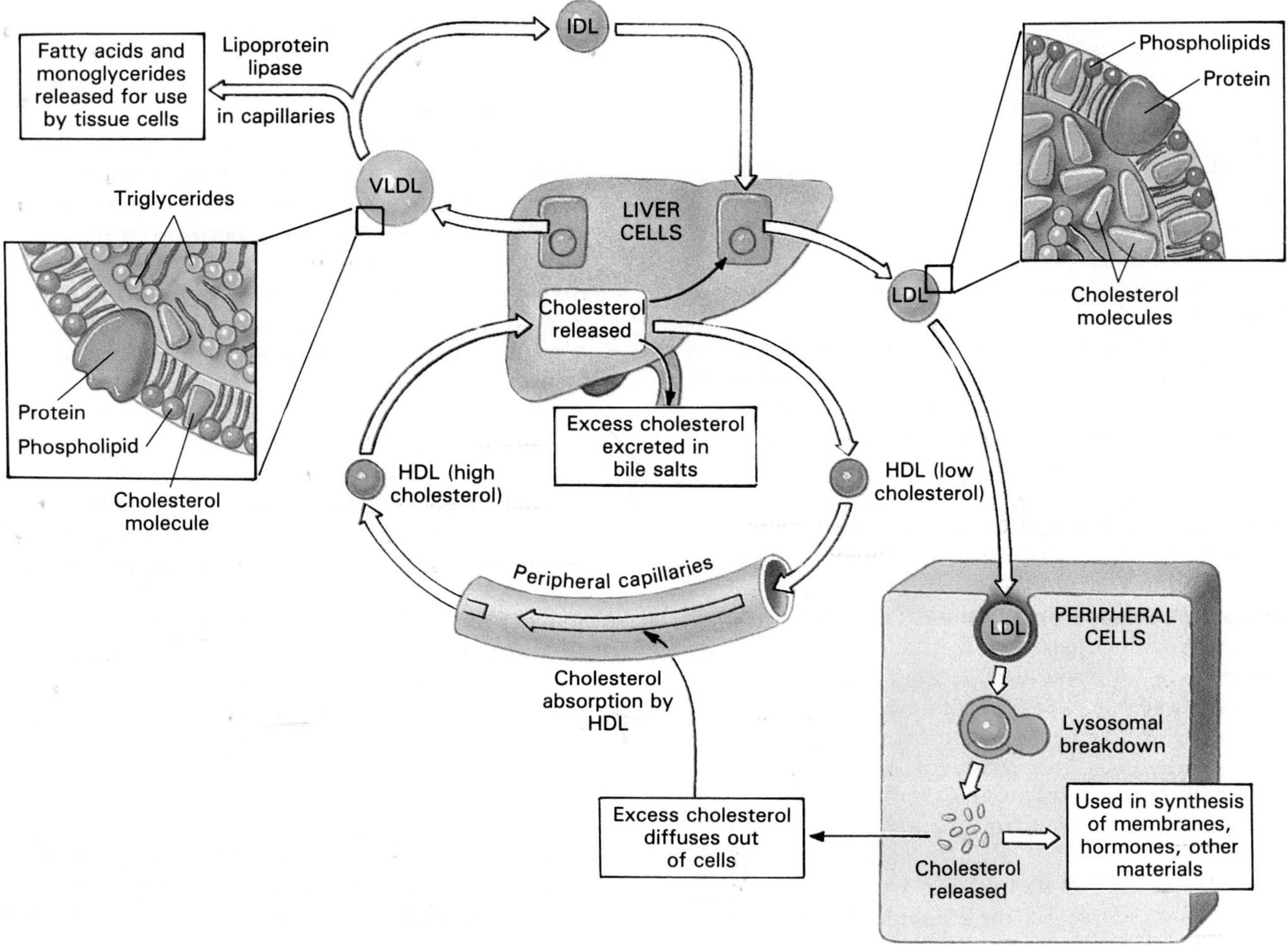

FIGURE 25-5
Lipid Transport and Utilization. Liver cells synthesize a VLDL that delivers triglycerides to peripheral tissues. Lipoprotein lipase in endothelial cells breaks down these triglycerides and releases fatty acids and monoglycerides that diffuse into the surrounding tissues. The IDL that remains returns to the liver, where it is absorbed and converted to an LDL that contains cholesterol. The LDL circulates to peripheral tissues, crosses the endothelium, and is absorbed by cells through endocytosis. The cholesterol is used in cellular processes; the excess diffuses back into the circulation. In the plasma, the cholesterol is absorbed by an HDL produced by the liver. On returning to the liver, the HDL is absorbed and the cholesterol extracted. Some of the cholesterol will be exported once again, in an LDL. Excess cholesterol will be excreted in bile salts.

the synthesis of lipid membranes. Excess cholesterol diffuses out of the cell and into the circulation, where it is absorbed by high-density lipoproteins and returned to the liver. On arrival in the liver, the HDLs are absorbed, and their cholesterol extracted. Some of the cholesterol recovered will be used in the synthesis of LDL, and the rest will be excreted in bile salts.

Free Fatty Acids **Free fatty acids (FFA)** are water-soluble lipids that can diffuse easily across cell membranes. Free fatty acids in the blood are usually bound to albumin, the most abundant plasma protein. Sources of free fatty acids include:

- fatty acids not used in the synthesis of triglycerides, which diffuse out of the intestinal epithelium and into the blood
- fatty acids that diffuse out of lipid stores, such as the liver and adipose tissue, when triglycerides are broken down.

Liver cells, cardiac and skeletal muscle fibers, and many other body cells can metabolize free fatty acids. They are an important energy source during periods of starvation, when glucose supplies are limited.

PROTEIN METABOLISM

There are roughly 100,000 different proteins in the human body, with varied forms, functions, and structures. All contain varying combinations of the same 20+ amino acids. (Appendix III details the structures of these amino acids.) Under normal conditions, there is a continual turnover of cellular proteins. Peptide

Diagnostics: Analysis of Cholesterol and Lipoprotein Levels

Standard blood tests report the total cholesterol, triglyceride, and lipid content of the plasma. Normal cholesterol values range from 150–250 mg/dℓ (see Health News: Dietary Fats and Cholesterol). Triglycerides are usually present at levels of 40–150 mg/dℓ, and the total lipid content of the blood may be from 400 to 1000 mg/dℓ.

When cholesterol levels are high, or when an individual has a family history of atherosclerosis or CAD, further tests may be performed to determine the relative amounts of cholesterol circulating in LDLs and HDLs. A high total cholesterol value linked to high LDL levels spells trouble. In effect, an unusually large amount of cholesterol is being exported to peripheral tissues. Problems can also exist if the individual has high total cholesterol—or even normal total cholesterol—and low HDL levels (below 35 mg/dℓ). In this case excess cholesterol delivered to the tissues cannot easily be returned to the liver for excretion. In either event the amount of cholesterol in peripheral tissues, and especially in arterial walls, is likely to increase. A standard guideline is that the ratio of total cholesterol to HDL cholesterol should be less than 4.5 to 1. If it is 4.5 or more, the individual is at risk for developing atherosclerosis.

bonds are broken, and the free amino acids are used to manufacture new proteins. This recycling occurs in the cytosol.

If other energy sources are inadequate, mitochondria can break down amino acids in the TCA cycle to generate ATP. Not all amino acids enter the TCA cycle at the same point, so the ATP benefits vary. However, the average yield is comparable to that of carbohydrate catabolism.

Amino Acid Catabolism

The first step in amino acid catabolism is the removal of the amino group. This requires a coenzyme derivative of **vitamin B$_6$** (*pyridoxine*). The amino group may be removed by **transamination** (trans-am-i-NĀ-shun) or **deamination** (dē-am-i-NĀ-shun). Figure 25-6 illustrates transamination and deamination reactions.

Transamination Transamination (Figure 25-6a) attaches the amino group of an amino acid to a **keto acid**. A keto acid resembles an amino acid except that the second carbon binds an oxygen atom rather than an amino group. Transamination produces a "new" amino acid that can enter the cytosol and be used for protein synthesis. It also converts the original amino acid to a keto acid that can enter the TCA cycle. Many different tissues perform transaminations. These reactions enable a cell to synthesize many of the amino acids needed for protein synthesis. Cells of the liver, skeletal muscles, heart, lung, kidney, and brain, which are particularly active in protein synthesis, perform large numbers of transaminations. [**CM:** *Phenylketonuria*]

Deamination Deamination (Figure 25-6b) is performed in preparing an amino acid for breakdown in the TCA cycle. Deamination is the removal of an amino group in a reaction that generates an ammonia molecule. Ammonia molecules are highly toxic, even in low concentrations. The liver, the primary site of deamination, has the enzymes needed to deal with the problem of ammonia generation. Liver cells take the ammonia and convert it to **urea**, a relatively harmless, water-soluble compound that is excreted in the

Health News: A Matter of Perception

The patient is an extremely thin adolescent woman brought to the clinic by her parents. Over the past few months her weight has declined by roughly 30 percent. Her skin is dry, there is some peripheral edema, and her heart rate and blood pressure are unusually low. She has trouble concentrating on schoolwork, and her short-term memory is poor. She is uncooperative, at least initially, and when questioned in detail she reveals her *real* problem: she's unhappy because she is just too fat!

The condition is called **anorexia nervosa**, and the cause is unknown. The pattern of physical symptoms resembles the patterns of severe starvation or protein deficiency diseases, but the problem occurs in relatively advanced societies where access to food is not a problem. Estimates of the incidence of anorexia range from 0.3 to 13.4 per 100,000 population, and the numbers are increasing. Affected individuals are obsessed with losing weight and perceive themselves as fat even when emaciated. Treatment is difficult, and only 50-60 percent of patients remain at normal weight for 5 years or more. Death rates from severe anorexia nervosa range from 5 to 10 percent.

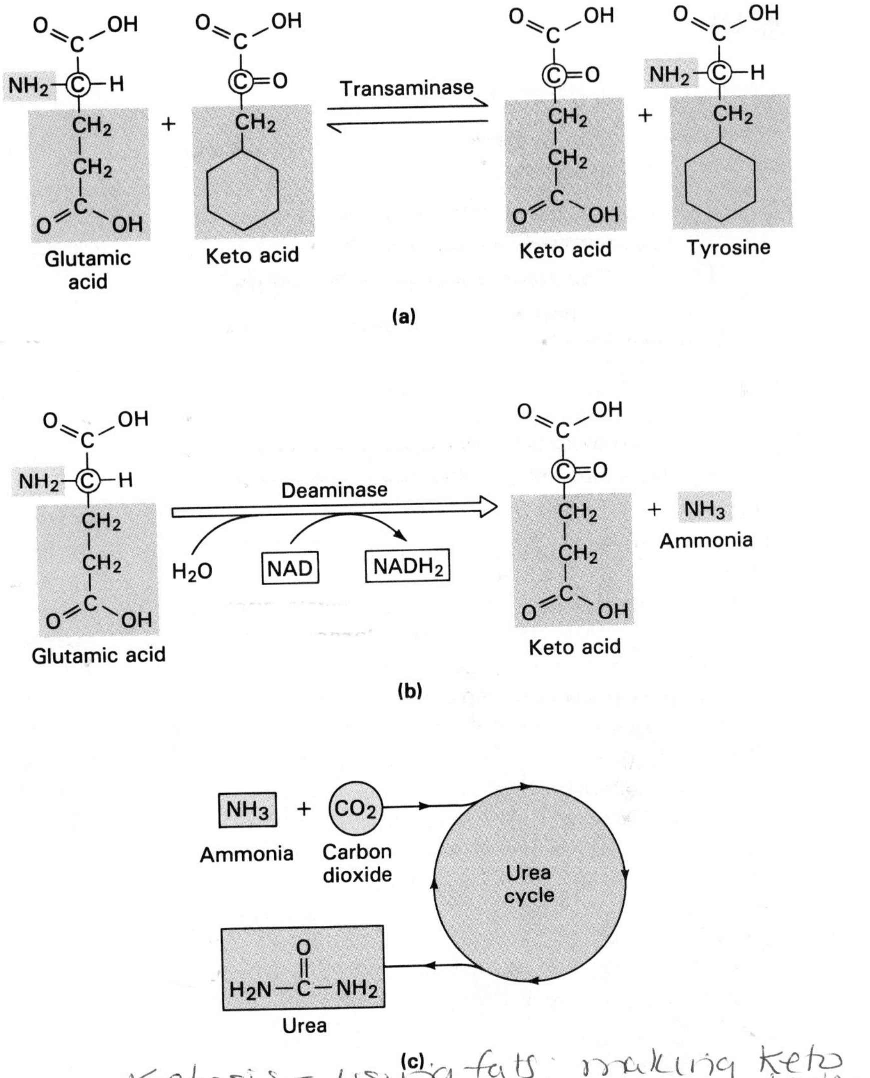

Glutamic acid + Keto acid ⇌ (Transaminase) Keto acid + Tyrosine

(a)

Glutamic acid → (Deaminase; H_2O, NAD → $NADH_2$) Keto acid + NH_3 Ammonia

(b)

NH_3 Ammonia + CO_2 Carbon dioxide → Urea cycle → Urea ($H_2N-C(=O)-NH_2$)

(c)

FIGURE 25-6
Amino Acid Catabolism. (a) During transamination an enzyme removes the amino group from one molecule and attaches it to a keto acid. (b) During deamination an enzyme strips the amino group from an amino acid and produces a keto acid and ammonia. (c) The urea cycle takes two metabolic waste products, carbon dioxide and ammonia, and produces a molecule of urea. Urea is a relatively harmless, soluble compound that is excreted in the urine.

urine. The **urea cycle** is the reaction sequence involved (Figure 25-6c).

When glucose supplies are low and lipid reserves are inadequate, liver cells break down internal proteins and absorb additional amino acids from the blood. The amino acids are deaminated, and the carbon chains are broken down to provide ATP.

Protein Synthesis

The basic mechanism for protein synthesis was detailed in Chapter 4 (Figures 4-12 and 4-13). (pp. 110, 112). The human body can synthesize roughly half of the different amino acids needed to build proteins. There are ten **essential amino acids**. Eight of them[1] cannot be synthesized at all; the other two[2] can be synthesized in amounts that are insufficient for growing children. The carbon frameworks of the amino acids not in the "essential" group can be synthesized readily, and a nitrogen group can be attached through transamination or **amination**. Amination is the attachment of an amino group, as indicated in Figure 25-7. Because the body can make these amino acids on demand, they are called the **nonessential amino acids**. [CM: *Protein Deficiency Diseases*]

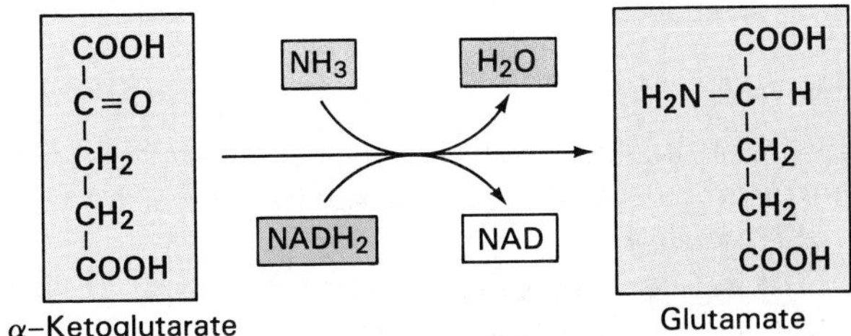

FIGURE 25-7
Amination. Amination attaches an amino group to a keto acid. This is an important step in the synthesis of nonessential amino acids. Amino groups can also be attached through transamination (see Figure 25-6).

[1] Isoleucine, leucine, lysine, threonine, tryptophan, phenylalanine, valine, and methionine.

[2] Arginine and histidine.

Proteins and Energy Production

Several factors make protein catabolism an impractical source of quick energy:

1. Proteins are more difficult to break apart than are complex carbohydrates or lipids.
2. Their energy yield is less than that of lipids.
3. One of the byproducts, ammonia, is a toxin that can damage cells.
4. Proteins form the most important structural and functional components of any cell. Extensive protein catabolism therefore threatens homeostasis at the cellular and systems levels.

NUCLEIC ACID METABOLISM

Living cells contain both DNA and RNA. Chapters 3 and 4 considered the replication of DNA, the mechanics of cell division, and the importance of DNA in regulating the structural and functional characteristics of the cell. ∞ (pp. 91–92, 109) The RNA in the cell is involved with protein synthesis.

Nucleic Acid Catabolism

RNA molecules are broken down and replaced on a regular basis. The bonds between nucleotides break first (see Figure 25-8). The nucleotides are usually recycled into new nucleic acids. However, the nucleotides can be catabolized to simple sugars and nitrogen bases. The sugars can enter the glycolytic pathways. Pyrimidines (cytosine, thymine, and uracil) are converted to acetyl-CoA and metabolized via the TCA cycle.

The genetic information contained in the DNA of the nucleus is absolutely essential to the long-term survival of the cell. As a result, the DNA in the nucleus

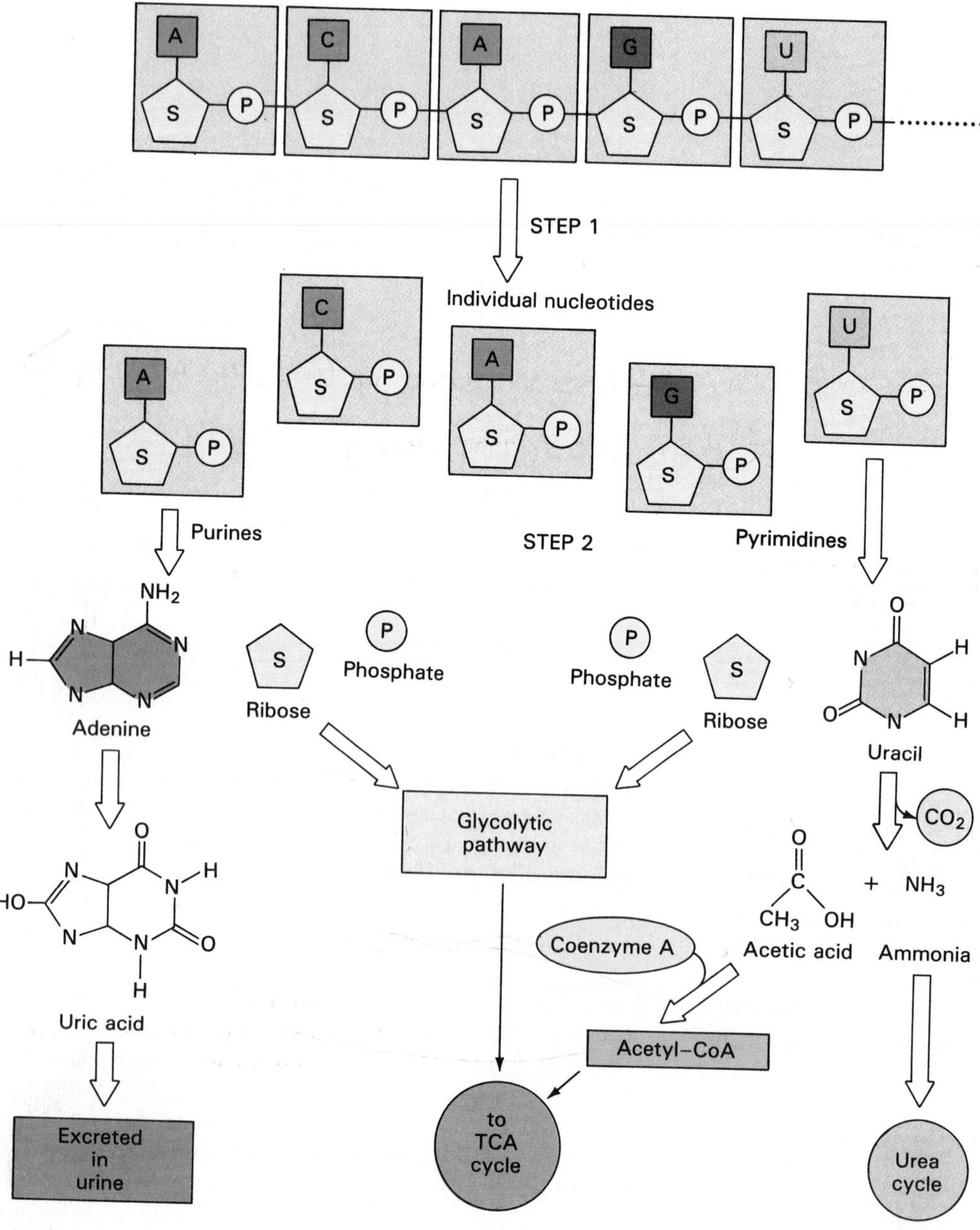

FIGURE 25-8 Nucleic Acid Catabolism. STEP 1: The RNA strand is broken down into its component nucleotides. STEP 2: Disassembling the nucleotides yields a phosphate group, a sugar (ribose), and a nitrogen base. The nitrogen base may be a purine or a pyrimidine. Purines cannot be used to provide energy. They are converted to uric acid and excreted in the urine. Pyrimidine catabolism produces acetyl-CoA, carbon dioxide, and ammonia.

is never catabolized for energy, even if the cell is dying of starvation.

Nucleic Acids and Energy Production

Nucleic acids are relatively insignificant contributors to the total energy reserves of the cell. Proteins account for 30 percent of the weight of the cell, and much more energy can be provided through the catabolism of nonessential proteins. Even when they are broken down, only the sugars and pyrimidines provide energy. The purines (adenine and guanine) cannot be catabolized at all. Instead they are deaminated and excreted as **uric acid**. Uric acid is another relatively nontoxic waste product, but it differs from urea in that it is far less soluble. Urea and uric acid are called **nitrogenous wastes** because they contain nitrogen atoms. [**CM:** *Gout*]

Nucleic Acid Synthesis

All cells synthesize RNA, but DNA synthesis only occurs in cells that are preparing for mitosis (cell division) or meiosis (gamete production). The process of DNA replication was described in Chapter 3. (p. 91) Messenger RNA, tRNA, and rRNA are transcribed by different forms of RNA polymerase. Messenger RNA is manufactured as needed, when specific genes are activated. Although several ribosomes can be reading the same message at any given moment, a strand of mRNA has a lifespan measured in minutes or hours. Molecules of ribosomal and transfer RNA in the cytosol are also broken down and replaced. Ribosomes are more durable than mRNA strands—the half-life of a ribosome is just over 5 days. Because each cell contains roughly 100,000 ribosomes, their replacement involves a considerable amount of synthetic activity.

How would a diet that is deficient in vitamin B_6 affect protein metabolism?

Elevated levels of uric acid in the blood could be an indicator of increased metabolism of what macromolecule?

AN OVERVIEW OF CELLULAR METABOLISM

Figure 25-9 summarizes the major pathways of cellular metabolism as described in this section. You should refer to this figure from time to time as you proceed into the next section dealing with metabolism at the tissue and organ levels.

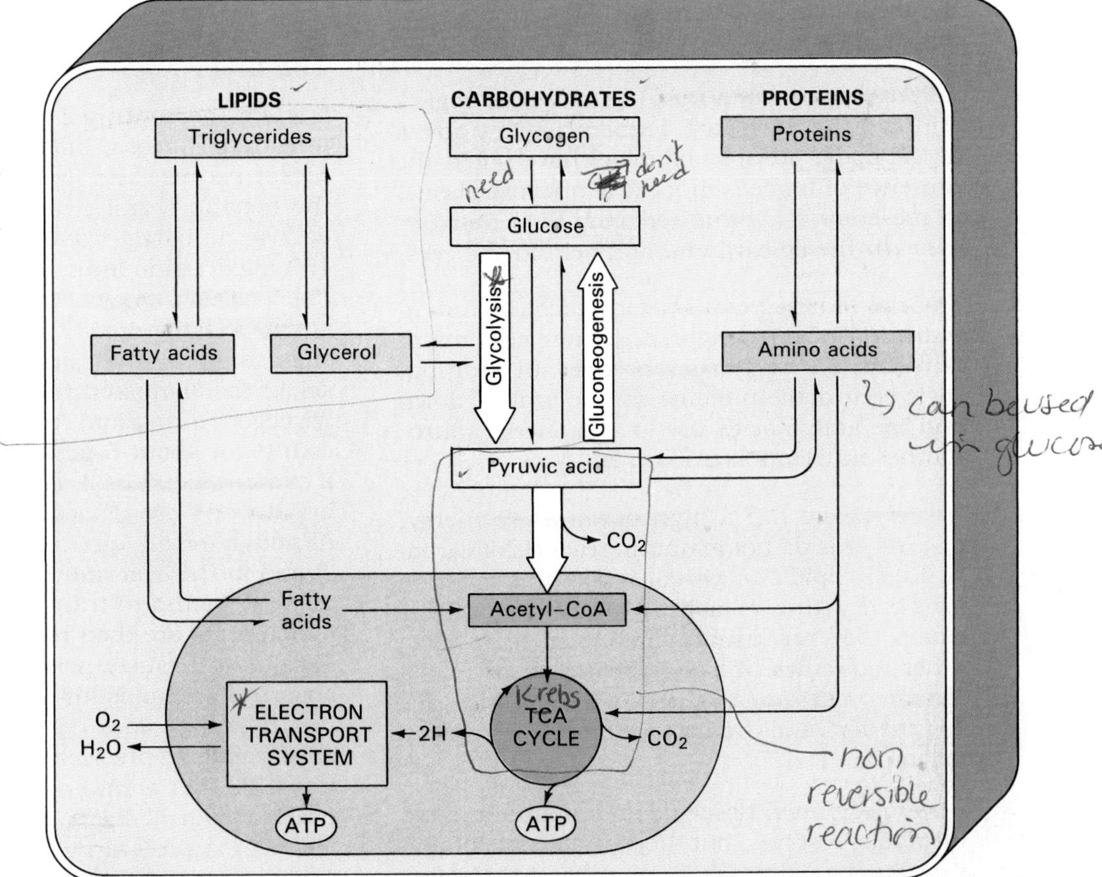

FIGURE 25-9 A Summary of the Pathways of Catabolism and Anabolism. This diagram provides an overview of major catabolic and anabolic pathways.

This diagram follows the reactions in a "typical" cell. Yet no one cell can perform all of the anabolic and catabolic operations and interconversions required by the body as a whole. As differentiation proceeds each cell type develops its own complement of enzymes, and this enzyme complement determines the cell's metabolic capabilities. In the presence of such cellular diversity, homeostasis can be preserved only when the metabolic activities of tissues, organs, and organ systems are coordinated.

Metabolic Interactions

The nutrient requirements of each tissue vary depending on the types and quantities of enzymes present in the cytoplasm. From a metabolic standpoint, the body can be considered in terms of five distinctive components: the *liver, adipose tissue, skeletal muscle, neural tissue,* and *other peripheral tissues*.

1. The liver represents the focal point for metabolic regulation and control. Liver cells contain a great diversity of enzymes, and they can break down or synthesize most of the carbohydrates, lipids, and amino acids needed by other cells in the body. Because of their extensive circulatory supply, liver cells are in an excellent position to monitor and adjust the nutrient composition of the circulating blood. The liver also contains significant energy reserves in the form of glycogen deposits.

2. Adipose tissue stores lipids, primarily as triglycerides. Adipocytes are found in many areas of the body; previous chapters have noted the presence of fat cells in loose connective tissue, in mesenteries, within red and yellow marrow, beneath the epicardium, and behind the eyes.

3. Skeletal muscle accounts for almost half of an individual's body weight, and these cells maintain substantial glycogen reserves. In addition, their contractile proteins can be broken down and the amino acids used as an energy source if other nutrients are unavailable.

4. Neural tissue has a high demand for energy, but the cells do not maintain reserves of carbohydrates, lipids, or proteins. Neurons must be provided with a reliable supply of glucose, because they are usually unable to metabolize other molecules. If blood glucose becomes too low, neural tissue in the CNS cannot continue to function, and the individual becomes unconscious.

5. Other peripheral tissues do not maintain large metabolic reserves, but they are able to metabolize glucose, fatty acids, or other substrates. Their preferred source of energy varies, depending on the instructions provided by the endocrine system.

The interrelationships among these five components can best be understood by considering the events that occur over a 24-hour period.

THE ABSORPTIVE STATE

After you have eaten, the absorption of nutrients continues for around 4 hours. If you are fortunate enough to eat three meals a day you spend 12 hours out of every 24 in the absorptive state. A typical meal contains proteins, lipids, and carbohydrates in varying proportions, and during the absorptive period the intestinal mucosa busily absorbs these nutrients. Glucose and amino acids enter the circulation, and the hepatic portal vein carries them to the liver. Most of the absorbed fatty acids enter the lacteals packaged in chylomicrons.

Some of these carbohydrates, lipids, and amino acids will be broken down at once to provide the energy needed to support cellular operations. The remainder will be stored, lessening the impact of future shortages. We will now consider the activities under way in specific sites, with particular reference to Figure 25-10.

The Liver

Despite the continual absorption of glucose at the intestinal mucosa, blood glucose levels do not skyrocket because the liver cells remove glucose from the hepatic portal circulation. Blood glucose levels do rise, but only from about 90 mg/dℓ to perhaps 150 mg/dℓ, even after a meal rich in carbohydrates. The liver uses some of the absorbed glucose to generate the ATP required to perform synthetic operations, such as glycogen formation, plasma protein synthesis, or the manufacture of proenzymes. Glycogenesis (glycogen formation) continues until glycogen accounts for about 5 percent of the total liver weight. If excess glucose still remains in the circulation, the hepatocytes use glucose to synthesize triglycerides. Although small quantities of lipids are normally stored in the liver, most of the synthesized triglycerides are bound to transport proteins. The resulting lipoproteins are then released into the bloodstream. Peripheral tissues, primarily adipose tissues, then absorb these lipids for storage.

The liver does not control circulating levels of amino acids as precisely as it does glucose concentrations. Plasma amino acid levels normally range between 35 and 65 mg/dℓ, but they may become elevated following a protein-rich meal. The absorbed amino acids are used to support the synthesis of proteins,

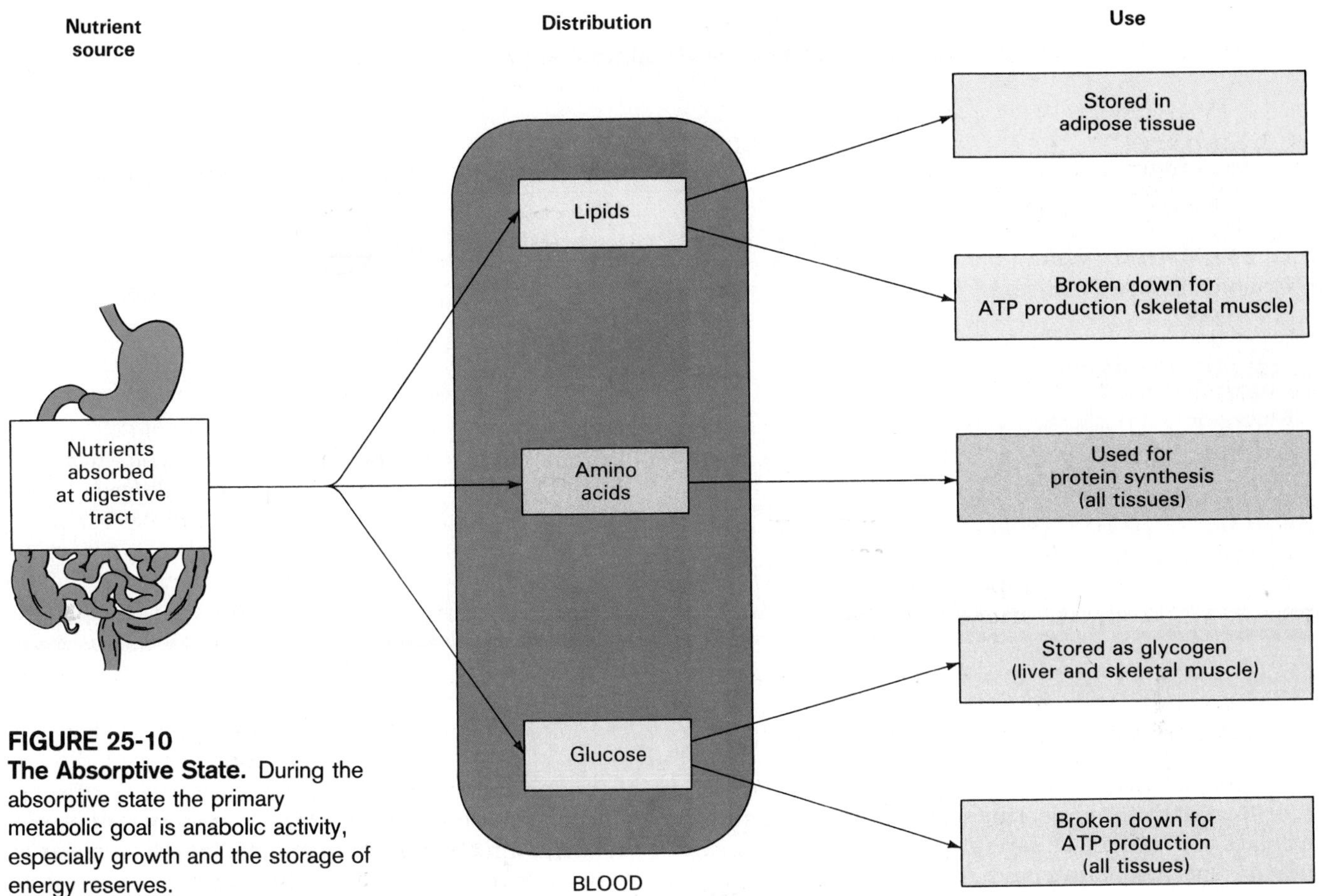

FIGURE 25-10
The Absorptive State. During the absorptive state the primary metabolic goal is anabolic activity, especially growth and the storage of energy reserves.

including plasma proteins and the proenzymes of the clotting system. Liver cells are also capable of synthesizing many amino acids, and an amino acid present in abundance may be converted to another, less common type, and released into the circulation.

Most of the lipids absorbed by the digestive tract do not reach the liver. Triglycerides, cholesterol, and large fatty acids bypass the liver, reaching the general venous circulation via the thoracic duct. Most of these lipids will be absorbed by other tissues rather than by the liver.

Adipose Tissue

During the absorptive state adipocytes remove fatty acids and glycerol from the circulation. Removal of lipids from the blood continues for 4–6 hours following a fatty meal. Over this period the presence of chylomicrons may give the plasma a milky appearance, a characteristic called **lipemia** (līp-Ē-mē-a).

Adipocytes are particularly active in absorbing these lipids and synthesizing new triglycerides for storage. At normal blood glucose concentrations any glucose entering these cells will be catabolized to provide the energy needed for lipogenesis (lipid synthesis). Adipocytes also absorb amino acids as required for protein synthesis. Although these cells can use glucose or amino acids to manufacture triglycerides, they do so only if circulating concentrations are unusually high. [**CM:** *Obesity*]

Skeletal Muscle, Neural Tissue, and Other Peripheral Tissues

When blood glucose and amino acid concentrations are elevated, all of the body's tissues increase their rates of absorption and utilization. Glucose molecules are catabolized for energy, and the amino acids are used to build proteins. Glucose is normally retained in the body because the kidneys reabsorb any glucose molecules entering the urine. This ability to conserve glucose breaks down only when blood glucose concentrations are extraordinarily high, somewhere in excess of 180 mg/dℓ. Amino acids are not reclaimed as readily, and amino acids often appear in the urine after a protein-rich meal.

When blood glucose levels are elevated, most cells ignore the circulating lipids, and the adipocytes have little competition. In resting skeletal muscles, a significant portion of the metabolic demand is met through the catabolism of fatty acids, and glucose molecules are used to build glycogen reserves that may account for 0.5–1 percent of the weight of each muscle fiber.

TABLE 25-1 Hormones and Their Effects on Peripheral Metabolism

State/Hormone	*Effect on General Peripheral Tissues*	*Effect on Liver*	*Effect on Adipose Tissue*	*Effect on Skeletal Muscle*
THE ABSORPTIVE STATE				
Insulin	Increases glucose uptake and utilization	Glycogenesis	Lipogenesis	Glycogenesis
Insulin and growth hormone	Increase amino acid uptake and protein synthesis			Fatty acid catabolism
THE POSTABSORPTIVE STATE				
Glucagon		Glycogenolysis		
Glucocorticoids	Most decrease use of glucose; increase reliance on ketone bodies and fatty acids	Glycogenolysis, gluconeogenesis	Lipolysis	Glycogenolysis, protein breakdown, and amino acid release
Growth hormone	Complements effects of glucorticoids			

The activities of the absorptive state are largely coordinated by the endocrine system. Table 25-1 summarizes the effects of several prominent hormones during the absorptive and postabsorptive periods.

THE POSTABSORPTIVE STATE

The remaining 12 hours of each day are spent in the postabsorptive state, although a person who is skipping meals or is on a hunger strike can extend it considerably. Attention in the postabsorptive state is focused on the mobilization of energy reserves and the maintenance of normal blood glucose levels. We will now examine the events under way in specific tissues during this period, as diagrammed in Figure 25-11.

The Liver and Gluconeogenesis

As the absorptive state ends, the intestinal cells stop providing glucose to the portal circulation. At first the peripheral tissues continue to remove glucose from the blood, and blood glucose levels begin to decline. The liver responds by reducing its synthetic activities, and when plasma concentrations fall below 80 mg/dℓ, liver cells begin breaking down glycogen reserves and releasing glucose into the circulation. As glycogen reserves decline and plasma glucose levels fall to around 70 mg/dℓ, liver cells begin to manufacture glucose in an attempt to stabilize blood glucose concentrations.

Through gluconeogenesis, liver cells synthesize glucose molecules from smaller carbon fragments. In effect, any carbon fragment that can be converted to pyruvic acid or one of the three-carbon compounds involved in the cytoplasmic reactions of glycolysis can be used to synthesize glucose. (The conversion of lactic acid to glucose in the liver was discussed in Chapter 10.) ∞ (pp. 320–321) With glucose already in short supply, lipids and amino acids must be catabolized to provide the ATP needed for these synthetic operations.

Utilization of Lipids During the postabsorptive state the liver absorbs fatty acids and glycerol from the blood. The three-carbon glycerol molecules are converted to glucose. Beta-oxidation of the fatty acids produces large quantities of acetyl-CoA, but because the conversion from pyruvic acid to acetyl-CoA cannot be reversed, these carbon chains cannot be used to synthesize glucose. Many of these two-carbon fragments are ultimately broken down in the TCA cycle. This breakdown provides most of the ATP needed for gluconeogenesis, as well as for the other synthetic operations carried out by liver cells over this period. In addition, some of the molecules of acetyl-CoA are converted to special compounds known as *ketone bodies.*

Ketone Bodies Ketone bodies are organic acids produced when lipid or amino acid catabolism is under way. These molecules form when the concentration of acetyl-CoA rises within the mitochondria. The first step involves the combination of two molecules of acetyl-CoA to form **acetoacetate** (as-ē-tō-AS-e-tāt). **Acetone** (AS-e-tōn) and **betahydroxybutyrate** (be-ta-hī-droks-e-BŪ-te-rāt) may be created by subsequent reactions. During the postabsorptive state the liver breaks down both lipids and amino acids, producing large amounts of acetyl-CoA. As the intracellular concentration of acetyl-CoA rises, ketone bodies form in significant numbers. The liver does not metabolize ketone bodies, and they diffuse through the cytoplasm and into the general circulation. Cells in peripheral tissues then absorb the ketone bodies and reconvert them to acetyl-CoA, for introduction into the TCA cycle.

The Utilization of Amino Acids Before an amino acid can be used for either gluconeogenesis or energy production via the TCA cycle, the amino group (NH_2) must be removed. The structure of the remaining carbon chain determines its subsequent fate. After deamination, some amino acids can be converted to molecules of pyruvic acid and then used for gluconeogenesis. These are known as the **glucogenic amino acids**. Other amino acids can be converted only to acetyl-CoA, and must be broken down further or converted to ketone bodies. These are called the **ketogenic amino acids**. All of the essential amino acids are ketogenic.

In the liver glucogenic and ketogenic amino acids are broken down, and the ammonia generated by the deamination reactions is converted to urea. This relatively harmless, water-soluble compound is later excreted in the urine. The urea concentration in the blood rises during the postabsorptive period, due to an increase in the rate of amino acid catabolism.

In summary, during the postabsorptive state the liver attempts to stabilize blood glucose concentrations, first by the breakdown of glycogen reserves, and later by gluconeogenesis. Over the remainder of the postabsorptive state, the combination of lipid and amino acid catabolism provides the necessary ATP and generates large quantities of ketone bodies that diffuse into the circulation.

FIGURE 25-11
The Postabsorptive State. In the postabsorptive state, energy reserves are mobilized and peripheral tissues (except neural) shift from glucose catabolism to fatty acid or ketone body catabolism to obtain energy.

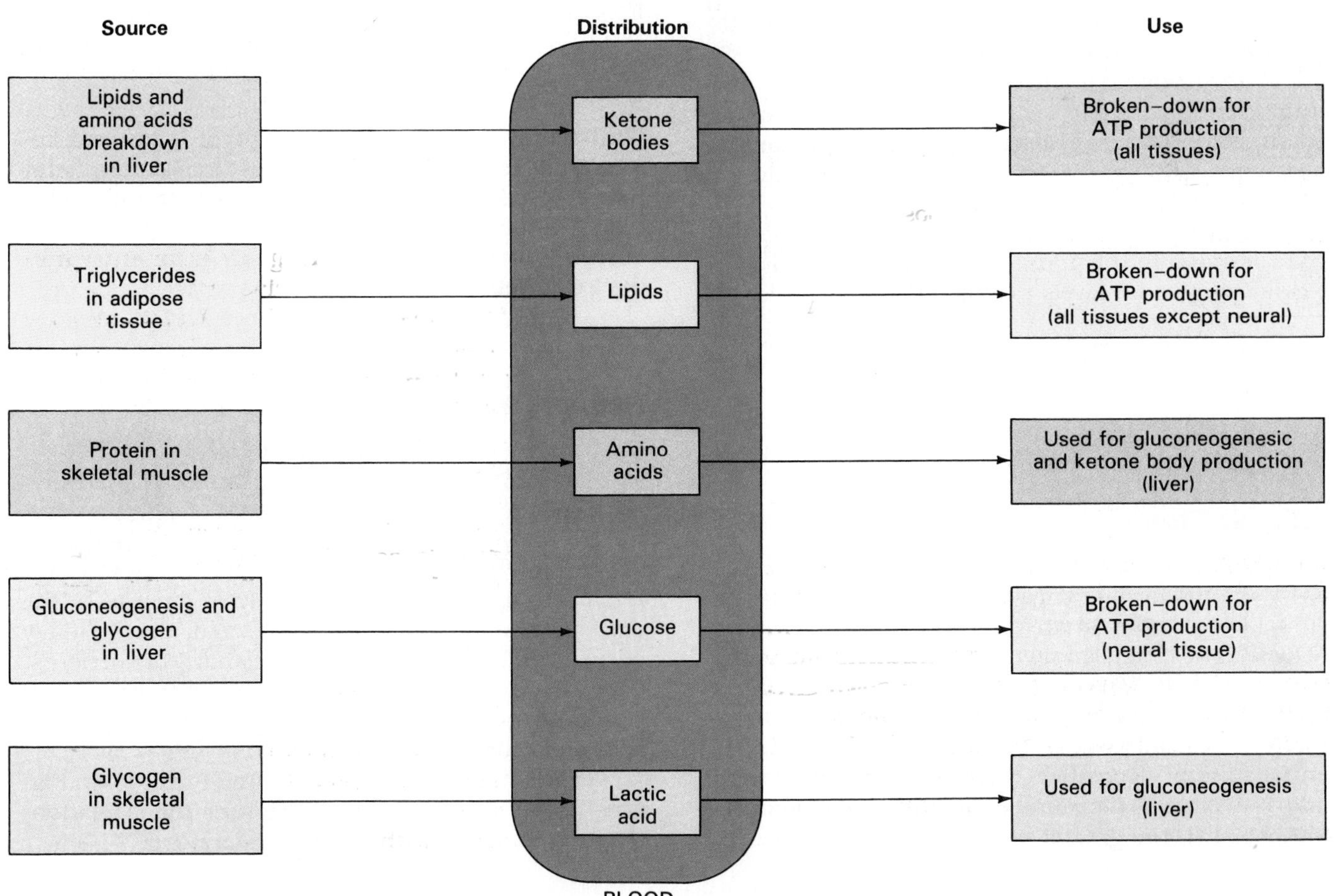

Clinical Comment: Ketosis and Ketoacidosis

The normal concentration of ketone bodies in the blood is around 30 mg/dℓ, and very few appear in the urine. During even a brief period of fasting, the increased production of ketone bodies results in **ketosis** (kē-TŌ-sis), a high concentration of ketone bodies in body fluids. **Ketonemia** (kē-tō-NĒ-mē-a) is an elevated concentration of ketone bodies in the blood, and **ketonuria** (kē-tō-NU-rē-a) is an elevated concentration in the urine. Both conditions indicate that catabolic activities are under way. Acetone, which diffuses into the alveoli very readily, may be smelled on the breath.

A ketone body is also called a "keto-acid" because it dissociates in solution, releasing a hydrogen ion. As a result, the appearance of ketone bodies in the circulation presents a threat to the plasma pH that must be controlled by buffers. During prolonged starvation, buffering capacities are exceeded, and a dangerous drop in pH occurs. This acidification of the blood is called **ketoacidosis** (kē-tō-as-i-DŌ-sis). In ketoacidosis the circulating concentration of ketone bodies can reach 200 mg/dℓ, driving the pH down to 7.05 or even lower. This lowered pH can disrupt normal tissue activities and cause coma, cardiac arrhythmias, and death.

In *diabetes mellitus* (Chapter 18) an inability to utilize glucose leads to lipid and protein catabolism to meet normal energy demands. ∞ (p. 594) The resulting **diabetic ketoacidosis** is the most common and life-threatening form of ketoacidosis. Such patients are starving in the midst of plenty, for their circulating glucose levels may be extremely high.

Adipose Tissue

Adipose tissue contains a tremendous storehouse of energy in the form of triglycerides. The average individual owes approximately 15 percent of his or her body weight to fat, enough to provide a 1–2 month reserve of ATP. Although some areas, including the eyelids, nose, and the backs of the hands and feet, rarely contain adipose tissue, other regions are preferential sites of deposition. Typically, an individual's adipose tissue is distributed among the hypodermis (50 percent), the greater omentum (10–15 percent), between muscles (5–8 percent), and packed around the kidneys (12 percent) and reproductive organs (15–20 percent).

As blood glucose levels decline, the rate of triglyceride synthesis falls. Under the stimulation of growth hormone and glucocorticoids, the adipocytes soon begin breaking down their lipid reserves, releasing fatty acids and glycerol into the bloodstream. This process, called *fat mobilization*, continues for the duration of the postabsorptive state.

Skeletal Muscle

At the start of the postabsorptive state, skeletal muscles obtain energy by breaking down their glycogen reserves and catabolizing the glucose released. As the concentrations of fatty acids and ketone bodies in the circulation increase, these substrates become increasingly important as an energy source. Glycogenolysis and glycolysis continue, but not all of the glucose molecules are ultimately broken down in the mitochondria. Instead, many of the pyruvic acid molecules generated are converted to lactic acid, which promptly diffuses into the circulation. As the postabsorptive state continues, muscle proteins are hydrolyzed by special enzymes called **cathepsins** (ka-THEP-sinz), and the amino acids released diffuse into the blood.

Other Peripheral Tissues

With rising plasma concentrations of lipids and ketone bodies and falling blood glucose levels, peripheral tissues gradually decrease their reliance upon glucose. Circulating ketone bodies and fatty acids are absorbed and converted to acetyl-CoA, for entry into the TCA cycle.

Neural Tissue

Neurons are unusual in that they continue "business as usual" during the postabsorptive state. Neurons are dependent on a reliable supply of glucose, and the alterations in the activity of liver, adipose tissue, skeletal muscle, and other peripheral tissues are intended to ensure that the supply of glucose to the nervous system continues unaffected, despite daily or even weekly changes in the availability of nutrients. Only after a prolonged period of starvation will neural tissue begin to metabolize ketone bodies and lactic acid molecules as well as glucose.

You should now take the time to compare Figures 25-10 and 25-11 until the functional relationships in each state are clearly understood.

ADJUSTMENTS TO STARVATION

The metabolic reserves of a typical 70-kg individual are indicated in Figure 25-12a. Because of its high energy content, the adipose tissue represents a disproportionate percentage of the total reserve in the form of triglycerides. Most of the available protein reserve is found in the contractile proteins of skeletal muscle. Carbohydrate reserves are sufficient for a typical postabsorptive period, between meals or overnight, but they do not provide a buffer against longer fasting periods.

Figure 25-12b shows changes in these metabolic stores during starvation. Carbohydrate utilization declines almost immediately, as the stores are depleted. The liver contains 75–100 g of glycogen which is readily available, and this reserve would be adequate to maintain blood glucose levels for about 4 hours. Skeletal muscle contains twice that amount of glycogen, but spread over a much greater total volume of tissue. These glucose reserves are not directly available to other tissues because the lack of a key enzyme makes skeletal muscle cells unable to release glucose molecules into the circulation. But even when shuffled to the liver as lactic acid, the combined carbohydrate reserves amount only to enough to get a person through a good night's sleep.

As blood glucose levels decline, gluconeogenesis accelerates, using glycerol, amino acids, and lactic acid. The glycerol is provided by adipocytes; the amino acids and lactic acid are provided primarily by skeletal muscle. At this point the kidneys begin to assist the liver by deaminating amino acids and generating additional glucose molecules.

Gluconeogenesis is accompanied by an increase in circulating ketone bodies, some derived from ketogenic amino acids, others from the catabolism of fatty acids. As the duration of the stress continues, peripheral tissues further restrict their glucose utilization, and the ketone bodies generated through fatty acid catabolism become the primary energy source.

The fasting individual gradually becomes weak and lethargic as peripheral systems are weakened by protein catabolism and stressed by pH changes. Buffer systems are challenged by the circulating amino acids, lactic acid, and ketone bodies, and ketoacidosis becomes a potential problem. Under these circumstances most tissues begin catabolizing ketone bodies almost exclusively, and in extreme starvation over 90 percent of the daily energy demands are met through the oxidation of ketone bodies. At this stage even neural tissue relies upon them to supplement declining glucose supplies. Table 25-2 presents a summary of the pattern of resource utilization during a prolonged fast.

Structural proteins are the last to be mobilized, with other forms, such as the contractile proteins of skeletal muscle, more readily available. When peripheral tissues catabolize proteins, the amino acids are exported to the liver, where they can be safely deaminated. The carbon fragments are then catabolized to provide ATP or used to manufacture glucose molecules or ketone bodies that can be broken down by peripheral tissues.

FIGURE 25-12
Metabolic Reserves and the Effects of Starvation. (a) Estimated metabolic reserves of a 70-kg individual. (b) Projected effects of prolonged fasting on the metabolic reserves of the same individual.

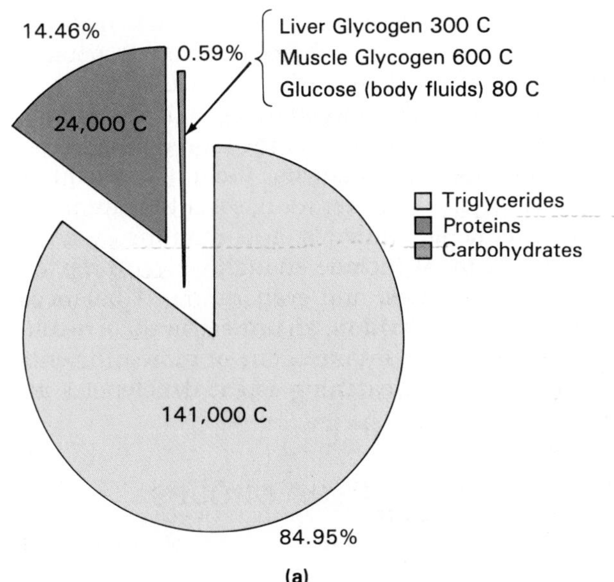

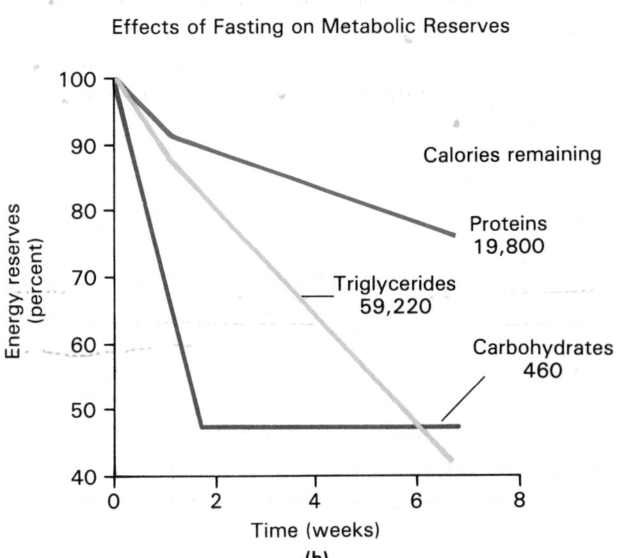

TABLE 25-2 The Maintenance of Plasma Glucose During Fasting

Duration of Fast	*Source of Glucose*	*Mechanism*	*Tissue Metabolism*	*Brain Metabolism*
3–4 hours	Diet	Intestinal absorption	Glucose	Glucose
4–16 hours	Liver	Liver glycogen	Glucose, but muscle and adipose tissue use declines	Glucose
16–48 hours	Liver	Gluconeogenesis	Glucose use reduced, ketone body formation increases	Glucose
2–24 days	Liver, kidney	Gluconeogenesis	Most tissues use ketone bodies as an energy source	Glucose, some ketone bodies
Over 4 weeks	Liver, kidney	Gluconeogenesis	Same	Increased dependence on ketone bodies

When lipid reserves are exhausted, crises follow relatively swiftly. On a gram-for-gram basis, cells must catabolize almost twice as much protein as lipid to obtain the same energy benefits. Making matters worse, by this time most of the easily mobilized proteins have already been broken down. As structural proteins are disassembled, a variety of unpleasant effects may appear. Accelerated protein catabolism causes problems with fluid balance, because the nitrogenous and acid wastes must be excreted in the urine. Urinary water losses climb, and the combination of dehydration and acidosis can damage the kidneys. When glucose concentrations can no longer be sustained above 40–50 mg/dℓ, the individual becomes disoriented and confused. The eventual cause of death may be kidney failure, ketoacidosis, protein deficiency, or hypoglycemia.

How long does it take to reach this critical state? It essentially depends on the size of the person's lipid reserves. Prolonged starvation for most people would be about 8 weeks, but the truly obese can hold out far longer. With adequate water and vitamin supplements, an 8-month fast has been used as a weight-loss technique. There are inherent risks involved, most notably the possibility of damage to the kidneys or severe ketoacidosis, but when one is that obese the risks of *not* doing anything are even greater.

Metabolic adjustments during starvation are coordinated primarily by the endocrine system. The glucocorticoids produced by the adrenal cortex are the most important hormones, aided by growth hormone from the pituitary. The effects of these hormones on peripheral tissues were included in Table 25-1.

What process in the liver would you expect to increase following a meal that is high in carbohydrates?

Why is there an increase in the amount of urea in the blood during the postabsorptive state?

Why is a person who is starving more susceptible to infectious diseases?

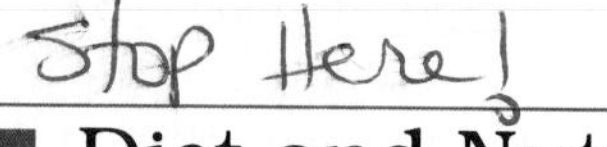

Diet and Nutrition

Metabolic adjustments during starvation can maintain the postabsorptive state for a considerable period, but a prolonged postabsorptive condition is abnormal and eventually fatal. Homeostasis can be maintained indefinitely only if the digestive tract absorbs fluids, organic substrates, minerals, and vitamins on a regular basis, keeping pace with cellular demands. The absorption of nutrients from food is called **nutrition**.

The individual requirement for each nutrient varies from day to day and from person to person. *Nutritionists* attempt to analyze a diet in terms of its ability to meet the needs of a specific individual. A **balanced diet** contains all of the ingredients necessary to maintain homeostasis, including adequate substrates for energy generation, essential amino acids and fatty acids, minerals, and vitamins. In addition, the diet must include enough water to replace losses in urine, feces, and evaporation. A balanced diet prevents **malnutrition**, an unhealthy state resulting from inadequate intake of one or more nutrients that becomes life-threatening as the deficiencies accumulate.

THE FOUR BASIC FOOD GROUPS

The traditional American method of avoiding malnutrition is to include members of each of the four **basic**

■ **TABLE 25-3 The Four Basic Food Groups**

Group	*Typical Foods*	*Nutrients Supplied*	*Deficiencies*	*Recommended Daily Intake*
Milk group	Milk, cheese, yogurt, other dairy products	Complete proteins, fats, carbohydrates, calcium, potassium, magnesium, sodium, phosphorus, vitamins A, B_{12}, pantothenic acid, thiamine, riboflavin	Dietary fiber, vitamin C	2 cups
Meat group	Meat, fish, poultry, eggs	Complete proteins, fats, potassium, phosphorus, iron, zinc, vitamins E, thiamine, B_6	Carbohydrates, dietary fiber, several vitamins	1 egg, and 4 oz of meat, fish, or poultry
Vegetable and fruit group	Potatoes, carrots, spinach, oranges, apples	Carbohydrates, vitamins A, C, E, folacin, dietary fiber, potassium	Often low in fats, calories, and protein	4 oz vegetables, 2 fruit, 1 potato
Bread and cereal group	Pasta, cereal, bread, margarine	Carbohydrates, vitamins E, thiamine, niacin, folacin, calcium, phosphorus, iron, sodium, dietary fiber	Fats (except in margarine)	3 oz cereal, 3 slices of bread, margarine

food groups in the diet. These are summarized in Table 25-3. Each differs from the others in the typical balance of proteins, carbohydrates, and lipids contained, as well as in the amount and identity of vitamins and minerals.

It should be realized that such groupings are rather artificial at best, and downright misleading at worst. What is important is to obtain nutrients in sufficient *quantity* (adequate to meet energy needs) and *quality* (including essential amino acids, fatty acids, vitamins, and minerals). How these are packaged is of no concern whatever. There is nothing magical about the number four, as long as intelligent choices are made. Although no single group can fulfill all dietary needs, a pair can do quite nicely, and a person could survive indefinitely without ever seeing an egg or a leafy vegetable. Conversely, there are often significant differences between the members of a particular group, so the wrong selections can lead to malnutrition even if all four groups are represented.

For example, consider the case of the essential amino acids. The liver cannot synthesize any of these ketogenic amino acids, and they must be obtained from the diet. Some members of the meat and milk groups, such as beef, fish, poultry, eggs, and milk, contain all of the essential amino acids in sufficient quantities. They are said to contain **complete proteins**. Many plants contain adequate *amounts* of protein, but they contain **incomplete proteins** that are deficient in one or more of the essential amino acids. True vegetarians, who restrict themselves to the fruit and vegetable group (with or without the bread and cereal group), must become adept at juggling the constituents of their meals to include a combination of ingredients that will meet all of their amino acid requirements. Even with a proper balance of amino acids, the vegetarian faces a significant problem, since vitamin B_{12} is obtained only from animal products.

Efforts are currently under way to expand the four groups to six by adding a fruit group and a fats, oils, and sweets group. These groups are then arranged in a *food pyramid* with the bread and cereal group at the bottom (Figure 25-13). The aim of this display is to emphasize the need to restrict dietary fats, oils, and sugar and to increase consumption of breads and cereals, which are rich in complex carbohydrates (polysaccharides such as starch).

NITROGEN BALANCE

A variety of important compounds in the body contain nitrogen atoms. These **N compounds** include:

1. Amino acids that are part of the framework of all proteins and protein derivatives, such as glycoproteins and lipoproteins;
2. Purines and pyrimidines, the nitrogen bases of RNA and DNA;
3. Creatine, important in energy storage in muscle tissue (as creatine phosphate);

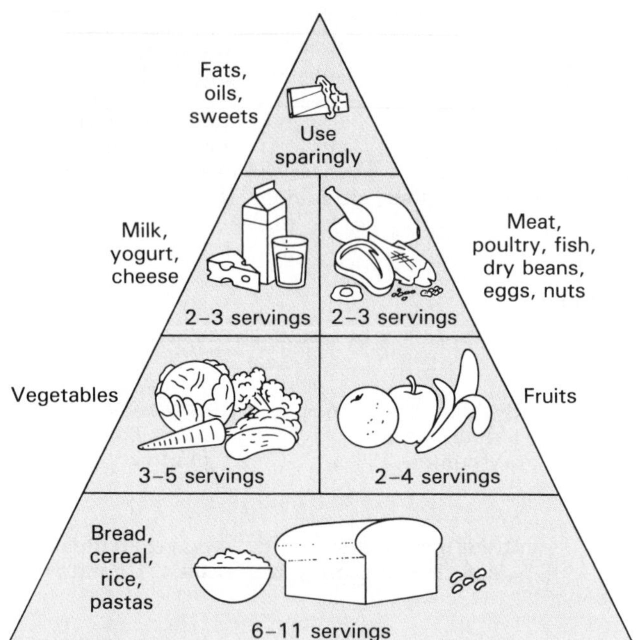

FIGURE 25-13
The Food Pyramid

4. Porphyrins, complex ring-shaped molecules that bind metal ions, and are essential to the function of hemoglobin, myoglobin, and the cytochromes.

N compounds are essential components of living systems, for they play key roles in determining the direction and rates of intracellular processes, and form structural proteins. When a person consumes an adequate diet of fats, carbohydrates, and proteins, the fats and carbohydrates are catabolized, and the amino acids are incorporated into proteins or converted to other N compounds. As a result, dietary fats and carbohydrates are often called **protein sparers**.

Despite the importance of nitrogen to these compounds, the body neither stores nitrogen nor maintains reserves of N compounds as it does carbohydrates (glycogen) or lipids (triglycerides). The carbon chains of the N compounds can be synthesized in the body, but the nitrogen atoms must be obtained either by recycling N compounds already in the body or by absorbing nitrogen from the diet. The individual is in **nitrogen balance** if the amount of nitrogen absorbed from the diet balances the amount lost in the urine and feces. This is the normal condition, and it means that the rates of N compound synthesis and breakdown are equivalent.

Growing children, athletes, and pregnant women are actively synthesizing N compounds, so they must absorb more nitrogen than they excrete. Such individuals are in a state of **positive nitrogen balance**. When excretion exceeds ingestion, **negative nitrogen balance** exists. This is an extremely unsatisfactory situation; the body contains only around a kilogram of nitrogen tied up in N compounds, and a decrease of one-third can be fatal. Even when energy reserves are mobilized, as during starvation, carbohydrates and lipid reserves are broken down first, and N compounds are conserved.

Such conservation is relative, not absolute. Proteins are the most abundant organic constituents of living cells, accounting for roughly 20 percent of total body weight. If periods of energy shortage are prolonged, protein catabolism becomes increasingly important as other reserves are exhausted. For example, during the first week of starvation a person may catabolize 1–1.5 kg of proteins, but in the eighth week the same person may be catabolizing the same amount each *day*.

MINERALS, VITAMINS, AND WATER

Minerals, vitamins, and water are essential components of the diet. The body cannot synthesize minerals, and our cells can generate only a small quantity of water and very few vitamins.

Minerals

Minerals are inorganic ions released through the dissociation of electrolytes. Minerals are important because:

1. Ions such as sodium and chloride determine the osmolarities of body fluids. Potassium is important in maintaining the osmolarity of the cytoplasm inside body cells.
2. Ions in various combinations play major roles in important physiological processes discussed, including
   - the maintenance of transmembrane potentials (Chapters 3, 10, and 12)
   - action potential generation (Chapter 12)
   - neurotransmitter release (Chapters 10 and 12)
   - muscle contraction (Chapters 10 and 20)
   - the construction and maintenance of the skeleton (Chapter 7)
   - the transport of respiratory gases (Chapter 23)
   - buffer systems (Chapters 2 and 27)
   - fluid absorption (Chapter 24)
   - waste removal (Chapters 26 and 27)
3. Ions are essential *cofactors* in a variety of enzymatic reactions. For example, the calcium-dependent ATPase in skeletal muscle also requires the presence of magnesium ions, and another ATPase required for the conversion of glucose to pyruvic acid needs both potassium and magnesium ions. Carbonic anhydrase, important in CO_2 transport, buffering systems,

TABLE 25-4 Minerals and Mineral Reserves

Mineral	*Significance*	*Total Body Content*	*Primary Route of Excretion*	*Recommended Daily Intake*
BULK MINERALS				
Sodium	Major cation in body fluids; essential for normal membrane function	110 g, primarily in body fluids	Urine, sweat, feces	1.1–3.3 g
Potassium	Major cation in cytoplasm; essential for normal membrane function	140 g, primarily in cytoplasm	Urine	1.9–5.6 g
Chloride	Major anion in body fluids	89 g, primarily in body fluids	Urine, sweat	1.7–5.1 g
Calcium	Essential for normal muscle and nerve function, structural support of bones	1.36 kg, primarily in skeleton	Urine, feces	0.8–1.2 g
Phosphorus	As phosphate in high-energy compounds, nucleic acids, and structural support of bones	744 g, primarily in skeleton	Urine, feces	0.8–1.2 g
Magnesium	Cofactor of enzymes, required for normal membrane functions	29 g, 17 in skeleton and the rest in cytoplasm and body fluids	Urine	0.3–0.4g
TRACE MINERALS				
Iron	Component of hemoglobin, myoglobin, and cytochromes	3.9 g, 1.6 stored (ferritin or hemosiderin)	Urine (traces)	10–18 mg
Zinc	Cofactor of enzyme systems, notably carbonic anhydrase	2 g	Urine, hair (traces)	15 mg
Copper	Required for hemoglobin synthesis, as cofactor	127 mg	Urine, feces (traces)	2–3 mg
Manganese	Cofactor for some enzymes	11 mg	Feces, urine (traces)	2.5–5 mg

and gastric acid secretion, requires the presence of zinc ions. Finally, the components of the electron transport system each require an iron atom, and the terminal cytochrome must bind a copper ion as well.

The major minerals and a summary of their functional roles are presented in Table 25-4. The body contains significant reserves of several important minerals, and this helps to reduce the effects of dietary variations in supply. The reserves are often relatively small, however, and chronic dietary reductions can lead to a variety of clinical problems. Alternatively, because storage capabilities are limited, a dietary excess of mineral ions can prove equally dangerous. [**CM:** *Iron Deficiencies and Excesses*]

Vitamins

Vitamins are needed in very small amounts, and they affect certain tissues more than others. As a result, two individuals who differ in body size, genetic programming, or activity patterns may have very different vitamin requirements.

Vitamins can be assigned to either of two groups, depending on their chemical structure and characteristics: *fat-soluble vitamins* and *water-soluble vitamins.*

Fat-Soluble Vitamins Vitamins A, D, E, and K are the **fat-soluble vitamins**. These vitamins are absorbed primarily from the digestive tract along with the lipid contents of micelles, but the skin can synthesize small amounts of vitamin D when exposed to sunlight.

There is considerable uncertainty over the mode of action of these vitamins. Vitamin D binds to cytoplasmic receptors within the intestinal epithelium and somehow promotes an increase in the rate of intestinal uptake of calcium and phosphorus. Vitamin A has long been recognized as a structural component of the visual pigment retinal, but its more general metabolic effects are not well understood. Vi-

TABLE 25-5 The Fat-Soluble Vitamins

Vitamin	Significance	Sources	Daily Requirement	Effects of Deficiency	Effects of Excess
A	Maintains epithelia; required for synthesis of visual pigments	Leafy green and yellow vegetables	1 mg	Retarded growth, night blindness, deterioration of epithelial membranes	Liver damage, skin peeling, CNS effects (nausea, anorexia)
D	Required for normal bone growth, calcium and phosphorus absorption at gut, and retention at kidneys	Synthesized in skin exposed to sunlight	None[a]	Rickets, skeletal deterioration	Calcium deposits in many tissues, disrupting functions
E (tocopherols)	Prevents breakdown of vitamin A and fatty acids	Meat, milk, vegetables	12 mg	Anemia; other problems suspected	None reported
K	Essential for liver synthesis of prothrombin and other clotting factors	Vegetables; production by intestinal bacteria	0.7–0.14 mg	Bleeding disorders	Liver dysfunction, jaundice

[a] Unless there is poor exposure to sunlight for extended periods. Alternative sources are provided in fortified milk products.

tamin K appears to be a necessary participant in a reaction essential to the synthesis of several proteins, including at least three of the clotting factors. Vitamin E probably stabilizes intracellular membranes. Current information concerning the fat-soluble vitamins is summarized in Table 25-5.

Because they dissolve in lipids, fat-soluble vitamins normally diffuse into cell membranes and other lipids in the body, including the lipid inclusions in the liver and adipose tissue. The body therefore contains a significant reserve of these vitamins, and normal metabolic operations can continue for several months after dietary sources have been cut off. As Table 25-5 points out, *too much* of a vitamin may produce effects just as unpleasant as *too little*. **Hypervitaminosis** (hī-per-vī-ta-min-Ō-sis) occurs when the dietary intake exceeds the abilities to store, utilize, or excrete a particular vitamin. This condition most often involves one of the fat-soluble vitamins because the excess is retained and stored in body lipids. [**CM:** *Hypervitaminosis*]

Water-Soluble Vitamins Most of the **water-soluble vitamins**, detailed in Table 25-6, are components of coenzymes. For example, NAD is derived from niacin, FAD from riboflavin, and coenzyme A from pantothenic acid.

Water-soluble vitamins are rapidly exchanged between the fluid compartments and the circulating blood, and excessive amounts are readily excreted in the urine. For this reason hypervitaminosis involving water-soluble vitamins is relatively uncommon.

Because these vitamins are not stored in large quantities, insufficient intake may lead to initial symptoms of vitamin deficiency within a period of days to weeks. The condition that results is termed a **deficiency disease**, or **avitaminosis** (ā-vī-ta-min-Ō-sis). Avitaminosis involving either fat-soluble or water-soluble vitamins can be caused by a variety of factors other than dietary deficiencies. An inability to absorb a vitamin from the digestive tract, inadequate storage, or excessive demand may each produce the same result. Interestingly enough, the bacterial inhabitants of our intestines help prevent deficiency diseases by producing five of the nine water-soluble vitamins, in addition to fat-soluble vitamin K.

Vitamin B_{12} has several unusual features. All of the water-soluble vitamins except B_{12} can be easily absorbed by the intestinal epithelium. The B_{12} molecule is large, and as you will recall from Chapter 24 it must be bound to the *intrinsic factor* from the gastric mucosa before absorption can occur. (p. 824) Vitamin B_{12} is also unusual because the liver contains reserves as large as those of any fat-soluble vitamin.

Water

Daily water requirements average 2500 mℓ, or roughly 40 mℓ/kg body weight per day. The specific requirement varies with environmental and metabolic activity. For example, exercise increases metabolic energy requirements, and it also accelerates water losses due to evaporation and perspiration. The

TABLE 25-6 The Water-Soluble Vitamins

Vitamin	*Significance*	*Sources*	*Daily Requirement*	*Effects of Deficiency*	*Effects of Excess*
B_1 (thiamine)	Coenzyme in decarboxylation reactions	Milk, meat, bread	1.9 mg	Muscle weakness, CNS and cardiovascular problems including heart disease; called *beriberi*	Hypotension
B_2 (riboflavin)	Part of FMN and FAD	Milk, meat	1.5 mg	Epithelial and mucosal deterioration	Itching, tingling sensations
Niacin (nicotinic acid)	Part of NAD	Meat, bread, potatoes	14.6 mg	CNS, GI, epithelial, and mucosal deterioration; called *pellagra*	Itching, burning sensations, vasodilation, death after large dose
B_6 (pyridoxine)	Coenzyme in amino acid and lipid metabolism	Meat	1.42 mg	Retarded growth, anemia, convulsions, epithelial changes	CNS alterations, perhaps fatal
Folacin (folic acid)	Coenzyme in amino acid and nucleic acid metabolism	Vegetables, cereal, bread	0.1 mg	Retarded growth, anemia, gastrointestinal disorders	Few noted except at massive doses
B_{12} (cobalamin)	Coenzyme in nucleic acid metabolism	Milk, meat	4.5 μg	Impaired iron absorption causing *pernicious anemia*	Polycythemia
Biotin	Coenzyme in decarboxylation reactions	Eggs, meat, vegetables	0.1–0.2 mg	Fatigue, muscular pain, nausea, dermatitis	None reported
Pantothenic acid	Part of acetyl-CoA	Milk, meat	4.7 mg	Retarded growth, CNS disturbances	None reported
C (ascorbic acid)	Coenzyme; delivers hydrogen ions, antioxidant	Citrus fruits	60 mg	Epithelial and mucosal deterioration; called *scurvy*	Kidney stones

temperature rise accompanying a fever has a similar effect, and for each degree the temperature rises above normal the daily water loss increases by 200 mℓ. Thus the advice "drink plenty of fluids" when one is sick has a definite physiological basis.

Most of the daily water ration is obtained by eating or drinking. The food consumed provides roughly 48 percent, and another 40 percent is obtained by drinking fluids. But a small amount of water is actually produced in the mitochondria by the activities of the electron transport system. Water produced in mitrochondria is called *metabolic water*. Table 25-7 indicates the amount of water produced during the catabolism of fats, carbohydrates, and lipids. The actual amount produced per day would therefore vary depending on the composition of the diet. A typical mixed diet in the United States contains 46 percent carbohydrates, 40 percent lipids, and 14 percent protein. This diet would produce roughly 300 mℓ of water per day, about 12 percent of the average daily requirement.

TABLE 25-7 Water Balance

Daily Input	
Water content of food	1000
Water consumed as liquid	1000
Metabolic water during catabolism[a]	300
Total	2300 mℓ
Daily Output	
Urine	1000
Evaporation at skin	750
Evaporation at lungs	400
Lost in feces	150
Total	2300 mℓ

[a] Each gram of lipid catabolized generates 1.7 mℓ of water; for proteins and carbohydrates the values are 0.41 mℓ and 0.55 mℓ.

DIET AND DISEASE

Diet has a profound influence on general health. We have already considered the effects of too many or

Health News: Water and Weight Loss

The safest way to lose weight is to reduce the intake of food while ensuring that all dietary essentials are available in adequate quantities. Water must be included on the list of essentials along with the amino acids, fatty acids, vitamins, and minerals. Because nearly half of our normal water intake comes from food (see Table 25-7), a person who eats less becomes more dependent on drinking fluids and whatever water is generated metabolically. At the start of a fast, the system conserves water and catabolizes lipids. That is why the first week of dieting may seem rather unproductive. Thus starving individuals actually need to increase their fluid intake or risk dehydration. As ketogenesis accelerates, more water must be lost at the kidneys to remove the accumulating waste products. Dehydration under these conditions can be especially dangerous because the concentrations of ketone bodies and nitrogenous wastes in the circulation will increase as the volume of the plasma declines.

too few nutrients, hypervitaminosis or avitaminosis, and above or below normal concentrations of minerals. More subtle, long-term problems may be encountered when the diet includes the wrong proportions or combinations of nutrients. The average American diet contains too many calories, and too great a proportion of those calories are provided by lipids. Only protein consumption figures are close to ideal values. This diet increases the incidence of obesity, heart disease, atherosclerosis, hypertension, and diabetes in the U.S. population. A summary of U.S. government recommendations for improving our diet is included in Table 25-8. [**CM:** *Alcohol: A Risky Diversion*]

The goal of a proper nutritional program is to provide a diet containing all essential nutrients that will also meet the energy needs of the individual. Most readers will be familiar with the term calorie counting; the next section considers the meaning and significance of the practice. [**CM:** *Nutrition and Nutritionists*]

- **Would an athlete in intensive training try to maintain positive or negative nitrogen balance?**
- **How would a decrease in the amount of bile salts in the bile affect the amount of vitamin A in the body?**

TABLE 25-8 Dietary Changes Advocated by the U.S. Government

	Percentage of Total Daily Energy Intake	
Dietary Component	*Average American Diet*	*Dietary Goals*
Carbohydrates	46	58
Lipids	38	30
Proteins	16	12

Summary of Recommendations:
1. Increase consumption of fruits, vegetables, and whole grains.
2. Decrease consumption of refined sugars.
3. Decrease consumption of fats, replace saturated with unsaturated fats.
4. Decrease consumption of animal fats, by selecting lean meats, poultry, and fish.
5. Decrease consumption of high-cholesterol foods, such as butter and eggs.
6. Decrease consumption of salt and foods high in salt content.
7. Decrease caloric intake to maintain desirable weight.

Bioenergetics

When chemical bonds are broken, energy is released. Inside cells, some of that energy may be captured as ATP, but much of it is lost to the environment as heat. The process of **calorimetry** (kal-o-RIM-e-trē) determines the total amount of energy released when the bonds of organic molecules are broken. The unit of measurement is the **calorie** (KAL-o-rē), defined as the amount of energy required to raise the temperature of 1 gram of water one degree centigrade. One gram of water is not a very practical measure when you are interested in the metabolic operations that keep a 70-kg human alive, so the **kilocalorie** (KIL-o-kal-o-rē) (kc), or simply **Calorie** (with a capital C), is used instead. Each Calorie represents the amount of energy needed to raise the temperature of 1 *kilo*gram of water one degree centigrade. When you turn to the back of a dieting guide to check the caloric value of various foods, the numbers indicate Calories, not calories.

FOOD AND ENERGY

In living cells, organic molecules are oxidized to carbon dioxide and water. Oxidation also occurs when something burns, and this process can be experimentally controlled. A known amount of material is placed in a chamber, called a **calorimeter** (kal-o-RIM-e-ter), that is filled with oxygen and surrounded by a known volume of water. Once the material is inside, the chamber is sealed and the substance is electrically ignited. When it has completely oxidized, and only ash remains in the chamber, the number of Calories released can be determined by comparing the water temperatures before and after the test.

When considering the energy potential of food, the results are usually expressed in Calories per gram (C/g). The catabolism of lipids entails the release of a considerable amount of energy, roughly 9.46 C/g. The catabolism of carbohydrates and proteins is not as rewarding, since many of the carbon and hydrogen atoms are already bound to oxygen. Their average yields are comparable: 4.18 C/g for carbohydrate and 4.32 C/g for protein. Most foods are mixtures of fats, proteins, and carbohydrates, and the values in a "Calorie counter" vary as a result.

The energy yield at the cellular level is usually expressed in terms of Calories per mole of substrate. For each mole of glucose catabolized aerobically, 686 Calories are released. That's a lot of energy, but glycolysis occurs in a series of small steps. Converting a mole of ADP to ATP takes roughly 7.7 C, and if biological processes were 100 percent effective a mole of glucose could produce 89 moles of ATP. In actuality, our cells are unable to capture even half of the energy content of glucose or any other nutrient.

It is interesting to compare the energy provided by the cytoplasmic and mitochondrial stages of glycolysis. The anaerobic conversion of a mole of glucose to two moles of pyruvic acid nets the cell two moles of ATP, representing an energy capture of 15.5 C. Because the reaction sequence releases a total of 56 C, anaerobic glycolysis has a rather low efficiency of 27.5 percent (15.5/56). The other 40.5 C are lost as heat, and this explains why you warm up so rapidly when exercising at peak levels. The mitochondrial reactions are far more efficient; the TCA cycle and the electron transport chain capture 277 C out of the 627 C released when the pyruvic acid molecules are completely broken down. Thus, aerobic glycolysis has an efficiency of 44 percent, and for the complete catabolism of glucose, start to finish, the average slips only slightly, to about 42 percent (292.5/686).

METABOLIC RATE

It is possible to examine the metabolic state of an individual and determine how many Calories are being utilized. The result can be expressed as Calories per hour, Calories per day, or Calories per unit of body weight per day, but what is actually measured is the sum total of all of the varied anabolic and catabolic processes occurring in the body. This value represents the **metabolic rate** of the individual at that time. The metabolic rate will change according to the activity under way—sprinting and sleeping measurements are quite different. In an attempt to reduce the variations, the testing conditions are standardized so as to determine the **basal metabolic rate** (**BMR**). Ideally, the BMR would represent the minimum, resting energy expenditures of an awake, alert person.

A direct method of determining the BMR simply monitors respiratory activity, for in resting subjects energy utilization is proportional to oxygen consumption. Assuming that average amounts of carbohydrates, lipids, and proteins are being catabolized, the ratio works out to 4.825 Calories per liter of oxygen consumed.

An average individual has a BMR of 70 C per hour or about 1680 C per day. Although the test conditions are standardized, there are many uncontrollable factors that can influence the BMR. These include age, sex, physical condition, body weight, and genetic differences such as variations among ethnic groups.

Because measuring the BMR is technically difficult, and circulating thyroid hormone levels have a profound effect on the BMR, clinicians usually monitor the concentration of thyroid hormones rather than the actual metabolic rate. The results are then compared with normal values to obtain an index of metabolic activity. One such test, the **T_4 assay**, measures the amount of thyroxine in the blood.

Daily energy expenditures for a given individual vary widely depending on the activities undertaken. For example, a person leading a sedentary life may have near-basal energy demands, but a single hour of swimming can increase the daily caloric requirements by 500 C or more. If the daily energy intake exceeds the total energy demands of the individual, the excess will be stored, primarily as triglycerides in adipose tissue. If the daily caloric expenditures exceed the dietary supply, there will be a net reduction in the body's energy reserves and a corresponding loss in weight. This accounts for the significance of calorie-counting and exercise in a weight-control program.

THERMOREGULATION

The BMR estimates the rate of energy use by monitoring heat production. Although energy is lost as heat due to biological inefficiency, heat loss serves an important homeostatic purpose. Humans are subject to vast changes in environmental temperatures, but our complex biochemical systems have a major limitation: the enzyme systems will only operate over a relatively narrow range of temperatures. Our bodies have anatomical and physiological mechanisms that keep body temperatures within acceptable limits, regardless of the environmental conditions. This homeostatic process is called **thermoregulation**. Failure to

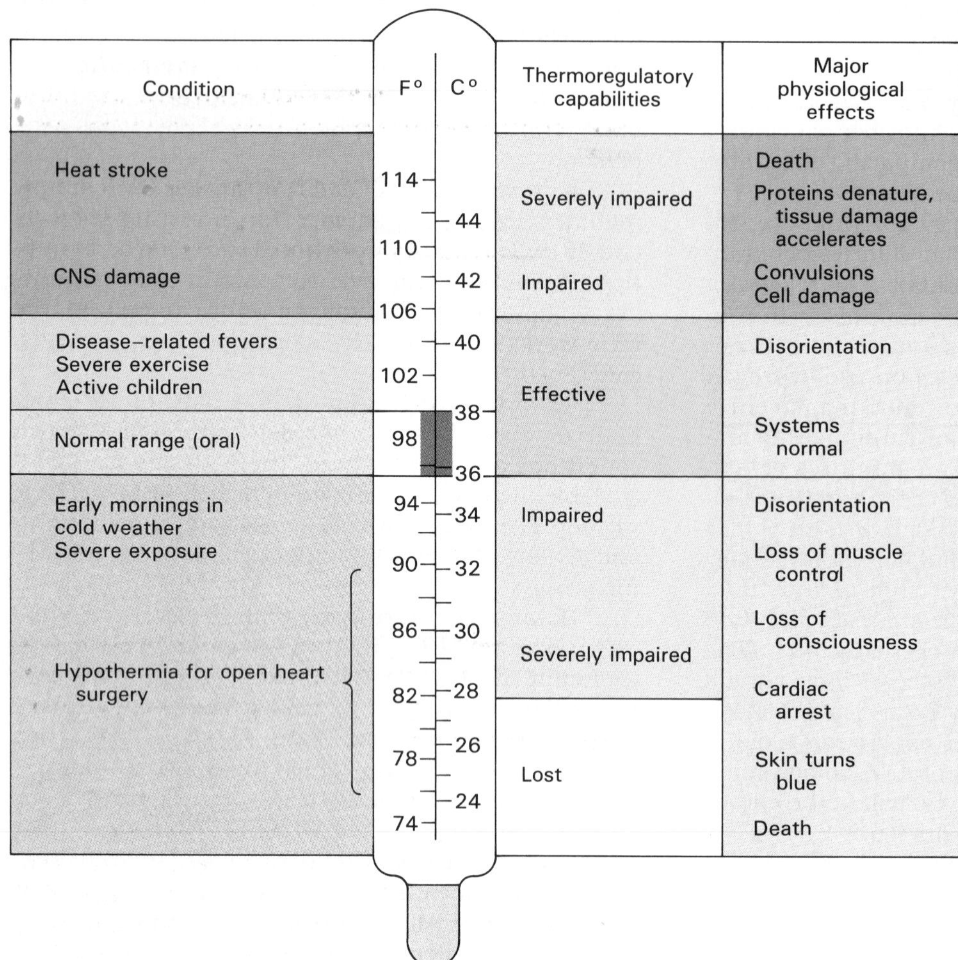

Condition
F°
C°
Thermoregulatory capabilities
Major physiological effects
Heat stroke
CNS damage
Disease-related fevers
Severe exercise
Active children
Normal range (oral)
Early mornings in cold weather
Severe exposure
Hypothermia for open heart surgery
114
110
106
102
98
94
90
86
82
78
74
44
42
40
38
36
34
32
30
28
26
24
Severely impaired
Impaired
Effective
Impaired
Severely impaired
Lost
Death
Proteins denature, tissue damage accelerates
Convulsions
Cell damage
Disorientation
Systems normal
Disorientation
Loss of muscle control
Loss of consciousness
Cardiac arrest
Skin turns blue
Death

FIGURE 25-14
Normal and Abnormal Variations in Body Temperature

control body temperature can result in a series of physiological changes, as indicated in Figure 25-14.

We are continually producing heat as a byproduct of metabolism, and that heat must be lost to the environment at the same rate if body temperature is to remain constant. When the environmental conditions vary from "ideal," becoming too warm or too cold, the gains or losses must be controlled to maintain homeostasis.

Mechanisms of Heat Transfer

Heat exchange with the environment involves four basic processes: *radiation, conduction, convection*, and *evaporation*. These mechanisms are illustrated in Figure 25-15.

Radiation Warm objects lose heat energy as infrared radiation. When we feel the heat from a stove top or from the sun, we are experiencing the radiant heat being given off by very hot objects. Our bodies lose heat the same way, but in proportionately smaller amounts. Over half of our heat loss is attributable to radiation; the exact percentage varies, depending on both body temperature and skin temperature. We can of course gain heat by radiational mechanisms as well as lose it—as any sun worshiper can attest.

Conduction Conduction refers to the direct transfer of energy through physical contact. When you arrive in an air-conditioned classroom and sit down on a cold plastic chair, you are immediately aware of this process. While occasionally startling, conduction is usually not an effective mechanism for gaining or losing heat.

Convection Convection is the result of conductive heat loss to the air that overlies the surface of the body. Warm air is lighter than cooler air, and so it rises. That is why your face gets blasted with hot air when you open an oven door at waist level. As the body conducts heat to the air adjacent to the skin, that air warms and rises, moving away from the surface. Cooler air replaces it, and as it in turn becomes warmed, the cycle repeats. Convection accounts for roughly 15 percent of our heat loss; this heat loss can be reduced by restricting air movement, as by bundling up with clothing. Convection is insignificant as a mechanism for heat gain.

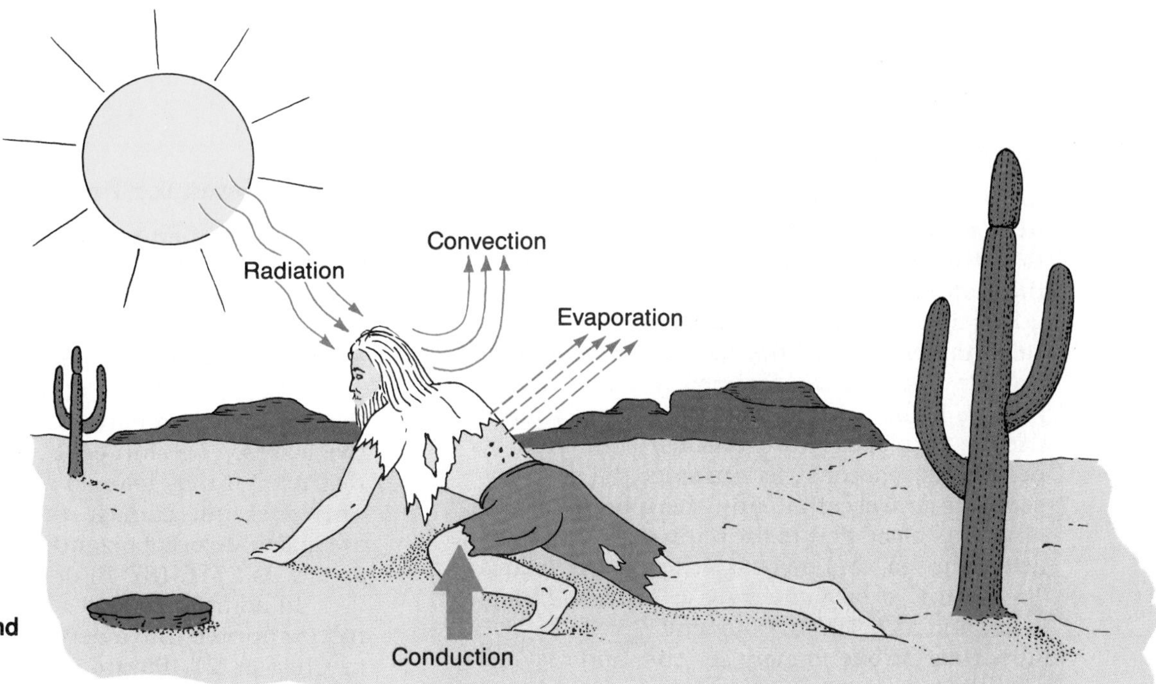

FIGURE 25-15
Routes of Heat Gain and Loss

Evaporation When water evaporates, it changes from a liquid to a vapor. This process absorbs energy, roughly 0.58 C per gram of water evaporated. When evaporation occurs at the surface of the skin, a significant amount of heat can be lost, which is one reason why the expression "cold and wet" is so often appropriate.

The rate of evaporation occurring at the skin is highly variable. Each hour, 20–25 mℓ of water crosses epithelia to be evaported from the alveolar surfaces and the surface of the skin. This **insensible perspiration** remains relatively constant; at rest it accounts for roughly one-fifth of the average heat loss. The sweat glands responsible for **sensible perspiration** have a tremendous range of activity, ranging from virtual inactivity to secretory rates of 2-4 liters per hour. This is equivalent to an entire day's resting water loss in under an hour. A maximal secretion rate would, if it were completely evaporated, remove 2320 C per hour!

Such levels of perspiration could not be sustained for much more than an hour, due to problems with fluid balance, but the figure should give you an idea of the adaptability of this mechanism. Its efficiency varies with the environmental conditions, especially the "relative humidity" of the air. If the air is saturated (100 percent humidity), it already holds as much water vapor as it will accept at that temperature. Under these conditions, evaporation is ineffective as a cooling mechanism. This is why humid, tropical conditions can be so uncomfortable—people perspire continually but remain warm and wet.

Biochemical reactions produce heat, and that heat is retained in the water that accounts for nearly 66 percent of the body weight. Water is an excellent conductor of heat, so the heat produced in one region of the body is rapidly distributed by diffusion, as well as via the circulation. The greater the number of biochemical reactions under way, the more heat will be produced. The relative efficiency of the reactions also plays a role, as noted in our comparison of anaerobic and aerobic glycolysis. As a result, heat production can be quite variable, and the maintenance of a constant body temperature requires a finely tuned and adaptable homeostatic mechanism. Thermoregulatory control mechanisms were discussed earlier, in Chapters 1 and 14. ∞ (pp. 8, 445)

Heat loss and heat gain involve the coordinated activity of many different systems, and the **heat loss center** and **heat gain center** of the preoptic area modify the activities of other hypothalamic nuclei. The heat-loss center exerts its principal effects via the parasympathetic division of the autonomic nervous system, and the heat-gain center directs many responses via the sympathetic division. The overall effect is to control temperature by influencing two events—the rate of heat production and the rate of heat loss to the environment. These may be further supported by behavioral modifications.

Mechanisms for Increasing Heat Loss

Mechanisms for increasing heat loss include *physiological mechanisms* and *behavioral modifications.*

Physiological Mechanisms When the temperature at the preoptic nucleus exceeds its thermostat setting, the heat-loss center is stimulated. Stimulation of this center has three major effects:

1. Inhibition of the vasomotor center causes peripheral vasodilation, and warm blood flows to

Clinical Comment: Heat Exhaustion and Heat Stroke

Heat exhaustion and *heat stroke* represent malfunctions of the thermoregulatory system. In **heat exhaustion**, also known as heat prostration, the individual experiences difficulties with the maintenance of blood volume. The heat-loss center is stimulating sweat glands that dump salts and water over the surface of the skin to provide evaporative cooling. As fluid losses mount, blood volume decreases. The resultant decline in blood pressure is not countered by peripheral vasoconstriction, since the heat-loss center is actively stimulating peripheral vasodilation. As blood flow to the brain declines, headache, nausea, and eventual collapse will follow. Treatment is straightforward: provide fluids, salts, and a cooler environment.

Heat stroke is more serious, and may occur following an untreated case of heat exhaustion. Predisposing factors include any preexisting condition, such as heart disease or diabetes, that affects peripheral circulation. The thermoregulatory center ceases to function, the sweat glands are inactive, and the skin becomes hot and dry. Unless the situation is recognized in time, body temperature may climb to 41°–45° C (106°–113° F). Temperatures in this range will quickly destroy brain, liver, skeletal muscle, and kidney cells, and disrupt a variety of vital physiological systems. Proper treatment involves lowering body temperature as rapidly as possible.

the surface of the body. The skin takes on a reddish color, skin temperatures rise, and radiational and convective losses increase.

2. As integumentary blood flow increases, sweat glands are stimulated to increase their secretory output. The perspiration flows across the body surface, and evaporative losses accelerate.
3. The respiratory centers are stimulated, and the depth of respiration increases. Often the individual begins respiring through an open mouth, rather than through the nasal passageways, increasing evaporative losses through the lungs.

Behavioral Modifications Behavioral alterations also assist in lowering body temperatures. The person becomes uncomfortably aware of the warm temperatures and may seek a shady spot, thereby decreasing radiative gains. A particularly resourceful individual may take the opportunity to lie down in some cool grass or slide into the pool or ocean in an effort to increase conductive losses.

Mechanisms for Promoting Heat Gain

When the temperature at the preoptic nucleus drops below acceptable levels, the heat-loss center is inhibited, and the heat-gain center is activated.

Heat Conservation The sympathetic vasomotor center decreases blood flow to the dermis of the skin, thus reducing radiational, convective, and conductive losses. The skin cools, and with the circulation restricted it may take on a bluish or pale coloration. This does not damage the epithelial cells, as they are able to tolerate extended periods at temperatures as low as 25° C (87° F) or as high as 49° C (110° F).

In addition, blood returning from the extremities is shunted into a network of deep veins described in Chapter 21. Figure 25-16 shows the changes in circulation that occur. In warm weather, blood flows in a superficial venous network (Figure 25-16a). In cold weather, blood is diverted to a network of deep veins that lie beneath an insulating layer of subcutaneous fat. This venous network wraps around the deep arteries (Figure 25-16b). Heat diffuses from the warm blood flowing outward to the limbs into the cooler blood returning from the periphery. This arrangement traps the heat close to the body core and restricts heat loss by reducing the temperature gradient between the arterial blood and the outside world. Diffusion between fluids moving in opposite directions is called *countercurrent exchange* (Figure 25-16c). (We will return to this topic in Chapter 27.)

A less effective response occurs when the sympathetic system orders the contraction of the arrector pili muscles, and "goose bumps" appear. For mammals naturally blessed with a fur coat, piloerection increases the insulation value of the pelt by causing the hairs to stand erect, rather than lie flat. Fur works like the thermopane windows on a house, by trapping a "dead space" of air next to the skin. Convection is blocked, and conduction through air is very slow. Although the response does little to warm a cold human, it does provide an amusing reminder of our evolutionary history.

Heat Generation The mechanisms available to generate heat can be divided into two broad categories. In **shivering thermogenesis** (ther-mō-JEN-e-sis), muscle tone is gradually increased. This increases the energy consumption of skeletal muscle tissue throughout the body. Both agonists and antagonists are involved, and the degree of stimulation varies, depending on the demand. If the heat-gain center is extremely active, muscle tone increases to the point where stretch receptor stimulation will produce brief, oscillatory contractions of antagonistic muscles. In

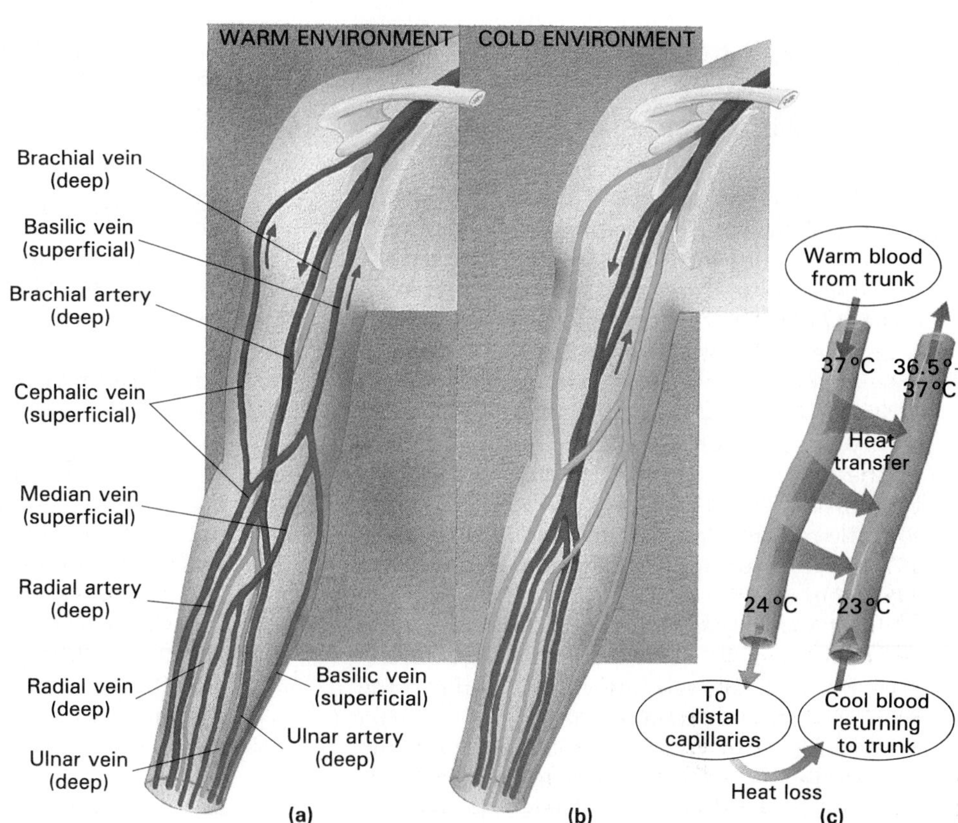

Figure 25–16 Countercurrent Heat Exchange. (a) Circulation through the blood vessels of the forearm in a warm environment. Blood enters the limb in a deep artery and returns to the trunk in a network of superficial veins. These veins radiate heat into the environment through the overlying skin. (b) Circulation through the blood vessels of the forearm in a cold environment. Blood now returns to the trunk via a network of deep veins that flow around the artery. This reduces the amount of heat loss, as indicated in (c). (c) Countercurrent heat exchange occurs as heat radiates from the warm arterial blood into the cooler venous blood flowing in the opposite direction. By the time the arterial blood reaches distal capillaries, where most of the heat loss to the environment occurs, it is already 13° C cooler than it was when it left the trunk. This mechanism reduces the rate of heat loss while conserving body heat. In effect, the countercurrent exchange traps heat near the trunk.

other words, the person begins to **shiver**. Shivering increases the workload of the muscles, and further elevates oxygen and energy consumption. The heat that is produced warms the deep vessels, to which blood has been shunted by the sympathetic vasomotor center. Shivering can elevate body temperature quite effectively since it increases the rate of heat generation by as much as 400 percent.

In **nonshivering thermogenesis** hormones are released that increase the metabolic activity of all tissues.

- The heat-gain center stimulates the adrenal medulla, via the sympathetic division of the ANS, and epinephrine is released. Epinephrine increases the rates of glycogenolysis in liver and skeletal muscle and the rate of aerobic respiration in most tissues.
- The preoptic area itself directs the activity of the thyroid gland. When temperatures are below normal the thyroid gland releases the hormone thyroxine into the blood. Thyroxine not only increases the rate of carbohydrate catabolism, but also the rate of catabolism of all other nutrients. This is why the BMR is so indicative of abnormal thyroid function.

Behavioral Modifications If we feel cold, in the summer we may get into the sun (increasing radiational heating), lie on warm sand (increasing conduction), or change the environmental temperature by resetting the thermostat on the air conditioner. In the winter we may reduce the radiational surface area exposed and increase insulation by adding a sweater, donning gloves, or putting up the hood on a jacket. We may decide on some strenuous activity to boost our energy consumption, although unfortunately shivering thermogenesis stops when the exercising begins. We may also take the easy way out, and head for a warm house.

In any case, we are likely to find that we are hungry. In addition to replenishing energy supplies reduced by shivering or through a hormonally directed increase in metabolic activity, the digestion and absorption of a meal involve appreciable energy expenditures, and therefore increased heat generation. On the other hand, the traditional stiff belt of brandy is not a particularly good idea, for alcohol promotes peripheral vasodilation. This reaction stimulates the thermoreceptors, so the individual feels warmer, but the rate of heat loss actually increases. [**CM:** *Hypothermia*]

Sources of Individual Variation

The timing of thermoregulatory responses may differ from individual to individual. A person may undergo **acclimatization** (a-klī-ma-ti-ZĀ-shun), making phys-

Health News: Fad Diets

Dieting has become a national pastime, and magazines and bookstores are flooded with "how-to" guides for losing weight. Despite the unusual, astounding, and often preposterous claims made on the covers, none has proved to be ideal and many have turned out to be actually dangerous. Now that you have a basic familiarity with metabolic processes and interconversions, we can briefly review several dietary myths. Analyzing the most popular fad diets is beyond the scope of this text, but you might find the information summarized in Figure 25-17 useful if you are considering an intensive diet program.

Identifying fad diets can be quite easy, once you have learned the warning signs. They invariably promise almost immediate results, and most claim to be virtually painless. The "immediate results" turn out to be temporary water losses, and if dieting in any form were easy, Americans would not have a weight problem in the first place.

Carried to extremes, some fad diets can produce fatal alterations in blood chemistry and physiological systems. Many of the trendy diets advocate the elimination of most or all of the members of one or more food groups, making it difficult to obtain dietary essentials. Several of the diets intentionally produce ketosis, a condition that can cause several unpleasant and even disastrous side effects. Equally significant, such restrictions are artificial and usually are intended to be followed only while dieting. Because the eating patterns that caused the original problem are not addressed, once the diet has ended the individual immediately begins putting on weight. This weight gain leads to another dieting cycle 6 months later, often using a different fad diet. This cycle of dieting delights publishing companies, who thrive on a combination of diet books and cookbooks, but it can be very frustrating for the individual. Only about 5 percent of dieters sustain their weight loss for 1-5 years or more.

The advocacy of "secret ingredients" should also ring warning bells, and you should be on the alert for exaggerated claims concerning either the amount of weight lost or the potential fringe benefits of the program. Paid testimonials from successful dieters are highly suspect and hardly a suitable ground for decision making. Such testimonials often take the place of scientific evidence, and pseudoscientific terms are often used to convince the public that scientific proof exists to support the inflated statements. In some cases the "scientific justifications" are totally unrelated to the weight loss experienced by those following a particular diet. For example, eating fruits and vegetables, coupled with reduced quantities of beef, will lower caloric intake and bring the diet in closer agreement with the recommendations listed in Table 25-8. Most people will lose weight on such a diet regardless of whether it is accompanied by complicated mumbo-jumbo about acupressure, cosmic contemplation, or toxic combinations in the stomach.

The most effective weight-loss program does not involve a crash diet, miracle drugs, elimination of all foods beginning with the letter B, or subscriptions to the diet-of-the-month club.

- The weight loss should be gradual, rather than sudden.

iological adjustment to a particular environment over time. For example, Indians in Tierra del Fuego once lived naked in the snow, but Hawaii residents unpack their sweaters when the temperature plummets below 70° F.

Another interesting source of variation is body size. Although heat *production* occurs within the mass of the body, heat *loss* must occur across a body surface. This situation introduces an interesting series of complications stemming from the geometric fact that as an object gets larger, the surface area increases at a much slower rate than does the total volume. The effects of such geometric principles on heat exchange are familiar to us all. When attempting to eat a scalding bowl of soup, we blow across a spoonful, not across the bowl, because the spoonful cools much more quickly. It has a high *surface-to-volume ratio.*

Thermoregulatory Problems of Infants Infants have problems with thermoregulation because of their relatively high surface-to-volume ratios. During embryonic development, temperature regulation is no concern of theirs, as the maternal surroundings are quite reliable. At birth, the temperature-regulating mechanisms are not fully functional. With such high surface-to-volume ratios, newborns must be dried quickly and kept bundled up; for those born prematurely, a thermally regulated incubator is uti-

Diet	High cost	Exercise not advocated strongly	Long-term eating habits unchanged	Low fat, low calorie	Low carbohydrate	Low protein	High protein	High fat	Nutrient deficiencies	Health risks reported	Potential risks and remarks
Formula diets: Cambridge, Nutralife, Liquid protein, Herbalife	■	■	■	■		■			■	■	Ketosis, ketoacidosis, muscle wasting, cardiac arrhythmias
Packaged diets: Nutri/ System, Genesis TM	■		■						■	■	High sodium levels in packaged meals
Pritikin diet (maximum weight loss)			■			■				■	Intestinal gas, diarrhea, muscle wasting
Beverly Hills diet		■	■						■	■	
Dr. Stillman, Water, or Quick Loss diet					■		■	■	■	■	Ketosis, kidney stones, inadequate calcium, increased serum cholesterol
Scarsdale, Atkin, or Air Force diet					■		■	■	■	■	Ketosis, ketoacidosis; inadequate calcium, vitamins
Zen macrobiotic diet			■	■					■	■	Ketosis
Weight Watchers				■							Not a diet as such; provides long-term individual planning and education under professional supervision

FIGURE 25-17
A Critical Review of Popular Fad Dieting Methods. Data sources: "Fad Diet Summary," *Kaiser Permanente*; "Rating the Diets," *Consumer Reports*; "A Practical Guide to Fad Diets," *Clinical Sports Medicine* 3(3):723-729; "The Sense and Nonsense of Best-Selling Diet Books," *Can. Med. Assoc. Journal* 126(6):696-701 (1982).

- Dietary modifications should be moderate, not sweeping, and intended to alter long-term eating habits and ensure that the weight loss will be permanent.

Given an opportunity, the human body will preserve homeostasis. By following the advice "everything in moderation," we can avoid testing our homeostatic limits, but many people tend to follow the "more must be better" philosophy when it comes to meeting nutritional needs. For example, it has been said that Americans have the most valuable urine in the world, thanks to our tendency to overdose ourselves on water-soluble vitamins. *It should be kept in mind that no amount of vitamins, minerals, dietary restrictions, or attention to nutritional recommendations can counteract the negative effects of even a single serious health risk such as smoking, excessive alcohol intake, or drug abuse.*

lized. Infants' body temperatures are more unstable than those of adults, and their metabolic rates decline when they are sleeping, then rise following arousal.

Infants cannot shiver, but they have a different mechanism for raising body temperature rapidly. The adipose tissue between the shoulder blades, around the neck, and possibly elsewhere in the upper body is histologically and functionally different from most of the adipose tissue in the adult. The tissue is highly vascularized, and the individual adipocytes contain numerous mitochondria. Together these characteristics give the tissue a deep, rich color responsible for the name **brown fat**. The individual adipocytes are innervated by sympathetic autonomic fibers. When these nerves are stimulated, lipolysis accelerates in the adipocytes. The cells do not capture the energy released through fatty acid catabolism, and it radiates into the surrounding tissues as heat. This heat quickly warms the blood passing through the surrounding network of vessels, and it is then distributed through the body.

In this way an infant can accelerate metabolic heat generation by 100 percent very quickly, while nonshivering thermogenesis in the adult will raise heat production by only 10–15 percent after a period of weeks. With increasing age and size, body temperature becomes more stable, and the importance of this thermoregulatory mechanism declines. There is little if any brown fat in the adult; with increased body size, skeletal muscle mass, and insulation, shivering thermogenesis is significantly more effective in elevating body temperature.

Because of their relatively large surface area, infants and small children may undergo dangerous alterations in body temperature quite rapidly should their regulatory mechanisms be unable to cope with a thermal stress. Thus for an infant, a cold swimming pool or a steaming jacuzzi (whirlpool bath) can be extremely hazardous.

Thermoregulatory Variations among Adults Adults of the same body weight may differ in their thermal responses if their weight is distributed differently such that they have different surface-to-volume ratios—just consider two 150-pound individuals, one 6 feet 6 inches, and the other 5 feet 4 inches. What tissues account for the weight also plays a factor. Adipose tissue is an excellent insulator, conducting heat at only about one-third the rate of other tissues. As a result, individuals with a more substantial layer of subcutaneous fat may not begin to shiver until long after their more slender companions.

The principal thermoregulatory problem of an obese individual is how to lose metabolic heat across a relatively undersized surface area. Obese persons perspire during modest activity and may become extremely uncomfortable in weather that seems quite delightful to thinner individuals. In cold weather, the surface-to-volume ratio works in their favor, and they do not have to dress as warmly as those around them.

In addition to hormone levels, environmental acclimatization, body weight, age, tissue distribution, and surface-to-volume ratios, our hypothalamic thermostats also affect our thermoregulatory responses. Two otherwise similar individuals may differ in their response to temperature changes because their hypothalamic thermostats are at different settings. There are daily oscillations in body temperature, with temperatures falling 1°–3° F at night and peaking sometime during the day or early evening. Individuals vary in terms of their time of maximum temperature setting, and some have a series of peaks, with an afternoon low. The origin of these patterns is uncertain, since it is not the result of daily activity regimens—people who work at night still show their temperature peaks over the same range of times as the rest of the population.

Finally, two otherwise similar individuals may differ because of variations in their diets. Following a meal consisting of a mixture of carbohydrates and lipids, the BMR rises about 4 percent. But after a meal rich in proteins the BMR may rise as much as 30 percent and remain elevated for 3–12 hours. This metabolic effect, termed the **specific dynamic action** of foods, is the reason why BMR tests are made on fasting subjects.

Fevers An elevated body temperature, or **pyrexia** (pī-REK-sē-a), may occur for a variety of reasons, not all of them pathological. In young children, transient fevers may result from exercise in warm weather, with no ill effects. Similar exercise-related elevations were rarely encountered in adults until running marathons became a national pastime. Temperatures ranging from 39–41° C (103°–106° F) may result, and it is for this reason that competitions are usually held when the air temperature is below 28° C (82° F).

Fevers were discussed when we examined nonspecific defenses in Chapter 22. (p. 725) A fever is the maintenance of a body temperature greater than 37.2° C (99° F). Fevers may result:

- from abnormalities affecting the entire thermoregulatory mechanism, such as heat exhaustion or heat stroke
- from clinical problems that restrict circulation, such as congestive heart failure
- from conditions that impair sweat gland activity, such as drug reactions and some skin conditions
- from the resetting of the hypothalamic thermostat by circulating *pyrogens*, most notably interleukin-1. When the thermostat setting is raised, the heat-gain center is activated. The individual feels cold and may curl up in a blanket. Shivering may begin and continue until the temperature at the preoptic area corresponds with the new setting. When the fever passes, the thermostat is reset to normal. The **crisis phase** then ensues, as the heat-loss center is stimulated. The individual feels unbearably warm, the blanket is discarded, the skin is flushed, and the sweat glands work furiously to bring the temperature down. Repeated cycles of this type constitute the "chills and fever" pattern of many illnesses.

Fevers may be classified as *chronic* or *acute*. Chronic fevers may persist for weeks or months as the result of infections, cancers, or thermoregulatory disorders. Sometimes a discrete cause cannot be determined, leading to a classification as an **FUO**, or *fever of unknown origin*. Acute fevers, such as those seen during heat stroke, in certain diseases, or in an occasional marathon runner, are life-threatening. Immediate treatments may involve cooling the individual in an ice bath (increasing conduction) or giving alcohol rubs (increasing evaporation) in combination with the administration of **antipyretic drugs** such as aspirin or acetaminophen.

√ **How would the BMR (basal metabolic rate) of a pregnant woman compare with her BMR in the nonpregnant state?**

√ **What effect would vasoconstriction of peripheral blood vessels have on body temperature on a hot day?**

√ **Why do infants have greater problems with thermoregulation than adults?**

CHAPTER REVIEW

Review of Selected Clinical Terms

ketosis (kē-TŌ-sis) (*p. 842*)
Abnormally high concentration of ketone bodies in body fluids.

ketoacidosis (kē-tō-as-i-DŌ-sis) (*p. 842*)
Acidification of the blood due to the presence of ketone bodies.

heat exhaustion (*p. 854*)
A malfunction of the thermoregulatory system caused by excessive fluid loss in perspiration.

heat stroke (*p. 854*)
A condition in which the thermoregulatory center stops functioning and body temperature rises uncontrollably.

Additional Terms of Clinical Importance

A MATTER OF PERCEPTION (*p. 834*): **anorexia nervosa**

KETOSIS AND KETOACIDOSIS (*p. 842*): **ketonemia, ketonuria, diabetic ketoacidosis**

Study Outline

Related Key Terms

Introduction (pp. 827–828)

1. Cells in the human body are chemical factories that break down organic substrates to obtain energy.

Cellular Metabolism (pp. 828–838)

1. In general, cells will break down excess carbohydrates first, then lipids, while conserving amino acids.

CARBOHYDRATE METABOLISM (p. 828)

2. Glycolysis and aerobic respiration provide most of the ATP used by typical cells. Glycogen can be broken down to glucose molecules. In glycolysis, each molecule of glucose yields two molecules of pyruvic acid and two molecules of ATP.
3. In the presence of oxygen the pyruvic acid molecules enter the mitochondria, where they are broken down completely. The carbon and oxygen atoms are lost as CO_2, and the hydrogen atoms are passed to coenzymes that deliver them to the electron transport system. The electron transport system generates ATP and the hydrogens combine with oxygen to form water.
4. **Gluconeogenesis,** the synthesis of glucose, enables a cell to create glucose molecules from other carbohydrates, glycerol, or some amino acids. **Glycogenesis** is the process of glycogen formation (glycogen is an important energy reserve when the cell cannot obtain enough glucose from the extracellular fluid).

LIPID METABOLISM (pp. 828–833)

5. During **lipolysis** (lipid catabolism), lipids are broken down into pieces that can be converted into pyruvic acid or channeled into the TCA cycle.
6. Triglycerides are the most abundant lipids in the body. Triglycerides are split into glycerol and fatty acids. The glycerol enters the glycolytic pathways, and the fatty acids enter the mitochondria.
7. **Beta-oxidation** is the breakdown of a fatty acid molecule into two-carbon fragments that can be used in the TCA cycle. The steps of beta-oxidation cannot be reversed, and the body cannot manufacture all of the fatty acids needed for normal metabolic operations.
8. Lipids cannot provide large amounts of ATP in a short amount of time. However, cells can shift to lipid-based energy production when glucose reserves are limited.
9. Lipids circulate as **lipoproteins** (lipid-protein complexes that contain large glycerides and cholesterol) and as **free fatty acids (FFA)** (water-soluble lipids that can diffuse easily across cell membranes). The liver contains **lipoprotein lipase,** an enzyme that breaks down complex lipids.

lipogenesis • linoleic acid
arachidonic acid • linolenic acid
essential fatty acids
chylomicrons
very low-density lipoproteins (VLDLs)
intermediate-density lipoproteins (IDLs)
low-density lipoproteins (LDLs)
high-density lipoproteins (HDLs)

Chapter Review

PROTEIN METABOLISM (pp. 833–836)

10. If other energy sources are inadequate, mitochondria can break down amino acids in the TCA cycle to generate ATP. In the mitochondria the amino group may be removed by **transamination** or **deamination,** and the carbon skeleton converted to one of the compounds involved in glycolysis.
11. Protein catabolism is an impractical source for quick energy.

Related Key Terms

vitamin B_6 • keto acid • urea
urea cycle
essential amino acids
amination
nonessential amino acids

NUCLEIC AND METABOLISM (pp. 836–837)

12. DNA in the nucleus is never catabolized for energy. RNA molecules are broken down and replaced regularly; usually they are recycled as new nucleic acids, but the nucleotides can be catabolized to simple sugars and nitrogen bases. In general, nucleic acids do not contribute significantly to the cell's energy reserves.

uric acid • nitrogenous wastes

AN OVERVIEW OF CELLULAR METABOLISM (pp. 837–838)

13. No one cell can perform all of the anabolic and catabolic operations necessary to support life. Homeostasis can be preserved only when metabolic activities of different tissues are coordinated.

Metabolic Interactions (pp. 838–844)

1. The body has five metabolic components: the liver, adipose tissue, skeletal muscle, neural tissue, and other peripheral tissues.
2. The liver represents the focal point for metabolic regulation and control. Adipose tissue stores lipids, primarily as triglycerides. Skeletal muscle contains substantial glycogen reserves, and the contractile proteins can be mobilized and the amino acids used as an energy source. Neural tissue, which does not contain energy reserves, depends on aerobic respiration for energy production. Other peripheral tissues are able to metabolize glucose, fatty acids, or other substrates under the direction of the endocrine system.

THE ABSORPTIVE STATE (pp. 838–840)

3. For around four hours after a meal, nutrients enter the blood as intestinal absorption proceeds.
4. The liver closely regulates the glucose content of blood arriving via the hepatic portal vein and the circulating levels of amino acids.
5. **Lipemia** (milky appearance of the plasma due to the presence of lipids) often marks the absorptive state. Adipocytes remove fatty acids and glycerol from the circulation and synthesize new triglycerides for storage.
6. During the absorptive state glucose molecules are catabolized and amino acids are used to build proteins. Skeletal muscles may also catabolize circulating fatty acids, and the energy is used to increase glycogen reserves.

THE POSTABSORPTIVE STATE (pp. 840–842)

7. The postabsorptive state extends from the end of the absorptive state to the next meal.
8. When blood glucose falls, the liver begins breaking down glycogen reserves and releasing the glucose into the circulation. As the duration of the fast increases, liver cells synthesize glucose molecules from smaller carbon fragments and from glycerol molecules. Fatty acids undergo beta-oxidation; the fragments enter the TCA cycle or combine to form ketone bodies.

acetoacetate • acetone
betahydroxybutyrate

9. **Glucogenic amino acids** can be converted to pyruvic acid and used for gluconeogenesis. **Ketogenic amino acids** can be converted to acetyl-CoA and catabolized or converted to ketone bodies.
10. The average individual carries a 1–2 month energy reserve in adipose tissue. During the postabsorptive state lipolysis increases and the fatty acids are released into the circulation for catabolism.
11. Skeletal muscles metabolize ketone bodies and fatty acids. Their glycogen reserves are broken down to yield lactic acid which diffuses into the bloodstream. After a prolonged fast, **cathepsins** (proteolytic enzymes) begin breaking down contractile proteins.
12. Neural tissue continues to be supplied with glucose as an energy source until blood glucose levels become extremely low.

ADJUSTMENTS TO STARVATION (pp. 843–844)

13. During starvation carbohydrate reserves are exhausted, and as blood glucose levels decline, gluconeogenesis accelerates, using glycerol, amino acids, and lactic acid. As the fast continues, protein catabolism becomes increasingly important. When lipid reserves

are exhausted, crises follow as contractile and structural proteins are mobilized.

Related Key Terms

Diet and Nutrition (pp. 844–850)

1. **Nutrition** is the absorption of nutrients from food. A **balanced diet** contains all of the ingredients necessary to maintain homeostasis; it prevents **malnutrition.**

nutritionists

THE FOUR BASIC FOOD GROUPS (pp. 844–845)

2. The four **basic food groups** are the Milk; Meat; Vegetable and Fruit; and Bread and Cereal groups.

complete proteins
incomplete proteins

NITROGEN BALANCE (pp. 845–846)

3. Amino acids, purines, pyrimidines, creatine, and porphyrins are **N compounds** that contain nitrogen atoms. An adequate dietary supply of nitrogen is essential, since the body does not maintain large nitrogen reserves.

protein sparers
nitrogen balance
positive nitrogen balance
negative nitrogen balance

MINERALS, VITAMINS, AND WATER (pp. 846–849)

4. **Minerals** act as cofactors in a variety of enzymatic reactions. They also contribute to the osmolarity of body fluids, and they play a role in transmembrane potentials, action potentials, construction and maintenance of the skeleton, transport of gases, buffer systems, fluid absorption, and waste removal.
5. Vitamins are needed in very small amounts. Vitamins A, D, E, and K are the **fat-soluble vitamins;** taken in excess, they can lead to **hypervitaminosis. Water-soluble vitamins** are not stored in the body; lack of adequate dietary supplies may lead to **deficiency disease (avitaminosis).**
6. Daily water requirements average about 40 mℓ/kg body weight. Water is obtained from food, drink, and metabolic generation.

DIET AND DISEASE (pp. 849–850)

7. A balanced diet can improve general health. Most Americans eat too much and consume too much fat, sugar, and sodium.

Bioenergetics (pp. 850–858)

1. The energy content of food is usually expressed as **Calories** per gram (C/g). Less than half of the energy content of glucose or any other nutrient can be captured by our cells.

calorimetry • **calorie**
kilocalorie • **calorimeter**

2. The efficiency of glycolysis averages 28 percent, while the entire process of aerobic respiration averages about 42 percent.
3. The total of all the anabolic and catabolic processes underway in the body represents the **metabolic rate** of an individual. The **basal metabolic rate (BMR)** is the rate of energy utilization at rest.

T_4 assay

THERMOREGULATION (pp. 851–858)

4. The homeostatic regulation of body temperature is **thermoregulation.** Heat exchange with the environment involves four processes: radiation, conduction, convection, and evaporation.

insensible perspiration
sensible perspiration

5. The **preoptic area** of the hypothalamus acts as the body's thermostat, affecting the **heat-loss center** and the **heat-gain center.**
6. Mechanisms for increasing heat loss include both physiological mechanisms (peripheral vasodilation, increased perspiration, and increased respiration) and behavioral modifications.
7. Responses that conserve heat include decreased blood flow to the dermis, countercurrent exchange, and piloerection. Heat may be generated by **shivering thermogenesis** and **nonshivering thermogenesis.**

shiver

8. Thermoregulatory responses differ between individuals. One important source of variation is **acclimatization** (adjusting physiologically to an environment over time).

brown fat
specific dynamic action

9. **Pyrexia** (fever), a body temperature above 37.2° C (99° F), can result from problems with the thermoregulatory mechanism, circulation, or sweat gland activity, or from the resetting of the hypothalamic thermostat by circulating pyrogens.

crisis phase • **FUO**
antipyretic drugs

Chapter Review

Review Planner

	Objective	Level 1
1	Compare the mechanisms of carbohydrate, lipid, protein, and nucleic acid metabolism.	**1 2 3 7 8 12 19 20 23 27 29 30 33**
2	Explain how the metabolic activities of tissues, organs, and organ systems are coordinated.	**22 31 32 43**
3	Describe the energy reserves of an average person.	**1 2 4 7**
4	Describe how we utilize our energy reserves to maintain homeostasis.	**1 2 5 6 7 21 22 25 37 41 43**
5	Explain what constitutes a balanced diet, and why it is important.	**8 10 12 13 14 15 17 24 39 40**
6	Discuss the functions of vitamins, minerals, and other important nutrients.	**9 13 14 16**
7	Discuss the homeostatic mechanisms that maintain a constant body temperature.	**11 18 21 22 25 26 28 34 35 36 37 38 41 42**

Level 2: **31 32 33 37 39 40 43 44**

Level 3: **45 46**

C M:
Phenylketonuria
Protein Deficiency Diseases
Gout
Iron Deficiencies and Excesses
Nutrition and Nutritionists
Hypothermia

NEWS C M:
Obesity • Hypervitaminosis
Alcohol: A Risky Diversion

ABC NEWS:
Hypertension and Black Americans

Review Questions

MULTIPLE CHOICE

1. In general, a cell with adequate reserves will break down __________ first to obtain energy: a) carbohydrates; b) lipids; c) proteins; d) RNA
2. In the preceding example, if the cell has exhausted its "first choice" for energy, it will begin to break down __________: a) carbohydrates; b) lipids; c) proteins; d) RNA.
3. Glycogenesis produces: a) glycogen; b) pyruvic acid; c) glycerol; d) glucose.
4. Which of the following uses energy, but does not maintain its own energy reserves?: a) hepatocyte; b) adipocyte; c) skeletal muscle fiber; d) neuron.
5. The absorptive state: a) begins 4 hours after a meal; b) provides glucose, amino acids, and lipids for storage; c) supports peripheral tissues through ketogenesis; d) can be extended through starvation.
6. Which of the following does *not* characterize the postabsorptive state?: a) blood glucose levels decline; b) glycogenesis accelerates; c) gluconeogenesis accelerates; d) ketone bodies increase.
7. During a starvation diet: a) levels of circulating ketone bodies decline; b) carbohydrate reserves become the dominant energy source; c) the amount of adipose tissue in the body increases; d) protein catabolism weakens the body.
8. Complete proteins contain: a) all of the essential and nonessential amino acids; b) all of the circulating lipoproteins; c) the entire range of incomplete proteins; d) all of the essential amino acids, but not necessarily the nonessential amino acids.
9. Fat-soluble vitamins: a) cannot be stored in the body; b) can cause hypervitaminosis; c) include the B vitamins; d) are rapidly excreted in the urine.
10. Even with a good balance of nutrients, someone who follows a strict vegetarian diet will be unable to: a) maintain normal bone growth; b) maintain a positive nitrogen balance; c) obtain a coenzyme required for nucleic acid metabolism; d) synthesize visual pigments.
11. Which of the following statements applies to shivering thermogenesis?: a) the thyroid gland stimulates heightened carbohydrate catabolism; b) inhibition of the vasomotor center causes peripheral vasodilation; c) epinephrine increases the rate of glycogenolysis in skeletal muscles; d) muscle tone increases until stretch receptors are stimulated.
12. Essential fatty acids are called "essential" because: a) our bodies are unable to synthesize them; b) our bodies can break them down easily to obtain nutrients; c) they are "protein sparers" which conserve amino acids to form vital N compounds.

DIAGRAMS

13. Create a table that lists the primary function(s) of the following minerals: sodium, potassium, chloride, calcium, phosphorus, magnesium, iron, zinc, copper, and manganese.
14. Create a table that presents the primary function(s), effects of deficiency, and effects of excess of the fat-soluble and water-soluble vitamins.

TRUE/FALSE

(Your instructor may wish to have you correct the statement if it is false.)

15. **T/F:** It is possible to eat representative selections from all four of the basic food groups and yet suffer from malnutrition.
16. **T/F:** Our bodies cannot synthesize any minerals at all.
17. **T/F:** Reducing water intake is a safe and effective way to lose weight.

MATCHING QUESTIONS

(Match each term with the most appropriate statement.)

Statements:

18. A heat transfer mechanism involving secretions from sweat glands that evaporate from the skin surface.
19. Lipoproteins that carry absorbed lipids from the intestinal tract to the circulation.
20. Removal of an amino group by attaching it to a keto acid, producing a new amino acid.
21. A homeostatic process that keeps body temperatures within acceptable limits.
22. The control center in the hypothalamus that regulates body temperature.
23. Removal of an amino group in a reaction that generates an ammonia molecule.
24. Consumption of all the necessary nutrients in order to maintain homeostasis.
25. The body's minimum resting energy expenditure.
26. An elevated body temperature.
27. Lipoproteins that deliver cholesterol to peripheral tissues.
28. A heat transfer mechanism involving continuous, uncontrolled evaporation from the skin surface.
29. The breakdown of a fatty acid molecule into two-carbon fragments.
30. Lipoproteins that transport excess cholesterol from peripheral tissues to the liver.

Terms:

a) Beta-oxidation
b) Chylomicrons
c) Low-density lipoproteins
d) High-density lipoproteins
e) Basal metabolic rate (BMR)
f) Balanced diet
g) Thermoregulation
h) Sensible perspiration
i) Insensible perspiration
j) Preoptic area
k) Pyrexia
l) Deamination
m) Transamination

SHORT ANSWER/ESSAY

31. Discuss the roles of the digestive, endocrine, nervous, cardiovascular, and respiratory systems in the metabolism of the human body.
32. Distinguish between the absorptive and postabsorptive states.
33. Compare the metabolism of carbohydrates, lipids, proteins, and nucleic acids. Which is normally the most important source of energy for cells?
34. It's a hot, sunny afternoon. You're lying on a sandy beach. Describe the ways in which you are exchanging heat with your environment.
35. In the previous example, suppose that you start to feel uncomfortably warm. Describe the thermoregulatory mechanisms initiated by your CNS to cool you off.
36. You return to the same beach a few months later; it's a chilly day and you're getting uncomfortably cool. What steps does your CNS take to conserve and generate heat?
37. What does a person's basal metabolic rate represent? How can it be determined, and what factors can affect it?
38. Why do infants have more thermoregulatory problems than adults? What mechanism(s) can raise infants' body temperatures rapidly when necessary?
39. What are N compounds, and why is negative nitrogen balance undesirable?
40. What are the four basic food groups? Analyze this method as a means of maintaining a balanced diet.
41. What is a Calorie? Explain its significance.
42. One person feels quite comfortable, while someone else in the same room shivers and dons a sweater. Explain how these two adults can have such different responses to the same environment.
43. What hormones are important in coordinating the physiological responses to starvation? How do they help maintain homeostasis?
44. From a metabolic standpoint, our bodies have five distinctive components. Identify these components, and discuss their metabolic interactions.

CRITICAL THINKING/APPLICATIONS

45. Articles in popular magazines sometimes refer to "good cholesterol" and "bad cholesterol." To what types and functions of cholesterol might these terms refer? Explain your answer.
46. You're visiting with friends at an outdoor barbecue on a hot summer day. One friend, who is drinking beer and sweating profusely, begins to complain of a headache and indigestion; suddenly he faints. What has happened, and what steps would you take? How could you determine whether this is a life-threatening condition? Did the beer play a role in his collapse?

Urinating in public is considered bad form these days. But this whimsical fountain in Brussels, Belgium, depicts what is, after all, a very natural and necessary process. After our digestive system assimilates a meal and our body harvests and utilizes the nutrients it needs, we must somehow discharge the resulting waste products. The urinary system performs this service, while efficiently conserving water and other valuable substances. It also, as we shall see, carries out a variety of other vital but less obvious functions (none of which has been immortalized in sculpture, so far as we know).

CHAPTER 26

The Urinary System

Chapter Objectives

After reading this chapter, you will be able to:

1 Identify the components of the urinary system and their functions.
2 Describe the structure and functions of the kidneys.
3 Describe the processes involved in the formation of urine and the changes that occur in the filtrate as it moves through the nephron.
4 Discuss the factors that regulate the composition of the urine.
5 Describe the structures and functions of the ureters, urinary bladder, and urethra.
6 Discuss the process of urination and how it is controlled.
7 Describe the effects of aging on the urinary system.
8 Explain how the urinary system interacts with other body systems to maintain homeostasis in body fluids.

Introduction

The human body contains trillions of cells bathed in extracellular fluid. Previous chapters compared these cells to factories that burn nutrients to obtain energy. Imagine what would happen if *real* factories were built as close together as cells in the body. What a mess they would make! Each would generate piles of garbage, and the smoke they produced, together with the depletion of oxygen, would drastically reduce air quality. In short, there would be a serious pollution problem.

The coordinated activities of the digestive, cardiovascular, respiratory, and urinary systems prevent the development of similar pollution problems inside the body. The digestive tract absorbs nutrients from food, and the liver adjusts the nutrient concentration of the circulating blood. The cardiovascular system delivers these nutrients and oxygen from the respiratory system to peripheral tissues. As blood leaves these tissues it carries the carbon dioxide and waste products to sites of excretion. The carbon dioxide is eliminated at the lungs, as described in Chapter 23. Most of the organic waste products are removed and excreted by the urinary system, the focus of this chapter.

The urinary system performs vital excretory functions and eliminates the organic waste products generated by cells throughout the body. But it also has a number of other essential functions that are often overlooked. A more complete list of urinary system functions includes:

1. Regulating plasma concentrations of sodium, potassium, chloride, calcium, and other ions by controlling the quantities lost in the urine;
2. Regulating blood volume and blood pressure by (a) adjusting the volume of water lost in the urine, (b) releasing erythropoietin, and (c) releasing renin;
3. Contributing to the stabilization of blood pH;
4. Conserving valuable nutrients by preventing their excretion in the urine;
5. Eliminating organic waste products, especially nitrogenous wastes such as urea and uric acid;
6. Assisting the liver in detoxifying poisons and, during times of starvation, deaminating amino acids so that they can be catabolized by other tissues.

These activities are carefully regulated to keep the composition of the blood within acceptable limits. A disruption of any one of these functions will have immediate and potentially fatal consequences.

This chapter considers the functional organization of the urinary system and describes the major regulatory mechanisms that control urine production and concentration. The next chapter takes a closer look at how the activities of different organ systems affect plasma electrolyte composition, fluid volume, and body fluid pH.

An Overview of the Urinary System

The components of the urinary system are indicated in Figure 26-1. The excretory functions are performed by the **kidneys**. These organs produce **urine**, a fluid containing water, ions, and small soluble compounds. Urine production occurs in several stages and involves filtration, reabsorption, and secretion. Urine leaving the kidneys travels along the **ureters** (ū-RE-terz) to the **urinary bladder** for temporary storage. When **urination**, or **micturition** (mik-tu-RI-shun), occurs, contraction of the muscular bladder forces the urine through the **urethra** and out of the body. The kidneys perform all of the filtration and transport operations involved in urine production. The other components of the system are responsible for storage and conduction of the urine to the exterior.

The kidneys are extremely active organs. They receive approximately 20 percent of the cardiac output, a phenomenal amount for organs with a *combined* weight of less than 300 g (10.5 oz).

The Kidneys

The kidneys are situated between the last thoracic and third lumbar vertebrae. The right kidney often sits slightly lower than the left, as indicated in Figure 26-2a. These organs lie between the muscles of the dorsal body wall and the peritoneal lining (Figure 26-2b). This position is called **retroperitoneal** (re-trō-per-i-tō-NĒ-al; *retro-*, behind) because they are behind the peritoneum.

The kidneys are protected and anchored by three

FIGURE 26-1
Components of the Urinary System

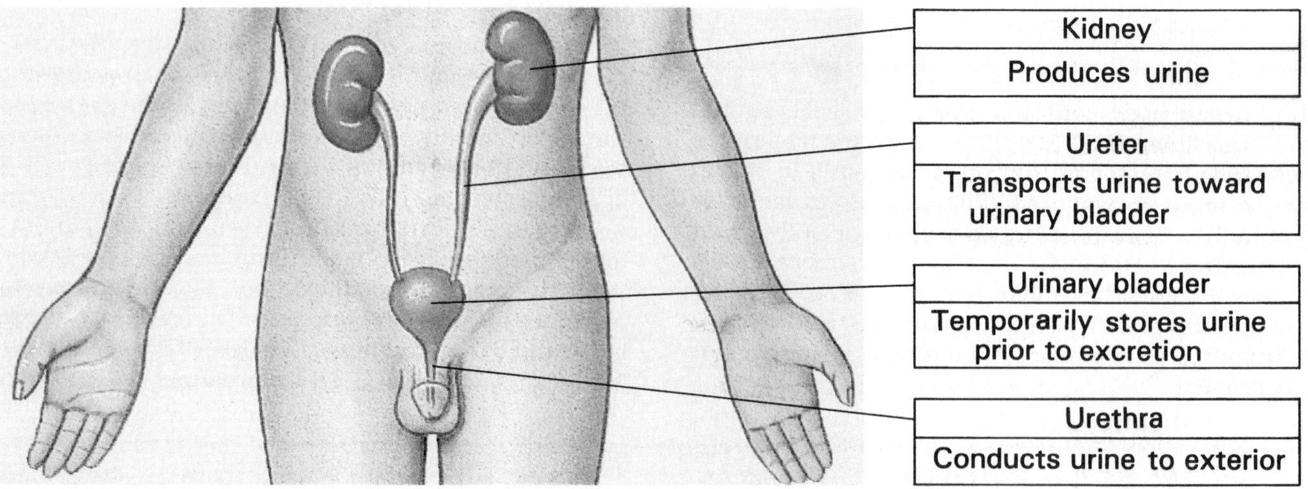

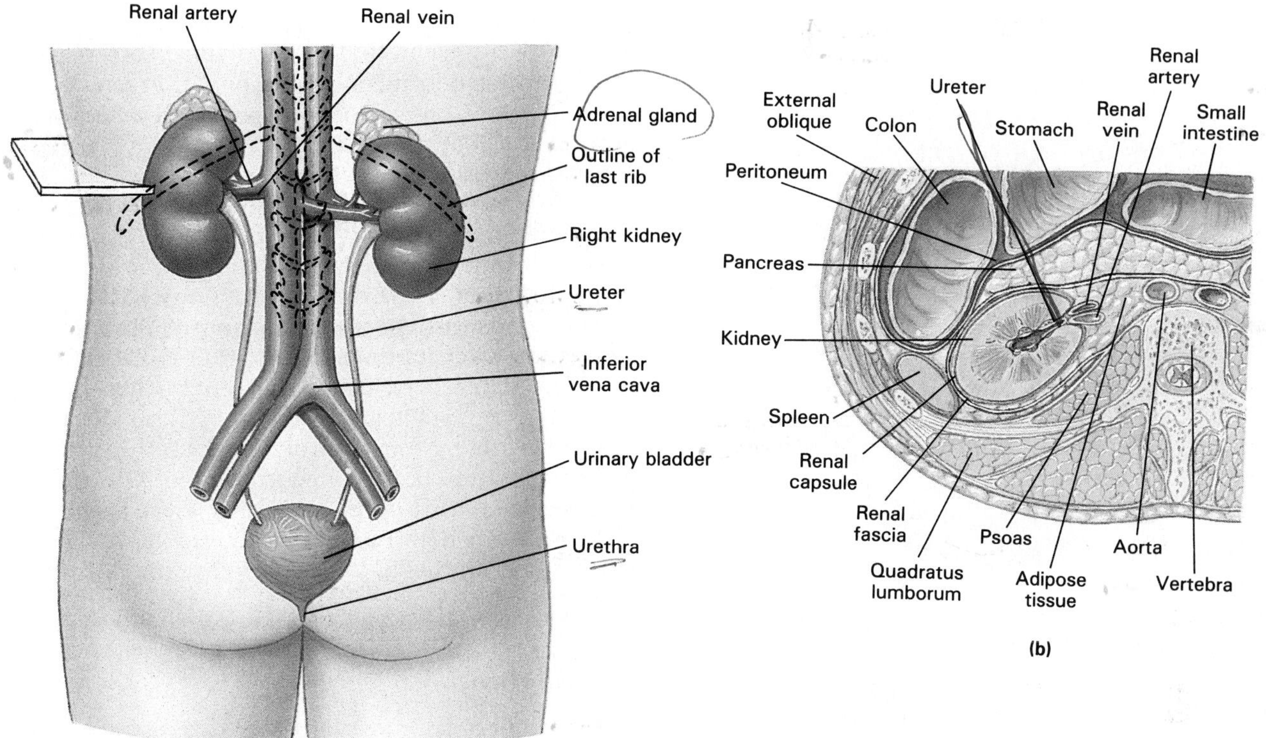

FIGURE 26-2
An Introduction to Renal Anatomy. (a) Posterior view of the trunk, showing the positions of the kidneys and other components of the urinary system. (b) Sectional view at the level indicated in (a).

concentric layers of connective tissue:

- A layer of collagen fibers covers the outer surface of the entire organ. This layer, the **renal capsule**, is also known as the *fibrous tunic* of the kidney.
- A layer of adipose tissue called the **adipose capsule** surrounds the renal capsule. This layer can be quite thick, so that the adipose capsule often obscures the outline of the kidney.
- Collagen fibers extend outward from the renal capsule through the adipose capsule to anchor the kidney to surrounding structures. These fibers create a dense outer layer known as the **renal fascia**. Posteriorly the renal fascia fuses with the deep fascia surrounding the muscles of the body wall. Anteriorly the renal fascia forms a thick fibrous layer that is covered by the peritoneum.

In effect the kidney hangs suspended by these collagen fibers and packed in a soft cushion of adipose tissue. This arrangement prevents the jolts and shocks of day-to-day existence from disturbing normal kidney function. If the suspensory fibers break or become detached, a slight bump or jar may displace the kidney and stress the attached vessels and ureter. This condition, called a **floating kidney**, can be especially serious in malnourished individuals. If little adipose tissue exists between the kidney and the renal fascia, the range of possible movement increases, and the ureters or renal blood vessels may become twisted or kinked.

SUPERFICIAL AND SECTIONAL ANATOMY OF THE KIDNEY

Each kidney has the shape of a lima bean, with a prominent indentation, or **hilus**, facing medially. A typical adult kidney (Figure 26-3) measures approximately 10 cm (4 in.) in length, 5.5 cm (2.2 in.) in width, and 3 cm (1.2 in.) in thickness. The surface of the organ is covered by a dense fibrous capsule, and in sectional view the inner portion of the capsule folds inward at the hilus to line an internal cavity, the **renal sinus**. Renal vessels and the **ureter** draining the kidney pass through the hilus and branch within the renal sinus. A thickened, external portion of the capsule extends across the hilus and stabilizes the position of these structures.

Seen in section, the kidney can be divided into

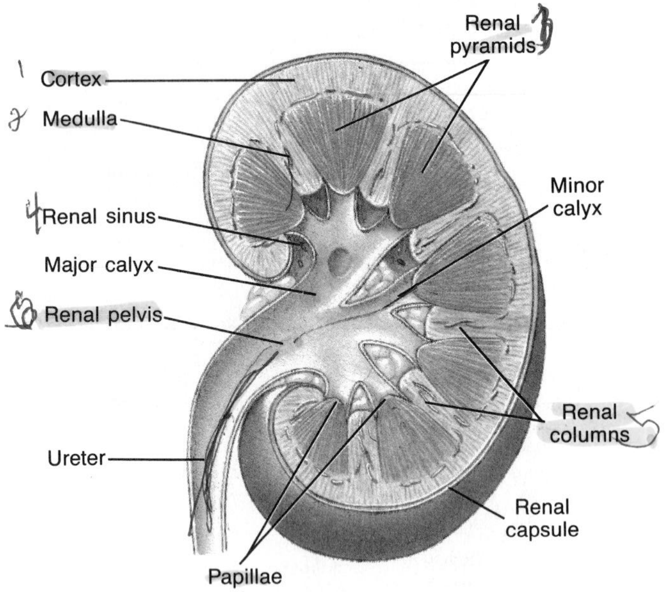

FIGURE 26-3
Structure of the Kidney. Frontal section through the left kidney, showing major anatomical landmarks.

an outer **cortex** and an inner **medulla**. The medulla contains 6 to 18 conical **renal lobes**, or **pyramids**, whose tips, or **papillae**, project into the renal sinus. Each pyramid has a series of fine grooves that radiate from the papillae. **Renal columns** extend from the cortex inward toward the renal sinus between adjacent renal pyramids. The columns have a distinctly granular texture, similar to that of the cortex.

Where the ureter enters the renal sinus it expands to form an enlarged chamber called the **renal pelvis**. The pelvis branches to form two **major calyces** (KĀL-i-sēz), and each major calyx gives rise to four or five cup-shaped **minor calyces**. Each minor calyx surrounds the exposed papilla of a single renal pyramid. Ducts within the papilla connect to the wall of the calyx and discharge the urine produced in the cortex and medulla. As production continues, the urine passes through the calyces and into the ureter.

HISTOLOGICAL ORGANIZATION

Figure 26-4 introduces the histological organization of the **nephron** (NEF-ron), the basic functional unit in the kidney. A single nephron consists of an elongate **renal tubule** that has straight and convoluted segments. For convenience, the nephron shown in this figure has been shortened and straightened out. The nephron begins at an expanded chamber, called a **renal corpuscle** (KŌR-pusl), that contains a capillary knot, or **glomerulus** (glo-MER-ū-lus). A typical glomerulus consists of around 50 intertwining capillaries. Filtration across the walls of the glomerulus produces a protein-free solution known as a **filtrate**. From the renal corpuscle the filtrate enters a long tubular passageway that is subdivided into regions with varied structural and functional characteristics. Major subdivisions include the **proximal convoluted tubule**, the **loop of Henle** (HEN-lē), and the **distal convoluted tubule**.

Each nephron empties into the **collecting system**. A **collecting tubule** connected to the distal convoluted tubule carries the filtrate toward a nearby **collecting duct**. The collecting duct leaves the cortex and descends into the medulla, carrying fluid toward a **papillary duct** that drains into the renal pelvis.

The urine arriving at the renal pelvis is very different from the filtrate produced at the glomerulus. As it travels along the nephron and collecting system, the composition and concentration of the filtrate change. The nature of the changes and the properties of the urine are determined by the activities and properties of the epithelial cells exposed to the filtrate.

The Nephron

Nephrons perform most of the vital functions of the kidneys. To be specific, they are responsible for

1. The production of filtrate, and
2. All of the reabsorption of organic substrates, and over 80 percent of the reabsorption of water and ions that occurs in the production of urine.

We will now take a closer look at the structural and functional specializations along each segment of the nephron. Descriptions of the histological appearance of the various sections should be followed in Figure 26-4; Table 26-1 summarizes important information concerning the regions of the nephron and collecting system.

The Renal Corpuscle Figure 26-5a on p. 865 provides a more realistic view of nephron organization. The proximal and distal convoluted tubules are highly coiled and packed tightly together around the renal corpuscle (Figure 26-5b). The renal corpuscle consists of the expanded initial segment of a nephron, sometimes called *Bowman's capsule*, and the glomerulus. This oval structure has a diameter averaging 150–250 μm. The outer wall of the capsule is lined by a simple squamous **capsular epithelium** continuous with the **glomerular epithelium** that covers the enclosed capillary network. The connection between the capsular and glomerular epithelium lies at the **vascular pole** of the renal corpuscle. At the vascular pole the glomerular capillaries are connected to the bloodstream. Blood arrives via the **afferent arteriole** and departs in the **efferent arteriole**. (This unusual circulatory arrangement will be discussed further in a later section.)

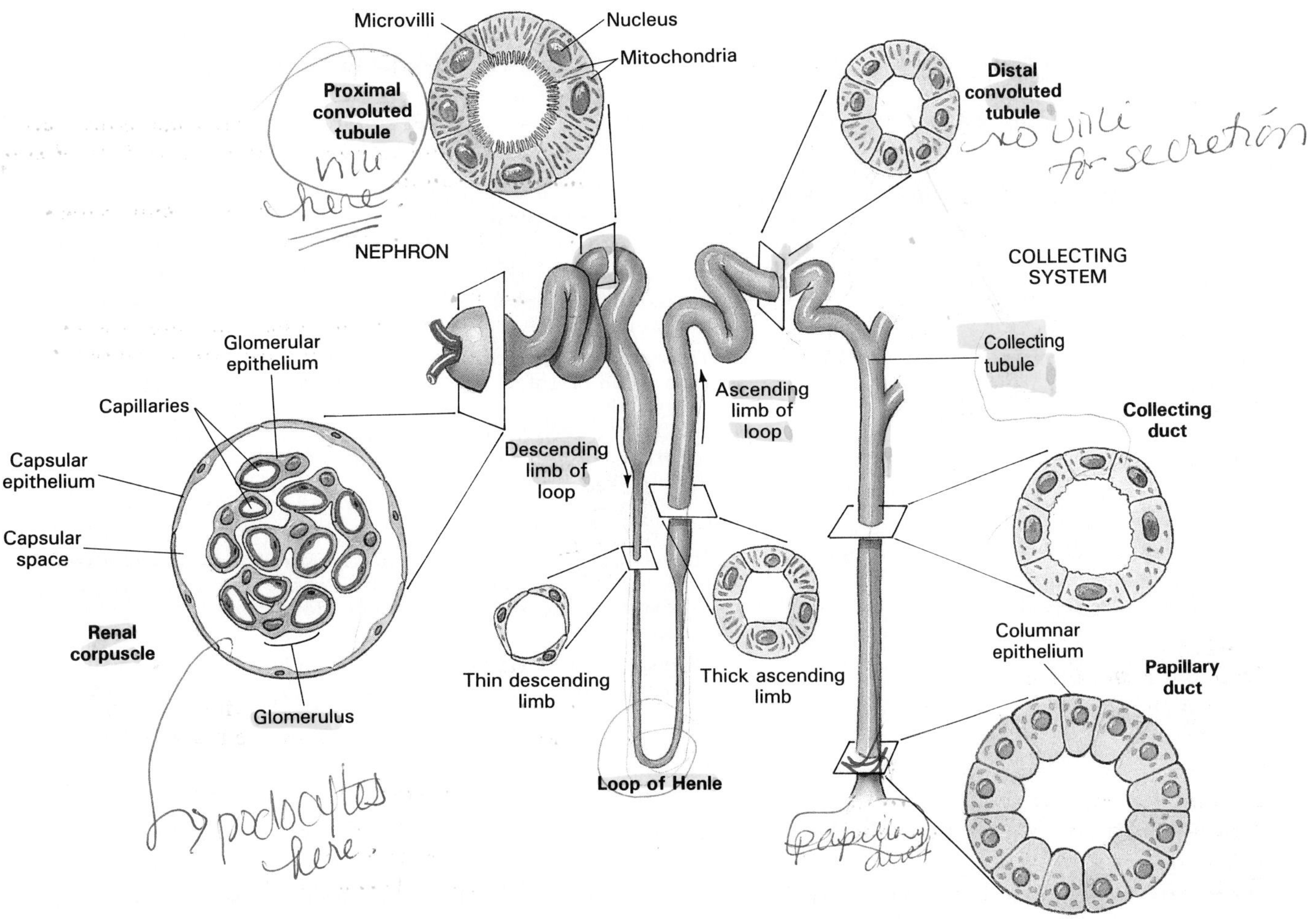

FIGURE 26-4
Histology of a Typical Nephron

The process of urine formation begins as fluid and dissolved materials leave the bloodstream at the glomerulus and enter the **capsular space** that separates these epithelial layers. The filtrate entering the capsular space is very similar to plasma with the blood proteins removed.[1] The filtration process involves passage across three physical barriers, illustrated in Figure 26-5c:

1. *The capillary endothelium*: The glomerular capillaries are fenestrated capillaries (Figure 21-3b). (p. 667) The diameters of the pores range from 60 to 100 nm (0.06–0.1 μm). These openings are small enough to prevent the passage of blood cells, but they are too large to restrict the diffusion of dissolved compounds, even those the size of plasma proteins.
2. *The basement membrane*: The basement membrane that surrounds the capillary wall has several times the density and thickness of a typical basement membrane. This layer, called the **lamina densa**, restricts the passage of the larger plasma proteins, but permits the movement of smaller molecules, including albumin, various nutrients, and ions.
3. *The glomerular epithelium*: Cells of the glomerular epithelium, called **podocytes** (PŌ-do-sīts, "foot cells"), have long cellular processes that wrap around the outer surfaces of the basement membrane. These delicate "feet," or **pedicels** (PED-i-celz), can be clearly seen in Figure 26-5d. Adjacent pedicels are separated by narrow spaces called **slit pores**. These slit pores are only 6–9 nm in width, small enough to block the passage of most of the smaller protein molecules. As a result, under normal circumstances few albumin molecules (average diameter 7 nm) enter the capsular space.

The diffusion barriers within the renal corpuscle primarily limit the upper size of the materials gaining access to the capsular space. This kind of size-selec-

[1] There are slight differences in ionic concentration due to charge interactions across the filtration barriers; positively charged ions reach the filtrate more quickly than negatively charged ions of the same size.

TABLE 26-1 The Organization of the Nephron and Collecting Systems in the Kidney

Region	*Length (mm)*	*Diameter (μm)*	*Primary Function*	*Secondary Function*	*Histological Characteristics*
Renal corpuscle	Spherical	150–250	Filtration of plasma to initiate urine formation		Glomerulus and podocytes in capsule
Proximal convoluted tubule (PCT)	14	60	Reabsorption of ions, organic molecules, vitamins, water	Secretion of acids, wastes	Cuboidal cells with microvilli
Loop of Henle	30	15	Descending limb: reabsorption of water from filtrate	Reabsorption of salts, water	Low cuboidal or squamous
		30	Ascending limb: reabsorption of ions; assists in creation of the medullary concentration gradient		
Distal convoluted tubule (DCT)	5	30–50	Reabsorption of sodium ions; secretion of acids, ammonia, drugs	Reabsorption of water (terminal portion only)	Cuboidal cells without microvilli
Collecting tubule	Variable	50	Reabsorption of water, sodium ions	Secretion of hydrogen ions	Cuboidal cells without microvilli; pale compared with DCT
Collecting duct	15	50–100	Reabsorption of water, sodium ions	Secretion of hydrogen ions	Cuboidal to columnar cells
Papillary duct	5	100–200	Conduction of urine to minor calyx		

tion process has obvious functional limitations. In addition to metabolic wastes and excess ions, compounds such as glucose, free fatty or amino acids, vitamins, and other solutes enter the capsular spaces. *These useful materials must be reclaimed*, and much of the burden falls upon the adjacent portion of the nephron, the proximal convoluted tubule (**PCT**). [**CM:** *Glomerulonephritis*]

The Proximal Convoluted Tubule The entrance to the proximal convoluted tubule, or **PCT**, lies almost directly opposite the vascular pole of the glomerulus, in a region called the **tubular pole**. The term "convoluted" fits quite well, for the path of the PCT twists and turns repeatedly. The lining consists of a simple cuboidal epithelium whose exposed surfaces are blanketed with microvilli. These cells actively absorb nutrients, plasma proteins, and ions from the filtrate as it flows along the length of the PCT. As these materials are absorbed, osmotic forces pull water across the wall of the PCT and into the surrounding interstitial spaces. Although absorption represents the primary function of the PCT, the epithelial cells can also secrete substances into the lumen.

The Loop of Henle The distal portion of the proximal convoluted tubule bends sharply and descends into the medulla. This bend marks the start of the loop of Henle. This portion of the nephron can be divided into a **descending limb** that travels toward the renal pelvis and an **ascending limb** that returns toward the cortex. Each limb contains a **thick segment** and a **thin segment**. (The terms "thick" and "thin" refer to the thickness of the surrounding epithelium, not the diameter of the lumen.) Thick segments are found closest to the cortex, whereas a thin squamous epithelium lines the deeper medullary portions.

The thin segments are freely permeable to water but relatively impermeable to ions and other solutes. The thick segments contain active transport mechanisms that pump materials out of the filtrate. These epithelial activities are essential to the reabsorption of water and the concentration of the filtrate. The thick ascending limb delivers fluid to the last of the segments of the nephron, the **distal convoluted tubule** (**DCT**).

The Distal Convoluted Tubule The distal convoluted tubule begins near the vascular pole of the glomerulus, between the afferent and efferent arterioles. The ascending limb ends here as it forms a sharp angle that places the tubular wall in close contact with the glomerulus and its attendant vessels. The tubule cells adjacent to the vascular pole are taller

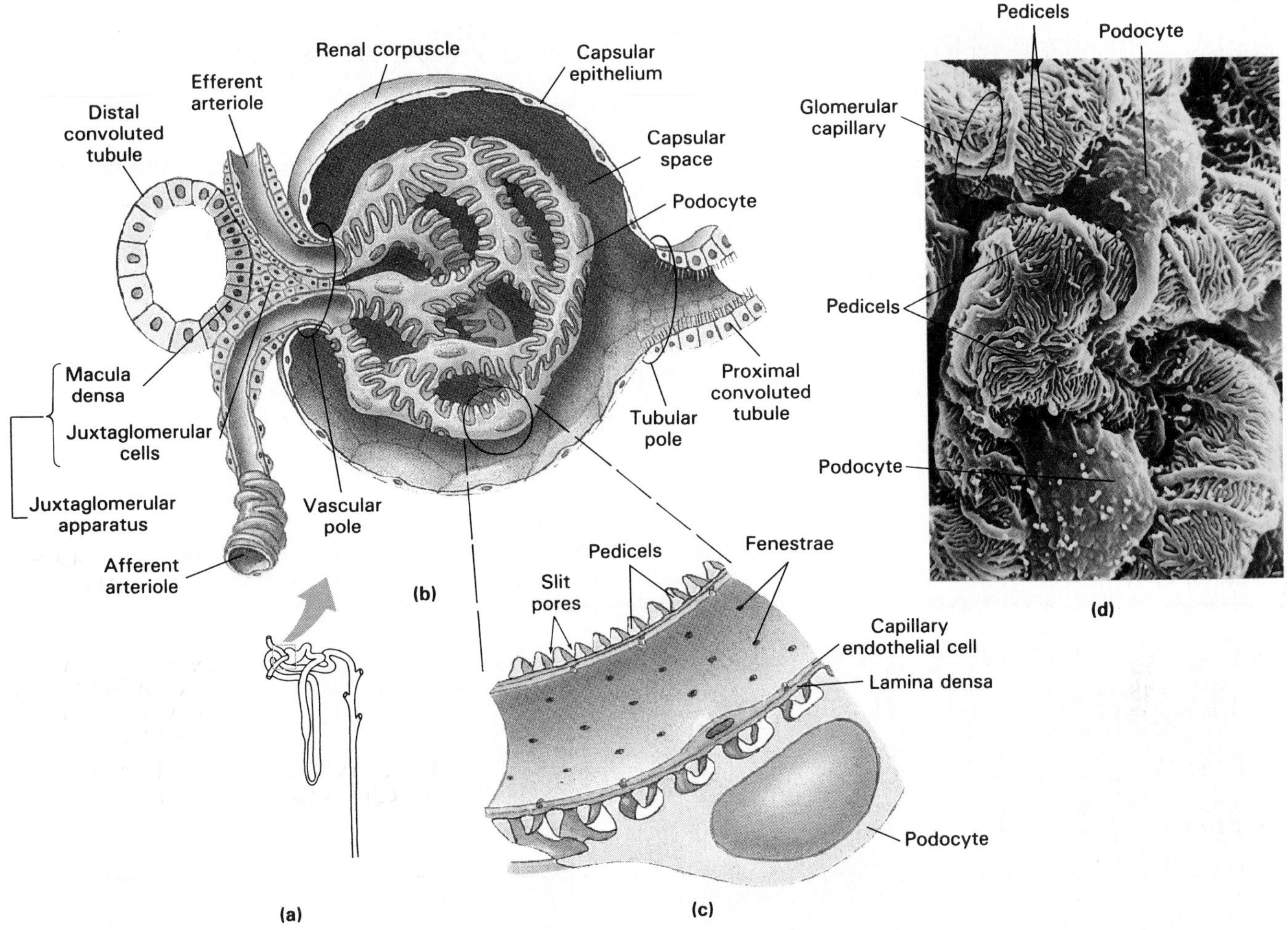

FIGURE 26-5
The Renal Corpuscle. (a) A more realistic view of a nephron, showing its three-dimensional structure. (b) The renal corpuscle, showing the structure of the glomerulus and the filtration apparatus. (c) Diagrammatic view of a podocyte with pedicels covering the adjacent surfaces of the lamina densa. (d) Electron micrograph of the glomerular surface, showing individual podocytes and their processes. (SEM, ×3563)

than those seen elsewhere along the DCT, and their nuclei are clustered together. This region, diagrammed in Figure 26-5b, is called the **macula densa** (MAK-ū-la DEN-sa).

The cells of the macula densa are closely associated with unusual smooth muscle fibers, the **juxtaglomerular cells**, in the wall of the afferent arteriole. Together the macula densa and juxtaglomerular cells form the **juxtaglomerular apparatus**. The juxtaglomerular apparatus is an endocrine structure that secretes two hormones, *renin* and *erythropoietin*, introduced in Chapter 18. (p. 588)

The epithelium lining the rest of the DCT consists of cuboidal cells. Viewed in section, the distal convoluted tubule differs from the proximal convoluted tubule in that it has a smaller diameter and the epithelium lacks the microvilli characteristic of the PCT. The DCT is an important site for the active secretion of ions and other materials, and it also reabsorbs sodium ions from the urine. In the final portions of the DCT an osmotic flow of water assists in concentrating the filtrate.

The Collecting System

The distal convoluted tubule opens into the collecting system, which consists of *collecting tubules, collecting ducts*, and *papillary ducts* (Figure 26-4). The collecting system is responsible for making final adjustments to the sodium ion concentration and volume of urine.

Collecting tubules and the proximal portions of the collecting ducts are lined by a simple cuboidal epithelium. Like the terminal segments of the DCT, these areas are involved in the reabsorption of sodium ions and water. Collecting ducts are also capable of actively secreting hydrogen ions into the filtrate. The epithelium gradually becomes more columnar as adjacent collecting ducts unite to form the papillary ducts

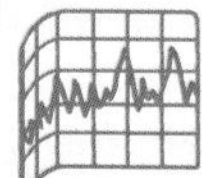

Diagnostics: Examination of the Urinary System

The functional anatomy of the urinary system can be examined using a variety of sophisticated procedures. Figure 26-6 presents several views of this system that should be compared with the diagrammatic view in Figure 26-1.

Administering a radiopaque compound that will enter the urine permits the creation of a **pyelogram** (PĪ-el-o-gram) by taking an X-ray of the kidneys (Figure 26-6a). Pyelography permits detection of unusual kidney, ureter, or bladder structures and masses.

Computerized tomography (CT) scans may also provide useful information concerning localized abnormalities (Figure 26-6b). The blood flow through the kidney can be checked by angiography (Figure 26-6c), as described earlier for the heart.

FIGURE 26-6
Images of the Urinary System. (a) A pyelogram of the urinary system, taken after the introduction of a radiopaque dye that was filtered into the urine. (b) A CT scan showing the position of the kidneys in a transverse section through the trunk. (c) An angiogram of the kidneys, showing their extensive blood supply.

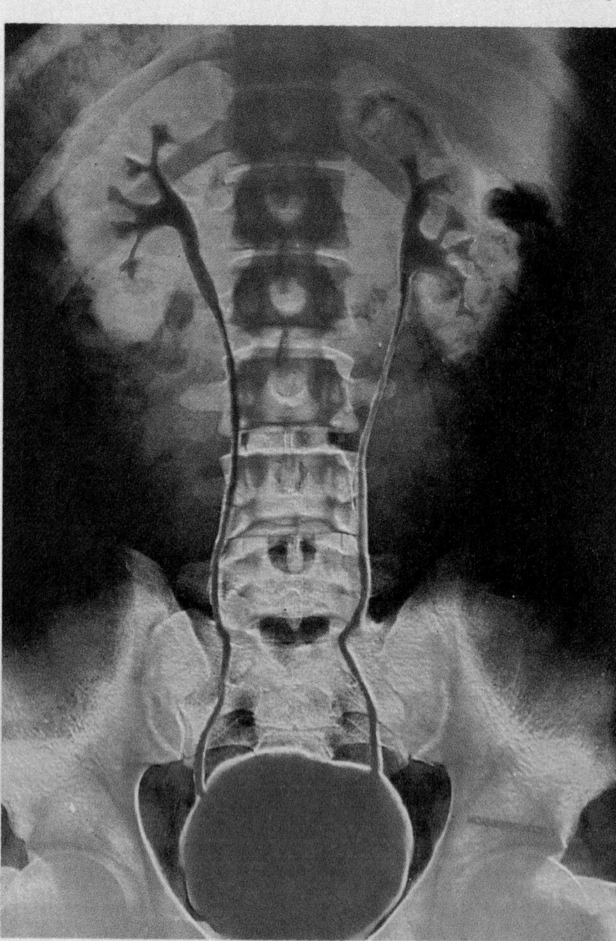

(a)

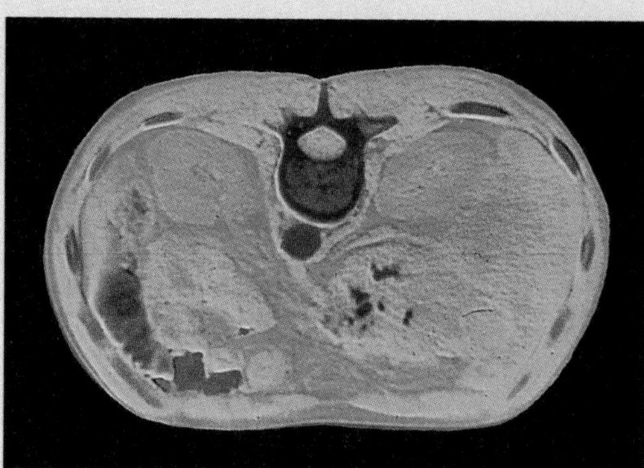

(b)

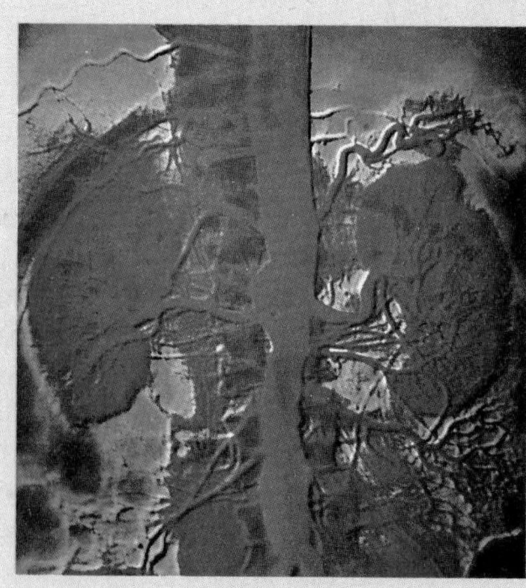

(c)

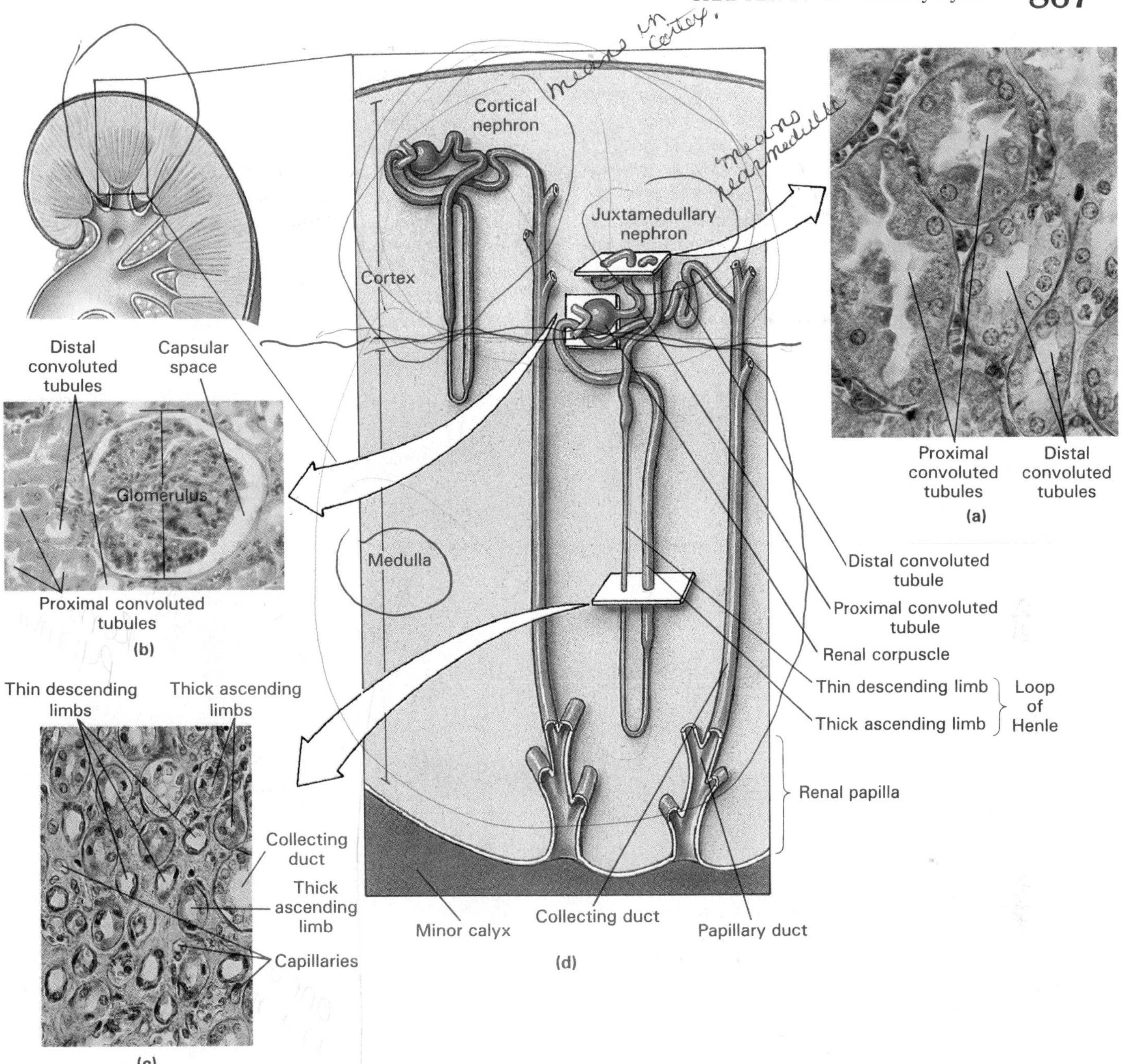

FIGURE 26-7
Cortical and Juxtamedullary Nephrons. The light micrographs show the tubules in section.

that open into the minor calyx at the apex of the lobe.

Before continuing, you should review the basic functions of each of the regions of the nephron as detailed in Table 26-1. This table also contains additional information concerning the nephron and collecting system. [**CM:** *Polycystic Kidney Disease*]

Cortical and Juxtamedullary Nephrons

There are approximately 1.25 million nephrons in each kidney, with a combined length of around 85 *miles*. There are slight differences in nephron structure depending on their location. Roughly 85 percent of the nephrons are called **cortical nephrons** because they are found in the superficial cortex of the kidney. The loops of Henle of cortical nephrons do not extend deeply into the medulla. The remaining 15 percent of nephrons, the **juxtamedullary** (juks-ta-MED-u-la-rē) **nephrons**, are located closer to the medulla, and they extend deep into the renal pyramids. Cortical and juxtamedullary nephrons are shown in Figure 26-7.

The Blood Supply to the Nephron

Each kidney receives blood from a *renal artery* that originates along the lateral surface of the abdominal aorta near the level of the superior mesenteric artery (see Figure 21-10). (p. 675) As the renal artery enters the renal sinus, it divides into a series of **interlobar arteries** that radiate outwards between the lobes (Figure 26-8a). They then turn, arching along the boundary lines between the cortex and medulla as the **arcuate** (AR-kū-āt) **arteries**. Each of the arcuates gives rise to a number of **interlobular arteries** supplying portions of the adjacent lobe. A comparable arrangement exists for the distribution of **interlobular**, **arcuate**, and **interlobar veins**, shown in Figure 26-8b.

Blood reaches the vascular pole of each glomerulus through an afferent arteriole and leaves in an efferent arteriole. Blood travels from the efferent arteriole to the **peritubular capillaries** that supply the proximal and distal convoluted tubules and the **vasa recta**, a capillary that accompanies the loop of Henle into the medulla. This circulatory arrangement, shown in Figure 26-8c, provides a route for the pickup or delivery of materials that are reabsorbed or secreted by the kidney tubules. In either event, diffusion occurs between the capillaries of the vasa recta and the tubular cells through the interstitial fluid, or **peritubular fluid**, that surrounds the nephron. **[CM:** *PAH and Analysis of Renal Blood Flow*]

FIGURE 26-8
Blood Supply to the Kidneys. (a) Sectional view, showing major arteries and veins; compare with Figure 26-3. (b) Circulation in the cortex. (c) Circulation to an individual nephron.

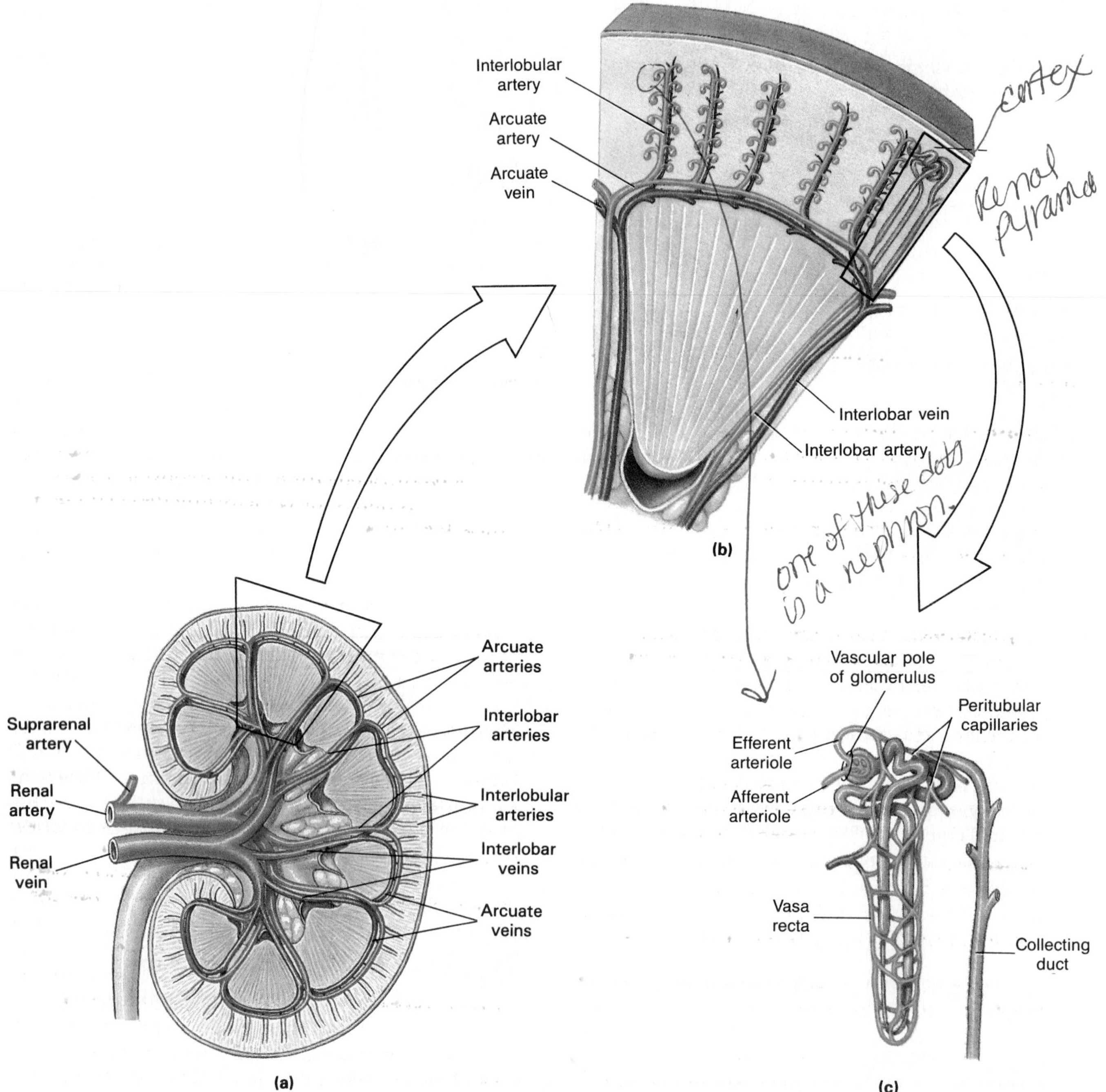

Innervation of the Nephron

The kidneys and ureters are innervated by **renal nerves**. Most of these nerve fibers are sympathetic postganglionic fibers originating in the superior mesenteric ganglion. Those innervating the kidneys enter at the hilus and follow the branches of the renal arteries and veins to reach individual nephrons. Known functions of sympathetic innervation include (1) regulation of glomerular blood flow and pressure, through innervation of afferent and efferent arterioles; (2) stimulation of renin release; and (3) direct stimulation of water and sodium ion reabsorption. (These functions will be discussed further in a later section.)

√ **Why don't plasma proteins pass into Bowman's capsule under normal circumstances?**

√ **Damage to what part of the nephron would interfere with the control of blood pressure?**

Principles of Tubular Function

Now that the basic organization of the kidney has been described we can examine kidney function in more detail. The formation of urine involves three processes:

1. **Filtration**: In filtration, hydrostatic pressure forces water across a membrane. Solute molecules small enough to pass through membrane pores are carried along by the surrounding water molecules. Filtration in the kidney occurs at the glomeruli. The filtrate formed then travels along the rest of the nephron toward the collecting system.
2. **Reabsorption**: Reabsorption is the removal of water and solute molecules from the filtrate. Many different mechanisms are involved. Water reabsorption occurs passively, through osmosis. Solute reabsorption may involve simple diffusion or the activity of carrier proteins in the tubular epithelium. Materials reabsorbed from the filtrate pass into the peritubular fluid, from which they eventually reenter the blood.
3. **Secretion**: Secretion is the transport of solutes across the tubular epithelium and into the filtrate. Secretion usually involves the activity of carrier proteins in the tubular epithelium.

Together these processes create a fluid that is very different from other body fluids. Table 26-2 provides an indication of the efficiency of the renal system by comparing the composition of the urine and plasma. All segments of the nephron and collecting system participate in the process of urine formation. Most regions of the nephron perform a combination of reabsorption and secretion, but the balance shifts from one region to another. For example, reabsorption dominates along the proximal convoluted tubule, whereas secretion is the primary function of the distal convoluted tubule. Other examples of the division of labor along the nephron will be encountered later in this chapter.

TABLE 26-2 Significant Differences between Urinary and Plasma Solute Concentrations

Component	*Urine*	*Plasma*
Ions (mEq/ℓ)		
Sodium (Na^+)	147.5	138.4
Potassium (K^+)	47.5	4.4
Calcium (Ca^{2+})	9.6	4.9
Magnesium (Mg^{2+})	8.9	1.8
Chloride (Cl^-)	153.3	106
Bicarbonate (HCO_3^-)	1.9	27
Phosphate (PO_4^{3-})	0.01 g/ℓ	0.1 g/ℓ
Sulfate (SO_4^{2-})	37.5	62
Metabolites and Nutrients (mg/dℓ)		
Glucose	0.009	90
Lipids	0.002	600
Ketone bodies	0.017	9
Amino acids	0.188	4.2
Proteins	0.000	7.5 g/dℓ
Nitrogenous Wastes (mg/ℓ)		
Urea	1800	305
Creatinine	150	8.6
Ammonia	60	0.2
Uric acid	40	3
Bilirubin	0.2	0.5
Urobilinogen	1.25	Trace

Before considering the functional specializations of the nephron and collecting system, we need to identify the mechanisms responsible for the movement of solutes and water across the tubular epithelium. Chapter 3 introduced the primary factors responsible for the movement of solutes and water across cell membranes. The passive processes of *diffusion*, *osmosis*, and *filtration* were considered in detail at that time. (Because this discussion assumes a familiarity with this material you may want to review Figures 3-17 through 3-20.) (pp. 77, 79, 80, 81)

Two types of carrier-mediated transport, *facilitated diffusion* (Figure 3-23) and *active transport* (Figure 3-24), were also introduced in Chapter 3. A third transport mechanism, *cotransport* (Figure 24-28), was presented in Chapter 24, and a fourth, *countertransport*, is discussed below. Because all four carrier-mediated transport mechanisms are involved in tubular reabsorption and tubular secretion, a closer look at these processes will prove helpful.

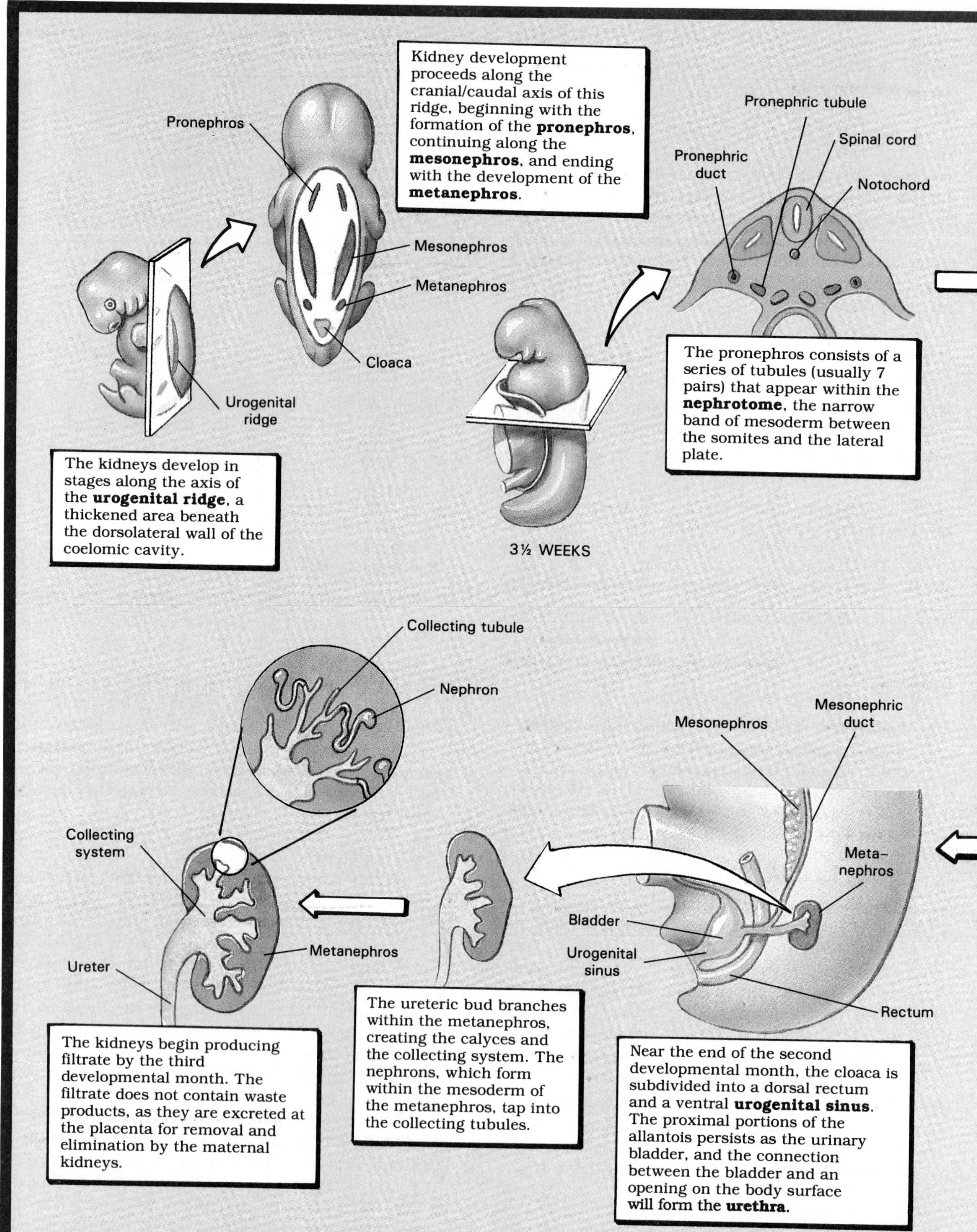

Kidney development proceeds along the cranial/caudal axis of this ridge, beginning with the formation of the **pronephros**, continuing along the **mesonephros**, and ending with the development of the **metanephros**.
Pronephros
Mesonephros
Metanephros
Cloaca
Urogenital ridge
The kidneys develop in stages along the axis of the **urogenital ridge**, a thickened area beneath the dorsolateral wall of the coelomic cavity.
Pronephric tubule
Spinal cord
Pronephric duct
Notochord
3½ WEEKS
The pronephros consists of a series of tubules (usually 7 pairs) that appear within the **nephrotome**, the narrow band of mesoderm between the somites and the lateral plate.
Collecting tubule
Nephron
Mesonephros
Mesonephric duct
Collecting system
Meta-nephros
Bladder
Urogenital sinus
Metanephros
Ureter
Rectum
The kidneys begin producing filtrate by the third developmental month. The filtrate does not contain waste products, as they are excreted at the placenta for removal and elimination by the maternal kidneys.
The ureteric bud branches within the metanephros, creating the calyces and the collecting system. The nephrons, which form within the mesoderm of the metanephros, tap into the collecting tubules.
Near the end of the second developmental month, the cloaca is subdivided into a dorsal rectum and a ventral **urogenital sinus**. The proximal portions of the allantois persists as the urinary bladder, and the connection between the bladder and an opening on the body surface will form the **urethra**.

Embryology Summary: Development of the Urinary System

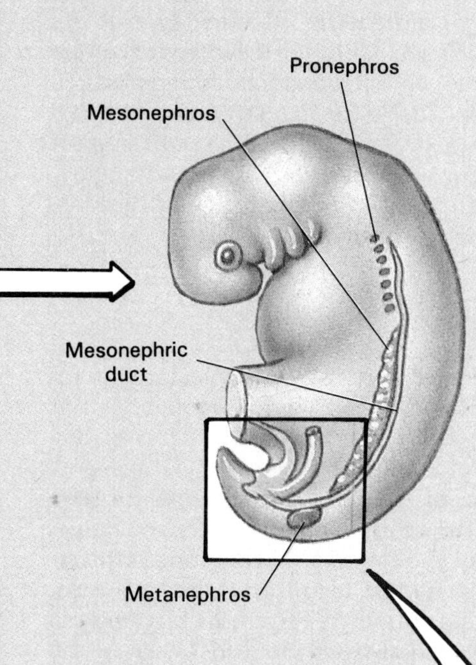

The pronephric kidneys are very small and nonfunctional, and they disintegrate almost at once. The only significant contribution of the pronephros is the formation of a pair of **pronephric ducts** that grow caudally until they connect to the *cloaca* (p. 805).

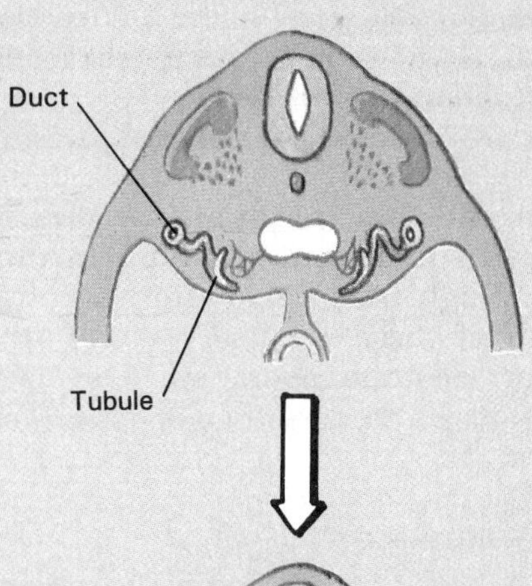

After approximately 4 weeks of development the mesoderm midway along the urogenital ridge begins organizing into the mesonephros. On either side of the midline, approximately 70 tubules develop within these segments. These tubules grow toward the adjacent pronephric duct and fuse with it. From this moment on the duct will be called the **mesonephric duct**.

Aorta

In each segment a branch of the aorta grows toward the nephrotome, and the tubules form large nephrons with enormous glomeruli. Like the pronephros, the mesonephros does not persist, and when the last segments of the mesonephros are forming the first are already beginning to degenerate.

Nephrotomal mesoderm of the metanephros forms a dense mass without a trace of segmental organization. This will become the functional adult kidney.

Mesonephros

Mesonephric duct

Cloaca

Metanephros

A **ureteric bud** forms in the wall of each mesonephric duct, and this blind tube elongates and branches within the adjacent metanephros. Tubules developing within the metanephros then connect to the terminal branches.

Most of the metabolic wastes produced by the developing embryo are passed across the placenta to enter the maternal circulation. The small amount of urine produced by the kidneys accumulates within the cloaca and the *allantois* (p. 805), and endodermal sac that extends into the umbilical stalk.

CARRIER-MEDIATED TRANSPORT MECHANISMS

Facilitated Diffusion

In facilitated diffusion (Figure 3-23) a carrier protein transports a molecule across the cell membrane without an expenditure of energy. (p. 84) The transport always follows the concentration gradient for the ion or molecule transported. Facilitated transport is important in the reabsorption of glucose and amino acids when their concentrations in the filtrate are relatively high. (At lower filtrate concentrations, reabsorption of these substances continues via active transport or cotransport, as these carrier mechanisms can operate despite an opposing concentration gradient.)

Active Transport

Carrier protein activity in active transport is driven by the hydrolysis of ATP to ADP on the inner membrane surface. The sodium-potassium exchange pump (Figure 3-24), which ejects three intracellular sodium ions in exchange for two extracellular potassium ions, is a familiar example. (p. 85)

Exchange pumps are just one of several types of ion pump active along the kidney tubules. Many other carrier proteins transport only one type of ion across the membrane. Examples include the active transport mechanisms for calcium, magnesium, chloride, iodide, and iron.

Cotransport

In cotransport, carrier protein activity is not directly linked to the hydrolysis of ATP. In this process, two substrates (ions, molecules, or both) cross the membrane while bound to the carrier protein. Both substrates are then released on the inner surface of the membrane. The movement always occurs following the concentration gradient *for at least one of the transported substances*. Chapter 24 introduced one important example of cotransport, sodium-linked glucose cotransport, along the digestive tract (see Figure 24-28). (p. 822) Comparable cotransport mechanisms remove glucose, other simple sugars, amino acids, lactic acid, phosphate, chloride ions, and hydrogen ions from the filtrate.

Countertransport

The fourth carrier-mediated transport mechanism involved in the modification of the filtrate, one that has not previously been discussed, is called **countertransport**. Countertransport resembles cotransport in all respects except that the two transported ions move in *opposite* directions. In **sodium-linked countertransport** one of the two is a sodium ion. **Sodium-calcium countertransport** is an important example. In this process, diagrammed in Figure 26-9, a calcium ion is exchanged for a sodium ion. Most cells in the body use sodium-calcium countertransport to keep intracellular calcium concentrations low. Along the kidney tubules, carriers on the epithelial surface bring calcium ions *into* the cell, while those on the surface attached to the basement membrane pump them into the peritubular fluid. The parathyroid hormone, *parathormone*, stimulates these carrier proteins to reduce the urinary loss of calcium ions.

A second important example of countertransport activity involves the exchange of chloride ions,

FIGURE 26-9
Sodium-Calcium Countertransport. In countertransport two ions, both carrying positive or negative charges, move across the membrane in opposite directions. In this example of sodium-linked cotransport, the cell must expend ATP to remove the sodium ions it gains when moving calcium ions out of the cytoplasm. Other countertransport mechanisms do not bring sodium ions into the cell, and no energy expense is involved.

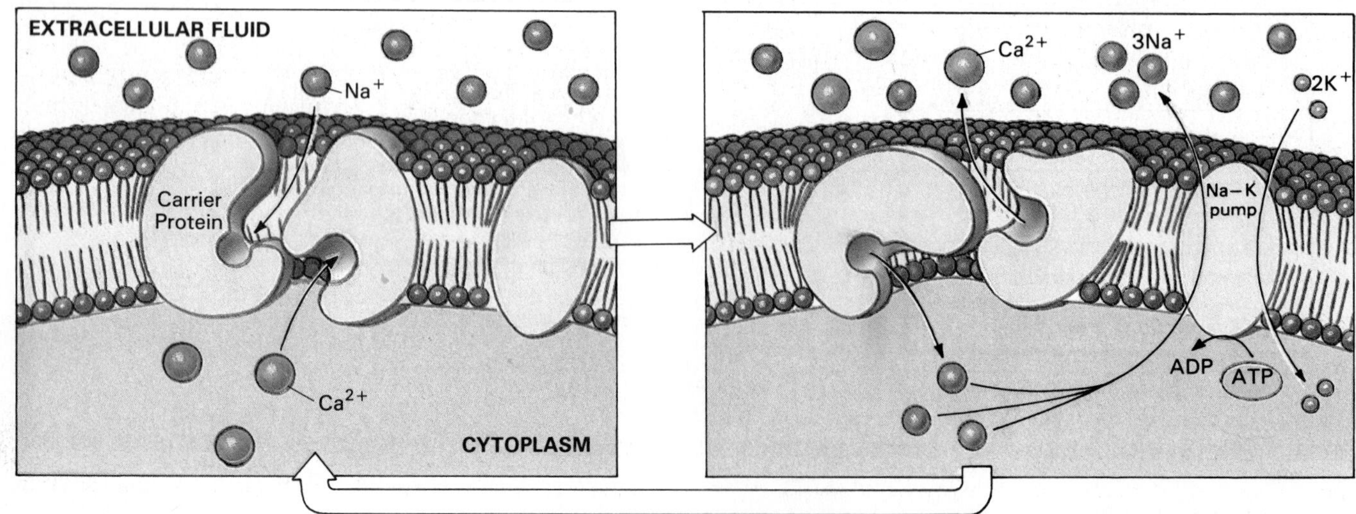

Cl^-, for bicarbonate ions, HCO_3^-. Bicarbonate ions are important buffers, and their reabsorption from the filtrate can help prevent dangerous changes in the pH of the blood.[2]

FEATURES OF CARRIER-MEDIATED TRANSPORT

All four of the carrier-mediated processes just discussed have certain common features that are important for an understanding of kidney function:

1. *In carrier-mediated transport, a specific substrate binds to a carrier protein that facilitates its movement across the membrane.*
2. *A given carrier protein normally works in one direction only.* In facilitated diffusion, that direction is determined by the concentration gradient of the substance being transported. In active transport, cotransport, and countertransport, it is the location and orientation of the carrier proteins that determine whether a particular substance will be reabsorbed or secreted. For example, the carrier protein that transports amino acids from the filtrate to the cytoplasm will not carry amino acids back into the filtrate.
3. *The distribution of carrier proteins can vary from one portion of the cell surface to another.* Transport between the filtrate and the interstitial fluid involves two steps, because the material must cross both the inner (attached) and outer (lumenal) surfaces of the tubular cells. In essence, the transport activities of the tubular cells are comparable to those of the intestinal epithelial cells diagrammed in Figures 24-28 through 24-30. (pp. 822, 824) For example, in the case of amino acids, glucose, or other nutrients, carrier proteins facing the basement membrane accept substrates from the cytoplasm and deliver them to the interstitial fluid.
4. *The membrane of a single tubular cell contains more than one type of carrier protein.* Each cell can have multiple functions, and the same cell that reabsorbs one compound can secrete another.
5. *Carrier proteins, like enzymes, have saturation limits.* When all the available carrier proteins are occupied at one time, the system of carriers cannot work any faster regardless of the size of the concentration gradient. The carriers are then said to be *saturated*. For any substance, the concentration at saturation is called the **transport maximum** (T_m), or *tubular maximum*. Although both carriers involved with reabsorption and those responsible for secretion have saturation limits, the concept is most often applied to tubular reabsorption.

The T_m and Renal Threshold

Normally any plasma proteins and nutrients, such as amino acids and glucose, are removed via active transport, cotransport, or facilitated diffusion. If the concentrations of these nutrients in the filtrate rise, the rates of reabsorption increase until the carrier saturation limits are reached. *A concentration higher than the tubular maximum will exceed the reabsorptive abilities of the nephron, and some of the material will remain in the filtrate and appear in the urine.* The transport maximum thus determines the **renal threshold**, the plasma concentration at which a specific compound or ion will begin appearing in the urine.

The renal threshold varies, depending on the substance involved. The renal threshold for glucose is approximately 180 mg/dℓ. When plasma glucose concentrations are higher than this, glucose concentrations in the filtrate exceed the T_m of the tubular cells, and glucose appears in the urine. Plasma concentrations seldom reach this level, and the appearance of glucose in the urine, a condition called *glucosuria*, is very unusual.[3] The renal threshold for amino acids is lower, and they appear in the urine when plasma concentrations exceed 65 mg/dℓ. Plasma levels often reach these levels after a protein-rich meal.

Formation and Modification of the Filtrate

The following general principles were introduced earlier in the chapter:

1. Filtration occurs exclusively in the renal corpuscle, across the walls of the glomerular capillaries.
2. The proximal convoluted tubule is the primary site of nutrient reabsorption.
3. The distal convoluted tubule is the primary site for secretion of substances into the filtrate.
4. The loop of Henle and the collecting system interact to regulate the amount of water and the number of sodium and potassium ions lost in the urine.

[2] The exchange of chloride for bicarbonate serves different purposes in different cells of the body. In red blood cells this process, called the chloride shift, is involved in the transport of carbon dioxide (Chapter 23). (p. 775) In the stomach a comparable process, termed the alkaline tide, is essential to the secretion of hydrochloric acid (Chapter 24). (p. 797)

[3] Glucosuria is a classic symptom of *diabetes mellitus*, a condition often resulting from inadequate insulin secretion. (For additional details review the discussion in Chapter 18.) (p. 594)

We will now examine each of these steps in greater detail.

GLOMERULAR FILTRATION

Glomerular filtration occurs as fluids move across the wall of the glomerulus and into the capsular space. The major forces acting across capillary walls were detailed in Chapters 21 and 22. (You may find it helpful to review Figures 21-23 and 21-24 before proceeding.) (pp. 693, 694) Primary factors involved in glomerular filtration are diagrammed in Figure 26-10.

Glomerular Hydrostatic Pressure

The hydrostatic (blood) pressure in the glomerular capillaries—the **glomerular hydrostatic pressure**, or $\mathbf{G_{hp}}$—provides the outward pressure forcing water and solute molecules across the glomerular wall. This pressure is unusually high, due to the arrangement of vessels at the glomerulus.

Blood pressure is low in typical systemic capillaries because the capillary blood flows into the venous system, whose resistance is relatively low. At the glomerulus, however, blood leaving the glomerular capillaries flows into an efferent arteriole whose diameter is *smaller* than that of the afferent arteriole. This vessel offers considerable resistance, and relatively high pressures are needed to force blood into it. As a result, glomerular pressures are comparable to those of small arteries, averaging 60 mm Hg instead of the 30 mm Hg typical of peripheral capillaries.

FIGURE 26-10
Glomerular Filtration. (a) In filtration, hydrostatic pressure forces water and dissolved solute molecules across a membrane. The size of the membrane pores controls what substances can or cannot cross the membrane. (b) An analysis of forces acting across the filtration apparatus of the glomerulus.

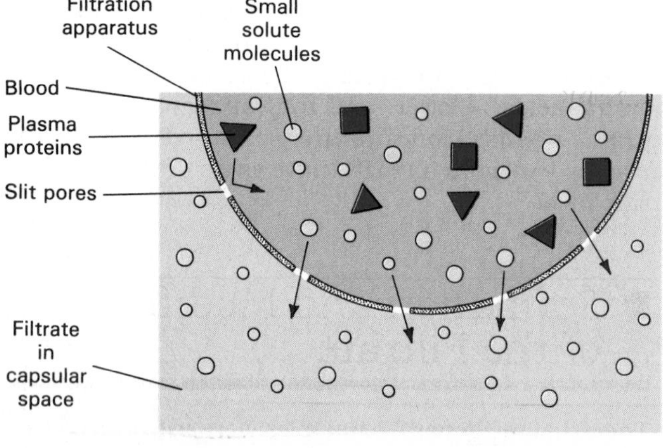

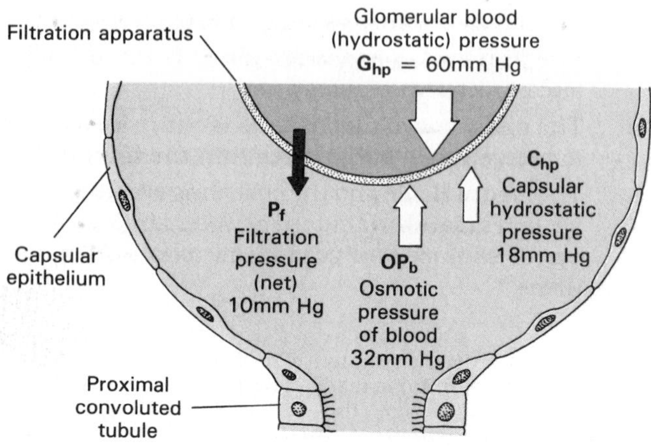

Capsular Hydrostatic Pressure

Glomerular hydrostatic pressure tends to force water and dissolved materials out of the bloodstream and into the capsular space. That movement is opposed by the **capsular hydrostatic pressure,** $\mathbf{C_{hp}}$. This pressure represents the resistance to flow along the nephron and conducting system. (Before additional filtrate can enter the capsule, some of the filtrate already present must flow into the proximal convoluted tubule.) The capsular hydrostatic pressure averages around 18 mm Hg.

Blood Osmotic Pressure

Filtration will also be opposed by the osmotic pressure of the blood, $\mathbf{OP_b}$. This value rises as fluid leaves the glomerular capillaries, because the plasma proteins, which cannot cross the glomerular wall, become increasingly concentrated. Over the entire length of the glomerular capillary bed the osmotic pressure averages about 32 mm Hg.

Filtration Pressure

The **filtration pressure** ($\mathbf{P_f}$) at the glomerulus is the difference between the blood pressure and the opposing capsular and osmotic pressures. This relationship, illustrated in Figure 26-10, can be summarized as:

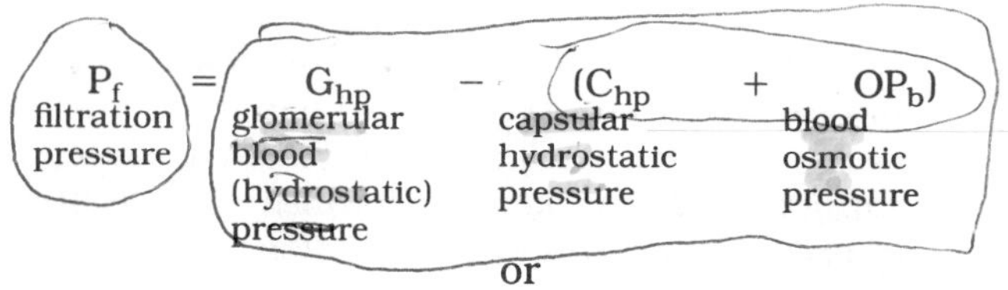

or

$$P_f = 60 \text{ mm Hg} - (18 \text{ mm Hg} + 32 \text{ mm Hg}) = 10 \text{ mm Hg}$$

This represents the average pressure forcing water and dissolved materials out of the glomerular capillaries and into the capsular spaces.

Glomerular Filtration Rate

The **glomerular filtration rate** (**GFR**) is the amount of filtrate produced in the kidneys each minute. Each kidney contains around 6 square meters of filtration surface, and the GFR averages an astounding *125 mℓ per minute*. This means that almost 20 percent of the plasma delivered to the kidneys by the renal arteries leaves the bloodstream and enters the capsular spaces. In the course of a single day the glomeruli generate about 180 liters (50 gal) of filtrate, roughly 70 times the total plasma volume. But as the filtrate passes through the renal tubules, about 99 percent of it is reabsorbed. The significance of tubular reabsorption should now be apparent! An inability to reclaim the water entering the filtrate, as in diabetes insipidus, can quickly cause death by dehydration. (This condition, caused by inadequate ADH secretion at the posterior pituitary, is discussed in a Clinical Comment in Chapter 18.) [**CM:** *Estimating the GFR*]

The glomerular filtration rate depends on the filtration pressure acting across the glomerular capillaries. Any factor that alters the filtration pressure will therefore alter the GFR and affect kidney function. One of the most significant factors is a decline in renal blood pressure.

Despite the large volume of filtrate generated, the filtration pressure is relatively low. Whenever the mean arterial pressure falls by 10 percent (from a normal value of about 100 mm Hg to 90 mm Hg), the GFR is severely restricted. If blood pressure at the glomeruli drops by 16 percent, from 60 to 50 mm Hg, kidney filtration will cease altogether. The kidneys are therefore sensitive to changes in blood pressure that have little or no effect on other organs. Hemorrhaging, shock, and dehydration are relatively common clinical conditions that can cause a dangerous decline in the GFR.

Glomerular filtration is the vital first step essential to all kidney functions. If filtration does not occur, waste products are not excreted, pH control is jeopardized, and an important mechanism for blood volume regulation is eliminated. A variety of regulatory mechanisms exist to ensure that the GFR remains within normal limits; these mechanisms will be detailed in a later section. [**CM:** *Conditions Affecting the Filtration Process*]

CHANGING THE COMPOSITION OF THE FILTRATE

Glomerular filtration produces a filtrate with a composition similar to that of blood plasma without the plasma proteins. For convenience, filtrate modification will be broken down into two phases: (1) changes in the composition of the filtrate through selective reabsorption and secretion and (2) changes in the total osmotic concentration of the filtrate.

Tubular Reabsorption and the Proximal Convoluted Tubule

The cells of the proximal convoluted tubule (PCT) normally reabsorb 60 percent of the volume of filtrate produced in the renal corpuscle.

Nutrient Reabsorption Under normal circumstances virtually all of the glucose, amino acids, and other nutrients are reabsorbed before the filtrate leaves the PCT. This reabsorption involves a combination of facilitated transport and cotransport mechanisms.

Ion Reabsorption The PCT actively reabsorbs ions, including sodium, potassium, calcium, magnesium, bicarbonate, phosphate, and sulfate ions. The ion pumps involved are individually regulated and may be influenced by circulating ion or hormone levels. For example, the presence of parathormone stimulates calcium and phosphate ion reabsorption.

In terms of total effort, the PCT is particularly active in the reabsorption of sodium ions (Figure 26-11). These ions diffuse into the tubular cell and move through the cytoplasm toward its base. The cell membrane adjacent to the basement membrane contains an ion pump that ejects the sodium ions into the peritubular fluid. The positively charged sodium ion is usually accompanied by a negatively

FIGURE 26-11
Directional Sodium Transport. Sodium ions diffuse from the filtrate across the tubular epithelium and into the cytoplasm. Chloride ions accompany them, drawn by the attraction between opposite charges. At the inner membrane surface, sodium ions are actively pumped into the peritubular fluid in exchange for potassium ions.

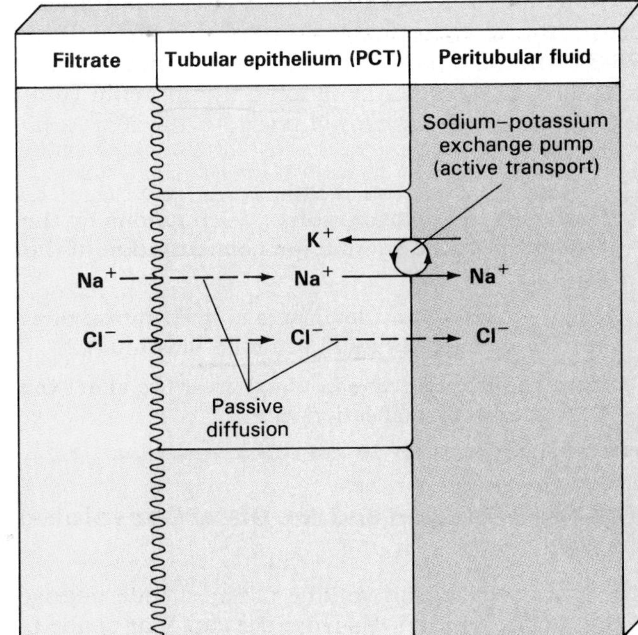

charged ion, often a chloride ion (Cl^-). No energy must be expended to move the anion, for its movement results from the electrical attraction between positive and negative charges.

Water Reabsorption The reabsorptive processes have a direct effect on the solute concentration of the filtrate and on the concentration in the peritubular fluid. The effects are diagrammed in Figure 26-12.

STEP 1: The filtrate entering the PCT has the same osmotic concentration as the peritubular fluid.

STEP 2: As transport activities proceed, the solute concentration of the filtrate decreases and that of the peritubular fluid and adjacent capillaries increases.

STEP 3: Osmotic forces then pull water out of the filtrate and into the peritubular fluid.

STEP 4: An osmotic equilibrium is restored, but the remaining filtrate now contains a relatively high concentration of urea.

Passive Urea Reabsorption Urea filtered into the renal corpuscle is not actively reabsorbed by the PCT, but the tubular epithelium is freely permeable to urea. As the urea concentration in the filtrate rises, urea molecules diffuse across the tubular epithelium and into the peritubular fluid. This further reduces the osmolarity of the filtrate, leading to additional water reabsorption.

Secretion A relatively small amount of active secretion also occurs along the PCT. The mechanisms involved resemble those of the distal convoluted tubule, which are discussed below. A number of other compounds in the filtrate are completely ignored by the tubular cells and by other portions of the nephron. As water and other compounds are removed, the concentration of these materials gradually rises. Table 26-3 contains a partial listing of substances actively reabsorbed, secreted, or ignored by the renal tubules. [**CM:** *Inherited Problems with Tubular Function*]

√ **How does the reabsorption of calcium ions by the kidney affect the sodium ion concentration of the filtrate?**

√ **What occurs when the plasma concentration of a substance exceeds its T_m (tubular maximum)?**

√ **How would a decrease in blood pressure affect the GFR (glomerular filtration rate)?**

Tubular Secretion and the Distal Convoluted Tubule

The composition and volume of the filtrate change dramatically as it travels from the capsular space to

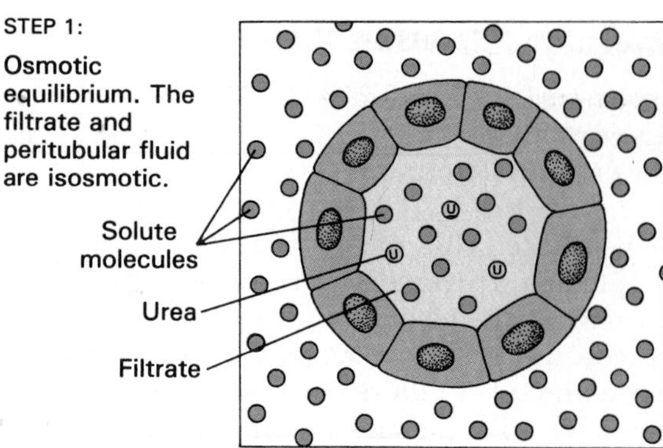
STEP 1:
Osmotic equilibrium. The filtrate and peritubular fluid are isosmotic.
Solute molecules
Urea
Filtrate

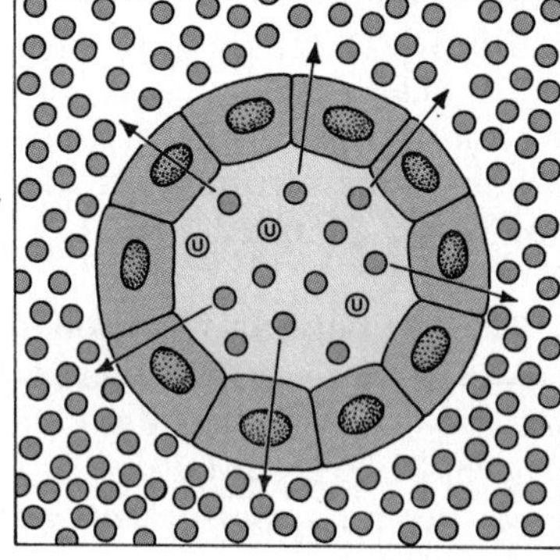
STEP 2:
Absorption via active transport of glucose, sodium, chloride, proteins, etc. The osmolarity of the peritubular fluid increases.

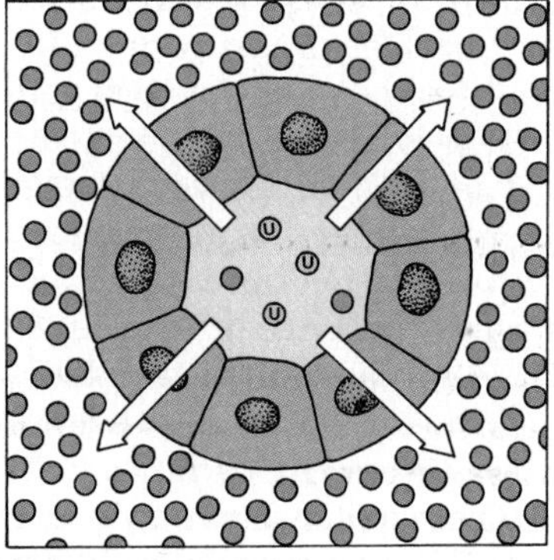
STEP 3:
Osmotic water flow from the filtrate into the peritubular fluid.

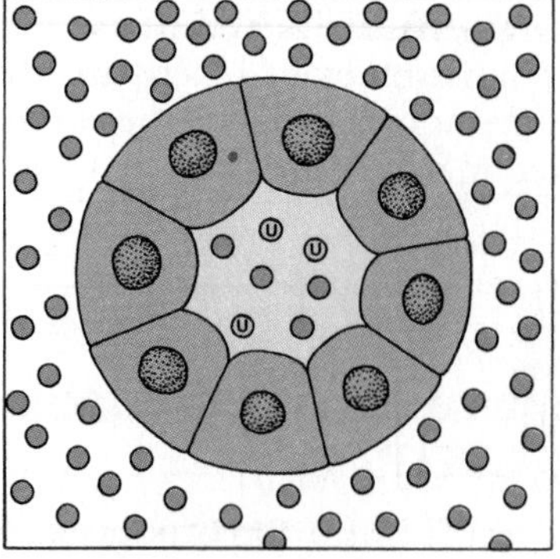
STEP 4:
Osmotic equilibrium is restored, filtrate volume is reduced.

FIGURE 26-12
Tubular Water Reabsorption. Although shown in four steps, osmosis occurs as long as the tubular cells continue to extract solute molecules from the filtrate.

TABLE 26-3 Tubular Absorption and Secretion

Reabsorbed	*Secreted*	*No Active Transport*
Ions Sodium (Na^+) Chloride (Cl^-) Potassium (K^+) Calcium (Ca^{2+}) Magnesium (Mg^{2+}) Phosphate (PO_4^{3-}) Sulfate (SO_4^{2-}) Bicarbonate (HCO_3^-) **Metabolites** Glucose Amino acids Proteins Vitamins	**Ions** Potassium (K^+) Hydrogen (H^+) Calcium (Ca^{2+}) **Wastes** Creatinine Ammonia Organic acids and bases **Miscellaneous** Neurotransmitters (ACh, NE, E, DOPA) Histamine Drugs (penicillin, atropine, morphine, numerous others) PAH	Urea Water Urobilinogen Bilirubin Alcohol

the distal convoluted tubule. Roughly 60 percent of the water and 65 percent of the solutes are reabsorbed in the PCT, and another 20 percent of the water and 25 percent of the dissolved materials, principally sodium and chloride ions, enter the peritubular fluid of the medulla along the loop of Henle. Selective reabsorption or secretion, primarily along the distal convoluted tubule, then performs the final adjustments in filtrate composition.

Filtration does not force all of the dissolved materials out of the plasma, and blood entering the peritubular capillaries still contains a number of potentially undesirable substances. In most cases their presence is not significant, for the remaining concentrations are too low to cause physiological problems. If the concentration of specific ions or compounds in the peritubular capillaries remains unacceptably high, the tubules may actively excrete these materials into the filtrate. A partial listing of substances secreted into the filtrate by the DCT can be found in Table 26-3.

The rate of potassium ion and hydrogen ion secretion rises or falls in response to changes in their concentration in the peritubular capillaries. The higher the concentration, the greater the secretory activity. These ions merit special attention because the concentration of potassium ions and hydrogen ions in the blood must be maintained within relatively narrow limits.

Potassium Ion Secretion Figure 26-13a diagrams the mechanism of potassium ion secretion along the distal convoluted tubule. Potassium ions are removed from the peritubular fluid by the tubular cells in exchange for intracellular sodium ions. At the tubular lumen, these potassium ions diffuse into the filtrate via potassium channels. In effect, the tubular cells are trading sodium ions in the filtrate for excess potassium ions in the body fluids.

The ion pump and the potassium ion channels are controlled by the hormone *aldosterone*, produced

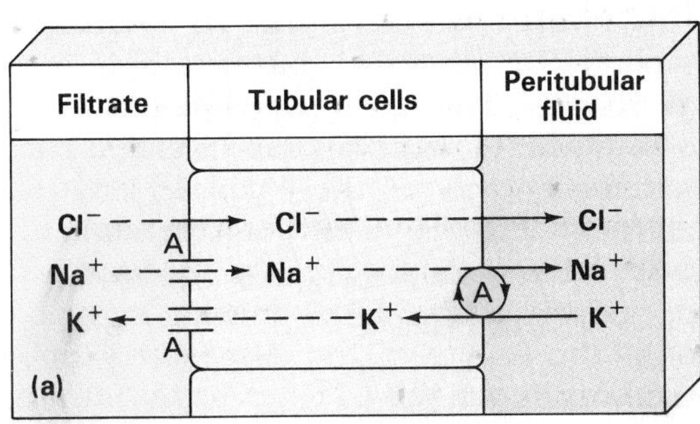

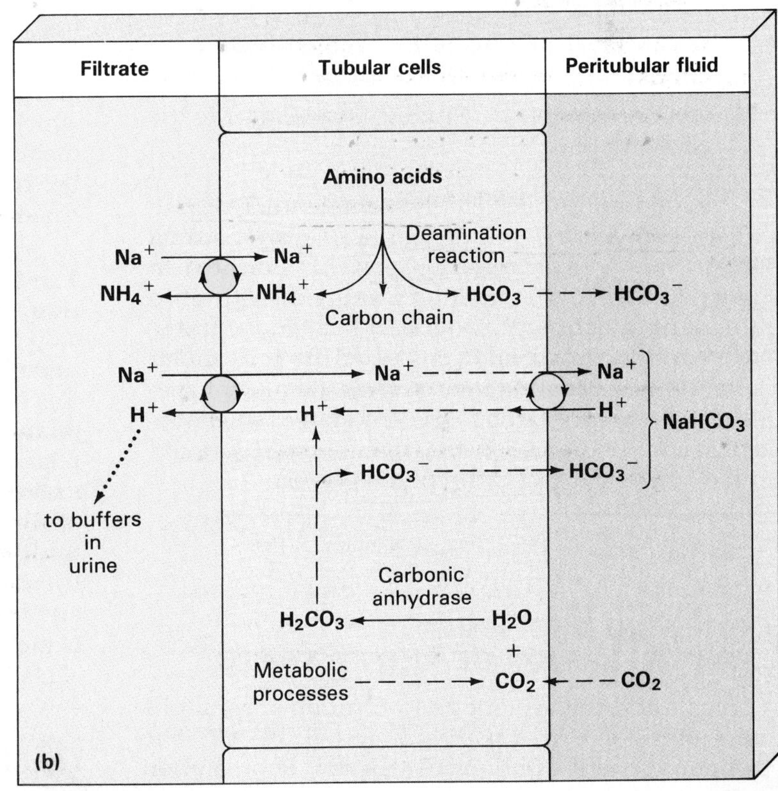

FIGURE 26-13
Tubular Secretion of Potassium and Hydrogen Ions. (a) Basic pattern for the active secretion of potassium. The process is linked to the removal of sodium ions; the exchange pumps involved are stimulated by aldosterone. (b) Hydrogen ion secretion occurs via several different routes. The central theme is the exchange of hydrogen ions in the cytoplasm for sodium ions in the filtrate.

by the adrenal cortex. Aldosterone stimulates ion pumps along the DCT, the collecting tubule, and the collecting duct, thus reducing the number of sodium ions lost in the urine. However, sodium ion conservation is associated with potassium ion loss. Prolonged aldosterone stimulation can therefore produce a dangerous reduction in the plasma concentration of potassium ions (hypokalemia).

Hydrogen Ion Secretion Hydrogen ion secretion is also associated with the reabsorption of sodium, and the ion pumps involved are stimulated by aldosterone. Hydrogen ion secretion lowers the pH of the urine while elevating that of the blood. Two secretory possibilities are summarized in Figure 26-13b. Both involve a familiar reaction sequence, the generation of carbonic acid by the enzyme *carbonic anhydrase* (see Figures 23-22 and 24-12). (pp. 774, 798) Hydrogen ions generated by the dissociation of the carbonic acid are secreted by means of sodium-linked countertransport, in exchange for sodium ions in the filtrate. The bicarbonate ions then diffuse into the peritubular fluids and into the bloodstream, where they help prevent changes in plasma pH.

Hydrogen ion secretion accelerates when the pH of the blood falls, as in lactic acidosis or ketoacidosis. The combination of hydrogen ion removal and bicarbonate ion production by the DCT plays an important role in the control of blood pH, a topic we will pursue in detail in Chapter 27.

Hydrogen ion secretion also accelerates during starvation, for several reasons:

- muscle fibers breaking down glycogen reserves release lactic acid;
- the number of circulating ketone bodies increases, and ketone bodies are acids;
- adipose tissues are releasing fatty acids into the circulation.

In addition, during starvation the DCT deaminates amino acids in reactions that yield **ammonium ions**, NH_4^+, and bicarbonate ions. As indicated in Figure 26-13b, the ammonium ions are then pumped into the filtrate through sodium-linked countertransport, and the bicarbonate ions enter the peritubular fluid. Tubular deamination thus has two major benefits: (1) it provides carbon chains suitable for catabolism, and (2) it generates bicarbonate ions that add to the buffering capabilities of the plasma.

CHANGING THE OSMOTIC CONCENTRATION OF THE FILTRATE

Osmosis and the osmotic concentration of solutions were introduced in Chapter 3. (pp. 80–82) The osmotic concentration, or osmolarity, of a solution is the total number of dissolved particles in each liter. Osmolarity is usually expressed in terms of **osmoles** or **milliosmoles** per liter.[4] If each liter of a fluid contains a mole of dissolved particles, it is a 1 **osmolar (Osm)** or 1000 **milliosmolar (mOsm)** solution. Our body fluids have an osmotic concentration of about 300 mOsm. By comparison, seawater has an osmolarity of about 1000 mOsm, and fresh water about 5 mOsm.

The normal kidneys are capable of producing a concentrated urine with an osmotic concentration of 1200–1400 mOsm, more than four times that of plasma. Because the urine can be highly concentrated, waste products are eliminated with minimal water loss. *Without this ability to concentrate the filtrate produced by glomerular filtration, fluid losses would lead to fatal dehydration in hours.*

Filtrate concentration involves an interplay between the loop of Henle, the vasa recta, and the collecting ducts. Their interaction produces unusually high concentrations of a number of ions and molecules in the peritubular fluid of the renal medulla. Near the cortex, the osmotic concentration of the interstitial fluid approaches 300 mOsm, but deep in the medulla that concentration rises to about 1200 mOsm. This **osmotic gradient** is responsible for concentrating the filtrate as it travels through the collecting system toward the renal pelvis.

Formation and Maintenance of the Osmotic Gradient

The high osmotic concentrations found in the medulla are primarily due to the presence of sodium ions, chloride ions, and urea. Two different mechanisms are involved in the formation of the osmotic gradient in the medulla. Sodium and chloride ions, which account for roughly two-thirds of the total osmolarity, are transported into the peritubular fluid by the loop of Henle. Urea enters by diffusion from the distal portions of the collecting system.

The Loop of Henle and Countercurrent Multiplication The descending and ascending limbs of the loop of Henle are very close together, but they have different permeability characteristics. The descending limb is permeable to water, but relatively impermeable to solutes. The thick ascending limb is relatively impermeable to both water and solutes. The exchange that occurs between these parallel lengths of the loop of Henle has been called **countercurrent multiplication**. *Countercurrent* refers to the fact that the exchange occurs between fluids moving in opposite directions—the filtrate in the descending limb is moving toward the renal pelvis, whereas the filtrate

[4] For methods of reporting solute concentrations, see Chapter 2 and Appendix IV.

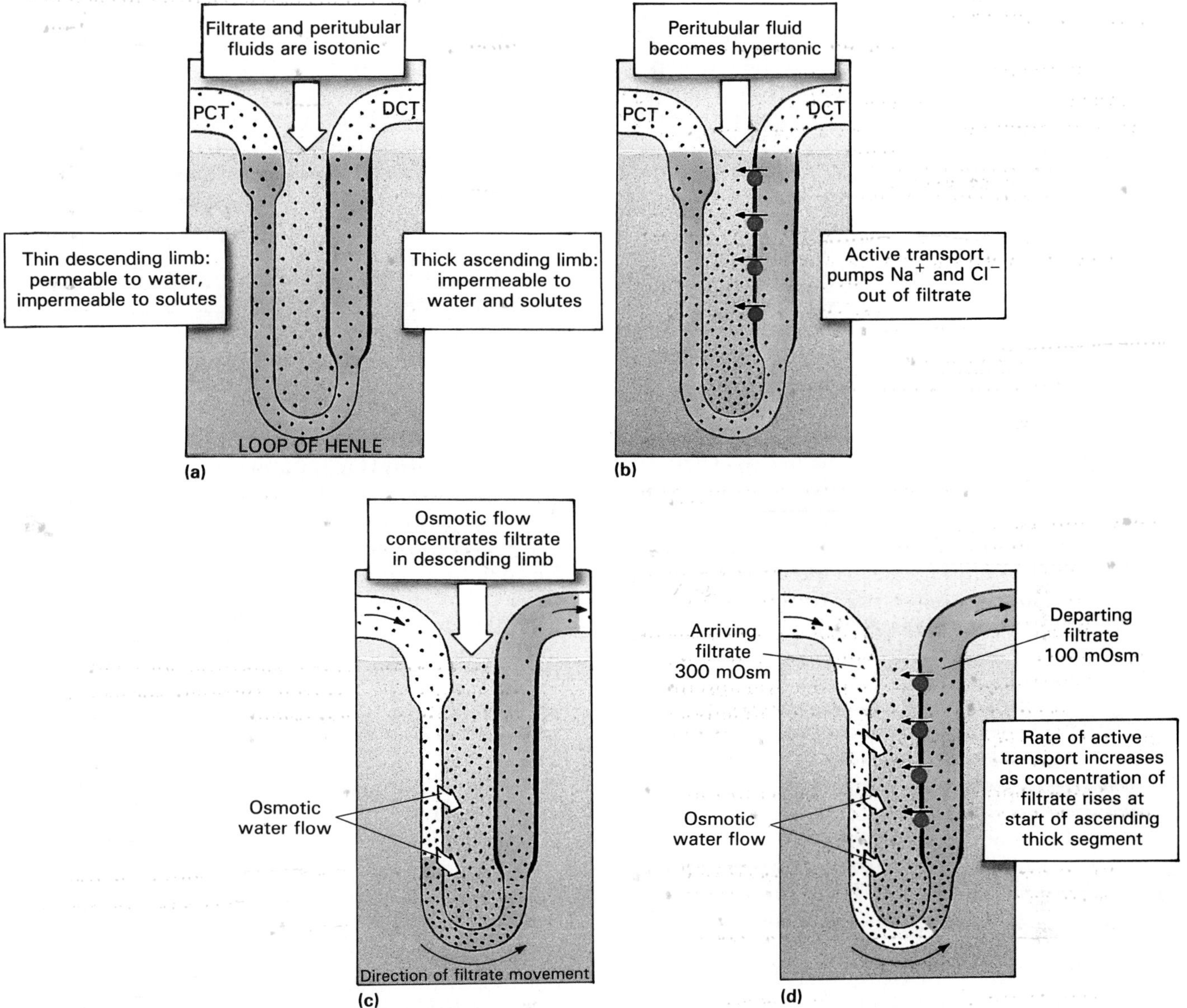

FIGURE 26-14
Countercurrent Multiplication. The operation of the countercurrent multiplier in the loop of Henle. At (a), the tubule contains a filtrate having the same osmolarity as other body fluids. Note the differences in permeability between the descending and ascending limbs. The thick ascending limb actively transports sodium and chloride ions into the peritubular fluid, elevating the osmotic concentration around the descending limb (b). This increase causes an osmotic flow of water out of the filtrate in the descending limb (c). Notice that filtrate is now shown to be moving along the tubule. The term "multiplication" refers to the fact that the osmotic gradient continues to increase as filtrate movement continues. The water loss concentrates the filtrate, and when the concentrated filtrate reaches the thick ascending limb additional Na^+ and Cl^- ions are pumped into the peritubular fluid. This leads to increased water loss, and the cycle continues. The filtrate arriving in the DCT has a lower osmolarity than the filtrate entering the loop of Henle. The missing ions have been trapped in the peritubular fluid of the medulla.

in the ascending limb is moving toward the cortex.[5] *Multiplication* refers to the fact that the effect increases as fluid movement continues. The basic concept is straightforward:

- *Sodium and chloride are pumped out of the filtrate in the ascending limb and into the peritubular fluid.*
- *This pumping elevates the osmotic concentration in the peritubular fluid around the descending limb.*
- *The result is an osmotic flow of water out of the filtrate held in the descending limb and into the peritubular fluid.*

The countercurrent exchange mechanism is diagrammed in Figure 26-14. For demonstration pur-

[5] Countercurrent exchange was introduced in Chapter 25; countercurrent heat exchange between veins and arteries was diagrammed in Figure 25-16. (p. 855)

poses, part (a) shows the loop of Henle containing a filtrate isotonic with other body fluids (roughly 300 mOsm). Parts (b) through (d) follow the changes in osmotic concentration that occur as active transport begins in the ascending limb and filtrate moves through the loop of Henle.

Activate transport along the thick ascending limb elevates sodium and chloride levels in the interstitial (peritubular) fluid of the medulla. As filtrate passes along this portion of the nephron, ion pumps transport sodium chloride (NaCl) into the surrounding interstitial fluid. The removal of ions from the filtrate in the ascending limb immediately elevates the osmotic concentration of the peritubular fluid around the descending limb, as shown in Figure 26-14b. The pumping mechanism is very effective, and almost two-thirds of the sodium and chloride ions are removed before the filtrate reaches the DCT.

Because the descending limb is permeable to water, as the filtrate travels along the descending limb, water moves out of the tubule by osmosis. Solutes remain behind, and the filtrate reaching the turn of the loop therefore has a higher osmolarity than it did when it left the PCT (Figure 26-14c).

When this relatively concentrated filtrate enters the ascending limb, additional sodium and chloride ions are pumped out. The rate of transport is proportional to the concentration in the filtrate. As a result, more sodium and chloride enter the medulla at the start of the thick ascending limb, where NaCl concentrations are highest, than near the cortex, where the concentrations are relatively low.

The result is the creation of a concentration gradient for sodium and chloride ions, with the highest concentrations deep in the medulla. As filtrate movement continues, the concentration gradient gets larger. This triggers additional water losses from the filtrate in the descending limb (Figure 26-14d).

This interaction essentially removes sodium and chloride ions from the filtrate and traps them in the medulla. The mechanism is very effective in restricting the urinary losses of sodium chloride and water. Approximately 40 percent of the filtrate produced at the glomeruli enters the descending limb of the loop of Henle (the rest is reabsorbed by the PCT). The filtrate enters the loop of Henle with an osmolarity of 300 mOsm. Roughly half of that volume reaches the distal convoluted tubule, and it arrives in the DCT with an osmolarity of 100 mOsm.

Tubular Permeability and Urea Concentrations Figure 26-15 diagrams the primary factors involved in the formation and maintenance of the osmotic gradient in the medulla. Figure 26-15a summarizes the countercurrent mechanism responsible for elevated sodium and chloride concentrations in the medulla. A separate mechanism, diagrammed in Figure 26-15b, accounts for the elevated urea concentrations in the medulla.

As water is lost by osmosis from the descending loop of Henle and the filtrate volume decreases, the urea concentration rises. The epithelia lining the thick ascending limb, the distal convoluted tubule, and collecting tubules and ducts are relatively impermeable to urea. As further water reabsorption occurs, the urea concentration continues to rise, so that the filtrate entering the papillary duct contains a relatively high concentration of urea. This region is permeable to urea, so under normal conditions urea diffuses out of the filtrate and into the peritubular fluid of the deepest portions of the medulla. The urea concentration in this region averages 450 mOsm, roughly equivalent to the concentration in fresh urine.

Function of the Vasa Recta Once established, the osmotic gradient is maintained by exchange between the renal tubules and the capillaries of the vasa recta. As it travels through the medulla, the vasa recta absorbs and transports solutes and water. Blood flow through these vessels and filtrate movement through the loop of Henle occur in opposite directions, as indicated in Figure 26-15c. As blood flows parallel to the ascending limb, it travels through the osmotic gradient of the medulla. Passing through regions of increasing osmotic concentration, the plasma becomes hypotonic with respect to the surrounding peritubular fluid, and large numbers of solute molecules diffuse into the plasma. (A small amount of water leaves the plasma at the same time, but the presence of plasma proteins severely restricts this water loss.)

At the bottom of the capillary loop, deep in the medulla, plasma osmolarity is identical to that of the surrounding peritubular fluid. This highly concentrated blood then flows around the loop of Henle and proceeds toward the cortex alongside the descending limb. As it enters regions where the osmolarity of the adjacent peritubular fluid is lower, the plasma becomes hypertonic with respect to its surroundings, and solutes diffuse into the peritubular fluid. At the same time, there is an osmotic flow of water from the surrounding fluid into the blood.

Blood enters the vasa recta with an osmolarity of approximately 300 mOsm. The blood flowing into the medulla increases in osmolarity through the gain of solutes; water losses are limited by the presence of plasma proteins. Blood flowing toward the cortex gradually decreases in osmolarity until it departs with an osmolarity of 300 mOsml. But this decrease in osmolarity occurs primarily through gain of water rather than loss of solutes. The net result is that

1. Fewer solute molecules diffuse *out* of the vasa recta as it ascends to the cortex than diffused *in* when it descended into the medulla;
2. The volume of blood leaving the vasa recta is larger than the volume that entered this capillary loop. *Under normal conditions the removal*

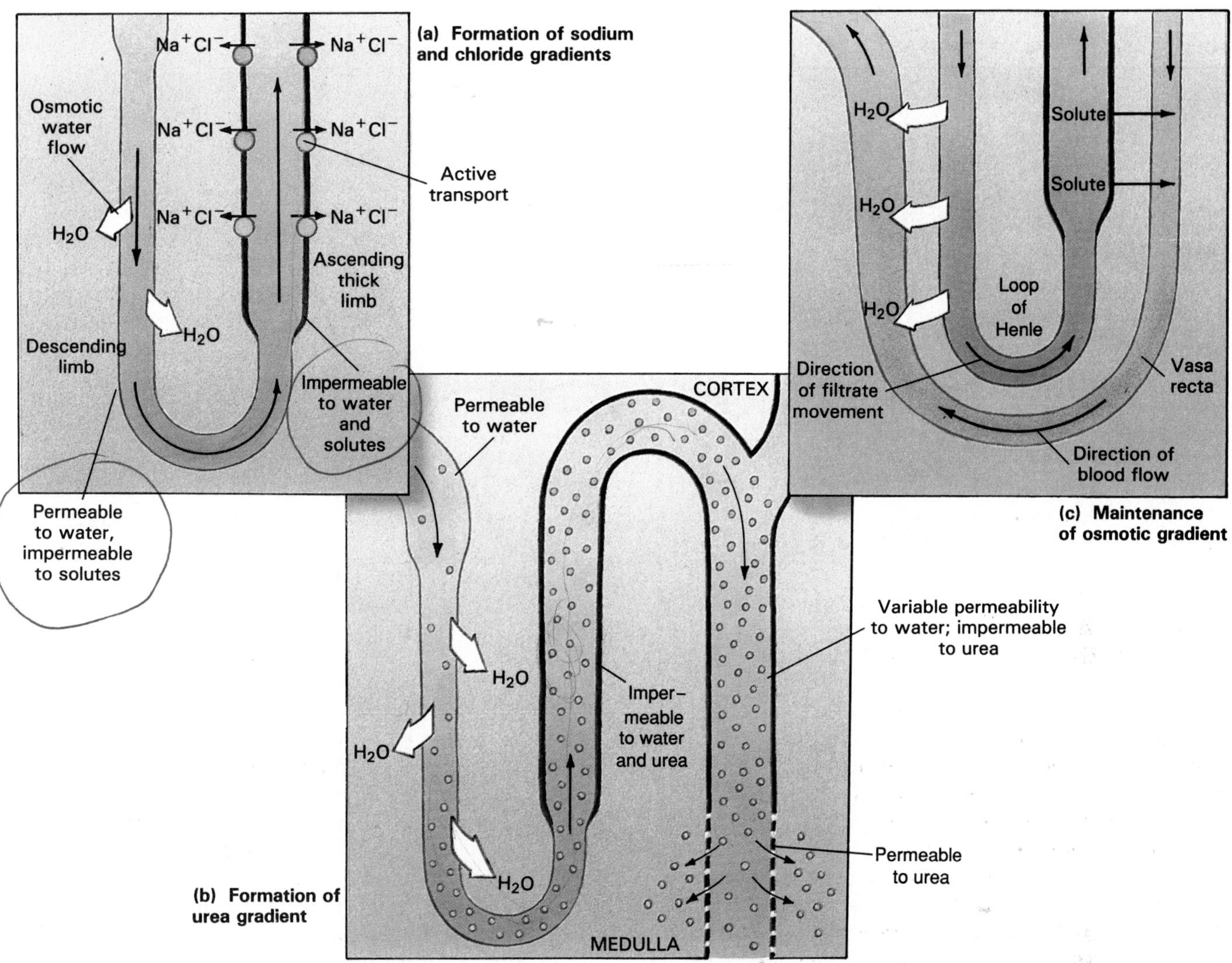

FIGURE 26-15
Fluid and Solute Movements Along the Loop of Henle. (a) Active transport of NaCl along the ascending thick limb results in the movement of water from the descending limb (see Figure 26-14). (b) The permeability characteristics of the loop and the collecting duct tend to concentrate urea in the filtrate and in the medulla. (c) Capillaries of the vasa recta maintain the osmotic gradient by removing the water and solutes reclaimed from the filtrate.

of solutes and water by the vasa recta precisely balances the rates of solute reabsorption and osmosis in the medulla. This mechanism effectively maintains the osmotic gradient.

Production of Hypertonic Urine

Figure 26-16 provides a functional overview that summarizes the major steps involved in the reabsorption of water and the production of hypertonic urine.

STEP 1: The filtrate produced at the renal corpuscle has the same osmolarity as the plasma, about 300 mOsm.

STEP 2: In the proximal convoluted tubule, the active removal of ions and other solutes produces a continual flow of water out of the filtrate. This reduces the volume of filtrate, but keeps the solutions inside and outside of the tubule isotonic. Roughly 60 percent of the volume of filtrate has been reabsorbed before the filtrate reaches the descending limb of the loop of Henle.

STEP 3: In the proximal convoluted tubule and descending limb of the loop of Henle, water moves into the surrounding interstitial fluids, leaving a small volume (roughly 20 percent of the original filtrate) of highly concentrated filtrate.

To this point the tubules are freely permeable to water, and changes in the internal or external solute concentration result in an immediate osmotic movement of water out of the filtrate. Because this permeability cannot be altered, this process is called **obligatory water reabsorption.**

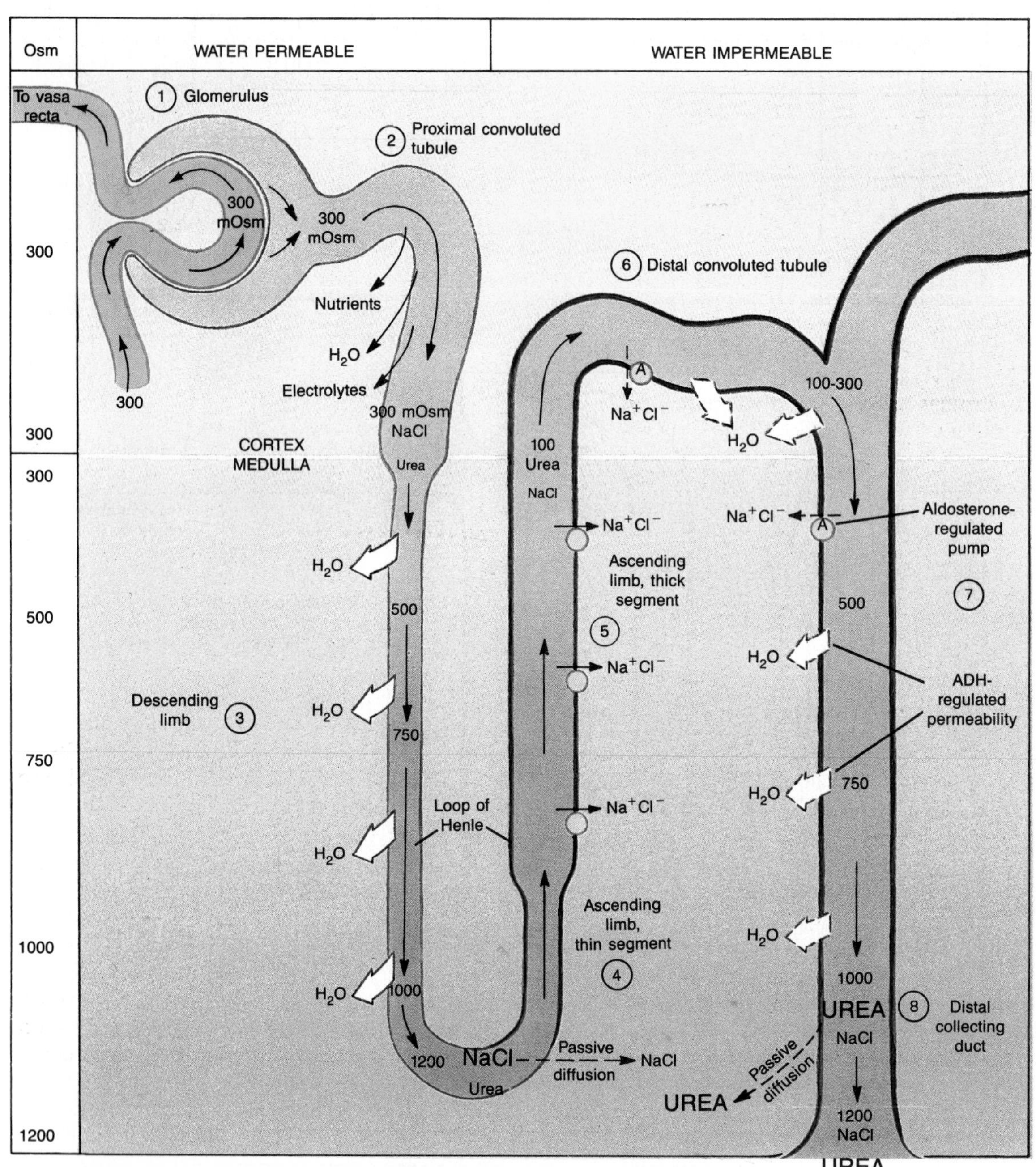

FIGURE 26-16
Major Steps in Urine Production

STEP 4: The filtrate then enters the ascending limb, where the epithelium is relatively impermeable to water and urea. Although the total osmolarities are equal, the filtrate now contains relatively fewer urea molecules and a higher concentration of sodium and chloride ions than does the surrounding interstitial fluid. Because the thin ascending limb is permeable to sodium and chloride ions, some of the ions diffuse out of the filtrate and into the medulla at this point.

STEP 5: As detailed earlier, the thick ascending limb actively transports sodium and chloride ions out of the filtrate. This transport lowers the osmotic concentration of the filtrate *without affecting filtrate volume.* The filtrate reaching the distal convoluted tubule is hypotonic relative to the peritubular fluid, with an osmolarity of only about 100 mOsm. Because only sodium and chloride ions are removed, urea now accounts for a significantly higher proportion of the total osmotic concentration at the end of the loop than it did at the start.

STEP 6: The final composition and concentration of the filtrate will be determined by the events under way in the DCT and the collecting system. Although the DCT, collecting tubule, and collecting duct are generally impermeable to solute molecules, the osmolarity of the filtrate can be adjusted through active transport. One important process, the aldosterone-stimulated reabsorption of sodium ions, has been shown in the figure.

STEP 7: The osmotic concentration of the urine is controlled by variations in the water permeabilities of the distal portions of the DCT, the collecting tubules, and the collecting ducts. These segments are impermeable to water unless exposed to antidiuretic hormone (ADH) from the posterior pituitary. In the absence of ADH, no water reabsorption occurs, and the individual excretes virtually all of the filtrate entering the DCT. Such a person produces relatively large quantities of dilute urine—up to 24 liters per day. (This problem, called *diabetes insipidus*, is discussed in a Clinical Comment in Chapter 18.) ∞ (p. 576)

At high concentrations of ADH the distal portions of the DCT and the collecting tubules and ducts become freely permeable to water. The filtrate entering the collecting ducts then has an osmolarity of about 300 mOsm, similar to that of the surrounding renal cortex. As the collecting duct descends toward the renal papilla through the medulla, it travels through regions of gradually increasing osmotic concentration. Because the duct is now permeable to water, there is an osmotic flow of water out of the filtrate. Under these conditions the urine entering the minor calyx has an osmolarity approaching 1200 mOsm.

Because the amount of water reabsorption in the DCT, collecting tubules, and collecting ducts can be regulated, the process just described is called **facultative water reabsorption.**

STEP 8: As the filtrate becomes increasingly concentrated, the urea concentration rises accordingly. In the final segments of the collecting duct and papillary duct the urea concentration of the filtrate exceeds that in the surrounding medulla. Because this portion of the collecting system is permeable to urea, urea molecules diffuse out of the filtrate and into the peritubular fluid.

Figure 26-17 provides a different perspective on the reabsorptive and secretory functions of the nephron and the collecting system. This graph follows the reabsorptive and secretory activities under way for each substance by recording the percentage remaining in the filtrate as one travels along the nephron. A value of 100 percent is assigned to the amount entering the nephron by filtration at the glomerulus. This method of presentation highlights several important points about kidney function that were discussed in this section. [**CM:** *Diuretics*]

What effect would increased amounts of aldosterone have on the potassium ion concentration of the urine?

What effect would a decrease in the sodium ion concentration of the filtrate have on the pH of the filtrate?

If the nephrons lacked a loop of Henle, how would this affect the volume and osmolarity of the urine they produced.

Regulation of Kidney Function

Out of the 180 liters or more of filtrate produced each day at the glomerulus, more than 99 percent is absorbed before reaching the renal pelvis. General characteristics of normal urine are listed in Table 26-4. However, the composition of the excreted 1.8 liters of urine varies, depending upon the metabolic and hormonal events under way.

The composition and concentration of the excreted urine are two integrated but distinct properties. The *composition* of the urine reflects the filtration, absorption, and secretion activities of the nephrons. Some compounds, such as urea, are neither actively excreted nor reabsorbed along the nephron. In contrast, nutrients and vitamins are usually completely absorbed, and other compounds, such as creatinine, that are missed by the filtration process are actively secreted into the filtrate.

These processes determine the identities and

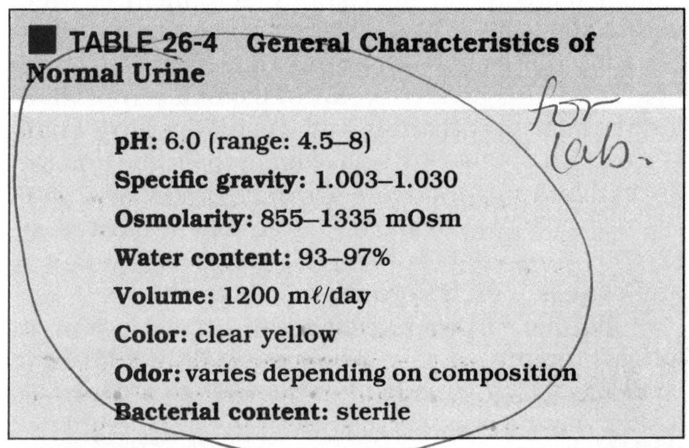

TABLE 26-4 General Characteristics of Normal Urine

pH: 6.0 (range: 4.5–8)
Specific gravity: 1.003–1.030
Osmolarity: 855–1335 mOsm
Water content: 93–97%
Volume: 1200 mℓ/day
Color: clear yellow
Odor: varies depending on composition
Bacterial content: sterile

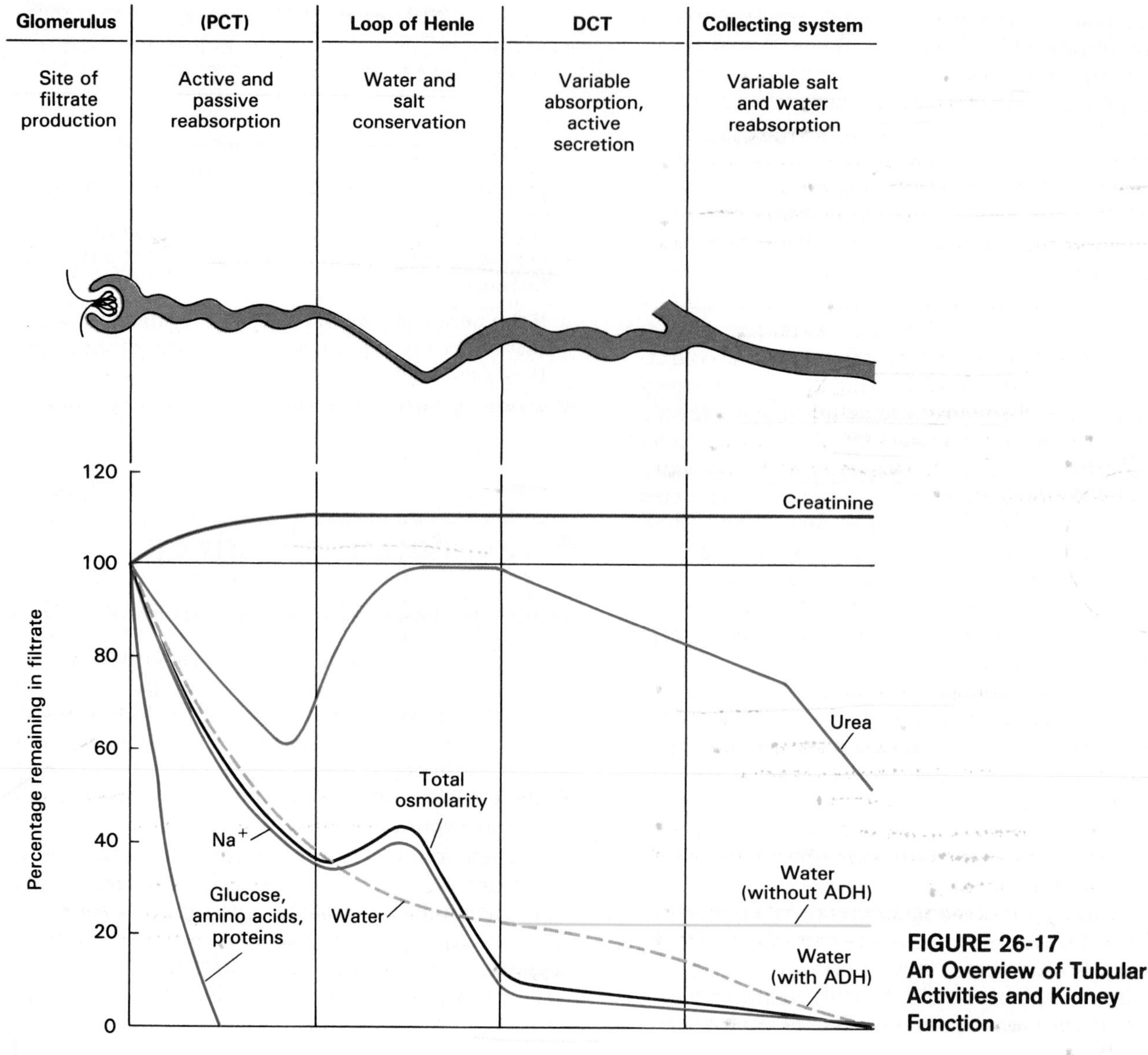

FIGURE 26-17 An Overview of Tubular Activities and Kidney Function

amounts of materials eliminated in the urine. The *concentration* of these components in a given urine sample depends on the osmotic movement of water across the walls of the tubules and collecting ducts. Because the composition and concentration of the urine vary independently, an individual can produce a small quantity of concentrated urine or a large quantity of dilute urine and still excrete the same amount of dissolved materials. As a result, physicians often request a 24-hour urine collection rather than a single sample. This enables them to assess both quantity and composition accurately.

Normal kidney function can continue only as long as the processes of filtration, reabsorption, and secretion function within relatively narrow limits. To a significant extent these processes can be regulated independently. A disruption in kidney function has immediate effects on the composition of the circulating blood. If both kidneys are affected, death will occur within a few days unless medical assistance is provided.

Many factors involved in the regulation of tubular reabsorption and secretion have already been described. In most instances changes in the concentration of specific solutes in the peritubular fluid or filtrate have a direct effect on the tubular rates of secretion or reabsorption. In the case of sodium ions and calcium ions, the rates of reabsorption or secretion can be adjusted by circulating hormones.

Regulatory mechanisms that alter the rate of glomerular filtration are much more complex and have a more widespread effect on renal function. Filtration is the key to urine production, for without it the rest of the nephron would have nothing to do.

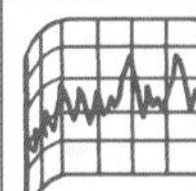

Diagnostics: Urinalysis

Normal urine is a clear, sterile solution with a yellow color. The color results from the presence of urobilinogen, generated by intestinal bacteria and absorbed in the colon. The evaporation of small compounds accounts for the characteristic odor; the presence of compounds, such as ammonia or ketone bodies, may also affect the smell. The analysis of a urine sample is a diagnostic tool of considerable importance even in modern high-technology medicine. A standard **urinalysis** involves an assessment of urine color and odor, two characteristics that can be determined without specialized equipment. Another easily performed test, classifying the taste of the urine as sweet, salty, and so on, is no longer popular, but in the seventeenth century this procedure also provided useful information. Quantitative chemical tests have since replaced the taste bud assay. Average values for urinalysis are presented in the Clinical Manual.

The total volume produced in a 24-hour period may also be of interest. **Polyuria** (pol-ē-Ū-rē-a) refers to excessive production of urine, well over 2 liters per day. **Oliguria** (o-li-GŪ-rē-a) refers to inadequate urine production, under 500 mℓ/day. In **anuria** (a-NŪ-rē-a) no urine is produced, and a crisis is imminent.

Many other tests involve recording changes in the color of test strips that are dipped in the sample. Urine pH and urinary concentrations of glucose, ketones, bilirubin, urobilinogen, plasma proteins, and hemoglobin can be monitored using this technique. In addition, the density or *specific gravity* of the urine is usually determined, using a simple device known as a **urinometer** (ū-ri-NOM-e-ter), or **densitometer** (den-si-TOM-e-ter). The sample may also be spun in a centrifuge and any residue examined under the microscope. Mineral crystals, bacteria, red or white blood cells, and deposits, known collectively as *casts*, can be detected in this way. During a urinary tract infection, bacteria may be cultured to determine their specific identities.

More comprehensive analyses can determine the total osmolarity of the urine and the concentration of individual electrolytes and minor metabolites, metabolic wastes, vitamins, and hormones. A test for one hormone in the urine, human chorionic gonadotrophin (HCG), provides an early and reliable indication of pregnancy.

The information provided by urinalysis can be especially useful when correlated with the data obtained from blood tests. The term **azotemia** (a-zō-TĒ-mē-a) refers to the presence of excess metabolic wastes in the blood. This condition may result from overproduction of urea or other waste products by the liver. In **uremia** (ū-RĒ-mē-a), by contrast, all normal kidney functions are adversely affected. (The symptoms of uremia, an important sign of kidney failure, are discussed in the Clinical Comment: Renal Failure.)

CONTROL OF GLOMERULAR FILTRATION RATES

Filtration depends on adequate circulation to the glomerulus and the maintenance of normal filtration pressures. (Factors affecting filtration pressure were discussed earlier in this chapter, and you may wish to refresh your memory before proceeding.)

Autoregulation of GFR

The nephron is capable of a considerable amount of autoregulation through changes in the diameter of the afferent arterioles, the efferent arterioles, and the glomerular capillaries. These adjustments occur in response to changes in local blood pressure and blood flow. For example, a reduction in blood flow and a decline in glomerular pressure trigger the dilation of the afferent arteriole and glomerular capillaries and the constriction of the efferent arteriole. This combination increases blood flow and elevates glomerular blood pressure to normal levels. As a result, filtration rates remain relatively constant over a broad range of systemic arterial pressures.

Hormonal Regulation of the GFR

The hormones renin, erythropoietin, and ADH (Chapters 18–21), which regulate blood pressure and volume, all affect the GFR.

Renin and Erythropoietin If the glomerular pressures remain low because of a decrease in blood volume, a fall in systemic pressures, or a blockage in

Clinical Comment: Renal Failure

Renal failure occurs when the kidneys become unable to perform the excretory functions needed to maintain homeostasis. When kidney filtration slows for any reason, urine production declines. As the decline continues, symptoms of renal failure appear because of the retention of water, ions, and metabolic wastes. The uremia that develops affects virtually all other systems in the body. For example, there are disturbances in fluid balance, pH, muscular contraction, metabolism, and digestive function. The individual usually becomes hypertensive, anemia develops due to a decline in erythropoietin production, and central nervous system problems may lead to sleeplessness, seizures, delirium, and even coma.

Acute renal failure occurs when filtration suddenly slows or stops altogether due to exposure to toxic drugs, renal ischemia, urinary obstruction, or trauma. Sensitized individuals may also develop acute renal failure after exposure to antibiotics or anesthetics. In these cases recovery may occur if the individual survives the incident. In *chronic renal failure* kidney function deteriorates gradually, and the associated problems accumulate over time. The condition usually cannot be reversed, only delayed, and symptoms of acute renal failure eventually develop. Chronic and acute renal failure may be treated by renal transplantation or the use of dialysis equipment (see the Health News box later in this chapter).

the renal artery or its tributaries, the juxtaglomerular apparatus releases *renin* and *erythropoietin* into the circulation. Because these compounds were considered in Chapters 18 and 21, only a brief summary will be presented at this time. (pp. 588, 701)

Erythropoietin stimulates red blood cell production in the bone marrow, thereby increasing the oxygen-carrying capacity of the blood. Renin converts the inactive protein angiotensinogen to angiotensin I, and a converting enzyme in the capillaries of the lungs subsequently converts this to angiotensin II, a potent hormone. Figure 26-18 diagrams the important hormonal effects. These include:

1. *In peripheral capillary beds,* angiotensin II causes a brief but powerful vasoconstriction, elevating pressures in the renal arteries and their tributaries.
2. *At the nephron,* angiotensin II causes the constriction of the efferent arteriole, further elevating glomerular pressures and filtration rates.
3. *In the CNS,* angiotensin triggers the release of ADH, stimulating the reabsorption of water and sodium ions, and causes the sensation of thirst.
4. *At the adrenal gland,* angiotensin II stimulates the secretion of aldosterone by the cortex and epinephrine by the medulla. The heart responds by increasing its rate and force of contraction.

The juxtaglomerular apparatus will also respond to changes in the osmolarity of the filtrate arriving at the DCT. If the rate of glomerular filtration declines, so does the rate of filtrate movement along the nephron. Because the filtrate then spends more time in the ascending limb of the loop of Henle, the concentration of sodium and chloride ions becomes abnormally low. When this occurs, the cells of the macula densa release renin, triggering responses that ultimately increase the glomerular filtration rate.

ADH Antidiuretic hormone release occurs under angiotensin II stimulation. ADH release also occurs when hypothalamic neurons are stimulated by a fall in blood pressure or an elevation in the osmolarity of the circulating blood. The nature of the receptors involved has not been determined, but these specialized hypothalamic neurons have been named **osmoreceptors**.

Autonomic Regulation of GFR

Sympathetic activation has three primary effects on the nephron:

1. It produces a powerful vasoconstriction of the afferent arterioles, decreasing the GFR and slowing the production of filtrate. The sympathetic activation triggered by an acute fall in blood pressure or a heart attack will override the hormonal regulatory mechanisms, at least over the short term. As the sympathetic tone decreases, the filtration rate gradually returns to normal levels. Sustained sympathetic stimulation at high levels may lead to renal ischemia and kidney damage.
2. Sympathetic activation alters the GFR by changing the regional pattern of blood circulation. For example, the dilation of superficial vessels in warm weather shunts blood away from the kidneys, and glomerular filtration declines temporarily. The effect becomes especially pronounced during periods of strenuous exercise. As the blood flow increases to the skin and skele-

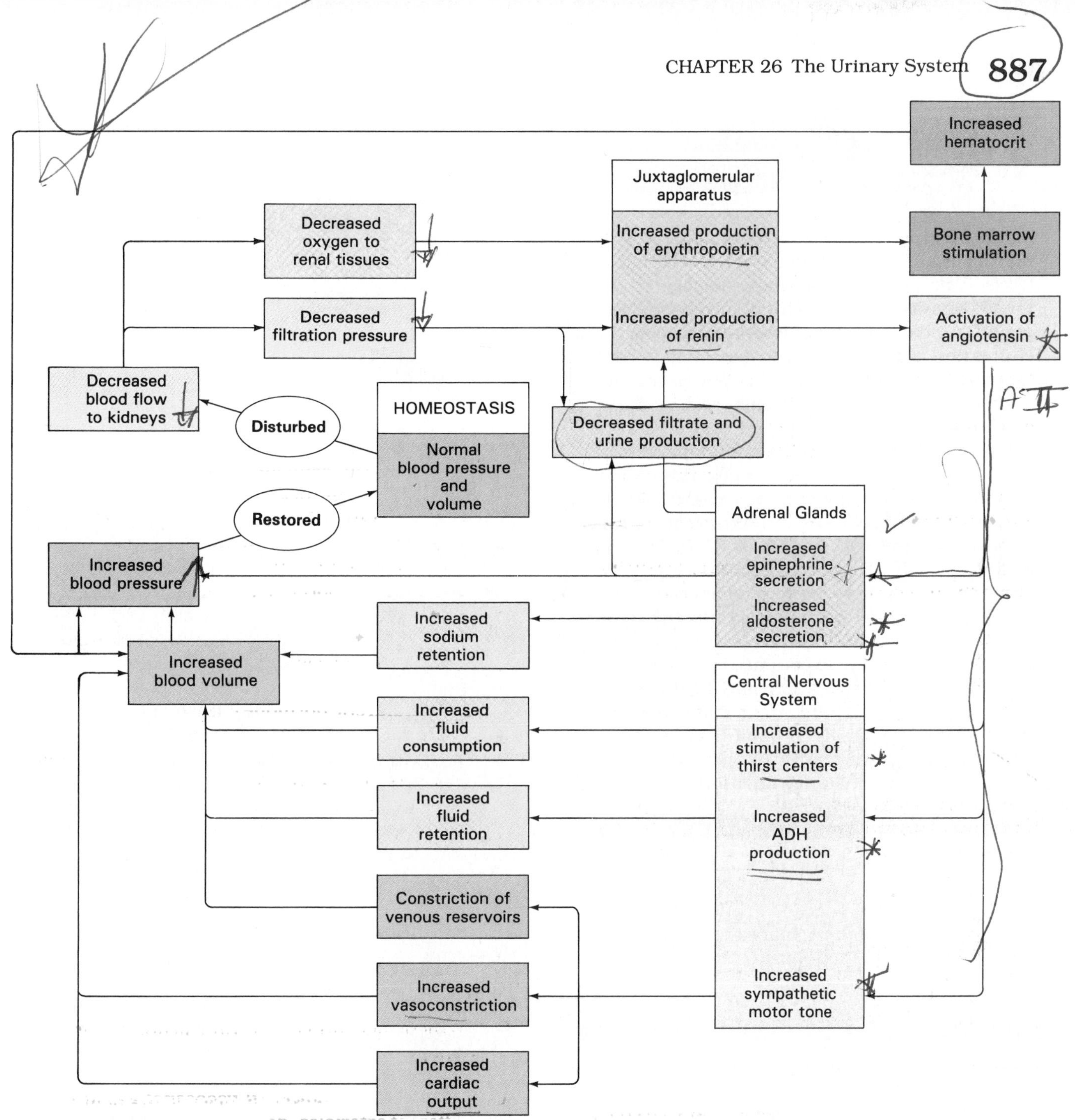

FIGURE 26-18
Role of the Kidneys in the Homeostatic Regulation of Blood Pressure and Blood Volume. This system is based on the maintenance of normal blood flow to the kidneys and the maintenance of normal glomerular filtration pressures.

tal muscles, kidney perfusion gradually declines.

3. Sympathetic activation stimulates the release of renin by the juxtaglomerular apparatus. Cell membranes in the JGA contain beta receptors. The binding of catecholamine to these receptors activates the cells, causing them to release renin.

Urine Transport, Storage, and Elimination

Filtrate modification and urine production end when the fluid enters the renal pelvis. The rest of the urinary system is responsible for the transport, storage, and elimination of the urine. We will now examine the

Sports and Fitness: Kidney Problems and Strenuous Exercise

At maximal levels of exertion renal blood flow may be less than one-quarter of normal resting levels. This can create special problems for distance swimmers and marathon runners because potentially dangerous conditions develop as the circulating concentration of metabolic wastes increases and peripheral water losses mount. If you read the papers carefully after a major marathon you will usually find that several of the participants developed acute kidney disorders during or following the event. Commonly encountered problems include elevated serum concentrations of K^+, lowered serum concentrations of Na^+, and a decrease in the GFR. Proteinuria may occur after the event, and 18 percent of marathon runners develop hematuria. Usually these symptoms disappear within 48 hours. Some marathon and ultramarathon participants, however, have experienced renal failure, complicated by dehydration and heat stroke. Such incidents require immediate medical treatment. Even with prompt intervention, damage to nephrons may result in some permanent impairment of kidney function.

other components of the urinary system and consider their functional interactions.

THE URETERS AND URINARY BLADDER

The ureters (Figure 26-19) extend caudally from the kidneys for about 30 cm (12 in.) before reaching the urinary bladder. Like the kidneys, the ureters are retroperitoneal, and they penetrate the posterior wall of the bladder without ever entering the peritoneal cavity.

In the male the base of the urinary bladder lies between the rectum and the symphysis pubis, as indicated in Figure 26-19a. In the female (Figure 26-19b) the urinary bladder sits inferior to the uterus and anterior to the vagina. Its dimensions vary, depending on the state of distension, but the full urinary bladder can contain up to a liter of urine.

The superior surfaces of the urinary bladder are covered by a layer of peritoneum, and several peritoneal folds assist in stabilizing its position. The **middle umbilical ligament**, or **urachus** (U-ra-kus), extends from the anterior and superior border toward the umbilicus (Figure 26-19c). The **lateral umbilical ligaments** pass along the sides of the bladder and also

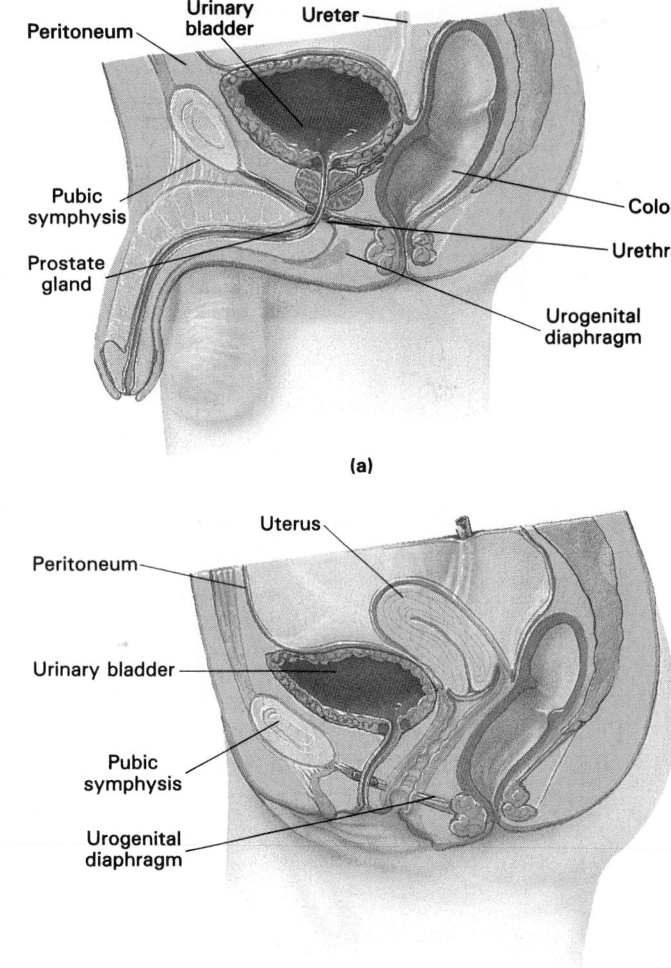

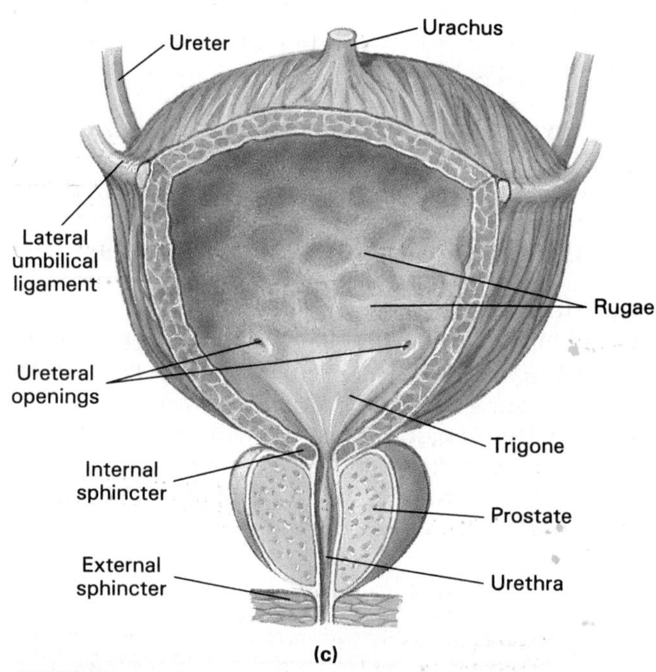

FIGURE 26-19
Organs Responsible for the Conduction and Storage of Urine. (a) Position of the ureter, urinary bladder, and urethra in the male. (b) Position of the same organs in the female. (c) Anatomy of the urinary bladder.

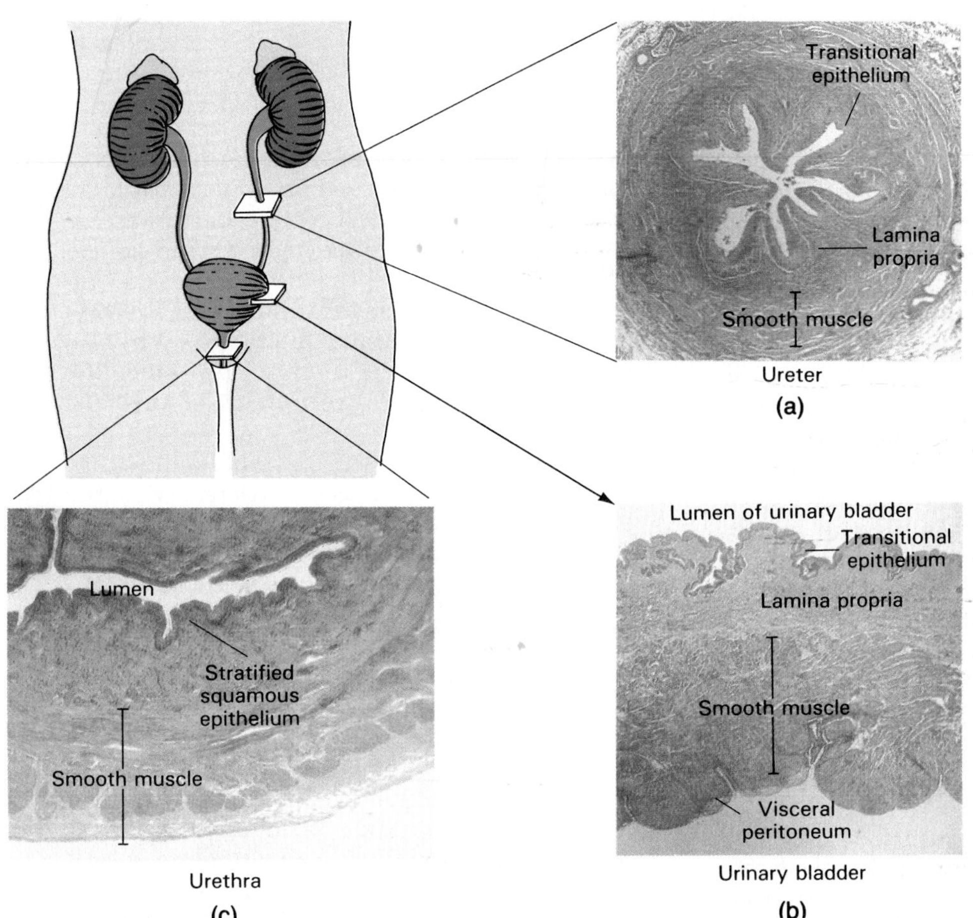

FIGURE 26-20
Histology of the Collecting and Transport Organs. (a) A ureter seen in transverse section. Note the thick layer of smooth muscle surrounding the lumen. (For a closeup of transitional epithelium, review Figure 5-6.) (LM, ×16) (b) The wall of the urinary bladder. (LM, ×17) (c) A transverse section through the urethra. (LM, ×42)

reach the umbilicus. These fibrous cords contain the vestiges of the two *umbilical arteries* that serviced the placenta during embryonic development. (For further information concerning the development of this system, see the Embryology Summary on p. 870.) The posterior, inferior, and anterior surfaces lie outside the peritoneal cavity, and in these areas tough ligamentous bands anchor the bladder to the pelvic and pubic bones.

In sectional view the triangular area bounded by the openings of the ureters and the entrance to the urethra constitutes the **trigone** (TRĪ-gōn) of the bladder. The urethral entrance lies at the apex of this triangle, at the lowest point in the bladder. The area surrounding the urethral entrance, called the **neck** of the urinary bladder, contains a muscular sphincter that also extends along the proximal portions of the urethra. This **internal sphincter** provides involuntary control over the discharge of urine from the bladder.

Sectional views of the ureter and urinary bladder can be found in Figure 26-20a,b. A *transitional epithelium* lines the renal pelvis, the ureter, and the urinary bladder. This stratified epithelium can tolerate a considerable amount of stretching. In the relaxed state the epithelium is several cell layers in thickness, and the exposed cells bulge into the lumen. When stretched, the epithelium thins to only two or three cell layers, with outer squamous and inner cuboidal cells. These changes in epithelial structure were detailed in Figure 5-6. (p. 132)

The underlying lamina propria is surrounded by longitudinal and circular muscle layers. Under normal conditions peristaltic contractions begin at the apex of the renal papillae and sweep along the minor and major calyces toward the renal pelvis. Similar contractions move urine out of the renal pelvis and along the ureter to the bladder.

The lamina propria of the mucous membrane contains abundant elastin fibers, and in the relaxed condition the epithelium forms a series of prominent folds, called **rugae**. These folds disappear as the bladder fills with urine and the walls stretch. The surrounding smooth muscle layers differ from those of the digestive tract in possessing inner and outer longitudinal muscle layers, with a circular layer sandwiched between. These layers form the powerful **detrusor** (de-TROO-sor) muscle of the bladder. Contraction of this muscle compresses the urinary bladder and expels its contents into the urethra.

THE URETHRA

In the female the urethra is very short, extending 25–30 mm from the bladder to the vestibule. In the male the urethra may be 18–20 cm (7–8 in.) in length, and it can be subdivided into regions as it passes specific landmarks and transits several reproductive organs. These divisions are described in Chapter 28.

Figure 26-20c shows a sectional view of the urethra. The urethral lining consists of a stratified epithelium that varies from transitional close to the bladder, through stratified columnar at the midpoint, to stratified squamous near the urethral opening. The lamina propria is thick and elastic, the mucous membrane is thrown into longitudinal folds, and mucus-secreting cells are found in the epithelial pockets. In the male the epithelial mucous glands may form tubules that extend into this region, and the connective tissues of the lamina propria anchor the urethra to surrounding structures. In the female the lamina propria contains an extensive network of veins, and the entire complex is surrounded by concentric layers of smooth muscle. In both sexes, as the urethra passes through the urogenital diaphragm a circular band of skeletal muscle forms the **external sphincter**. The external sphincter consists of skeletal muscle fibers, and its contractions are under voluntary control. [**CM:** *Problems with the Conducting System*]

THE MICTURITION REFLEX AND URINATION

Urine reaches the urinary bladder by the peristaltic contractions of the ureters. The process of urination is coordinated by the **micturition reflex**. Components of this reflex are diagrammed in Figure 26-21.

Stretch receptors in the wall of the urinary bladder are stimulated as it fills with urine. Afferent fibers in the pelvic nerves carry the impulses generated to the sacral spinal cord. Their increased level of activity (1) facilitates parasympathetic motor neurons in the sacral spinal cord and (2) stimulates interneurons that relay sensations to the cerebral cortex. As a result, we become consciously aware of the fluid pressure in the urinary bladder.

The urge to urinate usually appears when the bladder contains about 200 mℓ of urine. The micturition reflex begins to function when the stretch receptors have provided adequate stimulation to the parasympathetic motor neurons. At this time activity in the motor neurons generates action potentials that reach the smooth muscle in the bladder wall. These efferent impulses travel over the pelvic nerves, producing a sustained contraction of the urinary bladder.

This contraction elevates fluid pressures in the urinary bladder, but urine ejection does not occur unless both the internal and external sphincters are relaxed. Relaxation of the external sphincter occurs under voluntary control. When the external sphincter relaxes, so does the **internal sphincter**. If the external sphincter does not relax, the internal sphincter remains closed. Stimulation of the smooth muscles in the urinary bladder soon subsides, and the bladder gradually relaxes.

Clinical Comment: Urinary Tract Infections

Urinary tract infections, or **UTIs**, result from the colonization of the urinary tract by bacterial or fungal invaders. The intestinal bacterium *Escherichia coli* is most often involved, and women are particularly susceptible to urinary tract infections because of the proximity of the urethral orifice to the anus. Sexual intercourse may also push bacteria into the urethra and—since the female urethra is relatively short—toward the bladder.

The condition may be asymptomatic (without symptoms), but it can be detected by the presence of bacteria and blood cells in the urine. If inflammation of the urethral wall occurs, the condition may be termed **urethritis**, while inflammation of the lining of the bladder represents **cystitis**. Many infections affect both regions to some degree. Urination becomes painful, a symptom known as **dysuria** (dis-Ū-rē-a), and the bladder becomes tender and sensitive to pressure. Despite the discomfort produced, the individual feels the urge to urinate frequently. Urinary tract infections usually respond to antibiotic therapies, although subsequent reinfections may occur.

In untreated cases the bacteria may proceed along the ureters to the renal pelvis. The inflammation of the walls of the renal pelvis produces symptoms of **pyelitis** (pī-e-LĪ-tis), and if the bacteria invade the renal cortex and medulla as well, **pyelonephritis** (pī-e-lō-nef-RĪ-tis) results. Signs include a high fever, intense pain on the affected side, vomiting, and diarrhea. Blood cells and pus are present in the urine.

A further increase in bladder volume begins the cycle again, usually within an hour. Each increase in urinary volume leads to an increase in stretch receptor stimulation that makes the sensation more acute. Once the volume of the urinary bladder exceeds 500 mℓ, the micturition reflex may generate enough pressure to force open the internal sphincter. This leads to a reflexive relaxation in the external sphincter, and urination occurs despite voluntary opposition or potential inconvenience.

Voluntary urination involves relaxing the external sphincter and descending facilitation of the micturition reflex. Tensing of the abdominal and expiratory muscles increases abdominal pressures and assists in compressing the bladder. At the end of a normal micturition, less than 10 mℓ of urine remain in the bladder. [**CM:** *Problems with the Micturition Reflex*]

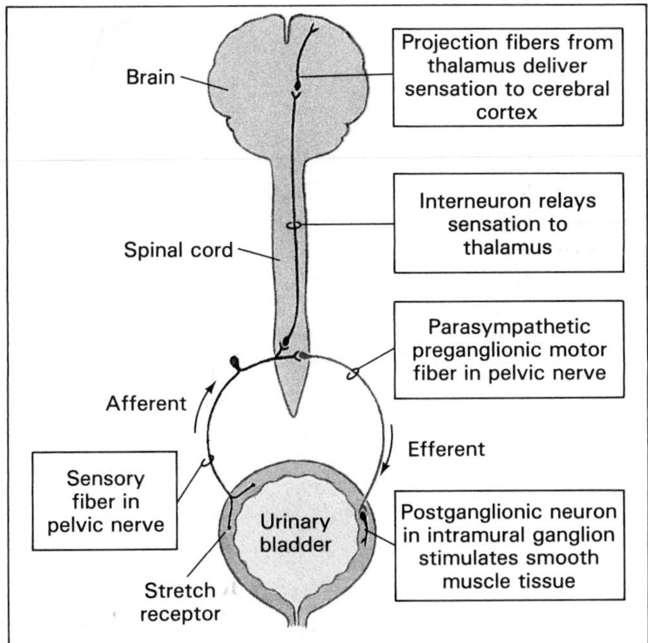

FIGURE 26-21
The Micturition Reflex. Basic components of the reflex arc involved in the micturition reflex.

Aging and the Urinary System

In general, aging is associated with an increased incidence of kidney problems. Many examples, such as *nephrolithiasis* (kidney stones), were described in Clinical Comment boxes earlier in the chapter. Other age-related changes in the urinary system include:

1. *A decline in the number of functional nephrons*: The total number of kidney nephrons drops by 30–40 percent between ages 25 and 85.
2. *A reduction in the GFR*: This results from decreased numbers of glomeruli, cumulative damage to the filtration apparatus in the remaining glomeruli, and reductions in renal blood flow.
3. *Reduced sensitivity to ADH*: With age the distal portions of the nephron and collecting system become less responsive to ADH. Less reabsorption of water and sodium ions occurs, and more potassium ions are lost in the urine.
4. *Problems with the micturition reflex*: Several factors are involved in such problems.
   a. The sphincter muscles lose muscle tone and become less effective at voluntarily retaining urine. This leads to problems with incontinence, often involving a slow leakage of urine.
   b. The ability to control micturition is often lost following a stroke, Alzheimer's disease, or other CNS problems affecting the cerebral cortex or hypothalamus.
   c. In males, **urinary retention** may develop secondary to chronic inflammation of the prostate gland. In this condition swelling and distortion of surrounding prostatic tissues compress the urethra, restricting or preventing the flow of urine.

Integration with Other Systems

The urinary system is not the only organ system concerned with excretion. The urinary, integumentary, respiratory, and digestive systems are sometimes considered to form an anatomically diverse **excretory system**. The components of this system perform all of the excretory functions of the body that affect the composition of body fluids. Examples of excretory activities discussed in earlier chapters include:

1. *Integumentary system*: Water and electrolyte losses in sensible perspiration can affect plasma volume and composition. The effects are most apparent when losses are extreme, as in maximum sweat production. Small amounts of metabolic wastes are also excreted in perspiration.
2. *Respiratory system*: The lungs excrete the carbon dioxide generated by living cells. Small amounts of other compounds, such as acetone and water, evaporate into the alveoli and are eliminated during exhalation.
3. *Digestive system*: Small amounts of metabolic waste products are excreted in liver bile, and a variable amount of water is lost in feces.

These excretory activities have an impact on the composition of body fluids. The respiratory system, for example, is the primary site of carbon dioxide excretion. But the excretory functions of these systems are not regulated as closely as are those of the kidneys, and under normal circumstances the effects of integumentary and digestive excretory activities are minor compared with those of the urinary system.

Figure 26-22 summarizes the functional relationships between the urinary system and other systems. Many of these relationships will be explored further in the next chapter, which considers major aspects of fluid, pH, and electrolyte balance.

√ **Why does a decrease in the amount of sodium ions in the distal convoluted tubule lead to an increase in blood pressure?**

√ **An obstruction of the ureters by a kidney stone would interfere with the flow of urine between what two points?**

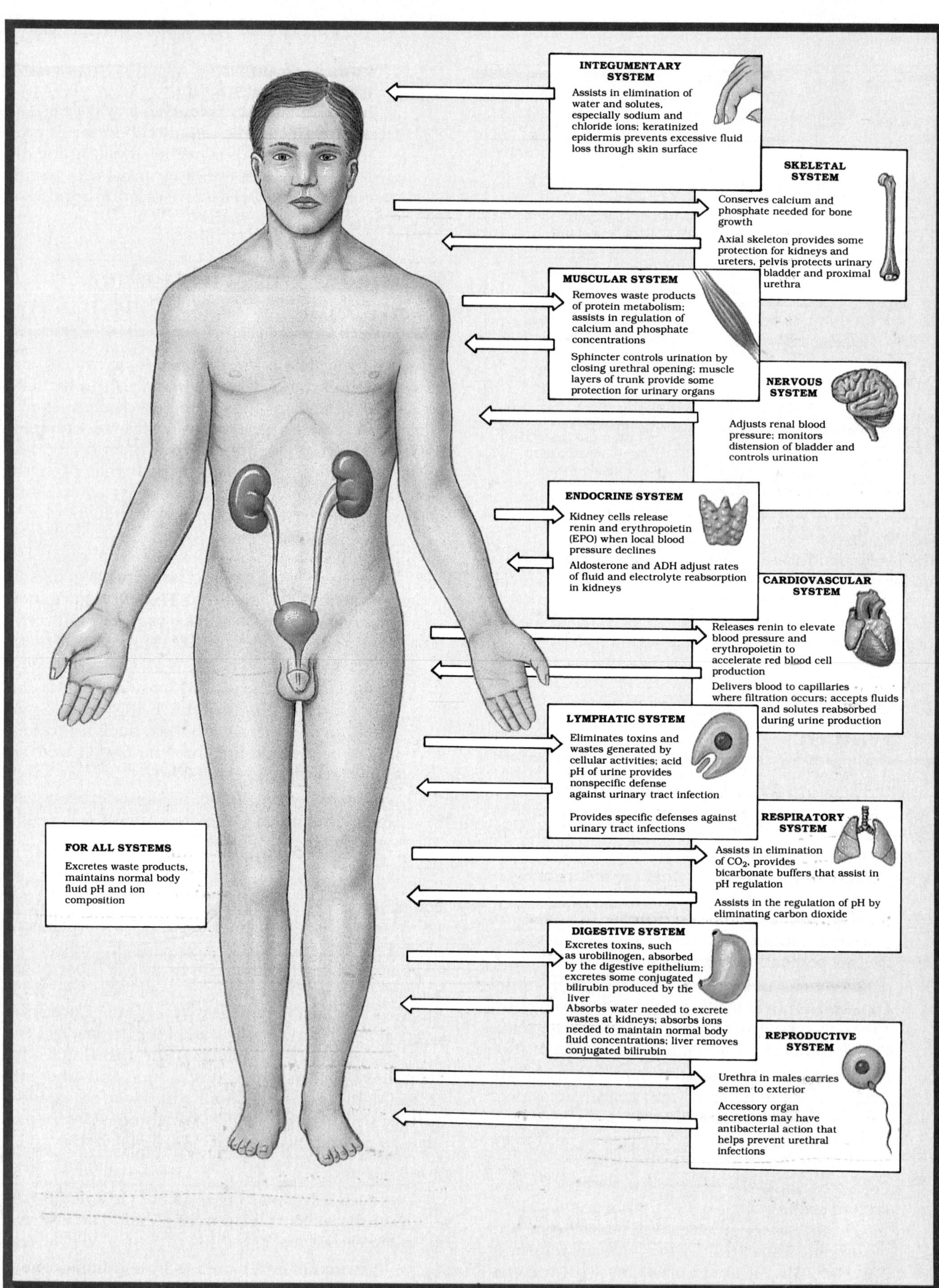

INTEGUMENTARY SYSTEM
Assists in elimination of water and solutes, especially sodium and chloride ions; keratinized epidermis prevents excessive fluid loss through skin surface
SKELETAL SYSTEM
Conserves calcium and phosphate needed for bone growth
Axial skeleton provides some protection for kidneys and ureters, pelvis protects urinary bladder and proximal urethra
MUSCULAR SYSTEM
Removes waste products of protein metabolism; assists in regulation of calcium and phosphate concentrations
Sphincter controls urination by closing urethral opening; muscle layers of trunk provide some protection for urinary organs
NERVOUS SYSTEM
Adjusts renal blood pressure; monitors distension of bladder and controls urination
ENDOCRINE SYSTEM
Kidney cells release renin and erythropoietin (EPO) when local blood pressure declines
Aldosterone and ADH adjust rates of fluid and electrolyte reabsorption in kidneys
CARDIOVASCULAR SYSTEM
Releases renin to elevate blood pressure and erythropoietin to accelerate red blood cell production
Delivers blood to capillaries where filtration occurs; accepts fluids and solutes reabsorbed during urine production
LYMPHATIC SYSTEM
Eliminates toxins and wastes generated by cellular activities; acid pH of urine provides nonspecific defense against urinary tract infection
Provides specific defenses against urinary tract infections
RESPIRATORY SYSTEM
Assists in elimination of CO_2, provides bicarbonate buffers that assist in pH regulation
Assists in the regulation of pH by eliminating carbon dioxide
DIGESTIVE SYSTEM
Excretes toxins, such as urobilinogen, absorbed by the digestive epithelium; excretes some conjugated bilirubin produced by the liver
Absorbs water needed to excrete wastes at kidneys; absorbs ions needed to maintain normal body fluid concentrations; liver removes conjugated bilirubin
REPRODUCTIVE SYSTEM
Urethra in males carries semen to exterior
Accessory organ secretions may have antibacterial action that helps prevent urethral infections
FOR ALL SYSTEMS
Excretes waste products, maintains normal body fluid pH and ion composition

FIGURE 26-22
Functional Relationships between the Urinary System and Other Systems

Health News: Advances in the Treatment of Renal Failure

Many different conditions can result in renal failure. Management of chronic renal failure typically involves restricting water and salt intake and reducing caloric intake to a minimum, with few dietary proteins. This combination reduces strain on the urinary system by (1) minimizing the volume of urine produced and (2) preventing the generation of large quantities of nitrogenous wastes. Acidosis, a common problem in patients with renal failure, can be countered with infusions of bicarbonate ions.

If drugs, infusions, and dietary controls cannot stabilize the composition of the blood, more drastic measures are taken. In one technique, called **hemodialysis** (hē-mō-dī-AL-i-sis), an artificial membrane is used to regulate the composition of the blood. The basic principle involved in this process, called **dialysis**, involves passive diffusion across a semipermeable membrane. The patient's blood flows across an artificial *dialysis membrane* that contains pores large enough to permit the diffusion of small ions, but small enough to prevent the loss of plasma proteins. On the other side of the membrane flows a special **dialysis fluid**.

The composition of dialysis fluid is indicated in Table 26-5. As diffusion takes place across the membrane, the composition of the blood changes. Potassium ions, phosphate ions, sulfate ions, urea, creatinine, and uric acid diffuse across the membrane into the dialyzing fluid. Bicarbonate ions and glucose diffuse into the bloodstream. In effect, diffusion across the dialysis membrane takes the place of normal glomerular filtration, and the characteristics of the dialysis fluid ensure that important metabolites remain in the circulation, rather than diffusing across the membrane.

In practice, silastic tubes, called *shunts*, are inserted into a medium-sized artery and vein. (The usual location is in the forearm, although the lower leg is sometimes used.) The two shunts are then connected as shown in Figure 26-23a. The connection acts like a "short circuit" that does not impede blood flow, and the shunts can be used like taps in a wine barrel, to draw a blood sample or to connect the individual to a *dialysis machine* (Figure 26-23b).

When connected to the dialysis machine, the individual sits quietly while blood circulates from the arterial shunt, through the machine, and back via the venous shunt. Inside the machine, the blood flows across a dialyzing membrane, where diffusion occurs.

TABLE 26-5 The Composition of Dialysis Fluid

Constituent	*Normal Plasma*	*Dialyzing Fluid*
ELECTROLYTES (mEq/ℓ)		
Potassium	5	3
Bicarbonate	27	36
Phosphate	3	0
Sulfate	0.5	0
NUTRIENTS (mg/dℓ)		
Glucose	100	125
NITROGENOUS WASTES (mg/dℓ)		
Urea	26	0
Creatinine	1	0
Uric acid	0.3	0

Note: Only the significant variations are noted; values for other electrolytes are usually similar. Although these values are representative, the precise composition can be tailored to meet the specific clinical needs. For example, if plasma potassium levels are too low, the dialyzing fluid concentration can be elevated to remedy the situation. Changes in the osmolarity of the dialyzing fluid can also be used to adjust an individual's blood volume, usually by adjusting the glucose content of the dialyzing fluid.

Use of a dialysis machine is suggested when a patient's *BUN* (*blood urea nitrogen*) exceeds 100 mg/dℓ (the normal value is 30 mg/dℓ). Dialysis techniques are useful because they can maintain patients awaiting a transplant or those whose kidney function has been temporarily disrupted. Hemodialysis does have a number of drawbacks, however: (1) the patient must sit by the machine about 15 hours per week; (2) between treatments the symptoms of uremia will gradually develop; (3) hypotension can develop as a result of fluid loss during dialysis; (4) air bubbles in the tubing can cause embolism formation in the bloodstream; (5) anemia often develops; and (6) the shunts can serve as sites for recurring infections.

One alternative to the use of a dialysis machine is *peritoneal dialysis*. In **peritoneal dialysis** the peritoneal lining is used as a dialysis

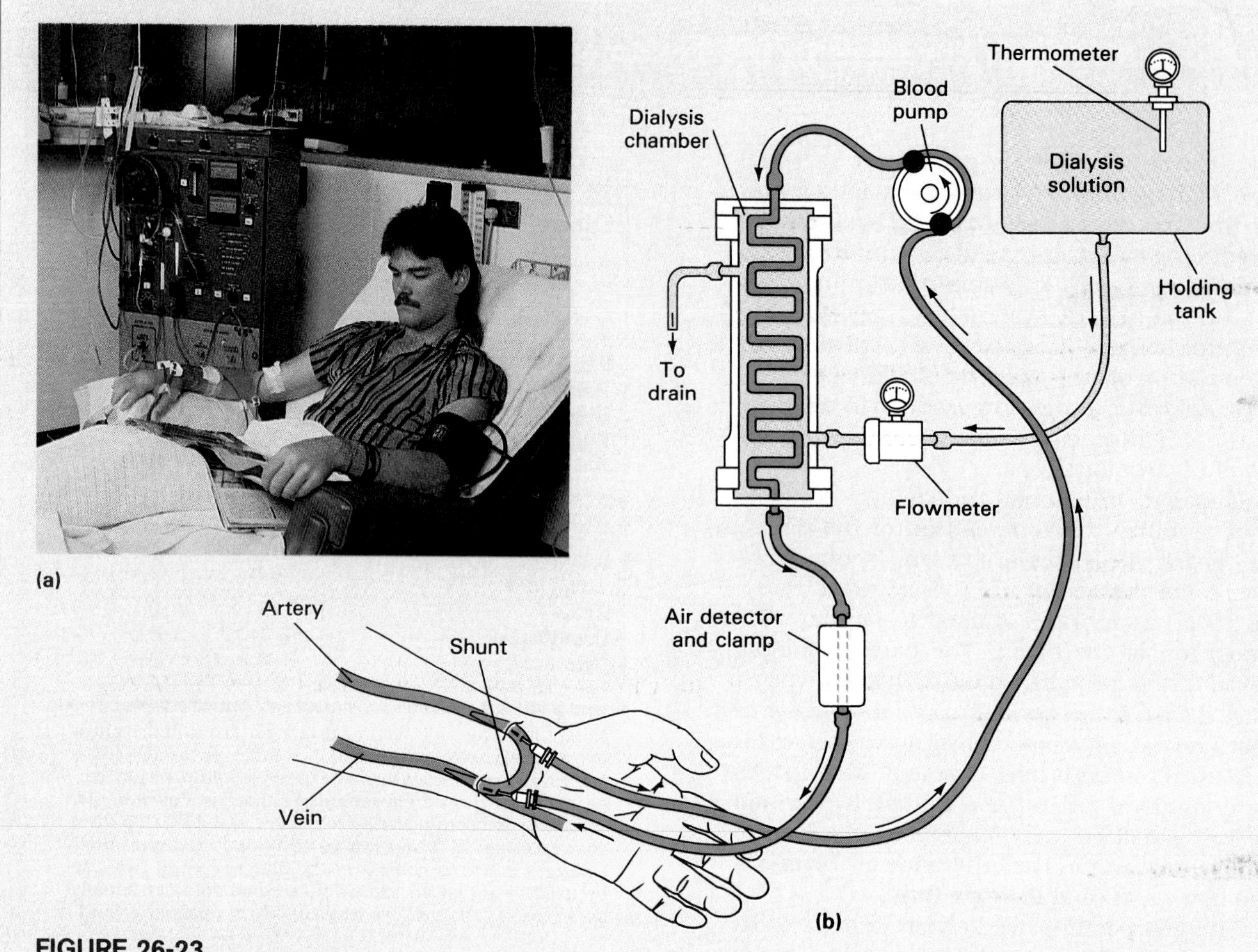

FIGURE 26-23
Hemodialysis. (a) Preparation for hemodialysis typically involves implantation of a pair of shunts connected by a loop that permits normal blood flow when the patient is not hooked up to the dialysis machine. (b) A diagrammatic view of the dialysis procedure.

membrane. Dialyzing fluid is introduced into the peritoneum through a catheter in the abdominal wall, and at intervals the fluid is removed and replaced. For example, one procedure involves cycling 2 liters of fluid in an hour—15 minutes for infusion, 30 minutes for exchange, and 15 minutes for fluid reclamation. This process is usually performed in a hospital. An interesting variation on this procedure is called **continuous ambulatory peritoneal dialysis** (**CAPD**). In this procedure the patient administers 2 liters of dialyzing fluid through the catheter and then continues with life as usual until 4–6 hours later, when the fluid is removed and replaced.

Probably the most satisfactory solution, in terms of overall quality of life, is *kidney transplantation.* This procedure involves the implantation of a new kidney obtained from a living donor or a cadaver. One-third of the approximately 8000 kidneys transplanted last year were obtained from living, related donors. The damaged kidney is usually removed. When left in place, an arterial graft is inserted to carry blood from the iliac artery or the aorta to the transplant, located in the pelvis or lower abdomen.

The success rate for this procedure varies, depending on how aggressively the recipient's T cells attack the donated organ and whether or not an infection develops. The 1-year success rate for implantation is now 85–95 percent. The use of kidneys taken from close relatives significantly improves the chances for a successful transplant. Immunosuppressive drugs are administered to reduce tissue rejection, but unfortunately this treatment also lowers the individual's resistance to infection.

CHAPTER REVIEW

Review of Selected Clinical Terms

pyelogram (PĪ-el-o-gram) (*p. 866*)
An image obtained by taking an X-ray of the kidneys after a radiopaque compound has been administered.

urinalysis (*p. 885*)
A physical and chemical assessment of urine.

uremia (ū-RĒ-mē-a) (*p. 885*)
A change in the composition of the blood indicating that all kidney functions are abnormal.

urethritis (*p. 890*)
Inflammation of the urethral wall.

cystitis (*p. 890*)
Inflammation of the urinary bladder lining.

dysuria (dis-Ū-rē-a) (*p. 890*)
Painful urination.

pyelitis (pī-e-LĪ-tis) (*p. 890*)
Inflammation of the walls of the renal pelvis.

pyelonephritis (pī-e-lō-nef-RĪ-tis) (*p. 890*)
Inflammation of the kidney tissues.

hemodialysis (hē-mō-dī-AL-i-sis) (*p. 893*)
A technique in which an artificial membrane is used to regulate the composition of the blood.

Additional Terms of Clinical Importance

URINALYSIS (*p. 885*): **polyuria, oliguria, anuria, urinometer (densitometer), azotemia**

RENAL FAILURE (*p. 886*): **renal failure**

AGING AND THE URINARY SYSTEM (*p. 891*): **urinary retention**

ADVANCES IN THE TREATMENT OF RENAL FAILURE (*p. 893*): **dialysis, dialysis fluid, peritoneal dialysis, continuous ambulatory peritoneal dialysis (CAPD)**

Study Outline

Related Key Terms

Introduction (pp. 859–860)

1. The digestive, cardiovascular, respiratory, and urinary systems work together to excrete wastes and regulate the water and electrolyte composition of body fluids. The functions of the urinary system include: (1) eliminating organic waste products; (2) regulating plasma concentrations of ions; (3) regulating blood volume and pressure by adjusting the volume of water lost, and releasing erythropoietin and renin; (4) helping to stabilize blood pH; (5) conserving nutrients; and (6) assisting the liver in detoxifying poisons and, during starvation, deaminating amino acids so they can be catabolized by other tissues.

An Overview of the Urinary System (p. 860)

1. The urinary system includes the **kidneys,** the **ureters,** and **urinary bladder,** and the **urethra.** The kidneys produce **urine** (a fluid containing water, ions, and soluble compounds); during **urination** urine is forced out of the body.

Related Key Terms: **micturition**

The Kidneys (pp. 860–869)

SUPERFICIAL AND SECTIONAL ANATOMY OF THE KIDNEY (pp. 861–862)

1. The **ureter** communicates with the **renal pelvis.** This chamber branches into two **major calyces,** each connected to four or five **minor calyces** that enclose the renal papillae.

Related Key Terms: **renal capsule** • **retroperitoneal** **adipose capsule** • **floating kidney** **renal fascia** • **pyramids** **hilus** • **renal sinus** • **cortex** **medulla** • **renal lobes** **papillae** • **renal columns**

Chapter Review

HISTOLOGICAL ORGANIZATION (pp. 862–869)

Related Key Terms

2. The **nephron** (the basic functional unit in the kidney) includes the **renal corpuscle** and a **renal tubule** that empties into the **collecting system** via a **collecting tubule,** a tributary of a **collecting duct.** From the **renal corpuscle** the **filtrate** travels through the **proximal convoluted tubule,** the **loop of Henle,** and the **distal convoluted tubule.**

papillary duct

3. Nephrons are responsible for: (1) production of filtrate; (2) reabsorption of nutrients; and (3) reabsorption of water and ions. The capsular epithelium lines the outer wall of the renal corpuscle. Blood arrives via the **afferent arteriole** and departs in the **efferent arteriole.**

glomerular epithelium
vascular pole

4. At the **glomerulus, podocytes** cover the **lamina densa** of the capillaries that project into the **capsular space.** The **pedicels** of the podocytes are separated by narrow **slit pores.**
5. The proximal convoluted tubule (PCT) actively reabsorbs nutrients, plasma proteins, and electrolytes from the filtrate. The loop of Henle includes a **descending limb** and an **ascending limb;** each limb contains a **thick segment** and a **thin segment.** The ascending limb delivers fluid to the **distal convoluted tubule (DCT),** which actively secretes ions and reabsorbs sodium ions from the urine.

tubular pole • **macula densa**
juxtaglomerular cells
juxtaglomerular apparatus

6. Roughly 85 percent of the nephrons are **cortical nephrons** found within the cortex; the **juxtamedullary nephrons** are closer to the medulla with their loops of Henle extending deep into the renal pyramids.
7. The vasculature of the kidneys includes the **interlobar, arcuate,** and **interlobular arteries,** and the **interlobar, arcuate,** and **interlobular veins.** Blood travels from the efferent arteriole to the **peritubular capillaries** and the **vasa recta.** Diffusion occurs between the capillaries of the vasa recta and the tubular cells through the **peritubular fluid** that surrounds the nephron.

renal nerves

Principles of Tubular Function (pp. 869–873)

1. Urine formation involves **filtration, reabsorption,** and **secretion.**

CARRIER-MEDIATED TRANSPORT MECHANISMS (pp. 872–873)

2. In carrier-mediated transport, a specific substrate binds to a carrier protein that facilitates its movement across the membrane. The saturation limit of a carrier protein is its **transport maximum.** Four types of **carrier-mediated transport** (facilitated diffusion, active transport, cotransport, and countertransport) are involved in modifying the filtrate. During **countertransport** the two trans-

renal threshold
sodium-linked countertransport
sodium-calcium countertransport

Formation and Modification of the Filtrate (pp. 873–883)

GLOMERULAR FILTRATION (pp. 874–875)

1. **Glomerular filtration** occurs as fluids move across the wall of the glomerulus into the capsular space, in response to the hydrostatic (blood) pressure in the glomerular capillaries. This movement is opposed by the **capsular hydrostatic pressure** (**C_{hp}**) and by the osmotic pressure of the blood (**OP_b**). The **filtration pressure (**P_f**)** at the glomerulus is the difference between the blood pressure and the opposing capsular and osmotic pressures.
2. The **glomerular filtration rate (GFR)** is the amount of filtrate produced in the kidneys each minute. Any factor that alters the filtration pressure acting across the glomerular capillaries will change the GFR and affect kidney function.

BP − (capsular + osmotic) = FP

Related Key Terms

CHANGING THE COMPOSITION OF THE FILTRATE (pp. 875–878)

3. Glomerular filtration produces a filtrate with a composition similar to blood plasma, but without the plasma proteins.
4. The cells of the PCT normally reabsorb 60 percent of the volume of filtrate produced in the renal corpuscle. The PCT generally reabsorbs sodium and other ions, water, and almost all of the nutrients in the filtrate. It also secretes various substances.
5. The DCT performs final adjustments by actively secreting or absorbing materials. Sodium ions are actively absorbed, in exchange for potassium or hydrogen ions discharged into the filtrate. Aldosterone secretion increases the rate of sodium reabsorption and potassium loss. The production of ammonia by DCT cells accelerates when the pH of the blood declines. **Ammonium ions** then appear in the blood and bicarbonate ions enter the circulation.

CHANGING THE OSMOTIC CONCENTRATION OF THE FILTRATE (pp. 878–883)

6. An **osmotic gradient** in the medulla encourages the osmotic flow of water out of the filtrate. The **countercurrent multiplication** between the ascending and descending limbs of the loop of Henle helps create the osmotic gradient in the medulla. As water is lost by osmosis and the filtrate volume decreases, the urea concentration rises.

osmolar (Osm)
milliosmolar (mOsm) • **osmoles**
milliosmoles

7. The vasa recta helps maintain the osmotic gradient by opposing the diffusion of solutes out of the medulla and by removing water and solutes retrieved from the filtrate. Normally the removal of solutes and water by the vasa recta precisely balances the rates of reabsorption and osmosis in the medulla.
8. Hypertonic urine is produced by a combination of **obligatory water reabsorption** and **facultative water reabsorption**.

Regulation of Kidney Function (pp. 883–887)

1. More than 99 percent of the filtrate produced each day is reabsorbed before reaching the renal pelvis. The composition of the excreted urine reflects the filtration, absorption, and secretion activities of the nephrons. The concentration depends on the osmotic movement of water across the walls of the tubules and collecting ducts.

CONTROL OF GLOMERULAR FILTRATION RATES (pp. 885–887)

2. The glomerulus shows autoregulation through alterations in the diameter of the afferent and efferent arterioles. Dropping filtration pressures stimulate the juxtaglomerular apparatus to release renin and erythropoietin. ADH production and release occur after stimulation by angiotensin II, or after stimulation of hypothalamic **osmoreceptors.**
3. Sympathetic activation (1) produces a powerful vasoconstriction of the afferent arterioles, decreasing the GFR and slowing the production of filtrate; (2) alters the GFR by changing the regional pattern of blood circulation; and (3) stimulates the release of renin by the juxtaglomerular apparatus.

Urine Transport, Storage, and Elimination (pp. 887–891)

1. Filtrate modification and urine production end when the fluid enters the renal pelvis. The rest of the urinary system is responsible for transporting, storing, and eliminating the urine.

THE URETERS AND URINARY BLADDER (pp. 888–889)

2. The ureters extend from the renal pelvis to the urinary bladder. The bladder is stabilized by the **urachus (middle umbilical ligament)** and the **lateral umbilical ligaments.** Internal features include the **trigone,** the **neck,** and the **internal sphincter.** The mucosal lining contains prominent **rugae** (folds). Contraction of the **detrusor** muscle compresses the bladder and expels the urine into the urethra.

Chapter Review

THE URETHRA (pp. 889–890)

3. In both sexes, as the urethra passes through the urogenital diaphragm a circular band of skeletal muscles forms the **external sphincter,** which is under voluntary control.

THE MICTURITION REFLEX AND URINATION (p. 890)

4. The process of urination is coordinated by the **micturition reflex,** which is initiated by stretch receptors in the bladder wall. Voluntary urination involves coupling this reflex with the voluntary relaxation of the external sphincter, which allows the opening of the **internal sphincter.**

Related Key Terms

renal threshold
sodium-linked countertransport
sodium-calcium countertransport

Aging and the Urinary System (p. 891)

1. Aging is usually associated with increased kidney problems. Age-related changes in the urinary system include: (1) declining number of functional nephrons; (2) reduced GFR; (3) reduced sensitivity to ADH; (4) problems with the micturition reflex (**urinary retention** may develop in men whose prostate glands are inflamed).

Integration with Other Systems (pp. 891–894)

1. The urinary, integumentary, respiratory, and digestive systems are sometimes considered as an anatomically diverse **excretory system.** The system's components work together to perform all of the excretory functions that affect the composition of body fluids.

■ Review Planner

	Objective	Level 1
1	Identify the components of the urinary system and their functions.	**1 2 8 9 17 19 35 37 41**
2	Describe the structure and functions of the kidneys.	**1 9 19 22 32 33 34 42 43 44**
3	Describe the processes involved in the formation of urine and the changes that occur in the filtrate as it moves through the nephron.	**7 10 11 12 14 15 20 23 24 25 28 36 38**
4	Discuss the factors that regulate the composition of urine.	**3 4 5 6 10 13 21 31**
5	Describe the structures and functions of the ureters, urinary bladder, and urethra.	**2 35 37 41**
6	Discuss the process of urination and how it is controlled.	**2 16 26 35 37 45**
7	Describe the effects of aging on the urinary system.	**27**
8	Explain how the urinary system interacts with other body systems to maintain homeostasis in body fluids.	**3 4 6 18**

Level 2: **18 19 24 29**

Level 3: **46 47**

C M: **Glomerulonephritis • Polycystic Kidney Disease • Conditions Affecting the Filtration Process Inherited Problems with Tubular Function • Diuretics • Problems with the Conducting System Problems with the Micturition Reflex • Average Values for Urinalysis**

C M: **PAH and Analysis of Renal Blood Flow • Estimating the GFR**

Review Questions

MULTIPLE CHOICE

1. Identify all of the structures that are responsible only for producing and modifying urine: a) urinary bladder; b) urethra; c) kidneys; d) ureters.
2. Identify all of the structures that are responsible only for storing urine and conducting it to the exterior: a) urinary bladder; b) urethra; c) kidneys; d) ureters.
3. A patient had developed a plaque that constricts the renal artery. Which of the following substances would the juxtaglomerular apparatus secrete into the circulation? (may be more than one answer): a) erythropoietin; b) ADH; c) renin; d) angiotensin II.
4. Which of the following would trigger the release of ADH?: a) osmoreceptors detect a rise in blood pressure; b) the osmolarity of the circulating blood declines; c) levels of angiotensin II fall; d) blood pressure falls.
5. Inadequate aldosterone levels will: a) produce hypercalcemia; b) conserve too much sodium; c) conserve hydrogen ions; d) produce hypokalemia.
6. All of the following would result from sympathetic activation, *except*: a) the afferent arterioles dilate; b) the regional pattern of blood circulation changes; c) the juxtaglomerular apparatus releases renin; d) production of filtrate slows down.
7. High concentration of ammonium ions in the filtrate indicate: a) deamination of amino acids; b) catabolism of nucleic acids; c) a surplus of ammonia in the diet.

DIAGRAMS

8. Cover the labels in Figure 26-1 and identify the components of the urinary system.
9. Cover the labels in Figure 26-3b. Identify the following: renal corpuscle, glomerulus, collecting tubule, proximal convoluted tubule, collecting duct, distal convoluted tubule, loop of Henle, renal tubule. What is being described?

MATCHING QUESTIONS

(*Match each term with the most appropriate statement.*)

Statements:

10. The blood pressure in the glomerular capillaries.
11. When all of the carrier enzymes are at saturation limits.
12. A pressure that represents the resistance to flow along the nephron and conducting system.
13. The amount of filtrate produced in the kidneys each minute.
14. The plasma concentration at which a specific substance will begin to appear in the urine.
15. The difference between the blood pressure and the opposing capsular and osmotic pressures.
16. Another term for urination.

Terms:

a) Micturition
b) T_m
c) Renal threshold
d) GFR
e) G_{hp}
f) C_{hp}
g) P_f

SHORT ANSWER/ESSAY

17. Discuss the function(s) of the urinary system.
18. Distinguish between the urinary system and the excretory system. Identify the components of the excretory system and their respective roles in preserving homeostasis.
19. What is a nephron? Explain its significance.
20. Compare the four types of carrier-mediated transport that are involved in modifying the filtrate. Name some substances that are transported by each method.
21. What factors determine the GFR?
22. Fill in the blanks: Blood reaches the vascular pole of each glomerulus through a(n) __________ arteriole and leaves in a(n) __________ arteriole.
23. Discuss the process of urine formation.
24. Explain how the composition of the filtrate changes as the fluid moves through the urinary system. How does the filtrate that first enters the capsular space differ from the urine that is excreted from the body?
25. Discuss how the osmolarity of the filtrate changes as it travels from the capsular space to the urethra.
26. Describe the micturition reflex. What is its function?
27. Mr. Casey is a healthy 67-year-old man. Explain how his urinary system would differ from that of his 10-year-old grandson.
28. You know that two unmarked but full test tubes contain either urine or plasma. Tube A contains higher concentrations of glucose and lipids; substance B is higher in chloride and potassium ions. Which is which?
29. Compare the roles of the PCT, the loop of Henle, and the DCT in maintaining homeostasis.
30. Mrs. Woodward's doctor asks her for urine samples collected over a 24-hour period, rather than a single sample. Why?
31. Fill in the blanks: Hormones involved in regulating the GFR include: __________, __________, and __________.

Identify and give a function for each of the following:

32. Collecting system.
33. Juxtaglomerular apparatus.
34. Peritubular capillaries.
35. Urethra.
36. Osmotic gradient.
37. Urinary bladder.
38. Countercurrent multiplication.
39. Slit pores.

Chapter Review

40. Vasa recta.
41. Ureter.

Distinguish between the following:

42. Renal sinus and renal pelvis.
43. Podocytes and pedicels.
44. Cortical and juxtamedullary nephrons.
45. Internal sphincter and external sphincter.

CRITICAL THINKING/APPLICATIONS

46. An 18-year-old college student is struck by a drunk driver who leaves the scene of the accident. When the student arrives at the emergency room, she is unconscious and suffering from an internal hemorrhage. How would her condition affect her GFR? Why?
47. For the past week Susan has felt a burning sensation in the urethral area when she urinates. She checks her temperature and finds she has a low-grade fever. Diagnose Susan's condition. What unusual substances are likely to be present in her urine? Explain why Susan is more likely to suffer from this condition than her husband is.

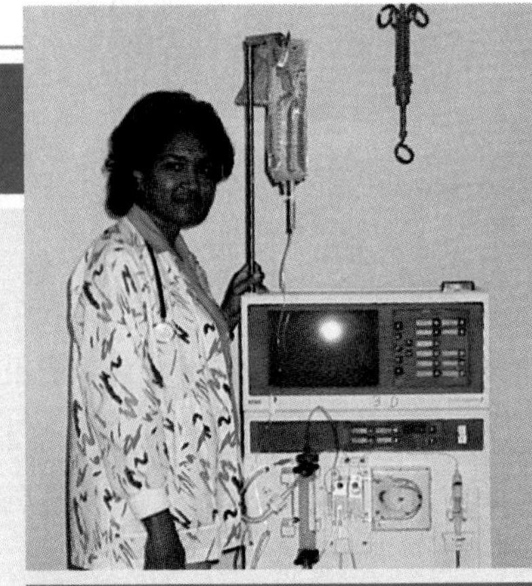

Career Close-up: Hemodialysis Nurse

Anita McDonald always wanted to be a nurse—as she says, "To be in a helping profession, involved with people." She went to a college that had a nursing program, and earned her BS and RN degrees concurrently. On graduation (Magna Cum Laude) she took a job at the St. Louis Veterans Administration Medical Center, where she had been working part time already. That was 17 years ago, and Anita is now a hemodialysis staff nurse there, in charge of the hemodialysis home training program.

The St. Louis Veterans Administration Medical Center treats patients with kidney disease from causes such as hypertension, diabetes, and drug abuse. Although they are supposed to be trained to be dialyzed in their homes, most of the men on Anita's unit would rather go to the hemodialysis unit than be treated at home. That is, the ones who come for treatment at all; noncompliance is a major problem for the hemodialysis staff.

"We have some people who just don't want to come in for treatment. Many of our patients are tired . . . washed out. We had one young man who was noncompliant. He lived with his mother. She would drop him off for treatment at the front door, and he'd leave through the back door. His mother thought he was coming here, until his disease got worse and he began having seizures. He was admitted to the intensive care unit in much worse shape than he had ever been."

Anita likes working on the hemodialysis unit because it's interesting and challenging; there's always something new to deal with. She credits her education for her ability to handle the variety of patient complications she encounters. "Even when we were being trained in the mechanics of hemodialysis, we also reviewed our anatomy and physiology," notes Anita. "We have to be aware of normal electrolytes, creatinine levels, and BUN, for instance. We have to be able to look at creatinine clearance rates and know how they affect our patients." For example, many of her patients have hematocrits as low as 20. Until recently, these patients would have had to receive transfusions. But now there's erythropoietin, which will raise the hematocrit if iron binding capacity and serum ferritin values are normal.

But, Anita cautions, "You have to know your A and P. We start with 10,000 units—some as high as 20,000 units of erythropoietin—and we have a sliding scale. As the hematocrit goes up, the erythropoietin administration is reduced. We do weekly hematocrits on everybody, so we monitor them closely."

The nurses also have to be on top of their patients' condition when they return from a weekend. The patients are dialyzed three times a week, with two days off. And anything could happen on those two days. A patient could forego his medication, or not adhere to his diet, or just simply not come in to the unit.

Fresh fruit . . . iced tea . . . tomato sauce . . . these everyday foods can wreak havoc on a body that's already lost its delicate balance. Just a couple of pieces of watermelon can throw off potassium levels, for example. Changes in calcium and phosphorus levels will also alert Anita to a noncompliance problem.

Nurses on the unit know to be suspicious when their patients come in with pulmonary or abdominal edema. Fluid restrictions are often violated; in a city where summer temperatures often hover around 100 degrees, it's easy to see why. It's not unusual for a man to come back to the unit weighing 10 to 12 pounds more than he did two days previously.

Anita explains that it's difficult to remove that fluid: patients cramp up when they lie down, so they must remain standing for hours. Or they might become hypotensive and nauseated. Yet these men will continually behave the same way, week after week, no matter how punishing the results of their noncompliance.

But for every noncompliant patient, there's another who is cooperative and appreciative. "Everyone is an individual," Anita says, "and people come to us with individual problems. We do a lot of education on a daily basis. We talk with out patients and their families. Sometimes it's successful, and sometimes it isn't. But the rewards are there. To have people be healthy because of what you do is the best reward."

We have already seen that water is vital to life. But it may surprise you to learn that you not only need water, you are largely water—some 66 percent of your body weight. In this chapter we will explore further the role of water in the human body and why it is so crucial to homeostasis. Water is the body's chief transport medium and solvent. The amount of water in the body, the precise concentrations of various ions in your blood and tissue fluids, the pH: all of these are critical, and a slight variation in any one of them can trigger profound physiological changes, some of them disastrous.

CHAPTER 27

Fluid, Electrolyte, and Acid-Base Balance

Chapter Objectives

After reading this chapter, you will be able to:

1. Describe how water and electrolytes are distributed within the body.
2. Explain the homeostatic mechanisms involved in the control of the fluid and electrolyte composition of body fluids.
3. Discuss the ways in which the body compensates for sudden losses or gains of water and/or electrolytes.
4. Explain the homeostatic mechanisms that stabilize the pH of intracellular and extracellular fluids.
5. Identify the most frequent threats to acid-base balance, and explain how the body responds when the pH varies outside normal limits.

■ Introduction

The next time you see a small pond, take a moment to think about the fish it contains. They live out their lives totally dependent on the quality of that isolated environment. Of course, polluting the pond with toxic substances will kill the fish, but more subtle changes can have equally grave effects. Changes in the volume of the pond, for example, can be quite important. If evaporation removes too much of the pond water, the fish become overcrowded; oxygen and food supplies run out, and the fish suffocate or starve. The ionic concentration of the pond water is also crucial. Most of the fish in a freshwater pond will die if the water becomes too salty; those in a saltwater pond will be killed if their environment becomes too dilute. The pH of the pond water, too, is a vital factor—that is why acid rain is such a problem.

The cells of our bodies live in a pond whose shores are the exposed surfaces of the skin. Most of the weight of the human body is water. Water accounts for up to 99 percent of the volume of extracellular fluid (ECF), and it is an essential ingredient of cytoplasm. All of a cell's operations rely on water as a diffusion medium for the distribution of gases, nutrients, and waste products. If the water content of the body changes, cellular activities are jeopardized. For example, when the water content of the body reaches very low levels, proteins denature, enzymes cease functioning, and cells die.

The ionic concentrations and pH of the body water are as important as its absolute quantity. If concentrations of calcium or potassium ions in the ECF become too high, cardiac arrhythmias develop, and the individual's life is in jeopardy. A pH outside of the normal range can lead to a variety of dangerous problems. Low pH is especially dangerous because hydrogen ions break chemical bonds, change the shapes of complex molecules, disrupt cell membranes, and impair tissue functions.

This chapter considers the dynamics of exchange between the various body fluids and between the body and the external environment. Maintenance of normal volume and composition in the extracellular and intracellular fluids is an important, multifaceted process. Several different but interrelated types of homeostasis are involved:

- A person is in **fluid balance** when the amount of water gained each day is equal to the amount lost to the environment. The maintenance of normal fluid balance involves regulating body water content and distribution.
- **Electrolyte balance** exists when there is neither a net gain nor a net loss of any ion in body fluids.
- **Acid-base balance** exists when the production of hydrogen ions precisely offsets their loss. While acid-base balance exists, the pH of body fluids remains within normal limits.

Much of the material in this chapter was introduced in earlier chapters. Those chapters focused on limited aspects of fluid, electrolyte, or acid-base balance as they related to specific systems. This chapter provides an overview that integrates those discussions to highlight important functional patterns. Few other chapters have such wide-ranging clinical importance: *treatment of any serious illness affecting the nervous, cardiovascular, respiratory, urinary,*

FIGURE 27-1
Body Fluid Compartments. The major body fluid compartments in a normal individual.

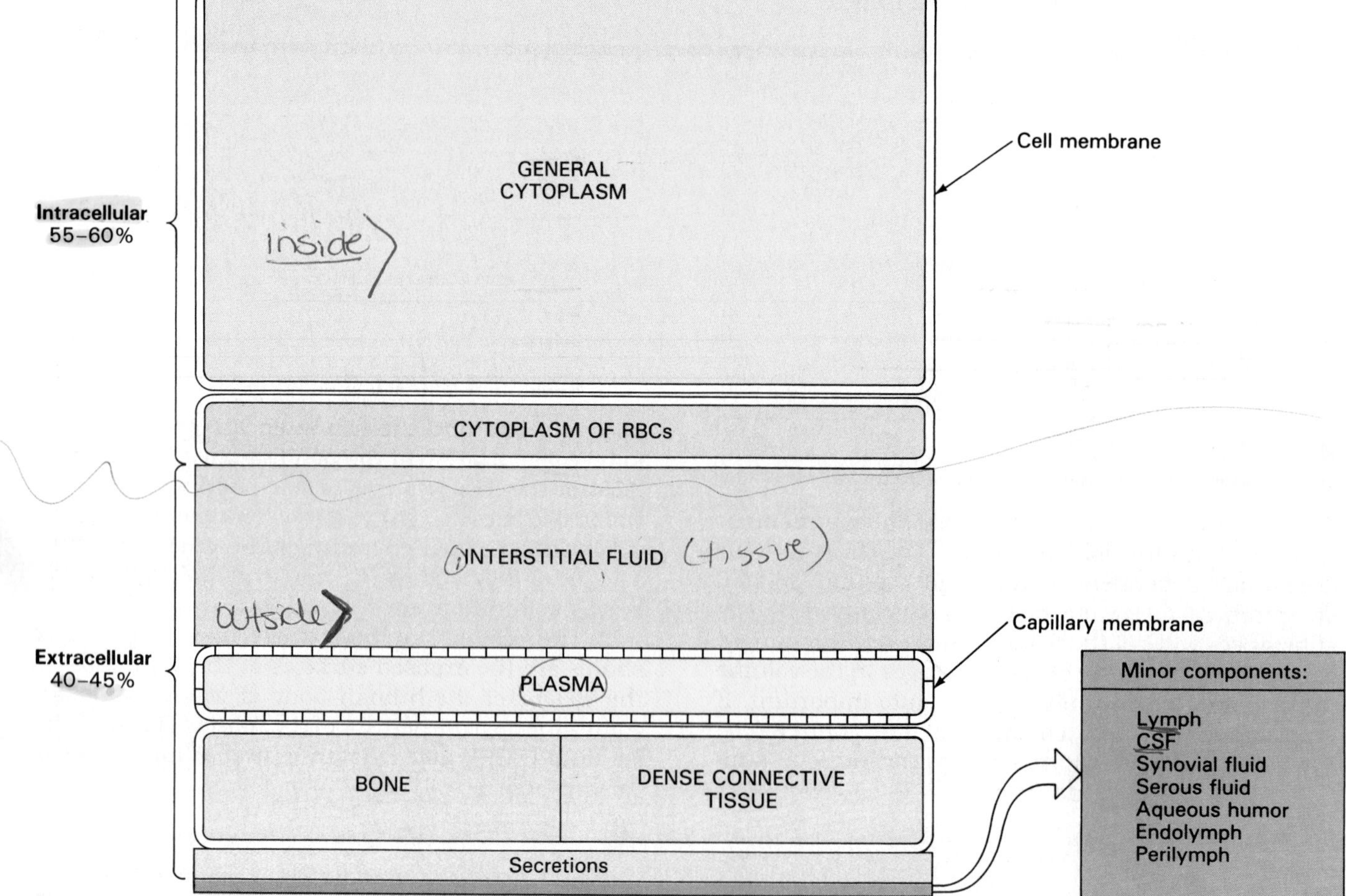

or digestive system must include steps to restore normal fluid, electrolyte, and acid-base balance.

Fluid and Electrolyte Balance

Figure 27-1 details the distribution of water in the body. Nearly two-thirds of the total body water content is found inside living cells, as the fluid medium of the **intracellular fluid (ICF)**, introduced in Chapter 3. (p. 63) The **extracellular fluid (ECF)** contains the rest of the body water. The largest subdivisions of the ECF are (1) the *tissue fluid*, or *interstitial fluid*, in peripheral tissues and (2) the *plasma* of the circulating blood. Minor components of the ECF include lymph, cerebrospinal fluid (CSF), synovial fluid, serous fluids (pleural, pericardial, and peritoneal fluids), aqueous humor, perilymph, and endolymph.

Exchange between these subdivisions of the ECF occurs primarily across the endothelial lining of capillaries. Fluid may also travel from the interstitial spaces to the plasma via the channels of the lymphatic system. There are regional variations in the identity and quantity of dissolved electrolytes, proteins, nutrients, and waste products within the ECF.[1] Yet these variations are relatively minor compared with the compositional differences *between* the ECF and the ICF.

In many respects the ICF and ECF behave as distinct entities, and they are often called **fluid compartments**. Exchange between the ECF and ICF occurs across selectively permeable cell membranes, by processes discussed in Chapter 3. (For a review of the mechanisms involved, see Table 3-7.) (p. 87) These processes enable cells to maintain internal environments with a composition different from that of their surroundings. As noted in earlier chapters, the principal ions in the ECF are sodium, chloride, and bicarbonate. The ICF contains an abundance of potassium, magnesium, and phosphate ions, plus large numbers of negatively charged proteins. (A comparison between the ICF and the two major subdivisions of the ECF can be found in Figure 27-2.)

Despite these differences in the concentration of specific substances, the intracellular and extracellular osmolarities are identical. The cell membranes

[1]For an analysis of the composition of ECF compartments, see Appendix V.

FIGURE 27-2
Cations and Anions in Body Fluids. Note the differences in cation and anion concentrations in the various body fluid compartments. For information concerning the composition of other body fluids, see Appendix V.

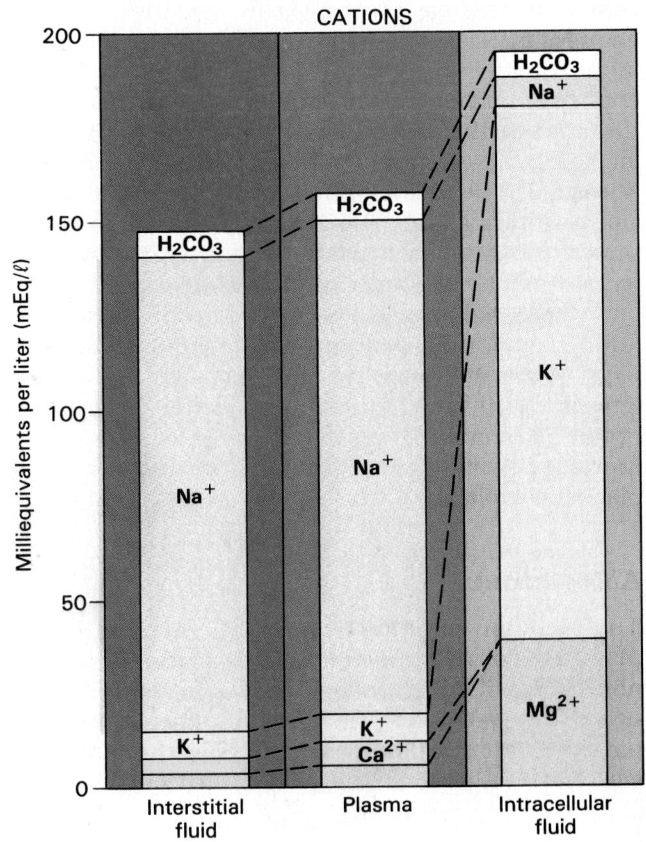

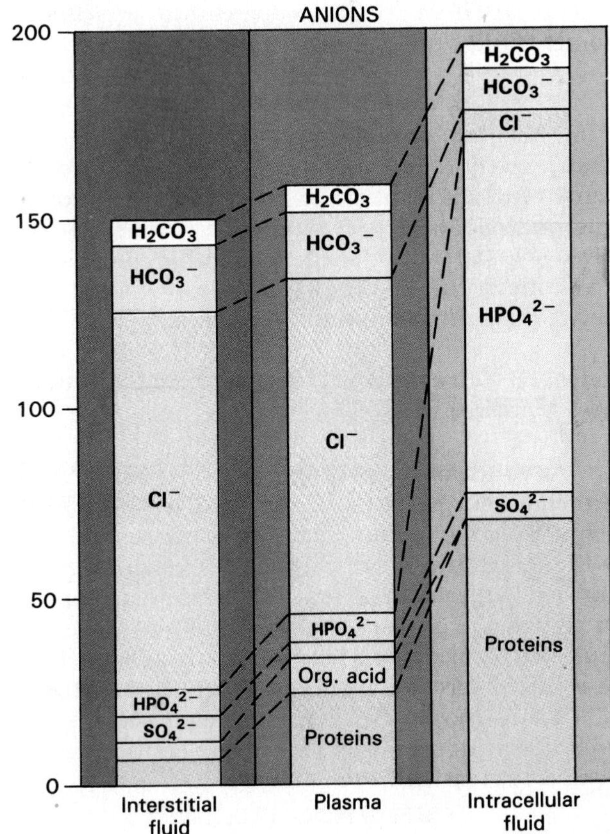

are freely permeable to water, and osmosis eliminates any minor concentration differences almost at once.

BASIC CONCEPTS PERTAINING TO FLUID AND ELECTROLYTE REGULATION

Four basic principles must be understood before we can proceed to a discussion of fluid and electrolyte balance.

Concept 1: All of the homeostatic mechanisms that monitor and adjust the composition of body fluids respond to changes in the ECF, not in the ICF.

Receptors monitoring the composition of two key components of the ECF, the plasma and the cerebrospinal fluid, detect significant changes in composition or volume and trigger appropriate neural and endocrine responses. This arrangement makes functional sense because a change in one component of the ECF will spread rapidly throughout the extracellular compartment and will affect all the cells in the body. By contrast, no receptors directly monitor the ICF. The ICF is contained within trillions of individual cells that are physically and chemically isolated from one another by their cell membranes. Thus changes in the intracellular fluid in one cell will have no direct effect on the composition of the intracellular fluid in distant cells and tissues *except via the extracellular fluid.*

Concept 2: There are no receptors that can directly monitor fluid and electrolyte balance.

Receptors cannot detect how much water, sodium, chloride, or potassium is present in the body at any given time. Important homeostatic adjustments occur in response to changes in *plasma volume* or *osmolarity* (the total concentration of dissolved solutes). Minor alterations in the composition or volume of the interstitial fluids are ignored until they affect plasma osmolarity or volume.

Concept 3: Our cells are unable to move water molecules by active transport.

All water movements across cell membranes and epithelia occur passively, in response to osmotic gradients. These gradients, however, can be established by the active transport of specific ions, such as sodium and chloride. You may find it useful to remember the simple phrase: *water follows salt.* When sodium and chloride ions (or other solutes) are actively transported across a membrane or epithelium, water follows by osmosis. This basic principle accounts for water absorption across the digestive epithelium and water conservation in the kidneys.

Concept 4: The body content of water or electrolytes will rise if intake exceeds outflow and fall if losses exceed gains.

Homeostatic mechanisms affect the rates of dietary absorption and urinary excretion. Homeostatic responses may involve both changes in behavior and physiological changes in specific systems. For example, a person may experience a sensation of thirst that leads her to drink more fluids or a craving for salt that prompts her to consume heavily salted foods. Because the rates of water and electrolyte uptake are proportional to the dietary supply, these changes in behavior have a direct effect on fluid or electrolyte balance. Physiological adjustments affecting fluid and electrolyte balance are mediated primarily by three hormones, *antidiuretic hormone* (ADH), *aldosterone*, and *atrial natriuretic peptide* (ANP). We will consider each of these in the following section.

AN OVERVIEW OF THE PRIMARY REGULATORY HORMONES

The hormones involved with fluid and electrolyte regulation were detailed in earlier chapters. Figure 18-17 includes flow charts that summarize the regulation and action of each of these hormones. ∞ (p. 589)

Antidiuretic Hormone

The hypothalamus contains special cells known as *osmoreceptors* that monitor the osmotic concentration of the plasma. These cells are sensitive to subtle changes in plasma osmolarity. A 2 percent alteration (approximately 6 mOsmol) is sufficient to trigger a change in osmoreceptor activity. These osmoreceptors include the neurons that secrete ADH. They are located in the anterior hypothalamus, and their axons release ADH near fenestrated capillaries in the posterior pituitary gland. The rate of ADH release varies directly with the osmolarity: the higher the osmolarity, the greater the amount of ADH released.

Increased release of ADH has two important effects: (1) it stimulates water conservation at the kidneys, reducing urinary water losses, and (2) it stimulates the thirst center to promote the drinking of fluids. The combination of decreased water loss and increased water gain gradually restores normal plasma osmolarity.

Aldosterone

The secretion of aldosterone by the adrenal cortex plays a major role in determining the rate of sodium absorption along the distal convoluted tubule and collecting system of the kidneys. The higher the plasma concentration of aldosterone, the more efficiently the kidney will conserve sodium ions. Because

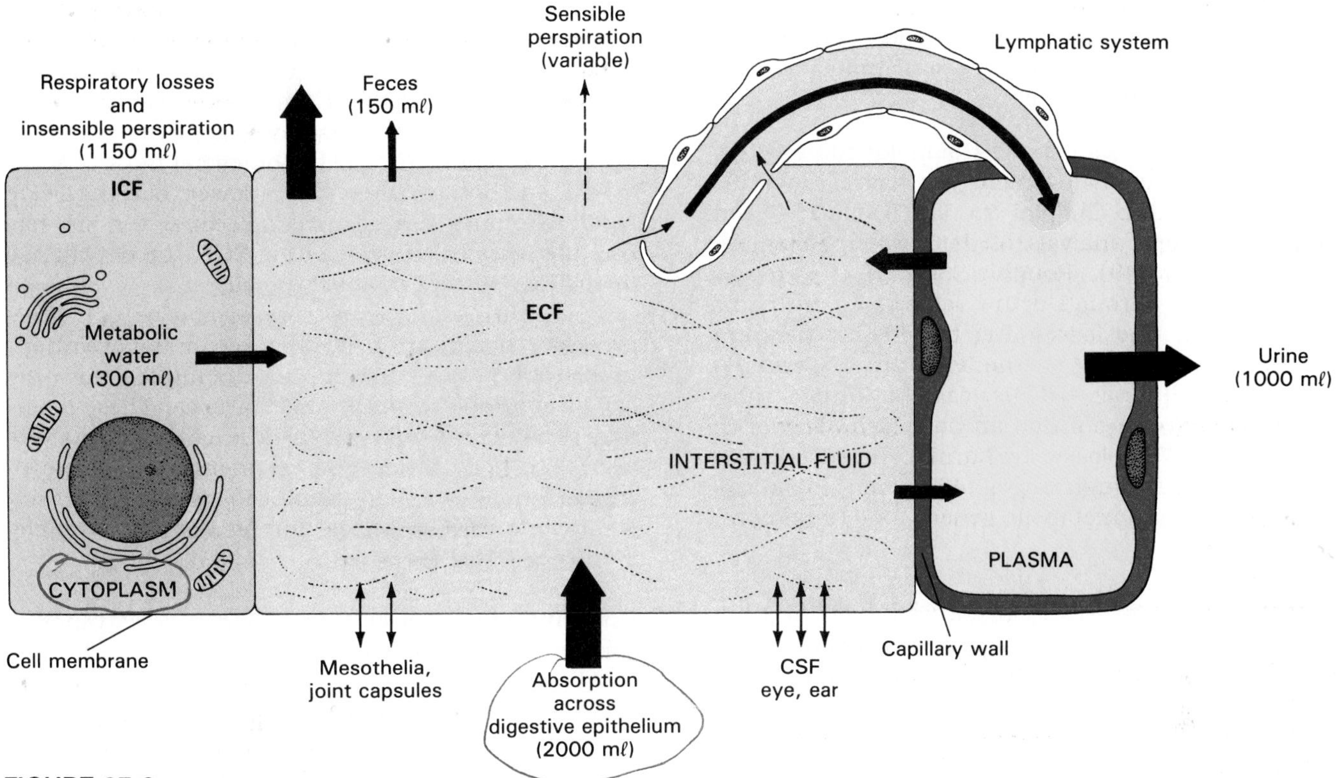

FIGURE 27-3
Fluid Exchanges. A diagrammatic representation of fluid exchanges between the fluid compartments of the body and the external environment. (The volumes are not to scale.)

"water follows salt," the conservation of sodium ions also stimulates water retention.

Secretion of aldosterone is not directly influenced by sodium ion concentrations. Rather, it occurs in response to a fall in plasma volume or blood pressure at the juxtaglomerular apparatus of the nephron, and also (as we will see later in this chapter) to potassium ion concentrations. (For a more detailed analysis of the regulation of aldosterone secretion, see Chapter 18.) ∞ (p. 587)

Atrial Natriuretic Peptide

Atrial natriuretic peptide is released by cardiac muscle fibers in response to abnormal distension of the atrial walls. This hormone (1) reduces thirst and (2) blocks the release of ADH and aldosterone that might otherwise lead to water and salt conservation. The resulting diuresis lowers the plasma volume and eliminates the source of the stimulation.

INTERPLAY BETWEEN FLUID AND ELECTROLYTE BALANCE

At first glance, it can be very difficult to distinguish between water balance and electrolyte balance. For example, when an individual loses body water, plasma volume decreases and electrolyte concentrations rise. Conversely, when excess electrolytes are gained or lost, there is an associated water gain or loss due to osmosis. However, because the regulatory mechanisms involved are quite different, it is often useful to consider fluid balance and electrolyte balance as distinct entities. This distinction is absolutely vital in a clinical setting, where problems with fluid balance and electrolyte balance must be identified and corrected promptly.

FLUID BALANCE

Water circulates freely within the extracellular fluid compartment. At capillary beds throughout the body, hydrostatic pressure forces water out of the plasma and into interstitial spaces. Some of that water is reabsorbed along the distal portion of the capillary bed, and the rest circulates into lymphatic vessels for transport to the venous circulation. This circulation is diagrammed in Figure 27-3, which indicates two additional important relationships between components of the ECF:

1. Water moves back and forth across the mesothelial surfaces lining the peritoneal, pleural, and pericardial cavities and through the synovial membranes lining joint capsules. The flow rate is significant; for example, roughly 7 liters of peritoneal fluid are produced and reabsorbed each day.

2. Water also moves between the blood and the cerebrospinal fluid, the aqueous and vitreous humors of the eye, and the perilymph and endolymph of the inner ear.

Figure 27-3 also shows the major routes of exchange with the environment. (These relationships were discussed in Chapter 25, and Table 25-7 contains a detailed analysis of daily water gains and losses.) (p. 849) Roughly 2300 mℓ of water are lost each day through urine, feces, and insensible perspiration. The losses due to sensible perspiration vary, depending on the activities undertaken, but the additional deficits can be considerable, reaching well over a gallon an hour (see Chapter 6). (p. 197) Water losses are normally balanced by the gain of fluids through eating (48 percent), drinking (40 percent), and metabolic generation (12 percent).

Fluid Exchange between the ECF and ICF

As noted earlier, there are significant differences in composition between the ICF and the ECF in terms of the major anions and cations. The characteristic differences persist because each cell expends energy by pumping ions into or out of the cytoplasm, as discussed in Chapter 3 (see Figure 3-24). (p. 85) Because the cell membranes are freely permeable to water, the ECF and ICF remain isotonic (of equal osmolarity) despite the differences in ionic distribution.

Changes in the osmolarity of the extracellular fluid can have a direct effect on the distribution of water within these compartments. Because the volume of the ICF is much greater than that of the ECF, the ICF acts as a "water reserve." When the osmolarity of the ECF changes, water movement between the ICF and ECF tends to oppose the change.

- If the ECF becomes more concentrated (hypertonic) with respect to the ICF, water will move from the cells into the ECF until osmotic equilibrium is restored.
- If the ECF becomes more dilute (hypotonic) with respect to the ICF, water will move from the ECF into the cells, and the volume of the ICF will increase accordingly.

In effect, instead of a large change in the composition of the ECF there are smaller changes in both the ECF and ICF. Water movement between the ECF and ICF is called a **fluid shift**. Fluid shifts occur relatively rapidly, reaching equilibrium within a period of minutes to hours.

Disturbances of Fluid Balance

Disturbances of fluid balance are very important and potentially deadly. Two examples will demonstrate the dynamic exchange between the ECF and ICF.

Allocation of Water Losses When water is lost but electrolytes are retained, the osmolarity of the ECF rises. As indicated in Figure 27-4a, osmosis then moves water out of the ICF and into the ECF until the two solutions are again isotonic. At this point both the ECF and ICF will be somewhat more concentrated, and both volumes will be lower than they were before the fluid loss occurred. Because the ICF has roughly twice the volume of the ECF, the net change in the ECF is kept relatively small.

Conditions that cause severe water losses include excessive perspiration (exercising in hot weather), inadequate water consumption (being lost in a desert), repeated vomiting, and diarrhea. These disorders promote water losses far in excess of electrolyte losses, so body fluids become increasingly concentrated. Responses that attempt to restore homeostasis include ADH secretion and (as soon as possible) an increase in fluid intake.

Distribution of Water Gains When pure water is consumed, or when hypotonic solutions are administered through intravenous infusion, the water content of the body increases without a corresponding increase in the concentration of electrolytes (Figure 27-4b). As a result, the extracellular fluid becomes hypotonic with respect to the ICF. A fluid shift then occurs, and the volume of the ICF increases at the expense of the ECF. Once again, the larger volume of the ICF enables it to limit the amount of osmotic change. After the fluid shift the ECF and ICF have slightly larger volumes and slightly lower osmolarities than they did originally.

Under normal circumstances this situation will be promptly corrected. The reduced plasma osmolarity depresses the secretion of ADH, discouraging fluid intake and increasing water losses in the urine. If the situation is *not* corrected, a variety of clinical problems develop. [**CM:** *Water Excess and Depletion*]

ELECTROLYTE BALANCE

An individual is in electrolyte balance when the rates of gain and loss are equal for each of the individual electrolytes in the body. Electrolyte balance is important because:

- total electrolyte concentrations have a direct effect on water balance, as detailed above;
- the concentrations of individual electrolytes can have an effect on a variety of cell functions. Many examples were encountered in earlier chapters, including the effect of abnormal sodium ion concentrations on nerve cell activity and the effects of high or low calcium and potassium ion concentrations on cardiac muscle tissue.

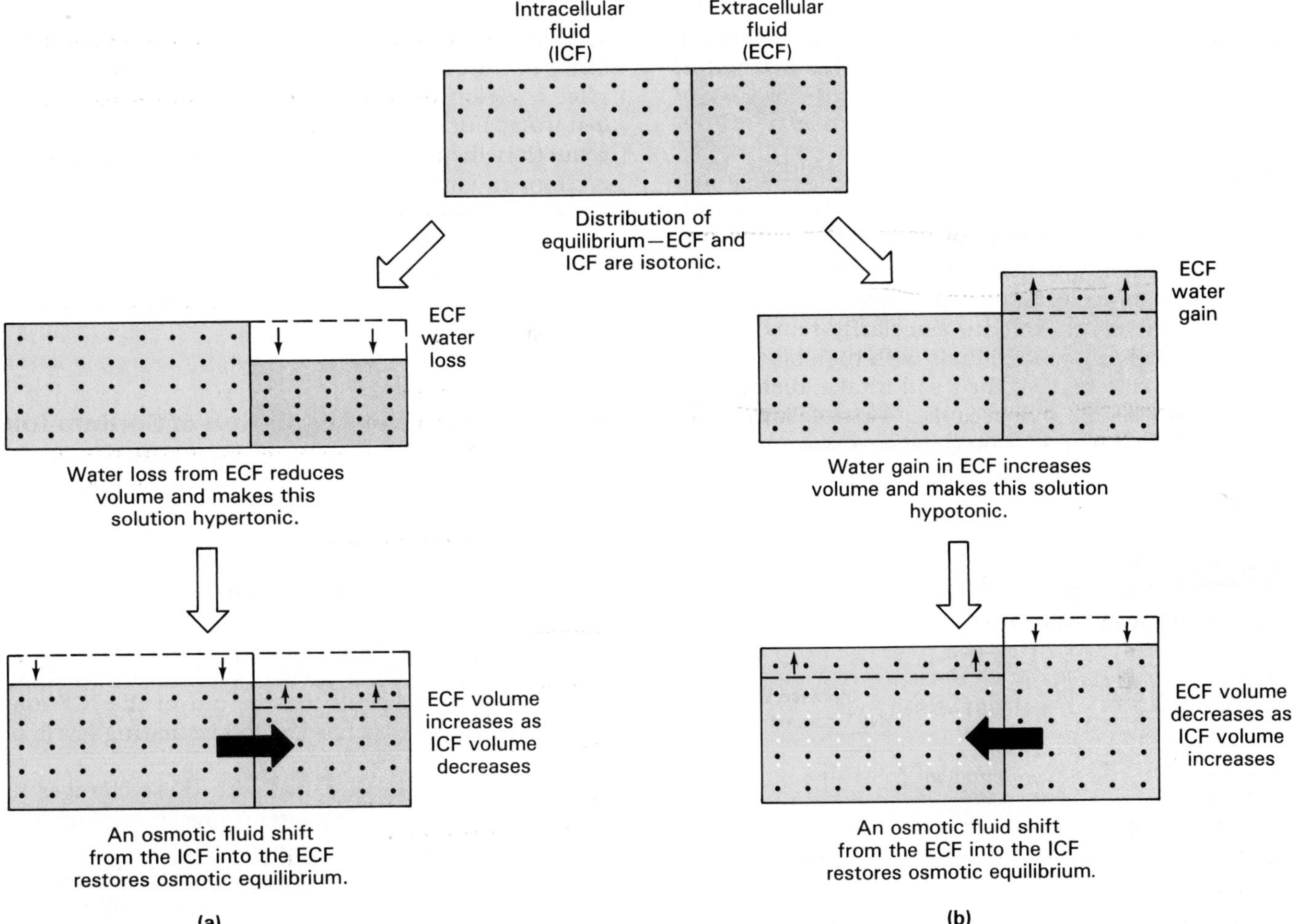

FIGURE 27-4
Osmosis and Fluid Shifts between the ECF and ICF. (a) A fluid shift into the ECF occurs when a water loss makes the ECF hypertonic with respect to the ICF. (b) A fluid shift into the ICF occurs when a water gain makes the ECF hypotonic with respect to the ICF.

Two cations, sodium and potassium, merit special attention because (1) they are major contributors to the osmolarities of the ECF and ICF, and (2) they have direct effects on the normal functioning of living cells.

Sodium is the dominant cation within the extracellular fluid. Because more than 90 percent of the osmolarity of the ECF results from the presence of sodium salts, principally sodium chloride (NaCl) and sodium bicarbonate ($NaHCO_3$), alterations in the osmolarity of body fluids usually reflect changes in the concentration of sodium ions. Potassium is the dominant cation in the intracellular fluid; extracellular potassium concentrations are normally low. Two general rules are worth noting:

1. The most common problems with electrolyte balance are caused by an imbalance between sodium gains and losses.
2. Problems with potassium balance are less common but significantly more dangerous than those related to sodium balance.

Factors That Affect Sodium Ion Concentrations in the ECF

The total amount of sodium in the ECF represents a balance between two factors:

1. *Sodium ion uptake across the digestive epithelium.* Sodium ions enter the ECF by crossing the digestive epithelium via diffusion and active transport mechanisms. The rate of uptake varies directly with the amount included in the diet.
2. *Sodium ion excretion at the kidneys and other sites.* Sodium losses occur primarily in the urine and through perspiration. The kidneys are the most important sites of sodium ion regulation. The mechanisms for sodium reabsorption at the kidneys were detailed in Chapter 26. (p. 875)

When sodium gains exceed sodium losses, the total sodium ion content of the ECF goes up; when losses exceed gains, the sodium ion content declines.

However, a change in the sodium ion *content* of the ECF does not produce a change in sodium ion *concentration. Whenever the rate of sodium intake or output changes, there is a corresponding gain or loss of water that tends to keep the sodium concentration constant.*

For example, eating a heavily salted meal will not elevate the osmolarity of the ECF, for as sodium chloride crosses the digestive epithelium, osmosis brings additional water into the ECF. Thus when sodium gains exceed sodium losses, the volume of the ECF increases, but its osmolarity remains the same. (This is why individuals with high blood pressure are told to restrict their salt intake; dietary salt will be absorbed, and because "water follows salt" the blood volume and blood pressure will increase.)

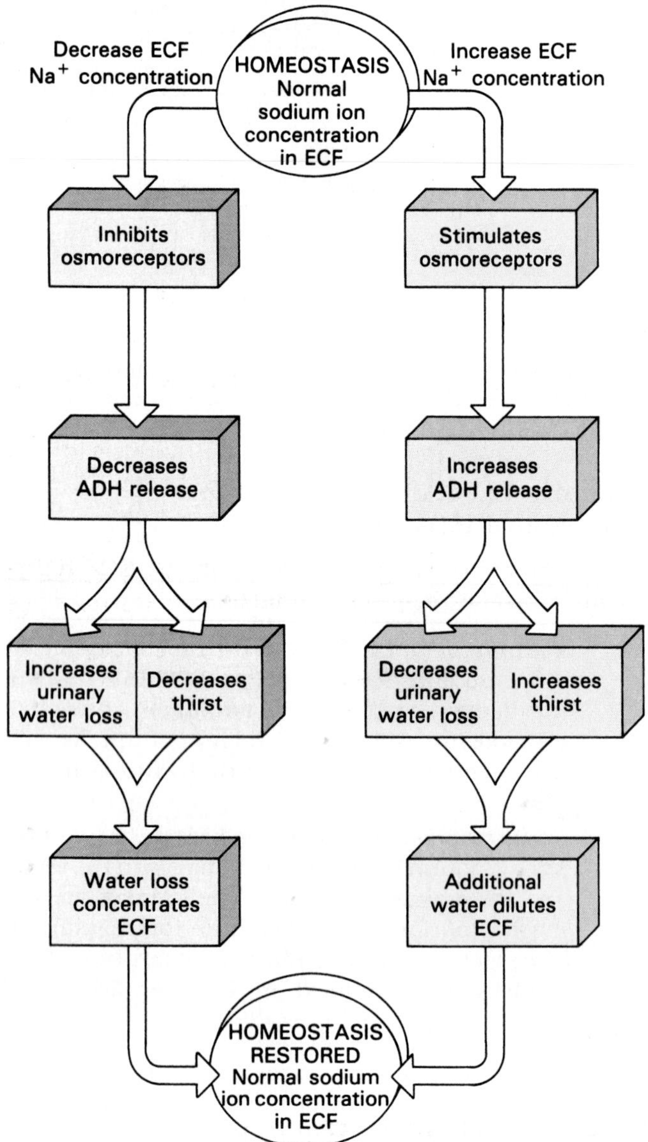

FIGURE 27-5
Homeostatic Regulation of Normal Sodium Ion Concentrations in Body Fluids

Conversely, consuming a large glass of tap water, a hypotonic solution, will not decrease the osmolarity of the ECF for very long. Drinking pure water after a period of excessive perspiration replaces the lost water but not the missing salts. The consequent reduction in plasma osmolarity depresses ADH secretion, causing an increased water loss at the kidneys. As water leaves the extracellular compartment, the osmolarity returns to normal. Thus when sodium losses exceed gains, the extracellular fluid volume decreases—but again, without a change in osmolarity.

A Summary of the Regulation of Sodium Ion Concentrations

When the mechanisms responsible for maintaining normal fluid balance are fully operational, *alterations in sodium balance result in the expansion or contraction of the extracellular fluid compartment.* The ECF volume changes, but the sodium ion concentration remains relatively stable. The sequence of events is diagrammed in Figure 27-5. Two steps are involved: (1) an initial fluid shift into or out of the ICF and (2) "fine tuning" via changes in circulating levels of ADH.

How does the body deal with these changes in ECF volume produced by alterations in sodium balance? As was pointed out previously (concept 2), there are no receptors sensitive to the water content of the body: the regulatory mechanisms cannot detect changes in ECF *volume*, only changes in ECF *concentration* (osmolarity). When the volume changes involved are small, it does not matter—minor fluctuations in ECF volume have no adverse physiological effects. Large variations in the ECF volume are corrected by the homeostatic mechanisms triggered by changes in blood volume. The regulatory steps are detailed in Figure 27-6. If the plasma volume is inadequate:

- the fall in blood pressure along the afferent arterioles of the kidneys stimulates the release of renin by the juxtaglomerular apparatus of each nephron;
- renin initiates a chain of events leading to the activation of angiotensin II (for additional details, see Figure 21-28); (p. 700)
- angiotensin II then produces a coordinated elevation in the extracellular fluid volume by stimulating thirst, causing the release of ADH, and triggering the adrenal production and secretion of aldosterone;
- thirst and ADH secretion lead to increased ingestion and retention of water, while aldosterone promotes the retention of sodium ions along the distal portions of the kidney nephrons.

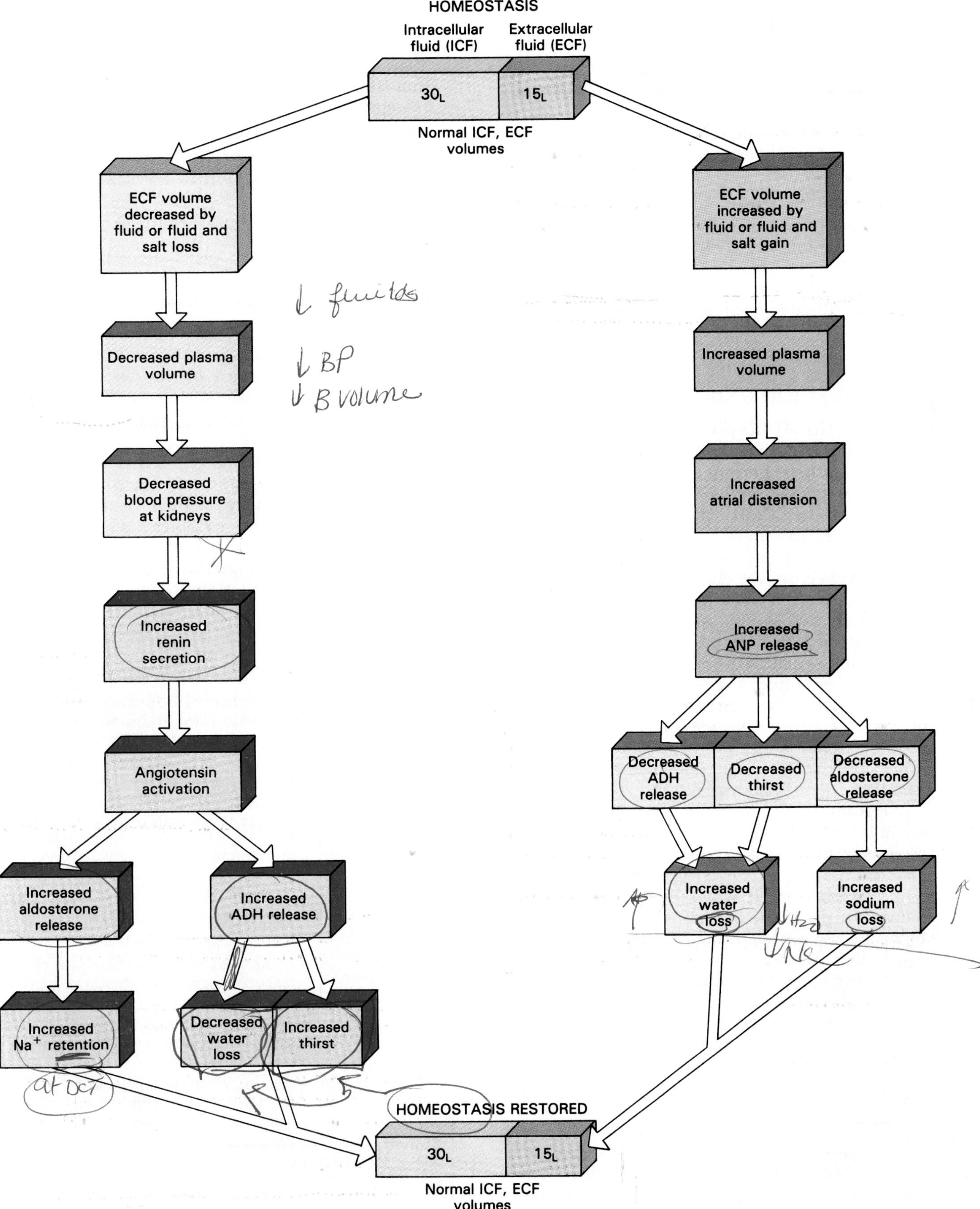

FIGURE 27-6
The Integration of Fluid Volume Regulation and Sodium Ion Concentrations in Body Fluids

As a result of this mechanism, water and sodium ion losses are reduced, gains are increased, and the volume of the ECF increases. The concentration of sodium ions, however, remains unchanged.

If the plasma volume becomes abnormally large:

- venous return increases, stretching the atrial wall and stimulating the release of atrial natriuretic peptide;
- ANP reduces thirst and blocks the secretion of ADH and/or aldosterone, which might promote water or salt conservation.

This mechanism causes increased salt and water loss at the kidneys, and the volume of the ECF declines.

POTASSIUM ION REGULATION

Potassium ions are the primary cations of the intracellular fluid, usually found in concentrations of 160 mEq/ℓ. *Roughly 98 percent of the potassium content of the body lies within the ICF.* The potassium concentration of the ECF is relatively low, ranging from 3 to 5 mEq/ℓ. The concentration in the ECF at any given moment represents a balance between (1) the rate of entry across the digestive epithelium and (2) the rate of loss into the urine.

Urinary losses are normally limited to the amount gained by absorption across the digestive epithelium, normally 50–150 mEq per day. (The potassium losses in feces and perspiration are negligible by comparison.) The concentration of potassium in the ECF is controlled by adjustments in the rate of active secretion along the distal convoluted tubule of the nephron. The rate of tubular secretion of potassium ions changes in response to:

1. **Alterations in the potassium ion concentration in the ECF:** In general, the higher the extracellular concentration of potassium, the higher the rate of secretion along the distal convoluted tubule.

2. **Changes in pH:** When the pH of the ECF falls, so does the pH of the peritubular fluid. The rate of potassium ion secretion then declines, because hydrogen ions—instead of potassium ions—are secreted in exchange for sodium ions in the filtrate.

3. **Aldosterone levels:** The rate at which potassium is lost in the urine is strongly affected by aldosterone, because the ion pumps sensitive to this hormone reabsorb sodium ions from the filtrate in exchange for potassium ions from the peritubular fluid. Aldosterone secretion is stimulated by angiotensin II as part of the regulation of blood volume. High plasma potassium concentrations also stimulate aldosterone secretion. Either way, under the influence of aldosterone *there is a direct relationship between the amount of sodium conserved and the amount of potassium excreted in the urine.* [**CM:** *Hyperkalemia and Hypokalemia*]

√ **What effect would drinking a pitcher of distilled water have on your level of ADH?**

√ **How would eating a meal high in salt affect the amount of fluid in the ICF (intracellular fluid) compartment?**

√ **What effect would being lost in the desert for a day without water have on your blood osmolarity?**

■ Acid-Base Balance

Chapter 2 introduced the topic of pH and the chemical nature of acids, bases, and buffers. Table 27-1 reviews key terms important to the discussion that follows; if you need a more detailed review, refer to the appropriate sections of Chapter 2 before proceeding. (pp. 40–42)

The pH of body fluids represents balance between the acids, bases, and salts in solution. The pH of body fluids normally remains within relatively narrow limits, usually from 7.35 to 7.45. Any deviation outside the normal range is extremely dangerous because changes in hydrogen ion concentrations disrupt the stability of cell membranes, alter protein

TABLE 27-1 A Review of Important Terms Relating to Acid-Base Balance

pH	The negative exponent (negative log) of the hydrogen ion concentration
Neutral	A solution with a pH of 7, which contains equal numbers of hydrogen and hydroxyl ions
Acidic	A solution with a pH below 7, in which hydrogen ions predominate
Basic or alkaline	A solution with a pH above 7, in which hydroxyl ions predominate
Acid	A compound that dissociates to release hydrogen ions, shifting the pH toward acidity
Base	A compound that dissociates to liberate hydroxyl ions or tie up hydrogen ions and increase pH
Salt	A compound whose dissociation does not release hydrogen or hydroxyl ions
Buffer	A compound that tends to oppose changes in the pH of a solution by removing or replacing hydrogen ions; in body fluids, buffers maintain pH within normal limits (7.35–7.45)

structure, and change the activities of important enzymes.

When the pH falls below 7.35, a state of **acidosis** exists. **Alkalosis** exists if the pH increases above 7.45. These conditions affect virtually all systems, but the nervous system and cardiovascular system are particularly sensitive to pH fluctuations. For example, severe acidosis can be deadly because (1) CNS function deteriorates, and the individual becomes comatose; (2) cardiac contractions grow weak and irregular, and symptoms of heart failure develop; and (3) peripheral vasodilation produces a dramatic drop in blood pressure, and circulatory collapse may occur.

The control of pH is therefore an extremely important homeostatic process of great physiological and clinical significance. Although acidosis and alkalosis are both dangerous, in practice *problems with acidosis are much more common than problems with alkalosis*. This is the case because several different types of acids are generated by normal cellular activities.

ACIDS IN THE BODY

There are three general categories of acids in the body: *volatile acids*, *fixed acids*, and *organic acids*.

Volatile Acids

A **volatile acid** is an acid that can leave solution and enter the atmosphere. Carbon dioxide is an important volatile acid found in body fluids, and *carbon dioxide concentration is the most important factor affecting the pH in body tissues*. In solution, carbon dioxide interacts with water to form molecules of carbonic acid (H_2CO_3). As noted in Chapter 23, the carbonic acid molecules then dissociate to produce hydrogen ions and bicarbonate ions. (p. 775) The complete reaction sequence is:

$$CO_2 + H_2O \rightleftharpoons H_2CO_3 \rightleftharpoons H^+ + HCO_3^-$$

This reaction occurs spontaneously in body fluids, but it occurs very rapidly in the presence of *carbonic anhydrase*, an intracellular enzyme found in red blood cells, liver and kidney cells, parietal cells of the stomach, and in many other cell types.

Because most of the carbon dioxide in solution is converted to carbonic acid, and most of the carbonic acid dissociates, there is a direct relationship between the pCO_2 and the pH (Figure 27-7). When carbon dioxide concentrations rise, additional hydrogen ions and bicarbonate ions are released, and the pH goes down.[2] At the alveoli, CO_2 diffuses into the atmosphere, the number of hydrogen ions and bicarbonate ions declines, and the pH rises. (This process, which effectively removes hydrogen ions from solution, will be considered in more detail later in the chapter.)

Fixed Acids

Fixed acids are acids that do not leave solution; once produced, they remain in body fluids until excreted at the kidneys. **Sulfuric acid** and **phosphoric acid** are the most important fixed acids. They are gener-

[2] Remember that because the pH is a *negative* exponent, when the concentration of hydrogen ions goes up, the pH goes down.

FIGURE 27-7
Basic Relationships between the pCO_2 and Plasma pH. The pH is inversely related to the pCO_2; when the pCO_2 increases, the pH declines.

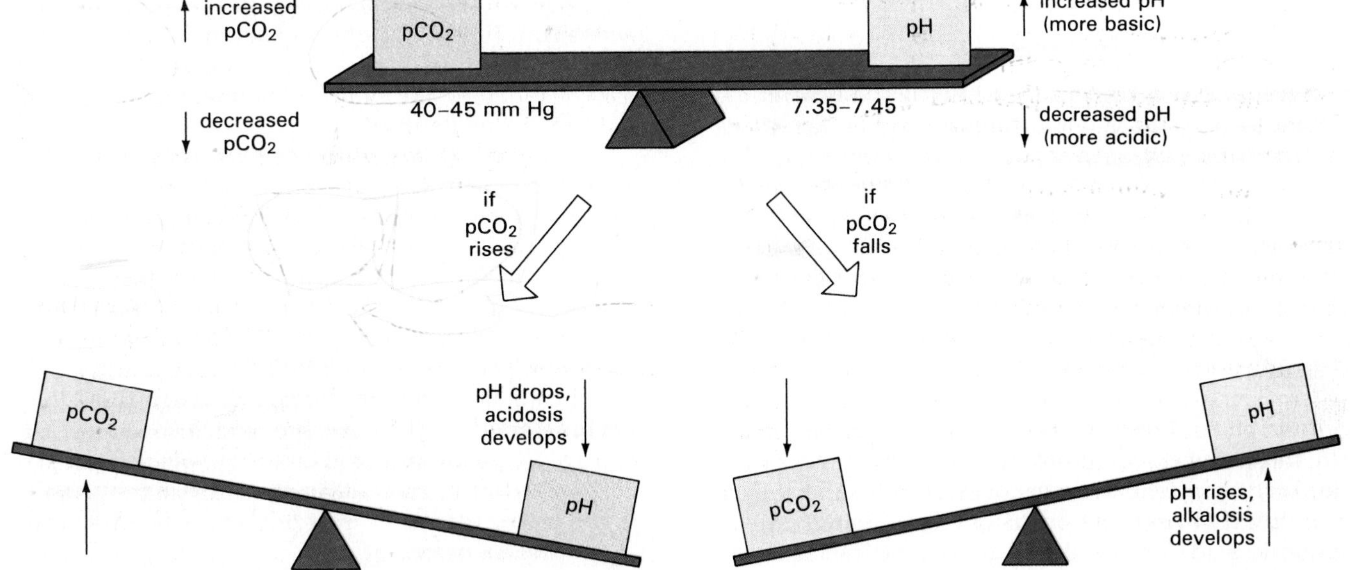

ated in small amounts during the catabolism of amino acids and compounds containing phosphate groups, including phospholipids and nucleic acids.

Organic Acids

Organic acids are acid participants in or byproducts of cellular metabolism. Lactic acid and ketone bodies are important examples of organic acids considered in earlier chapters. Under normal conditions, most organic acids are metabolized rapidly, and significant accumulations do not occur. But relatively large amounts of organic acids are produced during periods of anaerobic respiration, starvation, or excessive lipid catabolism.

BUFFERS AND BUFFER SYSTEMS

The acids discussed above, produced in the course of normal metabolic operations, must be controlled by the buffers and buffer systems in body fluids. *Buffers*, introduced in Chapter 2, are dissolved compounds that can provide or remove hydrogen ions and thereby stabilize the pH of a solution. (p. 41) Buffers include *weak acids* that can donate hydrogen ions and *weak bases* that can absorb them. A **buffer system** consists of a combination of a weak acid and its dissociation products: a hydrogen ion (H^+) and an anion.

There are three major buffer systems, each with slightly different characteristics and distribution. *Protein buffer systems* contribute to the regulation of pH in the ECF and ICF. These buffer systems interact extensively with the other buffer systems. The *carbonic acid-bicarbonate buffer system* is an important buffer system in the ECF. The *phosphate buffer system* has an important role in buffering the pH of the intracellular fluids. The functional relationships between these buffer systems are diagrammed in Figure 27-8.

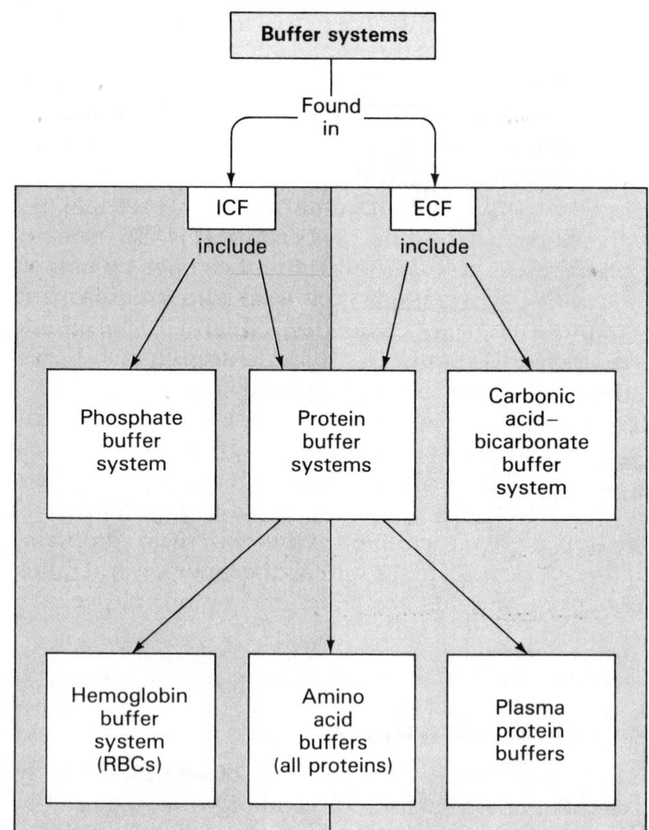

FIGURE 27-8
Buffer Systems in Body Fluids. Phosphate buffers are found primarily in the ICF, whereas the carbonic acid-bicarbonate buffer system is found primarily in the ECF. Protein buffer systems are found in both the ICF and the ECF. Extensive interactions occur between these buffer systems.

Protein Buffer Systems

Protein buffer systems depend on the ability of amino acids to respond to alterations in pH by accepting or releasing hydrogen ions. The underlying mechanism can be seen in Figure 27-9. If the pH climbs, the carboxyl group (COOH) of the amino acid can dissociate, releasing a hydrogen ion. If the pH drops, the amino group can accept an additional hydrogen ion, forming an NH_3^+ group.

The plasma proteins contribute to the buffering capabilities of the blood. In the interstitial fluids, there are extracellular protein fibers and dissolved amino acids that also assist in the regulation of pH. In the intracellular fluids of active cells, structural and other proteins provide an extensive buffering capability that prevents destructive pH changes when organic acids are produced by cellular metabolism.

The protein buffer systems inside cells are very important in stabilizing the pH of the extracellular fluid. For example, when the pH of the ECF declines, cells absorb hydrogen ions from the ECF and buffer them internally. When the pH of the ECF rises, cells exchange intracellular hydrogen ions for extracellular potassium. However, buffering activities that involve hydrogen ion transport between compartments occur slowly, and they are not able to make rapid, large-scale adjustments in pH.

The **hemoglobin buffer system** is an intracellular buffer system that has a much more immediate effect on extracellular pH. This buffering mechanism was described in the discussion of carbon dioxide transport in the cytoplasm of red blood cells (Figure 23-22). (p. 774) *The hemoglobin buffer system helps prevent drastic alterations in pH when the plasma pCO_2 is rising or falling.* Carbonic anhydrase in the cytoplasm rapidly converts carbon dioxide to carbonic acid. As the carbonic acid dissociates, the hydrogen ions are buffered by hemoglobin molecules and the bicarbonates are transported into the plasma. At the lungs the entire reaction sequence proceeds in the reverse direction.

In alkaline medium acts like an acid

Neutral pH

In acid medium acts like a base

FIGURE 27-9
Amino Acid Buffers. Amino acids can either accept a hydrogen ion or donate one, depending on the pH of their surroundings.

The Carbonic Acid-Bicarbonate Buffer System

Carbon dioxide generation occurs in all living tissues. As detailed above, most of the carbon dioxide is converted to carbonic acid, which then dissociates into a hydrogen ion and a bicarbonate ion. The carbonic acid and its dissociation products form the **carbonic acid-bicarbonate buffer system**. Its buffering properties are diagrammed in Figure 27-10a. Because the reaction is freely reversible, changing the concentrations of any one of the participants will alter the equilibrium concentrations of the others. Thus if hydrogen ions are removed, they will be replaced through the dissociation of carbonic acid molecules; if hydrogen ions are added, most will be removed through the formation of carbonic acid.

Because carbonic acid is produced by the reaction of carbon dioxide and water, this buffer system cannot provide protection against pH changes that accompany rising or falling pCO_2. However, this is not usually a problem because carbon dioxide levels are monitored by chemoreceptors and controlled by varying the respiratory rate. *The primary role of the carbonic acid-bicarbonate buffer system is in preventing pH changes caused by organic acid and fixed acids in the ECF.* The mechanism is detailed in Figure 27-10b. The hydrogen ions released through the dissociation of organic acids combine with bicarbonate ions, elevating carbonic acid levels and the pCO_2. The additional CO_2 is then lost at the lungs.

This system is very effective, but buffering activity can continue only as long as there are adequate numbers of bicarbonate ions in solution. Every time a hydrogen ion is removed, a bicarbonate goes with it, and once all of the bicarbonates have been tied up, buffering capabilities are lost. Fortunately, body fluids contain a large reserve of bicarbonate ions, primarily in the form of dissolved molecules of *sodium bicarbonate*, $NaHCO_3$. This readily available supply of bicarbonate ions is known as the **bicarbonate reserve**. When hydrogen ions enter the ECF, the bicarbonate ions tied up in carbonic acid molecules are replaced by bicarbonates from the bicarbonate reserve.

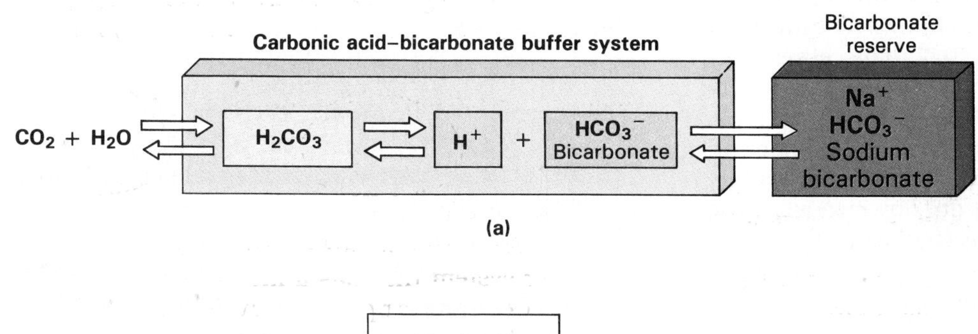

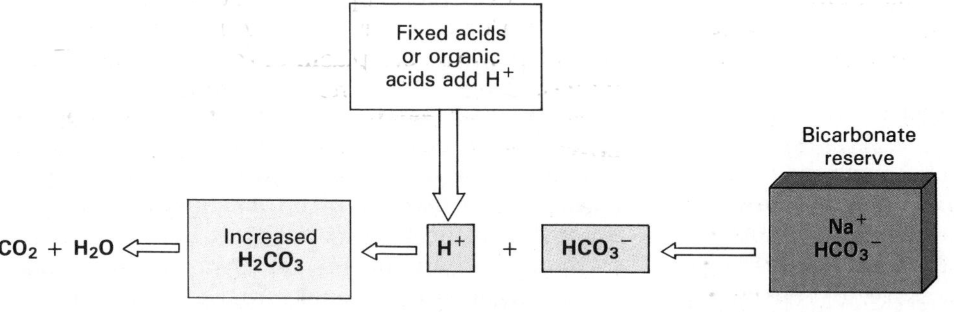

FIGURE 27-10
The Carbonic Acid-Bicarbonate Buffer System. (a) Basic components of the carbonic acid-bicarbonate buffer system, showing their relationships to CO_2 and the bicarbonate reserve. (b) Response of the carbonic acid-bicarbonate buffer system to hydrogen ions generated by fixed or organic acids in body fluids.

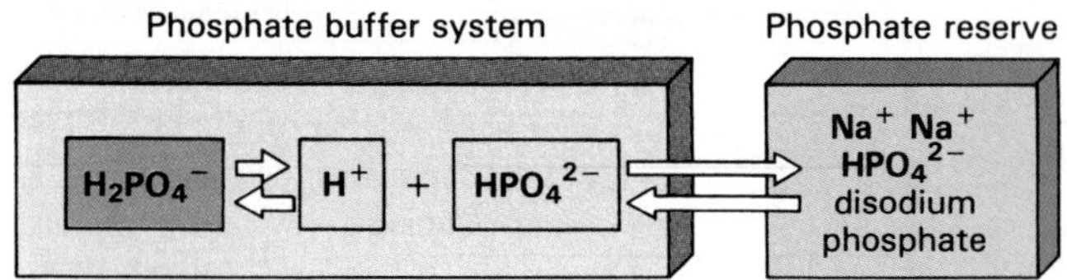

FIGURE 27-11
The Phosphate Buffer System. This buffer system functions primarily in the cytosol.

The Phosphate Buffer System

The **phosphate buffer system** consists of an anion, $H_2PO_4^-$, that is a weak acid. In solution, it reversibly dissociates into a hydrogen ion and HPO_4^{2-}. The basic operation of the phosphate buffer system resembles that of the carbonic acid-bicarbonate buffer system. The reactions involved are diagrammed in Figure 27-11. In the ECF the phosphate buffer system plays only a supporting role in the regulation of pH, primarily because the concentration of bicarbonate ions far exceeds that of phosphate ions. However, the phosphate buffer system is quite important in buffering the pH of the intracellular fluids, where the concentration of phosphate ions is relatively high.

MAINTENANCE OF ACID-BASE BALANCE

Maintaining acid-base balance involves more than the stabilization of pH. In the long run, gains and losses of hydrogen ions must be equal. Although buffers can tie up excess hydrogen ions, they provide only a temporary solution, because the hydrogen ions have not been eliminated, merely rendered harmless. The underlying problem is that the supply of buffer molecules is limited. For example, suppose that a buffer molecule prevents a pH change by binding a hydrogen ion that enters the ECF. This essentially ties up that buffer molecule and reduces the capacity of the ECF to cope with additional hydrogen ions. Eventually, all of the buffer molecules will be bound to hydrogen ions, and pH control becomes impossible. The situation can be resolved only by removing the hydrogen ion from the ECF, thereby freeing the buffer molecule, or by replacing the buffer molecule. Similarly, if a buffer provides a hydrogen ion to maintain normal pH, either that hydrogen ion or the buffer must eventually be replaced.

The maintenance of acid-base balance thus includes regulating hydrogen ion losses and gains. This regulation involves coordinating the actions of buffer systems with pulmonary mechanisms and renal mechanisms. The pulmonary and renal mechanisms support the buffer systems by (1) secreting or absorbing hydrogen ions, (2) controlling the excretion of acids and bases, and, when necessary, (3) generating additional buffers. It is the *combination* of buffer systems and these pulmonary and renal mechanisms that maintains pH within narrow limits.

Pulmonary Contributions to pH Regulation

The lungs contribute to pH regulation by their effects on the carbonic acid-bicarbonate buffer system. *Increasing or decreasing the rate of respiration can have a profound effect on the buffering capacity of body fluids by lowering or raising the pCO_2.* Changes in pCO_2 affect the concentration of hydrogen ions in the plasma (Figure 27-7) via the equilibrium depicted in Figure 27-10.

Mechanisms responsible for the control of respiratory rate were detailed in Chapter 23, and only a brief summary will be presented here. (Those seeking a more detailed review should refer to Figures 23-25 and 23-26.) (pp. 778, 779) Chemoreceptors of the carotid and aortic bodies are sensitive to the pCO_2 of the circulating blood. Other receptors located on the ventrolateral surfaces of the medulla oblongata monitor the pCO_2 of the cerebrospinal fluid. A rise in pCO_2 stimulates the receptors; a fall in the pCO_2 inhibits them. Stimulation of the chemoreceptors leads to an increase in the respiratory rate. As the rate of respiration increases, more CO_2 is lost at the lungs, and the pCO_2 returns to normal levels. When the pCO_2 of the blood or CSF declines, respiratory activity becomes depressed and the breathing rate falls. This causes an elevation of the pCO_2 in extracellular fluids.

Changes in respiratory rate affect pH because when the pCO_2 rises, the pH declines, and when the pCO_2 decreases, the pH increases (Figure 27-7). A change in the respiratory rate that helps stabilize pH is called **respiratory compensation**. Respiratory compensation occurs whenever the pH strays outside of normal limits. There are two reasons for this respiratory response:

1. Whenever the pH goes down, the equilibrium represented in Figure 27-10 is shifted to the left. More CO_2 is formed, and the pCO_2 goes up, stimulating the respiratory chemoreceptors.
2. The same chemoreceptors that are sensitive to pCO_2 levels are also directly sensitive to pH. When the pH drops, respiration is stimulated; when the pH rises, respiration is inhibited.

Renal Contributions to pH Regulation

Glomerular filtration puts hydrogen ions, carbon dioxide, and the other components of the carbonic acid-bicarbonate and phosphate buffer systems into the filtrate. The kidney tubules subsequently modify the pH of the filtrate by secreting hydrogen ions or reabsorbing bicarbonate ions. Under normal conditions, the body generates enough organic and fixed acids to add around 100 mEq of hydrogen ions to the extra-

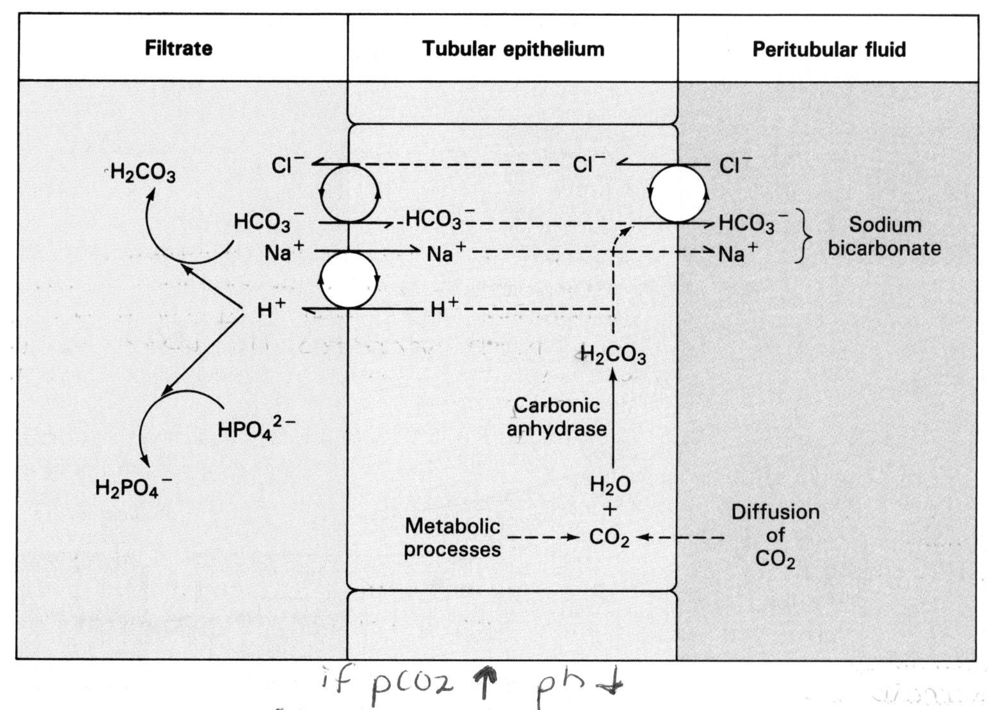

FIGURE 27-12
Kidney Tubules and pH Regulation. Response of the kidney tubules to acidosis. Note (1) active secretion of hydrogen ions, (2) removal of CO_2 from peritubular fluids, (3) reabsorption of sodium bicarbonate, and (4) the activity of other buffers in the filtrate.

cellular fluids each day. An equivalent number of hydrogen ions must therefore be excreted in the urine to maintain acid-base balance.

Tubular hydrogen ion secretion involves carbonic anhydrase activity within the cells of the proximal convoluted tubule and distal convoluted tubule. As indicated in Figure 27-12, this process results in the diffusion of sodium bicarbonate into the ECF. Bicarbonate ions are also recovered from the filtrate by countertransport in exchange for chloride ions. In this way urinary bicarbonate losses are minimized. A change in the rates of hydrogen ion and bicarbonate ion secretion or absorption in response to changes in plasma pH is called **renal compensation**.

The pH of the filtrate must be kept above 4.5 because hydrogen ion secretion cannot continue against a large concentration gradient. If there were no buffers in the filtrate, the kidney tubules could secrete only around 1 percent of the acid produced each day before the pH reached this limit. Buffers in the filtrate are therefore important, because they keep the pH within the acceptable range.

Buffer Activity in the Filtrate Components of the phosphate and carbonic acid-bicarbonate buffer systems enter the urine by filtration at the glomerulus. Together, they manage to tie up most of the secreted hydrogen ions.

Renal Response to Acidosis When the normal buffer mechanisms are stressed by excessive numbers of hydrogen ions, acidosis results. The kidney tubules are not able to distinguish among the various acids that may cause this condition. Whether the fall in pH results from the production of volatile, fixed, or organic acids, the renal contribution remains limited to the secretion of hydrogen ions and the simultaneous generation or reabsorption of bicarbonates. The tubular cells thus bolster the capabilities of the bicarbonate buffer system by increasing the concentration of bicarbonate ions in the ECF, replacing those already used to remove hydrogen ions from solution. During starvation, the tubular cells break down amino acids, yielding carbon chains for catabolism and bicarbonates to help buffer ketone bodies in the blood (see Figure 26–13). (p. 877)

Renal Response to Alkalosis When the pH rises, alkalosis occurs. Under these conditions the rate of hydrogen ion secretion declines, and the kidneys do not reclaim bicarbonates entering the filtrate. As a result, the concentration of bicarbonate ions in the plasma declines. This decline promotes the dissociation of carbonic acid molecules, and the hydrogen ions released help to return the pH to normal levels.

■ Disturbances of Acid-Base Balance

Figure 27-13 summarizes the interactions between buffer systems, respiration, and renal function in normal acid-base balance. In combination, these mechanisms are usually able to control pH very precisely, so that the pH of the extracellular fluids seldom varies more than 0.1 pH units, from 7.35 to 7.45. When buffering mechanisms are severely stressed, the pH wanders outside of these limits, producing symptoms of alkalosis or acidosis.

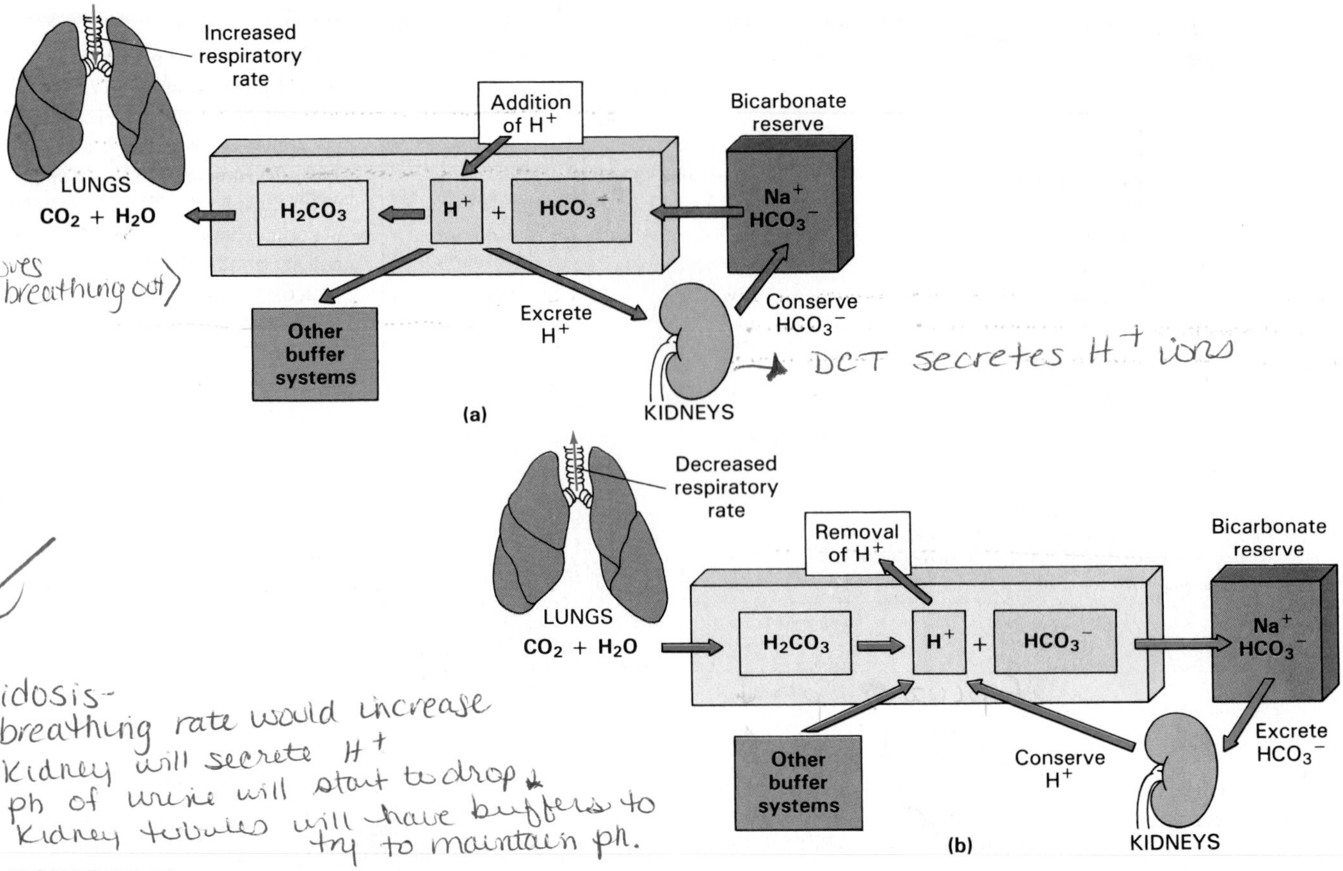

Increased respiratory rate

Addition of H^+

Bicarbonate reserve

LUNGS

$CO_2 + H_2O$

H_2CO_3

$H^+ + HCO_3^-$

Na^+ HCO_3^-

Other buffer systems

Excrete H^+

Conserve HCO_3^-

KIDNEYS

(a)

Decreased respiratory rate

Removal of H^+

Bicarbonate reserve

LUNGS

$CO_2 + H_2O$

H_2CO_3

$H^+ + HCO_3^-$

Na^+ HCO_3^-

Other buffer systems

Conserve H^+

Excrete HCO_3^-

KIDNEYS

(b)

FIGURE 27-13
The Central Role of the Carbonic Acid-Bicarbonate Buffer System in the Regulation of Plasma pH. The carbonic acid-bicarbonate buffer system plays a central role in pH regulation. This diagram summarizes the interactions between this buffer system and other buffer systems and compensation mechanisms. Part (a) shows the response to a fall in pH, part (b) the response to a rise in pH.

Readers considering a career in any health-related field will find that understanding acid-base dynamics is essential for clinical diagnosis and management of a wide variety of conditions. Disturbances of acid-base balance can result from any factor affecting one of the principal regulatory mechanisms.

- Any disorder affecting circulating buffers, respiratory performance, or renal function will potentially disrupt acid-base balance. Several conditions described in earlier chapters, including emphysema, chronic obstructive pulmonary disease (COPD), and renal failure, are associated with dangerous changes in pH. **[CM:** *Bronchitis, Emphysema, and COPD*] (p. 886)
- Cardiovascular conditions, such as heart failure or hypotension, can also affect the pH of internal fluids by causing fluid shifts, changing the rates of glomerular filtration, and altering respiratory efficiency.
- Conditions affecting the CNS will also disrupt normal acid-base balance. For example, neural damage or CNS disease will affect the respiratory and cardiovascular reflexes that are essential to normal pH regulation.

Temporary alterations in the pH of body fluids occur frequently. Rapid and complete recovery occurs through a combination of buffer system activity and the respiratory and renal responses. Serious abnormalities in acid-base balance typically show an initial *acute phase*, in which the pH moves rapidly away from the normal range. If the condition persists, physiological adjustments occur and the individual enters the *compensated phase*. *Unless the underlying problem is corrected, compensation cannot be complete, and the blood chemistry will remain abnormal.* The pH often remains outside normal limits, even after compensation has occurred.

The primary source of the problem is usually indicated by the name given to the resulting condition. **Respiratory disorders** result from abnormal carbon dioxide levels in the ECF. These conditions are usually directly related to an imbalance between the rate of carbon dioxide removal at the lungs and its generation in other tissues. **Metabolic disorders** are caused by the generation of organic or fixed acids or by conditions affecting the concentration of bicarbonate ions in the extracellular fluids. The problems facing individuals with respiratory disorders are quite different from those with metabolic disorders.

Pulmonary compensation alone can often restore normal acid-base balance in individuals suffering from respiratory disorders. In contrast, compensation mechanisms for metabolic disorders may be able to stabilize pH, but other aspects of acid-base balance (buffer system function, bicarbonate levels, and pCO_2) remain abnormal until the underlying metabolic cause is corrected.

The respiratory and metabolic categories can be subdivided, creating four major classes of acid-base disturbances: *respiratory acidosis*, *respiratory alkalosis*, *metabolic acidosis*, and *metabolic alkalosis*. Each of these will be discussed separately.

RESPIRATORY ACIDOSIS

Respiratory acidosis develops when the respiratory system is unable to eliminate all of the CO_2 generated by peripheral tissues. The primary symptom is low plasma pH due to **hypercapnia**, an elevated plasma pCO_2. Carbonic acid concentrations rise in parallel, and as these molecules dissociate, the hydrogen and bicarbonate ion concentrations begin rising. Other buffer systems can tie up some of the hydrogen ions, but once the combined buffering capacity has been exceeded the pH begins to fall rapidly. The effects are diagrammed in Figure 27-14a.

FIGURE 27-14
Respiratory Disorders of Acid-Base Regulation. Respiratory acidosis (a) and alkalosis (b) result from inadequate or excessive pulmonary ventilation. The rate of pulmonary ventilation is closely regulated by mechanisms described in Chapter 23. In normal individuals these responses, combined with those of other buffer systems and the kidneys, are usually able to restore normal acid-base balance.

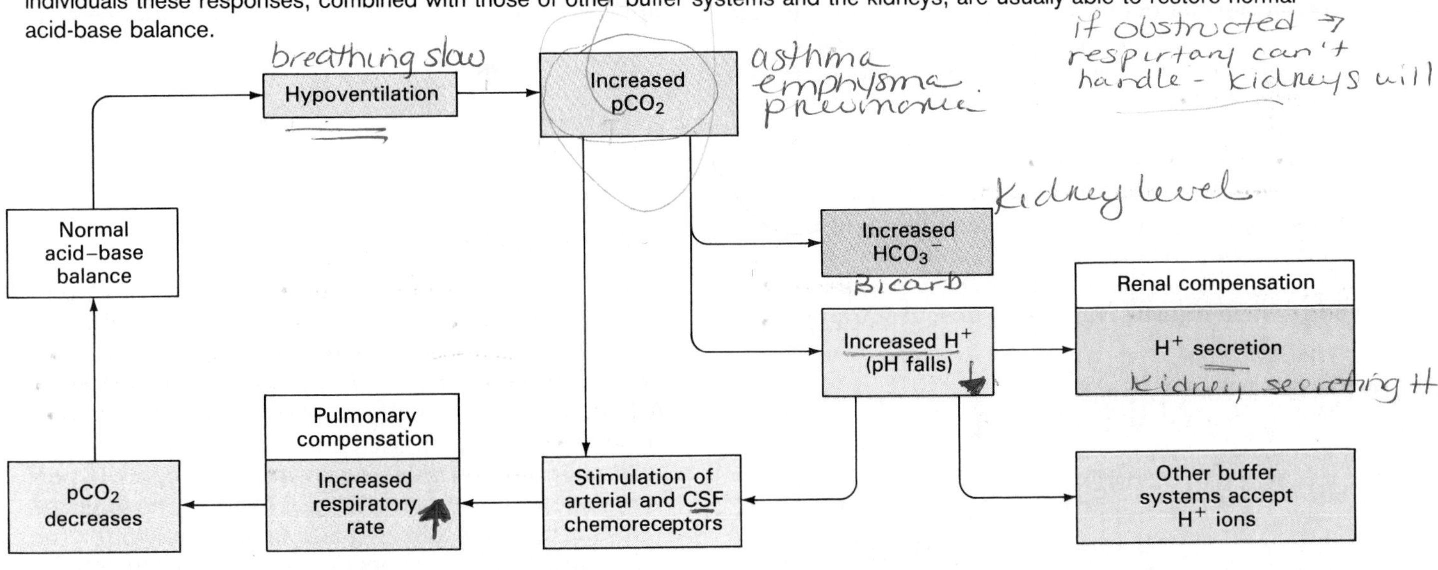

(a)

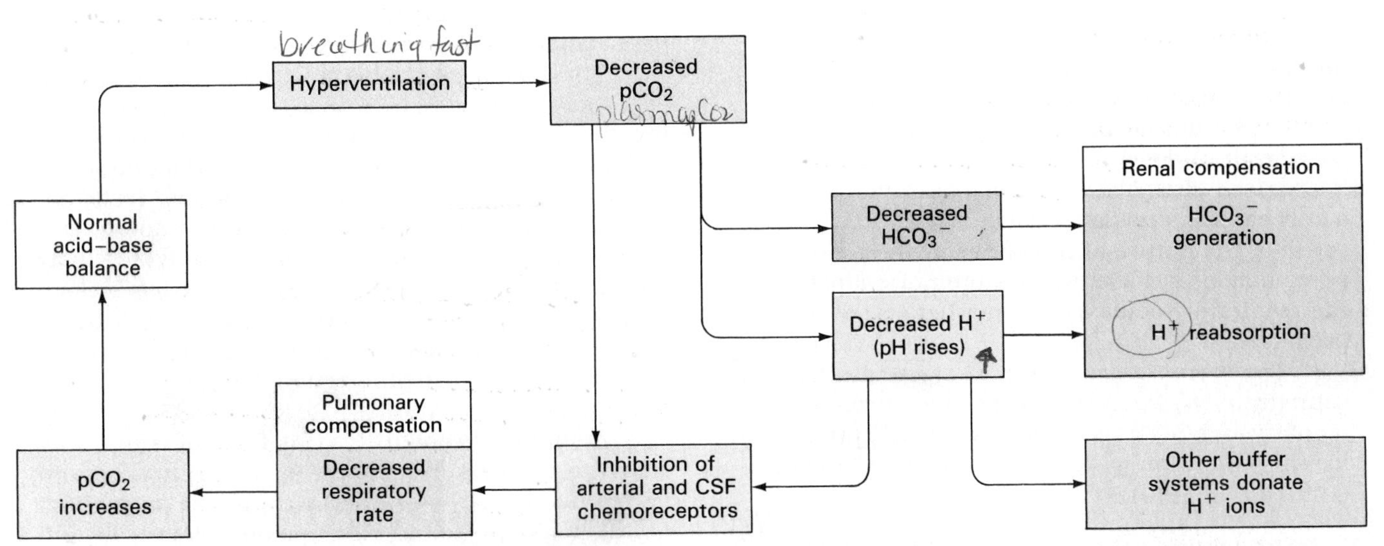

(b)

Respiratory acidosis represents the most frequent challenge to acid-base equilibrium. The usual cause is *hypoventilation*, an abnormally low respiratory rate. (Hypoventilation and hyperventilation, and their effects on pCO_2, were discussed in Chapter 23.) (p. 779) Our tissues generate carbon dioxide at a rapid rate, and even a few minutes of hypoventilation can cause acidosis, reducing the pH of the ECF to as low as 7.0. Under normal circumstances the chemoreceptors monitoring the pCO_2 of the plasma and CSF will eliminate the problem by calling for an increase in pulmonary ventilation rates.

If the chemoreceptors fail to respond, if pulmonary ventilation cannot be increased, or if the circulatory supply to the lungs is inadequate, the pH continues to decline. If the decline is severe, **acute respiratory acidosis** will develop. Acute respiratory acidosis is an immediate, life-threatening condition. It is especially dangerous in persons whose tissues are generating large amounts of CO_2 or who are incapable of normal respiratory activity. For this reason reversing acute respiratory acidosis is probably the major goal in the resuscitation of cardiac arrest or drowning victims. Thus first aid, CPR, and life-saving courses always stress the "ABCs" of emergency care: *Airway*, *Breathing*, and *Circulation*.

Chronic respiratory acidosis develops when normal respiratory function has been compromised but the compensatory mechanisms have not failed completely. Individuals suffering from central nervous system injuries, or those whose respiratory centers have been desensitized by drugs such as alcohol or barbiturates, do not respond to the warning signals from the chemoreceptors. As a result, these people are prone to develop acidosis due to chronic hypoventilation. Even with intact and functional respiratory centers, increased pulmonary exchange may be prevented by damage to some components of the respiratory system. Examples of conditions fostering chronic respiratory acidosis include emphysema, congestive heart failure, and pneumonia (in which alveolar damage or blockage typically occurs). Pneumothorax and respiratory muscle paralysis have a similar effect, for they too limit the ability to maintain adequate pulmonary ventilation rates.

When a normal pulmonary response does not occur, the kidneys respond by increasing the rate of hydrogen ion secretion into the filtrate. This response slows the rate of pH change, but renal mechanisms alone cannot return the pH to normal until the underlying respiratory or circulatory problems are corrected.

Clinical Comment: Treatment of Respiratory Acidosis

In respiratory acidosis, the primary problem is that the rate of pulmonary exchange is inadequate to keep the arterial pCO_2 within normal limits. The efficiency of pulmonary ventilation can often be improved temporarily by inducing bronchodilation or using mechanical aids that provide air under positive pressure. If breathing has ceased altogether, artificial respiration or a mechanical ventilator will be required. Because of the nature of the problem, these measures may be sufficient to restore normal pH if the respiratory acidosis was neither severe nor prolonged.

Treatment of acute respiratory acidosis is complicated by the fact that it causes a complementary metabolic acidosis. (Treatment options for this condition are discussed in the Clinical Comment on p. 915.)

RESPIRATORY ALKALOSIS

Problems with **respiratory alkalosis** (Figure 27-14b) are relatively uncommon. Respiratory alkalosis develops when respiratory activity lowers plasma pCO_2 to below normal levels, a condition called **hypocapnia**. A temporary hypocapnia can be produced by hyperventilation, when increased respiratory activity leads to a reduction in the arterial pCO_2. Continued hyperventilation can elevate the pH to levels as high as 7.8–8. This condition usually corrects itself, for the reduction in pCO_2 removes the stimulation for the chemoreceptors, and the urge to breathe fades until carbon dioxide levels have returned to normal. *Respiratory alkalosis caused by hyperventilation seldom persists long enough to cause a clinical emergency.*

Physical or psychological stresses or conscious effort may produce an increased respiratory rate. During the period of hyperventilation the pH rises and CNS function becomes affected. The initial symptoms involve tingling sensations in the hands, feet, and lips. A light-headed feeling may also be noted, and if the hyperventilation continues the individual may lose consciousness. When this occurs, the psychological stimuli are removed, and the breathing rate declines. The pCO_2 then rises until the pH returns to normal. (There is more than enough O_2 in the exhaled air to prevent hypoxia.)

A simple treatment for this form of respiratory alkalosis consists of having the victim breathe and rebreathe the air contained in a small paper bag. As the pCO_2 in the bag rises, so do the alveolar and arterial carbon dioxide concentrations. This eliminates the problem and restores the pH to normal levels. Other problems with respiratory alkalosis are rare and primarily involve patients on mechanical respirators or those with injuries to the brainstem who are incapable of responding to alterations in plasma CO_2 concentrations.

METABOLIC ACIDOSIS

Metabolic acidosis is the second most common type of acid-base imbalance. It has three major causes:

1. The most frequent cause of metabolic acidosis is an impaired ability to excrete hydrogen ions at the kidneys. For example, conditions marked by severe kidney damage, such as glomerulonephritis, often result in a severe metabolic acidosis. [**CM:** *Glomerulonephritis*] Metabolic acidosis may also be caused by diuretics that turn off the sodium-hydrogen transport system in the kidney tubules. The secretion of hydrogen ions is linked to the reabsorption of sodium, and when sodium reabsorption stops so does hydrogen secretion.
2. A less common cause of metabolic acidosis is the production of a large number of fixed and/or organic acids. As diagrammed in Figure 27-15a, the hydrogen ions liberated by these acids overload the bicarbonate buffer system, and the pH begins to decline. Two important examples

FIGURE 27-15
Factors Causing Metabolic Acidosis. (a) Increased acid production or decreased acid secretion leads to a buildup of H^+ in body fluids and a fall in pH. Pulmonary and renal compensation mechanisms can stabilize pH, but the blood chemistry remains abnormal until acid production or secretion returns to normal levels. (b) A loss of bicarbonate ions inactivates the carbonic acid-bicarbonate buffer system, leading to a fall in pH.

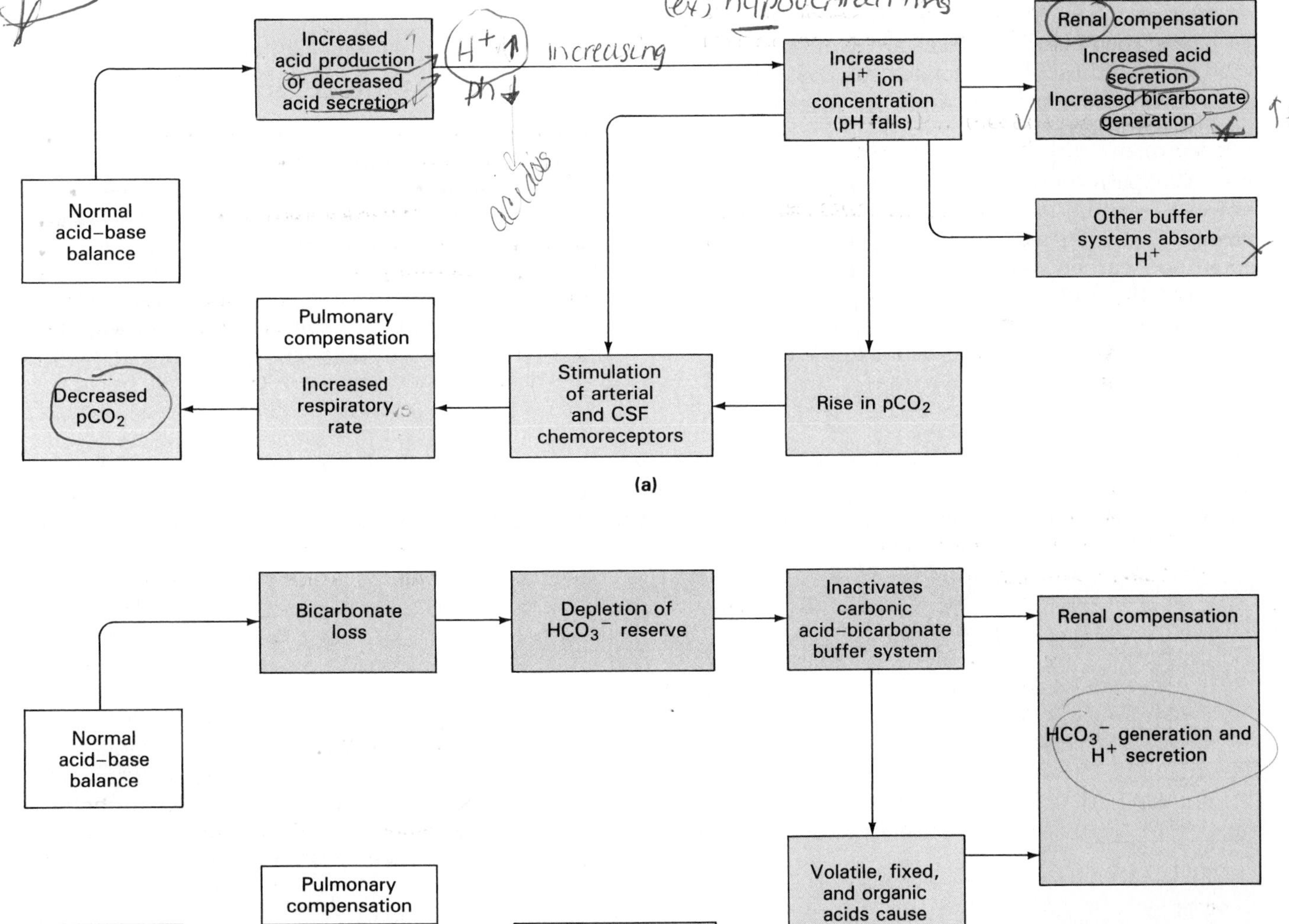

of metabolic acidosis were introduced in earlier chapters. **Lactic acidosis** may develop following severe exercise or prolonged tissue hypoxia (oxygen starvation) as active cells rely upon anaerobic respiration. (For a more detailed discussion, see Chapter 10, especially Figures 10-19 and 10-20.) (pp. 319, 321) **Ketoacidosis** results from the generation of large quantities of ketone bodies during the postabsorptive state. (For more details, see the related discussion in Chapter 25.) (p. 842)

3. Metabolic acidosis can occur following a severe bicarbonate loss (Figure 27-15b). The carbonic acid-bicarbonate buffer system relies on bicarbonate ions to balance hydrogen ions that threaten pH balance. A decline in the bicarbonate concentration in the ECF thus reduces the effectiveness of this buffer system, and acidosis soon develops. The most common cause of bicarbonate depletion is chronic diarrhea. Under normal conditions most of the bicarbonate ions secreted into the digestive tract in pancreatic, hepatic, and mucous secretions are reabsorbed before the feces are eliminated. In diarrhea these bicarbonates are lost, and the bicarbonate concentration in the ECF drops accordingly.

Compensation for metabolic acidosis usually involves a combination of respiratory and renal mechanisms. Hydrogen ions interacting with bicarbonate ions form carbon dioxide molecules that are eliminated at the lungs, while the kidneys excrete additional hydrogen ions into the urine and generate bicarbonate ions that are released into the ECF.

Clinical Comment: Combined Respiratory and Metabolic Acidosis

Metabolic and respiratory acidosis are frequently associated because sustained hypoventilation leads to decreased arterial pO_2, and oxygen-starved tissues generate large quantities of lactic acid. The problem can be particularly serious in cases of near-drowning, where the body fluids have high pCO_2, low pO_2, and large amounts of lactic acid generated by the muscles of the struggling victim. Prompt emergency treatment is essential, and the usual procedure involves some form of artificial or mechanical respiratory assistance coupled with the intravenous infusion of an isotonic solution containing sodium lactate, sodium gluconate (another weak base), or sodium bicarbonate.

METABOLIC ALKALOSIS

Metabolic alkalosis occurs when bicarbonate ion concentrations become elevated, as in Figure 27-16. The bicarbonate ions then interact with hydrogen ions in solution, forming carbonic acid. The reduction in H^+ concentrations then causes symptoms of alkalosis.

Cases of metabolic alkalosis are relatively rare, but one particularly interesting cause was noted in Chapter 24. (p. 797) The secretion of hydro-

FIGURE 27-16
Metabolic Alkalosis. Metabolic alkalosis most often results from the loss of acids, especially stomach acids lost through vomiting. During generation of replacement gastric acids, the alkaline tide introduces large numbers of bicarbonate ions into the bloodstream (see Figure 24-12). This leads to an increase in pH.

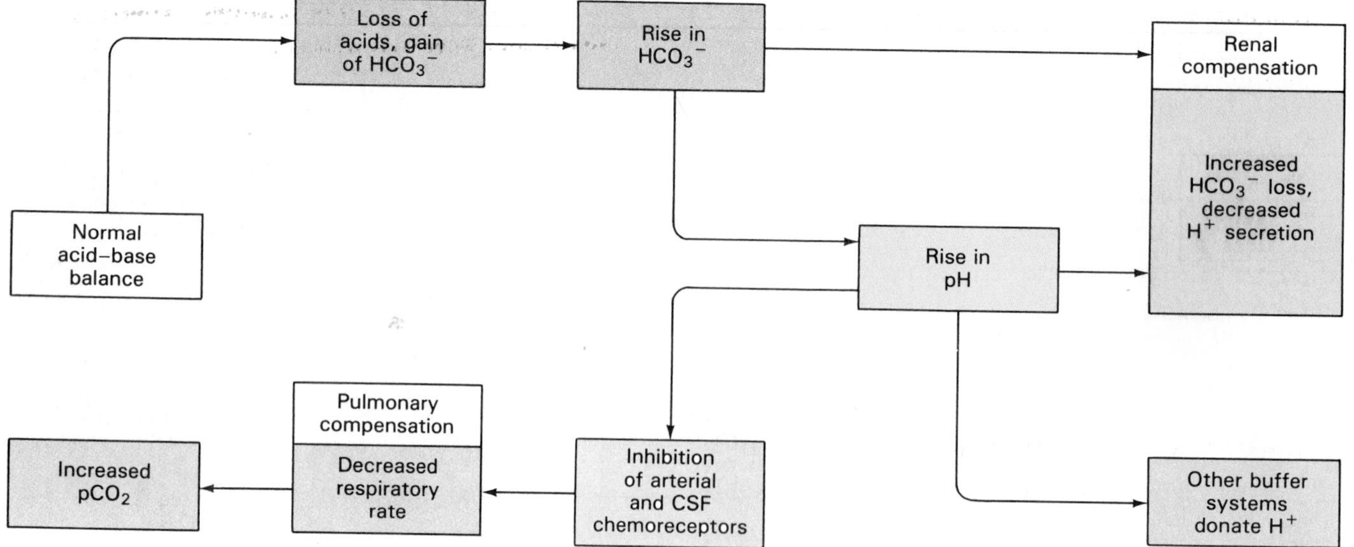

TABLE 27-2 Changes in Blood Chemistry Associated with the Major Classes of Acid-Base Disorders

Disorder	*pH*	*[HCO_3^-]*	*pCO_2*	*Remarks*	*Treatment*
Respiratory acidosis	**decreased**	**increased**	**increased**	**Usually caused by hypoventilation and CO_2 buildup in tissues and blood**	**Improve ventilation, sometimes with bronchodilation and mechanical assistance (PEEP, resuscitators)**
Metabolic acidosis	**decreased**	**decreased**	**decreased (as H^+ stimulates hyperventilation)**	**Caused by organic or fixed acid buildup, impaired H^+ excretion at kidneys, or bicarbonate loss in urine or feces**	**Bicarbonate administration (gradual) with other steps as needed to correct primary cause**
Respiratory alkalosis	**increased**	**decreased**	**decreased**	**Usually caused by hyperventilation and reduction in plasma CO_2 levels**	**Reduce respiratory rate, allow rise in pCO_2**
Metabolic alkalosis	**increased**	**increased**	**increased**	**Usually caused by prolonged vomiting and associated acid loss**	**pH below 7.55, no treatment; pH above 7.55, may require ammonium chloride administration**

chloric acid by the gastric mucosa is associated with the influx of large numbers of bicarbonate ions into the extracellular fluid. This phenomenon, known as the *alkaline tide*, temporarily elevates the bicarbonate concentration in the ECF at mealtimes. An individual who begins vomiting repeatedly will continue to generate stomach acids to replace those lost, and the ECF bicarbonate concentration will therefore continue to rise. Compensation for metabolic alkalosis involves a reduction in pulmonary ventilation, coupled with the increased loss of bicarbonates in the urine. [**CM:** *Classifying Acid-Base Disorders*]

The alterations in blood chemistry characteristic of the major types of acid-base disorders are summarized in Table 27-2.

- **What effect would a decrease in the pH of the body fluids have on the respiratory rate?**
- **Why does the renal filtrate need to be buffered.**
- **How would a prolonged fast affect the body's pH?**
- **Prolonged vomiting can lead to a condition known as alkalosis. How does this occur?**

Clinical Comment: Treatment of Metabolic Alkalosis

Metabolic alkalosis is a rarity, but it can be a serious problem following prolonged vomiting and the loss of gastric acids. Treatment of mild cases addresses the primary cause, in this case controlling vomiting. Treatment of acute cases may involve the administration of ammonium chloride, NH_4Cl. Metabolism of the ammonium ion in the liver results in the liberation of a hydrogen ion, so in effect the introduction of ammonium chloride leads to the internal generation of hydrochloric acid, HCl, a strong acid. As the HCl diffuses into the bloodstream, the pH falls toward normal levels.

CHAPTER REVIEW

Review of Selected Clinical Terms

respiratory acidosis (*p. 912*): Acidosis resulting from inadequate respiratory activity, characterized by elevated levels of carbon dioxide (hypercapnia) in body fluids.

respiratory alkalosis (*p. 912*): Alkalosis due to excessive respiratory activity, which depresses carbon dioxide levels and elevates the pH of body fluids.

metabolic acidosis (*p. 913*): A type of acidosis caused by the inability to excrete hydrogen ions, the production of numerous fixed and/or organic acids, or a severe bicarbonate loss.

metabolic alkalosis (*p. 914*): A rare form of alkalosis resulting from high concentrations of bicarbonate ions in body fluids.

Additional Terms of Clinical Importance

RESPIRATORY ACIDOSIS (*p. 912*): **acute respiratory acidosis, chronic respiratory acidosis**

METABOLIC ACIDOSIS (*p. 913*): **lactic acidosis, ketoacidosis**

Study Outline

Introduction (pp. 895–897)

1. All of our cells' operations depend on water as a diffusion medium for dissolved gases, nutrients, and waste products. Maintenance of normal volume and composition in the extracellular and intracellular fluids is vital to life. Three types of homeostasis are involved: **fluid balance**, **electrolyte balance**, and **acid-base balance**.

Fluid and Electrolyte Balance (pp. 897–904)

1. The **intracellular fluid (ICF)** contains nearly two-thirds of the total body water; the **extracellular fluid (ECF)** contains the rest. Exchange occurs between the ICF and ECF, but the two **fluid compartments** retain their distinctive characteristics.

BASIC CONCEPTS PERTAINING TO FLUID AND ELECTROLYTE REGULATION (p. 898)

2. Homeostatic mechanisms that monitor and adjust the composition of body fluids respond to changes in the ECF, not the ICF.
3. Receptors involved in fluid and electrolyte balance respond to changes in plasma volume or osmolarity
4. Our cells cannot move water by active transport; all water movements across cell membranes and epithelia occur passively in response to osmotic gradients.
5. The body content of water or electrolytes rises if intake exceeds outflow, and falls if losses exceed gains.

AN OVERVIEW OF THE PRIMARY REGULATORY HORMONES (pp. 898–899)

6. ADH encourages water resorption at the kidneys and stimulates thirst. Aldosterone increases the rates of sodium resorption at the kidneys. ANP opposes these actions and promotes fluid and electrolyte losses in the urine.

FLUID BALANCE (pp. 899–900)

7. Water circulates freely within the ECF compartment. At capillary beds hydrostatic pressure forces water from the plasma into the interstitial spaces. Water moves back and forth across the mesothelial lining of the peritoneal, pleural, and pericardial cavities; through synovial membranes lining joint capsules; and between the blood and cerebrospinal fluid, the aqueous and vitreous humors of the eye, and the perilymph and endolymph of the inner ear.
8. Water losses are normally balanced by gains through eating, drinking, and metabolic generation.
9. If the ECF becomes hypertonic relative to the ICF, water will move from the ICF into the ECF (a **fluid shift**) until osmotic equilibrium has been restored. If the ECF becomes hypotonic compared to the ICF, water will move from the ECF into the cells and the volume of the ICF will increase accordingly.

Related Key Terms

10. Gaining water without electrolytes will lower the osmolarities of the ECF and ICF. Losing water without electrolytes will increase the osmolarities of both compartments. Compensation occurs primarily through changes in the rate of ADH secretion.

ELECTROLYTE BALANCE (pp. 900–904)

11. Electrolyte balance is important because total electrolyte concentrations affect water balance, and because the levels of individual electrolytes can affect a variety of cell functions.
12. Problems with electrolyte balance generally result from an imbalance between sodium gains and losses. Problems with potassium balance are less common but more dangerous.
13. The rate of sodium uptake across the digestive epithelium is directly proportional to the amount of sodium in the diet. Sodium losses occur mainly in the urine and through perspiration.
14. Alterations in sodium balance result in the expansion or contraction of the ECF. Large variations in the ECF volume are corrected by the homeostatic mechanisms triggered by changes in blood volume. If the volume becomes too low, ADH and aldosterone are secreted; if the volume becomes too high, ANP is secreted.

POTASSIUM ION REGULATION (p. 904)

15. Potassium ion concentrations in the ECF are very low, and not as closely regulated as sodium ion concentrations. Potassium excretion increases as ECF concentrations rise, under aldosterone stimulation, and when the pH rises. Potassium retention occurs when the pH falls.

Acid-Base Balance (pp. 904–909)

1. The pH of normal body fluids ranges from 7.35 to 7.45; variations outside this relatively narrow range produces **acidosis** or **alkalosis**.

ACIDS IN THE BODY (pp. 905–906)

2. **Volatile acids** can leave solution and enter the atmosphere; **fixed acids** remain in body fluids until excreted at the kidneys; **organic acids** are acid participants in or byproducts of cellular metabolism.
3. Carbon dioxide, a volatile acid, is the most important factor affecting body tissue pH. In solution CO_2 reacts with water to form carbonic acid; the dissociation of carbonic acid releases H^+ ions.
4. **Sulfuric acid** and **phosphoric acid**, the most important fixed acids, are generated during the catabolism of amino acids and compounds containing phosphate groups.
5. Organic acids include metabolic products such as lactic acid and ketone bodies.
6. A **buffer system** consists of a weak acid and its dissociation products. There are three major buffer systems: (1) **protein buffer systems** in the ECF and ICF; (2) the **carbonic acid-bicarbonate buffer system**, most important in the ECF; and (3) the **phosphate buffer system** in the intracellular fluids and urine.

bicarbonate reserve

7. The **hemoglobin buffer system** is a protein buffer system that helps prevent drastic changes in pH when the pCO_2 is rising or falling. The carbonic acid-bicarbonate buffer system prevents pH changes due to organic acids and fixed acids in the ECF. The phosphate buffer system plays a supporting role in regulating the pH of the ECF, but is important in buffering the pH of the ICF.

MAINTENANCE OF ACID-BASE BALANCE (pp. 908–909)

8. Pulmonary and renal mechanisms support the actions of buffer systems by secreting or absorbing hydrogen ions, controlling the excretion of acids and bases, and generating additional buffers.
9. The lungs help regulate pH by affecting the carbonic acid-bicarbonate buffer system; changing the respiratory rate can raise or lower the pCO_2 of body fluids, affecting the buffering capacity. This process is called **respiratory compensation**.
10. In the process of **renal compensation** the kidneys vary their rates of hydrogen ion secretion and bicarbonate ion resorption depending on the pH of extracellular fluids.

Disturbances of Acid-Base Balance (pp. 909–915)

1. **Respiratory disorders** result when abnormal respiratory function causes an extreme rise or fall in CO_2 levels in the ECF. **Metabolic disorders** are caused by the generation of organic or fixed acids,

or by conditions affecting the concentration of bicarbonate ions in the ECF.

Related Key Terms

RESPIRATORY ACIDOSIS (pp. 911–912)

2. **Respiratory acidosis** results from excessive levels of carbon dioxide in body fluids. **Chronic respiratory acidosis** develops when compensatory mechanisms have not completely failed. If normal homeostatic adjustments do not occur, **acute respiratory acidosis** develops. — **hypercapnia**

RESPIRATORY ALKALOSIS (pp. 912–913)

3. **Respiratory alkalosis** is a relatively uncommon condition associated with hyperventilation. — **hypocapnia**

METABOLIC ACIDOSIS (pp. 913–914)

4. **Metabolic acidosis** results from depletion of the bicarbonate reserve. It is caused by an inability to excrete hydrogen ions at the kidneys, the production of large numbers of fixed and organic acids, or the bicarbonate loss that accompanies chronic diarrhea. — **lactic acidosis • ketoacidosis**

METABOLIC ALKALOSIS (pp. 914–915)

5. **Metabolic alkalosis** occurs when bicarbonate ion concentrations become elevated, as from extended periods of vomiting.

Review Planner

Objective	Level 1
1 Describe how water and electrolytes are distributed within the body.	**1 22 24**
2 Explain the homeostatic mechanisms involved in the control of the fluid and electrolyte composition of body fluids.	**2 4 10 11 14 18 21 30**
3 Discuss the ways in which the body compensates for sudden losses or gains of water and/or electrolytes.	**2 3 4 5 10 14 18 21 32**
4 Explain the homeostatic mechanisms that stabilize the pH of intracellular and extracellular fluids	**6 12 13 15 16 17 19 20 23 26 28 29 31**
5 Identify the most frequent threats to acid-base balance and explain how the body responds when the pH varies outside normal limits.	**6 7 8 9 33 34 35**

Level 2	**25 27 30**
Level 3	**36–37**
C M	Water Excess and Depletion Hyperkalemia and Hypokalemia
C M	Classifying Acid-Base Disorders

Review Questions

MULTIPLE CHOICE

1. About two-thirds of the body's water is found in the: a) intracellular fluid; b) extracellular fluid; c) blood plasma; d) interstitial fluid.
2. Plasma osmolarity rises when: a) even a small number of sodium ions are absorbed by the digestive epithelium; b) volume depletion occurs; c) water is lost without accompanying electrolytes; d) water is gained without accompanying electrolytes.
3. The most common problems with electrolyte balance result from an imbalance between gains and losses of: a) potassium; b) sodium; c) calcium; d) hydrogen ions.
4. When sodium ions are lost in excess: a) the ECF volume increases; b) the osmolarity decreases dramatically; c) the ECF volume decreases; d) the ICF volume decreases.
5. Most potassium is lost from the body: a) in sensible

perspiration; b) through absorption across the digestive epithelium; c) through interaction with a buffer system; d) in the urine.

6. A body fluid pH of 7.46 indicates: a) acidosis; b) alkalosis; c) normal acid-base balance.
7. Which of the following would *not* cause metabolic acidosis?: a) impaired sodium reabsorption at the kidneys; b) chronic diarrhea; c) prolonged hypoventilation; d) increased ketone body production.
8. "Hypercapnia" is typically seen in: a) respiratory acidosis; b) metabolic alkalosis; c) abnormally low pH in body fluids; d) respiratory alkalosis.

DIAGRAMS

9. Create a table that summarizes the following information for the four major categories of acid-base disorders: how each disorder affects pH, pCO_2 and HCO_3^- levels; frequent causes of each disorder; and usual treatments.

TRUE/FALSE

(*Your instructor may wish to have you correct the statement if it is false.*)

10. **T/F:** All of the homeostatic mechanisms that monitor fluid balance respond to changes in the ICF, not the ECF.
11. **T/F:** Our cells cannot move water molecules by active transport.
12. **T/F:** Carbon dioxide is an important organic acid found in the body.

MATCHING QUESTIONS

(*Match each term with the most appropriate statement.*)

Statements:

13. An acid that remains in body fluids until excreted at the kidneys.
14. A hormone that reduces thirst and blocks the release of two other hormones involved in regulating fluid and electrolyte balance.
15. The negative exponent of the hydrogen ion concentration in a solution.
16. This homeostatic mechanism draws on the body's reserve of $NaHCO_3$.
17. An acid that can leave solution and enter the atmosphere.
18. A hormone that reduces urinary water losses and stimulates the thirst center.
19. This homeostatic mechanism depends on the ability of amino acids to accept or release hydrogen ions as needed.
20. An acid that is a participant in or byproduct of cellular metabolism.
21. A hormone that stimulates sodium conservation.
22. Its principal ions are sodium, chloride, and bicarbonate.
23. It plays a relatively minor role in regulating the pH of the ECF, but is important in adjusting the pH of the ICF and urine.
24. Its principal ions are potassium, magnesium, and phosphate.

Terms:

a) ECF
b) ICF
c) ADH
d) ANP
e) Aldosterone
f) Protein buffer system
g) Carbonic acid-bicarbonate buffer system
h) Phosphate buffer system
i) pH
j) Volatile acid
k) Fixed acid
l) Organic acid

SHORT ANSWER/ESSAY

25. Define fluid, electrolyte, and acid-base balance, and explain why they are important to homeostasis.
26. Why does a loss in bicarbonate ions have the same effect on pH as a gain in hydrogen ions?
27. Define and give an example of a volatile acid, a fixed acid, and an organic acid. Which represent(s) the greatest threat to acid-base balance? Why?
28. Describe the effect(s) of adding or removing small numbers of hydrogen ions to this buffer system:

$$CO_2 + H_2O \rightleftharpoons H_2CO_3 \rightleftharpoons H^+ + HCO_3^-$$

29. Describe what would happen if we added or removed small numbers of bicarbonate ions to the buffer system in the previous question.
30. What are fluid shifts? What is their function, and what factors can cause them?
31. What are the three major buffer systems found in body fluids? How does each system work?
32. You play a brisk game of tennis on a hot day. Afterward, you're perspiring heavily and feel very thirsty, so you down a large glass of water. Describe the hormone(s) that were involved in regulating your fluid balance.
33. Which is more common, respiratory acidosis or respiratory alkalosis? Explain why.
34. Describe the pulmonary and renal responses to acid-base imbalance. Why are these important, even though buffer systems are present in body fluids?
35. Distinguish between metabolic and respiratory alkalosis. What can cause these conditions?

CLINICAL APPLICATIONS

36. Mr. Zaloudek, 67, has complained all afternoon of difficulty breathing. His family calls an ambulance, but he is unconscious by the time it reaches the hospital. An arterial blood gas test measures a plasma pH of 7.15 and a pCO_2 of 160. Identify Mr. Zaloudek's condition and discuss its cause.
37. Would Mr. Zaloudek's situation in the previous question affect his potassium balance? Explain.

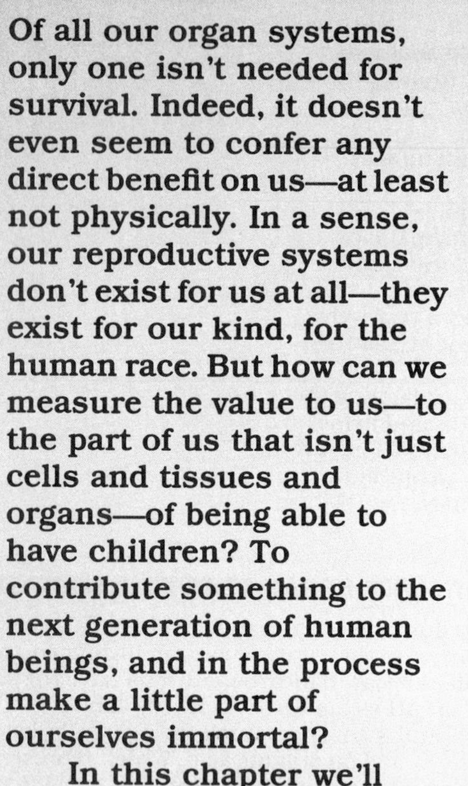

Of all our organ systems, only one isn't needed for survival. Indeed, it doesn't even seem to confer any direct benefit on us—at least not physically. In a sense, our reproductive systems don't exist for us at all—they exist for our kind, for the human race. But how can we measure the value to us—to the part of us that isn't just cells and tissues and organs—of being able to have children? To contribute something to the next generation of human beings, and in the process make a little part of ourselves immortal?

In this chapter we'll study the system that has enabled human beings to exist on this earth for a very long time—and, with luck, to stick around for at least a while longer.

CHAPTER 28

The Reproductive System

Chapter Objectives

After reading this chapter, you will be able to:

1 Summarize the functions of the human reproductive system and its principal components.
2 Describe the components of the male and female reproductive systems.
3 Detail the physiological processes involved in the ovarian and menstrual cycles.
4 Describe the hormonal mechanisms that regulate reproductive activities.
5 Discuss the production, storage, and transport of sex cells.
6 Describe the anatomical, physiological, and hormonal changes that accompany pregnancy.
7 Discuss the changes in the reproductive system that occur at puberty and with aging.
8 Explain how the reproductive system interacts with other body systems.

Introduction

An individual lifespan can be measured in decades, but the human species has survived for hundreds of thousands of years through the activities of the reproductive system. The entire process of reproduction seems almost magical; many primitive societies even failed to discover the basic link between sexual activity and childbirth and assumed that cosmic forces were responsible for producing new individuals. Although our society has a much clearer view of the reproductive process, a sense of wonder remains. Sexually mature males and females produce individual reproductive cells that are brought together by the sexual act. Their fusion starts a chain of events leading to the appearance of an infant that will mature as part of the next generation.

The next two chapters will consider the mechanics of this remarkable process. We will begin by examining the anatomy and physiology of the reproductive system. The human reproductive system produces, stores, nourishes, and transports functional male and female reproductive cells, or **gametes** (GAM-ēts). The next chapter begins with **fertilization**, the fusion of a **sperm** contributed by the father and an *egg*, or **ovum** (Ō-vum), from the mother. All of the cells in a human body are the mitotic descendants of a single fertilized egg, or **zygote** (ZĪ-gōt). Over the next 15–20 years that zygote will become a functional adult. This gradual transformation, the main focus of Chapter 29, is the process of *development*.

An Overview of the Reproductive System

The reproductive system includes:

- reproductive organs, or **gonads** (GŌ-nads), that produce gametes and hormones,
- ducts that receive and transport the gametes,
- accessory glands and organs that secrete fluids into these or other excretory ducts, and
- perineal structures associated with the reproductive system. These perineal structures are collectively known as the **external genitalia** (jen-i-TĀ-lē-a), or the *pudendum* (pū-DEN-dum; *pudere*, to be ashamed of).

The functional roles of the male and female reproductive systems are quite different. In the adult male the gonads, or **testes** (TES-tēz; singular *testis*), produce large numbers of sperm, perhaps as many as a half a billion per day. After storage, mature sperm travel along a lengthy duct system where they are mixed with the secretions of accessory glands. This creates the mixture known as **semen** (SĒ-men). During **ejaculation** (e-jak-ū-LĀ-shun) the semen is expelled from the body.

The gonads, or **ovaries**, of a sexually mature female usually produce only one mature egg per month. This gamete travels along short **uterine tubes** (*oviducts*) that terminate in a muscular chamber, the **uterus** (Ū-ter-us). A short passageway, the **vagina** (va-JĪ-na), connects the uterus with the exterior. During the sexual act the male ejaculation introduces semen into the vagina, and the sperm cells ascend the female reproductive tract. If a single sperm contacts the cell membrane of an egg, the two may fuse. This process, called fertilization, produces a zygote that subsequently undergoes repeated mitoses. After entering the uterus the resulting cell mass attaches to the uterine wall, developing over the next 9 months into a fetus containing billions of specialized cells. At birth the fetus travels down the vaginal canal and enters the world as a newborn infant.

We will now examine the anatomy of the male and female reproductive systems in detail and consider the physiological and hormonal mechanisms responsible for the homeostatic regulation of reproductive function. The anatomical reference points used in the following discussions were introduced in earlier chapters, and you may find it helpful to review the figures on the pelvic girdle (Figures 9-11 and 9-12), perineal musculature (Figure 11-10), pelvic innervation (Figure 13-10), and the regional blood supply (Figures 21-13 and 21-17). (pp. 281, 282, 344, 416, 678, 683)

FIGURE 28-1
Components of the Male Reproductive System. The male reproductive organs as seen in sagittal section.

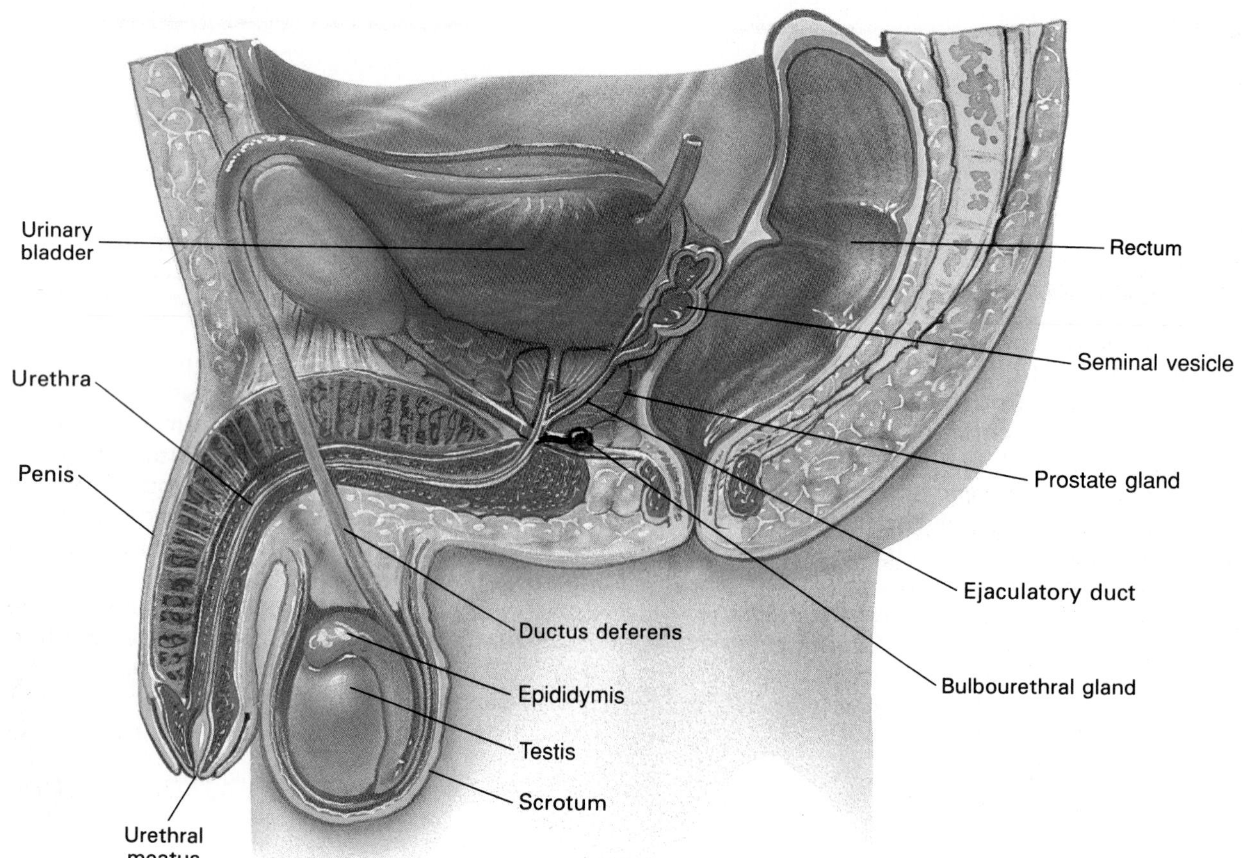

The Reproductive System of the Male

The principal components of the male reproductive system are indicated in Figure 28-1. Proceeding from the testes, the sperm cells, or **spermatozoa** (sper-ma-tō-ZŌ-a), travel along the **epididymis** (ep-i-DID-i-mus), the **ductus deferens** (DUK-tus DEF-e-renz), or *vas deferens*, the **ejaculatory** (ē-JAK-ū-la-tō-rē) **duct**, and the **urethra** before leaving the body. Accessory organs, notably the **seminal** (SEM-i-nal) **vesicles**, the **prostate** (PROS-tāt) **gland**, and the **bulbourethral** (bul-bō-ū-RĒ-thral) **glands** secrete into the ejaculatory ducts and urethra. Externally visible structures include the **scrotum** (SKRŌ-tum), which encloses the testes, and the **penis** (PĒ-nis), an erectile organ that surrounds the distal portions of the urethra. Together the scrotum and penis constitute the external genitalia of the male.

THE TESTES

Each testis has the shape of a flattened oval roughly 5 cm (2 in.) in length and 2.5 cm (1 in.) in width, with a weight of 10–15 g (0.35–0.53 oz.). The testes hang within the scrotum, a fleshy pouch suspended below the perineum anterior to the anus.

Descent of the Testes

During development, the testes form inside the body cavity adjacent to the kidneys (see the Embryology Summary on page 946). The relative positions of these organs change as the fetus enlarges, and the testes gradually move caudally toward the anterior abdominal wall. Usually by the end of the seventh developmental month they have moved through the abdominal musculature, accompanied by small pockets of the peritoneal cavity. This process, known as the **descent of the testes**, is diagrammed in Figure 28-2. [**CM**: *Cryptorchidism*]

As it moves through the body wall, each testis is accompanied by the ductus deferens and the blood vessels, nerves, and lymphatics that service the organ. Together these structures form the body of the **spermatic cord**. The narrow canals linking the scrotal chambers with the peritoneal cavity are called the **inguinal canals**. These passageways usually disappear, but their formation creates weak points in the abdominal wall that remain throughout life. As a result *inguinal hernias*, discussed in Chapter 11, are relatively common in males. (p. 345)

Position and Orientation of the Testes

Figure 28-3a details the orientation and position of the testes within the scrotum. The scrotum is divided into two separate chambers, and the boundary between the two is marked by a raised thickening in the scrotal surface known as the **raphe** (RA-fē). Each testis lies in a separate compartment, with a narrow space separating the inner surface of the scrotum from the outer surface of the testis. A mesothelial lining reduces friction between the opposing surfaces.

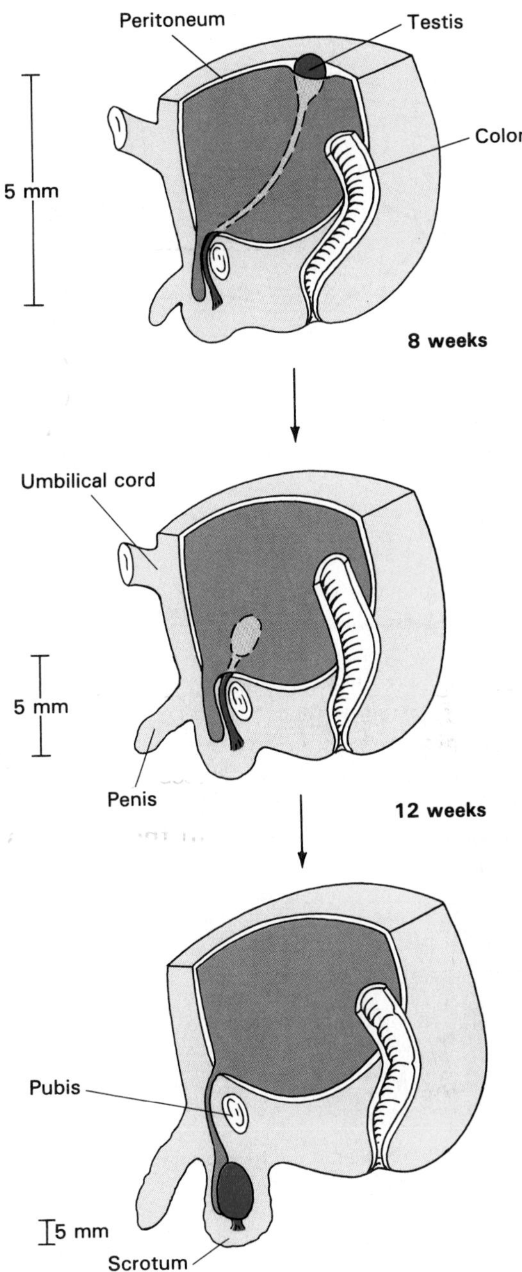

FIGURE 28-2
The Descent of the Testes

The scrotum consists of a thin layer of skin and the underlying superficial fascia. The dermis contains a layer of smooth muscle, the **dartos** (DAR-tōs), and tonic contraction of the dartos muscle causes the

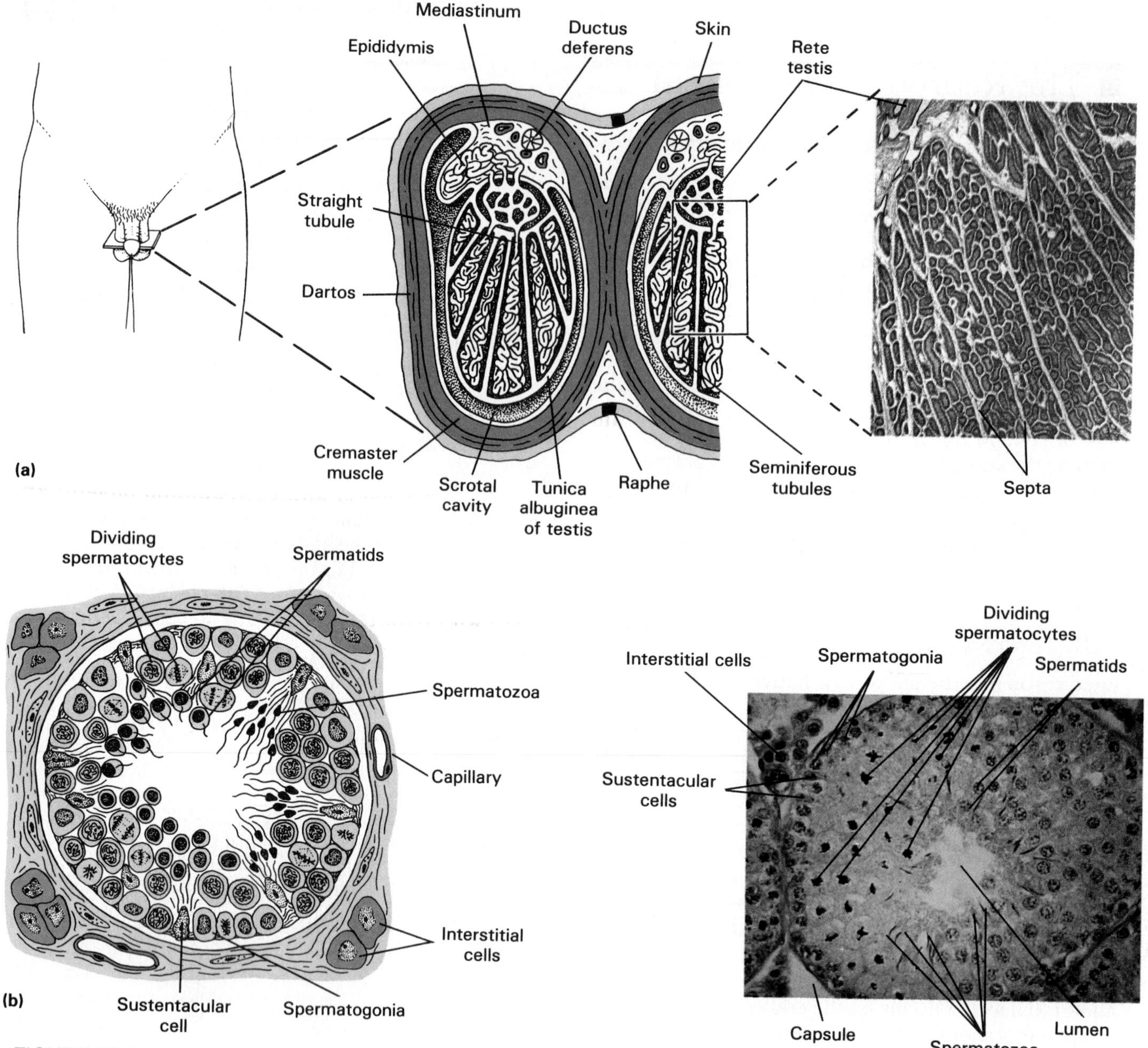

FIGURE 28-3
Structure of the Testes. (a) Diagrammatic sketch and anatomical relationships of the testes. (LM, ×15) (b) A single seminiferous tubule seen in cross section. (LM, ×756)

characteristic wrinkling of the scrotal surface. A layer of skeletal muscle, the **cremaster** (kre-MAS-ter) **muscle**, lies beneath the dermis. Contraction of the cremaster tenses the scrotum and pulls the testes closer to the body. The degree of tension in the cremaster muscle is closely regulated, for normal sperm development in the testes requires temperatures around 2°F lower than those found elsewhere in the body. Cooling the scrotum, as when entering an icy swimming pool, results in cremasteric contractions that pull the testes closer to the body and keep testicular temperatures from falling. When environmental temperatures rise, the cremaster relaxes. The testes then move away from the body, and excess heat is lost across the surface of the scrotum. A mesothelial layer (the *tunica vaginalis*) lines the inner surface of the cremasteric muscle layer.

Structure of the Testes

Beneath the mesothelium covering the testis lies the **tunica albuginea** (TŪ-ni-ka al-bū-JIN-ē-a), a dense layer of connective tissue rich in collagen fibers. The fibers of this network are continuous with those surrounding the adjacent *epididymis*, a highly convoluted tube where sperm remain while they complete their functional maturation. The collagen fibers of the tunica albuginea also extend into the substance of the testis, forming fibrous partitions, or *septa*. These septa converge toward the **mediastinum** of the testis, the region adjacent to the epididymis.

The septa subdivide the testis into a series of **lobules**. Roughly 800 slender, tightly coiled **seminiferous** (se-mi-NIF-e-rus) **tubules** are distributed among the lobules. Each tubule averages around 80 cm (31 in.) in length, and a typical testis contains nearly half a mile of seminiferous tubules. Sperm production occurs within these tubules.

Each seminiferous tubule has the form of a U connected to a single **straight tubule** that enters the mediastinum of the testis. Within the mediastinum these tubules are extensively interconnected, forming a maze of passageways known as the **rete** (RĒ-tē) **testis**. Fifteen to twenty large **efferent ducts** connect the rete testis to the epididymis.

Because the seminiferous tubules are tightly coiled, histological preparations most often show them in transverse section. Each tubule is surrounded by a delicate capsule, and loose connective tissue fills the spaces between the tubules. Within those spaces are found numerous blood vessels and large **interstitial cells** (*cells of Leydig*). The interstitial cells are responsible for the production of male sex hormones, called *androgens*. The steroid *testosterone* is the most important androgen. (Testosterone and other sex hormones were introduced in Chapter 18.) (p. 596)

Sperm cells, or spermatozoa, are produced through the process of **spermatogenesis** (sper-ma-tō-JEN-e-sis). Spermatogenesis begins at the outermost layer of cells in the seminiferous tubules. Stem cells called **spermatogonia** (sper-ma-to-GŌ-nē-a) divide to produce cells that are gradually pushed toward the lumen of the tubule. Over the next 3 weeks the relatively large **spermatocytes** (sper-MA-to-sīts) produced through spermatogonial divisions enter *meiosis*, a form of cell division that produces gametes containing one-half the normal complement of chromosomes. (Meiosis will be detailed in Chapter 29.) In this process spermatocyte divisions generate smaller **spermatids** (SPER-ma-tidz), and each spermatid matures into a single **spermatozoon** (sper-ma-tō-ZŌ-on), or sperm cell. These cell types are shown in Figure 28-3c,d.

There are no blood vessels inside the seminiferous tubules, and all nutrients must enter by diffusion from the surrounding interstitial fluids. Large **sustentacular** (sus-ten-TAK-ū-lar) **cells** (*Sertoli cells*) are attached to the tubular capsule and extend toward the lumen between the spermatocytes and developing spermatids. Sustentacular cells have four important functions:

1. *Maintenance of the blood-testis barrier*: The seminiferous tubules are isolated from the general circulation by a **blood-testis barrier** comparable to the blood-brain barrier described in Chapters 12 and 14. (pp. 374, 453) Extensions of sustentacular cells form a layer that surrounds the seminiferous tubule beneath the spermatogonia. Tight junctions between adjacent sustentacular cells prevent free diffusion from the interstitial fluid to the seminiferous tubule. Transport across the sustentacular cells is tightly regulated so that conditions inside the tubule remain very stable.

2. *Support of spermiogenesis*: **Spermiogenesis** is the physical transformation of a spermatid to a spermatozoon. Spermiogenesis requires the presence of sustentacular cells because these cells surround and enfold the spermatids, providing nutrients and chemical stimuli that promote their development.

3. *Secretion of inhibin*: Sustentacular cells secrete a hormone, **inhibin** (in-HIB-in), across the capsular wall of the tubule. Inhibin, introduced in Chapter 18, depresses the pituitary production of follicle-stimulating hormone (FSH) and gonadotropin-releasing hormone (GnRH). (p. 577) The faster the rate of spermatogenesis, the greater the amount of inhibin secreted. Inhibin thus represents a component of a negative feedback control mechanism that determines the maximum rate of sperm production. (This mechanism will be described in more detail in a later section.)

4. *Secretion of androgen-binding protein*: **Androgen-binding protein** (**ABP**) binds androgens (primarily testosterone) in the fluid contents of the seminiferous tubules. This protein is thought to be important in elevating the concentration of androgens within the tubules and stimulating spermiogenesis.

ANATOMY OF A SPERMATOZOON

The goal of this activity in the testes is the production of physically mature spermatozoa. The other portions of the male reproductive system are concerned with the functional maturation, nourishment, and transport of spermatozoa. Figure 28-4 presents several views of a typical spermatozoon. There are three distinct regions to each sperm cell: the *head*, the *middle piece*, and the *tail*. The **head** is a flattened oval. Densely packed chromosomes occupy most of the volume of the head. The tip forms the **acrosomal** (ak-rō-SŌ-mal) **cap**, which contains an enzyme that plays a role in fertilization. A very short **neck** attaches the head to the **middle piece**. The neck contains both centrioles of the original spermatid, oriented at right angles. The microtubules of the distal centriole (the one farthest from the head) are continuous with those of the middle piece and tail. Throughout the middle piece these microtubules are surrounded by an inner layer of dense fibers and an outer layer of mitochondria. The spirally arranged mitochondria provide the energy needed to move the tail.

The fat middle piece ends where the mitochondrial spiral stops, but the dense fibers extend almost to the tip of the tail. The entire **tail** is surrounded by a fibrous sheath. The spermatozoon tail is the only flagellum in the human body. In its internal organization the tail resembles an elongate cilium, but while cilia move materials past an immobile cell, a flagellum moves a cell from one place to another. Although cilia beat in a predictable, waving fashion, the tail of a spermatozoon has a complex, corkscrew motion.

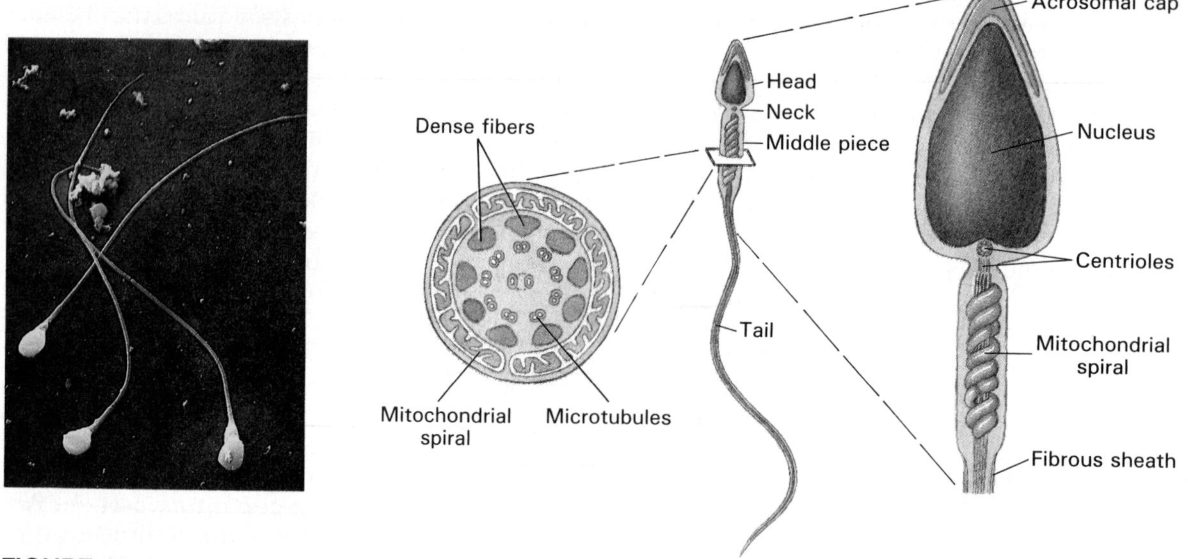

FIGURE 28-4
Spermatozoon Structure. The basic structure of a single spermatozoon as seen with the electron microscope is compared with diagrammatic views highlighting aspects of internal organization described in the text. (SEM, ×8750)

The entire streamlined package measures only 60 μ m in total length. Unlike other, less specialized cells, a mature spermatozoon lacks an endoplasmic reticulum, Golgi apparatus, lysosomes, peroxisomes, and inclusions, among other structures. Because the cell does not contain glycogen or other energy reserves, it must absorb nutrients, primarily fructose, from the surrounding fluid.

It takes about 3 weeks for an individual spermatozoon to complete its development within the seminiferous tubules. To reach the exterior, the spermatozoa produced in each testis must travel along a duct system that averages 8 meters in length. We will now examine each of the various subdivisions of that duct system.

THE EPIDIDYMIS

Late in their development the spermatozoa become detached from the sustentacular cells and lie within the lumen of the seminiferous tubule. Although they have most of the physical characteristics of mature sperm cells, they are still functionally immature and incapable of coordinated locomotion. Fluid currents then transport the cell along the straight tubule, through the rete testis, and into the **epididymis**.

As indicated in Figure 28-5, the epididymis lies along the posterior border of the testis. It has a firm texture and can be felt through the skin of the scrotum. The epididymis consists of an elongate tubule, almost 7 meters (23 ft.) long, so twisted and coiled that it actually takes up very little space. The epididymis can be divided into a *head*, *body*, and *tail*. The superior **head** includes the portions of the efferent ducts as they leave the testis. The **body** begins distal to the last of the efferent ducts and continues to the inferior tail. In the **tail** the number of convolutions decreases, and the histological organization changes until it becomes indistinguishable from that of the attached ductus deferens.

The epididymis has three important functions:

1. *It monitors and adjusts the composition of the tubular fluid.* The columnar epithelial lining of the epididymis (Figure 28-5b,c) bears distinctive elongate microvilli, or *stereocilia*, that increase the surface area available for absorption and secretion into the tubular fluid.
2. *It acts as a recycling center for damaged spermatozoa.* Cellular debris and damaged spermatozoa are absorbed, and the products of enzymatic breakdown are released into the surrounding interstitial fluids for pickup by the circulation.
3. *It is the site of physical maturation of spermatozoa.* It takes about 2 weeks for a spermatozoon to transit the epididymis, and during this period it completes its physical maturation. The spermatozoa in the epididymis remain immotile, and transport along the epididymis probably involves some combination of fluid movement and peristaltic contractions of smooth muscle layers. After passing along the tail of the epididymis the spermatozoa arrive at the ductus deferens.

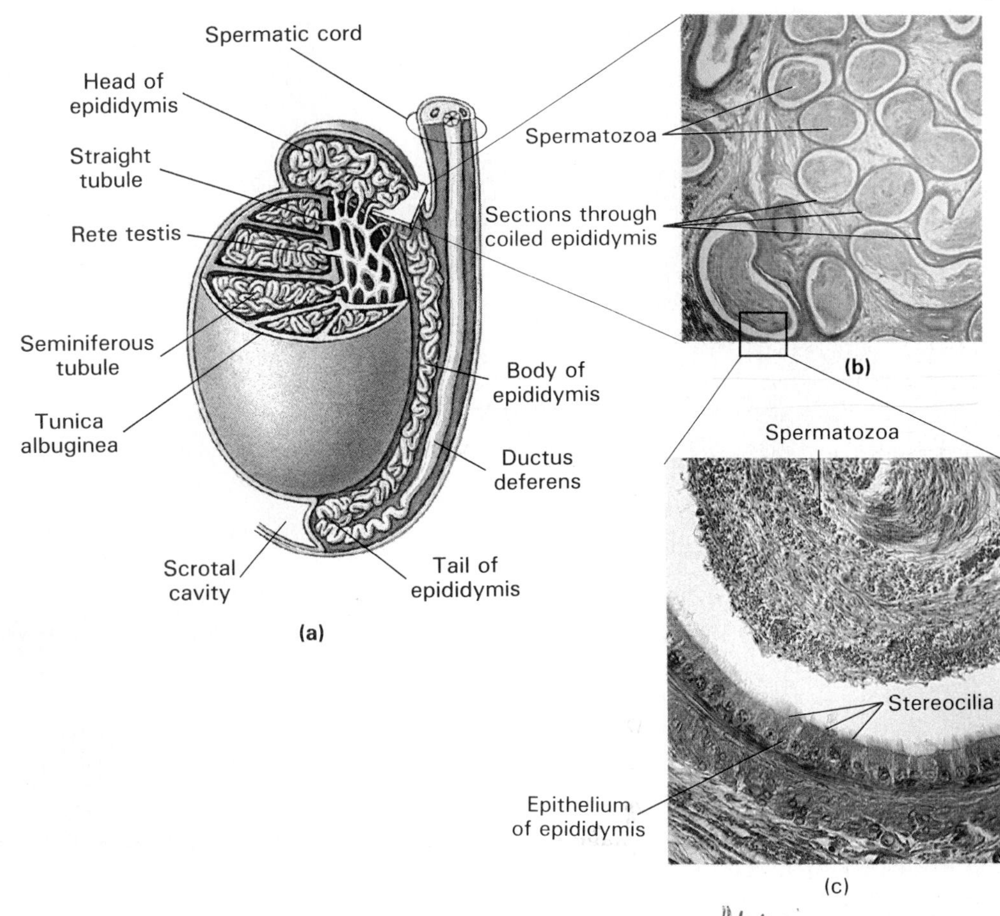

FIGURE 28-5
The Epididymis. (a) Appearance of the epididymis on gross dissection, showing the sectional plane of (b). (b) A light micrograph showing the organization of tubules and the surrounding connective tissues. (LM, ×62) (c) A micrograph showing epithelial characteristics, especially the elongate stereocilia characteristic of the epididymis. (LM, ×648)

THE DUCTUS DEFERENS

The ductus deferens, also known as the *vas deferens*, is 40–45 cm (16–18 in.) in length. It begins at the tail of the epididymis and ascends within a connective tissue sheath. That sheath also encloses the blood vessels, nerves, and lymphatics servicing the testis as well as the proximal portions of the cremaster muscle. The entire complex, called the spermatic cord, is formed during the descent of the testis. [**CM**: *Testicular Torsion*]

After passing through the inguinal canal, the components of the spermatic cord go their separate ways. The ductus deferens passes posteriorly, curving downward along the lateral surface of the urinary bladder toward the superior and posterior margin of the prostate gland. Just before it reaches the prostate and seminal vesicles, the ductus deferens becomes enlarged, and the expanded portion is known as the **ampulla** (am-POOL-a). The path of the ductus deferens can be seen in Figure 28-1, and its various subdivisions are outlined in Figure 28-6.

As the sectional views presented in Figure 28-6b indicate, the walls of the ductus deferens contain an astonishing amount of smooth muscle. Peristalsis in this dense muscular coat provides the propulsion that moves spermatozoa and fluid along the length of the duct. The epithelial lining resembles that of the epididymis, but the tension in the smooth muscles and an abundance of elastic fibers tend to throw the mucosa into small folds. In addition to providing propulsion, the ductus deferens can serve as a storage site for spermatozoa for up to several months. During this period the spermatozoa are in a state of suspended animation, remaining inactive with low metabolic rates.

The junction of the ampulla with the base of the seminal vesicle marks the start of the **ejaculatory duct**. This relatively short (2 cm, or less than 1 in.) passageway penetrates the muscular wall of the prostate where it fuses with the ejaculatory duct from the other side. The common ejaculatory duct then empties into the urethra.

THE URETHRA

The urethra of the male extends from the urinary bladder to the tip of the penis, a distance of 15–20 cm (6–8 in.). It can be divided into three regions, illustrated in Figure 28-7:

1. The **prostatic urethra** passes through the center of the prostate gland. Prostatic glands secrete into the urethra as it passes through the body of the prostate, and the ejaculatory ducts on

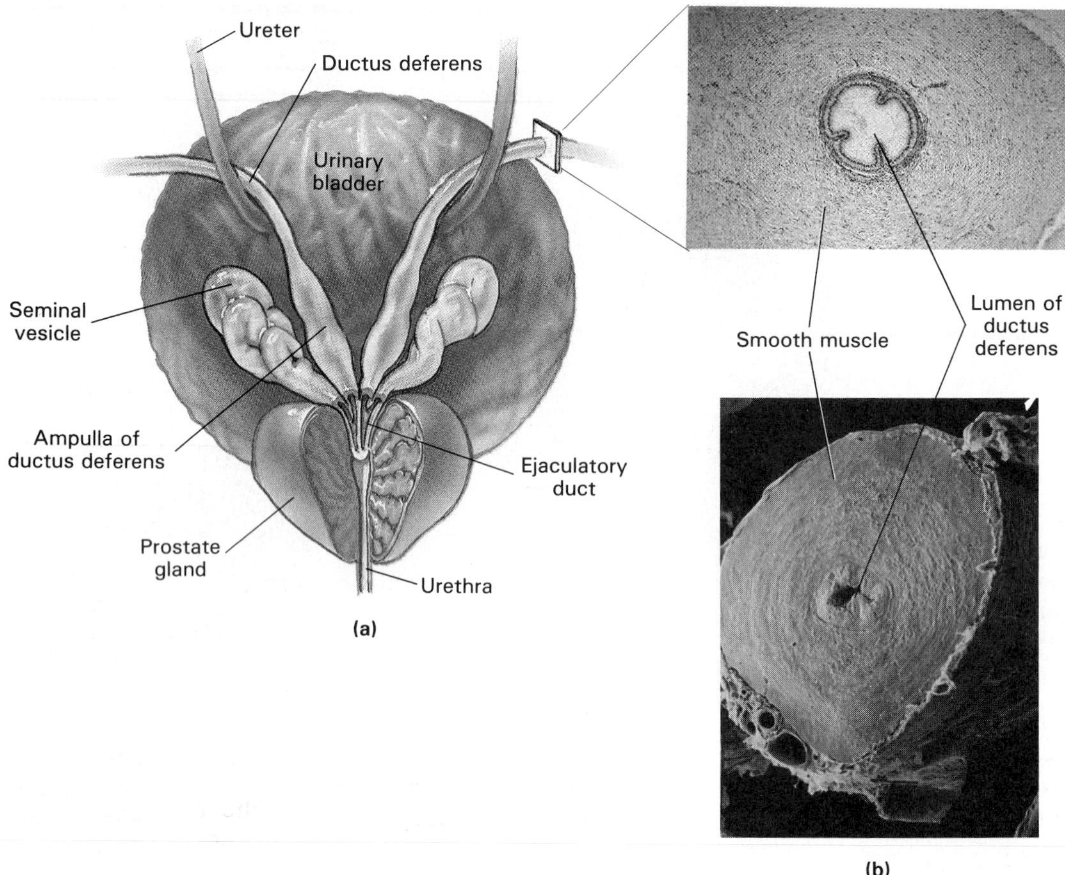

FIGURE 28-6
The Ductus Deferens. (a) A posterior view of the prostate, showing subdivisions of the ductus deferens in relation to surrounding structures. (b) Light and scanning electron micrographs showing extensive layering with smooth muscle around the lumen of the duct. (LM, ×30)

either side communicate with the prostatic urethra midway along its length.

2. The **membranous urethra** includes the short segment that penetrates the urogenital diaphragm, the muscular floor of the pelvic cavity. This is the narrowest portion of the urethra, and its position makes it relatively inelastic. As a result, perineal injuries may tear the urethra at this site.
3. The **penile** (PĒ-nīl) **urethra** extends from the distal border of the urogenital diaphragm to the **external urethral meatus** (*orifice*) at the tip of the penis. The penile urethra receives the secretions of numerous mucous glands, most notably the *bulbourethral glands* (*Cowper's glands*) located near the base of the penis.

THE PENIS

Figure 28-7 also details the structure of the **penis**. The skin overlying the penis resembles that of the scrotum. The dermis contains a layer of smooth muscle, and the underlying loose connective tissue allows the thin skin to move without distorting underlying structures. The subcutaneous layer also contains superficial arteries, veins, and lymphatics.

A fold of skin, the **prepuce** (PRĒ-pūs), or *foreskin*, surrounds the tip of the penis. The prepuce attaches to the relatively narrow **neck** of the penis and continues over the **glans** that surrounds the external urethral meatus. There are no hair follicles on the opposing surfaces, but **preputial** (prē-PŪ-shal) **glands** in the skin of the neck and the inner surface of the prepuce secrete a waxy material known as **smegma** (SMEG-ma). Unfortunately, smegma can be an excellent nutrient source for bacteria. Mild inflammation and infections in this region are common, especially if the area is not washed thoroughly and frequently. One way of avoiding trouble is to perform a **circumcision** (ser-kum-SIZH-un) and surgically remove the prepuce. In Western societies this procedure is usually performed shortly after birth. Recently the medical value of circumcision has been questioned, but there are still strong cultural biases favoring the practice.

Beneath the loose connective tissue, a dense

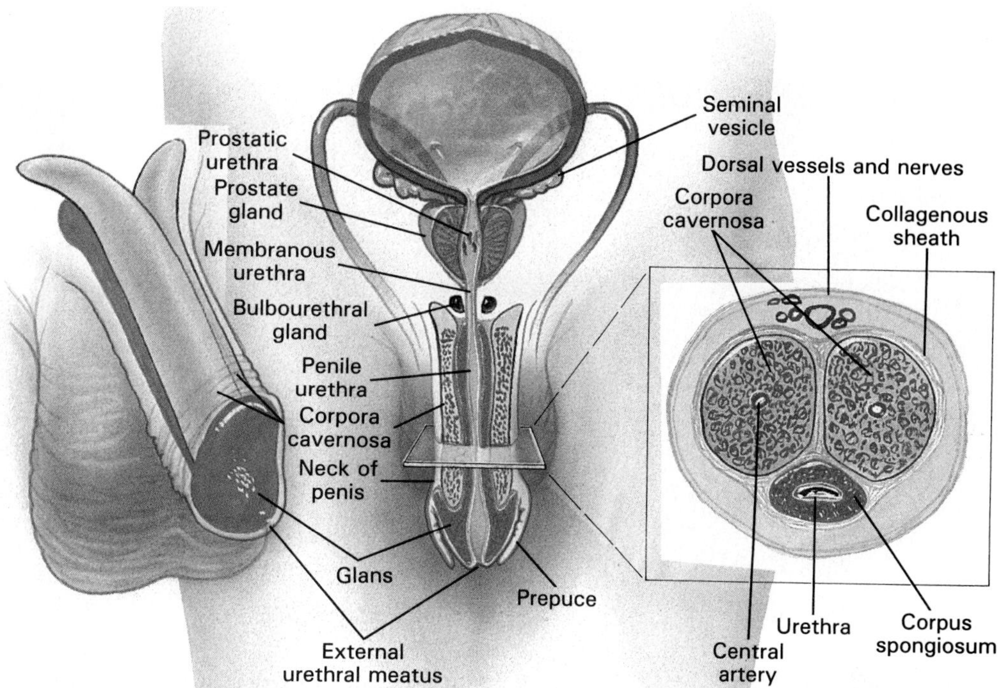

FIGURE 28-7
The Penis

network of elastic fibers encircle the internal structures of the penis. Most of the body of the penis consists of three masses of **erectile tissue**. Erectile tissue consists of a three-dimensional maze of vascular channels incompletely separated by partitions of elastic connective tissue and smooth muscle fibers.

On the anterior surface of the flaccid penis, the two cylindrical **corpora cavernosa** (KŌR-po-ra ka-ver-NŌ-sa) are separated by a thin septum and encircled by a dense collagenous sheath. At their bases the corpora cavernosa are bound to the pubic and ischial bones of the pelvic girdle. They extend along the axis of the penis as far as the neck, and the erectile tissue within each corpus cavernosum surrounds a central artery. Branches from the central arteries travel through partitions of smooth muscle and elastic tissue before emptying into the vascular channels.

The relatively slender **corpus spongiosum** (spon-jē-Ō-sum) surrounds the urethra. This erectile body extends from the superficial fascia of the urogenital diaphragm to the tip of the penis, where it expands to form the glans. The sheath surrounding the corpus spongiosum contains more elastic fibers than does that of the corpora cavernosa, and the erectile tissue contains a pair of arteries. In the resting state, the arterial branches are constricted, and the muscular partitions are tense. This combination reduces blood flow into the erectile tissue. Veins draining the erectile tissue are located between the corpora cavernosa and the corpus spongiosum and superficial to the corpora cavernosa. The smooth muscles relax under parasympathetic stimulation. When this occurs: (1) the vessels dilate, (2) blood flow increases, (3) the vascular channels become engorged with blood, and (4) **erection** of the penis occurs. The flaccid penis hangs beneath the pelvic symphysis anterior to the scrotum, but during erection the penis stiffens and assumes a more upright position. Most of the rigidity of the erect penis results from the blood pressure within the corpora cavernosa; high pressures in the corpus spongiosum could squash the urethra and prevent the passage of semen.

THE ACCESSORY ORGANS

The fluids contributed by the seminiferous tubules and the epididymis account for only about 5 percent of the final volume of semen. The seminal fluid is a mixture of the secretions of many different glands, each with distinctive biochemical characteristics. Important glands include the *seminal vesicles*, the *prostate gland*, and the *bulbourethral glands* (Fig. 28-8). Major functions of these glands include:

1. Activating the spermatozoa;
2. Providing the nutrients spermatozoa need for motility;
3. Propelling spermatozoa and fluids along the reproductive tract, primarily through peristaltic contractions; and
4. Producing buffers that counteract the acidity of the urethral and vaginal contents.

The Seminal Vesicles

The ductus deferens on each side ends at the junction between the ampulla and the duct draining the **seminal vesicle**. Each seminal vesicle is a tubular gland, with a total length of around 15 cm (6 in.). The body

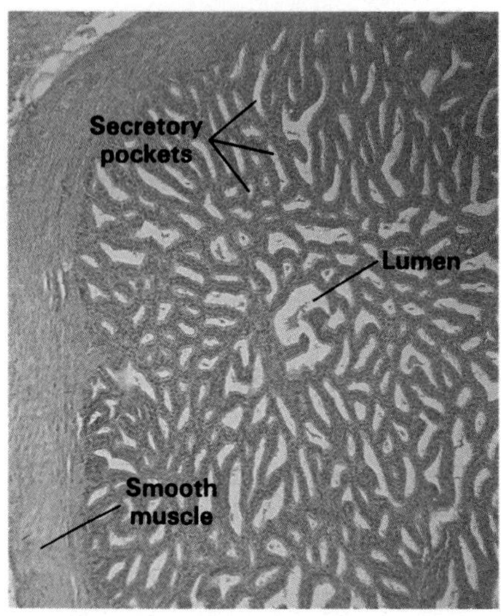

(a) Seminal vesicle

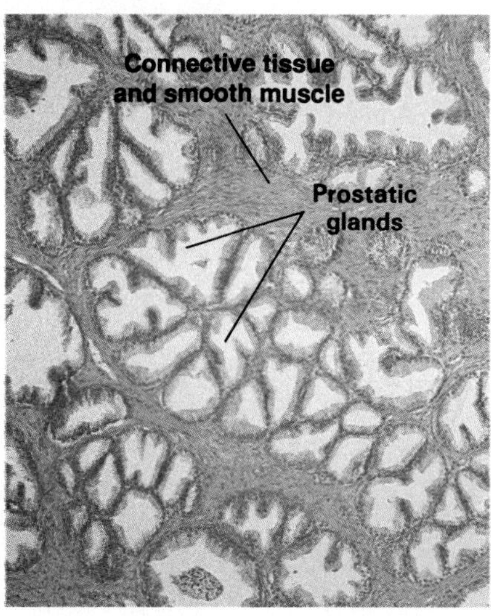

(b) Prostate gland

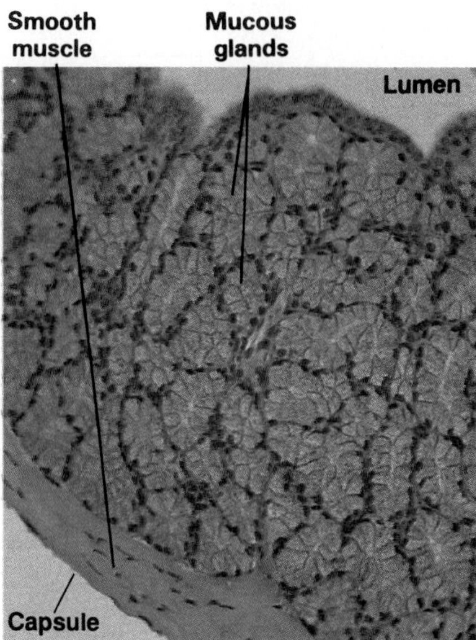

(c) Bulbourethral gland

FIGURE 28-8
Accessory Organs of the Male Reproductive Tract. (a) Light micrograph showing the appearance of the seminal vesicles. Note the extensive glandular surface area here; these organs produce most of the volume of seminal fluid. (LM, ×51) (b) The glands of the prostate. The tissue between the individual glandular units consists largely of smooth muscle. Contractions of this muscle tissue help move the secretions into the ejaculatory duct and urethra. (LM, ×51) (c) The bulbourethral glands, which secret a thick mucus into the penile urethra. (LM, ×206)

of the gland has many short side branches, and the entire assemblage is coiled and folded into a compact, tapered mass roughly 5 cm x 2.5 cm (2 in. x 1 in.). The location of the seminal vesicles can be seen in Figures 28-1 (lateral view), 28-6 (posterior view), and 28-7 (anterior view).

The seminal vesicles are extremely active secretory glands, with an epithelial lining that contains extensive folds. The appearance in this section can be seen in Figure 28-8a. The seminal vesicles contribute about 60 percent of the volume of semen, and although the vesicular fluid usually has the same osmotic concentration as blood plasma, the composition is quite different. In particular, the secretions of the seminal vesicles contain relatively high concentrations of fructose, a six-carbon sugar easily metabolized by spermatozoa. When mixed with the secretions of the seminal vesicles, previously inactive but mature spermatozoa begin beating their flagellae and become highly mobile.

The Prostate Gland

The **prostate gland** is a small, muscular, rounded organ with a diameter of about 4 cm (1.6 in.). As indicated in Figure 28-6, the prostate gland encircles the proximal portion of the urethra as it leaves the bladder. The prostatic wall contains roughly two dozen branching glands producing alkaline secretions that contribute about 30 percent of the volume of semen. The alkalinity assists in neutralizing the acids normally found in the urethra, as well as those encountered within the vagina. In addition to several other compounds of uncertain significance, prostatic secretions contain an unidentified compound with antibiotic properties that may help to prevent urinary tract infections in males. Figure 28-8b shows the appearance of the prostatic glands. [**CM**: *Prostatitis*]

The Bulbourethral Glands

The paired **bulbourethral glands**, or *Cowper's glands*, are situated at the base of the penis, covered by the fascia of the urogenital diaphragm. The glands are round, with diameters approaching 10 mm (less than 0.5 in.). The duct of each gland travels alongside the cavernous urethra for 3–4 cm (1.2–1.6 in.) before emptying into the urethral lumen. These glands, seen in Figure 28-8c, secrete a thick, sticky, alkaline mucus that has lubricating properties.

SEMEN

A typical ejaculation releases 2–5 mℓ of semen. This volume of fluid, called an **ejaculate**, contains:

1. *Spermatozoa*: A normal **sperm count** ranges from 20 to 100 million spermatozoa per milliliter.
2. *Seminal fluid*: **Seminal fluid**, the fluid component of semen, is a mixture of glandular secretions with a distinctive ionic and nutrient composition; a partial description can be found

TABLE 28-1 The Average Characteristics and Composition of Semen

Characteristic/Component	*Value*
Ejaculatory volume	3.4 mℓ
Color	opalescent white
Water content	91.8%
Specific gravity	1.028
pH	7.19
Sperm count	20–600 million/mℓ
Electrolytes (mEq/ℓ)	
Sodium	117
Chloride	43
Potassium	31
Calcium	12
Magnesium	12
Zinc	3
N-compounds (mg/dℓ)	
Ammonia	2
Urea	72
Amino acids	130
Uric acid	6
Proteins	4.5 g/dℓ
Metabolites (mg/dℓ)	
Fructose	224
Phospholipids	84
Cholesterol	103
Vitamins	
C	4.3 mg/dℓ
E	1 mg/dℓ
B_{12}	0.45 μg/dℓ

SOURCES: See Appendix V.

in Table 28-1. It also contains at least one prostaglandin of uncertain function. (Prostaglandins were first identified and named from prostatic secretions.) In terms of total volume, the seminal fluid contains the combined secretions of the seminal vesicles (60 percent), the prostate (30 percent), the sustentacular cells and epididymis (5 percent) and the bulbourethral glands (< 5 percent).

3. *Enzymes*: Several important enzymes are present in the seminal fluid. For example, semen includes (1) a protease that may help to dissolve mucus secretions in the vagina and (2) an enzyme with antibiotic properties. This type of enzyme, called **seminalplasmin** (se-mi-nal-PLAZ-min) has been isolated from bull semen. It kills a variety of bacteria including *Escherichia coli*, a normal resident of the large intestine. It has been suggested that this enzyme could enhance the survival of the spermatozoa within the vagina, but at present, further research is needed to confirm the existence of seminalplasmin in human semen and to determine the relationship between seminalplasmin and the unidentified antibiotic secretion of the prostate.

Within a few minutes after ejaculation semen coagulates, liquifying again after a variable period. The function of this clotting is unknown.

MALE SEXUAL FUNCTION

Male sexual function is coordinated by complex neural reflexes that are incompletely understood. The reflex pathways utilize the sympathetic and parasympathetic divisions of the autonomic nervous system. During **arousal** erotic thoughts or the stimulation of sensory nerves in the genital region lead to an increase in the parasympathetic outflow over the pelvic nerves. Under parasympathetic stimulation the smooth muscle fibers in the erectile tissues relax, blood enters the cavernous spaces, and erection occurs. The integument covering the glans of the penis contains numerous sensory receptors, and erection tenses the skin and increases their sensitivity. Subsequent stimulation may initiate the secretion of the bulbourethral glands, lubricating the penile urethra and the surface of the glans.

During the sexual act, called **coitus** (KŌ-i-tus), **intercourse**, or **copulation** (kop-ū-LĀ-shun), the sensory receptors are rhythmically stimulated. This stimulation eventually results in the coordinated processes of *emission* and *ejaculation*. **Emission** occurs under sympathetic stimulation. The process begins with the peristaltic contractions of the ampulla, pushing fluid and spermatozoa into the prostatic urethra. The seminal vesicles then begin contracting, and the contractions increase in force and duration over the next few seconds. Peristaltic contractions also appear in the walls of the prostate gland. The combination moves the seminal mixture into the membranous and penile portions of the urethra. While the contractions are proceeding, sympathetic commands also cause the contraction of the sphincter guarding the entrance to the urinary bladder. This contraction effectively prevents the passage of semen into the bladder.

Ejaculation occurs as powerful, rhythmic contractions appear in the *ischiocavernosus* and *bulbocavernosus* muscles, two superficial skeletal muscles of the pelvic floor. The ischiocavernosus muscles insert along the sides of the penis, and their contractions primarily serve to stiffen that organ. The bulbocavernosus muscle wraps around the base of the penis, and its contraction pushes semen toward the external urethral orifice. These contractions are controlled by somatic motor neurons in the lower lumbar and upper sacral segments of the spinal cord. (The positions of these muscles can be seen in Figure 11-12.) (p. 347)

Ejaculation is associated with intense pleasurable sensations, an experience known as male **orgasm** (ŌR-gazm). Several other noteworthy physiological changes occur at this time, including pronounced but temporary increases in heart rate and blood pressure. After ejaculation has been completed, blood begins to leave the erectile tissue, and the erection begins to subside. This subsidence, called **detumescence** (de-tū-MES-ens), is mediated by the sympathetic nervous system.

In summary, arousal, erection, emission, and ejaculation are mediated by a complex interplay between the sympathetic and parasympathetic divisions of the autonomic nervous system. Higher centers, including the cerebral cortex, can facilitate or inhibit many of the important reflexes, thereby modifying the patterns of sexual function. Any physical or psychological factor that affects a single component of the system can result in male sexual dysfunction, also called **impotence**. The most common condition is an inability to achieve or maintain an erection. There may be a variety of possible physical causes, for erection involves vascular changes as well as neural commands. For example, low blood pressure in the arteries servicing the penis, due to a circulatory blockage such as a plaque, will affect the ability to obtain an erection. Drugs, trauma, or illnesses that affect the autonomic nervous system or the CNS may have the same effect. But male sexual performance can also be strongly affected by the psychological state of the individual, and the majority of clinical cases of impotence probably reflect psychological rather than anatomical problems. Temporary periods of impotence are relatively common in normal individuals experiencing severe stresses or emotional problems. Depression, anxiety, and fear of impotence are examples of emotional factors that may result in sexual dysfunction.

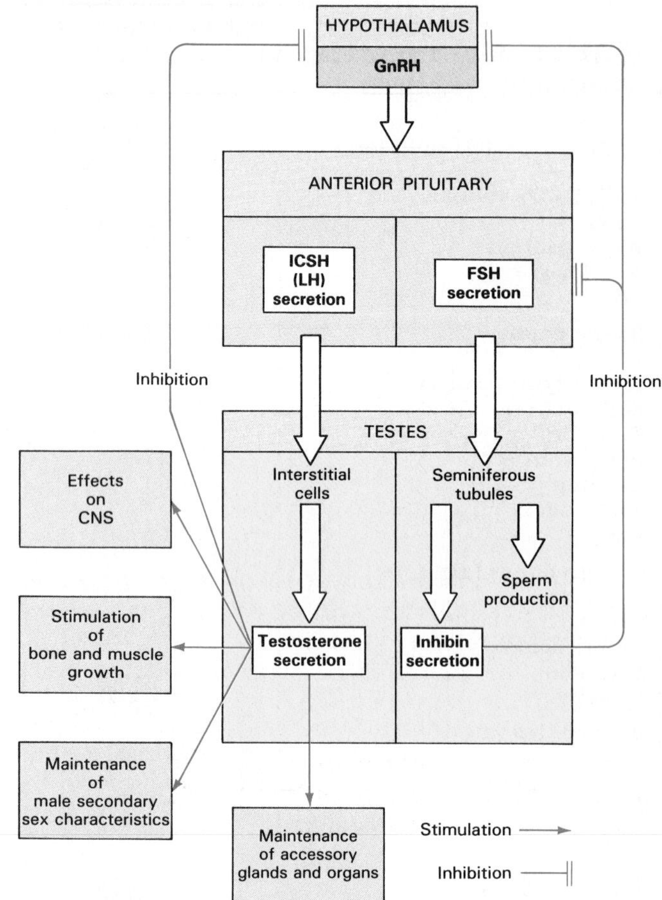

FIGURE 28-9
Hormonal Feedback and the Regulation of Male Reproductive Function

HORMONES AND MALE REPRODUCTIVE FUNCTION

Major reproductive hormones were introduced in Chapter 18, and the hormonal interactions in the male are diagrammed in Figure 28-9. (p. 596) The anterior pituitary releases *follicle-stimulating hormone* (**FSH**) and a second peptide hormone that is called *interstitial cell-stimulating hormone* (**ICSH**) in the male, but that is identical to *luteinizing hormone* (**LH**) in females. The pituitary release of these hormones occurs in the presence of *gonadotrophin-releasing hormone* (**GnRH**), a compound synthesized in the hypothalamus and carried to the anterior pituitary in the hypophyseal portal system.

In the male, ICSH causes the secretion of testosterone and other androgens by the interstitial cells of the testes. Testosterone, the most important androgen, has numerous functions, including:

1. Promoting the functional maturation of spermatozoa,
2. Maintaining the accessory organs of the male reproductive tract,
3. Determining the secondary sexual characteristics such as the distribution of facial hair, increased muscle mass and body size, and the quantity and location of characteristic adipose tissue deposits,
4. Stimulating metabolic operations throughout the body, especially those concerned with protein synthesis and muscle growth, and
5. Influencing brain development by stimulating sexual behaviors and sexual drive.

Testosterone production accelerates markedly at puberty, initiating sexual maturation and the appearance of secondary sexual characteristics. The rate of spermatogenesis is regulated by a negative feedback mechanism involving GnRH, FSH, and inhibin. Under GnRH stimulation, FSH promotes spermatogenesis along the seminiferous tubules. However, as spermatogenesis accelerates so does the rate of inhibin secretion by the sustentacular cells of the testes. As noted earlier in the chapter, inhibin acts to reduce the rate of GnRH and FSH production by the anterior pituitary. The net effect is that when FSH levels become elevated, inhibin production increases until the FSH levels return to normal. If FSH levels decline, inhibin production falls, and the rate of FSH production accelerates.

- √ **On a warm day would the cremaster muscle be contracted or relaxed? Why?**
- √ **If a spermatozoan lacked an acrosomal cap, how would this affect its ability to fertilize an ovum?**
- √ **What effect does a vasectomy have on sperm production?**
- √ **What will occur if the arteries serving the penis dilate?**

The Reproductive System of the Female

The location and orientation of the reproductive organs of the female are indicated in Figure 28-10. The principal organs include the *ovaries*, the *uterine tubes* (*Fallopian tubes* or *oviducts*), the *uterus* (womb), the *vagina*, and the components of the external genitalia. As in the male, a variety of accessory glands secrete into the reproductive tract. Physicians specializing in disorders of the female reproductive tract are called **gynecologists** (gī-ne-KOL-o-jists).

AN OVERVIEW OF FEMALE REPRODUCTIVE ANATOMY

The ovaries, uterine tubes, and uterus are enclosed within an extensive mesentery known as the **broad ligament**, indicated in Figure 28-11. The reproductive organs and their attendant blood vessels, lymphatics, and nerves lie within this mesentery. The uterine tubes run along the superior border of the mesentery and open into the pelvic cavity lateral to the ovaries. A thickened fold of mesentery, the **mesovarium** (mes-ō-VAR-ē-um), supports and stabilizes the position of each ovary. The broad ligament attaches to the sides and floor of the pelvic cavity, its epithelium becoming continuous with that of the parietal peritoneum. The broad ligament thus subdivides the pelvic cavity. The pocket formed between the posterior wall of the uterus and the anterior surface of the rectum is the **rectouterine** (rek-tō-Ū-te-rin) **pouch**, while that between the uterus and the posterior wall of the bladder is the **vesicouterine** (ves-i-kō-Ū-ter-in) **pouch**. These subdivisions are most apparent when you consider a sagittal section through the pelvic cavity, as in Figure 28-10.

Several other ligaments assist the broad ligament in supporting and stabilizing the position of the uterus and associated reproductive organs. These ligaments travel within the mesentery sheet of the

FIGURE 28-10
The Female Reproductive System

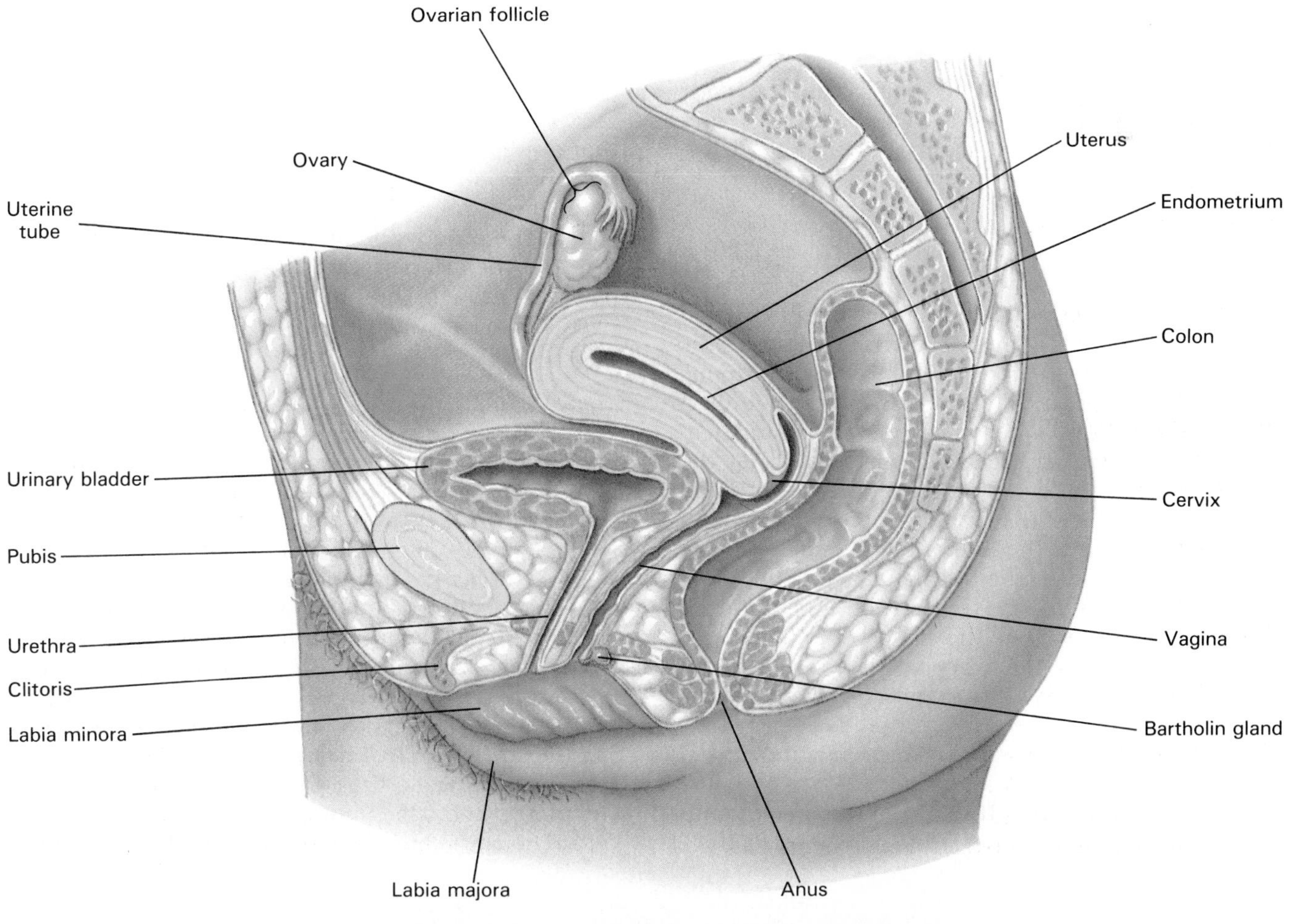

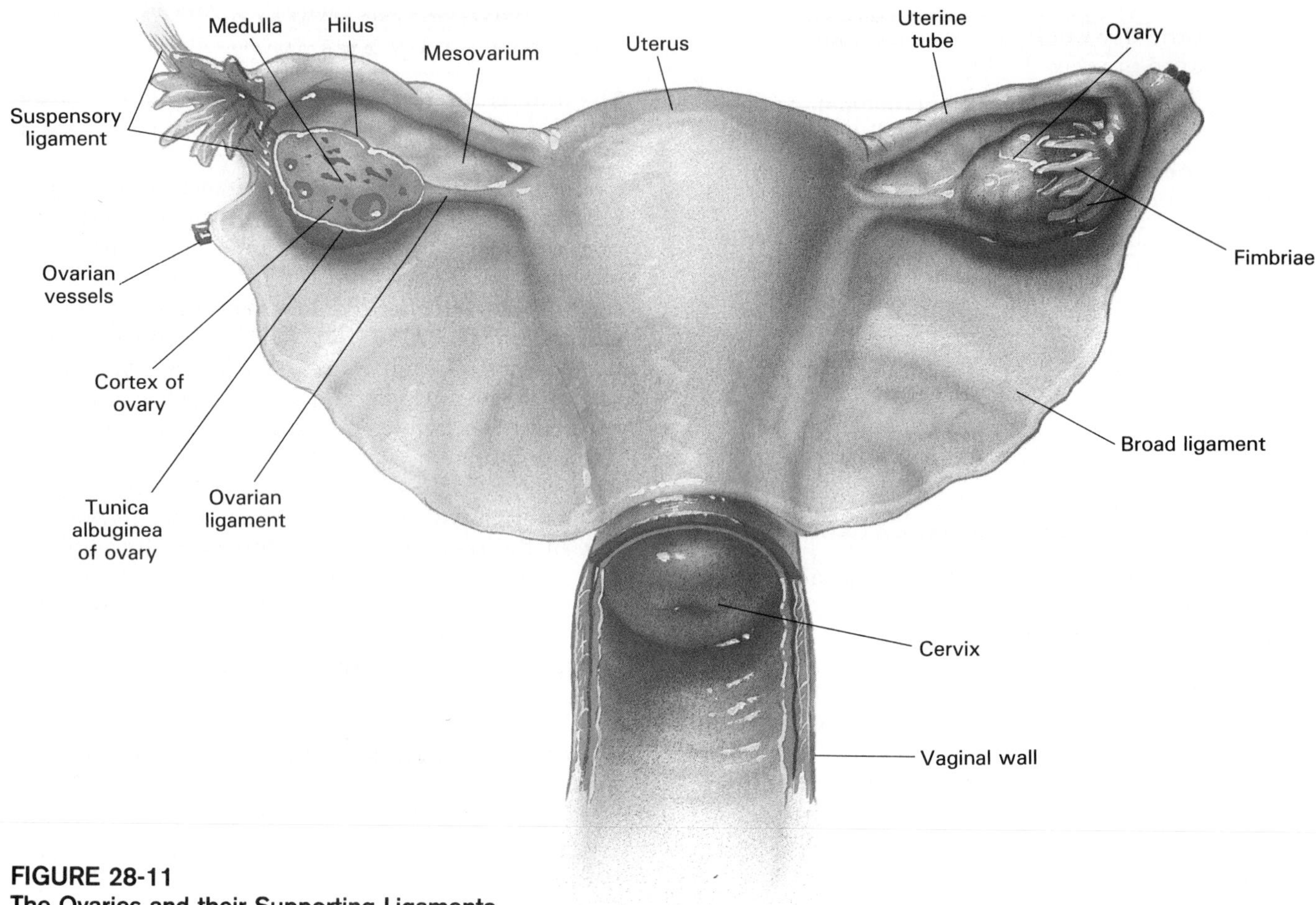

FIGURE 28-11
The Ovaries and their Supporting Ligaments

broad ligament on the way to the ovaries or uterus. The broad ligament limits side-to-side movement and rotation, and the other ligaments (described below) prevent superior-inferior movement.

THE OVARIES

The ovaries are small, almond-shaped organs located near the lateral walls of the pelvic cavity. These organs are responsible for egg production and sexual hormone secretion. The position of each ovary is stabilized by the mesovarium and by a pair of supporting ligaments. The **ovarian ligament** extends from the uterus, near the attachment of the uterine tube, to the medial surface of the ovary. The **suspensory ligament** extends from the lateral surface of the ovary past the open end of the uterine tube to the pelvic wall. Major blood vessels traveling to and from the ovary are found within the suspensory ligament. These vessels enter the ovary at the **ovarian hilus**, where the edge of the ovary attaches to the mesovarium.

A typical ovary measures approximately 5 cm x 2.5 cm (2 in. x 1 in.) and weighs 6–8 g (roughly 0.25 oz). It has a pale white or yellowish coloration and a nodular consistency that resembles cottage cheese or lumpy oatmeal. The exposed surfaces of each ovary are covered by a thickened mesothelium that overlies a layer of dense connective tissue called the **tunica albuginea**. The interior of the ovary can be divided into a superficial *cortex* and a deep *medulla*. The production of gametes occurs in the cortex, and the arteries, veins, lymphatics, and nerves passing through the hilus branch within the relatively narrow medulla.

THE UTERINE TUBES

Each **uterine tube** measures roughly 13 cm (5 in.) in length. The end closest to the ovary forms an expanded funnel, or **infundibulum** (in-fun-DIB-ū-lum), with numerous fingerlike projections that extend into the pelvic cavity. The projections, called **fimbriae** (FIM-brē-ē), and the inner surfaces of the infundibulum are carpeted with cilia that beat toward the entrance to the **ampulla**, the expanded initial segment of the uterine tube. The elongate ampulla leads to the **isthmus** (IS-mus), a short segment adjacent to the uterine wall. The **intramural** (in-tra-MU-ral) **portion** of the uterine tube passes through the wall of the uterus and opens into the uterine cavity.

The epithelium lining the ampulla has numer-

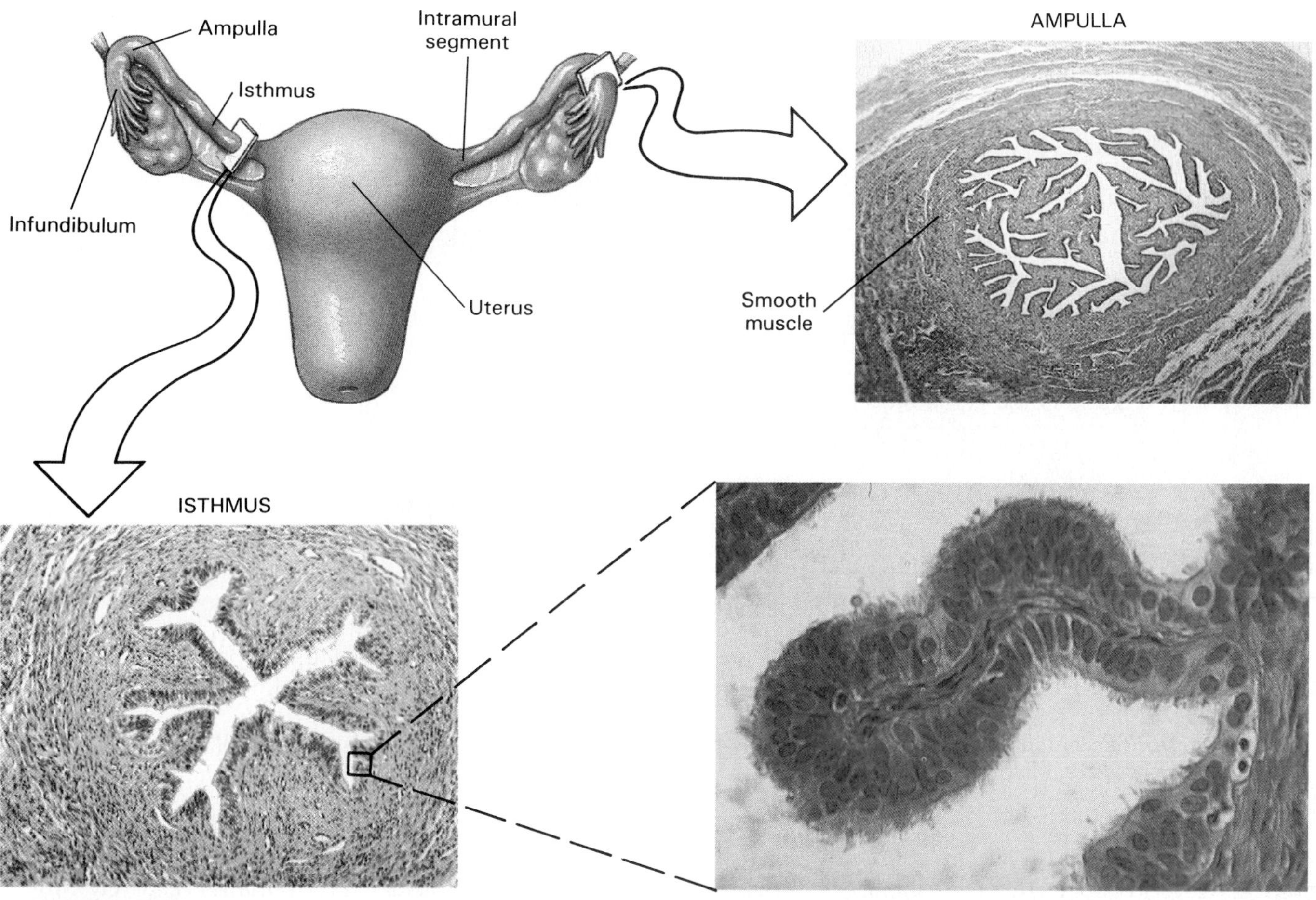

FIGURE 28-12
The Uterine Tubes (LMs, Ampulla ×45, Isthmus ×89)

ous pockets and grooves, and the mucosa is surrounded by concentric layers of smooth muscle (Figure 28-12b,c). There are cilia on the exposed epithelial surfaces (Figure 28-12d), and ovum transport presumably involves a combination of ciliary movement and peristaltic contractions in the walls of the uterine tube. It normally takes 3-4 days for an ovum to travel from the infundibulum to the uterine chamber. *If fertilization is to occur, the ovum must encounter spermatozoa during the first 12–24 hours of its passage.* Unfertilized eggs degenerate in the terminal portions of the uterine tubes or within the uterus.

THE UTERUS

The **uterus** provides mechanical protection and nutritional support to the developing embryo. The position of the uterus within the pelvic cavity can be seen in Figure 28-10, and additional details are presented in Figure 28-13. The typical uterus is a small, pear-shaped organ about 7.5 cm (3 in.) in length with a maximum diameter of 5 cm (2 in.). It weighs 30–40 g (1–1.4 oz). In its normal position the uterus bends anteriorly near its base, a condition known as **anteflexion** (an-tē-FLEK-shun). In this position the body of the uterus lies across the superior and posterior surfaces of the urinary bladder. If instead the uterus bends backwards toward the sacrum, the condition is termed **retroflexion** (re-trō-FLEK-shun). Retroflexion, which occurs in about 20 percent of adult women, has no clinical significance.

Suspensory Ligaments of the Uterus

In addition to the mesenteric sheet of the broad ligament, three pairs of suspensory ligaments stabilize the position of the uterus and limit its range of movement. The **uterosacral** (ū-te-rō-SĀ-kral) **ligaments** extend from the lateral surfaces of the uterus to the anterior face of the sacrum, keeping the body of the uterus from moving down and forward. The **round ligaments** arise on the lateral margins of the uterus just below the uterine tubes. They extend anteriorly, passing through the inguinal canal before ending in the connective tissues of the external genitalia. These ligaments primarily restrict posterior movement of the uterus. The **lateral** (*cardinal*) **ligaments** extend from the base of the uterus and vagina to the lateral walls of the pelvis. These ligaments also tend to prevent the downward movement of the

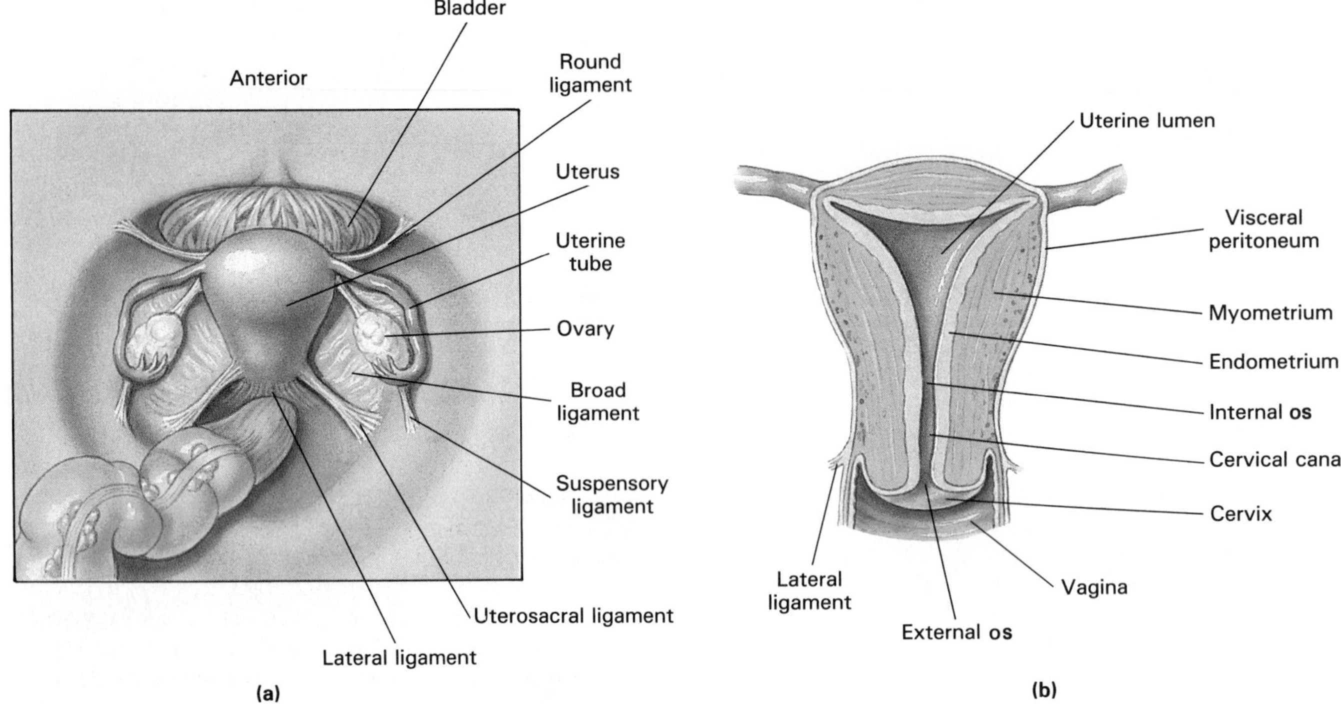

FIGURE 28-13
The Uterus. (a) The uterus and ligaments stabilizing its position in the pelvic cavity. (b) Details of uterine structure; for histological details, see Figure 28-19.

uterus. Additional mechanical support is provided by the muscles and fascia of the urogenital diaphragm.

Sectional Anatomy of the Uterus

The uterus can be divided into an expanded, superior **body**, a constricted **isthmus**, and an inferior **cervix** (SER-viks). The cervix projects a short distance into the vagina, and the uterine cavity opens into the vagina at the **external orifice**, or *cervical os*. The space within the uterus is shown in Figure 28-13b. The **uterine cavity** lies within the uterine body; it is connected to a narrow **cervical canal** via the constricted **internal orifice**, or *internal os*.

In histological section the thick uterine wall can be divided into an inner **endometrium** (en-dō-MĒ-trē-um), a muscular **myometrium** (mī-ō-MĒ-trē-um), and a superficial **serosa** continuous with the mesothelium of the broad ligament. The endometrium, or *mucosa*, of the uterus includes the epithelium lining the uterine chambers and the underlying connective tissues of the lamina propria. Uterine glands opening onto the endometrial surface extend deep into the lamina propria almost all the way to the myometrium. The myometrium consists of a thick mass of interwoven smooth muscle fibers. [**CM**: *Uterine Tumors*]

The endometrium can be divided into a superficial *functional zone* and a deeper *basilar zone*. The structure of the basilar layer remains relatively constant over time, but that of the functional zone undergoes cyclical changes in response to sexual hormone levels. These alterations produce the characteristic features of the *menstrual cycle*, the topic of a later section.

THE VAGINA

The **vagina** is a muscular tube extending between the uterus and the external genitalia. It has an average length of 7.5–9 cm (3–3.5 in.), but because the vagina is highly distensible its length and width are quite variable. The position and structure of the vagina and external genitalia can be seen in Figure 28-14 on page 934.

At the proximal end of the vagina, the cervix projects into the **vaginal canal**. The shallow recess surrounding the cervical protrusion is known as the **fornix** (FŌR-niks). The vagina lies parallel to the rectum, and the two are in close contact. After leaving the urinary bladder the urethra turns and travels along the superior wall of the vagina.

In sectional view the lumen of the vagina appears constricted, forming a rough H. The vaginal walls contain a network of blood vessels and layers of smooth muscle, and the lining is moistened by the secretions of the cervical glands and by the movement of water across the permeable epithelium. Prior to the onset of sexual activity a thin epithelial fold, the

Clinical Comment: Pelvic Inflammatory Disease (PID)

Pelvic inflammatory disease (PID) is a major cause of sterility in women. This condition, an infection of the uterine tubes, affects an estimated 850,000 women each year in the United States alone. Sexually transmitted pathogens are often involved, and as many as 80 percent of all first cases may be due to infection by *Neisseria gonorrhoeae*, the organism responsible for symptoms of **gonorrhea** (gon-ō-RĒ-a), a sexually transmitted disease discussed in the Clinical Manual. PID may also result from invasion of the region by bacteria normally found within the vagina. Symptoms of pelvic inflammatory disease include fever, lower abdominal pain, and elevated white blood cell counts. In severe cases the infection may spread to other visceral organs or produce a generalized peritonitis.

Sexually active women in the 15–24 age group have the highest incidence of PID. Although oral contraceptive use decreases the risk of infection, the presence of an intrauterine device (IUD) may increase the risk by 1.4–7.3 times. Treatment with antibiotics may control the condition, but chronic abdominal pain may persist. In addition, damage and scarring of the uterine tubes may cause infertility by preventing the passage of a zygote to the uterus. In this case an embryo may begin to develop in the uterine tube or elsewhere outside of the uterus, a condition known as an ectopic pregnancy. Recently, another sexually transmitted bacterium, *Chlamydia*, has been identified as the probable cause of up to 50 percent of all cases of PID. Despite the fact that women with this infection may develop few if any symptoms, scarring of the uterine tubes may still produce infertility.

hymen (HI-men), partially or completely blocks the entrance to the vagina. The two bulbocavernosus muscles pass on either side of the vaginal orifice, and their contractions constrict the entrance.

The vagina has several reproductive functions:

1. It serves as a passageway for the elimination of menstrual fluids;
2. It receives the penis during coitus and holds spermatozoa prior to their passage into the uterus;
3. In childbirth it forms the lower portion of the birth canal through which the fetus passes on its way to an independent existence.

The vagina normally contains resident bacteria supported by the nutrients found in the cervical mucus. As a result of their metabolic activities the normal pH of the vagina ranges between 3.5 and 4.5, and this acid environment restricts the growth of many pathogenic organisms. An infection of the vaginal canal, known as **vaginitis** (va-jin-Ī-tis), may be caused by fungal, bacterial, or parasitic organisms. In addition to any discomfort that may result, the condition may affect the survival of sperm and thereby reduce fertility. [**CM**: *Vaginitis*]

The External Genitalia The region enclosing the female external genitalia is usually called the **vulva** (VUL-va), or female pudendum. The components of the vulva are indicated in Figure 28-14a. The vagina opens into the **vestibule**, a central space bounded by the **labia minora** (LĀ-bē-a mi-NOR-a). The labia minora are covered with a smooth, hairless skin. The urethra opens into the vestibule just anterior to the vaginal entrance. Anterior to the urethral opening, the **clitoris** (KLI-tō-ris) projects into the vestibule. The clitoris is the female equivalent of the penis, derived from the same embryonic structures (see the Embryological Summary later in this chapter). Internally the clitoris contains erectile tissues that become engorged with blood during arousal. A small erectile *glans* sits atop the organ, and extensions of the labia minora encircle the body of the clitoris, forming the *prepuce*.

A variable number of small **lesser vestibular glands** discharge their secretions onto the exposed surface of the vestibule, keeping it moistened. During arousal a pair of ducts discharges the secretions of the **greater vestibular glands** (*Bartholin's glands*) into the vestibule near the posterolateral margins of the vaginal entrance. These mucous glands resemble the bulbourethral glands of the male.

The outer limits of the vulva are established by the *mons pubis* and the *labia majora*. The prominent bulge of the **mons pubis** is created by adipose tissue beneath the skin anterior to the pubic symphysis. Adipose tissue also accumulates within the fleshy **labia majora** that encircle and partially conceal the labia minora and vestibular structures. The outer margins of the labia majora are covered with the same coarse hair that carpets the mons pubis, but the inner faces are relatively hairless. Sebaceous glands and scattered apocrine sweat glands secrete onto the inner surface of the labia majora, moistening them and providing lubrication.

THE MAMMARY GLANDS

At birth the newborn infant cannot fend for itself, and several key systems have yet to complete their development. Over the initial period of adjustment to an independent existence, the infant gains nour-

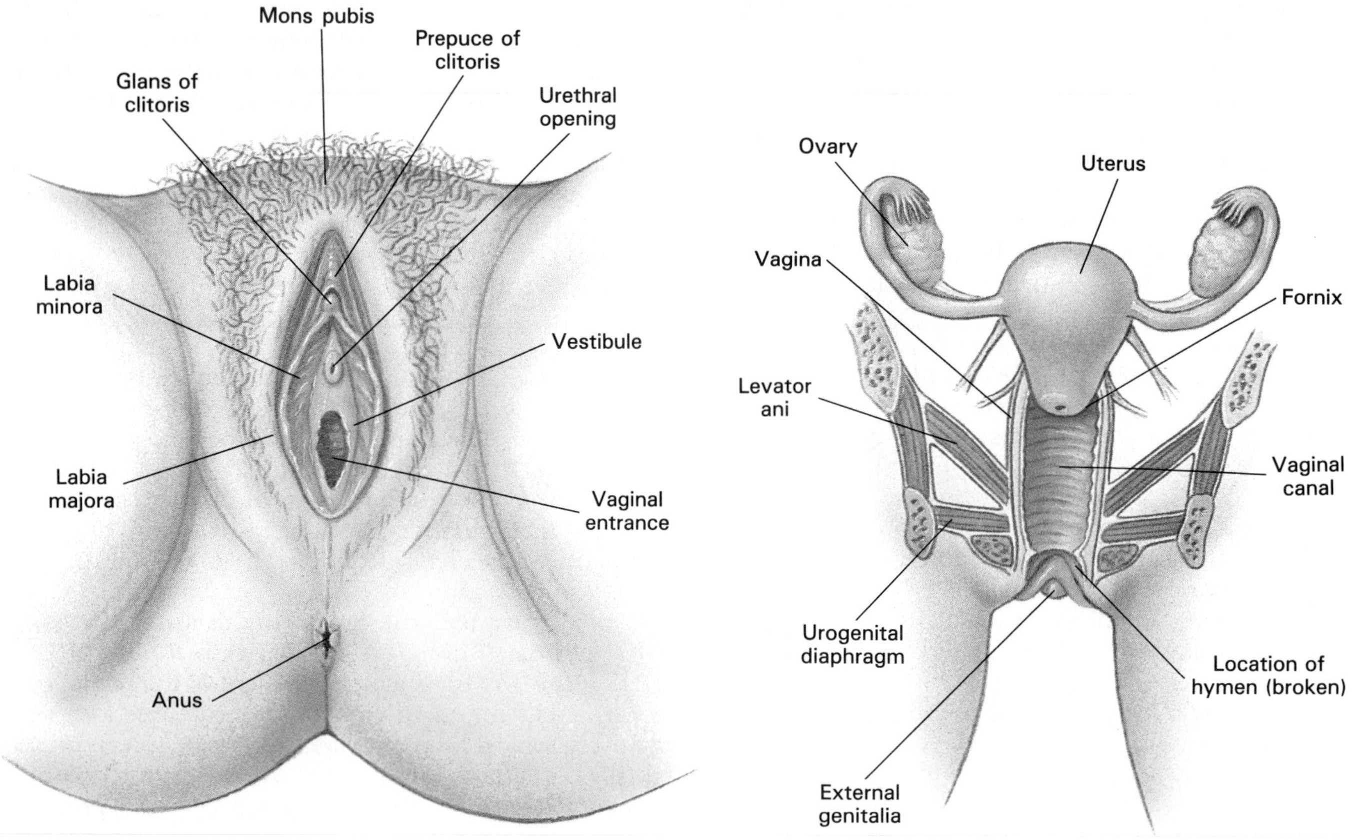

FIGURE 28-14
The Female External Genitalia. (a) An external view of the female perineum. (b) A sectional view showing the relative positions of the internal and external reproductive structures.

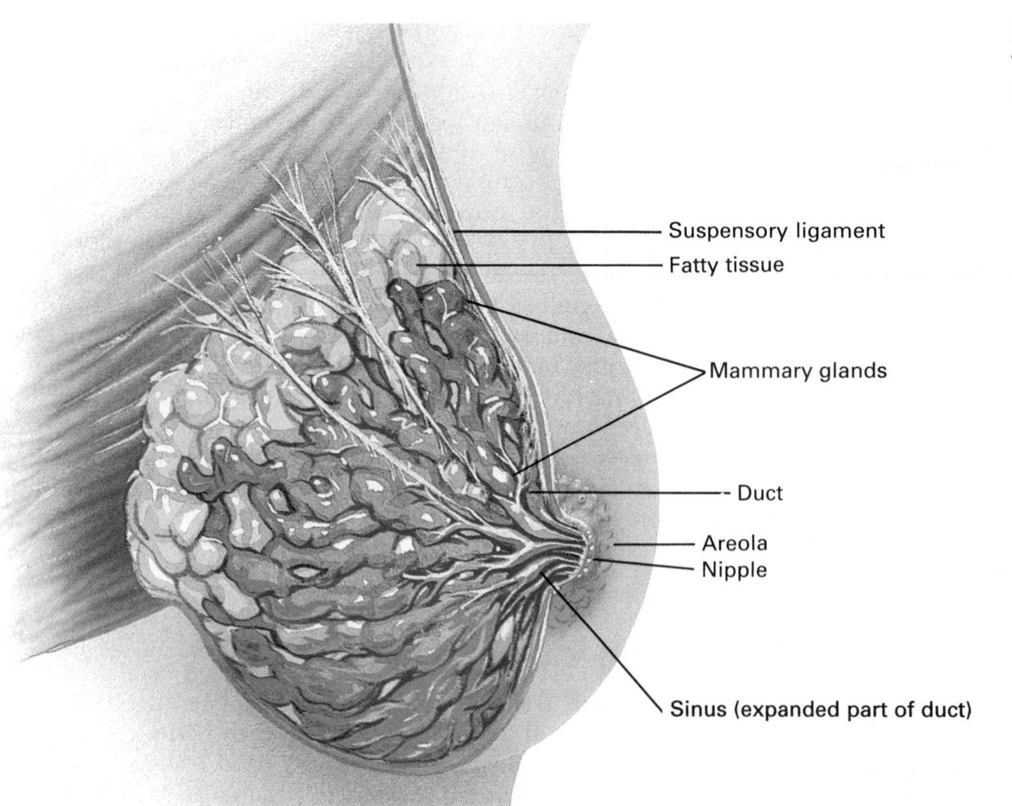

FIGURE 28-15
The Mammary Glands of the Female Breast

ishment from the milk secreted by the maternal **mammary glands**. Milk production, or **lactation** (lak-TĀ-shun), occurs in the mammary glands of the breasts, specialized accessory organs of the female reproductive system. The general anatomy and orientation of a breast can be seen in Figure 28-15.

The mammary glands lie in the subcutaneous layer beneath the skin of the chest. Each breast bears a small conical projection, the **nipple**, where the ducts of underlying mammary glands open onto the body surface. The skin surrounding each nipple has a reddish brown coloration, and this region is known as the **areola** (a-RĒ-ō-la). Large sebaceous glands beneath the areolar surface give it a granular texture.

Health News: Breast Cancer

The mammary glands are cyclically stimulated by the changing levels of circulating reproductive hormones that accompany the menstrual cycle. Usually the effects go unnoticed, but there can be occasional discomfort and inflammation of breast tissues late in the cycle. If inflamed lobules become walled off with scar tissue, **cysts** are created. Clusters of cysts can be felt in the breast as discrete masses, a condition known as **fibrocystic disease**. Because the symptoms are similar, biopsies may be needed to distinguish between this benign condition and breast cancer.

Breast cancer is the primary cause of death for women between the ages of 35 and 45, but it actually becomes even more common after age 50. There were approximately 45,000 deaths in the United States from breast cancer in 1989, and nearly 115,000 new cases reported. An estimated 7–9 percent of women in the United States will develop breast cancer at some point in their lifetimes. The incidence is highest among Caucasians, somewhat lower in blacks, and lowest in Asians and American Indians. Notable risk factors include (1) a family history of breast cancer, (2) a pregnancy after age 30, and (3) early menarche (first menstrual period) or late menopause (last menstrual period). Despite repeated studies (and rumors), there are no proven links between oral contraceptive use, estrogen therapy, breastfeeding, fat consumption, or alcohol use and breast cancer. It appears likely that multiple factors are involved; most women never develop breast cancer, even women in families with a history of this disease.

Early detection of breast cancer is the key to reducing mortalities. *Most breast cancers are found through self-examination*, but the use of clinical screening techniques has increased in recent years. **Mammography** involves the use of X-rays to examine breast tissues; the radiation dosage can be restricted because only soft tissues must be penetrated. This procedure gives the clearest picture of conditions within the breast tissues. Ultrasound can provide some information, but the images lack the detail of standard mammograms. **Thermography** maps the surface temperatures on the skin of the breasts. Because cancer cells have abnormally high metabolic rates and increased vascularization, tumors are significantly warmer than the surrounding tissues. The heat can be detected with this technique, but unfortunately the results are subject to considerable variation.

For treatment to be successful the cancer must be identified while it is still relatively small and localized. Once it has grown larger than 2 cm (0.78 in.) the chances for long-term survival worsen. A poor prognosis also follows if the cancer cells have spread through the lymphatic system to the axillary lymph nodes. If the nodes are not yet involved, the chances of 5-year survival are about 82 percent, but if four or more nodes are involved, the survival rate drops to 21 percent.

Treatment of breast cancer begins with the removal of the tumor. Because the cancer cells usually begin spreading before the condition is diagnosed, surgical treatment involves the removal of part or all of the affected breast.

- In a **segmental mastectomy**, or "lumpectomy," only a portion of the breast is removed.
- In a **total mastectomy** the entire breast is removed, but other tissues are left intact.
- In **radical mastectomy** the pectoralis muscles, the breast, and the axillary lymph nodes are removed. In a *modified radical mastectomy*, the most common operation, the breast and nodes are removed but the muscular tissue remains intact.

A combination of chemotherapy, radiation treatments, and hormone treatments may be used to supplement the surgical procedures.

The glandular tissue of the breast consists of a number of separate lobes, each containing several secretory lobules. Within each lobe the ducts leaving the lobules converge, giving rise to a single **lactiferous** (lak-TIF-e-rus) **duct**. Near the nipple, that lactiferous duct expands, forming an expanded chamber called a **lactiferous sinus**. There are usually 15–20 lactiferous sinuses opening onto the surface of each nipple. Dense connective tissue surrounds the duct system and forms partitions that extend between the lobes and lobules. These bands of connective tissue, known as the **suspensory ligaments of the breast**, originate in the dermis of the overlying skin. A layer of loose connective tissue separates the mammary complex from the underlying muscles, and the two can move relatively independently.

FEMALE SEXUAL FUNCTION

The phases of female sexual function resemble those of the male:

1. During arousal, parasympathetic activation leads to an engorgement of the erectile tissues of the clitoris and increased secretion of the greater vestibular glands.
2. During coitus, rhythmic contact with the clitoris and vaginal walls provides stimulation that eventually leads to orgasm.
3. Female orgasm is accompanied by peristaltic contractions of the uterine and vaginal walls and rhythmic contractions of the bulbocavernosus and ischiocavernosus muscles. The latter contractions give rise to the pleasurable sensations experienced.

[**CM**: *STDs*]

- ✓ **As the result of infections such as gonorrhea, scar tissue can block the lumen of the uterine tube. How would this affect a woman's ability to conceive?**
- ✓ **What is the advantage of the normally acidic pH of the vagina?**
- ✓ **Would blockage of a singe lactiferous sinus interfere with delivery of milk to the nipple? Explain.**

Hormones and the Female Reproductive System

As in the male, the activity of the female reproductive tract falls under hormonal control involving an interplay between pituitary and gonadal secretions. But the regulatory pattern is much more complicated, for a woman's reproductive system does not just produce functional gametes. It must also be prepared to protect and support a developing embryo and to nourish the newborn infant. Accomplishing these goals requires the coordination of the ovarian and uterine activities and the timely preparation of the mammary glands. Because the processes are complex and difficult to study, many of the biochemical details still elude us, but the general patterns are reasonably clear.

THE OVARIAN CYCLE

Ovum production, or **oogenesis** (ō-ō-JEN-e-sis), occurs on a monthly basis, as part of the **ovarian cycle**. Important features of this cycle are diagrammed in Figure 28-16. Ovum development occurs in specialized structures called **ovarian follicles** (ō-VAR-ē-an FOL-i-klz). In the outer portions of the cortex, just beneath the tunica albuginea, there are scattered clusters of immature eggs, or **oocytes** (Ō-o-sīts). Each oocyte within one of these **egg nests** has pale cytoplasm and a large, round nucleus with a prominent nucleolus. It is encircled by a simple squamous layer of follicular cells, and the combination is known as a **primordial** (prī-MŌR-dē-al) **follicle**. At monthly intervals some of those primordial follicles become active, and this marks the start of the ovarian cycle.

STEP 1: Formation of Primary Follicles

The cycle begins as the activated follicles develop into **primary follicles**. In a primary follicle the follicular cells enlarge and undergo repeated cell divisions. This division creates several layers of follicular cells around the oocyte. As the wall of the follicle thickens further, a space opens up between the developing oocyte and the innermost follicular cells. Within this space, called the **zona pellucida** (ZŌ-na pel-LŪ-si-da), microvilli originating at the surface of the oocyte interdigitate with those of the follicular cells. These microvilli increase the surface area available for absorption by roughly 35 times, and the follicular cells are continually providing the developing oocyte with nutrients. In addition to diffusion, facilitated diffusion, and cotransport, pinocytosis occurs at the bases of the microvilli.

STEP 2: Formation of Secondary Follicles

Although many primordial follicles develop into primary follicles, usually only a few will take the next step. The transformation begins as the wall of the follicle thickens and the deeper follicular cells begin secreting small amounts of fluid. This fluid accumulates in small pockets that gradually expand and separate the inner and outer layers of the follicle. At this stage the complex is known as a **secondary follicle**. Although the oocyte continues to grow at a slow pace, the follicle as a whole now enlarges rapidly due to this accumulation of fluid.

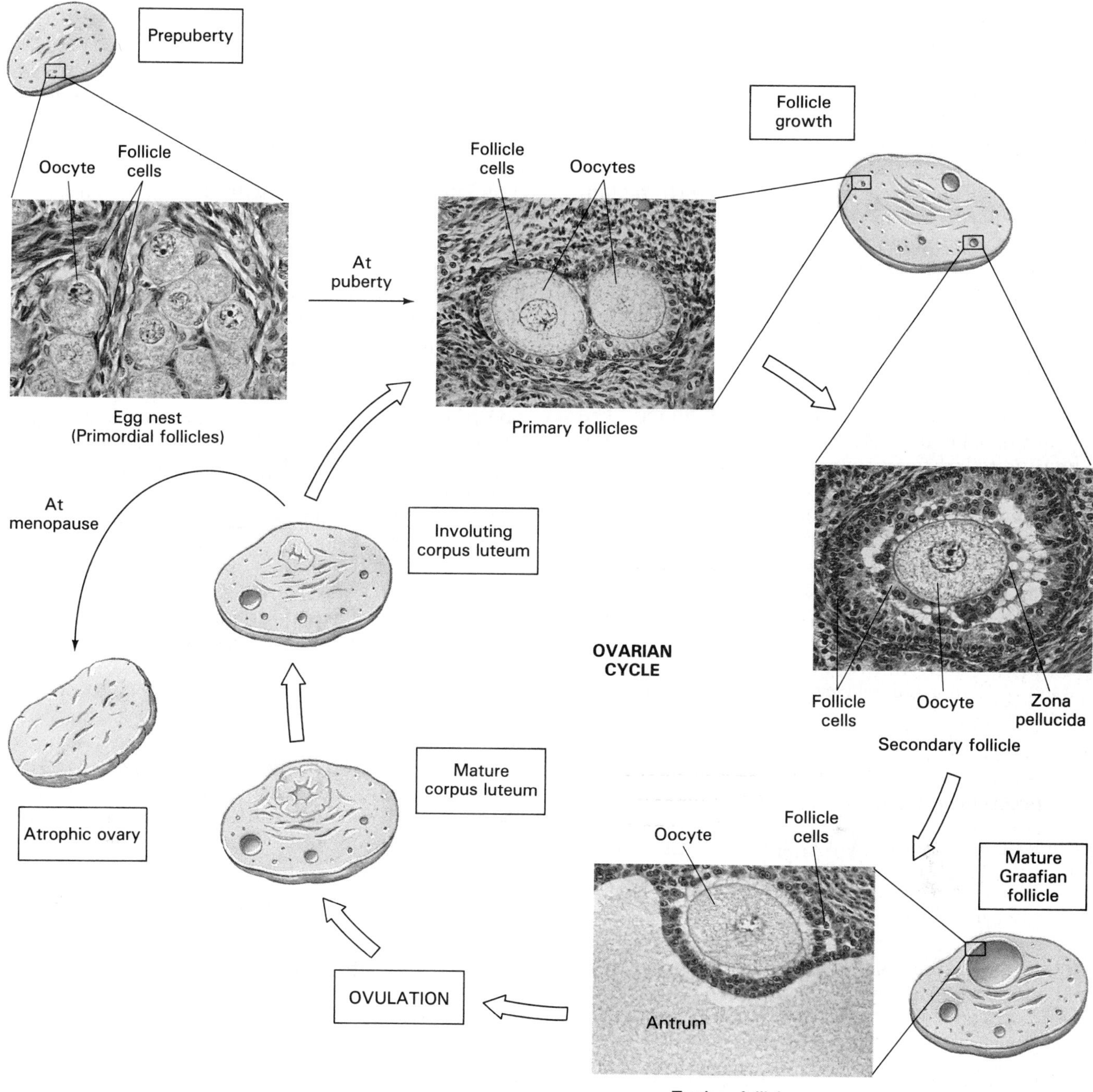

FIGURE 28-16
The Ovarian Cycle (LMs, ×486, Egg nest ×648)

STEP 3: Formation of a Tertiary Follicle

Eight to ten days after the start of the ovarian cycle, the ovaries usually contain only a single secondary follicle destined for further development. By the tenth to fourteenth days of the cycle it has formed a mature **tertiary follicle**, or *Graafian* (GRAF-ē-an) *follicle*, roughly 15 mm in diameter. This complex spans the entire width of the cortex and stretches the ovarian capsule, creating a prominent bulge in the surface of the ovary. The oocyte projects into the expanded central chamber, or **antrum** (AN-trum), surrounded by a mass of follicular cells.

STEP 4: Ovulation

As the time of egg release, or **ovulation** (ōv-ū-LĀ-shun), approaches, the oocyte and its follicular associates lose their connections with the follicular wall and drift free within the antrum. This event usually occurs at day 14 of a 28 day cycle. The follicular cells surrounding the oocyte are now known as the **corona radiata** (ko-RŌ-na rā-dē-A-ta). The distended follicular wall then ruptures, releasing the follicular contents, including the oocyte, into the pelvic cavity. The corona radiata has a sticky surface, so the oocyte usually attaches to the ovarian surface near the ruptured wall of the follicle. Direct contact with the fimbriae or fluid currents established by the ciliated epithelium of the fimbriae then transfer the oocyte to the uterine tube that will carry it toward the uterus.

STEP 5: Formation and Degeneration of the Corpus Luteum

The empty follicle initially collapses, and ruptured vessels leak blood cells into the lumen. The remaining follicular cells then invade the lumen, proliferating to create an endocrine structure known as the **corpus luteum** (LŪ-tē-um), named for its yellow color (*lutea*, yellow). Unless pregnancy occurs, after about 12 days the corpus luteum begins to degenerate. Fibroblasts then invade the region, producing a knot of pale scar tissue called a **corpus albicans** (AL-bi-kanz). The disintegration, or *involution*, of the corpus luteum marks the end of the ovarian cycle, but this event is followed by the activation of another cluster of primordial follicles and the start of another ovarian cycle.

Age and Oogenesis

Although many primordial follicles may have developed into primary follicles, and several primary follicles converted to secondary follicles, usually only a single oocyte will be released into the pelvic cavity at ovulation. The rest degenerate, a process termed **atresia** (a-TRĒ-zē-a). At puberty a woman has approximately 200,000 primordial follicles in each ovary. Forty years later few if any follicles remain, although a relatively small number will have actually completed their development into functional gametes over the interim.

Hormones and the Preovulatory Period

Follicular development begins under FSH stimulation, and each month some of the primordial follicles begin their development into primary follicles. As the follicular cells enlarge and multiply, they release steroid hormones collectively known as *estrogens*. The hormone **estradiol** (es-tra-DĪ-ol) is the most important estrogen. Estrogens (1) stimulate bone and muscle growth, (2) maintain female secondary sex characteristics such as body hair distribution and the location of adipose tissue deposits, (3) affect CNS activity, including sex-related behaviors and drives, (4) maintain functional accessory reproductive glands and organs, and (5) initiate repair and growth of the endometrium. Small quantities of estrogens are also contributed by interstitial cells scattered within the ovarian cortex. Changes in circulating estrogen concentration are the primary mechanism for coordinating female sexual function, and the relationships are summarized in Figure 28-17.

The upper portion of Figure 28-17 summarizes the hormonal regulation of the ovarian cycle. As follicular development proceeds the concentration of circulating estrogens rises, for the follicular cells are increasing in number and secretory activity. Rising estrogen levels, combined with rising inhibin production (Chapter 18), inhibit both the hypothalamic secretion of GnRH and the pituitary production and release of FSH. Estrogen also has an effect on the rate of LH secretion. Although the synthesis of LH occurs under GnRH stimulation, the rate of release into the bloodstream depends on the circulating concentration of estrogens. Thus as the follicles develop and estrogen concentrations rise, the pituitary output of LH gradually increases. Despite a slow decline in FSH concentrations, the combination of estrogens, FSH, and LH continues to support follicular development and maturation. (p. 596)

Estrogen concentrations take a sharp upturn in the second week of the ovarian cycle, as this month's tertiary follicle enlarges in preparation for ovulation. At about day 14 estrogen levels peak, accompanying the maturation of that follicle. The high estrogen concentration then triggers a massive outpouring of LH from the anterior pituitary. That sudden surge in LH concentration causes the rupture of the follicular wall and ovulation.

Hormones and the Postovulatory Period

After ovulation LH stimulates the remaining follicular cells to form the corpus luteum, and the yellow color of this mass results from the presence of lipid reserves. These compounds are used to manufacture steroid hormones known as **progestins** (prō-JES-tinz), principally the steroid **progesterone** (prō-JES-ter-ōn). Although moderate amounts of estrogens are also secreted by the corpus luteum, progesterone is the principal hormone of the postovulatory period. Its primary function is to prepare the uterus for pregnancy. It also stimulates metabolic activity, leading to a rise in basal body temperature.

Luteinizing hormone levels remain elevated for only two days, but that is long enough to stimulate the formation of the functional corpus luteum. Progesterone secretion continues at relatively high levels for the next week, but unless pregnancy occurs the corpus luteum then begins to degenerate. Roughly 12 days after ovulation, the corpus luteum becomes nonfunctional, and progesterone and estrogen levels fall markedly.

The decline in progesterone and estrogen levels stimulates the hypothalamic receptors, and GnRH

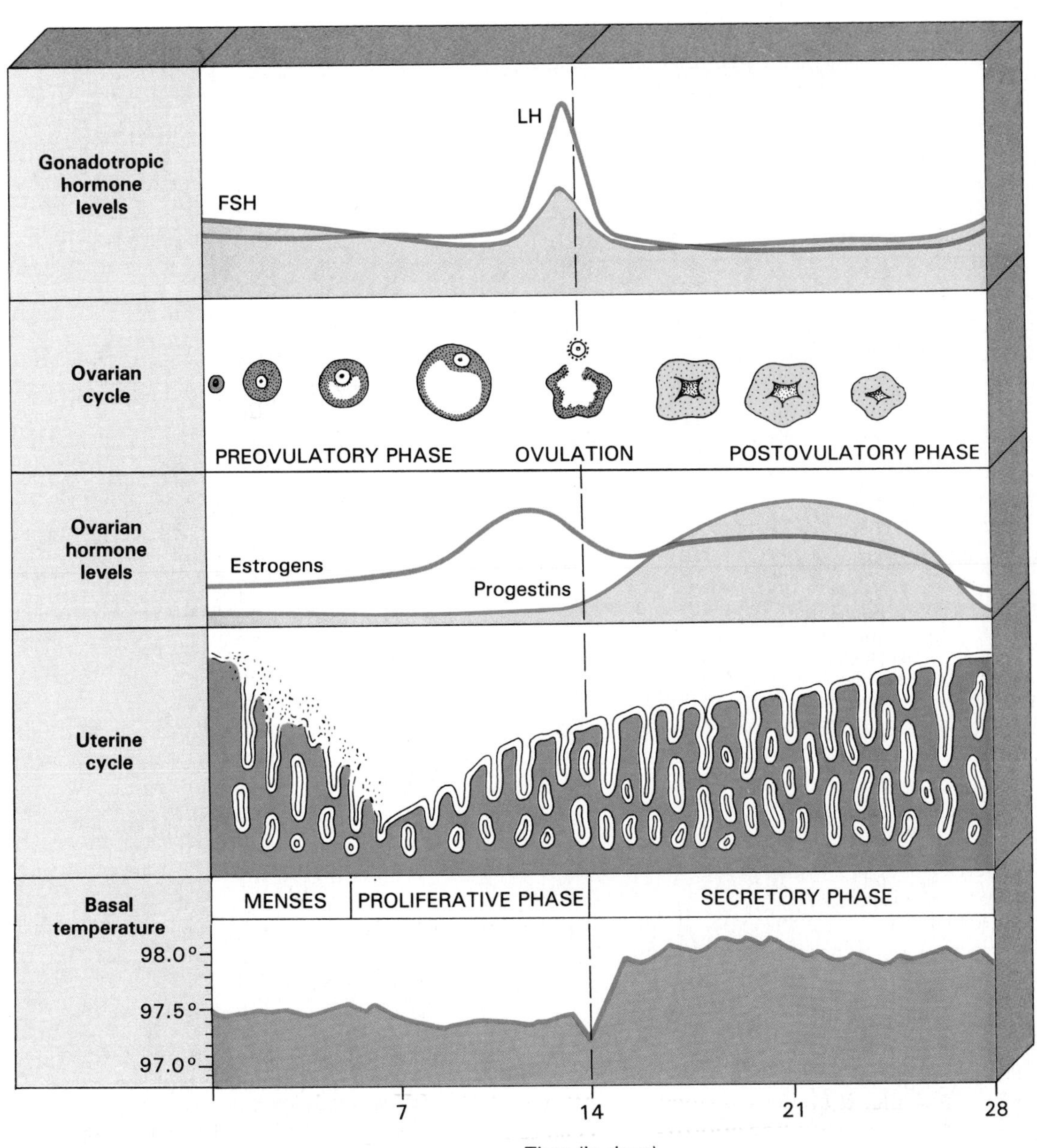

FIGURE 28-17
The Hormonal Regulation of Female Reproductive Function

production increases. This increase leads to an increase in FSH and LH production in the anterior pituitary, and the entire cycle begins again. Figure 28-18 reviews the functional relationships among the hormones involved in regulating the ovarian cycle.

Secondary Effects of the Ovarian Cycle

The hormonal changes involved with the ovarian cycle in turn affect the activities of other reproductive tissues and organs. At the uterus, the hormonal changes are responsible for the maintenance of the *menstrual cycle.*

THE MENSTRUAL CYCLE

The **menstrual** (MEN-stru-al) **cycle** averages 28 days in length, but it can range from 21–35 days in normal individuals. The cycle can be divided into three stages: *menses*, the *proliferative phase*, and the *secretory phase.*

Menses

The menstrual cycle begins with the onset of **menses** (MEN-sēz), a period marked by the wholesale destruction of the functional zone of the endometrium. The

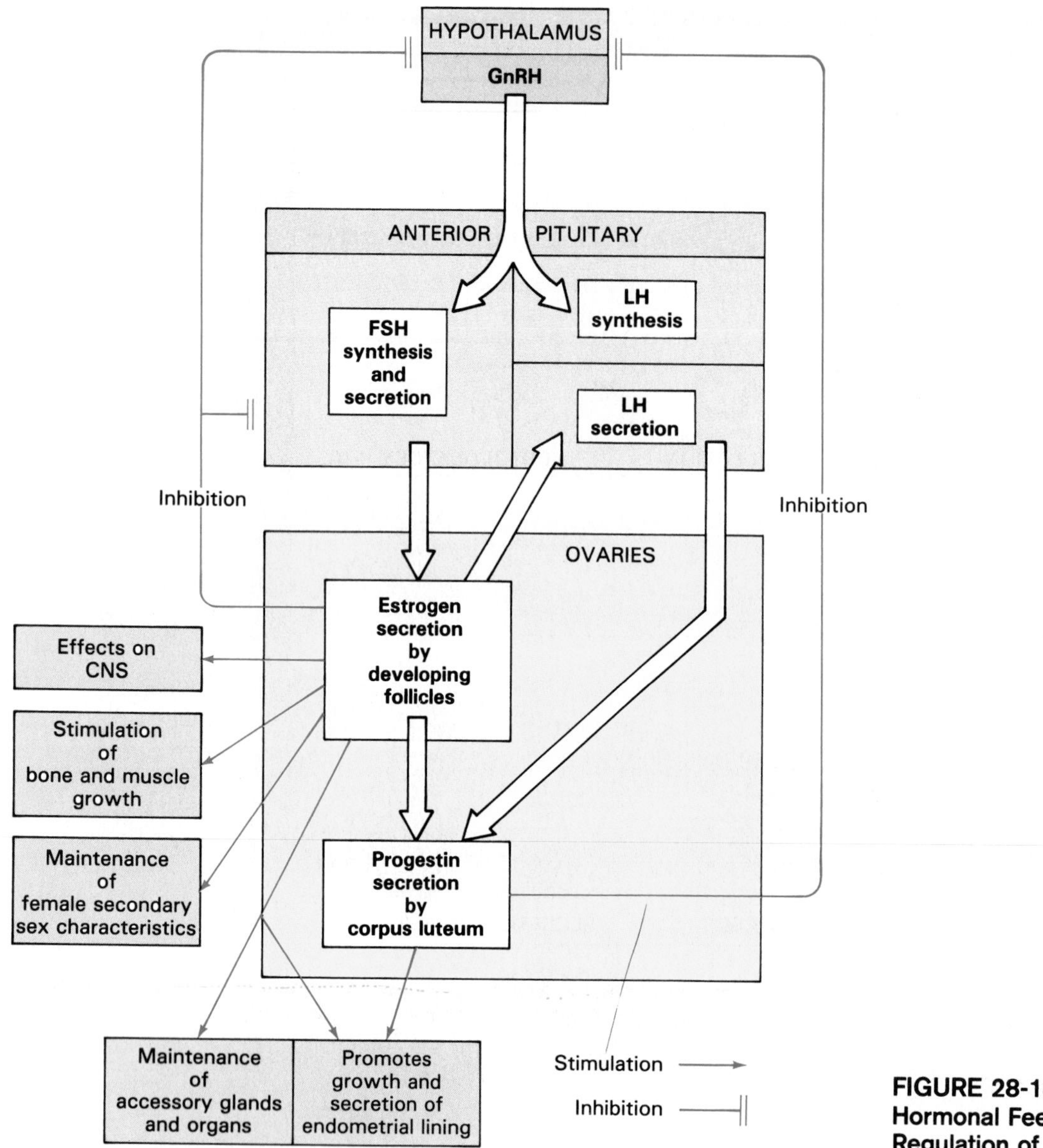

FIGURE 28-18
Hormonal Feedback and the Regulation of Ovarian Activity

arteries begin constricting, reducing blood flow to this region, and the secretory glands, epithelial cells, and other tissues of the functional zone begin to die of oxygen and nutrient deprivation. Eventually the weakened arterial walls rupture, and blood pours into the connective tissues of the functional zone. Blood cells and degenerating tissues break away and enter the uterine lumen, to be lost by passage through the external orifice and into the vagina. This sloughing of tissue, which continues until the entire functional zone has been lost (Figure 28-19a), is called **menstruation** (men-stru-Ā-shun). Menstruation usually lasts from 1 to 7 days, and over this period roughly 35/mℓ of blood are lost. Painful menstruation, or **dysmenorrhea**, may result from uterine inflammation and contraction or from conditions involving adjacent pelvic structures. [**CM**: *PMS*]

The Proliferative Phase

The basilar zone, including the basal portions of the uterine glands, survives menses intact because its circulatory supply remains constant. In the days following the completion of menses, the epithelial cells of the glands multiply and spread across the endometrial surface, restoring the integrity of the uterine epithelium. An early stage of this process is shown in Figure 28-19b. Further growth and vascularization will result in the complete restoration of the functional zone. As this reorganization proceeds, the endometrium is said to be in the **proliferative phase**. Because this occurs at the same time as the enlargement of primary and secondary follicles in the ovary, it is also known as the *preovulatory phase* or the *follicular phase* of the menstrual cycle.

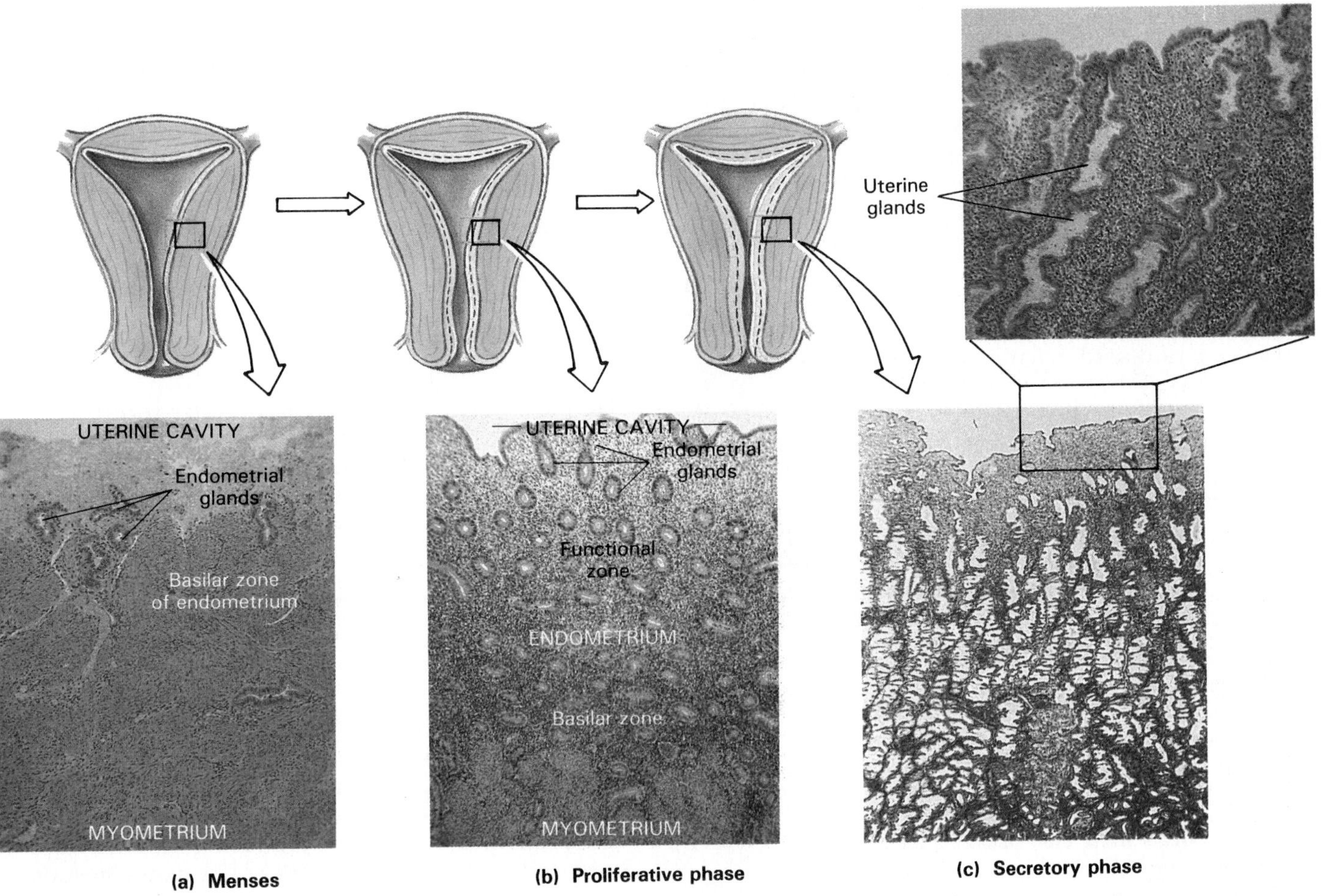

FIGURE 28-19
Uterine Changes in the Menstrual Cycle. These micrographs show the appearance of the endometrium at menstruation and during the proliferative phase and the secretory phase of the cycle. [LMs, (a)(b) and (c) inset ×57, (c) ×45]

By the time ovulation occurs, the functional zone is several millimeters thick, and prominent mucous glands extend to the border with the basilar zone. At this time the endometrial glands are manufacturing a mucus rich in glycogen. The entire functional zone is highly vascularized, with small arteries spiralling toward the inner surface from larger trunks in the myometrium.

The Secretory Phase

During the secretory phase of the cycle the glands enlarge, accelerating their rates of secretion, and the arteries elongate and spiral through the tissues of the functional zone. Because this phase begins at the time of ovulation, and it persists as long as the corpus luteum remains intact, it can also be called the *postovulatory phase*, or *luteal phase*, of the menstrual cycle.

Secretory activities peak about 12 days after ovulation, and Figure 28-19c shows the uterine lining at this time. Over the next day or two the glandular activity declines, and the menstrual cycle comes to a close. A new cycle then begins with the onset of menses and the disintegration of the functional zone. The postovulatory phase usually lasts 14 days, and individual variations are uncommon. As a result, the date of ovulation can be determined after the fact by counting backward 14 days from the first day of menses.

This fascinating cycle of events begins with the **menarche** (me-NAR-kē), or first menstrual period at puberty, typically at age 11–12. The cycles continue until age 45–50 when **menopause** (MEN-ō-paws), the last menstrual cycle, occurs. Over the intervening three and a half to four decades the regular appearance of menstrual cycles will be interrupted only by unusual circumstances, such as illness, stress, starvation, or pregnancy.

Clinical Comment: Amenorrhea

If menarche does not appear by age 16, or if the normal menstrual cycle of an adult becomes interrupted for six months or more, the condition of **amenorrhea** (ā-men-ō-RĒ-a) exists. Primary amenorrhea may indicate developmental abnormalities, such as nonfunctional ovaries or even the absence of a uterus, or an endocrine or genetic disorder. A transient amenorrhea may be caused by severe physical or emotional stresses. In effect, the reproductive system gets "switched off" under these conditions. Examples of factors that can cause amenorrhea include drastic weight reduction programs, anorexia nervosa, and severe depression or grief. Amenorrhea has also been observed in marathon runners and others engaged in training programs that require sustained high levels of exertion.

Hormones and the Menstrual Cycle

The sudden declines in progesterone and estrogen levels that accompany the degeneration of the corpus luteum result in the endometrial breakdown of menstruation. The sloughing of endometrial tissue continues for several days, until rising FSH, LH, and estrogen levels stimulate the repair and regeneration of the functional zone of endometrium. The preovulatory phase continues until rising progesterone levels mark the arrival of the postovulatory phase. The combination of estrogen and progesterone then causes the enlargement of the endometrial glands and an increase in their secretory activities. [**CM**: *Endometriosis*]

The hormonal fluctuations also cause physiological changes that affect the core body temperature. During the preovulatory period, when estrogen is the dominant hormone, the resting, or "basal," body temperature measured upon awakening in the morning is about one-half a degree Fahrenheit lower than it is during the postovulatory period, when progesterone dominates the endocrine picture. At the time of ovulation, basal temperature declines sharply, and this makes the temperature rise over the following day even more noticeable. As a result, by keeping records of body temperature over a few menstrual cycles a woman can often determine the precise day of ovulation. This information can be very important for individuals wishing to avoid or promote a pregnancy, for such an event can occur only if an ovum becomes fertilized within a day of its ovulation.

- **What changes would you expect to observe in the ovulatory cycle if the LH surge did not occur?**
- **Blockage of progesterone receptors in the uterus would have what effect on the endometrium?**
- **What event occurs in the menstrual cycle when the levels of estrogen and progesterone decline?**

PREGNANCY, HORMONES, AND MATERNAL SYSTEMS

If fertilization occurs, the zygote (fertilized egg) undergoes a series of cell divisions, forming a hollow ball of cells known as a **blastocyst** (BLAS-tō-sist). Upon arrival in the uterine cavity, the blastocyst initially obtains nutrients by absorbing the secretions of the uterine glands. Within a few days it contacts the endometrial wall, erodes the epithelium, and buries itself in the endometrium. This process, known as **implantation** (im-plan-TĀ-shun), initiates the chain of events leading to the formation of a special organ, the **placenta**, that will support embryonic and fetal development over the next nine months.

The placenta provides a medium for the transfer of dissolved gases, nutrients, and waste products between the fetal and maternal bloodstreams. It also acts as an endocrine organ, producing the hormones indicated in Figure 28-20. These hormones include *human chorionic gonadotropin*, *relaxin*, and *human placental lactogen* as well as estrogens and progestins.

Human Chorionic Gonadotropin

The hormone **human chorionic** (kō-rē-ON-ik) **gonadotropin** (**HCG**) appears in the maternal bloodstream soon after implantation has occurred. The presence of HCG in blood or urine samples provides a reliable indication of pregnancy, and kits sold for the early detection of pregnancy are sensitive to the presence of this hormone.

In function HCG resembles LH, for in the presence of HCG the corpus luteum does not degenerate. Instead it remains functional and continues to produce progesterone. Because progesterone secretion continues, the endometrial lining remains perfectly functional, and menstruation does not occur. If it did the pregnancy would end, for the functional zone of the endometrium would disintegrate.

In the presence of HCG, the corpus luteum persists for about three months before degenerating. Its departure does not trigger the return of menstrual periods, because by then the placenta is actively secreting both estrogen and progesterone. Over the following months the placenta also synthesizes two additional hormones, **relaxin** and **human placental lactogen** (LAK-tō-jen), or **HPL**.

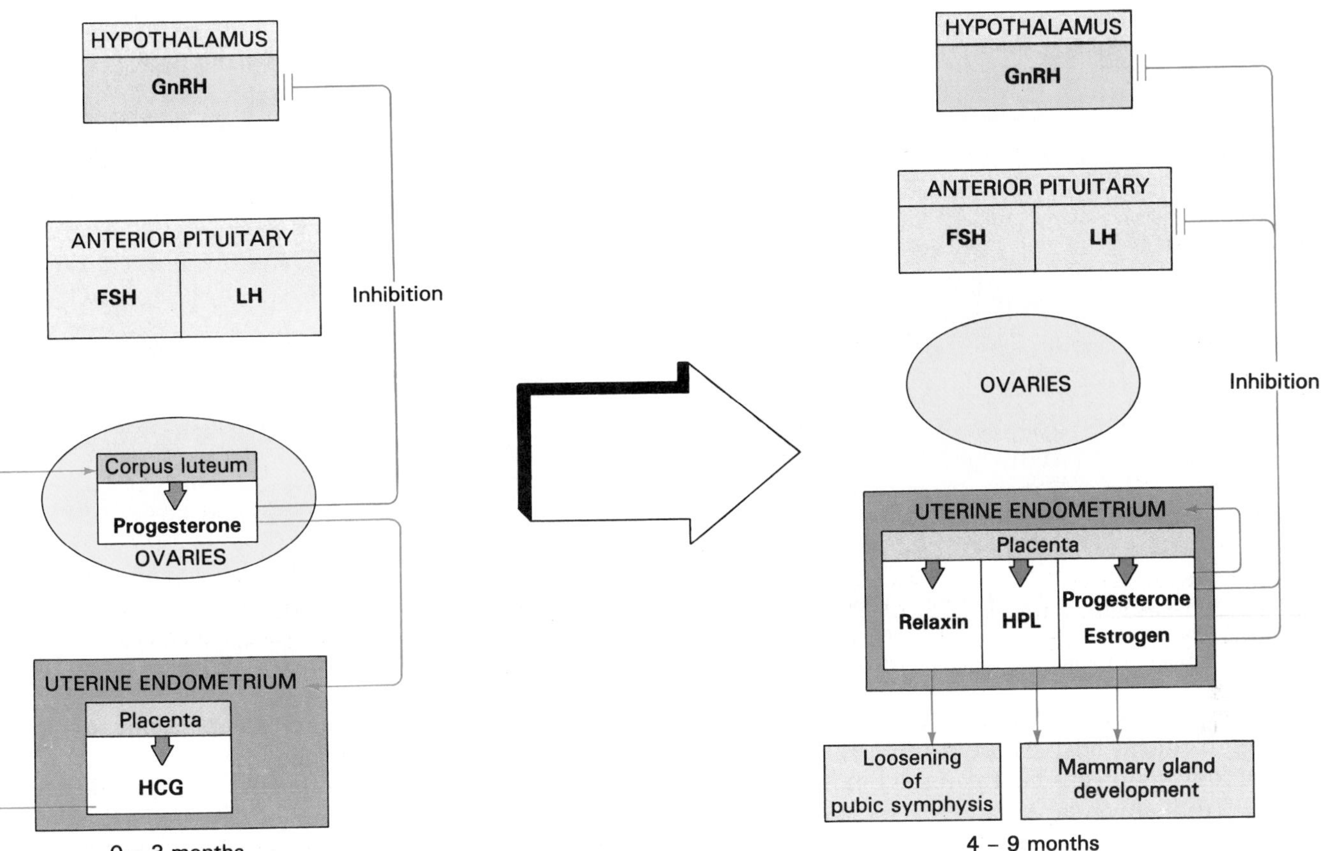

FIGURE 28-20
Hormones and the Effects of Pregnancy on the Maternal Reproductive System

Relaxin

Relaxin increases the flexibility of the symphysis pubis, permitting expansion of the pelvis during delivery. It also causes dilation of the cervix, making it easier for the fetus to leave the uterus and enter the vaginal canal.

Human Placental Lactogen

Human placental lactogen helps prepare the mammary glands for milk production. The resting mammary glands are relatively unaffected by the changes in estrogen and progesterone concentrations that accompany pregnancy. The conversion from resting to active status requires the presence of two additional hormones. One of them is HPL, and the other is the pituitary hormone *prolactin* (*PRL*). As noted in Chapter 18, prolactin is produced by cells in the anterior pituitary. (p. 577) But in the nonpregnant woman, the hypothalamus produces an inhibitory hormone (*PIH*) that keeps prolactin secretion at a minimum. During pregnancy the sustained, elevated levels of estrogens and progesterone depress the production of PIH, and the prolactin concentration in the blood begins to rise. The combination of PRL and HPL then prepares the mammary glands for their secretory roles.

Figure 28-21a,b compares the histological organization of the inactive and active mammary glands. The resting mammary gland is dominated by components of the duct system, rather than glandular cells. The secretory apparatus does not develop until pregnancy occurs. At that time the ducts become mitotically active, and gland cells begin to appear. By the end of the sixth month of pregnancy the mammary glands are fully developed, and the gland begins producing a secretion known as **colostrum** (ko-LOS-trum). Colostrum contains relatively more proteins and far less fat than milk, and it will be provided to the infant during the first two or three days of life. Many of the proteins are immunoglobulins that may help the infant ward off infections until its own immune system becomes fully functional.

As colostrum production declines, the mammary glands convert to milk production. Milk consists of a mixture of water, proteins, amino acids, lipids, sugars, and salts. It also contains large quantities of lysozymes, enzymes with antibiotic properties. Human milk provides roughly 750 calories per liter. The secretory rate varies, depending on the demand, but a 5–6 kg (11–13 lb) infant usually requires around 850 mℓ of milk per day.

The actual secretion of the mammary glands is triggered when the infant begins to suck on the nipple. Stimulation of tactile receptors at that site

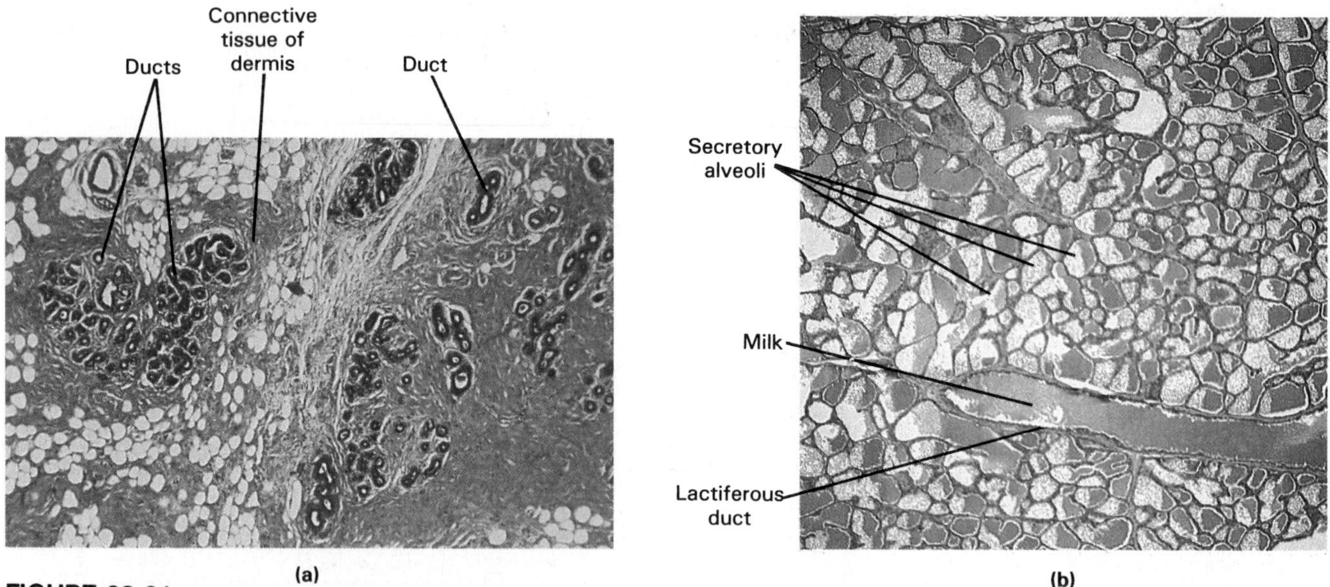

FIGURE 28-21
Functional Development of the Mammary Glands. Micrographs comparing the histological organization of (a) the resting and (b) active mammary glands. [LMs, (a) ×247, (b) ×110] These functional changes occur under the direction of several interacting hormones, diagrammed in Figure 28-20.

leads to the release of oxytocin at the posterior pituitary. When oxytocin reaches the mammary gland it causes the contraction of smooth muscles in the walls of the lactiferous ducts and sinuses. This results in the ejection of milk, and this milk ejection reflex, or milk let-down, was detailed in Chapter 18 and diagrammed in Figure 18-10. (p. 576) This reflex continues to function until weaning occurs, typically 1–2 years after birth. Milk production ceases soon after, and the mammary glands gradually return to a resting state.

Aging and the Reproductive System

The aging process affects the reproductive systems of men and women. The most striking age-related changes in the female reproductive system occur at menopause.

MENOPAUSE

Menopause is usually defined as the time that ovulation and menstruation cease. Menopause typically occurs at age 45–55, but in the years preceding the finale the regularity of the ovarian and menstrual cycles gradually fades. A shortage of primordial follicles is the underlying cause of these developments. It has been estimated that almost 7 million potential oocytes are found in fetal ovaries after five months of development, but the number drops to about 1 million at birth, and to a few hundred thousand at puberty. By age 50, there are no primordial follicles left to respond to FSH; in **premature menopause** this occurs before age 40.

Menopause is accompanied by a sharp and sustained rise in the production of GnRH, FSH, and LH, while circulating concentrations of estrogen and progesterone decline. The decline in estrogen levels leads to reductions in the size of the uterus and breasts, accompanied by a thinning of the urethral and vaginal walls. The reduced estrogen concentrations have also been linked to the development of osteoporosis, presumably because bone deposition proceeds at a slower rate. A variety of neural and cardiovascular effects are also reported, including "hot flashes," anxiety, and depression, but the hormonal mechanisms involved are not well understood. The symptoms accompanying and following menopause are sufficiently unpleasant that about 40 percent of menopausal women eventually seek medical assistance. Hormone replacement therapies involving a combination of estrogens and progestins can often prevent osteoporosis and the neural and vascular changes associated with menopause.

THE MALE CLIMACTERIC

Changes in the male reproductive system occur more gradually, over a period known as the **male climacteric**. Circulating testosterone levels begin to decline between ages 50 and 60, coupled with increases in circulating levels of FSH and LH. Although sperm production continues—many men can father children well into their eighties—there is a gradual reduction in sexual activity in older men. This may be linked to declining testosterone levels, and some clinicians are now tentatively suggesting the use of testosterone replacement therapy to enhance libido (sexual drive) in elderly men.

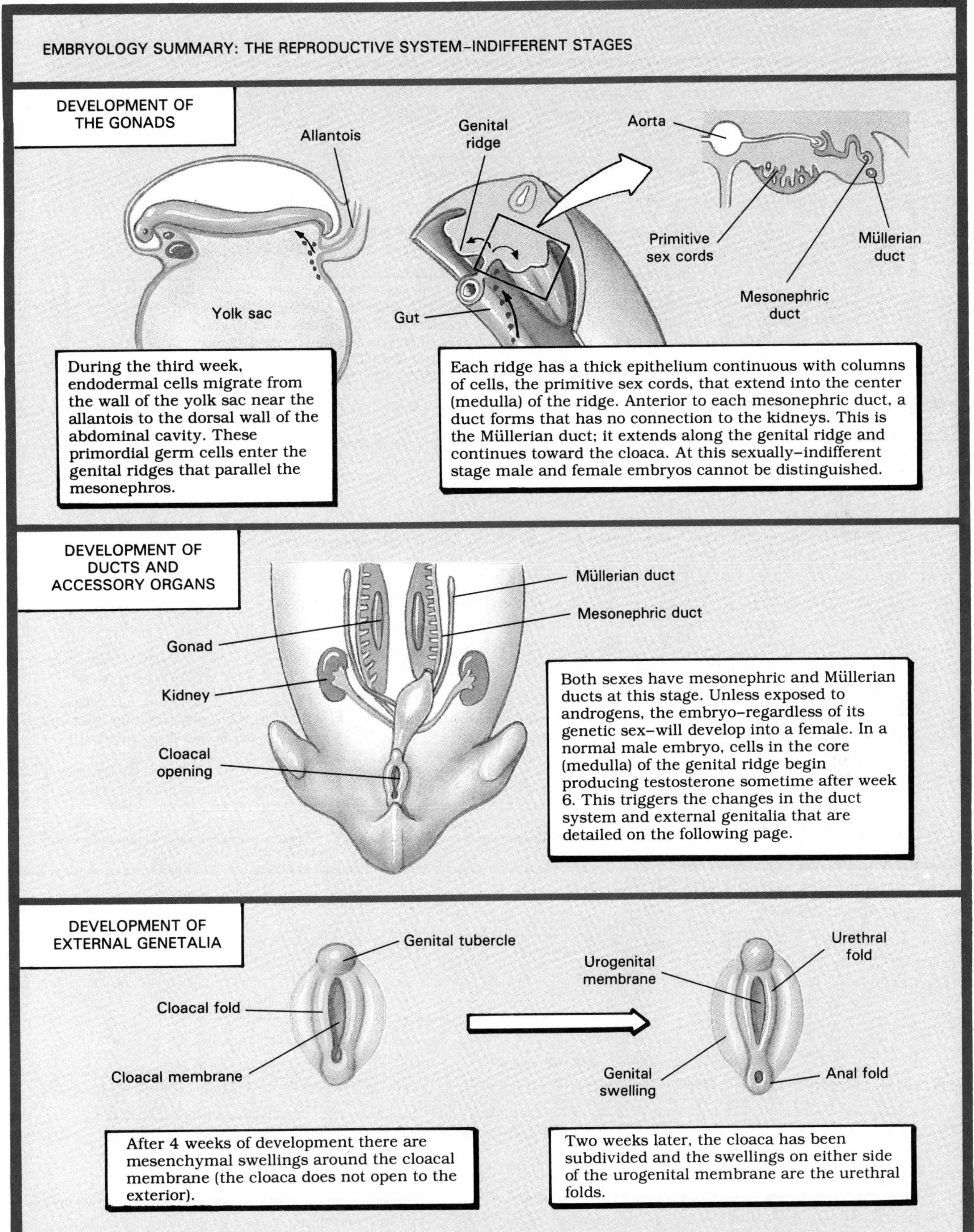

EMBRYOLOGY SUMMARY: THE REPRODUCTIVE SYSTEM–INDIFFERENT STAGES
DEVELOPMENT OF THE GONADS
Allantois
Genital ridge
Aorta
Primitive sex cords
Müllerian duct
Mesonephric duct
Yolk sac
Gut
During the third week, endodermal cells migrate from the wall of the yolk sac near the allantois to the dorsal wall of the abdominal cavity. These primordial germ cells enter the genital ridges that parallel the mesonephros.
Each ridge has a thick epithelium continuous with columns of cells, the primitive sex cords, that extend into the center (medulla) of the ridge. Anterior to each mesonephric duct, a duct forms that has no connection to the kidneys. This is the Müllerian duct; it extends along the genital ridge and continues toward the cloaca. At this sexually–indifferent stage male and female embryos cannot be distinguished.
DEVELOPMENT OF DUCTS AND ACCESSORY ORGANS
Müllerian duct
Mesonephric duct
Gonad
Kidney
Cloacal opening
Both sexes have mesonephric and Müllerian ducts at this stage. Unless exposed to androgens, the embryo–regardless of its genetic sex–will develop into a female. In a normal male embryo, cells in the core (medulla) of the genital ridge begin producing testosterone sometime after week 6. This triggers the changes in the duct system and external genitalia that are detailed on the following page.
DEVELOPMENT OF EXTERNAL GENETALIA
Genital tubercle
Cloacal fold
Cloacal membrane
Urogenital membrane
Urethral fold
Genital swelling
Anal fold
After 4 weeks of development there are mesenchymal swellings around the cloacal membrane (the cloaca does not open to the exterior).
Two weeks later, the cloaca has been subdivided and the swellings on either side of the urogenital membrane are the urethral folds.

EMBRYOLOGY SUMMARY: Development of the Reproductive System

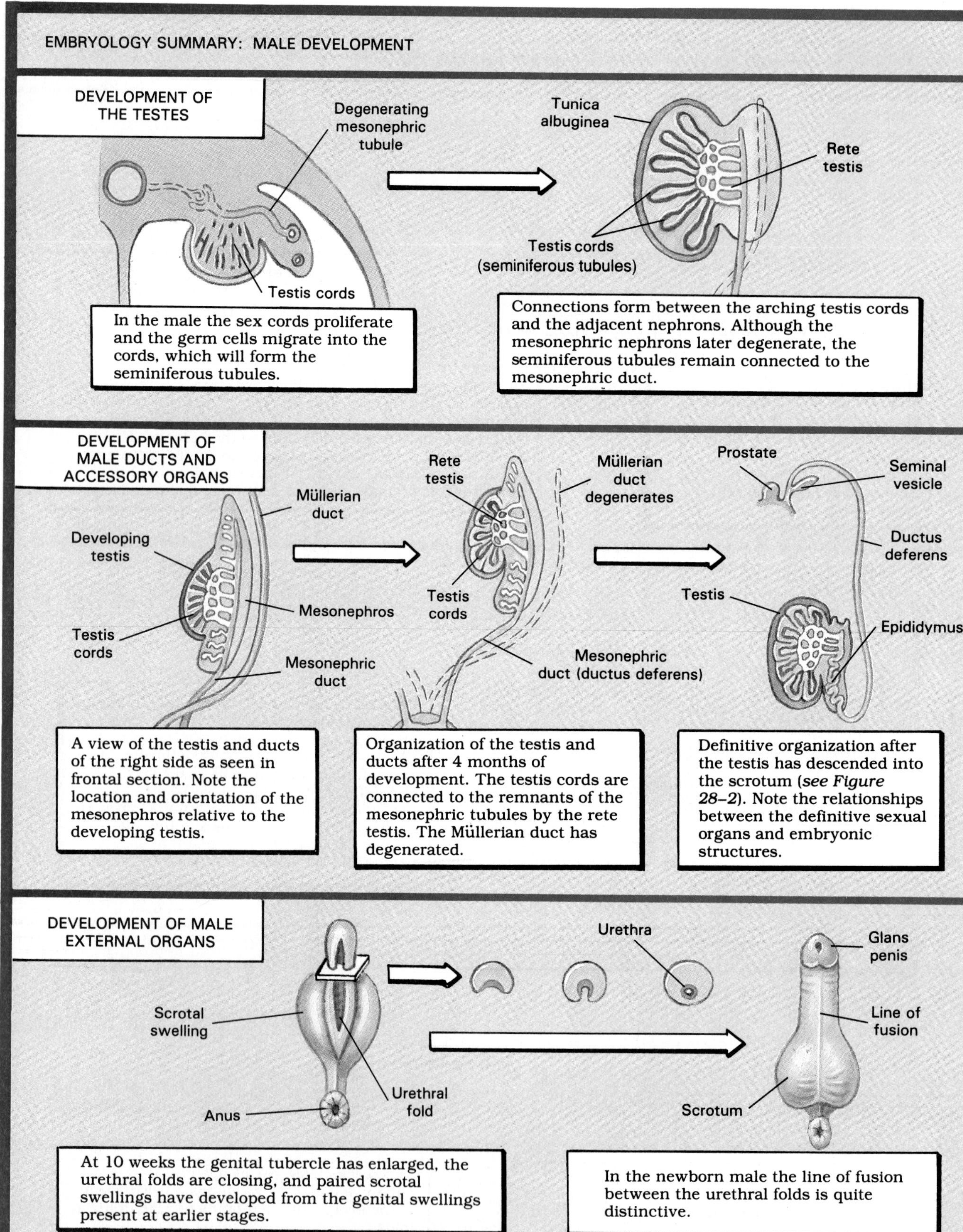

In the male the sex cords proliferate and the germ cells migrate into the cords, which will form the seminiferous tubules.

Connections form between the arching testis cords and the adjacent nephrons. Although the mesonephric nephrons later degenerate, the seminiferous tubules remain connected to the mesonephric duct.

A view of the testis and ducts of the right side as seen in frontal section. Note the location and orientation of the mesonephros relative to the developing testis.

Organization of the testis and ducts after 4 months of development. The testis cords are connected to the remnants of the mesonephric tubules by the rete testis. The Müllerian duct has degenerated.

Definitive organization after the testis has descended into the scrotum (*see Figure 28–2*). Note the relationships between the definitive sexual organs and embryonic structures.

At 10 weeks the genital tubercle has enlarged, the urethral folds are closing, and paired scrotal swellings have developed from the genital swellings present at earlier stages.

In the newborn male the line of fusion between the urethral folds is quite distinctive.

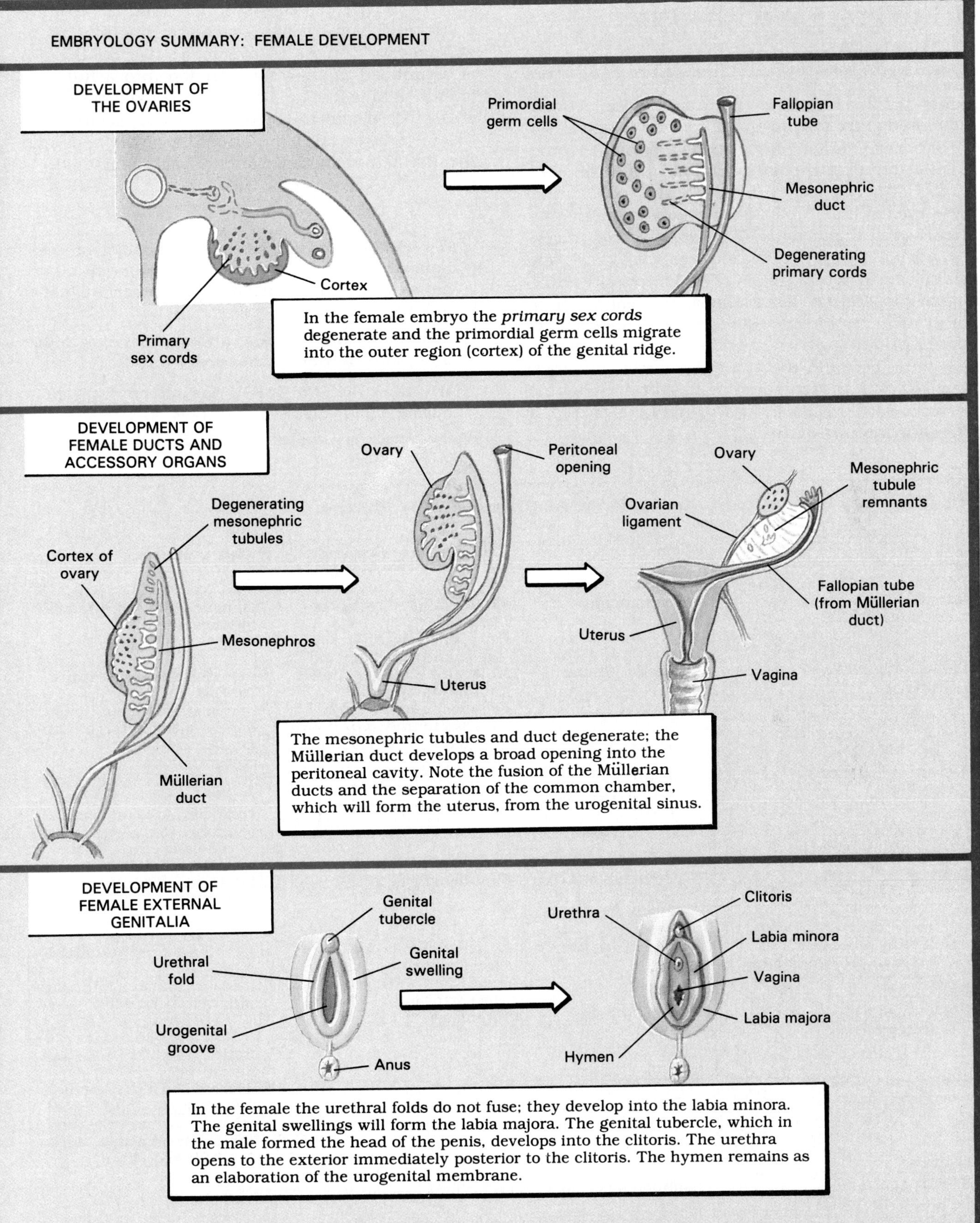

In the female embryo the *primary sex cords* degenerate and the primordial germ cells migrate into the outer region (cortex) of the genital ridge.

The mesonephric tubules and duct degenerate; the Müllerian duct develops a broad opening into the peritoneal cavity. Note the fusion of the Müllerian ducts and the separation of the common chamber, which will form the uterus, from the urogenital sinus.

In the female the urethral folds do not fuse; they develop into the labia minora. The genital swellings will form the labia majora. The genital tubercle, which in the male formed the head of the penis, develops into the clitoris. The urethra opens to the exterior immediately posterior to the clitoris. The hymen remains as an elaboration of the urogenital membrane.

Integration with Other Systems

Figure 28-22 summarizes the relationships between the reproductive system and other physiological systems. Normal human reproduction is a complex process that requires the participation of multiple systems. Hormones play a major role in coordinating these events, and Table 28-2 reviews the hormones discussed in this chapter. The reproductive process depends on a variety of physical, physiological, and psychological factors, many of which require intersystem cooperation. For example, the male's sperm count must be adequate, the semen must have the correct pH and nutrients, and erection and ejaculation must occur in the proper sequence. For these steps to occur, the reproductive, digestive, endocrine, nervous, cardiovascular, and urinary systems must all be functioning normally.

Even when all else is normal, and fertilization occurs at the proper time and place, a normal infant will not result unless the zygote, a single cell the size of a pinhead, manages to develop into a full-term fetus weighing 3–4 kg. Chapter 29 considers the process of development, focusing on the mechanisms that determine both the structure of the body and the distinctive characteristics of each individual.

- **Her doctor tells Sue that her pregnancy test indicates elevated levels of the hormone HCG (human chorionic gonadotropin). Is she pregnant or not?**
- **Sometimes while nursing, a woman may experience noticeable uterine contractions. Why?**
- **Why does the level of FSH rise and remain high during menopause?**

TABLE 28-2 Hormones of the Reproductive System and the Placenta

Hormone	*Source*	*Regulation of Secretion*	*Primary Effects*
REPRODUCTIVE SYSTEM			
Gonadotrophin-releasing hormone (GnRH)	Hypothalamus	*Male*: inhibited by testosterone *Female*: inhibited by estrogens and/or progestins	Stimulates FSH secretion, LH synthesis
Follicle-stimulating hormone (FSH)	Anterior pituitary	*Male*: stimulated by GnRH, inhibited by inhibin *Female*: stimulated by GnRH, inhibited by estrogens and/or progestins	*Male*: stimulates spermatogenesis *Female*: stimulates follicle development, estrogen production, and egg maturation
Estrogens (primarily estradiol)	Follicular and interstitial cells of ovaries	Stimulated by FSH	Stimulates LH secretion, maintains secondary sex characteristics and sexual behavior, stimulates repair of endometrium, inhibits secretion of GnRH
Inhibin	Sustentacular cells of testes and interstitial cells of ovaries	Stimulated by FSH	Inhibits secretion of GnRH and FSH
Luteinizing hormone/ interstitial cell-stimulating hormone (LH/ICSH)	Anterior pituitary	*Male*: stimulated by GnRH *Female*: production stimulated by GnRH, secretion by estrogens	*Male*: stimulates interstitial cells *Female*: stimulates follicular and interstitial cells
Progestins (primarily progesterone)	Corpus luteum	Stimulated by LH	Stimulates endometrial growth and glandular secretion, inhibits GnRH secretion
Androgens (primarily testosterone)	Interstitial cells of testes	Stimulated by ICSH	Maintains secondary sexual characteristics and sexual behavior, promotes maturation of spermatozoa, inhibits GnRH secretion
PLACENTA			
Human chorionic gonadotrophin (HCG)	Maintains corpus luteum for first 3 months of pregnancy		
Relaxin	Increases flexibility of pubic symphysis, dilates cervix		
Human placental lactogen (HPL)	Assists prolactin in preparing the mammary glands for milk production		
Estrogens, Progestins	Functions as noted above		

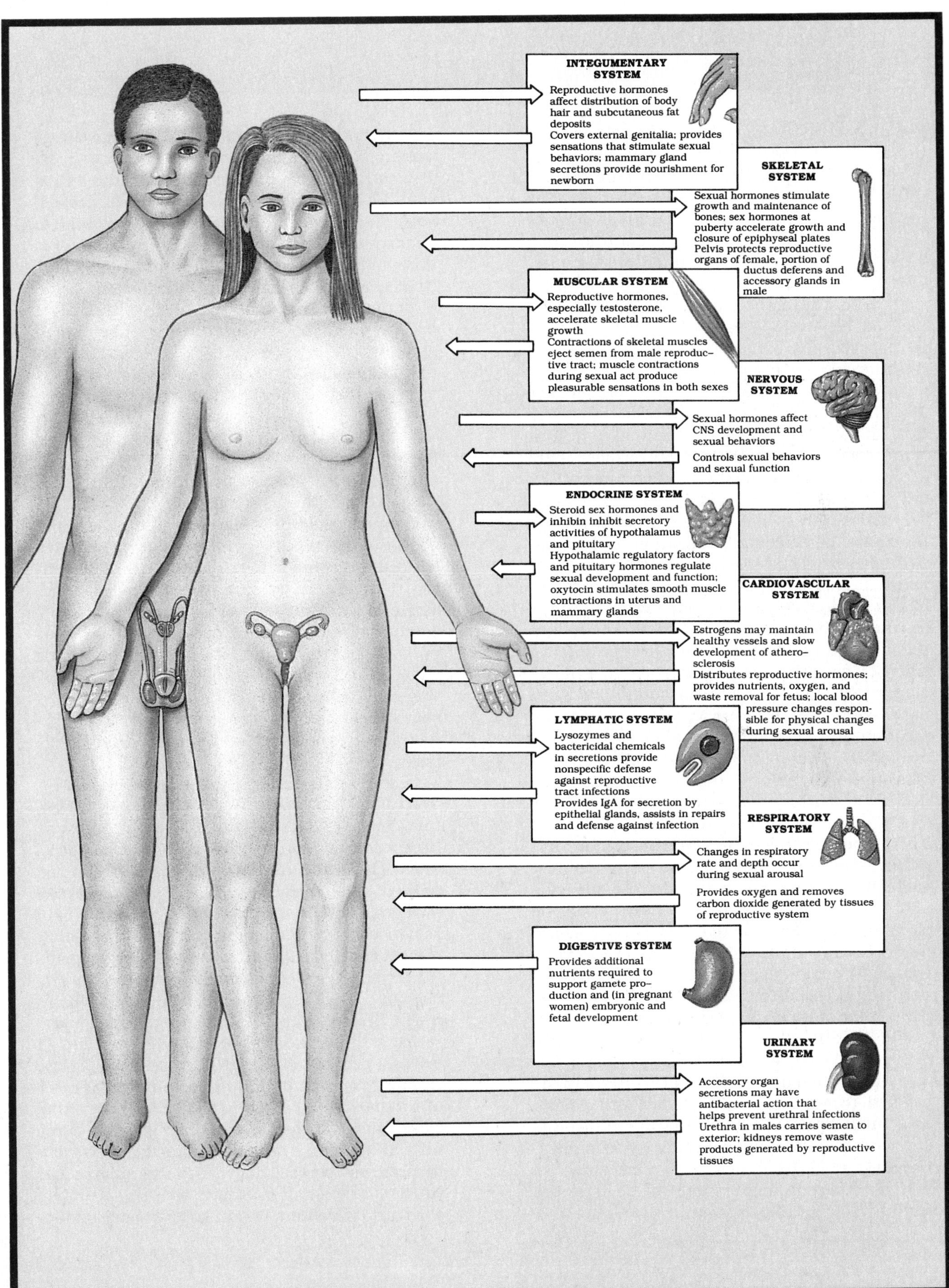
INTEGUMENTARY SYSTEM
Reproductive hormones affect distribution of body hair and subcutaneous fat deposits
Covers external genitalia; provides sensations that stimulate sexual behaviors; mammary gland secretions provide nourishment for newborn
SKELETAL SYSTEM
Sexual hormones stimulate growth and maintenance of bones; sex hormones at puberty accelerate growth and closure of epiphyseal plates
Pelvis protects reproductive organs of female, portion of ductus deferens and accessory glands in male
MUSCULAR SYSTEM
Reproductive hormones, especially testosterone, accelerate skeletal muscle growth
Contractions of skeletal muscles eject semen from male reproductive tract; muscle contractions during sexual act produce pleasurable sensations in both sexes
NERVOUS SYSTEM
Sexual hormones affect CNS development and sexual behaviors
Controls sexual behaviors and sexual function
ENDOCRINE SYSTEM
Steroid sex hormones and inhibin inhibit secretory activities of hypothalamus and pituitary
Hypothalamic regulatory factors and pituitary hormones regulate sexual development and function; oxytocin stimulates smooth muscle contractions in uterus and mammary glands
CARDIOVASCULAR SYSTEM
Estrogens may maintain healthy vessels and slow development of athero-sclerosis
Distributes reproductive hormones; provides nutrients, oxygen, and waste removal for fetus; local blood pressure changes responsible for physical changes during sexual arousal
LYMPHATIC SYSTEM
Lysozymes and bactericidal chemicals in secretions provide nonspecific defense against reproductive tract infections
Provides IgA for secretion by epithelial glands, assists in repairs and defense against infection
RESPIRATORY SYSTEM
Changes in respiratory rate and depth occur during sexual arousal
Provides oxygen and removes carbon dioxide generated by tissues of reproductive system
DIGESTIVE SYSTEM
Provides additional nutrients required to support gamete production and (in pregnant women) embryonic and fetal development
URINARY SYSTEM
Accessory organ secretions may have antibacterial action that helps prevent urethral infections
Urethra in males carries semen to exterior; kidneys remove waste products generated by reproductive tissues

FIGURE 28-22
Functional Relationships between the Reproductive System and Other Systems

Health News: Conception Control—Principles and Prospects

For physiological, logistical, financial, or emotional reasons most adults practice some form of conception control during their reproductive years. When the simplest and most obvious method, sexual abstinence, is unsatisfactory for some reason, another method of contraception must be used to avoid unwanted pregnancies. The selection process can be quite involved, for there are many methods available. Because each has specific strengths and weaknesses, the potential risks and benefits must be carefully analyzed on an individual basis.

Well over 50 percent of U.S. women age 15–44 are practicing some method of contraception; in 1990 an estimated 50 million American women were taking oral birth control pills. There are many different methods of contraception; only a few will be considered here.

Sterilization makes one unable to provide functional gametes for fertilization. Either sexual partner may be sterilized with the same net result. In a **vasectomy** (vaz-EK-to-mē) a segment of the ductus deferens is removed, making it impossible for spermatozoa to pass from the epididymis to the distal portions of the reproductive tract. The surgery can be performed in a physician's office in a matter of minutes. The spermatic cords are located as they ascend from the scrotum on either side, and after each cord is opened the ductus deferens is severed. After a 1-cm section is removed, the cut ends are usually tied shut. With the section removed, the cut ends do not reconnect; in time, scar tissue forms a permanent seal. A more recent vasectomy procedure often makes it possible to restore fertility at a later date. In this procedure the cut ends of the ductus deferens are blocked with silicone plugs that can later be removed. After a vasectomy the man experiences normal sexual function, for the epididymal and testicular secretions normally account for only around 5 percent of the volume of the semen. Spermatozoa continue to develop, but they remain within the epididymis until they degenerate.

In the female the uterine tubes can be blocked through a surgical procedure known as a **tubal ligation**. Since the surgery involves entering the abdominopelvic cavity, complications are more likely than with vasectomy. As in a vasectomy, attempts may be made to restore fertility after a tubal ligation.

Oral contraceptives manipulate the female hormonal cycle so that ovulation does not occur. The contraceptive pills produced in the 1950s contained relatively large amounts of progestins. These concentrations were adequate to suppress pituitary production of GnRH, so FSH was not released and ovulation did not occur. Unpleasant side effects included endometrial bleeding, and most of the oral contraceptive products developed subsequently added small amounts of estrogens. Current combination pills differ significantly from the earlier products because the hormonal doses are much lower, with only one-tenth the progestins and less than half the estrogens. The hormones are administered in a cyclic fashion, beginning five days after the start of menses and continuing for the next three weeks. Over the fourth week the woman takes placebo pills or no pills at all. Low-dosage combination pills are sometimes prescribed for women experiencing irregular menstrual cycles, for they create a 28-day cycle. There are now at least 20 different brands of combination oral contraceptives available, and over 200 million women are using them worldwide. In the United States, 25 percent of women under age 45 use the combination pill to prevent conception. The progestin-only "minipill" has proven to be less effective in preventing pregnancy.

The **condom**, also called a *prophylactic*, or "rubber," covers the body of the penis during intercourse and keeps spermatozoa from reaching the female reproductive tract. **Vaginal barriers** such as the *diaphragm*, *cervical cap*, or *vaginal sponge* rely on similar principles. A diaphragm, the most popular form of vaginal barrier in use at the moment, consists of a dome of latex rubber with a small metal hoop supporting the rim. Because vaginas vary in size, women choosing this method must be individually fitted. Before intercourse the diaphragm is inserted so that it covers the cervical os, and it is usually coated with a small amount of spermicidal jelly or cream, adding to the effectiveness of the barrier. The cervical cap is smaller and lacks the metal rim. It, too, must be fitted carefully, but unlike the diaphragm it may be left in place for several days. The vaginal sponge consists of a small synthetic sponge saturated with a *spermicide* a sperm-killing foam or jelly.

An **intrauterine device (IUD)** consists of a small plastic loop or a T that can be inserted into the uterine chamber. The mechanism of action remains uncertain, but it is known that IUDs stimulate prostaglandin production in the uterus. The net result is an alteration in the chemical composition of uterine secretions, and the changes in the intrauterine environment lower the chances for fertilization and subsequent implantation. IUDs are in limited use today in the U.S. but they remain popular in many other countries.

The **rhythm method** involves abstaining from sexual activity on the days ovulation might be occurring. The timing is estimated based on previous patterns of menstruation and sometimes by following changes in basal body temperature.

Sterilization, oral contraceptives, condoms, and vaginal barriers are the primary contraception methods for all age groups. But the relative proportion of the population using a particular method changes with age. Sterilization is most popular in older women, who may already have had children. Relative availability may also play a role. For example, a sexually active female under age 18 can buy a condom more easily than she can obtain a prescription for an oral contraceptive. But many of the observed changes occur because the relationship between risks and benefits varies for each age group.

When attempting to make a decision concerning the use and selection of contraceptives, many people simply examine the list of potential complications and make the "safest" choice. For example, media coverage of the potential risks associated with oral contraceptives made many women reconsider their use. Women taking oral contraceptives are at increased risk for venous thrombosis, strokes, pulmonary embolism, and (for women over 35) heart disease. But complex decisions should not be made on such a simplistic basis, and the risks associated with contraceptive use must be considered in light of their relative efficiencies. Pregnancy, although a natural phenomenon, has its risks, and the mortality rate for pregnant women averages around 8 deaths per 100,000 pregnancies. That average incorporates a broad range; in 1983 the rate was 5.4 per 100,000 in women under 20, and 27 per 100,000 for women over 40. Although these risks are small, for pregnant women over age 35 the chances of dying from complications related to pregnancy are almost twice as great as the chances of being killed in an automobile accident, and many times greater than the risks associated with the use of oral contraceptives.

By combining the data for failure rate with the known risks of pregnancy, and comparing this with the risks of the contraceptive procedure itself, you can assess the methods more precisely. This has been done in Table 28-3.

Before age 35, *any contraceptive method is safer than pregnancy*. In general over this period the risks are proportional to the failure rates of each method. The notable exception involves individuals taking the pill who also smoke cigarettes. Younger women are more fertile, so despite a lower mortality rate for each pregnancy they are likely to have more pregnancies. As a result birth control failures imply a higher risk in the younger age groups.

After age 35 the risk of complications associated with oral contraceptive use increases,

TABLE 28-3 Relative Risks of Mortality for Women Using Contraceptives

Contraceptive Technique	*Mortality Rate in Each Age Group*[a]		
	15–24	25–34	35–44
None[b]	7.2	12.9	26.9
Oral contraceptives			
Smokers	3.0	10.3	84.5
Nonsmokers	0.6	1.6	23.1
IUD	1.2	1.3	2.0
Barrier methods[b]	1.5	0.9	4.5

[a] Deaths per 100,000 women.
[b] Deaths directly related to conception with or without contraceptive failure.
Howard W. Orý, *Mortality Associated with Fertility and Fertility Control: 1983*, in Family Planning Perspectives, Vol. 15, No. 2 (1983).
Benjamin P. Sachs, Peter M. Layde, George S. Rubin, and Roger W. Rochat, *Reproductive Mortality in the United States*, in JAMA, Vol. 247, No. 20 (1982), p. 2789–2792.

while those of other methods remain relatively stable. Women over age 35 (smokers) or 40 (nonsmokers) are therefore often advised to seek other forms of contraception.

Several interesting contraceptive methods are now being tested. Vaginal rings and the Norplant system involve silicone rubber structures impregnated with steroid hormones that leak out at a constant rate. The vaginal rings are placed in the vagina, while the Norplant system involves the insertion of small rods beneath the skin. Both methods have undergone clinical trials, and the Norplant system has shown long-term effectiveness comparable to that of combined oral contraceptives, at lower steroid levels. Although the Norplant system is now available by prescription, a relatively high cost has to date limited the use of this contraceptive method.

Male contraceptives are also under development. *Gossypol*, a yellow pigment extracted from cottonseed oil, produces a dramatic decline in sperm count and sperm motility after 2 months. It can be administered topically, as it is readily absorbed through the skin. Fertility returns within a year after treatment is discontinued.

If contraceptive methods fail, options exist to either prevent implantation or terminate the pregnancy. The "morning-after pills" contain estrogens or progestins. They may be taken within 72 hours of intercourse, and they appear to act by altering the transport of the zygote or preventing its attachment to the uterine wall. **Abortion** refers to the termination of a pregnancy. Most clinicians discriminate between *spontaneous*, *therapeutic*, and *induced abortions*. **Spontaneous abortions**, or *miscarriages*, occur naturally, due to some developmental or physiological problem. **Therapeutic abortions** are performed when continuing the pregnancy represents a threat to the life and health of the mother.

Induced abortions ("elective abortions") are performed at the request of the individual. Each year there are approximately 1.5 million induced abortions in the United States, roughly one abortion for every three births. Most involve unmarried and/or adolescent women. The ratio between abortions and deliveries for married women averages 1:10, while for unmarried women and adolescents there are nearly twice as many abortions as deliveries. Induced abortions are currently legal during the first three months after conception, and many states permit abortions, sometimes with restrictions, until the fifth or sixth developmental month.

These operations are now the focus of considerable controversy. In simple terms, the pro-choice groups feel that each woman has a right to control her reproductive behavior and that abortion may be preferable to delivery of another unwanted infant into an already crowded world. The anti-abortion ("pro-life") groups feel that abortion at any stage is ending a human life and that there are couples waiting for the chance to adopt newborn infants. Opinions concerning the morality of abortion and current abortion laws must be left to the individual reader.

CHAPTER REVIEW

Review of Selected Clinical Terms

pelvic inflammatory disease (PID) (*p. 933*)
An infection of the uterine tubes.

gonorrhea (gon-o-RĒ-a) (*p. 933*)
A sexually transmitted disease.

mammography (*p. 935*)
The use of X-rays to examine breast tissues.

mastectomy (*p. 935*)
Surgical removal of part or all of a cancerous breast.

vasectomy (vaz-EK-to-mē) (*p. 950*)
Surgical removal of a segment of the ductus deferens, making it impossible for spermatozoa to reach the distal portions of the reproductive tract.

Additional Terms of Clinical Importance

BREAST CANCER (*p. 935*): **cyst, fibrocystic disease, breast cancer, thermography, segmental mastectomy, total mastectomy, radical mastectomy**

AMENORRHEA (*p. 942*): **amenorrhea**

CONCEPTION CONTROL—PRINCIPLES AND PROSPECTS (*p. 950*): **tubal ligation, oral contraceptives, intrauterine device (IUD), condom, vaginal barrier, rhythm method, abortion, spontaneous abortion, therapeutic abortion, induced abortion**

Study Outline

Related Key Terms

Introduction (p. 917)

1. The human reproductive system produces, stores, nourishes, and transports functional **gametes** (reproductive cells). **Fertilization** is the fusion of a **sperm** from the father and an **ovum** from the mother to create a **zygote** (fertilized egg).

An Overview of the Reproductive System (p. 918)

1. The reproductive system includes **gonads**, ducts, accessory glands and organs, and the **external genitalia**.
2. In the male the **testes** produce sperm, which are expelled from the body in **semen** during **ejaculation**. The **ovaries** (gonads) of a sexually mature female produce an egg that travels along **uterine tubes** to reach the **uterus**. The **vagina** connects the uterus with the exterior.

The Reproductive System of the Male. (pp. 919–928)

1. The **spermatozoa** travel along the **epididymis**, the **ductus deferens**, the **ejaculatory duct**, and the **urethra** before leaving the body. Accessory organs (notably the **seminal vesicles**, **prostate gland**, and **bulbourethral glands**) secrete into the ejaculatory ducts and urethra. The **scrotum** encloses the testes, and the **penis** is an erectile organ.

THE TESTES (pp. 919–921)

2. The **descent of the testes** through the **inguinal canals** occurs during development. The testes remain connected to internal structures via the **spermatic cord.** The **raphe** marks the boundary between the two chambers in the scrotum.
3. The **dartos** muscle gives the scrotum a wrinkled appearance; the **cremaster muscle** pulls the testes closer to the body. The **tunica albuginea** surrounds each testis. Septa extend from the tunica albuginea to the **mediastinum**, creating a series of **lobules**. **Seminiferous tubules** within each lobule are the sites of sperm production. From there sperm pass through a **straight tubule** to the **rete testis**. **Efferent ducts** connect the rete testis to the epididymis. Between the seminiferous tubules there are **interstitial cells** that secrete sex hormones. Seminiferous tubules contain **spermatogonia**, stem cells involved in **spermatogenesis**.

spermatocytes • spermatids spermatozoon • sustentacular cells
blood-testis barrier
spermiogenesis

ANATOMY OF A SPERMATOZOON (pp. 921–922)

4. Each spermatozoon has a **head**, **middle piece**, and **tail**.

acrosomal cap • neck

THE EPIDIDYMIS (p. 922)

5. From the testis the spermatozoa enter the **epididymis**, an elongate tubule with **head**, **body**, and **tail** regions. The epididymis monitors and adjusts the composition of the tubular fluid, and serves as a recycling center for damaged spermatozoa.

Chapter Review

Related Key Terms

THE DUCTUS DEFERENS (p. 923)

6. The ductus deferens begins at the epididymis and passes through the inguinal canal as one component of the spermatic cord. Near the prostate it enlarges to form the **ampulla**. The junction of the base of the seminal vesicle and the ampulla creates the **ejaculatory duct**, which empties into the urethra.

THE URETHRA (pp. 923–924)

7. The urethra extends from the urinary bladder to the tip of the penis. It can be divided into three regions: **prostatic urethra**, **membranous urethra**, and **penile urethra**.

external urethral meatus

THE PENIS (pp. 924–925)

8. The skin overlying the **penis** resembles that of the scrotum. Most of the body of the penis consists of three masses of **erectile tissue**. Beneath the superficial fascia there are two **corpora cavernosa** and a single **corpus spongiosum** that surrounds the urethra. Dilation of the erectile tissue with blood produces an **erection**.

prepuce • **glans**
preputial glands • **smegma**
circumcision

THE ACCESSORY ORGANS (pp. 925–926)

9. Each **seminal vesicle** is an active secretory gland that contributes about 60 percent of the volume of semen; its secretions contain fructose that is easily metabolized by spermatozoa. The **prostate gland** secretes alkaline fluids that help neutralize the acids normally found in the urethra and vagina. Alkaline mucus secreted by the **bulbourethral glands** has lubricating properties.

SEMEN (pp. 926–927)

10. A typical ejaculation releases 2–5 ml of semen (an **ejaculate**), which contains 20 to 100 million sperm per milliliter.

seminal fluid • **seminalplasmin**

MALE SEXUAL FUNCTION (pp. 927–928)

11. During **arousal** erotic thoughts or sensory stimulation lead to parasympathetic activity that produces erection. Stimuli accompanying **coitus** (**intercourse** or **copulation**) lead to **emission** and **ejaculation**. Strong muscle contractions are associated with **orgasm**.

detumescence • **impotence**

HORMONES AND MALE REPRODUCTIVE FUNCTION (p. 928)

12. Important regulatory hormones include **FSH** (follicle-stimulating hormone), **ICSH** (interstitial cell-stimulating hormone; identical to **LH** [luteinizing hormone] in females), and **GnRH** (gonadotrophin-releasing hormone). Testosterone is the most important androgen.

The Reproductive System of the Female (pp. 929–936)

1. Principal organs of the female reproductive system include the ovaries, uterine tubes, uterus, vagina, and external genitalia.

gynecologists

AN OVERVIEW OF FEMALE REPRODUCTIVE ANATOMY (pp. 929–930)

2. The ovaries, uterine tubes, and uterus are enclosed within the **broad ligament** (an extensive mesentery). The **mesovarium** supports and stabilizes each ovary.

rectouterine pouch
vesicouterine pouch

THE OVARIES (p. 930)

3. The ovaries are held in position by the **ovarian ligament** and the **suspensory ligament**. Major blood vessels enter the ovary at the **ovarian hilus**. Each ovary is covered by a **tunica albuginea**.

THE UTERINE TUBES (pp. 930–931)

4. Each **uterine tube** has an **infundibulum** with **fimbriae** (projections), an **ampulla**, an **isthmus**, and an **intramural portion** that opens into the uterine cavity. For fertilization to occur, the ovum must encounter spermatozoa during the first 12–24 hours of its passage from the infundibulum to the uterus.

THE UTERUS (pp. 931–932)

5. The uterus provides mechanical protection and nutritional support to the developing embryo. Normally the uterus bends anteriorly near its base (**anteflexion**). It is stabilized by the broad ligament, **uterosacral ligaments**, **round ligaments**, and the **lateral ligaments**.

retroflexion

6. Major anatomical landmarks of the uterus include the **body**, **isthmus**, **cervix**, **external orifice**, **uterine cavity**, **cervical canal**, and **internal orifice**. The uterine wall can be divided into an inner **endometrium**, a muscular **myometrium**, and a superficial **serosa**.

THE VAGINA (pp. 932–933)

7. The **vagina** is a muscular tube extending between the uterus and external genitalia. Prior to the onset of sexual activity a thin epithelial fold, the **hymen**, partially blocks the entrance to the vagina.

fornix • **vaginal canal** • **vaginitis**
mons pubis

8. The components of the **vulva** include the **vestibule**, **labia minora**, **clitoris**, **labia majora**, and the **lesser** and **greater vestibular glands**.

THE MAMMARY GLANDS (pp. 933–936)

9. At birth a newborn infant gains nourishment from milk secreted by maternal **mammary glands**.

FEMALE SEXUAL FUNCTION (p. 936)

10. The phases of female sexual function resemble those of the male, with parasympathetic arousal and muscular contractions associated with orgasm.

Related Key Terms

lactation • **nipple** • **areola**
lactiferous duct
lactiferous sinus
suspensory ligaments of the breast

Hormones and the Female Reproductive System (pp. 936–944)

11. Hormonal regulation of the female reproductive system is more complicated than that of the male.

THE OVARIAN CYCLE (pp. 936–939)

12. **Oogenesis** (ovum production) occurs monthly in **ovarian follicles** as part of the **ovarian cycle**. As development proceeds one finds **primordial**, **primary**, **secondary**, and **tertiary follicles**. At **ovulation** an **oocyte** and the surrounding follicular walls of the **corona radiata** are released through the ruptured ovarian wall.

egg nests • **zona pellucida**
antrum

13. The follicular cells remaining within the ovary form the **corpus luteum** that later degenerates into a **corpus albicans** of scar tissue. The hypothalamic secretion of GnRF triggers the pituitary secretion of FSH and the synthesis of LH. FSH initiates follicular development, and activated follicles and ovarian interstitial cells produce estrogens. **Progesterone**, one of the steroid hormones called **progestins**, is the principal hormone of the postovulatory period. Hormonal changes are responsible for the maintenance of the menstrual cycle.

atresia • **estradiol**

THE MENSTRUAL CYCLE (pp. 939–942)

14. A typical 28-day **menstrual cycle** begins with the onset of **menses** and the destruction of the functional zone of the endometrium. This process of **menstruation** continues from one to seven days.

dysmenorrhea

15. After menses, the **proliferative phase** begins and the functional zone undergoes repair and thickens. Menstrual activity begins at **menarche** and continues until **menopause**.

PREGNANCY, HORMONES, AND MATERNAL SYSTEMS (pp. 942–944)

16. If fertilization occurs, **implantation** of the **blastocyst** occurs in the endometrial wall. The **placenta** that develops produces several hormones that modify the ovarian and menstrual cycles. **Human chorionic gonadotropin (HCG)** maintains the corpus luteum for several months. The placenta also produces **relaxin** and **human placental lactogen (HPL)**. HPL and **PRL** (prolactin) are primarily responsible for preparing the mammary glands for milk production. For the first few days after delivery the mammary glands produce **colostrum**, a secretion that contains immunoglobulins that may help the infant ward off infections.

Aging and the Reproductive System (p. 944)

MENOPAUSE (p. 944)

1. Menopause (the time that ovulation and menstruation cease in women) typically occurs around age 50. Production of GnRH, FSH, and LH rise, while circulating concentrations of estrogen and progesterone decline.

premature menopause

THE MALE CLIMACTERIC (p. 944)

2. During the **male climacteric**, between ages 50 and 60, circulating testosterone levels decline, while levels of FSH and Lh rise.

Chapter Review

Review Planner

		Level 1	Level	
1	Summarize the functions of the human reproductive system and its principal components.	**1 2 3 29 44**	Level 2	**29 30 39**
2	Describe the components of the male and female reproductive systems.	**1 2 3 4 11 12 15 19 20 23 34 37 41 45 48**	Level 3	**49–51**
3	Detail the physiological processes involved in the ovarian and menstrual cycles.	**5 6 7 8 10 21 22 39**	CM	**Cryptorchidism Testicular Torsion • Prostatitis Uterine Tumors • Vaginitis STDs • PMS • Endometriosis**
4	Describe the hormonal mechanisms that regulate reproductive activities.	**5 6 7 8 9 10 32 43 46 47**		
5	Discuss the production, storage, and transport of sex cells.	**1 2 3 4 16 17 20 23 25 27 28 31 33 38 44**	ABC NEWS	**Chlamydia**
6	Describe the anatomical, physiological, and hormonal changes that accompany pregnancy.	**9 13 14 18 25 40 42**		
7	Discuss the changes in the reproductive system that occur at puberty and with aging.	**24 26 35**		
8	Explain how the reproductive system interacts with other body systems.	**10 36**		

Review Questions

MULTIPLE CHOICE

1. Which of the following is a function of the epididymis? (*May be more than one answer*): a) monitoring the composition of the tubular fluid; b) secreting inhibin; c) recycling damaged sperm; d) secreting an alkaline fluid that helps neutralize the urethra.
2. If the seminal vesicles were blocked, which of the following would result?: a) the production of smegma would stop; b) erection of the penis could not occur; c) the environment of the urethra would be too acid for sperm; d) sperm would not obtain adequate nourishment.
3. If the prostate gland were blocked, which of the following would result?: a) the production of smegma would stop; b) erection of the penis could not occur; c) the environment of the urethra would be too acid for sperm; d) sperm would not obtain adequate nourishment.
4. Semen: a) is manufactured by the seminal vesicles; b) contains buffers, enzymes, nutrients, and prostaglandins; c) has an alkaline pH; d) contains sustentacular cells as well as spermatozoa.
5. During the preovulatory period of the ovarian cycle: a) the corpus luteum forms; b) the concentration of circulating estrogens rises sharply; c) the functional zone of the endometrium sloughs off; d) basal body temperature is roughly one-half degree Fahrenheit higher than during the postovulatory period.
6. Throughout most of the postovulatory period of the menstrual cycle: a) the functional zone is prepared for the arrival of a blastocyst; b) progesterone remains the dominant hormone; c) LH production remains extremely high; d) secondary follicles are maturing into tertiary follicles.
7. At the time of ovulation: a) estrogen levels have been steadily rising, and LH levels suddenly increase dramatically; b) estrogen levels have been falling, and FSH levels suddenly decrease; c) progesterone and LH levels decrease; d) GnRH production increases.
8. During menses: a) the basilar zone of the endome-

trium is completely destroyed; b) the functional zone prepares for implantation; c) estrogen and progesterone levels are relatively low; d) ovulation occurs.

9. Ejection of milk occurs: a) under estrogen secretion alone; b) when HCG levels decline; c) after the mammary glands have been activated by PRL and HPL, and stimulated by oxytocin; d) following each menstrual cycle.

DIAGRAMS

10. Create a table that lists the source, regulation of secretion, and primary effects (in both men and women) of the following hormones: GnRF, FSH, estrogens, inhibin, LH/ICSH, progestins, androgens, HCG, relaxin, and HPL.
11. Cover the labels in Figure 28-1 and identify the components of the male reproductive system.
12. Cover the labels in Figure 28-10 and identify the components of the female reproductive system.

TRUE/FALSE

(If the statement is false, your instructor may wish to have you correct it.)

13. **T/F:** Fertilization normally occurs in the uterus.
14. **T/F:** Fertilization must occur during the first 3–4 days after an ovum is released from the ovary.
15. **T/F:** Bacteria are normally present in the vagina.

MATCHING QUESTIONS

(Match each term with the most appropriate statement).

Statements:

16. Female gamete.
17. Cells containing one-half the normal complement of chromosomes that develop into mature male sex cells.
18. A hollow ball of cells formed after the fertilized egg has undergone several cycles of cell division.
19. Sites of lactation.
20. Spermatogenesis begins when these stem cells divide.
21. A progestin-secreting mass of follicle cells that develops in the ovary after ovulation.
22. The sloughing of tissue, resulting from the destruction of the functional zone of the endometrium.
23. Reproductive organs that produce sex cells and hormones.
24. The last menstrual cycle.
25. A fertilized egg.
26. The first menstrual cycle.
27. Male gamete.
28. The release of an egg from an ovary.

Terms:

a) Spermatozoa
b) Ovum
c) Zygote
d) Gonads
e) Spermatogonia
f) Spermatid
g) Mammary glands
h) Corpus luteum
i) Ovulation
j) Menopause
k) Menarche
l) Menstruation
m) Blastocyst

SHORT ANSWER/ESSAY

29. Compare the respective functions of the male and female reproductive systems.
30. Why are inguinal hernias more common in men than in women?
31. Distinguish between emission and ejaculation.
32. Which is the most important androgen? What are its functions?
33. What is spermatogenesis, and where does it take place? Summarize the process.
34. What are sustentacular cells, and why are they important?
35. Describe the changes that occur with aging in the male and female reproductive systems. How do hormone levels change?
36. Discuss at least four ways in which the reproductive system interacts functionally with other body systems.

Distinguish between the following:

37. Anteflexion and retroflexion
38. Primordial, primary, and secondary follicles.
39. Ovarian cycle and menstrual cycle.
40. Zona pellucida and corona radiata.

Identify and give a function for each of the following:

41. Cremaster muscle
42. Colostrum
43. Inhibin
44. Ovaries
45. Interstitial cells of the testis
46. Human chorionic gonadotropin
47. Estradiol
48. Broad ligament

CRITICAL THINKING/APPLICATIONS

49. A combination oral birth control pill contains both estrogens and progestins. How do these hormones prevent pregnancy?
50. Mrs. Elliott, 38, comes to the doctor because she discovered a lump in one breast. What disorders could this indicate? What questions could you ask Mrs. Elliott, and what tests could you recommend, to distinguish between these disorders?
51. Which contraceptive methods do you think are most effective? Least effective?

The physiological processes we have studied thus far have been brief ones. Many last only a fraction of a second; others may take hours at most. But some important processes of life are measured in months, years, or decades. A human being develops in the womb for nine months, grows to maturity in 15 or 20 years, and may live the better part of a century. During that whole span of time, he or she will never cease to change. Birth, growth, maturation, aging, and death are all parts of a single, continuous process. And that process does not end with the individual, for human beings can pass at least some of their characteristics on to their offspring. Thus each generation gives rise to a new generation that will repeat the same cycle. . . . In this chapter we will explore the continuity of life, from conception to death.

CHAPTER 29
Development and Inheritance

Chapter Objectives

After reading this chapter, you will be able to:

1 Explain the differences between ordinary cells and gametes and relate the specializations of gametes to their functions.
2 Describe how gametes are formed.
3 Relate basic principles of genetics to the inheritance of human traits.
4 Describe the process of fertilization.
5 Discuss the stages of embryonic and fetal development.
6 Explain how pregnancy affects the mother's body systems and describe the processes of labor and delivery.
7 Explain how developmental processes are regulated.
8 Discuss the major stages of life after delivery.
9 Summarize the processes and factors involved in aging.

Introduction

Time refuses to stand still; today's infant will be tomorrow's adult. The process of **development** is the gradual modification of physical and physiological characteristics during the period from conception to physical maturity. The changes are truly remarkable—what begins as a single cell slightly larger than the period at the end of this sentence becomes an individual whose body contains trillions of cells organized into tissues, organs, and organ systems. The creation of different cell types during development is called **differentiation**. The basic mechanism involved was introduced in Chapter 4: *differentiation occurs through selective changes in genetic activity.* (p. 121) As development proceeds, some genes are turned off and others turned on. The identities of these genes vary from one cell type to another.

A basic understanding of development provides a framework that greatly enhances the understanding of anatomical and physiological processes. In addition, many of the mechanisms of development and growth are similar to those responsible for the repair of injuries. This chapter will focus on major aspects of development and consider highlights of the developmental process rather than describe the events in great detail. We will also consider the regulatory mechanisms and how developmental patterns can be modified—for good or ill. Few topics in the biological sciences hold such fascination, and fewer still confront the investigator with so dazzling an array of technological, moral, and logistical challenges.

An Overview of Topics in Development

Development involves the division and differentiation of cells and changes in genetic activity that produce and modify anatomical structures and physiological processes. In essence, the identities of the available genes ultimately determine the potential range of developmental variations. Although every human being goes through the same developmental stages, the differences in genetic makeup produce distinctive individual characteristics. **Inheritance** refers to the transfer of genetically determined characteristics from generation to generation. **Genetics** is the study of the mechanisms responsible for inheritance. This chapter considers basic genetics as it applies to the appearance of inherited characteristics such as sex, hair color, and various diseases.

Development, which begins at fertilization, or **conception**, can be divided into periods characterized by specific anatomical and physiological changes. **Embryology** (em-brē-OL-o-jē) considers the developmental events that occur in the first 2 months after fertilization. **Fetal development** begins at the start of the ninth week and continues up to the time of birth. Embryological and fetal development are sometimes referred to collectively as **prenatal development**. **Postnatal development** commences at birth and continues to maturity, when another process, **senescence**, or aging, begins.

We will begin our examination of development with a look at how gametes (sperm and ova) are produced. Although development actually begins at fertilization, the production of functional gametes establishes important guidelines for future development.

The discussion of gamete formation, or **gametogenesis** (ga-mē-tō-JEN-e-sis), brings us to a consideration of how the genetic material contained in the sperm and egg determines the characteristics of the individual. In the process we will take a brief look at the basic regulatory mechanisms that guide development.

With this background we then proceed to a description of fertilization and zygote formation and follow the sequence of events under way during prenatal and postnatal development.

Meiosis and the Formation of Gametes

Meiosis (mī-Ō-sis) is a special form of cell division leading to the production of sperm or eggs. Mitosis (Figure 3-32) and meiosis differ significantly in terms of the events that take place in the nucleus. (p. 93) A typical cell in the human body contains 46 chromosomes. If a gamete resembled other cells, at fertilization the fusion of an egg with a sperm would produce a zygote containing 92 chromosomes. The next generation would contain 184, the one after that 368, and so on. This would hardly be practical. Even ignoring the potential problems arising from excess copies of every gene, the extra chromosomes would take up space needed for things such as mitochondria, endoplasmic reticulum, and so forth.

These problems are avoided because our cells contain 23 chromosome *pairs* rather than 46 individual chromosomes. The normal chromosome complement of a typical **somatic** (sō-MAT-ik), or body, cell is therefore described as **2N**, or **diploid** (DIP-loid). At fertilization one of the chromosomes in each pair was contributed by the sperm, and the other by the egg. Gametes are therefore different from ordinary somatic cells *because they contain only half the normal number of chromosomes*. As a result a gamete can be described as **1N**, or **haploid** (HAP-loid). N stands for the normal number of chromosome pairs, and in humans that number is 23. There is nothing magical about the number 23 unless you are human. A cat produces gametes containing 18 chromosomes, while a rooster manufactures sperm with 40.

During gamete formation, meiosis (Figure 29-1) splits the chromosome pairs, producing haploid gametes. During the interphase preceding meiosis all of the chromosomes replicate themselves just as if they were about to undergo mitosis. As the prophase of the first meiotic division (**meiosis I**) arrives, the chromosomes condense and become visible. Each chromosome remains attached to its duplicate, and you see 46 doubled chromosomes. As in mitosis, the attached duplicates are called **chromatids** (KRO-ma-tidz). The corresponding maternal and paternal chromosomes pair off, an event known as **synapsis** (sin-AP-sis). This produces 23 pairs of chromosomes, each member of the pair consisting of two identical chromatids. This combination of four chromatids represents a **tetrad** (TET-rad; *tetras*, four).

As metaphase I begins, the nuclear envelope disappears and the tetrads line up along the metaphase plate. At the end of metaphase I the tetrads break up, and the maternal and paternal chromosomes separate.

As anaphase proceeds, the maternal and paternal components are randomly distributed, as indicated in Figure 29-1. As a result, telophase I ends with the formation of two daughter cells containing

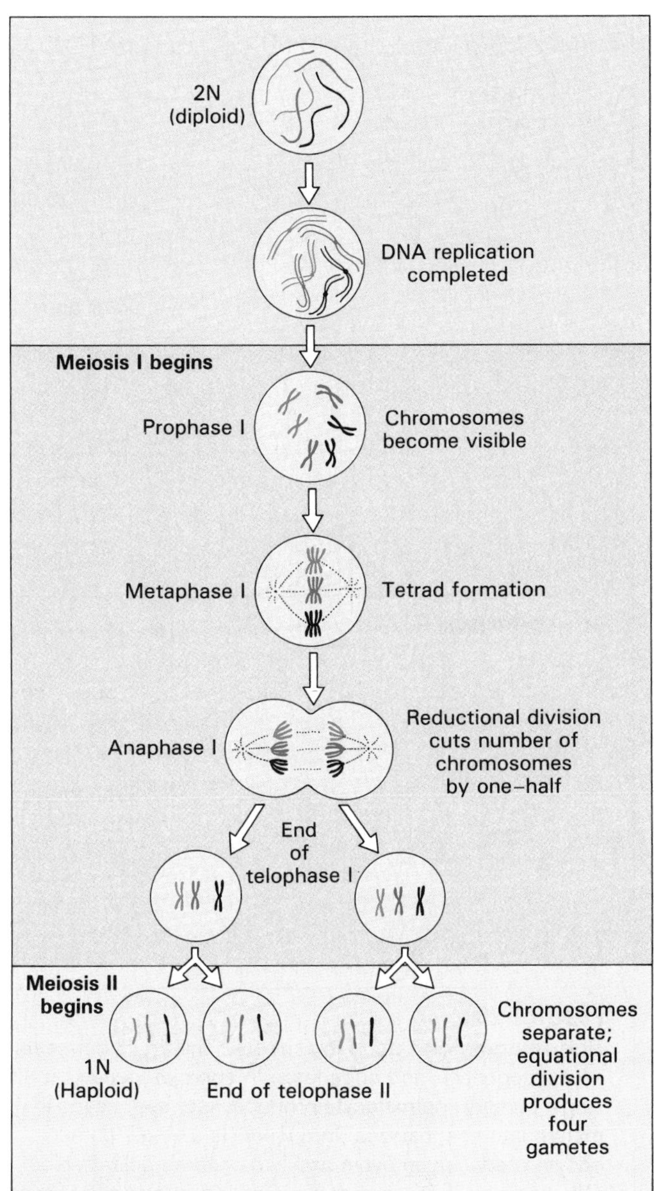

FIGURE 29-1
Meiosis and the Formation of Gametes

unique combinations of maternal and paternal chromosomes. Each daughter cell has 23 chromosomes. Because the first meiotic division reduces the number of chromosomes from 46 to 23 it is called a **reductional division**. Yet you will notice that each chromosome remains attached to the duplicate chromatid that formed before the start of prophase I. The duplicate chromatids are separated during **meiosis II**.

The interphase separating meiosis I and II is very brief, and the daughter cells proceed into prophase II without further growth or DNA replication. The completion of metaphase II, anaphase II, and telophase II produces four gametes, each containing 23 chromosomes. Because the chromosome number has not changed, meiosis II represents an **equational division**.

This description serves equally well as a model for either spermatogenesis (sperm formation) or oogenesis (ovum production). The cytoplasmic events of spermatogenesis were introduced in Chapter 28. ∞ (p. 921) *Spermatogonia* (stem cells) in the seminiferous tubules of the testes undergo mitoses throughout adult life. One of the daughter cells remains as an undifferentiated stem cell, while the other prepares to enter meiosis I. In doing so it differentiates into a *primary spermatocyte*. Meiosis I produces a pair of *secondary spermatocytes*, and meiosis II yields four functional *spermatids*. Spermatogenesis thus produces four functional spermatids for every primary spermatocyte undergoing meiosis. These spermatids are then transformed into sperm through the process of *spermiogenesis*. (You may find it helpful to review Figure 28-3 at this time.) ∞ (p. 920)

In the female, the **oogonia** (ō-o-GŌ-nē-a) complete their mitotic divisions before birth, and at puberty the ovaries contain **primary oocytes** (Ō-o-sīts) awaiting the hormonal signal to complete meiosis I. Although the nuclear events under way during meiosis I and II are identical to those described for spermatogenesis, the cytoplasm of the original oocyte is not evenly distributed at each step along the way. Oogenesis produces one functional **ovum**, containing most of that cytoplasm, and two or three nonfunctional **polar bodies** that later disintegrate. Polar bodies are little more than encapsulated chromosomes, and the first polar body often fails to complete meiosis II. Spermatogenesis and oogenesis are diagrammed in Figure 29-2.

The differences between spermatogenesis and oogenesis reflect the contrasting functional roles of sperm and eggs. A spermatozoon simply carries the paternal chromosomes to the site of fertilization, but the ovum must provide all of the nourishment and programming to support the development of an embryo for nearly a week after conception. The volume of the ovum is therefore much greater than that of the spermatozoon, and each primary oocyte entering meiosis contains only enough cytoplasm to produce a single functional ovum. At the time of fertilization, the diameter of the ovum is over twice the entire length of the spermatozoon. The relationship between the egg and sperm volumes is even more striking, on the order of 2000:1.

Genetics, Development, and Inheritance

Every somatic cell in the body carries copies of the original 46 chromosomes present in the fertilized egg or *zygote*. Those chromosomes and their component

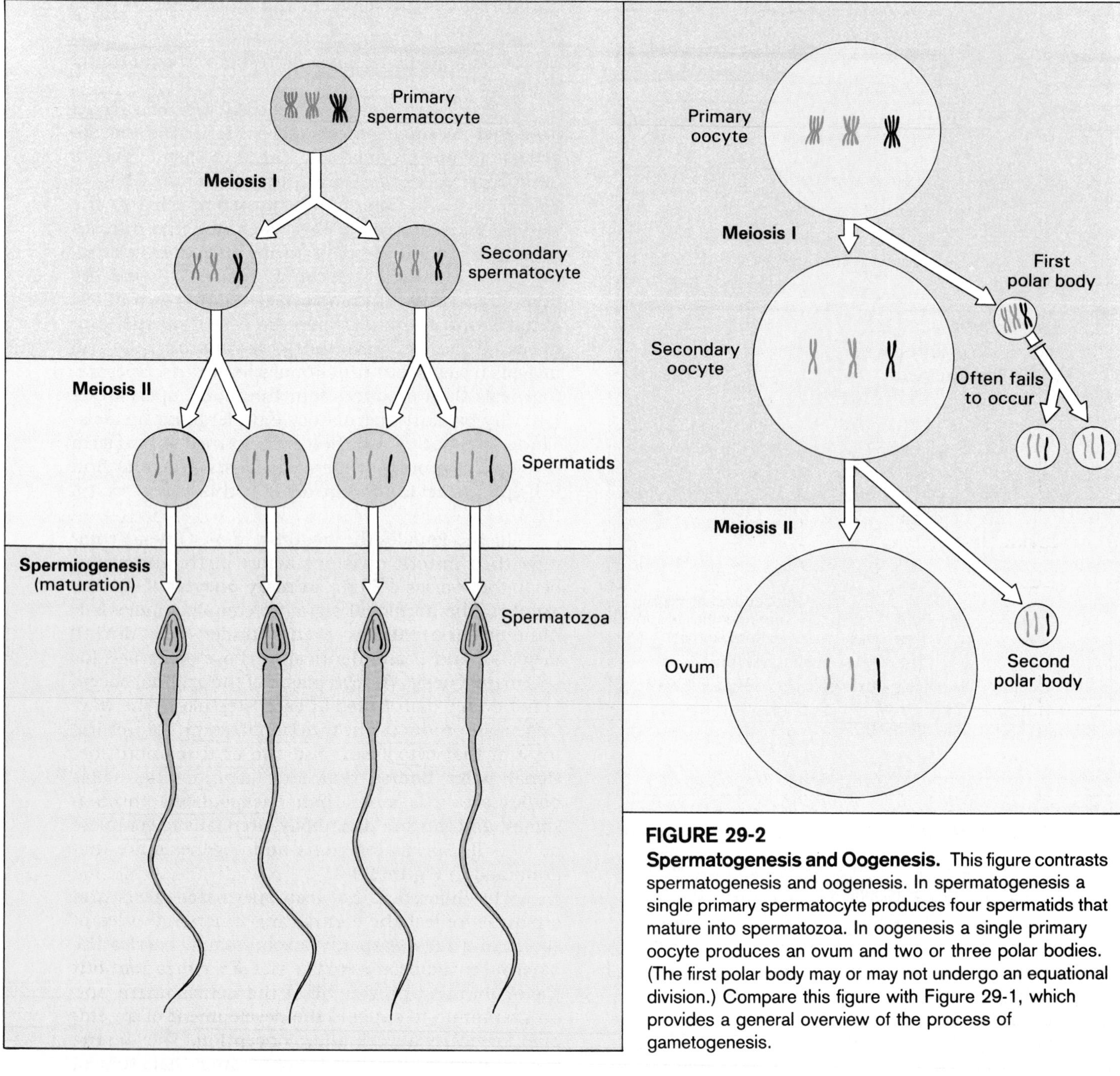

FIGURE 29-2
Spermatogenesis and Oogenesis. This figure contrasts spermatogenesis and oogenesis. In spermatogenesis a single primary spermatocyte produces four spermatids that mature into spermatozoa. In oogenesis a single primary oocyte produces an ovum and two or three polar bodies. (The first polar body may or may not undergo an equational division.) Compare this figure with Figure 29-1, which provides a general overview of the process of gametogenesis.

genes represent the individual's **genotype** (JĒN-ō-tīp). Through development and differentiation, the instructions contained within the genotype are expressed in many different ways. No single living cell or tissue makes use of all of the information and instructions contained within the genotype. For example, in muscle fibers the genes important for excitable membrane formation and contractile proteins are active, while a different set of genes is operating in the cells of the pancreatic islets. But the instructions contained within the genotype determine the anatomical and physiological characteristics of each individual. Those visible characteristics are known as the individual's **phenotype** (FĒN-ō-tīp).

Differences in genotype lead to distinctive variations in phenotype, but the relationships are not always predictable. You are not an exact copy of either parent, nor are you an easily identifiable mixture of their characteristics. Our discussion will begin with the basic patterns and their implications and then will examine the mechanisms responsible for regulating the activities of the genotype during subsequent prenatal development.

GENES AND CHROMOSOMES

The basic structure of genes and chromosomes was introduced in Chapters 3 and 4. (For a quick review of the important concepts, take a moment to examine Figures 3-11, 3-31, 4-12, and 4-13 before proceeding with this section.) (pp. 72, 91, 110, 112)

Autosomal Chromosomes

Every somatic cell contains 23 pairs of chromosomes. One member of each pair was contributed by the sperm, and the other by the ovum. The members of each pair are known as **homologous** (hō-MOL-o-gus) chromosomes. Twenty-two of those pairs are known as **autosomal** (aw-to-SŌ-mal) **chromosomes**, for their genes affect only somatic characteristics such as hair color or skin pigmentation. Each chromosome in an autosomal pair has the same structure and carries genes that affect the same traits. If one member of the pair contains three genes in a row, with number 1 determining hair color, number 2 eye color, and number 3 skin pigmentation, the other chromosome will carry genes affecting the same traits, and in the same sequence.

The various forms of any one gene are called **alleles** (a-LĒLS). If both chromosomes of a homologous pair carry the same allele of a particular gene, the individual is **homozygous** (hō-mō-ZĪ-gus) for that trait. For example, if a zygote receives a gene for curly hair from the sperm and one for curly hair from the egg, the individual will be homozygous for curly hair. *If you are homozygous for a particular trait, your phenotype will have that characteristic.* Usually about 80 percent of an individual's genetic complement consists of homozygous alleles. Because the chromosomes of a homologous pair have different origins, one paternal and the other maternal, they do not *have* to carry the same alleles. When an individual has two different alleles carrying different instructions, the individual is **heterozygous** (het-er-ō-ZĪ-gus) for that trait. In the case of a heterozygous genotype, the phenotype will be determined by the interactions between the corresponding alleles.

If an allele is **dominant** it will be expressed in the phenotype *regardless of any conflicting instructions carried by the other allele.* If both alleles must agree on the outcome before they can affect the phenotype, the alleles are said to be **recessive**. For example, Chapter 6 described the albino condition, characterized by an inability to synthesize a yellow-brown pigment, *melanin*. A single dominant allele determines normal skin coloration; two recessive alleles must be present to produce an albino individual.

Not every characteristic sorts out neatly into dominant or recessive categories, but a number of examples are included in Table 29-1. If you restrict attention to these alleles it is possible to predict the characteristics of individuals based on those of their parents. Dominant traits are traditionally indicated by capitalized abbreviations, while recessives are abbreviated in lower case. For a given trait, the possibilities are indicated by *AA* (homozygous dominant), *Aa* (heterozygous), or *aa* (homozygous recessive). The gametes involved in fertilization each contribute a single allele for a given trait. That allele must be one of the two contained by all other cells in the parental body. Consider, for example, the offspring of an albino

TABLE 29-1 The Inheritance of Phenotypic Characters

DOMINANT TRAITS
One allele determines phenotype, and the other is suppressed
normal skin coloration
bradydactyly (short fingers)
ability to taste phenylthiocarbamate (PTC)
ability to roll the tongue
free earlobes
curly hair
presence of Rh factor on red blood cell membranes
Both dominant alleles may be expressed (codominance)
presence of A or B antigens on red blood cell membranes
structure of serum proteins (albumins, transferrins)
structure of hemoglobin molecule
RECESSIVE TRAITS
albinism
blond hair
red hair (only expressed if individual is also homozygous for blond hair)
lack of A, B agglutinogens (type O blood)
SEX-LINKED TRAITS
color blindness
POLYGENIC TRAITS
eye color
hair colors other than pure blond or red

Note: For a listing of inherited clinical conditions, see Table 29-2.

mother and a normal father. Because albinism is a recessive trait, the maternal alleles can be abbreviated *aa*. No matter which of her ova gets fertilized, it will carry the recessive *a* gene. The father has normal coloration, and this is a dominant trait. He may be homozygous or heterozygous for this trait, since *AA* or *Aa* will have the same visible effect.

A simple box diagram known as a **Punnett square** lets us predict the characteristics of the children by showing us the genetic combinations that can occur. In the square shown in Figure 29-3, the maternal alleles are listed along the horizontal axis, and the paternal ones along the vertical axis. The possible combinations are indicated in the small boxes. Figure 29-3a considers the possible offspring of an *aa* mother and an *AA* father. All of the children must have the genotype *Aa*, and they will all have normal skin coloration. Compare these results with those of Figure 29-3b, for a heterozygous father (*Aa*). The heterozygous individual produces two types of gametes, *A* and *a*, and the egg may be fertilized by either one. As a result, half of the children born will have normal skin color with the phenotype *Aa*, and the other half will be albino, with the genotype *aa*. The Punnett square can also be used in reverse, to draw conclusions about the identity and genotype of a parent. For example, a man with the genotype *AA* cannot be the father of an albino child (*aa*). [**CM**: *Down Syndrome*]

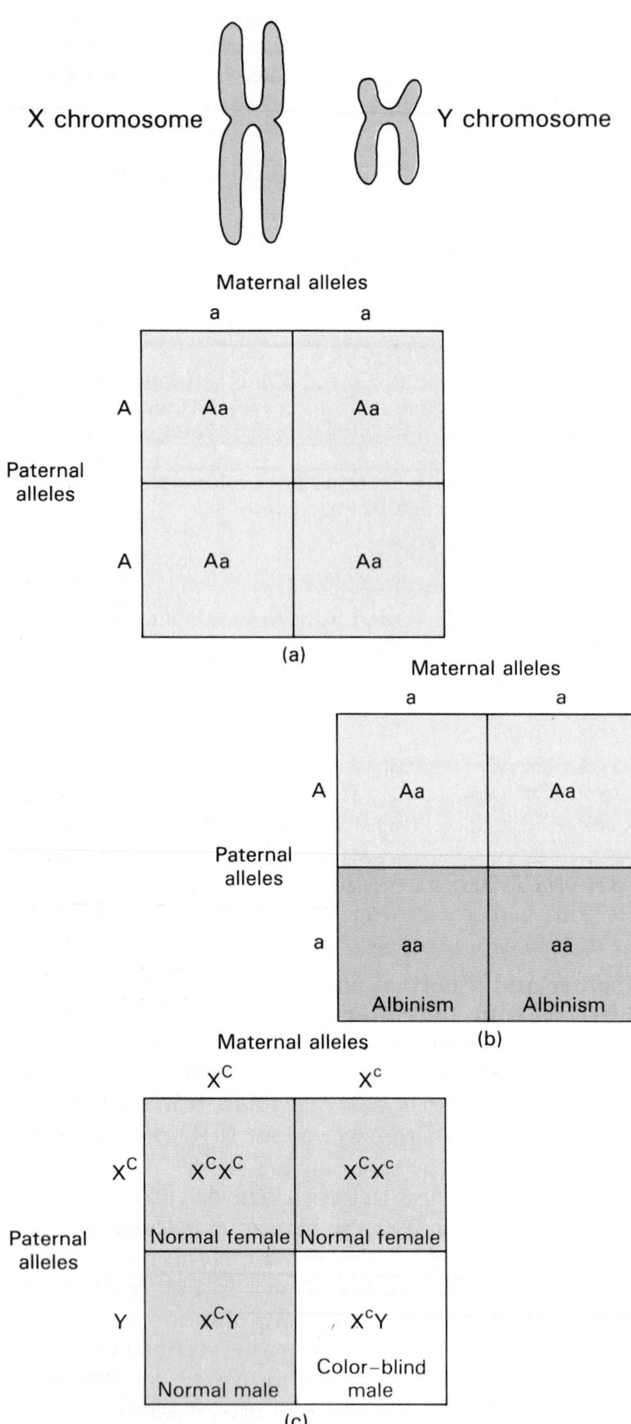

FIGURE 29-3
Predicting Phenotypic Characteristics. Top: a diagrammatic view of X and Y chromosomes. Note the differences in size and shape. (a) The offspring of a homozygous-dominant father and a homozygous-recessive mother will all be heterozygous for that trait. Their phenotype will be the same as that of the father. (b) The offspring of a heterozygous father and a homozygous-recessive mother will either be heterozygous or homozygous for the recessive trait. In this example, half of the offspring will have normal skin coloration and the other half will be albinos. (c) The inheritance of color blindness, a sex-linked trait. A color-blind father and a heterozygous mother will produce daughters with normal vision, but half of the sons will be color-blind.

Sex Chromosomes

The chromosomes of the twenty-third pair are called the **sex chromosomes** because they determine the genetic sex of the individual. Unlike other chromosomal pairs, the sex chromosomes are not necessarily identical in appearance and gene content. There are two different sex chromosomes, an **X chromosome** and a **Y chromosome.** The Y chromosomes are considerably smaller than X chromosomes, and they contain fewer genes. But among those genes are dominant alleles that specify that an individual with that chromosome will be a male. The normal male genotype is *XY*; females lack the Y chromosome, and their genotypes are *XX*.

The ova produced by a woman will always carry *X*, and sperm may carry *X* or *Y*, so if you make a Punnett square you can show that the male:female sex ratio in the offspring should be 1:1. The birth statistics differ slightly from that prediction, with 106 males born for every 100 females. It has been suggested that more males are born because the sperm carrying the Y chromosome can reach the oocyte first, since they do not have to carry the extra weight of the larger X chromosome.

The X chromosome also carries genes that affect somatic structures. These characteristics are called **X-linked** because in most cases there are no corresponding alleles on the Y chromosome. The inheritance of characteristics regulated by these genes does not follow the pattern of alleles on autosomal chromosomes. The best known of these single-allele characters are those associated with noticeable diseases or functional deficits.

The inheritance of color blindness, a condition discussed in Chapter 17, exemplifies the differences between sex-linked and autosomal inheritance. ∞ (p. 562) A relatively common form of color blindness is associated with the presence of a dominant or recessive gene on the X chromosome. Normal color vision is determined by the presence of a dominant gene, *C*, and color blindness results from the presence of the recessive gene *c*. A woman, with her two X chromosomes, can be either homozygous, *CC*, or heterozygous, *Cc*, and still have normal color vision. She will be color-blind only if she carries two recessive alleles, *cc*. But a male has only one X chromosome, so whatever that chromosome carries will determine whether he has normal color vision or is color-blind. A Punnett square for an X-linked trait, as in Figure 29-3c, reveals that the sons produced by a normal father and a heterozygous mother will have a 50 percent chance of being color-blind, while the daughters will all have normal color vision. A number of other clinical disorders noted earlier in the text are X-linked traits, including certain forms of *hemophilia*, *diabetes insipidus*, and *muscular dystrophy*. In several instances advances in molecular genetic techniques have permitted the localization of the specific genes on the X chromosome. This technique provides a relatively direct method of screening for the presence of a particular condition before the symptoms appear, and even before birth.

At present, relatively few of the multitude of phenotypic characteristics can be linked to individual alleles located on specific chromosomes. A particular pair of alleles specifies the identity of a protein that can be manufactured in the cytoplasm, and unless that protein is involved in the production of a visible pigment, as in the case of hair color, it will probably affect biochemical operations that do not produce an obvious phenotypic change. But over time, with several genes altering physiological operations, visible changes will often appear. Thus many human character traits reflect the combined, integrated activities of a large number of genes on several different chromosome pairs. For this reason geneticists attempting to trace inheritance prefer to examine details of enzyme activity and protein structure that reflect the action of individual genes, rather than relatively complex physical characteristics. [**CM:** *Sex Chromosome Abnormalities*]

■ **TABLE 29-2 Relatively Common Inherited Disorders**

Disorder	*Chapter in text or CM*
Autosomal dominants	
Adult polycystic kidney disease	26 CM
Marfan's syndrome	5 CM
Huntington's disease	15 CM
Autosomal recessives	
Deafness	17 CM
Albinism	6, 29
Sickle cell anemia	19
Cystic fibrosis	23
Phenylketonuria	25 CM
Tay-Sachs disease	12 CM
X-linked	
Duchenne's muscular dystrophy	10 CM
Hemophilia (one form)	19 CM
Testicular feminization	29 CM
Color blindness	17

INHERITANCE AND GENETIC DISORDERS

Few of the genes responsible for inherited disorders have been identified or even localized to a specific chromosome. But these conditions can be broadly categorized as involving *simple inheritance* or *polygenic (multifactorial) inheritance.*

Simple Inheritance

In **simple inheritance** phenotypic characters are determined by interactions between a single pair of alleles. The frequency of appearance of an inherited disorder resulting from simple inheritance can be predicted using a Punnett square. Although they are rare disorders in terms of overall numbers, more than 1200 different inherited conditions have been identified that reflect the presence of one or two abnormal alleles for a single gene. A partial listing is included in Table 29-2, along with the location where additional information can be obtained.

Polygenic Inheritance

Polygenic inheritance involves interactions between alleles on several genes. Because multiple alleles are involved, the frequency of occurrence cannot easily be predicted using a simple Punnett square. Several important adult disorders, including hypertension and coronary artery disease, fall within this category. Many of the developmental disorders responsible for fetal mortalities and congenital malformations also result from multiple genetic interactions. In these cases the particular genetic composition of the individual does not by itself determine the onset of the disease. Instead, the conditions regulated by these genes establish a susceptibility to particular environmental influences. This means that not every individual with the genetic tendency for a particular condition will actually develop it. It is therefore difficult to track polygenic conditions through successive generations. However, because many inherited polygenic conditions are *likely* but not *guaranteed* to occur, steps can be taken to prevent a crisis. For example, hypertension may be prevented or reduced by controlling diet and fluid volume, and coronary artery disease may be prevented by lowering serum cholesterol concentrations.

√ **Why is an ovum so much larger than a spermatozoan?**

√ **The ability to roll your tongue is an autosomal dominant trait. What would be the phenotype of a person who is heterozygous for this trait?**

√ **Joe has three daughters and complains that it's his wife's fault that he doesn't have any sons. What would you tell him?**

■ Fertilization

Normal fertilization occurs in the upper one-third of the uterine tube, usually within a day of ovulation. Over this period the oocyte has traveled a few centimeters, but the spermatozoa must cover the distance between the vagina and the upper portions of the

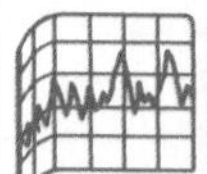

Diagnostics: Chromosomal Analysis

In **amniocentesis** a sample of amniotic fluid is removed and the fetal cells that it contains are analyzed. This procedure permits the identification of at least 20 different congenital conditions. The needle inserted to obtain a fluid sample is guided into position using ultrasound. Unfortunately, amniocentesis has two major drawbacks:

1. Because the sampling procedure represents a potential threat to the health of the fetus and mother, amniocentesis is performed only when there are known risk factors present. Examples of risk factors would include a familial history for specific conditions, or in the case of Down syndrome, a maternal age over 35.
2. Sampling cannot safely be performed until the volume of amniotic fluid is large enough that the fetus will not be injured during the sampling process. The usual time for amniocentesis is at a gestational age of 14–15 weeks. It may take several weeks to obtain results once samples have been collected, and by the time the results are received, the option of therapeutic abortion may no longer be available.

A relatively new technique known as **chorionic villi sampling** analyzes cells collected from the villi during the first trimester. Although this procedure cannot detect as many conditions, it may replace amniocentesis as a testing method for some disorders, including Down syndrome.

female reproductive tract. The sperm arriving in the vagina are already motile, but they cannot fertilize an egg until they have undergone **capacitation** (ka-pas-i-TĀ-shun) within the female reproductive tract. The precise mechanism behind capacitation remains uncertain, but it appears that a component normally present in semen prevents capacitation while the sperm remain within the male reproductive tract.

It is not known how spermatozoa move from the vagina to the fertilization site in the time available. An individual spermatozoon can propel itself at speeds of only about 34 μm per second, roughly equivalent to 5 inches per hour, and in theory it should be several hours before any spermatozoa arrive in the upper portions of the uterine tubes. In fact, the passage time may be considerably shorter, with estimates ranging from 30 minutes to 2 hours. Evidence from other mammals indicates that dead sperm travel as fast as live sperm, and contractions of the uterine musculature and ciliary currents in the uterine tubes have been suggested as likely transport mechanisms.

Even with transport assistance and available nutrients this is not an easy passage. Of the 200 million spermatozoa introduced into the vagina in a typical ejaculate, only around 10,000 make it past the uterus, and fewer than 100 actually reach the egg. A male with a sperm count below 20 million per milliliter will usually be sterile, because too few sperm survive to reach the egg. One or two spermatozoa cannot accomplish fertilization, because of the condition of the oocyte at ovulation.

THE OOCYTE AT OVULATION

Ovulation occurs prior to the completion of oocyte maturation, and the egg leaves the follicle with the nucleus still in metaphase II. Other metabolic operations have also been discontinued, and the oocyte drifts in a sort of suspended animation, awaiting the stimulus for further development. If fertilization does not occur, the oocyte disintegrates without completing meiosis II.

Fertilization is complicated by the fact that when it leaves the ovary the oocyte is surrounded by a layer of follicle cells, the *corona radiata*. The events that follow are diagrammed in Figure 29-4. These cells protect the gamete as it passes through the ruptured follicular wall and into the infundibulum of the uterine tube. Although the physical process of fertilization requires only a single sperm in contact with the oocyte membrane, that spermatozoon must first penetrate the corona radiata. The acrosomal cap contains **hyaluronidase**, an enzyme that breaks down the intercellular cement between adjacent follicle cells. Apparently there must be at least a hundred or more sperm thrashing around, bumping into the corona and releasing hyaluronidase, before the connections between the follicular cells begin to break down. No matter how many sperm slip through the gap, only a single spermatozoon will accomplish fertilization and activate the oocyte. When that sperm contacts the oocyte their plasma membranes fuse, and the sperm enters the ooplasm. This event irrevocably alters conditions inside the oocyte, and **activation** follows.

OOCYTE ACTIVATION

Activation involves a series of changes in the metabolic activity of the oocyte. The most dramatic changes are:

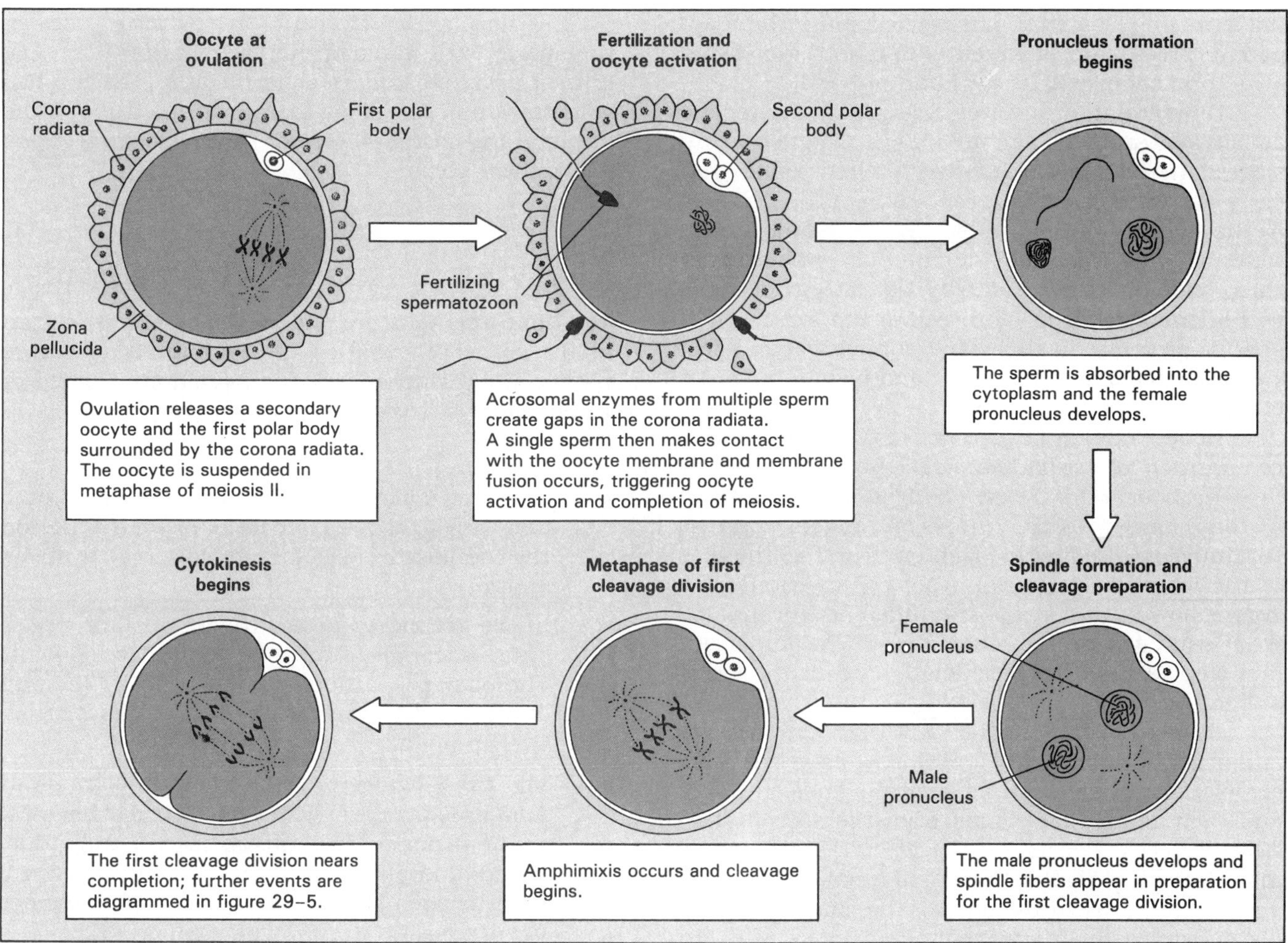
Oocyte at ovulation
Corona radiata
First polar body
Zona pellucida
Ovulation releases a secondary oocyte and the first polar body surrounded by the corona radiata. The oocyte is suspended in metaphase of meiosis II.
Fertilization and oocyte activation
Second polar body
Fertilizing spermatozoon
Acrosomal enzymes from multiple sperm create gaps in the corona radiata. A single sperm then makes contact with the oocyte membrane and membrane fusion occurs, triggering oocyte activation and completion of meiosis.
Pronucleus formation begins
The sperm is absorbed into the cytoplasm and the female pronucleus develops.
Spindle formation and cleavage preparation
Female pronucleus
Male pronucleus
The male pronucleus develops and spindle fibers appear in preparation for the first cleavage division.
Metaphase of first cleavage division
Amphimixis occurs and cleavage begins.
Cytokinesis begins
The first cleavage division nears completion; further events are diagrammed in figure 29–5.

FIGURE 29-4
Fertilization

1. The metabolic rate of the oocyte increases rapidly, and meiosis II is completed.
2. Vesicles just beneath the surface of the oocyte fuse with the cell membrane and discharge their contents through exocytosis. This process, called a **cortical reaction**, is important in preventing penetration by additional sperm. Fertilization by more than one sperm, a condition known as **polyspermy**, produces a nonfunctional zygote.

PRONUCLEUS FORMATION AND AMPHIMIXIS

After oocyte activation and the completion of meiosis, the nuclear material remaining within the ovum reorganizes as the **female pronucleus**, and the centrosome forms an aster with spindle fibers. While these changes are under way, the nucleus of the spermatozoon swells in size, becoming the **male pronucleus**. The male pronucleus then migrates toward the center of the cell, accompanied by its own aster and spindle fibers, and the two pronuclei fuse in a process called **amphimixis** (am-fi-MIK-sis). Fertilization is now complete, with the formation of a zygote containing the normal complement of 46 chromosomes.

These events set the stage for the development process that will, over the next 9 months, produce a fully formed infant. The factors that regulate development are complex and only partially understood. A brief discussion of the basic mechanisms will prove useful, however, because it can explain a great deal about the appearance of *congenital disorders*, those present at the birth of an individual.

INDUCTION AND THE REGULATION OF DEVELOPMENT

During prenatal development a single cell forms a 7-pound infant that in postnatal development grows through adolescence and maturity toward old age and eventual death. One of the most fascinating aspects of development is its apparent order and simplicity. A continuity exists at all levels and at all times. Nothing leaps into existence, unheralded and without apparent precursors; differentiation and increasing structural complexity occur hand in hand.

Differentiation involves changes in the genetic activity of some cells and not others. Chapter 4 considered the exchange of information between the nucleus and cytoplasm in a cell. (p. 110) Activity in the nucleus varies in response to chemical messages arriving from the surrounding cytoplasm. In turn, ongoing nuclear activity will alter conditions within the cytoplasm by directing the synthesis of specific proteins. In this way the nucleus can affect enzyme activity, cell structure, and membrane properties.

In development, *differences in the cytoplasmic composition of individual cells trigger alterations in genetic activity.* These changes in turn lead to further changes in the cytoplasm, and the process continues in a sequential fashion. But if all the cells of the embryo are derived from cell divisions of a zygote, how do the cytoplasmic differences originate? What sets this process in motion? The important first step occurs before fertilization, while the egg is still in the ovary.

Prior to ovulation, the growing oocyte accepts amino acids, nucleotides, and glucose as well as more complex materials such as phospholipids, mRNAs, and even complete proteins from the surrounding follicular cells. Because the follicle cells are not all manufacturing and delivering the same nutrients and instructions to the oocyte, the contents of the oocyte cytoplasm are not evenly distributed. After fertilization, subsequent divisions subdivide the cytoplasm of the zygote into ever smaller cells that differ from one another in terms of their cytoplasmic composition. These differences alter genetic activity, creating cell lines with increasingly diverse fates.

As development proceeds, some of these cells will release chemical substances, such as RNAs, polypeptides, and small proteins, that affect the differentiation of other embryonic cells. This type of chemical interplay between developing cells is called **induction** (in-DUK-shun). Induction can work over very short distances, as when two different cell types are in direct contact, or it may operate long-distance, in which case the inducing chemicals are functioning as hormones. [**CM**: *Induction and Sexual Differentiation*]

Such an arrangement is very efficient, but the appearance of an abnormal or inappropriate inducer can throw the entire system off track. During the 1960s the European market was strong for **thalidomide**, a drug effective in promoting sleep and preventing nausea. Thalidomide was often prescribed for women in early pregnancy, with disastrous results. The drug crossed the placenta and entered the fetal circulation, where (among other effects) it interfered with the induction process responsible for limb development. As a result, many infants were born without limbs, or with drastically reduced ones.

In normal development, the pattern is set as the sperm and ova complete their maturation. The critical first step occurs at fertilization, as the two gametes unite. The steps then continue through the prenatal and postnatal periods. The next two sections follow these steps.

A PREVIEW OF PRENATAL DEVELOPMENT

The time spent in prenatal development is known as the period of **gestation** (jes-TĀ-shun). For convenience, the gestation period is usually considered as three integrated **trimesters**, each three months in duration:

- The **first trimester** is the period of embryonic and early fetal development. During this period the rudiments of all of the major organ systems appear.
- In the **second trimester** the organs and organ systems complete most of their development. The body proportions change, and by the end of the second trimester the fetus looks distinctively human.
- The **third trimester** is characterized by rapid fetal growth. Early in the third trimester most of the major organ systems become fully functional, and an infant born 1 month or even 2 months prematurely has a reasonable chance of survival.

■ The First Trimester

At the moment of conception, the fertilized ovum is a single cell with a diameter of around 0.135 mm (0.005 in.). In the first month, its weight increases by a factor of 140,000. By the end of the first trimester (12th developmental week) the embryo is almost 75 mm in length (3 in.), weighing perhaps 14 g (0.5 oz).

The events that occur in the first trimester are complex and vital to the survival of the embryo. Because accidents often happen, *the first trimester is the most dangerous period in prenatal or postnatal life.* Only about 40 percent of conceptions produce embryos that survive the first trimester, and an additional number of fetuses enter the second trimester already doomed or deformed by some developmental mistake. For this reason pregnant women are usually warned to take great care to avoid drugs or other disruptive stresses during the first trimester, in the hopes of preventing an error in the delicate processes under way.

Many important and complex developmental events occur during the first trimester. We will focus attention on four general processes: *cleavage, implantation, placentation,* and *embryogenesis.*

- **Cleavage** (KLĒV-ij) is a sequence of cell divisions that begins immediately after fertilization and ends at the first contact with the uterine wall. Over this period the zygote develops into a multicellular complex known as a **blastocyst**. (Cleavage and blastocyst formation were introduced in the Embryology Summary in Chapter 5.) (p. 125)
- **Implantation** begins with the attachment of the blastocyst to the endometrium and continues as the blastocyst invades the maternal tissues. During the time implantation is under way, a number of other important events take place that set the stage for the formation of vital embryonic structures.
- **Placentation** (pla-sen-TĀ-shun) occurs as blood vessels form around the periphery of the blastocyst, and the **placenta** appears. The placenta is a vital link between maternal and embryonic systems, and it will support the fetus during the second and third trimesters.
- **Embryogenesis** (em-brē-ō-JEN-e-sis) is the formation of a viable embryo. This process establishes the foundations for all major organ systems.

CLEAVAGE AND BLASTOCYST FORMATION

Cleavage is a series of cell divisions that subdivide the cytoplasm of the zygote. At the start of cleavage, the zygote is an enormous cell with a relatively tiny nucleus. The ratio between cytoplasmic volume and nuclear volume is about 550:1. At the end of cleavage that ratio is 6:1, about the same as the ratio for normal somatic cells in the adult. Figure 29-5 follows the changes in appearance that occur as cleavage progresses.

The first cleavage division produces two identical cells, called **blastomeres** (BLAS-tō-mērz). The first division is completed roughly 30 hours after fertilization, and subsequent cleavage divisions occur at intervals of 10–12 hours. The second division occurs in the same plane as the first, increasing the number

FIGURE 29-5
Cleavage and Blastocyst Formation

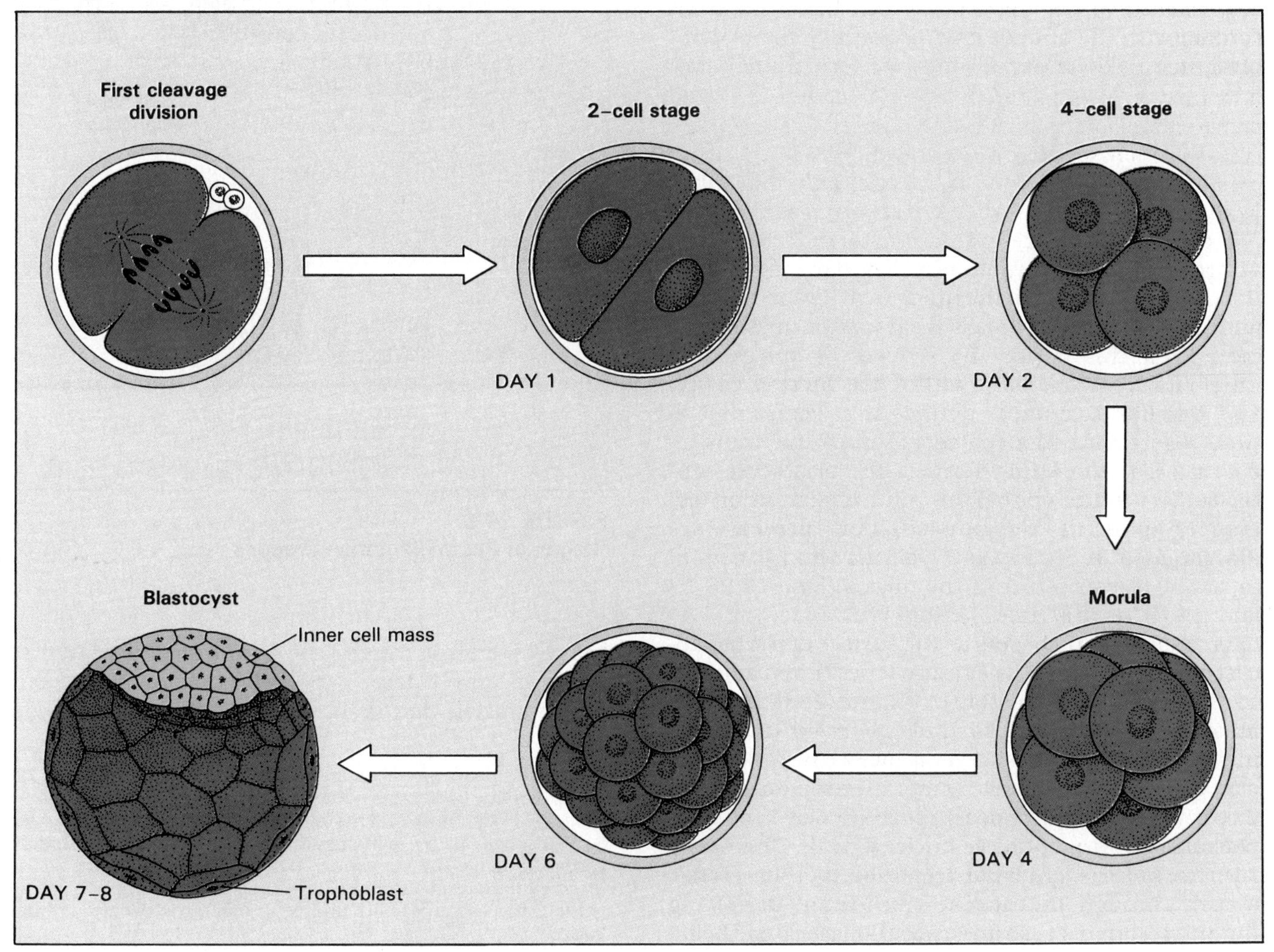

of blastomeres to four. The third cleavage division occurs at right angles to the first two, separating the blastomeres along the equatorial axis. During the initial cleavage divisions all of the blastomeres undergo mitosis simultaneously, but as the number of blastomeres increases the timing becomes less predictable.

Daughter cells maintain contact with their neighbors, some of which are their antecedents, and for a time the embryo is a solid ball of cells, resembling a mulberry. This stage is called the **morula** (MOR-ū-la; *morula*, mulberry). After 5 days of cleavage the blastomeres form a hollow ball, the **blastocyst**, with an inner cavity known as the **blastocoele** (BLAS-tō-sēl). At this stage you can begin to see differences between the cells of the blastocyst. Some of those variations result from the unequal distribution of cytoplasm, but in addition the blastocyst now has a complex three-dimensional shape. As a result, the blastomeres are no longer exposed to identical environments. The outer layer of cells, separating the outside world from the blastocoele, is called the **trophoblast** (TRŌ-fō-blast). The function is implied by the name: *tropho*, food + *blast*, precursor. These cells will be responsible for providing food to the developing embryo. A second group of cells, the **inner cell mass**, lies clustered at one end of the blastocyst. These cells are exposed to the blastocoele, but insulated from contact with the outside environment by the trophoblast. In time the inner cell mass will form the embryo.

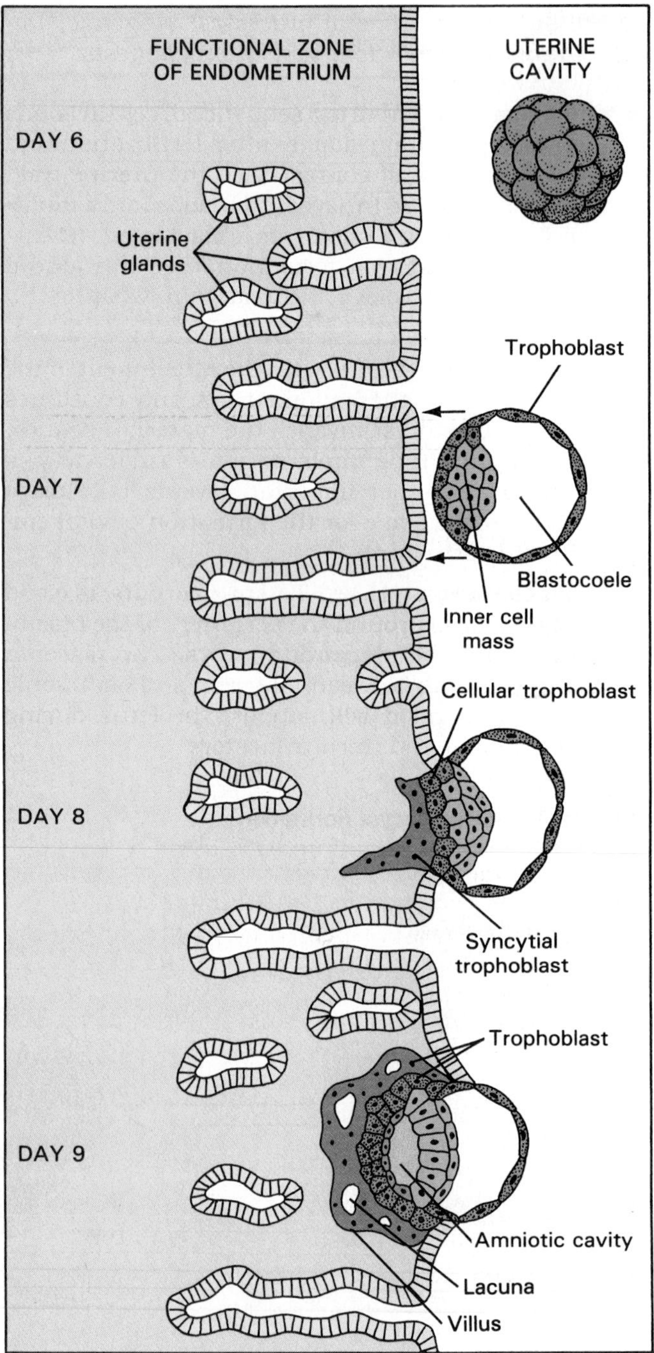

FIGURE 29-6
Stages in the Implantation Process

IMPLANTATION

At fertilization the zygote is still 4 days away from the uterus. It arrives in the uterine cavity as a morula, and over the next 2–3 days blastocyst formation occurs. Over this period the active cells are gaining nutrients from the fluid within the uterine cavity, and that fluid contains abundant glycogen due to the progesterone-induced secretion of the endometrial glands. When fully formed, the blastocyst contacts the uterine epithelium, and implantation occurs. Stages in the implantation process are diagrammed in Figure 29-6; you may find it helpful to review the structure of the uterus (Figures 28-13 and 28-19) at this time. ∞ (pp. 932, 941)

Implantation begins as the surface of the blastocyst closest to the inner cell mass touches and adheres to the uterine lining (Day 7, Figure 29-6). In this area the superficial cells undergo rapid divisions, making the trophoblast several layers thick. Near the endometrial wall the cell membranes separating the trophoblast cells disappear, creating a layer of cytoplasm containing multiple nuclei (Day 8). This outer layer, called the **syncytial trophoblast**, then erodes a path through the uterine epithelium, dissolving the intercellular cement between adjacent epithelial cells by secreting the enzyme *hyaluronidase*.[1] At first this erosion creates a gap in the uterine lining, but the migration and divisions of epithelial cells soon

[1] Hyaluronidase was also mentioned in Chapters 5 and 22, because many bacteria produce this enzyme to ease their spread through body tissues. This effective but destructive enzyme is also found in the acrosomal cap of sperm. Hyaluronidase cannot damage the blastocyst, however, because there are no intercellular connections holding the syncytial trophoblast together, and the enzyme is unable to reach the deeper, cellular layers.

repair the surface. When the repairs are completed, the blastocyst loses contact with the uterine cavity, and development occurs entirely within the functional zone of the endometrium.

The syncytial trophoblast continues to enlarge as implantation proceeds, and soon the trophoblastic surface resembles an interwoven carpet whose fibers spread into the surrounding endometrium (Day 9). The digestion of uterine gland cells releases quantities of glycogen and other nutrients that are absorbed by the syncytial trophoblast and distributed by diffusion across the underlying **cellular trophoblast** to the inner cell mass. These nutrients provide the energy needed to support the early stages of embryo formation. Trophoblastic extensions grow around endometrial capillaries, and as the capillary walls are destroyed, maternal blood begins to percolate through trophoblastic channels known as **lacunae**. Fingerlike **villi** extend away from the trophoblast into the surrounding endometrium, and these extensions gradually increase in size and complexity as development proceeds. Over the next few days the trophoblast begins breaking down larger endometrial veins and arteries, and blood flow through the lacunae accelerates. [**CM**: *Problems with the Implantation Process*]

The trophoblast establishes the basic physical structure of the placenta that will support development for the next 8–9 months. But completion of the placenta requires the participation of embryonic tissues originating at the inner cell mass. So before we can discuss the formation and anatomy of the placenta we need to consider what has happened to the inner cell mass while implantation was under way.

Formation of the Blastodisc

In the early blastocyst the inner cell mass has little apparent organization. But by the time of implantation, the inner cell mass is separating from the trophoblast. The separation gradually increases, creating a fluid-filled chamber called the **amniotic** (am-nē-OT-ik) **cavity**. The amniotic cavity can be seen in Day 9 of Figure 29-6, and additional details from Days 9–10 are shown in Figure 29-7. At this stage the cells of the inner cell mass are organized into an oval sheet that is two cell layers thick. This oval, called a **blastodisc** (BLAS-tō-disk), initially consists of an epithelial layer, or **epiblast** (EP-i-blast), facing the amniotic cavity and an underlying **hypoblast** (HĪ-pō-blast) confronting the blastocoele.

Gastrulation and Germ Layer Formation

A few days later, a third layer begins forming through the process of **gastrulation** (gas-troo-LĀ-shun) (Day 12, Figure 29-7). During gastrulation cells in specific areas of the epiblast move toward the center of the blastodisc, toward a line known as the **primitive streak**. At the primitive streak these migrating cells leave the surface and move between the epiblast and hypoblast. This creates three distinct embryonic layers with markedly different fates. Once gastrulation begins the layer remaining in contact with the amniotic cavity is called the **ectoderm**, the hypoblast is known as the **endoderm**, and the intervening, poorly organized layer is the **mesoderm**. The formation of mesoderm and the developmental fates of these **germ layers** were summarized in the Embryology Summary

FIGURE 29-7
Blastodisc Organization and Gastrulation

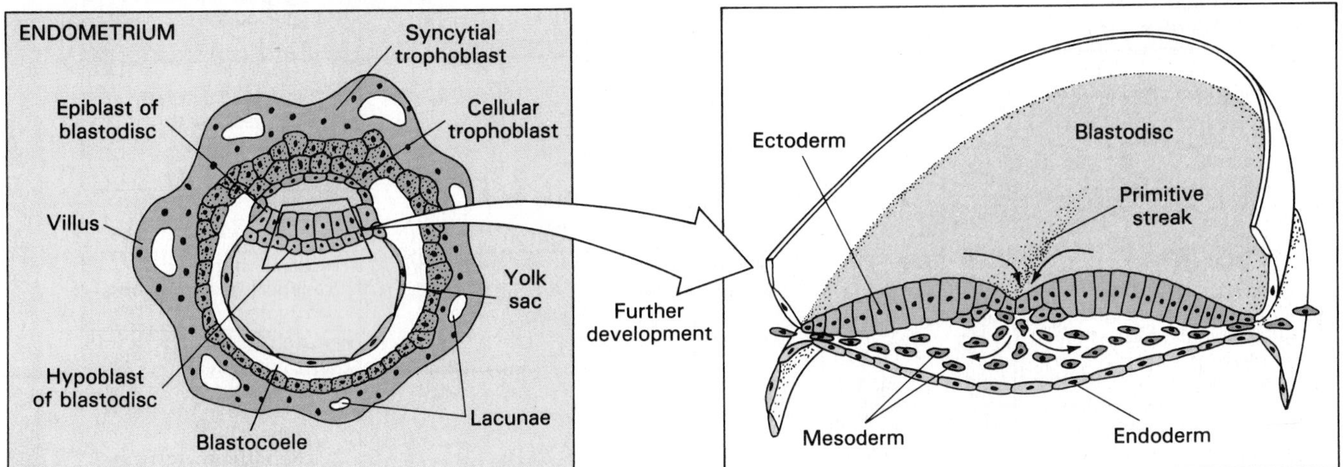

Day 10: The blastodisc begins as two layers, an epiblast facing the amniotic cavity and the hypoblast exposed to the blastocoele. Migration of epiblast cells around the amniotic cavity is the first step in the formation of the chorion. Migration of hypoblast cells creates a sac that hangs below the blastodisc. This is the first step in yolk sac formation.

Day 12: Migration of epiblast cells into the region between epiblast and hypoblast gives the blastocyst a third layer. From the time this process, called gastrulation, begins the epiblast is called *ectoderm*, the hypoblast *endoderm*, and the migrating cells *mesoderm*.

Health News: Technology and the Treatment of Infertility

Infertility, or the inability to have children, has been the focus of media attention for the past 5 years. The reason is simple: problems with infertility are relatively common. An estimated 10–15 percent of U.S. marriages are infertile, and another 10 percent are unable to have as many children as desired. With 20–30 percent of all marriages as potential customers, it is not surprising that reproductive physiology has become a popular field, and the treatment of infertility has become a major medical industry.

An infertile, or sterile, woman is unable to produce functional eggs and/or support a developing embryo. An infertile man is incapable of providing a sufficient number of motile sperm for successful fertilization. Because sterility of either sexual partner will have the same result, diagnosis and treatment of infertility must involve evaluation of both sexual partners. Approximately 60 percent of infertility cases can be attributed to problems with the female reproductive system.

Recent advances in our understanding of reproductive physiology are providing a number of new solutions to fertility problems. These approaches are diagrammed in Figure 29-8.

In cases of male infertility due to low sperm counts, semen from several ejaculates can be pooled, concentrated, and introduced into the female reproductive tract. This technique, known as *artificial insemination*, may lead to normal fertilization and pregnancy. If the husband cannot produce functional sperm, sperm can be obtained from a "sperm bank" that stores

FIGURE 29-8
Strategies to Treat Infertility

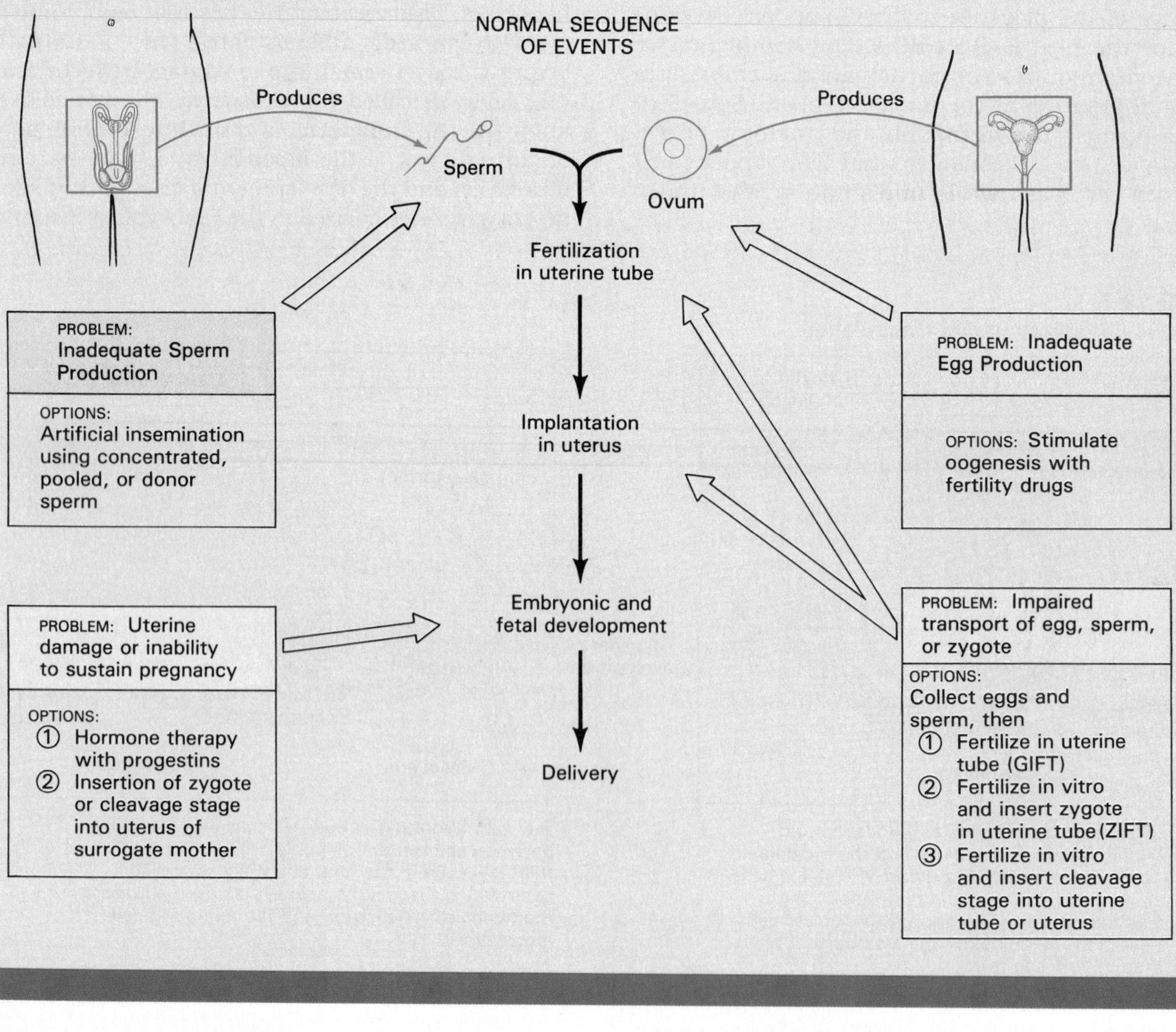

NORMAL SEQUENCE OF EVENTS

Produces

Sperm

Produces

Ovum

Fertilization in uterine tube

Implantation in uterus

Embryonic and fetal development

Delivery

PROBLEM: Inadequate Sperm Production

OPTIONS: Artificial insemination using concentrated, pooled, or donor sperm

PROBLEM: Inadequate Egg Production

OPTIONS: Stimulate oogenesis with fertility drugs

PROBLEM: Uterine damage or inability to sustain pregnancy

OPTIONS:
① Hormone therapy with progestins
② Insertion of zygote or cleavage stage into uterus of surrogate mother

PROBLEM: Impaired transport of egg, sperm, or zygote

OPTIONS: Collect eggs and sperm, then
① Fertilize in uterine tube (GIFT)
② Fertilize in vitro and insert zygote in uterine tube (ZIFT)
③ Fertilize in vitro and insert cleavage stage into uterine tube or uterus

donor sperm. If the problem involves the woman's inability to ovulate due to low gonadotrophin or estrogen levels, or her ability to maintain adequate progesterone levels after ovulation, these hormones can be provided. So-called *fertility drugs*, such as clomiphene (*Clomid*), stimulate ovarian egg production. Clomiphene works by blocking the feedback inhibition of estrogen on the hypothalamus and pituitary gland. As a result, circulating FSH levels rise, and more follicles are stimulated to complete their development. The chance of a single egg being fertilized through normal sexual intercourse is around 1 in 3. By increasing the number of eggs released, the odds of a pregnancy are increased. Unfortunately, it is not easy to determine just how much ovarian stimulation will be required. As a result, multiple births have often resulted from treatment with fertility drugs.

When there are problems with the transport of the egg from the ovary to the uterine tube, due to scarring of the fimbriae or other problems, a procedure called **GIFT** can be used. GIFT is short for *gamete intrafallopian tube transfer* (*Fallopian tube* is another name for the uterine tube or *oviduct*). Eggs are removed from the ovaries, placed in the uterine tubes, and exposed to high concentrations of sperm from the husband or donor. A large "crop" of mature eggs is collected after stimulating the ovaries with injected hormones, and the individual eggs are examined for defects before insertion into the uterine tubes. The success rate for this procedure is less than that of natural fertilization (33 percent), and not every pregnancy produces an infant. The cost of a single procedure (successful or not) averages $4000–5000.

In the GIFT procedure, fertilization occurs in its normal location, within the uterine tube. This site is not essential, and fertilization can also take place in a test tube or petri dish. This process is called **in vitro fertilization** (*vitro*, glass). If a carefully controlled fluid environment is provided, early development will proceed normally. One variation on the GIFT procedure, called **ZIFT** (*zygote intrafallopian tube transfer*), exposes selected eggs to sperm outside the body and inserts zygotes or early cleavage-stage embryos, rather than eggs, into the uterine tubes. If multiple zygotes are available, some can be frozen and stored for later insertion in case the initial procedure fails to produce a successful pregnancy. The cost for a single ZIFT procedure ranges between $5000 and $7500.

Alternatively, the zygote can be maintained in an artificial environment through the first 2–3 days of development. This procedure is often selected if the uterine tubes are damaged or blocked. The cleavage-stage embryo is then placed directly into the uterus, rather than into one of the uterine tubes. The cost of this procedure ranges from $4500 to $6500.

If fertilization and transport occur normally, but the uterus cannot maintain a pregnancy, the problem may involve low levels of progestin secretion by the corpus luteum. Hormone therapy may solve this problem. If the maternal uterus simply cannot support development, the zygote or cleavage-stage embryo can be introduced into the uterus of a substitute mother, or *surrogate mother*. If the embryo survives and makes contact with the endometrium, development will proceed normally despite the fact that the mother has no genetic relationship with the embryo. This procedure, which sounds relatively simple and straightforward, has proven to be one of the most explosive in terms of ethics and legality. Over the past 5 years, several court cases have resulted from disputes over surrogate motherhood and who merits legal custody of the infant.

Interpreting the rights and obligations of surrogate mothers has not been the only problem for lawyers and judges that can be attributed to new reproductive technologies. Legal battles have also broken out over a variety of complex questions, and some of them will take years to sort out. To understand the problem, just take a moment and try to decide:

- Do parents share property rights over frozen and stored zygotes? Can a husband have any of the stored zygotes implanted into the uterus of his second wife without the consent of his first wife, who provided the eggs?
- If the husband provided the sperm that fertilized the egg of a donor who is not his wife, for implantation into a surrogate mother, can the wife, the surrogate mother, or the egg donor sue for custody of the child after a divorce?

If you use your imagination, you can probably think of even more complex problems, many of which will probably be debated in a courtroom within the next decade.

■ **TABLE 29-3 The Fates of the Primary Germ Layers**

Ectodermal contributions
- *Integumentary system*: epidermis, hair follicles and hairs, nails, and glands communicating with the skin (apocrine and eccrine sweat glands, mammary glands, and sebaceous glands)
- *Digestive system*: mucous epithelium of mouth and anus, salivary glands
- *Respiratory system*: mucous epithelium of nasal passageways
- *Nervous system*: all neural tissue, including brain and spinal cord
- *Endocrine system*: posterior pituitary gland and the adrenal medullae

Mesodermal contributions
- *Muscular system*: all components
- *Skeletal system*: all components
- *Lymphatic system*: all components
- *Cardiovascular system*: all components, including bone marrow
- *Urinary system*: the kidneys, including the nephrons and the initial portions of the collecting system
- *Reproductive system*: the gonads and the adjacent portions of the duct systems
- *Endocrine system*: adrenal cortex, endocrine tissues in the gonads
- *Miscellaneous*: the lining of the body cavities (thoracic, pericardial, peritoneal) and the connective tissues supporting all organ systems

Endodermal contributions
- *Digestive system*: mucous epithelium (except mouth and anus), exocrine glands (except salivary glands), liver, and pancreas
- *Respiratory system*: respiratory epithelium (except nasal passageways) and associated mucous glands
- *Urinary system*: urinary bladder and distal portions of the duct system
- *Reproductive system*: distal portions of the duct system, stem cells that produce gametes
- Endocrine system: thymus, thyroid, pancreas, and anterior pituitary gland

on p. 180, which should be reviewed at this time. Table 29-3 contains a more comprehensive listing of the contributions each germ layer makes to the body systems described in earlier chapters.

Formation of Extraembryonic Membranes

Germ layers also participate in the formation of four **extraembryonic membranes**: the *yolk sac* (endoderm and mesoderm), the *amnion* (ectoderm and mesoderm), the *allantois* (endoderm and mesoderm), and the *chorion* (mesoderm and trophoblast). Although these membranes support embryonic and fetal development, they leave few traces of their existence in adult systems. Figure 29-9 details stages in the development of the extraembryonic membranes.

The Yolk Sac The first of the extraembryonic membranes to appear is the **yolk sac**. The yolk sac begins as the migration of hypoblast cells spreads out around the outer edges of the blastocoele to form a complete pouch suspended below the blastodisc. This pouch is already visible 10 days after fertilization, and it is shown in Figure 29-7. As gastrulation proceeds, mesodermal cells migrate around this pouch and complete the formation of the yolk sac (Figure 29-9a). Blood vessels soon appear within the mesoderm, and the yolk sac becomes an important site of blood cell formation.

Although the human yolk sac has no other function, in the blastocysts of primitive mammals such as the Australian platypus or spiny anteater the yolk sac cavity contains a yolk mass comparable to that of a chicken egg. The yolk proteins and lipids provide the energy required for fetal development of these mammals, which do not develop placental connections. In humans nourishment is provided via the placenta, which absorbs nutrients from the maternal bloodstream, and a yolk supply is not essential to survival. The yolk sac remains, however, reminding us of our evolutionary history.

The Amnion The ectodermal layer also undergoes an expansion, and ectodermal cells spread over the inner surface of the amniotic cavity. As in the case of the yolk sac, mesodermal cells soon follow, creating a second, outer layer. This combination of mesoderm and ectoderm is the **amnion** (AM-nē-on). As the embryo and later the fetus enlarges, this membrane continues to expand, increasing the size of the amniotic cavity. The amnion encloses fluid that surrounds and cushions the developing embryo and fetus (Figure 29-9c–e).

The Allantois The third extraembryonic membrane begins as an outpocketing of the endoderm near the base of the yolk sac. As indicated in Figure 29-9b,c, the free tip grows toward the wall of the blastocyst surrounded by a mass of mesodermal cells. This sac of endoderm and mesoderm is the **allantois** (a-LAN-tō-is), and the base of the allantois later gives rise to the urinary bladder. (The formation of the allantois and its relationship to the urinary bladder is detailed in the Embryology Summary in Chapter 26.) (p. 870)

Throughout embryonic and fetal development metabolic wastes generated by the fetus are eliminated by transfer to the maternal circulation. The developing kidneys produce a relatively dilute urine that is discharged directly into the amniotic cavity. A full-term fetus generates around 500 mℓ of urine each day and swallows a comparable amount of amniotic fluid for absorption and eventual recycling.

The Chorion The mesoderm associated with the allantois spreads out, once it reaches the wall of the blastocyst. Soon it extends completely around the inside, forming a mesodermal layer underneath

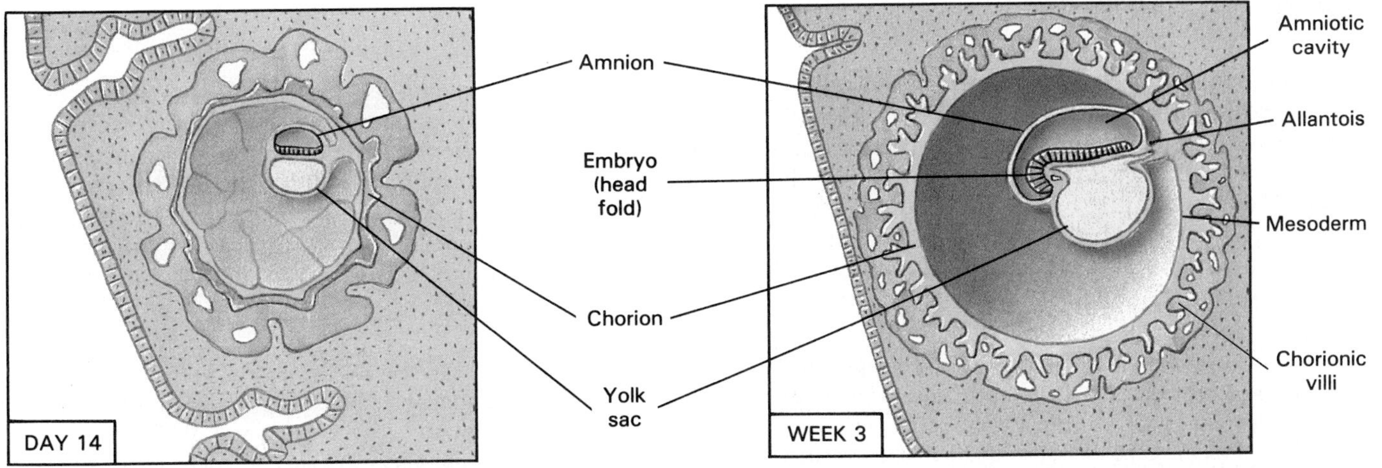

Migration of mesoderm around the inner surface of the trophoblast creates the chorion. Mesodermal migration around the outside of the amniotic cavity, between the ectodermal cells and the trophoblast, creates the amnion. Mesodermal migration around the endodermal pouch below the blastodisc creates the definitive yolk sac.

The embryonic disc bulges into the amniotic cavity at the head fold. The allantois, an endodermal extension surrounded by mesoderm, extends toward the trophoblast.

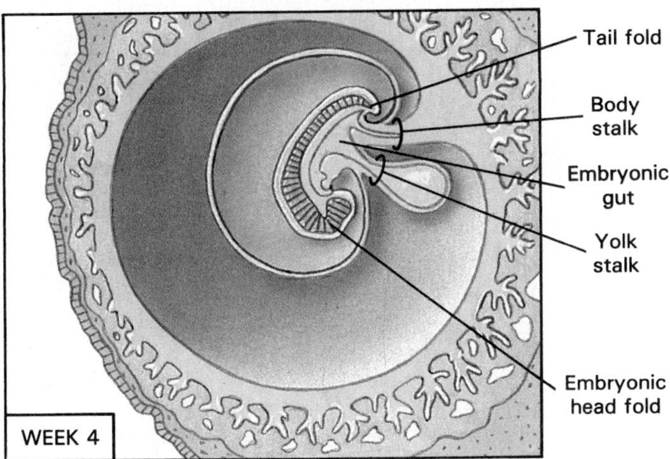

The embryo now has a head fold and a tail fold. Constriction of the connection between the embryo and the surrounding trophoblast constricts the yolk stalk and body stalk.

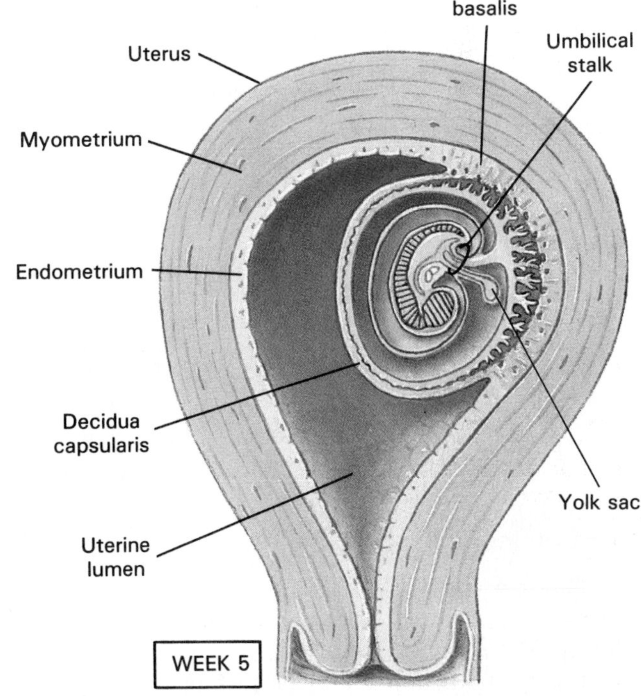

The developing embryo and extraembryonic membranes bulge into the uterine cavity. The trophoblast pushing out into the uterine lumen remains covered by endometrium, but no longer participates in nutrient absorption and embryo support. The embryo moves away from the placenta, and the body stalk and yolk stalk fuse to form an umbilical stalk.

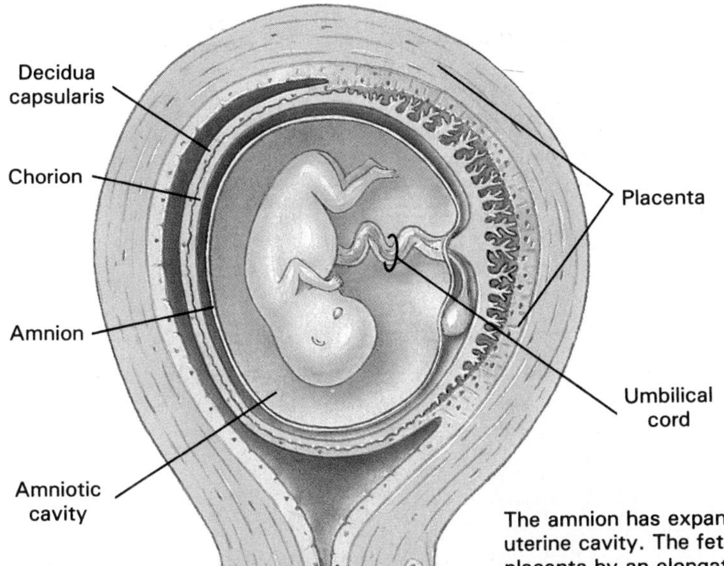

The amnion has expanded greatly, filling the uterine cavity. The fetus is connected to the placenta by an elongate umbilical cord that contains a portion of the allantois, blood vessels and the remnants of the yolk stalk.

FIGURE 29-9
Embryonic Membranes and Placenta Formation

the trophoblast. This combination of mesoderm and trophoblast is the **chorion** (KOR-ē-on) (Figure 29-9a). The significance of the chorion becomes apparent when you consider the problems facing the developing embryo as it continues to enlarge. When implantation first occurs, the nutrients absorbed by the trophoblast can easily reach the blastodisc by simple diffusion. But as the embryo and the trophoblastic complex enlarges, the distance between the two increases and diffusion alone can no longer keep pace with the demands of the developing embryo. The chorion solves this problem, for blood vessels developing within the mesoderm provide a rapid transit system linking the embryo with the trophoblast. Circulation

FIGURE 29-10

A Three-Dimensional View of Placental Structure. For clarity the uterus is shown after the embryo has been removed and the umbilical cord cut. Blood flows into the placenta through ruptured maternal blood arteries. It then flows around chorionic villi that contain fetal blood vessels. Fetal blood arrives over paired umbilical arteries and leaves over a single umbilical vein. Maternal blood reenters the venous system of the mother through the broken walls of small uterine veins. Note that no actual mixing of maternal and fetal blood occurs. The inset micrograph is a cross section through the vilus, showing the syncitial trophoblast exposed to the maternal blood space. (LM, × 2120)

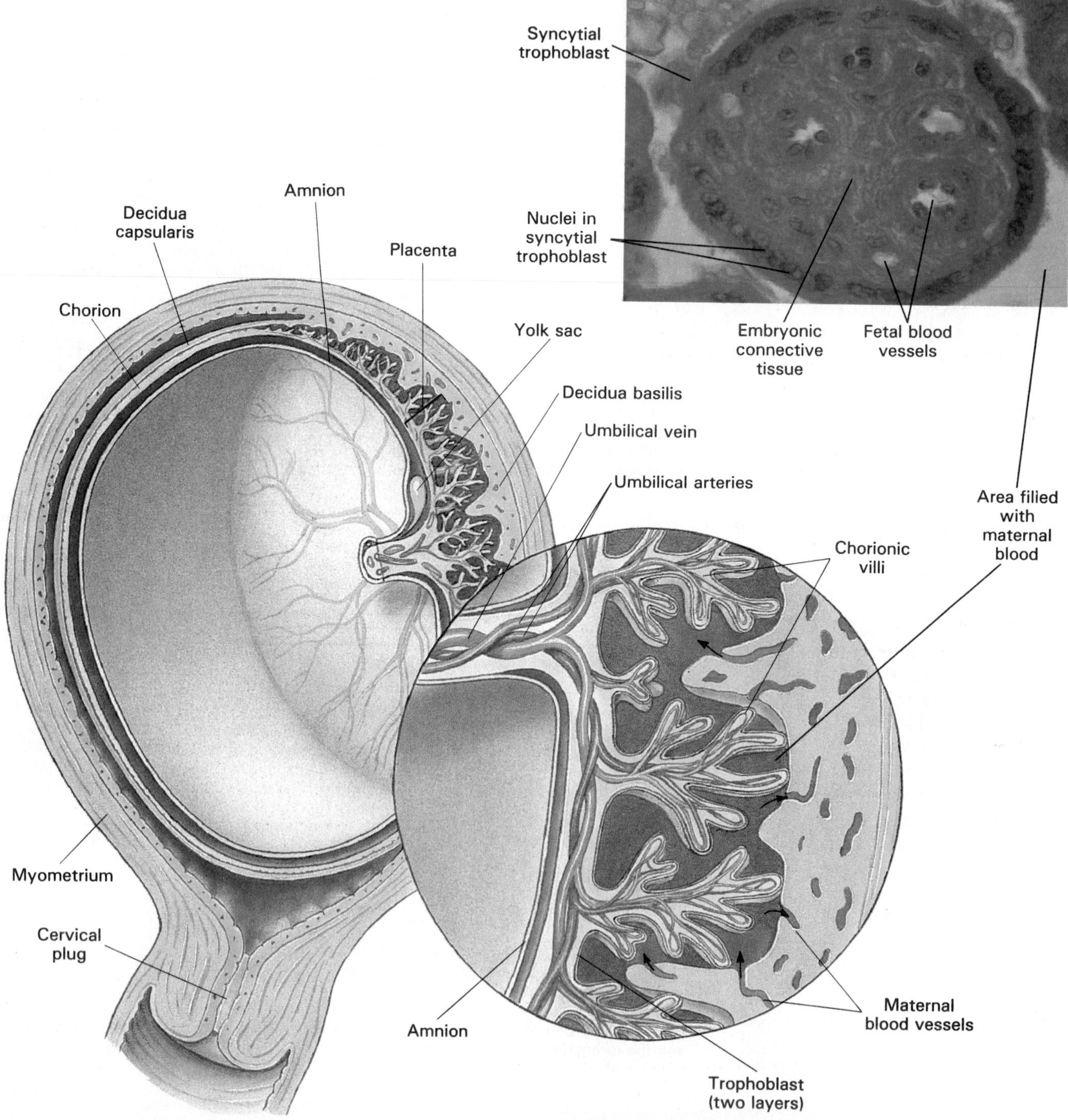

through those chorionic vessels begins early in the third week of development, when the heart starts beating.

PLACENTATION

The appearance of blood vessels in the chorion is the first step in the creation of a functional placenta. Figure 29-9 also provides details concerning the development and growth of the placenta. By the third week of development (Figure 29-9b), mesoderm extends along the core of each of the trophoblastic villi, forming **chorionic villi** in contact with maternal tissues. These villi continue to enlarge and branch, forming an intricate network within the endometrium. Blood vessels continue to be eroded, and maternal blood slowly percolates through lacunae lined by syncytial trophoblast. Chorionic blood vessels pass close by, and exchange between the embryonic and maternal circulations occurs by diffusion across the syncytial and cellular trophoblast layers.

At first the entire blastocyst is surrounded by chorionic villi. The chorion continues to enlarge, expanding like a balloon within the endometrium, and by the fourth week the embryo, amnion, and yolk sac are suspended within an expansive, fluid-filled chamber (Figure 29-9c). The only communication between the embryo and the chorion occurs at the site where mesoderm cells first came into contact with the inner wall of the trophoblast. The intervening connection, which contains the distal portions of the allantois and blood vessels carrying blood to and from the placenta, is known as the **body stalk**. The narrow connection between the endoderm of the embryo and the yolk sac is called the **yolk stalk**. (The formation of the yolk stalk and body stalk were detailed in the Embryology Summary in Chapter 24.) (p. 804)

The placenta cannot continue to enlarge indefinitely, for there are physical limits imposed by the structure of the uterus. Regional differences in placental organization begin to develop as placental expansion creates a prominent bulge in the endometrial surface. This relatively thin portion of the endometrium, the **decidua capsularis** (dē-SID-ū-a kap-sū-LA-ris), no longer participates in nutrient exchange, and the chorionic villi disappear (Figure 29-9d). Placental functions are now concentrated in a disc-shaped area situated in the deepest portion of the endometrium, a region called the **decidua basalis** (ba-SA-lis). As the end of the first trimester approaches, the fetus moves farther away from the placenta (Figure 29-9d,e). It remains connected by the **umbilical cord**, or *umbilical stalk*, that contains the allantois, blood vessels, and the yolk stalk.

Placental Circulation

Figure 29-10 indicates the extent of the fetal circulation at the placenta near the end of the first trimester. Blood flows to the placenta through the paired **umbilical arteries** and returns in a single **umbilical vein**. It has been estimated that the chorionic villi provide around 90 square meters of surface area for active and passive exchange between the fetal and maternal bloodstreams. As noted in Chapter 28, the placenta also synthesizes important hormones that affect maternal as well as embryonic tissues. (p. 942) Human chorionic gonadotrophin production begins within a few days of implantation, and during the second and third trimesters the placenta also releases progesterone, estrogens, human placental lactogen, and relaxin. These hormones are synthesized and released into the maternal circulation by the trophoblast. [**CM**: *Problems with Placentation*]

EMBRYOGENESIS

Shortly after gastrulation begins, folding and differential growth of the embryonic disc produce a bulge that projects into the amniotic cavity as indicated in Figure 29-10b. This projection is known as the **head fold**, and similar movements lead to the formation of a **tail fold** (Figure 29-10c). The embryo proper is now physically as well as developmentally separated from the blastodisc and the extraembryonic membranes. The definitive orientation of the embryo can now be seen, complete with dorsal and ventral surfaces and left and right sides.

These changes in orientation and form are illustrated in more detail in the Embryology Summary dealing with the formation of the basic body plan. (p. 14) Figure 29-11 on page 974 presents a visual summary of the changes in proportions and appearance that occur between the fourth developmental week and the end of the first trimester. The developmental history of structures labeled in Figure 29-11a can be followed in more detail by referring to the Embryology Summaries in earlier chapters.

The first trimester is a critical period for development because events in the first 12 weeks establish the basis for organ formation, a process called **organogenesis**. Embryology Summaries in earlier chapters described major features of organogenesis in each organ system. Important developmental milestones are indicated in Table 29-4; those interested in additional details should refer to the Embryology Summaries cross-referenced in the table. [**CM**: *Teratogens and Abnormal Development*]

- √ **What would happen if a spermatozoan did not become capacitated in the vagina?**
- √ **What is the fate of the inner cell mass of the blastocyst?**
- √ **Improper development of which of the extraembryonic membranes would affect the circulatory system?**
- √ **What are two important functions of the placenta?**

TABLE 29-4 An Overview of Prenatal Development

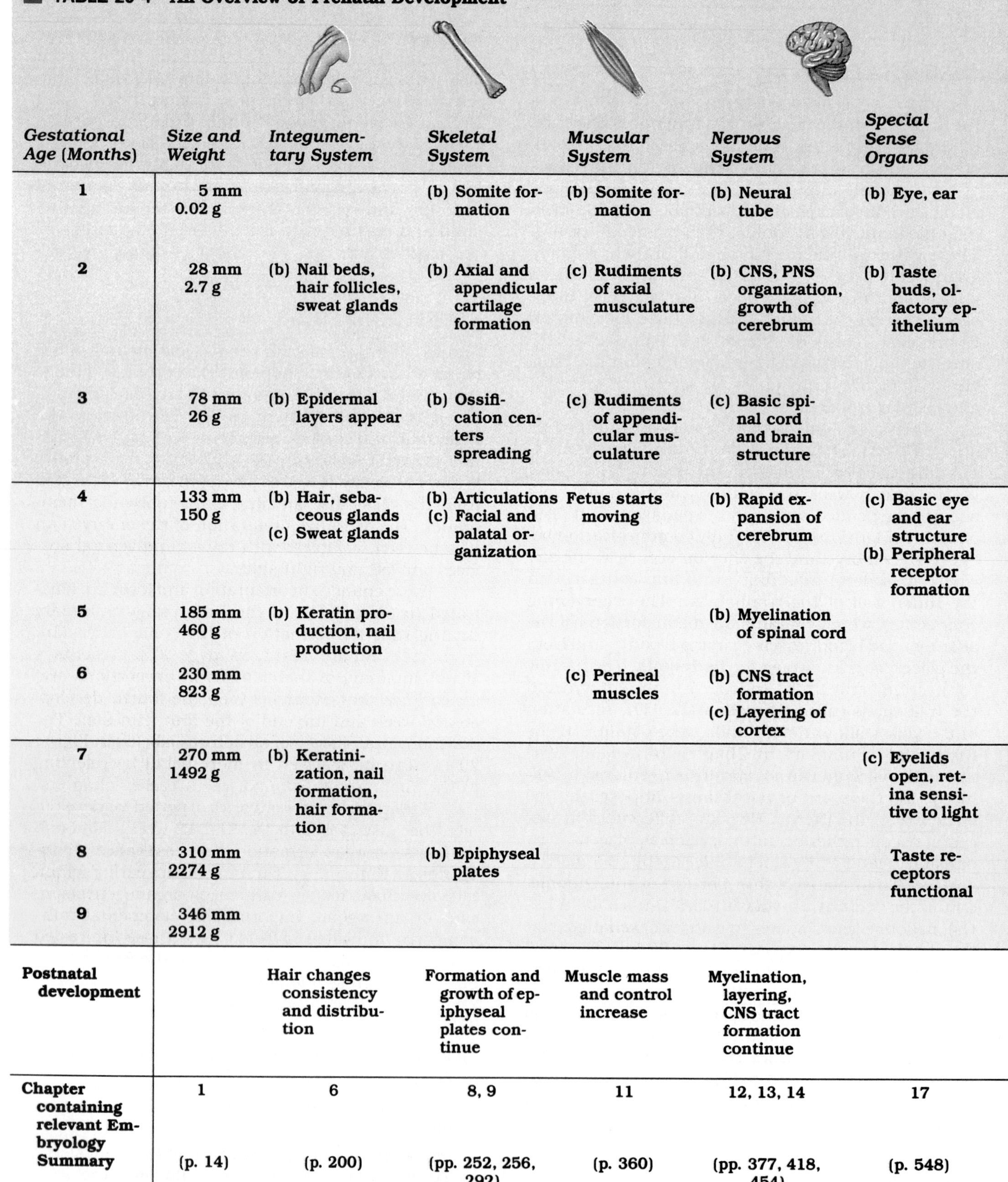

Gestational Age (Months)	*Size and Weight*	*Integumentary System*	*Skeletal System*	*Muscular System*	*Nervous System*	*Special Sense Organs*
1	5 mm 0.02 g		(b) Somite formation	(b) Somite formation	(b) Neural tube	(b) Eye, ear
2	28 mm 2.7 g	(b) Nail beds, hair follicles, sweat glands	(b) Axial and appendicular cartilage formation	(c) Rudiments of axial musculature	(b) CNS, PNS organization, growth of cerebrum	(b) Taste buds, olfactory epithelium
3	78 mm 26 g	(b) Epidermal layers appear	(b) Ossification centers spreading	(c) Rudiments of appendicular musculature	(c) Basic spinal cord and brain structure	
4	133 mm 150 g	(b) Hair, sebaceous glands (c) Sweat glands	(b) Articulations (c) Facial and palatal organization	Fetus starts moving	(b) Rapid expansion of cerebrum	(c) Basic eye and ear structure (b) Peripheral receptor formation
5	185 mm 460 g	(b) Keratin production, nail production			(b) Myelination of spinal cord	
6	230 mm 823 g			(c) Perineal muscles	(b) CNS tract formation (c) Layering of cortex	
7	270 mm 1492 g	(b) Keratinization, nail formation, hair formation				(c) Eyelids open, retina sensitive to light
8	310 mm 2274 g		(b) Epiphyseal plates			Taste receptors functional
9	346 mm 2912 g					
Postnatal development		Hair changes consistency and distribution	Formation and growth of epiphyseal plates continue	Muscle mass and control increase	Myelination, layering, CNS tract formation continue	
Chapter containing relevant Embryology Summary	1 (p. 14)	6 (p. 200)	8, 9 (pp. 252, 256, 292)	11 (p. 360)	12, 13, 14 (pp. 377, 418, 454)	17 (p. 548)

Note: (b) = begin formation; (c) = complete formation.

Endocrine System	*Cardiovascular and Lymphatic Systems*	*Respiratory System*	*Digestive System*	*Urinary System*	*Reproductive System*
	(b) Heartbeat	(b) Trachea and lung formation	(b) Intestinal tract, liver, pancreas (c) yolk sac	(c) Allantois	
(b) Thymus, thyroid, pituitary, adrenal glands	(c) Basic heart structure, major blood vessels, lymph-nodes and ducts (b) Blood formation in liver	(b) Extensive bronchial branching into mediastinum (c) Diaphragm	(b) Intestinal subdivisions, villi, salivary glands	(b) Kidney formation (adult form)	(b) Mammary glands
(c) Thymus, thyroid gland	(b) Tonsils, blood formation in bone marrow		(c) Gallbladder, pancreas		(b) Definitive gonads, ducts, genitalia
	(b) Migration of lymphocytes to lymphatic organs, blood formation in spleen			(b) Degeneration of embryonic kidneys	
	(c) Tonsils	(c) Nostrils open	(c) Intestinal subdivisions		
(c) Adrenal glands	(c) Spleen, liver, bone marrow	(b) Alveolar formation	(c) Epithelial organization, glands		
(c) Pituitary gland			(c) Intestinal plicae		(b) Testes descend
		Complete pulmonary branching and alveolar formation		Complete nephron formation at birth	Descent complete at or near time of delivery
	Cardiovascular changes at birth; immune system becomes operative thereafter				
18	20, 21	23	24	26	28
(p. 590)	(pp. 647, 670)	(p. 760)	(p. 804)	(p. 870)	(p. 945)

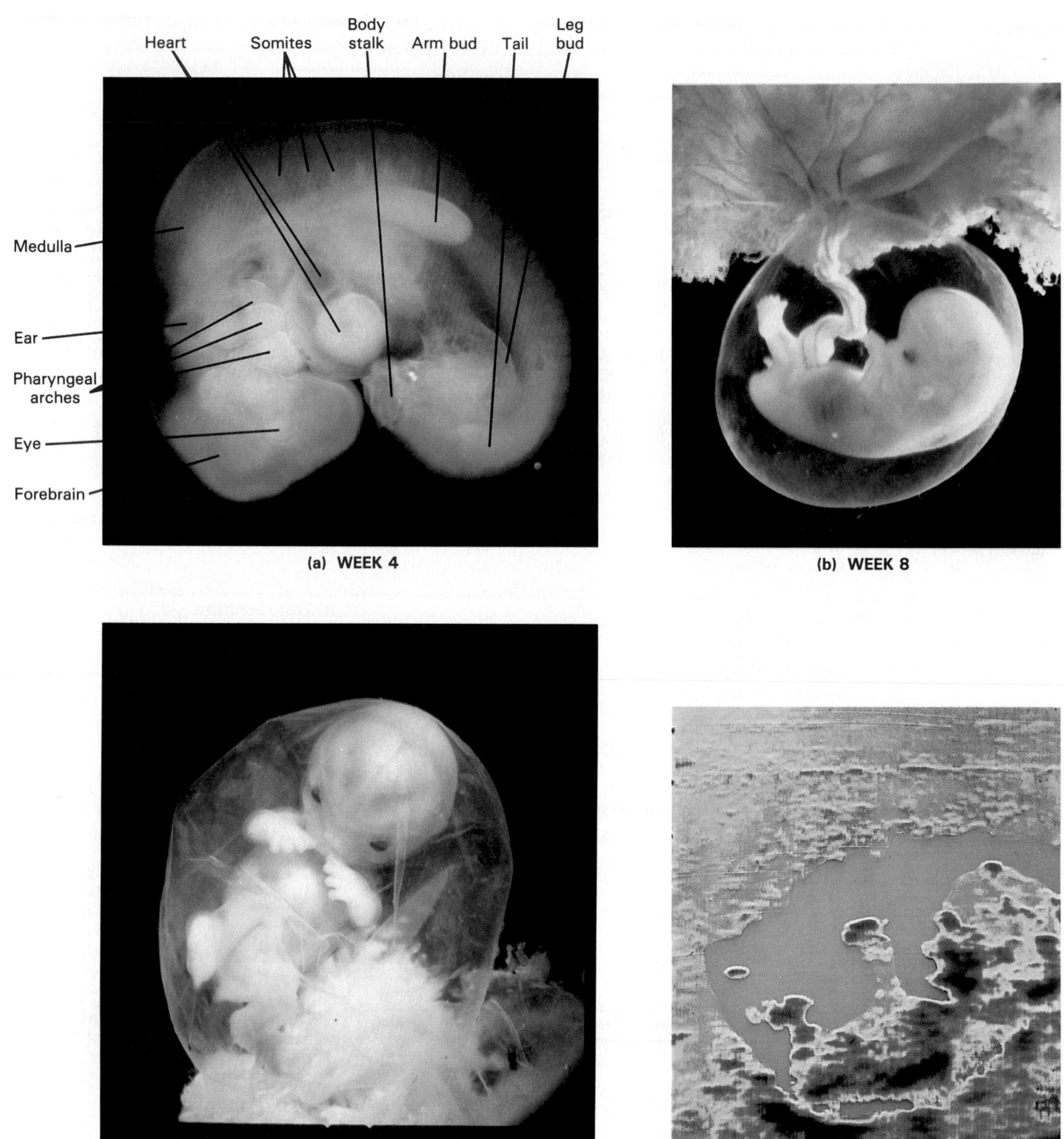

(a) WEEK 4

(b) WEEK 8

(c) WEEK 12

(d)

FIGURE 29-11
The First Trimester. Parts (a-c) are fiberoptic views of actual human embryos. View (d) is an ultrasound image that should be compared with (c).

■ The Second and Third Trimesters

By the start of the second trimester the rudiments of all of the major organ systems have formed. Over the next 3 months those systems near functional completion, and the fetus grows rapidly, tripling its length and increasing in weight by a factor of 50, ending the second trimester at around 0.64 kg (1.4 lb). Over this period the fetus, encircled by the amnion, grows faster than the surrounding placenta. Soon the mesodermal outer covering of the amnion fuses with the inner lining of the chorion. The new relationships between the various extraembryonic membranes can be seen in Figure 29-10, and representative views of second-trimester and third-trimester fetuses are shown in Figure 29-12.

During the third trimester, all of the organ systems become functional. The rate of growth begins to decrease, but in absolute terms this trimester sees the largest weight gain. In 3 months the fetus puts on around 2.6 kg (5.7 lb), reaching a full-term weight of somewhere near 3.2 kg (7 lb). The Embryology Summary in Chapter 1 shows the changes in body form that occur during the second and third trimesters. (p. 14) Important events in organ system development in the second and third trimesters were included in the Embryology Summaries in earlier chapters, and highlights are noted in Table 29-4.

PREGNANCY AND MATERNAL SYSTEMS

The developing fetus is totally dependent on maternal organ systems for nourishment, respiration, and waste removal. These functions must be performed by maternal systems in addition to their normal operations. For example, the mother must absorb enough oxygen, nutrients, and vitamins for herself *and* her fetus, and she must eliminate all of the generated wastes. Although this is not a burden over the initial weeks of gestation, the demands placed upon the mother become significant as the fetus grows larger. To survive under these conditions the maternal systems must make major compensatory adjustments. [**CM**: *Problems with the Maintenance of a Pregnancy*]

In practical terms the mother must breathe, eat, and excrete for two.

- *The respiratory rate goes up and the tidal volume increases.* As a result, the lungs obtain the extra oxygen required and remove the excess carbon dioxide generated by the fetus.

FIGURE 29-12
The Second and Third Trimesters. (a) A 4-month fetus seen through a fiberoptic microscope. (b) A 6-month fetus seen with ultrasound equipment.

(a)

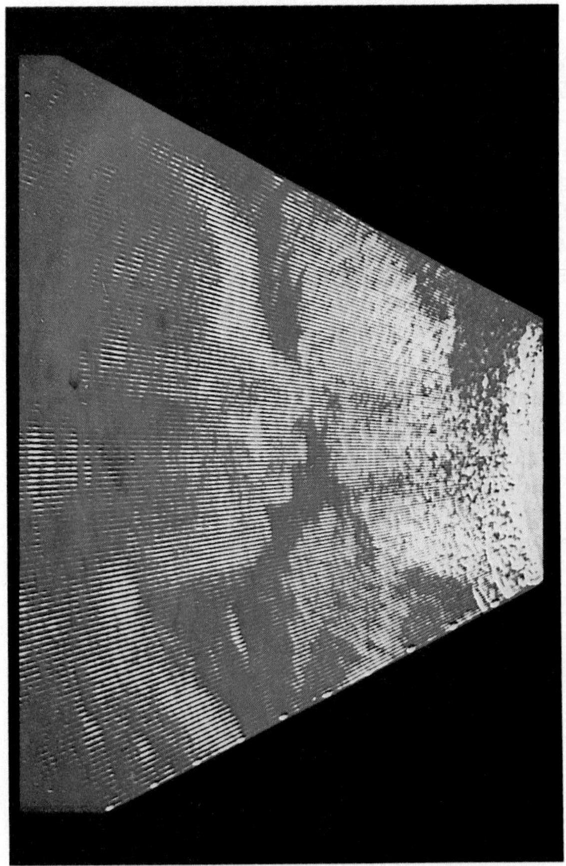
(b)

- *The maternal blood volume increases.* This occurs because (1) blood flowing into the placenta reduces the volume in the rest of the systemic circuit, and (2) fetal activity lowers the blood pO_2 and elevates the pCO_2. The combination stimulates the production of renin and erythropoietin (EPO), leading to an increase in maternal blood volume through mechanisms detailed in Figure 19-15a. (p. 630) By the end of gestation the maternal blood volume has increased by almost 50 percent.
- *The maternal requirements for nutrients and vitamins climb 10–30 percent.* Pregnant women, who must "eat for two," are often hungry.
- *The glomerular filtration rate increases by roughly 50 percent.* This corresponds to the increase in blood volume, and it accelerates the excretion of metabolic wastes generated by the fetus. Because (1) the volume of urine produced increases and (2) the weight of the uterus presses down on the urinary bladder, pregnant women need to urinate frequently.
- *The uterus undergoes a tremendous increase in size.* Structural and functional changes in the expanding uterus are so important that we will discuss them in a separate section.

STRUCTURAL AND FUNCTIONAL CHANGES IN THE UTERUS

At the end of gestation a typical uterus will have grown from 7.5 cm (3 in.) in length and 60 g (2 oz) in weight to 30 cm (12 in.) in length and 1100 g (2.4 lb) in weight. It may then contain almost 5 liters of fluid, giving the organ with contents a total weight of roughly 10 kg (22 lb). This remarkable expansion occurs through the enlargement and elongation of existing cells (especially muscle fibers), rather than by an increase in the total number of cells in the uterus.

The tremendous stretching of the myometrium is associated with a gradual increase in the rates of spontaneous smooth muscle contractions. In the early stages of pregnancy the contractions are weak, painless, and brief in duration. There are indications that the progesterone released by the placenta has an inhibitory effect on the uterine smooth muscle, preventing more extensive and powerful contractions.

Three major factors oppose the calming action of progesterone:

1. Estrogens, also produced by the placenta, increase the sensitivity of the uterine smooth muscles and make contractions more likely. Throughout pregnancy progesterone exerts the dominant effect, but as the time of delivery approaches estrogen production accelerates, and the myometrium becomes restless.
2. Oxytocin can trigger an immediate increase in the force and frequency of uterine contractions. Oxytocin release occurs at the posterior pituitary when the hypothalamus is stimulated by high estrogen levels and/or mechanical stimulation of the cervix.
3. In addition to estrogens and oxytocin, uterine tissues late in pregnancy produce prostaglandins that stimulate smooth muscle contractions.

After 9 months of gestation, multiple factors interact to produce **labor contractions** in the uterine wall. Once they have begun, a positive feedback mechanism operates to ensure that the contractions continue until delivery has been completed. Figure 29-13 diagrams important factors that stimulate and sustain labor.

1. *Relaxin* produced by the placenta softens the symphysis pubis, allowing it to move relatively easily, and the weight of the fetus then deforms the cervical orifice.
2. Deformation of the cervix and rising estrogen levels promote the release of oxytocin, and estrogen makes the already critically stretched smooth muscles even more excitable.

Late in pregnancy some mothers experience oc-

FIGURE 29-13
Interacting Factors in Labor and Delivery

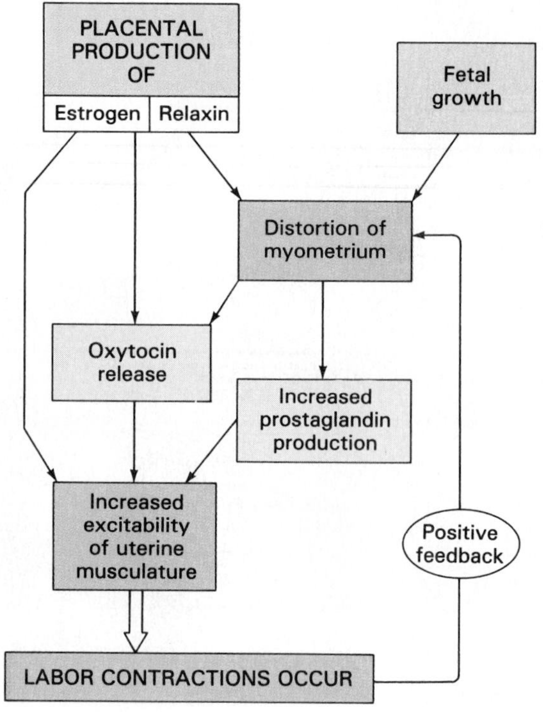

casional spasms in the uterine musculature, but the contractions are neither regular nor persistent. These contractions are called **false labor**. **True labor** begins when the biochemical and mechanical factors reach the point of no return.

■ Labor and Delivery

The goal of labor is the forcible expulsion of the fetus, a process known as **parturition** (par-tū-RISH-un), or birth. During true labor, each labor contraction begins near the top of the uterus and sweeps in a wave toward the cervix. These contractions are strong and occur at regular intervals. As parturition approaches the contractions increase in force and frequency, changing the position of the fetus and moving it toward the cervical canal.

STAGES OF LABOR

Labor has traditionally been divided into three stages, the *dilation stage*, the *expulsion stage*, and the *placental stage*. These stages are illustrated in Figure 29-14.

The Dilation Stage

The **dilation stage** begins with the onset of true labor, as the cervix dilates completely and the fetus begins

FIGURE 29-14
The Stages of Labor

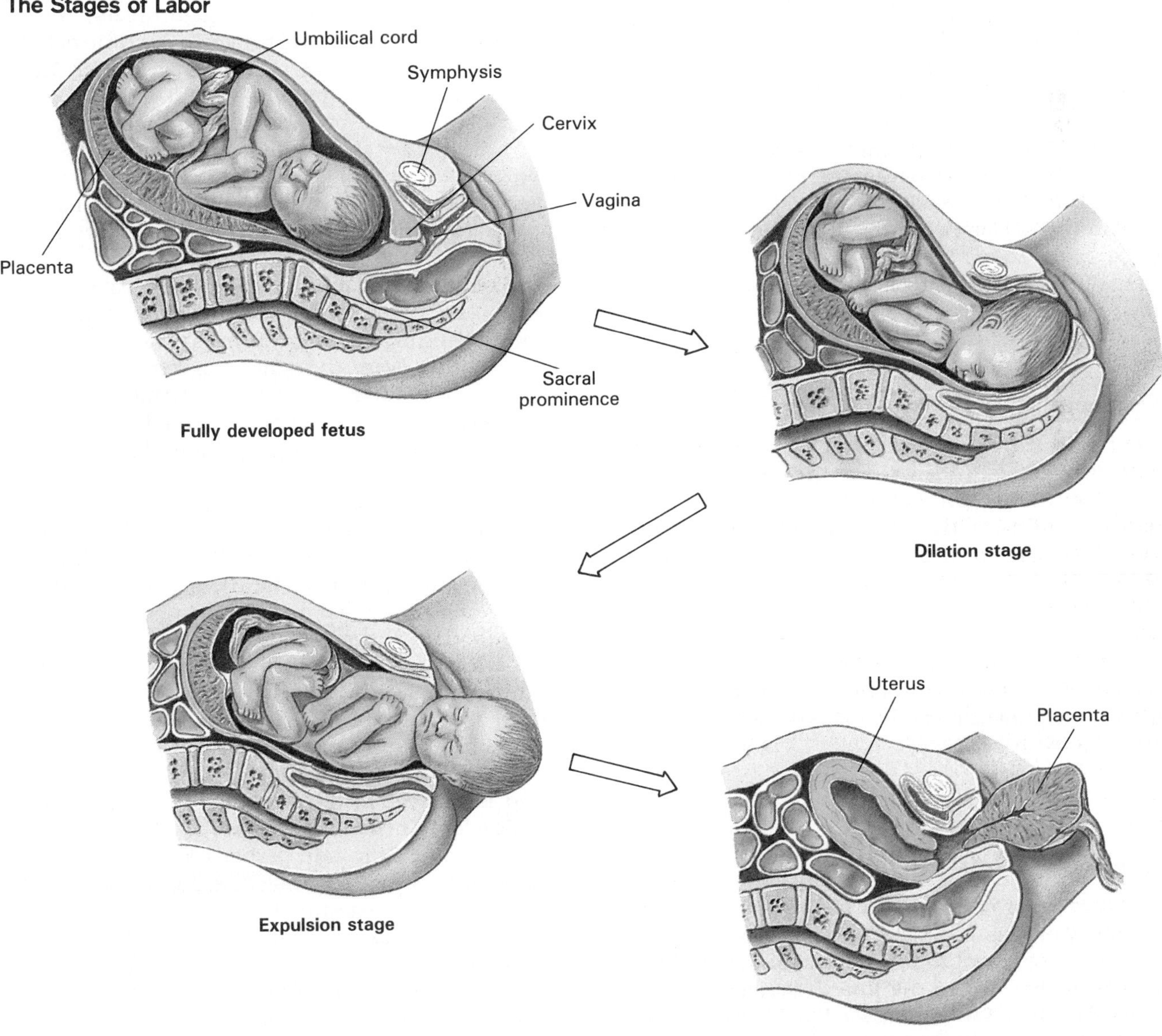

to slide down the cervical canal. This stage may last 8 or more hours, but during this period the labor contractions occur at intervals of once every 10–30 minutes. Late in the process the amnion usually ruptures, an event sometimes referred to as "having the water break."

The Expulsion Stage

The **expulsion stage** begins as the cervix dilates completely, pushed open by the approaching fetus. Expulsion continues until the fetus has completed its emergence from the vagina, a period usually lasting less than 2 hours. The arrival of the newborn infant into the outside world represents the birth, or **delivery**.

If the vaginal entrance is too small to permit the passage of the fetus, the entryway may be temporarily enlarged by making an incision through the perineal musculature (otherwise the tissues may tear uncontrollably). After delivery this **episiotomy** (e-pēz-ē-OT-o-mē) can be repaired with sutures, a much simpler procedure than dealing with a potentially extensive perineal tear. If unexpected complications arise during the dilation or expulsion stages, the infant may be removed by **caesarean section**. In such cases an incision is made through the abdominal wall, and the uterus is opened just enough to allow passage of the infant's head. This procedure is performed during 15–20 percent of the deliveries in the United States.

The Placental Stage

During the third, or **placental**, stage of labor the muscle tension builds in the walls of the partially empty uterus, and the organ gradually decreases in size. This uterine contraction tears the connections between the endometrium and the placenta. Usually within an hour after delivery the placental stage ends with the ejection of the placenta, or "afterbirth." The disruption of the placenta is accompanied by a loss of blood, perhaps as much as 500–600 mℓ, but because the maternal blood volume has increased during pregnancy the loss can be tolerated without difficulty. [**CM**: *Common Problems with Labor and Delivery*]

PREMATURE LABOR

Premature labor occurs when true labor begins before the fetus has completed normal development. This is a major threat to the survival of the newborn infant. Roughly half of the infant deaths following delivery are linked to premature labor and delivery.

The chances of newborn survival are directly related to body weight at delivery. Even with massive supportive efforts, infants born weighing less than 400 g (14 oz) will not survive, primarily because the respiratory, cardiovascular, and urinary systems are unable to support life without the aid of maternal systems. As a result, the dividing line between *spontaneous abortion* and **immature delivery** is usually set at 500 g (17.6 oz), the normal weight near the end of the second trimester.

In practical terms, infants delivered before completing 7 months of gestation (weight under 1 kg) have less than a 50:50 chance of survival and most survivors suffer from severe developmental abnormalities. A **premature delivery** produces a newborn weighing over 1 kg (35.2 oz), and its chances of survival range from fair to excellent depending on the individual circumstances.

MULTIPLE BIRTHS

Multiple births (twins, triplets, quadruplets, and so forth) may occur for several reasons. The ratio of twin to single births in the U.S. population is roughly 1:89. In "fraternal," or **dizygotic** (dī-zī-GOT-ik), twins two separate eggs were ovulated and subsequently fertilized. Because of the shuffling of chromosomes that occurs during meiosis, the odds against any two zygotes from the same parents having identical genes exceeds 1:8.4 million. Seventy percent of twins are dizygotic.

"Identical," or **monozygotic**, twins may result from the separation of blastomeres early in cleavage or by splitting of the inner cell mass prior to gastrulation. In either event the genetic makeup of the pair will be almost identical, for both formed from the same pair of gametes. Triplets, quadruplets, and larger multiples can result from multiple ovulations, blastomere splitting, or some combination of the two. For unknown reasons the statistics for naturally occurring multiple births fall into a pattern: twins occur at a rate of 1:89, triplets at a rate of $1{:}89^2$, quadruplets at $1{:}89^3$, and so forth. The incidence of multiple births can be increased by exposure to fertility drugs that stimulate the maturation of abnormally large numbers of oocytes (see the related Health News box for a more detailed discussion of fertility and infertility).

Complete splitting of the embryonic portion of the blastodisc can produce identical twins. If the separation is not complete, **conjoined** (*Siamese*) **twins** may develop. These infants often share a portion of the liver, some skin, and perhaps other internal organs as well. When the fusion is minor the infants can be surgically separated with some success, but those with more extensive fusions usually fail to survive delivery.

Even normal multiple pregnancies pose special problems for maternal systems, for the strains are multiplied proportionately. The incidence of toxemia and premature labor are both elevated, and the maternal mortality rate is higher than for single births.

The risks for the fetus are also increased, both during gestation and after delivery, for even at full term the newborn infants have a lower average birth weight. Multiple pregnancies also lead to complications at delivery; in the case of twins, more than half of the deliveries involve one or both fetuses entering the vaginal canal in an abnormal position.

- √ **Why does a mother's total blood volume increase during pregnancy?**
- √ **What effect would a decrease in progesterone have on the uterus during late pregnancy?**

Postnatal Development

Developmental processes do not cease at delivery, for the newborn infant has few of the anatomical, functional, or physiological characteristics of the mature adult. In the course of postnatal development each individual passes through a number of **life stages**, each typified by a distinctive combination of characteristics and abilities.

These stages are a familiar part of human experience. You could probably identify the features and functions associated with **infancy**, **childhood**, **adolescence**, and **maturity** (Figure 29-15). Although each stage has distinctive features, the transitions between them are gradual and the boundaries often indistinct. Once maturity has arrived, development ends and the process of aging, or **senescence**, begins. We will now consider highlights of each of the stages of life.

THE NEONATAL PERIOD, INFANCY, AND CHILDHOOD

The **neonatal period** extends from the moment of birth to 1 month thereafter. **Infancy** then continues to 2 years of age, and **childhood** lasts until puberty commences. Two major events are under way during these developmental stages.

FIGURE 29-15
Stages of Life

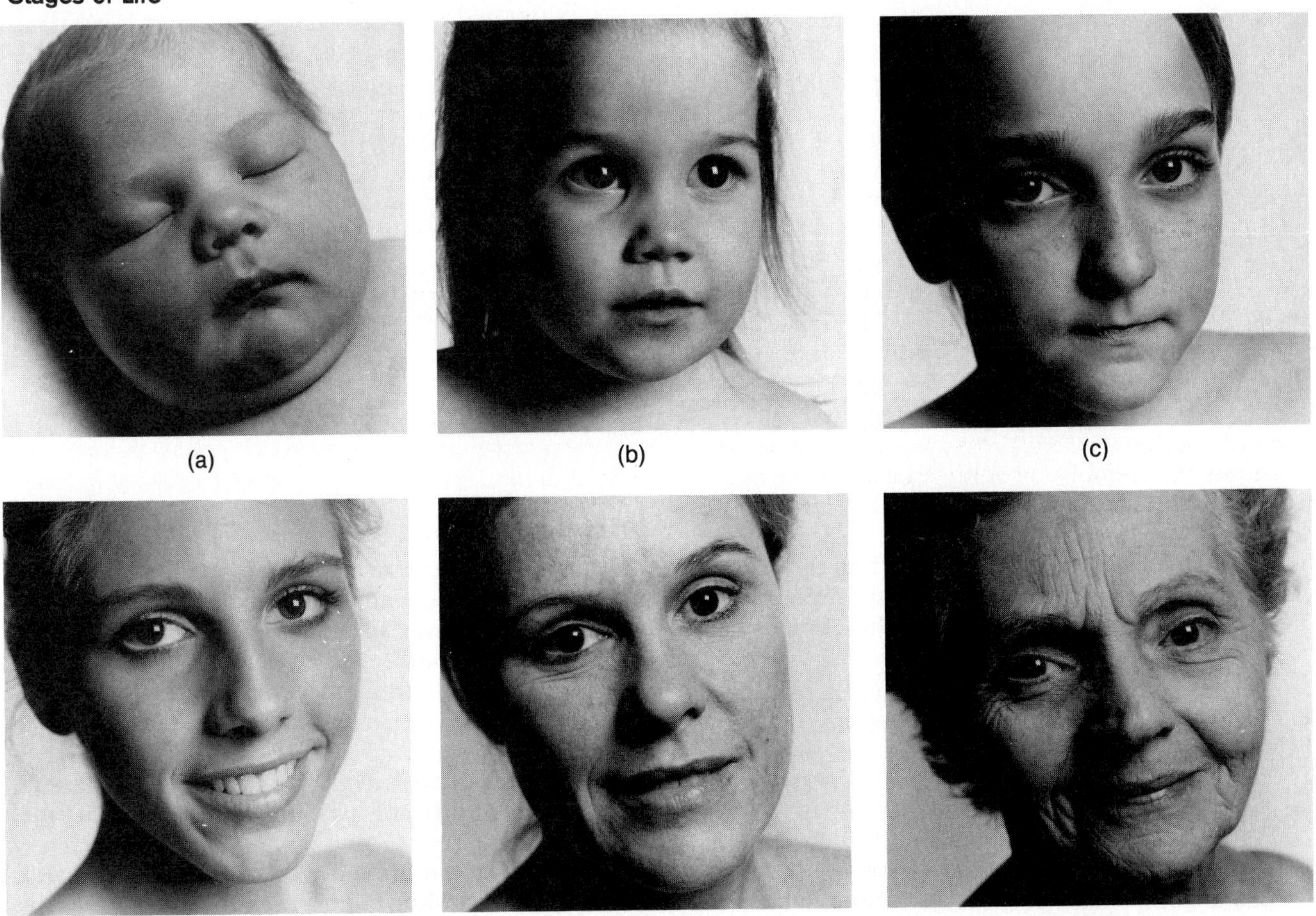

(a) (b) (c) (d) (e) (f)

Health News: Complexity and Perfection: An Improbable Dream

The expectation of prospective parents that every pregnancy will be idyllic and every baby a perfect specimen reflects deep-seated misconceptions about the nature of the developmental process. These misconceptions lead many to believe that when serious developmental errors occur someone or something is at fault, and blame might be assigned to maternal habits, such as smoking, alcohol consumption, or improper diet, maternal exposure to toxins or prescription drugs, or the presence of other disruptive stimuli in the environment. The prosecution of women giving birth to severely impaired infants for "fetal abuse" (exposing a fetus to known or suspected risk factors) represents an extreme example of this philosophy.

Although environmental stimuli may indeed lead to developmental problems, such factors are only one component of a complex system normally subject to considerable variation. Even if every pregnant woman were packed in cotton fluff and locked in her room from conception to delivery, developmental accidents and errors would continue to appear with regularity.

Spontaneous mutations are the result of random errors in the replication process, and such incidents are relatively common. At least 10 percent of fertilizations produce zygotes with abnormal chromosomes, and because spontaneous mutations usually fail to produce visible defects, the actual number of mutations must be far larger. Most of the affected zygotes die

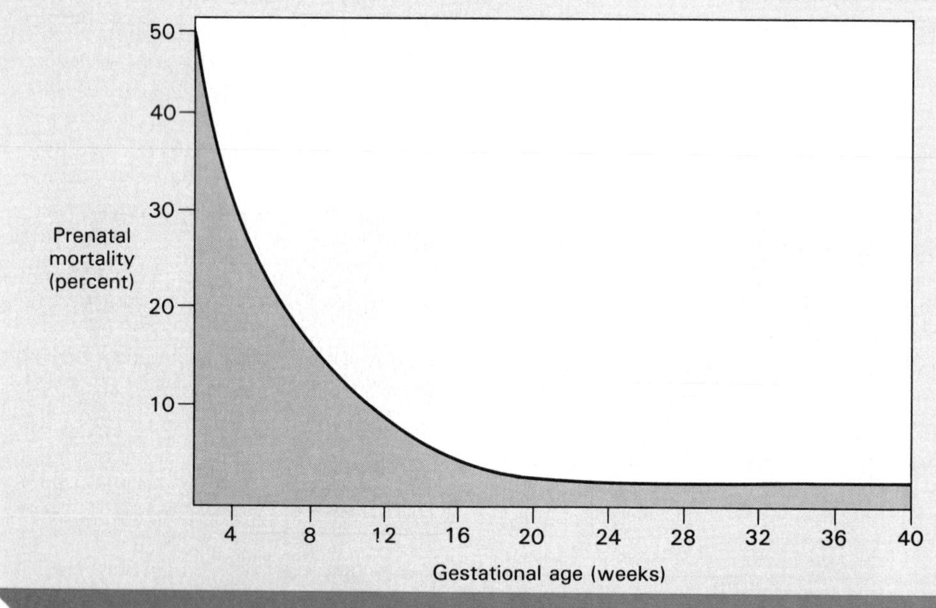

FIGURE 29-16
Prenatal Mortality

1. The major organ systems other than those associated with reproduction become fully operational and gradually acquire the functional characteristics of adult structures.
2. The individual grows rapidly, and there are significant changes in body proportions.

The Neonatal Period

A variety of physiological and anatomical alterations occur as the fetus completes the transition to the status of a newborn infant, or **neonate**. Prior to delivery transfer of dissolved gases, nutrients, waste products, hormones, and immunoglobulins occurred across the placental interface. At birth the newborn infant must become relatively self-sufficient, with the processes of respiration, digestion, and excretion performed by its own specialized organs and organ systems.

Specifics have been noted in previous chapters, so they will receive only passing mention at this time.

- The lungs at birth are collapsed and filled with fluid, and filling them with air involves a massive and powerful inspiratory movement.

before completing development, and only about 0.5 percent of newborn infants show chromosomal abnormalities resulting from spontaneous mutations.

Because of the nature of the regulatory mechanisms, prenatal development does not follow precise, predetermined pathways. For example, considerable variation exists in the pathways of blood vessels and nerves, because it doesn't matter how the blood or neural impulses get to their destination, as long as they do get there. If the variations fall outside acceptable limits, however, the embryo or fetus fails to complete development. Very small changes in heart structure may result in the death of a fetus, while large variations in venous distribution are extremely common and relatively harmless. Virtually everyone can be considered abnormal to some degree, because no one has characteristics that are statistically "average" in every respect. An estimated 20 percent of your genes differ from those found in the majority of the population, and minor defects such as extra nipples or birthmarks are quite common.

Current evidence suggests that as many as half of all conceptions produce zygotes that do not survive the cleavage stage. These disintegrate within the uterine tubes or uterine cavity, and because implantation never occurs there are no obvious signs of pregnancy. These instances of preimplantation mortality are often associated with chromosomal abnormalities. Of those embryos that implant, roughly 20 percent fail to complete 5 months of development, with an average survival time of 8 weeks. Severe problems affecting early embryogenesis or placenta formation are usually responsible. A curve for embryonic and fetal mortality is included in Figure 29-16.

Prenatal mortality tends to eliminate the most severely affected fetuses. Those with less extensive defects may survive, completing full-term gestation or arriving via premature delivery. **Congenital malformations** are structural abnormalities, present at birth, that affect major systems. Spina bifida, hydrocephaly, anencephaly, cleft lip, and Down syndrome are among the most common congenital malformations; these conditions were described in earlier chapters of the text and the Clinical Manual. The incidence of congenital malformations at birth averages around 6 percent, but only one-third of them are categorized as severe. Only 10 percent of these congenital problems can be attributed to environmental factors in the absence of chromosomal abnormalities or genetic factors, including a family history of similar or related defects.

Medical technology continues to improve our abilities to understand and manipulate physiological processes. Genetic analysis of potential parents may now provide estimates concerning the likelihood of specific problems, although the problems themselves remain outside our control. But even with a better understanding of the genetic mechanisms involved, it will probably never be possible to control every aspect of development and thereby prevent spontaneous abortions and congenital malformations. There are simply too many complex, interdependent steps involved in prenatal development, and malfunctions of some kind are statistically inevitable.

- When the lungs expand, the pattern of cardiovascular circulation changes due to alterations in blood pressure and flow rates. The ductus arteriosus closes, isolating the pulmonary and systemic trunks, and the closure of the foramen ovale separates the atria of the heart completing separation of the pulmonary and systemic circuits.
- Typical heart rates of 120–140 beats per minute and respiratory rates of 30 breaths per minute in neonates are considerably higher than those of adults.
- Prior to birth the digestive system remains relatively inactive, although it does accumulate a mixture of bile secretions, mucus, and epithelial cells. This collection of debris is excreted in the first few days of life. Over that period the newborn infant begins to nurse, obtaining nourishment from the milk produced by the mother's mammary glands.
- As waste products build up in the arterial blood they are filtered into the urine at the kidneys. Glomerular filtration is normal, but the urine cannot be concentrated to any significant degree. As a result, urinary water losses are high,

and neonatal fluid requirements are relatively greater than those of adults.

- The neonate has little ability to control body temperature, particularly in the first few days after delivery. For this reason newborn infants are usually kept bundled up in warm coverings. As the infant grows larger and increases the thickness of its insulating adipose "blanket," its metabolic rate also rises, and thermoregulatory abilities become more pronounced. Nevertheless, daily and even hourly alterations in body temperature continue throughout childhood.

The most rapid growth occurs during prenatal development, and after delivery the relative rate of growth continues to decline. Postnatal growth during infancy and childhood occurs under the direction of circulating hormones, notably growth hormone from the pituitary, adrenal steroids, and thyroid hormones. These hormones affect each tissue and organ in specific ways, depending on the sensitivities of the individual cells. As a result, growth does not occur uniformly, and the body proportions gradually change. These alterations were summarized in the Embryology Summary in Chapter 1 (p. 14).

Clinical Comment: Evaluating the Newborn Infant

Each newborn infant gets a close scrutiny after delivery. This procedure checks for the presence of anatomical or physiological abnormalities and provides baseline information useful in assessing postnatal development. In addition to general appearance, the pulse, respiratory rates, weight, length, and other physical dimensions are noted. Newborn infants are also screened for genetic and/or metabolic disorders, such as *phenylketonuria (PKU)* or congenital *hypothyroidism*.

The **Apgar rating** considers heart rate, respiratory rate, muscle tone, response to stimulation, and color at 1 and 5 minutes after birth. In each category the infant receives a score ranging from 0 (poor) to 2 (excellent), and the scores are then totaled. An infant's Apgar rating (0–10) has been shown to be an accurate predictor of newborn survival and the presence of neurological damage. For example, newborn infants with *cerebral palsy* usually have a low Apgar rating.

MONITORING POSTNATAL DEVELOPMENT

Pediatrics is a medical specialty focusing on postnatal development from infancy through adolescence. Because infants and young children cannot clearly describe the problems they are experiencing, pediatricians and parents must be skilled observers. A number of standardized testing procedures, such as the Apgar test (see the Clinical Comment: Evaluating the Newborn Infant), are used to assess an individual's developmental progress relative to normal values. In the **Denver Developmental Screening Test (DDST)** infants and children are checked repeatedly during their first 5 years. The test checks gross motor skills, such as sitting up or rolling over, language skills, fine motor coordination, and social interactions. The results are compared with normal values determined for individuals of similar age. These screening procedures assist in identifying children who may need special teaching and attention.

Too often parents tend to focus on a single ability or physical attribute, such as the age at first step or the rate of growth. This kind of one-track analysis has little practical value, and the parents may become overly concerned with how their infant compares with the norm. *It should be realized that normal values are statistical averages, not absolute realities.* For example, an infant usually begins walking between 11 and 14 months of age. But around 25 percent start before then, and another 10 percent do not start walking by the fourteenth month. Walking early does not indicate true genius, and walking late does not mean that the infant will need physical therapy. The questions on such screening tests are intended to determine if there are *patterns* of developmental deficits. Such patterns appear only when a broad range of abilities and characteristics are considered.

ADOLESCENCE AND MATURITY

Adolescence begins at puberty, when three events interact to promote increased hormone production and sexual maturation. These factors can be summarized as follows:

1. The hypothalamus increases its production of gonadotrophin-releasing hormone (GnRH).
2. The anterior pituitary becomes more sensitive to the presence of GnRH, and there is a rapid elevation in the circulating levels of FSH and LH (ICSH).
3. Ovarian or testicular cells become more sensitive to FSH and LH. These changes initiate gametogenesis and the production of male or female sex hormones that stimulate the appearance of secondary sexual characteristics and behaviors.

In the years that follow, the continued background secretion of estrogens or androgens maintains these sexual characteristics. In addition, the combination of sex hormones and growth hormone, adrenal steroids, and thyroxine leads to a sudden acceleration in the growth rate. The timing of the increase in size varies between the sexes, corresponding to different ages at the onset of puberty. In girls the growth rate is maximum between ages 10 and 13, whereas boys grow most rapidly between ages 12 and 15. Growth continues at a slower pace until ages 18–21, when most of the epiphyseal plates close.

The boundary between adolescence and maturity is very hazy, for it has physical, emotional, behavioral, and legal implications. Maturity is often associated with the end of growth in the late teens or early twenties. Although growth normally ceases at maturity, physiological changes continue. These changes are part of the process of senescence, or aging. Aging reduces the efficiency and capabilities of the individual, and even in the absence of other factors senescence will ultimately lead to death.

SENESCENCE

The characteristic physical and functional alterations under way as part of the aging process affect all organ systems. Examples include:

- A loss of elasticity in the skin that produces sagging and wrinkling.
- A decline in the rate of bone deposition, leading to weak bones, and degenerative changes in joints that make them less mobile.
- Reductions in muscular strength and ability.
- Impairment of coordination, memory, and intellectual function.
- Reductions in the production of and sensitivity to circulating hormones.
- Appearance of cardiovascular problems, and a Reduction in peripheral blood flow that can affect a variety of vital organs.
- Reduced sensitivity and responsiveness of the immune system, leading to problems with infection and cancer.
- Reduced elasticity in the lungs, leading to decreased respiratory function.
- Decreased peristalsis and muscle tone along the digestive tract.
- Decreased peristalsis and muscle tone in the urinary system, coupled with a reduction in the glomerular filtration rate.
- Functional impairment of the reproductive system, which eventually becomes inactive when the menopause or climacteric occurs.

These changes have been detailed in chapters dealing with specific systems, and readers interested in a more comprehensive review should consult appropriate sections in those chapters.

Aging does not affect all systems equally, and the age-related changes do not occur simultaneously. For example, peak muscular strength is found between ages 20 and 30, but auditory sensitivity reaches a peak in adolescence. Each system ages at its own rate, and if you compare any two adults you will encounter significant differences in the observed patterns of aging. The variations reflect differences in the genetic activity and programming of the individuals involved.

Factors Involved in Senescence

Aging appears to involve four distinct processes that influence the genetic programming of individual cells. These processes can be summarized as follows:

1. **Some cell populations grow smaller throughout your life.** Specialized cell types including neurons, cardiac muscle fibers, and skeletal muscle fibers have extremely long lives, but they are unable to undergo cell division. When one of these cells is injured or killed it cannot be replaced, and as the years go by accidents and disease processes take their toll.
2. **The ability to replace other cell populations decreases.** Most cells have lifespans ranging from a matter of hours to several years. Their destruction must be balanced by the divisions of **stem cells** that usually reside in the same region. There is some evidence that stem cell activity diminishes after a predetermined number of divisions has occurred. The rates of division must keep pace with the rate of destruction, so wherever cells lead short lives the *mitotic rates* are very high. These areas, including the skin, hair, and lining of the digestive tract, are often the first to show the effects of aging.
3. **Genetic activity changes over time.** Different genes may be activated, or the existing genes may be damaged by mutation (see point 4), but in any case the result is a change in protein synthesis and cellular metabolism. Several of the alterations seen during aging can be produced by hormones, making this an interesting topic of research.
4. **Mutations occur and accumulate. Somatic mutations** are those affecting somatic cells, those cells that are not concerned with meiosis and the production of gametes. Somatic mutations occur throughout life, and as they accumulate they affect cellular functions.

Taken together, these alterations reduce the functional abilities of the individual. They also affect homeostatic mechanisms. As a result, the elderly are less able to make homeostatic adjustments in response to internal or environmental stresses. As immune function deteriorates, the risks of contracting a variety of bacterial or viral diseases are proportionately increased. This deterioration leads to drastic physiological alterations that affect all internal systems. Death finally occurs when some combination of stresses cannot be countered by existing homeostatic mechanisms. Physicians attempt to forestall death by adjusting homeostatic mechanisms or removing the sources of stress. **Geriatrics** is a medical specialty concerned with the mechanics of the aging process, and physicians trained in geriatrics are known as **geriatricians**. Geriatricians are usually involved in the treatment of patients over 70 years of age. Problems commonly encountered by geriatricians include infections, cancers, heart disease, strokes, arthritis, and anemia; these conditions can be directly related to the age-induced changes listed above.

TABLE 29-5 The Five Major Causes of Death in the U.S. Population

	Ages 1–14		*Ages 15–34*	
Rank	**Male**	**Female**	**Male**	**Female**
1	Accidents		Accidents	
2	Cancer		Homicide	Cancer
3	Congenital anomalies		Suicide	
4	Homicide		Cancer	Homicide
5	Pneumonia, influenza		Heart disease	

Note: CVD = cerebrovascular disease (strokes); COPD = chronic obstructive pulmonary disease.

FIGURE 29-17
Major Causes of Postnatal Mortality

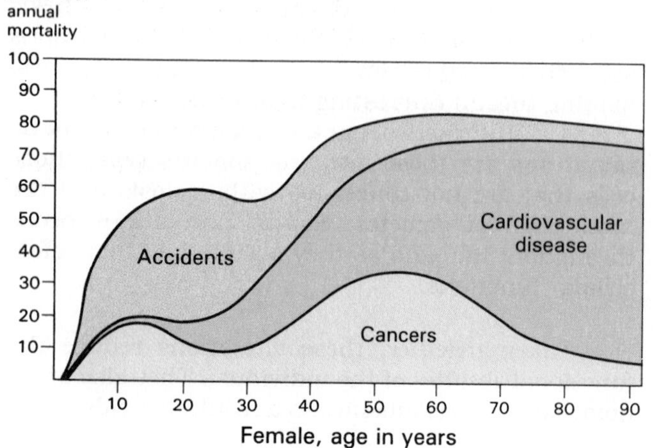

DEATH AND DYING

Despite exaggerated claims, there have been no substantiated cases of individuals living more than 120 years. Estimates for the lifespan of individuals born in the United States during 1984 are 71.1 years for males and 78.3 years for females. Interestingly enough, the causes of death vary depending on the age group under discussion. Consider the graphs shown in Figure 29-17, indicating the mortality statistics for various age groups. Accidents are the major cause of death in young people, and cardiovascular diseases in those over 40-45. More specific information concerning the major causes of death can be found in Table 29-5. Many of the characteristic differences in mortality figures result from changes in the functional capabilities of the individuals linked to development or senescence. It should also be realized that these figures would differ significantly if tabulated for countries and cultures with different genetic and environmental pressures.

The differences in mortality figures for male and female are related to differences in the accident rates for young people and in the rates of heart disease and cancer for older individuals. The upswing in female cancer rates reflects a rising breast cancer incidence for those over age 34, while lung cancer is the primary cancer killer of older men. In women the incidence of lung cancers and related killers, including pulmonary disease, heart disease, and pneumonia, has been steadily increasing as the number of women smokers has increased. This change has narrowed the difference between male and female life expectancies.

Experimental evidence and calculations suggest that the human lifespan has an upper limit of around 150 years. As medical advances continue, research

Ages 35–54		Ages 55–74		Ages 75 +	
Male	Female	Male	Female	Male	Female
Heart disease	Cancer	Heart disease		Heart disease	
Cancer	Heart disease	Cancer		Cancer	CVD
Accidents		CVD		CVD	Cancer
Cirrhosis of the liver	CVD	Accidents	Diabetes	Pneumonia, influenza	
Suicide	Cirrhosis of the liver	COPD	Accidents	COPD	Arterio-sclerosis

must focus on two related issues: (1) extending the average lifespan towards that maximum and (2) improving the functional capabilities of long-lived individuals. The first objective may be the easiest from a technical and moral standpoint. It is already possible to reduce the number of deaths attributed to specific causes. For example, new treatments promote remission in a variety of cancer cases, and anticoagulant therapies may reduce the risks of death or permanent damage following a stroke or heart attack. Many defective organs can be replaced with functional transplants, and the use of controlled immunosuppressive drugs will increase the success rates for these operations. Artificial hearts have been used with limited success, and artificial kidneys and endocrine pancreases are under development.

The second objective poses more of a problem. Few people past their mid-90s lead active, stimulating lives, and most would find the prospect of living another 50 years rather horrifying unless the quality of their lives could be significantly improved. Our abilities to prolong life now involve making stopgap corrections in systems on the brink of complete failure. Reversing the process of senescence would entail manipulating the biochemical operations and genetic programming of virtually every organ system. Although investigations continue, breakthroughs cannot be expected in the immediate future.

Over the interim, we are left with some serious ethical and moral questions. Now that we can postpone the moment of death almost indefinitely, how do we decide when it is appropriate to do so? How can medical and financial resources be fairly allocated? Who gets the limited number of hearts, livers, kidneys, and corneas available for transplant? Who should be selected for experimental therapies of potential significance? Should we take into account that care of an infant or child may add decades to a lifespan, while the costly insertion of an artificial heart in a 60-year old will add only months to years? How shall we allocate the costs for sophisticated procedures that may run to hundreds of thousand dollars over the long run? Are these individual or family responsibilities? In either case, will only the rich survive? Should the funds be provided by the government? If yes, what will happen to tax rates as the baby boomers become elderly citizens? And what about the role of the individual involved? If you decline treatment, are you mentally and legally competent? Can your survivors bring suit because you are forced to survive, or because you were allowed to die? These and other difficult questions will not go away, and in the years to come we will have to find answers we are content to live and die with.

√ **An increase in the levels of GnRH, FSH, LH, and sex hormones in the blood mark the onset of what stage in development?**

√ **What effect does a cell's lifespan have on the process of aging?**

CHAPTER REVIEW

Review of Selected Clinical Terms

amniocentesis (*p. 960*)
An analysis of fetal cells taken from a sample of amniotic fluid.

chorionic villi sampling (*p. 960*)
An analysis of cells collected from the cherionic villi during the first trimester.

infertility (*p. 966*)
The inability to have children.

in vitro fertilization (*p. 967*)
Fertilization outside the body, usually in a test tube or petri dish.

congenital malformation (*p. 981*)
A severe structural abnormality, present at birth, that affects major systems.

Additional Terms of Clinical Importance

TECHNOLOGY AND THE TREATMENT OF INFERTILITY (*p. 966*): **gamete intrafallopian tube transfer (GIFT), zygote intrafallopian tube transfer (ZIFT)**

COMPLEXITY AND PERFECTION: AN IMPROBABLE DREAM (*p. 980*): **spontaneous mutation**

EVALUATING THE NEWBORN INFANT (*p. 982*): **Apgar rating**

Study Outline

Related Key Terms

Introduction (pp. 953–954)

1. **Development** is the gradual modification of physical and physiological characteristics from conception to maturity. The creation of different cell types is **differentiation**.

An Overview of Topics in Development (p. 954)

1. **Inheritance** refers to the transfer of genetically determined characteristics from generation to generation. **Genetics** is the study of the mechanisms of inheritance. **Prenatal development** occurs before birth; **postnatal development** begins at birth and continues to maturity, when **senescence** (aging) begins.

Related Key Terms: **conception** • **fetal development** **embryology** • **gametogenesis**

Meiosis and the Formation of Gametes (pp. 954–955)

1. **Meiosis** is a special form of cell division leading to the production of sperm or eggs containing one-half of the normal chromosome number of **somatic** cells. **Meiosis I**, the first meiotic division, is a **reductional division,** while **meiosis II** is an **equational division.**
2. In the male, spermatogonia differentiate into primary spermatocytes before undergoing meiosis I, which produces a pair of secondary spermatocytes. In meiosis II each secondary spermatocyte divides to yield spermatids that will mature into spermatozoa. In the female, the **oogonia** differentiate into **primary oocytes.** Each primary oocyte that enters meiosis I generates a secondary oocyte and a nonfunctional **polar body.** Subsequent division of the secondary oocyte produces a mature **ovum** and two or three polar bodies.

Related Key Terms: **2N (diploid)** • **1N (haploid)** **chromatids** • **synapsis** **tetrad**

Genetics, Development, and Inheritance (pp. 955–959)

1. Every somatic cell carries copies of the original 46 chromosomes in the zygote; these represent the individual's **genotype**. The visible characteristics of the individual are the **phenotype**.

GENES AND CHROMOSOMES (pp. 956–959)

2. Every somatic human cell contains 23 pairs of chromosomes; each pair consists of **homologous** chromosomes. 22 pairs are **autosomal** chromosomes that affect somatic characteristics only. The various forms of a gene are called **alleles**. If both homologous chromosomes carry the same allele of a particular gene, the individual is **homozygous**; if they carry conflicting instructions, the individual is **heterozygous**.
3. The twenty-third pair of chromosomes are the **sex chromosomes** because they determine the sex of the individual. There are two

Related Key Terms: **dominant** • **recessive**

Related Key Terms

different sex chromosomes, an **X chromosome** and a **Y chromosome.** The normal male genotype is XY; that of females is XX. The X chromosome carries **X-linked** genes that affect somatic structures but have no corresponding alleles on the Y chromosome.

INHERITANCE AND GENETIC DISORDERS (p. 959)

4. In **simple inheritance** phenotypic characters are determined by interactions between a single pair of alleles. **Polygenic (multifactorial) inheritance** involves interactions between alleles on several genes.

Fertilization (pp. 959–962)

1. Fertilization normally occurs in the uterine tube within a day after ovulation. Sperm cannot fertilize an egg until they have undergone **capacitation.**

THE OOCYTE AT OVULATION (p. 960)

2. The acrosomal caps of the spermatozoa release **hyaluronidase**, an enzyme that separates cells of the corona radiata and exposes the oocyte membrane. When a single spermatozoon contacts that membrane, fertilization occurs and **activation** follows.

OOCYTE ACTIVATION (pp. 960–961)

3. During activation the oocyte completes meiosis II, and a **cortical reaction** occurs that prevents the penetration of additional sperm.

polyspermy • **female pronucleus**
male pronucleus • **amphimixis**
thalidomide

INDUCTION AND THE REGULATION OF DEVELOPMENT (pp. 961–962)

4. During development, differences in the cytoplasmic composition of individual cells trigger changes in genetic activity. The chemical interplay between developing cells is **induction.**

A PREVIEW OF PRENATAL DEVELOPMENT (p. 962)

5. The 9-month **gestation** period can be divided into three **trimesters.**

The First Trimester (pp. 962–975)

1. **Cleavage** subdivides the cytoplasm of the zygote in a series of mitotic divisions; the zygote becomes a **blastocyst.** During **implantation** the blastocyst burrows into the uterine endometrium. **Placentation** occurs as blood vessels form around the blastocyst and the **placenta** appears.

embryogenesis

CLEAVAGE AND BLASTOCYST FORMATION (pp. 963–964)

2. The **blastocyst** consists of an outer **trophoblast** and an **inner cell mass.**

blastomeres • **morula**
blastocoele

IMPLANTATION (pp. 964–971)

3. As the trophoblast enlarges and spreads, maternal blood flows through open **lacunae**. After **gastrulation** the **blastodisc** contains an embryo composed of **endoderm**, **ectoderm**, and an intervening **mesoderm**. These **germ layers** help form four **extraembryonic membranes:** the yolk sac, amnion, allantois, and chorion.
4. The **yolk sac** is an important site of blood cell formation. The **amnion** encloses fluid that surrounds and cushions the developing embryo. The base of the **allantois** later gives rise to the urinary bladder. Circulation within the vessels of the **chorion** provides a rapid transit system linking the embryo with the trophoblast.

syncytial trophoblast
cellular trophoblast
amniotic cavity • **epiblast**
hypoblast • **primitive streak**

PLACENTATION (p. 971)

5. **Chorionic villi** extend outward into the maternal tissues, forming an intricate, branching network through which maternal blood flows. As development proceeds the **umbilical cord** connects the fetus to the placenta. The trophoblast synthesizes HCG, estrogens, progestone, HPL, and relaxin.

body stalk • **yolk stalk**
decidua capsularis
decidua • **basalis**
umbilical arteries
umbilical vein

EMBRYOGENESIS (p. 971)

6. The first trimester is critical because events in the first 12 weeks establish the basis for **organogenesis** (organ formation).

head fold • **tail fold**

The Second and Third Trimesters (pp. 975–977)

1. In the second trimester the organ systems near functional completion. During the third trimester the organ systems become functional.

PREGNANCY AND MATERIAL SYSTEMS (pp. 975–976)

2. The developing fetus is totally dependent on maternal organs for nourishment, respiration, and waste removal. Maternal adaptations include increased blood volume, respiratory rate, tidal volume, nutrient intake, and glomerular filtration.

Chapter Review

STRUCTURAL AND FUNCTIONAL CHANGES IN THE UTERUS (pp. 976–977)

3. Progesterone produced by the placenta has an inhibitory effect on uterine muscles; its calming action is opposed by estrogens, oxytocin, and prostaglandins. At some point multiple factors interact to produce **labor contractions** in the uterine wall.

Related Key Terms

false labor • **true labor**

Labor and Delivery (pp. 977–979)

1. The goal of labor is **parturition** (forcible expulsion of the fetus).

STAGES OF LABOR (pp. 977–978)

2. Labor can be divided into three stages: the **dilation stage**, **expulsion stage**, and **placental stage**.

PREMATURE LABOR (p. 978)

3. Complications of labor and delivery include **forceps delivery**, **breech birth**, and **premature delivery**.

MULTIPLE BIRTHS (pp. 978–979)

4. Twin births may be **dizygotic** ("fraternal") or **monozygotic** ("identical").

delivery • **episiotomy**
caesarean section
immature delivery
conjoined twins

Postnatal Development (p. 979–985)

1. Postnatal development involves a series of **life stages**, including **infancy**, **childhood**, **adolescence**, and **maturity**. **Senescence** begins at maturity and ends in the death of the individual.

THE NEONATAL PERIOD, INFANCY, AND CHILDHOOD (pp. 979–982)

2. The **neonatal period** extends from birth to 1 month of age. **Infancy** then continues to 2 years of age, and **childhood** lasts until puberty commences. During these stages major organ systems (other than reproductive) become operational and gradually acquire adult characteristics, and the individual grows rapidly.
3. In the transition from fetus to **neonate** the respiratory, circulatory, digestive, and urinary systems begin functioning independently. The newborn must also begin thermoregulating.

MONITORING POSTNATAL DEVELOPMENT (p. 982)

4. **Pediatrics** focuses on postnatal development from infancy through adolescence.

Denver Developmental Screening Test (DDST)

ADOLESCENCE AND MATURITY (pp. 982–983)

5. Adolescence begins at puberty when: (1) the hypothalamus increases its production of GnRH; (2) circulating levels of FSH and LH (ICSH) rise rapidly; and (3) ovarian or testicular cells become more sensitive to FSH and LH. These changes initiate gametogenesis, production of sex hormones, and a sudden acceleration in growth rate.

SENESCENCE (pp. 983–984)

6. Aging affects all organ systems and involves four processes: (1) some cell populations grow smaller throughout life; (2) the ability to replace cells decreases; (3) genetic activity changes over time; and (4) mutations occur and accumulate.

somatic mutations
geriatricians

DEATH AND DYING (pp. 984–985)

7. Mortality patterns vary with age as the result of physiological and behavioral factors, but research suggests that the human lifespan has an upper limit of around 150 years.

Review Planner

	Objective	Level 1
1	Explain the differences between ordinary cells and gametes and relate the specializations of gamates to their functions.	**7 10 13 14 16 23**
2	Describe how gametes are formed.	**1 4 5 10 13 14 21 23**
3	Relate basic principles of genetics to the inheritance of human traits.	**2 6 7 14 22 29 44 47 48 49 51**
4	Describe the process of fertilization.	**18 20 31**
5	Discuss the stages of embryonic and fetal development.	**3 11 15 19 24 32 33 34 36 37 45**
6	Explain how pregnancy affects the mother's body systems and describe the processes of labor and delivery.	**9 25 26 27 28 30 34**
7	Explain how developmental processes are regulated.	**15 17 18 19 32**
8	Discuss the major stages of life after delivery.	**8 12 15 38 39 40 41**
9	Summarize the processes and factors involved in aging.	**8 12 42 43**

Level 2 **23 26 46**

Level 3 **52–54**

C M

Down Syndrome
Sex Chromosome Abnormalities
Induction and Sexual Differentiation
Problems with the Implantation Process
Problems with Placentation
Teratogens and Abnormal Development
Problems with the Maintenance of a Pregnancy
Common Problems with Labor and Delivery

Helping Premature Infants Breathe • Chlamydia

Review Questions

MULTIPLE CHOICE

1. Nonfunctional structures that form during oogenesis are called: a) secondary ova; b) oogonia; c) polar bodies; d) teratogens.
2. The members of each pair of autosomes are called: a) heterozygous; b) homologous; c) alleles; d) homozygous.
3. Cleavage: a) produces blastomeres of increasing size; b) produces a morula and then a blastocyst; c) occurs immediately following implantation; d) produces a secondary oocyte and a polar body.

DIAGRAMS

4. Draw a diagram showing the stages of spermatogenesis. Label the primary spermatocyte, secondary spermatocyte, spermatids, and spermatozoa. Show which are diploid and which are haploid, and where meiosis I and meiosis II occur.
5. Draw a diagram showing the stages of oogenesis. Label the primary and secondary oocytes, ovum, and polar bodies. Show which are diploid and which are haploid, and where meiosis I and meiosis II occur.
6. Draw a Punnett square that predicts the genotypes of the offspring of parents who are both heterozygous for dominant trait "B" (the trait is located on an autosome). What phenotype(s) will the offspring exhibit?

TRUE/FALSE

(If a statement is false, your instructor may wish to have you correct it).

7. **T/F:** The normal male genotype is XX.
8. **T/F:** Typically, our body systems age at a constant rate.
9. **T/F:** Pregnancy causes a slight decline in maternal blood volume.

MATCHING QUESTIONS

(Match each term with the most appropriate statement).

Statements:

10. A haploid cell.
11. Birth.
12. The process of aging.
13. Four chromatids.
14. The transfer of

Terms:

a) Development
b) Differentiation
c) Inheritance
d) Gamete
e) Somatic cell

Chapter Review

genetically determined characteristics from one generation to the next.
15. The gradual modification of physiological characteristics between conception and physical maturity.
16. A diploid cell.
17 Release of chemicals that affect the differentation of developing cells.
18. Acquisition of the ability to fertilize an egg.
19. The creation of different cell types as an individual grows.
20. Changes in the oocyte following fertilization.

f) Tetrad
g) Parturition
h) Senescence
i) Capacitation
j) Activation
k) Induction

SHORT ANSWER/ESSAY

21. Summarize the process of meiosis.
22. What would you conclude about a trait in each of the following situations?:
    a) Children who exhibit this trait have at least one parent who exhibits the same trait;
    b) Children exhibit this trait, even though neither of the parents exhibits it;
    c) The trait is expressed more frequently in sons than in daughters;
    d) The trait is expressed equally in both daughters and sons.
23. Relate differences between spermatogenesis and oogenesis to the functional roles of the gametes.
24. What is the umbilical cord, and what structures are found in it?
25. Note at least four important changes in maternal systems that occur during pregnancy. Why are these changes functionally significant?
26. Discuss the changes that occur in the uterus during pregnancy. How do these changes affect uterine tissues, and what hormones are involved?
27. Identify the three stages of labor, and describe the events that characterize each stage.
28. Describe four common complications that occur with labor and delivery.
29. Explain why more men than women are colorblind. What type of inheritance is involved?
30. What factors initiate labor contractions?
31. Describe the changes that occur in the sperm and egg immediately after fertilization.
32. What is induction? Explain its significance.
33. Summarize the developmental changes that occur during the first, second, and third trimesters.
34. What are the four extraembryonic membranes, how do they form, and what are their functions?
35. Why is it important for pregnant women to monitor their eating and drinking carefully? When is this especially crucial?
36. Identify the trophoblast, and describe three of its major functions.
37. Place these terms in the correct sequence and describe each: blastodisc, blastocyst, morula, zygote. Indicate when each appears during development.
38. Describe the physiological adjustments that an infant must make during the neonatal period in order to survive.
39. Identify three life stages that occur between birth and approximately age 10. Describe the characteristics of each stage and when it occurs.
40. What hormonal events are responsible for puberty? What life stage does it initiate?
41. What occurs during the life stage of maturity?
42. What is senescence? Explain how it affects human organ systems throughout the body.
43. Describe the factors involved in senescence that affect the genetic programming of individual cells.

Distinguish between the following:

44. Genetics and inheritance.
45. Decidua basalis and decidua capsularis.
46. Monozygotic and dizygotic.
47. Mitosis and meiosis.
48. Genotype and phenotype.
49. Heterozygous and homozygous.
50. Blastocyst and blastocoele.
51. Simple and polygenic inheritance.

CRITICAL THINKING/APPLICATIONS

52. Hemophilia A ("h"), a condition in which the blood does not clot properly, is a recessive trait located on the X chromosome. Draw a Punnett square which predicts the genotypes of the male and female offspring of a heterozygous mother and a father who is not a hemophiliac. How many of their daughters are likely to be hemophiliacs? How many of the sons?
53. Elaine is overjoyed when she finds out she's pregnant. She tries hard to do everything "right"—eating nutritious meals, exercising, avoiding unnecessary drugs, and getting prenatal checkups. Unfortunately, at 10 weeks gestation she starts to bleed from her uterus and has severe cramps. What could be happening?
54. A baby is born prematurely at 32 weeks gestational age. Its Apgar ratings are 4 after 1 minute and 4 after 5 minutes. Explain what the Apgar rating measures. Are these good or bad scores? Why is premature birth dangerous for the baby?

Career Close-up: Nurse/Companion

It is a picture-postcard scene: sun sparkling on water, palm fronds waving gently in warm breezes, people relaxing on lounges by a beautiful, turquoise pool. It's difficult to believe that a registered nurse is practicing right here, right now. . . .

And yet, if you were to watch the two women in the shallow section of the pool, you would see Norma Leone Gardner, RN, helping her patient, Mrs. W., do exercises for her severe arthritis. The warm water, buoyant and free of friction, is the perfect medium in which to move swollen, painful joints. After the exercise session, Mrs. W.—who has great difficulty walking on land—glides across the width of the pool several times in a relaxed, rhythmic side stroke, while Norma practices the kicking that she has learned from Mrs. W.

For in addition to being her nurse, Norma is also Mrs. W.'s companion. And as such she and her patient benefit from each other's consideration and expertise. When Norma told Mrs. W. that she would like to learn to swim, Mrs. W. taught her. Similarly, when Mrs. W. needs to go shopping or to the doctor, Norma drives her and does much of the legwork.

Norma, born in Jamaica to Panamanian parents, went to nursing school in England at the age of 18.

After three years of schooling and one year of postgraduate midwifery training, Norma became a Queen's Nurse—what we would call a visiting nurse. For the next nine years, she served as a District Nurse, delivering and caring for babies, making her rounds on a bicycle. She then came to New York City, where she worked, first as a maternity nurse, and then, after special training, as a psychiatric nurse.

Norma was offered the job of nurse/companion to Mrs. W. when she was about to retire from psychiatric nursing. Mrs. W., a widow in her late 70s, suffers from hypertension and painful osteoarthritis. An esophageal spasm makes eating difficult, she suffers from migraine headache almost daily, and has difficulty reading due to macular degeneration. Although she employs a full-time housekeeper, she needed the hands-on care that a private nurse can give. Norma took the job.

"I do everything," Norma, explains. "My job is total responsibility for Mrs. W. I watch for signs and symptoms. I administer her medications, often using my judgment as to which ones she needs at a given time. I am responsible for her diet and exercise. I note any change in her condition so I can notify her doctor. I am responsible for her total well-being."

Norma doesn't usually make any medication decisions without consulting the doctor, but nursing decisions can be made. The nurse keeps a daily record, which she takes to all medical appointments. Mrs. W. always wants Norma at her appointments, to tell the physician what has been going on, what medications she is taking, and how she reacts to them.

Her anatomy and physiology training always comes into play, notes Norma. "Working in the home, you're responsible for being aware of all drug interactions. You have to know when to give each medication so the patient can tolerate it better. When the patient presents a new sign or symptom, I immediately try to ascertain the cause. Knowledge of anatomy and physiology are very important on this job."

Norma distinguishes home nursing from institutional nursing, "This nursing is completely different in all ways. There is no doctor here, as in a hospital. I have a 24 hour responsibility here, as opposed to the eight hour shift in a hospital. I work a five-day week, including nights. Another nurse covers for me on my days off.

"Often I am more a companion than a nurse. I make many decisions for my patient, including those about what she should wear. I bathe her . . .. I help her shower. I have to be there in case she falls. Her meals are my responsibility too.

"Sometimes this patient can be demanding when it comes to medication. I say no. She says yes. I don't want to upset my patient, so I usually explain that my license will be in jeopardy if I don't follow medication orders exactly. Then Mrs. W. suggests we call the doctor, and we do. The doctor usually substantiates what I've been saying."

Norma works from 8:00 AM, when her patient awakens, to 12:30 AM, when she gives Mrs. W. her nighttime medications and bids her good night. Although the job encompasses many hours, the day is broken up so that Norma has frequent periods of free time during the day—when her patient is resting, telephoning, reading, watching television or doing other things. But Norma is always there for her patient, who can call her on an intercom at any time, day or night.

Although all of Mrs. W.'s family get-togethers and photos include Norma, "I am still her nurse. Sometimes I will go places with my patient as part of the family, but I am always the nurse.

"This is my patient, and my patient comes first."

Appendix I
Weights and Measures

Accurate descriptions of physical objects would be impossible without a precise method of reporting the pertinent data. Dimensions such as length and width are reported in standardized units of measurement, such as inches or centimeters. These values can be used to calculate the volume of an object, a measurement of the amount of space it fills. **Mass** is another important physical property. The mass of an object is determined by the amount of matter it contains; on earth the mass of an object determines its weight.

Most U.S. readers describe length and width in terms of inches, feet, or yards; volumes in pints, quarts, or gallons; and weights in ounces, pounds, or tons. These are units of the **U.S. system** of measurement. Table A-1 summarizes the familiar and unfamiliar terms used in the U.S. system. For reference purposes, this table also includes a definition of the "household units," popular in recipes and cookbooks. The U.S. system can be very difficult to work with, because there is no logical relationship between the various units. For example, there are 12 inches in a foot, 3 feet in a yard, and 1760 yards in a mile. Without a clear pattern of organization, converting feet to inches or miles to feet can be confusing and time consuming. The relationships between ounces, pints, quarts and gallons, or ounces, pounds, and tons are no more logical.

In contrast, the **metric system** has a logical organization based on powers of 10, as indicated in Table A-2. For example, a **meter** (m) represents the basic unit for the measurement of size. When measuring larger objects, data can be reported in terms of **dekameters** (*deka*, ten), **hectometers** (*hekaton*, hundred), or **kilometers** (**km**; *chilioi*, thousand); for smaller objects, data can be reported in **decimeters** (0.1 m; *decem*, ten), **centimeters** (**cm** = 0.01 m; *centum*,

TABLE A-1 The U.S. System of Measurement

Physical Property	*Unit*	*Relationship to Other U.S. Units*	*Relationship to Household Units*
Length	inch (in.)	1 in. = 0.083 ft	
	foot (ft)	1 ft = 12 in.	
		= 0.33 yd	
	yard (yd)	1 yd = 36 in.	
		= 3 ft	
	mile (mi)	1 mi = 5,280 ft	
		= 1,760 yd	
Volume	fluidram (fl dr)	1 fl dr = 0.125 fl oz	
	fluid ounce (fl oz)	1 fl oz = 8 fl dr	= 6 teaspoons (tsp)
		= 0.0625 pt	= 2 tablespoons (tbsp)
	pint (pt)	1 pt = 128 fl dr	= 32 tbsp
		= 16 fl oz	= 2 cups (c)
		= 0.5 qt	
	quart (qt)	1 qt = 256 fl dr	= 4 c
		= 32 fl oz	
		= 2 pt	
		= 0.25 gal	
	gallon (gal)	1 gal = 128 fl oz	
		= 8 pt	
		= 4 qt	
Mass			
	grain (gr)	1 gr = 0.002 oz	
	dram (dr)	1 dr = 27.3 gr	
		= 0.063 oz	
	ounce (oz)	1 oz = 437.5 gr	
		= 16 dr	
	pound (lb)	1 lb = 7000 gr	
		= 256 dr	
		= 16 oz	
	ton (t)	1 t = 2000 lb	

hundred), **millimeters** (**mm** = 0.001 m; *mille*, thousand), and so forth. Notice that the same prefixes are used to report weights, based on the **gram** (**g**), and volumes, based on the **liter** (**ℓ**). This text reports data in metric units, usually with U.S. equivalents. You should use this opportunity to become familiar with the metric system, because most technical sources report data only in metric units, and most of the rest of the world uses the metric system exclusively. Conversion factors are included in Table A-2.

Pharmacies at one time used the **apothecary system**, a relatively specialized system of measurement borrowed from England when America was still a colony. This system has been largely replaced by the metric system, and we will ignore apothecary units in this text. For reference purposes Table A-3 includes its units of measurement and the relationships between apothecary, metric, and U.S. units. The apothecary system deals only with volumes and weights, and the volumetric units are comparable to those of the U.S. system. The two systems differ, however, in terms of the definitions of mass units.

The U.S. and metric systems also differ in their methods of reporting temperatures; in the U.S., temperatures are usually reported in degrees Fahrenheit (°F), whereas scientific literature and individuals in most other countries report temperatures in degrees Centigrade or Celsius (°C). The relationship between temperatures in degrees Fahrenheit and those in degrees Centrigrade has been indicated at the bottom of Table A-2.

TABLE A-2 The Metric System of Measurement

Physical Property	*Unit*	*Relationship to Standard Metric Units*	*Conversion to U.S. Units*	
Length	nanometer (nm)	1 nm = 0.000000001 m (10^{-9})	= 4 × 10^{-8} in.	25,000,000 nm = 1 in.
	micrometer (μm)	1 μm = 0.000001 n (10^{-6})	= 4 × 10^{-5} in.	25,000μm = 1 in.
	millimeter (mm)	1 mm = 0.001 m (10^{-3})	= 0.0394 in.	25.4 mm = 1 in.
	centimeter (cm)	1 cm = 0.01 m (10^{-2})	= 0.394 in.	2.54 cm = 1 in.
	decimeter (dm)	1 dm = 0.1 m (10^{-1})	= 3.94 in.	0.25 dm = 1 in.
	meter (m)	standard unit of length	= 39.4 in.	0.0254 m = 1 in.
			= 3.28 ft	0.3048 m = 1 ft
			= 1.09 yd	0.914 m = 1 yd
	dekameter (dam)	1 dam = 10 m		
	hectometer (hm)	1 hm = 100 m		
	kilometer (km)	1 km = 1000 m	= 3280 ft	
			= 1093 yd	
			= 0.62 mi	1.609 km = 1 mi
Volume	microliter (μl)	1 μℓ = 0.000001 ℓ (10^{-6})		
		= 1 cubic millimeter (mm^3)		
	milliliter (mℓ)	1 mℓ = 0.001 ℓ (10^{-3})	= 0.03 fl oz	5 mℓ = 1 tsp
		= 1 cubic centimeter (cm^3 or cc)		15 mℓ = 1 tbsp
				30 mℓ = 1 fl oz
	centiliter (cℓ)	1 cℓ = 0.01 ℓ (10^{-2})	= 0.34 fl oz	3 cℓ = 1 fl oz
	deciliter (dℓ)	1 dℓ = 0.1 ℓ (10^{-1})	= 3.38 fl oz	0.29 dℓ = 1 fl oz
	liter (ℓ)	standard unit of volume	= 33.8 fl oz	0.0295 ℓ = 1 fl oz
			= 2.11 pt	0.473 ℓ = 1 pt
			= 1.06 qt	0.946 ℓ = 1 qt
Mass	picogram (pg)	1 pg = 0.000000000001 g (10^{-12})		
	nanogram (ng)	1 ng = 0.000000001 g (10^{-9})		
	microgram (μg)	1 μg = 0.000001 g (10^{-6})	= 0.000015 gr	66,666 μg = 1 gr
	milligram (mg)	1 mg = 0.001 g (10^{-3})	= 0.015 gr	66.7 mg = 1 gr
	centrigram (cg)	1 cg = 0.01 g (10^{-2})	= 0.15 gr	6.7 cg = 1 gr
	decigram (dg)	1 dg = 0.1 g (10^{-1})	= 1.5 gr	0.67 dg = 1 gr
	gram (g)	standard unit of mass	= 0.035 oz	28.35 g = 1 oz
			= 0.0022 lb	453.6 g = 1 lb
	dekagram (dag)	1 dag = 10 g		
	hectogram (hg)	1 hg = 100 g		
	kilogram (kg)	1 kg = 1000 g	= 2.2 lb	0.453 kg = 1 lb
	metric ton (kt)	1 mt = 1000 kg	= 1.1 t	
			= 2205 lb	0.907 kt = 1 t

Temperature	*Centigrade*	*Fahrenheit*	
Freezing point of pure water	0°		32°
Normal body temperature	36.8°		98.6°
Boiling point of pure water	100°		212°
Conversion °C → °F:	°F = (1.8 × °C) + 32	°F → °C:	°C = (°F − 32) × 0.56

TABLE A-3 The Apothecary System of Measurement

Physical Property	*Unit*	*Relationship to Other Apothecary Units*	*Relationship to Metric Units*	*Relationship to U.S. Units*[a]
Volume	minim (min)	1 min = 0.017 fl dr	= 0.06 mℓ	no comparable unit
	fluidram (fl dr)	1 fl dr = 60 min = 0.125 fl oz	= 3.7 mℓ	1 fl dr ap = 1 fl dr
	fluidounce (fl oz)	1 fl oz = 480 min	= 29.6 mℓ	1 fl oz ap = 1 fl oz
	pint (pt)	1 pt = 128 fl dr = 16 fl oz	= 473 mℓ = 0.47 ℓ	1 pt ap = 1 pt
Mass	grain (gr)	1 gr = 0.05 s = 0.002 oz	= 0.06 g	1 gr ap = 1 gr
	scruple (s)	1 s = 20 gr = 0.33 dr	= 1.3 g	no comparable unit
	dram (dr)	1 dr = 60 gr = 3 s = 0.01 lb	= 3.9 g	1 dr ap = 2.2 dr
	ounce (oz)	1 oz = 480 gr = 24 s = 8 dr = 0.08 lb	= 31.1 g	1 oz ap = 1.1 oz
	pound (lb)	1 lb = 5760 gr = 288 s = 96 dr = 12 oz	= 373 g = 0.37 kg	1 lb ap = 0.82 lb

[a] 1 dr ap = 1 dram in apothecary units.

Appendix II
The Periodic Table

Table A-4 presents the known elements in order of their atomic weights. Each horizontal row represents a single electron shell; the number of elements in that row is determined by the maximum number of electrons that can be stored at that energy level. The element at the left end of each row contains a single electron in its outermost electron shell; the last element in the row has a filled outer electron shell. Organizing the elements in this fashion highlights similarities that reflect the composition of the outer electron shell, and these relationships are evident when you examine the vertical columns. Helium, neon, argon, krypton, xenon, and radon all have full electron shells, they are all gases at normal atmospheric temperature and pressure, and they do not react readily with other elements. These are known as the *noble* or *inert* gases. In contrast, lithium, sodium, potassium, and so forth are silvery, soft metals that are so highly reactive that pure forms cannot be found in nature. The fourth and fifth electron levels can hold 18 electrons; table inserts are used to save space, as higher levels may store up to 32 electrons. Elements of particular importance to our discussion of human anatomy and physiology are highlighted in the table.

TABLE A-4 The Periodic Table of the Elements

Key: 1 — Atomic Number; H — Chemical Symbol; Hydrogen — Element; 1.0079 — Atomic Weight

1 **H** Hydrogen 1.0079																	2 **He** Helium 4.00260
3 **Li** Lithium 6.941	4 **Be** Beryllium 9.01218											5 **B** Boron 10.81	6 **C** Carbon 12.011	7 **N** Nitrogen 14.0067	8 **O** Oxygen 15.9994	9 **F** Fluorine 18.99840	10 **Ne** Neon 20.179
11 **Na** Sodium 22.98977	12 **Mg** Magnesium 24.305											13 **Al** Aluminum 26.98154	14 **Si** Silicon 26.086	15 **P** Phosphorus 30.97376	16 **S** Sulfur 32.06	17 **Cl** Chlorine 35.453	18 **Ar** Argon 39.948
19 **K** Potassium 39.098	20 **Ca** Calcium 40.08	21 **Sc** Scandium 44.9559	22 **Ti** Titanium 47.90	23 **V** Vanadium 50.99414	24 **Cr** Chromium 51.996	25 **Mn** Manganese 54.9380	26 **Fe** Iron 55.847	27 **Co** Cobalt 58.9332	28 **Ni** Nickel 58.71	29 **Cu** Copper 63.546	30 **Zn** Zinc 65.38	31 **Ga** Gallium 69.72	32 **Ge** Germanium 72.59	33 **As** Arsenic 74.9216	34 **Se** Selenium 78.96	35 **Br** Bromine 79.904	36 **Kr** Krypton 83.80
37 **Rb** Rubidium 85.4678	38 **Sr** Strontium 87.62	39 **Y** Yttrium 88.9059	40 **Zr** Zirconium 91.22	41 **Nb** Niobium 92.9064	42 **Mo** Molybdenum 95.94	43 **Tc** Technetium 98.9062[b]	44 **Ru** Ruthenium 101.07	45 **Rh** Rhodium 102.9055	46 **Pd** Palladium 106.4	47 **Ag** Silver 107.868	48 **Cd** Cadmium 112.40	49 **In** Indium 114.82	50 **Sn** Tin 118.69	51 **Sb** Antimony 121.75	52 **Te** Tellurium 127.60	53 **I** Iodine 126.9045	54 **Xe** Xenon 131.30
55 **Cs** Cesium 132.9054	56 **Ba** Barium 137.34	57 **La** Lanthanum 138.9055	72 **Hf** Hafnium 178.49	73 **Ta** Tantalum 180.994.79	74 **W** Tungsten 183.85	75 **Re** Rhenium 186.2	76 **Os** Osmium 190.2	77 **Ir** Iridium 192.22	78 **Pt** Platinum 195.09	79 **Au** Gold 196.9665	80 **Hg** Mercury 200.59	81 **Tl** Thallium 204.37	82 **Pb** Lead 207.2	83 **Bi** Bismuth 208.9804	84 **Po** Polonium (210)[a]	85 **At** Astatine (210)[a]	86 **Rn** Radon (222)[a]
87 **Fr** Francium (223)[a]	88 **Ra** Radium 226.0254[b]	89** **Ac** Actinium (227)[a]	104 **Rf** Ruther-fordium 261	105 **Ha** Hahnium 262	106 **Unh** Unnil-hexium 263	107 **Uns** Unnil-septium 262	108 **Uno** Unnil-octium 265	109 **Une** Unnil-hexium 266									

*	58 **Ce** Cerium 140.12	59 **Pr** Praseo-dymium 141.9077	60 **Nd** Neo-dymium 144.24	61 **Pm** Promethium (145)[a]	62 **Sm** Samarium 150.4	63 **Eu** Europium 151.96	64 **Gd** Gadolinium 157.25	65 **Tb** Terbium 158.9254	66 **Dy** Dysprosium 162.50	67 **Ho** Holmium 164.9304	68 **Er** Erbium 167.26	69 **Tm** Thulium 168.9342	70 **Yb** Ytterbium 173.04	71 **Lu** Lutetium 174.97
**	90 **Th** Thorium 232.038[b]	91 **Pa** Pro-tactinium 231.0359[b]	92 **U** Uranium 238.029	93 **Np** Neptunium 237.0482[b]	94 **Pu** Plutonium (242)[a]	95 **Am** Americium (243)[a]	96 **Cm** Curium (247)[a]	97 **Bk** Berkelium (249)[a]	98 **Cf** Californium (251)[a]	99 **Es** Einsteinium (254)[a]	100 **Fm** Fermium (253)[a]	101 **Md** Mende-levium (256)[a]	102 **No** Nobelium (254)[a]	103 **Lr** Lawren-cium (257)[a]

[a] Mass number of most stable or best-known isotope.

[b] Mass of most commonly available, long-lived isotope.

Appendix III
Condensed Structural Formulas for the Amino Acids

Amino acids with acidic or basic side chain functional groups

Aspartic acid (asp)

Glutamic acid (glu)

Tyrosine (tyr)

Lysine (lys)

Arginine (arg)

Histidine (his)

Amino acids with polar, neutral side chains

Serine (ser)

Threonine (thr)

Methionine (met)

Cysteine (cys)

Tryptophan (trp)

Asparagine (asn)

Glutamine (gln)

Amino acids with hydrocarbon side chains

Glycine (gly)

Alanine (ala)

Valine (val)

Leucine (leu)

Isoleucine (ile)

Phenylalanine (phe)

Proline (pro)

Appendix IV
Biological Solutions and Concentrations

Physiologists and clinicians pay particular attention to ionic distributions across membranes, and the electrolyte composition of body fluids. Standard values for physiological tests are provided throughout the text, and summarized in Appendix V. Data must be analyzed from several different perspectives, and physiological values may be reported in several different ways. One method is to report the concentration of atoms, ions, or molecules in terms of weight per unit volume of solution. Although grams per liter (g/ℓ) may be used, values are most often expressed in terms of grams (g), milligrams (mg), or micrograms (μg) per 100 mℓ. Since 100 mℓ is 0.1 liter, or 1 deciliter (**dℓ**), the abbreviations most often used in this text are **g/dℓ** and **mg/dℓ**.

Osmolarity depends on the total number of individual atoms, ions, and molecules in solution, without regard to molecular weight, electrical charge or molecular identity. As a result, if fluid balance and osmolarity are being monitored, concentrations are usually reported in terms of moles per liter (M/ℓ) or millimoles per liter (mM/ℓ or mmol/ℓ), rather than in terms of g/dℓ or mg/dℓ. To convert from g/dℓ to M/ℓ, multiply by 10 and divide by the atomic weight of the element. For example, a sample of plasma (blood with the cells removed) contains sodium ions at a concentration of roughly 0.32 g/dℓ (320 mg/dℓ). This value can be converted to M/ℓ as follows:

$$\frac{\text{g/d}\ell \times 10}{\text{atomic weight}} = \frac{0.32 \times 10}{22.99} = 0.140 \text{ M/}\ell \text{ (140 mM/}\ell\text{)}$$

Moles or millimoles per liter can also be used to indicate the concentration of molecules in solution; the same conversion can be performed by substituting molecular weight for atomic weight in the above equation.

Because electrolyte concentrations have profound effects on living cells, it is often important to know how many positive and negative charges are present in a biological solution. In this case the important question is not just how many ions or molecules are present, but how many positive or negative charges they bear. For example, a single ion of calcium (Ca^{2+}) has twice the electrical charge of a single sodium atom (Na^{+}), although the two are identical in terms of their effects on osmolarity. One **equivalent (Eq)** represents a mole of positive or negative charges; physiological concentrations are often reported in terms of **millequivalents per liter (mEq/ℓ)**. You should become familiar with both methods of expression, and fortunately the conversion from millimoles to millequivalents is relatively easy to perform. For **monovalent ions**, those with a +1 or −1 charge, millimole and millequivalent values are identical, and no calculation is needed. For **divalent ions**, with +2 or −2 charges, the number of charges (mEq) is twice the number of ions (mM); if an ion had a +3 or −3 charge, the number of millequivalents would be 3 times the number of millimoles. To reverse the process, and convert mEq to mM, simply divide by the ionic valence (number of charges).

Table A-5 compares the different methods of reporting the concentration of major electrolytes in plasma in terms of weight, moles, and equivalents; the tables included in Appendix V provide data in terms currently accepted for clinical laboratory reports.

There is no doubt that physiologists and clinicians would benefit from the use of standardized reporting procedures; it can be very frustrating to consult three references and find that the first reports electrolyte concentrations in mg/dℓ, the second in mM/ℓ, and the third in mEq/ℓ. In 1984, the American Medical Association House of Delegates endorsed a plan to standardize clinical test results through the use of metric **SI** (Systeme Internationale) units, with a target date of July 1, 1987 for the switchover. Unfortunately there was no mechanism for enforcing compliance, and the plan may not prove any more successful than the ill-fated attempt to drop the U.S. system of measurement in favor of the metric system.

The major problem is that the relationships to current normal values are difficult to remember. Electrolyte concentrations, now most often indicated in mEq/ℓ, will be reported in SI units that represent mM/ℓ. That means that the values for sodium or potassium concentrations remain unchanged, but the normal values for calcium or magnesium are reduced by 50 percent. The situation becomes more confusing when metabolite concentrations are considered. Cholesterol and glucose concentrations are now reported in terms of mg/dℓ, but the SI units represent mM/ℓ (mmol/ℓ). However, total lipid concentrations, also currently listed as mg/dℓ, and total protein concentrations, now shown as g/dℓ, will be reported in terms of g/ℓ. To be useful in a clinical setting, physicians must not only remember the definition of each SI unit, but convert and relearn the normal ranges. As a result, it appears unlikely that the conversion to SI units will be completed in the immediate future.

TABLE A-5 A Comparison of Methods for Reporting Concentrations of Solutes in the Blood

Solute	*mg/dℓ*	*mM/ℓ*	*mEq/ℓ*	*SI Units*
Electrolytes				
Sodium (Na^+)	320	140	140	140 mmol/ℓ
Potassium (K^+)	16.4	4.2	4.2	4.2 mmol/ℓ
Calcium (Ca^{2+})	9.5	2.4	4.8	2.4 mmol/ℓ
Chloride (Cl^-)	354	100	100	100 mmol/ℓ
Metabolites				
Glucose	90	5	nr	5 mmol/ℓ
Lipids, total	600	nr	nr	0.6 g/ℓ
Proteins, total	7 g/dℓ	nr	nr	70 g/ℓ

Note: For the purposes of this table values were selected from within normal ranges. nr = standard tests do not report results in these units.

Appendix V
Normal Physiological Values

Tables A-6 and A-7 present normal averages or ranges for the chemical composition of body fluids. These should be considered approximations, rather than absolute values, as test results vary from laboratory to laboratory due to differences in procedures, equipment, normal solutions, and so forth. Blanks in the tabular data appear where data were not available; sources used in the preparation of these tables are indicated below. Additional information concerning body fluid analysis can be found at the following locations in the text:

- Table 19-2 (p. 609) compares the compositions of blood plasma, interstitial fluid, and cytoplasm.
- Table 19-3 (p. 610) presents data on the cellular composition of whole blood.
- Table 23-1 (p. 770) details the partial pressures of gases in the blood.
- Table 24-5 (p. 816) contains a chemical analysis of bile.
- Table 26-2 (p. 869) compares the average compositions of plasma and urine.
- Table 26-4 (p. 883) gives the general characteristics of normal urine.
- Table 28-1 (p. 927) provides a detailed analysis of semen composition.

Sources

Ballenger, John Jacob. 1977. *Diseases of the Nose, Throat, and Ear.* Philadelphia, Pa.: Lea and Febiger.

Braunwauld, Eugene, Kurt J. Isselbacher, Robert G. Petersdorf, Jean D. Wilson, Joseph B. Martin, and Anthony S. Fauci, eds. 1987. *Harrison's Principles of Internal Medicine*, 11th ed. New York: McGraw-Hill.

Lentner, Cornelius, ed. 1981. *Geigy Scientific Tables*, 8th ed. Basel, Switzerland: Ciba-Geigy Limited.

Davidsohn, Israel, and John Bernard Henry, eds. 1969. *Todd-Sanford Clinical Diagnosis by Laboratory Methods*, 14th ed. Philadelphia, Pa.: W. B. Saunders Company.

TABLE A-6 The Composition of Minor Body Fluids

Test	*Normal Averages or Ranges*					
	Perilymph	*Endolymph*	*Synovial Fluid*	*Sweat*	*Saliva*	*Semen*
pH			7.4	4–6.8	6.4[a]	7.19
Specific gravity			1.008–1.015	1.001–1.008	1.007	1.028
Electrolytes (mEq/ℓ);						
Potassium	5.5–6.3	140–160	4.0	4.3–14.2	21	31.3
Sodium	143–150	12–16	136.1	0–104	14[a]	117
Calcium	1.3–1.6	0.05	2.3–4.7	0.2–6	3	12.4
Magnesium	1.7	0.02		0.03–4	0.6	11.5
Bicarbonate	17.8–18.6	20.4–21.4	19.3–30.6		6[a]	24
Chloride	121.5	107.1	107.4	34.3	17	42.8
Proteins (mg/dℓ)						
Total	200	150	1.72 g/dℓ	7.7	386[b]	4.5 g/dℓ
Metabolites (mg/dℓ)						
Amino acids				47.6	40	1.26 g/dℓ
Glucose	104		70–110	3.0	11	224 (fructose)
Urea				26–122	20	72
Lipids, total	12[c]		20.9	[d]	25–500[c]	188

[a] Increases under salivary stimulation.
[b] Primarily alpha-anylase, with some lysozymes.
[c] Cholesterol.
[d] Not present in eccrine secretions.

TABLE A-7 The Chemistry of Blood, Cerebrospinal Fluid, and Urine

Test	Normal Ranges		
	Blood[a]	*CSF*	*Urine*
pH	S: 7.38–7.44	7.31–7.34	4.6–8.0
Osmolarity (mOsm/ℓ)	S: 280–295	292–297	500–800
Electrolytes	(mEq/ℓ unless noted)		(urinary loss per 24-hour period[b])
Bicarbonate	P: 21–28	20–24	
Calcium	S: 4.5–5.3	2.1–3.0	6.5–16.5 mEq
Chloride	S: 100–108	116–122	120–240 mEq
Iron	S: 50–150 μg/ℓ	23–52 μg/ℓ	40–150 μg
Magnesium	S: 1.5–2.5	2–2.5	4.9–16.5 mEq
Phosphorus	S: 1.8–2.6	1.2–2.0	0.8–2 g
Potassium	P: 3.8–5.0	2.7–3.9	35–80 mEq
Sodium	P: 136–142	137–145	120–220 mEq
Sulfate	S: 0.2–1.3		1.07–1.3 g
Metabolites	(mg/dℓ unless noted)		(urinary loss per 24-hour period[c])
Amino acids	P/S: 2.3–5.0	10.0–14.7	41–133 mg
Ammonia	P: 20–150 μg/dℓ	25–80 μg/dℓ	340–1200 mg
Bilirubin	S: 0.5–1.2	<0.2	0.02–1.9 mg
Creatinine	P/S: 0.6–1.2	0.5–1.9	1.01–2.5 g
Glucose	P/S: 70–110	40–70	16–132 mg
Ketone bodies	S: 0.3–2.0	1.3–1.6	10–100 mg
Lactic acid	WB: 5–20[d]	10–20	100–600 mg
Lipids (total)	S: 400–1000	0.8–1.7	0–31.8 mg
Cholesterol (total)	S: 150–300	0.2–0.8	1.2–3.8 mg
Triglycerides	S: 40–150	0–0.9	
Urea	P/S: 23–43	13.8–36.4	12.6–28.6 g
Uric Acid	S: 2.0–7.0	0.2–0.3	80–976 mg
Proteins	(g/dℓ)	(mg/dℓ)	(urinary loss per 24-hour period[c])
Total	S: 6.0–7.8	20–45	47–76.2 mg
Albumin	S: 3.2–4.5	10.6–32.4	10–100 mg
Globulins (total)	S: 2.3–3.5	2.8–15.5	7.3 mg (average)
Immunoglobulins	S: 1.0–2.2	1.1–1.7	3.1 mg (average)
Fibrinogen	P: 0.2–0.4	0.65 (average)	

[a] S = serum, P = plasma, WB = whole blood
[b] Because urinary output averages just over 1 liter per day, these electrolyte values are comparable to mEq/ℓ.
[c] Because urinary metabolite and protein data approximate mg/ℓ or g/ℓ, they must be divided by 10 for comparison with CSF or blood concentrations.
[d] Venous blood sample.

Diem, K., and C. Lenter, eds. 1970. *Scientific Tables*, 7th ed. Basel, Switzerland: Ciba-Geigy Limited.

Halsted, James A. 1976. *The Laboratory in Clinical Medicine: Interpretation and Application*. Philadelphia, Pa.: W. B. Saunders Company.

Harper, Harold A. 1987. *Review of Physiological Chemistry*. Los Altos, Calif.: Lange Medical Publications.

Appendix VI
Pathways of Aerobic Respiration

FIGURE A-1 The reactions of the TCA cycle

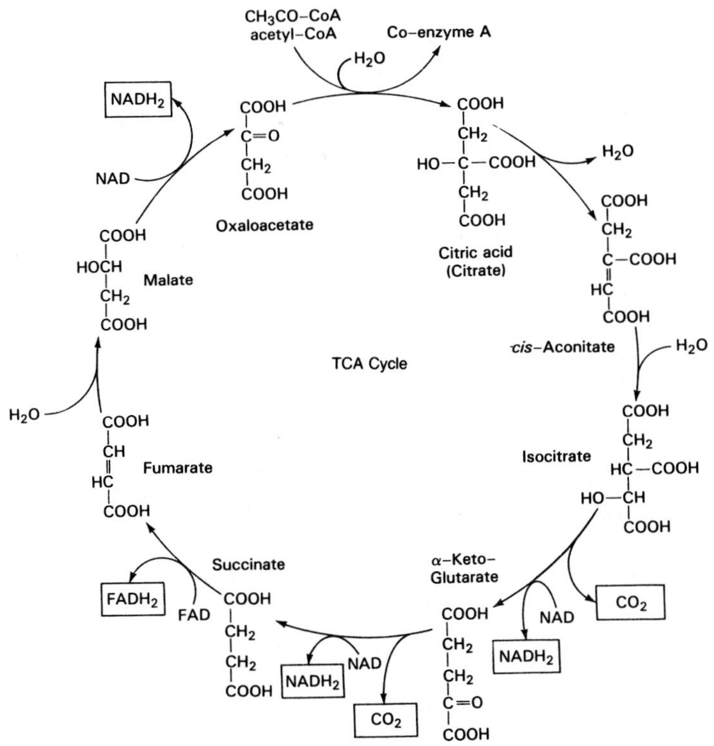

CH_3CO–CoA
acetyl–CoA
Co–enzyme A
H_2O
$NADH_2$
NAD
Oxaloacetate
Citric acid (Citrate)
H_2O
cis–Aconitate
H_2O
TCA Cycle
Malate
H_2O
Fumarate
Isocitrate
Succinate
$FADH_2$
FAD
$NADH_2$
NAD
CO_2
α–Keto–Glutarate
NAD
$NADH_2$
CO_2

FIGURE A-2 The glycolytic pathways

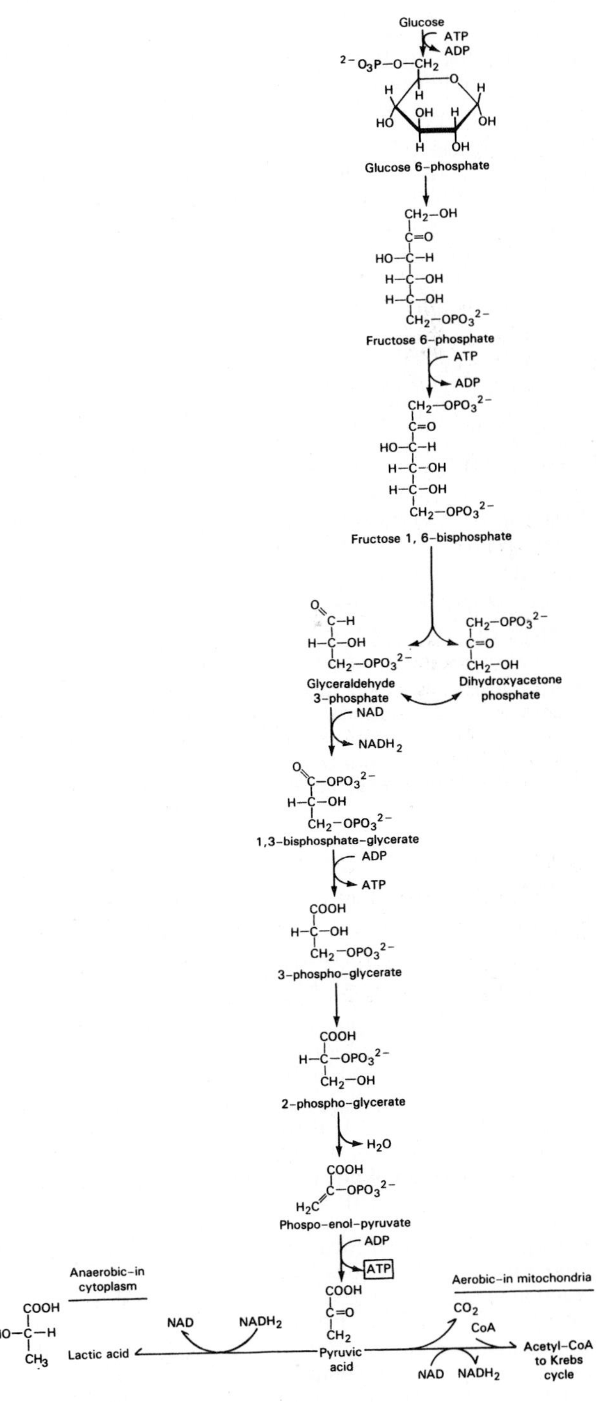

Appendix VII
Answers to Concept Checkpoint Questions

Chapter 1

Page 6

1. A *histologist* investigates the structure and properties of tissues.

2. Study of the physiology of specific organs is called *special physiology*. In this particular case the field of study is *cardiac physiology* (study of heart function). Since heart failure is often caused by disease, this specialty would overlap or be closely related to *pathological physiology*.

Page 10

1. Physiological systems can function normally only under carefully controlled conditions. Homeostatic regulation serves to prevent potentially disruptive changes in the body's internal environment.

2. When homeostasis fails, organ systems function less efficiently or begin to malfunction. The result is the state that we call *disease*. If the situation is not corrected, death may result.

3. Positive feedback is useful in processes that must move quickly to completion once they have begun, such as blood clotting. It is harmful in situations where a stable condition must be maintained, because it will serve to increase any departure from the desired condition. For example, positive feedback in the regulation of body temperature would cause a slight fever to spiral out of control, with fatal results. For this reason physiological systems usually exhibit negative feedback, which tends to oppose any departure from the norm.

Page 22

1. The two eyes would be separated by a *midsagittal section*.

2. The body cavity inferior to the diaphragm is the *abdominopelvic* (or *peritoneal*) *cavity*.

3. MRI scans are better at showing structural details of soft tissues (such as make up the brain) than ordinary X-rays, which are best for visualizing dense materials such as bone. Moreover, because computers are used to reconstruct the images, this technique is superior for depicting three-dimensional relationships.

Chapter 2

Page 35

1. Atoms combine with each other so as to gain a complete set of eight electrons in their outer energy levels. Oxygen atoms do not have a full outer energy level and so will readily react with many other elements to attain this stable arrangement. Neon already has a full other energy level and thus has little tendency to combine with other elements.

2. Hydrogen can exist as three different isotopes: hydrogen-1, with a mass of 1; deuterium, with a mass of 2; and tritium, with a mass of 3. The heavier sample must contain a higher proportion of one or both of the heavier isotopes.

Page 38

1. Since this reaction involves a large molecule being broken down into two smaller ones, it is an example of a *decomposition* reaction. Because energy is released in the process, the reaction can also be classified as *exergonic*.

2. Removing the product of a reversible reaction would keep its concentration low compared to the concentration of the reactants. Thus the formation of product molecules would continue but the reverse reaction would slow down, resulting in a shift in the equilibrium *toward the product*.

Page 41

1. When salt dissolves in water it dissociates into charged particles called ions that are capable of conducting an electric current. Sugar molecules are held together by covalent bonds and do not dissociate in solution; thus there are no ions to carry a current.

2. Stomach discomfort is often the result of excess stomach acidity ("acid indigestion"). Antacids contain a weak base that neutralizes the excessive acid.

3. The normal pH range for body fluids is 7.35–7.45. Fluctuations in pH outside this range can break chemical bonds, alter the shapes of molecules, and affect the functioning of cells, thereby causing harm to cells or tissues.

Page 57

1. A C:H:O ratio of 1:2:1 would indicate that the molecule is a *carbohydrate*. The body uses carbohydrates chiefly as an energy source.

2. The heat of boiling will break bonds that maintain the protein's tertiary &/or quaternary structure. The resulting change in shape will affect the ability of the protein molecule to perform its normal biological functions. These alterations are known as *denaturation*.

Chapter 3

Page 69

1. The finger-like projections on the surface of the intestinal cells are *microvilli*. They serve to increase the cells' surface area so that they can absorb nutrients more efficiently.

2. Since the flagellum is an organelle of locomotion, sperms cells lacking a flagellum would be unable to move. (As a result, they would probably be unable to reach and fertilize an egg.)

Page 76

1. The function of mitochondria is to produce energy for the cell in the form of ATP molecules. A large number of mitochondria in a cell would indicate a high demand for energy.

2. SER functions in the synthesis of lipids such as steroids. Ovaries and testes would be expected to have a great deal of SER because they produce large amounts of steroid hormones.

Page 88

1. In order to transport H^+ ions against their concentration gradient—that is, from a region where they are less concentrated (the cells lining the stomach) to a region where they are more concentrated (the interior of the stomach)—energy must be expended. An *active transport* process must be involved.

2. The 10 percent salt solution would be hypertonic with respect to the cells of the nasal lining, because it contains a higher concentration of salt than do the cells. The hypertonic solution would draw water out of the cells, causing the cells to shrink and loosening the mucus, thus relieving the congestion.

Page 92

1. If the cell membrane were freely permeable to sodium ions, more of these positively charged ions would move into the cell and the transmembrane potential would move closer to zero.

2. If spindle fibers failed to form during mitosis, the cell would not be able to separate the chromosomes into two sets. If cytokinesis occurred, the result would be one cell with two sets of chromosomes and one cell with none.

Chapter 4

Page 102

1. When the body temperature rises above 105°F for any length of time, proteins—including enzymes—begin to denature. When an enzyme becomes denatured it loses its normal structure and can no longer bind its substrate. As a result, the metabolic pathway in which the enzyme is involved will shut down. If that pathway is a vital one, death will result.

2. The body "burns" sugars and fats for energy through multistep processes with the aid of many enzymes. These enzymes lower the activation energy for the reactions involved, so that they can take place at body temperatures. Without enzymes, the activation energy for burning sugars and fats is so high that these compounds will not burn unless ignited by a hot flame.

Page 109

1. ATP is the most important energy compound in living cells. Since muscle cells require a great deal of energy to power the process of contraction, we would expect to find large quantities of ATP in muscle tissue.

2. NAD is required for an important step in the pathway of glycolysis. Decreasing the amount of NAD in a cell would thus decrease the rate of glycolysis.

3. During oxidative phosphorylation, hydrogen ions are pumped across the inner mitochondrial membrane into the space surrounding it, lowering the pH.

Page 117

1. Twenty different amino acids are commonly found in proteins, but there are only four different nucleotide bases. A code in which a single base was used to specify each amino acid could therefore "name" only four different amino acids. A code in which two bases were used to specify each amino acid could "name" only sixteen different amino acids. A three-base code is therefore needed.

2. A change in a single nucleotide can alter a codon so that it specifies a different amino acid in a cellular protein. Changing even one amino acid can sometimes alter a protein enough to destroy its ability to function. If the affected protein plays a vital role in the body, the result could be fatal.

3. Proteins are synthesized from amino acids. During starvation the body will catabolize amino acids for energy rather than use them to make new proteins.

Chapter 5

Page 136

1. No. A simple squamous epithelium does not provide enough protection against infection, abrasion, and dehydration and is not found in the skin surface.

2. The process described is *holocrine secretion*.

Page 152

1. The tissue is probably *fascia*, a type of dense regular connective tissue that attaches muscles to skin and bones.

2. Collagen fibers add strength to connective tissue. We would therefore expect vitamin C deficiency to result in the production of connective tissue that is weaker and more prone to damage.

Page 158

1. This is an example of a *mucous membrane*.

2. Since cardiac and skeletal muscle are both striated (banded), this must be *smooth muscle* tissue.

3. All of these regions are subject to mechanical trauma and abrasion—by food (pharynx and esophagus), feces (anus), and intercourse or childbirth (vagina).

Page 162

1. Antihistamines block the release of histamine from mast cells. Histamine is responsible for the increased blood flow and capillary permeability that causes swelling.

Chapter 6

Page 186

1. Cell are constantly shed from the outer layers of the *stratum corneum*.

2. The splinter is lodged in the *stratum granulosum*.

Page 187

1. Because fresh water is hypotonic with respect to skin cells, water will move into the cells by osmosis, causing them to swell.

2. Sanding the tips of ones fingers will not permanently remove fingerprints. Since the ridges of the fingerprints are formed in layers of the skin that are constantly regenerated, they will eventually reappear. The actual pattern of the ridges is determined by arrangement of tissue in the dermis, which is not affected by sanding.

Page 190

1. When exposed to the ultraviolet radiation in sunlight or tanning lamps, melanocytes in the epidermis and dermis synthesize the pigment melanin, darkening the color of the skin.

Page 195

1. When the dermis is stretched excessively, the elastic fibers are overstretched and are not able to recoil. The skin then forms folds or wrinkles, called stretch marks, in the affected areas.

2. Contraction of the arrector pili muscles pull the hair follicles erect, depressing the area at the base of the hair and making the surrounding skin appear higher. The result is known as "goose bumps" or "goose pimples."

3. Hair is a derivative of the epidermis, and if the epidermis is destroyed by the injury there would be no hair follicles to produce new hair.

Page 205

1. If the duct of a sebaceous gland is blocked by infection, the result is a *furuncle* or *boil*.

2. Apocrine sweat glands produce a secretion containing several kinds of organic compounds. Some of these have an odor and others produce an odor when metabolized by skin bacteria. Deodorants are used to mask the odor of these secretions.

3. As a person ages, the blood supply to the dermis decreases and merocrine sweat glands become less active. These changes make it more difficult for the elderly to cool themselves in hot weather.

Chapter 7

Page 214

1. If the ratio of collagen to hydroxyapatite in a bone increased, the bone would be more flexible and less strong.

2. Concentric layers of bone around a central canal is indicative of an Haversian system. Haversian systems make up compact bone. Since the ends (epiphyses) of long bones are primarily cancellous (spongy) bone, this sample most likely came from the shaft (diaphysis) of a long bone.

3. Since osteoclasts function in breaking down or demineralizing bone, the bone would have less mineral content and as a result it would be weaker.

Page 219

1. Long bones of the body, like the femur, have a plate of cartilage, called the epiphyseal plate, that separates the epiphysis from the diaphysis as long as the bone is still growing lengthwise. An X-ray would indicate whether or not the epiphyseal plate was still present. If it was, then growth was still occurring, and if not the bone had reached its adult length.

2. The increase in the male hormone, testosterone, that occurs at puberty, contributes to an increased rate of bone growth and the closure of the epiphyseal plates. Since the source of testosterone, the testes, is removed in castration, we would

expect these boys to have a longer, though slower, growth period and be taller than they would have been if they had not been castrated.

3. Women who are pregnant need large amounts of calcium to support the needs of the developing fetus for bone growth. If the expectant mother does not include enough calcium in her diet, her body will mobilize the calcium reserves of her skeleton to provide for the needs of the fetus, resulting in weakened bones and an increased risk of fracture.

Page 224

1. Bones increase in thickness in response to physical stress. One common type of stress that is applied to a bone is that produced by muscles. We would expect the bones of a body builder to be thicker after the addition of the extra muscle mass because of the greater stress that the muscle would apply to the bone.

2. Since the parathyroid hormone PTH causes an increase in the levels of blood calcium, we would expect to see elevated blood levels of the calcium ion and problems associated with excess calcium (such as tetanic muscle contractions which are covered in chapter 10). We would also expect to see bones that have less mass and that are more brittle, because PTH increases osteoclastic activity.

3. As an individual ages osteoclastic activity exceeds osteoblastic activity, and bones become more brittle. Eventually, the weight of the body causes compression of the vertebrae which results in a slight loss of height.

Page 236

1. Originally, the joint is a type of syndesmosis. When the bones fuse, the bones along the suture represent a synostosis.

2. a. abduction
   b. supination
   c. flexion

Chapter 8

Page 243

1. The bone that forms the superior portion of the orbit is the frontal bone and the maxilla forms the inferior portion of the orbit. These would be the two bones fractured by the ball.

2. The mastoid and styloid processes are projections found on the temporal bones of the skull.

3. The bones that articulate with the vomer are the ethmoid bone, sphenoid bone, maxillae, and palatine bones.

Page 247

1. The internal jugular veins pass through openings in the temporal bones.

2. The sella turcica contains the pituitary gland and is located in the sphenoid bone.

3. Nerve fibers to the olfactory bulb which deals with the sense of smell pass through the cribriform plate from the nasal cavity. If the cribriform plate failed to form these sensory nerves could not reach the olfactory bulbs and the sense of smell (olfaction) would be lost.

Page 254

1. The paranasal sinuses function to make some of the heavier skull bones lighter and to produce mucus.

2. The most powerful muscles that are involved in closing the mouth attach to the mandible at the coronoid process. A fracture of the coronoid process would make it difficult for these muscles to function properly and close the mouth.

3. Since many muscles that move the tongue and the larynx are attached to the hyoid bone, you would expect a person with a fractured hyoid bone to have difficulty moving their tongue and swallowing.

Page 265

1. The odontoid process is found on the second cervical vertebra or axis which is located in the neck.

2. Improper compression of the chest during CPR could and frequently does result in a fracture of the sternum or ribs.

3. In adults, the five sacral vertebrae fuse to form a single sacrum.

Page 276

1. The clavicle attaches the scapula to the sternum and thus restricts the scapula's range of movement. If the clavicle is broken, the scapula will have a greater range of movement and will be less stable.

2. The two rounded prominences on either side of the elbow are parts of the humerus (the lateral and medial epicondyles).

3. The radius is in a lateral position when the arm is pronated and in a medial position when the arm is supinated.

Page 279

1. Since the subscapular bursa is located in the shoulder joint, and inflammation of this structure (bursitis) would be found in the tennis player. The condition is associated with repetitive motion that occurs at the shoulder, such as swinging a tennis racket. The jogger would be more at risk for injuries to the knee joint.

2. Mary has most likely fractured her ulna.

Page 286

1. The three bones that make up the coxa are the ilium, ischium, and pubic bones.

2. Although the fibula is not part of the knee joint nor does it bear weight, it is an important point of attachment for many leg muscles. When the fibula is fractured, these muscles cannot function properly to move the leg and walking is difficult and painful. It also helps stabilize the ankle joint.

3. Joey has most likely fractured the calcaneus (heel bone).

Page 294

1. The iliofemoral, pubofemoral, and ischiofemoral ligaments would all be found in the hip joint.

2. Damage to the menisci in the knee joint would result in a decrease in the joint's stability. The individual would have a harder time locking the knee in place while standing and would have to use muscle contractions to stabilize the joint. When standing for long periods, the muscles would fatigue and the knee would "give out." We would also expect the individual to experience pain.

Chapter 10

Page 304

1. Since tendons attach muscles to bones, severing the tendon would disconnect the muscle from the bone so that when the muscle contracted nothing would happen.

2. Skeletal muscle appears striated when viewed under the microscope because it is composed of the myofilaments actin and myosin which are arranged in such a way as to produce a banded appearance in the muscle.

3. You would expect to find the greatest concentration of calcium ions to be located in the cisternae of the sarcoplasmic reticulum of the muscle.

Pages 310–311

1. Since the ability of a muscle to contract depends upon the formation of cross bridges between the myosin and actin myofilaments, a drug that would interfere with cross bridge formation would prevent the muscle from contracting.

2. Because the amount of cross bridge formation is proportional to the amount of available calcium ion, increased permeability of the sarcolemma to calcium ion would lead to an increased intracellular concentration of calcium and a greater degree of contraction. In addition, since relaxation depends on decreasing the amount of calcium in the sarcoplasm, an increase in the permeability of the sarcolemma to calcium could result in a situation in which the muscle would not be able to relax completely.

3. Without acetylcholinesterase, the motor endplate would be continuously stimulated by the acetylcholine and the muscle would be locked into contraction.

Page 317

1. The ability of the muscle to contract depends on the ability to form cross bridges between the actin and myosin. If the myofilaments overlap very little, then very few cross bridges are formed and the contraction is weak. If the myofilaments to not overlap at all, then no cross bridges form and the muscle cannot contract.

2. A motor unit with 1500 fibers is most likely from a large muscle involved in powerful, gross body movement. Muscles that control fine and/or precise movements such as movement of the eye or the fingers have only a few fibers per motor unit whereas muscles of the legs, for instance, that are involved in powerful contractions have hundreds of fibers per motor unit.

3. There are two types of muscle contractions, isometric and isotonic. In an isotonic contraction, tension remains constant and the muscle shortens. In isometric contractions, however, the same events of contraction occur, but instead of shortening, the tension in the muscle increases.

Page 325

1. The sprinter requires large amounts of energy for a relatively short burst of activity. To supply this demand for energy, the muscles switch to anaerobic respiration. Anaerobic respiration is not as efficient in producing energy as aerobic respiration and the process also produces acidic waste products. The combination of less energy and the waste products contribute to fatigue. Marathon runners, on the other hand, derive most of their energy from aerobic respiration which is more efficient and does not produce the level of waste products that anaerobic respiration does.

2. We would expect activities that require short periods of strenuous activity to produce a greater oxygen debt because this type of activity relies heavily on energy production by anaerobic respiration. Since lifting weights is more strenuous over the short term we would expect this type of exercise to produce a greater oxygen debt than swimming laps which is an aerobic activity.

3. Individuals who are naturally better at endurance types of activities such as cycling or marathon running have a higher percentage of slow twitch muscle fiber which are physiologically better adapted to this type of activity than the fast twitch fibers which are less vascular and fatigue faster.

Chapter 11

Page 335

1. The joint between the occipital bone and the first cervical vertebra would be an example of a first class lever system. The joint between the two bones is the fulcrum, which lies between the skull which is the resistance, and the neck muscles which are the force.

2. The main antagonist of the biceps brachii muscle is the triceps brachii muscle.

3. The name flexor carpi radialis longus tells you that this is a long muscle that lies next to the radius and functions to flex the hand.

Page 348

1. Contraction of the masseter muscle raises the mandible, while the mandible depresses when the muscle is relaxed. These movements are important in the process of chewing or mastication.

2. Damage to the external intercostal muscles would interfere with the process of breathing.

3. A blow to the rectus abdominis would cause the muscle to contract forcefully resulting in flexion of the torso. In other words, you would "double up".

Page 355

1. When you shrug your shoulders you are contracting your levator scapulae muscles.

2. The rotator cuff muscles include the supraspinatus, infraspinatus, subscapularis, and teres minor. The tendons of these muscles help to enclose and stabilize the shoulder joint.

3. Injury to the flexor carpi ulnaris would impair the ability to flex and adduct the hand.

Page 365

1. Injury to the obturator muscle would interfere with your ability to rotate your leg laterally.

2. The hamstring refers to a group of five muscles that collectively function in flexing the leg. These muscles are the biceps femoris, semimembranous, semitendinosus, gracilis, and sartorius.

3. The achilles (calcaneal) tendon attaches the soleus and gastrocnemius muscles to the calcaneus (heel bone). When these muscles contract, they cause extension of the foot. A torn achilles tendon would make extension of the foot difficult and the opposite action, flexion, would be more pronounced as a result of less antagonism from the soleus and gastrocnemius.

Chapter 12

Page 379

1. The afferent division of the nervous system is composed of nerves that carry sensory information to the brain and spinal cord. Damage to this division would interfere with a person's ability to experience a variety of sensory stimuli.

2. Sensory neurons of the peripheral nervous system are usually unipolar, thus this tissue is most likely associated with a sensory organ.

3. Microglial cells are small phagocytic cells that are found in increased number in damaged and diseased areas of the CNS.

Page 386

1. Depolarization of the neuron membrane involves the opening of the sodium channels and the rapid influx of sodium ions into the cell. If the sodium channels were blocked, a neuron would not be able to depolarize and conduct an action potential.

2. If the extracellular concentration of potassium ion decreased, more potassium ion would leave the cell and the electrical difference across the membrane (transmembrane potential) would be greater. This condition is called hyperpolarization.

3. Because of saltatory conduction, myelinated fibers conduct action potentials much faster than nonmyelinated fibers, so the axon conducting at 150 m/sec is myelinated.

Page 398

1. When an action potential reaches the presynaptic terminal of a cholinergic synapse, calcium channels are opened and the influx of calcium triggers the release of acetylcholine into the synapse to stimulate the next neuron. If the calcium channels were blocked, the acetylcholine would not be released and transmission across the synapse would cease.

2. In an adrenergic synapse, the neurotransmitter norepinephrine brings about depolarization of the postsynaptic membrane by way of the second messenger cAMP. If the enzyme that converts ATP to cAMP is blocked, then the norepinephrine could not bring about depolarization at the postsynaptic terminal and transmission at that synapse would be blocked.

3. A neurotransmitter that opens the potassium channels but not the sodium channels would cause a hyperpolarization at the postsynaptic membrane. The transmembrane potential would be greater and it would be more difficult to bring the membrane to threshold. This would result in an inhibitory postsynaptic potential (IPSP).

Page 404

1. In order for a severed axon to heal, it must come into contact with and grow into the new cord of Schwann cells that forms

distal to the site of the injury. If the axon fails to make this connection, the column of Schwann cells degenerates and the axon stops growing and the connection will not be reestablished. By closely aligning the two ends of the axons after an injury, there is a better chance that the connection will be made and innervation reestablished.

2. Interneurons are found in the CNS and are responsible for analyzing sensory input and coordinating motor output. Without the interneurons, the nervous system would not be able to process sensory information or make appropriate motor responses.

Chapter 13

Page 410

1. The ventral root of spinal nerves is composed of visceral and somatic motor fibers. Damage to this root would interfere with motor function.

2. The cerebrospinal fluid that surrounds the spinal cord is found in the subarachnoid space which lies beneath the epithelium of the arachnoid layer and on top of the pia mater.

Page 417

1. Since the polio virus would be located in the somatomotor neurons, we would find it in the anterior gray horns of the spinal cord where the cell bodies of these neurons are located.

2. The dorsal rami of spinal nerves innervates the skin and muscles of the back. In this case we would expect the skin and muscles of the back of the neck and shoulders to be affected.

3. The phrenic nerves that innervate the diaphragm originate in the brachial plexus. Damage to this plexus or more specifically to the phrenic nerves would greatly interfere with the ability to breathe and possibly result in death.

Page 431

1. The suckling reflex is an example of an innate reflex.

2. When the stretch receptors are stimulated by the gamma motor neurons, the spindles become narrower and less sensitive to stretch. As a result it would take more force to get the muscles of the leg to contract for the knee jerk reflex, and as a result the reflex would be slower.

3. A positive Babinski reflex is abnormal for an adult and indicates possible damage of descending tracts in the spinal cord.

Chapter 14

Page 442

1. If one of the interventricular foramina became blocked, cerebrospinal fluid would not be able to flow from the first or second ventricle into the third. Since cerebrospinal fluid would continue to be formed, the blocked ventricle would swell with fluid, a condition known as hydrocephalus.

2. The primary motor cortex is located in the precentral gyrus of the frontal lobe of the cerebrum.

3. The extrapyramidal system is composed of pathways that control muscle tone and coordinate learned movement patterns. A person suffering an injury to these tracts would exhibit difficulty in walking and in fluid precise movements of the arms and hands.

Page 451

1. The lateral geniculate nuclei are involved with processing visual information. Damage to these nuclei would interfere with the sense of sight.

2. Changes in body temperature would stimulate the preoptic area of the hypothalamus a division of the diencephalon.

3. The vermis and arbor vitae are structures associated with the cerebellum.

4. Even though the medulla is small, it contains many vital reflex centers including those that control breathing and regulate the heart and blood pressure. Damage to the medulla can result in a cessation of breathing, or changes in heart rate and blood pressure that are incompatible with life.

Page 469

1. Diffusion across the arachnoid villi is the means by which cerebrospinal fluid re-enters the blood stream. If this process decreased, then excess fluid would start to accumulate the ventricles and the volume of fluid in the ventricles would increase.

2. Damage to the vagus nerve (cranial nerve X) can result in death, since it has motor fibers to regulate breathing, heart rate, and blood pressure.

3. The glossopharyngeal nerve (cranial nerve IX) controls swallowing muscles and provides sensory information from the tongue.

4. Since the abducens nerve (cranial nerve VI) controls lateral movements of the eyes, we would expect an individual with damage to this nerve to not be able to move their eyes laterally.

Chapter 15

Page 483

1. The fasciculus gracilis in the posterior column of the spinal cord is responsible for carrying information about touch and pressure from the lower part of the body to the brain.

2. The anatomical basis for opposite side motor control is that crossing over (decussation) occurs, and the pyramidal motor fibers innervate lower motor neurons on the opposite side of the body.

3. The superior portion of the motor cortex exercises control over the hand, arm, and upper portion of the leg. An injury to this area would affect the ability to control the muscles in those regions of the body.

Page 490

1. We would expect Tina's brain waves to be beta waves which are characteristic of adults who are experiencing stress and/ or tension.

2. An inability to comprehend the written or spoken word indicates a problem with the general interpretive area of the brain, which in most individuals is located in the left temporal lobe of the cerebrum.

3. Recalling facts for an A & P test involves declarative memory, specifically secondary memory (long-term memory).

Page 497

1. The reticular activating system (RAS) is responsible for rousing the cerebrum to a state of consciousness. If a sleeping individual's RAS were stimulated, she would certainly wake up.

2. We would expect a drug that increases serotonin levels to produce a heightened perception of certain sensory stimuli such as auditory or visual, and hallucinations.

3. Some possible reasons for slower recall and loss of memory in the elderly include a loss of neurons (possibly those involved in specific memories), changes in synaptic organization of the brain, changes in the neurons themselves, and decreased blood flow which would affect the metabolic rate of neurons and perhaps slow the retrieval of information from memory.

Chapter 16

Page 508

1. The neurons that synapse in the collateral ganglia originate in the lower thoracic and upper lumbar portion of the spinal cord and pass through the chain ganglia to the collateral ganglia.

2. Since acetylcholine is the neurotransmitter that is released by all of the preganglionic fibers of the sympathetic nervous system, a drug that stimulates acetylcholine receptors would stimulate the postganglionic fibers of the sympathetic nerves resulting in increased sympathetic activity.

3. When alpha-1 receptors are stimulated, calcium channels are opened and the influx of calcium ions brings about the change associated with the stimulation of those receptors. Alpha-2 receptors activate a second messenger system that uses cAMP. Therefore, a change in extracellular calcium would have a greater affect on cells with alpha-1 receptors.

4. Blocking the beta-receptors on cells would decrease or prevent sympathetic stimulation of those tissues. This would result in decreased heart rate and force of contraction and relaxation of the smooth muscle in the walls of blood vessels. These changes would contribute to lowering a person's blood pressure.

Page 521

1. Muscarinic receptors are a type of acetylcholine receptors found in the postganglionic synapse of the parasympathetic nervous system. Stimulation of these receptors at the heart would cause an opening of more potassium channels resulting in hyperpolarization of the membrane and a decreased heart rate.

2. Since most blood vessels receive mainly sympathetic stimulation, a decrease in sympathetic tone would lead to a relaxation of the muscles in the walls of the vessels and vasodilation (the vessels would increase in diameter). This in turn would result in increased blood flow to the tissue.

3. A patient who is anxious about impending root canal would probably exhibit some or all of the following changes: a dry mouth, increased heart rate, increased blood pressure, increased rate of breathing, cold sweats, an urge to urinate or defecate, change in motility of the digestive tract (i.e. "butterflies in the stomach"), and dilated pupils. These changes would be the result of anxiety of stress causing an increase in sympathetic stimulation.

Chapter 17

Page 536

1. By the end of the lab period adaptation has occurred. In response to the constant level of stimulation, the receptor neurons have become less active partially as the result of synaptic fatigue.

2. Since nociceptors are pain receptors, if they are stimulated, you would perceive a painful sensation in your affected hand.

3. Proprioceptors relay information about limb position and movement to the central nervous system, especially the cerebellum. Lack of this information would result in uncoordinated movements and the individual probably would not be able to walk.

4. The taste receptors (taste buds) are only sensitive to molecules and ions that are in solution. If you dry the surface of the tongue, there is no moisture for the sugar molecules or salt ions to dissolve in and they will not stimulate the taste receptors.

Page 547

1. The role of the pharyngotympanic tube (Eustachian tube) is to allow for equalizing pressure on both sides of the tympanic membrane (eardrum). If this tube is blocked, there will be greater pressure on the inside of the tympanic membrane forcing it outward and producing pain.

2. Without the movement of the round window, the perilymph would be moved by the vibration of the stapes at the oval window, and there would be little or no perception of sound.

3. Loss of stereocilia (as a result of constant exposure to loud noises for instance) would reduce hearing sensitivity and could eventually result in deafness.

Page 565

1. The first layer of the eye to be affected by inadequate tear production would be the conjunctiva. Drying of this layer would produce an irritated, scratchy feeling.

2. When the lens is round you are looking at something closer to you.

3. Even with a congenital lack of cone cells in the eye you would still be able to see as long as you had functioning rod cells. Since cone cells function in color vision, you would see only black and white.

4. A deficiency or lack of vitamin A in the diet would affect the quantity of retinal that the body could produce and thus interfere with night vision.

Chapter 18

Page 579

1. Phosphodiesterase is the enzyme that converts cAMP to AMP thus inactivating it. If this enzyme were blocked, the effect of the hormone would be prolonged.

2. Dehydration increases the osmotic pressure of the blood. The increase in blood osmotic pressure would stimulate the neurohypophysis to release more ADH.

3. Somatomedins are the mediators of growth hormone action. If the level of somatomedins is elevated, we would expect to see the level of growth hormone elevated as well.

4. Increased levels of cortisol would inhibit the cells that control ACTH release from the pituitary, therefore the level of ACTH would decrease. This is an example of a negative feedback mechanism.

Page 588

1. If an individual lacked iodine in their diet, they would not be able to form the hormone thyroxine. As a result we would expect to see the symptoms associated with thyroxins deficiency, such as decreased rate of metabolism, decreased body temperature, poor response to physiological stress, and an increase in the size of the thyroid gland (goiter).

2. Most of the thyroid hormone in the blood is bound to a protein called thyroid-binding globulins. This represents a large reservoir of thyroxine that guards against rapid fluctuations in the level of this important hormone. Because there is such a large amount stored in this way, it takes several days to deplete the supply of hormone, even after the thyroid gland has been removed.

3. Removal of the parathyroid glands would result in a decrease in the blood levels of calcium ion. This could be counteracted by increasing the amount of vitamin D and calcium in the diet.

4. One of the functions of cortisol is to decrease the cellular use of glucose while increasing the available glucose by promoting the breakdown of glycogen and the conversion of amino acids to carbohydrates. The net result is an elevation in the level of glucose in the blood.

Page 597

1. One of the functions of the hormone atrial natriuretic peptide is to increase the rate of sodium excretion in the kidneys. As a result the sodium excreted in the urine would increase.

2. An individual with Type I diabetes has such elevated levels of glucose in the blood that the kidney cannot reabsorb all of the glucose, some is lost in the urine. The water lost with the glucose elevates blood osmotic pressure promotes the thirst.

3. Glucagon stimulates the conversion of glycogen to glucose in the liver. Increased amounts of glucagon would then lead to decreased amounts of liver glycogen.

4. The pineal gland receives neural input from the optic tracts and its secretion, melatonin, is influenced by light-dark cycles. Increased amounts of light inhibit the production and release of melatonin from the pineal gland.

Page 601

1. The type of hormonal interaction exemplified by the insulin and glucagon is antagonism. In this type of hormonal interaction, two hormones have opposite effects on their target tissues.

2. The hormones growth hormone, thyroid hormone, parathyroid hormone, and the gonadal hormones all play a role in formation and development of the skeletal system.

3. During the resistance phase of GAS, there is a high demand for glucose, especially by the nervous system. The GH-RH and CRH increase the levels of growth hormone and ACTH respectively. Growth hormone mobilizes fat reserves and promotes the catabolism of protein. ACTH increases cortisol which stimulates the conversion of glycogen to glucose as well as the catabolism of fat and protein.

Chapter 19

Page 616

1. The globulin protein fraction includes proteins called antibodies that defend our body against infection. A decrease in the amount of these proteins would make an individual more likely to catch a disease and it would make it more difficult for their body to rid itself of the infectious organism.

2. The hematocrit measures the amount of formed elements (mostly red blood cells) as a percentage of the total blood. In hemorrhage the loss of blood, especially red blood cells, would cause the hematocrit to be less.

3. The liver conjugates bilirubin, that is makes it more soluble by combining it with glucouronic acid so that it can be excreted more easily. Diseases that damage the liver such as hepatitis or cirrhosis would impair the livers ability to perform this function. As a result, the bilirubin would accumulate in the blood producing a condition known as jaundice.

4. The factors that are common to all forms of anemia are a low hematocrit and a low amount of hemoglobin.

Page 624

1. If a person with type A blood receives a transfusion of type B blood, the red cells would clump or agglutinate, potentially blocking blood flow to various organs and tissues.

2. In an infected cut we would expect to find a large number of neutrophils. Neutrophils are phagocytic white cells that are usually the first to arrive at the site of an injury and which specialize in dealing with infectious bacteria.

3. The type of white blood cell that produces circulating antibodies is the B lymphocyte, and these would be found in increased numbers.

4. Megakaryocytes are the precursors of platelets which play an important role in hemostasis and the clotting process. A decreased number of megakaryocytes would result in fewer platelets, which in turn would interfere with the ability to clot properly.

Page 630

1. The use of broad-spectrum antibiotics would lower the number of intestinal bacteria, and thus the amount of vitamin K produced. This decrease in vitamin K would lead to a decrease in the production of several clotting factors, most notably prothrombin. As a result, clotting time increases.

2. The kidneys release erythropoietin in response to low levels of oxygen. If the amount of oxygen reaching the kidneys is high, the kidneys would release less erythropoietin, and the level of erythropoietin in the blood would fall.

Chapter 20

Page 642

1. The semilunar valves on the right side of the heart guard the opening to the pulmonary artery. Damage to these valves would interfere with the blood flow through this vessel.

2. The most obvious characteristic that differentiates cardiac muscle tissue from skeletal muscle tissue is cardiac muscle fibers are small with a centrally placed nucleus. In addition, there are intercalated discs in the cardiac muscle and they have a different arrangement of T-tubules and sarcoplasmic reticulum.

Page 651

1. A decrease in the amount of calcium in the extracellular fluid (hypocalcemia) would result in a weaker contraction of the heart since fewer calcium ions would be available to bind to the troponin molecules to initiate the contraction sequence.

2. If these cells were not functioning, the heart would still continue to beat but at a slower rate.

3. If the impulses from the atria were not delayed at the AV node, they would be conducted through the ventricles so quickly by the bundle branches and Purkinje fibers that the ventricles would begin contracting immediately before the atria had finished their contraction. As a result the ventricles would not be as full of blood as they could be and the pumping of the heart would not be as efficient, especially during activity.

Page 657

1. When pressure in the left ventricle is rising, the heart is contracting but no blood is leaving the heart. During this initial phase of contraction, the AV valves and semilunar valves are both closed. The increase in pressure is the result of increased tension as the muscle contracts. When the pressure in the ventricle exceeds the pressure in the aorta, the aortic semilunar valves are forced open and the blood is rapidly ejected from the ventricle.

2. When the ventricles begin to contract, they force the AV valves to close which in turn pulls on the chordae tendineae which then pull on the papillary muscles. The papillary muscles respond by contracting, counteracting the force that is pushing the valves upward.

3. If the heart beats too quickly (tachycardia), there is not sufficient time for it to fill completely between the beats. Since the heart pumps blood proportionately to what enters, the less blood that enters, the less it will be able to pump. If it beats too fast, very little blood will enter circulation and tissues will suffer damage from lack of blood supply.

Page 661

1. Stimulating the acetylcholine receptors of the heart would cause the heart to slow down. Since the cardiac output is the product of stroke volume times the heart rate, if the heart rate decreases so will the cardiac output (assuming no change in the stroke volume).

2. The venous return fills the heart with blood stretching the heart muscle. According to Starling's law, the more the heart muscle is stretched, the more forcefully it will contract (to a point). The more forceful the contraction the more blood the heart will eject with each beat (stroke volume). Therefore, increased venous return will increase the stroke volume, assuming all other factors are constant.

3. Increased sympathetic stimulation of the heart will result in increased heart rate and increased force of contraction. The ESV represents the amount of blood that remains in a ventricle after a contraction (systole). The more forcefully the heart contracts the more blood it will eject and the lower the ESV will be. Therefore, increased sympathetic stimulation should result in a lower ESV.

Chapter 21

Page 672

1. The blood vessels are veins. Arteries and arterioles have a relatively large amount of smooth muscle tissue in a thick, well-developed tunica media.

2. Relaxation of the precapillary sphincters would increase the blood flow to a tissue.

3. Blood pressure in the arterial system pushes blood into the capillaries. Blood pressure on the venous side is very low, and other forces help keep the blood moving. Valves prevent the blood from flowing backward whenever the venous pressure drops.

Page 686

1. The left subclavian artery is the branch of the aorta that sends blood to the left shoulder and arm.

2. The common carotid arteries carry blood to the head. A compression of one of the common carotid arteries would result in decreased blood flow to the brain and loss of consciousness or even death.

3. Organs served by the celiac artery include the stomach, spleen, liver, and pancreas.

Page 695

1. In a normal individual, the pressure should be greatest in the aorta and least in the vena cavae. Blood, like other fluids, moves along a pressure gradient from high pressure to low pressure. If the pressure were higher in the inferior vena cava, the blood would flow backwards.

2. While standing for periods of time, blood tends to pool in the lower extremities. This decreases the venous return to the heart, and in turn the cardiac output decreases sending less blood to the brain, causing the light-headedness and fainting. A hot day adds to the effect, due to a loss of body water through sweating.

Page 706

1. In exercise (1) blood flow to muscles increases, (2) cardiac output increases and (3) resistance in visceral tissues increases.

2. Pressure at this site would decrease blood pressure at the carotidsinus, where the carotid baroreceptors are located. This causes a decreased frequency of action potentials along the glossopharyngeal nerve (IX) to the medulla, and more sympathetic impulses will be sent to the heart. The net result will be an increase in the heart rate.

3. Vasoconstriction of the renal artery will decrease both blood flow and blood pressure at the kidney. In response the kidney will increase the amount of renin that it releases, which in turn will lead to an increase in the level of angiotensin II. The angiotensin II will bring about increased blood pressure and increased blood volume.

4. In circulatory shock, there is a decreased venous return to the heart. As a result of the decrease in venous return, the cardiac output is decreased which accounts for the weak pulse. Because the cardiac output is decreased, the baroreceptors are stimulated and in turn there is increased sympathetic stimulation to the heart, causing the rapid heart rate. Although the heart is beating faster, there is less blood to pump, and pulse pressure remains low.

Chapter 22

Page 720

1. The thoracic duct drains lymph from the area beneath the diaphragm and the left side of the head and thorax. Most of the lymph enters the venous blood by way of this duct. A blockage of this duct would not only impair circulation of lymph through most of the body, it would also promote accumulation of fluid in the extremities (lymphedema).

2. The hormone thymosin from the thymus plays a role in the differentiation of stem lymphocytes into T lymphocytes. A lack of this hormone would result in an absence of T lymphocytes.

3. During an infection, the lymphocytes and phagocytes in the lymph nodes in the affected region undergo cell division better deal with the infectious agent. This increase in the number of cells in the node causes the node to become enlarged or swollen.

Page 726

1. A decrease in the number of monocyte forming cells in the bone marrow would result in a decreased number of macrophages in the body, since all of the different macrophages are derived from the monocytes. This would include the microglia of the CNS, the Kuppfer's cells of the liver, Langerhan's cells in the skin and digestive tract, alveolar macrophages as well as others.

2. A rise in interferon would indicate a viral infection. Interferon is released from cells that are infected with viruses. It does not help the infected cell, but "interferes" with the virus' ability to infect other cells.

3. Pyrogens stimulate the temperature control area of the preoptic nucleus of the hypothalamus. The result is an increase in body temperature or fever.

Page 739

1. Cytotoxic T cells function in cell-mediated immunity. A decrease in the number of cytotoxic T cells would interfere with the ability to kill foreign cells and tissues as well as cells infected by viruses.

2. Helper T cells promote B cell division, the maturation of plasma cells and the production of antibody by the plasma cells. Without the helper T cells the humoral immune response would take much longer to occur and would not be as efficient.

Page 742

1. The developing fetus is protected primarily by passive immunity, the product of IgG antibodies that cross the placenta from the mother's circulation. In addition, the fetus may show some degree of active cellular immunity by the third month of development.

2. Stress can interfere with the immune response by depressing the inflammatory response, reducing the number and activity of phagocytes, and inhibiting interleukin secretion.

Page 759

1. Increased tension in the vocal cord will cause a higher pitch in the voice.

2. The tracheal cartilages are C-shaped to allow room for esophageal expansion when large portions of food or liquid are swallowed.

3. Without surfactant, surface tension in the thin layer of water that moistens their surfaces would cause the alveolar to collapse.

4. Chronic smoking damages the lining of the air passageways. Cilia are seared off the surface of the cells by the heat, and the large number of particles that escape filtering in the nose are trapped in the excess mucus that is secreted to protect the irritated lining. This combination of circumstances creates a situation in which there is a large amount of thick mucus that is difficult to clear from the passages. The cough reflex is an attempt to remove this material from the airways.

Page 769

1. Since the rib penetrates the chest wall, air atmospheric air will enter the thoracic cavity. This space is normally at a lower pressure than the outside air, so when the air enters the natural elasticity of the lungs will not be compensated and the lung will collapse. This condition is called a pneumothorax.

2. As a result of emphysema the larger air spaces and lack of elasticity will increase compliance.

3. Since the fluid produced in pneumonia takes up space that would normally be occupied by air, the vital capacity will be decreased.

Page 775

1. On a hot, humid day, the air that we breathe contains more water vapor than on a cool dry day, and this means that the partial pressure of oxygen in the air is less. Since gases expand when heated, on a hot day the same volume of gas would contain fewer molecules. Thus an individual must breathe deeper or faster (or both) to gain the same amount of oxygen as compared to a cool dry day.

2. As skeletal muscles become more active they generate more heat and more acid waste products which lowers the pH of surrounding fluid. The combination of lower pH and higher temperature causes the hemoglobin to release more oxygen than it would under conditions of lower temperature and higher pH.

3. An obstruction of the airways would interfere with the body's ability to gain oxygen and eliminate carbon dioxide. Since most carbon dioxide is carried in the blood as bicarbonate ion that is formed from the dissociation of carbonic acid, an inability to eliminate carbon dioxide would result in an excess of hydrogen ions thus lowering the body's pH.

Page 780

1. The pneumotaxic center inhibits the inspiratory center and the apneustic center. Exciting this center in the brainstem would result in shorter breaths and a more rapid rate of breathing.

2. Chemoreceptors are more sensitive to carbon dioxide. When this gas dissolves it produces hydrogen ions that lower pH and alter cell or tissues activity.

Chapter 24

Page 795

1. Parasympathetic stimulation increases muscle tone and motility in the digestive tract. A drug that blocks this activity would decrease the rate of peristalsis.

2. The fauces is the opening between the oral cavity and the pharynx.

3. The process that is being described is deglutition or swallowing.

Page 807

1. The larger a meal (especially in terms of protein) the more stomach acid there is secreted. The hydrogen ions for the acid come from the blood that enters the stomach therefore, the blood leaving the stomach will have fewer than normal hydrogen ions and be decidedly alkaline, that is have a higher pH. This is referred to as the alkaline tide.

2. The vagus nerve contains parasympathetic motor fibers that can stimulate gastric secretions. This can occur even if food is not present in the stomach (caphalic phase of gastric digestion). Cutting the branches of the vagus that supply the stomach would prevent this type of secretion from occurring and decrease the chance of ulcer formation.

3. The small intestine has several adaptations that increase surface area to increase its absorptive capacity. First walls of the small intestine are thrown into folds called the plicae circularis. The tissue that covers the plicae forms fingerlike projections, the villi. The cells that cover the villi have an exposed surface that is covered by small fingerlike projections called the microvilli. In addition, the small intestine has a very rich blood and lymphatic supply to transport the nutrients that are absorbed.

4. It would increase.

Page 816

1. A narrowing of ileocecal valve would interfere with the flow of chyme from the small intestine to the large intestine.

2. A blockage of the parotid duct would interfere with carbohydrate digestion.

3. A decrease in the amount of bile salts would decrease the effectiveness of fat digestion and absorption.

Page 825

1. Lipids, proteins, and nucleic acids; carbohydrate digestion would be affected to a lesser extent.

2. Chylomicrons are formed from the fats that are digested in a meal. A meal this is high in fat would increase the number of chylomicrons in the lacteals.

3. Removal of the upper portion of the stomach would interfere with the absorption of vitamin B12. This vitamin requires a glycoprotein, called intrinsic factor (IF), that is produced by the parietal cells in the stomach.

Chapter 25

Page 837

1. Vitamin B6 (pyridoxine) is an important coenzyme in the processes of deamination and transamination, the first step in processing amino acids in the cell. A deficiency in this vitamin would interfere with the ability to metabolize proteins.

2. Uric acid is the product of purine degradation in the body. The macromolecules that contain purines are the nucleic acids. An increase in uric acid levels could indicate increased breakdown of nucleic acids.

Page 844

1. After a meal that is high in carbohydrate, we would expect increased glycogenesis (the formation of glycogen) to occur in the liver.

2. Urea is formed from by-products of protein metabolism. During the postabsorptive state, many amino acids are being metabolized and the ammonia produce by deamination is converted to urea in the liver, thus the amount of urea in the blood increases.

3. During starvation, the body must use fat and protein reserves to supply the necessary energy. Some of the protein that is metabolized for energy is the gamma globulin fraction of the blood that is mostly antibodies. This loss of antibodies coupled with a lack of amino acids to synthesize new ones, renders an individual more susceptible to contracting a disease and less likely to recover from it.

Page 850

1. We would expect a person who is in a positive nitrogen balance to be adding muscle mass.

2. Bile salts are necessary for the digestion and absorption of fats and fat soluble vitamins. Vitamin A is a fat soluble vitamin. A decrease in the amount of bile salts in the bile would result in a decreased ability to absorb the vitamin A from food and result in a vitamin A deficiency.

Page 858

1. The BMR of a pregnant woman should be higher than the BMR of the woman in a nonpregnant state because of increased metabolism associated with support of the fetus as well as the added effect of fetal metabolism.

2. Vasoconstriction of peripheral vessels would decrease blood flow to the skin and decrease the amount of heat that the body can lose. As a result, the body temperature would increase.

3. Infants have higher surface area to volume ratios than adults, and the temperature-regulating mechanisms of the body are not fully functional at birth. As a result they must expend more energy to maintain body temperature and they get cold more easily than a normal adult.

Chapter 26

Page 869

1. The slit pores created by the podocytes will only allow substances less than 6–9 nm in size to pass into the capsule.

2. Damage to the juxtaglomerular apparatus portion of the nephrons would interfere with the normal control of blood pressure.

Page 876

1. The absorption of calcium ion by the kidney is an example of a countertransport mechanism in which sodium ions are traded for calcium. As more calcium is reabsorbed more sodium enters the filtrate, increasing the sodium ion concentration of the filtrate.

2. When the plasma concentration of a substance exceeds its tubular maximum, the excess is not reabsorbed and is excreted in the urine.

3. Decreases in blood pressure would reduce the blood hydrostatic pressure within the glomerulus and decrease the GFR.

Page 883

1. Aldosterone promotes sodium retention and potassium secretion at the kidneys. In response to increases in aldosterone levels, the potassium ion concentration of the urine would increase.

2. The secretion of hydrogen ions by the nephron involves a countertransport mechanism with sodium. If the concentration of sodium in the filtrate decreased fewer hydrogen ions could be secreted. This would result in a urine with a higher pH.

3. If the nephrons lacked a loop of Henle, the kidneys would not be able to form a concentrated urine.

Page 891

1. When the amount of sodium ion in the filtrate passing through the DCT is low, the cells of the macula densa are stimulated to release renin. Renin activates angiotensin and this brings about an increase in blood pressure.

2. An obstruction of the ureters would interfere with the passage of urine from the renal pelvis to the urinary bladder.

Chapter 27

Page 904

1. Drinking a pitcher of distilled water would temporarily lower the blood osmolarity (osmotic pressure). Since ADH release is triggered by increases in osmolarity, a decrease in osmolarity would lead to a decrease in the level of ADH in the blood.

2. Consuming a meal high in salt would temporarily increase the osmolarity of the ECF. As a result some of the water in the ICF would shift to the ECF.

3. Fluid loss through perspiration, urine formation, and respiration, would increase the osmolarity of body fluids.

Page 915

1. A decrease in the pH of body fluids would have a stimulating effect on respiratory center in the medulla. The result would be an increase in the rate of breathing. This would lead to an elimination of more carbon dioxide which would tend to cause the pH to increase.

2. Hydrogen ion secretion ceases at a pH of 4.5 to 5.0. Since the body usually needs to eliminate more hydrogen ions than the amount necessary to lower the pH to that critical range, the filtrate needs to be buffered. The buffers allow the filtrate to take more hydrogen ions without decreasing the pH below the critical level.

3. In a prolonged fast, fatty acids are mobilized and large numbers of ketone bodies are formed. These molecules are acids that lower the body's pH. This would eventually lead to a condition known as ketoacidosis.

4. In vomiting, large amounts of stomach acid are lost from the body. This acid is formed by the parietal cells of the stomach by taking hydrogen ions from the blood. Excessive vomiting would lead to excessive removal of hydrogen ions from the blood to produce the acid thus raising the body's pH and creating a condition called metabolic alkalosis.

Chapter 28

Page 929

1. The cremaster muscle as well as the dartos muscle would be relaxed on a warm day, so that the scrotal sac could descend away from the warmth of the body and cool the testes.

2. The acrosomal cap contains enzymes necessary for penetrating the cell layers around the ovum. Without these enzymes, fertilization would not occur.

3. A vasectomy has no effect on spermatogenesis (sperm production).

4. Relaxation of the arteries serving the penis will result in erection.

Page 936

1. A blockage of the uterine tube would cause sterility.

2. The acidic pH of the vagina helps to prevent bacterial, fungal, and protozoal infections in this area.

3. Blockage of a single lactiferous sinus would not interfere with milk moving to the nipple because each breast usually has between 15 and 20 lactiferous sinuses.

Page 942

1. If the LH surge did not occur during an ovulatory cycle, ovulation and corpus luteum formation would not occur.

2. Progesterone is responsible for the functional maturation and secretion of the endometrial lining. Blocking progesterone receptors would inhibit endometrial development and make it unsuitable for implantation.

3. A sudden decline in the levels of estrogen and progesterone during the menstrual cycle signals the beginning of the menses.

Page 948

1. After fertilization, the developing trophoblasts and then later on the placenta produce and release the hormone HCG. She is pregnant.

2. The hormone oxytocin from the posterior pituitary gland is released during suckling and is responsible for the process of milk letdown. It can also cause contractions of the smooth muscle of the uterus.

3. At menopause, circulating estrogen levels begin to drop. Estrogen has an inhibitory effect on GN-RH and FSH and as the level of estrogen declines the levels of these two hormones rise and remain elevated.

Chapter 29

Page 959

1. The spermatozoan only has to carry a set of chromosomes to the ovum for fertilization. The ovum needs to provide all of the nutrients, cellular machinery, and regulatory molecules to support the development of the zygote for at least a week.

2. A person who is heterozygous for tongue rolling would have one dominant gene and one recessive gene. The person's phenotype would be "tongue roller."

3. Joe is misinformed about the determination of an offspring's sex. Sex is determined by a pair of sex chromosomes. Females have two X chromosomes whereas males have one X and one Y chromosome. The mother can only donate an X chromosome to her offspring. The father, however, can donate either an X or a Y chromosome. It is therefore the father, not the mother that determines the sex of the child.

Page 971

1. A sperm cannot fertilize an ovum unless it has undergone capacitation in the female reproductive tract.

2. The inner cell mass of the blastocyst eventually develops into the embryo.

3. The yolk sac would thus affect the development and function of the circulatory system.

4. The two important functions of the placenta are: (1) to provide a source of nutrition, gas exchange, and wastes elimination and (2) to produce hormones that regulate growth and development and adapt the mother to the demands of pregnancy.

Page 979

1. Because: (1) blood flow through the placenta reduces the volume of blood in the systemic circuit and this must be replaced and (2) the fetus uses oxygen and nutrients and produces carbon dioxide and wastes, stimulates and this increases in maternal blood volume.

2. Progesterone reduces uterine contractions. A decrease in progesterone at any time during the pregnancy can lead to uterine contractions and in late pregnancy, labor.

Page 985

1. An increase in blood levels of GnRH, FSH, LH, and sex hormones would signal the onset of puberty.

2. Long-lived cells accumulate mutations errors and damage from injury or disease. Short-lived cells are usually replaced by the divisions of long-lived stem cells. Over time these cells, and other long-lived cells, die and can no longer be replaced.

Foreign Word Roots, Prefixes, Suffixes, and Combining Forms

Many of the words we use in everyday English have their roots in other languages, particularly Greek and Latin. This is especially true for anatomical terms, many of which were introduced into the anatomical literature by Greek and Roman anatomists. This list includes some of the foreign word roots, prefixes, suffixes, and combining forms that are part of many of the biological and anatomical terms you will see in this text.

Each entry starts with the commonly encountered form or forms of the prefix, suffix, or combining form followed by the word root (shown in italics) with its English translation. One example is given to illustrate the use of the prefix, suffix, or combining form, but there are many others, and you will see them as you progress through the text.

a-, *a-*, without: avascular
ab-, *ab*, from: abduct
-ac, *-akos*, pertaining to: cardiac
ad-, *ad*, to, toward: adduct
aden-, adeno-, *adenos*, gland: adenoid
af-, *ad*, toward: afferent
-al, *-alis*, pertaining to: brachial
-algia, *algos*, pain: neuralgia
ana-, *ana*, up, back: anaphase
andro-, *andros*, male: androgen
angio-, *angeion*, vessel: angiogram
anti-, ant-, *anti*, against: antibiotic
apo-, *apo*, from: apocrine
arachn-, *arachne*, spider: arachnoid
arthro-, *arthros*, joint: arthroscopy
-asis, -asia, state, condition: homeostasis
astro-, *aster*, star: astrocyte
atel-, *ateles*, imperfect: atelectasis
baro-, *baros*, pressure: baroreceptor
bi-, *bi-*, two: bifurcate
blast-, -blast, *blastos*, precursor: blastocyst
brachi-, *brachium*, arm: brachiocephalic
brady-, *bradys*, slow: bradycardia
bronch-, *bronchus*, windpipe, airway: bronchial
cardi-, cardio-, -cardia, *kardia*, heart: cardiac
-centesis, *kentesis*, puncture: thoracocentesis
cerebro-, *cerebrum*, brain: cerebrospinal
chole-, *chole*, bile: cholecystitis
chondro-, *chondros*, cartilage: chondrocyte
chrom-, chromo-, *chroma*, color: chromatin
circum-, *circum*, around: circumduction
-clast, *klastos*, broken: osteoclast
coel-, -coel, *koila*, cavity: coelom
contra-, *contra*, against: contralateral
cranio-, *cranium*, skull: craniosacral
cribr-, *cribrum*, sieve: cribriform
-crine, *krinein*, to separate: endocrine
cyst-, -cyst, *kystis*, sac: blastocyst
desmo-, *desmos*, band: desmosome
di-, *dis*, twice: disaccharide
dia-, *dia*, through: diameter
diure-, *diourein*, to urinate: diuresis
dys-, *dys-*, painful: dysmenorrhea
-ectasis, *ektasis*, expansion: atelectasis
ecto-, *ektos*, outside: ectoderm
ef-, *ex*, away from: efferent
emmetro-, *emmetros*, in proper measure: emmetropia
encephalo-, *enkephalos*, brain: encephalitis
end-, endo-, *endos*, inside: endometrium
entero-, *enteron*, intestine: enteric
epi-, *epi*, on: epimysium
erythema-, *erythema*, flushed (skin): erythematosis
erythro-, *erythros*, red: erythrocyte
ex-, *ex*, out, away from: exocytosis
ferr-, *ferrum*, iron: transferrin
-gen, -genic, *gennan*, to produce: mutagen
genicula-, *geniculum*, kneelike structure: geniculates
genio-, *geneion*, chin: geniohyoid
glosso-, -glossus, *glossus*, tongue: hypoglossal
glyco-, *glykys*, sugar: glycogen
-gram, *gramma*, record: myogram
-graph, -graphia, *graphein*, to write, record: electroencephalograph
gyne-, gyno-, *gynaikos*, woman: gynecologist
hem-, hemato-, *haima*, blood: hemopoiesis
hemi-, *hemi-*, half: hemisphere
hepato-, *hepaticus*, liver: hepatocyte
hetero-, *heteros*, other: heterosexual
histo-, *histos*, tissue: histology
holo-, *holos*, entire: holocrine
homeo-, homo-, *homos*, same: homeostasis
hyal-, hyalo-, *hyalos*, glass: hyaline
hydro-, *hydros*, water: hydrolysis
hyo-, *hyoeides*, U-shaped: hyoid
hyper-, *hyper*, above: hyperpolarization
ili-, ilio-, *ilium*: iliac
infra-, *infra*, beneath: infraorbital
inter-, *inter*, between: interventricular
intra-, *intra*, within: intracapsular
ipsi-, *ipse*, itself: ipsilateral
iso-, *isos*, equal: isotonic
-itis, *-itis*, inflammation: dermatitis
karyo-, *karyon*, body: megakaryocyte
kerato-, *keros*, horn: keratin
kino-, -kinin, *kinein*, to move: bradykinin
lact-, lacto-, -lactin, *lac*, milk: prolactin
-lemma, *lemma*, husk: plasmalemma
leuko-, *leukos*, white: leukocyte
liga-, *ligare*, to bind together: ligase
lip-, lipo-, *lipos*, fat: lipoid
lyso-, -lysis, -lyze, *lysis*, dissolution: hydrolysis
mal-, *mal*, abnormal: malabsorption
mamilla-, *mamilla*, little breast: mamillary
mast-, masto-, *mastos*, breast: mastoid
mega-, *megas*, big: megakaryocyte
mero-, *meros*, part: merocrine
meso-, *mesos*, middle: mesoderm
meta-, *meta*, after, beyond: metaphase
mono-, *monos*, single: monocyte
morpho-, *morphe*, form: morphology
-mural, *murus*, wall: intramural
myelo-, *myelos*, marrow: myeloblast
myo-, *mys*, muscle: myofilament
natri-, *natrium*, sodium: natriuretic
neur-, neuro-, *neuron*, nerve: neuromuscular
oculo-, *oculus*, eye: oculomotor
oligo-, *oligos*, little, few: oligopeptide
-ology, *logos*, the study of: physiology
-oma, *-oma*, swelling: carcinoma
onco-, *onkos*, mass, tumor: oncology

-opia, *ops*, eye: optic
-osis, *-osis*, state, condition: neurosis
osteon, osteo-, *os*, bone: osteocyte
oto-, *otikos*, ear: otoconia
para-, *para*, beyond: paraplegia
patho-, -path, -pathy, *pathos*, disease: pathology
pedia-, *paidos*, child: pediatrician
peri-, *peri*, around: perineurium
-phasia, *phasis*, speech: aphasia
-phil, -philia, *philus*, love: hydrophilic
-phobe, -phobia, *phobos*, fear: hydrophobic
-phylaxis, *phylax*, a guard: prophylaxis
physio-, *physis*, nature: physiology
-plasia, *plasis*, formation: dysplasia
platy-, *platys*, flat: platysma
-plegia, *plege*, a blow, paralysis: paraplegia
-plexy, *plessein*, to strike: apoplexy
podo, *podon*, foot: podocyte
-poiesis, *poiesis*, making: hemopoiesis
poly-, *polys*, many: polysaccharide
presby-, *presbys*, old: presbyopia
pro-, *pro*, before: prophase
pterygo-, *pteryx*, wing: pterygoid
pulp-, *pulpa*, flesh: pulpitis
retro-, *retro*, backward: retroperitoneal
-rrhea, *rhein*, flow, discharge: amenorrhea
sarco-, *sarkos*, flesh: sarcomere
scler-, sclero-, *skleros*, hard: sclera
semi-, *semis*, half: semitendinosus
-septic, *septikos*, putrid: antiseptic
-sis, state or condition: metastasis
som-, -some, *soma*, body: somatic
spino-, *spina*, spine, vertebral column: spinodeltoid
-stomy, *stoma*, mouth, opening: colostomy
stylo-, *stylus*, stake, pole: styloid
sub-, *sub*, below; subcutaneous
syn-, *syn*, together: synthesis
tachy-, *tachys*, swift: tachycardia
telo-, *telos*, end: telophase
therm-, thermo-, *therme*, heat: thermoregulation
-tomy, *temnein*, to cut: appendectomy
trans-, *trans*, through: transudate
-trophic, -trophin, -trophy, *trophikos*, nourishing: adrenocorticotrophic
tropho-, *trophe*, nutrition: trophoblast
tropo-, *tropikos*, turning: troponin
uro-, -uria, *ouron*, urine: glycosuria

Eponyms in Common Use

Eponym	*Equivalent Terms*	*Individual Referenced*
The cellular level of organization (Chapter 3)		
Golgi apparatus		Camillo Golgi (1844–1926), Italian histologist; shared Nobel Prize in 1906
Krebs cycle	Tricarboxylic or citric acid cycle	Hans Adolph Krebs (1900–1981), British biochemist; shared Nobel Prize in 1953
The skeletal system (Chapters 7–9)		
Colles' fracture		Abraham Colles (1773–1843), Irish surgeon
Haversian canals	Central canals	Clopton Havers (1650–1702), English anatomist and microscopist
Haversian systems	Osteons	
Pott's fracture		Percivall Pott (1713–1788), English surgeon
Volkmann's canals	Perforating canals	Alfred Wilhelm Volkmann (1800–1877), German surgeon
Wormian bones	Sutural bones	Olas Worm (1588–1654), Danish anatomist
The muscular system (Chapters 10–11)		
Achilles' tendon	Calcaneal tendon	Achilleus, hero of Greek mythology
Cori cycle		Carl Ferdinand Cori (1896–) and Gerty Theresa Cori (1896–1957), American biochemists; shared Nobel Prize in 1947
The nervous system (Chapters 12–16)		
Broca's center	Speech center	Pierre Paul Broca (1824–1880), French surgeon
Foramina of Luschka	Lateral foramina	Hubert von Luschka (1820–1875), German anatomist
Foramen of Magendie	Median foramen	François Magendie (1783–1855), French physiologist
Foramen of Munro	Interventricular foramen	John Cummings Munro (1858–1910), American surgeon
Nissl bodies		Franz Nissl (1860–1919), German neurologist
Purkinje cells	Basket cells	Johannes E. Purkinje (1787–1869), Czechoslovakian physiologist
Nodes of Ranvier		Louis Antoine Ranvier (1835–1922), French physiologist
Island of Reil	Insula	Johann Christian Reil (1759–1813), German anatomist
Fissure of Rolando	Central sulcus	Luigi Rolando (1773–1831), Italian anatomist
Schwann cells		Theodor Schwann (1810–1882), German anatomist
Aqueduct of Sylvius	Mesencephalic aqueduct	Jacobus Sylvius (Jacques Dubois, 1478–1555), French anatomist
Sylvian fissure	Lateral sulcus	Franciscus Sylvius (Franz de le Boë, 1614–1672), Dutch anatomist
Pons varolii	Pons	Costanzo Varolio (1543–1575), Italian anatomist
Sensory function (Chapter 17)		
Organ of Corti		Alfonso Corti (1822–1888), Italian anatomist
Eustachian tube	Pharyngotympanic tube	Bartolomeo Eustachio (1520–1574), Italian anatomist
Golgi tendon organs	Tendon organs	*See* Golgi apparatus *under* The Cellular Level
Hertz (Hz)		Heinrich Hertz (1857–1894), German physicist
Meibomian glands		Heinrich Meibom (1638–1700), German anatomist
Corpuscles of Meissner		Georg Meissner (1829–1905), German physiologist
Merkel's discs		Friedrich Siegismund Merkel (1845–1919), German anatomist
Pacinian corpuscles		Filippo Pacini (1812–1883), Italian anatomist
Ruffini's corpuscles		Angelo Ruffini (1864–1929), Italian anatomist
Canal of Schlemm		Friedrich S. Schlemm (1795–1858), German anatomist
Glands of Zeis		Edward Zeis (1807–1868), German ophthamologist
The endocrine system (Chapter 18)		
Islets of Langerhans	Pancreatic islets	Paul Langerhans (1847–1888), German pathologist

Eponym	Equivalent Terms	Individual Referenced
Interstitial cells of Leydig	Interstitial cells	Franz von Leydig (1821–1908), German anatomist
The cardiovascular system (Chapters 19–21)		
Bundle of His		Wilhelm His (1863–1934), German physician
Purkinje cells		*See under* The Nervous System
Starling's law		Ernest Henry Starling (1866–1927), English physiologist
Circle of Willis		Thomas Willis (1621–1675), English physician
The lymphatic system (Chapter 22)		
Hassall's corpuscles		Arthur Hill Hassall (1817–1894), English physician
Kupffer cells		Karl Wilhelm Kupffer (1829–1902), German anatomist
Langerhans cells		*See* Islets of Langerhans *under* The Endocrine System
Peyer's patches		Johann Conrad Peyer (1653–1712), Swiss anatomist
The respiratory system (Chapter 23)		
Adam's apple	Thyroid cartilage	Biblical reference
Bohr effect		Niels Bohr (1885–1962), Danish physicist; won 1922 Nobel Prize
Boyle's law		Robert Boyle (1621–1691), English physicist
Charles' law		Jacques Alexandre César Charles (1746–1823), French physicist
Dalton's law		John Dalton (1766–1844), English physicist
Henry's law		William Henry (1775–1837), English chemist
The digestive system (Chapter 24)		
Plexus of Auerbach	Myenteric plexus	Leopold Auerbach (1827–1897), German anatomist
Brunner's glands	Duodenal glands	Johann Conrad Brunner (1653–1727), Swiss anatomist
Kupffer cells	Stellate cells	*See under* The Lymphatic System
Crypts of Lieberkuhn	Intestinal crypts	Johann Nathaniel Lieberkuhn (1711–1756), German anatomist
Plexus of Meissner	Submucosal plexus	*See* Corpuscles of Meisser *under* Sensory Function
Sphincter of Oddi	Hepatopancreatic sphincter	Ruggero Oddi (1864–1913), Italian physician
Peyer's patches		*See under* The Lymphatic System
Duct of Santorini	Accessory pancreatic duct	Giovanni Domenico Santorini (1681–1737), Italian anatomist
Stensen's duct	Parotid duct	Niels Stensen (1638–1686), Danish physician/priest
Ampulla of Vater	Duodenal ampulla	Abraham Vater (1684–1751), German anatomist
Wharton's duct	Submandibular duct	Thomas Wharton (1614–1673), English physician
Foramen of Winslow	Epiploic foramen	Jacob Benignus Winslow (1669–1760), French anatomist
Duct of Wirsung	Pancreatic duct	Johann Georg Wirsung (1600–1643), German physician
The urinary system (Chapter 26)		
Bowman's capsule	Glomerular capsule	Sir William Bowman (1816–1892), English Physician
Loop of Henle		Friedrich Gustav Jakob Henle (1809–1885), German histologist
The reproductive system (Chapters 28–29)		
Bartholin's glands	Greater vestibular glands	Casper Bartholin, Jr. (1655–1738), Danish anatomist
Cowper's glands	Bulbourethral glands	William Cowper (1666–1709), English surgeon
Follopian tube	Uterine tube/oviduct	Gabriele Falloppio (1523–1562), Italian anatomist
Graafian follicle	Tertiary follicle	Reijnier de Graaf (1641–1673), Dutch physician
Interstitial cells of Leydig	Interstitial cells	*See under* The Endocrine System
Glands of Littre	Lesser vestibular glands	Alexis Littre (1658–1726), French surgeon
Sertoli cells	Sustentacular cells	Enrico Sertoli (1842–1910), Italian histologist

Abbreviations Used in This Text

ACh	acetylcholine
AChE	acetylcholinesterase
ACTH	adrenocorticotrophic hormone
ADH	antidiuretic hormone
ADP	adenosine diphosphate
AIDS	acquired immunodeficiency syndrome
AMP	adenosine monophosphate
ANP	atrial natriuretic peptide
ANS	autonomic nervous system
AP	arterial pressure
ARC	AIDS-related complex
ARDS	adult respiratory distress syndrome
atm	atmospheric pressure
ATP	adenosine triphosphate
ATPase	adenosine triphosphatase
AV	atrioventricular
BCDF	B cell differentiation factor
BCGF	B cell growth factor
BMR	basal metabolic rate
B_{op}	blood osmotic pressure
bpm	beats per minute
C	kilocalorie; centigrade
CAD	coronary artery disease
cAMP	cyclic-AMP
CAPD	continuous ambulatory peritoneal dialysis
CCK	cholecystokinin
CHF	congestive heart failure
C_{hp}	capillary hydrostatic pressure
CNS	central nervous system
CO	cardiac output; carbon dioxide
CoA	coenzyme A
COMT	cathecol-O-methyltransferase
COPD	chronic obstructive pulmonary disease
CP	creatine phosphate
CPK or CK	creatine phosphokinase
CPM	continual passive motion
CPR	cardiopulmonary resuscitation
CRF	chronic renal failure
CRH	corticotrophin-releasing hormone
CSF	cerebrospinal fluid; colony-stimualting factors
CT	computed tomography; calcitonin
CVA	cerebrovascular accident
CVS	cardiovascular system
DC	Doctor of Chiropractory
DCT	distal convoluted tubule
DDST	Denver Developmental Screening Test
DIC	disseminated intravascular coagulation
DJD	degenerative joint disease
DMD	Duchenne's muscular dystrophy
DNA	deoxyribonucleic acid
DO	Doctor of Osteopathy
DOPA	dopamine
DPG	diphosphoglycerate
DPM	Doctor of Podiatric Medi-cine
DSR	dynamic spatial reconstruction
ECF	extracellular fluid
ECG	electrocardiogram
EDV	end-diastolic volume
EEG	electroencephalogram
EKG	electrocardiogram
ELISA	enzyme-linked immunoabsorbent assay
EPSP	excitatory postsynaptic potential
ESV	end-systolic volume
ETS	electron transport system
FAD	flavin adenine dinucleotide
FES	functional electrical stimulation
FMN	flavin mononucleotide
FSH	follicle-stimulating hormone
GABA	gamma aminobutyric acid
GAS	general adaptation syndrome
GC	glucocorticoids
GFR	glomerular filtration rate
GH	growth hormone
GH-IH	growth hormone-inhibiting hormone
G_{hp}	glomerular hydrostatic pressure
GH-RH	growth hormone-releasing hormone
GIP	glucose-dependent insulinotropic hormone
GnRH	gonadotrophin-releasing hormone
GTP	guanosine triphosphate
Hb	hemoglobin
HCG	human chorionic gonadotrophin
HCL	hydrochloric acid
HDL	high-density lipoprotein
HDN	hemolytic disease of the newborn
HIV	human immunodeficiency virus
HLA	human leukocyte antigen
HMD	hyaline membrane disease
HP	hydrostatic pressure
HPL	human placental lactogen
HR	heart rate
HTLV-III	human T cell lymphotrophic virus, type III (=HIV)
Hz	Hertz
ICF	intracellular fluid
ICSH	interstitial cell-stimulating hormone, also called LH
IH	inhibiting hormone
IM	intramuscular
IPSP	inhibitory postsynaptic potential
ISF	interstitial fluid
kc	kilocalorie
LDH	lactate dehydrogenase
LDL	low-density lipoprotein
L-DOPA	levdopa
LH	luteinizing hormone
LM	light micrograph
LSD	lysergic acid diethylamide
MAO	mono-amine-oxidase
MAP	mean arterial pressure
MC	mineralocorticoid
MD	Doctor of Medicine
MI	myocardial infarction
mm Hg	millimeters of mercury
mOsm	milliosmoles
MRI	magnetic resonance imaging
mRNA	messenger RNA
MS	multiple sclerosis
MSH	melanocyte-stimulating hormone
MSH-IH	melanocyte-stimulating hormone–inhibiting hormone
NAD	nicotinamide adenine dinucleotide
NE	norepinephrine
NRDS	neonatal respiratory distress syndrome
OP	osmotic pressure
Osm	osmoles
PAC	premature atrial contractions
PAT	paroxysmal atrial tachycardia
PBI	protein-bound iodine
PCT	proximal convoluted tubule
PCV	packed cell volume
PEEP	positive end-expiratory pressure
PET	positron-emission tomography
P_f	filtration pressure
PFC	perfluorochemical emulsions
PIH	prolactin-inhibiting hormone
PMN	polymorphonuclear leukocyte
PNS	peripheral nervous system
PR	peripheral resistance
PRL	prolactin
psi	pounds per square inch
PTA	post-traumatic amnesia; plasma thromboplastin antecedent

PTC	phenylthiocarbamide
PTH	parathormone
PVC	premature ventricular contraction
RAS	reticular activating system
RBC	red blood cell
REM	rapid eye movement
RER	rough endoplasmic reticulum
RH	releasing hormone
RHD	rheumatic heart disease
RNA	ribonucleic acid
rRNA	ribosomal RNA
SA	sinoatrial
SCID	severe combined immunodeficiency
SEM	scanning electron micrograph
SER	smooth endoplasmic reticulum
SGOT	serum glutamic oxaloacetic transaminase
SIADH	syndrome of inappropriate ADH secretion
SIDS	sudden infant death syndrome
SLE	systemic lupus erythematosus
SV	stroke volume
SVC	superior vena cava
T_3	triiodothyronine
T_4	tetraiodothyronine, also called thyroxine
TB	tuberculosis
TBG	thyroid-binding globulin
TEM	transmission electron micrograph
TIA	transient ischemic attack
T_m	transport (tubular) maximum
t-PA	tissue plasminogen activator
TRH	thryotrophin-releasing hormone
tRNA	transfer RNA
TSH	thyroid stimulating hormone
TX	thyroxine
U.S.	United States
UTI	urinary tract infection
UTP	uridine triphosphate
UV	ultraviolet
VF	ventricular fibrillation
VLDL	very low density lipoprotein
VPRC	volume of packed red cells
VT	ventricular tachycardia
WBC	white blood cell

Glossary of Key Terms

abdomen: Region of trunk bounded by the diaphragm and pelvis.
abdominopelvic cavity: portion of the ventral body cavity that contains abdominal and pelvic subdivisions.
abducens (ab-DŪ-senz): Cranial nerve VI, innervates the lateral rectus muscle of the eye.
abduction: Movement away from the midline.
abortion: Premature loss or expulsion of an embryo or fetus.
abruptio placentae (ab-RUP-shē-ō pla-SEN-tē): Premature loss of placental connection to uterus, leading to maternal hemorrhaging and shock.
abscess: A localized collection of pus within a damaged tissue.
absorption: The active or passive uptake of gases, fluids, or solutes.
accommodation: Alteration in the curvature of the lens to focus an image on the retina; decrease in receptor sensitivity or perception following chronic stimulation.
acetabulum (a-se-TAB-ū-lum): Fossa on lateral aspect of pelvis that accommodates the head of the femur.
acetyl group: $CH_3C{=}O$.
acetylcholine (ACh) (as-ē-til-kō-lēn): Chemical neurotransmitter in the brain and PNS; dominant neurotransmitter in the PNS, released at neuromuscular junctions and synapses of the parasympathetic division.
acetylcholinesterase (AChE): Enzyme found in the synaptic cleft, bound to the postsynaptic membrane, and in tissue fluids; breaks down and inactivates ACh molecules.
acetyl-CoA: An acetyl group bound to Coenzyme A, a participant in the anabolic and catabolic pathways for carbohydrates, lipids, and many amino acids.
achalasia (āk-a-LĀ-zē-a): Condition that develops when the lower esophageal sphincter fails to dilate, and ingested materials cannot enter the stomach.
Achilles tendon: Calcaneal tendon.
acid: A compound whose dissociation in solution releases a hydrogen ion and an anion; an acid solution has a pH below 7.0 and contains an excess of hydrogen ions.
acidosis (a-sid-Ō-sis): An abnormal physiological state characterized by a plasma pH below 7.35.
acinus/acini (AS-i-nī): Histological term referring to a blind pocket, pouch, or sac.
acne: Condition characterized by inflammation of sebaceous glands and follicles, commonly affects adolescents and most often involves the face.
acoustic: Pertaining to sound or the sense of hearing.
acquired immunodeficiency syndrome (AIDS): A disease caused by the **human immunodeficiency virus (HIV),** characterized by destruction of helper T cells and a resulting severe impairment of the immune response.
acromegaly: Condition caused by overproduction of growth hormone in the adult, characterized by thickening of bones and enlargement of cartilages and other soft tissues.
acromion (a-KRŌ-mē-on): Continuation of the scapular spine that projects above the capsule of the scapulohumeral joint.
acrosomal cap (ak-rō-SŌ-mal): Membranous sac at the tip of a sperm cell that contains hyaluronic acid.
actin: Protein component of microfilaments; form thin filaments in skeletal muscles and produce contractions of all muscles through interaction with thick (myosin) filaments; *see* **sliding filament theory.**
action potential: A conducted change in the transmembrane potential of excitable cells, initiated by a change in the membrane permeability to sodium ions; *see also* **nerve impulse.**
activation energy: The energy required to initiate a specific chemical reaction.
active transport: The ATP-dependent absorption or excretion of solutes across a cell membrane.
acute: Sudden in onset, severe in intensity, and brief in duration.
adaptation: Alteration of pupillary size in response to changes in light intensity; in CNS often used as a synonym for accommodation; physiological responses that produce acclimatization.
Addison's disease: Condition resulting from hyposecretion of glucocorticoids, characterized by lethargy, weakness, hypotension, and increased skin pigmentation.
adduction: Movement toward the axis or midline of the body as viewed in the anatomical position.
adenine: A purine, one of the nitrogen bases in the nucleic acids RNA and DNA.
adenohypophysis (ad-e-nō-hī-POF-i-sis): The anterior portion of the pituitary gland, also called the **anterior pituitary** or the **pars distalis.**
adenoid: The pharyngeal tonsil.
adenosine: A nucleoside consisting of adenine and a 5-carbon sugar.
adenosine phosphate (AMP): A nucleotide consisting of adenine plus a phosphate group (PO_4); also known as adenosine monophosphate.
adenosine diphosphate (ADP): Adenosine with two phosphate groups attached.
adenosine triphosphate (ATP): A high-energy compound consisting of adenosine with three phosphate groups attached; the third is attached by a high-energy bond.
adenylcyclase: An enzyme bound to the inner surfaces of cell membranes that can convert ATP to cyclic AMP.
adhesion: Fusion of two mesenterial layers following damage or irritation of their opposing surfaces.
adipocyte (AD-i-pō-sīt): A fat cell.
adipose tissue: Loose connective tissue dominated by adipocytes.
adrenal cortex: Superficial portion of adrenal gland that produces steroid hormones.
adrenal gland: Small endocrine gland secreting steroids and catecholamines, located superior to each kidney.
adrenal medulla: Core of the adrenal gland; a modified sympathetic ganglion that secretes catecholamines into the blood following sympathetic activation.
adrenergic (AD-ren-er-jik): A synaptic terminal that releases norepinephrine when stimulated.
adrenocortical hormone: Any of the steroids produced by the adrenal cortex.
adrenocorticotrophic hormone (ACTH): Hormone that stimulates the production and secretion of glucocorticoids by the zona fasciculata of the adrenal cortex; released by the anterior pituitary in response to CRF.
adventitia (ad-ven-TISH-a): Superficial layer of connective tissue surrounding an internal organ; fibers are continuous with those of surrounding tissues, providing support and stabilization.
aerobic: Requiring the presence of oxygen.
aerobic glycolysis: The complete breakdown of glucose into carbon dioxide and water, via pyruvic acid; a process that yields large amounts of ATP but requires mitochondria and oxygen.
afferent: Toward.
afferent arteriole: An arteriole bringing blood to the glomerulus of the kidney.
afferent fiber: Axons carrying sensory information to the CNS.
afterbirth: The distal portions of the umbilical cord and

placenta that are ejected from the uterus during the placental stage of labor.
agglutination (a-glōō-ti-NA-shun): Aggregation of red blood cells due to interactions between surface agglutinogens and plasma agglutinins.
agglutinins (a-GLŌŌ-ti-ninz): Immunoglobulins in plasma that react with antigens on the surfaces of foreign red blood cells when donor and recipient differ in blood type.
agglutinogens (a-GLOO-tin-o-jenz): Antigens on the surfaces of red blood cells whose presence and structure are genetically determined.
agonist: A muscle responsible for a specific movement.
agranular: Without granules; **agranular leukocytes** are monocytes and lymphocytes; the **agranular reticulum** is an intracellular organelle that synthesizes and stores carbohydrates and lipids.
AIDS: *see* **Acquired immunodeficiency syndrome.**
AIDS-related complex (ARC): Early symptoms of HIV infection, consisting chiefly of lymphadenopathy, fevers, and chronic nonfatal infections.
alba, albicans, albuginea (AL-bi-kanz) (al-bū-JIN-ē-a): White.
albinism: Absence of pigment in hair and skin caused by inability of body to produce melanin.
aldosterone: A mineralocorticoid (steroid) produced by the zona glomerulosa of the adrenal cortex that stimulates sodium and water conservation at the kidneys; secreted in response to the presence of angiotensin II.
aldosteronism: Condition caused by the oversecretion of aldosterone, characterized by fluid retention, edema, and hypertension.
alkalosis (al-kah-LŌ-sis): Condition characterized by a plasma pH of greater than 7.45, and associated with relative deficiency of hydrogen ions or an excess of bicarbonate ions.
allantois (a-LAN-tō-is): One of the extraembryonic membranes; it provides vascularity to the chorion and is therefore essential to placenta formation; the proximal portion becomes the urinary bladder.
alleles (a-LĒLZ): Alternate forms of a particular gene.
allergen: An antigenic compound that produces a hypersensitivity response.
alpha-blockers: Drugs that prevent stimulation of **alpha receptors.**
alpha cells: Cells in the pancreatic islets that secrete glucagon.
alpha receptors: Membrane receptors sensitive to norepinephrine or epinephrine; stimulation usually results in excitation of the target cell.
alveolar sac: An air-filled chamber that supplies air to several alveoli.
alveolus/alveoli (al-VĒ-o-lī): Blind pockets at the end of the respiratory tree, lined by a simple squamous epithelium and surrounded by a capillary network; gas exchange with the blood occurs here.
Alzheimer's disease: Disorder resulting from degenerative changes in populations of neurons in the cerebrum, causing dementia characterized by problems with attention, short-term memory, and emotions.
amacrine cells (AM-a-krīn): Modified neurons in the retina that facilitate or inhibit communication between bipolar and ganglion cells.
amenorrhea (ā-men-ō-RĒ-a): Failure to commence menstruation at adolescence or the cessation of menstruation.
amination: The attachment of an amine group to a carbon chain; performed by a variety of cells and important in the synthesis of **amino acids.**
amine group: NH_2, also called an amino group.
amino acids: Organic compounds whose chemical structure can be summarized as $R\text{-}CHNH_2COOH$.
amnesia: Temporary or permanent memory loss.
amniocentesis: Sampling of amniotic fluid for analytical purposes; used to detect certain forms of genetic abnormalities.
amnion (AM-nē-on): One of the extraembryonic membranes; surrounds the developing embryo/fetus.
amniotic fluid (am-nē-OT-ik): Fluid that fills the amniotic cavity; provides cushioning and support for the embryo/fetus.
amphiarthrosis (am-fē-ar-THRŌ-sis): An articulation that permits a small degree of independent movement.
amphicytes (AM-fi-sīts): Supporting cells that surround neurons in the PNS; also called satellite cells.
amphimixis (am-fi-MIK-sis): The fusion of male and female pronuclei following fertilization.
ampulla/ampullae (am-PUL-la): A localized dilation in the lumen of a canal or passageway.
amygdala/amygdaloid nucleus (ah-MIG-da-loid): A cerebral nucleus that is a component of the limbic system and acts as an interface between that system, the cerebrum, and sensory systems.
amylase: An enzyme that breaks down polysaccharides, produced by the salivary glands and pancreas.
anabolism (a-NAB-ō-lizm): The synthesis of complex organic compounds from simpler precursors.
anaerobic: Without oxygen.
anaerobic glycolysis: The cytoplasmic breakdown of glucose into lactic acid via pyruvic acid, with a net gain of 2 ATP.
analgesia: Relief from pain.
anal triangle: The posterior subdivision of the perineum.
anamnestic response (an-am-NES-tic): Sudden and exaggerated production of antibodies following second exposure to a specific antigen, due to the activation of memory B cells.
anaphase (AN-ā-fāz): Mitotic stage in which the paired chromatids separate and move toward opposite ends of the spindle apparatus.
anaphylaxis (a-na-fi-LAK-sis): Hypersensitivity reaction due to antigen binding to immunoglobulins (IgE) on the surfaces of mast cells; mast cell release of histamine, serotonin, and prostaglandins then causes widespread inflammation; a sudden decline in blood pressure may occur, producing a condition known as anaphylactic shock.
anastomosis (a-nas-to-MŌ-sis): The joining of two tubes, usually referring to a connection between two peripheral vessels without an intervening capillary bed.
anatomy (a-NAT-o-mē): The study of the structure of the body.
anatomical position: An anatomical reference position, the body viewed from the anterior surface with the palms facing forward; supine.
anaxonic neuron (an-ak-SON-ik): A CNS neuron that has many processes but no apparent axon.
androgen (AN-drō-jen): A steroid sex hormone primarily produced by the interstitial cells of the testis, and manufactured in small quantities by the adrenal cortex in either sex.
anemia (a-NE-mē-ah): Condition marked by a reduction in the hematocrit and/or hemoglobin content of the blood.
anencephaly (an-en-SEF-a-lē): Developmental defect characterized by incomplete development of cerebral hemispheres and cranium.
anesthesia: Total or partial loss of sensation from a region of the body.
aneurysm (AN-ū-rizm): A weakening and localized dilation in the wall of a blood vessel.
angiogram (AN-jē-ō-gram): An X-ray image of circulatory pathways.
angiography: X-ray examination of vessel distribution following the introduction of radiopaque substances into the bloodstream.

angiotensin I, II: Angiotensin II is a hormone that causes an elevation in systemic blood pressure, stimulates secretion of aldosterone, promotes thirst, and causes the release of ADH; a converting enzyme in the pulmonary capillaries converts angiotensin I to angiotensin II.
angiotensinogen: Blood protein produced by the liver that is converted to angiotensin I by the enzyme renin.
angstrom (Å): Unit of measure equivalent to 0.1 nanometers.
anion (AN-ī-on): An ion bearing a negative charge.
ankyloglossia (ang-kī-lō-GLOS-ē-a): Condition characterized by an overly robust and restrictive lingual frenulum.
annulus (AN-ū-lus): A cartilage or bone shaped like a ring.
anorectal canal (ā-nō-REC-tal): The distal portion of the rectum that contains the rectal columns and ends at the anus.
anorexia nervosa: An eating disorder marked by a loss of appetite and pronounced weight loss.
anoxia (a-NOKS-ē-a): Tissue oxygen deprivation.
antagonist: A muscle that opposes the movement of an agonist.
antebrachium: The forearm.
anteflexion (an-te-FLEK-shun): Normal position of the uterus, with the superior surface bent forward.
anterior: On or near the front or ventral surface of the body.
anterior pituitary: *See* **pituitary.**
anterograde amnesia: Inability to store memories of events that occur after a specific incident or time.
anthracosis (an-thra-KŌ-sis): "Black lung disease," deterioration of respiratory exchange efficiency due to chronic inhalation of coal dust.
antibiotic: Chemical agent that selectively kills pathogenic microorganisms.
antibody (AN-ti-bod-ē): A globular protein produced by plasma cells that will bind to specific antigens and promote their destruction or removal from the body.
anticholinesterase: Chemical compound that blocks the action of acetylcholinesterase and causes prolonged and intensive stimulation of postsynaptic membranes.
anticoagulant: Compound that slows or prevents clot formation by interfering with the clotting system.
anticodon: Triplet of nitrogen bases on a tRNA molecule that interacts with an appropriate codon on a strand of mRNA.
antidiuretic hormone (ADH) (an-tī-dī-ū-RET-ik): Hormone synthesized in the hypothalamus and secreted at the posterior pituitary; causes water retention at the kidneys, and an elevation of blood pressure.
antigen: A substance capable of inducing the production of antibodies.
antigen-antibody complex: The combination of an antigen and a specific antibody.
antigenic determinant site: A portion of an antigen that can interact with an antibody molecule.
antihistamine (an-ti-HIS-ta-mēn): A chemical agent that blocks the action of histamine on peripheral tissues.
antipyretic agents: Chemicals that reduce fever.
antrum (AN-trum): A chamber or pocket.
anuria (a-nŪ-rē-a): Cessation of urine production.
anus: External opening of the anorectal canal.
aorta: Large, elastic artery that carries blood away from the left ventricle, and into the systemic circuit.
aortic reflex: Baroreceptor reflex triggered by increased aortic pressures; leads to a reduction in cardiac output and a fall in systemic pressure.
Apgar: A test used to assess the neurological status of a newborn infant.
aphasia: Inability to speak.
apnea (AP-nē-a): Cessation of breathing.
apneustic center (ap-NŪ-stik): Respiratory center whose chronic activation would lead to apnea at full inspiration.
apocrine secretion: Mode of secretion where the glandular cell sheds portions of its cytoplasm.
aponeurosis/aponeuroses (ap-ō-nū-RŌ-sēz): A broad tendinous sheet that may serve as the origin or insertion of a skeletal muscle.
apoplexy: A stroke (cerebrovascular accident).
appendicitis: Inflammation of the appendix.
appendicular: Pertaining to the arms or legs.
appendix: A blind tube connected to the cecum of the large intestine.
appositional growth: Enlargement by the addition of cartilage or bony matrix to the outer surface.
aqueous humor: Fluid similar to perilymph or CSF that fills the anterior chamber of the eye.
arachidonic acid: One of the essential fatty acids.
arachnoid (a-RAK-noid): The middle meninges that encloses CSF and protects the central nervous system.
arachnoid villi: Processes of the arachnoid that project into the superior sagittal sinus; sites where CSF enters the venous circulation.
arbor vitae: Central, branching mass of white matter inside the cerebellum.
arcuate (AR-kū-āt): Curving.
areflexia (ā-re-flek-sē-a): Absence of normal reflex responses to stimulation.
areola (a-RĒ-ō-la): Pigmented area that surrounds the nipple of a breast.
areolar: Containing minute spaces, as in areolar connective tissue.
arrector pili (ar-REK-tor PI-li): Smooth muscles whose contractions cause piloerection.
arrhythmias (a-RITH-mē-as): Abnormal patterns of cardiac contractions.
arteriole (ar-TE-rē-ol): A small arterial branch that delivers blood to a capillary network.
artery: A blood vessel that carries blood away from the heart and toward a peripheral capillary.
arthritis (ar-THRĪ-tis): Inflammation of a joint.
arthroscope: Fiberoptic device intended for visualizing the interior of joints; may also be used for certain forms of joint surgery.
articular: Pertaining to a joint.
articular capsule: Dense collagen fiber sleeve that surrounds a joint and provides protection and stabilization.
articular cartilage: Cartilage pad that covers the surface of a bone inside a joint cavity.
articulation (ar-tik-ū-LĀ-shun): A joint; formation of words.
arytenoid cartilages (ar-i-TĒ-noid): A pair of small cartilages in the larynx.
ascending tract: A tract carrying information from the spinal cord to the brain.
ascites (a-SĪ-tēz): Overproduction and accumulation of peritoneal fluid.
aseptic: Free from pathogenic contamination.
asphyxia: Unconsciousness due to oxygen deprivation at the CNS.
aspirate: To remove or obtain by suction; to inhale.
association areas: Cortical areas of the cerebrum responsible for integration of sensory inputs and/or motor commands.
association neuron: *See* **interneuron.**
asthma (AZ-ma): Reversible constriction of smooth muscles around respiratory passageways, frequently caused by an allergic response.
astigmatism: Visual disturbance due to an irregularity in the shape of the cornea.
astrocyte (AS-trō-sīt): One of the glial cells in the CNS; responsible for the blood-brain barrier.
ataxia: Failure to coordinate muscular activities normally.
atelectasis (at-e-LEK-ta-sis): Collapse of a lung or a portion of a lung.

atherosclerosis (ath-er-ō-skle-RŌ-sis): Formation of fatty plaques in the walls of arteries, leading to circulatory impairment.
atom: The smallest stable unit of matter.
atomic number: The number of protons in the nucleus of an atom.
atomic weight: Roughly, the average total number of protons and neutrons in the atoms of a particular element.
atresia (a-TRĒ-zē-a): Closing of a cavity, or its incomplete development; used in the reproductive system to refer to the degeneration of developing ovarian follicles.
atria: Thin-walled chambers of the heart that receive venous blood from the pulmonary or systemic circuits.
atrial natriuretic factor (nā-tre-ū-RET-ik): Hormone released by specialized atrial cardiocytes when they are stretched by an abnormally large venous return; promotes fluid loss and reductions in blood pressure and venous return.
atrial reflex: Reflexive increase in heart rate following an increase in venous return; due to mechanical and neural factors; also called the **Bainbridge reflex.**
atrioventricular (AV) node (ā-trē-ō-ven-TRIK-ū-lar): Specialized cardiocytes that relay the contractile stimulus to the bundle of His, the bundle branches, the Purkinje fibers, and the ventricular myocardium; located at the boundary between the atria and ventricles.
atrioventricular valve: One of the valves that prevent backflow into the atria during ventricular systole.
atrophy (AT-rō-fē): Wasting away of tissues from lack of use or nutritional abnormalities.
auditory: Pertaining to the sense of hearing.
auditory ossicles: The bones of the middle ear: malleus, incus, and stapes.
autoantibodies: Antibodies that react with antigens on the surfaces of one's own cells and tissues.
autodigestion: The digestion of tissues by digestive acids or enzymes from the stomach or pancreas.
autoimmunity: Immune system sensitivity to normal cells and tissues, resulting in the production of autoantibodies.
autolysis: Destruction of a cell due to the rupture of lysosomal membranes in its cytoplasm.
automatic bladder: Reflex micturition following stimulation of stretch receptors in the bladder wall; seen in patients who have lost motor control of the lower body.
automaticity: Spontaneous depolarization to threshold, a characteristic of cardiac pacemaker cells.
autonomic ganglion: A collection of visceral motor neurons outside the CNS.
autonomic nerve: A peripheral nerve consisting of preganglionic or postganglionic autonomic fibers.
autonomic nervous system (ANS): Centers, nuclei, tracts, ganglia, and nerves involved in the unconscious regulation of visceral functions; includes components of the CNS and PNS.
autopsy: Detailed examination of a body after death, usually performed by a pathologist.
autoregulation: Alterations in activity that maintain homeostasis in direct response to changes in the local environment; does not require neural or endocrine control.
autosomal (aw-to-SŌ-mal): Chromosomes other than the X or Y chromosomes that determine the genetic sex of an individual.
avascular (ā-VAS-kū-lar): Without blood vessels.
avitaminosis (ā-vī-ta-min-Ō-sis): Condition caused by inadequate intake of one or more essential vitamins.
avulsion: An injury involving the violent tearing away of body tissues.
axilla: The armpit.
axolemma: The cell membrane of an axon, continuous with the cell membrane of the soma and dendrites and distinct from any glial cell coverings.
axon: Elongate extension of a neuron that conducts an action potential away from the soma and toward the synaptic terminals.
axon hillock: Portion of the neural soma adjacent to the initial segment.
axoplasm (AK-so-plazm): Cytoplasm within an axon.
azotemia (a-zo-TĒ-mē-a): Condition resulting from impaired kidney function and the retention of nitrogenous wastes, especially urea.
Babinski sign: Reflexive dorsiflexion of the toes following stroking of the plantar surface of the foot; positive reflex (Babinski sign) is normal up to age 1.5 yrs; thereafter a positive reflex indicates damage to descending tracts.
bacteria: Single-celled microorganisms, some pathogenic, that are common in the environment.
Bainbridge reflex: *See* **atrial reflex.**
baroreception: Ability to detect changes in pressure.
baroreceptor reflex: A reflexive change in cardiac activity in response to changes in blood pressure.
baroreceptors (bar-ō-rē-SEP-torz): Receptors responsible for baroreception.
basal metabolic rate: The resting metabolic rate of a normal fasting subject under homeostatic conditions.
base: A compound whose dissociation releases a hydroxyl ion (OH^-) or removes a hydrogen ion from the solution.
basement membrane: A layer of filaments and fibers that attach an epithelium to the underlying connective tissue.
basilar membrane: Membrane that supports the organ of Corti and separates the cochlear duct from the scala tympani in the inner ear.
basophils (BĀ-sō-filz): Circulating granulocytes (WBCs) similar in size and function to tissue mast cells.
B cells: Lymphocytes capable of differentiating into the plasma cells that produce antibodies.
benign: Not malignant.
beta cells: Cells of the pancreatic islets that secrete insulin in response to elevated blood sugar concentrations.
beta oxidation: Fatty acid catabolism that produces molecules of acetyl-CoA.
beta receptors: Membrane receptors sensitive to epinephrine; stimulation may result in excitation or inhibition of the target cell.
bicarbonate ions: HCO_3^-; anion components of the bicarbonate buffer system.
bicuspid (bī-KUS-pid): A sharp, conical tooth, also called a canine tooth.
bicuspid valve: The left AV valve, also known as the **mitral valve.**
bifurcate: To branch into two parts.
bile: Exocrine secretion of the liver that is stored in the gallbladder and ejected into the duodenum.
bile salts: Steroid derivatives in the bile, responsible for the emulsification of ingested lipids.
bilirubin (bil-ē-ROO-bin): A reddish pigment, a product of hemoglobin catabolism.
bioenergetics: The analysis of energy production and utilization by living cells.
biofeedback: Using artificial signals to provide feedback about unconscious, visceral motor activities.
biopsy: The removal of a small sample of tissue for pathological analysis.
bipennate: A muscle whose fibers are arranged on either side of a common tendon.
bladder: A muscular sac that distends as fluid is stored, and whose contraction ejects the fluid at an appropriate time; used alone, the term usually refers to the urinary bladder.
blastocoele (BLAS-tō-sēl): Fluid-filled cavity within a blastocyst.
blastocyst (BLAS-tō-sist): Early stage in the developing embryo, consisting of an outer trophoblast and an inner cell mass.

blastodisc (BLAS-tō-disk): Later stage in the development of the inner cell mass; it includes the cells that will form the embryo.
blastomere (BLAS-tō-mēr): One of the cells in the morula, a collection of cells produced by the division of the zygote.
blockers/blocking agents: Drugs that block membrane pores or prevent binding to membrane receptors.
blood-brain barrier: Isolation of the CNS from the general circulation; primarily the result of astrocyte regulation of capillary permeabilities.
blood clot: A network of fibrin fibers and trapped blood cells.
blood pressure: A force exerted against the vascular walls by the blood, as the result of the push exerted by cardiac contraction and the elasticity of the vessel walls. It is usually measured along one of the muscular arteries, with systolic pressure measured during ventricular systole, and diastolic pressure during ventricular diastole.
blood-testis barrier: Isolation of the seminiferous tubules from the general circulation, due to the activities of the sustentacular (Sertoli) cells.
Bohr effect: Increased oxygen release by hemoglobin in the presence of elevated carbon dioxide levels.
boil: An abscess of the skin, usually involving a sebaceous gland.
bolus: A compact mass; usually refers to compacted ingested material on its way to the stomach.
bone: *See* **osseous tissue.**
botulinus toxin (bot-ū-LĪ-nus): A toxin produced by the anaerobic bacterium *Clostridium botulinum*, sometimes a source of severe food poisoning.
bowel: The intestinal tract.
Boyle's law: The principle that in a gas, pressure and volume are inversely related.
brachial: Pertaining to the upper arm.
brachial plexus: Network formed by branches of spinal nerves C_5-T_1 en route to innervate the arm.
brachium: The upper arm.
bradycardia (brad-ē-KAR-dē-a): Slow heart rate, below 50 bpm.
brain stem: The brain minus the cerebrum and cerebellum.
brevis: Short.
Broca's center: The speech center of the brain, usually found on the neural cortex of the left cerebral hemisphere.
bronchial tree: The trachea, bronchi, and bronchioles.
bronchitis (bronq-KĪ-tis): Inflammation of the bronchial passageways.
bronchodilation: Dilation of the bronchial passages; may be caused by sympathetic stimulation.
bronchodilators (brong-kō-dī-LĀ-torz): Drugs that produce bronchodilation; some are used clinically in treating asthma.
bronchoscope: A fiberoptic instrument used to examine the bronchial passageways.
bronchus/bronchi: One of the branches of the bronchial tree between the trachea and bronchioles.
buccal (BUK-al): Pertaining to the cheeks.
buffer: A compound that stabilizes the pH of a solution by removing or releasing hydrogen ions.
buffer system: Interacting compounds that prevent increases or decreases in the pH of body fluids; includes the bicarbonate buffer system, the phosphate buffer system, and the protein buffer system.
bulbar: Pertaining to the brain stem.
bulbourethral glands (bul-bō-ū-RĒ-thral): Mucous glands at the base of the penis that secrete into the penile urethra; also called Cowper's glands.
bundle branches: Specialized conducting cells in the ventricles that carry the contractile stimulus from the bundle of His to the Purkinje fibers.
bundle of His (hiss): Specialized conducting cells in the interventricular septum that carry the contracting stimulus from the AV node to the bundle branches and thence to the Purkinje fibers.
bursa: A small sac filled with synovial fluid that cushions adjacent structures and reduces friction.
bursectomy: The surgical removal of an inflamed bursa.
bursitis: Painful inflammation of one or more bursae.
Caesarian section: Surgical delivery of an infant via an incision through the lower abdominal wall and uterus.
calcaneal tendon: Large tendon that inserts on the calcaneus; tension on this tendon produces plantar flexion of the foot; also called the Achilles tendon.
calcaneus (kal-KĀ-nē-us): The heelbone, the largest of the tarsal bones.
calcification: The deposition of calcium salts within a tissue.
calcitonin (kal-si-TŌ-nin): Hormone secreted by C cells of the thyroid when calcium ion concentrations are abnormally high; restores homeostasis by increasing the rate of bone deposition and the renal rate of calcium loss, and inhibiting calcium uptake at the digestive tract.
calculus/calculi (KAL-kū-lī): Concretions of insoluble materials that form within body fluids, especially the gallbladder, kidneys, or urinary bladder.
callus: A localized thickening of the epidermis due to chronic mechanical stresses; a thickened area that forms at the site of a bone break as part of the repair process.
calorie (c) (KAL-o-rē): The amount of heat required to raise the temperature of one gram of water 1°C.
Calorie (C): The amount of heat required to raise the temperature of one kilogram of water 1°C.
calorigenic effect: The stimulation of energy production and heat loss by thyroid hormones.
calvaria (kal-VAR-ē-a): The skullcap, formed of the frontal, parietal, and occipital bones.
calyx/calyces (KĀL-i-sēz): A cup-shaped division of the renal pelvis.
canaliculi (kan-a-LIK-ū-lī): Microscopic passageways between cells; bile canaliculi carry bile to bile ducts in the liver; in bone, canaliculi permit the diffusion of nutrients and wastes to and from osteocytes.
cancellous bone (KAN-sel-us): Spongy bone, composed of a network of bony struts.
cancer: A malignant tumor that tends to undergo metastasis.
cannula: A tube that can be inserted into the body; often placed in blood vessels prior to transfusion or dialysis.
canthus, medial and lateral (KAN-thus): The angles formed at either corner of the eye between the upper and lower eyelids.
capacitation (ka-pas-i-TĀ-shun): Activation process that must occur before a spermatozoon can successfully fertilize an egg; occurs in the vagina following ejaculation.
capillary: Small blood vessels, interposed between arterioles and venules, whose thin walls permit the diffusion of gases, nutrients, and wastes between the plasma and interstitial fluids.
capitulum (ka-PIT-ū-lum): General term for a small, elevated articular process; used to refer to the rounded distal surface of the humerus that articulates with the radial head.
caput: The head.
carbaminohemoglobin (kar-bam-ē-nō-hē-mō-GLŌ-bin): Hemoglobin bound to carbon dioxide molecules.
carbohydrase (kar-bō-HĪ-drāz): An enzyme that breaks down carbohydrate molecules.
carbohydrate (kar-bó-HĪ-drāt): Organic compound containing carbon, hydrogen, and oxygen in a ratio that approximates 1:2:1.
carbon dioxide: CO_2, a compound produced by the decarboxylation reactions of aerobic glycolysis.
carbonic anhydrase: An enzyme that catalyzes the reaction

$H_2O + CO_2 \rightleftharpoons H_2CO_3$; important in carbon dioxide transport, gastric acid secretion, and renal pH regulation.

carboxyl group (kar-BOKS-il): -COOH, an acid group found in fatty acids, amino acids, etc.

carboxypeptidase (kar-bok-sē-PEP-ti-dāz): A protease that breaks down proteins and releases amino acids.

carcinogenic (kar-SIN-ō-jen-ik): Stimulates cancer formation in affected tissues.

cardia (KAR-dē-a): The area of the stomach surrounding its connection with the esophagus.

cardiac: Pertaining to the heart.

cardiac cycle: One complete heartbeat, including atrial and ventricular systole and diastole.

cardiac glands: Mucous glands characteristic of the cardia of the stomach.

cardiac output: The amcunt of blood ejected by the left ventricle each minute; normally about 5 liters.

cardiac reserve: The potential percentage increase in cardiac output above resting levels.

cardiac tamponade: Compression of the heart due to fluid accumulation in the pericardial cavity.

cardiocyte (KAR-dē-ō-sīt): A cardiac muscle cell.

cardiomyopathy (kar-dē-ō-mī-OP-a-thē): A progressive disease characterized by damage to the cardiac muscle tissue.

cardiopulmonary resuscitation: Method of artificially maintaining respiratory and circulatory function.

cardiovascular: Pertaining to the heart, blood, and blood vessels.

cardiovascular centers: Poorly localized centers in the reticular formation of the medulla of the brain; includes cardioacceleratory, cardioinhibitory, and vasomotor centers.

cardium: The heart.

carina ka-RĪ-na): A ridge on the inner surface of the base of the trachea that runs anteroposteriorly, between two primary bronchi.

carotene (KAR-ō-tēn): A yellow-orange pigment found in carrots and in green and orange leafy vegetables; a compound that the body can convert to vitamin A.

carotid artery: The principal artery of the neck, servicing cervical and cranial structures; one branch, the internal carotid, represents a major blood supply for the brain.

carotid body: A group of receptors adjacent to the carotid sinus that are sensitive to changes in the carbon dioxide levels, pH, and oxygen concentrations of the arterial blood.

carotid sinus: A dilated segment of the internal carotid artery whose walls contain baroreceptors sensitive to changes in blood pressure.

carotid sinus reflex: Reflexive changes in blood pressure that maintain homeostatic pressures at the carotid sinus, stabilizing blood flow to the brain.

carpus/carpal: The wrist.

cartilage: A connective tissue with a gelatinous matrix and an abundance of fibers.

castration: Removal of the testes, also called bilateral orchiectomy.

catabolism (ka-TAB-ō-lizm): The breakdown of complex organic molecules into simpler components, accompanied by the release of energy.

catalyst (KAT-ah-list): A substance that accelerates a specific chemical reaction, but that is not altered by the reaction.

cataract: A reduction in lens transparency that causes visual impairment.

catecholamine (kat-e-KŌL-am-in): Epinephrine, norepinephrine, and related compounds.

cathepsins (ka-THEP-sinz): Enzymes present in the sarcoplasm of skeletal muscle cells that can break down contractile proteins, providing amino acids that can act as a supplemental energy source.

catheter (KATH-e-ter): Surgical instrument, a tube inserted into a body cavity or along a blood vessel or excretory passageway for the collection of body fluids, blood pressure monitoring, or the introduction of medications or radiographic dyes.

cation (KAT-ī-ons): An ion that bears a positive charge.

cauda equina (KAW-da ek-WĪ-na): Spinal nerve roots distal to the tip of the adult spinal cord; they extend caudally inside the vertebral canal en route to lumbar and sacral segments.

caudal/caudally: Closest to or toward the tail (coccyx).

caudate nucleus (KAW-dāt): One of the cerebral nuclei of the extrapyramidal system, involved with the unconscious control of muscular activity.

cavernous tissue: Erectile tissue that can be engorged with blood; found in the penis and clitoris.

cecum (SĒ-kum): An expanded pouch at the start of the large intestine.

cell: The smallest living unit in the human body.

cell-mediated immunity: Resistance to disease through the activities of sensitized T cells that destroy antigen-bearing cells by direct contact or through the release of lymphotoxins; also called cellular immunity.

cellulitis (sel-ū-LĪ-tis): Diffuse inflammation, usually involving areas of loose connective tissue, such as the subcutaneous layer.

cementum (se-MEN-tum): Bony material covering the root of a tooth, not shielded by a layer of enamel.

center of ossification: Site in a connective tissue where bone formation begins.

central canal: Longitudinal canal in the center of an osteon that contains blood vessels and nerves, also called the Haversian canal; a passageway along the longitudinal axis of the spinal cord that contains cerebrospinal fluid.

central nervous system (CNS): The brain and spinal cord.

central sulcus: Groove in the surface of a cerebral hemisphere, between the primary sensory and primary motor areas of the cortex.

centriole: A cylindrical intracellular organelle composed of 9 groups of microtubules, 3 in each group; functions in mitosis or meiosis by forming the basis of the spindle apparatus.

centromere (SEN-trō-mēr): Localized region where two chromatids remain connected following chromosome replication; site of spindle fiber attachment.

centrosome: Region of cytoplasm containing a pair of centrioles oriented at right angles to one another.

centrum: The vertebral body.

cephalic: Pertaining to the head.

cerebellum (ser-e-BEL-um): Posterior portion of the metencephalon, containing the cerebellar hemispheres; includes the arbor vitae, cerebellar nuclei, and cerebellar cortex.

cerebral cortex: An extensive area of neural cortex covering the surfaces of the cerebral hemispheres.

cerebral hemispheres: Expanded portions of the cerebrum covered in neural cortex.

cerebral nuclei: Nuclei of the cerebrum that are important components of the extrapyramidal system.

cerebral palsy: Chronic condition resulting from damage to motor areas of the brain during development or at delivery.

cerebral peduncle: Mass of nerve fibers on the ventrolateral surface of the mesencephalon; contains ascending tracts that terminate in the thalamus and descending tracts that originate in the cerebral hemispheres.

cerebrospinal fluid: Fluid bathing the internal and external surfaces of the CNS; secreted by the choroid plexus.

cerebrovascular accident (CVA): A stroke; occlusion of a blood vessel supplying a portion of the brain, resulting in damage to the dependent neurons.

cerebrum (SER-ē-brum): The largest portion of the brain, composed of the cerebral hemispheres; includes the cerebral cortex, the cerebal nuclei, and the internal capsule.

cerumen: Waxy secretion of integumentary glands along the external auditory canal.

ceruminous glands (se-RŪ-mi-nus): Integumentary glands that secrete cerumen.

cervical enlargement: Relative enlargement of the cervical portion of the spinal cord due to the abundance of CNS neurons involved with motor control of the arms.

cervix: The lower part of the uterus.

chalazion (kah-LA-zē-on): An inflammation and distension of a Meibomian gland on the eyelid; also called a sty.

chancre (SHANG-ker): A skin lesion that develops at the primary site of a syphilis infection.

charleyhorse: Soreness and stiffness in a strained muscle, usually involving the quadriceps group.

chemoreception: Detection of alterations in the concentrations of dissolved compounds or gases.

chemotaxis (kē-mō-TAK-sis): The attraction of phagocytic cells to the source of abnormal chemicals in tissue fluids.

chemotherapy: Treatment of illness through the administration of specific chemicals.

chloride shift: Movement of plasma chloride ions into RBCs in exchange for bicarbonate ions generated by the intracellular dissociation of carbonic acid.

cholecystitis (kō-lē-sis-TĪ-tis): Inflammation of the gallbladder.

cholecystokinin (CCK) (kō-lē-sis-tō-KĪ-nin): Duodenal hormone that stimulates the contraction of the gallbladder and the secretion of enzymes by the exocrine pancreas; also called pancreozymin.

cholelithiasis (kō-lē-li-THI-a-sis): The formation or presence of gallstones.

cholesterol: A steroid component of cell membranes and a substrate for the synthesis of steroid hormones and bile salts.

choline: Chemical compound, a breakdown product or precursor of acetylcholine.

cholinergic synapse (kō-lin-ER-jik): Synapse where the presynaptic membrane releases ACh on stimulation.

cholinesterase (kō-li-NES-te-rās): Enzyme that breaks down and inactivates ACh.

chondrocyte (KON-drō-sīt): Cartilage cell.

chondroitin sulfate (kon-DROI-tin): The predominant proteoglycan in cartilage, responsible for the gelatinous consistency of the matrix.

chordae tendineae (KOR-dē TEN-di-nē-ē): Fibrous cords that brace the AV valves in the heart, stabilizing their position and preventing back-flow during ventricular systole.

chorion/chorionic (KOR-ē-on) (ko-rē-ON-ik): An extraembryonic membrane, consisting of the trophoblast and underlying mesoderm, that forms the placenta.

choroid: Middle, vascular layer in the wall of the eye.

choroid plexus: The vascular complex in the roof of the third and fourth ventricles of the brain, responsible for CSF production.

chromatid (KRŌ-ma-tid): One complete copy of a DNA strand.

chromatin (KRŌ-ma-tin): Histological term referring to the grainy material visible in cell nuclei during interphase; the appearance of the DNA content of the nucleus when the chromosomes are uncoiled.

chromosomes: Dense structures, composed of tightly coiled DNA strands and associated histones, that become visible in the nucleus when a cell prepares to undergo mitosis or meiosis; normal human somatic cells contain 46 chromosomes apiece.

chronic: Habitual or long term.

chylomicrons (kī-lō-MĪ-kronz): Relatively large droplets that may contain triglycerides, phospholipids, and cholesterol in association with proteins; synthesized and released by intestinal cells and transported to the venous blood via the lymphatic system.

chyme (kīm): A semifluid mixture of ingested food and digestive secretions that is found in the stomach and small intestine as digestion proceeds.

chymotrypsin (kī-mō-TRIP-sin): A protease found in the small intestine.

chymotrypsinogen: Inactive proenzyme secreted by the pancreas that is subsequently converted to chymotrypsin.

ciliary body: A thickened region of the choroid that encircles the lens of the eye; it includes the ciliary muscle and the ciliary processes that support the suspensory ligaments of the lens.

cilium/cilia: A slender organelle that extends above the free surface of an epithelial cell, and usually undergoes cycles of movement; composed of a basal body and microtubules in a 9×2 array.

circulatory system: The network of blood vessels that are components of the cardiovascular system.

circumduction (sir-kum-DUK-shun): A movement at a synovial joint where the distal end of the bone describes a circle, but the shaft does not rotate.

circumvallate papilla (sir-kum-VAL-āt pa-PIL-la): One of the large, dome-shaped papillae on the dorsum of the tongue that form the V that separates the body of the tongue from the root.

cirrhosis (sir-RŌ-sis): A liver disorder characterized by the degeneration of hepatocytes and their replacement by connective tissue.

cisterna (sis-TUR-na): An expanded chamber.

citric acid cycle: *See* **Krebs cycle.**

cleavage (KLĒV-ij): Mitotic divisions that follow fertilization of the ovum and lead to the formation of a blastocyst.

cleavage lines: Stress lines in the skin that follow the orientation of major bundles of collage fibers in the dermis.

climacteric: Age-related cessation of gametogenesis in the male or female due to reduced sex hormone production.

clitoris (KLI-to-ris): A small erectile organ of the female that is the developmental equivalent of the male penis.

clone: The production of genetically identical cells.

clonus (KLŌ-nus): Rapid cycles of muscular contraction and relaxation.

clot: A network of fibrin fibers and trapped blood cells; also called a **thrombus.**

clotting factors: Plasma proteins synthesized by the liver that are essential to the clotting response.

clotting response: Series of events that result in the formation of a clot.

coccygeal ligament: Fibrous extension of the dura mater and filum terminale; provides longitudinal stabilization to the spinal cord.

coccyx (KOK-siks): Terminal portion of the spinal column, consisting of relatively tiny, fused vertebrae.

cochlea (KŌK-lē-a): Spiral portion of the bony labyrinth of the inner ear that surrounds the organ of hearing.

cochlear duct (KOK-lē-ar): Membranous tube within the cochlea that is filled with endolymph and contains the organ of Corti; also called the **scala media.**

codon (KŌ-don): A sequence of three nitrogen bases along an mRNA strand that will specify the location of a single amino acid in a peptide chain.

coelom (SĒ-lom): The ventral body cavity, lined by a serous membrane and subdivided during development into the pleural, pericardial, and abdominopelvic (peritoneal) cavities.

coenzymes (kō-EN-zīmz): Complex organic cofactors, usually structurally related to vitamins.

cofactor: Ions or molecules that must be attached to the active site before an enzyme can function; examples include mineral ions and several vitamins.

colectomy (ko-LEK-to-mē): Surgical removal of part or all of the colon.

collagen: Strong, insoluble protein fiber common in connective tissues.
colitis: Inflammation of the colon.
collateral ganglion (kō-LAT-er-al): A sympathetic ganglion situated in front of the spinal column and separate from the sympathetic chain.
Colles' fracture (KOL-lēz): Fracture of the distal end of the radius and possibly the ulna, with posterior and dorsal displacement of the distal bone fragments.
colliculus/colliculi (kol-IK-ū-lus): A little mound; in the brain, used to refer to one of the cortical thickenings in the roof of the mesencephalon; the superior colliculus is associated with the visual system, and the inferior colliculi with the auditory system.
colloid/colloidal suspension: A solution containing large organic molecules in suspension.
colon: The large intestine.
colonoscope (kō-LON-ō-skōp): A fiberoptic device for examining the interior of the colon.
colostomy (kō-LOS-to-mē): The surgical connection of a portion of the colon to the body wall, sometimes performed after a colectomy to permit the discharge of fecal materials.
colostrum (ko-LOS-trum): Secretion of the mammary glands at the time of childbirth and for a few days thereafter; contains more protein and less fat than the milk secreted later.
coma (kō-ma): An unconscious state from which the individual cannot be aroused, even by strong stimuli.
comedo (kō-MĒ-dō): An inflamed sebaceous gland.
comminuted: Broken or crushed into small pieces.
commissure: A crossing over from one side to another.
common bile duct: Duct formed by the union of the cystic duct from the gallbladder and the bile ducts from the liver; terminates at the duodenal ampulla, where it meets the pancreatic duct.
common pathway: In the clotting response, the events that begin with the appearance of thromboplastin and end with the formation of a clot.
compact bone: Dense bone containing parallel osteons.
compensation curves: The cervical and lumbar curves that develop to center the body weight over the legs.
complement: 11 plasma proteins that interact in a chain reaction following exposure to activated antibodies or the surfaces of certain pathogens, and which promote cell lysis, phagocytosis, and other defense mechanisms.
compliance: The ability of certain organs to tolerate changes in volume; a property that reflects the presence of elastic fibers and smooth muscles.
compound: A molecule containing two or more elements in combination.
concentration: Amount (in grams) or number of atoms, ions, or molecules (in moles) per unit volume.
concentration gradient: Regional differences in the concentration of a particular substance.
conception: Fertilization.
concha/conchae (KONG-kē): Three pairs of thin, scroll-like bones that project into the nasal cavities; the superior and medial conchae are part of the ethmoid, and the inferior are separate bones.
concussion: A violent blow or shock; loss of consciousness due to a violent blow to the head.
conducted change: An action potential; a change in the transmembrane potential that spreads across an excitable membrane.
condyle: A rounded articular projection on the surface of a bone.
cone: Retinal photoreceptor responsible for color vision.
congenital (kon-JEN-i-tal): Already present at the birth of an individual.
congestive heart failure (CHF): Failure to maintain adequate cardiac output due to circulatory problems or myocardial damage.
conjunctiva (kon-junk-TĪ-va): A layer of stratified squamous epithelium that covers the inner surfaces of the lids and the anterior surface of the eye to the edges of the cornea.
conjunctivitis: Inflammation of the conjunctiva.
connective tissue: One of the four primary tissue types; provides a structural framework for the body that stabilizes the relative positions of the other tissue types; includes connective tissue proper, cartilage, bone, and blood; always has cell products, cells, and ground substance.
contractility: The ability to contract, possessed by skeletal, smooth, and cardiac muscle cells.
contracture: A permanent contraction of an entire muscle following the atrophy of individual muscle cells.
contralateral reflex: A reflex that affects the opposite side of the body from the stimulus.
conus medullaris: Conical tip of the spinal cord that gives rise to the filum terminale.
convergence: In the nervous system, the term indicates that the axons from several neurons innervate a single neuron; this is most common along motor pathways.
coracoid process (ko-RA-kōid): A hook-shaped process of the scapula that projects above the anterior surface of the capsule of the shoulder joint.
Cori cycle: Metabolic exchange of lactic acid from skeletal muscle for glucose from the liver; performed during the recovery period following muscular exertion.
cornea (KOR-nē-a): Transparent portion of the fibrous tunic of the anterior surface of the eye.
corniculate cartilages (kor-NIK-ū-lāt): A pair of small laryngeal cartilages.
cornification: The production of keratin by a stratified squamous epithelium; also called **keratinization.**
cornu: Shaped like a horn.
corona radiata (ko-RŌ-na rā-dē-A-ta): A layer of follicle cells surrounding an oocyte at ovulation.
coronoid (kō-RŌ-nōid): Hooked or curved.
corpus/corpora: Body.
corpus callosum: Bundle of axons linking centers in the left and right cerebral hemispheres.
corpus cavernosum (KOR-po-ra ka-ver-NŌ-sum): Masses of erectile tissue within the body of the penis (male) or clitoris (female).
corpus luteum (LOO-tē-um): Progestin-secreting mass of follicle cells that develops in the ovary after ovulation.
corpora quadrigemina (KOR-po-ra quad-ri-JEM-i-na): The superior and inferior colliculi of the mesencephalic tectum (roof) in the brain.
corpus spongiosum (spon-jē-ō-sum): Mass of erectile tissue that surrounds the urethra in the male penis, and expands distally to form the glans.
cortex: Outer layer or portion of an organ.
Corti, organ of: Receptor complex in the scala media of the cochlea that includes the inner and outer hair cells, supporting cells and structures, and the tectorial membrane; provides the sensation of hearing.
corticobulbar tracts (kor-ti-kō-BUL-bar): Descending tracts that carry information/commands from the cerebral cortex to nuclei and centers in the brain stem.
corticospinal tracts: Descending tracts that carry motor commands from the cerebral cortex to the anterior gray horns of the spinal cord.
corticosteroid: A steroid hormone produced by the adrenal cortex.
corticosterone (kor-ti-kos-te-rōn): One of the corticosteroids secreted by the zona fasciculata of the adrenal cortex; a glucocorticoid.
corticotrophin: *See* **adrenocorticotrophic hormone (ACTH).**
corticotrophin-releasing hormone (CRH): Releasing hormone secreted by the hypothalamus that stimulates secretion of ACTH by the anterior pituitary.

cortisol (KOR-ti-sōl): One of the corticosteroids secreted by the zona fasciculata of the adrenal cortex; a glucocorticoid.
costa/costae: A rib.
cotransport: Membrane transport of a nutrient, such as glucose, in company with the movement of an ion, usually sodium; transport requires a carrier protein but does not involve direct ATP expenditure and can occur regardless of the concentration gradient for the nutrient.
countercurrent exchange: Diffusion between two solutions traveling in opposite directions.
countercurrent multiplication: Active transport between two limbs of a loop that contains a fluid moving in one direction; responsible for the concentration of the urine in the kidney tubules.
covalent bond (kō-VĀ-lent): A chemical bond between atoms that involves the sharing of electrons.
coxa/coxae: The bones of the hip.
cranial: Pertaining to the head.
cranial nerves: Peripheral nerves originating at the brain.
craniosacral division (krā-nē-ō-SAK-ral): *See* **parasympathetic division.**
craniostenosis (krā-nē-ō-sten-Ō-sis): Skull deformity caused by premature closure of the cranial sutures.
cranium: The brainbox; the skull bones that surround the brain.
creatine: A nitrogenous compound synthesized in the body that can bind a high-energy phosphate and serve as an energy reserve.
creatine phosphate: A high-energy compound present in muscle cells; during muscular activity the phosphate group is donated to ADP, regenerating ATP.
creatinine: A breakdown product of creatine metabolism.
crenation: Cellular shrinkage due to an osmotic movement of water out of the cytoplasm.
cribriform plate: Portion of the ethmoid bone of the skull that contains the foramina used by the axons of olfactory receptors en route to the olfactory bulbs of the cerebrum.
cricoid cartilage (KRĪ-koid): Ring-shaped cartilage forming the inferior margin of the larynx.
crista/cristae: A ridge-shaped collection of hair cells in the ampulla of a semicircular canal; the crista and cupula form a receptor complex sensitive to movement along the plane of the canal.
cross-bridge: Myosin head that projects from the surface of a thick filament, and that can bind to an active site of a thin filament in the presence of calcium ions.
cruciate ligaments: A pair of intracapsular ligaments (anterior and posterior) in the knee.
cryosurgery: A surgical technique that involves freezing and killing cells in a localized area.
cryptorchid testis: An undescended testis that is in the abdominopelvic cavity rather than the scrotum.
cuneiform cartilages (kū-NĒ-i-form): A pair of small cartilages in the larynx.
cupula (KŪ-pū-la): A gelatinous mass that sits in the ampulla of a semicircular canal in the inner ear, and whose movement stimulates the hair cells of the crista.
curare: A toxin that prevents neural stimulation of neuromuscular junctions.
Cushing's disease: Condition caused by oversecretion of adrenal steroids.
cutaneous membrane: The epidermis and papillary layer of the dermis.
cuticle: Layer of dead, cornified cells surrounding the shaft of a hair; for nails, *see* **eponychium.**
cutis: The skin.
cyanosis: Bluish coloration of the skin due to the presence of deoxygenated blood in vessels near the body surface.
cyst: A fibrous capsule containing fluid or other material.
cystic duct: A duct that carries bile between the gallbladder and the common bile duct.
cystitis: Inflammation of the urinary bladder.
cytochrome (SĪ-tō-krōm): A pigment component of the electron transport system; a structural relative of heme.
cytokinesis (sī-tō-ki-NĒ-sis): The cytoplasmic movement that separates two daughter cells at the completion of mitosis.
cytology (sī-TOL-ō-jē): The study of cells.
cytoplasm: The material between the cell membrane and the nuclear membrane.
cytosine: One of the nitrogen base components of nucleic acids.
cytoskeleton: A network of microtubules and microfilaments in the cytoplasm.
cytosol: The fluid portion of the cytoplasm.
cytotoxic: Poisonous to living cells.
cytotoxic T cells: Lymphocytes of the cellular immune response that kill target cells by direct contact or through the secretion of lymphotoxins; also called killer T cells.
daughter cells: Genetically identical cells produced by mitosis.
deamination (dē-am-i-NĀ-shun): The removal of an amine group from an amino acid.
decarboxylation (dē-kar-boks-i-LĀ-shun): The removal of a molecule of CO_2.
decerebrate: Lacking a cerebrum.
decomposition reaction: A chemical reaction that breaks a molecule into smaller fragments.
decubitis ulcers: Ulcers that form where chronic pressure interrupts circulation to a portion of the skin.
decussate: To cross over to the opposite side, usually referring to the crossover of the pyramidal tracts on the ventral surface of the medulla oblongata.
defecation (def-e-KĀ-shun): The elimination of fecal wastes.
deglutition (deg-loo-TISH-un): Swallowing.
degradation: Breakdown, catabolism.
dehydration: A reduction in the water content of the body that threatens homeostasis.
dehydration synthesis: The joining of two molecules associated with the removal of a water molecule.
delta cell: A pancreatic islet cell that secretes somatostatin.
dementia: Loss of mental abilities.
demyelination: The loss of the myelin sheath of an axon, usually due to chemical or physical damage to Schwann cells or oligodendrocytes.
denaturation: Irreversible alteration in the three-dimensional structure of a protein.
dendrite (DEN-drīt): A sensory process of a neuron.
denticulate ligaments: Supporting fibers that extend laterally from the surface of the spinal cord, tying the pia mater to the dura mater and providing lateral support for the spinal cord.
dentin (DEN-tin): Bonelike material that forms the body of a tooth; it differs from bone in lacking osteocytes and osteons.
deoxyribonucleic acid (dē-ok-sē-rī-bo-nū-KLĀ-ik): DNA strand: a nucleic acid consisting of a chain of nucleotides containing the sugar deoxyribose and the nitrogen bases adenine, guanine, cytosine, and thymine DNA molecule: two DNA strands wound in a double helix and held together by weak bonds between complementary nitrogen base pairs.
deoxyribose: A 5-carbon sugar resembling ribose but lacking an oxygen atom.
depolarization: A change in the transmembrane potential that moves it from a negative value toward 0 mV.
depression: Inferior (downward) movement of a body part.
dermatitis: Inflammation of the skin.
dermatome: A sensory region monitored by the dorsal rami of a single spinal segment.
dermis: The connective tissue layer beneath the epidermis of the skin.

detrusor muscle (de-TRŌŌ-sor): Smooth muscle in the wall of the urinary bladder.
detumescence (dē-tū-MES-ens): Loss of a penile erection in the male.
development: Growth and the acquisition of increasing structural and functional complexity; includes the period from conception to maturity.
diabetogenic effect: Elevation in blood sugar concentrations following the secretion of growth hormone or glucagon.
diabetes insipidus: Polyuria due to inadequate production of ADH.
diabetes mellitus (mel-LI-tus): Polyuria and glycosuria, most often due to inadequate production of insulin with resulting elevation of blood glucose levels.
dialysis: Diffusion between two solutions of differing solute concentrations across a semipermeable membrane containing pores that permit the passage of some solutes and not others.
diapedesis (dī-a-pe-DĒ-sis): Movement of white blood cells through the walls of blood vessels by migration between adjacent endothelial cells.
diaphragm (DĪ-a-fram): Any muscular partition; often used to refer to the respiratory muscle that separates the thoracic from the abdominopelvic cavities.
diaphysis (dī-A-fi-sis): The shaft of a long bone.
diarrhea (dī-a-RĒ-a): Abnormally frequent defecation, associated with the production of unusually fluid feces.
diarthrosis (dī-ar-THRO-sis): A synovial joint.
diastolic pressure: Pressure measured in the walls of a muscular artery when the left ventricle is in diastole.
diencephalon (dī-en-SEF-a-lon): A division of the brain that includes the epithalamus, thalamus, and hypothalamus.
differential count: The determination of the relative abundance of each type of white blood cell, based on a random sampling of 100 WBCs.
differentiation: The gradual appearance of characteristic cellular specializations during development, as the result of gene activation or repression.
diffusion: Passive molecular movement from an area of relatively high concentration to an area of relatively low concentration.
digestion: The chemical breakdown of ingested materials into simple molecules that can be absorbed by the cells of the digestive tract.
digestive system: The digestive tract and associated glands.
digestive tract: An internal passageway that begins at the mouth and ends at the anus.
dilate: To increase in diameter; to enlarge or expand.
diploid (DIP-lōid): Having a complete somatic complement of chromosomes (23 pairs in human cells).
disaccharide (di-SAK-ah-rīd): A compound formed by the joining of two simple sugars by dehydration synthesis.
dislocation: Forceful displacement of an articulating bone to an abnormal position, usually accompanied by damage to tendons, ligaments, the articular capsule, or other structures.
dissociation (di-sō-sē-Ā-shun): *See* **ionization.**
distal: Movement away from the point of attachment or origin; for a limb, away from its attachment to the trunk.
distal convoluted tubule: Portion of the nephron closest to the collecting tubule and duct; an important site of active secretion.
diuresis: Fluid loss at the kidneys; the production of urine.
divergence: In neural tissue, the spread of excitation from one neuron to many neurons; an organizational pattern common along sensory pathways of the CNS.
diverticulum: A sac or pouch in the wall of the colon or other organ.
diverticulitis (dī-ver-tik-ū-LĪ-tis): Inflammation of a diverticulum.
diverticulosis (dī-ver-tik-ū-LŌ-sis): The formation of diverticula.
dizygotic twins (dī-zī-GOT-ik): Twins that result from the fertilization of two different ova.
dominant gene: A gene whose presence will determine the phenotype, regardless of the nature of its allelic companion.
dopamine (DŌ-pah-mēn): An important neurotransmitter in the CNS.
dorsal: Toward the back, posterior.
dorsal root ganglion: PNS ganglion containing the cell bodies of sensory neurons.
dorsiflexion: Elevation of the superior surface of the foot.
Down syndrome: A genetic abnormality resulting from the presence of three copies of chromosome 21; individuals with this condition have characteristic physical and intellectual deficits.
duct: A passageway that delivers exocrine secretions to an epithelial surface.
ductus arteriosus (ar-te-rē-Ō-sus): Vascular connection between the pulmonary trunk and the aorta that functions throughout fetal life; normally closes at birth or shortly thereafter, and persists as the ligamentum arteriosum.
ductus deferens (DUK-tus DEF-e-renz): A passageway that carries sperm from the epididymis to the ejaculatory duct.
duodenal ampulla: Chamber that receives bile from the common bile duct and pancreatic secretions from the pan- creatic duct.
duodenal glands: *See* **submucosal glands.**
duodenal papilla: Conical projection from the inner surface of the duodenum that contains the opening of the duodenal ampulla.
duodenum (dū-A-dē-num): The proximal 1 ft of the small intestine that contains short villi and submucosal glands.
dura mater (DŪ-ra MĀ-ter): Outermost component of the meninges that surround the brain and spinal cord.
dynamic equilibrium: Maintenance of normal body orientation as sudden changes in position (rotation, acceleration, etc.) occur.
dynorphin (dī-NOR-fin): A powerful neuromodulator produced in the CNS that blocks pain perception by inhibiting pain pathways.
dyslexia: Impaired ability to comprehend written words.
dysmenorrhea: Painful menstruation.
dysmetria (dis-MET-rē-a): Difficulty in performing movements due to problems with the interpretation and anticipation of the distance to be covered.
dysuria (dis-Ū-rē-a): Painful urination.
eccrine glands (EK-rin): Sweat glands of the skin that produce a watery secretion.
echocardiography (ek-ō-kar-dē-OG-rafē): Examination of the heart using modified ultrasound techniques.
ectoderm: One of the three primary germ layers; covers the surface of the embryo and gives rise to the nervous system, the epidermis and associated glands, and a variety of other structures.
ectopic (ek-TOP-ik): Outside of its normal location.
effector: A peripheral gland or muscle cell innervated by a motor neuron.
efferent: Away from.
efferent arteriole: An arteriole carrying blood away from the glomerulus of the kidney.
efferent fiber: An axon that carries impulses away from the CNS.
ejaculation (e-jak-ū-LĀ-shun): The ejection of semen from the penis as the result of muscular contractions of the bulbocavernosus and ischiocavernosus muscles.
ejaculatory duct (e-JAK-ū-la-to-rē): Short ducts that pass within the walls of the prostate and connect the ductus deferens with the prostatic urethra.

elastase (ē-LAS-tāz): A pancreatic enzyme that breaks down elastin fibers.
elastin: Connective tissue fibers that stretch and rebound, providing elasticity to connective tissues.
electrical coupling: A connection between adjacent cells that permits the movement of ions and the transfer of graded or conducted changes in the transmembrane potential from cell to cell.
electrocardiogram (ECG, EKG) (e-lek-trō-KAR-dē-ō-gram): Graphic record of the electrical activities of the heart, as monitored at specific locations on the body surface.
electroencephalogram (EEG): Graphic record of the electrical activities of the brain.
electrolytes (ē-LEK-trō-līts): Soluble inorganic compounds whose ions will conduct an electric current in solution.
electron: One of the three fundamental particles; a subatomic particle that bears a negative charge and normally orbits around the protons of the nucleus.
electron transport system: Cytochrome system responsible for most of the energy production in living cells; a complex bound to the inner mitochondrial membrane.
eleidin (el-Ē-i-din): A protein that forms as a precursor of keratin.
element: All of the atoms with the same atomic number.
elephantiasis (el-e-fan-TĪ-a-sis): A lymphedema caused by infection and blockage of lymphatics by mosquitoborne parasites.
elevation: Movement in a superior, or upward, direction.
embolism (EM-bō-lizm): Obstruction or closure of a vessel by an embolus.
embolus (EM-bo-lus): An air bubble, fat globule, or blood clot drifting in the circulation.
embryo (EM-brē-o): Developmental stage beginning at fertilization and ending at the start of the third developmental month.
embryology (em-brē-OL-o-jē): The study of embryonic development, focusing on the first 2 months after fertilization.
emesis (EM-e-sis): Vomiting.
emmetropia: Normal vision.
emulsification (ē-mul-si-fi-KĀ-shun): The physical breakup of fats in the digestive tract, forming smaller droplets accessible to digestive enzymes; normally the result of mixing with bile salts.
enamel: Crystalline material similar in mineral composition to bone, but harder and without osteocytes, that covers the exposed surfaces of the teeth.
encephalitis: Inflammation of the brain.
endocarditis: Inflammation of the endocardium of the heart.
endocardium (en-dō-KAR-dē-um): The simple squamous epithelium that lines the heart and is continuous with the endothelium of the great vessels.
endochondral ossification (en-dō-KON-dral): The conversion of a cartilaginous model to bone, the characteristic mode of formation for skeletal elements other than the bones of the cranium, the clavicles, and sesamoid bones.
endocrine gland: A gland that secretes hormones into the blood.
endocrine system: The endocrine glands of the body.
endocytosis (EN-dō-sī-tō-sis): The movement of relatively large volumes of extracellular material into the cytoplasm via the formation of a membranous vesicle at the cell surface; includes pinocytosis and phagocytosis.
endoderm: One of the three primary germ layers; the layer on the undersurface of the embryonic disc, that gives rise to the epithelia and glands of the digestive system, the respiratory system, and portions of the urinary system.
endogenous: Produced within the body.
endolymph (EN-dō-limf): Fluid contents of the membranous labyrinth (the saccule, utricle, semicircular canals, and cochlear duct) of the inner ear.
endometrial glands: Secretory glands of the endometrium.
endometrium (en-dō-MĒ-trē-um): The mucous membrane lining the uterus.
endomysium (endō-MĪS-ē-um): A delicate network of connective tissue fibers that surrounds individual muscle cells.
endoneurium: A delicate network of connective tissue fibers that surrounds individual nerve fibers.
endoplasmic reticulum (en-dō-PLAZ-mik re-TIK-ū-lum): A network of membranous channels in the cytoplasm of a cell that function in intracellular transport, synthesis, storage, packaging, and secretion.
endorphins (en-DOR-finz): Neuromodulators produced in the CNS that inhibit activity along pain pathways.
endosteum: An incomplete cellular lining found on the inner (medullary) surfaces of bones.
endothelium (en-dō-THĒ-lē-um): The simple squamous epithelium that lines blood and lymphatic vessels.
enkephalins (en-KEF-a-linz): Neuromodulators produced in the CNS that inhibit activity along pain pathways.
enteritis (en-ter-Ī-tis): Inflammation of the intestinal tract.
enterocrinin: A hormone secreted by the duodenal lining when exposed to acid chyme; stimulates the secretion of the duodenal glands.
enteroendocrine cells (en-ter-ō-EN-dō-krin): Endocrine cells scattered among the epithelial cells lining the digestive tract.
enterogastric reflex: Reflexive inhibition of gastric secretion initiated by the arrival of acid chyme in the small intestine.
enterohepatic circulation: Excretion of bile salts by the liver, followed by absorption of bile salts by intestinal cells for return to the liver via the hepatic portal vein.
enterokinase: An enzyme in the lumen of the small intestine that activates the proenzymes secreted by the pancreas.
enzyme: A protein that catalyzes a specific biochemical reaction.
eosinophils (ē-ō-sin-ō-filz): A granulocyte (WBC) with a lobed nucleus and red-staining granules; participates in the immune response, and is especially important during allergic reactions.
ependyma (ep-EN-di-mah): Layer of cells lining the ventricles and central canal of the CNS.
epicardium: Serous membrane covering the outer surface of the heart; also called the visceral pericardium.
epidermis: The epithelium covering the surface of the skin.
epididymis (ep-i-DID-i-mus): Coiled duct that connects the rete testis to the ductus deferens; site of functional maturation of spermatozoa.
epidural block: Anesthesia caused by the elimination of sensory inputs from dorsal nerve roots following the introduction of drugs into appropriate regions of the epidural space.
epidural space: Space between the spinal dura mater and the walls of the vertebral foramen; contains blood vessels and adipose tissue; a frequent site of injection for regional anesthesia.
epiglottis (ep-i-GLOT-is): Blade-shaped flap of tissue, reinforced by cartilage, that is attached to the dorsal and superior surface of the thyroid cartilage; it folds over the entrance to the larynx during swallowing.
epimysium (ep-i-MĪS-ē-um): A dense investment of collagen fibers that surrounds a skeletal muscle, and is continous with the tendons/aponeuroses of the muscle, and with the perimysium.
epineurium: A dense investment of collagen fibers that surrounds a peripheral nerve.
epiphyseal plate (e-pi-FI-sē-al): Cartilaginous region between the epiphysis and diaphysis of a growing bone.

epiphysis (e-PIF-i-sis): The head of a long bone.
epistaxis (ep-i-STAK-sis): Nosebleed.
epithelium (e-pi-THĒ-lē-um): One of the four primary tissue types; a layer of cells that forms a superficial covering or an internal lining of a body cavity or vessel.
eponychium (ep-ō-NIK-ē-um): A narrow zone of stratum corneum that extends across the surface of a nail at its exposed base; also called the **cuticle.**
equational division: The second meiotic division.
equilibrium (ē-kwi-LIB-rē-um): A dynamic state, where two opposing forces or processes are in balance.
erection: Stiffening of the penis prior to copulation due to the engorgement of the erectile tissues of the corpora cavernosa and the corpus spongiosum.
erythema (er-i-THĒ-ma): Redness and inflammation at the surface of the skin.
erythrocyte (e-RITH-rō-sīt): A red blood cell; an anucleate blood cell containing large quantities of hemoglobin.
erythrocytosis (e-rith-rō-sī-TŌ-sis): An abnormally large number of erythrocytes in the circulating blood.
erythropoietin (e-rith-rō-POI-ē-tin): Hormone released by tissues, especially the kidneys, exposed to low oxygen concentrations; stimulates hematopoiesis in bone marrow.
Escherichia coli: Normal bacterial resident of the large intestine.
esophagus: A muscular tube that connects the pharynx to the stomach.
essential amino acids: Amino acids that cannot be synthesized in the body in adequate amounts, and must be obtained from the diet.
essential fatty acids: Fatty acids that cannot be synthesized in the body, and must be obtained from the diet.
estrogens (ES-tro-jenz): Female sex hormones, notably estradiol.
eupnea (ŪP-nē-a): Normal quiet breathing.
evaporation: Movement of molecules from the liquid to the gaseous state.
eversion (ē-VER-shun): A turning outward.
excitable membranes: Membranes that conduct action potentials, a characteristic of muscle and nerve cells.
excitatory postsynaptic potential: The depolarization of a postsynaptic membrane by a chemical neurotransmitter released by the presynaptic cell.
excretion: Elimination from the body.
exocrine gland: A gland that secretes onto the body surface or into a passageway connected to the exterior.
exocytosis (EK-sō-sī-tō-sis): The ejection of cytoplasmic materials by fusion of a membranous vesicle with the cell membrane.
expiration: Exhalation; breathing out.
expiratory reserve: The amount of additional air that can be voluntarily moved out of the respiratory tract after a normal tidal expiration.
extension: An increase in the angle between two articulating bones; the opposite of flexion.
external auditory canal: Passageway in the temporal bone that leads from the external auditory meatus to the tympanic membrane of the inner ear.
external auditory meatus: The entrance to the external auditory canal.
external ear: The pinna, external auditory meatus, external auditory canal, and tympanic membrane.
external nares: The nostrils; the external openings into the nasal cavity.
external respiration: Diffusion of gases between the alveolar air and the alveolar capillaries.
exteroceptors: Sensory receptors in the skin, mucous membranes, and special sense organs that provide information about the external environment and our position within it.
extracellular fluid: All body fluid other than that contained within cells; includes plasma and interstitial fluid.
extraembryonic membranes: The yolk sac, amnion, chorion, and allantois.
extrafusal fibers: Contractile muscle fibers, as opposed to the sensory intrafusal fibers (muscle spindles).
extrapyramidal system: Nuclei and tracts associated with the involuntary control of muscular activity.
extremities: The limbs.
extrinsic pathway: Clotting pathway that begins with damage to blood vessels or surrounding tissues and ends with the formation of tissue thromboplastin.
fabella: A sesamoid bone often found in the gastrocnemius muscle just behind the knee.
facilitated diffusion: Passive movement of a substance across a cell membrane via a protein carrier.
facilitation: Depolarization of a nerve cell membrane toward threshold, or making the cell more sensitive to depolarizing stimuli.
falciform ligament (FAL-si-form): A sheet of mesentery that contains the ligamentum teres, the fibrous remains of the umbilical vein of the fetus.
falx (falks): Sickle-shaped.
falx cerebri (falks ser-E-brē): Curving sheet of dura mater that extends between the two cerebral hemispheres; encloses the superior sagittal sinus.
fascia (FASH-a): Connective tissue fibers, primarily collagenous, that form sheets or bands beneath the skin to attach, stabilize, enclose, and separate muscles and other internal organs.
fasciculus (fa-SIK-ū-lus): A small bundle, usually referring to a collection of nerve axons or muscle fibers.
fatty acids: Hydrocarbon chains ending in a carboxyl group.
fauces (FAW-sēz): The passage from the mouth to the pharynx, bounded by the palatal arches, the soft palate, and the uvula.
febrile: Characterized by or pertaining to a fever.
feces: Waste products eliminated by the digestive tract at the anus; contains indigestible residue, bacteria, mucus, and epithelial cells.
fenestra: An opening.
fertilization: Fusion of egg and sperm to form a zygote.
fetus: Developmental stage lasting from the start of the third developmental month to delivery.
fibrillation (fi-bri-LĀ-shun): Uncoordinated contractions of individual muscle cells that impair or prevent normal function.
fibrin (FĪ-brin): Insoluble protein fibers that form the basic framework of a blood clot.
fibrinogen (fī-BRIN-o-jen): Plasma protein, soluble precursor of the fibrous protein fibrin.
fibrinolysis (fī-bri-no-LI-sis): The breakdown of the fibrin strands of a blood clot by a proteolytic enzyme.
fibroblasts (FĪ-brō-blasts): Cells of connective tissue proper that are responsible for the production of extracellular fibers and the secretion of the organic compounds of the extracellular matrix.
fibrocartilage: Cartilage containing an abundance of collagen fibers; found around the edges of joints, in the intervertebral discs, the menisci of the knee, etc.
fibrous tunic: The outermost layer of the eye, composed of the sclera and cornea.
fibula: The lateral, relatively small bone of the lower leg.
filariasis (fil-a-RĪ-a-sis): Condition resulting from infection by mosquitoborne parasites; may cause elephantiasis.
filiform papillae: Slender conical projections from the dorsal surface of the anterior two-thirds of the tongue.
filtrate: Fluid produced by filtration at a glomerulus in the kidney.
filtration: Movement of a fluid across a membrane whose pores restrict the passage of solutes on the basis of size.

filtration pressure: Hydrostatic pressure responsible for the filtration process.
filum terminale: A fibrous extension of the spinal cord that extends from the conus medullaris to the coccygeal ligament.
fimbriae (FIM-brē-ē): A fringe; used to describe the finger-like processes that surround the entrance to the uterine tube.
fissure: An elongate groove or opening.
fistula: An abnormal passageway between two organs or from an internal organ or space to the body surface.
flaccid: Limp, soft, flabby; a muscle without muscle tone.
flagellum/flagella (fla-JEL-ah): An organelle structurally similar to a cilium, but used to propel a cell through a fluid.
flatus: Intestinal gas.
flavin adenine dinucleotide: A coenzyme important in oxidative phosphorylation; cycles between the oxidized ($FADH_2$) and reduced (FAD) states.
flavin adenine mononucleotide: A coenzyme important in oxidative phosphorylation; cycles between the oxidized ($FMNH_2$) and reduced (FMN) states.
flexion (FLEK-shun): A movement that reduces the angle between two articulating bones; the opposite of extension.
flexor: A muscle that produces flexion.
flexor reflex: A reflex contraction of the flexor muscles of a limb in response to an unpleasant stimulus.
flexure: A bending.
fluoroscope: An instrument that permits the examination of the body with X-rays in real-time, rather than via fixed images on photographic plates.
folia (FŌ-lē-ah): Leaf-like folds; used in reference to the slender folds in the surface of the cerebellar cortex.
follicle (FOL-i-kl): A small secretory sac or gland.
follicle-stimulating hormone (FSH): A hormone secreted by the anterior pituitary; stimulates oogenesis (female) and spermatogenesis (male).
folliculitis (fo-lik-ū-LĪ-tis): Inflammation of a follicle, such as a hair follicle of the skin.
fontanel (fon-tah-NEL): A relatively soft, flexible, fibrous region between two flat bones in the developing skull.
foramen: An opening or passage through a bone.
forearm: Distal portion of the arm between the elbow and wrist.
forebrain: The cerebrum.
fornix (FOR-niks): An arch, or the space bounded by an arch; in the brain, an arching tract that connects the hippocampus with the mamillary bodies; in the eye, a slender pocket found where the epithelium of the ocular conjunctiva folds back upon itself as the palpebral conjunctiva.
fossa: A shallow depression or furrow in the surface of a bone.
fourth ventricle: An elongate ventricle of the metencephalon (pons and cerebellum) and the myelencephalon (medulla) of the brain; the roof contains a region of choroid plexus.
fovea (FŌ-vē-a): Portion of the retina providing the sharpest vision, with the highest concentration of cones; also called the **macula lutea.**
fracture: A break or crack in a bone.
frenulum (FREN-ū-lum): A bridle; *see* **lingual frenulum.**
frontal plane: A sectional plane that divides the body into anterior and posterior portions.
fructose: A hexose (simple sugar containing 6 carbons) found in foods and in semen.
fundus (FUN-dus): The base of an organ.
fungiform papillae: Mushroom-shaped papillae on the dorsal and dorsolateral surfaces of the tongue.
furuncle (FŪ-rung-kl): A boil, resulting from the invasion and inflammation of a hair follicle or sebaceous gland.
gallbladder: Pear-shaped reservoir for the bile secreted by the liver.
gametes (GAM-ēts): Reproductive cells (sperm or eggs) that contain one-half of the normal chromosome complement.
gametogenesis (ga-mē-tō-JEN-e-sis): The formation of gametes.
gamma aminobutyric acid (GABA) (GAM-ma a-MĒ-nō-bū-TIR-ik): A neurotransmitter of the CNS whose effects are usually inhibitory.
gamma motor neurons: Motor neurons that adjust the sensitivities of muscle spindles (intrafusal fibers).
ganglion/ganglia: A collection of nerve cell bodies outside of the CNS.
gangliosides: Glycolipids that are important components of cell membranes in the CNS.
gap junctions: Connections between cells that permit electrical coupling.
gaster (GAS-ter): The stomach; the body or belly of a skeletal muscle.
gastrectomy (gas-TREK-to-mē): Partial or total surgical removal of the stomach.
gastric: Pertaining to the stomach.
gastric glands: Tubular glands of the stomach whose cells produce acid, enzymes, intrinsic factor, and hormones.
gastrin (GAS-trin): Hormone produced by enteroendocrine cells of the stomach, when exposed to mechanical stimuli or vagal stimulation, and the duodenum, when exposed to chyme containing undigested proteins.
gastritis (gas-TRĪ-tis): Inflammation of the stomach.
gastroenteric reflex (gas-trō-en-TER-ik): An increase in peristalsis along the small intestine triggered by the arrival of food in the stomach.
gastroileal reflex (gas-trō-IL-ē-al): Peristaltic movements that shift materials from the ileum to the colon, triggered by the arrival of food in the stomach.
gastrointestinal (GI) tract: An internal passageway that begins at the mouth, ends at the anus, and is lined by a mucous membrane; also known as the **digestive tract.**
gastroscope: A fiberoptic instrument that permits visual inspection of the stomach lining.
gastrulation (gas-troo-LĀ-shun): The movement of cells of the inner cell mass that creates the three primary germ layers of the embryo.
gene: A portion of a DNA strand that functions as a hereditary unit and is found at a particular locus on a specific chromosome.
genetics: The study of mechanisms of heredity.
geniculate (je-NIK-ū-lāts): Like a little knee; the medial geniculates and the lateral geniculates are thalamic nuclei in the walls of the thalamus of the brain.
genitalia (jen-i-TĀ-lē-a): Reproductive organs.
genotype (JĒN-ō-tīp): The genetic complement of a particular individual. cess.
germinal centers: Pale regions in the interior of lymphatic tissues or nodules, where mitoses are underway.
gestation (jes-TĀ-shun): The period of intrauterine development.
genetic engineering: Term used to describe research and experiments involving the manipulation of the genetic makeup of an organism.
gingivae (JIN-ji-vē): The gums.
gingivitis: Inflammation of the gums.
gland: Cells that produce exocrine or endocrine secretions, derived from epithelia.
glans penis: Expanded tip of the penis that surrounds the urethra meatus; continuous with the corpus spongiosum.
glaucoma: Eye disorder characterized by rising intraocular pressures due to inadequate drainage of aqueous humor at the canal of Schlemm.
glenoid fossa: A rounded depression that forms the articular surface of the scapula at the shoulder joint.

glial cells (GLĒ-al): Supporting cells in the neural tissue of the CNS and PNS.
globular proteins: Proteins whose tertiary structure makes them rounded and compact.
glomerular capsule: Expanded initial portion of the nephron that surrounds the glomerulus.
glomerular filtration rate: The rate of filtrate formation at the glomerulus.
glomerulonephritis (glo-mer-ū-lō-nef-RĪ-tis): Inflammation of the glomeruli of the kidneys.
glomerulus (glo-MER-ū-lus): A ball or knot; in the kidneys, a knot of capillaries that projects into the enlarged, proximal end of a nephron; the site where filtration occurs, the first step in the production of urine.
glossopharyngeal nerve (glos-ō-fa-RIN-jē-al): Cranial nerve IX.
glottis (GLOT-is): The passage from the pharynx to the larynx.
glucagon (GLOO-ka-gon): Hormone secreted by the alpha cells of the pancreatic islets; elevates blood glucose concentrations.
glucocorticoids: Hormones secreted by the zona fasciculata of the adrenal cortex to modify glucose metabolism; cortisol, cortisone, and corticosterone are important examples.
glucogenic amino acids: Amino acids that can be broken down, converted to pyruvic acid, and used in the synthesis of glucose (gluconeogenesis).
gluconeogenesis (gloo-kō-nē-ō-JEN-e-sis): The synthesis of glucose from protein or lipid precursors.
glucose (GLOO-kōs): A 6-carbon sugar, $C_6H_{12}O_6$, the preferred energy source for most cells and the only energy source for neurons under normal conditions.
glucose-dependent insulinotrophic hormone (GIP): A duodenal hormone released when the arriving chyme contains large quantities of carbohydrates; triggers the secretion of insulin and a slowdown in gastric activity.
glycerides: Lipids composed of glycerol bound to 1–3 fatty acids.
glycogen (GLĪ-kō-jen): A polysaccharide the represents an important energy reserve; a polymer consisting of a long chain of glucose molecules.
glycogenesis: The synthesis of glycogen from glucose molecules.
glycogenolysis: Glycogen breakdown, and the liberation of glucose molecules.
glycolipids (glī-cō-LIP-idz): Compounds created by the combination of carbohydrate and lipid components.
glycolysis (glī-KOL-i-sis): The catabolism of glucose to pyruvic acid in the cytoplasm.
glycoprotein (glī-kō-PRŌ-ten): A compound containing a relatively small carbohydrate group attached to a large protein.
glycosuria (glī-cō-SŪ-rē-a): The presence of glucose in the urine.
goblet cell: A goblet-shaped, mucus-producing, unicellular gland found in certain epithelia of the digestive and respiratory tracts.
goiter: Enlargement of the thyroid gland.
Golgi apparatus (gol-jē): Cellular organelle consisting of a series of membranous plates that give rise to lysosomes and secretory vesicles.
Golgi tendon organ (gol-jē): *See* **tendon organ.**
gomphosis (gom-FŌ-sis): A fibrous synarthrosis that binds a tooth to the bone of the jaw; *See* **periodontal ligament.**
gonadotrophic hormones: FSH and LH, hormones that stimulate gamete development and sex hormone secretion.
gonadotrophin-releasing hormone (GnRH) (gō-nad-ō-TRŌ-fin): Hypothalamic releasing hormone that causes the secretion of FSH and LH by the anterior pituitary gland.
gonadotrophins: Gonadotrophic hormones.
gonads (GŌ-nads): Organs that produce gametes and hormones.
gout: Clinical condition resulting from elevated uric acid concentrations in the blood and peripheral tissues.
granulocytes (GRAN-ū-lō-sīts): White blood cells containing granules visible with the light microscope; includes eosinophils, basophils, and neutrophils; also called granular leukocytes.
gray matter: Areas in the central nervous system dominated by nerve cell bodies, glial cells, and unmyelinated axons.
gray ramus: A bundle of postganglionic sympathetic nerve fibers that go to a spinal nerve for distribution to effectors in the body wall, skin, and extremities.
greater omentum: A large fold of the dorsal mesentery of the stomach that hangs in front of the intestines.
greater vestibular glands: Mucous glands in the vaginal walls that secrete into the vestibule; the equivalent of the bulbourethral glands of the male.
greenstick fracture: A fracture most often affecting the long bones of young children.
groin: The inguinal region.
gross anatomy: The study of the structural features of the human body without the aid of a microscope.
growth hormone (GH): Anterior pituitary hormone that stimulates tissue growth and anabolism when nutrients are abundant, and restricts tissue glucose dependence when nutrients are in short supply.
guanine: A purine, one of the nitrogen bases found in nucleic acids.
gustation (GUS-tā-shun): Taste.
gynecologists (gī-ne-KOL-o-jists): Physicians specializing in the pathology of the female reproductive system.
gyrus (JĪ-rus): A prominent fold or ridge of neural cortex on the surfaces of the cerebral hemispheres.
hair: A keratinous strand produced by epithelial cells of the hair follicle.
hair cells: Sensory cells of the inner ear.
hair follicle: An accessory structure of the integument; a tube lined by a stratified squamous epithelium: that begins at the surface of the skin and ends at the hair papilla.
hair root: A thickened, conical structure consisting of a connective tissue papilla and the overlying matrix, a layer of epithelial cells that produces the hair shaft.
hallux: The big toe.
haploid (HAP-lōid): Possessing one-half of the normal number of chromosomes; a characteristic of gametes.
hapten: A partial antigen; one that can bind to an antibody but not stimulate antibody production; a foreign compound that has only one antigenic-determinant site.
hard palate: The bony roof of the oral cavity, formed by the maxillary and palatine bones.
Hassall's corpuscles: Aggregations of epithelial cells in the thymus whose functions are unknown.
haustra (HAWS-trē): Saclike pouches along the length of the large intestine that result from tension in the taenia coli.
Haversian system: *See* **osteon.**
heart block: A cardiac arrhythmia due to conduction delays that affect communication between the atria and ventricles.
heat exhaustion: Condition characterized by excessive perspiration and resulting in dangerous fluid and salt losses.
heat stroke: Condition resulting from failure of the normal temperature control mechanisms, characterized by a cessation of sweating and a potentially fatal elevation in body temperature.
Heimlich maneuver (HĪm-lik): A technique for removing an airway blockage by external compression of the abdomen and forceful elevation of the diaphragm.
helper T cells: Lymphocytes (T cells) whose secretions and

other activities coordinate the cellular and humoral immune responses.

hematocrit (hē-MAT-ō-krit): Percentage of the volume of whole blood contributed by cells; also called the packed cell volume (PCV) or the volume of packed red cells (VPRC).

hematoma: A tumor or swelling filled with blood.

hematuria (hē-ma-TŪ-rē-a): The presence of abnormal numbers of red blood cells in the urine.

heme (hēm): A porphyrin ring containing a central iron atom that can reversibly bind oxygen molecules; a component of the hemoglobin molecule.

hemiplegia: Paralysis affecting one side of the body (arm, trunk, and leg).

hemocytoblasts: Stem cells whose divisions produce all of the various populations of blood cells.

hemodialysis (hē-mō-dī-AL-i-sis): Dialysis of the blood.

hemoglobin (hē-mō-GLŌ-bin): Protein composed of four globular subunits, each bound to a single molecule of heme; the protein found in red blood cells that gives them the ability to transport oxygen in the blood.

hemolysis: Breakdown (lysis) of red blood cells.

hemophilia (hē-mo-FIL-ē-a): A congenital condition resulting from the inadequate synthesis of one of the clotting factors.

hemopoiesis (hē-mō-poi-Ē-sis): Blood cell formation and differentiation.

hemorrhage: Blood loss.

hemorrhoids (HEM-o-rōidz): Swollen, varicosed veins that protrude from the walls of the rectum and/or anorectal canal.

hemostasis: The cessation of bleeding.

hemothorax: The entry of blood into one of the pleural cavities.

heparin (HEP-a-rin): An anticoagulant released by activated basophils and mast cells.

hepatic duct: Duct carrying bile away from the liver lobes and toward the union with the cystic duct.

hepatic portal vein: Vessel that carries blood between the intestinal capillaries and the sinusoids of the liver.

hepatitis (hep-a-TĪ-tis): Inflammation of the liver, resulting from exposure to toxic chemicals, drugs, or viruses.

hepatocyte (he-PAT-ō-sit): A liver cell.

hernia: The protrusion of a loop or portion of a visceral organ through the abdominopelvic wall or into the thoracic cavity.

herniated disc: Rupture of the connective tissue sheath of the nucleus pulposus of an intervertebral disc.

heterotopic: Ectopic; outside of its normal location.

heterozygous (het-er-ō-ZĪ-gus): Possessing two different alleles at corresponding loci on a chromosome pair; the individual's phenotype may be determined by one or both of the alleles.

hexose: A 6-carbon simple sugar.

hiatus (hī-Ā-tus): A gap, cleft, or opening.

high-density lipoprotein: A lipoprotein with a relatively small lipid content, thought to be responsible for the movement of cholesterol from peripheral tissues to the liver.

hilus (HĪ-lus): A localized region where blood vessels, lymphatics, nerves, and/or other anatomical structures are attached to an organ.

hippocampus: A region beneath the floor of a lateral ventricle involved with emotional states and the conversion of short-term to long-term memories.

hirsutism (HER-sut-izm): Excessive hair growth in women that follows the distribution pattern typical of adult males; sometimes caused by the overproduction of androgens.

histamine (HIS-ta-min): Chemical released by stimulated mast cells or basophils to initiate or enhance an inflammatory response.

histology (his-TOL-ō-jē): The study of tissues.

histones: Proteins associated with the DNA of the nucleus, and around which the DNA strands are wound.

holocrine (HŌ-lō-krin): Form of exocrine secretion where the secretory cell becomes swollen with vesicles and then ruptures.

homeostasis (hō-mē-ō-STĀ-sis): The maintenance of a relatively constant internal environment.

homologous chromosomes (hō-MOL-o-gus): The members of a chromosome pair, each containing the same gene loci.

homozygous (hō-mō-ZĪ-gus): Having the same gene for a particular character on two homologous chromosomes.

hormone: A compound secreted by one cell that travels through the circulatory system to affect the activities of cells in another portion of the body.

human chorionic gonadotrophin (HCG): Placental hormone that maintains the corpus luteum for the first 3 months of pregnancy.

human immunodeficiency virus (HIV): The infectious agent that causes **acquired immunodeficiency syndrome (AIDS).**

human leukocyte antigen (HLA): Antigens on cell surfaces important to foreign antigen recognition and that play a role in the coordination and activation of the immune response.

human placental lactogen (HPL): Placental hormone that stimulates the functional development of the mammary glands.

humoral immunity: Immunity resulting from the presence of circulating antibodies produced by plasma cells.

hyaluronic acid: A proteoglycan in the matrix of many connective tissues that gives the matrix a viscous consistency; also functions as intercellular cement.

hyaluronidase: An enzyme that breaks down hyaluronic acid; produced by some bacteria and found in the acrosomal cap of a sperm cell.

hydrocephalus: Condition resulting from excessive production or inadequate drainage of cerebrospinal fluid.

hydrogen bond: Weak interaction between the hydrogen atom on one molecule and a negatively-charged portion of another molecule.

hydrolysis (hī-DROL-i-sis): The breakage of a chemical bond through the addition of a water molecule; the reverse of dehydration synthesis.

hydrophilic (hī-drō-PHIL-ik): Freely associating with water; readily entering into solution.

hydrophobic: Incapable of freely associating with water molecules; insoluble.

hydrostatic pressure: Fluid pressure.

hydroxyl group (hī-DROK-sil): OH^-.

hypercapnia (hī-per-KAP-nē-a): High plasma carbon dioxide concentrations, often the result of hypoventilation or inadequate tissue perfusion.

hyperglycemia: Elevated plasma glucose concentrations.

hyperkalemia (hī-per-kā-LĒ-mē-a): Abnormally high potassium concentrations in the extracellular fluid.

hypernatremia: Abnormally high sodium concentrations in the extracellular fluid.

hyperopia: The farsighted condition, characterized by an inability to focus on objects closeby.

hyperplasia: Abnormal enlargement of an organ due to an increase in the number of cells.

hyperpnea (hī-perp-NĒ-a): Abnormal increases in the rate and depth of respiration.

hyperpolarization: Movement of the transmembrane potential away from the normal resting potential and farther from 0 mV.

hyperreflexia: Abnormally exaggerated reflex responses to stimulation.

hypersecretion: Overactivity of glands that produce exocrine or endocrine secretions.

hypersensitivity: Overreaction to an allergen that results in tissue damage and inflammation.
hypertension: Abnormally high blood pressure.
hyperthermia: Excessively high body temperature.
hyperthyroidism: Excessive production of thyroid hormones.
hypertonic: A term used when comparing two solutions, to refer to the solution with the higher osmolarity.
hypertrophy (HI-per-trō-fē): Increase in the size of tissue without cell division.
hyperventilation (hī-per-ven-ti-LĀ-shun): A rate of respiration sufficient to reduce the plasma pCO_2 to levels below normal.
hypervitaminosis (hī-per-vī-ta-min-Ō-sis): Clinical condition caused by the excessive ingestion and uptake of vitamins.
hypesthesia: Abnormally decreased sensitivity to stimuli.
hypoblast (HĪ-pō-blast): The undersurface of the inner cell mass that faces the blastocoel of the early embryo.
hypocapnia: Abnormally low plasma pCO_2, usually the result of hyperventilation.
hypodermic needle: A needle inserted through the skin to introduce drugs into the subcutaneous layer.
hypodermis: The subcutaneous layer, a region of loose connective tissue also called the **superficial fascia.**
hypokalemia (hī-pō-kā-LĒ-mē-a): Abnormally low plasma potassium concentrations.
hyponatremia: Abnormally low plasma sodium concentrations.
hyponychium (hī-pō-NIK-ē-um): A thickening in the epidermis beneath the free edge of a nail.
hypophyseal portal system (hī-po-FI-sē-al): Network of vessels that carry blood from capillaries in the hypothalamus to capillaries in the anterior pituitary gland (hypophysis).
hypophysis (hī-POF-i-sis): The anterior pituitary gland, which can be further subdivided into the pars distalis and the pars intermedia.
hyporeflexia: Abnormally depressed reflex responses to stimuli.
hyposecretion: Abnormally low rates of exocrine or endocrine secretion.
hypothalamus: The floor of the diencephalon; region of the brain containing centers involved with the unconscious regulation of visceral functions, emotions, drives, and the coordination of neural and endocrine functions.
hypothermia (hī-pō-THER-mē-a): Abnormally low body temperatures.
hypothesis: A prediction that can be subjected to scientific analysis and review.
hypotonic: When comparing two solutions, used to refer to the one with the lower osmolarity.
hypoventilation: A respiratory rate insufficient to keep plasma pCO_2s within normal levels.
hypovitaminosis: Clinical condition resulting from inadequate vitamin ingestion and uptake; vitamin deficiency.
hypovolemic (hī-pō-vo-LĒ-mik): An abnormally low blood volume.
hypoxia (hī-POKS-ē-a): Low tissue oxygen concentrations.
ileum (IL-ē-um): The last 8 ft of the small intestine.
ileocecal valve (il-ē-ō-SĒ-kal): A fold of mucous membrane that guards the connection between the ileum and the cecum.
ileostomy (il-ē-OS-to-mē): Surgical creation of an opening into the ileum; the opening created when the ilium is surgically attached to the abdominal wall.
ilium (IL-ē-um): The largest of the three bones whose fusion creates a coxa.
immunity: Resistance to injuries and diseases caused by foreign compounds, toxins, and pathogens.
immunization: Developing immunity by the deliberate exposure to antigens under conditions that prevent the development of illness but stimulate the production of memory B cells.
immunodeficiency: An inability to produce normal numbers and types of antibodies and sensitized lymphocytes.
immunoglobulin (i-mū-nō-GLOB-ū-lin): A circulating antibody.
immunosuppression (i-mū-nō-su-PRE-shun): The suppression of immune responses by the administration of drugs.
implantation (im-plan-TĀ-shun): The erosion of a blastocyst into the uterine wall.
impotence: Inability to obtain or maintain an erection in the male.
inclusions: Aggregations of insoluble pigments, nutrients, or other materials in the cytoplasm.
incontinence (in-KON-ti-nens): Inability to voluntarily control micturition (or defecation).
incus (IN-kus): The central auditory ossicle, situated between the malleus and the stapes in the middle ear cavity.
inducer: A stimulus that promotes the activity of a specific gene.
inexcitable: Incapable of conducting an action potential.
infarct: An area of dead cells resulting from an interruption of circulation.
infection: Invasion and colonization of body tissues by pathogenic organisms.
inferior: A directional reference meaning below.
inferior vena cava: The vein that carries blood from the parts of the body below the heart to the right auricle.
infertility: Inability to conceive.
inflammation: A nonspecific defense mechanism that operates at the tissue level, characterized by swelling, redness, warmth, pain, and some loss of function.
inflation reflex: A reflex mediated by the vagus nerve that prevents overexpansion of the lungs.
infundibulum (in-fun-DIB-ū-lum): A tapering, funnel-shaped structure; in the nervous system, refers to the connection between the pituitary gland and the hypothalamus; the infundibulum of the uterine tube is the entrance bounded by fimbriae that receives the ova at ovulation.
ingestion: The introduction of materials into the digestive tract via the mouth.
inguinal canal: A passage through the abdominal wall that marks the path of testicular descent, and that contains the testicular arteries, veins, and ductus deferens.
inguinal region: The area near the junction of the trunk and the thighs that contains the external genitalia.
inhibin (in-HIB-in): A hormone produced by the sustentacular cells that inhibits the pituitary secretion of FSH.
inhibitory postsynaptic potential (IPSP): A hyperpolarization of the postsynaptic membrane following the arrival of a neurotransmitter.
initial segment: The proximal portion of the axon, adjacent to the axon hillock, where an action potential first appears.
injection: Forcing of fluid into a body part or organ.
inner cell mass: Cells of the blastocyst that will form the body of the embryo.
inner ear: *See* **internal ear.**
innervation: The distribution of sensory and motor nerves to a specific region or organ.
insertion: Point of attachment of a muscle that is more movable.
inspiration: Inhalation, the movement of air into the respiratory system.
inspiratory reserve: The maximum amount of air that can be drawn into the lungs over and above the normal tidal volume.
insoluble: Incapable of dissolving in solution.
insomnia: Sleep disorder characterized by an inability to fall asleep.
insulin (IN-su-lin): Hormone secreted by the beta cells of the pancreatic islets; causes a reduction in plasma glucose concentrations.

integument (in-TEG-ū-ment): The skin.
intercalated discs (in-TER-ka-lā-ted): Regions where adjacent cardiocytes interlock and where gap junctions permit electrical coupling between the cells.
intercellular cement: Proteoglycans, especially hyaluronic acid, found between adjacent epithelial cells.
intercellular fluid: *See* **interstitial fluid.**
interdigitate: To interlock.
interferons (in-ter-FĒR-ons): Peptides released by virally infected cells, especially lymphocytes, that make other cells more resistant to viral infection and slow viral replication.
interleukins (in-ter-LOO-kins): Peptides released by activated monocytes and lymphocytes that assist in the coordination of the cellular and humoral immune responses.
internal capsule: Term given to the appearance of the white matter of the cerebral hemispheres on gross dissection of the brain.
internal ear: The membranous labyrinth that contains the organs of hearing and equilibrium.
internal nares: The entrance to the nasopharynx from the nasal cavity.
internal respiration: Diffusion of gases between the blood and interstitial fluid.
interneuron: An association neuron; neurons inside the CNS that are interposed between sensory and motor neurons.
interoceptors: Sensory receptors monitoring the functions and status of internal organs and systems.
interosseous membrane: Fibrous connective tissue membrane between the shafts of the tibia and fibula or the radius and ulna; an example of a fibrous amphiarthrosis.
interphase: Stage in the life of a cell during which the chromosomes are uncoiled and all normal cellular functions except mitosis are under way.
intersegmental reflex: A reflex that involves several segments of the spinal cord.
interstitial fluid (in-ter-STISH-al): Fluid in the tissues that fills the spaces between cells.
interstitial cell-stimulating hormone: An alternative name for LH in the male; stimulates androgen production by the interstitial cells of the testes.
interstitial growth: Form of cartilage growth through the growth, mitosis, and secretion of chondrocytes inside the matrix.
interventricular foramen: The opening that permits fluid movement between the lateral and third ventricles.
intervertebral disc: Fibrocartilage pad between the centra of successive vertebrae that acts as a shock absorber.
intestinal crypt: A tubular epithelial pocket lined by secretory cells and opening into the lumen of the digestive tract; also called an intestinal gland.
intestine: Tubular organ of the digestive tract.
intracellular fluid: The cytosol.
intrafusal fibers: Muscle spindle fibers.
intramembranous ossification (in-tra-MEM-bra-nus): The formation of bone within a connective tissue without the prior development of a cartilaginous model.
intramuscular injection: Injection of medication into the bulk of a skeletal muscle.
intraocular pressure: The hydrostatic pressure exerted by the aqueous humor of the eye.
intrapleural pressure: The pressure measured in a pleural cavity; also called the intrathoracic pressure.
intrapulmonary pressure (in-tra-PUL-mo-ner-ē): The pressure measured in an alveolus; also called the intraalveolar pressure.
intrauterine: Within the uterus; used to refer to the period of prenatal development.
intrinsic factor: Glycoprotein secreted by the parietal cells of the stomach that facilitates the intestinal absorption of vitamin B_{12}.
intrinsic pathway: A pathway of the clotting system that begins with the activation of platelets and ends with the formation of platelet thromboplastin.
inversion: A turning inward.
in vitro: Outside of the body, in an artificial environment.
in vivo: In the living body.
involuntary: Not under conscious control.
ion: An atom or molecule bearing a positive or negative charge due to the acceptance or donation of an electron.
ionic bond (ī-ON-ik): Molecular bond created by the attraction between ions with opposite charges.
ionization (ī-on-i-ZĀ-shun): Dissociation; the breakdown of a molecule in solution to form ions.
ipsilateral: A reflex response affecting the same side as the stimulus.
iris: A contractile structure made up of smooth muscle that forms the colored portion of the eye.
ischemia (is-KĒ-mē-a): Inadequate blood supply to a region of the body.
ischium (IS-kē-um): One of the three bones whose fusion creates the coxa.
islets of Langerhans: See pancreatic islets.
isthmus (IS-mus): A narrow band of tissue connecting two larger masses.
isometric contraction: A muscular contraction characterized by rising tension production but no change in length.
isotonic contraction: A muscular contraction during which tension climbs and then remains stable as the muscle shortens.
isotonic: A solution having an osmolarity that does not result in water movement across cell membranes; of the same contractive strength.
isotope: Forms of an element whose atoms contain the same number of protons but different numbers of neutrons (and thus differ in atomic weight).
jaundice (JAWN-dis): Condition characterized by yellowing of connective tissues due to elevated tissue bilirubin levels; usually associated with damage to the liver or biliary system.
jejunum (je-JŪ-num): The middle portion of the small intestine.
joint: An area where adjacent bones interact; an articulation.
juxtaglomerular apparatus: The macula densa and the juxtaglomerular cells; a complex responsible for the release of renin and erythropoietin.
juxtaglomerular cells: Modified smooth muscle cells in the walls of the afferent and efferent arterioles adjacent to glomerulus and the macula densa.
karyotyping (KAR-ē-ō-tī-ping): The determination of the chromosomal characteristics of an individual or cell.
keratin (KER-a-tin): Tough, fibrous protein component of nails, hair, calluses, and the general integumentary surface.
keratinization (KER-a-tin-ī-za-shun): The production of keratin by epithelial cells.
keratohyalin (ker-a-tō-HĪ-a-lin): A protein precursor of keratin.
keto acid: A molecule that ends in -COCOOH; the carbon chain that remains after the deamination or transamination of an amino acid.
ketoacidosis (kē-tō-as-i-DŌ-sis): A reduction in the pH of body fluids due to the presence of large numbers of ketone bodies.
ketogenic amino acids: Amino acids whose catabolism yields ketone bodies rather than pyruvic acid.
ketone bodies: Keto acids produced during the catabolism of lipids and ketogenic amino acids; specifically acetone, acetoacetate, and beta-hydroxybutyrate.
ketonemia (kē-to-NĒ-mē-a): Abnormal concentrations of ketone bodies in the blood.
ketonuria (kē-to-NŪ-rē-a): Abnormal concentrations of ketone bodies in the urine.

ketosis (kē-TŌ-sis): Condition characterized by the abnormal production of ketone bodies.
kidney: A component of the urinary system; an organ functioning in the regulation of plasma composition, including the excretion of wastes and the maintenance of normal fluid and electrolyte balance.
killer T cells: *see* **cytotoxic T cells.**
kilocalorie (KIL-o-kal-o-rē): The amount of heat required to raise the temperature of a kilogram of water 1° C.
Krebs cycle: Aerobic reaction sequence occurring in the mitochondrial matrix; in the process organic molecules are broken down, carbon dioxide molecules are released, and hydrogen molecules are transferred to coenzymes that deliver them to the electron transport system.
Kupffer cells (KOOP-fer): Stellate cells of the liver; phagocytic cells of the liver sinusoids.
kyphosis (kī-FŌ-sis): Exaggerated thoracic curvature.
labia (LĀ-bē-a): Lips; labia majora and minora are components of the female external genitalia.
labrum: A lip or rim.
labyrinth: A maze of passageways; usually refers to the structures of the inner ear.
lacrimal gland (LAK-ri-mal): Tear gland on the dorsolateral surface of the eye.
lactase: An enzyme that breaks down milk proteins.
lactation (lak-TĀ-shun): The production of milk by the mammary glands.
lacteal (LAK-tē-al): A terminal lymphatic within an intestinal villus.
lactic acid: Compound produced from pyruvic acid during anaerobic glycolysis.
lactiferous duct (lak-TIF-e-rus): Duct draining one lobe of the mammary gland.
lactiferous sinus: An expanded portion of a lactiferous duct adjacent to the nipple of a breast.
lacuna (la-KŪ-nē): A small pit or cavity.
lambdoidal suture (lam-DŌID-al): Synarthrotic articulation between the parietal and occipital bones of the cranium.
lamellae (la-MEL-lē): Concentric layers of bone within an osteon.
lamina (LA-min-a): A thin sheet or layer.
lamina propria (LA-mi-na PRO-prē-a): Loose connective tissue that underlies a mucous epithelium and forms part of a mucous membrane.
laminectomy: Removal of the spinous processes of a vertebra to gain access and treat a herniated disc.
Langerhans cells (LAN-ger-hanz): Cells in the epithelium of the skin and digestive tract that participate in the immune response by presenting antigens to T cells.
laparoscope (LAP-a-ro-skōp): Fiberoptic instrument used to visualize the contents of the abdominopelvic cavity.
large intestine: The terminal portions of the intestinal tract, consisting of the colon, the rectum, and the anorectal canal.
laryngopharynx (la-rin-gō-FAR-inks): Division of the pharynx inferior to the epiglottis and superior to the esophagus.
larynx (LAR-inks): A complex cartilaginous structure that surrounds and protects the glottis and vocal cords; the superior margin is bound to the hyoid bone and the inferior margin is bound to the trachea.
latent period: The time between the stimulation of a muscle and the start of the contraction phase.
lateral: Pertaining to the side.
lateral apertures: Openings in the roof of the fourth ventricle that permit the circulation of CSF into the subarachnoid space.
lateral ventricle: Fluid-filled chamber within one of the cerebral hemispheres.
laxatives: Compounds that promote defecation via increased peristalsis or an increase in the water content and volume of the feces.
lens: The transparent body lying behind the iris and pupil and in front of the vitreous humor.
lesion: A localized abnormality in tissue organization.
lesser omentum: A small pocket in the mesentery that connects the lesser curvature of the stomach to the liver.
leukemia (loo-KĒ-mē-ah): A malignant disease of the blood-forming tissues.
leukocyte (LOO-kō-sīt): A white blood cell.
leukocytosis (loo-kō-sī-TŌ-sis): Abnormally high numbers of circulating white blood cells.
leukopenia (loo-kō-PĒ-nē-ah): Abnormally low numbers of circulating white blood cells.
ligament (LI-ga-ment): Dense band of connective tissue fibers that attach one bone to another.
ligamentum arteriosum: The fibrous strand found in the adult that represents the remains of the ductus arteriosus of the fetus.
ligamentum teres: The fibrous strand in the falciform ligament that represents the remains of the umbilical vein of the fetus.
ligate: To tie off.
limbic system (LIM-bik): Group of nuclei and centers in the cerebrum and diencephalon that are involved with emotional states, memories, and behavioral drives.
limbus (LIM-bus): The edge of the cornea, marked by the transition from the corneal epithelium to the ocular conjunctiva.
liminal stimulus: A stimulus sufficient to depolarize the transmembrane potential of an excitable membrane to threshold, and produce an action potential.
linea alba: Tendinous band that runs along the midline of the rectus abdominis.
lingual: Pertaining to the tongue.
lingual frenulum: An epithelial fold that attaches the inferior surface of the tongue to the floor of the mouth.
lipase (LI-pāz): A pancreatic enzyme that breaks down triglycerides.
lipemia (lip-Ē-mē-a): Elevated concentration of lipids in the circulation.
lipid: An organic compound containing carbons, hydrogens, and oxygens in a ratio that does not approximate 1:2:1; includes fats, oils, and waxes.
lipofuscin (li-po-FŪ-shun): A pigment inclusion of uncertain significance that is found in aging nerve cells.
lipogenesis (li-pō-JEN-e-sis): Synthesis of lipids from nonlipid precursors.
lipoids: Prostaglandins, steroids, phospholipids, glycolipids, etc.
lipolysis: The catabolism of lipids as a source of energy.
lipoprotein (li-po-PRŌ-tēn): A compound containing a relatively small lipid bound to a protein.
liver: An organ of the digestive system with varied and vital functions that include the production of plasma proteins, the excretion of bile, the storage of energy reserves, the detoxification of poisons, and the interconversion of nutrients.
lobule (LOB-ūl): The basic organizational unit of the liver at the histological level.
long-term memories: Memories that persist for an extended period.
loose connective tissue: A loosely organized, easily distorted connective tissue containing several different fiber types, a varied population of cells, and a viscous ground substance.
lordosis (lor-DŌ-sis): An exaggeration of the lumbar curvature.
lumbar: Pertaining to the lower back.
lumen: The central space within a duct or other internal passageway.
lungs: Paired organs of respiration, situated in the left and right pleural cavities.
luteinizing hormone (LH) (LOO-tē-in-ī-zing): Anterior pituitary hormone that in the female assists FSH in follicle

stimulation, triggers ovulation, and promotes the maintenance and secretion of the endometrial glands; in the male, stimulates spermatogenesis; also known as **interstitial cell-stimulating hormone.**

luxation (luks-Ā-shun): Dislocation of a joint.

lymph: Fluid contents of lymphatic vessels, similar in composition to interstitial fluid.

lymphadenopathy (lim-fad-e-NOP-a-the): Pathological enlargement of the lymph nodes.

lymphatics: Vessels of the lymphatic system.

lymphedema (lim-fe-DĒ-ma): Swelling of peripheral tissues due to excessive lymph production or inadequate drainage.

lymph nodes: Lymphatic organs that monitor the composition of lymph.

lymphocyte (LIM-fō-sīt): A cell of the lymphatic system that participates in the immune response.

lymphokines: Chemicals secreted by activated lymphocytes.

lymphopoiesis: The production of lymphocytes.

lymphotoxin (lim-fō-TOK-sin): A secretion of lymphocytes that kills the target cells.

lysis (LĪ-sis): The destruction of a cell through the rupture of its cell membrane.

lysosome (LĪ-so-sōm): Intracellular vesicle containing digestive enzymes.

lysozyme: An enzyme present in some exocrine secretions that has antibiotic properties.

macrophage: A phagocytic cell of the monocyte-macrophage system.

macula (MAK-ū-la): A receptor complex in the saccule or utricle that responds to linear acceleration or gravity.

macula densa (MAK-ū-la DEN-sa): A group of specialized secretory cells in a portion of the distal convoluted tubule adjacent to the glomerulus and the juxtaglomerular merular cells; a component of the juxtaglomerular apparatus.

macula lutea (LOO-tē-a): The fovea.

malignant cancer: A form of cancer characterized by rapid cellular growth and the spread of cancer cells throughout the body.

malleus (MAL-ē-us): The first auditory ossicle, bound to the tympanic membrane and the incus.

malnutrition: An unhealthy state produced by inadequate dietary intake of nutrients, calories, and/or vitamins.

mamillary bodies (MAM-i-lar-ē): Nuclei in the hypothalamus concerned with feeding reflexes and behaviors; a component of the limbic system.

mammary glands: Milk-producing glands of the female breast.

manus: The hand.

marrow: A tissue that fills the internal cavities in a bone; may be dominated by hemopoietic cells (red marrow) or adipose tissue (yellow marrow).

mass peristalsis: Powerful peristaltic contraction that moves fecal materials along the colon and into the rectum.

mass reflex: Hyperreflexia in an area innervated by spinal cord segments distal to an area of injury.

mast cell: A connective tissue cell that when stimulated releases histamine, serotonin, and heparin, initiating the inflammatory response.

mastectomy: Surgical removal of part or all of a mammary gland.

mastication (mas-ti-KĀ-shun): Chewing.

mastoid sinus: Air-filled spaces in the mastoid process of the temporal bone.

matrix: The ground substance of a connective tissue.

maxillary sinus (MAK-si-ler-ē): One of the paranasal sinuses; an air-filled chamber lined by a respiratory epithelium that is located in a maxillary bone and opens into the nasal cavity.

meatus (mē-Ā-tus): An opening or entrance into a passageway.

mechanoreception: Detection of mechanical stimuli, such as touch, pressure, or vibration.

medial: Toward the midline of the body.

mediastinum (mē-dē-as-TĪ-num): Central tissue mass that divides the thoracic cavity into two pleural cavities; includes the aorta and other great vessels, the esophagus, trachea, thymus, the pericardial cavity and heart, and a host of nerves, small vessels, and lymphatics.

medulla: Inner layer or core of an organ.

medulla oblongata: The most caudal of the five brain regions, also known as the **myelencephalon.**

medullary cavity: The space within a bone that contains the marrow.

medullary rhythmicity center: Center in the medulla that sets the background pace of respiration; includes inspiratory and expiratory centers.

megakaryocytes (meg-a-KAR-ē-ō-sīts): Bone marrow cells responsible for the formation of platelets.

meiosis (mī-Ō-sis): Cell division that produces gametes with half of the normal somatic chromosome complement.

melanin (ME-la-nin): Yellow-brown pigment produced by the melanocytes of the skin.

melanocyte (me-LAN-ō-sīt): Specialized cell found in the deeper layers of the stratified squamous epithelium of the skin, responsible for the production of melanin.

melanocyte-stimulating hormone (MSH) (me-LAN-ō-sīt): Hormone of the pars intermedia of the anterior pituitary that stimulates melanin production.

melanomas (mel-a-NŌ-mas): Dangerous malignant skin cancers that involve melanocytes.

melatonin (mel-a-tŌ-nin): Hormone secreted by the pineal gland; inhibits secretion of MSH and GnRF.

membrane: Any sheet or partition; a layer consisting of an epithelium and the underlying connective tissue.

membrane flow: The movement of sections of membrane surface to and from the cell surface and components of the endoplasmic reticulum, the Golgi apparatus, and vesicles.

membrane potential: *See* **transmembrane potential.**

membranous labyrinth: Endolymph-filled tubes of the inner ear that enclose the receptors of the inner ear.

memory: The ability to recall information or sensations; can be divided into short-term and long-term memories.

menarche (me-NAR-kē): The beginning of menstrual function.

meninges (men-IN-jēz): Three membranes that surround the surfaces of the CNS; the dura mater, the pia mater, and the arachnoid.

meningitis: Inflammation of the spinal or cranial meninges.

meniscectomy: Removal of a meniscus.

meniscus (mēn-IS-kus): A fibrocartilage pad between opposing surfaces in a joint.

menses (MEN-sēz): The first menstrual period that normally occurs at puberty.

merocrine (MER-o-krin): A method of secretion where the cell ejects materials through exocytosis.

mesencephalic aqueduct: Passageway that connects the third ventricle (diencephalon) with the fourth ventricle (meten cephalon).

mesencephalon (mez-en-SEF-a-lon): The midbrain.

mesenchyme: Embryonic/fetal connective tissue.

mesentery (MEZ-en-ter-ē): A double layer of serous membrane that supports and stabilizes the position of an organ in the abdominopelvic cavity and provides a route for the associated blood vessels, nerves, and lymphatics.

mesoderm: The middle germ layer that lies between the ectoderm and endoderm of the embryo.

mesothelium (mez-ō-THĒ-lē-um): A simple squamous epithelium that lines one of the divisions of the ventral body cavity.

messenger RNA (mRNA): RNA formed at transcription to direct protein synthesis in the cytoplasm.
metabolic turnover: The continual breakdown and replacement of organic materials within living cells.
metabolism (me-TAB-ō-lizm): The sum of all of the biochemical processes underway within the human body at a given moment; includes anabolism and catabolism.
metabolites (me-TAB-ō-līts): Compounds produced in the body as the result of metabolic reactions.
metacarpals: The five bones of the palm of the hand.
metalloproteins (met-al-ō-PRO-tēnz): Plasma proteins that transport metal ions.
metaphase (MET-ā-faz): A stage of mitosis wherein the chromosomes line up along the equatorial plane of the cell.
metaphysis (me-TA-fi-sis): The region of a long bone between the epiphysis and diaphysis, corresponding to the location of the epiphyseal plate of the developing bone.
metarteriole (met-ar-TĒ-rē-ōl): A vessel that connects an arteriole to a venule and that provides blood to a capillary plexus.
metastasis (me-TAS-ta-sis): The spread of a disease from one organ to another.
metatarsal: One of the five bones of the foot that articulate with the tarsals (proximally) and the phalanges (distally).
metencephalon (met-en-SEF-a-lon): The pons and cerebellum of the brain.
micelle (mī-SEL): A spherical aggregation of bile salts, monoglycerides, and fatty acids in the lumen of the intestinal tract.
microcephaly (mī-krō-SEF-a-lē): An abnormally small cranium, due to premature closure of one or more fontanelles.
microfilaments: Fine protein filaments visible with the electron microscope; components of the cytoskeleton.
microglia (mī-KROG-lē-a): Phagocytic glial cells in the CNS, derived from the monocytes of the blood.
microphages: Neutrophils and eosinophils.
microtubules: Microscopic tubules that are part of the cytoskeleton, and are found in cilia, flagella, the centrioles, and spindle fibers.
microvilli: Small, fingerlike extensions of the exposed cell membrane of an epithelial cell.
micturition (mîk-tū-RI-shun): Urination.
midbrain: The mesencephalon.
middle ear: Space between the external and internal ear that contains auditory ossicles.
midsagittal plane: A plane passing through the midline of the body that divides it into left and right halves.
mineralocorticoid: Corticosteroids produced by the zona glomerulosa of the adrenal cortex; steroids such as aldosterone, that affect mineral metabolism.
miscarriage: Spontaneous abortion.
mitochondrion (mī-tō-KON-drē-on): An intracellular organelle responsible for generating most of the ATP required for cellular operations.
mitosis (mī-TŌ-sis): The division of a single cell that produces two identical daughter cells; the primary mechanism of tissue growth.
mitral valve (MĪ-tral): The left AV, or bicuspid, valve of the heart.
mixed gland: A gland that contains exocrine and endocrine cells, or an exocrine gland that produces serous and mucous secretions.
mixed nerve: A peripheral nerve that contains sensory and motor fibers.
mole: A quantity of an element or compound having a mass in grams equal to its atomic or molecular weight.
molecular weight: The sum of the atomic weights of the atoms in a molecule.
molecule: A compound containing two or more atoms that are held together by chemical bonds.
monoclonal antibodies (mo-nō-KLŌ-nal): Antibodies produced by genetically identical cells under laboratory conditions.
monocytes (MON-o-sīts): Phagocytic agranulocytes (white blood cells) in the circulating blood.
monoglyceride (mo-nō-GLI-se-rīd): A lipid consisting of a single fatty acid bound to a molecule of glycerol.
monokines: Secretions released by activated cells of the monocyte/macrophage system to coordinate various aspects of the immune response.
monosaccharide (mon-ō-SAK-ah-rīd): A simple sugar, such as glucose or ribose.
monosynaptic reflex: A reflex where the sensory afferent synapses directly on the motor efferent.
monozygotic twins: Twins produced through the splitting of a single fertilized egg (zygote).
morula (MOR-ū-la): A mulberry-shaped collection of cells produced through the mitotic divisions of a zygote.
motor unit: All of the muscle cells controlled by a single motor neuron.
mucins (MŪ-sins): Proteoglycans responsible for the lubricating properties of mucus.
mucosa (mū-KŌ-sa): A mucous membrane; the epithelium plus the lamina propria.
mucous: An adjective referring to the presence or production of mucus.
mucous membrane: *See* **mucosa.**
mucus: Lubricating secretion produced by unicellular and multicellular glands along the digestive, respiratory, urinary, and reproductive tracts.
multifactorial trait: A phenotypic character that reflects the interactions of many different genes.
multipennate: A muscle whose internal fibers are organized around several different tendons.
multipolar neuron: A neuron with many dendrites and a single axon, the typical form of a motor neuron.
multiunit smooth muscle: Smooth muscle tissue whose muscle cells are innervated in motor units.
muriatic acid: Hydrochloric acid (HCl).
muscarinic receptors (mus-kar-IN-ik): Membrane receptors sensitive to acetylcholine (ACh) and to muscarine, a toxin produced by certain mushrooms; found at all parasympathetic neuroeffector junctions and at a few sympathetic neuroeffector junctions.
muscle: A contractile organ composed of muscle tissue, blood vessels, nerves, connective tissues, and lymphatics.
muscle tissue: A tissue characterized by the presence of cells capable of contraction; includes skeletal, cardiac, and smooth muscle tissue.
muscularis externa (mus-kū-LAR-is): Concentric layers of smooth muscle responsible for peristalsis.
muscularis mucosae: Layer of smooth muscle beneath the lamina propria responsible for moving the mucosal surface.
mutagens (MŪ-ta-jenz): Chemical agents that induce mutations and may be carcinogenic.
myalgia (mī-AL-jē-a): Muscle pain.
myasthenia gravis (mī-as-THĒ-nē-a GRA-vis): Muscular weakness due to a reduction in the number of ACh receptor sites on the sarcolemmal surface; suspected to be an autoimmune disorder.
myelencephalon (mī-el-en-SEF-a-lon): The medulla oblongata.
myelin (MĪ-e-lin): Insulating sheath around an axon consisting of multiple layers of glial cell membrane; significantly increases conduction rate along the axon.
myelination: The formation of myelin.
myenteric plexus (mī-en-TER-ik): Parasympathetic motor neurons and sympathetic postganglionic fibers located between the circular and longitudinal layers of the muscularis externa.
myocardial infarction (mī-ō-KAR-dē-al): Heart attack;

damage to the heart muscle due to an interruption of regional coronary circulation.

myocarditis: Inflammation of the myocardium.

myocardium: The cardiac muscle tissue of the heart.

myofibril: Organized collections of myofilaments in skeletal and cardiac muscle cells.

myofilaments: Fine protein filaments, composed of the proteins actin (thin filaments) and myosin (thick filaments).

myoglobin (mī-ō-glō-bin): An oxygen-binding pigment especially common in slow skeletal and cardiac muscle fibers.

myogram: A recording of the tension produced by muscle fibers on stimulation.

myometrium (mī-ō-MĒ-trē-um): The thick layer of smooth muscle in the wall of the uterus.

myopia: Nearsightedness, an inability to accommodate for distant vision.

myosepta: Connective tissue partitions that separate adjacent skeletal muscles.

myosin: Protein component of the thick myofilaments.

myositis (mī-ō-SĪ-tis): Inflammation of muscle tissue.

nail: Keratinous structure produced by epithelial cells of the nail root.

narcolepsy: A sleep disorder characterized by falling asleep at inappropriate moments.

nares, external (NA-rēz): The entrance from the exterior to the nasal cavity.

nares, internal: The entrance from the nasal cavity to the nasopharynx.

nasal cavity: A chamber in the skull bounded by the internal and external nares.

nasolacrimal duct: Passageway that transports tears from the nasolacrimal sac to the nasal cavity.

nasolacrimal sac: Chamber that receives tears from the lacrimal ducts.

nasopharynx (nā-zō-FAR-inks): Region posterior to the internal nares, superior to the soft palate, and ending at the oropharynx.

N compound: An organic compound containing nitrogen atoms.

necrosis (NEK-rō-sis): Death of cells or tissues from disease or injury.

negative feedback: Corrective mechanism that opposes or negates a variation from normal limits.

neonate: A newborn infant, or baby.

neoplasm: A tumor, or mass of abnormal tissue.

nephritis (nef-RĪ-tis): Inflammation of the kidney.

nephrolithiasis (nef-rō-li-THĪ-a-sis): Condition resulting from the formation of kidney stones.

nephron (NEF-ron): Basic functional unit of the kidney.

nerve impulse: An action potential in a nerve cell membrane.

neural cortex: An area where gray matter is found at the surface of the CNS.

neurilemma (nū-ri-LEM-ma): The outer surface of a glial cell that encircles an axon.

neuroeffector junction: A synapse between a motor neuron and a peripheral effector, such as a muscle, gland cell, or fat cell.

neurofibrils: Microfibrils in the cytoplasm of a neuron.

neurofilaments: Microfilaments in the cytoplasm of a neuron.

neuroglandular junction: A specific type of neuroeffector junction.

neuroglia (nū-ROG-lē-a): Non-neural cells of the CNS and PNS that support and protect the neurons.

neurohypophysis (NŪ-rō-hī-pof-i-sis): The posterior pituitary, or pars nervosa.

neuromodulator (nū-rō-MOD-ū-la-tor): A chemical that adjusts the sensitivities of another neuron to specific neurotransmitters.

neuromuscular junction: A specific type of neuroeffector junction.

neuron (NŪ-ron): A nerve cell.

neurotransmitter: Chemical compound released by one neuron to affect the transmembrane potential of another.

neurotubules: Microtubules in the cytoplasm of a nerve cell.

neurulation: The embyrological process responsible for the formation of the CNS.

neutron: A fundamental particle that does not carry a positive or negative charge.

neutropenia: An abnormally low number of neutrophils in the circulating blood.

neutrophil (NŪ-tro-fil): A phagocytic microphage (granulocyte, WBC) that is very numerous and usually the first of the mobile phagocytic cells to arrive at an area of injury or infection.

nicotinic receptors (nik-o-TIN-ik): ACh receptors found on the surfaces of sympathetic and parasympathetic ganglion cells, that will also respond to the compound nicotine.

nipple: An elevated epithelial projection on the surface of the breast, containing the openings of the lactiferous sinuses.

Nissl bodies: The ribosomes, Golgi, RER, and mitochondria of the perikaryon of a typical nerve cell.

nitrogenous wastes: Organic waste products of metabolism that contain nitrogen, such as urea, uric acid, and creatinine.

nociception (nō-sē-SEP-shun): Pain perception.

node of Ranvier: Area between adjacent glial cells where the myelin covering of an axon is incomplete.

nodose ganglion (NŌ-dōs): A sensory ganglion of cranial nerve X.

noradrenaline: Catecholamine secreted by the adrenal medulla, released at most sympathetic neuroeffector junctions, and at certain synapses inside the CNS; also called **norepinephrine.**

norepinephrine (nor-ep-i-NEF-rin): A catecholamine neurotransmitter in the PNS and CNS, and a hormone secreted by the adrenal medulla; also called **noradrenaline.**

nucleic acid (nū-KLĒ-ik): A polymer of nucleotides containing a pentose sugar, a phosphate group, and one of four nitrogenous bases that regulate the synthesis of proteins and make up the genetic material in cells.

nucleolus (nū-KLĒ-ō-lus): Dense region in the nucleus that represents the site of RNA synthesis.

nucleoplasm: Fluid content of the nucleus.

nucleoproteins: Proteins of the nucleus that are generally associated with the DNA.

nucleoside: A nitrogen base plus a simple sugar.

nucleotide: Compound consisting of a nitrogen base, a simple sugar, and a phosphate group.

nucleus: Cellular organelle that contains DNA, RNA, and proteins; a mass of gray matter in the CNS.

nucleus pulposus (pul-PO-sus): Gelatinous central region of an intervertebral disc.

nutrient: An organic compound that can be broken down in the body to produce energy.

nystagmus: Involuntary, continual movement of the eyes as if to adjust to constant motion.

obesity: Body weight 10–20 percent above standard values as the result of body fat accumulation.

occlusal surface (o-KLŪ-sal): The opposing surfaces of the teeth that come into contact when processing food.

ocular: Pertaining to the eye.

oculomotor nerve (ok-ū-lō-MŌ-ter): Cranial nerve III, that controls the extrinsic oculomotor muscles other than the superior oblique and the lateral rectus.

olecranon: The proximal end of the ulna that forms the prominent point of the elbow.

olfaction: The sense of smell.

olfactory bulb (ol-FAK-tor-ē): Two olfactory nerves that lie beneath the frontal lobe of the cerebrum.
olfactory tract: Tract over which nerve impulses from the retina are transmitted between the optic chiasma and the thalamus.
oligodendrocytes (o-li-gō-DEN-drō-sīts): CNS glial cells responsible for maintaining cellular organization in the gray matter and providing a myelin sheath in areas of white matter.
oligopeptide (ol-i-gō-PEP-tīd): A short chain of amino acids.
oncogene (ON-kō-jen): A gene that can turn a normal cell into a cancer cell.
oncologists (on-KOL-ō-jists): Physicians specializing in the study and treatment of tumors.
oocyte (Ō-o-sīt): A cell whose meiotic divisions will produce a single ovum and three polar bodies.
oogenesis (ō-o-JEN-e-sis): Ovum production.
oogonia (ō-o-GŌ-nē-a): Stem cells in the ovaries whose divisions give rise to oocytes.
oophorectomy (ō-of-o-REK-to-mē): Surgical removal of the ovaries.
oophoritis (ō-of-o-RĪ-tis): Inflammation of the ovaries.
ooplasm: The cytoplasm of the ovum.
opsin: A protein, one structural component of the visual pigment rhodopsin.
opsonization: An effect of coating an object with antibodies; the attraction and enhancement of phagocytosis.
optic chiasma (OP-tik ki-AZ-ma): Crossing point of the optic nerves.
optic nerve: Nerve that carries signals from the eye to the optic chiasma.
orbit: Bony cavity of the skull that contains the eyeball.
orchiectomy (or-kē-EC-to-mē): Surgical removal of one or both testes.
orchitis: Inflammation of the testes.
organelle (or-gan-EL): An intracellular structure that performs a specific function or group of functions.
organic compound: A compound containing carbon, hydrogen, and usually oxygen.
organogenesis: The formation of organs during embryological and fetal development.
organs: Combinations of tissues that perform complex functions.
origin: Point of attachment of a muscle that is less movable.
oropharynx: The middle portion of the pharynx, bounded superiorly by the nasopharynx, anteriorly by the oral cavity, and inferiorly by the laryngopharynx.
osmolarity (oz-mo-LAR-i-tē): The total concentration of dissolved materials in a solution, regardless of their specific identities, expressed in terms of moles.
osmoreceptor: A receptor sensitive to changes in the osmolarity of the plasma.
osmosis (oz-MŌ-sis): The movement of water across a semipermeable membrane toward a solution containing a relatively high solute concentration.
osmotic pressure: The force of osmotic water movement; the pressure that must be applied to prevent osmotic movement across a membrane.
osseous tissue: A strong connective tissue containing specialized cells and a mineralized matrix of crystalline calcium phosphate and calcium carbonate.
ossicles: Small bones.
ossification: The formation of bone.
osteoblast: A cell that produces the fibers and matrix of bone.
osteoclast (OS-tē-ō-klast): A cell that dissolves the fibers and matrix of bone.
osteocyte (OS-tē-ō-sīt): A bone cell responsible for the maintenance and turnover of the mineral content of the surrounding bone.
osteogenic layer (os-tē-ō-JEN-ik): The inner, cellular layer of the periosteum that participates in bone growth and repair.
osteolysis (os-tē-ŌL-ī-sis): The breakdown of the mineral matrix of bone.
osteon (OS-tē-on): The basic histological unit of compact bone, consisting of osteocytes organized around a central canal and separated by concentric lamellae.
otic: Pertaining to the ear.
otitis media: Inflammation of the middle ear cavity.
otoconia (otoliths) (ō-tō-KŌ-nē-a): Aggregations of calcium carbonate crystals in a gelatinous membrane that sits above one of the maculae of the vestibular apparatus.
oval window: Opening in the bony labyrinth where the stapes attaches to the membranous wall of the scala vestibuli.
ovarian cycle (ō-VAR-ē-an): Monthly chain of events that leads to ovulation.
ovary: Female reproductive gland.
ovulation (ōv-ū-LĀ-shun): The release of a secondary oocyte, surrounded by cells of the corona radiata, following the rupture of the wall of a tertiary follicle.
ovum/ova (Ō-vum): A gamete produced by the reproductive system of a female; an egg.
oxidation: The loss of hydrogen atoms or the acceptance of an oxygen atom.
oxidative phosphorylation: The capture of energy as ATP during a series of oxidation/reduction reactions; a reaction sequence that occurs in the mitochondria and involves coenzymes and the electron transport system.
oxytocin (oks-i-TŌ-sin): Hormone produced by hypothalamic cells and secreted into capillaries at the posterior pituitary; stimulates smooth muscle contractions of the uterus or mammary glands in the female, but has no known function in males.
pacemaker cells: Cells of the SA node that set the pace of cardiac contraction.
Pacinian corpuscle (pa-SIN-ē-an): Receptor sensitive to vibration.
palate: Horizontal partition separating the oral cavity from the nasal cavity and nasopharynx; can be divided into an anterior bony (hard) palate and a posterior fleshy (soft) palate.
palatine: Pertaining to the palate.
palpate: To examine by touch.
palpebrae (pal-PĒ-brē): Eyelids.
pancreas: Digestive organ containing exocrine and endocrine tissues; exocrine portion secretes pancreatic juice, endocrine portion secretes hormones, including insulin and glucagon.
pancreatic duct: A tubular duct that carries pancreatic juice from the pancreas to the duodenum.
pancreatic islets: Aggregations of endocrine cells in the pancreas.
pancreatic juice: A mixture of buffers and digestive enzymes that is discharged into the duodenum under the stimulation of the enzymes secretin and cholecystokinin.
pancreatitis (pan-krē-a-TĪ-tis): Inflammation of the pancreas.
Papanicolaou (Pap) test: Test for the detection of malignancies of the female reproductive tract, especially the cervix and uterus.
papilla (pa-PIL-la): A small, conical projection.
paralysis: Loss of voluntary motor control over a portion of the body.
paranasal sinuses: Bony chambers lined by respiratory epithelium that open into the nasal cavity; includes the frontal, ethmoidal, sphenoid, and maxillary sinuses.
parasagittal: A section or plane that parallels the midsagittal plane but that does not pass along the midline.
parasympathetic division: One of the two divisions of the autonomic nervous system; also known as the craniosacral division; generally responsible for activities that conserve energy and lower the metabolic rate.
parasympathomimetic drugs: Drugs that mimic the actions of parasympathetic stimulation.

parathormone: Hormone secreted by the parathyroid gland when plasma calcium levels fall below the normal range; causes increased osteoclast activity, increased intestinal calcium uptake, and decreased calcium ion loss at the kidneys.

parathyroid glands: Four small glands embedded in the posterior surface of the thyroid; responsible for parathor mone secretion.

parenchyma (par-ENG-ki-ma): The cells of a tissue or organ that are responsible for fulfilling its functional role.

paresthesia: Sensory abnormality that produces a tingling sensation.

parietal: Referring to the body wall or outer layer.

parietal cell: Cells of the gastric glands that secrete HCl and intrinsic factor.

Parkinson's disease: Progressive motor disorder due to degeneration of the cerebral nuclei.

parotid glands (pa-ROT-id): Large salivary glands that secrete a saliva containing high concentrations of salivary (alpha) amylase.

pars distalis (dis-TAL-is): The large, anterior portion of the anterior pituitary gland.

pars intermedia (in-ter-MĒ-dē-a): The portion of the anterior pituitary immediately adjacent to the posterior pituitary and the infundibulum.

pars nervosa: The posterior pituitary gland.

parturition (par-tū-RISH-un): Childbirth, delivery.

patella (pa-TEL-le): The sesamoid bone of the kneecap.

pathogenic: Disease-causing.

pathologist (pa-THO-lo-jist): An M.D. specializing in the identification of diseases based on characteristic structural and functional changes in tissues and organs.

pedicel (PED-i-sel): A slender process of a podocyte that forms part of the filtration apparatus of the kidney glomerulus.

pedicles (PE-di-k'ls): Thick bony struts that connect the vertebral body with the articular and spinous processes.

pelvic cavity: Inferior subdivision of the abdominopelvic (peritoneal) cavity; encloses the urinary bladder, the sigmoid colon and rectum, and male or female reproductive organs.

pelvis: A bony complex created by the articulations between the coxae, the sacrum, and the coccyx.

penis (PĒ-nis): Component of the male external genitalia; a copulatory organ that surrounds the urethra, and that serves to introduce semen into the female vagina.

pepsin: Proteolytic enzyme secreted by the chief cells of the gastric glands in the stomach.

peptidases: Enzymes that split peptide bonds and release amino acids.

perforating canal: A passageway in compact bone that runs at right angles to the axes of the osteons, between the periosteum and endosteum.

perfusion: The blood flow through a tissue.

pericardial cavity (per-i-KAR-dē-al): The space between the parietal pericardium and the epicardium (visceral pericardium) that covers the outer surface of the heart.

pericarditis: Inflammation of the pericardium.

pericardium (pe-ri-KAR-dē-um): The fibrous sac that surrounds the heart, and whose inner, serous lining is continuous with the epicardium.

perichondrium (pe-re-KON-drē-um): Layer that surrounds a cartilage, consisting of an outer fibrous and an inner cellular region.

perikaryon (per-i-KAR-ē-on): The cytoplasm that surrounds the nucleus in the soma of a nerve cell.

perilymph (PER-ē-limf): A fluid similar in composition to cerebrospinal fluid; found in the spaces between the bony labyrinth and the membranous labyrinth of the inner ear.

perimysium (per-i-MĪS-ē-um): Connective tissue partition that separates adjacent fasciculi in a skeletal muscle.

perineum (per-i-NE-um): The pelvic floor and associated structures.

perineurium: Connective tissue partition that separates adjacent bundles of nerve fibers in a peripheral nerve.

periodontal ligament (pe-rē-ō-DON-tal): Collagen fibers that bind the cementum of a tooth to the periosteum of the surrounding alveolus.

periosteum (pe-rē-OS-tē-um): Layer that surrounds a bone, consisting of an outer fibrous and inner cellular region.

peripheral nervous system (PNS): All neural tissue outside of the CNS.

peripheral resistance: The resistance to blood flow primarily caused by friction with the vascular walls.

peristalsis (per-i-STAL-sis): A wave of smooth muscle contractions that propels materials along the axis of a tube such as the digestive tract, the ureters, or the ductus deferens.

peritoneal cavity: *See* **abdominopelvic cavity.**

peritoneum (pe-ri-tō-NĒ-um): The serous membrane that lines the peritoneal (abdominopelvic) cavity.

peritonitis (per-i-to-NĪ-tis): Inflammation of the peritoneum.

peritubular capillaries: A network of capillaries that surrounds the proximal and distal convoluted tubules of the kidneys.

peroxisome: A membranous vesicle containing enzymes that break down hydrogen peroxide (H_2O_2).

permeability: Ease with which dissolved materials can cross a membrane; if freely permeable, any molecule can cross the membrane; if impermeable, nothing can cross; most biological membranes are selectively permeable.

perspiration, insensible: Evaporative water loss by diffusion across the epithelium of the skin or evaporation across the alveolar surfaces of the lungs.

perspiration, sensible: Water loss due to sweat gland secretion.

pes: The foot.

petrosal ganglion: Sensory ganglion of the glossopharyngeal nerve (N IX).

petrous: Stony, usually used to refer to the thickened portion of the temporal bone that encloses the inner ear.

Peyer's patches (PI-erz): Lymphatic nodules beneath the epithelium of the small intestine.

pH: The negative exponent of the hydrogen ion concentration, in moles per liter.

phagocyte: A cell that performs phagocytosis.

phagocytosis (FA-gō-sī-tō-sis): The engulfing of extracellular materials or pathogens; movement of extracellular materials into the cytoplasm by enclosure in a membranous vesicle.

phalanx/phalanges (fa-LAN-gez): A bone of the fingers or toes.

pharmacology: The study of drugs, their physiological effects, and their clinical uses.

pharyngotympanic tube: A passageway that connects the nasopharynx with the middle ear cavity; also called the Eustachian or auditory tube.

pharynx: The throat; a muscular passageway shared by the digestive and respiratory tracts.

phasic response: A pattern of response to stimulation by sensory neurons that are normally inactive; stimulation causes a burst of neural activity that ends when the stimulus either stops or stops changing in intensity.

phenotype (FĒN-ō-tīp): Physical characteristics that are genetically determined.

phonation (fō-NĀ-shun): Sound production at the larynx.

phosphate group: PO_4^{3-}.

phospholipid (fos-fō-LIP-id): An important membrane lipid whose structure includes hydrophilic and hydrophobic regions.

phosphorylation (fos-for-i-LĀ-shun): The addition of a high-energy phosphate group to a molecule.

photoreception: Sensitivity to light.

physiology (fiz-ē-OL-o-jē): Literally the study of function; considers the ways living organisms perform vital activities.

pia mater: The tough, outer meningeal layer that surrounds the CNS.

pigment: A compound with a characteristic color.

piloerection: "Goosebumps" effect produced by the contraction of the arrector pili muscles of the skin.

pineal gland: Neural tissue in the posterior portion of the roof of the diencephalon, responsible for the secretion of melatonin.

pinna: The expanded, projecting portion of the external ear that surrounds the external auditory meatus.

pinocytosis (PI-nō-sī-tō-sis): The introduction of fluids into the cytoplasm by enclosing them in membranous vesicles at the cell surface.

pituitary gland: The "master gland," situated in the sella turcica of the sphenoid bone and connected to the hypothalamus by the infundibulum; includes the posterior pituitary (pars nervosa) and the anterior pituitary (pars intermedia and pars distalis).

placenta (pla-SENT-a): A complex structure in the uterine wall that permits diffusion between the fetal and maternal circulatory systems; also called the **afterbirth.**

plantar: Referring to the sole of the foot.

plasma (PLAZ-mah): The fluid ground substance of whole blood; what remains after the cells have been removed from a sample of whole blood.

plasma cell: Activated B cells that secrete antibodies.

plasmalemma (plaz-ma-LEM-a): Cell membrane.

platelets (PLĀT-lets): Small packets of cytoplasm that contain enzymes important in the clotting response; manufactured in the bone marrow by cells called **megakaryocytes.**

pleura (PLOO-ra): The serous membrane lining the pleural cavities.

pleural cavities: Subdivisions of the thoracic cavity that contain the lungs.

pleuritis (ploor-Ī-tis): Inflammation of the pleura.

plexus (PLEK-sus): A network or braid.

plica (PLĪ-ka): A permanent transverse fold in the wall of the small intestine.

pneumotaxic center (nū-mō-TAKS-ik): A center in the reticular formation of the pons that regulates the activities of the apneustic and respiratory rhythmicity centers to adjust the pace of respiration.

pneumothorax (nū-mō-THO-raks): The introduction of air into the pleural cavity.

podocyte (PŌ-do-sīt): A cell whose processes surround the glomerular capillaries and assist in the filtration process.

polar bond: A form of covalent bond in which there is an unequal sharing of electrons.

polar body: A nonfunctional packet of cytoplasm containing chromosomes eliminated from an oocyte during meiosis.

polarized: Histological use indicates cells that have regional differences in organelle distribution or cytoplasmic composition along a specific axis, such as between the basement membrane and free surface of an epithelial cell.

pollex: The thumb.

polymer: A large molecule consisting of a long chain of subunits.

polymorph: Polymorphonuclear leukocyte; a neutrophil.

polypeptide: A chain of amino acids strung together by peptide bonds; those containing over 100 peptides are called proteins.

polyribosome: Several ribosomes linked by their translation of a single mRNA strand.

polysaccharide (pol-ē-SAK-ah-rīd): A complex sugar, such as glycogen or a starch.

polysynaptic reflex: A reflex with interneurons interposed between the sensory fiber and the motor neuron(s).

polyunsaturated fats: Fatty acids containing carbon atoms linked by double bonds.

polyuria (pol-ē-Ū-rē-a): Excessive urine production.

pons: The portion of the metencephalon anterior to the cerebellum.

popliteal (pop-LI-tē-al): Pertaining to the back of the knee.

porphyrins (POR-fi-rinz): Ring-shaped molecules that form the basis for important respiratory and metabolic pigments, including heme and the cytochromes.

porta hepatis: A region of mesentery between the duodenum and liver that contains the hepatic artery, the hepatic portal vein, and the common bile duct.

positive feedback: Mechanism that increases a deviation from normal limits following an initial stimulus.

postabsorptive state: A period that begins 4 hours after a meal, characterized by falling blood glucose concentrations and the mobilization of metabolic reserves.

postcentral gyrus: The primary sensory cortex, where touch, vibration, pain, temperature, and taste sensations arrive and are consciously perceived.

posterior: Toward the back; dorsal.

postganglionic neuron: An autonomic neuron in a peripheral ganglion, whose activities control peripheral effectors.

postovulatory phase: The secretory phase of the menstrual cycle.

potential difference: The separation of opposite charges; requires a barrier that prevents ion migration.

precentral gyrus: The primary motor cortex on a cerebral hemisphere, located rostral to the central sulcus.

prefrontal cortex: Rostral portion of each cerebral hemisphere thought to be involved with higher intellectual functions, predictions, calculations, and so forth.

preganglionic neuron: Visceral motor neuron inside the CNS whose output controls one or more ganglionic motor neurons in the PNS.

premolars: Bicuspids; teeth with flattened occlusal surfaces located anterior to the molar teeth.

premotor cortex: Motor association area between the precentral gyrus and the prefrontal area.

preoptic nucleus: Hypothalamic nucleus that coordinates thermoregulatory activities.

preovulatory phase: A portion of the menstrual cycle; period of estrogen-induced repair of the functional zone of the endometrium through the growth and proliferation of epithelial cells in the glands not lost during menses.

prepuce (PRĒ-pūs): Loose fold of skin that surrounds the glans penis (males) or the clitoris (females).

preputial glands (prē-PŪ-shal): Glands on the inner surface of the prepuce that produce a viscous, odorous secretion, called **smegma.**

presbyopia: Farsightedness; an inability to accommodate for near vision.

presynaptic membrane: The synaptic surface where neurotransmitter release occurs.

prevertebral ganglion: *See* **collateral ganglion.**

prime mover: A muscle that performs a specific action.

proenzyme: An inactive enzyme secreted by an epithelial cell.

progesterone (prō-JES-ter-ōn): The most important progestin secreted by the corpus luteum following ovulation.

progestins (prō-JES-tinz): Steroid hormones structurally related to cholesterol.

prognosis: A prediction concerning the possibility or time course of recovery from a specific disease.

projection fibers: Axons carrying information from the thalamus to the cerebral cortex.

prolactin (prō-LAK-tin): Hormone that stimulates functional development of the mammary gland in females; a secretion of the anterior pituitary gland.

prolapse: The abnormal descent or protrusion of a portion of an organ, such as the vagina or anorectal canal.

proliferative phase: *See* **preovulatory phase.**
pronation (prō-NĀ-shun): Rotation of the forearm that makes the palm face posteriorly.
pronucleus: Enlarged egg or sperm nucleus that forms after fertilization but before amphimixis.
properdin: Complement factor that prolongs and enhances nonantibody dependent complement binding to bacterial cell walls.
prophase (PRŌ-fāz): The initial phase of mitosis, characterized by the appearance of chromosomes, breakdown of the nuclear membrane, and formation of the spindle apparatus.
proprioception (prō-prē-ō-SEP-shun): Awareness of the positions of bones, joints, and muscles.
prostaglandin (pros-tah-GLAN-din): Lipoid secreted by one cell that alters the metabolic activities or sensitivities of adjacent cells; sometimes called "local hormones."
prostate gland (PROS-tāt): Accessory gland of the male reproductive tract, contributing roughly one-third of the volume of semen.
prostatectomy (pros-ta-TEK-to-mē): Surgical removal of the prostate.
prostatitis (pros-ta-TĪ-tis): Inflammation of the prostate.
prosthesis: An artificial substitute for a body part.
protease: *See* **proteinase.**
protein: A large polypeptide with a complex structure.
proteinase: An enzyme that breaks down proteins into peptides and amino acids.
proteinuria (prō-tēn-ŪR-ē-a): Abnormal amounts of protein in the urine.
proteoglycan (prō-tē-ō-GLĪ-kan): Compound containing a large polysaccharide complex attached to a relatively small protein; examples include hyaluronic acid and chondroitin sulfate.
prothrombin: Circulating proenzyme of the common pathway of the clotting system; converted to thrombin by the enzyme thromboplastin.
proton: A fundamental particle bearing a positive charge.
protraction: To move anteriorly in the horizontal plane.
proximal: Toward the attached, base of an organ or structure.
proximal convoluted tubule: The portion of the nephron between Bowman's capsule and the loop of Henle; the major site of active reabsorption from the filtrate.
pruritis (prū-RĪ-tus): Itching.
pseudopodia (sū-dō-PŌ-dē-a): Temporary cytoplasmic extensions typical of mobile or phagocytic cells.
pseudostratified epithelium: An epithelium containing several layers of nuclei, but whose cells are all in contact with the underlying basement membrane.
psoriasis (sō-RĪ-a-sis): Skin condition characterized by excessive keratin production and the formation of dry, scaly patches on the body surface.
psychosomatic condition: An abnormal physiological state with a psychological origin.
puberty: Period of rapid growth, sexual maturation, and the appearance of secondary sexual characteristics; usually occurs at ages 10–15.
pubic symphysis: Fibrocartilaginous amphiarthrosis between the pubic bones of the coxae.
pubis (PŪ-bis): The anterior, inferior component of the coxa.
pudendum (pū-DEN-dum): The external genitalia.
pulmonary circuit: Blood vessels between the pulmonary semilunar valve of the right ventricle and the entrance to the left atrium; the blood circulation through the lungs.
pulmonary ventilation: Movement of air in and out of the lungs.
pulp cavity: Internal chamber in a tooth, containing blood vessels, lymphatics, nerves, and the cells that maintain the dentin.
pulpitis (pul-PĪ-tis): Inflammation of the tissues of the pulp cavity.
pupil: The opening in the center of the iris through which light enters the eye.
purine: An N compound with a ring-shaped structure; examples include adenine and guanine, two nitrogen bases common in nucleic acids.
Purkinje cell (pur-KIN-jē): Large, branching neuron of the cerebellar cortex.
Purkinje fibers: Specialized conducting cardiocytes in the ventricles.
pus: An accumulation of debris, fluid, dead and dying cells, and necrotic tissue.
putamen (pū-TĀ-men): Thalamic nucleus involved in the integration of sensory information prior to projection to the cerebral hemispheres.
P wave: Deflection of the ECG corresponding to atrial depolarization.
pyelogram (PI-el-ō-gram): A radiographic image of the kidneys and ureters.
pyelonephritis (pī-e-lō-nef-RĪ-tis): Inflammation of the kidneys.
pyloric sphincter (pi-LOR-ic): Sphincter of smooth muscle that regulates the passage of chyme from the stomach to the duodenum.
pylorus (pī-LOR-us): Gastric region between the body of the stomach and the duodenum; includes the pyloric sphincter.
pyrexia (pī-REK-sē-a): A fever.
pyrimidine: An N compound with a ring-shaped structure; examples include cytosine, thymine, and uracil, nitrogen bases common in nucleic acids.
pyrogen (PĪ-ro-jen): A compound that promotes a fever.
pyruvic acid (pī-RŪ-vik): 3-carbon compound produced by glycolysis.
quadriplegia: Paralysis of the arms and legs.
quaternary structure: Three-dimensional protein structure produced by interactions between individual protein sub-units.
radiodensity: Relative resistance to the passage of X-rays.
radiographic techniques: Methods of visualizing internal structures using various forms of radiational energy.
radiopaque: Having a relatively high radiodensity.
rami communicantes: Axon bundles that link the spinal nerves with the ganglia of the sympathetic chain.
ramus: A branch.
raphe (RĀ-fē): A seam.
receptor field: The area monitored by a single sensory receptor.
recessive gene: An allele that will affect the phenotype only when the individual is homozygous for that trait.
recombinant DNA: DNA created by splicing together a specific gene from one organism into the DNA strand of another organism.
rectal columns: Longitudinal folds in the walls of the anorectal canal.
rectouterine pouch (rek-tō-Ū-te-rin): Peritoneal pocket between the anterior surface of the rectum and the posterior surface of the uterus.
rectum (REK-tum): The last 15 cm (6 in.) of the digestive tract.
rectus: Straight.
red blood cell: *See* **erythrocyte.**
reduction: The gain of hydrogen atoms or electrons, or the loss of an oxygen molecule.
reductional division: The first meiotic division, which reduces the chromosome number from 46 to 23.
reflex: A rapid, automatic response to a stimulus.
reflex arc: The receptor, sensory neuron, motor neuron, and effector involved in a particular reflex; interneurons

may or may not be present, depending on the reflex considered.
refractory period: Period between the initiation of an action potential and the restoration of the normal resting potential; over this period the membrane will not respond normally to stimulation.
relaxation phase: The period following a contraction when the tension in the muscle fiber returns to resting levels.
relaxin: Hormone that loosens the pubic symphysis; a hormone secreted by the placenta.
renal: Pertaining to the kidneys.
renal corpuscle: The initial portion of the nephron, consisting of an expanded chamber that encloses the glomerulus.
renin: Enzyme released by the juxtaglomerular cells when renal blood pressure or pO_2 declines; converts angiotensinogen to angiotensin I.
rennin: Gastric enzyme that breaks down milk proteins.
replication: Duplication.
repolarization: Movement of the transmembrane potential away from + mV values and toward the resting potential.
residual volume: Amount of air remaining in the lungs after maximum forced expiration.
respiration: Exchange of gases between living cells and the environment; includes pulmonary ventilation, external respiration, internal respiration, and cellular respiration.
respiratory minute volume: The amount of air moved in and out of the respiratory system each minute.
resting potential: The transmembrane potential of a normal cell under homeostatic conditions.
rete (RĒ-tē): An interwoven network of blood vessels or passageways.
reticular activating center: Mesencephalic portion of the reticular formation responsible for arousal and the maintenance of consciousness.
reticular formation: Diffuse network of gray matter that extends the entire length of the brain stem.
reticulospinal tracts: Descending tracts that carry involuntary motor commands issued by neurons of the reticular formation.
retina: The innermost layer of the eye, lining the vitreous chamber; also known as the neural tunic.
retinene (RET-i-nēn): Visual pigment derived from vitamin A.
refraction: The bending of light rays as they pass from one medium to another.
retraction: Movement posteriorly in the horizontal plane.
retroflexion (re-trō-FLEK-shun): A posterior tilting of the uterus that has no clinical significance.
retrograde flow (RET-rō-grād): Transport of materials from the telodendria to the soma of a neuron.
retroperitoneal (re-trō-per-i-to-NĒ-al): Situated behind or outside of the peritoneal cavity.
reverberation: Positive feedback along a chain of neurons, so that they remain active once stimulated.
rheumatism (ROO-ma-tizm): A condition characterized by pain in muscles, tendons, bones, or joints.
Rh factor: Agglutinogen that may be present (Rh-positive) or absent (Rh-negative) from the surfaces of red blood cells.
rhizotomy: Surgical transection of a dorsal root, usually performed to relieve pain.
rhodopsin (rō-DOP-sin): The visual pigment found in the membrane disks of the distal segments of rods.
rhythmicity center: Medullary center responsible for the basic pace of respiration; includes inspiratory and expiratory centers.
ribonucleic acid (rī-bō-nū-KLĀ-ik): A nucleic acid consisting of a chain of nucleotides that contain the sugar ribose and the nitrogen bases adenine, guanine, cytosine, and uracil.
ribose: A 5-carbon sugar that is a structural component of RNA.
ribosome: An organelle containing rRNA and proteins, that is essential to mRNA translation and protein synthesis.
right lymphatic duct: Lymphatic vessel delivering lymph from the right side of the head, neck, and chest to the venous system via the right subclavian vein.
rigor mortis: Extended muscular contraction and rigidity that occurs after death, as the result of calcium ion release from the SR and the exhaustion of cytoplasmic ATP reserves.
rod: Photoreceptor responsible for vision under dimly lit conditions.
rostral: Toward the nose; used when referring to relative position inside the skull.
rough endoplasmic reticulum (RER): A membranous organelle that is a site of protein synthesis and storage.
round window: An opening in the bony labyrinth of the inner ear that exposes the membranous wall of the scala tympani to the air of the middle ear cavity.
rubrospinal tracts: Descending tracts that carry involuntary motor commands issued by the red nucleus of the mesencephalon.
Ruffini corpuscles (ru-FĒ-nē): Receptors sensitive to tension and stretch in the dermis of the skin.
rugae (ROO-gē): Mucosal folds in the lining of the empty stomach that disappear as gastric distension occurs.
saccule (SAK-ūl): A portion of the vestibular apparatus of the inner ear, responsible for static equilibrium.
sagittal plane: Sectional plane that divides the body into left and right portions.
salivatory nucleus (SAL-i-va-tō-rē): Medullary nucleus that controls the secretory activities of the salivary glands.
salt: An inorganic compound created by a reaction between an acid and a base.
saltatory conduction: Relatively rapid conduction of a nerve impulse between successive nodes of a myelinated axon.
sarcolemma: The cell membrane of a muscle cell.
sarcoma (sar-KŌ-ma): A tumor of connective tissues.
sarcomere: The smallest contractile unit of a striated muscle cell.
sarcoplasm: The cytoplasm of a muscle cell.
scala media: The central, endolymph-filled chamber of the inner ear; *see* **cochlear duct.**
scala tympani: The perilymph-filled chamber of the inner ear below the basilar membrane; pressure changes here distort the round window.
scala vestibuli: The perilymph-filled chamber of the inner ear above the vestibular membrane; pressure changes here result from distortions of the oval window.
scar tissue: Thick, collagenous tissue that forms at an injury site.
Schlemm, canal of: Passageway that delivers aqueous humor from the anterior chamber of the eye to the venous circulation.
Schwann cells: Glial cells responsible for the neurilemma that surrounds axons in the PNS.
sciatic nerve (sī-A-tik): Nerve innervating the posteromedial portions of the thigh and lower leg.
sciatica (SĪ-AT-i-ka): Pain felt along the peripheral distribution of the sciatic nerve.
sclera (SKLER-a): The fibrous, outer layer of the eye forming the white area of the anterior surface; a portion of the fibrous tunic of the eye.
sclerosis: A hardening and thickening that often occurs secondary to tissue inflammation.
scoliosis (skō-lē-Ō-sis): An abnormal, exaggerated lateral curvature of the spine.
scrotum (SKRŌ-tum): Loose-fitting, fleshy pouch that encloses the testes of the male.

sebaceous glands (sē-BĀ-shus): Glands that secrete sebum, usually associated with hair follicles.
sebum (SĒ-bum): A waxy secretion that coats the surfaces of hairs.
secretin (sē-KRĒ-tin): Duodenal hormone that stimulates pancreatic buffer secretion and inhibits gastric activity.
secondary sex characteristics: Physical characteristics that appear at puberty in response to sex hormones, but that are not involved in the production of gametes.
semen (SĒ-men): Fluid ejaculate containing spermatozoa and the secretions of accessory glands of the male reproductive tract.
semicircular canals: Tubular components of the vestibular apparatus responsible for dynamic equilibrium.
semilunar valve: A three-cusped valve guarding the exit from one of the cardiac ventricles; includes the pulmonary and aortic valves.
seminal vesicles (SEM-i-nal): Glands of the male reproductive tract that produce roughly 60 percent of the volume of semen.
seminiferous tubules (se-mi-NIF-e-rus): Coiled tubules where sperm production occurs in the testis.
senescence: Aging.
septae (SEP-tē): Partitions that subdivide an organ.
serosa: *See* **serous membrane.**
serotonin (ser-ō-TO-nin): A neurotransmitter in the CNS; a compound that enhances inflammation, released by activated mast cells and basophils.
serous cell/secretion: A cell that produces a watery secretion containing high concentrations of enzymes.
serous membrane: A squamous epithelium and the underlying loose connective tissue; the lining of the pericardial, pleural, and peritoneal cavities.
serum: Blood plasma from which clotting agents have been removed.
sesamoid bone: A bone that forms in a tendon.
sigmoid colon (SIG-mōid): The S-shaped 8-inch portion of the colon between the descending colon and the rectum.
sign: A clinical term for visible evidence of the presence of a disease.
simple epithelium: An epithelium containing a single layer of cells above the basement membrane.
sinus: A chamber or hollow in a tissue; a large, dilated vein.
sinusoid (SĪ-nus-ōid): An extensive network of vessels found in the liver, adrenal cortex, spleen, and pancreas; similar in histological structure to capillaries.
sinusitis: Inflammation of a nasal sinus.
skeletal muscle: A contractile organ of the muscular system.
skeletal muscle tissue: Contractile tissue dominated by skeletal muscle fibers; characterized as striated, voluntary muscle.
sliding filament theory: The concept that a sarcomere shortens as the thick and thin filaments slide past one another.
small intestine: The duodenum, jejunum, and ileum; the digestive tract between the stomach and large intestine.
smegma (SMEG-ma): Secretion of the preputial glands of the penis or clitoris.
smooth endoplasmic reticulum: Membranous organelle where lipid and carbohydrate synthesis and storage occur.
smooth muscle tissue: Muscle tissue found in the walls of many visceral organs; characterized as nonstriated, involuntary muscle.
soft palate: Fleshy posterior extension of the hard palate, separating the nasopharynx from the oral cavity.
sole: The inferior surface of the foot.
solute: Materials dissolved in a solution.
solution: A fluid containing dissolved materials.
solvent: The fluid component of a solution.
soma (SŌ-ma): Body.
somatic (sō-MAT-ik): Pertaining to the body.
somatic nervous system: System of nerve fibers that run from the central nervous to the muscles of the skeleton.
somatomedins: Compounds stimulating tissue growth, released by the liver following GH secretion.
somatostatin: GH-IH, a hypothalamic regulatory hormone that inhibits GH secretion by the anterior pituitary.
somatotrophin: Growth hormone, produced by the anterior pituitary in response to GH-RH.
sperm: *See* **spermatozoa.**
spermatic cord: Spermatic vessels, nerves, lymphatics, and the ductus deferens, extending between the testes and the proximal end of the inguinal canal.
spermatids (SPER-ma-tidz): The product of meiosis in the male, cells that differentiate into spermatozoa.
spermatocyte (sper-MA-to-sīt): Cells of the seminiferous tubules that are engaged in meiosis.
spermatogenesis (sper-ma-to-JEN-e-sis): Sperm production.
spermatogonia (sper-ma-to-GŌ-nē-a): Stem cells whose mitotic divisions give rise to other stem cells and spermatocytes.
spermatozoa (sper-ma-to-ZŌ-a): Sperm cells; singlular form = spermatozoon.
spermicide: Compound toxic to sperm cells, sometimes used as a contraceptive method.
spermiogenesis: The process of spermatid differentiation that leads to the formation of physically mature spermatozoa.
sphincter (SFINK-ter): Muscular ring that contracts to close the entrance or exit of an internal passageway.
spinal nerve: One of 31 pairs of nerves that originate on the spinal cord from anterior and posterior roots.
spindle apparatus: A muscle spindle (intrafusal fiber) and its sensory and motor innervation.
spinocerebellar tracts: Ascending tracts carrying sensory information to the cerebellum.
spinothalamic tracts: Ascending tracts carrying poorly localized touch, pressure, pain, vibration, and temperature sensations to the thalamus.
spinous process: Prominent posterior projection of a vertebra, formed by the fusion of two laminae.
spleen: Lymphatic organ important for red blood cell phagocytosis, immune response, and lymphocyte production.
splenectomy (sple-NEK-to-mē): Surgical removal of the spleen.
sprain: Forceful distortion of an articulation that produces damage to the capsule, ligaments, or tendons but not dislocation.
sputum (SPŪ-tum): Viscous mucus ejected from the mouth after transport to the pharynx by the mucus escalator of the respiratory tract.
squama: A broad, flat surface.
squamous (SKWĀ-mus): Flattened.
squamous epithelium: An epithelium whose superficial cells are flattened and platelike.
stapedius (stā-PĒ-dē-us): A muscle of the middle ear whose contraction tenses the auditory ossicles and reduces the forces transmitted to the oval window.
stapes (STĀ-pez): The auditory ossicle attached to the tympanic membrane.
stenosis (ste-NŌ-sis): A constriction or narrowing of a passageway.
stereocilia: Elongate microvilli characteristic of the epithelium of the epididymis and portions of the ductus deferens.
steroid: A ring-shaped lipid structurally related to cholesterol.
stimulus: An environmental alteration that produces a change in cellular activities; often used to refer to events

that alter the transmembrane potentials of excitable cells.
stratum (STRĀ-tum): Layer.
stratified: Containing several layers.
stretch receptors: Sensory receptors that respond to stretching of the surrounding tissues.
stroma: The connective tissue framework of an organ, as distinguished from the functional cells (parenchyma) of that organ.
subarachnoid space: Meningeal space containing CSF; the area between the arachnoid membrane and the pia mater.
subclavian (sub-CLĀ-vē-an): Pertaining to the region under the clavicle.
subcutaneous layer: The layer of loose connective tissue below the dermis; also called the **hypodermis** or **superficial fascia.**
sublingual glands (sub-LING-gwal): Mucus-secreting salivary glands situated under the tongue.
submandibular glands: Salivary glands nestled in depressions on the medial surfaces of the mandible; salivary glands that produce a mixture of mucins and enzymes (salivary amylase).
submucosa (sub-mū-KŌ-sa): Region between the muscularis mucosae and the muscularis externa.
submucosal glands: Mucous glands in the submucosa of the duodenum.
subserous fascia: Loose connective tissue layer beneath the serous membrane lining the ventral body cavity.
substrate: A participant (product or reactant) in an enzyme-catalyzed reaction.
sulcus (SUL-kus): A groove or furrow.
summation: Temporal or spatial addition of stimuli.
superficial fascia: *See* **subcutaneous layer.**
superior: Directional reference meaning above.
superior vena cava: The vein that carries blood from the parts of the body above the heart to the right atrium.
supination (su-pi-NĀ-shun): Rotation of the forearm so that the palm faces anteriorly.
supine (SU-pīn): Lying face up, with palms facing anteriorly.
suppressor T cells: Lymphocytes that inhibit B cell activation and plasma cell secretion of antibodies.
suprarenal gland (sū-pra-RĒ-nal): *See* **adrenal gland.**
surfactant (sur-FAK-tant): Lipid secretion that coats alveolar surfaces and prevents their collapse.
sustentacular cells (sus-ten-TAK-ū-lar): Supporting cells of the seminiferous tubules of the testis, responsible for the differentiation of spermatids, the maintenance of the blood-testis barrier, and the secretion of inhibin.
sutural bones: Irregular bones that form in fibrous tissue between the flat bones of the developing cranium; also called **Wormian bones.**
suture: Fibrous joint between flat bones of the skull.
sympathectomy (sim-path-EK-to-mē): Transection of the sympathetic innervation to a region.
sympathetic division: Division of the autonomic nervous system responsible for "fight or flight" reactions; primarily concerned with the elevation of metabolic rate and increased alertness.
sympathomimetic drugs: Drugs that mimic the actions of sympathetic stimulation.
symphysis: A fibrous amphiarthrosis, such as those between adjacent vertebrae or between the pubic bones of the coxae.
symptom: Clinical term for an abnormality of function due to the presence of disease.
synapse (SIN-aps): Site of communication between a nerve cell and some other cell; if the other cell is not a neuron, the term neuroeffector junction is often used.
synaptic delay (sin-AP-tik): The period between the arrival of an impulse at the presynaptic membrane and the initiation of an action potential in the postsynaptic membrane.
synarthrosis (sin-ar-THRŌ-sis): A joint that does not permit relative movement between the articulating elements.
synchondrosis (sin-kon-DRŌ-sis): A cartilaginous synarthrosis, such as the articulation between the epiphysis and diaphysis of a growing bone.
syncope: A sudden, transient loss of consciousness; a faint.
syncytial trophoblast: Multinucleate cytoplasmic layer that covers the blastocyst; the layer responsible for uterine erosion and implantation.
syncytium: A multinucleate mass of cytoplasm, produced by the fusion of cells or repeated mitoses without cytokinesis.
syndesmosis (sin-dez-MŌ-sis): A fibrous amphiarthrosis.
syndrome: A discrete set of symptoms that occur together.
syneresis (si-NER-e-sis): Clot retraction.
synergist (SIN-er-jist): A muscle that assists a prime mover in performing its primary action.
synostosis (sin-os-TŌ-sis): A synarthrosis formed through the fusion of the articulating elements.
synovial cavity (si-NŌ-vē-ul) Fluid-filled chamber in a diarthrodial joint.
synovial fluid (sī-NŌV-ē-ul): Substance secreted by synovial membranes that lubricates joints.
synovial membrane: An incomplete layer of fibroblasts confronting the synovial cavity, plus the underlying loose connective tissue.
synthesis (SIN-the-sis): Manufacture; anabolism.
system: An interacting group of organs that performs one or more specific functions.
systemic circuit: Vessels between the aortic semilunar valve and the entrance to the right atrium; the circulatory system other than vessels of the pulmonary circuit.
systole (SIS-to-lē): The period of cardiac contraction.
systolic pressure: Peak arterial pressures measured during ventricular systole.
tachycardia (tak-ē-KAR-dē-a): Abnormally rapid heart rate.
tactile: Pertaining to the sense of touch.
taenia coli (TĒ-nē-a KŌ-lī): Three longitudinal bands of smooth muscle in the muscularis externa of the colon.
tarsus: The ankle.
T cells: Lymphocytes responsible for cellular immunity, and for the coordination and regulation of the immune response; includes regulatory T cells (helpers and suppressors) and cytotoxic (killer) T cells.
tears: Fluid secretion of the lacrimal glands that bathes the anterior surfaces of the eyes.
tectorial membrane (tek-TŌR-ē-al): Gelatinous membrane suspended over the hair cells of the organ of Corti.
tectospinal tracts: Descending extrapyramidal tracts carrying involuntary motor commands issued by the colliculi.
tectum: The roof of the mesencephalon of the brain.
telencephalon (tel-en-SEF-a-lon): The forebrain or cerebrum, including the cerebral hemispheres, the internal capsule, and the cerebral nuclei.
telodendria (te-lō-DEN-drē-a): Terminal axonal branches that end in synaptic knobs.
telophase (TEL-o-fāz): The final stage of mitosis, characterized by the disappearance of the spindle apparatus, the reappearance of the nuclear membrane and the disappearance of the chromosomes, and the completion of cytokinesis.
temporal: Pertaining to time (temporal summation) or pertaining to the temples (temporal bone).
tendinitis: Painful inflammation of a tendon.
tendon: A collagenous band that connects a skeletal muscle to an element of the skeleton.
tendon organ: Receptor sensitive to tension in a tendon.
tentorium cerebelli (ten-TŌR-ē-um ser-e-BEL-ē): Dural partition that separates the cerebral hemispheres from the cerebellum.
teratogen (TER-a-to-jen): Stimulus that causes developmental defects.

teres: Long and round.
terminal: Toward the end.
tertiary follicle: A mature ovarian follicle, containing a large, fluid-filled chamber.
tertiary structure: Protein structure that develops as the result of interactions between distant portions of the same molecule.
testes (TES-tēz): The male gonads, sites of gamete production and hormone secretion.
testosterone (tes-TOS-te-rōn): The principal androgen produced by the interstitial cells of the testes.
tetanic contraction: Sustained skeletal muscle contraction due to repeated stimulation at a frequency that prevents muscle relaxation.
tetanus: A tetanic contraction; also used to refer to a disease state resulting from the stimulation of muscle cells by bacterial toxins.
tetany: A tetanic contraction; also used to refer to abnormally prolonged contractions resulting from disturbances in electrolyte balance.
tetrad (TET-rad): Paired, duplicated chromosomes visible at the start of meiosis I.
tetraiodothyronine (tet-ra-ī-ō-dō-THĪ-rō-nēn): T_4, or thyroxine, a thyroid hormone.
thalamus: The walls of the diencephalon.
thalassemia (thal-ah-SĒ-mē-ah): A hereditary disorder affecting hemoglobin synthesis and producing anemia.
theory: A hypothesis that makes valid predictions, as demonstrated by evidence that is testable, unbiased, and repeatable.
therapy: Treatment of disease.
thermogenesis (ther-mō-JEN-e-sis): Heat production.
thermography: Diagnostic procedure involving the production of an infrared image.
thermoreception: Sensitivity to temperature changes.
thermoregulation: Homeostatic maintenance of body temperature.
thick filament: A myosin filament in a skeletal or cardiac muscle cell.
thin filament: An actin filament in a skeletal or cardiac muscle cell.
thoracoabdominal pump (tho-ra-kō-ab-DOM-i-nal): Changes in the intrapleural pressures during the respiratory cycle that assist the venous return to the heart.
thoracolumbar division (thō-ra-kō-LUM-bar): The sympathetic division of the ANS.
thorax: The chest.
threshold: The transmembrane potential at which an action potential begins.
thrombin (THROM-bin): Enzyme that converts fibronogen to fibrin.
thrombocyte (THROM-bo-sīt): *See* **platelet.**
thrombocytopenia (throm-bō-sī-tō-PĒ-nē-ah): Abnormally low platelet count in the circulating blood.
thromboembolism (throm-bō-EM-bō-lizm): Occlusion of a blood vessel by a drifting blood clot.
thromboplastin: Enzyme that converts prothrombin to thrombin; enzyme formed by the intrinsic or extrinsic clotting pathways.
thrombus: A blood clot.
thymine: A pyrimidine found in DNA.
thymosin (thī-MO-sin): Thymic hormone essential to the development and differentiation of T cells.
thymus: Lymphatic organ, site of T cell formation.
thyroglobulin (thī-rō-GLOB-ū-lin): Circulating transport globulin that binds thyroid hormones.
thyroid gland: Endocrine gland whose lobes sit lateral to the thyroid cartilage of the larynx.
thyroid hormones: Thyroxine (T_4) and triiodothyronone (T_3), hormones of the thyroid gland; hormones that stimulate tissue metabolism, energy utilization, and growth.
thyroid stimulating hormone (TSH): Anterior pituitary hormone that triggers the secretion of thyroid hormones by the thyroid gland.
thyroxine (TX) (thī-ROKS-in): A thyroid hormone (T_4).
tidal volume: The volume of air moved in and out of the lungs during a normal quiet respiratory cycle.
tissue: A collection of specialized cells and cell products that perform a specific function.
titer: Plasma antibody concentration.
tolerance: Failure to produce antibodies against antigenic compounds normally present in the body.
tonic response: An increase or decrease in the frequency of action potentials by sensory receptors that are chronically active.
tonsil: A lymphatic nodule beneath the epithelium of the pharynx; includes the palatine, pharyngeal, and lingual tonsils.
topical: Applied to the body surface.
toxic: Poisonous.
trabecula (tra-BEK-ū-la): A connective tissue partition that subdivides an organ.
trabeculae carnae (tra-BEK-ū-lē CAR-nē): Muscular ridges projecting from the walls of the ventricles of the heart.
trachea (TRĀ-kē-a): The windpipe, an airway extending from the larynx to the primary bronchi.
tracheal ring: C-shaped supporting cartilage of the trachea.
tracheostomy (trā-kē-OS-to-mē): Surgical opening of the anterior tracheal wall to permit airflow.
trachoma: An infectious disease of the conjunctiva and cornea.
tract: A bundle of axons inside the CNS.
tractotomy: The surgical transection of a tract, sometimes used to relieve pain.
transamination (trans-am-i-NĀ-shun): The enzymatic transfer of an amine group from an amino acid to another carbon chain.
transcription: The encoding of genetic instructions on a strand of mRNA.
transdermal medication: Administration of medication by absorption through the skin.
transection: To sever or cut in the transverse plane.
transfusion: Transfer of blood from a donor directly into the bloodstream of another person.
transient ischemic attack: A temporary loss of consciousness due to the occlusion of a small blood vessel in the brain.
translation: The process of peptide formation using the instructions carried by an mRNA strand.
transmembrane potential: The potential difference, in millivolts, measured across the cell membrane; a potential difference that results from the uneven distribution of positive and negative ions across a cell membrane.
transudate (TRANS-ū-dāt): Fluid that diffuses across a serous membrane and lubricates opposing surfaces.
treppe (TREP-e): "Staircase" increase in tension production following repeated stimulation of a muscle, even though the muscle is allowed to complete each relaxation phase.
triad (liver): The combination of branches of the hepatic duct, the hepatic portal vein, and the hepatic artery, found at each corner of a liver lobule.
triad (muscle cell): The combination of a T tubule and two cisternae of the sarcoplasmic reticulum.
tricarboxylic acid cycle (trī-kar-bok-SIL-ik): *See* **Krebs cycle.**
tricuspid valve (trī-KUS-pid): The right atrioventricular valve that prevents backflow of blood into the right atrium during ventricular systole.
trigeminal nerve (trī-GEM-i-nal): Cranial nerve V, responsible for providing sensory information from the lower portions of the face, including the upper and lower jaws, and delivering motor commands to the muscles of mastication.

triglyceride (trī-GLIS-e-rīd): A lipid composed of a molecule of glycerol attached to three fatty acids.
trigone (TRĪ-gōn): Triangular region of the bladder bounded by the exits of the ureters and the entrance to the urethra.
triiodothyronine: T_3, one of the thyroid hormones.
trisomy: The abnormal possession of three copies of a chromosome; trisomy 21 is responsible for the Down syndrome.
trochanter (trō-KAN-ter): Large processes near the head of the femur.
trochlea (TRŌK-le-a): A pulley.
trochlear nerve (TRŌK-lē-ar): Cranial nerve IV, controlling the superior oblique muscle of the eye.
trophoblast (TRŌ-fō-blast): Superficial layer of the blastocyst that will be involved with implantation, hormone production, and placenta formation.
troponin/tropomyosin (trō-PŌ-nin) (trō-pō-MĪ-ō-sin): Proteins on the thin filaments that mask the active sites in the absence of free calcium ions.
trunk: The thoracic and abdominopelvic regions.
trypsin (TRIP-sin): One of the pancreatic proteases.
trypsinogen: The inactive, proenzyme secreted by the pancreas and converted to trypsin in the duodenum.
T tubules: Transverse, tubular extensions of the sarcolemma that extend deep into the sarcoplasm to contact cisternae of the sarcoplasmic reticulum.
tuberculum (tū-BER-kū-lum): A small, localized elevation on a bony surface.
tuberosity: A large, roughened elevation on a bony surface.
tubulin: Protein subunit of microtubules.
tumor: A tissue mass formed by the abnormal growth and replication of cells.
tunica (TŪ-ni-ka): A layer or covering; in blood vessels: t. externa, the outermost layer of connective tissue fibers that stabilizes the position of the vessel; t. intima, the innermost layer, consisting of the endothelium plus an underlying elastic membrane; t. media, a middle layer containing collagen, elastin, and smooth muscle fibers in varying proportions.
turbinates: *See* **conchae.**
T wave: Deflection of the ECG corresponding to ventricular repolarization.
twitch: A single contraction/relaxation cycle in a skeletal muscle.
tympanic membrane (tim-PAN-ik): Membrane that separates the external auditory canal from the middle ear; membrane whose vibrations are transferred to the auditory ossicles and ultimately to the oval window; the "eardrum."
type A axons: Large myelinated axons.
type B axons: Smaller myelinated axons.
type C axons: Small unmyelinated axons.
ulcer: An area of epithelial sloughing associated with damage to the underlying connective tissues and vasculature.
ultrasound: Diagnostic visualization procedure that uses high-frequency sound waves.
umbilical cord (um-BIL-i-kal): Connecting stalk between the fetus and the placenta; contains the allantois, the umbilical arteries, and the umbilical vein.
umbilicus: The navel.
unicellular gland: Goblet cells.
unipennate muscle: A muscle whose fibers are all arranged on one side of the tendon.
unipolar neuron: A sensory neuron whose soma lies in a dorsal root ganglion or a sensory ganglion of a cranial nerve.
unmyelinated axon: Axon whose neurilemma does not contain myelin, and where continuous conduction occurs.
uracil: One of the pyrimidines characteristic of RNA.
uremia (ū-RĒ-mē-a): Abnormal condition caused by impaired kidney function, characterized by the retention of wastes and the disruption of many other organ systems.
ureters (ū-RĒ-terz): Muscular tubes, lined by transitional epithelium, that carry urine from the renal pelvis to the urinary bladder.
urethra (ū-RĒ-thra): A muscular tube that carries urine from the urinary bladder to the exterior.
urethritis: Inflammation of the urethra.
urinalysis: Analysis of the physical and chemical characteristics of the urine.
urinary bladder: Muscular, distensible sac that stores urine prior to micturition.
urination: The voiding of urine; micturition.
urobilinogen: A compound derived from the bilirubin excreted in the bile.
uterus (Ū-ter-us): Muscular organ of the female reproductive tract where implantation, placenta formation, and fetal development occur.
utricle (Ū-trē-k'l): The largest chamber of the vestibular apparatus; contains a macula important for static equilibrium.
uvea: The vascular tunic of the eye.
uvula (Ū-vū-la): A dangling, fleshy extension of the soft palate.
vagina (va-JĪ-na): A muscular tube extending between the uterus and the vestibule.
varicose veins (VAR-i-kōs): Distended superficial veins.
vasa vasorum: Blood vessels that supply the walls of large arteries and veins.
vascular: Pertaining to blood vessels.
vascularity: The blood vessels in a tissue.
vascular spasm: Contraction of the wall of a blood vessel at an injury site, a process that may slow the rate of blood loss.
vasoconstriction: A reduction in the diameter of arterioles due to contraction of smooth muscles in the tunica media; an event that elevates peripheral resistance, and that may occur in response to local factors, through the action of hormones, or from stimulation of the vasomotor center.
vasodilation (vaz-ō-dī-LĀ-shun): An increase in the diameter of arterioles due to the relaxation of smooth muscles in the tunica media; an event that reduces peripheral resistance, and that may occur in response to local factors, through the action of hormones, or following decreased stimulation of the vasomotor center.
vasomotion: Alterations in the pattern of blood flow through a capillary bed in response to changes in the local environment.
vasomotor center: Medullary center whose stimulation produces vasoconstriction and an elevation in peripheral resistance.
vein: Blood vessel carrying blood from a capillary bed toward the heart.
venae cavae (VĒ-na CĀ-va): The major veins delivering systemic blood to the right atrium.
ventilation: Air movement in and out of the lungs.
ventilatory rate: The respiratory rate.
ventral: Pertaining to the anterior surface.
ventricle (VEN-tri-k'l): One of the large, muscular pumping chambers of the heart that discharges blood into the pulmonary or systemic circuits.
ventricular escape: The initiation of ventricular contractions after a pause caused by impaired conduction of the contractile stimulus from the AV node.
ventricular folds: Mucosal folds in the laryngeal walls that do not play a role in sound production; the false vocal cords.
venule (VEN-ūl): Thin-walled veins that receive blood from capillaries.
vermis (VER-mis): Midsagittal band of neural cortex on the surface of the cerebellum.
vertebral canal: Passageway that encloses the spinal cord,

a tunnel bounded by the neural arches of adjacent vertebrae.
vertebral column: The cervical, thoracic, and lumbar vertebrae, the sacrum, and the coccyx.
vertebrochondral ribs: Ribs 8–10, false ribs connected to the sternum by shared cartilaginous bars.
vertebrosternal ribs: Ribs 1–7, true ribs connected to the sternum by individual cartilaginous bars.
vertigo: Dizziness.
vesicle: A membranous sac in the cytoplasm of a cell.
vestibular membrane: The membrane that separates the scala media from the scala vestibuli of the inner ear.
vestibular nucleus: Processing center for sensations arriving from the vestibular apparatus; located near the border between the pons and medulla.
vestibule (VES-ti-būl): A chamber; in the inner ear, the term refers to the utricle, saccule, and semicircular canals.
vestibulospinal tracts: Descending tracts of the extrapyramidal system, carrying involuntary motor commands issued by the vestibular nucleus to stabilize the position of the head.
villus: A slender projection of the mucous membrane of the small intestine.
virus: A pathogenic microorganism.
viscera: Organs in the ventral body cavity.
visceral: Pertaining to viscera or their outer coverings.
visceral smooth muscle: Smooth muscle tissue forming sheets or layers in the walls of visceral organs; the cells may not be innervated, and the layers often show automaticity (rhythmic contractions).
viscosity: The resistance to flow exhibited by a fluid due to molecular interactions within the fluid.
viscous: Thick, syrupy.
vital capacity: The maximum amount of air that can be moved in or out of the respiratory system; the sum of the inspiratory reserve, the expiratory reserve, and the tidal volume.
vitamin: An essential organic nutrient that functions as a coenzyme in vital enzymatic reactions.
vitreous humor: Gelatinous mass in the vitreous chamber of the eye.
vocal folds: Folds in the laryngeal wall containing elastic ligaments whose tension can be voluntarily adjusted; the true vocal cords, responsible for phonation.
voluntary: Controlled by conscious thought processes.
vulva (VUL-va): The female pudendum (external genitalia).
Wallerian degeneration: Disintegration of an axon and its myelin sheath distal to an injury site.
white blood cells: Leukocytes; the granulocytes and agranulocytes of the blood.
white matter: Regions inside the CNS that are dominated by myelinated axons.
white ramus: A nerve bundle containing the myelinated preganglionic axons of sympathetic motor neurons en route to the sympathetic chain or a collateral ganglion.
Wormian bones: *See* **sutural bones.**
xiphoid process (ZĪ-fōid): Slender, inferior extension of the sternum.
Y chromosome: The sex chromosome whose presence indicates that the individual is a genetic male.
yolk sac: One of the three extraembryonic membranes, composed of an inner layer of endoderm and an outer layer of mesoderm.
Zeis, glands of (ZĪS): Enlarged sebaceous glands on the free edges of the eyelids.
zona fasciculata (ZŌ-na) (fa-sik-ū-LA-ta): Region of the adrenal cortex responsible for glucocorticoid secretion.
zona glomerulosa (glo-mer-u-LŌ-sa): Region of the adrenal cortex responsible for mineralocorticoid secretion.
zona pellucida (pel-LŪ-si-da): Region between a developing oocyte and the surrounding follicular cells of the ovary.
zona reticularis (re-ti-kū-LAR-is): Region of the adrenal cortex responsible for androgen secretion.
zygote (ZĪ-gōt): The fertilized ovum, prior to the start of cleavage.

Index

A

B

C

D

E

F

G

I

J

K

L

M

N

O

P

Q

R

S

T

U

V

Illustration Credits

PHOTOGRAPHS

Chapter 1 **CO** David Pollack/The Stock Market **1–14a** AFIP/Science Source/Photo Researchers **1–14b** Agfa/ Science Source/Photo Researchers **1–14c** Biophoto Associates/Photo Researchers **1–15a,b** CNRI/Science Photo Library/Photo Researchers **1–15c** Diagnostic Radiological Health Sciences Learning Laboratory

Chapter 2 **CO** Tom Sanders/The Stock Market **2–3b** SIU/ Photo Researchers **2–3c** Mallinckrodt Institute of Radiology

Chapter 3 **CO** Four By Five **3–5a** Fawcett, Hirokawa, Heuser, Science Source/Photo Researchers **3–5b** M. Schliwa/ Visuals Unlimited **3–7a** David M. Phillips/Visuals Unlimited **3–8** Photo Researchers **3–9** CNRI/Science Photo Library/ Photo Researchers **3–10a** From *A Textbook of Histology*, by Bloom and Fawcett. Published by W. B. Saunders Company, 1968. Ninth edition **3–10b** Biophoto Associates/Photo Researchers**3–14a** Biophoto Associates/Photo Researchers **3–15b** Dr. Birgit H. Satir, Albert Einstein College of Medicine, Yeshiva University, Bronx, New York **3–21** Dr. Simon Watkins/Dana-Farber Cancer Institute **3–27b** Boerhinger-Ingelheim/Lennart Nilsson **3–29e** M. G. Farquhar and J. E. Palade, "Junctional Complexes in Epithelia" *Journal of Cell Biology*, 17: 379 (1963)

Chapter 4 **CO** Christopher Springmann/The Stock Market **4–14** Courtesy of Dr. Don Fawcett, S. Palay, E. Yamada, Photo Researchers **4–15b** Martin M. Rotker **4–15c** Michael and Elvan Habicht

Chapter 5 **CO** Tony Stone Worldwide **5–2** Dr. C. P. LeBlond/McGill University **5–3b** Dr. Fred Lightfoot, George Washington University **5–4** by the author **5–5b** by the author **5–6** by the author **5–7** by the author **5–8** Carolina Biological Supply Company **5–9** by the author **5–10a** S. Elem/Visuals Unlimited **5–10b** by the author **5–13a** Science Source/Photo Researchers **5–13b** by the author **5–13c** Biophoto Associates/Photo Researchers **5–14a** John D. Cunningham/Visuals Unlimited **5–14b** Bruce Iverson/Visuals Unlimited **5–14c** by the author **5–16a** Robert Brons/Biological Photo Service **5–16b** Science Source/Photo Researchers **5–16c** Ed Reschke/Peter Arnold **5–18** by the author **5–20a,c** by the author **5–20b** © Manfred P. Kage/Peter Arnold **5–21** by the author

Chapter 6 **CO** Robert Torrez/Tony Stone Worldwide **6–3** Custom Medical Stock Photo **6–4** by the author **6–5** From *Tissues and Organs: A Text-Atlas of Scanning Electron Microscopy* by Richard G. Kessel and Randy H. Kardon. Copyright © 1979 W. H. Freeman and Company. Used with permission. **6–7a** Carolina Biological Supply Co. **6–7b** Copyright © David Scharf **6–7c** Carolina Biological Supply Co. **6–9** Francois Lochon/Gamma-Liaison **6–10a** Manfred Kage/Peter Arnold, Inc. **6–10c** by the author **6–10d** by the author **6–13** by the author **6–14** by the author **6–18** C. Vergalee/Photo Researchers

Chapter 7 **CO** Kobal Collection/SuperStock **7–2a,b** by the author **7–2c** © Dr. Kessel/Dr. Kardon, Peter Arnold, Inc. **7–3** Biophoto Associates/Photo Researchers **7–5** by the author **7–8** by the author **Table 7–1** Custom Medical Stock Photo **7–15** Smith & Nephew Richards Inc.

Chapter 8 **CO** Georg Gerster/Comstock **Table 8–2** Frederick McDonald

Chapter 9 **CO** Alan Becker/The Image Bank **9–15c** SIU/ Photo Researchers **9–17** T. W. Henderson/Focus on Sports **9–19** Alexander Tsiaras/Photo Researchers **Profile box** Steven Caras

Chapter 10 **CO** NASA **10–2** Fred Hossler/Visuals Unlimited **10–3** From *Tissues and Organs: A Text-Atlas of Scanning Electron Microscopy* by Richard G. Kessel and Randy H. Kardon. Copyright © 1979 W. H. Freeman and Company. Used with permission. **10–8** J. J. Head, Carolina Biological Supply Co. **10–21a** by the author **10–21b** *Ham's Histology*, ninth ed. by Arthur W. Ham.; ed. David Cormack, © 1987, J. B. Lippincott Co., Philadelphia, p. 394.

Chapter 11 **CO** John P. Kelly/The Image Bank **11–24a** Focus on Sports **11–24b** Mickey Palmer/Focus on Sports **11–24c** David Lissy/Focus on Sports **11–24d** © 1987 Trafidlo

Chapter 12 **CO** Synapter/Science Photo Library/Photo Researchers **12–6** by the author **12–7** by the author **12–8a** Photo Researchers **12–8b** Biophoto Associates/ Photo Researchers **12–13** David Scott/Phototake

Chapter 13 **CO** **13–3a** Peres/Custom Medical Stock Photo **CO 13** (c) Lennart Nilsson, *Behold Man* **13–3b** From *Tissues and Organs: A Text-Atlas of Scanning Electron Microscopy* by Richard G. Kessel and Randy H. Kardon. Copyright © 1979 W. H. Freeman and Company. Used with permission. **13–5b** From *Tissues and Organs: A Text-Atlas of Scanning Electron Microscopy* by Richard G. Kessel and Randy H. Kardon. Copyright © 1979 W. H. Freeman and Company. Used with permission.

Chapter 14 **CO** Visuals Unlimited/G. Musil **14–5a** Pat Lynch/Photo Researchers **14–11** by the author **14–17** SIU/ Visuals Unlimited

Chapter 15 **CO** Gabe Palmer/The Stock Market **15–10e** Larry Mulvehill/Photo Researchers **15–11** Alan Gevins, EEG Systems Laboratory, San Francisco, California **15–13** Will & Deni McIntyre/Photo Researchers

Chapter 16 **CO** NASA/Comstock

Chapter 17 **CO** David Langley/The Stock Market **17–8** by the author **17–14b** Lennart Nilsson, from *Behold Man*, Little, Brown & Co, 1973 **17–26a** by the author **17–32b** Richmond Products, Boca Raton, Florida

Chapter 18 **CO** Ted Horowitz/The Stock Market **18–9b** Manfred Kage/Peter Arnold **18–13b** by the author **18–15** by the author **18–16** by the author **18–18** by the author

Chapter 19 **CO** Mark M. Lawrence/The Stock Market **19–2a** by the author **19–2b** Ed Reschke **19–2c** Dennis Kunkel/CNRI, Phototake **19–4** Bill Longcore/Photo Researchers **19–9** Ed Reschke **19–10** by the author **19–11** CNRI, Science Photo Library/Photo Researchers

Chapter 20 **CO** Gary Brettnacher/Tony Stone Worldwide **20–4b** Lennart Nilsson from *Behold Man*, © 1973, Little, Brown & Company **20–6c** L. V. Bergman & Associates, Inc. **20–7c** Manfred P. Kage/Peter Arnold **20–9a,b** Du Pont Merck Pharmaceutical Co. **20–9c** Peter Arnold, Inc. **20–13a,b** Mallinckrodt Institute of Radiology **20–13c** Courtesy of Picker International, Inc. **20–13d** Robert A. Levine, M.D., Harvard Medical School, Massachusetts General Hospital **20–13e** Science Photo Library/Photo Researchers

Chapter 21 **CO** Dick Luria/FPG **21–1** by the author **21–3a** Bolender, R. J., *J. Cell. Biol.* 61:269 **21–3b** *Bailey's Textbook of Microscopic Anatomy*, Kelly, Wood, and Enders eds., Williams and Wilkins, 1984 **21–5** Frank Sloop/W. Ober **21–16b** CNRI/Science Photo Library/Photo Researchers **21–22b** St. Bartholomew's Hospital/Photo Researchers

Chapter 22 **CO** Thomas Digory/Stockphotos Inc. **22–3a** by the author **22–6a** by the author **22–8d** by the author **22–9c** Astrid & Hanns Frider Michler/Photo Researchers **22–20** Centers for Disease Control **22–22a** Science Photo Library/Photo Researchers

Chapter 23 **CO** DiMaggio/Kalish/The Stock Market **23–3a** by the author **23–3b** CNRI/Science Photo Library/ Photo Researchers **23–5** CNRI/Phototake **23–7** John D. Cunningham/Visuals Unlimited **23–9b** Lester Bergman & Associates

Chapter 24 **CO** Nick Koudis/The Stock Market **24–3a** by the author **24–11a** From *Tissues and Organs: A Text-Atlas of Scanning Electron Microscopy* by Richard G. Kessel and Randy H. Kardon. © 1979 W. H. Freeman and Co. Used with permission. **24–11b** From *Histology* by Thomas S. Leeson and Roland C. Leeson. Philadelphia, W. B. Saunders Company, 1981, p. 338. **24–11c,d** by the author **24–15c** From *Tissues and Organs: A Text-Atlas of Scanning Electron Microscopy* by Richard G. Kessel and Randy H. Kardon. Copyright © 1979 W. H. Freeman and Company. Used with permission. **24–19** by the author **24–21** by the author **24–23** by the author **24–25c** by the author

Chapter 25 **CO** Richard Lowenberg/Science Source, Photo Researchers

Chapter 26 **CO** J.-C. Lozouet/The Image Bank **26–5d** David M. Phillips/Visuals Unlimited **26–6** CNRI Science Photo Library/Photo Researchers **26–20** by the author **26–23c** Science Source/Photo Researchers

Chapter 27 **CO** Comstock

Chapter 28 **CO** Dennis Hallinan/FPG **28–3** by the author **28–4** David M. Phillips/Visuals Unlimited **28–5** by the author **28–6b** (top) by the author; (bottom) from *Tissues and Organs: A Text-Atlas of Scanning Electron Microscopy* by Richard G. Kessel and Randy H. Kardon. Copyright © 1979, W. H. Freeman and Company. Used with permission. **28–8** by the author **28–12** (ampulla and isthmus) by the author **28–12** (epithelium) A. Copley/Visuals Unlimited **28–16** by the author **28–19a,b** by the author **28–19c** (top) by the author; (bottom) Biophoto Associates/ Photo Researchers **28–21a** Biophoto Associates/Photo Researchers **28–21b** by the author

Chapter 29 **CO** Ed Bock/The Stock Market **29–10** by the author **29–11a,b,c** Lennart Nilsson from *A Child Is Born*, © 1977 Dell/Delacorte **29–11d** CNRI/Science Photo Library/Photo Researchers **29–12a** Lennart Nilsson from *Behold Man*, © 1973 Little, Brown and Company **29–12b** CNRI/Science Photo Library, Photo Researchers **29–15** Nathan Benn/Woodfin Camp, Inc.

ART

Illustrations by William Ober and Claire Garrison

Figures 1-1, 1-2, 1-11, 1-13, 2-1, 2-4, 2-5, 2-6, 2-8, 2-9a, 2-10, 2-16, 2-19, 2-21, 2-22, 3-1, 3-2, 3-6, 3-7, 3-8, 3-9, 3-10, 3-11, 3-12, 3-13, 3-15, 3-16, 3-17, 3-19, 3-22, 3-23, 3-25, 3-29, 3-31, 3-32, 4-2, 4-3, 4-9, 4-12, 4-13, 5-2, 5-3, 5-4, 5-5, 5-6, 5-7, 5-8, 5-9, 5-10, 5-12, 5-13, 5-14, 5-16, 5-18, 5-19, 5-20, 5-21, 5-22, 5-23, 5-24, 5-25, 6-2, 6-4, 6-6, 6-7, 6-10, 6-13, 6-14, 6-15, 6-16, 6-17, 6-18, 6-19, 6-20, 7-1, 7-4, 7-5, 7-6, 7-7, 7-10, Table 7-1, 7-17, 7-18, 7-19, 7-20, 7-21, 8-18, 8-22, 9-7, 10-1, 10-4, 10-6, 10-7, 10-10, 10-11, 10-12, 10-13, 10-17, 10-19, 10-23, 11-5, 11-8, 11-14, 11-16, 11-17, 11-23, 12-2, 12-3, 12-4, 12-5, 12-9, 12-10, 12-12, 12-13, 12-14, 12-15, 12-17, 12-18, 12-19, 12-21, 12-23, 12-24, 12-25, 13-1, 13-2, 13-4, 13-6, 13-14, 13-17, 13-18, 13-19, 14-1, 14-2, 14-3, 14-4, 14-5, 14-6, 14-7, 14-8, 14-9, 14-11, 14-12, 14-13, 14-14, 14-16, 14-18, 14-19, 15-2, 15-5, 15-6, 15-7, 15-9, 15-12, 16-5, 16-10, 16-11, 17-1, 17-4, 17-5, 17-6, 17-7, 17-11, 17-12, 17-14, 17-15, 17-19, 17-20, 17-21, 17-22b, 17-23, 17-26, 17-29, 17-35, 18-1, 18-9, 18-11, 18-13a, 18-15, 18-16, 18-21, 19-3, 19-12, 19-14, 19-15, 20-17, 21-1, 21-6, 21-31, 22-1, 22-2, 22-4, 22-7, 22-22, 22-23, 22-24, 22-28, 23-4, 23-5, 23-7, 23-9, 23-10a and c, 23-15, 23-22, 23-28, 24-1, 24-9, 24-15, 24-18, 24-19, 24-21, 24-23, 24-25, 24-28, 24-31, 26-1, 26-2, 26-3, 26-4, 26-5, 26-7, 26-8, 26-9, 26-14, 26-15, 26-16, 26-19c, 26-22, 28-4, 28-5, 28-10, 28-12, 28-13, 28-16, 28-22, 29-7, 29-9, 29-10, 29-14, all Embryology Summaries, and Systems Overview Art.

Illustrations by Tina Sanders

Figures 8-3, 8-7a and b, 8-11, 8-13, 9-10, 9-12, 13-3, 13-5, 13-11, 14-10, 14-15, 14-16 (pullout), 15-9, 15-12, 15-14, 16-1, 16-3, 16-4, 16-7, 16-8, 16-9, 16-13, 17-14b and c, 17-16, 17-22a and c, 17-24, 17-25, 17-27, 17-34, 18-2, 18-3, 18-4, 18-5, 18-6, 18-8, 18-13c, 18-14, 20-1, 20-7, 20-14c, 21-2, 21-3, 21-4, 21-9, 21-10, 21-11, 21-12, 21-13, 21-14, 21-16, 21-17, 21-18, 21-19, 21-22, 22-5, 22-10, 22-12, 22-13, 22-14, 23-10b, 23-14, 24-3, 24-17, 25-16, and all orientation drawings.

Illustrations by MediVisuals

9-10, 9-20, 13-8, 13-9, 13-10, 14-19, 14-20, 14-21, 14-22, 14-23, 14-24, 14-25, 14-26, and 14-27.

Illustrations by Craig Luce

17-14a, 20-2, 20-3, 20-4, 20-5, 20-6, 20-11a, 20-14a and b, 22-9, 23-1, 23-2, 23-8, 24-2, 24-6, 24-7, 24-8, 24-22, 24-24, 26-19a and b, 28-1, 28-6, 28-7, 28-11, 28-14, and 28-15.

Illustrations by Ron Ervin

2-25, 3-14, 7-13, 8-1, 8-2, 8-4, 8-5, 8-6, 8-7c and d, 8-8, 8-9, 8-10, 8-12, 8-14, 8-15, 8-17, 8-19, 8-20, 8-21, 9-1, 9-2, 9-3, 9-4, 9-5, 9-6, 9-8, 9-9, 9-11, 9-13, 9-14, 9-15, 9-16, 9-18, 9-21, 11-4, 11-6, 11-7, 11-9, 11-10, 11-11, 11-12, 11-13, 11-15, 11-19, 11-20, 11-21, 11-22, 13-22, and 13-23.

Figure 1-6 by Ed Fisher; © 1990 *The New Yorker Magazine*, Inc.

All other drawings by Vantage Art, Inc.

Regional and Directional References

Region of Body	*Adjective*	*Directional Reference*	*Examples of Correct Usage*
Front	Ventral or anterior	Ventrally	The navel is on the ventral/anterior surface of the trunk.
Back	Dorsal or posterior	Dorsally	The dorsal body cavity encloses the brain and spinal cord; moving dorsally from the navel you find the muscles of the abdominal wall.
Head	Cranial	Cranially	The cranial border of the pelvis; moving cranially from the pelvis toward the ribs.
	Cephalic		Same as cranial.
	Superior		Same as cranial but you refer to the superior surface of the skull.
	Rostral	Rostrally	Same as cranial but refers to nose: the eyes are on the rostral surface of the head; moving rostrally from the back of the skull brings you to the face.
Tail (coccyx)	Caudal	Caudally	The hips are caudal to the waist; moving caudally from the shoulder brings you to the hips.
	Inferior	Inferiorly	Same as caudal but the soles are on the inferior surfaces of the feet.
Close to long axis of the body	Medial	Medially	The medial surfaces of the thighs may be in contact; moving medially across the surface of the chest, you arrive at the sternum.
Away from the long axis	Lateral	Laterally	The leg articulates with the lateral surface of the pelvis; moving laterally from the nose brings you to the eyes.
Close to the center of the body	Proximal	Proximally	The wrist is proximal to the fingers; moving proximally from the wrist brings you to the elbow.
Away from the center of the body	Distal	Distally	The fingers are distal to the palm; moving distally from the elbow brings you to the wrist.

Terms That Indicate Planes of Section

Orientation of Plane	*Adjective*	*Directional Reference*	*Examples of Correct Usage*
Parallel to long axis	Sagittal	Sagitally	A sagittal section separates right and left portions; you examine a sagittal section, but you section sagitally.
	Midsagittal		In a midsagittal section the plane passes through the midline, dividing the body in half and separating right and left sides.
	Parasagittal		A parasagittal section misses the midline, separating right and left portions of unequal size.
	Frontal or coronal	Frontally or coronally	A frontal or coronal section separates anterior and posterior portions of the body; coronal usually refers to sections passing through the skull.
Perpendicular to the long axis	Transverse or horizontal	Transversely or horizontally	A transverse or horizontal section separates superior and inferior portions of the body.

Boxes in the Text

CLINICAL COMMENTS

Supplemental Topics in the Clinical Manual

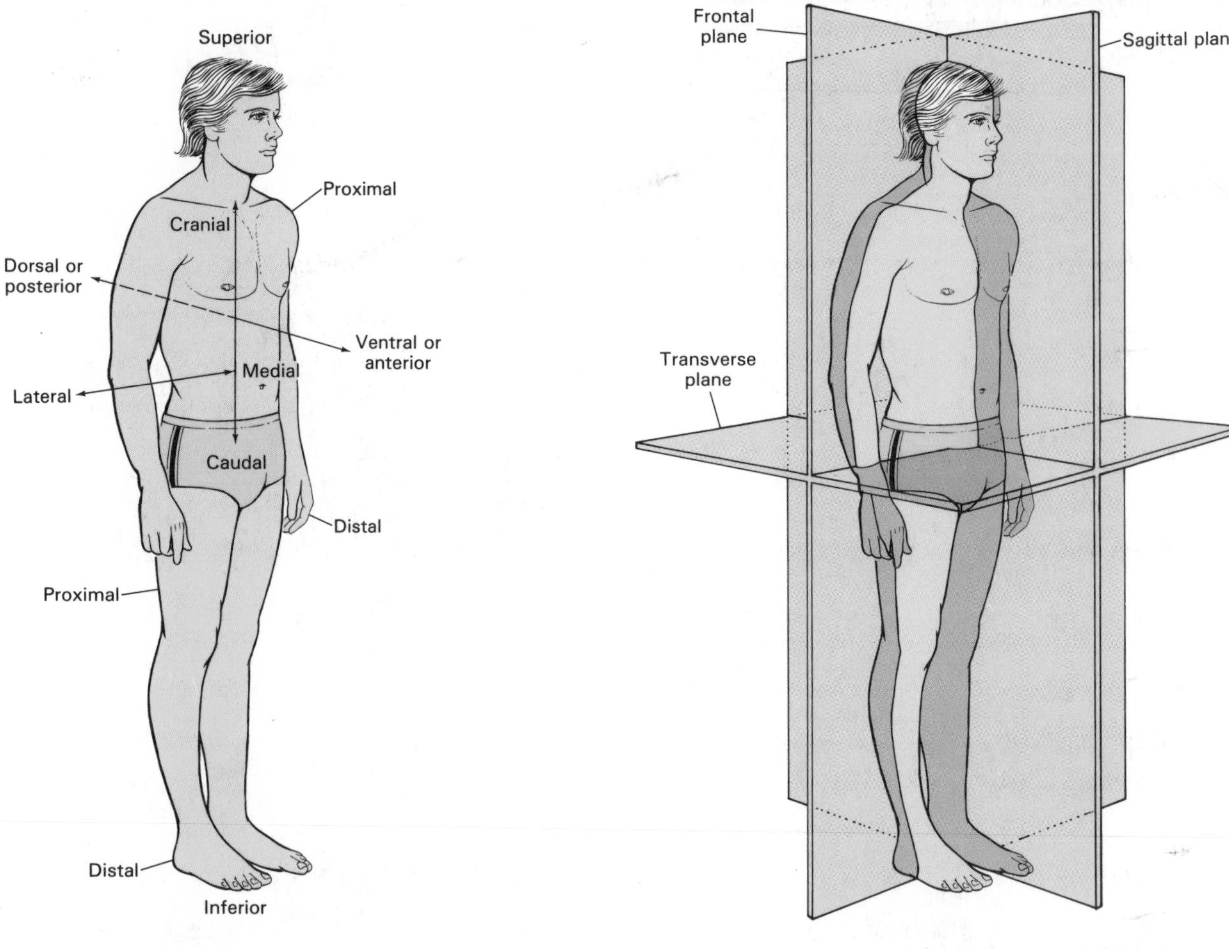

Some Useful Conversions

Measurement	*Metric to English*		*English to Metric*	
Volume	1 liter (ℓ)	= 1.057 quarts	1 gallon	= 3.785 ℓ
	1 milliliter (mℓ) 1 cubic centimeter (cc) 0.001 ℓ (10^{-3} ℓ)	= 0.034 fl oz approx. ⅛ teaspoon approx. 15 or 16 drops	1 quart	= 0.946 ℓ = 946 mℓ
			1 pint	= 0.473 ℓ = 473 mℓ
			1 fluid ounce	= 29.57 mℓ
			1 teaspoon	= approx. 5 mℓ
Length	1 meter (m) 100 centimeters (cm) 1000 millimeters (mm)	= 39.37 inches	1 foot	= 0.305 m = 30.5 cm
	1 cm	= 0.394 inch	1 inch	= 2.54 cm = 25.4 mm
	1 mm	= 0.039 inch		
Mass	1 kilogram (kg) 1000 grams (g)	= 2.205 pounds 35.28 ounces	1 pound 1 ounce	= 0.4536 kg = 28.35 g
	1 g 1000 milligrams (mg)	= 0.0353 ounce		

Temperature	*Centigrade*	*Fahrenheit*
Freezing point of pure water	0°	32°
Room temperature	20°	68°
Normal body temperature	36.8°	98.6°
Boiling point of pure water	100°	212°
Conversion	°C → °F: °F = (1.8 × °C) + 32	°F → °C: °C = (°F − 32) × 0.56